AF704247

WITHDRAWN

THE GEOCHEMISTRY OF GOLD AND ITS DEPOSITS
(together with a chapter on geochemical prospecting for the element)

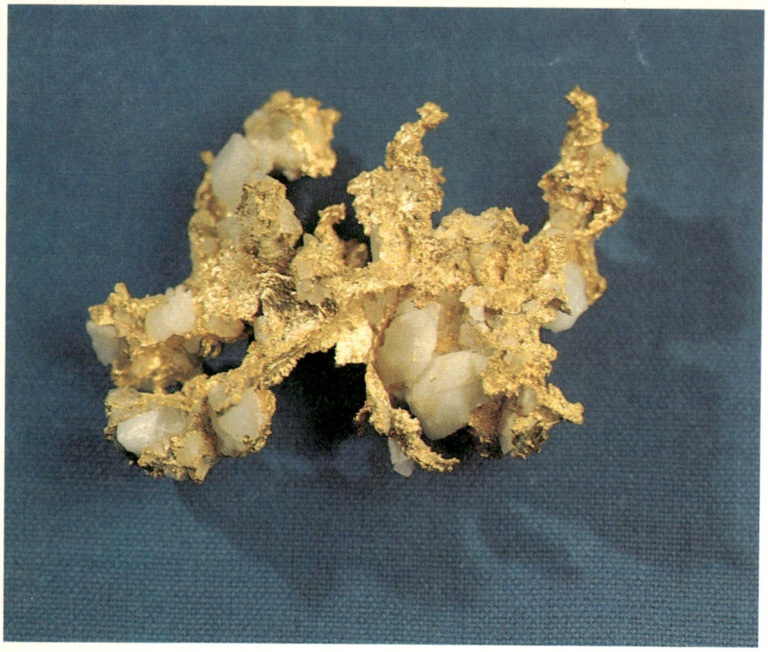

Native gold, Paymaster mine, South Porcupine, Ontario (Natural size).

Native gold nuggets, Goldrun Creek placer, Cassiar District, British Columbia (Natural size).

The gold cups of Vaphio: From the tholos tomb near Vaphio, Laconia, Greece. First Late Minoan, late 16th-early 15th century B.C., National Archaeological Museum, Athens.

GEOLOGICAL SURVEY
BULLETIN 280

THE GEOCHEMISTRY OF GOLD AND ITS DEPOSITS (together with a chapter on geochemical prospecting for the element)

R.W. Boyle

© Minister of Supply and Services Canada 1979

Available in Canada through

authorized bookstore agents
and other bookstores

or by mail from

Canadian Government Publishing Centre
Supply and Services Canada
Hull, Quebec, Canada K1A 0S9

and from

Geological Survey of Canada
601 Booth Street
Ottawa, Canada K1A 0E8

A deposit copy of this publication is also available
for reference in public libraries across Canada

Catalogue No. M42-280 Canada: $45.00
ISBN 0-660-01769-5 Other countries: $54.00

Price subject to change without notice.

QE
185
.A43
no. 280

Scientific Editor
R.G. BLACKADAR

Critical Readers
R.J.W. DOUGLAS
R.I. THORPE
I.R. JONASSON
R.G. GARRETT
J.J. LYNCH

Editor
MARK KENNEDY

Text printed on Kelmscott offset, smooth (brilliant white)
Set in Times Roman with News Gothic captions
by SOUTHAM MURRAY, Toronto

Artwork by CARTOGRAPHIC UNIT, GSC

3000-1978-6693-9

This book is dedicated to all those
who have moiled for gold in the rain
and cold and under the tropic sun,
and have used their winnings wisely.

Preface

Gold, the noblest of metals, has been utilized by man for more than 5000 years, first in artistic objects and jewellery, then largely in coinage and more recently in the industrial arts. The metal is today the principal medium for settling international accounts. Despite the great commercial and artistic value of gold, little was actually known about the geochemistry of the element until the middle of the present century; since then considerable research in various geochemical aspects of the metal has been carried out.

The author here presents a comprehensive compilation of the geochemistry of gold, partly from his own research, which has extended over 25 years, and partly from the literature. He describes the principal types of gold deposits and discusses their origin. The data here presented, in addition to being a fundamental contribution to the geochemistry of gold, will be found useful to those interested in geochemical prospecting for the element.

D.J. McLaren
Director General
Geological Survey of Canada

Ottawa, 25 August 1976

THE GEOCHEMISTRY OF GOLD AND ITS DEPOSITS

Abstract

Gold is a member of Group IB of the periodic table, which includes copper, silver and gold. In its chemical reactions gold resembles silver in some respects, but its chemical character is markedly more noble. The principal oxidation states of gold are Au(I)(aurous) and Au(III)(auric). These states are unknown as aquo-ions in solution, the element being present mainly in complexes of the type $[Au(CN)_2]^-$, $[AuCl_2]^-$, $[Au(OH)_4]^-$, and $[AuCl_4]^-$. There is only one naturally occurring isotope of gold: ^{197}Au.

In nature gold occurs predominantly in the native state or as a major constituent of various alloys containing mainly silver, copper or platinoid metals. Several gold and gold-silver tellurides are known of which the most common are sylvanite, calaverite, petzite, krennerite and nagyagite. The antimonide, aurostibite, $AuSb_2$, occurs as a hypogene mineral in some auriferous deposits, and there is also a selenide, fischesserite, Ag_3AuSe_2, and a bismuthide, maldonite, Au_2Bi, which is fairly well differentiated. The principal ore minerals of gold are the native metal, aurostibite and the various tellurides.

The abundance of gold in the upper lithosphere is about 0.005 ppm, and the Au/Ag ratio is about 0.1. The average gold content of igneous-type rocks in parts per million is – ultrabasic (0.004), gabbro-basalt (0.007), diorite-andesite (0.005) and granite-rhyolite (0.003). The average gold content of sedimentary rocks in parts per million is – sandstone and conglomerate (0.03), normal shale (0.004) and limestone (0.003). Certain graphitic shales, sulphide schists, phosphorites and some types of sandstones and conglomerates may contain up to 2.1 ppm Au or more.

The average gold content of soils is 0.005 ppm, and the average for natural fresh waters is 0.00003 ppm. Sea and ocean waters contain an average of 0.000012 ppm Au. Gold is a trace constituent of many plants and animals. Some coals are slightly enriched in gold with 0.05 to 0.1 ppm Au in the ash.

Gold is won both from deposits mined essentially for the element and as a byproduct of the mining and treatment of nickel, copper, zinc, lead and silver ores. The following types of primary deposits, exploited mainly for gold, can be distinguished:

1. Auriferous porphyry dykes, sills and stocks; coarse-grained granitic bodies, aplites and pegmatites.
2. Auriferous skarn-type deposits.
3. Gold-silver and silver-gold veins, stockworks, lodes, mineralized pipes and irregular silicified bodies in fractures, faults, shear zones, sheeted zones and breccia zones essentially in volcanic terranes.
4. Auriferous veins, lodes, sheeted zones and saddle reefs in faults, fractures, bedding plane discontinuities and shears, drag folds, crushed zones and openings on anticlines essentially in sedimentary terranes; also replacement tabular and irregular bodies developed near faults and fractures in chemically favourable beds.
5. Gold-silver and silver-gold veins, lodes, stockworks, silicified zones, etc. in a complex geological environment, comprising sediments, volcanics and igneous or granitized rocks.

Résumé

L'or est un élément du groupe 1B du tableau périodique, qui comprend le cuivre, l'argent et l'or. Du point de vue de sa réactivité chimique, l'or rappelle l'argent à certains égards, mais son caractère chimique est beaucoup plus 'noble'. Les principaux états d'oxydation de l'or sont Au (I) (aureux) et Au (III) (aurique). On ne trouve pas ces états d'oxydation sous forme d'aquo-ions en solution, mais surtout sous forme de complexes du type $[Au(CN)_2]^-$, $[AuCl_2]^-$, $[Au(OH)_4]^-$ et $[AuCl_4]^-$. Il n'existe à l'état naturel qu'un seul isotope de l'or: ^{197}Au.

Dans la nature, l'or se présente généralement à l'état natif, ou allié à d'autres métaux, surtout l'argent, le cuivre, ou les métaux du groupe du platine. On connaît plusieurs tellurures d'or et d'or et argent, dont les plus communs sont la sylvanite, la calavérite, la petzite, la krennerite et la nagyagite. L'antimoniure, l'aurostibite, $AuSb_2$, se présente sous forme de minéral hypogène dans certains gisements aurifères, et il existe aussi un séléniure, la fischesserite, Ag_3AuSe_2, et un bismuthiure, la maldonite, Au_2Bi, lequel est assez bien différencié. Les principaux minéraux d'or sont l'or natif, l'aurostibite et les divers tellurures.

L'abondance de l'or dans la partie supérieure de la lithosphère est d'environ 0.005 ppm et la proportion Au/Ag est d'environ 0.1. La teneur moyenne en or des roches ignées, exprimée en parties par million, est de 0.004 pour les roches ultrabasiques, de 0.007 pour les gabbros et les basaltes, de 0.005 pour la diorite et l'andésite, et de 0.003 pour le granite et la rhyolite. La teneur moyenne en or des roches sédimentaires, exprimée en parties par million, est de 0.03 pour les grès et les conglomérats, de 0.004 pour les schistes argileux normaux et de 0.003 pour le calcaire. Certains schistes graphiteux, certaines phosphorites, certains schistes cristallins sulfureux, et certains types de grès et de conglomérats peuvent contenir jusqu'à 2.1 ppm de Au ou davantage.

La teneur moyenne en or des sols est de 0.005 ppm et la teneur moyenne des eaux douces naturelles est de 0.00003 ppm. Les eaux marines et océaniques contiennent en moyenne 0.000012 ppm de Au. L'or se présente sous forme d'élément-trace chez de nombreux végétaux et animaux. Certains charbons son légèrement enrichis en or et leurs cendres contiennent de 0.05 à 0.1 ppm de Au.

On extrait l'or à la fois dans des gisements aurifères proprement dits et dans des gisements où il est un sous-produit de l'extraction et du traitement des minerais de nickel, de cuivre, de zinc, de plomb et d'argent. On peut distinguer les types suivants de dépôts primaires, que l'on exploite surtout pour l'or:

1. Les dykes, sills et stocks de porphyres granitiques aurifères, et les corps granitiques à grain grossier, les aplites et les pegmatites.
2. Les gisements aurifères de type skarn.
3. Les veines, stockwerks, filons, colonnes minéralisées et masses irrégulières silicifiées, aurifères-argentifères et argentifères-aurifères, situés dans les fractures, les failles, les zones de cisaillement, les zones à feuillets minéralisés et les zones de brèches, que l'on rencontre surtout dans les terrains volcaniques.
4. Les veines, filons, zones à feuillets minéralisés, et gîtes en selle aurifères, situés dans les failles, fractures, discordances stratigraphiques, et les zones de cisaillement, plis d'entraînement, zones broyées et voûtes anticlinales que l'on rencontre surtout dans les

6. Disseminated and stockwork gold-silver deposits in igneous, volcanic and sedimentary rocks.

(a) Disseminated and stockwork gold-silver deposits in igneous bodies.

(b) Disseminated gold-silver occurrences in volcanic flows and associated volcaniclastic rocks.

(c) Disseminated gold-silver deposits in volcaniclastic and sedimentary beds: deposits in tuffaceous rocks and iron formations and deposits in chemically favourable sedimentary beds.

7. Gold deposits in quartz-pebble conglomerates and quartzites.

Placers constitute the principal secondary type of gold deposit.

The quartz-pebble conglomerate deposits provide the bulk of the world's production of gold, some 56 per cent. The other deposits, mainly the vein and disseminated types, eluvial and alluvial placers and the various polymetallic veins, lodes, massive bodies and stockworks (byproduct gold) now provide the remaining 44 per cent of the production.

The epigenetic vein, lode, stockwork and disseminated types of gold deposits appear to have originated mainly by metamorphic secretion processes, the source rocks of the gold and its associated elements being mainly the enclosing volcanic and/or sedimentary piles. Modern gold placers are of sedimentary origin, the gold being winnowed into pay streaks as the result of both chemical (accretion) and physical (gravity) processes operating during weathering and subsequent sedimentation. The auriferous quartz-pebble conglomerate deposits appear to have originated as placers, the gold and many of its associated elements having undergone radical chemical reworking during subsequent diagenetic and metamorphic events.

The oxidation processes in gold deposits are complex and depend essentially on the Eh and pH. Colloidal and coprecipitation phenomena also play a large part. Iron and manganese minerals and carbonates in the gangue and ore greatly influence the reactions that lead to the secondary enrichment of the element. Gold is not easily solubilized in nature, and its soluble forms are readily reduced to the metal by a great variety of natural materials. The result of this behaviour is that the only common gold mineral found in the oxidized zones of auriferous deposits is the native metal. In this form it ultimately collects in both eluvial and alluvial placers, which have been exploited throughout the world since time immemorial.

Practically all the geochemical methods of prospecting are applicable in the search for auriferous deposits. The most effective methods appear to be those based on the sampling of stream and lake sediments and soils, analyzing these materials directly or analyzing heavy mineral separates obtained from them. The most specific indicator (pathfinder) elements for gold are Ag, As, Sb and Te.

terrains sédimentaires; et aussi, les gisements de substitution tabulaires et irréguliers, qui se sont formés à proximité de failles et fractures, dans des lits à caractères chimiques favorables.

5. Les veines, filons, stockwerks, zones silicifiées, aurifères-argentifères et argentifères-aurifères, situées dans un milieu géologique complexe, comprenant des sédiments, des roches volcaniques et des roches ignées ou granitisées.

6. Les disséminations et stockwerks aurifères et argentifères, dans les roches ignées, volcaniques et sédimentaires.

(a) Disséminations et stockwerks aurifères et argentifères dans les corps ignés.

(b) Disséminations aurifères et argentifères dans les coulées volcaniques et les roches associées volcaniclastiques.

(c) Disséminations aurifères et argentifères dans les lits volcaniclastiques et sédimentaires: les gîtes qui se sont formés dans les roches tufacées et les formations ferrifères; et, les gîtes qui se sont formés dans des lits sédimentaires de composition chimique favorables.

7. Les gîtes aurifères que l'on rencontre dans les bankets et les quartzites.

Les placers constituent le principal type de gisements secondaires d'or.

Les bankets (conglomérats contenant surtout des galets de quartz vitreux) fournissent la plus grande partie de l'or mondial, soit environ 56 pour cent. Les autres gisements, surtout les disséminations, veines filoniennes, et placers éluviaux et alluviaux, et les diverses veines filoniennes, filons, corps massifs et stockwerks polymétalliques (où l'or est un sous-produit) constituent actuellement les 44 pour cent restants de la production.

Les types de gisements aurifères épigénétiques, tels que les veines, filons, stockwerks, et disséminations semblent s'être formés surtout par des processus d'imprégnation diffuse au cours du métamorphisme, les roches mères de l'or et des éléments associés étant surtout les colonnes encaissantes volcaniques ou sédimentaires, ou les deux à la fois. Les placers aurifères modernes sont d'origine sédimentaire, l'or s'accumulant suivant des traînées, les 'pay streaks', sous l'effet de processus chimiques (accrétion) et physiques (gravité) qui agissent au cours de l'altération et de la sédimentation ultérieure. Les bankets aurifères semblent avoir été originellement des placers, où l'or et une grande partie des éléments qui l'accompagnent ont subi par la suite un remaniement chimique extrême, par diagénèse et métamorphisme.

Les processus d'oxydation qui ont lieu dans les gîtes aurifères sont complexes, et dépendent essentiellement du Eh et du pH. Des phénomènes colloïdaux et de coprécipitation jouent aussi un rôle important. Les minéraux riches en fer et en manganèse ainsi que les carbonates que l'on trouve dans la gangue et le minerai ont une influence profonde sur les réactions qui aboutissent à l'enrichissement secondaire en cet élément. Dans la nature, l'or est difficilement entraîné en solution, et ses formes solubles sont facilement réduites par un grand nombre de matériaux naturels pour donner l'or métallique. Il en résulte que le seul minerai aurifère que l'on rencontre habituellement dans les zones oxydées des gisements aurifères est le métal natif. C'est sous cette forme qu'il s'accumule finalement dans les placers éluviaux et alluviaux, que l'on exploite dans le monde entier depuis des temps immémoriaux.

Pratiquement, toutes les méthodes géochimiques de prospection conviennent à la recherche de gîtes aurifères. Les méthodes les plus efficaces semblent être celles fondées sur l'échantillonnage de sédiments et sols fluviatiles et lacustres, et sur l'analyse de ces matériaux eux-mêmes, ou des minéraux lourds obtenus par séparation. Les éléments indicateurs qui accompagnent le plus fréquemment l'or sont Ag, As, Sb et Te.

Contents

Chapter IV. Oxidation and secondary enrichment of gold deposits

Chapter V. Prospecting for gold deposits

Tables

Frontispiece

Plates

Figures

Au

The modern (upper left) and alchemical (lower right) symbols for gold, the most noble of metals.

The Argonauts examining the Golden Fleece at Colchis on the Euxine (Black Sea). (From Agricola, *De Re Metallica.*) According to Greek mythology, the Golden Fleece was taken from the ram on which Phrixus and Helle escaped from being sacrificed. It was hung up in the grove of Ares in Colchis and recovered from King Aeetes by the Argonautic expedition under Jason, with the help of the sorceress, Medea the king's daughter. In actual fact the Argonauts were early prospectors who sought the source of the ancient placers on the Black Sea. At that time (1200 B.C.) the workers of auriferous placers recovered the gold by trapping the metallic particles on sheep's fleeces placed in crude sluices. The fleeces were then hung up to dry in nearby trees and were later shaken to collect the gold.

☉

I have given thee the gold countries;
given thee what is in them of electrum,
lapis lazuli, and malachite.
–An epistle to Seti I from the Sun God
as given in the inscription in the
Temple at Edfu, Egypt, circa 1300 B.C.

Chapter I. Introduction

Research on the geochemistry of gold was begun by the writer some 25 years ago and has continued at intermittent intervals in the field and laboratory since that time. The work is now essentially complete, and it was thought that increasing world interest in the metal merited a report at this time. This volume is meant as a companion to that on silver (Boyle, 1968*b*) and should be read in conjunction with it.

Historical notes on gold

Of all elements gold has been the most esteemed by man since the earliest times. The desire for gold has markedly influenced his history and was a deciding factor in the development of chemistry. The cry 'gold' has lured men across oceans and continents, over the highest mountain peaks, into the Arctic tundras, into scorching deserts and through impenetrable jungles. Its gleam prompted the expeditions and conquests of Jason of Thessaly, Cyrus and Darius of Persia, Alexander of Greece, Caesar of Rome, Columbus of Genoa, Cortez and Pizzarro of Spain, Raleigh of England and many others down through history. Gold, according to Pliny the Elder (*Historia naturalis,* 79 A.D.), is the first of man's follies, silver the second. To make gold from baser metals was a major preoccupation of the alchemists as was also their ceaseless efforts to discover the elixir of life and the fountain of youth. In fact, the alchemists considered that they had indeed discovered the fountain of youth in the potable gold made by solution of the metal in aqua regia followed by treatment with ethereal oils. Roger Bacon certainly thought so for in urging the elixir "*aurum potable, oleum auri, quinta essentia auri*" on Pope Nicholas IV he related.

> An old man, when ploughing a field in Sicily, one day found some of the yellow potable gold in a golden phial, and, supposing it to be dew, drank up the liquor. He was thereupon transformed into a hale, robust, and accomplished youth. The youth was thereafter received into the service of the Sicilian king, where he served some eighty years.

A thousand and more years of alchemy and innumerable experiments to make gold by concocting mixtures and potions of every conceivable type gave us the basis of modern chemistry. We can truly conclude with Francis Bacon.

> Surely to alchemy this right is due, that it may be compared to the husbandman whereof Aesop makes the fable; that, when he died, to:• his sons that he had left unto them gold buried in his vineyard; and they digged all over the ground, and gold they found none; but by reason of their stirring and digging the mould about the roots of their vines, they had a great vintage the year following: so assuredly the search and stir to make gold hath brought to light a great number of good and fruitful inventions and experiments.

Cadmus, the Phoenician, is said by some early writers to have discovered gold, but this is surely legend. Others say that Thoas first found it in the Pangaeus Mountains in Thrace. The *Chronicum Alexandrinum* ascribes its discovery to Mercury, the son of Jupiter, or to Pisus, king of Italy, who quitting his own country went into Egypt. In actual fact the discovery of the element we call gold is lost in antiquity. Reference to the metal can be found in most of the ancient Hindu, Chinese and Hebrew manuscripts, and in some, such as the Old Testament, it is the first metal mentioned. Gold beads, gold-hafted flint knives, stone jars with gold-covered mouths, gold ornaments, gold jewellery and various golden decorations have been found in the excavations of the most ancient civilizations, in the Neolithic monuments of France, in the Celtic graves of Europe, among the Sumerian relics of Ur, in predynastic monuments and graves of Egypt and among the very ancient remains of Minoan Crete, India and China. The tombs of the Pharaohs, especially that of Tutankhamen (c.1371–c.1352 B.C.), contained many beautifully executed gold pieces, and representations of quartz-crushing and gold-refining processes dating back to at least 1350 B.C. have been reported from these tombs. In the Code of Menes, who reigned in Egypt about 3500 B.C., it was decreed that "one part of gold is equal to two and one-half parts of silver in value." By 1000 B.C. both gold and silver were probably in widespread use as coinage metals in all countries between the Indus and the Nile. Other uses for gold mentioned in old manuscripts from various civilizations include the fabrication of idols, shrines, altars, bowls, vases, flasks, drinking cups, funeral face masks, sarcophagi, mummy cases and ornamental weapons. The art and trade of the goldsmith and minter have, therefore, come down to us from antiquity. Their story is told in magnificent detail by Singer (1954), Sutherland (1969), Blakemore (1971) and Hobson (1971).

The chemists and archaeologists tell us that the earliest gold objects of all ancient civilizations were fashioned directly from native gold (Caley, 1964; Hartmann and Sangmeister, 1972). Later, around the middle of the first millennium B.C., a method of purifying native gold came into existence, and at about the same time the practice of alloying metals such as silver and copper with gold and thus debasing the metal came into use[1]. It was apparently Archimedes, in his bathtub, who devised a means of determining the amount of alloyed metals, although it seems probable that his principle was known long before the advent of Greek science.

Original manuscript submitted: 28 June 1975
Approval for publication: 29 June 1976

[1]The practice of alloying metals with gold was called *diplosis* by the early Egyptian goldsmiths, meaning the art of 'doubling' the weight of gold without materially changing its outward appearance. In Colombia where platinoids were abundant in the early gold placers, the Spanish treasurers encountered no end of troubles with 'doubling' of the gold by individuals by means of alloying with the platinoids, elements unknown in Europe at that time (1750). It is said that to avoid this debasing practice that the treasurers ordered all of the platinum won from the placers to be thrown into the sea or nonauriferous rivers (Weeks, 1968).

The early metallurgy of gold probably involved little more than separating the metal by gravity from the dross of the placers and the quartz and other gangue of veins. Later, probably prior to 1000 B.C., it was discovered that gold adhered to mercury, and this developed into the amalgamation process that still finds a use in the treatment of gold ores. In the latter part of the 18th century and the greater part of the 19th the chlorination process was used extensively. Chlorine was passed through the moistened ore forming gold chloride which was leached out with water, the gold was then precipitated by ferrous sulphate. In 1887 McArthur and Forrest invented the cyanide process of dissolving gold followed by its precipitation with zinc dust. This is the method now commonly used on a large scale for the extraction of the element from its ores.

The extraction of gold and silver from sulphide ores by the process of liquation[2] was probably known to the ancients, as the Greeks seem to have been familiar with it and the Romans are known to have practised the art.

The process of 'parting' gold and silver by cementation with salt, whereby the silver is converted to chloride and the gold is run off, was apparently known in the 6th century B.C. Parting, employing acids, was unknown to the Romans. As far as can be determined, the first reference to the parting of gold and silver by means of nitric acid is to be found in the treatises of the Moslem alchemists of the thirteenth century. Modern processes of separating silver and other metals from gold involve the treatment of molten gold with chlorine and precipitation of the metal by electrolysis. The latter produces a pure product with only mere traces of silver, copper and other metals.

Native gold is a relatively common mineral and is present in gold-quartz veins, in oxidized zones of many sulphide and other types of auriferous deposits and in placers in streams and rivers in most parts of the world. It would seem that much of the gold of the ancients came first from placers and later from primary deposits and their oxidized zones. One of the richest placer streams which supplied much electrum to the Middle East, and is said to have been the source of the wealth of Croesus, was the river Pactolus (Sarabat), a tributary of the Hermus (modern Gediz) after passing Sardis in Lydia (now Anatolia, Turkey). The Pactolus drains the auriferous region on the flanks of Mount Tmolus (the present Boz Sira Daglari); it is the river in which it is said that Midas, the mythical founder of the Phrygian Kingdom, on the advice of Bacchus, bathed in its water to rid himself of the fatal faculty of turning everything he touched into gold. The other two legendary sources of gold, mentioned repeatedly in the Old Testament, Havilah and Ophir, have not been precisely identified.

In Genesis 2:10–12 it is written:

"And a river went out of Eden to water the garden; and from thence it was parted, and became into four heads.

"The name of the first is Pison, that is it which compasseth the whole land of Havilah, where there is gold;

"And the gold of that land is good: there is bdellium and the onyx stone."

There has been much speculation as to the location of the land of Havilah, the most probable from a geological viewpoint being that the river Pison is the modern Coruh which drains into the Black Sea near Batumi and that Havilah is the Pontic goldfield in Turkey, near Trabzon. This field is also probably one of the sites where Jason and the Argonauts sought the Golden Fleece, since within historical times placer miners used sheep's fleeces in this and other fields to catch the gold in their crude sluices. This method of winning gold (Frontispiece) is recorded by Strabo (63 B.C.–A.D. 19?), the Greek geographer, who wrote "the mountain torrents are said to bring down gold, and these barbarians catch it in troughs perforated with holes and in fleecy skins."

Strabo's reference is probably to the river Phasis (the present Rioni in Georgian S.S.R., draining from the Caucasus) which drained the Colchis region, supposedly the goal of the Argonauts in the legend of Jason. Whether or not this river produced gold in ancient times is uncertain. If it did, it is another of the possible sites where the legendary Golden Fleece was sought.

Ophir, the fabulously rich land of gold from which King Solomon's Phoenician (Tharshish) navy brought large amounts of the metal (some 34 metric tons) to his kingdom, has led to much speculation as to its location. In Genesis X it is associated with Havilah, which as noted above was probably the Pontic goldfield on the Black Sea. This may account for the long period of time, some three years, to make the voyage from Ezion-geber at the head of the Gulf of Aqaba to Ophir and back (I Kings 10:22). The cargos mentioned, almug (sandalwood) trees, precious stones, ivory, apes and peacocks suggest circumnavigation of Africa. Tharshish or Tarshish (a region centred on Cadiz) suggests that the gold may have come from Spain, and specifically from the oxidized deposits of the Huelva region where the modern mining town of Tharsis often equated with Tarshish is located. Other possibilities are East Africa, principally Rhodesia, and specifically the Zimbabwe region, where some think King Solomon's mines and metallurgical plants were located. This may also have been the site of Punt, exploited by the Egyptians, and from whence in Queen Hatshepsut's time (1600 B.C.) and later great store of gold was brought by the Egyptian navy. Still other possibilities suggested for the site of King Solomon's mines are southern Turkey (Bolkar Mountains), northwest Saudi Arabia (the land of the ancient Midians and possibly the Eldorado of the Hebrews), Sudan (Ancient Nubia; *Nub* means 'gold' in ancient Egyptian), Altai (Purington, 1903), Ethiopia, India, Cuba, Peru, the Far East, particularly Japan, Arctic Canada and a hundred other places. The story of Ophir is well told by Rickard (1932) and Sutherland (1969). The history of ancient Rhodesian gold mines and of Zimbabwe is related in fascinating detail by Summers (1969).

The most ancient map known, the famous *Carte des mines d'or* at Turin, is a Rameside papyrus and fragments depicting a gold mining region in Egypt. On it are located roads, miner's houses, gold mines, quarries, auriferous mountains and so on. The papyrus is said to date back to about 1320 B.C. The exact site represented by the map is in some

[2]In the process of liquation the copper, arsenic, etc., ores were first smelted often with a siliceous flux, to yield an impure ingot. This ingot was then alloyed with lead if there was insufficient lead in the ores, and the alloy was heated to a temperature between the melting point of lead and copper, during which the lead would liquate out, carrying the silver (and gold) with it. The lead was then cupelled on bone ash.

doubt. Some have suggested the mines represented are those of the Wadi Kareim or the Wadi Hammamat, on the Qena-Qoseir road (Gardiner, 1914). Derry (1951) says that the area represented on the map is the Wadi Fawakhir in which the El Sid Gold Mine is situated.

Gold has a widespread occurrence in practically every country in the world. In Europe, Asia and Africa ancient gold mines are known in Spain, Great Britain, Greece, Turkey, Saudi Arabia, Iran, India, China, Japan, U.S.S.R. (Uzbek S.S.R., Armenian S.S.R. and elsewhere) and numerous other countries. Placers have yielded gold from the rivers Tagus, Guadalquiver, Tiber, Po, Rhone, Rhine, Hebrus (Maritsa), Nile, Zambezi, Niger, Senegal, Pactolus (Sarabat), Oxus (Amu Darya which flows through the golden land of Samarkand), Ganges, Lena, Aldan, Yangtze and hundreds of others too numerous to mention. The Egyptians mined gold extensively in Sinai, eastern Egypt and Sudan (Nubia) as far back as 4000 years ago. It was from them that the Persians, Greeks and Romans learned the techniques of gold prospecting, mining and metallurgy. The Greeks and Romans mined gold ores extensively in the metalliferous regions of their empires (Davies, 1935; Tylecote, 1962; Healy, 1978). Pliny the Elder (A.D. 23–79) in his *Historia naturalis*[3] written in the early years of our era, repeatedly mentions the mining and metallurgy of gold, and Agricola[4] and many others before him refer to the element often in some detail. Rickard (1932, 1934, 1944) and Forbes (1964) have traced the history of gold mining, metallurgy and usage in great detail.

Compared with the gold placers and mines of the Old World those in parts of the New may be just as ancient, although it would appear that the Aborigines of North and South America put little emphasis on gold beyond its use as ornaments, jewellery, sacrificial knives, etc. Columbus of Genoa found the natives in possession of gold nuggets, a fact which excited the Spaniards to later pursue their conquests of Mexico and South America. Their avarice for gold and their ceaseless search for Eldorado, irregardless of the circumstances and cruelty they wrought on the natives, convinced the latter that the white man's god was 'gold'. The letter of King Ferdinand of Spain to his colonists in America dated July 25, 1511 would seem to bear this out for it commanded them[5] "Get gold, humanely if you can; but at all hazards get gold". The Spaniards were afflicted by the *auri sacra fames*[6] as are many individuals of our present civilization. In the pursuit of golden treasure, gods as well as men are often destroyed as in Wagner's great opera "Der Ring der Nibelungen" of which "Das Rheingold" is the prelude.

It seems certain from the accounts of Peter Martyr, Bernal Díaz del Castillo, José de Acosta and others that the Aztecs, Incas and the various peoples preceding them washed gold from placers with the gourd and *batea* (wooden pan) far back in antiquity. They probably also obtained much gold from the oxidized zones of various types of primary sulphide deposits, since they mined these extensively for silver long before the arrival of the Spaniards.

Gold has influenced the exploration and settlement of the Soviet Union, United States, Australia, South Africa and Canada in many ways. Even theories of the origin of gold deposits have been a factor in the history of Canada. It will be recalled that an early theory about the origin of gold deposits postulated that they developed under the celestial influence of the sun. From this it was easy to draw the conclusion that the largest number of gold deposits would be generated in the regions bounded by the Tropics of Capricorn and Cancer where the sun's influence was greatest. Columbus, in fact, wrote in his journal as he approached Cuba in 1492, that "From the great heat which I suffer, the country must be rich in gold." It will be further recalled that a dispute (Nootka Sound Controversy) arose between Spain and Great Britain over the sovereignty of the lands bordering the northwest coast of America. Spain contended she possessed sovereignty by authority of the Papal Bull of Alexander VI in 1493, but Britain took the view that rights of sovereignty could be obtained only through trade and the establishment of colonies. Spain meanwhile had established a settlement on Nootka Sound and in 1789 seized four British ships in the sound. This act nearly led to war but was finally resolved in favour of the British viewpoint in a convention signed on October 28, 1790. It appears probable that among the factors that influenced the Spanish decision was the advice given the Spanish king, Charles IV, that gold was unlikely to occur in the northern regions of America because it was thought that the element was generated only in those regions most influenced by the sun.

The Soviet Union has long been a legendary source of gold. The land of Colchis drained by the river Phasis (the modern Rioni) in Georgian S.S.R. is reputed by legend to have provided great quantities of gold. Similarly the Persians are said to have obtained much gold from the Scythians, a polyglot group of tribes that inhabited the region north of the Black Sea, and from various Iranian and other tribes who inhabited the Ural-Uzbek-Altai region. The golden road to Samarkand was known centuries before Christ. With time the monopoly of gold mining became the sole preserve of the Imperial Czars who pursued extensive placer and lode mining first in the Urals, beginning about 1774, and later in many parts of Siberia, especially in the Altai region where alluvial deposits were exploited as early as 1820. In 1829 the placer deposits of the Lena were first exploited and in 1840 those of the Yenisei Ridge came into production; the placers in the drainage system of the Amur were apparently first worked around 1867, and those in the Far Eastern Maritime area appear to have been first exploited around 1870 or earlier.

Gold washing and mining have a relatively long history in the United States. In New Mexico, Arizona and probably California, the Spaniards operated mines from which they obtained silver and possibly some gold at least as far back as 1620. There are also old records mentioning early discoveries of gold nuggets in California in the period 1775–1780 (Clark, 1970). Gold was also known to the Indians, Spaniards and early settlers in the southern Appalachian region where mining began about 1792 (Becker, 1895). These deposits though rich were relatively small and soon depleted. The scene

[3]*Historia naturalis*, book XXXIII.
[4]Agricola, G., *De re metallica*.
[5]Helps, A., *The Spanish Conquest of America*, London, 1857.
[6]Quid non mortalia pectora cogis,
Auri sacra fames!

Accursed thirst for gold!
What dost thou not compel mortals to do.
–Virgil-Aeneid

then turned to California, to Coloma on the American River, where in the tail race of Sutter's Mill on the morning of January 24, 1848, James W. Marshall[7] plucked the first gold nugget from the sand that led to the great California gold rush of 1849–1850. There followed then in succession the discoveries of the Mother Lode and Grass Valley in California and the famous Comstock Lode in Nevada during the 1850's. The gold telluride deposits of Cripple Creek in Colorado were discovered in 1892, and by 1905 the Tonopah and Goldfield deposits in Nevada and the Alaskan placer deposits had been discovered.

Among the miners of the great California rush of 1849 was one Edward Hammond Hargraves, an Australian. After sojourning in California for some time and noting similarities in geology with areas where he had travelled in New South Wales he returned to the latter where, near Bathurst, on February 12, 1851 he discovered payable gold at the junction of Summer Hill Creek and Lewes Ponds (Lewis Pounds) Creek a tributary of the Macquarie River. This discovery led to the great gold rushes of Australia that have made that country famous as a gold producer. Previously (1823) gold was noticed by surveyor James McBrien in the Fish River between O'Connels Plain and Diamond Swamp north of the old Bathurst Road, near where the town of Oberon, New South Wales now stands. Many of the Australian miners stayed on to discover lead, zinc, copper, silver and tin deposits; others found a different 'Golden Fleece' on the grassy plains and hills of that fair continent. The history of the 'Midas Gullies' of Australia and of Australian mining is related in fascinating detail by Blainey (1969).

Gold was known to the Maoris long before the coming of the pakehas to New Zealand. However, there is little mention of the precious metal in the journals of the early explorers of New Zealand, and it was left mainly to the prospectors who had followed the great gold rushes to California and later to Australia to establish the presence of gold in commercial amounts. Alluvial gold was first discovered in 1852 at Coromandel (Hauraki Goldfield) North Island in Driving Creek by Charles Ring; later at Collingwood in Nelson, South Island in 1857; at Gabriels Gully in Otago by Gabriel Read in 1861; and in a number of sites along the western coastline of the South Island in 1864. Lode gold in the Hauraki Goldfield, the main centres being Waihi, Thames, Karangahake and Coromandel, was first exploited in the 1860's; the quartz deposits in Otago, South Island, were first worked in the 1870's; and the productive quartz lodes at Reefton produced their first gold in the early 1870's. The history of the discovery and mining of the gold deposits of New Zealand, of which there are many, ranging in age from Precambrian to Recent, is admirably told by Salmon (1963).

Gold and gold tellurides are widely distributed in the Fijian Islands, and the presence of these minerals was evidently known to the early Fijians. Baron A.B. de Este is said to have discovered gold in the Tavua area in 1872, but some 60 years were to pass before any serious prospecting was carried out. In 1932 Bill Borthwick and Jack Sinclair discovered payable gold on Vunisina Creek a small tributary of the Nasivi River. Further investigation of this prospect, which is associated with the Tertiary Vatukoula volcanic caldera, led to the development of a number of mines, of which the famous Emperor Mine has been in continuous production since about 1935. The history of prospecting in Fiji, and the development of the Emperor and other mines, is related in some detail by Fraser (1954).

Gold from West Africa found its way into Europe as early as the 10th century and probably before. Most of this gold came by Sahara caravan to Barbary and thence to Europe, the original sources being the Kingdoms of Ghana, Mali and Songhai. It is said that much of this gold came from a region known as Wangara (probably the basin of the Falémé, a tributary of the Senegal River and noted for its placers), but it seems more than probable that the gold had a much more widespread source considering the aurificity of West Africa. In any event one of the motives of the Portuguese voyages inspired by Henry the Navigator was to ascertain and exploit the west African gold. The Portuguese were soon followed by a host of English, French, Dutch and Spanish entrepreneurs. It is thought that annually more than a quarter of a million ounces of gold reached Europe during the 15th and 16th centuries from African sources.

In South Africa an historic event took place in 1834 when a Boer, Carel Kruger, while on a hunting expedition north of the Vaal River discovered gold on the Witwatersrand, or White Waters Ridge. Little attention was, however, paid to the find, the goldfields of Barberton and the DeKaap Valley being the chief focus of gold prospecting through 1885. In 1886 George Harrison, an Australian gold-digger, and George Walker, an Englishman, discovered payable gold reef on the Witwatersrand. This discovery soon led to the development of the great reefs which constitute the largest gold deposits known and which have made South Africa the foremost gold producer in the world. The story of the discovery of the remarkable deposits of the Witwatersrand, of the growth of Johannesburg and of the men who struggled for control, not only of the gold reefs, but of what is now South Africa is told in an admirable way by Cartwright (1962), Rosenthal (1970) and Robert Crisp in *The Outlanders* (Granada Publishing Limited, Mayflower Books Ltd.; Frogmore, St. Albans, Herts.; England, 1974).

There are a few references to gold in the annals of the voyages of Cabot, Jacques Cartier and Samuel de Champlain, and it is certain that these gentlemen explorers had an eye to the metal wherever they went. The references are, however, vague and the existence of gold in some of the sites mentioned appears doubtful. Jacques Cartier for instance in his second voyage[8] mentions that the Indians informed him that the people of the Kingdom of the Saguenay possessed a great store of gold and copper. Further, in the third voyage[9] he mentions diamonds and flakes of fine gold as thick as a man's nail near Cap-Rouge, southwest of Quebec City. In Roberval's[10] account of his voyage in 1542–1543, he records that Cartier brought to St. John's in Newfoundland diamonds and a quantity of gold ore which he found in the country of

[7]Del Mar (1880, p. 165) says that it was Marshall's daughter who first discovered gold in the tail race of Sutter's Mill.

[8]Biggar, H.P.; *The voyages of Jacques Cartier,* Publications of the Public Archives of Canada, No. 11, 1924, p. 201.
[9]*Ibid.*, p. 255.
[10]*Ibid.*, p. 264.

Canada. On being tried in the furnace, the ore was found to be good (i.e., it apparently contained gold)[11].

The initial quest for gold in Canada is a story of ignorance and knavery, involving the explorer Martin Frobisher, and one, a promoter and London merchant, Michael Lok by name (Low, 1906; Rickard, 1944; King, 1955; Cooke, 1964; Bruemmer, 1966; Blackadar, 1967; Kenyon, 1975). On the first voyage in 1576 Frobisher brought Lok a dark glittering stone from Hall Island which government assayers, among them the Queen's assayer of the Tower (the Mint), declared worthless but which an alchemist, the Italian John Baptista Agnello (or Angello ?), by 'coaxing nature' found the stone to be high grade gold ore. This was sufficient to promote a second voyage in 1577 by Frobisher, who on an island (Countess of Warwick Island) in Frobisher Bay "found good store of Ore, which in the washing held golde plainly to be seene." Two hundred tons of the ore were mined and the expedition returned to England where Commissioners for the Adventurers to the Northwest for the Discovery of a Northwest Passage and others examined the material and pronounced it to be worth as much as £67 and more in gold and silver. The next year Frobisher, under the auspices of the Company of Cathai, set sail with 15 ships and mined some 1350 tons of ore from the Countess of Warwick Island and other areas in Countess of Warwick Sound before returning to England. On their return they received the unpleasant news that the cargo was worthless, contained no gold, but only pyrite (marcasite ?) and a bronze mica, probably biotite. And so the venture ended: Lok found himself in debtors' prison; Frobisher remained free but in some disfavour with the Queen and others who had financed the mining venture. All that remains of this gold venture are a few rockcuts, small quarries and trenches in the rocks on and around what is now called Kodlunarn Island and a reminder that "all that glisters is not golde."

[11]For some years I have been interested in these references and have attempted to verify the sites and the presence of gold at them. The Kingdom of Saguenay was apparently bounded by the rivers Chamouchouan, Saguenay, St. Lawrence, Ottawa and Gatineau. There is gold in some of the deposits of this area, viz. Tetreault deposit near Quebec City, and the goldfields of Quebec and Ontario lie just to the north and west. I have, however, not been able to discover evidence of any ancient workings in the Kingdom of the Saguenay, nor have I seen any evidence of placer gold. There are, however some strange references to the presence of white men inhabiting this kingdom as related by the Indians to Cartier (*Ibid.*, p. 221), and one wonders if we are not here dealing with Norsemen or Spaniards who mined or washed gold somewhere in Ontario or Quebec. I suspect that the Wanapitei Lake and Vermilion River areas of Ontario where placer gold is known may have been the sites of some of the early Norse gold washings. That the Spaniards also penetrated parts of Ontario in search of gold also seems probable to me, and this may be the reason for the portrayal on one of the maps in Cartier's journal of men who resemble Spanish miners. According to legend, the Spaniards apparently called Canada 'Capa de Nada' (Cape with nothing) signifying that they found little if any gold in the country.

As for the diamonds at Cap-Rouge this is evidently an error in mineralogy. The minerals referred to are quartz crystals as has been suggested by others. I have also been unable to find any gold at this site nor can I find any reference to gold ever having been obtained anywhere along the North Shore of the St. Lawrence near Quebec City. Cartier could, however, have obtained gold from the Indians who may have picked it up along the Chaudière River which as mentioned below is gold-bearing. The gold ore that proved at St. John's to be good seems not to have stood the test of further analysis, otherwise one would have thought that Cartier or others would have been interested in examining and exploiting the occurrence or occurrences by further sampling and mining.

The first authentic discovery of gold in Canada is related by Lieutenant, later General, W.W. Baddeley of the Royal Engineers in the American Journal of Science, v. XXVIII, July 1835, p. 112–113, as follows:

> I have to inform you of two interesting facts, which have been lately ascertained respecting Quebec and its vicinity. The first is that native gold has been picked up, about thirty miles to the southward of this metropolis; it was met with, in a small stream running into the Chaudiere, and over a region, as I suspect, (for I have not visited the *precise* locality), of talcose slate. A similar specimen was found in the same neighbourhood several years ago. That in question is of a flat ovate form, weights 10.63 grains, and has a sp. gr. of 15.7. The geological associations of this ore appear to be analogous to those of the Russian and American localities and when we consider them all, additional probability is given to the conjecture, expressed by Professor Eaton, at page 52, Vol. 18 of the American Journal of Science.
>
> It is worthy of notice, that in the neighbourhood of the place where the gold was found, two or three Canadian peasants have been mining for several years past. About the year 1825, I visited the scene of their operations, and found a shaft, ten feet cube, sunk in talcose slate, the predominating rocks at hand being of serpentine. I have lately been informed that the depth of this shaft, is now upwards of fifty feet. At the period of my visit, silver was said to be the object of search, and presuming that they had mistaken the deceptive lustre of the silvery talc for that of this metal, I endeavored to dissuade them from so ruinous a pursuit. Nothing has yet transpired as to their success, which is generally considered not to have been encouraging, and yet it is scarcely conceivable that they would persevere through so many years, without being stimulated to do so by some substantial return, a consideration which joined to what is stated above, renders it not improbable that they have met with some small deposits of gold.

The exact locality appears to have been in the Seigniory of Rigaud-Vaudreuil, the property of the heirs of Charles Etienne Chaussegros de Léry, Esq., on a small stream called the Touffe des Pins, a tributary on the right bank of the Chaudière about 58 mi from Quebec (Geol. Surv. Can., Report of Progress, 1847–1848, p. 76–77).

From the Chaudière the scene shifted to British Columbia, and specifically to that great tempestuous river, the Fraser, named in honour of Simon Fraser, that intrepid Scottish explorer who descended it to a point near its mouth in 1808. Strange as it may seem, Fraser noticed no gold in the river despite the fact that it is gold-bearing from its mouth to its source. In any event as the years rolled on, the Hudson's Bay Company began buying gold from the Indians, the largest amount being purchased by Chief Trader McLean at Fort Kamloops in 1852. It seems that much of this gold may have come from the Thompson River for James Douglas in a memorandum dated 1860 remarked:

> Gold was first found on Thompson's River by an Indian, a quarter of a mile below Nicoamen. He is since dead. The Indian was taking a drink out of the river. Having no vessel he was quaffing from a stream when he perceived a shining pebble which he picked up and it proved to be gold. The whole tribe forthwith began to collect the glittering metal.

Rickard (1942) says that this initial discovery was probably in 1857, but it would appear to have been much earlier. The same author also relates that the earliest discovery of gold in British Columbia to arouse public interest was made around 1850 by an Indian on one of the Queen Charlotte Islands. Whatever the source of the gold purchased by Chief Trader McLean it appears that it was the sale of this gold, some say 800 oz, in San Francisco that attracted miners north to the Fraser. In 1855 flour or skim gold was discovered on the bars of the lower Fraser, but this attracted little attention. Moving northward coarse gold was found on the Nicoamen River above Lytton in 1857, and by 1858 the rich bars at Yale were being worked. The news of these discoveries was followed by an extraordinary migration from the goldfields of California to those of British Columbia. Following the golden stream of the Fraser ever northward, the miners reached the fabulously rich gravels of the Cariboo. The initial discovery of this famous camp in June, 1859 is described by Art Downs in his *Wagon Road North*[12] as follows:

> In the vanguard was a party of Americans, Peter Dunlevy, Jim Sellers, Ira Crow, Tom Moffit, and Tom Manifee. At the junction of the Chilcotin and Fraser Rivers they met a huge Indian who called himself "Tomaah". He wondered what the miners were doing and they showed him the flakes and nuggets from the bars. Tomaah scoffed and said he could show them a river where gold lay like beans in a pan. But, he explained in his pidgin English, the men must wait. He was currently employed by the Hudson's Bay Company and could not yet leave his job. In a few days he would meet them and show them the river.
>
> He returned as promised and took the party to his tribal encampment. Here he introduced them to another Indian, Long Baptiste, a majestic specimen over six feet tall with broad shoulders and narrow waist. It developed that Tomaah was in love and didn't want to leave, but Baptiste would act as their guide, and in June he led the miners through the wilderness to a river now called the Horsefly. Here they found the nuggets like beans in a pan. And although the excited men weren't concerned, they also became the first white men to pan gold in a region soon to become world famous – the Cariboo.

Others sought elsewhere along the golden stream and in 1860 richer gravels were located at Quesnel Forks and on Keithley and Antler creeks. In 1861 the celebrated Williams and Lightning creeks were discovered. The golden creeks of the Cariboo ultimately lured men to all parts of British Columbia and in the years that followed gold was discovered in a hundred streams, from the United States border to Yukon and beyond.

Meanwhile, in Nova Scotia lode mining of gold held the scene, the first indications of a profitable quartz lode having been found by Lieutenant C. L'Estrange while moose hunting in 1858 in the Mooseland area on the Tangier River. This discovery ultimately led to the mining of gold from numerous deposits of the gold-bearing Meguma Group. Previous to the discovery of lode gold, Heatherington[13] points out that the native metal was probably known to the early Acadians in Cape Breton Island, hence the French names – Bras d'Or, Cape d'Or and Jeu d'Or (now corrupted into Jeddore). This could well be true since gold deposits are known at Whycocomagh and Middle River in the central part of Cape Breton Island. One, John Campbell, a resident of Dartmouth discovered alluvial gold in 1857 in the sands of Fort Lawrence in Halifax Harbour, and it seems probable that discoveries of alluvial gold were made at Isaac Harbour, Sherbrooke, Ovens and elsewhere in Nova Scotia as early as 1830 (Malcolm, 1912, p. 3–4).

In the Canadian Shield, lode gold was first discovered in 1866 near Madoc, Hasting County, Ontario by a part-time prospector, Marcus Herbert Powell. This later became known as the Richardson mine from the name of the owner of the farm on which the gold was discovered.[14]

Other discoveries were made in the years following in the same area, and in northwestern Ontario near Port Arthur and in the Lake of the Woods area. Following the discovery of silver at Cobalt in 1903, prospectors ranged widely over the Precambrian areas of Ontario, Quebec and Manitoba. In Ontario and Quebec, Abitibi and Larder Lake were discovered in 1906; Porcupine, 1909; Swastika, 1910; Kirkland Lake, 1911; Matachewan, 1916; Rouyn (Noranda), 1924; and Red Lake, 1925. In Manitoba the Rice Lake district was discovered in 1911, and in Northwest Territories, the deposits in the sediments of the Yellowknife area were discovered in 1933 and those in the greenstones in 1935.

But we must go back more than 35 years to one of the greatest gold rushes of history – that of the Klondike in Yukon.

Discoveries of gold were reported on the Yukon River as early as 1869, and further important placers were located in 1878 on tributaries of this legendary river. It was not until 1896, however, that the fabulously rich placers of the Klondike River, a right bank tributary of the Yukon were located. Pierre Berton,[15] in his book *Klondike,* records the initial discovery thus:

> Carmack tried the prospects at Gold Bottom, but did not stake, and the trio headed back over the mountain almost immediately. The way was hard. They struggled over fallen trees and devil's clubs, a peculiarly offensive thorn, and they forced their way through interlaced underbrush, brier roses, and raspberry bushes. On the far side of the mountain they floundered into a niggerhead swamp that marked the headwaters of Rabbit Creek, and here they had to hop from clump to clump on their slippery moccasins or sink to their thighs in the glacial ooze. Hordes of gnats and mosquitoes rose about them as they stumbled on, unable to swat the insects for fear of losing balance.
>
> Thus they came wearily to the fork of Rabbit Creek once more, and pressed on for about half a mile before making camp for the night. It was August 16, the eve of a memorable day that is still celebrated as a festive holiday in the Yukon Territory.
>
> Who found the nugget that started it all? Again, the record is blurred. Years afterward Carmack insisted it was he who happened upon the protruding rim of bedrock from which he pulled a thumb-sized chunk of gold. But Skookum Jim and Tagish Charley always claimed that Carmack was stretched out asleep under a birch tree when Jim, having shot a moose, was cleaning a dishpan in the creek and made the find.
>
> At any rate, the gold was there, lying thick between the flaky slabs of rock like cheese in a sandwich. A single panful yielded a quarter of an ounce, or about four dollars' worth. In a country where a ten-cent pan had always meant good prospects, this was an incredible find. Carmack flung down the pan and let out a war whoop, and the three men began to perform a wild dance around it – a sort of combination Scottish hornpipe, Indian fox trot, syncopated Irish jig, and Siwash hula, as Carmack later described it. They collapsed, panting, smoked a cigarette apiece, and panned out some more gravel until Carmack had gathered enough coarse gold to fill an empty Winchester shotgun shell. Then they settled down for the night, the Indians chanting a weird song of praise into the embers of the fire while Carmack, staring at the dying flames, conjured up visions of wealth – of a trip around the world, of a

[12]Downs, Art; *Wagon Road North,* the story of the Cariboo gold rush in historical photos; Northwest Digest Ltd., Quesnel, B.C., 1960.

[13]Heatherington, A; *Goldfields of Nova Scotia,* John Lovell, Printer, Montreal, 1868, p. 20.

[14]For an interesting piece of Canadiana about prospecting for gold and other metals in the Madoc gold area and other places in Canada, United States and British Columbia see, *The Gold-Seekers Handbook and Practical Assayist,* published by S. Vivian, Bookseller, Napanee, Ontario, 1867.

[15]Berton, Pierre; *Klondike*, McClelland and Stewart, Limited, Toronto, 1958.

suburban mansion rimmed with flower borders, or a suitcase full of gilt-edged securities. In that instant of discovery something fundamental had happened to Siwash George: suddenly he had ceased to be an Indian. And he never thought of himself as an Indian again.

The following morning the trio staked claims on Rabbit Creek. Under Canadian mining law, no more than one claim may be staked in any mining district by any man except the discoverer, who is allowed a double claim. Carmack blazed a small spruce tree with his hand ax, and on the upstream side wrote with a pencil:

TO WHOM IT MAY CONCERN

I do, this day, locate and claim, by right of discovery, five hundred feet, running up stream from this notice. Located this 17th day of August, 1896.

– G.W. Carmack

While Carmack made the discovery which led to the historic rush to the Klondike he was not the true discoverer of the great gold placers. Rather, one Robert Henderson staked the first claims on the gold creeks of the Klondike, specifically on what is now called Gold Bottom Creek. The sequence of events is told in a letter to the Engineering and Mining Journal Press (February 17, 1923, p. 305–306) by A. E. Wills and corroborated by Henderson himself in a letter to the same journal (June 9, 1923, p. 1009) in which he states:

> George Carmack and the Indians were camped at the mouth of the Klondike, and I gave him the information, asked him to come up, and told him the easiest route to take to reach my discovery. They came, and it was on their way back to the mouth of the Klondike that they made their discovery on Rabbit Creek (now called Bonanza Creek). Carmack and his two Indians staked next to me on Gold Bottom Creek, but I believe they did not record the Gold Bottom staking.
>
> My discovery on Gold Bottom was made in July, 1896. I had no calendar, but as nearly as I could determine the date it was the fifteenth of July.

Green (1977) in his recent book *The gold hustlers* has chronicled the history of the Klondike from its discovery, through its hey-day, to the present. Today only small operators work the placers, the days of the great dredges being past, at least until there is a marked increase in the price of gold.

We see that gold has had a long history, and we shall see in what follows that gold is many things to many men. To the geochemist it is a rare metal, the geochemistry of which is intricate and complex; to the mining engineer and metallurgist it presents a challenge of extraction from the earth and from its ores; to the artist, goldsmith and jeweller it is a metal of superb and everlasting beauty; to the industrial artist it is a metal with unique properties useful in electronics and many other artifacts of man; to the numismatist it is a coinage metal with a long and interesting history; and finally to the economist a valuable standard against which wealth is measured and an imperishable medium for balancing international accounts. In literature the word gold and its derivatives appears more often than any other word, creating legends, tinging metaphor and simile and providing innumerable parables, analogies and proverbs.

The term *gold* is said by the scholars to come from the Sanskrit *jvalita* derived from *jval,* to shine. Our word *gold* derives from the Anglo-Saxon, *gold,* a word apparently corrupted from the Teutonic, *gulth,* the glowing or shining metal. The Latin term for gold, *aurum,* and the earlier Sabine *ausum* are said to be words of early Italian origin related to *aurora* meaning the glowing dawn. Another version has it that the Latin word, *aurum,* derives from the Hebrew, *aor,* meaning light. The Latin term is preserved in the chemical symbol for gold, Au, and in the terminology of its salts, *aurous* and *auric.*

The Egyptians used the most perfect of planar geometric figures, the circle, as the symbol for gold, the most perfect and noblest of the metals. The alchemists associated gold with *Sol* (the sun) or with the Greek sun-god Apollo and represented it by the symbol of perfection, the circle with a dot at the centre, or by the circle with a crown of rays to represent the king or Apollo of metals. To the early Hindu philosophers gold was the 'mineral light'; to the early Western philosophers the metal was the image of solar light and hence of the divine intelligence of the universe.

There is an extensive bibliography on the history of gold in Gmelin (1950).

Acknowledgments

Colleagues in the Geological Survey and students in Carleton University contributed various field and laboratory data on the geochemistry of gold. Particular thanks are due J.J. Lynch and his colleagues, S. Abbey and his colleagues and Ms. G. Aslin who provided much of the analytical data recorded in this bulletin. H. Steacy was particularly helpful in the selection and preparation of pure minerals from the National Mineral Collection in which gold (and silver) were determined, and A.G. Plant very kindly provided electron probe data on the constitution of native gold from numerous deposits in Canada and elsewhere. P.J. Lavergne carried out innumerable mineral separations during the research project. Critical field and analytical data were provided by T. Presant on soils, A.Y. Smith on sandstones and coal, S.S. Rajah on gossans and I.R. Jonasson on a number of types of earth materials. Samples of gold ores and hot spring precipitates from the various Recent and Tertiary auriferous deposits of New Zealand were provided by B.W. Robinson, Institute of Nuclear Sciences, Lower Hutt, New Zealand.

Many of the gold and silver analyses referred to in this bulletin were done in the laboratories of Bondar-Clegg and Co. Limited, Ottawa, Canada and the Canada Centre for Mineral and Energy Technology. The dedication of the personnel of these laboratories in the analytical work and their advice on precision and various other details concerning the analytical chemistry of the precious metals is greatly appreciated.

The extraction of data on gold from the world literature has been a formidable task which I have shared with many individuals through the years particularly with Mrs. P. Grove, Mrs. Alice P. Leech, Mrs. Eileen F. Foster, Rani Arora, Robin L. Monroe, A.G. Douma and B. O'Heare. Thanks are also due C. de Leuchtenberg who assisted greatly in the translation of a large number of Russian papers, and repeatedly brought to the writer's attention valuable information in the foreign literature which otherwise would have gone unnoticed. A number of references in the Russian literature, not available to me, were kindly provided by V.V. Shcherbina of the Vernadsky Institute of Geochemistry and Analytical Chemistry, Moscow; for these I am most grateful. D.R. Boyle, Applied Chemistry Research Group, Royal School of Mines, London, England brought to my attention numerous theses on geochemical prospecting for gold and its associated elements.

During my research on gold deposits I have visited and

worked in many mines throughout the world and in places have had access to various private company reports, assays, bullion returns, etc. Everywhere, I have been courteously treated for which I thank collectively the many mine managers and geologists with whom I have come in contact. I hope this book will serve them well in their further search for gold.

I wish also to record here the many communications both written and verbal that I have had through the years with numerous individuals who are particularly knowledgable about the geology and chemistry of gold, especially Frank Ebbut, H.S. Bostock, J.F. Henderson, C.F. Gleeson and I.R. Jonasson.

R.G. Garrett and A.G. Douma assisted in the statistical work recorded in this bulletin and helped in many other ways. To them I am most grateful. I wish also to thank R.J.W. Douglas, R.I. Thorpe, I.R. Jonasson, R.G. Garrett and J.J. Lynch who read parts of the manuscript and made numerous valuable suggestions for improvement. And finally I wish to express my gratitude to my wife, whose help in the field and in the referencing, reading and correction of the manuscript has been more valuable than fine gold itself.

The quotation about the discovery of the Cariboo is included with permission of Art Downs from his *Wagon Road North* published by Northwest Digest Ltd., Quesnel, B.C., and the quotation about the discovery of the Klondike is published by permission of Pierre Berton and McClelland and Stewart Limited, Toronto from the book *Klondike.* To both these men and their publishers I extend my sincere thanks. The golden cups of Vaphio are reproduced by permission of the Director, E. Thanos, National Archaeological Museum, Athens, Greece and the coloured photograph of the remarkable gold specimen from the Timmins (Porcupine) gold belt of Ontario was provided by the Mineralogy Department, Royal Ontario Museum, Toronto.

A note about the references: Malcolm Maclaren, in his classic treatise on gold written in 1908, remarked that the literature on gold was stupendous. Nearly 70 years later there is no superlative adjective adequate to describe the amount of literature available on the element. In the Selected bibliography I have attempted to include all of the papers and books published up to the end of 1977 that one would require in researching the details of the features of the gold-bearing belts of the world together with those on the geochemistry of the element. I extend my apologies to those whose papers could not be included mainly because of space considerations. Where possible, references to English abstracts of foreign publications appearing in Chemical Abstracts and other abstract journals are included with the original references.

Gold is characterized chemically by an extreme indifference to the action of all bodies usually met with in nature.
– T.K. Rose and W.A.C. Newman, 1937

Chapter II. Chemistry, mineralogy and geochemistry of gold

General chemistry

Gold, Au: atomic no. 79; atomic wt. 196.967; specific gravity 19.32 (20°C); melting point 1063.0°C; valence 1,3.

Comprehensive reviews of the chemistry and physical metallurgy of gold are contained in the works by Mellor (1923), Rose and Newman (1937), Gmelin (1950), Sidgwick (1950), Pascal (1957), Emery and Leddicotte (1961), Badone and Spence (1961), Pourbaix *et al.* (1963), Johnson and Davis (1973) and Myerscough (1973). The summary that follows deals principally with the chemical features of gold as they affect the geochemistry of the element.

Gold is a member of Group IB of the periodic system, which includes copper (Cu), silver (Ag), and gold (Au). Except that the so-called 'coinage metals' exhibit monovalency (oxidation state +1), there is little resemblance to the alkali metals of Group IA. In fact, the differences between the two subgroups are marked, as shown for instance by the general decrease in reactivity of the Group IB elements with increase in atomic number as contrasted with the reverse behaviour of the Group IA elements. As might be expected from electron orbital considerations, there is some similarity between Ni, Pd and Pt (Group VIII) and the corresponding elements of Group IB, namely Cu, Ag and Au. Thus, platinum has only one electron less than gold, and since the energy differences are small in the electronic constitution of the two elements, the differences in the chemical behaviour of the two elements in a few respects are slight. In natural environments, however, there is no close relationship between gold and platinum except that they tend to occur in the native state, and some platinum may be present in certain varieties of native gold and vice versa. There are no particular similarities between gold and mercury, the corresponding element in Group IIB of the periodic system. As is well known, however, gold and mercury form an amalgam, and well-defined compounds occur in the system Au-Hg; natural amalgam is known, but formal Au-Hg compounds do not appear to occur in nature, or if they do they have not been adequately characterized. There are marked differences in the chemistry of silver and gold, a consequence that is attributable to the diminishing effect of the lanthanide contraction between atomic numbers 57 (La) and 79 (Au).

Gold is considered by some to be a transition element of the third series, although it does not fulfil the necessary

Table 1. Energetic properties of copper, silver and gold

Property	Copper	Silver	Gold
Outer electron configuration	$3d^{10}4s^1$	$4d^{10}5s^1$	$5d^{10}6s^1$
Oxidation states (valence) (Those in brackets are uncommon)	1, 2, (3)	1, 2, (3)	1, 3
Electronegativity	1.9	1.9	2.4
Ionization potential, eV	Cu(I) 7.724 Cu(II)20.29	Ag(I) 7.574 Ag(II)21.48	Au(I) 9.223 Au(III) ~30
Lattice energy coefficient	Cu(I) 215 Cu(II)630	Ag(I)180	Au(I) ~183
Atomic radius, Å (co-ordination in brackets)	1.24 (8) 1.28 (12)	1.40 (8) 1.44 (12)	1.40 (8) 1.44 (12)
Ionic radius, Å (co-ordination in brackets)	Cu(I) 0.96 (6) Cu(I) 1.00 (8) Cu(II) 0.82 (6)	Ag(I) 1.13 Ag(I) 1.13 (10) Ag(II) 0.97 (6)	Au(I) 1.37 Au(I) 1.37 Au(III) 0.78 (4) sq. Au(III) 0.91 (6)
Miscellaneous radii, Å	Cu(I) (in fluorides) 1.00 Cu(I) (in oxides) 1.05 Cu(II) (in fluorides) 0.87 Cu(II) (in oxides) 0.97 Cu (tetrahedral covalent) 1.35 Cu(I) (hydrated) 1.7 Cu(II) (hydrated) 3.5	Ag(I) (in fluoride) 1.33 Ag(I) (in oxides) 1.35 Ag (tetrahedral covalent) 1.53 Ag(I) (hydrated) 1.7	Au(I) (hydrated) 1.4
Ionic potential, $\frac{Z}{r}$	Cu(I) 1.00 Cu(II) 2.44	Ag(I) 0.88	Au(I) 0.73 Au(III) 3.30

Sources: Szádeczky-Kardoss (1958); Green (1959); Samsonov (1968); Whittaker and Muntus (1970).

Note: The values given for the various radii are approximations at best. Those chosen appear to be the most satisfactory from the viewpoint of the geochemistry of gold, but as is well known all radii should be used with caution, since the radius of an element varies widely depending on the structure of the substance containing the element, the co-ordination configuration of the element and other factors.

electronic conditions[16] that characterize transition elements. Gold is a soft, yellow metal with a moderately high melting point (1063.0°C) and a high electrical and thermal conductivity exceeded only among the elements by silver and copper. It is the most malleable and ductile of metals; 1 oz of gold can be beaten into foil some 300 ft² in area. Of all the elements, gold in its pure state is undoubtedly the most beautiful, especially when it occurs in crystalline groupings or in dendritic and arborescent arrangements.

The energetic properties of gold, together with those of copper and silver, the two elements with which it is so often associated in nature, are given in Table 1. The thermodynamic properties of the element, its ions and compounds are given in Table 2. The Pourbaix (Eh-pH) diagrams for some of the common gold species are shown in Figure 1, and the cycle of gold interconversions under natural conditions is displayed in Figure 2. Additional data can be found in the publications by Kelley (1960), Wicks and Block (1963), Sillén and Martell (1964), Samsonov (1968), Gedansky and Hepler (1969), Hancock and Finkelstein (1971), Finkelstein and Hancock (1974), and Kubaschewski and von Goldbeck (1975). The vapour pressure of gold has been recently discussed in detail by Paule and Mandel (1972). The vapour pressure of the gold standard is 9.92×10^{-9} atm at 1300°K increasing to 1.87×10^{-3} atm at 2100°K. The heat of sublimation of gold is given as 87 720 ± 210 cal/mole at 298°K; the corresponding value for silver

[16]Strictly speaking, the transition elements are characterized by an inner *d* or *f* level, which is not filled to capacity, e.g., nickel. In gold the 5*d* orbital is completely filled.

Table 2. Thermodynamic data on gold

Formula	State	ΔH°	ΔF°	S°
Au (metal)	c	0.0	0.0	11.32
Au(I)	aq		39.0	
Au (III)	aq		103.6	
Au_2O_3	c	19.3	39.0	30.0
$H_2AuO_3^-$			–45.8	
$HAuO_3^{2-}$	aq		–27.6	
AuO_3^{3-}	aq		–5.8	
$Au(OH)_3$	c	–100.0	–69.3	29.0
$Au(OH)_3$	aq		–61.8	
AuCl	c	–8.4	–4.2	(24.)*
$AuCl_3$	c	–28.3	–11.6	(35.)
$AuCl_3 \cdot 2H_2O$	c	–167.7	–123.3	(54.)
$AuCl_4$	aq	–77.8	–56.2	61.
AuBr	c	–4.4	–3.7	(27.)
$AuBr_2^-$	aq		–27.1	
$AuBr_3$	c	–13.0	–5.9	(24.)
$AuBr_3$	aq	–9.2		
$AuBr_4^-$	aq	–45.5	–38.1	75.
AuI	c	0.2	–0.76	(28.5)
Au_2P_3	c	–24.1		
$Au(CN)_2^-$	aq	58.4	64.4	29.5
$Au(CNS)_2^-$	aq		57.7	
$Au(CNS)_4^-$	aq		130.1	

Sources: Latimer (1952); Handbook of chemistry and physics, 53rd ed.; Chemical Rubber Co., Cleveland, Ohio, 1972.

Notes: ΔH° – the standard heat of formation in kilocalories of the substance or ion from its elements at 25°C

ΔF° – the standard free energy of formation in kilocalories of the substance from its elements at 25°C

S° – the entropy of the substance or ion in cal/degree in its thermodynamic reference state at the reference temperature of 25°C

c – crystalline

aq – in aqueous solutions

*Values in parentheses are estimated.

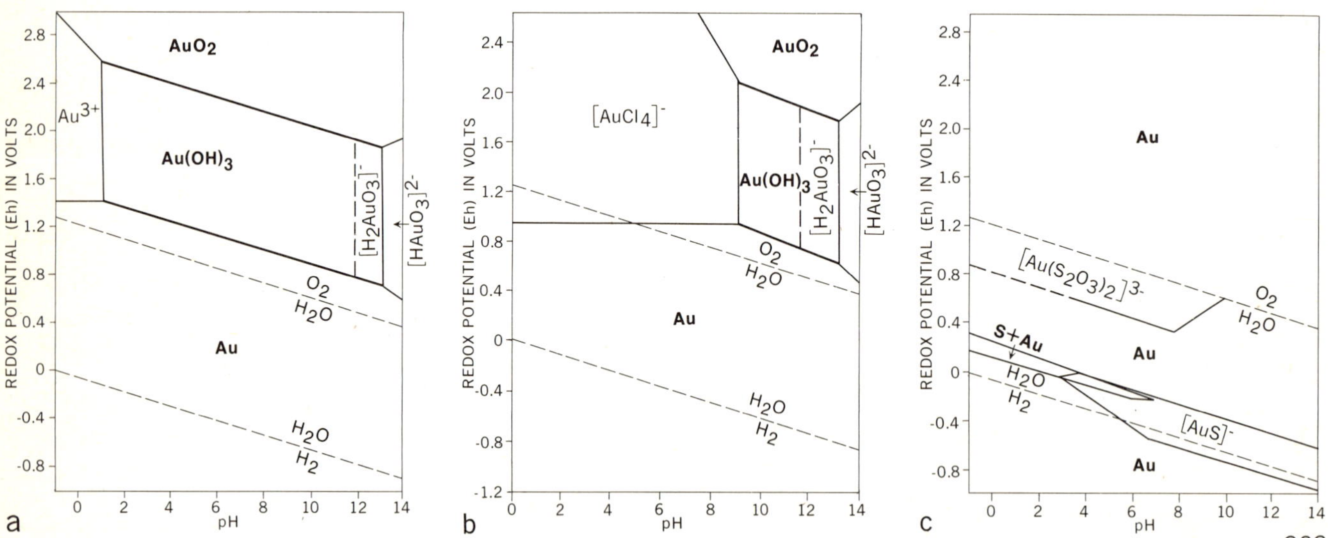

Figure 1. Pourbaix (Eh-pH) diagrams for gold. (a) System Au-H_2O at 25°C and 1 atm total pressure; concentrations of all gold species 10^{-4}M. (b) System Au-H_2O-Cl at 25°C and 1 atm total pressure; concentration of Au(III) = 10^{-2}M, $C1^-$ = 2M. (c) System Au-S at 25°C and 1 atm total pressure; concentration of all sulphur species = 10^{-1}M. [Solid species are shown in bold type, dissolved species in lighter type. For details of the thermodynamics of the solution chemistry of gold see the works by Latimer (1952), Pourbaix *et al.* (1963), Tyurin (1965*b*), Garrels and Christ (1965), Goleva *et al.* (1970*b*), Hancock and Finkelstein (1971) and Finkelstein and Hancock (1974).]

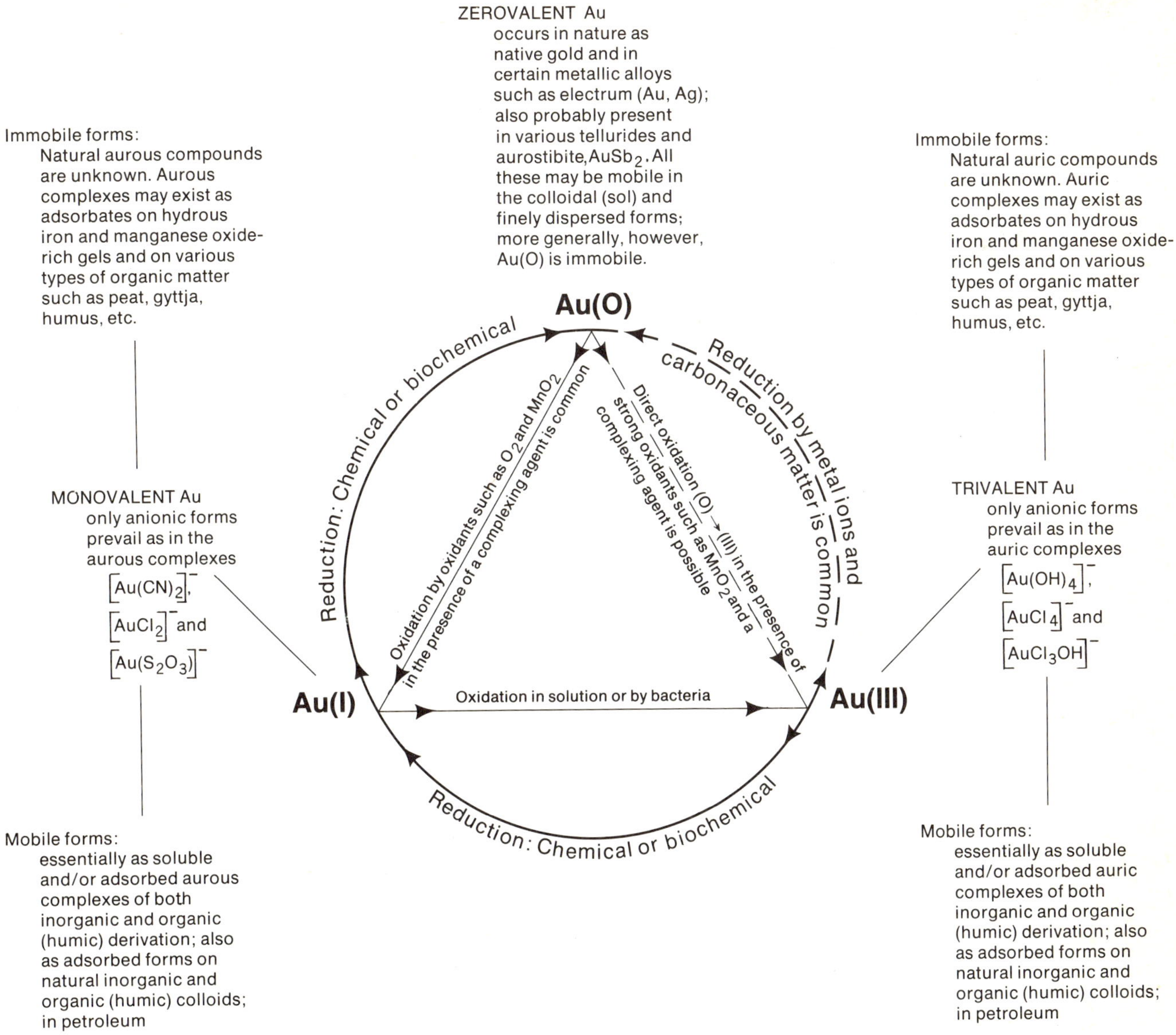

Figure 2. Cycle of gold interconversions in nature.

is 68 010 ± 300 cal/mole. These figures are of interest in natural systems in that they indicate that silver is the more mobile of the two elements at high temperatures.

Gold is usually found in one of three oxidation states: 0 (native), +1 (aurous) and +3 (auric) designated further as Au, Au(I) and Au(III) in this volume. Au(II) and Au(V) states are also known to occur in certain complex arrangments (Bergendahl, 1975). These appear to be relatively unstable states and probably do not occur in nature, except possibly in some organic-rich or chloride-rich natural waters. Gold in the native state is the most noble of metals. It is chemically unreactive and is not affected by water or most acids, nor attacked by oxygen or sulphur. The halogens react readily with gold as do also solutions that generate them such as aqua regia. Gold dissolves readily in alkali cyanide solutions in the presence of air, sparingly in alkali sulphide solutions, readily in selenic acid at about 225°C, readily in solutions containing telluric and sulphuric or phosphoric acid, readily in sulphuric and hydrochloric acid solutions containing an oxidizing agent such as MnO_2, and sparingly in alkali thiosulphate solutions. The element is probably sparingly soluble in solutions containing HCl and $CuCl_2$ in the presence of an oxidizing agent; also sparingly soluble in ferric chloride and ferric sulphate solutions containing an oxidizer. Finely divided gold is said by some investigations to be slightly soluble in alkali hydroxides. These and other reactions are discussed in the sections on natural waters, soils and oxidation processes in gold-bearing deposits. Mercury dissolves or amalgamates about 0.7 per cent by weight of gold at 25°C, and molten tellurium forms an intermetallic compound with the element.

Crystallographically gold belongs to a group that includes gold, silver, copper and lead. All of these metals are isometric, hexoctahedral – $4/m\ \bar{3}\ 2/m$ – and have the copper-type structure with the atoms lying at the points of a face-centred cubic lattice. This structure conforms to the space group F*m* 3*m*, and the unit cell contains 4 atoms. The cell edges for the four pure metals are as follows:

	Cu	Au	Ag	Pb
a_0Å	3.61496(18°C)	4.07825(25°C)	4.08548(20°C)	4.9496(20°C)

The atomic radius of gold is identical to that of silver (1.44Å), a feature that permits a continuous alloy series with a minimum cell edge at about 20 per cent silver. In nature, however, the series appears to be discontinuous. There is much less similarity in the atomic radii of gold (1.44Å) and copper (1.28Å), and as a consequence most samples of native gold contain relatively small amounts of copper (up to 20% Cu). Native copper, likewise, usually contains only traces of gold. In the laboratory gold forms alloys with a large number of metals, among which the most important are Ag, Cu, Al, Cd, Hg, Co, Pb, Fe, Ni, Pd, Pt, Sn, Ga, Ca, In and Zn. Some of these elements, such as Ag, Cu, Ni, Pd and Pt, are miscible with gold in all proportions. Details of these alloys are given by Badone and Spence (1961), Wise (1964), and Raynor (1976). Natural gold invariably contains some of the elements noted above, particularly Ag, Cu, Al, Fe, Bi, Sn, Pb, Zn, Pd and Pt. The colour of natural gold alloys is largely dominated by silver. With increasing content of silver the golden colour is attenuated until at about 65 per cent silver the colour of the alloy is barely distinguishable from pure silver.

The effect of mercury on gold has been known for centuries. It is usually said that mercury 'dissolves' gold, but this is an error that has crept into the literature and has been repeated for many years (Brown, 1960). In actual fact the solubility of gold in mercury is slight, only 0.13 atomic per cent at 20°C. The amalgamating effect of mercury is essentially the result of mechanical mixing, wetting and surface reactions between gold and mercury, about which we know little. On the other hand gold forms an alloy with mercury, the maximum solid solubility of Hg in Au being approximately 35 atomic per cent. At least three artificial gold-mercury compounds are known in the system Au-Hg – Au_3Hg, a second compound somewhat richer in gold than Au_3Hg and an indefinite compound near the composition Au_2Hg (Rolfe and Hume-Rothery, 1967; Berndt and Cummins, 1970). Lindahl (1970) also synthesized a phase Au_6Hg_5 at 25 and 80°C. Other phases, such as $AuHg_2$, Au_2Hg_3 and Au_5Hg_8, also seem to be stable in the system Au-Hg (Shklyar and Kakovskii, 1972). A compound close to Au_2Hg_3 and called gold amalgam occurs in nature. Its structure and composition require further work.

In its compounds gold as Au(I) (aurous) and Au(III) (auric) has the electronic configuration of d^{10}-type and d^8-type respectively and exhibits mainly a covalent character because of the high ionization potentials of the two states of oxidation. This is particularly so in its natural compounds such as the tellurides and the antimonide, $AuSb_2$ (aurostibite). There once was some slight evidence for the existence of Au(II) in certain gold-organic complexes, and compounds with empirical compositions such as AuS and $AuCl_2$ might suggest a (II) oxidation state. It is, however, now generally agreed that compounds such as AuS contain equal numbers of Au(I) and Au(III) species and should be characterized as mixed compounds written as $Au(I)_2S{\cdot}Au(III)_2S_3$. The compound $Cs_2Au_2Cl_6$ also appears to contain gold in both the (I) and (III) states rather than in the (II) state. The possible transitory (catalytic) existence of Au(II) in redox reactions involving Fe^{2+} is discussed by Rich and Taube (1954).

The aqueous chemistry of gold is essentially that of complex ions since the ions Au^{+1} and Au^{+3} are unstable in water because of their high oxidation potentials (Fig. 1). Both Au(I), d^{10}, and Au(III), d^8, are B-type metal ions, which means that the stability of their complexes tend to decrease with increasing electronegativity of the ligand donor atom. Thus, the stability order for the halogen complexes is $I^- > Br^- > Cl^- > F^-$ where the order of electronegativities is $F^- > Cl^- > Br^- > I^-$. Similarly the stability of the various complexes with the donor atoms in Groups VA and VIA fall in the order Bi > Sb > As > P and Te > Se > S > O. This explains why gold tellurides, gold bismuthide and the antimonide (aurostibite) are stable enough to occur as minerals whereas we do not find natural gold selenides (*sensu strictu*), sulphides, oxides and arsenides. Nevertheless there is a bond between gold and selenium and arsenic since the precious metal is often enriched in deposits containing abundant selenides, arsenides and their oxidized derivatives. Smirnov and Entin (1975) have also noted a bond between gold and phosphorus in genetically diverse phosphatic deposits.

Au(I) generally exhibits linear twofold co-ordination, although tetrahedral fourfold co-ordination is also observed in some aurous substances. Most aurous compounds that are not complexes are insoluble with the exception of AuCN. Aurous oxide, Au_2O, and aurous hydroxide, AuOH, are mentioned in the literature, but recent research throws some doubt on their existence. Aurous chloride, AuCl, aurous bromide, AuBr, and aurous iodide, AuI, can be prepared. The chloride slowly disproportionates in water to yield metallic gold and auric chloride. The bromide and iodide behave in a similar manner, although not as rapidly. Aurous fluoride has not been prepared, probably because it is unstable in the presence of water. Aurous chloride dissolves in alkali chloride solutions to form the chloroaurite (I) complex ion $[AuCl_2]^-$. Aurous bromide and iodide behave in a similar manner in solutions containing an excess of bromide or iodide. Aurous cyanide, AuCN, can be easily prepared. It dissolves in alkali cyanide solutions to form the aurous cyanide complex $[Au(CN)_2]^-$. Aurous sulphide, Au_2S, is the most stable of the sulphides of the element. It is insoluble in water ($K_{sp} = <10^{-13}$) but may form a colloid; it is insoluble in most acids, soluble in strong oxidizing agents such as aqua regia, soluble in alkali cyanide solutions, and soluble in alkali polysulphide solutions to give thioaurite(I) compounds of the type Na[AuS]. Aurous polysulphides characterized as $Na_4[Au_2S_3]$, $K_3[AuS_2]$ and $K_5[AuS_3]$ are obtained by the action of polysulphide solutions on aurous sulphide and other aurous and auric compounds in solution. These polysulphides are slightly soluble in water. Acidification of a solution containing thioaurite(I) compounds or aurous polysulphides yields aurous sulphide. Treatment of an acidified solution of $K[Au(CN)_2]$ with hydrogen sulphide, likewise, throws down Au_2S. Aurous thiosulphates of the type $Na[Au(S_2O_3)]$ can be prepared, but they are not as stable as the fourfold complex $Na_3[Au(S_2O_3)_2]$. Aurous thiosulphate is very soluble in water giving complex ions of the type $[Au(S_2O_3)]^-$ and $[Au(S_2O_3)_2]^{3-}$. The later complex is apparently of the open chain type commonly written as $[O_3S{\cdot}S\text{-}Au\text{-}S{\cdot}SO_3]^{3-}$. By loss of sulphur, complex sulphites of the form

$[O_3S\text{-}Au\text{-}SO_3]^{3-}$ are derived. Oxy-salts of gold(I) are uncommon. There is no simple carbonate or nitrate, and the existence of a sulphate is doubtful. The gold(I) thiocyanate, AuSCN, can be prepared. It is highly insoluble in water.

Summarizing, we observe that gold(I) is characterized by numerous complexes that are stable in aqueous solution. Most of these are monomeric in character, familiar examples being $[AuCl_2]^-$, $[Au(CN)_2]^-$ and $[Au(S_2O_3)]^-$. Crystalline compounds of the type $K[Au(CN)_2]$ can be obtained, and the acid $H[Au(CN)_2]$ can be isolated. Numerous complexes of Au(I) with substituted sulphur, arsenic and antimony ligands are also known. Some of these such as $[AuS]^-$, $[AuS_2]^{3-}$, $[Au_2S_3]^{4-}$, $[AuSb]^{2-}$ and the thioarsenito and thioantimonito complexes, of the type $[Au(AsS_3)]^{2-}$ and $[Au(SbS_3)]^{2-}$, are probably of great importance in the transport of gold in endogenic processes in nature.

Gold(I) also forms a number of organometallic compounds (alkyls, aryls, olefins, etc.) and chelates, but these are much too numerous and complicated to be discussed here. The interested reader should consult the summary by Johnson and Davis (1973) and the specialized reviews by Jack and Powell (1972), Nesmeyanov *et al.* (1974) and Johnson (1976). The organometallic aurous compounds are as a rule not as stable as the corresponding auric compounds. Some of the aurous organometallic compounds such as the phosphine, arsine, mercaptide and methylated complexes and chelates, may occur in natural settings, especially those rich in humus; none of these compounds have, however, been characterized.

Gold(III) commonly exhibits fourfold planar co-ordination, although higher co-ordination numbers, fivefold trigonal bipyramidal and sixfold octahedral, are possible. Some of the complexes of Au(III) are isomorphous with those of Cu(III) and Ag(III). Most of the auric compounds are powerful oxidizing agents. Gold(III) hydroxide, $Au(OH)_3$ is obtained by the addition of alkali hydroxide to solutions containing chloroaurate ion, $[AuCl_4]^-$. The hydroxide is probably not a true hydroxide but rather a hydrated acid hydrate of some form. The hydroxide can be dehydrated to give auric acid AuO(OH) and ultimately the oxide Au_2O_3. Both these oxides have amphoteric properties; in acidic solutions they give complexes of the type $[AuCl_4]^-$ and $[Au(SO_4)_2]^-$, and in basic solutions complexes characterized as $[Au(OH)_4]^-$ or $[AuO_2]^-_{aq}$ are stable. The last two are commonly referred to as aurate complexes, the corresponding salts such as, $KAuO_2 \cdot 3H_2O$, being designated aurates. Gold(III) hydroxide when treated with strong alkalies yields another series of aurates characterized by the ion AuO_3^{3-}; salts such as Na_2HAuO_3 and Na_3AuO_3 have been prepared.

Gold(III) forms compounds with all of the halides. Auric fluoride is not stable in water and is rapidly decomposed. The chemistry of the chloride, bromide and iodide is similar. Auric chloride and bromide are soluble in water, in which they are partly hydrolyzed to give species such as $[AuCl_3OH]^-$; excess chloride or bromide ion yields the complex chloraurate or bromoaurate ions, $[AuCl_4]^-$, $[AuBr_4]^-$. Chloroauric acid, $HAuCl_4 \cdot 4H_2O$ and the corresponding bromine compound can be isolated, and well defined salts such as $KAuCl_4$ and $Ca(AuCl_4)_2$ are known. The stability of the auric halide complexes increases in the order I>Br>Cl>F. The phase diagram of the gold-chlorine system is discussed by Janssen *et al.* (1976). Two compounds occur in the system – AuCl and Au_2Cl_6.

Gold(III) salts of oxyanions are not particularly stable, but solutions containing complex auric nitrates, sulphates, etc., e.g. $[Au(NO_3)_4]^-$, $[Au(SO_4)_2]^-$ can be prepared. Gold selenate, $Au_2(SeO_4)_3$, and gold selenite, $Au_2(SeO_3)_3$, are relatively stable.

Gold(III) cyanides of the type $K[Au(CN)_4] \cdot xH_2O$ are known. In solutions containing excess cyanide they are said to form the various complex ions $[Au(CN)_4]^-$, $[Au(CN)_5]^{2-}$ and $[Au(CN)_6]^{3-}$. Gold(III) thiocyanates are known; the most common of these $K[Au(SCN)_4]$ is insoluble in water.

Gold(III) sulphide, Au_2S_3, can be prepared by the action of H_2S on auric salts, but it is readily reduced in water, yielding Au_2S and sulphur. It is soluble in alkali sulphide and polysulphide solutions to give thioaurate(III) compounds, but these are unstable and degrade to the more stable thioaurite(I) species.

Gold(III) forms numerous inorganic complexes, some of which have been mentioned above. Most of these are anionic complexes. Relatively little is known about soluble cationic and neutral gold complexes but Beneš (1967) and Beneš and Riedel (1967) consider that in strongly acid chloride solutions gold exists prevailingly in the form of undissociated $HAuCl_4$. With increasing pH, hydrolysis occurs with the formation of positively charged hydrolytic products (pseudocolloids). Gold(III) also forms a host of organometallic compounds and chelates that are too numerous and complicated to be discussed here. They are reviewed by Jack and Powell (1972), Johnson and Davies (1973) and Nesmeyanov *et al.* (1974). Some of these compounds and complexes may be of importance in the geochemistry of gold in organic (humic) environments, but as yet none have been characterized.

Gold forms a large number of complex compounds in which the state of oxidation of the element is multiple or uncertain. Reference to the sulphide AuS has already been made; it appears to be a compound of Au_2S and Au_2S_3. The salt, $Na[AuS] \cdot H_2O$, has been isolated. There are a number of complex double sulphides, including Ag_3AuS_2, AgAuS and Ag_3AuS_2; a number of tellurides; two selenides AuSe and Au_2Se_3; seleno-tellurides of the type Au_2SeTe; and mixed selenides and tellurides of the types Ag_3AuSe_2 and Ag_3AuTe_2; various telluride halides of the type $AuXTe_2$ (where X = Cl, Br, I) are also known (Rabenau *et al.*, 1970). Of all these artificial compounds only the tellurides and the silver-gold selenide are known in nature; some of the double sulphides would seem to be stable enough to persist in nature and may occur as minerals. One of these, the double sulphide, Ag_3AuS_2, uytenbogaardtite has recently been described. The antimonide, $AuSb_2$, occurs in nature as aurostibite and the silver-gold selenide, Ag_3AuSe_2, as fischesserite. The arsenide is unknown as a natural mineral, and as far as the writer can ascertain it has not been synthesized.

Gold has a marked affinity for the elements tellurium and bismuth and less so for antimony. The reason for this is obviously that the various combinations Au-Te, etc. give a particularly stable electronic configuration. The atoms of Te, Bi and Sb have large radii, and this was noted by Goldschmidt (1954) who sought to explain the affinity of gold with these elements as due to their ability to form packings such that gold could be collected (captured) in their interstices. The phenomena are, however, mainly connected with the electronegativity of the donor atoms of the complexes as discussed above. There are a number of gold tellurides in nature, a poorly differentiated bismuthide and an antimonide. Numerous artificial Au-Sb compounds are known (Muller, 1974). Gold does not form a selenide (*sensu strictu*) in nature, but as noted above there is a silver-gold selenide (fischesserite) that occurs in places (*vide infra*). The various phases in the system gold-bismuth-tellurium have been investigated by Gather and Blacknik (1974).

Gold in solution is readily reduced to the metal by nearly all of the metals and metallic ions and by numerous compound substances. Of interest in this respect in natural environments are the ions Fe^{2+} and Mn^{2+} and a great variety of organic compounds including citric and acetic acid and their salts. This feature of the chemistry of gold is discussed in more detail in the sections on soils and natural waters and in the chapter on the oxidation of gold-bearing deposits.

Gold forms a colloid readily when a dilute gold solution is reduced. The particle size is apparently determined by the concentration, temperature and pH. A low pH favours large particles whereas a high pH yields small particles. Gold colloids are hydrophobic and generally carry a negative charge. In a recent study Turkevich *et al.* (1951) have shown that in the formation of gold colloids nucleating agents can be identified with reducing agents which form a mixed polymer with chlorauric ion before the reduction to the gold nucleus takes place. They also showed that the law of growth of the gold colloids is exponential. The average size, the deviation from the average size and the character of the particle size distribution curve are determined by the amount of gold, the nucleation process and the law of growth.

The possible role of gold colloids in the migration of gold in nature is discussed in a subsequent section.

Table 3 lists the solubility constants of some of the common salts and other compounds of gold likely to be encountered in nature.

Table 3. Solubility of some gold compounds

Compound	Solubility in water, g/100cc (25°C)	Solubility in other reagents
Au	Insoluble	Soluble in aqua regia, KCN, hot H_2SO_4; insoluble in acids and alkalies
AuBr	Insoluble	Dissociates in acids; soluble in NaCN
$AuBr_3$	Slightly soluble	
AuCl	Very slightly soluble	Soluble in HCl, HBr
$AuCl_3$	68	
AuCN	Very slightly soluble	Soluble in KCN, NH_4OH; insoluble in alkalies
$HAu(CN_4)\cdot 3H_2O$	Very soluble	
AuI	Very slightly soluble	Soluble in KI
AuI_3	Insoluble	Soluble in iodides
$H[Au(NO_3)_4]\cdot 3H_2O$	Soluble, dissociates	Soluble in HNO_3
Au_2O_3	Insoluble	Soluble HCl, concentrated HNO_3, NaCN
$Au_2O_3\cdot xH_2O$	5.7 x 10^{-11}	Soluble HCl, NaCN, concentrated HNO_3
Au_2P_3	Insoluble	Insoluble in HCl, dilute HNO_3
Au_2Se_3	Insoluble	
Au_2S	Insoluble	Soluble aqua regia, KCN; insoluble in acids
Au_2S_3	Insoluble	Soluble in Na_2S
$AuTe_2$	Insoluble	

Source: Handbook of chemistry and physics, 53rd ed.; Chemical Rubber Co., Cleveland, Ohio, 1972.

Isotopes

Natural gold consists of only one isotope, $^{197}Au_{79}$, whose half life is estimated to be greater than 3×10^{16} years (Strominger *et al.* 1958). Some 20 artificial radiosotopes are known; all have short half-lives, none exceeding 185 days. Mauer and Pose (1943) found no evidence of natural fission in gold and concluded that the half-life for the process is longer than 2×10^{15} years.

Abundance and geochemical character: gold in meteorites, tektites and lunar materials

The cosmic abundance of gold as given by Green (1959) from the data of Suess and Urey (1956) is 0.000534 ppm or 0.145 atom of Au per 10^6 Si atoms. More recent estimated cosmic abundances of gold as given by Cameron (1959), Clayton and Fowler (1961), Baedecker and Ehmann (1965) and Cameron (1968) are respectively 0.145, 0.13, 0.18 and 0.20 atom of Au per 10^6 Si atoms. These values can be compared with the corresponding value for silver – 0.5 – as given by Cameron (1968). Ehmann and Gillum (1972) give the cosmic atomic abundance of gold as 0.19 (Si = 10^6 atoms); that for platinum on the same basis is 1.37 giving an Au/Pt atom ratio for the primitive solar nebula equal to 0.14.

Gold is a trace constituent of meteorites, its abundance in the various types of these objects having a considerable range as shown in Table 4. The iron meteorites (siderites) and the metal phase of chondrites have the highest gold contents indicating the marked siderophilic character of gold. The gold content of the silicate phases of meteorites is low whereas the sulphide (troilite) phase is relatively high, indicating respectively a general poor lithophile tendency and a relatively high chalcophile habit for gold.

In the chondrites the Au/Ag ratio is consistently around 1, whereas in the siderites the average ratio is extremely high, at 40 or more. The ratio in the sulphide (troilite) phase is of the order of 0.05 if the available analyses for silver are reasonably correct. All these data indicate a weak siderophile character and a marked chalcophile nature for silver compared with gold, features that are well known to all chemists.

Some of the correlations of gold with other elements in meteorites are of interest. There is a reasonable correlation between nickel, cobalt and gold in some chondrites and siderites according to the data presented by Goldberg *et al.* (1951) and Vinogradov *et al.* (1973). Linn and Moore (1968), on the other hand, found that the positive correlation of gold concentrations in relation to nickel and to cobalt was not pronounced in their data on iron meteorites. The correlation between gold and iron is not marked in the data on meteorites, although some investigators have observed a positive

Table 4. Gold and silver contents of meteorites, tektites and the moon*

Type of meteorite, etc.	Element	Number of samples	Range	$\bar{X}$§	S§	G§	$\bar{X}g$§	Sg§
				(ppm)				
Chondrites (world)	Au	246	0.0170–1.6000	0.2225	0.1427	0.1998	-0.6994	0.1881
	Ag	34	0.0330–0.5700	0.1769	0.1464	0.1285	-0.8911	0.3542
	Au/Ag			1.2578		1.5549		
Chondrites,								
Abee†	Au	1	0.25					
Benton†	Au	1	0.96					
Holman Island†	Au	1	0.94					
Chondrites, metal phase (world)	Au	52	0.4480–2.9000	1.5085	0.6016	1.3846	0.1413	0.1888
Chondrites, nonmetal (silicate) phase (world)	Au	47	0.0004–0.0910	0.0199	0.0181	0.0125	-1.9030	0.4788
Achondrites (world)	Au	11	0.0005–0.2800	0.0310	0.0830	0.0053	-2.2775	0.7186
Siderolites (50% iron, 50% silicate) (world)	Au	7	0.0400–2.000	0.9143	0.7262	0.5318	-0.2743	0.6053
Siderites (irons) (world)	Au	291	0.0550–8.7440	1.2050	0.8999	0.9597	-0.0179	0.3102
	Ag	87	0.0050–0.1320	0.0301	0.0316	0.0174	-1.7597	0.4717
	Au/Ag			40.0332		55.155		
Siderites								
Annaheim†	Au	1	0.9					
	Ag	1	0.07					
	Au/Ag		12.85					
Skookum†	Au	1	0.5					
	Ag	1	0.15					
	Au/Ag		3.33					
Madoc†	Au	1	0.7					
	Ag	1	0.03					
	Au/Ag		23.33					
Troilite (sulphide phase) (world)	Au	16	0.0700–8.0000	0.9787	1.9217	0.4128	-0.3842	0.5405
	Ag			18‡				
	Au/Ag			0.054				
Tektites (world)	Au	15	0.0003–0.0242	0.0081	0.0080	0.0040	-2.4020	0.6296
	Ag	8	0.2500–0.5000	0.4063	0.1294	0.3856	-0.4139	0.1558
	Au/Ag			0.0199		0.0104		
Lunar rocks and soils	Au	529	0.0000025–0.08700	0.00485	0.00855	0.00173	-2.7613	0.8021
	Ag	299	0.0001–10.7800	0.0980	0.6689	0.0052	-2.2819	0.9077
	Au/Ag			0.04949		0.3327		
Lunar rocks	Au	292	0.0000025–0.0870	0.0047	0.0090	0.0011	-2.9603	0.9478
	Ag	166	0.0001–10.7800	0.0933	0.8440	0.0027	-2.5622	0.8602
	Au/Ag			0.0504		0.4074		
Lunar soils and fines	Au	237	0.000011–0.0737	0.0051	0.0080	0.0031	-2.5161	0.4716
	Ag	133	0.0002–3.0800	0.1039	0.3458	0.0117	-1.9321	0.8445
	Au/Ag			0.0491		0.2650		
Lunar metal grains	Au	58	0.050–19.000	1.5332	2.5940	0.7737	-0.1114	0.5564
	Ag	1	0.0730	0.0730		0.0730		
	Au/Ag			21.003		10.599		

* Sources of information: *see* the Selected bibliography. Doubtfully high values for gold and silver have been excluded from some of the published data. For details of individual meteorites and bibliographies on meteorites *see* Mason (1962), Middlehurst and Kuiper (1963), Hey (1966), Mason (1971) and Ramdohr (1973). For information on most of the lunar samples see the volumes containing the papers delivered at the 1st, 2nd, 3rd, 4th and 5th Lunar Science conferences held at Houston Texas.

† Analyses done in the Geological Survey of Canada laboratories by precise chemical methods (analyst, J.G. Sen Gupta). (For descriptions of the meteorites *see* Geol. Surv. Can., Pap. 70-66.)

‡ Estimated value.

§ In this and subsequent tables $\bar{X}$, is the arithmetic mean; S, the standard deviation; G, the geometric mean; and $\bar{X}g$ and Sg are respectively the mean and standard deviation of the data expressed in $\log_{10}$ units.

correlation in some varieties of meteorites; for instance Ehmann *et al.* (1970) found such a correlation with the total iron content in enstatite chondrites. The relationship of gold with the various platinoids appears to be rather erratic judging from the overall available data, although Ehmann and Gillum (1972) found a positive correlation for platinum and gold within each major chondrite group they investigated. The correlation coefficient was 0.70 which is significant at the 99.9 per cent confidence level. Hintenberger *et al.* (1973) observed an antipathetic correlation between gold and mercury in some Antarctic meteorites. Fouché and Smales (1966) noted a strong correlation between gold and arsenic in iron meteorites and a weak correlation between gold and antimony. Both these correlations extend to terrestrial materials as we shall see later. Similar correlation features were noted for gold, arsenic and antimony in chondritic meteorites by Fouché and Smales

(1967). Other correlations and relationships of gold in meteorites are briefly commented on by Ehmann (1971).

A number of investigators have suggested that an approximation of the average composition of the earth's mantle and even of the entire earth is provided by the composition of the chondrites. Others have suggested that the earth derived from an agglomeration of chondrites. If these theories are valid there is a strange difference, which requires an explanation, in the Au/Ag ratios between chondrites and ultrabasic and basic rocks supposedly derived from the mantle (Table 13). Perhaps primitive differentiation processes in the earth are the reason, the gold settling into the metallic phase (Ni-Fe core) and the silver remaining largely in the mantle. Such a differentiation is perhaps supported by the fact that the Au/Ag ratios in siderites are extremely high. All of this discussion is of course highly speculative to say the least, especially the supposition that the core of the earth is composed of nickel-iron; it may be that chondrites and the other meteorites bear no relationship whatever to the earth in any of its aspects.

The gold and silver data for tektites (Table 4) is similar to that for terrestrial igneous rocks (Table 13). The Au/Ag ratio is, likewise, similar. The significance of these facts is obscure.

The gold and silver data for the lunar rocks and fines (soils) (Table 4) are extensive and are discussed at length by a number of investigators (*see* the list of references in the Selected bibliography). To generalize one notes that the overall gold and silver contents of the lunar materials are not much different from those recorded for igneous rocks on the earth (Table 13). A number of investigators have remarked that the lunar soils are enriched in gold and silver relative to the rocks, the reason being that the soils have a micrometeorite component. The enrichment is, however, not great as can be seen from the statistics recorded in Table 4.

There are no data on the nature of gold in any of the extra-terrestrial materials just described. Presumably in meteorites much of the gold is present in the metallic (mainly Fe-Ni) phases and possibly in part in the sulphide (troilite) phase; also possibly in some of the iron oxides (magnetite, ilmenite) and chromite. In all of these minerals gold may be present as a submicroscopic dispersed phase or in lattice sites in the metal alloys, sulphides and oxides. Ramdohr (1973) mentions the presence of native gold (observed by A.B. Edwards) in the Wedderburn iron meteorite. He also thinks that some osbornite observed in meteorites could be mistaken for gold and vice versa. This possibility should be checked by electron probe.

There is reasonable agreement with respect to the value of the terrestrial abundance of gold. Goldschmidt (1954) gives 0.005 ppm for the lithosphere; Rankama and Sahama (1950) give 0.005 ppm as the average for igneous rocks; Green (1959) gives ~ 0.001 ppm as the average for igneous rocks; Vinogradov's (1962) estimate is 0.0043 ppm for the crust; and Taylor (1964) has given 0.004 ppm as the continental crustal abundance of the element, a mean of the basalt (0.004 ppm) and granite (0.004 ppm) averages. More recently Li and Yio (1966) have given the figures for the abundance of Au, Ag and Pt in the various parts of the earth's crust noted in Table 5. From their estimates the Au:Ag:Pt ratios are 1:20.4:13.2 and the Au/Ag ratio is 0.048. Horn and Adams (1966) computed the abundance of gold in igneous rocks to be 0.00357 ppm; in sedimentary rocks the range was 0.00179 to 0.00457 ppm. The writer estimates the terrestrial abundance of gold and silver to be respectively 0.005 and 0.05 ppm from analyses quoted in this work and a previous report on silver (Boyle, 1968*b*). This gives an Au/Ag ratio of 0.1.

Table 5. Abundance of gold, silver and platinum in domains of the earth's crust*

Source	Au	Ag	Pt
		(ppm)	
Deep oceanic regions	0.004	0.098	0.095
Suboceanic regions	0.0029	0.082	0.05
Continental shield regions	0.0034	0.067	0.031
Folded belt regions	0.0038	0.062	0.022
Oceanic crust	0.0035	0.091	0.075
Continental crust	0.0035	0.065	0.028
Earth's crust	0.0035	0.075	0.046
Crystalline rocks	0.0036	0.077	0.049
Sedimentary rocks	0.0051	0.065	–

*From Li and Yio (1966).

In the upper lithosphere gold occurs in the native state and as a constituent of a few natural alloys. It is also a trace element in various native metals and semi-metals such as iron, copper and arsenic. These features indicate the strong siderophile character of the element. The metal also combines readily with tellurium, bismuth and antimony, forming a number of mineral species, but its chalcophile character is weak since no natural gold sulphides are known as minerals. The element appears to have no oxyphile character, and there are no known natural gold oxides, salts or basic salts. Maclaren (1908) refers to silicate of gold, but there are no modern investigations to confirm the presence of such a compound. No silicates of gold are known in nature.

Gold combines with a large number of artificial organic compounds, but the occurrence of these in natural environments is largely uninvestigated. The element is a trace constituent of various terrestrial and marine organisms, but its biological role, if any, is unknown. Gold is also a trace constituent of some coals and is found as traces in some petroleum residues and in oil shales.

Of all metals gold is most
frequently found native.
–Richard Kirwan, 1799

Mineralogy of Gold

Gold minerals

The minerals of gold are given in Table 6. Of those listed the most common are native gold and the tellurides. For the details of the gold minerals the reader should consult Palache, *et al.* (1944, 1951), Ramdohr (1960), Hey (1962, 1963), Hey and Embrey (1974), Chukhrov (1960–1965) and Berry and Thompson (1962). Occurrences of Canadian gold minerals are noticed by Traill (1970).

Native gold crystallizes in a great variety of forms, the most common being parallel groups and twinned aggregates branching at 60 degrees parallel to the edges or the diagonals of the faces of (111); also commonly reticulated, dendritic, arborescent, filiform, wiry, mossy and spongy. Massive forms include rounded fragments (nuggets), flattened grains, scales,

Table 6. Gold minerals

Native elements, alloys and metallic compounds	
Gold	Au
Argentian gold (electrum)	(Au,Ag)
Cuprian gold (cuproauride)	(Au,Cu)
Palladian gold (porpezite)	(Au,Pd)
Rhodian gold (rhodite)	(Au,Rh)
Iridic gold	(Au,Ir)
Platinum gold	(Au,Pt)
Bismuthian gold	(Au,Bi)
Gold amalgam	Au_2Hg_3 (?)
Maldonite	Au_2Bi
Auricupride	$AuCu_3$
Palladium cuproauride	$(Cu,Pd)_3Au_2$
Sulphide	
Uytenbogaardtite	Ag_3AuS_2
Tellurides	
Calaverite	$AuTe_2$
Krennerite	$(Au,Ag)Te_2$
Montbrayite	$(Au, Sb)_2Te_3$
Petzite (antamokite)	Ag_3AuTe_2
Muthmannite	(Ag,Au)Te
Sylvanite	$(Au,Ag)Te_4$
Kostovite	$AuCuTe_4$
Nagyagite	$Pb_5Au(Te,Sb)_4S_{5-8}$
Antimonide	
Aurostibite	$AuSb_2$
Selenide	
Fischesserite	Ag_3AuSe_2
Tellurate	
Gold tellurate (?)	

spangles, sprigs and finely divided particles (flour gold, mustard gold and paint gold). All of these and other forms are figured by Weinig (1960).

The purity of native gold, gold bullion, gold coin, etc. is expressed in terms of fineness, which is commonly defined as the proportion of pure gold in the sample in parts per thousand.[17] Thus, native gold which is 900 fine contains 90 per cent of the element gold. In some deposits native gold is relatively pure, but more generally it contains some silver, copper, and iron and traces to minor amounts of one or several of the following elements: Li, Na, K, Be, Mg, Ca, Sr, Ba, Zn, Cd, Hg, B, Al, Ga, In, Sc, Si, Ge, Sn, Pb, Ti, Zr, As, Sb, Bi, V, Se, Te, Cr, Mo, W, Mn, Co, Ni, Rh, Pd, Ir, Pt, U, Th and rare-earths.

There is a vast literature on the composition of native gold in practically all types of deposits in which the mineral occurs. Some of the papers noticed during the compilation of this chapter include those by Warren and Thompson (1944), Ferguson (1950), Perelyaev (1953), Gay (1963, 1964), Denisov *et al.* (1966), Jones and Fleischer (1969), Antweiler and Sutton (1970), Lantsev *et al.* (1971), Desborough *et al.* (1971), Badalova *et al.* (1971), Petrovskaya *et al.* (1971), Berman *et al.* (1973), Semenenko and Latysh (1973), Kuznetsov *et al.* (1973*b*), Petrovskaya (1973), Davidenko (1973*a,b*), Igumnova (1974) and Petrovskaya *et al.* (1976). Some highlights from some of these papers may be mentioned.

There is a great variation in the trace and minor element content of primary and secondary native gold, although within some specific deposits the contents of the various elements appear to range within relatively narrow limits. Native gold in oxidized zones and in associated placers generally contains lesser amounts of silver and a number of other elements compared with the native gold in the corresponding primary ores. Within some specific deposits single crystals and particles of native gold are frequently relatively homogeneous with respect to the distribution of trace and minor elements; in other deposits the native gold is heterogeneous, a silver and/or other elemental zoning being present. In some deposits there is an increase in the silver and other elemental content from the centres of the crystals or particles outward; in other deposits the reverse is the case. In some gold lodes the crystals and particles of native gold contain numerous inclusions, generally associated gangue minerals such as pyrite, magnetite, chalcopyrite, tellurides, etc.; in other lodes the native gold has a scarce assortment of inclusions; and in still others inclusions are virtually absent. Within some single deposits the intergrain variations and differences in trace and minor element contents are often marked; in other deposits the grains and crystals of gold throughout the deposits are relatively homogeneous within narrow limits. An example of the latter is the gold in some of the Witwatersrand ore shoots. Placer gold commonly has thin rims of relatively high purity gold from which, according to some investigators, the silver and other trace elements have been leached. This feature of placer gold is discussed at some length in Chapter III. The gold in the oxidized zones of certain lodes may also exhibit the rim effect; it is uncommon in primary gold ores in most goldfields. Some investigators claim that the types and other features of gold deposits such as temperatures of deposition, nature of the depositing solutions, etc. can be divined from the analyses of native gold samples. Short resumes of some of the papers of the investigations will illustrate their contentions.

Warren and Thompson (1944) considered that by thorough and detailed work it may be possible, by means of spectrographic analysis; to determine the metallogenetic province or zone, and in some cases even the deposit, from which a particular sample of gold originates.

More recently Antweiler and Sutton (1970) report results for 174 quantitative spectrochemical analyses for silver and semiquantitative spectrochemical analyses for 32 other elements in samples of lode and placer gold obtained from 67 localities in eight western states of the continental United States, including Alaska, and from one locality in Brazil. Of the 24 elements detected (besides gold) in the samples, Ag, Cu, Fe and Ti were the most frequently detected. Pb, Pd, Bi, V, Ni and Co were identified more often in lode samples than in placer samples, but Ti, Zr, La, Y and Cr occurred more frequently in placer samples. They concluded that their data could possibly be used in prospecting for deposits of gold and other metals and in determining ore depositional conditions or genetic associations.

Lantsev *et al.* (1971) determined the contents of the following trace elements – As, Sb, Te, Pb, Cu, Fe, Bi, Sn, Ni, Co, Mn, Cr, Ti, Zn, Cd and Ga – in 120 gold samples from 40 endogenic deposits formed under various conditions and at various depths. They concluded that the concentration of trace elements in gold deposits increases as one ascends through ores formed at greater depths to those of near-surface genesis. This tendency was found for all elements, with the exception of Cu, during the passage from great depths to

[17]Fisher (1945) defines the true gold fineness as the ratio Au/(Au + Ag) × 1000. This definition eliminates the base metal factor.

medium depths and for Sb, Sn, Zn and Mn during the passage from medium to surface depths. This regularity was thought to be caused mainly by diffusion upwards of isomorphous and mechanical admixtures from gold during the process of recrystallization at high temperature.

Research based on polished sections and spectrographic, chemical and electron-probe analyses of native gold from primary deposits and placers carried out by the writer indicates that most of the metallic and semimetallic elements may be lattice constituents of native gold. However, many of the metals and other elements are often contaminants, being present in minute inclusions of the minerals commonly associated with native gold, particularly quartz, carbonates, sericite, sulphides, arsenides, tellurides, selenides, etc. In my experience the following elements exhibit a frequency which suggests that they are predominantly lattice constituents when they are present in pure samples of native gold: Ag, Cu, Fe, Ni, Hg, As, Sb, Bi, Te, Pd and Pt.

Several varieties of native gold are known which contain significant amounts of other metals. These have been given specific mineral names in some cases, but more generally they are referred to by a prefix indicating the contained element, e.g. argentian gold, palladian gold, etc.

Argentian gold. Normally all native gold contains some silver generally in the range 5–15 per cent Ag. Most of this silver is a lattice constituent, and there is a complete(?) substitutional series from gold through argentian gold to aurian silver (kustelite) to native silver. *Electrum* has been applied to gold containing 20 per cent or more silver. The name is an ancient one and has been used for centuries, apparently being first applied to the gold won from the famous placers of the Pactolus a small river of ancient Lydia which joined the Hermus (modern Gediz) after flowing through Sardis. Pliny, in his *Historia naturalis,* said "when the proportion of silver in gold is one-fifth, it is called electrum and possesses a pale yellow colour."

The range and significance of the Au/Ag ratio in native gold and in gold deposits is discussed in Chapter III.

Cuprian gold. Most samples of native gold contain only traces to minor amounts of copper probably substituting in the lattice in most cases. Cases are known, however, where the copper is apparently plated on the gold (Ferguson, 1950). Some samples of gold mentioned in the literature contain from 0.10 to 20 per cent Cu. One of the copper varieties has been named *cuproauride* and given the formula AuCu. According to Ramdohr (1960) it is rare and requires further investigation. The compound, $AuCu_3$, containing some 40 per cent gold, named *auricupride* is reported by Ramdohr to occur frequently. It has a high metallic lustre and a yellow colour with a reddish tint. Oen and Kieft (1974) describe a new copper-gold alloy with a formula approximating $CuAu_{2.6}$. The alloy occurs in the chromite-niccolite ore from Beni-Bousera, Morocco, as irregular grains mostly along grain boundaries of niccolite and loellingite. In the auriferous ores of the Karabashsk deposit Novgorodova and Tsepin (1976) observed the following phases among the auricuprides, Cu_3Au, CuAu and $CuAu_4$. The artificial isomorphous series Cu-Au is complete, but in nature the copper end members with high contents of gold (>3% Au) have not been recognized.

Palladian gold termed *porpezite* in the old literature is said to contain 5 to 10 per cent Pd in solid solution, but samples of this type of gold require restudy to ensure their homogeneity. It has been reported from Porpez, Taguaril and other localities in Minas Geraes and Goyaz, Brazil. Recently, *palladium gold* (Au,Pd) has been described from the Stillwater Complex in Montana by Cabri and Laflamme (1974). The mineral is face-centered cubic ($a=4.047(1)$Å) and is cream-coloured with a metallic pinkish cast under reflected light in air. It is associated with a number of other rhodium and platinum alloys and minerals including platinian rhodium, rhodian platinum, platinum-iron, and braggite/vysotskite.

Rhodian gold, also referred to as *rhodite,* is said to contain 34 to 43 per cent Rh. It requires restudy to determine its homogeneity. Rhodian gold is reported from Colombia and Mexico.

Iridic gold is reported in the old literature to contain up to 30 per cent Ir. This type of gold requires restudy to establish its homogeneity. Some of the reported occurrences may be intergrowths of gold and osmium-iridium and platinum as described by Syrovatskii *et al.* (1969).

Platinum gold is reported in the old literature to contain up to 10 per cent Pt. It requires reexamination. Some of the reported occurrences may be intergrowths of gold and osmium-iridium and platinum as described by Syrovatskii *et al.* (1969).

Gold amalgam is reported by Palache *et al.* (1944) to occur in grains and lumps of white to yellowish colour with metallic lustre in the Mariposa region, California, with platinum in Colombia and in the vicinity of Pleiari, Borneo. The composition of some of the samples analyzed corresponds well with the composition Au_2Hg_3, suggesting a compound homologous with the silver mineral, moschellandsbergite, Ag_2Hg_3.

Maldonite or bismuthic gold. Some samples of native gold contain traces to minor amounts of bismuth (up to 3% Bi) possibly in solid solution. The mineral maldonite was early reported to be an intermetallic compound with the formula Au_2Bi. However, microscopic examination of the maldonite from Maldon, Victoria, the type locality, shows a fine myrmekitic intergrowth of gold and bismuth, derived apparently from the breakdown of Au_2Bi, which is said to be stable only at high temperatures (Ramdohr, 1953*a*, 1960).

The maldonite from Salsigne (Aude) in France described by Boyer and Picot (1963) averages 64.58 per cent Au and 35.72 per cent Bi, approximating the formula Au_2Bi. The sample in this case appears to have been homogenous. Boyer and Picot confirm Ramdohr's opinion that the mineral is formed within a relatively narrow temperature interval and think the mineral is a reliable geothermometer. From the Au-Bi phase diagram they suggest a temperature of deposition below or equal to 373°C.

According to Razin *et al.* (1971) there is some doubt about the existence of *auricupride,* $AuCu_3$. These same authors describe a new mineral, *palladium cuproauride,* in the ores of the Talnakh copper-nickel deposit, U.S.S.R. The mineral is an intermetallic compound with formula $(Cu,Pd)_3Au_2$. It contains traces and minor amounts of Rh, Ag, Bi and Ni.

The physical properties of native gold, the morphology of gold crystals and particles in deposits and the relationship of these parameters to the depositional conditions in gold deposits have been extensively studied by many investigators prior to 1970. Papers reviewed during the compilation of this chapter since that date include those by Korobeinikov (1971), Koroleva *et al.* (1971), Petrovskaya *et al.* (1971*a–d*), Yablokova (1972), Berman *et al.* (1973) and Dzhantuganov (1974). Some of the features brought out by all of these studies, both pre- and post-1970, are summarized briefly below. Only the properties of epigenetic gold and that in the oxidized zones of auriferous deposits are reviewed here; placer gold is discussed in detail in Chapter III.

The morphology of native gold in epigenetic deposits is exceedingly variable, the most common shapes of the particles being nodular and lamellar, lumpy masses, amoeba-shaped masses, simple platy forms, ramifying platy and veinlike masses, dendritic masses, wires, sprigs, spear- and splinterlike forms, crystals, droplike inclusions in sulphides and other minerals, smears on shear planes, pulverulent aggregates (mustard gold), and spongy masses. There appear to be no clear-cut relationships between the shapes of the native gold particles and the types or age of the deposits. In general crystals are rare in the deposits formed at great depth (mainly deposits of Precambrian and Paleozoic age), whereas crystals are relatively common in younger deposits formed near the surface especially in those of Tertiary age. Crystals, wires, dendrites, pulverulent aggregates and spongy masses are especially characteristic of secondary gold found in the oxidized zones of auriferous deposits. Deep-seated deposits, which include most lodes of Precambrian and Paleozoic age, have much more uniformity in the morphology of their native gold particles than near-surface deposits of Mesozoic and Tertiary age. The morphology of the gold particles and crystals in the old (deep-seated) deposits is also generally much simpler than is that of the gold in the younger (near-surface) deposits.

The size of gold particles and crystals in epigenetic gold-bearing deposits varies widely, the usual range being from less than one thousandth of an ounce up to masses weighing a thousand ounces or more. Deposits formed under deep-seated conditions (most Precambrian and Paleozoic deposits) generally are marked by fair uniformity in the size of the gold particles, whereas deposits formed under near-surface conditions (many Tertiary deposits) are often characterized by great variability in the size of the native gold particles and crystals. This variability is said by some investigators to indicate fluctuating conditions during the deposition of the gold. In numerous deposits gold is present in an extremely finely divided form in sulphides, sulphosalts or gauge minerals such as quartz. The division is so fine in some deposits that the gold cannot be seen under the highest magnification in minerals such as arsenopyrite and pyrite. The nature and significance of this submicroscopic gold is discussed subsequently. There are no clear-cut relationships between the size of gold particles and crystals and depth in epigenetic deposits as far as this writer can ascertain from the literature or from his experience. It is of interest to note that the gold of the Witwatersrand is remarkably uniform in its size of particles (av. diameter – 80μ; 0.08 mm), a feature that is discussed in more detail in Chapter III.

Most investigators agree that the morphology and size of gold particles and crystals in most epigenetic deposits are essentially dependent upon the size and shape of the sites of deposition which in turn are predicated on the original or induced porosity of rocks and on structural events such as fracturing, formation of open spaces along faults, shearing etc. In addition there are also various chemical factors which seem to be involved, but about which we have only a superficial knowledge. These include replacement phenomena, exsolution reactions, coprecipitation phenomena and so on.

Native gold contains gaseous inclusions that have recently been studied in some detail by Petrovskaya *et al.* (1971*a*). They found that some 90 per cent of the gas was CO_2 and the remainder mainly nitrogen. The pressure in the inclusions studied was estimated at 5 to 20 atm. The authors concluded that the presence of CO_2 and N_2 (also found in gaseous inclusions in quartz in the same deposits) is due to the adherence of bubbles of CO_2 at the surface of hydrophobic gold grains during their growth. Their further conclusion was that carbon dioxide plays a significant role in the crystallization of gold from hydrothermal solutions.

The optical properties of gold have been extensively investigated by Hodgson (1968), and the determination of the silver content of natural gold alloys by reflected-light microscopy is described by Eales (1968) and Eales and Viljoen (1973). A criticism of the first paper has been published by Stumpfl (1969).

Gold is unique among the precious metals in the variety and weight of its placer nuggets. Some have been found that are beautifully metallic and crystallized, whereas others are inconspicuous, being rounded, pitted and covered with black manganese oxides, brown limonite, humic compounds, silica, clay or other nondescript compounds. In weight there are records of nuggets ranging from less than 1 oz up to masses weighing nearly 2400 oz (200 lb troy). Australia has produced many of the largest nuggets, the goldfields of Victoria exceeding all others. Of the nuggets officially recorded from Victoria 12 exceeded 1000 oz, and 33 were between 1000 and 500 oz. The Dunolly district, which is deeply weathered and oxidized, has surpassed all others in the number of nuggets found. The Welcome Stranger weighed slightly more than 2284 oz and was found in detrital material just below the surface at Black Lead, Mt. Moliagul, near Donolly, Victoria. Another nugget, the Welcome, weighed 2217 oz and was found in alluvial material at a depth of 180 ft at Bakery Hill, Ballarat, Victoria. Two others, also found in alluvial gravels in Victoria, the Viscount Canterbury and Viscountess Canterbury, weighed 1114 and 912 oz respectively. In the United States some very large nuggets have been discovered, the one from Carson Hill, Calaveras County, California weighing some 2340 oz (Gaarden, 1940). Another water-worn nugget, the Willard, Dogtown or Magalia nugget, weighing 54 lb troy, was found at Magalia in California in 1859. The discovery of this nugget was the signal for a celebration which has been held annually ever since (Clark, 1970).

Despite the large amount of gold taken from the Klondike, Cariboo and other placer areas of Canada few large nuggets have been discovered. Robinson (1935) mentions an 85 oz nugget from the Klondike, and in eastern Canada a nugget weighing 52 oz, 11 dwt, 6 grains, called the 'Kilgour Nugget' was taken from the Gilbert River (Chaudière), Beauce, County, Quebec. In the same area a nugget weighing 45 oz, 12 dwt called the 'McDonald Nugget' was found on Lot

16, De Lury Concession, Gilbert River, Quebec. Numerous other smaller nuggets weighing a few tens of ounces have been taken from the Klondike, Cariboo and other placers in Canada (Frontispiece).

Gold nuggets have been found in every conceivable geological environment, from birds' nests and crops to potholes in placers. The origin of gold nuggets is discussed in Chapter III.

Certain of the gold tellurides probably occur in minute amounts in most primary gold deposits judging by the frequency with which traces of tellurium are found in gold ores and the nearly universal presence of isolated pockets and veinlets of tellurides in most gold veins and other types of auriferous deposits. In some deposits the tellurides are important ore minerals of both gold and silver. Localities where gold (and silver) tellurides are particularly abundant include Nagyág and Brad, Romania; Kalgoorlie and Mulgabbie, Western Australia; Cripple Creek, Colorado; Kirkland Lake, Ontario; and a number of localities in Japan (Watanabe, 1952); Vatukoula, Fiji; Philippines (Callow and Worley, 1965); and a number of localities in U.S.S.R. (Chukhrov, 1960; Amiryan, 1960; and Pankov and Chukhrov, 1962).

Krennerite, calaverite, sylvanite, nagyagite, muthmannite and *petzite* appear to be the most common of the gold tellurides, having been recorded in a large number of places. *Montbrayite* has been recognized only in the Robb-Montbray Mine, Montbray Township, Quebec where it is associated with other tellurides and gold (Thompson, 1949). *Kostovite* is a new mineral and occurs in the copper ore deposit of Chelopech, Bulgaria where it is associated with tennantite, chalcopyrite, pyrite, native tellurium, native gold, tellurides and other minerals (Terziev, 1966). Tunell (1954) has reviewed the main features of the crystal structures of the various gold-silver tellurides.

Thompson (1949) has described most of the known gold and silver telluride occurrences in Canada.

Markham (1960) has discussed the synthetic and natural phases in the system Au-Ag-Te and also his findings with respect to the occurrence of tellurides at Kalgoorlie, Western Australia and Vatukoula, Fiji. He finds that the experimental data are insufficient to indicate a meaningful range of depositional temperatures for any given telluride, but thinks that when the calaverite-krennerite transition is experimentally determined, it should provide a useful geological thermometer. Cloke (1963) and Petersen (1964) have commented on some of the physico-chemical features of Markham's paper.

Shcherbina and Zar'yan (1964) have discussed the parageneses of gold and silver tellurides from Armenian sulphide deposits in terms of the experimentally determined parts of the system Ag-Au-Te. They concluded that the assemblages hessite-petzite-gold and hessite-petzite-krennerite were formed by deposition from hydrothermal solutions during the galena-sphalerite stage of mineralization, and by later replacements. Native silver and tellurium, formed by the decomposition of gold-silver tellurides, occur in some assemblages.

Cabri (1965*a*) recently carried out a systematic investigation of the phase relations in the Au-Ag-Te system and found that krennerite, rather than being a polymorph of $AuTe_2$, should be interpreted as a single mineral entity with the approximate formula Au_4AgTe_{10}. It is stable up to its incongruent melting point of 382 ± 5°C, and hence its presence in deposits is not necessarily indicative of a low temperature of formation. The incongruent melting temperature of sylvanite (354 ± 5°C) indicates a maximum temperature for its formation. Cabri also found that hessite and petzite are each characterized by three polymorphs. Continuous solid solution exists between the highest polymorphs and extends to a composition about 7 weight per cent more gold-rich than stoichiometric petzite. A new phase, termed the 'x' phase, was found to be stable along the petzite-hessite join from about 2.5 to 14.5 weight per cent gold within temperature limits of 50 to 415°C. The observations of Forsythe (1968) with respect to the occurrence of tellurides at Vatukoula lend support to some of Cabri's conclusions.

Howie and Veale (1966) have discussed the low temperature synthesis of gold telluride. They could not make gold sesquitelluride by low temperature coreduction and precipitation techniques; sometimes a metastable telluride of composition close to $AuTe_{1.75}$ was obtained. They state that the mineral montbrayite, Au_2Te_3, may represent an intermediate step in the ordering of this structure.

The relation between the composition and X-ray diffraction data of gold-silver tellurides is discussed by Cabri and Rucklidge (1968). Stumpfl (1970) has noted considerable compositional variations in the gold content of specimens of gold tellurides from different deposits.

Our semiquantitative and qualitative analyses of the various gold-silver tellurides indicate that most are relatively pure. Traces to minor amounts (tens to thousands of ppm) of S, Se, Sb and As are commonly present in most samples probably replacing Te. In addition to these elements the following are commonly found in traces to minor amounts: Cu, Fe, Ni, Co, Pb, Sn, Hg, W, Mo and Bi. Some of these elements evidently replace Au and Ag, although some of them may be largely in impurities. The platinum metals were not found in any of the samples.

The isotope ^{130}Te is radioactive and decays by double beta emission to yield ^{130}Xe. This feature of tellurium minerals permits age determinations of gold and silver deposits containing tellurium to be made. Srinivasan *et al.* (1972) have determined the half-life for double beta decay of ^{130}Te to be $2.83 \pm 0.30 \times 10^{21}$ years from an analysis of xenon and tellurium in the tellurides of Kalgoorlie, Australia. Three telluride samples from different locations (Vakukoula, Fiji; Facebaja, Transylvania, Romania; and Moctezuma, Sonora, Mexico) gave ages consistent with geologic age estimates. The authors conclude that the method is particularly useful in yielding the age of gold and silver mineralization directly.

Some of the facts concerning the occurrence of silver (and gold) tellurides may be noted here. The tellurides seldom if ever are associated with the selenides; for some reason they seem to be mutually exclusive. It may be that observers have failed to recognize the presence of the two types of minerals together, but this seems unlikely considering the amount of mineralogical work that has been done on gold ores. Even in the highly seleniferous gold ores of Salida, Indonesia there are apparently no selenides (or doubtful selenides, naumannite?) associated with the various tellurides (sylvanite, petzite, hessite, altaite). There, the tellurides are invariably associated with galena and the selenium is bound mainly in acanthite, galena and arsenpolybasite (Kieft and Oen, 1973). An alternative reason for the antipathy of selenides and tellurides in deposits may be that the solutions that deposit gold and silver tellurides are inhibited from carrying selenium and vice versa. This, likewise, seems improbable from a chemical viewpoint. There are a number of silver selenides, but no natural gold selenide *sensu strictu* is known. The silver-gold selenide, fischesserite, Ag_3AuSe_2, is however, stable. Artificially a stable phase α-AuSe can be prepared, and while tellurium does not substitute in the AuSe lattice to a detectable extent, selenium will substitute up to at least 15 atomic per cent of the tellurium atoms in the monoclinic $AuTe_2$ (calaverite) lattice at 400°C (Cranton and Heyding, 1968). There is also at least one ternary phase, Au_2SeTe, in the system Au-Se-Te.

The tellurides generally occur in deposits in volcanic assemblages, or in rocks closely associated with andesites, basalts, etc. Particularly notable is their occurrence in Precambrian greenstones, examples being the Kirkland Lake–Noranda–Malartic areas of Ontario and Quebec; Porcupine area, Ontario; Red Lake area, Ontario; Flin Flon, Manitoba; and the famous telluride-gold lodes of Kalgoorlie, Western Australia. In younger (generally Tertiary) volcanic rocks, notable occurrences of gold and silver tellurides are known at Cripple Creek, Colorado; Brad and Nagyag, Romania (Transylvania); Vatukoula, Fiji and elsewhere.

In detail, the gold tellurides occur mainly in quartz and quartz-carbonate veins and in pyritized, silicified and carbonated zones in volcanic rocks. Wall-rock alteration effects characterized by carbonatization, hydration, pyritization and silicification are generally marked in these deposits. Some massive sulphide deposits in volcanic assemblages, particularly those rich in copper, carry gold and silver tellurides but with less frequency than the veins and carbonatized-silicified zones. Skarn deposits do not seem to contain gold and silver tellurides in any abundance according to my investigations, although other tellurides, mainly those of bismuth, may occur in some quantity in such an environment. The minerals most commonly associated with gold and silver tellurides include pyrite, arsenopyrite, pyrrhotite, tetrahedrite-tennantite, native gold, native tellurium (rare in Precambrian but common in Tertiary deposits), other tellurides (particularly coloradoite, HgTe), tellurbismuth, various Cu, Pb, Zn, Ag and Bi sulphides and sulphosalts, tourmaline and fluorite. The paragenesis of deposits containing gold tellurides is often quartz and pyrite followed in turn by base metal sulphides, tellurides and native gold. This sequence is, however, not invariable, gold and tellurides apparently being deposited simultaneously in some deposits, or gold may precede the tellurides.

Aurostibite has been recognized in the gold deposits of the Yellowknife and Larder Lake areas in Canada (Graham and Kaiman, 1952), in the deposits of Krasna Hora and Milesov in Czechoslovakia (Sobotka, 1954), in the antimony veins of Costerfield, Victoria, Australia (Stillwell, 1952), in the Lone Hand and Jessie Gold mines, Gwanda district, Rhodesia (Eales, 1962), and in the Indarama Gold-Antimony Mine, Sebakwe area, Que Que, Rhodesia (Mehliss, 1968). It is probably more common than these few occurrences would suggest. According to my observations (Boyle, 1961*a*) aurostibite is generally closely associated with gold in deposits that contain stibnite and other antimony minerals. Aurostibite in these environments frequently rims some of the gold particles and appears to be younger than gold, but good relationships for establishing the paragenesis of the mineral are uncommon. In the Czechoslovakian occurrences aurostibite is said to be younger than the native gold. In the Victoria occurrences the mineral occurs in the quartz-gold-stibnite veins. According to Stillwell (1952) the mineral is best preserved in quartz and, where observed in stibnite, tends to appear as relict areas. Its occurrence in a gold grain containing a gold-aurostibite eutectic led Stillwell to conclude that the mineral formed above 360°C, a higher temperature than that which the younger stibnite was deposited. Stillwell states that aurostibite is a relatively unstable compound and has been largely dissociated in the later stages of vein formation, yielding the dull, rusty, cellular-textured gold that is a common feature of stibnite-bearing veins. The dissociation yields native antimony and gold. The aurostibite in the Gwanda district of Rhodesia contains of the order of 1 per cent nickel and nondetectable amounts of silver. Much of the aurostibite in these deposits rims gold and is probably late, but some of the purer gold in the ore of the Jessie Mine is younger than aurostibite. In the Indarama Mine in Rhodesia the aurostibite was deposited simultaneously with native gold, both minerals being late in the general mineral paragenesis.

The aurostibite in the Bestyube goldfield in northern Kazakhstan, U.S.S.R., forms reaction rims around native gold and is thus a late mineral formed by the reaction of the gold with antimony-bearing solutions in a reducing environment (Naz'mova *et al.,* 1975). Electron microprobe analyses showed the aurostibite to contain Au – 44.52; Ag – 0.12; Cu – 0.08; Sb – 52.73; and As – 1.73 weight per cent. In the Yakutian deposits aurostibite occurs in quartz-stibnite veins where it is intimately intergrown with or rims native gold (Indolev and Zhadanov, 1975). The paragenesis in some samples is native gold followed by aurostibite; in a few others stibnite followed successively by aurostibite and a late generation of native gold. The aurostibite is considered to have formed in the deposits mainly by reactions between early precipitated gold and late stage solutions containing antimony.

Barton (1971) has concluded that aurostibite is stable only under conditions of relatively low sulphur activity, and that the assemblage pyrite + aurostibite is stable only at low temperatures (below about 125°C, but the uncertainties are too large for this to be a useful point on the geothermometric scale). This temperature agrees with the finding of the writer (Boyle, 1954) that late quartz, probably of the same generation as gold and aurostibite, has a temperature of deposition (actually decrepitation temperature and liquid inclusion filling temperature) around 150°C.

Fischesserite, the silver-gold selenide, Ag_3AuSe_2, is isometric and isostructural with petzite. It occurs in carbonate-quartz veins at Predborice, Bohemia, Czechoslovakia associated with chalcopyrite, naumannite, clausthalite, berzelianite, gold and other selenides (Johan *et al.,* 1971). It appears to be a late mineral in the paragenetic sequence and is commonly situated between the grain boundaries of carbonate minerals. The mineral has also been recognized in certain unusual ores in the Texas Gulf massive sulphide body at Timmins, Ontario.

Gold tellurate is reported as a supergene mineral at the Daté Mine, Hokkaido, Japan (Watanabe, 1952). The mineral is evidently derived by the oxidation of gold tellurides. It requires confirmation.

A number of synthetic compounds of gold are relatively stable and may occur as minerals, although they have not yet been identified. These compounds include various selenides, seleno-tellurides, mixed Au-Ag sulphides, the phosphide, the chloride and various telluride halides of the type $AuXTe_2$ (X=Cl,Br,I). By analogy with $AuSb_2$ one might also suspect that the compound $AuAs_2$ should occur in nature, although a search by electron-microbe by the writer and his colleagues has failed to find such a mineral in auriferous ores carrying abundant native arsenic, arsenopyrite and other arsenic-bearing minerals. The reason for the nonappearance of $AuAs_2$ in nature may result from the relatively high instability of the mineral at low temperatures due to the unfavourable electronic (electronegativity) relationships discussed previously. (*See further* the discussion under gold in pyrite and arsenopyrite.)

Barton *et al.* (1978) have recently described a double sulphide of silver and gold, *uytenbogaardtite*, Ag_3AuS_2, from Tambang Sawah, Benkoelen district, Sumatra, Indonesia; the Comstock Lode, Nevada, U.S.A.; and Smeinogorski, Altai, U.S.S.R. At all three localities uytenbogaardtite occurs as small (up to 100μ) blebs and rims intimately associated with electrum, acanthite, quartz and other minerals.

Gold in other minerals

Gold is a minor and trace constituent of a great variety of minerals, a feature of the geochemistry of the element that has been extensively investigated for many years. A summary of the status of gold in minerals up to 1969 is contained in the paper by Jones and Fleischer (1969).

The writer's investigations, based on optical and electron microprobe work on polished and thin sections, indicates that the gold in many minerals is present in particulate form, generally as native gold but also as gold tellurides and aurostibite. In some minerals, however, the traces of gold are probably lattice constituents.

Gold in silicates

The older data (prior to 1950) recorded in the literature on the gold content of silicates are sparse and seemingly unreliable since the values recorded range from 0.5 to >0.5 ppm, amounts that are relatively high considering the abundance of gold in nature.

Table 7 records some of the more recent (post-1950) analyses of silicate minerals, a few from the National Mineral Collection of Canada. The latter were specially purified, and the gold in these samples was done by a fire assay–atomic absorption method the detection limit being 0.01 ppm. The silver in the samples was done by spectrographic and atomic absorption methods.

As shown by Table 7 the gold content of silicates ranges from about 0.0002 to 0.924 ppm. There is no strikingly apparent relationship in the gold content of silicates in the series quartz-feldspar-mica-amphibole-pyroxene-olivine, although there is a positive indication that minerals like biotite and amphibole contain a little more gold than the others. This is suggested from recent studies such as those carried out by Zvereva and Gavrilenko (1971). In the dark silicate minerals gold may substitute in the iron sites, although we suspect that much if not all of the gold may be present in microscopic

Table 7. Gold and silver contents of common silicate minerals

Mineral and locality	Au (ppm)	Ag (ppm)	Reference
Quartz			
U.S.S.R.			
Altai and Transbaikal	0.0006–0.0042 (av., 0.0008)	–	1
Altai-Sayan	0.011 (av.)	–	2
Kazakhstan	0.0011–0.012		4
Altai, Transbaikal and Urals	0.0016 (av.)		5
Urals	0.0017 (av.)		6
U.S.A.			
Helena, Montana (quartz and feldspar)	0.006–0.176 (av., 0.065)		3
Canada			
Buckingham, Que.	<0.01	<0.05	7
Lansdowne, Ont. (crystals)	<0.01	<1.0	7
Feldspars			
U.S.S.R.			
Altai and Transbaikal	0.0002–0.0005 (av., 0.0003)	–	1
Altai-Sayan	0.0040 (av.)		2
Altai-Sayan	0.018		2
Kazakhstan	0.0013–0.0052		4
Altai, Transbaikal and Urals	0.0016–0.0017		5
Urals	0.0016–0.0017		6
Canada			
Eau Claire, Ont. (albite)	<0.01	<0.5	7
Villeneuve Tp., Que. (albite)	<0.01	<0.5	7
Bathurst Tp., Ont. (microcline)	<0.01	<0.5	7
Portland Tp., Que. (microcline)	<0.01	<0.5	7
Nain Island, Lab. (Ca feldspar)	<0.01	<0.5	7
Feldspathoids			
Canada			
Davis Hill, Ont. (nepheline)	<0.01	<0.5	7
Nephton, Ont. (nepheline)	<0.005	0.9	7
Bancroft, Ont. (sodalite)	<0.01	<0.5	7
Micas			
Muscovite:			
U.S.S.R.			
Altai-Sayan	0.0038 (av.)	–	2
Urals	0.0109 (av.)		6
Canada			
Eau Claire, Ont.	<0.01	<0.5	7
Biotite:			
U.S.S.R.			
Altai-Sayan	0.0040 (av.)	–	2
Altai	0.0091 (av.)	–	2
Kazakhstan	0.001–0.0119	–	4
Altai, Transbaikal and Urals	0.0019 (av.)		5
Urals	0.0018 (av.)		6
U.S.A.			
Helena, Montana	0.002–0.924 (av., 0.076)		3
Canada			
Douglas, Ont.	<0.01	<0.5	7
Amphiboles			
U.S.S.R.			
Altai-Sayan	0.0059 (av.)	–	2
Kuznetsk, Ala-Tau	0.077 (av.)	–	2
Kazakhstan	0.0026–0.0201		4
Altai, Transbaikal and Urals	0.0020 (av.)		5
Urals	0.0016 (av.)		6
U.S.A.			
Helena, Montana (hornblende)	0.003–0.823 (av., 0.100)	–	3
Canada			
Wilberforce, Ont.	<0.01	<0.5	7
Bathurst Tp., Ont.	<0.01	<0.5	7

Table 7. (cont'd.)

Mineral and locality	Au	Ag	Reference
	(ppm)		
Pyroxenes			
U.S.S.R.			
Altai-Sayan	0.016 (av.)	–	2
Kazakhstan	0.0137–0.0240		4
Canada			
Yates Mines, Sandy Creek, Que.	<0.01	<0.5	7
Ball Mines, Haliburton, Ont.	<0.01	<0.5	
Chlorite			
U.S.S.R.			
Altai, Transbaikal and Urals	0.0030 (av.)		5
Urals	0.0025 (av.)		6
Olivine			
U.S.S.R.			
Altai-Sayan	0.014 (av.)	–	2
Canada			
Bigelow Tp., Que.	<0.01	<0.5	7
Epidote			
U.S.S.R.			
Altai, Transbaikal and Urals	0.0026 (av.)		5
Urals	0.0025 (av.)		6
Tourmaline			
U.S.S.R.			
Altai-Sayan	0.012 (av.)		2
Canada			
Portland Tp., Que.	<0.01	<0.5	7
Leduc Mine, Wakefield Tp., Que.	<0.01	<0.5	7
Villeneuve Mine, Villeneuve Tp., Que.	<0.01	<0.5	7
South March, Ont.	<0.005	0.1	7
Sphene			
U.S.S.R.			
Altai-Sayan	0.0039 (av.)		2
Altai, Transbaikal and Urals	0.0029 (av.)		5
Urals	0.0027 (av.)		6
Garnet			
Canada			
Labelle Mine, Que.	0.01	<0.5	7
Dana Tp., Ont.	<0.01	<0.5	7
U.S.S.R.			
Urals	0.0052 (av.)		6
Zircon			
Canada			
Saramac Mines, Ont.	0.01	<0.5	7

References: 1. Anoshin and Potap'yev (1966); neutron activation methods.
2. Shcherbakov and Perezhogin (1964); neutron activation methods.
3. Mantei and Brownlow (1967); neutron activation methods.
4. Zvereva and Gavrilenko (1971); neutron activation methods.
5. Anoshin *et al.* (1970); neutron activation methods.
6. Bushliakov (1971); neutron activation methods.
7. Geological Survey of Canada; assay and atomic absorption methods.

inclusions of pyrite and/or magnetite which these minerals often contain. Biotite and other ferromagnesian minerals devoid of pyrite and/or magnetite contain only very low amounts of gold (<0.01 ppm) in our experience.

Gold in native elements

Gold is a frequent trace and minor constituent in native elements as shown in Table 8. In general the metallic elements contain more gold than the semimetallic and nonmetallic varieties such as graphite, sulphur, etc.

Native arsenic occurs in some gold deposits as does also native antimony. Both minerals contain relatively high contents of gold. My investigations of native arsenic from Bau, Sarawak by polished sections and microprobe show that some of the gold is present as minute inclusions. This gold amounts to about 1/700 of the overall amount determined by assay. It is concluded, therefore, that the remainder of the gold is present as a lattice constituent of the arsenic. The same features prevail in the native antimony from West Gore, Nova Scotia, except that more of the gold is present in the native form.

Bismuth is notably low in gold content according to my analyses. The small amounts present are probably lattice constituents. Graphite contains a little gold, but most of the element is probably present in the small crystals and specks of pyrite, which are invariably present in graphite from veins. The values for diamond given by Mirzaev and Khabirov (1972) are very low.

Copper, silver and the various platinoid metals and alloys invariably contain gold, often in relatively large amounts. Pure native copper from some deposits, mainly in Ontario and Michigan, contains only traces of gold according to my experience. Other deposits, however, appear to have considerable amounts of gold in the native copper according to the published analyses. Probably most of the gold reported in native copper is a lattice constituent. Native silver from Cobalt, Ontario is notably low in gold, but that from Kongsberg, Norway in similar deposits, is relatively enriched in the element according to the published analyses. Aurian silver or kustelite is reported from a number of deposits. In this mineral Au substitutes for Ag, and a complete series extends through argentian gold (electrum) to gold. The various platinoids including native platinum, platiniridium, osmiridium, aurosmiridium, iridosmine, etc. generally contain gold, especially in placers. The quantity of gold varies widely as shown by the analyses in Table 8. In some deposits most of the gold is probably present as a lattice constituent in the native platinoids; in other deposits the platinoids are intergrown with free gold. In a few deposits such as those in the Kuznetsk, Alatau region of U.S.S.R the gold forms thin blooms and films, and more rarely thin crusts on nuggets of osmiridium (sysertskite) and similar platinoids. Small fractures may also contain this type of gold which according to Syrovatskii *et al.* (1969) is definitely secondary and formed within the placers.

Some authors record up to 2.7 per cent Au in native tellurium, but my research on the mineral indicates that little gold is partitioned into pure native tellurium. Samples from Cupra mines, Quebec show only traces of gold, whereas associated tellurides such as altaite and tellurbismuthite are greatly enriched in the element.

The sulphur of gossans on auriferous deposits commonly contains some particulate native gold. A sample of native sulphur from acidic sulphate gossan on cupriferous pyrite deposits on the island of Cyprus is reported to contain 8200 ppm Au (H.A. Tourtelot; U.S. Geol. Surv. Circ. 622, p. 5, 1969). The gold occurs as particles, 10 to 100μ across, and the smaller ones show poorly formed crystal faces. The larger particles are rounded, resembling framboidal pyrite which many geologists believe to be of bacterial origin. The sulphur associated with fumaroles, salt domes and other types of deposits is normally low in gold (<0.01 ppm).

Gold in silver and base metal tellurides and selenides

Gold is a common trace and minor element in practically all of the silver and base metal tellurides and selenides, and in

Table 8. Gold and silver contents of native elements

Native element and locality	Au	Ag	Au/Ag ratio	Reference
	(ppm)			
Arsenic, As				
Germany; Andreasberg, Harz	150	>1000	<0.1	1
Malaysia; Bau, Sarawak	20–40	30–50	0.75	3
Germany; Thüringia	5	1	5	4
Allemontite, (As, Sb)				
Canada; Atlin, B.C.	500	–	–	4
Canada; Atlin, B.C.	421	294	1.43	3
Antimony, Sb				
U.S.A.; Kern County, California	35.2	0.7	50.2	3
Canada; West Gore, N.S.	12.0	0.3	40.0	3
Bismuth, Bi				
Germany; Altenberg, Saxony	0.03	3.5	0.008	3
Canada; O'Brien Mine, Cobalt, Ont.	0.050	318	0.0001	3
Australia; New South Wales	1.8	3.5	0.51	3
Graphite, C				
Canada; Black Donald Mines, Calabogie, Ont.	0.035	<0.5	>0.07	3
Canada; Buckingham, Que.	0.020	<0.5	>0.04	3
Diamond, C				
U.S.S.R., various	0.00056–0.0031			4
Copper, Cu				
Norway; Kviteseid	200	>1000	<0.2	1
Canada; Indian Mine, Thunder Bay District, Ont.	0.02	41	0.0004	3
U.S.A.; Pewabic bed, Keweenaw Peninsula, Michigan	<0.01	73.5	<0.0001	3
Various (Chukhrov)	2–3%			4
Silver, Ag				
Norway; Kongsberg	40	98.45%	≪1	4
Canada; (Nova Scotia Mine, Cobalt, Ont.)	3.0	99+%	–	3
Canada; (Castle-Tretheway Mine, Gowganda, Ont.)	4.1	99+%	–	3
Canada; (Echo Bay Mine, Echo Bay, N.W.T.)	13.5	99+%	–	3
Aurian silver (kustelite)	0–30%	–	–	4
Iron, Fe				
Greenland; Ovifak, Disko	1–5	5–10	0.4	2
Germany; Bühl	0.5	5–10	0.06	2
Zinc, Zn				
Canada; Keno Hill, Y.T.	nd	tr	–	3
Lead, Pb				
Various	tr	tr	–	4
Mercury, Hg				
Various	tr–m	tr–m	–	4
Iridosmine, (Os,Ir)				
Australia; New South Wales	800	–	–	4
U.S.S.R.; Urals	>1000	600	>1.6	1
Various (Chukhrov)	9000	–	–	4
Osmiridium (Ir, Os)				
Various (Chukhrov)	19.3%	–	–	4
Platinum, Pt				
Brazil	>1000	>1000	–	1
U.S.S.R.; Urals	500	200	2.5	1
Various (Chukhrov)	up to 3%	–	–	4
Canada; Similkameen River, B.C.	300	40	7.5	3
Canada; Horsefly River, B.C.	240	10	24	3
Canada; North Saskatchewan River, Alta.	3600	1300	2.8	3
Platiniridium, (Pt,Ir)				
Brazil	200	–	–	1
Sulphur, S				
Mexico; San Felipe, Baja California	0.030	<0.5	>0.06	3
Italy; Cianciani	<0.01	<0.5	–	3
Tellurium, Te				
Canada; Cupra Mines, Quebec	<10	>50		3
Various (Chukhrov)	0.15–2.78%			4

References: 1. Analysis by Noddack and Noddack (1931).
2. Analysis by Goldschmidt and Peters (1932).
3. Analysis by Geological Survey of Canada, Mines Branch and Bondar-Clegg and Co., Ltd., Ottawa.
4. Other sources: Hintze (1889–1968), Doelter *et al.* (1911–1931); Palache *et al.* (1944); Fischer (1958–59); Chukhrov (1960–1965); Mirzaev and Khabirov (1972).

Notes: nd – not detected
tr – trace
m – minor element

some it occurs in considerable quantities (Table 9). From the available data the tellurides appear to contain more of the element than the selenides, but the two types of minerals rarely coexist in one setting, and the partition of gold between the two species is, therefore, relatively unknown. Since natural gold tellurides are common and natural gold selenides *sensu strictu* have not yet been reported one can assume that the bond of gold to tellurium is much more stable than that to selenium in nature. In the double selenide, fischesserite, Ag_3AuSe_2, the gold content is 27.4 per cent.

Table 9. Gold and silver contents of silver and base metal tellurides and selenides

Mineral and locality	Au	Ag	Au/Ag ratio	Reference
	(ppm)			
Hessite, Ag_2Te				
U.S.S.R.; Krasnyi Klyutsh	2000	62.45%	0.003	1
Altaite, PbTe				
Australia; Kalgoorlie	200	4300	0.05	1
U.S.S.R.; Altai	400	>1000	<0.40	2
Melonite, $NiTe_2$				
Australia	3220	770	4.2	2
Clausthalite, PbSe				
Germany; Tilkerode, Harz	250	>1000	<0.25	2
Berzelianite, Cu_2Se				
Sweden; Skrikerum	100	>1000	<0.10	2
Czechoslovakia; Bukov	0.1	1460	0.00007	3
Umangite, Cu_3Se_2				
Canada; Beaverlodge district, Sask.	0.3	8450	0.00003	3

References: 1. Hintze (1889–1968); Doelter *et al.* (1911–1931); Palache *et al.* (1944); Chukhrov (1960–1965).
2. Noddack and Noddack (1931).
3. Laboratories, Mines Branch and Geological Survey of Canada.

As shown in Table 9 silver tends to predominate over gold in the base metal tellurides and selenides, and the Au/Ag ratio is invariably less than 1, often by several orders of magnitude, especially in the selenides. In the tellurides and selenides gold may replace Ag, Pb, Ni, Bi and Cu, although in many of these minerals, especially the tellurides, small amounts of gold tellurides may be mixed with the base metal varieties.

Gold in sulphides, arsenides, antimonides, sulphosalts and similar minerals

There is a large literature on the gold content of these minerals. The data in Tables 10 and 11 have been compiled up to the end of 1975 from the most significant works on the subject and from the numerous analyses done in the laboratories of the Geological Survey. It is recognized that the data are incomplete, but the ranges shown give an idea of the gold content to be expected in the common sulphides, arsenides, etc.

One of the major problems in ascertaining the role of gold in sulphides, arsenides and allied minerals is the matter of purity of the samples. In a great many occurrences the sulphides, arsenides, etc. are intimately admixed with native gold and gold minerals or contain these minerals as blebs, specks and small crystals. In many cases the investigators do not state whether the minerals are homogeneous or not, and it is impossible to tell if the gold is in the lattices of the minerals or present in one or more admixed gold minerals. Geochemists doing spectrographic and other types of analytical work on sulphides, arsenides and similar minerals should examine their specimens for homogeneity by microscopic methods or better still by means of the electron probe. The data would then be of more significance in ascertaining the extent of replacement of one element by another.

The arsenides and antimonides commonly carry small amounts of gold, those containing the platinum metals (sperrylite) having the highest concentrations according to our analyses. Other arsenides, particularly the copper-bearing species, e.g., domeykite also appear to contain relatively high amounts of gold as shown by the analyses of others. In the arsenides and antimonides gold is probably a lattice constituent and substitutes for the platinum metals and copper and possibly also for cobalt, nickel and iron.

Gold is a common microconstituent in a large number of sulphides and sulphide-arsenides, being concentrated mainly in the copper, silver and antimony species and in pyrite and arsenopyrite. The lead and zinc sulphides and sulphosalts do not normally contain enrichments of gold, although in some deposits galena and sphalerite are indicators of high gold values. Where this occurs the gold is mainly in the native form occurring in close associationship with galena and sphalerite and frequently within these two sulphides as specks and blebs. This has been recently confirmed by the researches of Mincheva-Stefanova and Stefanov (1972) who analyzed some 65 sphalerites from various lead-zinc and polymetallic deposits in Bulgaria. Their values range widely (0.0012–4.08 ppm Au); most of the gold is present in the sphalerite as very finely dispersed inclusions of the native metal. In some sphalerites the dark coloured zones were relatively enriched in gold compared with the lighter zones. The greenish shade of a yellowish green cleiophane (greenish sphalerite with a gold content of 0.0792 ppm) was thought to be due to the presence of very finely dispersed gold throughout the mineral. According to my observations, the case for galena may, however, be somewhat different, since some rather pure galenas contain high contents of gold. In such galenas gold may replace lead, the electronic compensation being provided by tellurium in the sulphur positions. Alternatively these galenas may contain microscopic specks of gold tellurides. Whatever the mechanism it is common to find tellurium in most auriferous galenas.

In a recent paper Parilov and Netaliyeva (1973) carried out a number of leaching experiments on argentiferous sphalerite using NH_4OH. Their results indicate that much of the silver in sphalerite is present in the halide form (AgCl), possibly as epitaxitic inclusions because of the similarities in the crystal structures of ZnS and AgCl. Perhaps some of the gold in some sphalerites is, likewise, present in the halide form.

In the various copper, silver and antimony sulphides and sulphide-arsenides gold occurs native and apparently as a lattice constituent. Chalcocite, bornite, chalcopyrite and enargite frequently carry relatively high contents of gold, yet do not exhibit any evidence of the presence of the free metal under the highest magnifications possible with the ore microscope. In this case it seems probable that gold is a lattice constituent substituting in the copper sites. The same situation prevails for a variety of silver minerals, including argentite, freibergite and argyrodite, suggesting that gold in these

Table 10. Gold and silver contents of some common sulphides, arsenides, antimonides, bismuthides and sulphosalts

Mineral and locality	Au (ppm)	Ag (ppm)	Au/Ag ratio	Reference	Remarks
ARSENIDES AND ANTIMONIDES					
Niccolite, NiAs					
Various	1–10	10–100	<1	1	Purity unknown
Canada; Cobalt, Ont.	<0.1	7.2	<1	9	Pure
Smalite, [(Co,Ni)As_{3-x}]					
Canada; O'Brien Mine, Cobalt, Ont.	0.045	48 700	<1	9	Contains micro-inclusions of silver
Skutterudite, (Co,Ni)As_3					
Various	0.1–5	1–1000	<1	1	Purity unknown
Morocco	8.4	<0.5	>1	9	Pure
Canada; Cobalt, Ont.	<0.1	23 000	<1	9	Contains micro-inclusions of native silver
Safflorite, (Co,Fe)As_2					
Various	0.3–5	0.5–100	<1	1	Purity unknown
Rammelsbergite, $NiAs_2$					
Germany; Schneeberg	0.3	5	0.06	1	Purity unknown
Domeykite, Cu_3As					
Mexico; Paracatas	25	>1000	<1	2	Purity unknown
Sperrylite, $PtAs_2$					
Canada; Sudbury, Ont.	40	300	0.13	2	Purity unknown
Canada; Sudbury, Ont.	700	2300	0.3	9	Pure
Breithauptite, NiSb					
Canada; Cobalt, Ont.	<0.01	460	<1	9	Contains micro inclusions of silver
SULPHIDES AND SULPHIDE-ARSENIDES					
Chalcocite, Cu_2S					
U.S.A.; Kennecott Mine, Alaska	<0.02	265	<0.00007	9	Pure
U.S.A.; Butte, Montana	12	500	0.02	2	Purity unknown
U.S.A.; Butte, Montana	1.05	175	0.006	9	Pure
S.W. Africa; Tsumeb	0.075	93	0.0008	9	Pure
Canada; Alma, N.B.	7.0	68.2	0.10	9	Pure
Canada; Dorchester, N.B.	0.03	320	0.00009	9	Pure
Bornite, Cu_5FeS_4					
Germany; Mansfield	8	>1000	<1	2	Purity unknown
Canada; Copper Mountain, B.C.	0.475	94.7		9	Pure
Canada; Batachewan Bay, Ont	0.07	31.2		9	Pure
Galena, PbS					
U.S.S.R.; Almalyk	13	450	0.03	3	Purity unknown
U.S.S.R.; Karamazar deposit	3.0	1126	0.002	8	Purity unknown
Germany; Thuringia	0.01	20–100	<1	4	Purity unknown
South Africa	30	>1000	<1	2	Purity unknown
Canada; Keno Hill, Yukon	0.01–0.19	1650–6300	<1	9	Contains minor amounts of tetrahedrite
Canada; Walton, N.S.	<0.02	355	<1	9	Pure; assay
Canada; Carleton Place, Ont.	0.015	32	0.0004	9	Pure; assay
Canada; Hudson Bay Range, B.C.	trace	2257	<1	12	98% pure
Sphalerite, ZnS					
Germany; Silesia	3	150	0.02	2	Purity unknown
Germany; Bühl	0.5	5	0.10	1	Purity unknown
U.S.S.R.; Chukotka	500			5	Purity unknown
U.S.S.R.; Karamazar deposit	3.3	364	0.009	8	Purity unknown
U.S.S.R.; Uzbekistan	1.3	75	0.017	8	Purity unknown
Canada; Keno Hill, Y.T.	0.3–0.5	10–5000	<1	9	Some specimens contain tetrahedrite
Canada; Yellowknife, N.W.T.	<10	10–100		9	Pure; spectrographic analyses
Canada; Hudson Bay Range, B.C.	3.4	68.4–342.0	0.01–0.05	12	94–97% pure
Canada; Orchan Mine, Que.	0.55	38.3	0.014	10	Mill concentrate
Canada; Normetal Mine, Que.	0.41	50.95	0.008	10	Mill concentrate

Table 10. (cont'd.)

Mineral and locality	Au	Ag	Au/Ag ratio	Reference	Remarks
	(ppm)				
Chalcopyrite, $CuFeS_2$					
Germany; Clausthal	20	800	0.025	2	Purity unknown
Germany; Zwickau	0.2	100–1000	<1	1	Purity unknown
Germany; Thüringia	0.02	~10	<1	4	Purity unknown
U.S.S.R.; Almalyk	22	0.02	1100	3	Purity unknown
U.S.S.R.; Karamazar deposit	4	927	0.004	8	Purity unknown
U.S.S.R.; Uzbekistan	8	90–212	<1	8	Purity unknown
Canada; Yellowknife, N.W.T.	<10	10–1000	<1	9	Pure; spectrographic analyses
Canada; Keno Hill, Y.T	0.9	1280	0.0007	9	Pure; assay
Canada; Walton, N.S.	0.17	1300	0.0001	9	Pure; assay
Canada; Hudson Bay Range, B.C.	6.84	694.2	0.009	12	97% pure
Canada; Horne Mine, Noranda, Que.	22.26	56.08	0.39	10	Mill concentrate
Canada; Opemiska Mine, Que.	4.44	102.6	0.04	10	Mill concentrate
Canada; Orchan Mine, Que.	2.32	533.5	0.004	10	Mill concentrate
Canada; Campbell-Chibougamu Mine, Que.	20.00	85.50	0.233	10	Mill concentrate
Canada; Normetal Mine, Que.	8.2	399.11	0.020	10	Mill concentrate
Pyrrhotite, $Fe_{1-x}S$					
Norway; Frøysa	2	100	0.02	2	Purity unknown
Germany; Bühl	0.5	5	0.1	1	Purity unknown
U.S.S.R.; Chukotka	2	–	–	5	Purity unknown
East Greenland	0.003	–	–	6	Analysis by neutron activation
Canada; Buckingham, Que.	0.1	<1	<1	9	Pure; assay sample from pegmatite
Canada; Yellowknife, N.W.T.	<10	<1	–	9	Pure; spectrographic analysis
Canada; Hudson Bay Range, B.C.	22.2	44.4	0.5	12	98% pure
Pentlandite, $(Fe,Ni)_9S_8$					
Norway; Espedalen	6	140	0.04	2	Purity uncertain
Germany; St. Blasien	0.5	10–100	<1	1	Purity uncertain
Stannite; Cu_2FeSnS_4					
England; Cornwall	2–5	100	<1	1, 2	Purity uncertain
Czechoslovakia; Zinnwald	0.2	100–1000	<1	1	Purity uncertain
Argentite (Acanthite), Ag_2S					
Germany; Freiberg	200	–		2	Purity uncertain
Canada; Cobalt, Ont.	0.130	–		9	Pure
Canada; Thunder Bay district, Ont.	0.380	–		9	Pure
Canada; Echo Bay, N.W.T.	0.320	–		9	Pure
Cinnabar, HgS					
Spain; Almaden	0.020	<0.5		9	Pure
U.S.S.R.; Khaydarkan deposit	0.070	<0.5		9	Pure
U.S.A.; New Almaden, Calif.	0.180	<0.5		9	Pure
Stibnite, Sb_2S_3					
Germany; Thüringia	0.9	2	0.45	4	Purity unknown
U.S.S.R.; Chukotka	20			5	Purity unknown
Canada; Yellowknife, N.W.T.	10–20	100–1000	<1	9	Impure; contains native gold, aurostibite, tetrahedrite, etc.
Japan	0.025	0.6	0.04	9	Pure
Bismuthinite, Bi_2S_3					
Canada; Wakefield Tp., Que.	0.025	28	0.0009	9	Pure
Hungary; Bánát Mines	3300	–	–	11	Contains inclusions of native gold
Molybdenite, MoS_2					
Various	0.05–10	10–>100	<1	1	Purity unknown
Germany; Thüringia	0.02	>100	<1	4	Purity unknown
Canada; Yellowknife, N.W.T.	0.145	<1		9	Pure

Table 10. (cont'd.)

Mineral and locality	Au (ppm)	Ag (ppm)	Au/Ag ratio	Reference	Remarks
Pyrite, FeS_2					
South Africa (Rand)	0.18	0.10	1.8	9	Probably contains inclusions of gold
East Greenland	0.016			6	Purity unknown
Germany; Thüringia	0.2	5	0.04	4	Purity unknown
Norway; various	0.4	10–70	<1	2	Purity unknown
Italy; various	0.12–200			7	Purity unknown
U.S.S.R.; Almalyk	3.5	36	0.097	3	Purity unknown
U.S.S.R.; Chukotka	10			5	Purity unknown
U.S.S.R.; Karamazar deposit	27	141	0.191	8	Purity unknown
U.S.S.R.; Uzbekistan	40–187	37–232	–	8	Purity unknown
Canada; Hudson Bay Range, B.C.	4.8	78.6	0.06	12	80% pure
Alabandite, MnS					
Roumania; Nagyág	2	30	0.07	2	Purity unknown
Marcasite, FeS_2					
Germany; Westphalia	4	100	0.04	2	Purity unknown
Siegenite, $(Co,Ni)_3S_4$					
Germany	1	>1000	<1	1	Purity unknown
Arsenopyrite, FeAsS					
Norway; Skutterud	8	90	0.09	2	Purity unknown
Germany; Saxony	0.5	10	0.05	1	Purity unknown
Germany; Thüringia	0.5	15	0.03	4	Purity unknown
U.S.S.R.; Chukotka	200			5	Purity unknown
U.S.S.R.; Karamazar deposit	1.5	318	0.004	8	
U.S.S.R.; Uzbekistan	40–559	44–160	–	8	
Canada; Hudson Bay Range, B.C.	12.9–20.5	3.4–17.1	0.75–5.9	12	99% pure
Cobaltite, CoAsS					
Canada; Cobalt, Ont.	5	5	1	1	Purity unknown
Norway; Skutterud	10	200	0.05	2	Purity unknown
Norway; Skutterud	0.5	1	0.5	1	Purity unknown
Sweden; Tunaberg	0.5	10	0.05	1	Purity unknown
Germany; various	0.3–3	5–10	<1	1	Purity unknown
Germany; Thüringia	5	20	0.25	4	Purity unknown
Canada; Cobalt, Ont.	0.04	50 000	<1	9	Contains microinclusions of silver
Gersdorffite, NiAsS					
Germany; various	5–100	1–100		1	
Ullmannite, NiSbS					
Germany; various	0.5–20	10–1000	<1	1	
Germany; Thüringia	1	10	0.1	4	
SULPHOSALTS					
Tetrahedrite, $Cu_{12}Sb_4S_{13}$					
England; Cornwall	8	>1000	<1	2	Purity unknown
Austria; Tyrol	60	>1000	<1	2	Purity unknown
Germany; Thüringia	0.02	20–100	<1	4	Purity unknown
U.S.S.R.; Karamazar deposit	91	15 097	0.006	8	Purity unknown
Canada; Yellowknife, N.W.T.	10–100	10–>1000	<1	9	Pure
Canada; Keno Hill, Y.T. (freibergite)	10.9–44.8	5–17%	<1	9	Pure
Canada; Clyde Forks, Ont.	0.8	762	0.001	9	Pure
Canada; Hudson Bay Range, B.C.	71.8	25–27%	<1	12	
Tennantite, $Cu_{12}As_4S_{13}$					
Canada; Walton, N.S.	<0.03	785	<1	9	Pure
Boulangerite, $Pb_5Sb_4S_{11}$					
Canada; Keno Hill, Y.T.	0.5	230	0.002	9	Pure
Jamesonite, $Pb_4FeSb_6S_{14}$					
Canada; Dublin Gulch, Y.T.	0.7	240	0.003	9	Pure
Germanite, $Cu_3(Ge,Ga,Fe,Zn)(As,S)_4$					
S.W. Africa; Tsumeb	40	800	0.05	2	Purity unknown

References:
1. Goldschmidt and Peters (1932).
2. Noddack and Noddack (1931).
3. Badalov and Terekhovich (1966).
4. Fischer (1958–1959).
5. Sidorov (1966).
6. Vincent and Crocket (1960).
7. Minguzzi (1947).
8. Badalov and Badalova (1967).
9. Analyses by laboratories, Geological Survey of Canada, Mines Branch, and Bondar-Clegg and Co. Ltd., Ottawa.
10. Mine analyses.
11. Koch, S. (1948), Acta Mineral. Petrog. vol.2, p. 4.
12. Kirkham (1968).

Table 11. Gold and silver contents of pyrite, pyrrhotite and arsenopyrite from some Canadian mineral deposits and rocks

Mineral and locality	Au (ppm)	Ag (ppm)	Au/Ag ratio	As (ppm)	Sb (ppm)	Remarks
Pyrite:						
Buckingham, Quebec (composite sample)	<0.01	0.2	<1	16	5	In pegmatites
Richardson Mine, Eldorado, Ont.	0.5	2		56	<1	In skarn
Keno and Galena Hills, Y.T. (6 samples from Pb-Zn-Ag Deposits)	0.08–6.1	<10–200	<1	1100–6760	10–265	Not associated with arsenopyrite. No gold seen in polished sections. Minor tetrahedrite in some samples
Keno and Galena Hills, Y.T. (3 samples from gold-quartz deposits)	2.6–15.7	<10–200	<1	2900–6500	10–1020	Associated with arsenopyrite. No gold seen in polished sections. Minor tetrahedrite and other sulphosalts in some samples
Yellowknife, N.W.T.	1–5	5–20	<1	100–500	5–50	In gold-quartz deposits
Walton, N.S. (1 composite sample)	<0.01	172	<1	5600	40	Associated with chalcopyrite, galena, sphalerite and tennantite in Pb-Zn-Ag orebody
Blind River, Elliot Lake, Ont.	0.41	7.5	0.05	300	6	Pure, from uraniferous ore
Geco Mine, Manitouwadge, Ont.	1.52	400	0.003	–	–	Massive sulphide body
Flin Flon Mine, Man.	0.75	33	0.02	–	–	Massive sulphide body
Kidd Creek Mine (Texas Gulf Sulphur), Timmins, Ont.	0.01	8	0.001	–	–	Massive sulphide body
Nigadoo Mine, Bathurst, N.B.	0.005	140	<0.001	–	–	Massive sulphide body
Heath Steele Mine, N.B.	1.00	28	0.035	–	–	Massive sulphide body
Horne Mine, Noranda, Que.	2.00	20	0.10	–	–	Massive sulphide body
Faro Mine, Y.T.	1.00	58	0.017	–	–	Massive sulphide body
Weedon Pyrite Mine, Que.	0.15	5	0.03	–	–	Massive sulphide body
Louvem Mine, Val d'Or, Que.	0.23	5	0.046	–	–	Massive sulphide body
New Calumet Mine, Que.	1.36	750	0.002	–	–	Skarn
Cobalt area, Ont.	0.160	5	0.03	1400	10	In pyritiferous graphitic schist and slate interbedded with greenstone. Minor amounts of chalcopyrite, pyrrhotite and other sulphides
Keno-Galena Hills, Y.T.	0.06–1.9	2–14	<1	500–1100	15–34	In quartzite, phyllite and schist remote from veins
Walton, N.S. (10 composite samples)	<0.02–0.17	0.34–4.0	<1	230–940	26–124	In shales; well removed from mineralized zones
Walton, N.S. (3 composite samples)	0.06–0.37	1.7–112.8	<1	70–1600	5–28	In coalified plant zones near minor faults
Ovens, N.S. (2 composite samples)	0.15	1.16	0.13	200	6–21	In slates and greywacke
Rawdon area, N.S. (composite sample)	<0.01	0.5	<1	6400	7	In slates
Wabana, Nfld. (composite sample)	0.08	1.4	0.06	620	30	Pyritiferous zone in iron beds
Florence area, Ont.	0.02	0.17	0.1	1600	41	In Devonian shales
Port Hood, N.S.	0.04	0.17	0.23	190	9.6	In coal
Stonehaven, N.B.	0.02	12	0.001	1200	16	In pyritiferous coal-bearing sandstone. Minor amount of galena present
Champlain Lookout, Gatineau Park, Que.	0.025	1	0.02	–	–	Pyrite cubes in Grenville gneiss
Pyrrhotite:						
South of Bancroft, Ont.	0.012	6.1	0.001	300	7.5	In quartzites
Buckingham, Que.	0.01	<1	<1	–	–	In pegmatites
Yellowknife, N.W.T.	<10	<1	–	–	–	In gold-quartz veins (spectrographic analyses)
St. Stephen, N.B.	0.105	6.2	0.017	160	<15	In nickel deposit
Argenteuil Co., Que.	1.010	6.0	0.168	196	–	
Noranda, Que.	0.980	6.0	0.163	130	<15	In Cu-Zn deposit; minor chalcopyrite
Hastings County, Ont.	0.020	2.7	0.007	76	<15	Pegmatite
MacDonald Mine,						

Table 11. (cont'd.)

Mineral and locality	Au	Ag	Au/Ag ratio	As	Sb	Remarks
	(ppm)			(ppm)		
Pyrrhotite (cont'd.):						
Garson Mine, Sudbury, Ont.	0.220	3.6	0.061	61	<15	Massive sulphide deposit
Alexo Mine, Cochrane, Ont.	0.220	5.0	0.044	58	<15	Massive sulphide deposit
Leroi Mine, Rossland, B.C.	2.110	88.0	0.024	61	<15	Massive sulphide deposit
Arsenopyrite:						
Keno-Galena Hills, Y.T.	9.7–17.3	10–400	<1	>40%	380–1160	In gold-quartz veins
Dublin Gulch area, Y.T.	2.4	2–20	<1	>40%	170–190	In gold-quartz veins
Yellowknife, N.W.T.	10	5	2	>40%	10–100	In gold-quartz veins
West Lake Gold Mine, Mt. Uniacke, N.S.	>500	48	>10	>40%	330	In gold-quartz veins
Marmora Tp., Hastings County, Ont.	3.75	2.8	1.34	>40%	45	Gold-quartz prospect
Howry Creek Gold Mine, Curtin Tp., Sudbury district, Ont.	17.85	2.5	7.14	>40%	50	In gold-quartz veins
Ovens area, N.S.	2.05	0.5	4.1	>40%	12–200	In slates and greywacke
Molega area, N.S.	5.95	0.8	7.4	>40%	180	In slates and greywacke near gold-quartz veins

*Analyses by Laboratories, Geological Survey of Canada and Bondar-Clegg and Co. Ltd., Ottawa.

minerals occupies some of the silver sites in the lattice. The situation with stibnite is complex. Pure stibnite from Japan appears to contain relatively little gold and silver. In other deposits, however, the mineral is gold- and silver-bearing as at Yellowknife and Red Lake in Canada and in the Chukotka deposits, U.S.S.R. In the Canadian deposits examined by the writer the stibnite is intimately intergrown with tetrahedrite which is gold-bearing and with aurostibite and finely divided native gold which cannot effectively be separated from the antimony sulphide prior to analysis. Hand picked blades of stibnite from these deposits, however, generally contain only a little gold. These facts suggest that relatively little gold enters the stibnite lattice, and that if gold is present in stibnite it occurs predominantly in the native form, as aurostibite, or in admixed sulphosalts. It is of interest to note, however, that Dickson *et al.* (1975) found that synthetic stibnite, orpiment and realgar took up considerable amounts of gold from the gold tubes used in their synthetic studies: stibnite from 1000 to 2000 ppm Au; orpiment from 100 to 2000 ppm Au; and realgar from 50 to 200 ppm Au. The exact nature of the gold in the sulphides was not determined.

Of the various sulphosalts, the tetrahedrite-tennantite series is the most consistently enriched in gold. Only rarely can native gold or other gold minerals be detected in pure auriferous tetrahedrite-tennantite samples suggesting that most of the gold in these minerals is a lattice constituent substituting for copper. That gold prefers the copper sulphosalts has been shown by the writer at Keno Hill (Boyle, 1965*a*). There, tetrahedrite (freibergite) contains from 11 to 45 ppm Au whereas the closely associated lead sulphosalts, boulangerite and jamesonite, contain of the order of 0.5 ppm Au. In a qualitative way the same is true at Yellowknife, although here great difficulty is encountered in obtaining pure separates of each mineral.

Gold is commonly found in pyrite and arsenopyrite, especially in auriferous deposits, a feature of the element which has been known for a long time. Commonly associated elements in the pyrite of both host rocks and deposits are silver, arsenic and antimony and in arsenopyrite, silver and antimony, as shown in Table 11. With respect to pyrite there is no consistent relationship between the arsenic and antimony contents and the gold content; the same is true as regards the antimony and gold contents in arsenopyrite. In pure pyrite the Au/Ag ratio is generally less than 1 in all types of occurrences according to our analyses and those published in the literature. Only rarely have we found exceptions to this generalization, and these are obvious such as in samples where the pyrite contains visible gold as blebs or in late fractures. In samples of pyrite from the Rand ores we have analyzed we have noted an Au/Ag ratio generally greater than 1, but some samples give ratios less than 1. Those having the high ratios generally have specks of native gold which cannot be physically separated from the pyrite. Many investigators have commented upon the close association of gold and pyrite in the Rand among whom Pelletier (1940), Liebenberg (1955) and Saager (1969) should be mentioned as having made detailed studies. It is of interest to note that the Au/Ag ratio in the pyrite in the relatively nonauriferous Blind River–Elliot Lake quartz-pebble conglomerates have a low ratio (0.05) (Table 11). The Au/Ag ratio in arsenopyrite appears to be erratic judging from some of my data and those in the literature. However, in a number of Nova Scotia deposits where I have carefully examined coexisting pyrite and arsenopyrite, both devoid of secondary gold in fractures, the ratio is generally greater than 1 in arsenopyrite and less than 1 in pyrite. This suggests a tendency for arsenopyrite to concentrate gold whereas pyrite tends to take up silver.

The problem of the nature of gold (and silver) in pyrite and arsenopyrite is an old one that first appeared when metallurgists sought to deal with certain gold ores that are not directly amenable to cyanidation processes. Most of these ores are arsenical and usually contain abundant arsenopyrite, which contains most of the gold. Some, such as those in the Bau area, Sarawak, contain the bulk of the gold in native arsenic and minor arsenopyrite. Others are characterized mainly by gold-bearing pyrite, and still others contain gold-bearing arsenopyrite, sulphosalts and pyrite. Visible and microscopic native gold is present in most of these deposits but often makes up only a small proportion of the total amount of gold in the ores. The remainder of the gold is in an 'invisible' form and can only be won by roasting the ores, which liberates the gold in a state amenable to cyanidation. In some gold ores such as those in Fiji and southeastern Ontario practically all of the gold is 'invisible' in either arsenopyrite or pyrite.

Investigations on the nature of gold (and silver) in pyrite and arsenopyrite have been carried out in the past by Bürg (1930, 1935), Head (1934, 1935*a,b*), Haycock (1937), Van Aubel (1939), Kurauti (1941), Maslenitsky (1944) and Stillwell and Edwards (1946) and more recently by Edwards (1958), Clark (1960), Boyle (1961*a*, 1965*a*), Aubert *et al.* (1964), Tyurin (1965*a*), Schweigart (1965), McPheat *et al.* (1969), Korobushkin (1970), Kalitkina (1971), Badalov (1972), Sakharova *et al.* (1972), Wells and Mullens (1973), Klempert *et al.* (1973) and Gavrilov *et al.* (1974). Two views predominate, one that the gold is present in chemical combination in the pyrite and arsenopyrite, and the other that the gold is present in a finely divided (colloidal?), often submicroscopic state ($<0.1\mu$). The former view appears the more probable as will be shown in the following.

Examination of most polished sections of gold-bearing pyrite and arsenopyrite at high magnifications generally reveals the presence of minute blebs or stringers of native gold or other gold minerals that frequently appear to be of a secondary (exsolved) nature. Calculations of the amount of gold present in these forms generally agree with the assay values within the experimental errors inherent in such work. There are, however, a number of deposits where the assay values of pyrite and arsenopyrite, particularly the latter, greatly exceed the amount of gold minerals visible at the highest magnifications under the microscope. This gold is generally not amenable to cyanidation and, in my experience, can only be recovered, by complete breakdown of the pyrite or arsenopyrite lattice either by roasting or by attack with nitric acid or aqua regia. Under the microprobe, domains or points of high gold content in the pyrite and arsenopyrite can rarely be made out, although in some specimens from Châtelet, France, Aubert *et al.* (1964) were able to distinguish gold-rich areas in arsenopyrite. Microscopic and microprobe work, therefore, suggest that much of the 'invisible' gold in pyrite and arsenopyrite is present in a submicroscopic form, probably a lattice constituent of these two minerals. This view is confirmed by recent research using a microprobe and other techniques on auriferous pyrite (0.065 oz/ton) by McPheat *et al.* (1969). They found that 0.02 to 0.03 oz/ton gold was in solid solution and the balance of the metal was finely dispersed as grains throughout the pyrite. Wells and Mullens (1973), however, opted for the particulate gold theory in the Carlin and Cortez ores of Nevada. Using a microprobe they found that much of the gold in the deposits was present in a submicroscopic form ($<0.5\ \mu$) in tiny grains (<0.005 mm) of arsenian pyrite and in the thin rims of larger pyrite grains; also in small grains of arsenopyrite where it is present. They concluded that since the gold content varies widely within the interior of large pyrite grains, within pyrite rims and within the disseminated pyrite and arsenopyrite – some pyrite and arsenopyrite is barren – that the gold possibly occurs as submicron particles of native gold or of a gold mineral.

Some synthetic and detailed analytical studies have been done on the problem of the state of gold in pyrite and arsenopyrite. Burg (1930, 1935), who first investigated the problem, showed that when natural gold-bearing pyrite from Brad is heated to 600°C in a vacuum and the liberated sulphur vapour is removed, the pyrite is converted to pyrrhotite, and the gold aggregates into small visible particles. If the sulphur vapour is retained, the gold reacts with it to form a gold sulphide with variable stability features. Burg thought that his results suggested solid solution of gold in the pyrite. Kurauti (1941) and Maslenitsky (1944) synthesized gold-bearing pyrite in which the gold could not be detected microscopically. Maslenitsky found that his synthetic pryrite, containing some 340 ppm gold, when heated in vacuo at 600°C behaved in a similar manner to the natural pyrite used by Burg. He also observed inclusions of gold sulphide in his heated synthetic pyrite, and he was also able to synthesize Au_2S_3 and determine its optical properties and etching behaviour. Kurauti's results are especially significant since he found that up to 2000 ppm gold could be substituted in the pyrite lattice, and further that the lattice constant of pyrite decreased with increasing gold content, a feature that proves solid solution of gold in pyrite. Tyurin (1965*a*), on the other hand, found in some experiments that gold concomitantly precipitated with pyrite was present in discrete particles that did not exceed 3μ in size. Clark (1960) also found that the solubility of gold in arsenopyrite at 600°C was insufficient to measurably affect the 131 spacing of the mineral. He did not rule out, however, the possibility of a small gold solubility in arsenopyrite, for example 0.1 per cent (1000 ppm or approximately 30 oz Au/ton of arsenopyrite). In a survey of research for 1969 (U.S.G.S., Prof. Pap. 650-A, p. A108) it was reported that P.B. Barton has found that chalcopyrite can dissolve more than 1.6 weight per cent Au at 650°C and that much of this gold is exsolved on cooling. In the 600–700°C temperature range, galena dissolves between 0.1 and 0.5 weight per cent Au. Bismuth increases the solubility of gold, whereas silver decreases it. This suggests that the mechanism for the incorporation of gold in galena involves Au(I) rather than Au(III). In the same range of temperatures, the solubility of gold in pyrite, pyrrhotite and arsenopyrite is less, perhaps much less, than 0.1 weight per cent. Korobushkin (1970) determined by selective chemical methods that gold in gold-bearing pyrite and arsenopyrite in a deposit in western Kazakhstan was present in two forms – in a cationic form, the gold substituting isomorphously for iron, and in an anionic form, the gold substituting for atoms of arsenic–because of the similarlity of the atomic radii of the two elements (Au, 1.44Å; As, 1.39Å). This view has been reiterated by Badalov (1972). Finally, Kirillov *et al.* (1970) experimented with iron monosulphide and dithiosulphatoaurate, $Na_3[Au(S_2O_3)_2]$ in an attempt to synthesize auriferous pyrite at low temperatures. Their results indicate that the thiosulphate complex disproportionates in aqueous solution in the presence of FeS and that gold passes into the precipitate. The precipitate evidently consisted of FeS grains covered by films of auriferous FeS_2. They write equations for the reactions as follows:

$$2Na_3[Au(S_2O_3)_2] + 4H_2O \rightarrow 2Au + S + 3Na_2SO_4 + 3H_2S$$
$$FeS + S \rightarrow FeS_2$$

Extensive chemical and microscopic work on both auriferous pyrite and arsenopyrite by Haycock (1937), Stillwell and Edwards 1946, Edwards (1958) and McPheat *et al.* (1969) on Canadian, Fijian and Australian gold ores strongly suggest solid solution of gold in the lattice of both pyrite and arsenopyrite. The writer's research on gold-bearing pyrite and arsenopyrite from Yellowknife, N.W.T.; Keno Hill, Yukon; southeastern Ontario; and Nova Scotia suggests the same phenomenon.

There are theoretical reasons for suspecting that gold (and silver) enter the structure of pyrite and arsenopyrite. In pyrite (and the closely related arsenopyrite) each iron atom is covalently bonded to six sulphur atoms at the corners of a nearly regular octahedron. In addition there is also a moderate degree of metallic bonding (Fe-Fe) indicated by the metallic lustre, especially of arsenopyrite. The octahedral covalent radius of ferrous iron in pyrite is 1.23Å, and for gold in aurostibite, $AuSb_2$, a mineral isostructural with pyrite, the octahedral radius is 1.40Å. There is also a similarity in the covalent radius of arsenic (1.39Å) and gold (1.40Å). Gold will form three rather unstable artificial sulphides, AuS, Au_2S and Au_2S_3 indicating that the bond to sulphur although weak is nonetheless possible. These considerations suggest that some gold may substitute in the iron sites of pyrite and arsenopyrite mainly because of similar octahedral radii, a certain affinity for sulphur, and the isostructural relationship of the two minerals pyrite and aurostibite.[18] There may also be some substitution of gold for arsenic in arsenopyrite because of the similar radii and charges of the two elements. In addition the metallic binding in the sulphides, especially arsenopyrite, confers certain alloying properties on them, permitting other metals such as gold and silver to reside in the lattices without excessive distortion especially at high temperatures. Another reason why pyrite is commonly auriferous may be due to certain epitaxial relationships which take place when pyrite and gold are precipitated in veins. Thus, Amosov and Gureev (1970) noted that gold forms thin crystalline crusts on pyrite and concluded that the crystalline data for gold and pyrite suggest epitaxy as a possibility. Further investigations by Amosov *et al.* (1975) indicate the cocrystallization of pyrite and native gold. They suggest that the cocrystallization phenomenon is not the trapping of gold particles by pyrite but the deposition of gold on the surface of the growing pyrite crystals. Research by Sakharova *et al.* (1972) on gold-bearing sulphides (pyrite, arsenopyrite, chalcopyrite and pyrrhotite) from the Darasun deposit in East Transbaikal further confirms the exsolution hypothesis. They found that during heating tests of polished sections of the various sulphides that the finely dispersed ('invisible') gold aggregated into larger particles around 250 to 300°C. A disintegration of some of the larger 'visible' gold particles was also noted in some cases. They concluded that over a long period of time, gold in the sulphides could be redistributed at relatively low temperatures (150–250°C) during the late stage of hydrothermal processes in veins. Kalitkina (1971) also found experimentally that gold migrated relatively rapidly by diffusion in the solid state, preferably along fractures and grain boundaries, in both pyrite and arsenopyrite. During heating tests Klempert *et al.* (1973), likewise, observed exsolution of gold from the auriferous ('invisible' gold) pyrite of the Terek deposit in Uzbekistan. Narseev and Starova (*in* Petrovskaya, 1974) noted that some 95 per cent of the gold in the Kalba deposits of eastern Kazakhstan was dispersed in pyrite and arsenopyrite. After heating the pyrite for 20 min. at 250°C enlarged gold particles and a decrease in the cell parameter of the pyrite took place (from 5.4275 to 5.4095). They suggested that point defects in the lattice of pyrite occur at the sites of the dispersed gold. The decreased microhardness found for the auriferous pyrite was considered to be due to linear dislocations. For the auriferous arsenopyrite the investigators noted a tendency to fine acicular to finely prismatic habit, increased cell parameters *a* and *b* and decreased microhardness. Other details of the auriferous pyrite and arsenopyrite should be sought in the original paper.

Having regard to the above discussion one can postulate, therefore, that early formed high temperature pyrite and arsenopyrite probably take up gold and silver largely in solid solution, or as atomic layers on the growing faces of the sulphide minerals, features that appear to be common in some of the deposits studied by the writer (e.g., Yellowknife; Boyle, 1961*a*). These features are also commented on by Haycock (1937), Stillwell and Edwards (1946), Edwards (1958) and McPheat *et al.* (1969). At lower temperatures the presence of foreign constituents such as gold and silver distort the pyrite and arsenopyrite lattices, with a consequent rise in the free energy of the crystals. In order to reduce the free energy of the system to a minimum under the prevailing conditions gold and silver migrate to nearby low chemical potential sites such as fractures and grain boundaries where they crystallize as silver-bearing native gold. This appears to explain the frequent occurrence of much 'invisible' gold and silver in relatively unfractured and unrecrystallized pyrite and arsenopyrite; where reworking, recrystallization and fracturing are evident the gold is largely present in the native form, although small amounts may remain within the lattice of the two minerals. The exsolution phenomenon described above is probably largely the result of diffusion processes, a conclusion supported by the experimental work of Clark (1960) who found that gold diffused rapidly through arsenopyrite at temperatures of 660°C and higher.

Gold in halides, carbonates, sulphates, oxides and similar minerals

Data on the gold content of some halides, carbonates, sulphates, oxides, etc. are given in Table 12. The nature of gold in many of these minerals is difficult to determine, particularly in those of supergene derivation because of their finely divided, often pulverulent habit, and their common intermixture with a variety of other supergene minerals.

The common halides contain very little gold, generally less than 0.01 ppm. Seeland (1973*a,b*) noted the presence of gold in only one halite sample from Paleozoic rocks in Michigan. The average gold content in the halite was 0.06 ppm. There appear to be no published analyses of the gold content of the various silver halides, nor of the mercury, copper and lead-copper halides and complex carbonate or

[18]Note that practically all pyrite and arsenopyrite (Table 11) contains antimony – sufficient it would seem to bind the gold in the pyrite and arsenopyrite lattice, thus neutralizing any charge differences.

Table 12. Gold and silver contents of some halides, carbonates, sulphates, phosphates, arsenates, antimonates, nitrates, tungstates and oxides.

Mineral and locality	Au	Ag	Au/Ag ratio	Reference	Remarks
	(ppm)				
HALIDES					
Halite, NaCl					
England; Cheshire	0.103	nd	–	1	Purity unknown
Germany; Stassfurt	0.132	nd	–	1	Purity unknown
Canada; Pugwash, N.S.	<0.01	<0.5	–	21	Pure
U.S.A.; Michigan	0.06	–	–	20	
Sylvite, KCl					
Germany; Mecklenburg	0.003	nd	–	2	Purity unknown
Germany; various	0.0001–0.0034	nd	–	14	Potassium salts
Carnallite, $KMgCl_3{\cdot}6H_2O$					
Germany; Bernburg	0.012	nd	–	2	Purity unknown
Sylvite, KCl, *and Carnallite,* $KMgCl_3{\cdot}6H_2O$					
Canada; Ormiston, Sask.	<0.01	<0.5	–	21	Pure
Fluorite, CaF_2					
Canada; St. Lawrence, Nfld.	<0.01	<0.5	–	21	Pure
Canada; Sandy Creek, Que.	<0.01	<0.5	–	21	Pure
Canada; Madoc, Ont.	0.01	<0.5	–	21	Pure
Chlorargyrite, AgCl					
Chile; Chanarcillo	<10	75%	–	21	Pure
Australia; Broken Hill, N.S.W.	<10	75%	–	21	Pure
CARBONATES					
Siderite, $FeCO_3$					
Canada; Keno Hill, Y.T.	0.17	<1	–	21	Pure
Dolomite, $CaMg\ (CO_3)_2$					
Canada; Portage du Fort, Que.	<0.01	<0.5	–	21	Pure
Canada; Bancroft, Ont.	<0.01	<0.5		21	Pure
Canada; Actinolite, Ont.	<0.01	<0.5	–	21	Pure
Calcite, $CaCO_3$					
Canada; Pinks Lake, Que.	0.010	<0.5	–	21	Pure
Canada; Brompton Lake, Que.	0.015	<0.5	–	21	Pure
Mexico; Chihuahua	<0.01	<0.5	–	21	Pure
Calcite (marble)					
Canada; Carleton Place, Ont.	<0.01	<0.5	–	21	Pure
Malachite, $Cu_2(OH)_2(CO_3)$					
Zaire; Katanga	0.015	<0.1	–	21	Pure
U.S.A.; Bisbee, Ariz.	0.01	106	0.00009	21	Pure
Azurite, $Cu_3(OH)_2(CO_3)_2$					
Southwest Africa; Tsumeb	0.190	<0.1	–	21	Pure
U.S.A.; Bisbee, Ariz.	0.210	29	0.007	21	Pure
SULPHATES					
Anhydrite, $CaSO_4$					
Germany; Plömnitz	0.007	nd	–	2	Purity unknown
Gypsum, $CaSO_4{\cdot}2H_2O$					
United States; Michigan	0.02	–	–	20	
United States; Grand Rapids, Mich.	0.083	nd	–	3	Purity unknown
United States; Salina, N.Y.	0.083	nd	–	3	Purity unknown
Canada; Keno Hill, Y.T.	<0.005	<0.005	–	21	In lead-zinc-silver veins and lodes
Canada; Walton, N.S.	0.085	0.51	0.16	21	Evaporite beds, minor pyrite in some samples
Kainite, $KMg(SO_4)Cl{\cdot}3H_20$					
Germany; Plömnitz	0.003	nd	–	2	Purity unknown
Barite, $BaSO_4$					
Canada; Keno Hill area, Y.T.	0.08	<1–5	–	21	In veins (secondary?)
Canada; Madoc, Ont.	<0.01–0.01	<0.5	–		Pure
Canada; Walton, N.S.	<0.01	<0.5	–	21	Pure
Anglesite, $PbSO_4$					
Canada; Keno Hill, Y.T.	0.17–0.5	>1000	<1	21	Oxidation product of galena
Canada; Bathurst, N.B.	<0.01–0.01	44	<1	21	In oxidized zone of massive sulphide deposits
Jarosite, $KFe_3(SO_4)_2(OH)_6$					
Various localities	up to 7900	up to 18%	<1		Gold probably largely present in finely divided native form in most samples
Various sulphates of iron, zinc, etc.					
Canada; Keno Hill area, Y.T.	up to 0.08	41.72	<1	21	Derived from oxidation of pyrite, siderite, etc.

Table 12. (cont'd.)

Mineral and locality	Au	Ag	Au/Ag ratio	Reference	Remarks
	(ppm)				
PHOSPHATES, ARSENATES, ANTIMONATES					
Apatite, $Ca_5(PO_4)_3(F,Cl,OH)$					
Canada; Sandy Creek, Que.	<0.01	<0.5	–	21	Pure
Canada; Sebastopol Tp., Ont.	<0.01	<0.5	–	21	Pure
Canada; Cantley, Que.	<0.01	<0.5	–	21	Pure
U.S.S.R.; various	0.003–0.03	nd	–	15	Bone phosphate
Triplite, $(Mn,Fe)_2(PO_4)F$					
South West Africa; Sandamab	0.5	5	0.1	4	Purity unknown
Pyromorphite, $[Pb_5(PO_4,AsO_4)_3Cl]$					
Canada; Keno Hill area, Y.T.	0.17	100–1000	<1	21	Contains some cerussite, otherwise pure
Germany; Schwarzwald	18–21	200	0.1	16	Purity unknown
Scorodite, $(Fe,Al)(AsO_4){\cdot}2H_2O$					
Canada; Keno Hill area, Y.T.	0.5–10.26	1–50	<1	21	Oxidation product of gold-quartz-arsenopyrite veins
Beudantite, $PbFe_3(AsO_4)(SO_4)(OH)_6$					
Canada; Bathurst, N.B.	0.425	87	0.004	21	In oxidized zone of massive sulphide deposits, pure
Bindheimite, $Pb_2Sb_2O_6(O,OH)$					
Canada; Keno Hill area, Y.T.	0.7	100–1000	<1	21	Alteration product of galena and freibergite
NITRATES					
Soda-Niter, $NaNO_3$					
Chile	0.110	nd	–	1	Purity unknown
TUNGSTATES					
Wolframite, $(Fe,Mn)(WO_4)$					
Canada; Keno Hill area, Y.T.	0.17	<1	–	21	Pure
Scheelite, $Ca(WO_4)$					
Canada; Keno Hill area, Y.T.	0.03	<1	–	21	Pure
OXIDES					
Magnetite, $FeFe_2O_4$					
United States; Helena, Mont.	0.003–0.329 (av., 0.037)	nd	–	7	Range and average of 44 analyses
U.S.S.R.; Altai-Sayan	0.048	nd	–	8	Average of seven analyses
U.S.S.R.; Kazakhstan	<0.001–0.0120	–	–	13	In rocks
U.S.S.R.; Urals	0.0036	nd	–	17	Average in granitoid rocks
U.S.S.R.; Altai, Transbaikal and Urals	0.0034	nd	–	18	Average in granitoid rocks
Canada; Marmora, Crowe Lake, Ont.	0.03	<0.4	–	21	In skarn orebody
Canada; Hull, Que.	<0.01	<1	–	21	Composite of magnetite body in skarn
Ilmenite, $FeTiO_3$					
U.S.S.R.; Altai, Transbaikal and Urals	0.0019	–	–	18	Average in granitoid rocks
U.S.S.R.; Urals	0.0022	–	–	17	Average in granitoid rocks
Canada; St. Urbain, Que.	<0.01	0.4	<1	21	Pure
Hematite, Fe_2O_3					
Canada; Yellowknife, N.W.T.	0.005	<0.04	–	21	In giant quartz veins
Canada; Walton, N.S.	<0.01	7.5	–	21	In oxidized zones of barite deposits
Canada; Balderson, Ont.	<0.01	<0.5	–	21	Pure
Canada; Plummer Tp., Ont.	<0.01	<0.5	–	21	Pure
Pyrolusite, MnO_2					
Czechoslovakia; Platten	0.2	4	0.05	5	Purity unknown
Psilomelane, $BaMn^{2+}Mn_8^{4+}O_{16}(OH)_4$					
Germany: Harz	2	50	0.04	5	Purity unknown
Hausmannite, $Mn^{2+}Mn_2^{3+}O_4$					
Sweden: Langban	0.1	10	0.01	5	Purity unknown
Wad, Mn Oxides					
Germany; Thuringia	0.01	5	<1	6	Purity unknown
United States; various	<100	30–300	–	9	Purity unknown
Various (U.S.A.; Sumatra, etc.)	8–35	up to 1000 or more	1	10	Purity unknown
United States; Arkansas	0.7	23	0.03	11	Purity unknown
Canada; Keno Hill, Y.T.	0.08	100–1000	<1	21	In oxidized zones of Pb-Zn-Ag deposits

Table 12. (cont'd.)

Mineral and locality	Au	Ag	Au/Ag ratio	Reference	Remarks
	(ppm)				
Wad (cont'd.)					
Canada; Walton, N.S.	<0.01	0.8–1.4	<1	21	Nodules in limestone
Canada; Walton, N.S.	0.34	0.68	0.5	21	Spring precipitates (average)
Oceans, manganese nodules	0.003	–	–	12	Average
Canada; Nova Scotia lakes, manganese nodules	0.004	–	–	12	Average
Limonite, iron oxides					
Various	1–35	1–1000 or more	<1		General estimate from world literature. All in oxidized zones of polymetallic deposits
Canada; Keno Hill area, Y.T.	0.08	100–1000	<1	21	In oxidized zones of Pb-Zn-Ag deposits
Canada; Walton, N.S.	<0.01	0.5	<1	21	In oxidized zones of barite deposits
Canada; Bathurst, N.B.	<0.01–21.2	0.5–500	<1	21	In gossans of massive sulphide deposits
Cassiterite, SnO_2					
Various localities	0.2–0.5	1–100	<1	4	Purity unknown
Canada; Keno Hill area, Y.T.	<0.01	1–5	<1	21	Pure
Uraninite, UO_2					
Norway; Brevik	0.1	2	0.05	5	Purity unknown
Czechoslovakia; Jáchymov	0.03	5	0.006	5	Purity unknown
Pitchblende, UO_2					
Canada; Great Bear Lake, N.W.T.	0.025	5.0	0.005	21	Pure
Thorianite, ThO_2					
Ceylon	0.05	2	0.02	5	Purity unknown
Columbite, $(Fe,Mn)(Nb,Ta)_2O_6$					
Norway; Arendal	0.05	2	0.02	5	Purity unknown
Chromite, $FeCr_2O_4$					
Various localities	0.2	1	0.2	4	Purity unknown
Canada; Bird River, Man.	<0.01	3	<1	21	Pure
Samarskite, tantalite, ilmenorutile					
Various localities	0.02–0.65	0.2–0.3	–	19	Purity unknown

Note: nd – not determined

References:
1. Liversidge (1897*a*).
2. Friedrick (1906).
3. Lincoln (1911*b*).
4. Goldschmidt and Peters (1932).
5. Noddack and Noddack (1931).
6. Fischer (1958–1959).
7. Mantei and Brownlow (1967).
8. Shcherbakov and Perezhogin (1964).
9. Hewett, Fleischer and Conklin (1963).
10. Clevenger and Caron (1925).
11. Foley (1960).
12. Harriss *et al.* (1968).
13. Zvereva and Gavrilenko (1971).
14. Goubeau and Birckenbach (1938).
15. Yasyrev (1969).
16. Schmidt (1919).
17. Bushliakov (1971).
18. Anoshin *et al.* (1970).
19. Lunde (1927) and Lunde and Johnson (1928).
20. Seeland (1973*a,b*).
21. Laboratories, Geological Survey of Canada and Bondar-Clegg and Co., Ltd., Ottawa.

hydroxide halides. Analyses of chlorargyite, AgCl, in laboratories of the Geological Survey show gold to be generally less than 10 ppm, indicating that there is little enrichment of the precious metal in silver halides.

All of the common carbonates have a very low gold content, generally much less than 0.01 ppm. This is probably due to the fact that the bond of gold to the carbonate radical is very weak or nonexistent. The complex copper carbonates can apparently accommodate small amounts of gold, probably substituting in the copper sites.

Among the simple sulphates investigated the largest amounts of gold have been noted in anglesite from Keno Hill, Yukon. Its presence in the mineral from this area, however, appears to be spurious; the element is probably largely present in admixed scorodite, bindheimite and beudantite which cannot be physically separated from the anglesite. The anglesite from the Bathurst area is relatively pure and contains on the average only 0.01 ppm Au.

Hand-sorted concentrates of supergene sulphates of copper and iron from underground workings of mines in the Empire district, Colorado, are reported to contain as much as 11.7 ppm Au, 10 ppm Te and 0.60 ppm Hg, all well above background amounts for the area (Lakin, 1969*a*). Higher-than-background amounts of manganese, bismuth, cobalt, nickel and zinc were also detected in the samples. Seeland (1973*a,b*) noted the presence of gold in one sample of gypsum-anhydrite from Paleozoic rocks in Michigan. The amount recorded was 0.02 ppm Au.

As noted above halite and the common sulphates contain only minor traces of gold. Friend and Allchin (1939), however, report 49 ppm Au in pale blue celestite and 85 ppm Au in deep red celestite from Yate, Gloucestershire, England. Colourless crystals contained no detectable gold. In a later paper the same authors (Friend and Allchin, 1940) found up to 4 ppm Au in blue anhydrite from Cropwell Bishop, Nottinghamshire and an average of 23 ppm Au in deep blue Stassfurt halite. They attributed the various colours of these minerals to the presence of colloidal gold. The gold values recorded in these halite and sulphate samples are exceptionally high and should be confirmed. The writer has been unable to find more than a trace of gold (<0.01 ppm) in orange gypsum from Nova Scotia and in most coloured halite, celestite and anhydrite samples from various localities. However, one sample of blue halite from Heringen, Germany, and a sample of blue celestite from Wogau, Germany contained 0.125 and 0.110 ppm Au respectively. The corresponding silver contents were 2.8 and 0.4 ppm.

The basic sulphates of ferric iron and other elements, the jarosites, frequently contain relatively high amounts of gold. Knopf (1915) recorded up to 7900 ppm Au (234 oz/ton) in plumbojarosite from the Boss Mine, southern Nevada, and Head and Miller (1928) found up to 89 ppm Au (2.6 oz/ton) in jarosite from Beatty, Nevada and Marysvale, Utah. In U.S.S.R. Chukhrov *et al.* (1940) report gold- and silver-bearing jarosites from the Maikain deposit, central Kazakhstan, and Friedrich (1960) found gold in jarosite in the Rodalquilar deposit, Spain. In all of these deposits most of the gold is apparently present in the native state as finely divided particles. Despite this feature it would seem from my observations that the jarosites, particularly plumojarosite and argentojarosite, can accommodate some gold in their lattices. Argentojarosite may contain up to 18 per cent silver, a feature proving considerable substitution of the metal in the alkali sites of the jarosite lattice. One can suppose, therefore, that where mobile gold is present in the solutions from which jarosite is crystallized that there may be some substitution of Au(I) (1.37Å) for Ag(I) (1.13Å) because of somewhat similar radii and similar electronic configurations.

There is relatively little gold in any of the apatites investigated in laboratories of the Geological Survey, but the phosphates, pyromorphite and triplite, can apparently accommodate some gold in their lattices probably substituting in the lead or iron sites. Some of the sedimentary bone phosphates in U.S.S.R. carry small amounts of gold. Smirnov and Entin (1975) have noted a bond between gold and phosphorus in a number of genetically diverse phosphatic deposits in U.S.S.R. In certain apatites from these deposits they recorded up to 0.030 ppm Au.

Scorodite from gold-bearing arsenopyrite veins invariably contains gold, frequently in amounts up to 10 ppm or more. In my experience most of this is present in finely divided native gold, although some pure specimens of scorodite in which gold cannot be seen microscopically nor removed either physically or chemically (by cyanide solutions) appear to contain the metal in lattice sites, probably substituting for iron. Beudantite and bindheimite, likewise, invariably contain gold, some of which cannot be removed physically nor chemically. Some of this gold (together with silver) probably substitutes in the lead sites of the lattices of these minerals. It is interesting to note that gold tends to concentrate in the supergene arsenates and antimonates, a feature which suggests a definite coherence between the metal and arsenic and antimony. This coherence is of course well known in hypogene mineral assemblages.

Both of the tungstates, wolframite and scheelite, may contain small amounts of gold. The nature of the gold is unknown. Probably most of it resides in microscopic particles of the native metal. It is worth noting that there is a general association of tungsten, gold and silver in many types of gold deposits.

In rocks and in a few auriferous deposits magnetite carries gold. Some values are given in Table 12; others are quoted in subsequent sections on gold in rocks. The amount of gold in magnetites can range up to 0.05 ppm. The element apparently substitutes for iron in the magnetite lattice since free gold can rarely if ever be seen in the mineral. Platinum group metals accompany gold in some magnetites from rocks in the Urals (Fominykh and Znamenskii, 1974).

Of the various supergene oxides, limonite and wad tend to be the most common carriers of gold, especially where they occur either in or near gold deposits. The presence of gold (and especially silver) in wad (up to 1.3 oz/ton Au and 100 oz/ton Ag or more) is an old problem discussed in great detail in the years past by Clevenger and Caron (1925) and numerous others. The reasons for the extended study were mainly metallurgical since the manganiferous silver-bearing ores and some wad ores containing gold are highly refractory and only yield their silver to cyanide solutions after intensive reduction. After much study Clevenger and Caron and others concluded that much of the silver in wad is held as silver manganite of uncertain composition. More recent work concerning silver in these ores is discussed by Boyle (1968*b*). With respect to gold in wad-silver ores Clevenger and Caron concluded that the gold in some of these ores is amenable to direct cyanidation, and hence its extraction is not increased by reduction. In other ores the gold follows silver, and is more or less refractory; on reduction the recovery of gold is increased from such ores to an even higher percentage than the silver. More recent papers on the occurrence of gold in manganese oxides include those by Foley (1960) who found up to 0.02 oz/ton Au in the manganese ores of Polk County, Arkansas; Harriss *et al.* (1968) who recorded the gold contents in manganese nodules shown in Table 12; and Hrytsyk *et al.* (1969) who found native gold in the earthy manganese ores of the Chyvchyn Mountains, U.S.S.R.

The occurrence of gold in limonite and other iron oxides, especially in the gossans of precious and base metal deposits, is well known and has been repeatedly described in innumerable publications (*see also* Chapter IV).

In recent years the writer has investigated the nature and distribution of gold (and silver) in wad and limonite from a variety of gold-bearing deposits, mainly in Yukon, Northwest Territories, New Brunswick and Nova Scotia. The results of the research concerning the silver have been published (Boyle, 1968*b*). The principal results with respect to gold include the following:

1. Some samples of wad and limonite have the gold (and silver) uniformly distributed throughout the mineral complexes; in other samples the gold (and silver) is erratically distributed.

2. Most of the gold- and silver-bearing wads and limonites carry considerable quantities of silica, alumina, titanium, phosphorus, arsenic and antimony (*see*, for instance, Tables 42, 56 and 60 in Boyle, 1965*a*).

3. In most samples of wad and limonite gold is present in the native state and can be superpanned from samples of the minerals or leached by potassium cyanide. In some samples of wad and limonite the gold follows silver and is tightly bonded to the manganese and iron complexes and is not readily soluble in potassium cyanide or potassium thiosulphate. This

gold may be present as an adsorbed constituent or as a gold manganite, or ferrite although its constitution is apparently amorphous (colloidal?) since no distinct X-ray patterns can be discerned from those of the wad and limonite minerals. In certain samples with a relatively high organic content, the gold may be bound in organic (humic) complexes of manganese and iron. In still other samples, particularly in gossans, the gold (and silver) appear to be an integral part of admixed minerals such as plumbojarosite or argentojarosite, beudantite, bindheimite, scorodite and anglesite.

One of the reasons for the enrichment of gold (and silver) in wad and limonite, especially in gossans, is the fact that the simple soluble species of the two elements are readily reduced to the metals by ferrous and manganous salts. This feature is discussed further in Chapter IV.

Another reason for the enrichment of gold (and silver) in wad and limonite is probably due to the colloidal behaviour of hydrous manganese dioxide and hydrous ferric oxide. Hydrous manganese dioxide sols and gels carry a negative charge and strongly adsorb cationic gold (and silver) to their surfaces. With aging, the gold may partake of the structure of the manganese dioxide complex, giving a gold manganite or some other similar compound. Alternatively the gold may simply remain in an adsorbed state either at the grain boundaries of the microcrystals of the complex or perhaps in interstitial positions. On recrystallization, after long periods of aging, some of the gold may also be split out of the complex giving nuclei of native gold. Silica sols and gels and humic colloids, likewise, carry negative charges and tend to adsorb cationic gold to their surfaces. Hence, the presence of these colloids in wad would also tend to increase the amount of gold precipitated in the wad-silica-humic complex. Some gold thus precipitated could conceivably also be bound as gold-manganese-iron organic complexes. Under other conditions the manganese dioxide colloids may present a positively charged surface due to adsorption of a variety of cations. This then permits the adsorption and/or coprecipitation of negative anionic complexes of gold or negative gold colloids. Another possibility may also account for the adsorption of gold on manganese colloids. The hydrated manganous oxide (hydroxide), $MnO{\cdot}H_2O$, appears to form during the initial stage of precipitation of manganese under a low oxidation potential, and this gelatinous precipitate is later oxidized to hydrated manganese dioxide, $MnO_2{\cdot}H_2O$. The hydrated manganous oxide (hydroxide) adsorbs anions strongly and would, therefore, tend to adsorb and/or coprecipitate the available negative anionic complexes of gold or negative gold colloids.

Ferric oxide hydrosols and gels tend to carry a positive charge and thus adsorb and precipitate various negative ionic complexes of gold, e.g., $[AuCl_2]^-$ and the negatively charged colloids of gold. With aging of the gels the gold is commonly split out as the native metal, but in many cases the gold is an intimate constituent of the limonitic complex.

In a milieu where manganese, iron, lead, silica, alumina, sulphate, phosphate, arsenate, antimonate, etc. are present in solution in oxidizing zones, the colloidal reactions must be very complex indeed. There, normal adsorption phenomena undoubtedly play a part, but hydrolytic reactions, especially those involved in the hydrolysis of iron and manganese sulphates, soluble arsenates, antimonates and phosphates, etc., also come into play. During these reactions gold is coprecipitated with argentojarosite, plumbojarosite, beudantite, bindheimite and other hydrolytic products, all of which may be intimately associated with the wad and limonite gels. On aging the wad-limonite complex tends to segregate out native gold and other gold- and silver-bearing minerals.

As for the earth . . . it hath dust of gold.
–Job 28:5–6

Geochemistry of gold in terrestrial materials

Summary accounts of the various aspects of the geochemistry of gold in most terrestrial materials are contained in a large number of works including those by Maclaren (1908), Crane (1908), Emmons (1917), Mellor (1923), Clarke (1924), Dunn (1929), Goldschmidt and Peters (1932), Lindgren (1933), Van Aubel (1934), Emmons (1937), Fersman (1939), Zvyagintsev (1941), Gmelin (1950–1954), Bateman (1950), Rankama and Sahama (1950), Smirnov (1951), Friedensburg (1953), Goldschmidt (1954), Routhier (1963), Shcherbakov and Perezhogin (1964), Smirnov (1965), Foss (1966*a,b*), Ziauddin (1970), Uklonskii (1970), Machairas (1970*a*), Shcherbina (1972), Allmann and Crocket (1972), Petrovskaya (1973), Simons and Prinz (1973) and Smirnov (1974). There is a selected annotated bibliography on the geochemistry of gold by Cooper (1971).

In the following sections there are numerous tables on the content of gold in the various earth materials. During the compilation of these tables considerable difficulty was encountered in deciding which analytical data in the literature should be used in the calculations because of the different analytical procedures employed, rock nomenclature, possible presence of gold mineralization and so on. These problems are familiar to all geochemists who attempt to calculate normal or background abundance figures. I have been selective in compiling the tables, utilizing only those data which in my opinion have been obtained by reliable modern analytical procedures on earth materials, which seemed to me from the descriptions of the samples to be well removed from the effects of gold mineralization. Admittedly this is a rather subjective procedure, and I would stress the fact that the abundance figures given in the tables are only estimates at best. All of the data sources are given in the Selected bibliography at the end of the bulletin.

Gold in igneous-type[19] rocks

The gold content of igneous-type rocks is given in Table 13. The table has been compiled from analyses in the literature and from analyses done for the Geological Survey of Canada. Excessively high values have been arbitrarily omitted from some of the data as have also the gold contents of rocks that are highly altered or have closely associated sulphide and gold deposits.

[19]The term igneous-type is used to avoid the controversy on the origin of granites, ultrabasic rocks and various types of porphyries included in the tables.

Table 13. Gold content of igneous-type rocks

Rock type	Number of samples	Range	$\bar{X}$ (ppm)	S	G	$\bar{X}g$	Sg	Ag* $\bar{X}$ (ppm)	Au/Ag ratio
Ultrabasic rocks									
Dunite, peridotite, pyroxenite, kimberlite, anorthosite, etc.	1185	0.0002–0.7800	0.0114	0.0474	0.0045	–2.3424	0.4899	0.08	0.1425
Lamprophyres									
Camptonite, minette, etc.	15	0.0007–0.0020	0.0016	0.0004	0.0015	–2.8117	0.1235	–	–
Basic rocks									
Intrusives: gabbro, norite, diabase, etc.	1493	0.0003–0.6800	0.0230	0.0631	0.0068	–2.1686	0.5621	0.14	0.1643
Extrusives: basalt, etc.	1752	0.0001–0.2300	0.0174	0.0180	0.0076	–2.1215	0.6636	0.11	0.1582
Intermediate rocks									
Intrusives: granodiorite, syenite, monzonite, diorite, etc.	4710	0.0001–0.3500	0.0075	0.0126	0.0042	–2.3769	0.4170	0.05	0.1500
Extrusives: andesite, trachyte, dacite, latite, etc.	1359	0.0001–0.0650	0.0129	0.0126	0.0068	–2.1680	0.5476	0.08	0.1613
Acid rocks									
Intrusives: granite, aplite, etc.	3454	0.0002–2.9000	0.0114	0.0796	0.0028	–2.5518	0.5233	0.05	0.2280
Extrusives: rhyolite, obsidian, etc.	372	0.0001–0.1130	0.0037	0.0074	0.0017	–2.7699	0.5248	0.05	0.0740
Alkali-rich rocks									
Nepheline syenite, phonolite, etc.	99	0.0001–0.0135	0.0034	0.0032	0.0021	–2.6775	0.4764	0.10	0.0340
All ultrabasic rocks	1185	0.0002–0.7800	0.0114	0.0474	0.0045	–2.3424	0.4899	0.08	0.1425
All lamprophyres	15	0.0007–0.0020	0.0016	0.0004	0.0015	–2.8117	0.1235	–	–
All basic rocks	3245	0.0001–0.6800	0.0200	0.0449	0.0072	–2.1432	0.6193	0.12	0.1667
All intermediate rocks	6069	0.0001–0.3500	0.0087	0.0128	0.0047	–2.3302	0.4579	0.06	0.1450
All acid rocks	3826	0.0001–2.9000	0.0106	0.0757	0.0027	–2.5730	0.5274	0.05	0.2120
All alkali-rich rocks	99	0.0001–0.0135	0.0034	0.0032	0.0021	–2.6775	0.4764	0.10	0.0340
All igneous rocks	14 439	0.0001–2.9000	0.0119	0.0474	0.0044	–2.3564	0.5414	0.07	0.1700

* The arithmetic mean for silver is taken from a previous compilation by Boyle (1968*b*). It does not refer to the samples quoted in the column-number of samples.

Some of the variations in the data on gold analyses of igneous rocks are evidently due to analytical rather than natural causes, a feature that is difficult to assess. Most of the data from the literature compiled in Table 13 are based on neutron activation analyses, which in my experience are not always satisfactory for some types of rocks. Most of the Geological Survey analyses have been done by a combined fire assay–atomic absorption method, which uses several tens of grams of rock powder. While not as sensitive as activation analyses in some respects the method does give more uniform reproducible results and tends to smooth out errors due to sample inhomogeneities.

There is considerable uncertainty about trends in the concentration of gold in the sequence – ultrabasic-basic-intermediate-acidic rocks. Some investigators, including Shilin (1968) and Shcherbakov (1968), maintain that mafic rocks are enriched in gold relative to acidic rocks. In central Kamchatka, Shilin, for example, found the older magmatic rocks of basic composition to contain more gold than the younger acid rocks of the same Miocene magmatic epoch. Others, including De Grazia and Haskin (1964), Anoshin and Potap'yev (1966) and Voskresenskaya *et al.* (1970) find a near uniformity of concentration of gold in all igneous rock types. Gottfried *et al.* (1969) report that gold is not concentrated in residual liquids during magmatic differentiation. In general they found gold contents in felsic members of the batholiths sampled to be lower than in the related mafic members and noted that relatively higher gold concentrations are associated with earlier crystallizing phases (e.g., magnetite, mafic silicates, etc.). They remarked that this was consistent with the trend of progressive depletion of gold from mafic to silicic rocks of calc-alkalic suites. Volarovich and Shilin (1971) report that gold gradually increases in the differentiation series gabbro to aplite, although the distribution of the element in the various rock types is frequently irregular. Most of the gold is said to be dispersed in igneous rocks in the native state. They also claim an increase in gold with increase in potassium content during differentiation, and suggest that during the postmagmatic stage potassium-rich solutions appear enriched with gold, from which gold deposits are formed. In the Kurile-Kamchatka volcanic province, however, Anoshin and Kepezhinskas (1972) found statistically that the gold content is independent of the basicity of the volcanic rocks; elevated gold contents were found, nevertheless, in potassium-rich volcanics in the basalt-andesite-liparite series. The authors concluded that generally there is no significant difference in gold content among the groups of volcanic rocks related to a single differentiation series. In other words gold is inert during differentiation of basalt magma. In the central Kyzylkum, Khamrabaev *et al.* (1972) noted that there was a general tendency for Pb, U and Th to increase in the granitic series from older granodiorites to younger granites. Gold showed an opposite tendency. Ivensen *et al.* (1974) indicate that gold is fractionated in magmas during differentiation and crystallization. They found an inverse correlation between the gold content and the amount of alkali feldspars. The gold content was also found to increase with an increase in the basicity of the rock and the Na content. In a recent paper Kashirin *et al.* (1975) noted a significant increase in the gold content from the first to the last differentiates of the granitic intrusions in the Lena gold-bearing region. Higher gold contents were found in the endocontact and apical parts of the intrusions as well as in parts of the intrusions that were altered (beresitized and sericitized). The granitic bodies were also found to be surrounded by extensive exocontact aureoles, the belt with the maximum gold contents being located some 500 m from the granitic contact.

Shcherbakov and Perezhogin (1964) find that in the Altai-Sayan folded belt of the U.S.S.R. gold is more abundant in extrusive rocks than in their intrusive equivalents. This trend is also apparent in the data of other investigators. There is also more gold in oceanic basalts than in those of terrestrial occurrence according to some investigators (De Grazia and Haskin, 1964; Anoshin and Emel'yanov, 1969).

In recent years a number of investigations have been conducted on the gold content of layered basic intrusives and ultrabasic masses to define the differentiation trend of the element. Vincent and Crocket (1960) postulated that the initial magma from which the Skaergaard rocks crystallized contained 0.0046 ppm Au. All the rocks throughout the differentiation series, including the acid granophyres, are within a factor of 2 of this amount. The ferrogabbro phase with a relatively high copper sulphide content contained ten times the gold content (0.028 ppm) of the average rock; ferrogabbros with iron sulphides contained relatively little gold (0.0029 ppm). Rowe (1969) investigated the differentiated Great Lake tholeiitic dolerite sheet in Tasmania and found that the gold content of parts of the core is related to the mafic index [$(FeO + Fe_2O_3) \times 100/FeO + Fe_2O_3 + MgO$]. The variation of gold content with depth and mafic index was found to be similar to that of copper, indicating in Rowe's opinion that gold and copper probably crystallized concomitantly from the magma. Crocket and Chyi (1972) noticed considerable fractionation of gold between dunite, peridotite and their serpentinized equivalents in the Mount Albert pluton, Gaspé, Quebec. In the peridotites and dunites the gold content averaged 0.0011 ± 0.0004 ppm whereas in the serpentinites the average was 0.0024 ± 0.0007 ppm. Gold is concentrated (0.0034 ppm) in the accessory chrome spinels in peridotites and dunites. In late deuteric chromitite veinlets the gold content was only 0.00083 ppm.

Voskresenskaya *et al.* (1970) found that the distribution of gold in the layered Zlatogorsk intrusion of ultrabasic and basic rocks was relatively uniform around 0.0012 ppm. An exception was the leucocratic norites (0.0049 ppm Au). The Zlatogorsk intrusion is characterized by a substantially lower content of gold compared with the Skaergaard and Anakit intrusions. Shcheka *et al.* (1971) examined the gold contents of various intrusive basic and ultrabasic bodies in U.S.S.R., Korean People's Democratic Republic and the Indian Ocean and found that the rocks of the ultrabasic dunite-harzburgite clan were richest in the element (0.021 ppm). Rocks of the gabbro clan had considerably less gold (0.004 ppm). Their data led them to the conclusion that gold is fairly uniformly distributed throughout the rocks during differentiation and auto-metasomatic processes of ultrabasic and basic suites; a notable concentration of the element, however takes place during the Cu-Ni sulphide stage. Examination of rocks near contacts indicated that reworking by granitic magmas of basic and ultrabasic rocks leads to considerable loss of gold from the invaded rocks. Moiseenko and Fat'yanov (1972) developed this thesis further, pointing out that the primary source of gold in the earth's crust is derived from basic and ultrabasic magmatic formations arising in the mantle. During granitization and metamorphism of basic and ultrabasic rocks gold is released, is transferred to melts and hydrothermal solutions, migrates and is deposited in any available local structures.

Anoshin *et al.* (1971) examined the Upper Paleozoic to Lower Mesozoic traprock (tholeiite-basalt) intrusions in the Tungusian and Putoransk subprovinces of U.S.S.R. and found an average gold content of 0.007 ppm; the minimum and maximum contents were 0.002 and 0.074 ppm respectively. In one of the differentiated sills they found an average of 0.008 ppm, which they thought represented the initial content of the undifferentiated magma. In the gabbro-dolerite horizon of the sill, containing abundant sulphides and magnetite, and enriched in copper, they found an average of 0.020 ppm Au. Other rocks in the differentiated sequence contained relatively uniform contents of gold (~0.004 ppm Au). These facts led them to the conclusion that gold is concentrated during the sulphide and hydrothermal formation stages of the differentiation of traprock magmas. The study of Noril'sk type differentiated intrusives (Talnakh and Noril'sk) in U.S.S.R by Godlevskii *et al.* (1970) indicates a similar trend. Gold is concentrated in the layered phases of the intrusives impregnated with sulphides (up to 0.42 ppm); associated layered differentiates such as gabbros and dolerites have gold contents ranging from 0.03 to 0.06 ppm, an order of magnitude higher than that for basic rocks in general.

Certain ultrabasic rocks appear to be greatly enriched in gold. Borisenko and Uskov (1971) found the following values in the Gusevogorsk Massif, Urals, U.S.S.R.

Rock	Au content	
	Range	Average
	(ppm)	
Diallage and hornblende pyroxenite	0.004–1.3	0.14
Olivine pyroxenites	0.002–0.5	0.08
Wehrlites	0.004–0.6	0.08
Hornblendites	0.020–0.3	0.12

Some of the gold is in the native form and as electrum, both associated with native platinum. The highest gold contents were found in layers and zones of the ultrabasic rocks enriched in sulphides (pyrite, chalcopyrite, bornite) and characterized by increased contents of hornblende.

In another paper, Borisenko *et al.* (1972) state that the irregular distribution of the gold content of ultrabasic rocks is due to the presence in these rocks of discrete inclusions of native gold and electrum. The remainder of the gold, which is rather uniformly distributed, occurs in the form of neutral atoms adsorbed by rock-forming and ore minerals. In the Noril'sk type of differentiated intrusives (Talnakh and Noril'sk) in U.S.S.R., Godlevskii *et al.* (1970) state that the gold is present in both the native form and as an isomorphic replacement in the rocks and associated ores.

Dodin *et al.* (1973) examined the platinoid and gold contents of a number of basic intrusives in U.S.S.R., some nickel-bearing of the Norilsk-Talnakh type. They found that the content of platinoids in both rock-forming minerals and the rocks and their various ratios were indicative of economic nickeliferous intrusives; the gold contents on the other hand were independent of the type of intrusion and hence not indicative of economic nickel-bearing massifs.

Ultrabasic inclusions in a kimberlite pipe in northern Yakutia were analyzed by Ukhanov and Pchelintseva (1972).

Low gold contents were found in garnetiferous peridotite (av. 0.0012 ppm) and garnetiferous pyroxenite (av. 0.0015 ppm). Eclogites contained increased amounts (up to 0.0045 ppm). The authors noted an increase in the gold content with the ferruginous character of the eclogite but no correlation with the various oxides. A direct correlation of gold with vanadium was also indicated. They concluded from their preliminary data that the gold is largely contained in clinopyroxene.

In another study of kimberlite rocks and ultrabasic inclusions Rozhkov *et al.* (1973) found the gold content to range from 0.001 to 0.010 ppm. They concluded that none of the kimberlite rocks are gold concentrators. They further remarked that their values did not confirm the supposition that the gold in ultrabasic rocks is heterogeneous, nor the hypothesis of an irregular content of the noble metal in the upper mantle on a regional basis.

Serpentinization may reduce the gold content of ultrabasic rocks according to the data of Li and Kornev (1972). In the ultramafic rocks of the Yenisei Range the initial gold content averages about 0.0041 ppm; in the serpentinites the gold content is 0.0017 ppm, a moderate decrease.

The distribution of gold in batholiths and stocks of intermediate to granitic composition appears to be highly variable although some general trends are apparent. Mantei and Brownlow (1967) showed that the Marysville quartz diorite stock in Montana has an appreciably higher gold content in and near the periphery where it averages 0.071 ppm; elsewhere it averages about 0.01 ppm. Voskresenskaya and Zvereva (1968) observed small but statistically significant differences in the average contents (0.0011, 0.0016, 0.0021 ppm) of gold in three granitic complexes in northern Kazakhstan. Hybrid rocks of the endocontact zones of the complexes contained from 1.5 to 4.5 times as much gold as the granitic rocks. Davletov and Dzhakshibayev (1970) used neutron activation analysis to determine the distribution of gold between the minerals in various igneous rocks. Feldspars were found to be the main carriers of gold (50% of the metal in the rock), whereas biotite was also observed to be a consistent concentrator mineral. The carriers and concentrators showed a progressive increase in Au content from the centre of an intrusion to the periphery and from the deep parts to the apical parts. Bushliakov (1971) found that the gold content decreases with increasing rock acidity in the granitoids of the Verkh-Isetsk massif in the Urals. Uvarov *et al.* (1972) examined the gold content (0.0006–0.003 ppm) from three granitic complexes on the Kalba-Narymsk region of eastern Kazakhstan. They found a general uniformity of gold contents in single complexes but significant differences in the level of the gold values of the three complexes. Antipin *et al.* (1969), however, found no difference in the content of gold (and molybdenum) in granitoids of different ages and geological settings in the East Transbaikal region of U.S.S.R. From this they reasoned that gold is relatively inert during the processes of magmatic differentiation of multi-phase intrusions. A somewhat similar conclusion was reached by Li and Datsenko (1973) who investigated the gold content of a variety of metasomatic and palingenic granitoids of the South Yenisei Range, U.S.S.R. They noticed a slight enrichment (up to 0.150 ppm Au) in the endocontact and central zones of the Taraksk and Posolnensk granitoid massif. In the rocks of the Nizhnekansk massif they found no significant differences in the gold contents (~0.002 ppm) of the different phases. They interpreted this behaviour as due to the inert character of gold during palingenesis. Shcherbakov and Perezhogin (1963) examined the gold content of granites and other rocks and their accessory magnetites in the Altai-Sayan gold belt. They noted that the average gold content in the granitic masses was higher in the peripheries, in the apical parts and in the region of the exocontacts.

The data in the literature suggest regional variations in the gold content of igneous rocks, and there may well be distinct gold geochemical provinces. The variations are, however, small in most cases, and it is not possible to determine whether they are real or due to analytical variance between investigators. Nevertheless, the gold contents of some suites of igneous rocks within certain specific areas exceeds the norm by factors as high as 10. With respect to gold content there appears to be no universal difference between batholiths and stocks with associated gold deposits and those that are barren. Shcherbakov and Perezhogin (1964) could find no clear-cut difference between the gold content in 'gold bearing' and other intrusives, a feature that agrees with most of our data. Mantei and Brownlow (1967) and others have, however, indicated that some stocks with associated gold deposits have higher than average gold contents. Shilin (1968) noted that certain Miocene magmatic rocks in central Kamchatka have gold contents one order of magnitude greater than others in the same region. He postulated a Miocene magmatic hearth enriched in gold below central Kamchatka. In the Rice Lake–Beresford Lake area of southeastern Saskatchewan Stephenson and Ehmann (1971) found evidence to indicate that the mafic intrusions may be the source rocks of the gold deposits. Some varieties of monchiquite, camptonite, odonite and other lamprophyric rocks appear to be enriched in gold (and silver) in some areas. Buturlinov and Latysh (1970) found up to 0.08 ppm Au and 3.6 ppm Ag in some of these basic lamprophyric rocks in the Donets Basin, U.S.S.R. Our experience with lamprophyric rocks is that they may contain up to 0.05 ppm Au or more where they contain abundant pyrite or where they are albitized and altered.

Gottfried *et al.* (1969, 1972) and Tilling *et al.* (1973) have concluded, following an extensive study of gold in igneous rocks that:

> The available data provide little support for the view that gold is concentrated in the residual liquids during magmatic differentiation in any of the calc-alkalic or alkalic provinces. The variation trends indicate that gold either remains relatively constant or decreases during differentiation processes.
>
> The gold content in suites of unaltered igneous rocks spatially associated with gold mineralization from selected areas – Boulder batholith region, Montana; Cornwall, Pa.; Jerome district, Arizona; Bullfrog district, Nevada; Mother Lode, California – is virtually the same as petrographically equivalent rocks from provinces devoid of mineralization. The relatively high concentration of gold that exceeds background levels for a given region invariably occurs only locally in altered rocks and may be ascribed to remobilization or secondary introduction. Geochemical abundance data alone neither provide unique criteria in identifying broad segments of the crust favorable for gold mineralization nor specify processes that resulted in the economic concentration of gold. Thus, in any exploration program, geologic factors, such as petrologic, structural and tectonic setting, must be considered in conjunction with purely geochemical abundance factors.

Against this view we have the research by Kushmuradov (1970), who worked in a gold-bearing area in Uzbek, U.S.S.R. and found an average of 0.01 ppm Au in the granitoids of the North Nuratau Mountains, a value some 2 to 3 times higher than the average abundance of gold in such rocks. Altered (albitized and silicified) granitoids and sedimentary rocks near the Charmitan, Karaulkhon Latta and Koktonly gold deposits contained 2 to 3 times more gold (0.02 ppm) than the unaltered rocks. Dyke rocks, including lamprophyres, granite porphyries and diorite porphyries, were relatively enriched in gold averaging 0.026 ppm. However, a study of a number of intrusive granitoid bodies in U.S.S.R. with diverse associated ore mineralization, carried out by Anoshin *et al.* (1970) found no significant statistical difference in the gold content between bodies with associated gold ores and those that are barren. These researchers also found similar gold contents in different phases of granitoid bodies with associated gold mineralization. They suggested that the increased amounts of gold in some intrusions in gold belts was secondary, that is, the metal was introduced or rearranged during the hydrothermal stage. In the Chatkal zone of the Tien Shan, Davletov (1974) finds that relatively high contents of gold occur in igneous intrusives spatially associated with the gold mineralization of this zone.

Wolfe (1975) gives interesting data on barren and auriferous granitoid bodies from six Early Precambrian felsic intrusions in Ontario. These range in size from 4 to 88 km². Four – Dome stock, Red Lake area; Larder Lake stock; Lebel stock, Kirkland Lake area; and Cairo stock, Matachewan area – have associated gold deposits and two (Kabenung Lake and Coyle Lake) are barren. Wolfe observed that the gold levels in unmineralized stocks are consistently below the 5 ppb detection limit of the analytical method used. In contrast, analyses of samples from stocks with associated gold mines within their borders and in nearby metavolcanic-metasedimentary host rocks show a strongly skewed trace element gold population with a substantial percentage of gold values exceeding 5 ppb and ranging as high as 320 ppb. Further details on this research are given in Chapter V.

In the granitic rocks and their contact haloes in the auriferous Lena region, Kashirin *et al.* (1975) observed that the former carried from 0.0016 to 0.0047 ppm Au whereas the latter contained from 0.0007 to 0.0139 ppm Au. They noted that the concentration of gold increased in the late-stage differentiates of the granitic magma and in rocks altered by autometasomatic processes. An enrichment of gold was also observed in the endocontact and apical facies of the granitic bodies.

In the recent symposium edited by Petrovskaya (1974) there are a number of papers referring to the abundance of gold in igneous rocks and the presence of auriferous deposits. One by Trunilina and Yakovlev notes that the content of gold in unaltered magmatic rocks of the Charka River Basin (northeastern Yakutia) increases with the sulphur content and that these rocks have a high gold content (0.0086–0.0205 ppm) when auriferous mineralization is present in the area. An accumulation of gold in peripheral parts of granitic massifs and in selvages of dykes is also noted, features that are ascribed to magmatic differentiation.

The writer concludes from this diversity of findings and opinion that the relationship between the gold content of igneous rocks and the presence or absence of gold deposits is unresolved. Part of the problem would seem to be the difficulty in obtaining an adequate sample in that gold is distributed extremely nonuniformly in unaltered igneous-type rocks. Even those that are equigranular and massive were found by Rozhkov *et al.* (1970) to display this habit. The sampling problem is even more troublesome in slightly altered rocks where the gold content can range widely from sample to sample.

To summarize the facts on the general distribution of gold in igneous rocks as shown in Table 13 and from the more detailed data in the numerous publications and in my files one can say that the variation from the mean gold content of igneous rocks is never great. The data suggest that on the average the ultrabasic and basic rocks have slightly more gold than other igneous-type rocks, but the excess amount is not great as shown by the geometric mean. In places there is evidence of some trends in the sequence ultrabasic-basic-acidic rocks with a general depletion in gold toward the acid end of the sequence. This trend is, however, not marked and a general uniformity of gold content seems to be present in many sequences. Specific observations indicate few definite trends except in rocks which contain relatively high contents of sulphides, especially pyrite and chalcopyrite. Such rocks are generally enriched in gold regardless of type. We have also noted relatively high gold contents in some quartz-feldspar porphyries containing pyrite and other sulphides, but the origin of these porphyries and their contained sulphides is somewhat controversial. They may in some cases be of metamorphic rather than igneous origin. There is a suggestion that some of the alkali-rich syenites and similar rocks are higher than normal in gold but only by a small amount. On the basis of the available data nepheline syenites and phonolites appear to be normal within a range 0.0001 to 0.0135 ppm Au. The nepheline syenites of Ontario are low in gold (<0.005 ppm), and the alkali basalts investigated by Gottfried *et al.* (1972) appear to have a distinctly lower gold content (~ 0.0006 ppm) than most Hawaiian and continental basalts of the tholeiitic magma type with a gold content ranging from 0.002 to 0.0035 ppm. Certain highly alkaline Permian magmatic rocks investigated by Lisogor and Babakhodzhaev (in Petrovskaya, 1974) are, however, enriched in gold compared with others in the Karamazar region of U.S.S.R.

The distribution of gold in some regional suites of igneous rocks is lognormal in nature; in others lognormal with a marked skew; and in still others the distribution is polymodal. The last type of distribution suggests addition of gold to the rocks during mineralization processes.

There is considerable difference of opinion on the distribution and nature of gold in igneous rocks. Most investigators are agreed that gold is distributed in a nonuniform manner in igneous rocks; in fact in some rocks its distribution can be said to be quite erratic according to our investigations. Rozhkov *et al.* (1970) comment at length on this particular feature after applying statistical principles to their data. Three views seem to be prevalent with respect to the nature of the gold in igneous rocks. One holds that the gold is present in the native form or as electrum; another that the element is dispersed throughout the rocks and constituent minerals in the form of neutral atoms; and a third that the gold is present as an atomic and/or ionic substitution in the lattice sites of minerals. Probably all three are correct depending on the type of rock, mineral composition and other factors. Mineral separates from igneous rocks show gold to be present in practically all of the minerals although the amounts vary, but generally only within narrow limits. Data from the literature and the writer's analyses indicate the following ranges in ppm to be expected: quartz, 0.0008–0.002; feldspars, 0.0003–0.002; biotite, 0.002–0.007; hornblende, 0.0005–0.003; pyroxene, 0.0005–0.004; olivine, 0.001–0.01; sphene, 0.004–0.009; magnetite, 0.003–0.1; and pyrite and other accessory sulphides, up to 0.5 or more. Considerably higher values for biotite, hornblende and magnetite are reported in some igneous rocks by Mantei and Brownlow (1967) and others. In some suites of rocks the sialic minerals, mainly the feldspars, carry the highest contents of gold. Khairullin *et al.* (1968) found the highest amounts of gold in the heavy fractions (e.g., biotite and magnetite), and Davletov and Dzhakshibayev (1970) found that biotite is the principal mineral concentrator of gold, but the feldspars contain the bulk of the element in igneous rocks (up to 50% of the total amount of gold present). Mantei and Brownlow (1967) and Zvereva and Gavrilenko (1971) noticed that the dark minerals including pryoxene, hornblende, biotite and magnetite are enriched in gold compared with quartz and feldspar, although the latter investigators emphasize the point that the magnetite is not always richer in gold than other minerals and that its gold content varies within broad limits. Anoshin *et al.* (1970) for instance state that in granitoids the gold content of the rock-forming minerals is relatively uniform, and that the differences are statistically insignificant. Their table indicates, however, that quartz and feldspar (0.0016 ppm) have lower average gold contents than biotite, amphibole, magnetite, sphene, etc. (0.0019, 0.0020, 0.0034, 0.0029 ppm Au respectively). Bushliakov (1971) gives the average gold content in rock forming minerals in the granitoids of the Urals as determined by activation analyses as follows: quartz, 0.0017 ppm; plagioclase, 0.0017 ppm; microcline, 0.0016 ppm; hornblende, 0.0016 ppm; biotite, 0.0018 ppm; muscovite, 0.0019 ppm; magnetite, 0.0036 ppm; ilmenite, 0.0022 ppm; sphene, 0.0027 ppm; apatite, 0.0024 ppm; garnet 0.0052 ppm; epidote, 0.0025; ppm; and chlorite, 0.0025 ppm. Sobolev *et al.* (1974) record the following ranges in parts per million for the various minerals in basic and ultrabasic formations in the Urals: olivine, 0.003–0.010; pyroxene, 0.0006–0.001; plagioclase, 0.0005–0.003; and hornblende, 0.0004–0.0007. Gottfried *et al.* (1972) found by material balance calculations that in most igneous rocks, even the very freshest, the gold measured in the rock tended to be greater than that indicated by the summation of the gold contribution from the principal rock-forming minerals. This suggested to them that some of the gold may reside elsewhere in the rock, possibly in phases volumetrically small but comparatively rich in gold (for example, sulfides) or even as intergranular native gold. Keays and Scott (1976) found that the gold contents of the glass rims of oceanic pillow basalts (av. 0.0015 ppm) to be some 5 to 7 times greater than those of the associated crystalline interiors (av. 0.0005 ppm). They considered that the hosts for the bulk of gold in basalts are probably (1) early separated oxide and silicate phases such as Cr-spinel and olivine and (2) sulphide phases.

The gold (and platinum content) of magnetites in rocks can apparently be used to differentiate igneous rocks of different genesis. In the Urals the granitoids have two origins – one as differentiates of basaltic magma and the second derived from sialic material (by granitization). In an interesting paper on the subject Fominykh and Znamenskii (1974) observed an increased content of gold in the magnetites from granitoids of the granite (sialic) formation, with which are often associated hydrothermal gold-bearing deposits. The gold content of magnetites of this formation ranged from 0.018 to 0.035 ppm; in addition an increased gold content (0.035 ppm) was noted in magnetites from more basic varieties of rocks (quartz diorites and diorites). Magnetites in the granitoids of the gabbro (basaltic) formation differ in gold content from magnetites of the granite formation by an order of magnitude and contain the same amount of gold regardless of the rock composition (av. 0.006 ppm).

Platinum group elements are also distributed differentially in magnetites from the different groups of granitoids. A high platinum content was noted in magnetites from the gabbro group of granitoids, the maximum amounts being found in magnetites from diorites, quartz diorites, granosyenites and syenodiorites. In the magnetites of the granitoids derived from sialic material the platinum content was low and did not depend on the rock composition.

Finally, experiments carried out by Davletov *et al.* (1973) on the phase leaching of gold from intrusive rocks and mineral carriers of gold in these rocks indicates that gold is not distributed evenly in the mineral carriers as should be the case with isomorphic replacement. Rather, the metal occurs in microcracks and lattice discontinuities in the minerals and in microfractures and along intergrain boundaries in the rocks. These features are of interest in theories of metamorphic secretion since gold in microfractures and grain boundaries can be easily mobilized (extracted) from these sites. This circumstance is referred to later in the section on the origin of epigenetic deposits in Chapter III.

The writer's investigations generally confirm the findings of a number of workers that the dark minerals, magnetite, ilmenite and particularly pyrite and chalcopyrite contain the highest amounts of gold in igneous rocks. In our work on igneous rocks we have not been able to separate free gold from large quantities of rock nor have we seen free gold in polished sections of mineral concentrates of magnetite, pyrite, chalcopyrite and other sulphides from fresh (unaltered) igneous rocks. There are, however, numerous older references in the literature (e.g., Blake, 1897; Spurr, 1903, 1904; Weed, 1904*a*, *b*; Brock, 1904) of free gold in supposedly fresh (unaltered) granites, diabase, etc. Some of the descriptions given are, however, suspicious of alteration or impregnation since many are in or near gold mines or sulphide deposits.

Correlations of gold with other elements in igneous rocks are imperfect and seemingly erratic in many cases. Shcherbakov and Perezhogin (1964) claim to have found a correlation of gold with the following metals and metal oxides in decreasing order of correlation as follows: Cu, Fe_2O_3, MgO, V and FeO. They found no correlation with Zn, Pb and other metals. Rowe (1969), as noted above, found a correlation with the mafic index and with copper in the Great Lake tholeiitic dolerite sheet in Tasmania. Zvereva and Gavrilenko (1971) could find no correlation of gold with iron in the minerals examined in their study. Gottfried *et al.* (1969, 1972) concluded that their data suggest a negative correlation between gold content and abundance of some 'residual' trace elements (e.g., Th, U and Nb) but a gross positive correlation with chalcophilic elements such as copper. Shcherbakov *et al.* (1972) carried out an extensive series of gold analyses (over 2000) on a large number of rock types. They found a definite correlation between the iron and gold contents of igneous and sedimentary rocks, the higher gold contents being present in rocks with high iron contents. The main gold concentrator mineral in granitoids was found to be magnetite; in skarn, hornfels and schists the gold was contained mainly in the femic minerals. Hamrabaev and Urunbaev (1971) noted positive correlations of gold with ferrous iron and copper in the igneous rocks of Uzbekistan. A possible correlation of gold and sodium was also noticed. Keays and Scott (1976) analyzed a number of oceanic pillow basalts from the mid-Atlantic Ridges; in the fresh crystalline interiors of the pillows they noted correlations between gold and Cr, Ni and MgO.

Studies by the writer on the correlation of gold with other elements in igneous rocks suggest a positive correlation with Fe and possible positive correlations with Cu, Ag, S, As and Sb, but the relationships are often erratic. I think the positive correlation with sulphur is genuine, and I have noted a general tendency for gold to be enriched in igneous rocks rich in pyrite, pyrrhotite or chalcopyrite. The correlation with arsenic, likewise, seems highly probable most of the arsenic being present in the sulphides, especially pyrite (Boyle and Jonasson, 1973). In the basic-ultrabasic suites of rocks there appears to be a weak positive correlation between gold and the various platinoids, and perhaps with chromium, although at the time of writing my research on the platinum group is not complete. It further appears that certain ultrabasic rocks are enriched in both the platinoids and gold compared with more acidic suites of rocks.

Goldschmidt (1954) estimated the abundance of gold in igneous rocks (lithosphere) to be 0.005 ppm. Others (*see* the section above on abundance) have given abundances, based mainly on igneous rocks, ranging from 0.001 to 0.0043 ppm. Calculations based on the data given in Table 13 yield an abundance of 0.0044 ppm (geometric mean) for igneous rocks. Like all abundance calculations this figure should be taken as a rough estimate until more accurate figures are available on the quantity of each igneous rock type in the earth.

Gold in sedimentary rocks

The gold content of a variety of unconsolidated sedimentary rocks is given in Table 14, and similar data for consolidated sediments are contained in Table 15.

The data are not extensive for most types of unconsolidated sediments, but those available indicate that certain types of tuffs (volcanic ash) and organic sediments may be enriched in gold. Some of the sediments associated with oceanic ridges and rises appear also to contain slightly higher than average amounts of gold. Crocket *et al.* (1973) consider that the gold in the sediments of the East Pacific Rise had a submarine volcanic exhalative origin. Piper and Graef (1974) on the other hand think that much of the gold in the flank sediments of this rise had a lithogenic origin (not precipitated from seawater). In the crestal sediments, however, where the gold is significantly higher in content, they considered that a fraction of the gold had a volcanogenic (exhalative) source (precipitated from seawater).

The Red Sea metalliferous sediments (Hendricks *et al.*, 1969) are enriched in gold (and silver) (Table 14). These sediments exhibit a wide variation in physical appearance and in chemical composition, from oxidized red muds to dark sulphide-rich muds and a few light-coloured carbonate-rich samples. Hendricks *et al.* (1969) found extreme variations in composition between adjacent sample sites, and considerable variation was noted even within different splits from the same core. Extreme vertical variations in the cores were also recorded, sometimes within a few centimetres of depth.

The data provided by Koeberlin (1934) on the gold content of the volcanic ash in central Chile is interesting, in that there may be areas where such deposits (both unconsolidated and indurated) may be of economic interest. Goodchild (1918) also mentions the fact that certain ash flows associated with the volcano La Sufral in Colombia contain from 0.42 to 1.71 ppm Au. Silver, likewise, has been found to be enriched in certain falls of volcanic ash in Ecuador, South America (Mallet, 1887,1890). Taken together these two facts are of prime importance considering the frequent occurrence of gold-silver deposits in or associated with tuffs (sedimented volcanic ash) or porphyries (crystallized volcanic material) in the Canadian Shield, Tertiary terranes, etc. These types of rocks in the Canadian Shield are enriched in both gold and silver and a variety of other associated elements (Boyle, 1968*a,b*) (Table 15). Such rocks would seem to be an adequate source for the derivation of gold-silver veins and other deposits by a metamorphic secretion mechanism. (*See* the section on origin of gold deposits in Chapter III.)

Anoshin *et al.* (1969) carried out an extensive investigation of marine sediments in the Atlantic Ocean basin with the results given in Table 14. Their main findings can be summarized in point form as follows: (Details should be sought in the original paper.)

1. The average gold content in modern sediments is a little higher than the average content in felsic, mafic and ultramafic rocks, but considerably lower than the content in oceanic basalts. The most frequent occurring concentrations of gold (0.003–0.006 ppm) fall within the range of consolidated sedimentary rocks or are slightly higher (*see also* Table 15).

2. The highest gold contents were in coquinoid limestones from the Black Sea, the lowest in sands and gravels from the Baltic Sea and Atlantic Ocean and in some pyroclastic sediments from the vicinity of Iceland.

3. In the clastic sediments the gold content increases with decrease in grain size. The correlation of gold content with the pelitic (clay) fraction is weak in some types of sediment but significant in others.

4. The distribution of gold in the unconsolidated sediments coincides closely with the distribution of uranium, an element often closely associated with organic matter.

5. In carbonate sediments very low and very high values were encountered. The highest contents are apparently concentrated in molluscan shells.

6. In general the concentration of gold in biogenic carbonate sediments is not dependent on the content of

Table14. Gold content of some Recent unconsolidated sedimentary rocks

Rock Type	Au (range)	Au (av.)	Ag (av.)	Au/Ag ratio	Reference
		(ppm)			
Volcanic ash, Chile	0.1–5	–	–	–	1
Deep sea red clay, volcanic ash, etc.,					
Atlantic region	0.015–0.267	0.108	0.737	0.146	2
Red clay					
Brazilian Basin	0.0042–0.0310		–		3
Lutites					
Argentinian Basin	0.0031–0.0173		–		3
Recent carbonates					
Florida Coast		0.0039	–		3
Recent coral					
Florida Coast		0.0008	–		3
Terrigenous sediments (silt, clay, sand, gravel) (30% $CaCO_3$)					
Baltic Sea	0.0001–0.0054	0.003	–		4
Black Sea	0.002–0.024	0.006	–		4
Mediterranean Sea	0.0011–0.009	0.0038	–		4
Atlantic Ocean	0.0006–0.0042	0.0013	–		4
Biogenic and chemical sediments (coquinoids, sand, silt, clay) (>30% $CaCO_3$)					
Black Sea	0.0046–0.150	0.0204	–		4
Mediterranean Sea	0.0011–0.0065	0.0031	–		4
Atlantic Ocean	0.0007–0.0034	0.0015	–		4
Volcanic sediments (silts; 10% pyroclasts)					
Atlantic Ocean (near Iceland)	0.0005–0.015	0.0042	–		4
Mediterranean Sea (Santorini Lagoon)		0.0140	–		4
Oceanic sediments; Arctic, Antarctic,					
Pacific, and Indian Oceans	0.00064–0.0076	0.00274	–		5
Oceanic sediments: East Pacific Rise	0.00036–0.00282	0.00134			6
Red Sea metalliferous sediments	0.0210–5.6000	0.6594	28.3973	0.0232	7
Shelf sediments Carolina Coast, U.S.A.	0–0.006				8
Alluvial placers (stream, river and marine)	<0.01–200+	–	0.1–20+		9

References:
1. Koeberlin (1938).
2. Wagoner (1908).
3. De Grazia and Haskin (1964).
4. Anoshin *et al.* (1969).
5. Crocket *et al.* (1973).
6. Piper and Graef (1974).
7. Hendricks *et al.* (1969).
8. Pilkey and Bornhold (1970).
9. Estimated.

Table 15. Gold content of unconsolidated and consolidated (lithified) sedimentary rocks

Rock type	Number of Samples	Range	$\bar{X}$	S	G	$\bar{X}g$	Sg	Ag* $\bar{X}$	Au/Ag ratio
			(ppm)					(ppm)	
Unconsolidated sediments, oceans	299	0.0003–2.5000	0.0523	0.2761	0.0035	–2.4615	0.6792	–	–
Sandstone, arkose, conglomerate and									
sedimentary breccias	3474	0.0002–0.4300	0.0570	0.0499	0.0263	–1.5807	0.7139	0.2	0.2850
Greywacke and subgreywacke	38	0.0010–0.1000	0.0132	0.0192	0.0076	–2.1207	0.4303	0.2	0.0660
Witwatersrand and similar auriferous									
quartz-pebble conglomerates		0.4–230	8.0					0.8	10.0
Shale, mudstone, siltstone, argillite	1356	0.0001–0.8000	0.0080	0.0266	0.0039	–2.4061	0.4730	0.2	0.0400
Sulphidic-schists, pyritic greywacke,									
pyritic black shale, pyritic tuff, etc.	19	0.0025–2.1000	0.1320	0.4778	0.0143	–1.8442	0.7228	0.6	0.2200
Tuffs	97	0.0005–0.1120	0.0069	0.0163	0.0023	–2.6341	0.5737	0.3	0.0230
Limestone, dolomite	440	0.0002–0.0889	0.0070	0.0128	0.0034	–2.4706	0.4648	0.1	0.0700
Evaporites:									
1) Gypsum, anhydrite	18	0.0050–0.0850	0.0208	0.0169	0.0172	–1.7644	0.2683	<0.1	<0.*n*
2) Halite, sylvite, etc.	15	0.0005–0.0300	0.0238	0.0109	0.0178	–1.7505	0.4880	<0.1	<0.*n*
(See also Table 12)									
Chert	3	0.0050–0.0400	0.0167	0.0202	0.0100	–2.0000	0.5214	0.2	0.0835
Phosphorite	1		0.0072						
Manganese nodules	1		0.0002						
All sedimentary rocks	5761	0.0001–2.5000	0.0403	0.0831	0.0122	–1.9125	0.7632	0.3	0.1343

* The arithmetic mean for silver is taken from a previous compilation by Boyle (1968*b*). It does not refer to samples quoted in the column–number of samples.

$CaCO_3$ but probably on the organic content since there is a general positive correlation between gold and organic matter (C_{org}). This suggests that gold is probably absorbed by organisms (principally plankton) from seawater which on their death are sedimented in the clays and other marine sediments. Another process is concerned with the precipitation of gold in organo-metallic compounds absorbed on clays and clastic material.

7. The distribution of gold in some marine sediments appears to depend on the types of rocks in the provenance areas and on the mineral composition of the marine sediments.

The gold content of modern eluvial and alluvial (placer) deposits may be high enough to provide an economic source of the element. The lowest economic unit for exploitation of these deposits varies from region to region and from deposit to deposit. Roughly it is about 0.1 ppm Au. Most of the gold in eluvial and alluvial deposits is particulate. Its origin is probably due to both mechanical and chemical processes. These concepts, and descriptions of the deposits, are discussed in greater detail in the section on placers in Chapter III.

In normal types of consolidated sedimentary rocks the highest amounts of gold unquestionably occur in the conglomerate, sandstone, greywacke suite and in the black shales, carbonaceous argillites, alum shales, oil shales and some phosphorites. Only small amounts of gold occur in normal shales, limestone, evaporites and other sedimentary rocks. Certain consolidated sandstones and conglomerates are greatly enriched in gold. Most of these constitute fossil placers; they are discussed in some detail under this heading in Chapter III.

Minnitt *et al.* (1973) analyzed three suites of essentially argillaceous sediment from the Wolkberg Group and the Black Reef–Dolomite transition zone (Transvaal Supergroup, northeastern Transvaal) for gold, using instrumental neutron activation analysis. The average gold content of the argillaceous rocks was found to be 0.0025 ppm, while the more arenaceous sediments carried, on the average, some 0.022 ppm gold. Sediment samples from the old workings on the Crags had an average gold content of 0.061 ppm; a value of 0.131 ppm was determined for one sample from these workings. Minnitt *et al.* (1973) concluded from their investigation that the background gold concentration in argillites is, on average, about 10 times lower than in more arenaceous sediments. An important exception was found in a carbonaceous silt-shale rock from the Dolomite–Black Reef transition zone, which carried nearly 0.090 ppm gold. Such a sediment they thought could act as proto-ore for gold-quartz reefs of the Pilgrims Rest-type, which are very often intimately associated with the carbonaceous shale beds in the Dolomite.

Certain terranes of sedimentary and derived metamorphic rocks appear to be greatly enriched in gold. Thus Li and Shokhina (1974 *a,b*) noted that the Proterozoic (sedimentary and derived metamorphic) rocks of the Yenisey Range are characterized by higher than normal contents of gold, in some cases by an order of magnitude of the clarke value (0.004 ppm). Particularly enriched are certain sandstones and quartzites and carbonaceous and graphitic phyllites and schists.

Certain oil shales and other types of bituminous sediments contain small amounts of gold in excess of the average for normal shales. Varley (1922) has described an investigation involving recovery of gold from such rocks in Wyoming (Green River Formation) and Utah. He records values ranging from 0.3 to 0.6 ppm (0.01–0.02 oz/ton Au). Cook (1973) in a later paper on the Green River oil shale gives the gold content as <0.1 ppm. The nature of the gold in the shales could not be determined by Varley. Korotayeva and Polikarpochkin (1969) also noted enriched contents of gold in the organic-rich rocks of the Nerchinsk Zavod region of eastern Transbaikal, U.S.S.R. The enrichment was two- to threefold and in places ten- to fifteenfold compared with the light (organic free) shales and siltstones. In other papers by Polikarpochkin *et al.* (1969) and Gapon (1970) the maximum gold values recorded in organic (carbonaceous) rocks was 200 ppm with some 50 per cent of the samples containing 5 ppm and more. These authors ascribe the enrichment of gold in the carbonaceous rocks to physico-chemical processes and calculated that some 50 per cent of the gold in the rock was precipitated from solution. In samples of the Athabasca tar sands we found less than 0.005 ppm Au.

Phosphorites appear to be enriched in gold in places. Yasyrev (1969) found from <0.01 to 0.15 ppm in the Mesozoic nodular phosphate deposits of the Russian Platform. Some 10 to 35 per cent of the gold was in the free state; the remainder was combined in a chemical form. In fossil reptile bones and sawfish teeth the amounts of gold ranged from <0.01 to 0.085 ppm. Yasyrev found a correlation of gold content in phosphorites with a number of other trace elements including Ti, Zr, etc. This might suggest that detrital processes were the prime factors in the concentration of gold in these deposits, but our observations indicate that such elements are also subject to marked hydrolysis phenomena (as is also gold in solution) causing their precipitation. In the United States and Morocco the marine phosphorites appear to be low in gold. Samples of marine phosphate rock from Florida, Montana, North Carolina and Morocco contained only up to 0.003 ppm Au as determined by the neutron activation–fire assay method (Z.S. Altschuler, U.S. Geol. Surv. Circ. 622, p. 11).

Evaporites are notably low in gold and silver. Goubeau and Birckenbach (1938) found a general average of 0.0014 ppm total precious metals (Ag, Au, Pt, etc.) in various potassium salts (sylvanite, carnallite, etc.) in Germany. Our analyses show that all salt deposits including halite, sylvite, etc. generally have less than 0.001 ppm Au. Gypsum and anhydrite from the Walton area, Nova Scotia, likewise, generally has less than 0.001 ppm Au. Pyrite in evaporites may, however, be enriched in gold (up to 0.01 ppm). Seeland (1973 *a,b*) reports higher than average amounts of gold in certain samples of gypsum, anhydrite and halite from Paleozoic rock formations south of the Great Lakes in United States. Twelve per cent of the gypsum and anhydrite samples analyzed contained detectable gold, the mean being 0.020 ppm; a similar percentage of halite samples reported gold with a mean of 0.061 ppm.

Popov (1975) investigated the gold content in the Upper Jurassic Halogenous Formation in the southern part of Russian Turkestan. In all some 120 samples were analyzed. The average gold values found were: rock salt – 0.0012 ppm;

sylvanite – 0.0017 ppm; and sylvanite-carnallite – 0.0031 ppm. The content of gold in the associated limestones and anhydrite was lower. The order of increasing concentration of gold in the various rocks was found to be anhydrite–anhydrite-limestone–rock salt–potassium salts, indicating that gold was precipitated in increasingly larger amounts from waters of increasing salinity.

There appear to be only limited data on the gold content of oxidate sediments. The gold and silver contents of a number of iron-formations and associated sedimentary rocks in Canada are shown in Table 16. It will be noted that the sulphide-bearing members are the most enriched in the two elements. The Algoma (Archean) type iron-formations generally contain more gold than the Superior (Proterozoic) type. The Ordovician iron-formation in the Bathurst area of New Brunswick contains only low gold contents, from <0.005 to 0.01 ppm. Analyses of fossiliferous sedimentary magnetite beds (Lower Devonian) in the Nictaux Falls area of Nova Scotia indicate that the gold contents are less than 0.005 ppm. Hematite beds at the base of the Nepean sandstone (Cambrian) near Delta, Ontario generally contain less than 0.005 ppm Au. Seeland (1973 *a,b*) noted that certain glauconitic sandstones (Franconia Sandstone in Wisconsin) were slightly enriched in gold. Some 17 per cent of the samples analyzed reported gold with a mean content of 0.031 ppm.

Rozhkov *et al.* (1971*a*) have provided some data for the

Table 16. Gold and silver content of Canadian Precambrian* and Paleozoic iron-formations

Locality, type and age	Description	Au	Ag	Au/Ag ratio
		(ppm)		
Temagami, Ont.	Banded pyrite-magnetite chert	0.050	2.4	0.02
Algoma-type iron-formation†	Banded chert with conformable pyrite layer	0.020	0.7	0.03
Precambrian (Archean)	Banded chert with minor disseminated pyrite	0.060	1.1	0.05
	Jasper and chert with pyrite	0.020	2.0	0.01
	Pyritic chert-magnetite breccia	0.055	0.8	0.07
Balmer Lake, Ont.	Magnetic laminated grey chert	<0.005	0.6	<0.008
Algoma-type iron-formation				
Precambrian (Archean)				
Kukatush, Sudbury district, Ont.	Magnetite-carbonate-chert iron-formation	<0.005	0.9	<0.005
Algoma-type iron-formation	Thinly banded carbonate iron-formation	<0.01	1.6	<0.006
Precambrian (Archean)	Jasper-magnetite-hematite iron-formation	<0.01	0.4	<0.02
Geraldton, Ont.	Magnetite-carbonate-quartz iron-formation	0.01	1.0	0.01
Algoma-type iron-formation	Magnetite-carbonate-quartz-pyrite iron-formation	0.01	1.1	0.009
Precambrian (Archean)	Magnetite-carbonate-quartz-pyrite iron-formation	0.085	0.9	0.09
	Carbonate-quartz-pyrite iron-formation	1.030	2.0	0.51
	Oxide-carbonate facies iron-formation	<0.005	0.8	<0.006
Schreiber, Ont.	Massive pyrite in iron-formation	0.070	2.2	0.03
Algoma-type iron-formation	Massive pyrrhotite in iron-formation	0.040	3.6	0.01
Precambrian (Archean)	Massive pyrite, pyrrhotite, minor galena in iron-formation	0.050	4.7	0.01
Michipicoten, Ont.	Buff siderite	<0.01	2.1	<0.004
Algoma-type iron-formation	Black siderite, minor disseminated pyrite	<0.005	1.4	<0.003
Precambrian (Archean)	Black siderite with pyrite in quartz-ankerite vein	0.005	1.6	0.003
	Siderite-chert iron-formation, minor pyrite	<0.01	1.0	<0.01
	Magnetite-siderite-chert iron-formation	0.025	1.4	0.02
	Magnetite-siderite-pyrrhotite-chert iron-formation	<0.005	1.1	<0.004
	Siderite-pyrite-chert iron-formation	0.005	1.8	0.003
Hearne Lake, N.W.T.	Recrystallized chert-magnetite iron-formation	<0.005	0.5	<0.01
Algoma-type iron-formation	Silicate facies (quartz-chlorite-pyrite and minor chalcopyrite)	0.235	1.9	0.12
Precambrian (Archean)				
Kaminak Lake area, N.W.T.	Laminated magnetite iron-formation	<0.005	0.3	<0.016
Algoma-type iron-formation	Magnetite iron-formation, some pyrite	0.015	0.8	0.018
Precambrian (Archean)	Pyritic cherty iron-formation	0.010	1.4	0.007
	Massive pyrrhotite in iron-formation	0.015	2.8	0.005
	Pyritic interflow	0.065	2.3	0.028
	Massive carbonate, some pyrite and arsenopyrite	0.090	1.1	0.081
	Massive carbonate	0.025	1.5	0.016
	Pyritic interflow	0.430	1.3	0.330
	Pyritic graphitic-interflow	0.005	0.4	0.125
	Cherty carbonate iron-formation	0.005	1.3	0.003
	Cherty iron-formation	0.005	0.4	0.012
	Magnetite chert bands	0.005	0.4	0.012
	Pyritic interflow, some chalcopyrite	0.180	13.0	0.013
	Pyritic cherty carbonate interflow	0.150	2.4	0.062
	Carbonate-rich tuff	0.060	0.9	0.066
	Siderite	<0.005	0.8	<0.006
	Cherty pyritic interflow	<0.005	0.9	<0.005
	Pyritic bands in iron-formation	0.020	1.2	0.016
	Cherty pyritic bands	0.020	1.4	0.014
	Cherty pyritic bands silicate iron-formation	0.760	1.5	0.506

Table 16. (cont'd.)

Locality, type and age	Description	Au	Ag	Au/Ag ratio
		(ppm)		
Kirkland Lake, Ont.				
Algoma-type iron-formation	Magnetite-chert	0.031	0.5	0.06
Precambrian (Archean)	Chert and fine-grained quartz	<0.01	0.1	<0.1
Schefferville, Labrador-Quebec	Nonmagnetic silicate-carbonate iron-formation, Ferriman section	<0.01	0.4	<0.02
Superior-type iron-formation†	Silicate-carbonate member of iron-formation, Wishart section	<0.01	0.9	<0.01
Precambrian (Proterozoic)	Silicate-carbonate member of iron-formation, French section	<0.01	0.8	<0.01
	Oxide facies; Lower Red cherty member of iron-formation, French section	<0.01	0.9	<0.01
	Oxide facies; Pink Cherty member of iron-formation, French section	<0.01	0.6	<0.02
	Oxide facies; Grey Cherty member of iron-formation, French section	<0.005	0.5	<0.01
	Oxide facies; Brown Cherty member of iron-formation, French section	<0.015	0.7	<0.02
	Oxide facies; Upper Grey member of iron-formation, French section	<0.010	0.6	<0.02
Bathurst district, N.B.				
Austin Brook area	Magnetite-chert-chlorite (composite sample)	<0.01	3.7	0.003
Algoma-type iron-formation	Magnetite-chert-chlorite-pyrite (composite sample)	0.075	4.1	0.018
Ordovician				
Nictaux Falls area, N.S.	Magnetite-hematite (composite sample)	0.023	1.9	0.012
Nictaux-Torbrook Formation				
Metamorphosed Clinton-type iron-formation				
Upper Devonian				

*The samples in these suites of iron-formations were kindly provided by G. A. Gross, R. H. Ridler and J.B. Henderson, Geological Survey of Canada; and H. Lovell, Division of Mines, Ontario Ministry of Natural resources.

†For classification of the types of iron-formations *see* G.A. Gross (1965), Geology of iron deposits of Canada, vol. 1; Geol. Surv. Can., Econ. Geol. Rep. no. 22, p. 116.

Krivoi Rog ores of U.S.S.R. as shown in Table 17. It will be noted that the gold contents are an order of magnitude higher than the contents in most sedimentary rocks. The gold tenor in all rocks and ores of the iron-formations was found to be 0.0*n* ppm. Some of the iron ores are relatively rich in gold; for example martite ore – 0.7 ppm, magnetite concentrate – 0.4 to 0.8 ppm. This suggested to the authors that some ore horizons were enriched in gold by sedimentary or metamorphic processes. Data provided by Pletnev (1972) for the Krivoi Rog siliceous ferruginous formations of Lower Proterozoic age show that in vertical profile the rocks range in gold content from 0.02 to 2 ppm. Two types of gold were recognized – one related to original sedimentation processes and the other representing a superimposed mineralization present in thin crushed zones. The latter may be a redistribution of the former (?).

Table 17. Gold contents of Krivoi Rog iron ores, U.S.S.R.

Description	Range in gold content
	(ppm)
Magnetite-martite ores	0–0.00*n*–0.*n*
Ferruginous quartzites	0–0.0*n*–0.*n*
Shales	0.0*n*–0.*n*
Mineralized zones in ferruginous quartzites	0–0.0*n*–0.*n*–*n*

Source: Rozhkov *et al.* (1971*a*).

The various data on iron-formations are interesting in the light of the common occurrence of commercial gold deposits in or near these formations in a number of places in the Canadian Shield and elsewhere in Precambrian rocks. The oxidate sediments together with sulphidic slates and schists that commonly accompany iron-formations may be the source of the gold that was later concentrated by secretion processes into deposits. These and other speculations are considered further in Chapter III.

Enrichment of gold in certain red (oxidized) sediments is also notable mainly as a result of adsorption and/or coprecipitation by ferric oxide hydrate. Thus, certain red sandstones and conglomerates may contain higher than normal amounts of gold than equivalent grey (reduced) rocks of the same type. Polikarpochkin *et al.* (1969) refer to this situation in the Baleisky, Onokhovsky and Lukinsky grabens where they find maximum contents of gold in red conglomerates up to 0.5 ppm. I have found higher than normal gold contents in some Carboniferous, Tertiary and Recent conglomerates containing abundant iron oxides in various parts of Canada. These are discussed in the section on placers.

There are apparently no systematic analyses for gold in the bedded sedimentary manganese deposits of the world. The results reported for deep sea manganese nodules generally show the gold content to be <0.001 ppm; some wads, however, are richer in gold (Table 12) with values up to 0.7 ppm Au or more.

Gold is frequently concentrated under unusual conditions in some types of sediments. Laur (1906) describes remarkable concentrations of gold and silver in Triassic sediments in the Meurthe-et-Moselle region of France. Up to 29 ppm Au (and 245 ppm Ag) are recorded for samples of sandy dolomite. The nature of the gold is not specified. Palmer (1935) has described an interesting occurrence of gold and silver in petrified logs in the Aspen district of Nevada. One log assayed 18.26 oz of gold and 10.8 oz of silver to the ton, and native gold was readily observed under the microscope. Palmer thought that the carbonized wood produced the reducing environment that precipitated the gold and silver from circulating solutions that came into contact with the logs. The writer has seen similar occurrences of gold in petrified Sequoia trees in samples from Alaska (G.B. Robbe collection, Michigan Technological College, Houghton, Michigan).

An interesting occurrence of traces of gold (and silver) is that in the salt-dome calcitic and gypsiferous cap rocks of Texas and Louisiana mentioned by Hanna and Wolf (1941). In these rocks the noble metals appear to be associated with sulphides, including pyrite, hauerite (MnS_2), galena and sphalerite and in some places with barite, celestite and native sulphur. The origin of the sulphides and their contained gold and silver seems to be best explained by the hypothesis that as H_2S rose from underlying petroliferous layers it came into contact with groundwaters (or connate waters in some cases) carrying lead, zinc, iron, manganese, silver, gold, etc. in the region of the cap rocks and deposited the sulphides in the available porous zones.

Occurrences of auriferous sandstones and conglomerates are widespread throughout the world, the most famous and productive being those of the Witwatersrand, South Africa, Tarkwa, Ghana and Jacobina, Argentina. The Witwatersrand ores average 8 ppm Au and 0.8 ppm Ag; those at Tarkwa contain 10 ppm Au and 0.5 Ag, and those at Jacobina 15 ppm Au and about 1 ppm Ag. Other less well known occurrences occur in many parts of the world throughout the geological column. Only a few can be mentioned as examples. For details and descriptions of other similar auriferous rocks consult Chapter III, especially the section on placers.

In the United States Devereux (1882) describes the occurrence of gold in the Potsdam conglomerate-breccia-sandstone of the Black Hills near the famous Homestake deposit. Some streaks near the bedrock are said to run 85 ppm (2.5 oz/ton), but the overall grade of the conglomerate is low, probably of the order of 1 ppm or less. Devereux recognized two types of gold in the fossil placers – mechanically concentrated gold and chemically precipitated gold. The fineness of the placer gold was greater than that of the quartz veins from which it was said to have been derived, a feature that is normal in practically all placers.

Hastings (1908) described fossil placers near Antelope Springs, Park County, Colorado. These placers occur in a large area of unconsolidated lake beds that average about 0.01 oz/ton (0.3 ppm). The nature of the gold could not be determined. Dumble (1913) described somewhat similar deposits from the Eocene of Texas. The gold-bearing beds are clays and sands with interbedded lignite. The grade is low, but some samples ran as high as 1 oz/ton (34.2 ppm) and upward. In Oregon and California, Diller (1914) describes Cretaceous fossil marine placers formed about an old island. The gold-bearing rocks are conglomerates derived from an underlying auriferous slate bedrock series. They contained from 1 to 5 ppm Au on the average. More recently Antweiler and Love (1967) have described gold-bearing Cretaceous and Tertiary quartzite conglomerate in northwest Wyoming. The gold values run from 0.035 to 0.222 ppm. The gold is largely confined to sandstone layers and the sandy matrix of the conglomerates. Much of it is the particulate native variety, but some evidently occurs in an absorbed state or in the structure of a variety of minerals. In the Rocky Mountain region of the conterminous United States certain fossil black sandstones of Late Cretaceous age contain up to 1.3 ppm Au in Wyoming and Colorado according to R.S. Houston (U.S. Geol. Circ. 622, 1969, p. 8). In Wyoming the younger deposits contain the higher gold contents.

In Canada there are also numerous occurrences of gold-bearing conglomerates, sandstones and quartzites. The uraniferous conglomerates and quartzites of the Blind River–Elliot Lake area carry small noncommercial amounts of gold in some beds. The Edmonton Formation in Alberta is gold-bearing in places. It probably supplied most of the gold to the placers worked many years ago along the North Saskatchewan River (Tyrrell, 1915). Another example of gold-bearing conglomerate and sandstone is found in the basal Carboniferous of Nova Scotia in the Gays River area (Malcolm and Faribault, 1929). The gold content of the richer streaks was said to average 5.4 ppm (0.158 oz/ton); our analyses show much lower values, around 0.03 ppm. Yet another example of gold-bearing limonitic-cemented conglomerate is found in the Beauceville area of Quebec. This has been mined extensively in the past. The rocks may be early Pleistocene in age although some investigators have considered them to be of Tertiary age (*see also* the section on placers in Chapter III).

Throughout other parts of the world gold-bearing conglomerates, sandstones and quartzites are widespread, and many have been mined or washed for gold. The Witwatersrand, Tarkwa and Jacobina deposits have been mentioned previously. These are the most famous of all of the fossil gold placers. In U.S.S.R. there are numerous occurrences of gold-bearing sandstones and conglomerates. One of these in Cambrian sandstones is described by Pokrovskii (1968); others are mentioned by a number of other Russian investigators, the principal publication on their occurrence being that by Ivenson *et al.* (1969). In Switzerland, Hofmann (1965) describes quartzite-rich conglomerates of the Upper Marine Molasse (Miocene) and coarse Triassic sandstones that carry gold contents of the order of 1 to 2 colours per 100 kg of rocks. In South Island, New Zealand, conglomerates and quartzites ranging in age from Cretaceous to Recent have been washed for gold for many years. These conglomerates are of fluvial and glaciofluvial origin. They are discussed in greater detail in Chapter III.

The fossil gold-bearing placers typified by conglomerates, sandstones and quartzites are of two types. These are only briefly described here since more details are given in the section on placers in Chapter III. The ancient Precambrian types (Witwatersrand, etc.) are characterized by relatively high gold contents and an abundance of pyrite. The gold is mainly free, but some is contained in pyrite and other minerals. Uraniferous minerals are common, and in some there are also thorium and rare-earth minerals. Iron oxides are rare except in the Tarkwa deposits. The more recent types are often characterized by limonite or iron oxide cemented conglomerates in which the gold is mainly free with only small amounts in adsorbed or other forms. Uranium, thorium and rare-earth minerals are rare or present in only small amounts; pyrite tends to be rare, but in some deposits, it is present in small amounts in the basal parts of the sequences. While both types may represent similar sedimentary conditions of formation it is obvious that their subsequent history has been much different.

The writer has extensively investigated the nature of gold in sedimentary rocks, and while it is often difficult to determine precisely the location of parts per billion gold in some samples, concludes the following from his research:

1. In conglomerates, sandstones, quartzites and similar rocks most of the gold is present in particulate form as the native metal. The size of the particles ranges all the way from less than 1 μ to nuggets the size of peas. Much of this gold is of detrital or detrital-chemical origin. Some gold is also present in forms, the nature of which we have not been able to ascertain with certainty. Where pyrite is present there are small amounts of the element in this mineral as shown by the results from Elliot Lake (Table 11). Other minerals such as limonite, magnetite, pyrrhotite, etc. may also contain traces of gold, the form of which is unknown.

2. In pyritiferous black shales and schists most of the gold is present in pyrite and/or pyrrhotite as shown by the analyses in Table 11. It is of interest to note that in rocks containing coexisting sedimentary pyrite and arsenopyrite gold prefers the arsenopyrite whereas silver is concentrated mainly in the pyrite. This feature is particularly clear in samples from the Meguma slates in Nova Scotia (Ovens, N.S., Table 11). In some black shales, oil shales, etc. the gold is largely associated with the carbonaceous substance, probably in an adsorbed form or as some type of organometallic compound. Where fine-grained pyrite and/or pyrrhotite are present, however, the gold tends to be partitioned into these mineral phases. The same is true if other sulphides such as chalcopyrite and galena are present. From this it seems to follow that the degree of diagenesis and metamorphism of the black shales and schists more or less determines the location of gold (and silver). In black shales that have undergone little metamorphism the gold appears to be associated with both the carbonaceous substance and fine-grained nearly colloidal sulphides. With increase of metamorphism and migration of the sulphides into knots or crystals, the bulk of the gold (and silver) tends to collect either in the iron sulphides or in chalcopyrite and other sulphides that occur in blebs or in small fractures in the pyrite and/or pyrrhotite. It is of interest to note that Kucha (1973) observed that much of the gold in the copper-bearing shale of the Fore-Sudetic monocline in Poland is present in minerals containing native silver but also in organic compounds of the thioaurate and chelate types. Legedza (1967) has examined the problem of the accumulation of gold in sulphides in the sediments of the Russian Platform during diagenesis. He thinks that thiosulphates in groundwaters act as collectors and carriers of both sulphur and gold, and that during diagenesis the metal is largely precipitated in pyrite (apparently in the metallic state) according to the reactions:

$$FeO + H_2S + nH_2O \rightarrow FeS{\cdot}nH_2O$$

$$2FeS{\cdot}nH_2O + [Au(S_2O_3)_2]^{3-} \rightarrow 2FeS_2 + Au + 2SO_3 + nH_2O$$

In lithified sulphide-free sediments, gold apparently remains in its finely dispersed state.

3. I have not been able to determine the nature of gold in normal shales, carbonate rocks, evaporites, etc. with any degree of certainty. The amounts of gold are so low that the mineral separation and analytical problems are in many cases insuperable. My results suggest that the gold, in some fine-grained sediments such as shales, is present in particulate form (flour gold) or in the small amounts of pyrite, magnetite, limonite and other heavy minerals that are frequently present in such rocks. In carbonates and evaporites the gold is probably largely present in carbonaceous matter or in colloidal pyrite. In all fine-grained sedimentary rocks there is probably some gold present as an adsorbed element.

The average gold content of the sediments of the Russian Platform is given by Nikitin and Yasyrev (1974) as 0.005 ppm. In the minerals of these sediments the average gold content ranges from 0.0008 ppm in chlorites to 0.0137 in the iron oxides; the organic constituents contain an average of 0.0089 ppm Au.

Gold in metamorphic rocks

The available data on the gold content of metamorphic rocks are given in Table 18. The number of analyses in some of the categories are too few to draw sound conclusions, but the following indications are noted.

In metamorphic rocks gold appears to occur in the same minerals as those noted for igneous and sedimentary rocks. My research shows that much of the gold occurs in pyrite, pyrrhotite, etc. and in the ferromagnesian minerals, principally magnetite, biotite, pyroxenes and amphiboles. Where pyrite and pyrrhotite are present in the same rocks, gold generally prefers the pyrite. If arsenopyrite is present, this mineral is enriched compared with pyrite and pyrrhotite. In certain quartzites and metaconglomerates the element may be present as the native metal. These types of metasediments are discussed in greater detail in the section on auriferous quartz-pebble conglomerates and placers in Chapter III.

Regionally metamorphosed rocks appear to have gold contents in the same range as those of the igneous and sedimentary rocks from which they were derived. Certain skarns and other contact rocks, on the other hand, are often enriched in gold (and silver). This appears to be normal since such rocks often contain copper deposits and a few gold-silver deposits. These are discussed in further detail in Chapter III.

Certain metamorphic rocks seem to be exceptionally high in gold over large areas, an example being those in the Sob River basin in the Ukraine (Latysh *et al.*, 1974). Certain shales, containing high contents of magnetite, report up to 0.3 ppm Au in this area. Other rocks contain from a trace to 0.02 Au. It is not clear whether these are fresh rocks, or whether there is a superimposed mineralization present. Relatively enriched quantities of gold are also recorded by Usenko *et al.* (1973) in the garnetiferous quartzites, crystalline schists, skarns and magnetite ores from the Sea of Azov region, U.S.S.R. They report gold contents ranging from 0.3 to 1.0 ppm in the ferruginous quartzites. Other rocks contain from traces to 0.2 ppm Au. Certain of the Grenville marbles in the Adirondacks of New York appear to be enriched in gold in places. Seeland (1973 *a,b*) found that 24 per cent of the samples he analyzed ranged in gold content from 0.020 to 0.20 and averaged 0.053 ppm.

The migration trend of gold during metamorphism is indicated in a number of investigations:

Brammall and Dowie (1936) noted low gold contents (0.07 ppm) in basic igneous rocks, grey granites, gneisses and schists of the Malvern Hills, England. Higher values (up to

Table 18. Gold content of metamorphic rocks

Rock type	Number of samples	Range	$\overline{X}$ (ppm)	S	G	$\overline{X}g$	Sg	Ag* $\overline{X}$ (ppm)	Au/Ag ratio
Quartzites and indurated sandstones	289	0.0002–1.1500	0.0317	0.0836	0.0047	–2.3325	0.9282	–	–
Slates, phyllites, meta-argillites	225	0.0009–0.0150	0.0022	0.0022	0.0016	–2.7959	0.3126	0.2	0.0110
Gneisses, granulites	1929	0.0002–0.3000	0.0031	0.0071	0.0027	–2.5682	0.1806	–	–
Amphibolites, greenstones	425	0.0001–0.1000	0.0071	0.0110	0.0041	–2.3837	0.4399	0.25	0.0284
Schists	1969	0.0001–3.7000	0.0186	0.1545	0.0037	–2.4351	0.5388	0.2	0.0930
Marble and crystalline limestones	40	0.0002–0.1000	0.0132	0.0224	0.0036	–2.4385	0.7243	0.05	0.2640
Hornfels and meta-pelites	368	0.0003–0.3000	0.0064	0.0262	0.0018	–2.7341	0.5037	–	–
Eclogites	103	0.0008–0.0400	0.0036	0.0045	0.0024	–2.6132	0.3626	0.10	0.0360
Skarns	44	0.0076–0.0220	0.0086	0.0028	0.0083	–2.0789	0.0887	–	–
Undifferentiated metamorphic rocks (porphyroids, chloritized and epidotized tuffs, etc.)	12	0.0004–0.0084	0.0038	0.0022	0.0029	–2.5308	0.3812	–	–
All metamorphic rocks	5404	0.0001–3.7000	0.0109	0.0960	0.0031	–2.5092	0.4703	0.15	0.0727

*The arithmetic mean for silver is taken from a previous compilation by Boyle (1968*b*). It does not refer to the samples quoted in the column – number of samples.

2.06 ppm) were recorded in red granites, red granite pegmatites and mixed rocks affected by the reddened granites. They attributed the increased amounts of gold to disseminations of hematite attributable to pneumatolytic processes associated with the latest phase of granitic intrusion.

Boyle (1961*a*) observed decreases in the gold contents of chloritized zones and marked increases in carbonatized zones associated with the formation of quartz in the Yellowknife Greenstone Belt, Northwest Territories, Canada.

Buryak (1964) noted that the gold mineralization of the Lena goldfields, U.S.S.R. is closely related in time, space and possibly genetically to regional metamorphism.

In the Altai and Transbaikalian massifs Anoshin and Potap'yev (1966) found a two- to threefold increase of gold in an early stage of altered granites compared with the unaltered varieties. The gold content, thereafter, decreased with increased hydrothermal alteration. In other words gold was lost from rocks that have undergone metamorphism according to their data.

Shilin (1968) found that low temperature carbonatization processes reduced the content of gold in primary gold-enriched magmatic rocks in central Kamchatka.

Voskresenskaya and Zvereva (1968) noted a direct correlation between gold mineralization and the degree of metasomatism in northern Kazakhstan. In weakly propylitized rocks they recorded an average of 0.006 ppm Au whereas in highly propylitized varieties the values were as high as 0.06 ppm. Roslyakov (1969) noted that gold was concentrated in rocks formed at the terminal stages of metamorphism in the Darasun ore field.

Chaika and Nezhdanova (1969) examined the problem of the migration of gold during the metamorphism of the sulphide-bearing graphitic kyanite shales of the Keyv Series of the Kola Peninsula, U.S.S.R. They noted that pyrite contained the most gold (0.006–0.008 ppm), followed by graphite (0.004–0.006 ppm), pyrrhotite (0.003 ppm) and muscovite (0.003–0.004 ppm). During the metamorphism of the shales gold migrated together with silica (quartz) from the shales into nearby cavities. These contained gold in quantities up to 2.86 ppm. In the kyanite schists of the Vitim-Patom Highland, U.S.S.R. Kepezinskas *et al.* (1970) state that a complete redistribution of gold results from metamorphism. Petrov *et al.* (1972) noted a similar tendency during the metamorphism of the sedimentary rocks of the Patom Highlands, U.S.S.R. Gold was lost from the rocks of the amphibolite facies and accumulated in the epidote-amphibolite and greenschist facies. Uranium and thorium were relatively inert. Buryak *et al.* (1972) also found that the granitization of the Upper Proterozoic Patom Series of shales, schists, etc. resulted in significant losses of gold. This gold is evidently slightly concentrated in the youngest pegmatites.

During the autometasomatic amphibolitization of ultrabasic rocks gold is practically inert according to Shcheka and Moiseenko (1970). However, during serpentinization processes much gold is lost from these rocks; granitization and migmatization of ultrabasic rocks also results in significant losses of gold. Li and Kornev (1972) also noted appreciable losses of gold during serpentinization of dunites. In another paper Moiseenko *et al.* (1970) state that the main sources of gold in the earth's crust are basic and ultrabasic rocks. During the metamorphism of these rocks the gold is freed, transferred to melts and solutions and after migration is deposited in local faults and fractures. The same theme, greatly amplified and illustrated, is discussed by Moiseenko *et al.* (1971) and Moiseenko and Fat'yanov (1972).

In their study of the Rice Lake Gold Belt of Manitoba, Stephenson (1971), Stephenson and Ehmann (1971) and Weber and Stephenson (1973) concluded that the low gold values recorded for high-grade gneisses was the result of gold depletion during high-grade metamorphism of the belt.

Finally, Pchelintseva and Fel'dman (1973) examined the gold content of the metamorphic rocks in the Koktchetav uplift and concluded that the gold distribution in these rocks is determined by the composition of the initial rocks (sediments and volcanics), not by the grade of the metamorphism which ranges from greenschist to amphibolite facies.

A number of more recent studies suggest that gold is relatively mobile during metamorphic processes. Li and Shokhina (1974*a*) found that during regional metamorphism no substantial change took place in the Precambrain rocks of the Yenisei Mountain Ridge. However, they observed that the deep reformation of the mineral composition of some of the rocks under conditions of high temperature was attended by an internal redistribution and dispersion of gold. Palandzhyan and Razin (1974) working in the ultrabasic and basic massifs of the Sevan Ridge, U.S.S.R. observed that metamorphic transformation of rocks of the ophiolite association is accompanied by a decrease in their gold content, and Mushkin *et al.* (1974) found that eclogitization of granulite facies gneisses leads to upward movement of gold, together with Si, Na and K, into quartz veins. Gavrilenko *et al.* (1974) observed that the regional metamorphism of Lower Proterozoic rocks in the Kolmozero-Voron'ya complex (Kola Peninsula) was essentially isochemical, the overall concentrations of petrogenic and ore-forming elements, including gold, being the same as the concentrations in the parent rocks. The great dispersion of gold observed and the anomalous concentrations of the metal in hydrothermally altered rocks in fissured zones, at contacts between mafic and ultramafic intrusives, and in mineralized pegmatites and their enclosing rocks were considered to be due to redistribution of gold to some degree during metamorphism. During the formation of contact-metasomatic zones associated with granitic intrusions in the Altai-Sayan Korobeinikov (1976) observed that most of the gold in the original rocks was redistributed and concentrated in the hornfels and marbles, particularly in magnetite, epidote, amphibole and chlorite.

Despite all these indications there are still insufficient data to establish unequivocally the migration trend of gold (and silver) during metamorphic processes. Nevertheless, certain trends seem to be indicated from my research and that of others. During high grade metamorphic processes and granitization gold (and silver) appears to be liberated from the rocks but is not concentrated in pegmatities despite the fact that the migration of much silica is involved in the process. The element is, however, markedly enriched in some skarn zones, in various types of veins and in a variety of other deposits that appear to form during the late stages of the metamorphic and granitization history of a mineralized belt.

Gold is not a chalcophile element but has certain tendencies in this direction since it appears to follow sulphur, arsenic and antimony during regional and contact metamorphic processes. There is also considerable evidence to suggest that the migration characteristics and diffusion rates of gold are about the same as those of K, Na, Mo, W, Cu, Ag, Zn, Cd, Hg, Pb, As, Sb, Se and Te since these elements are commonly associated in deposits. The tendency of gold to follow copper and silver is common in some skarn deposits (Table 44); in others it follows arsenic and/or antimony. There is also a marked coherence in the migration tendencies of silica and gold, the two constituents being nearly always associated in auriferous deposits. An explanation of these and other features is reserved for the discussion on the origin of gold deposits in Chapter III.

> As a land surface is worn away in regions where conditions are unfavourable for solution, gold tends gradually to become concentrated at the surface.
> –W.H. Emmons, 1937

Gold in soils and other products of weathering

The average gold content of most soils is low, but the element is enriched in certain types of soils and in a variety of glacial and weathered products in the vicinity of gold-bearing rocks or auriferous deposits. Some of these weathered products constitute valuable deposits of the element; they are discussed in the section on placers in Chapter III.

Relatively little precise knowledge on the migration and concentration of gold during weathering is available despite the importance of the subject. Both physical and chemical processes are involved; the importance of each in the migrational history of the element depends on numerous factors discussed below.

Soils

Data on the gold content of soils are limited, and the calculation of an average value seems unwarranted at the present time. Swaine (1955) does not list an average value, and Vinogradov (1959) states that the data on gold in soils were insufficient at the time of his compilation to give values for the contents of normal soils.

Nemec *et al.* (1936) found an average gold content of 0.1 ppm in some Czechoslovakian soils formed from andesite and rhyolite, and Nemec *et al.* (1937) recorded gold values from 0.185 to 0.350 ppm in other soils of Czechoslovakia. Babicka (1943) records values from 0.2 to 6 ppm Au in some Czechoslovakian soils. Goldschmidt and Peters (1933) found 0.1 ppm Au in the humus of a birch forest in Germany.

Relatively few data on the gold content of soils well removed from gold deposits are extant in the literature. One particular study, however, carried out in Greece gives considerable information on the content and nature of gold in Grecian residual and alluvial soils, including one loessial soil from Illinois (Yassoglou and Nobeli, 1968). The values recorded range from 0.005 to 0.150 ppm Au. The results indicated that the silt clay fractions contain much higher amounts of gold than does sand. Parent materials, which have undergone one cycle of soil formation, appear to contain gold in the silt in the resistant metallic form. In other cases, the element was found to be associated with easily weatherable minerals and appeared to be progressively lost from the silt fraction during processes of soil genesis. The authors thought that gold could be used as an index of soil weathering, either as a resistant or as a removable constituent of the silt, depending on the composition and the origin of the parent material of the soil. They also felt that gold can be used as an index of the lithological continuity of soil profiles.

Crocket *et al.* (1973) record gold values for some Arctic Island soils: the Prince Patrick grey soils contain 0.001 ppm Au and the Melville Island grey-brown and red-brown soils 0.0003 and 0.0018 ppm Au respectively. The average for the Arctic soils is given as 0.001 ppm Au. The analyses were done by neutron activation methods.

The writer has carried out investigations of the gold content of soils far removed from gold deposits in a number of Canadian areas, with the results noted in Table 19. Summarizing we can say that the general background range of the soils in most parts of Canada is from <0.0005 to 0.005 ppm. There is a slight enrichment of gold in the A (humic) horizons compared with other horizons in some soil profiles.

Numerous investigations of the gold content of soils in the vicinity of gold-bearing deposits are extant in the literature. These, together with some of my data, are summarized in Table 20. Enrichment of the element in the A (humic) horizons of soils is again apparent in numerous places, although in some profiles there is a considerable concentration

Table 19. Gold content of Canadian soils well removed from gold-bearing rocks and deposits

Description	Range in gold content (ppm)
Southwestern Ontario, Port Lambton–Wallaceburg area	
Soils developed on glacial lake clay	
A horizon (virgin)	<0.005–0.01
B horizon (virgin)	<0.005–0.01
C horizon (virgin glacial clay)	<0.005 (av.)
A horizon (cultivated area)	<0.005–0.01
B horizon (cultivated area)	<0.005–0.01
C horizon (virgin glacial clay) (cultivated area)	<0.005 (av.)
Soils developed on glacial sand	
A horizon (virgin)	<0.005
C horizon (virgin)	<0.005
A horizon (cultivated)	<0.005
C horizon (cultivated)	<0.005
Central Ontario, Minden-Bobcaygeon area	
C horizon (till)	<0.005
Eastern Ontario, Carleton Place–Ottawa area	
C horizon (till)	0.01 –0.035
C horizon (Leda clay)	<0.005
Eastern Ontario, Lavant Township, Lanark County	
Soils developed on glacial till	
B horizon below pine needle mat	<0.005–0.01
Northern Ontario, Cobalt area	
A horizon (humic)	<0.005
A_2 horizon (sand)	<0.005
B horizon (sand)	<0.005
C horizon (sand)	<0.005
A horizon (humic)	<0.005
C horizon (glacial clay)	<0.005
C horizon (till)	<0.005
New Brunswick, Bathurst area	
Soils developed on glacial till, sand and gravel	
$A_{00}+A_0$ horizon (humic)	<0.005–0.03
A_1 horizon (reddish loam)	<0.005–0.017
A_2 horizon (leached layer)	<0.005–0.017
B_2 horizon (abundant iron oxides)	<0.005–0.017
C horizon (sand, till)	<0.005–0.017
Manitoba, Falcon Lake area	
Soils developed on sandy glacial till	
A horizon	<0.005–0.015
B + C horizon	<0.005
Northwest Territories, Yellowknife area	
Soils developed on glacial till, sand and gravel	
A horizon (humic)	0.005–0.10
B horizon	0.008 (av.)
C horizon (sandy till; glacial sand and gravel; clay locally)	0.009 (av.)
Northwest Territories, Regan Lake area	
Regosoils*	<0.005–0.005

*Data from Cameron, E.M. and Durham, C.C. (1974), Geol. Surv. Can. Pap. 74-1, Pt. A, p. 53–60.

of gold in the B and C horizons directly over deposits (viz., Crestaurum profiles). As a generalization one can say that gold is enriched in one or more of the soil profiles in the vicinity of auriferous deposits.

A number of studies have been carried out by the writer to determine the mode of binding of gold in soils. The results of these studies suggest that gold is bound in the following ways in most soils:

1. As a constituent of flour gold and small flakes and nuggets of native gold. These occur almost universally in the weathered products of gold deposits, auriferous sulphide deposits, etc. and in nearby soils, especially those of a residual nature. Similar flakes and nuggets have been noticed in glacial till, sand and gravel overlying, or in the vicinity of, gold-bearing deposits. This gold is readily extracted by aqua regia, KCN, and other gold solvents. Frequently, the gold is of much greater fineness than that in the underlying deposits. This feature is discussed in greater detail in Chapters III and IV.

2. As a minor constituent of a number of partly oxidized residuals of primary vein minerals, particularly pyrite, arsenopyrite, chalcopyrite, sphalerite, magnetite, etc. These minerals have been noticed in soils overlying all types of gold deposits throughout Canada.

3. As a minor constituent of a number of discrete secondary (supergene) minerals, particularly limonite, wad, beudantite, bindheimite, anglesite, jarosite, copper carbonates, etc. These minerals generally occur in soils quite close to the underlying deposits. Most occur in the C horizons; only a few have been noted in the A and B horizons, and their presence in these horizons is generally due to solifluction and other disturbances of the soil.

4. As a minor constituent of manganese oxide coatings on sand and silt; also in similar material cementing sand, silt, etc. in the B and C horizons of the soil. It has been noted that some soils enriched in manganese oxides are also enriched in gold (and silver). The enrichment of gold is presumably due to the presence of foreign colloids such as positively charged hydrous iron oxides, alumina, etc., which impart a net positive charge to the initially negative hydrous manganese oxide gels. These in turn attract and coprecipitate the negatively charged gold complexes and colloids. (*See also* the section on gold in oxides in the first part of this chapter.) The gold in manganese oxides is evidently present mainly as an adsorbed phase or in a very finely divided native form. Some wads in oxidized auriferous deposits contain wiry and powdery forms of native gold.

5. As a minor constituent of various hydrous iron oxide coatings in sand and silt; also in similar material cementing sand, silt, etc. in the B and C horizons of the soil. Hydrous iron oxides in soils in the vicinity of gold-bearing deposits are invariably enriched in gold, the probable cause being that negatively charged gold complexes and hydrosols are strongly adsorbed and precipitated by positively charged hydrous iron oxide sols and gels. The gold in hydrous iron oxides is evidently present mainly as an adsorbed phase or in a very finely divided native form. Some limonites in oxidized auriferous deposits contain wiry and powdery forms of native gold.

6. As a minor constituent of organic matter (humus) in soils. We have found an enrichment of gold in some of the A (humic) horizons of soils in the vicinity of gold-bearing deposits and also in normal soils. The nature of the gold in this material is difficult to ascertain. Some is evidently a constituent of the undecayed plant and animal matter (*see* below under gold in plants and animals.) The remainder is probably present as a chelate or some other organometallic gold compound. Only small amounts of gold (and silver) are

Table 20. Gold content of soils in the vicinity of auriferous deposits

Locality	Material	Range or average (ppm)	Reference	Remarks
Région de l'Afema en Côte d'Ivoire	Argillaceous lateritic soils	0.02–5.4	1	
Minerai Banora en Guinée	Argillaceous lateritic soils	0.5–30	1	
Aldan Shield, U.S.S.R.	Soils	0.1–5.7	2	Kuranakh-type (karst) gold deposits
	A_0 horizon	0.1–5.5	2	
Mont Flotouo (Ity, Côte d'Ivoire)	Argillaceous lateritic soil	0–>8	3	
Northwest Wyoming	Soil on/or derived from Harebell and Pinyon conglomerate-sandstone formations	0.036–0.048	4	Rock underlying soil contains 0.0–0.43 ppm Au
Silver Reef, Utah	Soil	<0.05–0.35	5	Underlying rocks are sandstones, shales, etc.
Southeastern, U.S.A.	Near surface soil	0.02–0.52	6	0–5.5 ft deep
	Soil from a pit	0.03–0.23		0–12 ft deep
Ity, Ivory Coast	Lateritic ores	up to 18.3	7	
British Guiana	Laterites and ironstones	0–>17	8	Precipitated ironstones (limonite) are often rich in gold
Mill Canyon area, Nevada	Mull (forest humus)	0.8 (av.)	9	Composite of 2 samples
Empire district, Clear Creek County, Colorado	Mull (forest humus)	1.4 (av.)	9	Composite of 8 samples
Empire district, Clear Creek County, Colorado	Mull (forest humus)	0.6 (av.)	9	Composite of 9 samples
Empire district, Colorado	Mull layer	<0.04–>0.60	10	About 70% of values cluster in the range <0.04–0.59 and 30% in range >0.60
	Soil (6–12 in. depth)	<0.04–>0.60	10	About 90% of values cluster in the range <0.04–0.59 and about 10% in range >0.60
Auriferous areas in Colorado, Utah, Nevada and Arizona	Mull layer	<0.04–5.1	11	
	Other soil horizons	<0.04–29	11	High concentrations of gold commonly found near the bottoms of the profiles
Cros Gallet, Guyane française	Soil	0.053–1.10	12	
Rhodesia; near Leopardess Mine	Topsoils and subsoils	0.042–0.049	13	Kalahari sand
Lavant, Ont. (Joe Lake Mine)	B horizon below pine needle mat	0.01–0.05	14	Podsol developed on till
Falcon Lake, Man.	A horizon	<0.01–0.025	14	
(Falcon Lake Mine)	B + C horizon (till)	<0.01	14	
Yellowknife, N.W.T.				
Profile 1 (Crestaurum Mine area)	A (humic) horizon	0.85	14	Arctic brown soils developed on till. Permanently frozen in places
	B + C (rusty till)	24		
Profile 2 (Crestaurum Mine area)	A (humic) horizon	1.1	14	
	B + C (clay)	0.04		
Profile 3 (Lynx area)	A (humic) horizon	0.04	14	
	B horizon (clay)	0.02		
	C horizon (reddish till and clay)	9.0		
Near and over gossan zone, Regan Lake area, N.W.T.	Soils and disintegrated gossanous material	<0.005–0.810	15	Regosoils
Bathurst area, N.B.				
Profile 1 (Nigadoo vein) (For geological description *see* Davies *et al.* (1969), Geol. Surv. Can. Pap. 67-49)	A_1 (humic layer)	<0.01	14	Podsols developed on glacial till
	A_2 (leached zone)	<0.01		
	B_2 (rich in iron oxides)	<0.01		
	C (till	0.01		
	C (till and disintegrated gossan)	0.045		
	Gossan	0.41		
	Sulphide ore	0.01–0.20		
Profile 2 (Bathurst No. 6 deposit) (For geological description *see* Boyle and Davies (1964), Geol. Surv. Can. Pap. 63-24)	A_0 (humic layer)	0.025	14	
	A_2 (leached layer)	0.015		
	B_1 (reddish loam)	0.120		
	B_2 (yellowish-brown loam-till)	0.220		
	B_3 (reddish loam-till)	0.465		
	Gossan-limonite cemented conglomerate	0.017–0.684		
	Sulphide ore at interface of gossan and orebody	2.8		
	Sulphide ore	0.30		

References:
1. Martinet and Cuper (1961).
2. Razin and Rozhkov (1963).
3. Granier *et al.* (1963).
4. Antweiler and Love (1967).
5. Cornwall *et al.* (1967).
6. Kinkel and Lesure (1968).
7. Jones (1969).
8. Harrison (1908).
9. Curtin *et al.* (1970).
10. Curtin *et. al.* (1968).
11. Lakin *et al.* (1974).
12. Machairas (1970*a*).
13. Watson (1972).
14. Boyle (unpublished).
15. Cameron, E. M. and Durham, C. C. (1974) Geol. Surv. Can., Pap. 74-1, pt. A, p. 53–60.

removed by cyanide leaches of humic material. Ashing is necessary to effectively liberate the gold (and silver).

Zhukova (1974) found that up to 50 per cent of the gold in the organic horizons of soils over auriferous deposits in eastern Transbaikal, U.S.S.R. was water soluble. The remainder was bonded to organic substances, such as resins, waxes, fats, etc., extractable with an alcohol-benzene-chloroform mixture. Only small amounts of gold were bonded to humic and fulvic acids according to the data presented.

7. As a minor constituent of the clay fraction. The precise mode of binding of the gold in this fraction is uncertain. Small amounts are liberated by aqua regia and cyanide leaches indicating that some of the gold is present in an adsorbed state. The remainder of the gold appears to be tightly bound in the clay complex or clay-humate complexes. Roslyakov *et al.* (1971*a*) have noted that gold is relatively immobile where kaolin-type weathering predominates; on the other hand the metal is highly mobile in regions of lateritic weathering.

In addition to the above mechanisms, one can postulate that there is probably a minor amount of gold in the soil solutions of all soils. Roslyakov *et al.* (1971) found mobile gold in the form of ionic complexes – $[AuCl_2]^-$, $[Au(OH)_2]^-$ and $[Au(OH)_4]^-$ – in the pore solutions of soils and weathered debris in the vicinity of gold deposits. There are probably also small amounts of gold in bacteria and other organisms in the soil. These occurrences would seem to complete the possible modes of binding of gold in most soils.

Our investigations indicate that the mobility and fixation of gold in soils is influenced by the following factors:

1. Topography: in the mountainous parts of the Yukon and British Columbia, the soils tend to be consistently disturbed by solifluction and downhill migration. This results in poor profile development and mixing of material from different horizons. Both these features make systematic work difficult with respect to the migration and concentration tendencies of gold and other elements within the soil horizons. It is necessary to carry out the work on a local basis, and I have found that broad generalizations do not apply. On the other hand in the Canadian Shield and eastern Canada, where elevations are subdued and where profile development is fair to good, certain generalizations on the geochemistry of gold in the soils are permissible.

2. Climate: my investigations were too restricted to draw adequate conclusions about the effect that climate may have on the chemistry of gold in the soil. In the far northern regions where permafrost prevails, the soils are greatly disturbed by frost-boiling, solifluction, etc. This tends to disrupt the profiles and brings particles of unweathered material to the surface. This is a favourable feature for geochemical prospecting, since particles of gold and gold-bearing secondary oxidation products occur near the surface in the vicinity of deposits, a circumstance that permits relatively shallow sampling. In the more temperate regions where podzols and other types of soils predominate, normal chemical conditions prevail, and the chemistry of gold in the soils can be traced with more assurance. Under conditions of tropical weathering gold appears to be relatively mobile in lateritic terranes according to the recent investigations of Roslyakov *et al.* (1971*a*). This feature has been noted by others earlier (*see* the next section on gold in natural waters). Despite the asserted high mobility of gold under tropical weathering conditions much of the metal remains behind in the weathered mantle as can be seen by perusing Table 20. (*See also* the section on placers in Chapter III and the pertinent discussion in Chapter IV.)

3. Parent materials: no major differences have been noted in the gold content of soils developed from different rock types in Yukon. In the Canadian Shield and in eastern Canada, where the soils are developed on glacial tills or mixtures of tills and residual soils, I have noted no radical variations in the gold content of the normal soils. There may be slight enrichments of gold (up to 0.01 ppm or more) in residual soils and weathered residuum developed on pyritic graphitic schists, phyllites, quartzites, etc. containing an abundance of quartz blows and boudins. Evidently, the gold derives from the quartz bodies and oxidized pyrite. Some tills record slightly higher than average contents of gold (0.005–0.010 ppm) for reasons that are obscure.

In all of my investigations I have found that the presence of auriferous deposits normally influences the gold content of the soils. In Yukon, Canadian Shield areas, New Brunswick and Nova Scotia the gold content of shallow soils is generally increased over or near these deposits. Where deep glacial deposits overlie gold-bearing veins, as in some parts of the Yukon, Canadian Shield and eastern Canada, this generalization does not hold. In such situations, however, the basal members of the glacial materials normally report high contents of gold near auriferous deposits.

4. Drainage conditions: well-drained soils tend to have lower contents of gold (and silver) compared with poorly-drained soils in the same area. This is probably due to the fact that all gold (and silver) compounds are highly soluble and are gradually removed in time from the various soil horizons.

5. Presence of complexing agents: the free ions, Au^+ and Au^{3+}, are unknown in aqueous media. Both are highly complexed or hydroloysed, and the mobility of gold is hence largely dependent on suitable available complexing agents. Complexing agents such as $(S_2O_3)^{2-}$ or $(CN)^-$ or an excess of Cl^- give soluble complexes of the type $[Au(S_2O_3)]^-$, $[Au(CN)_2]^-$ and $[AuCl_2]^-$ which render gold highly mobile. Some organic (humic) complexes evidently have the same facility. Anions, cations and complexes which inhibit the mobility of gold are numerous. Cations and anions such as OH^-, Fe^{3+}, Mn^{2+}, etc. precipate insoluble compounds that coprecipitate or adsorb soluble gold thus decreasing the mobility of the element. The presence of H^+ in solution causes the dissociation of many gold complexes stable only in alkaline environments; OH^- has the same effect on the acid complexes. The presence of H_2S or S^{2-}, due to the bacterial reduction of protein complexes in organic environments, may precipitate the sulphide or the free metal, thus greatly reducing the mobility of gold.

6. Eh of soils: soluble forms of gold are easily reduced, generally to the metal, by practically all substances as shown in the next section on natural waters. The migration of gold in soluble forms in soils is, therefore, temporal and generally not extensive. Complexing, however, may increase the mobility of the element in some environments. Indirectly, the Eh is also of major importance in the migration of gold in soils since it

plays a part in the hydrolysis of iron and manganese, the precipitation of hydrous iron and manganese oxides, the oxidation of organic complexes, etc., and these exert a major influence on the mobility of gold as described below.

7. pH of soils: the pH of soils affects the mobility of gold in many ways, some of them directly and some indirectly. These are discussed below. Normally, gold is more soluble under acidic conditions such as where pyrite, pyrrhotite, and other sulphides are undergoing oxidation. However, in alkaline environments where suitable complexing agents are present gold is highly mobile.

The chloride complex $[AuCl_4]^-$ and other acid complexes are stable only under acidic conditions. On neutralization these complexes dissociate and the oxide, hydroxide or metal are precipitated. Similarly the alkaline complexes of gold viz. $[Au(CN)_2]^-$ and $[Au(S_2O_3)]^-$ are stable only under alkaline conditions, the oxide, hydroxide or metal being precipitated on neutralization. Changes in pH are, therefore, important, in the mobility of gold in soluble complex forms in soils. However, the chemistry of gold in soils is rarely dependent alone on the effect of pH on gold ions and complexes since the presence of iron, manganese, alumina and organic matter modify the course of events in a number of ways.

In an environment where decreasing acidity and a relatively high Eh prevail, iron if present in the ferrous state, is oxidized to the ferric state, undergoes hydrolysis and may be precipitated as a basic salt or as the oxide hydrate, $FeO(OH)\cdot nH_2O$. In some profiles where the pH ranges from 3 to 5, jarosite, $KFe_3(SO_4)_2(OH)_6$, and beudantite, $PbFe_3(AsO_4)(SO_4)(OH)_6$, two complex basic sulphates of ferric iron, may be present, and these tend to scavenge much gold, rendering the element immobile. In other profiles it is evident that much of the gold is tied up in the hydrous iron oxides. Normally ferric oxide hydrate is a good carrier of gold, since its colloid bears a positive charge, which tends to coprecipitate the negative gold complexes and the negatively charged gold colloids. Where PO_4^{3-}, AsO_4^{3-}, hydrated MnO_2, silica, humic colloids, etc. are present these are attracted to the ferric oxide hydrate and may give the combined colloidal complex a negative charge, a feature which permits the adsorption or coprecipitation of cationic gold as well as numerous other cations.

Decreasing acidity, likewise, causes the precipitation of hydrated manganese oxides and since these carry a negative charge they adsorb and coprecipitate any available hydrous iron oxides, alumina, etc., which impart a net positive charge to the gels. These in turn adsorb and coprecipitate the negatively charged gold complexes and hydrosols. Clay minerals have a similar effect. (*See also* the discussion on adsorption and coprecipitation of gold complexes and colloids in the section on gold in natural waters and their precipitates.)

These features serve to emphasize the fact that gold is relatively immobile in soils where the pH lies between 4 and 8. The mobility of the element is restricted in part by the dissociation of alkaline gold complexes in an acid medium, the dissociation of acidic gold complexes in an alkaline medium, but mainly by the hydrolytic and colloidal reactions that characterize the chemistry of iron, manganese and alumina in the soils.

8. Presence of organic compounds: organic matter, particularly the humic substances, adsorb gold strongly and apparently build the element into their structures, perhaps forming chelated complexes with the metal. Whatever the mode of binding, it is evident that much gold is immobilized where abundant organic matter is present in the soil. This is shown by the marked enrichments of gold in the A_0 (humic) horizons of many soils, and its presence in relatively increased amounts in some peat and half bog soils and in decomposed organic debris in marshes. Soils containing charcoal layers (from ancient forest and peat fires) may exhibit marked enrichments of gold (and silver) in these layers.

The pH seems to have little effect on the mobility of gold in the organic horizons of soils or in peats, etc. The pH may be as low as 3.0 with no particular decrease in the content of gold in the humic components.

Another point of interest is the fact that as the humification processes proceed the large co-ordinated humic groups dissociate and pass into a soluble (colloidal) form in the soil solutions and are ultimately removed by the ground waters. As the soluble groups appear to retain much of their gold during the disintegration process, the mobility of gold is increased in the soils. (*See also* the section on gold in natural waters in this chapter.)

Finally, under a high oxidation potential the humic complexes may be oxidized yielding CO_2, H_2O, etc. and liberating any bound gold. This gold is then in an extremely mobile state, although in most cases it is generally seized quickly by the remaining humic complexes. This process repeated *ad infinitum* leads to the enrichment of gold (and silver) noted in some of the organic layers of soils, in peats, etc.

9. Fe^{2+} – Fe^{3+} content of soil solutions: the acid reactions[20,21]

$$Fe^{2+} + Au^{+} \leftrightarrows Au + Fe^{3+}$$
$$3Fe^{2+} + Au^{3+} \leftrightarrows Au + 3Fe^{3+}$$

are reversible, the direction depending on the Fe^{2+}/Fe^{3+} ratio in solution. Where a high Fe^{2+} concentration prevails, gold is precipitated, and the mobility of the element is reduced. Where a high Fe^{3+} concentration is present gold is retained in the soluble and hence mobile state. It should be particularly noted that the reactions apply only to iron in solution; once hydrated oxides of hydroxides are precipitated these no longer play a part in the oxidation-reduction reactions. Soluble manganous salts probably also play a role similar to those of ferrous salts in slightly acid, neutral, and mildly alkaline soils.

$$2Au^{+} + Mn^{2+} + 4H^{+} + 2O_2 \leftrightarrows 2Au + MnO_2 + 2H_2O$$
$$2Au^{3+} + 6H_2O + 3Mn^{2+} \rightleftarrows 2Au + 3MnO_2 + 12H^{+}$$

[20] These reactions are some of the most important as regards the geochemistry of gold (and silver) in near surface processes. The first has been discussed in detail by Tananaeff (1925) as regards silver.

[21] The oxidation-reduction couples for ferrous, manganous, aurous and auric ions and their potentials as given by Latimer (1952) are as follows:

$Fe^{2+} = Fe^{3+} + e^-$	$E^0 = -0.771$
$2H_2O = O_2 + 4H^+ + 4e^-$	$E^0 = -1.229$
$Mn^{2+} + 2H_2O = MnO_2 + 4H^+ + 2e^-$	$E^0 = -1.23$
$Au = Au^{3+} + 3e^-$	$E^0 = -1.50$
$Au = Au^+ + e^-$	$E^0 = -1.68$

Note: Both ferrous and manganous ion can reduce gold from its ionic state to the metal.

10. Formation of colloids: practically nothing is known about natural gold colloids in soils. Gold colloids carrying a negative charge are easily prepared in the laboratory, and if these occur in nature they would be expected to increase the mobility of the element in soils. Because the gold hydrosols carry a negative charge the precipitation of these colloids would take place readily where positively charged colloids and electrolytes are present in the soil solutions. *(See also* the discussion on colloids in the section on natural waters.)

11. Influence of organisms: there are no direct data on the part that organisms such as bacteria and fungi play in the mobility of gold in the soil. Indirectly, one can assume with some confidence that the effects are important in two ways—some bacteria and fungi solubilize and hence mobilize gold, whereas others precipitate the element. Williams (1918) noted that *Penicillium glaucum* and *Oidium lactis* develop in colloidal gold solutions containing tannin and that they remove the metal from solution, but no explanation of the process could be suggested at that time.Parès and Martinet (1964) noted that *Bacillus subtilis, Serratia marcescens* and *Pseudomonas fluorescens* were capable of solubilizing gold to a marked degree, the effect diminishing and precipitation increasing with the age of the cultures. Lyalikova and Mokeicheva (1969) have also found that many heterotrophic bacteria have the ability to dissolve gold, and that sulphur compounds (thiosulphates and polythionates) which accumulate during the oxidation of sulphides by sulphur bacteria contribute to the solution of gold. A culture isolated from a gold deposit, similar to *Bacillus alvei,* was found to dissolve 500 to 600 ppb Au in 3 weeks. In an extended study on the solubilization of gold by micro-organisms, Doxtader (*in* Lakin *et al.,* 1974) found that the microbial strains utilized showed various abilities to solubilize the noble metal. *(See also* the section on gold in plants and animals.)

Bacteria and other micro-organisms also break down and decompose the various organic complexes and hence would liberate any complexed gold in a reactive and relatively soluble form. The production of H_2S by bacteria on the other hand would reduce the solubility of gold as mentioned above.

Finally, burrowing animals such as gophers, groundhogs and ants may physically transport gold particles from depth to the surface profiles of soils. An interesting case of the action of termites in Kalahari sand, Rhodesia is described by Watson (1972). The termites in excavating galleries to the water table have evidently brought some particles of gold to the surface from gold-bearing fissures and deposited them in their mounds. *(See also* the section on biogeochemical methods in geochemical prospecting in Chapter V.)

The above reminds one of the wonderful tale of the gold digging ants that has been passed from antiquity down through the generations to modern times. Herodotus had one version, Strabo another, Pliny yet another. In his *Historia naturalis* (Book II, Chapter 31) Pliny records:

"In the countrey of the Northerne Indians, named Dardae, the Ants do cast up gold above ground from out of the holes and mines within the earth: these are in colour like to cats, and as big as the wolves of Aegypt. This gold beforesaid, which they worke up in the winter time, the Indians do steale from them in the extreme heate of Summer, waiting their opportunitie when the Pismires (ants) lie close within their caves under the ground, from the parching sun. Yet not without great danger: for if they happen to wind them and catch their sent, out they go, and follow after them in great hast, and with such fury they fly upon them, that oftentimes they teare them in pieces; let them make away as fast as they can upon their most swift camels, yet they are not able to save them. So fleet of pace, so fierce of courage are they, to recover gold that they love so well."

As can be imagined various interpretations have been placed on this quaint tale which we cannot pursue here. Evidently the site of these gold-digging ants was the Gobi Desert, but there have been various other theories (Adams, 1938).

Other products of weathering

There are few data on the occurrence and content of gold (and silver) in weathered products such as bauxites, laterites, bog-iron ores, bog manganese deposits, etc.

Some laterites tend to be enriched in gold especially in the vicinity of underlying gold deposits. Some values for these laterites are shown in Table 20. Normally laterites contain only small amounts of gold, generally less than 0.01 ppm. There appear to be only limited investigations of the gold content of typical bauxites.

Tenyakov *et al.* (1970) determined the gold content of bauxites from various localities in U.S.S.R. and Hungary. They concluded from their results that:

1. Gold appears to be always present in bauxite since the element was detected in all samples analyzed from all deposits.

2. The gold concentration in bauxite exhibits an extremely wide range from 0.0004 ppm to 0.0290 ppm, a variation of two orders of magnitude. The average (arithmetic mean) gold content of bauxite is 0.0045 ppm.

3. There is some difference in the average content of gold in geosynclinal bauxite and platform bauxite; the latter is somewhat richer in gold (0.0049 ppm compared with 0.0036 ppm).

4. The gold content of bauxites is clearly related to their parent rock type; the metal is generally enriched in bauxite presumably formed from igneous or eruptive rocks but is much lower in that formed from clay and shale.

Tenyakov *et al.* (1970) thought that the gold in bauxites is probably mainly present in a mechanical form (as native gold).

The gold content of bog iron and manganese deposits is not well known. My analyses of most of the bog iron and maganese deposits of Ontario and Quebec show that such deposits contain only low amounts of gold (<0.005 ppm). A bog iron-manganese deposit in Canoe Lake, New Brunswick was found to contain 0.015 ppm Au and 2.2 ppm Ag; another in Bass Brook, near Walton, Nova Scotia had 0.34 ppm Au and 0.68 ppm Ag; and a third near Long Lake, New Brunswick had up to 1.4 ppm Au and 0.8 ppm Ag. The last bog is near stringers of quartz that carry a little gold.

During the weathering of gold- and silver-bearing rocks and deposits the two metals tend to concentrate in particles of native gold in the weathered detritus. Such auriferous blankets or eluvial deposits are common in Brazil, the Guianas, Mexico, Africa, Asia and elsewhere. They are composed principally of laterite–a mixture of ferric oxides, hydrated ferric oxides, clay and sand. Some contain essentially gravel, sand and clay and may be partly cemented by caliche and iron oxides. The origin of the gold-silver particles in these materials has been a long-standing problem. It is certain that some of the particles are of residual origin and represent the flakes and nuggets weathered from gold-quartz veins. Some of the gold and silver is probably also of chemical origin; that is, the particles are either enlarged or individually accreted by additions of gold and silver from solutions that derived the metals both from the weathered constituents of the rocks and from various gold- and silver-bearing sulphides in veins and other types of deposits.

As the weathering cycle proceeds the material in deposits may be further decomposed, eroded, transported and redeposited. This can be effected by wind, streams, rivers or by the surf of the sea, resulting in aeolian, stream, river, delta and beach placers. During these movements the gold-silver particles may be comminuted and reduced to a flour or, as is more often the case, especially in an aqueous environment, the small particles may be recrystallized, partly dissolved and the gold reprecipitated on the larger particles. Since silver is more mobile (more soluble) than gold the chemical processes of solution and reprecipitation generally lead to an increase in the fineness of the gold flakes and nuggets with increasing distance from the source. Furthermore, the outside parts of nuggets often have a significantly greater fineness than the inner parts, a feature that adds convincing support to the theory of accretion of gold from solutions as the particles are moved along by mechanical processes. Other evidence of the solution and precipitation of gold and silver in placers is witnessed by the occurrence of gold crystals in places, wire gold, sprigs of gold, replacement of deer horns by film gold, agatelike structures of certain nuggets and the fact that some investigators have found increased amounts of gold and silver in stream waters in the vicinity of gold veins and polymetallic deposits. (*See* the section on gold in natural waters.) Further details on the interesting problems of the geochemistry of the origin of gold in oxidized zones and placers is given in Chapters III and IV.

The most frequent associates of gold-silver particles in placers, in addition to the common silicates, are magnetite, ilmenite, rutile, monazite, garnet, zircon, tourmaline, pyrite, native bismuth, cassiterite and cinnabar. Nuggets of native silver are rare, although they do occur in some of the placers in Alaska and in those of Burwash and Arch creeks (Kluane and Donjek rivers district), Yukon. Arquerite, the silver-gold-mercury amalgam, occurs in the placers of Silver, Vital, Kenny and Kwanika creeks in the Fort St. James area, British Columbia and in Burwash Creek in the Kluane Lake district, Yukon.

Certain types of eluvial and alluvial fans and similar deposits, in addition to gold and silver, also contain galena, anglesite, bindheimite, mimetite and other oxidation products of lead-zinc-silver and gold veins. Such deposits are known in many parts of the world, particularly in the unglaciated regions. Those associated with a mineralized porphyritic granite stock near Guadalcázar in Mexico have received detailed attention (Foshag and Fries, 1942; Fries and Schmitter, 1948). Few of these deposits contain in excess of 0.10 ppm Au. In general the average gold content is about 0.025 ppm. Other types associated with sink-holes in limestone are known in Burma and China (Rundall, 1927–1928; Searls, 1952) and have been worked in the past. Somewhat similar deposits developed in karsts are mined in the Yakutia region of the U.S.S.R. There the deep oxidation of auriferous metasomatic rocks developed in limestones has led to concentrations of gold in karsts in the limestones. These deposits are called 'Kuranakh type' by the U.S.S.R. geologists; they are described in more detail in Chapter III.

Undoubted fossil gold-silver placers have been described from several countries. They occur in conglomerates, sandstones, and arkoses of various ages. Examples are known in the Black Hills, South Dakota; Gays River, Nova Scotia; Indian River, Yukon (Maclean, 1914); and elsewhere. (*See also* the section on placers in Chapter III.) The auriferous Witwatersrand quartz-pebble conglomerates, the greatest of all gold deposits, are considered to be modified fossil placers. These deposits, and others in the same category, are discussed at length in Chapter III.

Mountains big with mines, whence
many a bursting stream auriferous plays
–Thomson, *Summer*

Gold in natural waters and their precipitates

Gold is widely diffused in ground, spring, stream, river and lake waters, but generally only in small amounts. The average for all fresh waters is about 0.00003 ppm (0.03ppb) Au. Marked variations above and below this average have been reported for a number of regions.

Most of the gold in natural fresh waters is derived from gold present in rocks, mineral deposits and their weathered products. Infinitesimally small amounts of gold (and silver) in some fresh waters may arise from industrial and domestic usage.

Several investigations of the gold content of hot spring, ground, stream, river and lake waters have been carried out in a number of countries. Data from these investigations and from my own limited studies are recorded in Table 21.

Only a few detailed investigations of the gold content of hot springs and thermal waters are extant in the literature. Some of the more comprehensive studies are summarized below.

Koga (1960, 1961) carried out an extensive investigation of the gold, silver and numerous other trace elements in 27 hot springs in the Beppu area of Japan. The range in the gold contents was 0.0 to 2.2 ppb, and the average for all hot springs was 0.53 ppb Au and 10.7 ppb Ag, giving an average Au/Ag ratio of 0.05. Ninety per cent of the springs had less than 1 ppb Au. Hot springs with low pH values (2.0–3.8) contained the highest amounts of gold, especially those enriched in sodium sulphate and sodium chloride. The silver content was low in NaCl springs and high in sodium bicarbonate ones. Gold showed no particular relationship to sodium bicarbonate. The temperature of the springs influenced the gold content; in general the higher the temperature the higher the gold content. Most spring precipitates were found to contain gold, the highest value being 56.5 ppb (0.056 ppm). No

Table 21. Gold content of natural waters

Description	No. of samples	Au (range) (ppb)	pH (range)	Remarks
Hot springs	35	0.004–2.2	<1–14	65% of the values lie in the range 0.04–0.7 ppb
Groundwaters and cold springs	8	<0.005–0.2	6.3–7.3	H_2S carbonate springs at Matzesta (Sochi) U.S.S.R. are enriched in gold; up to 35 ppb according to Zverev *et al.*, (1947)
Groundwaters, cold spring waters and mine waters in vicinity of gold deposits	84	<0.0008–80	2–8	84% of the values lie in the range 0.001–1.0 ppb
Stream and river waters	24	0.002–4.7	6.5–7.7	73% of the values lie in the range 0.002–0.1 ppb
Stream and river waters in vicinity of gold deposits	24	0.0048–35	6.5–7.6	91% of the values lie in the range 0.0048–0.150
Lake waters				
Fresh	21	0.12–7.7		Lake Baikal (Jones, 1970) and Greek lakes (Grimanis *et al.*, 1965)
Saline	4	<0.1–32		Great Salt Lake, Utah and Owens Lake, California
Oceans and seas	73	0.001–46		71% of the values lie within the range 0.001–0.50 ppb

Compiled from data by: Albul and Miller (1959), Baur (1942), Borovitskii *et al.* (1966), Brooks (1960), Browne (1969), Chernyayev *et al.* (1969), Cushing and Rancitelli (1972), Fischer (1966), Gebler and Stramkovskaya (1957), Goldberg (1957), Goleva (1970), Goleva *et al.* (1970*a,b*), Gosling *et al.* (1971), Gottfried *et al.* (1972), Grange (1937), Grimanis *et al.* (1965), Haber (1927, 1928), Hummel (1957), Ikeda (1955*a,b*), Jenne *et al.* (1968), Jones (1970), Kaspar *et al.* (1972), Koga (1960, 1961), Kropachev (1935), Lincoln (1911*b*), Lombardi (1964), Noddack and Noddack (1939), Oka *et al.* (1964), Popova (1971), Putnam (1953), Razin and Rozhkov (1963), Rosenbaum *et al.* (1969), Schiller *et al.* (1971*a,b*), Sharma (1972), Stark (1943), Talipov *et al.* (1972), Vorotnikov *et al.* (1973), Weiss and Lai (1963), Weissberg (1969, 1970), White *et al.* (1963*a,b*), Yasuda (1927), Zverev *et al.* (1947), Zvyagintsev (1941)

detailed correlation between gold and silver was found in the spring water, but there was a positive correlation between the two elements in the spring precipitates. The average Au/Ag ratio in the spring precipitates was about 0.05, much the same as the average ratio for the water. Positive correlations of gold with chromium and manganese and a negative correlation with titanium were found in the spring waters, and a positive correlation with silver and arsenic and a negative correlation with manganese obtained in the spring precipitates. The positive correlation of gold with chromium is of interest considering the common occurrence of chromium-rich alteration minerals (fuchsite) in certain types of gold deposits.

Weissberg (1969, 1970) described red-orange amorphous sulphides with opaline silica, precipitating from thermal waters along the eastern margin of the Taupo volcanic belt in New Zealand; these contained a number of ore elements up to the following concentrations:

As	2%	Tl	1000 ppm
Sb	10%	Pb	50 ppm
Au	85 ppm	Zn	200 ppm
Ag	500 ppm	W	3%
Hg	2000 ppm	Au/Ag ratio 0.17	

The transporting waters carry these elements in the following maximum values:

As	8 ppm	Tl	0.007 ppm (7 ppb)
Sb	0.3 ppm	S (as sulphide)	120 ppm
Au	0.00004 ppm (0.04 ppb)	Au/Ag ratio 0.06	
Ag	0.0006 ppm (0.6 ppb)		

These thermal waters are essentially alkaline (pH 6–7) Na-K-Cl-HCO_3-CO_3-SO_4-SiO_2 waters, with temperatures ranging from 200 to 290°C at a depth of 1 km.

Browne (1969) records 0.0001 to 0.001 ppm Au and 0.002 Ag in water separated at atmospheric pressure from the Broadlands geothermal drillhole, Taupo Volcanic Zone, New Zealand. The pH of the water was 8.55 and the type – Na-K-Cl-HCO_3-CO_3 – carrying small amounts of As (9 ppm), Sb (1.5 ppm), W (0.03 ppm), Zn (0.015 ppm), Cu (0.012 ppm), Tl (0.01 ppm) and a number of other elements. This water has deposited base metal sulphides (pyrite, galena and sphalerite with silver telluride, cobaltite and Ni-glaucodot) in tuffs, tuffaceous sediments and rhyolites, which it penetrates.

Goleva *et al.* (1970*a,b*) found 0.8 to 1.6 ppb Au in strongly acid (pH <1, Eh >0.8 V) HCl fumaroles of the Ebeko volcano on Paranushir Island, Kamchatka, U.S.S.R. They concluded that the gold was present mainly as halide complexes.

Gottfried *et al.* (1972) report that the gold content of the thermal waters in Yellowstone National Park, U.S.A. is very low, generally 0.004–0.040 ppb, with a maximum of 0.1 ppb. The sinters on the other hand, are relatively enriched in gold, 5 to 50 ppb with maximum values up to 5000 ppb. Sinters having high gold concentrations also contain correspondingly higher amounts of arsenic, antimony and mercury. The pH of the waters ranges from 2 to 9, but most have a neutral to slightly alkaline reaction.

Summarizing the available data on hot springs one can say that the pH of the water seems to have little effect on the transfer of gold. All springs, including those with acid, neutral and alkaline reactions may carry the element, although there is a tendency for it to be enriched in neutral and alkaline waters. Temperature evidently has an effect, the hotter the water the higher the gold content in some spring systems. Correlations with other elements are difficult to establish with certainty mainly because of insufficient data. Gold is generally accompanied by silver, and in nearly all auriferous springs there is a relatively high content of As, Sb and SiO_2. Nearly all hot springs carry sufficient Cl, S^{2-}, As and Sb to complex and render mobile the small amounts of gold present.

There are insufficient data on the gold content of normal groundwaters and cold springs to draw satisfactory conclusions about the transport of the element in these waters. From my experience I conclude that the average gold content of most normal groundwaters and cold springs is about 0.005 ppb. My compilation of the gold content of normal stream and river waters shows that the values range from 0.002 to 4.7 ppb with some 73 per cent of the values in the range from 0.002 to 0.1 ppb. In my experience most streams and river waters have <0.005 ppb total gold including that dissolved and associated with particulate matter. In a recent study Cushing and Rancitelli (1972) found that the gold and silver contents of the Columbia River water, as determined by neutron activation analyses, varied from time to time as shown below.

Element	Date				
	11-4-68	2-12-69	4-24-69	6-11-69	8-14-69
			(ppb)		
Au	0.030	<0.005	<0.005	<0.005	<0.005
Ag	0.50	<0.01	<0.01	3.03	<0.01

There are insufficient data on the gold content of freshwater lakes to draw any conclusions. Grimanis *et al.* (1965), using activation analysis methods, reported gold contents in 11 Greek lakes to range from 0.2 to 7.7 ppb (av. 1.5 ppb). The corresponding silver contents were less than 10 ppb.

In salt lakes Lincoln (1911*b*) reported up to 32 ppb Au in Great Salt Lake, Utah; in the Nevada salt marshes the same author reported no gold present. Carpenter (1941) found no gold in the salts and other materials of the dry lakes in Nevada. In the summer brine of Owens Lake, California Lombardi (1964) reports the presence of gold in the 10 to 100 ppb range.

Most of the work on the gold content in cold springs, groundwaters and stream and river waters has been done in the vicinity of gold-bearing deposits. As shown in Table 21, the values tend to be higher than in similar bodies of water remote from gold deposits. A number of detailed investigations are extant in the literature, some of which merit brief reviews.

Kropachev (1935) seems to have been the first to examine the gold content of normal fresh waters as well as those in gold belts. He found from 40 to 60 ppb Au in the rivers of nonauriferous regions and some five times as much gold in the river water of auriferous regions. Still higher quantities of gold were found in the drainage waters of gold mines (500–1300 ppb). While his analytical methods were crude they did show that the waters in auriferous regions are commonly enriched in gold.

Borovitskii *et al.* (1966) examined the gold content in the waters of the Kuranakh gold-bearing area of the Aldan massif, U.S.S.R. The gold ores are mainly oxidized disseminated pyritic ores with pyrite, magnetite, hydrogeothite, hematite and gold. (*See also* Chapter III.) Associated elements are Ag, Cu, Pb, Zn, As, V and Ga. Three types of groundwaters are distinguished in the area – those in unconsolidated recent materials above the permafrost (mainly organic and calcium bicarbonate and sodium bicarbonate waters); those above- and inter-permafrost in the Lower Cambrian rocks (mainly calcium bicarbonate waters); and those below the permafrost in Lower Cambrian and Archean rocks. The results of the investigation showed that: the most common companions of gold in the groundwaters were As, Sb and Tl (and to a lesser extent Ag, Ti, Sn and Te); the inter-permafrost waters contained the highest gold contents; the gold contents in groundwaters near the gold deposits fluctuated from 0.03 to 10.0 ppb; the maximum gold concentrations occur in groundwaters with a high content of dissolved iron (up to 300 ppb) and manganese (up to 100 ppb); and that the gold contents in stream waters near the gold deposits were generally less than 0.05 ppb. An earlier investigation of the waters associated with the Kuranakh deposits by Razin and Rozhkov (1963) indicated that the more acid the waters the higher the gold content. Gold appears to be much more mobile than silver in the area since the ratio Au/Ag in the groundwaters as calculated from the data of Borovitskii *et al.* (1966) is generally high, in some cases up to 450.

Chernyayev *et al.* (1969) reviewed the literature on gold in natural water and carried out work on the natural waters of northern Kazakhstan and southern Urals. In the near-surface groundwaters of the Kokchetav uplift gold was found in 40 per cent of the analyzed samples (altogether 180 samples were analyzed). Its content varied from 0.1 to 4 ppb. The distribution of gold in the groundwaters was lognormal. Its average content in the samples for which positive results were obtained is 0.4 ppb, or if all analyzed samples are considered (including those in which gold was not detected), 0.15 ppb. Little difference was found in the gold content of waters circulating through different rock types (e.g., gabbro, 0.58 ppb; granodiorite, 0.5 ppb; shales and sandstones, 0.522 ppb). In the Shalkarsk area the mode was 0.2 ppb and the mean content (all samples) 0.154 ppb. Analyses of groundwaters from one of the Kokchetav gold deposits showed gold contents ranging from 0.4 to 1 ppb. These waters always contain As, Mo, La, Y, Yb, Li, Cu and a number of other elements. The content of silver in these waters is low. There was no well defined correlation between the content of gold in the waters and any of the trace elements nor between such constituents as carbonate, sulphate, chloride and hydrogen ion (pH). The hydrogeochemical profiles plotted across the gold deposits, however, show that directly over the orebodies there are peak concentrations of Au, Ag, As and Bi. The peaks for Ag and As are sharp; those for Bi are broad.

Goleva *et al.* (1970*a,b*) examined both ground and surface waters in the vicinity of a number of gold deposits in various geological and geographical settings in U.S.S.R., mainly in East Transbaikal, Central Asia, Southern Urals and Kamchatka. The contents they found ranged from 0.05 to 10 ppb Au. From their studies they concluded that:

1. The commoner types of natural water (pH 5–8 and Eh 0.1–0.5 V) most probably contain gold as uncharged particles and organic complexes. Gold in the presence of thiosulphate ions forms the very stable $[Au(S_2O_3)_2]^{3-}$ ion.

2. The halide complexes, $[AuCl_2]^-$, $[AuCl_4]^-$ and $[AuBr_4]^-$, can be formed only in media with Eh >0.9 V and pH <1 at appropriately elevated concentrations of chloride and bromide ions ($Cl^- \geq$ 2 mM, Br^- >0.6 μM). Such waters are exotic

under natural conditions and occur only in the HCl fumaroles of active volcanoes (Ebeko, etc.).

3. Waters containing H_2SO_4 are unfavorable for the migration of gold. Thus, waters in rapidly oxidizing zones of gold-sulphide deposits have only 0.1–0.5 μg/L.

4. The commonest coprecipitants of gold in natural waters are: iron hydroxides, calcium carbonate, disulphides and colloidal silica.

5. The gold content of groundwaters can be used as a specific means of prospecting for gold-bearing deposits.

In the Balei gold district of eastern Transbaikal, U.S.S.R., Goleva (1970) found background contents of <0.07 ppb Au and anomalous contents >0.15 ppb in the groundwaters. In alluvial and surface waters the background and anomalous contents were respectively <0.05 and >0.1 ppb Au. Most of the waters are weakly mineralized calcium, magnesium and sodium bicarbonate types with small amounts of sulphate. The main indicators of gold deposits in the hydrologic system were found to be Au, Ag, As and Zn and in some areas Bi and Sb.

Gosling *et al.* (1971) used neutron activation methods to investigate the gold content of stream, spring and mine water in both gold-bearing and barren areas underlain by sedimentary, igneous and metamorphic rocks of the Colorado Front Range. The total gold content ranged from nondetectible (<0.001 ppb) to 0.150 ppb. The 'solute' (dissolved) gold concentrations were generally higher by a factor of 2 to 6 than the gold held on or in particulates, and no relationship was found between both these values (solute and particulate) and rock or water types. Their data indicated a spring flush-out of dissolved gold in groundwater during the snow-melt recharge period. They concluded that direct hydrogeochemical prospecting using gold as an indicator of its deposits is unpromising.

Talipov *et al.* (1972) determined the Au, Ag and As content of the surface waters of the central Kyzylkum gold belt and found the gold content to range from 0.01 to 0.95 ppb; the corresponding arsenic contents were 10 to 378 ppb. The gold content was not related to the chemical composition of the water.

Kasper *et al.* (1972) found from 0.09 to 0.20 ppb in waters flowing from old gold mine tunnels in Austria (Imhoff and Bockhart) and Czechoslovakia (Jilové). They noted a general development of inhomogeneity in the waters after 21 days of storage.

Volkov and Shakhbazova (1973) determined the gold content of the surface and groundwaters of the Beskes auriferous copper ore field. The gold content was found to range from 0.0 to 5.2 ppb. The distribution of gold in the waters obeyed the lognormal law with a mean gold content of 0.09 ppb and a dispersion factor of 0.474. Within these parameters the minimum anomalous content was 0.3 and the maximum 2.5 ppb, the last value being a positive indication of the presence of gold in the ores. No relation was observed between the chemical composition and the distribution of gold in the waters although the highest concentration of the precious metal was noticed in $Ca(HCO_3)_2$ waters.

In the west Siberian artesian basin of U.S.S.R. Shvartsev (1974) found gold in all saline groundwater samples from the petroleum fields in the northern Tomsk region, the contents ranging from 0.02 to 0.49 μg/L. Shvartsev concluded that all the gold in the saline groundwaters was of lithological origin.

The available data on the gold content of ocean water is given in Table 21. The average gold content of the oceans appears to be about 0.0117 ppb and the average silver content about 0.3 ppb giving an Au/Ag ratio of 0.04.

The French chemist, J.L. Proust (1754–1826) was the first to report the presence of gold in seawater, and the English chemist S. Sonstadt (1872) carried out the first quantitative analysis finding less than a grain of gold per ton (about 65 ppb). Many years later Fritz Haber carried out over a period of some 7 years an extensive investigation of the gold content of seawater. Reviews of these and other investigations are given by Mellor (1923), Jaenicke (1935), Gmelin (1950), Purkayastha and Das (1965), Jones (1970) and Wood (1971).

The gold content of the oceans appears to be variable as will be readily apparent from the following short review of some of the significant investigations. This variation appears to be real, although some variations reported in the literature may be due to different methods of analysis.

Haber (1927, 1928) found that seawater did not contain nearly as high contents of gold as had been earlier reported. His values ranged from 0.003 to 4.8 ppb. He noted large variations in the sea both on an areal and local basis and attributed them to a variety of causes mainly to the presence of gold in various forms – as the chloroauric ion, in colloidal form and adsorbed by both organic and inorganic matter.

Yasuda (1927) found relatively large amounts of gold (3–20 ppb) in the water of bays in various parts of Japan. The equivalent silver content was 20 to 70 ppb.

Putnam (1953) gives an interesting account of the investigation of gold in seawater from Sonstadt and Haber up to the 1950's. He, himself, carried out a number of investigations of the gold content of seawater and concluded from these and analyses in the literature that the concentrations could vary through the range of 0.003 to 44 ppb Au.

Hummel (1957) found by radioactivation analysis that the gold content of the oceans depended on the distance from the shore. He recorded 0.40 ppb Au for English coastal waters and 0.015 ppb Au for water from the northwestern limit of the Bay of Biscay.

Brooks (1960) using an ion exchange enrichment technique obtained 0.009 ppb Au in seawater at Sea Point, Cape Town, South Africa. Weiss and Lai (1963) used cocrystallization methods to determine the gold content of seawater 40 mi due west of San Francisco. Their average was 0.068 ± 0.003 ppb. Oka *et al.* (1964) found from 0.006 to 0.429 ppb Au in surface ocean samples and samples taken at a depth of 100 to 200 m off the Japanese coast. Schutz and Turekian (1965) found the gold content of seawater from the Atlantic and Indian oceans to range from 0.004 to 0.027 ppb with an average of 0.011 ppb. The silver content in all oceans ranged from 0.055 to 1.5 ppb and the average was 0.29 ppb. The average Au/Ag ratio in the sea as calculated from their data is, therefore, 0.04.

Wood (1971) carried out an extensive investigation of the geochemistry of gold in the marine environment to determine possible natural concentrating processes which may be used to extract gold. Radioactive gold tracers ^{195}Au and ^{198}Au were used in the investigations. The solubility of gold in seawater was determined to be of the order of 4µg Au/L (4 ppb) about the same order of magnitude as previously predicted by Sillén.

Differences in the ion change properties of Au (I) and Au (III) were determined by experiments with electrolytically produced mixed solutions of aurous and auric chloride which were passed through strong anion exchange columns. Approximately 10 per cent of the soluble gold in equilibrated seawater solutions was found to be nonionic. The soluble character of the gold in these solutions was unchanged when mixed with sulphide waters. This indicated to Wood (1971) that gold would not be concentrated in anoxic basins. The character of the soluble gold was also studied using dialysis and electrophoresis.

Soluble gold was rapidly adsorbed on several clays followed by a release of gold within about 24 h of the mixing time. This is about the time for the equilibrium shift of auric to aurous gold to be established.

$$AuCl_4^- + 2\,Au + 2\,Cl^- \rightleftharpoons 3\,AuCl_2^- \quad K_{eq} = 1.8 \times 10^{-8}$$

Gold was found to be easily reduced and to be adsorbed on many different materials. Most materials used in sampling and analysis for gold in aqueous media adsorbed gold least from acidified solutions. Rubber and cellulose ester fiber filters were notable exceptions.

Neutron activation methods were devised by Wood (1971) to analyze marine samples for gold. Analyses of samples from southeast Alaska indicated that gold in seawater is predominantly in a soluble nonionic form, with ionic and particulate forms making up only about 20 per cent of the total. The average gold content in these waters was found to be 3 ng Au/L (0.003 ppb) with a range of 1 to 11 ng Au/L. Analyses of sediment in Gold Creek and Gastineau Channel at Juneau, Alaska indicate that the transport of gold is mainly by physical means with the majority of the gold appearing in the 4ϕ size range in and near Gold Creek.

Cationic forms of gold were found by Wood (1971) to be absent in seawater, and gold was not associated to any great extent with dissolved organic matter which is extractable in either chloroform or ethyl acetate.

The gold concentrations in samples from the east Pacific found by Wood (1971) showed a heterogeneous distribution. The average concentration was 0.004 ppb with a range of 0.001 to 0.025 ppb Au.

Wood (1971) estimates that approximately 10 per cent of the gold in seawater is ionic, probably Au(I)chlorohydroxy anions, e.g., $[AuClOH]^-$. Another 10 per cent is associated with particulate matter, and the remaining 80 per cent is less than 0.45 µ, nonionic, not extractable in ethyl acetate nor chloroform and probably colloidal.

Ryabinin *et al.* (1974) determined the gold content in 89 water samples collected with bathometers from different levels of the Atlantic Ocean tropical zone at some 19 sites. The concentrations ranged from 0.004 to 3.4 ppb. The average concentration for all samples was 0.202 ppb, a value considerably higher than that found by other investigators for ocean water. The distributions were very irregular, and no particular pattern was discernible.

Polyakov (1959) observed that the amount of gold and silver in foam and slime of Kara-Bogas-Gol Bay (Caspian Sea) U.S.S.R. was considerably greater than that in the seawater. Dudey *et al.* (1969) found 7.3 × 10^{-7} µg/m^3 Au in marine aerosols over the Atlantic Ocean.

The investigations of the gold content in ocean water indicate that there are major areal variations. The data are not extensive enough to draw any conclusions about the variations with depth. Most studies seem to indicate that nearshore samples are higher in gold than those in the open oceans, a feature that is probably due to the greater amount of inorganic and organic particulate matter, and hence more adsorbed and particulate gold, in nearshore waters compared with those at a great distance from the shore. Major local variations in the gold content of seawater are probably due to the presence of variable amounts of particulate matter. Future research on the gold content of the oceans should be directed to determining precisely the quantity of 'solute', adsorbed, and particulate gold in the samples.

The source of the gold in ocean water is speculative and problematical. Presumably most of the precious metal is the result of terrestrial weathering, and there may be a small contribution from the weathering of marine sediments. Along the ridges and rises of the oceans some of the gold in oceanic water may be of volcanic exhalative origin.

The oceans are an enormous reservoir of gold (and silver). A calculation using the average gold content of seawater (0.02 ppb) and the total mass of the oceans (143 × 10^{16} metric tons) indicates some 28.6 × 10^6 metric tons of the metal in the oceanic system. Other estimates in the literature of the total gold content of the oceans range from 5 × 10^6 to 6 × 10^{10} metric tons. The presence of so much gold in the oceans has whetted the scientific interest of many men, especially as regards commercial recovery of the metal. Fritz Haber (1927, 1928), as mentioned previously, had this in mind and spent some 7 years dealing with the problem, finally abandoning any hope during his lifetime of developing a feasible commercial recovery process. Others have taken up the problem with the hope of alleviating the economic ills of their nations (Purkayastha and Das, 1965). Perhaps in the future a highly specific resin for gold such as that described by Koster and Schmuckler (1967) will form the basis for commercially extracting the metal from the vast reservoir of the oceans. Lancaster (1973), however, considers that the extraction of gold from seawater by any known method is an unrealisable dream.

Many of the natural precipitates from hot springs, cold springs and groundwaters contain small amounts of gold (and silver) as shown in Table 22. Some precipitates are greatly enriched in the element, especially those near gold deposits. Others, such as those probably from submarine hot springs as in the Red Sea, contain up to 5.6 ppm Au (Table 14).

Manganese oxides (wad) frequently concentrate gold and certain varieties of limonite, especially if they are admixed with or contain manganese oxides, phosphate, arsenate, etc., tend to carry considerable amounts of the element. Organic compounds (humic complexes) also seem to be major concentrators of gold in some areas. Sulphate efflorescences and stalactites in the drifts and workings of gold-silver mines of the Tintic district, Utah and Empire district, Colorado were found by Morris and Lovering (1952) and Lakin (1969*a*) to be

Table 22. Gold content of natural precipitates (dry matter)

Location	Description of precipitate	Au (ppm)	Ag (ppm)	Au/Ag ratio	Reference
New Guinea	Wad and other manganese oxides; mangenese content 7.15%	35.9	4801.7	0.007	1
	Wad and other manganese oxides; manganese content 12.15%	28.7	147.0	0.2	1
New Guinea	Dark brown to greenish, pin point organic (iron humate) compound along seepages	27 920	15 660	1.8	1
Guiana (British Guiana)	Precipitated ironstones (limonite) on sides of hills near gold-bearing rocks and deposits	0–>17			24
U.S.A.; Tintic Standard Mine, Tintic district, Utah	Various sulphates such as chalcanthite, epsomite, pisanite, etc. Value given is maximum value in chalcanthite. Minerals on shale walls were highest in gold	2.7	1162	0.002	2
U.S.A.; Southeastern states	Transported limonite	0.14–2.9	–	–	17
U.S.A.; (various areas)	Iron oxides and hydroxides (crusts and stains)	0.5–1.7	–	–	16
U.S.A.; Empire district, Colorado	Supergene sulphates of copper and iron	11.7	–	–	27
U.S.A.; Arkansas	Manganese ore leached from country rock	0.7	23	0.03	4
U.S.S.R.; Kuznetski-Alatau gold deposits	Ocherous and nodular transported limonite	<10–100	<10–100	–	15
U.S.S.R.; Chatkal-Kurama region	Iron hydroxide precipitates from mine drainage	0.06	20	0.003	26
Canada; Holman Mine, N.S.	Limonite	1.1	–	–	3
Canada; Keno Hill, Y.T.	Limonite, residual	0.08	500	0.0001	5
	Limonite and cerussite (transported)	0.7	500	0.001	5
	Wad and minor limonite (transported)	0.08	50	0.002	5
	Limonite, scorodite and other oxidation products	0.5	500	0.001	5
	Scorodite and limonite (residual in part)	10.26	–	–	5
	Anglesite (transported)	0.5	>1000	<0.0005	5
	Bindheimite (residual in part)	0.7	500	0.001	5
	Pyromorphite and cerussite (residual in part)	0.17	500	0.0003	5
	Various sulphates of iron and zinc	0.08	–	–	5
Canada; Walton, N.S.	Manganese oxides; wad; precipitate from springs	0.3	0.7	0.4	6
	Iron oxide precipitates (limonite)	<0.08	18.5	<0.004	6
Canada; Yellowknife, N.W.T.	Limonite, Lynx area (transported)	0.17			7
Canada; Ross Mine, Holtyre, Ont.	Limonite precipitate, 450 ft level	0.06	2.0	0.03	7
	Hydrous manganese oxide precipitate, 150 ft level	3.04	7.8	0.39	7
Canada; Annapolis Royal, N.S.	Bog manganese oxides (wad)	<0.01	1.4	<0.007	7
Canada; Bathurst area, N.B.	Bog iron and manganese, Canoe L.	0.01	2	0.005	7
	Transported limonite and other Fe-Mn-Si precipitates associated with massive sulphide deposits	<0.01–0.77	<0.8–26.6	<0.05	7
Canada; Long Lake area, N.B.	Bog manganese in peat	1.4	0.8	1.75	7
U.S.A.; Ojo Caliente, N. Mex.	Calcareous tufa (calcite and limonite)	0.047	2.73	0.017	22
U.S.A.; Anaconda Hot Springs, Mont.	Limonite and travertine	1.7	–	–	8
U.S.A.; Steamboat Springs, Nev.	Sinters, muds, carbonates, sulphides, etc.	0.2–12	39–42	0.15	11
	Metal-bearing siliceous muds	up to 200	–	–	19
	Siliceous sinters	0.11	–	–	19
U.S.A.; Casa-Diablo hot spring area, Mono County, Calif.	Hot spring sinter	trace	up to 17	–	12
U.S.A.; Niland, Calif.	Residue from high temperature steam well	3.7	13 030	0.0003	13
U.S.A.; Yellowstone National Park, Wy.	Siliceous sinters	0.005–0.05 (up to 5.5)	–		18, 19
New Zealand; Rotorua	Siliceous sinter	0.855	25.8	0.331	9
	Sulphurous sinter	1.99	138.1	0.014	9
New Zealand; Whakarewarewa and other areas	Siliceous sinter deposited mainly from alkaline springs; maximum values	5.3	138	0.04	10

Table 22. (cont'd.)

Location	Description of precipitate	Au	Ag	Au/Ag ratio	Reference
		(ppm)			
New Zealand; Taupo volcanic zones	Siliceous and sulphide precipitates associated with hot springs	85	500	0.17	14, 23
New Zealand; Puhipuhi, North Island	Hot spring sinter (opal). Sinter also enriched in Hg. *See* paper by Hoggins and Brooks (1973).	0.03–1.30	1.6–3.0	0.02–0.43	7
Japan; Motoyu spring, Nasu	Yellow precipitate	1–10			21
Columbia, South America; La Sufral volcano	Native sulfur	0.42–1.71	–	–	25

Compiled from data by: 1. Blanchard (1933) 2. Morris and Lovering (1952) 3. Knight (1911) 4. Foley (1960) 5. Boyle (1965*a*) 6. Boyle (1972) 7. Boyle (unpublished) 8. Weed (1905) 9. Maclaren (1906) 10. Grange (1937) 11. Brannock *et al.* (1948) 12. White (1955) 13. White *et al.* (1963*a,b*) 14. Weissberg (1970) 15. Mikov (1970) 16. U.S. Geol. Surv. (1968), Prof. Pap. 600-A, p. 7 17. Kinkel and Lesure (1968) 18. Gottfried *et al.* (1972) 19. U.S. Geol. Surv. (1969*c*), Prof. Pap. 650-A, p. 121 20. Lincoln (1911*b*) 21. Ikeda (1955*a,b*) 22. Lindgren (1910) 23. Browne (1969) 24. Harrison (1908) 25. Goodchild (1918) 26. Skryabin (1968) 27. H. Lakin (1969*a*), U.S. Geol. Surv., Circ. 622, p. 4.

gold-bearing. Of the precipitates associated with hot springs, the siliceous sinters (usually with some sulphides) deposited from alkaline waters tend to concentrate more gold (and silver) than the carbonate and limonitic sinters.

In the natural precipitates gold is most frequently accompanied by enrichments of SiO_2, As, Sb, Cu, Ag, Zn, Cd, Hg, Ti, Zr, Mn, Fe, B, Al, P, Te and S and less frequently by Li, K, Ca, Mo, W, Co, Ni, Tl, Bi, Se, V, Cr, Sn, Th and U. These are also the common associates of gold in natural waters. Judging from the data in the literature (Table 22) the Au/Ag ratio of hot spring deposits such as siliceous sinter, travertine, etc. is generally less than 1. There is considerable variability in the Au/Ag ratios of cold spring deposits, precipitates from groundwaters, organic precipitates, etc., but they are commonly much less than 1. The highest contents of gold and silver generally occur in the siliceous sinters precipitated from hot alkaline saline waters. This circumstance is particularly marked at Whakarewarewa and other thermal areas in New Zealand (Grange, 1937; Browne, 1969; Weissberg, 1970). Such natural data suggest that some gold-silver-quartz deposits were probably precipitated from alkaline, saline siliceous waters or from diffusion currents in an alkaline environment.

Numerous other occurrences of gold in spring precipitates are recorded in the literature. Many of these are associated with pyrite, which partly replaces rotting wood and other organic debris in the spring waters. References as far back as Sterry Hunt (1879) and Liversidge (1893*c*) draw attention to this mode of occurrence of gold and other metals. There are a number of references in the old literature attesting to the solubility of gold in mine waters and its precipitation (or plating) on iron particles such as rails, shovels, mine cars, etc.

The mode of occurrence of gold in natural precipitates is exceedingly variable. In manganese oxides and limonite much of the gold is probably present in an adsorbed form or as an integral part of the oxide complexes. Some varieties of auriferous limonites, particularly those associated with sulphide and gold-quartz deposits, contain the element in a number of ways: in minute specks, wire and sprigs of native gold; in a great variety of secondary minerals such as jarosite (plumbojarosite and argentojarosite), bindheimite, beudantite, anglesite, pyromorphite, malachite, azurite, etc.; and in an adsorbed or bound form in the hydrous iron, manganese, aluminum and silicon oxides. Limonite and wad, intimately mixed with organic substances, probably hold the element partly as some form of gold humate or other gold-organic compound. In precipitates containing secondary sulphides such as pyrite, melnikovite and marcasite, most of the gold is present in these sulphides in an unknown form. More details on the form of gold in natural precipitates and secondary oxides, sulphates, carbonates, etc. are given in a previous section on gold in other minerals in this chapter.

There are only scattered references to the possible presence of gold in the various products of volcanoes and fumaroles. Recent quantitative data appear to be lacking, but the element is probably present in the precipitates and sublimates associated with these high temperature phenomena since Goleva *et al.* (1970*a,b*) detected up to 1.6 ppb Au in the waters of the HCl fumaroles of the volcano, Ebeko, in the Kuril Islands, U.S.S.R., Goodchild (1918) mentions the presence of up to 1.7 ppm Au in the sublimated sulphur of La Sufral volcano in Columbia. Gold has also been detected in traces in the pyrite associated with the sulphur "blowers" in the Sulphur Bank at Borax Lake, California (Liversidge, 1893*c*).

The extremely small amounts of gold in natural waters (circa 0.03 ppb) make the problem of the determination of the state of the element most difficult. There has been much theorizing on the state of gold in waters, but as yet we have no satisfactory data on which to base solid conclusions, nor have we yet available any experimental techniques to decide exactly how gold exists in natural waters. About all that we can truthfully say is that gold migrates in waters by some process of natural alchemy as yet undiscovered. This statement notwithstanding we can suggest from the known chemistry of gold that the element may be present in natural waters in one or all of the following forms:

1. As finely divided particles of native gold: gold is slightly 'soluble' in water, especially in a very finely divided state. The form of this 'dissolved' gold is, however, unknown. In some cases the particles are probably colloidal in size and the solution is, therefore, in the nature of a colloid (*see* 13 below). In other cases the finely divided particles may be hydrated, coated with a variety of natural substances which

stabilize the particles in 'solution' or adsorbed to fine silt, organic compounds or a variety of colloids such as the hydrous iron oxides. In some surface waters much gold is transported as fine particles that appear to float or are carried along as silt. This gold is the familiar 'flour gold' or 'float gold' frequently referred to in the literature. Goni *et al.* (1967) have shown by experimental methods that gold can be made to form a stable suspension of very fine (colloidal) particles by prolonged abrasion by quartz grains. This suspension is stabilized by the presence of carbonate in small amounts (up to 1 per cent). Above this concentration the carbonate acts as a flocculent. The colloidal silica also formed by abrasion, assists in maintaining the gold in suspension, probably acting as a protective colloid. Machairas (1967, 1970*a*) refers to particles of gold in iron oxide stalactites in the Cros Gallet Tunnel in France which he thinks have been transported in suspension in groundwaters. These particles have a maximum size of 20μ. Chernyayev *et al.* (1969) have also concluded that gold migrates in the supergene zone of gold and sulphide deposits partly as very fine particles in suspension in the oxidizing solutions.

In a recent paper Goleva *et al.* (1970*b*) claim that gold in the common types of natural waters (pH 5–8 and Eh 0.1–0.5 V) is present mainly in the form of Au(0) (uncharged particles) and as organic complexes.

At high temperatures and pressures metallic film gold precipitated on quartz and porcelain is relatively soluble in water according to the experiments carried out by Vilor and Sarapulova (1971). At 100°C they noted a solubility of 6.6 ppb of the film gold; at 300°C, 67.5 ppb; and at 350°C a decrease in solubility to 34–45 ppb. Lower values were recorded when film gold precipitated on porcelain was used. They concluded that gold dispersed in sedimentary rocks during sedimentation may be leached out by dilute aqueous solutions in the course of progressive metamorphism, transported and then precipitated on encountering a change in pressure and temperature in the rocks through which filtration of the solutions took place.

2. As the ions Au^+ and Au^{3+}: these states have been suggested by some early investigators, but their presence as ions in any normal natural water appears unlikely. Gold has an extremely high affinity for electrons, hence its character as the most noble of metals. The ionization potentials for both ions are (Au^+–9.2 eV; Au^{3+}– ~30 eV) and hence their interaction with water is marked. Aurous ion decomposes to form metallic gold and auric ions in aqueous solutions. It can, however, be stabilized by ligands such as $(CN)^-$. Auric ion is stable in very acid waters (pH <2) (Fig. 1); at higher pH it is strongly hydrated or is stabilized in water as the numerous complexes noted in 3 and 4 below. As a rule gold is highly inert in nature, and strong oxidizing agents such as MnO_2, Fe^{3+}, As^{5+}, Sb^{5+}, Cu^{2+}, O_3 or O_2 are necessary to oxidize the metal to the (I) and (III) states. Only when complexing agents such as $(S_2O_3)^{2-}$, $(CN)^-$ and Cl^- are present can gold be maintained in a mobile (soluble) state in the presence of these oxidizing substances.

3. As gold hydroxide, $Au(OH)_3$: gold hydroxide is slightly soluble in water (Lenher 1918; Johnston and Leland, 1938), the quantity being about 3.1×10^{-6} mole/L or approximately 600 ppb. Being amphoteric the hydroxide exhibits increased solubility in acid solutions especially those containing the halogens or halogen acids – $4HCl + Au(OH)_3 = HAuCl_4 + 3H_2O$ – and in alkaline solutions in which it forms the complex ions $[AuO_2]^-$, $[AuO_3]^{3-}$ and their hydrated equivalents, e.g., $[H_2AuO_3]^-$ and $[HAuO_2]^{2-}$. Tyurin and Kakovskii (1960), from thermodynamic calculations, consider that gold may be present as the hydroxide or its complexes in some supergene waters. Vorotnikov *et al.* (1973) also consider that some of the gold in these waters is present as $[Au(OH)_2]^-$.

Lenher (1918) carried out a number of experiments on the hydrothermal transport of gold as the hydroxide. At relatively low temperatures and pressures (110°C, 1.5 atm) the hydroxide is relatively stable in solution; at higher temperatures and pressures decomposition to the metal takes place to a variable degree. Vilor and Shkarupa (1971) investigated the system Au–SiO_2–H_2O within the temperature interval 100 to 400°C and at H_2O pressures of up to 1000 atm. Below 200°C concentrations of Au and SiO_2 are equal and independent of pressure. Above 200°C they increase with temperature and pressure. The gold is said to be present as $[AuO_2]^-$ and colloidal particles. The colloidal gold system is stabilized by dissolved and polymerized SiO_2. Narseev (*in* Petrovskaya, 1974), likewise, advocates the transport of gold in hydrothermal solutions in the form of hydroxy complexes and chloroaurates stabilized by polymeric silica compounds when high concentrations of silica are present in the solutions.

4. As soluble halogen salts and halogen complexes: when salts such as $AuCl_3$, $AuBr_3$ and AuI_3 dissolve in water they do so with the formation of complex ions of the type $[AuCl_4]^-$, $[AuCl_3(OH)]^-$, $[AuCl_2(OH)_2]^-$, $[AuCl(OH)_3]^-$ and so on. There are analogous bromine and iodine complexes whose stability increases in the order I >Br >Cl. Actually the formation of these complexes is a very intricate matter, but it is apparent that the auric ion can probably form every state of the fourfold complex from $[Au(OH)_4]^-$ to $[AuCl_4]^-$ depending on the pH of the solution and the concentration of chloride ions. This is evidently because an aqueous solution of auric chloride is in reality a solution of the acid $H[AuCl_3OH]$. There is also an unusual series of addition products of auric chloride with the halides of P, Se, Si, Sb, etc. These evidently give soluble complex ions but relatively little is known about their chemistry. Some, such as those with silicon and antimony, may have geochemical significance in the migration of gold since the three elements are invariably found together.

Further discussion of the various aspects of the halide complexes of gold is beyond the scope of this book, but four recent papers dealing with traces of gold in chloride solution are of interest. Beneš (1967) concluded from dialysis and electomigration experiments that traces of gold (10^{-7}–10^{-8}M) (2–20 ppb) in chloride solutions between the pH range 6 to 13 exist predominantly as complex anions or molecules, probably of the hydroxoaurate type, $[Au(OH)_xCl_y]^{(x+y-3)-}$ with ($x + y \leq 4$). In the pH range 0 to 4 Beneš and Riedel (1967) and Beneš and Singer (1968) concluded from adsorption experiments that the trace amounts of gold in chloride solutions were present predominantly as undissociated molecules of $HAuCl_4$ at low pH and positively charged hydrolytic products of this acid at higher pH. With increasing pH (>4) the negatively charged hydroxo complexes noted above are dominant. On the mechanism of the dissolution of gold in chlorine water Svistunov (1970) concluded from thermodynamic data that the gold is present mainly as $HAuCl_4$ and that the metal reacts with atomic rather than molecular chlorine as is generally considered to be the case.

Complex aurous halides, bromides and iodides of the type $[AuCl_2]^-$ are known, but they are relatively unstable, the aqueous solutions rapidly decomposing with separation of metallic gold and the formation of auric complexes.

Numerous investigations have considered that gold in surface waters leaching gold-bearing deposits is largely present as a complex halide. Lenher (1909, 1912) was one of the first to deal with this form of transport in detail, and he summarized the earlier literature on the subject. Brokaw (1910) and Emmons (1917) pointed out that gold could be dissolved in acid supergene solutions as the soluble auric chloride complex when a strong oxidizing agent such as manganese dioxide was present to oxidize the gold to the Au(III) state. This view is supported experimentally by Miller and Fisher (1974). Chukhrov (1947) considered that the solution of gold in the oxidation zone of mineral deposits was due to the formation of soluble alkali gold bromides and iodides. Shcherbina (1956*b*), Krauskopf (1951), Kelly and Cloke (1961) and Shabynin (1966) have postulated that the auric chloride complex, $[AuCl_4]^-$, is one of the important forms in which gold is transported in supergene waters. Kelly and Cloke (1961) concluded that if an activity of 10^{-5} moles/L (2ppm) is chosen as the lower limit of the geological solubility of gold, the element is soluble in acid chloride solutions (pH < 5.5), oxidation potentials greater than 0.9 V, and chloride ion activities greater than $10^{-3.2}$ moles/L. In the waters of the southern Enisei region, U.S.S.R. Vorotnikov *et al.* (1973) concluded from dialysis studies that gold migrated in an ion-dispersed state in the form of negatively charged anion complexes, principally as $[AuOH_2]^-$ and $[AuCl_3OH]^-$. Roslyakov *et al.* (1972) found that the $[AuCl_2]^-$ complex prevailed in the pore solutions of oxidized ores, eluvial layers and soils overlying gold deposits. It was commonly accompanied by gold in the form of the $[Au(OH)_2]^-$ complex.

The views that supergene waters transport gold mainly as $[AuCl_4]^-$ and as other halide complexes has been questioned by a number of investigators particularly Boydell (1924), Tyurin and Kakovskii (1960), Goni *et al.* (1967), Goleva (1968) and Goleva *et al.* (1970*a,b*). They point to the relative instability of the auric chloride complex at a pH greater than 4 (the normal range of most supergene waters), the relatively easy reduction of the complexes by metallic minerals, organic compounds etc., the paucity of chloride, bromide and iodide in most supergene waters, and the general absence or late stage appearance of oxidants such as MnO_2 in many deposits.

Boydell (1924) and Goni *et al.* (1967) consider colloidal transport more probable, and Tyurin and Kakovskii think that the effective soluble complex is a thiosulphate one (*see* 5 below). Lakin *et al.* (1974) stress the relative immobility of gold chloride in geologic environments because of the easy reduction of the various complexes to the metal by most common geological materials. They also note that while bromide and iodide form stronger complexes with gold they are rarely if ever in sufficient natural concentrations to be of importance as a supergene agent of transfer of the precious metal. The writer's observations on the halide content of supergene waters leaching gold-bearing deposits in temperate climates confirms the low content of these elements. Only in rare cases do the supergene waters contain more than a few, a few tenths or a few hundredths parts per million respectively of Cl, Br or I. Higher concentrations can be expected in arid regions where some supergene transport of gold may take place as halide complexes.

Transport of gold in hypogene waters (hydrothermal solutions) as chloride complexes has also appealed to a number of investigators, particularly Ogryzlo (1935), Krauskopf (1951), Helgeson and Garrels (1968) and Rabenau and Rau (1968). Krauskopf stated that gold may be transported at high temperatures and pressures as the ion $[AuCl_4]^-$ in acid solutions provided that reducing agents are absent. He thought that at the temperature and pressure of most hydrothermal solutions that the hydrogen ion of the acid was a sufficient oxidizing agent to keep the gold in the oxidized state. Helgeson and Garrels concluded from thermodynamic considerations that gold is carried in acid hydrothermal solutions primarily in the aurous state as chloride complexes. The writer (Boyle 1969*c*) has criticized this proposed acid-chloride transport mechanism from a number of angles, pointing to the general low content of chloride now present in gold veins and their wall-rock alteration zones, the alkaline character of the wall-rock alteration associated with gold deposition, the low transfer percentages of alumina in alteration zones associated with gold deposits, the general presence in gold deposits of carbonates which are difficult to precipitate from highly acid solutions and so on. Because of these and other features of gold deposits it seems probable that near neutral to alkaline sulphide solutions also carrying As, Sb and Ag are the most likely gold-transporting media. Recently Weissberg (1970) has shown experimentally and by examination of thermal waters precipitating gold in the Broadlands, Ohaki and Waiotapu areas, New Zealand that the $[AuS]^-$ complex is more than adequate to account for the transport of gold in these near-neutral low salinity hydrothermal solutions (*see* 12 below).

More recent views on the transport of gold as chlorides are found in a number of papers. Vilor and Shkarupa (1971) examined the system Au-SiO_2-NaCl-H_2O between the range 100 to 400°C and at H_2O pressures up to 1000 atm. They found that the gold and silica contents increase with temperature, pressure and acidity as compared to neutral solutions. At NaCl concentrations higher than 50 g/L the disperse Au phase was found to be stable within the whole temperature range. They concluded that their data confirmed the possibility of the formation of gold deposits by the leaching of gold from various rocks by metamorphogenic hydrothermal solutions. In a further study of the system Au-SiO_2-H_2O-NaCl-HCl-NaOH in the range 200 to 400°C, Vilor and Kazmin (1973) concluded that the basic gold complex in chloride solutions is $[AuCl_2]^-$ supporting the view of Helgeson and Garrels (1968) that the oxidation state of gold is low at high temperatures and pressures.

Kolonin and Aksenova (1971) studied the effects of temperature on complex formation of gold in chloride solutions by spectrophotometric methods. They observed that for $[AuCl_4]^-$, hydrolysis was greatly increased with increase in temperature in the interval 21 to 90°C.

Henley (1973) determined the solubility of gold in chloride solutions in the range 300 to 500°C. The solutions were buffered with respect to HCl by a K-feldspar–muscovite–quartz assemblage and to oxygen by the assemblage hematite-magnetite. Solubilities were found to increase rapidly with temperature from about 10 ppm at 300°C to 500 and 1000 ppm at 500°C at 1 and 2 kbar respectively. Based on this evidence Fyfe and Henley (1973) have concluded that water of dehydration (metamorphic water) in which chlorides are present could move gold from piles of rocks into available shear zones and faults where it precipitates at lower temperatures and pressures.

Rytuba and Dickson (1977) investigated the reaction of pyrite + pyrrhotite + quartz + gold with NaCl-H_2O solutions within a temperature range 300 to 500°C and pressure range from 500 to 1500 bars. Contrary to previous investigations they found no evidence of gold-chloride complexing under the near-neutral and reducing conditions of the experiments. They concluded that the appreciable solubility of gold in their experiments was due to the pyrite-pyrrhotite equilibrium, which sets the S^0, Fe^{2+} and S^{2-} activities; the hydrolysis of S^0, which forms HS^-, required for complexing; hydrolysis of Cl^-, which buffers the acidity; oxidation of gold by S^0 and H^+; and the formation of $[Au(HS)_2]^-$. Genetically, processes capable of depositing gold and gangue minerals from NaCl solutions initially saturated at 500°C and 1500 bars, were postulated to be: decreased temperature, decreased pressure, dilution to lower NaCl concentrations or combination of the three processes.

Gold chloride is volatile as Au_2Cl_6 (Landsberg and Hoatson, 1970; Visnapuu *et al.* 1971), and this form has appealed to some as a mode of transport during endogenic processes. Thus, Anoshin and Potap'yev (1966) have concluded that gold moves in the gaseous phases Au_2Cl_6 and Au_2Cl_2 said to be stable over a wide temperature interval (300–800°C). There are also a number of other volatile species of gold, e.g., $AuFeCl_6$ and $AuAlCl_6$, that could conceivably occur in nature. These have considerable volatility in the range 200 to 500°C (Heinen and Eisele, 1974).

A number of early investigators have concluded that the dissolved (solute) gold in seawater is present mainly as the $[AuCl_4]^-$ and $[AuO_2]^-$ complexes, the latter according to Krauskopf (1951) being about 100 times the concentration of the former. Peshchevitskii *et al.* (1965, 1970) have investigated the problem of gold in seawater carefully from the viewpoint of the thermodynamic stabilities of various complex forms of gold having regard to the composition of seawater, the redox potentials of the possible gold complexes and the redox potentials of seawater. They concluded that gold was present mainly as the aurous complex $[AuCl_2]^-$ and a few per cent as $[Au(Cl,Br)]^-$. The other relative contents of the complex forms of gold such as $[AuCl_4]^-$, $[AuBr_4]^-$, $[AuBr_2]^-$, $[AuI_2]^-$, $[AuS]^-$, $[AuCl_2OH]^-$, $[AuCl_2(OH)_2]^-$, $[AuCl(OH)_3]^-$, $[Au(OH)_4]^-$, $[AuOHCl]^-$, and $[Au(OH)_2]^-$ are small.

In summary we are probably justified in saying that the transport of gold as chloride complexes in hydrothermal and supergene oxidizing waters is minimal compared with other modes of migration. Certain acid chloride thermal springs may transport some gold, but in general these do not seem to be related to gold deposits. In arid regions oxidation of gold deposits may be attended by transport of gold in the form of halogen complexes. In the sea the soluble gold appears to be present mainly as the aurous chloride complex, $[AuCl_2]^-$, but at the concentration present (0.02 ppb) it will remain difficult to prove that this is in fact so.

5. As thiosulphate complexes of the type $[Au(S_2O_3)_2]^{3-}$: both gold and silver are soluble in thiosulphate solutions containing oxidants, gold forming the complex noted above and silver the complex $[Ag(S_2O_3)_2]^{3-}$. Both complexes are relatively stable in weakly acid, neutral and alkaline solutions (pH 4–9) but decompose in acid solutions. The silver complex yields S, Ag_2S and some Ag_2SO_4; the gold complex yields a variety of indefinite products according to my investigations. Depending on the concentration of thiosulphate these appear to be S, native gold and probably Au_2S. The presence of any reducing substance gives free gold and some sulphur in the precipitate. The general stability of the two metal complexes is also considerably different in my experience. Silver thiosulphate solutions are relatively stable to natural agents of oxidation and reduction, and silver is relatively difficult to precipitate from the solutions in any form. Gold on the other hand is relatively easily precipitated as the metal both by oxidation and reduction of the complex by a variety of agents. In a more general sense thiosulphate complexes of both gold and silver are relatively stable in the presence of ferrous sulphate (in gossans, etc.) and are not readily reduced by this compound. This is especially so for silver and less so for gold. The varied habit of the decomposition of the thiosulphate complexes of silver and gold are important in the oxidation of gold-silver deposits containing sulphides since with respect to silver, a soluble sulphate is formed whereas with the gold complex the metal is precipitated. These circumstances permit a ready separation of the two metals, and a natural refining, therefore, takes place – gold is concentrated mainly as the native metal whereas silver is leached out, a phenomenon consistently found in eluvial and alluvial placers (*see* Chapter III).

Other oxidized sulphur species can arise as intermediate products between sulphide and sulphate during the oxidation of sulphides according to the scheme: $S^{2-} \rightarrow [S_2O_3]^{2-} \rightarrow [SO_3]^{2-} \rightarrow [S_{2-5}O_6]^{2-} \rightarrow [SO_4]^{2-}$. Sulphite and polythionate form soluble complexes with gold. These can be characterized as $[Au(SO_3)_2]^{3-}$ and $[Au(S_{2-5}O_6)_2]^{3-}$. These complexes and other soluble species of the oxygenated sulphur compounds are fairly stable in alkaline solutions. They are not readily reduced by pyrite and other sulphides, and hence if formed endow gold with considerable mobility. On long contact with sulphides however, alkali species are ultimately reduced to native gold. In acid solutions the sulphite and polythionate complexes of gold and silver are decomposed to give the metals, sulphur, the metal sulphides or the metal sulphates (silver only) depending on the Eh and pH of the decomposition process.

Summarizing the above we see that all of the thio complexes can form only where the solutions are only slightly acid, neutral or alkaline, a condition that may obtain where oxidizing auriferous sulphide deposits have carbonate wall-rocks or gangue.

Information on the presence of thiosulphates and other thio species in waters leaching gold deposits is limited. My observations indicate that these constituents are present in neutral and alkaline environments but only in small amount (up to 5 ppm), most of the sulphur being present as sulphate. Thiosulphates are, however, formed in some quantity in various metallurgical practices involving oxidation of sulphides (Forward and Mackiw, 1955) and various thiosalts are produced during the alkaline oxidation of a variety of sulphides (Schmidt and Conn, 1971). It can be concluded, therefore, that thio complexes of both gold and silver may be instrumental in the transport of gold where the supergene solutions acting on sulphide-bearing gold deposits are only slightly acid, neutral or alkaline.

Tyurin and Kakovskii (1960) have postulated that gold (and silver) migrates in the supergene environment mainly as thiosulphate complexes. They find these complexes to be relatively stable and that the dissolution of gold and silver as thiosulphate complexes is greatly accelerated by the presence of cupric ion, which catalyzes the various reactions. Listova *et al.* (1966) showed experimentally, using natural Fe, Pb and Zn sulphides, that the oxidation of these minerals in the presence of carbonates yielded Ca thiosulphate and polythionate. They observed an increased solubility of gold when these sulphur compounds were present, and postulated that they were the main solvents of gold during oxidation processes in gold deposits. They concluded that gold could have a marked migration capacity in the supergene environment when thio salts were present. Cherepnin *et al.* (1973) have also suggested that the enrichment of gold in the oxidized zones of the sulphide-quartz veins of the Kuznetsk Alatau is the result of the deposition of gold from solutions of its thiosulphate salts. Lakin *et al.* (1974) concluded that thiosulphate ion may be present in enviroments high in sulphur but generally in very low concentrations. Above pH 5 gold may be transported as the thiosulphate complexes in these environments.

Lenher (1912) considered that gold could be transported as soluble thiosalts in hydrothermal solutions, and Tyurin (1963,1965*b*) set out to prove that thiosulphates transport gold in hydrothermal solutions. Shcherbina (1964), however, disagrees with this view pointing out that no thiosulphate minerals are known, that thiosulphate has not been detected in hot springs, and that liquid inclusions do not contain the compound. He also notes that at high temperatures (>100°C) the thiosulphate complexes are unstable and break down. Kim (1972) essentially agreed with this view, stating that the migration of gold in hydrothermal solutions in the form of the thiosulphate complex $[Au(S_2O_3)_2]^{3-}$ is insignificant. Tyurich (*in* Petrovskaya, 1974), on the other hand, maintains that the formation of complex compounds of gold in sulphide solutions such as the oxygenated species $Au_2S_2O_3$ and $[AuS_2O_3]^-$ must be taken into account during hydrothermal processes. Ellis (1967) states that thiosulphate and polythionate ions do not persist in natural waters at temperatures of 200 to 300°C but break down according to the reaction types:

$$S_2O_3^{2-} + H_2O \rightarrow H_2S + SO_4^{2-}$$
$$2S_5O_6^{2-} + 8H_2O \rightarrow 5SO_4^{2-} + 6H^+ + 5H_2S$$

A search of the literature revealed relatively little information on the abundance of thiosulphate in hot springs. In the great memoir on the hot springs of Yellowstone by Allen and Day (1935) values for $(S_2O_3)^{2-}$ in the waters range from 0 to 21 ppm. A similar range is found in a few other hot springs in various parts of the world according to scattered data in the literature and personal communications; values up to 37 ppm are recorded from hot springs according to the summary given by Lakin *et al* (1974). If these values are actually the contents in the thermal waters and not due to oxidation of the waters near the surface, it would appear that some low temperature hydrothermal solutions may carry enough thiosulphate to complex and transport the small amounts of gold that give rise to deposits.

6. As the soluble sulphate: gold is said to be soluble in H_2SO_4 solutions as the sulphate in the presence of an oxidizing agent such as O_2, MnO_2, etc. Gold (III) hydroxide is also said to be soluble in H_2SO_4 probably with the formation of the complex sulphate $[Au(SO_4)_2]^-$. These views are reiterated many times in the old literature (Lenher, 1904) and sometimes appear in the newer literature on gold compounds. Mellor (1923) characterized aurosoauric, $AuSO_4$, and auric, $Au_2(SO_4)_3$, sulphates and mentions complex alkali gold sulphates of the type $K[Au(SO_4)_2]$. These on solution may give rise to the complex ion $[Au(SO_4)_2]^-$. Relatively little is known about the solubilities or the stabilities of these various sulphates and complexes. They would appear to be relatively unstable, especially in the presence of reducing agents. In water, gold sulphates are extensively hydrolyzed, another feature that makes them relatively unstable. On thermodynamic grounds Roslyakov *et al.* (1972) consider that the solubility of gold as the sulphate in oxidizing waters is infinitesimal, the standard electrode potential for the $[Au(SO_4)_2]^-$ complex being unfavourable.

It has long been noted that gold exhibits some mobility where sulphides are undergoing oxidation, and it has frequently been postulated that the agent aiding its solution in such environments is ferric sulphate, $Fe_2(SO_4)_3$. One also notes that in some places the content of gold in oxidizing waters of sulphide deposits is greatest when the waters are acidic and charged with iron, manganese, arsenic and sulphate.

Ferric sulphate as an aid in the solution of gold in oxidizing waters has been questioned by a number of investigators in the past, particularly Stokes (1906) and Emmons (1917) who stated that the solvent action of ferric sulphate alone is negligible. Kreiter *et al.* (1959), on the other hand, found by experiment that solutions of ferric sulphate and sulphuric acid are good solvents of gold.

To resolve this problem I have carried out a number of experiments using atomic absorption methods to detect gold in solution. The results (Boyle *et al.*, 1975) indicate that amorphous gold is weakly soluble in ferric sulphate solutions alone, and more so when free H_2SO_4 is present. The presence of NaCl in the solutions appears in some cases to slightly reduce the solubility of the precious metal for reasons that are not clear.

7. As soluble gold arsenate or gold arsenate complexes: gold is soluble in arsenic acid, H_3AsO_4 or $As_2O_5 \cdot H_2O$, in the presence of MnO_2 and other strong oxidants. The nature of the solute is unknown, but presumably it is gold arsenate, a complex molecule of the type $[AuAsO_4]$, or a complex ion of the type $[Au(AsO_4)_2]^{3-}$ analogous to the sulphate.

The common occurrence of arsenic in gold deposits ensures a supply of dissolved arsenic in oxidizing supergene waters since the oxidation of arsenic minerals such as arsenopyrite yields arsenic acid or some derivative of this acid in solution as shown by the following simplified reaction:

$$4FeAsS + 13O_2 + 6H_2O \rightarrow 4FeSO_4 + 4H_3AsO_4$$

There are no detailed investigations to determine the extent of migration of gold as arsenate or arsenate complexes from oxidizing sulphide-arsenide deposits, but we have noted that some waters rich in soluble arsenic are, likewise, enriched in gold.

8. As soluble gold antimonite or antimonate complexes: relatively little is known about these. By analogy with arsenic it appears probable that gold will form complex antimonites of the form $[Au(SbO_2)_4]^-$ and possibly also complex antimonates. There are also a number of double complexes of antimony and chloride which may have geochemical significance in the migration of gold. The common occurrence of antimony in gold deposits ensures a supply of dissolved antimony since the oxidation of antimony minerals such as stibnite or tetrahedrite yields soluble antimonites and/or antimonates depending on the redox potential. We have noted increased amounts of gold migrating in waters enriched in antimony.

9. As soluble nitrates, nitrites and other soluble inorganic species: auric nitrate and a variety of complex nitrates and nitrites are known, but these appear to hydrolyze so extensively that they are not stable in water. In natural settings, however, some of these complexes may be stabilized by the organic constituents of humus, and gold may migrate in the form of combined organic nitrate, nitrite or ammine complexes, especially where a source of nitrates, nitrites or ammonium ion is available as in decaying vegetation, humus, etc.

Gold is soluble in phosphoric acid, H_3PO_4, presumably with the formation of gold phosphate. The stability of this compound in water is unknown, and nothing is known about its occurrence in natural situations. It may occur in some natural waters where phosphate is abundant.

Gold is said to be soluble in hot selenic acid, H_2SeO_4, at high temperatures (300°C) and in hot solutions of telluric acid containing H_2SO_4 or H_3PO_4. The transport of gold presumably as selenates and tellurates at high temperatures could conceivably take place, but there are no data by which to judge the efficiency of the mechanism. Transport of gold during low temperature supergene processes as selenates, etc. would not appear to be important, but again no data are available for judgement.

Colloidal gold is said to be soluble in dilute alkalis apparently with the formation of the hydroxide, hydroxide complexes or oxide complexes (*see* 3 above). Colloidal gold and very finely divided gold is said to exhibit some solubility in alkali carbonate solutions, although Ogryzlo (1935) found that sodium carbonate has no solvent effect on gold at high temperatures and pressures. The reported solubility of gold in carbonate solutions may be due to the formation of very slightly soluble alkali aurate species such as $KAuO_2$ (*see also* 3 above). The reaction may first involve the formation of the hydroxide or some hydrous complex followed by formation of the aurate.

$$2Au(OH)_3 + K_2CO_3 = 2KAuO_2 + CO_2 + 3H_2O$$

The writer has experimented with gold and various carbonate solutions and finds that the normal carbonates are poor solvents of gold. Solutions of alkali bicarbonates on the other hand dissolve small amounts of amorphous gold (Boyle *et al.*, 1975).

Two other mechanisms involving the solution of gold in potassium or sodium carbonate solutions concern the presence of H_2S in these solutions. The alkali sulphide or bisulphide formed are well known solvents of gold (*see also* 12 below).

$$Na_2CO_3 + H_2S = Na_2S + H_2O + CO_2$$
$$NaHCO_3 + H_2S = NaHS + H_2O + CO_2$$

These mechanisms may be important in the transfer of gold during both hypogene and supergene processes as shown by our experiments (Boyle *et al.*, 1975).

Liversidge (1893*c*) reported that gold is dissolved in a solution of potassium or sodium silicate at a pressure of 90 psi. The form in which the dissolved gold was present is not mentioned. It may have been as a colloid or as some complex, perhaps an aurate. Cumenge and Robellaz (1898) also refer to alkali auro-silicates and think that they play an important part in the formation of auriferous quartz deposits. The experiments by Von Veimarn (1913), the result of which led him to suggest the presence of an unstable gold silicate when $NaAuCl_2$ and Na_2SiO_3 are mixed, are also of interest in the formation of gold quartz veins. These experiments should be repeated since they have a direct bearing on the transfer of gold and silica during hypogene processes (*see also* the reference to Frondel and others in 13 below).

The writer has experimented with the solubility of gold in various solutions of sodium and potassium silicate (saturated – 10 per cent saturated, pH 10–12). The nature of the gold used in the experiments was the same as that described in Boyle *et al.* (1975). In saturated solutions of sodium metasilicate gold solubilities up to 40 μg/mL were noted, but in similar solutions of potassium silicate the solubility was much lower (up to 1.2 μg/mL). When H_2S was bubbled through the various alkali silicate solutions considerable amounts of gold were taken into solution (>100 μg/mL). On completion of the H_2S treatment the solutions gelled but still retained the gold in the gel. It can be assumed that AsH_3 and SbH_3 would have a similar effect on the solubility of gold judging from my previous work noted above.

Stokes (1906) found that gold is dissolved by cupric chloride and HCl, the reaction being as follows:

$$Au + 3CuCl_2 = AuCl_3 + 3CuCl$$

He remarks that addition of a chloride to cupric sulphate would probably also have the same solvent effect on gold. The same author found that ferric sulphate has no solvent effect on gold. Addition of sodium chloride to the ferric sulphate solutions, however, led to the solution of some gold. This is the oxidation and complexing effect on gold referred to in 4 above. Stokes (1906) also found that gold is soluble in a solution of ferric chloride and HCl. Again this is due to the oxidation and complexing effect. Experimentally, Miller and Fisher (1974) found that gold exhibits some solubility in $FeCl_3$ solutions in a dynamic system when (Fe^{2+}) is continuously removed. The reason for this is obvious, Fe^{2+} being a strong reductant of gold in solution. The gold dissolved in the ferric chloride solutions was completely sorbed by pyrite. The presence of bromide in the solution aided the dissolution of gold and impeded its reduction and precipitation on pyrite. All of these experiments are of interest in understanding the transport of gold during supergene processes as noted in

Chapter IV. Their significance in the hypogene transfer of gold is problematical.

Lenher (1909) found that boiling HCl under pressure dissolves gold readily, and that hot solutions of sodium chloride and HCl under pressure have a similar effect. High temperature sodium chloride and other alkali chloride solutions under high pressure, even in the absence of HCl, also have a solvent effect on gold. These mechanisms for the solution of gold may be significant in hypogene environments.

Babička (1943) mentions work by J. Barvir and C. Doelter that indicated that sodium chloride is a weak solvent for gold, but in my experiments I have not been able to confirm this (Boyle *et al.,* 1975). Babička (1943) also mentions that Egleston found that magnesium sulphate also dissolves the metal, but this seems improbable. Vilor and Shkarupa (1971) experimented with the Au-SiO_2-NaCl-H_2O system at temperatures between 100 and 400°C and H_2O pressures up to 1000 atm. They found that the gold and silica contents increase with temperature, pressure and acidity as compared to neutral solutions. At NaCl concentrations higher than 50 g/L the disperse gold phase decreased while the ionic form increased. The chloride complexes were found to be stable within the whole temperature range. The Au^{3+}/Au^{+} ratio increased with increase in temperature in the range from 200 to 400°C. They considered that silicon-chloride complexes play a major role in the transfer of SiO_2.

Morris and Lovering (1952) noticed an affinity of gold for cupric sulphate solutions as indicated by the presence of the precious metal in chalcanthite and pisanite efflorescences in the Tintic District, Utah (Table 22). They also found gold in copiapite, which suggested to them an affinity of gold for ferric sulphate solutions with iron just commencing to hydrolyze. Along the same line Lakin (1969*b*) mentions the fact that gold is relatively soluble in solutions leaching dumps of porphyry copper deposits (ferric and cupric sulphate solutions at pH 1.5) when sodium chloride and MnO_2 are added and air is bubbled through the solutions. Two aliquots of 200 mL each, one containing NaCl and MnO_2 and the other free of these constituents, contained 400 and 4 μg of gold respectively.

10. As a soluble cyanide or thiocyanate complex: gold forms very soluble cyanide and thiocyanate complexes in an oxidizing environment. The cyanide complexes are $[Au(CN)_2]^-$ and $[Au(CN)_4]^-$, and there are also a number of multiple complexes of the type $[Au(CN)_2Cl_2]^-$. The thiocyanate complex can be characterized as $[Au(CNS)_4]^-$. The dissolution of gold in cyanide and thiocyanate solution is usually written as:

$$4Au + 8CN^- + O_2 + 2H_2O = 4[Au(CN)_2]^- + 4OH^-$$
$$Au + 4[CNS]^- = [Au(CNS)_4]^-$$

The various gold cyanide and thiocyanate complexes are stable, but only in mildly acid, neutral and alkaline solutions.

Numerous plants and seed kernels, some fungi and certain insects contain cyanide in the form of cyanogenic glycosides, which are composed of one or more types of sugar with CN and other groups bonded through hydroxyl (OH) (Armstrong and Armstrong, 1931; McIlroy, 1951; Kingsbury, 1964; Conn, 1969). A familiar cyanogenic glycoside is amygdalin, the bitter essence in almonds. *In vitro,* cyanogenetic glycosides are hydrolyzed in acidic aqueous solutions and split out HCN, thiocyanates, sugars and other products generally aldehydes or ketones. *In vivo,* enzymatic reactions produce similar products. Plants, fungi and insects on their death undergo decomposition during which HCN may be split out of the organic complexes as a result of hydrolysis or bacterial action. Interactions of HCN with alkalis in soils, humus, etc. in neutral and alkaline environments may produce alkali cyanides and thiocyanates which, as noted above, solubilize gold (and silver). From what has been said it is evident that gold may be transported in the exogene cycle as cyanide or thiocyanate complexes in an environment where humification processes are proceeding. Such complexes would, however, seem to be transitory since their stability during humification processes is relatively low as a result of oxidation and bacterial activity which break them down to simpler compounds such as CO_2, H_2S, SO_4^{2-} and nitrates.

The presence of available cyanides and thiocyanates in or about oxidizing gold deposits where humification processes are progressing is uncertain. Biochemically they should exist. My research on this matter has failed, however, to definitely establish their presence, but work is continuing on the problem. Certain identification of traces of cyanide is difficult in natural waters as shown by Goulden *et al.* (1972).

Lungwitz (1900*a,b*) thought that cyanide derived from the decomposition of plants could be a solvent of gold but dismissed the idea because he thought this active agent could only be formed at high temperatures. He also noted that CN had not yet been found in the decay products of vegetation. Shcherbina (1956*b*) postulated that gold forms soluble complexes with humic acids, one of the complexes being $(Au(CNS)_4)^-$. Curtin *et al.* (1970) leached three gold-bearing samples of mull (forest humus layer) from Nevada and Colorado with demineralized water, obtaining gold in solution in the range 27 to 120 parts per trillion (1×10^{-12}). They suggested that most of the gold in the leachates was probably present as extremely small colloidal particles (<0.05 μ diameter), with much smaller amounts in the form of gold cyanide and thiocyanate ions.

In later papers Lakin *et al.* (1971; 1974) explained that at a pH of 5 to 8, $(CNS)^-$ and $(CN)^-$ may be expected to dissolve gold. They found that thiocyanates were relatively sparse in soils, but that cyanides could be abundant as a result of the hydrolysis of cyanogenetic glycosides. More than 1000 species of plants yield HCN on hydrolysis; many arthropods produce HCN; and some animals and fungi also yield appreciable amounts of the compound. Macerated aqueous suspensions of 16 species of native plants collected in Colorado, Utah, Nevada and Arizona were found to dissolve leaf gold. Lakin *et al.* (1974) consider that gold cyanide offers the most feasible form of soluble gold in soils.

Cyanide- and cyanate-bearing minerals exist in places. A number of cyanide minerals are recorded by Povarennykh and Rusakova (1973) and Rusakova and Povarennykh (1973) in various gold belts of U.S.S.R. One mineral kafehydrocyanite, $K_4[Fe(CN)_6]\cdot 3H_2O$, has been characterized. Cyanides were found in the waters from which these minerals were precipitated. According to the investigators, cyanides and thiocyanates are significant in the migration and redistribution of gold in the supergene zone of gold deposits. The source of the cyanides is considered to be decaying organic matter. A thiocyanate mineral, julienite, $Na_2Co(SCN)_4\cdot 8H_2O$ occurs in small amounts at Chamibumba, Katanga, Zaire (Belgian Congo).

It is usually considered that cyanides and thiocyanates play no role in the primary endogenic migration of gold. Our views in this respect may require reassessment. Recently, Mukhin (1974) has authenticated the presence of thiocyanates and hydrogen sulphide in the thermal springs in the caldera of the Golovnin volcano (Kunashir Island, Kuriles). From these and other data the author concludes that hydrocyanic acid (HCN) is present in volcanic gases. If so, a ready solvent for gold is available in deep-seated processes. Similarly, gold may have migrated as a soluble complex thiocyanate in endogenic processes.

11. As soluble organic complexes: gold forms a great variety of organic compounds. Gold (I) alkyls, isonitriles, carbenes and mercaptides have long been known and are quite stable, and there are also a great number of gold (III) alkyls and chelated diarsine, phosphine and other complexes. In addition there are the various cyanide and thiocyanate complexes noted in 10 above. The facility with which gold binds with organic molecules is, therefore, well established.

In natural environments organic compounds are extremely complex especially those occurring as plant and animal residues and in humus. The latter substance is a most complicated material with no well established composition. In addition to humic acid and fulvic acid components it apparently consists of a great variety of relatively unknown constituents derived from the breakdown of dead vegetable and animal matter. Furthermore, these constituents are exceedingly variable depending upon the degree of humification. Various types of humus appear to react differently toward the metals, and no one component can be effectively tagged as being responsible for the strong binding action on metals displayed by some varieties of humus; thus, the subject can only be discussed in general terms.

When available the metals of Group IB are generally strongly bound by humus. Copper forms a strong bond with humus, a feature that has been discussed at great length by numerous investigators and summarized by Manskaya and Drozdova (1968), Ong *et al.* (1970) and Boyle (1977). The nature of the copper bond to humus is uncertain, but the experimental evidence suggests some kind of internal or chelated complex. Silver, likewise, is strongly bound by humus probably as a chelate or some other such organometallic complex (Boyle, 1968*b*). Gold is no exception. The writer's observations and those of others (*see* the section on soils) show that gold is markedly enriched in the humic horizons of some soils, and that much of this gold is bound in the humic component. An important question in the transport of gold is not whether gold can be bound by humic materials – this much seems proved – but whether it can migrate in natural solutions in this manner. The latter has given rise to considerable controversy.

Lungwitz (1900*a,b*) working in Guyana, postulated that gold chloride may be formed during the oxidation of gold deposits by supergene waters containing chloride ion and nitric and sulphuric acids derived from the decomposition of organic matter. The gold chloride, he thought, could form double salts with water-soluble organic compounds such as those in the brown colouring matter (humus) in streams. This feature would give gold considerable mobility in the hydrologic regime and may account for the high contents of gold found in trees and other plants. Harrison (1908) concurred in the view that much of the secondary gold in Guyana was transported in organic-rich waters. Freise (1931) experimented with the solution of gold by humic acids extracted from lignite and black river water generated in virgin forests in Brazil that was rich in dissolved organic compounds. He reported relatively rapid dissolution of gold as a humate by these compounds when oxygen was excluded. Blanchard (1933) found certain humic materials near gold deposits in New Guinea (*see* Table 22) greatly enriched in gold. He thought that at least some of the gold was dissolved by and migrated in organic solutions but that the conditions were complicated by the presence of much dissolved manganese. Shcherbina (1956*b*) suggested that gold may possibly form soluble compounds with humic acids in organic environments, and Steelink (1963) thought it probable that gold may react with humic acids, in the presence of oxygen, to form soluble compounds. He considered the solution and precipitation reactions to be as follows:

Solution: $Au + H^+ + \text{humic acid} + O_2 = [\text{Au humate}]^{3+} + H_2O$

Precipitation: $[\text{Au humate}]^{3+} + R = Au + \text{humic acid} + R^+$

(R = a reducing agent such as Fe^{2+} or organic compounds)

Goleva (1968) mentions the fact that up to 70 per cent of the gold in certain natural waters is extracted by chloroform, a feature which suggests considerable organo-metallic binding of gold in these waters. Finally, Udodov *et al.* (*in* Petrovskaya, 1974) find that gold exhibits some solubility in humic and fulvic acids.

Investigations by Garces (1942) and Fetzer (1934, 1946) oppose the concept that humic acids dissolve gold. Both have found experimentally that gold (and silver) are not dissolved by humic acids extracted from peat and lignite.

Recent studies have tended to confirm the transport of gold as natural organic complexes, often colloidal in nature. Ong and Swanson (1969) have shown that gold is neither dissolved, complexed nor oxidized by organic (humic) acids, but rather that the element in solution as gold chloride is reduced by the acids to a negative colloid of metallic gold. This colloid is then stabilized by a protective layer of organic molecules around the gold thus permitting a relatively high mobility of the element in soil and other natural waters. Curtin *et al.* (1970) and Lakin *et al.* (1971) in the papers summarized in 10 above noted that gold is mobile in an organic environment mainly as a colloid and in minor amounts as gold cyanide and gold thiocyanate ions.

A recent study of the interaction of natural organic acids with gold by Fisher *et al.* (1974) indicates that humic acids alone do not dissolve gold, but if the gold occurs in a colloidally dispersed state, containing adsorbed oxygen, the humic and fulvic acids solubilize the element. Furthermore, these acids interact with dissolved $[AuCl_4]^-$ forming stable auriferous humate and fulvate complexes. Experimentally a gold-fulvate complex was shown to be very stable under most natural supergene conditions.

We have carried out a number of investigations involving the 'solubility' of gold in humic-rich solutions (Boyle *et al.*,

1975). These show that gold is slowly 'dissolved' in these solutions; the metal is intimately associated with the humic matter being either chelated, organometallically bonded or adsorbed. When the humic matter is removed by ultra filtration the solutions are essentially free of gold.

There seems to be little doubt that gold is 'solubilized' in one form or another by a variety of organic compounds. In some cases the compounds or complexes actually assist in the dissolution of gold; in other cases they act as protective agents (colloids) for gold colloids. Such mechanisms are undoubtedly important in the migration of gold under supergene conditions in organic-rich terranes. .

Under hypogene conditions there is some evidence to support the contention that gold is often transported as an organometallic complex. The common occurrence of gold in thucholite (an indefinite C-H-O-N-S-U-Th compound) as in the Rand and at the Richardson deposit in southeastern Ontario suggests that gold may migrate in natural waters combined with bitumen or other petroliferous substances, perhaps specifically as a porphyrin which can be synthesized. (*See also* the section on biogeochemistry.) Further details on the association of gold and hydrocarbon substances are given in the section on gold and its associated minerals in Chapter III.

12. As complex sulphide, polysulphide, telluride and sulphide-arsenic-antimony complexes: it has long been known that gold is soluble in alkali hydrogen sulphide (e.g., NaHS) solutions at low temperatures and that gold sulphide is soluble in solutions of alkali sulphide and polysulphide in excess at low temperatures. At high temperatures gold is also soluble in concentrated alkali sulphide solutions. Egleston (1881), Liversidge (1893*c*), Becker (1887), Lenher (1904, 1912, 1918), Mellor (1923), Gruner (1933), Ogryzlo (1935), Zviagincev and Paulsen (1940), Smith (1943), Krauskopf (1951), Kim (1972) and Seward (1973) have dealt in one way or another with the various aspects of the experimental and thermodynamic problems in the various aqueous systems involving gold, gold sulphide and alkaline sulphides. There is some agreement that the principal complex ion is $[AuS]^-$, although some $[AuS_2]^-$ may also be present in the systems. In a recent paper Seward (1973), determined the solubility of gold in aqueous sulphide solutions from pH 4 to 9.5 in the presence of a pyrite-pyrrhotite redox buffer at temperatures from 160 to 300°C and 1 kbar. Maximum solubilities were obtained in the neutral region of pH. It was concluded that three gold complexes contributed to the solubility: $[Au_2(HS)_2S]^{2-}$ predominated in alkaline solution, $[Au(HS)_2]^-$ in the near neutral pH region and $[Au(HS)]^0$ with less certainty in the acid pH region. The stabilities of the first two complexes are much greater than chloroaurate (I) species, a feature which suggests that the chloride complexes play only a subsidiary part in the transport of gold in neutral and alkaline hydrothermal solutions.

The sulphide and polysulphide complexes of gold are stable in aqueous solutions between a pH of 6 up to at least 10. The $[AuS]^-$ complex is also relatively stable to metallic sulphides and a number of other reductants. Since many hot spring waters are alkaline, and other effects such as certain types of wall-rock alteration suggest alkaline solutions, transport of gold as soluble alkaline sulphide complexes has appealed to a number of investigators as a transport mechanism for gold during the formation of hypogene deposits.

Smith (1943) carried out an extensive series of experiments involving the solution and transport of gold, silver and tellurium in alkali sulphide solutions. He was able to show experimentally that gold is soluble in, and can be crystallized from, such solutions, and that two other common gold minerals, electrum (Au,Ag) and calaverite ($AuTe_2$), can be similarly synthesized. His conclusions on the transport and deposition of gold, silver and tellurium merit quotation in full because of their probable importance in the formation of hypogene gold ores:

> Gold is much more soluble than most of the common metals in alkali sulphide solutions. It is included in the group of metals whose sulphides are very soluble in dilute aqueous solutions of alkali sulphide at ordinary temperature. This group also contains mercury, bismuth, antimony, arsenic, and tellurium. In the group of metals whose sulphides are soluble in aqueous solutions of alkali sulphide only at elevated temperature, the most soluble metal appears to be silver. Thus the natural association of minerals containing gold which are deposited late in the ore sequence with minerals containing silver, bismuth, and tellurium is explained. When the soluble double sulphides of gold and alkali are broken down by a lowering of the sulphide concentration, gold deposits as the metal, since its sulphides are very unstable above 40°C. In solutions containing both silver and gold, if the sulphide concentration is reduced to the point where metallic gold separates, it does so along with considerable silver in solid solution, forming argentiferous gold or electrum.
>
> If tellurium is present along with gold in natural alkali sulphide solutions, gold tellurides may or may not precipitate as the sulphide concentration is reduced, depending on the state of the tellurium in the solutions. If the free sulphur concentration in the polysulphide form is high, then the tellurium is present as alkali thiotellurites, and when in this form metallic gold precipitates first and much later the thiotellurite complex is broken down and metallic tellurium is precipitated. If the free sulphur concentration is low, then the tellurium is present in part as polytelluride ion and gold tellurides are first precipitated, followed later by the excess of metallic gold. These reactions have been duplicated by experiment, the order of solubility in alkali sulphide solutions being found to be, starting with the most insoluble – gold telluride, gold, tellurium. This explains the natural paragenesis of gold tellurides, gold, and tellurium, which apparently deposit in the same order.

In recent years Weissberg (1970) has reinvestigated the solubility of gold in alkaline sulphide solutions and finds that the solubility of gold experimentally determined at elevated temperatures in alkaline Na_2S solutions ranges from 1 to 6 ppm Au in solutions containing from 0.25 to 0.9 moles Na_2S/kg solution at temperatures between 150 and 290°C. In less alkaline solutions, where the HS^- ion predominates, the experimentally determined solubility of gold ranges from 100 to 200 ppm Au in solutions containing from 0.2 to 0.3 moles NaHS/kg solution at temperatures between 150 and 250°C, in good agreement with earlier workers. Having done the necessary calculations he concluded that gold solubility due to the formation of a gold sulphide complex tentatively formulated as $[AuS]^-$ is more than adequate to account for the transport of gold in some natural near-neutral low salinity hydrothermal solutions, such as in the Broadlands, Ohaki and Waiotapu areas, New Zealand, where currently forming surface precipitates contain up to 85 ppm Au (*see* Table 22). This makes a postulated near-neutral gold sulphide transport mechanism a particularly effective means for the formation of epithermal gold deposits.

One of the problems with the alkali sulphide theory of gold transport is that pyrite, one of the ubiquitous sulphide minerals in gold deposits, is insoluble or only sparingly soluble in alkali sulphide solutions. There are a number of other sulphides that also show this phenomenon to some degree, some of them relatively abundant in gold deposits, viz. molybdenite, argentite, pyrrhotite, sphalerite, etc. To circumvent these difficulties it is usually said that gold is deposited later than pyrite in deposits. This is true in many deposits, but late pyrite accompanied by gold is common in most deposits. Early pyrite, with gold as a lattice constituent or in a very finely divided submicroscopic form, is also common. Perhaps we are not dealing with pyrite as such in alkali sulphide solutions but with stable species of iron sulphide complexes, which are ultimately partitioned out of the solutions or diffusion currents as the solid phases pyrite and/or pyrrhotite. The same may be true for the other metals such as Mo, Ag, Zn, etc. With respect to pyrite we may also not be dealing with the transport of iron at all, this element being largely derived from the nearby host rocks as McKinstry (1957) has observed. The derivation of the other metals is more difficult to explain by this secretion hypothesis; they appear to have been transported some distance according to my investigations.

Gruner (1933) has referred to the problem of the low solubility of gold in hydrosulphide hydrothermal solutions charged with silica. Zvyagintsev and Paulsen (1940), however, doubt the earlier data and maintain that silica in fact promotes the solution of gold in hydrosulphide solutions. They also state that sulphides and quartz are not precipitants of gold from hydrosulphide solutions as some investigators have suggested. On the contrary, hydrosulphide solutions may pass through a sulphide zone without depositing gold and may in fact dissolve gold, which is contained in the sulphides. To precipitate the gold they call on decreased solubility as a result of falling temperature. Gold in the sulphides of gold deposits results from coagulation processes involving gold sols and the surface properties of sulphides.

Gold may also be dissolved in the oxidized zone of auriferous deposits principally as $[AuS]^-$. The formation of the H_2S necessary to form this complex is most probably provided by the action of sulphate-reducing bacteria active in the oxidized zones. In most oxidized zones the amount of H_2S is minimal but is invariably present. In their study of a number of oxidized zones, eluvial layers and soils overlying gold deposits in eastern Transbaikal and the Kuznetskii Alatau, Roslyakov *et al.* (1972) found that the sulphide complex of gold, $[AuS]^-$, prevailed over all other forms (chloride, hydroxy, etc.) in the aqueous extracts of the materials in and associated with the oxidized auriferous zones. From this data they reasoned that when H_2S is present in oxidized zones most of the gold migrates as a sulphide complex.

Arsenic and antimony are nearly universal associates of gold in its deposits, as discussed at length in the section on associated elements. These two elements as sulphides are also readily soluble in aqueous alkali sulphide solutions. These facts have led the writer (Boyle, 1969*c*) and others (Seward, 1973; Sorokin, 1973) to postulate that arsenic and antimony complexes are probably important agents in the hypogene transport of gold. The solution of the various arsenic and antimony sulphides in sodium sulphide gives the arsenites, thioarsenates, thioantimonites and thioantimonates as follows:

$$As_2S_3 + 3\,Na_2S = 2\,Na_3AsS_3 \text{ (thioarsenite)}$$
$$As_2S_5 + 3\,Na_2S = 2\,Na_3AsS_4 \text{ (thioarsenate)}$$
$$Sb_2S_3 + 3\,Na_2S = 2\,Na_3SbS_3 \text{ (thioantimonite)}$$
$$Sb_2S_5 + 3\,Na_2S = 2\,Na_3SbS_4 \text{ (thioantimonate)}$$

The complexes formed in these solutions are of the type $[AsS_3]^{3-}$. Complexing of these with gold could yield double complexes of the type $[Au(AsS_3)]^{2-}$ and $[Au(SbS_3)]^{2-}$ and others such as $[Au(AsS_2)]^0$, $[Au(Sb_2S_4)]^-$, etc. depending on the pH and other factors. These complexes may be responsible for the transport of gold (and silver) in typical hypogene environments. Experimentation with carbonate solutions containing various combinations of H_2S, AsH_3 and SbH_3 has shown that amorphous gold is relatively soluble under these conditions (Boyle *et al.*, 1975). The soluble species formed in the solutions are evidently $[AuS]^-$, $[Au(AsS_3)]^{2-}$, $[Au(SbS_3)]^{2-}$, $[AuAs]^{2-}$, $[AuSb]^{2-}$ or more complicated polymerized and hydrolyzed species.

In the supergene environment transport of gold as the $[AuS]^-$ complex or as the complexes of arsenic and antimony mentioned above may be a feature of the lower parts of oxidized zones where the solutions are neutral or alkaline and carry sulphide ion, H_2S, or dissolved arsenide and stibnide complexes. The enrichment of gold in the secondary (supergene) sulphide zones of some deposits suggests this mode of transport.

The complex $[AuS]^-$ is stable in seawater, but the general absence of S^{2-} and HS^- ions makes its presence in this environment practically negligible (Peshchevitskii *et al.* 1965, 1970).

Relatively little is known about complex telluride species and the migration of gold as tellurides. A sodium thiotellurate (IV), Na_2TeS_3, is reported to be formed by dissolution of Te in solutions of Na_2S_2 and Na_2S_3 but not Na_2S. The corresponding complex is probably $[TeS_3]^{2-}$ from which the double complex $[AuTeS_3]^-$ may be derived. In addition there may be a variety of other complexes such as $[Au(Te_2)]^-$, $[Au_2(Te_2)]^0$ and $[Au(Te_2)_2]^{3-}$ derived from the stable reduced ditelluride ion $(Te_2)^{2-}$ as suggested by Seward (1973). The attempts by Smith (1943) to synthesize gold tellurides from alkali sulphide solutions suggest that the systems are complex. He concluded that in alkali sulphide solutions, gold telluride is less soluble than gold sulphide, which in turn is less soluble than tellurium sulphide, and also that gold telluride appears in alkali sulphide systems only when the free sulphur concentration is low. For the extension of these data to natural systems see the quotation from Smith above.

13. As various colloids or sols: metallic gold readily forms a colloid, the 'Purple of Cassius' being known for centuries, and the red gold sol, obtained by reduction of a dilute gold solution with tannin and other organic compounds, was extensively investigated by Faraday. The metallic gold colloid carries a negative charge in the pH range 4 to 8. The charge can be reversed by the addition of ferric chloride and other salts (Weiser, 1933). Gold sols are, therefore, similar in most respects to the metallic silver colloids. Other gold compounds can apparently also form stable colloids. Beneš (1967) mentions positively charged hydrolytic products formed as the result of the hydrolysis of gold solutions in the pH range 1 to 3.

These products are apparently hydroxides, e.g. $Au(OH)_3$. Hydrous auric oxide, $Au_2O_3{\cdot}xH_2O$, and aurous oxide, $Au_2O{\cdot}xH_2O$, also have colloidal tendencies.

Gold colloids may arise in nature in a number of ways, some of which are: by comminution of gold particles by repeated abrasion; by reduction of a dissolved (ionic) gold compound or complex by numerous inorganic and organic substances, e.g., Fe^{2+}; by hydrolysis of dissolved gold compounds or complexes; as a result of the dissociation of a variety of dissolved gold complexes; and by release of gold held in the lattices of gold minerals such as tellurides and in solid solution in various sulphides and arsenides, particularly pyrite and arsenopyrite, as a result of oxidation processes. The gold (mustard gold) so released is extremely finely divided and reactive. Stabilization or peptization of gold colloids is accomplished by numerous organic and inorganic compounds, especially humic compounds, colloidal silica, hydrous iron oxides, etc. Gold colloids are stable at least up to temperatures of 100°C or more; at higher temperatures their stability decreases, but in the presence of suitable peptizers such as silica they are stable up to relatively high temperatures (350°C) (Frondel, 1938). Sulphide sols probably have a similar stabilizing effect at elevated temperatures. Flocculation and precipitation of gold from colloidal solutions is accomplished by various electrolytes, positive colloids, organic compounds and many other substances, all of which are mentioned subsequently.

Most of the work on gold colloids in the past (Zsigmondy and Thiessen, 1925) has been done on relatively enriched solutions. More recently some work on traces of gold ($<10^{-8}$M; 2 ppb) in solution has been carried out by Schweitzer and Bishop (1953) who showed by centrifugation studies that the gold exhibited colloidal behaviour in the pH range 2 to 13. Beneš (1967) has found, however, that traces of gold (10^{-8}–10^{-5}M) form only colloids (pseudocolloids) between pH 1 to 3. Above pH 3 the gold species is present mainly as a complex hydroxo anion (*see* 4 above). The pseudocolloids are evidently positively charged hydrolytic products, $Au(OH)_3$(?).

The transport of gold as a colloid in surface waters and supergene oxidizing groundwaters has long been advocated. Maclaren (1908) and Boydell (1924) and others before them considered this mode of transport as probable, and Lindgren (1933) inclined to the view that gold was transported as a colloid in certain oxidized zones of gold deposits.

Bastin (1915) experimented with gold colloids and was able to produce the dull brown finely divided gold often found in the oxidized zones of gold deposits. He suggested the possibility of gold being transported in colloidal solution in the oxidized zones of gold deposits in the presence of protective colloids such as silica sol.

Mering *et al.* (1953) observed that negative gold sols were fixed by kaolinite as a result of surface cation exchange processes (Au for Na, Ca, H, etc.).

Goni *et al.* (1967) investigated experimentally the stability, transport and flocculation of gold colloids in the surficial cycle. They observed that ionic and metallic gold give stable colloidal suspensions that permit migration over great distances. Chemically the colloidal suspensions were formed by reduction of dissolved gold by both humic and fluvic acids in acid and alkaline solutions respectively. Mechanically the gold colloids were formed by prolonged abrasion of gold particles. In both cases the colloids were found to be stabilized by the presence of colloidal iron hydroxide and colloidal silica. A high content of carbonate (as K_2CO_3) depressed the stability of the gold colloid, but low amounts (up to 0.2%) had only a minor effect. The authors postulated that the gold colloids could be flocculated in a number of ways, including changes in pH and Eh, interaction with organic matter in swamps and bogs, intermingling of waters with different salinities, adsorption by clay minerals, etc. Goni *et al.* (1967) were also able to make small artificial gold nuggets morphologically similar to those found in nature by compaction of colloidal particles due to turbulence. (*See also* the section on placers in Chapter III.)

A number of other investigators have suggested colloidal transport of gold under a variety of supergene conditions. Chernyayev *et al.* (1969) considered that some of the gold in the natural waters of northern Kazakhstan, U.S.S.R. was present as a colloid. Curtin *et al.* (1970) reasoned from their experimental work on the leaching of gold-bearing mull (forest humus layer) by demineralized water that most of the mobile ('soluble') gold was present in the leachate as extremely small colloidal particles ($<0.05\mu$ diameter). Machairas (1970*a*) thinks that some of the gold in the exogene cycle travels in (colloidal) suspension and gives some natural examples of where he thinks this process has been active (*see* 1 above). Gosling *et al.* (1971) concluded that some of the gold in the natural waters in Colorado was present in a fine colloidal form with an organic acid protective coating, the complex particles being capable of passing the 0.1 μ membrane filter used in the determination of 'solute' and 'particulate' gold. During oxidation processes of gold-quartz deposits Kraft *et al.* (1974) consider that gold migrates mainly in colloidal forms (*see also* Chapter IV).

Ong and Swanson (1969) and Ong *et al.* (1970) carried out an extensive investigation of the effect of organic acids in stabilizing gold colloids in the surficial environment. To summarize their findings they stated that:

> Interactions of various forms of gold, as plates, coarse particles, colloids, and ions $[AuCl_4]^-$, with different types of natural organic acids show that gold is not oxidized and complexed by the organic molecules as reported by some investigators. To the contrary, organic concentrations in the range of 3 to 30 ppm have the capacity to reduce gold chloride solutions to negatively charged colloids of metallic gold. For the 30-ppm organic acid solutions, the reduction process is accomplished by the formation of a protective coating of hydrophilic organic molecules around the hyrophobic gold sol making the gold very stable for at least 8 months and not easily coagulated by cations. The gold sols so formed are less than 10 mμ in size. This protective layer is also formed when colloidal gold is mixed with organic acids. For the 3 ppm organic acid, the organic matter concentration is too low to form the protective coating and the colloidal gold precipitates.
>
> Two of the organic acids used in this study were extracted from peat and humate-cemented sand, and one was collected from a brown lake water. All three are similar in composition and in physical properties to organic acids isolated from soils and surface waters. These natural organic acids can be considered as hydrophilic colloids having a particle size in the 2.4–10 mμ range and having negative charges due to the dissociations of the carboxylic and phenolic groups.
>
> The significances of this study in geochemical processes involving the mobility of gold are: (1) in a very acidic solution, for example near an ore deposit, where gold is soluble as its chloride and organic acids are not soluble, the presence of solid organic

matter causes the reduction of gold chloride, and metallic gold of colloidal size will be precipitated and intimately associated with the organic matter; (2) in slightly acidic to basic waters, for example in the general range of composition of natural waters, gold can only be transported as stable organic protected colloids; and (3) these organic-protected gold colloids are precipitated when they enter a different chemical environment, such as sea water or brackish water with their abundant ions or clay colloids, or when they enter an acid environment ($pH<3$).

The hydrothermal hypogene transport of gold in a colloidal form received some attention in the early years of the present century, but relatively little work has been done on the problem in recent years. Hatschek and Simon (1912) experimented with silica gels throughout which were uniformly dispersed soluble gold salts in an effort to produce some of the features common in gold-quartz veins. By using a number of reducing agents, viz. oxalic acid, ferrous sulphate, sodium sulphite, graphite, charcoal, bituminous shale and crude petroleum, they found that in a gel consisting of silicic acid in which a gold salt is uniformly distributed, a reducing or precipitating agent has the following effects:

(a) If the reducing agent is hypotonic (undersaturated) in relation to the salts in the gel the whole of the gold will leave the gel and will deposit in aggregates, probably of crystalline form, at the point or points of contact of the reducing agent with the gel. The gel, as a result of losing part of its soluble gold content, may cause the reducing agent to become hypertonic (supersaturated) and the gold still remaining in the gel would then be precipitated as below.

(b) If the reducing agent is hypertonic in relation to the salts in the gel the gold will be precipitated within the gel, not, however, necessarily uniformly throughout the gel, but often in distinct layers, the layers being parallel to the surface of the gel in contact with the precipitating agent, and the distance from one layer to the next one increasing with the distance from this surface.

From these experimental results they sought to explain the banding and ribbon structure so common in gold-quartz veins and to explain the distribution of gold in the quartz of the veins. They concluded that such phenomena could originate when a structure such as a fracture was filled with gelatinous silica into which or through which gold salts had diffused, precipitating native gold mainly as a result of reduction by a variety of common substances. Goni *et al.* (1967) repeated some of the diffusion experiments of gold chloride through silica gel and obtained gold films and some textures frequently present in gold deposits.

In a classic work on the role of colloidal solutions in the formation of mineral deposits Boydell (1924) referred to the occurrence of much fine-grained quartz, chalcedony and opal in gold deposits and to the presence of sulphides and gold in many deposits in an extremely fine state of division. He thought that these phenomena indicated colloidal transport of gold and silica, flocculation of the colloids to form gels, and finally recrystallization of the constituents of the gels to form the quartz and gold as it is now found. Lindgren (1933) advocated a somewhat similar origin for certain types of gold-quartz deposits. More recently Zhirnov (1972) has described spherical brownish-coloured colloidal-sized particles of gold (0.1 μ and smaller) in colloform bands in the Kaul'da gold deposit of the Almalyk ore field U.S.S.R. The fineness of these particles was 794. Zhirnov concluded that these particles initially represented colloidal gold protected by silica sols, and that they were deposited from solutions by a rapid drop of temperature at the margins of the veins where pyrite assisted in the coagulation of the gold colloids. Wright (1969) figures numerous textures of ore and gangue minerals in the auriferous Tennant Creek (Australia) deposits, which he thinks proves that the minerals were at one time in the colloidal state. Subsequent to deposition they have crystallized, and some have been partly mobilized. He suggests that the mobilization process took place by thixotropic reliquefaction of the gelatinous parent materials of the minerals now present.

Frondel (1938) seems to have been one of the last investigators to experiment seriously with colloidal gold under hydrothermal conditions. He found that gold sols were protected by silica colloids against electrolytes (NaCl) and spontaneous coagulation by increase of temperature. He also noticed an increase in the stability of protected gold sols with increase in temperature, and found that the protected sols were stable up to 350°C. From the experimental data Frondel concluded that in acid hydrothermal solutions gold is very probably transported in true solution; part or all of the gold salt, however, may be carried as an adsorbed electrolyte on colloidal silica. With approaching neutrality or alkalinity of the originally acid solution, the gold is precipitated out as sol, protected by colloidal silica. The protected gold sol may be coagulated in various ways – by decrease of temperature, dilution, changes in pH, mixing of the gold sol with other solutions carrying different kinds and concentrations of electrolytes and adsorption of the sol on mineral surfaces and gel surfaces.

More recently Vilor and Shkarupa (1971), experimenting with the Au-SiO_2-H_2O system at 100 to 400°C and at H_2O pressures up to 1000 atm, observed that the gold and silica contents in the solution were about equal and independent of pressure up to 200°C. At higher temperatures they increase with temperature and pressure, and the gold is present as $[AuO_2]^-$ and colloidal particles. The colloidal particles are stabilized by dissolved and polymerized SiO_2.

Finally, Fijii and Haramura (1976) have noted that a solution of $HAuCl_4$ can be reduced to a gold colloid in silica solution without adding reducing agents. In their experiments at room temperature and pressure they used (1) silica solution + $HAuCl_4$ solution + silica gel and (2) silica solution + $HAuCl_4$ solution. The concentration of the silica solution was about 50 ppm SiO_2. In the first experiment deposition of aggregates of colloidal purple gold and quartz crystals were observed. Colloidal purple gold was also absorbed in silica gel giving the gel a darker colour. Dissolution of the silica gel by alkali solution revealed the presence of fine purple gold and growth of gold and quartz crystals. In the second experiment, deposition of aggregates of fine purple gold, quartz crystals and a small amount of silica gel were observed. The colloidal purple gold did not adhere to the quartz crystal surfaces. The authors concluded that although the experiments were simple and at room temperature and pressure, they suggest that both colloidal and true solution processes are involved in the origin of some gold-quartz veins.

Despite all of the favourable experiments and speculations indicating the colloidal transport of gold it should be pointed out that no one has yet shown that gold is present in natural waters as a colloid. My investigations using dialysis methods have all been negative, but this is not surprising since gold in natural waters is present in such low amounts (circa 0.005 ppb) that one can probably not hope to attain success with present day apparatus. I have, however, noted that gold travels with other sols, e.g., hydrous ferric oxides, manganese sols, humic colloids, etc., but its state is unknown. It is probably largely present in an adsorbed form.

14. As aerosols and vapours in water gas (steam): the constitution of the aerosols would be varied. In some cases they would probably be colloidal gold particles or gold compounds of various types; in others the gold would probably be adsorbed to hydrated silica, $SiO_2{\cdot}nH_2O$, hydrated alumina, $Al_2O_3{\cdot}nH_2O$, etc. Gold aerosols are probably present in volcanic emanations or in steam impregnated rocks near the surface. There are few data on these phenomena: Goleva *et al.* (1970*a,b*) found from 0.8 to 1.6 ppb Au in the HCl fumaroles of the Ebeko volcano on Paranushir Island, Kurile Islands, Kamchatka, U.S.S.R. They concluded that the gold was present mainly as the chloride complex $[AuCl_4]^-$. There are a number of other volatile compounds of gold that could conceivably exist as aerosols or vapours under natural conditions, e.g., $AuFeCl_6$ and $AuAlCl_6$. These have considerable volatility in the range 200 to 500°C (Heinen and Eisele, 1974).

15. As an integral part of or adsorbed to the various humic complexes (colloids) that commonly occur in surface waters and in some underground waters: this is probably an important mode of transport in regions where decaying organic matter is present. Further details are given in 11 and 13 above.

16. As an integral part of or adsorbed to suspended particulate matter: in some places this is a major mode of transport of gold (and silver) in surface waters and also in waters leaching porous deposits undergoing oxidation. We have repeatedly found gold (and silver) in colloidal and particulate hydrous iron oxides, hydrous manganese oxides and fine silt, and in these substances with an abundant humic component in both surface and underground waters. The nature of the gold in these materials is discussed in the section on gold in other minerals in this chapter (e.g., limonite, wad, etc.) and in 13 above. The content of gold in these colloidal and particulate substances is variable. Where the so-called 'flour gold' is present (*see* 1 above) the contents are very high; in more normal environments as in the Keno Hill area and elsewhere the fine silt and other particulate matter in the streams averages about 0.05 ppm.

Fischer (1966) reports gold in amounts up to 0.1 ppm in the suspended silt of the Saale River in Germany. In the Aldan Shield of U.S.S.R. Razin and Rozhkov (quoted in Trofimov, 1969) found that 'floating gold' ('flour gold') and gold in suspended matter predominates in ground and creek waters. Their values for these materials range up to 0.037 ppm. In their paper on gold migration in the vicinity of the 'Kuranakh type' of gold deposit in Yakutia, Razin and Rozhkov (1963) found an average of 0.001 ppm dissolved gold and 0.009 ppm in suspended matter in waters rich in dissolved iron and manganese. Gosling *et al.* (1971), in their research on the gold content of natural waters in Colorado, found that the particulate solids in the waters ranged from nondetectible to 9.6 ppm, with most of the values falling between 0.1 and 0.7 ppm.

17. As an integral part of or adsorbed to various free swimming or floating micro-animal organisms and plants: Krauskopf (1956) showed experimentally that plankton adsorbs silver strongly. Gold probably behaves in a similar manner. (*See also* the section on the biogeochemistry of the element.)

The migration of gold in natural waters can be summarized in the following statements:

In ground, spring and surface waters gold is present in only small amounts generally of the order of 0.03 ppb. Much of this gold is apparently present as a colloid or adsorbed to organic, inorganic or organo-inorganic colloids and particulate matter. The remainder of the gold is in a soluble form in a variety of complexes depending on the environment of migration of the metal. In hypogene (hydrothermal) solutions the gold is evidently largely present in a complex form mainly as some species combining sulphide, antimony-arsenic-sulphide or telluride-sulphide. Under acidic condition the metal may also be complexed as some form of the chloro complex. Judging from the elemental associates of gold in its epigenetic deposits (*see* the section on epigenetic gold deposits in Chapter III), the nature of the wall-rock alteration zones associated with gold deposits (Chapter III), and the nature of thermal waters now depositing gold, a reasonable estimate can be made of the type of solution or diffusion current from which epigenetic gold deposits have been or are being precipitated. This solution is probably relatively dilute containing no more than 1 ppm Au, generally much less. Its pH is nearly neutral, and its constitution is varied but in most cases it contains the principal elements in the following forms:

- Alkalis (Na, K, etc.) as carbonates, bicarbonates and sulphates
- Alkaline earths (Mg, Ca, Sr, Ba) as bicarbonates
- Boron as borates and boric acid
- Silicon as colloidal silica, alkali silicate or $Si(OH)_4$
- Arsenic and antimony as As-Sb-sulphide complexes
- Sulphur as H_2S, various sulphide complexes and sulphates
- Tellurium as telluride-sulphide complexes
- Fluorine as alkali fluorides
- Chloride as alkali chlorides
- Iron and manganese as bicarbonates
- Base metals (Cu, Zn, Hg, etc.) as sulphide, bisulphide and polysulphide complexes
- Gold and silver as sulphide, bisulphide, As-Sb-sulphide and tellurium-sulphide complexes

The alkali carbonates and H_2S appear in many cases to control the migration and solution of gold mainly through formation of alkali sulphide and hydrogen sulphide complexes which render gold soluble. The formation of sulphide and bisulphide ion in solutions of alkali carbonates and H_2S probably follows reaction such as:

$$K_2CO_3 + H_2S = K_2S + H_2O + CO_2$$
$$KHCO_3 + H_2S = KHS + H_2O + CO_2$$

When arsenic and antimony are present in the system gold may be rendered soluble as complexes of these elements.

Changes in temperature, pressure and concentration of solutions and gases containing dissolved gold species commonly lead to the precipitation of gold. In near surface (supergene) processes changes in temperature, pressure, etc. are usually small and hence have only minimal effects, although they may be important at the orifices of springs where evaporation of solutions takes place, or where sublimation phenomena prevail at the vents of fumaroles. Under deep seated primary vein forming conditions, however, changes in temperature, pressure and concentration have marked effects. When highly heated water gas under high pressure and charged with K, Na, SiO_2, Cl, CO_2, H_2S, As, Sb, Te and gold in the form of various complexes, enters dilatant fractures or other structures it undergoes a decrease in pressure with consequent precipitation of various minerals including gold. Similar effects are produced as these gases (or solutions) rise toward the surface where they come into contact with cooler rocks, where they mingle with cool, circulating groundwaters or where they react with the wall-rocks.

In addition to changes in temperature, pressure and concentration there are a number of other mechanisms by which gold may be precipitated from natural waters and vapours. These are:

1. Decomposition of gold complexes due to changes in pH and other parameters: most dissolved gold species are complexes, some stable only in acid solutions, viz. the halide complexes such as $[AuCl_4]^-$, others stable only in neutral and alkaline waters, viz. the thiosulphate complexes such as $[Au(S_2O_3)_2]^{3-}$ and the thioaurite complex $[AuS]^-$. Changes in pH lead to the precipitation of gold from these complexes depending on the availability of complexing ions and other factors. Considering the acid halide complexes, neutralization of solutions containing these leads to the decomposition of the complexes and the precipitation of native gold where ions capable of complexing the element under neutral and alkaline conditions are absent. Similar conditions prevail during the neutralization or acidification of alkaline solutions containing complex ions of the type $[Au(S_2O_3)_2]^{3-}$ and $[AuS]^-$. In this case the complexes are decomposed and native gold is precipitated where ions capable of complexing gold under acidic conditions are absent.

Shifts in the pH of gold-bearing solutions through the neutral region into the acid and alkaline ranges must be common in nature. Acidic gold-bearing chloride solutions coming into contact with carbonates either under supergene or hypogene conditions generally rapidly and quantitatively precipitate their gold as one can readily see by simply adding powdered calcite to a solution of dilute gold chloride in a test tube. The effect in this case seems to involve an initial precipitate of gold hydroxide or oxide which is unstable, and which is ultimately reduced to native gold. Similarly, neutralization of alkaline waters containing dissolved gold in the form of alkaline complexes leads to precipitation of the metal where such waters come into contact with acid waters in oxidized zones or in humic environments whose waters are invariably buffered on the acid side of the pH scale. In hypogene environments loss of alkalis such as K, Na, Rb and Cs by reaction with wall rocks (sericitization, albitization, etc.) leads to a shift in the pH toward the acid range, the decomposition of the alkaline complexes and the precipitation of native gold or its compounds. Similarly, withdrawal of sulphide and arsenide ions by reaction with iron in the host rocks (formation of pyrite, aresenopyrite, etc.) leads to the breakdown of the gold sulphide and arsenide complexes and the consequent precipitation of gold. Dilution of solutions containing $[AuS]^-$, likewise, leads to a reduction of the sulphide ion concentration and the precipitation of gold. Oxidation (change in Eh) results in the precipitation of gold from sulphide, arsenide and antimonide complexes.

Decomposition of gold complexes in solution due to changes in pH, Eh and a variety of reactions involving the withdrawal of alkali and alkaline earth metals (the balancing cations of the complexes) and sulphide ion are probably the most important modes of precipitation of gold in both hypogene and supergene environments.

2. Adsorption and/or coprecipitation of soluble gold complexes and colloids: this is probably the second most important mode of precipitation of gold in natural settings. Concomitant with these phenomena there are invariably a variety of reduction reactions, which lead to the precipitation of metallic gold in the systems. Flocculation invariably takes place in the case of the gold colloids.

In the supergene environment gold (and silver) is strongly adsorbed and/or coprecipitated by natural gels, particularly hydrous iron oxides (limonite), hydrous manganese oxides (wad), silica-alumina gels, clay minerals and humic gels. Some of these gold-bearing substances are mentioned in the section on mineralogy, and analyses of these materials are given in Tables 12 and 22. Other data are given in Chapter IV.

The reasons for the adsorption and/or coprecipitation of gold by the various natural gels, clay minerals, etc. is a particularly complicated problem, mainly because experimentation with these natural substances is difficult, the results obtained being often strange and nonreproducible. The reason for this is that these natural precipitates, gels and resultant mineral complexes are never pure, and their composition varies widely as one can see by a glance at the analyses given in Boyle (1965*a*, 1972) and Dyck (1971). In addition to a great variation in the major constituents such as Fe, Mn, Si, Al, Mg, Ca, etc., there are also marked variations in the content of the humic constituents, which are inextricably mixed with the gels of Fe, Mn, Si, etc. Dyck (1971), experimenting with the adsorption of silver on these natural substances in our laboratories, found that in general they adsorb much larger quantities of silver than do laboratory preparations of hydrous oxides of iron and manganese under similar conditions of pH and equilibrium silver concentrations. He attributes this to the presence of ions in the colloidal complexes, which form insoluble compounds with silver. Gold behaves in a similar manner judging from natural and experimental data obtained by the writer. However, gold does not form many stable compounds, and ions associated with the gels do not have the same precipitating effect as found by Dyck (1971) for silver. On the other hand gold is readily reduced to the metal by practically any organic compound and by a great variety of ions such as Fe^{2+}, Mn^{2+}, etc. Once in the metallic form it is not easily desorbed (dissolved) by natural solutions, except where relatively high concentrations of complexing anions such as $[S_2O_3]^{2-}$ and $[Cl]^-$ are present.

The metallic gold colloid carries a negative charge in the

pH range 4 to 8, which is that found in most natural waters. Reversal of the charge is possible by the addition of various salts (Weiser, 1933). Most of the common gold complexes carry a negative charge, e.g., $[AuCl_4]^-$, $[Au(S_2O_3)_2]^{3-}$. Some of the natural colloids and gels carry positive charges under certain conditions. Below a pH of approximately 7.5 (the isoelectric point) hydrous ferric oxide carries a positive charge; above this pH the sol is negative. From this it follows that hydrous ferric oxide would adsorb and/or coprecipitate and carry colloidal and anionic gold extensively in natural solutions with an acidic and neutral reaction. Under alkaline conditions hydrous ferric oxides would tend to repel and disperse colloidal and anionic complexed gold. These features seem to account for the common enrichment of gold in many limonitic gossans and for the transport of gold adsorbed to colloidal ferric oxides in acidic and neutral surface and groundwaters during the oxidation of gold deposits.

Aluminum hydroxide and hydrous alumina behave in much the same way as hydrous ferric oxide, the isoelectric point being approximately at pH 8.4. Aluminosilicates behave similarly, the isoelectric point shifting to a more acid pH as silica is substituted in the system Al_2O_3-SiO_2. With 57 per cent Al_2O_3 the isoelectric point is at a pH of 7.0 whereas with 34 per cent Al_2O_3 the isoelectric point is at 5.0 (Johansen and Buchanan, 1957). Much more complex are adsorption phenomena involving kaolinite and other clay minerals. In kaolinite the negative charge on the 001 face is always greater than the positive charge on the edge faces and the aluminum mineral invariably carries a net negative charge down to a pH of 3.6 (Street and Buchanan, 1956). One would expect, therefore, that gold colloids and anionic complexes would be repelled by the clay minerals. Such, however, is not the case as shown by Mering *et al.* (1953). They found that gold sols were extensively adsorbed by kaolinite, particularly on the edge faces (i.e., the sites of positive potential). These features explain why gold is frequently concentrated in clay mineral aggregates and mixed limonite–clay mineral–silica aggregates in the oxidized zones of some gold deposits (*see also* Chapter IV). The phenomena may also account for the rich primary shoots of gold in certain highly kaolinized and argillized rocks, which owe their condition to sedimentation processes or hydrothermal alteration. Foster (1971*a*) considers this probable at Carlin, Nevada, but Radtke and Scheiner (1971) disagree and state that there is no apparent correlation between the amount of gold deposited in these disseminated orebodies and the total clay or illite content. The reason why certain clay minerals such as kaolinite do not always adsorb negative gold colloids or complexes probably has something to do with the fact that as the adsorptive sites of clay minerals become saturated with phosphate, arsenate and other negative anionic complexes they become negatively charged and hence repel the gold species. This has been shown experimentally as far as phosphated edges are concerned (Marshall, 1964).

Many other gels and sols carry negative charges, examples being silica, hydrous manganese dioxide, humic substances and various sulphide sols. Theoretically these should repel and disperse the metallic gold colloid and anionic gold complexes. In nature, however, the theory does not in general apply for we find that many of these substances are avid for gold and tend to precipitate the element extensively as one can see by a glance at Table 22. The reasons for this are complicated and only imperfectly known. The answer probably lies in the fact that seldom are these gels found in a pure state. Most are highly contaminated by hydrous ferric oxide and alumina gels, and these are probably mainly responsible for the adsorption and/or coprecipitation of gold in the composite gels. Other possibilites also exist: the hydrous MnO or $Mn(OH)_2$ gel is positively charged and adsorbs anions such as sulphate strongly. It would be expected, therefore, that the gold colloid and anionic complexes would also be adsorbed. Hydrous MnO is probably one of the initial manganese precipitates in neutral and alkaline environments characterized by a low oxidation potential, such as gossans, stream sediments, etc. During maturation of the gels the MnO or $Mn(OH)_2$ gel is oxidized, ultimately yielding MnO_2 and other manganese oxides. The problem is also concerned with the adsorption of gold by humic substances and the formation of gold organic compounds. Since humus is invariably a constituent of the composite sols in the surface environment, this mechanism may also be important in the enrichment of gold in gels. Finally, natural sols and gels tend to adsorb a great variety of substances which change their effective surface charges. Adsorption of a variety of cationic complexes by negative sols and gels may give them an effective positive charge, which permits them in turn to adsorb the negative gold sols and anionic complexes.

3. Changes in oxidation-reduction potential: gold is the most noble of metals and hence the lowest member of the electrochemical series, the potentials in acid solutions being as follows (Latimer, 1952):

$Cu = Cu^{2+} + 2e^-$	$E^o = -0.337$
$Fe^{2+} = Fe^{3+} + e^-$	$E^o = -0.771$
$Ag = Ag^+ + e^-$	$E^o = -0.7991$
$2H_2O = O_2 + 4H^+ + 4e^-$	$E^o = -1.229$
$Mn^{2+} + 2H_2O = MnO_2 + 4H^+ + 2e^-$	$E^o = -1.23$
$Au = Au^{3+} + 3e^-$	$E^o = -1.50$
$Au = Au^+ + e^-$	$E^o = -1.68$

From a consideration of these and other redox couples it can be seen that all the metals and most of their ions can reduce gold from its ionic state to the metal. The effect produced by iron and manganese is marked. Thus, an increase in the redox potential with the consequent production of ferric ion and MnO_2 renders gold more soluble and mobile if complexing agents are available. Alternatively a decrease in the redox potential permits greater concentrations of ferrous and manganous ions, which precipitate gold. Increase in the oxidation potential in solutions containing iron has other effects. Ferric iron undergoes hydrolysis and is precipitated as hydrous oxide gel which carries a positive charge. This results in the strong adsorption of anionic gold complexes and gold colloids as noted in 2 above.

Changes in the oxidation-reduction potential are effective in precipitating gold from natural solutions in a variety of supergene environments – in the sea, in freshwater bodies, in manganese and iron bogs, in precipitates at spring orifices and in the zones of oxidation of gold deposits. Similarly, these changes greatly affect the mobility of gold during hypogene processes. Where the oxidation potential is high, gold is maintained in the oxidized (ionized) state, permitting its transfer for considerable distances. Where the oxidation potential is low or where reducing substances are present the mobility of the metal is restricted. Finally, changes in the oxidation potential have a marked effect on the solubility of gold as sulphide, arsenide and antimonide complexes. Oxidation of these complexes results in the rapid and nearly quantitative precipitation of the metal.

The effects of various anions in solution may influence the coprecipitation of gold and silver on sulphides as shown by the researches of Sakharova *et al.* (1976). They found that the influence of the anions Cl^-, F^-, CO_3^{2-}, HCO_3^- and PO_4^{3-} upon the precipitation of gold and silver in the system $HAuCl_4 + AgNO_3 + NH_4OH$ was marked. The presence of Cl^- and F^- lowers the overvoltage of the silver reduction reaction and raises the overvoltage of gold reduction leading to a decrease in the fineness of the gold-silver precipitates. The influence of CO_3^{2-} and HCO_3^- ions results in a sharper rise of the overvoltage of the process of silver reduction than that of gold and consequently the formation of precipitates with a high fineness. The PO_4^{3-} ion exerts no appreciable influence upon the coprecipitation of Au and Ag.

4. Decrease in H_2S, S^{2-} or Cl^- concentration of vapours, gases or solutions: where gold is carried as the sulphide or chloride complexes, reaction of the vapours, gases or solutions with the wall rocks or gangue minerals may reduce the concentration of H_2S, S^{2-} or Cl^- as a result of the formation of pyrite, chloride-bearing minerals, etc. This breaks up the sulphide and chloride complexes causing the precipitation of gold in one form or another. Similar effects probably result when gold is transported as complex S-Se-Te and S-Sb-As compounds and the S^{2-}, HS^- or As ion concentration is reduced by reactions with iron (formation of pyrite and arsenopyrite) in the wall rocks.

Reactions such as those noted above appear to be partly responsible for the primary precipitation of gold (and silver) in a number of epigenetic deposits in basic rocks and in certain types of sedimentary rocks. These reactions may also operate in a modified way in the zones of reduction (secondary sulphide zones) of oxidized gold deposits.

5. Precipitation of gold as a result of the reaction of gold-bearing solutions, gases and colloids with various natural materials: the process is evidently one of reduction although other mechanisms such as adsorption and colloidal coagulation may also be involved. Whatever the cause it has been shown by a great number of investigators (Liversidge, 1893*c*; Rickard 1896; Lenher, 1909; Palmer and Bastin, 1913; Morris 1918; Sakharova and Lobacheva 1967; Radtke and Scheiner, 1970*a*; Shabynin, 1970; Zaitseva *et al.*, 1973; Lakin *et al.*, 1974; Sakharova *et al.*, 1975; and others) that most minerals and rocks precipitate gold from its solutions. The most efficient precipitants appear to be the sulphides, arsenides, tellurides, sulphosalts, native metals including gold, calcite, clay minerals, finely divided chlorite, graphite, carbon, carbonaceous matter, organic carbon compounds, pyritiferous carbonaceous slate and schist, carbonate rocks, sapropel and so on. Charcoal has long been known to be an active precipitant of gold (and silver) mainly by adsorption (Gross and Scott, 1927). Morris (1918) was able to grow beautiful dendritic crystals of gold on calcite in hydrothermal bombs charged with gold chloride. Shilo *et al.* (1974) have shown that gold and silver can be precipitated as a solid solution on pyrite. Limonite, wad, jarosite, scorodite and a veritable host of other supergene minerals precipitate, adsorb or otherwise incorporate gold from solution. Hydrogen in serpentine (mainly in magnetite) is considered by Coveney (1971) to have been the effective precipitant of gold from aqueous gold species in the Oriental Mine, Alleghany, California. The stability of the various soluble gold complexes in the presence of the various earth materials noted above varies widely and depends on many parameters. Generally speaking the halide complexes are the most readily reduced followed by the thiosulphate, thiocyanate, cyanide and various S-As-Sb complexes in about that order. Considering the great variety of natural precipitants of gold one wonders how the element is capable of migrating in any environment.

6. Coagulation of colloids: colloidal gold, colloidal gold compounds and gold adsorbed to various inorganic and organic colloids can be precipitated by a great variety of chemical changes such as a decrease or increase in pH, coprecipitation by other colloids and precipitation by electrolytes. Pyrite, arsenopyrite, galena and various other sulphides, arsenides and sulphosalts also coagulate gold on their surfaces as shown by Zvyagintsev and Paulsen (1940) and a number of earlier workers. The coagulating effect of minerals like quartz, mica, barite and calcite is less marked than that of the sulphides, etc., but is effective over long time intervals.

There are no data on the presence of gold colloids in natural hydrothermal solutions, and hence the effect of coagulation processes in primary vein formation is a matter of speculation. In natural fresh- and seawater it is probable that much gold is precipitated as a result of the coagulation of colloids.

7. Sublimation: this process is probably effective mainly in volcanic and fumarolic processes. The sublimates incorporate the available gold either in their structures or in an adsorbed form.

8. Decomposition of organic (humic) complexes due to oxidation: humic complexes, carrying gold in an adsorbed state or as an integral part of their structure, on oxidation yield CO_2, H_2O, etc. and liberate the gold. This gold may then be precipitated by a variety of mechanisms.

9. Sedimentation of particulate matter carrying gold: this process acts in all sedimentary basins in lakes, seas and in the oceans. The gold is incorporated in the sediments in a variety of forms.

10. Death and sedimentation of gold-bearing plants and organisms: numerous organisms and plants in natural fresh and saline waters absorb gold into their vital parts, or like plankton adsorb the element on their surfaces. On their death their remains are deposited on the bottoms of lakes, seas and the oceans. There, further reactions attendant on the decay of the organisms and plants fix the gold in the sediments. One of the most common of these reactions in a reducing environment is the bacterial degradation of the protein complexes of the organisms with the consequent production of H_2S, which precipitates iron and other sulphides which incorporate the gold. Legedza (1967) considered that sedimentary pyrite absorbs gold dissolved as the thiosulphate during sedimentation and diagenesis in the following manner:

$$FeO + H_2S + nH_2O \rightarrow FeS{\cdot}nH_2O$$

$$2\,FeS{\cdot}nH_2O + Au(S_2O_3)_2^{3-} \rightarrow 2\,FeS_2 + Au + 2SO_3 + nH_2O$$

11. Precipitating action of organisms: certain types of bacteria precipitate gold directly from natural waters. (*See* the section on biogeochemistry.) In addition iron and manganese bacteria play an indirect part since they oxidize Mn^{+2} to Mn^{+4} and Fe^{2+} to Fe^{3+}, which then undergo hydrolysis and are precipitated as hydrous manganese and iron oxides. During these reactions gold may be coprecipitated or adsorbed by the hydrous oxides.

> Gold is elaborated by vegetable growth in soils where there are no pretended geological indications; it is found occasionally in rain water; it may, for anything I know to the contrary, exist in the air, vapourized and afloat, as reguline.
> –W.B. Clarke, (Southern Gold Fields, Sydney, 1860)

Biogeochemistry of gold

Gold is a microconstituent of many plants and animals, including man. The literature on the biochemistry of the element is extensive, but is scattered through a great variety of medical and biological journals, many not readily accessible to the geologist and geochemist. Only a few of the most important papers have been consulted during the preparation of the following compilation.

Gold has been used in dentistry and as a medicine for centuries. The potable gold known to the alchemists was a dilute solution of the metal in aqua regia. Many medicinal virtues were ascribed to this potion, including those of the elixir of life. In more recent times gold solutions (thiosulphates) were used in the treatment of syphilis; today gold compounds still find a medicinal use in the treatment of rheumatoid arthritis.

Gold has long been known to be a constituent of plants and animals. The Czechoslovakian alchemist Paterson Hain Janos is said to have found gold in Hungarian grapes, and he wrote a treatise on the occurrence of gold in vegetation entitled *"De Auro vegetabili et vitibus Hungariae aureis"*. V.I. Vernadsky cites the researches of Rouelle and d'Arcet who found gold in French wines in 1779, and C.-L. Berthollet the famous French chemist is said to have determined the content of gold in the ashes of vegetables. In more recent times a number of voluminous compilations dealing with gold in plants and animals have appeared. Among these the works by Berg (1928) and Bertrand (1932) summarize their own work and that of those before them. Other exhaustive reviews, replete with many references, covering some 10 years of research and about all that was known concerning gold in plants and animals at the time, are the works by Babička (1943) and Babička *et al.* (1945). More recently Jones (1970) has compiled most of the data available on the gold content of plants and animals and added a commentary. The reader will also find a valuable discussion on the biogeochemistry of gold (and silver) in the recent compilation by Manskaya and Drozdova (1968) on the geochemistry of organic matter.

Gold in plants

The gold content of plants varies over a wide range, and hence average values are practically meaningless. It is difficult to comment on the analyses quoted in the literature since they have been done by many different investigators and by a great variety of different methods over a period of some 50 years. Table 23 shows the ranges in various types of plants. For particulars the reader should consult the extensive compilation by Jones (1970) and the various references noted in the table.

Table 23. Gold content of plants*

Type of plant	Au in ash (range) (ppm)
Nonvascular	
Bacteria	no data
Algae (marine)	0.00015–1.7
Algae (terrestrial)	no data
Fungi (terrestrial)	trace–11.6
Lichens	1–1.3
Mosses	0–19.2
Vascular	
Herbs	
Grasses and sedges	0–8.6
Forbs	0–610
Shrubs	0–600
Trees	
Coniferous	0–6
Deciduous	0–10
Miscellaneous	
Honey	not detected
Wine (U.S.S.R.)	0.030–0.758 μg/L (ppb)

*Compiled from data by: Aferov *et al.* (1968)
Aripova and Prikhid'ko (1965)
Aripova and Talipov (1966)
Babička (1943)
Beridze *et al.* (1969)
Boyle (unpublished)
Cannon *et al.* (1968)
Curtin *et al.* (1968)
Erämetsä (1954)
Fukai and Meinke (1962)
Kaspar *et al.* (1972)
Khotamov *et al.* (1966)
Korotayeva and Polikarpochkin (1970)
Lungwitz (1900*a,b*)
Razin and Rozhkov (1963, 1966)
Schiller *et al.* (1971*a,b*, 1973)
Stark (1943)
Warren and Delavault (1950)
Zvyagintsev (1941)

There are insufficient data on the gold content of nonvascular plants to draw any satisfactory conclusions about the capacity of these plants to accumulate gold. Certain species of fungi (*Boletus*) seem to accumulate gold. Fungi (mushrooms) also appear to accumulate silver in places (*see* the compilation by Boyle, 1968*b*). Razin and Rozhkov (1966) analyzed five samples of various lichens and found an average of 1.0 ppm Au in the ash, about twice that found in their soil substrates. Lichens tend to concentrate copper, silver and other elements where they are available in the substrate (Lounamaa, 1956; LeRoy and Koksoy, 1962), and may well be accumulators of gold where it is available either in the substrate or in surface waters.

Shacklette (1965) did not detect gold in his study of various liverworts and mosses, but Razin and Rozhkov (1966) found up to 19.2 ppm Au in various species of bryophytes. The amount of gold in the ashes of the mosses generally exceeded that present in their substrates.

Fungi and bacteria may have an important role in the

cycle of gold in nature. Hatschek (1919) and Williams (1918) noted that colloidal preparations of gold produced by the addition of tannic acid tend to become mouldy, and that the mould, in the course of time, removed the metal from solution. Williams' (1918) experiments with colloidal gold solutions and fungi showed the following:

1. Conidia of *Penicillium glaucum* and *Oidium lactis* developed in colloidal gold solutions which contained tannin or gum arabic.

2. The living fungus, during growth, removed the metal from the solution. Retention occurred in such cell walls as are not cuticularized.

3. The process goes on more irregularly when masses of dead fungus were introduced into solutions.

4. Solutions with a higher rate of diffusion coloured the fungus more quickly than those with a lower rate. In all cases the accumulation of gold finally produced a blue coloration.

5. No explanation of the process could be given.

There are also a number of other scattered references to the capability of fungi to extract gold from solution. In one interesting reference (selected abstracts from the technical literature on gold by Johnson, Matthey and Co. Ltd. for 1969, p. 1) it is reported that the Novosti Information Service in Moscow states that mould fungi extract gold from solution. Active fungi extracted 98 per cent of the metal in 15 to 20 h. Babička (1943) mentions that *Penicillium glaucum* is capable of depositing colloidal gold on its membranes and in its vicinity, and suggested that fungi in the surface parts of gold deposits and mines probably have a great influence in the transition of gold from the ionized to the colloidal state. The same author cites a number of references, which bear on the deposition of gold mainly as a colloid by various micro-organisms. Barbieri (1935) describes various gold-bacterium complexes, which are formed when bacteria are treated with solutions of $AuCl_3$.

Parès and Martinet (1964) carried out an extensive investigation of the role bacteria can play in the cycle of gold. They found that gold from laterites and other gold-bearing materials could be solubilized by autotrophic bacteria, such as the nitrogen fixing species, and to a much smaller extent by various heterotrophic bacteria existing in the soil, water and air. The solubilization of gold by bacteria can be marked, but several factors influence the phenomenon, such as the pH, the age of the cultures, the quality of inoculum, the concentration of gold in the medium and the nature of the nitrogenous and carbonaceous nutritive substances. These and other factors discussed by the authors should be referred to in the original paper. It is of particular interest to note that the solubilization activity of the various bacteria was strongest with strains obtained from soils over gold-bearing deposits.

In a Canadian patent, A. A. Mayling (Chemical Abstracts, v. 66, 13120b, 1967) describes the use of bacteria to keep aerated ferric chloride solutions oxidized for leaching ores containing Au, Ag, Cu and Pb. Suitable bacteria are *Ferrobacillus ferrooxidans* and *Thiobacillus ferrooxidans*. At acid pH (1–3) and temperatures ranging from 15 to 40°C these Fe-oxidizing bacteria depress the normal reduction of ferric chloride resulting from reaction with the metals or sulphide in the ore. Presumably the gold taken into solution is complexed as chloride. The implication of the action of such bacteria in the mobilization of gold in the oxidized zones of gold deposits containing abundant iron minerals is obvious, considering the fact that the Fe-oxidizing bacteria are of common occurrence and that complexing substances such as Cl and organic compounds are, likewise, common in some environments.

Lyalikova and Mokeicheva (1969) examined the role of bacteria in the migration of gold during the oxidation of gold deposits. They found that numerous heterotrophic bacteria have the ability to dissolve gold, and that various sulphur compounds (e.g., thiosulphates) that accumulate in solutions during the oxidation of sulphides by sulphur bacteria contribute to the solution of gold. A culture isolated from the Zodsk gold deposit, similar to *Bacillus alvei* was the most active and dissolved 500 to 600 μg/L (ppb) Au in 3 weeks. The largest amounts of gold solubilized in the experiments were obtained when stibnite was present – an interesting fact and one that supports the writer's contention that gold is frequently transferred in nature as some form of antimony compound. The authors concluded from their laboratory experiments and microbiological examination of the Zodsk gold deposit that bacteria play a significant role in gold migration.

In a recent report K.G. Doxtader (U.S.Geol. Surv., Circ. 622, p. 13) summarized his work involving the testing of the ability of various soil bacteria and fungi to grow in liquid media containing different concentrations of the metals including gold, silver, mercury, copper, lead, manganese, cobalt, chromium, molybdenum, tin, cadmium, uranium, vanadium, tungsten, zinc and palladium. He found that one fungus could grow in a concentration of 25 ppm Au, and 12 species of bacteria and fungi could grow in cultures containing 10 ppm Au. On the other hand, cultures containing >5 ppm Ag or Hg were found to be lethal to all organisms tested; some organisms could grow in saturated solutions of copper, manganese, molybdenum, lead, uranium, tungsten and zinc. Two bacteria species (not named in the report) previously reported to be capable of solubilizing native gold (up to 1.5 ppm in solution) were also tested for a period of 3 weeks, with negative results.

In a more detailed account of his work Doxtader (*in* Lakin *et al.*, 1974) concluded as follows:

> In the microbiological studies, elemental gold was added to liquid nutrient media which were then inoculated with pure cultures of bacteria and fungi either obtained from stock collections or isolated from soil. In a glucose-peptone medium containing colloidal gold, significant gold solubilization occurred. In other media or with powdered gold, metal solubilization was slight. The microbial strains showed various abilities to solubilize the noble metal. In sterile glucose-peptone medium gold solubilization occurred, but, in the same medium inoculated with various microorganisms, the solubilization of gold increased.

Doxtader (*in* Lakin *et al.*, 1974) also tested soils and stream sediments from Cripple Creek, Colo., to determine whether specific bacteria or fungi occurred in materials containing gold. No significant differences in numbers or kinds of micro-organisms present were found between gold-bearing and barren samples.

During recent investigations of the effect of organisms on the geochemical cycle of gold Mineev (1976) concluded that the dissolution, colloidal formation and precipitation of gold take place in nature with the participation of various micro-organisms.

Of the vascular plants the herbs have received the most attention with respect to their gold content. Razin and Rozhkov (1966) analyzed a number of samples of various grasses and sedges, including entire plants and the stems, leaves and fruit of the plants. Their values range from 0.005 to 8.6 ppm in the ash. Over some of the placer deposits in the Beauceville area of Quebec the writer has found an average of 2.4 ppm Au in the ash of various grasses (whole plants, excluding roots; 9.3% ash, mainly silica). This compares with an average of 9.2 ppm Au in the ash of *Equisetum* in the same area (whole plants, excluding roots; 26% ash, mainly silica). It would appear from these values that there is a relationship between the gold and silica content of plants, the higher the silica the greater the gold content, all other things being equal. This feature is also mentioned by Babička (1943) and Thyssen-Bornemisza (1942).

The gold contents of grasses and maize (*Zea mays*) have been investigated recently by Schiller *et al.* (1973) using neutron activation analysis with a view to determining the influence of seasonal variations of samples in a gold-bearing area. The contents in the ash of maize ranged from 0.008 to 0.103 ppm Au; in the ash of grasses the values were 0.009 to 0.040 ppm Au. Marked seasonal variations were noted between the three seasons, spring, summer and autumn. These are important data for biogeochemical prospecting purposes. The investigators recommend that sampling should be done preferably in the spring since this is the time when the maximum amount of gold is taken up by the plants due probably to higher rainfall and movement of elements in soil waters, etc.

Considerable work on the gold content of various forbs has been done in the last 40 years. The range in the values is from 0 (not detected) to 610 ppm Au. Babička (1943) and his colleagues carried out investigations on the gold content of a variety of forbs for some 10 years, paying particular attention to plants such as corn, beans and horsetails (*Equisetum*). In the ash of *Equisetum palustre* (marsh horsetail) he found 610 ppm Au in the andesitic region of Oslany, Czechoslovakia, where the gold content of the soil averaged about 0.2 ppm. In two other samples of *Equisetum arvense* contents of 16 and 63 ppm Au were reported in the ash. These results seemed to indicate that *Equisetum* is a gold accumulator. However, Cannon *et al.* (1968) found that *Equisetum* averages only 0.17 ppm Au in the ash in the samples they investigated from a number of mineralized areas in the conterminous United States; in samples from Alaska they found up to 0.5 ppm Au in the ash. These values are about the same as those reported by Warren and Delavault (1950) and Razin and Rozhkov (1966) in various varieties of *Equisetum*. Cannon *et al.* (1968) concluded that the high values quoted by Babička and other workers may be erroneous since a study of the analytical procedures that were used suggest that other metals in the plants, some of which are relatively high (e.g., zinc, copper), may have been measured and reported as gold.

Our research on *Equisetum* seems to indicate that horsetails may be accumulators of gold. Over some of the placers of the Beauceville area, Quebec the ash of horsetails contained up to 9.2 ppm Au (whole plants, excluding roots; 26% ash). *Equisetum* ash from the Ramore area of Ontario, well removed from gold mineralization, contained 0.095 ppm Au; the corresponding silver value was 2.4 ppm. *Equisetum* growing on old tailings piles on the Hollinger property at Timmins, Ontario contain considerable amounts of gold according to a personal communication to the writer from Mr. C.D. MacKenzie, Chief Geologist of Hollinger Gold Mines (1970). The gold content of the ash of these plants ranged from 0.01 to 0.25 oz Au/ton (0.3 to 8.5 ppm).

The gold content of shrubs ranges from 0 to 600 ppm. Babička (1943) found 600 ppm in the ash of the seeds of *Clematis vitabla* (virgins' bower) in the andesitic region of Oslany, Czechoslovakia. The soil and weathered debris contained 0.2 ppm Au. In the wood and fruit of the same shrub and of *Corylus avellana* (hazel nut) and *Rosa Canina* (dog rose) from Oslany and other areas in Czechoslovakia the same author found gold contents ranging from trace to 110 ppm in the ash. It is interesting to note that relatively large amounts of gold occur in seeds. This is also true for silver as pointed out by the writer (Boyle, 1968*b*). Warren and Delavault (1950) found only small amounts of gold (nil and <0.03 ppm) is the ash of *Juniperus communis* (dwarf juniper) in the North Fork of Watson Bar Creek in the Clinton Mining Division of British Columbia. Gold occurs in quartz veinlets in this area. Razin and Rozhkov (1966) analyzed a number of shrubs in the vicinity of the Kuranakh gold deposits, Yakutia. Their values range from 0 (not detected) to a maximum of 10 ppm Au in the ash; this can be compared with a maximum of 8 ppm Au in the soil. However, they found that in some 65 per cent of the soil and shrub ash samples that the maximum gold contents in the soil exceeded those in the ashes of the shrubs. Aripova and Talipov (1966) investigated the gold content of a number of shrubs and herbs in the gold belts of Uzbekistan. The species sampled were wormwood (*Artemisia terrae-alba*), salsola (*Salsola* sp. and ferula (*Ferula* sp.). The largest amounts of gold were found in the ash of wormwood (up to 125 ppm Au). The average gold content in the plants over auriferous zones ranged from 20 to 50 ppm Au and greater. The largest amounts of gold were found in the plants during springtime and early summer (May-June), the blooming and fruiting period. Khotamov *et al.* (1966) used neutron activation methods to determine the gold content of a number of shrubs and their organs in one of the ore fields of the gold-bearing Kyzyl-Kum district, U.S.S.R. Their values ranged from 0.3 to 36 ppm in the ash. The soil substrates contained from 0.2 to 1.5 ppm Au. The leaves of some shrubs contained more gold than the stems and roots, whereas in other plants the gold content was relatively uniform in all organs. The largest amounts of gold (and arsenic and antimony) in the plants were observed during the period of intensive growth and the smallest amounts in dried (dead) plants. No relationship was detected between the ash content and the overall content of gold (and arsenic and antimony). Arsenic and antimony were also determined by activation methods on the vegetation and soil. A direct correlation between gold and antimony was found in the soil and plant ash, whereas a correlation between gold and arsenic in the soil and plant ash was not always observed.

The gold content of trees ranges from 0 (not detected) to 10 ppm in the ash. Lungwitz (1900*a,b*) appears to have been one of the first to report the presence of gold in trees, specifically in what he called 'iron wod' from the Omai Valley in British Guiana. The work was done mainly to prove that gold migrates in surface waters. Babička (1943) records up to 10 ppm Au in the ash of the blossoms of *Tilia parvifolia* (linden). Warren and Delavault (1950) found from 0 to 1.02 ppm Au in the ash of *Pinus contorta* (lodgepole pine), *Pinus ponderosa* (western pine), *Pseudotsuga taxifolia* (Douglas fir), *Salix* sp (willow) and *Populus tremuloides* (aspen) in the North Fork of Watson Bar Creek, British Columbia. In areas unmineralized with gold (negative areas) the gold contents in the ash of these trees ranged from nil to 0.03 ppm. In the vicinity of quartz veinlets carrying gold (positive areas) the contents in the ash ranged from nil to 1.02 ppm. Silver was also determined in the ash of the trees. Calculations based on the values given indicate that the Au/Ag ratio in the ash is about 0.06. Other investigations of the gold content of trees include those by Zvyagintsev (1941), Razin and Rozhkov (1963, 1966), and Khotamov *et al.* (1966). Their values range from 0 (not detected) to 6 ppm in the ash of various species.

Concerning plant products Berg (1928) did not detect gold in honey. In wines from the U.S.S.R., Beridze *et al.* (1969) by activation analysis found from 0.030 to 0.758 μg/L (ppb) Au. The level of the gold content in red wines was considerably higher than that in white wines. Most of the gold was thought to have come from the soil with minor contributions from the apparatus used in the processing of the vineous spirit.

Seasonal variations in the gold content of plants have been noted by some investigators, but the data are insufficient to draw adequate conclusions. Aripova and Talipov (1966) report that the ash of Artemisia (wormwood) contains from 10 to 20 times more gold (and copper, lead and chromium) in May than in October, and that these metals tend to be highly concentrated in plants in desert regions. Khotamov *et al.* (1966) found that the largest amounts of gold (and arsenic and antimony) occur in plants during the period of intensive growth and the least in plants that have died and dried up. As noted above Schiller *et al.* (1973) noted marked seasonal variations in the gold content of plants, the maximum concentrations being present during springtime.

With respect to the content of gold in the various organs of plants, Khotamov *et al.* (1966) state that the largest amounts of gold are concentrated in the above ground portions of plants, particularly in the leaves of some species. Seeds may be concentrators of gold in a manner similar to that found for silver (Boyle, 1968*b*).

There appears to be no particular relationship between the gold content and the ash content of plants. There may, however, be a relationship between the silica content of plants and their gold content, but more work is required to confirm this. Khomatov *et al.* (1966) found a definite direct correlation between gold and antimony in the ash of plants, a correlative feature that also holds in most gold deposits. The correlation with arsenic was less definite.

The data with respect to the ratio – Au in soil/Au in plant ash – are conflicting. Some investigators have found this ratio to be less than 1 whereas others have found the reverse. It is well known that this ratio depends on a veritable host of factors, and one would expect wide variations and reversals to be general, even locally, as some of the data in the literature clearly show. A detailed study by Razin and Rozhkov (1963) over the Kuranakh (karst) type of gold deposits in Yakutia showed that there was a direct relationship between the content of gold in the ashes of the plants and the content in the soil substrate.

The uptake and translocation of gold by plants has been studied by a number of investigators including Zvyagintsev (1941), Babička (1943), Razin and Rozhkov (1963), Shacklette *et al.* (1970) and Schiller *et al.* (1971*a*).

Zvyagintsev (1941) speculated that the solutions secreted by the root hairs of plants were capable of solubilizing gold in their vicinity, and that the gold entered the plant in an ionized form.

Babička and his colleagues after more than 10 years of study on the gold content of plants concluded the following:

1. Only a few plants can take up gold in large amounts.
2. The concentration of gold is characteristic for certain plants; it is not, however, definitely known whether gold has a favorable effect on the growth and the life functions of these plants.
3. The gold taken up by the roots of a plant spreads throughout the plant and becomes concentrated in certain organs. In the case of seed-bearing plants, this is usually in the seeds. With increasing age of a plant there is a progressive increase in the gold content.
4. Plants absorb ionized gold derived from solid particles of gold, particularly when the metal is in the colloidal state. In nature, colloidal gold is concentrated mostly in the rhizosphere (root layer), perhaps by the actions of plants and organisms generally. If gold occurs in the soil in the form of small nuggets, tablets, fine wires, etc. it is practically unavailable to plants.

Razin and Rozhkov (1963) concluded from their data on gold in plants and soils over the Kuranakh (karst) type of gold deposits in Yakutia that the gold is assimilated by plants in the ionized form.

Shacklette *et al.* (1970) experimented with the uptake and translocation or radioactive (^{198}Au) and nonradioactive gold (^{197}Au) by impatiens (*Impatiens holstii*), garden balsam (*I. balsamina*) and onion (*Allium cepa*). They found that colloidal gold was not absorbed by the roots of these plants, and that of the solutions of gold used (chloride, cyanide, bromide, iodide, thiocyanate, thiosulphate and gold cyanide solution *via* flax) the gold cyanide is the most readily absorbed by roots and transported in largest amounts to the leaves. They concluded that if gold is present in the soil and if cyanogenic plants are rooted in this soil, a mechanism is present for the entrance of gold into the biogeochemical cycling process.

Schiller *et al.* (1971*a*) experimented with the uptake of radioactive gold by barley from solutions containing in one case aurothiomalic acid-Au-198-sodium salt and in another case colloidal gold (Au-198) (particle size range up to 200 Å) stabilized with gelatin. In both cases the uptake of gold was almost quantitative, from which the authors concluded that gold accumulation by plants, both in colloidal and ionized form, is possible.

In recent papers Talipov *et al.* (1975, 1976) have noted that the accumulation of gold (and a number of other elements) in plants in the Karamin Range, U.S.S.R., depends to a large degree on the concentration of the element(s) in the groundwaters and soils. In particular they noted a linear correlation between the Au, Ag, Cu, Zn and Pb contents of plants and that of soils and waters. They also observed that some elements have an antagonistic effect on the uptake of certain elements in plants; thus high calcium contents in groundwaters decreased the migration of gold and hence its accumulation in plants.

The nature of gold in plants is not known. It may be present largely in the chlorophyll (a porphyrin), since a gold (III) porphyrin, which is relatively stable, has been synthesized (Fleischer and Laszlo, 1969). It may also be present as an organometallic compound, perhaps as a protein complex, since the metal, like silver, is apparently concentrated in seeds which are especially high in protein. The role of gold, if any, in plant metabolism is unknown.

According to Babička (1943) small quantities of gold stimulate the growth of plants and micro-organisms, whereas large amounts are toxic. He cites a number of researches to support this contention. Felippe and Martins (1926) found that native gold had an oligodynamic (growth inhibiting) action on *B. dysenteriae* Flexner, and Nadson and Stern (1934) observed that native gold placed at a distance of 2 to 3 mm from white mustard seeds retarded the germination of the seeds and the growth of the sprouts. Shacklette *et al.* (1970) found that concentrated solutions of radioactive gold cyanide were extremely toxic to all *Impatiens* whereas low concentrations of the compound were not toxic. Radioactive gold bromide solutions were very toxic to *Impatiens*, and radioactive gold iodide solutions were found to be extremely toxic to all plants experimented with. Radioactive gold thiocyanate was toxic to unrooted cuttings of *Impatiens*, whereas the nonradioactive solutions were less toxic. Gold thiosulphate solutions were only slightly toxic to all *Impatiens.* The toxic effects manifested by gold solutions on plants are principally wilting of leaves and stems, necrosis and chlorotic symptoms. None of these effects have been observed by the writer in plants growing in soils and placers rich in gold in Canada, nor are there any references in the literature to such effects attributable to the presence of gold in other countries. The writer has also not observed any stimulative effects on plants that could be related to the presence of small amounts of gold in soil and soil solutions.

Some plants appear to be gold accumulators. The case for *Equisetum* needs verification, and further careful studies should be carried out on this plant, especially where it grows on gold-enriched soils. Grasses also appear to accumulate gold in places. Fungi and lichens are probably accumulators of gold where the element is readily available, and some mosses also appear to concentrate gold according to the data provided by Razin and Rozhkov (1966). Khotamov *et al.* (1966) state that *Lagochilus intermedius, Artemisia terrae-alba* and *Cousinia hamadae* are concentrators of gold in the Kyzyl-Kum region of U.S.S.R. Certainly some of the wormwood shrubs (*Artemisia*) are greatly enriched in gold (up to 125 ppm Au in the ash) in the arid regions (Tamdynsk Mountains) of Uzbek, U.S.S.R. according to the data of Aripova and Talipov (1966). The average for these shrubs over gold-bearing zones in many cases ranges from 20 to 50 ppm Au. According to a later communication by Talipov and Khotamov (1974) a number of desert and arid land plants in the Tamdynsk Mountains (central Kyzyl-Kum) concentrate large amounts of gold, including *Lagochilus intermedius* (36 ppm in the ash), some *Salsola* (saltwort) species, *Artemisia* and *Stipa* grasses. Certain plants such as *Spiraea* and honeysuckle accumulate large amounts of gold when growing on or near auriferous deposits in the Chatkal-Kurama region of U.S.S.R. (Talipov *et al.* 1976). In India Dekate (1971) has noted that plants producing hydrocyanic acid such as *Sorghum saccharatum* are capable of preferentially absorbing gold from the soil over auriferous deposits, at times, to the extent of depleting gold in the soil in the proximity of plant roots. (*See also* the section on biogeochemical prospecting in Chapter V.)

Direct gold indicator plants appear to be rare if in fact such plants actually exist. Dorn (1937) says that "Gold indicator plants are up until the present only known from Brazil where *Cecropia laetivirens Hub., C. palmate Willd, C. lyratiloba, Alpinia speciosa Schum., Typha dominguensis Kunth,* and *Typha latifola L.* not only contain gold in their ashes but also according to F. Freise are reliable indicators of gold placers." Babička (1943) mentions that prospectors searching for gold and silver deposits in Queensland, Australia, looked for the shrub *Lonicera confusa DC,* which grows preferentially on gold- and silver-bearing soils. A shrub in California is also said by R.H. Stretsch to indicate the presence of gold, but no details of this plant are given.

It is interesting to note that the early Chinese considered some plants to be indicators of gold and silver deposits. Needham (1959) quotes Tuan Chlêng-Shịh in the *Yu-Yang Tsa Tsu* written *circa* 800 A.D. as saying:

"When in the mountains there is the *tshung* plant (ciboule onion; *Allium fistulosum*), then below silver will be found. When in the mountains there is the *hsiai* plant (shallot; *Allium bakeri*), then below gold will be found."

There are numerous other similar statements pertaining to gold in the old Chinese literature on geological and mining subjects, but space precludes their mention here. One is of interest, however, namely that the Chinese noted the presence of the *chhi* (essence) of gold in the rape-turnip (*Brassica rapa-depressa*) (Needham, 1959).

It is worth noting that the coincidence of occurrence of the onion species (*Allium* sp.) and the precious metals, silver and gold, is probably an indirect one, the relationship being more dependent on the presence of sulphur. The writer has noted that various species of wild onions and the meadow garlic (*Allium canadensis*) often proliferate where an abundance of sulphate is available from the oxidation of sulphide deposits, some of which may be auriferous. It is also well known that the *Allium* species, especially garlic, is generally rich in sulphur, and some species seem to take up more than average amounts of arsenic, an element universally associated with gold and silver. The turnip is likewise rich in sulphur.

According to a recent report (Engineering and Mining Journal, v. 177, p. 32, 1976) (Chem. Abstr., v. 34, 111234h) the water hyacinth (*Eichhornia crassipes*) is an accumulator of gold and silver. If experiments are successful it is planned to use the plant to recover the two precious metals from solutions leaching old Cripple Creek mining tailings. In the article it is stated that "under a regular harvesting program, an acre of hyacinths can yield silver at a rate of a pound every four days, and similar results are anticipated for gold."

Gold in animals

Gold has been determined in only a few animals, the data available being shown in Table 24. Reviews of the gold content of animals have been published by Babička (1943), Vinogradov (1953) and Jones (1970).

Noddack and Noddack (1939) found from 7 to 30 ppb Au in the dry matter of a number of marine organisms, contents similar to those found by neutron activation methods in a variety of sea animals by Fukai and Meinke (1962) and Bowen (1968). Their values range from 0.0012 to 126 ppb Au in the dry matter. The average gold content for all marine organisms done by activation methods is 23.84 ppb. Compared with the average gold content of the sea (0.02 ppb) the enrichment factor of gold in the dry part of marine organisms is 1200. Enrichment factors similar to this figure are not unusual in marine organisms.

Babička (1943) and his colleagues analyzed june bugs from the auriferous Oslany region of Czechoslovakia finding an average of 25 ppm Au in their ash. In another location, which is nongold bearing, near Stříbrná Skalice (Bohemia) they found no gold in the ash of these insects. They reasoned that the june bugs feeding on the leaves of gold-bearing plants accumulate gold, whereas those feeding on nongold-bearing plants do not have the opportunity of concentrating the metal. To prove this they starved the gold-bearing june bugs for a time, then fed them on leaves which contained no gold. A reduction of the gold content from 25 to 18.5 ppm was observed in their ash. Razin and Rozhkov (1966) did not detect gold in the water beetle (*Dytiscus* sp.) or in the carpenter ant (*Camponotus vagus*) but found 400 ppb Au in the ash of a bee (*Vispidae* sp.).

Table 24. Gold content of animals

Type of animal	Au	Reference	Remarks
Marine animals	(ppb)		
Sponge	10	1	In the dry matter
Sponge (various)	18–126	2	In the dry matter
Sea urchin	7	1	In the shell
Starfish	30	1	In the dry matter
Sea cucumber	24	1	In the dry matter
Jelly fish	7	1	In the dry matter
K. fische (Lippfische)	10	1	In the dry matter
Clam (*Tapes japonica*)	5.7	3	In the dry matter of the soft parts
Prawn (*Pandalus* sp.)	0.28	3	In the dry matter of the soft parts
Mackerel (*Pneumatophorus japonicus*)	0.12	3	In the dry matter of the muscle
Terrestrial animals			
Crustaceans (crab)	0	4	In the shell
Insects			
June bugs	25000	4	In the ash
Bee (*Vispidae* sp.)	400	5	In the ash
Water beetle	0	5	In the ash
Carpenter ant	0	5	In the ash
Birds			
Various	0–1500	5	In the ash
Animals			
Various	0–14000	4,6	In wet and dry matter

Reference: 1. Noddack and Noddack (1939.)
2. Bowen (1968).
3. Fukai and Meinke (1962).
4. Babička (1943).
5. Razin and Rozhkov (1966).
6. Jones (1970).

Razin and Rozhkov (1966) analyzed a number of birds, including crossbills, buntings, pipits and a thrush in the vicinity of the Kuranakh gold deposits, southern Yakutia. The gold contents in their ashes ranged from 0 (not detected) to 1.5 ppm. There are numerous references to the occurrence of gold nuggets in the crops of birds.

Gold has been detected in a number of animals and in man. Babička (1943) and his colleagues detected relatively high contents of gold in the ash of the antlers of deer and in the hair. Contents from 0 (not detected) up to 68.6 ppm Au were found in the antlers and other parts of deer in the gold-bearing Oslany region of Czechoslovakia. The highest contents of gold were found in the tips of the antlers. No gold was found in the antlers of deer from the barren Malacky area in western Slovakia or in the eastern Carpathian Mountains. A wild boar bagged in the Oslany area contained no detectable gold. Jones (1970) found only 0.5 to 28.3 ppb Au in the ashes of the antlers of 15 mule deer (*Odocoileus hemionus*) harvested from four gold-mining regions in the United States. The largest amounts of gold occurred in the tips of the antlers. The contents found by Jones are much less than those recorded by Babička; they approximate the contents found in common rocks.

Bertrand (1932) detected gold in cow liver and brains, and in human blood, feces and urine. His values range from 0.2 to 14 ppm Au, the largest amount being in the cow brains. In wet human liver Parr and Taylor (1963) found from 0.013 to 0.79 ppb. In human blood and its components the gold content ranges from 0.06 to 0.8 ppb (Bagdavadze *et al.*, 1965). Human hair appears to be enriched in gold according to the data given by Bate and Dyer (1965). They recorded means ranging from 2.7 to 43 ppb Au in people from Tennessee and New Zealand.

The data on the gold content of animals is inadequate, scattered, unsystematic and often conflicting. Conclusions seem unwarranted except to note that marine animals seem to have slightly lower amounts of the element than terrestrial animals. Gold seems to be concentrated in relatively large amounts in the proteinaceous substances of animals (hair, brains, muscle of fishes), which suggests that the element may be largely present as gold-protein complexes in animals, probably bound to the thiol groups in the complexes. In this respect gold behaves much like mercury, its adjacent element in the periodic table. The role of gold, if any, in animal metabolism is unknown.

There appears to be no auriferous animals as far as the data goes. Animals in gold-bearing regions may contain larger amounts of the metal, but the data provided by Babička (1943) and Jones (1970) are conflicting with respect to the antlers of deer. The subject should be further investigated.

It is interesting to note that the old literature is replete with references of unusual occurrences of gold in animals. Gold nuggets and particles of auriferous quartz have been found in the crops of partridges, hens, ducks and geese in many gold-bearing regions of the world, but particularly in Rumania and Hungary. Gold, plating the teeth of goats, sheep and cattle grazing in auriferous regions, is also recorded in some old manuscripts. Buried human skulls in certain auriferous streams have been observed to contain particles of attached gold.

Homo sapiens is a concentrator of gold and silver but in ways other than biochemical. Since the beginning of early gold mining (circa 4000 B.C.) man has accumulated an estimated hoard of some 3 billion (3×10^9) oz in his vaults, a small amount (0.00008%), however, compared with the total amount of gold in the earth's crust (3.8×10^{15} oz). By contrast man has won about 35 billion (35×10^9) oz of silver from the earth. The ratio of extraction of Au/Ag is, therefore, 3/35 or 0.08, a value near that for the ratio found for the two elements in the earth's crust (0.1).

Gold in humus, peat bogs, muskegs and marshes

Gold, in common with many of the metals, tends to become enriched in humus and humic deposits. The reasons are the strong adsorptive and absorptive capacity of humus for metals and the presence in humic deposits of various reducing substances and H_2S.

Goldschmidt (1935) found 0.2 ppm Au in the ash from beech humus in the Sababurg area near the Weser River in Germany. Deul (1958) noted the presence of gold in the ash of humic acid fractions separated from peat. He did not detect the element in plant ashes, in the whole peat ash, nor in any other fraction separated from the peat. This suggests that gold in the peat is mainly associated with the humic components. Curtin *et al.* (1970) and Lakin *et al.*, (1974) found from <0.04 to 5.1 ppm Au in samples of mull (forest humus layer) taken from gold-bearing areas in U.S.A. (*see* Table 20). They were able to leach small amounts of gold (27–120 parts per trillion, 10^{12}) from the mull with demineralized water and concluded that much of the gold in the leachates was present mainly in the form of extremely small colloidal particles (<0.05 μ diameter) with minor amounts of gold in the ionized state as thiocyanate and cyanide. My research on gold in soil profiles indicates that gold is commonly enriched in the humic layer of the soil (*see* Table 20 and the section on gold in soils in this chapter), and I think that much of the gold in the humic horizon of soils is bound mainly as a humate, perhaps in a chelated or other organometallic configuration.

The organic matter in bogs, muskegs and marshes is enriched in gold in certain areas. The copper bog near Sackville, New Brunswick, which contains up to 4 per cent copper, is also enriched in gold in amounts up to 2.3 ppm and in silver in amounts up to 5 ppm in the dry matter (Boyle, 1977). Bog manganese in peat in some places in New Brunswick carries from 0.01 to 1.4 ppm Au and from 0.8 to 2 ppm Ag in the dry matter (Table 22).

Other types of humic deposits also seem to be high in gold and silver in places. Blanchard (1933) mentioned organic (probably humic) substances with relatively large amounts of manganese near seepages in New Guinea that carry up to 27 920 ppm Au and 15 660 ppm Ag (Table 22). Decaying wood tends to concentrate gold from solution. There are numerous references in the older literature to the occurrence of relatively high amounts of gold in pit props and mine timbers. Babička (1943) also refers to the facility with which decaying wood concentrates gold from solution. Palmer (1935) found marked enrichments of gold and silver in petrified (silicified and carbonized) logs in the Aspen district of Nevada. Samples taken of one log contained 18.26 oz Au/ton and 10.8 oz Ag/ton (624.5 and 369.3 ppm respectively). He concluded that the carbonized wood had precipitated the gold from circulating solutions.

In U.S. Geological Survey, Circular 560, p. 17 (1968) it is reported that 12 samples of peat, muck and driftwood from streams and wet areas near the Front Range mineral belt in Colorado were analyzed for gold on a dry sample basis. Muck from Elk Park and peat from Rudolphs Bog contained 0.3 ppm gold; driftwood from Clear Creek and soft peat from Caribou Peat Bog contained 0.2 ppm. All other samples had less than 0.1 ppm.

Bensusan (1942) records data on an interesting concentration of gold in Lower Burma, which is probably at least partly due to organic agencies. He states:

> Between Tenasserim and Mergui, and impinging on the latter, there exists a very remarkable auriferous deposit in which it is estimated there is upwards of £200,000,000 worth of gold in one concession alone of about 120 square miles. This statement might appear to be exaggerated, but it is a fact, nevertheless. The writer had to investigate this occurrence as well as the deposit at Tenasserim and he was asked to find a means of recovering the gold. Research work was carried out both in Burma and in London with the help of skilled chemists, metallurgists and geologists for nearly a year working on a ton sample sent for the purpose. While hundreds of assays on large samples show without any doubt that the material carries an average of about 4.0 grains of fine gold per ton (0.285 ppm Au), no means could be found for recovering it by any straking or amalgamation method. The gold appears to be in colloidal form as repeated examinations of sample and concentrate under high power microscope failed to show any free gold. In spite of all that was done the tailings from the tests always assayed as much as the heads. No satisfactory theory has yet been put forward as to the origin of the gold nor as to the form in which it exists.
>
> It has been shown that the deposit, which is mangrove swamp mud, could be exploited by means of a suction dredge at a cost of 3d. or 4d. per cubic yard, i.e., as far as excavation and delivery of alluvium to strakes or other appliances used on dredges for alluvial gold recovery are concerned: and since the average value of the mud can be very conservatively taken at 1/6 per yard, there is a fortune waiting for anyone who can devise an economical method of extracting the gold.
>
> A number of boreholes were put down in the mud in various places to depths of about 30 ft. and every sample taken showed gold by fire assay. One borehole was put down to 90 ft. without meeting bed rock and samples roughly taken from the bank of the river always showed values.
>
> In all the straking experiments a small amount of concentrate was recovered which consisted of pyriticised foraminifera, very beautiful microscopic shells of innumerable shapes, but this concentrate yielded only a few grains of gold per ton as well as traces of platinum and iridium.

The manner of occurrence of gold in humic deposits is relatively unknown. In some deposits a small part of the gold appears to be in an adsorbed form, since it can be removed by weak cyanide and aqua regia solutions. The remainder, generally the greater part, is tightly bound in the brown humic complexes. Ong and Swanson (1969) consider that gold is present mainly as a colloid stabilized by a protective layer of organic molecules around the gold in an organic environment (*see also* the section on natural waters in this chapter). Shabynin (1970) observed experimentally that gold is absorbed by peat in about the same amounts as by quartz. He concluded that organic gold compounds are formed, part of which remain in the peat and part of which are soluble and render the gold mobile.

Gold in coal

Gold is a frequent microconstituent in a number of coal seams, but is generally present in amounts less than 0.05 ppm in the ash (Table 25). Certain coal seams, however, appear to be greatly enriched in gold and silver. Lock (1882) and Maclaren (1908) mention the presence of gold in the pyrite of coal, in coaly matter or in sandstones and conglomerates of coal measures in Borneo; Gippsland, Victoria; Tallawang, New South Wales; New Zealand; Tasmania; Deadwood, South Dakota; and the Witwatersrand. The last is probably not coal but thucholite which carries abundant gold in the Rand (*see* the section on the Witwatersrand in Chapter III). Rose and Newman (1937) mention the presence of a rich deposit of gold in lignite in Japan but give no details. There is also a reference to gold in coal in the Batrina Mine in Verespatak, Transylvania in the old literature but again no details. In a summary on gold in coal Lincoln (1911*b*) and Gibson and Selvig (1944) mention the presence of the element in a number of Cretaceous coals in the western United States. The ashes of some of these contain up to 0.5 oz Au/ton. Details are given in the papers by Chance (1900), Jenny (1903), Parsons (1907) and Stone (1912). Jones (1969) mentions, however, that more recent analyses have not confirmed these high figures. Thus, it is reported in U.S. Geol. Survey, Circular 560, p. 17 (1968) that some 400 samples of coal and associated rocks from Colorado, Wyoming, Utah and Idaho contain <0.020 ppm Au. Samples from the Cambria coal field near Newcastle, Wyoming reported in the early literature to contain up to 0.5 oz Au/ton (17 ppm) were found to contain less than 0.1 ppm Au. Some western coals, however, were found to contain small amounts of gold, particularly in coal from Kemmerer, Wyoming and Wales, Utah. No values are given for these coals since the analyses were nonreproducible, suggesting that the gold may be particulate. In U.S. Geol. Survey, Circular 622, p. 10 (1969) it is further stated by J.D. Vine that 100 samples of coal ash from U.S. Appalachian and interior coal fields all showed less than 0.05 ppm Au.

In Canada, Tyrrell (1915) mentions the presence of traces of gold in the coal of the Cretaceous Edmonton Series on the North Saskatchewan River.

At the Tyozin Mine in Japan Tertiary coal beds intercalated with sandstones are exceptionally rich in gold (Watanabe, *in* Newhouse 1942). Similarly in the Coromandel Peninsula, New Zealand, thin coal seams are rich in gold where the auriferous Tertiary veins cross them. In these cases, the gold is apparently hydrothermal in origin.

The gold content of the Carboniferous coals of the Maritime Provinces of Canada is relatively low. The coalified plant fragments in the Carboniferous sandstones are greatly enriched in silver in places, but as a general rule the gold content rarely exceeds 0.08 ppm Au, despite the presence of abundant pyrite, chalcocite and other sulphides.

There seem to be no detailed investigations on the gold content of the different ranks of coal or of the different components, viz. fusain, vitrain, clarain and durain. Analyses for gold of specialized types of coal such as torbanite, cannel, boghead, jet and natural coke are lacking as are also those for the various resinous substances such as amber associated with

Table 25. Gold content of coal

Locality	Remarks	Au (range)	Ag (range)	Au/Ag ratio	Reference
		(ppm)			
Germany and England	In ash	0.2–1.0	0.5–10	~0.1	1
U.S.A.; Wyoming and South Dakota	Ashes of Cambria coal	17.1 (0.5 oz/ton)	–	–	2
	Sandstone roof of Cambria coal bed (pyritiferous)	0.1	–	–	2
	Coke made from Cambria coal	4.2 (av.)	14.7 (av.)	0.28	2
Western U.S.A.	Coal	<0.020	–	–	3
Eastern U.S.A.	In ash	<0.05	–	–	4
Yugoslavia; Timok Basin	In ash of lignite, bituminous and semianthracite coals	0.117–0.573	2.56–6.07	–	5
Canada; Gaspereau River area, Cape Breton Island, N.S.	In ash; Pennsylvanian coal	<0.01	0.78	<0.01	6
Canada; Harbour seam, Sydney, N.S.	In coal; Pennsylvanian coal	<0.001	1.2	<0.0008	6
Canada; Stonehaven, N.B.	In the ash; Pennsylvanian coal	<0.01–0.015	2.3	~0.004	6
Canada; Shubenacadie, N.S.	In the coal; pyritized Cretaceous lignite	<0.005	1.3	<0.004	6
Canada; Walton, N.S.	In the coal; pyritized coal fragments near faults in Mississippian sandstones	0.034	5.47	0.006	6
Canada; Walton, N.S.	Pyrite and other sulphides separated from pyritized coal fragments near faults in Mississippian sandstone	0.6	77.3	0.0008	6
Canada; Port Hood, N.S.	Pyrite; Pennsylvanian coal	0.04	0.17	0.23	6
Canada; Dorchester Mine near Sackville, N.B.	Coal fragments containing chalcocite in Pennsylvanian sandstones	<0.01–0.03	2–190	<0.005	6
Canada; Shippegan Island, N.S.	Coal fragments containing chalcocite in Pennsylvanian sandstones	0.01–0.08	27.5	~0.002	6

References: 1. Goldschmidt and Peters (1933).
2. Stone (1912).
3. U.S. Geological Survey Circular 560 (1968).
4. U.S. Geological Survey Circular 622 (1969*b*).
5. Popovic (1954).
6. Boyle (unpublished).

coal or derived from it. These substances would seem to provide an interesting field of investigation.

The occurrence of gold in coal seams is generally erratic. Judging from the writer's experience with silver (Boyle, 1968*b*) and the limited analyses for gold in coal, the distribution of gold in the seams can be summarized as follows:

1. Higher contents of gold (and silver) are generally recorded from coal seams low in ash and rich in other trace elements such as copper, lead and zinc.

2. Higher contents of gold (and silver) often accompany parts of the coal seams or their overlying or underlying rocks that are enriched in sulphides particularly pyrite. Splits containing abundant pyrite are sometimes greatly enriched in silver and gold as well as in various other chalcophile elements.

3. A parallel enrichment of the following elements has been noted in some coal seams in Nova Scotia: gold, silver, copper, zinc, lead, cobalt, nickel, tin, molybdenum, arsenic, antimony and barium. This relationship is marked in coalified fragments in the Walton area in Nova Scotia.

4. Higher gold and silver contents tend to occur in the coal near the top and the bottom of individual seams. There are also a number of references in the literature to higher contents of silver, gold and other trace elements in the rocks immediately overlying the seams.

5. In some coal seams, and in sandstones containing coal fragments, higher gold and silver contents tend to occur in the coal or in coalified fragments near faults, fractures or crushed zones.

The mode of occurrence of gold in coal and coalified fragments has not been extensively investigated. Some general observations from the literature and the writer's research follow:

1. In some coal seams the gold is present as macroscopic and microscopic forms. Much of this gold is commonly of hydrothermal origin.

2. Some gold is probably associated with the coaly substance of coal and may be bound as an organometallic complex. Bouška *et al.* (1963) concluded that gold is bound principally to the organic part of the coal.

3. A very small amount of the gold in coal may be adsorbed to the coaly substances.

4. Much gold (nature unknown) is present in or associated with pyrite (Tables 11 and 25), and some gold is generally observed in the small amounts of galena, chalcopyrite, arsenopyrite, bornite, sphalerite and marcasite that are found in some coal seams. Popovic (1954) noted that the noble metal content of the coals he studied varied with the sulphur content, especially the chemically bound sulphur – the higher the bound sulphur content the higher the noble metal content.

5. The gold in coalified fragments in most sandstones and shales is either in or associated with pyrite, chalcocite, bornite and galena. Its nature is unknown.

6. In the weathered parts of coal seams gold (and silver) may occur in jarosites and iron oxides; also in the native form.

Gold may be concentrated in coal by one or all of the following processes:

1. Goldschmidt (1933*a*, 1935) originally suggested that gold and other trace elements were first concentrated by the coal plants, and these in turn contributed the elements to the humic substances that are the mother of coal: we have seen in the section on the biogeochemistry of the element that gold is a microconstituent of numerous plants, and hence it seems logical to conclude that some of the gold in coals came directly from the accumulated plant material.

2. Precipitation of gold from surface waters flowing into or through the ancient coal swamps: humus in bogs and swamps tends to adsorb and absorb the available gold from surface waters and thus becomes relatively enriched in the metal. Some gold may also be precipitated by bacteria. During humification and coalification processes this gold is incorporated in the coal.

3. Precipitation of gold from circulating groundwaters that course through the coal seams during and after coalification processes: in this case the source of the gold may be the coal itself, and the element is simply redistributed within the seam. More generally, however, much of the gold probably comes from the enclosing sedimentary rocks. Some investigators maintain, however, that the gold is of hydrothermal origin, especially where the element is precipitated in zones containing coal fragments in sandstones.

In coal seams and zones of coalified fragments gold circulating in aqueous systems is precipitated by adsorption processes, probably by chelation and other organometallic binding processes, and by reduction of the complex gold ions to the metal. Gebler and Stramkovskaya (1957) experimented with the adsorption properties of lignite for Cu, Ag and Au. For copper and silver they found a 4.31 and 5.2 per cent copper and silver ion absorption by weight of lignite. For gold they found no absorption but a strong reduction of the gold in solution to the metal.

The enrichment of gold in most coal seams and in coalified fragments in sandstones seems to be best explained by the circulating water hypothesis.

Gold in organic shales, petroleum and other natural hydrocarbons

Shales and other rocks containing organic matter, bitumen, kerogen and their degradation products tend to be slightly enriched in gold compared with normal shales. The distribution of gold in these rocks is, however, commonly erratic. Commercial recovery of gold and silver from oil shales and tarry residues is suggested in a patent by Petty (1968). For further details see the section on gold in sedimentary rocks.

There are relatively few data on the gold content of petroleum. Colombo *et al.* (1964) and Shah *et al.* (1970) determined the gold content by neutron irradiation methods in oil and asphalt samples from a number of fields in Italy, U.S.A. and Libya. Their values range from <0.1 to 3.012 ppb in the oil (Table 26). In the asphalt the values ranged from 0.1 to 1.0 ppb. Al-Shahristani and Al-Atyia (1973) report the presence of gold in some Iraqi crude oils but give no quantitative data.

Hitchon *et al.* (*in* Yen, 1975) report the presence of gold in four samples from a total of 88 crude oils from the Alberta fields of Canada. The average concentration of gold in these samples was 0.438 ppb; the maximum was 1.32 ppb.

The mode of occurrence of gold in petroleum is unknown. It seems probable, however, that the element is combined as some type of organo-gold complex or as a chelated complex, perhaps a porphyrin, since the element is capable of forming porphyrins easily (Fleischer and Laszlo, 1969). Shah *et al.* (1970) also analyzed the 10 crude samples shown in Table 26 for Sc, Ni, Fe, Co, Zn, Se, Sb, Cr, As, Eu, Hg and U. A study of the correlation coefficients between these elements and gold indicates a significant correlation between Au and Sb and also between Au and Sc, Fe, Co and Cr.

Table 26. Gold content of crude oil and asphalt

Sample number and type	Locality	Gold content (ppb)
Data from Colombo *et al.* (1964)		
Crude oil	Alanno, Italy	<0.1
Crude oil	Chieuti, Italy	<0.1
Crude oil	Cercemaggiore, Italy	<0.1
Crude oil	Avanella, Italy	<0.1
Asphalt	Gela, Italy	0.5±0.2
Asphalt	Noto, Italy	1.0±0.1
Asphalt	Vittoria, Italy	0.2±0.1
Asphalt	Ragusa, Italy	0.2±0.1
Data from Shah *et al.* (1970)		
Crude oil, 010	California, U.S.A.	1.238±0.032
Crude oil, 011	California, U.S.A.	0.723±0.039
Crude oil, 012	California, U.S.A.	1.174±0.06
Crude oil, 013	California, U.S.A.	1.911±0.34
Crude oil, 014	California, U.S.A.	3.012±0.40
Crude oil, 015	Libya	0.666±0.032
Crude oil, 016	Libya	1.365±0.076
Crude oil, 017	Libya	1.242±0.261
Crude oil, 018	Louisiana, U.S.A.	0.668±0.06
Crude oil, 019	Wyoming, U.S.A.	1.489±0.051

The gold in petroleum may be derived from the following sources.

1. From the original plants and animals whose remains gave rise to the petroleum: in the present section on biogeochemistry we have reported that many marine plants and animals concentrate gold, and it seems logical to suppose that some gold in petroleum came from this source.

2. By adsorption and absorption processes from sedimentary rocks: petroleum migrating through sedimentary rocks on its way to reservoirs probably adsorbs and absorbs some gold from these rocks.

3. By contact with connate and other waters: these waters probably carry small amounts of gold, which may be absorbed by petroleum when the two solutions come into contact.

There are relatively few data on the gold content of the natural bitumens and asphalts and the various solid hydrocarbons such as gilsonite, ozokerite, grahamite, anthraxolite, thucholite, etc. Thucholite is commonly greatly enriched in gold in the Witwatersrand quartz-pebble conglomerates of South Africa (Liebenberg, 1955) (*see also* Chapter III). The thucholite in the skarn-type gold deposits of the Richardson Mine near Madoc, Ontario is also phenomenally enriched in gold. Other solid hydrocarbons and bituminous materials analyzed by the writer vary widely in their gold content as shown in Table 27. With respect to the possible migration of gold as a complex hydrocarbon species see the section on the origin of the Witwatersrand in Chapter III.

Table 27. Gold content of bituminous materials and solid hydrocarbons

Description	Location	Au (ppm)	Ag (ppm)	Au/Ag ratio	Remarks
Thucholite*	Richardson Mine, Eldorado, Ont.	>1000	>2	~500*	In the solid hydrocarbon
Thucholite	Witwatersrand	>1000	>10	12	In the solid hydrocarbon (*see also* the discussion on the Rand in Chapter III)
Albertite	Albert County, N.B.	<0.005	<0.3	~0.01	In the ash of the solid hydrocarbon
Vanadium-bearing hydrocarbon-shale	Quadra Island, B.C.	0.25	47	0.005	In the solid hydrocarbon and shale
Antraxolite	Sudbury, Ont.	0.03	2.5	0.012	In the ash of the solid hydrocarbon

*By electron probe the ratio Au/Ag of the visible native gold is 99.8/0.2. *See* the papers by Boyle and Steacy (1973), Steacy *et al.* (1973) and Steacy *et al.* (1974*a,b*).

Gold in the atmosphere

Relatively little data are available on the gold content of the atmosphere. The element is mentioned as being present in rainwater in the old literature, but no quantitative data are given. Peirson *et al.* (1973) have measured the gold content of rainwater and air by neutron activation at various sites in the United Kingdom. They give values of 0.01 μg/L for rainwater and an average of less than 0.003 ng/kg for air.

Nearly all of the gold in the atmosphere is present in particulates; some may be present as volatile methylgold as Wood (1974) has suggested, or as some other volatile gold-organic compound arising from microbiological activity. The source of the particulates of which gold is either an integral constituent or present in an adsorbed form is probably multiple and includes: terrestrial aerosols raised from the surface of the land and ocean by winds; volcanic aerosols; industrial pollutant aerosols; and extraterrestrial particles (meteoritic dust) from sources unknown. There are no data on the contribution, which each of these makes to the gold content of the atmosphere.

Gold is a constituent of certain terrestrial gases and aerosols, especially in volcanic areas. These are discussed in the section on natural waters.

Chapter III. Gold Deposits

Introduction

The primary gold content of igneous bodies of rock is low (0.004 ppm), and bodies with economic concentrations of the element are unknown except where later shearing and fracturing has taken place permitting the development of veins, stockworks and replacement lodes. In carbonatites gold is found only in traces, but some of these bodies contain higher than normal amounts of silver. My analyses of the various rocks in carbonatite complexes show gold to be generally present in the range 0.005 ppm; the silver contents cluster around 3 ppm. Most of the gold is in pyrite, which may contain up to 0.075 ppm Au. Some pegmatites contain enrichments of gold, but few of these bodies are economic sources of the element. Certain skarn deposits contain economic concentrations of gold, and some have been mined solely as auriferous deposits; from others considerable amounts of gold are also won as a byproduct of copper, lead or zinc mining and smelting.

In hypogene veins, lodes, stockworks, mantos and other types of similar deposits, gold is frequently strongly enriched, and these deposits are one of the main sources of gold production in the world. It is of interest to note, however, that veins greatly enriched in silver, especially the Ni-Co arsenide–native silver type are notably low in gold. Analyses by the writer of such veins from Cobalt and elsewhere in Canada show gold to be present in amounts clustering around 0.05 ppm in the ores. Disseminated gold deposits, many with a bedded (stratabound) aspect, developed in limestones, limy shales and schists, iron formations and other types of sedimentary rocks have been exploited in recent years.

The most productive gold deposits are those found in quartz-pebble conglomerates. Some of these deposits also yield uranium, rare-earths and small amounts of platinum metals.

Gold has been won extensively through the centuries from ancient and recent placers, both eluvial and alluvial. Unusual types of placers developed in karst terranes produce gold in some countries.

Finally, gold is won as a byproduct of many other types of metalliferous deposits including massive and disseminated Ni-Cu sulphide ores associated with basic rocks, massive Pb-Zn-Cu sulphide ores in volcanic and sedimentary terranes, various types of polymetallic deposits and the so-called porphyry Cu-Mo deposits.

This chapter deals with the classification of gold deposits, an outline of the metallogenic epochs of gold mineralization, the elements associated with gold in its deposits, the common minerals associated with gold in its deposits, a discussion of the Au/Ag ratio in deposits, an outline of wall-rock alteration processes in epigenetic gold deposits, descriptions of typical gold deposits including placers and ends with a section on the origin of epigenetic gold deposits and the controversial auriferous quartz-pebble conglomerates.

Classification of gold deposits

The following summary of gold deposits differs somewhat from the commonly accepted classifications given in most textbooks on economic geology. This departure may or may not be justified, but it is the writer's opinion that a classification based on the geological and geochemical setting of gold deposits is much more valuable than any based on their origin be it magmatic, hydrothermal, sedimentary or otherwise. It seems best to state objectively in what types of rocks and structural settings the deposits occur. With this clearly set down and a knowledge of the geochemistry of the deposits, it may be possible to discern their origin and, more importantly, to predict the environment where one may prospect for similar deposits. In the last 50 years many investigators have advocated a magmatic hydrothermal origin for all gold deposits – an extreme view that is fundamentally erroneous, because it fails to consider many of the geochemical features of gold, especially the behaviour of the element in the cycle of weathering and the part played by metamorphic processes in concentrating the element.

The classification suggested by the writer considers gold deposits under nine main categories, as follows:

1. Auriferous porphyry dykes, sills and stocks; auriferous coarse-grained granitic bodies, aplites and pegmatites.

2. Auriferous skarn-type deposits.

3. Gold-silver and silver-gold veins, stockworks, lodes, mineralized pipes and irregular silicified bodies in fractures, faults, shear zones, sheeted zones and breccia zones essentially in volcanic terranes.

4. Auriferous veins, lodes, sheeted zones and saddle reefs in faults, fractures, bedding-plane discontinuities and shears, drag folds, crushed zones and openings on anticlines essentially in sedimentary terranes; also replacement tabular and irregular bodies developed near faults and fractures in chemically favourable beds.

5. Gold-silver and silver-gold veins, lodes, stockworks, silicified zones, etc. in a complex geological environment, comprising sediments, volcanics and various igneous intrusive and granitized rocks.

6. Disseminated and stockwork gold-silver deposits in igneous, intrusive, volcanic and sedimentary rocks. (a) Disseminated and stockwork gold-silver deposits in various igneous intrusive bodies. (b) Disseminated gold-silver occurrences in volcanic flows and associated volcaniclastic rocks. (c) Disseminated gold-silver deposits in volcaniclastic and sedimentary beds: Deposits in tuffaceous rocks and iron-formations and deposits in chemically favourable sedimentary beds.

7. Gold deposits in quartz-pebble conglomerates and quartzites.

8. Placers. (a) Eluvial, including those developed in karst terranes. (b) Alluvial. (c) Fossil eluvial and alluvial.

9. Miscellaneous sources of gold.

L'or est de tous les âges
—Proust, 1920

Metallogenetic epochs and provinces of gold mineralization

Lindgren (1933) defines a 'metallogenetic' epoch as the time interval favourable for the deposition of certain useful substances and a 'metallogenetic' province as an area in which mineralization has been active at one or more periods.

The literature of economic geology is replete with discussions of the metallogenic epochs of gold mineralization, and numerous world maps have been published outlining metallogenic gold provinces. The auriferous metallogenic epochs are summarized in Table 28 and the geological distribution of gold deposits are generalized in Figures 3, 5, 6 and 7. It should be emphasized that both the table and the figures are no more than general guides.

The following books, papers and maps have greatly assisted the writer in compiling the data on the metallogenesis of gold: Curle (1905), Maclaren (1908), De Launay (1913), Lindgren (1909, 1917, 1933), Redmayne (1922), Miller and Singewald (1919), Dunn (1929), International Geological Congress (1930), Kittl (1931–1932), Finch (1933), Bruce (1933), Chibber (1934), Emmons (1937), Cooke (1946), Juan (1946), Hills (1947), Gmelin (1950–1954), Bateman (1950), Bilibin (1947, 1950, 1955), Ho (1953), Freidensburg (1953), Edwards (1953), Turneaure (1955), Jenks (1956), Brinck (1956), Smirnov (1959), Burnham (1959), Petrovskaya (1960), Radkevich (1961), Bishopp (1962), Routhier (1963), Roy (1963), Ch'ang ta (1963), Rankama (1963–1970), Pelletier (1964), Laffitte and Rouveyrol (1964), McAndrew (1965), McLeod (1965), De Kun (1965), Campbell (1965), Sher (1966, 1972), Musylev (1966), Ogura (1967, 1969), Il'jin (1968, 1974), Rozhkov (1968*a,b*), Laffitte (1968–1970), Ridge (1968, 1972), Ramovic (1968), Koschmann and Bergendahl (1968), King (1969), Ianovici *et al.* (1969), Gümüs (1970), Petersen (1970), Folinsbee (1971), Kim (1971), Guild (1972), Magakian (1972), Galkiewicz (1972), Shilo (1971, 1972), Warren (1972), Goossens (1972), Abreu (1973), Hughes (1973), Tsypukov and Lombo (1973), Takashima and Igarashi (1973), Fisher (1973), Cordani *et al.* (1973), Konstantinov (1974), Itsikson (1974), Choubert (1974), Smirnov (1974), Williams (1974), Anhaeusser (1976*a,b*) and Knight (1975); also the various geological maps and reports published by geological societies and surveys throughout the world.

The geological history of the earth has been marked by a number of epochs of epigenetic gold mineralization, the greatest of which took place in Precambrian time. Younger epochs occurred in Paleozoic, Mesozoic and Cenozoic times. Gold placers are known in sedimentary rocks of practically all ages, particularly in those of Tertiary and Quaternary age. If it is accepted that the Witwatersrand deposits are placers, or modified placers, the greatest epoch of formation of gold placers occurred in Precambrian time.

Precambrian

Gold deposition took place extensively during Precambrian time (Fig. 3), the typical auriferous deposits formed being mainly gold-quartz veins and silicified zones and modified placer quartz-pebble conglomerates. Gold was also deposited during this time in skarn deposits, polymetallic lodes, massive sulphide bodies, disseminated and massive Ni-Cu sulphide deposits, and other types of metalliferous lodes. Most of the typical auriferous quartz deposits occur in volcanic rocks and their associated sediments, mainly greywacke and slate; the modified placers are in pyritiferous or hematitic quartz-pebble conglomerates and associated quartzites.

In North America most of the Precambrian auriferous deposits occur in the Canadian Shield and in the Black Hills of South Dakota. Other deposits, probably of Precambrian age, are known in the Appalachian fold belt of the United States and Canada, and in folded and metamorphosed rocks in Wyoming, New Mexico and Arizona. In all of these widely separated areas the principal deposits belong to the gold-quartz and disseminated types, developed mainly in volcanic and sedimentary terranes. Precambrian skarn-type deposits are of minor importance as gold producers. Considerable quantites of gold are won from a variety of polymetallic lodes, massive Cu-Zn sulphide bodies, Ni-Cu ores associated with basic rocks and other types of metalliferous bodies. The known pyritiferous quartz-pebble conglomerates in the Precambrian (Lower Proterozoic; Aphebian) of North America carry only minor amounts of gold and silver. Because of extensive glaciation only small and economically insignificant gold placers occur in the Canadian Shield; in the United States placers developed from Precambrian gold deposits have been of some importance. More details on the metallogenic epochs of gold mineralization in the Canadian Shield are discussed below.

The Early, Middle and Late Precambrian formations of South America, occurring mainly in the Brazilian and Guyanian shields, contain numerous gold deposits principally in the Brazilian states of Bahia and Minas Geraes, in Uruguay, and in the Guyanas and Venezuela. Most of the primary auriferous deposits are of the gold-quartz type or of the pyrite impregnation types as in Surinam (Brinck, 1956) and Guyana (Fernandes, 1961). In Venezuela the auriferous deposits are of three types (Pasquali and Bisque, 1975) – massive quartz veins with native gold; quartz veins containing abundant pyrite; and zones of gold-bearing pyrite in sheared lavas, without quartz. There are also auriferous quartz-pebble conglomerates in the Serra de Jacobina in Brazil. In the Cordillera of South America numerous areas are underlain by undifferentiated Precambrian rocks that are apparently difficult to separate from those of Paleozoic age. A number of gold deposits occur in these rocks, but their age is uncertain. Most investigators consider them to be Paleozoic or younger, and hence they are discussed in the sections that follow. Placers developed from the Precambrian deposits are or were important in the Guyanas, Venezuela and Brazil.

Precambrian gold deposits, ranging in age from Archean to Proterozoic, occur in many parts of Africa. In fact, the oldest of all gold mines, those in Egypt (Nubia), seem to have been worked more than 4000 years ago. From Egypt and Sudan, east of the Nile, great belts of Precambrian rocks

Table 28. Characteristics of the principal auriferous metallogenic epochs of the world

Era	Precambrian (3000–570)*	Paleozoic (570–225)*	Mesozoic (225–65)*	Cenozoic (65–0)*
Age of host rocks	Gold-quartz deposits mainly in Archean rocks (>2500 m.y.); extensive deposits in Proterozoic quartz-pebble conglomerates; some large gold-quartz and varied stockwork deposits in Proterozoic rocks	Mainly Paleozoic; some deposits in Precambrian or both Paleozoic and Precambrian rocks	Mainly Mesozoic but also Paleozoic and Precambrian	Mainly Tertiary but also all older rocks
Principal types of host rocks	Gold-quartz deposits mainly in meta-volcanics, iron – formation and meta-sediments of greywacke-slate type; also in associated quartz-feldspar porphyries and other granitic rocks. Largest deposits in quartz-pebble conglomerates and associated quartzites (modified placers)	Gold-quartz and skarn-type deposits mainly in metasediments, especially in greywacke-slate sequences; similar deposits in meta-volcanics, in porphyries and in granitic rocks	Gold-quartz and skarn-type deposits in meta-sediments and meta-volcanics	Andesites, dacites, tuffs, rhyolites and a great variety of meta-volcanic and meta-sedimentary rocks
Principal types of deposits	Modified placer (?) deposits in quartz-pebble conglomerates; gold-quartz veins, lenses, saddles, silicified zones, etc.; skarn types; auriferous polymetallic veins, lodes and massive sulphide bodies	Gold-quartz veins, lodes, lenses, silicified zones; saddle reefs and bedding-plane veins are characteristic in many areas. Skarn and hornfels types. Auriferous polymetallic veins, lodes and massive sulphide bodies	Gold-quartz veins, lodes, lenses, silicified zones; skarn-type deposits; auriferous polymetallic veins, lodes and massive sulphide bodies	Gold-quartz veins, lodes, lenses and silicified zones; auriferous polymetallic veins, lodes and massive sulphide bodies
Principal gangue minerals	Quartz-pebble conglomerates – pyrite, quartz, hematite, sericite. Gold-quartz type – quartz, pyrite, arsenopyrite, pyrrhotite, carbonates	Quartz, pyrite, arsenopyrite and carbonates	Quartz, pyrite, arsenopyrite and carbonates	Quartz, pyrite, arsenopyrite carbonates, adularia, alunite, fluorite, barite; rhodochrosite and rhodonite are characteristic
Principal metallic minerals	Quartz-pebble conglomerates – pyrite, hematite. Gold-quartz type – pyrite, pyrrhotite, arsenopyrite, galena, sphalerite, chalcopyrite, stibnite and sulphosalts	Pyrite, arsenopyrite, pyrrhotite, chalcopyrite, galena, sphalerite, stibnite and sulphosalts	Pyrite, arsenopyrite, pyrrhotite, chalcopyrite, galena, sphalerite, stibnite and sulphosalts	Pyrite, arsenopyrite, chalcopyrite, galena, sphalerite, argentite, sulphosalts, stibnite and silver selenides
Nature of the gold	Quartz-pebble conglomerates – free gold, average size 80 microns. Gold-quartz type – free gold and as a lattice or submicroscopic constituent of pyrite, arsenopyrite, sulphosalts, etc; in tellurides	Generally free; in some deposits much of the gold is present as a lattice or submicroscopic constituent of pyrite, arsenopyrite, sulphosalts, etc; in tellurides	Generally free; in some deposits much of the gold is present as a lattice or submicroscopic constituent of pyrite, arsenopyrite, sulphosalts, etc; in tellurides	Generally free; in some deposits much of the gold is present as a lattice constituent or submicroscope constituent of pyrite and various sulphosalts; in tellurides
Characteristic associated elements†	K, Na, Cu, Ag, Zn, Cd, B, Th, U, rare earths, SiO_2, Pb, As, Sb, Bi, Te, (Se), Mo, W, Cr, Mn, Fe, (Co), (Ni), (Pt metals). U, Th, and rare earths are characteristic of quartz-pebble conglomerates	K, Na, (Sr), (Ba), (Mo), W, (Co), (Ni), Cu, Ag, Zn, Cd, (Hg), B, (Sn), Pb, As, Sb, Bi, (Se), Te, (F)	K, Na, (Sr), (Ba), (Mo), W, (Co), (Ni), Cu, Ag, Zn, Cd, (Hg), B, (Sn), Pb, As, Sb, Bi, (Se), (Te), (U), (F)	K, (Sr), Ba, (U), Mo, W, Mn, (Co), (Ni), Cu, Ag, Zn, Cd, Hg, (B), T1, (Sn), Pb, As, Sb, Bi, Se, Te, F
Characteristic wall-rock alteration processes	Pyritization, arsenopyritization, carbonatization, sericitization, biotitization, tourmalinization, chloritization, silicification, albitization and development of skarn minerals. Development of pyrite, sericite, pyrophyllite in quartz-pebble conglomerates	Pyritization, arsenopyritization, carbonatization, sericitization, tourmalinization, chloritization, silicification, listvenitization, beresitization, albitization and development of skarn and hornfels	Pyritization, arsenopyritization, carbonatization, sericitization, tourmalinization, chloritization, silicification, albitization, listvenitization, beresitization, (alunitization), development of adularia; development of skarn and hornfels	Pyritization, carbonatization, sericitization, chloritization, silicification, alunitization; propylitization is widespread in most auriferous areas.

*Approximate duration of the geological eras (in millions of years).

†Excluding the common and nearly universal associates such as Si, Fe, S, etc.; elements in brackets are rare or localized.

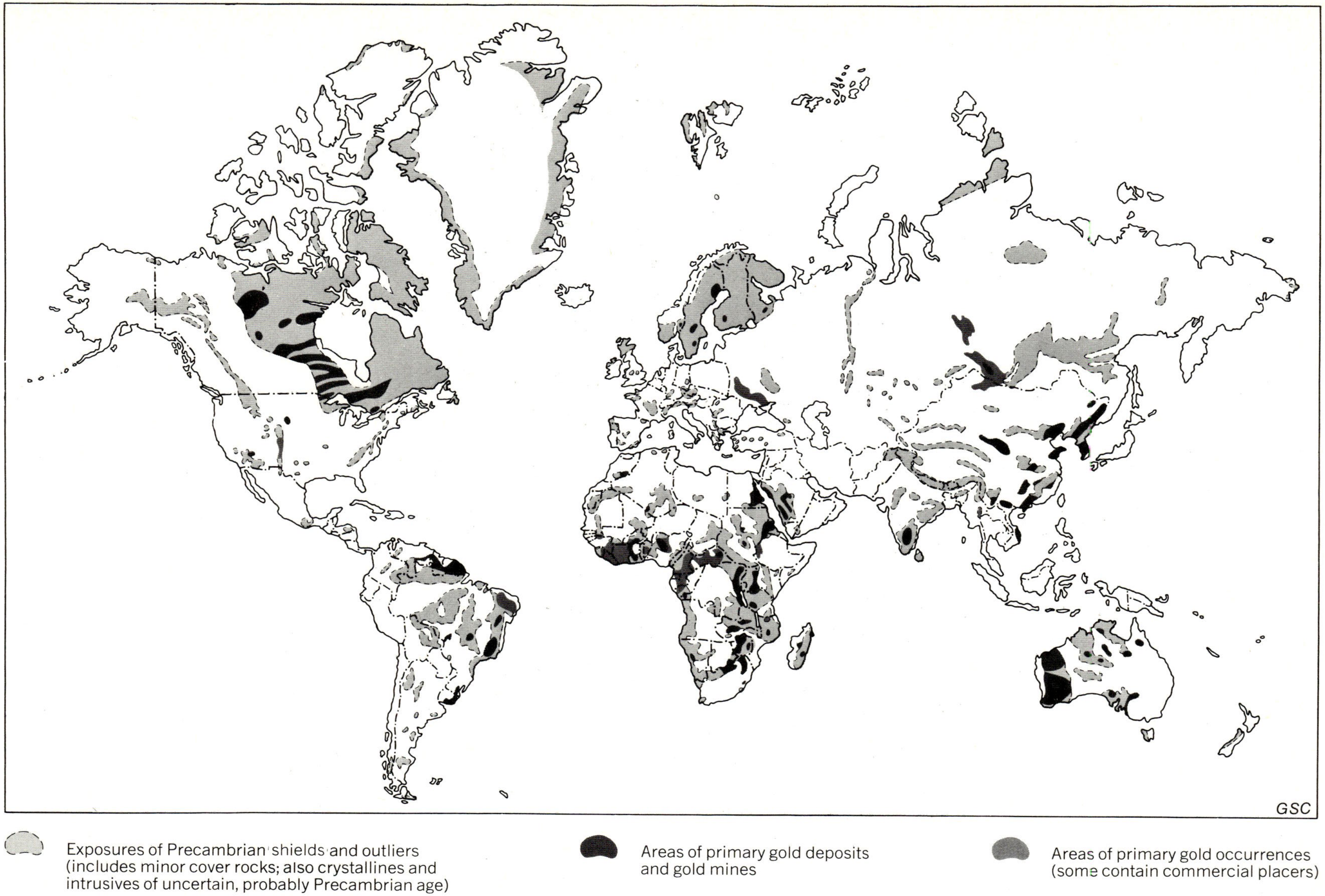

Figure 3. World distribution of primary gold deposits and occurrences of Precambrian age.

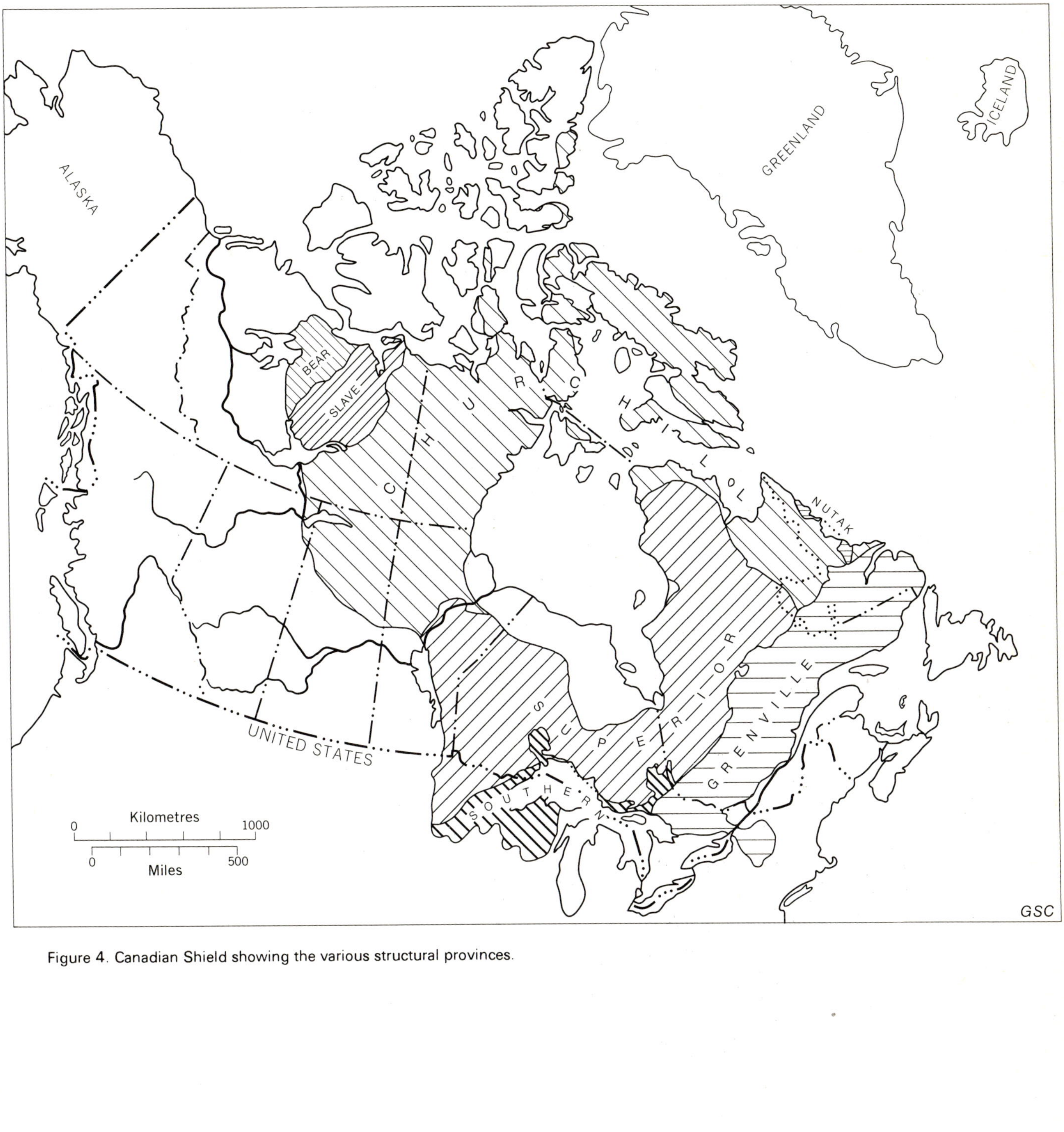

Figure 4. Canadian Shield showing the various structural provinces.

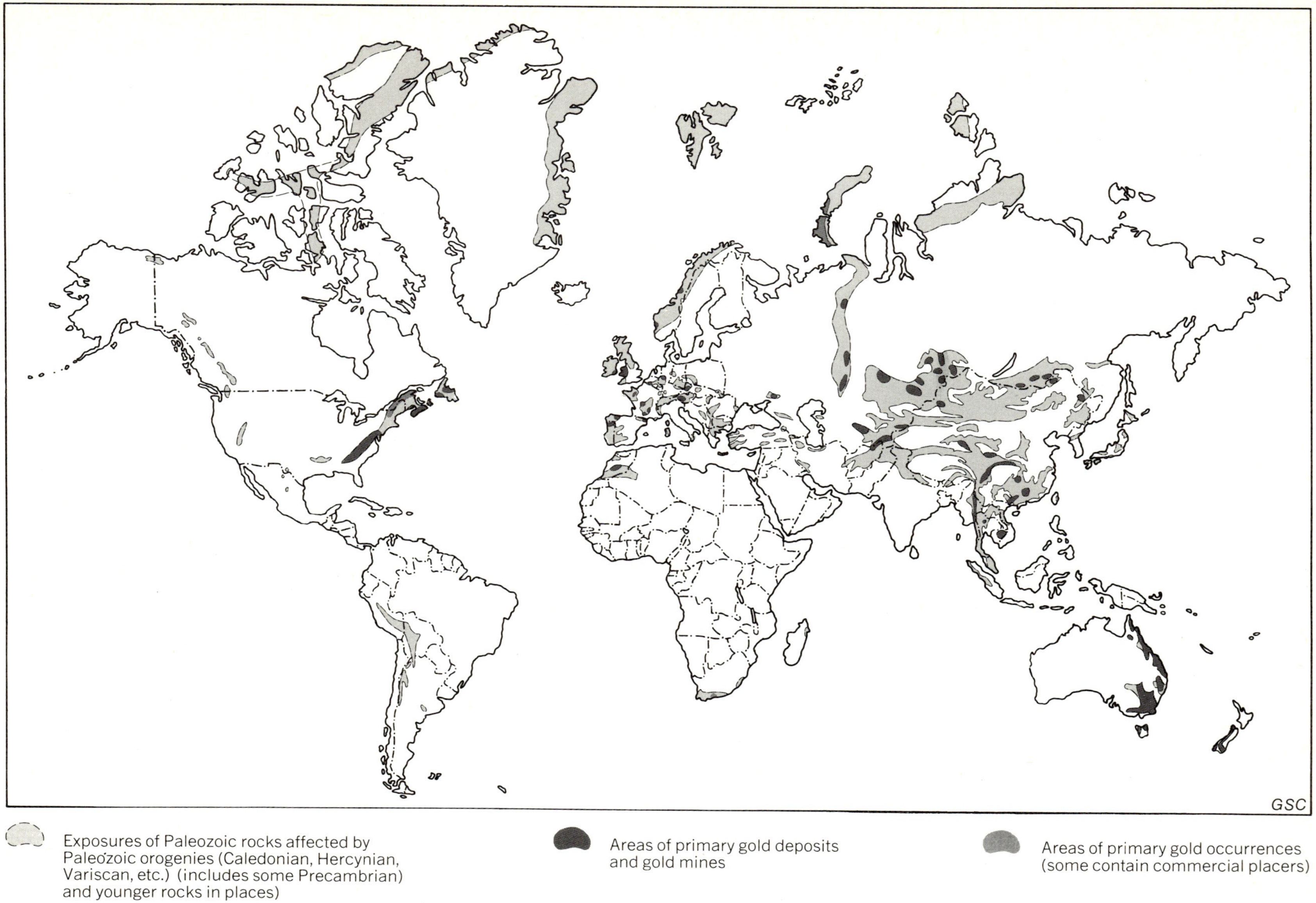

Figure 5. World distribution of primary gold deposits and occurrences of Paleozoic age.

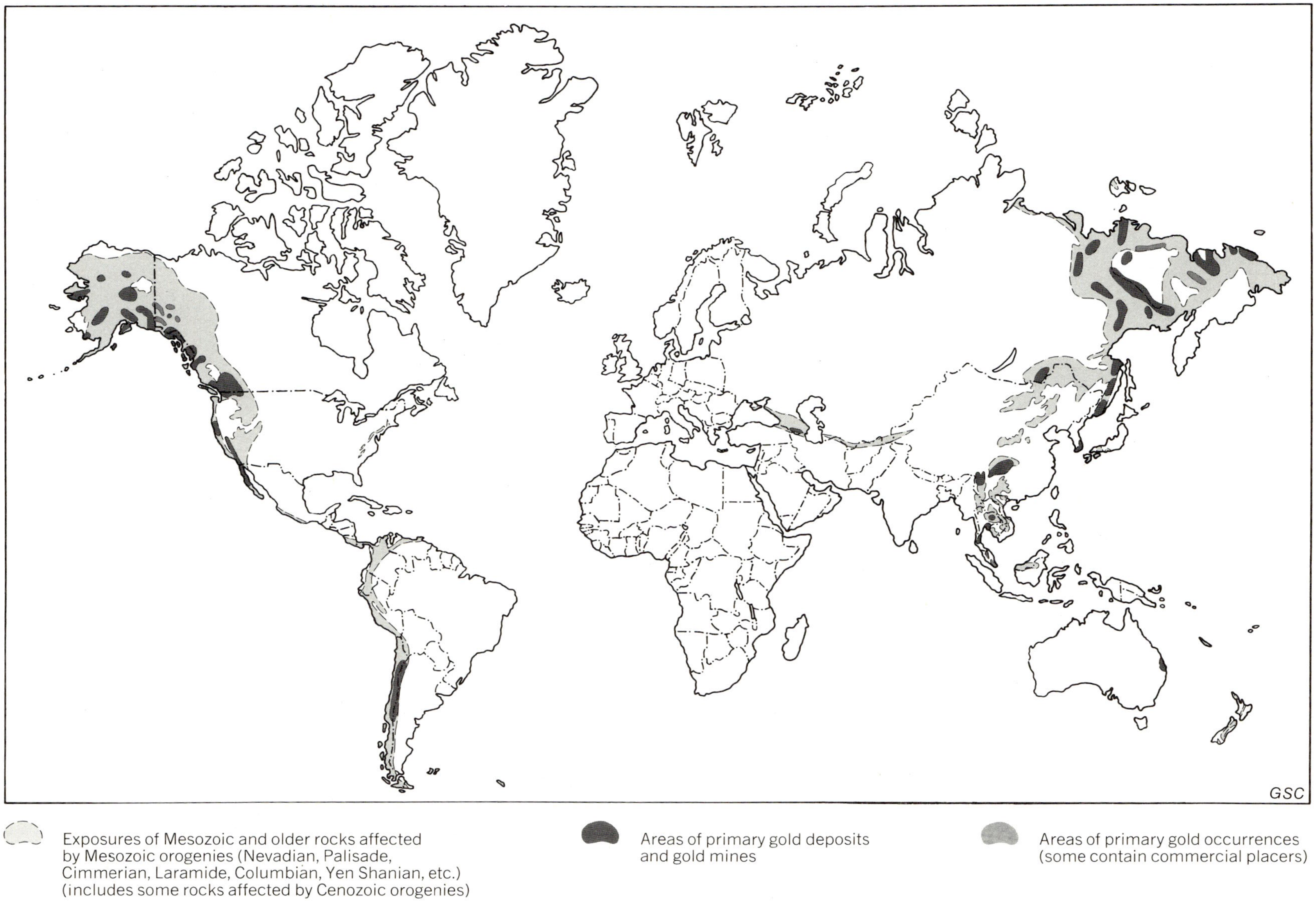

Figure 6. World distribution of primary gold deposits and occurrences of Mesozoic age.

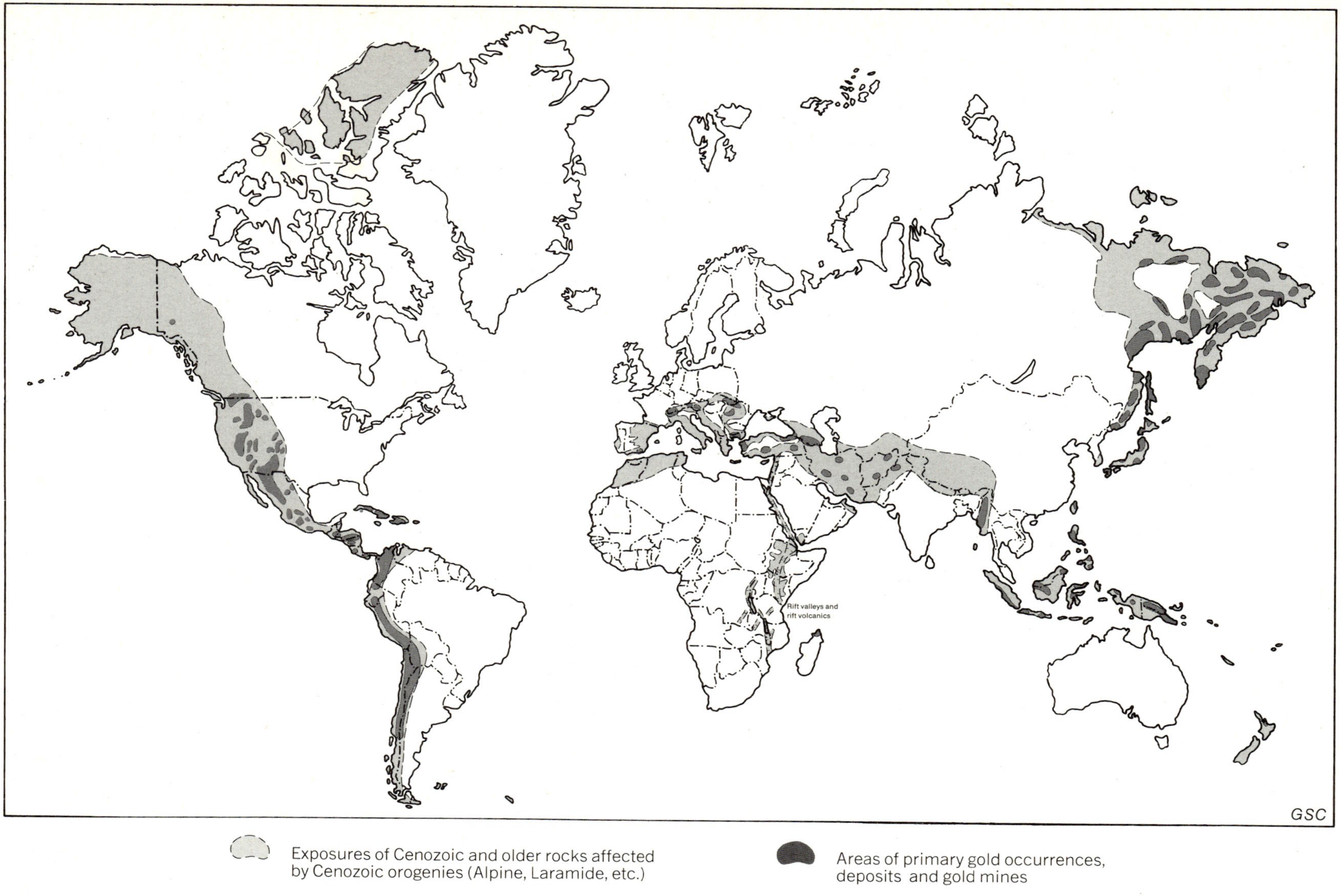

Figure 7. World distribution of primary gold deposits and occurrences of Cenozoic age (may include some deposits older than Tertiary).

stretch longitudinally to the Cape Province of South Africa and transversely from the Gulf of Aden almost to the western extremity of the continent. Precambrian rocks also underlie most of the Malagasy Republic (Madagascar Island) and occur in a narrow belt adjoining the west coast of the African continent from the Gulf of Guinea southward. Throughout these rocks there are a vast number of deposits of the gold-quartz type in veins and silicified zones developed in volcanic, sedimentary and granitic terranes. Many of these are small, but a number of deposits have yielded large amounts of gold. Disseminated types in banded iron-formations and polymetallic lodes of Precambrian age also yield some gold. Of special interest, principally in South Africa, are the Precambrian pyritiferous quartz-pebble conglomerates (Witwatersrand system), which contain the world's largest known concentrations of gold. Similar types of deposits, but with hematite instead of pyrite in the matrix, occur in quartz-pebble conglomerate in Ghana (Tarkwa field). Placers, developed mainly from the Precambrian gold-quartz veins and lodes, are important in many countries in Africa, particularly in Mauritania, Guinea, Senegal, Mali, Sierra Leone, Liberia, Ivory Coast, Upper Volta, Ghana, Togo, Dahomey, Nigeria, Cameroun, Angola, Uganda, Zaire, Malawi, Mozambique, Ethiopia and Malagasy.

In Europe, Precambrian rocks are exposed mainly in the Baltic (Fennoscandian) Shield, which underlies much of Norway, most of Sweden, Finland and northwestern U.S.S.R. west of the White Sea. Similar Precambrian rocks occur in the core of the Urals and underlie the Orel-Kursk region and a belt in the Ukraine, extending from a point east of Rovno to the sea of Azov (Ukranian Shield). Other areas of Precambrian rocks occur in the Hebrides, in northern Scotland and Ireland and in scattered areas in Wales and England (Anglesey, Malvern Hills, Charnwood Forest); also in some of the massifs of western Europe, viz. in France (Armorican), in Czechoslovakia (Bohemian) and in central Germany (Schwarz-wald). The Precambrian of Europe, especially in the Baltic Shield, is marked by four cycles of sedimentation separated by periods of granitic formation. The most intense and widespread tectonic and metamorphic events occur in the third or Karelidic cycle that ended about 1800 million years ago. The youngest Precambrian rocks (Middle and Late Proterozoic) are relatively fresh and are referred to the Riphean in U.S.S.R; in Norway these rocks are Eocambrian. Few important gold deposits of Precambrian age occur in the Precambrian rocks of Europe, although representatives of the skarn and pyrite-arsenopyrite-gold-quartz types are known in the Baltic Shield and have yielded some gold principally in the Adelfors and Skelleftea districts of Sweden and the Tampere district of western Finland (Haveri Mine). There are also gold-bearing deposits in eastern Karelia, U.S.S.R. (Voitovich and Khenkina, 1971). There, the Voitsk and Shuezersk deposits are mainly veins and lodes in strike-slip fault zones cutting phyllites, quartzites and graphitic schists, intruded by post-middle Proterozoic gabbro, diabase, diorite porphyries, granite porphyries and granosyenites. The principal ore minerals are chalcopyrite, pyrite, bornite, scheelite, molybdenite, galena, native gold, native copper, malachite and azurite, the last three probably of supergene origin. In the Kola Peninsula the productive gold deposits are mainly quartz veins and quartz-sulphide lodes (Lobanov, *in* Petroskaya, 1974). Some of the deposits in the Bohemian massif have also been mined extensively at Jilové, Roudny and elsewhere. Most of these deposits were auriferous quartz veins, some with relatively large amounts of antimony (stibnite), which was also won as a byproduct. Although found mainly in Precambrian rocks (gneisses, schists, phyllites and greenstones), these deposits are of Hercynian age (Morávek, 1971). Most of the gold won from the Precambrian of Europe has come from ores mined principally for their copper, zinc and lead content. Examples of these deposits are the massive copper sulphide deposits of Falun and Boliden in Sweden. The latter contains massive concentrations of pyrite, chalcopyrite, pyrrhotite, arsenopyrite, boulangerite, jamesonite, and bismuth- and selenium-bearing minerals. Some of the ores averaged over 0.5 oz Au/ton and 1.5 oz Ag/ton. Because of extensive glaciation only small placers of relatively little economic value occur in the Precambrian terranes of Europe.

In Asia large areas are underlain by Precambrian rocks in the Arabian Peninsula, U.S.S.R. (Siberia), Mongolia, China, Korea and India. These rocks are generally subdivided into two great groups of Archean and Proterozoic age. The Precambrian rocks in Siberia and the Asian part of U.S.S.R. underlie part of the Pamirs; an area southeast of Kyzyl (Sangilen massif); areas west, northwest and northeast of Irkutsk (Baikalides); west of the Sea of Okhotsh (Aldan Shield); north of Vladivostok (Khanka massif); west of Birobidzhan (Bureya massif); southeast of Khatanga (Anbar Shield); part of the Cherskiy mountain range (Kolyma massif); and other scattered areas east to Bering Strait. A review of the Precambrian of U.S.S.R. is given by Salop (1964). In Mongolia, Precambrian rocks occur in the western part of the country and in scattered areas elsewhere. North Korea and parts of South Korea are underlain by Precambrian schists of sedimentary origin intruded by a variety of granitic rocks. In China, most of the Precambrian rocks occur in Manchuria and in a number of belts in the central, western and southern part of the country. Three great systems are differentiated: T'ai Shan (Archean?), Wu-t'ai and Sinian (Proterozoic?). The first consists of highly metamorphosed sedimentary and igneous rocks, mostly gneiss and schist, and the second comprises less metamorphosed sediments. The Sinian consists mainly of sediments, including arkosic clastics below grading up into thick cherty limestones. The Sinian also includes sedimentary iron ore, tillites and local lava flows. These are unmetamorphosed except where involved in much later mountain-forming disturbances and generally resemble the overlying Paleozoic formations. In India a great belt of Precambrian rocks underlies the central, eastern and southern part of the peninsula. Similar rocks underlie much of Sri Lanka (Ceylon). There are no definite Precambrian rocks in Japan, although certain groups of metamorphics (Hida series) may have a Precambrian component.

Numerous gold deposits occur in the Precambrian of Asia. Most belong to the gold-quartz type developed in ancient (Archean) volcanic and sedimentary terranes. Typical examples of these are the famous deposits of Mysore, India (Kolar and other fields) in metavolcanics and associated sediments. Numerous ancient gold mines were worked in metavolcanic and metasedimentary rocks in Saudi Arabia as far back as King Solomon's time. Most workings were in the oxidized zones of gold-quartz-pyrite veins and lodes in fissures and shear zones; one deposit, the Mahd Ad-Dahab, now works hypogene quartz-pyrite ores in rhyolite breccias (Kingdom of Saudi Arabia, 1965, 1969, 1970). According to some geologists these deposits are of Riphean age (Proterozoic) related to granitic bodies of the same age. In Korea gold-quartz veins are mined or were formerly mined at Unsan, Nurupi, Sak Ju, Sen Sen and elsewhere. Numerous gold deposits of Precambrian age are also known in China, but in recent years there has been little information available on their nature and exploitation. Judging from the older reports (Wong, 1919; Emmons, 1937; Juan, 1946) most of the deposits are gold-quartz veins, gold-bearing skarns and gold-quartz-sulphide lodes related to orogenic events at the end of the Algonkian (Proterozoic). The host rocks are mainly greenstones (amphibolites), schists and gneisses invaded by a variety of granitic rocks. Despite their considerable extent and similarity in rock types to shields in other countries containing large gold deposits, the shields and areas of Precambrian rocks in the Asiatic as well as those in the European part of U.S.S.R., including the Baltic, Ukranian, Aldan and Anabar shields apparently contain few commercial deposits of Precambrian age mined exclusively for gold (Rozhkov and Zhadnova, 1972; Borodaevskaya and Rozhkov 1974; Buryak *et al.,* 1974). There are, however, numerous gold occurrences in these various areas of Precambrian rocks, some apparently of Precambrian age (Buryak, 1967 *a,b,c*; Buryak *et al.,* 1974) and some of later age (Zakharov, 1971). The gold mineralization described by Buryak *et al.* (1974) in the Sayan-Baikal folded belt consists of three types: slightly auriferous polymetallic sulphide desposits, gold-quartz veins and impregnation sulphide deposits containing some gold. These deposits occur mainly in phyllites, shales, carbonate rocks and basic volcanic flows. The mineralization is older than the Early Proterozoic granitic rocks and hence may be of Archean age. According to Perelygina (1973) most of the gold deposits in the Aldan Shield, some of commercial grade, are associated with the rejuvenated parts of the shield marked by the Aldan volcano-plutonic complex of Late Mesozoic age. The deposits include gold-sulphide bodies and veins, oxidized gold-pyrite-quartz bodies (Kuranakh-type), auriferous silicified and limonitized zones, auriferous quartz veins and pyritized zones and auriferous zones characterized by potassic feldspar-pyrite-carbonate metasomatites. Much gold is won from a variety of polymetallic deposits in Precambrian Proterozoic rocks in U.S.S.R. (Smirnov, 1959). Similar polymetallic deposits in Precambrian rocks in China, Korea and other countries in Asia yield some gold. The age of these deposits is uncertain. So far as is known economic gold-bearing quartz-pebble conglomerates have not been found in the Precambrian of Asia, although a number of occurrences have been reported (*see* the section on fossil placers). Numerous large gold placers derived in part from lean deposits in Precambrian rocks are worked in the headwaters of the Lena River (Aldan Shield), in a number of rivers north of Manchuria and elsewhere in U.S.S.R. (*see* Fig. 87). Similar, but smaller gold placers derived from Precambrian deposits have been exploited in Mongolia, China, Korea, India and other countries in Asia.

The Precambrian of Australasia is concentrated principally in the great shield area of Western Australia and parts of Northern Territory and South Australia with relatively smaller zones in the cores of mountains or highlands in Queensland, New South Wales and Tasmania. New Zealand has no definitely dated Precambrian rocks, but certain metasediments and gneisses in Westland and South Nelson that contain gold-quartz lodes with pyrite, arsenopyrite and stibnite in shears and crushed zones (Westland, Paparoa, Lyell and Reefton goldfields) may be of this age (Williams, 1974). Elsewhere in Australasia rocks of undoubted Precambrian age appear to be rare.

In Western Australia and Northern Territory numerous gold deposits of the gold-quartz type have been mined principally in the Norseman, Kalgoorlie, Coolgardie, Wiluna, Lawlers, Marble Bar, Brocks Creek, Paterson Range and Tennant Creek areas (Edwards, 1953; Blockley, 1974; Knight, 1975). Most of the deposits occur in ancient (Archean) volcanics (greenstones) and their associated sediments and granitic intrusives; in places also in iron-formation. The Tennant Creek gold-copper-bismuth deposits are in lower Proterozoic greywacke and shale interbedded with iron-formation; the oxidized gold quartz orebodies of the Telfer Mine in the Paterson Range are in pyritic slates and quartzites of Middle Proterozoic age. Elsewhere in the Precambrian of Australia gold is won from gold-quartz veins and a variety of massive sulphide and polymetallic bodies, including those at Broken Hill and in the Duchess, Selwyn and Kuridala areas southeast of Mount Isa. The Etheridge goldfield in northern Queensland comprises mainly quartz veins associated with the Forsayth Granite and Cobbold Dolerite, which are intrusive into Archean and Lower Proterozoic schist and gneiss. The mineralization is probably of Proterozoic age (Woodal, 1975) and comprises quartz, pyrite, galena, chalcopyrite and sphalerite. Most of the oreshoots averaged more than 1 oz/ton but were shallow in depth (300 ft). Deposits in quartz-pebble conglomerates are known in the basal members of the Proterozoic Nullagine Series, Marble Bar district, Western Australia, but they are only of marginal economic importance. A few small gold placers, mostly of Tertiary age, derived from the Precambrian deposits, have been worked in Western Australia and elsewhere in Australia. The Precambrian lodes in Westland and South Nelson in New Zealand also gave rise to productive alluvial gold concentrations.

Precambrian rocks probably underlie much of Antarctica as is witnessed by abundant exposures along the margins of the continent, particularly East Antarctica, which is probably a shield area. Most of the Precambrian rocks dated so far in Antaractica appear to be relatively young, 1800 million years or so. No gold deposits have yet been discovered in these rocks.

Gold deposition during Precambrian time has been a long and complex process differing in details in the various shields and other Precambrian terranes of the earth. The writer has studied only the Canadian Shield in detail with respect to gold deposition in Precambrian time. Some of the features of the gold deposits in this vast terrane of rocks are similar to those found elsewhere in Precambrian rocks, but it should also be borne in mind that there are significant differences as well. One striking similarity is the frequent development of deposits of the gold-quartz type in ancient volcanics (greenstones) and their associated sediments, particularly the iron-formations and greywacke-slate sequences. A marked difference is the general absence of economic quantities of gold in the Early Proterozoic uraniferous quartz-pebble conglomerates of the Canadian Shield (e.g., Blind River–Elliot Lake) similar to those developed in South Africa (Witwatersrand), Ghana (Tarkwa) and Brazil (Jacobina). The Canadian Shield is also relatively poor in economic Proterozoic deposits of the gold-quartz type, compared to the Australian, Arabian, African and other shields.

Gold deposition took place in the Canadian Shield during or following all of the four major orogenies as defined by Stockwell (1970, 1973) (Table 29), but the intensity has varied greatly. The greatest and most productive period is related to the Kenoran with minor and weak periods of mineralization associated with the Hudsonian, Elsonian(?) and Grenvillian. Most of the productive gold deposits in the Canadian Shield are in Archean rocks with only minor deposits in Proterozoic rocks. With respect to the structural provinces of the Canadian Shield (Fig. 4), as defined by Stockwell (1970), nearly all of the highly productive gold deposits occur in the Superior and Slave provinces; only a few, relatively small, rich or large low-grade deposits are known in the Churchill, Bear, Southern and Grenville; and no economic deposits are known in the Nutak Province. A summary of the principal characteristics of the gold deposits of the Canadian Shield is given in Table 30.

The Archean rocks of the Canadian Shield are characterized by thick sequences of volcanic flows and various intrusives of basic, intermediate and acidic composition, all generally referred to in the older literature as Keewatin. These are interbedded with or overlain often conformably by great thicknesses of greywackes and allied rocks commonly referred to in the older literature as Temiskaming-type sediments. The Archean volcanic terranes have been much studied over the last 75 years by many geologists. In the writer's opinion they differ little from what one sees in volcanic terranes of more recent vintage. One feature is of great importance in the writer's opinion – the presence, in places, of considerable thicknesses of sulphide-bearing slates, tuffs, graphitic schists and iron-formation (Boyle, 1968*a*, 1976*a*). These are commonly enriched in gold, silver, base metals and other elements found in the epigenetic deposits of the volcanic terranes (Tables 15 and 16). They were probably one of the major sources of the metallic elements, sulphur, arsenic, etc. in the epigenetic deposits of the Archean. Similar enrichments of gold and other metals occur in bands and beds of graphitic pyritiferous rocks in the Timiskaming-type sediments. Considered overall, the Archean piles of volcanic and sedimentary rocks initially contained sufficient enrichments of gold and silver and other metals to provide a source for all of the epigenetic deposits found within them (Boyle, 1976*a*). All of the Archean rocks were folded, sheared, faulted, metamorphosed, granitized and intruded by granitic rocks during the late Archean Kenoran orogeny and possibly older orogenies. The Archean rocks are best preserved in the Superior and Slave provinces but have also been recognized in parts of the Grenville, Churchill, Southern and Nutak provinces.

Table 29. Precambrian time-stratigraphic classification in relation to orogenies of the Canadian Shield

Eon	Era	Sub-Era	Orogeny (mean K-Ar mica age, m.y.)
Proterozoic			
	Hadrynian		
			Grenvillian (955)
	Helikian	Neohelikian	
			Elsonian (1370)
		Paleohelikian	
			Hudsonian (1735)
	Aphebian		
Archean			Kenoran (2480)

Source: Stockwell (1970, 1973).

The Proterozoic rocks of the Canadian Shield are a varied lot, and few generalizations can be made about them. In most regions they comprise conglomerate, quartzite, greywacke, arkose, shale, stromatolitic limestone and other allied sedimentary rocks. In some regions these rocks are present in addition to basic volcanic flows and sills, iron-formation, limestone (marble) and other allied rocks. Some Proterozoic rocks are relatively flat-lying and little disturbed except by faulting and minor folding as in the Cobalt area of Ontario; others are highly folded, faulted, sheared, metamorphosed, granitized and intruded by granitic rocks during the Hudsonian, Elsonian and Grenvillian orogenies.

The gold deposits related to the Kenoran orogeny are developed mainly in shear zones and faults in the Archean volcanic and sedimentary rocks of the Superior and Slave provinces, but some important deposits (e.g., Kirkland Lake) occur in faults in associated intrusive granitic rocks and porphyries in these provinces. Most of the highly auriferous deposits belong to the gold-quartz and disseminated types, but considerable amounts of gold and silver have been won as byproducts from the massive and disseminated sulphide orebodies mined essentially for copper and zinc. The main characteristics of the auriferous deposits related to the Kenoran are shown in Table 30. The principal gangue is quartz with some carbonate, often ankerite. The main metallic minerals are pyrite, arsenopyrite and pyrrhotite, with minor amounts of galena, sphalerite, chalcopyrite, stibnite, molybdenite, scheelite and various sulphosalts. The gold in most deposits is free or combined as a telluride or antimonide (aurostibite). In some deposits the gold is intimately associated with arsenopyrite, pyrite or sulphosalts as a lattice or submicroscopic constituent, and these ores frequently have to be roasted before cyanidation. The characteristic elements accompanying gold vary from place to place, but a generalized association includes Cu, Ag, Zn, Cd, B, Pb, As, Sb, Se, Te, Cr, Mo and W. Not all of these are found in any one deposit. Barite is rare as a major gangue mineral, and fluorite is met with only occasionally. Cobalt and nickel occur in the ores only as traces or minor amounts, and cobalt-nickel minerals are rare in most deposits. Platinum metals are not found except as traces, and then only sporadically. Bismuth is relatively rare in the ores, and uranium and thorium are lacking in more than traces. The rare-earths, likewise, occur in the ores only as traces. Niobium and tantalum are not found, and tin occurs generally only as traces in the ores. Mercury is generally only a trace constituent of the gold, sulphide minerals and sulphosalts. Cinnabar has not been recorded in

Table 30. Characteristics of gold deposits related to principal orogenies of the Canadian Shield

	Kenoran (2480)*	Hudsonian (1735)*	Elsonian (1370)*	Grenvillian (955)*
Structural province affected	Superior, Southern and Slave; parts of Churchill and Grenville	Churchill, Bear and Southern	Nain Subprovince, Grenville Province	Grenville
Age of host rocks	Archean – Keewatin and Timiskaming types of volcanics and sediments and their associated granitic rocks and porphyries	*Churchill Province* Archean – Keewatin and Timiskaming types of volcanics and sediments, their granitized equivalents and associated granitic bodies and porphyries *Southern Province* Proterozoic – Huronian quartzite, greywacke, slate and their metamorphosed equivalents. Post-Huronian diabase		Archean and Proterozoic
Principal types of host rocks	Basic, intermediate and acidic volcanics (greenstones) Greywacke and allied rocks. Granitic rocks and porphyries	*Churchill Province* Basic, intermediate and acidic volcanics and their metamorphosed equivalents. Greywackes and allied rocks and their metamorphosed equivalents including gneisses and schists. Grantitic rocks and porphyries *Southern Province* Quartzite, greywacke, slate and their metamorphic equivalents. Diabase		Amphibolites (greenstones), gneiss, schist, skarn, syenite and gabbro
Principal types of deposits	Gold-quartz veins and silicified zones. Massive and disseminated Cu-Zn sulphide bodies	*Churchill Province* *Flin Flon–Herb Lake area* Gold-quartz; massive and disseminated Cu-Zn sulphide bodies *Lake Athabasca area* Quartz stockwork in granitic sill; disseminated quartz-carbonate-sulphide bodies *Southern Province* Gold quartz veins and stockworks; massive Ni-Cu ores	No deposits known	Gold-quartz veins. Disseminated and vein like bodies in skarn. Cu-Zn-Pb sulphide bodies in skarn
Principal gangue minerals	Quartz, carbonates (mainly ankerite, dolomite and calcite	Quartz, carbonates (mainly ankerite, dolomite and calcite). Quartz and siderite in places in the Southern Province		Quartz, carbonates (mainly ankerite, dolomite and calcite); various skarn minerals
Principal metallic minerals	Prryite, arsenopyrite and pyrrhotite; minor galena, sphalerite, chalcopyrite, stibnite, molybdenite, scheelite and various sulphosalts	*Churchill Province* *Flin Flon–Herb Lake area* Pyrite, arsenopyrite and pyrrhotite; minor sphalerite, chalcopyrite and galena *Lake Athabasca area* Pyrite, arsenopyrite, chalcopyrite, Ni-Co arsenides, sphalerite and pitchblende *Southern Province* Pyrite pyrrhotite, arsenopyrite; very minor chalcopyrite, sphalerite and galena		Pyrite, arsenopyrite, pyrrhotite, magnetite, ilmenite; variable amounts of sphalerite, chalcopyrite, galena and tetrahedrite
Nature of the gold	Free or combined as telluride or antimonide (aurostibite). Some intimately associated with pyrite, arsenopyrite and sulphosalts	Generally free. In the Flin Flon–Herb Lake area and in the Southern Province some gold is intimately associated with arsenopyrite		Generally free but in some deposits intimately associated with arsenopyrite

Table 30. (cont'd.)

Characteristic associated elements†	Cu, Ag, Zn, Cd, B, Pb, As, Sb, Se, Te, Cr, Mo, W	*Churchill Province* *Flin Flon–Herb Lake area* Cu, Ag, Zn, Cd, B, Pb, As, Sb, Cr *Lake Athabasca area* Cu, Ag, Zn, Cd, Pb, As, Sb, Se, Mo, U, Rare earths, Hg (Pt), (Te) *Southern Province* Cu, Ag, Zn, Cd, Pb, As, Sb, W	Cu, Ag, Zn, Cd, B, Pb, As, Sb, Co, Bi, F; U; Th and rare earths in some deposits
Characteristic wall-rock alteration processes	Chloritization, silicification, sericitization, carbonatization, tourmalinization, albitization, pyritization and arsenopyritization	Chloritization, sericitization, silicification, carbonatization, tourmalinization, albitization, pyritization and arsenopyritization. Hematitization in Lake Athabasca region	Chloritization, silicification, carbonatization, tourmalinization, pyritization and arsenopyritization. Development of skarn

*Mean K-Ar mica age in millions of years.
†Excluding the common and nearly universal associates such as Si, Fe, S, etc.; elements in parentheses are rare or localized.

the deposits as far as the writer is aware, but mercury tellurides occur in some ores. The Au/Ag ratio of these ores is high as a rule, averaging about 8. Local reversals of the ratio yielding values of 1 or less may occur in ore shoots where silver tellurides, silver-bearing tetrahedrite or a secondary mineralization of silver minerals are present.

The Hudsonian orogeny is manifest mainly in the Churchill, Bear and Southern provinces. Gold deposits in these provinces tend to be small or have large tonnages with very low grades; some of the smaller deposits are exceptionally rich. In Churchill Province most of the gold deposits belong to the gold-quartz type and are clustered in the Flin Flon–Herb Lake belt of volcanics and sediments which may be of Archean age. Those at Nor-Acme, Laguna, Gurney and Squall Lake are thought by some geologists to be related to the Kenoran orogeny, which has been affected later by the Hudsonian. On the north shore of Lake Athabasca the Box Mine was developed in a granitic sill that appears to intrude or was derived by granitization from the Tazin Group of sediments, which are thought to be of Archean age. In the same area the Nicholson Mine, originally prospected for gold, occurs in dolomite, calcareous quartzite and ferruginous quartzite of the Tazin Group intruded by an amphibolite sill. South of Lake Athabasca in the Carswell structure one of the Cluff Lake uranium deposits is enriched in gold and gold tellurides. Elsewhere in the Churchill Province numerous gold-quartz veins are known, but none are of importance economically. Considerable amounts of gold have been won from massive and disseminated Cu-Zn ores in volcanic and sedimentary terranes in the Churchill Province.

The character of the gold-quartz deposits in the Churchill Province are essentially the same as those mentioned for the Superior and Slave provinces. There are, however, a few minor differences in their elemental contents in the Flin Flon–Herb Lake area, and in the Athabasca region. The gold ores in the Flin Flon–Herb Lake area appear to be low in tellurides and hence tellurium, although this element occurs in the Flin Flon massive Cu-Zn sulphides and in other similar bodies in the area. Molybdenum and tungsten, likewise, appear to be particularly low in abundance. In the Lake Athabasca area the orebodies of the Box Mine contain some pitchblende, and those at the Nicholson Mine contain pitchblende, thucholite, Co-Ni arsenides, a variety of silver minerals and tiemannite (HgSe). Platinum has also been recorded in the ores. The Cluff Lake deposit, enriched in gold, carries abundant pitchblende, clausthalite (PbSe), Ni-Co arsenides and copper sulphides. There is some platinum in the deposit. All of these mines occur in the Athabasca uraniferous belt, and it is difficult to determine the relative ages of the gold and uranium mineralization. The gold mineralization may be earlier than the uranium one, although examination of samples suggests they are often contemporaneous; in some samples the gold is late.

There are no known economic gold deposits in the Bear Province. Lord (1951) mentions only one significant occurrence of gold in this province, on Norris Lake about 145 mi north-northwest of Yellowknife (Doris Yellowknife Gold Mines Ltd.). The occurrence is mainly in Proterozoic black slate and argillite of the Snare Group. The mineralization occurs in contorted shear zones and consists of quartz stringers and lenses with some carbonate and abundant pyrite, galena, sphalerite, arsenopyrite, pyrrhotite and chalcopyrite. According to Broughton (1973) there are about 81 000 tons of ore containing 0.228 oz Au/ton in this deposit. Numerous other quartz veins and other types of deposits carry traces of gold in other areas of the Bear Province including the Oro, Bud and Thompson occurrences on Great Bear Lake (Lord, 1951). None of these were of economic interest as gold deposits when Lord examined them.

The gold deposits and occurrences in the Southern Province occur in highly folded and contorted Huronian (Proterozoic) quartzites, greywackes and slates, their metamorphic equivalents, and in associated granitic rocks; also in (Nippissing?) diabase and gabbro, which intrudes the Huronian rocks throughout the area north of Lake Huron. There are also some gold-quartz veins in greenstones that are probably of Huronian age southwest of the early Middle Proterozoic Sudbury irruptive. The Sudbury ores, associated with norite, gabbro and associated rocks, contain some gold and silver, the average contents being about 3.5 ppm Au and 45 ppm Ag in the massive sulphides and 0.08 ppm Au and 2.3 ppm Ag in the ore as mined. One deposit south of the irruptive (the Vermilion Mine), essentially a massive pyrite-chalcopyrite body, was enriched in gold, silver and platinum minerals. Analyses of massive pyrite and chalcopyrite taken from the old dumps of this mine contained the following amounts of precious metals in ppm: Au 1.9, Ag 100, Pt 35 and Pd 108. Quartz veins in the near vicinity of this deposit, mineralized with small amounts of pyrrhotite and chalcopyrite, are low in the precious metals, representative analyses being as follows in ppm: Au 0.055, Ag 1.0, Pt<0.005, and Pd 0.008.

The gold deposits in the Southern Province belong mainly to the gold-quartz type. The principal gangue is quartz with some carbonate, often ankerite, dolomite or siderite (Crystal Mine on Lake Wanapitei). Barite is present in a few deposits; fluorite appears to be absent. The principal metallic minerals are pyrite, arsenopyrite and pyrrhotite, with very minor amounts of chalcopyrite, sphalerite, galena and native arsenic and/or native copper in some deposits. The gold is mainly free, but some is intimately associated with arsenopyrite. The characteristic elements accompanying gold vary from place to place, but a generalized association includes Cu, Ag, Zn, Cd, Pb and As. Sb, Te, Mo, Co, Ni, W, Ba, Sr and Se are present in trace and minor amounts in some deposits. Pt, Pd, U and Sn were not detected in analyses of ores done for the writer from various veins in the area west of the Sudbury irruptive to Sault Ste. Marie. The deposits include those on Lake Wanapitei (Crystal Mine); the Havilah (Ophir) north of Thessalon; prospects on Howry Creek (Bousquet Mine); and occurrences in the Whiskey Lake area. The Bruce Copper Mines, which are in Keweenawan diabase, were fairly regular quartz veins containing chalcopyrite and small quantities of calcite and barite; they also carried a little gold in places, up to 2.5 oz/ton in some pockets. Descriptions of these various deposits are given by Hopkins (1922), Bruce (1933) and Kindle (1936).

The Southern Province also contains the extensive pyritiferous quartz-pebble conglomerates of the Matinenda Formation of the Elliot Lake Group (Early Proterozoic) in the Blind River–Elliot Lake district of Ontario. These contain large tonnages of uranium ore, but their gold tenor, unlike those of the Witwatersrand is measured only in minor and trace amounts. The average gold and silver contents of typical ore are 0.1 and 2.7 ppm respectively. The pyrite may contain up to 0.4 ppm Au and 9 ppm Ag.

The Elsonian orogeny appears to have affected principally the Nain Subprovince of the eastern Churchill Province and extended through much of the Grenville Province. So far as is now known there are no economic gold deposits related to this orogeny. Even gold occurrences seem to be absent or sparse.

The Grenvillian orogeny is manifest in the Grenville Province. This province is a complex mixture of rocks of the various Precambrian eras including both Archean and Proterozoic types. Greenstones (now mainly amphibolites and granulites) occur in some areas, but the most characteristic rocks are paragneiss, schist, marble and skarn. There are also large areas of quartzite, slate, metamorphosed iron-formation and allied rocks. All of these rocks are intruded by pegmatites, granites, syenites, nepheline syenites and anorthosites, the origin of which has long been problematical. Most of the granitic types and the ordinary syenites seem to have been derived by a process of intense granitization of the sediments. The others may have been derived by some uncharacterized process related to deep-seated differentiation.

Four types of gold-bearing deposits occur in the rocks of the Grenville Group: gold-quartz veins and stringers in gabbro (Cordova Mine); similar bodies in carbonatized amphibolite, epidotized rock, hornblende schist, greywacke and dolomite; skarn-type bodies (Richardson and other deposits); and auriferous skarn deposits mined essentially for their Cu, Pb and Zn contents (New Calumet and Tetreault deposits).

The gold-quartz and skarn-type deposits of eastern Ontario occur in the belt of Grenville rocks extending from Belmont Township, Peterborough County eastward across the counties of Hastings, Addington and Frontenac into the western part of Lanark County, in all a distance of about 70 mi. Gold was first found in Ontario in this belt near Eldorado (Richardson Mine). A large number of deposits are known, but most are small and low grade, although some were exceptionally rich near the surface. Descriptions of the deposits can be found in the reports of Vennor (1870), Miller (1902), Miller and Knight (1914), Hopkins (1922), Hurst (1927), Rickaby (1932) and Cooke (1946). Most deposits consist mainly of quartz veins, silicified zones or irregular skarn-type bodies, some with abundant arsenopyrite with which the gold is often closely associated. Other common minerals include pyrite, ankerite, pyrrhotite, tourmaline, feldspar, fluorite, chalcopyrite, tetrahedrite, galena, magnetite and ilmenite. The Richardson Mine is interesting from a scientific viewpoint in that some of the gold occurs in thucholite or is associated with uraninite and brannerite. The Au/Ag ratio in all of these deposits is generally greater than 1 and appears to average about 10 according to my analyses.

The auriferous skarn-type deposits mined essentially for Cu, Pb and Zn, including New Calumet and Tetreault in Quebec, contain essentially pyrite, galena, sphalerite, chalcopyrite and tetrahedrite in a gangue of skarn minerals. Free gold and electrum occur in the ores. The Au/Ag ratio is invariably less than 1.

Paleozoic

The Paleozoic is that era of geologic time between the Precambrian and Mesozoic eras and comprises rocks belonging to the Cambrian, Ordovician, Silurian, Devonian, Mississippian, Pennsylvanian and Permian systems. The Paleozoic era is generally considered to have been closed by orogenic disturbances known by various names throughout the world – the Appalachian in eastern North America, the Hercynian (Variscan) in Europe and so on (Fig. 5). There are also a number of other orogenies that have affected Paleozoic rocks at various times, e.g., the Acadian in eastern North America, which occurred in Devonian time, and the Caledonian in Europe, which is of Siluro-Devonian age. Space does not permit discussion of the various Paleozoic orogenies throughout the world, but reference will be made to those that are of interest with respect to gold mineralization.

Paleozoic rocks, mainly sediments with local centres of volcanics, were laid down over much of the North American continent. Those in the central parts of the continent suffered relatively little disturbance during Paleozoic time, and no epigenetic gold deposits are known in these rocks. In the Cordillera there is evidence for a widespread orogeny in Devonian time and possibly also in late Permian time. Evidence for these orogenies is most clearly seen in parts of

British Columbia, in the Antler Peak region of Nevada, and elsewhere in western United States and Mexico. It appears, however, that the effects of the Paleozoic orogenies in the western Cordillera are not readily separated from those that took place during Mesozoic and Cenozoic times. In any case there are no epigenetic gold deposits known to be associated with these supposed Paleozoic orogenies of the western Cordillera. Gold deposits do occur, however, in the Paleozoic rocks in places in the western Cordillera, but they are related to Mesozoic and Tertiary orogenies and are discussed below. In the northern part of the continent, in the Arctic Islands, the Paleozoic rocks of the Innuitian orogen suffered several periods of orogenic disturbance but were not extensively granitized or injected by granitic rocks. No gold deposits are known in these rocks. The Paleozoic rocks flanking the eastern coast of the continent were highly folded, faulted, granitized and intruded by granitic rocks during two main orogenic events, the Acadian and Appalachian. Gold deposition is associated principally with the Acadian orogeny of Devonian age.

In the southern Appalachians of the United States a number of gold belts extend from the Potomac River near Washington, D.C. to Alabama (Becker, 1895; Lindgren, 1933; Pardee and Park, 1948). The deposits belong mainly to the gold-quartz type and are veins or irregular silicified and pyritized bodies. They occur in slates, schists, gneisses, amphibolites, altered volcanic tuffs and breccias and a variety of granitic rocks, all of which range in age from Precambrian to middle Paleozoic (Devonian?). Some think the deposits are of late Paleozoic age, but the results of geochemical and geological studies by Bell (1960) in the Concord area suggest two periods of mineralization. The gold is apparently related to the earlier period which is associated with small granite plutons, possibly older than Devonian(?). The characteristic gangue is quartz with carbonates, biotite, chlorite, magnetite, ilmenite, tourmaline, albite, gahnite, garnet and minor barite. Fluorite is rare. The metallic minerals are pyrite, arsenopyrite, pyrrhotite, molybdenite and less commonly ferberite, scheelite, galena, sphalerite, tetrahedrite and chalcopyrite. The gold is generally free; a few tellurides (tetradymite, altaite and nagyagite) have been recorded. The principal elements enriched in the deposits in addition to gold are Cu, Ag, Ba, Zn, Cd, B, Pb, As, Sb, (Bi), (Te), Mo and W. Those in brackets are relatively rare. The Au/Ag ratio of the ores appears to range from 5 to 8. Placers derived from these deposits have been worked in the past.

Another belt of auriferous deposits occurs throughout the length of the mainland of Nova Scotia mainly in Ordovician rocks (Meguma Group). These rocks are mainly greywacke, quartzite, arkose, and slate which have been metamorphosed, granitized and intruded by Devonian granitic rocks related to the Acadian orogeny. The gold deposits belong to the gold-quartz type and are developed mainly as veins and saddle reefs (Bendigo type) (Malcolm 1912; Malcolm and Faribault, 1929; Douglas, 1948; Bell, 1948). The principal gangue is quartz with minor amounts of carbonates; the metallic minerals are pyrite, arsenopyrite, chalcopyrite, galena, sphalerite, stibnite, tetrahedrite, scheelite and molybdenite. The gold is generally free, but some is closely associated with arsenopyrite. No tellurides are recorded; only traces of Te and Se have been found in the ores by the writer. The Au/Ag ratio in these deposits ranges from 1.5 to 5.5 in most deposits, but values as high as 50 or more are recorded from some veins. The elements enriched in the deposits in addition to gold include Cu, Ag, Zn, Cd, B, Pb, As, Sb, Mo and W.

Gold-quartz deposits similar to those in Nova Scotia and the southern Appalachians of the United States are widespread throughout the Paleozoic rocks of Newfoundland (Snelgrove, 1935), New Brunswick, Gaspé and Eastern Townships of Quebec, Green Mountains of Vermont and the New England states. There are also gold-quartz veins and gold-bearing silicified zones in Precambrian rocks in Cape Breton Island (Baddeck area), southern New Brunswick and elsewhere in the Appalachians. As noted previously, these are probably of Precambrian age. All of these deposits in both Precambrian and Paleozoic rocks are small and generally low grade according to the writer's observations. Some gold is won as a byproduct of massive and disseminated Cu-Zn-Pb ores in Paleozoic rocks in Newfoundland, New Brunswick, Quebec and the Appalachian regions of U.S.A. In these deposits the Au/Ag ratio is invariably less than 1 by a large factor.

Tertiary and Quaternary placers derived from the Paleozoic deposits just described have been worked in the past in Quebec and Nova Scotia. Most were small, but those in the Chaudière valley of Quebec were of moderate size and relatively rich. Ancient (fossil) placers in Carboniferous rocks derived from gold deposits in rocks of Devonian age or older are known in Nova Scotia at Gays River and at Nauwigewauk in New Brunswick. These are small and low grade where observed, but they should receive further investigation since larger and richer bodies may lie beneath the thick drift cover. (*See also* the section on placers.)

The continent of South America is not marked by gold deposition related to Paleozoic orogenies. Paleozoic rocks are, however, widespread in South America and are metamorphosed, granitized and intruded by granitic rocks in places, particularly in Colombia. No gold deposits of economic interest seem to be definitely related to Paleozoic orogenies in the Cordillera as far as the writer can ascertain. The rich gold deposits of Colombia, so avidly sought by the Conquistadores, while in rocks as old as Precambrian, are mainly of Tertiary age or placers derived from epigenetic deposits of that age.

In northwestern Argentina, from the Bolivian border to a point south of Mendoza, there is a belt of gold deposits of uncertain age in schists, gneisses and various types of sedimentary and volcanic rocks. A number of the deposits are said to be related to Paleozoic or older granites, and some are associated with Tertiary igneous rocks. Probably most of the gold deposits are of Tertiary age (Whiting, 1959). A number of the deposits are quartz veins and stockworks; others are silicified zones and stockworks. The principal gangue minerals are quartz and carbonates with pyrite, arsenopyrite, stibnite, galena, sphalerite, chalcopyrite, tetrahedrite, enargite, famatinite and chalcocite. The gold is free in most deposits. Many of the deposits are deeply oxidized and enriched in both gold and silver. Placers derived from the deposits have been worked in some districts.

Africa is underlain in places by a variety of Paleozoic rocks, and these are folded, faulted, granitized and intruded by granitic bodies related to the Hercynian and earlier

Paleozoic disturbances, particularly in the Atlas and Anti-Atlas ranges which suffered their greatest deformation during the great Alpine (Tertiary) orogeny. The gold deposits of Morocco are mainly in Precambrian rocks but may be of Hercynian age according to Jouravsky (1952*a*). Most of the deposits are hematitic gold-quartz veins and lodes with pyrite, arsenopyrite, pyrrhotite, galena, sphalerite, chalcopyrite, chalcocite and native gold. Those in the Bou Azzer region are unusual in that the native gold is in or associated with brannerite, chalcopyrite, pyrite and a great assemblage of cobalt-nickel arsenides in a gangue of calcite, dolomite, quartz, talc and chlorite. In the Cape System of South Africa gold-quartz veins with minor amounts of galena, chalcopyrite and marcasite occur in the much disturbed Paleozoic terrane of Knysna and in a broad zone of crushing at Hell's Gate in Alfred County, Natal.

In Europe, the Caledonian and Hercynian orogenies are manifest in Paleozoic rocks in three great belts. On the northwest, the older Lower Paleozoic Caledonian orogenic belt extends southwestward through Scandinavia into the northern and western parts of the British Isles. On the south the Hercynian orogenic belt appears in a great series of massifs, the intervening parts being buried by younger rocks, which extends from southern Ireland and Cornwall through France and the Iberian peninsula eastward to the region north of the Sea of Azov. The third belt occurs along the site of the Urals and extends northward to Novaya Zemlya. In these belts the Paleozoic rocks are highly folded, faulted, granitized and intruded by granitic rocks. Tin, lead, zinc, copper, silver, arsenic and antimony predominate among the metals deposited during the orogenies that built these belts. Gold deposits figure only in places, particularly in the Urals, but also in Wales and elsewhere in Europe.

In Wales, the Romans mined gold from a number of gold-quartz veins and lodes near Llandovery (Ogofau workings) (Nelson, 1944). The most important Welsh deposits, now exhausted, are, however, in Merionethshire in the Mawddach Valley, north of Dolgellau (Andrew, 1910*b*; Thomas, 1961; Collins, 1975; Hall, 1977). The deposits are gold-quartz veins and lodes related to the Caledonian orogeny. They occur in highly folded and faulted Cambrian black slates, quartzites, phyllites and schists, highly pyritiferous in places, and intruded by diorite, dolerite and diabase dykes, sills and plugs. The principal gangue is quartz with minor carbonates (mainly calcite). The metallic minerals include pyrite, arsenopyrite, pyrrhotite, sphalerite, galena and chalcopyrite. Tetradymite, the bismuth telluride, is reported as occurring in some veins. The gold is generally free, and the Au/Ag ratio of the deposits is about 9.5. Copper deposits occur in the same area, and a porphyry-type (disseminated) copper deposit has been recently discovered near some of the old gold workings.

Gold has been won from veins and placers in the Leadhills district and elsewhere in Scotland and in several places in Ireland, mainly County Wicklow (Atkinson, 1619; Reeves, 1971; Collins, 1975), the gold being probably derived from scattered gold-bearing quartz veins and sulphide deposits in Paleozoic rocks affected by the Caledonian orogeny. Other gold-quartz veins and base metal lodes, containing a little gold and related to the Hercynian orogeny, are known in various places in western Europe – in Cornwall and Devon in England and in Portugal, Spain, France, Czechoslovakia and Germany. Most of these deposits are small, but they probably contributed much gold to the Tertiary and Quaternary placers worked by the Romans and described by Strabo, Diodorus and Pliny.

A few deposits in Norway and France have provided a moderate amount of gold. In Norway gold-quartz deposits in a variety of Paleozoic rocks occur on Bömmel Island, southern Norway, and in the Bindalen district (Vokes, 1958). In France gold has been won from deposits at La Lucette, La Bellière, Châtelet, Cheni, La Gardette and in the district of Salsigne (Aude). Most of the deposits are quartz veins and lodes in schists and gneisses intruded by granitic rocks. The gold deposits of the Salsigne district (Tollon, 1970) are in complexly folded Paleozoic limestones, shales and grits. They constitute veins and replacement bodies containing quartz, arsenopyrite, pyrite, pyrrhotite and chalcopyrite. Among a number of other minerals present are bismuth, bismuthinite, galena and sphalerite. The deposits of the St. Yrieix district in the Massif Central (Cheni, La Fagassière-Les Gareillas) are veins in schists, gneisses, tonalites and granites (Sevensma 1941; Normand and Phan, 1968). The mineralization is quartz, arsenopyrite, pyrite, galena, sphalerite, jamesonite-boulangerite, chalcopyrite, pyrrhotite, stibnite, tetrahedrite, bournonite and native gold. Some of the apophyses of the granitic bodies contain specks of free gold in tiny quartz veinlets. This gold is associated with cassiterite, scheelite, pyrite, arsenopyrite and a few other sulphides. At Châtelet the veins contain quartz, pyrrhotite, pyrite, marcasite, arsenopyrite, berthierite, stibnite and carbonates. The gold is never native in this deposit as has been shown by microprobe analysis (Aubert *et al.*, 1964; Burnol *et al.*, 1974). Rather the gold apparently occurs in layers of the arsenopyrite either in lattice sites or in an extremely finely divided form.

Some of the gold deposits in the Armorican Massif (Brittany) may be of Precambrian age according to Guigues and Machairas (1972). This may also be the case for some of the other (generally minor) gold deposits of eastern Europe in the various massifs.

Numerous gold deposits occur in the region southwest of Prague, Czechoslovakia, and these have been mined extensively in the past. The richest occur in the Jilové district where veins and stockworks cut mainly Precambrian greenstones and sediments but are of Hercynian age related to the central Bohemian pluton (Morávek, 1971). The principal gangue is quartz and carbonates mineralized with pyrite, arsenopyrite, pyrrhotite, molybdenite, galena, chalcopyrite, sphalerite, tetrahedrite, meneghinite, bournonite, boulangerite, cubanite, native bismuth, bismuthinite, kobellite, cosalite, native arsenic, scheelite, native gold and tellurides, including tetradymite, tellurobismuthite, calaverite, petzite, hessite, altaite and coloradoite. The veins exhibit several stages of mineralization. The fineness of the gold ranges from 850 to 900.

In Austria deposits of a similar nature and probably of Hercynian age occur in the Hohe Tauern, a region of crystalline gneisses, schists and quartzites invaded by the Ankogl gneissic granitic mass, some 45 km long and 20 km wide. Most of the deposits occur near Rauris, Fusch, Heiligenblut, Döllach, Gastein and Murwinkel. They comprise vein systems and impregnation lodes in gneiss, quartzites and various schists, and contain mainly quartz, dolomite, tourmaline, arsenopyrite, loellingite, pyrite, chalcopyrite, galena, stibnite, argentite, sphalerite, molybdenite, scheelite and gold. The ores, now exhausted averaged about 0.25 oz Au/ton. The silver content was high, averaging 2 oz or more per ton. Much of the gold was free, but some was closely associated with arsenopyrite. The oxidized parts of the lodes were probably very rich; old reports mention up to 500 g/ton or more. Old descriptions of these deposits and others in Austria are given by Beyschlag *et al.* (1914, 1916) and Emmons (1937).

The characteristic elements concentrated in the gold deposits related to the Caledonian and Hercynian orogenies in western Europe are essentially similar and comprise Cu, Ag, Zn, Cd, B, Pb and As. Co and Ni occur in small amounts in some of the deposits in Wales and Switzerland. Te is relatively rare, but Sb and Bi are abundant in places, especially in some of the deposits in Portugal and France (Le Châtelet, La Lucette, Salsigne). Mo, W and Sn are abundant in other types of deposits related to the Paleozoic orogenies but relatively rare in the typical gold deposits.

Gold occurs at many places along the Ural chain, mainly on its eastern flank (Mushketov, 1930; Sirin, 1963; Smolin, 1970; Ovchinnikov, 1972; Borodaevskaya and Rozhkov, 1974). Placers were formerly the main type of deposit, being derived mainly from a great variety of gold-quartz veins and pyritized silicified zones in highly folded and faulted gneisses, schists, phyllites, quartzites, listwänites (quartz-sericite-talc-calcite-dolomite rocks, derived from the alteration of serpentinites and associated intrusives including serpentinites, gabbro, norite and diabase), granite, syenite, various types of porphyries and highly altered (sericitized and pyritized) fine-grained granitic rocks called beresites. The endogene gold deposits are of Hercynian age. In the northern and central part of the Urals the gold deposits appear to show a close relationship to the acidic types of dykes and other intrusions. At Berezovski many of the gold-quartz deposits are ladderlike veins in the beresitic rocks. In the southern part of the Urals there appears to be a closer relationship with the intermediate and basic dykes and other intrusives. The Kochkar deposit is a series of quartz veins and lenses associated with highly altered 'tabashki' dykes, which are meso- and melanocratic rocks consisting principally of biotite, amphibole, feldspars, epidote, quartz and carbonates characterized by a blastic structure. The veins and lenses are most productive where they occur in shears, fractures and altered zones in the 'tabashki' dykes. The principal gangue in most of the typical gold deposits of the Urals is quartz with some carbonates; tourmaline is present in places. The metallic minerals include pyrite, arsenopyrite, pyrrhotite, chalcopyrite, galena, tennantite-tetrahedrite, boulangerite, bournonite, stibnite, aikinite, tetradymite, bismuthinite, etc. Cinnabar is recorded from a few deposits, and scheelite is mentioned as a fairly abundant mineral in one district. The gold is generally free and the Au/Ag ratio of most deposits ranges from 7 to 9. Tellurides appear to be relatively rare. The grade of the deposits is about 10 g/ton (0.34 oz Au/ton). The alteration processes are chloritization, sericitization, silicification, carbonatization, pyritization, beresitization and listwänitization. The characteristic elements concentrated in the gold deposits in addition to gold are Cu, Ag, Ba, Zn, Cd, Hg, B, Pb, As, Sb, Bi, (W), (Ni) and (Co). Those in parentheses are minor or rare.

Prior to the Russian revolution numerous placers and hypogene gold deposits were worked in the Urals. Following the revolution there was a general decline, and little gold mining was done in the area. In the last 30 years a number of new mines have been developed mainly in the Berezovski, Kochkar, Plast-Miass and Dzhetygara areas. The central and southern Urals are also noted for their Hercynian polymetallic and massive and disseminated pyritic Cu-Zn-Pb bodies (Sverdlovsk, Karabash, Bajmak-Sibai, Mednogorsk-Orsk and Kurosan areas). Gold is won as a byproduct from some of these deposits.

The gold placers of the Urals, derived from the deposits of Paleozoic age, were formed during succeeding eras. They include fossil Mesozoic placers and Cenozoic (Tertiary and Quaternary) placers. Many of them produce platinum metals in addition to gold. (*See also* the section on placers.)

In Asia the Caledonian, Hercynian (Variscan) and other minor Paleozoic orogenies were attended by widespread gold mineralization. The Caledonian orogeny is manifest mainly in western Kazakhstan and neighbouring Kirghiz S.S.R., particularly in the regions south and west of Alma-Ata and north and northwest of Lake Balkhash in the Altai-Sayan ranges of Siberia, Mongolia and northwest China and in the Ulan Ude-Chita region (Transbaikal). Hercynian (Variscan) orogenic events are manifest in great belts that extend through Uzbekistan, and the Lake Balkhash and Lake Zaysan regions in Kazakhstan, U.S.S.R., eastward and southward across Mongolia and China into parts of the Malay Peninsula. These great belts of Caledonian and Hercynian disturbances are marked by a large number of productive gold deposits of many types.

Gold was won from placers and mines in Uzbek, S.S.R. long before the Christian era, the mines being mentioned by Strabo and Herodotus in their writings. Today, the birthplace of one of the first of the geological fraternity, Abu Ali Ibn Sina (Avicenna) (980–1037) is one of the foremost gold-producing areas in the Soviet Union. Many deposits are known in the great belts of Paleozoic rocks that stretch from the southern Urals to the T'ien Shan mountains and include the Nuratau Mountains, the Kyzyl-kum desert and Hissar range in Uzbek, S.S.R. The deposits belong mainly to the gold-quartz type and are generally enriched in arsenic. Prospecting in the late 1950's using geochemical methods and arsenic as indicator, led to the discovery of the great Muruntau deposit in the Muruntau Mountains. This deposit is localized in gently dipping metamorphosed sediments on the south limb of an anticline between two large faults. The deposit is a vast stockwork of quartz-pyrite and quartz-arsenopyrite stringers and veins in interbedded metamorphosed carbonaceous sandstones, siltstones, shales, phyllites and a variety of schists some of which are chloritic, all invaded by swarms of granitoid dykes, including various porphyries, granite, syenite and monzonite, apparently all of Hercynian age. The wall-rock alteration is unusual for gold deposits of this type. It includes essentially the development of a hornfelslike (cherty) rock consisting of quartz, biotite, feldspar, sericite, chlorite, actinolite, with pyroxene and hornblende in places. The principal gangue minerals are quartz, feldspar and tourmaline. The main metallic minerals are pyrite, arsenopyrite, pyrrhotite and chalcopyrite. The average sulphide content of the ore is about 1 per cent. Minor and rare minerals include marcasite, scheelite, sphalerite, galena, bismuthinite, native bismuth, cinnabar, miargyrite, tetrahedrite, pyrargyrite and others. The gold is generally free; gold tellurides are suspected. The Au/Ag ratio of the native gold is 94/4; in the deposit as a whole the ratio is 4:1. The grade of the deposit is not stated clearly in the various publications but is probably about 3.4 ppm (0.1 oz/ton). The elements enriched in the deposit include the following in addition to gold: Cu, Ag, Sr, Ba, Zn, Cd, Hg, B, Pb, As, Sb, Bi, Se, Te, Mo, W, Co and Ni. An extended description of the deposit is given in the book edited by Khamrabaev (1969).

In addition to the deposit just described there are various other types of gold deposits in Uzbekistan. According to Petrovskaya (1968) and Khamrabaev (1969) these include quartz-gold-telluride deposits, pyritic quartz deposits with gold and tellurides, quartz-carbonate-adularia-gold (electrum) and skarn-chalcopyrite-gold deposits. The elements concentrated in these deposits are essentially the same as those mentioned for the Muruntau deposit. In places Sb and Te figure more prominently, whereas in others Cu, Mo and W are relatively abundant in the ores.

Northeastward in the Paleozoic rocks of Kazakhstan and in the region west and east of Lake Baikal there are a large number of gold deposits belonging to the following types – skarn, gold-quartz-pyrite-arsenopyrite veins and lodes and various polymetallic deposits enriched in gold and silver. The characteristic elements associated with gold in these deposits include: Cu, Ag, Ba, Zn, Cd, Hg, B, Pb, As, Sb, Bi, Se, Te, Mo and W. Extended descriptions of the deposits in northern Kazakhstan are contained in the papers by Esenov *et al.* (1968, 1970) and the book edited by Abdulkabirova *et al.* (1971).

Gold deposits – mainly of the gold quartz type, but also including the skarn type and gold-bearing polymetallic deposits – occur in Paleozoic rocks in Mongolia, in Manchuria and in the Caledonian and Hercynian belts in China (mainly in Szechwan, Hunan and Kwang-si provinces). There are also a few gold-quartz and skarn-type gold deposits and gold-bearing polymetallic deposits in Paleozoic rocks in Korea and Japan that are probably of Hercynian age. These deposits are, however, relatively insignificant compared with those formed in Japan during the Tertiary.

In Australasia there were two periods of Paleozoic orogeny confined mainly to Australia (Tasman Geosyncline). One marked the period between the close of the Ordovician and the close of the Devonian and appears in places to have affected Silurian rocks as well (the Siluro-Devonian orogeny of Hills, 1947). In Tasmania the age of the granitic intrusives of this period ranges from 375 to 335 m.y. (Late Devonian to Early Carboniferous) (Williams *et al., in* Knight, 1975). In Victoria and New South Wales the age of the granitic intrusives is mainly Siluro-Devonian (Spencer-Jones and Vandenberg, *in* Knight, 1975; Pogson, *in* Knight, 1975). The other period of orogeny and granitic intrusion is Permo-Triassic. To the first belong the rich gold-quartz deposits of Victoria (Bowen and Whiting, *in* Knight, 1975) and adjacent parts of New South Wales (Markham, *in* Knight, 1975) and those in Tasmania (Green, *in* Knight, 1975). Here belong the great gold-quartz lodes of Bendigo, Castelmaine, Ballarat, Daylesford, Hill End, Hill Grove and others. Bendigo and Ballarat are the leading fields. Much gold is also won from the copper sulphide deposits of Cobar, New South Wales (Rayner, 1969; Brooke, *in* Knight, 1975) and from other polymetallic deposits related to the Siluro-Devonian orogeny.

At Bendigo and Castelmaine the rocks are Ordovician slates and greywackes intruded by Devonian granite. The deposits are mainly quartz saddle reefs and bedding plane veins. At Ballarat the deposits are mainly persistent quartz veins in fissures, shears and faults that cut Ordovician slates and greywackes. They are exceedingly rich where they cross indicators (carbonaceous and pyritiferous beds). The deposits at Daylesford are veins and lodes in closely folded Ordovician slates, quartzites and greywacke. In the belt between Walhalla and Woods' Point quartz veins form ladders in diorite dykes or occur in fractures and faults in an area of closely folded Silurian sediments intruded by granodiorite. The principal gangue in all these deposits is quartz with minor carbonates. The main metallic minerals are pyrite, pyrrhotite, arsenopyrite, chalcopyrite, galena, sphalerite and minor molybdenite, wolframite, native bismuth, stibnite, bournonite and other sulphosalts. Some of the deposits are relatively rich in sulphides (up to 60% of the ore). There is little or no wall-rock alteration. The gold is free and no tellurides have been reported. The Au/Ag ratio is high, ranging from 7 to 9. The characteristic elements concentrated in the veins in addition to gold include: Cu, Ag, Zn, Cd, Pb, As, Sb, Mo and W.

At Hill End, north of Bathurst in New South Wales, lenticular quartz veins occur in Silurian slates and tuffs intruded by quartz porphyry. At Hargraves the deposits are saddle reefs. In the Hill Grove area in the northeastern part of New South Wales gold-quartz veins occur in slates and greywackes. The gangue is invariably quartz and minor carbonate in all of these deposits with subordinate amounts of pyrite, arsenopyrite, pyrrhotite, galena, sphalerite, stibnite and minor scheelite. The characteristic elements concentrated, in addition to gold, include Cu, Ag, Zn, Cd, Pb, As, Sb and W.

Most of the gold-quartz deposits of Tasmania are related to the Siluro-Devonian period of orogeny and are located in the northeast part of the island (Green, *in* Knight, 1975). The deposits at Lisle-Golconda, Mathinna, Mangana and Beaconsfield are quartz veins and pyritiferous quartz veins mainly in fissures, faults and shears in folded Ordovician conglomerates, limestones, quartzites, slates and various schists and serpentinites intruded by granite, granodiorite and a variety of porphyrites. Elsewhere in Tasmania there are other gold-quartz deposits and polymetallic deposits carrying small amounts of gold. The famous Mount Lyell and nearby deposits, 15 mi northeast of Strahan, are some of the world's greatest copper-gold deposits (Reid, *in* Knight, 1975). The Mount Lyell deposit was a massive sulphide body, greatly enriched at the surface in gold, contained within sericite schists (acid lavas and pyroclastics) near a conglomerate contact. The primary ore consisted principally of pyrite, chalcopyrite, a little bornite, enargite, galena, sphalerite, molybdenite and tennantite with a little quartz, fluorite and barite. In summary, the characteristic elements concentrated with gold in the deposits related to Paleozoic orogenies in Tasmania include Cu, Ag, Ba, Zn, Cd, Pb, As and Sb. Sn and

W occur in some gold-bearing lodes, and F is recorded in places.

Paleozoic rocks, highly deformed and intruded by granitic batholiths of Devonian and Permo-Triassic age characterize the mountain ranges of Queensland and northeastern New South Wales and extend along the coast for some 1400 mi (Murray, *in* Knight, 1975). Associated with the Paleozoic orogenies are a number of important gold districts some still productive, including Capetown (Charters Towers), Mount Morgan, Gympie, Mount Coolon, Mount Taylor, Palmerville, Ravenswood, Croydon, Cracow, Hodgkinson and others. The first three have produced the largest amounts of gold. Most of the deposits are gold-quartz veins and lodes in slates, quartzites, limestone, various schists, volcanics, porphyries and granitic rocks. The gangue is quartz with minor carbonates, and the metallic minerals are mainly pyrite, pyrrhotite, arsenopyrite, sphalerite, galena, chalcopyrite, stibnite and tennantite-tetrahedrite. Gold also occurs in some quantity in a variety of skarn deposits that contain essentially the same minerals noted above in addition to bornite, bismuthinite, native bismuth, magnetite, etc. A variety of polymetallic deposits also contain considerable gold, which is won as a byproduct. The great quartz-pyrite-chalcopyrite stockwork pipe with massive and disseminated sulphide zones in quartz porphyries (volcanics) at Mount Morgan, highly enriched in gold in the surface oxidation zones, is an example of these types of deposits. Most of the gold in the Queensland deposits is free, but some is intimately associated with fine pyrite and arsenopyrite. Gold and silver tellurides occur at Charters Towers, Gympie and Golden Plateau at Cracow. Some of the Queensland gold deposits were exceedingly rich, averaging 1 oz/ton in many deposits. The characteristic elements concentrated with gold in the deposits related to the Paleozoic orogenies in Queensland and New South Wales are Cu, Ag, Zn, Cd, B, Pb, As, Sb, Bi and Te.

In New Zealand there are a number of gold districts associated with Paleozoic orogenies and in places with younger (Jurassic) granites – those along or near the west coast of South Island in early Paleozoic rocks, including the Gold Blocks, Aorere, Mount Arthur, Mount Owen, Wangapeka and Preservation Inlet districts; and those in the Marlborough and Alpine schists and in the Otago schists near the southeast end of South Island (Williams, 1974). The Otago schists, shales and greywackes are thought to be at least partly of Paleozoic age by some and of Jura-Triassic age by others. The deposits in these rocks are mainly quartz-pyrite lodes and veins at Bendigo, Macetown, Glenorchy, Skippers, Carrik and Macraes. The deposits at Reefton and Lyell are in the Greenland and Waiuta groups, both consisting of sequences of greywackes, slates and phyllites. These deposits are mainly gold-quartz lodes in shears and faults. The age of both the host rocks and deposits is uncertain. They may be Paleozoic although some have considered them to be Precambrian (Proterozoic) in age (Williams, 1974).

In the Paleozoic gold districts of South Island, New Zealand, the gold deposits belong mainly to the gold-quartz type. The gangue is invariably quartz with pyrite, arsenopyrite, tetrahedrite, stibnite, chalcopyrite, galena, sphalerite, pyrrhotite and abundant scheelite in places. The gold is mainly free, and in most deposits the Au/Ag ratio is greater than 8. The characteristic elements associated with gold in the Paleozoic gold districts of South Island are Cu, Ag, Zn, Cd, Pb, As, Sb and W.

Elsewhere in Australasia there are a few small primary deposits of the gold-quartz type that may be related to Paleozoic orogenies, although some investigators consider them to be of Precambrian or Mesozoic age. These deposits occur in Borneo and New Guinea and as far as is known they are not being worked (International Geological Congress, 1930; Emmons, 1937).

Gold placers derived mainly from Paleozoic lodes, were formerly of considerable importance in Eastern Australia in Queensland, Victoria, New South Wales and Tasmania; in New Zealand, and in parts of Borneo and New Guinea. Some are still worked in a small way. Many of these placers are fossil and are of Tertiary age; others are of Quaternary (Recent) age.

Mesozoic

The Mesozoic is that era of geologic time between the Paleozoic and Cenozoic eras and comprises rocks belonging to the Triassic, Jurassic and Cretaceous systems. The orogenic history, and consequently the metallogeny, of the era is complex (Fig. 6).

Mesozoic rocks are confined mainly to the Western Cordillera and Arctic Islands of North America. A few Triassic rocks, including volcanics and sediments, are known along the eastern seaboard of the continent and these are flexed and faulted in places (Palisade disturbance). There are no known gold deposits in the eastern Mesozoic rocks nor in those of the Arctic Islands.

Two major orogenies are manifest in the Cordilleran region, the late Jurassic–early Cretaceous Nevadan (Columbian) and the late Cretaceous–early Tertiary Laramide. The effects of the former, together with earlier, local, Mesozoic orogenies are most pronounced in the western part of the continent whereas those of the latter are more evident along the mountain front. The rocks deformed during these orogenies range in age from Precambrian to Tertiary, but the mineralization is mainly late Jurassic to early Tertiary.

Gold deposits related to the Mesozoic orogenies of the Western Cordillera of North America are widespread from Alaska to Central America. Most belong to the gold-quartz type, but there are also numerous gold-bearing skarn-type deposits, and a veritable host of gold- and silver-rich polymetallic deposits. Placers of Cenozoic (Tertiary and Quaternary) age derived mainly from deposits related to the Mesozoic orogenies were formerly important in the western United States – mainly in California, in British Columbia, in Yukon and in Alaska. Only a few of the hypogene deposits can be mentioned here.

Many of the great gold deposits of the western United States, Canada and Alaska are related to Mesozoic orogenies. Here belong the extensive gold-quartz deposits of California and southwestern Oregon, including the Mother Lode, Grass Valley and Alleghany districts, numerous gold-quartz deposits in Washington, Idaho and other western states, the gold-quartz veins of the Zeballos, Tulsequah, Stewart, Hazelton, Nelson-Ymir-Salmo, Bridge River, Cariboo (Barkerville), Atlin and other districts in British Columbia, gold-quartz veins in Yukon, and gold-quartz veins and lodes in Alaska, particularly at Juneau, Fairbanks and in nearby areas. The host rocks of these gold deposits are extremely varied comprising slates, greywackes, quartzites, argillites, schists, various volcanic rocks, serpentinites and granitic rocks, ranging in age from Precambrian to Cretaceous, and younger in places. In all of the deposits the gangue is quartz with minor carbonates, often ankerite. The metallic minerals are pyrite, arsenopyrite, pyrrhotite, chalcopyrite, galena, sphalerite, tetrahedrite, stibnite and molybdenite. Scheelite is common in some veins. The gold is free in most of the deposits, but in some it is intimately associated with pyrite and/or arsenopyrite. In a few deposits gold-silver tellurides are reported. The Au/Ag ratio of the deposits is variable; in some the ratio is as high as 9, in others as low as 5. The grade averages about 0.5 oz/ton Au, lower in some and higher in others. The characteristic elements concentrated in addition to gold include Cu, Ag, Zn, Cd, B, Pb, As, Sb, Bi, Se, Te, Mo and W.

In addition to the typical gold-quartz veins related to Mesozoic orogenies in the Western Cordillera there are also a few skarn-type gold deposits of the same period, exemplified by the Nickel Plate and French mines at Hedley, British Columbia. Considerable amounts of gold are also won as a byproduct of a variety of Mesozoic polymetallic deposits, mainly those mined for copper, silver, lead and zinc.

Gold deposits associated with Mesozoic orogenies in the Cordillera of South America are minor and of little commercial importance compared with those of Tertiary age. Some gold-quartz veins and impregnation zones in Chile and neighbouring Argentina are thought to be related to granites and diorites of late Cretaceous and early Tertiary age. More probably the granitic rocks are entirely Tertiary according to some geologists. In Chile the auriferous deposits related to intrusions of Late Jurassic to Late Cretaceous age are mainly veins and breccia pipes (Ruiz and Ericksen, 1962). The veins comprise gold-pyrite types containing quartz, pyrite and chalcopyrite; they are low grade except in the oxidized zones. Another vein type is marked by the presence of specularite (hematite), magnetite and chalcopyrite; they are medium to high grade deposits. The pipes are circular or lenticular breccia bodies cemented by quartz and coarse-grained tourmaline. They are developed in granitic rocks of various types and grade into or are associated with chalcopyrite-tourmaline pipes. Much of the gold occurs as fine blebs in the chalcopyrite of these deposits.

In Mesozoic time extensive diastrophism and igneous activity did not take place in Africa and Europe, and no gold deposits of definite Mesozoic age are known in these continents. In western Asia the effects of Mesozoic orogeny and igneous activity are confined mainly to the Caucasus, and particularly to the Lesser Caucasus where gold-quartz deposits in ophiolites and other rocks are developed such as at Zodsk (Konstantinov and Grushin, 1972; Borodaevskaya and Rozhkov, 1974). A number of auriferous polymetallic deposits are also worked in the Lesser Caucasus. Some of the gold deposits cut rocks of Upper Cretaceous age and hence may actually be of Tertiary age. In eastern Asia a great belt of Mesozoic and older rocks are folded, faulted, granitized and intruded by a variety of granitic rocks from southwest Borneo, through the Malayan peninsula, across China, and into the far eastern part of U.S.S.R. The Verkhoyansk range and the regions flanking the Kolyma massif (Precambrian and Paleozoic) and stretching to Bering Strait reflect this orogeny as do also the Sikote-Alin range in U.S.S.R., parts of South Korea and the southern part of Honshu in Japan. This orogenic epoch, commonly referred to as Yen Shanian, is mainly of late Cretaceous age, but its effects are also manifest in early Tertiary rocks in some districts.

A number of gold-quartz veins and lodes are known in Mesozoic rocks in southwest Borneo, Malaya, Thailand, Cambodia, Laos and Vietnam, but these are mainly small and are not worked. Derived placers still provide a little gold in places. In many places the primary lode gold and vein deposits thought to be of Mesozoic age in these countries may actually be late Cretaceous–Tertiary in age. In China gold deposits, mainly of the gold-quartz type, related to post-Jurassic orogenic events, have been productive mainly in Szechwan Province. In the eastern Transbaikal region and in the far eastern U.S.S.R., in rocks affected by Mesozoic diatrophism and granitic intrusion, there are a large number of productive gold deposits, mainly of the gold-quartz type, but also including gold-bearing pyritized, silicified zones, skarns and gold-bearing dykes, in the Yana-Kolyma folded belt, in the Okhotsk Mesozoic folded regions, in the Chukotka region, on the north slope of the Aldan Arch, in the Sikhote Alin range and elsewhere (Kunaev, 1963; Ditmar, 1966; Sidorov, 1966; Konstantinov *et al.*, 1967; Apel'tsin *et al.*, 1968; Gorzhevskii *et al.*, 1968; Shilo *et al.*, 1969; Gamyanin and Solov'ev, 1969; Kogen, 1969; Obolenskii and Obolenskaya, 1972; Perelygina, 1973; Berger and Kuz'min, 1975; and others). According to Zagruzina (1972) the sequence of gold mineralization, as determined by absolute age dating, in the northeastern U.S.S.R. from the Upper Jurassic to the Paleogene was as follows: early (mainly Mesozoic) gold-quartz veins, stockworks, silicified zones, etc.; gold-rare metals deposits; and late (mainly Paleogene) gold-silver deposits. Large eluvial, eluvial-talus, alluvial and beach placers, which have derived much of their gold from the hypogene Mesozoic deposits, are also worked in all of these areas.

A general review of the nature of the gold-quartz and other gold-bearing deposits in the Mesozoic and older rocks of Asia and related to late Mesozoic diastrophism and associated granitic activity indicates that quartz is the principal gangue with carbonates and minor barite in some districts. The metallic minerals include pyrite, pyrrhotite, arsenopyrite, galena, sphalerite, chalcopyrite and tetrahedrite. A number of deposits contain stibnite, various sulphosalts often silver-bearing, molybdenite, bismuthinite, scheelite and rarely cassiterite. The gold is generally free, but some is intimately associated with arsenopyrite and pyrite or with stibnite and sulphosalts in some deposits. Tellurides occur in some deposits (Darasun). Wall-rock alteration effects (beresitization, chloritization, tourmalinization, carbonatization, etc.) are marked in many of the deposits. Mineral zoning and pulsation are also common in some of the deposits. Many of the deposits are deeply oxidized and enriched (e.g., the karstified Kuranakh deposits) (*see* Chapter IV). The Au/Ag ratio commonly ranges from 7 to 9, but in some deposits the ratio is 1 or less. Deposits with low Au/Ag ratios generally belong to late orogenic events that span Cretaceous–early Tertiary time.

Most seem to belong to the Tertiary suite of gold deposits mentioned in the next section. The characteristic elements concentrated with gold in the Mesozoic orogenic belts of Asia include: Cu, Ag, (Ba), Zn, Cd, (Hg), B, (Sn), Pb, As, Sb, (Bi), Se, Te, Mo, W, F, (Co), (Ni) and (U). Those noted in parentheses are occasional or rare.

Gold deposits related to Mesozoic orogeny and igneous activity appear to be rare in Australasia. In the Tasman geosynclinal zone in southeastern Queensland, Australia, the Au, Cu, Sb and Hg mineralization near Maryborough is probably of Upper Cretaceous age, related to the Maryborough orogeny. The gold deposits occur at North Arm and are apparently associated with Late Cretaceous intrusions of granite, syenite and diorite (Jones, 1947; Hills, 1947). Warren (1972) records that plugs and bosses of granite were intruded during Cretaceous time in places in northeastern Australia. They have associated small gold lodes west of Mackay. Small gold lodes also occur in Cretaceous peralkaline plugs at Mount Dromedary south of Sydney and at Cygnet, south of Hobart, Tasmania.

Cenozoic

The Cenozoic is the latest of the four periods of geologic time, extending from the close of the Mesozoic era to and including the present. The Cenozoic era includes the rocks formed during the Tertiary (Paleogene and Neogene) and Quaternary periods.

The Cenozoic represents one of the major periods of gold concentration in the geological history of the earth (Fig. 7). The Tertiary (Paleocene to Pliocene) period is notable for the extensive deposition of hypogene gold deposits in many parts of the world, and both the Tertiary and Quaternary represent two periods when many of the great gold placers of the world were formed. Even today in the Recent (Holocene) epoch hypogene gold deposits are still forming (in New Zealand and elsewhere), and gold is being concentrated in stream, river and other types of sediments. Finally, the oxidation of gold deposits during Tertiary and Quaternary time has led to secondary enrichment of gold (and silver) in many deposits throughout the world.

In the Cordillera of North America the early Tertiary period was marked by the extensive injection of sills and stocks of monzonite, granodiorite and their related porphyries into older rocks ranging in age from Precambrian to Cretaceous. This period was accompanied, or followed in the early and late Tertiary, by marked orogenic disturbances, which resulted in uplift, warping and faulting of large regions of the Cordillera, and the outpouring of andesitic and rhyolitic lavas on a vast scale. While this general scheme of events may hold in some districts, when the details of the geological features in particular areas are examined there are many complexities. For instance it seems that in some areas of the eastern parts of the Cordillera the early Tertiary magmatic events are simply a continuation or an eastward extension of those that affected Cretaceous and older rocks in the western parts of the Cordillera. In addition, the sequence of sedimentation, volcanic activity and magmatic activity during the Tertiary is extremely complex and not readily correlated from district to district. Finally, in many districts it is difficult to sort out mineral deposits which belong to Tertiary orogenic events from those that were formed during the Mesozoic, Paleozoic or Precambrian. Increasing use of radioactive age-dating methods may resolve some of these difficulties in the future.

The above complexities notwithstanding, it can be said that Tertiary orogenic events, manifest by widescale diastrophism, volcanism and intrusion by granitic rocks and porphyries, can be traced from Alaska, through British Columbia, where they are less pronounced, southward with increasing intensity through the western United States into Mexico and Central America where they attain their greatest development. Gold deposits of great importance are related to these Tertiary events particularly in the western United States, Mexico and Central America. The gold-bearing pyrrhotite-chalcopyrite-pyrite ores of Rossland, and a few other gold-bearing deposits in the Boundary district of British Columbia, are now considered to be of early Tertiary age (Fyles *et al.*, 1973), but Thorpe and Little (1973) caution against accepting this age for the mineralization. Scattered deposits in Alaska (Unalaska Island, Unga Island) are probably also of Tertiary age. The Dusty Mac gold-silver deposit in silicified Eocene volcanic rocks near Okanagan Falls, British Columbia is Cenozoic in age (Church, 1973). It is described in more detail subsequently. These seem to be the only gold deposits of definite Tertiary age in British Columbia and Alaska, but there are probably others which may prove to be of this age. The Mount Nansen gold-silver deposit in the Yukon, described by Saager and Bianconi (1971), is most probably of late Cretaceous–early Tertiary age. This deposit is a system of narrow, steeply dipping veins in rocks ranging in age from Precambrian to Tertiary. The mineralization consists essentially of quartz veins with pyrite, arsenopyrite, sphalerite, galena, freibergite and native gold. The Au/Ag ratio of the deposit is low, averaging about 0.03; the fineness of the native gold is about 800.

The characteristic hypogene gold deposits of the Tertiary in western North America are predominantly gold-quartz and allied types often with an abundance of silver and base metal sulphides and sulphosalts. Here belong the deposits of Republic in Washington; Tonopah, Rawhide, Goldfield, Comstock Lode, Tuscarora and Jarbidge in Nevada; Bodie in California; the Black Mountains in Arizona; De Lamar in Idaho; the San Juan region (Telluride, Ouray, Silverton, Lake City, Rico, Needle Mountains, La Plata and Creede) and Cripple Creek in Colorado; Pachuca, El Oro and Guanajuato (Veta Madre) in Mexico; and a number of deposits in Central America. The disseminated Carlin, Cortez, Getchell and Gold Acres deposits in Nevada also belong in the listing, according to the latest dating investigations.

The host rocks of the Tertiary deposits of the Cordillera of North America are varied and include andesites, trachytes, rhyolites and various porphyries of Tertiary age; Tertiary shales, limestones and sandstones; and older Mesozoic, Paleozoic and Precambrian rocks of sedimentary, extrusive volcanic and intrusive igneous origin. The alteration processes are, likewise, varied depending on the rock types – characteristic, however, is the marked propylitization (chloritization, carbonatization and pyritization) and sericitization manifest in the Tertiary andesites. In other rocks silicification, pyritization, sercitization and alunitization are common. The gangue of the veins is invariably quartz, often chalcedonic, banded or crustified, with abundant calcite in some places, rhodochrosite in others and rhodonite, adularia, fluorite and barite in a few deposits. The metallic minerals include pyrite, chalcopyrite, galena, sphalerite, argentite, polybasite, pyrargyrite, proustite, tetrahedrite, pearceite, silver selenides (naummanite), stibnite, molybdenite, bismuthinite and wolframite. Arsenopyrite and pyrrhotite are relatively uncommon. The gold is generally free, although some is intimately associated with the sulphosalts. Much of the gold is extremely fine grained giving a yellow hue to the quartz in some deposits. Tellurides are common as at Cripple Creek in Colorado. The Au/Ag ratio is relatively high in some deposits ranging from 5 to 40 with an average of about 10 at Cripple Creek; elsewhere and more generally the ratio is 1:1 or as low as 1:250 (Tonopah, Pachuca, etc.).

The disseminated deposits of Carlin, Cortez, Getchell and Gold Acres in Nevada are described in more detail in a later section. Briefly they are bedded and irregular disseminated deposits in altered and mineralized calcareous siltstones and limestone or in brecciated and fractured zones in these rocks. The mineral assemblage is pyrite, arsenopyrite, quartz, realgar, stibnite, cinnabar, sphalerite, galena and gold.

The elements associated with gold in the Tertiary deposits of the Cordillera of North America are variable – generally they include: Cu, Ag, Ba, Zn, Cd, Hg, (B), Pb, As, Sb, Bi, Se, Te, Mo, W, F, Mn, (Co), (Ni) and (U). Those in parentheses are rare or occasional.

The Tertiary rocks and the orogenic events, which have affected them and the older rocks of the Cordillera of South America, are closely analagous to those described for North America. The Tertiary in South America was a time of intensive mineralization, particularly for copper, silver and tin. Gold accompanied this mineralization over a widespread area, but most of the strictly gold deposits are small and low grade with the exception of those in Colombia and a few in Chile and Peru.

Most of the Colombian gold deposits belong to the gold-quartz type and are found in fractures, faults and stockworks that cut Tertiary and older rocks of both sedimentary and volcanic derivation. The most productive gold deposits lie in the Central Cordillera mainly in the departments of Antioquia, Caldas, Tolima and Chocó. The gangue is invariably quartz with carbonates, and the metallic minerals are mainly pyrite, arsenopyrite, pyrrhotite, galena, sphalerite, argentite, chalcopyrite, bournonite, jamesonite, tetrahedrite, stibnite, stephanite, pyrargyrite and cinnabar. The gold is generally free, but in some deposits it is intimately associated with arsenopyrite and various sulphosalts. Tellurides are reported in some of the deposits. The Au/Ag ratio exceeds 1 in some deposits; in others it is much less than 1. Placers derived from the hypogene Tertiary deposits were extensively worked in some of the departments. Some of these contained relatively large amounts of platinum minerals. The elements associated with gold in the hypogene Tertiary deposits of Colombia and other countries in South America include: Cu, Ag, Ba, Zn, Cd, Hg, (B), Pb, As, Sb, (Bi), Se, Te, (Mo), (W), (F), (Co) and (Ni). Those in parentheses occur only in some deposits as far as the writer can ascertain. Boron is especially abundant in some of the quartz-chalcopyrite-tourmaline-breccia pipes in granitic rocks in Chile, as at the El Chivato deposit (Ruiz and Ericksen, 1962).

Tertiary (Alpine) orogenic events affected parts of the North African coastal region and certain zones in Ethiopia, Uganda, Kenya, Tanganyika and Mozambique included in the Great Rift Valley system of Africa. There are no gold deposits of importance associated with these orogenic events in North Africa, although gold occurs in small quantities in some of the copper, lead-zinc, antimony and mercury deposits in Morocco, Algeria and Tunisia. Post-Triassic (probably Tertiary) gold-quartz veins and stockworks occur in Malagasy (Madagascar) near Andavakorera in a hot spring area. The mineralization of these deposits is essentially quartz, dolomite, calcite, barite, chalcopyrite, galena, sphalerite, pyrite, marcasite and native gold. The fineness of the gold averages about 700. These deposits appear to be the only manifestation of gold mineralization associated with Tertiary orogenic events in southern Africa.

Tertiary rocks are widespread in Europe and comprise extensive thicknesses of sediments and volcanics, particularly in a belt that flanks the Mediterranean Sea (the Tethys Geosyncline) and extend from Portugal and Spain to Turkey in Asia and eastward through the Black Sea region into the Caucasus. These rocks were profoundly affected during the great Alpine orogeny of the Tertiary, which extended into the Quaternary. Parts of the belt are still tectonically active. Mineralization of various types occurs all along this belt but only in four areas – in Spain (Rodalquilar goldfield), in Romania (Transylvania), in eastern Czechoslovakia and adjacent Ukrainian S.R.R. and in a few places in the Caucasus – were gold deposits of economic significance deposited. The Spanish, Romanian and Czechoslovakian deposits are typical and are briefly described.

The Rodalquilar goldfield is located in the Tertiary volcanic complex of the Iberian Peninsula, east of Almeria in southeastern Spain. The deposits are quartz veins, stockworks, breccia fillings and silicified bodies mainly in dacites, andesites, rhyolites and pyroclastics. Near the deposits and in places regionally, the volcanic rocks are extensively propylitized and argillitized, the principal alteration minerals being quartz, alunite, pyrite, illite, kaolinite and chlorite. Near the veins the associated alteration is quartz-alunite-pyrite-illite-kaolinite, and this grades outward into chloritic rocks. The mineralization of the deposits is essentially crystalline and colloform quartz, pyrite, alunite, jarosite, gold tellurides and a variety of sulphides including marcasite, pyrrhotite, covellite and chalcocite. At a depth of 200 m the gold mineralization is said to change sharply to a zone of complex sulphides. The age of the mineralization is Tortonian-Andalucian (Miocene). Details of the deposits are given by Friedrich (1960) and Vivaldi *et al.* (1971).

In Romania, particularly in the Apuseni Mountains (Brád) gold belt, the basement rocks are folded Mesozoic limestones, shales, conglomerates and sandstones on which are laid Tertiary sediments and tuffs, now folded. Intruding and covering the Tertiary sediments and tuffs are rhyolites, dacites, andesites and basalts. Volcanic necks composed of rhyolite, andesite, dacite and breccias of these rocks invade the Tertiary sediments and tuffs at many points. Faults, fractures and brecciated zones of Tertiary age cut the various Tertiary volcanic and sedimentary rocks, and in places these are filled with quartz accompanied by calcite, rhodochrosite and barite. The metallic minerals are mainly pyrite, arsenopyrite, sphalerite, stibnite, tetrahedrite, bournonite, galena, argentite, pyrargyrite, stephanite, cinnabar and gold. Most of the gold is free and tellurides occur in some of the deposits. In the typical gold deposits the Au/Ag ratio ranges from 1:2 to 1:4; silver-rich deposits have ratios in the range 1:100 to 1:200. The wall-rock alteration includes extensive propylitization, kaolinization, sericitization and silicification. Many of the veins occur within the confines of the volcanic necks, lying parallel to the long axes of the necks; others rim the necks. There are similar veins in or associated with Tertiary rhyolite, andesite and dacite plugs in the Baia Mare district some 120 mi north Brád. These veins show a marked hypogene zoning from the surface downward as follows: a zone rich in gold-bearing sulphides, argentiferous sulphosalts and silver; a zone rich in argentiferous galena and sulphosalts; a zone rich in sphalerite; a zone rich in copper sulphides; and a zone rich in pyrrhotite. There is considerable similarity both in the structural setting and mineralization of the veins in the Brád and Baia Mare gold belts to those in the Cripple Creek volcanic centre in Colorado. The elements associated with gold in the Transylvanian deposits include Cu, Ag, Ba, Zn, Cd, Hg, (B), Pb, As, Sb, (Bi), Mn, Se and Te. Those in parentheses occur only in some deposits as far as the writer can ascertain.

In the Western Carpathians of Czechoslovakia is situated Banskä Stiavnica, one of the oldest mining towns in Europe. Near this town a number of gold-silver deposits have been worked in the Kremnicko-Štiavnické Rudohorie Mountains since the 15th century, and it is said before this by the Celts. The area is underlain by Paleozoic schists, conglomerates and sandstones and Triassic shales, quartzites, limestones and dolomites, intruded by granodiorite, quartz diorite and dacite (Mesozoic and Tertiary?). All of these rocks are overlain and intruded by Tertiary andesites. The veins developed in faults and fractures, cross all of the rocks of the district and are probably Miocene in age or younger. The mineralization is of a polymetallic type, lead and zinc prevailing over copper. In the Štiavnica district the gangue material consists essentially of quartz, in places with hematite (the so-called cinopel). Other gangue minerals, present usually in small amounts, are carbonates, rhodonite and barite. The upper portions of the veins were rich in silver and gold. In the Hodruša district the gangue is mainly quartz and carbonate with negligible amounts of pyrite and the base metal sulphides. The wall-rock alteration consists of extensive propylitization, some of it regional in character, pyritization, sericitization and carbonitization. The metallic minerals include pyrite, chalcopyrite, galena, sphalerite, tetrahedrite, stephanite, argentite and polybasite. Some of the veins, especially in the Štiavnica district, show a vertical zonation from the surface downward as follows: a zone rich in gold and silver (secondary?); a zone rich in galena and sphalerite with abundant rhodonite; a zone rich in sphalerite and galena with considerable chalcopyrite; and a zone rich in chalcopyrite and pyrite. Scheelite appears in the chalcopyrite zone and increases in content with depth. The ores are generally low in gold, the ratio Au/Ag being about 1:45 according to old reports. A modern description of the veins is given by Koděra (1963). Somewhat similar gold-bearing deposits occur in western Ukrainian S.S.R. in the Beregovo mining district (Boev, 1971). The Begansk deposit is a quartz-alunite-barite-polymetallic lode, and the Beregova deposit is characterized by the presence of quartz, alunite, gold and various polymetallic sulphides. The host rocks are Tertiary volcanics and sediments mainly of Miocene age. The elements concentrated with gold in the Carpathian deposits are Cu, Ag, Ba, Zn, Cd, (Hg?), Pb, As, Sb, Se, Mn and W.

The southern and eastern parts of Asia are marked by two great Tertiary orogenic belts that seemingly coalesce in the East Indies. The southern belt, usually designated as the Alpine Himalayan belt, passes through Turkey, Iran, Afghanistan, the Himalayan region and thence into Burma. Parts of Thailand seem also to have been affected by Tertiary and Quaternary orogenic and magmatic events since numerous hot springs are still active in places. The eastern belt includes the mountainous volcanic islands of Sumatra, Java, Borneo and Sarawak, New Guinea, the island arcs of the Celebes, the Philippines, Taiwan, Ryukyu Islands, Japan, Sakhalin Island, the Kuriles and Kamchatka and the region to the northeast. Bordering the sea of Okhotsk and extending into the region of northeast Chukotka there is another volcanic fold belt, designated by some Russian writers as the East Asiatic tectonic-magmatic belt, that was active in the Tertiary and is still tectonically active in places. These great belts of Tertiary and Quaternary orogeny affect rocks of many ages, ranging in some places from Precambrian to Recent. The eastern belt in particular was the site of great volcanic activity in the Tertiary and is still the site of such activity in places today.

The Alpine-Himalayan orogenic belt contains gold deposits related to Tertiary orogenic events, but most of the hypogene veins and lodes mined essentially for gold appear to have been small. The Anatolia goldfield in Turkey extends south from the Dardanelles (and the site of ancient Troy) to the valley of the present Gediz, the river a tributary of which was known in antiquity as the Pactolus and from which the treasure of Croesus is said to have come. Still farther south in the Boz Dagh (Mount Tmolos) area and in southern Anatolia in the Bolkar-Dag (Taurus Mountains), Tertiary hypogene gold deposits are also known. Most of the gold in the Anatolia goldfield was won from Tertiary and Quaternary placers or from the oxidized zones of gold-quartz and polymetallic deposits. These deposits occur in mica schists, liparites, augite andesites and basalts of both Tertiary and older age. The veins and lodes contain essentially quartz, pyrite, argentiferous galena, sphalerite and chalcopyrite. The Au/Ag ratio appears to range from 1:10 to 1:5.

The Pontic goldfield, on the south shore of the Black Sea in the Trabzon (Izpir) district of Turkey, is in an area of Tertiary volcanics and sediments. One theory has it that this is the field in which Jason sought the 'Golden Fleece' since the ancient miners there used sheep fleeces to catch the gold during placer operations. Yet another theory holds that this is the site of the golden land of Havilah mentioned in Genesis 2(10–12) (*see also* Chapter I). Few hypogene gold deposits actually occur in this area. Most of the gold appears to have been derived from the oxidation zones of massive and disseminated Pb-Zn-Cu sulphide bodies.

Descriptions of the gold deposits and occurrences in Turkey are given by Karajian (1920), Bayramgil (1945), Nahai (1958), Ryan (1960) and Gümüs (1970).

Gold-bearing deposits occur on the eastern extension of

the Turkish Pontic belt in Armenian S.S.R., east of Yereven and in other parts of the Trans-Caucasus (Konstantinov, 1973*a*, 1974; Borodaevskaya and Rozhkov, 1974). There, the deposits are apparently of Oligocene to Early Miocene age and occur in intrusive rocks (gabbro, diorite, granodiorite) and effusive-sedimentary rocks (rhyolites, quartz albitophyres, andesite porphyries, tuffs, etc.). The orebodies are gold-quartz-sulphide zones, impregnations and veins (Zod, Tutkun, Lalikh, etc.), and fracture-filled veins (Megradzor, Armanis, Agduzdag) in great fault zones. The mineralization is essentially quartz, carbonates, pyrite, arsenopyrite, chalcopyrite, tetrahedrite, various other sulphosalts, stibnite, gold, tellurides and a host of other minerals in small amounts. Listwänitization of gabbros and serpentinites, beresitization of rhyolites and silicification, chloritization, pyritization and carbonitization of various other rocks constitute the principal wall-rock alteration effects. In Iran there are also gold-quartz veins and polymetallic deposits that are mainly of Tertiary age (Wilson, 1948; Bariand and Pelissier, 1972).

In Afghanistan gold-quartz lodes, probably of Tertiary age, were exploited near Kandahar, and a number of gold placers have been worked in the northeast part of the country. The gold in the placers is evidently largely derived from extensive, lean gold-bearing conglomerates derived from the erosion of the Pamirs (Levat, 1903).

In Burma, gold-quartz veins mainly in Tertiary andesite near Nankan have been worked in the past. The gangue is quartz with carbonates, pyrite, chalcopyrite, franklinite and tellurides, mainly altaite. The great Bawdwin lead-zinc-silver mine produces a little gold. Placers derived mainly from lean Tertiary deposits occur at various places along the great Irrawaddy River and its tributaries: also along a number of other rivers.

The gold mines in Indonesia (Sumatra, Java, Celebes and Borneo) and Sarawak are essentially similar and can be described in general. All are in Tertiary andesites, dacites, trachytes and rhyolites, Tertiary sedimentary rocks including tuffs, conglomerate, sandstone, limestone and shale, or in older rocks most of which are of Mesozoic or Paleozoic age.

The deposits tend to occur close to the Tertiary andesites and other eruptive rocks. All are essentially gold-quartz veins attended by extensive propylitization of the andesites, often on a regional scale. The gangue is quartz with much calcite, rhodochrosite, rhodonite, adularia and zeolites in places. The metallic minerals are variable amounts of pyrite, arsenopyrite, sphalerite, galena, chalcopyrite, tetrahedrite, argentite, pearceite and polybasite.

Many of the veins are crustified and have abundant cockade structures. Much of the gold is free, but some is intimately associated with the various sulphides. Electrum is common in many of the veins, and gold-silver-tellurides occur in a few. The Au/Ag ratio of the ores range from 1:1 to 1:40 in most deposits. The Radjang Lebong deposits contain considerable amounts of selenium. The deposits at Bau in Sarawak are replacements and disseminations in Upper Jurassic to Lower Cretaceous limestone, sandstone and shale, intruded by Tertiary (?) dykes, stocks and sills of dacite and granodiorite porphyry (Wolfenden, 1965; Pimm, 1967). They are marked by considerable zones of silicification (quartz and calc-silicate minerals) in which variable amounts of pyrite, arsenopyrite, galena, sphalerite, proustite, cinnabar, native arsenic, native antimony, realgar, orpiment, stibnite and jamesonite occur. Most of the free gold is microscopic, and there is seemingly much gold in the lattice of the native arsenic according to the investigations by Lau (1970). Placers derived from the hypogene Tertiary deposits are worked in places in Indonesia (Sigit *et al.*, 1969).

The gold deposits of the Philippines occur mainly in Tertiary rocks in the Baguio, Aroroy, Paracale (Mambulao) and Suyoc districts. The host rocks are mainly andesites and associated sediments intruded by porphyritic dacites and quartz diorites. The deposits are mainly gold-quartz veins and polymetallic deposits in fissures and faults. The gangue is essentially quartz with manganiferous carbonate, anhydrite and calcite. The quartz is frequently opaline. The metallic minerals include pyrite, sphalerite, galena and chalcopyrite. Crocoite ($PbCrO_4$) occurs in some of the veins near Manila. The gold is mainly free. A recent description of the Philippine gold deposits is given by Bryner (1969).

The gold deposits of Taiwan are mainly gold-quartz veins in Tertiary andesites and associated sediments intruded by andesite, dacite and various porphyries. The mineralization is typical of many Tertiary deposits and includes quartz, barite, rhodochrosite, alunite, pyrite, enargite, chalcopyrite, sphalerite, galena and stibnite. Much of the gold is free.

The gold deposits of Japan are mainly of Tertiary (Neogene) age, although some (Omidaui, Asahi and Kabutoyama in southern Honshu) are of Cretaceous age. Those of Neogene age are described by Nishiwaki *et al.* (1971).

The Miocene and Pliocene geology of Japan is characterized by a volcanic and associated sedimentary complex extending from Northern Hokkaido to southern Kyushu, comprising andesites, dacites and rhyolites interbedded with tuffs, breccias and terrestrial and marine shales, limestones, arkose, etc. The andesites are highly propylitized as are also many of the tuffaceous rocks giving rise to the general term 'green tuff' used by many Japanese geologists. This term is somewhat equivalent to 'greenstone' used by Canadian and other geologists for volcanic terranes in Precambrian rocks. Nearly all of the economic gold-silver deposits of Japan occur in these 'green tuff' rocks, as do also many of the massive sulphide deposits called 'kurokô' by the Japanese. The latter deposits also yield considerable amounts of silver and gold.

The principal types of Japanese gold-silver deposits include gold-quartz fissure veins, gold-quartz stockworks and massive gold-quartz stockwork deposits in which the intervening host rock is intensely silicified. The gangue is invariably quartz often with adularia, barite, calcite, rhodochrosite, rhodonite, inesite, johannsenite, ankerite, sericite, kaolinite and other clay minerals. The quartz is commonly chalcedonic with a rhythmic banded crustification of white to grey quartz alternating with grey to black fine-grained sulphides greatly enriched in gold and silver (the 'ginguro' of the Japanese prospectors). The metallic minerals occur in variable amounts and include pyrite, chalcopyrite, sphalerite, galena, arsenopyrite, stibnite, ruby silver, argentite, pearceite, polybasite, stephanite, miargyrite, native silver and tetrahedrite-tennanite. The gold minerals are mainly gold-electrum usually ranging from 10 to 20μ in diameter. Gold-silver tellurides are rare as are also the selenide minerals. In some veins the polybasite and argentite are rich in selenium. Enargite and luzonite occur in some deposits, particularly in the large silicified Kasuga and Akeshi deposits in southern Kyushu. The latter deposit also contains abundant native sulphur. The Au/Ag ratio in some deposits is 1:1 or less, in others 1:10 to 1:40 or more. The elements accompanying gold in the Tertiary deposits of Japan include: Cu, Ag, Ba, Zn, Cd, (Hg), (B), Pb, As, Sb, (Bi), Se, Te, (Mo), (W) and Mn. Those in parentheses are rare or occasional.

The Tertiary terranes and orogenic belts of the far eastern U.S.S.R. including the Kuril Island arc, Sakhalin Island and Kamchatka, and the great volcanic fold belt bordering the Sea of Okhotsk and extending into Chukotka contain a number of hypogene gold deposits similar in many respects to those found in Japan and elsewhere in the eastern Tertiary orogenic belt. Placers derived from these deposits are widespread (Fig. 87). Summary descriptions of the gold-bearing areas and deposits are given by Nikol'skii (1963), Radkevich *et al.* (1965, 1969), Sidorov (1966), Yugai (1967), Aprelkov and Kharchenko (1968), Shilo *et al.* (1969), Rozhkov (1971*b,c,d*), Bolarovich (1972), Zagruzina (1972), Borodaevskaya and Rozhkov (1974) and Roslyakova and Roslyakov (1975).

The gold-bearing deposits of the far eastern U.S.S.R. can be classified into four categories: gold-quartz, gold-rare metal, gold-silver and gold sulphide (Shilo *et al.,* 1969).

The gold-quartz deposits are mainly veins and irregular bodies generally related to folding. They are particularly common in the Mesozoic rocks of the northeastern U.S.S.R. Their mineralization consists essentially of quartz, carbonates, pyrite, arsenopyrite, galena, sphalerite, sulphosalts and native gold. The age of many of these deposits is apparently middle to late Mesozoic (Zagruzina, 1972).

The gold-rare metal deposits are mainly veins and lodes associated with faults and fracture systems and are said to be related mainly to granitoid intrusions of the volcanoplutonic gabbro-andesite-granodiorite clan. The mineralization occurred in at least two stages – first greisenization with the development of wolframite and molybdenite followed by quartz, carbonates, disseminated base metal sulphides, cassiterite (wood tin in places), wolframite, scheelite, bismuth tellurides, gold and gold-silver tellurides. The age of this mineralization is Late Cretaceous–Paleogene.

The gold-silver deposits are veins and stockworks associated mainly with deep faults. Their mineralization comprises essentially quartz, commonly colloform banded, rhythmically crystalline banded, drusy or platy, adularia, rhodonite, rhodochrosite, base metal sulphides, hematite, stibnite, argentite, silver sulphosalts and gold. Some deposits are transitional to the gold-rare metal-polymetallic types of deposits. Propylitization is characteristically developed in the volcanic host rocks of many of the gold-silver deposits. Most of the gold-silver deposits are Late Cretaceous to Paleogene in age.

The gold-sulphide deposits are complex mineralogically, comprising essentially quartz, albite, carbonate, chlorite, hydromica, pyrite, chalcopyrite, galena, sphalerite, tetrahedrite, silver sulphosalts and gold. The sulphides may make up 30 to 40 per cent of the ores. Propylitization of the andesite host rocks of the gold-sulphide deposits is common and widespread. The age of the deposits is apparently Late Cretaceous–Paleogene, mainly the latter.

From the descriptions of various deposits in the far eastern U.S.S.R. it would appear that propylitization is not as common as in other Tertiary deposits, but in some areas of Tertiary andesites and older basic volcanics this type of alteration is well developed. Alunitization is common in some deposits in Kamchatka and elsewhere. Argillization is distinctive in a number of deposits, and hydromicas occur abundantly in many veins. Adularia is a frequent mineral in numerous deposits. Some deposits have an early phase marked by quartz, cassiterite and tourmaline. Bismuth minerals, molybdenite and wolframite are present in many of the deposits but rarely in commercial amounts. Tellurides are common in some veins, selenides in others. Zeolites occur in places and manganese minerals (rhodonite, rhodochrosite) are a general feature of many of the deposits. The gold is commonly finely divided and usually low in fineness ($<$500–800). The elements accompanying gold in the Tertiary deposits of the far eastern U.S.S.R. include Cu, Ag, Ba, Zn, Cd, Hg, B, Pb, Sn, As, Sb, Bi, Se, Te, Mo, W, U and Mn.

Tertiary rocks are relatively widespread in Australasia, particularly in Australia, New Zealand and New Guinea. The Tertiary of Australia is characterized by sandstone, shale, marl, oil shales, clays, lignite, limestone, chert, tuffs and thick basalt flows in some areas. The small gold deposits of the Cygnet area of Tasmania were thought to be of Tertiary age (Jones, 1947), but are now considered to be Cretaceous (Warren, 1972). With the possible exception of the Cygnet area, hypogene gold deposits of Tertiary age are unknown in Australia, but gold placers, both of eluvial and alluvial character, were formed on a vast scale in Victoria and New South Wales in Tertiary and later times. Many of these were covered with sediments and basalt flows often several hundred feet thick. These are the 'deep leads', which produced much of the gold of Australia before the turn of the century.

The 'auriferous series' of the Coromandel Peninsula of New Zealand in the north part of North Island are composed mainly of Tertiary andesites, dacites, breccias and tuffs, all much propylitized, in places on a regional scale (Williams 1974). These rest on folded and faulted Paleozoic and Mesozoic sediments and volcanics. The deposits are gold-quartz veins that lie chiefly in the Tertiary andesites and other volcanics, but at places some extend into the Mesozoic sediments. The gangue minerals are essentially quartz, calcite, ankerite, adularia, barite, rhodonite and rhodochrosite. The metallic minerals are pyrite, arsenopyrite, chalcopyrite, galena, sphalerite, cinnabar, tetradymite, proustite and argentite. The gold is generally free although in some veins it is not visible and is closely associated with pyrite, arsenopyrite and other sulphides. Most of the native gold is of low fineness, the average being probably about 700 or slightly less; gold-silver tellurides occur in some veins. The Au/Ag ratio in the deposits averages about 1:7. Some of the veins in the Thames and other districts were fabulous bonanzas, averaging thousands of ounces of gold and silver to the ton. The veins were all deeply oxidized. The elements concentrated with gold in the Tertiary deposits of New Zealand include Cu, Ag, Ba, Zn, Cd, Hg, (B), Pb, As, Sb, (Bi), Se, Te, (Mo), (W) and Mn. Those in parentheses are rare or occasional.

Manifestations of gold mineralization still persist in New Zealand along the Taupo Volcanic Zone of North Island

(Weissberg, 1969). This zone comprises a prism of Quaternary and upper Tertiary volcanic rocks striking northeast in a volcanic-tectonic graben. Hot springs predominate along the eastern margin of the volcanic zone at Waimangu, Waiotapu, Ohahi-Broadlands and Rotokawa. The sinters of these springs and the drillhole discharge precipitates contain abundant opaline silica and amorphous sulphides greatly enriched in Fe, As, Sb, Au, Ag, Hg, Tl, Pb, Zn, P and W. The Au and Ag contents range from 1 to 85 and 5 to 500 ppm respectively. The wall-rock alteration of the rhyolites, ignimbrites and associated lacustrine sediments near the thermal springs and their conduits is marked by the presence of adularia, sericite, quartz, montmorillonite, illite, pyrite, calcite and chlorite.

In New Guinea, Papua and New Irian, gold-bearing deposits are known in several areas. Most are of Tertiary age. Those in New Guinea (Morobe Goldfield, Kainantu Goldfields; Porgera gold deposits) are apparently of late Tertiary (3.5 m.y.) age (Page and McDougall, 1972; Rebec *et al., in* Knight, 1975). The deposits occur in a variety of settings, most being in Tertiary volcanics and sediments. Some are quartz lodes and veins in volcanics and sediments and others are quartz stockworks in highly altered porphyry stocks and sills (gold porphyry deposits). The characteristic metallic minerals are pyrite, chalcopyrite, bornite and molybdenite in the auriferous porphyries. In the banded quartz-carbonate (rhodochrosite) veins pyrite, chalcopyrite, galena, silver sulphides and sulphosalts and minor cinnabar occur. The average primary gold content of the lodes in the Morobe goldfield is about 0.5 oz/ton, and the gold-silver ratio probably averages about 0.04. Many of the near surface parts of the veins, vein zones and auriferous stockworks were greatly enriched in both gold and silver. Both eluvial and alluvial placers have been worked all over New Guinea, Papua and New Irian. Those in the Morobe Goldfield have been highly productive (Fisher, 1945).

The Fiji Islands, lying north of New Zealand, are underlain mainly by Tertiary rocks in which the Vatukoula goldfield on the island of Viti Levu is situated. Blatchford (1953), Cohen (1962) and Denholm (1967) describe the ores of this field as quartz veins in fissures cutting basalt, volcanic ashbeds and andesite dykes of Miocene age. The gangue minerals are quartz and calcite with pyrite, sphalerite, chalcopyrite, pyrrhotite, marcasite and tetrahedrite-tennantite. Gold-silver and other tellurides, including krennerite, sylvanite, empressite, petzite, coloradoite, calaverite and native tellurium, are abundant in the ores (Stillwell, 1949; Markham, 1960; Forsythe, 1968). Native gold when present is mainly free, but some occurs in a submicroscopic form intimately associated with pyrite, arsenopyrite and tetrahedrite-tennantite. Similar deposits occur in Vanua Levu at the former Mount Kasi Gold Mine. The gold-silver ratio in the Fijian ores appears to average about 3:1. The submicroscopic 'native gold' is estimated to be 940 fine by Stillwell (1949).

Remarks on the metallogenesis of gold

As we look back through the pages of geological history, spanning some 3 billion years, we see few distinct patterns with respect to the metallogenesis of gold as related to the great orogenic events that have punctuated the history of the earth. We see dimly a possible relation of gold deposits to particular rock types such as volcanics and their associated sediments, and an indistinct pattern in the elemental constituents of gold deposits, not so much with regard to the elements that are present in the deposits but mainly with respect to those that are not present, or occur only in trace amounts.

In discussing the metallogenesis of hypogene gold deposits we must consider three fundamental aspects: (1) the source of the elements in the deposits; (2) the pathways, including conductive structures and permeable rocks, by which the elements in the deposits reached their sites of deposition; and (3) the sites of deposition or traps containing the deposits, including favourable structures and favourable host rocks for replacement or impregnation. The last two aspects require little discussion here, except to say that during orogenesis there are generally sufficient structures formed, and these are usually extensive enough both in their strike and dip dimensions to tap enormous volumes of rock that could have supplied the constituents of the veins and other types of deposits. In addition, in most orogenic districts there are sufficient structural traps and favourable replaceable rocks for the formation of epigenetic gold deposits. The problem therefore, is one mainly of the source of the gold and accompanying elements. Did these elements come from very deep sources, for example the mantle, from magmas which are generated in most orogenic belts or from the piles of volcanic and sedimentary rocks which constitute the zones affected by orogenesis? Some of the answers seem relatively clear in the light of the preceding discussion of gold deposits in the orogenic belts of the world and the geochemical data given in Chapter II.

When we look in detail at the successive orogenies in the geological history of the earth we note that none are devoid of gold deposits. There appear, however, at first glance to be two general maxima in the number of gold deposits of all types (epigenetic, placer, modified placer, etc.) formed – one maxima characterizes Precambrian time and the other Tertiary time. The orogenies of the Paleozoic and Mesozoic have a number of associated significant economic deposits in many parts of the world, but the total number of such deposits in both epochs is far overshadowed by those of the Precambrian and Tertiary. Perhaps this is an allusion – future prospecting and development may bring to light many deposits that are related to the Paleozoic and Mesozoic orogenies. However, the gold deposits of the Witwatersrand are so overwhelming in their size that one cannot but conclude that the Precambrian was the greatest period of formation of gold deposits in the history of the earth. The Tertiary, on the other hand, may be rivalled or superseded by deposits of Paleozoic and Mesozoic age yet to be discovered. Such seems to be the case as the new gold deposits in Paleozoic and Mesozoic terranes are discovered, developed and brought into production in the U.S.S.R. and elsewhere. We are, therefore, probably only justified in concluding that the Precambrian was the greatest period of gold deposition, and that later periods have about equal numbers of gold deposits. But a closer analysis narrows our conclusion still further. The Precambrian spanned an enormous period of time, some 2.5 billion years, nearly four times that of all of the later eras combined. Furthermore, as the geological histories of the shields and other Precambrian areas in the world are carefully documented, it becomes evident that the number of major orogenies in Precambrian time about equal and may exceed those of later time. It will be recalled that there are at least four major orogenies in the Canadian Shield, and perhaps a number of minor orogenies within these major diastrophic manifestations that remain to be worked out. In Canada, gold deposits are associated with three of the main Precambrian orogenies; in other countries a similar pattern is revealed, numerous gold deposits being associated with the orogenic events that closed the Archean as well as with those that occurred in the Proterozoic. The age of the Witwatersrand system, containing the world's largest gold deposits, was at one time thought to be Proterozoic but is now considered to be Archean (2900 m.y.) on the basis of radioactive age determinations. However, the stratigraphic position of the Witwatersrand system *vis-a-vis* Archean-type greenstones and granitic rocks and numerous other features mentioned in a later section of this chapter suggest to the writer that this great auriferous quartz-pebble conglomerate system is of Aphebian (Early Proterozoic) age. If so, the Early Proterozoic was the greatest period of gold deposition in the history of the earth rivalled only by the Archean. The orogenies that followed in Late Proterozoic and the following eras appear to have about equal numbers of gold deposits, at least to a first approximation.

Those who advocate deep sources, such as the mantle, for the gold and its associated elements in epigenetic deposits have relatively few points in their favour. It is conceivable that some sort of ichor may have delivered gold and its associated elements to favourable structures via fault and fracture systems that penetrate deeply into the earth's crust. It is even conceivable that the generation of gold was marked in Archean time because the source (mantle?) was much nearer the contemporaneous surface than at later times during which the crust of the earth gradually thickened. There are, however, a number of facts that seem to contradict the deep source hypothesis. If deep fault and fracture zones were the conduits up which gold and its associated elements migrated one would expect to find a marked distribution of gold deposits along or near these structures. Such is not the case – few gold deposits are actually localized in the great fault and fracture zones of the earth. Most are in or near relatively small and local structures within a geological terrane, a particularly good example being the formation of gold-quartz lenses and veins in dilatant zones on the crests of anticlines in sedimentary rocks (Bendigo type). Many more examples could be cited such as quartz lodes in drag folds in sediments, in local fracture systems in volcanic necks and dykes, and so on. Furthermore, the formation of most gold deposits seems to be keyed to metamorphic, granitization and magmatic events within a particular area and not to great dislocations in the crust. This relationship is clearly shown by many of the gold deposits in the Canadian Shield. In the Yellowknife district, for example, early shear zones and fractures formed essentially in the volcanic (greenstone) belt, and a variety of local dilatant zones restricted mainly to a greywacke-slate suite of sediments, were extensively mineralized with gold and silver during and following metamorphism, granitization and injection of granitic rocks. Later fault and fracture systems, some extending for distances of 50 mi or more are barren even though they contain many large quartz bodies in places. The writer (Boyle, 1961*a*) postulated that the fundamental differences between the two periods of mineralization in the Yellowknife district can be related to metamorphic events. The productive gold deposits were formed while regional metamorphism, granitization and intrusion of granites was operative; when the late fault and fracture systems were mineralized the wave of regional metamorphism, granitization, etc. had long since subsided. This theme will be further developed in the last section of this chapter.

Numerous investigators have related the occurrence of gold deposits in Archean and Tertiary times, as well as those in other periods, to volcanism and associated sedimentation. This correlation is valid for many gold belts in the world, but a direct relationship to magmatism during volcanic periods is improbable on several counts. As shown in Chapter II, basalts, andesites, rhyolites and other magmatic rocks characteristic of volcanic periods are not enriched in gold. On the other hand the sediments associated with volcanic piles frequently show marked enrichments in gold and silver and their associated elements. In the Canadian Shield these features are quite evident when the analyses of the magmatic volcanic rocks, which average 0.007 ppm Au, are compared with those of their associated sediments that may contain up to 0.1 ppm Au or more. (*See also* Table 16 on the content of gold and silver in iron formations.) One can logically conclude from these data that the main enrichment of gold in volcanic piles is in the sediments, not in the basalts, andesites, etc.

The timing of the formation of gold deposits in volcanic terranes is a most important consideration. Throughout the world gold-quartz veins and other types of gold deposits postdate the volcanic events by a period of time. How long is uncertain, but certainly after consolidation of the lavas and sediments and the injection of igneous dykes and stocks, after folding or block-faulting of these rocks and in most areas after extensive fracturing, faulting and shearing, not only of the volcanic rocks and their associated sediments, but also of the granitic stocks, dykes and batholiths that invade them at a late stage in the orogenic cycle. One can only conclude from these features that there is no direct relationship between volcanic magmatism (i.e., outpouring of lavas, injection of dykes, etc.) and the gold deposits in volcanic terranes.

Tertiary epigenetic gold deposits throughout the world exhibit an intimate relationship to volcanic centres often marked by andesite and dacite flows, dykes, stocks, volcanic necks and tuffaceous sediments. At first glance one might conclude that these deposits are directly related to volcanic magmatism, and some investigators have indeed asserted this view. A close examination of the geological data, however, indicates no direct relationship between the formation of the gold deposits and the magmatic events. An indirect connection is, nevertheless, probable as will be shown presently.

We need not repeat the same arguments about the gold content of andesites, etc. and the paragenesis of the veins with respect to consolidation of the rocks, dyke and stock intrusion, and faulting and fracturing. One sees exactly the same phenomena with respect to epigenetic Tertiary gold deposits as that present in older rocks. Tertiary gold deposits postdate all of the magmatic events except the formation of some late postore dykes in places. Most Tertiary gold deposits appear to be related to thermal waters, which pervade the volcanic and sedimentary piles and today in places appear as hot springs, the sinters of which, in New Zealand particularly, contain gold. All of the evidence from isotopic and other studies suggests that these waters are of meteoric or metamorphic origin. They appear to have collected their gold and other associated elements from the rocks through which they pass. The source rocks are, therefore, of considerable importance.

There are few detailed studies on the source rocks that provided the gold in Tertiary terranes. Judging from the relatively low contents of gold in andesites and dacites these rocks would seem to be poor sources for the metal. More probable source rocks would seem to be the tuffs and sediments in the Tertiary terranes. Even more probable source rocks are those older rocks that form the basement on which the Tertiary rocks were laid down. In many areas these are sediments, often of the carbonaceous pyritic variety, which are considerably enriched in gold wherever they occur. During the dying phases of volcanism in Tertiary times these rocks were evidently heated and permeated by meteoric and metamorphic waters, which mobilized the gold and carried it upward to be deposited in available fractures and faults in the overlying Tertiary andesites and other related volcanics. The Tertiary rocks, mainly andesites, are, therefore, not the main source of the gold but rather the chemically reactive receptacle rocks, which provided the necessary dilatant zones and favourable chemical environment for the deposition of auriferous orebodies. These concepts are discussed in more detail in the last section in this chapter on the origin of gold deposits.

Certain metallogenetic facts should be noted with respect to the relationship of epigenetic gold deposits to volcanism. Considering the Canadian Shield it is apparent that those provinces marked by an abundance of Archean greenstone belts are rich in epigenetic gold deposits, e.g., the Superior and Slave provinces. Those provinces containing relatively few greenstone belts of Archean or younger age are poor in economic gold deposits, e.g., the Bear, Churchill, Southern, Grenville and Nain provinces. On a worldwide basis the same features are evident. Precambrian shields marked by extensive greenstone belts are highly auriferous (e.g., Canadian, Indian, Australian, South African–Rhodesian and Guianian–Brazilian), whereas those shields with few greenstone belts are poor in epigenetic gold deposits (e.g., Baltic, Ukranian, Taimyr, Anabar and Aldan). On a more detailed basis greenstone belts that are complex, containing abundant Algoman-type iron-formations, tuffs, sulphidic interflow sediments and discordant (intrusive) porphyry bodies, and exhibiting extensive shearing, carbonatization and well developed metamorphic facies, especially an inner greenschist facies, are rich in economic epigenetic gold deposits; those composed mainly of piles of volcanics, exhibiting only one facies of metamorphism, tend to be relatively barren. Similarly in Tertiary volcanic terranes, and even in those of Paleozoic and Mesozoic age, complexity seems to produce epigenetic gold deposits. Tertiary and older volcanic terranes marked by quartz porphyry and dacitic plugs, widescale faulting and fracturing, and extensive propylitization are auriferous (e.g., western United States, New Zealand, Romania), whereas those piles of Tertiary volcanics that are relatively simple and unaltered are barren (e.g., northern British Columbia, western Scotland, Ireland, Iceland).

Numerous investigators have attempted to relate gold deposits to orogenic magmatic events, involving either granitization followed by intrusion of the mobile material as granitic bodies, or intrusion of granitic bodies derived by differentiation of mantle or lower crustal materials (Hillebrand, 1954). Gold deposits are supposed to represent one of the hydrothermal stages of magmatic differentiation, an hypothesis that is familiar to all. A detailed criticism of this hypothesis is given at the end of this chapter. It is sufficient to state here that some gold-quartz veins and other auriferous deposits exhibit a spatial relationship to intrusive granitic bodies in orogenic zones throughout the world; other gold-quartz veins and gold-bearing deposits show no such relationship. Nearly all types of epigenetic gold deposits, however, are restricted to rocks that exhibit a low to moderate degree of metamorphism (greenschist to amphibolite facies). One does not find epigenetic gold deposits in rocks that do now show the effects of considerable recrystallization and alteration such as regional propylitization. From this fact the logical conclusion seems to follow that epigenetic gold deposits are more closely related to metamorphic than to magmatic phenomena. In the grand scheme of orogenesis the occurrence of epigenetic gold deposits in the vicinity of intrusive granitic rocks is not fortuitous. The reason for this, however, may not be that the granitic bodies provided the gold, but that these bodies are simply one in the series of products of intense metamorphism. These bodies also include the gold deposits, the gold being derived from piles of sedimentary and volcanic rocks and concentrated as a result of granitization and later metamorphic processes that continued long after the emplacement, crystallization and consolidation of batholiths, stocks and dykes of granite, granodiorite, etc. This theme is developed further in the last section of this chapter on the origin of gold deposits.

The distribution of the great gold belts of the world, ranging in age from Precambrian to Cenozoic, are shown on Figures 3, 5, 6 and 7. Any connection between the belts on one continent and those on another is tenuous to say the least. No relationship appears to exist between the Appalachian (Paleozoic) belt and the Caledonian and Hercynian belt in Europe (or North Africa) as far as the writer can ascertain, although apparently such a possibility exists as regards the European belt (Rast and Grant, 1973). The Precambrian belts in eastern South America and those in western Africa, likewise, appear to bear no particular relationship to each other. Hills (1947) has also drawn attention to the absence of any correspondence between the Precambrian belts in India (Kolar) and those in western Australia. Crawford (1970) has, however, attempted to relate these fields by their age relationships, but his argument on this basis is not convincing. The reader is left to decide how all of this negative evidence is to be resolved in the light of the current wide acceptance of the existence of continental drift.

Pereira and Dixon (1964–1965) discuss evolutionary trends in ore deposition and present a curve that purports to show that there has been an exponential decrease in the grade of gold-bearing deposits with time. This is a difficult problem to deal with because of the lack of relevant data. However, from the writer's experience with gold deposits of all ages this curve cannot be correct. One has only to recall the phenomenal gold bonanzas of the Tertiary. Few such high grade gold deposits occur in earlier periods. Even Tertiary polymetallic deposits seem to show a greatly increased gold content, as in Japan and elsewhere. By comparison, polymetallic deposits of Precambrian, Paleozoic and Mesozoic age are in general much less enriched in gold. The quantitative evaluation of evolutionary trends in gold deposition by Travis (1966), likewise, does not support the conclusions of Pereira and Dixon.

Laznicka (1973) discussed the development of nonferrous metal deposits through geological time and concluded that:

> The sequence of depositional ages based on ore deposits presently outcropping on the earth's surface is, from the oldest to the youngest, Cr,Ni-Au-Cu, Zn-Pb, Ag-Sn, W, Sb, Mo-Hg. Although this sequence parallels several evolutionary trends involved in the formation of the earth's crust, it is believed and semiquantitatively demonstrated that depth of denudation rather than evolution is the more important factor; in other words, the conditions of preservation have had greater influence on the age relationships of ore deposits outcropping on the present earth's surface than have the conditions of formation and their progressive evolution in the earth's history.

The vagaries of the Au/Ag ratio in gold-bearing deposits is considered in some detail in a later section. Here we shall mention only generalities. Overall, Precambrian deposits mined essentially for gold show a high Au/Ag ratio generally greater than 1 and averaging about 5. Paleozoic and Mesozoic deposits, likewise, have ratios generally greater than 1 but probably a lower average, about 3.5 according to the writer's data. Tertiary gold deposits commonly have ratios less than 1, averaging about 0.05. These data seem to suggest that with time silver has been precipitated in increasing amounts compared to gold in strictly gold deposits. In polymetallic deposits of practically all ages, however, silver generally always exceeds gold in amount and the Au/Ag ratio is less than 1, averaging about 0.025. These general statements are negated in a number of gold and polymetallic deposits. One need only recall that some orebodies of Precambrian age (e.g., Yellowknife, Red Lake) have Au/Ag ratios equal to or less than 1. Similarly, some highly productive orebodies of Tertiary age (e.g., Cripple Creek, Fiji) have Au/Ag ratios averaging 3 or more.

The elemental associates of gold in its deposits are discussed at length in the next section. As a preamble some significant facts are presented for consideration. Copper, silver and zinc are invariable associates of gold in epigenetic deposits of all ages; lead is relatively rare in Precambrian deposits occurring generally only as trace amounts in deposits of that age. In younger deposits lead is frequently a minor or abundant element. Mercury is relatively rare in most Precambrian deposits occurring mainly as low traces in the ores. The element appears in trace to minor amounts in younger deposits and may be relatively abundant in those of Tertiary age. Barium is usually present as traces in Precambrian deposits but often appears in minor amounts in those of a younger age. Thallium seems to be enriched in deposits younger than Paleozoic. Tin is relatively rare in practically all epigenetic gold deposits except those rich in sulphides or in stanniferous provinces such as the eastern U.S.S.R. Arsenic is an invariable and intimate associate of gold in all types and ages of epigenetic deposits. Antimony is, likewise, a common associate of gold and is found in practically all types and ages of epigenetic deposits. No particular abundance relationships of arsenic in gold-bearing deposits with respect to type or age of epigenetic deposits are evident. Antimony appears in greatest abundance in vein and lode deposits; it is relatively rare in skarn-type and quartz-pebble conglomerate deposits. Bismuth is relatively rare in most Archean deposits and appears to increase in abundance in those of Proterozoic, Paleozoic and Mesozoic age. In Tertiary deposits its abundance is usually low. Niobium and tantalum are generally only present in trace amounts in all types of epigenetic deposits as are also zirconium and hafnium. Selenium is usually present as traces and minor amounts in epigenetic deposits of all types and ages; in some Tertiary deposits the element is greatly enriched. Tellurium is one of the most frequent associates of gold occurring in practically all types of epigenetic deposits. No age relationships are evident. Deposits in or near volcanic rocks appear to contain the most tellurium; those in sedimentary rocks generally contain only small amounts of tellurium, but some deposits in skarns may be enriched in the element. Molybdenum and tungsten are common in most types and practically all ages of epigenetic gold deposits. The abundance of both elements is usually low. Manganese is found in all epigenetic gold-bearing deposits; in those of Tertiary age the element is frequently abundant as witnessed by the presence of rhodochrosite, rhodonite, inesite, etc. in many deposits throughout the world. Cobalt and nickel are rarely found in any abundance in epigenetic gold deposits, but in some places Co-Ni arsenides and other minerals are notable. Platinum metals are infrequent associates of gold except in the quartz-pebble conglomerate deposits (Witwatersrand) and in placers. Thorium and uranium accompany gold in the quartz-pebble conglomerate and in some skarn, vein and stockwork deposits in uraniferous areas. Most gold deposits containing the two radioactive elements are of Precambrian (Lower Proterozoic) age, although some epigenetic deposits of Paleozoic, Mesozoic and especially Tertiary age are uraniferous.

The auriferous quartz-pebble conglomerates and quartzites are characterized by relatively large amounts of iron in the form of pyrite or the oxides, depending on the particular deposits. The content of sulphur varies almost directly with the amount of pyrite in the ores. The trace and minor element suite (Table 35) is of interest in the origin of these enormous deposits, a point that will be elaborated upon in the last part of this chapter. Briefly, a number of deposits exhibit enrichments of U, Th and rare earths in addition to Au and Ag. There are also enhanced contents of the various base metals in some deposits, including Cu, Zn, Pb, Ni and Co. The As content appears to vary with the content of pyrite in most deposits, and in certain of these deposits the content of gold appears to vary directly with the arsenic content. Deposits with abundant pyrite and other sulphides seem to be enriched in Hg. In general the following elements are usually present in low amounts: In, Tl, Sn, Sb, Bi, Nb, Ta, Se, Te, Mo and W. The general low content of tellurium in these gold deposits is rather remarkable. Some reefs in the Rand (e.g., Carbon Leader) are greatly enriched in platinoids, a feature that is unusual for auriferous deposits.

> Gold – Found in nature as the free metal and in tellurides. Very widely distributed and under a great variety of conditions, but almost invariably associated with quartz and pyrite.
> –F.W. Clarke, 1924

Associated elements in gold deposits

The elements with which gold is most commonly associated in hypogene and supergene deposits are mainly chalcophile and include S, Se, Te, As, Sb, Bi, Cu, Ag, Zn, Cd, Hg, Sn, Pb, Mo, W, Fe, Pt, Pd, Co and Ni. There are in addition numerous other elemental associates that are of interest. The author has made a laboratory and literature study of the elemental associations of gold in all types of its deposits, the results of which are given in the various tables accompanying this section and in the discussion that follows. Particularly helpful in this study have been the condensed data given by Lincoln (1911*b*), Cooper (1923), Schwartz (1944), Liebenberg (1955), Petrovskaya (1955, 1973), Radtke *et al.* (1972*a*), Feather and Koen (1975), and Marakushev (1977). Some remarks concerning the association of gold with each of the elements of the periodic table follow. Tables 31, 32 and 33 give a synoptic view of the elements associated with gold in the various types of deposits. The zoning relationships of gold and various elements in epigenetic deposits is discussed in the last part of this chapter and shown in Figures 94 and 95.

Group IA: H, Li, Na, K, Rb, Cs

Hydrogen requires little comment. The element is an invariable constituent of water and hydroxyl in the various micaceous minerals found in the zones of chloritization, sericitization, etc. associated with many types of epigenetic gold deposits, particularly veins, silicified zones and disseminated deposits. The element is similarily disposed in the sericite, chlorite, chloritoid and pyrophyllite, which forms much of the matrix of the quartz-pebble conglomerate deposits. In the supergene environment hydrogen is a constituent of the water and hydroxyl in hydrous and hydrated oxides of iron, manganese, etc. that invariably occur with gold in the oxidized zones of deposits and in eluvial and alluvial placers.

The role of hydrogen (water) in endogenic wall-rock alteration processes associated with gold deposition is discussed at length in a subsequent section.

Lithium is a typical pegmatite element occurring as a major constituent of minerals such as spodumene, lepidolite, petalite and amblygonite. It is also frequently enriched in certain greisen and micaceous alteration zones associated with tin, tungsten and molybdenum deposits in granitic rocks, gneisses and metasediments. In hypogene gold deposits and their alterations zones lithium is rarely present in amounts greater than 400 ppm. I have not observed lithium minerals in

Table 31. Elements associated with gold in skarn-type, gold-quartz and disseminated deposits. Gangue elements are excluded except if present in abnormally high concentrations.

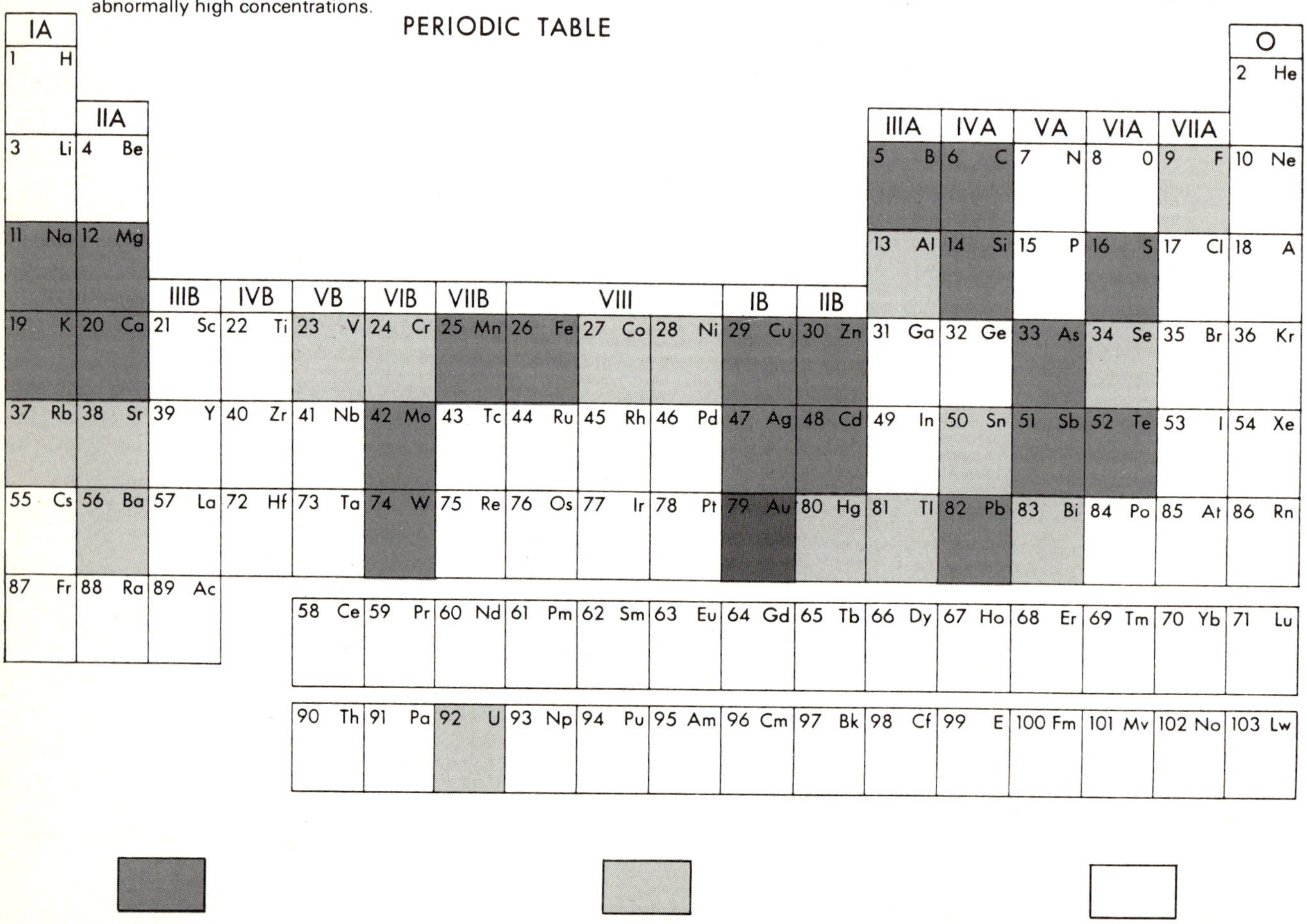

Elements exhibiting a high frequency of enrichment in deposits

Elements exhibiting a low or occasional high frequency of enrichment in deposits

Elements rarely enriched in deposits

Table 32. Elements associated with gold in quartz-pebble conglomerate deposits. Gangue elements are excluded except if present in abnormally high concentrations. (For key *see* Table 33.)

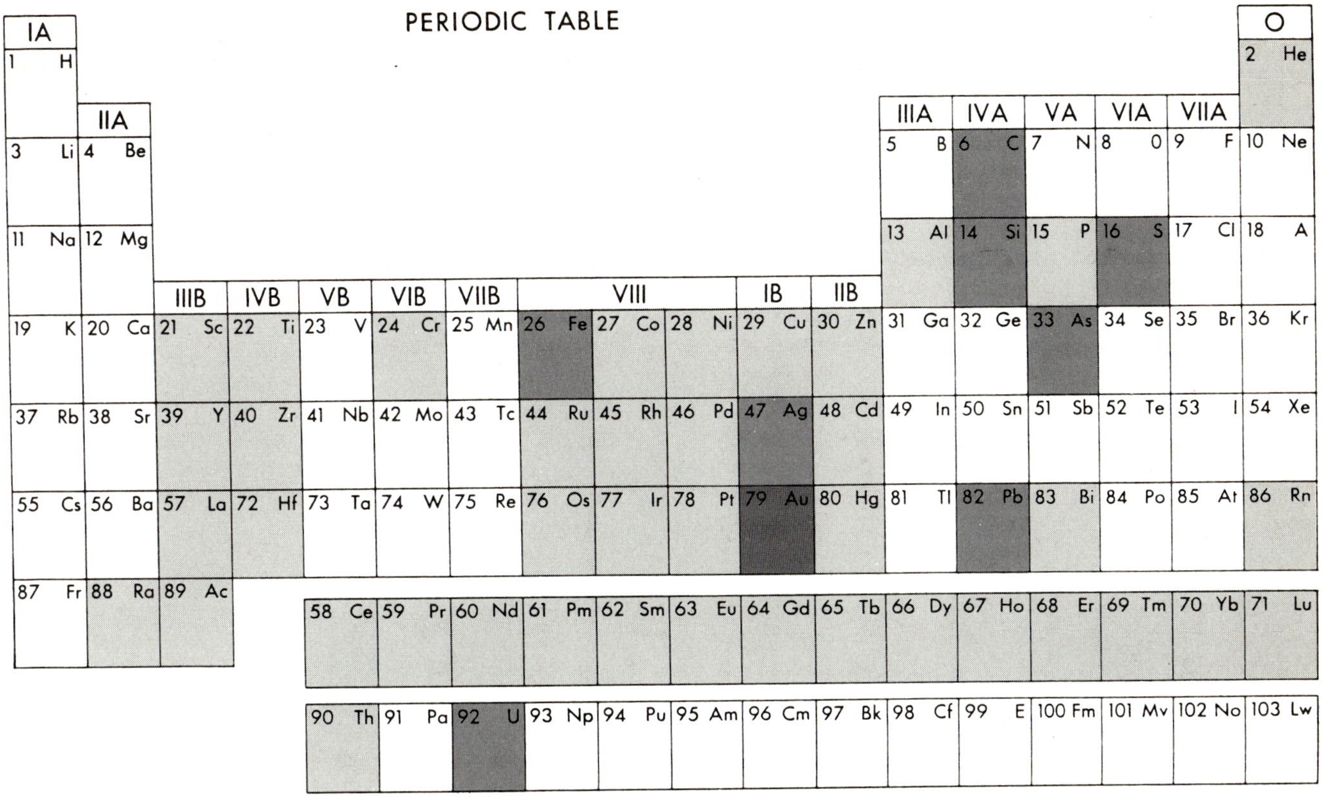

Table 33. Elements associated with gold in eluvial and alluvial placers. Gangue elements are excluded except if present in abnormally high concentrations.

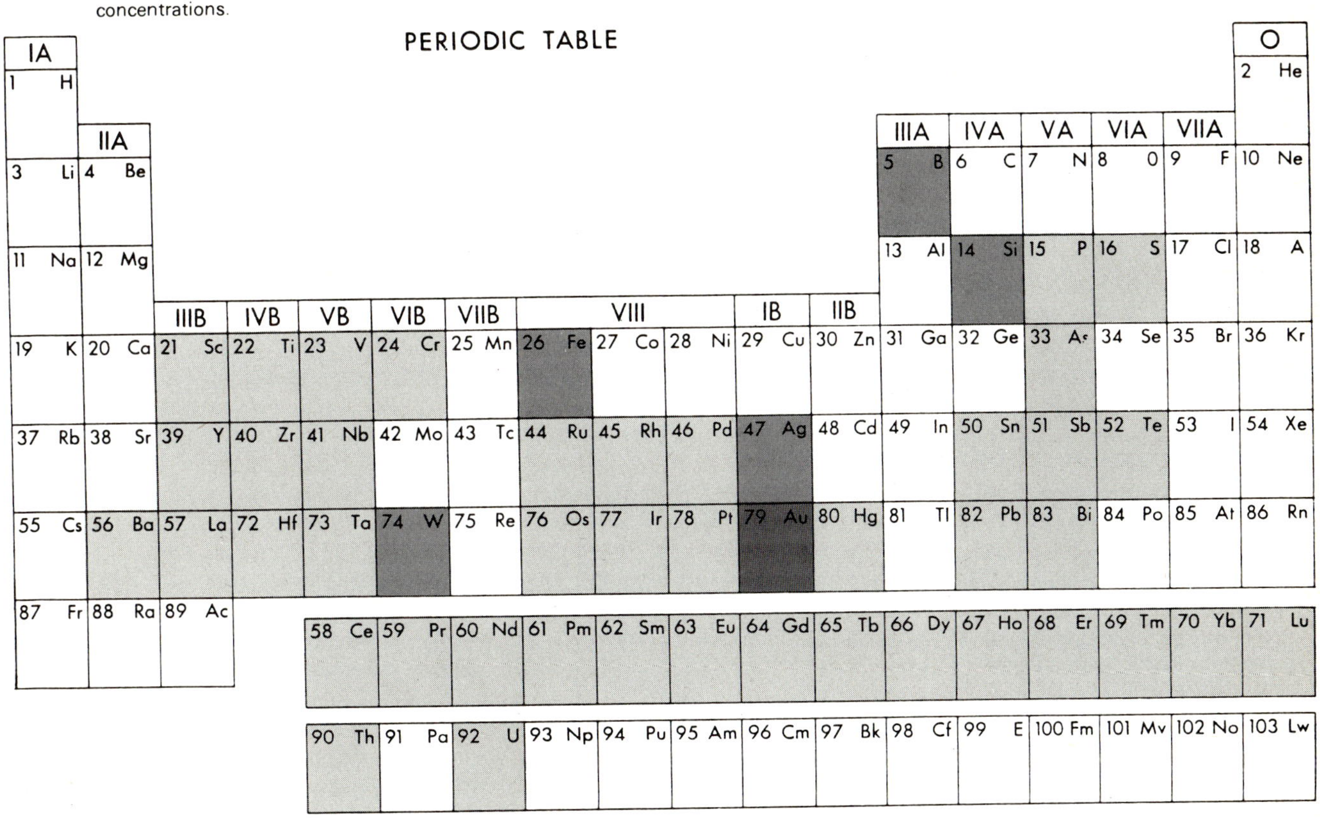

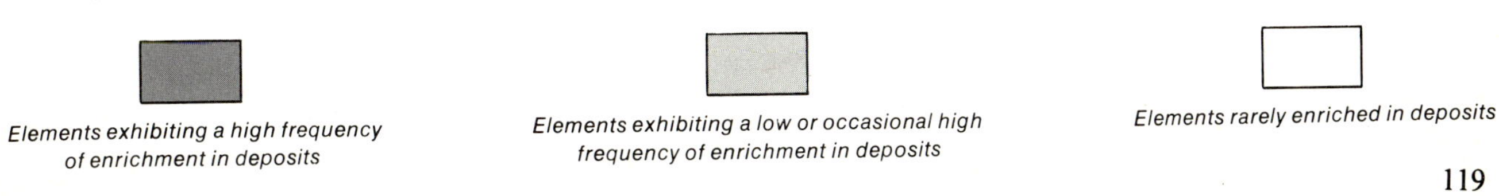

the gold deposits that I have investigated, and there are relatively few references to the presence of such minerals in gold deposits described in the literature.

Lithium minerals are occasionally found in skarn deposits, but the gold-bearing sulphide skarn deposits appear to be relatively low in the element according to my analyses. In auriferous skarn in Yukon, at the New Calumet Mine in Quebec and elsewhere the lithium contents ranged from 6 to 17 ppm. Most of the lithium is present in Mg-Fe amphiboles and pyroxenes, tourmaline and micas. There is no correlation between the lithium and gold contents of skarn deposits as far as I can ascertain.

Gold-quartz veins and silicified bodies and their alteration zones generally contain only small amounts of lithium as shown in Table 34. (*See also* the tables accompanying the section on wall-rock alteration processes.) There is a general

Table 34. Alkali element content of some gold deposits and their alteration zones

Description	Li	Rb	Cs	K_2O	Na_2O	MgO	Remarks
		(ppm)			(%)		
*Yellowknife district, N.W.T., Canada**							
Greenstone Belt:							
Amphibolite (greenstone)	16	11	0.3	0.39	1.85	6.53	Country rock; composite sample of epidote amphibolite facies
Chlorite-carbonate schist	49	12	0.4	0.41	0.90	8.43	Main constituent of shear zones
Carbonate-sericite schist	12	106	4.0	2.82	0.29	5.77	Zone of intense alteration adjacent to quartz lenses
Gold-quartz ore	5	5	<1	0.1	<0.1	0.3	Mainly quartz with some pyrite, arsenopyrite, etc.
Quartz-feldspar porphyry	5	58	2.0	1.63	4.15	0.37	Country rock
Sericite schist derived from above	6	171	6.7	3.94	0.63	0.77	Zone of alteration adjacent to quartz veins
Gold-quartz ore	5	5	<1	<0.1	<0.1	0.3	Mainly quartz with some pyrite, arsenopyrite, etc.
*Sedimentary area (Ptarmigan Mine)**							
Greywacke and minor slate	40	63	4	2.2	3.3	3.0	Country rock
Greywacke and minor slate	76	67	6	2.2	3.3	3.0	2–6 ft from vein
Greywacke and minor slate	79	227	47	4.0	2.0	3.1	6–12 in. from vein
Altered greywacke and slate	66	87	10	2.7	2.8	3.0	0–12 in. from vein
Quartz vein	2	2	<1	0.08	0.01	0.03	Mainly quartz with minor pyrite, sphalerite and galena
Caribou Mines, Nova Scotia, Canada							
Greywacke and quartzite	27	65	2.6	1.56	2.8	0.8	Host country rock
Greywacke	20	90	3.6	2.19	2.7	1.1	Host rock, 10 in. from vein
Greywacke	19	88	3.4	2.17	2.6	1.1	Host rock, 7 in. from vein
Greywacke	19	89	3.7	2.30	2.6	1.1	Host rock, 5 in. from vein
Greywacke	22	89	3.7	2.15	2.7	1.1	Host rock, 3 in. from vein
Greywacke	23	90	3.6	2.21	2.7	1.1	Host rock, 1–2 in. from vein
Hector-Calumet Mine, Galena Hill, Yukon, Canada†							
Gabbro (greenstone)	95	32	3.4	0.22	0.04	7.62	Host rock
Chlorite-carbonate zone	94	144	5.6	1.31	0.53	5.32	Altered zone 1–6 in. from lodes
Carbonate-chlorite zone	24	310	11.6	3.00	0.21	1.85	Altered zone 1 in. from lodes
Bralorne Mine, British Columbia, Canada‡							
Diorite	16	5	0.6	0.3	4.0	4.9	Host country rock
Chloritized diorite	38	15	3.5	0.6	2.7	7.2	10–25 ft from veins
Chlorite-sericite schist	52	46	3.5	2.6	1.6	5.7	0–5 ft from veins
Carbonate-sericite schist	47	64	4.2	3.6	2.3	2.4	Adjacent to veins
Gold-quartz ore	<5	<5	<1	0.3	0.2	0.3	Veins
Kushikino Mine, Japan§							
Andesite (relatively fresh)	49	144	6.5	3.7	2.1	4.0	Host country rock
Altered andesite (propylite)	28	293	11.7	6.6	0.9	1.8	Host country rock
Silicified and chloritized andesite	33	252	5.3	6.5	1.8	1.6	Near quartz veins
Highly silicified and chloritized andesite	130	148	8.5	2.8	0.1	0.4	Adjacent to quartz veins
Witwatersrand quartz-pebble conglomerate deposits, South Africa‖							
Giant Reefs, Project 68/61/A	20	100	3.5	2.53	0.1	3.0	Black (carbonaceous) reef
Sub Nigel-Kimberley reef	12	17	<0.1	0.49	0.05	0.6	Quartz-pebble conglomerate, some fuchsite
Welkom basal reefs	2	12	1.5	0.39	0.06	0.06	Quartz-pebble conglomerate

*For description of deposits *see* Boyle (1961*a*).
†For description of deposits *see* Boyle (1965*a*).
‡For description of deposits *see* Joubin (1948).
§For description of deposits *see* Taneda and Mukaiyama (1970).
‖ For description of deposits *see* Liebenberg (1955).

enrichment of lithium in most alteration zones, excepting those in granitic rocks where there is commonly a depletion. No particular pattern is evident with respect to distance from the veins and lodes. Pampura and Trufanova (1968) noted, however, that the inner zones of the argillic alteration halos were enriched in lithium (200 ppm) compared with the zones remote from the mineralization (130 ppm) in the epithermal Balei gold deposit in eastern Transbaikal. Radanova and Karadzhova (1971) also noted concentrations of lithium in the dickite-quartz facies of the propylites in the Tertiary Spahievo Pb-Zn field in Bulgaria. Much of the lithium in the alteration halos and gold ores is present in micas and clay minerals such as sericite and kaolinite, and in Mg-Fe minerals, particularly chlorite and talc; also in tourmaline where this mineral is present in the alteration mineral suite. In some alteration zones lithium tends to follow magnesium and/or iron, a feature that is in agreement with the general geochemistry of the element. There is generally no particular correlation with K_2O such as that shown by rubidium and cesium.

Lithium probably exhibits a behaviour in auriferous polymetallic deposits similar to that noted above for gold-quartz deposits, judging from the results obtained from analyses of the alteration zones associated with the Pb-Zn-Ag lodes at Keno Hill, Yukon (Table 34).

There appear to be no data on the lithium contents of the disseminated gold deposits such as those that occur in Nevada (Cortez, Carlin, Gold Acres) and elsewhere.

There are no data in the literature on the lithium contents of gold-bearing quartz-pebble conglomerates. Analyses done in laboratories of the Geological Survey on samples from some of the reefs of the Witwatersrand system indicate a range in the lithium content from 2 to 20 ppm (Tables 34 and 35). The black (carbonaceous) reef material appears to contain the highest amounts of the element.

Lithium is a common trace element in most hot spring waters. Those in the Taupo volcanic zone of New Zealand, from which gold-silver ore-grade materials are being precipitated, contain 3.2 to 12 ppm Li (Weissberg, 1969).

The behaviour of lithium during the oxidation of gold deposits appears to be erratic (Table 36). In some cases the element is leached from the oxidized zones; in other cases there may be a slight enrichment. I have noted that lithium may be considerably enriched where wad (hydrous manganese oxides) is abundant in gossans.

Sodium is most commonly associated with hypogene gold deposits, being concentrated in the alteration zones of these deposits, particularly in those in basic to intermediate igneous rocks but also in places in those in granites, shales, slates, skarn, gneisses and similar rocks. Most of the sodium is concentrated in the alteration (metasomatic) feldspars, mainly albite and adularia, with small amounts in sericite (paragonite), chlorite and other alteration minerals. Sodium feldspars (albite or adularia) and micas may also occur in the vein quartz or silicified zones of gold deposits. In places skarn-type gold deposits show an enrichment in sodium, the element being mainly in feldspars and other silicates. Gallagher (1940) and Firsov (1956) have remarked on the frequent occurrence of albite-rich rocks in the vicinity of gold deposits. Reid (1941), Bruce (1941*a*), and Wisser (1944) have commented on this association and have provided a number of exceptions, especially in wall-rock alteration phenomena. The behaviour of sodium in gold deposits is discussed in greater detail in the section on alteration processes.

Potassium is enriched in the alteration zones of many types of hypogene gold deposits, the element occurring mainly in sericite, clay minerals, feldspar (orthoclase, albite and adularia) or alunite. The element is particularly characteristic

Table 35. Trace and minor element content of some auriferous quartz-pebble conglomerate deposits

Element	Giant Reefs, Witwatersrand, South Africa	Sub-Nigel-Kimberley Reef, Witwatersrand, South Africa	Welkom Basal Reefs, Witwatersrand, South Africa	Tarkwa, Ghana	Elliot Lake, Ontario
			(ppm)		
Li	20	12	2	6.6	13
Rb	100	17	12	28	100
Cs	3.5	<0.1	1.5	2.0	2
Cu	137	110	86	10	100
Ag	1.6	2.7	8.6	0.2	2.3
Au	4.35	300	98	11.0	0.090
Be	3.5	<2	<2	<2	<2
Sr	136	13	21	105	18
Ba	330	<50	<50	1050	410
Zn	300	330	132	11	10
Cd	2	0.7	0.3	<0.5	1
Hg	0.130	3.28	1.95	0.040	5.0
B	150	20	10	35	25
Ga	5	5	2	5	8
In	<1	<1	<1	<1	<1
Tl	nf	nf	nf	<100	<100
Sc	16	0.7	10	10	10
Y	10	14	25	9	70
La	26	15	21	45	130
Ce	nf	nf	nf	nf	5
Th	2	15	130	27	435
U	0.6	167	1500	0.7	2000
Ge	0.8	1.3	<0.7	<0.5	<0.5
Sn	3.5	5.5	5.0	2.0	3.0
Pb	51	96	450	7	535
Ti	2847	629	839	2877	1468
Zr	47	80	320	220	15
As	13600	108	216	3	28
Sb	2	3	3.5	2	1
Bi	nf	2.2	1.7	nf	43
V	96	18	15	22	2
Nb	12	18	15	4	6
Ta	<3	<3	<3	<3	<3
Se	2	1.5	1	nf	3
Te	<0.2	0.4	0.2	<0.2	0.7
Cr	585	710	705	29	4
Mo	3	2.5	2.5	1	11
W	nf	nf	nf	nf	2
F	195	10	5	25	35
Cl	250	45	15	10	60
Mn	590	108	55	1400	2
Co	37	59	102	16	57
Ni	111	106	158	8	45
Pt	<0.005	<0.005	<0.005	0.020	0.005
Pd	0.025	<0.025	0.062	0.005	0.006
P	7500	100	170	87	300

Note: nf–not found.

Table 36. Alkali element content of some oxidized gold deposits

Description	K_2O (%)	Li	Rb (ppm)	Cs	Remarks
*Yellowknife Greenstone Belt, N.W.T., Canada**					
Gold-quartz ore, Lynx prospect	1.62	7	53	<1	Contains essentially quartz, pyrite, arsenopyrite and sulphosalts
Gossan, Lynx prospect	0.01	1	<1	<0.1	Mainly limonite with highly leached schist particles
Keno Hill, Yukon, Canada†					
Gold-quartz ore, No. 6 vein	0.16	2	6	0.2	Contains essentially quartz, pyrite and arsenopyrite
Oxidized zone, No. 6 vein	4.5	8	231	18.5	Mainly limonite, quartz and jarosite minerals
Keno Hill, Yukon, Canada†					
Pb-Zn-Ag ore (Hector-Calumet Mine)	0.17	1	2	1.2	Composite sample oxidized ore, 100-foot level, No. 3 vein
Pb-Zn-Ag ore (Hector-Calumet Mine)	0.08	1	7	0.0	Composite sample oxidized ore, 300-foot level, No. 3 vein
Pb-Zn-Ag ore (Hector-Calumet Mine)	0.12	1	14	0.0	Composite sample unoxidized ore, 400-foot level, No. 3 vein
Pb-Zn-Ag ore (Hector-Calumet Mine)	0.09	1	8	0.0	Composite sample unoxidized ore, 525-foot level, No. 3 vein
Pb-Zn-Ag ore (Hector-Calumet Mine)	0.07	1	6	0.0	Composite sample unoxidized ore, 650-foot level, No. 3 vein
Pb-Zn-Ag ore (Hector-Calumet Mine)	0.05	1	2	0.0	Composite sample unoxidized ore, 775-foot level, No. 3 vein

*For description of deposits *see* Boyle (1961*a*).
†For description of deposits *see* Boyle (1965*a*).

of gold-quartz deposits marked by sericitization and alunitization and the development of potassic feldspar, micas and adularia in the vein quartz or silicified zones. Skarn-type gold deposits may also exhibit an enrichment of potassium in places. The disseminated gold deposits (Carlin, Nevada) exhibit a marked enrichment of potassium in the primary ore-bearing parts (1.6%) compared with the host carbonate country rocks (0.7%) (Radtke *et al.*, 1972*a*). Some of the auriferous beds of the quartz-pebble conglomerates have a relatively high potash content (Table 34). Most of the potassium in these deposits is present in micas (sericite). Thermal waters carrying and precipitating gold are invariably enriched in potassium and sodium. Those in New Zealand have potassium and sodium contents ranging from 49 to 224 ppm and 545 to 1785 ppm respectively (Weissberg, 1969). The wall rocks of the fissures carrying the thermal waters are also enriched in potassium (adularia) according to the data provided by Steiner (1970). The oxidized zones of gold deposits frequently show an enrichment in potassium (mainly in jarosite) in gold-rich sections; sodium, on the other hand, is commonly leached out.

The behavior of potassium and sodium in hypogene gold deposits is discussed in greater detail in the section on alteration processes. There, it is shown that the K_2O/Na_2O ratio commonly increases as gold mineralization is approached, a useful feature that can be employed during detailed prospecting for gold-quartz and other types of epigenetic gold deposits (Boyle, 1974*b*).

Rubidium and *cesium* follow potassium closely in gold deposits being concentrated mainly in sericitized and feldspathized alteration zones. The bulk of both elements is concentrated in sericite, potassic feldspars or alunite. I have not observed pollucite or any other cesium or rubidium minerals in gold deposits.

Skarn deposits enriched in potassium may have elevated contents of both rubidium and cesium, but the gold-bearing varieties analyzed in the Geological Survey laboratories were all relatively low in both elements, rarely containing more than 45 ppm Rb and 1 ppm Cs.

Gold-quartz veins and silicified bodies and their alteration zones contain only small amounts of rubidium and cesium as shown in Table 34. In the alteration zones of gold-quartz veins both elements tend to be sympathetically enriched with potassium, increases in the elements being evident as the veins are approached. In the Balei gold-quartz deposits, eastern Transbaikal, Pampura and Trufanova (1968) noted marked enrichments of cesium (average 45 ppm) in the alteration zones. The chief host minerals were found to be kaolinite and hydromica, mainly the latter.

Rubidium and cesium probably exhibit a behaviour in auriferous polymetallic deposits similar to that noted above for the gold-quartz deposits, judging from the results obtained from analyses of the alteration zones associated with the Pb-Zn-Ag lodes at Galena Hill, Yukon (Table 34). Radanova and Karadzhova (1971) noted high concentrations of rubidium and cesium in the propylite facies (13–250 and 0–50 ppm respectively) associated with the Tertiary Pb-Zn deposits in the Spahievo ore district in Bulgaria. Cesium tended to be concentrated in the epidote-chlorite-albite subfacies.

There are no extensive data in the literature on the rubidium and cesium contents of the auriferous quartz-pebble conglomerates. Analyses done in the Geological Survey laboratories on samples from some of the reefs of the Witwatersrand system (Tables 34 and 35) indicate a range for rubidium of 12 to 100 ppm and for cesium <0.1 to 3.5 ppm. The black (carbonaceous) reef material appears to contain the highest amounts of the elements. The range for rubidium given by Feather and Koen (1975) is from <5 to 56 ppm.

There appear to be no data on the rubidium and cesium contents of the disseminated gold deposits such as those that occur in Nevada (Cortez, Carlin, Gold Acres) and elsewhere.

Rubidium and cesium are common trace elements in some hot spring waters, but there are no data on these two elements in gold-bearing hot springs. The data provided by White *et al.* (1963*b*) indicate that up to 0.2 ppm Rb may be present in thermal waters closely associated with certain epithermal mineral deposits.

The behaviour of rubidium and cesium during the oxidation of gold deposits is similar to that of potassium. In some deposits both elements are leached from the oxidized zones; in others there is an enrichment. The latter takes place mainly in gossans and oxidized zones where jarosite minerals are developed, such as in some of the gold-bearing veins at Keno Hill (Table 36). Gold and silver may also be concentrated in these jarosite-rich zones (*see* Chapter IV).

Group IB: Cu, Ag

Copper, varying in amount from traces to a few per cent, is a constant associate of gold in all types of deposits. The element may from a suite of its own minerals and/or may occur as a constituent in a variety of sulphides and sulphosalts and in native gold and its various minerals. The concentration of copper in native gold is highly variable (*see* Chapter II). The most common copper minerals in gold deposits are chalcopyrite, enargite, chalcocite and tetrahedrite-tennantite; in some skarn zones bornite is abundant. The chalcocite zones of certain supergene enriched copper deposits are commonly rich in gold. Native copper occurs in some gold-quartz veins (e.g., Windpass Mine, B.C.; and Ptarmigan Mine, Yellowknife, N.W.T.). In these deposits the mineral is primary in origin. The mineral may also occur in certain oxidized zones of gold deposits but generally only in small amounts.

Some auriferous skarn deposits are greatly enriched in copper; others such as the Nickel Plate Mine at Hedley, British Columbia contain only minor amounts of the element. Both gold and silver tend to follow copper in auriferous skarns, but the relationships are often complex (Table 37).

Most Precambrian gold-quartz deposits and silicified deposits are low in copper as are also most of those in younger epochs. In some of the latter, however, copper may appear in some abundance, especially in Tertiary deposits.

The disseminated gold deposits (Carlin) are enriched in copper (up to 200 ppm; average 35 ppm) in the primary mineralized zones (Radtke *et al.*, 1972*a*). Most of the copper is in chalcopyrite and probably also in sulphosalts.

The quartz-pebble conglomerates usually carry copper, mainly in chalcopyrite; some reefs also contain small amounts of cubanite, bornite, chalcocite, covellite, neodigenite, chalcopyrrhotite (?) and tennantite-tetrahedrite, (Liebenberg, 1955; Saager, 1968; Feather and Koen, 1975). Analyses of a number of the reefs done in the Geological Survey laboratories show copper contents ranging from 10 to 137 ppm (Table 35). On the basis of only a few samples it appears that the carbonaceous reefs are the richest in the metal.

Hot springs carrying and precipitating gold invariably carry some copper. The amounts are usually small (0.002–0.2 ppm) in the water.

Gold is positively correlated with copper in some hypogene deposits; in others there is no particular relationship. The old Chilean proverb "If thou findest copper, thou has gold" is, therefore, not universally true. The zoning relationships of copper with respect to gold are discussed in the last part of this chapter and shown in Figures 94 and 95. Near vein and other types of epigenetic gold deposits, copper commonly increases in content as the veins are approached. (*See* the section on wall-rock alteration processes and the accompanying tables.)

In oxidized copper deposits, and in some gold deposits, the secondary chalcocite and other supergene sulphide zones are greatly enriched in gold (and silver). In the gossans and oxide zones of many polymetallic, massive sulphide and gold deposits copper and gold tend to part company – the gold largely remaining behind and the copper being removed to lower levels or dispersed in the groundwaters. In a few oxidized gold deposits azurite, malachite, chrysocolla and other secondary copper minerals may accompany rich auriferous parts of gossans. (*See also* Chapter IV.)

Silver, varying from traces to minor amounts, is a constant associate of gold in all types of deposits. The element may form its own suite of minerals and/or may occur as a trace constituent of various sulphides and sulphosalts and in minor or major amounts in native gold and its minerals. The most common silver-bearing mineral in gold deposits is probably tetrahedrite-tennantite. Argentite (acanthite) is common in some vein deposits younger than Precambrian, as are also proustite and pyrargyrite. Other silver sulphosalts tend to be rare. Silver tellurides are common in deposits in volcanic rocks in places. The selenide, naumannite and other silver selenides occur sparingly in certain Tertiary deposits. Silver-bearing minerals in the quartz-pebble conglomerates of the Rand, exclusive of gold, include tennantite, stromeyerite, proustite and rarely native silver and dyscrasite (Saager, 1968; Feather and Koen, 1975). Native silver is relatively rare in the hypogene parts of most gold deposits, although it has been recorded in a few deposits of all ages. In most places it appears to represent a stage in the mineralization that is quite

Table 37: Gold, silver, copper, K_2O and Na_2O content of skarn, Whitehorse Copper Belt, Yukon

Rock type	Au	Ag	Au/Ag	Cu	Cu/Ag	K_2O	Na_2O
	(ppm)			(ppm)		(%)	
Diorite-granodiorite	<0.005	0.2	<0.02	15	75	1.2	2.8
Skarn adjacent to above	<0.005	0.45	<0.01	22	48.8	1.2	4.4
Normal skarn	<0.005	0.2	<0.02	12	60	0.1	0.2
Skarn adjacent to magnetite-bornite deposit	0.05	1.5	0.03	3900	2600	~0.1	~0.2
Magnetite-bornite deposit	5.0	60	0.08	6.4%	1066.6	~0.1	~0.1
Crystalline limestone adjacent to above	0.05	0.1	0.5	135	1350	<0.1	<0.1
Crystalline limestone	<0.005	~0.05	<0.1	5	~100	<0.1	<0.1

Note: Traverses across granodiorite, skarn, copper deposit and limestone at Arctic Chief Mine. For geology *see* Kindle (1964) and Figure 10.

separate from that of gold. Native silver is common in some of the oxidized zones of auriferous deposits.

The Au/Ag ratio of both native gold and of gold deposits as a whole is variable depending on the many factors discussed in a later section. The silver and gold contents of some hypogene deposits exhibit a definite positive correlation; in other deposits there is no particular relationship. The zoning relationships of silver with respect to gold are discussed in the last part of this chapter and shown in Figures 94 and 95. With respect to the loci of gold (and silver) mineralization in veins and similar deposits, silver generally exhibits a consistent increase in content as the deposits are approached. (*See* the section on wall-rock alteration processes and the accompanying tables.)

Silver and gold are frequently concentrated together in supergene sulphide (mainly chalcocite) zones of oxidized polymetallic, massive sulphide, porphyry copper and gold deposits. In the gossans and oxide zones of these deposits the chemistry of gold and silver is complex and is discussed in Chapter IV.

Group IIA: Be, Mg, Ca, Sr, Ba, Ra

Beryllium frequently exhibits a slight enrichment in the alteration zones of certain types of gold deposits, particularly the vein types where abundant sericite and tourmaline are developed. In the sericite phases of the hypogene alteration in the greenstone belt at Yellowknife the beryllium content ($\sim$ 2ppm) appears to vary directly with the amount of sericite present. In other deposits the element appears to be present mainly in tourmaline (up to 15 ppm), and the content of the element varies with the abundance of this mineral. The author has observed no beryllium minerals in hypogene gold vein deposits, but Ol'shevskii (1973) has noted the occurrence of beryl (aquamarine) in some gold-quartz veins of the Karalveem deposit, and beryllium minerals, including beryl, phenakite and bavenite have been reported in some of the auriferous veins of the Hohe Tauren in Austria (Ertl *et al.*, 1975). In skarn deposits containing enrichments of gold, beryllium may exhibit a slight concentration, the element being present mainly in various lime silicates and in minerals such as vesuvianite, helvite, humite, etc. Radtke *el al.* (1972*a*) record <1 to 3 ppm Be in the primary ores of the Carlin, Nevada disseminated ores and <1 to 1 ppm in the host carbonate rocks indicating a slight enrichment of the element during mineralization. In the 'oxidized ores' they found <1 to 3 ppm indicating no particular enrichment during oxidation.

The beryllium content of the Witwatersrand reefs range from <2 to 3.5 ppm (Table 35). The location of the beryllium is probably in the micas and tourmaline. Other quartz-pebble conglomerate deposits show a similar range in the content of beryllium.

The writer has not traced the fate of beryllium during the oxidation of gold deposits in detail. Some analyses of gossans suggest that the element is concentrated in clay minerals and aluminous limonite; the beryllium in tourmaline goes with the mineral into the resistates, which may collect in the soils and nearby streams.

Beryl and other berylliferous minerals may accompany native gold in certain eluvial and alluvial placers, but in most deposits of this nature these minerals are relatively rare.

Magnesium is a frequent associate of gold particularly in skarns and in vein deposits in intermediate and basic rocks. In skarn the element forms a large number of magnesium silicates and spinels; in vein deposits the element is present mainly in chlorite, talc, serpentine, biotite and similar minerals and in magnesian calcite, dolomite and ankerite. All of these minerals may also occur in the vein quartz and in silicified zones. Magnesite appears to be rare in gold deposits, although it has been reported in some according to the old literature. Much of the magnesium in gold deposits appears to have been derived from the host rocks. (*See further* under the section on wall-rock alteration processes.)

During oxidation processes in gold deposits large amounts (commonly more than 75%) of magnesium are leached. Some magnesium is, however, retained in the gossans and oxidized zones mainly in the forms of supergene carbonates and secondary chloritic minerals.

Spinels may appear in the heavy concentrates of some gold placers. A host of other magnesium minerals may also be present including garnets, amphiboles, pyroxenes, etc.

Calcium is universally present in all types of hypogene gold deposits. In skarn deposits the element forms a number of calc-silicates and may be present in minerals such as fluorite, apatite, sphene and scheelite. In vein deposits calcium is a major constituent of the carbonates of the alteration zones and is frequently present in this form in the quartz veins and silicified zones. The principal calcium carbonates in gold deposits in their order of abundance are ankerite, dolomite and calcite. Anhydrite and gypsum are found in a great variety of gold deposits, mainly vein types, often in some abundance. These two minerals occur both in the alteration zones and in the vein quartz and silicified zones. Large amounts of anhydrite and gypsum are a feature of some of the deposits in the Timmins (Porcupine) area, Ontario, and the two minerals have been noted by the writer in a number of gold deposits in various greenstone belts in Canada.

Much of the calcium in gold deposits appears to have been derived from the host rocks. (*See further* under the section on alteration processes.) In some deposits, however, there is evidence to suggest that the calcium has been introduced from distant sources.

During the oxidation of gold deposits much calcium is lost, commonly more than 50 per cent. Some calcium is, however, retained in the gossans and oxidized zones mainly in the form of supergene carbonates, gypsum and anhydrite.

In placer deposits the most persistent calcium minerals associated with native gold appear to be sphene and garnet. A host of other calcium minerals may also be present in varying amounts, including various silicates, apatite, scheelite, etc.

Strontium tends to follow calcium in most hypogene gold deposits, being concentrated in small amounts mainly in the various hypogene carbonates and alteration zones. At Yellowknife, and in other gold deposits in greenstone belts in Canada, the strontium content of the carbonates rarely exceeds 500 ppm. In the carbonate alteration zones in basic and intermediate igneous rocks the amounts of strontium usually range from 20 to 50 ppm Sr. Strontianite appears to be rare in gold deposits as is celestite, although small amounts of the mineral have been recorded in the Kirkland Lake deposits, at Cripple Creek in Colorado and elsewhere. Both strontianite and celestite may occur in small amounts in gold-bearing polymetallic deposits. Most of the strontium in hypogene gold deposits appears to be locally derived, mainly from the nearby host rocks. In the quartz-pebble conglomerate deposits the contents of strontium range from 13 to 136 ppm (Table 35). Much of this strontium is probably in feldspar, the remainder in minerals such as mica. In the disseminated ores at Carlin, Nevada, Radtke *et al.* (1972*a*) recorded the following strontium contents in ppm in the host carbonate rocks, 'primary ores' and 'oxidized ores', respectively: 150 to 2000 (av. 625); 20 to 1500 (av. 150); and 20 to 2000 (av. 140). The averages indicate a depletion of strontium during primary mineralization and practically no change during oxidation processes.

During the oxidation of most auriferous deposits, strontium generally follows calcium and is commonly extensively leached. Some remains behind in the gossans and oxidized zones, mainly in the supergene carbonates.

Barium tends to follow potassium and calcium in hypogene gold deposits being concentrated in small amounts in the various potash feldspars, sericite and carbonates in the wall-rock alteration zones of vein-type deposits. In skarn deposits the element is evidently concentrated mainly in the calc-silicates but again only in very small amounts. At Yellowknife, the carbonates in the wall-rock alteration zones and in the veins rarely contain more than 50 ppm Ba and the sericite-carbonate alteration zones average about 150 ppm. Similar conditions prevail in deposits in greenstone in other volcanic belts of the Canadian Shield. Where the veins occur in intermediate igneous rocks (syenite, granodiorite), however, as at Kirkland Lake and Red Lake, Ontario (Red Lake Gold Shore Mine) barium contents up to 1000 ppm may be recorded in the alteration zones. Barium is apparently also more abundant in certain wall-rock alteration zones in other parts of the world as can be readily seen by perusing the various analyses given by Lindgren (1933) and others. These analyses quote contents up to 1200 ppm Ba, principally in altered intermediate rocks such as monzonite and diorite. The association and correlation of gold with barium in auriferous deposits and their alteration zones is also noted by a number of investigators in U.S.S.R. (Roslyakova and Roslyakov, 1975). In most gold deposits studied by the writer strontium generally exceeds barium in the carbonates by a factor of 2 or more. In the wall-rock alteration zones strontium either equals or exceeds barium where carbonatization is marked. This appears to be a generality judging from analyses quoted in the literature, although there may be some exceptions. (*See also* the section on alteration processes.)

Barite occurs as a minor gangue mineral in some gold deposits, mainly the low and medium temperature types (the epithermal and mesothermal deposits of Lindgren). Witherite and other barium minerals have not been seen by the writer in gold deposits, and there appears to be no reference in the literature to the occurrence of these minerals in auriferous deposits.

The disseminated gold ores of Carlin, Nevada are greatly enriched in barium, present mainly in barite. Radtke *et al.* (1972*a*) report the following barium contents in ppm in the host carbonate rocks, 'primary ores' and 'oxidized ores', respectively: 27 to 1000 (av. 150); 150 to 70 000 (av. 2200); and 200 to 100 000 (av. 2300). The averages indicate only a slight enrichment during oxidation processes.

In the quartz-pebble conglomerate deposits barium occurs in amounts ranging from <50 to 1050 ppm (Table 35). Most of this barium appears to be present in feldspars, the remainder in micas.

Barite, having a relatively high specific gravity and general resistance to weathering and erosion, may accompany the heavy resistate suite of minerals in certain eluvial and alluvial gold placers of recent and ancient derivation. Under conditions of intense weathering and leaching much barium may go into solution and be removed from the gossans and oxidized zones of auriferous deposits. Considerable amounts usually remain behind, however, adsorbed or combined with manganese oxides (wad) and manganiferous limonite. Some is also retained in various supergene sulphates and basic sulphates.

Radium accompanies gold only in those deposits containing uranium and/or thorium. The classic example of such deposits is the Witwatersrand quartz-pebble conglomerates of South Africa. Similar gold deposits in other parts of the world, such as those at Jacobina in Brazil, also carry some radium. Uraninite, pitchblende and thorium minerals, and hence radium, are relatively rare in vein and stockwork gold deposits and in gold-bearing silicified zones and skarn. Exceptions to this generally are met in uranian and thorian metallogenic provinces. Thus, the Richardson Gold Mine in skarn at Eldorado, Ontario carries uraninite, brannerite and thucholite and contains radium; similarly some gold deposits (e.g., the Box and Nicholson mines) in the Beaverlodge uraniferous province of Saskatchewan contain pitchblende and hence carry radium. Other hypogene gold deposits, frequently of Tertiary age and localized in porphyries and granitic rocks, in various parts of the world, carry radium if present in uranium and thorium provinces. (*See also* the section on uranium and thorium.)

During the oxidation of uraniferous and thorium-bearing gold deposits radium behaves much like barium. It tends to be strongly absorbed by wad and manganiferous limonite and may be concentrated in secondary sulphates and basic sulphates. The gossans and oxidized zones may, therefore, be highly radioactive. Under conditions of intense weathering and leaching much radium is removed from the gossans and oxidized zones in a soluble form.

Eluvial and alluvial placers, both recent and fossil, are commonly marked by the occurrence of various uranium- and thorium-bearing resistate minerals such as zircon, monazite, thorite, etc. These normally contain traces of radium.

Group IIB: Zn, Cd, Hg

Zinc is a nearly universal associate of gold in all types of hypogene gold deposits, occurring mainly in sphalerite and a great variety of zinc-bearing sulphosalts; occasionally in franklinite as in some veins in Burma. In some deposits zinc may be present in percentage amounts, in others only as traces or minor quantities. In polymetallic and some skarn deposits for instance the element is invariably abundant; in gold-quartz veins and silicified deposits zinc is a minor element; and in the quartz-pebble conglomerate type of gold deposits is it present generally only in minor amounts (10–330 ppm) according to our analyses (Table 35), mainly in sphalerite (Liebenberg, 1955; Saager, 1968; Feather and Koen, 1975). In some hypogene gold deposits there is frequently a positive correlation between the quantities of zinc and gold in the ores, and the zinc minerals (sphalerite and sulphosalts) are direct indicators; in other deposits no such relationships are evident. The zoning relationships of zinc with respect to gold are discussed in the last part of this chapter and shown in Figures 94 and 95.

In the oxidized zones of gold deposits, gold and zinc tend to part company, the gold remaining behind in the gossans and oxidized rubble of the deposits and the zinc migrating away in the oxidizing meteoric waters. In some deposits, however, the situation is more complex: Zinc may be concentrated together with gold in the supergene sulphide zones, and in the oxidized zones zinc silicates (calamine) and carbonates (smithsonite), and minerals like adamite ($Zn_2(OH)AsO_4$) may be developed where gold is enriched. This seems to take place mainly where the wall rocks are carbonates or where an abundance of dissolved arsenates or colloidal silica was present in the supergene solutions.

The writer has no extensive data on the occurrence of zinc in recent and fossil eluvial and alluvial gold placers. I have, however, noted higher than average amounts of zinc in the heavy mineral fractions of some of these deposits. Most of the zinc is present mainly in magnetite, in residual grains of sphalerite and in particles of wad and limonite. In the clay fractions of eluvial gold deposits I have noted slight concentrations of zinc in places; presumably this zinc is present in an adsorbed form or perhaps in some kind of a zinc clay mineral like sauconite.

Cadmium follows zinc closely in hypogene gold deposits and is generally quantatively contained in sphalerite and zinc-bearing sulphosalts, particularly tetrahedrite and tennantite. The galena and chalcopyrite of some gold deposits may also contain up to 200 ppm cadmium. In the writer's experience gold ores rarely contain more than 500 ppm Cd; more generally these ores contain only from 5 to 60 ppm Cd. Those of the quartz-pebble conglomerate deposits contain from <0.5 to 2 ppm Cd (Table 35). The author has no data on the correlation tendencies of Cd and Au in gold ores. The geochemistry of cadmium in various types of ores, including those bearing gold, is discussed at length in Vlasov (1966).

In the oxidized zones of gold deposits gold and cadmium tend to part company, the gold remaining behind in the gossans and oxidized parts of the deposits and the cadmium migrating away in the oxidizing meteoric waters. In some deposits, however, the situation is more complex. I have noted the presence of relatively high amounts of cadmium in the lower parts of some oxidized zones. Most of this cadmium is precipitated as hawleyite (βCdS) or greenockite (αCdS) on pyrite. Where supergene zinc and lead minerals such as smithsonite, anglesite, beudantite, bindheimite and pyromorphite are present cadmium may be significantly enriched. Limonite, wad and secondary clay minerals may also concentrate small amounts of cadmium. Cadmium may also exhibit a slight concentration in the supergene sulphide zones of certain auriferous deposits. Further details on the fate of cadmium during oxidation processes in mineral deposits can be found in the papers by Boyle (1965*a,b*) and Goleva and Chitaeva (1973).

The writer has no extensive data on the content of cadmium in recent and fossil eluvial and alluvial gold placers. The data available suggest that the element is normally very low in these deposits since only traces have been found in the heavy mineral concentrates. In these, most of the cadmium appears to be in oxidized residuals of galena, sphalerite and sulphosalts and in particles of wad and limonite.

Mercury is a common associate of gold in practically all types of hypogene gold deposits, but the element is invariably present only in small amounts, often much less than 1 ppm in the gold ores as a whole. In some deposits, however, mercury may be more abundant. Many of these are near-surface (epithermal) veins that occur in known mercuriferous belts in various parts of the world. Most of the gold deposits enriched in mercury are young, often of Tertiary age. However, deposits containing marked amounts of mercury are known in Precambrian rocks in Rhodesia (Robertson, 1972), in eastern Ontario and elsewhere. Cinnabar, metacinnabar, livingstonite and mercurian tetrahedrite and tennantite may appear in gold deposits, but they are generally not abundant. Examples of deposits where cinnabar and minor amounts of the other mercurian minerals accompany gold include Mercur, Utah (Spurr, 1923); White Caps Mine at Manhattan, Nevada (Spurr, 1923); a number of Paleozoic deposits in Uzbekistan and the southern Tien-Shan, U.S.S.R. (Finkelshtein, 1970; Rakhmatullaev, 1970); in certain deposits in Azerbaijan S.S.R. (Nasibov *et al.,* 1972); in some deposits in Yakutia (Obolenskii and Obolenskaya, 1972); in a number of mines along the 'Antimony Line' in the Murchison Range in the northeastern Transvaal, South Africa (Sahli, 1961); and in the Carlin deposit in Nevada (Radtke *et al.,* 1972*a,b*). In all of these deposits the mercury minerals are invariably late and commonly associated with stibnite and arsenic minerals such as orpiment and realgar. Coloradoite, HgTe, appears to have a relatively widespread geographical and age distribution in gold-quartz veins, occurring in the Precambrian deposits of Kalgoorlie (kalgoorlite) in Western Australia and at Kirkland Lake, Ontario and in the Tertiary veins of Boulder County, Colorado and elsewhere. Tiemannite, HgSe, may occur in deposits enriched in selenium; Robinson (1955) records its presence in the Nicholson Mine in the Goldfields area, Saskatchewan. Gold accompanies cinnabar and metacinnabar in the active solfatara fields of the Mendeleev volcano in the Kurile Islands (Ozerova *et al.,* 1969).

The general range in the mercury content of some gold deposits is shown in Table 38. Several features are evident from the data. The first is that the age of the deposits seems to have relatively little control on the presence of mercury since certain Precambrian deposits have as much as or more mercury than the younger Tertiary deposits. The deciding factor seems to be the presence of sphalerite and/or late sulphosalt minerals such as tetrahedrite-tennantite. This is especially notable at Yellowknife (greenstone belt), Campbell

Table 38. Mercury content of some gold ores

Description	Hg content (range or average) (ppm)	Remarks
Yellowknife, N.W.T., Canada		
Deposits in greenstone belt	0.025–11.5	Precambrian (Archean) gold-quartz deposits with pyrite, arsenopyrite, stibnite, various sulphosalts, sphalerite, galena, etc.
Deposits in sedimentary area	0.013–0.050	Precambrian (Archean) gold-quartz veins containing pyrite, arsenopyrite, galena and sphalerite
Contwoyto Lake, N.W.T., Canada	0.020	Precambrian (Archean) disseminated native gold deposits in arsenopyritic and pyritic sediments
Red Lake, Ont., Canada		
Red Lake Gold Shore Mine	0.109–0.183	Precambrian (Archean) quartz veins in granodiorite and granite
Campbell Red Lake Mine	0.234–15.4	Precambrian (Archean) quartz bodies in greenstones. Some of the ores are rich in stibnite and sulphosalts
Cochenour Willans Mine	0.04	Precambrian (Archean) silicified bodies in greenstones and iron-formation
Kirkland Lake–Larder Lake district, Ont., Canada		
Macassa Mine	0.020–0.040	Precambrian (Archean) gold-quartz veins in syenite and metasediments
Upper Canada Mine	2.4–25.0	Precambrian (Archean) tabular veins, lenses and silicified zones in volcanics and tuffs
Kerr Addison Mine	0.066–0.180	Precambrian (Archean) gold-quartz veins, stockworks and silicified zones in chlorite schists (greenstones)
Ramore area, Ont., Canada		
Ross-Holtyre Mine	25.3	Precambrian (Archean) gold-quartz veins with abundant sulphides in greenstones
Cordova, Ont., Canada		
Cordova Mine	0.017–0.040	Precambrian (Grenvillian) gold-quartz veins in gabbro
Goldfields, Sask., Canada		
Box Mine	0.036–0.048	Precambrian (Archean) gold-quartz stockwork in granite sill
Sudbury, Ont., Canada		
Vermilion Mine	0.033–0.053	Precambrian (Proterozoic) massive sulphide ore greatly enriched in Au and platinoids
Keno Hill, Y.T., Canada		
No. 6 vein, Keno Hill	0.40	Mesozoic(?) gold-quartz veins with abundant arsenopyrite and sulphosalts in quartzites and phyllites
Dublin Gulch, Potato Hills	0.026–0.368	
Bridge River district, B.C., Canada		
Bralorne Mine	0.072–1.9	Mesozoic gold-quartz veins in diorite
Nova Scotia, Canada		
Various mines in Meguma Formation	0.040–0.319	Paleozoic gold-quartz veins and saddles in greywacke and slate
Japan		
Kushikino Mine	0.150	Tertiary gold-quartz veins and lenses in silicified andesite
Hauraki Goldfield, New Zealand		
Martha Mine	2.5	Tertiary gold and sulphide deposits in andesites and associated rocks
Maratoto Mine	0.122	
Tui Mine	0.060	
Waihi Mine	11.69	
Puhipuhi, New Zealand		
Puhipuhi hot springs	360–760	Slightly auriferous siliceous sinter
Witwatersrand, South Africa	0.130–3.28	Precambrian (Archean?) quartz-pebble conglomerates
Dublin Gulch, Y.T., Canada	765–5350	Gold nuggets in Tertiary(?) placer

Red Lake and the Ross-Holtyre Mine. At Yellowknife, orebodies with little sphalerite and few sulphosalts contain low contents of mercury (0.025–0.10 ppm) whereas those rich in sulphosalts are generally enriched in mercury (>5 ppm). The Ross-Holtyre Mine is unusual in that it has one of the highest mercury contents of any gold deposit examined in the Canadian Shield. This mine is also unusual in other respects – it has an Au/Ag ratio much less than 1 for some of its orebodies, a feature that is anomalous compared with most gold deposits in the Canadian Shield of Ontario. Two periods of mineralization are present in places, the first a gold-quartz one and the second quartz-dolomite veins with sphalerite, tennantite, gold, niccolite and native silver (Jones, 1948*b*). The last two minerals are reminiscent of Cobalt, Ontario where the mineralization, especially the native silver, is enriched in mercury. Another feature of the data on mercury is that deposits in volcanics appear on the whole to contain more of the element than those in sedimentary terranes. This is exemplified again at Yellowknife when a comparison between deposits in the greenstones and those in the sedimentary area are considered. In this case sphalerite is not a factor since both types of deposits contain about equal amounts of the mineral. The deposits in the sedimentary area, however, contain no sulphosalts, and this may be the controlling factor. Yet another feature is the general enrichment of mercury in the ores of the Witwatersrand. This is paralleled by similar enrichments of mercury in the Elliot Lake deposits (Table 35). The final feature of note is the great enrichment of mercury in some gold nuggets compared with the amount of the metal in the primary deposits as in the Dublin Gulch area of Yukon. This feature is discussed in more detail in the section on placers in this chapter.

To summarize, one can say that the content of mercury in gold deposits is relatively independent of age, the controlling factors being the presence of sphalerite and/or sulphosalts and cinnabar. In general, however, Tertiary deposits tend to be enriched in mercury compared with older deposits. Deposits in volcanics tend to contain more mercury than these in sedimentary terranes. From the literature it is evident that primary gold deposits, and the gold in placers, in mercuriferous belts contain much more mercury than those outside these belts, although there are exceptions to this generalization.

Sears (1971) examined the distribution of mercury in selected base metal and gold deposits in Quebec, Canada. He found that mercury was more abundant in zinc sulphide ores than in iron, copper or lead sulphides and that, generally, there was an increase in mercury with increasing zinc content of the ore.

Two base metals mines from the Grenville Province of the Precambrian Shield (New Calumet and Montauban) were found to have the highest tenors of mercury of all ores sampled, followed by the base metal ores of the Superior Province of the Precambrian Shield, the Appalachian region and the St. Lawrence Lowlands.

Samples from three gold deposits (Camflo, Sigma and Lamaque mines) in the Superior Province showed that contrary to expectations the gold ores were extremely low in mercury (0.001–0.125 ppm). Sears (1971) commenting on this feature noted that this was the reverse to that found for gold deposits in the United States and Sweden (James, 1962, 1964; Akright *et al.,* 1969).

Tsonev (1972) examined the relationships of gold and mercury in the Radka deposit, Bulgaria and found that the late stage native silver contained the highest amounts of the element.

Radtke *et al.* (1972*a,b*) found an enrichment of mercury in the disseminated Tertiary gold ores at Carlin, Nevada. The fresh carbonate rocks contain from 0.02 to 0.24 ppm Hg (av. 0.08); the mineralized carbonate rocks or 'primary ores' from 0.4 to 453 ppm Hg (av. 25 ppm); and the oxidized mineralized rocks or 'oxidized ores' from 0.2 to 130 ppm Hg (av. 20 ppm). These contents compare with average zinc values as follows: fresh carbonate rocks (14 ppm), 'primary ores' (185 ppm) and 'oxidized ores' (110 ppm).

In a recent paper Jonasson and Sangster (1975) report the mercury contents of a large number of samples of sphalerite from Canadian sulphide deposits, some of which carry gold. They noted that the significant factors which influence the magnitude of the mercury levels include age and type of deposit, bulk composition of the ore (proportion of zinc and perhaps lead) and proximity to postore activation or reactivation of regional fault structures. With respect to the age of the host rocks, mercury in sphalerite in sulphide deposits in volcanic terranes varied in its mean relative abundance as follows: Proterozoic (99 ppm)>Phanerozoic (50 ppm)>Archean (33 ppm). In carbonate-hosted deposits, the order of abundance was similar: Proterozoic (~10 ppm)>Phanerozoic (3.7 ppm). The mercury contents of sphalerite in Zn- and Pb-rich deposits in volcanic terranes were generally higher than those in the Cu-rich ones. Sphalerites in deposits which contain sulphosalt minerals (generally Pb-Zn, Ag ores), e.g., tetrahedrite, were found to be high in mercury. Mercury contents, up to 450 ppm were found in sphalerite in the sulphide deposits in volcanic terranes in the Flin-Flon–Snow Lake–Lynn Lake belts – the highest found for deposits of this kind in the Canadian Shield.

In most gold deposits the bulk of the mercury is present principally as a trace constituent in the various sulphides, arsenides and sulphosalts, particularly sphalerite and tetrahedrite-tennantite. Analyses performed at the Geological Survey show great differences in the amount of mercury present in these minerals in gold deposits, the range in the sphalerite being 0.1 to 200 ppm and in the tetrahedrite-tennantite 10 to 1000 ppm. We have noticed that mercurian tetrahedrite (with Hg>1%) generally tends to be very low in silver and gold content, and the reverse is also true that argentian tetrahedrite-tennantite with as much as 5 per cent Ag and 45 ppm Au tends to be remarkably low in mercury, containing amounts usually less than 5000 ppm Hg. The reason for this is not clear. It should be noted, however, that a new variety of argentian mercurian tetrahedrite has been described by Atanasov (1975) from the Chiprovtsi lead-silver deposit, western Stara-Planina mountains, Bulgaria. This tetrahedrite contains more than 16 weight per cent Ag and more than 18 weight per cent Hg; the gold content is not given. The mineral has an extremely large unit-cell edge, $a = 10.64 \pm 0.01$ Å. The argentian mercurian tetrahedrite is associated with a late stage pyrite-cinnabar mineralization.

Native gold, in most of the gold deposits examined by the author, contains mercury, but the amounts are highly variable ranging from mere traces in some deposits to several hundreds and even thousands of parts per million in others. The gold, both hypogene and placer, in the mercuriferous belts of the earth's crust seems to contain the highest contents of mercury. Where native silver occurs in auriferous deposits it may be greatly enriched in mercury.

Relatively little is known about the fate of mercury during the oxidation of gold deposits. Only generalizations can be made. The author has found that during the weathering and oxidation of deposits containing mercury, the element enters the exogene (surficial) cycle as compounds that are both of a physical and chemical nature (Jonasson and Boyle, 1972). Both cinnabar and native mercury are relatively stable to oxidation during weathering processes and may pass into the soil and ultimately into stream and lake sediments as finely divided particles of these two minerals. In its chemical forms mercury passes into the surficial cycle as the metal, Hg^0, and its two ions, Hg_2^{2+} and Hg^{2+}. Mercury metal is relatively soluble (6×10^{-5} g/L; 60 ppb) under practically all Eh and pH conditions found in nature. The ions exhibit variable solubilities depending upon the anionic species available and the Eh-pH regime. The author's experience suggests that mercury in most surficial waters can migrate as sulphate, nitrate, various chloride complexes, carbonate complexes, hydroxide complexes, ammonia complexes, methylated forms, e.g., $Hg(CH_3)_2$, and as a variety of chelated and other organometallic complexes resulting from the decay of plant and animal matter and the maturation of humus. Precipitation of the small amounts of mercury in downward percolating waters in ore deposits takes place mainly by adsorption on inorganic materials such as limonite, wad, etc. and by adsorption and/or chelation or other forms of organic binding by humic constitutents if present in the oxidized zones. Production of H_2S by inorganic processes or by bacteria in the vein systems precipitates the element as the sulphide in the secondary sulphide zones of some deposits. Under extremely arid conditions or where sodium chloride or other chlorides are available from sea winds, salt pans, saline waters, etc. the minerals montroydite, HgO, calomel, Hg_2Cl_2, and a number of oxychlorides such as terlinguaite, Hg_2OCl, and eglestonite, Hg_4OCl_2, may be developed in the oxidized zones. Mercurian tetrahedrite in some deposits oxidizes to give pulverulent cinnabar where the oxidation potential is moderately low as in the lower parts of the oxidized zones. Native gold (and silver) in oxidized zones tends to collect mercury and may become enriched in the element in some deposits.

Under most conditions mercury is lost from the oxidized zones of gold deposits especially where the oxidation has been intense and prolonged. Where the conditions are less severe mercury may exhibit an enrichment. At Yellowknife the gold ores in the greenstone belt average about 0.375 ppm Hg; gossans where developed contain up to 1.2 ppm Hg, indicating a general enrichment. At Clyde Forks, Lanark County, Ontario the gold-bearing tetrahedrite-cinnabar occurrence averages about 5000 ppm Hg and the gossanous material, where developed, averages about 500 ppm. At Carlin, Radtke *et al.* (1972*a*) report values of 0.4 to 453 (av. 25) and 0.2 to 130 (av. 20) ppm Hg respectively for the mineralized carbonate rocks (primary ores) and oxidized mineralized rocks (oxidized ores) respectively, indicating some loss of mercury during oxidation processes. In the oxidized ores the mercury is present mainly as remnant cinnabar and secondary schuetteite, $Hg_3(SO_4)O_2$, montroydite, HgO, and minor amounts of the native metal.

Mercury is present in relatively large amounts in some gold placers; the element occurring usually as cinnabar but also occasionally in the form of secondary minerals such as native mercury and calomel. Mercury is also a trace or minor constituent of most samples of placer gold, especially in mercuriferous terranes.

Group IIIA: B, Al, Ga, In, Tl

Boron is commonly associated with gold in practically all types of its deposits. The most frequent boron mineral in auriferous deposits is tourmaline, much less frequent are danburite, datolite, axinite, dumortierite, kotoite and ludwigite. Tourmaline is particularly characteristic of certain gold-quartz veins and silicified zones in greywacke and pelitic sedimentary rocks and their metamorphic and granitized equivalents including most granitic rocks; it is usually less abundant in veins in diorite and greenstone. Axinite and dumortierite tend to occur most frequently in skarn-type deposits, although these two minerals and danburite and datolite, sometimes occur in near-surface (epithermal) and deeper seated gold-quartz deposits. Kotoite and ludwigite are practically restricted to skarn-type deposits, particularly those enriched in copper. Boron is enriched in the disseminated gold deposits (Carlin type) in amounts ranging from <7 to 200 ppm (av. 79 ppm) as compared with the carbonate host rocks from <7 to 70 ppm (av. 15 ppm) (Radtke *et al.*, 1972*a*; Harris and Radtke, 1976). The nature of the boron minerals is not stated. In the auriferous quartz-pebble conglomerates, tourmaline tends to be disseminated and relatively scarce except locally in the reefs. To some investigators the tourmaline in these deposits is detrital, to others introduced by hydrothermal action. The content of boron in these deposits rarely exceeds 150 ppm (Table 35).

In gold-quartz deposits tourmaline tends to be developed in the greatest amounts in the wall-rock alteration zones, particularly those immediately adjacent to the quartz veins or silicified zones. In some veins and silicified zones there is also frequently a development of disseminated crystals and sunbursts of tourmaline in the quartz; also in vugs and cross-cutting veinlets. This tourmaline and that in the wall-rock alteration zones generally carries a large number of trace elements, some present in considerable quantity. An example from the Ptarmigan Mine at Yellowknife will suffice to illustrate this point. This tourmaline contains enrichments of the following elements: Li – 23 ppm; Cs – 2 ppm; Zn – 1700 ppm; Cd – 31 ppm; Ga – 50 ppm; Ge – 5.4 ppm; As – 150 ppm; and Sn – 110 ppm.

The distribution of boron in the wall-rock alteration zones of some typical gold-quartz deposits is shown in Tables 44 to 59 inclusive. Enrichment with proximity to the veins is general and most marked in basic rocks (greenstones). Some veins in sedimentary rocks are marked by a zone enriched in boron, others show little, if any, concentration of the element near the veins.

The writer has not been able to trace the fate of boron during the oxidation of gold deposits with any degree of accuracy, mainly because its habit appears to be extremely erratic in the deposits that have been investigated. Tourmaline and most other boron silicates have a marked resistate tendency, although in places they alter to micaceous and other products, and boron is set free when subjected to supergene waters that have been percolating through sulphides and carbonates. At Keno Hill, where we have examined the fate of tourmaline in gold-quartz and lead-zinc silver veins, we have found that a considerable amount of the mineral passes into the resistates in the oxidized zones, then into the soils, and ultimately into the stream systems and placers of the area. Some boron, however, enters solution during oxidation processes and is carried away in the groundwaters and ultimately into the streams. This statement is supported by the presence of boron in small amounts in the underground waters and in some of the spring precipitates of the area (up to 50 ppm B). Much boron is, however, also retained in the oxidized zones particularly in limonite and wad (up to 5000 ppm B). The nature of this boron has not been characterized, although some of it is in tourmaline.

Boron exhibits remarkably little enrichment during oxidation of the disseminated (Carlin type) gold deposits. Radtke *et al.* (1972*a*) report 7 to 300 ppm B (av. 80 ppm) in the oxidized mineralized rocks ('oxidized ores') at Carlin, Nevada as compared with <7 to 200 ppm B (av. 70 ppm) in the mineralized carbonate rocks ('primary ores').

Tourmaline, and in places other resistant boron minerals, are common associates of gold in recent and ancient placers. The content of tourmaline varies widely, some of the mineral being contributed by the auriferous deposits and some coming from sediments, tourmaline granites and other boriferous rocks. The deposits in some places appear to be the main contributors (or contribute an excess of boron over the general background) since the pattern of the distribution of boron (mainly in tourmaline) in the stream sediments and placers is directly correlative with the abundance of gold in the stream sediments and placers and with the location of the gold veins.

Such patterns are clearly evident in the Dublin Gulch–Keno Hill area of Yukon as one can see by comparing the boron map published by Gleeson *et al.* (1968) and the map of gold in heavy concentrates of stream sediments by Boyle and Gleeson (1972).

Aluminum is a constant associate of gold in its deposits being present in the silicates of the wall rocks of deposits and in minerals such as sericite and feldspars in the veins and silicified zones and in the reefs of the quartz-pebble conglomerate type of deposit (pyrophyllite). The habit of aluminum during wall-rock alteration processes associated with gold deposits is discussed in the section on alteration.

Some gold-quartz deposits are marked by the development of alunite, $KAl_3(SO_4)_2(OH)_6$, the basic sulphate of aluminum, and occasionally also by the formation of aluminous minerals such as diaspore and gibbsite. Such deposits commonly occur in Tertiary acid and intermediate volcanic rocks, thought by some investigators to be altered by sulphuric acid solutions or vapours. The type deposits are at Goldfield, Nevada (Ransome, 1909) and the Kasuga Mine, southern Kyushu, Japan (Taneda and Mukaiyama, 1970). Other occurrences of this nature occur in Jurassic and Triassic lavas on Vancouver Island (Clapp, 1915) and elsewhere.

During the oxidation of gold deposits the aluminous minerals in the ore materials are generally highly altered and much of the aluminum is carried away especially where the solutions are highly acid or alkaline. Much alumina, however, remains behind, and the bulk of this is bound in the limonite and wad complexes. In some deposits abundant kaolinite and other aluminous clay mineral complexes are formed, and some aluminum enters supergene minerals such as scorodite and jarosite. In these deposits aluminum often exhibits an enrichment in the oxidized zones over that present in the primary ores. Alunite may also be formed in the oxidized zones where acid sulphate solutions derived from the oxidation of pyrite and other sulphides are widespread, and numerous other aluminum-iron sulphates such as alunogen and halotrichite may occur, but they are generally rare. Wavellite, the aluminum phosphate, is a notable mineral in the oxidized zones of some gold deposits. In some oxidized zones gold is notably concentrated in the highly aluminous parts particularly in kaolinite and in the aluminum-bearing jarosite and scorodite. Lindgren (1933) remarks on the enrichment of gold in kaolinized zones, Schwartz (1944) mentions rich gold concentrations in kaolin in the San Juan district of Colorado, and Petrovskaya (1955) states that kaolinite is a direct indicator of gold in the oxidized zones of some of the gold districts of eastern Transbaikal, U.S.S.R. The frequent occurrence of gold in scorodite and jarosite is noted in the section on gold in other minerals. Kaolin and minerals such as jarosite and scorodite appear to be avid precipitators of gold especially when in the colloidal form as noted in the section on gold in natural waters.

Gallium occurs only in the parts per million range in most gold deposits. It is concentrated principally in the aluminum silicates (sericite, pyrophyllite, etc.) and other aluminous minerals (up to 200 ppm) and in sphalerite (up to 4000 ppm) in which it evidently substitutes for aluminum and zinc respectively. These features have long been known and are reviewed in detail by Shaw (1957), Vlasov (1966–1968) and Borisenok (1970). According to the statistics provided by Ivanov and Meituv (1970) the average gallium content of sphalerite in lead-zinc deposits is highest (100 ppm) in low temperature (carbonate-barite-fluorite) deposits, from 3 to 35 ppm in medium temperature quartz-silicate-sulphide assemblages, and relatively low (7 ppm) in skarn deposits. Deposits in carbonate rocks are invariably enriched in gallium compared with other types of host rocks. Gallium also occurs in magnetite, pyrite, galena, chalcopyrite and tetrahedrite-tennantite but generally only in the range <10 to 100 ppm. This is the general range of the element in the pyrite of the gold ores of the Yellowknife district (Boyle, 1961*a*). In a sulphide millieu I have noted that gallium and cadmium have marked chemical similarities, and the two elements appear to have similar substitutional capacities in sulphides and sulphosalts. The ionic covalent radii of the two elements are respectively (Ga^{3+}, 0.62; Ga, 1.22) and (Cd^{2+}, 1.03; Cd, 1.49), and the considerable differences in size and change would seem to preclude the sympathetic replacement of the two elements in sulphide structures. The answer may, however, lie in the fact that in the sulphide GaS, gallium is not present as Ga^{+} or Ga^{3+}, but is combined in Ga-Ga units in a layer lattice, each gallium atom being tetrahedrally co-ordinated by one Ga and three S atoms. Somewhat similar arrangements with tetrahedral co-ordination are found for Zn, Cd, Cu and S in sphalerite, wurtzite, greenockite and tetrahedrite-tennantite. Co-ordination number 4 predominates in sulphide structures containing Zn, Cd and Ga, and this appears to be the most energetically stable arrangement (the lowest free energy for the system). From this it follows that gallium and cadmium tend to show sympathetic substitutional habits in some sulphide systems. This explanation, however, does not apply to the frequent occurrence of Cd and Ga in galena in which the co-ordination number of the metal with sulphur is 6.

I have noted no particular coherence or correlation between gold and gallium in auriferous deposits nor has any been mentioned in the literature. The Tsumeb deposits of Namibia (Southwest Africa) appear to be particularly rich in gallium; a number of gallium sulphides such as gallite, $CuGaS_2$, occur there, but there is no marked enrichment of gold.

In the contact skarn copper deposits at Whitehorse, Yukon the magnetite in the magnetite-hematite-bornite chalcopyrite-chalcocite bodies is greatly enriched in gallium in amounts up to 600 ppm (Kindle, 1964). The gold content of these ores is low (0.005–0.135 oz Au/ton; the silver values are relatively high, up to 3.4 oz/ton).

Much of the gallium in the pyrite in the Yellowknife gold deposits appears to have come from the country rocks (Boyle, 1961*a*, p. 119). In the Keno Hill area, Yukon, increases in the content of gallium (and indium) were noted in the wall rocks as some of the lead-zinc veins (some containing gold) were approached. In other veins a depletion of gallium was noted in the altered zones near the veins (Boyle, 1965*a*).

Radtke *et al.* (1972*a*) found <2 to 15 ppm Ga in the 'primary ores' at Carlin, Nevada and <2 to 30 ppm in the 'oxidized ores', indicating a slight enrichment of the element during oxidation processes. The carbonate host rocks contained <2 to 15 ppm Ga.

The gallium content of the auriferous quartz-pebble conglomerates rarely exceeds 5 ppm (Table 35). Most of the gallium appears to be present in the sulphides although there is probably some in the aluminosilicates.

Gallium hydroxide, $Ga(OH)_3$, like that of aluminium, is amphoteric and the metal is soluble in acid and base to give $[Ga(H_2O)_6]^{3+}$ and gallate $[Ga(OH)_4]^-$, ions respectively. The aquo ions undergo ready hydrolysis and hydrated forms of gallia such as GaO·OH and $Ga(OH)_3$ are precipitated between about pH 4 and 9. Hydrous iron and manganese oxides and aluminum hydroxide and hydrated oxides coprecipitate and carry gallium. In these the gallium is probably present as the hydroxide (Razenkov and Galaktionova, 1962). During the oxidation of gold deposits gallium tends to migrate in acid or strongly alkaline solutions and may be largely removed from the deposits. Some remains behind, however, in the limonite, wad and aluminous complexes and hence may be concentrated to a moderate degree with gold in some gossans and oxidized zones. Gallium may also be enriched in the secondary sulphide zones of some deposits, although there are no extensive investigations on this aspect of the chemistry of the element. Its coprecipitation with secondary lead, zinc, and cadmium minerals such as anglesite, calamine, smithsonite and adamite also seems probable, but minerals of this type analyzed by Papish and Stilson (1930) showed no detectable Ga. Kulikova (1962*a*, 1964, 1966), however, found the element in small amounts in smithsonite (1–10 ppm) and hydrozincite (2–3 ppm) but none in hemimorphite. In sauconite, the secondary zinc clay mineral, the element was enriched (up to 1000 ppm). Only low amounts were found in supergene lead minerals (traces–50 ppm). The hydrous iron and manganese oxides (limonite wad) contained up to 44 ppm Ga.

From her extensive research on the fate of gallium during the oxidation of sulphide deposits Kulikova (1966) concluded that the element is impoverished in the oxidized zones and is generally dispersed, especially during the initial stages of oxidation when the solutions are acidic in character. She also concluded that the principal minerals that concentrate gallium (and indium) are the hydrous iron oxides (limonite). To generalize we can say that in this respect gallium behaves somewhat like gold, but in its general ease of dispersion during the oxidation of sulphide deposits it differs from the noble metal.

There are no extensive data on the occurrence of gallium in recent and ancient eluvial and alluvial gold placers. The writer has noted the presence of the element (up to 20 ppm) in the weathered clay material of the eluvial gold placers of Dublin Gulch, Yukon, and from this presumes that in placers gallium is mainly present in the aluminous complexes and clay minerals which these deposits usually contain. There are probably also traces of gallium in oxidized residuals of sphalerite, galena, etc. and in minerals such as magnetite in most placers. The cassiterite concentrate from the Dublin Gulch placer was found to contain up to 100 ppm Ga.

Indium is present in most gold deposits in the parts per million range where the ores contain sphalerite, galena, chalcopyrite, tetrahedrite-tennantite and tourmaline. In typical gold-quartz deposits where sulphides are minimal the indium contents are invariably much less than 1 ppm. In the Witwatersrand quartz-pebble conglomerates the indium content in all samples analyzed in the Geological Survey laboratories was less than 1 ppm (Table 35). The data given by Shaw (1957), Vlasov (1966–1968) and Kulikova (1966) in their reviews of the geochemistry of indium indicate the maximum contents of the element in the following minerals: sphalerite (7000 ppm), galena (22 ppm), chalcopyrite (1500 ppm) and tetrahedrite-tennantite (10 ppm). Pyrite and arsenopyrite tend to be low in indium, although Auger (1941) noted its presence in pyrite at Noranda, and Boorman and Abbott (1967) found 100 ppm and 50 ppm in pyrite and arsenopyrite respectively in the Mount Pleasant, New Brunswick tin ores. The pyrite in the Yellowknife gold deposits contains up to 10 ppm In (Boyle, 1961*a*). Wolframite, cassiterite and other tin minerals such as stannite tend to be enriched in indium (up to 1000 ppm in cassiterite), but these minerals are not generally found in gold deposits except in placers. Certain gold deposits in the far eastern part of U.S.S.R. are relatively rich in cassiterite, wolframite and tourmaline and are hence relatively enriched in indium (up to 10 ppm) (Meshcheryakov, 1963). Scheelite is said to contain some indium in places, but I did not detect the element in this mineral in the gold-quartz veins at Yellowknife.

Three independent indium minerals have been identified: indite, $FeIn_2S_4$; roquesite $CuInS_2$; and dzhalindite, $In(OH)_3$. The first was found in a cassiterite deposit in U.S.S.R., the second in a copper-tin deposit in France, and the third is an oxidation product found in an indium-rich deposit in U.S.S.R. All of these minerals are extremely rare and have not yet been identified in gold deposits.

I have noted no particular coherence or correlation between indium and gold in typical hypogene gold deposits, nor has any been reported in the literature. In this respect gold differs somewhat from silver, which frequently shows a geochemical affinity with indium in a number of deposits (Badalov, 1961). In the gold-bearing waters of the Taupo volcanic zone in New Zealand, Browne (1969) reports the indium content to be less than 0.00005 ppm. By contrast the thallium content is 0.01 ppm.

The oxide and hydroxide of indium are essentially basic yielding the $[In(H_2O)_6]^{3+}$ ion. A number of indium complexes in sulphate solutions are possible, including such species as $[InSO_4]^+$ and $[In(SO_4)_2]^-$. The indium aquo ions are readily hydrolyzed, and hydrous iron and manganese oxides and aluminum hydroxide and hydrous oxide coprecipitate and carry indium mainly as the hydroxide, $In(OH)_3$, or as a variety of basic salts. We should, therefore, expect that during the oxidation of indium-bearing sulphides in gold deposits that acidic solutions remove much indium and disperse it into the water regime of an area. Where neutralization of the solutions takes place the aquo ions of indium are hydrolyzed and largely coprecipitated in the wad, limonite and aluminous complexes of the gossans and oxidized zones. Basic salts such as beudantite and jarosite should also take up some indium. On passing downward some indium is probably also precipitated in the secondary sulphide zones. Secondary zinc and lead minerals such as calamine, smithsonite, cerussite, anglesite and bindheimite may also incorporate available indium. A perusal of the literature suggests that this does in fact occur. Prokopenko (1941) found indium in jarosite and cerussite, and Anderson

(1953) noted its presence in cerussite and pyromorphite. A series of studies on indium in the oxidized zones of polymetallic deposits have been published by Vishnevsky (1958) and Kulikova (1962*b*, 1964, 1966). In general they find that indium is largely dispersed during oxidation especially when the solutions are acid. In the oxidized zones some indium is retained by limonite and wad and by hydrocassiterite if it is present. An enrichment of indium may occur in beudantite (up to 25 ppm), plumbojarosite (up to 90 ppm) and in bindheimite (up to 5000 ppm) in some oxidized zones. Kulikova thinks that indium arsenate, $InAsO_4 \cdot H_2O$, may form in the presence of pentavalent arsenic in acid solutions and may be coprecipitated with bindheimite and beudantite.

The above suggests that gold and indium may be enriched together in some gossans; also in the deeper zones where secondary arsenates and antimonates are developed. At Keno Hill a slight enrichment of gold was noted in bindheimite, beudantite and scorodite, but relatively little indium was present in these minerals. This may be because most of the sphalerite in these deposits contains only small amounts of indium (<100 ppm).

There are no data on the presence of indium in gold placers. Presumably the element is present in traces, particularly in residuals of sphalerite, pyrite, etc. and in cassiterite where it is present.

Thallium is a trace constituent of most gold deposits containing sulphides and sulphosalts. According to Ivanov (1959) the largest amounts of thallium are concentrated in the gold-bearing polymetallic deposits containing abundant lead, zinc and copper sulphides and sulphosalts. These deposits contain on the average from 1 to 5 ppm Tl with some having contents ranging from 10 to 50 ppm. Ivanov also states that the sulphide minerals containing thallium are: galena (50–100 ppm) with the highest amounts, and lead sulphosalts, sphalerite, chalcopyrite, pyrite and marcasite (1–5 ppm) generally with the lowest amounts. In the Karamazar region of U.S.S.R. Badalov and Rabinovich (1966) found that tetrahedrite and chalcopyrite were the richest thallium minerals with up to 15 and 5.6 ppm respectively. In the alteration zones of these deposits and in others, thallium may also be concentrated in sericite and alteration feldspars but only in the parts per million range (Kurbanayev, 1966). This is a consequence of the close coherence of thallium and potassium (and rubidium) in nature and the well known fact that thallium as Tl^{+} replaces K^{+} in micas, feldspars and other potassium minerals. A review of these and other phenomena involving thallium is given by Shaw (1957), Vlasov (1966–1968), and Voskresenskaya (1967). A number of polymetallic deposits contain thallium minerals such as lorandite, $TlAsS_2$, vrbaite $Tl(As,Sb)_3S_5$, crookesite, $(Cu,Tl,Ag)_2Se$ and hutchinsonite, $(Pb,Tl)_2As_5S_9$. These are, however, rare in ores with a relatively high gold content, but lorandite has been identified in a few gold-quartz and disseminated-type deposits. In some polymetallic gold deposits enriched amounts of thallium have been noted in pyrite (up to 1000 ppm or more) and in other sulphides and sulphosalts.

Typical gold-quartz deposits appear to have very low contents of thallium, except those with an abundance of sulphides. I did not detect the element in the Yellowknife ores by spectrographic methods, and other gold-quartz deposits I have investigated in Canada have contents less than 1 ppm in the ores. The quartz-pebble conglomerates of the Witwatersrand have less than 1 ppm Tl according to my analyses (Table 35). Skarn-type gold deposits also appear to be relatively low in thallium except where they contain abundant lead, zinc and copper minerals. Even in these the contents are low according to Ivanov and Meituv (1970), who give a statistical average of 1 ppm Tl for such deposits.

My review of the literature on the occurrence of thallium and gold confirms the view of Ivanov (1959) and Ivanov and Meituv (1970) that thallium tends to be concentrated in medium and low temperature deposits. In this respect it is interesting to note that thallium is enriched in some present day gold-bearing precipitates from hot springs and drillhole discharges, as witnessed by the evidence given by Weissberg (1969), who found up to 5000 ppm Tl in red-orange amorphous sulphides with opaline silica in the Taupo volcanic zone of New Zealand. The alkaline waters of this zone contain up to 0.01 ppm Tl (Browne, 1969). The disseminated deposits (Carlin type) are somewhat enriched in thallium according to the data provided by Radtke *et al.* (1972*a*). Lorandite occurs in barite veins in some of the Carlin orebodies (Callahan, 1974). The mineral occurs intergrown with realgar, orpiment and an unknown Tl-Hg-As sulphide mineral. The orpiment is thallium-rich according to Radtke *et al.* (1974*a*). In the ores of the Getchell Mine, galkaite, $(Hg,Cu,Tl,Zn)(As,Sb)S_2$, has been identified by Botinelly *et al.* (1973), and in the ores of the Carlin deposit a new mineral carlinite, Tl_2S, has been described by Radtke and Dickson (1975). Carlinite occurs as small subhedral to anhedral grains with fine-grained quartz and hydrocarbon compounds in carbonaceous limestones in the east pit of the deposit. In contrast to these relatively low temperature types of deposits, fumarolic sublimates are relatively low in thallium according to the data compiled by White and Waring (1963). In those of Vesuvius the values range from not found to 10 ppm Tl, and those of Papandajan in Indonesia have 80 ppm Tl.

It is difficult to establish a direct coherence and correlation between thallium and gold in auriferous deposits, because no detailed study of this feature has been done. Having examined the literature carefully it is apparent that gold and thallium have a geochemical affinity or coherence in deposits with abundant Pb, Zn, As, Sb and Hg; in other deposits, particularly the gold-quartz and skarn-type the situation is less certain. By anology with silver, which exhibits a definite coherence with certain elements of the second long period of the periodic table, viz. Pd, Ag, Cd, In, Sn, Sb, we should expect that gold would show a marked coherence with thallium both of which are members of the third long period, viz. Pt, Au, Hg, Tl, Pb and Bi. Support for this view is found in the following: the oxidation states of both gold and thallium are the same, Au(I), Au(III); Tl(I), Tl(III), and the radii of the thallous and aurous ions are similar, Tl(I), 1.44Å; Au(I), 1.37Å. The thallous state is more stable in aqueous solutions than the thallic, and both are markedly stabilized in solutions by complexing (e.g., with Cl) in a manner analogous to the Au(I) and Au(III) states. Thallic and auric salts are quite similar in their chemical reactions. In general Tl(I) and Tl(III) are much more stable toward reducing compounds than the corresponding gold states, but the ions of both metals are precipitated by H_2S to yield the sulphides Tl_2S, Au_2S and Tl_2S_3. Thallous sulphide is relatively stable and its analogues occur in nature; thallic sulphide is a very unstable compound. Gold sulphides are markedly unstable, yielding the free metal readily. Brooks (1961) has noted a coherence between thallium and bismuth, and as explained later, gold also has a marked affinity for bismuth.

It should also be mentioned that thallium shows a marked coherence with silver and lead, a feature that has been remarked on by the writer in a previous publication on silver (Boyle, 1968*b*). The coherence between these three elements is a consequence of the chemical similarity of the precipitation of their sulphides and the chemistry of their complexes in solution. Their ionic radii are also similar, Tl(I), 1.44Å; Ag(I), 1.31Å; and Pb(II), 1.32Å.

Thallium occurs in two oxidation states (I) and (III). The latter probably does not occur in nature except where the oxidation potential is very high. Thallous hydroxide, TlOH, resembles the alkali metal hydroxides in being a soluble strong base. Thallic hydroxide, $Tl(OH)_3$, is purely basic. Thallous sulphate is moderately soluble in water; the halide is sparingly soluble. Thallium forms a number of soluble complexes in aqueous solutions; with chloride the complexes are of the type $[TlCl_6]^{3-}$. The thallous ion is not particularly sensitive to pH (like K^+), but the thallic ion is extensively hydrolyzed to $[TlOH]^{2+}$ and the colloidal hydrous oxide at a pH as low as 2. Thallic hydroxide and the hydrous oxide are extensively coprecipitated by limonite, wad and various natural alumina complexes; like K^+ the thallous ion is probably strongly adsorbed by clay minerals and other colloidal complexes.

With the above features in mind one can predict the fate of thallium during the oxidation of thallium-bearing gold deposits containing sulphides. Released from the sulphides and minerals like sericite thallium is probably largely in the Tl^+ state and under both acid and alkaline conditions would possess a relatively high mobility, much of the element being dispersed into the water regime of the area. Any thallic salts or complexes formed would be quickly hydrolyzed and coprecipitated with the limonite, wad and other oxidation products of the gossans and upper parts of oxidized zones of the deposits. Under such conditions one might expect an enrichment of both gold and thallium. More generally, however, the thallium as Tl^+ passes downward in solution where some is adsorbed and incorporated by the hydrous iron and manganese oxides, supergene clay minerals and particularly by basic salts, such as jarosite, containing K^+ and probably also by numerous supergene lead minerals such as cerussite, anglesite, beudantite, bindheimite, etc. In the supergene sulphide zones some thallium may also be precipitated with the various sulphides. Since gold is also precipitated in all of these zones during oxidation processes one might expect a sympathetic enrichment of the two metals.

Studies of the fate of thallium during the oxidation of deposits have produced mixed results. Ivanov (1959) states that in certain rich thallium-bearing deposits the highly oxidized zones are depleted in thallium compared with the primary ores, and that the greatest enrichments of the element occur in the zones of secondary sulphides, apparently mainly in pyrite and marcasite. Mogarovskii (1961) found thallium phenomenally enriched in jarosite (up to 2.04%). As limonite and other iron oxides succeed this basic salt, thallium tends to be strongly leached except where wad is formed. In this complex thallium tends to be enriched in psilomelane and pyrolusite. It may also form the independent mineral, avicennite, Tl_2O_3, under conditions of high oxidation potential. Kulikova (1962*a*, 1964) found that thallium was generally leached, especially during the acid stages of oxidation. During the later neutral and alkaline stages the element was retained in small amounts of wad, limonite and other supergene minerals. In the Maykain gold-barite polymetallic deposits, Kurbanayev (1966) found that much thallium was retained in the weathered products, particularly those derived from sericite. These various studies indicate that gold and thallium may concentrate together during oxidation especially where basic salts such as jarosite are formed and where considerable adsorption and/or coprecipitation by hydrous iron and manganese oxides and clay minerals takes places. Retention of much thallium in oxidized zones is suggested by the sporadic occurrence of the element in spring and mine waters in mineralized belts containing gold and polymetallic deposits. I did not find thallium (or indium) in the natural precipitates at spring orifices in the Keno Hill area, Yukon, and Borovitskii *et al.* (1966) found only sporadic concentrations of the element in amounts up to 1 ppb in the mine waters of the gold regions in the Aldan region, U.S.S.R.

There are few data on the thallium content of recent and ancient eluvial and alluvial placers. Presumably the element is present in exceedingly low amounts in these deposits being present probably only as traces to minor amounts in oxidized residuals of galena, sphalerite, etc.; also in clay minerals. The last seem to be important concentrators in places since Kurbanayev (1966) found that the eluvium and alluvium derived from the weathering of sericite contained more thallium (av. 0.1 ppm) than the bedrock (av. 0.02 ppm) containing the Maykain gold-barite polymetallic deposit.

Group IIIB: Sc, Y, La, Ce and other rare-earths, Th, U

Scandium appears to be a relatively constant associate of gold in its deposits judging from the work done on the element in the Yellowknife district. There, scandium appears in amounts ranging from 10 to 100 ppm in the tourmaline in vein quartz and wall rocks in greywacke and slate; also in about the same amounts in late carbonates (mainly calcite) in the quartz and silicified bodies in greenstones. I did not detect it in scheelite or other minerals in the deposits. I have noted the presence of scandium in other gold deposits at Red Lake where the element occurs mainly in tourmaline in amounts up to 50 ppm. These values are not much different from what I find in the tourmaline of pegmatites at Bernic Lake, Manitoba and Buckingham, Quebec (100 and 30 ppm Sc respectively). There is probably also a slight enrichment of scandium in auriferous skarn deposits that are relatively rich in W, Sn, Mo and rare-earths. Auriferous deposits carrying uranium also seem to be slightly enriched in scandium in places. In the Carlin disseminated deposits Radtke *et al.* (1972*a*) report an average of 8 ppm Sc in the 'primary ores' and 10 ppm Sc in the 'oxidized ores'. The carbonate host rocks contained an average of 3 ppm Sc. In the uraniferous conglomerates at Elliot Lake the scandium content averages about 10 ppm. In the quartz-pebble conglomerates of the Witwatersrand the scandium contents range from 0.7 to 16 ppm. There is no relationship between the uranium and scandium contents of the ore in either deposits (Table 35).

In most gold deposits there is invariably some scandium in the wall-rock alteration zones usually in tourmaline, apatite and carbonates; the habit with respect to proximity to gold ore, is commonly erratic, but in some places there is a slight increase as ore is approached (Tables 44–59 inclusive).

According to the literature on scandium in hypogene mineral deposits, the element is concentrated principally in deposits enriched in tungsten, niobium, tin, zirconium, rare-earths, beryllium and uranium, principally in pegmatites, greisen veins, skarn and carbonatites (Shcherbina, 1960; Borisenko, 1961; Park, 1966; Vlasov, 1966–1968; Frondel and Ito, 1967). There is no mention of enrichments of scandium in gold (and silver) deposits in these reviews, and no instance of any concentration of scandium minerals in typical gold deposits could be found in the world literature. The principal minerals carrying scandium in the above mentioned deposits are wolframite, cassiterite and a great variety of rare-earth minerals such as euxenite, xenotime, allanite, etc. The scandium minerals thortveitite and sterrettite are very rare. In ferromagnesium silicates and other minerals, Frondel and Ito (1967) find that scandium largely replaces Al and Fe in the 6-co-ordinated (Al, Fe^{3+}) positions.

Scandium is exclusively trivalent in nature and its solution chemistry is similar to that of the rare-earths and aluminum, but its oxide is decidedly more basic than the latter but less so than the oxides of the rare-earths. The scandium ion, being smaller, has a greater tendency to hydrolyze than the rare-earth ions, and complex formation is more pronounced. In alkaline carbonate solutions soluble complexes of the type $[Sc(CO_3)_4]^{5-}$ are stable. Scandium forms a hydrous oxide, $Sc_2O_3 \cdot nH_2O$, and a definite hydrated oxide, ScO(OH), similar in structure to AlO(OH). The oxide is amphoteric. The sulphate is soluble; the phosphate, carbonate and fluoride, like those of the rare-earths, are insoluble. Hydrous iron and manganese oxides and aluminum hydroxides adsorb and/or coprecipitate scandium to a marked degree.

Relatively little is known about the fate of scandium during the oxidation of gold deposits. As might be expected from the above summary of its solution chemistry, scandium tends to be slightly concentrated in limonite and wad and in aluminous clay minerals. At Keno Hill I found scandium contents ranging from 10 to 100 ppm in some of these materials. One might expect some concentration of scandium in supergene carbonates and phosphates, but this was not found at Keno Hill, probably because of the large amount of sulphate in the waters that renders the element highly soluble. In general, it is probable that scandium is largely leached from deposits during oxidation processes, especially where the waters are acidic in nature. This was the conclusion of Kalenov (1958) regarding wolframite and allanite deposits. In this respect scandium tends to follow calcium, strontium and the rare-earths.

Scandium is present as traces in recent and ancient gold placers mainly in cassiterite, monazite, wolframite, baddeleyite, allanite, other resistant rare-earth minerals, and probably in wad, limonite and the clay complexes. Its presence in placers, together with gold, is therefore, largely incidental and dependent mainly on the occurrence of scandium-bearing resistate minerals.

Yttrium, lanthanum, cerium and the other *rare-earths* appear to have a fairly widespread distribution in gold deposits of all types and exhibit slight enrichments in some skarn and quartz-pebble conglomerate deposits. There is a vast literature (Ginzburg *et al.*, 1959; Vlasov, 1966–1968) on the rare-earths, very little of which pertains to gold deposits. I have, however, investigated the distribution of yttrium, lanthanum and the rare-earths in a semiquantitative way in a number of types of gold deposits from which some general conclusions can be drawn.

In auriferous skarn deposits yttrium, lanthanum and the various other rare-earths are bound in carbonates, garnet, sphene, epidote, apatite, allanite, monazite, brannerite, uraninite, fluorite and scheelite. Excepting the carbonates, garnet and epidote these minerals are relatively minor in amount, and as a whole most skarns generally only contain from 10 to 100 ppm total rare-earths. There appears to be no particular relationship between gold and yttrium, lanthanum and the rare-earths in skarns, but the problem requires more detailed study than I have been able to devote to it.

Recently we have noted native gold in brannerite in the tourmaliniferous skarn at the Richardson Mine, Eldorado, Ontario (Steacy *et al.*, 1974*a,b*). The native gold is later than the brannerite and may belong to a different period of mineralization.

In gold-quartz deposits there is a slight but often erratic enrichment of yttrium, lanthanum and the other rare-earths in the wall-rock alteration zones. The enrichment tends to follow the calcium content and appears to be largely dependent on the abundance of calcium-bearing carbonates. In the quartz bodies yttrium, lanthanum and the other rare-earths prefer the carbonates with a high calcium content and are, therefore, generally enriched in the younger carbonates in the deposits. The older carbonates such as ankerite tend to contain only low amounts of yttrium, lanthanum and the other rare-earths (<10 ppm) whereas the young calcite, especially if well crystallized may contain up to 500 ppm of these elements. The general enrichment of the yttrium earth group of metals in the carbonates (Boyle, 1961*a*), in contrast to low and erratic contents of the cerium earth group, is probably due to the easier capture of the yttrium earths to replace Ca^{2+} because of the closer similarity in radii as shown below.

$$Ca^{2+} - 0.99 \text{Å}$$
$$Y^{3+} - 0.92 \text{Å}$$
$$Yb^{2+} - 1.03 \text{Å}$$
$$Ce^{3+} - 1.07 \text{Å}$$
$$La^{3+} - 1.07 \text{Å}$$

Some gold-quartz and polymetallic deposits carrying gold may be enriched in yttrium, lanthanum and the other rare-earths mainly in minerals such as uraninite, brannerite and thucholite. Examples of these deposits are present in the Goldfields area of Saskatchewan (Robinson, 1955) and in the Cluff Lake deposit in the Carswell structure, Saskatchewan (Little *et al.*, 1972), where native gold occurs in complexly mineralized veins containing pitchblende, uraninite, clausthalite, Ni-Co arsenides and thucholite; in the La Gardette (Isère) area of France (Geffroy, 1963) where brannerite occurs in the gold veins; and at the Little Nell occurrence Plumas County, California where brannerite is also present with gold in the veins and in fractures in an albitite dyke (Pabst and Stinson, 1960). Jouravsky (1952*b*) also mentions the association of brannerite with gold and molybdenite in the Ni-Co arsenide deposits of Bou-Azzer in Morocco. The bastnaesite deposits at Mountain Pass, San Bernardino County, California are apparently not enriched in gold, but auriferous sulphides occur in the immediate vicinity of these deposits (Sulphide Queen Mine) (Olson *et al.*, 1954). The Sulphide Queen orebody apparently contained some rare-earth carbonate and thorium-bearing minerals.

Aleksandrov *et al.* (1975) noted the presence of thorian florencite, Ce $Al_3(PO_4)_2(OH)_6$, in the carbonate-sulphide alteration zones of the Bodaibo deposits, eastern Siberia. They considered the mineral to be an indicator for gold mineralization in the deposits.

Radtke *et al.* (1972*a*) detected a general enrichment of Ce, La, Y and Yb in the 'primary ores' at Carlin, Nevada compared with the host carbonate rocks. The 'oxidized ores' also show a slight enrichment of these elements compared with the 'primary ores' when the averages are considered.

Auriferous quartz-pebble conglomerates and associated quartzites, etc. (Witwatersrand type) are frequently enriched in rare-earths and particularly in yttrium and lanthanum (Table 35). Minerals containing these elements in the deposits include principally brannerite, uraninite, zircon, allanite and thucolite. Brannerite is the most common rare-earth (mainly Y) mineral in the Elliot Lake–Blind River deposits. In the Rand, uraninite carries the bulk of the rare-earths; other rare-earth–bearing minerals include monazite, betafite, euxenite, brannerite and uranothorite. The rare-earth–bearing mineral in the Jacobina, Brazil conglomerates appears to be uraninite, although uncertainties about this (Gross, 1968) make one wonder if brannerite or some other such mineral is not present.

The relationship between gold, yttrium, lanthanum and the other rare-earths (also U and Th) in the quartz-pebble conglomerate type of deposit is not entirely clear. Gold and these elements may be cogenetic although there has been much dispute and discussion about this point, the gist of which is reviewed in the section on these deposits in this chapter. In the Rand, Liebenberg (1955), in a classic study, found a general sympathetic relationship between gold and uranium (and hence probably between gold and rare-earths). This led him to conclude that the gold and uraninite were deposited simultaneously regardless of their ultimate origin.

Uranium (actually the radioactivity) and gold also vary sympathetically in a number of conglomerate beds in the Jacobina deposits (Gross, 1968). This probably means that rare-earths and gold also vary sympathetically since uranium and rare-earths are invariably associated in such deposits. A number of economic gold-bearing conglomerates in the Jacobina area are, however, virtually barren of radioactivity, a feature that has led Gross (1968) to conclude that there is no reason to assume that gold and uranium are cogenetic.

Sulphide-rich parts of the Blind River–Elliot Lake conglomerates are enriched in gold and in many cases in uraninite and brannerite and hence in yttrium and rare-earths. It is not known, however, how gold (and silver) vary overall with yttrium and the rare-earths within the deposits because few gold analyses with this purpose in mind have been done.

The solution chemistry of yttrium, lanthanum and the rare-earths is relatively simple in general, but somewhat complex when each of the elements is considered in detail. The oxidation state for most of these elements is 3, and the ions tend to be relatively mobile especially where the waters are acid and abundant sulphate is present in the solutions. All of the rare-earths and yttrium and lanthanum form oxides of the type M_2O_3. The hydroxides, $M(OH)_3$, are definite compounds, and their basicity decreases with increasing atomic number as their ionic radii decrease. The hydroxides are not amphoteric. The fluorides are insoluble as are also the carbonates and phosphates. In alkaline carbonate solutions soluble complexes of the type $[M^{3+}(CO_3)_3]^{3-}$ are stable. Cerium can exist in the Ce^{4+} state in natural aqueous solutions under conditions of a high oxidation potential. Its aqueous chemistry is similar to zirconium, hafnium and the quadrivalent actinides. Except at very low pH (<1) hydrolysis and polymerization of ceric ion, $[Ce(H_2O)]^{4+}$, is extensive. A number of the rare-earths can also exist in the +2 state of oxidation. Their aqueous chemistry is similar to barium although there are some major differences. The pH of hydrolysis of the various ions of the rare-earths and yttrium show considerable spread from about 2 for Ce^{4+} to 8 for La^{3+}. Because of this they tend to be separated in supergene processes, but the details are complex and not well understood.

Not much is known about the fate of yttrium, lanthanum and the other rare-earths during the oxidation of gold-quartz deposits. My general studies indicate that the elements are relatively mobile and are largely leached in deposits where the strong oxidation of pyrite gives acid sulphate solutions. Small amounts of most of the rare-earth elements and yttrium are, however, retained in the wad and limonite and secondary carbonates. Supergene phosphates such as pyromorphite may collect the rare-earths as I have found at Keno Hill. With increasing neutralization of the solutions hydrolysis phenomena apparently lead to coprecipitation of many of the rare-earths in limonite and secondary carbonates. Under alkaline conditions it seems probable that yttrium, lanthanum and the other rare-earths have considerable mobility in complex forms, particularly as the relatively stable $[M^{3+}(CO_3)_3]^{3+}$ ion as Krainov and Loginova (1968) have suggested.

There are no detailed studies of the fate of the rare-earths during the oxidation of pyritiferous gold-bearing conglomerates. In the near-surface parts of the deposits at Blind River and Elliot Lake acid conditions prevail because of the formation of sulphates and sulphuric acid derived from the oxidation of pyrite. Uranium and thorium are partly leached and dispersed, and yttrium, lanthanum and the other rare-earths appear to suffer the same fate although small amounts are retained in the limonite and other oxidation products. Gold appears to be more abundant in the oxidized parts of the outcrops than in the primary ore, but the writer's work is too limited to indicate that this is the general case. If an enrichment of gold is a general one as I suspect it is, it would appear that yttrium, lanthanum and the other rare-earths part company with gold during the oxidation of pyritiferous conglomerates, the gold remaining behind and yttrium, lanthanum and the other rare-earths moving downward or being dispersed in the groundwaters.

Yttrium, lanthanum and the rare-earths invariably accompany gold in eluvial and alluvial placers, being concentrated mainly in monazite, allanite, zircon, scheelite and a number of other rare-earth–bearing resistate minerals. Most of these minerals come from the country rocks. The association of gold with yttrium, lanthanum and the other rare-earths in placers is, therefore, largely incidental, being predicated on the fact that gold and some of the rare-earth–bearing minerals have a relatively high specific gravity and resistance to weathering and comminution during the formation of placers.

Uranium and *thorium* are infrequent associates of gold in its deposits, although the association is marked in the world's largest deposits, namely the pyritiferous quartz-pebble conglomerate (Rand) type. Some uranium is found in a few gold-quartz deposits, and both uranium and thorium may accompany gold in skarn type deposits, but generally only in traces and minor amounts.

Uranium and thorium are relatively rare elements in gold-quartz deposits in Canada. Most analyses of the ores of these deposits show only a few parts per million (generally <5 ppm) combined uranium and thorium. An exception to this is the Box Mine in the Beaverlodge area of Saskatchewan where Geological Survey analyses show from 5 to 55 ppm U and 20 to 34 ppm Th in the wall rocks and mineralized zones. The gold-bearing polymetallic deposits are, likewise, devoid of any concentration of uranium and thorium in Canada with the exceptions of the Opemiska Copper Mine in Quebec, the Consolidated Nicholson Mine in the Beaverlodge area of Saskatchewan where pitchblende and uraninite occur (Lang *et al.,* 1962), and the unusual auriferous uranium deposits at Cluff Lake in the Carswell structure, Saskatchewan where native gold and gold tellurides are associated with abundant pitchblende, uraninite, clausthalite, chalcopyrite and Ni-Co sulpharsenides (Little *et al.,* 1972). Digenite also occurs abundantly in parts of these deposits.

In foreign deposits uranium and thorium are encountered in relatively few gold-quartz and polymetallic deposits. The veins of the La Gardette (Isère) district in France contain some brannerite (Geffroy, 1963) and those of the Little Nell Mine in Plumas County, California carry the same mineral in small amounts (Pabst and Stinson, 1960). In some veins from Guadalupe, Chihuahua, Mexico, uraninite and gold are associated (Krieger and Hagner, 1943). In some of the Tertiary cupriferous veins of Mitterberg, Salzburg, Austria, gold occurs in nodules of brannerite and uraninite associated in places with organic matter (Bernhard, 1966; Bauer and Schermann, 1971; Siegl, 1972; Paar, 1976). The gold is a late mineral in these deposits ramifying through the uraniferous nodules. Pitchblende accompanies gold in some of the Tertiary auriferous quartz veins and stockworks in the Central City district of Colorado. The pitchblende is an early mineral in the paragenesis of the veins (Sims *et al.,* 1963).

A study of the distribution of gold-quartz and gold-bearing polymetallic deposits in Canada suggests that these deposits only contain concentrations of uranium and thorium in metallogenic provinces characterized by the two actinide elements (viz. Box Mine, Nicholson Mine and Cluff Lake deposits in the Athabasca U-Th province). Similar conditions probably prevail in other parts of the world, but the data are too scanty to draw adequate conclusions. The exact relationships (paragenesis) of gold and uranium in the Nicholson, Box and Cluff Lake deposits are uncertain. In the Nicholson, where carbonates, nickel-cobalt arsenides, pyrite, chalcopyrite, galena, tiemannite, dyscrasite etc. occur, some of the gold may be cogenetic with the uranium, but some of the gold is definitely later than the pitchblende and thucholite (Robinson, 1955); in the Box the pitchblende appears to be related to a later stage of mineralization not necessarily connected with the gold.

Gold-bearing skarn-type deposits may contain traces to minor amounts of uranium and thorium, present mainly in minerals like apatite, monazite, zircon and allanite; uraninite and similar U-Th minerals are rare in most of these deposits. In the Keno Hill area we noted only 1.6 to 4.5 ppm U in sulphide-scheelite skarns containing up to 0.10 oz Au/ton. The uranium appears to vary directly with the tungsten content. In the Eastern Ontario gold belt, however, uranium-thorium minerals, including uraninite, brannerite and thucholite, occur in some quantity in the gold-bearing tourmaliniferous skarn zone at the Richardson Mine near Eldorado. In some samples from this deposit gold is later than the uraninite and brannerite; in other samples the gold is disseminated through the thucholite in a manner such as to suggest exsolution from a hydrocarbon gel (Boyle and Steacy, 1973; Steacy *et al.,* 1973, 1974*a,b*).

Many of the pyritiferous gold-bearing quartz-pebble conglomerate deposits are characterized by enrichments of uranium and in places by thorium (and rare-earths). These elements occur principally in uraninite, pitchblende and brannerite; also in smaller amounts in thucholite (or uraniferous carbon), uranothorite, betafite, euxenite, monazite, zircon and similar minerals. Details of these deposits and speculations on their origin are considered at greater length in a later part of this chapter.

The relationships between gold and uranium and thorium in the quartz-pebble conglomerate type of deposit are not entirely clear. The three elements may be cogenetic although there has been much controversy about this point. In the Rand deposits, Liebenberg (1955), after an extensive investigation, found that the uranium and gold contents varied sympathetically in the majority of the reefs and concluded that gold and uraninite, the principal U-Th mineral, were deposited simultaneously, regardless of their ultimate origin. In a later study Von Rahden and Urli (1969) showed statistically that there is a strong sympathetic relation and a high correlation between gold and uranium in a number of the reefs. In the volume edited by Haughton (1964*a*) on the Rand there is agreement among a number of authors that gold and uranium exhibit a sympathetic relationship. In the Virginia section of the Orange Free State Goldfield, Winter (*in* Haughton 1964*a*), however, states that while the uranium values follow the trend of gold values, the relationship varies widely and is very erratic. Feather and Koen (1975), summarizing the relationship between gold and uranium in the Rand, state that some investigators find a strong correlation between the two elements, others only a low degree of positive correlation, and a few no correlation at all.

Some sulphide-rich parts of the Elliot Lake–Blind River

conglomerates are slightly enriched in gold and in many cases in uraninite and brannerite. It is not known, however, how gold (and silver) vary overall with uranium and thorium within the deposits because few gold analyses with this purpose in mind have been done. Lang *et al.* (1962) state that in general high uranium contents are related to abundant pyrite. Since the latter appears to carry or is associated with most of the gold (and silver) (*see* Table 11) a positive correlation between U, Th, Au and Ag seems probable.

In the Jacobina conglomerates in Brazil, Gross (1968) found that the radioactivity tends to increase with the gold content, this relationship being particularly obvious for samples containing either very high or very low quantities of gold. However, there are a number of economic gold-bearing conglomerates in the Jacobina area that are virtually barren of radioactivity, a feature that has prompted Gross (1968) to conclude that there is no reason to assume that gold and uranium are cogenetic.

Certain uranium deposits may be enriched in gold. The Jabiluka deposits in Northern Territory, Australia contain significant gold values (15 ppm), which can be won when the deposits are mined (Rowntree and Mosher, *in* Knight, 1975). These deposits occur in fractured and crushed zones in Lower Proterozoic metasediments of the Koolpin and Fisher Creek formations invaded in places by granitic pegmatites. The mineralization, including pitchblende and uraninite with some pyrite, chalcopyrite, chalcocite, covellite, galena and native gold, is most intense where the rocks are chloritized and hematitized. The gold is late and is associated with tellurobismuthite and tellurides, mainly altaite and melonite. The Cluff Lake deposits in Saskatchewan, referred to above, also contain significant gold values. Most of the uranium deposits that are enriched in gold appear to be of Lower Proterozoic age. In this respect it is of interest to note that some of the auriferous magnetite bodies in Lower Proterozoic rocks at Tennant Creek, Northern Territory, Australia have a uraniferous halo which ranges in magnitude from 20 to 280 ppm U (Large, 1975). In the writer's experience this is not unusual for many Proterozoic gold deposits, especially those in sedimentary terranes.

The oxidation of uranium deposits has been investigated intensively, and most of the supergene features of the element are well known. Much less is known about the fate of thorium, and there appears to be no extensive studies of the oxidation phenomena in gold deposits containing uranium and thorium, particularly the quartz-pebble conglomerate type.

In aqueous solutions the thorium ion Th^{4+} is extensively hydrolyzed at pH higher than 2.5; the resultant soluble species are numerous complexes depending on the nature of the anions, their concentrations, etc. The sulphate and the nitrate are soluble in acid solutions, and the complex ion $[Th(CO_3)_5]^{6-}$ is thought to be the principal species in alkaline carbonate solutions. The fluoride and phosphate are insoluble. In organic environments thorium is readily transported as thorium-humates or absorbed to (colloidal) humus. Thorium is strongly adsorbed and/or coprecipitated and carried by hydrous iron and manganese oxides, and various alumino-silicate gels probably as hydrous thoria, $ThO_2{\cdot}nH_2O$. The pH of precipitation of the hydroxide is about 6. Uranium differs from thorium in having two principal oxidation states in natural solutions, U(IV) and U(VI). The U(III) and U(V) states are also possible making the aqueous chemistry of uranium particularly intricate when it is realized that a veritable host of complexes and polymeric species are formed depending on the anions present, their concentration, etc. The ions of uranium are usually characterized as U^{3+}, U^{4+}, UO_2^+ and UO_2^{2+}. The uranyl carbonate complexes $[UO_2(CO_3)_2]^{2-}$ and $[UO_2(CO_3)_3]^{4-}$ are probably the most important in natural alkaline waters. Uranium (IV) sulphate and nitrate are soluble; the phosphate, arsenate, molybdate and fluoride are insoluble, precipitates of these compounds being formed when U^{4+} ion is mixed with PO_4^{3-}, AsO_4^{3-}, MoO_4^{3-} and F^- in acid solutions. Uranyl (VI) sulphate and nitrate are soluble. In organic environments uranium is readily transported as uranium-humates or adsorbed to (colloidal) humus. A veritable host of supergene complex uranium (uranyl) minerals may form during the oxidation of uranium deposits, depending upon the presence of phosphate, molybdate, vanadate, silicate, carbonate, arsenate, selenite, tellurite and sulphate anions and cations such as Ca, Mg, Pb, Ba, Bi, Cu and K. Many of these appear to form by reaction with $UO_3{\cdot}H_2O$, the hydrous uranyl oxide gel, a hydrolysis product of uranyl ion. Solutions containing U^{4+} hydrolyze when neutralized, and the hydroxide $U(OH)_4$ or hydrous oxide $UO_2{\cdot}2H_2O$ is precipitated from which water can be split out to yield the oxide, UO_2 (uraninite). Alternatively, acidification of alkaline uranyl carbonate solutions yields the hydroxide $UO_2(OH)_2$ or hydrous oxide $UO_3{\cdot}H_2O$ from which uranium oxide (uraninite) may form by reduction. These reactions appear to be responsible for the occurrence of much supergene uranium black (uraninite). Finally uranium is adsorbed and/or precipitated and carried by hydrous iron and manganese oxides, silica-gels, etc.

With the above discussion in mind it is possible to predict the fate of Th, U and Au during the oxidation of deposits. Thorium is relatively immobile except under extremely acid conditions or when alkali carbonates are abundant in alkaline solutions. In these cases some thorium may be leached. More generally, however, considerable thorium is retained in the gossans as a result of adsorption and coprecipitation by limonite, etc., and the element is, therefore, enriched with gold. Supergene phosphates and a number of other minerals such as the arsenates may also retain much thorium. Under acid conditions uranium is extensively leached especially where pyrite is undergoing oxidation, the element being highly mobile as the sulphate. As the solutions are neutralized a great variety of supergene minerals may be formed, principally secondary uraninite and various phosphates, vanadates, carbonates, etc. Under alkaline conditions where the stable uranyl carbonate complex is formed uranium may be relatively mobile. Under most conditions where carbonates are minor gangue and wall-rock minerals, and where the deposits contain much pyrite as in the quartz-pebble conglomerates, uranium and gold tend to part company, the uranium being largely removed as the soluble sulphate from the oxidized zones and the gold remaining behind. Such phenomena appear to have taken place in some of the oxidized outcrops of the Blind River–Elliot Lake area.

Where abundant humic constituents are present in the surface waters and groundwaters U, Th and Au may be rendered mobile, and the three elements may be leached from deposits in which they occur. Alternatively, in gossans where humus and iron are strongly concentrated (humic limonite) Au may be enriched with U and Th evidently mainly as the result of adsorption phenomena.

Uranium and thorium are minor and trace elements in most recent and ancient alluvial and eluvial gold placers, but the presence of the elements with gold is largely incidental being predicated on the circumstances that gold and many of the uranium- and thorium-bearing minerals such as zircon, monazite and apatite, which derive mainly from the country rocks, are resistates.

Group IVA: C, Si, Ge, Sn, Pb

Carbon accompanies gold in all of its deposits being present as amorphous carbon, graphite, carbonaceous matter, various solid and semisolid hydrocarbons, and in a great variety of carbonates. The last are discussed in the next section on associated and host minerals of gold.

Black and grey quartz in gold-quartz deposits have long been recognized as a favourable indication of high gold values, and this is essentially true in many Precambrian gold belts in both sedimentary and volcanic terranes. In deposits in younger rocks, black and grey quartz are not particularly abundant except in some sedimentary terranes containing the familiar black slates or schists. Black carbonates may occur in the carbonated wall rocks or be intergrown with the black and grey quartz in some deposits, but they are relatively rare. Only in one or two deposits, in the writer's experience, have black carbonates been observed cutting older generations of black, grey and white quartz. The nature of the colouring of black and grey quartz is discussed at length in the next section on minerals associated with gold in its deposits. There, it is stated that the colouring is due mainly to amorphous carbon and graphite. The gold in black quartz may have been transported as an auriferous hydrocarbon the carbon and graphite being the polymerized residues of the hydrocarbon, or alternatively the gold may have been selectively precipitated from solutions or diffusion currents by the carbon or graphite.

Amorphous carbon and graphite accompany gold in some skarn deposits, but the relationships between the elements (C and Au) are not clear in most places. The carbon and graphite in some deposits appear to be residual from carbonate rocks after replacement by silicates; in others there is considerable evidence to suggest that the carbon minerals were deposited from some type of mobile carbon compound. It follows, therefore, that gold and carbon in skarn bodies may be associated as a result of the alternate processes suggested for their relationship in black quartz veins. Support for the transport of gold into skarn bodies as an auriferous hydrocarbon is found in the recent discovery of gold in thucholite in the Richardson Mine near Eldorado, Ontario. There, the gold appears as though it exsolved from a thucholite gel (Boyle and Steacy, 1973).

Carbonaceous matter, graphite and amorphous carbon in phyllites, slates and schists have long been regarded as favourable precipitants for gold and silver where they are in contact with or enclosed within veins, saddle reefs and silicified zones. These are the black leaders, beds or indicators of the old gold miners. In places, these indicators are phenomenally enriched in gold, some of which is difficult to extract by cyanidation. The reason for the enrichment is obviously the marked reducing action which the carbonaceous matter has on dissolved gold, a feature demonstrated long ago by Rickard (1896) with black bituminous Rico shale.

Carbon, bound mainly in authigenic thucholite (or uraniferous carbon) and what is called amorphous hydrocarbon (i.e., nongraphitic), is present in some quantity in the Rand conglomerates (Davidson and Bowie, 1951; Liebenberg, 1955). In the Blind River–Elliot Lake deposits thucholite is present mainly along slips and fractures. Some of the gold is intimately associated with thucholite in many of the Rand conglomerates, but the association is not always invariable. Liebenberg (1955) has concluded from his studies that some of the gold was deposited later than the thucolite, being in some cases moulded around the grains of the hydrocarbon, present as specks in the material, or in crosscutting veinlets and seams. In these cases the thucholite probably had a precipitating effect on the migrating gold. In the Black Reef, characterized in places by amorphous hydrocarbon, gold is not closely associated with this substance. According to Liebenberg (1955) this type of hydrocarbon was introduced at a later stage and did not affect the deposition of the (remobilized) gold.

Mention of 'carbon' frequently appears in the older literature on the Rand. For instance Graton (1930, p. 143) says "In those mines where it is at all common the 'carbon' is generally recognized, I believe, as the most significant and reliable of all indicators of good gold values." In a recent contribution on the carbon of the Rand, Hallbauer and van Warmelo (1974) have concluded that much of the thucholite represents fossilized plant (lichen) material. (*See* the further discussion on this finding in the section on the Rand in this chapter.)

On casual inspection the gold values in the Blind River–Elliot Lake conglomerates do not appear to bear any relationship to thucholite and other carbonaceous substances.

Thucholite and carburan are relatively rare in most gold-quartz-sulphide deposits, but examples are known in the auriferous Boliden ores of Sweden, in the gold-quartz ores of the Box Mine in the Uranium City district of Saskatchewan and in auriferous ores in other uraniferous belts.

Carbon and carbonaceous matter are common constitutents of the disseminated type gold deposits, being particularly abundant in those in carbonate rocks as at Carlin, Nevada. Radtke and Scheiner (1970*a*,*b*) consider that the carbonaceous matter at Carlin and Cortez played a large part in the precipitation of the gold from acid hydrothermal solutions. Subsequent oxidation of the gold-organic complexes so formed precipitated the gold in a microscopic form. Wells (1971) on the other hand, using electron microprobe methods, noted no particular association of gold with carbonaceous matter in these deposits, but a close relationship to arseniferous minerals such as pyrite was found.

Silicon: 'Quartz is the mother of gold', so has run the prospector's adage since the time of Pliny and probably before. It is axiomatic to say that there is no hypogene gold deposit (including the conglomerate type) from which quartz in one form or another is absent. The correlative that all quartz veins or silicified zones contain economic quantities of gold is, however, not true, although the writer has not seen a quartz vein which on analysis does not contain a slightly higher amount of gold than that present in normal rocks (0.004 ppm).

On a statistical basis gold favours a quartz gangue whereas silver tends to occur in veins marked by a carbonate and/or barite gangue. These associations have been known for centuries, but the precise reasons for them have, however, escaped us to the present. The association of gold and silicon (silica) also extends to the plant kingdom as noted in the section on biogeochemistry. Such a marked relationship between gold and silicon must reflect some fundamental parallel in the chemistries of the two elements, either in their solution or colloidal characteristics. Further remarks on these phenomena are contained in the sections on gold in natural waters in Chapter II and on the origin of gold deposits in this chapter. The precise relationship of gold to quartz, chalcedony, silicification, etc., an important consideration in any study of gold, is reviewed in the next section on host and associated minerals. The chemistry of silica in the wall-rock alteration zones of gold deposits is discussed in the section on wall-rock alteration processes in this chapter.

The behaviour of silica during the oxidation of gold deposits is variable and complicated depending on the composition of the deposits and a number of other factors that are discussed in some detail in Chapter IV. Briefly, in some deposits quartz is particularly resistant to chemical attack and largely remains behind often with much gold in the rubble of the gossans and oxidized zones; in other deposits there is considerable solution and redeposition of silica in the gossans and oxidized zones; and in still others much silica is leached and dispersed into the groundwater systems. Generally speaking, gold and silica (quartz) often show a parallel behaviour during oxidation processes, due to two circumstances – their resistate nature and their solution and colloidal characteristics.

Quartz is a persistent mineral in gold placers. The 'white leads' of many placer fields and the 'white channel gravels' of others emphasize the importance of the mineral in these secondary deposits. Resistate silicates of many kinds also occur in all gold placers. They are listed and discussed in the section on placers.

Germanium is usually present in gold deposits containing chalcopyrite, tetrahedrite, enargite and magnetite, minerals that commonly contain from 5 to 50 ppm Ge. Higher values are found in cassiterite, stannite, topaz and in low temperature sphalerite (up to 3000 ppm); also in cinnabar (up to 2000 ppm or more), pyrargyrite and in the sulpho-germanates, germanite, renierite, argyrodite-canfieldite (up to 7% or more). All of these minerals, with the exception of sphalerite, are rare in gold deposits. In most gold deposits sphalerite is not noticeably germanium-bearing, containing only amounts up to 50 ppm according to the available data in the literature. Those we have investigated in gold deposits invariably contain less than 50 ppm Ge.

Reviews of the geochemistry of germanium can be found in the works by Fleischer and Harder (1955), Ginzburg *et al.* (1959) and Vlasov (1966).

Relatively little is known about the germanium content of skarns containing economic amounts of gold. In most of these bodies the element is present in low amounts (<50 ppm) according to the writer's investigations. Miroshnichenko *et al.* (1971) found amounts up to 90 ppm in rare metal and polymetallic skarns associated with granitic contacts in central Kazakhstan. Those near intermediate intrusives had lower contents. Skarn-greisen deposits were found to be especially rich in germanium, much of the element being present in topaz and muscovite (50 ppm each). All of the calc-silicates contained germanium, some garnets containing up to 180 ppm Ge.

In gold-quartz deposits with small amounts of sulphides the content of germanium is low, in the parts per million range or less; larger amounts can be expected in the pyritic, copper pyritic and polymetallic deposits where sphalerite and copper minerals are present. In the polymetallic deposits the contents may range up to 100 ppm judging from the literature. If bornite is present and germanium minerals are developed, some polymetallic deposits may contain up to 1000 ppm Ge or more in the ore. Fleischer and Harder (1955) mention that two gold mines in South Dakota produced concentrates, which contained 140 ppm Ge. Argyrodite has been identified in some of the silver-gold ores of the Hauraki goldfield, New Zealand (Ramsay and Kobe, 1974). There appear to be no extensive data on germanium in the Rand ores. Liebenberg (1955) reports only traces in the uraninite concentrates, and we found <2 ppm in suites of the ores (Table 35). In the Blind River–Elliott Lake ores the germanium content is <0.5 ppm.

There appears to be no particular relation between germanium and gold in its deposits, although the data are not extensive enough to substantiate this generalization in all cases. Germanium tends to be associated more with copper and silver than with gold, among the members of Group IB. The ores of the Tsumeb Mine in Namibia (South-West Africa), noted for their high germanium content, yield essentially Cu, Pb, Zn, Cd and Ag; the gold content is very low. Some germanium compounds (e.g., GeS) have a relatively high volatility, hence the low content of the element in deep, high temperature deposits. In near-surface (low temperature?) deposits a relationship between gold and germanium may exist. Two facts suggest this – the presence of argyrodite in the silver-gold ores of Great Barrier Island, Hauraki goldfield, New Zealand (Ramsay and Kobe, 1974) and a concentration of up to 0.004 ppm Ge in some of the thermal gold-bearing waters in the Taupo volcanic zone in New Zealand (Browne, 1969).

Germanium has oxidation states (II) and (IV). The (II) state is stable in nature where the oxidation potential is low and is probably the most important state in sulphides and some sulphosalts. In minerals like argyrodite, Ag_8GeS_6, however, germanium is apparently in the (IV) state. The quadrivalent state is stable under high oxidation potentials and is generally the state found in minerals in the supergene environment. The divalent state forms an amphoteric hydroxide characterized as $Ge(OH)_2$, although its exact nature is unknown. It may be a hydrous oxide of the type $GeO{\cdot}nH_2O$. The divalent hydroxide is readily oxidized in air to $GeO_2{\cdot}nH_2O$ from which water can be split out to give GeO_2. This oxide is not markedly amphoteric, but Ge(IV) is soluble in acids such as HCl. The germanates are well characterized compounds forming the ions $[GeO(OH)_3]^-$, $[GeO_2(OH)_2]^{2-}$ and $[(Ge(OH)_4)_8(OH)_3]^{3-}$, as well as other more complex forms. It is probably in these forms that much of the germanium migrates in the supergene environment.

There are no data *per se* on the fate of germanium during the oxidation of gold deposits. Data in the literature on the transport of the element in natural waters and in sulphide deposits containing germanium, however, give the general outline (Frondel and Ito, 1957*b*; Ershov and Shcheglova, 1958; Pazenkova, 1967; Goleva and Vorob'yeva, 1967; Krainov, 1967; and Lyalikova and Kulikova, 1969). From these studies it is evident that germanium is relatively mobile under both acid and alkaline conditions. The forms of migration are speculative, but in acidic solutions it can be assumed that cationic complexes are present whereas in alkaline solutions the anionic complexes noted above are probably predominant. According to some investigators organic complexes are also important transfer agents. Bacteria aid the oxidation of sulphides containing germanium in both an acid and alkaline environment.

As initially released from sulphides germanium is probably largely in the (II) state; some released from minerals like argyrodite would be in the (IV) state. Where the oxidation potential is low germanium may migrate in the divalent state, but more generally this state is oxidized to the quadrivalent state, and germanium migrates largely as various germanate ions. In a milieu where abundant iron and manganese is present in solution, where hydrolysis phenomena are marked, and where limonite and wad are being precipitated, much germanium is absorbed and/or coprecipitated mainly in the limonite, perhaps as an iron germanate as Pazenkova (1967) has suggested. Passing downward through the gossans and oxidized zones some germanium as Ge(II) may substitute for Pb(II) in minerals such as anglesite, cerussite and a new basic lead sulphide, which Frondel and Ito (1957*b*) have described. Much germanium is, however, oxidized to the (IV) state, and some is taken up by various heavy metal arsenates such as bayldonite, mimetite, duftite, etc. where Ge(IV) is thought to substitute for As(V) in the (AsO_4) tetrahedra. Supergene minerals such as smithsonite, malachite, calcite, etc. contain relatively little germanium, but secondary silicates such as hemimorphite, the zinc silicate, may take up some germanium. Some germanium on reaching the secondary sulphide zone is incorporated in minerals such as chalcocite and covellite. The remainder is dispersed in the groundwater system, which may contain up to 1.5 ppb in the vicinity of sulphide and other deposits containing germanium.

We can see from the above outline that germanium and gold would tend to be concentrated together in various phases of the oxidation zones – in the oxide or gossan facies, in the basic salt and arsenate-sulphate facies, and in the secondary sulphide facies.

There are no data on the content of germanium in recent and ancient eluvial and alluvial gold placers. Presumably, the element is present as traces in these deposits, being present in limonite and wad particles, magnetite, garnet, cassiterite and in oxidized residuals of sphalerite, chalcopyrite and a variety of other minerals.

Tin is associated with gold in a number of its deposits but generally only in minor or trace amounts. The classic tin deposits are notably low in gold, and the reverse is also true that deposits enriched in gold seldom carry much tin; this despite the fact that both gold and tin deposits are marked by extensive silicification and the development of quartz, arsenic, antimony and bismuth minerals, tourmaline and other common mineral associates of gold. Silver on the other hand is commonly enriched with tin in a number of deposits, particularly those containing stannite and the rarer sulphosalts such as canfieldite, cylindrite, teallite, etc. An exception to these generalizations is found in the stanniferous provinces of the far eastern U.S.S.R. and other parts of eastern Asia; also in certain stanniferous regions in Australia (New South Wales, Tasmania). In the deposits of the far eastern U.S.S.R. Nekrasov (1973) emphasizes the fact that tin and gold minerals are not antagonistic. There, gold in two forms – native and dispersed in sulphides and arsenides – occurs in typical Au-Sn deposits. These are skarns with glaucodot-cobaltite, safflorite-loellingite and magnetite mineralization; quartz-gold-cassiterite and quartz-molybdenite-gold veins; and silicate-sulphide veins, stockworks and zones of greisenization mineralized with quartz, tourmaline, chlorite, arsenopyrite, pyrrhotite, scheelite, galena, sphalerite and gold. In all types the tin minerals are earlier than those containing gold.

Auriferous skarn deposits in most places contain only traces to minor amounts of tin, generally well below 50 ppm. Most of the tin is in the sulphides such as chalcopyrite, sphalerite, galena and various micas and the calc-silicates, probably largely as specks of cassiterite and other tin minerals. Well crystallized cassiterite, stannite and other tin minerals are relatively rare in the gold-bearing skarns of most mineralized belts.

In typical gold-quartz deposits tin invariably accompanies gold but only in trace amounts. At Yellowknife the gold ores in the greenstone belt carry less than 10 ppm Sn; at Kirkland Lake the range is 2 to 11 ppm Sn. Most of the tin at Yellowknife is present in stannite, which has been identified in polished sections. Traces (10–100 ppm) occur in pyrite, galena and sphalerite in most auriferous deposits. Gold veins and saddles in sedimentary rocks usually contain only traces of tin mainly in pyrite, galena and sphalerite. In the gold-quartz veins of the Balei mineralized belt U.S.S.R., Gapontsev and Polikarpochkin (1969) found tin to range from 2 to 30 ppm. In some ore shoots there was an upward increase in the tin content. In stanniferous regions throughout the world small amounts of cassiterite are common in auriferous veins as in New South Wales, Australia and in the Far Eastern tin belt of the U.S.S.R.

In the disseminated gold ores (Carlin type) only small amounts of tin are recorded (<7–20 ppm) (Radtke *et al.*, 1972*a*). There is no appreciable change with oxidation of these types of ores according to the available data.

Polymetallic deposits enriched in gold may carry higher amounts of tin than the typical gold-quartz deposits. In these deposits the tin may occur in stannite, canfieldite and other tin minerals, including minor amounts of cassiterite in places, but more generally the element occurs in amounts ranging from 10 to 1000 ppm in pyrite, chalcopyrite, galena, sphalerite and a variety of sulphosalts. The ores as a whole generally carry less than 100 ppm Sn.

Tin occurs only in traces in the auriferous quartz-pebble conglomerates (Table 35). Liebenberg (1955) described detrital cassiterite with a strong pleochroism from the Dominion Reef of the Rand. In most reefs only traces (3–6 ppm) of tin are present in the Rand ores according to our analyses.

During oxidation processes tin is relatively immobile due to the resistate nature of its oxide and the ease with which the element is precipitated from solution. Tin exists in two states of oxidation, (II) and (IV). The stannous state is probably not stable in nature except under a very low oxidation potential or when the ion is extensively complexed under very acid conditions. At the normal pH found in nature (5–8) stannous ion is readily oxidized by air giving stannic solutions. Stannous ion is, therefore, relatively rare in most supergene environments.

Stannic oxide or cassiterite, SnO_2 is amphoteric. The stannic ion in acid solutions is often characterized as Sn^{4+}, but this is a simplification since the ion is extensively hydrolyzed and complexed. Dilution or addition of base to a stannic solution leads to extensive hydrolysis and the precipitation of the hydrous oxide, $SnO_2 \cdot nH_2O$. From this water can be split out to give cassiterite. The oxide is soluble in excess base giving stannate ion usually written $[Sn(OH)_6]^{2-}$. Tin may, therefore, migrate in strongly acid and alkaline solutions in a cationic or anionic form respectively. Between pH 5 and 8 both forms are extensively hydrolyzed and the oxide is precipitated. Hydrous iron and manganese oxides (limonite and wad), the various silica-alumina gels (clay minerals), and hydrous titanium oxide coprecipitate, adsorb and carry tin in relatively large amounts.

In oxidizing zones cassiterite is relatively unaffected except under very acid and alkaline conditions when some tin may enter solution. More generally the cassiterite remains as a resistate in the gossans and oxidized zones, utlimately passing into eluvial and alluvial deposits. Stannite and the other sulphostannates on oxidation yield stannic or stannate ion to the solutions depending on the pH. These ions, however, have relatively little mobility – hydrolysis due to neutralization of the solutions or dilution precipitates the oxides as cassiterite (needle tin and wood tin). Coprecipitation and/or absorption of the soluble tin by limonite, wad and other gellike substances further restrict the mobility of the ions. Finally, much tin is captured and incorporated in a variety of basic salts and other supergene minerals precipitated during the various hydrolytic reactions in oxidation zones. At Keno Hill tin was found in anglesite (100–1000 ppm), smithonsite (200 ppm), beudantite (1000 ppm), bindheimite (100–1000 ppm) and jarosite (100–1000 ppm). Overall the lodes at Keno Hill showed relatively little change in their tin content with depth through the oxidized zones into the primary ore (Boyle, 1965*a*). In this respect the element behaved much the same as gold.

In some recent and ancient eluvial and alluvial placers, tin, mainly in the form of cassiterite, is concentrated with gold. Such placers, however, tend to be relatively rare, few tin placers containing economic concentrations of gold and vice versa. In some placers traces of tin accompany gold in the heavy concentrates. The element is present mainly in residuals of sphalerite, pyrite, galena, sphene, rutile and other common minerals that contain tin in primary deposits.

Lead is a universal associate of gold in all types of its deposits, the largest amounts being present in gold-bearing polymetallic and skarn deposits and the lowest in the gold-quartz and quartz-pebble conglomerate types. Galena is the principal lead mineral in all primary epigenetic gold deposits; some deposits carry appreciable amounts of various lead sulphosalts; and a few contain lead tellurides, selenides, bismuthides, etc. Tetrahedrite-tennantite, and various other sulphosalts, arsenides, sulphides and tellurides may contain minor to trace amounts of lead in gold deposits. The zoning relationships of lead with respect to gold are discussed in the last part of this chapter and shown in Figures 94 and 95.

Lead is present only in trace to minor amounts (up to 100 ppm in the ore) in some auriferous skarn deposits, particularly those marked by the presence of abundant copper sulphides or arsenopyrite (e.g., Nickel Plate and French mines, British Columbia). Other gold-bearing skarn deposits (e.g., New Calumet Mine, Quebec) contain considerable amounts of lead (up to 3%). There appears to be no particular correlation between the lead and gold contents in skarn deposits, although in some orebodies the lead-rich parts tend to be enriched in gold and particularly in silver.

Most gold-bearing polymetallic deposits are generally enriched in lead, but there appears to be no particular correlation between the gold and lead contents in these types of deposits in Canada according to the writer's investigations, which are, however, limited in their scope. Judging from the available data in the literature, correlations between the gold and lead contents in polymetallic deposits in other parts of the world appear to be weak or nonexistant, although further detailed research is desirable on this particular aspect of the geochemistry of gold.

The lead content of most gold-quartz deposits is low and often erratic. Deposits in Precambrian greenstones (meta-basalts, meta-andesites) and intermediate and acidic rocks generally contain less than 200 ppm Pb in the ore, mainly in galena. Higher amounts (up to 5000 ppm Pb) occur in some of these deposits (e.g., Yellowknife, Red Lake) where sulphosalts such as jamesonite are abundant in the ore. Some of the Tertiary gold-quartz ores in andesites and other volcanics are exceptionally rich in lead containing percentage amounts of the element. Gold-quartz deposits in sedimentary rocks of all ages tend to carry more lead than those in volcanic and other igneous rocks. The amounts of lead are highly variable in the deposits in sediments, ranging from mere traces to 2 per cent or more in the ore.

Schwartz (1944) mentions that galena is a particularly good indicator of high grade gold shoots in a relatively large number of gold-quartz and other types of auriferous deposits. He attributes this to the fact that galena and gold are frequently the last minerals in the paragentic sequence, and hence the two often occur together. In the writer's experience the presence of altaite, PbTe, commonly signals the presence of gold tellurides in deposits.

The lead content of the disseminated gold deposits (Carlin Type) ranges from <7 to 1500 ppm in the 'primary ores' and from <7 to 200 ppm in the 'oxidized ores' at Carlin according to Radtke *et al.* (1972*a*). The carbonate host rocks contain <7 to 15 ppm Pb.

Lead is a trace to minor constituent of the quartz-pebble

conglomerate gold ores of the Rand. The content of lead in the ores appears to vary widely in sympathy with the abundance of galena. In local areas galena is abundant, but overall the lead content of the ores in samples analyzed at the Geological Survey ranges from 51 to 450 ppm (Table 35). This seems to agree relatively well with the analyses given by Saager (1969) who states that in the Basal Reef the chalcopyrite, galena and sphalerite on the average do not form more than 0.05 per cent of the total ore. Some of the lead is radiogenic occurring as specks and veinlets of galena in uraninite and thucholite. Ordinary (i.e., largerly nonradiogenic lead-bearing) galena occurs throughout the reefs associated with pentlandite, pyrrhotite, sphalerite, chalcopyrite, cobaltite, linnaeite and arsenopyrite. This galena, characterized as secondary and pseudohydrothermal by Liebenberg (1955) and Saager (1969) respectively, occurs mainly interstitially in the matrix of the conglomerates where it replaces the matrix materials to a varying extent. Gold is apparently closely associated with galena and the other sulphides in some parts of the reefs, but Liebenberg (1955) concluded from mineragraphic work that most of the gold is genetically unrelated to the sulphide mineralization. The relationship of gold and lead (as well as elements in the various sulphides such as Zn, Co, Ni, Cu, As, etc.) in the Rand deposits is obviously complex as one can readily see by reading the various papers on the origin of these deposits. It would be interesting to investigate the detailed correlation of gold with these various chalcophile elements in the conglomerate orebodies. Such information could help in the solution of the origin of these controversial deposits.

Lead can exist in two oxidation states (II) (plumbous) and (IV) (plumbic). The plumbous compounds are prevailingly basic, and their salts ionize in solution. The plumbic compounds are few, are largely covalent, require the presence of high oxidation potentials, and are relatively unstable in natural situations. Only the plumbous compounds are important in nature forming the sulphide (PbS) and a variety of sulphosalts, tellurides, etc. in hypogene deposits. In the oxidized zones of lead deposits the most common plumbous minerals are the sulphate (anglesite), the carbonate (cerussite) and minerals like bindheimite (an antimonate), beaudantite (a basic sulphate) and pyromorphite-mimetite (a phosphate-arsenate).

Plumbous salts have a strong tendency to hydrolyze and form the insoluble hydroxide $Pb(OH)_2$ when base is added. This hydroxide is amphoteric dissolving in acid to give the plumbous ion and in excess base to form the plumbite ion $[Pb(OH)_3]^-$. The nitrate and acetate are soluble, the other plumbous salts such as the sulphate and carbonate are insoluble or only sparingly soluble. Soluble complexes of the type $[PbCl_3]^-$ are formed in the presence of excess chloride ion, and soluble carbonate complexes are stable in solutions enriched in CO_2 or hydrogen carbonates. Judging from our observations the soluble species of lead are relatively rare under most oxidizing conditions.

During the oxidation of gold deposits containing galena and sulphosalts, the lead is largely immobilized as the sulphate (anglesite) which in time is converted to the carbonate (cerussite) the most stable phase in oxidized zones. Some lead is also precipitated in the basic sulphate-arsenate, beudantite; in the pyroantimonate, bindheimite; in the phosphate-arsenate minerals, pyromorphite and mimetite; and in the basic sulphate, plumbojarosite. The oxide minerals, massicot, minium and plattnerite are rare in most oxidized zones. Supergene galena is also rare in most deposits. Limonite and manganese oxides in gossans tend to adsorb and/or coprecipitate considerable amounts of lead.

All of the supergene lead minerals are relatively insoluble especially under slightly acid and neutral conditions. At Keno Hill and in the vicinity of other lead deposits I have noted that only traces of lead (<0.005 ppm) migrate in the groundwaters (Boyle, 1965*a*). The general tendency, therefore, is for marked enrichments of lead to take place in the oxidized zones of deposits compared with the primary zones. In this respect gold and lead behave in a similar manner, and it is common to find both elements enriched in gossans and their underlying leached zones. In addition I have noted that a number of the supergene lead minerals in gossans and oxidized zones tend to adsorb, coprecipitate or otherwise accumulate gold. The following ranges in ppm for the gold content of supergene lead minerals from a number of gold-bearing deposits in Canada can be quoted: anglesite (0.01–0.5); cerussite and limonite (0.7); bindheimite (0.7); pyromorphite (0.17); cerussite (0.2); and beudantite (0.01–0.4).

Most recent and fossil eluvial and alluvial gold placers contain only traces of lead mainly in unoxidized residuals of galena and sulphosalts, and in minerals such as anglesite and cerussite. A few eluvial placers containing gold and silver may have relatively high contents of lead minerals. (*See* the section on gold in other products of weathering in Chapter II.)

Group IVB: Ti, Zr, Hf

These three elements show relatively little association with gold except in some placers.

Titanium accompanies gold mainly in minerals such as leucoxene, rutile, ilmenite, sphene and brannerite, in some skarn and gold-quartz deposits. The titanium in some of the minerals appears to represent an introduction of the element; in others the titanium minerals are residuals from alteration, or the element was derived from the wall rocks of the deposits. There appears to be no particular correlation between gold and titanium in these deposits according to the writer's investigations. Leucoxene (rutile, sphene, etc.) is a frequent alteration product in the alteration zones of some gold deposits, particularly those in basic and intermediate volcanic and intrusive rocks. The behaviour of titanium in these alteration zones is discussed in more detail in the section on alteration processes. For the association of brannerite with gold see the section on uranium and thorium.

The titanium content of the 'primary ores' at Carlin, Nevada ranges from 200 to 5000 ppm; in the 'oxidized ores', the range is 300 to 5000 ppm indicating no particular enrichment during oxidation of these gold ores (Radtke *et al.*, 1972*a*). The host rocks contain from 70 to 2000 ppm Ti, indicating an enrichment of titanium during primary mineralization.

Titanium, mainly in minerals such as leucoxene, rutile, sphene, ilmenite and brannerite, is a common minor and trace constituent of the quartz-pebble conglomerate deposits (Table 35). Some of the leucoxene in the Rand is radioactive (Liebenberg,

1955); brannerite is one of the principal uraniferous minerals in the Elliot Lake–Blind River ores. The relationship of titanium and titanium minerals to gold in the conglomerate deposits is uncertain. According to the placer theory, which is favoured by many, the minerals are mainly detrital, like the gold, but subsequently reconstituted or altered in part. Other ideas on the origin of these minerals in the quartz-pebble conglomerates are discussed later in this chapter.

Rutile, sphene, leucoxene, ilmenite and a variety of other titanium minerals accompany gold in recent and fossil eluvial and alluvial placers. Their association with gold in these deposits is mainly a consequence of their resistate character and high specific gravity. Brannerite occurs as a resistate mineral in the gold placers of Stanley Basin, Idaho.

Zirconium and *hafnium* exhibit no particular association with gold in hypogene deposits. Zirconium minerals are relatively rare in gold-bearing skarn deposits and in typical gold-quartz veins. Our analyses show only traces of zirconium in these deposits (<50 ppm). In the quartz-pebble conglomerate type of gold deposit zircon and a few other rarer zirconium-hafnium–bearing minerals are present in the matrix of the quartz conglomerate. Some of the zircons in the Rand are uraniferous (Liebenberg, 1955).The relationship of zircon and other zirconium-hafnium–bearing minerals to gold in the quartz-pebble conglomerate type of deposit is not clear. Some of the zircons may be detrital. The ores of the Rand contain from 47 to 320 ppm Zr (Table 35). Von Rahden and Urli (1969) noted a high correlation between Zr, Ti and Cr in some of the reefs in the Rand. Radtke *et al.* (1972*a*) record the following zirconium contents in the disseminated gold ores of Carlin, Nevada: carbonate host rocks – 30 to 150 ppm; 'primary ores' – 30 to 300 ppm; 'oxidized ores' – 30 to 500 ppm. Primary mineralization appears to have introduced some zirconium; oxidation has enriched the ores only slightly in the element.

Zircon commonly accompanies gold in recent and ancient eluvial and alluvial placers. Baddeleyite, ZrO_2, and the other rarer zirconium-hafnium–bearing minerals tend to be rare or of local occurrence in these deposits. All accompany gold as a consequence of their resistate character.

Group VA: N, P, As, Sb, Bi

Nitrogen exhibits no particular association with gold in any of its deposits. Phosphorus may accompany gold in some types of deposits. The other three elements in this group are closely associated with gold in many types of hypogene deposits.

Phosphorus, mainly in apatite, occurs in some gold deposits, principally in the skarn type but also in some gold-quartz veins. Hulin (1930) describes an occurrence of apatite as an important vein mineral in the Kennedy Mine of the Mother Lode district in California, and Schwartz (1944) states that apatite is present in a few other gold deposits. The writer has observed that apatite occurs in very minor amounts in most skarn and gold-quartz deposits, but there appears to be no particular correlation or association of phosphorus with gold in any of these deposits. In the Carlin, Nevada, disseminated gold ores Radtke *et al.* (1972*a*) report <0.01 to 0.5 per cent P in the 'primary ores' and 0.01 to 0.9 per cent in the 'oxidized ores' indicating some enrichment during oxidation processes. The carbonate host rocks contain <0.01 to 0.05 per cent P. In the quartz-pebble conglomerate type of gold deposit, apatite, monazite and some of the rarer phosphates occur sparingly in the matrix of the conglomerates. Geological Survey analyses show phosphorus contents ranging from 100 to 7500 ppm in the Witwatersrand ores (Table 35). Those recorded by Feather and Koen (1975) for the Rand show a range from <40 to 390 ppm P. There appears to be no particular association of phosphate minerals with gold in the conglomerate ores.

Smirnov and Entin (1975) have noted a bond between gold and phosphorus in a number of genetically diverse phosphatic deposits (magmatic, hydrothermal, sedimentary, etc.) in U.S.S.R. The contents of gold recorded in these deposits ranged from 0.0017 to 0.470 ppm with an average of 0.023 ppm. In certain apatite separates from the deposits up to 0.030 ppm Au was recorded. Smirnov and Entin point to the fact that the first three ionization potentials of gold and phosphorus are similar and suggest that complexation through sodium involving compounds of the type Na_3AuO_3 and Na_3PO_4 may be the reason for the bond between the two elements.

Supergene phosphates occur in some oxidized zones of gold deposits. Wavellite occurs in the oxidized parts of the veins at Cripple Creek, Colorado and elsewhere. Vivianite and ludlamite are also frequent accessory minerals in oxidized gold ores and may be primary in places. Pyromorphite-mimetite is common in gold deposits enriched in lead minerals. All of these phosphates originate by downward migration of soluble phosphate in the oxidized zones followed by precipitation where cations such as Fe, Al and Pb, bind the phosphate as insoluble minerals. Some gossans are relatively enriched in phosphorus due evidently to the strong adsorption and binding of phosphate by hydrous ferric oxides and wad. This was noted at Keno Hill and elsewhere in the oxidized zones of gold-quartz deposits. Unoxidized quartz-arsenopyrite gold ore in the No. 6 vein at Keno Hill contained 0.02 per cent P_2O_5 whereas the highly oxidized material ran 0.24 per cent P_2O_5 (Boyle, 1965*a*).

Because of their resistate character apatite, monazite and the rarer phosphates tend to occur in some abundance in gold placers of recent and fossil origin. My observations suggest that there is no particular relationship between the quantity of phosphate minerals and gold in placers, an anticipative circumstance considering that most of the phosphates are derived from the country rocks and not from the gold deposits.

Arsenic exhibits a nearly universal enrichment in most types of hypogene gold deposits, and there is a marked coherence between gold and arsenic during both hypogene and supergene processes. The most common arsenic minerals in hypogene gold deposits are arsenopyrite and tennantite-tetrahedrite. Other arsenic-bearing sulphosalts are common in certain deposits. Native arsenic is relatively uncommon, but occurs in relatively large amounts in some auriferous deposits, examples being the Engineer Gold Mine, Yukon (allemontite); Nagyág, Transylvania and Bau, Sarawak. Enargite is common in a number of the Japanese gold deposits; also in a few of the straight gold deposits and extensively in some of the copper deposits containing low gold values in the Cordillera of North and South America, in U.S.S.R., and elsewhere. Niccolite, cobaltite, gersdorffite, smaltite, safflorite and other Ni-Co-Fe arsenides are relatively rare in gold deposits, but a few gold-quartz veins and skarn deposits containing some of these minerals are known in Mexico; in the Middleburg district, northern Transvaal; in the French Mine, Hedley, British Columbia; in some of the gold deposits at Ramore, Ontario and elsewhere. The Ni-Co-Fe arsenide veins (Cobalt, Ontario type) with abundant native silver generally contain only traces of gold (<0.1 ppm). Hypogene orpiment and realgar appear to be rare in gold deposits, but the oxidized and secondary sulphide zones of some deposits may contain these minerals, usually in restricted amounts. Examples of the latter are found at Nagyág, in some Japanese deposits and elsewhere. Hypogene realgar and orpiment occur in the disseminated deposits of Carlin and Getchell in Nevada. When arsenic minerals are absent in gold deposits the element is largely camouflaged in pyrite or marcasite. Some average arsenic values for pyrite where arsenic minerals are rare or absent in the gold ore shoots of deposits, are given below:

Location	As in pyrite (ppm)
Keno Hill	2500
Yellowknife (Greenstone Belt)	500
Kirkland Lake	500
Timmins	500
Red Lake	300
Cordova Mine, Cordova, Ontario	5000
Nova Scotia (Meguma Group)	1000
Elliot Lake quartz-pebble conglomerate ores	250
Rand quartz-pebble conglomerate ores	2000

Arsenic, mainly in arsenopyrite, is highly enriched in certain gold-bearing skarn deposits. An excellent example of this type of deposit was the Nickel Plate Mine in British Columbia where the arsenic content was up to 25 per cent and more in the high grade ore. Other gold-bearing skarn deposits have much lower contents of arsenic, usually in the range 100 to 1000 ppm. Most of this is in pyrite, disseminated arsenopyrite and other sulphides and sulphosalts.

Gold-bearing polymetallic deposits are invariably enriched in arsenic as are the typical gold-quartz veins. In these two types of deposits the arsenic content ranges from a few tens of parts per million in the ore to percentage amounts. Some of these deposits are so rich in arsenic minerals that great difficulty is encountered in winning the gold. In fact, in some deposits the gold is so intimately associated with arsenopyrite and arsenic-bearing sulphosalts that the element would appear in many cases to be a lattice constituent of these minerals. This problem is discussed in more detail in a section on gold in other minerals in Chapter II. Arsenic is also commonly found in the alteration zones of both polymetallic deposits and gold-quartz veins in relatively large amounts in some deposits. Our studies show that most of this arsenic is present in disseminated pyrite and/or arsenopyrite. There is frequently a consistent increase in the arsenic content of the wall rocks and alteration zones as the deposits are approached as shown in Tables 44 to 59 inclusive.

The disseminated gold deposits of north-central Nevada, spatially related to the Roberts Mountains thrust fault, including Carlin, Cortez, Getchell and Gold Acres, are all enriched in arsenic (Radtke *et al.*, 1972*a*). The most widely distributed arsenic mineral in the hypogene ores appears to be arsenopyrite, but realgar, as noted above occurs in some abundance in the Getchell and Carlin deposits. The host carbonate rocks 'primary ores' and 'oxidized ores' at Carlin contain respectively in ppm As: <10 to 10; 50 to $>10\,000$; and 40 to >3000. According to the averages given some arsenic is lost during oxidation processes. In the primary ores of the Getchell Mine, Berger (1975) noted a fair statistical relationship between the gold tenor and the arsenic content.

The gold-bearing hot spring precipitates are characterized by high contents of arsenic, some of which is present as orpiment and realgar in certain areas. Weissberg (1969) found up to 2 per cent As in the metal-rich gold-bearing precipitates from hot springs and drillhole discharges in the Taupo volcanic zone of New Zealand. Some of the arsenic is probably present as orpiment. The As/Sb ratio in the precipitates is highly variable ranging from 0.003 to 37.5 in the different springs and drillholes.

The auriferous quartz-pebble conglomerates contain relatively high amounts of arsenic, mainly in the pyrite, according to the writer's investigations. Other arsenic-bearing minerals such as arsenopyrite, cobaltite, gersdorffite and skutterudite appear to be omnipresent in most reefs in the Rand, but their abundance is relatively low judging from the literature on these deposits. Saager and Mihálik (1967) have described two intergrown types of pyrite, forming porous buckshot pyrite in the Rand deposits. The isotropic variety contained 2010 ppm As and the anisotropic variety 8770 ppm As. The respective Ni and Co contents were 1140 and 490 for the isotropic pyrite and 4600 and 2330 for the anisotropic variety. Our analyses show that the arsenic content of the pyrite in the Rand and Elliot Lake conglomerate deposits ranges from 250 to 2000 ppm. In the pyritiferous conglomerate ores as a whole the arsenic appears to range from 28 to 13 600 ppm As (Table 35). The precise relationship of arsenic and gold in these deposits is not known. Correlative studies of the two elements could prove useful in genetic studies of these deposits.

The fate of arsenic during the oxidation of arsenic-bearing deposits has been extensively investigated by the writer and his colleagues (Boyle, 1965*a*, Boyle and Dass, 1971*a*; Boyle, 1972; Boyle and Jonasson, 1973). The details are complex and varied, and only the salient points can be discussed in the space available here.

Arsenic exists in three principal oxidation states in nature, As(0), As(III) and As(V). The first two states are those found in native arsenic, in the sulphide As_2S_3 and in the various arsenides and sulpharsenides. The pentavalent state is the one most stable in an oxidizing environment, being a constituent of a veritable host of arsenates and basic sulphate-arsenates.

Oxidation of native arsenic, arsenopyrite and the various other arsenides and sulpharsenides yields a variety of supergene arsenates and basic arsenate-sulphates the most common being scorodite, $Fe(AsO_4)\cdot 2H_2O$; erythrite, $Co_3(AsO_4)_2\cdot 8H_2O$;

annabergite $Ni_3(AsO_4)_2 \cdot 8H_2O$; mimetite, $Pb_5(AsO_4)_3Cl$; pharmacosiderite, $Fe_3(AsO_4)_2(OH)_3 \cdot 5H_2O$; adamite, $Zn_2(AsO_4)(OH)$; olivenite, $Cu_2(AsO_4(OH)$; and beudantite, $PbFe_3(AsO_4)(SO_4)(OH)_6$. Of these, the most important in gold deposits is scorodite; the others occur only where there are high contents of Ni, Co, Cu, Pb or Zn. The oxide, As_2O_3 (arsenolite), may form in oxidized zones where water circulation is restricted; being relatively soluble the mineral is not stable in most oxidized zones. The sulphides, orpiment and realgar, and the silver sulphosalt, proustite, may be of supergene origin, but they are relatively rare in the oxidized zones of most gold deposits.

Oxidation of primary arsenic minerals yields soluble arsenic acid, H_3AsO_4, and its derivatives the soluble alkali and alkaline-earth arsenates. In these forms arsenic is relatively mobile, and under certain conditions considerable amounts of the element may be removed from the deposits during oxidation. More generally, however, much of the arsenic, in the form of arsenates, is coprecipitated and/or adsorbed by hydrous iron and other oxides or reacts with cations such as Fe, Cu, Pb, Zn, Co, Ni, etc. to give a variety of insoluble arsenates and basic arsenate-sulphates during various hydrolytic and colloidal reactions in the oxidized zones. These reactions take place mainly between the pH range 4 to 7. In gold deposits the arsenate most commonly formed is scorodite, although in some deposits pharmacosiderite, beudantite and other similar minerals may be present in some abundance. All of these minerals tend to concentrate gold. In Yukon we have found some varieties of scorodite in the oxidized zones of gold deposits to contain as much as 10.3 ppm Au, mostly in the free state. In the gossans of the massive and vein sulphide deposits in the Bathurst area, New Brunswick, scorodite is, likewise, commonly enriched in gold in amounts up to 0.6 ppm. This gold is present in an extremely finely divided form, and some may be in an adsorbed state or in the lattice of the mineral. Beudantite may also be enriched in gold. Samples from the Bathurst gossans contain from 0.4 to 3 ppm Au.

Gold and arsenic exhibit a sympathetic enrichment in the oxidized zones of some gold deposits. In others the oxidized zones are leached of both elements to a considerable degree. In still others gold is markedly enriched and arsenic is strongly leached. The last situation seems to prevail where the wall rocks and gangue contain relatively little carbonate to neutralize the downward moving solutions. Under these conditions the arsenates are not stable, or those that are formed, are gradually dissolved, and the arsenic is carried away in solution. Intense leaching by meteoric waters also leads to the alteration of scorodite to limonite. In such oxidized zones about all that remains is gold-enriched limonite and wad with traces to minor amounts of adsorbed or chemically combined arsenic.

Gold and arsenic may show a sympathic enrichment in some supergene sulphide zones, the gold occurring principally in the free state or in the sulphides and the arsenic being concentrated in secondary pyrite, marcasite and a variety of other sulphides.

Arsenic tends to be enriched in recent and ancient eluvial and alluvial gold placers, especially those containing a relatively high content of limonite particles or limonite-cemented gravels. This is a consequence of the marked adsorptive and coprecipitative capacity of hydrous ferric oxides for arsenate. In some of the limonitic particles and cemented gravels of the Beauceville district, Quebec I have found up to 180 ppm As; in the Carboniferous placers of Gays River, Nova Scotia the limonitic material contains up to 125 ppm As. In the eluvial placers of Dublin Gulch, Yukon the arsenic content in the limonite-clay materials may exceed 400 ppm As. These derive from a terrane containing gold-quartz-arsenopyrite veins. I have also noted arsenopyrite and arsenic-bearing pyrite in nearly all of the heavy mineral concentrates of the gold placers in Yukon and elsewhere in Canada.

Antimony exhibits a nearly universal enrichment in most types of hypogene gold deposits, and there is a marked coherence between gold and antimony during both hypogene and supergene processes. This relationship also extends to the plant kingdom (*see* the section on Biogeochemistry). Gold forms a mineral with antimony, aurostibite, $AuSb_2$. This mineral is generally hypogene in origin, usually formed late in the paragenetic sequence and probably at relatively low temperatures.

The common antimony minerals in hypogene gold deposits are stibnite, tetrahedrite-tennantite and a variety of other sulphosalts of which the most persistent are jamesonite, boulangerite, bournonite, meneghinite, jordanite, chalcostibite, zinkenite, berthierite and semseyite. Native antimony is not particulary common in gold deposits, although it has been found in small amounts in the Engineer Gold Mine, Yukon (allemontite), in the antimony-rich Yellowknife ores, at West Gore, Nova Scotia and elsewhere. The silver sulphosalts, pyrargyrite, polybasite and stephanite are relatively uncommon minerals in most gold deposits, although in some gold-bearing polymetallic deposits and in some of the great (mainly Tertiary) Au-Ag bonanzas of the world these minerals may be relatively abundant. When antimony minerals are absent in gold deposits, the element is largely camouflaged in pyrite, marcasite or arsenopyrite. Some average antimony values for pyrite and arsenopyrite, where antimony minerals are rare or absent in the gold ore shoots of deposits, are given below:

Location and mineral	Sb (ppm)
Keno Hill (pyrite)	100
Keno Hill (arsenopyrite)	500
Yellowknife Greenstone Belt (pyrite)	100
Yellowknife Greenstone Belt (arsenopyrite)	400
Kirkland Lake (pyrite)	8
Timmins (pyrite)	20
Red Lake (pyrite)	10
Cordova Mine, Cordova, Ontario (pyrite)	20
Nova Scotia (Meguma Group; pyrite, arsenopyrite)	2–100
Elliot Lake quartz-pebble conglomerate ores (pyrite)	2
Rand quartz-pebble conglomerate ores (pyrite)	4

Antimony is not particularly enriched in gold-bearing skarn deposits in contrast to arsenic. Antimony minerals are relatively rare in skarn deposits, although tetrahedrite has been noted in small amounts in a number of these deposits (e.g., New Calumet Mine, Quebec). Where pyrite and arsenopyrite are abundant in the ores the antimony content may be relatively high, up to 50 ppm Sb or more.

Gold-bearing polymetallic deposits are invariably enriched in antimony, as are some of the typical gold-quartz veins. In these two types of deposits the antimony ranges from a few tens of parts per million in the ore to percentage amounts. Some of the gold-quartz deposits are so rich in antimony minerals, mainly stibnite and sulphosalts, that great difficulty is encountered in winning the gold. It is in these deposits that the gold antimonide, aurostibite, frequently makes its appearance. Gold also appears to substitute in the lattice of a number of the antimony sulphosalts in these deposits, principally in tetrahedrite-tennantite. Antimony is also commonly found in the alteration zones of both polymetallic deposits and gold-quartz veins, in relatively large amounts in some deposits. Our studies show that most of this antimony is present in disseminated pyrite and/or arsenopyrite. There is frequently a consistent increase in the antimony content of the wall rocks and alteration zones as the deposits are approached as shown in Tables 44 to 59 inclusive.

The disseminated gold deposits of north-central Nevada, including Carlin, Cortez, Getchell and Gold Acres, are all enriched in antimony (Radtke *et al.*, 1972*a*). The antimony mineral in these deposits is mainly stibnite and antimoniferous orpiment. The host carbonate rocks, 'primary ores', and 'oxidized ores' at Carlin contain respectively in ppm Sb: <0.5 to 3; 5 to 450; and 5 to 450. According to the averages given there is relatively little change in antimony content during oxidation processes. Berger (1975) found no significant statistical relationship between the gold and antimony contents in the primary ores of the Getchell Mine.

The gold-bearing hot-spring precipitates are characterized by high contents of antimony, some of which is present as stibnite in certain areas. Weissberg (1969) found up to 30 per cent Sb in the metal-rich gold-bearing precipitates from hot springs and drillhole discharges in the Taupo volcanic zone of New Zealand. Some of the antimony is probably present as stibnite. The As/Sb ratio in the precipitates is highly variable ranging from 0.003 to 37.5 in the different springs and drillholes.

The auriferous quartz-pebble conglomerate deposits contain some antimony mainly in the pyrite according to our investigations. Other antimony-bearing minerals are relatively rare in these deposits; Saager (1968) recorded the presence of tennantite, proustite and dyscrasite in the Basal Reef in the Free State Geduld Mine, near Welkom in the Orange Free State Goldfield; Feather and Koen (1975) record the presence of stibnite locally in some of the Rand reefs. Galena is relatively common in some of the auriferous conglomerates of the Rand, and it may contain some antimony. There may also be more silver-antimony sulphosalts in the ores than is generally thought judging from the research by Saager (1969) on the gold-silver relationships. Geological Survey analyses show that the antimony content of the pyrite in the Rand and Elliot Lake conglomerates ranges from 2 to 4 ppm. In the pyritiferous conglomerate ores of the Rand as a whole the antimony content appears to average about 3 ppm (Table 35). The relationship of antimony and gold in the conglomerates is not known. Correlative studies of the two elements could prove useful in genetic studies of these deposits.

The fate of antimony during the oxidation of antimony-bearing deposits has been extensively studied by the writer and his colleagues (Boyle, 1965*a*; Boyle and Dass, 1971*a*; Boyle, 1972). The details are particularly complex and only the salient points can be discussed in the space available here.

Antimony exists in three principal oxidation states in nature: Sb(0), Sb(III) and Sb(V). The first two states are those found in native antimony, in the sulphide, Sb_2S_3, and in the various antimonides and sulphantimonides. The pentavalent state is the most stable in an oxidizing environment, being a constituent of a number of the natural oxides (cervantite, stibiconite) and antimonates.

Oxidation of native antimony, stibnite and the various other antimonides and sulphantimonides yields a variety of supergene antimony minerals the most common being cervantite, $Sb_2O_3{\cdot}Sb_2O_5$, senarmontite, Sb_2O_3, valentinite, Sb_2O_3, stibiconite, $Sb(III)Sb(V)_2(O,OH,H_2O)_7$, kermesite, Sb_2S_2O, bindheimite, $Pb_{2-y}Sb_{2-x}(O,OH,H_2O)_{6-7}$, romeite, $(Ca,Na)_2Sb_2(O,OH)_7$, and numerous other antimonites and antimonates. Yellow and white ocherous antimony compounds that are amorphous to X-rays characterize the oxidation zones of many gold deposits containing abundant hypogene antimony minerals.

Oxidation of primary antimony minerals yields the various oxides which are only sparingly soluble in solutions with pH ranging from 4 to 8 and hence these minerals have a relatively high degree of stability in most oxidized zones. The trivalent oxide is amphoteric yielding $[Sb(OH)_2]^+$ ions in acid solutions and antimonite anions usually written as $[Sb(OH)_4]^-$ in basic solutions. The cationic complex is particularly prone to hydrolysis on dilution or neutralization of acidic solutions and the hydrous trioxide is precipitated. Antimonites, precipitated as a result of the reaction of various cations such as Ca and Fe with antimonite-bearing solutions, are relatively rare. The pentavalent oxide is acidic dissolving in base to give antimonate anion, usually written as $[Sb(OH)_6]^-$. Reactions of this complex and/or the various derivatives of antimonic acid with cations, particularly lead and iron, yield a number of insoluble minerals of which bindheimite and tripuhyite are the most common.

Under very acid ($pH<3$) and highly alkaline ($pH>8$) conditions much antimony is removed from deposits during oxidation processes. Under more normal pH conditions (4–8) much of the antimony is retained in the oxidized zones in the form of oxides and various antimonites and antimonates and in an adsorbed or chemically combined state in limonite and wad. There may also be considerable amounts of antimony camouflaged in the various arsenates and a variety of other supergene minerals. In the Keno Hill deposits a depletion of antimony and gold was noted in the oxidized zones of some deposits and a sympathetic enrichment of both elements in others compared with the primary ores (Boyle, 1965). At Yellowknife the writer observed a general enrichment of both elements in the shallow oxidized zones of some veins. In some of the gossans overlying the massive sulphide deposits of the Bathurst district, New Brunswick both gold and antimony are considerably enriched. Under conditions of intense leaching by groundwaters antimony is largely removed from the gossans and oxidized zones. This effect is particularly marked where the wall rocks and gangue contain relatively little carbonate to neutralize the downward moving acidic solutions. In such oxidized zones about all that remains is gold-enriched limonite and wad with traces to minor amounts of adsorbed or chemically combined antimony.

Gold and antimony may show a sympathetic enrichment in some supergene sulphide zones, the gold occurring principally in the free state or in the sulphides and the antimony being concentrated in secondary pyrite, marcasite and a variety of other sulphides. Aurostibite in some of the near surface parts of the Yellowknife veins may be of supergene origin, although this is difficult to verify.

Antimony tends to be enriched in recent and ancient eluvial and alluvial gold placers, especially those containing a relatively high content of limonite particles or limonite-cemented gravels. This is a consequence of the marked adsorptive and coprecipitative capacity of hydrous ferric and manganese oxides for antimonate. In some of the limonite particles and cemented conglomerates of the Carboniferous placer of Gays River, Nova Scotia I have found from 2 to 12 ppm Sb. In the limonite-cemented conglomerate of the Beauceville district, Quebec, on the other hand, only <2 ppm Sb was obtained. These differences evidently reflect the provenance of the element in the parent terrane of the placer deposits. In the case of Gays River, antimony and arsenic are relatively abundant in the Meguma Group whereas these elements are relatively rare in the rocks and deposits of the Beauceville district. In the eluvial placers of Dublin Gulch, Yukon the antimony content of the limonite-clay materials may exceed 10 ppm. These derive from a terrane containing gold-quartz-arsenopyrite-jamesonite veins. I have also noted jamesonite, boulangerite and antimony-bearing pyrite in many of the heavy mineral concentrates of the gold placers in Yukon and elsewhere in Canada.

Bismuth is a common associate of gold in some hypogene deposits, although it is generally present in only small amounts (<5 ppm) in most gold ores. There is a coherence between gold and bismuth exemplified by the fact that gold bismuthide (maldonite, Au_2Bi) occurs in nature.

The common bismuth minerals in hypogene gold deposits are bismuthinite, native bismuth, galenobismutite, bismuthian tetrahedrite-tennantite (goldfieldite), wittichenite and various bismuth tellurides and selenides including joseite, hedleyite, tellurbismuth, wehrlite and guanajuatite. Tetradymite, cosalite and other bismuth minerals are relatively rare in gold deposits. When bismuth minerals are absent in gold deposits, the element may be present in small amounts in a great variety of sulphides and sulphosalts. At Yellowknife the writer has noted the presence of bismuth in samples of arsenopyrite (~10 ppm), pyrite (<10 ppm), stibnite (~10 ppm), various sulphosalts, particularly tetrahedrite (~100 ppm), sphalerite (<10 ppm), chalcopyrite (~10 ppm) and galena (~100 ppm). In the Keno Hill area some of the galena in the gold-quartz veins is rich in bismuth (up to 8000 ppm) as are also jamesonite and boulangerite (up to 500 ppm).

Bismuth may be slightly enriched in gold-bearing skarn zones, the principal bismuth minerals being native bismuth, bismuthinite and bismuth tellurides. The tellurides frequently exhibit a close relationship with native gold as in the French Mine, Hedley district, British Columbia. Most of the ores in the gold-bearing skarn category rarely contain more than 10 ppm Bi.

Some gold-bearing polymetallic deposits and typical gold-quartz veins may be enriched in bismuth. Typical examples occur in Queensland (Burnett district), Northern Territory (Tennant Creek) and New South Wales, Australia; Norway (Svartdal); Japan (Nishizawa and Nakanosawa mines); the United States (Goldfield); U.S.S.R. (Darasun and Sredniy Golgotay goldfields); Canada (Hazelton and other areas in British Columbia); France (Salsigne-Aude) and elsewhere. In the Tennant Creek Juno deposit the most abundant bismuth-bearing mineral is junoite, $Bi_8Pb_3Cu_2(S,Se)_{16}$, a new mineral described by Large and Mumme (1975) and Mumme (1975). It is accompanied by the bismuth-selenium sulphosalt, wittite and a variety of other selenium-bearing sulphosalts. One of the best known examples of the gold-silver-bismuth type of deposits is at Goldfield, Nevada, where bismuthinite is found with pyrite, marcasite, sphalerite, enargite, tellurium-bismuth tetrahedrite (goldfieldite), arsenical famatinite, tennantite, various gold-silver tellurides and native gold, all in irregular silicified and alunitized zones referred to as "ledges" by Ransome (1909). Lindgren (1933) gives an analysis of rich ore from the Mohawk Mine which indicates 0.35 per cent Bi (3500 ppm). This is a high assay; most gold-bismuth deposits contain bismuth in the range 10–100 ppm in the ore. At Kirkland Lake for instance the ores of the Upper Canada Mine contain from 5 to 11 ppm Bi.

There are no data on the bismuth content in the disseminated gold deposits of north-central Nevada (e.g., Carlin, Cortez). Likewise, data are particularly sparse on the bismuth content of gold-bearing precipitates associated with present day hot springs. Browne (1969) records the presence of bismuth (0.003 ppm) and gold (0.0001–0.001 ppm) in the water of the Broadlands Geothermal drillhole, Taupo volcano zone, New Zealand. Base metal sulphides have been deposited in rocks penetrated by this drillhole, and enriched gold-bearing precipitates are associated with the hot springs and steam wells in the area (*see also* under arsenic and antimony).

There are no published data on the bismuth content of the gold-bearing conglomerate ores, nor, excepting the rare occurrence of michenerite, (Pd,Pt)BiTe, in the Rand, are there any references to the presence of bismuth minerals in these ores. Our analyses (Table 35) show that the bismuth content of the Rand ores runs from not found to 2.2 ppm; in the Elliot Lake ores the content averages about 43 ppm Bi. The pyrite in the Elliot Lake ores is also rich in bismuth (40 ppm). Some of the bismuth in these ores may be of radiogenic origin.

The fate of bismuth during the oxidation of bismuth-bearing deposits has been investigated by the writer and his colleagues (Boyle, 1965*a*; Boyle and Dass, 1971*a*). Only a few of the salient points can be discussed here.

The principal supergene bismuth minerals in gold deposits are bismite-sillenite Bi_2O_3, bismutite, $(BiO)_2CO_3$, bismoclite-daubreeite, BiO(OH,Cl) and montanite, $Bi_2TeO_6{\cdot}2H_2O$. There are also supergene bismuth arsenates, vanadates, molybdates and complex uranates, but these are relatively rare in gold deposits. Bismuth ochres, which are probably mixtures of the oxide and carbonate but are amorphous to X-rays, are common in some oxidized zones.

Bismuth exists in two oxidation states in nature, Bi(0) and Bi(III). The pentavalent state, Bi(V), is apparently not reached under the oxidation potentials existing in nature, since no mineral 'bismuthates' are known. The first two states are those found in native bismuth, in the sulphide, Bi_2S_3, in the oxide Bi_2O_3 and in the various bismuthides and complex bismuth sulphosalts. The oxide, Bi_2O_3, is basic and is practically insoluble in water. It does dissolve however, in acidic solutions yielding hydrolized bismuth (bismuthyl) ion that can be characterized as $[BiO]^+$, $[BiOH]^{2+}$ or $[Bi(OH)_2]^+$. It is probably in this form that the small amounts of bismuth migrate in the oxidized waters of gold deposits.

Studies by the writer indicate that native bismuth oxidizes slowly in oxygenated water, forming a hydrous oxide from which water is split out to give the yellow oxide, Bi_2O_3, bismite. Bismuthinite, Bi_2S_3, bismuthian tetrahedrite, bismuth tellurides and the various other primary bismuth minerals also yield the oxide directly, or hydrolyzed bismuth ion, which follows a number of courses depending on the pH and the presence of other constituents. If the pH is low some bismuth is leached from the deposits. On neutralization of the solutions the bismuth ion undergoes rapid hydrolysis with the precipitation of the oxide or hydroxide. If CO_2 is present in the solutions bismuthyl carbonate (bismutite) may form in two ways, one by direct precipitation and the other by reaction of CO_2 with the hydroxide. The two reactions can be expressed as follows:

$$4(BiO)^+ + O_2 + 2CO_2 = 2(BiO)_2CO_3 \text{ (bismutite)}$$
$$2Bi(OH)_3 + CO_2 = (BiO)_2CO_3 \text{ (bismutite)} + 3H_2O$$

Bismutite may also be found where carbonates are present in the gangue or wall rocks. The simple reactions in this case involve the formation of hydrogen carbonate ion, which fixes the bismuthyl ion as the carbonate:

$$CaCO_3 + CO_2 + H_2O = Ca^{2+} + 2(HCO_3)^-$$
$$2(BiO)^+ + (HCO_3)^- = (BiO)_2CO_3 \text{ (bismutite)} + H^+$$

Bismite and bismutite are the most common supergene bismuth minerals, but where other constituents are present in the oxidizing solutions a variety of secondary bismuth minerals may be formed in the pH range 4 to 8. Arsenate precipitates bismuthyl ion as insoluble arsenates; tellurate as montanite; chloride as bismoclite or daubreeite; vandate as pucherite, $BiVO_4$; and so on. In addition bismuthyl ion is adsorbed and/or coprecipitated probably as the oxide or hydroxide by hydrous ferric oxides (limonite), hydrous manganese oxides (wad) and hydrous silica-alumina complexes. Much bismuth is also adsorbed and/or absorbed in some manner or other by minerals such as senarmontite, bindheimite, anglesite and jarosite. In these secondary minerals I have found bismuth contents ranging from 100 to 1000 ppm. As noted previously some of these minerals also tend to concentrate gold.

From the above discussion it will be seen that bismuth is enriched in oxidation zones where the pH of the waters lies in the range 4 to 8, that is, where carbonate gangue of wall rocks is available to neutralize the acid waters generated during the oxidation of sulphides such as pyrite. Under these conditions gold is also enriched in the gossans and underlying oxidized parts of the deposits. In some highly oxidized deposits native bismuth encrusted with bismite and/or bismutite and large particles of gold may be present in the rubble. These are largely residuals. Where the oxidizing waters retain their acidity some bismuth is leached, and hence gold and bismuth tend to part company since gold tends to remain behind in the gossans and oxidized zones under such conditions. The amount of bismuth passing downward in oxidizing waters in most deposits is relatively low, generally in the range 0.0001 to 0.001 ppm where bismuth minerals are present in the ores. Some of this bismuth may be precipitated together with gold in the secondary sulphide zones of deposits (Naumov *et al.*, 1971).

There are few data on the presence of bismuth in recent and ancient gold placers. Native bismuth and bismuthinite are relatively resistant to oxidation when coated with bismite and bismutite. In this condition the element is a persistent constituent of the gold placers of the Dublin Gulch area, Yukon. Tetradymite has also been found in these placers. There are scattered references in the literature noting the occurrence of native bismuth, bismuthinite and bismuth tellurides in gold placers in many parts of the world. Various other bismuth-bearing minerals have also been found in small amounts in the heavy concentrates of gold placers, including galena, various sulphosalts, bismutotantalite and bismuthian gold.

Group VB: V, Nb, Ta

Vanadium is present in some hypogene gold deposits principally in minerals like sericite, roscoelite, pyrite, apatite, magnetite, ilmenite and leucoxene. At Yellowknife the element was found in amounts up to 100 ppm in all of these minerals in the wall-rock alteration zones and in the gold ores. Most of the vanadium appears to have been contributed by the host rocks (meta-andesites). There is no correlation of vanadium with gold in the Yellowknife deposits nor in any other auriferous vein deposits investigated by the writer. For other vein deposits there may be a correlation between the two elements as noted below.

Vanadium is present in the gold-bearing conglomerates of the Rand in amounts ranging from 15 to 96 ppm (Table 35), apparently mainly in the pyrite and micas (sericite). There appears to be no particular correlation between gold and vanadium in these ores as far as the writer is able to ascertain.

Roscoelite, $K(V,Al)_3(AlSi_3)O_{10}(OH)_2$, is reported to occur in some abundance in a few of the gold veins of the Mother Lode system northwest of Placerville, El Dorado County, California; also at Cripple Creek, Colorado, and at Kalgoorlie, Western Australia. According to Lindgren (1933) the roscoelite in the Mother Lode system often accompanies native gold. At Cripple Creek, according to the same authority, roscoelite is often abundant in rich ore. In the Kalgoorlie gold belt vanadium mica is apparently common in rich shoots of gold and gold tellurides according to some authors. It should be mentioned that gold ores rich in tellurides frequently seem to have higher than average amounts of vanadium either in pyrite or in the vanadiferous micas. The reason for the relationship between the three elements, Te, V and Au, is obscure to the writer, but should be borne in mind by those carrying out further investigations on these ores. Perhaps the relationship is connected with the fact that gold and gold tellurides commonly occur in host rocks that are relatively rich in vanadium (e.g., basic to intermediate volcanics, serpentinites, etc.)

Vanadium minerals may be present in gold ores containing uranium and rare earth minerals. The Nicholson deposits in the Goldfields area, Saskatchewan, contains nolanite, $Fe_3V_7O_{16}$, in some abundance (Robinson, 1955). The relationship of the gold values to this mineral is unknown.

Nolanite has also been observed in nonuraniferous ores, particularly in the South Kalgoorlie Mine along the Golden Mile at Kalgoorlie, Australia (Taylor and Radtke, 1967). There, the mineral is closely associated with native gold and numerous Au-Ag tellurides. In the paragenetic sequence coloradoite and native gold are younger than the nolanite. In further research on the 'green leader' at Kalgoorlie, Nickel (1977) recognized sericitic vanadium muscovite, nolanite and an unidentified titanium vanadate. The color of the 'green leader' is due to vanadium muscovite in which the vanadium is present in the trivalent state.

In the disseminated gold ores of Carlin, Nevada, Radtke *et al.* (1972*a*) report the following vanadium contents for host carbonate rocks, 'primary ores', and 'oxidized ores' respectively: 10 to 500, 15 to 500 and 50 to 1500 ppm. The oxidized ores are apparently not enriched according to the averages given for the 'primary ores' (140 ppm V) and 'oxidized ores' (140 ppm V).

Concentrations of *niobium* and *tantalum* have not been detected in any of the hypogene gold deposits examined by the writer. This was expected since deposits containing Nb and Ta are commonly found in quite different environments than those characterized by the presence of gold, a feature also remarked upon by Sadovskii *et al.* (1971). The contents of niobium in the skarn, gold-quartz and polymetallic ores investigated were all less than 5 ppm; similarly the tantalum contents never exceeded 2 ppm. In the quartz-pebble conglomerate ores the niobium contents range from 4 to 18 ppm with the highest amounts in the Rand ores; the tantalum contents were <3 ppm in all occurrences (Table 35). In the Carlin, Nevada disseminated ores Radtke *et al.* (1972*a*) recorded the following ranges in the host carbonate rocks, 'primary ores' and 'oxidized ores' respectively: <7 to 7; <7 to 15; and <7 to 20 ppm Nb. Tantalum was not found.

The heavy resistate minerals of vanadium, niobium and tantalum frequently accompany gold in recent and fossil placers. The most common vanadium-bearing minerals in these surficial deposits are magnetite, ilmenite, rutile and apatite, which are usually present in some abundance. Tantalite-columbite, pyrochlore, fergusonite, euxenite and a variety of other niobates and tantalates may occur in gold placers, but they are usually rare or present in only small amounts. There appears to be no particular correlation between gold and the vanadium-, niobium- and tantalum-bearing resistates in placers, a circumstance to be expected since these minerals are largely contributed by the host rocks and deposits such as pegmatites that are either devoid of gold or are unrelated genetically to gold deposits.

Group VIA: O, S, Se, Te

Oxygen does not combine with gold in nature, and there are no known natural gold oxides. Numerous other oxygen compounds occur with gold in all of its deposits. These are discussed above and below.

Sulphur is a constant associate of gold in all types of deposits, being present mainly in pyrite, arsenopyrite, chalcopyrite, sphalerite, galena and numerous other sulphides and sulphosalts. Hypogene sulphates are relatively uncommon gangue minerals in most hypogene gold deposits, but some deposits are notably enriched in some of these minerals. Barite is a gangue mineral in a number of gold deposits, principally the polymetallic and gold-quartz type. Anhydrite and gypsum are much more common in gold deposits than is generally realized, being present in both the polymetallic and gold-quartz type of all ages, often in some abundance. Alunite, the basic hydrous potassium aluminum sulphate, is a characteristic mineral in the gold-quartz-alunite assemblage exemplified by Goldfield, Nevada and other deposits.

Gold and its various minerals exhibit a very close association with a number of sulphides, sulphide-arsenides and sulphosalts. Of these the most common associates of gold are pyrite and arsenopyrite. In some cases it is evident that the element is a lattice constituent of these minerals as discussed in Chapter II; in other cases the gold is of later generation and may have exsolved from these two minerals. The other sulphur minerals with which gold and its minerals show an intimate association are chalcopyrite, galena, sphalerite, tetrahedrite-tennantite and stibnite. Their relationship with gold is discussed further in the section on mineral associates of gold.

The most common sulphide in the auriferous quartz-pebble conglomerate deposits is pyrite. Other sulphur-bearing minerals present in trace and minor amounts in the Rand ores include the following: arsenopyrite, stibnite, marcasite, glaucodot, pyrrhotite, sphalerite, galena, chalcopyrite, cubanite, bornite, chalcocite, covellite, chalcopyrrhotite, bravoite, molybdenite, mackinawite, niccolite, millerite, pentlandite, cobaltite, linnaeite, gersdorffite, skutterudite, safflorite, tetrahedrite-tennantite, stromeyerite and proustite (Saager, 1968; Feather and Koen, 1975). Not all the quartz-pebble conglomerate deposits are enriched in sulphur-bearing minerals. The Tarkwa conglomerate ores in Ghana carry mainly hematite; pyrite and other sulphides are very rare.

During oxidation processes the sulphides, sulphide-arsenides and sulphosalts generally yield either normal sulphates, acid sulphates or basic sulphates depending on the pH, Eh and various complicated hydrolytic phenomena. Thiosulphate, sulphite and a variety of other polymerized oxidized sulphur species may form as intermediate products during the oxidation of sulphides and sulphosalts as noted in Chapters II and IV. These species are, however, probably only transitory, being oxidized on to sulphate in most natural environments. Most of these species are also only stable under alkaline conditions. When present some are good solvents of gold and silver as mentioned in Chapter II. Relatively large amounts of gold may, therefore, be transported in oxidized zones where species such as thiosulphate are formed. As noted in Chapter II, gold is not readily soluble as the sulphate, and it seems probable that relatively little of the metal migrates as a simple sulphate. Some may migrate as complex sulphates, but little is known about these species.

Most sulphates are soluble (>1 g/100 mL) except those of Sr, Ba, Ra and Pb (<0.1 g/100 mL). The sulphates of Ca,

Ag and Hg are sparingly soluble. The oxidation and leaching of gold deposits containing sulphides results in most cases in a strong depletion in the amount of sulphur in the oxidized zones. This depletion varies from place to place but is rarely less than 25 per cent in most oxidized zones and may exceed 95 per cent in some. The sulphur, which is set free as sulphate or oxidized polymerized species in the oxidized zones of gold deposits containing sulphides, may follow three main courses. Some is precipitated as insoluble sulphates, mainly anglesite and gypsum, or as basic sulphates or basic sulphate-arsenates such as jarosite and beudantite. Gold is frequently concentrated in these minerals, particularly in the basic sulphate-arsenates. In some jarosites for instance gold (mainly in the native form) may occur in amounts up to 7900 ppm; in minerals like beudantite we have found up to 0.7 ppm Au. Under certain conditions a considerable portion of the oxidized sulphur species is reduced and precipitated as various sulphides in the zones of reduction (secondary sulphide zones). Gold may be enriched in these zones. The remaider of the sulphur, mainly as sulphate, passes into the groundwater system and ultimately through springs into the stream systems of the area.

The fate of gold and sulphur during oxidation processes can be summarized as follows: in the highly oxidized and leached zones sulphur and gold tend to part company; much of the gold remains behind in the gossans and the sulphur (mainly as sulphate) is leached. Where basic sulphates and basic arsenate-sulphates are formed during the neutralization of acid sulphate solutions some gold may be concentrated in these minerals. Gold and sulphur passing downward are both precipitated in the secondary sulphide zones of the deposits where the appropriate conditions for reduction of sulphate and soluble gold compounds are present. Under certain conditions native sulphur may form in the oxidized zones of auriferous deposits, evidently as the result of the oxidation of H_2S. When this occurs finely divided gold, wires and mossy aggregates of the metal are commonly concentrated in the sulphur.

Sulphur accompanies gold in recent and ancient eluvial and alluvial placers, being present mainly in pyrite, arsenopyrite, galena and a variety of other sulphides and sulphosalts. The total quantity of these minerals is never great, rarely more than a few per cent of the heavy mineral suite. Pyrite is by far the most abundant of the sulphides, etc. in placers. Barite is present in a few gold placers, often in some abundance, but its distribution is generally local. Anglesite may occur in some gold placers near veins containing lead minerals. Gypsum crystals are common in the clays and weathered residuum of eluvial placers but are relatively uncommon in alluvial placers.

Selenium follows sulphur closely in nature and is hence a nearly universal associate of gold in its deposits. In some gold-bearing deposits selenium exhibits a marked enrichment; in others only traces of the element are present. Gold selenide *sensu strictu* has not yet been verified in nature, although there are a number of reports suggesting its occurrence, especially in the selenium-rich gold deposits. A silver-gold selenide, fischesserite, Ag_3AuSe_2, has recently been described.

Practically all sulphides, sulphosalts, sulphates and other species of sulphur minerals contain selenium, but the amounts are highly variable. The element is frequently enriched in minerals such as pyrite, chalcopyrite, bismuthinite, tetrahedrite and some of the other sulphosalts, various sulphates, etc. in which selenium replaces or substitutes for sulphur. Selenium combines readily with Cu, Ag, Pb, Bi and Hg forming a number of individual hypogene minerals of which the most common are umangite, Cu_3Se_2, klockmannite, CuSe, penroseite, $(Ni,Cu)Se_2$, naumannite, Ag_2Se, aguilarite, Ag_4SeS, clausthalite, PbSe, tiemannite, HgSe, guanajuatite, $Bi_2(Se,S)_3$, and junoite, $Bi_8Pb_3Cu_2(S,Se)_{16}$. Supergene species include the oxide, selenolite, SeO_2, and a number of selenites and selenates, viz. chalcomenite, $CuSeO_3{\cdot}2H_2O$, ahfeldite, $NiSeO_3{\cdot}2H_2O$, kerstenite, $PbSeO_4{\cdot}2H_2O$, selenojarosite, and others. All of these are relatively rare minerals.

A number of gold deposits are greatly enriched in selenium. Most of these occur within relatively restricted metallogenic zones that seem to be characterized by the presence of the selenides. Examples occur in the Tonapah area of Nevada; the De Lamar area of Idaho; in the Republic Mine, Republic, Washington; in the Radjang-Lebong goldfield, Sumatra, Indonesia; in the Waihi, Maratoto, Tui and Great Barrier Island lodes of New Zealand; and in the Goldfields area of Saskatchewan. Selenium (0.05%) is present in the auriferous magnetite ores of the Juno Mine in the Tennant Creek Goldfield, Northern Territory, Australia. Certain auriferous uranium-bearing deposits may be enriched in selenium (e.g., Nicholson Mines, Goldfields, Saskatchewan). At Tennant Creek the element occurs in a variety of selenium-bearing sulphosalts including wittite, heyrovskyite and the new mineral junoite, $Bi_8Pb_3Cu_2(S,Se)_{16}$ (Large and Mumme, 1975). In additon to the presence of selenides in auriferous deposits it is commonly noted that the other sulphides and sulphosalts are greatly enriched in selenium, particularly galena, tetrahedrite, tennanite, polybasite, argyrodite and so on. It is of interest to note that these selenium-rich deposits contain few if any tellurides; in fact tellurides and selenides are rather rare companions in most gold deposits.

Excepting the selenium-rich gold deposits, some of which contain up to 200 ppm Se in the ores, most gold-quartz and skarn-type deposits contain relatively little selenium, usually in the range from 1 to 10 ppm in the ores. With increase in sulphides in these deposits there is generally an increase in the selenium content. Polymetallic-type gold deposits usually report relatively high selenium contents, in the range from 10 to 400 ppm or more. The quartz-pebble conglomerates of the Rand contain from 1 to 2 ppm Se (Table 35). The element is mainly in pyrite. In the disseminated gold ores of Carlin, Nevada, Radtke *et al.* (1972*a*) found <1 to 20 ppm Se in the 'primary ores' and <1 to 40 ppm in the 'oxidized ores'. There is no apparent correlation between selenium and gold values in auriferous deposits according to our analytical data.

Berzon and Karyshev (1973) have investigated the Te/Se ratio in the gold deposits of the Urals and have grouped the deposits according to the value of this ratio. (*See* the discussion in the following section on tellurium.)

Selenium occurs in nature in the minus(II), (0), (IV) and (VI) oxidation states. Of these, the (IV) and (VI) states are of interest during oxidation processes. The (IV) state is that contained in the oxide, SeO_2, and various selenites, and the (VI) state prevails in the selenate ion, $[SeO_4]^{2-}$. Chemically, selenium is like sulphur in most respects and similar to arsenic in a few features. A review of the geochemistry of selenium is given by Vlasov (1966–1968).

The oxidation of seleniferous pyrite, selenides and other hypogene selenium-bearing minerals in gold deposits is complex and not particularly well understood. Only a general outline is given here since our research on selenium is not yet complete. Under conditions of low Eh, oxidation of selenium minerals yields selenious acid (H_2SeO_3) and selenite ions to the meteoric waters, and these may migrate considerable distances. Native selenium (and seleniferous sulphur or sulphurous selenium) may also be formed in some oxidized zones by a variety of mechanisms, but these minerals are generally rare. Selenolite, SeO_2, may form in dry places, but this mineral is also rare because of its high solubility. Selenite ions and selenious acid tend to be adsorbed by a variety of colloidal substances, especially hydrous iron oxides with which they form very insoluble basic ferric selenite in the pH range 3.5–9. Various cations, principally Ca, Na, K, Cu, Pb, Ni, Co and Fe, and hydrolytic products of these ions, may combine with selenite ion to form a number of insoluble products, poorly differentiated as minerals. Under conditions of higher oxidation potential selenium is oxidized to the VI state and migrates as selenic acid (H_2SeO_4) or various soluble selenates. Selenates and selenic acid are also strongly adsorbed and/or coprecipitated by hydrous iron oxides in the pH range 5 to 7. Selenate ion is also readily incorporated in various normal sulphates and in basic sulphates such as jarosite. Numerous cations and their hydrolytic products, particularly Fe, Pb, Ni and Co, precipitate insoluble compounds, poorly differentiated as minerals. Where soluble selenium species pass into the zone of reduction of gold-bearing sulphide deposits, some of the element as selenide is incorporated in supergene sulphides such as pyrite, marcasite, covellite and chalcocite.

Comparing the migration tendencies of sulphur and selenium it is evident from a consideration of the appropriate oxidation potentials that the latter element is much less mobile in some environments than the former. This is because selenium requires a higher oxidation potential to mobilize it as soluble species, and these when formed are much more easily reduced than the soluble sulphur species. It would also appear from our studies that selenium in its various soluble states is much more susceptible to adsorption by hydrous ferric oxides and other similar colloidal substances. Because of these features sulphur and selenium may be separated in some oxidized zones, the sulphur being largely removed whereas considerable amounts of selenium remain behind.

In some oxidized zones gold and selenium may exhibit a sympathetic enrichment, especially where secondary sulphates and basic sulphates such as jarosite are present. In others, the two elements part company, the selenium being largely removed whereas much of the gold remains behind in the gossans. This condition prevails particularly where the waters have an acid reaction or where large volumes of water, with either an acid or alkaline reaction, pass through the oxidized zones. In the zones of reduction (secondary sulphide zones) gold and selenium may again exhibit a sympathetic enrichment.

Generally speaking selenium tends to be concentrated in the gossans and weathered rubble of gold-bearing deposits in Canada. For example, at Yellowknife and Sudbury the enrichments shown in Table 39 in the gossans compared with the primary ores have been noted.

There are no data on the occurrence of selenium in fossil and recent eluvial and alluvial placers. Presumably traces of the element are present in the sulphides and other sulphur minerals that frequently accompany gold in these deposits.

Table 39. Selenium, tellurium and gold contents of primary ores and gossans of some Canadian gold-bearing deposits

Location	Se in ore	Se in gossan	Te in ore	Te in gossan	Au in ore	Au in gossan
			(ppm)			
Lynx deposit, Yellowknife, N.W.T. (gold-quartz type)	1	33	<0.2	4.0	0.034	0.17
Vermilion deposit, Sudbury, Ont. (polymetallic type)	165	250	115	160	1.9	0.54

Tellurium appears to be a nearly constant associate of gold in all types of deposits, according to the writer's investigations. The element is particularly abundant in some types of gold-quartz deposits, e.g., Kirkland Lake, Ontario; Kalgoorlie, Australia; Nagyág, Romania; and Cripple Creek, Colorado. All of these deposits are in or near volcanic rocks of widely differing ages. Tellurium is present in most gold-bearing skarn deposits but generally only in small amounts. Auriferous polymetallic deposits invariably contain some tellurium, and certain auriferous uranium-bearing deposits may be slightly enriched in the element (e.g., Jabiluka deposit, Northern Teritory, Australia). The conglomerate type of gold deposit seems to be relatively low in the element, but my data are not extensive for these deposits (Table 35). The only tellurides in the Rand ores are michenerite; (Pd,Pt)BiTe, and moncheite, $PtTe_2$ (Feather and Koen, 1975).

Analyses by the Geological Survey laboratories of gold-ores show the following general ranges for tellurium, in ppm:

Gold-quartz deposits	0.2–2200
Skarn deposits	0.2–0.5
Polymetallic deposits	0.2–10
Quartz-pebble conglomerate deposits (Rand)	<0.2–0.7
Disseminated deposits (Carlin, Nevada)	<0.2–0.6 (Radtke *et al.*, 1972*a*) (av. 0.02) (Harris and Radtke, 1976)

Gold combines with tellurium to form a number of hypogene tellurides and telluride-sulphides of which the most common are sylvanite, calaverite, krennerite, montbrayite, petzite and nagyagite. Native tellurium and a variety of other Pb, Ni, Co, Fe, Ag, Hg and Bi tellurides and telluride-sulphides, including tetradymite, tellurbismuth, joseite, hedleyite, altaite, hessite, aguilarite, frohbergite, melonite, rickardite and coloradoite, commonly accompany gold in all types of its deposits. When tellurium forms no minerals in gold deposits the element is camouflaged in a number of sulphides and sulphosalts, particularly galena, stibnite, bismuthinite, lead sulphosalts and tetrahedrite-tennantite. At Yellowknife up to 1000 ppm Te was found by the writer in the galena of the gold ores, up to 100 ppm in stibnite and up to 100 ppm in the lead sulphosalts. The maximum amount of tellurium recorded in the literature for tetrahedrite (goldfieldite) is 17 per cent. Goldfieldite, however, appears to be a mixture, according to some authors and hence requires further study. Native gold frequently contains some tellurium, commonly as admixed tellurides, but probably also as a lattice constituent in some samples.

Native gold is commonly closely associated with the various tellurides and telluride-sulphides in hypogene deposits, but the paragenesis of these minerals is often complex. Generally speaking, however, the tellurides and gold are usually the latest minerals to be deposited in the mineralization sequence. Forsythe (1971) noted a decrease in the Au/Ag ratio with increase in depth in the main ore shoot of the Crown-Crescent lode in the Emperor Gold Mine, Fiji, which he attributed to vertical zoning of gold-silver tellurides caused by a complex deposition sequence, resurgence of mineralization and replacement phenomena. The details of his scheme should be consulted in the original paper.

Berzon and Karyshev (1973) have investigated the Te/Se ratios in the gold deposits of the Urals and have grouped the deposits as follows: (1) gold-sulphide-quartz and gold-quartz-sulphide veins with high Se and Te in the sulphides and an overall ratio of 0.6; (2) gold-selenium-sulphide-quartz veins in which the Se and Te are mainly in the late sulphides; the Te/Se ratio in these deposits is 0.02; and (3) gold-telluride-sulphide-quartz veins, stockworks and silicified zones in schistose rocks; the Se and Te are in late minerals, particularly tetradymite, tellurbismuth and Au-Ag tellurides. The Te/Se ratio is 4. The last are the principal auriferous deposits in the Urals.

Tellurium generally shows a positive correlation with gold (and silver) in most auriferous deposits, and there is commonly a positive correlation between Te, Bi, Au and Ag. In some deposits the correlations are erratic apparently because of the occurrence of abundant native tellurium and Hg, Pb, Fe and Ni tellurides containing little if any gold. In some deposits in volcanic terranes there may be a positive correlation between Te and V. The reason for this circumstance is obscure. There does not seem to be any relationship between the tellurium content of auriferous deposits and their age. Many Precambrian deposits are telluriferous, as are also numerous Tertiary deposits; the latter appear, however, to be more generally telluriferous than the former according to my data.

The supergene minerals of tellurium include the native element; tellurite and paratellurite, TeO_2; montanite, $Bi_2TeO_6{\cdot}2H_2O$; teineite, $CuTeO_3{\cdot}2H_2O$); emmonsite, $Fe_2Te_3O_9{\cdot}2H_2O$; rodalquilarite (chlorotellurite of iron); quetzalcoatlite, $Cu_4Zn_8(TeO_3)_3(OH)_{18}$; and a large number of poorly differentiated Pb, Bi, Cu, Hg, Zn and Mn tellurites and tellurates that occur as ochers and crusts, often intermixed with a great variety of other supergene minerals.

The oxidation states of tellurium of interest in supergene processes are minus (II), (0), (II), (IV) and (VI). The minus (II) state prevails in the tellurides, the (0) state is represented by native tellurium, the (IV) state by the oxide and various tellurites and the (VI) state by the tellurates. Chemically tellurium is similar to selenium in some respects and somewhat like antimony in a few features, particularly its tendency to combine readily with gold and silver. The geochemistry of tellurium, especially its behaviour during supergene processes in gold deposits, is very poorly known (Vlasov, 1966–1968). The writer's research on the element can be outlined as follows:

The common oxides of tellurium, TeO_2 and TeO_3, are both insoluble in water but are amphoteric dissolving in acids and bases. The acids are tellurous, H_2TeO_3, and orthotelluric, H_6TeO_6 or $Te(OH)_6$, all soluble in water. Orthotelluric acid is much different in its ionization properties from those of the corresponding sulphur and selenium acids, and it has a strong tendency to polymerize to form metatelluric acid often characterized as $(H_2TeO_4)_{(n)}$. The salts of tellurous acid are the tellurites and those of telluric acid the tellurates. The dissolved species in acid solution are H_2TeO_3 and H_6TeO_6 and their ionization products. In strong acid solutions TeO_2 forms the positive ion $[TeOOH]^+$. In basic solutions the ions are TeO_3^{2-}, TeO_4^{2-} and their hydrolysis products.

Tellurium forms a sulphate, $TeSO_4$, which is soluble in solutions containing H_2SO_4. The element also readily forms an elemental colloid and may be present in some natural near surface waters in this state. In addition, a number of soluble inorganic complexes are known, and metal-organic and chelate complexes are a feature of the organic chemistry of tellurium, suggesting that the element may migrate in some near surface waters bound to humic and other organic compounds.

The behaviour of tellurium during the oxidation of gold-bearing deposits is varied. In some places the tellurides are relatively resistant to oxidation and pass into the gossans, weathered debris and soils and ultimately into placers as minute corroded crystals and masses. In other places intensive oxidation yields native tellurium, tellurite or a variety of relatively rare Bi, Cu, Pb, Hg, Fe, Zn and Mn tellurites and tellurates. The details of the formation of these minerals are not well known. It would seem from our investigations that oxidation of hypogene native tellurium, tellurides and other tellurium-bearing minerals yields mainly tellurous acid and soluble tellurites where the oxidation potential is relatively low. Under conditions of higher oxidation potential telluric acid and soluble tellurates are produced. Hydrolysis of solutions containing tellurous acid and soluble tellurites appears to precipitate tellurite or paratellurite, TeO_2, close to the primary tellurides or native tellurium, and fixation by ferric ion, hydrous ferric oxide and a number of cations such as Cu, Zn and Mn precipitates insoluble tellurites of which the most common are emmonsite, $Fe_2Te_3O_9{\cdot}2H_2O$; mackayite, $FeTe_2O_5(OH)$; and blakeite, $Fe_2(TeO_3)_3$. Soluble tellurate is also fixed by ferric iron, hydrous ferric oxide, or cations such as Bi, Cu, Pb, Hg, Zn and Mn as various tellurates that are poorly differentiated as minerals. Adsorption of tellurite and tellurate by hydrous ferric oxides and hydrous manganese oxides, and coprecipitation and/or adsorption and absorption by a variety of secondary minerals such as jarosite and beudantite, are also features that tend to restrict the mobility of tellurium in zones of oxidation. Where chloride is present in the system the chlorotellurite of iron, rodalquilarite, may form in the oxidized zones of gold deposits (Lopez *et al.*, 1968). In a number of gold deposits oxidation of tellurium-bearing minerals appears to result in relatively little migration of tellurium, and in this respect the gold and tellurium contents may remain relatively constant or show only a slight sympathetic enrichment or depletion. This is evidently what took place at Cripple Creek, Colorado according to Lindgren (1933). There, the oxidation of the tellurides yielded dark brown powdery gold, colourless tellurite, and ferric tellurites like durdenite and emmonsite; some of the tellurium was carried away in solution. At Carlin, Nevada Radtke *et al.* (1972*a*) noted no particular change in the tellurium content between the 'primary ores' and the 'oxidized ores'. At Yellowknife, Sudbury and elsewhere in Canada the gossans commonly show an enrichment in tellurium as shown by the analyses in Table 39. Where highly acidic conditions prevail or large volumes of water, regardless of pH, pass through the oxidized zones, much tellurium is removed. Under these conditions gold and tellurium appear to part company. Downward migrating tellurium on entering the zone of reduction may be incorporated in the secondary sulphides, in which it may again be enriched with gold. Supergene tellurides are rare in most oxidized auriferous deposits, but in some oxidized telluride-bearing veins small amounts of secondary (probably supergene) tellurides have been noted in local zones of reduction.

As noted above tellurides tend to resist oxidation and may pass into eluvial and alluvial placers together with gold. A variety of tellurides, including joseite, hessite, petzite, sylvanite, krennerite, tellurbismuth, wehrlite, hedleyite, tetradymite, altaite and nagyagite, have all been found in small amounts in the heavy concentrates of various placers in Canada, U.S.A., U.S.S.R. and elsewhere.

Group VIB: Cr, Mo, W

Chromium is a minor and trace constituent of a great variety of minerals found in the various types of gold deposits. In the skarn-type most of the chromium is present in chlorite, biotite, magnetite, ilmenite, pyrite, garnet and numerous other Ca-Mg-Fe silicates. In gold-quartz and polymetallic deposits most of the chromium occurs in minerals such as chlorite, biotite, sericite, pyrite, magnetite and leucoxene. In the conglomerate type, chromium is present mainly in the micas (sericite) and pyrite, with small amounts and traces in a variety of other minerals including chromite which appears to be an original resistate (placer) mineral in the Rand ores (Liebenberg, 1955; Feather and Koen, 1975). The chromium content of the Rand ores appears to range from 585 to 710 ppm (Table 35). Von Rahden and Urli (1969) found a high correlation between Cr, Zr and Ti in some of the reefs of the Rand. In the disseminated gold ores of Carlin, Nevada, chromium occurs in amounts ranging from 7 to 200 ppm (Radtke *et al.*, 1972*a*). This is a considerable enrichment over that found in the carbonate host rocks (7–70 ppm Cr).

Much of the chromium in gold-quartz and gold-bearing polymetallic deposits appears to be inherited from the host rocks, although in places there may be an introduction or remobilization of the element. In skarn-type deposits some of the chromium appears to have been introduced or remobilized from the host rocks. As far as I have been able to ascertain there is no particular correlation between chromium and gold in any of the auriferous hypogene deposits.

Mariposite, the chromium-bearing variety of phengite, and fuchsite, the greenish chromium-bearing variety of muscovite (sericite), are commonly developed in the wall-rock alteration zones and ores of gold-quartz and polymetallic deposits, generally where they traverse serpentine, peridotite and basic volcanic and plutonic rocks. Examples are widespread in the Mother Lode system of California; in Canada at Yellowknife, Timmins, in the Kerr Addison Mine, Kirkland Lake district, Ontario, in British Columbia and elsewhere; in Australia at Kalgoorlie; in a number of deposits in U.S.S.R. in amphibolites and other basic rocks and in some of the auriferous deposits of Taiwan (Chen and Lee, 1974). Other occurrences and various details of the two minerals are discussed by Whitmore *et al.* (1946). The development of mariposite and fuchsite is always intimately associated with carbonatization, especially ankeritization. During this type of alteration the chromium released by carbonatization of chromium-bearing minerals such as serpentine, amphibole, pyroxene, biotite and chlorite is taken up mainly by the alteration micas (sericite) and to a lesser extent by pyrite and arsenopyrite. Some chromium is also transferred to the veins and lodes where it enters pyrite, arsenopyrite and the micas. Studies by the writer on chromium in the Yellowknife and Kerr Addison deposits show that pyrite and arsenopyrite may contain up to 100 ppm Cr; the fuchsite averages about 1600 ppm Cr. Analyses in the literature indicate that some varieties of fuchsite may contain up to 5 per cent Cr_2O_3 and mariposite up to 1 per cent Cr_2O_3.

Fuchsite and mariposite are frequently closely associated with rich gold shoots, and the two green micas can often be considered as good indicators. However, the association is not invariable, there being large stretches of shear zones and altered basic rocks in many deposits with plenty of fuchsite or mariposite but only traces of gold.

During wall-rock alteration processes the behaviour of chromium is variable as shown by Tables 44 to 59 inclusive. In some places there is an enrichment of chromium in the highly altered phases; in others no change or a slight decrease. In basic and ultrabasic rocks the changes can be followed with a fair degree of accuracy. As the rocks containing chromiferous silicates, magnetite, ilmenite, chromite, etc. are subjected to carbonatization and hydration, much chromiun is released since the element cannot be accommodated in the carbonates. It finds a home, however, in chlorite where this mineral is formed or in fuchsite and/or mariposite where these minerals appear in the equilibrium assemblage. The chromium that is left over is accommodated in part in leucoxene, but some appears to migrate upward, together with Al, SiO_2, K and Na, in the vein systems where they may form an abundance of chromiferous micas in the higher level silicified zones and quartz veins.

In nature the most stable states of oxidation of chromium are (III) and (VI); the former state is present in chromite and in the various chromiferous silicates including fuchsite and mariposite; the latter state is present predominantly as chromate (CrO_4) groups in the relatively rare chromates.

During oxidation of chromium-bearing gold deposits

much of the chromium in minerals such as fuchsite, mariposite and pyrite appears in the hydrolysis products (limonite, wad, kaolin, etc.) probably as the hydrous sesquioxide, $Cr_2O_3{\cdot}nH_2O$, which is readily coprecipitated and/or adsorbed by hydrous iron, alumina and other hydrolysates. The chromium contents in the limonite of gossans are highly variable ranging from 50 to 1000 ppm or more according to my investigations. In the gossans of the Vermilion Mine at Sudbury 125 ppm Cr was found compared with 1250 ppm in the primary ore, indicating leaching of the element during oxidation. On the other hand, at Yellowknife in the Lynx prospect gossan 50 ppm Cr was found compared with 10 ppm in the primary ore. Radtke *et al.* (1972*a*) found chromium contents ranging from 15 to 700 ppm (av. 60 ppm) in the oxidized mineralized ores of the Carlin deposit in Nevada. This indicates only a slight enrichment, if any, over the primary ores which range from 7 to 200 ppm Cr (av. 60 ppm). Under conditions of high oxidation potential some chromium is delivered to the oxidizing solutions probably as soluble alkali or calcium chromates. Where lead minerals are present in the ores, or dissolved lead occurs in the supergene solutions, the mineral crocoite, $PbCrO_4$ may appear, but it is rare. Other species of mineral chromates and dichromates with K, Na, Mg, Pb and Zn, are, likewise rare. Species of phosphate, arsenate and vandate minerals with chromium as an essential element are also known, but these occur only locally. More generally one finds traces to minor amounts of chromium in the common phosphates and arsenates. At Keno Hill, for instance, the pyromorphite in the oxidized zones contains up to 1000 ppm or more Cr. Greenish and purplish coloured chromium-bearing sulphates may occur in the dry parts of the oxidized zones of some gold deposits containing abundant fuchsite, mariposite and other primary chromium minerals. These sulphates are relatively rare and appear to be temporal, dissolving readily with any influx of water.

Few chromium minerals have been recorded in the oxidized zones of auriferous deposits. In some of the Beresovsk veins, Urals, U.S.S.R., crocoite, $PbCrO_4$, occurs with limonite in cavities in quartz near the surface. Pyromorphite is asociated with the chromate in some of the veins.

Few details are available on the concomitant behaviour of gold and chromium during oxidation processes. The writer's limited investigations suggest little change or a slight sympathetic enrichment of the two elements in some gossans. In others, gold is enriched and much chromium is leached, and in still others there is a general loss of both elements as at the Vermilion Mine at Sudbury. There are no recorded instances of a sympathetic enrichment of chromium and gold in the secondary sulphides in the zones of reduction.

Chromite, chromiferous magnetite and leucoxene, chrome micas and a variety of other chromium-bearing minerals accompany gold in some fossil and recent eluvial and alluvial placers. As far as I have been able to ascertain there is no particular correlation between gold and chromium in placers, a feature that one might suspect since most of the chromium-bearing minerals are contributed by the host rocks and not by the hypogene gold deposits.

Molybdenum is commonly associated with gold in most types of its deposits, but the element is generally present only in small amounts. Molybdenite is the only common hypogene molybdenum mineral found in gold deposits. Traces of molybdenum may also occur in some of the sulphides and in scheelite in gold deposits. Our analyses of ores from the various types of gold deposits show the following ranges in molybdenum content in ppm:

Skarn deposits	<1–500
Gold-quartz veins	2–840
Polymetallic deposits	2–840
Quartz-pebble conglomerates (Rand)	2.5–3

In the skarn, gold-quartz and polymetallic deposits, gold is only rarely found in direct association with molybdenite. The paragenesis of the two minerals is, therefore, often difficult to establish. In most deposits where relationships can be clearly ascertained the molybdenite is an early mineral, contemporaneous with quartz. Gold and gold tellurides are generally later than the quartz and, therefore, later than molybdenite. Only occasionally has the writer seen gold intergrown with molybdenite in gold deposits. This confirms Lincoln's (1911*b*) earlier observations. In a few auriferous deposits molybdenite appears to be a very late mineral occurring along post-ore fractures. This age of molybdenite may be younger than the last generation of gold and gold tellurides although the paragenetic relationships are difficult to establish.

The molybdenum content of the disseminated gold ores of Carlin, Nevada ranges from <2 to 100 ppm (Radtke *et al.* 1972*a*).

There are no references to enrichment of molybdenum in gold-bearing precipitates from thermal waters in New Zealand or elsewhere.

Saager (1968) and Feather and Koen (1975) mention the presence of molybdenite in the Rand ores. Our analyses of typical ores from the Rand (Table 35) show molybdenum to be present in the range of 2 to 3 ppm. Those of the Blind River–Elliot Lake district are a little higher in molybdenum content (11 ppm).

In some gold deposits there is a positive correlation between molybdenum and gold as can be seen from the table given by Thomson (1950, p. 97) for the Kirkland Lake ores. In other gold veins the correlation is weak, negative or nonexistent. For example, in one of the Crestaurum veins near Ryan Lake at Yellowknife the molybdenum (molybdenite) content increases as the vein progresses into the granite; concomitantly the gold values drop from a third of an ounce or more per ton (10.8 ppm) to very low values (0.145 ppm).

The common oxidation states of molybdenum are (II), (III), (IV), (V) and (VI). Only the (IV) and (VI) states are of importance in nature, the former occurring in molybdenite, MoS_2, and the latter in a number of rather rare molybdates, including powellite, wulfenite, koechlinite, lindgrenite and ferrimolybdite, all generally supergene minerals. An indefinite oxide, probably $Mo_3O_8{\cdot}nH_2O$, called ilsemannite or molybdenum blue, also occurs in the oxidized zones of certain molybdenite deposits.

The oxide MoO_3 is acidic and dissolves in basic solutions to form a complicated series of oxyanions called molybdates, the simplest of which is probably MoO_4^{2-} or some similar hydrated species. When alkaline molybdate solutions are acidified, and the pH lowered below 7, a most complicated series of condensed polymolybdate ions are formed. These have been variously categorized, but no particular consensus appears to prevail among investigators as to their true nature. The hydrates $MoO_3 \cdot 2H_2O$ and $MoO_3 \cdot H_2O$ can be precipitated from approximately neutral solutions of molybdates. These are not definitely known in nature, although they may be constituents of certain molybdenum ochers. When molybdate solutions or a suspension of MoO_3 are reduced, a deep blue oxide (ilsemannite?) of variable composition $MoO_{2.5\text{-}3.0}$ is formed. According to some investigators this oxide has colloidal properties and is characterized by a defect structure. MoS_2 is precipitated in a hydrated form when H_2S reacts with slightly acid solutions of molybdates.

The alkali and Mg molybdates are soluble; the Ca, Ba, Ag, Pb, Cu, Bi and Fe molybdates are insoluble or sparingly soluble. The complex chemistry of molybdenum is extremely intricate and fraught with many difficulties of interpretation. Most of the complexes are anionic and often polymeric. The element also forms a number of chelates in the laboratory, suggesting that this feature of its chemistry may be important in natural organic environments.

Molybdenite tends to oxidize slowly and in some deposits, especially those characterized by a quartz gangue and a paucity of pyrite and other sulphides, passes relatively unchanged into the eluvium and ultimately into the stream sediments. Under conditions of intense oxidization, particularly in the presence of pyrite and other sulphides, the molybdenum in molybdenite is oxidized to the hexavalent state and enters solution as molybdate or polymolybdate. In this form it probably migrates as the relatively soluble K, Na and Mg species or as the so-called molybdic acid H_2MoO_4. There is also the possibility of migration as a sulphate complex of the type $MoO_2 \cdot SO_4$.

In a few places the colloidallike hydrous oxide, natural molybdenum blue or ilsemannite, is precipitated close to concentrations of molybdenite. This defect oxide appears to form as the result of the hydrolysis of acidic solutions of soluble molybdates or molybdic acid. More generally the secondary minerals formed are ferrimolybdite where soluble iron is present, or powellite, wulfenite, koechlinite or lindgrenite where Ca, Pb, Bi or Cu are respectively present in the oxidizing solutions.

Ferrimolybdite forms in acid solutions through reaction of ferric iron with soluble molybdates. According to Vinogradov (1957) and others, ferrimolybdite has maximum stability between pH 3 to 4. Jones (1957) gives a pH of 2.7, and Sarafian and Furbish (1965) give a pH of 3.35. In a more acid environment it tends to dissolve and in a more alkaline environment it hydrolyzes according to the equation

$$Fe_2(MoO_4)_3 + 6H_2O \rightarrow 2Fe(OH)_3 + 3H_2MoO_4$$

This suggests that at the pH of most natural waters (5–8) ferrimolybdite is unstable, and much of the molybdenum present in the mineral passes into solution and is ultimately dispersed. Some molybdenum may, however, be retained as an adsorbed constituent on limonite, and some may be incorporated in secondary minerals such as jarosite, pyromorphite and wad.

Powellite has maximum stability in the pH range from 5 to 6 and wulfenite is probably also most stable in this range. Data on the solubility of the other natural molybdates are lacking.

From the above it is evident that under mildly acidic conditions in oxidized zones there is a relatively high retention of molybdenum in insoluble minerals. Under conditions of high acidity much molybdenum enters solution and is dispersed. The same is true where the waters are neutral or alkaline. Lindgren (1933, p. 64), for example, mentions that the water from a tunnel in the Gilpin County gold mines contained nearly 8 g/L of molybdenum oxide (5332 ppm Mo), and that the water had a deep greenish blue colour. Some of the soluble molybdenum migrating downward in auriferous deposits may be precipitated in the zones of reduction (secondary sulphide zones) probably as colloidal or very fine-grained MoS_2.

In some oxidized zones I have noted that the contents of gold and molybdenum show little change; in others gold is enriched and molybdenum is extensively leached. In the gossans of the Lynx prospect at Yellowknife the molybdenum content averages about 8 ppm as compared with the primary ore which averages about 11 ppm. The corresponding gold values are 0.175 and 0.045 ppm respectively. Radtke *et al.* (1972*a*) report a range of <2 to 50 ppm Mo (av. 7 ppm) in the oxidized ores of the Carlin deposit in Nevada. This indicates little change in the molybdenum content during oxidation processes, since the primary ores contain from <2 to 100 ppm Mo (av. 7 ppm). There is a slight enrichment of gold in the oxidized zones at Carlin. Gold and molybdenum may show a slight enrichment in the zones of reduction of some deposits where secondary sulphides are concentrated.

The writer has observed molybdenite in the heavy concentrates of stream sediments near molybdenite deposits, but has not found the mineral in the heavy concentratres of gold placers. Traces of molybdenum are, however, common in the heavy mineral concentrates of gold placers, the element being probably present in pulverulent molybdenite or in sulphides and other minerals such as pyrite, apatite, and monazite.

Tungsten is a nearly universal associate of gold in its hypogene deposits, a coherence considered by Foster (1977) to be due in part to the similar solubilities of scheelite and gold. I have found the element in practically all of the gold ores that I have examined, the contents ranging from a few parts per million to 5000 ppm or more in some gold-quartz veins.

The principal hypogene tungsten mineral in gold deposits is scheelite. Wolframite and its varieties, such as ferberite and hubnerite, are less frequent but in places may be relatively abundant.

The largest amount of tungsten appear to occur in the skarn-type gold deposits and in gold-quartz veins. Gold-bearing polymetallic deposits generally have low amounts of the element, but some deposits are notably enriched. The Geological Survey's analyses of ores from the various types of auriferous deposits show the following general ranges:

Skarn-type deposits	2–1000
Gold-quartz veins	2–5000
Polymetallic deposits	<2–1000

Rich shoots of scheelite or wolframite may occur in all of

these gold deposits. Some have been mined for their tungsten content. The grade of these shoots is generally low, ranging from 0.05 to 0.75 per cent WO_3. In some gold veins, ore shoots rich in scheelite are commonly relatively low in gold.

In the skarn, gold-quartz and polymetallic deposits gold is only rarely found in direct association with scheelite and wolframite. the paragenesis of the tungstates and gold is, therefore, often difficult to establish. In most deposits where relationships can be clearly ascertained the tungstates are early minerals, contemporaneous with quartz. Gold, in a few deposits seen by the writer, occurs as inclusions in scheelite and also along fractures in the mineral. Gold tellurides exhibit similar relationships in some deposits.

Tungsten is enriched in the disseminated gold ores of Carlin, Cortez, Getchell and Gold Acres in Nevada (Roberts *et al.*, 1971). The tungsten content of the Cortez deposit averages about 35 ppm (Wells *et al.*, 1969). Radtke *et al.* (1972*a*) record tungsten contents ranging from <20 to 400 ppm in primary ores of the Carlin deposit. The mineral forms of the element are not stated in the published accounts. Tungsten is, likewise, enriched in the gold-bearing precipitates from thermal waters in the Taupo volcanic zone, New Zealand (Weissberg, 1969). In these precipitates tungsten contents up to 3 per cent have been recorded.

Relatively little is known about the distribution of tungsten in the auriferous conglomerate deposits. Liebenberg (1955), Saager (1968) and others make no mention of the presence of scheelite or wolframite in the Rand ores and analyses of uraninite concentrates from amalgam barrel residues given by Liebenberg show only doubtful traces of tungsten. Analyses of typical Rand ores confirm that tungsten is present in nondetectable amounts (Table 35). In the Blind River–Elliot Lake ores tungsten is, likewise, present in low amounts (2 ppm).

The aqueous chemistry of tungsten is intricate because of the number of oxidation states that the element manifests, (II), (III), (IV), (V) and (VI). Of these only the (IV) and (VI) states are of interest, the former occurring in the mineral tungstenite, WS_2, and the latter in scheelite and other tungstates. During oxidation processes the hexavalent state predominates in the various reactions. This state, represented by the canary yellow oxide WO_3, has a most complicated chemistry many features of which are relatively uncharacterized. In general the chemistry of hexavalent tungsten is much like that of hexavalent molybdenum.

The oxide, WO_3, occurs as tungstite, $WO_3 \cdot H_2O$, in nature. It is acidic and dissolves in basic solutions to form a complicated series of oxyanions called tungstates, the simplest of which is probably $[WO_4]^{2-}$ or some similar hydrated species. The normal alkali metal and magnesium tungstates are relatively soluble in water; the other metal tungstates are insoluble or only sparingly soluble. These include a variety of supergene minerals of which the most common are stolzite. $Pb(WO_4)$, raspite, $Pb(WO_4)$, cuprotungstite, $Cu_2(WO_4)(OH)_2$, ferritungstite, $Ca_2Fe_2(WO_4)_7 \cdot 9H_2O$, tungsten-powellite, $Ca(Mo,W)O_4$, anthoinite, $AlWO_3(OH)_3$, sanmartinite, $ZnWO_4$ and thorotungstite, an indefinite thorium-aluminum tungstate. There are in addition a number of yellow tungsten ochers, often found in oxidized zones containing hypogene tungsten minerals. These are amorphous to X-rays but are probably various tungsten oxides, hydrous oxides and tungstates.

Scheelite and wolframite tend to resist weathering and solution and generally pass into the gossans and oxidized zones of gold deposits relatively unaffected. Ultimately these minerals pass from the eluvium to stream placers where they are concentrated with gold. Under some conditions, however, especially where the oxidizing waters contain abundant dissolved Na, K or Mg salts or where sulphuric acid is present in abundance from oxidizing pyrite both wolframite and scheelite are slowly dissolved, yielding probably the soluble tungstates and the so-called tungstic acid:

$$CaWO_4 + 2Na^+ = Ca^{2+} + Na_2WO_4$$
$$CaWO_4 + H_2SO_4 = CaSO_4 + H_2WO_4$$

Neither of these soluble compounds have much mobility, especially the acid. It hydrolyzes readily, precipitating tungstite, $WO_3 \cdot H_2O$. Where the solutions are alkaline the soluble tungstates possess some mobility. Acidification of these solutions, however, yields tungstite on hydrolysis. Where they come into contact with Pb, Cu, Zn, Al, Th and Fe minerals or ions in solution a variety of insoluble minerals noted above may be precipitated. Hydrous iron oxides also coprecipitate and/or adsorb tungstate forming the tungstate, ferritungstite.

The oxidation of gold deposits containing tungsten minerals, therefore, generally leads to accumulations of the resistate tungsten minerals, insoluble secondary tungstates or limonite enriched in tungsten. In this respect tungsten and gold behave in a similar manner and tend to show little change or some enrichment in the oxidized zones. In some deposits tungsten may be leached out of the oxidized zones leaving enrichments of gold. This is what obtains in the gossans of the Lynx prospect at Yellowknife where the tungsten and gold contents of the primary ore and gossans are as follows:

	Primary ore	Gossan
	(ppm)	
W	4	2
Au	0.045	0.175

Radtke *et al.* (1972*a*) noted no marked change in the tungsten content during the oxidization of the disseminated ores of Carlin, Nevada. There are no recorded cases of concomitant enrichment of gold and tungsten in the zones of reduction containing secondary sulphides (Gannett, 1919).

As noted above, scheelite and wolframite resist weathering and solution and pass with gold into eluvial and alluvial placers. Raspite, the lead tungstate, may also accumulate in gold placers. Generally the amounts of the tungsten minerals in gold placers are low and not economic. However, placers are known where both gold and tungsten minerals may be greatly enriched. The Dublin Gulch gold placers of the central Yukon provide such an example (Fig. 69). In the eluvial placers the tungsten content (mainly in scheelite with some wolframite) is 0.5 per cent or more. Parts of the alluvial placers have a similar tenor of tungsten, although these placers as a whole probably only contain about 1/10 as much tungsten. The source of the gold and tungsten in Dublin Gulch is multiple – most of the gold derives from arsenopyrite-pyrite-jamesonite-quartz veins; the wolframite and some of the scheelite comes from stringers of quartz in a nearby granitic stock; and the bulk of the scheelite is weathered mainly from skarn bodies surrounding the stock (Boyle, 1965*a*).

Group VIIA: F,Cl, Br, I

Fluorine accompanies gold in all types of deposits but generally only in small amounts, except in some deposits where fluorite or topaz are important gangue minerals. The most common hypogene fluorine mineral in gold deposits is fluorite. Other hypogene fluorine-bearing minerals include fluormicas, tourmaline, topaz, fluorapatite and a great variety of common silicates such as chlorite, apophyllite, sphene, amphibole, etc. with a small fluorine component. Minerals such as cryolite, and the various rare-earth, U, Nb and Ta fluoride-bearing minerals are extremely rare or nonexistent in gold deposits.

Table 40. Fluorine and chlorine contents of various gold ores

Type of deposit	F	Cl
	(ppm)	
Skarn	1150 (av.)	200 (av.)
Gold-quartz	10–2100	100–500
Polymetallic	10–1000	100–500
Quartz-pebble conglomerate (Rand)	5–195	15–250

Analyses of ores from the various types of gold deposits (Table 40) show the content of fluorine to be relatively low in most places. Where fluorite is a major gangue mineral, as in the Cripple Creek, Colorado deposits the amount of fluorine may exceed 5 per cent (10% fluorite). Similarly, where topaz is an important gangue (alteration) mineral as in the Brewer Gold Mine, Chesterfield County, South Carolina (Peyton and Lynch, 1953) the fluorine contents may run up to 3.5 per cent or more with long core lengths of 1 per cent or more.

Fluorite is not mentioned in the literature as a gangue mineral in the disseminated gold ores of Carlin and elsewhere in Nevada, and Radtke *et al.* (1972*a*) give no analyses for the element.

Analyses of gold-bearing thermal waters in the Taupo volcanic zone in New Zealand contain from 4 to 9 ppm F (Weissberg, 1969).

The fluorine content of the quartz-pebble conglomerates appears to range widely from about 5 to 195 ppm F (Table 35). Most of the fluorine is in the micas and apatite; also in monazite and other Th, U and rare-earth minerals in the Elliot Lake ores.

The behaviour of fluorine in the wall-rock alteration zones of a number of auriferous deposits is shown in Tables 44 to 59 inclusive. The wall-rock alteration zones in deposits in sediments and granites commonly show increases in their fluorine content as the auriferous mineralization is approached; deposits in greenstones, diorites and similar rocks show no change or only slight increases toward the veins.

Fluorite is relatively insoluble in water (16 ppm at 18°C) as are also the fluorapatites (20 ppm at 18°C). Carbonated solutions and mineral acids such as H_2SO_4 dissolve fluorite and apatite slowly yielding fluoride ion, which migrates mainly as the alkali species. The same agents release fluoride ion from fluor-micas and other fluoride-bearing minerals.

In some gold deposits, especially those containing abundant calcium-bearing carbonates, or where Pb, Ca or Al are present in solution, insoluble fluorides, chloride-fluorides and basic fluorides such as secondary fluorite, matlockite, creedite, etc. may be precipitated. Some fluoride may also be adsorbed and/or coprecipitated by hydrous ferric oxides, wad or hydrous alumina complexes, particularly the last. More generally, however, much fluoride is leached from gossans and oxidized zones leaving only trace or minor amounts in these materials. Gold and fluoride, therefore, tend to part company during the oxidation of deposits, gold becoming enriched in the gossans and oxidized zones and the fluorine being largely removed. This is the situation at Cripple Creek, Colorado as described by Lindgren and Ransome (1906). The gossans at the Lynx prospect, Yellowknife exhibit a slight depletion in their fluoride content (30 ppm) compared with the primary gold-bearing quartz-pyrite ore (40 ppm).

Fluorite and fluorapatite occur as minor constituents in some eluvial and alluvial gold placers, and fluorine is generally present in tourmaline, topaz, micas and other silicates in these deposits. Fluoriferous rare-earth, Y, U, Nb and Ta minerals such as yttrofluorite, microlite and pyrochlore may occur in some gold placers, but they are rare.

Chlorine is a constituent of most gold deposits, but the element is invariably present only in small amounts as shown by the analyses in Table 40. Most of the chlorine in gold deposits occurs in chlorapatite and as NaCl and KCl in liquid inclusions in a great variety of minerals, particularly quartz and carbonates. Chlorine-bearing minerals such as the sodalite group, the scapolites and eudialite are rare or nonexistent in most gold deposits. Scapolite occurs in a few skarn-type gold deposits, but it is generally a minor mineral in the gangue.

The writer (Boyle, 1961*a*) made an extensive investigation of the distribution of chlorine (chloride) in the Yellowknife gold-quartz deposits. There, the contents of Cl in the host rocks, alteration zones, and gold ores were about the same (200 ppm Cl, av.). Much of the chloride occurs in carbonates as shown in the following:

Carbonate	Cl
	(ppm)
Ankerite in carbonate-sericite schist (principal gold stage)	50
Calcite and ankerite in cross-cutting quartz-carbonate veins (minor gold in places)	900
Scalenohedral calcite in late fractures (barren of gold)	100

Much of the chloride ion in the carbonates is extractable by pulverizing the minerals and leaching with distilled water. This suggests that the chloride ion is contained in minute inclusions of sodium or potassium chloride in the carbonates, or in these salts in crystal cavities (liquid inclusions).

The behaviour of chlorine (chloride) in the wall-rock alteration zones of a number of gold-bearing deposits is shown in Tables 44 to 59 inclusive. Some of the profiles show little change in the chloride content through the alteration zones, others show some leaching of chloride in certain alteration phases, and still others exhibit small enrichments in some phases.

Gold is soluble as chloride complexes $[AuCl_2]^-$, $[AuCl_4]^-$, etc. as discussed in the section on natural waters in Chapter II. This has led a number of investigators to assume that gold migrates in this form during hydrothermal processes. As noted previously these complexes are stable only in acid solutions below a pH of about 6. At neutral and alkaline pH the concentration of the aurous and auric chloride complexes is insignificantly small. Helgeson and Garrels (1968) have assumed that gold is transferred in acid (pH 2–4) hydrothermal sodium chloride solutions as the aurous chloride complex, but this mechanism has been criticized by Boyle (1969*c*) on several grounds, some of which are mentioned in Chapter II. Gold may be transported in acid sodium chloride solutions, and some gold-quartz veins particularly those containing alunite may have been precipitated from such solutions, but they must be rare considering the small amounts of chloride present in most gold deposits (Table 40) and the general abundance of carbonates in gold veins, minerals that are not readily precipitated from acid solutions with pH 2 to 4. More pertinent, however, is the fact that most present day thermal waters carrying and depositing gold, while containing relatively large amounts of Cl (up to 3000 ppm), are neutral or slightly alkaline. These waters also contain considerable amounts of S, Sb and As, the elements with which gold (and silver) are most commonly associated in nature. It would seem on thermodynamic evidence that gold in such waters is transferred mainly as the $[AuS]^-$ complex or as various other complexes of S, As and Sb as postulated in Chapter II. Weissberg (1970) has concluded from experimental work and from the natural gold-bearing water systems in the Taupo volcanic zone in New Zealand that "The experimental evidence indicates that the most effective gold transport in aqueous sulfide solutions occurs in near-neutral pH solutions containing a high proportion of sulfide as HS^- ions. In more strongly acid or alkaline solutions, where the sulfide species present are dominantly H_2S or S^{2-} respectively, the solubility of gold is much lower and these solutions also seem less likely to play an important role in the hydrothermal transport of gold from the geological point of view."

Seward (1973) concluded, however, that gold is probably transported as both thio and chloro complexes. In near-neutral and alkaline media he thought that three thio complexes were largely responsible for the solubility of gold – the complex $[Au_2(HS)_2S]^{2-}$ in alkaline media, the complex $[Au(HS)_2]^-$ in neutral media and possibly the complex $[Au(HS)]^0$ in acid media.

Before leaving the hypogene processes of gold deposition from chloride solutions it is interesting to note that present day waters in faults in many auriferous regions are charged with chlorides, mainly NaCl. At Yellowknife such waters contain up to 11 383 ppm Cl and occur at a depth of 2300 ft and more (Boyle, 1961*a*, p. 168). Others have recorded similar chloride-rich waters in Precambrian rocks at Timmins, Ontario (Langford and Hancox, 1936), in the Sturgeon River Gold Mine near Nezah in Ontario (Bruce, 1941*b*); in the Kalgoorlie goldfields, Australia (Simpson and Gibson, 1912); and elsewhere. All of these waters have a neutral to alkaline pH. They have been considered to be connate waters by some investigators, but they differ substantially from sea water in having low magnesium contents. In fact their Ca/Mg ratios are usually the reverse to what one finds in sea water. Their origin is enigmatic; they may represent fossil hydrothermal solutions of metamorphic derivation, but this is only a guess.

The alkali and alkaline-earth chlorides are soluble, and their migration is relatively independent of pH and Eh conditions such as those that prevail during oxidation of gold deposits. Oxidation of most types of gold deposits, therefore, generally leads to a marked removal of chloride. In certain types of gold deposits, however, especially those containing an abundance of primary Ag, Pb, Bi, Hg and Cu minerals a number of relatively insoluble chlorides, oxychlorides, complex chlorides and compound chlorides may develop in the oxidized zones and gossans, viz. chlorargyrite, AgCl, calomel, HgCl, cotunnite, $PbCl_2$, eglestonite, Hg_4OCl_2, mendipite, $Pb_3O_2Cl_2$, matlokite, PbFCl, bismoclite, BiOCl, atacamite, $Cu_2(OH)_3Cl$, boleite, $Pb_9Cu_8Ag_3Cl_{21}(OH)_{16}{\cdot}H_2O$, and kleinite, $Hg_2N(Cl,SO_4){\cdot}nH_2O$. Most of these minerals are developed in oxidized zones in hot arid regions or where special conditions prevail such as where the groundwaters are highly saline or where deposits occur near the sea.

Under most conditions in a humid temperate climate, oxidation of gold deposits generally leads to removal of chloride from gossans and oxidized zones. The gold content may show little change, enrichment or depletion depending on the chemical system (*see* Chapter IV). In a hot arid climate oxidation may lead to a sympathetic enrichment of gold, silver and chloride in the gossans and oxidized zones of deposits containing primary Ag, Pb, Bi, Hg and Cu minerals.

Bromine and *iodine* are probably present in most gold deposits, but the amounts would seem to be very low. I have no data on the distribution of these elements in gold ores, but investigations described in the literature indicate that the general levels of abundance of the two elements in mineral deposits are respectively: Br – 0.3 to 5 ppm; I – 0.1 to 1.5 ppm.

Both bromine and iodine occur in some of the gold-bearing thermal waters of the Taupo volcanic zone in New Zealand. Weissberg (1969) records the following ranges in these waters: Br – 3 to 7.2 ppm; I – 0.4 to 2.4 ppm.

The chemistry of bromine and iodine is essentially the same as that discussed for chlorine in oxidation processes. In most humid temperate climates bromine and iodine are extensively leached from gossans and oxidized zones; in hot arid climates Br and I may replace Cl in the chlorides, and the bromide and iodide analogues of some of the chloride minerals mentioned above may form where the primary ores contain Ag, Pb, Bi, Hg and Cu. Of these the most common are bromargyrite, AgBr, and iodargyrite, AgI. Bromates are not known in nature, but in environments where a very high oxidation potential prevails iodates, hydrous iodates, basic iodates, iodate-oxychlorides and iodate-chromates of Cu, Ca and Pb may develop. They are rare.

Group VIIB: Mn, Re

Manganese generally accompanies gold in hypogene deposits mainly in the form of manganiferous carbonates, particularly ankerite. Rhodochrosite and occasionally rhodonite, johannsenite and inesite are important gangue minerals in some gold deposits, principally the gold-quartz and polymetallic types of Tertiary age. Various other manganiferous minerals accompany gold in its deposits including sphalerite in vein and polymetallic deposits, magnetite in a variety of deposits and various manganiferous silicates in skarn deposits. In the auriferous quartz-pebble conglomerate deposits only minor amounts of manganese are present (Table 35), principally in pyrite, sphalerite, chlorite and other minerals.

Certain auriferous quartz deposits are marked by the presence of amethyst, especially those of Tertiary age, which also contain an abundance of manganese minerals. The colour of this amethystine quartz is disputed, but a number of investigations, including those of the writer, suggest that the colouring is due at least in part to manganese ions in the lattice of the quartz.

The behaviour of manganese during wall-rock alteration processes in gold deposits is discussed in a later section. Most of the manganese in the alteration zones and in gold ores appears to have been contributed by the nearby host rocks, although in some deposits contributions from more distant sources are indicated.

The supergene manganese minerals are varied and frequently difficult to determine because they occur in earthy forms and are commonly intermixed with a great variety of other supergene minerals. Pyrolusite, MnO_2, and a great variety of other oxides, hydroxides, basic oxides and oxide-silicates, all intermixed in an aggregate usually called wad, are characteristic of gossans and oxidized zones. There are also a number of phosphates and sulphates that may appear in oxidized zones, but these are usually rare.

The aqueous chemistry of manganese is particularly intricate as befits an element with all oxidation states from (II) to (VII). Since space dictates brevity here, only a brief and relatively simple outline will be given of the fate of manganese during oxidation processes.

Oxidation of most hypogene manganese minerals such as the carbonates yields the aquo complex $[Mn(H_2O)_6]^{2+}$ to solutions. This ion is stable in acid and neutral solutions, and migrates mainly as the sulphate or hydrogen carbonate complex. In alkaline solutions the hydroxide is precipitated, then oxidizes rapidly in the presence of oxygen, water is split out, and MnO_2 (pyrolusite) is formed.

$$2\,Mn(OH)_2 + O_2 = 2\,MnO_2 + 2H_2O$$

Alternatively, the aquo manganous complex is oxidized to the Mn^{4+} state in solution and undergoes hydrolysis with the precipitation of a variety of oxides and basic oxides depending on the pH, Eh and presence of various cations. MnO_2 (pyrolusite) is often the most common of the oxides formed, but it is invariably mixed with a great variety of substances constituting wad.

The formation of gossans and oxidized zones containing supergene manganese minerals is a most complicated problem not readily summarized here. One can contrive Eh-pH diagrams and equations for the iron and manganese systems, but these are generally misleading in their simplicity since it is impossible to take into account all of the phases present, the nonstoichiometric substances formed, the colloidal reactions and the influence of SiO_2, Al_2O_3, PO_4^{3-}, SO_4^{2-}, AsO_4^{3-} and a host of cationic complexes that undergo a most intricate history of hydrolysis reactions. Nevertheless, some general features are evident. Some limonitic gossans are relatively free of large amounts of manganese minerals even though the primary ore and gangue are highly manganiferous. This is probably explained by the fact that comparing the Fe(II) and Mn(II) states the former is more easily oxidized at a lower pH than the latter. Thus, in waters leaching sulphide deposits most of the iron is precipitated as limonite well below a pH of 5 whereas the manganese still retains considerable mobility. Numerous cases of this have been observed by the writer. At Keno Hill, where the pH of the oxidizing waters leaching sulphides and manganiferous siderite averages 6.5, the iron content ranges from 1 to 5 ppm whereas the manganese content frequently exceed 25 ppm (Boyle, 1965*a*).

Under these conditions much manganese is leached, especially from the near surface parts of the oxidized zones. In other deposits, however, the oxidized zones are rich in both limonite and wad or rich in wad with considerable amounts of iron removed, despite the fact that the Eh and pH conditions are such that one might expect a high retention of iron and considerable loss of manganese. The reasons for this type of behavior are not at all clear to this writer. Obviously some uncharacterized chemical factor is at work, perhaps inorganic or organic complexing which renders iron more mobile than manganese.

The presence of supergene manganese minerals, particularly the various hydrous oxides and soluble species of manganese in the oxidized zones of deposits, are important in the migration of gold in three respects. Firstly, the oxides tend to adsorb and/or coprecipitate gold from solution (*see* also the discussion in the section on gold in other minerals in Chapter II). Secondly, Mn(IV) ion has sufficient potential to oxidize gold and render it mobile if complexing agents for gold are present. On the other hand Mn(II) ion reduces soluble gold to the metal. Thirdly, manganese dioxide, and some of the other oxides, aid in the formation of chlorine which complexes with gold and renders it mobile. Thus HCl (derived mainly by the reaction of H_2SO_4 with NaCl or other chlorides in oxidized zones) acting on MnO_2 yields Mn^{2+} and liberates Cl_2 which renders gold mobile as the chloride complex. The various reactions probably run as follows:

$$H_2SO_4 + 2NaCl \rightarrow 2HCl + Na_2SO_4$$
$$4HCl + MnO_2 \rightarrow MnCl_2 + 2H_2O + Cl_2$$
$$Au + Cl_2 \rightarrow \text{Soluble complexes such as } [AuCl_4]^-$$

We see, therefore, that in the absence of chlorides gold tends to be enriched in oxidized zones containing supergene manganese minerals. Where chlorides and an ample supply of H_2SO_4 are available, gold may be mobilized and may migrate downward to the zone of reduction where some of the element may be precipitated in the native state or incorporated in the secondary sulphides. The remainder of the element migrates in the groundwater system and is ultimately lost from the deposits.

It should also be mentioned that the presence of manganese in natural waters, especially surface waters, seems in places to be responsible for the relatively high migration capacity of gold. The precise reasons for this behaviour are unknown – one factor seems to be that gold is adsorbed to the various hydrous manganese oxide colloids and is carried in this manner; another may be adsorption and/or incorporation of gold in suspended humic manganese colloids.

Manganese is a constituent of a great many light and

heavy minerals that accompany gold in recent and fossil eluvial and alluvial placers. In some placers manganese is a major constituent of limonite-wad cemented gravels, sand and silt. These materials, as well as gravels and sands that contain an abundance of limonite-wad particles, are invariably enriched in gold, probably largely as a result of adsorption processes.

Rhenium is a very rare element that occurs in amounts of a few parts per million or less in a number of minerals, principally the sulphides, molybdenite, pyrite, chalcopyrite, bornite, sphalerite and galena. It is often enriched in molybdenite in which it may be present in amounts from <1 to 3000 ppm or more according to analyses quoted in the literature. Three poorly differentiated rhenium minerals are known, but these are extremely rare. Of these the sulphide, dzhezkazganite, ReS_2, is probably the most common. The geochemistry of the element is discussed by Vlasov (1966–1968) and Ivanov *et al.* (1972).

Speaking generally, it can be said that rhenium follows Mo and W in hypogene processes, and in places there appears to be a coherence with Cu, Pb, Cd, Zn, and perhaps with Se, Te and U.

There are only a few data on the content of rhenium in gold deposits. Ivanov *et al.* (1972) mention rhenium contents ranging from 0.02 to 0.86 ppm in pyrite from some gold-bearing deposits in U.S.S.R. The writer has detected the presence of rhenium in molybdenite in some gold deposits but only in small amounts (~ 10 ppm).

During oxidation processes rhenium appears to be more mobile than molybdenum (Morachevskii and Nechaeva, 1960). The element, once liberated from its hypogene host minerals, mainly molybdenite, probably migrates as the perrhenate ion, ReO_4^-, in the form of the soluble calcium or sodium salt. In chloride waters it may also migrate as the soluble ReO_3Cl compound (Mukanov and Aubakirova, 1969). Perrhenate ion does not form highly insoluble salts with the cations of Ca, Pb, Sr and Fe, as does the molybdate ion, a feature that further increases its mobility with respect to molybdenum. The perrhenate ion is to some extent adsorbed by hydrous iron oxides, but the effect of manganese oxides is uncertain; it is probably coprecipitated with manganese because of the chemical similarities of the perrhenate and permanganate ions. In most places wads are in general low in rhenium. Potassium-bearing supergene minerals such as jarosite probably concentrate rhenium because of the low solubility of potassium perrhenate.

One can only speculate that rhenium is retained in some degree in the gossans during the oxidation of gold deposits, but where extensive leaching has taken place probably much of the element is removed. Since rhenium forms a sulphide readily (ReS_2, dzhezkazganite), certain zones of reduction (secondary sulphides) may be enriched in the element, although there are no data available to suggest that this in fact does take place.

Group VIII: Fe, Co, Ni and Pt group metals

Iron is a constant companion of gold in all types of its deposits. There is no need to belabour the details of this marked association, but a few points should be stressed. The principal hypogene iron minerals accompanying gold in its deposits in their order of abundance are pyrite, arsenopyrite, pyrrhotite, ankerite, siderite, chalcopyrite, bornite, magnetite and ilmenite. To these can be added a veritable host of iron silicates, particularly biotite, chlorite, epidote, hornblende, amphibole, etc. In nearly all deposits gold, gold tellurides and aurostibite are later than these iron minerals in the paragenetic sequence.

The behaviour of iron during the wall-rock alteration processes that attend gold deposition is discussed in a later section of this chapter. As McKinstry (1957) has pointed out, much, if not all of the iron, bound in sulphides and carbonates in gold deposits, is derived locally from the host rocks. Some of the iron delivered to solutions through leaching of iron-bearing rocks may migrate considerable distances as witnessed by the fact that iron sulphides and other iron minerals are common constituents of the precipitates found at the orifices of auriferous hot springs.

The fate of iron during the oxidation of sulphides, carbonates, silicates, etc. in gold deposits is a most complex problem that can be treated here only in a general outline. The oxidation states of interest are Fe(II) and Fe(III), the former occurring in sulphides, carbonates and most silicates, and the latter in some silicates, in magnetite and in hematite, limonite, a variety of basic salts and a host of other supergene minerals.

When iron sulphides, carbonates, silicates, etc. are subjected to the action of meteoric waters, iron is initially released as the aquo ferrous ion $[Fe(H_2O)_6]^{2+}$ which migrates mainly as the sulphate, hydrogen carbonate complex, $[Fe(HCO_3)_2]$, chloride or as a number of organic complexes usually referred to collectively as humates for lack of exact characterization. The oxidation potentials are such that ferrous ion can be converted to ferric ion in acid solution by the oxygen of the air. The conversion is slow or inhibited in very acid solutions ($pH<2$). Change in pH through the neutral point into the alkaline range results in the precipitation of ferrous hydroxide, which is readily oxidized to the brown hydrous ferric oxide in air. Ferric iron in the absence of complexing agents is present as the aquo ion $[Fe(H_2O)_6]^{3+}$. The principal feature of this ion is its marked tendency to hydrolyze, to enter a variety of basic salts, and to form complexes. To prevent hydrolysis the pH of the solutions must be acid ($pH \sim 0$); with increasing pH (circa 2) extensive hydrolysis sets in, and at a pH of 2 to 3 condensed polynuclear species appear, and ultimately gelatinous $Fe_2O_3 \cdot nH_2O$ is precipitated. But this is a gross oversimplification of what actually takes place in nature when SO_4^{2-}, PO_4^{3-} and AsO_4^{3-} are found in solution in the presence of ferric ion, K, Na, Pb, Zn, Ag and other metal ions. In this situation as the pH increases a veritable host of basic salts may appear such as jarosite, pharmacosiderite, cacoxenite, beudantite, etc. These, however, tend to be unstable to extensive leaching by meteoric waters, and some of their constituents pass into solution leaving ferric iron as limonite. Under other conditions a number of normal ferric salts may form, some of which are relatively insoluble, examples being scorodite and a number of phosphates.

Oxidation of gold ores containing iron-bearing sulphides and other minerals invariably leaves a residue of limonite (gossan) in which scorodite and/or the basic salts may be developed. Pyrite, the iron carbonates and other iron-bearing minerals first yield ferrous iron to the solutions that may oxidize to ferric iron, and this in turn undergoes hydrolysis with the precipitation of hydrous iron oxides (limonite) or basic salts, or minerals such as scorodite, depending on the pH and the availability of various cations and anions. Alternatively, ferrous hydroxide may be precipitated, then readily oxidized, yielding hydrous ferric oxides (limonite). Iron passing downward in the deposits may be precipitated in the zone of reduction mainly as marcasite or pyrite. Gold may be enriched in the limonitic parts of gossans, in the zones of scorodite and/or basic salts and in the zone of reduction. More is said about these enrichments in Chapter IV.

The presence of supergene iron minerals, particularly the hydrous oxides, and soluble species of iron in the oxidized zones of deposits is important in the migration of gold in two respects. Firstly, the oxides tend to adsorb and/or coprecipitate gold from solution (*see also* the discussion in the section on gold in other minerals in Chapter II). Secondly ferric ion has sufficient potential to oxidize gold and render it mobile, provided complexing species such as chloride are present. On the other hand, in the absence of complexing species, ferrous iron reduces soluble gold to the metal. The interplay of all these reactions is complex and is discussed in more detail in Chapter IV.

It should also be mentioned that the presence of iron in natural waters, especially surface waters, seems in places to be responsible for the relatively high migration capacity of gold. The precise reasons for this behaviour are unknown – one factor seems to be that gold is adsorbed to the various hydrous iron oxide colloids and is carried in this manner; another may be adsorption and/or incorporation of gold in suspended humic iron colloids.

Iron is a constituent of a great many light and heavy minerals that accompany gold in recent and ancient eluvial and alluvial placers. The most abundant heavy iron mineral in gold placers is magnetite; ilmenite is common in some placers, and pyrite is ubiquitous but is generally present in small amounts. In some placers iron is a major constituent of limonite-wad cemented gravels, sand and silt. These materials, as well as gravels and sands that contain an abundance of limonite-wad particles, are frequently enriched in gold, probably largely as a result of adsorption processes.

Cobalt accompanies gold in most types of deposits but generally only in trace or minor amounts. In most deposits cobalt is a constituent of pyrite, pyrrhotite, pentlandite, magnetite, chalcopyrite, arsenopyrite and tennantite-tetrahedrite in which it is usually present in amounts up to 100 ppm. Smaltite, cobaltite, safflorite, linnaeite and other cobalt sulphides and arsenides are relatively rare in gold deposits, but a few gold-quartz veins and skarn deposits containing some of these minerals are known in the Middleburg district, northern Transvaal; in the French Mine, Hedley, British Columbia; in the Windpass Mine, British Columbia (Uglow and Osborn, 1926); in some of the gold-quartz deposits at Ramore, Ontario; and elsewhere. The Ni-Co-Fe arsenide veins (Cobalt, Ontario type) with abundant native silver generally contain only traces of gold (<10–100 ppb).

The cobalt content of the Rand ores appears to range from 37 to 102 ppm (Table 35). In these ores gold is often associated with cobaltite, gersdorffite, skutterudite and linnaeite which it either replaces or with which it is intergrown (Pelletier, 1940; Liebenberg, 1955; Feather and Koen, 1975). These sulphides are mainly secondary (remobilized) and late in the paragenetic sequence according to Liebenberg. According to Feather and Koen (1975) both cobaltite and gersdorffite occur in two distinct generations, one detrital and the other authigenic. The detrital variety of these two solid solution minerals is essentially cobaltite whereas the authigenic variety associated with the main period of gold remobilization is gersdorffite. Feather and Koen also showed that the order-disorder relation of the authigenic cobaltite-gersdorffite and the mixed crystal compositions in the system FeAsS-CoAsS-NiAsS can be used as a sensitive indicator to estimate the temperature at which the reefs have been elevated during the period of metamorphic activity that remobilized the gold and precipitated the associated secondary minerals. Their results indicated a temperature in the region of 600°C for most reefs.

The disseminated gold deposits of Nevada, U.S.A. (Carlin, Cortez, Getchell and Gold Acres) are not characterized by marked enrichments of cobalt (Wells *et al.*, 1969; Roberts *et al.*, 1971); in fact some of the highly mineralized limestones at Cortez are depleted in the element. At Carlin, however, Radtke *et al.* (1972*a*) found a slight enrichment (<2–50 ppm) in the primary ore as compared with the fresh carbonate rocks (<2–7 ppm). In the 'oxidized ores' they found from <2 to 150 ppm Co.

Weissberg (1969) makes no mention of the presence of cobalt in the gold-rich silica-sulphide precipitates from thermal waters in the Taupo volcanic zone, New Zealand.

The fate of cobalt during oxidation processes has been extensively studied by the writer and his colleagues (Boyle and Dass, 1971*a*). Only some of the details need be reiterated here.

The oxidation states of cobalt of interest are (II) (cobaltous) and (III) (cobaltic). The former occurs in a variety of sulphides and arsenides (e.g., linnaeite, Co_3S_4; smaltite, $CoAs_2$), in the carbonate, cobaltocalcite $CoCO_3$, and the arsenate, erythrite, $(Co,Ni)_3(AsO_4)_2{\cdot}8H_2O$. The (III) state occurs in various ill-defined cobaltic oxides and hydrous oxides of which stainierite (heterogenite), CoO(OH), is fairly well differentiated.

The aqueous chemistry of cobalt (II) is dominated by the formation of the aquo ion $[Co(H_2O)_6]^{2+}$ and its many ionic complexes. Unlike ferrous ion, the cobaltous ion is relatively stable to oxidation. Addition of a base yields $Co(OH)_2$ whose pH of precipitation is approximately 6.8. The hydroxide exhibits colloidal and gel properties. It is amphoteric yielding the cobaltous ion in acid solutions and the ion $[Co(OH)_4]^{2-}$ in concentrated alkali. Hydrated cobaltous salts with the most common anions are known, and many of these are relatively soluble, particularly the sulphate, chloride and nitrate. The carbonate is insoluble but exhibits increased solubility with increased amounts of CO_2 in solution, probably because of the formation of the hydrogen carbonate complex $Co(HCO_3)_2$. Arsenate ions precipitate the insoluble arsenate, erythrite, and sulphide ions or H_2S precipitate an insoluble sulphide from cobaltous solutions. Trivalent cobalt is unstable as the free aquo ion $[Co(H_2O)_6]^{3+}$, since it is reduced by water to the cobaltous state. The reduction is less marked in basic than in acid media. In its many complexed forms, however, cobaltic ion is relatively stable to reduction. Hydrolysis of cobaltic solutions leads to the precipitation of indefinite hydrous oxides; one species CoO(OH) appears to be a definite compound. Cobaltic oxide is also formed by the oxidation of freshly precipitated $Co(OH)_2$. The hydrous cobaltic and cobaltous-cobaltic oxides have colloidal and gel properties.

The oxidation of cobalt sulphides, arsenides and other primary cobaltian minerals yields cobaltous ions to meteoric waters, and these migrate mainly as the hydrogen carbonate, sulphate, chloride or as soluble organic complexes. Where arsenic is abundant in the environment complex hydrolytic reactions lead to the precipitation of erythrite and other arsenates, under midly acid, neutral or slightly basic conditions. Locally, where evaporation of the solutions takes place hydrated sulphates such as bieberite, moorhouseite and aplowite are precipitated. These are soluble and hence temporal. With increasing alkalinity or by reaction with alkali bicarbonate solutions cobaltocalcite may be precipitated although this secondary mineral is rare. The indefinite hydrous cobaltic and cobaltous oxides appear to originate by oxidation of $Co(OH)_2$, which is precipitated or coprecipitated with ferrous hydroxide, alumina and other oxides and hydroxides in weakly acid, neutral and basic solutions. Where hydrous manganese dioxide is precipitated cobaltian wad (asbolite) may be abundant. Strong adsorption of cobalt to hydrous ferric oxides (limonite) is also a feature of some oxidized gold-bearing sulphide-arsenide deposits containing cobalt. Despite all of these various mechanisms of precipitation cobalt retains a relatively high mobility, and much of the element is leached from the deposits. Some cobalt passing downward into the zones of reduction is precipitated in the secondary sulphides.

Gold and cobalt may be concentrated together under certain conditions in gossans, particularly in those containing abundant wad or limonite-wad aggregates. More generally, however, gold is enriched in the gossans, and cobalt is extensively leached as at the Lynx prospect, Yellowknife and elsewhere (Table 41). Gold and cobalt may again join company and be moderately enriched in secondary sulphides in the zone of reduction.

Cobalt is a trace to minor constituent of a number of heavy minerals that accompany gold in recent and fossil eluvial and alluvial placers. Chief among these are pyrite, arsenopyrite, pyrrhotite, magnetite and ilmenite. In some placers cobalt is a minor or trace constituent of limonite-wad cemented gravels, sand and silt. These materials, as well as gravels and sands that contain an abundance of limonite-wad particles, are frequently enriched in gold, probably largely as a result of adsorption processes.

Table 41. Cobalt, nickel and precious metal contents of some auriferous ores and gossans

	Co	Ni	Pt	Pd	Au
			(ppm)		
Lynx prospect, Yellowknife					
Quartz-pyrite ore	46	5	<0.02	0.03	0.045
Gossan	2	<2	<0.02	0.03	0.175
Vermilion Mine, Sudbury					
Chalcopyrite-pyrite ore	100	8850	35	108	1.9
Gossan	7	312	130	100	0.545

Nickel accompanies gold in most types of deposits but generally only in trace or minor amounts. In most deposits nickel is a constituent of pyrite, pyrrhotite, chalcopyrite, arsenopyrite, tennantite-tetrahedrite and other sulphosalts in which it is usually present in amounts up to 100 ppm. Pentlandite, maucherite, millerite, niccolite, breithauptite, chloanthite and other nickel sulphides and arsenides are relatively rare in gold deposits, but a few gold-quartz veins, polymetallic deposits and skarn deposits containing these minerals are known in British Columbia and Ontario; at Alistos, Mexico (Krieger and Hagner, 1943); at the Mackinaw Mine, Washington, U.S.A. (Milton and Milton, 1958); and elsewhere. A number of these deposits are in or near peridotites, serpentinites and other basic and ultrabasic rocks. An excellent example of this is found at the old Cuniptau Mine, near Goward, Ontario where the auriferous ore contained pyrrhotite, chalcopyrite, pentlandite, violarite, pyrite, sphalerite, magnetite, chromite, calaverite and sperrylite (Sandefur, 1942).

The Ni-Co-Fe arsenide veins (Cobalt, Ontario type) with abundant native silver generally contain only traces of gold (<10–100 ppb). The Ni-Co-Cu ores of the Sudbury type are, likewise, low in gold. Hawley (1962) mentions the extreme rarity of the metal in the Sudbury ores, and found native gold in only one of the deposits. According to our calculations and analyses the Sudbury ores as mined contain on the average about 0.08 ppm Au and 2.5 ppm Ag. The ores of the Norilsk region in U.S.S.R. have about the same content of gold and silver. In the Rand ores gold is associated with pentlandite and gersdorffite in some reefs and with pyrrhotite, arsenopyrite and various cobalt sulphides and arsenides, which contain some nickel in other reefs. These sulphides and arsenides are mainly secondary (remobilized) and late in the paragenetic sequence according to Liebenberg (1955) and Feather and Koen (1975). (*See also* the discussion under cobalt.) The nickel content of the Rand ores ranges from 106 to 158 ppm (Table 35). The disseminated gold deposit at Cortez, Nevada, exhibits a slight enrichment of nickel in the highly mineralized limestone according to the data provided by Wells *et al.* (1969) and Radtke *et al.* (1972*a*). The latter investigators give the range of nickel content as 7 to 200 ppm in the Carlin ores as compared to 1 to 30 ppm in the carbonate host rocks; in the 'oxidized ores' they found 1.5 to 500 ppm Ni indicating some enrichment during oxidation. Weissberg (1969) makes no mention of the presence of nickel in the gold-rich silica-sulphide precipitates from thermal waters in the Taupo volcanic zone, New Zealand.

The fate of nickel during oxidation processes has been extensively studied by the writer and his colleagues (Boyle and Dass, 1971*a*). Only some of the details of interest as they pertain to gold deposits need be mentioned here.

Nickel can assume five oxidation states, (0), (I), (II), (III) and (IV). Only the divalent state is of importance in nature, occurring in various sulphides, arsenides, the oxide, a few hydroxides, a complex carbonate, several arsenates and a number of silicates. The (0) oxidation state is represented by certain natural nickel-iron alloys.

The aqueous chemistry of nickel (II) is dominated by the formation of the aquo ion $[Ni(H_2O_6]^{2+}$ and its many ionic complexes. Unlike ferrous ion and cobaltous ion, the nickelous ion is quite stable to oxidation in natural environments. Addition of a base yields $Ni(OH)_2$, whose pH of precipitation is approximately 6.6. This hydroxide is basic with no amphoteric properties. It exhibits colloidal and gel properties, dissolves readily in acids and also in ammonia with the formation of the complex ion $[Ni(NH_3)_6]^{2+}$. Nickel hydroxides and compound nickel hydroxides are known in nature. Most occur with garnierite. Nickelous oxide, NiO, is formed when water is split out of the hydroxide. It is basic in character, insoluble in water and occurs as the mineral bunsenite.

Nickelous salts with most of the common anions are known, and many of these are relatively soluble, particularly the sulphate, chloride and nitrate. The carbonate, $NiCO_3{\cdot}6H_2O$, which precipitates when alkali bicarbonates are added to nickelous solutions, the phosphate, $Ni_3(PO_4)_2{\cdot}8H_2O$, and the arsenate, $Ni_3(AsO_4)_2{\cdot}8H_2O$, which precipitate under mildly acid, neutral and mildly basic conditions are insoluble. The carbonate, phosphate and arsenate are, however, soluble in strongly acid solutions and sparingly soluble in solutions with excess CO_2, the latter probably because of the formation of hydrogen carbonate complexes. The normal carbonate has not been recognized in nature, but a hydrated basic carbonate, zaratite, $Ni_3(CO_3)(OH)_4{\cdot}4H_2O$, is known. The basic carbonate results probably because carbonate solutions containing nickel hydrolyze readily and precipitate basic carbonates rather than the normal carbonate. The phosphate is not known in nature, but a number of arsenates, of which annabergite is the most common, occur frequently in the oxidized zones of nickel deposits. Two common hydrated sulphates (retgersite and morenosite) are known and a number of silicates of variable composition, collectively known as garnierite occur as oxidation products of nickel minerals.

Addition of H_2S or sulphide ion to nickelous solutions yields NiS, which is extremely insoluble in water.

All oxidation states of nickel are characterized by their ability to form complex ions and compounds of great variety and multiplicity. Some of these such as $[Ni(NH_3)_6]^{2+}$ and various chelates may occur in nature, although they have not yet been characterized.

The oxidation of gold deposits containing nickel minerals yields solutions containing the element mainly as the sulphate, hydrogen carbonate, or as soluble organic complexes. The resultant secondary minerals depend on the type of deposit. In deposits containing nickeliferous sulphides and arsenides, the oxide, basic carbonate, arsenates (mainly annabergite) and hydrated sulphates may appear in small amounts, but the usual product is a nickelian limonite and wad. Where hydrolysis and other reactions take place in a milieu dominated by dissolved silica, silicates or silica gel, the garnierite group of hydroxyl-bearing nickel silicates, may be developed. During the oxidation of most types of gold deposits containing nickel much of the element is removed; some, usually only small amounts, is precipitated in the zone of reduction in the secondary (supergene) sulphides. Comparing cobalt and nickel it is my experience that the former is the more mobile of the two elements, although in some types of deposits both elements exhibit a high mobility, especially in acid solutions (Boyle and Dass, 1971*a*).

Gold and nickel may be concentrated together under certain conditions in gossans as indicated by the data of Radtke *et al.* (1972*a*) mentioned above. The enrichment of nickel is usually greatest in those zones containing an abundance of limonite and wad. More generally, however, gold is enriched in the gossans, and nickel is leached, as shown by the analyses of primary ore and gossan at the Lynx prospect, Yellowknife (Table 41). Gold and nickel may again join company and be moderately enriched in secondary sulphides in the zone of reduction of certain deposits.

Nickel is a trace to minor constituent of a number of heavy minerals that accompany gold in recent and ancient eluvial and alluvial placers. Chief among these are pyrite, arsenopyrite, pyrrhotite, magnetite and ilmenite. In some placers pellets of the nickel-iron alloys may occur especially in the vicinity of serpentinites. Nickel is a minor or trace constituent of limonite-wad cemented gravels, sand and silt in some places. These materials, as well as gravels and sands that contain an abundance of limonite-wad particles, are frequently enriched in gold, probably largely as a result of adsorption processes.

The *platinum* family of elements comprises the light platinoids, ruthenium, rhodium and palladium, and the heavy platinoids, osmium, iridium and platinum. These elements form a number of alloys in nature and are also combined with As, S, Se, Te, Sb, Bi or Sn to form numerous minerals of which the most common appears to be sperrylite, $PtAs_2$. For general reviews of the occurrence of platinoids with gold see the works by Kemp (1902), Mertie (1969) and Cabri (1972).

Some varieties of native gold contain the platinoids, mainly in placers but also in the hypogene deposits and their oxidized zones. The amount and type of the various platinoids in native gold appears to be highly variable, and no generalizations can be made except to say that perhaps palladium is the most common of the platinoids in gold (*see also* the section on gold minerals in Chapter II). Excepting native gold there is no exact knowledge how the platinoids occur in gold-bearing deposits that are enriched in these metals. The elements seem to have an affinity for copper minerals and are frequently concentrated in chalcopyrite, bornite, tetrahedrite-tennantite and chalcocite. Pyrrhotite, pentlandite and various arsenides also contain the elements, especially in the massive Ni-Cu deposits. Pyrite is usually very low in the elements. The Geological Survey analyses, and those of Hawley (1962) and Keays and Crocket (1970), indicate that most sulphides, arsenides, sulphosalts, etc. rarely contain more than a few parts per million, and most have less than 1 ppm, total platinoids. Some of these platinoids may be present in the various sulphide, arsenide, etc. minerals as discrete platinoid minerals, although in many cases they probably substitute for Cu, Ni or Fe. Certainly in some deposits (e.g., Vermilion Mine, Sudbury, Ontario) they are present as minute grains of sperrylite, mitchenerite or froodite. The platinoids have a marked affinity for tellurium and at least three platinoid tellurides are known in nature. This suggests that the platinoids may be present in the gold and other tellurides in some gold deposits.

Concentrations of platinum metals are relatively rare in

gold deposits, and the reverse is generally true that gold is generally present only as a trace constituent in typical platinum metal deposits. Gold-quartz veins that we have investigated contain only traces of Pt (<0.005 ppm) and Pd (<0.01–0.05 ppm) as do also the various other types of auriferous deposits that have been studied at the Geological Survey. Mercer and Crockett (1972) found average gold contents of approximately 1.5 ppm and 0.4 ppm respectively for the orebody and adjacent metasediments in the Heath-Steele B-1 deposit, Bathurst-Newcastle district, New Brunswick. The corresponding palladium contents were 0.0014 and 0.0026 ppm.

There are reports that certain quartz veins in Columbia, Finland, U.S.S.R., South Africa, New Zealand and Ontario contain platinoids, but their gold content is not mentioned. Some of those in New Zealand are described by Bell (1906). Those in the Waterberg district of South Africa are described by Wagner (1926). Certain contact metamorphic and other types of unusual skarn deposits are enriched in platinoids and often contain some gold. Examples are known in Brazil and South Africa (Lindgren, 1933, p. 779); also in U.S.S.R. in central Asia where Timofeeva (1972) described auriferous skarns with chalcopyrite, pyrrhotite, arsenopyrite, molybdenite, joseite, tiemannite and Pt-Pd minerals, e.g., niggliite, cupropalladium and sperrylite. Most of the deposits are richer in palladium than in platinum. A number of gold-bearing copper deposits have unusual concentrations of platinoids in places. The Cuniptau Mine at Goward, Ontario (Sandefur, 1942) developed Cu-Ni-Pt-Au ore in a shear zone in Archean serpentinite. The ore consisted mainly of pyrrhotite, chalcopyrite, pentlandite, violarite, pyrite, sphalerite, tetrahedrite, magnetite, calaverite, sperrylite and chromite. The Salt Chuck Mine on Prince Edward Island, Alaska is another example (Mertie, 1921). This mine operated on an ore containing disseminated bornite, chalcocite, covellite and epidote in a pyroxenite. The concentrates averaged about 40 per cent Cu, 1.2 oz/ton Au, 5.3 oz/ton Ag and 3.1 oz/ton Pt and Pd. The ore averaged 0.063 oz/ton Au and 0.253 oz/ton Pt and Pd. Yet another example is cited by Emmons (1903) and Knopf (1915) in the Yellow Pine district, southern Nevada. In this deposit abundant plumbojarosite (a supergene mineral) carried on the average in ounces per ton: Au – 3.46; Ag – 6.4; platinum – 0.70; and palladium – 3.38. The primary ore was siliceous and composed of lead, copper and iron sulphides.

Most epigenetic auriferous vein, replacement, polymetallic and skarn deposits, enriched in platinoids, occur in or near peridotites, pyroxenites, serpentines and other types of ultrabasic rocks, judging from the geological descriptions, some of which are rather inadequate on which to base conclusions. However, it does appear as though these ultrabasic rocks have in some manner or other provided the platinoids; the concentration into the veins, etc. appears to have been largely by a secretion mechanism long after the consolidation of the ultrabasic rocks.

Small amounts of iridosmine containing mainly osmium and iridium and minute amounts of platinum, ruthenium, rhodium and gold are won from the gold ores of the Witwatersrand quartz-pebble conglomerate (Mertie, 1969). A total of about 6000 oz of platinoids is said to be recovered annually from the ores. Geological Survey analyses of the platinoid content of the Rand ores and other quartz pebble conglomerates are given in Table 35. Cousins (1973) gives the value of 0.0035 ppm for the average recovery content of platinoids in the reefs of the Rand and an average Au-platinoid ratio of 2000. The Au-platinoid ratio of the West Rand ranges from 100 to 8000 (Stumpfl, 1974). According to Liebenberg (1955) iridosmine and other platinoids are relatively common in the amalgam barrel residues, but grains of the mineral are rarely seen in the matrix of the bankets. Some of the grains of iridosmine are intergrown with small amounts of native platinum, and the grains of iridosmine are also sometimes partly replaced by gold. Other grains of native platinum with small amounts of Pd, Rh, Ru, Os and Ir are occasionally found in the matrix of the conglomerate. Liebenberg, (1955) and others consider that the platinoid minerals are detrital in origin, but those in the Carbon Leader would seem to have been introduced in a manner other than by sedimentation. (*See* the discussion on the origin of these deposits in the last part of this chapter.) In a recent contribution Feather and Koen (1975) list a great number of platinoid minerals for the Rand of which the most abundant are osmium, iridosmine, osmiridium and iridium. They consider most of these to be detrital, but some, especially the platinoid mineral coatings such as sperrylite and hollingworthite on platinoid alloy grains, are thought to be of secondary derivation. (*See further* the section on the association of platinoid minerals with gold in the next section.)

The gold content of platinoid-rich layers of ultrabasic rocks such as those associated with the Bushveld and other differentiated basic bodies is very low generally in the range of 0.05 to 0.5 ppm according to our analyses. This can be compared with the platinoid-bearing ores of the Merensky Reef in the Bushveld complex, which average about 8 ppm (0.25 oz/ton) total platinoids. The Sudbury and similar deposits associated with ultrabasic and basic rocks, enriched in Ni and Cu, carry very low gold contents, about an average of 0.08 ppm. The total platinoid content of the Sudbury ores appears to range from 0.65 to 9 ppm (0.019–0.26 oz/ton), but the actual tenor varies greatly from place to place and with type of ore (Hawley, 1962). Judging from production figures the respective percentages of platinoids present are Pt, 45 per cent; Pd, 40 per cent, and the remainder of platinoids, 15 per cent. Some deposits seem to depart from these abundances, palladium being more abundant than platinum. Keays and Crocket (1970) found positive correlations between Os-Ru, Os-Ir and Ru-Ir and weak negative correlations between Pd-Ir and Pd-Os, but no correlation between gold and any of the platinoids in some of the minerals and orebodies at Sudbury.

The fate of the platinum metals during the oxidation of deposits in which they are enriched is only imperfectly known. The writer's research on these metals in the surficial cycle is incomplete, and hence some of the statements made below are tentative.

The native platinoid metals, their alloys and native gold containing the platinoid metals do not appear to dissolve to any great extent in most supergene solutions derived from the oxidation of pyrite, pyrrhotite, chalcopyrite and other sulphides or arsenides. Exceptions to this may occur where complexing agents such as thiosulphate, cyanide, chloride and certain organic compounds are present, in which case the gold, silver and perhaps some of platinoids may be dissolved and leached from the deposits. Sperrylite, essentially $PtAs_2$, seems to be relatively resistant to oxidation, occurring in a number of the gossans at Sudbury and elsewhere. It also occurs in some placers. Laurite, $(Os,Ru)S_2$, also appears to resist oxidation in some deposits. The fate of other platinoid arsenides, sulphides, selenides, tellurides, antimonides, bismuthides and stannides during oxidation processes is unknown. Probably some are oxidized, and their constituent platinoids are released to the supergene solutions. The platinoids in substitutional sites in the various sulphides, arsenides, etc. are probably released to the supergene solutions, perhaps in some cases as soluble sulphates (e.g., platinic sulphate) and in others as a variety of soluble complex ions which are especially characteristic of the platinum metals. Palladium seems to be more mobile than platinum during oxidation processes since the Pd-Pt ratio appears to decrease in oxidized zones and gossans compared with the primary ore in some deposits (Table 41). Ruthenium and osmium, likewise, appear to possess considerable mobility during oxidation, probably because they form relatively soluble complex oxy anions. Numerous other factors appear to affect the mobility of the platinoids in a differential way. Thus Pt and Pd in solution appear to be relatively easily reduced to the metallic state by ferrous sulphate (similar to Au and Ag) whereas Ru, Os, Rh and Ir seem to be little affected. Soluble arsenites and arsenates precipitate complex compounds, especially from solutions of Pt and probably Pd. H_2S appears to precipitate all of the platinoids from their solutions as sulphides.

Because platinoids in solution are readily reduced to the metallic state by a variety of reductants they do not generally migrate far in gossans and oxidized zones. The exact nature of the reductants is unknown, but organic matter, ferrous sulphate and residual sulphides appear to be involved. The dissolved platinoids probably follow four main courses. The largest amounts appear to be precipitated on the residual particles of native platinoids in the gossans and in the nearby soils and weathered residuum. Small amounts are either adsorbed and/or coprecipitated by limonite, wad and other colloidal materials or precipitated as an exceedingly fine dust, perhaps composed of various platinoid minerals, with minerals such as limonite, wad, malachite and plumbojarosite. Certain supergene minerals may accommodate the platinoids in their structures although proof of this feature is lacking as yet. The occurrence of high platinoid assays in oxidized zones of some deposits containing plumbojarosite and other basic sulphates suggests that these minerals may carry some of the platinum metals as lattice constituents. The fact that solutions of Ru, Rh, Pd, and Pt sulphates can precipitate basic and complex oxy-sulphates tends to support this contention. Finally, small amounts of the dissolved platinoids are transported downward in oxidized zones and are concentrated at lower levels in secondary (supergene) sulphides, particularly chalcocite and covellite. The nature of the platinoids associated with these sulphides is unknown.

The resultant of all these processes is that the platinoids tend to be enriched in most gossans and oxidized zones in the form of the native metals and alloys, or as sperrylite, and probably other platinoid minerals. Assays of some of the gossans at Sudbury recorded in the early literature show up to 10 oz/ton of platinoids whereas the general average of fresh sulphide ore in this area is about 0.1 oz/ton of platinoids. Gold, and frequently silver, exhibit a parallel enrichment with the platinoids in some gossans and oxidized zones, particularly those rich in jarositic minerals, developed on Ni-Cu and other platinoid-bearing deposits. These two precious metals are, likewise, enriched with the platinoids in supergene sulphide zones, especially those rich in copper where chalcocite and covellite are developed. Part of the enrichment of gold with platinoids in supergene environments probably involves the potentials of free metals; thus platinum metal will reduce and/or coprecipitate gold from solution, as can be seen from their position in the electrochemical series.

Gold is frequently accompanied in placers by platinoid minerals, particularly the alloys, but also by sperrylite and occasionally other platinoid arsenides, bismuthides, etc. The platinoids generally appear with gold in placers principally in terranes where serpentinites and other ultrabasic rocks or platinoid deposits of the Bushveld or Sudbury type are present.

> Le quartz est le plus fidèle compagnon de l'or
> –Cumenge et Robellaz (1898)

Mineral associates of gold

Mention is made in the previous section of the great variety of minerals accompanying gold in its deposits. Here we shall consider the most common mineral associates of gold and attempt to show how they differ or remain constant in gold deposits of all types and different geological ages. The paragenetic relationships of the various gold minerals with respect to other metallic and gangue minerals are also discussed. Particularly helpful in the compilation of this section have been the condensed data given by Lincoln (1911*b*), Schwartz (1944), Petrovskaya (1955, 1971, 1973, 1974), Feather and Koen (1975), and Anhaeusser (1976).

A number of native elements and alloylike compounds accompany gold in its deposits. Most of these minerals are present only in small amounts, although a few may be relatively abundant in some deposits.

Graphite, amorphous carbon and *carbonaceous matter,* commonly referred to as 'carbon', are frequently encountered in gold deposits of practically all types. The presence of graphite and amorphous carbon in black and grey quartz and in carbonates is discussed in detail in the sections describing quartz. The carbon minerals also occur in seams and slips and irregular masses along faults and in brecciated parts of the orebodies. Frequently these minerals are greatly enriched in native gold and other gold minerals. It is common for instance to find native gold smeared along graphitic slips in many orebodies. Mention should also be made here of the common occurrence of rich shoots of gold where veins and other deposits intersect graphitic beds – the so-called black leaders or indicators of the old gold miners.

'Carbon' has been of considerable interest to gold miners and metallurgists for many years, principally because it causes severe difficulties in winning the gold from the ores as a result of its precipitating effect on gold in the cyanide circuits. Innumerable descriptions of carbonaceous material in gold deposits appear in the literature. Here we may mention the papers by Bell (1924) on the carbonaceous matter in the Porcupine (Timmins) mines of Ontario; Van Aubel (1936) on the graphitic matter in the auriferous veins of Kivu, Zaire; Boyle (1953) on the carbon and graphite in the quartz of the veins at Yellowknife; Liebenberg (1955) and Feather and Koen (1975) on the 'carbon' of the Witwatersrand, South Africa; Baskakova (1970) on the organic matter in the gold-quartz veins of the Kyzyl-Kum desert, U.S.S.R.; Roberts *et al.* (1971) and Wells (1971) on the carbonaceous matter in the Carlin disseminated deposit, Nevada; Scheiner *et al.* (1971) on various carbonaceous gold ores in United States; and Saxby and Stephens (1973) on carbonaceous matter in the sulphide ores from Mount Isa and McArthur River in Australia. 'Carbon' or graphite occurs in auriferous deposits in all types of rocks including granites. It seems to be most abundant in deposits of Precambrian age diminishing through time and being rare in Tertiary deposits. In the younger deposits, however, bituminous substances may be relatively abundant in some gold belts.

The nature of 'carbon' in gold quartz deposits, mainly of Precambrian age, has been studied in many places by the writer. To summarize, it can be said that much of the 'carbon' is amorphous to X-rays and gives no patterns. Chemically this type of 'carbon' is essentially pure carbon with little if any other constituents such as H,O,P, etc. In some veins, however, the 'carbon' is crystallochemically organized and gives fair to good X-ray patterns for graphite. This 'carbon' or graphite is generally present in zones of high grade metamorphism (amphibolite facies and higher). Other types of 'carbon' are also present in gold deposits. Baskakova (1970) recognized bituminous substances and humic acids in the gold-quartz veins of the Kyzl-Kum desert. The bituminous substances on analysis proved to be hydrocarbons. The 'carbon' in the Rand ores is of two types Liebenberg (1955) – a radioactive variety or thucholite and a nonradioactive variety probably best described as anthraxolite.

Thucholite, carburan and *anthraxolite* are not commonly found in skarn deposits, but the first mineraloid has been recognized in the skarn-type Richardson deposit at Eldorado, Ontario (Boyle and Steacy, 1973). There, the mineraloid contains micro-grains of gold and uraninite distributed throughout the groundmass in no particular pattern. In gold-quartz-sulphide veins thucholite, carburan and anthraxolite are relatively rare, but examples are known from the auriferous Boliden ores in Sweden and the auriferous-uraniferous Box Mine, Nicholson and other ores of the Goldfields (Uranium City) district of Saskatchewan. In the Goldfields occurrences the gold is younger than the thucholite. In the quartz-pebble conglomerate ores of the Rand thucholite (or uraniferous carbon) is abundant and has received considerable study by Davidson and Bowie (1951), Liebenberg (1955), and others. (*See* the section on the Witwatersrand in this chapter.) The thucholite is dull-black in appearance and confined mainly to the conglomerate bands and pyritic quartzites; in places seams and bands of thucholite often mark the lower bedding planes of the conglomerates. The thucholite occurs mainly as small warty granules with a roundish or oval form mainly in the phyllosilicate matrix; irregular patches and stringers are found in places, but they are rare; columnar and massif thucholite occur as streaks or 'pencil lines' to relatively thick layers as in the 'Carbon Leader'. Under the microscope the massive thucholite is composed of aggregates of granules in places separated by narrow rims of gold and gold-bearing sulphides. The columnar variety of thucholite is generally fairly homogeneous; in places gold and sulphides occupy partings between the parallel adjacent columns. The granular type of thucholite has a tendency to be round and oval, and granules often merge into each other to form stringers and irregular masses. Most granules have an authigenic appearance and occupy interstitial spaces in the matrix of the conglomerates. All of the varieties of thucholite contain small grains and minute inclusions of uraninite, up to 150 in a single granule 0.5 mm in diameter. This phenomenon is interpreted by Liebenberg as due to replacement of large grains of uraninite by the hydrocarbon. Some of the specks and veinlets of uraninite in the thucholite are interpreted as secondary. Gold and auriferous sulphides are closely associated with thucholite, occurring as minute inclusions and veinlets in the mineraloid, in bands around the granules in places and in the boundaries between adjacent granules.

Carbonaceous substances other than thucholite in the Rand ores consist mainly of brownish hydrocarbons that contain no radioactive inclusions. Dull, black, nonradioactive carbonaceous matter mixed with chlorite is also common in the Black Reef. Gold is not associated with this hydrocarbon; according to Liebenberg (1955) this carbonaceous matter was introduced at a late stage.

The paragenetic relationships of gold, thucholite, carburan and anthraxolite in auriferous deposits are of interest. In some cases small blebs of gold are scattered at random through the hydrocarbons and seem to be of exsolution origin; in others the gold is late and occupies small cracks and fissures (syneresis cracks) in the hydrocarbons. The late gold may represent exsolution from the hydrocarbons, or it may have been introduced from other parts of the deposits.

The origin of the 'carbon' in gold deposits has long been a subject of speculation. In some deposits, particularly those in carbonaceous and graphitic shales or schists some of the carbon is a residue after replacement as shown by Van Aubel (1936) and others. In other deposits, especially those in volcanic, granitic bodies and conglomerates the carbon has obviously been introduced. Its source is a problem, however, and there has been much speculation on the subject. Bell (1924) hypothesized that:

> The carbon may have been the product of the decomposition of carbonates into carbon dioxide, carbon monoxide and, finally, elementary carbon. No less may it be the product of carbon, precipitated from soluble carbides, or even the result of decomposition of hydrocarbons. Owing to the common close connection of carbon and carbonates in the vein material, one would be tempted to favour the derivation of the former from the latter, were it not for a realization of the difficulties presented by the dissociation of carbon monoxide into carbon and oxygen in ordinary atmospheric temperature and pressure. At the same time it is recognized that these difficulties may be far from insurmountable under geological conditions.

That this is the case has been recently shown by Salotti *et al.* (1971) who discussed the origin of graphite in deposits. They envisage a reaction of the type: carbon mineral + hydrogen → (methane to butane) + water + metal oxide or hydroxide. A subsequent reaction involving the pyrolitic dissociation of methane yields carbon:

$$CH_4 \rightarrow C + 2H_2$$

This mechanism is an attractive one for the derivation of carbon in gold deposits and is supported by some field evidence. Thus, in many deposits in the Canadian Shield and in younger rocks methane, and in places other heavier hydrocarbons, are dissolved in the deep waters that course along the faults and slips. Carbonaceous matter is precipitated with gold in places along these faults probably as a result of the pyrolitic processes mentioned above. The source of the hydrocarbons in young rocks is easily explained, but the presence of methane and other hydrocarbons in Precambrian rocks is a mystery. The writer (Boyle, 1961*a*) once considered them to have come from the surface, but their presence at depths of several thousand feet in some places seems to negate this hypothesis. On the contrary they appear to be coming from great depth, and they may represent some uncharacterized synthesis deep within the earth. They deserve further study, especially as regards the fundamental problem of the origin of petroleum. In the Witwatersrand reefs methane is a common gas, and this has led Liebenberg (1955) to conclude that much of the thucholite and other carbonaceous substances are the result of polymerization of hydrocarbon gases by radioactive emanations. He does not, however, rule out the possibility that oily hydrocarbons in addition to gaseous hydrocarbons took part in the formation of the thucholite. In the Rand the methane and oily hydrocarbons were probably derived from primitive bacterial or algal material that permeated the conglomerates in their depositional sites.

While the gaseous hydrocarbon theory is an attractive one for the origin of hydrocarbons and carbon in gold deposits, there is considerable evidence to suggest that liquid hydrocarbons are the more probable sources of the carbon. Rather than the hydrocarbons replacing uraninite in the Rand bankets, it seems more probable that the uranium and the gold were introduced into the sites by some sort of liquid hydrocarbon-gold-uranium compound. Certainly the distribution of the specks and minute inclusions of uraninite and gold in the thucholite, both in the Rand and in the Richardson deposit, suggests exsolution of these elements from a hydrocarbon gel. In the case of the Richardson deposit the organometallic compound evidently migrated in solution or by diffusion into the skarn where it was precipitated. With time, polymerization processes (actually dehydrogenation and decarboxylation), with splitting out of uranium and gold left an aggregate of thucholite in which uraninite and native gold were dispersed. In the Rand deposits the organometallic compounds evidently absorbed (dissolved) uranium and flour gold from the quartzites, shales and conglomerates, and carried them into porous sites where the aging of the hydrocarbon gels liberated gold and uranium which now appear in, or in the near vicinity of, the thucholite as native gold and uraninite.

In the Rand the story may not be as simple as that just expounded. In a recent contribution Hallbauer and van Warmelo (1974) have examined both the columnar and 'fly speck' types of thucholite from the Carbon Leader Reef and other auriferous reefs by scanning electron microscope. Their results are most interesting. In the columnar type, internal structures morphologically resembling filamentous, branched and apparently septate cells of obvious biological origin were identified. These were partly encrusted with gold. In addition silicified structures, which they identified as primitive fungi, were noted. From these observations they concluded that the organism was some kind of fossilized symbiotic lichenlike plant with no modern equivalent that formed carpetlike colonies of up to several square metres in extent. These organisms were capable of extracting gold and uranium from the environment and depositing them inter- and intracellularly. The 'fly speck' carbon constitutes spherical nodules with irregular and pitted surfaces showing a remarkable morphological resemblance to living fungal *sclerotia* (reproductive hyphae). The amorphous coal-like thucholite is envisaged by Hallbauer and van Warmelo as due to exogenous plant debris washed into its present position and there attacked by a fungal organism. (*See also* the section in the last part of this chapter on the origin of the Witwatersrand deposits.)

In gold-quartz veins and other similar deposits the organometallic (hydrocarbon) compound evidently migrated in solution or by diffusion into the deposits where it was precipitated. With time dehydrogenation and decarboxylation proceeded almost to completion leaving only carbon (graphite) and gold as the residuals.

The question 'What is the precise relationship of gold to graphite and carbonaceous matter in gold deposits?' has been partly answered above. In some deposits at least some of the gold is evidently contemporaneous with the carbon. In other deposits it is obvious that the carbonaceous matter has precipitated some of the gold – the famous black leaders or indicators of Bendigo, Australia and the black bands in the Meguma Group of Nova Scotia are prime examples of this phenomenon.

Diamond and *carbonado* may accompany gold in placers but generally not in economic quantities. Occurrences are widespread: in the gold-bearing gravels of California; in the gravels of Minas Geraes, Bahia and other states in Brazil; in the gold washings of Victoria, Queensland and South Australia; in some of the golden gravels of the Irrawaddy River in Burma; in the auriferous placers of Borneo; in the gold-platinoid placers of the Urals; and elsewhere in many parts of the world. Most occurrences of diamonds in gold placers appear to be in terranes containing kimberlites, lamprophyres and similar rocks or in terranes characterized by ancient conglomerates (fossil placers) which may have initially collected the gold and diamonds. Platinoids are common associates of gold in placers containing diamonds. In general diamonds and gold are not associated in endogenic deposits, kimberlites containing only background amounts of gold (0.005 ppm); gold-quartz and other types of gold deposits are similarly devoid of diamonds.

Sulphur in the native form is not an important mineral in most gold deposits, except in the oxidized zones where it frequently occurs in small amounts, being derived in most places by oxidation of H_2S and polysulphides formed during the oxidation of hypogene sulphides. Where native sulphur occurs in oxidized zones it is commonly accompanied by wires, tufts, hairs, mossy aggregates, spangles, minute crystals and finely divided particles of native gold. Supergene minerals associated with the auriferous sulphur often include gypsum (selenite) and alum. Hypogene sulphur occurs in a number of Tertiary (near surface) gold deposits. An example is the Kasuga deposit in southern Kyushu in Japan (Taneda and Mukaiyama, 1970). There, the sulphur occurs as small irregular blebs, patches and masses in the silicified orebody. Associated minerals are quartz, enargite, luzonite and native gold. The gold appears to be contemporaneous with the native sulphur. In the Chinkuashih gold-copper deposits of Taiwan native sulphur and cinnabar are the last minerals to be deposited (Huang, 1973).

Native *arsenic* is present as an ore mineral in some gold deposits, principally in those of Tertiary age. The mineral is particularly common in the Bau gold district in Sarawak. There, the native arsenic is auriferous, containing gold as microscopic blebs and apparently also as a lattice constituent. The latter feature was studied by Lau (1970) and the writer. Native arsenic also occurs in the ores of Nagyág in Romania, Hauraki in New Zealand, in certain veins in New South Wales, Australia and elsewhere.

Native *antimony* is relatively rare in most gold deposits, generally occurring in those containing abundant antimony minerals. No relationships with respect to the age of deposits have been observed. At Yellowknife (Precambrian) native antimony has been observed in microscopic grains and blebs in massive stibnite in the ores in greenstones. It occurs in a similar manner in the stibnite ores of West Gore, Nova Scotia (Paleozoic). No paragenetic relationships with gold and other gold minerals could be established. The native antimony in auriferous deposits at Lucknow, New South Wales is exceptionally rich in gold (>200 ppm).

Allemontite, AsSb, is a relatively rare mineral in most gold deposits. It has been reported from the Ophir Mine, Comstock Lode, Nevada and from the Engineer Mine at Atlin, British Columbia (Walker, 1921). Some of the pure specimens of allemontite from Atlin are highly auriferous (Table 8). Using the electron probe the nature of the gold in the allemontite could not be determined; it is apparently not present in discrete particles of native gold and hence is probably in the lattice of the mineral.

Native *bismuth* occurs in gold deposits in only small amounts. There appears to be no particular relationships with respect to the age of deposits, although the mineral is not particularly common in Archean ores as far as the writer can ascertain. Gold occurs in native bismuth in the deposits at Tennant Creek and Cobar, Australia (Stillwell and Edwards, 1941, 1942), and some of the Barberton gold deposits in South Africa contain metallic bismuth. At the Windpass Mine, British Columbia, gold and bismuth are associated, and Lowell (1942) says that significant amounts of native bismuth and gold occurred in the ores of the Jewett and Irwin mines in southwestern Oregon. In France, native bismuth is common in the gold-quartz ores of the Salsigne-Aude district (Tollon, 1970). Native bismuth is present in skarn-type gold deposits in many parts of the world. The paragenesis of native bismuth and gold in deposits is often difficult to ascertain. In some deposits their deposition appears to overlap; in others native bismuth is frequently a late mineral, in places later than gold.

Maldonite, the bismuthic gold (Au_2Bi), is probably a mixture of gold and native bismuth, since the compound Au_2Bi is metastable at normal temperatures and hence tends to exsolve or partition into two phases. The mineral occurs at Maldon, Victoria, Australia where it is associated with bismuth minerals, native gold, scheelite, and apatite in veins and greisen veins; also in garnetiferous skarn at Redsbanya, Romania and rarely in the gold-quartz veins of the Salsigne-Aude district in France where it is associated with bismuthinite.

Native *tellurium* is commonly mentioned as being present in gold deposits. It occurs in very small amounts closely associated with various tellurides, pyrite, galena, alabandite, etc. in a number of gold deposits in Romania (Transylvania) and Western Australia (Kalgoorlie). Also in the mines of Balia, Asia Minor, in several gold districts in Japan (Kawazu Mine, Izu Peninsula) and in a number of gold and silver districts in the western United States, mostly in Colorado (Cripple Creek, Magnolia, Goldhill, Ballarat and Central districts, Boulder County; Vulcan, Garfield County; Gunnison County) and at Delamar, Lincoln County, Nevada. The analyses of the mineral quoted by Palache *et al.* (1944) and Chukhrov (1960) all show gold present in minor amounts (0.15–2.78%), probably as an admixture of the metal or as various gold tellurides.

Native *silver* is a rare mineral in the hypogene ores of gold deposits. I have seen it only occasionally in the deposits of the Timmins area and in the Ross Mine, near Ramore; also in the veins of the Berens River area, northwestern Ontario. In the Ross Mine it may belong to a separate stage of mineralization entirely divorced from the gold stage. In the Kidd Creek deposit at Timmins, native silver is abundant in the great massive sulphide orebody. This silver contains only a trace of gold. The massive sulphide ore is, likewise, almost devoid of gold. In the ores of Cobalt, Ontario, native silver is phenomenally enriched, but gold is present only in minute amounts (<10–100 ppb). Native silver is reported in the literature to be present in numerous auriferous deposits in U.S.S.R. and elsewhere. Most of this silver is in the oxidized zones, but some may be in the primary zones and may be of hypogene origin, although the descriptions do not always make this clear.

Native silver, as wires, foils, dendrites, scales, spangles and irregular nodules and masses, is common in the oxidized zones of some gold-silver deposits, especially the 'bonanza' type where the silver is largely present in tetrahedrite-tennantite, argentite and argentiferous galena in the hypogene ores. The chemistry and origin of this native silver is discussed at length by Boyle (1968*b*) and summarized in Chapter IV.

Electrum or *argentian gold* with a ratio of gold to silver of 1:1 corresponds to 36 per cent silver; 2:1 to 21 per cent, etc. Electrum is common in many gold deposits, particularly those of Tertiary age. (*See also* the next section on Au/Ag ratios in deposits and the section on the mineralogy of gold in Chapter II.)

Dyscrasite, the silver antimonide, appears to be very rare in gold deposits so far as this writer has been able to ascertain. Saager (1968) mentions its presence in small amounts in the Rand ores. He thinks it is a detrital (allogenic) mineral introduced with detrital sulphides into the bankets at the time of their formation.

Native *copper* occurs only infrequently as a hypogene

mineral in gold deposits. The mineral is occasionally present as small nuggets, irregular masses and sprigs in the quartz veins of the Ptarmigan Mine at Yellowknife, Northwest Territories. Its relationship to thc native gold could not be determined. Native copper has also been observed in some veins in the Beardmore area of Ontario, and there are scattered reports of the presence of hypogene native copper in various types of gold deposits in many parts of the world.

Supergene native copper is commonly encountered in the oxidized parts of gold deposits enriched in copper minerals.

The native *platinoids* appear to be uncommon hypogene minerals in auriferous deposits. They probably occur in microscopic amounts in some of the deposits enriched in platinoids, as discussed in the last part of the previous section on elemental associates of gold. Small amounts of iridosmine occur in the ores of the Witwatersrand. According to Liebenberg (1955) some of the grains of iridosmine are intergrown with small amounts of native platinum, and grains of iridosmine are sometimes replaced by gold. Liebenberg considers the platinoids to be of detrital origin, but it would seem that those in the Carbon Leader, including platiniridium, platinum and osmiridium, had an origin other than sedimentation. (*See* the discussion on the origin of the auriferous quartz-pebble conglomerates in the last part of this chapter.)

In a recent contribution to the occurrence, distribution and origin of platinoids in the Rand, Feather and Koen (1975) list the following platinoid minerals as being present in the reefs: osmium, iridosmine, osmiridium and iridium are the most abundant with lesser amounts of isoferroplatinum, ruthenosmiridium, ruthenian iridium, iridian ruthenium, Rh-Pt alloy, sperrylite, hollingworthite, iridarsenite, ruthenarsenite, geversite, michenerite, moncheite, laurite, braggite, cooperite, a palladium telluride and an unnamed mineral with composition RhS_2. They consider the alloys and the isoferroplatinum to be essentially of detrital origin, modified slightly by coatings and overgrowths of secondary platinoid minerals such as sperrylite and hollingworthite.

In recent placers gold is commonly accompanied by small to moderate amounts of native platinoids, particularly in terranes characterized by ultrabasic intrusives or by basic intrusives with associated Ni-Cu-Co sulphide deposits.

Native *mercury* appears to be extremely rare as a hypogene mineral in auriferous deposits. The metal may, however, occur in small amounts in the oxidized zones of gold-bearing deposits containing cinnabar and other mercury minerals. In these situations the metal has been observed occasionally in the gold deposits of Kalgoorlie, Australia, California and the U.S.S.R. Gold amalgam may accompany native mercury in somc deposits.

Native mercury, gold amalgam and a variety of indefinite amalgams principally of the platinoids may accompany gold in recent placers. In some cases these minerals seem to have resulted from the escape of mercury from amalgam barrels during former placer mining; in other cases they may be natural.

In oxidized zones and placers gold may be accompanied by a number of metals and unusual alloys, none of which occurs in any quantity. Native *lead* has been reported from the gold placers of the Urals and Altai, from those in Colombia, from certain auriferous stream sediments and soils in the Keno Hill area, Yukon (Boyle, 1976*b*) and from scattered diggings in many parts of the world. Most of these occurrences seem to be authentic, although in some placers the lead may be hunter's shot. Krylova (1975) mentions the presence of native lead in a Mesozoic-Cenozoic gold-silver deposit in the northeastern U.S.S.R. The native metal is associated with pyrite, galena, sphalerite and silver sulphosalts. Native *zinc* occurs both in the oxidized zones of some gold deposits and in various placers, but only sporadically and generally in small amounts. Its origin is somewhat of an enigma (Boyle, 1961*b*). Pellets, grains and plates of *nickel-iron* are relatively common in gold placers, particularly where ultrabasic rocks and basic rocks with associated nickeliferous sulphides are present in the terrane. Occurrences are widespread – in the Fraser and other rivers in British Columbia and Yukon; in the platinum placers of the Urals; in the gold washings of the Gorge River, New Zealand; and in various old placers in western U.S.A. Native *tin* is rare but has been noted in the gold-platinum sands of the Aberfoil and Sam rivers near Oban, New South Wales and elsewhere. Mustart (1965) noted an AuSnPb compound of unspecified name in the gold placers of Bonanza Creek, Klondike district, Yukon. Native *tantalum* is very rare but has been reported from the gold washings of the Urals and Altai Mountains, U.S.S.R. The mineral requires confirmation, since not all investigators agree as to its authenticity.

Quartz is the most common mineral in all types of gold deposits, a fact mentioned in the very earliest writings on the precious metal. Practically all varieties of this mineral are found in gold deposits, including crystalline, massive, cryptocrystalline or chalcedonic and amorphous (silica) (*see* pls. 1–16). The colours cover the spectrum ranging from colourless, white, black, grey, mottled black and white, blue, amethyst, to rose. Citrine (yellow quartz) does not seem to have been recorded in gold deposits. In some deposits the quartz is highly crushed, fractured, brecciated, recrystallized and recemented by later generations; in other deposits the mineral is little affected by shearing, fracturing or recrystallization.

In the writer's experience quartz crystal (rock crystal) occurs in gold deposits of all ages; those of Tertiary age tend to have more of this variety of quartz than do deposits of older epochs. However, some Tertiary deposits are markedly devoid of quartz crystal. The quartz in deposits of Precambrian, Paleozoic and Mesozoic age tends to be massive and highly crushed, fractured, recrystallized, etc. This is evidently due to repeated tectonic activity, although it should be noted that some Precambrian veins examined by the writer are nearly completely devoid of the effects of structural deformation. Cryptocrystalline and chalcedonic quartz, often beautifully banded and alternating with quartz crystals, is characteristic of Tertiary gold deposits wherever they are found. This type of quartz also occurs in some older deposits but is generally relatively rare. Amorphous quartz (silica), concretionary quartz, and sintery quartz is a feature of gold-bearing precipitates at the orifices of hot springs. Wood opal (petrified wood) containing gold occurs in some tuffaceous deposits (*see* page 47). Common opal and agate are rare in gold deposits in the writer's experience. Jasper, prase and chrysoprase are rare in gold deposits but appear in small amounts in places, the first where iron minerals such as hematite are relatively abundant and the last two in rocks rich in nickel minerals such as serpentines.

The colour of quartz in gold deposits is an interesting subject, studied by the writer for many years. In general most of the quartz in gold veins is colourless or white, often milky. The latter when coarsely crystalline with a greasy aspect is known to the prospector as 'bull quartz' or 'buck quartz' and often a sign of poor gold values. In Precambrian rocks this is certainly frequently the case; in younger rocks the criterion does not hold. In fact, some of the richest veins in Paleozoic and Mesozoic rocks are essentially 'bull quartz', but usually highly fractured.

Black and grey quartz owes its colour to two principal causes – electronic darkening due to bombardment by particles or rays emanating from radioactive minerals and colouring due to black impurities such as graphite, amorphous carbon, tourmaline and finely divided sulphides or sulphosalts. The first, electronically darkened or smoky quartz, is relatively rare in gold deposits mainly because of the general absence of uraninite and other radioactive minerals in the deposits. In some deposits, however, such as the Witwatersrand the darkening of the quartz pebbles in some reefs is partly due to this process and partly due to the presence of carbon according to the writer's investigations[22]. Other gold deposits containing pitchblende and other uranium minerals may have smoky quartz, but they are relatively rare. Most black and grey quartz and most of the greyish blue varieties are coloured by amorphous carbon or graphite which forms oriented trains and disseminations along wavy planes or occurs as randomly distributed fine dust throughout the quartz grains (Boyle, 1953). Such quartz is particularly common in Precambrian deposits in practically all types of rocks. It also appears, but not so frequently, in deposits of younger age mainly in those in greywacke-carbonaceous slate-argillite sequences. The exact relationship of the carbon with gold is not clear, but it has been repeatedly observed that when black quartz and white quartz occur in the same specimen, the former tends to contain much more gold. With some reservations it appears as discussed above under the section on graphite, etc. that the gold and carbon are intimately related for one or other of two reasons – the gold was transported by the mother substances (hydrocarbons) of the carbon or the carbonaceous matter acted as a precipitant for the gold.

The source of the carbon in many varieties of quartz is an enigma. In some veins the carbon is obviously relict from the replacement of carbonaceous and graphic sediments; in other veins, especially those in porphyries and volcanics such an origin seems improbable since these rocks rarely contain much free carbon. The carbon may have had a number of possible origins as discussed above under graphite and amorphous carbon. It may represent reduced carbon dioxide, which was obviously abundant during the formation of the veins and which commonly occurs in the liquid inclusions in the quartz. How this was accomplished is, however, not clear because of the high reduction potentials required. The carbon could conceivably represent diffusion or hydrothermal transport of the element in some manner or other into the veins, but we have no exact knowledge of this process. Graphite veins in metamorphic rocks suggest such a process, although there are alternate explanations. One of these, suggested by Salotti *et al.* (1971), involves the interaction of hydrogen with carbonate minerals to produce methane followed by the pyrolitic dissociation of the methane to yield free carbon. Finally, the carbon may represent the final decomposition product of a great variety of hydrocarbons of uncertain origin. Gold may also be transported with the hydrocarbons as some kind of organometallic or chelated compound as suggested by two lines of evidence. Firstly, black quartz is commonly highly enriched in gold, and secondly, gold is intimately associated with thucholite (or uraniferous carbon) in the Witwatersrand (Liebenberg, 1955) and has recently been found in a form suggesting exsolution from colloidal hydrocarbon aggregates in the skarn deposit at the Richardson Mine, Eldorado, Ontario (Boyle and Steacy, 1973). The source of the hydrocarbons, whose residuals after extensive polymerization are amorphous carbon and graphite in many black quartz veins and thucholite aggregates in the Witwatersrand and Richardson Mine, is conjectural as already discussed above. Bituminous materials and polymerized hydrocarbons are frequent constituents of inclusions in quartz and other minerals in many veins (not generally gold bearing) and pegmatites of practically all ages. In the writer's experience, most of these deposits occur in or near sedimentary rocks, and the hydrocarbons may have come from these rocks by means of metamorphic secretion processes. Some investigators think the hydrocarbons, including aromatic, olefinic, paraffin, polar and other phases, are biogenic; others consider them to be abiogenic. We are really dealing here with the origin of petroliferous substances and the nature of early life, subjects beyond the scope of this work. A recent interesting paper discussing the origin of a variety of organic spheres and particles in quartz is that by Mueller (1971).

While amorphous carbon and graphite are common colouring agents for goldbearing quartz, hydrocarbons appear to be relatively rare. None was observed in samples examined by the writer by UV and other methods. Personal communications, however, suggest that quartz and other minerals from deposits younger than Precambrian age may contain hydrocarbons, but at the time of writing no samples were available for testing. Others examining gold deposits should test the constituent minerals for hydrocarbons, because these substances may be important academically in explaining the transport and precipitation of gold in its various deposits. Recently Baskakova (1970) has observed bituminous substances, which on analysis proved to be hydrocarbons and humic acids, in the gold-quartz veins of the Kyzyl-Kum desert, U.S.S.R. These deposits are mainly of Paleozoic age, and the bituminous matter probably came from the organic (carbonaceous) matter in the host carbonate rocks and shales. According to Baskakova the bituminous matter acted as a precipitant of the gold and sulphides. From the descriptions it appears more probable that some of the gold and probably

[22]It is of interest to note that Horwood (1917) examined the problem of the colour of the quartz in the Witwatersrand by chemical and heating tests but did not come to a specific conclusion about the origin of the black colour of some of the pebbles. He figured numerous fine examples of black-edged pebbles that are obviously due to radioactive darkening a process that was not understood in Horwood's time. Horwood's research, however, is an excellent example of reporting the exact facts so that those who follow similar investigations can utilize the facts in the light of new knowledge.

part of the sulphur were carried into the deposits by the bituminous substances.

The nature of the carbon in quartz veins and other deposits appears to vary with the grade of metamorphism of the rocks in which the deposits occur. In the greenschist and sericite facies the carbon is invariably amorphous; with increasing metamorphism the carbon becomes more organized in three dimensions and is commonly graphite in the amphibolite and higher grade facies.

Black and grey quartz coloured by tourmaline or finely disseminated sulphides is common in quartz veins of all ages. No age differences with respect to abundance of these types of quartz have been noted by the writer.

Amethyst quartz is said to owe its colour to electronic processes (colour centres) based on impurity atoms or ions such as iron, manganese, titanium, nickel, etc. The exact type and concentration of these atoms and ions in the lattice of the mineral has not yet been satisfactorily characterized, but some progress has been made (Frondel, 1962; Dennen and Puckett, 1972). Amethyst is uncommon in gold veins in Precambrian rocks in the writer's experience. If present, it is of restricted or local occurrence in the veins and is generally not auriferous.

In younger deposits, especially those of Tertiary age, amethyst is common in some gold belts. The Veta Madre (Tertiary) of the Guanajuato region in Mexico is noted for its pale coloured amethyst, as are also a number of the Tertiary gold-bearing polymetallic veins in the Banská Štiavnica region of Czechoslovakia and the Paleozoic veins of the Berezovsk (Ural) area of U.S.S.R. (Avdonin and Vertushkov, 1955).

Rose quartz is exceedingly rare in gold deposits despite its common occurrence in pegmatites. Only minor amounts of rose quartz have been observed in the thousands of gold veins seen by the writer and even that present is of exceedingly poor quality and probably would not be considered rose quartz by fanciers of this beautiful mineral. At Yellowknife a variety of flesh coloured quartz is developed in some ladder and ramifying veins. This quartz and accompanying carbonates are late; in places both minerals carry phenomenal quantities of native gold with freibergite (tetrahedrite), chalcopyrite, bournonite and other sulphosalts.

Quartz, highly stained and impregnated by hematite, occurs in a number of gold-bearing polymetallic deposits in various parts of the world. It is common in the Tertiary Banská Štiavnica district of Czechoslovakia and is called 'Cinopel' by the Czech mining geologists. Greenish quartz (not prase), coloured mainly by disseminated patches and wisps of chlorite and other greenish silicates such as fuchsite, is especially common in gold-bearing deposits in greenstones (metavolcanic basic rocks). Golden (yellow) quartz is a feature of certain Tertiary deposits, the colour being due to extremely finely divided gold dispersed throughout the crystals and grains of the mineral.

Ribbon quartz and book structure (parallel quartz sheets sandwiched between slabs of wall rock) are common features of gold-quartz deposits throughout the world being particularly abundant in Precambrian deposits and less frequent in those of Paleozoic and Mesozoic age. They are not common in Tertiary deposits. There are numerous theories concerning the origin of these types of quartz, all discussed by McKinstry and Ohle (1949), Chace (1949), Stillwell (1950, 1953*b*) and Boyle (1961*a*). Most concern accretion and replacement processes or combinations of these. Farmin (1941), however, postulated that vein deposition through inflation of partings in the country rock by an injected, quasi-magmatic fluid was the probable *modus operandi.* He thought the ribbon structure was due to inclusions of parallel septa of country rock and of schistose elongation of more angular inclusions during strike faulting. Further discussion of the origin of ribbon quartz is given in the last section of this chapter dealing with the origin of epigenetic gold deposits.

Exhaustive studies of the microscopic character of gold-bearing quartz have been carried out at the University of Toronto mainly under the supervision of Professor F. G. Smith in connection with geothermal and other related research. The literature is voluminous and cannot be summarized in any satisfactory manner here. The reader is, however, referred to the following publications where summaries and bibliographies are given: Adams (1920), White (1943), Smith (1953) and Boyle (1954, 1961*a*).

At Yellowknife the microscopic description of the gold-bearing and barren quartz is typical of that seen in deposits in rocks of practically all ages in Canada. Elsewhere in the world there are some minor differences that are notable and are referred to later. Quartz deposits at Yellowknife are many and varied (Boyle, 1961*a*) (Figs. 17 and 18). Briefly they comprise, the oldest first:

Greenstone belt

1. *Early barren quartz-filled fractures ('bull quartz' with small amounts of feldspar and pyrite (Pl. 1):* These are cut and offset by shear zones and faults containing gold-bearing quartz.

The quartz lenses in the early fractures rarely exceed 25 ft in length and depth, and a few inches in width. They occur principally in the greenstones, but a few are present in the granitic rocks flanking the greenstone belt. The contacts of the lenses with the host rock are knife-sharp in nearly all localities, but some lenses exhibit a few wisps of unreplaced greenish chlorite adjacent to the slightly chloritized wall rock. The quartz in most lenses is coarse grained, but in some it is medium grained with a sugary texture. The colour is usually milky white, and the lustre is vitreous. Most specimens are brittle and break with an uneven splintery fracture.

Thin sections of the early quartz contain coarse- to medium-sized grains that exhibit a moderate degree of undulose extinction and are cut by small ramifying patches of clear quartz grains possessing sharp extinction. Liquid inclusions are present in most of the larger grains and include two types – two-phase inclusions consisting of a large bubble and a liquid, and three-phase inclusions consisting of a large bubble, a liquid and microscopic crystals. Some inclusions are randomly distributed throughout the grains and may be primary. Others mark rotated parts of the quartz grains, not in optical continuity with other parts of the same grain and are probably of secondary origin. Still other inclusions are randomly distributed along variously oriented planes which form a reticulating pattern. These planes of inclusions cut across grain boundaries and must, therefore, be secondary. All quartz grains contain unidentifiable microscopic acicular crystals and other solid inclusions.

Plate 1.
Early quartz-filled fracture, Yellowknife Greenstone Belt, Yellowknife, Northwest Territories.

2. *Gold-bearing quartz lenses and veins in shear zones and fractures:* The quartz in these bodies is mainly massive (Pl. 2) or ribboned (Pl. 3). Some exhibit vague comb structure (Pl. 4). Two or more ages are frequently present. In some veins the oldest quartz is black or grey, the younger is invariably white or milky (Pl. 5). Reversals of this sequence are notable in a few veins (Pl. 6). Mottled varieties (Pl. 7) are the result of brecciation of black and grey quartz and infilling by white and milky quartz, followed by widescale recrystallization of the mass. All ages of quartz contain carbonates, pyrite, arsenopyrite, stibnite, various sulphosalts, pyrrhotite, sphalerite, chalcopyrite, galena, scheelite, gold and aurostibite.

The macroscopic features of the quartz in the composite lenses (Pl. 2) is essentially the same in all occurrences. The colour of the quartz varies from dark to light grey to white depending upon the amount of graphite or carbon, grey sulphides, sulphosalts and liquid inclusions present. In parts of the Campbell System a light flesh-coloured variety occurs in some lenses. Ribbon structues are ubiquitous in all lenses, but much massive quartz occurs locally. Vugs, containing comb quartz, are rare. Most of the quartz has a vitreous, slightly schistose appearance and is brittle; some varieties, however, have a pronounced cherty appearance, are tough and break with a conchoidal fracture.

The microscopic features of the quartz in the composite lenses are similar in all occurrences. In thin sections the quartz appears as medium-sized grains and microcrystalline grains. The medium-sized grains generally show extreme undulose extinction and are crowded with innumerable solid and liquid inclusions. These features, as well as the presence of abundant

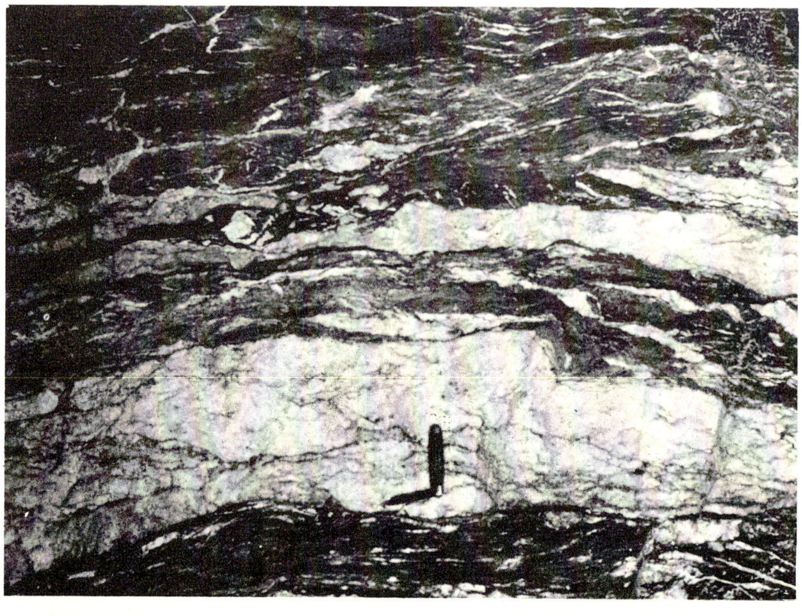

Plate 2.
Typical composite gold-quartz lenses composed essentially of massive quartz, Campbell shear zone system, Negus Mine, Yellowknife, Northwest Territories.

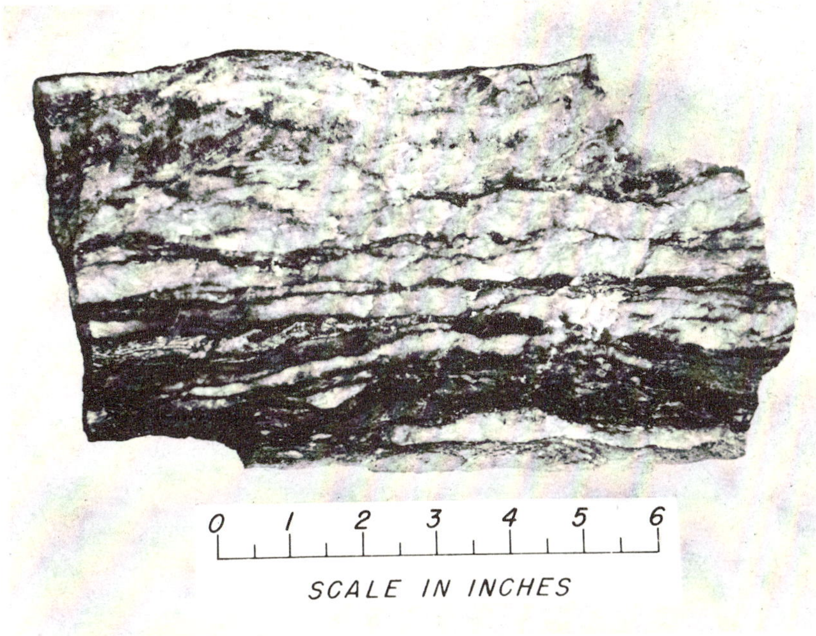

Plate 3.

Ribbon quartz, Campbell shear zone system, Negus Mine, Yellowknife, Northwest Territories.

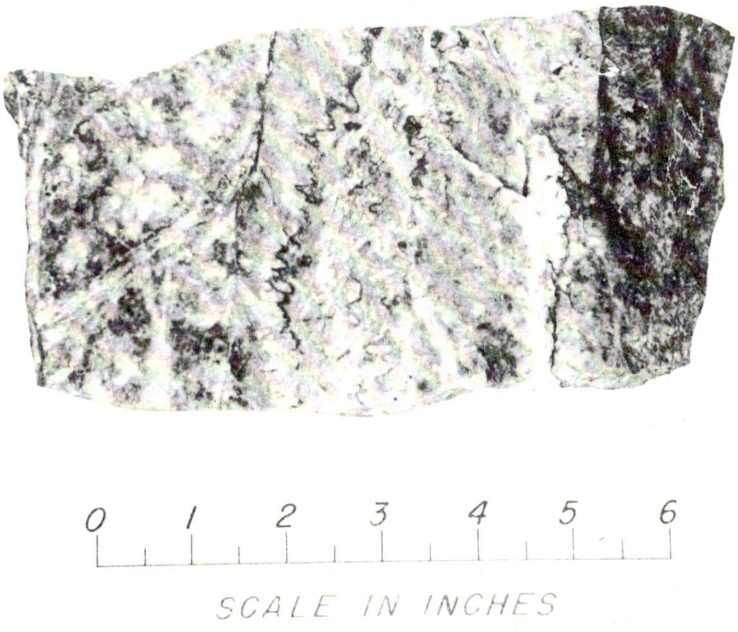

Plate 4.

Grey quartz exhibiting comb structure, Negus-Rycon shear zone system, Yellowknife, Northwest Territories.

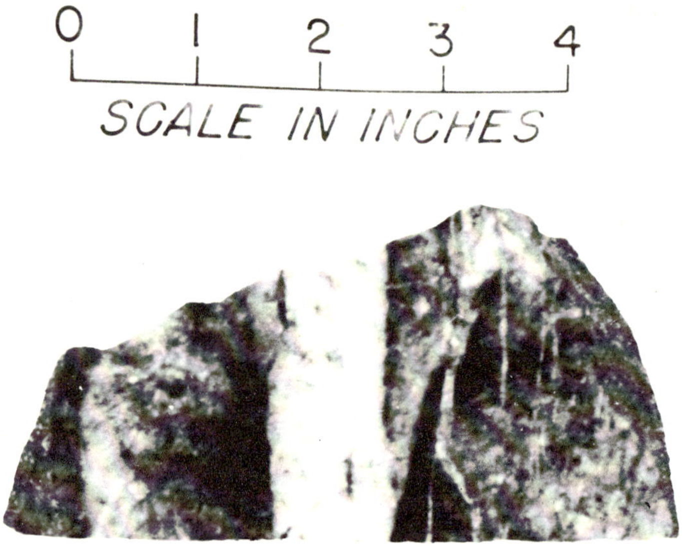

Plate 5.

Early black quartz cut by younger generation of white quartz, Negus-Rycon shear zone system, Yellowknife, Northwest Territories.

Plate 6.

Late generation of black quartz flanked by grey quartz, Negus-Rycon shear zone system, Yellowknife, Northwest Territories.

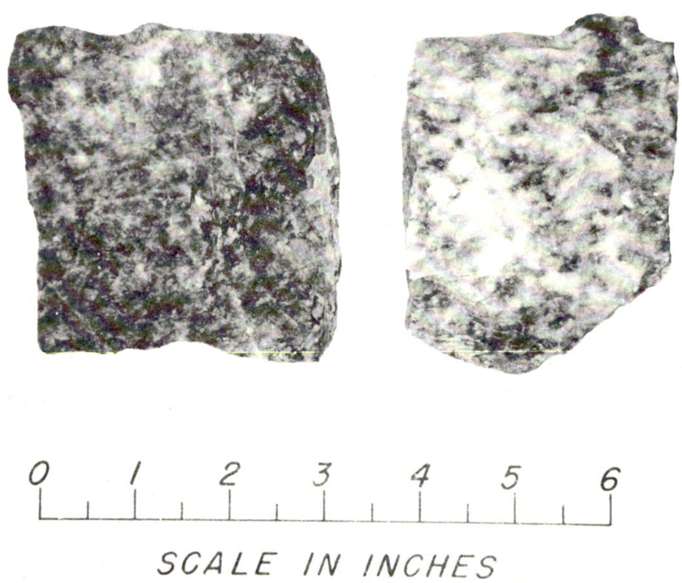

Plate 7.

Mottled varieties of quartz, Negus-Rycon shear zone system, Yellowknife, Northwest Territories.

microcrystalline clear quartz, are indicative of gold-bearing quartz and serve to differentiate it from the quartz in granites, pegmatites and late faults. The microcrystalline quartz is clear with sharp extinction and contains only a few liquid and solid inclusions. It marks the grain boundaries of medium-grained quartz and forms irregular patches, seams and phantom veinlets throughout this quartz.

In most lenses medium-grained quartz predominates and generally constitutes some 70 per cent of the quartz present; in other lenses the microcrystalline type may predominate and a cherty quartz is developed.

The liquid inclusions generally have two phases, liquid and gas. They are distributed at random through the quartz grains, along planes restricted to individual grains, or along diffuse planes transecting the boundaries of several grains. In some medium-sized quartz grains, beady lines of inclusions mark parts of the crystals slightly rotated out of optical continuity. Solid inclusions are distributed throughout the medium-sized grains and the fine-grained clear patches and veinlets. In most cases, where the identity of the solid inclusions can be established, they are fragments of carbonate, sericite, apatite, rutile or tourmaline. Some of these inclusions may be unreplaced fragments of the wall-rock minerals, and others may have crystallized with the quartz.

Some shear zones in the Negus-Rycon System and in the Shaft zone of the Con System contain gold-bearing quartz veins, which exhibit features somewhat different from the lenses just described. In these shear zones the veins are flanked by a slight schistose, dense dykelike alteration zone. The quartz veins are markedly vuggy and vague comb structures are common (Pl. 4). The vuggy parts are filled with carbonate minerals, and some contain sulphosalts.

Some shear zones of the Negus-Rycon System contain two ages of early quartz. The first is black to grey in colour and is partly replaced and veined by milky quartz (Pl. 5). In some veins, black and grey quartz abuts the vein walls, and the milky quartz occupies a medial position; in other veins the reverse is true (Pl. 6). Many veins contain a mottled variety of quartz that consists of small nuclei of black and grey quartz in a matrix of white quartz (Pl. 7).

Thin sections of the quartz from the above two shear-zone systems show many of the quartz grains to be euhedral or subhedral in outline. Growth lines, marked by primary liquid inclusions, are abundant, and secondary inclusions along planes cutting grain boundaries are common. Microscopic trains of carbon particles occur in the black and grey quartz and are primarily responsible for its colour.

3. *Quartz-filled fractures that form ladder veins and ramifying patterns in the shear zones:* These veins (Pls. 8 and 9) cut the auriferous lenses, veins and schist zones and are essentially barren in an ore sense. Most contain carbonates (mainly calcite or dolomite), and some contain small amounts of pyrite, arsenopyrite, chalcopyrite, sphalerite, galena, tetrahedrite, other sulphosalts and gold.

The macroscopic features of this age of quartz differ from those of the early gold-bearing quartz. In most occurrences the quartz is coarse grained, exhibits euhedral to subhedral crystals and is vuggy. Quartz in the stringers that have a strong development of bulbous masses is fine grained, slightly schistose, and may be cherty locally. The vuggy parts of the stringers and lenses contain calcite and ankerite and locally tetrahedrite, chalcopyrite and some gold. Lenses, which exhibit replacement features, include abundant unreplaced fragments of sericite schist containing pyrite and arsenopyrite, chlorite schist and early quartz. The contacts of the quartz with the various types of schists and early quartz is generally sharp and alteration zones are absent. The colour of the quartz ranges from white to faintly flesh coloured. Some irregular crosscutting stringers in the Giant System contain black or grey quartz and calcite due to dispersed microscopic particles of carbon in the two minerals.

Microscopic examination of this age of quartz shows

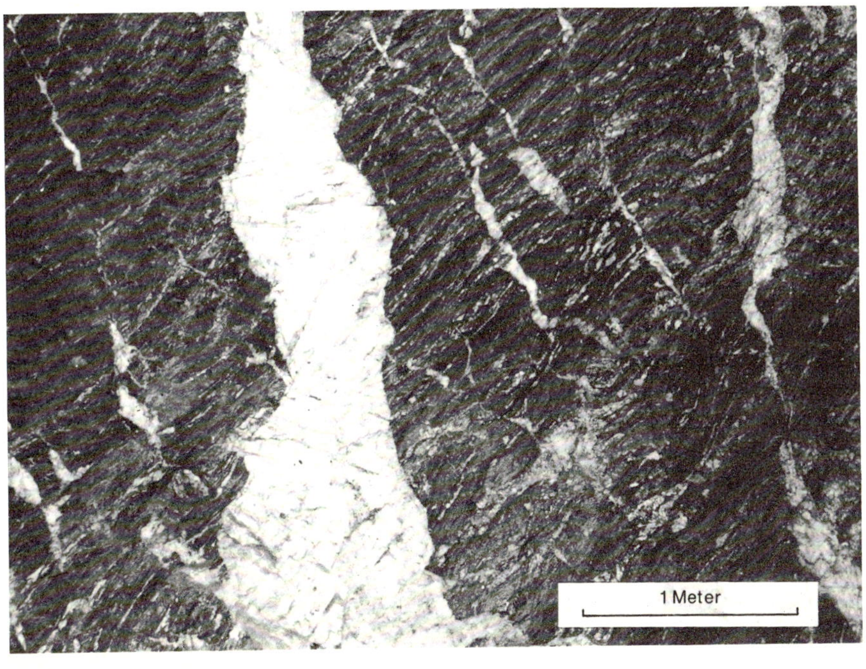

Plate 8.

Late quartz veins and lenses cutting silicified and mineralized schist zone, Con shear zone system, Yellowknife, Northwest Territories.

Plate 9.
Late quartz-carbonate veins arranged in en echelon bulbous masses, Campbell shear zone system, Negus Mine, Yellowknife, Northwest Territories.

many of the grains to be relatively clear and to have a general subhedral to euhedral habit with growth planes marked by primary liquid inclusions. In addition, randomly oriented solid and liquid inclusions, and a few planes of secondary liquid inclusions, are generally present. The quartz in the bulbous masses is commonly highly crushed, brecciated and recrystallized with a development of much fine-grained clear quartz. Grains, which escaped mylonitization, exhibit extreme anomalous extinction. These facts suggest that the bulbous masses are the result of structural contortion of the veins and stringers with consequent recrystallization in place.

Sedimentary area (Yellowknife Supergroup sediments)

Three types of quartz deposits occur in these rocks. Their age relationships could not be established with certainty when the first investigation was carried out in 1950. More recent detailed investigations in 1970 suggest that the deposits are probably contemporaneous, or nearly so, their differences being due only to the metamorphic facies in which they occur. The types of quartz deposits are:

1. *Quartz-bearing pegmatities in granite (Prosperous Lake granite) and adjacent quartz-mica schist and hornfels:* These carry no gold; assays of many samples all yield gold in amounts less than 0.01 ppm. The quartz is colourless to white or milky. Some varieties, mainly in the granitic areas have a faintly smoky colour.

Thin sections of the pegmatitic quartz show it to be composed of interlocked, clear, coarse grains with a general sharp extinction. Large irregular liquid inclusions with stationary bubbles and negative crystals with similar bubbles are a distinctive feature of the quartz. Solid inclusions include apatite, tourmaline, garnet, zircon and mica. Tourmaline tends to favour the quartz parts of the pegmatites, and garnets have a slight tendency to the same habit.

2. *Vitreous quartz of mainly one age in lenses and veins in greywacke, argillite, slate and their metamorphic equivalent, the knotted quartz mica schists:* The quartz lenses in the inner zone of the knotted quartz-mica schist aureole adjacent to the Prosperous Lake granite are regular with short and narrow dimensions; others are highly contorted ptygmatic veins (Pl. 10). The quartz in the lenses and veins has sharp contacts with the enclosing schist and hornfels, and fragments of wall rock in the lenses and veins are rare. Alteration haloes about the veins and lenses are not pronounced. Many lenses contain crystals and chalky weathering feldspar, and tourmaline, andalusite, muscovite and garnet are common in lenses cutting schists that contain an abundance of these minerals. Thin sections taken from samples of these quartz veins display a groundmass of clear grains possessing undulose extinction and containing a few liquid and solid inclusions. Patches and irregular veinlets of microcrystalline quartz intersect the groundmass of clear grains in most sections.

Most of these quartz veins are barren and contain no more than 0.1 ppm Au and 0.5 ppm Ag in most occurrences. On Outpost Island in Great Slave Lake, however, similar quartz bodies are auriferous (Lord, 1951). Some of the veins and lenses in the Yelllowknife area resemble pegmatites in the quartz-mica schists and appear to be a transition phase between quartz veins and pegmatites. Despite extensive investigations they have not been observed to grade from one to the other along strike or dip. The quartz is generally glassy ranging in colour from dense black, through grey, to bluish grey, to colourless, white or milky.

The features of the quartz deposits in the relatively unaltered sediments are clearly shown in the lenses at the Burwash prospect. Here, the quartz occurs in short narrow lenses having extremely sharp contacts with the slightly altered. wall rocks. Ribbon structures are rare, but a few fragments of schist and chlorite occur in the quartz. The colour of the quartz in most lenses is bluish black as a result of disseminated graphite; locally a mottled variety composed of diffuse areas of white to light grey quartz in dark grey to black quartz is present. A few grains of albite may occur intimately intergrown with the quartz grains, and a few veinlets of albite and carbonate may transect the massive quartz in places. Sulphides and native gold are sporadic in their occurrence.

Thin sections of the bluish quartz show that it is composed of a groundmass of medium-sized grains exhibiting anomalous blocky extinction features and containing numerous ovoid spots of graphite. Other solid and liquid inclusions, arranged at random and along reticulating planes, are common.

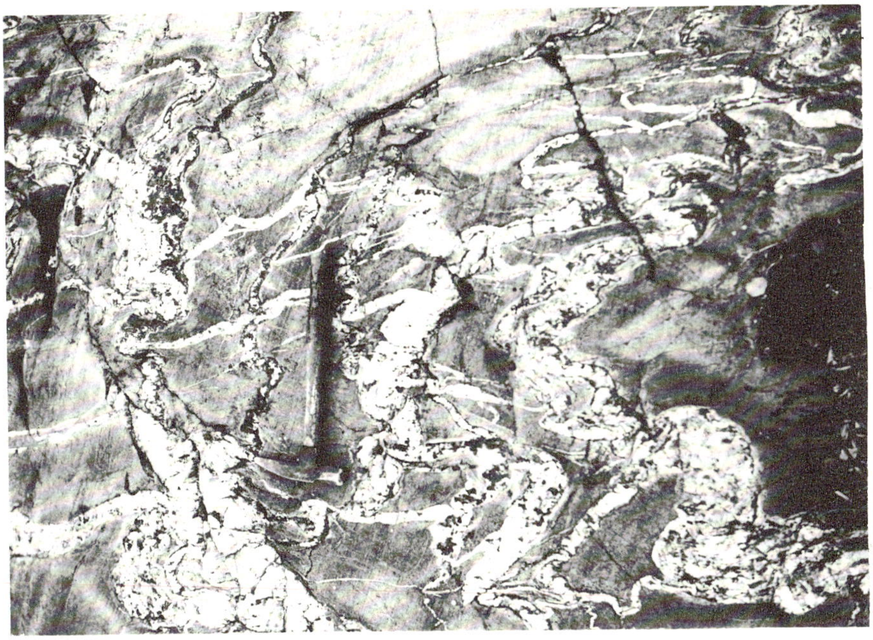

Plate 10.
Ptygmatic quartz veins in quartz-mica schist, Prosperous Lake, Northwest Territories.

The groundmass of medium-sized grains is cut and intersected by ramifying veinlets and patches of fine-grained clear quartz with sharp extinction and contains a few specks of graphite and other solid inclusions.

3. *Grey and white or milky quartz of two ages in veins and lenses mineralized with sulphides and gold:* These occur in fractures, faults, ruptured axes of folds, in contorted zones and as saddle reefs in greywackes, quartzites, argillites, slates, and their metamorphic equivalents, the quartz-mica schists. The early variety of quartz is black, grey or bluish grey, coloured by disseminated amorphous carbon in the argillites, greywackes, slates, etc. and by graphite in the quartz-mica schists. This quartz is cut or impregnated by a later white or milky variety (Pl. 11). Both varieties contain pyrite, sphalerite, galena, chalcopyrite, arsenopyrite, native copper, gold, tourmaline and minor scheelite.

Most of the quartz lenses and veins, which contain two ages of quartz and are mineralized with sulphides and gold, are localized in the outer part of the knotted quartz-mica schist aureole. Typical examples are the Ptarmigan series of lenses that occur in faults cutting quartz-mica schist (Fig. 37). The lenses are commonly arranged in an *en echelon* manner and may average 200 ft in length and 10 ft or more in width. In most places the lenses have sharp contacts with the slightly altered quartz-mica schist, and only a few inclusions of schist occur in the quartz near the walls. Ribbon structures are rare. Vugs containing euhedral quartz crystals, carbonate, pyrite, sphalerite, galena and gold are common; seams and patches of tourmaline, albite and carbonate are present in some part of the veins.

The Ptarmigan quartz lenses contain two ages of quartz. The first is mottled bluish grey, medium-grained quartz and makes up most of the lenses; the second is a limpid white, medium-grained quartz that cuts across and may replace the first age (Pl. 11).

Thin sections show the first age to be composed of medium-sized interlocked grains with a moderate degree of wavy extinction. The most characteristic inclusions in the grains are ovoid spots of graphite, which are mainly responsible for the grey colour of the quartz. Small numbers of random and planar liquid inclusions are generally present, together with some chlorite and sericite and a few unidentified acicular needles. The medium-sized grains are intersected by numerous veinlets and patches of fine-grained, clear quartz

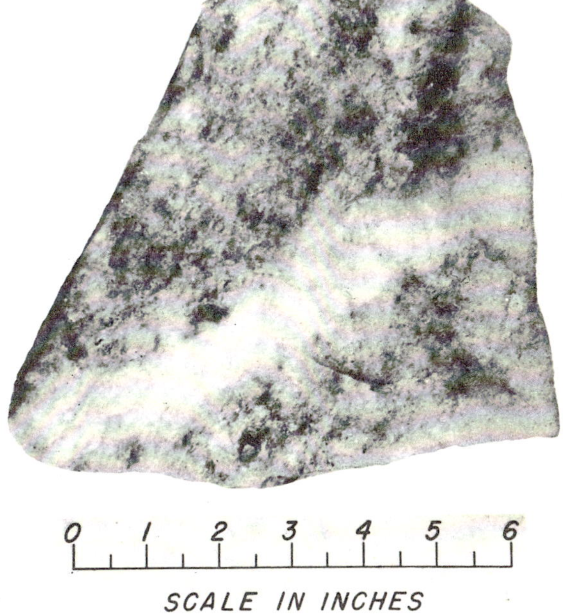

Plate 11.
Early grey quartz cut by later white quartz, Ptarmigan Mine, Yellowknife, Northwest Territories.

exhibiting sharp extinction and containing a few solid and liquid inclusions. Thin sections of the second age of quartz show that it is composed of an interlocked mass of relatively clear, medium-sized grains, which contain a few spots of graphite and some random liquid and solid inclusions.

Extensive faults cutting greenstone, sedimentary and granitic areas

Great faults extending for many tens of miles cut the greenstone, sedimentary and granitic rocks of Yellowknife and adjacent areas. Locally these faults contain enormous quartz veins and stockworks, locally called 'giant quartz veins'. As far as is known these bodies are essentially barren of gold, analyses revealing no more than 0.025 ppm Au and < 0.4 ppm Ag. In places they contain pyrite and chalcopyrite. Two general types of quartz bodies are present – relatively small quartz lenses and the 'giant quartz veins' noted above.

The smaller lenses consist of a groundmass of highly intergrown quartz crystals surrounding breccia fragments; in others the lenses are made up principally of prismatic comb quartz, which lines cavities in the faults. The quartz is commonly colourless or milky. In some lenses amethyst may occur.

In thin sections the intergrown quartz crystals of the small lenses exhibit a general euhedral to subhedral character, and relatively little mylonitization is present. Fracturing is common, however, and in most thin sections late carbonates and chlorite may fill the breaks. Primary liquid inclusions mark growth lines, and in some sections abundant secondary planar inclusions are present. Solid inclusions are abundant and include mica, chlorite and carbonates.

The comb quartz is composed of small to large prisms with milky bases and clear tips. In thin sections these crystals exhibit excellent growth zones, and all are crowded with large, two- and three-phase liquid inclusions. The surface of the crystals is etched in places, and solid inclusions such as chlorite, carbonates and sericite are present within growth zones and along growth planes. Many of the prismatic quartz crystals are fractured and healed by fine-grained clear quartz containing pyrite, chalcopyrite, chlorite and small amounts of carbonates.

The 'giant quartz veins' are remarkable structures and have been found in many of the late faults in the Northwest Territories. An excellent description of the structural characteristics and texture of these veins is given by Furnival (1935).

The 'giant quartz vein' at the junction of the West Bay and Akaitcho faults is a large stockwork that varies from 100 to 300 ft in width, and though not continuous, extends for nearly 2 mi along the West Bay Fault (Fig. 17).

The structure of the stockwork is complex and consists of several large lenses of massive white quartz intersected by innumerable veinlets of subhedral milky quartz containing many druses and elongated cavities lined with clear quartz prisms. Hematite is ubiquitous in the druses and cavities and forms a reticulating series of seams and veinlets throughout the massive quartz lenses and adjacent wall rocks.

Three distinct ages of quartz, each preceded by a period of fracturing, are developed in the stockwork. The first age of quartz is represented by large irregular lenses of massive white quartz, which developed after the initial fracturing and brecciation that formed the West Bay Fault. Only small amounts of hematite are associated with this age. The lenses appear to have been emplaced partly by replacement of the highly brecciated fault zone because most lenses contain abundant fragments of partly replaced breccia fragments. Thin sections of this age of quartz show it to contain medium- to small-sized anhedral interlocked grains enclosing some random liquid inclusions and numerous solid inclusions. Most grains exhibit anomalous extinction, and mylonitization of the larger grains is widespread.

After the massive lenses were formed, a second period of fracturing took place. This brecciated the early quartz lenses and adjacent wall rocks, especially the granodiorite, for distances up to 300 ft outward from the fault zone. Following this stage of brecciation quartz and hematite were deposited. The quartz is fine grained and anhedral; some is microcrystalline forming a red or black jasper as a result of occluded hematite.

The third stage of brecciation is local in places, but in others it extends outward from the fault plane for 100 ft or more. The fractures formed during this period are filled with comb quartz, and many druses and elongated cavities occur. In some cavities small amounts of red hematite encrust the small quartz prisms. In thin sections the third age of quartz exhibits marked euhedral characteristics: growth lines marked by liquid inclusions and a few solid inclusions are a common feature.

The chemical composition of 'pure' samples of all the quartz types at Yellowknife described above are given in the writer's original work (Boyle, 1961*a*) and need not be repeated here. The general conclusions reached by the writer are as follows:

> The contents of minor elements are extremely erratic and generally have a wide range from sample to sample in the same setting. It is very doubtful if the contents of minor elements reflect anything but the degree of contamination by sericite, sulphides, carbonates, etc. It is nearly impossible to remove these extraneous minerals by leaching and other treatment, because they are so intimately intergrown with the quartz. It follows, therefore, that contamination by the elements present in intergrown minerals masks any systematic variations in the substitution of elements in the quartz lattice or any variation in the elements in the salts present in the crystal cavities.

The problem of differentiating ore-bearing from barren quartz by the eye or microscope, is likewise, extremely difficult. At Yellowknife most of the ore-bearing quartz in the greenstone belt is intensely crushed and in part recrystallized, and there are often several generations of quartz probably in part representing remobilized silica from the confines of the veins and lenses. This feature has also been emphasized by Petrovskaya (1970) for deposits in U.S.S.R. Multiple generations or redeposition of quartz is less noticeable in the deposits in the sedimentary area. Ore quartz is invariably crowded with liquid and solid inclusions, and this is probably a good microscopic criteria for the differentiation of ore and barren quartz. Much of the gold in the quartz is late, occupying small fractures, crevices and seams in the quartz and other minerals. Where comb quartz is present the native gold and other gold minerals occupy interstices between the separate layers of crustified quartz. These features may be repeated, but invariably any one age of quartz is generally followed by a late deposition of gold and gold minerals.

The quartz in other gold deposits of Precambrian and

younger rocks exhibit features similar to those described for Yellowknife, and much of the native gold, tellurides and other gold-bearing minerals are invariably late. Some Tertiary deposits contain abundant rhythmically banded chalcedonic quartz (Pls. 12 and 13) and in this respect are somewhat different from the older deposits. The relationships of gold minerals to this quartz are, however, usually similar to those existing in older deposits.

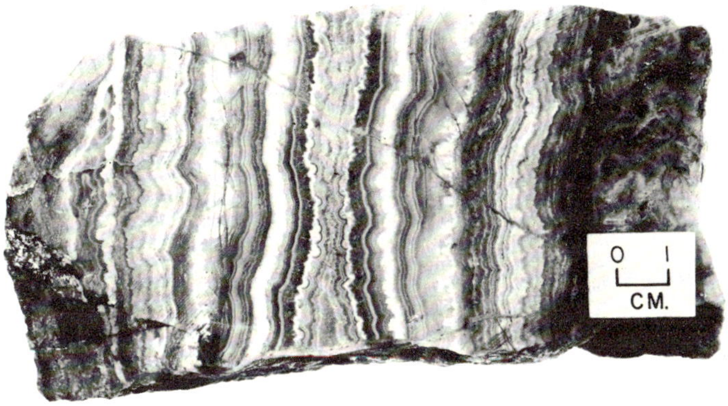

Plate 12.
Rhythmically banded chalcedony, Madjarovo veins, Eastern Rhodope Mountains, Bulgaria.

Plate 13.
Rhythmically banded quartz, chalcedony, carbonates and sulphides (galena, sphalerite, chacopyrite, pyrite), Tui Mine, Te Aroha, New Zealand.

The detailed relationship of gold to quartz in auriferous deposits is of considerable interest from the viewpoint not only of origin but also metallurgical practice in winning the precious metal from its ores. Mawdsley (1938), White (1943), Ebbutt (1948), Urashima (1956), Petrovskaya (1970, 1973) and Petrovskaya *et al.* (1971*b*) have dealt with some aspects of this subject. The summary to follow complements their work and extends the data to all types of deposits.

Gold-bearing pegmatites. In these deposits the relationships are essentially the same as those noted below for gold-quartz veins and silicified zones. The gold is mainly closely associated with sulphides or occurs in the free state in late fractures in the quartz and other minerals.

Skarn deposits. Gold occurs in these bodies in four forms. (1) In or closely associated with pyrite, arsenopyrite, tetrahedrite and other sulphides and sulphosalts as a lattice constituent or in microscopic form: The microscopic gold is usually distributed at random in the sulphides and sulphosalts. Paragenetically, it is usually a late mineral often occurring in minute fractures and seams. (2) In native arsenic as in the Bau deposits Sarawak: The gold is a lattice constituent and also appears in microscopic particles. (3) In tellurides: No good relationships can usually be established. Probably most are paragenetically late minerals. (4) As native gold: The metal is usually in contact with quartz, Ca-Mg-Fe-Mn–silicates, carbonates, etc.; the native gold is late, occupying tiny fractures and seams in quartz and these other minerals.

Gold-quartz veins, lodes and silicified zones. Gold occurs in a number of ways in these deposits. (1) In or closely associated with pyrite, pyrrhotite and arsenopyrite as a lattice constituent or in microscopic form (*see also* the section on gold in other minerals in Chapter II): The microscopic gold is usually distributed at random in the sulphides. Paragenetically, the gold is usually a late mineral often occurring in minute fractures or seams or rimming the sulphides. (2) In sulphosalts, mainly tetrahedrite or tennantite; in stibnite; in realgar; and in other sulphides and sulphosalts such as galena, sphalerite, chalcopyrite, boulangerite, jamesonite and enargite: The gold is a lattice constituent of some of these minerals; it also appears in microscopic form in a random fashion and in tiny fractures and seams in these minerals. (3) As native gold in contact with quartz and other gangue minerals: The gold is paragenetically late occurring in small faults, fractures, crushed areas, slips, chloritic and sericitic seams and small vuggy parts of the quartz (Pl. 14). In young deposits (Tertiary) the gold may be moulded around quartz crystals or occur in bands between successive combs or layers of quartz. (4) As aurostibite: Most of the aurostibite in gold deposits occurs in close proximity to gold and forms haloes and coatings about the gold particles or is present in narrow seams in the gold. It is also associated with gold as small diffuse aggregates scattered along the contacts of arsenopyrite, stibnite and sulphosalts. In most polished sections it appears to be later than the gold, but good age relationships are uncommon. Aurostbitie is usually restricted in its occurrence to deposits containing concentrations of antimony minerals and gold. (5) As tellurides: There is probably no gold deposit without gold-silver tellurides, although they may be present only in trace amounts. In nearly all deposits the gold tellurides accompany native gold and are late minerals in the paragenetic sequence occurring in fractures, slips and crushed zones in the quartz and other gangue minerals. Where crustification is prominent the gold and tellurides are invariably late following the deposition of quartz, carbonates, etc. In individual veins there may be several generations of crustified minerals, including native gold and gold tellurides, each superimposed on an earlier generation.

Auriferous polymetallic and massive sulphide deposits. In polymetallic deposits the relationships are essentially the same as those noted above for gold-quartz veins and silicified zones. In many of these deposits much gold is intimately associated with the sulphide and sulphosalt minerals being present either as lattice constituents or as microscopic particles. In the

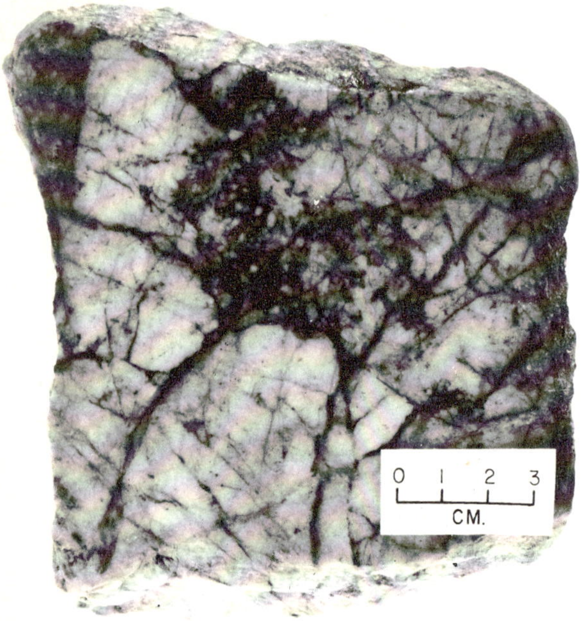

Plate 14.
Native gold associated with late generations of stibnite and sulphosalts (tetrahedrite, bournonite, jamesonite), Negus-Rycon shear zone system, Yellowknife, Northwest Territories.

massive sulphide deposits the distribution of the gold is varied, but invariably the bulk of the element is late in the paragenetic sequence. There is also a minor generation of gold which accompanies each stage of the sulphides and occurs in them as microscopic particles or in lattice sites. The classic account of the distribution of gold in massive sulphide bodies is that by Price (1934) who studied the rich auriferous Noranda ores. He identified the following main types of gold mineralization in the great massive sulphide bodies: (1) gold and tellurides (especially the former) forming minute veinlets, irregular areas and dots in idiomorphic pyrite crystals; (2) veinlets (predominantly gold) along the borders of pyrite crystals, between the pyrite and the adjacent sulphide; (3) gold-telluride areas replacing pyrrhotite, the free gold particles being surrounded or 'sheathed' by tellurides; (4) gold replacing chalcopyrite; (5) gold and tellurides in veinlets and angular patterns replacing minute chlorite aggregates in the massive sulphides; (6) quartz veinlets carrying gold and tellurides, cutting the earlier sulphides; and (7) type 5 repeated at the contact of highly chloritized syenite porphyry where such material occurs in the midst of a large sulphide mass.

Gold-bearing quartz-pebble conglomerates. An enormous amount of microscopic work has been done on these deposits. Early summaries of research are given by Graton (1930). Later work is described in detail by Liebenberg (1955) in his extensive paper on the Rand. Saager (1969) summarizes the recent research on the Rand in his paper on the relationship of gold and silver, and Feather and Koen (1975) have added further data. Gross (1968) describes the relationships of the gold to other minerals in the Jacobina conglomerates in Brazil, and Bray (1928) and Kitson (1930) give similar data for the Tarkwa conglomerates in Ghana. (*See also* the section on these deposits in a later part of this chapter.)

Detailed descriptions of the varieties of quartz in the quartz-pebble conglomerates are given in a subsequent section dealing with these deposits. Briefly, the quartz occurs in pebbles that are cemented by a matrix consisting of quartz grains, micaceous minerals, pyrite (Pl. 15) or iron oxides (Pl. 16).

Saager (1969) describing the occurrence of gold in the Rand says:

> The gold in the Basal Reef occurs in much the same way as in other ore-horizons of the Witwatersrand Goldfield. It varies

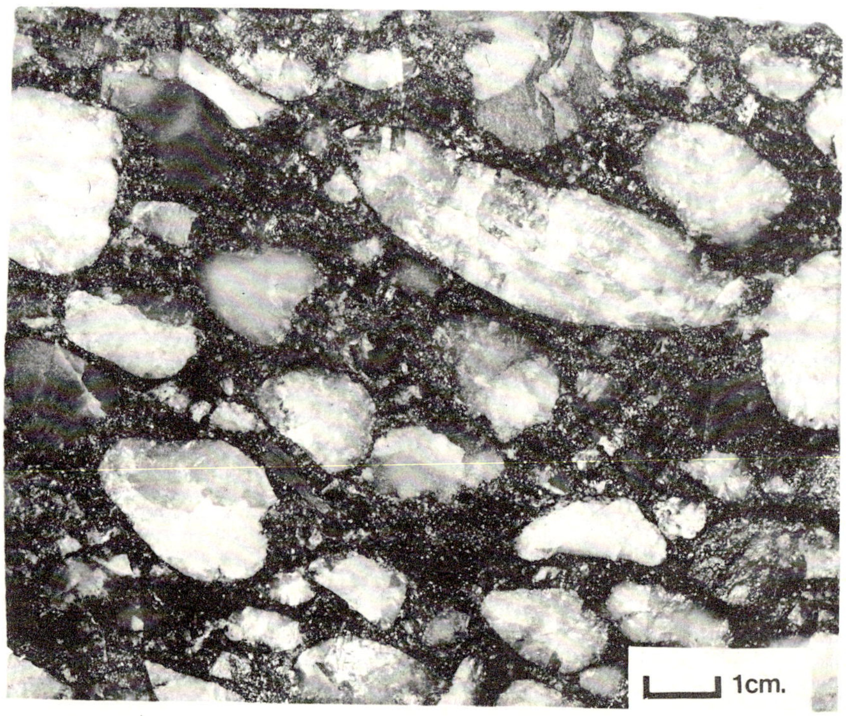

Plate 15.
Auriferous pyritic quartz-pebble conglomerate, Main Reef Leader, Robinson's Deep, Witwatersrand, South Africa.

Plate 16.
Auriferous, hematitic quartz-pebble conglomerate, Tarkwa goldfield, Ghana.

considerably in grain-size from 1 micron to over 1,000 microns, commonly averaging between 5 and 100 microns. Macroscopically visible gold is only very rarely encountered. Usually the gold occurs as irregular, jagged particles among the matrix constituents of the conglomerate and is intimately intergrown with secondary, reconstituted sulfides, or it fills fractures and cavities in earlier formed minerals. Frequently the gold during its reconstitution, has migrated into the marginal areas of porous pyrite grains where it now forms secondary inclusions. Primary inclusions and myrmekitic intergrowths of gold with detrital pyrite and linnaeite have occasionally been observed. Rarely, rounded gold grains are found in the matrix. According to a suggestion by Liebenberg (1955), these possibly represent detrital gold particles which still exhibit the abraded morphology obtained during transportation.

The close relationship between gold and thucholite in many Witwatersrand ores has been known for a long time, suggesting precipitation by the latter on gold taken into solution. The common close association also of uraninite with gold has led Liebenberg (1955) to suggest that the uraninite may have exerted a precipitating effect of gold from solution.

A vast array of incontrovertible macroscopic and microscopic evidence now accumulated from all areas of the Witwatersrand and associated subsidiary basins leaves no room for doubt that the gold is essentially of placer origin. The mode of occurrence of the gold and especially its relationship and association with other minerals clearly indicates subsequent mobilization and redeposition during the metamorphism of the Witwatersrand sediments.

No specific mention is made of quartz in this description, and it should perhaps be noted that occasionally gold is seen in small fractures in the quartz pebbles. There are also quartz veins in the Rand system that seem to receive little attention by the placerists. Graton (1930) refers to these and states that they carry the same minerals as the matrix of the conglomerate, viz. sulphides, sericite, chlorite, carbonates and gold. The gold is a late mineral in these veins.

According to Gross (1968) the bulk of the gold in the Jacobina conglomerates occurs in the sandy matrix. It is usually present as grains varying in size from 0.05 to 0.01 mm. Within individual conglomerate beds the gold increases markedly toward the tops of the beds. This suggested remobilization of (placer) gold to the author, who also points out that occasionally thin plates of gold are found in fractures in the pebbles and in the conglomerate. In some polished sections grains of pyrite occasionally exhibit crystal structure, which in places may embay or surround gold particles. Where ultramafic rocks form sills and steeply dipping dykes in the gold-bearing horizons, concentrations of gold occur near the contacts of these intrusions.

According to Bray (1928), Kitson (1930) and Rozhkov (1967) the matrix of the Tarkwa quartz-pebble conglomerate consists of quartz grains, sericite, chloritoid and much hematite and some magnetite in granular form. The gold is essentially all in the matrix, but some is found around the quartz pebbles and in little fractures in the matrix. Numerous quartz veins also traverse the Tarkwa conglomerate, and these carry pyrite, siderite, hematite and minor amounts of gold, the last a late mineral in the paragenetic sequence.

Summary of features of quartz in auriferous and barren deposits

It is difficult to summarize all of the above features of auriferous and barren quartz, but certain aspects seem to stand out in the writer's opinion.

Quartz deposited, reworked or recrystallized under deep-seated conditions in high grade metamorphic terranes (hypothermal deposits of some geologists) generally occurs in coarse-grained compact aggregates and is usually relatively clear and glassy to subvitreous and milky; some varieties are glassy, dark grey to black due to microinclusions of graphite. Vugs and prismatic crystals are rare; ribbon and book structures are uncommon. Liquid inclusions in the quartz tend to be relatively large but restricted in number. Native gold and gold minerals are invariably late occurring in microfractures or brecciated parts of the quartz.

Quartz precipitated, reworked or recrystallized under conditions of medium depth and medium to low grade metamorphism (mesothermal deposits of some geologists) is generally finer grained than the preceding type of quartz. Ribbon and book structures are especially common, and crude and coarse banding is a feature in some veins and lenses. Vugs and druses are common in some deposits but are usually not abundant. The quartz is usually milky to dull grey to black in colour, the last being due mainly to finely divided forms of carbon (carbonaceous matter). Liquid inclusions are abundant, some varieties of auriferous quartz being crowded with these objects. Liquid inclusions range in size from small to large and are of both primary and secondary derivation. Gold and gold minerals are invariably late occurring in microfractures, phantom veinlets or brecciated parts of the quartz.

Quartz deposited under near surface conditions (epithermal deposits of some geologists) generally in zones of low grade metamorphism (propylitic facies) is commonly fine grained, porcellanic, flinty, dull cherty, chalcedonic, to opaline. Some deposits are characterized by medium- to fine-grained quartz. Vugs are universal and cockade structures common; fine colloform banding is often characteristic, and developments of lamellar and hackly quartz are notable in some deposits. The quartz is clear to milky, only rarely dark grey to black. Liquid inclusions are very small. Gold and gold minerals may occur interbanded with successive generations of quartz, or these minerals may be late, filling small fractures and crushed zones in the quartz. Some varieties of quartz in Tertiary veins are heavily disseminated with microcrystalline gold such as to give the quartz a golden yellow colour.

Microscopically auriferous quartz is often characterized by a great abundance of liquid inclusions of both primary and secondary origin. This feature has been noted and commented on by numerous investigators including particularly Smith *et al.* (1949–1951), Boyle (1954) and Machairas (1970*b*). Machairas noted the presence of gold particles in some of the liquid inclusion cavities and observed that the content of CO_2 in the inclusions in the quartz was greatest in auriferous quartz. Finally he concluded that as a generality the concentrations of fluid inclusions and gold in quartz are proportional.

Despite all these favourable indications there is really no satisfactory way of unequivocally distinguishing barren from auriferous quartz as far as is known to the writer except by assay methods. The practised naked eye can sometimes distinguish quartz that is more 'kindly' to gold mineralization within a specific belt but may fail in other belts. Microscopically, auriferous quartz can sometimes be distinguished from barren quartz by the large degree of cataclasis, strain phenomena, recrystallization, phantom veinlets and secondary liquid inclusions, which characterize gold-bearing quartz in many belts. Chemically it is doubtful if auriferous quartz can be distinguished from barren quartz, although more work should be done on this aspect of the problem. The comment of W. M. Courtis (1890) made after a long study of auriferous quartz is appropriate at this point:

> After several years spent among the gold-mines of California with practical and professional gold-miners, I have come to the conclusion that very few care to risk an opinion on a piece of quartz until some of it has been "horned out." Nevertheless, there is what the miners call a "kindly" appearance, that is more often followed by good results than is a vitreous, water-wet-looking quartz. A dark, resinous quartz seems to be most in favor, but there are so many exceptions that even quartz as "white as a hound's tooth" produces gold in some mines.

Silicates of many types are common in the wall-rock alteration zones and ores of gold deposits of all categories. Most of these are discussed in the section on wall-rock alteration processes and the details can be omitted here. Among the feldspars the most common are *albite, oligoclase* and *adularia.* Much has been written on the association of gold and albite, a few of the principal references including those by Lindgren (1906), Gallagher (1940) and Ward (1958). In some cases there is an association with albite dykes or albitization of the wall rocks of the deposits; in other cases the albite occurs as a minor gangue mineral in quartz veins and silicified zones. Albite and oligoclase as vein minerals are relatively common in Precambrian deposits, diminishing in amount through deposits of Paleozoic, Mesozoic and Tertiary age. They occur most commonly in deposits in the higher grades of metamorphism (epidote amphibolite and amphibolite facies). Gold and gold minerals are generally later than albite and oligoclase in their paragenesis. In some places the gold can be seen along the cleavage planes of the feldspars. More details on the association of gold and albite are given in Chapter V. *Adularia* is usually found in the young (Tertiary) gold deposits mainly as a vein mineral. *Valencianite* is adularia; it occurs in some of the auriferous and argentiferous veins of Mexico and the western United States. Gold and gold minerals are generally later than adularia in their paragenesis.

Of the micaceous minerals in gold deposits the most common are *sericite (muscovite), biotite, talc* and *chlorite. Fuchsite* and *mariposite*, the chromium-bearing varieties of muscovite and phengite respectively, are generally restricted to the wall-rock alteration zones of gold deposits in basic and ultrabasic rocks. Some auriferous porphyries also contain fuchsite. *Chromian muscovite* is a common mineral in some of the auriferous quartz-pebble conglomerate bands at Jacobina in Brazil; also in the Rand quartz-pebble conglomerates. *Paragonite* is relatively rare compared with sericite in the wall-rock alteration zones of gold deposits; it occurs most commonly where there is a high degree of soda metasomatism and is often accompanied by albite. *Roscoelite*, the vanadium mica, occurs in a few of the gold quartz veins of the Mother Lode system in California, at Cripple Creek in Colorado and elsewhere. The mineral seems to occur most abundantly in deposits with basic and ultrabasic host rocks. According to Lindgren (1933) roscoelite is often indicative of rich auriferous shoots. In the Kalgoorlie gold belt of Western Australia *vanadium muscovite* (sericite) is particularly common and is responsible for the colour of the 'green leader'. It is closely associated with nolanite, native gold and tellurides (calaverite, petzite, altaite, melonite, coloradoite) (Nickel, 1977). *Pyrophyllite*, $Al_4(Si_8O_{20})(OH)_4$, is a common mineral in the micaceous matrix of the auriferous quartz-pebble conglomerates of the Witwatersrand. According to Liebenberg (1955) it is the main micaceous silicate in the matrix of some of the deposits. Gold is intimately associated with the various micaceous silicates, including pyrophyllite, replacing them completely or partially in places. Pyrophyllite occurs in the wall-rock alteration zones of certain (usually Tertiary) vein deposits. At Rodalquilar in southeast Spain the mineral occurs in the alteration zones, together with alunite, adjacent to veins with high concentrations of gold (Lodder, 1966). *Chloritoid* is not common in auriferous deposits but occurs in the quartz-pebble conglomerates of the Rand in moderate abundance (Liebenberg, 1955). It is commonly accompanied by chlorite and both minerals may be moulded on and around the grains of pyrite as observed in thin sections.

Serpentine is a common mineral in the alteration zones of auriferous deposits developed in ultrabasic rocks; also in certain skarn deposits. The mineral only rarely occurs in the quartz of gold-quartz deposits. In a few deposits native gold occurs in asbestos veinlets in serpentinized rocks; in some places the metal is present along the cross fibres of the asbestos, one characteristic occurrence being at Jones Creek, Gundagai, New South Wales.

The clay minerals, *kaolinite, montmorillonite*, etc., are common in auriferous deposits, primarily in those of Mesozoic and Tertiary age. They result mainly from endogene alteration of feldspars, feldspathoids and other silicates during wall-rock alteration processes. In some deposits the clay minerals are supergene products of oxidation and weathering processes. Gold is commonly intimately associated with kaolinite and the other clay minerals mainly as a result of adsorption processes. (*See also* the section on gold in natural waters in Chapter II.)

Zeolites, such as *stilbite, chabazite, laumontite* and *natrolite*, are not common minerals in auriferous deposits, but significant occurrences are reported in gold-quartz deposits in the Tertiary andesites, basalts and pyroclastics in New Zealand, Indonesia and Japan. Seiranyan *et al.* (1974) describe the wide distribution of calcic zeolites (laumontite, stilbite) in the auriferous sulphide ores of northern Armenia. These deposits are in volcaniclastic and effusive rocks of basaltic and andesitic composition. The zeolites occur both in the quartz-epidote gangue of the deposits and in veinlets and amygdules in the wall-rock alteration halos.

Amphiboles, pyroxenes, olivine, epidote, garnet, wollastonite, andalusite and *cordierite* are most abundant in auriferous skarn-type deposits, although garnet and andalusite have been noted in some abundance in gold-quartz veins in high grade metamorphic rocks. Hawley (1939) and Lord (1951) mention the presence of minerals such as staurolite, andalusite and corundum in the auriferous quartz bodies of Outpost Island, Great Slave Lake, Northwest Territories. These veins also contain ferberite, scheelite and molybdenite. The gold is later than the andalusite and staurolite, occurring frequently in tiny fractures in the former mineral. In the Kabin Mine in Thailand rich gold shoots occurred in quartz-grossularite veins.

All of the above mentioned minerals are commonly encountered in gold placers.

Nephrite (an amphibole) and *jadeite* (a pyroxene) collectively known as jade are frequently found as boulders and pebbles in certain auriferous stream gravels in Asia and elsewhere. This may partly explain why the Chinese have long associated gold with jade, the two most precious substances known to the early Chinese who worshipped both as immortal. The Chinese word for jade *jü* is employed in metaphor to symbolize nobility, beauty and purity much in the same manner as the English words 'gold' and 'golden'.

Tourmaline is one of the commonest minerals in auriferous deposits of practically all types. In some deposits such as the auriferous quartz-pyrite-tourmaline breccia pipes developed mainly in granitic rocks the borosilicate may be nearly as abundant as quartz. Tourmaline occurs both in the wall-rock alteration zones of gold-quartz deposits and in the quartz, carbonates and other gangue minerals. Gold and gold minerals are generally later than tourmaline in their paragenesis, although in some deposits the native gold often occurs along the elongated crystals of the borosilicate, and it is difficult to determine their respective ages. To some investigators the small amounts of tourmaline in the quartz-pebble conglomerates of the Rand are detrital; to others introduced by hydrothermal activity.

Tourmaline is a common associate of gold in auriferous placers.

Beryl is not commonly found in auriferous deposits, but minor amounts of the mineral may occur in skarn-type deposits and in certain veins carrying tin minerals and topaz. Ol'shevskiy (1973) records the presence of aquamarine in the gold-quartz veins of the Karalveem deposit in western Chukotka.

Some gold placers in granitic terranes contain small amounts of common beryl. In New South Wales emeralds are encountered in some gold washings. This gem is also found in auriferous placers in other gem-bearing parts of the world.

Rhodonite, inesite and *johannsenite* are common manganese silicates in young (Tertiary) gold deposits. In these they appear to be relatively early minerals and are commonly associated with manganese carbonates, especially rhodochrosite, manganiferous calcite and kutnahorite (Mn dolomite). *Bustamite* and a variety of other maganiferous silicates occur in small amounts in certain auriferous skarn deposits and high temperature vein and replacement deposits.

Topaz is relatively rare in gold deposits but occurs as an important alteration and gangue mineral in the Brewer Gold Mine. Chesterfield County, South Carolina (Peyton and Lynch, 1953). There, the topaz content may be as much as 17.5 per cent or more with long core lengths of 5 per cent or more. Topaz also occurs in auriferous deposits marked by tin and tungsten (wolframite) mineralization. It is evidently an early mineral in these deposits and may belong to a stage that is more or less divorced from the auriferous phase of mineralization.

Topaz is a minor constituent of some gold placers.

Sphene occurs only rarely in auriferous deposits and principally in skarn-type deposits. *Allanite*, the rare-earth silicate, is, likewise, uncommon in nearly all types of auriferous deposits. It may occur sporadically in skarn-type deposits and has been noted in certain quartz-pebble conglomerates. *Zircon* may occur in various type of auriferous deposits but only in traces; in the Rand the zircon in the matrix of the bankets is uraniferous. According to Liebenberg (1955) it is of detrital origin. In modern gold placers zircon is an invariable associate of gold.

Silicates, which seem not to have been recorded from auriferous deposits, include *pollucite, feldspathoids* and *spodumene*. In the supergene zone zinc silicate, *calamine*, and the copper silicate, *chrysocolla*, occur in the oxidized zones of some gold deposits containing abundant primary zinc and copper minerals.

Carbonates are ubiquitous minerals in most gold deposits being most abundant in the skarn, vein and polymetallic types and least abundant in the quartz-pebble conglomerates. The most common carbonates in gold deposits are *ankerite, dolomite, calcite, rhodochrosite* and *siderite. Magnesite* is mentioned as being present in some gold deposits in the old literature, and *kutnahorite*, the manganiferous dolomite, occurs in certain young (Cretaceous-Tertiary deposits). *Aragonite* does not seem to occur in gold deposits; at least there are no modern references to the presence of this carbonate in gold deposits in the literature, and the writer has not observed it in his investigations. *Strontianite* may occur in some gold-quartz deposits, but it is relatively rare as a gangue mineral. *Fluocarbonates* are rarely present in auriferous deposits, although *bastnaesite* and other similar carbonates apparently occur in the Sulphide Queen Mine at Mountain Pass, San Bernardino County, California. In Precambrian, Paleozoic and Mesozoic deposits, *ankerite, dolomite, calcite* and *siderite* predominate in about that order of abundance. Siderite is, however, uncommon in most gold-quartz deposits, but in some such as those in Proterozoic rocks north of Lake Huron, Ontario it is an important gangue mineral. In these the native gold is invariably later than both the quartz and siderite. Tertiary deposits are often characterized by *manganiferous calcite, rhodochrosite* and in places by *kutnahorite*, which are frequently accompanied by manganese silicates such as rhodonite, inesite and johannsenite.

Gold-quartz veins and silicified deposits, particularly those developed in basic volcanic rocks frequently exhibit two or more ages of carbonate. At Yellowknife, and in other gold-quartz deposits of the Canadian Shield, three ages of carbonate are commonly developed in deposits in the greenstone belts. The oldest, developed mainly in the wall-rock alteration zones, is ankerite with high magnesium and iron contents. The second age, present in crosscutting quartz-carbonate veins and lenses, is ankerite or calcite, characterized by a higher calcium content and lower iron and magnesium contents compared with the oldest carbonate. The youngest carbonate, in late crosscutting slips and small faults, is invariably calcite with low contents of magnesium and iron. In all three stages of carbonate, manganese remains relatively constant, with a slight increase in the youngest age.

The tendency for carbonates to become more calcitic with successive ages is apparently normal in most gold-quartz deposits of all geological ages in rocks of all types. The tendency is most marked in basic and intermediate igneous rocks and less marked in granitic rocks and sediments. Reversals of the trend have, however, been noted, but they are usually local.

Native gold, in contact with vein and wall-rock carbonates, is invariably late, occurring in seams, fractures, crushed zones and small slips. In some old (Precambrian) and young (Tertiary) deposits gold is sometimes found moulded on the scalenohedral prismatic and rhombohedral faces of carbonates, between crustified layers and in carbonate mineral vugs. Gold tellurides generally exhibit relationships to carbonate minerals similar to those for native gold.

Few colour relationships have been observed among the carbonates of gold deposits with respect to type of deposit or age of deposit. Black carbonates (mainly ankerite or calcite) are common in some Precambrian gold-quartz deposits. These carbonates are invariably associated with black and grey quartz and owe their colour to amorphous carbon in most places (Boyle, 1953). Reddish carbonates, stained by hematite, occur in certain gold deposits containing uranium minerals. The iron oxide appears to have been liberated from the carbonates in some manner or other by radioactive bombardment since the coloration often forms halos or streaks about the uranium minerals.

The microscopic features of vein carbonates in gold and other types of deposits are discussed in an excellent study by Grout (1946). Space does not permit a detailed summary of the results of this investigation here, but some general observations may be noted. Concerning sulphides in auriferous deposits Grout found a general trend (not without exceptions) for the presulphide carbonate to be mixed and complex, whereas the late carbonate is commonly calcite. The early carbonates are very commonly replaced by sulphides, whereas the late calcite fills minute fractures and only rarely replaces the sulphides. Grout noted the common occurrence of native gold with ankerite and along cleavage planes of calcite and other carbonates. In some deposits the carbonates associated with gold began forming early and continued even after the deposition of native gold. An exception to the general sequence of ankerite giving way to calcite during deposition in gold deposits was noted by Grout in the Howey Mine in the Red Lake district of Ontario where Mather (1937) observed that the early calcite deposited with the gold was calcite with ankerite increasing in proportion thereafter.

Smithsonite, the zinc carbonate, *cerussite*, the lead carbonate, and the two copper carbonates, *malachite* and *azurite*, are common in the oxidized zones of auriferous deposits containing primary zinc, lead and copper minerals. *Bismutite*, $(BiO)_2(CO_3)$, is common in the oxidized zones of auriferous deposits containing primary bismuth minerals. The carbonate is commonly associated with wires, hairs, spangles and finely divided particles of gold as in certain deposits in Honduras, Australia, U.S.S.R. and elsewhere.

Fluorite is a gangue mineral in some gold-quartz and polymetallic deposits. It is relatively rare in most of the skarn-type gold-bearing ores. Most Precambrian gold deposits are generally marked by their paucity of fluorite, but the mineral becomes increasingly abundant in younger deposits and is an important gangue mineral in some Tertiary deposits. Fluorite is not recorded as being present in the Witwatersrand bankets (Liebenberg, 1955; Saager, 1968; Feather and Koen, 1975). Various colours mark fluorite in gold deposits, from colourless to green, yellow, brown, reddish and violet-blue.

The veins of Cripple Creek, Colorado are exceptionally rich in violet fluorite which tends to crystalline forms (Lindgren and Ransome, 1906). There, as elsewhere in other gold deposits containing fluorite, quartz and carbonates are often highly intergrown or form remarkable crustified arrangements in the veins. Generally speaking fluorite, quartz, carbonates and sulphides are early minerals followed by gold and gold tellurides in depositional sequence. In crustified veins repetitions of the sequence are frequently present.

Halite occurs in a number of Precambrian deposits and in others where the underground waters are rich in sodium chloride. Occurrences in Canada are notable in the Con Mine, Yellowknife, in the Madsen Mine at Red Lake and in the Undersill Mine at Beardmore, Ontario. In most of these occurrences the halite is a late mineral localized in both crystalline and massive form along late slips and in small breccia faults. Both colourless and orange varieties have been observed; neither of these contain gold and silver (Au <0.005 ppm; Ag <0.1 ppm).

Other primary halide minerals are rare in auriferous deposits. The various silver halides including *chlorargyrite, bromargyrite* and *iodargyrite* may occur in the oxidized zones of gold-silver deposits in arid and semiarid climates or where there is an abundance of chloride, bromide or iodide in supergene waters. Copper halides such as *atacamite* have a similar distribution. The occurrence of these various halides with gold is considered in greater detail in Chapter IV.

Barite, celestite, anhydrite and *gypsum* are minor to major gangue minerals in some gold deposits, generally the gold-quartz type. Barite is rare in Precambrian deposits but is of increasing importance in younger deposits. Celestite is relatively rare, but the mineral occurs in some of the deposits at Kirkland Lake, Ontario, at Cripple Creek, Colorado and elsewhere. Anhydrite and gypsum appear to be relatively common in many gold-quartz deposits ranging in age from Precambrian to Tertiary. Both minerals are abundant in some of the veins in the Timmins (Porcupine) district of Ontario (Langford and Hancox, 1936). In the McIntyre Mine large zones of the quartz-feldspar porphyry are replaced by anhydrite. Some Tertiary gold veins are notable for the large amounts of barite and anhydrite they contain. In these deposits native gold and other gold minerals are paragenetically later than the two sulphates. In the paragenetic sequence of minerals in other gold deposits anhydrite and gypsum are generally later than both quartz and the first stage of carbonates, pyrite and arsenopyrite; in some wall-rock alteration zones, anhydrite appears to be about the same age as the replacement carbonates. With respect to gold and its minerals good relationships are rarely observed, but I have not seen gold in anhydrite or gypsum, which suggests that these two sulphates are probably very late minerals in some gold deposits.

Gypsum is a frequent mineral in the oxidized zones of auriferous deposits. Commonly gold is found in this mineral (selenite) in a wiry and dendritic form; also in bright spangles and as dust and mossy aggregates. Anhydrite is relatively rare in the oxidized zones of most auriferous deposits but has been noted in a few places.

Alunite, $KAl_3(SO_4)_2OH_6$, is a characteristic mineral of certain gold-quartz bodies and irregular silicified zones developed in volcanic sequences, usually dacite, latite, andesite and their equivalent volcanoclastics. Most of the deposits are of Tertiary age, although some occurrences are older, such as those in Mesozoic rocks on Vancouver Island (Clapp, 1915). Alunite is not known in deposits of Precambrian age.

Alunite occurs both in the wall rocks of the deposits as an alteration product and in the veins and irregular silicified zones. Characteristic examples occur at Goldfield, Nevada, at Rodalquilar in southeast Spain, and in the Kasuga Mine, southern Kyushu, Japan. The paragenetic relationships of alunite and gold minerals seem to be variable. In some cases the alunitization is paragenetically early and is followed by stages characterized by gold and silver; in a few deposits alunite, generally in small amounts, seems to have been deposited contemporaneously with native gold and gold tellurides. In some deposits alunite is commonly associated with *jarosite*, but the paragenesis of the latter is often difficult to ascertain. In most deposits it is probably of supergene origin, derived from the oxidation of pyrite.

The wall-rock alteration effects of alunitization are discussed in a following section.

Jarosites, including *argentojarosite*, are relatively common minerals in the oxidized zones of some auriferous deposits. *Anglesite*, the lead sulphate, and *brochantite* and *chalcanthite*, the copper sulphates, occur among the oxidized products of some auriferous deposits. These minerals are discussed in more detail in Chapter IV.

Apatite is often present in small amounts in various types and ages of gold deposits, but only in a few is it an important gangue mineral. Hulin (1930) mentions such a case in the Kennedy Mine of the Northern Lode System of California. Willimott (1884–1885) mentions gold in apatite near Péche Village, Wakefield Township, Gatineau County, Quebec. *Wavellite* is reported from some of the gold veins at Cripple Creek in Colorado and elsewhere. It may be a secondary (supergene) mineral in these deposits. Other phosphates are rare in gold deposits. Poznyak (1968) found *variscite* in a gold deposit in western Chukotka and Williams (1972) noted the presence of *embreyite*, $Pb_5(CrO_4)_2(PO_4)_2{\cdot}H_2O$, in specimens from the oxidized zones of the Beresov gold-quartz veins, Urals, U.S.S.R. These phosphates, in addition to *pyromorphite*, which occurs in certain plumbiferous gold deposits, are mainly supergene in origin. In the Bobrik auriferous lead-zinc deposits of the Donets Basin, U.S.S.R. Kuznetsov *et al.* (1973*a*) describe occurrences of *florencite*, a rare-earth–thorium phosphate. They claim that the rare-earth mineral can be used in prospecting when it is present in deposits. Aleksandrov *et al.* (1975) also found florencite in the carbonate-sulphide alteration zones of the Bodaibo deposits, eastern Siberia. *Monazite* is rare in most auriferous deposits particularly in the gold-quartz type. In the Rand the mineral occurs in only small amounts in the matrix of the conglomerates. According to Liebenberg (1955) the mineral is authigenic in origin in some bankets and detrital in others. Monazite is a universal associate of gold in recent placers.

Nolanite, $Fe_3V_7O_{16}$, appears to be relatively common in a number of gold deposits. The mineral occurs in the Nicholson deposits in the Goldfields area, Saskatchewan and in the deposits of the Golden Mile, Kalgoorlie, Western Australia. In the latter the nolanite is closely associated with native gold, various tellurides and vanadium muscovite (sericite) (Nickel, 1977). The vanadium is present in the trivalent state and is responsible for the green colour of the 'green leader'.

Numerous other phosphates, arsenates, vanadates and antimonates occur in gold deposits mainly in the oxidized zones. Among the most common of these are *pyromorphite-mimetite, vanadinite, scorodite, bindheimite* and *beudantite*. Where primary Ni-Co arsenides are present the supergene arsenates *annabergite* and *erythrite* may be present among the oxidation products. *Crocoite*, $PbCrO_4$, occurs in some of the oxidized parts of the veins in the Berezovsk goldfield, Urals, U.S.S.R. The occurrence of these minerals is discussed in more detail in Chapter IV.

Pyrite is the second most abundant mineral in gold deposits exceeded only by quartz. In some deposits pyrite may actually exceed quartz in quantity, as in certain skarn, polymetallic and massive sulphide deposits. All varieties of pyrite occur in gold deposits, including crystalline, massive, fine granular, radiating, reniform, globular ('buck shot') and stalactitic forms.

Gold occurs in pyrite in three forms – submicroscopic, perhaps in the lattice of the mineral; as native gold in small blebs, droplets and ramifying masses, generally at random; and as irregular masses, plates and grains along fractures, slips and seams or coating the faces of crystals. The first mode of occurrence is considered at length in the section on gold in other minerals in Chapter II. In many cases the second indicates exsolution; this mechanism is also discussed in Chapter II. The third indicates exsolution in some cases; in others introduction from sources outside the confines of the immediate pyrite crystal or pyritic mass is more probable. The source of this gold is considered in the section on the origin of gold deposits in this chapter.

Gold tellurides in contact with pyrite generally exhibit relationships similar to native gold. Most are late minerals occurring in fractures in individual pyrite crystals, in fractured parts of pyritic masses or perched on pyrite crystals.

In some auriferous ores the grain size of pyrite is indicative of the grade of the ore – the finer the grain size the higher the grade. Anhedral and subhedral pyrite are also commonly richer in gold than the euhedral sulphide. Large euhedral crystals of pyrite in deposits are generally low in gold, although, as usual, there are some exceptions to this observation. The reason for the enrichment of gold in fine-grained anhedral pyrite is not clear. In some cases it may be due to initial precipitation processes involving numerous sites for deposition of gold on pyrite nuclei; in other cases it may be related to crushing and mylonitization of the pyrite, followed by introduction of gold into the crushed and mylonitized zones.

Chesnokov and Popova (1971) studied the crystal morphology of pyrite from the Berezovo gold deposits of the Urals, U.S.S.R. The pyrite in the quartz veins, beresites and listwänites exhibited mostly a hexahedral, rarely a pentagonal-dodecahedral, crystal habit and formed essentially contemporaneously. Gold was found to be present in the form of small inclusions in the pyrite, the distribution of the metal increasing from the central to peripheral parts in large crystals (0.2–70 ppm Au). In undeformed crystals very small contents of gold were observed. In a somewhat similar study of the same deposits Chesnokov (1972) measured the paramater a_0 of pyrite. It was found to increase linearly with an increase in the Ni + Co content of the pyrite and also with an increase in the Ni content of the country rocks.

In the Kalba gold belt, Kazakhstan, U.S.S.R., Poltorikhin (1974*a,b*) noted six generations of fine-grained pyrite, the fifth generation (arsenic-gold-polymetallic stage) being the richest in gold. This pyrite tends to be complex and is often characterized by a pentagonal dodecahedral habit, outer layers ('jackets'), a zonal internal crystal structure, increased unit-cell parameters, the greatest mineral admixture (located along growth zones and microfractures), a lower microhardness and reflectivity and also a negative thermoelectric effect. In addition there is evidence of collective crystallization, the presence of several centres of crystallization and shifts in the angles of the growth zones, as well as cases of the recrystallization of globular pyrite into the crystalline variety. The gold-silver-rich pyrite is also said to exhibit the greatest number of defects in the crystal lattice.

In the Carlin and Cortez disseminated gold ores of Nevada, Wells and Mullens (1973) found by electron probe analysis that much of the gold in the unoxidized ores was concentrated together with arsenic in tiny pyrite grains (<0.005 mm) and in thin rims of larger pyrite grains scattered through the mineralized rock. No free gold was observed in the unoxidized ore, the particles being submicroscopic (<0.5 μ); in the oxidized ore, free gold is relatively abundant.

A number of investigations of the trace element content of pyrite in Canadian gold mines have been carried out, most with mixed results. Auger (1941) examined the minor element (Ag, Zn, Pb, Co, Ni, Cr, Pb, Ti, V, Sr and Ba) content of pyrite in a number of Canadian gold deposits. His results are too comprehensive to report in detail, but briefly he observed that (1) within one mine, one district or even one region, certain minor constituents show a consistent supremacy over other elements; (2) the type of ore deposit or the temperature of deposition has affected the nature of the accessory elements; (3) the nature of the wall rock has no apparent effect on the distribution of the minor elements in pyrite; (4) gold in the sulphide has a very erratic distribution; and (5) most of the elements found in pyrite except gold are part of the pyrite lattice.

Auger (1941) also made a systematic analysis of hundreds of sulphide samples from the Hollinger, Noranda and Siscoe mines. The samples taken at regular level intervals in the mines showed that: (1) there is a systematic variation with depth in the quantities of minor elements; (2) the type of variation does not agree with the usual zonal precipitation of minerals; and (3) the vein type of deposits gives more erratic values for the minor elements than does the massive sulphide types of deposits.

Hawley (1952) studied the distribution of Au, Ag, Te, Ni, Co, Pb, Zn, Sn, Mo, Mn, Cr, Ti, V, Bi and As in pyrite from four Canadian gold deposits in Ontario (Kirkland Lake area, Porcupine area) and Quebec (Powell-Rouyn Mine). Among other parameters variations in the concentration of the trace elements with depth, size of grain, host rock and matrix, vein type and temperature as determined with the pyrite geothermometer (Smith, 1947) were studied. The only persistent variation in composition of pyrite with depth noted was an increase in the cobalt content. Otherwise, excepting a few local variations, a uniformity in the trace elements to depths of 5000 ft were notable. Variations with other parameters appeared to be erratic, although there was a general consistency in the presence of tellurium in the pyrite from the Kirkland Lake area and its rarer occurrence in the other deposits.

Hawley and Nichol (1961) carried out a detailed study of trace elements (Co, Ni, Ag, Cr, Pb, Sn, Ti, V, Au, As, Bi, Mn and Mo) in pyrite from the Sudbury, Flin Flon, Chibougamau, Quemont, Noranda, Normetal, Manitouwadge and McIntyre deposits, all of which contain gold in minor or major amounts. Qualitatively the pyrite contained much the same suite of elements as pyrrhotite and chalcopyrite from the same deposits. Gold showed a greater preference for pyrite than pyrrhotite or chalcopyrite. Cobalt was preferentially concentrated in the order pyrite, pyrrhotite and chalcopyrite, the usual order of crystallization of these minerals in the ores.

The distribution of nickel was much less regular than cobalt in the three sulphides. With respect to silver in the three sulphides, chalcopyrite was always richer than pyrite or pyrrhotite. Other elements showed no regular or systematic differences in distribution in the three sulphides. Variations in the quantities of trace elements were noted in deposits of different as well as of similar types. Variations were also evident in the varieties of pyrite at McIntyre, in pyrite of different generations, in pyrrhotites of different generations, and in pyrrhotites of differently located orebodies, as at Sudbury. Variations also occurred with depth in certain ore shoots.

The writer has carried out trace element studies of pyrite in the gold deposits at Yellowknife, in those at Keno Hill and elsewhere. He has concluded that no particular patterns are indicated with respect to oreshoots within a single mineral belt or with depth. The reason for this is evidently due to the fact that pure pyrite cannot be separated from orebodies. It always contains wisps, blebs and patches of most of the ore minerals in the deposits in varying proportions. The trace element data, therefore, only indicate the degree of contamination of the pyrite by other ore minerals. Unless special procedures are taken, such as acid leaching, to remove these extraneous minerals the trace element data are erratic. Even these procedures are not satisfactory in most cases because pyrite cannot be ground fine enough. There are, however, some general qualitative relationships that are immediately observable in the trace element content of pyrite between different mineral belts as Auger, Hawley and Nichol and others have pointed out. Tellurium appears in greater abundance in some ores such as those at Kirkland Lake, and silver and gold are in general more abundant in pyrite in auriferous deposits than in those that are barren of the two metals. The nickel and cobalt contents and the Ni/Co ratio may also be indicative of certain types of orebodies and may also have depth significance. However, if these elements and their ratio are to be used effectively, pyrrhotite, arsenopyrite and minerals such as tetrahedrite and Ni-Co arsenides must be effectively removed completely from the pyrite before analysis.

My observations indicate that relatively high arsenic and antimony contents in pyrite are often an indication of enrichment of gold in auriferous deposits. This is a general statement, and not all deposits may show this effect. The reason may lie in the probability that much of the gold in endogenic processes migrates as an Au-As-S or Au-Sb-S complex as outlined in Chapter II.

Pyrite was used by Wanless *et al.* (1960) for isotopic studies of sulphur in the Yellowknife gold-quartz deposits. No systematic variations in the $^{32}S/^{34}S$ ratios were observed with depth in the mines, but some variations were noted in the pyrite from different ages and types of deposits and occurrences. These are discussed further in the section on origin of deposits in the last part of this chapter.

Pyrite is the most abundant metallic mineral in the gold-bearing quartz-pebble conglomerate reefs of the Witwatersrand in South Africa and Jacobina in Brazil; also in the uranium–thorium–rare-earth–bearing conglomerates of the Elliot Lake district, Ontario, Canada. In these three deposits one is justified in saying that with some slight modifications the nature and occurrence of pyrite is similar in all three deposits. The pyrite of the Witwatersrand has been extensively studied by Liebenberg (1955), Ramdohr (1958 *a,b*), Saager and Mihálik (1967) and Feather and Koen (1975). Saager (1970) has produced a table, which is self explanatory of the types of pyrite present in the Rand (Table 42). Most of the gold in the Rand bankets is not detrital in its present form. In its relationships with pyrite the authigenic or secondary gold is only rarely found as inclusions in the pyrite. Liebenberg (1955, p. 161) remarks that:

> Although gold is a relatively constant and widespread associate of the pyrite in the banket, this sulphide is only of secondary importance as a gold carrier. Mineragraphic work on specimens of the banket, and on pyrite concentrates, shows that amongst the tens of thousands of grains examined, only comparatively few of them contain included gold. Where present, the gold is found mainly as minute roundish and irregular inclusions in grains of buckshot pyrite, but relatively large irregular patches and veinlets of gold replacing the grains of pyrite were also observed. Replacement of pyrite by gold is sometimes indicated by minute stringers leading from the irregular inclusions of gold into the pyrite host. Cracks in the pyrite are occupied by gold. Occasional grains of pyrite are partly enveloped by thin films of gold from which fine stringers sometimes lead into the pyrite along microscopic fissures.

Saager (1969) mentions that:

> Frequently the gold, during its reconstitution, has migrated into the marginal areas of porous pyrite grains where it now forms secondary inclusions. Primary inclusions and myrmekitic intergrowths of gold with detrital pyrite and linnaeite have occasionally been observed.

Table 42. Different pyrite structures in the Basal Reef in the Orange Free State Goldfield, South Africa

Morphology	Structures	Allogenic	Authigenic
Rounded	Compact or porous	Detrital pyrite and black sands pyritized before their final deposition	Black sands pyritized *in situ*
	Pseudomorphic	Inhomogeneous black sands pyritized before their final deposition	Inhomogeneous black sands pyritized *in situ*
	Concretionary		Generally formed *in situ*; very rarely signs of short transportation
	Nodules containing 'mineralized bacteria'		Formed *in situ*, indicating presence of primitive life
Idiomorphic to hypidiomorphic	Compact or porous		Formed *in situ* by reconstitution of ferrous sulphide; rarely hydrothermal along dyke contacts
	Compact or porous encrustations		Formed *in situ* by reconstitution during metamorphism; generally overgrow older detrital pyrite
Xenomorphic	Veinlets and fracture fillings		Formed *in situ* by reconstitution during metamorphism

Source: Saager (1970).

Some varieties of buckshot pyrite in the Rand are relatively rich in arsenic (0.2–0.9%) and in nickel and cobalt (Ni – 0.1–0.4%; Co – 0.05–0.02%) (Saager and Mihálik, 1967). According to the present writer, investigations of all varieties of pyrite in the samples from the Rand and Elliot Lake that have been analyzed contain relatively high amounts of As, Sb, Ni, Co, Cu, Pb, Mo, Cd, Zn, W, Ag, Au, Hg, Te and Se. Some of these elements may be in the other sulphides and minerals that cannot be physically separated from the pyrite.

While pyrite is a fair indicator of the presence of gold in quartz-pebble conglomerates it is not an invariable one. Von Rahden and Urli (1969), who carried out a statistical treatment to determine the correlation of gold, uranium and pyrite in some of the reefs of the Rand, concluded that in certain reefs there is a significant correlation of the two elements with pyrite; in others the correlations are not significant.

Marcasite occurs in many types of auriferous deposits, but it is rarely differentiated from pyrite, and hence its exact paragenesis is relatively unknown. The mineral appears to be most common in the younger deposits, particularly those of Tertiary age. In these deposits marcasite is commonly an early mineral often accompanying or taking the place of pyrite. In all deposits marcasite continues to be precipitated from time to time during the mineralization process; in places it is a very late mineral lining fractures, small faults and slips that cut the gold orebodies. In these occurrences it is usually accompanied by calcite and is barren of gold.

Arsenopyrite accompanies gold in all types of its deposits. It is particularly common in Precambrian, Paleozoic and Mesozoic deposits, less so in Tertiary deposits. The relationships of arsenopyrite, gold, aurostibite and gold tellurides in deposits are essentially the same as those for pyrite and need not be repeated. In many deposits there is frequently a much closer association between gold and arsenopyrite than between the precious metal and pyrite, a feature emphasized by the frequent necessity to fine-grind and roast auriferous arsenopyrite ores to liberate the mineral in a form amenable to cyanide treatment. In some deposits even this treatment is not entirely satisfactory, since the gold is either submicroscopic or a lattice constituent of the arsenopyrite.

In some ores the grain size of arsenopyrite is indicative of the grade of the ore. At Yellowknife and elsewhere the finer the grain size of the arsenopyrite in some ore shoots the higher is the grade. The reason for this is not clear. In some cases it may be due to initial precipitation processes involving numerous sites for the deposition of gold on arsenopyrite nuclei; in other cases it may be related to crushing and mylonitization of the arsenopyrite, followed by introduction of gold into the crushed and mylonitized zones.

Arsenopyrite is a common mineral in gold-bearing quartz-pebble conglomerates although the amounts present are usually low (generally 1%). In the Witwatersrand Liebenberg (1955) mentions that gold particles occur in arsenopyrite more frequently than in pyrite. He has concluded that the arsenopyrite is a younger mineral, paragenetically speaking, than pyrite. Feather and Koen (1975) class arsenopyrite as an allogenic mineral in the Rand ores. They found up to 3 per cent Co and/or Ni in the mineral.

Some gold deposits contain massive shoots of arsenopyrite. Such deposits are known in the eastern Ontario gold belt (Marmora, Deloro, Actinolite), near Hedley British Columbia (Nickel Plate Mine), in the Bau gold district of Sarawak and elsewhere. In many of these deposits native gold is rarely visible, and considerable difficulty is encountered in winning the metal from the ores. In the Carlin and Cortez disseminated gold ores of Nevada, gold is intimately associated with arsenopyrite in an invisible form in the unoxidized ores (Wells and Mullens, 1973).

The presence of arsenopyrite or pyrite in deposits is not an invariable guide to gold mineralization, although it can be said that both minerals are favourable indicators, particularly in the gold-quartz and disseminated types of ores. There are, however, many of these deposits often with abundant pyrite that are practically devoid of gold. In the writer's experience, most deposits containing arsenopyrite appear to be enriched in gold, although the amounts may not be economic.

Both pyrite and arsenopyrite may be constituents of recent auriferous placers.

Pyrrhotite is present in nearly all types of gold deposits, being particularly common in those of Precambrian, Paleozoic and Mesozoic age. In Tertiary deposits, particularly those noted in the older literature as epithermal deposits, the mineral is relatively rare. The relationship of gold, aurostibite, gold tellurides and pyrrhotite are essentially the same as those for pyrite and need not be repeated.

According to Ebbutt (1948) pyrrhotite is much less commonly auriferous, a feature also noted by Dougherty (1935), Hawley and Nichol (1961) and others. Ebbutt thinks that this is due to the fact that pyrrhotite is prone to mash and recrystallize whereas pyrite, being brittle, holds fractures readily and allows the late gold to enter the fractures. Certainly, pyrrhotite exhibits a rather erratic behaviour with respect to its gold content in the writer's experience, a feature also commented on by Dougherty (1935) and Schwartz (1944).

In the Witwatersrand quartz-pebble conglomerates, pyrrhotite is present generally only in small amounts. Pelletier (1940) found a marked association between gold and pyrrhotite in these ores. Pyrrhotite concentrates show about twice as much gold as those of pyrite, which is much more abundant in the ores. The same author mentions that under the microscope gold also appears more commonly in pyrrhotite than in pyrite. According to Liebenberg (1955) pyrrhotite is a secondary (remobilized) sulphide, as are also galena, sphalerite, chalcopyrite, etc., and he agrees that gold is more commonly associated with these sulphides than with the paragenetically older pyrite. Feather and Koen (1975) state that some investigators have found no obvious relation to exist between pyrrhotite and gold.

Chalcopyrite generally accompanies gold in its deposits, the amount present being variable – minor quantities in gold-quartz deposits and minor to major amounts in polymetallic and skarn ores. In some gold-quartz deposits chalcopyrite is a good indicator of gold values; in others there is no particular relationship.

The relationship of gold to chalcopyrite is much the same as that described for pyrite and arsenopyrite. The gold may be present (1) in a submicroscopic form, probably in the lattice of the mineral, (2) in blebs and irregular masses in the mineral, the particles of gold having no particular orientation or relationships and (3) in seams, small cracks and fractures, and surrounding, penetrating and moulding the surfaces of the mineral grains. Gold occurs in its various forms in chalcopyrite for the same reasons as those given above for pyrite. The evidence is, however, stronger that gold (and silver) replace copper in the structure of the mineral because of the size of the atoms and other considerations discussed in Chapter II. This is commonly the relationship in chalcopyrite in some polymetallic, skarn and massive sulphide deposits.

Gold tellurides are usually paragenetically later than chalcopyrite. At Kirkland Lake the tellurides commonly occur in fractures in chalcopyrite and replace it extensively in places. The paragenesis of the tellurides in the Cripple Creek ores is complicated to say the least, but these minerals are generally later than the small amounts of chalcopyrite present in the ores (Lindgren and Ransome, 1906). In the complex seleniferous gold ores of Salida, Indonesia, the gold and other tellurides are later than chalcopyrite (Kieft and Oen, 1973).

In the Rand ores Liebenberg (1955) mentions that gold frequently occurs in chalcopyrite; however, usually only in small amounts. Feather and Koen (1975) state that chalcopyrite in the Rand is intimately associated with gold and sulphides, e.g., pyrrhotite, pentlandite and sphalerite, of the same paragenesis.

Sphalerite and *galena* are nearly universal minerals in gold deposits of all types, the amounts of the two minerals ranging from mere traces to several per cent. Sphalerite is the more abundant of the two minerals in most deposits.

Large amounts of galena are relatively rare in Precambrian deposits, especially those of Archean age. Younger deposits generally contain more galena, but the amounts are often highly variable from belt to belt and in different types of deposits.

Galena is generally paragenetically younger than pyrite, pyrrhotite and arsenopyrite in most types of gold deposits and often appears to have crystallized at or about the same time as gold. In many deposits the mineral is, therefore, a good indicator of gold values, a feature noted by numerous investigators (Schwartz, 1944). The same statements also apply to sphalerite, although this mineral is perhaps less indicative of good values in some deposits. One of the reasons why galena and gold are commonly associated in nature may be due to certain epitaxial relationships that exist between native gold and the lead sulphide.

The paragenetic and age relationships of gold, gold tellurides and aurostibite with respect to galena and sphalerite are not always clear-cut in gold-quartz, skarn and polymetallic deposits. In some ores the gold minerals occur as blebs, droplets and irregular masses in the two minerals and appear either to be exsolved constituents or to have crystallized together. The frequent occurrence of gold and other tellurides in or associated with galena is notable in places. The tellurides may have exsolved from the galena, replaced the galena or the minerals may have been coprecipitated together in some manner or other. All of these relationships probably are related to the fact that galena, PbS, and altaite, PbTe, are similar structurally and not greatly different from some of the bismuth, gold and silver tellurides in some of their parameters. There are, however, numerous references in the literature describing gold and gold tellurides in veinlets that occur in fractures in both galena and sphalerite, and gold frequently occurs along the cleavages of galena. Replacement of galena and sphalerite by gold is also commonly referred to, and veinlets of gold, gold tellurides, galena, sphalerite, chlorite, sericite and clear quartz commonly intersect pyrite and arsenopyrite. In crustified veins, gold and gold tellurides commonly occur along the crystal faces of galena and sphalerite, in crystal vugs of the two minerals and perched on crystals of the two sulphide minerals. In the complex seleniferous banded gold ores of Salida, Indonesia, the gold and other tellurides are closely associated with the bands of galena and fine-grained sphalerite (Kieft and Oen, 1973). In summary, most of the gold minerals in gold-quartz, skarn and polymetallic deposits are either contemporaneous with galena and sphalerite or later than these two minerals.

Liebenberg (1955) notes that gold frequently occurs as inclusions and replacement bodies in the small amounts of galena and sphalerite in the bankets of the Rand. Both of the base metal sulphides are paragenetically younger than the bulk of the pyrite. Feather and Koen (1975) class sphalerite as a consistent authigenic mineral in the reefs.

Stibnite is a relatively common mineral in gold deposits of the gold-quartz, polymetallic and disseminated type. The mineral is rare or absent in most skarn deposits seen by the writer. No age differences in the occurrence of stibnite have been noted by the writer, nor are any mentioned in the literature. Some Precambrian deposits (e.g., Yellowknife and Red Lake, Canada; Murchison Range, South Africa) are rich in stibnite, as are also some deposits in Paleozoic (West Gore, Nova Scotia; Costerfield, Victoria, Australia), Mesozoic and Tertiary rocks (Hauraki, New Zealand). Recent hot springs precipitating stibnite, also deposit gold in places. Some gold-bearing stibnite deposits have been mined mainly for antimony with gold as a byproduct (e.g., Murchison Range, South Africa).

Stibnite is not mentioned as being present among the ore minerals of the Witwatersrand (Saager, 1968), but Feather and Koen (1975) have recently identified the sulphide in a rough platinoid concentrate from Western Deep Levels, Limited.

In the disseminated ores of Carlin, Cortez and Gold Acres in Nevada, stibnite is the principal antimony mineral.

Stibnite is invariably a late mineral in gold deposits. At Yellowknife it occurs with gold in late fractures, as nests in crushed parts of the quartz veins and in vuggy parts of the veins. In the Red Lake deposits it is, likewise, late occurring in much the same manner as at Yellowknife. Horwood (1945) places it among the latest minerals deposited in the Cochenour Willans Gold Mine. Stibnite is also late at West Gore in Nova Scotia, in the Paleozoic auriferous deposits in France and in most of the other samples of stibnite-bearing gold ores seen by the writer. In young (Tertiary) deposits successive generations of stibnite and gold may occur in crustified veins.

The relationship of native gold and gold tellurides to stibnite are not always clear. Deposition of the gold minerals and stibnite in most deposits appears to be contemporaneous or nearly so. Native gold often occurs as blebs and droplets in massive stibnite, and some relationships suggest replacement of stibnite by gold. Occasionally one sees veinlets of native gold in broken blades of stibnite, and in some vugs gold is perched on stibnite crystals. Otherwise, stibnite and gold have mutual relationships and appear to have crystallized together.

The tellurides generally show the same relationships to stibnite as native gold when found together.

Aurostibite, $AuSb_2$, commonly occurs in stibnite-bearing gold veins. At Yellowknife the relationships of the mineral and stibnite are not clear. Most of the aurostibite rims or penetrates gold particles along small fractures. It appears to be later than the gold and stibnite, and probably also the sulphosalts. In some stibnite ores in the Canadian Shield where gold is present in close proximity to the stibnite, the particles of the metal have a peculiar brownish colour that may be due to incipient formation of aurostibite. Metallurgists believe that this gold resists amalgamation and cyanidation. Aurostibite is also known in the ores of the Chesterville Mine, Larder Lake, Ontario; in the gold deposits of Krasna Hora and Milesov in Czechoslovakia; in the antimony veins of Costerfield, Victoria, Australia; and in the Lone Hand and Jessie gold mines, Gwanda District, southern Rhodesia; also in a number of auriferous-antimoniferous belts in U.S.S.R. In all of these occurrences the aurostibite is late – it precedes, accompanies or follows gold in the paragenesis of the vein minerals. (*See also* the section on the mineral in Chapter II.)

Niccolite, cobaltite and *cobaltian loellingite* are relatively rare in gold-quartz, gold-bearing skarn and polymetallic deposits. In the Rand ores they occur together with minerals such as glaucodot, bravoite, millerite, pentlandite, linnaeite, gersdorffite, skutterdite and safflorite, in small amounts (Saager, 1968; Feather and Koen, 1975). Most of these minerals are authigenic (remobilized) minerals and are commonly associated with similarly reconstituted gold. Some investigators have noted a striking affinity between Ni-Co minerals, especially cobaltite, and gold in the Witwatersrand ores (Liebenberg, 1955; and others). Saager (1968), however, did not notice the intimate association of cobaltite and gold in the Basal Reef of the Orange Free State Goldfield and thinks that the cobaltite occurs mainly as water-worn detrital grains. Feather and Koen (1975) recognized two distinct generations of cobaltite and gersdorffite in the reefs – one allogenic and the other authigenic.

In some skarn bodies such as those of the French Mine at Hedley, British Columbia, safflorite and cobaltite occur in the ores, but the age relations of the gold and these minerals are not entirely clear. The gold appears to be paragenetically later than the Ni-Co sulphides and arsenides in most cases. Krieger and Hagner (1943) reported similar relationships with respect to the gold-bearing nickel minerals at Alistos, Kinaloa, Mexico. In the Soviet Union a number of the auriferous skarns carry small to moderate amounts of a number of nickel-cobalt minerals, including *gersdorffite, cobaltite, danaite, glaucodot* and *safflorite*. These are generally early minerals in the paragenetic sequence in the deposits. According to Byers (1940) cobaltite is a much earlier mineral than gold in the Nighthawk Peninsular Gold Mine in the Timmins area of Ontario. In the Ross Gold Mine at Ramore, Ontario, the age of the cobaltite and gold is not clear. The same is true where small amounts of Ni-Co arsenides occur in some of the gold mines (Algold Mine) in the Goudreau-Michipicoten area of Ontario. At the Mackinaw Mine, Snokomish County, Washington, the gold is later than niccolite and pentlandite and is always associated with maucherite ($Ni_{11}As_8$) (Milton and Milton, 1958).

It is of interest to note that veins with abundant native silver and Ni-Co arsenides and sulphides, such as those at Cobalt and elsewhere, contain relatively little gold. The ores at Cobalt contain only <10 to 100 ppb Au on the average. Certain similar Ni-Co arsenide deposits generally with low amounts of silver and commonly with pitchblende or uraninite may be enriched in gold. An example of this type of deposit is found at Cluff Lake in the Carswell structure, northern Saskatchewan. There, the veins contain abundant uraninite and colloform pitchblende, native gold, gold tellurides, clausthalite, chalcopyrite and nickel-cobalt sulpharsenides, mainly gersdorffite. The paragenesis of the gold with respect to these various minerals is uncertain; it appears in most cases to be a late mineral. Some of the Ni-Co arsenide veins at Bou Azzer in Morocco are auriferous, the gold being commonly intimately associated with maucherite and niccolite often in inclusions in the latter (Caillere and Dietrich, 1966).

Molybdenite is a common mineral in gold ores of all types, but is generally present only in very small amounts. It is noted among the ore minerals of the Rand by Saager (1968). The mineral appears in deposits of all ages from Precambrian to Cenozoic. It seems to be most frequent in those of Precambrian and Tertiary age.

Gold-bearing skarn bodies usually contain molybdenite, but the age relationships with native gold and gold tellurides are frequently difficult to ascertain, since the minerals are not often contiguous. Probably gold is later than molybdenite in most skarn bodies.

In gold-quartz and polymetallic ores, molybdenite is often late occurring along late slips and fractures. In this respect some molybdenite may be closely related paragenetically with gold and gold tellurides, although these minerals and the sulphide are rarely contiguous. At Yellowknife no good relationships between the gold minerals and molybdenite could be observed except in one vein where the gold appears to be contemporaneous or slightly younger than the sulphide. The molybdenite in the veins at Yellowknife investigated by the writer is predominantly the 2H type; only in the auriferous porphyry at the Con Mine at Yellowknife is the molybdenite of the 3R type (Traill, 1963; Boyle, 1968*c*). Another locality where 3R type molybdenite occurs is in some veins near Heron River, northern Ontario.

In some deposits molybdenite is commonly intergrown with pyrite, sphalerite or galena, indicating that it is perhaps paragenetically older than gold and gold tellurides. At the War Eagle Mine, Rossland, British Columbia, gold commonly impregnates molybdenite, but the age relationships are not entirely clear. Late generations of quartz seem, in some places, to bring molybdenite into various deposits. These appear to be younger than the last generation of gold and gold tellurides, since they do not contain gold minerals in most places.

Saager (1968) says that the molybdenite in the Rand always occurs as idiomorphic inclusions in rounded pyrite grains, which he thinks are allogenic components of the bankets.

Bismuth-bearing minerals, including *bismuthinite, matildite, cuprobismutite, tellurobismuthite, tetradymite, joseite, volynskite, cosalite* and a number of other bismuth sulphosalts, are recorded from many gold deposits, mainly the skarn, gold-quartz and polymetallic types. These minerals tend to occur more frequently in ores younger than Precambrian, although the element bismuth is generally recorded in all the Archean gold deposits, particularly in minerals such as galena and sulphosalts. Bismuth tellurides, native bismuth, bismuthinite and bismuth sulphosalts have all been recorded from Archean deposits. Some Proterozoic gold deposits are also notably enriched in bismuth.

In the ores of Goldfield, Nevada, (Tertiary) bismuthinite is relatively common and is accompanied by a bismuth-bearing variety of tetrahedrite (goldfieldite). The rich ores average 0.35 per cent Bi (Lindgren, 1933). The Tertiary ores of Boise Basin, Idaho, contained abundant galenobismutite, and this and other bismuth minerals were often intergrown with gold. According to Schwartz (1944) bismuth indicated good gold values in the Boise district. In the gold and silver-bearing sulphide of the Ohio mining district, Marysvale, Utah, bismuth minerals and cassiterite are relatively abundant. The principal bismuth minerals in these ores are cuprobismutite and bismuthinite. The native gold is paragenetically later than cuprobismutite (Radtke *et al.,* 1967).

Bismuth and bismuthinite accompany gold in the old Windpass Mine, British Columbia, and galenobismutite, bismuthinite and cosalite are common in the Cariboo Gold Quartz Mine at Wells, British Columbia (Warren, 1936). Bismuthinite occurs in some of the gold-quartz veins at Yellowknife, Northwest Territories.

In France, bismuth and bismuthinite are common in the gold ores of the Salsigne-Aude district (Tollon, 1970). In U.S.S.R. bismuth minerals accompany gold in a number of deposits in age from Paleozoic to Tertiary. Volynskite occurs in some of the gold deposits of Central Asia (Markova, 1967), and a number of bismuth minerals, including cosalite, native bismuth, bismuthinite, tellurobismuthite, tetradymite, wehrlite, galenobismutite, lillianite and emplectite are found in the Darasun gold field and Sredniy Gologtay deposits (Sakharova, 1969*b*, 1972; Sakharova and Krivitskaya 1972). In these deposits the bismuth minerals accompany the native gold, or the gold preferentially replaces the bismuth minerals. In any case there is a very close paragenetic relationship between gold and bismuth. Cosalite is reported from the gold ores of Berezovsk and Kochkarsh, Urals, U.S.S.R. (Vertushkov *et al.,* 1972).

Stillwell and Edwards (1941) report the occurrence of native bismuth in the gold ores of Tennant Creek (Precambrian) and native bismuth and galenobismutite in those at Cobar (Paleozoic) in Australia. Maldonite the black gold of some Australian mines (Maldon Mine, Victoria) is a mixture of gold and bismuth. The mineral also occurs in some greisen and skarn deposits. More recent research on the Tennant Creek ores has shown that the new mineral *junoite,* $Bi_8Pb_3Cu_2(S,Se)_{16}$, is the most abundant bismuth-bearing mineral at the Juno Mine (Large and Mumme, 1975; Mumme, 1975; Mumme and Watts, 1976). It is accompanied by the bismuth-selenium sulphosalt, *wittite,* and a variety of other selenium-bearing sulphosalts.

A few bismuth-bearing minerals have been recorded from the ores of the Witwatersrand (Feather and Koen, 1975). They include the platinoid minerals michenerite, moncheite and geversite. (*See also* the discussion on the native plantinoids in this section.)

These occurrences indicate the widespread distribution of bismuth minerals in gold deposits throughout the world. In general native gold and tellurides are later than or contemporaneous with most of the bismuth minerals. In some occurrences gold and native bismuth occur in an intergrown manner such as to suggest some sort of unmixing phenomena. Samples of native gold from numerous occurrences also contain traces to minor amounts of bismuth indicating that the two elements probably migrate together and are contemporaneously precipitated. There is also a frequent association of gold tellurides with various bismuth minerals, particularly bismuth tellurides and tetradymite.

In the oxidized zones of auriferous deposits containing hypogene bismuth minerals there is commonly an association of gold with supergene bismuth minerals such as bismite, bismoclite and bismutite. Such occurrences are reported from Honduras, Australia, U.S.S.R. and elsewhere. These minerals may also accompany gold in recent placers where they coat or are otherwise associated with native bismuth, bismuthinite and other bismuthian minerals.

Cinnabar is a relatively uncommon mineral in most gold deposits. It is particularly rare in skarn type deposits and has not been recorded in the Rand ores despite the fact that the latter are relatively high in mercury (Table 35).

Cinnabar is present in a number of gold-quartz and polymetallic deposits, some cutting rocks as old as Precambrian. Lindgren (1933) mentions the presence of the mineral in the veins of California (Mother Lode) and in those in central Idaho. Cinnabar is also common in some of the gold-bearing veins in Archean rocks in Rhodesia, mainly in carbonated andesitic lavas (greenstones) (Robertson, 1972); also in the Murchison Range, South Africa (Sahli, 1961). The mineral also occurs in some low-grade gold deposits in Grenville rocks near Clyde Forks, Ontario. Elsewhere, cinnabar is met with in small amounts in a number of Tertiary gold deposits throughout the world, examples being Cripple Creek in Colorado and a number of deposits in the far eastern U.S.S.R. Cinnabar occurs in small amounts in the disseminated ores of Carlin, Nevada (Radtke *et al.,* 1972*a,b*). Cinnabar, metacinnabar and gold are associated in the active solfataras of the Mendeelev volcano in the Kurile Islands (Ozerova *et al.,* 1969).

The age relationships of gold and cinnabar are not clear from most of the descriptions in the literature. Evidently, both gold and cinnabar are late minerals and may in part be contemporaneous. This is suggested by the common occurrence of mercury in native gold from many areas. In the Chinkuashih gold-copper deposits in Taiwan cinnabar and native sulphur are the last minerals to be deposited (Huang, 1973).

Supergene cinnabar is present in small amounts in the oxidized zones of certain auriferous deposits carrying mercuriferous tetrahedrite. The cinnabar evidently derives from the oxidation of the sulphosalt.

Cinnabar may accompany gold in certain placers, generally in mercuriferous belts.

Realgar and *orpiment* are rarely if ever found as hypogene minerals in gold deposits of Precambrian age. In younger deposits they appear with increased frequency and are common in some of the gold-quartz veins and disseminated deposits of the Tertiary. The realgar in the deposits at Nagyág in Romania is probably of hypogene origin, although some writers have considered it to be a supergene mineral.

In the disseminated ores of Carlin and Getchell in Nevada hypogene orpiment is relatively common, some of it being thallium-bearing (Radtke *et al.,* 1974*a*). Realgar is also present in these deposits. The auriferous Tertiary quartz vein at Zarehshuran in western Iran renowned for its orpiment, contains some realgar, stibnite, cinnabar, sphalerite, galena, jordanite, getchellite ($AsSbS_3$) and fine-grained black pyrite. The pyrite carries the bulk of the gold (Bariand and Pelissier, 1972). Present day hot springs depositing gold in the Taupo volcanic zone of New Zealand commonly contain up to 2 per cent arsenic, some of which is probably bound as orpiment (Weissberg, 1969).

The paragenetic relationships of realgar, orpiment and gold minerals are rarely mentioned in the literature. In general, the arsenic sulphides and gold minerals are late; in some deposits the minerals are essentially contemporaneous in age; in others gold is slightly younger than the arsenic sulphides. At Carlin, Nevada the orpiment and realgar together with stibnite and cinnabar are apparently later than the gold mineralization (Harris and Radtke, 1974).

Some of the realgar and orpiment in the zones of oxidation and reduction in auriferous deposits may be of supergene origin.

Argentite (acanthite), chalcocite and *stromeyerite* have a varied occurrence in gold deposits. *Argentite (acanthite)* is relatively rare in Precambrian deposits but common in younger deposits especially in the great gold-silver bonanzas of the Tertiary, classic examples being the celebrated Comstock Lode in Nevada and the veins at Tonopah, Nevada. Elsewhere in the world argentite-gold-quartz veins are or were mined extensively in Mexico, Japan, U.S.S.R. and New Zealand (Hauraki field). Argentite is generally a late mineral in auriferous deposits, and in many places it appears to have been deposited nearly contemporaneously with gold and its minerals. In some deposits gold is slightly later than argentite.

Primary *chalcocite* is not particularly common in any type of gold deposit. It is present in many auriferous skarn-type deposits, in various vein and porphyry-type deposits such as those at Butte, Montana and in a great variety of other types of deposits mined essentially for copper. The mineral is rare among the ore minerals of the Witwatersrand (Feather and Koen, 1975). The great supergene chalcocite bodies in many porphyry and other types of copper deposits are commonly enriched in both gold and silver. *Digenite* accompanies gold in the uraniferous deposits of Cluff Lake in the Carswell structure, northern Saskatchewan.

Stromeyerite, the silver copper sulphide, has about the same distribution in gold deposits as that mentioned for argentite and chalcocite. Some stromeyerite in deposits is a very late mineral and may actually be of supergene origin. Saager (1968) lists stromeyerite among the ore minerals of the Rand. It occurs near dykes crosscutting the Basal Reef in the Orange Free State Goldfield.

Bornite is not a particularly characteristic mineral of gold deposits, although it occurs in many types ranging in age from Precambrian to Mesozoic. It is not common in Cenozoic (Tertiary) deposits.

Bornite is probably most common in skarn-type deposits and gold-quartz veins. In the former it may be abundant, in the latter the mineral is generally an accessory, although in places, as in the auriferous disseminated copper orebody of the McIntyre Mine in the Timmins area, the sulphide is fairly abundant. The mineral is rare among the ore minerals of the Rand (Feather and Koen, 1975).

Cubanite has been recorded from a few gold deposits ranging in age from Precambrian (Morro Velho, Brazil) to Tertiary (far eastern U.S.S.R. gold belt). Paragenetically it appears at about the same stage as chalcopyrite. Cubanite is a very rare mineral in the reefs of the Rand (Feather and Koen, 1975).

Alabandite, MnS, is largely restricted to Tertiary gold deposits or to those of older age formed apparently at low temperatures. The mineral is invariably associated with veins rich in other manganese minerals such as rhodochrosite, rhodonite, johannsenite, inesite, etc. Alabandite occurs in the Tertiary gold and silver veins in Romania (Kapnik, Offenbánya and Nagyág), in Turkey, in Mexico and in Japan, and is relatively widespread in the western United States (Tombstone, Bisbee, Eureka Gulch and elsewhere). Some of these occurrences are described in more detail by Hewett and Rove (1930).

Albandite is generally earlier than native gold and its minerals in most deposits, but in some it appears to be nearly contemporaneous.

The *telluride* minerals are characteristic of nearly all types of gold deposits. They have not, however, been recorded in the disseminated deposits at Carlin and elsewhere in Nevada (Radtke *et al.,* 1972*a*). Deposits greatly enriched in base metal and precious metal tellurides include Kirkland Lake and Red Lake in Ontario (Precambrian); Brad and Nagyág, Romania (Transylvania) (Tertiary); Kalgoorlie in Western Australia (Precambrian); Vatukoula, Fiji Islands (Tertiary); and Cripple Creek in Colorado (Tertiary). The commonest silver tellurides in gold deposits are hessite, petzite, empressite, stuetzite and muthmannite in about that order. The various platinoid tellurides, including vincentite, kotulskite, merenskyite, temagamite and michenerite do not seem to have been recorded in gold deposits, except in the Rand where michenerite, moncheite and geversite have been noted (Feather and Koen, 1975). The most common base metal tellurides in gold deposits are altaite, PbTe, tetradymite, Bi_2Te_2S, joseite, $Bi_3Te(Se,S)$, coloradoite, HgTe, wehrlite, BiTe, hedleyite, Bi_7Te_3, and melonite, $NiTe_2$, in about that order. Frohbergite, $FeTe_2$, occurs in some of the veins of Sacaramb and Fata Baii in Romania (Ramdohr and Udubasa, 1973) and volynskite, $Ag(Bi,Sb)Te_2$, is present in the ores of the Zhana-Tyube goldfield in northern Kazakhstan (Spiridonov *et al.,* 1974). Tellurbismuth, Bi_2Te_3, often containing appreciable contents of antimony, has been recorded in a number of different types of auriferous deposits. Tellurantimony, Sb_2Te_3, is apparently rare in auriferous deposits.

Tellurides in skarn-type and gold-quartz deposits are invariably late minerals in the depositional sequence. In a number of auriferous deposits the tellurides are often closely associated with galena. Native gold and gold tellurides are commonly deposited contemporaneously, alternatively or clearly later than the base metal tellurides in most deposits. In others free gold and auriferous tellurides occur in association with the base metal tellurides but not in contact, and no particular paragenetic relationships can be established. At Cripple Creek, Colorado, the paragenesis of the mainly microscopic tellurides with respect to pyrite, galena, sphalerite, quartz and fluorite is complicated, but broadly speaking the tellurides are late (Lindgren and Ransome, 1906). In the Robb-Montbray gold property in Quebec, Thomson (1928) found the following genetic succession of sulphides and tellurides: pyrite, chalcopyrite and pyrrhotite, sphalerite, krennerite, tetradymite and altaite, petzite and coloradoite, native gold and chalcocite (supergene?). In the Kirkland Lake gold ores Thomson (1922) established the depositional sequence of minerals as follows: (1) quartz and pyrite; (2) carbonate stringers; (3) sphalerite followed in turn by chalcopyrite, tellurides and native gold. Thomson remarks repeatedly about the association of tellurides in the Ontario deposits with native gold, predominately in small late fissures and fractures in quartz and early sulphides. In the complex seleniferous gold ores of Salida, Indonesia the tellurides (hessite, sylvanite, petzite, altaite) are invariably associated with galena, which in the paragenesis of the deposits follows sphalerite, chalcopyrite and pyrite (Kieft and Oen, 1973). There are apparently no selenides in the deposit, the selenium being mainly in acanthite, galena and arsenpolybasite; hence, the paragenetic relationships of tellurides and selenides could not be determined. The general antipathy of telluride and selenide minerals in deposits has already been discussed in Chapter II.

The general distribution and habit of hypogene tellurides in auriferous deposits are of interest. Gold tellurides occur with greatest frequency in Tertiary deposits and in Precambrian deposits, mainly those of Archean age. Gold-quartz veins and polymetallic deposits, marked by the presence of quartz or extensive silicified zones, tend to be richer in auriferous tellurides than skarn deposits; the auriferous quartz-pebble conglomerates are notably poor in auriferous, argentiferous and base metal tellurides. In Precambrian deposits auriferous and argentiferous tellurides are generally accompanied by those of Hg, Pb, Cu and Bi; native tellurium is relatively rare in both Archean and Proterozoic deposits. Calaverite commonly predominates in Precambrian deposits, the other Au-Ag tellurides being subordinate. In Tertiary deposits tellurides of metals other than gold and silver are comparatively rare, native tellurium on the other hand is relatively abundant in some deposits. Krennerite tends to dominate over calaverite in a number of Tertiary deposits, but in some calaverite is the principle telluride. Native gold is commonly much more abundant in telluriferous Precambrian deposits (e.g., Kirkland Lake) than in most Tertiary deposits rich in tellurides (e.g., Cripple Creek). (*See also* the discussion on tellurides in Chapter II.)

Supergene tellurides appear to be relatively rare in auriferous deposits, although in the Central City district of Colorado some of the copper tellurides may be of this origin (Kelly and Goddard, 1969).

Some of the base meal tellurides accompany native gold and gold tellurides in certain recent placers but generally only in small amounts.

The *selenide* minerals, including naumannite, Ag_2Se, umangite, Cu_3Se_2, berzelianite, Cu_2Se, clausthalite, PbSe, aguilarite, $Ag_4(S,Se)$, penroseite, $(Ni,Cu)Se_2$, crookesite, $(Cu,Tl,Ag)_2Se$, guanajuatite, Bi_2Se_3, and tiemannite, HgSe, occur mainly in Tertiary gold-quartz veins. Examples are found at Republic, Washington; Tonopah, Nevada; Kushikino Mine, southern Kyushu, Japan; Radjang-Lebong, Sumatra, Indonesia; and the Hauraki belt (Waihi Mine) in New Zealand.

The Tennant Creek auriferous magnetite ores in Northern Territory, Australia are in Proterozoic rocks. They carry about 0.05 per cent selenium mainly partitioned into the new bismuth sulphosalt, junoite, $Bi_8Pb_3Cu_2(S,Se)_{16}$, wittite and heyrovskite, two other seleniferous bismuth-bearing, sulphosalts and a variety of other selenium-bearing sulphosalts (Large and Mumme, 1975). The galena in this deposit is also seleniferous containing up to 7.2 per cent Se.

Two facts about selenium in gold deposits are of interest: firstly, the element is a common trace element in the various sulphides and sulphosalts in nearly all types and ages of gold deposits, but selenide minerals appear only in abundance in deposits of Mesozoic-Cenozoic vintage; secondly, as mentioned above, tellurides and selenides are seldom associated with one another in gold deposits. The reason for this is obscure.

Sulphosalts are common in most types of gold deposits with the exception of the skarn type in which they are rare. *Tetrahedrite-tennantite* is probably the most abundant of the sulphosalts in gold deposits, occuring in veins and polymetallic lodes of all ages. The copper and lead sulphosalts such as *zinkenite, chalcostibite, jamesonite, cosalite, kobellite, semseyite, boulangerite, bournonite, meneghinite* and *jordanite* are most abundant in Precambrian, Paleozoic and Mesozoic deposits. They are relatively rare in Tertiary deposits. The silver sulphosalts, *proustite, pyrargyrite, matildite, miargyrite, smithite, stephanite* and *polybasite,* are uncommon in Precambrian deposits; they appear in increasing amounts in Paleozoic and Mesozoic deposits and reach their highest abundance and frequency in Tertiary deposits. There are some exceptions to this generalization. The Berens River deposit in northwestern Ontario for instance contained relatively abundant amounts of ruby silvers. In the Rand ores only tennantite-tetrahedrite, prousite and an unnamed $Ni_9Sb_2S_8$ phase have been identified (Saager, 1968; Feather and Koen, 1975). Several sulphosalt minerals are reported by Radtke *et al.* (1972*a*) in the disseminated ores at Carlin, Nevada.

The tin sulphosalts, *stannite,* Cu_2FeSnS_4, and *canfieldite,* Ag_8SnS_6, are restricted mainly to gold deposits in stanniferous regions.

Enargite, Cu_3AsS_4, and *famatinite,* Cu_3SbS_4, are particularly characteristic of deposits of Tertiary age. Locally they may appear in older deposits but rarely in any quantity; these minerals are very rare in Precambrian deposits. In the deposits located in the 'great circle of fire' that encompasses the Pacific, enargite is common in gold-bearing cupriferous polymetallic deposits. Here we may note its occurrence in many deposits in Chile, Peru, Argentina, western United States (particularly Butte, Tintic and Goldfield), Japan and the Phillipines (Luzon). In the siliceous gold ores of the Kasuga Mine in southern Kysushu, Japan, enargite, famatinite (luzonite) and native gold are the main ore minerals (Taneda and Mukaiyama, 1970). In the other great 'belt of fire' of the globe – the Mediterranean-Himalayan – enargite occurs in a number of auriferous deposits, but particularly in the quartz-sulphide ores at Bor, Yugoslavia.

Some of the sulphosalts in auriferous deposits, particularly tetrahedrite-tennantite, enargite and famatinite, commonly follow sphalerite and are more or less coeval with chalcopyrite in the paragenetic sequence of the veins. In these cases the minerals preceded the deposition of native gold and tellurides. Gold, however, accompanied the mineralization by tetrahedrite-tennantite in a number of deposits since these sulphosalts contain the precious metal as lattice constituents.

The remainder of the sulphosalts, including in places tetrahedrite-tennantite, are late minerals in most gold deposits, introduced at about the same time as gold. At Yellowknife, native gold and various sulphosalts are intimately associated, the latter being particularly good indicators of high grade shoots. A similar situation exists in all other gold deposits seen by the writer. Under the microscope the sulphosalts are commonly closely associated with gold in its ores, both forming mutual boundaries in many cases. In some sections one often sees gold apparently replacing the sulphosalts, and gold and sulphosalts often occur together in late veinlets with quartz and carbonates (Pl. 14). In banded and crustified ores the sulphosalts and gold are generally always intimately associated, sometimes alternating in their depositional sequence. In vugs and nests, gold crystals and dendrites can sometimes be seen perched on crystals and fibres of sulphosalts. Some of the gold in shoots containing sulphosalts is present as a lattice constituent of these minerals, particularly tetrahedrite-tennantite. Tiny grains of gold with no apparent orientation are also commonly present in sulphosalts. These may have been exsolved from the minerals, or may have been coprecipitated with them.

The rich silver-bearing sulphosalts, including pyargyrite, proustite and polybasite, commonly precede the deposition of native gold in most deposits, although in some the minerals commonly look to be coeval.

Gold tellurides apparently bear the same relationships to the sulphosalts as native gold, but I have rarely seen these two types of minerals in juxtaposition where unequivocal paragenetic conclusions could be drawn. One often sees, however, gold tellurides in very late veinlets and as projecting crystals in the latest vugs in gold veins. These are usually later than the sulphosalts. In the complex seleniferous gold ores of Salida, Indonesia, pearceite, polybasite, prousite, enargite (luzonite) and arsenpolybasite apparently followed the deposition of tellurides (Kieft and Oen, 1973). (For the relationship of aurostibite to sulphosalts *see* the discussion above under stibnite and the section on aurostibite in Chapter II.)

Supergene sulphosalts appear to be rare in auriferous deposits. In the writer's experience only the ruby silvers may occur as supergene minerals. In the zones of reduction of some auriferous deposits in Yukon pyrargyite is present but is very low in gold content (Boyle, 1965*a*). The relationships to supergene gold could not be determined.

Hematite, magnetite, ilmenite, franklinite, spinel and *gahnite* have all been recorded from gold deposits. Most of these minerals are relatively minor gangue constituents in gold-quartz and gold-bearing polymetallic deposits, but in some skarn and pipelike deposits they may be relatively abundant. In some skarns magnetite is a concentrator of gold together with arsenopyrite and chalcopyrite. Hematite occurs in deposits of all types and ages. The mineral is particularly abundant in the Tarkwa quartz-pebble conglomerates of Ghana. There, it occurs abundantly in the matrix and forms thin (sedimentary) layers in the conglomerate beds. In gold-quartz veins a dusting and impregnation by hematite is common in some districts, examples being Kirkland Lake, Ontario and Tennant Creek, Australia. Specular hematite also occurs in some of the veins at Kirkland Lake. At Tennant Creek, gold, bismuth and copper mineralization occurs in discordant magnetite- and hematite-rich lodes which have an ellipsoidal to pipelike shape (Large, 1974). These pipes cut felsic sediments and pyroclastics of Lower Proterozoic age. Massive magnetite and much chlorite form the core of the Juno lode, and gold is concentrated in the centre of this core. The other elements are vertically zoned.

Auriferous uranium deposits are commonly marked by hematitization, which consists of a pervasive reddening of the rock adjacent to the quartz-carbonate and pitchblende veins. The hematite appears in most cases to be locally derived by the oxidation of the ferrous iron in minerals such as amphiboles, pyroxenes and micas. Some ferric iron may be introduced into the sites of mineralization from sources in the nearby host rocks. Some auriferous uranium deposits have both intensely chloritized and hematitized zones that halo the mineralization sites. Such phenomena suggest introduction of both magnesium and iron, probably from relatively nearby rocks.

Kolosova and Onishchuk (1971) described gold deposits from eastern Transbaikal (Novinka, Magnitnyy and Ugol'noye in the Ust-Kara gold belt) that are rich in magnetite. The deposits are mainly magnetite veins and lodes in granitoids and various dykes and contain magnetite, hematite, actinolite and quartz with which the gold is for the most part associated. There are in addition a number of other minerals including bismuthite, tetradymite, chalcopyrite, arsenopyrite and very rare, isolated grains of marcasite, ilmenite, wolframite, galena, sphalerite, native bismuth, pyrrhotite and cobaltite. Feldspars occur often among the vein minerals, and tourmaline, chlorite, biotite, epidote and hornblende are found more rarely in the magnetite veins. Scheelite is present in relatively large amounts. Kolosova and Onishchuk state that part of the gold crystallized almost simultaneously with the magnetite; some gold was also precipitated before and after the iron oxide. Gold-bearing magnetite pebbles occur in the alluvium of a number of stream valleys in the area.

In Chile certain auriferous veins cutting granitic bodies are marked by an abundance of specularite and some contain magnetite (Ruiz and Ericksen, 1962).

Magnetite and ilmenite are rare in Tertiary deposits, and franklinite and gahnite are particularly rare in all deposits. In most of the quartz-pebble conglomerate ores (e.g., Rand and Jacobina) magnetite and ilmenite are rare. Geologists, who favour a placer origin for these deposits, maintain that this is because these minerals have been sulphidized and now appear mainly as pyrite.

Where relationships can be established gold and gold tellurides are generally later minerals than magnetite, ilmenite, hematite, franklinite and gahnite. In some deposits, however, hematite is a very late mineral, often related to a mineralization quite unrelated to the gold period. Such a situation occurs in places at Yellowknife where late faults cut the gold orebodies. Occasionally in some deposits one also sees veinlets of magnetite much later than the gold mineralization.

There are no particular indicator qualities of any of the iron and zinc oxide minerals for gold in any type of deposit as far as the writer can ascertain.

Hematite, magnetite, ilmenite, spinel and chromite are commonly encountered in auriferous placers, the first four minerals in various geological terranes and the last where ultrabasic bodies are a feature.

Corundum is an uncommon mineral in gold deposits. It may occur in the wall-rock alteration zones of certain auriferous quartz veins in highly aluminous rocks such as schists, gneisses, syenites, etc. Gem sapphires and rubies are commonly encountered in gold placers as in New South Wales; in the gold washings near Helena, Montana; and in the concentrates from Eldorado Bar on the Missouri River (sapphires); and elsewhere.

Rutile and *leucoxene* commonly occur as accessory minerals in gold deposits particularly those in basic rocks. Studies of the distribution of rutile and leucoxene in the Yellowknife deposits in greenstones (Boyle, 1961*a*) indicate that the minerals are formed mainly as a result of the carbonatization and sericitization of basic rocks. The titanium cannot be accommodated in any of the alteration minerals formed in the wall-rock alteration zones and veins and hence appears as rutile and/or leucoxene. This phenomena is discussed further in the section on alteration processes in this chapter.

Rutile and leucoxene are relatively common in the ores of the Witwatersrand. According to Ramdohr (1953*b*) and Liebenberg (1955) the rutile (and leucoxene) formed from titanium liberated during pyritization of black sand.

Rutile, brookite, anatase and leucoxene are common in recent gold placers.

Cassiterite and other tin minerals are relatively uncommon in most gold deposits, but in some mineral belts they (mainly cassiterite) are present. Examples are found in the stanniferous regions of the northeastern and eastern U.S.S.R. (Nekrasov, I. Ya., 1973; Radkevich *et al.,* 1965; Radkevich, 1969); in some of the auriferous deposits of the western U.S.A. as at Marysvale, Utah (Radtke *et al.,* 1967); in Bolivia, Malaya, New South Wales, Australia; and elsewhere. In most of these occurrences cassiterite is paragenetically older than native gold and other gold minerals.

Small amounts of apparently detrital cassiterite are present in some of the reefs of the Rand (Liebenberg, 1955), but in general the tin content of quartz-pebble conglomerates is very low (Table 35).

Cassiterite may accompany gold in certain recent placers.

The oxides and hydrous oxides of iron and manganese, mainly *limonite* and *wad,* are abundantly developed in the oxidized zones of most auriferous deposits. Their formation is examined in detail in Chapter IV.

Scheelite and *wolframite* are common minerals in gold deposits of various types mainly the skarn, gold-quartz and polymetallic type. Tungsten is present in only trace amounts in the Witwatersrand and other quartz-pebble conglomerates (Table 35). Scheelite and wolframite have not been recorded in the Rand ores (Liebenberg, 1955; Saager, 1968; Feather and Koen, 1975) – a fact which seems peculiar considering that the deposits are supposed to be modified placers, and that gold, especially that derived from Precambrian terranes, is usually accompanied by tungsten (scheelite).

Scheelite has a rather erratic distribution both locally and regionally in auriferous deposits. The mineral occurs in deposits of all ages from Precambrian to Tertiary, and there is seemingly no differnce in the abundance of the mineral with time. Some gold belts, however, are notable for their scheelite whereas others have little or none of the mineral. In the Porcupine (Timmins) district of Ontario for instance scheelite is rather common in the gold veins, whereas in the Kirkland Lake area the mineral is very sparse, although small amounts do occur in the Chesterville and Kerr Addison mines.

Wolframite is actually a rather rare mineral in gold deposits of all types. I have seen it in only two places (Outpost Island and Matthews Lake area, Northwest Territories) in Precambrian gold deposits and not at all in deposits of Paleozoic and Mesozoic age in Canada. Curiously enough, wolframite (hüberite) appears in very small amounts in some Tertiary deposits, particularly in those at Tonapah, Nevada and Cripple Creek, Colorado; also in some deposits in the far eastern U.S.S.R.

Scheelite is generally an early mineral in gold-quartz veins. Frequently it is fractured and seamed by late quartz veinlets carrying gold and gold tellurides. Some shoots in gold-quartz veins enriched in scheelite tend to be relatively low grade in gold. In the Meguma Group of Nova Scotia this seems to be due to the fact that some of the scheelite-rich veins appear to be younger than the gold veins.

Both scheelite and wolframite may accompany gold in recent placers.

Uranium and *thorium minerals* are relatively rare in most gold deposits throughout the world. In the Witwatersrand and other quartz-pebble conglomerates, however, they are ubiquitous and economic minerals. Gold-quartz veins, auriferous skarns and some auriferous polymetallic deposits in uraniferous provinces also carry uranium and thorium minerals in some districts. Examples are found in the Uranium City area of Saskatchewan; in the Chihuahua district, Mexico (Krieger, 1932); in the Central City district of Colorado (Sims *et al.,* 1963); in the Alligator rivers region, Northern Territory, Australia; in the Mitterberg district of Austria (Bauer and Schermann, 1971; Siegl, 1972; Paar, 1976); and in the Eldorado area of Ontario.

Detailed studies of the uranium minerals and their relationship to gold in the Witwatersrand bankets have been made by Liebenberg (1955), Ramdohr (1958*a,b*) and Feather and Koen (1975). The principal uranium species are uraninite and uraninite-bearing hydrocarbon (thucholite). Some of the uraninite is said to be detrital and some secondary or redistributed. The thucholite is interpreted as consisting of remnants of detrital uraninite replaced to varying degrees during thucholitization; some secondary uraninite may accompany the detrital grains in the thucholite. Gold is intimately associated with the uranium minerals, and there is a general sympathetic relationship between the uranium and gold contents, a feature also emphasized by Von Rahden and Urli (1969) who state:

At all three mines, the gold was found to occur in at least two generations. The older generation occurs in the form of small, generally rounded inclusions, totally enclosed within pyrite grains that show no evidence of healed cracks that could originally have permitted the entry of the gold. The second and younger generation of gold constitutes the major portion of the gold seen in polished sections. This younger gold occurs as stringer-like inclusions embedded in phyllosilicates as anhedral growths replacing pyrrhotite, pyrite, chalcopyrite and uraninite, and occasionally as small euhedral crystals showing bright facets and embedded in phyllosilicates or occurring within cracks in quartz grains. This second generation of gold has been described in considerable detail by Liebenberg, Von Rahden, and Schidlowski. These investigators attributed the form shown by this second generation of gold to the effects of dynamic and localized metamorphism of the conglomerates, when remobilization and recrystallization of the gold occurred. They postulated that it was during this period of remobilization of the gold that the uraninite and sulphides were replaced by gold. The movement of the gold did not exceed perhaps a few millimetres, and was not sufficient to upset or affect the original sedimentary distribution pattern to any significant extent.

The relationships of gold and uranium minerals in the Box Mine at Goldfields, Saskatchewan are not entirely clear. It would appear that the uranium minerals are later than the gold mineralization and that the two suites may be independent of each other. The same may also be true at the Nicholson Mine in the same area, although in some samples from this mine the gold, often containing high amounts of platinoids, occurs in calcite that veins and cements the pitchblende. In different samples from this deposit and in other deposits in the Goldfields area gold occurs as small inclusions in the pitchblende. Some of this gold appears to be late, although this observation is tenuous.

In the pyritic uraninite-gold-quartz veins in altered porphyries in the Chihuahua district in Mexico, the gold is paragenetically later than the uraninite occurring in fractures in the uraniferous mineral (Krieger, 1932). Samples from the Jabiluka deposits in the Alligator rivers region, Northern Territory, Australia show that the native gold and associated tellurides, including altaite and melonite, are later than pitchblende and uraninite in the paragenetic sequence.

At the Richardson Mine at Eldorado in eastern Ontario (Boyle and Steacy, 1973; Steacy *et al.*, 1973, 1974*a,b*), some of the native gold occurs in clots of thucholite, giving the impression that the gold exsolved from an organometallic gel. In the same mine native gold occurs in fractures in *brannerite.* Brannerite has also been noted in other gold and silver mines, notable at Bou Azzer (Jouravsky, 1952*b*), Plumas County, California (Pabst and Stinson, 1960) and La Gardette (Isère), France (Geffroy, 1963). At the Plumas County occurrence the gold is clearly later than the bannerite. In the La Gardette deposit the paragenesis is as follows: brannerite-quartz-native gold-galena-barite. Brannerite also occurs with uraninite in the veins of Mitterberg, Salzburg (Austria). Gold is an associated mineral and is later than brannerite and uraninite (Siegl, 1972). Brannerite is a constituent of some of the ores of the Rand, particularly the Vaal Reef (Feather and Koen, 1975). Its relationship with gold is not stated.

Excepting the Rand[23] and some deposits in the Lake Athabasca[23] region of Saskatchewan, uranium and thorium minerals are particularly rare or absent in deposits of Archean age. The reason for this is a problem in metallogenesis which remains unsolved.

Chromite is rarely encountered in auriferous deposits of the skarn, gold-quartz or polymetallic type. As rounded to euhedral grains the mineral is widely distributed in the reefs of the Witwatersrand but only in trace and minor amounts (Feather and Koen, 1975). It exhibits a closely sympathetic relation with zircon, and is generally considered by most investigators to be of detrital origin.

Chromite is a heavy mineral constituent of numerous recent gold placers, particularly those in terranes marked by the presence of ultrabasic-basic rock suites and Ni-Cu sulphide deposits.

Platinoid minerals constitute a number of natural alloys and a series of compounds with S, As, Sb, Bi, Te, Pb and Sn. The platinoid alloys and their association with gold have been mentioned in the first part of this section. The other platinoid compounds are relatively rare constituents of gold deposits especially the gold-quartz type. Certain auriferous skarn deposits associated with basic and ultrabasic rocks, or in calcareous rocks with associated serpentinites, may carry platinoid minerals, principally sperrylite, cooperite and stibiopalladinite. Deposits of this nature are known in South Africa and in Sumatra, Indonesia. Various auriferous polymetallic veins, rich in copper minerals, may carry minor amounts of platinoid minerals, principally sperrylite. The Cuniptau Mine at Goward, Ontario is an example of this type of deposit, others are known in the United States, Colombia and U.S.S.R. Without exception these deposits are either in or closely associated with ultrabasic-basic stocks, plugs or dykes. Certain unusual quartz veins and pegmatites in various parts of the world carry small amounts of platinoid minerals and traces of gold. Such veins occur in Colombia, U.S.S.R., South Africa, New Zealand and Ontario. The platinoid minerals are commonly associated with pyrite, specularite, chrome chlorite and sericite. Most of these veins occur in or near igneous ultrabasic-basic rocks. Certain gold-bearing uraniferous veins and others greatly enriched in selenium may contain small amounts of platinoids. The first is exemplified by the Nicholson Mine in the Beaverlodge area of Saskatchewan, the second by veins at Tilkerode, Harz, Germany and the Artonville Mine, Messina, South Africa.

The reefs of the Witwatersrand contain very minor amounts of platinoid minerals comprising various alloys, which have been previously described, and other compounds (Feather and Koen, 1975). The alloys are generally always associated with a number of platinoid sulphides, arsenides, antimonides and tellurides including sperrylite, hollingworthite, iridarsenite, ruthenarsenite, geversite, michenerite, moncheite, laurite, braggite, cooperite, a palladium telluride and an unnamed mineral with composition RhS_2. Feather and Koen (1975) consider the platinoid alloys to be essentially of detrital origin, modified subsequently by coatings and overgrowths of the secondary platinoid sulphides, arsenides, etc. Few of the

[23] The age of the Witwatersrand is uncertain; it may be Archean, although its stratigraphic position and other features suggest that it is of Lower Proterozoic age. The gold-uranium deposits of the Lake Athabasca region may be partly Archean in age. However, some of the uranium deposits cut Proterozoic rocks (Martin Formation), and age dating by Pb/U and Pb/Pb methods indicates that the uranium minerals have apparent ages ranging from 190 to 1850 m.y. (Robinson, 1955).

platinoid minerals of the Rand have been observed in polished sections, and hence their relationship to the gold is not known. The electron probe analyses of some of the platinoid minerals of the Rand indicate very low gold contents.

Platinoid minerals accompany gold in recent placers, particularly in terranes containing igneous ultrabasic-basic bodies and Ni-Cu sulphide deposits. Most of the platinoid minerals are alloys of the various elements, but sperrylite, arsenopalladinite, mertieite and numerous other platinoid compounds have been recorded from auriferous placers.

> In every species of gold there is a proportion of silver; in some one-tenth part, in others a ninth, and in others an eighth.
> —Pliny the Elder, 79 A.D.

Au/Ag ratios of auriferous deposits

The Au/Ag ratios of various types of gold deposits, especially with respect to metallogenic provinces and epochs, is a fascinating problem, but one that has received relatively little detailed research. In what follows the more recent significant papers dealing with Au/Ag ratios in gold deposits are reviewed, and some general conclusions are drawn from these and the writer's own research.

A clear distinction should be made between fineness of gold and the gold-silver ratio of deposits. The latter refers to the total gold and silver in the deposits as determined by analyses (assays) of the ores expressed as the ratio Au/Ag; the former is calculated in parts per thousand and refers to the proportion of gold present in the naturally occurring metal in the deposits. In general the minor element in native gold is mainly silver, but other base and precious metals may also occur, e.g., Cu, Fe, Pt metals. (*See also* the section on native gold in Chapter II.) Where the amount of base metals is high in the native gold or bullion the ratio (Au/Au + Ag) × 1000 or the *true fineness* should be used as suggested by Fisher (1945).

In the discussion that follows emphasis is placed on the Au/Ag ratio of deposits. Where appropriate, remarks will be made about the fineness of native gold in deposits. Before proceeding it should be noted that some of the investigations described in the literature are based on the fineness of bullion, which may not be an accurate measure of the Au/Ag ratio in deposits because the percentage recovery of the two precious metals from the ores is seldom stated. Where the quantity of argentiferous minerals is minimal in deposits the fineness of the bullion is a good estimate of the Au/Ag ratio; where argentiferous minerals are abundant the bullion results pertaining to the Au/Ag ratio of the deposits may be seriously in error since gravity and cyanide methods do not generally remove all of the silver from such minerals. Sharwood (1911*c*) has discussed some of the effects that metallurgical practise has on the Au/Ag ratio in bullion. His conclusions are valid to this day.

Pryor (1923–1924) studied the bullion assays of the Kolar goldfield, India. In addition to noting some differences in the fineness of the bullion from the different mines he found a general increase in the Au/Ag ratio of the bullion with depth in the deposits (surface fineness, 890; at depth, 930). He attributed this to thermal effects, the ores at depth being deposited at a higher temperature than those near the surface. This indicates a higher mobility for silver than gold, a feature that has been repeatedly emphasized for many years, and one that has some chemical support.

In the precious metal deposits of the western United States, Nolan (1933) observed that the Au/Ag ratios varied within wide limits. In some of the mines in the Silver City–De Lamar region the ratio is less than 1:100, but in a number of other camps it is almost 1:1. Nolan thought that the ratios were perhaps related to the temperature of deposition since he found an Au/Ag ratio of 1:105 in the (hot) central focus of mineralization whereas the ores more distant from the central focus, in the cooler zones, have a ratio of about 1:95. He also mentioned the fact that at Oatman, Arizona successive stages in the vein formation, five in number, have a progressively larger proportion of gold. Addie (1964) in a more recent paper, likewise, thought that Au/Ag ratios are a sensitive geothermometer. He interpreted the Au/Ag ratio to be an expression of exsolution of a gold-silver solid solution as the temperature changes. Thus, he maintained that high Au/Ag ratios represented higher temperatures and low Au/Ag ratios lower temperatures. A deposit that shows an increase in the Au/Ag ratio with depth would, therefore, indicate that the thermal source was being approached. Such a simplification seems too general in view of the data presented by Bruce mentioned below, although it agrees as noted above, with the chemical fact that silver is more mobile than gold.

Bruce (1943) investigated the Au/Ag ratios of the bullion from certain Ontario gold mines (Porcupine, Kirkland Lake, Malartic) and concluded that: (1) the Au/Ag ratio for some deposits seems to decrease with depth; (2) the ratio is different for different kinds of wall rocks even on the same mine level; and (3) silver shows less variation than gold.

Thomson *et al*, (1950) and Charlewood (1964) gave average Au/Ag ratios based on production figures for most of the gold mines along the main ore zone at Kirkland Lake. These range from 3.9 to 8.9 despite the fact that the ores are relatively uniform as regards their mineralogy. The average fineness of the bullion ranged from 798 to 899. The ratio Au/Ag, calculated by the writer from the yearly production figures, shows an erratic behaviour. In some mines the ratio decreases with time and in others an increase or erratic behaviour is evident. Whether these ratios can be keyed to depth is a question since there is seldom an orderly production depthwise with passing years in gold mines.

In the publication by Thomson *et al.* (1950) noted above, Hopkins discussed the Au/Ag ratios based on the annual production of the Wright-Hargreaves Mine. He noted that the Au/Ag ratio decreases with depth and remarked that this is due to an increase in the silver content rather than a decrease in the gold content. Hence, the trend in the Au/Ag ratio does not help to predict the depth at which the precious metal content will be below ore grade.

Mills (1954) studied bullion assays and mine production records finding a systematic variation in the purity of the gold recovered during the years 1936 to 1946 in the O'Brien Gold Mine, Kewagama, Quebec. From 500 to 1200 ft beneath the surface the ratio (Au/Au + Ag) × 1000 increased from 860 to 920; below 1200 ft the ratio was constant at an average value of 925. Between the surface and 500 ft the ratio was unexpectedly high (900–910); this was interpreted as possibly due to secondary enrichment. Mills also found that the proportion of gold recovered by cyanidation as against that recovered by amalgamation during 1939–1946 increased continuously suggesting that the particle size became smaller with depth.

Shcherbina (1956*a*) has given a number of Au/Ag ratios of various deposits throughout the world and drawn a few conclusions that merit repeating here. He found that:

1. Gold enrichment (a high Au/Ag ratio) is present mostly in higher temperature and deeper-seated ore deposits, whereas silver enrichment (a low Au/Ag ratio) is characteristic of low temperature ore deposits formed at intermediate depths or near the surface. This is correlative with the vapour pressures of the elements, viz. silver has a higher vapour pressure than gold at temperatures above 700°C. Silver is, therefore, more volatile than gold and hence migrates farther from the source of heat (magma) and tends to be enriched near the surface.

2. Deposits in older geological formations are frequently richer in gold (a high Au/Ag ratio), whereas in younger formations silver is more abundant (a low Au/Ag ratio).

3. Ore deposits formed by more alkaline hydrothermal solutions, as witnessed by the alteration of the wall rocks, are richer in gold, whereas those formed by weaker alkaline solutions are richer in silver. This circumstance is explained as due to the different modes of gold and silver in alkali sulphide solutions. Thus, silver is carried mainly as AgHS, whereas gold is transferred in the form of $[AuS_3]^{3-}$ and $[AuS_2]^-$ ions.

4. Gold is predominant in telluride ores, whereas silver is generally dominant in selenide ores. An analogous, but not a strict, relationship is observable with respect to As and Sb. Thus, gold prefers the arsenopyrite or smaltite mineral association, whereas silver favours antimony assemblages.

5. The Au/Ag ratio depends on regional metallogenic peculiarities, being low for the Circum-Pacific belt of America and high for the countries bordering the Pacific Ocean on the southwest.

6. During near-surface processes silver is much more mobile than gold.

Badalov and Badalova (1967) investigated the regularities of the distribution of gold and silver in the principle sulphide ore minerals of two different metallogenic provinces (Karamazar and western Uzbekistan) in U.S.S.R. Briefly their conclusions were:

1. The distribution of gold and silver in the sulphides of the various ore deposits depends on the content of sulphides, sequence of their deposition and the affinity of the two precious metals for certain ore minerals. Different minerals serve as concentrators and carriers of gold and silver depending on the character of ore-bearing generations, geochemical province and genetic type of ore deposit.

2. As a rule, the content of gold in the sulphides decreases from the older to the younger generations, while the content of silver increases.

3. The Au/Ag ratio in the ore minerals is characteristic of a given mineralization. In the Karamazar region the maximum ratio between the average contents of gold and silver is 1:28, while in the western Uzbekistan deposits it reaches 1:1, and even 2:1. Together with the fineness of gold, the Au/Ag ratio may be used in deciphering the geochemistry and genesis of the ores.

4. The sharp difference in the assemblage of ore-bearing generations and in the distribution of gold and silver in the sulphides of the Karamazar and western Uzbekistan mineralized zones may be explained by the difference in the role of the enclosing rocks during ore deposition. In western Uzbekistan, these rocks are sandstones and shales; in the Karamazar region, they are igneous rocks.

5. Investigation of the distribution of gold and silver in the endogene orebodies of a given region may be helpful in predicting the presence of gold deposits proper. Certain quantitative relationships have been established in the Karamazar region between the various gold-free but silver-rich sulphide deposits and the gold deposits of comparable size.

Fitzgerald *et al.* (1967*a,b*) examined the feasibility of using Au/Ag ratios in solving a fault displacement problem in the Bourlamaque district of Quebec (Lamaque and Sigma mines). The ore zones in these two mines are composed of a multiplicity of gold-bearing quartz veins extending vertically to depths greater than 4000 ft. The mines are about 2500 ft apart horizontally, and are separated by a steep-dipping fault zone. The investigators noted that the Au/Ag ratios at both Lamaque and Sigma, calculated from annual mint (bullion) returns over the past 30 years, represent very large samples. These values, when plotted against the annual mean depth of mining, were found to increase with depth in both mines at about the same rate, with Lamaque having the higher values. If it is assumed that both orebodies formed under similar conditions, the higher Au/Ag ratios found at Lamaque suggest that this orebody has been raised approximately 1300 ft with respect to Sigma by postore movement on the fault zone. Stope samples taken over a vertical range of 2285 ft at Lamaque were analyzed for Au and Ag by two independent methods. The Au/Ag ratios determined by conventional analysis (i.e., Au by fire assay; Ag by atomic absorption) were variable, but were in the same range as the ratios from the mint (bullion) returns. The investigators found no evidence that intrastope variations are caused by such geological features as the composition of the wall rock or the thickness or grade of the vein. The Au/Ag ratios obtained by electron probe microanalysis of gold grains taken from the stope samples were approximately three times higher than the ratios obtained by conventional analysis, and there appeared to be no change in gold fineness with depth. These results suggested to the investigators that both the intrastope variations and the increase in the Au/Ag values with depth are caused by the presence or absence of Au- and Ag-bearing minerals other than the native alloy. Silver was found to occur in petzite (25% Au, 42% Ag) and tetrahedrite (12% Ag), but these minerals are rare in the mines and could not adequately account for all the missing silver. Pyrite is the most common sulphide, making up about 2.5 per cent of the ore; it contains about 2.3 oz of silver per ton of pyrite, which is sufficient, it was thought, to account for the difference in silver content obtained from native gold and silver recovered either from individual stope samples or from the mill as reported in mint returns.

In the Balei ore field, U.S.S.R., (*see also* below) Gapontsev and Polikarpochkin (1969) observed significant differences in the Au/Ag ratios in various parts of the gold veins and their primary dispersion halos. Above the economic ore shoots the Au/Ag ratios ranged between 0.5 to 1.0; in the ore shoots the range was 1.25 to 5.0; below the ore shoots the range was 0.14 to 0.33, and in the peripheral parts of the primary halos the ratio fell to values as low as 0.02. To generalize, their results showed that there is both a lateral and vertical upward and downward increase in the Au/Ag ratios with proximity to ore. This is essentially the same as noted at Yellowknife (Table 48).

Fortin (1969) studied the distribution of gold and silver in the Sullivan orebody, Val d'Or, Abitibi-East County, Quebec. The orebody comprises auriferous quartz-tourmaline veins in the peripheral parts of the Bourlamaque batholith. Statistical studies showed the grade population to be bimodal, affected by an important particle (nugget) effect; each population was random. The Au/Ag ratio was found to increase systematically with depth in a linear manner.

Kitaev *et al.* (1970*c*) investigated in detail the Au/Ag ratio in the primary aureoles of the Balei goldfields in U.S.S.R. These deposits are veins with chalcedonic quartz and relatively few sulphides. The background content of gold in the country rocks is of the order of 0.001 ppm and silver 0.006 ppm. The Au/Ag ratio fluctuates in the vicinity of 0.166. In the halos the gold content ranges from 0.01 to 1 ppm or more; the respective silver values are 3 to 6 ppm. On the dip the Au/Ag ratio follows the curve shown in Figure 8; across the

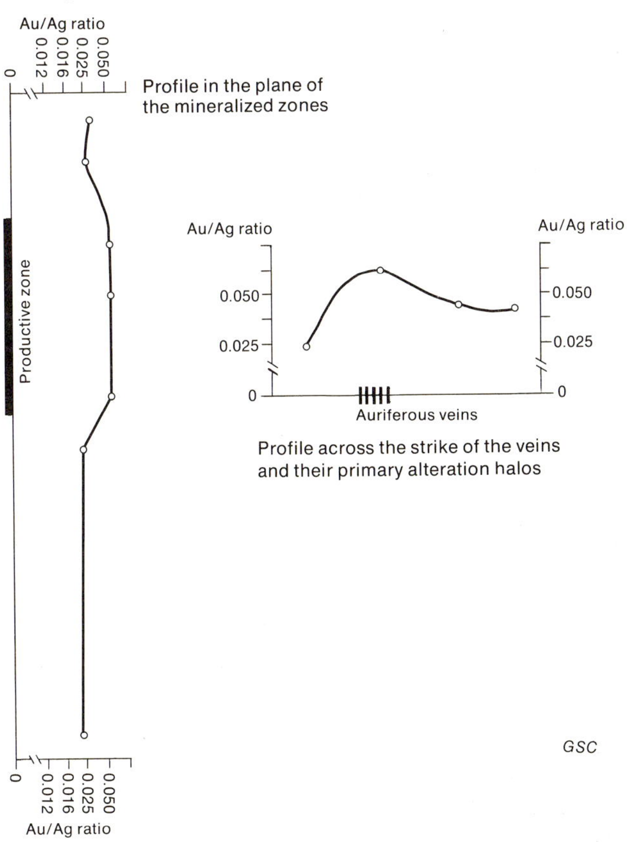

Figure 8. Generalized profiles of the Au/Ag ratio in the Balei goldfield, eastern Transbaikal, U.S.S.R. (after Kitaev *et al.*, 1970*c*).

strike the generalized situation is as shown in the right-hand part of the figure. The investigators concluded that a study of the silver-gold ratio could provide information on the paths of movement of hydrothermal solutions and details of the microstructure of aureoles. In addition, the ratio has considerable applied significance for the determination of the location of the zone of optimum mineralization and the level of erosion of the orebodies; and a comparative estimation of the degree of gold mineralization in the various orebodies.

Koroleva (1971) examined the Au/Ag ratios in the endogene gold-bearing deposits of middle Asia, U.S.S.R., which can be categorized as: (1) gold-pyrite-arsenopyrite-pyrrhotite-telluride deposits; (2) mixed polymetallic deposits containing gold and silver; and (3) polymetallic deposits containing admixed gold. Her conclusions are as follows:

1. The Au/Ag ratio decreases from the oldest mineral associations to the youngest. In all deposits the sulphide associations composed of minerals of iron, arsenic and copper, crystallizing during the oldest stages of the hydrothermal process, are the richest in gold. Silver accumulates in the younger minerals of a given stage, or at the end of the depositional process.

2. In the gold and gold-bearing deposits the Au/Ag ratio is determined by the dominance of the mineral with which gold is associated. The highest ratios are characteristic of ores in which the arsenopyrite-pyrite association predominates. With increase in pyrite content the ratio decreases, reaching a minimum in ores with predominant galena and silver minerals.

3. The Au/Ag ratio may be used to predict the mineralogical composition of an association exploitable for gold in little known ore deposits.

4. In the gold and gold-bearing deposits of middle Asia the Au/Ag ratio is generally higher than in the deposits of Altai, Caucasus and Mexico. This indicates a greater concentration of gold in middle Asia. The Au/Ag ratio may be used in preliminary evaluation of individual deposits and prospects.

5. There is a relation between the fineness of gold and its occurence in a given mineral association. Widespread variations in the fineness of gold in some deposits is due to the presence of several mineral associations.

In the Byngov deposit in the central Urals where the mineral assemblage includes a variety of base and precious metal tellurides and native gold in quartz veins, Berzon and Bushlyakov (1971) found that the Au/Ag ratio decreased sharply with depth from 1.5 to 1.1. They also found that the presence of disseminated pyrite and chalcopyrite in the veins decreased both the gold and silver contents.

Forsythe (1971) noted a decrease in the Au/Ag ratio with depth in the main ore shoot of the Crown-Crescent lode in the Emperor Gold Mine, Fiji. He attributed this to vertical zoning of gold-silver tellurides. The zoning apparently resulted from a complicated two stage mineral deposition sequence of ore minerals and reactions of second stage solutions with minerals deposited during the first mineralization stage.

Saager and Bianconi (1971), using statistical methods, examined the Au/Ag ratios in the Mount Nansen gold-silver deposit, Yukon, Canada. This deposit consists of narrow steeply dipping veins in rocks ranging from Precambrian to

Tertiary. The primary mineralization is complex with pyrite, arsenopyrite, various sulphosalts, stibnite, galena and sphalerite. The deposit may be of Cretaceous–early Tertiary age. A general down-dip decrease in silver content and a fairly uniform gold distribution was found in one of the mines, resulting in a general increase in the Au/Ag ratio with depth.

Stephenson (1971) noted that the silver content (~0.04 oz/ton) of the San Antonio Gold Mine in the Rice Lake district of Manitoba remained relatively constant during the period 1935–1968, whereas the gold tenor of the ores declined from about 0.30 to 0.20 oz/ton during the same period. He correlated these values with depth and observed that the Au/Ag ratio decreased from about 7.1 to 5.1 over a depth of 5380 ft. His curve shows a significant high Au/Ag ratio equal to 8 near the surface suggesting some supergene enrichment of gold, a feature also noted by Mills (1954) as stated above.

Randall (1973) found a rather complicated pattern in the Anglo-Rouyn Cu-Ag-Au Mine at La Ronge, Saskatchewan. The Au/Ag ratio changed from 0.18 at the start to 0.28 near the end of operations. With time, the copper grade decreased, the gold grade increased and the Cu/Ag ratio was nearly constant. Electron microprobe studies of the concentrate showed most of the silver to be present in chalcopyrite. Only half the gold was found to be so associated, the rest being in iron sulphides, sphalerite and silicates. Gold-silver assays from ore samples gave ratios from 0.003 to 0.83. Samples rich in quartz, chlorite or iron sulphides showed higher Au/Ag ratios than those rich in chalcopyrite. Ore with a high proportion of chalcopyrite gave ratios of 0.06 to 0.1. Metal contents and ratios were found to be spatially distributed in the shallow, elongate, northeast-southeast trending orebody. The centre of the body shows banded chalcopyrite ore with few apparent metamorphic effects and a relatively low Au/Ag ratio. Higher Au/Ag ratios were found in quartz-rich, nonbanded ore toward the ends of the orebody where ralative proportions of chalcopyrite, pyrrhotite and pyrite vary, and metamorphism is more apparent. The irregular distribution of gold was thought to be due to original deposition with copper, later deposition or selective redistribution during metamorphism.

The pattern of the various precious metals in the chalcopyrite ores of the southern Urals (Magnitogorsk) is likewise complicated (Novgorodova, 1976). She found that the distribution of Pd, Pt, Au and Ag in pyrite, pyrrhotite, chalcopyrite and sphalerite does not obey the lognormal law. The Pd/Pt ratio varies uniformly with changes in the Au/Ag ratio; the Pd/Pt ratio increases and the Au/Ag ratio decreases from the early- to late-stage mineral assemblages.

Berman and Gorelyshev (1974) concluded from their study of the Au/Ag ratio of an auriferous deposit in the northeastern part of the U.S.S.R., that the ratio is the result of many geological influences but particularly the character of the mineral associations.

Hargraves (1963), Von Rahden (1965), Addie (1964) and Saager (1969) have investigated the Au/Ag ratio in the ores of the Witwatersrand. Their conclusions are discussed subsequently.

Ol'shevskiy (1975) studied the Au/Ag ratio in 71 near-surface (mainly Tertiary) gold-silver deposits in various parts of the world, 25 of the deposits being in the eastern part of the U.S.S.R. He noted a dependence of the ratio on the types of wall rocks in which the deposits occurred. In general in the deposits in U.S.S.R., those in acidic rocks (rhyolites, etc.) have a low Au/Ag ratio (<0.05) whereas those in intermediate rocks (andesites) have a high value (>0.05). Similar results were noted in other (mainly Tertiary) deposits in various parts of the world. Ol'shevskiy thinks that the relationship between the Au/Ag ratio and the composition of the host rocks can be explained by the fact that the gold-silver mineralization is connected paragenetically with the plutonic-volcanic rock complexes. The more acid volcanic rocks are formed at near-surface horizons and consequently since silver is more mobile (or has a higher migration rate) more of the element accompanies the acid rocks than is the case for gold. The result is a low Au/Ag ratio in the deposits in near-surface acidic rocks and higher ratios in the deeper rocks of more basic composition. It will be recalled from the discussion above that Bruce (1943) and Badalov and Badalova (1967) noted a wall-rock effect on the Au/Ag ratio; others including Fitzgerald *et al.* (1967 *a,b*) have, however, noted no wall-rock effects on the ratio.

In some Canadian gold deposits the writer has noted a wall-rock effect on the Au/Ag ratio of the ores similar to that described by Ol'shevskiy (1975). A few examples will serve to illustrate this statement. Many gold deposits in (basic) greenstone rocks have relatively high ratios compared with similar deposits in (acidic) quartz-feldspar porphyries, granites and syenites regardless of age. Some basic examples are: Yellowknife, Northwest Territories (Archean greenstones – 5); Bralorne, British Columbia (Mesozoic greenstones and diorites – 5.5); Pickle Crow, Ontario (Archean greenstones and iron-formation – 11.5); Delnite Mine, Timmins, Ontario (Archean greenstones – 11); and Chesterville Mine, Larder Lake, Ontario (Archean greenstone – 18). Some acidic examples are: Box Mine, Lake Athabasca (Archean granite – 3); Howey Mine, Red Lake, Ontario (Archean quartz porphyry – 3); Young Davidson Mine, Matchewan, Ontario (Archean syenite – 4.3); and Kirkland Lake, Ontario (Archean syenite – 5.8). The effect is also notable in the many mines of the Timmins (Porcupine) area of Ontario. Mines such as the Moneta, Delnite, Aunor and Buffalo Ankerite, mainly in Archean greenstones, have high Au/Ag ratios for their ores (7–11.6), whereas those such as the Hollinger, McIntyre and Paymaster, which are partly in or near masses of quartz-feldspar porphyry, have lower ratios (3–5.2). An extreme example of this effect can be seen by comparing the general Au/Ag ratios in the various ores in the McIntyre Mine. The gold-quartz ores mainly in greenstones have a Au/Ag ratio equal to 5, whereas the cupriferous ores in altered porphyry have a ratio of 0.14 (Boyle, 1976*a*).

It should be remarked, that there are numerous exceptions to the basic and acidic wall-rock effect on the Au/Ag ratio in some gold belts, but statistically speaking it appears that there is some validity in the phenomenon. The reasons for the effect are, however, obscure. Ol'shevskiy (1975) attributes the effect to volcanic agencies, but this cannot be the case where granites are concerned. It seems more probable that the effect is due to differential reactions between the gold-bearing solutions or diffusion currents and the two kinds of wall rocks – basic and acidic. Basic rocks are marked by a much higher degree of carbonatization, sericitization and sulphidization than are acidic rocks in most gold deposits (*see* the next section on wall-rock alteration processes). This has involved a greater removal of CO_2,

K and S from the solutions or diffusion currents, a feature that has probably markedly affected the stability of the various gold and silver complexes. Gold complexes, such as those with S, As, Sb and Te, were probably rendered more unstable than similar silver complexes, thus leading to the precipitation of a greater proportion of gold in basic rocks than in those of acidic composition. This explanation is only a suggestion and seems to follow from our experiments (Boyle *et al.*, 1975) mentioned in Chapter II in the section on natural waters – actually, the whole problem of the wall-rock effect on the Au/Ag ratio requires much more detailed and sustained investigations. Finally, there may be some confirmatory evidence with respect to the acidity of the environment and the precipitation of gold from the recent research by Diman (1976). He observed that the purity (fineness) of gold in association with pyrite depends essentially on the sulphur regime and the acidity of solutions rather than on the temperature of deposition.

Konstantinov (1977) has characterized volcanogenic gold deposits on the basis of their Au/Ag ratios, which he notes range from 10:1 to 1:1000. Within this range he characterizes three types falling within the following ratios: 10:1 to 1:1, 1:1 to 1:5 and less than 1:20.

Those deposits with high Au/Ag ratios (10:1 to 1:1), including Cripple Creek in Colorado, Vatukoula in Fiji and certain deposits in the Soviet Union, comprise a gold-tellurium type. All have high tellurium contents, 5 to 20 times that of gold, the tellurium being combined mainly with Au, Ag, Bi and Hg and also in the native state. Most of these goldfields are rich but small in area (5–15 km^2). The native gold usually has a high fineness; in some deposits much or nearly all of the gold is present as telluride minerals. The productive gold-telluride shoots are usually shallow, localized within 1000 m from the surface and pass into gold-polymetallic shoots at greater depths. The deposits are mostly concentrated in local faults and fractures mainly in subvolcanic formations constituting necks, dikes and bodies of eruptive breccia that form part of the volcanic belts of island arcs (on an oceanic or transitional type of crust) or ophiolite zones. The host volcanic rocks are generally alkaline basalts and andesite-basalts of the calc-alkaline series with roughly equal proportions of sodium and potassium. Prior to the mineralization events an increase in alkalinity in the late volcanic phases is notable and is manifest by minor extrusions of dacite and rhyolite.

Volcanogenic deposits with ratios 1:1 to 1:5 such as those in Romania (Transylvania), Japan and eastern U.S.S.R. are marked by a strong development of adularia and relatively small amounts of sulphides compared to the previous type. The gold occurs in small nuggety masses and has a low fineness. Such deposits are commonly localized in volcano-tectonic depressions, calderas and grabens, frequently in radiating faults and fissures or in similar structures related to block faulting. Most of these deposits constitute large goldfields (30–50 km^2) and are confined to volcanic belts of continental interior type and less frequently to fragments of island arcs developed on a continental crust. The host volcanics are andesites, andesite-basalts and andesite-dacites. The subvolcanic facies are commonly more acidic in composition than the overlying rocks and are characterized by higher alkalinity with an increase in the relative role of potassium.

The third type, characterized by ratios of less than 1:20, includes the silver-rich Comstock Lode, various other argentiferous-auriferous deposits with ratios as low as 1:1000 in western United States, Mexico and the eastern regions of U.S.S.R. All are characterized by high amounts of argentite (acanthite), with native silver and/or electrum in some deposits. Base metals and occasionally tin occur in higher amounts in the associated sulphide ores. The deposits constitute goldfields covering considerable areas (70–100 km^2). Many of the deposits are localized in extensive tension faults and fissures related to tectonic uplifts and block faulting. The host volcanic rocks exhibit an increasing alkalinity with time resulting in preore acidic rocks of trachy-liparitic composition followed by deposition of potassium-rich gangue minerals such as adularia and hydromicas.

An attempt can now be made to systematize the above data together with that recorded in the numerous papers mentioned in the Selected bibliography. It should, however, be noted that when one attempts to compile the Au/Ag ratios of deposits on a worldwide basis one is beset with many inconsistencies and difficulties, too numerous to mention here. Some of these are related to the inadequacy or lack of suitable assays, others derive from the process of compiling from production data where the grade and percentage of recovery of the two precious metals are unknown. Despite these problems Table 43 has been compiled from the worldwide literature according to the type of deposit in which gold and silver occur. The following features are apparent:

1. Only three types of hypogene deposits have Au/Ag ratios consistently greater than 1. These are the auriferous quartz-pebble conglomerate deposits, certain skarn-type deposits and most, but not all, gold-quartz veins in Precambrian, Paleozoic and Mesozoic rocks. Tertiary deposits in certain belts also have ratios greater than 1, but they are relatively uncommon. All the other hypogene deposits generally have Au/Ag ratios less than 1. Gold placers always have Au/Ag ratios greater than 1.

2. The Au/Ag ratios in disseminated deposits in shales and sandstones (Kupferschiefer and 'red bed' types) are generally low, indicating a relatively high degree of mobility and concentration of silver during the formation of these particular deposits. Since these deposits probably are the result of low temperature aqueous processes, the ratios are what one would expect because silver is much more soluble and hence more mobile and more easily concentrated under such conditions.

3. In the auriferous quartz-pebble conglomerate deposits the range of the Au/Ag ratios is generally narrow, but there are some significant differences in some deposits such as the Witwatersrand, which have been intensively studied by Hargraves (1963), Von Rahden (1965), Saager (1969) and Viljoen (1971).

Hargraves (1963) found that low grade samples have high silver contents and vice versa, and that the silver content of bullion from individual reefs varies as a funciton of the relative elevation of the reef, i.e., the bullion of the shallow reefs has a higher silver content than deeper reefs of comparable grade. Hargraves (1963) interpreted these compositional gradients as the result of chemical rather than physical (detrital) processes and stated that in terms of the modified placer hypothesis for the origin of the gold, the gradients imply considerable redistribution of silver relative to gold during the 'solution' stage. This is, therefore, again a case of the greater mobility of silver compared to gold during relatively low temperature processes.

Table 43. Au/Ag ratios of a variety of mineral deposits

Type of deposit	Au/Ag ratio (range)	Remarks
Shale deposits (Kupferschiefer type)	0.006–0.025	
Disseminations, veins, etc, in sandstones (red-bed type)	0.003–0.01	Variable
Quartz-pebble conglomerates (Rand type)	5–20	Witwatersrand bullion ranges between (5.8–15.6) (Hargraves, 1963). On channel samples of the Ventersdorp Contact Reef and Main Reef the Au/Ag ratios range from 0.6–13.7 (Von Rahden, 1965)
Disseminated lead-zinc deposits in carbonate rocks (Mississippi Valley–Pine Point type)	0.001–0.1	Few data
Skarn-type deposits	0.005–10	Variable
Massive sulphides (Ni-Cu Sudbury type)	0.03–0.07	Shcherbina (1956*a*) gives 0.07 for Norilsk, and Hawley (1962) gives 0.03 for Sudbury ores
Porphyry copper type	0.001–0.1	Few data
Polymetallic massive sulphides (Flin Flon–Noranda–Bathurst type)	0.006–1	Average about 0.025
Polymetallic massive sulphides (Kurokô type)	0.005–0.05	Average about 0.02
Polymetallic veins, mantos, etc. (Keno Hill–Sullivan–Cœur d'Alene type)	<0.0001–0.02	
Gold-quartz veins, lodes, etc. (Precambrian, Paleozoic and Mesozoic age)	1.37–12.5	Average about 4.2
Gold-quartz veins, lodes, etc. (Tertiary age)	0.005–0.33	Average about 0.05
Native silver-Co-Ni-As-Bi-U veins (Cobalt-Jáchymov type)	<0.0001–0.01	
Hot spring siliceous sinters	<1	
Gold placers (all types and ages)	>1	

Van Rahden (1965) thought that his data, which show wide variations in the apparent fineness (and hence in the Au/Ag ratios) of channel samples from the same elevations and stopes in the Ventersdorp Contact Reef and Main Reef, could be ascribed to the presence of silver minerals (mainly native silver) in the ore. The silver minerals, he felt, were due to the redistribution of silver relative to gold during the solution stage of the modified placer theory suggested to explain the Rand deposits. Van Rahden further concluded that his results did not tend to confirm Hargraves' (1963) hypothesis that the silver content of the reefs varied with elevation. He recommended electron microprobe studies of the ore to resolve the problem.

Addie (1964), commenting on Hargraves' paper, considered that the increase in the Au/Ag ratio in the Witwatersrand deposits with depth is due to exsolution of a silver-gold solid solution as the temperature changes. By this he apparently meant that since the silver is more mobile it migrates out of the system. He thought that the increasing Au/Ag ratio signalled an approach to a thermal centre and postulated a thermal source for the mineralization.

Saager (1969) investigated the Au/Ag ratios of individual gold particles and of ore samples from the Basal Reef conglomerate horizon of the Upper Witwatersrand System in the Orange Free State Goldfield (South Africa). The values for individual gold grains were found to be exceptionally constant at about 12.3, whereas the Au/Ag ratios of the ore samples have markedly lower values and show a large variation. Saager concluded that these values indicate that the detrital gold has been homogenized during transportation and sedimentation and that part of the original silver content has been redistributed. He thought it probable that the first refining of detrital gold particles occurred in streams that eventually drained into the Witwatersrand basin. Further redistribution of the silver relative to the gold took place in the littoral zone of the basin, where the conglomerates were reworked by wave action and erosion. He further contended that supergene enrichment of gold might also have taken place close to the original shoreline. These processes, he believed, accounted for a decrease in the Au/Ag values of ore samples towards the northwest in the area investigated. Furthermore, Saager hypothesized that gold has probably been homogenized during the metamorphic period of the Witwatersrand System, when pseudohydrothermal transport and redeposition of certain constituents took place. Finally, he suggested that the main portion of the silver content of the original alloy that was redistributed during the various processes outlined above is now present as finely dispersed silver minerals. The hypothesis of an elevation control of the Au/Ag values, as postulated by Hargraves, was found to be improbable.

Viljoen (1971) carried out an electron microprobe analysis of gold grains in situ in samples from a variety of reefs from the Rand. The silver content of individual gold grains varied from 0.3 to 32.1 per cent. Variations were noted in one and the same sample, and some gold grains that were analyzed showed that they were considerably enriched in silver along the periphery. Contrary to the findings of other investigators no discrete silver-bearing minerals were encountered, but the silver content of galena in the Dominion Reef was up to 3 per cent and in the Vaal Reef up to 2 per cent. The data presented indicate an overall average of 10 per cent for the silver content of the gold particles.

4. There are few data on the gold content of the 'Mississippi Valley type' lead-zinc deposits, and hence a precise knowledge of the Au/Ag ratios is not obtainable. Our limited data on these deposits show that silver is generally greatly enriched compared to gold, and the ratios are, therefore, always low. This indicates a higher mobility for silver compared to gold, a feature that one would anticipate since these deposits are probably concentrated by low temperature connate or circulating groundwater.

5. The Au/Ag ratios in skarn deposits are exceedingly variable and apparently related to the mineralogy and hence

the chemistry of formation of these deposits. Taking some Canadian deposits as examples one notes that a deposit such as the Nickel Plate at Hedley, B.C., where arsenic is abundant, is characterized by a high Au/Ag ratio (5–10), and in this respect is similar to certain gold-quartz veins enriched in arsenic. Probably the mode of transport of the two noble metals during the formation of these two types of deposits has been the same, although the geological setting differs. It would seem that the migration of gold was greatly enhanced by the presence of arsenic in the diffusion or solution system. On the other hand the New Calumet and Tetreault lead-zinc deposits in the Grenville Province both have low Au/Ag ratios (<0.1) and simulate the normal type of silver-rich Pb-Zn deposits. These deposits are not particularly enriched in arsenic, but contain significant amounts of antimony present in tetrahedrite, polybasite, stibnite, etc. Hence, it seems probable that antimony has had a marked influence on the mobility of silver.

6. Massive nickel-copper sulphides of the Sudbury type, generally associated with basic igneous rocks, seem to have a narrow range of Au/Ag ratios from about 0.03 to 0.07, indicating a considerable enrichment of silver over gold at the relatively high temperatures at which these ores are supposed to have formed. Both the silver and gold appear to follow copper, although it would be interesting to know the precise relationships with arsenic and antimony. According to Hawley (1962) chalcopyrite is the principal argentiferous mineral at Sudbury.

7. The porphyry copper deposits tend to have a relatively low Au/Ag ratio judging from the few good data available. Certainly many of the high-grade copper and lead-zinc skarn and vein deposits associated with the porphyry copper bodies are greatly enriched in silver compared to gold. The situation in the Timmins area with respect to a porphyry copper deposit and the gold-quartz veins is of interest. There, silver is much more abundant in the McIntyre Cu-Mo orebody than in the gold-quartz veins that occur in the same environment. In the copper orebody the Au/Ag ratio is about 0.14, whereas in the gold-quartz deposits the ratio is about 5, in agreement with the general overall ratio (7) in the Porcupine (Timmins) gold belt.

In British Columbia the Au/Ag ratio of the various porphyry copper deposits shows a wide variation. The Island Copper orebodies on Vancouver Island have a ratio approaching 0.2 calculated from production records. Other deposits such as the Granisle, Bethlehem, Brenda and Lornex orebodies have low ratios, ranging from 0.002 to 0.095. The investigations of the Island Copper metallurgical staff and our work on the ores indicates that most of the gold in the Island Copper orebodies is present in the native state. The silver is largely in, or associated with, chalcopyrite.

Kesler (1973) investigated the gold abundance in porphyry copper deposits with respect to copper and molybdenum. He suggested that porphyry copper deposits are divisible into geologically real copper-molybdenum and copper-gold classes. The compositional variation, he observed, was thought to have been caused by primary compositional differences between mineralizing systems (that is, the genetically associated intrusive rock and its juvenile hydrothermal phase), or by differing chemical responses of similar systems to distinctive tectonic features (such as level of emplacement or permeability of wall rock) or wall-rock chemistry. Kesler concluded that the obvious important factor for such consideration is the possibility that copper-molybdenum and copper-gold porphyry copper deposits could represent mineralizing systems of fundamentally different compositions and, as such, could be analogous to petrologic variations among areas of different tectonic setting.

8. The massive polymetallic sulphide deposits (Flin Flon–Noranda–Bathurst type) nearly all have relatively low Au/Ag ratios, which average about 0.025. Hutchinson (1973) found the ratios in these types of deposits, mostly in volcanic terranes, to range from 0.0002 to 0.60. In some of these deposits silver follows lead and in others an association with copper is evident. Gold seems to follow arsenic and is concentrated in pyrite and arsenopyrite in some deposits; in others, gold appears to be associated mainly with chalcopyrite, galena and sphalerite. The occurrence of gold and silver in most deposits of this type is not well defined, and more research on this feature of the geochemistry of the deposits is desirable.

A number of deposits contain argentiferous tetrahedrite, and other silver-bearing sulphosalts, and hence there may be a relationship between silver and antimony. Other silver and gold minerals such as argentiferous tennantite, argentite, native silver, electrum, gold and sylvanite have been recorded from a number of deposits, and these may be responsible for part of the gold and silver content of the ore.

In the 'kurokô type' of polymetallic sulphide deposit silver is also greatly enriched with respect to gold, the ratio Au/Ag usually falling in the range 0.005 to 0.05. Much of the gold in these ores occurs as electrum, which is commonly about 770 fine. The electrum is most commonly associated with galena, sphalerite and tetrahedrite-tennantite in the ores. The silver in addition to being present in electrum is contained in tetrahedrite-tennantite, acanthite, stromeyerite, pearceite, polybasite and a number of other silver sulphosalts.

9. The polymetallic veins and the native silver – cobalt–nickel arsenide veins have the lowest ratios of all the various types of auriferous hypogene deposits, indicating an extreme mobility for silver and practically no mobility for gold during their formation. The reasons for the great enrichment of silver in these deposits are indeed obscure but must be related either to the source of the mineralizing media or to particularly good agents of transport for silver. In many of the polymetallic veins and mantos silver follows lead closely, and there appears to be a close association with antimony and sometimes with arsenic. In other deposits of this type silver and copper exhibit a close affinity. High contents of manganese in siderite are characteristic of a number of silver-rich polymetallic veins. In the Ni-Co arsenide type there is often an association of silver with antimony, as witnessed by the presence of antimonian silver, allargentum and dyscrasite in these deposits. The association of the native silver with the arsenides is, however, often so close as to suggest that the silver migrated as some highly mobile Ag-As complex. Why gold in quantity did not go along with the silver is an enigma since the element tends to follow arsenic in other types of deposits. An answer may lie in the complex composition of the solutions or diffusion currents that precipitated the silver and arsenides. These

solutions were evidently greatly enriched in carbonate as witnessed by the general abundance of calcite, dolomite and other carbonates in the veins. Silver carbonate is relatively soluble, especially where CO_2 is in excess, and this feature could account for the relatively great mobility of silver during the formation of the Ag-Ni-Co-As veins. On the other hand gold does not seem to be capable of forming a carbonate, and in fact there are indications from some of the natural associations of the element that solutions or diffusion currents highly enriched in CO_2 actually depress the solubility and mobility of gold.

10. Gold-quartz veins in Precambrian, Paleozoic and Mesozoic rocks generally have Au/Ag ratios greater than 2, and these average about 4.2. There are, however, some local exceptions to this generalization that seem to correlate with the mineralogy and hence chemistry of formation of the deposits. At Yellowknife, Northwest Territories, veins with arsenopyrite and pyrite tend to have normal ratios for this class whereas others with a high stibnite and sulphosalt (antimony) content tend to have ratios equal to or less than 1. This suggests that the mobility of silver was considerably increased when antimony was present in quantity in the chemical system. Another example in the Canadian Shield is the Ross Mine, near Ramore, Ontario. Some of the orebodies in this mine have an Au/Ag ratio of about 0.7, much lower than those in the nearby Porcupine and Kirkland Lake camps (~7). Two periods of mineralization are evident in the Ross Mine; the first consists of blue quartz veins with pyrite, chalcopyrite, sphalerite, galena and gold, and the second is marked by the development of quartz-dolomite veins containing pyrite, sphalerite, galena, tennantite, chalcopyrite, gold and native silver (Jones, 1944). In the first period of mineralization gold predominated over silver, and in the second, the reverse is true. Here again we note a preponderance of silver where carbonate was present in some abundance in the chemical system. Other gold mines in the Canadian Shield where unusual Au/Ag ratios prevail in the ores include the Ronda in the Matachewan area of Ontario where the ratio approximates 0.58, the Ardeen (Moss) in the Thunder Bay district of Ontario with a ratio of approximately 0.17 and the Berens River deposit where the ratio is highly anomalous being as low as 0.023. These deposits have an unusual mineralogy, most containing abundant argentiferous galena or other silver minerals especially sulphosalts and tellurides. The mineralogy of the Berens River deposit in northwestern Ontario is unusual for Archean greenstone belts. Its ores consisted of abundant sphalerite and galena, with pyrite, pyrrhotite, chalcopyrite, tetrahedrite, dyscrasite, native silver and ruby silvers in a gangue of quartz, carbonates and actinolite (Oliver, 1949). No native gold was observed despite the fact that the ores averaged 0.25 oz Au/ton. The Ardeen (Moss) Mine was characterized by abundant silver tellurides.

11. Gold-quartz veins, lodes and stockworks in Tertiary andesite, dacite, rhyolite and other associated volcanic rocks generally have ratios less than 1 in contrast to their counterparts in older rocks. Many of the young (Tertiary) deposits are, however, mineralogically different from the older deposits in that a number of them contain much larger amounts of galena, chalcopyrite, sphalerite, tetrahedrite-tennantite, argentite and silver sulphosalts. In this respect they are analogous to many of the polymetallic veins and have similar ratios. By contrast, veins with a simple mineralogy, containing essentially native gold, gold tellurides, pyrite and minor amounts of the base metal minerals, such as those at Cripple Creek, Colorado, have Au/Ag ratios greater than 1.

12. Siliceous sinters precipitated from present-day hot springs generally have Au/Ag ratios less than 1 (Table 22). In this respect they are similar to many of the Tertiary deposits.

13. Gold placers always have Au/Ag ratios greater than 1 regardless of the age of the source deposits. This is due to the natural refining action that takes place during weathering processes. It seems highly probable that at one time or another much of the gold in secondary zones and in placers was in solution, and that during this process most of the accompanying silver was removed. This seems to follow naturally from the chemistry of the two elements. Silver is much more soluble than gold in ground and surface waters and is not as readily reduced to the metal. A natural separation is, therefore, inevitable.

Some investigators would perhaps disagree with this interpretation. For instance, Mertie (1940) is of the opinion that the differences in the fineness between lode and placer gold are related to depth of erosion of the mother lodes. Thus, the placer gold, which is usually of great fineness, probably comes from the apical near-surface zones of lodes whose hypogene gold was of greater fineness than the deep-seated parts of the lodes that are now exposed to view. This opinion is certainly contrary to that held by MacKay (1944), who showed almost conclusively that the gold in the oxidized zones of a number of mines in Tanzania and Nigeria has been greatly increased in fineness as the result of oxidation and secondary enrichment processes.

14. Deposits in older geological formations are frequently richer in gold than those in younger formations. There are, however, numerous exceptions to this generalization in a number of mineral belts, Cripple Creek (Tertiary) being a good example.

15. The evidence that the Au/Ag ratio is an index of temperature of formation or recrystallation of deposits is conflicting. In some deposits the ratio increases with depth, in other deposits a decrease or erratic behaviour is evident. However, few precise studies of the variation of the Au/Ag ratios of single ore shoots or veins with increasing depth have been made, and judgment on the effectiveness of the ratio as a temperature index should be reserved. Furthermore, the assumption that the deep parts of the deposits were formed at temperatures higher than the near-surface parts may be in error. On a statistical basis, however, there is some evidence to support the contention that deep-seated (high temperature?) deposits have a higher Au/Ag ratio than those formed at intermediate depths or near the surface, presumably under conditions of lower temperatures.

16. There is considerable evidence to show that the Au/Ag ratio increases in primary halos with proximity to ore shoots in most types of epigenetic gold deposits. This increase is also manifest in a vertical sense in the vein mineralization, increasing downward toward ore shoots in some cases and upward toward the shoots in others. Both upward and downward increases in the Au/Ag ratio have been recorded from some deposits.

17. Shcherbina's (1956*a*) view that gold is predominant in telluride ores whereas silver is dominant in selenide ores appears to be true. The analogous situation that gold tends to follow arsenic, and that silver is associated with antimony, is likewise marked on a statistical basis. There are, however, some striking examples where gold is enriched in antimony-bearing ores and silver in arsenical ores.

18. Gold and silica (quartz) show a marked association whereas silver tends to be concentrated in an environment where carbonates are abundant. The association of silver with manganiferous siderite, manganiferous calcite or dolomite and rhodochrosite is striking in some districts. In Tertiary deposits gold shows a similar relationship to the manganiferous carbonates and silicates. (*See* the section on mineral associates.)

19. Evidence from many auriferous belts throughout the world indicates that there is a wall-rock effect on the Au/Ag ratio, the ratio being higher where the wall rocks are basic than where acidic rocks are the hosts. There are a number of exceptions to this generalization, but statistically the phenomenon seems to hold in a number of gold belts. The reasons for the effect are obscure but probably involve the stability of the complexes of gold and silver in the transporting solutions or diffusion currents. Basic rocks appear to have the greatest effect in rendering gold complexes unstable, thus permitting proportionally higher amounts of gold to be precipitated than is the case with acidic rocks. The wall-rock effect on the Au/Ag ratio merits further detailed investigation.

20. The Au/Ag ratio in deposits seems to depend on regional metallogenic peculiarities in a crude way, if only certain types of deposits are considered. When all deposits are brought into the picture only diffuse relationships emerge. There are perhaps as many gold-quartz deposits with ratios greater than 1 in the Circum-Pacific belt of America as there are other deposits with ratios less than 1. The same is true for the Appalachian belt and in some greenstone belts of the Canadian Shield. The ratios seem to depend more on the type of deposit (i.e., its mineral constitution and hence chemical history) than on its position in the earth's crust.

The purity or fineness of native gold in deposits is a complicated problem that has not received the attention it merits. With the development of the electron probe this problem is more amenable to a detailed attack, and it is to be hoped that we shall soon see some detailed research work on gold from all types of deposits, and also from single deposits with respect to the degree of fineness with increasing depth and other parameters.

The purity of native gold is considered in a general way in Chapter II. There, it was stated that a large number of metals may substitute in trace, minor or major amounts in gold. Of these Ag is frequently a minor or major constituent; Cu is commonly also present, and elements like Fe, Hg, Bi and Pt may be a minor or trace constituent of native gold.

The literature invariably emphasizes the fact that the fineness of gold in placers is higher than that in the deposits from which it came. This circumstance is discussed as some length in this chapter in the section on placers. Here, it is sufficient to state that this is due to a natural refining action predicated on the greater mobility (solubility) of silver compared with gold in the supergene environment. The silver is leached out in the general placer environment leaving the residual gold depleted in silver. Another feature repeatedly referred to in the literature is the commonly observed fact that young (Tertiary) deposits (some call these near-surface or epithermal deposits) have high contents of silver in their native gold. These deposits are the home of electrum. No hypothesis to explain this phenomenon seems to have been put forward. Perhaps it is metallogenetic in nature and cryptic from the present plane of observation. On the other hand the phenomenon may have something to do with the recycling of gold and silver in the crust with time. It is common to find late secretion veins in gold deposits greatly enriched in silver. Veins formed by what the writer has called a late extraction process from pre-existing sulphides in rocks and deposits are invariably rich in silver. The rich silver deposits at Cobalt may be an extreme example of this phenomenon (Boyle and Dass, 1971*b*). As mentioned later, the Tertiary deposits may have originated by this process of extraction of Au and Ag from underlying rocks – hence their native gold tends to be greatly enriched in silver.

Fisher (1945, 1950) has examined the fineness of gold with respect to the types of deposits. His general observations and conclusions, only slightly paraphrased are as follows.

Lodes formed under epithermal conditions at shallow depths typically contain gold from about 500 to 700 fine. In this zone the correspondence between gold fineness and temperature of formation appears to be close, and the gold fineness furnishes a sensitive measure of the conditions of deposition. Deposits formed in Late Tertiary times under shallow cover carry gold about 500 fine as a rule, while those formed near the bottom of the zone or in what Graton has called the leptothermal zone – transitional from the epithermal to the mesothermal – contain gold about 700 fine, possibly ranging up to as much as 800. Studies made by R. V. Gaines consequent upon the publication of Fisher's original paper (Fisher, 1945) furnish an illustration of this type of variation. Samples were collected at various horizons in the Timmins Ochali-Berlin Mine near Yaramul in the Antioquia region of Colombia, South America. This lode is considered to have been formed near the top of the mesothermal zone. Gold fineness in the primary ore was found to increase downwards from 680 to 760 over a vertical distance of 1300 ft. These figures provide a satisfactory confirmation of the deductions as to the probable variation in gold fineness with depth that were drawn from a general study of the subject, namely that the fineness of gold shows a general increase with depth of deposition, which implies high temperature and pressure, and appears to be correlated more closely with temperature.

Exceptions to the rule that epithermal deposits carry comparatively low grade gold occur only in deposits in which silver is conspicuously absent and insufficient silver was present in the ore-bearing solutions to form an alloy appropriate to the environment. As long as sufficient silver is present, even though it is only just sufficient, low-grade gold will be found in epithermal deposits.

If tellurium is present, it will combine with the gold and silver in proportions appropriate to the relative concentrations, so that the gold telluride, calaverite, will be formed in regions like Cripple Creek, where little silver is present in the ore, and silver-gold tellurides where silver is abundant. Gold appears to have a greater affinity for tellurium than for silver, and will combine with it in preference to silver. If excess gold and silver are present after all the tellurium has been taken into combination as tellurides, they will be deposited as an alloy of fineness typical of the environment, e.g., in epithermal deposits such as those of Karacs-Czebe, Transylvania, where low fineness gold is formed.

Deposits in the mesothermal zone normally carry gold from 750 to 900 fine, but the fineness may be higher in areas poor in silver. A very common fineness in lodes of the mesothermal zone is 850 to 870.

It is not impossible that under certain special chemical conditions, or in the presence of certain minerals, gold of lower than the typical fineness may be deposited in mesothermal deposits, but such cases, if they exist at all, are rare exceptions.

Deposits of the hypothermal zone always contain gold in excess of 800 fine, no matter how much silver may be present in the ore.

Gold of very high fineness, 990 or more, is generally the result of oxidation under conditions favourable for the complete removal of silver. Mount Morgan, Queensland and Kalgoorlie, Western Australia are excellent examples. High fineness is also often associated with the oxidation of telluride ores, e.g., Fiji and Cripple Creek, Colorado.

Alluvial gold, being derived from the oxidized ore, is of higher fineness than the average of the vein gold and shows an increase in value downstream due to surface refining action, as the size of the grains decreases.

Eales (1961, 1968) examined the fineness of gold in a number of Rhodesian mines (Olympus, Lonely, Horn and Turk). These deposits occur in Precambrian greenstone belts, some in iron-formations and sheared greenstones and others in granite gneiss, the principal mineralization consisting of quartz, pyrite, pyrrhotite, chalcopyrite, sphalerite, sulphosalts, gold tellurides and native gold. In the gold recovered from the four deposits he noted that high-grade ore generally contained silver-poor gold whereas silver-rich gold was more typical of low-grade ore. Mineragraphic studies indicated that gold of early crystallization, which occurs enclosed by sulphides such as chalcopyrite or sphalerite, contains more silver than that which crystallized late. Eales concluded that variations in fineness from place to place in a given deposit depend on the stage of the paragenesis during which the bulk of the gold was precipitated, and that the relationship between fineness and tenor arises out of the protracted crystallization of the gold in the richer sections. Accordingly, he reasoned that ore shoots are characterized by the association of late-stage minerals with a high proportion of late gold, which is poor in silver.

Sakharova (1969*b*) and Sakharova and Demidov (1972) examined the composition of native gold by X-ray spectral (electron probe) methods in the Darasun quartz sulphide deposit, Transbaikal, U.S.S.R. The elements investigated in the gold were Ag, Cu, Fe, Bi, Pb, Te, Sb, As and S. Their data indicated that gold of the same generation varied in composition both within the deposit and, in places, within the same polished section. For most grains the content of gold and silver was constant in the central part, but the concentration of gold decreased and that of silver generally increased in the marginal zone. The concentration of gold at the centre of grains ranged from 93 to 76.5 per cent, but it diminished at the edge to values ranging from 10.5 to as low as 2 per cent. The increase in silver content of the same interval equaled 2 to 10 per cent. The zones ranged in width from 3 to 15μ. The decrease in gold concentration in the marginal zones is clearest for grains in sulphides. For disseminations in quartz the gold content does not vary. For granules at the boundary between ore minerals and quartz the gold content does not vary on the side of quartz, but it diminishes by 2 to 2.5 per cent on the side of ore minerals (sulphides, tellurides).

Sakharova (1969*b*) concluded from the data that an exchange of atoms by diffusion is possible in natural mineral aggregates. She further postulated that there is interdiffusion between gold and the metals in its host sulphides. The metals that can diffuse into gold are iron and copper, i.e., elements that have small ionic radii and resemble gold in atomic structure in the metallic state. Elements of the anionic type, such as sulphur and tellurium, however, cannot diffuse into native gold, nor can high-valent cations like antimony and arsenic.

Finally, Sakharova and Demidov (1972) showed that the Au/Ag ratio in the ores of the Darasun deposit is 5 to 20 times higher than the average for the earth's crust. The Au/Ag ratio is lower in beresitized rocks than in the ore-bearing veins.

Yugai (1971) found that the purity (fineness) of gold increased with depth from 606 to 771 over a vertical range of 288 m in gold-quartz deposits in sheared sediments and Hercynian granites. He attributed this to the effects of supergene alteration processes.

An extensive investigation of the fineness and other characteristics of native gold in 15 gold deposits formed at shallow depth (mostly epithermal deposits of Cretaceous-Tertiary age, e.g., Belaya Gora, Agatovskoye) in the eastern U.S.S.R. by Berman *et al.* (1973) shows that the situation is extremely complex. They noted wide fluctuations in the fineness of the gold within single deposits but relatively uniform values in single ore shoots. In some cases the maximum difference was 600 units between different ore shoots. In some cases different parts of single gold particles showed a maximum difference of 120 units in fineness. Nevertheless average values generally indicated that the early generation of gold in most deposits was higher in fineness than later generations. The fineness of the native gold is also influenced by its host mineral. In some deposits the gold in the sulphides has a much higher fineness than that in the quartz and other gangue. Berman *et al.* (1973) think this may be typical only of gold-silver deposits with an excess of silver in the depositing solutions (shallow depth deposits), and mention investigations by N. L. Nikolayeva that indicate the native gold in sulphides in deeper mesothermal deposits has a lower fineness than that in the quartz and other gangue.

A number of recent data on the fineness of gold in deposits in U.S.S.R. is contained in the two volumes edited by Petrovskaya (1974). In one of the northern Kazakhstan deposits of Silurian age mineralized with quartz, arsenopyrite and sulphosalts, Sher found the average fineness to be 870$\pm$70. The main part of the gold with a fineness of 870 was deposited in the quartz gangue. In the eastern Yakutian gold-quartz-sulphide-sulphosalt deposits, Gamyanin and Leskova found a complex situation with respect to the composition (fineness) of the native gold. In many deposits the composition of the gold was erratic and changed with respect to the geological and physico-chemical conditions of its formation. In some oreshoots formed over a short dip or strike interval the composition of the gold was relatively constant. At the Karal'veemskoe deposit in western Chukotka the mineralization is essentially gold-quartz veins and irregular bodies with pyrite and arsenopyrite in Late Mesozoic diorite dykes and in greenstone diabase sills in Triassic phyllites, quartzites and schists. In the deposit Ol'shevskii and Kisterov found by histogram that the fineness of the gold is bimodal indicating two generations, one with an average fineness of 890 and the other 910. And finally, in the Muruntau deposit in Uzbek S.S.R. (For description *see* the section on deposits in this chapter.) Voronich *et al.* noticed that the gold and silver content of the deposit depends on the composition of the host rocks and is independent of depth. The deposit also lacks vertical zoning.

To summarize the data available on the variations in the fineness of gold in deposits one can say that the situation is generally complex, and in some cases each gold belt and commonly each deposit seems to have its own characteristics. In some auriferous deposits there is an indication that the early deposited gold is more silver-rich (lower in fineness) than that deposited later in the paragenetic sequence. At Kirkland Lake for instance the gold in early pyrite tends to be silver-rich (low in fineness) whereas the later generations of free gold are silver-poor (high fineness) (Blomfield *et al.,* 1936). Eales (1961, 1968), as noted above, also found a similar relationship in some of the gold mines in the Precambrian terrane of Rhodesia. Many of these investigations were carried out be reflectometry measurements about which there is some controversy as to their accuracy (Stumpfl, 1969). The author's investigations in this matter are equivocal, but on a statistical basis there is an indication that the early gold in sulphides is lower in fineness than the later generations of free gold. This probably only applies, to old (Precambrian and Paleozoic) deposits, since there is undoubted evidence that the late generations of gold in certain Mesozoic-Cenozoic deposits are exceptionally rich in silver compared with early generations.

There appears to be considerable variation of fineness with depth from deposit to deposit. In some cases there is an increase in the fineness of native gold with depth, a feature explained by some as due to an approach to a hot centre; silver being more mobile migrates farther from the hot centre than gold. In other deposits there is an indication that the fineness decreases with depth, and in still others there is no particular relationship, the fineness remaining relatively constant.

The data are equivocal concerning lateral zonations of the fineness of native gold in deposits with respect to thermal centres of intrusion or granitization. There are indications that the fineness of native gold in deposits near these centres is higher than in more remote deposits. Numerous aberrations seem, however, to complicate this picture and more data are required before definite statements can be made.

Within individual deposits there are often major differences in the fineness of native gold from oreshoot to oreshoot. Most individual oreshoots, however, generally record a relatively uniform fineness when their dip and strike dimensions are short.

In some deposits as revealed by microprobe studies the gold grains have a relatively uniform silver composition; in others they are markedly heterogeneous being zoned or having silver-rich or silver-poor sectors. Numerous deposits also seem to have marked intragrain compositional differences with respect to silver and other lattice constituents. These inhomogeneities obviously have a marked effect on systematic variations within deposits, features that must always be considered by those doing work on the fineness of gold in deposits.

There is obviously a wall-rock effect on the fineness of native gold, but the details are obscure and the whole problem requires detailed investigation.

The effect of temperature on the fineness of native gold in deposits should be carefully investigated, especially where deposits have a great vertical range and pass from low grade metamorphic facies into intrusives. Some of the data at hand suggest that there is a temperature effect, native gold near hot centres or fronts being finer than that in remote sites. This is most noticeable where intrusives such as dykes cut preexisting auriferous oreshoots. Adjacent to the dykes the native gold is often much finer than that at a distance from the dyke.

All of this discussion has centred essentially on the fineness (i.e., the silver content) of native gold. It seems probable that the contents of Cu, Te, Hg and the various other elements in lattice sites in native gold would also vary in certain systematic ways with depth, wall rock and other geological and geochemical features of the deposits. This problem has not been investigated in any detail within gold belts or deposits, although some preliminary work has been done as noted in the section on native gold in Chapter II. Particularly pertinent in these matters are the papers by Desborough *et al.* (1970, 1971) and Berman *et al.* (1973).

Oxidation processes have a marked effect on the fineness of gold in deposits. In many deposits the free gold in the oxidized zones is considerably finer than that in the hypogene zones. The range in fineness of hypogene gold is usually 700 to 925; that for native gold in oxidized zones is commonly 900 to 999. In placers the gold is normally finer than the hypogene gold in the original deposits, although there are situations where there is an equivalence in the fineness of the two types of gold.

> A study of the changes in rocks contiguous to orebearing fissures is essential to a thorough understanding of the genesis of such deposits.
> —W. Lindgren, 1901

Hypogene wall-rock alteration processes associated with epigenetic gold deposits

The term "wall-rock alteration" is difficult to define precisely since the phenomena accompanying it are extremely diverse, and the patterns developed vary with numerous geological and chemical parameters. In a general way wall-rock alteration can be defined as the mineralogical and chemical changes manifest in the host rocks of epigenetic deposits by the mineralizing media that may have been aqueous solutions, gases, melts or diffusion currents. Two general types can be differentiated –

hypogene wall-rock alteration developed contemporaneously or nearly so with the precipitation of the hypogene ore and gangue minerals and supergene wall-rock alteration produced as a result of supergene processes of oxidation and reduction. Here, only hypogene wall-rock alteration will be discussed in any detail. The supergene alteration processes in gold-bearing deposits are described in Chapter IV.

The type, degree and the lateral and vertical extent of wall-rock alteration attending epigenetic gold deposits depends on numerous interacting factors both geological and chemical. Briefly these are:

1. The nature of the host rock, specifically its mineralogical and chemical constitution, grain size, porosity and permeability, and the degree of shearing and fracturing adjacent to the sites of precipitation of the ore and gangue minerals.

2. The nature of the ore and gangue forming medium, specifically its state (liquid, gas or diffusion current), its chemical composition and its temperature and pressure.

Studies of wall-rock alteration associated with gold deposits are valuable in two respects. Firstly, they provide, together with chemical data on the ore and gangue elements and liquid inclusions in these minerals, an idea of the chemical constitution of the ore forming medium, and secondly they are useful in prospecting for ore. A few brief comments on the first of these features will be made in this section. The second aspect is dealt with in Chapter V.

The data used in the preparation of this section have come from many sources, principally published descriptions in the world literature together with data obtained by the writer during his long study of gold deposits.

The common mineral alterations in epigenetic mineral deposits are so well known that they hardly need mentioning. However, to avoid repetition, some of the common mineral alterations in all types of rocks are listed below.

Mineral	*Alteration products*
Quartz	Generally resists alteration. Quartz may be replaced by sericite, feldspar, carbonates and other minerals in some alteration zones.
Potash feldspar	Frequently resistant in alteration zones. In intense phases of alteration it yields sericite, hydromuscovite, clay minerals, alunite, epidote, zoisite, quartz and albite.
Plagioclase	Sericite, hydromuscovite, clay minerals, alunite, epidote, zoisite and quartz. Calcium component may yield calcite; aluminum component may give rise to chlorite or biotite. Frequently replaced by various sulphides. Calcic varieties of plagioclase often lose calcium and yield albite.
Muscovite	Hydromuscovite, clay minerals and quartz.
Biotite	Hydrobiotite, chlorite, sericite, hydromuscovite, clay minerals, epidote and quartz. Iron and magnesium components may yield carbonates (ankerite); iron and titanium components give rise to magnetite, rutile and leucoxene. Frequently replaced by various sulphides.
Hornblende	Biotite, hydrobiotite, chlorite, sericite, clay minerals, epidote and quartz. Calcium, iron and manganese components may yield carbonates, pyrite and/or magnetite. Titanium component gives rise to rutile and leucoxene. Frequently replaced by various sulphides. May yield serpentine minerals and talc under certain conditions.
Augite	Hornblende (uralite), biotite, chlorite, quartz and other products mentioned under biotite and hornblende. May yield serpentine minerals and talc under certain conditions.
Olivine	Serpentine minerals, talc, magnetite or hematite. Frequently carbonatized with development of ankerite. More rarely altered to chlorite, pyroxenes and amphiboles.
Magnetite	Frequently replaced by pyrite. Titanium component may yield leucoxene or rutile.
Ilmenite	Iron component may yield pyrite. Titanium component yields leucoxene or rutile.
Sphene	Leucoxene or rutile, calcite and quartz.
Zircon	Generally unaltered.
Apatite	Frequently unaltered. May be destroyed in intensely altered zones.
Carbonates	Various calcium-magnesium-manganese silicates, talc and serpentine minerals. Frequently entirely replaced by quartz.

Summary accounts of many of the features of wall-rock alteration, some pertaining to auriferous deposits can be found in the various papers and monographs by Lindgren (1901), Morey and Ingerson (1937), Schwartz (1947, 1950, 1955, 1959), Sales and Meyer (1948, 1950), Lovering *et al.* (1949), Lovering (1950), Stringham (1952), Khitarov (1957), Burnham (1962), Hawkins and Roy (1963), Korzhinskii (1965), Meyer and Hemley (1967) and Boyle (1970).

Roslyakova and Roslyakov (1975) give an extensive summary of the types of alteration associated with the various types of gold deposits in U.S.S.R. In addition they discuss the habit of gold and numerous other trace elements in the primary (endogenic) halos of these deposits (*see also* Chapter V).

Types of wall-rock alteration associated with epigenetic gold deposits

A number of types of hypogene wall-rock alteration are associated with epigenetic gold deposits. The most important of these, together with some brief general notes on their mineralogy and chemistry, are as follows. Details are discussed in subsequent sections.

Skarnification consists generally of the development of calcium-magnesium-manganese-iron silicate minerals, quartz, magnetite and a host of other minerals in limestones, dolomites or calcareous shales and schists. Gold deposition may or may not be associated with the process. The calcium-magnesium-manganese–iron silicates formed include olivine, wollastonite, garnet, pyroxenes, amphiboles, uralite and scapolite. Quartz, plagioclase and carbonates are generally present, and there may be a development of many minerals containing the mineralizers, S, H_2O, F, Cl and B, such as pyrite, pyrrhotite, serpentine, tourmaline, axinite, scapolite, vesuvianite, topaz, micas and fluorite. The process of skarnification may be essentially isochemical with removal of CO_2; more generally, however, it involves the introduction of SiO_2, Mg, Fe and the volatiles S, As, B, F, Cl and H_2O. Under these conditions removal of CO_2 is usually extensive.

In some skarn deposits the introduction of gold, economic sulphides, scheelite and various oxides appears to have taken place simultaneously with the formation of the principal skarn minerals. In other skarn deposits the introduction of the economic minerals appears to be later than the formation of the skarn. In this case the skarn is altered, the main effects being replacement of pyroxene by tremolite and actinolite and the development of quartz, epidote, calcite, sericite and chlorite or serpentine.

Amphibolitization and *pyroxenitization* involve the development of amphiboles and pyroxenes respectively in the wall rocks of auriferous deposits. Minerals accompanying these types of alteration include feldspar (commonly albite), epidote, tourmaline, biotite, magnetite, pyrite, pyrrhotite, garnet, biotite and occasionally andalusite and cordierite. Both amphibolitization and pyroxenitization are characteristic of certain deep-seated (high temperature?) gold-quartz veins and impregnation deposits mainly in igneous rocks of intermediate and basic composition but also in certain gneissic sedimentary terranes. Some of these deposits seem to be transitional to skarn bodies.

Feldspathization consists of the development of secondary feldspars in the host rocks of epigenetic gold deposits. Albitization is perhaps the most common, but the development of potassium feldspar (microcline, orthoclase or adularia) is common in certain types of deposits. Chemically the process may involve the introduction of K, Na and Al, although in some deposits these elements are simply rearranged, and the system is essentially isochemical. The genesis of hydrothermal K-feldspar (adularia), by replacement of andesine phenocrysts and by precipitation processes, in an active geothermal environment at Wairakei, New Zealand is discussed by Steiner (1970). In the Finish gold occurrence in the Okhotsk-Chukotka volcanic belt, U.S.S.R., Eremin (1970) noted that the near vein adularization of the liparite ignimbrites was superimposed on a preore propylitization stage. He also noted that the amount of adularization increased with depth.

Tourmalinization consists of the development of tourmaline in the wall rocks of certain types of epigenetic gold deposits. Chemically it involves the introduction of boron and in some cases Mg, Fe, Na, Ca, Al and Li. In zones of intense tourmalinization in shales and schists there is frequently a decrease in the SiO_2 and CO_2 content of the altered rock.

Fluoritization involves the introduction of fluorine into wall rocks with the development of fluorite. *Topazization* implies the development of topaz and an introduction of fluorine. *Greisenization* is an all-inclusive term involving the processes of fluoritization, topazization, sericitization and tourmalinization. These combined processes entail the conversion of feldspar, and other constituents of granitic and pelitic rocks in places, to aggregates of quartz, topaz, tourmaline, fluorite and commonly lepidolite or lithium-bearing muscovite. Judging from the available analyses there is a general introduction of fluorine, boron, lithium, tin and iron, and small abstractions of silica, calcium, sodium and potassium; alumina and magnesium generally remain relatively constant. All of these various fluoriferous alterations are generally associated with tin, tungsten and beryllium mineralizations. They are relatively rare in auriferous deposits. Fluoritization is, however, common in some Tertiary gold deposits, e.g., Cripple Creek, Colorado; topazization is notable at the Brewer Gold Mine, Chesterfield County, South Carolina; and greisenization is marked in some auriferous deposits bearing tin and tungsten minerals in the Mesozoic-Cenozoic tin-tungsten-gold belts of the far eastern U.S.S.R.

Silicification consists of the development of secondary quartz, jasper, chalcedony, chert, opal or other varieties of silica in the wall rocks of epigenetic gold deposits. By another terminology it also includes the development of various silicates in the host rocks, particularly in carbonate rocks or calcareous shales and schists. As noted above the latter process is called skarnification.

The chemistry of silicification is varied and depends essentially on the type of host rock, which is affected by the process. In carbonate rocks there is generally a major introduction of SiO_2 and a wholesale removal of Ca, Mg, Fe, CO_2 and other constituents. In silicate rocks SiO_2 may be redistributed within the host rocks, that is, leached in one place and added in another.

Biotitization is an uncommon type of alteration in gold deposits but occurs in certain places. It usually involves the development of biotite or hydrobiotite from various cafemic silicates. During the process H_2O and K are introduced, and some SiO_2 may be extracted. Less commonly the constituents of biotite, including K, Mg, Fe, Al and SiO_2, are all introduced into the rock.

Chloritization is one of the most common types of alteration in gold deposits. In many cases it develops directly by the hydrous alteration of cafemic silicates. During this process H_2O is introduced and some SiO_2 may be removed. In other cases constituents of chlorite, including Mg, Fe, Al and SiO_2, are introduced into the rock.

Sericitization is perhaps the most common of all of the types of wall-rock alteration in gold deposits developed in acidic, intermediate and basic igneous and metamorphic rocks. It consists of the development of potash mica, generally sericite or hydromuscovite, as a result of the hydration of feldspars or from the rearrangement of K, Al and SiO_2 within intensely altered wall rocks. Chemically, sericitization generally involves the introduction of K and H_2O into the rocks affected, and there is usually a removal of some SiO_2, Fe and Ca. The development of fuchsite and mariposite, the chromium-bearing micas, in basic and ultrabasic rocks adjacent to gold-quartz veins usually does not imply an introduction of chromium. Rather this element appears to be redistributed during the chemical breakdown of magnetite, pyroxenes and amphiboles, the principal chromium-bearing minerals in the rocks. The development of the mica, roscoelite, probably has a similar origin deriving its vanadium from magnetite, etc.

Granitic rocks, granite gneisses, quartz porphyries and certain types of schists heavily sericitized, albitized in places and impregnated with pyrite are called 'beresites' by the U.S.S.R. geologists. The term *beresitization* is commonly used in the Ural goldfields and in other gold-bearing districts of U.S.S.R. Numerous studies of the process have been carried out. According to Glukhov (1974) the wall rocks are generally enriched in K, CO_2 and Ca and exhibit depletions in Fe^{3+}, Al and SiO_2. Sericite and Ca-Mg carbonates are the principal resultant minerals. According to the same investigator the deposition of gold occurs during the late beresitization – early silicification (quartzitization) stages. Grabezhev (1973, 1974) states that beresites and greisens are formed within a similar temperature interval of 270 to 400°C. According to the same author fluorite is generally absent in beresites because the F concentration in the solutions from which gold was precipitated was lower than in those which gave rise to rare metal deposits (Be, Mo, W, Sn, Rb, Li). In the latter the F content often ranges from 937 to 8510 ppm, whereas in beresites the range is much smaller, 550–740 ppm. Most of the fluorine in beresites is in the micas (sericite).

Beresitization is particularly complex in certain deposits and may be zoned. Gavrikova *et al.* (1973) observed such a situation in the Itakinsk gold-sulphide deposits of eastern Transbaikal. There, they noted by analyses complex patterns with respect to Na and K as the veins were approached, although in a general way the K_2O/Na_2O ratio showed an increase towards the auriferous deposits. In addition they noted a vertical zonation over 1 km, the most outstanding feature being the development of seybertite (a high temperature brittle mica) at deep horizons and lower temperature chlorite in the upper horizons. They attributed this phenomenon partly to a temperature gradient along the direction of flow of solutions. Abdukhayumov (1973) and Babaev and Abdukhayumov (1974) also noted marked lateral zonations in beresitized rocks in the gold-sulphide deposits of Terekkan, U.S.S.R. The zoning comprised an external chlorite-albite-sericite-carbonate zone, an intermediate quartz-carbonate-sericite-pyrite zone and inner quartz-sericite-pyrite and quartz zones adjacent to the orebodies. During the beresitization process the content of Na, Mg and other bases decreased and S and CO_2 were introduced. The Si and K contents increased in the inner zone. In some deposits Al displayed an inert behaviour.

The development of the sodium mica, paragonite, is not common but occurs in some alteration zones where soda metasomatism is a marked feature of gold deposition. The process can be called *paragonitization;* it usually involves introduction or rearrangement of considerable amounts of Na and H_2O.

Carbonatization consists of the formation of secondary carbonates in the host rocks of epigenetic gold deposits. The phenomenon is particularly common in intermediate to basic rocks. *Dolomitization* consists of the formation of dolomite in the wall rocks of deposits usually at the expense of calcium carbonates, but also in places at the expense of magnesium silicates and other minerals. The process is relatively uncommon in gold deposits, but an analogous process *ankeritization,* which consists of the formation of ankerite, $Ca(Fe,Mg,Mn)(CO_3)_2$, is especially common.

The chemical processes involved in carbonatization are complex and depend essentially on the type of host rock affected. The ankeritization of basic and intermediate intrusive rocks and volcanics, such as frequently occurs in gold-quartz deposits, appears to require only the introduction of CO_2, the other constituents of ankerite such as Ca, Fe, Mg and Mn coming from the chemical breakdown of amphiboles, pyroxenes and feldspar. During this process much SiO_2 is removed and probably transferred to the veins where it crystallizes as quartz.

Sideritization and other types of carbonatization of sedimentary rocks, especially quartzites and sandstones, generally requires the introduction of nearly all of the components of the carbonates formed in the rocks.

Basic and ultrabasic rocks, heavily carbonated, sericitized and pyritized are called 'listwänites' by the Soviet geologists. The term *listwänitization* is commonly used in the Ural goldfields and in other auriferous districts of the U.S.S.R. Kashkai (1964) and Goncharenko (1970) have discussed the zoning and chemistry of listwänites in some detail. Briefly the metasomatic development of listwänitized rocks follows the sequence: (1) initial formation of serpentine-actinolite-chlorite rocks (with few carbonates) grading to serpentine-chlorite-carbonate rock; (2) intermediate stage of formation of serpentine-brucite-carbonate rocks containing hematite and quartz and grading to talc-chlorite-carbonate and talc-carbonate rocks; (3) listwänite stage of the mariposite listwänites and the quartz-carbonate listwänites and their varieties; (4) final stage of veinlet formation of quartz, carbonates and other components of listwänites and (5) a postlistwänite stage during which sulphides and other components, genetically related to younger hydrothermal activity, were precipitated in the listwänites.

The main characteristic of listwänitization is the conversion of serpentine into talc and/or carbonates. The chemical composition of listwänite is variable and is controlled by zonal factors and the composition of the host rocks. In general there is an introduction of K, Ca, Al, CO_2 and H_2O and an abstraction of SiO_2.

Pyritization and *arsenopyritization* are two of the most common types of alteration in gold deposits. They consist of the development of disseminated pyrite and arsenopyrite in the host rocks of epigenetic deposits. In many types of deposits only sulphur and arsenic are introduced, the iron coming from the breakdown of iron silicates and oxides as McKinstry (1957) has suggested. In other types of deposits all of the components of pyrite and arsenopyrite are introduced.

Propylitization is a paricularly common type of alteration associated with gold-quartz and polymetallic deposits in basic and intermediate intrusive rocks and volcanics. The principal secondary minerals developed are chlorite, epidote, zoisite, leucoxene, carbonates, sericite, feldspar, clay minerals, pyrite and arsenopyrite. Propylites commonly form the fringe zone of highly altered rocks where sericitization, carbonitization and pyritization are intense.

Chemically, much water is introduced during propylitization, and there may also be additions of CO_2, S and As. Some SiO_2 is generally extracted during the process, and there may be losses of Na, K and the alkaline earths in some districts; in others K and Na may be enriched.

Propylitization tends to pervade large volumes of rock in many districts and may, therefore, not be directly connected with epigenetic gold deposits. In places it appears to grade imperceptibly into the greenschist facies of regional metamorphism.

Rocks that are heavily propylitized exhibit several types of alteration including chloritization, carbonatization, pyritization and sericitization. The term propylitization is, therefore, a broad inclusive one.

Much has been written on the processes of propylitization in the past, especially with reference to the Mesozoic-Cenozoic gold deposits of Europe and the United States (Lindgren, 1933; Emmons, 1937), and numerous papers have appeared in recent years on the subject, most dealing with Tertiary rocks and deposits in which propylitization is especially marked (Szeky-Fux, 1964; Rusinov, 1968; Kazarinov *et al.,* 1970; Vasilevskii, 1972). There are many questions about the causes of propylitization, the source of the volatiles, such as CO_2 and H_2O, that take part in the regional phenomena, and the relationship of propylitization to mineral (gold) deposition. These problems lie more in the field of petrology and cannot be discussed here. About all that can be said with assurance respecting gold deposits is that many, especially those in Tertiary andesitic terranes, appear to be closely associated with extensive propylitization linked in some manner or other with a high mobility of potassium resulting in potash metasomatism of the propylitized rocks and the wall-rock alteration zones of the gold-quartz veins.

Argillic alteration consists of the development of clay minerals in the wall rocks of epigenetic gold deposits. The common clay minerals found include kaolinite, montmorillonite, dickite, illite and halloysite. The principal chemical effect appears to be the introduction of water, and there may be a removal of some SiO_2, Fe, K, Na, Ca and Mg, depending on the type of rock affected.

Alunitization is common in some low temperature (epithermal) gold deposits principally in acid and intermediate volcanic rocks. It generally consists of the development of alunite, $(K,Na)Al_3(SO_4)_2(OH)_6$, and quartz in feldspathic rocks, the alunite being largely the result of alteration of the feldspars. Chemically, the alteration generally involves an introduction of S and H_2O and removal of some SiO_2, Na, Ca, Mg and Fe.

Certain stability parameters for alunite were determined by Hemley *et al.* (1969). According to them a fairly high H_2SO_4 concentration is required in solutions before alunite can be produced in equilibrium with muscovite, kaolinite or K-feldspar. They also found that consistent with geologic observation alunite-kaolinite (including dickite or pyrophyllite) are the most likely products of strong hydrolytic (advanced argillic) types of alteration in either hot spring or higher temperature environments, whereas muscovite-alunite is favoured by higher temperature as well as higher potassium concentrations.

Some of the more recent studies of alunitization association with auriferous deposits are those by Harvey and Vitaliano (1964) for the Goldfield district, Nevada; Lodder (1966) for the gold-alunite deposits of Rodalquilar, southeast Spain; Takeuchi and Abe (1967) for a number of deposits in Japan; and Wang (1973*a,b*) for auriferous deposits in Taiwan.

In general, alunitization signifies acid attack by H_2SO_4 on the alkaline silicates of the wall rocks, particularly on the feldspars. In some deposits, however, the components of alunite, including K, Na, Al and SO_4 seem to have been introduced, at least locally, possibly as soluble sulphates.

Alunitization may be hypogene and/or supergene in origin in some gold deposits.

Hematitization is a type of alteration that frequently accompanies gold deposits, particularly those containing uranium. It is also found in a variety of other deposits often in syenitic rocks as at Kirkland Lake, Ontario. The presence of hematite generally indicates a low partial pressure of sulphur in the mineralizing solutions or diffusion currents.

The iron producing the hematitization may or may not be introduced. In some cases there is abundant chemical evidence that the element has been introduced as ferric iron into the wall rocks; in other cases where pervasive hematitization is present the iron is probably only redistributed while undergoing oxidation from the ferrous to ferric state. In the latter case some iron may actually be leached.

Serpentinization and the closely allied development of talc *(talcification)* generally consists of the formation of serpentine and talc in ultrabasic rocks such as dunite, peridotite and pyroxenite. Serpentinization and the development of talc may also occur in crystalline limestones and dolomites. The effect is marked in some gold-quartz and other types of deposits in ultrabasic rocks.

Serpentinization and the development of talc in ultrabasic rocks generally only involves the introduction of H_2O and the redistribution of the other constituents of the rocks. Where serpentine and talc are developed in limestones and dolomites there is generally an introduction of SiO_2 and H_2O and frequently some Mg.

Hydration is nearly universal in all types of wall-rock alteration in gold deposits. It indicates an introduction of water that is fixed either as hydroxyl or as water of hydration in the various alteration minerals. *Dehydration* involves the removal of water, such as takes place in the formation of certain types of wall-rock alteration zones.

Phosphatization is common in the alteration zones of some gold deposits. It involves the introduction or rearrangement of phosphate resulting mainly in the development of apatite. Some gold deposits exhibit this feature to a moderate degree.

Pyrophyllitization consists of the development of the hydrous aluminum silicate, pyrophyllite. In most types of epigenetic gold deposits this type of alteration is apparently rare. Pyrophyllite is, however, a very common mineral in the matrix of the auriferous quartz-pebble conglomerates of the Witwatersrand.

Zeolitization consists of the development of zeolites such as stilbite, natrolite, heulandite, chabazite, etc. The widespread process is frequently not associated with mineral deposition. The process, however, frequently accompanies native copper deposits in amygdaloidal basalts and is also associated with certain types of gold and silver veins. In all occurrences the zeolites are often accompanied by calcite, prehnite, pectolite, apophyllite and datolite.

Regularities of wall-rock alteration associated with gold deposits with respect to types of host rocks

A survey of the world literature of gold deposits, combined with studies at the Geological Survey indicates that wall-rock alteration, exhibits a certain regularity with respect to the types of host rocks. This regularity is discussed below, but it should be clearly understood that there are numerous exceptions to the generalizations given here.

For purposes of discussion the various types of host rocks are divided into the following categories:

1. Acidic rocks – include granites, rhyolites, quartz-feldspar porphyries and the acidic metamorphic equivalents of these rocks such as granite gneisses, mica schists, etc.

2. Intermediate and basic rocks – include granodiorite, syenite, nepheline syenite, monzonite, dacite, diorite, andesite, phonolite, gabbro, diabase, basalt and their metamorphic equivalents such as amphibolites, pyroxenites, etc.

3. Ultrabasic rocks – include dunite, peridotites, pyroxenites and their metamorphic equivalents.

4. Normal shales, slates, phyllites and derived schists.

5. Sandstone, conglomerate, quartzite, greywacke, arkose and their metamorphic equivalents.

6. Carbonate rocks.

Acidic rocks

In gold deposits the most prevalent types of wall-rock alteration in acid rocks are sericitization, argillization and the development of quartz and pyrite. The principal chemical features are the introduction of water and usually small amounts of CO_2 and the redistribution and loss of a number of constituents. Silica is generally liberated and may crystallize as quartz in the alteration zones or be transferred to the veins or other dilatant zones. Sodium and occasionally potassium are leached, and sulphur is generally introduced together with some iron. This type of wall-rock alteration is typical of many gold and base metal districts and of the copper-molybdenum porphyry deposits.

Feldspathization combined with the development of tourmaline and mica is common in the wall rocks of certain slightly auriferous pegmatites in granites, granite gneisses and schists. The phenomenon generally involves the introduction of some Na and K and variable amounts of B and H_2O. Feldspathization is only moderately common in the wall-rock alteration zones of gold-quartz and base metal deposits in acidic rocks. In some goldfields, however, there is often a marked development of albite or adularia.

Alunitization occurs in certain near-surface (epithermal Tertiary) gold deposits in granites, rhyolites and quartz-feldspar porphyries. It is, however, not as common as in intermediate rocks such as dacite, latite, etc. (*see* below).

Tourmalinization is marked in certain gold-quartz deposits in acidic rocks, especially in granodiorite. Biotitization, chloritization, carbonatization, zoisitization and epidotization are rather uncommon types of wall-rock alteration in acidic rocks, although there are many notable examples of these particular types in a few gold deposits.

Beresitization is a common alteration process in many parts of the world where gold deposits are localized in granite dykes, quartz-feldspar porphyry dykes and irregular bodies and other types of granitic rocks.

Intermediate and basic rocks

In gold deposits the most common types of wall-rock alteration in intermediate and basic rocks are chloritization, carbonatization, sericitization, pyritization, arsenopyritization, alunitization and propylitization. There is frequently a lateral zonation of these types of alteration with respect to the veins, lenses and lodes. Adjacent to the deposits the wall rocks exhibit an intense phase of sericitization, carbonatization and pyritization (also arsenopyritization) that grades imperceptibly into a phase characterized by chloritization, weak carbonatization and pyritization. In some types of deposits the latter phase grades into rocks that are areally propylitized and contain much chlorite, epidote, carbonate, etc.

Carbonatization, sericitization, pyritization and chloritization of intermediate and basic rocks generally involves an introduction of CO_2, H_2O, S and frequently As. There is usually a major extraction of SiO_2 that is probably transferred to the vein sites in most cases. Constituents like Fe, Ca, Mg and Mn remain more or less constant. Na and K exhibit a variable habit. In some cases there is a local redistribution of alkalis, especially K; some Na is leached from the more intensely altered zones in places.

Argillic alteration is comparatively rare in intermediate and basic rocks older than the Tertiary. In Tertiary and Quaternary volcanic zones, however, this type of alteration is common in gold deposits, particularly in those in andesites and their corresponding pyroclastics.

Alunitization is a frequent type of alteration in certain near-surface (epithermal Tertiary) deposits of gold, silver and base metals in dacites, latites and monzonites (e.g., Goldfield, Nevada). Alunite is generally developed together with much quartz at the expense of the feldspars. Chemically much S and H_2O are introduced, and there is a removal of some SiO_2, Na, Ca, Mg and Fe in intensely alunitized rocks. Carbonates are rare in alunitized rocks.

Zeolitization is common in certain vesicular basalts and other basic rocks containing native copper, gold and silver deposits. Minerals, such as orthoclase, chlorite, epidote, pumpellyite, prehnite, datolite, quartz and calcite, frequently accompany the zeolites. Analyses of altered zeolitized basalts appear to be lacking in the literature, and it is difficult to determine what changes take place during this particular type of alteration. Presumably H_2O, CO_2, B, Na, Ca and SiO_2 are introduced in most cases. Zeolitization is relatively rare in gold deposits older than those of Tertiary age. In the latter and in some Quaternary deposits this type of alteration may occur, but it is not common.

Feldspathization occurs in some intermediate host rocks but is relatively rare in basic rocks, although developments of albite are common in the alteration zones of certain gold-quartz and gold-bearing polymetallic deposits in some districts.

Tourmalinization is moderately abundant in some gold-quartz and other types of deposits developed in intermediate and basic rocks. Biotitization may also occur in places.

Ultrabasic rocks

The predominant type of alteration in ultrabasic rocks is serpentinization and the development of talc. In some gold districts serpentinites and other types of ultrabasic rocks show a marked carbonatization (ankeritization) and a development of magnesite and chromium-bearing fuchsite or mariposite. In such situations there is generally a marked loss of silica that is probably transferred to the veins where it crystallized as quartz. Listwänitization is common in some auriferous deposits in basic and ultrabasic rocks in U.S.S.R.

Normal shales, slates, phyllites and schists

Shales, slates and schists adjacent to gold deposits generally exhibit less tendency to alter compared with acidic, basic and ultrabasic igneous rocks. There are, however, certain well marked types of alteration in shales, slates and schists and these usually show some regularity.

In high grade metamorphic terranes and in the contact aureoles of granitic intrusions, calcareous shales may exhibit skarnification in the vicinity of gold, copper and other types of high temperature deposits.

Tourmalinization is particularly common where shales,

slates and schists are the host rocks for gold-quartz veins. Commonly a band of tourmaline borders both contacts of the quartz veins.

Feldspathization is generally not extensive in the wall rocks of most types of gold-quartz veins, although in some districts albitization is characteristic.

In many metalliferous districts normal shales, slates and schists tend not to exhibit extensive alteration zones in the vicinity of gold-quartz and other varieties of epigenetic gold and polymetallic deposits. In such districts the quartz veins are more or less frozen against the wall rocks. In other districts, however, such deposits may be marked by one or more of the following types of alteration: silicification, tourmalinization, biotitization, chloritization, sericitization, carbonatization, pyritization, arsenopyritization and argillic alteration. Of these tourmalinization, sercitization, chloritization, carbonatization and pyritization appear to be the most common.

Sandstone, conglomerate, quartzite, greywacke, arkose and their metamorphic equivalents

The so-called 'red bed' uranium, vanadium, copper, lead and silver deposits, usually devoid of gold, are generally, but not always, in bleached beds of sandstone, sandy limestone, sandy shale, conglomerate or arkose. The bleaching in most cases is said to be due to the reduction of ferric iron to the ferrous state. There is also generally a development of kaolinite and frequently some chlorite both within the deposits and in the enclosing rocks. The chemistry of the bleaching has not been investigated in sufficient detail in any of the deposits to draw adequate conclusions. Some analyses done for the writer on the rocks from Nova Scotia deposits suggest that the principal chemical changes in the bleached rocks, compared to the normal red sandstones and shales, are reduction of ferric iron and the introduction of some K, Al and H_2O.

Gold deposits in quartz-pebble conglomerates are excluded here because of their controversial origin. Most geologists consider these deposits to be modified placers. The various alterations that these deposits have undergone are considered in detail in the next section of this chapter.

Veins and other types of epigenetic gold deposits in sandstone, quartzite, greywacke and arkose are not generally marked by extensive alteration zones. Where alteration is present, as in some gold districts, the predominant types are tourmalinization, feldspathization, carbonatization, pyritization, arsenopyritization and choritization.

Carbonate rocks

The principal types of high temperature alteration in carbonate rocks are skarnification and silicification. These usually go hand in hand. In some gold deposits in skarns there may also be a weak to moderate development of tourmalinization, chloritization, serpentinization or the formation of talc. In the Muruntau deposits in Uzbekistan, U.S.S.R. the calcareous siltstones, shales and phyllites associated with the gold-bearing stockwork deposits are extensively altered in many stages during which formation of actinolite, pyroxene, feldspars, chlorite, biotite, tourmaline, sericite, quartz, pyrite, pyrrhotite, arsenopyrite and gold took place.

Skarnification is an extremely variable type of alteration that appears to depend on the initial composition of the carbonate rocks. Generally there is a major development of calcium-iron-magnesium-manganese silicates, quartz and magnetite. The chemical changes in the altered rock vary from district to district. A generalized compilation of the chemical data available in the literature indicates that the additions are mainly SiO_2, Fe, Mg, Ti, K, Al, S, B, F, Cl, P and H_2O, plus a number of metals of which the most significant are Cu, Mo, W, Sn, Pb, Zn, As, Au, Ag, Bi, Pt and Be. Not all of these additions occur in any one place. The principal subtraction is CO_2.

The alterations found in medium- to low-temperature deposits in carbonate rocks are dolomitization and silicification. The first is relatively rare in gold deposits. The chemical changes in dolomitization generally involve an increase in Mg and frequently some Mn and a little Fe. There may or may not be a loss of calcium. Silicification involves the introduction of SiO_2 and a general elimination of Ca, Mg, Mn, Fe and CO_2.

Timing of wall-rock alteration, open and closed systems and the nature of ore-forming solutions giving rise to auriferous deposits

The timing of the introduction and abstraction of the various constituents that participate in wall-rock alteration is a most complex problem – one that has not yet been satisfactorily solved. According to one school the various exchanges are an integral part of one surge of mineralization, the various reactions and exchanges taking place in a complex diffusion system within a relatively short geological interval. The other school advocates a number of successive but discontinuous surges or pulses, each producing characteristic effects that depend on the nature of the mineralizing medium, the nature of the wall rocks, the sequence of fracturing and other factors. One stage of wall-rock alteration may be superimposed on another, and the chemical and mineralogical relationships may be very complex indeed. The whole problem of the timing of wall-rock alteration is intimately associated with mineral paragenesis and zoning, two subjects that are too extensive and complex to be discussed at this point; they are, however, dealt with later in this chapter. The student of ore genesis and wall-rock alteration should always be aware of the two possibilities and should strive to obtain unequivocal criteria for the solution of the chemical and mineralogical relationships.

The problem of open and closed chemical systems in wall-rock alteration processes associated with epigenetic gold deposits remains essentially unsolved and will remain so until detailed material balance calculations are carried out throughout all parts of the various types of auriferous deposits, including the ores and the lateral, sub and supra alteration haloes. Similarly the problem of equilibrium in the various chemical reactions in wall-rock alteration is unanswered. Probably most reactions are not equilibrium reactions in the commonly accepted chemical sense. It is more fruitful to approach wall-rock alteration from a dynamic (kinetic) viewpoint, since diffusion seems to play such a large part in all of the processes.

One final aspect of the chemical and mineralogical regularity in wall-rock alteration in gold deposits that needs mentioning is the tendency of dissimilar rocks to converge to a uniform type of alteration. This feature has been noted by numerous investigators and is evidently due to prolonged attack by the mineralizing medium. There are for instance numerous references in the literature of interbedded slates, greywackes, amphibolites and other quite dissimilar rocks being all converted to carbonate-sericite rocks and of quartz porphyries, tuffs and shales intensely altered to an aggregate composed of quartz, sericite and clay minerals. Silicification of carbonate rocks, shales and sandstones, likewise yields a convergent type of alteration in some deposits that is indistinguishable with respect to the host rock except by very detailed studies.

It has long been recognized that the ore-forming medium may be either gaseous, an aqueous solution, meltlike or in the nature of a diffusion current of mobile ions or complexes. The features of wall-rock alteration in auriferous deposits do not seem to give unequivocal criteria for any of these mechanisms of transport, although there are certain trends that may be indicative.

The features of auriferous skarn zones suggest extensive diffusion of SiO_2 and other constituents into the carbonate rocks and CO_2 out of them. It also seems probable that diffusion of mineralizers, such as H_2O, S, As, F, Cl and B, played a large part in the formation of auriferous skarns.

The formation of the various types of auriferous veins and lodes probably took place mainly by diffusion in the deep zones of the earth and by precipitation from aqueous solutions under near-surface conditions. In certain cases, such as in the slightly auriferous massive nickel-copper ores, the sulphides may have crystallized from melts, since many have little or no associated wall-rock alteration. It seems impossible from studies of wall-rock alteration to decide which process – diffusion or precipitation from aqueous solutions – was the important agent in the formation of auriferous veins and lodes. Other criteria (Boyle, 1963) must be brought to bear on this problem. Whatever the agent or mechanism, it is certain that water played a large part in the formation of auriferous veins. Other volatiles, such as CO_2, S, As, Se, Te, Sb, B and F, have played a leading role in the formation of certain types of gold deposits. Wall-rock alteration zones associated with auriferous deposits rarely show the effects of chlorine metasomatism, a feature that might lead one to suspect that chlorides are not as important in the transport of gold and other metals as some investigators would have us believe. On the other hand sodium chloride is particularly abundant in some hot springs and thermal waters and is found in liquid inclusions. The reason for the low abundance of chlorides in wall-rock alteration zones and in auriferous veins may be the great solubility of most of these compounds. In other words, they are largely removed from the deposits at a late stage by the mineralizing solutions or diffusion currents.

Judging from the chemistry of wall-rock alteration, the constituents of most auriferous veins and lodes were deposited from an aqueous medium that was highly fluxed with CO_2 and S in most cases and with As, Sb, Se, Te, Bi, B, F and Cl in some cases. The gold was probably transported mainly as sulphide, arsenide, antimonide or telluride complexes, and most of the accompanying metals were carried as complex bicarbonates, sulphides, arsenides, antimonides, selenides, tellurides, bismuthides, fluorides or chlorides. Quartz and many of the constituents of the other gangue minerals, such as Fe, Ca, Mg and Mn, may have been largely derived from the adjacent wall rocks.

The pH of the ore forming medium is difficult to assess from wall-rock alteration studies of auriferous deposits. In some cases there is evidence that the solutions or diffusion currents were acid and that they gradually became alkaline by reaction with the wall rocks. In other cases they evidently were neutral or slightly alkaline. In most cases where direct measurement of or calculation from hot spring data have been done the solutions are at or near the neutral point.

Further discussion of the origin of epigenetic gold deposits is reserved for the last section of this chapter. Some typical examples of wall-rock alteration in various types of gold deposits are now discussed.

Examples of wall-rock alteration associated with gold deposits

The examples to be described follow the general scheme of types of deposits mentioned at the beginning of this chapter. Some of the data are taken from the published literature; the remainder were obtained during the study of gold deposits by the writer in Canada and other countries.

Auriferous porphyry dykes, sills and stocks; auriferous coarse-grained granitic bodies, aplites and pegmatites

Deposits in granitic rocks containing indigenous gold are few and most are not of economic importance. Some pegmatites carry small amounts of gold that may have crystallized with the pegmatitic minerals. The common types of wall-rock alteration associated with the formation of these pegmatites are feldspathization, tourmalinization, silicification and development of mica (frequently Li-bearing); development of beryl, apatite and garnet; and biotitization of amphibolites in places, indicating an introduction of K. These processes are described in some detail by Cameron *et al.* (1949) and Jahns (1955).

Where gold occurs in porphyry dykes and other granitic rocks as a late mineralizing phase, the granitoids are commonly fissured and silicified. There is often an introduction of tourmaline, and the feldspars are sericitized and frequently carbonated. Pyrite, arsenopyrite and other sulphides fill the fissures in part and invade the granitoid walls up to distances of 1 ft or more. These manifestations of the mineralization are similar to those described subsequently under granitic rocks and need not be further dealt with here.

Auriferous skarn-type deposits

These deposits are characterized by the development of various silicates of calcium, magnesium, manganese, iron and aluminum, especially garnet, epidote, vesuvianite, diopside, tremolite and wollastonite. There is in addition some quartz, abundant carbonate and minerals such as pyrite, pyrrhotite, molybdenite, arsenopyrite and magnetite. Typical examples of these types of deposits are the Nickel Plate deposit at Hedley, British Columbia that is marked by abundant auriferous arsenopyrite, often in massive lenses, and the French deposit in the same area that is mainly a skarn zone with little arsenopyrite.

Skarn-type gold deposits commonly exhibit minor to major secondary alteration effects superimposed on the skarn.

Where these are present the pyroxene is frequently replaced by tremolite and actinolite, and there are developments of quartz, epidote, calcite, sericite, and chlorite or serpentine.

The chemistry of skarnification is well illustrated by the copper deposits near Whitehorse, Yukon that carry, however, only small amounts of gold and silver (0.25 oz Au/ton and 6 oz Ag/ton of concentrates). There, the relationships are exceptionally clear as shown in Figure 9. Further details of the geology are outlined by Kindle (1964).

The analyses (Table 44) and profiles (Fig. 10) show the normal relationships commonly observed during skarnification. As silica, alumina, iron, titania and manganese are introduced into the limestone, there is a moderate extraction of lime and a wholesale removal of carbon dioxide. MgO shows an erratic behaviour. Na_2O and K_2O are enriched only in the rocks near the granodiorite, and P_2O_5 exhibits an enrichment near the central part of the skarn zone. Water tends to be enriched in the skarn compared with the original limestone. Sulphur exhibits an erratic behaviour being concentrated, however, in the skarn adjacent to the orebody.

It will be noted that lithium tends to follow the major alkalis rather than magnesium as is common in some rocks. Rubidium follows potassium closely, and barium behaves in a similar manner. Strontium exhibits a mixed behaviour, following the major alkalis in the zone next to the granodiorite and calcium in the limestone.

Copper shows a marked increase as the orebody is approached, a feature also exhibited by silver, gold, mercury and arsenic.

Considering various ratios it is apparent that there are some consistencies. Approaching the orebody from the granodiorite the data in Table 44 show that the K_2O/Na_2O ratios increase and the SiO_2/CO_2 and SiO_2/total volatiles ratios exhibit a decrease, although for the last two an erratic behaviour is apparent along parts of the traverse. As the orebody is approached from the limestone side the K_2O/Na_2O ratios show no change, whereas the SiO_2/CO_2 and SiO_2/total volatiles ratios register an increase.

Veins, lodes, stockworks, mantos and bedded disseminated deposits

There are several types of these deposits as noted at the beginning of this chapter. Here we shall deal successively with those dominantly in intermediate and basic intrusive and extrusive rocks; those dominantly in intrusive and extrusive acidic (granitic) rocks; and those dominantly in sedimentary rocks.

Gold-quartz veins and silicified zones dominantly in intermediate and basic extrusive and intrusive rocks (Yellowknife type)

Deposits of this type commonly occur in volcanic terranes, the principal host rocks being andesites and basalts, now largely greenstones. Similar types of deposits also occur in syenite, dacite, granodiorite, monzonite, diorite, diabase and gabbro plugs or stocks in various terranes. Their alteration zones are similar and are included here.

Deposits of this type range in age from Precambrian or Tertiary. Characteristic examples of Precambrian age in Canada include the deposits of Yellowknife, Northwest Territories, Squall Lake, Manitoba; San Antonio, Manitoba; various deposits in the Red Lake area of Ontario; various deposits in the Timmins area of Ontario; the Kerr Addison deposit in the Larder Lake district of Ontario; the Cordova Mine in eastern Ontario near Marmora; and a number of deposits in western Quebec including Arntfield, Wasa Lake, Senator-Rouyn, Stadacona, East Malartic; and others. Elsewhere in the world Precambrian members of this type occur in Kalgoorlie and Coolgardie districts of Western Australia, in the Barberton district of South Africa and in the Kolar goldfields of India.

Deposits of Mesozoic age of this type are widespread in the Western Cordillera of North America. Here belong the great deposits of the Mother Lode of California; a number of deposits in British Columbia, including the Bralorne Mine; and a number of deposits in Alaska including some of the orebodies of the Alaska Juneau Gold Mine.

Tertiary deposits of this type are, likewise, widespread throughout the world occurring in Romania, Japan, New Zealand, Fiji, U.S.S.R., the western United States and elsewhere.

Yellowknife area, Northwest Territories (Precambrian: Archean age). The alteration zones associated with the Yellowknife gold deposits in the greenstone belt are typical examples of gold-quartz veins and silicified zones in intermediate to basic volcanic and igneous rocks. They have been described in detail by the writer (Boyle, 1961*a*). Only the significant facts need be considered here.

Three metamorphic facies characterize the Yellowknife Greenstone Belt, which is composed essentially of meta-andesite and meta-basalt flows and diorite dyke swarms. An amphibolite facies is developed adjacent to the western granodiorite-granite, which grades inward through an intermediate epidote amphibolite facies into a poorly defined greenschist facies. The deposits are quartz veins and silicified zones in great shear zones and subsidiary faults. The principal minerals in the veins and silicified zones are quartz, carbonates, pyrite, arsenopyrite, pyrrhotite, stibnite, sulphosalts, gold and aurostibite.

Most shear zones in the amphibolite facies of the greenstone belt contain quartz lenses and stringers, which are flanked by characteristic alteration zones. The wall rocks are predominantly amphibolites, containing hornblende, oligoclase-andesine, quartz, magnetite, sphene, ilmenite and small amounts of epidote, pyrite and pyrrhotite. These minerals are destroyed in the shear zones, and a weakly schistose aggregate of chlorite, albite, sericite and leucoxene is present. Adjacent to quartz lenses, the chlorite schist of the shear zones grades imperceptibly into a chlorite-albite-sericite schist containing quartz, leucoxene and/or sphene and small amounts of calcite. Pyrite, pyrrhotite and arsenopyrite are present and may make up 0.5 per cent of the schistose rock. The shear zones are rarely more than 3 ft wide, and the alteration zones adjacent to quartz lenses are 1 ft or so wide.

The rocks of the epidote amphibolite and greenschist facies are cut by the highly productive shear zones of four ore-bearing systems (Fig. 17). There is relatively little difference in the internal arrangement of the alteration haloes in these four systems, and the following description applies to all shear zones in these two facies.

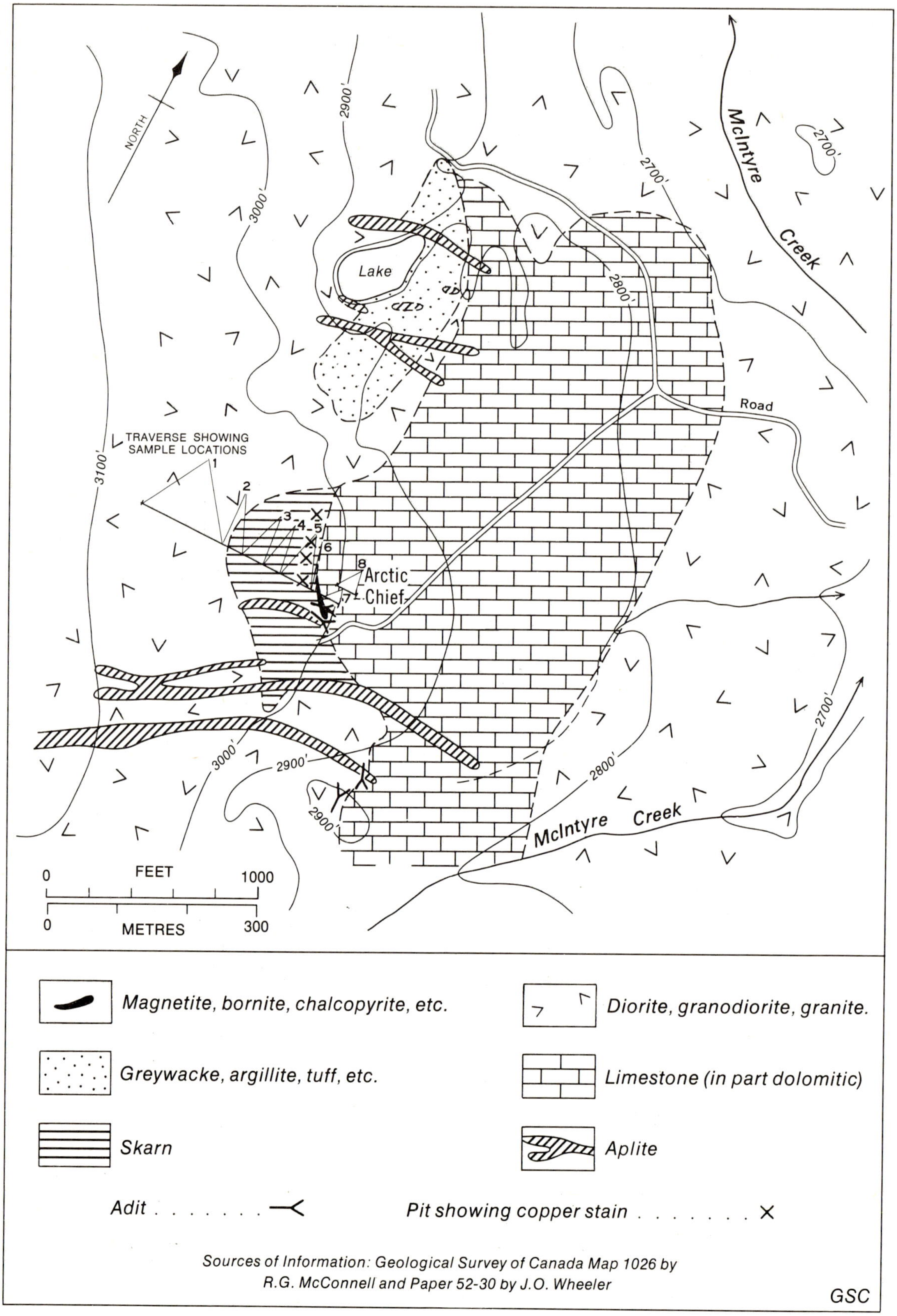

Figure 9. Surface geology at the Arctic Chief Mine, Whitehorse Copper Belt, Yukon Territory.

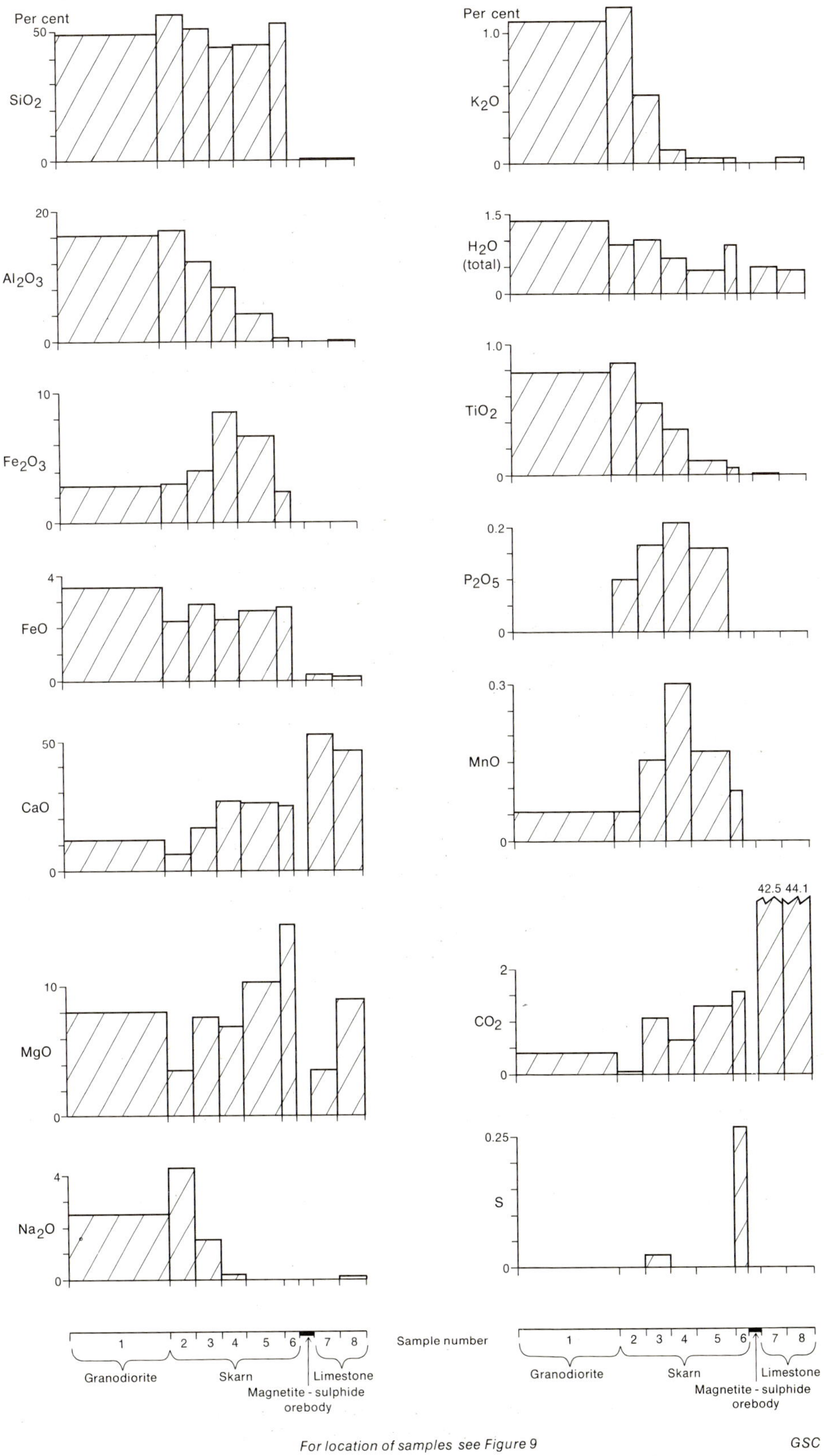

Figure 10. Profiles of elements along traverse across skarn, Arctic Chief Mine, Whitehorse Copper Belt, Yukon Territory (*see* Table 44).

Table 44. Analyses of rocks along traverse (Fig. 9) Whitehorse Copper (skarn) Belt, Whitehorse, Yukon*

Rock type	Granodiorite	Granodiorite in contact zone	Skarn	Skarn	Skarn	Skarn near ore	Limestone near ore	Limestone country rock
Sample number on traverse	1	2	3	4	5	6	7	8
				%				
SiO_2	49.6	57.8	50.9	44.2	45.8	51.3	0.81	0.16
Al_2O_3	16.3	17.8	12.2	8.7	4.4	0.89	0.00	0.08
Fe_2O_3	3.1	3.3	4.1	8.5	6.7	2.3	0.00	0.00
FeO	3.58	2.23	2.94	2.33	2.75	2.78	0.10	0.04
CaO	12.1	7.1	17.4	27.7	27.3	24.6	52.8	45.3
MgO	8.0	3.6	7.5	6.9	10.2	15.2	3.7	8.9
Na_2O	2.6	4.3	1.6	0.20	0.00	0.00	0.00	0.02
K_2O	1.1	1.2	0.52	0.10	0.02	0.02	0.00	0.02
H_2O (total)	1.42	0.90	1.0	0.66	0.44	0.84	0.48	0.42
TiO_2	0.79	0.85	0.57	0.37	0.10	0.05	0.01	0.00
P_2O_5	0.00	0.10	0.17	0.21	0.16	0.00	0.00	0.00
MnO	0.06	0.06	0.15	0.30	0.17	0.09	0.00	0.00
CO_2	0.48	0.06	1.1	0.68	1.28	1.56	42.50	44.10
S	nf	nf	0.02	nf	nf	0.27	nf	nf
C			0.04					
O≡S			0.01			0.11		
Total	99.1	99.3	100.2	100.9	99.3	99.8	100.4	99.0
				(ppm)				
Li	8	5	<1	<1	1	<1	<1	<1
Rb	24	23	24	3	<1	<1	<1	<1
Cs	<1	<1	<1	<1	<1	<1	<1	<1
Be	<2	<2	<2	<2	<2	<2	<2	<2
Sr	620	740	760	100	31	25	820	500
Ba	330	540	540	18	10	13	10	13
B	nf	nf	nf	<50	<50	nf	nf	nf
Sc	42	20	15	nf	nf	nf	nf	nf
U	0.3	0.8	1.1	2.8	1.7	1.0	1.4	1.9
Pb	4	7.5	4	<4	4	<4	<4	5
Cu	42	34	65	25	42	3380	140	9
Ag	<0.4	<0.4	<0.4	0.5	<0.4	0.6	<0.4	<0.4
Au	<0.005	<0.005	<0.005	<0.005	0.005	0.055	0.045	<0.005
Zn	63	50	52	49	60	47	17	10
Cd	<0.4	0.5	<0.4	<0.4	<0.4	<0.4	<0.4	<0.4
Hg (ppb)	127	113	293	170	45	637	488	428
Zr	60	72	150	120	57	nf	nf	nf
As	<6	<6	<6	<6	<6	<6	6	<6
Sb	<1	<1	<1	<1	<1	<1	<1	<1
Bi	nf	nf	nf	nf	nf	nf	nf	nf
V	210	230	240	200	140	60	nf	nf
Se	<1	<1	<1	1	<1	<1	<1	<1
Te	<0.2	<0.2	<0.2	<0.2	<0.2	<0.2	<0.2	<0.2
Cr	200	27	17	nf	nf	nf	nf	nf
Mo	1.2	1.2	0.7	1.0	0.5	0.5	0.5	0.5
W	2	3	<2	<2	<2	<2	<2	<2
Co	31	18	18	14	14.5	17.5	<2	<2
Ni	56	17	17.5	16	18.5	21	8	8.5
Pt(ppb)	<15	<15	<15	<15	<15	<15	<15	<15
Pd(ppb)	<2	<2	<2	<2	<2	<2	2	<2
F	400	500	650	300	425	300	650	350
Cl	600	350	400	300	250	300	<100	150
Ratios								
K_2O/Na_2O	0.42	0.28	0.33	0.50	>4	>4	1	1
SiO_2/CO_2	103.3	963.3	46.3	65.0	35.78	32.88	0.02	0.004
SiO_2/H_2O+CO_2+S	26.1	60.2	24.0	33.0	26.6	19.2	0.02	0.004

Note: nf–not found.

*For location of samples *see* Figure 9. Samples comprised a large aggregate of chip samples across a 10 ft width for distance noted along traverse.

Table 45. Phases of alteration in the shear zones of the Giant-Campbell, Con, Negus-Rycon and Crestaurum systems, Yellowknife, Northwest Territories

Phase	Minerals present (in order of relative abundance)	Remarks
Country rock	Fibrous amphibole, oligoclase, epidote, quartz, magnetite, leucoxene, sphene, chlorite, carbonates, pyrite, pyrrhotite, apatite, minor reticulating epidote-carbonate veinlets	Massive to slightly schistose greenstone country rock
1.	Chlorite, fibrous amphibole, oligoclase, epidote, quartz, magnetite, ilmenite, leucoxene, sphene, carbonates, pyrite, pyrrhotite, apatite, reticulating epidote-carbonate veinlets, hematite on slips in some places	Wall rock adjacent to shear zones; rock is weakly schistose
2.	Chlorite, albite, ankerite, quartz, leucoxene, pyrite, small amounts of fibrous amphibole, apatite and epidote in some places, pyrrhotite locally, rare reticulating epidote-carbonate veinlets	Shear zone; chlorite schist; marked foliation
3.	Chlorite, ankerite, albite, quartz, leucoxene, pyrite, colourless to faintly greenish mica, rare occurrences of tourmaline, apatite and pyrrhotite	Shear zone; chlorite-carbonate schist; strong foliation
4.	Ankerite, chlorite, sericite, quartz, albite, leucoxene, pyrite, arsenopyrite, rare occurrences of tourmaline and apatite	Shear zone; carbonate-chlorite-sericite schist; strong foliation
5.	Ankerite, sericite, quartz, chlorite, pyrite, arsenopyrite, rutile needles, rare occurrences of tourmaline and apatite	Shear zone; carbonate-sericite schist; strong foliation
6.	Quartz, sericite, ankerite, pyrite, arsenopyrite, sulphosalts, gold, sphalerite, etc., rutile needles, rare occurrences of tourmaline	Mineralized zones; orebodies in some cases
Quartz bodies	Quartz, arsenopyrite, pyrite, sulphosalts, sericite, ankerite, gold, rare occurrences of tourmaline and apatite, scheelite in some quartz lenses	Quartz lenses and veins; orebodies

In most of the shear zones of the four systems (Giant-Campbell, Con, Negus-Rycon and Crestaurum) the six alteration phases listed in Table 45 can be recognized. The widths of the phases vary extensively throughout the systems; in some places individual phases may be tens and even hundreds of feet wide. All phases are not necessarily present within a shear zone. All shear zones are marked by a development of phases 1 and 2 and locally by a mild development of phase 3. Where orebodies and mineralized zones are present, all phases are generally present, but some orebodies lack an adjacent pronounced sericite-carbonate phase.

The location of the more intensely altered phases (3–6) of the shear zone is controlled in most places by the same structural features that have localized the quartz veins and lenses, namely, shear-zone junctures and flexures and dragged parts of schist zones.

The country rocks are predominantly epidote amphibolites. The amphibolites contain fibrous amphibole (actinolite), albite-oligoclase, epidote, quartz, magnetite, ilmenite, leucoxene, chlorite, carbonate, pyrite and pyrrhotite in that order of abundance. These minerals are highly intergrown with one another, and the rock has a poorly developed schistose texture in thin sections. In hand specimens the rock is massive and breaks with a blocky to irregular fracture.

Phase 1 marks a transition between the epidote amphibolite country rock and the shear zones, and only a slight schistosity has been impressed on the rocks. The main feature of this phase is the occurrence of reticulating pinkish buff and pistachio-green epidote-carbonate veinlets. In most occurrences these veinlets contain a dense intergrowth of epidote, calcite or ankerite, quartz, albite, tremolite, leucoxene, chlorite, hematite and pyrite. In other occurrences quartz-carbonate stringers form a central core flanked by a narrow epidote-carbonate alteration zone containing the minerals listed above. Seams and joints coated with hematite occur in this phase in many places especially in the shear zones of the Giant System.

Phase 2 constitutes the chlorite schist phase of the shear zones. The rock is dark green and is highly foliated both in hand specimens and thin sections. Constituent minerals include, in order of abundance, green chlorite, albite, ankerite, quartz, leucoxene, pyrite and epidote. In some places a few small reticulating carbonate-epidote veinlets are present. The albite and quartz commonly occur in small elongated lenses and in bands in the highly foliated chlorite masses. The leucoxene occurs in small buff to white, optically opaque masses with a linear arrangement along the foliated chlorite masses. Narrow quartz and carbonate veinlets cut across the schistosity in most specimens.

Phase 2 grades imperceptibly by increase in ankerite into the chlorite-carbonate schist of phase 3. In thin sections the gradation is marked by a bleaching of part of the chlorite to give a colourless to faintly greenish mica that has low first order interference colours and contains small leucoxene knots. The leucoxene is clearer than in the chlorite schist, and in some thin sections small diffuse crystals of rutile can be discerned in the irregular leucoxene masses. Albite, quartz and pyrite are generally present, and in a few localities apatite and small amounts of tourmaline may occur. In thin sections the rock shows a banded structure. Some bands contain green chlorite, light green mica and albite. Locally ankerite occurs in small crenulations and seams, which cut the bands. Quartz is commonly present in augenlike masses around pyrite crystals or in seams and veinlets parallel to or transecting the bands of other minerals. The rock is green to greenish buff in colour and is strongly foliated in hand specimens and thin sections.

In the vicinity of quartz lenses and mineralized zones the chlorite-carbonate schist phase grades, by increase of ankerite, sericite and quartz, into the carbonate-chlorite-sericite schist of phase 4. This phase in turn grades, by increase in sericite, carbonate and quartz and decrease in chlorite, into the carbonate-sericite schist of phase 5. The latter grades into either mineralized zones (phase 6) or quartz veins and lenses, by increase in quartz and sulphides.

The macroscopic and microscopic features of phases 4, 5 and 6 are nearly the same in all occurrences. The rock is grey to buff and highly foliated in hand specimens and thin sections. Mashing and extreme contortion of the schist occurs in most shear zones, especially in those of the Giant-Campbell System. The rock contains variable amounts of ankerite, sericite, chlorite, quartz, pyrite, arsenopyrite and leucoxene depending upon the phase in which it occurs. In phases 4 and 5 the minerals occur in bands in some localities; in others, especially in parts of the Negus-Rycon System, all minerals occur as a dense intergrowth. Toward the mineralized zones and quartz bodies there is a definite increase in pyrite and arsenopyrite. These minerals occur principally in sericite bands, but may be scattered throughout the general groundmass. Trains of leucoxene masses, in which rutile needles are readily distinguishable, occur along all the sericite bands. The leucoxene is a marked feature of these phases of alteration and gives the rock its distinctive buff colour. Small quartz and carbonate veinlets intersect the highly altered phases throughout their extent.

The alteration phases are general in most of the shear zones of the four major systems. In a few shear zones, however, there are a few differences, mainly in the textures of the zones. In these shear zones alteration phases 3 to 6, are compressed into a zone rarely more than 1 ft wide. This composite zone consists of a dense intergrowth of microcrystalline ankerite, bleached chlorite, quartz, albite, sericite, chlorite, trains and disseminated masses of leucoxene and abundant pyrite and arsenopyrite. The rock of this zone is only slightly foliated and appears as a narrow buff to grey-coloured, dykelike zone adjacent to the quartz lenses. The transition to the chlorite schist phase, which also shows only a limited degree of foliation, is sharp. In many places there is a zoning of the sulphides outward from the quartz. Arsenopyrite occurs adjacent to the quartz veins, pyrite farther out but within the highly altered phase and pyrrhotite in the chlorite schist phase. In all occurrences small quartz-carbonate and albite-quartz veinlets ramify through the highly altered phase.

Chemical analyses have been made of the typical alteration phases in the shear zones of the greenstone belt where they intersect epidote amphibolite rocks.

Tables 46 and 47 record the percentage of each constituent, the adjusted percentage after volume calculations, the gains and losses and the percentage gains and losses for samples taken across well-defined alteration zones in the two shear-zone systems of the greenstone belt. Table 48 shows the results of analyses of large composite samples from the two major phases of alteration in shear zones that cut greenstones. These samples were selected with great care from all shear zones investigated and probably give the best data on the general overall trend of the chemical changes.

The alteration zoning in the Con, Negus-Rycon and Giant-Campbell systems was studied in detail by thin sections and chemical and spectographic analyses. The alteration features of these systems, while slightly different in a few respects from those of other systems, are typical and illustrate the trend in all shear zones cutting the greenstones.

For purposes of discussion the various phases given in Table 45 may be grouped into three general phases – the country rock (greenstone), chlorite and chlorite-carbonate schist phase and the carbonate-sericite schist phase.

The microscope reveals that the amphibole of the country rock was altered first to chlorite and ankerite; in the more intense phases the chlorite in turn was altered to sericite. The plagioclase may have lost calcium to form albite in the chlorite phase, but it has been generally obliterated in the sericite phase, and sericite and ankerite have formed at its expense. Epidote has been altered to chlorite or to carbonate depending upon the intensity of alteration. Ilmenite, magnetite and sphene have been generally obliterated, and pyrite, arsenopyrite, rutile and leucoxene have formed from some of the components of these minerals. Quartz and carbonate minerals have formed throughout the highly altered phases.

It is probable that the carbon dioxide, water and sulphur were derived from the greenstone country rocks, and that these compounds diffused into the dilatant shear-zone systems (Boyle, 1959). Here they reacted with the minerals present producing new minerals stable under the existing conditions. To maintain the equilibrium, various chemical exchanges, as shown by the profiles in Figures 11, 12 and 13, have taken place. For purposes of discussion let us examine the results as shown by the profiles drawn from the analyses of the composite samples (Fig. 13). Local variations will be reflected in the profiles from the individual shear zones.

The profile for SiO_2 shows a marked decrease in the chlorite-carbonate and sericite schist phases of the shear zones compared with the greenstone. This decrease is attended by an increase in CO_2 in both phases, by H_2O in the chlorite-carbonate phase and by S in the carbonate-sericite phase. This suggests that these gaseous components displaced the liberated SiO_2 to maintain the chemical equilibrium. The profiles from all shear zones show similar features and we may, therefore, conclude that as the volatiles were added to the system, silica was liberated and migrated into dilatant zones to form the quartz veins and lenses. More will be said about this in the sections to follow and in the section on the origin of gold deposits in this chapter.

The content of Al_2O_3 in the composite samples shows a decrease in the chlorite-carbonate phase and an increase in the carbonate-sericite phase as compared with the greenstone. Similar features are shown by the profile of the Negus-Rycon System, but in the Campbell System the Al_2O_3 content of the carbonate-sericite phase is lower than both the greenstone and chlorite-carbonate phase. This anomaly can perhaps be explained by the degree of carbonatization. The Campbell System shows much higher percentage gains in CO_2 than the Negus-Rycon System, and more carbonate has formed at the expense of the aluminous silicates. Hence, the greater migration of Al_2O_3 to maintain the chemical equilibrium.

The Fe_2O_3, FeO, CaO, MgO, TiO_2 and MnO contents of the composite samples show individual trends in the alteration phases, but their combined content shows an overall decrease in both alteration phases compared to the greenstone. This tendency is also shown by the analyses from the individual shear zones and appears to be general.

The Na_2O content decreases consistently in the two alteration phases compared with the greenstone. This loss of soda is also shown by the analyses from the two specific shear-zone systems.

Compared with the greenstone, the K_2O content of the composite samples is relatively constant in the chlorite-carbonate phase but increases many times in the carbonate-sericite phase. In specific shear zones the K_2O content may

Table 46. Chemical changes produced by alteration of greenstone, Negus-Rycon System, Yellowknife, Northwest Territories

Constituent	YC-16*	YC-17†	Adjusted§	Gains and losses in constituents	Percentage gains and losses	YC-16*	YC-18‡	Adjusted§	Gains and losses in constituents	Percentage gains and losses
		(%)					(%)			
SiO_2	49.18	45.40	43.54	−5.64	−11.45	49.18	34.96	35.30	−13.88	−28.20
Al_2O_3	10.59	7.35	7.05	−3.54	−33.50	10.59	12.21	12.33	+1.74	+16.45
Fe_2O_3	7.20	3.56	3.41	−3.79	−52.65	7.20	3.01	3.04	−4.16	−57.80
FeO	9.87	9.16	8.78	−1.09	−11.05	9.87	8.93	9.02	−0.85	−8.60
CaO	11.01	10.99	10.54	−0.47	−4.25	11.01	10.42	10.52	−0.49	−4.45
MgO	5.63	4.18	4.01	−1.62	−28.80	5.63	4.40	4.44	−1.19	−21.16
Na_2O	1.16	0.50	0.48	−0.68	−58.75	1.16	1.11	1.12	−0.04	−3.44
K_2O	0.24	1.02	0.98	+0.74	+308.00	0.24	2.48	2.50	+2.26	+941.00
H_2O^+	1.46	3.26	3.13	+1.67	+114.50	1.46	0.80	0.81	−0.65	−44.50
H_2O^-	0.07	0.14	0.13	–	–	0.07	0.31	0.31	–	–
TiO_2	0.84	1.05	1.01	+0.17	+20.20	0.84	1.79	1.81	+0.97	+115.75
P_2O_5	0.18	0.38	0.36	+0.18	+100.00	0.18	0.70	0.71	+0.53	+294.50
MnO	0.22	0.11	0.10	−0.12	−55.50	0.22	0.08	0.08	−0.14	−63.60
CO_2	1.39	9.56	9.17	+7.78	+560.00	1.39	12.13	12.30	+10.91	+785.00
S	0.85	2.17	2.08	+1.23	+145.00	0.85	6.47	6.55	+5.70	+670.00
As_2O_3	trace	1.57	1.50	+1.50	–	trace	1.93	1.95	+1.95	–
Total	99.89	100.30	–	–	–	99.89	101.73	–	–	–
Specific gravity	2.97	2.85	–	–	–	2.97	3.01	–	–	–

*YC-16–greenstone, chip sample across 80 ft, 11th level, Negus Mine, Negus-Rycon System.
†YC-17–chlorite and chlorite-carbonate schist, chip and channel sample across 2 ft, 11th level, Negus Mine, Negus-Rycon System.
‡YC-18–carbonate-chlorite and carbonate-sericite schist, chip and channel sample across 1 ft, 11th level, Negus Mine, Negus-Rycon System.
§–Adjusted % = percentages of YC-17 and YC-18 multiplied by the factors 2.85/2.97 and 3.01/2.97 respectively, to reduce to a common volume.

Table 47. Chemical changes produced by alteration of greenstone, Campbell System, Yellowknife, Northwest Territories

Constituent	YC-19*	YC-20†	Adjusted§	Gains and losses in constituents	Percentage gains and losses	YC-19*	YC-21‡	Adjusted§	Gains and losses in constituents	Percentage gains and losses
		(%)					(%)			
SiO_2	51.53	43.91	40.92	−10.61	−20.59	51.53	49.29	49.29	−2.24	−4.35
Al_2O_3	16.31	14.72	13.71	− 2.60	−15.90	16.31	13.10	13.10	−3.21	−19.65
Fe_2O_3	0.94	0.30	0.28	− 0.66	−70.10	0.94	1.21	1.21	+0.27	+28.75
FeO	9.65	6.87	6.40	− 3.25	−33.70	9.65	5.67	5.67	−3.98	−41.25
CaO	8.59	9.43	8.78	+0.19	+2.22	8.59	7.99	7.99	−0.60	−7.00
MgO	6.23	7.69	7.17	+0.94	+15.10	6.23	3.81	3.81	−2.42	−38.95
Na_2O	2.49	1.63	1.52	− 0.97	−38.90	2.49	0.33	0.33	−2.16	−86.60
K_2O	0.11	0.70	0.65	+0.54	+491.00	0.11	1.59	1.59	+1.48	+1345.50
H_2O^+	2.11	3.55	3.31	+1.20	+56.80	2.11	1.45	1.45	−0.66	−31.30
H_2O^-	0.10	0.46	0.43	–	–	0.10	0.10	0.10	–	–
TiO_2	0.84	0.58	0.54	− 0.30	−35.75	0.84	0.26	0.26	−0.58	−69.00
P_2O_5	0.22	0.33	0.31	+0.09	+40.95	0.22	0.14	0.14	−0.07	−31.80
MnO	0.16	0.22	0.20	+0.04	+25.00	0.16	0.08	0.08	−0.08	−50.00
CO_2	0.55	9.60	8.95	+8.40	+1529.00	0.55	10.48	10.48	+9.93	+1800.00
S	0.28	0.43	0.40	+0.12	+42.80	0.28	3.67	3.67	+3.39	+1210.00
As_2O_3	–	–	–	–	–	–	2.87	2.87	+2.87	–
Total	100.11	100.42	–	–	–	100.11	102.04	–	–	–
Specific gravity	2.95	2.75	–	–	–	2.95	2.94	–	–	–

*YC-19–Greenstone wall rock, chip sample across 250 ft, 13th level, Negus Mine, Campbell System.
†YC-20–Chlorite and chlorite-carbonate schist, chip sample across 70 ft, 13th level, Negus Mine, Campbell System.
‡YC-21–Carbonate-chlorite and carbonate-sericite schist, quartz stringers and small lenses. Chip and channel sample across 8 ft, 13th level, Negus Mine, Campbell System.
§Adjusted % = percentages of YC-20 and YC-21 multiplied by the factors 2.75/2.95 and 2.94/2.95 respectively, to reduce to a common volume.

Table 48. Chemical changes produced by alteration of greenstone, composite samples of all shear-zone systems, Yellowknife, Northwest Territories

Constituent	A-76*	A-251†	Adjusted§	Gains and losses in constituents	Percentage gains and losses	A-76*	A-250‡	Adjusted§	Gains and losses in constituents	Percentage gains and losses
		(%)					(%)			
SiO_2	50.45	40.36	37.12	−13.33	−26.42	50.45	38.19	38.30	−12.15	−24.08
Al_2O_3	13.96	12.87	11.84	−2.12	−15.19	13.96	16.26	16.31	+2.35	+16.83
Fe_2O_3	2.00	1.06	0.97	−1.03	−51.50	2.00	0.38	0.38	−1.62	−81.00
FeO	9.61	12.00	11.04	+1.43	+14.88	9.61	3.49	3.50	−6.11	−63.58
CaO	10.23	8.94	8.22	−2.01	−19.65	10.23	11.02	11.05	+0.82	+8.02
MgO	6.53	8.43	7.75	+1.22	+18.68	6.53	5.77	5.79	−0.74	−11.33
Na_2O	1.85	0.90	0.83	−1.02	−55.14	1.85	0.29	0.29	−1.56	−84.32
K_2O	0.39	0.41	0.38	−0.01	−2.56	0.39	2.82	2.83	+2.44	+622.64
H_2O^+	2.36	5.20	4.78	+2.42	+102.54	2.36	1.16	1.16	−1.20	−50.85
H_2O^-	0.11	0.13	0.12	+0.01	+9.09	0.11	0.10	0.10	−0.01	−9.09
TiO_2	1.02	1.05	0.97	−0.05	−4.90	1.02	0.82	0.82	−0.20	−19.61
P_2O_5	0.13	0.11	0.10	−0.03	−23.08	0.13	0.21	0.21	+0.08	+61.54
MnO	0.21	0.22	0.20	−0.01	−4.76	0.21	0.17	0.17	−0.04	−19.05
CO_2	0.68	8.03	7.39	+6.71	+986.76	0.68	15.57	15.62	+14.94	+2197.06
S	0.22	0.12	0.11	−0.11	−50.00	0.22	3.26	3.27	+3.05	+1386.36
Cl	0.02	0.00	0.00	−0.02	−100.00	0.02	0.01	0.01	−0.01	−50.00
Cr_2O_3	0.03	0.03	0.03	0.00	0.00	0.03	0.00	0.00	−0.03	−100.00
C	0.04	0.12	0.11	+0.07	+175.00	0.04	0.04	0.04	0.00	0.00
Total	99.84	99.98				99.84	99.56			
Less O≡S,Cl	0.10	0.05				0.10	1.22			
Net total	99.74	99.93				99.74	98.34			
Specific gravity	2.99	2.75				2.99	3.00			
				(ppm)						
Li	16	49				16	12			
Rb	11	12				11	106			
Cs	0.3	0.4				0.3	3.9			
B	15	93				15	150			
As	2	74				2	6900			
Sb	<1	26				<1	23			
Au	0.008	0.017				0.008	0.632			
Ag	0.7	0.256				0.7	1.13			
Ratios										
K_2O/Na_2O	0.21	0.46				0.21	9.72			
SiO_2/CO_2	74.19	5.03				74.19	2.45			
SiO_2/H_2O+CO_2+S	15.48	3.02				15.48	1.85			
Au/Ag	0.01	0.06				0.01	0.56			

*A-76—composite sample from epidote-amphibolite facies (greenstone country rock).
†A-251—composite sample of chlorite-carbonate schist of shear zones. Widths vary from 1–70 ft or more.
‡A-250—composite sample of carbonate-sericite schist of shear zones. Widths vary from 0.5 to 10 ft or more.
§Adjusted%=percentages of A-251 and A-250 multiplied by the factors 2.75/2.99 and 3.00/2.99 respectively, to reduce to a common volume.

increase significantly in the chlorite-carbonate phase.

The profiles of H_2O as shown by the analyses of both the composite samples and those from individual systems behave consistently. The water content increases in the chlorite-carbonate phase and decreases in the carbonate-sericite phase compared to the greenstone.

In all cases the profiles for CO_2 clearly indicate an increase in this constituent in all the alteration phases of the shear zones as compared with the greenstone. The greatest amounts are present in the carbonate-sericite phase with lesser amounts in the chlorite-carbonate phase.

The sulphur profiles of the alteration phases of the shear zones provide some interesting data on the origin of the sulphides in the orebodies and carbonate-sericite phase. As shown by the analyses of composite samples, the content of sulphur is less in the chlorite-carbonate phase than in the greenstone. In the carbonate-sericite phase the content is high, a feature that is common to all shear zones. Considering the profiles of the two specific shear-zone systems, however, it will be noted that the sulphur content is higher in both phases of the shear zones than in the greenstones. This apparent anomaly in the two results may be due to different sampling procedures. In the case of the individual shear zones, continuous samples were taken across the shear zones. Since the chlorite-carbonate zones always contain small seams of carbonate-sericite-pyrite schist these were unavoidably included with the chlorite-carbonate schist samples. In other words no separation on strictly petrographic grounds was made. In the case of the composite samples petrographic control was rigidly adhered to, and the samples of the two phases of alteration were prepared in such a way as to exclude mutual contamination. It is evident from the results of the composite samples that where chlorite-carbonate schist was formed from greenstone the process led to a decrease in the sulphur content of the rock. The liberated sulphur probably migrated together with other components into the carbonate-sericite phase or to

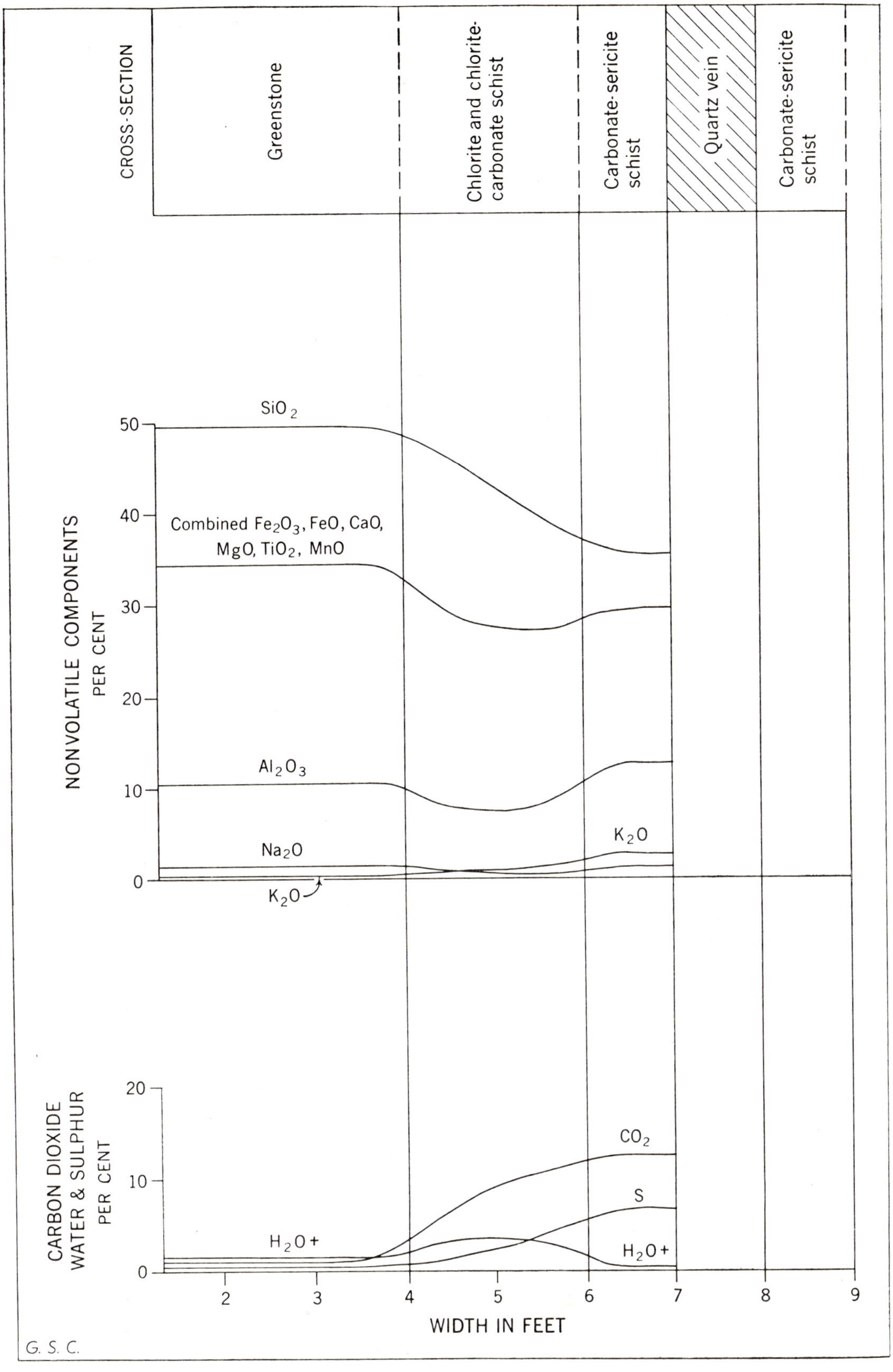

Figure 11. Chemical changes produced by alteration of greenstone, Negus-Rycon System, Yellowknife, Northwest Territories (*see* Table 46).

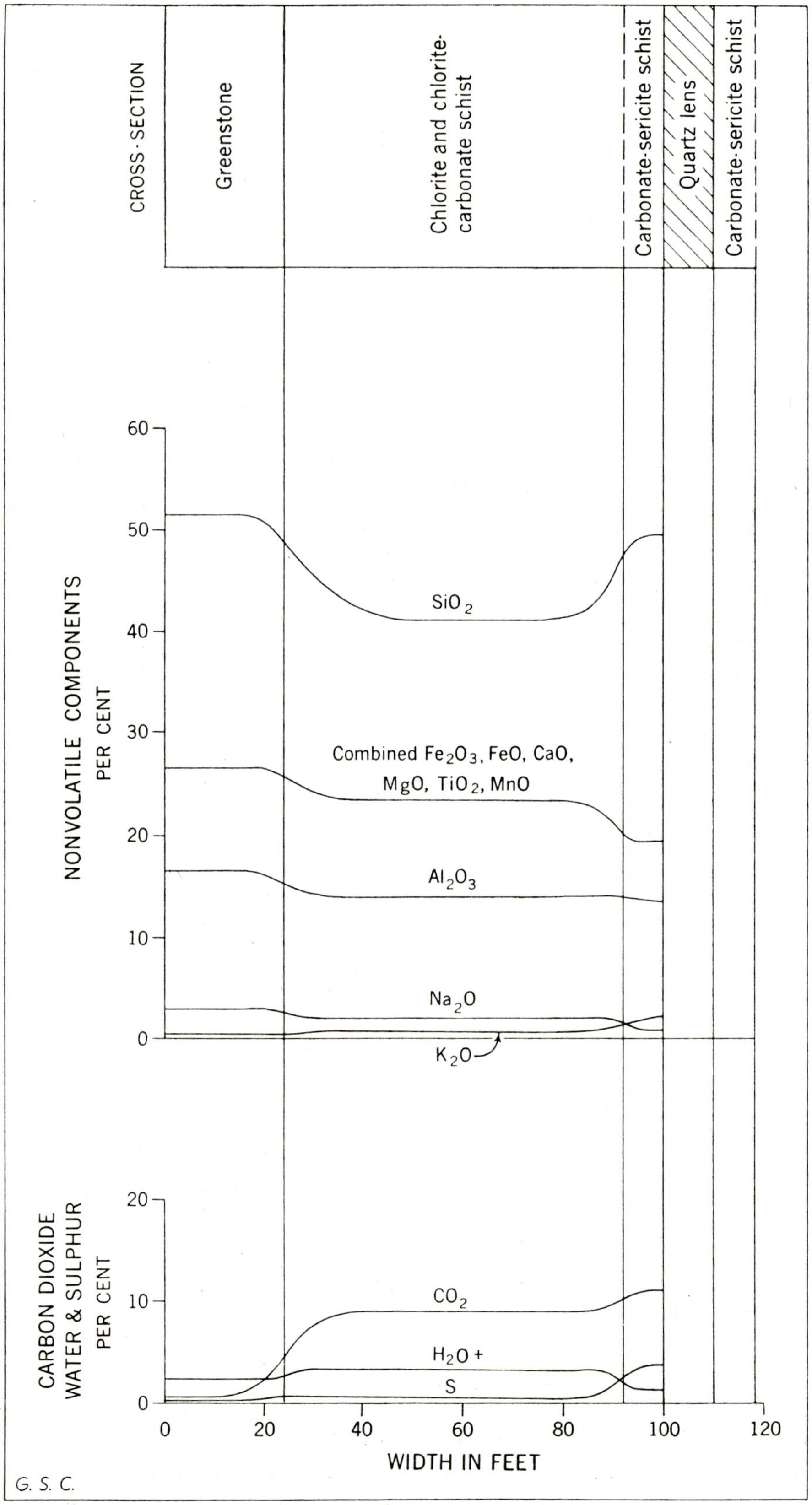

Figure 12. Chemical changes produced by alteration of greenstone, Campbell System, Yellowknife, Northwest Territories (*see* Table 47).

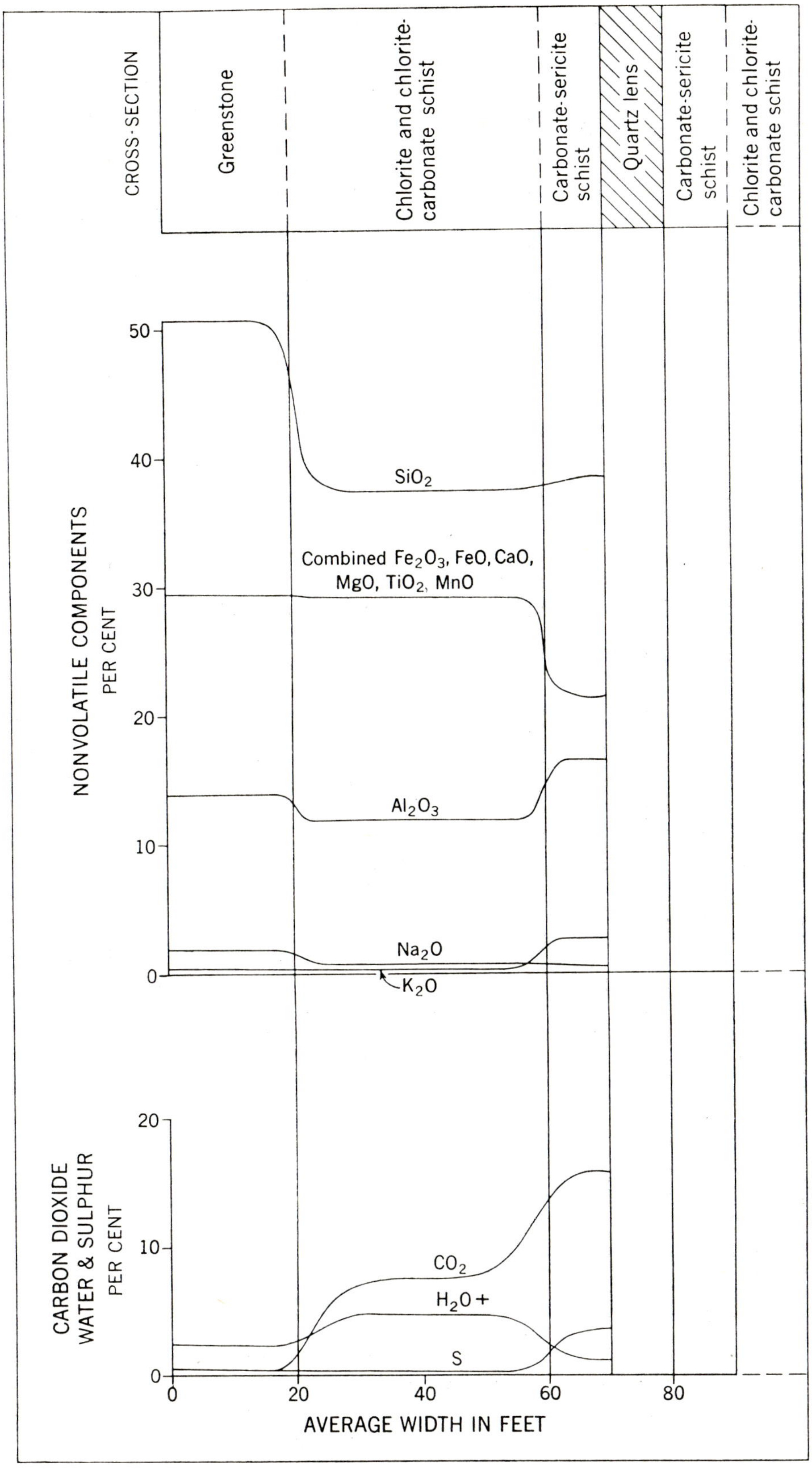

Figure 13. Chemical changes produced by alteration of greenstone, composite samples from all shear zones, Yellowknife, Northwest Territories (*see* Table 48).

dilatant zones where it was precipitated as pyrite and other sulphides. This concept is amplified further in the section on origin of gold deposits.

To summarize the chemical features displayed by the profiles, it can be said that the principal agents responsible for the alteration were carbon dioxide and water, and to a lesser extent sulphur and arsenic. These volatiles reacted differently with the various mineral constituents of the greenstone rocks, and the alteration minerals formed reflect the chemical equilibrium attained under the conditions of pressure, temperature and concentration that prevailed during the formation of the shear zones. As carbon dioxide and water were added, silica was abstracted together with some iron, calcium, magnesium, titanium and manganese. Soda was leached from all alteration phases, and potash was enriched in the sericite phase. Sulphur was evidently mobilized in the chlorite-carbonate phase and migrated into the carbonate-sericite phase or into the site of the quartz lenses.

The mineral transformations and elemental migrations inferred from thin sections and chemical and spectographic analyses provide much data on the geochemistry and relative mobilities of the elements during the alteration processes. These are interpreted in the following sections according to well-established crystallochemical principles.

The amphiboles of the greenstone in the presence of CO_2 and H_2O have yielded either ankerite or chlorite or both depending upon the concentration of the two volatiles. To form chlorite the elemental constituents of the amphibole must have undergone a major structural rearrangement. During this process some of the Mg, Fe, Mn, Al_2O_3 and SiO_2 was bound by OH into the chlorite structure. The Ca and the remainder of the Fe, Mg and Mn were liberated and bound by CO_2 into the ankerite structure. The K and Na contained in the original amphibole were liberated during the process.

In the more highly carbonated parts of the shear zones the process has gone further, and the chlorite was attacked by the CO_2, and its Mg, Fe and Mn were bound into ankerite. This has liberated SiO_2 and Al_2O_3.

The plagioclase has undergone a similar alteration – its Ca content was bound into ankerite, and the K, Na, Al_2O_3 and SiO_2 formed albite in certain circumstances or migrated. In the same manner the Ca, Fe, Mn, Al_2O_3 and SiO_2 of the epidote minerals were incorporated into chlorite that was ultimately attacked by CO_2 to form ankerite with the liberation of SiO_2 and Al_2O_3.

Sulphur and arsenic diffusing through the shear zones have bound some iron liberated by the alteration of amphibole, chlorite and epidote into pyrite and arsenopyrite. Polished sections also show that the pyrite and arsenopyrite contain irregular unreplaced nuclei of magnetite and ilmenite suggesting that the iron in these minerals was also bound by sulphur and arsenic. Titanium was liberated during the alteration of the magnetite and ilmenite, the carbonatization of the amphibole and chlorite and the sericitization of the chlorite. This element was not readily accommodated in the carbonate, sericite or sulphide structures and hence formed an independent mineral phase, which now appears as irregular white leucoxene patches or aggregates of yellow rutile needles.

Potassium is enriched in the shear zones especially in the carbonate-sericite phase, adjacent to quartz lenses. It is often assumed that such a potash metasomatism is due to the introduction of potassium by magmatic solutions. This assumption neglects the pertinent facts that a ready source of potassium exists in the greenstones and that the element has a high mobility where differential pressure and shearing prevails and where water and carbon dioxide are present. The greenstone contains an average content of 0.25 per cent K_2O, held principally in the plagioclase and amphiboles. During the carbonatization and hydration of these minerals, the K^+ was liberated because large amounts could not be readily accommodated in the chlorite or carbonate structures. In the shear zones enormous masses of rock have been carbonatized and hydrated, and it follows that large amounts of potassium must have been set free.

Potassium shows large percentage gains in most of the alteration phases and is strongly enriched in the sericite schist adjacent to quartz lenses. From thin section studies it is apparent that the sericite is formed at the expense of both chlorite and plagioclase. These observations taken together with the chemical considerations discussed above suggest that the geochemical history of potassium during the alteration processes was as follows:

1. Release as K^+ or as a hydrated ion from its initial host minerals such as plagioclase and amphibole during carbonatization and hydration.

2. Diffusion of K^+ ions from zones of intense shearing into zones of lower pressure where reactions with chlorite and plagioclase or combination with alumina and silica took place to form sericite.

Sodium is bound principally in the plagioclase, but small amounts are also present in the amphiboles. During the carbonatization of these two minerals most of the sodium was liberated. Recombination may have taken place with alumina and silica to form albite locally, but much sodium has been leached and probably migrated to higher levels or was ultimately dispersed at the surface.

The percentage gains and losses for sodium are lower than for potassium, suggesting that sodium has the lower mobility of the two elements.

Alumina is fixed in the plagioclase, amphibole and epidote of the greenstone. During alteration of these minerals the alumina was redistributed among the chlorite, plagioclase and sericite of the alteration zones. Some parts of the shear zones are leached and others enriched, but the percentage gains and losses are low, suggesting that Al_2O_3 had a relatively low mobility during the alteration processes.

During the alteration of the greenstone most of its Ca, Mg, Mn and Fe content has been bound by carbon dioxide into ankerite; some iron has been bound by sulphur and arsenic into pyrite and arsenopyrite. The remainder of this group of elements, as shown by profiles of their combined oxides, has been leached suggesting migration into low-pressure zones such as the vein sites. There, much of the Ca, Mg, Mn and Fe was bound by CO_2, S and As into the ankerite, pyrite and arsenopyrite of the quartz lenses. Some of these elements, however, may have migrated upward along the shear zones and were dispersed at the surface.

The behaviour of a number of trace elements in the alteration zones in the deposits in greenstones at Yellowknife has been discussed previously in the section on associated elements. Of particular interest (Table 48) is the behaviour of Li, Rb, Cs, B, As, Sb, Au and Ag.

Lithium tends to follow magnesium, rather than the alkalis, and there is no increase in the element as the veins are approached, the largest amounts of lithium being present in the chlorite schists rather than in the sericite schists. Rubidium and cesium follow potassium and are enriched consistently as the auriferous veins are approached. Boron, arsenic, antimony, gold and silver, likewise, increase as the veins are approached.

Regarding ratios, it will be noted that there is a marked increase in the K_2O/Na_2O ratios and consistent and marked decreases in the SiO_2/CO_2 and SiO_2/total volatiles ratios as auriferous mineralization is approached. Similarly there is a marked increase in the Au/Ag ratio. These combined features are useful in predicting proximity to ore ahead of the bit during drilling as suggested by the writer in another paper (Boyle, 1974*b*).

Kirkland Lake, Ontario (Precambrian: Archean age). A number of gold deposits in the Kirkland Lake area occur in fracture zones that transect syenite porphyry. The wall-rock alteration attending these deposits has been examined by a number of investigators and is summarized by Thomson (1950) as follows:

> Wall-rock alteration has had a two-fold effect in that the original mineral constituents have been largely obliterated and at the same time a colour change has taken place in the syenite rock. The disappearance of the granitoid texture in the intrusives is the result of rock movement, whereas the colour change is produced by hydrothermal alteration.
>
> The rock adjacent to a vein-fault in the intrusives is changed from a massive holocrystalline rock to a banded, brecciated, or streaky variety. This may be reddened, bleached, or silicified, or there may be a mixture of all phases. The cherty-looking, silicified material has undergone the most complete granulation. Greenish-coloured alteration is prominent along some of the branch veins in the Teck-Hughes mine and certain veins in the sediments in the Sylvanite and Toburn mines. Much of the very pale green alteration is due to sericitization, but in a few places deep-green chrome mica (fuchsite) is developed over areas seldom more than an inch in diameter.
>
> Microscopic study shows that the crystals are first cracked or sheared along parallel lines, then progressively broken, granulated, and pulverized, depending upon the amount of alteration involved. Chlorite, sericite, and leucoxene have developed along narrow seams in the rock, and carbonate minerals and secondary quartz and pyrite have been introduced. The fine-grained bleached rock thus produced has often been further shattered to form a breccia, with one type of alteration embedded in a slightly different phase. Even this late breccia shows later fracturing. The resulting product is called mylonite in the camp. It is most commonly found along the stronger vein faults. Sometimes the mylonite bands have sharply defined contacts and resemble dikes or dikelets in appearance. The widths of the mylonite zone vary greatly and may be only a few inches or many feet.
>
> Microscopic examination of the mylonite shows that it is a microbreccia with a cataclastic structure. Where it has been formed by the alteration of syenite porphyry, the original feldspar phenocrysts may be seen in all stages of fracturing, slicing, and granulation. Sericite and chlorite have been developed along the fracture planes. In many thin sections secondary carbonates (chiefly iron-bearing dolomite) make up about half of the mineral constituents; there is also much secondary quartz and some pryite crystals. The colour is due to the carbonates, ferric oxide, and leucoxene.
>
> Carbonatization may affect large areas of rock and is not necessarily associated with the ore. It may be observed in the vicinity of any slip, whether pre-ore or post-ore, and the extent of the alteration seems to be dependent upon the amount of fracturing in the rock. For example, in the Toburn mine the syenite porphyry shows reddish alteration for a distance of about 600 feet north of the South vein. At the other end of the camp, irregular areas of alteration occur throughout the system of hanging-wall veins in the Macassa mine which attain a width of 325 feet. It would appear that the period of intense carbonatization preceded the period of ore deposition and that zones of carbonatization and gold-bearing veins do not necessarily coincide, although in most cases they have followed the same channelways. The same conclusions would apply to the Larder Lake fault zone lying to the south of the Kirkland Lake camp, where there is a much great development of carbonates. It implies widespread carbonatization throughout the country along lines of weakness as a forerunner of the gold mineralization.
>
> Silicification and sericitization are generally limited to areas immediately adjacent to the veins and may show a rather sharp contact with the country rock. The H vein in the Teck-Hughes mine and certain veins in the Sylvanite show varying degrees of silicification along the vein walls. Silicification is seen in breccia fragments of syenite or syenite porphyry within the quartz veins where very fine quartz partly replaces the fragments. Microscopic examination of the vein material along the K vein in the Teck-Hughes mine shows a very high degree of sericitization.
>
> The alteration of the different types of wall rock may show a slight difference in appearance, but in many places it only makes the identification of the original country rock more difficult. The syenite generally becomes a more pronounced red or pink; the dark-coloured augite syenite becomes more mottled in appearance, the ferromagnesian minerals being bleached and the feldspars reddened. In many places the syenite porphyry changes to a buff-coloured crush breccia or mixture of buff-coloured, reddish, and cherty phases. The sediments and tuffs do not show the same degree of alteration but tend to become red, brown, or green adjacent to innumerable tiny fractures. The various colours are due to disseminated dusty particles of red hematite, brown limonite, or goethite, and to varying quantities of ferriferous carbonate and sericite.

Thomson (1950) gives an extended list of analyses from which Table 49 was compiled. In general the alteration of the syenite porphyry, syenite and augite syenite has involved an increase in CaO, MgO, CO_2 and S. Some varieties of the altered syenitic rocks exhibit gains in SiO_2; others are characterized by significant losses.

Cordova deposit, Ontario (Precambrian: Grenville age). The Cordova Mine, near Marmora, Ontario was developed in a number of shear zones cutting a gabbro stock of variable composition and texture near a large mass of diorite. The age of the rocks is Grenville, and the age of the deposit is probably the same. In the vicinity of the old mine workings the gabbro is coarse grained and relatively equigranular. In the shear zones the minerals of the diorite are largely destroyed and replaced by chlorite and chlorite-carbonate schist. The width of these schist zones is relatively narrow, a few inches to 1 ft in most places. The veins and lenses are composed essentially of quartz with minor carbonate, small amounts of pyrite, ilmenite and native gold. The grade of the deposit is 0.3 oz Au/ton and 0.03 oz Ag/ton. The gold is mainly free.

In thin sections the plagioclase, hornblende and pyroxene of the gabbro are seen to be altered to a schistose mass of carbonate, quartz, biotite and/or chlorite, with some pyrite and minor ilmenite close to the quartz veins. As seen from the analyses in Table 50 there is a decrease in SiO_2 content as the alteration increases and the vein is approached. Concomitently an increase in CO_2, S and P_2O_5 is noted. K_2O is enriched in the chlorite-biotite-carbonate phase. H_2O shows a major increase in the biotite-chlorite phase and only a slight increase

Table 49. Chemical changes produced by alteration of syenite porphyry, Kirkland Lake area, Ontario

Constituent	Augite syenite	Syenite	Syenite porphyry	Wall rock and mylonite									Ore		
	1	2	3	4	5	6	7	8	9	10	11	12	13	14	15
								(%)							
SiO_2	50.24	57.02	62.19	63.61	59.59	49.92	46.00	58.83	67.84	69.91	67.24	49.24	48.67	58.76	66.23
Al_2O_3	16.43	18.30	14.85	14.96	12.46	15.44	12.48	14.46	12.73	10.77	12.27	12.64	11.64	12.51	13.31
Fe_2O_3	2.77	0.98	1.38	1.08	0.94	0.97	3.00	1.37	1.37	1.51	0.12	0.41	1.49	0.91	(Fe)2.58
FeO	4.48	2.67	2.03	1.16	2.41	4.78	7.10	2.23	1.73	0.74	1.49	4.96	6.07	3.00	
MgO	3.80	1.04	2.33	1.85	2.80	3.79	6.45	3.69	1.71	0.75	1.80	3.14	5.88	3.04	1.68
CaO	5.84	1.99	3.88	3.36	5.51	5.82	6.90	4.77	3.13	4.58	3.90	10.36	6.69	4.76	3.64
Na_2O	3.54	4.25	4.92	4.30	3.82	4.49	2.85	5.44	5.34	4.00	4.60	4.82	1.86	2.78	7.06
K_2O	5.84	7.41	3.76	4.12	2.80	4.52	2.89	2.12	1.58	1.37	1.69	0.80	4.36	4.63	
H_2O (total)	1.90	0.95	1.05	0.91	0.88	1.40	2.32	0.71	0.74	0.89	0.68	0.77	1.84	1.31	
CO_2	3.35	3.06	2.79	4.08	7.68	7.22	8.20	5.96	3.10	3.64	5.06	11.84	8.86	5.91	3.10
TiO_2	0.93	1.90	0.44	0.42	0.39	0.98	0.81	0.45	0.28	0.25	0.28	0.65	0.86	0.55	
P_2O_5	0.51	0.23	0.37	0.26	0.35	0.42	0.55	0.06	0.10	none	0.12	0.15	0.13	0.22	
S								0.45	0.12	1.10	0.19	0.29			0.70
Cr_2O_3								none	0.08	none	trace	trace			
MnO	0.16	0.17	0.13	0.10	0.06	0.09	0.21	0.09	0.02	0.12	0.11	0.14	0.13	0.09	
FeS_2					0.28	0.30							1.81	1.59	
Total	99.79	99.97	100.12	100.21	99.97	100.14	99.76	100.63	99.87	99.63	99.55	100.21			
Specific gravity	2.819	2.681	2.730	2.699	2.689	2.750	2.805								

Source: Thomson (1950).

Notes: 1. Average of rock type based on 5 analyses of least altered material.
2. Average of rock type based on 2 analyses of least altered material.
3. Average of rock type based on 4 analyses of least altered material.
4. Altered porphyry from the north wall of the South vein, 1500-foot level, Wright-Hargreaves Mine.
5. Altered porphyry from the wall of an ore shoot on the main (north) vein, 600-foot level, Lake Shore Mine.
6. Altered syenite from the wall of a vein on the 2350-foot level, Kirkland Lake Gold Mine.
7. Altered augite syenite (lamprophyre) from the south wall of the main vein, 2475-foot level, Kirkland Lake Gold Mine.
8. Mylonite from No. 4002 drift, south vein, 4000-foot level, Wright-Hargreaves Mine.
9. Mylonite from No. 6102 drift, south vein, 6150-foot level, Wright-Hargreaves Mine.
10. Cherty phase of mylonite, No. 4801 west drift, 4825-foot level, Lake Shore Mine.
11. Buff-coloured mylonite, main south break, southeastern section, 1000-foot level, Sylvanite Mine.
12. Mylonite on No. 5 strike fault, shaft station, 5400-foot level, Sylvanite Mine.
13. Sample representing the mill run for 1 month from Kirkland Lake Gold Mine.
14. Sample representing 10 849 tons of ore milled at the Lake Shore Mine.
15. Analysis of representative mill feed, Wright-Hargreaves Mine.

in the highly altered phase near the veins. Al_2O_3, Fe_2O_3, FeO, CaO, MgO and MnO show a few changes, but these are not marked.

The trace and minor elements exhibit a varied habit. Those showing increases as the veins are approached include, Li, Rb, Cs, Cu, Ag, Au, Ba, B, Sn, Zr, As, Bi and V; those marked by decreases comprise Sr, Zn, Ga and Ni. Elements such as As, Sb and Bi show a definite correlation with the increase in pyrite as the veins are approached. Consistent increases in the K_2O/Na_2O and Au/Ag ratios and decreases in the SiO_2/CO_2 and SiO_2/total volatiles as the veins are approached are notable.

Bralorne Mine, Bridge River district, British Columbia, (Mesozoic age). The Bralorne deposit in the Bridge River district of British Columbia consists of gold-bearing quartz veins and silicified zones, mainly in a diorite stock of Mesozoic age. The diorite varies considerably in composition from an augite to a hornblende diorite. The veins are quartz-filled fissures that strike across the diorite. Adjacent to the veins the diorite is altered to chlorite and chlorite-carbonate-sericite schist over widths of several feet. The principal vein minerals are quartz, calcite, mariposite, arsenopyrite and pyrite, with minor amounts of scheelite, sphalerite, pyrrhotite, chalcopyrite, stibnite, galena and molybdenite. Native gold is the economic mineral. Joubin (1948) has described the deposit in some detail.

As shown by the analyses in Table 51, there is a decrease in SiO_2 as the alteration increases and the vein is approached. Attending this decrease there is a marked increase in CO_2, H_2O, S, K_2O and P_2O_5. Na_2O shows a general decrease as the ore is approached. Al_2O_3 remains nearly constant throughout the alteration sequence; there is a general decrease in Fe_2O_3 in the heavily chloritized zones and an increase in FeO, CaO, MnO, TiO_2 and MgO. The total iron content remains relatively constant, increasing slightly in the chlorite phases.

There is a general increase in the contents of the following trace elements as the veins are approached: Li, Rb, Cs, Ag, Au, Be, Ba, Hg, B, Sc, Pb, As, Sb, V, Cr, Mo, W, Cl, Co and Ni. The smooth increase is commonly broken in the highly altered carbonate-sericite schist, some elements such as Cu, Be, Sc, Cl and Ni being depleted in this zone compared with the original diorite. The element zinc shows a general decrease towards the vein, and cadmium, zirconium and fluorine exhibit an erratic behaviour. Strontium follows calcium closely, whereas barium follows potassium in the alteration sequence.

The data on the ratios show changes similar to those noted for the Yellowknife deposits. There is a marked increase in the K_2O/Na_2O ratios and a consistent decrease in the SiO_2/CO_2 and SiO_2/total volatiles as the gold-quartz veins are approached. The Au/Ag ratio shows a general increase towards the veins.

Table 50. Chemical changes produced by alteration of gabbro, Cordova Mine, Cordova, Ontario

Constituent	Gabbro*	Biotite-chlorite schist†	Adjusted§	Gains and losses in constituents	Percentage gains and losses	Gabbro*	Chlorite-biotite-carbonate schist‡	Adjusted§	Gains and losses in constituents	Percentage gains and losses
		(%)					(%)			
SiO_2	45.5	42.2	38.32	−7.18	−15.78	45.5	39.0	38.53	−6.97	−15.32
Al_2O_3	13.8	13.1	11.90	−1.90	−13.77	13.8	12.5	12.35	−1.45	−10.51
Fe_2O_3	3.0	1.55	1.41	−1.59	−53.00	3.0	4.55	4.50	+1.50	+50.00
FeO	10.6	10.8	9.81	−0.79	− 7.45	10.6	8.55	8.45	−2.15	−20.28
CaO	10.5	9.05	8.22	−2.28	−21.71	10.5	10.5	10.37	−0.13	−1.24
MgO	7.35	6.75	6.13	−1.22	−16.60	7.35	6.30	6.22	−1.13	−15.37
Na_2O	2.5	2.80	2.54	+0.04	+1.60	2.5	1.00	0.99	−1.51	−60.40
K_2O	0.35	0.30	0.27	−0.08	−22.86	0.35	1.00	0.99	+0.64	+182.86
H_2O (total)	2.00	4.20	3.81	+1.81	+90.50	2.00	2.10	2.08	+0.08	+4.00
TiO_2	2.67	1.69	1.54	−1.13	−42.32	2.67	2.03	2.01	−0.66	−24.72
P_2O_5	0.25	0.14	0.13	−0.12	−48.00	0.25	2.05	2.03	+1.78	+712.0
MnO	0.23	0.19	0.17	−0.06	−26.09	0.23	0.23	0.23	0.00	0.00
CO_2	0.20	7.1	6.45	+6.25	+3125	0.20	7.65	7.56	+7.36	+3680
S	0.14	0.22	0.20	+0.06	+42.86	0.14	1.57	1.55	+1.41	+1007.14
O≡S	0.06	0.09				0.06	0.59			
Total	99.1	100.0				99.1	98.6			
Specific gravity	3.114	2.826				3.114	3.076			
				(ppm)						
Li	8	10.5				8	12			
Rb	15	13.5				15	22			
Cs	0.3	<0.1				0.3	2.0			
Cu	39	36				39	82			
Ag	1.0	1.8				1.0	1.3			
Au	0.015	0.03				0.015	1.8			
Sr	230	220				230	190			
Ba	180	180				180	305			
Zn	91	85				91	84			
Cd	1	1				1	<1			
Hg (ppb)	54	58				54	52			
B	nf	nf				nf	490			
Ga	5.4	3.3				5.4	3.1			
Sc	54	27				54	53			
Y	32	<30				32	34			
Yb	<10	<10				<10	<10			
U	0.1	0.1				0.1	1.1			
Th	3	3				3	1			
Sn	<0.4	<0.4				<0.4	2.5			
Pb	5.0	2.9				5.0	5.1			
Zr	87	60				87	108			
As	12	8				12	29 600			
Sb	1	2.5				1	2.5			
Bi	<0.5	<0.5				<0.5	3.2			
V	500	315				500	885			
Se	<2	<2				<2	<2			
Te	<0.2	<0.2				<0.2	1.0			
Cr	44	45				44	<20			
Mo	1	2				1	2			
W	3	5				3	7			
F	600	550				600	620			
Cl	300	100				300	500			
Co	43	44				43	42			
Ni	47	22				47	22			
Pt (ppb)	<5	<5				<5	nf			
Pd (ppb)	<5	<5				<5	8			
Ratios										
K_2O/Na_2O	0.14	0.11				0.14	1.0			
SiO_2/CO_2	227.5	5.94				227.5	5.10			
SiO_2/H_2O+CO_2+S+As	19.43	3.66				19.43	2.73			
Au/Ag	0.015	0.017				0.015	1.38			

Note: nf–not found.

*Composite sample of fresh gabbro.

†Composite sample of biotite-chlorite schist derived from gabbro (18–6 in. from vein).

‡Composite sample of chlorite-biotite-carbonate schist derived from gabbro (6–0 in. from vein).

§Adjusted % = percentages of biotite-chlorite schist† and chlorite-biotite-carbonate schist‡ multiplied by the specific gravity factors 0.908 and 0.988 respectively to reduce to a common volume.

Other deposits in intermediate and basic intrusive and extrusive rocks. Throughout the world numerous gold deposits in intermediate and basic intrusive and extrusive rocks exhibit wall-rock alteration features similar to those just described. A brief review of some significant examples follows.

Lindgren (1905–1906, p. 539), after a study of some gold ores from Western Australia where the lodes occur mainly in amphibolites (greenstones), concluded that the alteration processes were as follows:

> The metasomatic processes consist then in the introduction of carbon dioxide and sulphur with gold, mercury and tellurium and the attendant development of carbonates at the expense of the lime, magnesia, and iron contained in the chlorite and amphibole, and of pyrite at the expense of the same minerals and the black iron ores. The silica set free by the decomposition of the silicate has been deposited as quartz where it was not needed for the formation of the new silicates sericite and albite.

Ransome (1909) describes the characteristic alteration of Tertiary dacites, latites, rhyolites and andesites associated with the gold-silver lodes of Goldfield, Nevada. The dacites, containing some 60 per cent SiO_2 and 2.5 and 3 per cent respectively of K_2O and Na_2O, are highly altered in the vicinity of the lodes to a comparatively soft rock containing essentially quartz, kaolinite, alunite and pyrite. During the alteration the dacite has been depleted in some of its silica, much of its Al_2O_3, MgO and CaO and some of its Na_2O and K_2O. Gains are represented by H_2O, S and SO_4. The iron present in the original rock was more than sufficient to form all of the pyrite. The alunitization process is extensive within the dacite and other rocks, expanding over widths of a mile in places and extending over a considerable area (*see* Fig. 32). The known productive gold areas are within the alunite halo but are relatively small compared with the area covered by the halo.

In a more recent investigation of the wall-rock alteration associated with the gold deposits of the Goldfield district, Harvey and Vitaliano (1964) distinguished the following mineralogical zones, which flank the veins in the Tertiary dacites, latites and andesites:

1. A propylitic zone farthest from the veins and characterized by the development of chlorite, chlorite-vermiculite, calcite and antigorite.

2. A bleached argillic zone, divided into a montmorillonite subzone in the outermost part and an illite-kaolinite subzone in the innermost part.

3. An alunite-quartz zone immediately adjacent to the veins in which alunite, quartz, hematite, jarosite, halloysite, gypsum, epidote and sulphides are present. The investigators postulate that alunite was formed by the reaction of sulphuric acid and K-mica. Thermodynamic data are given for the reaction.

Clapp (1915) describes alunitized and pyrophyllitized dacites and andesites of Triassic and Jurassic age from Kyuquot Sound, Vancouver Island. These rocks, now mainly quartz-sericite-chlorite, quartz-sericite, quartz-pyrophyllite and quartz-alunite (natroalunite) rocks, are pyritized in places and contain small amounts of gold (0.05 oz/ton) and silver (0.5 oz/ton). The analyses given by Clapp indicate that alunitization and sericitization of the rocks lead to an increase in S, SO_4 and H_2O. Decreases were noted for iron oxides, MgO and CaO, as well as Na_2O. SiO_2, Al_2O_3 and K_2O have

Table 51. Chemical changes produced by alteration of diorite, Bralorne Mine, Bridge River district, British Columbia

Constituent	Fresh diorite*	Chloritized diorite†	Adjusted[ll]	Gains and losses	Percentage gains and losses	Fresh diorite*	Chlorite-sericite schist‡	Adjusted[ll]	Gains and losses	Percentage gains and losses	Fresh diorite*	Carbonate-sericite schist§	Adjusted[ll]	Gains and losses	Percentage gains and losses
		(%)					(%)					(%)			
SiO_2	56.9	43.6	44.91	−11.99	−21.07	56.9	41.4	41.23	−15.67	−27.54	56.9	42.9	43.54	−13.36	−23.48
Al_2O_3	15.6	16.3	16.79	+1.19	+7.63	15.6	16.8	16.73	+1.13	+7.24	15.6	18.1	18.37	+2.77	+17.76
Fe_2O_3	1.8	1.1	1.13	−0.67	−37.22	1.8	0.9	0.896	−0.904	−50.22	1.8	1.8	1.83	+0.03	+1.67
FeO	5.9	8.7	8.96	+3.06	+51.86	5.9	7.8	7.77	+1.87	+31.69	5.9	5.5	5.58	−0.32	−5.42
CaO	6.2	9.7	9.99	+3.79	+61.13	6.2	8.7	8.67	+2.47	+39.84	6.2	6.4	6.50	+0.30	+4.84
MgO	4.9	7.2	7.42	+2.52	+51.43	4.9	5.7	5.68	+0.78	+15.92	4.9	2.4	2.44	−2.46	−50.20
Na_2O	4.0	2.7	2.78	−1.22	−30.50	4.0	1.6	1.59	−2.41	−60.25	4.0	2.3	2.34	−1.66	−41.50
K_2O	0.3	0.6	0.62	+0.32	+106.67	0.3	2.6	2.59	+2.29	+763.3	0.3	3.6	3.65	+3.35	+1116.7
H_2O (total)	2.7	4.6	4.74	+2.04	+75.56	2.7	5.1	5.08	+2.38	+88.15	2.7	4.4	4.47	+1.77	+65.56
TiO_2	0.47	0.72	0.74	+0.27	+57.45	0.47	0.59	0.588	+0.118	+25.11	0.47	0.78	0.79	+0.32	+68.09
P_2O_5	0.03	0.02	0.021	−0.009	−30.00	0.03	0.23	0.229	+0.199	+663.3	0.03	0.13	0.132	+0.102	+340.0
MnO	0.14	0.15	0.16	+0.02	+14.29	0.14	0.13	0.129	−0.011	−7.86	0.14	0.18	0.183	+0.043	+30.71
CO_2	0.2	2.8	2.88	+2.68	+1340.	0.2	6.7	6.67	+6.47	+3235.	0.2	9.1	9.24	+9.04	+4520.
C	0.1	0.1				0.1	nf				0.1	nf			
S	0.16	0.18	0.185	+0.025	+15.63	0.16	0.46	0.458	+0.298	+186.3	0.16	1.44	1.46	+1.30	+812.5

Cs [?]	0.6	3.3 [?]	0.6	3.5	0.6	4.2
Cu	50	39	50	83	50	25
Ag	1.3	1.5	1.3	1.6	1.3	1.8
Au	0.015	0.01	0.015	0.4	0.015	0.8
Be	2	4	2	5	2	<2
Sr	220	360	220	140	220	190
Ba	86	270	86	160	86	570
Zn	68	59	68	54	68	47
Cd	1.3	2.5	1.3	1.3	1.3	2.5
Hg (ppb)	17	8	17	46	17	285
B	53	69	53	150	53	150
Sc	46	56	46	60	46	33
U	<0.1	<0.1	<0.1	<0.1	<0.1	<0.1
Pb	10	15	10	20	10	15
Zr	60	nf	60	nf	60	52
As	13	19	13	63	13	17 000
Sb	<1	<1	<1	2	<1	16
Bi	<0.2	<0.2	<0.2	<0.2	<0.2	<0.2
V	240	390	240	440	240	300
Se	<2	<2	<2	<2	<2	<2
Te	<0.2	<0.2	<0.2	<0.2	<0.2	<0.2
Cr	46	140	46	180	46	75
Mo	1	1	1	2	1	4
W	<2	<2	<2	8	<2	28
F	200	<200	200	200	200	<200
Cl	<200	<200	<200	400	<200	<200
Co	42	59	42	78	42	57
Ni	41	62	41	83	41	34
Pt (ppb)	nf	<5	nf	nf	nf	nf
Pd (ppb)	<5	<5	<5	<5	<5	nf
Ratios						
K_2O/Na_2O	0.08	0.22	0.08	1.63	0.08	1.57
SiO_2/CO_2	284.5	15.6	284.5	6.2	284.5	4.7
SiO_2/H_2O+CO_2+S+As	18.6	5.8	18.6	3.4	18.6	2.6
Au/Ag	0.012	0.007	0.012	0.25	0.012	0.44

Note: nf–not found.
*Composite sample of fresh diorite (country rock) (25–50 ft from vein).
†Composite sample of chloritized diorite (chlorite schist) (10–25 ft from vein).
‡Composite sample of chlorite-sericite schist (1–5 ft from vein).
§Composite sample of carbonate-sericite schist (0–1 ft from vein).
‖Adjusted % = percentages of chloritized diorite†, chlorite-sericite schist‡ and carbonate-sericite schist§ multiplied respectively by the specific gravity factors 1.030, 0.996 and 1.015 to reduce to a common volume.

increased in some rocks and decreased or remained constant in others.

Simpson and Gibson (1912) give a series of analyses of typical Kalgoorlie rocks in which the propylites (chlorite-carbonate schist) and carbonated parts show substantial losses of silica and gains of carbon dioxide, sulphur and potash when compared with the amphibolites, gabbros and epidiorites from which they were derived. The analyses given by Feldtmann (1916) for the East Coolgardie goldfield exhibit similar features. A suite of analyses of specimens from Ora Banda (40 mi northwest of Kalgoorlie) given by Clarke and Ellis (1939), however, show an increase in the silica content as the veins are approached. The specimens were, however, not in a complete sequence from one crosscut.

In a more recent contribution to the subject of wall-rock alteration associated with the auriferous lodes of Kalgoorlie, Bartram and McCall (1971) note extensive regional carbonatization of the greenstones (dolerite) with the development of chlorite, albite, quartz, carbonates and minor muscovite. Weak carbonatization resulted in the formation of calcite whereas intense carbonatization produced both ankerite and calcite. All of the gold lodes occur within these altered zones and are flanked by zones enriched in sericite. No relationship was found between the gold values and the type of intensity of wall-rock alteration. Considering the overall chemical features at Kalgoorlie there is a general similarity to the alteration processes at Yellowknife. In both auriferous belts CO_2, H_2O, S and K have been introduced, and there has been a marked leaching of SiO_2. The exchange reaction between K and Na resulting in significant loss of Na does not appear to have been as pronounced at Kalgoorlie as at Yellowknife. In both belts there is little or no change in the total iron content from country rock to pyritic lode, indicating that most of the iron component was contributed by the host rocks and bound by the introduced sulphur.

Bartram and McCall (1971) seem to think that the regional carbonatization and lode formation are distinct processes, despite the fact that the lodes are always found in the zones of carbonatization. They say "Several lines of evidence suggest that the carbonatising solutions invaded the greenstone after regional metamorphism, that these solutions were active over a long period, and that their activity commenced before the major lode formation was completed." Commenting on the progressive changes in the nature of the solutions they state: "Either the hydrothermal solutions changed progressively, with time, becoming richer in potassium and poorer in carbon dioxide, or else there were two distinct phases of alteration, comprising an earlier carbon dioxide-rich phase and a later potassium-rich phase."

This view has been prevalent in the literature for many years. The writer can find no satisfactory evidence to substantiate changing solutions or pulsation effects in the alteration zones of epigenetic gold deposits. On the contrary it seems more logical from the field evidence and the chemistry involved that the formation of quartz lodes was concomitant with carbonatization. This statement is amplified further in the section on the origin of gold deposits at the end of this chapter.

McCann (1922), in his study of gold-quartz veins at the Bridge River district, British Columbia, where the veins occur in augite diorite, found a strong desilication of the adjacent wall rocks and concomitant increases in CO_2, H_2O, S, As and K_2O. He made comparisons with the Grass Valley and Juneau, Alaska gold deposits and found similar phenomena. About the silica he says (p. 364), "the silica which has been lost has probably contributed to vein filling . . ." Na_2O was leached from the highly altered carbonate sericite schist, as were also the oxides of iron, aluminum and magnesium. These results are essentially the same as those noted in the Bralorne Mine as described above.

Knopf (1929, p. 47) in his description of the Mother Lode System in California observed, after considering the alteration adjacent to the veins in greenstones, amphibolites and serpentinites that, "the carbon dioxide liberated immense quantities of silica from the wall rocks, and this silica was delivered to the vein channels, where it was in part precipitated as quartz."

At Porcupine, Ontario, Hurst (1935) found that during the alteration processes of the dacite, pillow lava (greenstone) and quartz porphyry in which the orebodies occur, the content of silica was decreased substantially in the altered equivalents of these rocks adjacent to the quartz bodies. K_2O and S showed marked gains, and CO_2 increased in the altered dacite. Na_2O exhibited an increase in the altered dacite and decreases in the altered pillow lava and porphyry.

Bateman (1940), in his description of the character of the wall-rock alteration in the Uchi Lake greenstones in Ontario, gives analyses and remarks on the desilication of the alteration zones adjacent to the gold veins and orebodies. His analyses show a marked increase in the K_2O, S and CO_2 contents of the altered greenstones and a slight increase in the Na_2O content.

In the Nighthawk Peninsular Gold Mine in the Porcupine Camp of Ontario the orebodies are carbonate zones with quartz stringers and disseminated sulphides in or near a stock of albite syenite intruding Keewatin lavas (andesites). Byers (1940) found that during the alteration of the syenite to a carbonate-sericite and carbonate-chlorite rock there was a decrease in SiO_2, Na_2O, Al_2O_3 and an increase in MgO, CaO, K_2O, FeO and H_2O.

Davidson and Banfield (1944), describing the alteration associated with the Beattie gold deposits that occur in syenite porphyry, give analyses showing a desilication of the syenite porphyry as the vein is approached. There is also a concomitant increase in K_2O, CO_2 and S, and a decrease in Na_2O. The andesite greenstone, however, shows an increase in silica as the orebody is approached. Only two analyses are given and it is difficult to draw conclusions. The andesite (dark greenstone) does not appear to be a normal sample of greenstone country rock, since it contains 29.70 per cent SiO_2 and 18.75 per cent $CaCO_3$. It probably represents a phase of the alteration zones.

Tooker (1963) made an extensive study of the primary wall-rock alteration of the auriferous base metal veins in the central part of the Front Range mineral belt, Gilpin and Clear Creek counties, Colorado. The host rocks are varied and include Proterozoic granite, granodiorite, granite gneiss, quartz diorite, amphibolite, garnet-gneiss and biotite schist, and Tertiary quartz monzonite and bostonite porphyry dykes and plutons (Fig. 48). The alteration zones flanking the veins range from a few tens of centimetres to several metres and consist essentially of an inner zone marked by extensive sericitization that grades outward through successive zones of strongly argillized rock and weakly argillized rock to fresh rock. The alteration zones are said to have been developed largely prior to the gold-base metal mineralization. Extensive mineralogical and geochemical data are given to which the interested reader should refer for details. Briefly, Tooker considered the distribution of elements in the altered wall rocks to be largely the result of the movement of ions in and out of the interstices of a silicon-oxygen or aluminum-oxygen framework in which the oxygen ion positions are the most abundant and largest. Calculations from chemical analyses of samples of representative rocks showed that, in the alteration of equal volumes of rock, small amounts of K^+, Fe^{2+}, C^{4+}, H^+, S^{2-}, and at places Al^{3+} were added, and Si^{4+}, Na^+, Ca^{2+}, Fe^{3+} and Mg^{2+} were removed. In a chemical sense, this is interpreted as a hydrolysis process in which there is a replacement of chemical equivalents of metal cations by H^+ ions, which enter the structure and produce polarized hydroxyl groups. The hydrothermal solutions that altered the rocks, as inferred by Tooker from observable surface and underground waters and the mineralogical and chemical changes in the host wall rocks, were relatively dilute, slightly acid and initially carried K and subsequently CO_2 and S ions or complexes. As the rocks were altered, these solutions were modified by constituents replaced in the wall rocks and by vein-forming ore metals.

In the Tertiary (Neogene) deposits in the Rodalquilar area of southeast Spain Lodder (1966) recognized the following zones from the ignimbrite wall rocks inward toward the veins: (1) least altered ignimbrite characterized by mixed-layer chlorite-vermiculite, illite, chlorite and carbonate; (2) a zone characterized by mixed-layer smectite-chlorite, illite, smectite, chlorite, carbonate, sulphides, iron oxides and minor gypsum; (3) a zone marked by mixed-layer illite-smectite, illite, smectite, quartz, adularia, plagioclase, sulphides, alunite, jarosite and minor gypsum, iron oxides and strongly leached biotite; (4) a kaolinite-illite zone marked by kaolinite, illite, quartz, adularia, plagioclase, jarosite, alunite and minor gypsum, sulphides and iron oxides; (5) a zone in which dickite is abundantly developed together with quartz, alunite, jarosite, illite, locally pyrophyllite, sulphides and iron oxides; and (6) alunite-quartz which comprises the veins. The principal mineral constituents are fine-grained banded vein quartz, alunite, jarosite, barite, gypsum, zircon, apatite, rutile, iron oxides, sulphides and gold. The widths of the zones are variable; the outer zones may range from 70 to 150 m in width whereas the inner dickite zone is rarely more than 0.5 m except where pyrophyllite is present in which case this zone may range up to 5 m wide. The alunite-quartz (vein) zone varies in width from 1 cm to 1.7 m.

The chemical analyses presented by Lodder (1966) show significant decreases in silica content in some of the outer zones with a gradual increase toward the alunite-quartz zone. Alumina contents show a reverse relationship to those for silica. There is a gradual decrease in the potassium and sodium contents of the rocks as the veins are approached. The K_2O/Na_2O ratios across the alteration zones show an unusual erratic behaviour, probably because the samples are not from the same traverse across the veins.

Taneda and Mukaiyama (1970) discussed the wall-rock alteration processes at the Kushikino Mine, southern Kyushu, Japan. The host rocks are Tertiary pyroxene andesites, which are propylitized; in the hanging walls of the veins sericitization is marked, and in the footwalls silicification is predominant, the silicified rocks consisting mainly of quartz, sericite, adularia, calcite and pyrite. Analyses of the altered footwall rocks show marked increases in the content of silica and decreases in Ti, Al, Fe, Mn, Ca, Mg, Na and K.

Weissberg and Wodzicki (1970) investigated the wall-rock alteration at the Tui Mine, Te Aroha, New Zealand. The deposit in this mine is a fault-controlled, open-space filling type, containing essentially lead-zinc-copper sulphides, and is typical of the epithermal gold-silver and base metal ore deposits that occur as lodes in fissures and faults in extensively altered Tertiary andesite lavas and pyroclastic rocks.

They observed that the hydrothermal alteration is zonally arranged around the lodes, increasing in intensity towards them. It is characterized by the following sequential mineralogical changes: (1) pyroxene to chlorite (plus carbonate and epidote), magnetite to pyrite; (2) plagioclase to K-feldspar (plus carbonate and epidote); (3) K-feldspar to sericite; and (4) intense silicification and obliteration of original textures. Interlayered montmorillonite-illite and minor kaolinite occur throughout most of the altered sequence and are probably of retrograde origin. These mineralogical changes are accompanied by a small gain of sulphur, carbon dioxide, water, potassium, manganese and possibly silica, and a loss of calcium and possibly minor magnesia and alumina.

The textural evidence indicated that at a given location K-feldspar preceded sericite, and kaolinite was the last mineral to crystallize, suggesting that the temperature and/or the ratio K^+/H^+ decreased with time. Similar hydrothermal minerals occur in the active hydrothermal fields of the Taupo volcanic zone, where their different spatial distribution probably reflects present day variations in physical and chemical conditions rather than changes in these conditions through time.

They further noted that the intensity of hydrothermal alteration and the concentration of As, Sb, Cu, Pb, Zn and Hg in the wall rocks increased with proximity to the lodes and other permeable zones. From this they concluded that the zoned alteration pattern is genetically related to ore emplacement and that the ore metals were added to the lodes and wall rocks by hydrothermal transport along fissures.

Farther north in the Coromandel Peninsula of New Zealand the host rocks of the Hauraki auriferous deposits are extensively propylitized over great widths, some zones exceeding 1000 ft or more. These zones occur mainly in the First Period volcanics, which consist mainly of Eocene andesitic breccias and flows with minor amounts of associated dacitic and rhyolitic rocks. Finlayson (1909) studied the alteration in the andesites in great detail and gives analyses of fresh and highly altered (propylitized) hornblende andesite. Petrographically the fresh andesite passes through a chloritized phase into the propylitic phase (actually a zone consisting essentially of quartz, sericite, carbonate and pyrite) as the auriferous quartz veins are approached. The analyses of the fresh and highly propylitized andesite indicate that during chloritization and carbonatization processes silica is lost in the chloritized parts of the alteration zones and increased slightly in the rocks nearest the quartz veins. Carbon dioxide is gained in all altered rocks, and there is considerable loss of soda and a gain of potash and sulphur in the intensely altered sericite-adularia-pyrite phases next to the veins. Iron shows little change being bound mainly by sulphur into pyrite. Constituents like alumina, titania and phosphorus are reduced in the rocks adjacent to the veins; magnesia is also reduced with a corresponding increase in lime.

Bragg (1943) described the wall-rock alteration at the San Antonio Mine in southeastern Manitoba. The orebodies are veins and pyritized zones in a diabase sill. The alteration types recognized were epidotization, chloritization, carbonatization, albitization, silicification and sericitization. Orebodies were mainly characterized by carbonatization, pyritization, albitization and sericitization. More recently Stephenson (1971) has given an extensive series of analyses of wall rocks adjacent to the various types of gold deposits in the Rice Lake–Beresford Lake greenstone belt, southeastern Manitoba, which includes the San Antonio deposit. He noted depletions of SiO_2, Al_2O_3, CaO, MgO and Na_2O in the altered greenstones (quartz diorite, gabbro) and enrichments of K_2O, CO_2, H_2O, S and B. The total iron content changed only slightly. He concluded that silica removed from the alteration zones contributed to the quartz vein filling. Similar features were noted by Bailes (1971) for dacitic and diabasic rocks in the Pilot-Smuggler shear zone in the Rice Lake region. Amukun and Turnock (1971), however, concluded that there is no evidence that the material in the quartz veins of the Bissett area of Manitoba originated from the country rock.

In Taiwan there are a number of auriferous deposits in Tertiary rocks of which the Chinkuashih gold-copper lodes in dacites and Neogene sediments are the most important. Wang (1973) describes the wall-rock alteration attending these lodes in some detail. Outward from the great Penshan quartz sulphide vein in dacite there is a marked zonal arrangement consisting of a well developed silicified zone that grades through an argillized zone into a wide propylitized zone and then into unaltered dacite. In places the propylitized zone is several hundred metres wide. The other zones are much narrower, although the silicified zone approaches 100 m in near-surface parts of the vein.

The alteration processes associated with mineralization in Tertiary volcanic areas can be complex as shown by the work of Stanciu (1973) in the Gutîi Mountains, Romania. There, the volcanism occurred in three stages – the first, a rhyodacite (Tortonian-Buglovian) stage lacks hydrothermal activity; the second, an andesitic stage (Volhynian-Pontian) is marked by intensive hydrothermal effects and mineralization; and the third, a late andesitic stage (Upper Pliocene) has been slightly affected by hydrothermal alteration. Three hydrothermal phases mark the second volcanic stage, which is also the main metallogenic stage; the first is associated with the eruption of Sarmatian pyroxene andesites, the second is associated with the eruptions of Pannonian quartz andesites, and the third is associated with eruptions of Pontian pyroxene andesites. The first and second phases are characterized by alteration processes in succession as follows: propylitization, chloritization, development of adularia, sericitization, argillization and carbonatization. Adjacent to the veins the development of adularia is most persistent with sericitization and argillization notable in places. The propylitization and chloritization are commonly regional in aspect. During the third phase adularization was strongly diminished, but propylitization, sericitization and argillization were strongly manifest. Stanciu (1973) states that the cupiferous ore deposits are marked by chloritic zones; polymetallic veins mainly by adularization and subordinately by propylite zones; and gold-silver veins by widely developed adularia zones.

In a further contribution to wall-rock alteration effects in Tertiary andesites at the auriferous Hanes deposit in Romania Borcos and Stanciu (1974) note that the propylitization and sericitization of the andesites took place under pressure and at temperatures greater than 365°C in the presence of acidic hypervolatile emanations. They noted a marked replacement of sodium by potassium during alteration processes and concluded that when the K_2O content is greater than 3 per cent it can serve as an indicator of economic gold-silver mineralization.

Little has been said about the wall-rock alteration effects associated with certain gold-quartz deposits formed in the intermediate and basic rocks under deep seated conditions in high grade metamorphic terranes. The reason for this is that there are relatively few of these deposits, and only a small number have been studied intensively from the viewpoint of wall-rock alteration phenomena.

Briefly most of the wall-rock alteration effects in these deposits are minimal, the quartz bodies either grading into or being frozen against amphibolite or pyroxene schists and gneisses and other high grade intermediate-basic rocks. Commonly a zone of biotitization, chloritization and feldspathization occurs adjacent to the quartz bodies as at the Cordova Mine described above.

Summary of wall-rock alteration effects adjacent to gold deposits in intermediate and basic intrusive and extrusive rocks. The principal types of wall-rock alteration attending gold deposits in intermediate and basic intrusive and extrusive rocks are chloritization, carbonatization and sericitization. Silicification, present in some types of deposits, is relatively uncommon. Alunitization is restricted mainly to Tertiary deposits. Where chloritization, carbonatization and sericitization is marked there is generally a significant loss of SiO_2 and marked gains of CO_2, K_2O and S. Na_2O is commonly leached. Fe, Mn, Ca, Mg and Ti often show little change in most deposits, but in some there are significant additions or subtractions. H_2O is increased in the chloritized zones and generally decreased in the carbonate-sericite zones compared with the original rocks. Silicification is usually marked by a large increase in SiO_2 and decreases in elements such as Al, Fe, Mn, Ca, Mg, Ti, Na and K. Alunitization is commonly accompanied by a loss of silica in the wall rocks adjacent to the veins, and there is frequently also a loss of Na in some of the highly altered phases of the alteration zones. Marked increases in S always accompany this type of alteration. The behaviour of the other elements is variable.

Where carbonatization is marked in the alteration zones of epigenetic gold deposits in intermediate and basic rocks the ratio K_2O/Na_2O generally shows a consistent increase as the veins are approached. Similarly, there is usually a marked decrease in the SiO_2/CO_2 and SiO_2/total volatiles ratios with proximity to the veins. Where silicification and alunitization prevail the behaviour of these ratios tends to be variable.

Gold-quartz veins and silicified zones dominantly in acidic (granitic) intrusive and extrusive rocks. Deposits of this type commonly occur in volcanic terranes, the principal host rocks being rhyolites and effusive quartz-feldspar porphyries, now largely quartz-mica schists. Similar types of deposits also occur in granitic stocks and plugs and in intrusive quartz-feldspar bodies in various terranes.

Deposits of this type range in age from Precambrian to Tertiary. Characteristic examples of the Precambrian type in Canada include certain veins in the Yellowknife area; the Box Mine in northern Saskatchewan; a few veins in the Red Lake area of Ontario; the Renabi Mine in Ontario; and numerous veins and silicified zones in the Timmins area, Ontario. Elsewhere in the Precambrian, members of this type occur in the Kalgoorlie and Coolgardie districts of Western Australia, and in the Barberton district of South Africa.

Paleozoic and Mesozoic deposits of this type are widespread in Europe, Asia and the Western Cordillera of North America. A number of deposits of this type occur in the Ural goldfields; other examples include some of the veins of Grass Valley and Nevada City in California, and a number of deposits in British Columbia, including the Premier Mine at the head of Portland Canal, and others near Nelson.

Tertiary deposits of this type are, likewise, fairly widespread throughout the world, occurring in the gold districts of Romania, Japan, New Zealand, U.S.S.R. and United States (De Lamar, Idaho; Rawhide, Nevada; and Cripple Creek, Colorado). In all of these districts the acidic host rocks are mainly rhyolites and rhyodacites that accompany andesites and basalts. At Cripple Creek some of the host rocks in the mineralized volcanic neck are latites and phonolites; the country rocks are granite.

Yellowknife district, Northwest Territories (Precambrian: Archean age). Some of the veins and silicified zones in the Yellowknife Greenstone Belt cut quartz-feldspar porphyries that form dykes and regular bodies in the meta-andesites. The alteration of the porphyries adjacent to the veins is mainly sericitization and hydration.

The fresh quartz-feldspar porphyry contains essentially quartz, plagioclase, chloritized biotite and muscovite. Accessory minerals are pyrite, pyrrhotite, leucoxene, epidote and carbonate. Thin sections show that alteration of the rock results in the development of sericite from the plagioclase and biotite and a general recrystallization of the quartz and sulphides. In places an increase in sulphides is apparent.

The chemical analyses of the fresh and altered porphyry are given in Table 52 and graphed in Figure 14. They indicate relatively little change in the SiO_2, Fe_2O_3, TiO_2, P_2O_5 and MnO during the alteration process. Much calcium and carbon dioxide have migrated out of the sericite zone together with sodium and some iron. Potassium, alumina, water and sulphur were added.

The alteration processes, therefore, were mainly an exchange of potassium for sodium during hydration, and to balance the equilibrium, alumina was added and calcium abstracted. This calcium together with carbon dioxide probably formed carbonates in the quartz lenses. Some of the iron was fixed by sulphur in the alteration zones, and the remainder migrated to the site of the quartz lenses where it was probably fixed as pyrite.

Table 52. Chemical changes produced by alteration of quartz-feldspar porphyry, Negus-Rycon System, Yellowknife, Northwest Territories

Constituent	A-23*	A-24†	Adjusted‡	Gains and losses in constituents	Percentage gains and losses
		(%)			
SiO_2	73.09	72.46	73.80	+0.71	+ 1.06
Al_2O_3	14.01	16.26	16.56	+2.55	+18.30
Fe_2O_3	0.32	0.34	0.35	+0.03	+9.36
FeO	0.89	0.45	0.46	-0.43	-48.50
CaO	2.05	0.94	0.96	-1.09	-53.20
MgO	0.37	0.77	0.78	+0.41	+111.00
Na_2O	4.15	0.63	0.64	-3.51	-84.50
K_2O	1.63	3.94	4.01	+2.38	+146.00
H_2O+	0.72	1.93	1.97	+1.25	+174.00
H_2O-	0.05	0.12	0.12	+0.07	+140.00
TiO_2	0.20	0.25	0.26	+0.06	+30.00
P_2O_5	0.05	0.10	0.10	+0.05	+100.00
MnO	0.02	0.01	0.01	-0.01	-50.00
CO_2	1.80	0.97	0.99	-0.81	-45.00
S	0.80	1.27	1.29	+0.49	+61.20
Cl	trace	0.01	0.01	–	–
Cr_2O_3	0.04	trace	–	–	–
Total	100.19	100.45	–	–	–
Less O≡S,Cl	0.34	0.55	–	–	–
Net total	99.85	99.90	–	–	–
Specific gravity	2.69	2.74	–	–	–
	(ppm)				
Li	5	6			
Rb	58	171			
Cs	2.0	6.7			
B	28	34			
As	142	140			
Sb	6.2	22			
Ratios					
K_2O/Na_2O	0.39	6.25			
SiO_2/CO_2	40.61	74.70			
SiO_2/H_2O+CO_2+S	22.02	17.38			

* A-23–composite sample of massive quartz-feldspar porphyry, Con Mine.
† A-24–Composite sample of altered quartz-feldspar porphyry adjacent to quartz veins, Negus-Rycon System.
‡ Adjusted % = percentages in A-24 multiplied by the factor 2.74/2.69 to reduce a common volume.

Considering some of the trace elements it is apparent that there is an increase in the Rb, Cs, B and Sb content as the veins are approached. Li and As remain relatively constant.

There is a marked increase in the K_2O/Na_2O ratio as the gold-quartz veins are approached. The SiO_2/CO_2 ratio shows an increase that reflects the silicification of the alteration zone. There is a small decrease in the SiO_2/total volatiles ratio.

Red Lake area, Ontario (Precambrian: Archean age). The Red Lake Gold Shore Mine and the Howey and Hasaga mines occurred in granitic rocks in the Red Lake district of Ontario, the first in a granitic stock and the last two in a quartz porphyry dyke in greenstones. At the Howey and Hasaga mines the normal grey porphyry is altered to a buff waxy rock rich in sericite. The analyses (Table 53) given by Horwood

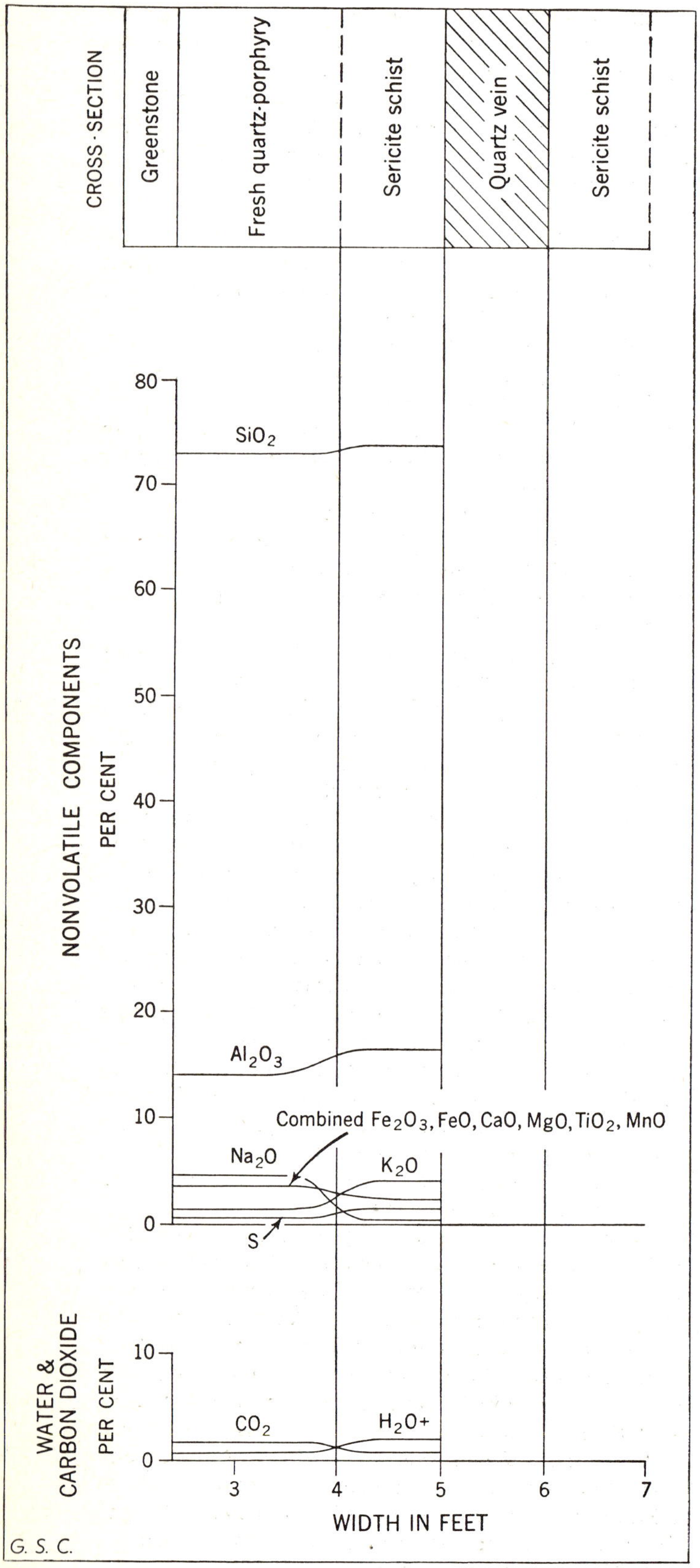

Figure 14. Chemical changes produced by alteration of quartz-feldspar porphyry, Negus-Rycon System, Yellowknife, Northwest Territories (*see* Table 52).

(1945) show a slight decrease in silica in the altered porphyry as the veins are approached, an increase in Al_2O_3, TiO_2, K_2O, CO_2 and H_2O and a decrease in Na_2O. The iron species exhibit an erratic behaviour; sulphur exhibits an increase in the slightly sericitized porphyry.

Table 53. Chemical changes produced by alteration of quartz porphyry, Howey Mine, Red Lake, Ontario

Constituent	Normal grey quartz porphyry	Partly sericitized porphyry	Highly sericitized porphyry
		(%)	
SiO_2	72.39	71.82	70.06
Al_2O_3	13.87	16.00	16.61
Fe_2O_3	1.45	0.10	1.80
FeO	1.13	0.36	0.85
MgO	0.46	0.21	0.51
CaO	1.24	1.07	1.10
Na_2O	4.75	4.62	3.32
K_2O	2.79	2.14	3.22
TiO_2	0.64	0.66	0.77
H_2O	0.43	1.27	1.34
CO_2	0.52	0.36	0.67
FeS_2	0.48	2.06	trace
Total	100.15	100.67	100.25

Source: Horwood (1945).

At the Red Lake Gold Shore Mine the host rocks of some of the quartz veins are granites and granodiorites with inclusions of greenstones. Near the veins the granitic rocks are silicified, sericitized, chloritized and moderately carbonated and pyritized.

The analyses given in Table 54 show a slight increase in SiO_2 and moderate to marked decreases in Al_2O_3, Fe_2O_3, FeO, CaO, MgO, TiO_2, P_2O_5 and MnO in the carbonate-sericite schist compared with the granite. Na_2O exhibits a small decrease in the sericite-carbonate schist, and K_2O remains relatively constant although some samples from this zone show a moderate increase of a percent or so. The major additions during sericitization and carbonatization are H_2O, CO_2 and S.

Lithium exhibits a decrease in the alteration zone, as does also cesium. Rubidium remains relatively constant. Increases as the veins are approached are apparent for the following trace elements: Cu, Au, B, Zr, As and F; decreases are registered by Sr, Ba, Zn, Cl and Co.

The K_2O/Na_2O ratio exhibits a slight increase as the gold-quartz veins are approached. The SiO_2/CO_2 and SiO_2/total volatiles both show significant decreases as alteration becomes more intense.

Box Mine, Goldfields, Saskatchewan (Precambrian: Archean age). The Box Mine at Goldfields, Saskatchewan is a quartz stockwork localized in a granite sill chiefly in quartzites. Adjacent to and between individual quartz lenses and stringers the granitic rock is highly altered to a quartz-sericite schist.

The analyses shown in Table 55 indicate slight decreases for SiO_2, Na_2O and CO_2 as the veins are approached. Increases are registered by FeO, total Fe, MnO, MgO, K_2O and S. The other constituents show little change or an erratic behaviour.

Table 54. Chemical changes produced by alteration of granite, Red Lake Gold Shore Mine, Red Lake, Ontario

Constituent	Granite*	Carbonate-sericite schist†	Adjusted‡	Gains and losses	Percentage gains and losses
	(%)				
SiO_2	68.1	72.3	72.1	+4.0	+5.9
Al_2O_3	15.1	13.8	13.8	–1.3	–8.6
Fe_2O_3	1.2	0.63	0.63	–0.57	–47.5
FeO	1.8	1.2	1.20	–0.60	–33.3
CaO	2.2	0.83	0.83	–1.37	–62.3
MgO	0.93	0.29	0.29	–0.64	–68.8
Na_2O	4.1	3.0	3.00	–1.10	–26.8
K_2O	3.0	3.0	3.00	+0.00	+0.00
H_2O (total)	0.8	1.2	1.20	+0.40	+50.00
TiO_2	0.42	0.38	0.38	–0.04	–9.5
P_2O_5	0.09	0.03	0.030	–0.06	–66.7
MnO	0.08	0.06	0.060	–0.02	–25.0
CO_2	0.1	0.6	0.60	+0.5	+500.0
S	0.08	0.20	0.20	+0.12	+150.0
C	0.02	0.02			
O≡S	0.03	0.08			
Total	98.0	97.5			
Specific gravity	2.703	2.696			
	(ppm)				
Li	27	12			
Rb	84	81			
Cs	4.4	1.5			
Cu	15	31			
Ag	<0.5	<0.5			
Au	<0.03	0.6			
Be	<20	<20			
Sr	300	100			
Ba	1000	940			
Zn	60	52			
Cd	<0.5	<0.5			
B	15	38			
U	2	2			
Sn	<15	<15			
Pb	<5	<5			
Zr	150	200			
As	<5	6.5			
Sb	<2	<2			
Bi	<10	<10			
V	30	<10			
Nb	<30	<30			
Se	<2	<2			
Te	0.2	0.2			
Cr	<10	<10			
Mo	1	1			
W	2	2			
F	700	800			
Cl	100	nf			
Co	5	<5			
Ni	5	5			
Pt (ppb)	nf	nf			
Pd (ppb)	<5	<5			
Ratios					
K_2O/Na_2O	0.73	1.0			
SiO_2/CO_2	681	120.5			
SiO_2/H_2O+CO_2+S	69.5	36.2			

Note: nf–not found.
*Composite sample of granite.
†Composite sample of carbonate-sericite schist (0–2 ft from vein).
‡Adjusted % = the percentages of carbonate sericite schist† multiplied by the specific gravity factor 0.997 to reduce to a common volume.

Among the trace elements the following are marked by significant increases as the veins are approached: Cu, Ag, Au, Ba, Zn, Cd, B, Pb, As, Bi, W, Co and Ni. Decreases are registered by Ga, U, Ge, V, Cr and Mo.

There is a gradual increase in the K_2O/Na_2O ratio with increasing intensity of alteration. The SiO_2/CO_2 ratio exhibits a general increase whereas the SiO_2/total volatiles shows a marked decrease in the highly altered zone.

Other deposits in acidic intrusive and extrusive rocks. Numerous gold deposits in granitic and rhyolitic rocks exhibit wall-rock alteration features similar to those just described. A brief review of some significant examples follows:

Moore (1912), in a study of the alteration of a Precambrian granite to quartz-sericite (paragonite) schist at the St. Anthony Gold Mine, Sturgeon Lake, Ontario, found a decrease in the amount of silica as the quartz veins were approached and concluded from the analyses, ". . . a comparison of the analyses shows a regular decrease in the silica content of the granite indicating a transfer to the veins."

An increase in the Na_2O, Al_2O_3 and iron contents and decreases in CaO, K_2O and S were also evident. The increased soda content was attributed to the formation of the soda mica, paragonite.

At Porcupine, Ontario, Hurst (1935) noted the changes during alteration of quartz porphyry (Pearl Lake porphyry) to quartz-mica schist given in Table 56. Of particular interest is the general loss in SiO_2 and Na_2O and the increase in K_2O, H_2O and S (pyrite) in the altered phase of the porphyry.

Laipanov and Sher (1971) record an extensive investigation of the wall-rock alterations associated with the gold-quartz veins in the Byeryozovsky gold deposit in the Urals. There, the veins containing quartz and sulphides occur in granite dykes, amphibolites and serpentinites of Paleozoic age.

In the granite dykes the alteration zones range from 10 to 25 cm wide and are composed of ankerite, pyrite, sericite and quartz. The alteration zones are characterized by increased contents of SiO_2, Al, Fe, Mg, Mn, K_2O, S, H_2O and decreased contents of Na_2O and Ca as compared to the fresh granite. The minor element content (W, Sn, Bi, Cu, Zn, Pb, Ag, As, Sb, Rb, Li, F) increased as the vein is approached. The gold content also increased from 0.01 to 0.5 ppm. Mica in the altered rocks is characterized by decreasing amounts of Fe^{3+}, Fe^{2+}, and Mg content towards the vein. No general trend in the ankerite composition was observed.

In the amphibolites, consisting of hornblende, oligoclase and sphene, the alteration zones are up to 10 m wide and can be divided into several zones, from an outer to inner as follows: a hornblende-albite-epidote-chlorite-rutile zone; an albite-chlorite-dolomite-rutile-quartz-pyrite zone; and an ankerite-Fe-muscovite-rutile-quartz-pyrite zone. The altered zones are characterized by an increased content of CO_2, H_2O, K_2O, S and a decreased content of Al, Fe, Mg, Ca, Na and SiO_2. The minor element content (Cu, Zn, Pb, Co) decreases. The gold content in the outer zones decreases from 0.015 to 0.008 ppm; that in the inner zones increases up to 0.03 ppm. The iron content in the amphiboles and mica decreases toward the vein; that of the carbonates increases.

In the serpentinites, consisting of serpentine and chromite, the alteration zones are the widest. From the outer to the

Table 55. Chemical changes produced by alteration of granite, Box Mine, Goldfields, Saskatchewan

Constituents	Granite*	Slightly altered granite†	Adjusted§	Gains and losses of constituents	Percentage gains and losses	Granite*	Highly altered granite‡	Adjusted§	Gains and losses of constituents	Percentage gains and losses
		(%)					(%)			
SiO_2	76.2	74.2	74.4	−1.8	−2.4	76.2	74.3	75.6	−0.6	−0.79
Al_2O_3	12.9	13.7	13.7	+0.8	+6.2	12.9	11.8	12.0	−0.9	−7.0
Fe_2O_3	0.4	0.1	0.1	−0.3	−75.0	0.4	0.4	0.4	–	–
FeO	0.6	1.0	1.0	+0.4	+66.7	0.6	2.7	2.8	+2.2	+366.7
CaO	0.2	0.2	0.2	–	–	0.2	0.2	0.2	–	–
MgO	0.9	0.5	0.5	−0.4	−44.4	0.9	1.4	1.4	+0.5	+55.6
Na_2O	3.8	3.5	3.5	−0.3	−7.9	3.8	3.0	3.1	−0.7	−18.4
K_2O	3.4	4.5	4.5	+1.1	+32.4	3.4	3.9	4.0	+0.6	+17.7
H_2O (total)	0.5	0.3	0.3	−0.2	−40.0	0.5	0.6	0.6	+0.1	+20.0
TiO_2	0.12	0.13	0.13	+0.01	+8.3	0.12	0.12	0.12	–	–
P_2O_5	<0.02	0.02	0.02	–	–	<0.02	<0.02	–	–	–
MnO	<0.02	<0.02	–	–	–	<0.02	0.02	0.02	–	–
CO_2	0.2	<0.1	–	–	–	0.2	<0.1	–	–	–
S	0.14	0.34	0.34	+0.20	+142.9	0.14	1.51	1.54	+1.40	+1000.0
C	<0.1	<0.1				<0.1	<0.1			
O≡S,Cl,F	0.05	0.13				0.05	0.57			
Total	99.5	98.4				99.5	99.4			
Specific gravity	2.640	2.646				2.640	2.687			
				(ppm)						
Li	4	2				4	3			
Rb	105	132				105	115			
Cs	0.5	0.5				0.5	0.4			
Cu	175	25				175	810			
Ag	0.9	<0.3				0.9	5			
Au	0.22	0.02				0.22	16.5			
Be	<2	<2				<2	<2			
Sr	22	30				22	20			
Ba	225	360				225	285			
Zn	12.5	15				12.5	106			
Cd	0.3	<0.3				0.3	1.4			
Hg (ppb)	45	40				45	48			
B	14	88				14	22			
Ga	7.2	5.5				7.2	2.8			
Tl	<2	<2				<2	<2			
Sc	<5	<5				<5	<5			
Y	28	22				28	28			
La	nf	<50				nf	<50			
Th	28	31				28	28			
U	50	6				50	7			
Ge	1.3	1.0				1.3	1.1			
Sn	2.3	1.3				2.3	2.3			
Pb	45	75				45	242			
Zr	215	190				215	207			
As	<10	<10				<10	10			
Sb	<2	<2				<2	<2			
Bi	0.87	<0.5				0.87	4.5			
V	24	<20				24	<20			
Se	<2	<2				<2	2			
Te	<0.2	0.2				<0.2	0.35			
Cr	13	<10				13	<10			
Mo	5	1				5	3.2			
W	3	2				3	3.5			
F	nf	100				nf	100			
Cl	100	100				100	100			
Co	3	4				3	5			
Ni	<1	<1				<1	2			
Pt (ppb)	nf	nf				nf	nf			
Pd(ppb)	nf	nf				nf	nf			
Ratios										
K_2O/Na_2O	0.89	1.29				0.89	1.30			
SiO_2/CO_2	381	>742				381	>743			
SiO_2/H_2O+CO_2+S	90.71	107.54				90.71	34.4			

Note: nf–not found.
*Granite host rock.
†Slightly altered granite (sericite schist) 6 in. from quartz stringers and veins.
‡Highly altered granite (sericite schist) adjacent to quartz stringers and veins (1 in. from veins).
§Adjusted % = percentages of slightly altered granite† and highly altered granite‡ multiplied by the specific gravity factors 1.002 and 1.018 respectively to reduce to a common volume.

Table 56. Chemical changes produced by alteration of quartz porphyry, Porcupine gold area, Ontario*

Constituent	Fresh porphyry†	Altered porphyry‡
	(%)	
SiO_2	63.53	61.56
Al_2O_3	15.68	18.74
Fe_2O_3	1.30	trace
FeO	4.08	0.87
CaO	1.19	2.48
MgO	1.08	1.18
K_2O	2.07	4.00
Na_2O	4.27	2.66
H_2O	1.02	1.85
CO_2	4.42	3.83
TiO_2	0.15	0.41
P_2O_5	0.12	0.18
MnO	0.13	trace
FeS_2 (pyrite)	0.47	2.89
Total	99.51	100.65
Specific gravity	2.760	2.752

Source: Hurst (1935).
*Samples taken in the vicinity of No. 218 vein, 200-foot level, Hollinger Mine.
†Sample typical of the quartz porphyry mass, taken 30 ft from the vein.
‡Waxy phase of altered porphyry from the walls immediately adjoining the vein.

inner zone the following mineral associations can be distinguished: serpentine, talc, magnesite, chromite, dolomite, talc, quartz, pyrite, chromite; and dolomite, quartz, fuchsite. In the altered serpentine rocks the CO_2, Ca and K_2O content decreases, while that of H_2O, Mg, Fe, SiO_2, Al and Na_2O increases. The minor element content (Cu, Zn, Pb) in the outer zones decreases, that in the inner zones increases. Gold content in the altered serpentine decreased from 0.015 to 0.004 ppm.

Laipanov and Sher (1971) concluded that during the process of wall-rock alteration at the Byeryozovsky deposit various exchange reactions take place. Thus, as CO_2, S and K_2O are fixed, SiO_2, Al, Mg and Fe are carried out of the basic and ultrabasic rocks and introduced and precipitated in the altered zones in the granite. H_2O is removed from the serpentinites and supplied to the alteration zones in the granite and amphibolites. Ca is introduced into the altered serpentinites. During the alteration of the basic and ultrabasic rocks Au, Cu, Zn and Pb are removed and partly redistributed. An increase in the concentration of these elements takes place during alteration of the granite. The gold content of the quartz veins can to a considerable degree be related to the extraction of gold from the wall-rocks.

Present day alteration of acidic rocks in the thermal areas of New Zealand has been extensively studied. Some of the waters carry gold in addition to various other elements such as arsenic, antimony and thallium (*see* Chapter II). The thermal waters and their effects on the rocks in the Waiotapu geothermal field have been studied by Healy *et al.* (1963). There, the country rocks are mainly of rhyolitic composition. The chemistry of the waters (steam) is complicated and details should be sought in the original paper. Briefly the waters (steam) are near neutral sulphate-chloride, chloride and bicarbonate-chloride waters containing considerable amounts of constituents such as K, Na, F, B, As, CO_2 and H_2S. At the surface the waters are acid sulphate waters due to surficial oxidation of H_2S. During the alteration of the rhyolitic rocks the quartz is largely inert whereas the magnetite, pyroxene, amphibole and biotite tend to be completely altered. Plagioclase is altered in some sections and relatively unaltered in others. The susceptibility to alteration in decreasing order is: magnetite, pyroxene and hornblende, biotite, plagioclase and quartz. The alteration products are varied, including quartz, opal, hydromica, potash feldspar, albite, wairakite (a zeolite), calcite, pyrite and alunite, the last occurring only in near surface rocks. To generalize, two types of hypogene alteration have been recognized – potassium silicate alteration and albitization associated with the formation of calcite and occasionally with wairakite. Hydromica and K-feldspar are the characteristic minerals of the potassium silicate alteration; these minerals replace the primary soda-lime plagioclase. In more basic rocks, such as andesite, calcite may be associated with the hydromica. The albitization results from the conversion of primary soda-lime plagioclase into albite; this process is commonly associated with the crystallization of calcite and occasionally with the formation of wairakite.

At Waiotapu the zoning of alteration is complex depending on rock types, fissuring and other features. Details should be sought in the original paper by Healy *et al.* (1963). An alteration zoning was also observed by Steiner (1953) at Wairakei in the rhyolitic tuffs and associated sedimentary rocks. Four zones of progressive alteration were distinguished. A superficial zone of sulfuric acid leaching is followed at depth by a zone of argillization. On the zone of argillization is superimposed a zone of zeolitization at shallow depth, and a zone of feldspathization lies at deeper levels.

Summary of wall-rock alteration effects adjacent to gold deposits in acidic (intrusive and extrusive) rocks. The principal types of wall-rock alteration associated with gold deposits in acidic intrusive and extrusive rocks are sericitization, silicification, alunitization and pyritization. Where sericitization and alunitization are marked, there is generally a marked loss of SiO_2 and appreciable gains of H_2O and S. Where silicification is a feature there is often, but not always, an addition of SiO_2. An increase in K_2O and a decrease in Na_2O is common in altered zones in acidic rocks. The Al_2O_3, CaO, MgO and iron oxides show a varied habit. CO_2 is commonly added, especially where carbonatization is extensive. In some wall-rock alteration zones in acidic rocks, however, the intensively sericitized zones next to the quartz veins and lodes are depleted in CO_2.

The ratio K_2O/Na_2O commonly shows an increase with increasing alteration and proximity to gold-quartz veins in acidic rocks, reflecting the extensive development of sericite, potash feldspar, potash-bearing clay minerals or alunite. Where albitization or paragonitization is a feature of the alteration processes the K_2O/Na_2O ratio may remain constant or show a slight decrease. The SiO_2/CO_2 ratio exhibits a variable habit but commonly increases with increase in the intensity of alteration. The SiO_2/total volatiles usually shows a consistent decrease as the veins are approached.

Gold-quartz veins and silicified zones dominantly in sedimentary rocks. Most deposits of this type are quartz veins and lenses dominantly in greywacke quartzite, shale, argillite, phyllite and slate sequences. Some deposits are localized in calcareous shales, argillites and slates. These include two types – stockworks and bedded disseminated deposits.

Gold deposits in sedimentary rocks range in age from

Precambrian to Tertiary. Characteristic examples of the Precambrian type occur in the Yellowknife district, Northwest Territories. There, the deposits are saddle reefs and quartz veins in slates, argillites and greywackes. Similar deposits occur in a number of other Archean gold belts in Canada and elsewhere throughout the world. Paleozoic examples are widespread in Victoria, Australia, particularly in the Bendigo district where the classic deposits are saddle reefs and legs in folded slates and argillites of Ordovician age. Similar deposits occur in the Meguma Group (Ordovician) of Nova Scotia and elsewhere throughout the gold-belts of the world where shales, argillites, slates and greywackes are the host rocks.

Cenozoic gold deposits in sedimentary rocks are relatively common in regions where Tertiary volcanism and hot-spring activity is marked, as in Japan, U.S.S.R., New Zealand and western U.S.A. The bedded disseminated deposits in calcareous shales at Carlin and Cortez in Nevada are of Tertiary age.

Most gold deposits in shales, argillites, phyllites, slates, quartzites and greywackes lack well-defined alteration zones, the quartz veins being more or less frozen against the unaltered rock. Some deposits exhibit a minimal degree of alteration, and a few are marked by characteristic alteration zones.

Yellowknife district, Northwest Territories (Precambrian: Archean age). The majority of quartz veins and pegmatites enclosed by argillites, slates and greywacke in the Yellowknife sedimentary area have no well-defined alteration zones. Some pegmatites and quartz veins, however, are marked along their borders by a development of tourmaline indicating an introduction of boron. A few quartz veins such as those at the Burwash prospect have a thin alteration zone marked by slight bleaching of greywacke on weathered surfaces, caused by the development of sericite and secondary feldspar in the fresh rock. An analysis of a composite sample from this type of alteration zone is given in Table 57. Figure 15 shows the differences graphically.

The alteration of the greywacke has resulted in the introduction of some silica, water, potash and sulphur and the removal of some alumina, calcium, ferrous iron and carbon dioxide.

At the Ptarmigan deposit, where gold-bearing quartz veins and lenses up to 20 ft thick occur in greywacke, quartzite and slate, the alteration is minimal and not immediately obvious except where narrow zones of black tourmaline border parts of the quartz veins.

The analyses given in Table 58 and graphed in Figure 16 indicate that the chemical effects are localized close to the vein, generally within the zone ranging from 1 to 6 in from the contact. Slight decreases are registered by SiO_2 and Na_2O, and there are small to significant increases in Al_2O_3, total Fe, CaO, MgO, K_2O, H_2O, TiO_2, P_2O_5, CO_2 and S.

Among the trace elements nearly all show an increase as the veins are approached. Marked increases are notable for Cs, Cu, Ag, B, Ga, Sc, As, V, Te, Co, Ni, Zn, Cd, Sn, Pb and W. F shows a significant increase toward the vein; Cl registers a decrease.

It will be noted that there is a general increase in the K_2O/Na_2O ratio and a general decrease in the SiO_2/CO_2 and SiO_2/total volatiles as the gold-quartz veins are approached. It also should be pointed out that the SiO_2/CO_2 ratios in deposits of this type are commonly erratic, and wide fluctuations are frequently encountered, presumably because of the small amounts of CO_2 (carbonatization) present in most deposits.

Table 57. Chemical changes produced by alteration of greywacke, Burwash prospect, Yellowknife, Northwest Territories

Constituent	Composite sample of greywacke	Composite sample of alteration zone	Gains and losses in constituents
		(%)	
SiO_2	62.61	64.48	+1.87
Al_2O_3	16.78	16.22	−0.56
Fe_2O_3	0.38	0.84	+0.46
FeO	5.81	4.95	−0.86
CaO	2.80	0.50	−2.30
MgO	2.99	3.09	+0.10
Na_2O	3.79	3.69	−0.10
K_2O	1.56	2.12	+0.56
H_2O+	1.94	2.77	+0.83
H_2O-	0.12	0.15	+0.03
TiO_2	0.68	0.66	−0.02
P_2O_5	0.16	0.14	−0.02
MnO	0.08	0.06	−0.02
CO_2	0.10	0.00	−0.10
S	0.14	0.20	+0.06
Cl	0.02	0.02	+0.00
Cr_2O_3	0.01	0.03	+0.02
C	0.13	0.10	−0.03
Total	100.10	100.02	–
Less O≡S,Cl	0.06	0.09	–
Net total	100.04	99.93	–
Specific gravity	2.73	2.73	–

Argosy Gold Mines, Casummit Lake area, Ontario (Precambrian: Archean age). The Argosy Gold Mine exploited quartz veins in Archean greenstones and a band of greywackes and slates on the north shore of Casummit Lake, Ontario. The minerals present in the veins were essentially quartz with variable amounts of albite, carbonate, arsenopyrite, pyrrhotite, pyrite, chalcopyrite, sphalerite, galena and native gold. Horwood (1938), who studied the wall-rock alteration at this mine, concluded that the alteration took place at two different times: carbonatization related to an early generation of quartz, and sericitization associated with a second generation of quartz. He gives analyses of greywacke, slate and seriticized slate which indicate that during the alteration of the sediments as much as 18 per cent of SiO_2 was lost in the alteration zones some of which are up to 1 ft wide flanking the quartz veins. Other constituents exhibit a variety of changes – potash shows a marked gain, soda a slight increase, sulphur and arsenic marked increases, and carbon dioxide and water decrease.

Caribou Mine, Nova Scotia (Meguma Group: Devonian age). The Caribou Mine is in the Caribou gold district of Nova Scotia about 45 mi northeast of the city of Halifax. The mine was worked by the Consolidated Mining and Smelting Company (Cominco) from 1934 until 1946. The geology is described by Bell (1948).

The orebodies are leg reefs and irregular quartz lenses and stringers in a drag folded zone on the limbs of a fold in quartzites, greywackes and slates of the Meguma Group. The quartz is frozen against the sedimentary rocks, and there are few visible signs of any wall-rock alteration except the

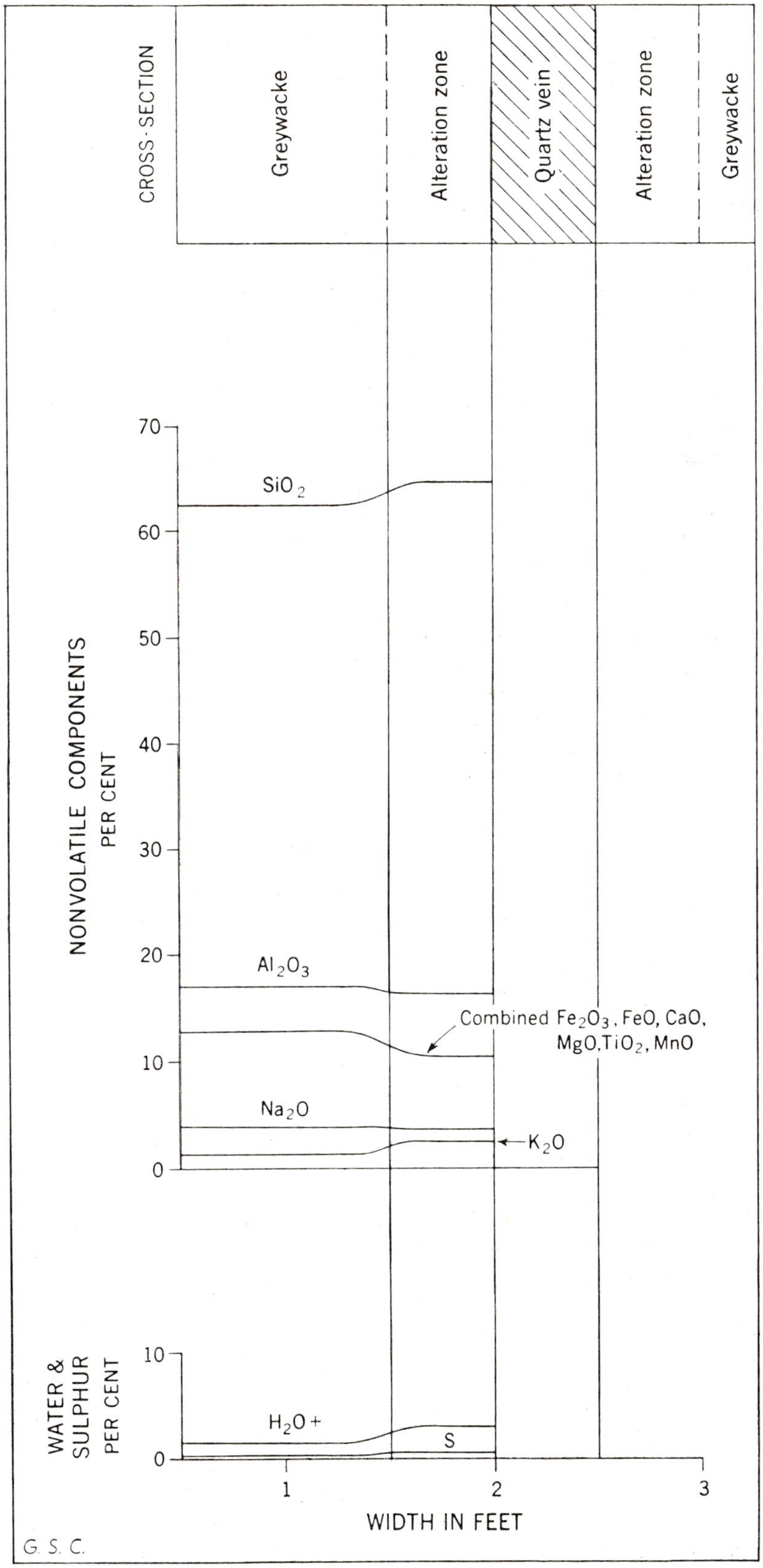

Figure 15. Chemical changes produced by alteration of greywacke, Burwash prospect, Yellowknife, Northwest Territories (*see* Table 57).

Table 58. Chemical changes produced by alteration of greywacke, Ptarmigan Mine, Yellowknife, Northwest Territories

Constituents	Grey-wacke*	Grey-wacke†	Adjusted''	Gains and losses	Percentage gains and losses	Grey-wacke*	Grey-wacke‡	Adjusted''	Gains and losses	Percentage gains and losses	Grey-wacke*	Grey-wacke§	Adjusted''	Gains and losses	Percentage gains and losses
		(%)					(%)					(%)			
SiO_2	62.9	62.4	63.02	+0.12	+0.191	62.9	53.1	54.91	−7.99	−12.70	62.9	51.3	51.97	−10.93	−17.38
Al_2O_3	16.4	17.8	17.98	+1.58	+9.63	16.4	20.9	21.61	+5.21	+31.77	16.4	22.2	22.49	+6.09	+37.13
Fe_2O_3	1.6	1.3	1.31	−0.29	−18.13	1.6	4.1	4.24	+2.64	+165.0	1.6	1.5	1.52	−0.08	−5.00
FeO	4.8	5.1	5.15	+0.35	+7.29	4.8	4.2	4.34	−0.46	−9.58	4.8	5.7	5.77	+0.97	+20.21
CaO	2.0	1.7	1.72	−0.28	−14.00	2.0	2.7	2.79	+0.79	+39.50	2.0	3.9	3.95	+1.95	+97.50
MgO	3.0	3.1	3.13	+0.13	+4.33	3.0	3.0	3.10	+0.10	+ 3.33	3.0	3.1	3.14	+0.14	+4.67
Na_2O	3.3	2.0	2.02	−1.28	−38.79	3.3	2.8	2.90	−0.40	−12.12	3.3	3.1	3.14	−0.16	−4.85
K_2O	2.2	4.0	4.04	+1.84	+83.64	2.2	2.7	2.79	+0.59	+26.82	2.2	4.5	4.56	+2.36	+107.27
H_2O (total)	1.4	1.9	1.92	+0.52	+37.14	1.4	2.0	2.07	+0.67	+47.86	1.4	2.3	2.33	+0.93	+66.43
TiO_2	0.66	0.69	0.697	+0.037	+5.61	0.66	0.95	0.982	+0.322	+48.79	0.66	1.09	1.10	+0.44	+66.67
P_2O_5	0.16	0.17	0.172	+0.012	+7.5	0.16	0.33	0.341	+0.181	+113.13	0.16	0.16	0.162	+0.002	+1.25
MnO	0.07	0.06	0.061	−0.009	−12.86	0.07	0.10	0.103	+0.033	+47.14	0.07	0.07	0.071	+0.001	+1.43
CO_2	<0.1	<0.1				<0.1	0.10	0.103			<0.1	0.1	0.101		
S	0.06	0.42	0.424	+0.364	+606.7	0.06	0.96	0.993	+0.933	+1555	0.06	1.56	1.58	+1.52	+2533
Total	98.6	100.7				98.6	97.7				98.6	100.1			
Specific gravity	2.756	2.783				2.756	2.850				2.756	2.792			
							(ppm)								
Li	52	45				52	51				52	54			
Rb	68	33				68	72				68	80			
Cs	4	4				4	21				4	14			
Cu	47	69				47	157				47	237			
Ag	<0.4	<0.4				<0.4	<0.4				<0.4	1.6			
Au	0.02	0.07				0.02	0.13				0.02	8.5			
Be	<2	<2				<2	<2				<2	<2			
Sr	340	210				340	280				340	390			
Ba	440	1000				440	520				440	720			
Zn	81	148				81	316				81	474			
Cd	1.3	1.3				1.3	1.9				1.3	14.4			
Hg (ppb)	4	4				4	7				4	6			
B	30	50				30	1300				30	20			
Ga	14	17				14	18				14	19			
In	<0.5	<0.5				<0.5	<0.5				<0.5	<0.5			
Tl	<1	<1				<1	<1				<1	<1			
Sc	13	15				13	24				13	31			
Y	<20	<20				<20	<20				<20	<20			
Yb	<4	<4				<4	<4				<4	<4			
U	0.7	1.2				0.7	0.8				0.7	1.7			
Ge	<0.5	0.55				<0.5	0.92				<0.5	<0.5			
Sn	1.2	18				1.2	16				1.2	7.6			
Pb	17	76				17	195				17	424			
Zr	160	170				160	190				160	220			
As	6	88				6	3500				6	50			
Sb	1	2				1	2				1	3			
Bi	<0.5	<0.5				<0.5	<0.5				<0.5	1.8			

[illegible]	325	599	325	827	325	713
Cl	300	100	300	200	300	100
Co	23	26	23	31	23	25
Ni	66	65	66	98	66	79
Pt (ppb)	<30	<20	<30	<20	<30	<60
Pd (ppb)	<4	<2	<4	4	<4	<8
Ratios						
K_2/Na_2O	0.7	2.0	0.7	1.0	0.7	1.5
SiO_2/CO_2	>629	>624	>629	531	>629	513
SiO_2/H_2O+CO_2+S+As	41.6	26.2	41.6	15.6	41.6	13.0

Note: nf-not found. Numbers in parentheses refer to sample in Figure 16.
*Composite chip sample of greywacke 2 to 35 ft from vein, Ptarmigan Mine (1).
†Composite chip sample of greywacke 6 to 18 in. from vein, Ptarmigan Mine (2).
‡Composite chip sample of greywacke 1 to 6 in. from vein, Ptarmigan Mine (3).
§Composite chip sample of greywacke at vein contact, Ptarmigan Mine (4).
||Adjusted % = percentages of greywacke‡ and greywacke§ multiplied by the specific gravity factors 1.010, 1.034 and 1.013 respectively to reduce to a common volume.

development of a selvage of chlorite in places and a general dissemination of arsenopyrite and pyrite outward from the vein for 1 ft or so. This dissemination of arsenopyrite near gold-bearing saddle reefs, legs and irregular lenses of quartz is common in all of the gold deposits of the Meguma Group in Nova Scotia and in fact is common in practically all gold-quartz veins in sequences of greywackes, quartzites, slates and argillites.

As shown by the analyses in Table 59 there is a slight loss of SiO_2, H_2O and MnO as the veins are approached and significant gains of Al_2O_3, Fe_2O_3, FeO, total Fe, CaO, MgO, TiO_2, P_2O_5, CO_2 and S. Na_2O exhibits little change whereas K_2O shows a significant increase.

The trace elements have a rather erratic behaviour, but general increases are registered by Rb, Cs, Ag, Au, Sr, Ba, Zn, Sc, Y, U, Zr, As, V, Cr and Mo. Li and Co show a slight decrease as the vein is approached. Cu, Cd, Hg and Ni are particularly erratic. F increases toward the vein; Cl shows no tendency in this direction.

Lena goldfields, Bodaibo synclinorium, U.S.S.R. (Upper Proterozoic age). In this field the country rocks are Upper Proterozoic polymict sandstones and schistose siltstones falling into the greenschist facies. For the auriferous quartz veins in sandstones Konovalov (1973) found the alteration zonation to be as follows: (1) virtually unaltered rocks composed of quartz, carbonates (dolomite, siderite), plagioclase (andesine, albite), sericite and chlorite and lesser amounts of microcline, pistacite (epidote) and hornblende; (2) slightly altered rocks (2–5 m from vein contacts) with the same mineralogy as above with the omission of chlorite, pistacite and hornblende; (3) highly altered rocks composed mainly of quartz, ankerite, calcite and sericite; and (4) auriferous quartz veins.

Material balance calculations show that during alteration SiO_2 and Na_2O have been leached and K_2O and CO_2 introduced. The other components registered little change. Konovalov (1973) concluded that the silica migrated into the fissure cavities where it crystallized as quartz.

Muruntau deposit, Uzbek, S.S.R. (Paleozoic age). The Muruntau deposit is an en echelon system of thick gold-bearing quartz veins, bordered by stockworks of gold-bearing quartz-sulphide stringers and veins. The host rocks are calcareous siltstones, shales and various types of schists, all of Paleozoic age. The wall-rock alteration associated with this deposit is discussed in detail by Rakhmatullaev and Sher (1969) and Baimukhamedov *et al.* (1975). A short description of the orebodies is given in the descriptions of deposits in this chapter.

The alteration in the Muruntau deposit is considerably different than that found in most gold deposits in sedimentary rocks. Briefly, it is a skarn type of alteration on which is superimposed a quartz-microcline metasomite.

An early preore alteration of the country rocks is the most widespread. During this stage the chlorite-sericite aggregate of the matrix of the siltstone was converted into a mass of fine flaky biotite, and rocks originally rich in carbonate are marked by developments of actinolite and less often of pyroxenes. The altered rocks have more compact structures than the primary ones and are marked by a conchoidal fracture. Also typical of these rocks is a banded coloration due

Table 59. Chemical changes adjacent to gold-quartz orebodies, Caribou Mine, Caribou gold district, Nova Scotia

Constituent	Quartzite*	Quartzite†	Adjusted**	Gains and losses	Percentage gains and losses	Quartzite*	Quartzite‡	Adjusted	Gains and losses	Percentage gains and losses	Quartzite*	Quartzite§
		(%)					(%)					(%)
SiO_2	77.8	71.4	72.3	−5.5	−7.1	77.8	72.7	73.0	−4.8	−6.2	77.8	73.0
Al_2O_3	10.8	11.0	11.1	+0.3	+3.1	10.8	11.2	11.2	+0.4	+3.7	10.8	11.6
Fe_2O_3	0.5	0.6	0.6	+0.1	+21.4	0.5	0.5	0.5			0.5	0.5
FeO	1.7	2.9	2.9	+1.2	+70.6	1.7	2.8	2.8	+1.1	+64.7	1.7	2.9
CaO	0.4	1.8	1.8	+1.4	+350.	0.4	1.8	1.8	+1.4	+350.	0.4	1.8
MgO	0.8	1.1	1.1	+0.3	+37.5	0.8	1.1	1.1	+0.3	+37.5	0.8	1.1
Na_2O	2.8	2.7	2.7	−0.1	−3.6	2.8	2.6	2.6	−0.2	−7.1	2.8	2.6
K_2O	1.5	2.2	2.2	+0.7	+46.7	1.5	2.2	2.2	+0.7	+46.7	1.5	2.3
H_2O (total)	1.4	1.2	1.2	−0.2	−14.3	1.4	1.2	1.2	−0.2	−14.3	1.4	1.3
TiO_2	0.61	0.64	0.65	+0.04	+6.6	0.61	0.63	0.63	+0.02	+3.3	0.61	0.66
P_2O_5	0.05	0.27	0.27	+0.22	+440.	0.05	0.14	0.14	+0.09	+180.	0.05	0.28
MnO	0.17	0.13	0.13	−0.04	−23.5	0.17	0.13	0.13	−0.04	−23.5	0.17	0.13
CO_2	0.1	2.5	2.5	+2.4	+2400.	0.1	2.7	2.7	+2.6	+2600.	0.1	2.7
S	<0.02	0.18	0.18			<0.02	0.18	0.18			<0.02	0.12
C												
O≡Cl,F,S,	0	0.08	0.08			0.	0.08	0.08			0	0.06
Total	98.6	98.6				98.6	99.8				98.6	101.0
Specific gravity	2.706	2.738				2.706	2.718				2.706	2.945
							(ppm)					
Li	27	20				27	19				27	19
Rb	65	90				65	88				65	89
Cs	2.6	3.6				2.6	3.4				2.6	3.7
Cu	36	4				36	23				36	25
Ag	0.7	0.9				0.7	1.0				0.7	1.1
Au	<0.01	0.01				<0.01	0.01				<0.01	0.01
Sr	130	160				130	170				130	170
Ba	280	390				280	400				280	410
Zn	37	43				37	45				37	41
Cd	1.9	1.4				1.9	1.2				1.9	1.4
Hg (ppb)	25	15				25	19				25	22
B	15	10				15	16				15	16
Ga	16	15				16	14				16	15
Sc	12	14				12	13				12	12
Y	<20	<20				<20	21				<20	30
Yb	<4	<4				<4	<4				<4	<4
U	0.1	0.1				0.1	0.3				0.1	<0.1
Pb	20	20				20	20				20	20
Zr	380	290				380	340				380	360
As	15	1250				15	1250				15	1500
Sb	<1	<1				<1	<1				<1	<1
Bi	<0.2	<0.2				<0.2	<0.2				<0.2	<0.2
V	43	51				43	53				43	60
Se	nd	nd				nd	nd				nd	nd
Te	<0.5	<0.5				<0.5	<0.5				<0.5	<0.5
Cr	43	62				43	55				43	48
Mo	1	1				1	1				1	3
W	<2	<2				<2	<2				<2	<2
F	<200	300				<200	200				<200	200
Cl	<200	<200				<200	<200				<200	200
Co	20	20				20	18				20	18
Ni	26	26				26	26				26	26
Pt	nd	nd				nd	nd				nd	nd
Pd	nd	nd				nd	nd				nd	nd
Ratios												
K_2O/Na_2O	0.54	0.81				0.54	0.85				0.54	0.88
SiO_2/CO_2	778.0	28.56				778.0	26.93				778.0	27.04
SiO_2/H_2O+CO_2+S+As	51.47	17.83				51.47	17.29				51.47	17.10

Note: nd–not detected.
*Composite chip sample of quartzite and greywacke, 1–50 ft from vein.
†Composite chip sample of quartzite and greywacke, 10–12 in. from vein.
‡Composite chip sample of quartzite and greywacke, 7–10 in. from vein.
§Composite chip sample of quartzite and greywacke, 5–7 in. from vein.
‖Composite chip sample of quartzite and greywacke, 3–5 in. from vein.
¶Composite chip sample of quartzite and greywacke, 1–3 in. from vein.
** Adjusted % = percentages of quartzite†, quartzite‡, quartzite§, quartzite‖, and quartzite¶ multiplied by the specific gravity factors 1.012, 1.004, 1.088, 1.014 and 1.017 respectively to reduce to a common volume.

Adjusted**	Gains and losses	Percentage gains and losses	Quartzite*	Quartzite‖	Adjusted**	Gains and losses	Percentage gains and losses	Quartzite*	Quartzite¶	Adjusted**	Gains and losses	Percentage gains and losses
(%)				(%)					(%)			
79.4	+1.6	+2.1	77.8	71.3	72.3	−5.5	−7.1	77.8	70.3	71.5	−6.3	−8.1
12.6	+1.8	+16.7	10.8	10.8	11.0	+0.2	+1.9	10.8	11.9	12.1	+1.3	+12.0
0.5			0.5	0.6	0.6	+0.1	+20.0	0.5	0.2	0.2	−0.3	−60.0
3.2	+1.5	+88.2	1.7	2.9	2.9	+1.2	+70.6	1.7	2.9	3.0	+1.3	+76.5
2.0	+1.6	+400.	0.4	1.9	1.9	+1.5	+375.	0.4	1.9	1.9	+1.5	+375.
1.2	+0.4	+50.0	0.8	1.1	1.1	+0.3	+37.5	0.8	1.1	1.1	+0.3	+37.5
2.8			2.8	2.7	2.7	−0.1	−3.6	2.8	2.7	2.8		
2.5	+1.0	+66.7	1.5	2.2	2.2	+0.7	+46.7	1.5	2.2	2.2	+0.7	+46.7
1.4			1.4	1.1	1.1	−0.3	−21.4	1.4	1.1	1.1	−0.3	−21.4
0.72	+0.11	+18.0	0.61	0.65	0.66	+0.05	+8.2	0.61	0.65	0.66	+0.05	+8.2
0.31	+0.26	+520.	0.05	0.32	0.32	+0.27	+540.	0.05	0.32	0.33	+0.28	+560.
0.14	−0.03	−17.6	0.17	0.13	0.13	−0.04	−23.5	0.17	0.13	0.13	−0.04	−23.5
2.9	+2.8	+2800.	0.1	2.7	2.7	+2.6	+2600.	0.1	3.0	3.1	+3.0	+3000.
0.13			<0.02	0.20	0.20			<0.02	0.18	0.18		
0.07			0	0.09	0.09			0	0.08	0.08		
			98.6	98.5				98.6	98.5			
			2.706	2.743				2.706	2.753			
						(ppm)						
			27	22				27	23			
			65	89				65	90			
			2.6	3.7				2.6	3.6			
			36	<4				36	8			
			0.7	1.4				0.7	1.3			
			<0.01	0.01				<0.01	0.01			
			130	170				130	180.			
			280	420				280	480.			
			37	41				37	43			
			1.9	2.1				1.9	1.7			
			25	18				25	20			
			15	13				15	15			
			16	15				16	15			
			12	14				12	11			
			<20	<20				<20	21			
			<4	<4				<4	<4			
			0.1	0.5				0.1	0.5			
			20	25				20	20			
			380	380				380	550			
			15	1500				15	1500			
			<1	<1				<1	<1			
			<0.2	<0.2				<0.2	<0.2			
			43	50				43	56			
			nd	nd				nd	nd			
			<0.5	<0.5				<0.5	<0.5			
			43	60				43	40			
			1	2				1	2			
			<2	<2				<2	<2			
			<200	300				<200	200			
			<200	<200				<200	<200			
			20	16				20	19			
			26	24				26	29			
			nd	nd				nd	nd			
			nd	nd				nd	nd			
			0.54	0.81				0.54	0.81			
			778.0	26.41				778.0	23.43			
			51.47	17.18				51.47	15.87			

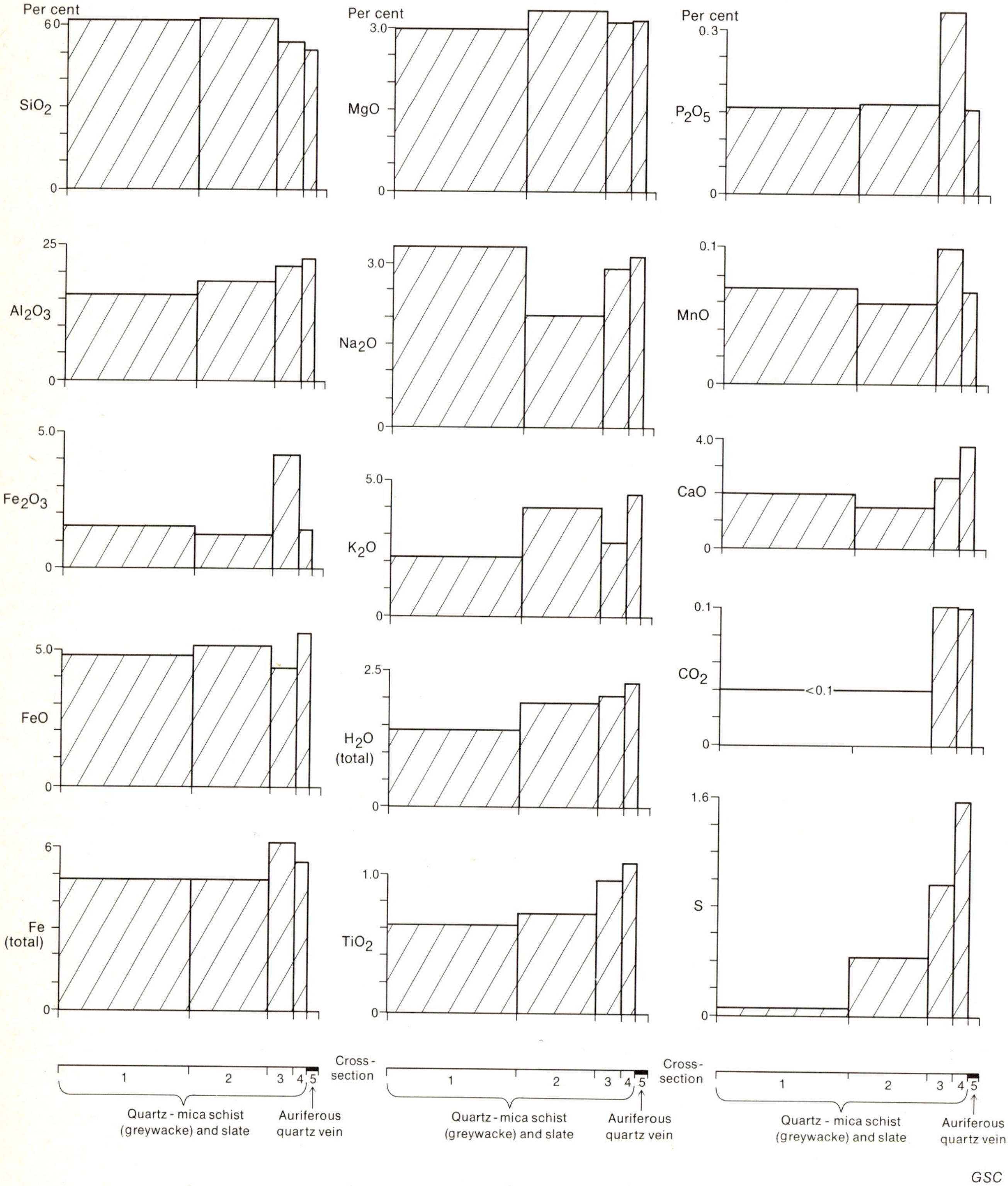

Figure 16. Chemical changes produced by alteration of greywacke, Ptarmigan Mine, Yellowknife, Northwest Territories. (Numbers on cross-section refer to samples described in Table 58.)

to the presence of brown biotite-rich, green actinolite-rich, and grey quartz-rich bands.

This skarn-type alteration has been interpreted as a contact metamorphic phenomenon (conversion of rocks into hornfels) related to an assumed granitoid intrusive body concealed at some depth.

A comparison of the mineral and chemical compositions of the initial and altered rocks reveals that the latter are marked by gains in the contents of magnesium, iron, alkali metals and partly aluminum. Carbon dioxide and calcium were lost from the calcareous rocks during the alteration processes. Silica was apparently redistributed; it shows losses in the initial aluminous rocks and gains in the altered limy rocks. The composition of newly formed pyroxenes are interesting. The pyroxene at shallow depths is diopside; at depths of 150 to 200 m, salite; and from 400 m downward, clinoenstatite. This indicates that the Fe and Mg contents of pyroxene increase depthwise. The only ore element introduced during this period was tungsten, precipitated as scheelite.

Superimposed on the hornfels are quartz-microcline metasomatites developed principally along the contacts of most gold-bearing and sulphide-bearing quartz veins and veinlets. This bordering alteration zone is 2 to 3 cm wide in most places and wider locally.

The quartz-microcline metasomatite is a fine-grained (0.02–0.04 mm) light pink rock consisting of quartz (35–50%), fine-grained microcline (25–50%), commonly replacing biotite, variable amounts of calcite (0.5%; occasionally up to 10%) and apatite (up to 9%). Relict minerals include rutile, leucoxene and zircon. The apatite is haloid free (hydroxylapatite). The enhanced apatite content of the metasomatites is attributed to the presence of phosphorite beds in the host rocks.

A comparison of the compositions of biotitized and actinolitized siltstone (hornfels) with that of the quartz-microcline metasomatites shows that during the formation of the metasomites, SiO_2, Ca, K and CO_2 were introduced and Al, Fe, Mg, Na and H_2O were removed from the hornfels.

The quartz-microcline metasomatites are associated with numerous stringers of biotite, amphibole and pyroxene, all cutting the quartz veins. They have a length of a few metres and their thickness ranges from 1.0 to 1.5 cm. Commonly they have a very irregular shape. The mineral composition of these stringers is the same as the metamorphic rocks they intersect.

Reefton gold lodes, New Zealand (Precambrian or Paleozoic). The Reefton lodes in South Island, New Zealand are quartz veins and stockworks in faults, sheared and crushed zones, and fractures cutting Waiuta Series greywacke, quartzite and phyllite. The mineralization of the quartz bodies includes native gold, pyrite, stibnite, arsenopyrite, chalcopyrite, molybdenite, calcite and rare galena (Gage, 1948).

Near the veins and in the stockworks the greywacke and phyllite are altered to a soft light-coloured rock composed mainly of quartz, sericite, pyrite, arsenopyrite and carbonate. The unaltered greywacke consists chiefly of quartz, feldspar and chlorite; of these minerals the feldspar and chlorite are replaced mainly by sericite and carbonate in the alteration zones. Henderson (1917) gives analyses of the altered and unaltered greywacke and argillite (phyllite), which show little change as alteration proceeds. There is a general loss of silica, in altered zones in greywacke as much as 7 per cent and a general increase in potash as the veins are approached.

Carlin gold deposit, Nevada (Tertiary age). This deposit is one of a number of disseminated bedded deposits in the Carlin district of Nevada. It serves as an example of the type of alteration present in these deposits.

The Carlin deposit is characterized by large disseminated replacement-type ore bodies in the upper beds of the Silurian Roberts Mountains Formation. (*See* the more detailed description of this and similar orebodies in the section on description of deposits in this chapter.) Radtke *et al.* (1972*a*) have described the alteration effects attendant upon mineralization in the Carlin deposit in some detail.

The host rocks for the orebodies are dark- to medium-grey, thin bedded, siliceous, argillaceous, dolomitic limestones containing widely varying amounts of calcite, dolomite, illite and quartz, plus minor kaolin, montmorillonite(?), chlorite, K-feldspar, plagioclase, pyrite, zircon, barite, rutile, sphene and carbonaceous materials.

The primary gold ores are composed of widely varying amounts of quartz, illite, calcite, dolomite and barite, the same minor accessory minerals as are present in the host rocks, plus pyrite, pyrrhotite, arsenopyrite, realgar, orpiment, stibnite, chalcopyrite, sphalerite, galena, cinnabar, molybdenite and several sulphosalt minerals.

The analyses given by Radtke *et al.* (1972*a*) show that the most important change in the rocks during mineralization was the removal of Ca, Mg, CO_3 and the introduction of SiO_2. Aluminum exhibits an enrichment as does also iron, the latter reflecting the introduction of pyrite and smaller amounts of pyrrhotite and arsenopyrite. In addition to SiO_2 and Fe the following elements exhibit significant enrichments in the ores compared with the host rock: Ba(2200 ppm), As(480 ppm), Zn(185 ppm), Sb(130 ppm), B(70 ppm), Cu(35 ppm), Pb(30 ppm), La(30 ppm), Hg(25 ppm), W(18 ppm?), Au(11 ppm), Mo(7 ppm), Se(2 ppm?) and Ag(0.4 ppm?). Approximate enrichment factors for these elements in the primary ores *vs.* the fresh host rocks are Au(>550X), Hg(>300X), Sb(160X), As(>50X), Ba(15X), Zn(13X), B(5X), Mo(>4X), Pb(>4X) and Se(>2X). Ag, W and La are also enriched, but the data sets available preclude the calculation of meaningful enrichment factors.

Other deposits in sedimentary rocks. There are a number of important gold deposits in sedimentary rocks that are especially difficult to classify with respect to wall-rock alteration effects. Some of these have skarn characteristics; others have high-grade metamorphic features. Examples are the Homestake Mine, Madsen Mine, Morro Velho Mine and others.

Gustafson (1933) and Slaughter (1968) describe the wall-rock alteration effects attending the great auriferous quartz lodes in the Homestake Mine at Lead, South Dakota. There, the host rocks (Homestake Formation) are cummingtonite schists bearing garnet, biotite and chlorite, probably derived from a sandy, clayey iron-magnesium carbonate sediment (sideroplesite). Near the quartz orebodies the cummingtonite schists, and in places the sideroplesite schists, are converted to rocks containing essentially chlorite, ankerite, arsenopyrite, pyrrhotite and pyrite. The strongly chloritized rock is generally auriferous. The mineralization stages are complex and are described in more detail in the section on description of deposits in this chapter.

The orebodies of the Madsen Mine in the Red Lake district of Ontario are in silicified, sulphide-bearing (pyrite, pyrrhotite and arsenopyrite) tuffs in which abundant garnets are developed near the orebodies but seldom in the ore. At depth there are orebodies in actinolite schists, which are remarkably low in quartz. Further details are given in the section on descriptions of deposits in this chapter.

The Morro Velho Gold Mine and other similar deposits in the Quadrilátero Ferrífero, Minas Gerais, Brazil are characterized by unusual types of wall-rock alteration that have been variously interpreted. The alteration at the Morro Velho has been called "Lapa Sêca" and consists of an uneven envelope of rock composed essentially of ankerite, albite and quartz with pyrite and pyrrhotite. Some investigators consider this rock to be part of a recrystallized iron-formation. Further details are given in the section on description of deposits in this chapter.

Coarse-grained gold-quartz veins with a pegmatitic aspect in high-grade metamorphic sedimentary schists and gneisses commonly grade imperceptibly into these metamorphic rocks or are frozen against walls of schist and gneiss. In many places a selvage or diffuse zone of tourmaline and/or mica is characteristic; in other places garnet, andalusite, staurolite, cordierite and even corundum may be developed both in the veins and adjacent wall rocks. Such veins are common in the Yellowknife Group sediments and in similar rocks on Outpost Islands, Great Slave Lake, Northwest Territories (Hawley, 1939; Lord, 1951).

Summary of wall-rock alteration effects adjacent to (gold deposits in sedimentary rocks). In greywacke, quartzite, argillite and slate sequences the wall-rock alteration effects are minimal. Commonly there is a slight desilication close to the vein and a development of chlorite and some sericite. Na_2O is usually leached and K_2O added. As the veins are approached increases in As, S, P_2O_5, B and CO_2 are characteristic.

Where gold deposits are developed in limestones, and calcareous quartzite, greywacke and slates marked alteration effects are noted – at Muruntau a skarn-type alteration with a superimposed quartz-microcline metasomite has developed, and in the Carlin-type deposits an extensive silicification, removal of CO_2 and characteristic additions of S, As, Sb, Hg and a host of other chalcophile elements are evident.

> Auriferous veins or deposits may be of any form, may occur in any rock, and may have received their gold from various sources.
> – J. M. Maclaren, 1908

Descriptions of typical gold deposits

The following descriptions treat gold deposits according to the classification outlined at the beginning of this chapter. Typical examples are described, but the descriptions are not detailed. Sufficient references are given, however, to enable the interested reader to pursue the details of each type of deposit in depth. Supplementary references to many of the deposits mentioned are included in the Selected bibliography.

Auriferous porphyry dykes, sills and stocks; auriferous coarse-grained granitic bodies, aplites and pegmatites

The indigenous gold content of these granitic rocks is invariably low as shown by analyses of a variety of types done in our laboratories (Table 60).

Table 60. Gold and silver content of fresh (unmineralized) granitic rocks

Rock type	Au average	Au average	Ag average	Ag
	(ppm)	(oz/ton)	(ppm)	(oz/ton)
Pegmatites	0.003	0.00009	0.05	0.0015
Coarse-grained granitic rocks	0.003	0.00009	0.05	0.0015
Quartz-feldspar porphyry	0.017	0.0005	0.1	0.003
Quartz-feldspar porphyry with indigenous pyrite	0.10	0.003	1.0	0.03

We conclude from these data that normal pegmatites, coarse-grained granites and porphyries cannot be considered potential economic sources for gold and silver. Such bodies may, however, be greatly enriched in both elements where they are sheared, faulted, fractured and otherwise structurally prepared, permitting the localization of quartz veins, silicified and pyritized zones and disseminations. Such granitic bodies occur throughout the world, and some constitute large gold deposits. They are described in category 6 below. A few pegmatites, coarse-grained granites and porphyries contain gold in commercial quantities that the earlier investigators attributed mainly to magmatic segregation processes within the bodies. A few of these are described briefly below.

Weed (1904*a*) summarizes numerous observations made with respect to the presence of gold as a primary constituent of eruptive (igneous) rocks. These include aplites, diorites, basalts and diabase in which he thinks the gold may have been of primary origin. In a later paper he (Weed, 1904*b*) refers to auriferous and argentiferous basaltic (monchiquite) dykes from the Krynka River area in eastern Russia. These dykes consist essentially of augite, biotite, hornblende and olivine in a glassy base. The auriferous dykes are especially rich in fluorite (up to 2.5%) and abundant pyrite. Brock (1904) also records the presence of gold, apparently primary, in a porphyry dyke from the North Fork of the Salmon River in British Columbia.

Lacroix (1922–1923) mentions several occurrences of gold, some with associated hematite, others in quartz in pegmatites in Madagascar, and Spurr (1906, 1923) asserts that gold occurs in quartz pegmatites (alaskites) in Alaska, at Silver Peak, Nevada and elsewhere. Butler *et al.* (1920) describe an unusual veinlike deposit from Gold Hill, Utah that contains native gold in an intergrown mass of quartz, orthoclase, fibrous amphibole, epidote, magnetite, carbonate, tourmaline, powellite, malachite, scheelite, molybdenite and apatite. Knopf (1924) mentions quartz-microcline-tourmaline-mica dykes or veins near Rochester, Nevada that carry coarse native gold, arsenopyrite and pyrite.

The Natas Mine in South-West Africa (Reuning, 1925) is in pegmatite dykes and quartz veins in the apex of a granitic stock. The gold in the pegmatite dykes is probably related to the quartz vein stage rather than to the pegmatite stage. It is associated with scheelite, molybdenite and copper sulphides.

Brammall (1926) mentions the presence of gold (and silver) in two places in the Dartmoor granite, southwest England. Judging from the occurrence of the visible metal as inclusions in both quartz and feldspar, and from the presence of both gold and silver in biotite, he inferred that the two metals were of primary origin. There are, however, quartz stringers in the vicinity of the occurrences suggesting that some of the gold may be introduced. Native gold has also been observed in pegmatites on Dartmoor (Brammall and Harwood, 1924). In the Bittleford pegmatite it occurs as specks and spangles embedded in feldspar and quartz. Associated minerals of interest are tourmaline, ilmenite, apatite and pyrrhotite.

An unusual type of gold deposit in an igneous body occurred on the farm Waaikraal, about 25 mi northeast of Rustenburg. According to Wagner (1927) the gold ore is a magmatic segregation in a sill belonging to the Bushveld Complex. The sill is a pyroxenitic quartz gabbro merging into a pegmatitic quartz gabbro and pegmatitic quartz monzonite enclosing slivers of a banded magnetite-augite rock, which is probably a recrystallized ironstone of sedimentary origin (iron-formation). Having regard to the common occurrence of gold and gold deposits in iron-formations one wonders if this is not simply a case of reconstitution and concentration of gold from the iron-formation induced by magmatic activity.

Shenon and Reed (1934) describe fractured and brecciated pegmatitic material in a shear zone at Orogrande, Idaho County, Idaho that contains quartz, pyrite and gold. Gallagher (1940) mentions a number of occurrences of gold in pegmatite and in quartz bodies containing feldspars.

In the Canadian Malartic Mine in the Malartic district of Quebec, Cooke (1946) mentions that pegmatite veins carrying gold have been found within a porphyry mass. These are commonly much richer than the average siliceous ore of the mine. The veins (pegmatites) consist of quartz intergrown in places with albite, and contain a large assortment of other minerals such as biotite, muscovite, fluorite, rutile, tourmaline, molybdenite, pyrite, chalcopyrite and native gold. Cooke thought that occurrences of such veins (pegmatites) indicated that the gold deposits originated in part from the known bodies of porphyry, and perhaps in part from similar bodies consolidating at greater depths.

Indigenous pyrite, pyrrhotite and arsenopyrite from pegmatites generally contain only traces of gold and silver according to Geological Survey analyses. There are, however, references in the literature to the fact that these sulphides as well as others are often auriferous, values up to 0.1 oz Au/ton and more being reported. It is generally not stated whether these sulphides were original or not; in most places they appear to have been introduced after the consolidation of the pegmatitic bodies.

The descriptions of many gold lodes in pegmatites and granitic rocks indicate that the gold mineralization is usually unrelated to the (magmatic) crystallization stage. The sequence of events is generally the introduction and consolidation of the granitic body followed much later by a number of stages of fracturing, faulting, shearing, alteration and introduction of sulphides, quartz and native gold. A number of examples exhibiting this sequence of events are described in a later section.

Auriferous skarn-type gold deposits

Gold is a frequent constituent of skarn deposits, in which it is commonly more abundant than the literature would indicate. Most skarn deposits yield gold as a byproduct of copper and lead-zinc mining, but there are a number of these deposits that are greatly enriched in gold and silver and are mined essentially for the two precious metals.

The general features of skarn deposits are well known and need not be described in detail here. Most of the deposits occur in highly metamorphosed terranes in which there has been much granitization and injection of granitic rocks. Some deposits occur near the contacts of granitic rocks and have long been called 'contact metamorphic' (Fig. 9); others are developed in favourable reactive beds or zones some distance from granitic contacts.

Most of the gold deposits occur in irregular skarn zones or are associated with skarn minerals developed in limestones, dolomites, carbonate-bearing shales or carbonate-bearing schists. The deposits are irregular and bunchy in form, but many are tabular, following beds that have been selectively replaced. The gold ore shoots in most deposits are commonly highly irregular, their limits being generally only determined by assay methods.

The deposits usually contain a characteristic suite of relatively high temperature Ca-Mg-Fe-silicates such as garnet, epidote, vesuvianite, diopside, tremolite, scapolite, wollastonite and axinite. A lower temperature suite of minerals is also commonly present including serpentine, phlogopite, muscovite, quartz, calcite, dolomite and various sulphides, mainly chalcopyrite, bornite, chalcocite, pyrite, pyrrhotite, arsenopyrite, galena, sphalerite, molybdenite and locally various Ni-Co sulphides and arsenides. Magnetite and hematite occur in many of the deposits and scheelite in a few. Stibnite is rare in most deposits, but occurs in quantity in those of the Bau mining district in Sarawak. The gold minerals include native gold, rarely electrum, and in places various tellurides. Aurostibite has not been reported from skarns.

The elements commonly accompanying gold in skarn deposits are Fe, S, Cu, Ag, Zn, Cd, B, Ga, Ge, Sc, Y, rare-earths, Th, U, Pb, Ti, P, As, (Sb), Bi, Se, Te, Mo, W, (F), (Cl), (Co), (Ni) and (Pt metals). Those in parentheses are relatively rare in deposits enriched in gold. The elements most frequently enriched with gold in skarn deposits are Fe, S, Cu, Ag, Zn, Pb, Mo, As, Bi and Te. There is commonly a positive correlation between Au and Cu in some skarn deposits. Tungsten is a common trace element in gold-bearing skarn deposits. The element belongs to the early phase of skarnification, whereas gold tends to be precipitated late in the mineralization processes. The two elements may, therefore, be negatively correlated, for it is common to find skarn deposits that are rich in tungsten (scheelite) practically devoid of gold and vice versa.

Skarn-type gold deposits have a widespread distribution throughout the world especially in belts of carbonate rocks invaded by diorites, monzonites, granodiorites and granites. Among these gold is particularly enriched in those containing relatively large amounts of arsenopyrite and pyrite. Skarn-type deposits rich in copper, or in copper, lead and zinc, generally carry a little gold and silver. Skarn-type iron deposits containing magnetite, titan-magnetite, and hematite, and the various skarn deposits enriched in tungsten (scheelite), tin (cassiterite), graphite or platinum, generally carry only traces of gold and silver. A few typical examples of skarns enriched in gold are described below:

In the Hedley district of British Columbia two large gold deposits were developed in skarn (Camsell, 1910; Billingsley and Hume, 1941; Dolmage and Brown, 1945) – The Nickel Plate Mine (Lamb *et al.*, 1957) and the French Mine (Lamb, 1957). These two mines worked orebodies developed in skarn near the contact of Cretaceous granodiorite and Triassic limestone, limy argillite, quarztite and tuffs and volcanic flows. In the Nickel Plate area numerous porphyritic gabbro, dioritic and other dykes and sills cut the country into blocks; this effect was not notable in the French Mine. The orebodies occurred all within the skarn and were tabular, pipelike and irregular. In the Nickel Plate Mine the principal mineral was gold-bearing arsenopyrite, often present in massive shoots. Other minerals included chalcopyrite, pyrrhotite and tetradymite. The ore in the French Mine was mainly skarn impregnated with small amounts of gold, bornite, chalcopyrite, pyrrhotite, molybdenite, chalcocite, covellite, safflorite, cobaltite, native bismuth and two tellurides, hedleyite and joseite. The grade of the two mines was approximately 0.4 Au oz/ton (13.7 ppm) and 0.035 Ag oz/ton (1.19 ppm), and the Au/Ag ratio ranged from 4 to 10 in the ores. The copper content averaged 0.04 per cent. In addition to gold and silver some copper was won from the deposits. Some of the ores carried up to 0.6 per cent cobalt.

The highly metamorphosed Precambrian Grenville Group in Canada contains a number of lead-zinc deposits carrying both silver and gold. The best known of these are the Tetreault deposit near Quebec City and the New Calumet deposit on Calumet Island in the Ottawa River. The Tetreault deposit produced considerable zinc, lead, gold and silver from orebodies containing abundant skarn minerals, sphalerite, galena, pyrrhotite, pyrite, chalcopyrite, tetrahedrite, stibnite, arsenopyrite, and grading 7 to 10 per cent Zn, 1 to 3 per cent Pb, 0.09 oz Au/ton and 5 to 9 oz Ag/ton (Dresser and Denis, 1949). The New Calumet deposit, now exhausted, was a tabular body of massive and disseminated sulphides in a highly altered impure limestone now essentially composed of pyroxene, tremolite, feldspar, chlorite, garnet and calcite (Dresser and Denis, 1949). The principal sulphides were pyrite, pyrrhotite, sphalerite and galena. The silver content ranged between 5 and 10 oz Ag/ton and there was about 0.05 oz Au/ton. The silver was closely associated with the galena and was present in various silver-bearing minerals, including tetrahedrite, argentite and polybasite. The gold was mainly in the native form associated with all the sulphides and as minute grains in the silicate gangue. One variety of ore, characterized by fine-grained pyrite and pyrrhotite in coarse calcite with feldspar and tremolite, was relatively rich in native gold. This native gold appeared to have been introduced late in the mineralization history of the deposit. It carried minor amounts of silver.

In the Richardson Mine at Eldorado, Ontario, the first gold mine discovered in Ontario (Vennor, 1870), ore was obtained from a skarn zone developed in carbonate rocks near the contact of a granite-syenite stock (Wilson, 1965). The gold occurs in pockets and small seams in skarn composed mainly of diopside, amphibole, epidote, mica, chlorite, talc, coarse calcite, tourmaline and pyrite. In some parts of the deposit brannerite is accompanied by gold, the precious metal being younger, occurring mainly in seams and small fractures in the uraniferous mineral (Steacy *et al.*, 1974*a,b*). In other parts of the deposit the gold occurs in thucholite as finely dispersed particles in a manner such as to suggest exsolution from a hydrocarbon gel (Boyle and Steacy, 1973).

The Richardson Mine is one of many in the eastern Ontario gold belt (Miller and Knight, 1914; Hurst, 1927; Cooke, 1946). Most of these deposits are irregular quartz veins or lodes with considerable amounts of arsenopyrite and pyrite in high grade amphibolites, gneisses and schists. Just north of the Richardson Mine at Bannockburn, auriferous quartz-carbonate veins with much pyrite and small amounts of galena, sphalerite and altaite are present (Rickaby, 1932). The gold occurs in the native form and is also present in small amounts of krennerite.

The ages determined for the deposition of brannerite in the Richardson Mine as given by lead isotopes are as follows: Pb^{206}/U^{238}, 936±28 m.y.; Pb^{207}/U^{237}, 957±29 m.y.; and Pb^{207}/Pb^{206}, 1001±30 m.y. These may be compared with ages carried out on the nearby Deloro stock by Wanless and Loveridge (Rubidum-strontium isochron age studies, Report 1; Geol. Surv. Can. Pap. 72-23). A Rb-Sr wholerock isochron age gave 1059±46 m.y. The K-Ar ages of two riebeckites from the stock gave 875 and 989 m.y.

Gold bearing skarn deposits are common in other parts of the world. Emmons and Calkins (1913) describe a rich gold-bearing contact metamorphic deposit at Cable in Montana. There, granite (granodiorite) contains a block of limestone in which skarn and gold-bearing orebodies are developed. The principal minerals in the ore are calcite, quartz, diopside, actinolite, garnet, biotite, sericite, chlorite, magnetite, specularite, pyrite, pyrrhotite, chalcopyrite, arsenopyrite, and rarely galena and sphalerite. The gold is native, frequently occurring in masses of several pounds. It also occurs finely disseminated in the iron sulphides, in calcite and in quartz. The silver content of the bullion was low.

At Ouray, Colorado, Burbank (1930) mentions gold-bearing contact metamorphic deposits in the Wanakah Mine. The ores include different mineralogical varieties ranging from low-grade ores containing magnetite, garnet and other lime silicates, to massive pyritic ores and galena ores.

Roberts *et al.* (1971) refer to the Battle Mountain (Copper Canyon and Copper Basin) pyrometasomatic deposits in north-central Nevada. The country rocks are silicic and volcanic rocks of the upper plate of the Roberts Mountains thrust (Scott Canyon, Valmy and Harmony formations), which was overlapped by late Paleozoic rocks (Battle Formation, Antler Peak Limestone and the Edna Mountain Formation) and overridden by the Golconda thrust plate. The orebodies at Copper Canyon were mainly in calcareous conglomerate and hornfels of the Battle Formation; adjacent to the Copper Canyon stock these rocks were metamorphosed to calc-hornfels. During late stage metallization processes the silicates were altered to chlorite and clay minerals, and sulphides including pyrite, pyrrhotite, arsenopyrite, chalcopyrite and minor sphalerite and galena were introduced. This ore ranged in grade from 0.5 to 1 per cent Cu, and contained 0.10 to 0.25 oz Au/ton and about 1.5 oz Ag/ton. At Copper Basin, gold ore occurred in metamorphosed limestone beds of the Harmony Formation and in the overlying Battle Formation adjacent to intrusive bodies. This ore contained remnants of garnet, diopside, epidote, pyrite, pyrrhotite and chalcopyrite in a groundmass of chlorite and clay minerals. The copper in the ore occurred mainly as supergene cuprite and chalcocite. The ore mined averaged 2.96 per cent Cu, 0.49 oz Au/ton and 2.15 oz Ag/ton.

In Nicaragua, Central America the Rosita Mine obtained gold-copper ore from a deeply weathered garnet-epidote skarn developed in Cretaceous marbles near a small Tertiary dioritic intrusion (Bevan, 1973). A description of the oxidized and primary orebodies is given in Chapter IV. The La Luz Mine, some 23 mi southwest of the Rosita Mine (Plecash *et al.*, 1963) was somewhat similar. The ore was in silicified epidote-rich skarn and contained much disseminated pyrite, chalcopyrite and fine-grained gold. The average grades were about 0.107 oz Au/ton, 0.032 oz Ag/ton and 0.4 per cent Cu.

In the Soviet Union a number of skarn deposits are mined essentially for gold, particularly in the Altai-Sayan. These have been described in detail by Vakhrushev (1972). The Sinyukhinskoe in the Gorny Altai, the Lebedskoe in the Gornaya Shoriya and the Natal'evskoe deposits in the Kuznetskii Alatau are in Paleozoic limestones, limy shales and argillites, and dolomites intruded by lower Paleozoic intrusions of the granodiorite-diorite clan. The deposits all occur in skarns containing mainly wollastonite, garnet and diopside mineralized with magnetite, bornite, chalcocite, chalcopyrite, arsenopyrite, pyrite, pyrrhotite, galena, tetrahedrite-tennantite, sphalerite, gold, the tellurides, melonite and altaite, and a number of rarer minerals. The quantity of sulphides is low in the deposits, generally less than a few per cent. Native bismuth occurs in the Natal'evskoe deposit. In a similar deposit, the Tardanskoe, in Tuva, U.S.S.R. also described by Vakhrushev, the gold is rich in the magnetite parts of the skarn especially where serpentine and phlogopite are present. In most of the deposits the gold is relatively fine averaging about 930; the gold in the Natal'evskoe deposit is relatively rich in silver. Electrum was also noted in this deposit. Details of the distribution of gold in the Tardanskoe and Natal'evskoe deposits are given by Kudryavteseva (1969) and Vasil'ev (1970) respectively. Other skarn deposits in U.S.S.R. in Kirghizia, Gornyi Altai and elsewhere are described by Andreev *et al.* (1970), Davletov *et al.* (1972) and Shcherbakov *et al.* (1972).

Small deposits of siliceous gold ores with high temperature calc-silicates occur in the Bau mining district, Sarawak, Malaysia (Wolfenden, 1965; Pimm, 1967). The ores occur as irregular pods or lenses along joints and fractures in Upper Jurassic marbles invaded by mid-Miocene acidic porphyry stocks, dykes and sills. The predominant calc-silicate minerals are wollastonite, grossularite, vesuvianite, andradite, diopside, epidote and rarely plagioclase; other gangue minerals include quartz, calcite, pyrite, arsenopyrite, sphalerite, chlorite and apatite. Stibnite is abundant in some of the lenses, in places in commercial quantities. The gold is coarser in some of the lenses, whereas in others it is finely disseminated. The tenor of the ores is about 0.5 oz Au/ton, with richer ores running 1 to 4 oz/ton.

Skarns constitute the most predominant type of mineral deposit in the Suian district of Korea (Watanabe, 1943). Most are found near the contact of the Suian granite. The orebodies are irregular in form, consisting of lime-magnesian-silicate gangue minerals, pyrite, chalcopyrite, pyrrhotite, arsenopyrite, loellingite, bismuthinite, tetradymite, sphalerite, galena, molybdenite and gold. The greater part of the gold and copper production of the Suian mining district has come from orebodies of this type. In some skarn deposits borate minerals, such as kotoite, ludwigite, etc., are abundantly found as gangue minerals. The age of the mineralization is thought to be late Mesozoic.

Gold-silver and silver-gold veins, stockworks, lodes, mineralized pipes and irregular silicified bodies, in fractures, faults, shear zones, sheeted zones and breccia zones essentially in volcanic terranes

Representatives of this type of deposit are widespread throughout the folded and relatively flat-lying volcanic terranes of the earth. The deposits occur in rocks of all ages, but the largest number occur in those of Precambrian and Tertiary age.

The favourable host rocks are commonly basalts, andesites, latites, trachytes and rhyolites. In Precambrian rocks such assemblages are usually referred to as 'greenstones'. Many deposits occur in tuffs, agglomerates and sediments interbedded with the volcanic flows, particularly in banded iron-formations. In the older terranes, the rocks are generally regionally metamorphosed and exhibit the characteristic regional metamorphic facies outward from igneous or granitized centres. The younger rocks generally show the effects of chloritization (propylitization) over broad areas, but locally some of the andesites and rhyolites may be relatively fresh.

In the older rocks the deposits are veins, lodes, stockworks, pipes and irregular mineralized masses generally in extensive fracture and shear zone systems. Some occur in drag folds. The deposits in the younger rocks are usually confined to fissures, fractures, faults and brecciated zones that often have a limited horizontal and vertical extent. Others, however, are associated with fracture and fault systems that extend for many miles.

The structure of the veins and lodes in the older rocks is generally compact, and the minerals are commonly intimately intergrown with one another. Recrystallization and a coarse-grained character are common. Replacement features abound, and book and ribbon structures are common in the quartz of the veins. The structure of the younger, generally Tertiary, deposits is characteristic. Open cavities are abundant, and filling, crustification, and comb and cockade structures are conspicuous. The vein walls are often irregular, and the vein matter is generally frozen to the walls. Banding and repeated surges of mineralization, witnessed by numerous overgrowths, are characteristic.

Quartz is the most important gangue in these deposits. In the older rocks the quartz is colourless, white, milky, grey or black, and forms coarse granular aggregates, intimately intergrown with the other vein minerals. Most of the quartz gives the appearance of being considerably recrystallized, and some varieties have a definite schistose and gneissic appearance. In silicified zones the quartz is usually microcrystalline. Vugs and crusts of quartz crystals occur in some deposits, but they are rare. The quartz in the younger rocks is characteristically white or light grey, and in some deposits an amethyst colour is often noticeable. It is usually fine grained and often chalcedonic. Some varieties exhibit a delicate colloform banding suggesting colloidal deposition. Several generations are often present, and these cement older brecciated quartz aggregates, ore minerals and wall-rock fragments.

Other gangue minerals in the older deposits include calcite, dolomite and ankerite. In the younger deposits, barite and fluorite are also common, and rhodochrosite and rhodonite are characteristic. Adularia is commonly developed in some Tertiary deposits, and kaolinite, dickite and other clay minerals, sometimes in abundant amounts, are present in the veins. Most are of hypogene origin, but some may be a product of descending surface waters. Zeolites occur in a few of the younger deposits.

The metallic minerals are varied and seem, in part, to be dependent on the age of the deposits. Native gold is the most common. In the older deposits it tends to be relatively pure with only a little silver, the ratio of Au/Ag ranging between 5:1 and 50:1. The gold of the younger deposits is usually much richer in silver, the ratio Au/Ag approaching 1:1 or less in some districts. Gold and silver tellurides commonly occur in all ages of deposits, but are particularly abundant in the younger deposits. Both the native gold and gold-silver tellurides tend to be late minerals in the paragenetic sequence of the veins. Silver selenides occur in some of the younger deposits, but these minerals are generally rare.

The principal silver mineral in the deposits is acanthite (argentite). In addition there may be proustite, pyrargyrite, miargyrite, stephanite, polybasite and tetrahedrite-tennantite. Native silver is common, but is often an oxidation product. These silver minerals are rare in the older (Precambrian) deposits; but a little silver-bearing tetrahedrite-tennantite and other silver sulphosalts may occur in some veins. Most of the deposits that are rich in silver minerals are in Paleozoic, Mesozoic or Tertiary rocks.

Among the other metallic minerals, pyrite, arsenopyrite, argentiferous galena, sphalerite and chalcopyrite are ubiquitous in all ages of deposits, although the last three minerals are generally present in only small amounts in the Precambrian deposits. Pyrrhotite occurs in some of the older deposits, but is rare in the younger. Stibnite and a variety of sulphosalts are most common in the younger deposits, but also occur in some abundance in certain Precambrian ones. Alabandite is usually restricted to the younger deposits. Molybdenite and scheelite are present in certain Precambrian deposits but are relatively rare in younger veins, although there are exceptions to this generalization.

Characteristic types of wall-rock alteration are generally developed adjacent to and in the vicinity of nearly all deposits in this class. In the old Precambrian rocks the most common types of alteration are chloritization, carbonatization, sericitization, pyritization, arsenopyritization and silicification. In the younger rocks propylitization (chloritization and pyritization) is especially characteristic, and there may also be a development of adularization, silicification, kaolinization, sericitization and more rarely alunitization.

Veins in Precambrian greenstones and in certain younger andesites and trachytes are generally flanked by zones of carbonate-sericite schist, which grade imperceptibly into zones of chlorite-carbonate and chlorite schist and then into the country rock. In general these alteration zones represent a loss of silica and some of the cafemic elements, and major increases in water, carbon dioxide, sulphur and potassium. Sodium is generally leached. The reactions to produce the alteration zones are complex and have been discussed in the previous section. The most significant feature is the transfer of silica to the vein sites where it probably crystallized as quartz. By means of detailed chemical profiles it can be shown that most of the vein quartz owes its origin to this process.

Where rhyolites or dacites are present the alteration effects are less marked. Sericitization, pyritization and silicification are, however, characteristic. Sediments, interbedded with the volcanic rocks, are not markedly affected in most places, although there may be a development of pyrite, sericite and some carbonate minerals near the veins.

Propylitization is characteristic of many Tertiary and younger deposits. The mineralogical changes include the abundant development of chlorite and pyrite, sometimes also epidote, from the ferromagnesian minerals in andesite and similar rocks. There is also a development of carbonates and some sericite in certain areas. The chemical changes consist of a moderate leaching of potassium and sodium, a decrease of silica and cafemic elements, and additions of water, sulphur and carbon dioxide in some areas. The propylitization is pervasive, being developed usually adjacent to the veins as well as in great volumes of rock in the general vicinity of the mineralized zones.

The alteration in the younger rhyolites and dacites is usually sericitization and silicification. Alunitization, the development of adularia and carbonatization may affect all younger rock types. Silicification, without the development of lime silicates, may take place in limestones and limy shales. Sandstones, shales and argillites are generally little altered.

The Precambrian deposits of this class exhibit remarkably little vertical elemental zonation; the younger deposits, on the other hand, especially those of Tertiary age are commonly markedly vertically zoned. From the surface downward the sequence is frequently Au-Ag, Pb-Zn, Cu. With greater depth the copper zone tends to pass into barren zones of quartz, carbonates and pyrite with only scattered base metal sulphides. In some Tertiary deposits there is a marked decrease and often an abrupt cessation of silver sulphosalts with depth. Tellurides may also exhibit this effect in some deposits. Increase in the contents of pyrite and the sulphides of Pb, Zn and Cu are common. At great depth the Precambrian deposits commonly pass into barren quartz and carbonates with scattered pyrite, pyrrhotite and minor base metal sulphides, although in some cases the structures appear to fail and mineralized rock no longer occurs. Pyrrhotite takes the place of pyrite at depth in some of the gold belts of the Canadian Shield, and there is commonly a decrease in the stibnite and sulphosalt content of the ores with increasing depth. In these belts, however, the gold and silver values are usually maintained at or near the amounts encountered near the surface.

Some deposits exhibit an increase in the Au/Ag ratio with depth (*see also* the section on Au/Ag ratios in this chapter). In some belts e.g., Kolar, India, there is apparently little diminution in gold values to depths as great as 10 200 ft, but the fineness of the gold is said to increase with depth, from 800 in the near-surface parts to 900 at depth.

The elements commonly concentrated in this class of deposits include Cu, Ag, Zn, Cd, Hg, B, Tl, Pb, As, Sb, Bi, V, Se, Te, S, Mo, W, Mn, Fe, (Co), (Ni), CO_2 and SiO_2; less commonly Ba, Sr, U, Th, Sn, Cr and F. Hg and Sb are particularly characteristic of the younger deposits. The source of these elements is probably the volcanic pile within which the deposits are enclosed. Interbedded sediments may also have made a major contribution of elements to some of the deposits. Tertiary gold-silver and silver-gold base metal deposits in relatively flat-lying volcanics such as andesites and dacites may have derived many of their elements from underlying older rocks in the basement. This and other speculations on the origin of the Tertiary deposits are discussed in a subsequent section on the origin of gold deposits.

Numerous deposits belong in this category. At least two general types can be recognized – one characterized by the presence of native gold as the principal economic mineral and the other by the presence of economic amounts of gold tellurides and/or native gold with or without selenides. Short descriptions and examples of these two general types follow.

Gold-silver and silver-gold deposits

Native gold, alloyed with variable though generally small amounts of silver, is the principal economic mineral in the gold-silver type. In some deposits gold-silver tellurides and aurostibite may contribute to the economic value of the deposits. The Au/Ag ratio of the deposits is 1 or more, in some deposits as high as 12 or more. Quartz is the principal gangue usually with subordinate amounts of carbonates. Other characteristic minerals include pyrite, pyrrhotite, arsenopyrite, chalcopyrite, and minor galena and sphalerite. Small amounts of silver minerals such as acanthite (argentite), tetrahedrite-tennantite, pyrargyrite, proustite, etc. occur in some deposits, usually younger than Precambrian.

The silver-gold deposits contain native gold (commonly electrum), gold-silver tellurides and a veritable host of silver minerals as the economic minerals. The principal primary silver-bearing minerals are tetrahedrite-tennantite, acanthite (argentite), pyrargyrite, proustite, pearceite, stromeyerite, naumannite, miargyrite and matildite. The Au/Ag ratio of the deposits is invariably much less than 1.

Deposits of the gold-silver and silver-gold types range in age from Precambrian to Tertiary. Most are of Precambrian, Paleozoic and Tertiary age.

Numerous examples occur in the greenstone belts of the Canadian Precambrian Shield, including those in the following districts: Yellowknife, Northwest Territories; Rice Lake, Manitoba; Crow River, Lake of the Woods, Red Lake, Porcupine and Kirkland Lake, Ontario; and Belleterre, Noranda-Rouyn, Val d'Or and Chibougamau, in western Quebec.

The deposits of this type at Yellowknife (Boyle, 1961*a*; Henderson and Brown, 1966) are lenses, veins, stockworks and silicified zones in extensive chlorite schist shear zones that intersect meta-andesites, basalts and dacites with interbedded pyritiferous slates and tuffs (Fig. 17). These rocks are cut by quartz-feldspar porphyries in places and by granodiorite along their western limit. The deposits are younger than both the quartz-feldspar porphyries and granodiorite. The alteration of the volcanic rocks includes chloritization, carbonatization, arsenopyritization, pyritization and sericitization, the last four especially intensive near the quartz bodies and silicified zones (Fig. 18). The principal minerals in the quartz deposits are ankerite, pyrite, arsenopyrite, stibnite, chalcopyrite, sphalerite, galena, tetrahedrite, a number of other Pb, Cu and Fe sulphosalts, tellurides in places, molybdenite and scheelite. The economic minerals are native gold and aurostibite. The sulphides and sulphosalts carry considerable amounts of gold in a finely divided form. The grade of the ores ranges from 0.5 to 1 oz Au/ton or more. The Au/Ag ratio of most ore shoots ranges from 3 to 6, but in some orebodies, especially where silver minerals such as tetrahedrite are common, the ratio is 0.7 to 0.9.

The San Antonio Mine in the Rice Lake district of Manitoba is within a sill of diabase that lies in tuff, arkose and conglomerate of the Rice Lake Group. (Gibson and Stockwell, 1948; Stephenson, 1971). The deposit consists of en echelon quartz veins and lenses flanked by zones marked by chloritization, carbonatization, pyritization and sericitization (Fig. 19). The principal minerals are ankerite, chlorite, tourmaline, pyrite, arsenopyrite, sphalerite, galena, chalcopyrite, pyrrhotite, magnetite, petzite and tellurobismuth. The grade of the ore is about 0.35 oz Au/ton.

Numerous gold-quartz veins and lodes occur in the highly metamorphosed greenstone belts and gneisses intruded by granitic rocks between Thunder Bay and the Manitoba border (Cooke and Johnston, 1932; Bruce, 1933; Thomson, 1936; Beard and Garratt, 1976). Most are concentrated in the Lake of the Woods area where mining and exploration were extensive at the turn of the century. The deposits, now exhausted, produced about 100 000 oz of gold from four principal mines, the Mikado, Regina, Sultana and Laurentian. The orebodies were mainly quartz veins, quartz stockworks and silicified zones in shear zones and fractures in amphibolites (greenstones), gneiss, schist, felsic intrusives and extrusives and granitic rocks. The mineralization was essentially quartz with ankerite and small amounts of pyrite, arsenopyrite, chalcopyrite, bismuthinite, molybdenite, free gold and minor tellurides. Tourmaline was common in some of the veins. The gold was coarse in many of the veins, especially near the surface, and the grade exceeded 1 oz/ton in many cases. Grade decreased markedly at depth, however, and few veins were economic below 750 ft. Opportunities for further discoveries of economic deposits in the region seem good.

The gold ore deposits of the Red Lake area, northwestern Ontario are varied (Horwood, 1945). Some occur in granodiorite (Red Lake Gold Shore); others are disseminations in quartz-feldspar porphyry (Howey-Hasaga) or in tuffs (Madsen). A number of deposits occur in shear zones, fractures and faults in meta-andesites, basalts and tuffs (greenstones). The Cochenour Willans Mine (Kuryliw, 1957) is in a volcanic sequence ranging from rhyolite to andesite, now highly altered as a result of chloritization, carbonatization, albitization, silicification and pyritization. The orebodies are quartz and quartz-carbonate lenses in shears and fractures of considerable extent, mainly in tuffs and altered volcanic rocks. The mineralization is primarily quartz and ankerite with free gold. Associated minerals are pyrite, arsenopyrite, sphalerite, pyrrhotite, chalcopyrite, stibnite and tellurides.

The Central Patricia Mine in the Crow River district of Ontario is in the greenstone belt north of Lake St. Joseph (Barrett and Johnston, 1948). The gold deposits are quartz veins and sulphide stringers mainly in fractures in banded iron-formation (Fig. 20). The principal minerals are pyrrhotite, arsenopyrite, green chlorite and ankerite. The gold is mainly free and averages about 0.33 oz/ton in the orebodies. At the Pickle Crow Mine (Bothwell, 1938; Corking, 1948) in the same district ore was mined from one quartz vein (Howell vein) mainly in iron-formation and another quartz vein in sericitic quartz porphyry. There are also veins in greenstones (Fig. 21). The principal minerals are pyrite, chalcopyrite, pyrrhotite, arsenopyrite, tourmaline and chlorite. The average grade of the ore was 0.6 oz Au/ton and 0.05 oz Ag/ton; the Au/Ag ratio was 12.

The deposits of the Porcupine (Timmins) district, Ontario, (Dunbar, 1948; Goodwin, 1965*a*; Ferguson, 1966; Davies, 1977) are mainly extensive quartz veins, silicified zones and silicified, sericitized and pyritized zones in sheared, fractured and highly altered (carbonatized and chloritized) Keewatin lavas and pyroclastics that range in composition from rhyolite to andesite. Some quartz veins, stockworks, silicified zones and pyritized zones produce much gold from quartz-feldspar porphyry and Temiskaming-type sediments. The vein systems are extremely complex as shown by the plans of the Hollinger orebodies (Fig. 22) and the descriptions of Ringsleben (1935), Buffam (1948*a,b*), Jones (1948*a*), Furse (1948), Carter (1948), Taylor (1948), Kinkel (1948), Longley and Lazier, (1948), Hawley and Hart, (1948), Holmes (1948), Bell (1948), Backman (1948), Price and Bray (1948), Byers (1948) and Griffis (1962). The age of the mineralization at the Preston East Dome Mine as calculated from lead isotope data on galena closely associated with gold is given as 2961 m.y. (Machairas, 1970*a*). The alteration is essentially chloritization, carbonization, sericitization, silicification and pyritization. The principal minerals are quartz, ankerite, calcite, tourmaline, anhydrite, pyrite, pyrrhotite, galena, sphalerite, chalcopyrite, arsenopyrite, native gold, tellurides and scheelite. The grade of the ore ranges from 0.1 to 0.4 oz Au/ton; the silver content ranges from 0.03 to 0.07 oz/ton; and the Au/Ag ratio averages about 5.5 with some mines having ratios as high as 10 or more, especially from near surface veins. Some parts of the deposits contained considerable amounts of tungsten. The Hollinger veins in places averaged about 0.33% WO_3, all in scheelite.

Many of the gold deposits of the Kirkland Lake–Larder Lake district (Thomson, 1950; Thomson *et al.*, 1950) are in rocks other than volcanics and are described in a later section. Some of the deposits in the eastern part of the district, particularly the Kerr Addison Mine and Chesterville Mine are in highly sheared, altered and mineralized volcanics and tuffs ranging in composition from dacite to andesite. The orebodies

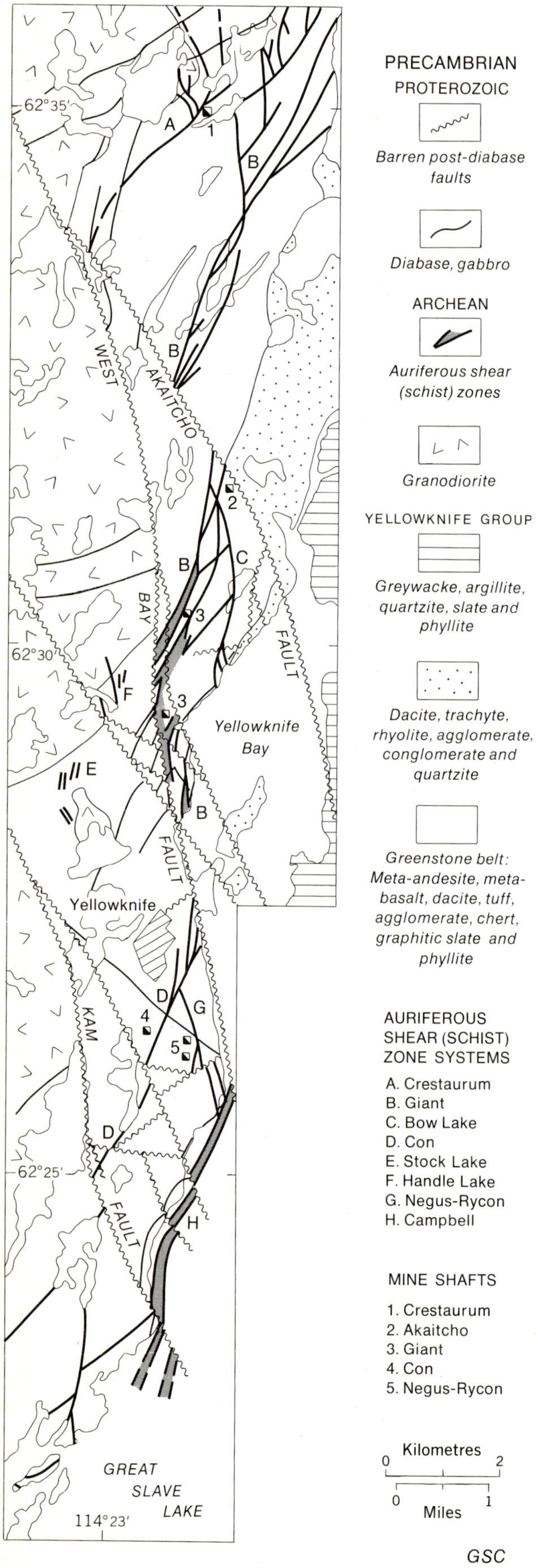

Figure 17. Yellowknife Greenstone Belt showing principal auriferous shear zone systems (after Boyle, 1961*a*).

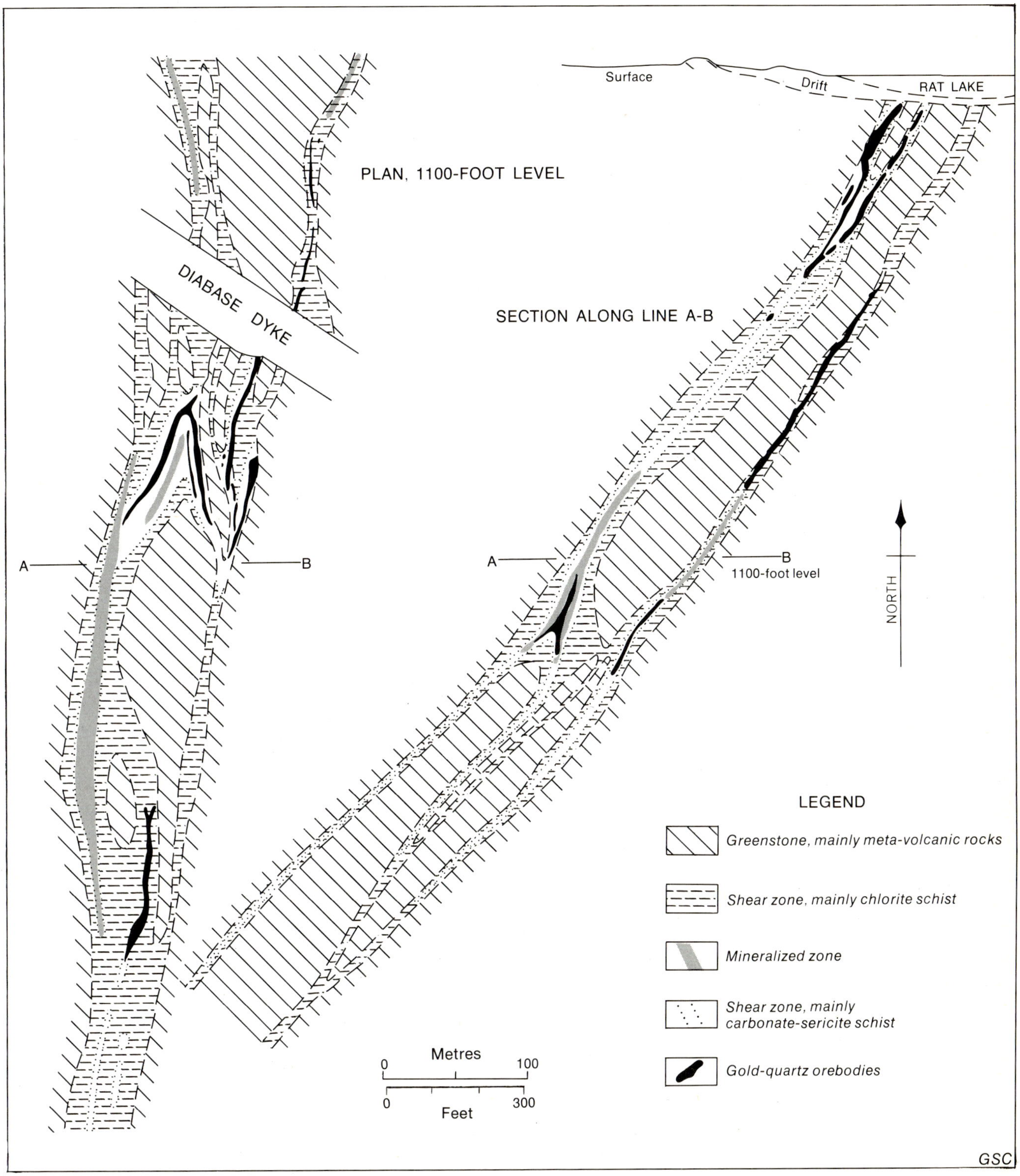

Figure 18. Generalized plan and section of the Con Mine (Con shear zone system), Yellowknife, Northwest Territories (after Boyle, 1961*a*).

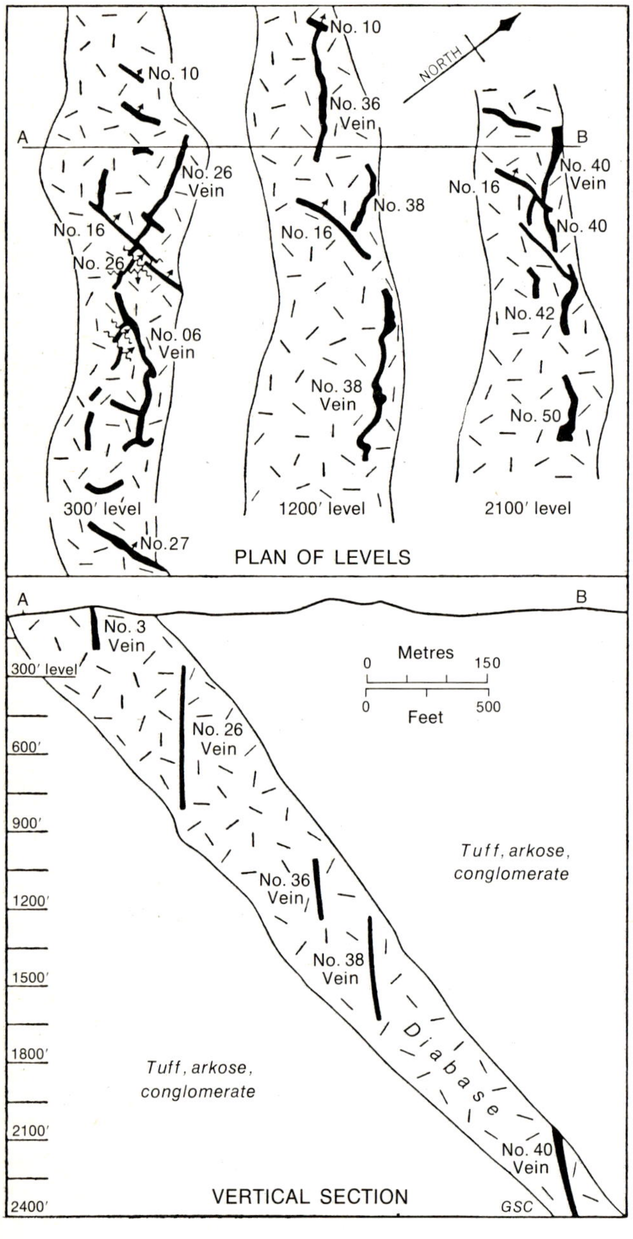

Figure 19. Generalized plans and section, San Antonio Mine, Bissett, Manitoba (after Gibson and Stockwell, 1948).

at the Kerr Addison Mine are great quartz-carbonate stockworks and disseminated pyritic bodies in intensely sheared and altered zones (Baker *et al.,* 1957). Those at the Chesterville Mine are similar. Figure 23 illustrates the great complexity of the orebodies in the carbonated and chloritized schist zones. Most of the gold is free. Other minerals include pyrite, arsenopyrite, chalcopyrite, millerite, sphalerite, tetrahedrite, galena and scheelite. The average gold and silver grades are 0.20 and 0.01 oz/ton respectively, and the Au/Ag ratio is 20, exceptionally high for this type of deposit.

Many of the deposits in the greenstone belts of northwestern Quebec occur in altered volcanics ranging in composition from rhyolite to basalt (Wilson, 1948; Norman, 1948*a, b*; Dresser and Denis, 1949; Dugas, 1966; Latulippe, 1966; Prochnau, 1971). Those in the intermediate and basic varieties of volcanic rocks include among others the Belleterre Mine in Guillet Township; the Francoeur, Arntfield, Wasa Lake, Senator Rouyn and Stadacona in the Noranda-Rouyn district; the O'Brien, East Malartic and Malartic goldfields, in the Lake Malarctic–Val d'Or district; and the Agnico-Eagle and other mines and prospects in Joutel Township, Harricana River area. These deposits are essentially similar. They occur in shear zones, faults, fractures or drag folds and comprise quartz veins, quartz-ankerite replacement zones, silicified zones and disseminated and massive sulphide bodies in highly silicified and carbonated volcanic rock. The principal minerals in the deposits are quartz, ankerite, calcite, arsenopyrite, pyrite, chalcopyrite, galena, tourmaline, free gold and gold-silver tellurides. Hematite occurs in some deposits (Arntfield, Francoeur), and scheelite is present in places but is rarely abundant. The gold grade of the ores ranges from 0.15 to 0.5 oz/ton and the silver grade averages about 0.05 oz/ton. The Au/Ag ratio ranges from 3 to 10 or more.

According to Price (1934) and Dugas (1966) the gold deposits of the Rouyn-Noranda area are younger than the massive base metal sulphide bodies.

Many examples of this type of deposit occur in other parts of the world. Here belong the gold deposits of the Mother Lode System of California where the ores are in greenstones; numerous deposits in the Kalgoorlie and adjacent districts of Australia; the ores of the Kolar goldfield in India; and gold deposits of the Barberton Mountain Land of South Africa. Many Tertiary examples are known in Roumania (Transylvania), New Zealand, Mexico, United States, Japan, Fiji and elsewhere.

The Mother Lode System of California (Fig. 24) comprises two-score mines or more in a belt some 120 mi long and about 1 mi wide on the western foothills of the Sierra Nevada (Knopf, 1929; Clark, 1970). The deposits are quartz veins and bodies of pyritized and mineralized country rock in a system of linked and anastomosing faults, fractures and shear zones cutting a complex of Jurassic and Paleozoic (Carboniferous) greenstones, serpentinites, slates, schists and granodiorite, the last probably of Cretaceous age. The principal minerals are coarse-grained milky quartz, ankerite, calcite, mariposite, pyrite, arsenopyrite, galena, sphalerite, chalcopyrite, tetrahedrite, scheelite, molybdenite, native gold and various tellurides. Alteration is intense in the greenstones and serpentinites, being mainly carbonatization (ankeritization), pyritization and sericitization. The average grade of the ore was 0.3 oz Au/ton, and the gold-silver ratio averaged about 10. Fabulous bonanzas were found in some of the veins of the system.

The Kalgoorlie district of Western Australia (Fig. 25) features the famous 'Golden Mile' along which and within the general vicinity of which numerous gold lodes occur in shear zones, fractures and pipes (intersections of shears) in a greenstone belt intruded by porphyries and granodiorite, all of Precambrian age (Campbell, 1953, 1965; Finucane and Jenson, 1953; Utting, 1953; Stillwell, 1953*a*; Woodall, 1965, 1975; Finucane, 1965; Travis *et al.,* 1971; Travis and Woodall, 1975; Tomich, 1974, 1976).

Two types of orebodies are present: (1) gold and gold telluride mineralization in the lodes of the Golden Mile zone localized mainly within a sill (Golden Mile Dolerite). These are the most productive and comprise steeply-dipping pyritic

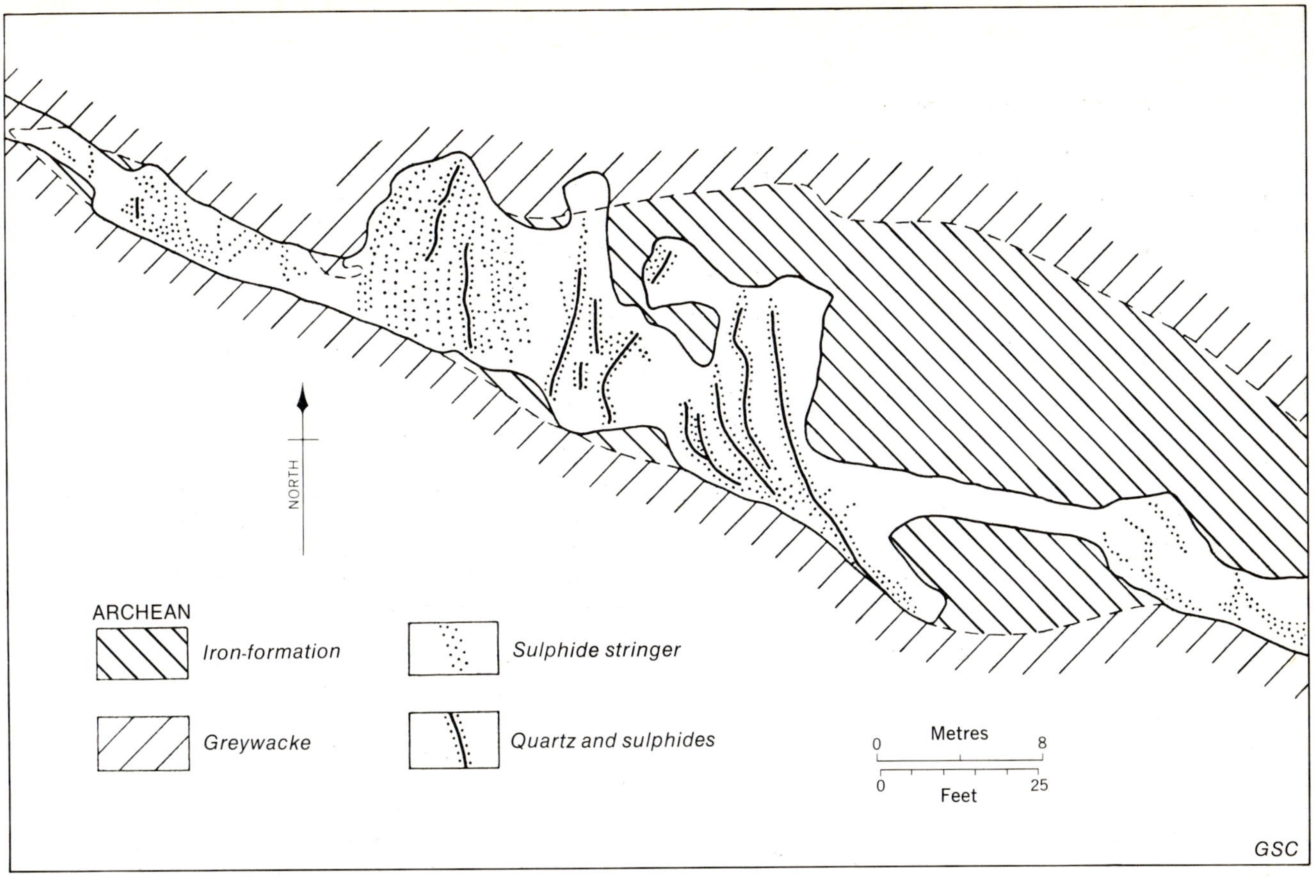

Figure 20. Plan of F-10-3 orebody, 1000-foot level, Central Patricia Mine, Crow River district, Ontario (after Barrett and Johnston, 1948).

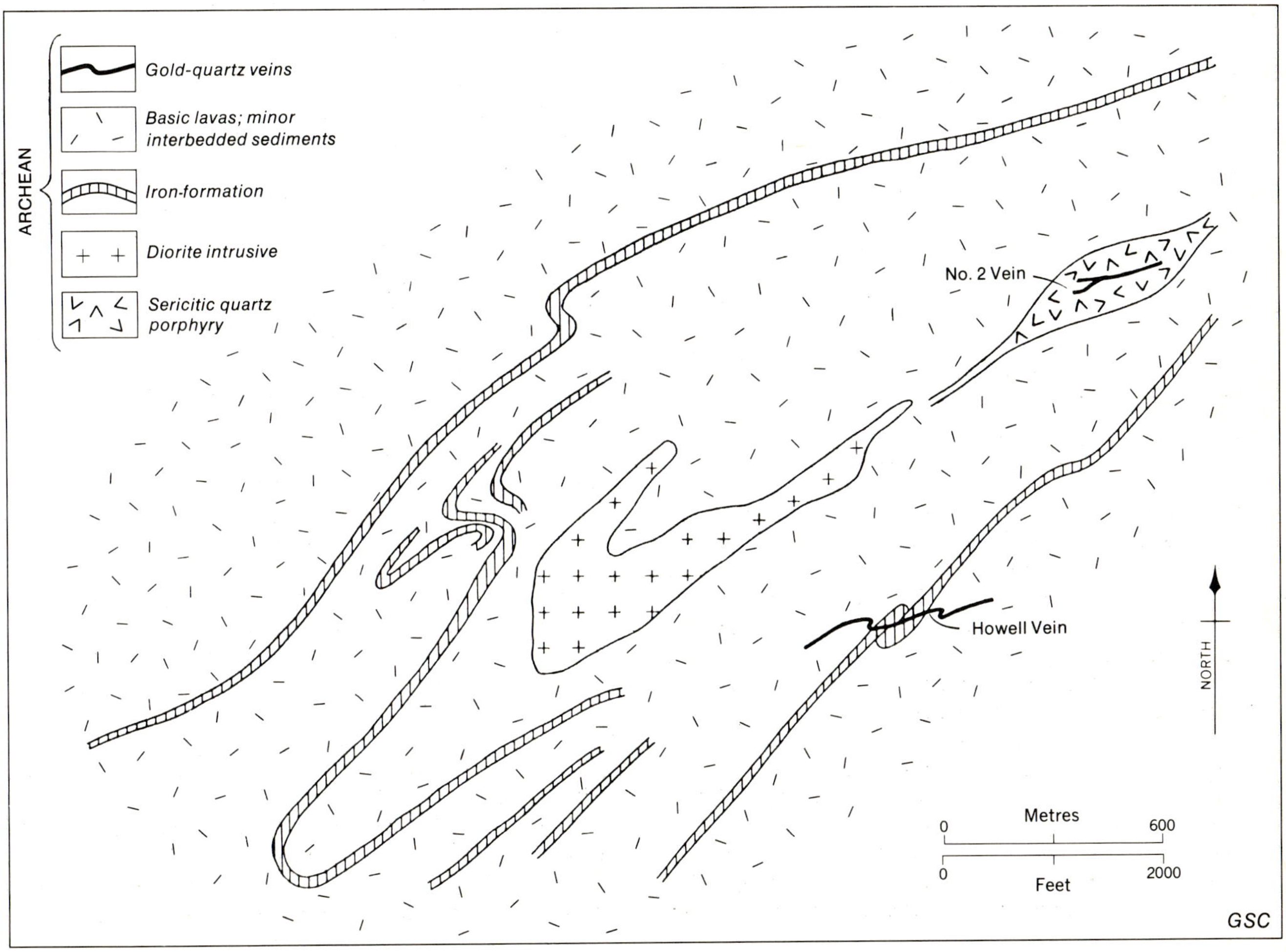

Figure 21. Major structure and auriferous quartz veins, Pickle Crow Mine, Crow River district, Ontario (after Corking, 1948).

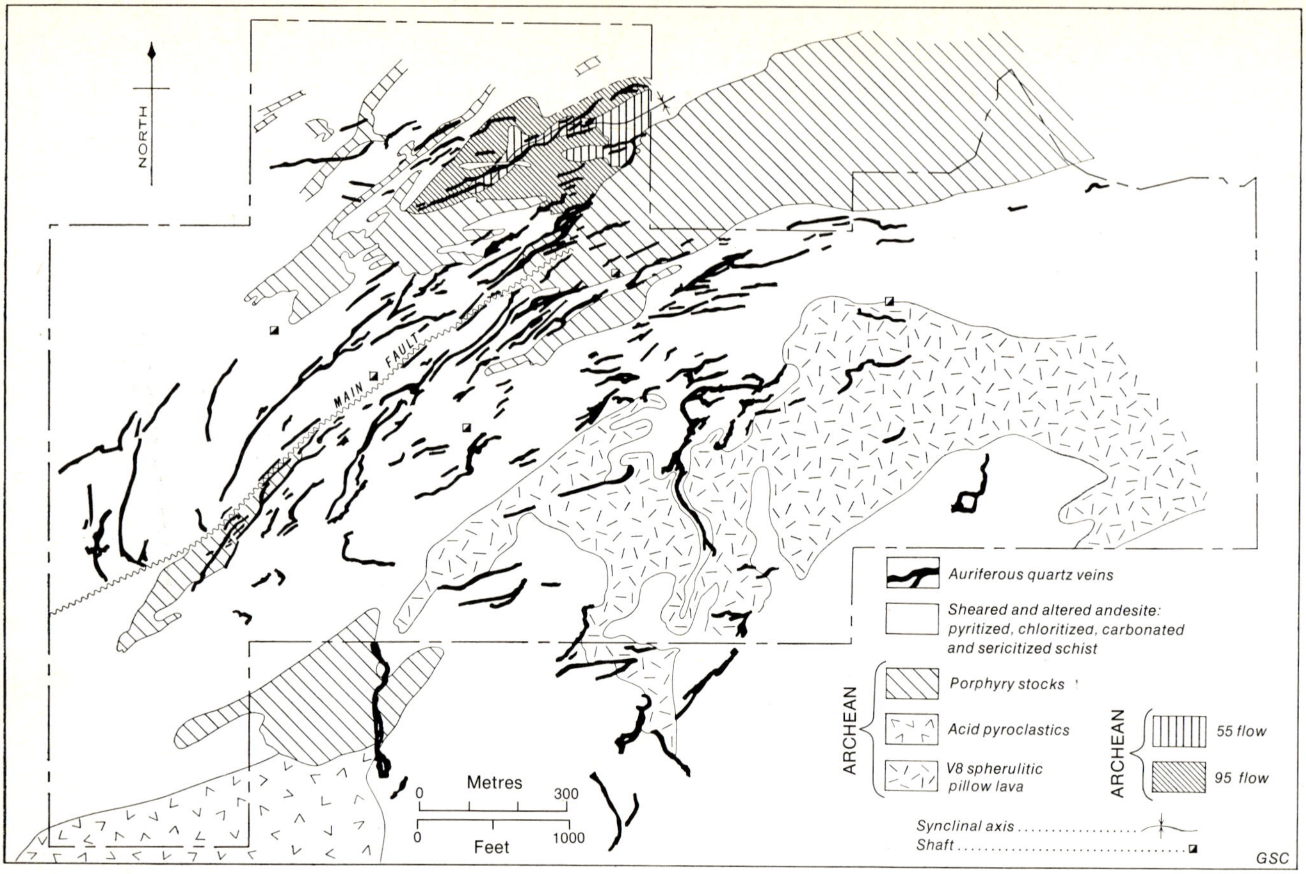

Figure 22. Plan of 425-foot level, Hollinger Mine, Timmins, Ontario, showing complex vein systems (after Jones, 1948*a*).

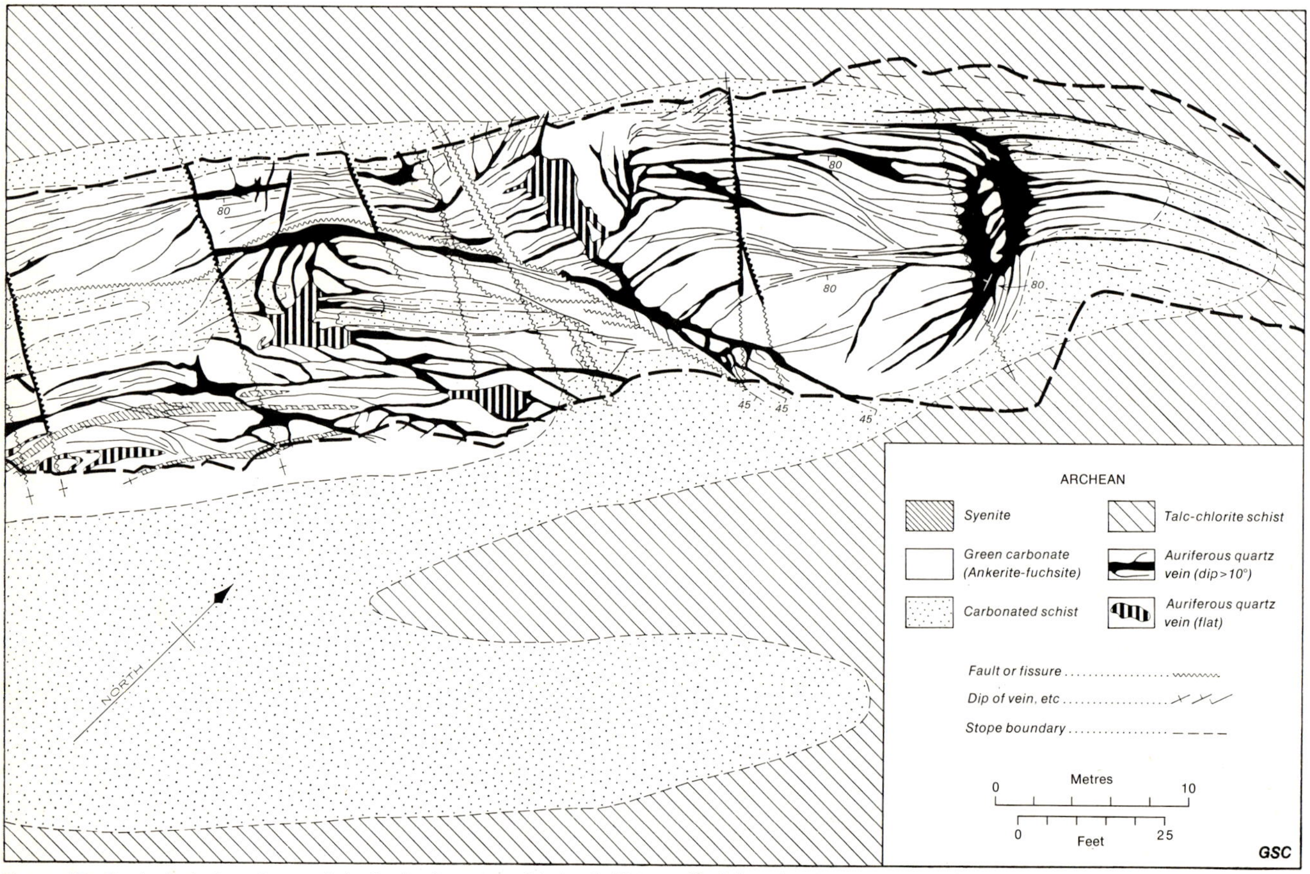

Figure 23. Geological plan of part of the back of a stope, 9th level, Chesterville Mine, Larder Lake, Ontario, illustrating the great complexity of the quartz bodies in the chloritized and carbonated schist zones (modified from Buffam and Allen, 1948).

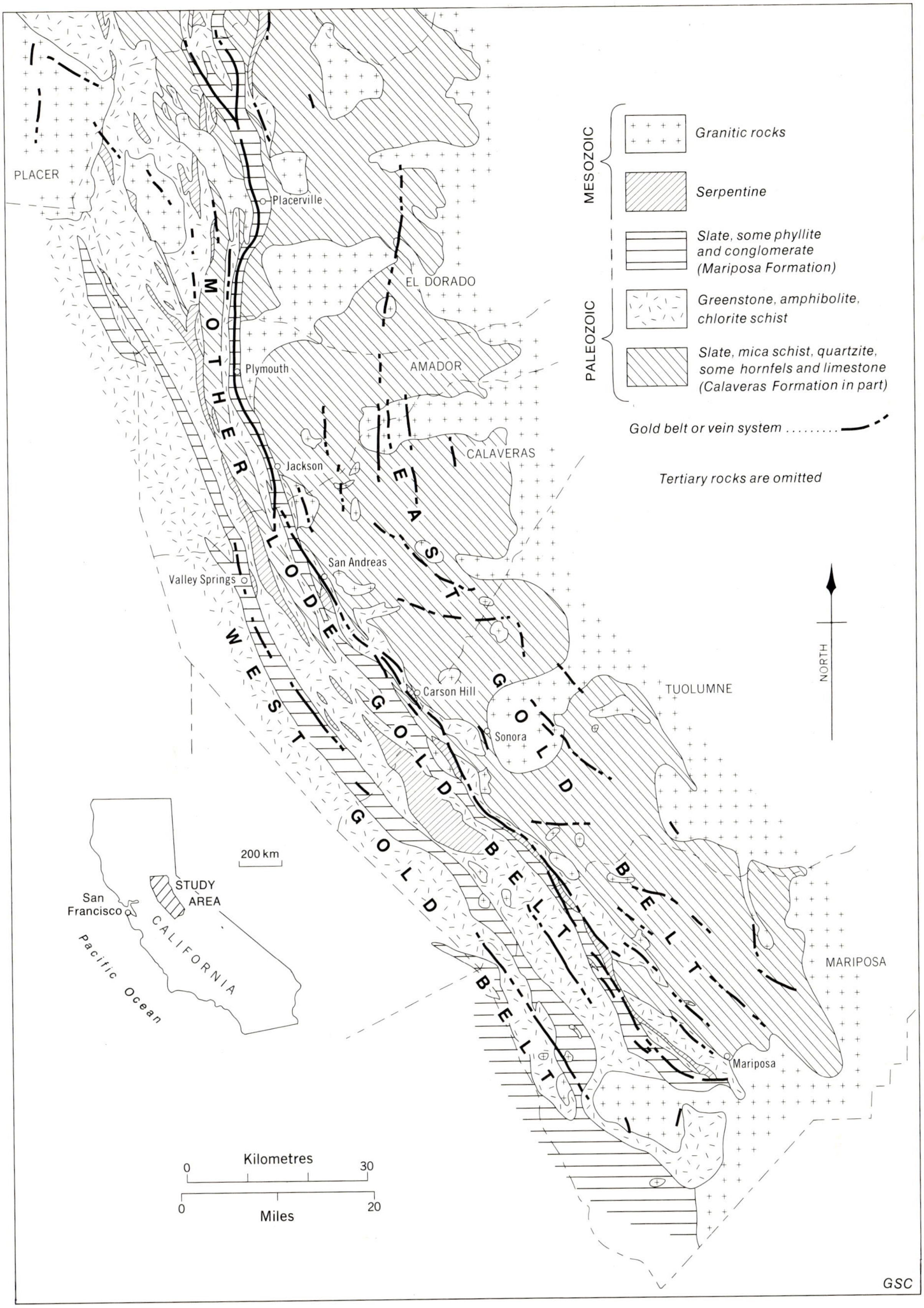

Figure 24. The 'Mother Lode System', Central Sierra Nevada, California. Also shown are the East and West gold belts (after Clark, 1970).

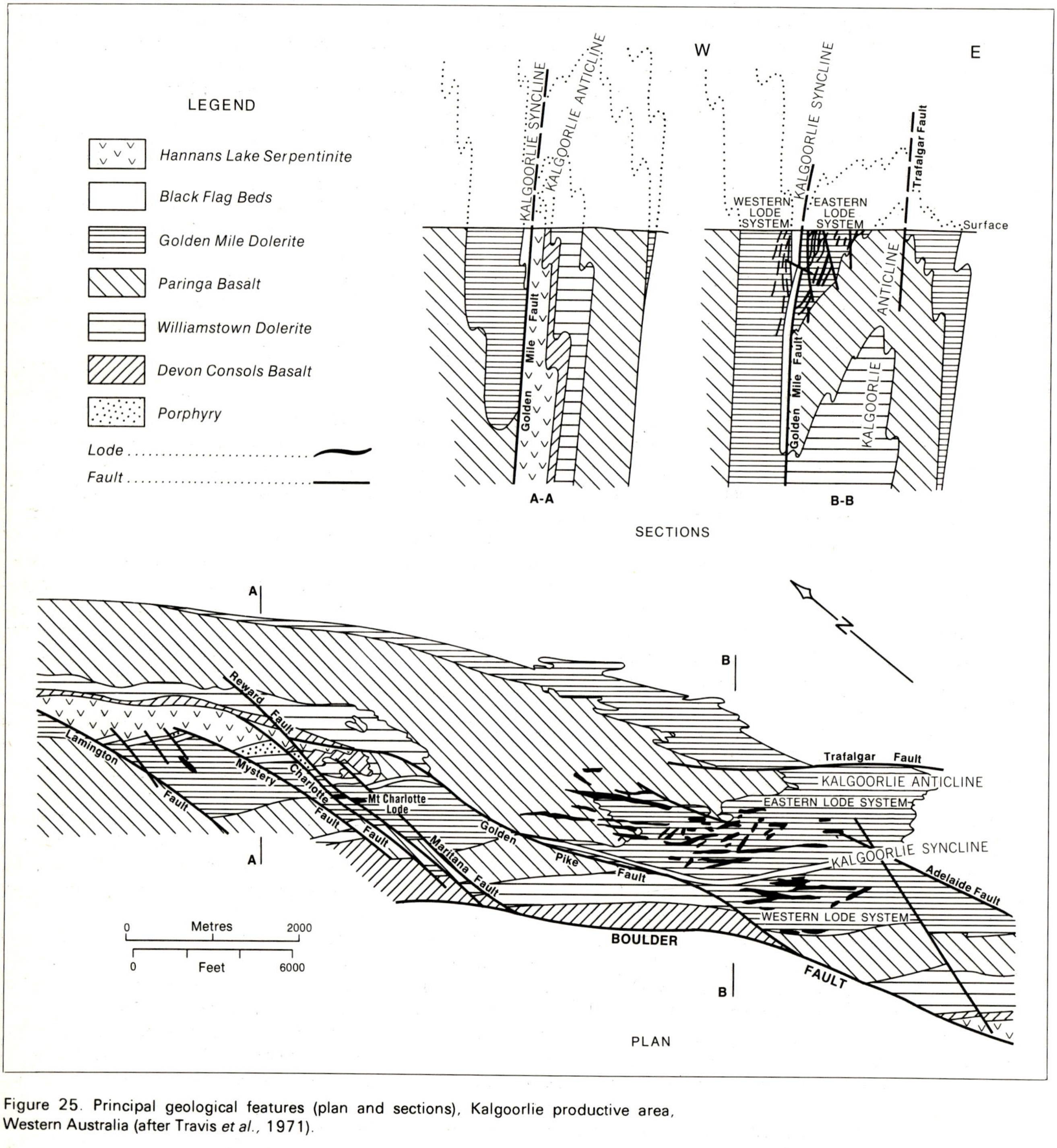

Figure 25. Principal geological features (plan and sections), Kalgoorlie productive area, Western Australia (after Travis *et al.*, 1971).

and silicified replacement bodies along narrow shears and fractures, and (2) gold-quartz mineralization that occurs as replacement bodies along shears or as stockworks of quartz veins (Mount Charlotte Mine). Tellurides are rare in this type of deposit. The principal minerals in the deposits are quartz, ankerite, calcite and pyrite, with small amounts of chalcopyrite, galena, sphalerite, tetrahedrite, stibnite, enargite, pyrargyrite, loellingite, specularite, magnetite, fluorite, tourmaline, native gold and several tellurides. Alteration effects are marked in the greenstones and include chloritization, carbonatization, pyritization and sericitization. The grade of the ore averages about 0.4 oz Au/ton and the Au/Ag ratio is about 8. Deposits somewhat similar to those at Kalgoorlie occur in the Coolgardie greenstone belt (McMath *et al.,* 1953) and in the other greenstone belts of Western Australia and Northern Territory (Hills, 1953; Woodall, *in* Knight, 1975).

The Kolar goldfield of India, worked in a primitive way for centuries and in modern times since 1880, is localized in a sliver of Precambrian Dharwar metavolcanics some 50 mi long and 2 to 3 mi wide in the eastern part of the Mysore Plateau, India. The most recent description of this famous goldfield is by Narayanaswami *et al.* (1960). Other similar fields are in narrow greenstone belts to the northwest at Ramajiri and Hutti (Murthy, 1964; Ghosh *et al.,* 1970). Sikka *et al.* (1964) have discussed the setting of the Kolar and other goldfields in the Precambrian terranes of India and compared the deposits with those of the Porcupine (Timmins) area of Ontario.

The deposits of the Kolar field are quartz lodes and quartz-sulphide bodies in north-striking subparallel schist zones in amphibolites (metavolcanics; greenstones) that can be traced for many miles along their strike (Fig. 26). Some also have great dip dimensions, a few of the lodes being mined down to vertical depths of 10 000 ft or more. The eastern part of the Kolar field is marked by the Muscoom, the famous Champion and Mundy's quartz lodes; in the central and western parts of the field sulphide-bearing lodes that are in places graphitic predominate. Most of these subparallel lodes follow the general strike of the greenstones and exhibit a stratigraphic control. The quartz lodes occur as en echelon bodies within well marked schistose zones. They contain essentially quartz, calcite, native gold, with minor amounts of magnetite, ilmenite, galena, pyrrhotite, pyrite, arsenopyrite, chalcopyrite, tourmaline and scheelite, often mixed with pyroxene, hornblende, cummingtonite, tremolite, epidote, biotite, garnet, wollastonite, dumortierite and minor chlorite. Gold tellurides are mentioned as being present by Rozhkov (1966). The quartz sulphide lodes are similar except that they contain more sulphides (about 10%) including pyrrhotite, arsenopyrite, pyrite, chalcopyrite and minor galena. Both the quartz and sulphide lodes are enveloped by a zone of micaceous schist representing alteration of the greenstones. The schist consists of alternating layers of brown biotite and granular quartz together with minor layers of dark fibrous cummingtonite. Adjacent to individual quartz veins and stringers there is a successive alteration sequence as follows: (1) a thin casing of pale-green diopside at the immediate contact of the vein quartz, followed outward by (2) thick or thin layers of dark hornblende and admixed epidote, and finally by (3) a thick layer of brown mica. Where the lode is composed of numerous parallel veins and stringers of quartz, there are symmetrically alternating layers of quartz, diopside, hornblende-epidote and brown mica. These layers are in places highly drag folded, contorted and sheared as a result of postmineralization shearing or probably premineralization drag folding of highly sheared and fractured wall rock. Streaky layers and lenses of diopside, hornblende, epidote and micaceous material are also seen at places in the highly sheared quartz, which is commonly dark gray to black at these localities.

The gold content of the Kolar lodes varies widely but averages about 0.4 oz/ton in the quartz lodes and 0.3 in the sulphide bodies; the Au/Ag ratio ranges from 8 to 10.

The Hutti gold deposits, Raichur district, Mysore are similar to those in the Kolar gold belt. They lie in the broadest part of the Precambrian Hutti-Maski schist (greenstone) belt (Dharwars). According to Vasudev and Naganna (1973) the deposits are mainly gold-quartz veins and lenses in narrow shear zones in basic metavolcanic rocks (amphibolites to chlorite schists) intruded by granites and aplites. Six reefs are productive; these contain quartz, chlorite, biotite, sericite, actinolite, hornblende, tourmaline, calcite, siderite and ankerite as gangue minerals in which are found native gold, pyrite, arsenopyrite, pyrrhotite, chalcopyrite, sphalerite, galena, specularite, rutile and ilmenite. No tellurides are present. Vasudev and Naganna (1973) recognized three distinct stages in the mineralization history of the deposits. Minerals formed during the first stage were rutile (residual in wall rocks), pyrite (with traces of gold), arsenopyrite, specularite, quartz and carbonate. Minerals of the second stage were pyrrhotite, chalcopyrite, sphalerite, galena and gold. Minerals of the third stage were pyrite, calcite and quartz developed along numerous minor faults (slips) and major faults. Based on textural characteristics and associations, three generations of gold can be distinguished as follows: (1) gold-pyrite association; (2) gold-arsenopyrite-quartz association; and (3) gold-sphalerite-quartz association. Only the gold of the second and third generations are important in the economics of the deposits.

The Barberton Mountain Land of South Africa, some 175 mi east of Pretoria, represents an ancient Precambrian (Swaziland System) greenstone belt, composed of a basal volcanic assemblage overlain by sedimentary successions and invaded by granitic rocks (De Villiers, 1957; Davies and Hunter, 1964; Gribnitz, 1964; Anhaeusser *et al.,* 1968; Viljoen *et al.,* 1969; Anhaeusser, 1976). The gold deposits, comprising quartz veins and lenses in greenstones, sediments and granodiorite, pyritic ores in fractured cherts and other sediments, and complex sulphide ores in porphyry, are localized in faults, fractures and shears. Some of the deposits in the Steynsdorp goldfield show a well-defined stratigraphic control. The most abundant minerals in the deposits are quartz, siderite, ankerite, calcite, pyrite, arsenopyrite, pyrrhotite, chalcopyrite, sphalerite, galena, stibnite, tetrahedrite, berthierite, bournonite, argentite, pyrargyrite, millerite, linnaeite, skutterudite, cobaltite, native bismuth and native gold. A number of other sulphosalts and sulphides are mentioned by De Villiers (1957). The Au/Ag ratio of the deposits is high, averaging about 33.

Somewhat similar desposits are present in the Murchison Range of the Northeastern Transvaal, some 200 mi northeast of Pretoria (Hall, 1930; Van Eeden *et al.,* 1939; Sahli, 1961). There, the Swaziland System comprises greenstone and other

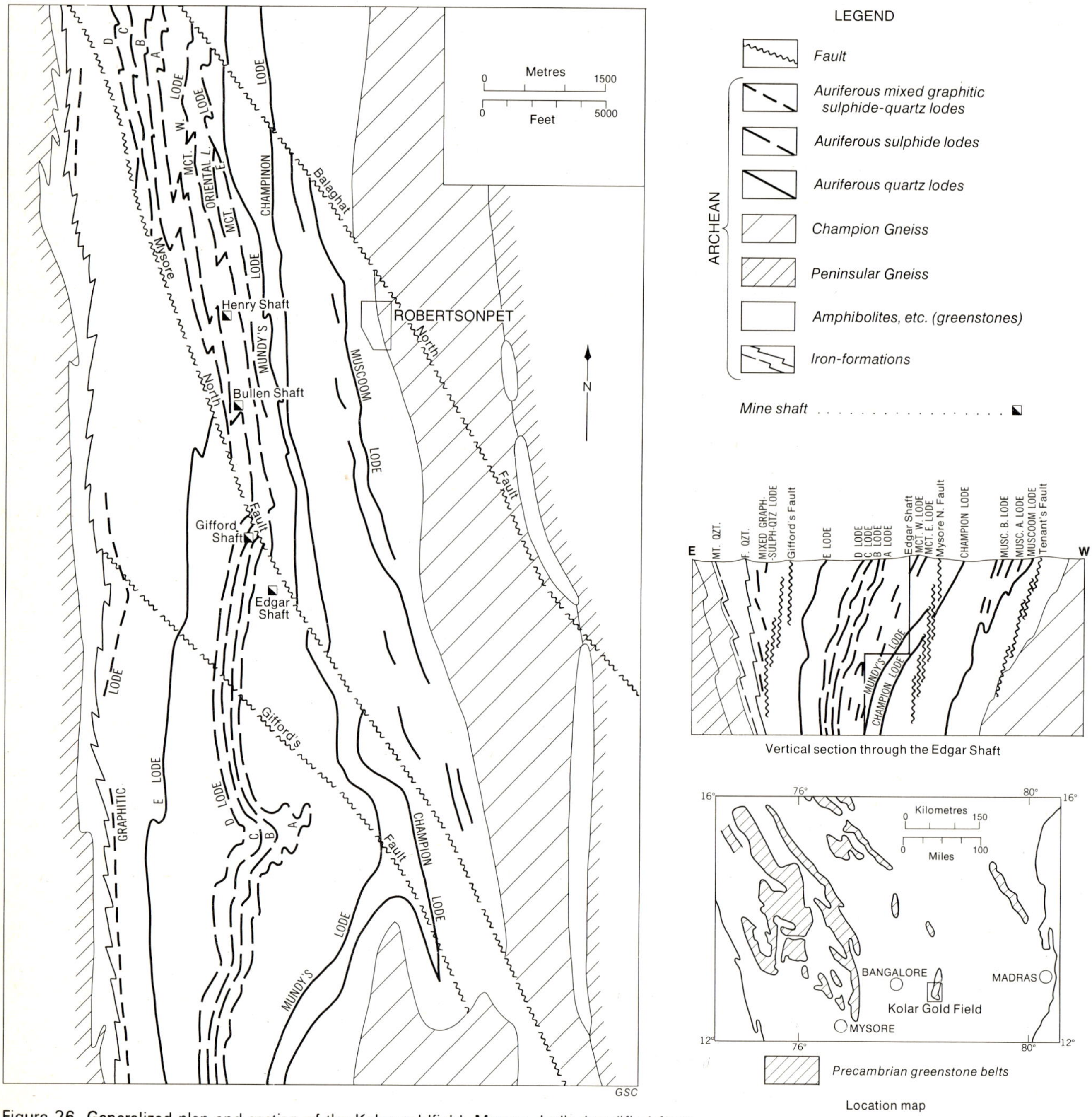

Figure 26. Generalized plan and section of the Kolar goldfield, Mysore, India (modified from Narayanaswami *et al.*, 1960).

schists invaded and surrounded by granitic rocks. The deposits are mainly quartz veins, sheeted zones and mineralized silicified zones in faults, fractures and shears along a prominent break (The Antimony Line). The deposits carry quartz, pyrite, carbonates, tetrahedrite, chalcopyrite and free gold. Stibnite is abundant in some of the deposits; these produce both gold and antimony. Some deposits also carry cinnabar as a very late mineral.

Northward in the African Shield gold deposits in or closely associated with ancient volcanic rocks have been mined extensively in Rhodesia (Lightfoot, 1930; Phaup, 1964; Goldberg, 1964; Visser, 1968; Anhaeusser, 1976), Zaire (Woodtli, 1961*a, b*), and in Sudan and Egypt (El Boushi, 1972; Neubauer, 1962).

Lightfoot (1930) classified the gold deposits of Rhodesia into five categories: (1) pyritic impregnations along shatter belts; (2) antimonial (stibnite) impregnations along shatter belts; (3) pyritic gold-quartz veins; (4) antimonial (stibnite) gold-quartz veins; and (5) cupriferous gold-quartz veins, containing mainly gold, auriferous pyrite and chalcopyrite. Greenstone schist is said to be the most kindly rock, followed by iron-formation, dolomitic schist, quartzites and granite.

The Cam and Motor Mine (Collender, 1964), the largest gold producer in Rhodesia, wins gold from quartz veins and quartz stockworks in lavas (greenstones), along lava-sediment contacts and in beds of sediments interbedded with lavas. The mineralization is characterized by the presence of ankerite, pyrite, arsenopyrite, stibnite, sphalerite, scheelite and gold. The gold varies in fineness from 900 to 930. The Dalny Mine (Leigh, 1964) is in an extensive shear zone in greenstones which are highly altered near the quartz lenses and silicified bodies. The alteration is typical of gold deposits in greenstone exhibiting intense sericitization and carbonatization adjacent the veins and chloritization farther out. CO_2 increases as the quartz bodies are approached. K_2O increases in a similar manner and Na_2O decreases. The mineralization is essentially pyrite, arsenopyrite and minor amounts of chalcopyrite, sphalerite and scheelite. Gold is the only element recovered; it occurs free and in intimate association with arsenopyrite. The grade of the ore averages 0.2 oz Au/ton. The Dawn Mine orebodies are in quartz reefs (veins) in shears in pillow lavas and greenstones that are carbonated, sericitized, chloritized and pyritized near the quartz bodies (Mehliss, 1964). The metallic minerals in the quartz are mainly pyrite and gold with minor amounts of sphalerite and galena. The Patchway Gold Mine is in a quartz vein cutting Archean pillow lava, the principal minerals being native gold, pyrite, galena, sphalerite, chalcopyrite and scheelite (Ward, 1968). The Indarama Mine at Sebakwe (Mehliss, 1968) is of interest because it contains two distinct types of mineralization – an earlier arsenical generation, which impregnates the greenstone walls of the fissures, is followed by a later antimonial generation, which, along with black and white quartz and calcite, forms the fissure fillings. The early generation is represented by pyrite, pyrrhotite, loellingite and arsenopyrite, and the late generations by aurostibite, gudmundite, tetrahedrite, berthierite and stibnite. Gold spans the two generations having been precipitated with both the early and late generations of sulphides and sulphosalts. The mineralization history is very similar to that at Yellowknife, Northwest Territories, Canada.

The deposits in Zaire (Woodtli, 1961*a, b*) are mainly quartz veins in shear and faulted zones in amphibolites, itabirite (iron-formation) and granitoid rocks at Kilo, and gold impregnation deposits in schistose rocks containing carbonate, quartz and albite at Moto.

In Egypt and the Sudan the deposits are varied. At the Gebeit Mine in Sudan (El Boushi, 1972) the orebodies are in quartz-pyrite-arsenopyrite veins in highly sheared, mylonitized and pseudo-schistose silicic volcanics interbedded with tuffs. The wall rocks are highly pyritized, and there is also a development of chlorite, epidote and calcite. Elsewhere in Sudan there are quartz veins, none of which is of economic value, in a variety of Precambrian schists of volcanic derivation near granitic masses. Much gold was won in early times from gold-quartz veins in or associated with Precambrian volcanics and associated rocks, now largely, schists (Emmons, 1937). In Egypt, the El Sid Mine, now largely exhausted, is in quartz veins in basic intrusives, sedimentary schists and gneisses, hornblende migmatite and granite-granodiorite near the contact of the granitic rocks with the schists and gneisses (Derry, 1951). The mineralization comprises quartz, pyrite, galena, sphalerite, chalcopyrite, arsenopyrite and traces of stibnite. The gold is free and grades about 1 oz/ton. Most of the gold in the Egyptian deposits ranged from 730 to 930 fineness.

The details of the metallogenesis of gold in the southeastern desert of Egypt are summarized by Ivanov *et al.* (1973). There appears to be some controversy regarding the age of the auriferous quartz veins with respect to other types of deposits. The writers conclude that at least one phase of gold mineralization is closely related to the Old Volcanics (Archean?), especially their propylitized (chloritized) members. According to El Shazly (1957) the gold veins occur in rocks ranging from early Precambrian schists to late granites. He assigns a late Precambrian age to the gold mineralization.

Tertiary volcanic terranes contain numerous famous gold districts. As examples of these may be mentioned the various deposits in Romania, New Zealand, Fiji, Indonesia, Japan, U.S.S.R., United States and South America.

The gold deposits of Romania occur mainly in faults and fractures that cut Tertiary (Neogene) volcanics in the districts of Brad (Apuseni Mountains) and Baia Mare (Gutăi Mountains) both in Transylvania (Von Palfy, 1911–1912; Streckeisen and Ghitulescu, 1930; Petrulian, 1934; Helke, 1934; Emmons, 1937; Ghitulescu and Socolescu, 1941; Bujor, 1948; Gorzhevskii *et al.*, 1971; Giuscă *et al.*, 1968; Pomârleanu *et al.*, 1972; Giuscă *et al.*, 1973; Cioflică *et al.*, 1973). The volcanics include flows of andesite, dacite and rhyolite, various tuffs, volcaniclastics and interbedded sediments, lying mainly on Mesozoic sediments and crystalline schists (Fig. 27). Volcanic necks of dacite, rhyolite and andesite, and their breccias are favourite sites for the localization of vein clusters (Fig. 28). In this respect the deposits are somewhat analogous to those at Cripple Creek in Colorado. The principal minerals are fine-grained quartz, jasper, chalcedony, calcite, fluorite, rhodochrosite, barite, hematite, pyrite, arsenopyrite, chalcopyrite, galena, sphalerite, alabandite, cinnabar, stibnite, tetrahedrite, bornite, chalcocite, enargite, stephanite, bournonite, argentite, pyrargyrite, proustite, native gold and gold and other tellurides, including sylvanite, nagyagite, petzite, krennerite, hessite

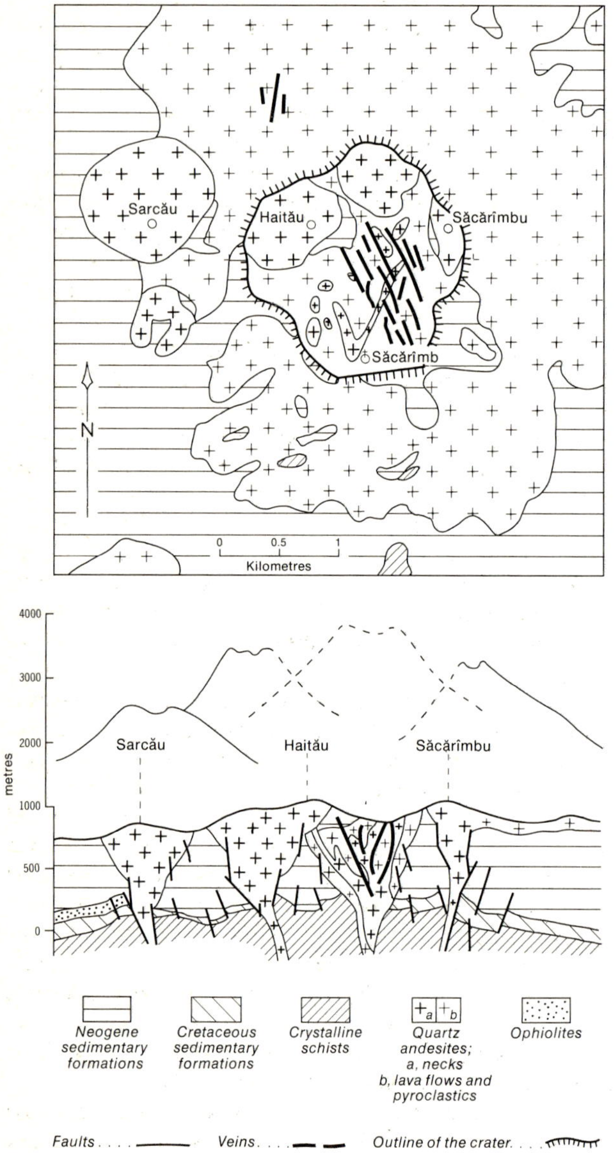

Figure 27. Săcărîmb (Nagyág) volcanic edifice, Apuseni Mountains, Romania (after Cioflica *et al.*, 1973).

Figure 28. Generalized plan and section of geological features in the vicinity of the Franz Tunnel level, Săcărîmb (Nagyág), Romania (after Von Palfy, 1912).

and altaite. The propylitic alteration of the volcanics is pervasive and regional in extent. Near the veins and stockworks there is a narrow alteration zone marked by the development of pyrite, carbonates, adularia, sericite, locally alunite and clay minerals. The texture of the veins is usually massive and commonly brecciated; banding is relatively rare and vugs are present but not large. 'Glauch' veins barren of gold are a feature of some of the districts. These veins are older than the ore-bearing veins and are interpreted as having been filled by ascending liquid muds containing fragments of shale and igneous rocks. These unusual veins are sometimes found in other Tertiary volcanic areas in many parts of the world.

The mineralization history of the economic deposits is complex. In the Apuseni Mountains, Cioflică *et al.* (1973)

recognize three metallogenic phases, oldest to youngest as follows:

1. A gold-silver phase characterized by pyrite, arsenopyrite, sphalerite, chalcopyrite, galena, alabandite, tetrahedrite, pearceite, polybasite, argentite, native gold and marcasite in that paragenetic order. These minerals occur in a quartz, carbonate and rhodocrosite gangue.

2. A gold-silver, base metal sulphide, copper and mercury phase. These deposits are varied and include gold deposits proper, gold deposits with base metal sulphides, porphyry copper mineralization and cinnabar mineralization. The gold deposits proper are characterized by pyrite, marcasite, arsenopyrite, sphalerite, galena, chalcopyrite, alabandite, gold tellurides, tetrahedrite and various other sulphosalts, stibnite, realgar, native arsenic and native gold in that paragenetic order in a gangue essentially composed of quartz, carbonates, rhodochrosite, barite and gypsum.

3. A subordinate base metal sulphide phase characterized in places by tetrahedrite and chalcopyrite, enargite, famatinite, bismuth and germanium minerals, and in places by gold-silver tellurides (sylvanite, hessite) and rarely native gold.

Some of the deposits, especially those with a polymetallic character are markedly vertically zoned as follows from the surface downwards: Au-Ag, Pb-Zn, Cu. In some deposits there is considerable telescoping and several generations of minerals. The geothermometric work by Giuscǎ *et al.* (1968) shows that mineralization proceeded through a large temperature range – 165–269°C for the gold stage and 154–316°C for the polymetallic stage. The grade of the Romanian ores is about 10 ppm Au (0.3 oz/ton). The native gold is relatively rich in silver containing on the average about 25 per cent Ag. The Au/Ag ratio in the deposits as a whole is highly variable especially where silver minerals are present; on the average it is low, about 1:2 to 1:4 judging from the available data. Where the deposits are silver-rich the ratio Au:Ag is in the range of 1:100 to 1:200.

The Tertiary gold deposits of New Zealand occur in the Coromandel Peninsula (Hauraki goldfield) and adjacent regions where andesite and dacite flows of Eocene and Miocene age overlain by Pliocene rhyolites lie on Mesozoic sediments. Several areas have been highly productive, including Waihi, Thames, Coromandel, Kuaotunu, Waitekauri, Komata-Maratoto, Karangahake, Te Aroha and Te Puke. There is an extensive amount of literature on the deposits in these areas, most of which is summarized by Henderson (1930), Emmons (1937) and Williams (1974). Of particular interest are the reports and papers by Park (1897), Fraser (1910), Fraser and Adams (1907), Finlayson (1909), Bell and Fraser (1912), Henderson and Bartrum (1913), Morgan (1924) and Williams (1974).

The deposits at Waihi occurred in a conjugate system of fissures in highly propylitized dacite and andesite. The principal fissure, the Martha, strikes northeast, dips steeply southeast and has been traced for about 1 mi. The subparallel Empire and Royal, and most of the numerous other lodes and fissures, lie to the northeast; they dip steeply northwest and join but, so far as known, do not cross the Martha in depth. Most of the ore-shoots became poorer or terminated altogether at from 600 to 800 ft below sea level.

The orebodies consisted essentially of quartz veins well mineralized with valencianite (orthoclase), calcite, rhodochrosite, pyrite, sphalerite, galena, chalcopyrite, pyrargyrite, acanthite and minor stibnite, native arsenic and molybdenite. The ores were banded and the quartz was colloform (chalcedonic) in some places and lamellar or platy in others. This lamellar or platy quartz is characteristic of some types of Tertiary gold deposits in various parts of the world. Some investigators have suggested that this type of quartz is due to replacement of calcite, but Morgan (1924) thinks this is erroneous. Much of the gold in the Waihi ores was free and there were some tellurides. The grade averaged 0.5 oz Au/ton or more and the Au/Ag ratio averaged about 0.3.

In the Thames area of New Zealand (Fig. 29) the deposits are quartz veins in highly propylitized andesites in a comparatively small downthrow fault block. There was practically no payable ore below a depth of 500 ft from the outcrop of any vein. Fabulous bonanzas characterized the Thames veins, some containing 1 to 6 oz electrum/lb of ore. The principal minerals were quartz, calcite, rhodochrosite, rhodonite, barite, ankerite, pyrite, chalcopyrite, galena, sphalerite, stibnite, molybdenite, cinnabar, argentite, pyrargyrite, hessite, petzite, native arsenic and native gold, the last containing up to 40 per cent silver.

The numerous veins at Coromandel were of the Thames bonanza type. At Te Puke the veins were similar to those at Waihi, but no large bodies of secondarily enriched ore were found as in the latter district. The quartz lodes in the Te Aroha area are mainly in extensive fissures in andesites and dacites. The main fissure or Waiorongomai Buck reef was followed for some 3 mi on strike. It dipped east and consisted of a zone of crushed silicified rock cut by numerous longitudinal and transverse fissures. In places it was 60 ft wide. The mineralization consisted of quartz, calcite, ankerite, rhodochrosite, valencianite (orthoclase), pyrite, chalcopyrite, galena, sphalerite, cinnabar, acanthite, pyrargyrite and native gold, the last ranging in fineness from 645 to 900. The ores of the Karangahake area are similar to those at Te Aroha.

A recent contribution to our knowledge of the Hauraki goldfield by Ramsay and Kobe (1974) is of interest. They studied the silver-gold ores on Great Barrier Island. There, the deposits comprise a number of east-striking, quartz-filled, mineralized fissures that transect andesites and an unconformably overlying siliceous sinter deposit of upper Pliocene to Pleistocene age. The wall-rock alteration exhibits a crude zoning consisting of a general propylitization in the andesites with argillic assemblages and silicification developed close to the veins. The endogene ore minerals comprise pyrite, marcasite, arsenopyrite, chalcopyrite, sphalerite, rare galena and tetrahedrite, acanthite, electrum, silver, chlorargyrite, pyrargyrite, proustite, polybasite-pearceite, argyrodite, aguilarite, naumannite, realgar and stibnite. The principal supergene minerals are limonite, hematite, covellite, native silver, kermesite and cervantite. The electrum is approximately 500 fine, while the Au/Ag ratio of the deposits as a whole is 0.033. The ore is seleniferous, some of the polybasite containing up to 3 per cent Se. The argyrodite is, likewise, selenium-bearing, as is also the galena. The two selenides, aguilarite and naumannite, are relatively rare. The presence of the germanite and argyrodite is of interest, considering that certain thermal waters in

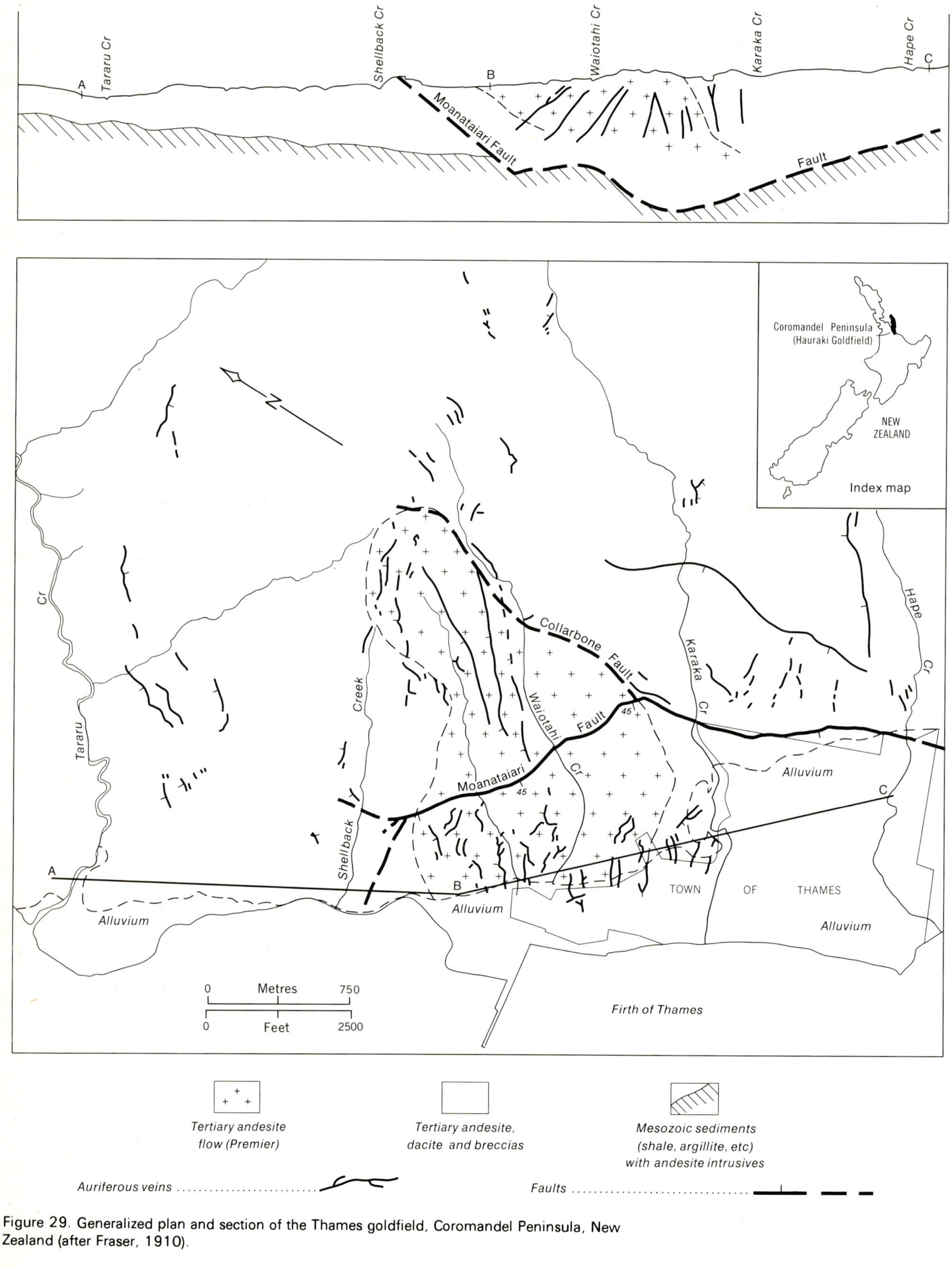

Figure 29. Generalized plan and section of the Thames goldfield, Coromandel Peninsula, New Zealand (after Fraser, 1910).

North Island, New Zealand (Taupo zone) are germanium bearing (Browne, 1969). Ramsay and Kobe (1974) considered that the maximum depth of deposition was of the order of 500 m and regarded the mineralization as subvolcanic and of Upper Pliocene or Pleistocene age. The paragenetic sequence of ore minerals involved a number of episodes that overlapped one another. Initial deposition was characterized by pyrite and marcasite, which continued intermittently through the subsequent base metal, silver sulphosalt, selenide and stibnite stages. A very late hypogene episode resulted in the deposition of chlorargyrite and widespread kaolinite. Deposition of electrum appears to have commenced with the silver-sulphosalt stage and increased with the closing stages of mineralization.

From the descriptions of the various mining regions both recent and past the Hauraki deposits appear to be crudely zoned with cinnabar mineralization near the surface, gold-silver mineralization also near the surface and passing successively downward into lead-zinc zones and then into copper-rich zones.

The Tertiary gold deposits of Fiji are all veins and lodes (Ibbotson, 1962; Houtz and Phillips, 1963; Rickard, 1966). Those in the Vatukoula field (Emperor Gold Mine) in northern Viti Levu (Blatchford, 1953; Markham, 1960; Cohen, 1962; Ibbotson, 1967; Denholm, 1967) occur in numerous narrow shear and shatter zones that vary in attitude from subhorizontal flatmakes to vertical shears. These cut gently dipping basaltic flows, tuff beds and ash beds intruded by numerous andesite dykes, all of which form part of a Tertiary volcanic caldera. The productive veins appear to favour the basalts on the western flank of the caldera, although many of the shears straddle the contact of the caldera and the underlying basalts. The andesite dykes are heavily propylitized, but the basalts and other volcanics only exhibit marked propylitic effects near the quartz lodes. The intervening basalt between shears is generally quite fresh. The lodes of the Emperor Mine are composed mainly of crustified and vuggy quartz accompanied by some carbonate. The metallic minerals include pyrite and arsenopyrite, both of which are auriferous, pyrrhotite, marcasite, chalcopyrite, galena, sphalerite, tetrahedrite-tennantite, native gold, native tellurium and a number of Au-Ag tellurides. Native gold is relatively rare and when present is dendritic and leafy. The native tellurium contains up to 0.1 per cent Au. The tellurides, native tellurium and native gold are late minerals in the paragenetic sequence. The gold content of the ores ranges from 0.25 to 1 oz/ton, and the Au/Ag ratio of the ores averages about 3. The fineness of the hypogene microscopic gold is about 940 according to Stillwell (1949). The Emperor Mine produces both gold and tellurium. The lodes when first found exhibited only shallow oxidation zones in which fine wire gold of great fineness (960) was encountered. As in New Zealand there were no significant placers associated with the Vatukoula goldfield. Deposits similar to those at Vatukoula occur on Vanua Levu at the former Mount Kasi Gold Mine (Green, *in* Fisher, 1973). There, the basaltic and andesitic pyroclastics are considerably altered by propylitization, silicification and sericitization, adjacent to the quartz lodes and veins that contain essentially quartz, carbonate, pyrite, barite, native gold and minor tellurides.

The most productive gold deposits in Indonesia are in Tertiary rocks. The deposits include those at Paleleb, Soemalta, Totok and Bolang Mongondou in the north arm of Celebes, those in the Lebong district in Sumatra and those in western Java, among which are the Tjikotok and Tjipitjoeng districts (De Jongh, 1930; Van Bemmelen, 1970). The deposits are mainly quartz veins and lodes and silicified zones in faults, fractures and sheeted zones in Tertiary andesites, porphyritic diorites, rhyolite, trachyte, and their associated tuffs and slaty sediments. The mineralization is mostly quartz and carbonates commonly manganiferous, with variable amounts of arsenopyrite, pyrite, pyrrhotite, sphalerite, galena, chalcopyrite, acanthite, pearceite, polybasite, tennantite, enargite (luzonite), native gold (commonly electrum) and gold-silver tellurides (hessite, sylvanite, petzite). Rhodonite, rhodochrosite and manganese-silver compounds are common in some veins; zeolites, some manganiferous, have been noted in a few veins; and silver and other selenides in the Lebong district. The gold content of the veins is variable but is commonly 0.5 oz/ton or more. The Au/Ag ratio ranges from 0.16 to 1. Some of the ores are highly seleniferous, the selenium being largely concentrated in sulphides, particularly in acanthite, galena and arsenpolybasite (Kieft and Oen, 1973).

The gold deposits in the Philippines have recently been described by Bryner (1969). They include the enargite-luzonite-telluride deposits at Lepanto, Luzon, localized along the intersection of an unconformity and a fault zone; the quartz-telluride-gold vein deposits of the Baguio district, Luzon; and a quartz–base metal sulphide deposit in the Paracale district, Camarines Norte. Lepanto and some of the Baguio district deposits are spatially related to centres of Tertiary explosive volcanic activity.

The Lepanto deposit is a pencil-shaped polymetallic body encased in a silicified and argillized envelope in late Miocene to early Pliocene dacitic pyroclastics and Cretaceous-Paleogene metavolcanics. The economic minerals of interest are enargite, luzonite, tennantite and Au-Ag tellurides. The Au/Ag ratio is about 1:4. The Baguio deposits are mainly veins in fractures and are composed essentially of quartz, calcite, rhodonite, rhodochrosite, pyrite, base metal sulphides, Au-Ag tellurides, native gold, stibnite and cinnabar. Some of the veins contained considerable amounts of gypsum and anhydrite. Many of the deposits are in andesites and dacites cut by numerous dioritic dykes and plutons. The Paracale-Gumaus Mine works veins containing pyrite, chalcopyrite, sphalerite, galena and gold in a quartz gangue. Many of the veins are in granodiorite beneath an ultramafic body. A pale greenish bleaching is widespread in the vicinity of the more productive veins.

The gold deposits of Taiwan are mainly in Tertiary volcanics and sediments (Huang, 1955, 1960, 1972, 1973). The gold-copper deposits of the Chinkuashih Mine occur as irregular lenses or pipes in silicified Miocene sandstone or dacite. Fissure-filling and impregnation or replacement are the principal processes of ore formation. Sheeted and brecciated structures are common. The ore minerals are pyrite, enargite, luzonite, bournonite, sphalerite, cinnabar and native gold, accompanied by the gangue minerals quartz, barite, alunite, kaolinitic clay, and minor amounts of diaspore and sulphur. The ores are characterized by crustification, ringore and drusy structures. Wall-rock alteration is marked. From the orebodies outwards there are three zones – a silicified zone, an argillized or alunitized zone and a chloritized and carbonated zone. Supergene enrichment is not present except for superficial enrichment of native gold of residual origin.

The Wutanshan area, adjacent to the Chinkuashih Mine, is underlain chiefly by Miocene sediments intruded by dacite, both greatly fractured by faulting. The ore deposits in this area occur as fissure-fillings of thin, lenticular form in dacite, and partly in sandstone and shale. The ore is composed mainly of quartz and pyrite, accompanied by minor barite, calcite, manganocalcite, kaolinitic clay, galena, sphalerite, marcasite and native gold. Crustification and drusy structures are common. In contrast with the Chinkuashih deposits, enargite, luzonite and alunite are absent, and quartz gangue is abundant. Three alteration envelopes associated with the deposits are defined: from the veins outward, there are zones of silicification, argillization and chloritization-carbonatization. Alunitization is absent and silicic alteration is not as common as in the Chinkuashih area. The deposits are of epithermal origin and were probably formed in late Pleistocene time. They show the same genetic and age relationships as the Chinkuashih deposit.

In Japan, gold deposits of Tertiary age are widespread (Tatsumi, 1970; Nishiwaki *et al.,* 1971; Kitami, 1973), being localized in the zone of andesite, basalt, rhyolite, dacite and associated tuffs and sediments known as the Neogene 'Green Tuff Region' by the Japanese geologists. Nearly all of these rocks are gently dipping and highly faulted. In addition they are extensively altered on a regional scale, the principal alteration effects being those of propylitization and argillization. Nishiwaki *et al.* (1971) recognized nine types of Neogene gold-silver deposits in Japan. These occur as fissure veins, stockworks and stockworks in intensely silicified rock. The types are as follows:

(1) Those rich in base metals with an Au/Ag ratio equal to about 1. Their mineralogy comprises pyrite, sphalerite, chalcopyrite, galena, hematite and gold (electrum);

(2) Those rich in chalcopyrite and mined mainly for copper;

(3) The argentite (acanthite) type typified by sphalerite, galena, pyrite, chalcopyrite and gold. Gangue minerals are quartz, adularia, sericite, clay minerals and calcite. The Au/Ag ratio ranges from 0.025 to 0.1;

(4) The manganese mineral type characterized by abundant rhodochrosite and rhodonite plus the mineral assemblage and Au/Ag ratios described in (3) above;

(5) Deposits with high amounts of silver sulphosalts such as polybasite and pearceite and a mineral assemblage as in (3) above;

(6) An intermediate type between types (5) and (7);

(7) Deposits with sulphosalts having a low silver content;

(8) The luzonite-enargite type characterized by the presence of luzonite-enargite, pyrite and quartz. These deposits are very poor in silver minerals and generally lack galena and sphalerite;

(9) Telluride-bearing deposits. These are rare. The mineral assemblage is usually tellurides, luzonite-enargite, tetrahedrite, pyrite, quartz and barite.

Vertical zoning is a conspicuous feature in some of these deposits, the general downward zonation being Au-Ag, Pb-Zn, Cu. A number of producing gold mines occur in southern Kyushu that are representative of some of these types and will serve as examples (Matsukuma, 1962; Taneda and Mukaiyama, 1970; Wakabayashi *et al.,* 1973).

The Kushikino deposit near the city of the same name is an anastomosing series of quartz veins in fractures in gently dipping propylitized andesite (Fig. 30). The footwalls of the veins are highly silicified whereas the hanging walls exhibit extensive sericitization. Both types of alteration are also associated with a development of small amounts of adularia, calcite and pyrite. Two types of ores are present in the Kushikino deposit; one contains mainly milky medium-grained quartz with native gold (electrum) and small amounts of scattered sulphides; the other is a fine-grained chalcedonic quartz with thin black bands (Ginguro ores) of fine-grained sulphides, sulphosalts, selenides and electrum, mainly pyrite, chalcopyrite, marcasite, sphalerite, galena, acanthite, pyrargyrite, stephanite, polybasite, naumannite, tetrahedrite and electrum. The medium-grained quartz ores contain up to 1 oz Au/ton and the Au/Ag ratio ranges from 0.25 to 1. The Ginguro ores are rich and usually contain >1 oz Au/ton; their Au/Ag ratio ranges from 0.03 to 0.1.

The Kasuga Mine near Makurazaki City represents a somewhat different type of Tertiary gold deposit. It consists of a stockwork of quartz veins and stringers in a porous silicified zone developed in andesite tuff, tuff breccia and tuffaceous shale. The core of the mineralized area is represented by a porous mass of fine-grained quartz. Surrounding this there is a dickite-kaolin zone, which grades imperceptibly through kaolinized rock into unaltered tuffs and other rocks (Fig. 31). The minerals in the quartz veins and stringers and silicified zone are mainly enargite and native gold. Native sulphur occurs in places. The grade of the quartz vein stringer zones of the deposits is about 0.6 oz Au/ton. The highly siliceous part of the deposits averages about 0.1 oz Au/ton and is used as a flux in copper smelters.

The Tertiary gold deposits of the U.S.S.R. occur mainly in the far east (Radkevich, 1966; Radkevich *et al.,* 1965; Shilo *et al.,* 1969; Rozhkov, 1971*b,c,d*; Zagruzina, 1972; Anoshin and Kepezhinskas 1972; Magakian, 1972; Kuznetsov and Panifilov, 1974) particularly in the mobile belt and areas underlain by Tertiary volcanics and other rocks in Kamchatka (Aprelkov and Kharchenko, 1968), the lower reaches of the Amur River (Yugai, 1967) and the Okhotsk-Chukotsk volcanic belt (Sidorov, 1965). The Tertiary rocks in this vast territory are varied and include basalts, andesites, dacites, trachytes, rhyolites and the corresponding tuffs. Some of the deposits are in Mesozoic rocks but are of Neogene age. Other deposits are confined solely to the Tertiary rocks, large stretches of which show the characteristic propylitization and argillization common to Tertiary rocks throughout the world. The deposits are, likewise, varied, but most are systems of quartz veins or stockworks in andesites and dacites and silicified zones with quartz veins in tuffs. To generalize it can be said that the chief minerals are quartz, carbonates, pyrite, arsenopyrite, the various base metal sulphides, tetrahedrite, stibnite, various silver minerals, minor scheelite, native gold and electrum. A few of the deposits contain a little cassiterite and others have small amounts of cinnabar. Adularia figures in the wall-rock

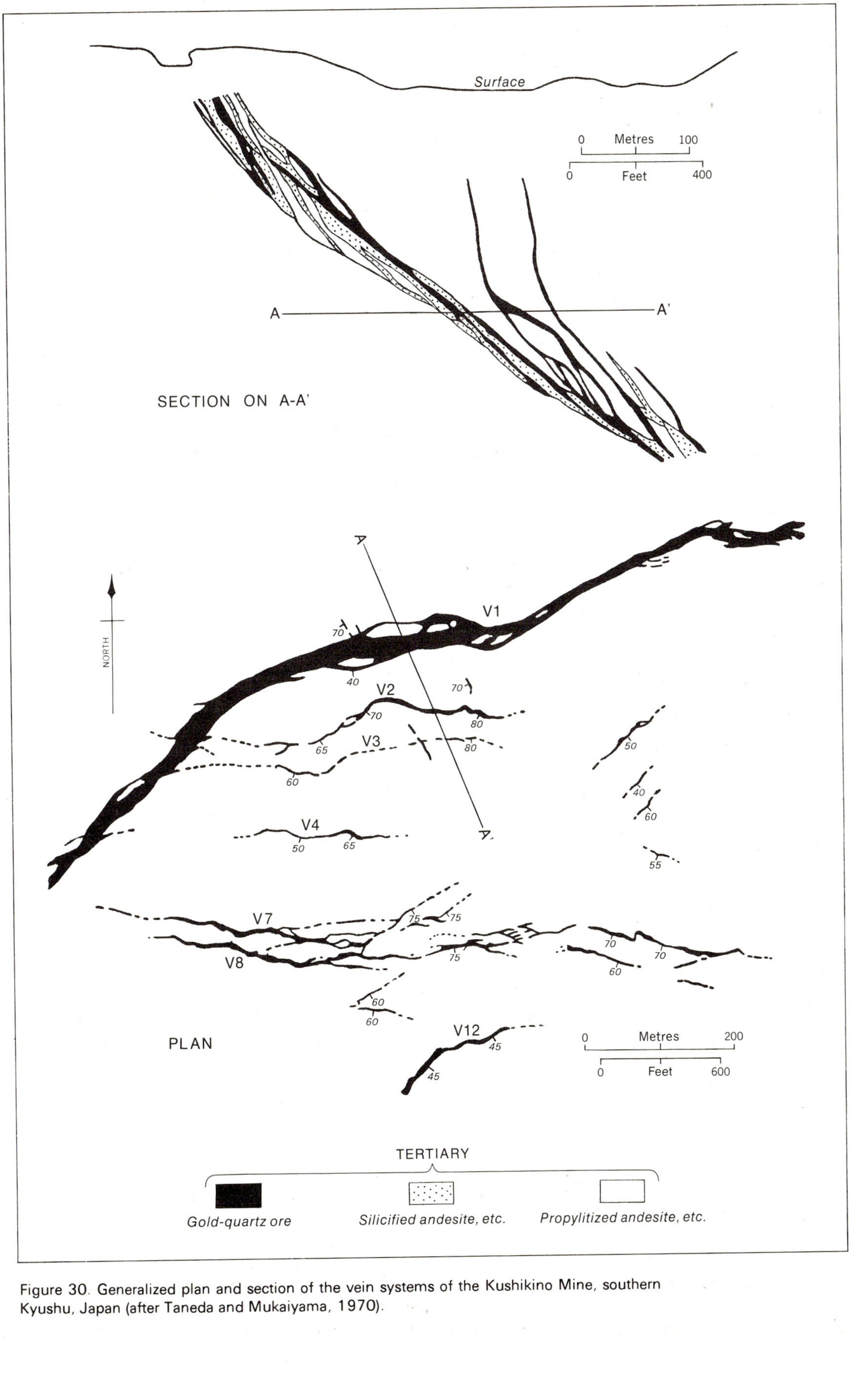

Figure 30. Generalized plan and section of the vein systems of the Kushikino Mine, southern Kyushu, Japan (after Taneda and Mukaiyama, 1970).

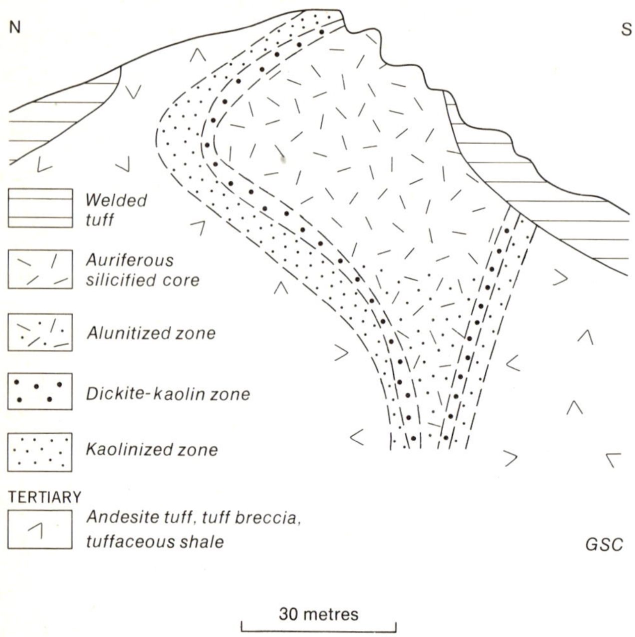

Figure 31. Generalized section showing the ore zone and alteration halos of the Honko deposit of the Kasuga Mine, southern Kyushu, Japan (after Taneda and Mukaiyama, 1970).

alteration zones and vein minerals in some veins and alunite in others. Braunite, the manganese silicate, is present in some veins and zeolites and epidote are characteristic of some veins and their alteration halos. The deposits are generally silver rich, the fineness of the gold averaging about 700.

In detail, the auriferous deposits of the far eastern U.S.S.R. can be classified into four categories: gold-quartz, gold-rare metal, gold-silver and gold-sulphide (Shilo *et al.*, 1969).

The gold-quartz deposits are mainly veins and irregular bodies generally related to folding. They are particularly common in the Mesozoides of the northeastern U.S.S.R. Their mineralization consists essentially of quartz, carbonates, pyrite, arsenopyrite, galena, sphalerite, sulphosalts and native gold. The age of many of these deposits is apparently middle to late Mesozoic (Zagruzina, 1972).

The gold-rare metal deposits are mainly veins and lodes associated with faults and fracture systems and are said to be related mainly to granitoid intrusions of the volcanoplutonic gabbro-andesite-granodiorite clan. The mineralization occurred in at least two stages – greisenization with the development of wolframite and molybdenite followed by quartz, carbonates, disseminated base metal sulphides, cassiterite (wood tin in places), wolframite, scheelite, bismuth tellurides, gold and gold-silver tellurides. The age of this mineralization is late Cretaceous-Paleogene.

The gold-silver deposits are veins and stockworks associated mainly with deep faults. Their mineralization comprises essentially quartz, commonly colloform banded, rhythmically crystalline banded, drusy, or platy, base metal sulphides, hematite, stibnite, acanthite and gold. Some deposits are transitional to the gold-rare metal polymetallic types of deposits. Propylitization is characteristically developed in the volcanic host rocks of many of the gold-silver deposits. Most of the gold-silver deposits are late Cretaceous to Paleogene in age.

The gold-sulphide deposits are complex mineralogically, consisting essentially of quartz, albite, carbonate, chlorite, hydromica, pyrite, chalcopyrite, galena, sphalerite, tetrahedrite and gold. The sulphides may make up 30 to 40 per cent of the ores. Propylitization of the andesite host rocks of the gold-sulphide deposits is common and widespread. The age of the deposits is apparently late Cretaceous–Paleogene, mainly the latter.

As many as four stages may be distinguished in the formation of some of the deposits. The first stage was one of silicification, chloritization and epidotization of the host rocks. Among the ore minerals, pyrite and hematite were introduced at this stage. During the second stage, the greater part of the quartz veins was formed, and sulphides of lead, zinc and copper were deposited along with silver and gold. The adularia also appeared during this stage. Thereafter, tectonic movements reopened the fissures, fractured the previously deposited ores and vein material and new quartz veins were deposited. During this third stage, some galena, sphalerite, chalcopyrite and probably also gold were deposited. During the fourth and last stage there was no influx of ore minerals – only the carbonates were precipitated in the central parts of the earlier veins.

One of the deposits, the Belaya Gora, in the gold-bearing region of Nizhyni Priamur, has been extensively investigated (Mel'nikov and Mel'nikova, 1970; Petrovskaya *et al.*, 1971*d*). The deposit is composed of impregnations and stockworks of minerals in the vent and near-vent facies of Oligocene volcanics comprising felsite porphyries, liparites, trachytes and dacites. These break through relatively flat-lying andesite and basalts. The mineralization consists principally of fine-grained quartz and clay minerals, adularia, gold and small amounts of sulphides including pyrite, arsenopyrite, chalcopyrite, sphalerite, galena, hematite, pyrargyrite, proustite, native silver, tetradymite, cinnabar and metacinnabar. There are also small amounts of wood tin. The Au/Ag ratio of the deposit is 1:4. The host rocks are extensively altered. There is a more or less regional propylitization; near the deposits there is much silicification, pyritization, argillization, some adularization, sericitization, carbonatization and chloritization dependent on the rock type affected.

The elements enriched in the Tertiary deposits of the Soviet far east include: Cu, Ag, Au, Ba, Zn, Cd, Hg, (B), Ga, La, Sn, Pb, As, Sb, Bi, (Se), Te, Mo, W, Mn, Fe and S. Those in parentheses are relatively rare or local.

Tertiary gold deposits in the Western Cordillera of North America are confined mainly to Mexico and the United States. Few deposits in Mexico are essentially gold deposits. Most, such as those at Pachuca, Real del Monte and Guanajuato (Veta Madre), are rich silver deposits with a little gold as a byproduct (Echegoyén *et al.* (1972)). Their oxidized zones in places, however, were apparently rich in gold.

El Oro district in Mexico, some 70 mi northwest of the capital, is noted for its Tertiary gold lodes in fault fissures (Flores, 1920). These cut Mesozoic pyritic shales intruded by Tertiary andesite sills. The veins were oxidized and then covered by later flows of andesite. The mineralization comprises fine-grained, often banded and crustified quartz, calcite, pyrite, sphalerite, acanthite and other sulphides. The ores average about 0.5 oz Au/ton and 10 oz Ag/ton. The main deposits are now exhausted. Somewhat similar deposits in Tertiary andesites and dacites were mined near Hostotipaquillo and elsewhere in Mexico (Ridge, 1968; Albers and Kleinhampl, 1970; Roberts *et al.,* 1971).

In the western United States there are numerous famous gold camps closely associated with Tertiary volcanics; their rich gold-silver ores are now largely exhausted. As examples, we may mention deposits in the Black Mountains of Arizona; De Lamar, Idaho; Rawhide, Jarbidge, Tonopah, Goldfield and Virginia City (Comstock Lode), Nevada; Cripple Creek and the San Juan region, Colorado; and Republic in Washington. These exhibit a varied geology and mineralogy that can be touched upon only briefly here. All deposits are veins, lodes or stockworks in faults, fractures, sheeted zones or shattered zones in Tertiary volcanic terranes. Most are silver-rich with a low to very low Au/Ag ratio. Fabulously rich gold-silver bonanzas were features of many of the deposits. Many of the veins were attended by extensive wall-rock alteration effects including pyritization, sericitization, chloritization, silification, carbonatization, and the development of adularia and alunite.

The deposits in the Black Mountains of Arizona near Oatman were veins and lodes along fault fissures, mainly in Tertiary andesite heavily chloritized and sericitized near the veins. The veins were beautifully crustified and filled with chalcedonic quartz, calcite, fluorite, adularia, minor base metal sulphides, acanthite, silver sulphosalts and finely disseminated gold. At De Lamar, Idaho, the veins in fissures and faults cut mainly rhyolite that exhibits silicification, development of adularia and pyritization near the veins. Mineralization consisted essentially of quartz, calcite, adularia, fluorite, minor base metal sulphides, acanthite and naumannite (Ag_2Se). The grade of the ore was about 0.5 oz Au/ton and the Au/Ag ratio is 0.05.

The gold veins at Jarbidge, Rawhide, Gold Circle, Round Mountain, Beatty and National, Nevada are essentially similar. They occurred in faults and fractures that cut rhyolite or, in places, more basic Tertiary volcanics and tuffs and contained essentially quartz, adularia, calcite and small amounts of pyrite, arsenopyrite, base metal sulphides, acanthite, tetrahedrite, pyrargyrite, other silver sulphosalts and naumannite. At National, stibnite is abundant. Some veins contained small amounts of barite, fluorite and clay minerals, mainly kaolinite and halloysite. The native gold was silver rich, often only 500 to 600 fine, and the Au/Ag ratios of the deposits as a whole ranged from 0.02 to 0.3. The grade of most deposits was about 0.5 oz Au/ton.

The silver-gold deposits at Tonopah, Nevada are replacement veins and lodes occupying faults and fractures that cut a complex series of Tertiary tuffs, breccias, rhyolite, trachyte and andesite flows injected by rhyolite plugs (Spurr, 1905; Bastin and Laney, 1918; Nolan, 1930, 1935; Bonham and Garside, 1974). The wall rocks are extensively altered near the veins as the result of silicification, sericitization and the development of adularia. This alteration grades outward into a kaolinite-montmorillonite-chlorite-carbonate-albite alteration phase and finally into the fresh wall rock. The veins were defined mainly by assay and in the case of the famous Mizpah vein were as much as 40 ft wide. The chief hypogene minerals in the veins and lodes were quartz, adularia, sericite, rhodonite, barite, manganiferous calcite, electrum, argentite, polybasite, pyrargyrite, selenides, huebnerite, scheelite, cinnabar, pyrite, sphalerite and galena. The Au/Ag ratio changed from about 1:85 in the central part of the district to less than 1:100 in the western and eastern extremities of the district. The ores, especially the oxidized ones, were exceedingly rich running hundreds of ounces of silver and tens of ounces of gold to the ton in some shoots.

Supergene enrichment of the lodes was notable in the Tonopah district reaching depths of 700 ft or more in places. The supergene minerals developed included limonite, kaolin, hyalite, turquoise, wavellite, dahllite, gypsum, jarosite, wulfenite, native gold and silver, covellite and a variety of silver halides, chlorargyrite in the upper parts of the veins, bromargyrite (embolite) in the middle and iodargyrite in the lower parts (Burgess, 1911). Some of the argentite, polybasite and pyrargyrite were considered by early investigators to be of supergene origin.

The upper limit for the age of the Tonopah mineralization is given as 20.4 m.y. and the lower limit as 17 m.y. by Bonham and Garside (1974).

At Goldfield, some 30 mi south of Tonopah, an intrusive sill of dacite invades Tertiary trachyandesite, rhyodacite, latite and rhyolite flows and tuffs, overlying Mesozoic quartz monzonite that intrudes Ordovician black argillite and shale (Ransome, 1909). Searls (1948) considers that the dacite, in which a high proportion of the gold veins occur, is actually a flow in the volcanic series. Ashley (1974) interprets the geology of the Goldfield district as a volcanic centre marked by a ring fracture that probably delineates a small caldera of Oligocene age (Fig. 32*a* and *b*). Alteration of the volcanics is extensive on a regional scale consisting mainly of silicification, argillization and alunitization. The age of the mineralization is Tertiary (20.6–21.1 m.y.) as determined by potassium-argon dating of primary alunite (Silberman and Ashley, 1970) and fission track methods (Ashley and Silberman, 1976). The orebodies were irregular sheets, pods and pipes with indefinite boundaries within linear silicified zones in the southwestern corner of the extensive alteration zones (Fig. 32*c* and *d*). These silicified zones and their contained ore sheets called "ledges" by Ransome (1909, 1910) consisted of quartz, pyrite, alunite, kaolinite, marcasite, sphalerite, bismuthinite, famatinite, tennantite, goldfieldite (Te-bearing tetrahedrite) and various gold tellurides (petzite, hessite, sylvanite). The average grade of the ore near the end of the life of the deposits was about 0.3 oz Au/ton, but much higher grades, up to 100 oz Au/ton or more, were common in the near surface parts of the lodes. The Au/Ag ratio of the ores was about 3 judging from the old records.

The famous Comstock Lode at Virginia City some 20 mi southeast of Reno, Nevada was one of nature's largest concentrations of gold and silver (Becker, 1882; Reid, 1904–1906; Bastin, 1922). The lode, which developed in a great fault fissure with upward projecting spurs, is of Tertiary age and cut mainly andesite (Fig. 33).

The wall rocks were intensely propylitized and sericitized. Most of the orebodies were bonanza shoots and consisted of intensely crushed quartz, some calcite, pyrite, chalcopyrite,

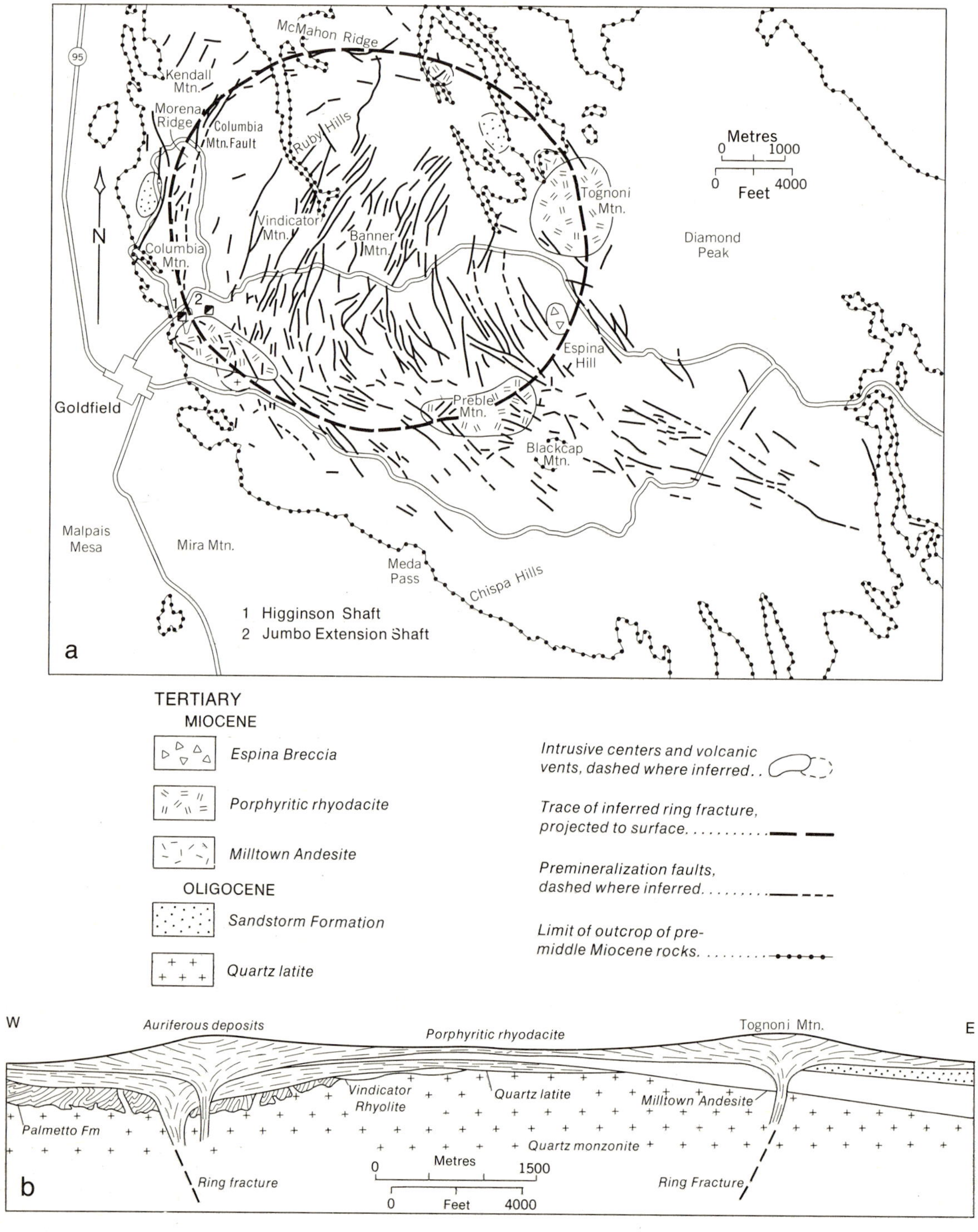

Figure 32. (a) Plan showing premineralization faults and intrusive centres for premineralization volcanic units, Goldfield district, Nevada (modified from Ashley, 1974). (b) Diagrammatic cross-section through the Goldfield volcanic centre prior to shingle faulting and ore deposition (modified from Ashley, 1974).

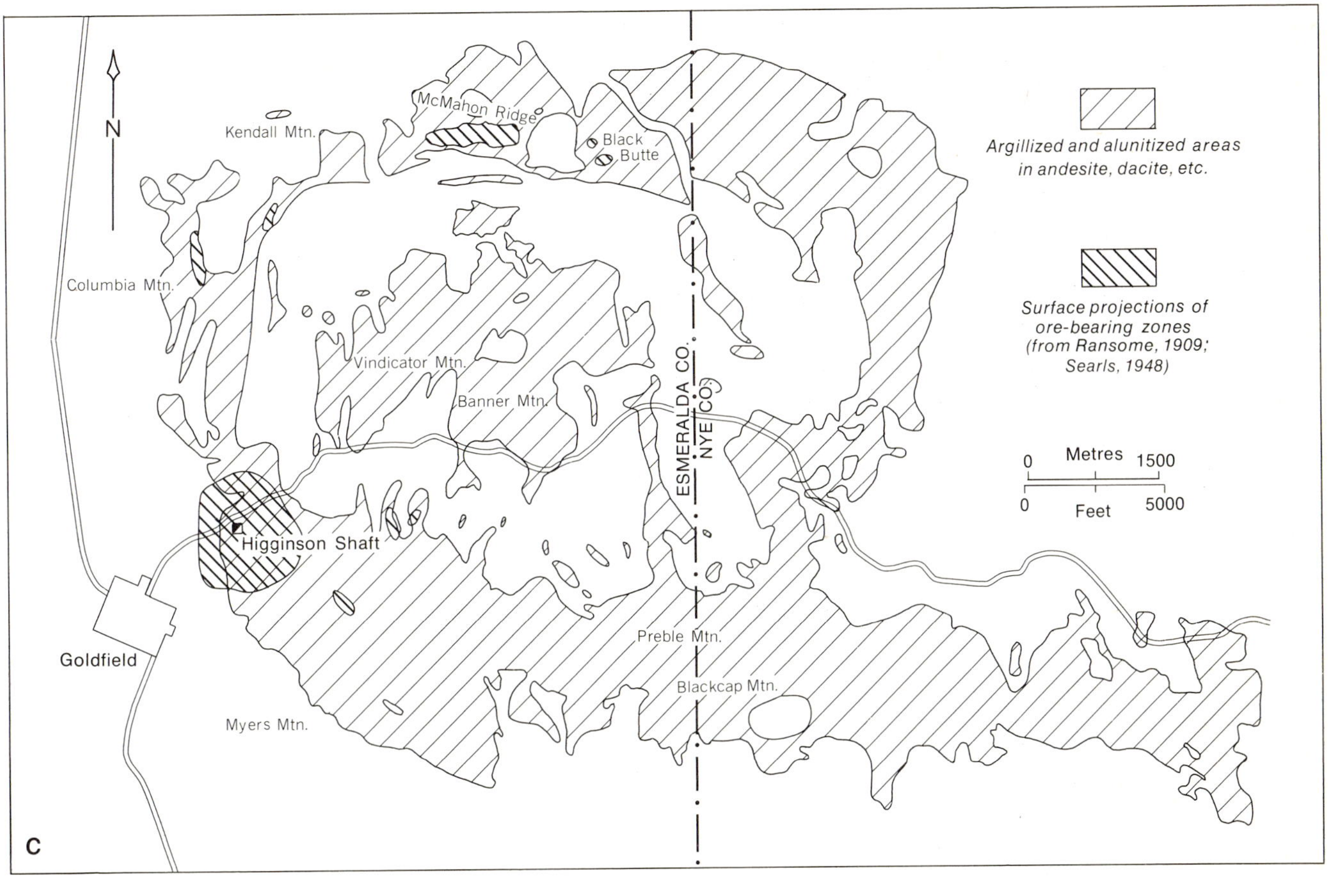

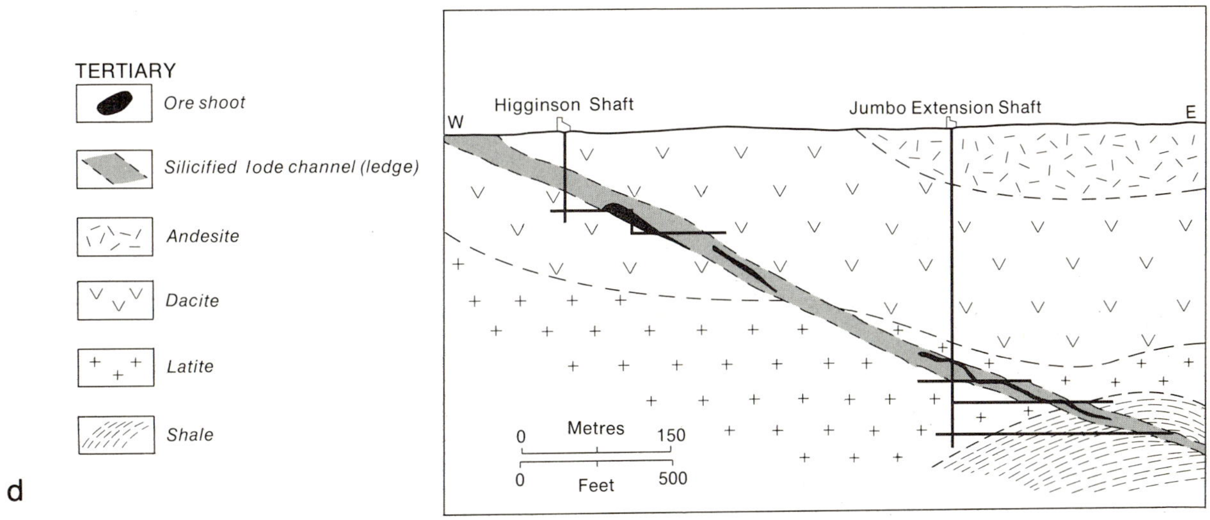

Figure 32. (c) Plan of Goldfield district, Nevada showing mineralized and altered zones (modified from Jensen *et al.*, 1971). (The argillized and alunitized areas enclose the productive and unproductive auriferous ledges.) (d) Vertical section of the Mohawk-Jumbo lode, Goldfield, Nevada (modified from Turner, 1915).

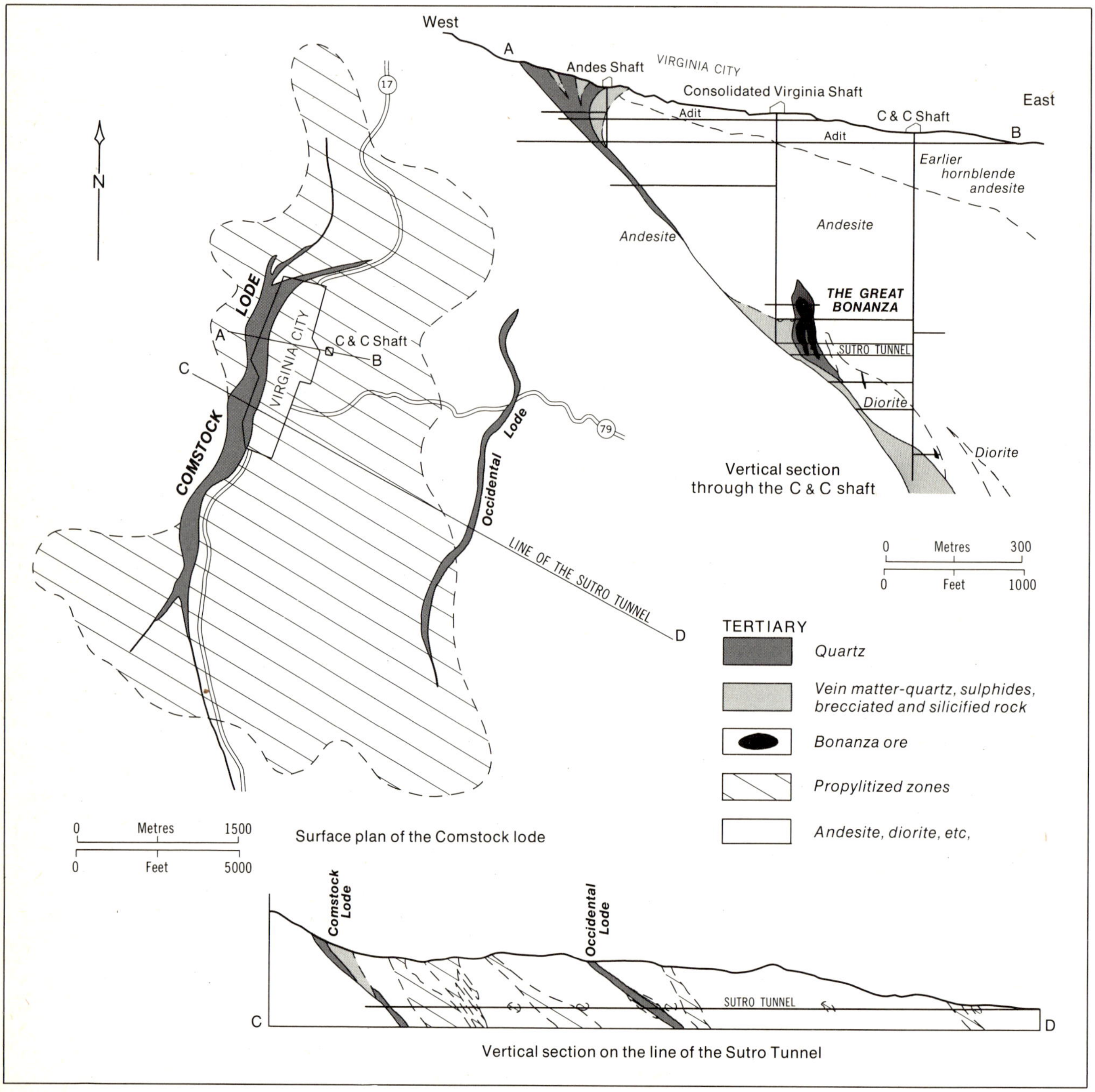

Figure 33. Generalized geological plan and sections of the Comstock Lode, Virginia City, Nevada (modified from Becker, 1882, and others).

sphalerite, galena and abundant gold, electrum, acanthite, polybasite and other silver sulphosalts. The Au/Ag ratio of the ores was about 0.025. In the deeper levels, flows of hot (170°F) calcium sulphate waters and rock temperatures of 115°F made mining of the lode difficult.

The San Juan region in southwest Colorado contains a number of important mineral districts, many of which contain polymetallic deposits with only small amounts of gold and silver. A number of deposits, however, mostly small but exceptionally high grade and now largely exhausted, were essentially gold-silver producers, including Telluride, Ouray, Silverton, Rico, Creede, La Plata, Bonanza, Lake City and Needle Mountains. The rugged San Juan Mountains are underlain chiefly by Tertiary tuffs, andesite, latite and rhyolite poured out on a basement of folded Mesozoic and Paleozoic sediments lying on Precambrian gneisses and other rocks (Steven, 1968; Burbank and Luedke, 1968; Steven and Lipman, 1976). The deposits are veins, sheeted zones and replacements mostly in the volcanics and tuffs, mineralized with quartz, rhodochrosite, rhodonite, barite, fluorite, galena, sphalerite, tetrahedrite, enargite, pearceite, gold tellurides and native gold. Some veins contain a little bismuthinite, molybdenite and huebnerite. Propylitization and sericitization are marked close to the veins. The ores in the Telluride district were in fissures in tuffs, andesite and rhyolite and contained crustified and drusy quartz, calcite, siderite, rhodochrosite, adularia, barite, fluorite, pyrite, galena, sphalerite, chalcopyrite, polybasite and native gold. The ores averaged about 0.3 oz Au/ton and the Au/Ag ratio of the ores was about 0.03. In the Ouray district the famous Camp Bird lode averaged about 1 oz Au/ton. The lode intersected tuff and andesite and consisted of beautifully banded and crustified quartz, rhodochrosite, calcite, fluorite, native gold, gold tellurides and a few per cent of galena, pyrite and sphalerite. The Camp Bird lode is one of many gold, silver and base metal deposits in the Silverton district in faults, fissures and brecciated zones related to the Silverton Cauldron (Fig. 34). The ores of the Creede district (Steven, 1968, 1969) were veins in fault fissures in rhyolite and consisted of banded amethystine quartz, hematite, barite, rhodochrosite, chlorite, argentiferous galena, sphalerite, chalcopyrite, pyrite and acanthite. The gold content averaged about 0.1 oz/ton and the silver about 12 oz/ton, giving an Au/Ag ratio of 0.008. The age of the mineralization (24.6 ± 0.3 m.y.) is latest Oligocene-Miocene (Steven, 1969; Bethke *et al.*, 1976).

In a recent contribution to the mineralization history of the western San Juan Mountains, Colorado Lipman *et al.* (1976) state that the ore deposits formed intermittently in middle to late Tertiary time (from about 30 to 10 m.y. ago) during essentially the same period as that of associated igneous activity. The mineralization occurred recurrently during the waning stages of evolution of several precaldera central volcanoes and also after the formation of the Uncompahgre, San Juan, Silverton and Lake City calderas. The rich deposits of the Silverton Caldera were formed later than the caldera and are closely related in space and time to volumetrically minor intrusions of quartz-bearing silicic porphyry. Lipman *et al.* (1976) stress the point that the time relationships of the volcanism and mineralization seem to preclude any significant genetic tie between the major mineralization and implacement of the caldera-forming magmas. The calderas, therefore, served mainly as structural controls for the mineralization, which seems to be closely related in some as yet unknown manner with minor intrusions of the quartz porphyry type.

The Bodie mining district lies near the eastern margin of the Bodie Hills in California, close to the Nevada border. The country rocks consist mainly of Tertiary andesite, dacite, rhyolite and welded tuffs lying on Mesozoic and Paleozoic sediments and late Mesozoic granitic rocks (Clark, 1970). The mineralization is Tertiary (7.2–8.6 m.y.) as determined by potassium-argon dating on adularia (Silberman *et al.,* 1972) and consists of quartz veins containing adularia, some pyrite, minor base metal sulphides and abundant finely disseminated gold, rich in silver. Most of the veins were unproductive below 1000 ft.

Numerous other gold-silver deposits associated with Tertiary volcanic centres occur in the western United States. Two, Cripple Creek, Colorado and Republic, Washington, are briefly described below; short descriptions of others can be found in Emmons (1937).

Auriferous mineralization of Tertiary age is relatively uncommon in British Columbia and Yukon although as noted elsewhere some of the deposits in the Boundary District of British Columbia and the Mount Nansen deposit in Yukon may be of Tertiary age. Another deposit of undoubted Tertiary age is the Dusty Mac prospect located about 1 mi east of Okanagan Falls, British Columbia (Church, 1973).

The Dusty Mac deposit is a lenslike zone in highly fractured silicified Eocene volcanic rocks and sedimentary debris (White Lake Formation) marked in places by the development of small quartz veins. The mineralization consists mainly of quartz, of at least three ages, disseminated pyrite, native gold and native silver, the last probably supergene in part. Some of the quartz veins carry minor bornite and chalcopyrite. The deposit is marked by a limonitic gossan. A published statement by Dusty Mac Mines quoted by Church (1973) indicates 67 790 tons of ore averaging 0.23 oz Au/ton and 4.97 oz Ag/ton. In a more recent report (Northern Miner, Oct. 9, 1975, v. 61, no. 3, p. 1) the reserves are quoted as 102 413 tons averaging 0.24 oz Au/ton and 4.29 oz Ag/ton with 10 per cent dilution.

Church (1973) considers that the following events led to the formation of the deposit:

1. Development of dilations in major shears.
2. Filling of the dilations with quartz, accompanied by gold and silver mineralization.
3. Late-stage movement in the shear zones resulting in brecciation of the quartz and intermixing of the quartz with crushed andesite wall rocks.

In South America there are numerous gold districts associated with Tertiary volcanic terranes. As an example one may mention the Zaruma mining area in Ecuador described by Billingsley (1926) and Goossens (1972). There the veins cut the Tertiary Portovelo andesite formation, which is intruded by small bodies of dacite, rhyolite and granodiorite. The mineralization is essentially quartz and calcite with pyrite, argentiferous galena, sphalerite, chalcopyrite and native gold. The ore ran 0.25 to 1.0 oz Au/ton and 1 to 3 oz Ag/ton.

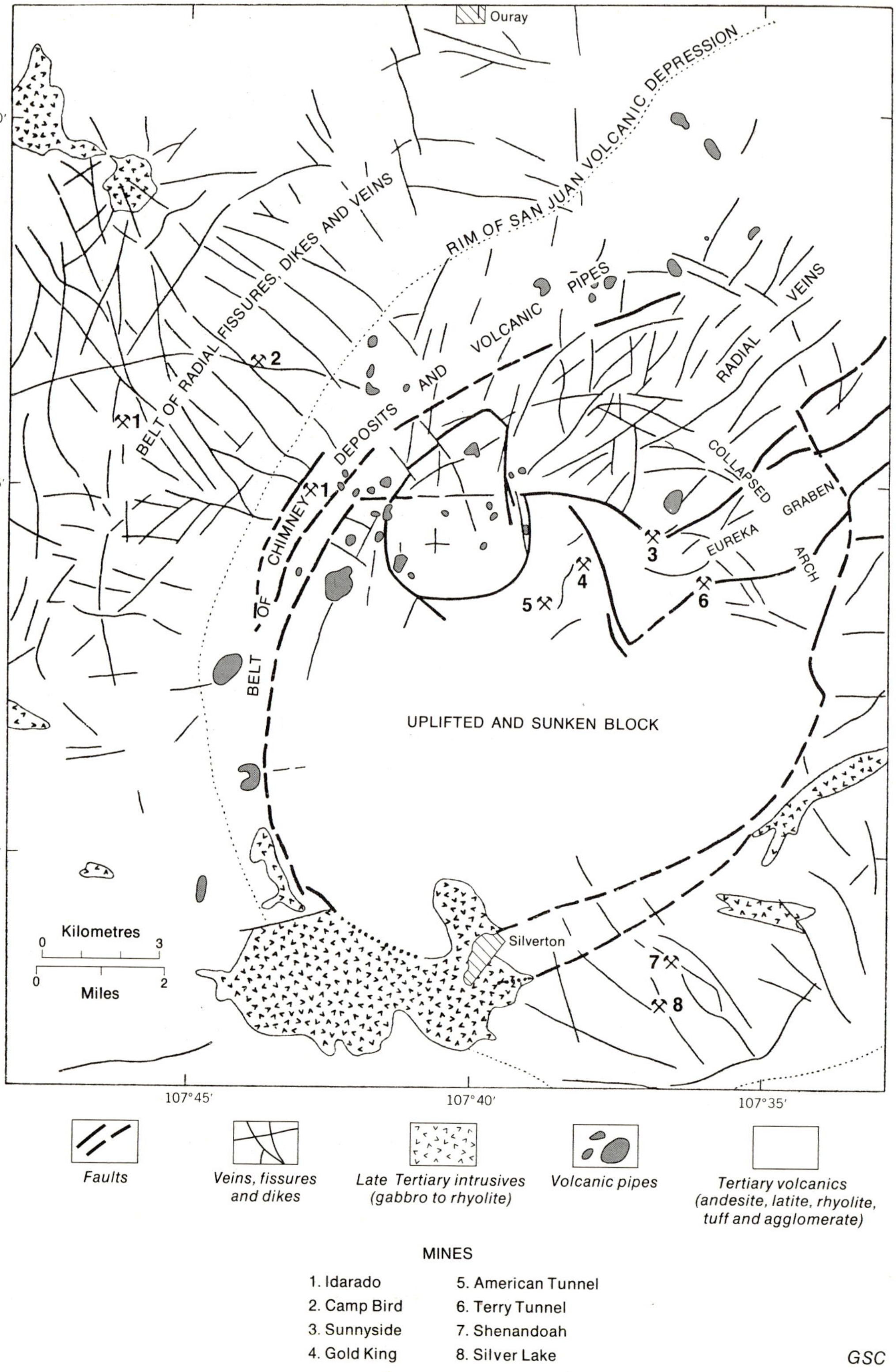

Figure 34. Structural map of Silverton Cauldron showing relationship of veins, faults and mines (modified from Burbank and Luedke, 1968).

Since the time of the Conquistadores, Colombia has produced more than a billion dollars in gold (50 million ounces), two thirds from placers and the remainder from lodes. Most of the lodes are probably of Tertiary age and occur principally in the Department of Antioquia but also in Chocó, Caldas, Nariño, Cauca and Tolima. The country rocks of the lodes, stockworks, mantos, impregnation zones and veins include granites and other granitic rocks, porphyries, sediments and volcanics ranging in age from Precambrian to Tertiary. In the Titiribí district of Antioquia some of the deposits occur in or near highly propylitized and pyritized andesites (Sierra Veta andesite) that are slightly auriferous. The mineralization in the various deposits in Colombia consists essentially of quartz, calcite, dolomite, pyrite, arsenopyrite, sphalerite, galena, bournonite, chalcopyrite, jamesonite, stephanite, ruby silver, cinnabar, argentite, tetrahedrite, stibnite and free gold. Alteration associated with the auriferous bodies comprises mainly sericitization, carbonitization and pyritization. The native gold varies in fineness from 340 to 920; the average gold grade of the deposits as a whole was about 0.3 oz/ton (10 ppm). Some deposits averaged 1 oz Au/ton or more. The silver content ran up to 10 oz/ton (340 ppm) or more in some shoots. The Au/Ag ratios of many of the deposits averaged about 0.35. The average fineness of the placer gold is about 800. A summary of the gold deposits of Colombia with a bibliography is given by Singewald (1950).

Gold-telluride and gold-selenide deposits

These deposits are veins, stockworks, sheeted zones and mineralized pipes commonly in Tertiary volcanic rocks, although some are known in older rocks. The economic minerals in the telluride deposits are mainly gold-silver tellurides, frequently with relatively little free gold except in the oxidized zones. The economic minerals in the selenide deposits are native gold closely associated with various selenides, some silver-bearing and others silver-gold-bearing (fischesserite). Both types of deposits are relatively rare.

Cripple Creek, Colorado is the best known example of the gold telluride deposits (Lindgren and Ransome, 1906; Loughlin and Koschmann, 1935). The country rocks are Precambrian granites and gneisses broken through by a great mass of Tertiary volcanic rocks representing a large volcano (Fig. 35). The core of this volcano is composed of tuffs and breccia of latite-phonolite, which are cut by dykes and masses of phonolite, syenite, monchiquite and vogesite. The deposits are veins, mineralized sheeted zones, replacements in breccia and along fissures and irregular pipes in mineralized breccia (Fig. 36). These cut both the Precambrian and Tertiary rocks,

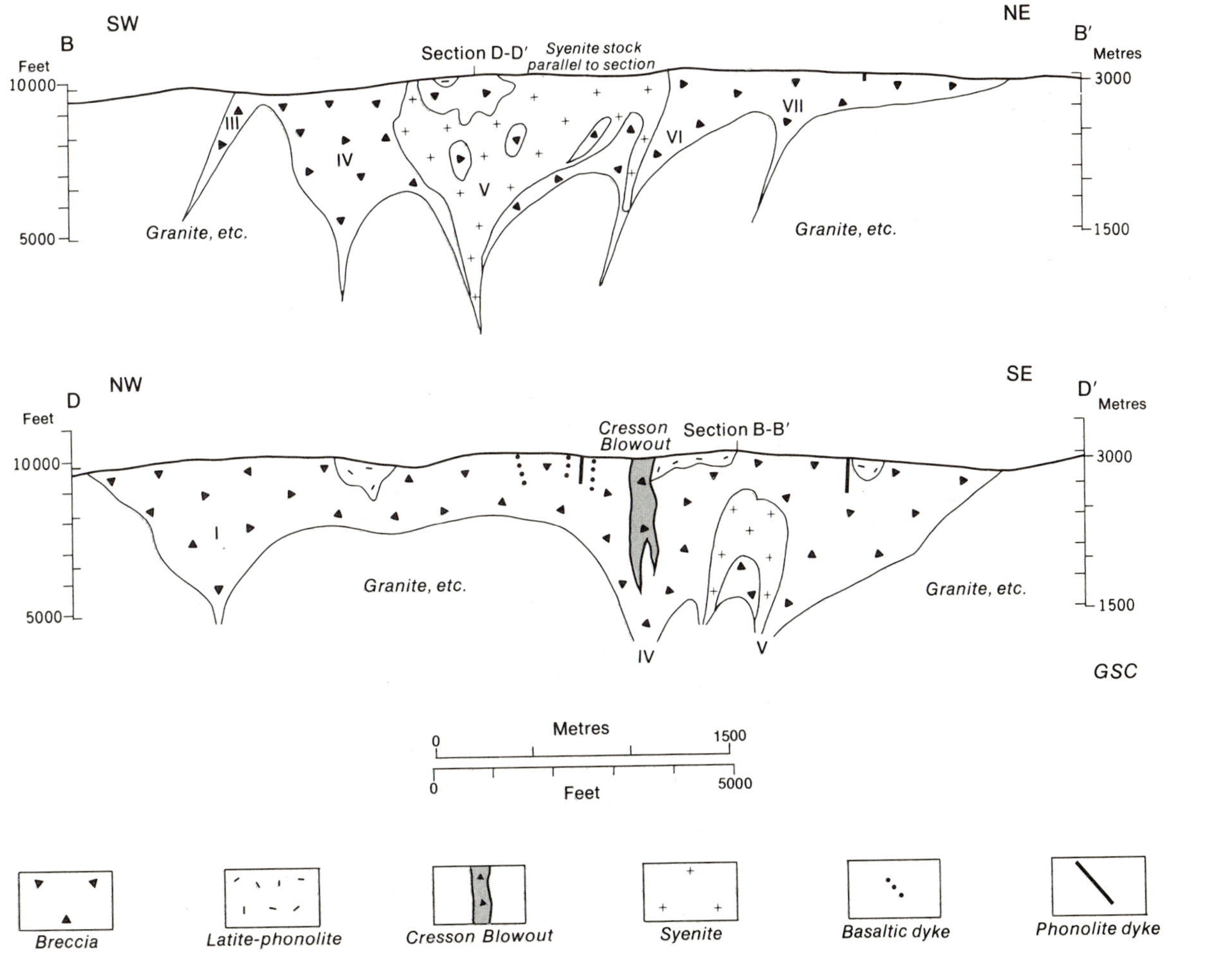

Figure 35. Generalized cross-sections through the Cripple Creek Crater, Cripple Creek, Colorado. Section BB' is approximately at right angles to Section DD'. Roman numerals indicate subcraters (modified from Loughlin and Koschmann, 1935).

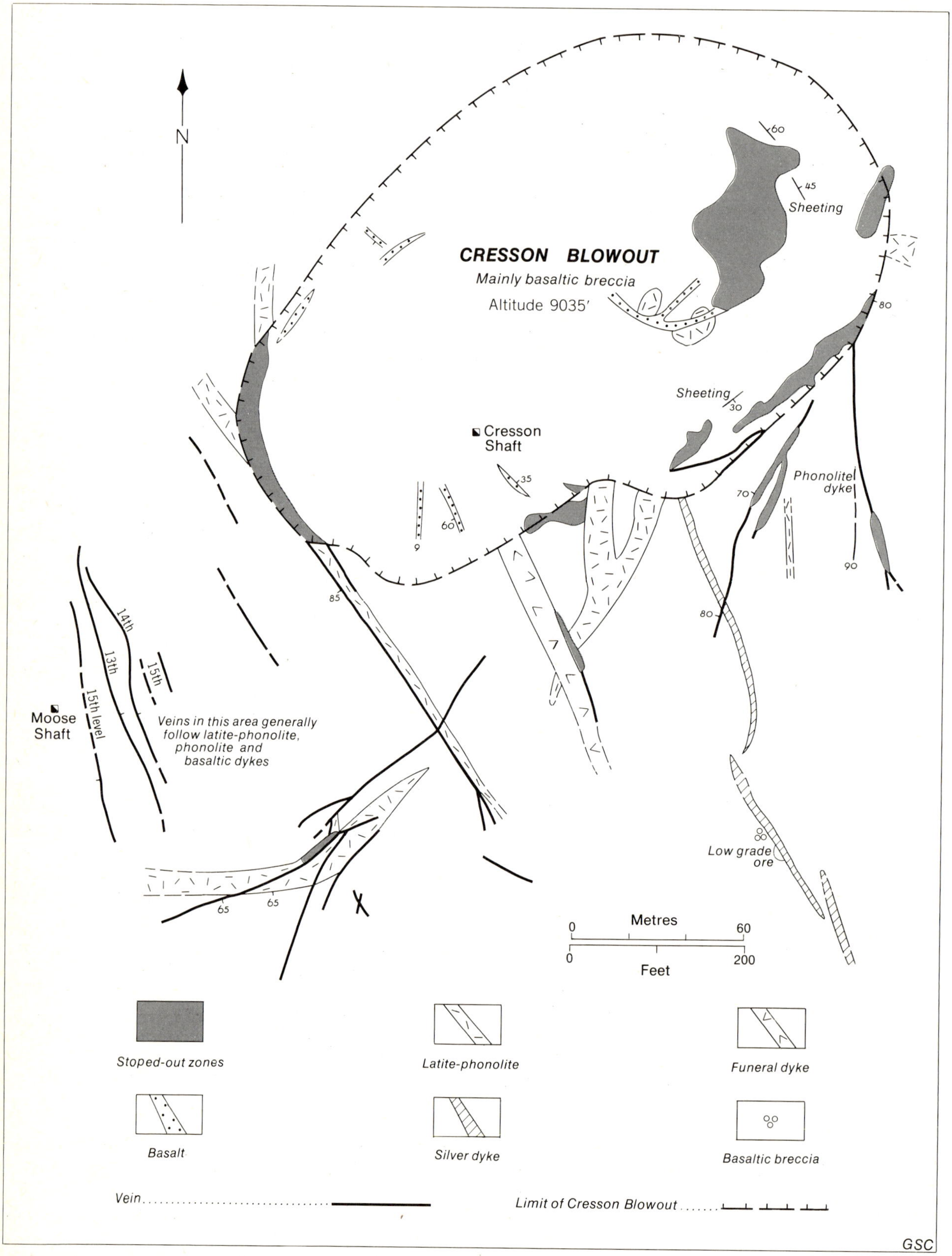

Figure 36. Plan of level 10 of the Cresson Mine and of the lower levels of the Moose Mine, showing relations of Cresson Blowout to the breccia and dykes of the main crater and to veins and irregular orebodies (after Loughlin and Koschmann, 1935).

but are best developed in the latter. The principal ore mineral is calaverite, $AuTe_2$, with a silver content generally less than 4 per cent. There is practically no hypogene native gold. Associated with the calaverite are small amounts of sylvanite, petzite, pyrite, sphalerite, galena, tetrahedrite, stibnite, cinnabar, molybdenite and minor amounts of wolframite (huebnerite). Coloradoite has been identified in the primary ore, and acanthite, native gold and jarosite in the oxidized zone (Hildebrand and Gott, 1974). The gangue is quartz, fluorite, carbonate and roscoelite (the vanadium mica). The vein structure is drusy. The alteration is pyritization, carbonatization and propylitization. Adularia is developed in some veins.

Loughlin and Koschmann (1935) recognized three stages in the formation of the Cripple Creek orebodies. The first stage was characterized mainly by the deposition of jasper and porous aggregates of quartz and adularia and fine-grained, massive mixtures of fluorspar and quartz accompanied by some coarsely crystalline pyrite. The characteristic minerals of the second stage are quartz, fluorspar, fine-grained pyrite, dolomite or ankerite, celestite, roscoelite, sphalerite, galena, tetrahedrite, tellurides of gold and in some places tellurides of silver and copper. The tellurides are late minerals generally following all of the other minerals of the second stage. The third general stage is represented mainly by openings containing clear to smoky quartz, chalcedony, fine-grained and radiating pyrite crystals, calcite and locally cinnabar.

The deposits of Cripple Creek are now largely exhausted. During their lifetime they produced some 20 000 000 oz of gold or more and 2 000 000 oz of silver. The fabulously rich calaverite 'vugs' or 'caverns' in the Cresson 'blowout' yielded more than 1 000 000 oz of gold.

Gold-silver tellurides are important ore minerals in a number of other deposits throughout the world. In these, however, the tellurides are generally subsidiary to native gold. Examples include Kirkland Lake, Ontario, Canada; Kalgoorlie, Western Australia; Brad, Romania; and Vatukoula, Fiji. Most of these deposits are described above or in the following sections.

The gold-selenide deposits are rather rare. There are, however, examples at Republic in Washington (Umpleby, 1910; Lindgren and Bancroft, 1914; Wright, 1947; Full and Grantham, 1968), in the Radjang-Lebong goldfield in Sumatra, Indonesia and at the Kushikino Mine, Kagoshima Perfecture, Japan (Mukaiyama and Izawa, 1964). Most of the deposits are veins, stockworks and pipes in Tertiary or younger andesites. The principal gangue is quartz and calcite, sometimes adularia, usually in banded and colloform arrangements or in fine-grained intergrown aggregates. The principal ore minerals are native gold, often high in silver, electrum, various silver, lead and copper selenides such as aguilarite, naumannite and berzelianite, pyrite, tetrahedrite, chalcopyrite, galena and sphalerite. These minerals are often fine grained and form distinctive black streaks, wisps, crusts and bands in the quartz. These bands are rich in gold and silver, often assaying hundreds and thousands of ounces of gold and silver to the ton. The alteration attending the veins is generally silicification, pyritization or propylitization. The Au/Ag ratio of the deposits in the Radjang-Lebong field varies from 0.01 to 1.

Auriferous veins, lodes, sheeted zones and saddle reefs in faults, fractures bedding plane discontinuities and shears, drag folds, crushed zones and openings on anticlines essentially in sedimentary terranes; also replacement tabular and irregular bodies developed near faults and fractures in chemically favourable beds

These deposits are widespread throughout the world and have produced a large amount of gold and silver. They are developed predominantly in sequences of shale, sandstone and greywacke dominantly of marine origin. Such sequences are invariably folded, often in a complex manner, metamorphosed, granitized and invaded by granitic rocks, forming extensive areas of slate, argillite, quartzite, greywacke and their metamorphic equivalents. Near the granitic bodies various types (kyanite, andalusite, cordierite) and quartz-mica schists and hornfels are developed, and these grade imperceptibly into relatively unmetamorphosed slates, argillites, quartzites and greywackes marked by the development of sericite, chlorite and other low-grade metamorphic minerals. Most of the gold deposits are developed in the lower grade facies. Few economic deposits occur in the granitic batholiths or large stocks that invade the greywacke-slate sequences.

The principal gangue mineral in these deposits is quartz, frequently ribbonned, with a glassy, white, grey, to black colour. Feldspar, micas, chlorite and minerals such as rutile are subordinate. Among the metallic minerals, pyrite and arsenopyrite are much the commonest, but galena, chalcopyrite, sphalerite and pyrrhotite also occur. Molybdenite, bismuth minerals and tungsten minerals are local. Stibnite occurs in abundance in a few deposits but is relatively rare in most deposits. Acanthite, tetrahedrite-tennantite, and other sulphosalts are not common in these deposits. Carbonates, mainly calcite and ankerite, are common but not abundant. The valuable ore minerals are native gold, generally low in silver, auriferous pyrite and auriferous arsenopyrite. Tellurides are relatively rare, and aurostibite is an uncommon mineral in these deposits.

A few deposits in this category are tabular or irregular replacement (disseminated) bodies developed in carbonate rocks or limy argillites and shales. The principal minerals developed in these deposits include quartz, fluorite, pyrrhotite, pyrite, arsenopyrite, sphalerite, galena, chalcopyrite and stibnite.

As a general rule the wall-rock alteration effects associated with these deposits are minimal and the quartz veins, saddle reefs and irregular masses are frozen against the slate, argillite or greywacke wall rocks. In places thin zones of mild chloritization, sericitization and carbonatization are present. Some veins are marked by thick black zones (up to 6 in.) of tourmalinized rock. Disseminated pyrite and arsenopyrite are common in the wall rocks of most of these deposits. This pyrite and arsenopyrite is commonly auriferous.

The elements exhibiting a high frequency of occurrence in this type of gold deposit include Cu, Ag, Mg, Ca, Zn, Cd, (Hg), B, (In), (Tl), Si, Pb, As, Sb, (Bi), S, (Se), (Te), (Mo), W, (F), Mn, Fe, (Co) and (Ni). Those in parentheses have a low to very low frequency of occurrence.

The sedimentary piles in which these deposits occur are marked by extensive developments of shales, argillites and slates, all containing abundant pyrite. These are invariably enriched in all of the elements contained in the deposits. In addition the clastic sediments in the sedimentary piles including conglomerate, greywacke and sandstone commonly have higher than normal amounts of gold, probably of detrital origin. These, together with the black pyritiferous sediments, are probably the source beds for the elements in the deposits. As the various structures developed in the sedimentary piles, especially along fault and fracture systems and on the noses of anticlines (saddle reefs), their dilatancy induced the diffusive flow of silica, gold and other elements into these sites where they were partitioned into the series of minerals we now observe in the veins and lodes. In places chemically receptive rocks have precipitated quartz, gold and other minerals from diffusion currents moving gold and gangue elements through the rocks.

In detail the precise relationships of the gold deposits in this category to granitic bodies invading the sedimentary terranes are often not entirely clear in many auriferous belts. In many sedimentary terranes the metamorphic, granitization and vein-forming processes appear to be coeval or nearly so. In other regions, however, particularly in greywacke-slate terranes, some of the quartz deposits appear to be truncated by the granitic bodies, and xenoliths of the quartz veins containing arsenopyrite and pyrite occur in the granitic rocks. In a few places (e.g., Yellowknife) some of the auriferous quartz veins are cut by pegmatites. Such relationships are often observed where there are two ages of granitic rocks in the terrane as in the Precambrian Yellowknife Supergroup and in the Ordovician Meguma Group of Nova Scotia. Commonly the first phase is a biotite granite whereas the second is a muscovite granite usually with a plethora of associated pegmatites. Where these geological features are present the sequence of events seems to have been:

1. Initial stage of folding of greywacke and slate with the concomitant effects noted in 2.

2. Rise of the geotherms through the sedimentary pile due to metamorphism and granitization at depth. This is attended by the migration of silica, sulphur, arsenic, gold and other constituents into available dilatant zones to form the saddle reefs, leg reefs and interbedded veins with their contained auriferous sulphides and native gold.

3. Continued folding resulting in the corrugation of the quartz veins, formation of quartz boudins, ptygmatic veins and a general crushing and recrystallization of the quartz. During this period remobilization and reconcentration of the initial gold in dilatant sites (shoots), commonly as the native metal, may take place in the quartz bodies. In addition further increments of gold, sulphur, arsenic, silica and other constituents may be provided by the prevailing diffusion currents.

4. Continued granitization at depth followed by the injection of high level granitic stocks and small batholiths. The development of pegmatites commonly takes place during this stage.

5. Late stage minor folding and faulting.

Deposits in sedimentary terranes are widespread in Canada, United States, France, U.S.S.R., Australia and South Africa. Only a few characteristic examples will be described.

The Precambrian sediments of the Yellowknife Supergroup in the Yellowknife district, Northwest Territories comprise mainly slate, argillite, quartzite and greywacke, and their metamorphic equivalents, various quartz-mica schists, near granitic batholiths and stocks. All of the sediments were intricately folded in an isoclinal manner, then cross folded, yielding basin and dome structures and finally faulted. The gold deposits occur as:

1. Quartz veins in contorted schist zones and faults that transect the sedimentary beds.

2. Quartz veins and small stockworks in schist zones and fractures that parallel the sedimentary beds. These comprise quartz veins and stockworks in ruptured and sheared beds and bedding planes, dragged slate beds, contacts and saddle reefs on the noses of folds.

3. Quartz veins and sheeted zones in ruptured and sheared axes of isoclinal folds.

4. Quartz veins and stringer zones in other irregular fractures and openings without any continuity or special relationships.

The Ptarmigan (Fig. 37) and Burwash deposits are quartz veins and lenses in small faults that cut greywacke and slate beds at a large angle (Boyle, 1961*a*). Wall-rock alteration is minimal (*see* page 240). The gangue is colourless to grey massive quartz with native gold and minor amounts of tourmaline, pyrite, sphalerite and galena. The grade of the ore was about 0.5 oz/ton, and the Au/Ag ratio of the ores ranged from 3 to 8.

The Thompson-Lundmark Gold Mine some 40 mi northeast of Yellowknife obtained its ore from large quartz veins lying parallel to the bedding of the knotted quartz-mica schists and slates (Fortier, 1946, 1947; Lord, 1941*a,b*, 1951). The wall-rock alteration is minimal, and the mineralization consists mainly of quartz and abundant native gold with small amounts of tourmaline, feldspar, pyrite, arsenopyrite, pyrrhotite, chalcopyrite, galena and sphalerite. The grade of the various ore-shoots ranged from 0.4 to 1.5 oz Au/ton and the Au/Ag ratio averaged about 5. Some of the veins in the vicinity of the Thompson Lundmark Mine are of interest, since they are traversed by pegmatites.

The ore shoots of the Camlaren Mine at Gordon Lake (Henderson and Fraser, 1948; Lord, 1951) are in saddle reefs and legs on the nose of a pitching anticline (Fig. 38). The grade of the ore was about 0.75 oz Au/ton and the Au/Ag ratio of the ores ranged from 5 to 8. The mineralization consisted essentially of quartz with some ankerite, native gold and small amounts of pyrite, chalcopyrite, galena and sphalerite. Literally thousands of narrow quartz veins, saddle reefs and legs occur in the slaty beds between more massive greywacke and quartzite in the Gordon Lake basin. All contain gold, but few are of economic grade and tonnage.

The Consolidated Discovery Yellowknife Mine in the Giauque Lake area of Northwest Territories developed and mined orebodies mainly in highly folded and crenulated metasediments near the nose of steeply dipping volcanics (Wiwchar, 1957). One of the orebodies (the north zone vein) had a saddle shape (Fig. 39); other zones were irregular. The vein mineralization was essentially quartz with small amounts of oligoclase, pyrrhotite, pyrite, muscovite, sericite, actinolite, chlorite and rare occurrences of scheelite, sphalerite, galena, chalcopyrite and andalusite. The gold was free and coarse, and the last mineral to crystallize. The wall-rock alteration consisted mainly of a band a few inches wide next to the vein

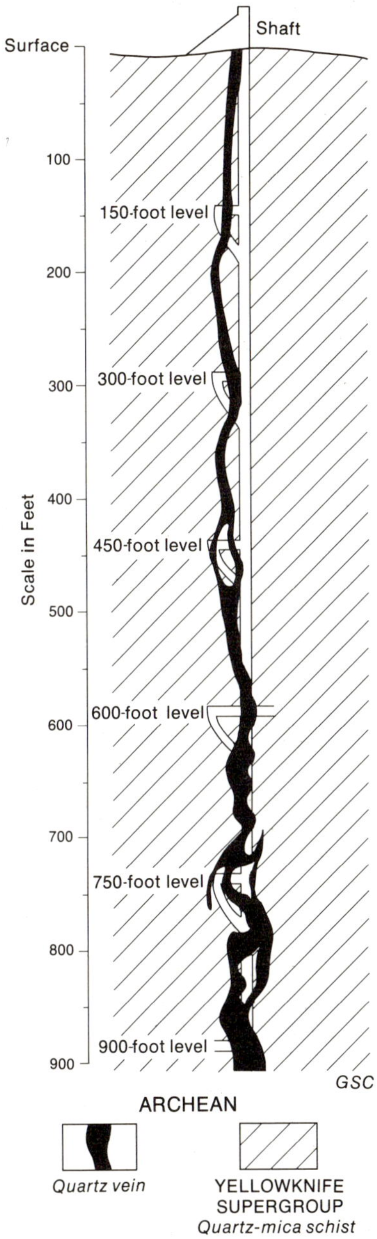

Figure 37. Vertical section through Ptarmigan shaft, Yellowknife, Northwest Territories, looking northwesterly along the strike of Ptarmigan No. 1 vein (after Boyle, 1961*a*).

enriched in biotite and chlorite. The grade of the veins was about 1 oz/ton Au.

The Cariboo Gold Quartz Mine at Wells, British Columbia (Fig. 78) obtained its ore from a myriad of quartz veins and bedded pyritic deposits in faults and fractures in the Paleozoic Cariboo series consisting mainly of quartzites, black argillites and schists (Hanson, 1935; Richards, 1948; Skerl, 1948; Sutherland Brown, 1957). Most of the quartz veins cut the sedimentary beds at large angles, but some lie nearly parallel with the strata. Their mineralization is essentially quartz with some calcite and ankerite, native gold, and small amounts of pyrite, cosalite, galena, sphalerite, galenobismutite, pyrrhotite and scheelite. The pyritic bodies are small and follow the strike and dip of the sediments. They replace limestone and limy sedimentary beds. Most consist of lenses of auriferous pyrite containing small amounts of quartz, pyrrhotite and scheelite. The average grade of the ore in the various types of deposits is 0.4 oz Au/ton. The Au/Ag ratio of the ores is about 8.

The Ordovician Meguma Group of Nova Scotia contains innumerable gold-quartz veins, some of which have been productive in the past. The Meguma Group is composed essentially of massive greywacke, quartzite, black argillite, slate and their metamorphic equivalents near Devonian granitic masses. The sedimentary rocks are folded into long pitching anticlines and synclines that yield basin and dome structures in many areas. The quartz deposits occur in the same structural locations as those mentioned for the Yellowknife Supergroup in Northwest Territories. Most of the economic deposits were saddle reefs, quartz legs and stockworks on the flanks of anticlines, or quartz masses on the limbs of anticlines in dragged and contorted slates and argillites between competent greywackes and quartzites (Figs. 40 and 41). The structural setting of the deposits has been discussed in detail by Rickard (1912), Malcolm and Faribault (1929), Newhouse (1936), Douglas (1948), Bell (1948) and Keppie (1976). The mineralization is essentially white or dark grey quartz, frequently ribboned, with small amounts of ankerite or calcite, some feldspar in places and minor to small amounts of pyrite, pyrrhotite, arsenopyrite, chalcopyrite, sphalerite and galena. Abundant stibnite is present in veins at West Gore, and molybdenite, scheelite and tourmaline occur in small amounts in a number of the deposits. Quartz veins and lenses similar to the gold veins carry small amounts of scheelite and wolframite in some areas. These are essentially barren of gold. Some of the tungsten veins appear to be younger than the gold veins. Native gold is the only economic mineral in the gold veins; gold tellurides and aurostibite have not been recognized in the deposits. Wall-rock alteration is absent to minimal in all of the deposits. The grade of the deposits ranges from 0.15 to 1 oz Au/ton or more. The fineness of the native gold spans the range 800 to 980, most samples being in excess of 900. The Au/Ag ratios of the deposits varies widely according to my analyses, from <1 to 52. Most of the ratios, however, are around 10.

Newhouse (1936) recognized a zonal arrangement of the gold deposits in Nova Scotia and related it to the granitic intrusives. Near the granitic bodies in a belt 1¼ mi wide he noted that the quartz was coarse and that the following minerals occurred in the veins with great frequency: biotite, muscovite, oligoclase, feldspar, hornblende, garnet, tourmaline, epidote, ilmenite and pyrrhotite and less frequently andalusite and molybdenite. Pyrite and arsenopyrite were universal; carbonates were rare. In an intermediate zone, 1½ mi or more from the granite contacts, carbonate and chlorite were abundant and arsenopyrite and pyrite were particularly abundant. The quartz was coarse grained but not as marked as in the zone next to the granitic bodies. In an outer zone stibnite and native antimony, as at West Gore, made their appearance. Newhouse recorded one interesting fact, the bulk of the gold was localized in the lower grade facies, that is in

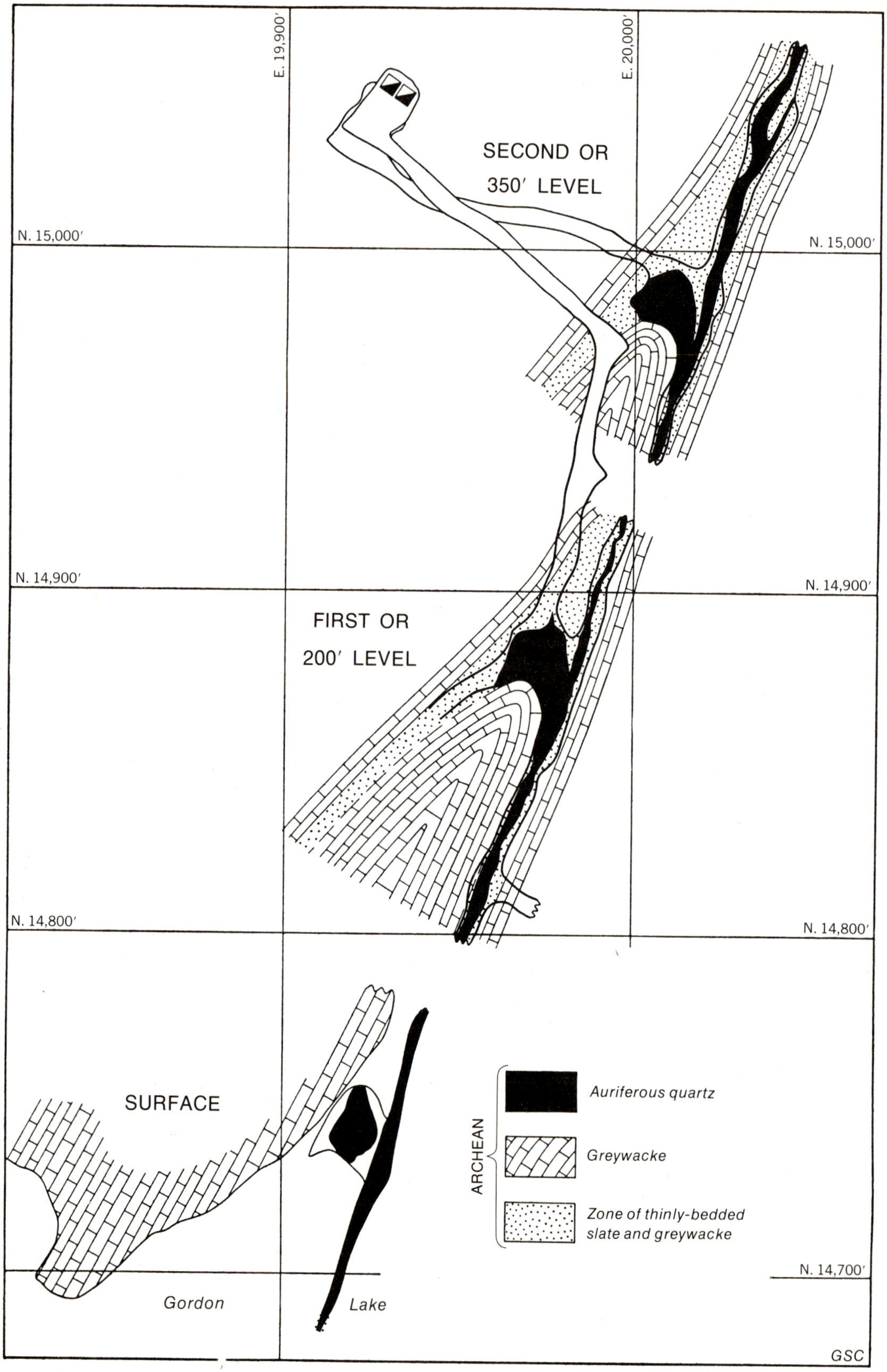

Figure 38. Plan of surface and part of underground workings, Camlaren Mine, Gordon Lake, Northwest Territories, showing geology of the Hump vein (after Henderson and Fraser, 1948).

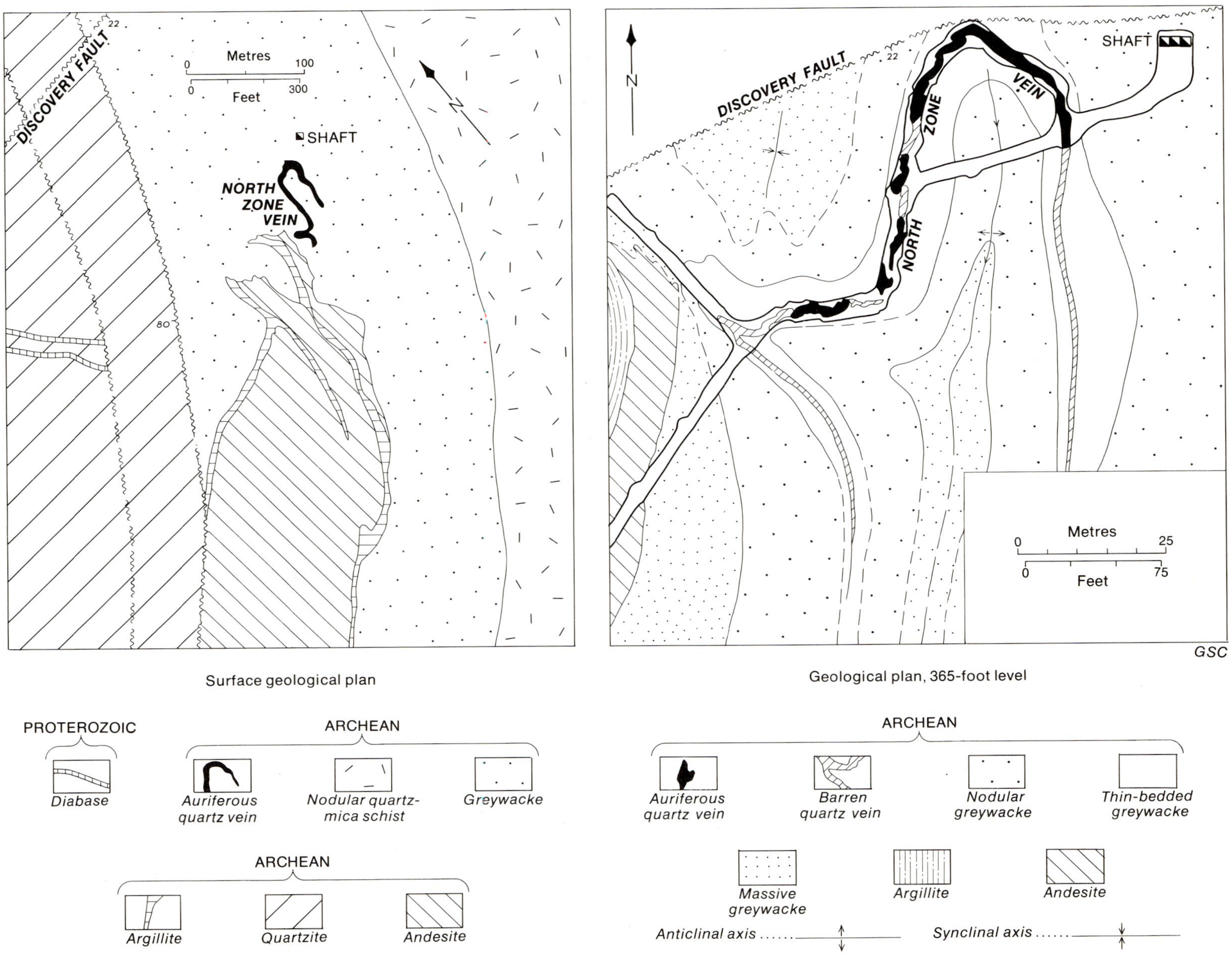

Figure 39. Generalized geological surface and underground plans of the Consolidated Discovery Yellowknife Mine (after Wiwchar, 1957).

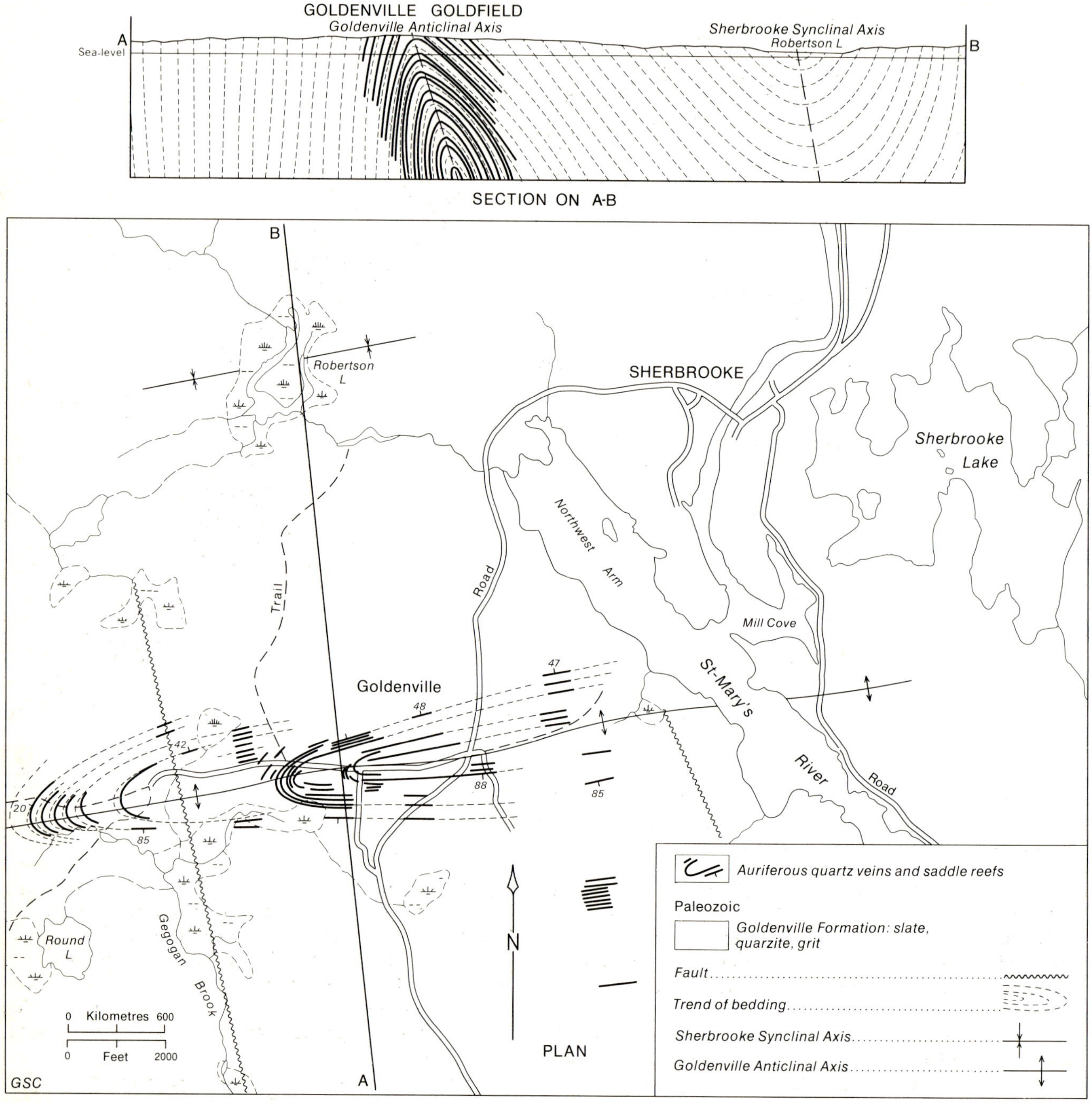

Figure 40. Generalized geological plan and section of the Goldenville gold district, Guysborough County, Nova Scotia (modified from Malcolm and Faribault, 1929).

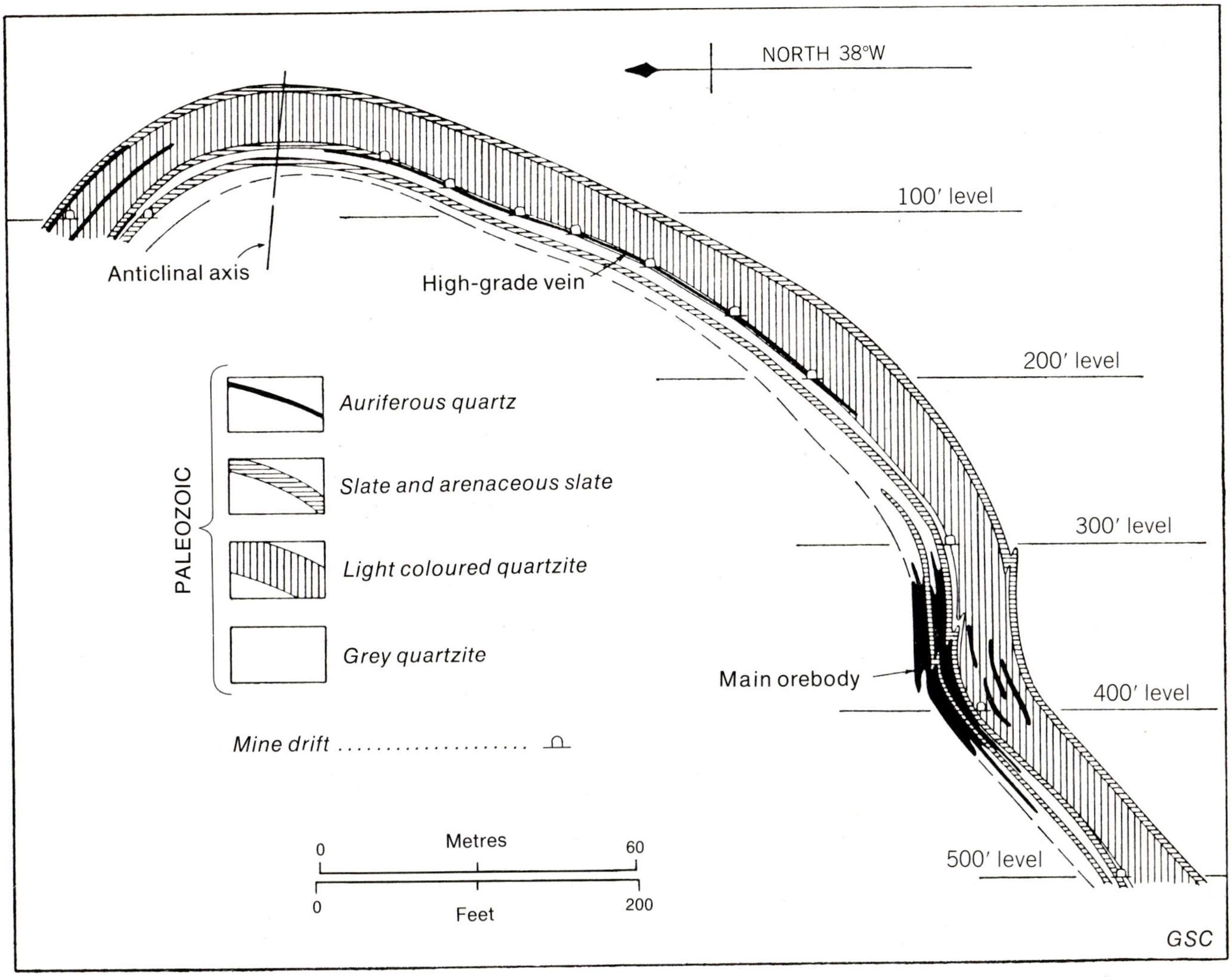

Figure 41. Vertical section through the southeastern part of the Caribou deposit, Caribou gold district, Halifax County, Nova Scotia (after Bell, 1948).

the intermediate and outer zones. Taylor and Schiller (1966), likewise, observed that most of the gold-quartz veins in the Meguma Group are confined to areas underlain by the greenschist zone of regional metamorphism. Other sedimentary gold belts in various parts of the world exhibit a similar pattern.

In the United States some of the best examples of gold deposits in sedimentary terranes include those in the Appalachian 'slate belt' that extends from Alabama, Georgia, South and North Carolina and Virginia, into Maryland (Becker, 1895; Pardee and Park, 1948; Bell, 1960). The 'slate belt' is a complex series of rocks ranging in age from Precambrian to Ordovician and including a great variety of volcaniclastic rocks, greenstones, slates, argillites, quartzites, greywackes and thin limestones here and there intruded by large and small bodies of granitic rocks of post-Devonian age (?). The deposits are varied. Some are quartz veins and lenses in shear zones, faults and fracture zones; others are conformable silicified zones in tuffaceous sedimentary rocks; and still others are replacement bodies in limestone as in the Kings Mountain area, North Carolina. The mineralization consists essentially of quartz, carbonates, tourmaline, pyrite, pyrrhotite, arsenopyrite and small amounts of galena, sphalerite, tetrahedrite, chalcopyrite, molybdenite, ferberite and scheelite. Native gold is the sole economic mineral, but gold-silver tellurides are mentioned as being present in some lodes. The grade of the ore is 0.25 oz Au/ton or less and the Au/Ag ratio of the deposits as a whole appears to average about 10 judging from the old records.

Worthington and Kiff (1970) have suggested a volcanogenic origin for the gold deposits in the slate belt of the North Carolina Piedmont. They note that many of the deposits have a spatial affinity to the Uwharrie volcanic formation, which consists of felsic pyroclastic rocks and tuffs. They envisage deposition of the deposits from hydrothermal emanations of hot springs and fumaroles associated with the waning stages of Uwharrie volcanism. Quartz vein deposits, somewhat similar to those in the Appalachians, occur in ignimbrites near Rodalquilar in southeast Spain and have been interpreted by De Roever and Lodder (1967) as related to fumarolic activity immediately following deposition of the ignimbrite.

The Salsigne Gold Mine, approximately 15 km north of

Carcassone, 90 km east of Toulouse, France obtains its ore from bodies in sedimentary rocks of Paleozoic age, folded and faulted during the Hercynian orogenic period (Reynolds, 1965; Tollon, 1970). The country rocks are limestone, dolomitic limestone, phyllite, grit and grès (phyllitic quartzite). The orebodies are sulphide bodies and quartz veins in faults mainly in phyllites, and irregular replacement bodies in dolomitic limestones and grès where intersected by faults (Fig. 42). The mineralization of the sulphide bodies consists of coarse-grained arsenopyrite, pyrite and chalcopyrite. Pyrrhotite occurs in some vein systems and minor amounts of wolframite and scheelite have been noted. The gangue is minor quartz, dolomite and siderite. The copper and silver values are relatively high, averaging 0.6 per cent Cu and 85 g Ag/ton. The gold content averages about 10 g/ton yielding an Au/Ag ratio of 0.12. The quartz veins contain essentially quartz, dolomite, siderite, chalcopyrite and minor arsenopyrite, pyrite and pyrrhotite. The grades are 12 and 28 g/ton respectively for Au and Ag yielding an Au/Ag ratio of 0.43. The mineralogy of the limestone replacements consists essen-

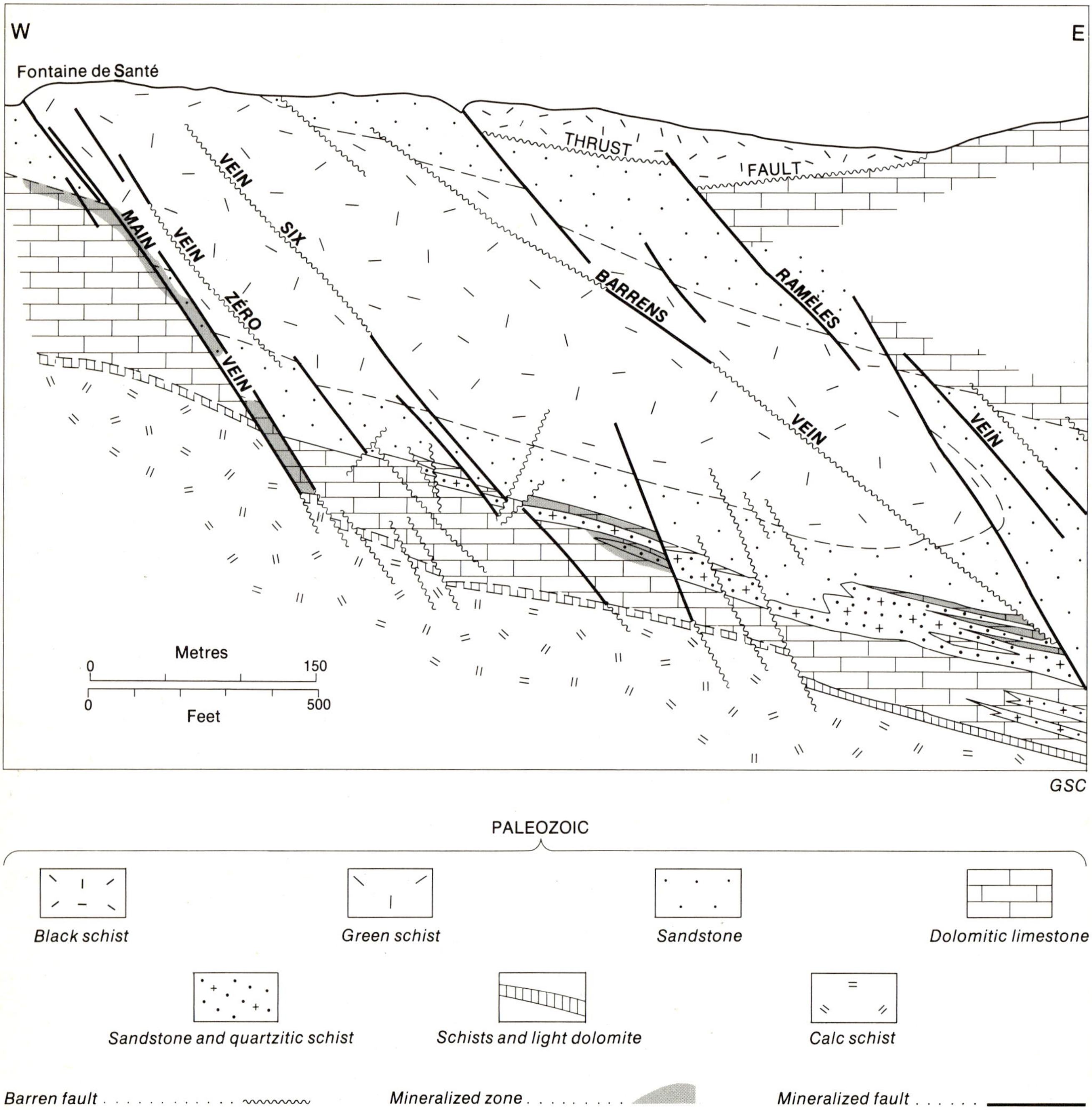

Figure 42. Generalized geological east-west section on 100 north, Salsigne Gold Mine, Montagne Noire, France (modified from Tollon, 1970).

tially of massive pyrite with much arsenopyrite and pyrrhotite, small amounts of chalcopyrite and minor amounts of galena and sphalerite. The grades are 11 g/ton Au and 25 g/ton Ag; the Au/Ag ratio, 0.44. The grès replacements are different in mineralogy and grade from the other types of ores. The mineralization consists mainly of small amounts of arsenopyrite with chlorite and quartz. Gold values may be as high as 3700 g/ton, and the bismuth contents range up to 1 per cent or more. Antimony contents are also as high as 1 per cent. The average grade of the ore is 18 g Au/ton and 10 g Ag/ton giving a Au/Ag ratio of 1.8. Wall-rock alteration effects are minimal in the deposit; in the grès replacements a little biotite is developed in the ore zones.

Tollon (1970) has recorded the presence of a large number of rarer minerals in the ores of the Salsigne Mine. These include relatively abundant bismuthinite and native bismuth, and rare pearceite, polybasite, matildite, kobellite, proustite, native silver, pyrargyrite, maldonite, cosalite, stannite, scheelite and wolframite.

The Salsigne orebodies as a whole yield gold, silver and small amounts of copper, bismuth, arsenic trioxide and sulphuric acid, the last from pyrite and pyrrhotite. In 1973 the mine produced 49 675 oz gold, 161 146 oz silver, 372 short tons copper, 41.9 short tons bismuth, 7305 short tons arsenious trioxide and 21 799 short tons of sulphuric acid (Northern Miner, Nov. 28, 1974, p. 9).

The Sovetskoe deposit described by Borodaevskaya and Rozhkov (1974) is located in the Eniseisk (Yenisey) auriferous region, which borders the Siberian platform on the west. The deposit is in Proterozoic sediments pierced by Upper Proterozoic granitoid intrusions, including granitic bodies, injection gneisses, aplite, pegmatite and diabasic porphyrite dykes. The mineralization is said to be linked with the Upper Proterozoic magmatic events. The principal stages in the geological development of the region are as follows: (1) accumulation of mainly terrigenous sediments during Lower and Middle Proterozoic time followed by carbonates in Late Proterozoic time; (2) intrusion of diabase bodies beginning during Lower Proterozoic time; (3) folding and fracturing with development of schistosity; (4) intrusion of granitic bodies accompanied by local and regional metamorphism; (5) a prolonged interval of erosion followed by the transgression of Lower Paleozoic seas and concomitant sedimentation; (6) appearance of grabens filled with Lower Paleozoic deposits and flanked by tectonic fractures that were initially activated during Upper Proterozoic time; and (7) prolonged continental conditions attended by the formation of auriferous placers in the Quaternary period (*see* Fig. 87).

The deposit is confined to uniform beds of Upper Proterozoic phyllite that has been highly sheared and drag folded, often in a complicated manner over widths of several dozens of metres and along a length of several kilometres. This shearing is related to a major thrust zone more than 10 km in extent, the ore-bearing shears being subsidiary to this so-called 'main thrust'. The orebodies are irregular and constitute stockworks of quartz veins of variable length and width that ramify hither and thither through the sheared rock. Quartz is the principal gangue mineral, and the main wall-rock alteration is tourmalinization, appearance of apatite and zircon and replacement of ilmenite by rutile; albitization, chloritization, sericitization, silicification, arsenopyritization and pyritization are marked adjacent to some of the quartz veins and stringers.

The ore minerals (about 5% of the ore) are represented by pyrite and arsenopyrite (predominant), chalcopyrite, galena, sphalerite and in places pyrrhotite and marcasite. Carbonates (calcite, ankerite), albite, sericite and chlorite are encountered among the vein minerals in addition to quartz.

The history of ore formation has been long and has included many stages. The following principal mineral associations, oldest to youngest, can be distinguished: (1) relict minerals: chlorite, muscovite, sericite, albite, ankerite, tourmaline, rutile, apatite, etc., derived from the country rocks; (2) semimilky quartz, with slight traces of muscovite and sericite; (3) arsenopyrite and pyrite; (4) polymetallic sulphide-pyrite, comb quartz, siderite, pyrrhotite, sphalerite, galena, chalcopyrite, marcasite and very rarely, bismuthinite, native silver, freibergite, maucherite and violarite; (5) gold with a small quantity of calaverite; and (6) carbonates and collomorphic pyrite. N. Petrovskaya (*in* Borodaerskaya and Rozhkov, 1974) has suggested that the main body of vein quartz was probably deposited from highly concentrated hot silicate solutions and has further demonstrated that the quartz underwent intensive metamorphism and recrystallisation before the mineral associations next in age were deposited.

The mineral associations enumerated above are unevenly distributed in the various bodies, with the result that the following ore types can be distinguished: homogeneous quartz ores, quartz ores with relicts from the surrounding shales, quartz-pyrite ores (with impregnated or pockety mineralization) and quartz-arsenopyrite and quartz-sulphide ores mixed with traces of the polymetallic sulphide group. The latter ore type has the highest gold content. The orebodies near the walls of the sheared zones, where breccia structures predominate, are the richest; the poorest are in the inner parts of the sheared zones; these are composed of massive, low sulphide-bearing quartz.

The gold in the ores is present mainly in the native form as small angular, fibrous and nodular forms, the grains ranging from 1μ to several millimetres in cross-section. The fineness of the native gold is rarely less than 850. The purest gold comes from quartz ores low in sulphides; in the polymetallic sulphide ores the gold contains traces of zinc, lead and copper; traces of selenium and tellurium are also noted in some samples.

The Sovetskoe deposit is a large tonnage proposition; it resembles in many respects the low grade Alaska Juneau deposit at Juneau described in the next section. The grade of the Sovetskoe deposit is not given, but it probably averages about 0.1 oz/ton.

The Muruntau deposit in Uzbec S.S.R. is an en echelon system of thick steeply dipping auriferous quartz veins, intersected by stockworks of gently dipping gold-bearing quartz and quartz-sulphide stringers and veins (Fig. 43). The host rocks are Paleozoic calcareous siltstones, shales and various types of schists. The geology and mineralization of this deposit is discussed in detail by Nesterova and Chebotarev (1969), Yudin (1971), Borodaevskaya and Rozhkov (1974) and Baimukhamedov *et al.* (1975*b*).

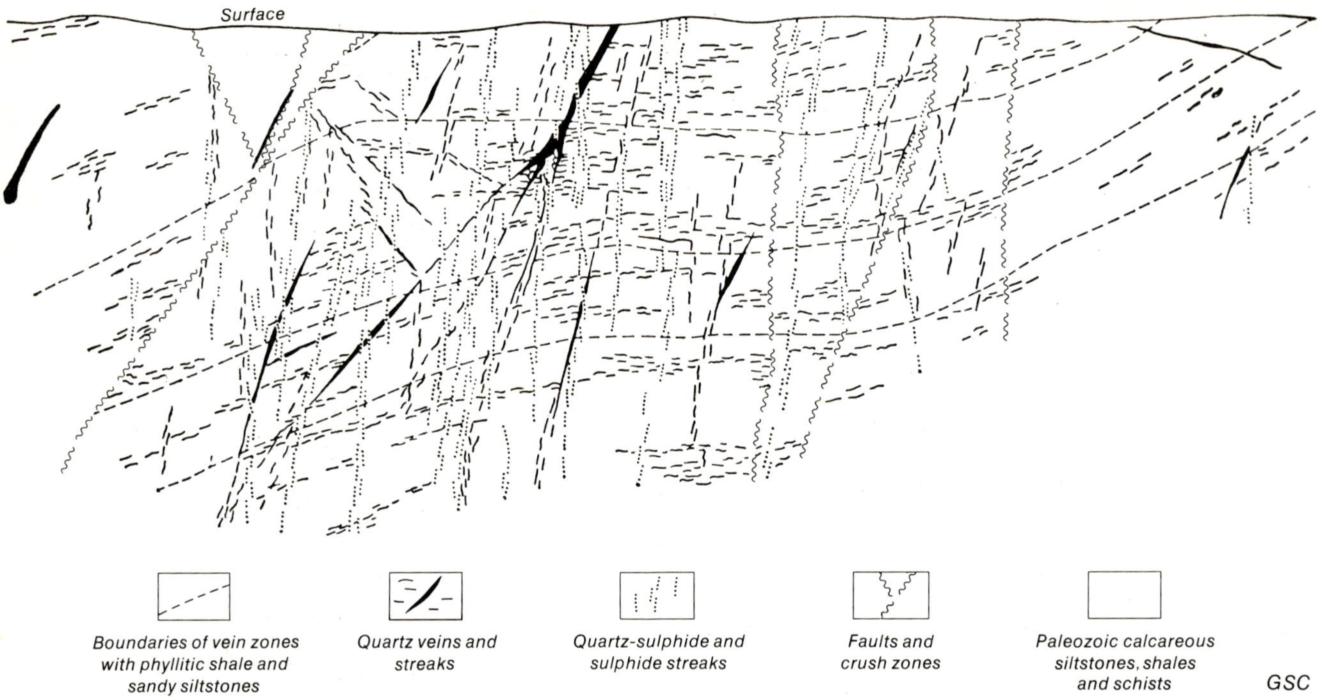

Figure 43. Geological section of the Muruntau gold deposit, Uzbec S.S.R. showing the interrelationship of gently and steeply dipping vein zones (after S. Sher, *in* Borodaevskaya and Rozhkov, 1974). (No scale given on original.)

Three stages of alteration and mineralization are recognized as follows: preore, ore and postore.

The preore stage includes three phases. The early mineral associations related to the metasomatic alteration of the wall rocks originated during the first phase. The new minerals formed were pyroxene, actinolite, feldspars, chlorite, biotite, sericite, graphite, ferruginous tourmaline (schorl) and pyrrhotite, all of which originated mainly as the result of the rearrangement of the components of the wall rocks. These minerals jointly with metasomatic quartz are distributed in the shales as accumulations along the bedding planes, as lenticular stringers, or as 'shadows' associated with quartz grains. The second phase, closely related to the first, produced a system of concordant and discordant veins and stringers filled with coarse-grained quartz, the constant associates of which are orthoclase, scheelite and wolframite. The third phase is represented by monomineralic veins of coarsely-crystalline columnar quartz.

The ore stage began with the intrusion of dykes of diverse composition. Next followed the formation of latitudinal fractures, brecciation of the coarse-grained quartz veins and stringers formed during the second phase of the preore stage and deposition of minerals of the second stage of the mineralization. This stage comprises four phases of mineral deposition. During the first phase quartz-albite metasomatites were superimposed upon the altered zones formed during the preore stage. The second phase of mineralization was repeatedly interrupted by fracturing during which some quartz-arsenopyrite-pyrite stringers were deposited first, followed by the main period of pyrite-arsenopyrite mineralization. Chalcedony and rosettelike quartz aggregates were deposited during the third phase, which was followed by the fourth or the main quartz-gold-polymetallic phase of ore deposition. This phase resulted in the deposition of two distinct associations of minerals, an early generation including pyrrhotite, molybdenite, pyrite, sphalerite, galena, chalcocite and tetrahedrite, and a late one marked by native gold, native bismuth, bismuthinite, gold tellurides (?), altaite and miargyrite.

The initiation of the postore stage was characterized by the precipitation of a peculiar quartz-tourmaline-pyrite association, localized on the whole in lenslike bodies of quartz-tourmaline breccia. Next followed the deposition of veins of columnar quartz with small amounts of pyrite and calcite. The last minerals formed were carbonates, represented mainly by calcite and subordinated by dolomite stringers oriented at random and intersecting all earlier stages of mineralization.

In a recent contribution on the geochemistry of the Muruntau gold deposit Petrov (1972) states that the ore is localized in quartz and quartz-mica schists containing lenticular lenses of limestones. The schists are said to be enriched in gold (100–1000 ppm), As ($<$100 ppm), W ($<$1000 ppm), Bi (10 ppm), Cu (4–10 ppm), Zn (2–3 ppm) and Co (1.5–2 ppm). He proposes a syngenetic origin for the deposit.

The grade of the Muruntau deposit is not stated in most publications. The fineness of the native gold ranges from 737 to 953 with an average of 889. In a recent report by Bendik and Nesterova (1971) the grade is given as 0.058 oz Au/ton (2 ppm) and 2.9 to 8.8 oz Ag/ton (100–300 ppm). The gold content stated seems low considering the values quoted by Petrov above. Baimukhamedov *et al.* (1975*b*) give the grade of the metasomatites as 0.5 to 3.5 g/ton Au.

The rich hypogene gold deposits of Victoria, Australia, now largely exhausted, are mainly in Lower Ordovician sedimentary sequences composed essentially of slate, argillite, greywacke, grit and quartzite intruded in places by Devonian granitic masses (Thomas, 1953; various papers in Edwards, 1953; McAndrew, 1965; Woodall, 1975; Bowen and Whiting, *in* Knight, 1975). The chief gold mining centres were Bendigo and Ballarat, about 80 mi northwest of Melbourne and Castlemaine, Clunes, Berringa, Scarsdale, Daylesford and Maldon. The sedimentary rocks are closely folded into series of pitching anticlines and synclines giving dome and canoe structures. The gold deposits are mainly quartz bodies along bedding planes, in faults cutting the sedimentary beds at a high angle and in saddle reefs on anticlines and trough reefs in synclines. Some saddle reefs are stacked and have well-developed spurs as shown in Figure 44. The Bendigo and Ballarat fields are well-known for their classic saddle reefs, and these two fields have been the most productive. The mineralization in the saddle reefs and other deposits is essentially coarse-grained quartz, some feldspar (albite), dolomite, ankerite or calcite, chlorite, pyrite, arsenopyrite, rarely pyrrhotite, galena, chalcopyrite, sphalerite, and locally molybdenite, bismuth, stibnite, bournonite and other sulphosalts. The only economic minerals are gold and stibnite. Arsenic has been recovered as a byproduct from some of the gold deposits. The native gold usually averages about 850 fine. The grade of the deposits ranged from 0.4 to 0.9 oz Au/ton.

The Telfer Mine developed by Newmont Pty. Ltd. is in the Paterson Range of Western Australia and comprises quartz and quartz-limonite reefs with pyrite boxworks and free gold mainly concordant with, but locally cutting at a small angle, a sequence of metamorphosed stable shelf sediments including sandstone, siltstone, shale and dolomite correlated with the Middle Proterozoic Bangemall Group (Blockley, 1974). Granite dated at 600 m.y. intrudes the sediments some 15.5 mi west of the deposits. Most of the auriferous quartz deposits occur in domes or anticlines, and many are saddle reefs. The average grade of the limonitic deposits (oxidized surface parts) is about 0.5 oz/ton Au. At depth in the primary zone the quartz contains mainly pyrite and gold, and the average grade is lower. Auriferous gossans mark many of the deposits.

The Reefton goldfield, the most productive of the gold deposits in Precambrian(?) rocks in the northwestern part of South Island, New Zealand, occurred in an extensive set of narrow, steeply dipping crushed zones, shears and small faults developed near the axes of folds in Waiuta Group greywackes, argillites and phyllites (Henderson, 1917; Gage, 1948; Williams, 1974). The lodes in the Blackwater Mine, the principal producer, were a series of quartz veins and stringers in crushed rock and clay. The mineralization was essentially quartz, minor carbonates, pyrite, arsenopyrite, stibnite, minor chalcopyrite, galena, minor silver and molybdenite. The average grade was about 0.5 oz Au/ton. Some 80 per cent of the gold was free though rarely visible. Secondary enrichment was unimportant.

The gold deposits of South Africa appear to belong to three metallogenic periods. These include an early Precambrian (Archean) period in which most of the deposits are in or associated with greenstones of volcanic origin in the Swaziland System; a later Archean or early Proterozoic period when the great deposits of the Witwatersrand were formed; and a late Precambrian period wherein the deposits appear to be

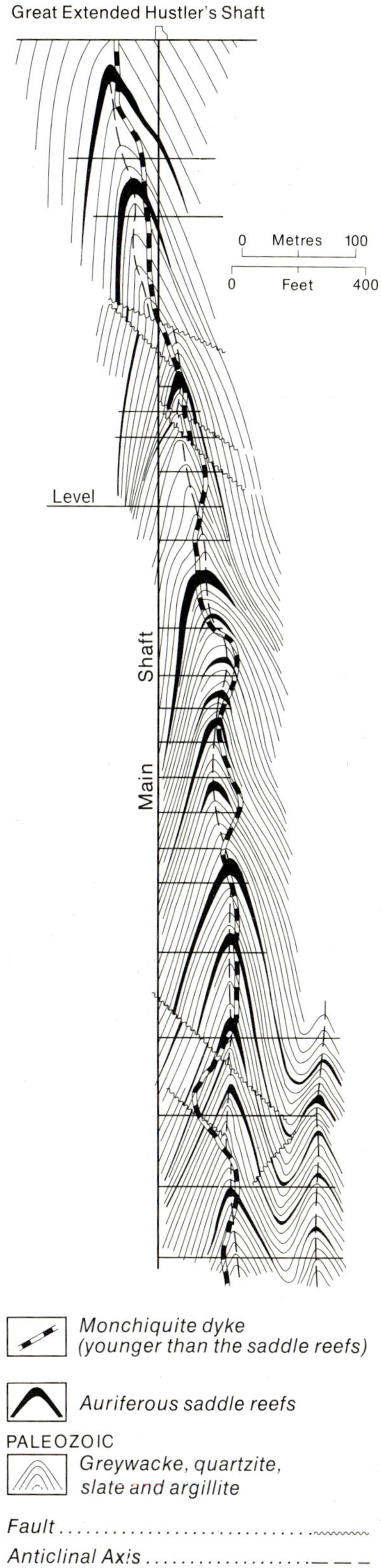

Figure 44. Stacked saddle reefs, Bendigo goldfield, Victoria, Australia (modified from Herman, 1914).

related to granitic and other rocks of the Bushweld Complex. The last period is marked by important deposits in the Pilgrim's Rest and Sabie districts. Most of these deposits are in sediments (Reinecke and Stein, 1930; Wybergh, 1930; Geol. Surv. S. Afr., 1959; Pelletier, 1964).

The gold deposits of the Pilgrim's Rest and Sabie goldfields lie mainly in the Dolomite and Black Reef series of the Transvaal System (*see* also Table 61). The Black Reef Series, composed essentially of sandstone and conglomerate, lies directly on the older gneiss, schist and granite (Swaziland System) and is followed upward by the Dolomite Series, which is essentially constituted of thick beds of dolomite with thin shale beds and bands of chert. Lying on top of the Dolomite Series is the Pretoria Series consisting of shales and interbedded quartzites. This series contains a few small gold deposits. The sedimentary beds are penetrated by a number of granitic, dioritic and gabbroic sills, stocks and dikes. The gold mineralization cuts these igneous bodies in places.

The more important deposits (reefs) are the Theta, Beta, Portugese, Vaalhoek and Glynn's reefs in the Dolomite Series and the Sandstone Reef in the Black Reef Series. These are mainly bedding plane quartz veins generally 1 to 3 ft or more thick. Cross-cutting quartz veins that join the bedding plane veins occur in some deposits, and a 'Blow' occurring just below Glynn's Reef in the Elandsdrift Mine near Sabie consisted of a steeply dipping fragmented pipelike mass containing pieces of dolomite, chert, shale and quartz cemented by quartz and pyrite. The mineralization in all of the deposits consisted mainly of quartz, both fine-grained laminated and coarse-grained massive, carbonates, chalcopyrite, arsenopyrite, sphalerite, galena, bismuthinite, considerable amounts of pyrite, frequently in massive layers in some reefs and native gold. The gold was commonly finely divided and invisible to the eye. The grade of the deposits ranged from 0.35 to 0.75 oz Au/ton.

Gold-silver and silver-gold veins, lodes, stockworks and silicified zones in a complex geological environment, comprising sediments, volcanics and various igneous intrusive and granitized rocks

Deposits in this category combine nearly all the epigenetic features described for the previous categories. The origin of their metallic and other components may lie in the combined group or groups of rocks in which they lie, particularly interbedded sedimentary and volcanic piles. The deposits are best illustrated by specific examples. Those chosen include the Alaska Juneau Mine at Juneau Alaska; Grass Valley, California; the Rossland gold-copper belt in British Columbia; the Little Long Lac–Sturgeon River district, Ontario; the Renabie Mine in Leeson Township, Ontario; the Kirkland Lake district, Ontario; and the Central City district, Colorado. There are hundreds of other deposits throughout the world that have similar features.

Most of the deposits in this category are mainly quartz veins or lodes, stockworks or silicified zones in faults, fractures, shear zones, sheeted zones and breccia zones. Many of the deposits occur in volcanic rocks and pass into intruding granitic bodies indicating that they are later than the major periods of granitization and igneous injection. Other deposits occur in sediments, their metamorphic equivalents and in the associated granitized zones or intruded granitic rocks. As in the volcanic terranes the deposits are later than the granitization and granitic injection periods.

Some deposits are in rocks as old as Precambrian but were deposited during Tertiary time. The Central City district is a good example of this phenomenon. There are many other examples of younger deposits in much older host rocks in U.S.S.R., Japan and elsewhere.

The Alaska Juneau Gold Mine, now exhausted, was well known for its large tonnage mining operation (10 000 tons/day) and gold production from gold orebodies averaging 0.04 oz Au/ton. The orebodies (Fig. 45) were localized in an extensive stockwork of quartz veins and stringers, some 1000 to 2000 ft wide, 3½ mi long and extending downward at least 3000 ft, ramifying through a belt of contorted and altered greenstone, sills and dykes of gabbro, slate, tuffs and schists of Mesozoic age (Wernecke, 1932). The stockwork zone was extensively altered, biotitization, sericitization and silicification being the principal processes. The mineralization consisted essentially of quartz, albite, ankerite, calcite, pyrite, galena, pyrrhotite, chalcopyrite, sphalerite and native gold. The gold grade averaged about 0.04 oz/ton; the silver values were low.

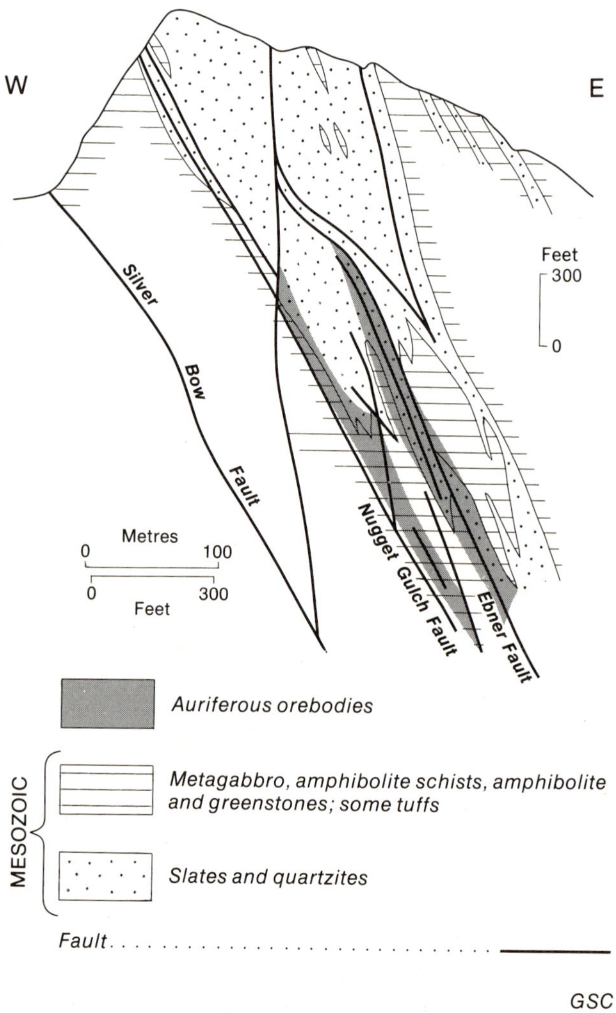

Figure 45. Typical cross-section, Alaska Juneau Gold Mine, Juneau, Alaska (after Wernecke, 1932).

The Grass Valley gold district and the Nevada City deposits some 4 mi apart formed one of the greatest lode-gold centres in east central California. The deposits are described in some detail by Lindgren (1896), Howe (1924), Farmin (1938), Johnston (1940) and Clark (1970). The deposits are persistent quartz veins and lodes and quartz stockworks and sheeted zones developed in faults, fractured zones and brecciated zones that cut Paleozoic and Mesozoic amphibolites (greenstones), slates, schists, etc. intruded by a granodiorite pluton of Mesozoic age. The deposits occur in all rock types, and the wall rocks are strongly sericitized and carbonated and moderately chloritized, epidotized, pyritized and arsenopyritized. The mineralization is essentially ribboned quartz, with variable amounts of ankerite and calcite, pyrite, arsenopyrite, galena, sphalerite, tetrahedrite, chalcopyrite, molybdenite, minor pyrargyrite, some scheelite, gold and small quantities of various tellurides. The grade averaged about 0.5 oz Au/ton, and the Au/Ag ratio of the ores was about 3.7. Johnston (1940) thought that the quartz veins grew by accretion from aqueous solutions; Farmin (1938, 1941) on the other hand considered that the veins were intruded as a highly concentrated magmatic solution of quartz (vein-dykes).

The Rossland gold-copper camp in the West Kootenay district of British Columbia produced during its lifetime some 6 200 000 tons of ore averaging 0.47 oz Au/ton, 0.6 oz Ag/ton and 1 per cent Cu from veins in a complex geological environment (Drysdale, 1915; Gilbert, 1948; Little, 1960; Thorpe, 1967).

The country rocks in the Rossland camp are interbedded greenstone flows, sills and sediments (Rossland volcanics) of Lower Jurassic age, all cut by a succession of offshoots or satellites of the Nelson batholith, including monzonite, granodiorite, alkaline syenite, pulaskite, granite, granite porphyry and lamprophyre. The orebodies were replacements along fissures chiefly in augite porphyry and monzonite. The mineralization consisted essentially of pyrrhotite, chalcopyrite, some pyrite and arsenopyrite, small amounts of gersdorffite, galena, sphalerite, stibnite, bismuthinite, molybdenite and marcasite, in a gangue of altered wall rock containing variable amounts of quartz and calcite. The alteration consisted of the development of minerals such as wollastonite, actinolite, garnet, epidote, prehnite, laumontite, chabazite, tourmaline, apophyllite, muscovite, biotite, chlorite and serpentine. Many of the veins were nearly massive sulphides; others were composed of disseminated sulphides. Some of the gold was free, and some was combined as telluride; a large part of the gold is present in chalcopyrite, pyrite, arsenopyrite and other sulphides as a finely divided or submicroscopic (lattice?) constituent. There was a close association between bismuthinite and gold in some stopes, and between molybdenite and gold in others. In most samples the gold is a late mineral in the paragenesis, impregnating arsenopyrite, bismuthinite, molybdenite and other sulphides.

Fyles *et al.* (1973) claim on the basis of K-Ar data that the mineralization at Rossland is of Tertiary age, but Thorpe and Little (1973) caution against accepting this age until U-Pb dating of zircon and whole-rock Rb-Sr dating are done.

The Little Long Lac–Sturgeon River gold district centred on Geraldton, Ontario had five major producing mines (Leitch, Northern Empire, Jellicoe, Bankfield and Tombill, and Hard Rock). The country rocks of the district are Archean (Keewatin) volcanics overlain unconformably by (Timiskaming) sediments and cut by diorite, feldspar porphyry and granodiorite (Horwood, 1948*a*; Pye, 1952). Most of the deposits are quartz veins and lodes in shear zones, drag folds, fracture zones and faults in folded greywacke, slate and iron-formation in the (Timiskaming) sediments. Other deposits are quartz veins and lodes similarly structurally located in Keewatin greenstones and their associated interbedded pyritiferous slates, tuffs and iron-formation. A few veins are in quartz porphyry, diorite and granodiorite. The mineralization was quartz with small amounts of pyrite, pyrrhotite, arsenopyrite, chalcopyrite, stibnite, tetrahedrite, other sulphosalts, tourmaline and native gold in most of the deposits in the sediments. In the deposits in the greenstones ankerite was common. Some of the gold-bearing bodies in iron-formation in the Hard Rock Mine were nearly massive pyrite and arsenopyrite with pyrrhotite in places (Matheson and Douglas, 1948). Alteration in the sediments was minimal; in the greenstones and porphyries carbonatization and sericitization predominate. The grade of the orebodies ranged from 0.3 to 0.5 oz Au/ton and the Au/Ag ratio of the ores was about 8 to 10.

In the Northern Empire Mine in the Little Long Lac district a thick, flat diabase sill cut out some 500 ft of the vein (Fig. 46). On the upward and downward extension of the vein for some 50 ft or more from the diabase the gold was extensively leached from the vein (Benedict and Titcomb, 1948). This mobilization and rearrangement of gold is common where late igneous bodies cut through gold-quartz deposits. At Yellowknife the late diabase dykes commonly become enriched in gold in the vicinity of the gold ore shoots; in other cases gold is concentrated in the orebodies adjacent to the diabase. In this case, as opposed to the situation at the Northern Empire Mine, the diffusion was, apparently toward the hot source, not away from it as one might expect on a cursory examination of the phenomenon. In the Rand of South Africa where igneous bodies cut through the gold-bearing bankets both enrichments and impoverishments of gold near dykes and sills have been noted (*see* page 324).

The Renabie Mine in Leeson Township, Ontario developed orebodies mainly in granite and tonalite gneiss and massive granite that intrudes andesitic lavas (greenstones) of Archean age (Bruce 1944, 1948; Rice, 1949). Four bodies of gold-bearing quartz were exploited including a pipe-shaped mass and three zones of quartz lenses and stockworks. Most of the quartz in the four orebodies is a granular white variety with brownish streaks, cut by veins of a younger white quartz. Only small amounts of metallic minerals occur in the older quartz and practically none in the younger. Pyrite is the most abundant sulphide; galena is sparingly present and erratically distributed. The native gold is microscopic in character; it is associated with carbonaceous substances (graphite) in some of the veins. The grade of the ore averages 0.2 oz Au/ton and the Au/Ag ratio is 2.7.

The Kirkland Lake gold belt in northern Ontario during its heyday boasted 10 major gold producers – Amalgamated Kirkland, Macassa, Kirkland Lake Gold, Teck Hughes, Lake Shore, Wright Hargreaves, Sylvanite, Toburn, Bidgood Kirkland and Upper Canada. After more than 50 years of mining

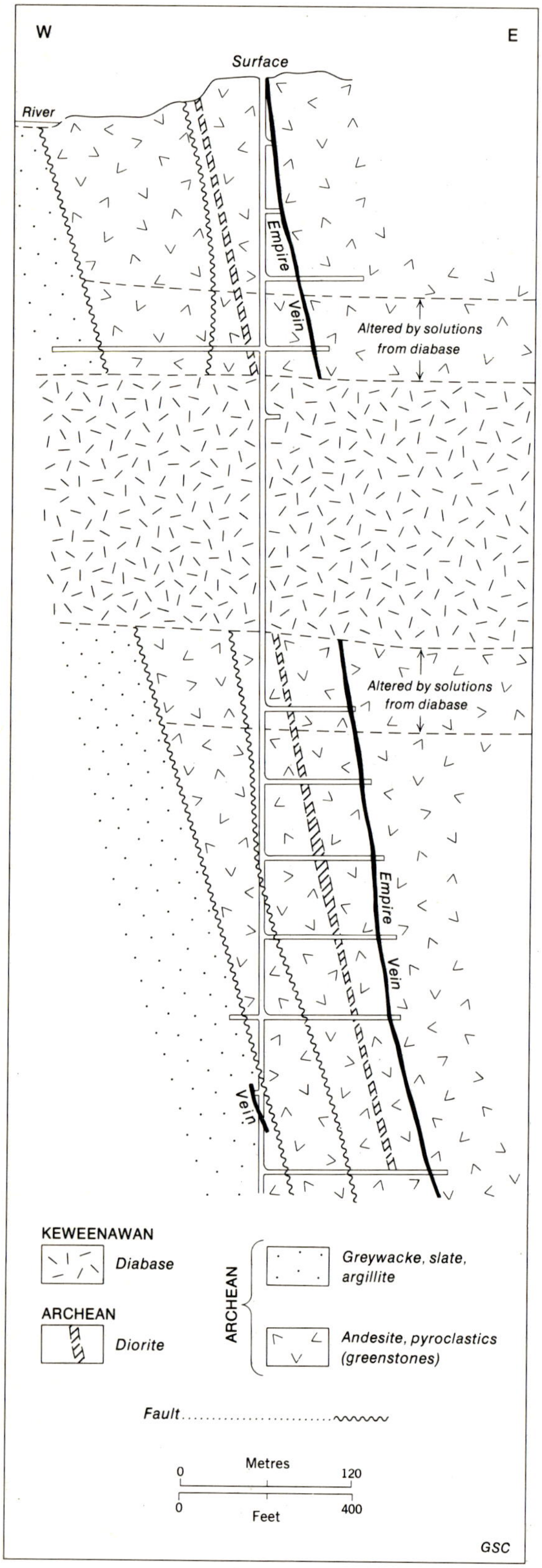

Figure 46. Vertical cross-section through shaft, Northern Empire Mine, Beardmore, Ontario (after Benedict and Titcomb, 1948).

most of the orebodies of these mines are now exhausted. The country rocks are steeply folded Precambrian sediments, including conglomerate, greywacke, slate, iron-formation, trachyte flows and tuffs lying in an east-west synclinal structure the flanks of which are marked both north and south by Keewatin greenstones and their associated tuffs and iron-formations. The greenstones and sediments are intruded by granodiorite, syenite and syenite porphyry, the last, one of the most favorable for ore shoots at Kirkland Lake. The structure and mineralization of the Kirkland Lake gold belt is described in great detail by Thomson *et al.* (1950).

Most of the gold deposits in the Kirkland Lake gold belt are in a thrust fault locally called the 'main break' or the Kirkland Lake Fault, which cuts through the sedimentary rocks and syenite and syenite porphyry plugs invading them (Fig. 47). Associated with the 'main break' are a number of bifurcating and parallel faults, all heavily mineralized in places. The orebodies were mainly fissure veins, but some were stockworks, breccia veins and gash veins. All were cross-faulted by postore movements. The principal gangue was quartz, but in some veins the mineral was relatively sparse, the oreshoots being silicified, mineralized mylonite along the fault. Other gangue minerals in small amount were calcite, albite, orthoclase, ankerite, barite, tourmaline, actinolite and minor apatite. The metallic minerals were mainly pyrite, chalcopyrite, galena, sphalerite, molybdenite, hematite, graphite, native gold and a number of tellurides, including altaite, coloradoite, calaverite, petzite and melonite. The grade of the ores was about 0.5 oz Au/ton and the average Au/Ag ratio of the deposits as a whole ranged from 4 to 9. The fineness of the native gold averaged about 940 and that of the bullion about 860. As Hopkins (1950) has pointed out this means that some 9 per cent of the silver in the bullion must have come from sources other than native gold. Some of the excess silver was provided by some of the tellurides, some by galena, but the bulk probably by pyrite in which silver tends to exceed gold in content according to the writer's analyses. This appears to be normal in a number of gold mines in Precambrian rocks.

The Central City district, Colorado, some 30 mi west of Denver, is of great interest as a gold-bearing province since the mineralization is of Tertiary age whereas the host rocks are mainly Precambrian. The geology of the district has been investigated in great detail and is described by Lovering and Goddard (1950) and Sims *et al.* (1963).

The Central City district is part of a larger zoned belt of mineralization that transects the grain of the Front Range at a small angle, stretching northeast from Breckenridge to Jamestown (*see* insert in Fig. 48). The Central City district is underlain mainly by Precambrian rocks comprising sedimentary gneisses and schists, migmatites, pegmatitic masses and a granodiorite body. The only younger rocks are a variety of Tertiary porphyries (Fig. 48).

The mineral deposits are mainly veins and lodes localized in a great anastomosing series of fractures and small faults that trend northeast and generally dip more than 60°. There is also a large oval stockwork (The Patch) that is believed to be a mineralized pipe of explosion breccia. All of the veins are younger than the Tertiary dykes and irregular bodies. Most veins and the stockworks contain essentially quartz, minor

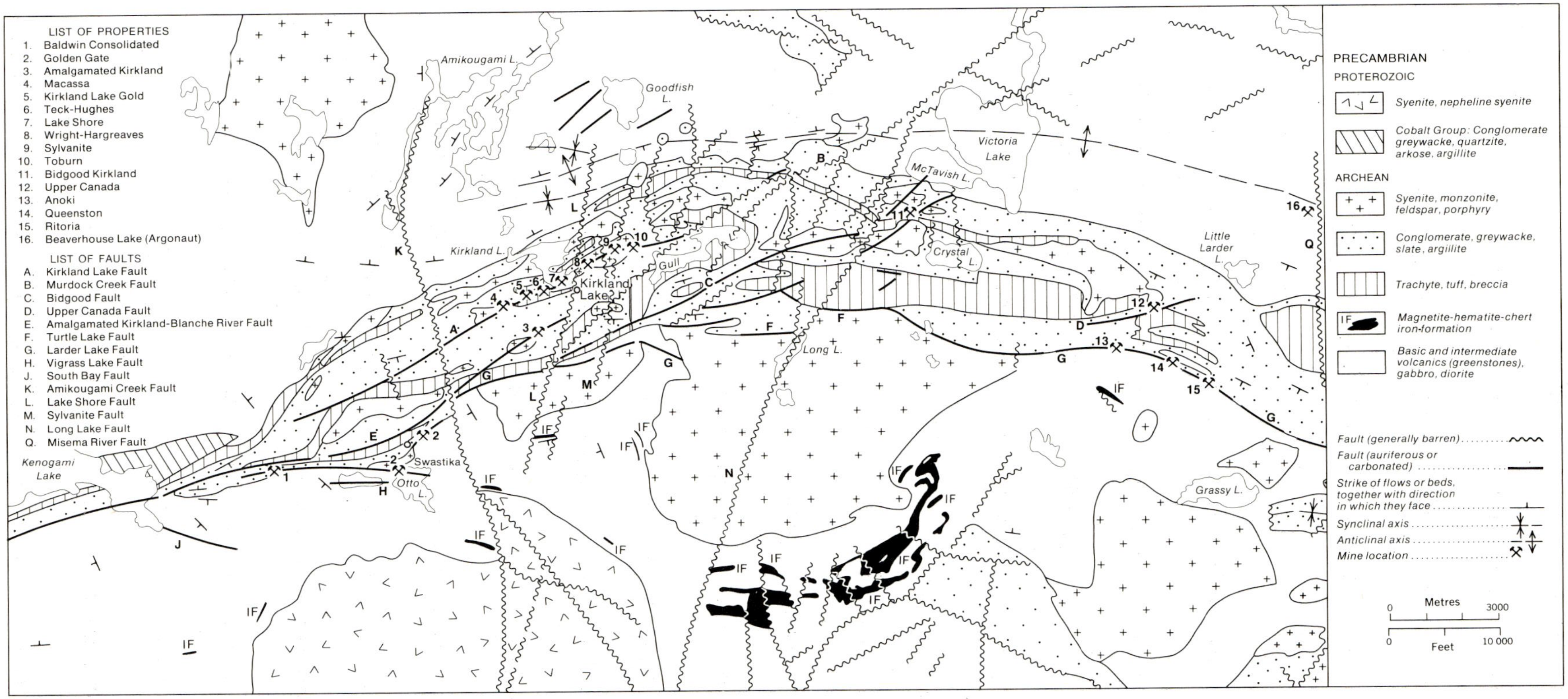

Figure 47. Generalized geological map showing the setting of the Kirkland Lake gold deposits (after Thomson, 1950).

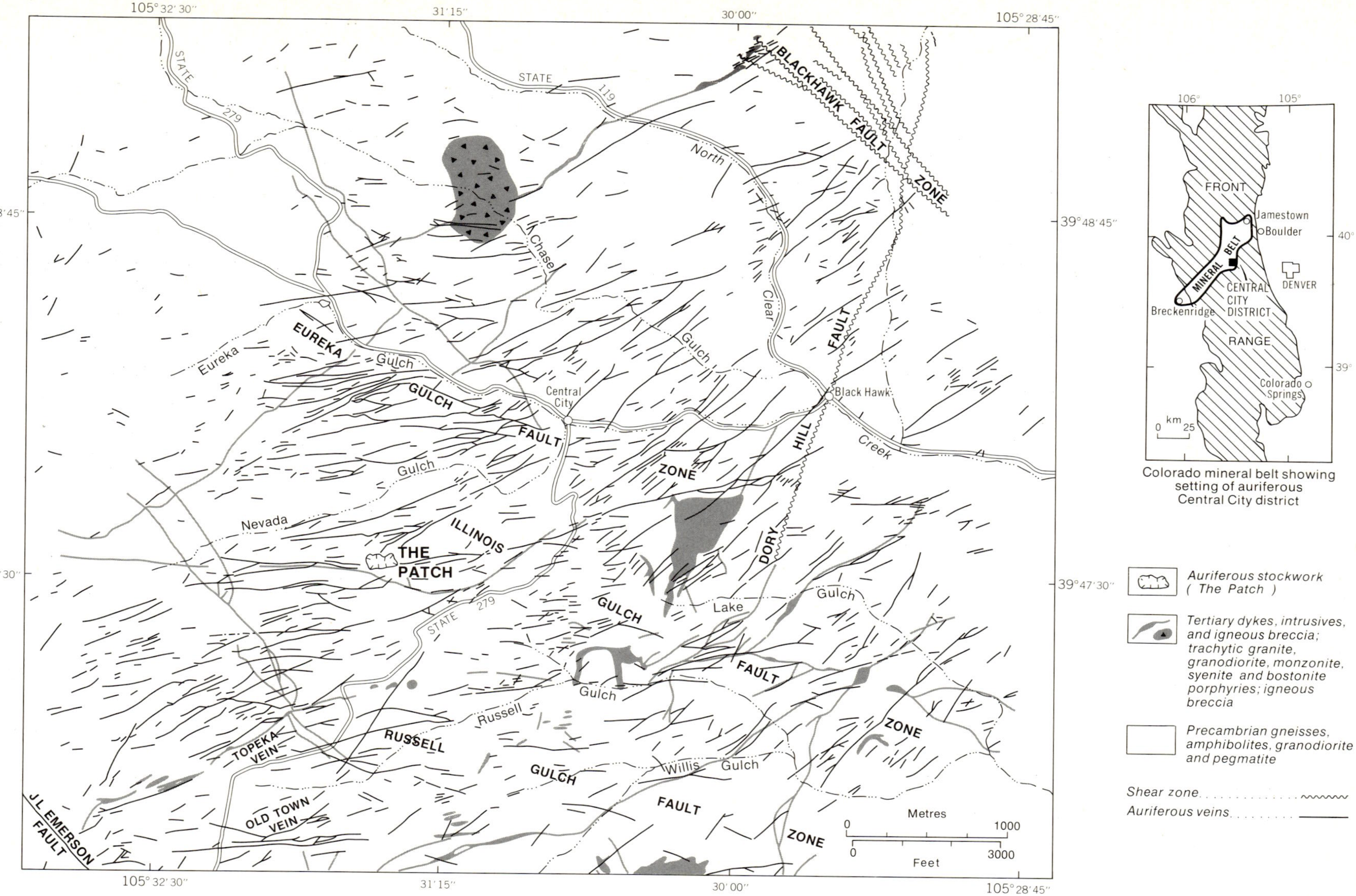

Figure 48. Generalized geological setting of the gold and other veins of the Central City district, Colorado (after Sims *et al.*, 1963).

fluorite, pyrite, sphalerite, chalcopyrite, tetrahedrite-tennantite, galena, native gold, and small amounts of bismuth, molybdenite, enargite, ruby silver, ferberite and Au-Ag tellurides. Pitchblende is a local constituent of some veins and has been mined as an ore mineral. The average grade of the smelting ore from the pyrite veins was 1.78 and 5.9 oz/ton of gold and silver respectively. The Au/Ag ratios of these veins ranged from 0.065 to 0.62. The average grade of the galena-sphalerite veins was 0.39 and 40.18 oz/ton of gold and silver respectively. The Au/Ag ratios ranged from 0.008 to 0.08. The highest silver concentrations were associated with galena and copper minerals. Pyrite in all types of deposits contained low contents of silver as did also most samples of sphalerite. Much of the gold was associated with chalcopyrite and tennantite, and less so with pyrite, galena and sphalerite.

According to Sims *et al.* (1963) the veins are distributed according to a well-defined, concentric zonal arrangement. A core of pyrite veins (central zone) is surrounded by a peripheral zone of galena-sphalerite veins; an intermediate zone contains transitional veins that consist of the minerals characteristic of both the central and peripheral zones. Most of the important mines are in the intermediate zone.

Sims *et al.* (1963) also state that although the veins were formed during a single period of mineralization, the major vein minerals were deposited in three stages, from oldest to youngest – a uranium, a pyrite and a base metal stage. The pyrite and base metal stages of mineralization were of broad areal extent and produced the district-wide zoning, whereas the uranium-stage mineralization was of local extent.

The ores were formed during the Laramide orogeny and were, according to Sims *et al.* (1963), derived from the magmas that consolidated to form the porphyries of the district. Lead-uranium isotope ages of uraninite from the veins indicate an absolute age of mineralization of about 60 m.y.

At the northeast extremity of the Colorado mineral belt, in the Jamestown, Gold Hill and Magnolia mining districts of Boulder County, there are a large number of gold telluride deposits that have been extensively mined in the past. These are described in detail by Kelly and Goddard (1969).

The predominant country rocks are Precambrian granites, gneisses and schists intruded here and there by various Tertiary intrusives mainly of biotite latite. The veins are of Early Tertiary age and are composed essentially of an interlacing network of pyritic or marcasitic horn quartz in which the ore minerals are sparse and irregularly distributed. The chief ore minerals are sylvanite, petzite, hessite, calverite, krennerite and native gold. Native tellurium and a variety of tellurides also occur. The gangue minerals are quartz, pyrite, marcasite, sphalerite, galena, chalcopyrite, various sulphosalts, roscoelite, ankerite, calcite, fluorite and barite. The gold and precious metal tellurides are late in the paragenetic sequence.

Kelly and Goddard (1969) consider that the tellurium, gold and other hypogene vein constituents were transported into their deposition sites as soluble chloride complexes.

The veins exhibit shallow oxidation and weathering with the development of jarosite, limonite, tellurium oxides, supergene native tellurium, mercury, hessite, copper tellurides, native silver, silver halides and fine spongy gold in limonite ('rusty gold').

Disseminated and stockwork gold-silver deposits in igneous intrusive, volcanic and sedimentary rocks

These deposits cover a wide spectrum. Some occur in igneous plugs, stocks, dykes and sills; others occur in diffuse altered zones in volcanic flows principally of Tertiary and Mesozoic age; and still others are disseminated bodies in various types of sedimentary rocks.

The principal economic element in these deposits in gold with small amounts of silver. A few deposits yield the base metals, but they are generally not known as base metal deposits. The grade of the deposits is highly variable. Most are relatively low grade (<0.5 oz Au/ton) but are characterized by large tonnages. The Au/Ag ratio is variable being generally high in the older deposits and low in the younger deposits. Most deposits are characterized by the presence of quartz or by silicification, often on a wide scale. Some deposits, particularly those containing massive pyrite, pyrrhotite and arsenopyrite, have little quartz and are not marked by the effects of silicification. The elements commonly concentrated in these deposits, neglecting those in the common gangue minerals such as quartz, silicates and carbonates, are: Cu, Ag, Au, (Ba), (Sr), Zn, Cd, Hg, B, (Sn), Pb, As, Sb, Bi, V, S, Se, Te, Mo, W, (F), Fe, Co and Ni. Those in parentheses are infrequent or occur only in certain deposits.

These deposits can be classified into three categories. Those in igneous bodies, those in volcanic flows and those in specific types of sedimentary beds. Each type is best described by examples.

Disseminated and stockwork gold-silver deposits in igneous intrusive bodies

These deposits occur in igneous plugs, stocks, dykes and sills that have been intensely fractured or shattered and infiltrated by quartz, gold and other minerals. Most of the deposits are stockworks or diffuse irregular impregnations. The alteration processes vary with the type of rock. In granitic (acid) rocks sericitization, silicification, feldspathization (development of albite, adularia, etc.) and pyritization are predominant; in intermediate and basic rocks carbonatization, sericitization, serpentinization and pyritization prevail. Alunitization may occur in both acid and basic rocks in places.

Seven Canadian examples, one from Chile, two from U.S.S.R. and one from Zaire will illustrate this type of deposit – the Howey and Hasaga mines at Red Lake, Ontario; some of the orebodies in the Matachewan district, Ontario; the Box Mine at Goldfields, Saskatchewan; the Sunbeam Kirkland Mine in the Falcon Lake stock, southeastern Manitoba; the Baton Lake dyke near Indin Lake, Northwest Territories; the Lamaque Mine, Val d'Or, Abitibi County, Quebec; the Camflo Mine near Malartic Quebec; the El Chivato (Talca) Mine in Chile; the Berezovsk deposits in the Urals of U.S.S.R.; and the Twangitza deposit in Zaire.

The orebodies of the Howey and adjacent Hasaga Mine at Red Lake, both now exhausted, were developed in a quartz porphyry dyke intruding Keewatin volcanic breccia (Horwood, 1945, 1948*e*). The orebodies were sections of the porphyry dyke extensively fractured and infiltrated by gold-bearing quartz veins and stringers (Fig. 49). These zones of veins and stringers were marked by considerable alteration the

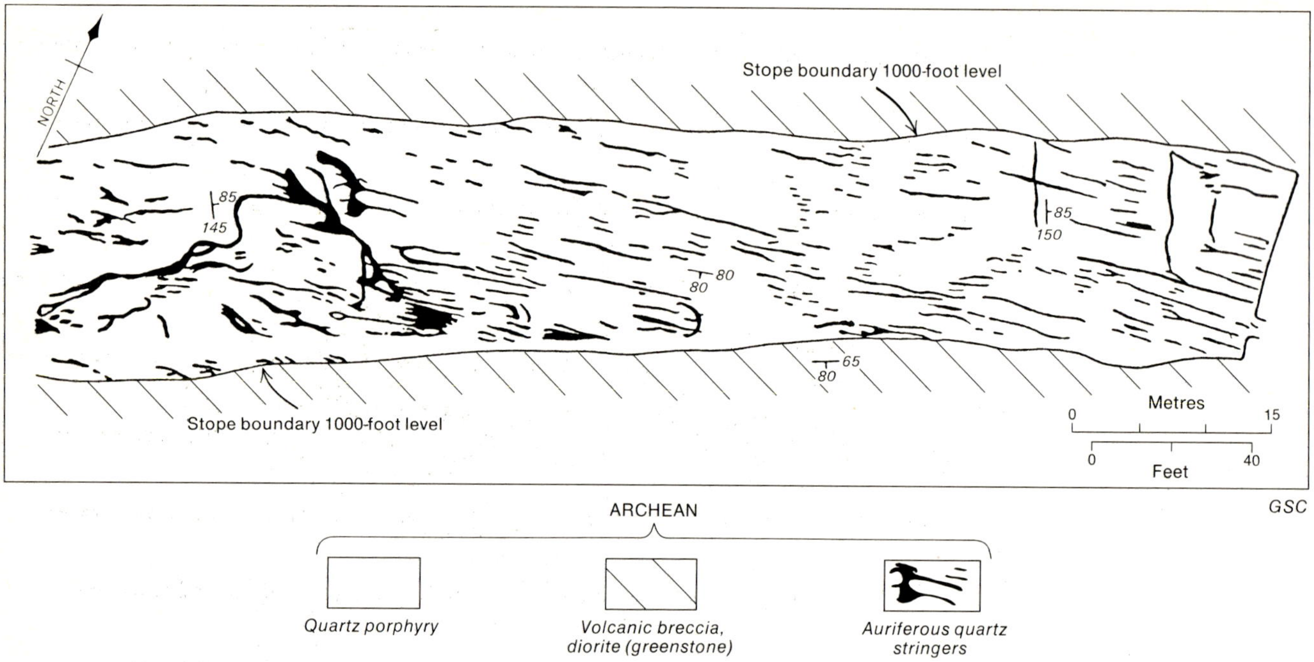

Figure 49. Sketch showing the distribution of auriferous quartz stringers in a portion of the quartz porphyry dyke, Howey Mine, Red Lake, Ontario (modified from Horwood 1945, 1948*e*).

principal processes being sericitization, albitization, carbonatization and pyritization. The principal minerals within the quartz veins and stringers included carbonates, pyrite, sphalerite and small amounts of galena, chalcopyrite, altaite and polybasite. The economic gold minerals were the native metal and small amounts of tellurides, mainly sylvanite. The grade of the orebodies averaged 0.10 oz Au/ton and 0.05 oz Ag/ton, giving an Au/Ag ratio of 2.

Some of the gold orebodies in the Matachewan district of Ontario are in syenite porphyry (Dyer, 1936). At the Young-Davidson and Matachewan Consolidated mines, now exhausted, porphyritic syenite bodies (mainly dykes) intrude Keewatin volcanics, including pillow lavas and agglomerate and Timiskaming conglomerate and greywacke (North and Allen, 1948; Derry *et al.*, 1948). In the Young-Davidson Mine (Fig. 50) the brick-red phase of the syenite contained 2 per cent of disseminated auriferous pyrite accompanied by minor amounts of chalcopyrite, galena, molybdenite, scheelite and specularite. Superimposed on this phase of the syenite are two sets of fractures, one containing quartz and carbonate veins usually less than 2 in. thick, and the other fine and hairlike containing visible gold. The average grade of the ore was 0.1 oz Au/ton and 0.02 oz Ag/ton, giving an Au/Ag ratio of 5. The ore in the prophyry of the Matachewan Consolidated Mine was similar to that in the Young-Davidson.

The orebody of the Box Mine at Goldfields, Saskatchewan, now abandoned, was somewhat similar to that described above for the Howey and Hasaga mines. The focus of mineralization was a granitic sill intruded into gneisses of the Tazin Group of possibly Archean age (Jewitt and Gray, 1940). The sill is marked by a stockwork of innumerable quartz veins and stringers developed in fractures and mineralized with small amounts of pyrite, galena, sphalerite and free gold. At least two and possibly three ages of quartz stringers occur. The wall-rock alteration in the heavily mineralized sections included sericitization and pyritization (*see* page 236). The grade of the ore averaged about 0.05 oz Au/ton, and the Au/Ag ratio of the orebody as a whole was about 5. In the general area there are a number of other similar deposits in granitic rocks (Alcock, 1936; Christie, 1953).

The Sunbeam Kirkland Mine, now abandoned, was developed in the quartz monzonite phase of the Falcon Lake stock (Fig. 51*a*) in the West Hawk Lake area of southeastern Manitoba (Brownell, 1941; Haugh, 1962; Gibbins, 1970). The deposit was a pipelike body (Fig. 51*b*) rudely circular in cross-section and plunging steeply to the northwest. The material in the pipe was highly fractured quartz monzonite that has been moderately to intensely sericitized, carbonated, silicified and impregnated with sulphides and gold. The sulphides, localized mainly along thin fractures, were pyrite, pyrrhotite, arsenopyrite, sphalerite, galena, tennantite and chalcopyrite. The gold was mainly free. Late fractures also carried some free gold. The average grade of the orebody was 0.23 oz/ton, and the Au/Ag ratio appears to be 1 according to analyses of ore samples carried out at the Geological Survey.

The Baton Lake dyke deposit some 30 mi north-northeast of the north end of Indin Lake, Northwest Territories is owned by Johnsby Mines Ltd. The deposit is a quartz stockwork in a quartz-albite (quartz diorite) dyke in andesites (greenstones) of the Yellowknife Supergroup. The dyke has been traced for some 20 000 ft and has an average width of about 110 ft. A host of glassy quartz veinlets and stringers occur in silicified parts of the dyke, follow definite fracture patterns, and range in width from a fraction of 1 in. to 2 ft or

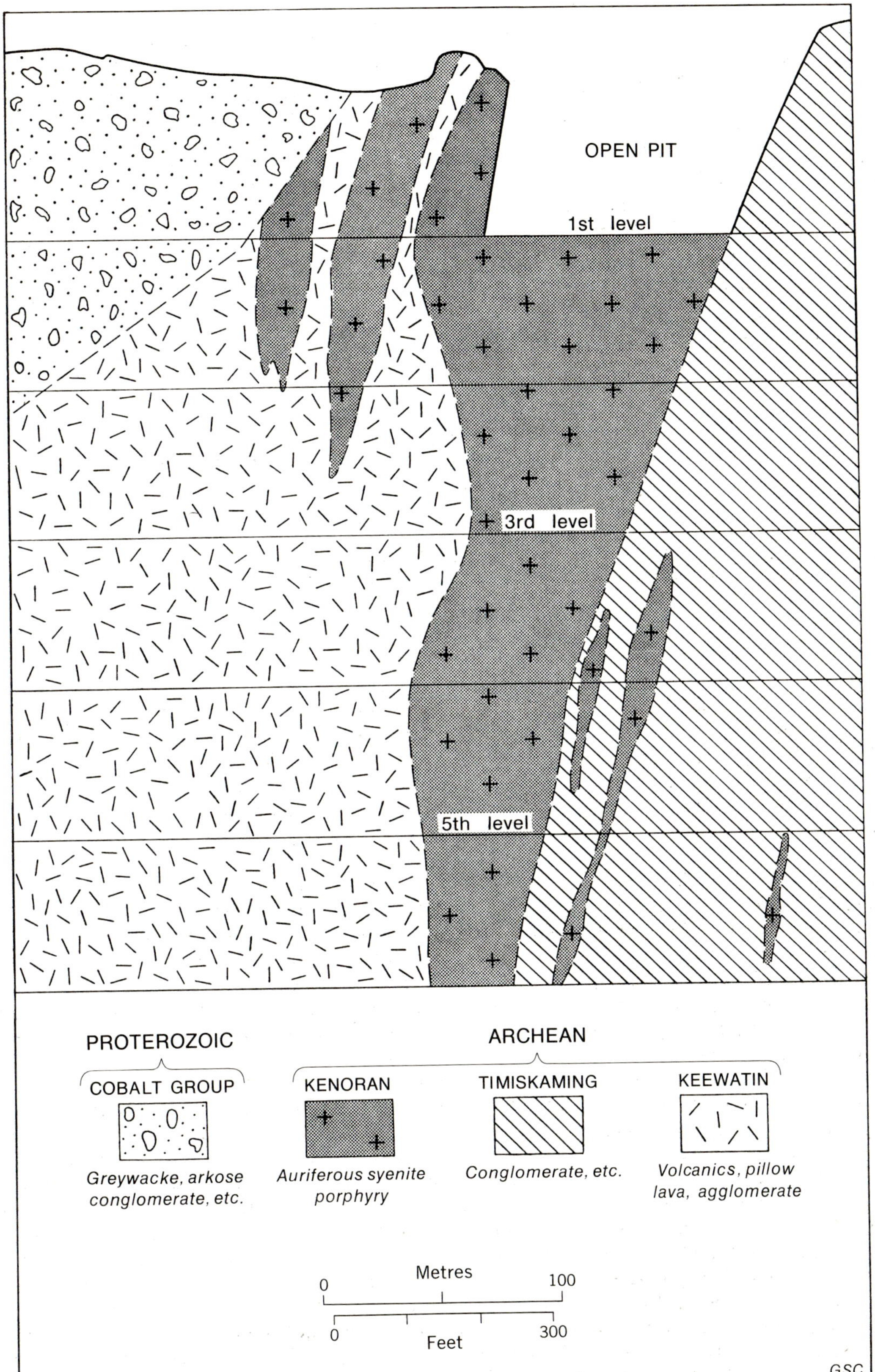

Figure 50. Typical north-south vertical section, Young-Davidson Mine, Matachewan, Ontario (after North and Allen, 1948).

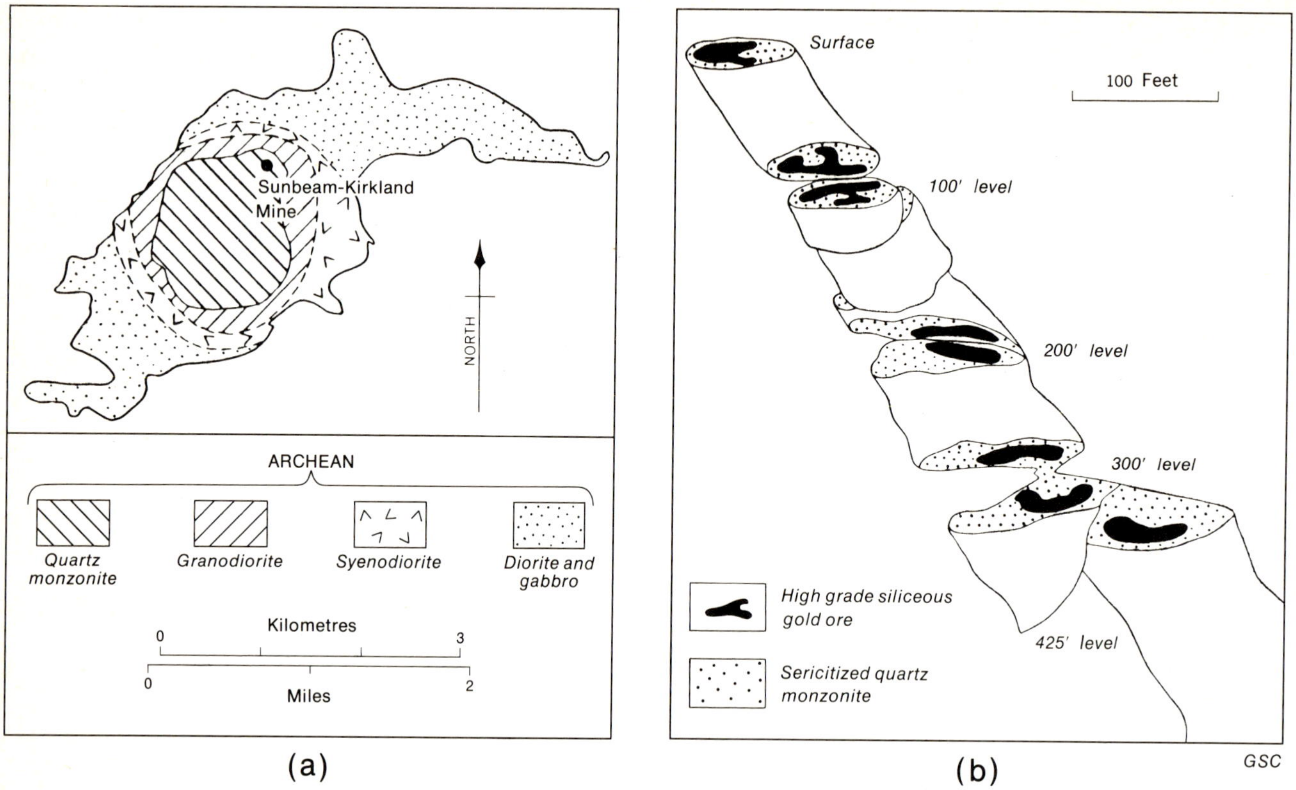

Figure 51. (a) General geology of the Falcon Lake stock, southeastern Manitoba. (b) Sunbeam-Kirkland auriferous pipe (after Brownell, 1941).

more. Most of the gold occurs free in the glassy quartz and is commonly associated with pyrite, pyrrhotite, arsenopyrite and sphalerite. In places there is some tourmaline in the quartz stringers. Pyrrhotite is abundant throughout the dyke and appears to be indigenous. It is reported that the deposit contains 15 million tons of ore averaging 0.068 oz Au/ton (Northern Miner, Aug. 3, 1972, p. 19). Assays quoted by Lord (1951) indicate an Au/Ag ratio of 1.1.

The Lamaque Mine at Val d'Or, Quebec is essentially in a granodiorite-diorite stock in Archean agglomerates and tuffs intruded by dykes and irregular masses of diorite porphyry (Wilson, 1948). The orebodies are veins or stockworks of quartz stringers in faults with little displacement (Fig. 52). The mineralization in the quartz consists of tourmaline, ankerite, scheelite, and small amounts of pyrite, chalcopyrite, native gold and several tellurides. Adjacent to most of the veins the wall rock is carbonatized, pyritized and tourmalinized for distances of a few inches to several feet. The average grade of the oreshoots is 0.32 oz Au/ton.

The geology and mineralization of the Lamaque orebodies are remarkably similar to those obtaining at the Morning Star Mine, Woods Point, Victoria, Australia (Clappison, 1953). There, the orebodies are in quartz veins in fissures in a diorite dyke that intrudes Upper Silurian slate and sandstone (Fig. 53). The quartz veins vary in width from a few inches up to 4 ft and are marked by flanking altered zones in which a fair amount of pyrite and quartz are developed. The mineralization in the quartz veins consists mainly of pyrite, with small amounts of galena, sphalerite, chalcopyrite and free gold, the last being commonly present in amounts of several ounces to the ton.

The Camflo Mine located between Malartic and Val d'Or, Quebec is in a highly fractured, moderately altered porphyritic syenite (monzonite) stock that intrudes Archean (Malartic and Kewagama groups) conglomerate, iron-formation, greywacke, phyllite and various volcanic rocks (tuff, andesite, diorite) (Meikle, 1970). The orebodies are zones marked by myriad white-grey quartz stringers and considerable amounts of disseminated pyrite. Calcite and fluorite are widespread in the ore zones, and minor amounts of scheelite are common. Small specks of visible gold occur in the quartz stringers, along minute fractures in the porphyry and coating the faces of pyrite crystals. Minor amounts of tellurides occur in places. Both the native gold and tellurides are late in the paragenetic sequence of minerals. One thousand tons per day are mined from the orebodies which average about 0.25 oz/ton Au. The Camflo orebodies resemble those of the Young-Davidson and Matachewan Consolidated mines described above.

Several other occurrences of large tonnage disseminated gold orebodies in shattered, sheared and fractured igneous bodies could be mentioned. One averaging in width from 60 to 100 ft, containing some 5 million tons carrying about 0.15 oz Au/ton occurs in shattered granodiorite at the Cochenour Willans Mine in the Red Lake area, Ontario (Northern Miner, Jan. 31, 1974, v. 59, no. 46, p. 1). Other possibilities of developing large low grade auriferous orebodies in granodiorite, syenite or other igneous type rocks would seem to be good in the Canadian Shield. Parts of the Kirkland Lake area

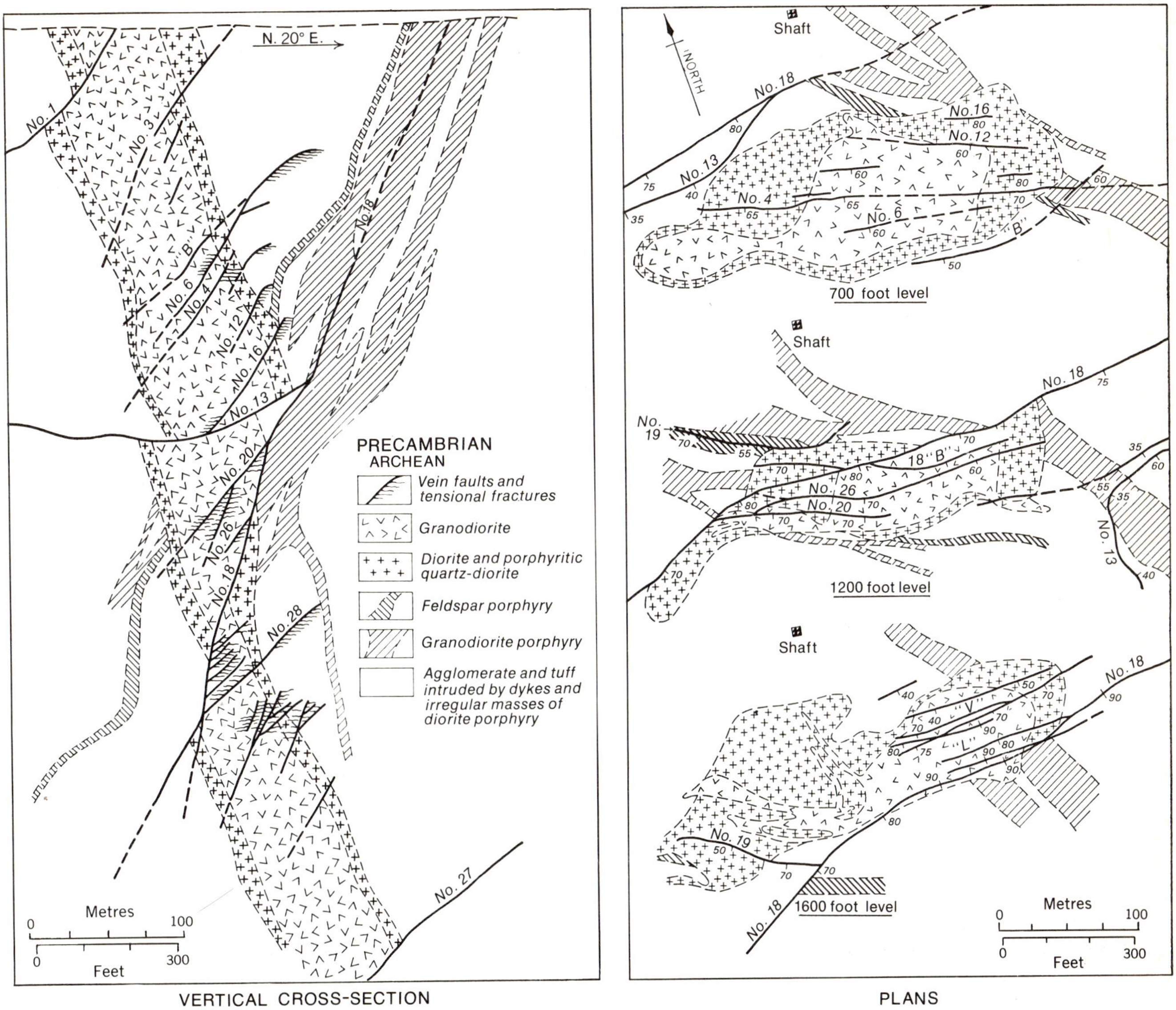

Figure 52. Plan and vertical cross-section transverse to veins, Lamaque Mine, Val d'Or, Quebec (after Wilson, 1948). (Compare with Figure 53.)

should be looked at in this respect since some of the syenites and porphyries are mineralized over considerable widths in places (*see* Burrows and Hopkins, 1925, p. 22).

Interesting and productive auriferous deposits of infrequent occurrence are certain gold-tourmaline-chalcopyrite circular and lenticular pipes developed mainly in small granitic stocks. The El Chivato (Talca) Mine in Chile about 150 mi south of Santiago is a striking example (Ruiz and Ericksen, 1962). The ore is altered and replaced granitic breccia cemented by quartz, carbonates and coarsely crystalline tourmaline. The brecciated granitic rock is of Late Jurassic to Late Cretaceous age. The sulphide minerals are pyrite, chalcopyrite, galena, sphalerite and tetrahedrite. Much of the gold is finely disseminated and commonly occurs as blebs in chalcopyrite. In Chile this type of deposit grades into or is associated with rich copper pipes characterized by abundant chalcopyrite and tourmaline with only small amounts of gold. According to Geier (1933) the orebodies are generally inverted cones. Most are highly oxidized and enriched in gold down to 200 ft or more.

Somewhat similar tourmaliniferous deposits occur in the eastern Transbaikal gold-molybdenum belt of the U.S.S.R. (Borodaevskaya and Rozhkov, 1974). The auriferous shoots of the Klyuchi gold deposit in this belt are developed in pipes, veins and irregular bodies localized in explosive (breccia) pipes at the intersection of fractures cutting mainly granites of unspecified Mesozoic age and Upper Cimmeric (Jurassic) minor intrusions. The host rocks are marked by an early phase of sericitization, chloritization, pyritization and silicification followed by an intense phase of auriferous quartz-tourmaline and quartz-sulphide veins and impregnations. The minerals include essentially pyrite, chalcopyrite, arsenopyrite and gold

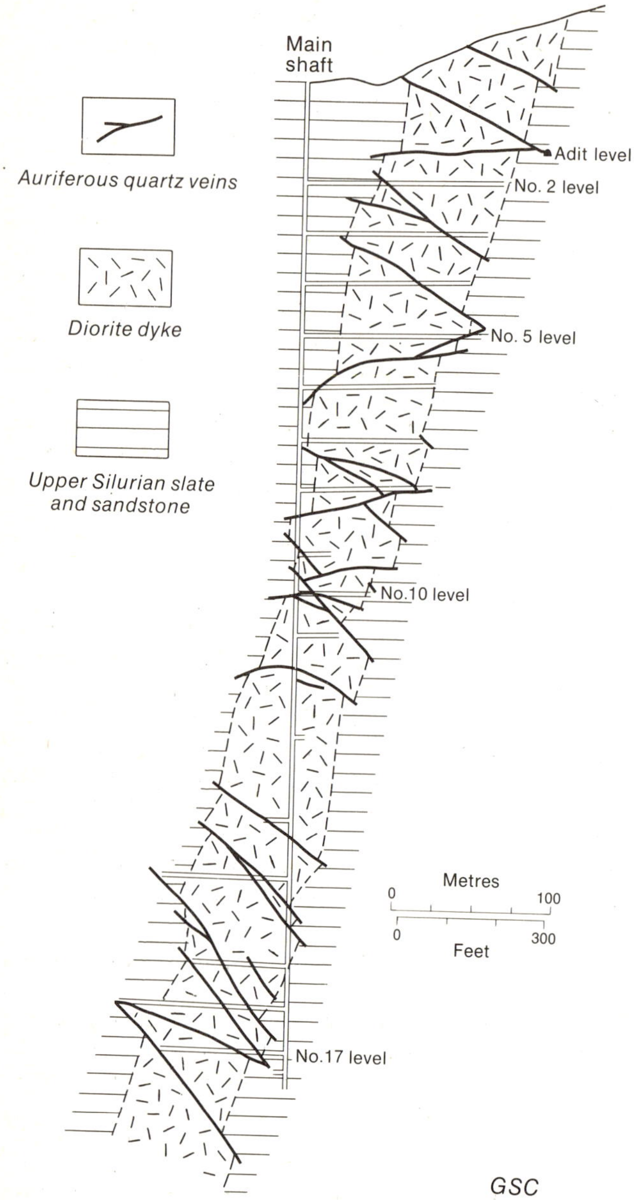

Figure 53. Typical north-south cross-section through the main shaft, Morning Star Mine, Woods Point, Victoria, Australia (after Clappison, 1953). (Compare with Figure 52.)

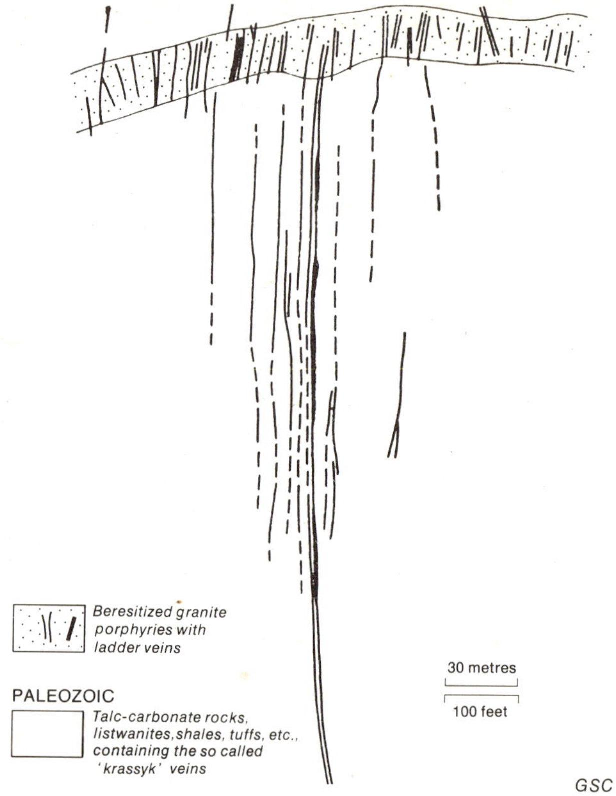

Figure 54. Generalized geological plan of the Molokova deposit, Berezovsk, Urals, U.S.S.R. showing auriferous ladder veins in beresitized granite porphyry dykes and the so-called gold-bearing 'krassyk' quartz veins in sedimentary, volcanogenic and basic-ultrabasic intrusives (after Borodaevskaya and Borodaevskii *in* Borodaevskaya and Rozhkov, 1974).

with minor amounts of freibergite, enargite, molybdenite, bismuthinite, native bismuth and a number of other sulphides and sulphosalts.

The Beresovsk deposits in the Urals, U.S.S.R., discovered in 1745 and now largely exhausted, are typical examples of gold deposits in igneous dykes and stocks. The Beresovsk district some 10 mi northeast of Sverdlovsk is underlain by Paleozoic mica schists and serpentinites intruded by Hercynian granitic bodies (Mushketov, 1930; Borodaevskii and Kytyukin, 1939; Boradaevskaya and Rozhkov, 1974). Intruding the schists in a fairly well-defined zone are a large number of imbricating quartz porphyry and microgranite dykes 10 to 150 ft wide, with a northerly strike and mostly vertical to steep dips, some of which are intersected by a network or ladder-vein system of quartz veins and stringers, a few inches to 3 ft or more wide (Fig. 54). Adjacent to the quartz veins and commonly throughout the quartz porphyry a type of alteration called beresitization occurs. This alteration is a combination of sericitization and pyritization with feldspathization in places. The veins consist of quartz, sericite, tourmaline, auriferous pyrite, galena, chalcopyrite, aikinite, covellite, magnetite, sphalerite, tetrahedrite and gold. The grade of the deposits ranges from 0.25 to 0.50 oz Au/ton. The Au/Ag ratio is about 8. At Pyschminsk some 5 mi northeast of Beresovsk similar quartz veins occur in a complex of sedimentary schist, serpentine and quartz porphyry. Most of the ladder-vein quartz zones lie within the quartz porphyry, but others are in the serpentinite, which is altered to listwänite, actually a carbonated (magnesite, dolomite), sericitized, pyritized, talcose serpentinite with quartz. Associated with the ladder veins in the Berezovsk goldfield are zones of so-called 'krassyk' veins in the sedimentary, volcanogenic and basic-ultrabasic country rocks. These are similar in strike and dip to the ladder veins in the dykes; their mineralization comprises drusy quartz with subordinate amounts of ankerite, calcite, dolomite, pyrite, tetrahedrite, scheelite, aikinite, galena and chalcopyrite.

The Twangitza gold deposit in Kivu Province, Zaire is in highly carbonated albitites (Safiannikoff, 1972). The mineralization is localized in an anticlinal zone in which black schists and quartzites of the Burundian system (Precambrian) were intruded by albitites. The albitites are composed essentially of albite and 25 to 30 per cent ankerite. There is practically no quartz. Most of the gold is associated with pyrite and arsenopyrite that is disseminated through the albitites. Some monazite is also disseminated through the albitite and adjacent rocks. Ankerite-pyrite-arsenopyrite-quartz veinlets form a few stockworks in fractured zones outside the albitites. The gold is very fine grained, and very little responds to amalgamation. The grade of the orebodies averages about 3 to 5 ppm (0.08–0.14 oz Au/ton). Safiannikoff likens the deposit to those at Beresovsk in U.S.S.R. and some of those at Kilo-Moto.

In closing this section it should be pointed out that gold deposits in igneous bodies are common. There is a marked relationship in many gold belts between albitites, albite porphyries, quartz-feldspar porphyries and gold. It is not always clear just what this relationship is. In most places the localization of the gold deposits in or near the porphyries has been ascribed to the structural competence of these rocks relative to the host rock. In other words they fractured more readily and formed structural sites for the orebodies. The origin of the various porphyries in gold belts is itself a major problem. Some seem to be of magmatic origin and are perhaps a differentiate either of a granitic magma or of granitization processes. Others appear to have formed by replacement from solutions or by diffusion. These various hypotheses have been discussed by Henderson and Brown (1966) and by Boyle (1961*a*, 1976*a*). In many places the pyrite, ankerite and gold appear to have been largely indigenous; in others they appear to have been partly redistributed during shearing and fracturing. Moorhouse (1942) mentions a number of these gold-bearing albitite dykes and porphyries in the gold belts of Canada; he thinks that the gold mineralization of some of the bodies is a differentiate more or less *in situ* of the bodies themselves. In the western San Juan Mountains, Colorado, Lipman *et al.* (1976) emphasize the relationship in space and time of the gold and base metal mineralization to volumetrically minor intrusions of quartz-bearing silicic porphyry, which post-dates the magmatic events that gave rise to the various volcanic caldera.

Disseminated gold-silver occurrences in volcanic flows and associated volcaniclastic rocks

Commercial deposits of this type have not been worked. Most occurrences are very low grade commonly less than 0.01 oz Au/ton (0.3 ppm). Silver contents are higher in places, averaging in some cases 3.5 oz Ag/ton (120 ppm).

The disseminated occurrences in this category are in reality large irregular and diffuse zones of alteration manifest mainly in rhyolites, andesites, basalts and their associated tuffaceous rocks. These zones of alteration constitute silicification, sericitization, epidotization, argillization or alunitization, commonly associated with pyritization and carbonatization. In the basic and intermediate rocks the effects are commonly collectively called propylitization. Large volumes of the volcanic country rocks are affected, giving them a bleached and altered aspect. Locally diffuse silicified zones, quartz veins, alunite veins and pyrite veins and segregations ramify through the altered rocks.

Such extensively altered zones of rock, in places carrying small amounts of gold and silver, are widespread in Mesozoic-Tertiary volcanic terranes throughout the world. They are particularly common in the western United States, in Japan, Indonesia, New Zealand (Coromandel Peninsula) and elsewhere. Some are closely associated with gold deposits as at Goldfield, Nevada (Fig. 32), Virginia City (Comstock Lode) (Fig. 33) and Titiribí, Colombia (Hoffmann, 1931). Most are of Tertiary age, but some such as those on Vancouver Island described by Clapp (1915) are of Mesozoic age. Others in British Columbia such as the Mount Edziza (Souther, 1967) and other volcanic complexes are of Tertiary age. One occurrence near Barstow, California is said to contain some 50 million tons containing an average of 3.5 oz Ag/ton (E.A. Schiller, pers. commun., 1968). The gold content is not known. The silicified and altered zones at Goldfield, Nevada range in gold content from about 0.02 to 3 ppm, the corresponding silver values are 0.3 to 10 ppm (Ashley and Keith, 1976). Deposits of this type mentioned in the Russian literature (Petrovskaya, 1974) appear to range widely in their gold and silver contents (Au – 0.02–3 ppm; Ag – 0.5–8.6 ppm).

Little can be said about the details of these occurrences since they have not been extensively studied. Most contain considerable amounts of fine-grained quartz, chlorite, epidote, sericite, adularia in places, clay minerals, carbonates and pyrite. Some are marked by developments of alunite as at Goldfield, Nevada (Jensen *et al.*, 1971; Ashley, 1974; Ashley and Keith, 1976). Those rich in pyrite appear in many cases to contain the most gold and silver. These altered and mineralized zones require detailed sampling and analysis before any conclusions can be drawn as to their commercial value.

Disseminated gold-silver deposits in volcaniclastic and sedimentary beds

These deposits are usually conformable with the sedimentary and volcaniclastic beds although in some cases their limits may infringe irregularly on overlying or underlying rocks. Some are large and of relatively high grade; others are commonly too low grade or not of sufficient tonnage to merit attention.

Two general categories of these deposits can be recognized – those developed in tuffaceous rocks and iron-formations within volcanic terranes and those resulting from extensive infiltration or replacement of chemically favourable beds such as limestones. The first is especially common in Precambrian terranes, although there is no reason why they should not occur in rocks of younger age; however, examples of the latter are rare to date. The second can apparently occur in rocks of any age. These deposits are best illustrated by specific examples.

Disseminated deposits in tuffaceous rocks and iron-formations

Gold deposits in tuffs and other volcaniclastic rocks are common in the greenstone belts of the Canadian Shield and in other greenstone belts of Precambrian age throughout the world. The Madsen Mine at Red Lake, Ontario will serve as an example of an extensive gold deposit in tuff (Horwood, 1945; McIntosh, 1948; Harriman, 1965).

The country rocks in the vicinity of the Madsen deposit

are predominantly Keewatin volcanics consisting of meta-andesite pillow lavas with interbedded tuffs, agglomerate and breccia of acid to intermediate composition. Included in the volcanic pile are quartz and feldspar porphyries, which probably represent flows although the quartz porphyry that forms the upper contact of the ore-bearing tuff may be a dyke or sill.

The orebodies at the Madsen Mine all lie mainly within tuff beds and were localized by the action of rolls in the quartz porphyry on the hanging wall and in the talc schist and andesite on the footwall. The ore zones (Fig. 55) consist of silicified and heavily mineralized lenses in sheared tuff. The silicification in some places may be original (that is the rock was initially a cherty tuff), although there are narrow stringers of quartz and greenish quartz-carbonate containing visible gold that cut through the lenses in places. The lenticular ore zones, defined by assay, consist mainly of silicified tuff, abundant pyrite and small to minor amounts of magnetite, pyrrhotite, arsenopyrite, sphalerite and chalcopyrite. The average grade of the orebodies is 0.23 oz Au/ton and the Au/Ag ratio is about 5.5.

Gold-silver deposits in the various facies of iron-formations are particularly numerous in Precambrian rocks, especially those in greenstone belts of Archean age. Numerous examples can be cited including one of the orebodies in the Hard Rock Gold Mine near Geraldton, Ontario; orebodies in the Central Patricia Mine, Crow River district, Ontario; possibly the orebodies of the Homestake Mine, Lead, South Dakota; probably the orebodies of the Morro Velho, Raposos and Passagem de Mariana deposits in the 'Quadrilátero Ferrífero'), Belo Horizonte area, Minas Gerais, Brazil; and possibly some of the conformable gold-bearing sulphide orebodies in the Kolar goldfield, India. Certain gold deposits in the Precambrian of Africa also belong in this category, e.g., some of those in Rhodesia and the Moto deposit in Zaire.

The orebodies of the Central Patricia Mine in the greenstone belt in the Crow River area north of Lake St. Joseph in northwestern Ontario were in iron-formation (Thomson, 1939*a*; Barrett and Johnston, 1948). The iron-formation is banded and composed of layers of magnetite, chert (quartz) and siderite. The orebodies consisted of quartz and sulphide stringers and narrow veins that crossed the banded iron-formation at a large angle but were confined almost entirely to the iron-formation (Fig. 20). The gold-bearing stringers were mineralized heavily with pyrrhotite and arsenopyrite, and minor amounts of pyrite, chalcopyrite, siderite and chlorite. Nearly all of the gold was present as native metal in a finely divided form (<200 mesh) in minute fractures and on the crystal faces of the sulphide minerals. The individual stringers assayed up to 11 oz Au/ton, but the stockwork orebodies averaged about 0.33 oz Au/ton.

Numerous other gold deposits occur in iron-formation and related rocks in the Canadian Shield, particularly in Ontario. Here may be mentioned deposits and occurrences in the Pickle Lake, Casummit Lake, Michipicoten and Goudreau, Beardmore, Kirkland Lake and Emerald Lake (Timagami Forest) areas.

An interesting example of auriferous tuffs in a sedimentary terrane is provided by the occurrences at the Andrew Yellowknife claims some 70 mi northwest of Yellowknife, Northwest Territories (Lord, 1951). The country rocks are metasedimentary strata of the Archean Yellowknife Group comprising impure quartzites, quartz-mica schists, andalusite schists, sericite schists and biotite hornfels. Interbedded with these are beds up to 1600 ft in strike length of dark grey to black amphibolite, biotite-garnet schists and biotite-chlorite-garnet schists containing abundant pyrite in streaks, blebs and irregular masses. Vein quartz occurs irregularly along the beds but appears to bear no particular relationship to the pyrite. These pyritic schists appear to have originally been sulphidic tuffs and sediments (possibly silicate and sulphide facies of iron-formation?), now metamorphosed to pyritic amphibolites and pyritic schists during granitization processes, the manifestation of which is evidenced by a granitic batholith and pegmatites nearby. Some of the pyritic amphibolites contain significant gold assays, as high as 0.22 oz/ton across 7 ft. Native gold occurs sparingly in parts of the beds, but most of the gold is intimately associated with the pyrite in an invisible form. Beds, probably sulphidic tuffs and sediments (iron-formation?), similar to those at the Andrew Yellowknife property are common in other parts of the Yellowknife Supergroup. All are indicated by their rusty brown (limonitic) outcrops; most contain some gold and silver but rarely in economic amounts.

The famous Homestake Mine at Lead, South Dakota obtains its ore from a complex of inclined mineralized pipes, pods, veins, saddles and lenses developed along the crests of plunging folds within the Homestake Formation (McLaughlin, 1931; Gustafson, 1933; Noble, 1950; Slaughter, 1968; Norton, 1974). This formation is of Precambrian age and is said to have been originally an Fe-Mg carbonate bed some 60 ft thick in the early description by McLaughlin. Slaughter also characterizes the formation as originally an Fe-Mg carbonate sediment, now a sideroplesite-quartz schist with some banding due to differing amounts of graphite in the outer metamorphic zone. In the high grade metamorphic zones the sideroplesite schist has been converted to cummingtonite schist. The formation has been isoclinally folded and then cross-folded resulting in numerous contorted zones. Dilation in zones of cross-folding, mainly on the crests of folds, has been postulated as the cause of ore localization (Fig. 56).

According to Slaughter (1968) chloritization of the cummingtonite or sideroplesite schists is the most widespread effect of the mineralization, which he divided into four stages. The principal minerals of the first stage are quartz, chlorite and arsenopyrite; in the second stage, quartz, ankerite and pyrrhotite. In the third stage only pyrrhotite was deposited. The fourth stage followed the intrusions of Tertiary dykes, which occur throughout the area. It consisted mostly of pyrite and calcite. Gold accompanies all stages and is always the last mineral to be deposited. It is most closely associated with arsenopyrite of the first stage but can be interpreted as having been deposited during the fourth stage and superimposed on the products of the earlier stages. Noble (1950) mentions the presence of galena, sphalerite, chalcopyrite, marcasite, tetrahedrite, magnetite and specularite in the ores. The last two are fairly widespread.

The average grade of the Homestake orebodies is about 0.4 oz Au/ton, and their Au/Ag ratio is about 4. The average fineness of the native gold, as given by Slaughter, is 820.

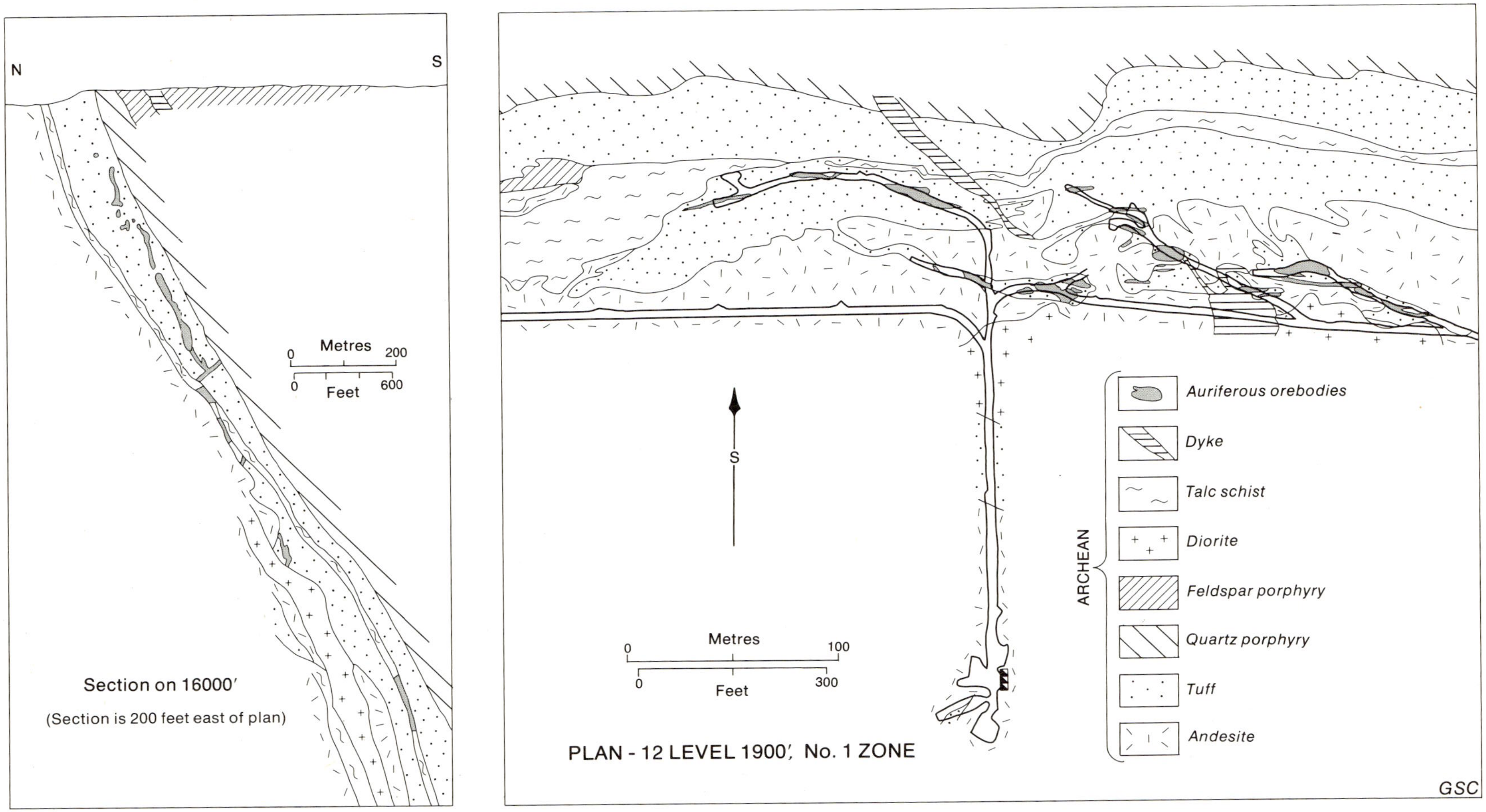

Fig. 55. Characteristic plan and section, Madsen Mine, Red Lake, Ontario (after Harriman, 1965).

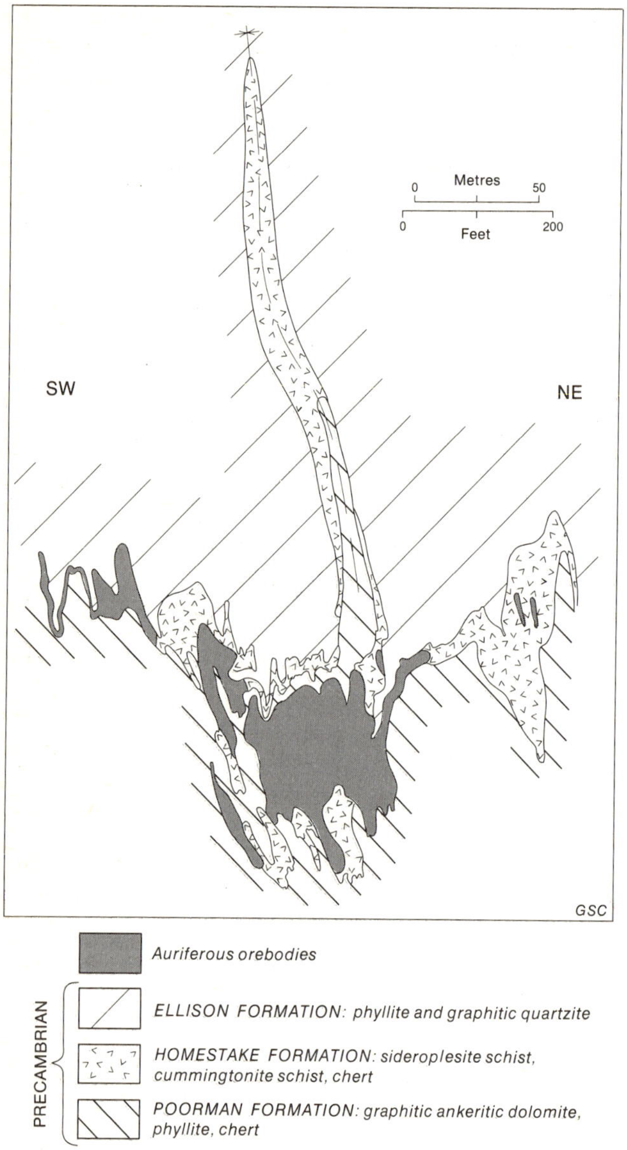

Figure 56. Vertical cross-section of the Main Orebody, Homestake Mine, South Dakota (after Noble, 1950). The isoclinal axis of the Pierce anticline is strongly deformed by a cross-fold, and the orebodies are localized in the zone of cross-folding.

Some investigators consider the deposit to be of Precambrian age; others have considered that the gold mineralization is Tertiary, since deposits of this age occur in the near vicinity. The orebodies in the Homestake Mine are, however, cut by Tertiary intrusives and faults, and the sulphur isotopic ratios of the sulphides in the Homestake are fundamentally different from those of the Tertiary deposits (Rye and Rye, 1974). It thus appears from these data that the gold was deposited during Precambrian time. The fineness of the gold is, however, low for Precambrian deposits, a feature that requires explanation.

Recently, on the basis of lead isotopes in galena from quartz veins, Rye *et al.* (1974*a*) concluded that the ore in the Homestake Mine was emplaced by metamorphic secretion from a 2.5 billion-year-old source (i.e., the Homestake Formation) during regional metamorphism and intrusion that occurred 1.6 billion years ago. The lead isotopes in Tertiary galenas from Paleozoic and Precambrian host rocks were found to have had different growth histories. The Tertiary galenas in Paleozoic rocks did not derive their lead entirely from Tertiary intrusive rocks in the area, but rather the lead may have been obtained mainly or in part from the Paleozoic rocks. Tertiary galenas in Precambrian rocks, on the other hand, seem to have obtained their lead largely from the transected metasedimentary host rocks or from the underlying basement, with little or no contribution from other sources.

The Homestake deposit is difficult to classify. Some have considered it to be simply a hydrothermal replacement deposit of favourable sedimentary carbonate rocks; other views now current consider that the Homestake deposit is confined to beds of carbonate iron-formation and that the gold and associated sulphides, original constituents of the iron-rich sediments, were later concentrated in dilatant zones during metamorphism (Sawkins and Rye, 1971, 1974). Sulphur, oxygen, carbon and lead isotopic data support this hypothesis (Rye and Rye, 1974; Rye *et al.* 1974*a*).

Deposits somewhat similar to those at the Homestake Mine occur in the vicinity of Contwoyto Lake, Northwest Territories, Canada (McConnell, 1964*b*; Tremblay, 1966; Bostock, 1967, 1968). There, layers and lenses of pyrrhotite and pyrite-bearing amphibolites (cummingtonite gneisses) are interbedded with Archean (Yellowknife Supergroup) greywacke and argillite and their metamorphic equivalents, nodular (cordierite and andalusite) biotite schist and gneiss. These layers and lenses of amphibolites and sulphide-bearing amphibolites are generally less than 60 ft thick and are thought by Bostock (1967, 1968) to be the metamorphic equivalents of an oxide-silicate-sulphide facies of iron-formation. An alternate explanation is that they are metamorphic replacements of carbonate layers or bands in an argillite-greywacke assemblage. To the writer the iron-formation hypothesis seems more probable.

The richest gold-bearing parts of these sulphide bands and lenses are in highly contorted areas, especially on the crests of folds where some secondary quartz stringers and small lenses are developed. The principal minerals in these sites are garnet, cummingtonite, quartz, pyrite, pyrrhotite, arsenopyrite, loellingite, chalcopyrite and native gold. Samples of this type of material are said to run from 0.07 to 0.65 oz Au/ton and 0.04 to 0.24 oz Ag/ton. The highest gold values appear to be concentrated in the arsenopyrite and loellingite-rich zones.

The gold deposits within the Quadrilátero Ferrífero, Minas Gerais, Brazil have been the subject of study for many years, and their origin has been variously interpreted. The principal mines are the Morro Velho, Raposos and Passagem de Mariana. The mineralogy of the orebodies in these mines is essentially similar. The economic mineral is native gold accompanied by pyrite, pyrrhotite, arsenopyrite, chalcopyrite and small to minor amounts of tetrahedrite, bornite, sphalerite, galena, stibnite, wolframite and scheelite. The gangue is quartz and ankerite, whose origin has been variously interpreted as original, redistributed during metamorphism, or introduced by hydrothermal solutions. In the Morro Velho and Raposos mines the sulphide content of the ore ranges from low amounts up to 50 per cent or more. Tourmaline is especially abundant in the Passagem deposit. The grade of the ores ranges from 0.25 to 0.5 oz Au/ton and the Au/Ag ratio is about 5 for most mines.

Recent descriptions of the country rocks in the "Quadrilátero Ferrífero" are given by Gair (1962), Dorr and Barbosa (1963), Barbosa (1968), Dorr (1969) and Fleischer and Routhier (1973). The rocks of interest are the Rio das Velhas Series and Minas Series, which include a great pile of Precambrian sediments comprising quartzite, phyllite, chert, black graphitic schist, chlorite schist, ferruginous quartzite, quartz dolomite-ankerite rock (Lapa Seca) and what seem to be from the descriptions and samples several facies of iron-formation including oxide, carbonate, silicate and sulphide facies. The terminology is somewhat confusing to say the least and is not helped by differences of opinion as to what constitutes itabirite (oxide facies of iron-formation) and other weathered products of iron-formations (Park, 1970; Dorr 1971). It appears from the literature that there are also serious problems in unravelling the stratigraphy and differences of opinion on the succession of rocks, facies changes, etc. These controversies, notwithstanding, there is considerable evidence to indicate that the three major gold deposits of the district, and numerous other smaller deposits, are either in or closely associated with one or other facies of iron-formation.

The best descriptions of the Morro Velho Mine is by Matheson (1956) and Gair (1962). The orebodies occur within a highly contorted siliceous banded dolomite-ankerite rock (Lapa Seca) that is probably the equivalent of the carbonate facies of iron-formation (Fig. 57). The rocks enclosing the Lapa Seca are quartz-carbonate-sericite-chlorite and graphitic schists. The main orebody is a flattened pipe that stands edgewise and plunges eastward to a depth exceeding 8000 ft. The other orebodies are ramifying, irregular, contorted and sinuous. The ore is particularly rich in sulphides, especially pyrrhotite and arsenopyrite. Some parts of the orebodies are nearly massive sulphides. The grade averages about 0.4 oz Au/ton.

The Raposos Mine about 3 mi northeast of the Morro Velho Mine has been described by Gair (1962) and Tolbert (1964). The orebodies are long, narrow sinuous bodies composed mainly of quartz, ankerite, pyrite, pyrrhotite and arsenopyrite in iron-formation, mainly the siderite facies. The gold occurs mainly in the native form, and the grade ranges from 9 to 16 g Au/ton (0.27 to 0.47 oz/ton).

The Passagem de Mariana gold deposit, in the southeastern part of the "Quadrilátero Ferrífero", has been described by Barbosa *et al.* (1948) and more recently by Fleischer and Routhier (1973). The orebodies, composed essentially of quartz, carbonates, tourmaline, arsenopyrite and gold, occur in a sedimentary horizon averaging some 5 m in thickness and extending over a distance of 15 km. The ore-bearing horizon (Batatal Formation according to Fleischer and Routhier) consists of graphitic phyllites, ferruginous dolomite and metacherts. The hanging wall is iron-formation (itabirite) and the footwall sericitic quartzite. The orebearing horizon has been considerably deformed, and in detail some of the orebodies are composed of boudins of tourmalinite containing arsenopyrite and gold, whereas others consist of tourmalinite with gold and sulphides in quartz and coarse dolomite.

The origin of the gold deposits of the "Quadrilátero Ferrífero" have been variously interpreted. Lindgren (1933),

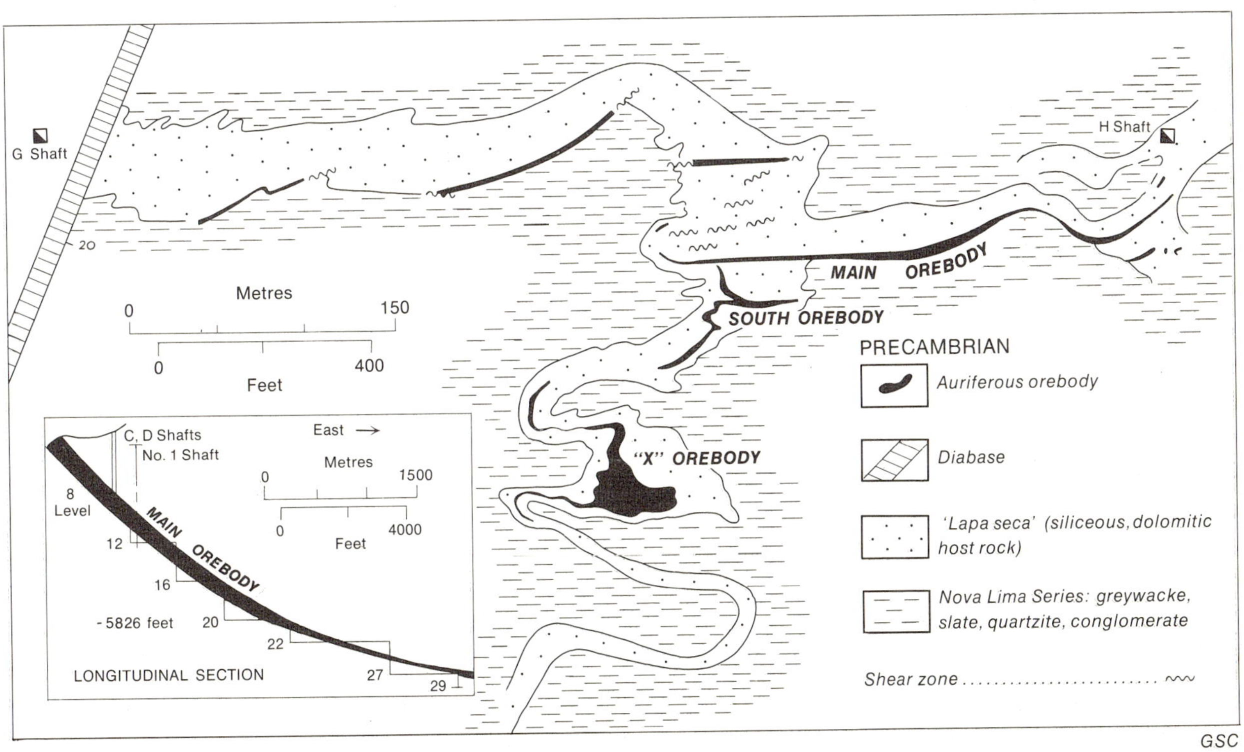

Figure 57. Geological plan 20 level (5826 ft below collar) and generalized longitudinal section, Morro Velho Mine, Minas Geraes, Brazil (after Matheson, 1956).

Matheson (1956), Gair (1962), Dorr (1969), and Park and MacDiarmid (1970) considered that the Morro Velho deposit was typically hydrothermal, the gold and other constituents presumably being derived from some underlying granitic body, although most of the authors are rather vague on the exact magmatic source of the elements in the deposit. Fleischer and Routhier (1973), on the other hand, think that the Morro Velho deposit is consanguineous-syngenetic, the gold and other elemental constituents being syngenetic with the chemical sediments and later remobilized and concentrated in favourable sites during structural deformation and metamorphism. This view has been disputed by Barbosa *et al.* (1974).

Tolbert (1964) considered the Raposos Mine to be a hydrothermal deposit but found no evidence for the ultimate origin of the gold. He considered the enclosing Precambrian rocks and a granitic body as possible sources of the precious metal. He thought that the iron, sulphur, arsenic and small amounts of chromiun in chromium sericite associated with the orebodies may have been derived from a local source.

A wide variety of origins has been suggested for the Passagem de Mariana deposit. Hussak (1898) considered the deposit to be a mineralized pegmatite, and Derby (1911) and others following him considered the ores to be of hydrothermal origin. Fleischer and Routhier (1973) think the deposit is a remobilized-syngenetic one, the gold and other constituents of the ores being an integral part of the chemical sediments in which the deposit is localized. This view has been disputed by Barbosa *et al.* (1974) who apparently think that the deposit is a hydrothermal bedding vein.

Numerous other examples of gold deposits in or near iron-formation occur in the Precambrian shields of the world. In Africa, Woodtli (1961) describes gold impregnation deposits in the Moto area of Zaire. The orebodies are in a Precambrian schistose rock containing ankerite, quartz and albite near itabirite (iron-formation) and graphitic schist. The mineralogy of the orebodies comprises quartz, albite, ankerite, siderite, magnetite, hematite, pyrite, pyrrhotite and native gold, the last rarely visible. The gold-bearing shoots are irregular and are outlined by assay. In Rhodesia, Lightfoot (1930), Pelletier (1964) and Anhaeusser (1976) mention that a number of the gold-quartz-sulphide deposits are in Precambrian banded ironstones. Pyrrhotite figures prominently in these deposits as it does in most auriferous orebodies in iron-formation throughout the world. The Wanderer Mine was a prolific producer until its exhaustion in 1952. More recently Fripp (1976) has described the stratabound gold deposits in Archean banded iron-formations in Rhodesia. He thinks that the gold-bearing sulphide-rich layers in the deposits predate the metamorphism and deformation of the iron-formation and the enclosing aquagene lavas and tuffs, and proposes that the sulphides, carbonates, chert and gold represent primary (syngenetic) sedimentary accumulations on the sea floor. These constituents were precipitated from geothermal brines during fumarolic-stage volcanic activity in Sebakwian (early Precambrian) times. These brines contained the gold as dissolved complex species with hydrogen and sulphur, and possibly with arsenic. The depth of the sea water, temperature and pH largely controlled the nature and gold content of the iron-formation.

In Tanzania some of the gold deposits in the Iramba-Sekenke goldfield are extensive impregnations of pyrite, pyrrhotite and gold in Precambrian banded quartzite and iron-formation; similar auriferous impregnation deposits are worked in the Southwest Mwanza goldfield, in the Musoma goldfield, in the Kahama-Nzega-Shinyanga area and elsewhere (Harris, 1961). Many of the deposits in the Lupa goldfield, likewise, appear to occur in sedimentary banded iron-formations (Grantham, 1928), although there has been a general dispute about the origin of some of these ironstones (Kondakov and Perunov, 1973).

In Australia gold-quartz veins and disseminations of gold are known in banded iron-formation in the Lennonville, Boogardie (Finucane, 1953) and north of Lake Austin goldfields, Western Australia and in the Tennant Creek district of Northern Territory (Sullivan and Ivanac, 1953; Crohn, 1965; Baxter, *in* Knight 1975; Woodall, 1975; Ryan, 1976). The ironstones and jaspers in the Tennant Creek area are apparently not iron-formations in the ordinary sense. Some appear to be replacement bodies in faults and shear zones. The gold-bismuth-copper Juno lode for instance is a discordant magnetite- and hematite-rich ellipsoidal to pipelike body cutting Lower Proterozoic felsic sediments and pyroclastics. Massive magnetite and abundant chlorite constitute the core of the lode, and gold is concentrated therein. Both the ore minerals and gangue minerals are vertically zoned within the lode (Large, 1974, 1975). Most are closely associated with hematitic shale and may have originated by mobilization and concentration of iron from these sediments. Crohn (1965), however, considered this improbable basing his conclusion on trace element analyses of the hematite from the two types of occurrences. Actually, the origin of the Tennant Creek lodes has been a controversial topic for years. Recently Dunn (1976) considered that the auriferous deposits of the Tennant Creek field are examples of the "massive oxide" facies of volcanogenic mineral deposits. In reply Large (1976) agreed that the ores could be considered as massive oxide equivalents of volcanogenic massive sulphide ores, but maintained that the Juno orebody is a subsurface replacement deposit and has significant differences from typical proximal massive sulphides deposited at the seawater-rock interface. In the Yilgarn block of Western Australia the Nevoria Mine was developed mainly in strongly deformed iron-formation (Williamson and Barr, 1965). The ore shoots are flat-lying quartz veins carrying much pyrrhotite and gold; some are interrupted by pegmatite dykes, but the exact age of the gold mineralization with respect to these dykes is not stated.

Some of the Precambrian gold-bearing sulphide lodes of the Kolar goldfield in India as described by Narayanaswami *et al.* (1960) may be in the sulphide facies of iron-formation. These are described as containing abundant graphite in places and grading into graphitic lodes (Fig. 26). The sulphide lodes are also marked by an abundance of pyrrhotite, a sulphide mineral that is particularly diagnostic of gold lodes in iron-formation.

The principal features of gold deposits in iron-formations can be summarized briefly. Nearly all deposits have an epigenetic aspect, that is the gold-bearing quartz stringers, lenses and irregular pods, the gold-bearing sulphide stringers and the massive sulphide lenses and irregular bodies are clearly in dilatant zones such as fractures, small faults, drag folds and contorted parts of the iron-formations.

In some deposits there is considerable evidence for replacement of the carbonate facies (siderite and dolomite) of the iron-formations by quartz, silicates, sulphides, etc. Most deposits are characterized by relatively marked amounts of pyrrhotite rather than pyrite, although the latter is usually present in the ores. Graphite is present in a number of deposits. The gold is intimately associated with the sulphides in many deposits and is rarely visible to the naked eye. The Au/Ag ratio in nearly all deposits is greater than 1. The grade of the deposits and occurrences ranges all the way from <0.01 to 0.5 oz Au/ton.

The source of the gold, silver and other elements such as arsenic and sulphur is probably the sedimentary beds comprising the iron-formations, particularly the graphitic sulphide facies and the pyritiferous chlorite facies. During metamorphism, granitization and injection of granitic bodies, dilatant zones appeared in competent beds of iron-formation, and these were filled with quartz, sulphides, gold, etc. extracted from the pile of sediments comprising the iron-formations and in places also from andesites, pyritiferous shales, pyritic tuffs, etc. Where carbonate facies constituted the bulk of the iron-formations replacement of the carbonates by diffusing constituents such as SiO_2, As, S, etc. predominated. More is said about these processes in the section on the origin of gold deposits at the end of this chapter.

Disseminated deposits in chemically favourable sedimentary beds

Gold deposits of this type occur mainly in carbonate-bearing rocks such as limestone, dolomite, calcareous quartzites and calcareous shales. They may be of any age, but the most productive to date are of Tertiary age. The orebodies are commonly irregular, but some are tabular following specific beds or series of beds. Most of the deposits occur in highly faulted terranes, and the orebodies generally spread out into beds laterally from faults or fractured zones. The replaced beds or series of beds are often disturbed but not generally highly brecciated or contorted. The mineralization is highly variable, but the development of quartz, pyrite, arsenopyrite, and small amounts of other sulphides and sulphosalts is universal. Silicification is the main process in their formation. The principal gold mineral in most of these deposits is the native metal, commonly low in silver. The native gold is usually widely disseminated through the deposits and is generally microscopic. Ore shoots usually have to be outlined by assay.

Numerous examples of this type of deposit occur throughout the world. Some of the replacement orebodies in dolomitic limestone and phyllitic quartzite of the Salsigne Gold Mine in France (Fig. 42) belong in this category. The protores of the Kuranakh type residual deposits in southern Yakutia, U.S.S.R. are also of this type (Fig. 70). The best known examples are in north-central Nevada, where four deposits – Getchell, Cortez, Carlin and Gold Acres – have been extensively investigated.

The literature on the Nevada disseminated deposits is extensive, the papers by Joralemon (1951), Hausen and Kerr (1968), Wells *et al.* (1969), Radtke and Scheiner (1970*a,b*), Wells (1971), Roberts *et al.* (1971) and Radtke and Dickson (1976) being particularly informative, especially the last, which includes an extensive bibliography.

The four deposits – Getchell, Cortez, Carlin and Gold Acres – differ in their general morphology but are remarkably similar in their mineralization and geochemistry. All are thought to be of Tertiary (Oligocene?) age. Silberman *et al.* (1974), however, state that Getchell has been dated at approximately 90 m.y., Gold Acres has a probable age of 92 to 94 m.y., Cortez is younger than 35 m.y., and the age of Carlin is uncertain. This suggests a broad range for the age of mineralization from late Mesozoic to middle Tertiary. The age of 80 m.y. for the Getchell Mine has been disputed by Joralemon (1975), who considers the deposit to be younger than 20 m.y. In a later discussion, however, Berger *et al.* (1975) maintain that the older age is most compatible with the geologic, geochemical and isotopic data available. The gold in the deposits is extremely fine grained (microscopic) in most parts of the deposits, and the Au/Ag-ratios range from 9 to 30. Such ratios are unusual but not unique for Tertiary deposits as witness the high ratios at Cripple Creek, Colorado (~12). The geochemistry of the various deposits is essentially similar. There has been widescale replacement of carbonates by silica which is manifest by a marked increase in the latter constituent in the orebodies and a decrease in CO_2 and some of the other constituents of the carbonates, especially calcium. Other elements introduced include Fe, Ba, As, Hg, Sb, Pb, Zn, Cu, W, Au, Ag and S. At Carlin and Cortez, Radtke and Scheiner (1970*a*) consider that carbonaceous matter played a large part in the precipitation of the gold from acid hydrothermal solutions at 175 ± 25°C. Subsequent oxidation of the gold-organic complexes so formed precipitated the gold in a microscopic form. Wells (1971) on the other hand, using electron microprobe methods, noted a close association of gold and arsenic with tiny grains of arsenic-bearing pyrite and no particular association with carbonaceous matter as suggested by Radtke and Scheiner (1970*a*).

The four disseminated deposits are spatially related to the Roberts Mountains Thrust Fault, the orebodies lying either within the thrust zone or within carbonate rocks a few hundred feet below it, especially where the thrust has been domed. Three prerequisites for ore are emphasized by Roberts *et al.* (1971), namely: (1) a source for gold-bearing solutions; (2) fractured and permeable ground to permit access of solutions; and (3) precipitants such as carbonate and/or organic carbon. Each of the deposits differs somewhat in structure and mineralogy and will be described briefly. Details should be sought in the references cited.

The gold orebodies in the Getchell deposit are sheetlike masses that lie along the various strands of the Getchell fault zone. They extend at least 7000 ft horizontally and 800 ft down the dip, varying in width from a few feet to more than 200, averaging about 40 ft. The ore is in sheared and mineralized argillite and limestone, granodiorite and andesite porphyry dykes cut by quartz, calcite and dolomite veins and containing a plastic 'gumbo' that is rich in gold. The 'gumbo' is evidently an alteration product of carbonaceous shale or argillite and chert. The mineralogy of the ores is low temperature, comprising quartz, gold, realgar, orpiment, stibnite, pyrite, marcasite, arsenopyrite and minor amounts of other sulphides. According to Silberman *et al.* (1974), K-Ar ages of the granodiorite, andesite, porphyry, tungsten-bearing tactites and altered granodiorite in or near the Getchell Mine indicate that emplacement, alteration and mineralization are all part of a magmatic-thermal episode that took place approximately 90 m.y. ago.

The Cortez deposit is mainly in altered calcareous siltstone and limestone of the Silurian Roberts Mountains Formation near an intrusive porphyry body (Fig. 58). No obvious structural feature controls the gold mineralization except perhaps the Roberts Mountains thrust. The hypogene mineralization comprises quartz, gold, pyrite and small amounts of other sulphides found in the other disseminated deposits. Based on hydrogen isotopic data Rye *et al.* (1974*b*) have concluded that the age of the mineralization in the Cortez deposit is Tertiary. They further concluded from hydrogen isotope data that meteoric water was the dominant component of the ore-forming fluids. The lead and sulphur isotopic data were considered to be consistent with the idea that the sulphur and gold were locally derived, probably from the Roberts Mountains Formation.

The gold orebodies of the Carlin deposit are in the upper part of the Roberts Mountains Formation, several hundred feet below the Roberts Mountains thrust (Fig. 59). The orebodies are diffuse irregular sheets that follow the stratigraphy of the carbonate beds but also cut across the bedding. In places a relationship to faults is evident. The mineralization comprises quartz, barite, realgar, pyrite, stibnite, cinnabar, sphalerite and galena in variable amounts. The gold is present as the native metal and as a gold-organic compound according to Roberts *et al.* (1971). Wells (1971) did not find the gold-organic compounds, however, using electron probe methods. Rather, he observed that most of the gold is associated with fine-grained arseniferous pyrite. In a recent contribution Harris and Radtke (1974) using statistical procedures have concluded that the paragenetic sequence at Carlin was probably: (1) early (?) barite; (2) disseminated quartz, pyrite, Au, Hg, As, Sb; (3) As, Sb, Hg sulphides, mainly as realgar, orpiment, stibnite and cinnabar accompanying late barite; and (4) overlapping Pb, Zn, Cu, Mo sulphides. They also observed that primary ores with low organic carbon contents show high correlations between the elements Au-Hg-As-Sb. In high carbon ores only Au to Hg show a high correlation reflecting the mutual associations between them and organic materials. Poor correlations within certain parts of the East and Main orebodies confirm the theory of a late superimposed As, Sb, Hg sulphide stage. Correlations among Au, Ba, Cu, Mo, Pb, Zn, in the West orebody are highest between base metals, followed by Ba to base metals and lowest between Au to base metals. Correlations between Au and Ba are consistently negative over the entire deposit, reflecting their paragenetic separation. B and W, present in anomalous amounts, together with Se and Te, which exhibit only very small concentrations in mineralized rocks, show exceedingly low correlations to each other and to Au. In a later publication Harris and Radtke (1976) have slightly modified the paragenetic sequence. They now consider most of the barite to be a late stage mineral. They also noted that linear regression analyses of gold with the independent variables boron, tellurium, selenium, tungsten suite showed that data for these elements fit a linear model poorly. The only significant

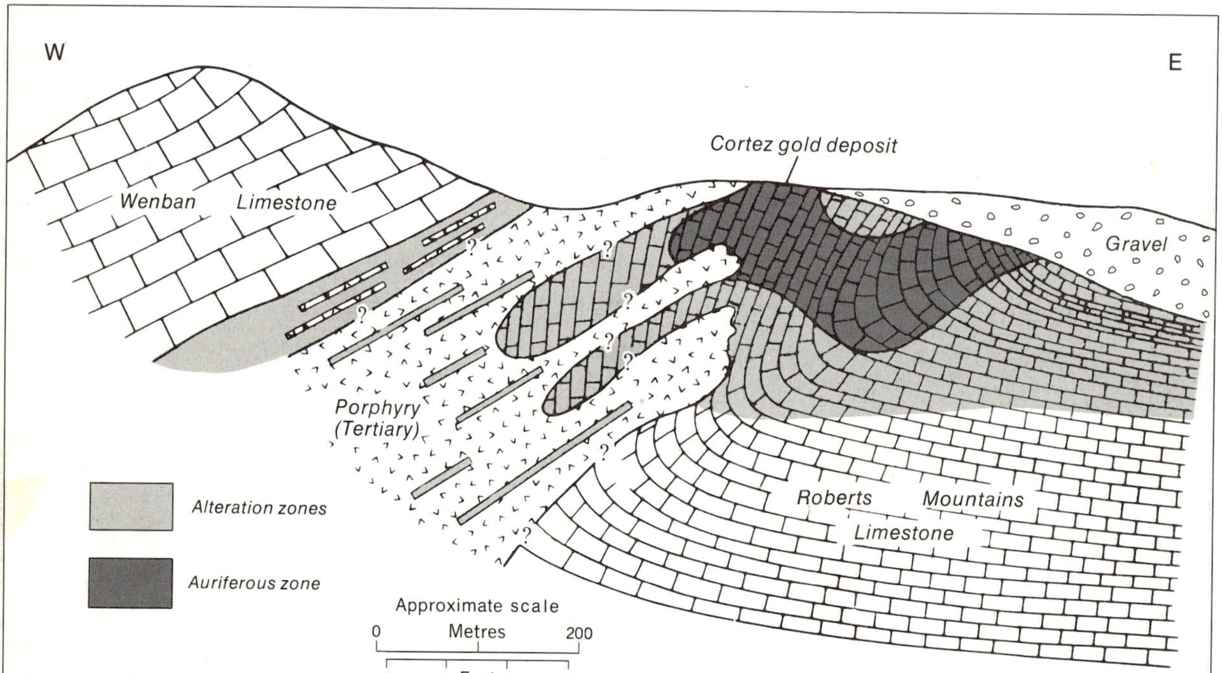

Figure 58. Idealized cross-section through the Cortez gold deposit from west to east showing rock units, gold deposit, alteration zones (the lower extent of which is unknown) and overlying gravels (after Wells *et al.*, 1969).

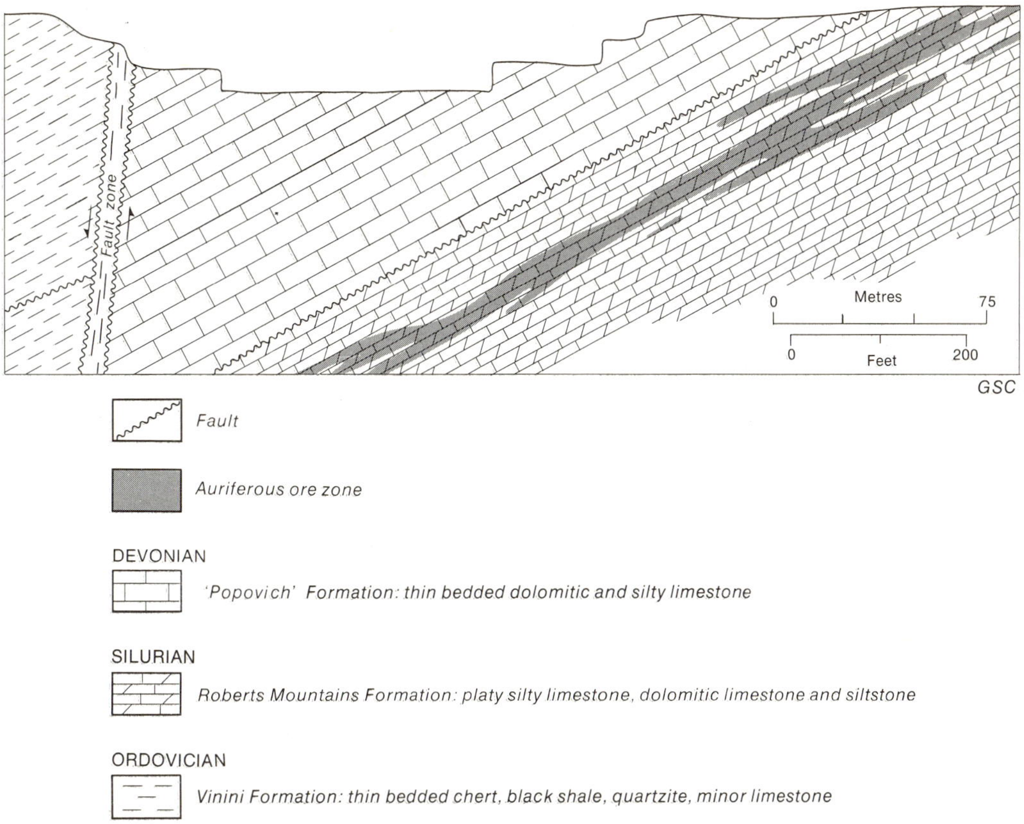

Figure 59. Cross-section of Carlin orebody, Carlin, Nevada (after Hausen and Kerr, 1968).

correlation found was between gold and tellurium; this correlation is strongest in the East orebody and becomes progressively weaker through the Main and West orebodies. The mode of tellurium occurrence was not firmly established, but the authors considered that tellurium and possibly, selenium could be present in hydrothermal pyrite or could coat pyrite grains together with gold, mercury, arsenic and antimony. Other details should be sought in the original paper.

The grade at the Carlin deposit ranges from 0.24 to 0.31 oz Au/ton according to the published reports.

The ore deposits at Gold Acres are in the brecciated zone of the Roberts Mountains thrust and in rocks above the thrust. The host rocks are sheared and brecciated chert, shale, limestone and altered felsitic intrusive rock. Tactite zones with pods and veinlets of pyrite, arsenopyrite, sphalerite, galena and low gold values are also present. Disseminated gold ore is erratically distributed within the thrust zone and along fracture zones and in felsite.

In a recent contribution Radtke and Dickson (1976) state that disseminated gold deposits of the Carlin type occur in many areas of the Basin and Range Province of the United States. The formation of these deposits, characterized by the elemental association, Au, As, Sb, Hg and Tl, was a part of the events that shaped the Basin and Range Province. Most deposits are located along the margins of ranges or near the interior of ranges along high-angle normal faults. They are not genetically related to thrust faults such as the Roberts Mountains thrust.

A deposit similar to that at Carlin in Nevada has recently been described by Richards *et al.* (1976) in Tertiary rocks in the Queen Charlotte Islands, British Columbia. This deposit (Specogna) is reported to contain an indicated reserve of more than 50 million tons averaging 0.06 oz/ton gold and 0.1 oz/ton silver. Other metallic elements present in anomalous amounts include Hg, As, Sb and Te. Possible major controls of mineralization include proximity to an unconformity, proximity to a major fault, and permeability of the host Skonun Formation of Miocene to Pliocene age. Quartz-feldspar porphyry dykes, although volumetrically insignificant, are thought to be important in the mineralizing process. Alteration effects include strong silica metasomatism, pervasive pyritization and intense clay alteration of feldspar in sandstones and in dykes. Although the gold mineralization is central to a much larger zone of silica, clay and pyrite, zoning of alteration is obscure. Dispersion of Hg in the rocks forms a halo about the better-grade gold. Other metallic elements form weak patterns being peripheral or central to the focus of mineralization.

Some brief remarks about disseminated gold deposits in chemically favourable rocks will serve to close this section. These deposits invariably have an epigenetic aspect and are most commonly developed in carbonate rocks in or near fault and fracture systems; commonly also near porphyry intrusions. Silicification is the predominant process in their formation. Most have been interpreted as being of low temperature hydrothermal origin. Thus, Radtke and Scheiner (1970*b*) postulated a low temperature (175±25°C) hydrothermal origin from acid solutions for the Carlin and Cortez deposits. The source of the gold and other constituents in the orebodies has not been considered in detail by any investigators studying these deposits, although a few allude to granitic rocks, porphyries, etc. as a possible source. It seems more probable that the pile of sediments in which these deposits occur is the source of the gold and other epigenetic elements, an hypothesis that should be tested in the areas where these deposits occur. Carbonates and possibly organic carbonaceous matter seem to have been the effective precipitants of silica, various sulphides and gold.

Gold deposits in quartz-pebble conglomerates and quartzites

These are the largest known gold deposits and have provided over 60 per cent of the world's gold production in the last 50 years. In addition some of these deposits contain economic quantities of uranium. Other deposits in this category are characterized by economic concentrations of uranium, thorium and rare-earths and only traces of gold and silver. There is an enormous amount of literature on these deposits, and no attempt will be made here to review any of it in any detail. While the deposits are remarkably similar in many respects, in detail there are numerous differences, some small and some major, that serve to stir up controversies about their origin. In fact, perhaps no metallic deposit has exercised the minds of economic geologists as regards origin as much as the gold-bearing quartz-pebble conglomerates since their discovery on the Witwatersrand of South Africa in 1886. By the mid-1940's the protracted controversies between the placerists, modified placerists and hydrothermalists had subsided, only to flare up again with renewed vigour and acrimony among a new generation when uranium was discovered or rediscovered in some of the deposits.

The general characteristics of these deposits will be dealt with first, some typical examples will be described briefly, and finally the various views on their origin will be outlined. Details about individual deposits or clusters of deposits are not given; these can be consulted in the references cited.

Quartz-pebble conglomerate deposits bearing gold, uranium, thorium and rare-earths are widely distributed. The best known examples are the Witwatersrand in South Africa, the Tarkwa conglomerates in Ghana, the Jacobina conglomerates in Brazil and the Elliot Lake–Blind River conglomerates in Ontario, Canada. In Rhodesia, gold-bearing conglomerates somewhat similar to those of the Rand occur in the Bulawayo area (Mennell, 1905). Some of these were worked for gold at the Eldorado Mine and elsewhere. The grade at the Eldorado Mine averaged 0.5 to 1 oz Au/ton or more. In Tanganyika near Tabora, a gold-bearing conglomerate has been explored, and in Zaire the Ruwe Mine, some 80 mi west of Kambove yielded both gold and platinum from (Permo-Triassic?) sandstones and quartzites. Old records indicate the grade at this mine to average about 10 ppm Au, 11 ppm Pt and 2 ppm Pd. Relatively little work seems to have been done on the last two conglomerates in recent years. Early reports and references are given in Emmons (1937). Numerous other occurrences of these types of conglomerate are known in Finland, U.S.S.R., Australia and elsewhere (Davidson, 1964–1965; Rozhkov and Pisemskey, 1966; Radkevich, 1966; Volodin, 1966; Nozhkin *et al.*, 1972; Pisemskii and Pletnev, 1972; Andrade Ramos and Fraenkel, 1974; Robertson, 1974).

The productive quartz-pebble conglomerate deposits are all of Precambrian age. Phanerozic examples are known, but none of these have yet proven to be of economic interest. The details of the ages of the Precambrian quartz-pebble conglomerates are not entirely clear. If we can believe the isotopic dating, the Witwatersrand deposits are 2.3 to 2.8 billion years old, the Blind River–Elliot Lake deposits (usually considered as Aphebian in age) 2.3 to 2.5 billion years and the Jacobina 2.5 billion years. According to some age scales this places at least some of the deposits well into the Archean. Stratigraphically speaking, however, all of these deposits lie on old basement rocks (greenstones, greywacke, slate, gneisses, schists, granites, etc.) that have been traditionally referred to as Archean in many countries. The absolute ages of the deposits does not really concern us here, and we shall not consider the problems further. The superincumbent stratigraphic relationship of the conglomerate deposits to the older (Archean) rocks, which throughout the world contain a great variety of epigenetic gold deposits, deposited before the conglomerates were laid down, is important.

The mineralized quartz-pebble conglomerates commonly occur as interbedded thin beds and elongated lenses in thick sedimentary sequences composed of clastic sediments, mainly quartzites, grits and microconglomerates with some hydrolysates, principally shales. Commonly the conglomerates are restricted to specific members and not scattered indiscriminately throughout the sedimentary sequence. In some sequences (e.g., Witwatersrand, Ventersdorp System) volcanic rocks are present, generally near the top of the stratigraphic section. The ore-bearing conglomerate beds or sequences of beds vary greatly in thickness, but most are only a few feet thick; ore-bearing horizons thicker than 10 ft are uncommon; some of the richest conglomerate beds are only inches thick. In some districts the mineralized conglomerate beds and conglomerate horizons are particularly persistent, extending for tens and even hundreds of miles.

The quartzites, shales and argillites in the sedimentary sequence appear in most places to be relatively normal for these types of rocks. The quartzites generally consist essentially of quartz grains, feldspar, small chert fragments, sericite, chlorite, pyrophyllite in places and chloritoid. The shales and argillites are similarly constituted but with much less quartz and more micas. Some of the shales and argillaceous quartzites are red due to the presence of hematite. Others carry magnetite. The quartzites generally exhibit a moderate degree of crossbedding, a feature that is relatively uncommon in the conglomerates, although in some deposits rude crossbedding is present in these coarse rocks. The shales are seldom highly graphitic, although examples of carbonaceous shales and conglomerates are known in the Black Reef in the Witwatersrand Basin. In both the quartzites and conglomerates erosion channels and 'cut and fill' structures are present but are not a marked feature in most districts. Ripple marks are present on the shales and quartzites of some districts.

The origin of the conglomerates and their associated sediments has been variously interpreted by a host of sedimentary experts, few of whom seem to agree on either the major or minor details. A whole spectrum of broad origins has been suggested including terrestrial, littoral and marine. Among these, deposition in alluvial fans, on piedmonts and flood plains of large rivers, in enclosed basins, in large deltas, along marine shorelines and on gently sloping continental shelves have all had their proponents. Normal weathering and erosion processes, sometimes drastic, have appealed to many as the agency producing the sedimentary materials, but some have invoked glaciation followed by sedimentary milling and winnowing of the products as the responsible mechanism. Particularly perplexing to many theorists is how persistent beds of thin oligomictic conglomerates (sheet gravels) could have accumulated over such vast areas. One constantly wonders whether these conglomerates and their associated sediments, and indeed most Precambrian sediments, can be interpreted in terms of uniformitarian principles.

The paleoclimatic conditions under which the quartz-pebble conglomerates and their associated quartzites were laid down can only be conjectured. These rocks suggest intensive weathering, erosion and sedimentary milling. Intensive weathering infers the presence of considerable amounts of oxygen, carbon dioxide and water in the paleoatmosphere. There is little doubt that the last two compounds were present since the conglomerates and quartzites were water lain, and volcanism in Archean times produced abundant carbon dioxide (Boyle, 1976*a*), much of which was probably carried over into the Proterozoic atmosphere to persist through various cyclic processes to the present day, being augmented from time to time during younger periods of volcanism. The availability of oxygen during early Proterozoic time has, however, been questioned, a number of investigators claiming that the quartz-pebble conglomerates were laid down under anoxogenic conditions. This appears improbable since hematitic iron-formations are relatively abundant in the Archean and that molecular oxygen was probably one of the gaseous products of the early earth, its origin being due to radiolysis and other early geochemical processes (Boyle, 1976*a*). There is a strong possibility that the quartz-pebble conglomerates and quartzites of Proterozoic age represent well winnowed accumulations of quartz and micas from glacial tills, in which case the weathering and erosion were essentially mechanical. This is suggested by the fact that the Aphebian (Lower Proterozoic) is marked by glacial periods, the tillites of the Gowganda Formation and the Witwatersrand (Wiebols, 1955) being manifestations. Having regard to these speculations it seems probable that the Proterozoic quartz-pebble conglomerates and their associated quartzites were deposited during a warming postglacial period under an atmosphere with a relatively high oxygen content.

The detailed constitution of the conglomerates is remarkably similar, but there are a few marked differences between certain deposits. The pebbles in nearly all of the ore-bearing conglomerates are predominantly quartz with occasional occurrences of chert, quartzite, porphyry and slate. Other types of pebbles are rare according to the literature and this writer's examination of the Elliot Lake–Blind River deposits and samples from other deposits in many parts of the world. The quartz and other pebbles exhibit a relatively high degree of sorting in most deposits and show evidence of thorough abrasion, two features suggesting extensive travel or long sedimentary milling. Initially, most of the pebbles were probably well-rounded, but in many deposits they show flattening and roughly ellipsoidal forms. Some resemble large almonds hence the name 'banket', which is Dutch for almond layer cake. The flattening and stretching of the pebbles is evidently due to tectonic activity during the folding and metamorphism of the beds. The bulk of the matrix of the conglomerates is composed essentially of fine-grained quartz, sericite, chlorite, chloritoid and minor amounts of a variety of other minerals which need not concern us at this point. In some of the Witwatersrand conglomerates, pyrophyllite is relatively abundant (Liebenberg, 1955).

With respect to iron minerals there are two types of conglomerate deposits – those containing pyrite and pyrrhotite and those with hematite or more rarely ilmenite or magnetite. The iron sulphides and oxides seem to be mutually exclusive in the conglomerates of particular fields. Iron oxides are absent or extremely rare in the conglomerates of the Witwatersrand, Jacobina and Blind River–Elliot Lake deposits, and pyrite and pyrrhotite are rare in the Tarkwa conglomerates. It is interesting, however, that some of the interbedded shales in the Rand and the other fields commonly carry considerable amounts of magnetite. Conglomerates rich in pyrite and pyrrhotite carry uranium, thorium and rare-earths; those with the iron oxides do not carry these elements in more than trace amounts in the deposits known to date. The pyrite in the conglomerates is predominantly of two types – a rounded 'buckshot' type and an euhedral to subhedral type. These types have been variously interpreted as detrital or formed during or after sedimentation by a number of processes. Some pyrite occurs in veinlets and fracture fillings and is obviously epigenetic. The hematite in the Tarkwa deposits is present mainly as grains that do not appear to be detrital in their present form. They seem to be more like crystallized grains derived from a chemical precipitate. More will be said about the iron minerals in the descriptions of the principal deposits, but it should perhaps be mentioned here that Robertson (1974) places considerable emphasis on the point that all of the uraniferous quartz-pebble conglomerates are younger than the first Proterozoic red beds, a feature that he thinks points to an anoxygenic atmosphere for the earth during the sedimentary deposition of detrital uraninite in the quartz-pebble conglomerates. This view is open to serious doubt not only because of the existence of the hematitic Tarkwa deposits but also because of the presence of abundant iron-formations (Algoman-type) in the Archean, a fact that strongly indicates that oxygen was relatively abundant in the very early atmosphere (Boyle, 1976*a*).

The above description covers the principal components of the quartz-pebble conglomerate deposits. Those of economic interest, however, commonly contain a number of base metal sulphides, uraninite, brannerite, solid hydrocarbons, platinoids and other minerals in addition to gold. These constituents are best described in the context of specific deposits where the details are more meaningful.

The source of the most abundant and specific components of the conglomerates and their associated sediments is a difficult problem, in some areas probably insoluble, because the source rocks have been either completely removed by erosion or deeply buried by younger rocks. There is evidence in some regions such as the Rand, Jacobina and Tarkwa for an Archean hinterland marked by the presence of gold-quartz deposits, gneisses and granitic rocks, the last two a possible source of uranium, thorium and rare-earths. In other regions such as Blind River–Elliot Lake the hinterland appears to have been mainly granitic gneisses, intrusive granitic bodies and only minor belts containing gold-quartz deposits. This difference may partly account for the general paucity of gold in the Blind River–Elliot Lake ores.

The precise source of the quartz in the quartz pebbles and matrix of the conglomerates is an interesting problem. All investigators agree that the pebbles are vein quartz, but few specify from what type of veins the quartz was derived. Pretorius (1964), describing the vein quartz pebbles in the Central Rand conglomerates, mentions the spectrum of colours present, including white, cream, grey, mottled grey and black, jet black, opalescent blue and emerald green, the last relatively rare. These colours are typical of nearly all quartz-pebble conglomerates throughout the world. The writer has examined the macroscopic and microscopic characteristics of the quartz pebbles and matrix quartz from various quartz-pebble conglomerate deposits from various parts of the world in an attempt to relate the quartz to Archean deposits of the mineral. He concludes that the white to milky variety of quartz in a number of the deposits is essentially similar to quartz present in many of the gold-quartz veins in greenstones and Archean sediments. Many of the white quartz pebbles in a number of the deposits also match the so-called barren 'bull quartz' commonly present in volcanics and sediments in Archean gold belts. Some of the glassy colourless to white varieties of quartz pebbles in some deposits resembles pegmatitic quartz and were probably derived from these granitic bodies. The grey, mottled grey and black, and jet black varieties of quartz pebbles are characteristic of Archean gold deposits as mentioned in a previous section. These varieties of quartz could be derived from deposits either in greenstones or sediments. The opalescent blue to dusky blue quartz can be matched with that present in certain gold deposits in Archean greywackes and graphitic slates. The dark colour of some of the quartz requires comment. Part of the colour in some cases may be due to radioactivity (electronic darkening) associated with uranium and thorium now present in the conglomerates. Alternatively the quartz may have been initially or partly darkened in radioactive pegmatites. The latter appears the least probable since many of the quartz pebbles carry a dark outer border, which appears to have been induced in situ. Horwood (1917), the first geologist to investigate the black-edged pebbles, included illustrations of many fine examples of the phenomenon in his book on the Witwatersrand; the phenomenon is also notable in the Elliot Lake deposits, particularly in some of the ore shoots of the Denison Mine.

One of the striking features about the quartz pebbles is their general lack of original particles of gold, sulphides, uraninite, etc. Where these minerals are present in the quartz pebbles, they are obviously of secondary origin, since they commonly cut both the matrix and the pebbles or show undoubted replacement textures. These features have been repeatedly mentioned in the literature and are generally accepted by most investigators. Along the same line the writer has noted that the quartz pebbles do not generally show the marked cataclastic features of many gold-quartz veins. How are these features to be explained? The answer appears to lie in the extreme weathering and sedimentary milling conditions that the pebbles have endured. In the weathering and disintegration of gold-quartz veins containing sulphides the writer has noted that the vein quartz tends to disaggregate first along the seams and zones bearing sulphides, gold and carbonates and finally along the zones of cataclasis cemented by secondary quartz, carbonates, sericite, etc. The final product is dominantly quartz practically free of any other mineral, particularly gold and pyrite, and relatively free of abundant zones, seams and patches of cataclasis. Quartz of this type appears first in eluvial placers and ultimately in alluvial placers. The 'White Channel' gravels of the Klondike in Yukon, Canada are a case in point. Careful examination by the writer of the white quartz pebbles, which constitute a large proportion of these gravels, revealed only a few pebbles with pyrite and other sulphides and none with visible gold. Examination of numerous gold concentrates from these gravels and indeed from others in various placers in Yukon, British Columbia and elsewhere in Canada revealed only rare nuggets or pieces of gold with attached quartz gangue or pyrite. These observations agree with the observations of many investigators of gold placers as described in the literature for the last 70 years. It is thus evident that vein quartz and gold generally part intimate company but retain their general association during weathering processes and placer accumulation. This probably accounts for the fact that the quartz pebbles in the quartz-pebble conglomerates rarely contain gold or sulphides that can be proven to be of primary origin.

The source and manner of formation of the matrix of the conglomerates is of interest with respect to the origin of certain elements of a hydrolysate nature as will be mentioned later in the discussion on origin. Microscopic examination suggests that the quartz in the matrix is probably only partly detrital, a large part having been initially present in a dissolved or colloidal state. The same is probably true of the components that now constitute sericite, chlorite, chloritoid and pyrophyllite. Regardless of whether these observations are valid or not, there is no doubt that the matrix of both the conglomerates and their associated quartzites has undergone thorough compaction, cementation and reconstitution leaving minerals characteristic of the sericite-chlorite facies of metamorphism.

Krendelev *et al.* (1970) make reference to a number of the features just mentioned in their extensive study of sulphide-bearing quartz-pebble conglomerates of many types from various parts of the world. They note a general reconstitution of the matrix of these conglomerates and claim that even the quartz pebbles are often fully recrystallized in some deposits. They investigated gas-fluid inclusions in the quartz for their gaseous content and homogenization temperatures. They concluded that the gold and uraniferous ores in Precambrian sulphide-bearing conglomerates resulted from the activity of solutions with temperatures about 220–400°C and containing gases such as CO_2, O_2 and NH_3 (Rand).

One other general feature of the mineralized quartz-pebble conglomerates that requires mention is the common presence both in the conglomerates and their associated rocks of mineralized gash veins, fractures and faults. In the Rand many of the gash veins are filled with quartz, sulphides and gold. In the Blind River–Elliot Lake area the faults and fractures contain thucholite and sulphides. Elsewhere, these mineralized veins and fractures are present in or near the quartz-pebble conglomerate deposits but have not been described in detail. The origin of these mineralized structures and their relationship to the mineralized conglomerates has led to much controversy, a brief outline of which is given below.

Witwatersrand Basin, South Africa

The Witwatersrand, or more simply the Rand as it is generally called, was discovered in 1886; since then its many mines have produced more than 1 billion oz Au, 100 million oz Ag, considerable uranium and some platinoids. The annual production is about 30 million oz Au/year making the Rand the largest producer of the precious metal by far in the world. Some of its mines are the deepest in the world exceeding 10 000 ft in places. The reserves of ore averaging 0.1 to 0.3 oz Au/ton are said to be large.

The literature on the geology, mineralogy and geochemistry of the Witwatersrand is voluminous. An extensive bibliography covering the years up to 1959 has been compiled by Hargraves and Winter (1964) to which the interested reader is referred. Later references of interest are mentioned in the text that follows. The definitive textbook on the Rand is that edited by Haughton (1964*a*) in which the papers by Borchers (1964) and Brock and Pretorius (1964*a,b*) provide excellent summaries of the history, exploration, geology and mineralization of this unique goldfield. There is a recent review of the problems of the origin of the Rand by Pretorius (1975).

The Witwatersrand Basin, some 250 mi long and 90 mi wide, lies southwest of Johannesburg and has major orebodies along its northeastern, eastern, western and southern edges (Fig. 60). The general structure of the basin is a synclinal trough with peripheral faults in places and extensive cross faults in others.[24] The rocks of particular economic interest are all sediments of Precambrian age resting on a basement of granite, gneiss and schists. The generalized geological column for the basin is given in Table 61. The various systems are fairly readily correlated throughout the Rand Basin; correlation of the various individual reefs and reef systems from area to area is relatively difficult because of the discontinuous and lenselike nature of the reefs. The geological history of the basin is complex, involving several stages of extensive sedimentation and volcanic activity; only a brief outline of the salient features will be attempted here.

The oldest rocks, the Swaziland System, underlying the basin are Precambrian amphibolites, schists and gneisses intruded by or granitized to a variety of granitic rocks. Rocks equivalent to these in South Africa (Barberton Mountain Land), Rhodesia and elsewhere in Africa are the hosts for a variety of gold-quartz deposits. On the deeply eroded surface of the basement complex the Dominion Reef System, some 3000 ft thick, was laid down apparently only in some areas. The upper part of this system is composed of lavas, a lower sequence of amygdaloidal andesites followed upwards by felsite and rhyolite; the lower part of the system, some 300 ft thick, comprises grits, quartzites and conglomerates. The auriferous-uraniferous Dominion Reef conglomerates at or near the base have been worked in the Klerksdorp area. The Dominion Reef System was extensively folded before the deposition of the Witwatersrand System.

Unconformably overlying the Dominion Reef System, where it is present, and the basement complex generally is the Witwatersrand System, which is divisible into Lower and Upper divisions. The generalized stratigraphic column of both these divisions is given in Table 62. The Lower Division consists mainly of shales, quartzites, grits and conglomerates, only two of the last being sufficiently auriferous to be mined (Government Reef, and reefs in the Jeppestown Series in the Klerksdorp area). The Upper Division comprises mainly quartzites and conglomerates with minor shale. This division contains the most productive auriferous and uraniferous conglomerates of the Rand, including the Main-Bird Series, the Kimberley Reefs and the less persistent but payable Elsburg Reefs.

After deposition of the Upper Division of the Witwatersrand System there followed a major volcanic period during which lavas of dominantly basaltic to andesitic composition, quartz porphyries and felsites with minor intercalated banded tuffs, shales, cherts and breccias of the Ventersdorp System were laid down. Near the base in some areas there are quartzites, grits and conglomerates. The system varies greatly in thickness, exceeding 10 000 ft in some districts of the Orange Free State. The auriferous Ventersdorp Contact Reef lies at the base of the system and is mined on the West Wits Line, in the West Rand areas, and at one mine in the Klerksdorp area. It is generally assumed that the folded and eroded beds of the Witwatersrand System were the source of the gold and pebbles of this reef.

Following the outpouring of the Ventersdorp lavas, sedimentary and volcanic members of the Transvaal and Karroo systems were laid down; the latter now covers more than half of the Witwatersrand Basin. The Transvaal System is of interest since the Black Reef lies in quartzites, conglomerates and carbonaceous shales at the base. This reef occurs as conglomerate or channel fill and transgresses rocks of the Basement Complex, Witwatersrand System and Ventersdorp System. Gold values tend to be erratic, and uranium values are consistently poorer than in the underlying reefs. It is mined in the Klerksdorp, West Wits, West Rand, Central and East Rand areas. Some geologists consider that the gold and other minerals in the Black Reef represent the redistribution products of eroded reefs in the Upper Division of the Witwatersrand System.

The dip of the sediments and volcanics in the Rand Basin is variable as might be expected for rocks covering such a large area. In general the dips are less than 45°. Locally they may be very steep and even overturned as in the region of the peripheral faults of the basin and the Vredefort Dome.

Igneous bodies of several ages penetrate the rocks of the Witwatersrand Basin. Most of these are sills and steeply dipping dykes of basic and intermediate composition related to the Ventersdorp, Bushveld Complex and Karroo periods. Most of the rocks comprising the sedimentary pile of the Witwatersrand Basin fall into the greenschist (sericite-chlorite)

[24]The synclinal trough idea may require modification as further work proceeds. Folding of a complex nature may account for the configuration of the Witwatersrand basin according to Dr. D. A. Pretorius (lecture, Ottawa, October 2, 1973).

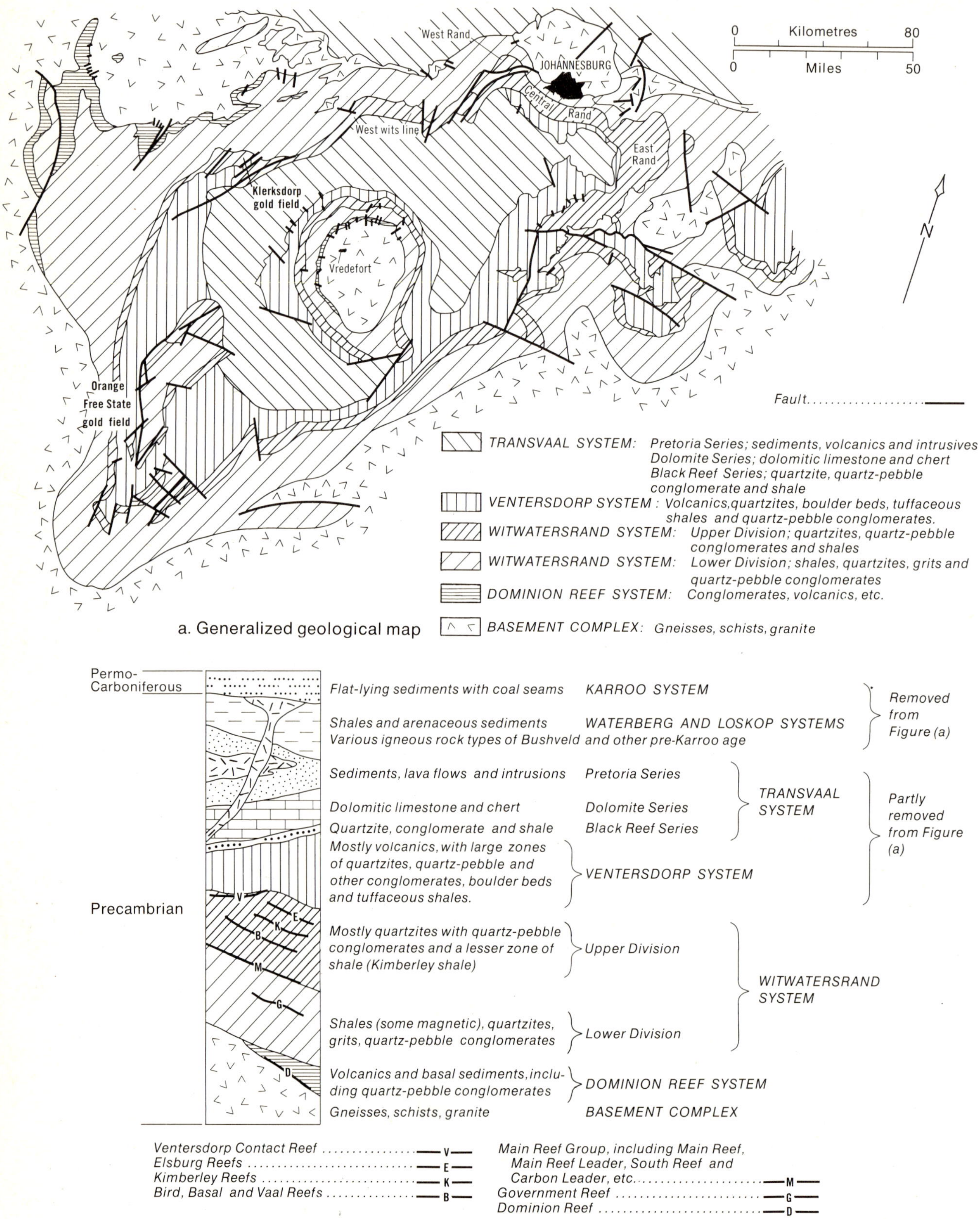

Figure 60. (a) Generalized geological map of the distribution of the Witwatersrand System beneath the younger covering rocks (after Borchers, 1964; Koppel and Saager, 1974; and others). (b) Generalized geological column showing the various geological relationships in the Witwatersrand Basin (after Borchers, 1964; and others).

Table 61. Generalized geological column, Witwatersrand Basin, South Africa

Era	System		Lithology and remarks
Permo-carboniferous	Karroo System		Flat lying shales, grits, sandstones, etc. with coal seams
			Unconformity
Precambrian	Transvaal System		Shales, quartzites, dolomite, chert, conglomerate, lava flows, etc. Black Reef at base
			Unconformity
	Ventersdorp System		Mainly andesitic and basaltic lavas with porphyries, felsites, tuffs, shales, cherts and breccia some quartzites and conglomerates. Ventersdorp Contact Reef at base
			Conformity in some areas; unconformity in others
	Witwatersrand System	Upper Division	Mostly quartzites and conglomerates, and minor shale. Contains the auriferous Elsburg, Kimberley, Bird and Main Reef groups
			Conformity in places; unconformity in others
		Lower Division	Amygdaloidal lava at top, tuff, shales, quartzites, grits, conglomerate. Contains the auriferous Government Reef and other gold-bearing reefs near the middle of the division
			Unconformity
	Dominion Reef System		Conglomerate, quartzite, grit, arkose, amygdaloidal andesites, felsites and rhyolite. Auriferous Dominion Reef at base
			Great unconformity
	Swaziland System (Basement rocks)		Amphibolites, actinolite and hornblende schists, talc schists, serpentinites, gneisses and intrusive granitic rocks

Table 62. Generalized stratigraphical column for the Witwatersrand System in the Central Rand

Division and total thickness	Series and total thickness	Main stratigraphical marker horizons	Thickness of marker horizons
	(ft)		(ft)
Upper division Witwatersrand System (9400)	Kimberley-Elsburg Series 6100	Elsburg Reefs	1500
		Kimberley Reefs	600
		Kimberley Shales	550
Mainly quartzites and conglomerates with minor shale	Main-Bird Series 3300	Bird Reef Marker	100
		Bird Reefs	300
		Main Reef Group	100
Lower division Witwatersrand System (15 200)	Jeppestown Series 3800	Jeppestown Amygdaloid	100
Mainly shales, quartzites, grits, and conglomerates	Government Reef Series 6100	Government Reef	5
		Coronation Reef	5
		Coronation (or West Rand) Shales	500
		Promise Reef	5
	Hospital Hill Series 5300	Hospital Hill Quartzites	1300
		Contorted Bed	150
		Speckled Bed	5
		Ripple-marked Quartzites	20
		Water Tower Slates	800
		Orange Grove Quartzites	550

Source: Brock and Pretorius (1964*b*).

facies of regional metamorphism. The thermal effect of the various igneous bodies is local and overall minimal.

The rocks of the Witwatersrand Basin have suffered several periods of tilting, subsidence and erosion followed at the end of the Transvaal period by extensive folding and thrusting. In the centre of the basin the basement was thrust up, doming and in places overturning the overlying sediments to give a pronounced structure known as the Vredefort Dome. The sediments have been partly eroded from the top of the dome thus exposing a concentric pattern of outcrop of the rocks of the Witwatersrand, Ventersdorp and Transvaal systems. The idealized outcrop pattern of the Witwatersrand System, as well as the other systems, as shown by Borchers (1964) is an outer elliptical annulus of rocks dipping inward with low to moderate dips and a more or less circular annulus of rocks in the centre having variable dips. Extensive, and in places intricate, fault systems disrupt this ideal pattern throughout the basin.

The environment of deposition of the sediments and volcanics of the Witwatersrand Basin has been the subject of much discussion, and a number of hypotheses have been proposed including; the littoral (marine shoreline) hypothesis (Young, 1917); the sea floor hypothesis involving cyclothem deposition (Sharpe, 1949) or littoral source of pebbles (Graton, 1930); the deltaic or estuarine hypothesis (Mellor, 1916); the pediment hypothesis with a desert connotation (Antrobus, 1956); the alluvial hypothesis with deposition on a piedmont, flood plain or region of alluvial fans associated with large rivers and their tributaries (Reinecke, 1930; Bain, 1960); and the glacial hypothesis (Wiebols, 1955) involving the winnowing and sedimentation of materials from extensive till sheets and other glacial deposits. Brock and Pretorius (1964*a*) combine a number of these concepts into a four-delta hypothesis, the sedimentary materials being derived from a hinterland of red granite with inset synclines of gold-bearing Archean schists undergoing continuous uplift along a fault at the periphery of the basin. Sedimentary milling of the weathered products along the littoral zone is envisaged followed by movement and sedimentation of pebbles and other materials along the sea floor by wave-bottom effects assisted by gravity (Fig. 61).

In a more recent contribution to the problem of the depositional conditions that produced the Witwatersrand Pretorius (1975) attempts to resolve the various problems that have arisen since the 1964 synthesis. His remarks about this famous goldfield deserve to be quoted in full:

> It would now appear that the Witwatersrand depository was an intermontane, intracratonic, yoked basin with a fault-bounded northwestern edge and a gently downwarping more passive southeastern boundary. The enclosed basin was at least 350 km long in an east-northeasterly direction, and 200 km wide in a north-northwesterly direction. The structural environment resembled that of the Basin and Range Province, but a far wetter climate prevailed. The basin was a shallow-water lake or inland sea, no connexion to an open ocean having yet been found. The depository became structurally more unstable with time, and a pattern of interference folding produced structural depressions and culminations both on the rim and within the depository. The various goldfields developed in downwarps between basement domes. The northwestern side was episodically but continuously rising, causing the basin-edge to advance progressively farther towards the depositional axis. The final depository was smaller than the original, so that, overall, the sediments were laid down in a shrinking basin. Conditions were generally transgressive in the Lower Witwatersrand Division and generally regressive in the Upper Division. Second-order transgressions and regressions were superimposed on these primary trends. Between the base of the Lower Division and the base of the Upper Division, the edge of the basin moved southeastwards by 60 km, and the depositional axis by about 10 km.
>
> A high-energy transfer system from the source to the depository took the form of a relatively short, linear fluvial array. From the areal geometry of the different stratigraphic horizons, the patterns of facies variations, the trends in the changes of grain-sizes of sediments, the directions and patterns of paleoflow, the nature of the environmental indicators, and the distribution of heavy minerals, it would appear that a goldfield is a fluvial fan or fan-delta that was formed where the river system debouched into the lake via a canyon cut through the high ground to the northwest of the peripheral faults. In this type of environment, there were far greater amounts of water on a typical alluvial fan and the energy-level was higher than that on an oceanic delta. After emerging from the canyons, the rivers flowed short distances over a piedmont plain and then dispersed through a braided-stream pattern into the basin. The fluvial fans were restricted to the northwestern margin of the depository, and some of them coalesced in their more distal parts, leading to the impression of extensive sheets of uniform gravel. The largest of the fans was that constituting the East Rand goldfield, which measured 40 km in length down the central section from the apex to the fanbase, 50 km wide in width in the midfan section, and 90 km in width in the fanbase section. The western margin of this fan was 45 km long and the eastern flank 60 km. Six fluvial fans have so far been discovered – those constituting the Orange Free State, Klerksdorp, West Wits, West Rand, East Rand, and Evander goldfields. The original Central Rand, on which the Main Reef conglomerates were discovered, would seem to be a geographical entity only, since, geologically, it represents the coalescence of the eastern part of the West Rand fan and the western part of the East Rand fan.
>
> The gold and uranium mineralization has been recognized to occur in five forms: (i) in the matrices of the conglomerates, (ii) in heavily pyritic sands filling erosion channels, (iii) on quartzites along a plane of unconformity between successive cycles of sedimentation, (iv) on shales along planes of unconformity, and (v) in carbon seams on, or adjacent to, planes of unconformity. The three last-mentioned types of reefs were formed in the terminal stages of one cycle of sedimentation, and the first-mentioned two in the initial stages of a succeeding cycle. The gold and uranium were transported as detrital particles and in solution as chloride- and cyanide-complexes. Concentration took place physically, through gravity settling and subsequent winnowing by wave- and current-action, and biochemically through interaction between the gold and uranium and the algal or lichen colonies that preferentially developed about the mouths of the major rivers, in the quieter water conditions on the margins and fanbases of the fluvial fans, and at the end of certain cycles of sedimentation.
>
> The conceptual model of a typical Witwatersrand goldfield is portrayed in Figure 62 (this volume). The apex of the fluvial fan was located along the tectonically unstable basin-edge where repeated uplift of the source-area side took place along longitudinal faults. The fanheads of earlier fans were thus uplifted and reworked into later fans, while the midfan and fanbase sections were structurally depressed and thereby preserved. The downward displacement of the midfan and fanbase also caused transgression of the lake waters, producing winnowing of the fines and lag concentrations of the heavier minerals. Longshore currents moved the finer sediments farther away from the entry-points to form asymmetrical fans owing to the clockwise movement of the water in the depository. The typical fluvial fan had two main lobes in which were located a larger number of braided-stream channels, thicker and coarser clastic sediments, and higher concentrations of detrital gold and uranium. The material that was laid down between the lobes took the form of sands, silts, and muds, similar to that which accumulated on the fan margins and base. Conditions under these lower-energy regimes at times provided the optimum environment for the growth of algae or lichens, which took the form of thin algal mats. The gold was of too fine a grain to settle in the fanhead facies. The highest concentrations took place in the midfan lobes, with the peak of uranium mineralization a little farther down the slope than the peak of the gold. The energy-level dropped too low to permit the transportation of detrital gold to the fanbase section. However, the gold and uranium that were in solution did interact with the biogenic material that was present in the low-energy environments.
>
> A fluvial fan was built up in a series of pulses of sedimentation, which started with progradation during regression, went through aggradation during transgression, and ended with degradation during stillstand. These three stages constituted a single cycle of sedimentation. A new cycle was initiated through tectonic adjustment along the longitudinal faults. Such adjustment produced a steepening of the paleoslope, with the result that the increased competency of the streams brought greater amounts of coarse debris onto the fan. The higher energy-level caused progradation and a consequent regressive relationship with the earlier sediments. The first pulse laid down an openwork gravel, and the next pulse, the sand matrix. Heavy minerals were brought in with the sand phase and not with the gravels. Thereafter, as the energy-level dropped, transgression took place, with the deposition of finer-grained material, until a state of equilibrium was reached and deposition came to a standstill. End-of-cycle winnowing by the waters of the streams and the lake produced a greater concentration of residual heavy minerals on the erosion surface. Incipient tectonic activity caused tilting of the erosion surface, thereby producing the unconformable relationships between successive cycles. On the tilted surface, degradation was enhanced, winnowing intensified, and lag concentration brought to an optimum. Continued tectonic adjustment culminated in the prograding sedimentation of the next cycle. The turbulent gravels broke up the depositional floor and incorporated the thin streaks of lag gold and uranium of the unconformity. Thus, these minerals could have been introduced into the gravels in two processes – picking-up from the footwall sediments and downward infiltration during the sand pulse that succeeded the laying down of the pebbles.

Not having seen the Rand the writer should perhaps not make comparisons. However, the voluminous detailed descriptions of the rocks in the literature, provide a fairly reliable

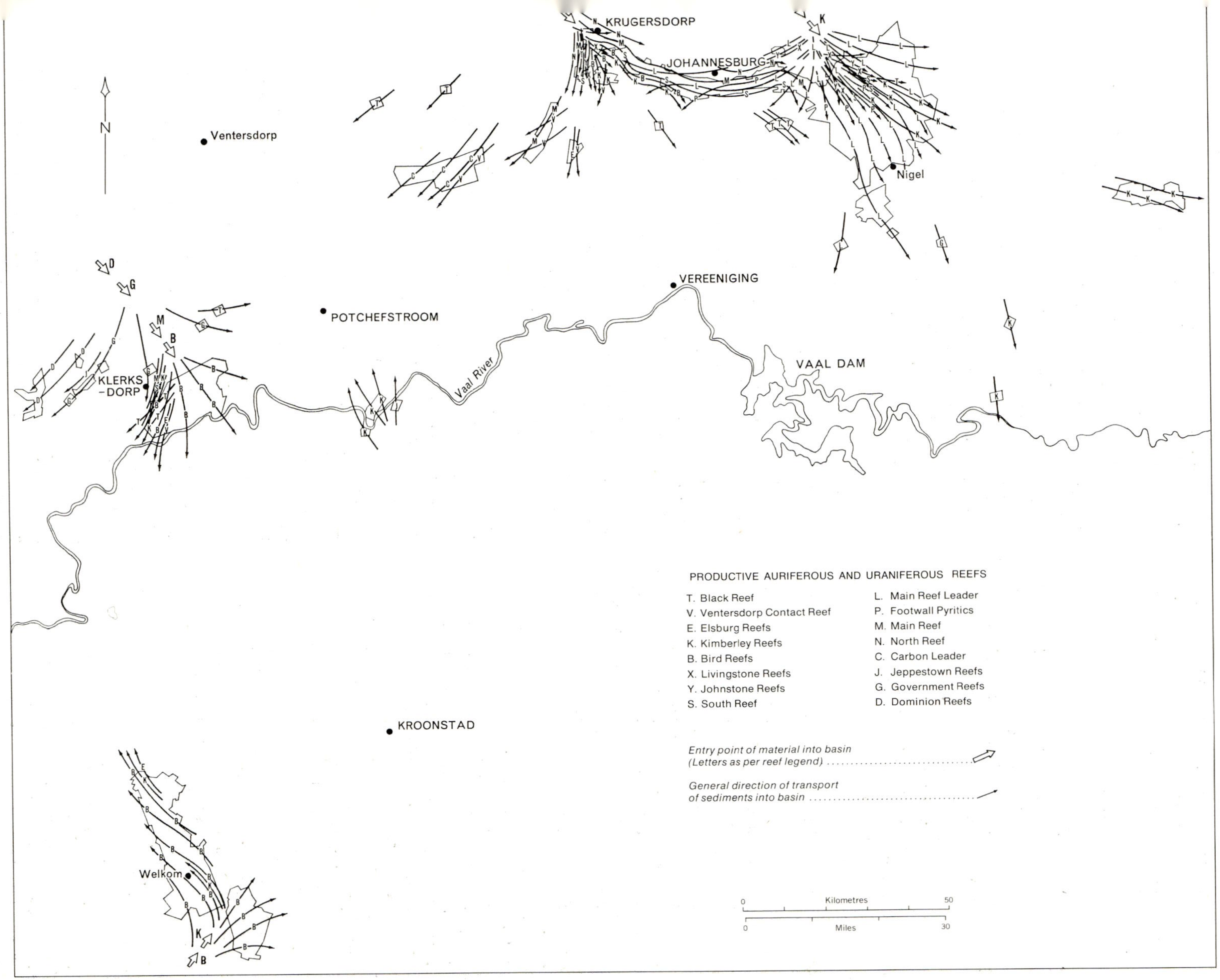

Figure 61. Sketch map of deltas and points of entry of sediment into the Witwatersrand Basin (after Brock and Pretorius, 1964*a,b*).

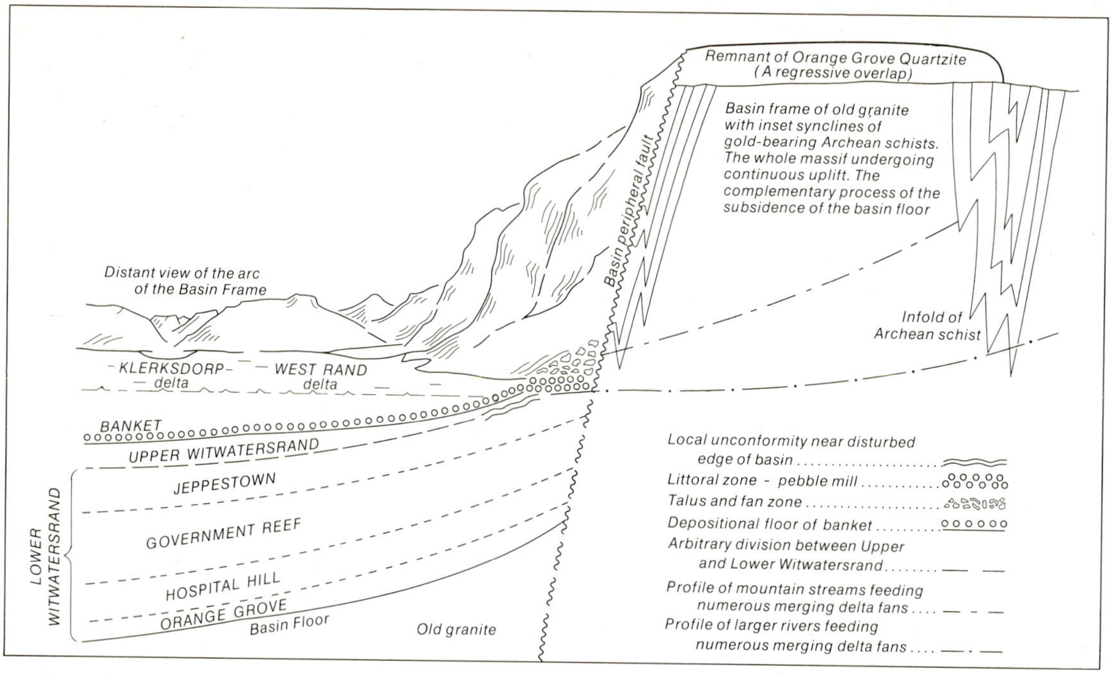

General concept-the banket, on a smooth depositional floor, is fed by ellipsoidal pebbles accumulated on the littoral 'pebble mill' released under exceptional but not cataclysmic conditions. The pebbles were moved by wave-bottom effects assisted by gravity (after Brock and Pretorius, 1964).

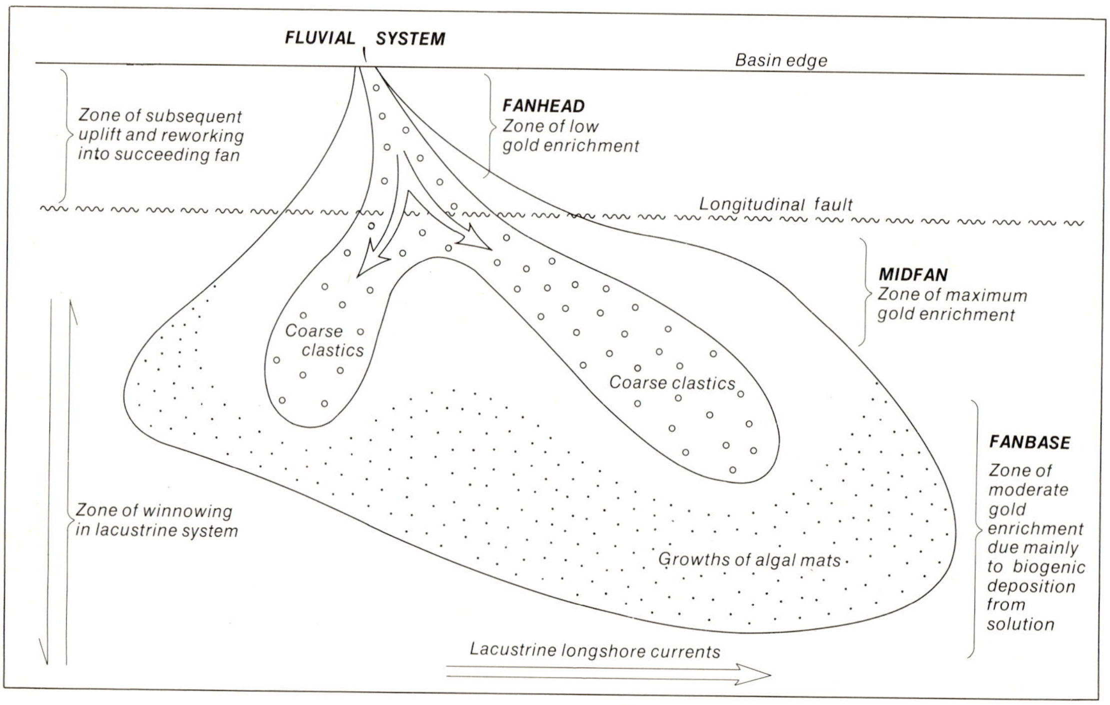

Detailed conceptual model of a Witwatersrand goldfield as a fluvial fan developed at the mouth of a major river flowing from a source-area in the northwest and debouching into a shallow-water, intermontane, intracratonic lake over a peripheral hinge-line consituted by faults parallel to the depositional axis. The general geometry of the fan, the various fan facies of deposition, the locations of coarse clastics and algal mats, the arrangement of varying zones of enrichment in gold, the portion of the fan subjected to transgression and winnowing and the clockwise flow direction of the longshore currents in the lake are shown (after Pretorius, 1975).

Figure 62. Diagrammatic sketches of the 'Basin Edge' concept of the formation of the Rand.

substitute for actual observations and permit some general comparisons.

There are a number of sedimentary basins that are analogous to the Rand and resemble it in many ways with respect to sedimentation phenomena, volcanism and tectonism. Most of these are in grabens or in subsidence zones bounded by profound fault systems. Here, we may mention the Michigan Copper Basin (Keweenaw Peninsula), the Sudbury–Blind River Basin, and the Bay of Fundy Basin in the Maritime Provinces of Canada. All are characterized by conglomerates and successions of grits, sandstones and quartzites interbedded or overlain by lavas commonly basaltic and frequently amygdaloidal. The mineralization in all of these basins differs considerably. Some are cupriferous, others uraniferous and still others uraniferous and auriferous as in the Rand. There are also analogies between the Rand and the so-called continental 'red beds' with their interbedded green beds. These constitute great basins filled with conglomerates, grits and sandstones. Some are cupriferous, others argentiferous, still others uraniferous; none are now known that are auriferous. Why should these major differences in mineralization occur in rocks that seem to have accumulated under similar conditions? We do not know the answer. Perhaps the parent rocks from which the sediments were derived were the controlling factor, or perhaps the degree of metamorphism and hence the extent of the redistribution and concentration of the various elements in the sedimentary and volcanic pile is the responsible factor. The latter seems the more probable.

Summaries of the types of gold (and uranium) mineralization in the Rand Basin are given by Pretorius (1964, 1975), and Brock and Pretorius (1964*b*). They distinguish the following types of occurrences of gold and uranium: (1) in the matrices of the conglomerates, (2) in banded pyritic quartzites, (3) in quartzites along a plane of unconformity between successive cycles of sedimentation, (4) on shales along planes of unconformity, (5) in carbon seams and thin films on parting planes between unconformities, or disconformities, (6) in quartz veins, and (7) in dykes.

Of these, only the first five constitute exploitable orebodies; the other two types have contributed to the production of gold on rare occasions and to a very limited extent. Those occurrences in categories (3), (4) and (5), according to Pretorius (1964, 1975), were formed in the terminal stages of one cycle of sedimentation, and those in categories (1) and (2) in the initial stages of a succeeding cycle. Categories (1), (2), (5), (6) and (7) are of most interest and are described in some detail below. Categories (3) and (4) are discussed mainly under category (2).

Conglomerate beds: These constitute the main orebodies of the Rand and are commonly referred to as 'banket' or 'reef'. Here belong the oreshoots or 'pay streaks' in the Dominion Reef; the Government, Coronation and Promise reefs in the Lower Division of the Witwatersrand System; the Elsburg, Kimberley, Bird and Main reefs of the Upper Division of the Witwatersrand System; the Ventersdorp Contact Reef; and the Black Reef at the base of the Transvaal System.

The conglomerates are composed essentially of pebbles of vein quartz with a few pebbles of quartzite and chert, all set in a matrix composed mainly of quartz, muscovite, sericite, chlorite, chloritoid and pyrophyllite (Pl. 15). In the matrix there is a veritable host of minerals occurring in variable amounts from place to place. Detailed descriptions of these minerals and their manner of occurrence have been given by Young (1917), Liebenberg (1955), Ramdohr (1955), Saager and Mihálik (1967), Saager (1968), Feather and Koen (1975), Barrass (1975) and a number of others (Haughton, 1964*a*). The list of minerals in the matrix includes:

Pyrite	Pentlandite	Chromite
Marcasite	Cobaltite	Leucoxene
Arsenopyrite	Linnaeite	Monazite
Glaucodot	Gersdorffite	Apatite
Leucopyrite	Skutterudite	Xenotime
Loellingite	Safflorite	Zircon
Pyrrhotite	Tennantite	Sphene
Sphalerite	Tetrahedrite	Garnet
Galena	Stomeyerite	Corundum
Chalcopyrite	Proustite	Diamond
Cubanite	Dyscrasite	Dolomite
Bornite	Bismuth	Calcite
Chalcopyrrhotite	Tellurium	Tourmaline
Chalcocite	Magnetite	Uraninite
Covellite	Hematite	Uranothorite
Neo-diginite	Ilmenite	Brannerite
Bravoite	Goethite	Thucholite
Molybdenite	Rutile	Graphite and carbon
Stibnite	Anatase	Native silver
Mackinawite	Brookite	Platinoid alloys and minerals
Niccolite	Ilmeno-rutile	Gold
Millerite	Cassiterite	Gold telluride(?)

Of these minerals pyrite is the most abundant constituting from 2 to 20 per cent or more of the matrix. It has been studied by Ramdohr (1958*a,b*) and classified in great detail by Saager (1970) (*see* Table 42). He distinguishes two general types – allogenic and authigenic[25] – with rounded, idiomorphic to hypidiomorphic (euhedral to subhedral) and xenomorphic (anhedral) morphologies. The xenomorphic pyrite occurs in veinlets and fracture fillings and is attributed to redistribution during metamorphism. The rounded (buckshot) pyrite is said to be partly detrital and partly due to pyritization of black sands and to concretionary processes, both having taken place in situ. Nodules containing 'mineralized bacteria' are thought to be due to the activity of primitive life. The idiomorphic to hypidiomorphic pyrite is thought to have been formed in situ as a result of reconstitution processes operating mainly during metamorphism. The pyrite is, therefore, considered to be partly detrital (allogenic) and partly due to syngenetic and metamorphic reconstitution processes (authigenic).

[25]These two terms, originally introduced by Young (1917), for the minerals of the Rand are particularly appropriate. Allogenic minerals are those that have a detrital origin; authigenic minerals include all those that grew within the conglomerates. However, it is difficult to make a clear-cut differentiation between these two suites as anyone who has studied these ores will recognize. The classification depends on subjective interpretations made by eye or microscope using criteria that not all investigators agree are specific. How the authigenic minerals grew and the source of their constituents (syngenetic or introduced from distant sources) is also a matter of interpretation and has led to an enormous amount of controversy.

Hoefs *et al.* (1968) have investigated the sulphur isotopic composition of the allogenic (detrital) and authigenic (recrystallized or reconstituted) pyrite. Both types were slightly heavier than meteoric sulphur by only a few parts per thousand and are similar to what is thought to be magmatic sulphur (equivalent to the sulphur in the Palabora deposit). Since the samples showed no major shift in enrichment of ^{32}S the authors concluded that there was little biologic activity involved in the formation of the authigenic pyrite. On the contrary, they thought that the primary source of the pyrite was of magmatic rather than sedimentary origin. Their data, however, do show a slight shift toward enrichment of the lighter isotope in the reconstituted pyrite compared with the detrital variety. To the writer this may mean a fractionation in terms of a preferential mobilization of the lighter sulphur isotope during the reconstitution process as the investigators have suggested; the shift may, however, also be due to the activity of primitive organisms operating on sulphur of early Precambrian age. Thus, only slight shifts in the sulphur in Archean sediments in the Yellowknife area where bacterial (?) activity has been active have been noted (Wanless *et al.,* 1960). It would seem from these data that early Archean organisms had only a limited capacity to fractionate the sulphur compounds compared with their evolutionary descendents.

Pyrrhotite, arsenopyrite, the various base metal sulphides and the various sulphosalts and other metallic minerals are generally considered by most investigators to be authigenic minerals. Most of these minerals are present in only small amounts, although some such as pyrrhotite, galena, sphalerite and some of the Ni-Co minerals occur in some reefs in moderate amounts, in places in a massive, intergrown manner. Pelletier (1940) noted that high gold values were frequently associated with pyrrhotite and considered the mineral to be an enriching factor in Rand ores. The detailed relationship of gold (and uranium) to the various other base metal sulphides, sulphosalts, etc., and particularly the correlation with their elemental components, such as Sb, As, Co, Ni, Cu, etc. is relatively unknown, a gap in the knowledge of these deposits that should receive attention from the viewpoint of genesis of the metals. Microscopically, Liebenberg (1955) found that much of the gold was later (younger) than the base metal sulphides. In most of the reefs the authigenic gold commonly replaces pyrrhotite, chalcopyrite, pentlandite, sphalerite, cobaltite, linnaeite, galena and arsenopyrite.

The so-called 'carbon' of the Rand conglomerates and other types of closely associated deposits is a hydrocarbon-uraninite mixture probably best referred to as thucholite or uraniferous carbon. It was observed by the early workers, has been studied in detail by Macadam (1931), Davidson and Bowie (1951) and Liebenberg (1955) and is referred to at some length by De Kock (1964), McKinney *et al.* (1964) and others in the book edited by Haughton (1964*a*). The mineraloid occurs in a granular, columnar and massive form and is jet black to dull black in colour. Occasional thin sheets, stringers and veinlets are also encountered in some reefs. The granular variety occurs in the matrix of most reefs as small (0.1–4 mm) black spheroidal or nodular forms with a warty outer appearance. The ovoid granules frequently extend beyond the ends of the conglomerate lenses and are the only indication of the position of conglomerate horizons. The granules are generally distributed throughout the reefs, but in places there is often a concentration along the parting planes between the hanging wall and footwall of the reefs; in some areas the concentrations of granules are localized along the parting planes on the footwalls of the reefs. The massive thucholite resembles vitrinite and the columnar variety occurs as vitreous fibres normal to the bedding. The thickness of the seams of columnar and massive thucholite is generally only a few millimetres, rarely more than 1 cm. Both varieties are commonly associated with one another, or they may occur alone. Both may occur in the conglomerates, although they are more commonly found along the bedding planes of conglomerate bands and along the basal contacts of reefs. In the latter situation it is extensively developed in the "Carbon Leader" reef of the West Wits Line (De Kock, 1964).

The replacement, displacement and other growth features of the thucholite indicate that it is an authigenic mineraloid. Associated minerals include uraninite, gold, platinoids and sulphides such as pyrrhotite, chalcopyrite, pentlandite, sphalerite, cobaltite, linnaeite, galena and arsenopyrite. Microscope examination of the thucholite indicates that the uraninite commonly occurs as a large number of tiny inclusions in the mineraloid. Gold occurs as veinlets, specks and small irregular patches within the body of the thucholite; bands of gold also surround the granules of thucholite, and thin films and veinlets cut across the columns or occur with phyllosilicates between the columns of the columnar variety of the mineraloid. The various sulphides noted above commonly occur in the interstitial spaces between granules of thucholite and also appear in veinlets and patches in cracks and cavities in the mineraloid. Inclusion of sulphides in thucholite are also observed in places.

The intimate association of gold and thucholite in the Rand has been known and commented on for nearly 50 years. In and near the mineraloid, visible and occasionally coarse gold can be frequently seen, a feature that is unusual in the Rand ores. Thucholite is also rich in other metals in the "Carbon Leader"; besides gold there is native silver, a variety of platinoid minerals (sperrylite, braggite, cooperite, platiniridium, platinum and osmiridium), chalcopyrite, a number of Ni-Co arsenides and sulphides, sphalerite and galena (De Kock, 1964).

The origin of the thucholite is unresolved. Davidson (1964–1965) considers the mineraloid a carburan polymerized radiogenically from the methane and other hydrocarbon gases so prevalent in the South African mines. De Kock (1964) thinks the thucholite is possibly the organic remains of a primitive form of algae, which existed in the waters of the basin in which the reefs were deposited. Actually, the methane may be similarly derived, although it may have come from a very deep source as the result of abiogenic processes. Schidlowski (1966) considered that the carbonaceous material (thucholite) in the Rand formed by radiolytic polymerization of originally mobile (preferably gaseous) hydrocarbons percolating through the conglomerates, the prerequisite for its formation being the presence of detrital uraninite within the heavy fraction of the individual reefs. The writer is of the opinion that nearly all of these solid hydrocarbons are polymerized residues of petroleumlike materials derived in some manner, as yet unknown, from the humification of algae, bacteria and other low-order forms of life. There is a strong suggestion that these petroleumlike substances are capable of transporting gold, uranium, thorium and various other metals, probably in a chelated form. As they set (polymerize) in a gellike mass, resulting in the ovoid and columnar form of many of these hydrocarbons, the gold, uranium and other components are split out, appearing as minute inclusions or along cracks in the hardened gel.

The origin of the thucholite has been variously related to biogenic sources. Schidlowski (1965) found cell-like structures in the Witwatersrand rocks and Prashnowsky and Schidlowski (1967) identified amino acids and monosaccharides in the carbonaceous substances (thucholite). Hoefs and Schidlowski (1967) concluded from carbon isotope data of the thucholite that its constituents were of biogenic origin. The sulphur isotopic data on the allogenic and authigenic pyrite presented by Hoefs *et al.* (1968), however, showed only a small enrichment in ^{32}S in the recrystallized authigenic pyrite, suggesting only slight, if any, activity of organisms operating on the sulphur cycle in the conglomerates.

In a recent contribution to the origin of the carbon of the Rand, Hallbauer and van Warmelo (1974) have examined both the columnar and 'fly speck' types of thucholite from the Carbon Leader Reef and other auriferous reefs by scanning electron microscope. In the columnar type internal structures morphologically resembling filamentous, branched and apparently septate cells of obvious biological origin partly encrusted with gold were identified. In addition silicified structures that were identified as primitive fungi were noted. From these observations they concluded that the organism was some kind of a fossilized symbiotic lichenlike plant with no modern equivalent that formed carpetlike colonies of up to several square metres in extent. These organisms were capable of extracting gold and uranium from the environment and depositing them inter- and intracellularly. The 'fly speck' carbon constitutes spherical nodules with irregular and pitted surfaces showing a remarkable morphological resemblance to living fungal *sclerotia* (reproductive hyphae). The amorphous coallike thucholite is envisaged by Hallbauer and van Warmelo (1974) as due to exogenous plant debris washed into its present position and there attacked by a fungal organism.

Carbonaceous substances other than thucholite are found in the reefs and other rocks of the Rand Basin but generally only in small amounts. Liebenberg (1955) mentions the presence of nonradioactive bituminous matter in the reefs and quartz veins. The Black Reef conglomerate in particular contains appreciable quantities of a nonradioactive carbonaceous substance that together with chlorite is responsible for the colour of the reef. The gold in the reef is not closely associated with this carbonaceous matter.

Uraninite[26] was first noted in the Rand ores by Cooper (1923); its occurrence has been studied in detail by Liebenberg (1955). As already stated the uraninite is commonly present in thucholite as minute scattered particles. In addition the mineral occurs in the matrix, commonly as small rounded masses, isolated from the thucholite. The majority of this uraninite is said by Liebenberg (1955) to be detrital, an interpretation that a number of investigators have disputed at length. The uraninite in thucholite is, likewise, thought to be detrital, replaced to varying degrees by the hydrocarbon. Some of the microscopic uraninite is secondary occurring as tiny veinlets and irregular patches in cracks and cavities in the matrix. Most of the uraninite is thorium poor ($\sim 2\%$ ThO_2); the rare-earth content is of a similar order. Gold is closely associated with uraninite in most occurrences.

Most of the gold of the Rand conglomerate is extremely finely divided averaging about 80μ (0.08 mm) in diameter. Some is visible to the eye especially where the metal is closely associated with or in veinlets and patches in thucholite. Much of the gold occurs as tiny ragged particles or clusters of particles in the fine-grained matrix of the banket where it is generally associated with quartz, sericite and chlorite. It also occupies tiny fissures and cavities in most allogenic and authigenic minerals and may occur enclosed in or as replacements of secondary sulphides, especially pyrrhotite, pyrite, sphalerite, galena, etc. The general paragenetic relationship between gold and these sulphides is the same as that commonly seen in veins, namely that the gold is generally the last mineral to crystallize. Only minor amounts of gold occur in small rounded grains, and there are no descriptions of nuggets, large or small, such as one might find in Tertiary and Quaternary placers. This is confirmed by the recent work of Hallbauer and Joughin (1972) who examined the distribution and size of gold particles in the reefs. They found that all gold particles after release by hydrofluoric acid treatment passed through a sieve with apertures of 2.4 mm. All of the gold particles from the Kimberley Reef passed through a sieve with apertures of 0.075 mm. On the other hand, less than half the gold from the Vaal and Ventersdorp Contact Reefs passed through the same sieve. The separated gold had various shapes, the two most prominent forms being platelike particles with irregular edges and crystalline particles consisting of elongated octahedrons. Some of the crystalline particles were more complicated in their shape with branchlike extensions. Hallbauer and Joughin (1972) state that such similarly shaped particles have been observed in placer deposits.

The average grade of Rand gold ore is 0.25 oz Au/ton, the fineness of the native gold averages about 925, and the general Au/Ag ratio in the conglomerate deposits is approximately 9. According to Myers (1971) the fineness of the gold in the Rand is exceptionally uniform from particle to particle. The Ag/Au ratios in the various reefs of the Rand and the relationship of gold and silver have been intensively studied by Hargraves (1963), Von Rahden (1965), Schidlowski (1968), Saager (1969) and Viljoen (1971).

Hargraves (1963) found that low grade samples have high silver contents and vice versa, and that the silver content of bullion from individual reefs varies as a function of the

[26]There has been much argument, mainly academic, on the use of the terms uraninite and pitchblende in the quartz-pebble conglomerates and other types of uranium deposits. In fact, pitchblende is a cryptocrystalline variety of uraninite, but some investigators insist on distinguishing between the varieties on the basis of their origin, composition (U/Th ratio), or habit. Neither of these bases is really satisfactory, but the composition of the mineral may be significant in some contexts. Thus, it appears that in the Dominion Reef the uraninite carries relatively large amounts of thorium whereas in the other reefs the mineral (pitchblende) is relatively low in thorium. This fact would seem to be significant in the origin of the uraniferous mineral in the various reefs. Similarly in pegmatites the thorium content of uraninite is commonly high (about 25%) whereas in vein deposits the thorium content is generally low, features that must reflect significantly different processes in the origin of these deposits.

relative elevation of the reef, i.e., the bullion of the shallow reefs has a higher silver content than deeper reefs of comparable grade. Hargraves interpreted these compositional gradients as the result of chemical rather than physical (detrital) processes and stated that in terms of the modified placer hypothesis for the origin of the gold, the gradients imply considerable redistribution of silver relative to gold during the 'solution' stage.

Von Rahden (1965) thought that his data, which show wide variations in the apparent fineness (and hence in the Au/Ag ratios) of channel samples from the same elevations and stopes in the Ventersdorp Contact Reef and Main Reef, could be ascribed to the presence of silver minerals (mainly native silver) in the ore. The silver minerals, he felt, were due to the redistribution of silver relative to gold during the solution stage of the modified placer theory suggested to explain the Rand deposits. Von Rahden (1965) further concluded that his results did not tend to confirm Hargraves' (1963) hypothesis that the silver content of the reefs varied with elevation. He recommended electron microprobe studies of the ore to resolve the problem.

Addie (1964), commenting on Hargraves' (1963) paper, considered that the increase in the Au/Ag ratio in the Witwatersrand deposits with depth is due to exsolution of a silver-gold solid solution as the temperature changes. By this he apparently meant that since the silver is more mobile it migrates out of the system. He thought that the increasing Au/Ag ratio signalled an approach to a thermal centre and postulated a thermal source for the mineralization.

Schidlowski (1968) investigated the gold fraction of the Witwatersrand conglomerates from the Orange Free State goldfield and confirmed that the bulk of the native gold is concentrated within the bottom parts of the 'reefs', mostly near the footwall contacts. This, together with other criteria, suggested to him that the distribution of the metal was governed by sedimentary processes. On the other hand, the shape of the gold grains and the general fabric of the gold-enriched layers often lack conspicuous sedimentary characteristics, a fact that he related to subsequent metamorphism, which produced marked structural modifications and a limited textural rearrangement of the gold and its host-rock, thus obliterating to a certain degree the primary alluvial character of the metallization. Schidlowski postulated that the large-scale reconstitution of the gold fraction was effected either by a mechanical reshaping of the malleable and ductile gold particles or by an intermediate solution process followed by reprecipitation. These processes he thought took place more or less in situ, the original sedimentary distribution pattern not being profoundly affected, whereas the individual allogenic gold grains were transformed into authigenic ones or sometimes even into major gold aggregates. Schidlowski found by electron-probe microanalysis that the gold contains on the average between 9.9 and 12.4 per cent silver, which he considered also suggestive of an alluvial origin. Schidlowski concluded that an integration of all available data strongly supports a modified placer theory for the origin of the Witwatersrand gold.

Saager (1969) studied the Au/Ag ratios of individual gold particles and of ore samples obtained from the Basal Reef conglomerate horizon of the Upper Witwatersrand System in the Orange Free State Goldfield. He found that the values of individual gold grains were exceptionally constant at about 12.31, whereas the Au/Ag ratios of the ore samples had markedly lower values and showed a large variation. He thought that these values indicated that the detrital gold was homogenized during transportation and sedimentation and that part of the original silver content was redistributed. Saager further postulated that the first refining of detrital gold particles occurred in streams that eventually drained into the Witwatersrand basin. A redistribution of the silver relative to the gold took place in the littoral zone of the basin, where the conglomerates were reworked by wave action and erosion. Supergenic enrichment of gold might also have taken place close to the original shoreline. These processes he believed accounted for a decrease in the Au/Ag values of ore samples towards the northwest in the area investigated. Finally, the gold was homogenized during the metamorphic period of the Witwatersrand System, when pseudohydrothermal transport and redeposition of certain constituents took place. The main portion of the silver content of the original alloy, which was redistributed during the various processes outlined above, is now present as finely dispersed silver minerals. The hypothesis of an elevation control of the Au/Ag values, as suggested by Hargraves (1963), was found to be improbable.

Viljoen (1971) compared the composition of gold from the Witwatersrand with that from the Barberton Mountain Land. The elements most frequently found in the gold of both types of deposits are silver and copper with occasional nickel. The silver content of the Witwatersrand gold was found to vary considerably (0.3–32.1%) compared with the smaller variation (3.6–21.6%) of the Barberton ores.

Summarizing, we can say that nearly all investigators agree that the gold in the Rand conglomerates is authigenic. Some postulate hydrothermal processes as its source and others a placer origin with later redistribution of the metal as a result of metamorphic processes.

The ore shoots, pay shoots or pay streaks in the conglomerate beds are thin bodies with extensive dimensions on both their strike and dip. Most shoots are only a few feet thick, and some are only inches or fractions of inches such as the Carbon Leader. In places a conglomerate band may thin to a single line of pebbles; in others the reef is manifest only by a parting plane along which gold, hydrocarbon and uraninite occur. On their dip and strike dimensions the shoots may be hundreds to several thousand feet long. There appear to be some differences as to the exact location of the pay streaks in the conglomerate beds. As noted above some investigators find that the bulk of the gold is concentrated in the bottom parts of the reefs in some areas. In other areas evidently this is not always the case since there are numerous statements in the literature that the gold shows no preference for the bottoms of the conglomerate bands, being distributed randomly throughout the reefs or in some cases even concentrated at the top of the conglomerate beds (Pretorius, 1964, p. 97). The longer axes of the streaks tend to be parallel to the longer axes of the pebbles in a number of areas and trend roughly in the same direction. Some streaks are braided. Reinecke (1927), who carried out an extensive study on the outline of the assay and stope plans in the Main Reef of the East Rand, found that the gold distribution radiated southeastward from a northwest source and coincided with well-graded pebbles, large pebbles, aligned pebbles and relatively wide sections of the reefs. He interpreted these features to indicate deposition of gold in stream channels and concluded that the gold was originally of placer origin. Brock and Pretorius (1964*a,b*) reiterate these general views:

> In any given goldfield the total amount of gold present per unit volume of strata in the Upper Witwatersrand Division is greater near the basin's edge than it is farther away. As the sediments fan out towards the centre of the basin so the amount of gold becomes progressively less. The number of conglomerate bands and the overall coarseness of the sediments also vary proportionately with the mineralization, indicating that the pebbles, the gold and the uranium were introduced into the basin contemporaneously and at the same point by the same transporting agencies.
>
> Paystreaks, representing better developed conglomerates and higher values, radiate outwards from the same points of entry. The East Rand affords the best documented example, but almost identical conditions exist in the Central Rand, West Rand, West Wits Line and Orange Free State. Although conglomerates are also present between the paystreaks, here the pebbles are usually not closely packed and have not been "milled" to the same degree of homogeneity, and the conglomerates carry uneconomic amounts of gold. It has been observed that the coarseness of the banket type reef, up to certain size limits, generally varies directly with the gold content. The higher velocities of the transporting media in the pay channels brought more and larger pebbles and more gold and uranium.
>
> With the development of empirical rules regarding the amount of gold with relation to the edge of the basin, all the readily available data pertaining to sedimentational trends of pay reefs were collected and reviewed. These include paystreaks, footwall channels, increase in pebble size, increase in number of pay reefs, and increase in total gold per unit area of cross-section. In each of four goldfields a shoreward convergence was demonstrated. The gold-bearing zones are essentially fan-shaped with a well-defined apex in each case. They can hardly be called anything but deltas.
>
> The four apices, representing four points of entry (northwest of Krugersdorp; north of Benoni; northwest of Klerksdorp; and south of Welkom) (*see* Figure 61 in this bulletin), must represent points on the ultimate rim of the basin at that particular time. Thus we have confirmation regarding the essential correctness of our outlining of the basin and its basin edge ramifications, and confirmation too that the loss by erosion was not excessive.

The general scheme of formation of the Rand banket suggested by Brock and Pretorius (1964*a,b*) and Pretorius (1975) combines the earlier work of a number of investigators; it has been termed the basin edge concept and is shown diagramatically in Figure 62 and described in previous pages.

Banded pyritic quartzites. The banded pyritic quartzites occur in erosion channels, scoured in the shales, quartzites and conglomerates, beneath major auriferous horizons in the Main Reef Group, the Bird Reef and the Kimberly Reef Group of the Upper Witwatersrand Division and in the Black Reef Series of the Transvaal System. The thickness of these quartzites varies from thin streaks to many feet, and the channels containing the banded quartzites are up to 100 ft deep. The economic mineralized beds consist of dense, dark quartzites alternating with layers of pyrite and barren quartzite. The layers of pyrite follow the bedding in the quartzites. The mineralization of the pyritic quartzites is essentially the same as that described above for the conglomerates and need not be repeated. According to Pretorius (1964) thucholite and uraninite are present in the pyritic quartzite but are not necessarily associated with the gold. There is, however, a more definite relationship in the quartzites than in the conglomerates between a high pyrite content and sympathetic gold content. The quartzites normally contain only small ore shoots of gold and uranium, but some are relatively rich with average grades higher than those in the conglomerates.

In some parts of the Witwatersrand gold is concentrated on quartzites and shales along planes of unconformity. In these cases the thickness of the pay zone may be only fractions of an inch. The gold is usually very finely divided but may be visible in some specimens. According to Pretorius (1964) and others this gold is essentially detrital, formed in the terminal stages of one cycle of sedimentation.

Carbon seams and thin films. The best example of this type of orebody is the Carbon Leader described in detail by Swiegers (1940) and De Kock (1964). This horizon is generally a carbon-rich streak or seam in quartzites, grits or conglomerates, or a mixture of these. It commonly forms a thin footwall seam and is heavily mineralized with pyrrhotite and pyrite. On the seam rests a variable layer of pebbles of vein quartz and some angular chert fragments. The pebble band itself is only 2 to 3 in. thick and the carbon seam varies from a mere streak to a solid band of carbon up to 3 in. thick. The minerals composing the Carbon Leader include hydrocarbon (thucholite), gold, native silver, uraninite, various platinoids, pyrite, pyrrhotite, chalcopyrite, a variety of Ni-Co arsenides and sulphides (mainly skutterudite, gersdorffite, niccolite, millerite, pentlandite and cobaltite), sphalerite, galena, chromite, muscovite, sericite, pyrophyllite, chlorite, chloritoid, biotite, and minor amounts of cassiterite, rutile, leucoxene, brannerite, tourmaline and zircon.

The gold in the Carbon Leader horizon occurs as veinlets, specks and anhedral grains, and is predominantly associated with the phyllosilicates, which occur as infilling between the cracks and cavities of the columnar type of thucholite. The remainder of the gold is directly included in the hydrocarbon or occurs in association with the pyrite and quartz. The very wide fluctuation in the Au/Ag ratios in assays of the Carbon Leader is due to the presence of native silver.

The platinoid content of the Carbon Leader is high, up to 0.75 oz/1000 tons of ore milled (0.00075 oz/ton or 0.025 ppm) the various platinoids present include the arsenide, sperrylite; the sulphides braggite and cooperite; and the metallic minerals platiniridium, platinum and osmiridium.

Quartz veins. Gold-bearing quartz veins are relatively common in some parts of the Rand Basin. They appear to be more prevalent in the shales of the Lower Division of the Witwatersrand System than in the quartzites and other rocks of the Upper Division (Pretorius, 1964). Their presence is usually minimized by the placerists and emphasized by the hydrothermalists of whom Graton (1930) gives the most extensive description of these bodies.

The quartz bodies occur as definite veins and veinlets cutting the reefs and associated rocks, as flat seams and veins along parting and bedding planes, as large veins and irregular masses or 'blows' near dykes and faults and as irregular patches or blotches in the reefs. The mineralization of the quartz bodies comprises native gold, which may be coarse in places, massive pyrite or pyrrhotite, chalcopyrite and more rarely galena, sphalerite and arsenopyrite. Some of the quartz veins have been mined in the past, but in general most of these deposits are small and uneconomic.

Several investigators have noted that the ore-bearing conglomerates near the auriferous quartz veins are depleted in gold, indicating that the conglomerates have been impoverished to supply the gold in the veins. Furthermore, it has been noted that these quartz veins become consistently less auriferous and ultimately barren with increasing distance from the gold-bearing conglomerates and quartzites. These features suggest that the quartz veins are of late secretion origin, unrelated to the initial formation of the gold in the bankets.

Dykes. Low grade gold mineralization is present in certain dykes, particularly where these intrusives cut across the reef horizons (Pretorius, 1964). As an example Reinecke (1927) mentioned that the "South Rand Dike", in Crown Mines averaged 0.06 oz Au/ton over a length of 380 ft. The highest values recorded reached 0.225 oz Au/ton.

The interpretation of the origin of gold in dykes crossing gold horizons is frequently difficult. In some places gold increases as the dyke is approached; in others the reverse is true – the gold appears to have been swept out of the rocks adjacent to the dyke. Such phenomena are repeatedly reported in gold deposits. With respect to the Rand two examples need only be cited – one by Sharpe (1955) where gold is highly concentrated adjacent to the Van Ryn Deep dyke where it crosses the Main Reef Leader and the other by De Kock (1964) where the gold in the Carbon Leader was depleted where crossed by a basic dyke. The writer has observed similar phenomena where dykes cross gold deposits and has also observed that some diabase dykes become slightly enriched in gold where they intersect gold-bearing ore shoots. These differences in the migration habits of gold are difficult to explain in generalities. Each occurrence has to be studied in detail. In some cases fluids are released in the rocks adjacent to the dyke, which assist in the migration of the gold in the pre-existing deposits away from the hot source; in other cases thermal diffusion and dilitancy (fracturing, etc.) are involved and the gold moves toward and into the dyke especially during the late cooling stages.

Origin of the Rand gold deposits. No type of gold deposit has prompted as much discussion and argument with respect to the origin of its gold (and uranium) as the Rand. Four general theories have been advanced:

Early workers considered the gold to be strictly of placer origin without modification. A somewhat similar theory says the ores are syngenetic, but the result of chemical precipitation of gold rather than mechanical deposition of the metal (Penning, 1888; De Launay, 1913; Garlick, 1953). Miholić (1954) considered that the gold was a diagenetic addition to chemically precipitated uranium-hydrocarbon deposits, all later modified during metamorphism. He envisaged initial marine gravel deposits along a steep coast in which a rich vegetation of uranium concentrating organisms developed under anaerobic conditions giving rise to the formation of thucholite and pyrite. At a later period gold in an ionic form was reduced and precipitated by the organic matter from percolating thermal waters that invaded the conglomerates.

In a recent paper Reimer (1975) has postulated yet another variation on the syngenetic chemical theory for the origin of the gold and uranium in the Rand. Having analyzed the published age data from the western part of the Transvaal he concluded that the auriferous sediments of the Witwatersrand System were deposited over a period of about 30 to 40 m.y. during the time range of 2480 to 2370 m.y. The age relationship between these sediments and the over- and underlying formations he thought invalidated the concept of a "Witwatersrand Triad", which has been visualized as comprising the Dominion Reef, Witwatersrand and Ventersdorp systems, covering a time span of about 600 m.y. Reimer further opined that the age of the Witwatersrand System as deduced from his study has important implications for the origin of the gold in the conglomerates. Thus, doubt is cast on the simple model of its derivation as clastic grains direct from Archean greenstone belts. On the contrary it would appear that most of the gold was dissolved in trace amounts from rocks in the source area and transported in solution to the depository. Here it was precipitated, possibly aided by primitive organisms and then concentrated by normal sedimentary processes. Reimer also rejects derivation of the uraninite from pegmatitic sources and suggests that most of the uranium was transported to the depository in a dissolved form. It was then precipitated under strong organic influences to form nodular grains that, like the gold, were concentrated by normal sedimentary processes in the conglomerate horizons.

The most recent contribution to the origin of the uranium mineralization of the Witwatersrand and Dominion Reef systems is that by Simpson and Bowles (1977), who assume the presence of a depositional basin overlying an Archean craton enriched in oxyphile metals including uranium. In the basin they consider that uranium was deposited in two ways, as detrital allogenic uraninite and by precipitation from solution. The detrital components were concentrated in the higher energy conglomerate environment of the Dominion Reef and proximal parts of the Witwatersrand Basin. The solution components including uranium and gold were precipitated under reducing conditions through the agency of decaying organic matter in the lower energy or distal regions of the basin, giving rise through diagenesis and metamorphism to such uraniferous horizons as the Carbon Leader of the West Wits, which are not directly related to known paleogeographical channels. They further state that it can now be clearly demonstrated that relatively small amounts of thorium increase the resistance to attrition and oxidation to the extent that a thorian uraninite with 1 per cent or more thorium oxide can survive as river detritus under present-day atmospheric conditions. They find that all of the uraninite grains from the Witwatersrand Basin considered to be of detrital origin contain 1 per cent thorium oxide or more; hence, it is unnecessary to postulate a reducing atmosphere, indeed the retention of sulphate and uranyl ions in solution would seem to require the existence of an oxidizing atmosphere in the model proposed for the deposition of uranium from solution.

Most South African geologists writing on the Rand adhere to the modified placer theory. They argue that the gold was deposited as placer grains, the metal being later mobilized and recrystallized during the metamorphism of the rocks. Uraninite has suffered a similar fate. Advocates of this theory include Mellor (1916, 1931), Reinecke (1927, 1930), Young (1931), Macadam (1936), De Kock (1940), Du Toit (1940), Frankel (1940), Liebenberg (1955), Ramdohr (1958*a,b*), Bain (1960), Nel (1960), Brock and Pretorius (1964*a*), Schidlowski

(1968), Saager (1969), Viljoen *et al.* (1970), Minter (1970), Myers (1971) and Pretorius (1974*a,b,* 1975).

The third theory argues that the ores are of hydrothermal origin, although some investigators are hazy about the source of the gold. Presumably they consider that it derived from a cooling magma. The supporters of this theory include Hatch and Corstorphine (1909), Horwood (1917), Graton (1930) and Davidson (1953, 1957). The publication of Graton's treatise brought a veritable blizzard of criticism from the placerists most of which is published as an annex to the Transactions of the Geological Society of South Africa v. XXXIV, 1931, p. 1–92.

The fourth theory suggested by Davidson (1964–1965) states that the gold and uranium were leached from overlying volcanics by heated saline waters. These waters sank to the lowest permeable horizons mainly along the conglomerates and unconformities where they deposited their metals toward the cooler outer margins of the basins. This idea brought forth another storm of criticism, much of which was unfavourable to the mechanism (Bowie, 1964–1965; Collender, 1964, 1965; Dowie, 1964–1965; Rice, 1964–1965; Taylor, 1964–1965; Bishopp, 1964–1965; Gillett, 1964–1965; Way, 1965; Greenberg, 1964–1965). It was pointed out that the volcanics were generally impermeable and showed no evidence of leaching. There is also no evidence of saline beds ever having been present, none are present now, and there is no evidence of collapse structures indicating their former presence.

A considerable number of features are invoked to support the modified placer theory among which the following are the most important.

1. The great persistence of the mineralization in individual beds of conglomerate or sequences of beds.

2. A ready source of placer gold in the Swaziland System known to contain numerous gold-quartz deposits.

3. The presence of abundant quartz pebbles that is a feature of many placers including the Klondyke (White Channel gravels) and the Victoria placers in Australia (White Leads).

4. The presence of undoubted detrital minerals that commonly accompany gold in placers an example being the platinoid metal compounds, chromite, garnet, etc. Uraninite is, likewise, said by some investigators to be detrital, but this has been disputed.

Viljoen *et al.* (1970) and Saager (1973, 1975) have postulated a granite-greenstone provenance model for the gold and a number of other minerals in the Rand. They claim that all the available evidence suggests that the sediments of the two basins (Dominion Reef, Witwatersrand System) were largely derived from the erosion of a granite-greenstone terrane similar to the Barberton region of the eastern Transvaal. They showed that a correlation exists between the expected placer minerals derived from successive erosional levels of such a terrane, and the actual placer minerals encountered in the Dominion Reef and Witwatersrand systems. More recently Köppel and Saager (1973, 1974) have advanced lead isotopic data of sulphides (galena, tetrahedrite) in the auriferous deposits of the greenstones of the Swaziland Sequence, which compares well with that of some of the data from the sulphides (pyrite, galena) of the Rand deposits. From this they concluded that the greenstone belts of the Swaziland System and their numerous gold deposits contributed detritus to the Witwatersrand depository.

Modification involving redistribution, migration and recrystallization of gold is admitted by the placerists, who attribute these phenomena to metamorphism evidenced by the presence of abundant sericite, chlorite and secondary quartz.

The principal evidence advanced in favour of the hydrothermal theory is:

1. The gold as it is now found is not in the form of nuggets or dust; on the contrary it is similar in its various forms and aspects to epigenetic gold. The silver content is also high, a feature that is not generally found in placer deposits.

2. Much of the gold is not at the base of the conglomerate beds as is normal in placers; rather the gold is commonly scattered throughout the bankets; some is even at the top of the bankets.

3. The gold is in shoots or pay streaks that are not identifiable with true pay streaks, leads, bars, channels, etc. in modern placers.

4. The bulk of the pyrite is certainly not detrital; some pyrite is like that found in epigenetic deposits.

5. The bulk of the uraninite is probably not detrital; it is in veins and veinlets and scattered at random through thucholite.

6. The general paucity of detrital minerals, especially magnetite. The placerists claim, however, that the magnetite has been sulphidized.

7. The presence of elements such as Cu, Pb, Zn, Ag, etc. bound in various sulphides and sulphosalts. These minerals, all investigators agree, exhibit epigenetic features.

8. The presence of sericite, chlorite, carbonate, tourmaline and other hydrothermal minerals in the deposits. In places these show an intimate relationship to gold and seem to have been precipitated with it.

9. The presence of quartz veins in faults, fractures and gash fractures. These veins carry gold and a variety of epigenetic sulphides.

10. Similarity in some respects to copper deposits in conglomerates such as those in Michigan. The copper is obviously introduced into the conglomerates; why not the gold? The geological setting is somewhat similar as well. There are conglomerate and quartzites in a basin interbedded with volcanics and considerable evidence of igneous activity in both places (e.g., Duluth Batholith and Bushveld Complex).

When all of the factors are taken into consideration most of the South African geologists on the Rand favour a modified placer origin. But the story is probably not yet finished, for new information is forthcoming every year. As Saager (1968) has pointed out, the extremely complex and complicated ore paragenesis of the Rand can probably only be explained by a polygenetic origin of the different ore components, "With certain reservations, it is possible to obtain specimens from the Witwatersrand ores that exhibit classic hydrothermal, placer, or even syngenetic-sedimentary characteristics. 'Proof' can thus be obtained for any genetic interpretation it is wished to construe on the origin of the Witwatersrand conglomerate ores."

For one not familiar with the Rand it would seem that the overall geological and geochemical pattern is the key to the origin problem as it is in any other gold deposit. Specific criteria may be found that support the overall theory, but these may be overshadowed in some places by later effects that may be irrelevant but serve to confuse the pattern. After the description of the other conglomerate deposits in this category an attempt is made in the section on the origin of gold deposits to formulate a general theory on the origin of the conglomerate type.

The Tarkwa goldfield, Ghana

The Tarkwa goldfield in southwestern Ghana near the town of the same name is in a Precambrian terrane consisting mainly of a thick series of argillaceous and arenaceous sediments with two well-defined gold-bearing quartz-pebble conglomerate beds. These lie in a long narrow northeast striking syncline. The geology of the field has been described in detail by Hirst (1938) and Junner *et al.* (1942).

The geological column for the Tarkwa goldfield is given in Table 63, and a generalized geological section is shown in Figure 63. The oldest rocks, the Birrimian System, are essentially greenstone belts isolated by granites. They contain a number of gold deposits throughout their extent in Ghana. At Abuasi, some 70 mi northwest of Tarkwa, a number of gold quartz lodes occur in the Birrimian schists and phyllites. The Ashanti lode, the most productive, is a steeply dipping series of quartz veins in sericitic and graphitic schists and phyllites, containing essentially quartz, carbonates, sericite and small amounts of pyrite, arsenopyrite, sphalerite, galena and gold, the last mostly invisible to the eye. The average grade of the ore is about 1 oz Au/ton (Kitson, 1930). The Birrimian rocks and their deposits may have been the principal source of alluvial gold in the Banket Series of the Tarkwa according to a number of investigators.

The Banket Series consists of a conglomerate bed at its base known as the Basal or Main Reef, another series of conglomerates and breccia conglomerates termed the Middle Reef Series some 15 to 40 ft above the Basal Reef, and an upper Breccia Reef Series lying from 40 to 75 ft or more above the Middle Reef. In general the reefs strike northeast and dip from 30 to 50° toward the synclinal axis. The reefs tend to lense out along strike according to some investigations; others attribute this feature to strike faulting. A controversy has, therefore, arisen as to which is the more important (Hobson and Robinson, 1943; Junner and Hirst, 1944). The Basal or Main Reef carries the best gold values; in most places its oreshoots average about 0.27 oz Au/ton and the Au/Ag ratio ranges from 4 to 49. The Breccia Reef Series is low grade, averaging less than 0.05 oz Au/ton. The ore shoots in the Middle Reef tend to be short and low grade, averaging about 0.1 oz Au/ton.

The gold-bearing shoots of the Basal and Middle reefs are in beds or series of beds of quartz-pebble conglomerates (Pl. 16) interbedded with quartzites. The Breccia Reef consists of alternating bands of coarse to fine breccia and grit containing a large amount of Birrimian phyllite and schist debris in the form of squeezed pebbles, fragments and dust. Crossbedded seams of hematite occur throughout the series, particularly in the coarse members. The Basal and Middle Reef series consist of relatively persistent conglomerate beds interbedded with

Table 63. Generalized geological column, Tarkwa goldfield, Ghana

	System	Group, etc.	Lithology
Precambrian		Post Tarkwaian Intrusives	Epidiorite, norite, gabbro, amphibolite, porphyry, diabase and epidiorite
	Intrusive contact		
		Huni Sandstone	Sandstone, grit, quartzite and phyllite
	Tarkwaian	Tarkwa Phyllite	Phyllite and chloritoid phyllite with subsidiary arenaceous beds
		Banket Series	Quartzites, grits, breccias and banket conglomerates. Three reefs are recognized: a lower main or Basal Reef, a Middle Reef and an upper Breccia Reef
		Kawere Group	Sandstones, quartzites, grits, breccias and conglomerates
	Great unconformity		
	Birrimian	Upper Birrimian	Volcanics (greenstones), pyroclastics, phyllite, greywacke and manganiferous phyllite, intruded by and granitized in places to granites and porphyries. Also intruded by epidiorites in places

Source: Junner *et al.* (1942).

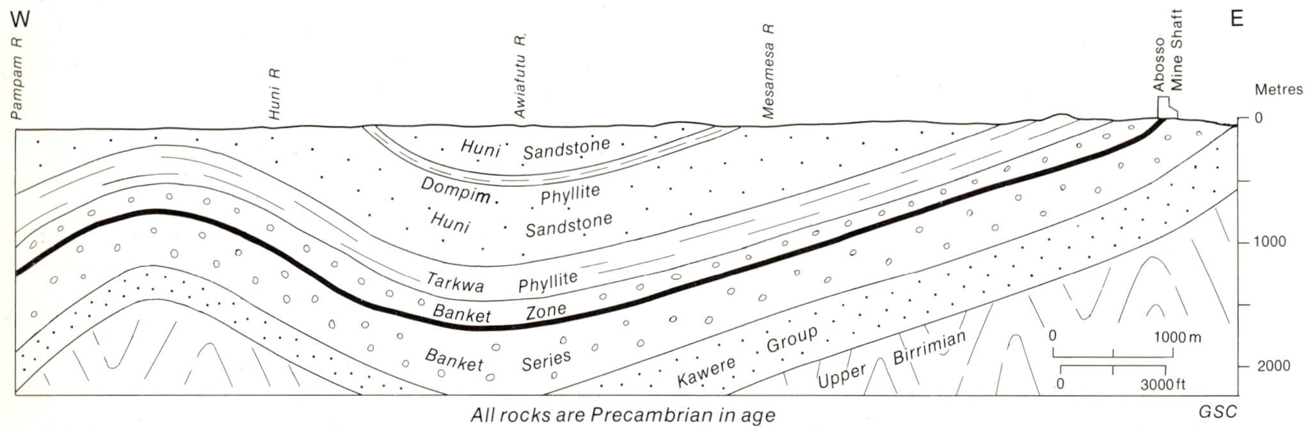

Figure 63. Generalized geological section across the Tarkwa goldfield, Ghana (after Hirst, 1938).

quartzites. The conglomerates are composed essentially of vein quartz pebbles and some quartzite and schist pebbles. The matrix of the conglomerate consists of quartz and black sand (mainly hematite admixed with ilmenite, magnetite and rutile) together with variable but minor amounts of sericite, chlorite, chloritoid, tourmaline, rutile, zircon, garnet and gold. Epidote, carbonate and pyrite are rare except near later dykes, faults and quartz veins. The black sand occurs as disseminated grains and aggregates of grains in the interstices between the pebbles of quartz and also in the form of bedded and crossbedded seams. Most investigators agree that the hematite and other constituents of the black sand are undoubtedly of detrital origin. The conglomerates and the associated quartzites are shallow water deposits deposited in a slowly sinking basin as witnessed by the presence of crossbedding, erosion channels and the occurrence of other features such as ripple marks in the Banket Series and in underlying and overlying sediments. Junner *et al.* (1942) state that the deposits are not lacustrine, glacial or aeolian deposits; they think they may have been either deltaic or fluvioglacial deposits.

The Banket Series and associated rocks have undergone regional metamorphism up to the chlorite-sericite stage. In addition a number of narrow quartz veins intersect the reefs, and small lenticular quartz veins are scattered throughout the Banket Series and other rocks of the Tarkwa System. In the mines the most common are those following footwall shears and veins with steep dips that traverse the reefs. The latter are commonly associated with dykes. Both types carry sulphides, mainly pyrite, chalcopyrite and galena. The gold values of the reefs are said to be unaffected by proximity to dykes or quartz veins.

The gold in the conglomerates is very fine grained and generally not visible to the naked eye. In general the gold values are highest and most persistent in well sorted and closely packed hematite-rich quartz conglomerates with a moderate thickness (1–6 ft). The gold is segregated into definite pay shoots separated by unprofitable intervals, and in the pay shoots the gold is segregated into definite zones. The wall-rock quartzites carry little or no gold, but some sporadic high assays are obtained where they contain an abundance of hematite. In several of the mines the gold is concentrated in the bottom foot of individual conglomerate beds. The bulk of the gold is intimately associated with the detrital hematite containing water-worn crystals of zircon and rutile. Individual particles of gold average about 60μ (0.06 mm) in diameter. This is approximately the size of the gold in alluvial placers in a number of Ghanian rivers according to Junner *et al.* (1942). The particles of gold as seen in polished and thin sections tend to be irregular and hackly, and they are commonly associated with a green talcose mineral. They are certainly not now in the form of placer nuggets or dust, a feature that indicates either remobilization of placer gold or deposition from hydrothermal solutions.

As with the Rand two general theories have been suggested to explain the occurrence of gold in the Tarkwa conglomerates, Junner *et al.* (1942) consider the gold to be of placer origin, later remobilized and recrystallized. Rozhkov (1967) agrees with this theory. Bray (1928), however, attributes the mineralization to ore-bearing solutions, but does not state from whence came the gold except to suggest a deep-seated source.

The arguments used in favour of each theory for the origin of the gold in the Tarkwa deposits are essentially the same as those for the Rand. Junner *et al.* (1942) have summarized the principal evidence for each theory as follows:

For the placer theory:

1. Gold occurs in varying amounts in the folded and metamorphosed Banket Series conglomerates throughout the Gold Coast. On the other hand it is rare in the underlying Kawere conglomerate, which is also of Tarkwaian age, and is also rare in quartz veins and igneous intrusions cutting the Tarkwaian.
2. Auriferous quartz veins and lodes are abundant in the Birrimian – the next oldest formation to the Tarkwaian – and there is abundant evidence that the Banket Series beds were largely derived from the Birrimian.
3. The gold in the veins and lodes in the Birrimian and also in the streams draining Birrimian rocks is of similar size to that in the banket conglomerates.
4. Although intrusive igneous rocks are abundant in the Tarkwaian in the Tarkwa goldfield and there is evidence of hydrothermal action accompanying them, the auriferous conglomerates are little altered. The detrital black sand in the banket and adjacent wall rocks has not been replaced by sulphides except in a very few places alongside intrusions and quartz veins, and carbonates and sulphides are absent or rare in these rocks. Elsewhere in the world carbonates and sulphides almost invariably accompany gold deposits of hydrothermal origin.
5. The distribution of the gold in the banket in the Tarkwa goldfield is clearly related to sedimentary features.
6. The fineness of the gold (950–1000) is consistently high and is appreciably greater than the average of the fineness of hydrothermal gold.

For the hydrothermal theory:

1. The gold is not water worn, nuggets are absent, and the gold is more finely divided than normal alluvial gold.
2. The occurrence of most of the visible gold and of some of the invisible gold appears to be controlled by minute fractures and shear planes, particularly along the footwall contacts of the conglomerates.
3. The presence of tourmaline, chlorite, sericite and secondary quartz, and rarely some carbonate and sulphides, in the conglomerates and in quartz veins and dykes cutting them.
4. Gold occurs only occasionally in the quartz pebbles.

A recent sedimentological study of the Tarkwaian gold deposits by Sestini (1970, 1973) suggested that rather than a plain with a very low gradient, the site of deposition was an east-sloping pediment surface formed by coalescing alluvial fans. Most of the gold was derived from primary quartz veins in the Birrimian, to the west of Tarkwa; a small contribution may have come from gold-quartz-tourmaline veins in the Birrimian east of Tarkwa.

Jacobina auriferous-uraniferous conglomerates, Brazil

The auriferous-uraniferous Jacobina conglomerates of Brazil occur about 170 mi northwest of Salvador, Bahia, Brazil. They are of Precambrian age and have been described in some detail by White (1956), Bateman (1958), Leo *et al.* (1964), Cox (1967), Lemos (1974), Robertson (1974) and Andrade Ramos and Fraenkel (1974). The generalized geological column for the Jacobina Series is shown in Table 64 and their distribution in Figure 64.

The Jacobina Series lies in a long narrow fault block lying within the predominant gneissic terrane of the Brazilian Shield. In general the sediments of the series strike north and dip steeply (55–80°). They are cut by a number of steep angle transverse faults, some with considerable throw. According to Cox (1967) the grade of metamorphism imposed on the sediments corresponds to the amphibolite facies of regional metamorphism.

The rocks of interest are the auriferous conglomerates, which occur as lenses and beds up to 1 m thick separated by thicker beds of quartzites. These lenses and beds exhibit scour and fill structures at their base, and crossbedding and more rarely ripple marks are prominent in the conglomerates and quartzites indicating shallow water conditions. The conglomerates are composed essentially of pebbles of vein quartz set in a matrix of quartz, sericite, fuchsite and chlorite, with small amounts of magnetite, tourmaline and zircon. Two types of conglomerates are recognized, Piritoso and Chabu. The former, as the name implies, is heavily mineralized with pyrite, the latter contains little pyrite, the quartz pebbles being cemented by an arenaceous matrix. The gold and uraninite are concentrated in the Piritoso type.

Gold has been won from a number of mines deriving ore from the Piritoso type of conglomerates. These include Serra Branca, north of Jacobina, and the Canavieiras, Morro do Vento and João Belo mines to the south (Fig. 64). The production from these mines is small, and only the Canavieiras is operating at present (1975), milling about 50 tons/day of ore. The grade is variable ranging from 0.25 to 0.45 oz Au/ton and the uranium (U_3O_8) content appears to range from 0.008 to 0.05 per cent. Thorium is present only in trace amounts.

Table 64. Generalized geological column, Jacobina Series, Serra de Jacobina, Brazil

	Formation	Thickness (m)	Lithology
			Ultramafic sills and dikes; amphibolite lenses and dykes in places
	Intrusive contact		
Middle Precambrian(?)	Cruz das Almas	2100(+)	Pelitic schists, micaceous quartzite and minor conglomerate
Middle Precambrian(?)	Rio do Ouro	2000–2300	Well bedded white to pale green quartzites with locally prominent ripple marks and crossbedding
Middle Precambrian(?)	Serro do Córrego	1800–2000	Interbedded quartzite and auriferous conglomerate
Middle Precambrian(?)	Bananeira	1000(+)	Massive coarse-grained pelitic schist; interbedded quartz muscovite schist and micaceous quartzite
	Lower Precambrian of Brazilian Shield (?)	?	Gneisses and granitic rocks, skarn, etc.

Source: Leo *et al.* (1964) and Cox (1967).

The ore-bearing Piritoso type of conglomerate contains from 2 to 5 per cent pyrite in the matrix of the conglomerate. There are in addition native gold, uraninite (pitchblende), chalcopyrite, pyrrhotite, sphalerite, ilmenite, molybdenite, rutile, brannerite and zircon (Ramdohr, 1958*b*; Gross, 1968). Thucholite is not mentioned as being specifically present in the Jacobina deposits, but in other similar deposits in Brazil (Belo Horizonte) small rounded balls of the mineraloid carry inclusions of uraninite (Robertson, 1974). Chromian muscovite is widespread in the quartzites and conglomerates, especially near the zone of the Jacobina fault (Leo *et al.*, 1965). According to Leo *et al.* (1965), the mineral is not of magmatic origin. Instead, the probable sources of the chromium are detrital chromite in original quartz sandstone beds, and chromium leached from ultramafic rocks by hydrothermal solutions in the Jacobina fault zone and introduced into adjacent quartzite layers. The gold is extremely finely divided varying in size from 0.05 to 0.1 mm. It is concentrated in the Piritoso conglomerates in ore grade levels but has a widespread distribution in nearly all conglomerates being present in average amounts of 0.015 oz/ton (0.5 ppm) according to Gross (1968). In the pyritic reefs the gold is included in pyrite grains, and minor amounts occur as thin sheets in fracture zones that cut the conglomerate layers.

The distribution of the gold shoots is of interest. White (1961) noted that the pyrite-gold-uraninite zone in the Morro do Vento followed a fracture that cut across all lithologic units. Similar features are apparent at the Serra Branca Mine according to Cox (1967). Gross (1968) mentions that there is no particular relationship between primary sedimentary structures such as size of pebbles, texture of pebbles (glassy or sugary), degree of sorting or rounding, or matrix-to-pebble ratio. He further states that in a number of beds in the Canavieiras Mine the gold is concentrated in the tops of the conglomerate beds. He does not think that the intrusive ultramafic rocks, which invade the reef horizons in places, have had any marked effect on the distribution of the gold. There is also no evidence to suggest that the dykes introduced the gold. Cox (1967), however, suggested that gold migration and deposition seems in places to have been controlled in part by the dykes of ultramafic rocks.

The same theories for the origin of gold at Jacobina have been suggested as for those in the Rand and Tarkwa, namely the hydrothermal and modified placer theories. White (1961), noting that the mineralization is crosscutting as mentioned above, considered that the gold and uraninite were emplaced by hydrothermal solutions introduced along a north-south fracture or fracture zone, apparently parallel to the bedding. This conclusion, he felt, was strengthened by the presence of quartz stringers and veins and the extensive sericitization and chloritization. Cox (1967) concluded that a placer origin for the deposits is probable but that there has been considerable redistribution of the minerals during metamorphism. Gross (1968) also suggested a modified placer origin for the auriferous conglomerate of the Canavieiras Mine. He quoted the following evidence in support of the theory:

1. Gold is found in the matrix of practically all the conglomerates in the mine area.
2. Economic concentrations of gold occur in a foreset system of cross-beds.

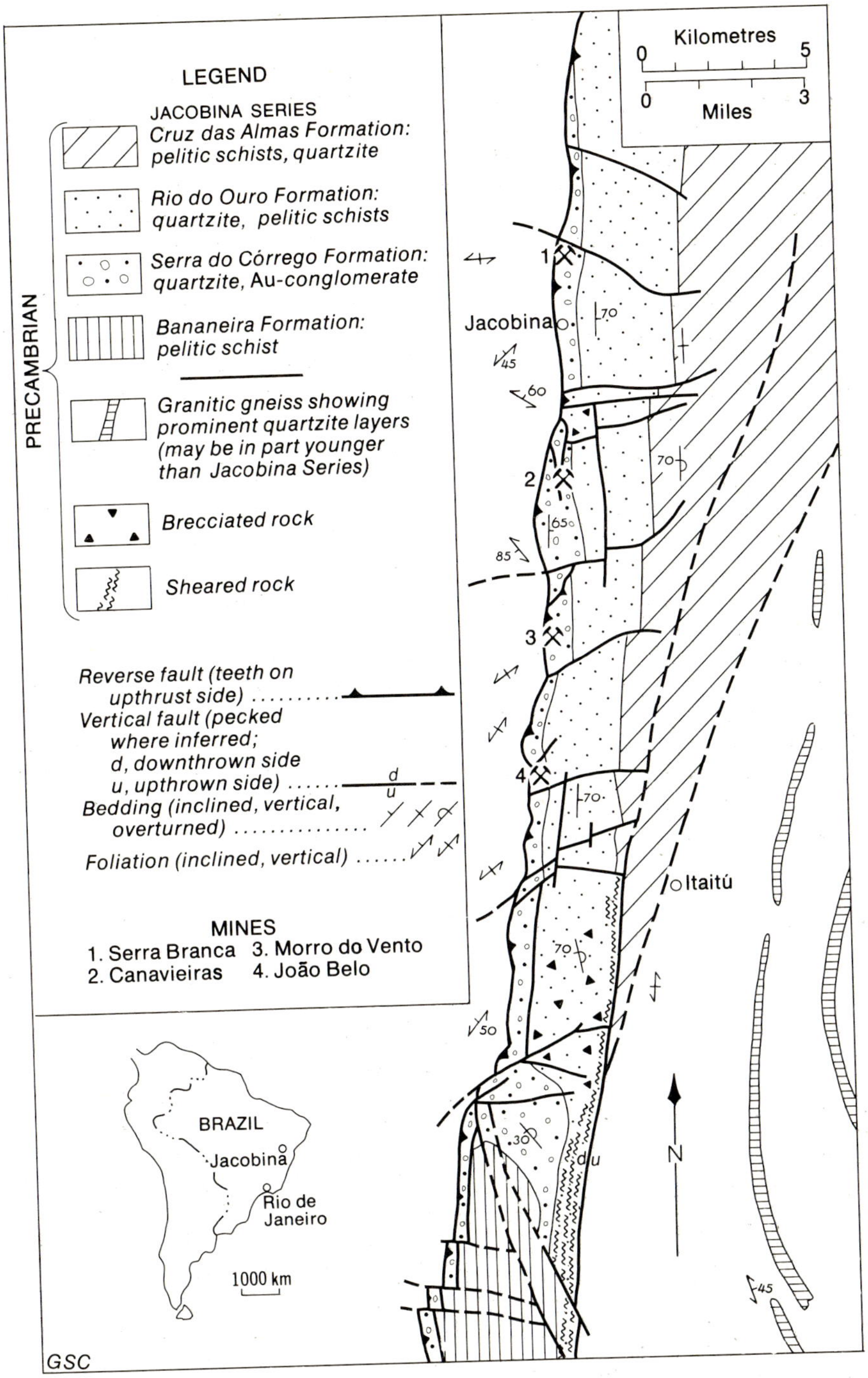

Figure 64. Generalized geological map of the southern part of the Serra de Jacobina, Brazil (after Cox, 1967).

Evidence suggesting that the placer gold was redistributed within the conglomerates during folding, metamorphism and intrusion is as follows:

1. Gold tends to concentrate towards the top of conglomerate beds.
2. Occasionally gold is found in tiny fractures that cut pebbles of the conglomerates.
3. Pitchblende occurs with high-grade gold ore.
4. Concentrations of gold occur near the contacts of ultramafic dyke intrusions.

Blind River–Elliot Lake uraniferous conglomerates
The Blind River–Elliot Lake conglomerates contain uranium, thorium and rare-earths (mainly yttrium earths). Gold values are erratic and of no consideration in the economic value of the ores. The deposits are briefly described solely for the purposes of comparison.

The Blind River–Elliot Lake area is north of Lake Huron some 95 mi west of Sudbury. The geology of the area and mineralization of the uraniferous belt has been described in detail by Collins (1925), Joubin and James (1957), Hart *et al.* (1957), Robertson and Steenland (1960), Lang *et al.* (1962), Pienaar (1963) and Roscoe (1969), all of whom also give extensive bibliographies.

The geological column for the Blind River-Elliot Lake district is shown in Table 65 and a generalized plan and section is shown in Figure 65. The rocks of interest are the Matinenda Formation at the base of the Proterozoic Huronian Supergroup. These contain all of the uraniferous orebodies.

Table 65. Generalized geological column, Blind River–Elliot Lake district, Ontario

Era	Group	Formation	Lithology
Precambrian – Proterozoic			Nipissing quartz diabase; granite; olivine diabase
	Intrusive contact		
Precambrian – Proterozoic – Huronian	Cobalt		Conglomerate, arkose, quartzite, greywacke and argillite
	Quirke Lake	Serpent	Subarkose, argillite and conglomerate
		Espanola	Limestone, siltstone, argillite, calcareous sandstone and conglomerate
		Bruce	Paraconglomerate, greywacke, subarkose
	Hough Lake	Mississagi	Subgreywacke, subarkose, siltstone, shale
		Pecors	Argillite, siltstone, greywacke and subarkose
		Ramsey Lake	Paraconglomerate, subarkose and greywacke
	Elliot Lake	McKim (300 ft)	Argillite, siltstone, greywacke and quartzite
		Matinenda (10–800 ft)	Quartzite, subarkose and uraniferous quartz pebble conglomerates; Residual palaeosoil in places
	Great uncomformity		
Precambrian – Archean	Basement complex		Volcanics (greenstones), cherty iron formation, granodiorite, pegmatites and other granitic rocks

Adapted from Table IV-4 *in* Douglas (1970).

The Matinenda Formation is composed essentially of oligomictic conglomerates interbedded with beds of quartzite and subarkose, all metamorphosed to the sericite facies. The conglomerates are composed mainly of well rounded vein quartz pebbles with a few pebbles of chert, feldspar and acidic volcanics in a matrix of quartz, feldspar and sericite. The conglomerate beds range in thickness from 1 in. to 10 ft. Crossbedding is common in the quartzites and conglomerate beds and scour channels are frequent. Locally the conglomerates and quartzites are intersected by irregular albitite dykes, and there is considerable albitization of the conglomerates in places, particularly at the Pronto Mine (Fig. 66). Many conglomerate beds are heavily mineralized with pyrite (av. 3–5%; max. 25%) and some carry economic quantities of uranium. There are in addition considerable concentrations of thorium and rare-earths (mainly yttrium earths). The average uranium content of the ores throughout the district is about 0.12 per cent U_3O_8, but some sections of conglomerates and quartzites are exceptionally rich containing percentages of U_3O_8 according to the writer's observations. These frequently occur where there is extensive crossbedding. The average thorium content appears to be about 0.05 per cent ThO_2 and the average rare-earth (mainly yttrium earths) content about 0.25 per cent. The gold content is exceptionally low averaging about 0.003 oz/ton (0.09 ppm) according to analyses done for the writer. The reserves of uranium and thorium ore in the Blind River–Elliot Lake district approach 300 000 000 tons averaging 0.1 per cent U_3O_8 and 0.05 per cent ThO_2. Enormous tonnages of grades at 0.01 to 0.05 per cent U_3O_8 and 0.005 per cent ThO_2 are available. The mineralogy of the ores has been extensively studied by Traill (1954), Nuffield (1954), Roscoe (1959), Roscoe and Steacy (1958) and Robertson and Steenland (1960). They list the following minerals as being present:

Pyrite	Pitchblende	Anatase	Grunerite
Pyrrhotite	Monazite	Rutile	Tourmaline
Chalcopyrite	Zircon	Chromite	Chlorite
Galena	Uranothorite	Spinel	Amphibole
Sphalerite	Coffinite	Epidote	Diopside
Molybdenite	Thucholite	Sericite	Apatite
Marcasite	Ilmenite	Biotite	Cassiterite
Cobaltite	Hematite	Chamosite	Fluorite
Brannerite	Magnetite	Garnet	Barite
Uraninite	Sphene	Greenalite	Carbonate

The most abundant metallic mineral is pyrite, which is disseminated in small grains and aggregates of grains in the matrix and around the borders of the quartz pebbles. The grains are anhedral to euhedral, and there is much spheroidal or 'buck-shot' pyrite. Some of the pyrite contains leucoxene cores suggestive of sulphidization of titaniferous magnetite. The principal uranium minerals are uraninite and brannerite. The brannerite tends to occur in the matrix in ovoid aggregates about 0.3 mm in diameter that also carry anatase or rutile. Uraninite is commonly present as anhedral to subhedral grains about 0.1 mm in diameter. It normally contains about 5 per cent ThO_2. In pyrite streaks in both conglomerates and quartzites there is usually considerable uraninite in aggregate form, and certain crossbedded parts of the beds may

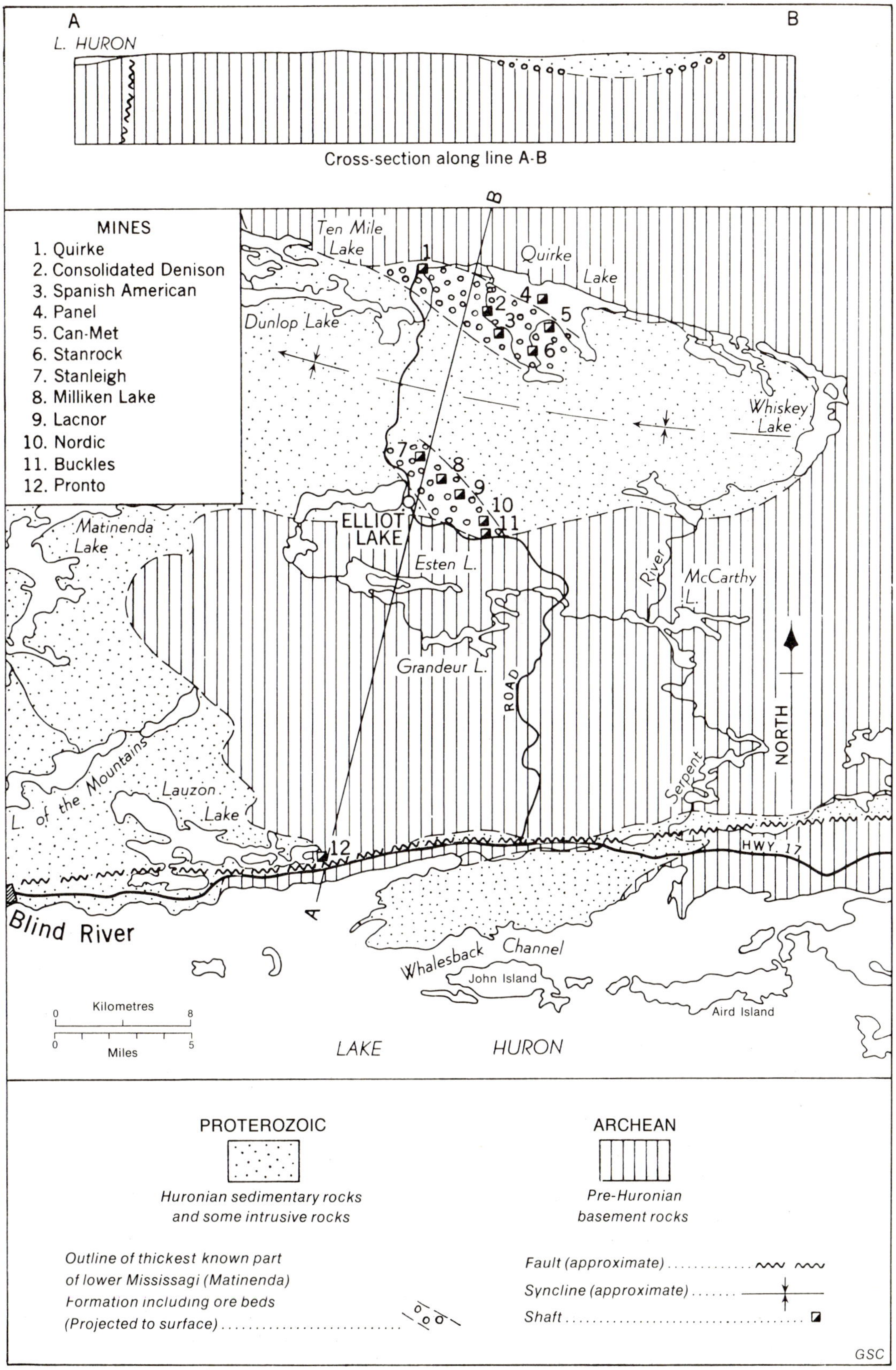

Figure 65. Diagrammatic geological plan and cross-section, Blind River–Elliot Lake area, Ontario (after Lang *et al.*, 1962).

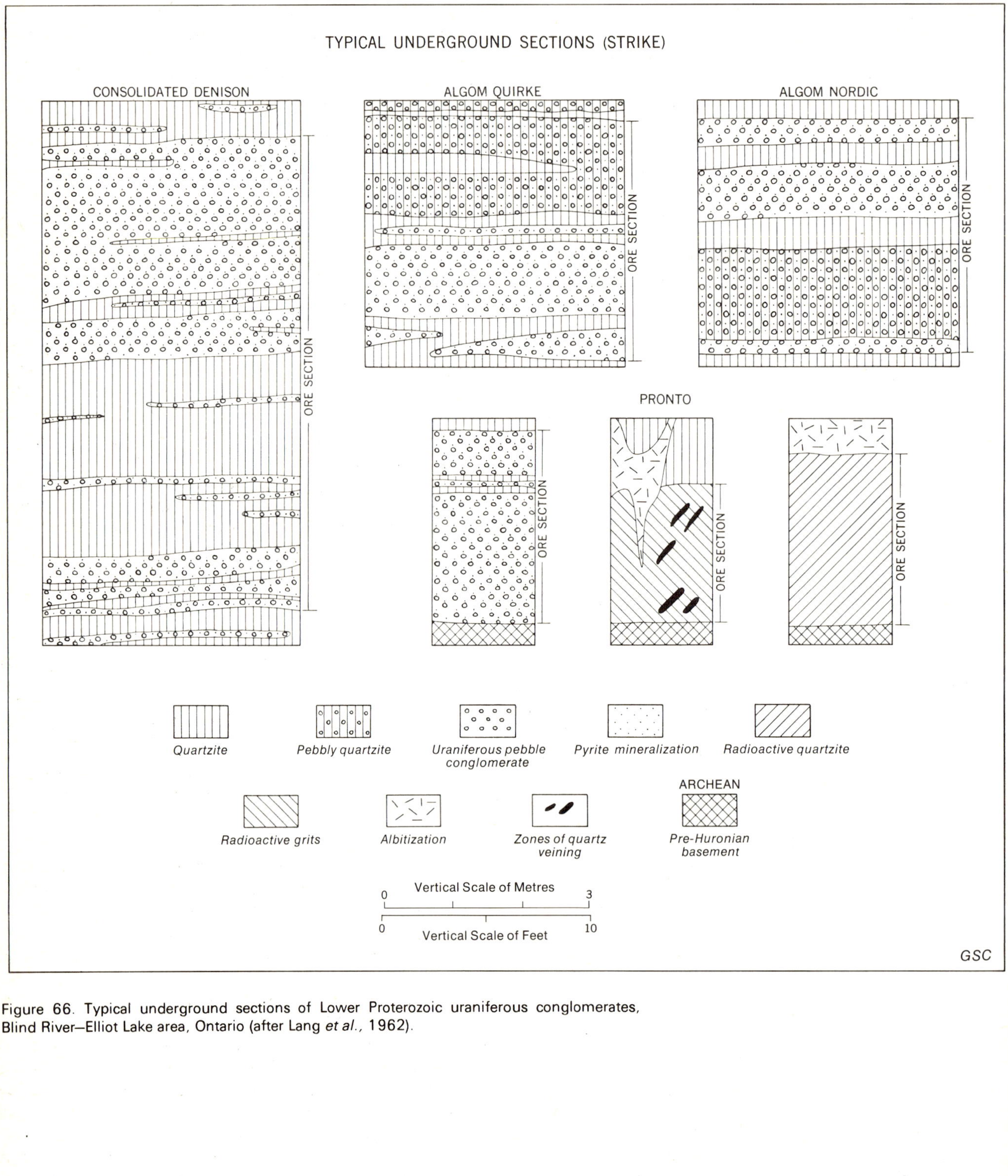

Figure 66. Typical underground sections of Lower Proterozoic uraniferous conglomerates, Blind River–Elliot Lake area, Ontario (after Lang *et al.*, 1962).

be made up of nearly massive aggregates of uraninite. There is commonly a higher content of uraninite in pyrite-rich parts of the beds. Thucholite was not seen in the matrix of the conglomerate or quartzite by the writer; it occurs, however, as warty aggregates along faults and fractures in places and as thin columnar seams with uraninite and pyrrhotite along bedding planes. Ruzicka and Steacy (1976) describe some of the latter occurrences in detail.

Gold and silver occur in the Blind River–Elliot Lake ores only in small amounts, the average gold content for most samples being 0.09 ppm (0.003 oz Au/ton). Locally, high assays can be obtained particularly where sulphides such as galena, sphalerite, cobaltite, etc. occur. Most of the gold and silver are in the pyrite and sulphide concentrates as indicated by assays of these materials in our laboratories. The average for the pyrite and mixed sulphide concentrates is about 0.64 ppm (0.019 oz Au/ton). No native gold was seen in these concentrates by the writer, but the metal has been reported as occurring sporadically in some of the ores.

The oreshoots in the Blind River–Elliot Lake mines are usually gently dipping tabular bodies composed of lenses and interfingering beds of pyritic quartz-pebble conglomerates and quartzites as shown in Figure 66. In most of the stopes the pyritic mineralization of both the conglomerates and quartzites commonly shows a wispy to banded appearance, and there is undoubted evidence of replacement of crossbedded features by pyrite and uraninite. One can also see seams of pyrite along bedding planes, veinlets of pyrite in the quartz pebbles in places and little trains or veinlets of uraninite (pitchblende) in the matrix of the conglomerates. All conglomerates and quartzites rich in uranium are generally always relatively rich in pyrite, but the converse is not necessarily true. Sericite-rich areas seem to be indicative of high uranium contents in most samples. Finally, in some areas quartz veining is evident. These and isolated fractures contain pyrite, chalcopyrite, pyrrhotite, galena, sphalerite and occasionally cobaltite. They are not radioactive.

On the question of origin the two general theories – hydrothermal and modified placer – common to these types of deposits prevail. The early workers (Joubin and James, 1957; Holmes, 1957) supported the modified placer theory and later workers have tended to agree (Roscoe and Steacy, 1958; Robertson and Steenland, 1960; Bain, 1960; Pienaar, 1963; Robertson, 1974). Davidson (1957, 1964–1965), however, maintains that the uraniferous ores are hydrothermal or due to downward leaching by saline waters as explained in the section on the Rand.

Other Precambrian occurrences of uraniferous and auriferous quartz-pebble conglomerates in Canada

Quartz-pebble conglomerates are developed in a number of other districts in the Canadian Shield; many contain uranium and thorium with minor amounts of gold, others are considerably enriched in gold as well as in uranium and thorium. East of the Blind River–Elliot Lake district, some 35 mi west of Sudbury, the basal Huronian rocks yield uranium from the Agnew Lake Mine. The pyritic oligomictic conglomerates comprising the ore horizons are steeply dipping and extend to depths of 3000 ft or more. The principal ore mineral is uranothorite, and the Th/U ratio in the ores is about 3:1 compared to 1:2 for the Elliot Lake deposits. Gold is found only in traces. Somewhat similar pyritic conglomerates and quartzites occur elsewhere along the base of the Huronian Supergroup west and east of Sudbury, one particular occurrence of note being present at the southwest corner of Lake Wanapitei. This conglomerate contains essentially monazite and leucoxene as the radioactive minerals; gold occurs only in traces.

Uraniferous occurrences in the Montgomery Lake–Padlei area of Northwest Territories are in Aphebian pyritic oligomictic conglomerates in the Montgomery Lake Group, which underlies the predominantly clastic Hurwitz Group and rests on rocks of Archean age. The occurrences contain abundant pyrite with uranothorite, zircon, leucoxene and notable amounts of gold (up to 0.1 oz Au/ton).

Occurrences of highly deformed pyritic uraniferous conglomerate, probably of Aphebian age, and belonging to the Sakami Formation, lie on Archean greenstones and greywackes on the west side of Sakami Lake in northern Quebec. Locally they carry in excess of 10 per cent pyrite and contain disseminated uraninite (with 10% ThO_2) and a thorium silicate, either thorite or allanite (Robertson, 1974). Gold occurs only in traces as far as is now known. Other basal quartz-pebble conglomerate lenses occur in quartzites and arkoses in the Papaskwasati Basin north of Lake Mistassini in Quebec. These belong to the Papaskwasati Formation of the Mistassini Group of Aphebian age, contain minor pyrite, are radioactive and have received some attention. Their gold content as far as is now known is low.

> The large nuggets found in the drift are simply the reliquiae of the chief masses of gold which once occupied the uppermost parts of the reefs, and that like the blocks of many an ancient conglomerate, they have been swept from the hilltops into adjacent valleys by former great rushes of water.
>
> –Murchison, Siluria, 5th ed., 1872

Placer deposits

Placer deposits provided early man with the first samples of gold and since that time have accounted for a large production of the metal. If we include the Witwatersrand and other quartz-pebble conglomerates as fossil placers or modified fossil placers, then the placer type of deposit has provided more than two-thirds of man's store of gold.

Before proceeding further certain terms with respect to placers should be defined.

The term 'placer' is evidently of Spanish derivation and was used by the early Spanish miners in both North and South America as a name for gold deposits found in the sands and gravels of streams. Originally, it seems to have meant 'sand bank' or 'a place in a stream where gold was deposited'. While many other terms have been coined for deposits in weathered residuum and alluvium none is quite as succinct and expressive as 'placer'.

The terminology of the zone or stratum containing an economic concentration of gold in eluvial and alluvial placers is varied. We shall use the miner's term 'pay streak', which is commonly used in Canada and the United States. Other

English terms in use include 'pay gravel', 'pay sand', 'pay dirt', 'pay wash', 'pay channel', 'pay lead', 'run of gold', 'gutter' and 'wash dirt'. There are a host of Spanish terms used mainly in Mexico and South America, and there are also miner's terms for the rich gold zones in practically all languages where eluvial and alluvial placers are worked. Rickard, many years ago, advocated the use of "gold-bearing channel" for all these terms, but it has not been generally accepted.

The tenor of pay streaks or of placer gold gravels and sands, in general, is referred to by the value (in ounces, pennyweights, or in any unit of currency) per cubic yard, per running foot of channel or per square foot of surface; also occasionally in bonanzas by dollars or some other unit of currency per pan. It is of interest to note that placer deposits can be worked whose gold content is as low as 0.1 ppm.

The pay streaks of placer deposits may rest on or near bedrock or on some stratum above bedrock. The bedrock in placer deposits is commonly referred to as the 'true bottom', although the term is little used today. When the streaks rest on a well-defined stratum of sand, gravel, shingle or clay above the bedrock they are said to be on a 'false bottom'.

Placers have been variously categorized, but here we shall use a simple nomenclature based upon whether the placers are formed by concentration of gold in situ over or in the immediate vicinity of primary deposits, namely 'residual' or 'eluvial placers', or by agencies that have concentrated the gold in the near vicinity or at some distance from the primary source. In the latter category we recognize 'alluvial', 'beach' and 'aeolian placers'. The terms 'saprolite' or 'saprolitic placer' were formerly used for certain types of eluvial placers, mainly in the eastern United States.

Eluvial, alluvial, beach and aeolian placers may become buried after their formation and are sometimes referred to as 'buried placers'. These placers may be buried under (1) volcanic deposits as in California and Australia; (2) glacial deposits as in Canada and U.S.S.R.; (3) talus and other slope deposits; (4) aeolian deposits as in Australia; (5) alluvial sands and gravels; and (6) marine and lacustrine deposits.

There are fossil (lithified) equivalents of eluvial, alluvial, aeolian and beach placers. These are described subsequently.

The gold in auriferous placers may come from one or more of the following sources:

1. Auriferous quartz veins and other types of gold-bearing deposits mentioned in the section on epigenetic deposits.

2. Auriferous sulphide impregnation zones, porphyry copper deposits, etc.

3. Auriferous polymetallic deposits.

4. Slightly auriferous quartz stringers, blows and veins in schists, gneisses and various other rocks.

5. Various slightly auriferous minerals such as pyrite and other sulphides in graphitic schists and other rocks.

6. Slightly auriferous conglomerates, quartzites and other rocks.

7. Old placers.

It should be noted that the geological history of productive placers is frequently complex, much more so than the sequence: primary vein or lode source → eluvial placer → alluvial placer. Often an intermediate collector of gold is involved, mainly auriferous conglomerates, quartzites, etc. Fayzullin (1968) emphasizes this point in his papers. He recognizes a number of variants in the lode-placer sequence as follows: (a) lode → deluvial placer → interceptor → alluvial placer, (b) lode → interceptor → alluvial placer, and (c) interceptor → alluvial placer.

The primary agency that produces gold placers is weathering; a process that involves numerous complex chemical reactions that are explained in detail in Chapter IV. Three things may happen to gold in primary deposits: (1) the gangue minerals may be disintegrated and leached away, leaving the gold relatively untouched; the gold may remain in situ in the oxidized zones or pass into eluvial and alluvial placers; (2) the gold may be dissolved and carried far away from the deposits in which case no placers are formed; or (3) the dissolved gold may be wholly or partly reprecipitated on nuclei of gold in the residuum or on similar nuclei as they are moved along in the alluvium of streams, rivers, beaches, etc. The last process is largely responsible for nuggets. The formation of placers is, therefore, a combination of both mechanical and chemical processes working in consort over long periods of time. We shall return to this theme later in the present section.

The mechanical agencies that assist in the transport and winnowing of gold into placers are gravity, the running water of streams and rivers, the agitation of waves along the shores of lakes, seas and oceans, the wind, and glaciers. Gravity is a force that operates in the formation of all types of placers; it is the principal agent that concentrates gold in eluvial placers. The other agents give rise respectively to the following types of placers: stream and river (alluvial) placers, deltaic placers, beach placers and aeolian placers. There are few authenticated aeolian gold placers of any size, and we shall not consider them further except to note that near the outcrops of some of the primary deposits in Australia the wind had blown away the finer detritus exposing coarser material in which gold was enriched in places (Hoover, 1899, Rickard, 1899). Auriferous sand dunes have been recorded by Spurr (1906) in the Silver Peak quadrangle, Nevada. These apparently contain only traces of gold and silver. Placers due entirely to glaciation are, likewise, uncommon, and none of economic value are known to this writer in moraines, kames or eskers. Certain outwash plains, where worked by glacial and postglacial streams may, however, contain low grade placers in some glacial terrains. Winnowing of gold by present-day streams and rivers from glacial sands and till and working of these materials along certain lakes has also led to minor concentrations of gold in many places in Canada and other glaciated countries. All of these types of glacial deposits are mentioned briefly in the alluvial category.

The minerals concentrated in placers include two types – those with low to medium specific gravity (the light minerals) and those with a medium to high specific gravity (the heavy minerals). All of these minerals have three features in common – great physical resistance to mechanical abrasion and comminution, great chemical resistance to solution in surficial waters and a general equidimensional character in their form. The last is of importance since flaky minerals such as molybdenite, scaly gold and specularite are difficult to concentrate in spite of their high specific gravity. Another factor is the degree to which a mineral can exist in a subdivided form and yet remain relatively chemically stable. This is a factor with respect to gold since the metal can exist in almost infinitely subdivided form giving the so-called 'flour', 'float', 'flood' or 'skim' gold. Such gold does not sink readily in sand and gravel and hence is not concentrated to any extent in placers; on the contrary it may be transported hundreds of miles in running water to be deposited over broad deltas or on the floors of lakes and oceans where it is commonly disseminated throughout great thicknesses of sediment. This particular feature is of importance in discussing the origin of the very finely divided gold of deposits such as the Witwatersrand. (*See* the final section of this chapter on the origin of deposits.) Commonly much flour gold accumulates temporarily in streams and rivers on bars on the inside of curves or meanders or in other areas of slack water ('skim bars'). Unless these are recognized they may give an inflated impression of the potential of a placer stream or river. Skim bars are also misleading during geochemical prospecting within a restricted area since they give false anomalies. This aspect is discussed further in Chapter V.

'Moss gold' is flour gold trapped by mosses and other plants along river and stream beds during high water. 'Moss miners' have collected this gold for centuries along the auriferous rivers throughout the world. It is easily recovered by burning the moss.

With skim and moss gold a new 'crop' appears each year after high water in auriferous water courses. This is partly the reason for the adage that 'gold grows in placers'. It is, however, probably not the whole story as will be seen from a later discussion.

The most common minerals with low to medium specific gravity in placers are quartz (S.G.=2.65), muscovite, amphibole, pyroxenes, tourmaline, garnet, diamond, chromite, rutile, barite, corundum, wad, limonite and zircon (S.G.=4.5); those with medium to high specific gravities in placers include monazite (S.G.=5), magnetite, ilmenite, cassiterite, wolframite, scheelite, cinnabar, gold and platinum (S.G.=22 when pure). A great variety of sulphides and sulphosalts may accompany gold in placers. This writer has seen pyrite (often abundant), galena, sphalerite, arsenopyrite, boulangerite and jamesonite in placers in Yukon Territory and elsewhere. Native bismuth is present in some placers; native mercury in others; and more rarely native arsenic, native silver, arquerite (Ag-Hg amalgam), native copper, native lead (not shot), native zinc, cinnabar, realgar, sperrylite, molybdenite, chalcopyrite, hematite, carbonates, feldspars, kyanite, topaz, spinel, allanite, epidote, sphene, tantalite-columbite and apatite. Some gold placers contain diamonds and other gems such as rubies, sapphires, emeralds, topaz, garnets, etc. Commonly the gems are secondary in value to gold, and in some placers they are so sporadic as to be a curiosity. Certain placer sands and gravels are cemented or coated by limonite, wad or mixtures of these mineral aggregates. These aggregates commonly contain much silica, alumina, titania and other hydrolysate oxides and frequently large amounts of humic material. Placer gravels or sands strongly cemented by limonite and other iron oxides are often referred to as 'cement gravels' in English speaking countries or 'cangalli' in South American countries. A few placer sands and gravels have been cemented by carbonates, silica or clay, and most placers in the permafrost of Canada, Alaska and U.S.S.R. are frozen solid, requiring a complicated procedure to thaw them out in places.

The gold of placers has varied characteristics. The most common habit is as dust that comprises particles ranging from specks the size of the tip of a needle or less to those the size of a flaxseed (in quantitative terms, <0.1–2 mm). Also common are small scales and spangles; less common are nuggets, crystals, wires, leaves, tufts and hairs and arborescent, reticulated, dendritic, filiform, mossy and spongy forms. Extremely finely divided gold (flour, float or skim gold), the bane of the placer miner and geochemist, is common in some placer districts. Some of the largest natural specimens of the metal have come from placers, frequently of the eluvial type. From these have come some of the largest nuggets of gold won by man. The 'Welcome Stranger' found by accident in a cart rut in the eluvium near Ballarat, Victoria, Australia weighed 2516 oz troy and the 'Blanch Barkley' also from Victoria weighed 1743 oz. The Carson Hill nugget found in California topped the scales at 1296 oz. More generally the gold of placers occurs in small flattened scales or grains averaging a few millimetres in diameter or as moderately finely divided particles known as dust averaging a few tenths of a millimetre in diameter. The fineness of placer gold varies greatly and has been studied by many investigators. A survey of the literature indicates that for most placer deposits the fineness ranges from 500 to 999, that is from electrum to nearly pure metal; most placer gold is above 850 fine. The other elements in placer gold are mainly silver, copper and iron (*see also* Chapter II). Vein gold generally has a fineness that ranges from about 500 to 900. There are innumerable references in the literature that refer to the fact that in any particular district the fineness of the placer gold will usually be higher than that in the veins from which the metal presumably came. Lindgren (1933) for instance states that in California the vein gold averages 850 fine whereas the transported placer gold in the Tertiary channels averages 930 to 950. Similarly, the placer gold of Manhattan, Nevada is finer than the lode gold (Ferguson, 1917). In the Carolinas and Georgia the placers and eluvial deposits commonly have gold with a fineness greater than 900 whereas the vein gold is generally much less than this value, in places being as low as 500. Distance travelled and size of the placer gold also seem to be factors in the fineness, the further from the source and the smaller the size of the particles the higher the fineness. These features are difficult to assess in the writer's experience, but they are evidently true according to a number of placer miners who were consulted on the matter. Certainly, flour gold generally has a high fineness. Hill (1915) and Hite (1933*a,b*) gave an average of about 950 as the fineness of the Snake River flour gold. There is much evidence to suggest that most of the gold in gossans and oxidized zones is of greater fineness than that in the primary ores. Fisher (1945), who made a worldwide study of the fineness of gold in various environments, concluded that gold in the oxidized zone is nearly always higher in grade than the primary gold, and this difference varies largely according to chemical conditions and the facilities that exist for removal of the silver in solution, the gold being redeposited close by, or merely left enriched by the removal of the silver. Under conditions suitable for taking gold into solution the gold redeposited, if the concentration of silver remains high, will not necessarily be of high fineness. He went on to state that very high gold fineness is usually the result of oxidation under conditions favourable for the complete removal of silver, e.g., Mt. Morgan, Queensland, and that high fineness is often associated with the oxidation of telluride ores, e.g., Fiji and Cripple Creek, Colorado. Finally, he noted that alluvial gold, being derived from the oxidized ore, is of higher quality (fineness) than the average of the vein gold, and shows an increase in value downstream, due to surface refining action, as the size of the grains decreases.

In Yukon Territory and at Yellowknife, Northwest Territories some of the supergene gold examined by the writer is of great fineness (990), whereas the primary gold is in the range 900 to 950. Another feature of placer gold, first examined by McConnell (1907) in the Klondike, is that the outer parts of nuggets commonly have a higher fineness compared with the inner parts. The writer has examined some of McConnell's samples and others from a number of placers by electron probe, and found that the rim of greater fineness is rarely more than 0.03 mm (30 μ). A rim of relatively pure gold on nuggets seems to be rather general since Johnston and Uglow (1926) found the effect on the gold of the Cariboo of British Columbia; Fisher (1935) in the Morobe goldfield, New Guinea; and Petrovskaya and Fastalovich (1955) in much of the placer gold of U.S.S.R. Ramdohr (1965) also noted the phenomenon in Rheingold; Mustart (1965) in much of the placer gold of Yukon; and Desborough *et al.* (1970) in placer gold from various parts of the United States. Stumpfl and Clark (1965), however, rightly point out that the rind of purer gold on nuggets, flakes and dust is not always present. They noted its absence in gold-platinoid concentrates from placers in southeast Borneo. The rim effect has generally been interpreted as the result of solution of silver, but the writer is of the opinion that it represents the precipitation of gold (and some silver) on gold nuclei. An argument in favour of the latter process is given in a later section on the origin of placers.

It should be noted that the rim effect in placer gold evidently begins during oxidation of the primary deposits, because it is found in gold in the oxidized zones of auriferous deposits and in eluvial deposits. Gold with enriched rims has been observed by a number of investigators in near-surface shallow deposits (mainly Tertiary) in a number of places, and the phenomenon is noticeable in some kurokô ores in Japan. Enriched rims are rare in deep-seated gold deposits in the writer's experience. In the oxidized ores and placer deposits the rims are generally much wider and better developed than in the primary deposits.

Placer gold commonly contains many inclusions usually small in size (0.005–0.50mm). The inclusions as seen in polished sections and under the microprobe include various sulphides, particularly pyrite, galena and chalcopyrite; also arsenopyrite, various sulphosalts, tellurides and quartz, sericite, rutile, etc., most of the minerals normally found in gold deposits. Many of these minerals are probably original constituents of the gold nuggets since they are commonly near the core of the gold particles. Some often appear as though they were the nuclei about which gold precipitated in the oxidized zones of the deposits. These inclusions should receive detailed study during geochemical prospecting surveys since one might be able to divine the exact source of the gold, especially if material from a few primary sources in the district under investigation is available for control purposes.

Other internal features of placer gold have been extensively studied. In the classic paper by Fisher (1935) the high grade rims referred to above and the general granular structure of the nuggets and flakes are mentioned. Petrovskaya and Fastalovich (1955), who have carried out a long term study of the placer gold of U.S.S.R., also note the nearly universal occurrence of high grade rims and reentrants on gold nuggets and flakes and found that these have a fine-grained polyhedral texture rather than a laminated one as thought by the earlier investigators. They also noted that much of the placer gold retains the internal granular structure and other microscopic features of the primary deposits of gold, but found that much placer gold, especially that held in placers for long periods of time, exhibits marked internal deformation and recrystallization textures. They also observed the presence of intergranular stringers of gold in nuggets and flakes. These were found in the placer gold and in the gold from oxidized zones of auriferous deposits but not in the gold from the primary ores.

The external form, colour, lustre and other features of placer gold visible to the naked eye are commonly characteristic. The miners and bankers in a placer district can generally tell from which creek the gold originated simply by visual inspection. It is said that one banker in Dawson in its heyday could tell precisely from which of the many creeks and gulches of the Klondike a specimen of gold originated, and I am inclined to believe the story having seen gold from all of the creeks either in the old gold room of Yukon Consolidated, Ltd. or in the National Mineral Collection of Canada.

Placer gold usually has an entirely different appearance from that found in veins and other deposits. The high lustre of the vein gold is replaced by a subdued lustre in the placer, due it appears to incipient crystallization on the surface. Some varieties of gold are coloured due to a number of reasons. Black and deep brown gold are commonly due to the presence of coatings of manganese and iron oxides or manganese and iron humates. Alternatively some black and brownish gold is due to a coating of very finely divided colloidal or microcrystalline gold. Whitish and dull greyish varieties of gold owe their colour generally to thin coatings of calcium carbonate, colloidal silica or fine-grained (colloidal) clay. In most placers the gold has the characteristic golden colour, but in others foreign coatings are quite common. 'Black gold' (ouro preto) is very characteristic of a number of placer districts in Brazil and other tropical areas.

The writer has either seen, heard about or read about practically every form of gold in placers and the closely associated oxidized zones of gold deposits. The most common forms are scales, plates and nuggets. In addition some placers contain crystals of gold, hopper-shaped crystallized particles, masses having filiform, reticulated and dendritic shapes and films, wire and mossy gold. The growth forms of large nuggets are commonly fairly regular, many looking like nuts or potatoes; other nuggets are highly irregular and gnarled in shape. Some partly enclose vein quartz fragments or rounded quartz and other types of pebbles and mineral fragments. Some nuggets have unusual growth proturberances that may appear as distorted crystals. A few nuggets have been found that have secondary crystalline outlines.

Various terms are used in the following descriptions to give a semi-quantitative estimation of the size of the gold particles in placers. A 'colour' has no exact meaning; it is used by placer miners to refer to a small piece of gold that usually varies from 1/16 in. (1.5mm) upward. Estimation of the value of gold colours in a pan and hence of the value of placer grounds can only be gained by experience since the value varies with the size of the colours, their thickness, and their purity. A good estimate can be made by picking out the colours from a number of pans, weighting them on a small pocket microbalance, and calculating the value per yard. There are about 50 pans (18 in. diameter) per yard. Coarse gold (nuggets) is usually considered to be that which remains on a 10-mesh screen (>1.5mm); medium gold will pass a 10-mesh screen and be held on a 20-mesh screen; fine gold passes a 20-mesh screen and is held on a 40-mesh screen (value 1 colour per cent); and very fine gold passes a 40-mesh screen. Flour gold is very much finer than the last; in some places 500 to 1000 colours or more are required to equal the value of a cent!

In general it can be said that the farther that gold has travelled from its source the smaller in size the particles become. Near the source the gold particles are usually rough and relatively large; with distance the particles become smooth, much rounded and smaller. The particles also apparently lose mass with distance from the source. Antweiler and Lindsey (U.S. Geol. Surv. Circ. 622, 1969*b*, p. 6) noted that particles of the Snake River gold, near its source below the outcrops of auriferous conglomerates in Jackson Hole, with diameters approximating 0.125 mm, averaged 10μg in weight, whereas near Twin Falls, Idaho, some 180 mi downstream, particles of the same diameter weighed only half as much. The cause of the mass loss was not discussed.

To summarize this introduction we can say that for the development of placers of any type, four requisites are necessary:

1. The occurrence of gold in deposits, in widespread quartz veins and blows or in a disseminated form in pyritic shales or other country rocks.

2. A fairly long period of deep secular chemical and mechanical weathering on a surface of submature to mature topography, during which time the gold is set free from the deposits or country rocks.

3. Concentration of the gold by some agency, generally water.

4. Absence of extensive glaciation: glaciation does not entirely preclude the occurrence of placers since both eluvial and alluvial types of placers may be overriden and little disturbed by the glaciers in some cases and buried by their deposits of till, clay, etc. in others. Such placers occur in British Columbia, Alaska, Yukon, Quebec and the Lena district of U.S.S.R.

Country kindly to the occurrence of extensive placers is readily recognized. The topography is subdued and marked by broad, often terraced entrenched valleys and rounded deeply weathered hills commonly with nearly accordant summit levels. Few extensive placers are found in terrains marked by sharp alpine features and high gradient V-shaped valleys; similarly, excessively flat terrains far from mountain systems and their foothills yield few productive placers. Like all generalities in geomorphology and geology there are some exceptions to these observations.

The literature on placers is voluminous. Works containing discussions of general principles include those by Liversidge (1893*c*), Weatherbe (1907), Crane (1908), Tyrrell (1912), Longridge (1913), MacKay (1921), Raeburn and Milner (1927), Park (1927), Idriess (1933), Boericke (1933), Lindgren (1933), Gardner and Johnson (1934–1935), Fisher (1935), Crampton (1937), Bilibin (1938), Mertie (1940), Averill (1946), Grutterink (1950), Bateman (1950), Shilo (1956, 1960), Gorbunov (1959), Griffith (1960), Harrison (1962), Vasyunina (1963), Barkovskaya (1963), Trofimov (1964), Fayzullin (1968), Ivenson *et al.* (1969), Shilo and Shumilov (1970, 1976), Valpeter and Davidenko (1970), Romanowitz *et al.* (1970), Volarovich and Shokhor (1970), Kartashov (1971), West (1971), Kazakevich (1972, 1976), Sigov *et al.* (1972), Wells (1973), Denisov (1973), Tayurskii (1973), Kogen *et al.* (1974), Kushnarev and Kashcheev (1974), Tishchenko and Tishchenko (1974), Segerstrom and Ryberg (1974), McLellan *et al.* (1974), Saks (1974, 1975), Vorob'yev and Krapivner (1975), Clift (1975), Prokuronov (1975), Smirnov (1976), Kisterov (1976), Fricker (1976) and Zhelnin and Travin (1976).

The references by Gardner and Johnson (1934–1935), Romanowitz *et al.* (1970), West (1971), Wells (1973) and Fricker (1976) contain up-to-date summaries of the procedures for the prospecting, sampling, evaluation and working of placers. It need hardly be emphasized here that the processes involved in the concentration of gold in placers are extremely complex, and the conditions that control them so variable that even under the most favourable circumstances observable, it is generally not possible to estimate the potential of a placer even roughly without detailed sampling, either by pitting, test holes, or deep cuts or trenches.

The distribution of the principal eluvial and alluvial auriferous placer districts of the world is shown in Figure 67. Many of the districts have long been exhausted, some before the beginning of the Christian Era.

Papers and other works describing placer areas in various parts of the world and methods whereby they are worked are included in the Selected bibliography.

Eluvial placers

Eluvial placers[27] are formed in the weathered residuum over or in the immediate vicinity of primary gold deposits of all types. Commonly these placers develop downhill from the outcrop of the primary deposits (Fig. 68). Some are of the nature of talus deposits or fanglomerate facies on mountain slopes or in the debris at the foots of hills and mountains. An unusual type is formed in the depressions of deeply and irregularly weathered terrane (karst topography). Eluvial placers are closely associated with alluvial placers in many districts, the former giving birth to the latter in many places.

The principal mechanism for the concentration of heavy minerals in eluvial placers is the winnowing action of gravity and downhill creep, the latter being essentially dependent on

[27]Some geologists, particularly in U.S.S.R., classify these placers as eluvial, deluvial and proluvial. By this classification eluvial placers are those whose outlines coincide more or less with those of the primary deposits. Deluvial (scree or talus) placers are those whose upper limit is at or near the primary source and whose downhill front lies at the foot of a slope. Proluvial placers form in the disintegrated debris at the foot of hills or mountains.

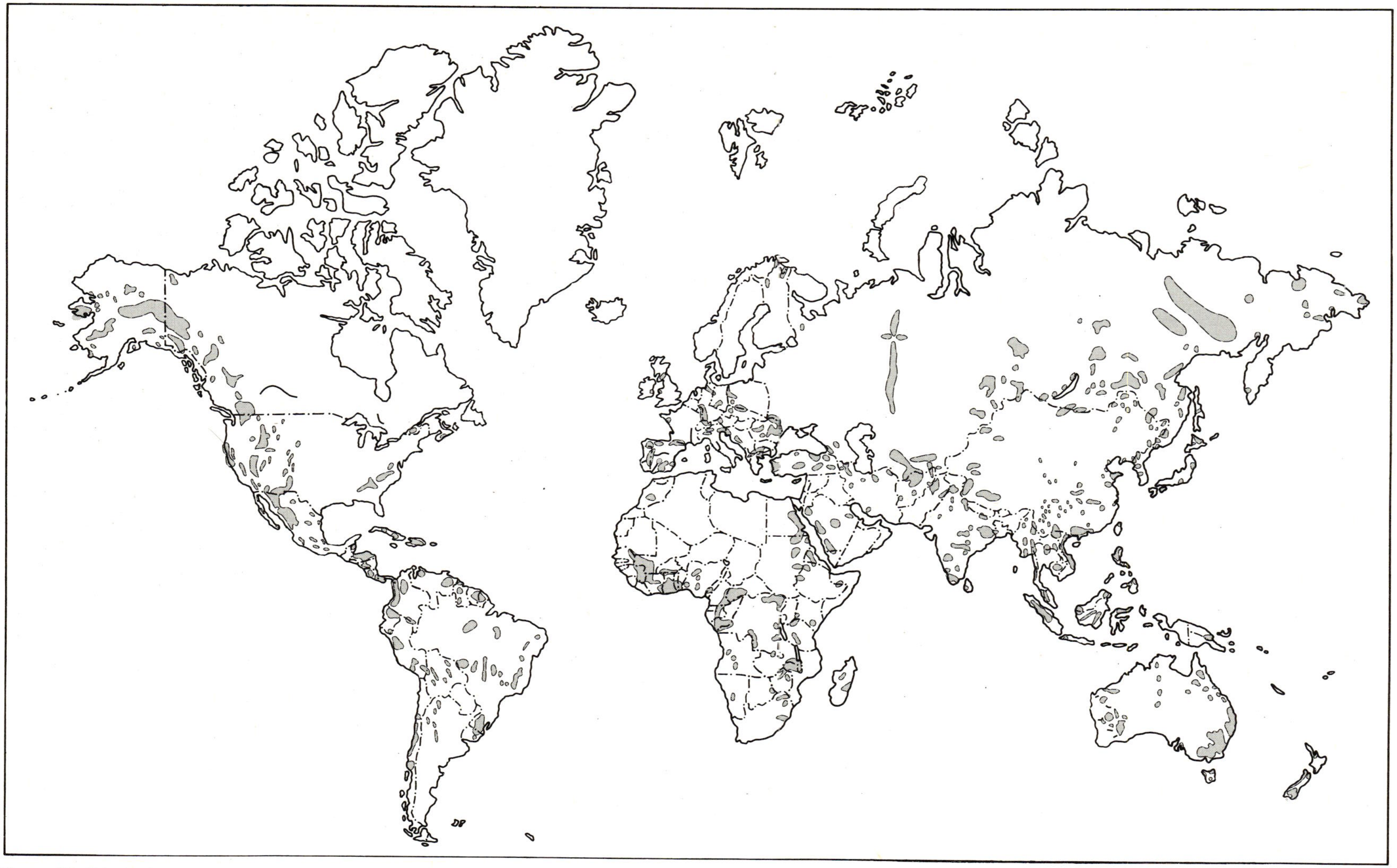

Figure 67. Principal unconsolidated (nonlithified) eluvial and alluvial auriferous placer districts of the world. (Many of the districts shown have long been exhausted.)

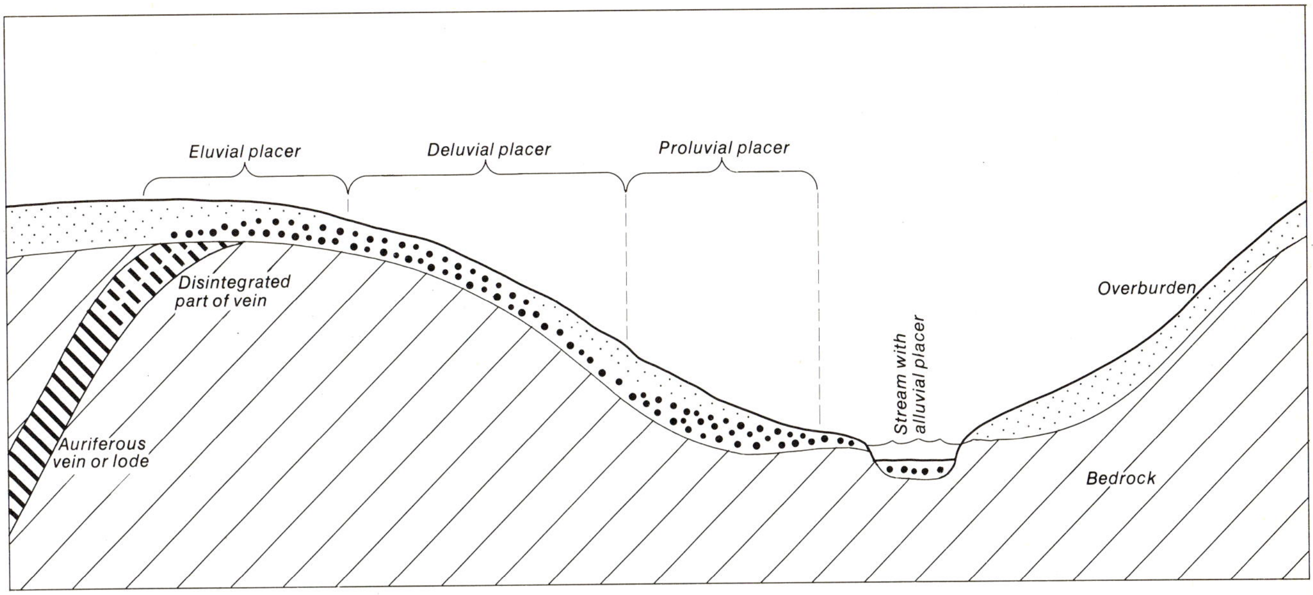

Figure 68. Sketch illustrating outcrop of a gold-quartz vein supplying material to form eluvial and alluvial placers. By one terminology, an eluvial placer embraces all materials not transported by streams. By another, only the placer materials over weathered and disintegrated deposits constitute an eluvial placer; downslope and other placer materials are classified as deluvial and proluvial as shown.

the angle of slope or gradient where the placers are formed on the sides of hills or mountains. Secondary factors include the thickness of the slope materials (scree, talus, residuum), the size and specific gravity of the weathered particles in the residuum, the coefficients of thermal expansion and contraction of residual particles, the coefficient of friction, the movement of ice and snow (glaciers) and the annual and daily variation of temperature. The last governs the freezing and thawing of the residuum in temperate zones, and is particularly important in permafrost regions where solifluction is widespread. Continuous downhill creep appears to be the main way by which gold reaches gulches and creeks. Moving water plays only a small part in the concentration of heavy minerals in eluvial placers, and it is common to find that these placers are largely independent of water courses either in the detritus or on the surface.

Auriferous eluvial placers commonly exhibit considerable gradation and variation in the content and nature of their gold and associated heavy and light minerals. Near the primary deposits the particles of gold are invariably larger and commonly of less fineness than those farther down the dispersion fans and trains on the slopes. During the downward creep of minerals and rock fragments on slopes the lighter fractions tend to gravitate to the top layers and move relatively faster downhill than the heavier fractions. This is why one commonly finds a greater concentration of light minerals and rock fragments such as quartz floats at the foots of hills and mountains rather than farther uphill and at the outcrop of auriferous quartz veins. Finally, nonuniformity and erratic behaviour commonly prevail in the continuity and value of the pay streaks in most eluvial placers, a feature that dictates that most of these deposits must be carefully pitted, trenched, drilled and evaluated before exploitation.

Eluvial gold placers were widely worked in many parts of the world prior to the turn of the century. Most were residual blanketlike deposits developed on a relatively flat terrain, or fanlike accumulations formed on gentle slopes. The materials of these deposits varied greatly depending principally on the types of bedrocks and deposits and the type and degree of weathering. Most had a high iron (lateritic) content although some were relatively highly enriched in aluminum (bauxite or clay). Most consisted of a residuum of quartz, sand, clay, iron oxides, aluminum oxides and manganese oxides in which the particles of gold, commonly with a number of other chemically resistant heavy minerals, were concentrated on or near the bedrock. Some of these deposits are highly cemented with clay, limonite or caliche (carbonates); more rarely by silica. Most eluvial placers are low grade in gold, but some are of enormous yardage (tonnage). Much of the gold in eluvial placers is rough and irregular in form, and the fineness is commonly only slightly higher than that in the primary deposits. Eluvial placers are known for their large nuggets or masses of gold, some weighing several hundred to a thousand ounces or more; alluvial placers only rarely contain large nuggets (>100 oz). Some eluvial placers yield silver, lead, cassiterite, cinnabar, diamonds and platinoids in addition to gold. A few descriptions of typical examples of eluvial deposits follow.

Extensive eluvial placers are unknown in Canada, but there are a number of small examples mainly in Yukon that are of interest from an academic viewpoint. In the Dublin Gulch area, Yukon (Boyle, 1965*a*), small eluvial placers are developed on the sides of the hills above the economic gold placer in the gulch (Fig. 69). The eluvial placers contain mainly scheelite with small amounts of wolframite and gold, these minerals being derived from a complex of scheelite-

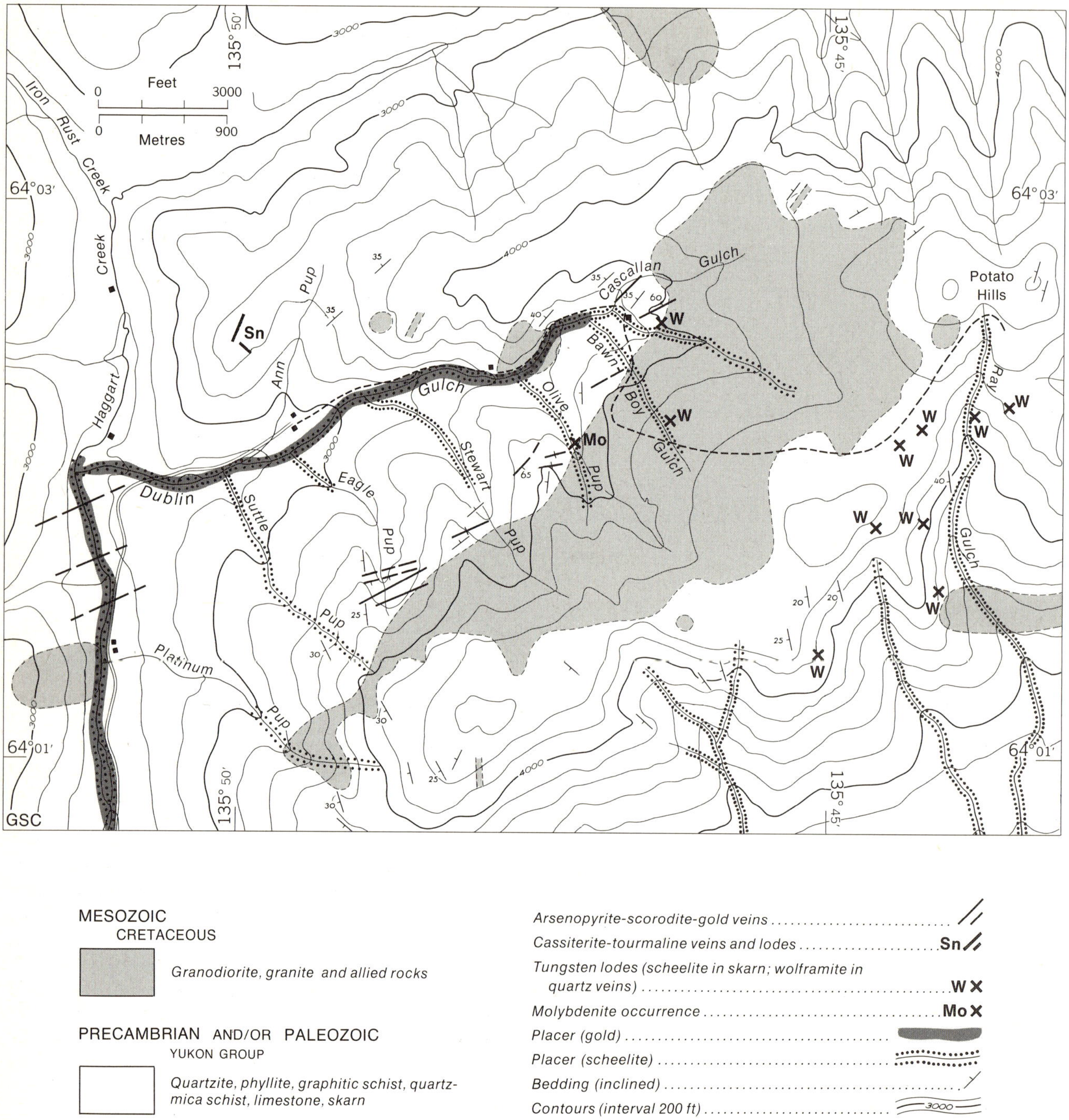

Figure 69. Generalized geology of Dublin Gulch area, Yukon showing gold and scheelite gulch placers (after Boyle, 1965*a*).

bearing skarn, quartz-wolframite veins and northeast striking quartz-arsenopyrite-pyrite-scorodite-gold veins, all lying slightly uphill from the eluvial placers. The constitution of the eluvial material is varied depending on the type of bedrock. Over schists the materials are mainly mixtures of sand, clay, limonite and disintegrated schist; over granodiorite the materials are principally a limonitic clay mixture with highly disintegrated feldspar and quartz. The principal heavy minerals in the eluvium are gold, scheelite, wolframite, extrinsic hematitic nodules of iron-formation, garnet, cassiterite, magnetite, bismuth, pyrite, arsenopyrite, jamesonite, siderite, nodules of oxidized galena, limonite, wad and scorodite. Eluvial material containing gold and scheelite is greatly enriched in arsenic, antimony and tungsten a feature that can be used in geochemical prospecting for this type of deposit.

In the western United States eluvial placers have been worked in the past in California, Oregon, Nevada and Montana, but few rivalled the alluvial types in these states in their economic importance. In the Appalachians of the United States eluvial placers were worked on a fairly large scale before the turn of the century, especially in Georgia, (Becker, 1895; Pardee and Park, 1948; and Lesure, 1971). There, in the Dahlonega gold belt in Lumpkin County, the bedrock is a complex of mica schist, mica gneiss, quartzite, amphibolite, migmatites and granite. The schists contain abundant gold-bearing quartz veins and stringers of various ages. All rocks and deposits are deeply weathered forming at the surface what has been called a saprolite by Becker (1895). Saprolite is in reality only a form of laterite; it is brownish in colour, composed of soft, earthy, clay-rich materials derived from thoroughly decomposed schist and other metamorphic rocks. Zones of disorganized and disintegrated quartz veining in the saprolitic schists were mined by deep open cuts on a large scale near Dahlonega, the gold being won partly by washing and partly by crushing of the quartz in stamp mills. The enrichment in the saprolite was evidently due to considerable leaching and precipitation of gold as suggested by the geochemical results quoted by Lesure (1971), who found up to 2.9 ppm in freshly precipitated limonite. The gold distribution in the fresh vein quartz is low and spotty, and the primary deposits are apparently not attractive mining ventures.

Certain gossans on stockworks or massive sulphide deposits and their attendant eluvial deposits have been mined or are being mined for gold. At the Greenhorn Mine in Shasta County, California an auriferous gossan was worked profitably for many years (Huttl, 1940). The gossan was developed on a copper sulphide orebody and consisted of a mass of limonitic material of varying thickness containing localized zones of rhyolite. The principal values in the gossan were gold and silver with some flakes of native copper. The gold particles were extremely fine, almost microscopic. The combined gold and silver content averaged about 0.20 oz/ton. Similar gossans have been mined in many countries, e.g., Spain (Rio Tinto), Australia (Mt. Morgan) and elsewhere. Some of these are described in more detail in Chapter IV.

Eluvial deposits have been worked in many places in the lateritic residuum in Mexico, Central America, Cuba, Haiti, Puerto Rico and the countries of South America.

In Mexico eluvial-alluvial placers in three large fans of gravels downslope from a mineralized porphyritic granite stock and its limestone contact zone are found near Guadalcázar in the central part of the State of San Luis Potosi (Fries and Schmitter, 1948). The placers were derived from the weathering and erosion of the mineralized parts of the granitic body and its contact zone and contain tin as cassiterite, mercury as cinnabar and minor amounts of gold and silver. The gold in the finer fraction of the gravels and sands is free; in the coarser fractions it is chiefly in sulphide minerals (pyrite, arsenopyrite, galena, sphalerite, stibnite, silver sulphosalts) and their oxidation products. Part of the silver is in sulphides and sulphosalts and a part is alloyed with the native gold. Average assays of some of the placer materials show the following: 84 ppm Sn, 15 ppm Hg, 9.44 ppm Ag and 0.058 ppm Au. Some 550 million m^3 of gravel are present in the fans.

In South America most of the eluvial deposits were blankets and eluvial fans in the lateritic materials near zones of quartz veining in Archean greenstones and associated sediments. In actual fact gold in the laterites of the Guianas is widespread ranging from 0.003 to 0.135 ppm over hundreds of square miles in some places. Such areas seem to be centred on auriferous Precambrian greenstone and sedimentary belts intruded by granitic rocks and porphyries. During weathering, therefore, gold has apparently been concentrated in the laterites both from the rocks and from epigenetic gold-bearing deposits. Extensive natural winnowing of the laterites has evidently given rise to the important alluvial placers, which are a feature of the Guianas. One interesting deposit at Omai, Guiana (British Guiana) was a deeply weathered aplite dyke highly decomposed to a depth of more than 100 ft (Maclaren, 1908). The primary dyke material was shot through by slightly auriferous quartz stringers and heavily pyritized, the pyrite being auriferous. The decomposed material consisted essentially of limonite-stained sand and clay, in which well crystallized free gold was abundant.

Other interesting types of eluvial deposits occur in Brazil mainly in the Quadrilátero Ferrífero, Minas Gerais. All of these are associated with gold-bearing iron-formations (itabirites). General descriptions of these deposits are given by Derby (1884, 1903), Bensusan (1929), De Oliveira (1930, 1932), Gair (1962) and Dorr (1969) (*see also* p. 304 of this chapter). One type of secondary deposit, not necessarily eluvial, but often grading into eluvial deposits is locally known as "jacutinga" (De Oliveira, 1932). The jacutinga occurs as thin (inches to a few feet) lines or bands in itabirite and is a decomposition, or more accurately a surficial chemical disintegration product of the iron-formation. It is composed essentially of powdery ferric oxide (limonite and hematite) manganese oxides, clay minerals and talc in which nuggets, plates and threads of native gold are present. Evidently the gold originated by chemical solution and reprecipitation of the metal, from low grade gold-bearing iron-formation. From the descriptions, some of the jacutinga were rich, up to 0.5 oz Au/ton or more, and some were mined to depths of 700 ft. Some of the gold was rich in palladium.

The other type of deposit associated with iron-formations in Brazil is the gold-bearing 'Tapanhoancanga' or 'Canga' for short. This is an irregular layer or blanket up to 10 ft or more thick of limonite-cemented fragments of iron-formation (itabirite). It is commonly developed on all iron-formations throughout the world that are extensively oxidized. Much of the limonite appears to derive from the oxidation of iron silicates and pyrite in the iron-formation. Where the iron-formations are also gold-bearing the canga is commonly enriched in gold, the metal occurring in small flakes, wires, specks and also in a submicroscopic form associated in some manner or other with the limonite. In the past, deposits of this type were worked in a small way in many parts of the ferriferous quadrangle in Minas Gerais.

Eluvial placers have not been important sources of gold in Europe although they have been worked in the past, some by the Romans, in Portugal, Spain, France, Switzerland, Germany and Romania. Important eluvial deposits were once worked in the Berezovsk, Kochkar and other districts in the U.S.S.R. In Africa a few eluvial placers have been worked mainly in Egypt, Sudan, Ethiopia, Zaire, Ghana and elsewhere in the western gold belt of Africa. Most were small but have mothered extensive alluvial placers in a number of countries. It is of interest to note that the great Rand deposits of South Africa gave rise neither to extensive eluvial nor to alluvial deposits, a feature that is difficult to explain. The reason for the paucity of placer gold associated with these enormous deposits probably has much to do with the size of the gold in the reefs (50 μ) and the presence of abundant pyrite, which weathers readily to give soluble iron salts and colloidal hydrous iron oxides. Both of these features would tend to render the gold soluble or mobile as an absorbed component, thus ensuring its broad dispersion.

Eluvial gold was relatively common in Australia before the turn of the century, but most of these placers are now exhausted. Fabulous nuggets were found in some districts in the eluvium or weathered rubble of the oxidized deposits. At Big Nugget Hill in the Hargraves goldfield of New South Wales a nugget weighing 106 lb was found in the eluvium by an aboriginal shepherd. At Lucknow, also in New South Wales, the eluvium was of great richness at and near the outcrop of the veins. In Victoria even larger nuggets were found in the eluvium and disintegrated country rock near the veins of Ballarat, Tarnagulla and other places, including the Welcome Stranger (2516 oz), Welcome (2195 oz), Blanch Barkley (1743 oz), Canadian (1319 oz), Dunolly (1364 oz) and Sarah Sands (755 oz). In Western Australia the eluvial placers were small, rich and soon worked out in the Pilbara, Kimberly, Coolgardie, Kalgoorlie and other Yilgarn fields.

In Asia eluvial gold has been won from soils and weathered debris near outcroppings of gold in Japan, China, Burma, India, U.S.S.R. and elsewhere. Most of these deposits were small and have long been exhausted. One type of eluvial deposit in U.S.S.R. is of considerable interest both economically and scientifically – the Kuranakh type. The Kuranakh gold deposits found in 1959–1962 are mainly in Lower Cambrian bituminous limestones and dolomites in the Aldan anteclise of southern Yakutia (Razin and Rozhkov, 1963, 1966; Kazarinov, 1969; Borodaevskaya and Rozhkov, 1974). There is considerable similarity between the primary gold occurrences and those described in this chapter for the disseminated deposits of north-central Nevada (Carlin, Cortez, etc.). The primary occurrences in the Kuranakh district are mainly potash feldspar (adularia) and quartz metasomatites with about 6 ppm gold occurring as irregular replacements of the limestones and dolomites in places in stratiform bodies up to 3 km in length, 300 to 500 m in breadth and 5 to 10 m in thickness. The gold in the adularia zones is very finely-divided (2–5 μ) and mainly in pyrite. Thin kersantite dykes and sills of late Mesozoic age are also accompanied by adularization along their contacts both in the Cambrian strata and in Lower Jurassic sandstones, and these igneous rocks have a comparable tenor of gold within the ore zones. Oxidation of the primary gold-bearing zones and karstification of the Lower Cambrian limestones in post-Jurassic and Recent times has given rise to economic eluvial deposits consisting essentially of highly weathered material composed mainly of oxidized ore, limonite, clay and sandy materials in the karst cavities (Fig. 70). The gold in the eluvial material occurs as very finely-divided flakes ranging in size from 5 to 20 μ. The ratio of concentration is not given, but presumably it is about 2 making the grade of the deposits about 12 ppm (0.35 oz Au/ton).

In Luzon, Philippines, eluvial placers are common and were extensively mined in the past. Gilbert (1940) describes a number of these from the Paracale-Mambulao district where the primary auriferous material is mainly veins and stockworks of quartz and pyritic quartz veins in granite intruding serpentinites. The auriferous eluvial material consisted mainly of weathered debris and disintegrated bedrock carrying free gold.

Certain eluvial auriferous deposits take the form of talus accumulations. A typical example of this type of placer occurs near the Belaya Gora deposit in the Lower Amur region of U.S.S.R. (Borodaevskaya and Rozhkov, 1974). The bedrock deposit occurs near the summit of Belaya Gora in a strongly fractured, kaolinized and silicified Oligocene volcanic neck. The gold is mainly free and accompanied by about 0.5 per cent combined pyrite, arsenopyrite, sphalerite and argentiferous sulphosalts. The talus placer associated with the primary deposit lies below the deposit on a gentle slope and averages some 5–6 m in thickness. In composition the talus material is largely variegated clay and sandy clay and oxidized rubble with a heavy mineral suite composed mainly of limonite, magnetite, ilmenite, chromite, epidote and small amounts of zircon, sulphides, manganese oxides and other minerals. The gold is disseminated through the talus material and also forms short streaks and enriched lenses. The gold in the talus is not much different from that in the bedrock. The gold grains range in size from 0.05 to 1.5 mm. Some adsorbed and very finely divided gold also occurs in concretions of kaolinite and quartz. Downhill from the talus placer there is a colluvial and alluvial placer system associated with springs and the river fed by the springs.

Alluvial placers

Alluvial placers are those formed in present and past water courses in gulches, creeks, rivers, flood plains and deltas. Reworking of some of these deposits together with others formed as a result of sedimentation or glacial processes by wave action may yield beach placers, which are treated separately.

Alluvial placers have been worked since ancient times in practically every country (Fig. 67) and have produced probably about one-quarter of man's store of gold. If we include the

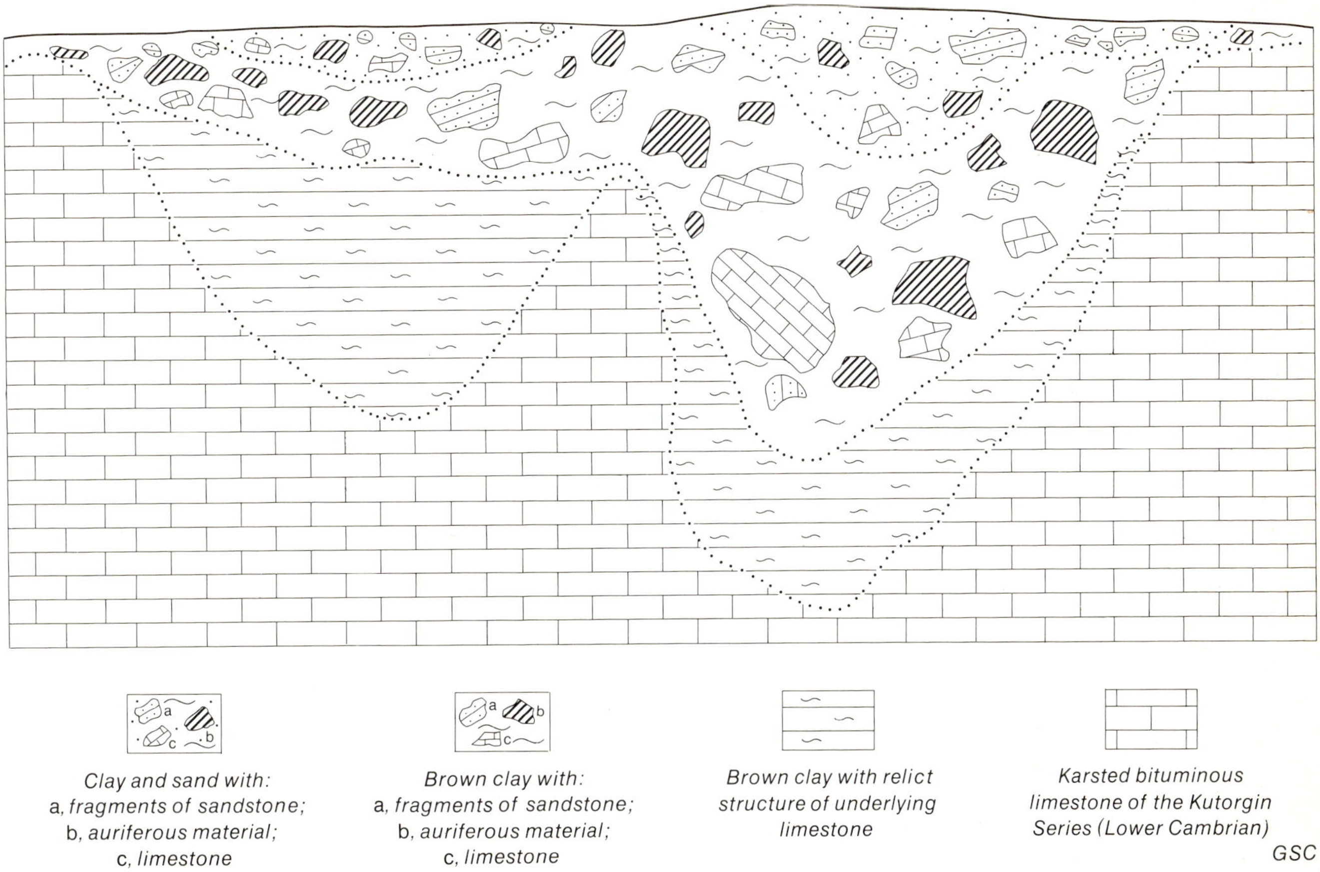

Figure 70. Schematic cross-section of an auriferous karst, Kuranakh district, southern Yakutia, U.S.S.R. (after Razin and Rozhkov, 1963). (No scale given on original.)

Witwatersrand deposits as fossil alluvial placers, the amount of gold produced from these types of placers probably approaches two-thirds of man's store of the precious metal.

Alluvial placers can be classified into two general categories – modern and fossil. The distinction between the two is commonly difficult to make in the field. Placers formed in present day water courses and most of those of Pleistocene and Tertiary age fall into the modern category. Those of greater age, commonly buried deeply by superincumbent sediments or volcanics and generally lithified we shall call fossil. Fossil placers occur throughout the geological column and are described subsequently.

There is an enormous amount of literature on alluvial gold placers, dealing with descriptions of the placer fields, the principles of placer formation and the methods of panning, rocking, sluicing, hydraulicking and dredging. The interested reader is referred to the general references mentioned in the introduction to the section on placers.

There are some general characteristics of alluvial gold placers that will serve as a basis for the descriptions and discussion that follow. These include: alluvial placer gold in pay streaks near its source is invariably coarse and is found in the lower layers of the alluvium either on bedrock, in a zone a few feet above the bedrock or in crevices, fractures, etc. in the bedrock within a few feet of the surface. An exception to this is where the so called 'false bottoms' or 'false bedrocks' occur in thick beds of alluvium. These false bottoms may be clay layers (hard pans) within the gravels, compacted sands or more rarely limonite-cemented sands and gravels (conglomerates). Alluvial placer gold far from its source is generally finely divided, and while part of the gold may be on or near bedrock or false bottoms most of it is dispersed throughout great thicknesses of the sediments. This is especially true of extensive flood plain and deltaic deposits. Two other features are characteristic of most alluvial gold placers. The first, as intimated above, is that the further from the source the more finely divided the gold, and the second is that placer gold is finer in value than its source lode gold and that with increasing distance from the source the finer is the gold. These two statements are general. The first is invariable, but there are exceptions to the second in some districts.

Alluvial placers are composed of loose unconsolidated gravels and sands that are commonly relatively clean. The terms 'white channel gravels', 'white leads' and 'white bars' reflect the latter circumstances with respect to quartz. In places, however, the pebbles and the gold may be coated with limonite, wad and other precipitates. Some alluvial gravels and sands are heavily impregnated with limonite and wad forming the 'cement gravels' or 'cangalli' of the alluvial miner. Such gravels invariably occur where the primary deposits and wall rocks are rich in pyrite, siderite, chlorite and other iron- and manganese-bearing minerals. Secondary siliceous and calcareous cements are rare in alluvial gravels except in the vicinity of siliceous springs and lime-bearing waters; some alluvial deposits may be cemented by caliche in arid regions. In permafrost zones the alluvial gravels are frozen solid and have to be thawed out before dredging or hydraulicking.

The heavy mineral suite accompanying gold in alluvial placers differs depending on the host rocks and types of primary deposits. Magnetite and ilmenite are the most common, and these may be accompanied by varying amounts of monazite, pyrite, arsenopyrite, cassiterite, wolframite, scheelite, cinnabar, native bismuth, bismuthinite, galena, sulphosalts, platinoids, tourmaline, garnet, chromite, rutile, barite, corundum, zircon, wad and limonite.

Most alluvial deposits in gulches, streams and rivers are characterized by a lack of regular and persistent bedding or stratification, but pseudobedding, laminations, current or false bedding may be developed in some accumulations. In deltaic deposits bedding and stratification is poorly- to well-developed depending upon the rate of sedimentation. Fractures, fissures and even faults can be seen in some modern alluvial placers. The last may throw the pay streaks several tens of feet in places.

Gold placers can be classified in a number of ways based upon their location or genesis. Kartashov (1971) gives a classification that is current among some placer geologists in the Soviet Union. He distinguishes two types of alluvial placers – autochthonous and allochthonous (Table 66). In the authochthonous variety he lists those placers formed essentially near the primary or more rarely the secondary source(s) of their gold; the allochthonous variety implies considerable transport of the gold and deposition far from the primary and/or secondary source(s). This classification is quite satisfactory but requires a considerable knowledge of the details of placer deposits and the dynamics of their formation before they can be adequately categorized. Another classification of placers is provided by Kazakevich (1972) (Table 67).

The simplified classification of placers adopted here follows essentially the scheme suggested by Lindgren (1933) and others based mainly on the geological location and the tectonic history (uplift and depression) as shown in Table 68. The latter factor is emphasized by Trofimov (1964), who considers it to be the chief control on the formation of placers.

The formation of alluvial placers depends upon many interacting physical and chemical factors, our general knowledge of which is fairly well established. However, our understanding of many of the chemical and physical details of the accumulation of gold in alluvial placers is still remarkably poor, especially with respect to the formation of nuggets. There are also many differences of opinion as to why gold is concentrated on or near bedrock in alluvial placers. These problems are discussed subsequently.

The physical and chemical factors involved in the formation of alluvial gold placers are: an adequate source of primary gold, favourable oxidation processes for its release

Table 66. Characteristics of the kinds of alluvial gold placers

Notes	Autochthonous: Bottom*	Autochthonous: Above-bottom*	Allochthonous
1. Are represented by	Channel, valley, terrace and watershed placers	Valley, terrace and watershed placers	Point-bar delta, river-plain, valley, terrace and water-shed placers
2. Occur	Adjacent to their gold-bearing sources		More or less far from their gold-bearing sources, being separated from them and from autochthonous placers by zones of dispersion of 'placer' minerals
3. Are concentrated	At the base of an instrative† or substrative† alluvium and in the crevices of the bedrock	At the base of a perstrative† alluvium and within constrative† strata, in the same parts of valleys as bottom placers	In surficial horizons of a perstrative alluvium and within constrative strata, downstream from autochthonous placers
4. Consist of mineral grains	Derived directly from gold-bearing sources or redeposited from older placers and not carried away by rivers from concentration zones		Brought by rivers into concentration zones
5. Accumulate during	Entire time of destruction of primary gold-bearing deposits, embracing, as a rule, many stages of river development	The last equilibrium and/or aggradation stages of river development	
6. An enclosing alluvium is formed during	Downcutting stages or transition from them to equilibrium stages	Equilibrium and/or aggradation stages	
7. An enclosing alluvium, being reworked during a downcutting stage, is	Not destroyed but displaced to the level of a new bedrock bottom	Displaced to the level of a new bedrock bottom and added to bottom placers	Completely destroyed
8. The mechanism of concentration of 'placer' minerals	Does not essentially depend upon the hydrodynamic properties of flowing water		Depends to a large extent upon the hydrodynamic properties of flowing water

Adopted from Kartashov (1971).
*These terms refer to the position of the pay streaks with respect to the bedrock.
†Instrative, perstrative and constrative correspond to the stages of river downcutting, dynamic equilibrium and aggradation. Refer to Figure 73 and the original article by Kartashov (1971).

Table 67. Genetic classification of gold placers

Type	Sub-type
I.Eluvial	1. Zones of weathering of gold-quartz deposits 2. Zones of oxidation of auriferous sulphide deposits
II.Slope	1. Solifluction and land slide placers 2. Solifluction and deluvial placers 3. Deluvial and landslide placers
III.Watercourse	1. Alluvial placers 2. Proluvial placers 3. Lacustrine placers
IV.Glacial	1. Placers of main and lateral moraines 2. Placers related to interglacial and glacial streams and other water courses of glacial origin
V.Eolian	Placers in eolian sands
VI.Marine	1. Beach placers 2. Placers on underwater slopes 3. Placers in near shore still water (lagoons, coves)

Source: Kazakevitch (1972).

Table 68. Classification of alluvial placers according to their geological location and tectonic history (uplift or depression)

Present form	Elevated equivalents	Depressed equivalents
1. Gulch, canyon and creek gravels	High level gulch and creek gravels	Deeply buried gulch and creek gravels
2. River and bar gravels	Bench gravels High level river and bar gravels	Deeply buried river and bar gravels
3. River flood-plain gravels	High level flood-plain gravels	Depressed (deeply buried) flood plain gravels
4. Deltaic gravels	Elevated deltaic gravels	Depressed deltaic gravels
5. Beach and shore line gravels	Elevated beach and shore line gravels	Depressed beach and shore line gravels

and favourable concentration sites, all of which depend essentially on the tectonic environment, which controls the dynamics of streams and rivers. Each of these is discussed below.

1. Nature of the oxidation processes in the primary gold-bearing deposits: several factors are involved, including climatic and groundwater conditions and the nature of the primary gold, gangue and wallrocks. These are discussed in detail in Chapter IV. The nature of the primary gold and the chemical system operating during oxidation of the deposits are two critical factors in the formation of alluvial placers.

The size of the primary gold particles is a major consideration. Where the particles of native gold in a primary deposit exceed 100 μ there appears to be relatively little solution or flotation of the gold in the natural waters. During oxidation processes the particles tend to go directly into the eluvium and creep from there into the alluvial placers. They may, however, accrete gold from the soil solutions and stream waters as they proceed on their journey.

In some districts, however, where the primary gold is extremely finely divided (<50 μ) or is a microscopic or lattice constituent of auriferous pyrite, arsenopyrite, tellurides, etc., alluvial placers are not formed to any extent even though other conditions are favourable. One striking example of this is the paucity of alluvial placers associated with the great Witwatersrand deposits. In such districts the gold does not seem to aggregate to form large particles. On the contrary it appears to be dispersed as flour gold or is dissolved and dispersed throughout the ground and surface water systems. In other districts the reverse situation exists, and alluvial deposits abound even though the gold is finely divided or a constituent of pyrite, etc. in the primary deposits. In these the gold accretes mainly in the oxidized zones and eluvial deposits to form relatively large particles, which ultimately find their way into the alluvial placers. To explain the difference in behaviour between the two extremes is not simple because of the complex chemical systems involved. In the writer's experience one important factor appears to be the purity of the finely divided primary gold. If its silver and/or copper content is high the gold appears to be more readily dissolved. This is understandable from etching procedures in microscopic work on gold – pure gold etches with great difficulty whereas alloyed gold is readily etched. Once dissolved the gold may be dispersed or reprecipitated (accreted) depending on the chemical system, especially on the Eh and pH of the environment and the presence of various colloids and precipitants (*see* Chapter IV). The relatively pure, finely divided gold, on the other hand, enters the erosion cycle chemically unaffected and is mechanically dispersed far and wide. Finely divided gold contained within pyrite and arsenopyrite and lattice gold in these minerals and in tellurides, aurostibite, etc. tends to accrete and form large particles, wires, etc. for reasons that are obscure but which may involve the solvent effects of thiosulphates, cyanides and ferric sulphate, interaction of the dissolved gold with natural reductants and various colloidal reactions. The surface characteristics of finely divided gold also seem to be important in its mechanical transport in water and probably also in its dissolution in oxidizing solutions. Some finely divided gold particles appear to have a relatively small contact angle due probably to films of organic substances, which permit air bubbles to attach themselves readily, a feature that promotes their easy flotation downstream. In the case of the Witwatersrand gold the film in some cases may be the hydrocarbon (thucholite) with which the gold is intimately associated. Films of all kinds, including both organic and inorganic substances, commonly inhibit the solution of finely divided gold in natural solutions. The most common inhibitors appear to be the various hydrous iron and manganese oxides and silica.

2. Nature of the bedrock or false bottoms: since alluvial gold is invariably found on or near the bedrock surface or on false bottoms the characteristics of these substrates are commonly decisive in the formation of alluvial placers. Few 'pay streaks' or 'runs of gold' occur on smooth bedrock bottoms for obvious reasons. The most favourable bedrocks are those that form natural riffles perpendicular to the stream or river course as in a gold sluice. Alternating thin beds of soft schist and hard quartzite or slate and quartzite are particularly favourable. It is a surprising fact, however, that in some placer areas bedrock riffles parallel to the direction of the stream flow are

more effective in trapping the gold than those crossing it at a large angle. Limestone bedrocks that are pitted and pinnacled provide particularly good settling places for gold, as do potholes and nests of boulders in some placer streams. Partly weathered schists and gneisses in which the leaves of the schist and bands of the gneiss are slightly separated are also ideal rocks for the concentration of alluvial gold. The little nuggets, flakes and specks of gold dust work their way into such rocks often to depths of 10 ft or more. Similar phenomena are noted where the bedrocks are sheared, fractured or intensely shattered. Limestones in karst terrane are also good traps for gold, the metal having worked its way down into the rock to depths of 30 or 40 ft in some placer areas. Mining of these zones of enriched bedrock is common in many alluvial gold districts. As for false bottoms, indurated clay and hardpan layers appear to be particularly favourable, although rough gravels and sands cemented by limonite, wad or carbonate provide suitable surfaces for the accumulation of gold in some districts. Compacted sands with rough surfaces are favourable in others. Some placer areas are characterized by stacked pay streaks due to alternating bands of clay, compacted sand, etc. in the sedimentary section.

3. Stream and river dynamics: this subject is complex involving hydrodynamics, sedimentation phenomena, geomorphology and isostasy (uplift and subsidence), subjects that can be treated only in a general manner here.

It is assumed that the reader is familiar with the general rudiments of the principles of sedimentation of falling bodies in static water. Briefly, the controlling factors involved are mainly the differences in specific gravity, size and shape of particles. It is axiomatic that of two spheres of the same weight but of different size, the smaller, with its lesser surface area, and hence lesser friction to water, sinks more rapidly in a static water medium. This is part of the reason why gold as small nuggets is commonly associated with quartz pebbles 1 in. or more in diameter.[28] Furthermore, the shape of particles is a factor – a spherical mineral has less surface area than a platy mineral of the same weight and hence sinks more rapidly. Finally, during jigging of bodies of variable specific gravity in water, those with the higher specific gravity sink to the lowest level. During this motion, round, shotty particles of gold will also find the bottom quicker than scaly or spongy particles.

We are, however, not generally dealing with static bodies of water in the formation of placers but with bodies of water that move and exhibit turbulent flow. The principal parameters in stream and river dynamics are the gradient and the volume of water. The gradient is determined by many secondary factors in an area but fundamentally by isostasy, which involves uplift (or depression) of one segment of the earth with respect to the other. The volume of water delivered to a stream or river, likewise, depends on many secondary factors involving the runoff and groundwater systems but primarily on the rainfall of the district. In a district with a pluvial climate once a gradient has been established stream and river systems develop that erode the land and yield a pattern of gulches, canyons and valleys.

The dynamics of the movement of sedimentary particles in streams and rivers are of interest in the formation of placers. The factors involved are many, including the velocity of the water, which depends essentially on the gradient; the degree of turbulent flow, the specific gravity of the minerals being transported and the nature of the stream or river bed. Two types of solid load transported by streams and rivers can be distinguished – the suspended load and the bed load. The suspended load is maintained in the body of water mainly by turbulence, although flotation of sedimentary particles attached to air bubbles may be important in places. The bed load consists of material that is rolled, pulled, slid or otherwise swept along the bed of the stream by traction or saltation (skipping of pebbles along the bottom in a series of leaps). In addition to these movements there is a constant creep of the sediment down gradient. This effect is small where the slope is minor, but may be marked where high gradients prevail. As an approximation it is said that the transporting power of a stream varies as the sixth power of the velocity. Calculations show that a stream running 2 mph (3 ft/sec) will carry a stone or nugget weighing about 3 oz; at 4 mph one weighing 45 lb; at 10 mph one weighing 1.5 tons; and so on. It is easy to understand how a spate of water moving down a gulch at 20 mph, as during flood times, can move boulders weighing 100 tons and more.

The evolution of a landscape following a general uplift above sea level follows definite patterns depending essentially on the amount of rainfall received. In districts with humid climates most rivers flow into the oceans, although interrupted in places by lakes. Four normal stages of evolution of drainage systems in humid areas can be recognized – initial, youthful, mature and old age. During the initial stage the gradients are low, depressions become filled with water, and streams lazily follow various gulleys and flaws in the rock. The rate of erosion is minimal, and few if any placers are formed. As the youthful stage sets in an integrated pattern of major streams with numerous tributaries develops by downward and headward up-gradient erosion, most lakes disappear and V-shaped valleys, and in places steep gulches, canyons and gorges, mark the landscape. With time the normal evolutionary pattern develops further to maturity. Gulch, creek, stream and some river placers are characteristically developed during these stages. As the rivers erode down toward their base level, lateral erosion develops wider and wider valleys with gently sloping sides, flood plains are formed, and meandering courses that wander hither and thither from channel to channel are characteristic. Some creek and stream placers continue to develop, but river, flood plain and deltaic placers are the mark of this stage. Continued slow downcutting of rivers and streams or uplift may leave the remnants of earlier valley floors as terraces in which remnants of the river flood plain placers are preserved. In old age the level of the whole land is reduced to a peneplain characterized by a low relief with low rounded hills, shallow valleys, rolling landscapes and sluggish meandering rivers, marked by oxbow lakes that flow in various directions following the very low gradients that prevail. Few if any placers are formed during this stage.

[28] In heavy mineral studies, mineral grains deposited at the same time are considered as hydraulically equivalent or equi-settling, under the conditions prevailing during deposition. Various factors, principally specific gravity and grain shape, are of importance in determining the relative size of hydraulically equivalent grains. The theoretical and practical aspects of the hydraulic equivalence of grains, with a bibliography, are discussed by McIntyre (J. Geol., v. 67, p. 278–301, 1959).

In arid climates the evolutionary pattern is significantly different, mainly because of reduced rainfall, the absence of thick soil cover and the lack of vegetation. Wind erosion is more important in such climates; extensive downhill creep along valley sides is reduced because of the lack of moisture, and numerous basins of interior drainage develop generally with no connections to oceans. The infrequent precipitation during the youthful stage gives rise to numerous nonintegrated streams in steep-walled valleys that cut deeply back into the highlands. With increasing maturity a poorly integrated drainage develops mostly into closed basins that are marked by pediments on which alluvial fans may form. In old age the drainage pattern is generally completely disintegrated, there being in most districts no streams of any length and commonly no interconnection between those that do exist during a pluvial period. Arid conditions are not particularly favourable for the formation of placers, although some may form in gulches, in canyons and in alluvial fans on pediments where the conditions are right. Arid areas are, however, often characterized by great spates and walls of water that come off the highlands or mountains during cloud bursts, rush down the dry washes, gulches, canyons and valleys carrying everything before them, and finally debouch onto the alluvial fans on pediments or into the closed basins. Such violent activity is not conducive to the formation of placers; on the contrary any that formed would be sluiced down the gulches, creeks and canyons and the gold spread indiscriminately over the lower valley bottoms and inland basins.

Desert placers as a whole are small and the pay streaks are often erratic and commonly scattered. 'Bajada' or 'pediment placers' in desert terrains may be more productive, but they are commonly difficult to work. Descriptions of some typical desert placer areas in the southwestern United States and Mexico may be found in the publications by Haley (1923), Webber (1935), Vanderburg (1936), Wilson and Fansett (1961) and Johnson (1972*a,b,* 1973*a,b,* 1974).

The above descriptions characterize the normal evolution of drainage patterns and the formation of placers under the extremes of humid and arid conditions. But many abnormal conditions prevail in any particular district. Sudden uplifts at the mature or old age stages cause renewed downcutting of channels and destruction or deep burial of placers already formed. Similarly, great or sudden increases in rainfall may produce floods that sluice out the gulches, creeks and canyons, change the courses of the rivers, and greatly modify, destroy or bury the most recent placers. Depressions of valleys may result in basins in which lakes develop or inlets are formed, which are invaded by the ocean. These ultimately become filled with sediment burying deeply any placers formed in the valleys. Finally, glaciers forming in high mountains and extending down gulches and valleys may plough out all sediments and scatter the placer materials far and wide.

The location of the pay streaks in placers is of prime interest to the placer miner. Given an adequate primary gold source pay streaks in general are fairly uniform and have considerable continuity in moderately hilly country where uniform rainfall has prevailed and where deep secular decay has been followed by a gradual restricted uplift. Any aberrations in this ideal pattern invariably cause marked variations in the tenor and continuity of the pay streaks. The richest pay streaks are those produced by reworking of preexisting auriferous gravels.

The law of the pay streak in placer deposits is variable depending on whether the placer is formed in gulches, in river channels, on flood plains or in deltaic deposits. Lindgren (1911, p. 66) said of the pay streak referring to the deposits of the Sierra Nevada of California, some of which are buried beneath Tertiary lava flows:

> It has become almost an axiom among miners that the gold is concentrated on the bedrock and all efforts in placer mining are generally directed toward finding the bedrock in order to pursue mining operations there. It is well known to all drift miners, however, that the gold is not equally distributed on the bedrock in the channels. The richest part forms a streak of irregular width referred to in the English colonies as the "run of gold" and in the United States as the "pay streak" or "pay lead". This does not always occupy the deepest depression in the channel and sometimes winds irregularly from one side to the other. It often happens that the values rapidly diminish at the outside of the pay lead, but again the transition to poorer gravel may be gradual. An exact explanation of the eccentricities of the pay lead may be very difficult to furnish. Its course depends evidently on the prevailing conditions as to velocity of current and quantity of material at the time of concentration.

Tyrrell (1912) while admitting the fickle nature of the pay streak concluded that "the pay-streak is a feature in the structure and growth of the valley in which it occurs, its formation is governed by certain geological laws, and those laws should be recognizable without great difficulty if the growth of the valley can be traced with reasonable accuracy."

The features of the evolution of valleys is discussed above, but this knowledge is only of general value in studying placers. To accurately determine the location of their pay streaks, especially in the buried types and those on flood plains and deltas, requires detailed profiling by trenching or overburden drilling. Geophysical methods may assist in some terranes. Where the cover is light and the bedrock magnetically low, a magnetometer survey may outline the position of the magnetic 'black sands' so commonly accumulated with gold. Where the bedrock has a higher magnetic intensity than the gravel, the position of magnetic lows may indicate the position of the pay streak channels. Electrical methods may also assist where there are conductivity or other differences between the bedrock and the basal alluvium. Hammer seismic methods under favourable conditions can also be used for profiling and outlining the notches or gutters containing pay streaks. The pay streak in gulches and youthful streams is generally easily located. It is commonly in or near the notch or gutter of the V-shaped valley or gulch on and just above bedrock (Fig. 71).

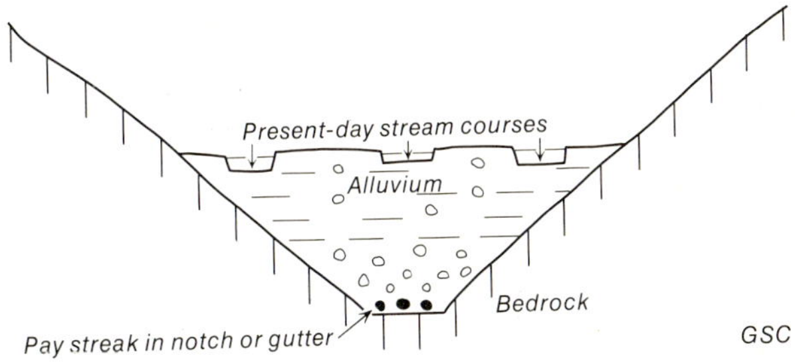

Figure 71. Sketch showing the location of an auriferous pay streak in the notch or gutter of a V-shaped valley or gulch.

As the valleys mature, widen out and become filled with sediments the streams shift their location from time to time, and the pay streaks may no longer be in the lowest depression or gutter. Furthermore, pay streaks long since buried may bear no relation to the present streams (Fig. 72). Pay streaks in wide valleys, flood plains and deltas are extremely irregular in their distribution and are the product of an environment where shifting channels and migrating meanders hold sway (Fig. 73). In most streams and rivers where placers are forming, gold collects along bars due to some obstruction, diversion or slackening of the water course in the slower moving water on the inside curves of fast-flowing meanders, in the main stream near the mouths of tributaries and elsewhere where slack water prevails (Fig. 74). These guides are useful in the study of both recent and ancient water courses. With respect to the latter, however, care must be taken to ascertain the direction of the paleogradients. For instance the Tertiary channels of the Sierra Nevada have been tilted by movements initiated by block faulting so that some of the gold channels now 'run up hill' as compared with the present gradient of the country.

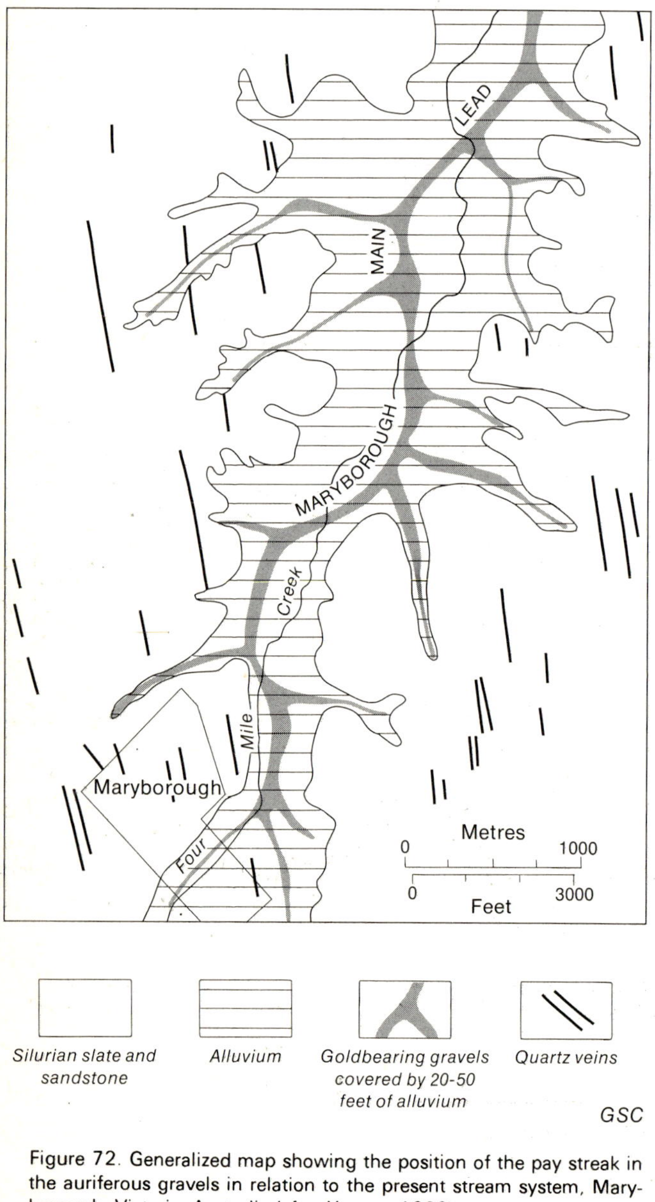

Figure 72. Generalized map showing the position of the pay streak in the auriferous gravels in relation to the present stream system, Maryborough, Victoria, Australia (after Hunter, 1909).

Pay streaks may be reworked by subsequent streams in the same or different courses into exceedingly rich concentrations (Fig. 75). In a broader areal sense the reworking of former placers may also be economically important in places. Thus, the lean placer concentrations of gold and platinum formed during the Mesozoic and early Tertiary in the Urals, U.S.S.R. were uplifted and crossed by a Pliocene-Quaternary drainage system in which rich new placers were concentrated (Sigov *et al.*, 1972).

Deltas may form in rivers, lakes, seas and oceans, but few gold placers in deltas in these locations have been described, probably because they are relatively rare. The principles governing the pay streaks in deltas are the least understood. Deltas in rivers tend to concentrate gold in fairly well-defined streaks just downstream from the mouths of the streams, mainly in topset and foreset beds if these are developed. Deltas in lakes commonly receive only fine-grained to very fine-grained (flour) gold, and the metal may be dispersed more or less throughout the sedimentary pile. In some cases, however, the fine-grained gold may collect in streaks on the foreset beds, and these may have relatively high angles of inclination. Deltas in seas and oceans do not seem to develop well-defined streaks, probably because of the extremely fine subdivision of most of the gold reaching the oceanic environment. If streaks do develop, they are usually erratic and seem to be mainly located in the topset and foreset beds.

An unusual type of (alluvial) placer is described by Bensusan (1942) in Burma. This placer, near Tenasserim, consists of gold in mangrove swamp mud. The gold appears to be present in a colloidal form since repeated examinations of samples and concentrates under a high power microscope failed to show any free gold. From the description it would seem that this is a case of concentration of gold in a humic environment, perhaps by chelation or other organic bonding (*see* also Chapter II). The swamp mud averages about 0.285 ppm Au and is said to contain some £200 million worth of the metal.

At this point a few words should be said about alluvial placers originating from the reworking of glacial debris by glacial and later streams. Some examples of these are given later in the descriptions that follow, and hence only generalities need concern us here.

Valley glaciers give rise to vast terminal moraines and during wasting may clog the valleys with great accumulations of glacial till, clay and sand. If these glaciers have scoured valleys containing gold placers or the sides of valleys containing oxidized or unoxidized gold deposits, the glacial materials may be slightly auriferous. Reworking of these materials by meltwaters and subsequent streams may produce stream and river placers in the normal manner. Many of these placers are often buried, difficult to recognize, and their pay streaks are frequently erratic. Nevertheless, a number of these types of placers have been worked in the past particularly in New Zealand (Park, 1969; Williams, 1974), and there are others in the Rocky Mountains of Canada and the United States, and elsewhere.

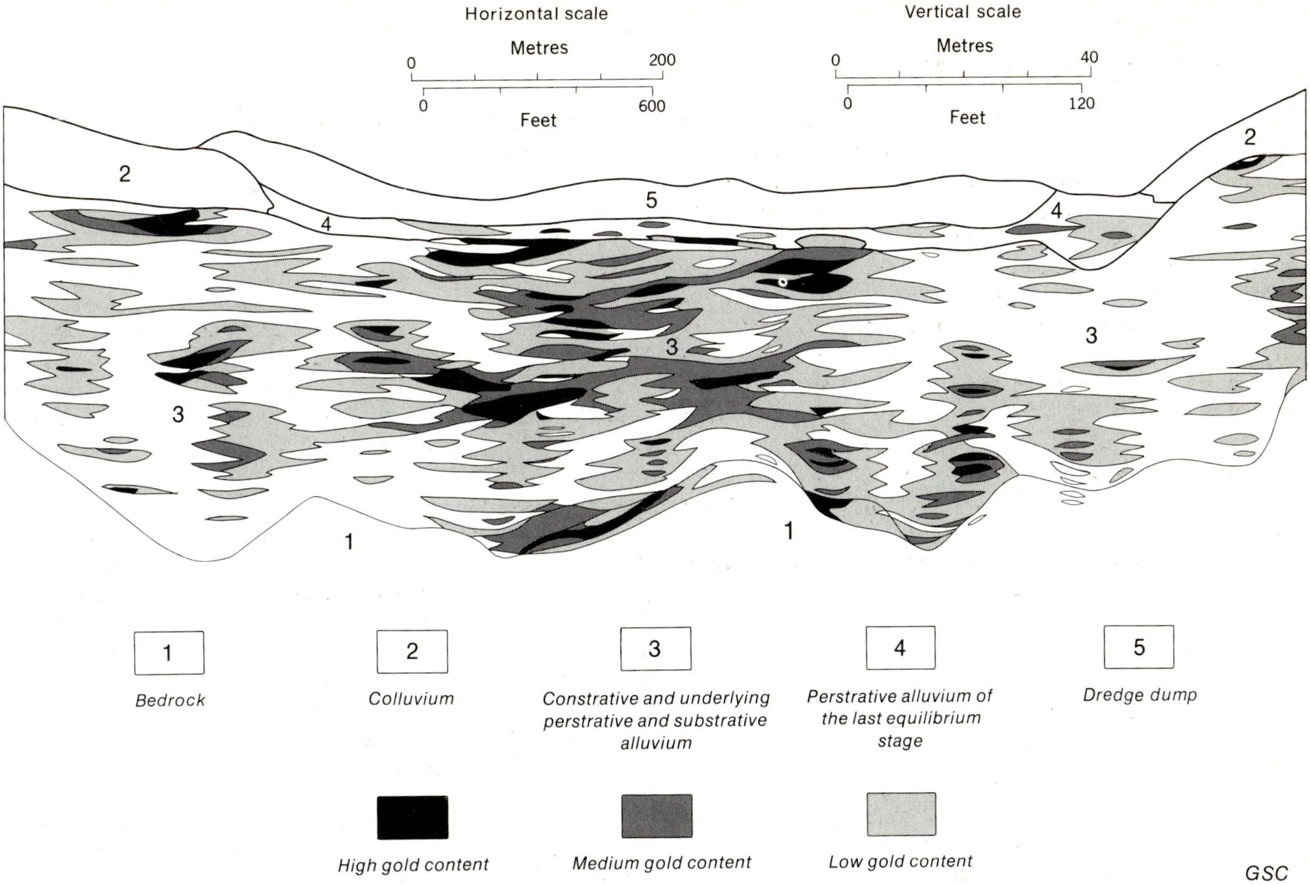

Figure 73. Cross-section of an allochthonous gold placer in the valley of the Kuranakh River, southern Yakutia, U.S.S.R. (after Kartashov, 1971).

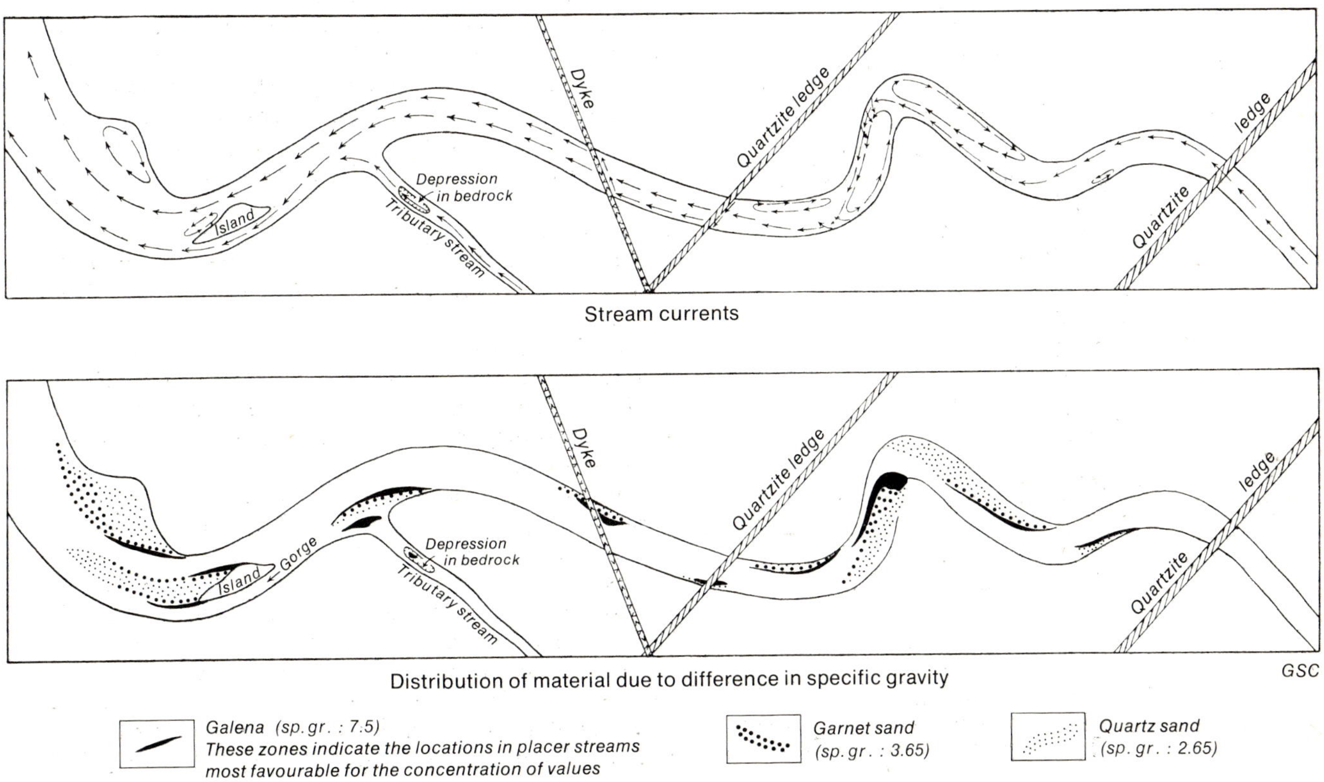

Figure 74. Diagram of a laboratory stream showing currents and distribution of sedimentary materials of different specific gravity (after MacKay, 1921). Galena was employed as the mineral of high specific gravity. The analogy with the placer conditions found in most rivers and streams with moderate gradient is excellent.

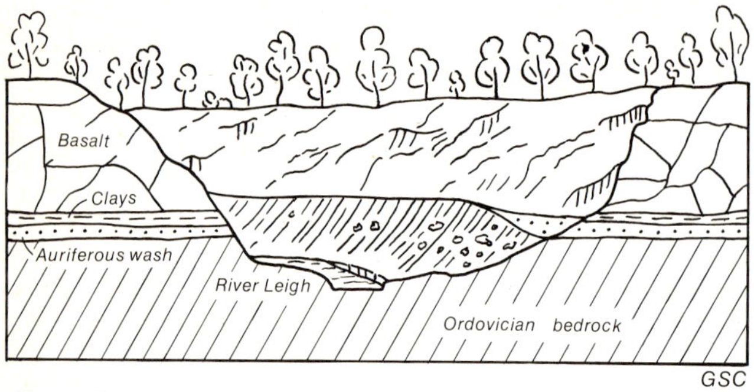

Figure 75. Sketch showing reworking by subsequent stream of an older pay streak, River Leigh, Victoria, Australia (after Hunter, 1909). An old auriferous sub-basaltic river is crossed by the present stream in which some payable gold was found downstream from the old pay streak.

The action of piedmont glaciers and ice sheets (continental glaciers) has generally been disruptive and destructive of preexisting placers. In places, however, there have been reconcentrations by glacial streams from the disrupted placers as in Yukon, British Columbia, Quebec, Siberia and elsewhere. Some of these are described subsequently.

From the general considerations discussed above, and others that will be mentioned subsequently in the text, some general conclusions about the concentration of gold and other heavy minerals in alluvial placers can be stated.

1. Coarser gold is generally deposited in the upper reaches of a placer gulch, stream or river, and the finer grained gold is generally deposited in the lower reaches of placer water courses.

2. The richest and coarsest gold and nuggets are commonly deposited in the layers with comparatively coarse sediment, and the finer gold particles are commonly deposited in the finer, sandy fractions of the sediment.

3. Gold should normally occur in streaks with other heavy minerals such as magnetite, scheelite, barite, etc.

4. Gold found on a bottom with a steep gradient will commonly be coarser than that found on a flat-lying level bottom.

5. In narrow gulches and gorges mainly coarse gold will be deposited.

6. Little gold will lodge in places where scouring action is marked.

7. Pot holes do not generally form effective lodgements for gold mainly because of the centrifugal grinding action and ejection of fine gold from such sites.

8. On a favourable bottom the gold generally collects on the downstream side of natural riffles.

9. Zones of slack water in water courses are favourable sites for the accumulation of gold particles.

10. Glaciation does not preclude the occurrence of placers.

Beach placers

Beach placers can be subdivided into two categories – modern and fossil. They are formed by the winnowing action of waves, undertow and alongshore currents along present and past shorelines bordering lakes, seas and oceans where a source of primary gold has been available for concentration. Tidal movements and strong winds may accentuate the concentrating processes in some places. Beach placers tend to form predominantly along rectilinear shorelines; few are known along markedly broken and rocky shores. The source of the gold may be in the rocks along the shores; in sediments deposited by streams upon the coast; and in slightly auriferous stream or river terraces, sea terraces and gravel plains that border the coasts.

The constituents of beach placers are essentially the same as those in the alluvial category. Quartz pebbles and sand predominate, but on some beaches there may be considerable shingle composed of the pebbles and stones of the country rock of the adjacent region. Magnetite and ilmenite form the bulk of the heavy minerals in most deposits. Clay beds and hard pans are common in some beach deposits especially in those of a raised character. These may form false bottoms for pay streaks. Coquina accumulations may be present in some instances and may mark the lines along which the richest pay streaks occur in modern beach placers. In fossil beach placers coquina may be invaluable as marker horizons.

Most of the gold in beach placers is fine grained and generally of high fineness, commonly greater than 900. The richest pay streaks usually follow the strand lines; most are a few inches to a few feet in thickness and generally only a few tens to several hundred feet in width (Figs. 88 and 89). Their continuity along the strandline is frequently erratic, and numerous cutouts are encountered. Elevation or depression of the land leads respectively to the formation of raised or deeply buried beach placers.

Descriptions of typical auriferous alluvial and beach placers and districts

Alluvial placers were the first type of gold deposits worked by man. Ancient Chinese and Hindu writings mention these deposits, and we have the story in Greek mythology of Jason, commander of the good ship *Argo*, who sought the Golden Fleece at Colchis in Asia Minor on the far shore of the Euxine (Black) Sea (Frontispiece). The metal gold also figures extensively in the Old Testament, and there are many allusions to its source in placers. Herodotus and Strabo, the Greek historian and geographer respectively, repeatedly mention the working of alluvial placers in Thrace, the Aegean Islands and Asia Minor. Pliny, writing at the beginning of our era, describes many of the Roman placer operations, which seem to have stretched from Asia Minor to Spain, France and Wales. In particular he mentions gold in the stream beds of the Tagus in Spain, the Po in Italy, the Hebrus in Thracia, the Pactolus in Asia Minor and the Ganges in India, the last probably from heresay. Pliny mentions booming and sluicing and collection of gold on ulex, a rough, prickly plant that was burned and the gold washed out of the ashes. In the Dark and Middle Ages some gold was won from the old placers of the Mediterranean area and western Europe, but by the time of the rediscovery of America by Columbus in 1492 nearly all of these placers were exhausted.

When Columbus made his first landfall in the New World he found the natives in possession of gold, and later exploring in Hispaniola (Haiti) in 1493 on the second voyage he observed crude placer operations in the streams and rivers of the interior. In 1494 Alonso de Ojeda discovered the rich

Cibao placers, and a year later Pablo Belvis arrived from Spain with a large quantity of mercury for amalgamation purposes. The first gold won was sent immediately to the King of Spain, who donated it to Pope Alexander VI in Rome where it was dedicated to the service of Christianity in the gilding of a cathedral dome. Thus began the *auri sacra fames* of the Spaniards in America that was to wreak such havoc on the natives and was also to introduce the Negro slave trade to the New World. The sordid story has been told many times and need not be repeated here; it is one of the black pages in the history of gold mining.

A more pleasant story can be written about the great gold rushes of the last century. Alluvial gold gravels were worked in the Altai of Siberia as early as 1820, but some 9 years later extensive alluvial deposits were found in the Lena Basin, probably the largest alluvial gold deposits known. These discoveries led ultimately to the colonization of southern Siberia. These and other deposits in eastern Siberia are still worked extensively today. The great gold rush of 1849 to California can be said to have opened up the American west. The California gold rush was followed in 1851 by the Australian rushes to New South Wales and Victoria, an impetus to mining that has kept Australia in the forefront of world mineral production ever since. The golden gravels of the Fraser River in British Columbia were known as far back as 1852, but it was not until 1858 that the great stampede up the Fraser began ultimately in 1861 reaching Williams and Lightning creeks in the Cariboo, the most celebrated of all gold creeks in the province. Gold placers in the Yukon Basin were worked as early as 1880, but it was not until 1896, and perhaps 2 years earlier according to some accounts, that gold was discovered on the tributaries of the Klondike River. The great rush took place in 1897–1898 to Dawson – one of the greatest gold rushes in history and certainly the most colourful, made immortal by Robert W. Service in his novel *The Trail of '98*, and in his poems, *Songs of a Sourdough* and *Ballads of a Cheechako*.

Alluvial gold placers in streams and rivers have been the richest and of most interest to the prospector. But some beach placers have also caught his eye and have been extensively worked. Gold was discovered in Alaska as far back as 1865–1866 and alluvial prospecting in the streams and rivers near Nome was in full swing in 1899, the year that gold was discovered on the beaches by a soldier and a prospector. There followed a frenzied digging and grubbing along the coast for many miles, more than one million dollars having been won by hand rockers in less than 2 months (Collier *et al.*, 1908).

The methods of the alluvial placer miner have changed much since Jason's time. Originally, early man probably plucked by hand the heavy shining nuggets from the gravels of the streams. This is the stage at which Columbus and his mining engineers found some of the natives working the placers of Haiti. Later the gravels of the streams were stirred up by the workers using crude booming (hushing) operations and sluiced over the fleeces of sheep and goat pelts, the gold remaining mainly trapped within the wool and goat's hair. This method is apparently still employed in some of the placer streams of Asia Minor, Afghanistan and Mongolia. Panning is an old technique certainly known to the ancients of the Old World and the natives of Africa, who used the calabash (gourd) and the natives of Central and South America who employed the batea. The rocker and the sluice were known to the Greeks and Romans, and they were adept at booming (hushing) operations. The dry washer or dry blower has long been known to those who have sought gold under desert conditions. The water monitor-, bulldozer- and dragline-sluice operations and the great mechanical dredges are modern mechanical adaptations of age-old techniques of separating the gold from the dross.

Gulch and creek placers

Gulch and creek placers are common in nearly all placer districts of the world. They lie in the small valleys, guts and gulches that are tributary to the main stream systems of a district. Literally thousands of these deposits have been worked, nearly all in a small way. The main characteristics of these types of placers are:

1. Most occur in moderately hilly country exhibiting the effects of protracted weathering and denudation. A few occur in regions of alpine topography, but these are generally marked by spotty pay streaks.

2. The gradients of the present (and past) stream systems are moderate.

3. The source of the gold and associated heavy minerals is generally close at hand, either at the heads of the gulches and creeks or along their valley sides. The source of the gold is commonly auriferous quartz veins or gold-bearing sulphide bodies; disseminated gold in quartz blows and stringers in the country rocks and in pyrite and other sulphides in graphitic shales and other rocks is the principal source in some instances.

4. The heavy mineral constituents accompanying the gold are mainly those found in the primary gold deposits, in closely associated deposits or in the enclosing country rocks.

5. The gold is usually coarse and commonly higher in fineness than that in the primary deposits. Large nuggets, wires and crystals are a feature, and nodules of vein quartz or sulphides with veinlets of gold or containing disseminated gold are common.

6. The pay streaks are rich and generally on the bedrock or in the top few feet of the bedrock. Most pay streaks are well defined and fairly regular. False bottoms may occur but are not a common feature.

7. The overburden covering the pay streaks is generally not deep except in depressed areas or where glacial deposits complicate the picture.

Gulch and creek placers are common in the Yukon and a number occur in the Keno Hill–Galena Hill area. One in particular, Dublin Gulch, is typical and will serve as an example of what one might expect to encounter with gulch and creek placers the world over.

The Dublin Gulch and Haggart Creek placers (Fig. 69), discovered and first worked in 1898, have been described in some detail by Keele (1905), Bostock *in* Little (1959), Boyle (1965*a*) and Boyle and Gleeson (1972). The topography is characterized by relatively low rounded hills and numerous small streams tributary to the main water courses, all marks of a deeply dissected upland. The bedrock of the area is folded and faulted quartzite, phyllite, graphitic schist, limestone and quartz-mica schist of the Yukon Group intruded by a body of granodiorite and granite probably of Cretaceous age. Skarn, carrying scheelite, is developed in places near the granitic rocks. In addition there are numerous quartz stringers carrying wolframite and scheelite in the granitic rocks and their contact zone, a cassiterite-tourmaline lode and several northeast striking gold-bearing quartz-arsenopyrite-pyrite-sulphosalt veins. Primary vein material in these veins averages about 0.2 oz Au/ton and 1.5 oz Ag/ton. The gold is largely submicroscopic to microscopic and is mainly in the arsenopyrite, pyrite and sulphosalts; only rarely is free gold seen in the quartz and sulphides.

The country rocks and the gold-bearing veins are deeply weathered, the latter down to at least 20 ft. The stable minerals in the oxidized zones of the veins are mainly scorodite, limonite, residuals of pyrite and arsenopyrite and some anglesite and beudantite. Assays of the oxidized material are highly variable in gold content, ranging from 0.015 to 0.30 oz Au/ton; the silver content is, likewise, highly variable averaging about 0.5 oz/ton. The gold in the oxidized zones consists of fine dust, small flakes and abundant wires. Some of the gold is closely associated with the scorodite and limonite and is probably in a chemically bound or adsorbed form. Most of the gold in the oxidized zones is about 850 fine, although there are marked differences from vein to vein.

The area immediately around Dublin Gulch is beyond the limit of the last well-marked Pleistocene glaciation, but evidence of an early glacial incursion is present as witnessed by what seem to be remnants of weathered tills and also by the appearance in the placer concentrates of nodules of hematite (iron-formation), which probably came from the Snake River deposit in northern Mackenzie Mountains, Yukon. The streams in Dublin Gulch and Haggart Creek have entrenched parts of their courses in deep overburden forming terraces with old modified profiles. The exact nature of the original profile in the lower part of Dublin Gulch is difficult to ascertain because of the extensive stirring and digging since 1898. What appears to have been present at the surface was about 6 to 8 ft of gravels, schist particles, soil and granite boulders, probably mainly due to creep and slope wash. This was succeeded in places by 2 to 3 ft of bluish clay that overlay 3 to 4 ft of yellowish (limonite coated) gravels and weathered debris. The last materials represent the detritus arising from deep secular weathering of the bedrock of the area and the contained deposits during a period of uplift in late Tertiary time.

According to old reports fine colours of gold occurred in the surface materials where washed by the stream in the gulch. Most of the gold, however, is on the bedrock in the bottom foot or so of the lower gravels and weathered debris. The pay streak begins at Bawn Bay Gulch and continues with only a few breaks downstream. It is about 100 ft wide near the mouth of Dublin Gulch, narrowing gradually upstream. The grade is about 0.03 oz/yd.

The gold is accompanied by scheelite (about 1 lb/yd^3), some wolframite and a little cassiterite. In addition there are a number of other heavy minerals, including magnetite, hematite nodules (iron-formation), arsenopyrite nodules, jamesonite nodules, bismuth, galenobismutite, rarely bismuth tellurides, pyrite, tourmaline and garnet. The gold is present as fine dust, scales, rough wires and sprigs, occasional crystals and small nuggets ranging from pea size up to about the size of hickory nuts. Most of the nuggets are worn and pitted. The writer did not observe any free gold in the quartz or sulphide pebbles and so far as is known none of this type of gold has been reported. Mustart (1965), however, found angular arsenopyrite fragments in some of the nuggets. The principal minor and trace metals in the gold are Ag, Cu, Fe, Hg, Bi and Sn. All of these occur in the deposits from which the gold came. The average fineness of the Dublin Gulch gold is about 900.

The Dublin Gulch area provides an excellent example where gold can be traced from primary deposit to stream placer. The primary gold is mainly microscopic to submicroscopic and associated with pyrite, arsenopyrite and sulphosalts. Particles of gold greater than 100 μ are rare in the writer's experience. On the oxidation of the sulphides and sulphosalts the gold was released and formed flour or mustard gold in some cases, but there was also a nucleation and growth of gold on fine rough wires, flake and small nuggets in the oxidized zones. All of these types of gold passed gradually into the eluvium where they probably accreted more gold on their journey to the gulch. Here, they passed slowly down gradient, some particles possibly accreting more gold but most being rolled along and hammered by the pebbles of the mills of nature. The fineness has changed during the journey from oxidized zone to placers in Haggart Creek a distance of some 3 mi. In the oxidized zones the fineness averages about 890; in Dublin Gulch 900 and in Haggart Creek 925.

Stream and river placers

The main features of these placers are discussed in some detail above. Briefly their characteristics are: (1) most occur in districts with a subdued topography marked by broad, often terraced, entrenched valleys developed in a terrain of rounded, deeply weathered hills; (2) the present (and past) gradients of the streams are moderate to low; (3) the pay streaks are generally not as rich as those in gulches and creeks, but they are longer, commonly wider and more uniform. Multiple pay streaks may occur at the same elevation or they may be stacked at different elevations. Most of the rich pay streaks are on the bedrock; (4) the gold is generally finer grained than that in creek and gulch gravels. Large nuggets are relatively rare in stream and river placers; (5) the fineness of the gold is usually higher than that in creek, gulch and eluvial placers; (6) the overburden is commonly tens of feet to hundreds of feet deep; and (7) most stream and river placers are amenable to large scale hydraulicking and dredging operations.

There are many famous districts where stream and river placers have been extensively worked. Some provide good examples and will be described briefly. They include the Klondike district in Yukon, the Cariboo and other districts in British Columbia, the Chaudière and other areas in southeastern Quebec, the Sierra Nevada of California; the Chocó in Columbia, the deep leads of Victoria, Australia, and the extensive placers of the Lena, Aldan, Amur and other drainage systems in Siberia, U.S.S.R., Mongolia and northern China.

Klondike district, Yukon. The stream and river placers of the Klondike district near Dawson, Yukon were discovered in 1896 or earlier and have yielded some 10 million ounces, more than half of the placer gold production of the Western Cordillera of Canada. Their heyday is past, although there are still some small hydraulic and bulldozer operations on some of the streams. Extensive descriptions of the placers of the Klondike are given by McConnell (1905, 1907) and more recently by Gleeson (1970).

The most productive creeks were Bonanza, Hunker, Dominion, Gold Run, Sulphur and Quartz (Fig. 76). The underlying bedrock is mainly the Klondike Schist (Klondike Series) comprising folded and faulted sedimentary quartz-mica schists, chlorite schists, sericite schists, quartzites, phyllites, pyritic graphitic schist and highly sheared quartz porphyry sills or flows sandwiched between the sediments. The age of these rocks is not precisely known; they are probably Precambrian.

Throughout the Klondike Schist there are myriad primary quartz bodies, most carrying traces to small amounts of gold. Briefly these bodies include:

1. Quartz boudins, blows, stringers and bedded veins sandwiched between the sedimentary beds: These carry small amounts of pyrite and all are auriferous carrying however, only small amounts of gold, generally in the range of 0.01 to 0.05 oz/ton.

2. Quartz-barite veins containing minor amounts of galena: Most of these veins are slightly auriferous containing approximately 0.02 oz Au/ton and 0.03 oz Ag/ton. The gold is said to occur in the native state in these deposits.

3. Quartz veins in the chloritic phase of the Klondike Schist: Some of these veins and their adjacent wall rocks contain coarse cubes of pyrite and minor amounts of galena. Spectacular specimens of free gold are said to have been obtained from these veins. On the average, however, the quartz and pyrite are only feebly auriferous.

4. Stringers, veins and irregular lenses of quartz in extensive, wide northwest-trending shear zones that exhibit marked sericitization and pyritization: The metallic minerals are mainly pyrite, galena and sphalerite. Some barite is present. Gold values are spotty and appear to be concentrated in pockets with galena, sphalerite and pyrite. Some of these assay in the ounces per ton; the quartz zones as a whole are very low grade (0.01–0.05 oz Au/ton). The gold is mainly free as small grains and ragged particles.

The pyritic graphitic schists of the Klondike Schist are also feebly auriferous. They contain about 0.005 ppm Au, mostly in the pyrite, samples of which contain up to 2 ppm Au according to our analyses.

The Klondike district is a thoroughly dissected upland, a part of the Yukon Plateau, marked by rounded hills and a multitude of small streams tributary to the main water courses. The valleys are flat and wide in their lower reaches, but gradually narrow towards their heads into steep-sided narrow gulches ending abruptly in broad amphitheatres. Many of the lower slopes of the valleys are conspicuously terraced. The district has not been glaciated, but is underlain by permafrost, in places to a depth of 200 ft. This necessitates thawing the ground out before dredging operations can begin.

The placers of the Klondike owe their existence to a general uplift in late Tertiary time followed by deep secular weathering of the bedrock and the contained gold-bearing quartz bodies. V-shaped valleys were formed first, and these were gradually widened and filled at maturity with extensive gravel deposits. A later period of uplift, probably of Recent or Late Pleistocene age, has resulted in the streams channelling their valleys deeply, almost to bedrock in places, and leaving the characteristic terraces as benchmarks in their downward migration.

McConnell (1905) classified the auriferous Klondike gravels into three categories as follows:

- High level gravels
  - River gravels
  - White Channel gravels
    - Yellow gravels
    - White gravels
- Gravels at intermediate levels
  - Terrace gravels
- Low level gravels
  - Gulch gravels
  - Creek gravels
  - River and stream gravels

Their stratigraphic relationships are shown in Figure 77, and their description as given below is paraphrased only slightly from McConnell's classic report.

The 'low level creek gravels' are the most important gravels in the Klondike. These gravels floor the bottoms of all the valleys to a depth of from 4 to 10 ft. They rest on bedrock usually consisting of decomposed and broken schists and are overlaid by a sheet of black frozen muck ranging in thickness from 2 to 30 ft or more. They are local in origin and consist entirely of the schists and other rocks outcropping along the valleys. The schist pebbles are usually flat round-edged discs measuring 1 to 2 in. in thickness and 2 to 6 in. in length. They constitute the greater part of the deposit but are associated with a varying proportion of rounded and subangular quartz pebbles and boulders, and less frequently, with pebbles derived from the later eruptive rocks of the region. The pebbles are loosely stratified, are usually embedded in a matrix of coarse reddish sand and alternate in places with thin beds of sand and muck.

The creek gravels frequently enclose leaves, roots and other vegetable remains and also the bones of various extinct and still existing northern animals, such as the mammoth, buffalo, bear, musk-ox and mountain sheep and goat.

The 'low level gulch gravels' occupy the upper portions of the main creek valleys and small tributary valleys. They differ from the creek gravels in being coarser and more angular. A considerable proportion of their material consists of almost unworn fragments of schist washed down from the adjacent slopes. They contain the same vegetable and animal remains as the creek gravels.

The only 'low level river gravels' of the district proven, so far, to contain gold in paying quantities occur in the wide flats bordering the lower portion of the Klondike River below the mouth of Hunker Valley. The river gravels consist of quartzite, slate, chert, granite and diabase pebbles largely derived from the western slopes of the Ogilvie Range. They are harder and better rounded than the creek gravels, a result of the greater distance travelled.

The 'intermediate level terrace gravels' lie on rock terraces that occur at various points cut into the steep slopes of the present valleys. The terraces were produced during the

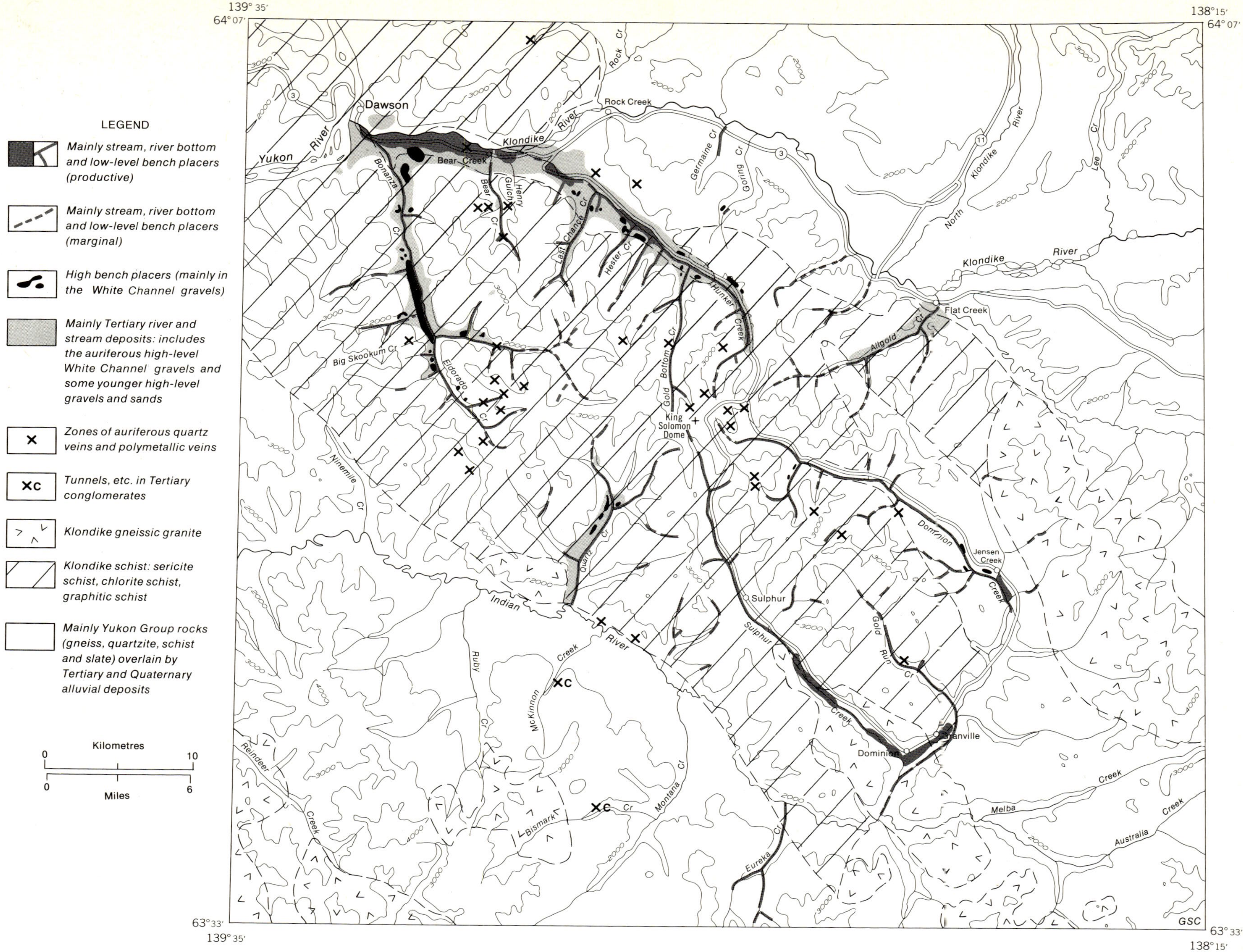

Figure 76. Generalized geology of the Klondike (Dawson) district, Yukon showing the relation-

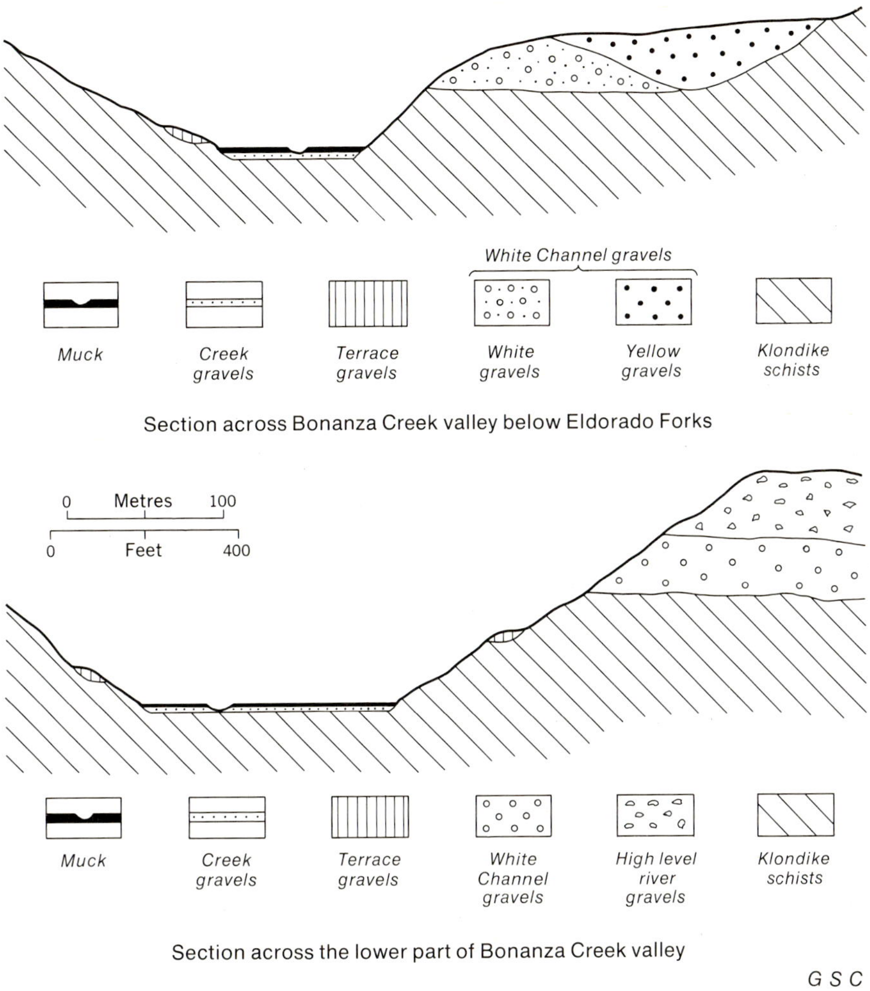

Figure 77 Generalized sections across the placer deposits of Bonanza Creek, Klondike (Dawson) district, Yukon (after McConnell, 1907).

deepening of the valleys and are simply remnants of former valley-bottoms. Most are small, seldom exceeding a few yards in width and a few hundred yards in length, irregular in distribution and occur at all elevations up to the bottoms of the old valleys. The beds of gravel supported by the terraces are usually from 6 to 15 ft in thickness, very similar to those in the creek bottoms but showing somewhat more wear. The terrace gravels, like the creek gravels, are overlain, as a rule, with muck, and at one point on Hunker Creek were found buried beneath 100 ft of this material.

'High level river gravels' are extensively distributed along Bonanza and Hunker creeks and some of their tributaries and also occur on Eldorado, Bear, Quartz, Nine Mile and Allgold creeks. They consist, principally, of ancient creek deposits overlain near the mouths of some of the valleys by gravels laid down by the Klondike River when it ran at a much higher level than at present and occupied a somewhat wider valley.

These gravels occur at various points along the Klondike River. In the Klondike district they are found covering the small plateaux in which the ridges separating Bonanza and Hunker creeks from the Klondike River terminate. They rest, in both places, on high level creek gravels at an elevation of about 450 ft above the present valley bottoms. They have a thickness of from 150 to 175 ft, and consist principally of well-rolled pebbles of quartzite, slate, chert, granite, diabase and conglomerate embedded in a matrix of grey sand and derived, like those in the present stream, from the western part of the Ogilvie Range. The high level river gravels are reported to contain gold in paying quantities at Acklens farm, a name given to a portion of the bench on the right limit of the Klondike, 2 mi above its mouth; they are, however, generally of little economic importance.

The 'high level creek gravels' or 'White Channel gravels' are ancient creek deposits laid down in the wide, flat-bottomed valleys, which characterized the region before the last general uplift. After their deposition, the country was elevated 600 to 700 ft, and the increased grades acquired by the streams enabled them to cut down through their old gravel beds into the bedrock beneath, and to excavate the steep-sided troughlike valleys in which they now run. The old gravels now occur on wide benches bordering the present valleys at elevations of from 150 to 300 ft above them, the elevation generally increasing downstream. Their distribution along the valleys is irregular, as a large portion of the deposit was destroyed during the deepening of the main valleys and the tributary valleys and gulches.

The general character of the White Channel gravels is remarkably similar in the various Klondike creeks but differs considerably from the ordinary type of stream deposits in other parts of Yukon. They consist of a compact matrix of small, clear, little-worn and often sharply angular grains of quartz and scales of sericite thickly packed with rounded quartz pebbles and rounded and subangular and wedge-shaped quartz boulders often 2 to 3 ft in diameter. Flat and subangular pebbles of sericite schist, the principal rock of the district, are also present but in much smaller numbers than the quartz constituents. The schist pebbles are usually decomposed and crumble rapidly when thawed out. The deposit is always stratified, but, except in rare instances, there has been no sorting of the various constituents into separate beds, and the composition is very uniform throughout. The colour is characteristically white or light grey due to the preponderance of the quartz constituents and the leaching out of the greater part of the iron. The colour is darker and the sands are noticeably coarser towards the limit of the deposit on the upper part of the creeks.

The White Channel gravels vary in thickness from a few feet to 150 ft and in width from 100 ft to ½ mi or more. The deposit increases in volume descending the streams, and attains its greatest development near their mouths.

The deposits, unlike the creek and gulch gravels, appear to be destitute of vegetable and animal remains. None were found by McConnell (1905) and the few reported discoveries by miners lack confirmation.

On Gold, Adams and other hills on Bonanza Creek the typical compact white variety of the White Channel deposit is replaced towards the sides of the old valley by flat rusty coloured gravels, more loosely bedded and containing a smaller proportion of quartz than the ordinary white variety. These probably represent flood plain deposits. They have the appearance of overlying the white variety and were formerly, in the absence of sections, considered by McConnell to be younger. However, the long exposures, now available for study in the various hydraulic cuts, show that the two varieties pass gradually one into the other both horizontally and vertically and in places are interbedded, evidence of nearly contemporaneous deposition. The loose yellow variety seldom carries much gold.

McConnell (1905) concluded that the White Channel gravels were probably deposited by winding streams with easy grades and comparatively slack currents. The preponderance of vein quartz pebbles and boulders, the most resistant rock in the district, gives them the character of a residual deposit. They were built up slowly and in the long process the softer rocks were mostly destroyed and carried away.

The great length of the White Channel period is indicated by the enormous gold accumulations, all derived from the slow breaking up of auriferous quartz veins that took place in it. Since the close of the period the additions to the supply have been trifling, although a sufficient time has elapsed to enable small streams to excavate channels, mostly through hard schists from 300 to 600 ft in depth. Practically all the gold in the present low level valley flats is of secondary origin and derived from the partial distribution of the older gravels.

The age of the White Channel gravels has not been determined, but they must date back to the Pliocene at least. They were certainly deposited before the advent of the present severe climatic conditions, as the white coloration is largely due to the leaching out of the greater portion of the iron by circulating surface waters, and this must have taken place before they were permanently frozen.

Summarizing we can say that the gold in the Klondike placers has passed through at least four stages of concentration as follows: (1) primary concentration in quartz veins and other bodies; (2) secondary concentration in the oxidized zones of the quartz veins and other auriferous bodies; (3) tertiary concentration in the placers of the high level White Channel gravels; and (4) quaternary concentration in the low level gulch, creek and stream gravels.

Each of these stages has resulted in a successive enrichment, the last stage being phenomenally rich in some creeks.

The rich pay streaks in the White Channel gravels invariably occur in the first few feet above the bedrock or in the upper foot of the bedrock. False bottoms are rare. The remainder of the gravels are slightly auriferous carrying a few cents per yard. In the low level creek, gulch and river gravels and in the terrace gravels the pay streaks are commonly on the bedrock in the first few feet of gravel and weathered debris. Where the bedrock is tight the thin stratum of gravel immediately above the bedrock is exceedingly rich. Where the bedrock is open the first few feet of the bedrock and the first few feet of the overlying gravels constitute pay material. In some places in a few creeks the shattered and frost wedged schist layers of the bedrock near the erosion surface contained all of the gold, the overlying gravels being practically barren. Most of the pay streaks in the Klondike were remarkably persistant in length, wide, some up to 1000 ft or more, and exceedingly rich. The one in the low level creek gravel of Eldorado Creek for instance ran some 65 oz per running foot for 4 mi.

The gold of the pay streaks is variable both in appearance and fineness. Near the heads of streams and in the gulches and draws the gold in the low level gravels occurs as coarse, commonly angular grains, flakes, sprigs and wires, and as nuggets that are fairly large. Farther downstream the gold particles are more rounded, flattened, finer grained, and large nuggets are uncommon. The gold in the White Channel gravels ranges from fine grained to relatively coarse, rough flakes, wires and particles. Nuggets are common but not generally very large. In both types of gravels, low and high level, the gold commonly has a crystallized character, and some of the nuggets enclose quartz or are partly enclosed by this mineral. The quartz is identical with that seen in the gold-bearing quartz veins, blows, etc. according to the writer's observation. Dendritic gold occurs in places on pebbles and stones in the placers, and crystals are not uncommon.

Mustart (1965) and Gleeson (1970) give spectrographic analyses of gold from a number of the creeks. All samples report Ag, Cu, Hg, Ti, Mg, Al and Fe, and some contain traces of Pb, As, Sb, V, Ba and Sn. Mustart noted an unidentified Au-Sn-Pb compound in gold from Monte Cristo Hill, Bonanza Creek.

The fineness of the Klondike gold is exceedingly variable not only on different creeks but also along different portions of the same creek. The lowest fineness recorded appears to be about 625 and the highest 890. McConnell (1907) concluded that the grade of the placer gold appeared to depend mainly on original differences in the grade of vein gold. He wrote:

> The variation in grade of the placer gold appears to depend mostly on original differences in grade of the vein gold from which it was derived. Creeks draining certain areas in the district carry low grade gold, while other areas supply high grade. An important centre of dispersion for low grade gold occurs west of the lower portion of Hunker Creek. Hester and Last Chance Creeks, Henry Gulch and Bear Creek all head in the same ridge within a comparatively short distance of each other and all carry low grade gold. Big Skookum Creek, a tributary of Bonanza Creek, heads in a low grade area and the gold brought down by it lowers appreciably the general grade of the Bonanza Creek gold for several claims. The Dome and surrounding region furnishes a good example of a high grade area. The streams flowing outwards from this centre, including Upper Dominion, Upper Hunker, Sulphur and Gold Bottom Creeks, all carry high grade gold although the values differ considerably.

McConnell (1907) also thought that there was an increase in the fineness of some of the gold due to leaching out of silver from the near surface parts of the nuggets. We shall return to this problem later.

Gleeson (1970) studied the minerals accompanying gold in the Klondike placers. The most abundant suite in the low level creek gravels included actinolite, barite, chlorite, epidote, hornblende, hypersthene, garnet, goethite, ilmenite, leucoxene, magnetite, pyrite, sphene, rutile, tremolite, tourmaline, zoisite and zircon; the less abundant to rare suite composed andalusite, anatase, apatite, biotite, cerussite, chromite, clinopyroxene, clinozoisite, dolomite, jarosite, micrometeorites, fluorite, kyanite, olivine, spinel, staurolite, topaz, monazite, scheelite, sphalerite and cassiterite. The heavy minerals in the White Channel gravels are the same but slightly more rounded. Highly resistant minerals such as garnet, magnetite and zircon are generally more plentiful than the softer minerals such as barite, goethite, pyrite and jarosite.

The origin of the gold in the Klondike placers has long been debated because few sizeable and rich primary gold-bearing deposits have ever been found in the district. McConnell (1907) originally considered that the gold came from the quartz deposits in the district, a view that is essentially correct in the writer's opinion.

There are a number of facts which support the view that most of the gold came from the quartz deposits, including the blows, stringers, etc. that are probably metamorphic secretion products and are younger than the quartz-barite and quartz sulphide veins mentioned above. These facts can be enumerated as follows:

1. All of the quartz-bearing deposits are auriferous. Some carry visible gold, occasionally in spectacular amounts.
2. The gold nuggets with quartz indicate that at least some of the gold came from quartz deposits. The quartz in the nuggets can be matched with that in the primary quartz deposits in the writer's opinion.
3. A number of the heavy minerals accompanying the gold in the placers evidently came from the gold-bearing veins. Barite is one of these diagnostic minerals, galena is another. Some of the pyrite in the placers matches that in the veins, and some came from the pyritic graphitic schists.
4. There is a general spatial relationship between the gold placers and the areal distribution of the Klondike Schist and the gold-bearing deposits it encloses. (Fig. 76). This suggests that the vein deposits, and probably also some of the rocks such as the pyritiferous graphitic schists, gave rise to the gold placers.

It seems probable, basing conclusions on these facts, that the gold came mainly from the quartz deposits enclosed within the Klondike Schist. The amount of quartz as a rough estimate in the Tertiary White Channel gravels as originally laid down probably exceeded 6×10^9 tons which at 0.01 oz Au/ton of quartz according to our average analyses would yield about 60 million ounces of gold of which say one half, or 30 million ounces were released to accumulate in the modern gulch and stream placers. If we add gold which would have been released during Tertiary time from the weathered pyrite in the country rocks, especially the graphitic schists, the amount of gold for placer concentration in the White Channel gravels would be greatly augmented.

The Cariboo, British Columbia. The Cariboo, with Barkerville its capital, was the premier placer gold district of British Columbia. It lies in the central part of the province within the intermontane plateau region bounded by the Coast Range on the west and Cariboo and other mountain ranges on the east. The placer gold production of the district is probably of the order of 2.5 to 3 million ounces.

The Cariboo is a deeply dissected region with low rounded hills and an irregular pattern of streams, creeks and gulches. The weathering and erosion that gave rise to the dissection of the country evidently originated in early Tertiary time and extended throughout that period. In Pleistocene time a stagnant ice sheet lay over the land, removing much of the weathered mantle at higher elevations but having little effect on the placer deposits in most of the valleys. The bedrock is folded and faulted phyllite, quartzite, argillite, slate and limestone of the lower Paleozoic Cariboo Group. These enclose a host of quartz bodies ranging from stringers, blows and boudins to fairly well-defined quartz-vein zones and siliceous replacements in the limestones. All these deposits are auriferous, some of the vein zones and siliceous replacements having been mined extensively (see page 281). The principal minerals in the auriferous deposits are pyrite, galena, chalcopyrite, sphalerite, cosalite, bismuthinite, arsenopyrite, pyrrhotite, scheelite and free gold. Details of the geology of the Cariboo district are given by Hanson (1935) and Sutherland Brown (1957); the placers are described at length by Johnston and Uglow (1926).

The placers of the Cariboo (Barkerville) district, discovered in 1860, are now largely exhausted after having yielded probably over 2.5 million ounces of gold. The principal productive creeks were Williams and Lightning, but a number of shorter creeks produced considerable quantities of gold (Fig. 78). Several types of auriferous gravels are found in the district, and the gold has had a complex history following its release from the deposits and rocks during extensive oxidation and erosion in Tertiary time. First there was an extensive accumulation of gravels and gold in the valleys during a

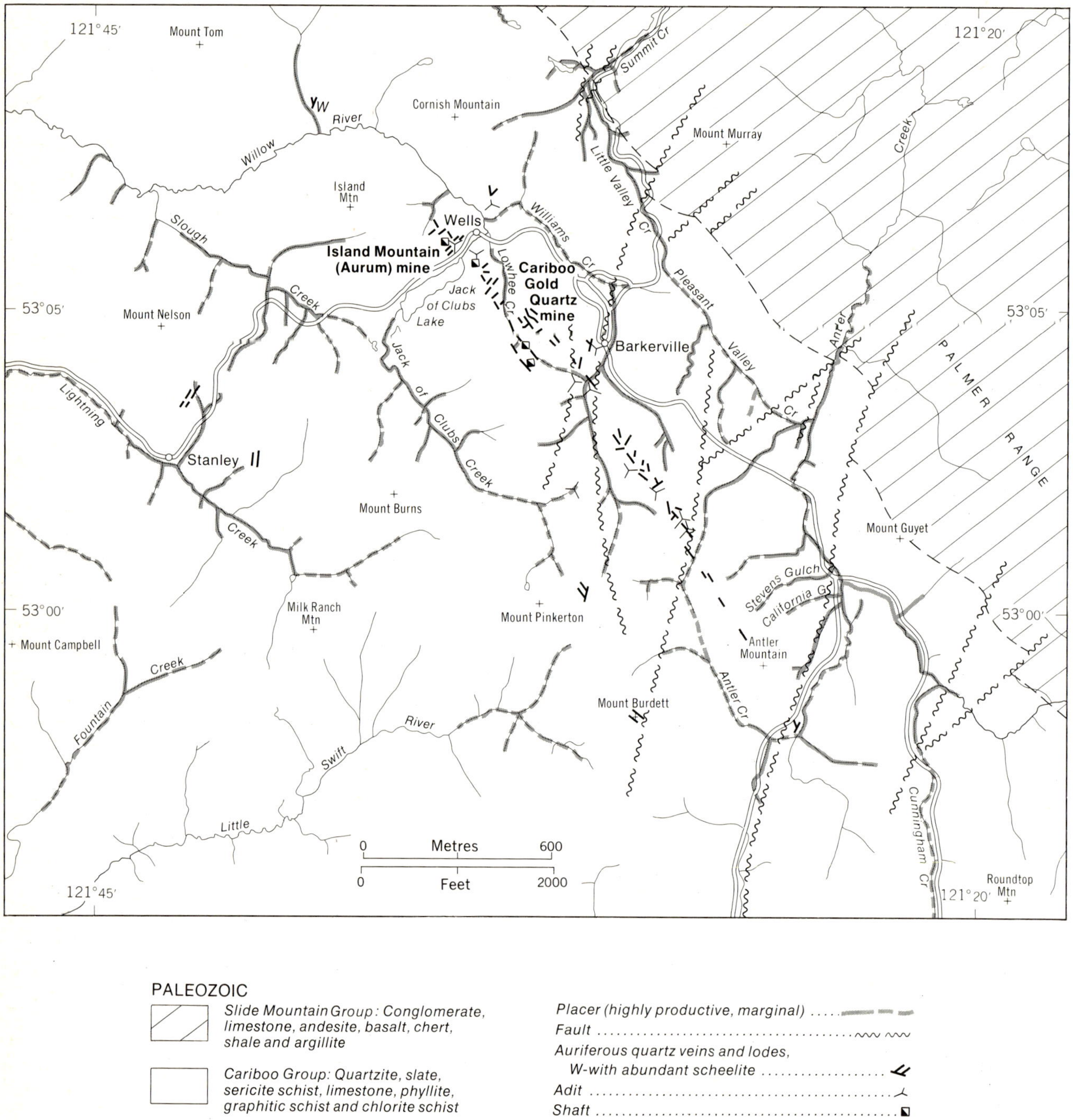

Figure 78. Generalized geological map of the Barkerville (Cariboo) district showing relationship of the auriferous placers to the bedrock gold veins (after Johnston and Uglow, 1926; Hanson, 1935; Sutherland Brown, 1957; with additions by the writer from field observations and airphotos).

gradual period of Tertiary uplift. This was followed in Pleistocene time by the development of valley glaciers and stagnant ice sheets, the latter largely protecting the Tertiary gravels from scouring and destruction. During interglacial periods some gold placers were formed probably from reworked gravels at the heads of valleys. In addition some outwash deposits with some gold were formed in places from the sluicing action of the meltwater on the glacial debris and old gravels. Slight uplift after the disappearance of the glaciers rejuvenated the streams, and some have cut down and eroded the drift deposits yielding postglacial gravels that are slightly auriferous in places. Briefly the types of auriferous gravels corresponding to the various periods are as follows with accompanying notes:

1. *Ancient stream gravels resting on bedrock in the valley bottoms and on rock benches or in old channels.* These averaged about 5 ft in thickness and were the richest and most important of the auriferous gravels. They consisted of weathered and water-worn pebbles of country rock and considerable amounts of old talus or scree. Quartz pebbles were comparatively rare. The pay streaks were relatively persistent in length, developed mainly in the narrow, deep parts of the creeks and were mainly on the bedrock or in the upper few feet of the weathered and shattered rock. The gold was mostly coarse, very little flour or fine scale gold being present. Flattened grains, pellets and small nuggets of gold predominated, but there were also some crystals, angular particles, mammillary nuggets and quartz nodules with veinlets of gold. The accompanying heavy minerals of interest were pyrite, galena, scheelite and barite. The gold in these gravels was concentrated during Tertiary time, and the gravels were covered by glacial drift, in places up to 250 ft deep, in Pleistocene time. During Pleistocene time some of the Tertiary gravels were reworked giving rise to those described below.

2. *Interglacial stream gravels.* These are generally thought not always underlain and overlain by glacial drift. In most valleys they occur above a false bottom of boulder clay well above the bedrock of the creek. Most of the gold appears to have come from the reworking of ancient (Tertiary) gulch and creek gravels near the heads of the valleys and from glacial drift (eluvium) scraped off the adjacent hills. The gold in these gravels is essentially the same as that mentioned for the ancient stream gravels. Most of the pay streaks were lean but fairly extensive.

3. *Auriferous glacial outwash gravels.* These occur in irregular masses included in the glacial drift and to some extent in glacial gravels filling parts of the stream valleys. The gold is highly disseminated through these gravels, and pay streaks of value do not seem to occur in them.

4. *Postglacial stream gravels.* These have been derived mainly from the erosion of the glacial drift and ancient (Tertiary) auriferous gravels. They were deposited in places in the beds and along the benches of the present streams. Some were productive, but the pay streaks were thin. Some of the gold in the surface gravels on the benches was fairly coarse, but as a general rule the gold in most of the postglacial stream gravels was fine grained and commonly scattered through the gravels.

The placer gold of the Cariboo varied in fineness from about 775 to about 950 as compared with the vein gold which varied from 850 to 910. Johnston and Uglow (1926) found no evidence that the placer gold was, as a rule, of greater fineness than the vein gold. They considered that most of the gold was detrital and supported their view with evidence similar to that recorded above for the Klondike. They noted gold-rich rims on the nuggets, which they thought was due to leaching of silver. They thought that perhaps some of the placer gold resulted by accretion processes in the gravels, but concluded that this mode of origin was probably of little importance in the overall formation of the placers.

The precise source of the gold in the Cariboo placers is uncertain. The northwesterly trending vein zones that cut the rocks west of Wells and Barkerville (Fig. 78) may have provided much of the gold in Williams Creek and its many tributaries and that in Stevens Gulch, California Gulch and other gulches draining the Antler Mountain area. The source of the gold in Lightning Creek, Jack of Clubs Creek and the left limit tributaries of Slough Creek remains undetected. There may be a series of northwest trending vein zones that underlie the heavily drift covered area Mount Pinkerton–Mount Burns–Mount Nelson.

Other British Columbia placers. A number of other placer areas in British Columbia (Fig. 79) are of interest since they exhibit features that differ in some respects from the normal placers of the province. In the Atlin district, a deeply dissected and glaciated plateau is underlain mainly by sedimentary, volcanic and intrusive rocks ranging in age from Precambrian to Pleistocene (Aitken, 1959). The placers are stream placers found in an area containing base metal sulphide deposits and myriad quartz veins carrying some gold. The auriferous stream gravels are of two kinds: (1) yellow, highly decomposed gravels, some probably of Tertiary age, others of pre-Wisconsin interglacial age and generally buried beneath glacial drift, and (2) postglacial gravels formed by stream erosion of glacial drift and ancient gold-bearing gravels. The first type includes the important economic gold placers. They also carry some wolframite in places (e.g., Boulder Creek). Some of the auriferous gravels occur beneath Tertiary and Pleistocene basalts and scoria and in this respect are similar to the deep leads of Victoria, Australia. In the Cassiar district stream and river gravels in the Dease Lake area are similar to those at Atlin. The main auriferous rivers and streams were Dease, Thibert, McDame and Goldpan (Hanson and McNaughton, 1936; Gabrielse *et al.*, 1962). In places east of Dease Lake and along the Stikine River above Telegraph Creek, slightly auriferous stream and river gravels underlie basaltic lavas that are apparently Pleistocene in age.

In the Princeton district of central British Columbia the underlying rocks are a succession of volcanics ranging in age from late Paleozoic to late Tertiary; sedimentary rocks, mostly in minor amounts, interbedded with the volcanic groups; and intrusive rocks ranging in composition from granite to peridotite and in age from Jurassic to late Cretaceous or early Tertiary (Camsell, 1913; Rice, 1960; Raicevic and Cabri, 1976). The most interesting placers of the area occur in the Similkameen and Tulameen rivers. These carry both gold and platinoid minerals accompanied by chromite, magnetite and occasionally native copper. The ratio of platinoids to gold is about 1 to 4 in the Similkameen River and lower reaches of

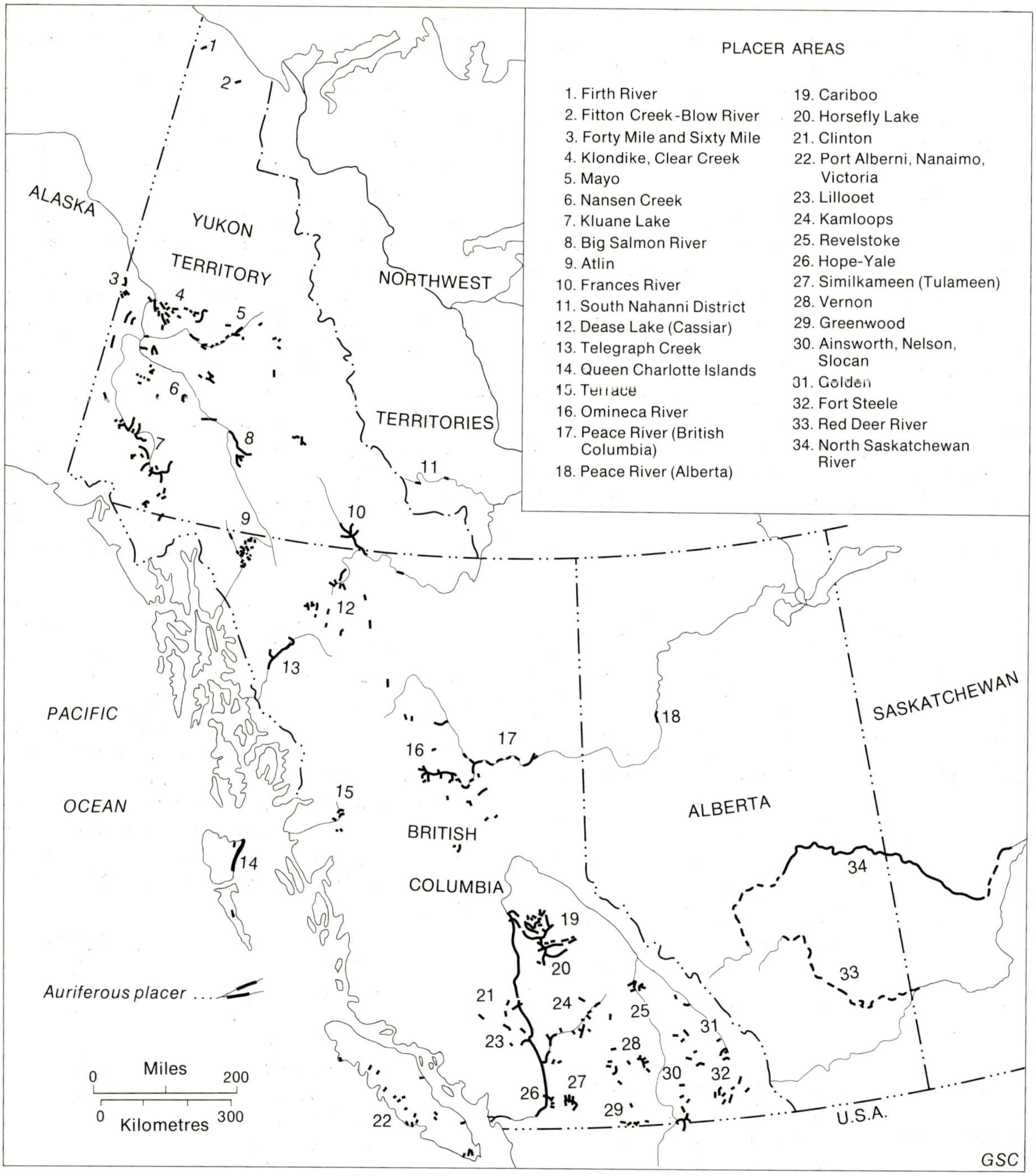

Figure 79. Placer gold districts in western Canada.

the Tulameen River, but increases upstream to a point where the platinoids equal and in places exceed the amount of gold. The platinoid minerals (some 19 in number; Raicevic and Cabri, 1976) are mainly isoferroplatinum, native platinum, ferroan platinum, tulameenite, iridosmine, osmium and ruтheniridosmine. The gold, as described by Camsell (1913) and Rice (1960), occurs in rough, angular or slightly flattened, rarely well-flattened nuggets. Flour gold is comparatively scarce. The largest nugget reported weighed about 8 oz, but nuggets over 1 oz in weight are rare. Quartz can be seen adhering to occasional nuggets, and this and the general lack of flattening suggest that the gold has not moved far from its source. The platinum nuggets are not generally flattened and rarely even subangular, but occur in small rounded grains resembling fine shot, and mostly of relatively uniform size. These nuggets are smaller than the gold on the average, but at the same time the platinum does not occur in fine, flaky particles. Most of the pelletlike nuggets have a rough pitted surface, many show adhering grains of chromite and magnetite, and olivine and pyroxene are occasionally attached to them.

Commenting on the age and origin of the gold and platinum placers Rice (1960) had the following to say:

> The original source of the gold and platinum in the placer deposits is believed to be the gold-bearing veins of Grasshopper Mountain and vicinity and the Olivine Mountain body of platinum-bearing ultrabasic rocks. Erosion of these bodies has released their metal content, but it is evident that only the residue of a large amount of such rocks could have produced the rich deposits of Tulameen and Similkameen Rivers. Since glacial time canyons have been cut in the floors of some of the valleys, but over most of the area erosion has not succeeded in removing the mantle of glacial material, much less eroding any quantity of the underlying rock. It is, therefore, clear that the gold and platinum of the placers must have been released from the parent rocks by erosion in pre-Glacial time and deposited in pre-Glacial placers. It also follows that these placers must have partly escaped the dissipating effect of glaciation, and that the present deposits are mainly either remnants of pre-Glacial placers or those only slightly disturbed by the ice. There has, of course, been some reconcentration of the earlier deposits in the beds of the present streams, but it seems most unlikely that Glacial and post-Glacial streams could have formed the rich placer deposits of the Tulameen except by reworking deposits that were already rich.

Chaudière River basin, Quebec. The gold placers of the Beauceville area, in the Chaudière River drainage area, 50 mi southeast of Quebec, were discovered circa 1830 and since that time have yielded about 100 000 oz of gold. By comparison these placers were small, but they are of interest since they occur in a glaciated terrain of eastern Canada.

The bedrocks are of Ordovician age, belonging to the Beauceville Formation and Caldwell Group and include folded and faulted argillites, quartzites, slates, tuffs and basic volcanics, all containing relatively large amounts of pyrite. Innumerable quartz bodies occur in all of these rocks especially in the quartzites, in every conceivable form – as contorted masses, as lenses and veins in fractures and small faults and as blows, boudins and irregular masses. Some carry carbonates and pyrite; other sulphides such as arsenopyrite, chalcopyrite and galena are rare. No gold was seen in any of the quartz bodies by the writer, but channel samples taken for analyses reported from 0.01 to 0.02 ppm Au. The sedimentary rocks reported less than 0.01 ppm Au, but pyrite separated from these rocks contained up to 0.4 ppm Au with an average of about 0.025 ppm Au.

Details of the general geology of the Beauceville area are given by Cirkel (1911), MacKay (1921), Tolman (1936), Gorman (1955) and Béland (1957). The surficial geology is reported on by Gadd (1964). The placers are described in detail by MacKay (1921).

The Beauceville area lies within the Chaudière basin, which is characterized by deeply dissected uplands with low rounded hills, wide valleys and rock terraces marking stages in the lowering of the valleys, all features denoting an area of mature topography. The final stages of the sculpturing of the terrain were evidently accomplished in Tertiary time and were followed during the Pleistocene by extensive glaciation, which has left a mantle of glacial till and other deposits over much of the area. Two directions of glaciation are evident, both resulting from a general south to southeastward movement of the ice. In valleys lying in the general direction of the ice movement the gold placers were destroyed, whereas in the smaller valleys with axes across the direction of movement of the ice the auriferous gravels were preserved and covered with glacial drift.

Considerable amounts of gold have been won from three rivers in the area, including in their order of importance Rivière Gilbert, Rivière du Moulin and Rivière des Plantes and in small amounts from the following water courses; rivières Chaudière, Famine, Linière, Metgermette and Cumberland and ruisseaux Bolduc, d'Ardoise and Stafford (Fig. 80). The surficial deposits in all of these watercourses consist of a lower series of gravels, richly auriferous in places, overlain by glacial sands, gravels, boulder clay, till, etc. that have been reworked and redeposited in most streams and rivers as shallow stream gravels. A section through the surficial deposits in Rivière Gilbert, as given by MacKay (1921), shows the following sequence:

	Thickness
Surface	(ft)
Loam and river gravel	4
Hard, blue boulder clay	36
Hard grey sand	23
Firm grey sand	10
Gravel and sand	14
Ferruginous sand	3
Auriferous gravel	7
Bedrock	

A section across Rivière du Moulin at the hydraulic pit is shown in Figure 81.

There are differences of opinion among investigators as to the origin and age of the rich gold gravels in the bottoms of the rivers and streams. MacKay (1921) considered them to be of Tertiary age derived largely from deep erosion during that time. He noted that some of the gravels were cemented by iron oxides (limonite), a criterion he thought indicated Tertiary derivation. Gadd (1964), however, considered that the gold-bearing gravels belong to the nonglacial sand and gravel sequence that are no greater than interglacial in age and probably interstadial. He noted that some of these gravels and sands rest on older till and in turn are capped by younger till

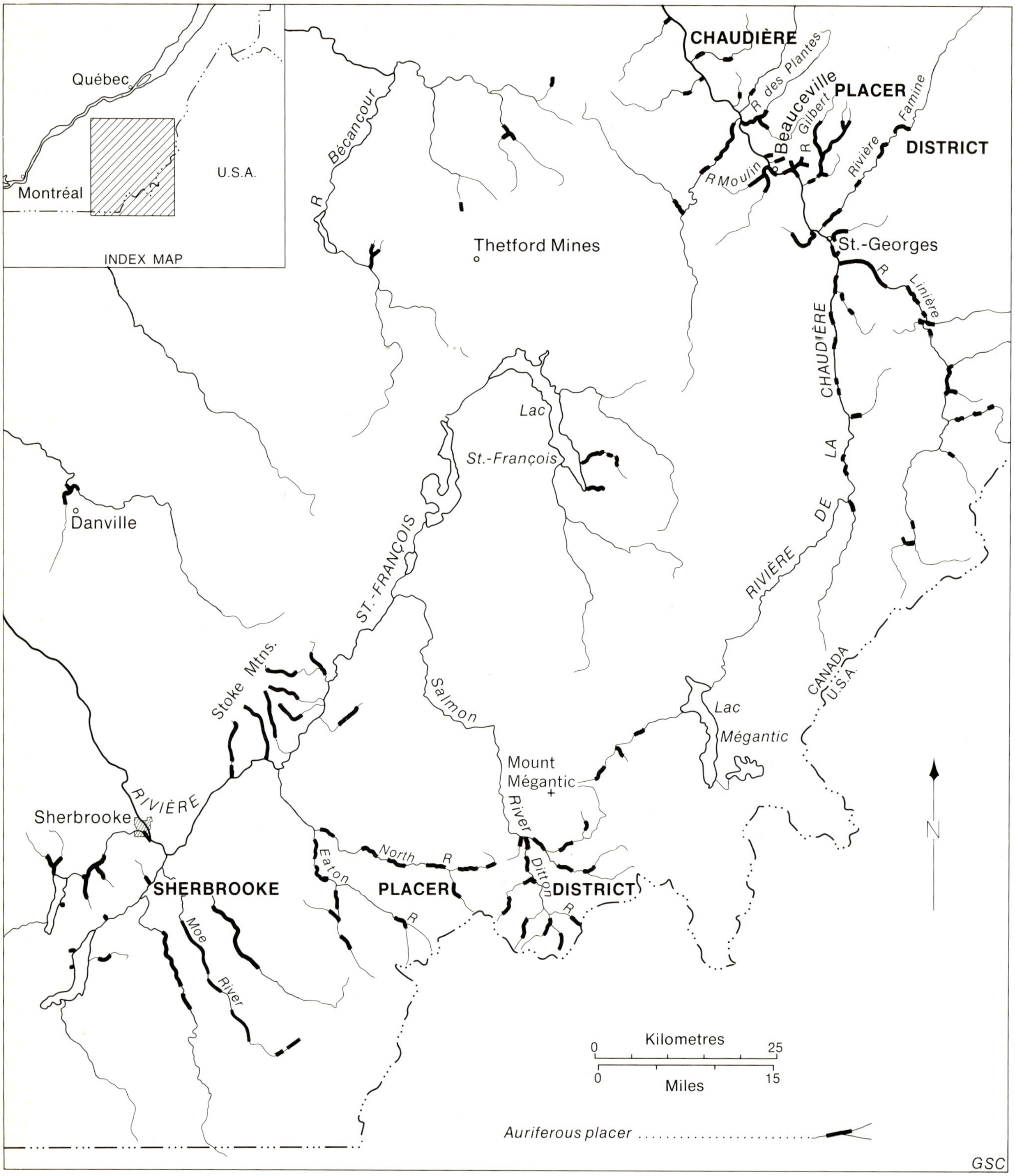

Figure 80. Auriferous placers of southeastern Quebec.

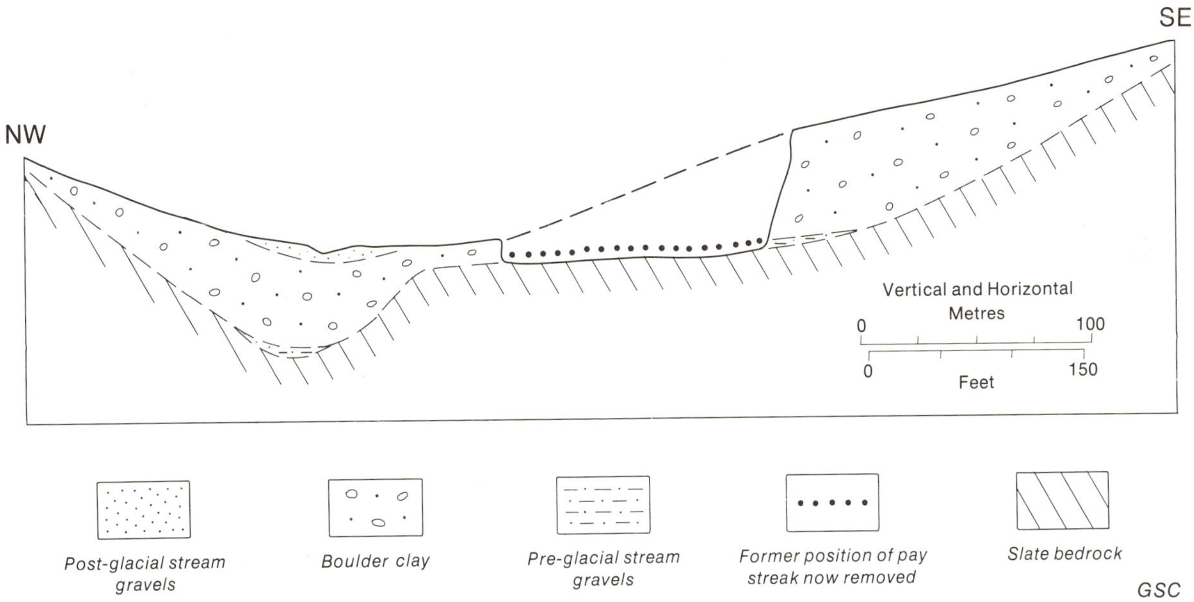

Figure 81. Cross-section of auriferous placer in Rivière du Moulin, Beauceville district, Quebec (after MacKay, 1921).

of the region. The cementation and iron oxide staining of some of the gravels were considered to be phenomena of secondary enrichment under groundwater conditions of present or relatively recent time.

It is difficult to reconcile these opposing views. The observations of the writer show that the cementation phenomena of the sands and gravels by iron oxides is proceeding today, being related to a number of springs at the heads and in the tributaries of many of the gold-bearing streams and rivers. Their waters carry iron, evidently derived from the weathering of the pyritiferous schists, slates and phyllites, which precipitates in the sands and gravels under slightly oxidizing conditions. Cementation phenomena cannot, therefore, be related exclusively to the Tertiary. On the other hand the materials of the gold-bearing gravels represent materials from a deeply weathered and eroded terrain, and these have taken time to form, much longer than an interglacial period. The evidence of the gold pay streaks suggest a very long period of accumulation and winnowing of the metal. The nuggets, especially, are highly knarled, rounded, abraded and pitted and have been a part of the dynamics of the erosion cycle for a long time. All this suggests that the gold, its accompanying heavy minerals and the weathered products with which it is associated were largely accumulated in the eluvium and drainage systems in pre-Pleistocene (probably Tertiary) time. During Pleistocene time these materials have been reworked in part and probably some of the gold has been mechanically redeposited.

MacKay (1921) divided the placers of the area into preglacial and postglacial categories. The former may now be considered as interglacial, at least in part. Brief descriptions of each follow.

Preglacial (interglacial) placers. These constitute the chief sources of gold and include stream bed placers, terrace or beach placers, river bar placers and alluvial plain placers. Only the stream bed and terrace placers have been productive. The average thickness of the auriferous gravels in these deposits was 5 ft, although some are over 30 ft thick. Most of these gravels are yellow, oxidized, compact and frequently cemented (limonite-cemented sands and gravels). Underlying the gravels in places is a yellow oxidized unctuous clay. The gravels contain stones and boulders up to 2 ft in diameter that are said by MacKay (1921) not to be glaciated and all of local origin. The writer would agree with this conclusion.

Analyses of the limonite-cemented gravel and sand and the puttylike clay from Rivière Gilbert are given in Table 69.

Table 69. Analyses of limonite-cemented conglomerate and sand and limonitic clay from Rivière Gilbert, Beauce County, Quebec

Constituent	Limonite-cemented conglomerate and sand	Limonitic clay
	(%)	
SiO_2	66.2	67.2
Al_2O_3	11.7	15.3
Fe_2O_3	9.4	4.1
FeO	1.5	0.7
CaO	0.7	0.4
MgO	1.6	1.2
Na_2O	2.4	2.0
K_2O	1.9	2.7
H_2O (total)	3.7	4.0
TiO_2	0.59	0.66
P_2O_5	0.13	0.09
MnO	0.07	0.06
CO_2	<0.1	<0.1
S	0.06	0.02
F	0.02	0.04
Cl	0.02	<0.02
C	0.23	0.11
O ≡ Cl, F, S	0.03	0.03
Total	100.2	98.6

Both the limonite cemented gravel and clay contain relatively small amounts of gold (<0.01 ppm). Heavy concentrates from these materials, however, show the following gold values:

	Au (ppm)
Limonite-cemented sand and gravel	0.02
Limonitic lower clay	0.30

The gold values are disseminated through the gravels in places, but more generally they are in the material immediately above the clay. Where the clay is missing the values are on the bedrock and/or in the upper few feet of bedrock. The gold occurs as fine dust and in nuggets ranging up to the size of goose eggs. Most nuggets are highly water worn, but a few of the smaller ones show angular outlines. Some contain vein quartz. The heavy minerals accompanying the gold are magnetite, specularite, hematite, chromite, ilmenite, pyrite, garnet, rutile, zircon, ottrelite, corundum and platinum. The platinum occurred in only minor amounts in Rivière des Plantes. It and the chromium probably derived from the ultrabasic rocks in the basin of this watercourse.

Postglacial placers. These include stream bed placers, terrace or bench placers, river bar placers and alluvial plain placers in the modern and postglacial sands and gravels. The first gold found in the area was in the river bar and alluvial plain placers of the Chaudière down river from Rivière Gilbert. The gold is generally fine grained in these deposits and commonly highly disseminated through the gravels. Only an occasional nugget is found. None of these placers were of importance except where the yellow (interglacial or preglacial) gravels were exposed to stream cutting, and while these were rich they were pockety. The heavy minerals in these placers are the same as those mentioned above for the preglacial and interglacial gravels.

MacKay (1921) and others thought that the gold in the Beauceville placers was concentrated during Tertiary time from relatively local sources. Logan and Hunt (*in* Logan *et al.*, 1863) writing earlier, however, removed the source to the crystalline schists of the Notre Dame range lying north of the Beauceville area. Cirkel (1911) considered that the source of the gold lay outside the placer area. The writer considers the source to be the rocks and quartz deposits of the Beauceville Formation and Caldwell Group within which the gold-bearing streams occur. There is certainly enough gold in the pyrite alone of these rocks (up to 0.4 ppm) to provide the gold in the placers.

In other parts of the Eastern Townships (Fig. 80) and Gaspé of Quebec small gold placers are known, and some have been worked in the past (Dresser and Denis, 1949). According to McGerrigle (1936) all the placers in the Eastern Townships, including those in the Chaudière Valley, lie southeast of the Sutton range and between St. Francis-Massawippi Valley on the southwest and the vicinity of the Chaudière River in the northeast. Areas, excluding the Chaudière Valley, where gold has been washed or recorded, include the Ditton River and Moe River basins in Compton County, the Stoke Mountain area in Richmond, Wolfe and Compton counties and others, described in some detail by McGerrigle (1936). He recognized three types of placer gold deposits: pre-Glacial, Glacial and Recent. The pre-Glacial deposits, exemplified by those in the Ditton River and Stoke Mountain areas, occur mainly in the stream bottoms in brownish weathered materials, often in a yellowish clay. They are probably of Tertiary age. The Glacial deposits are in sands and gravels filling parts of the stream and river valleys (e.g., Ascot Township). The Recent deposits were formed after the melting of the ice sheets and include those that have resulted from the reworking of older deposits by the present streams, as in gravel and sand bars along the rivers and streams (e.g., Moe River). Most of the gold in the deposits is well worn, fine grained, scaly (dust) and tarnished to bright in colour and lustre. A few small nuggets have been recovered, and some streams contain rough angular particles and dendrites. Some of the nuggets in the Ditton River contain inclusions of hessite. The principal heavy minerals encountered with the gold include magnetite, pyrite, galena and zircon.

McGerrigle (1936) considered that the gold in the placers was derived from quartz veins and probably partly from sedimentary bedrocks (conglomerates and quartzites). The fine-grained (dusty) nature of much of the gold suggests other possible sources. The district in which the placers occur is largely underlain by pyritiferous black phyllites and schists from which some of the gold probably originated during the course of weathering. Pyritic copper deposits, some auriferous, occur in a number of places in the Sherbrooke area. Weathering of these deposits may have given rise to some of the gold placers as is perhaps the case for those in Felton Brook, which heads near the Suffield, King, Howard and Silver Star copper sulphide deposits. Other gold placers whose source of gold is unknown may indicate similar sulphide deposits hidden beneath the heavy drift.

Sierra Nevada, California. The placers of the Sierra Nevada of California were discovered in 1848 after which followed the great rush of 1849. Since then some 55 million ounces of gold have been won from the placers that lie mainly on the southwestern flank of the Sierra Nevada along a distance of some 250 mi (Fig. 82).

The productive gravels of the Sierra Nevada are mainly of Tertiary and Quaternary age and are described at length by Lindgren (1911), Haley (1923), Pardee (1933), Jenkins and Wright (1934), Clark (1970) and Yeend (1974). Most of the rich placers are now exhausted and hydraulic mining of lean gravels was largely prohibited by law many years ago, because of environment problems in the basins of the Feather, Yuba, Bear, American and other rivers. Dredging along sections of the Yuba River (Hammonton district), American River (Folsom district) and elsewhere has, however, continued for some 70 years, terminating in 1968.

The general topography of California is simple – a mountain range along the ocean, the Coast Range; another along the eastern border, the Sierra Nevada; and between them closed in at both ends by their junctions a splendid valley in a tectonic trough, the Great Valley separated by the Sacramento River into the Sacramento and San Joaquin valleys. The part that concerns us is the northern part of the Sierra Nevada, the northern part of the San Joaquin valley and the southern part of the Sacramento valley (Fig. 83). The bedrock underlying this large area is mainly of Paleozoic and Mesozoic age consisting of highly folded and faulted sediments and volcanics intruded by ultrabasics (serpentinites) and granitic rocks.

Figure 82. Map showing the location of the principal placer districts of California.

Along extensive fault and fracture systems great gold-bearing quartz lodes were formed in late Mesozoic time; of these the Mother Lode system was the greatest. There followed then a long period of uplift and deep secular weathering and erosion that extended throughout most of the Tertiary. Much gold was accumulated in the streams that flowed southwestward off the Sierra Nevada. Near the end of Tertiary time volcanism was the dominant feature and most of the auriferous gravels were buried beneath rhyolite flows and tuffs that filled the valleys and in many places diverted the streams into new courses, some of which crossed the old courses. Some gold was deposited in these new courses. Near the end of Tertiary time vast eruptions of andesitic flows, tuffs and breccias took place filling all the valleys to their rims. New streams initiated as the result of marked uplift and tilting in late Pliocene and early Quaternary time cut deeply into these tuffs and other volcanic rocks, and into, and in many cases, below the auriferous Tertiary gravels. Great V-shaped canyons, some 3000 to 4000 ft deep, developed, and in many cases some of the gold was swept out into the valley at the foot of the Sierras, there to form large accumulations of gravels that have been extensively dredged in places (Fig. 83). Later in the Quaternary, mountain and valley glaciers developed, the latter mainly in the upper parts of the valleys. They had only a small effect on the redistribution of the Tertiary gold.

The present topography of the placer fields is a valley (the Great Valley) from which a foothills region rises abruptly, giving way first to a plateau or middle slopes, rugged in part and ultimately as one moves westward to the high Sierras with an average elevation of 6000 ft from which some high peaks and ridges rise well above the snowline. Through the foothills, middle slopes and high Sierras the streams have cut deep V-shaped canyons that widen out to U-shaped valleys in the high Sierras.

The placers first worked in the Sierra Nevada were in the auriferous gravels of the present streams, from which the gold was traced to the Tertiary river beds on the summits of the ridges and ultimately to the primary sources in the quartz veins of the Mother Lode and other systems. Here, we shall consider only the auriferous Tertiary gravels.

The nature and history of the Tertiary gravel deposits is best shown on Lindgren's classic section (Fig. 84), on which it is evident that considerable complexity exists. The location of the various courses of the Tertiary auriferous channels and the main intervolcanic (late Tertiary) channels of the Sierra Nevada are also shown in fascinating detail on the map by Lindgren (1911) from which generalized Figure 83 was largely compiled by Clark (1970). It will be seen that many of the Tertiary channels bear only a general relationship to the present channels, and that in many cases they cross the latter at large angles.

According to Lindgren (1911) the occurrence of gold in economic quantities in the Tertiary gravels depends essentially on the distribution of the primary deposits in the pre-Tertiary rocks of the Sierra Nevada. The Tertiary and Recent river courses are barren where traversing extensive granitic areas but carry values when they reach the flanking contact zones and sediments. The constituents of the Tertiary gravels are mainly pebbles and sands of quartz and metamorphic rocks, and the pay streaks are mainly on the bedrocks (Fig. 85). The gold is both coarse and fine, and large nuggets have been recovered from some channels. Lindgren (1911) did not think that chemical precipitation of gold was responsible for the formation of the nuggets. The fineness of the Tertiary channel gold is greater than that of the primary veins, an average of 900 for the former as compared with values ranging from 620 to 900 for the latter. The gravels in the richer channels ran from about 3.5 to about 25 oz Au/linear foot. The accompanying detrital heavy minerals in the Tertiary channels are magnetite, ilmenite, chromite, pyrite, monazite, zircon, garnet and small amounts of native copper, cinnabar, amalgam and platinoids, the last derived from the serpentinites of the gold belt. Detrital diamonds were recorded here and there, derived it is thought from the peridotites (now serpentinites). Among the authigenic minerals in the Tertiary placers Lindgren (1911) mentions calcite, dolomite, silica (opal and chalcedony) and pyrite, some auriferous. These minerals were evidently deposited by migrating groundwaters in the gravel channels, some probably heated during the volcanic periods. Fossils of many types lie buried in the gravels of the channels. They include Tertiary mammals, doubtful human remains and artifacts, fossil leaves, diatoms and abundant silicified fossil wood.

The Chocó, Colombia. The great placers of the Chocó and other districts in Colombia were first worked before the coming of the Conquistadores by native Indians using the calabash (gourd) and batea (pan). With the coming of the Spaniards gold was sought throughout Colombia, being found in practically every part of the country. Since mining began in 1550 more than 500 million dollars in gold has been won from the placer deposits. The Chocó, and the departments of Antioquia, Caldas and Tolima, were the most important New World sources of revenue for Spain during the 17th and 18th centuries.

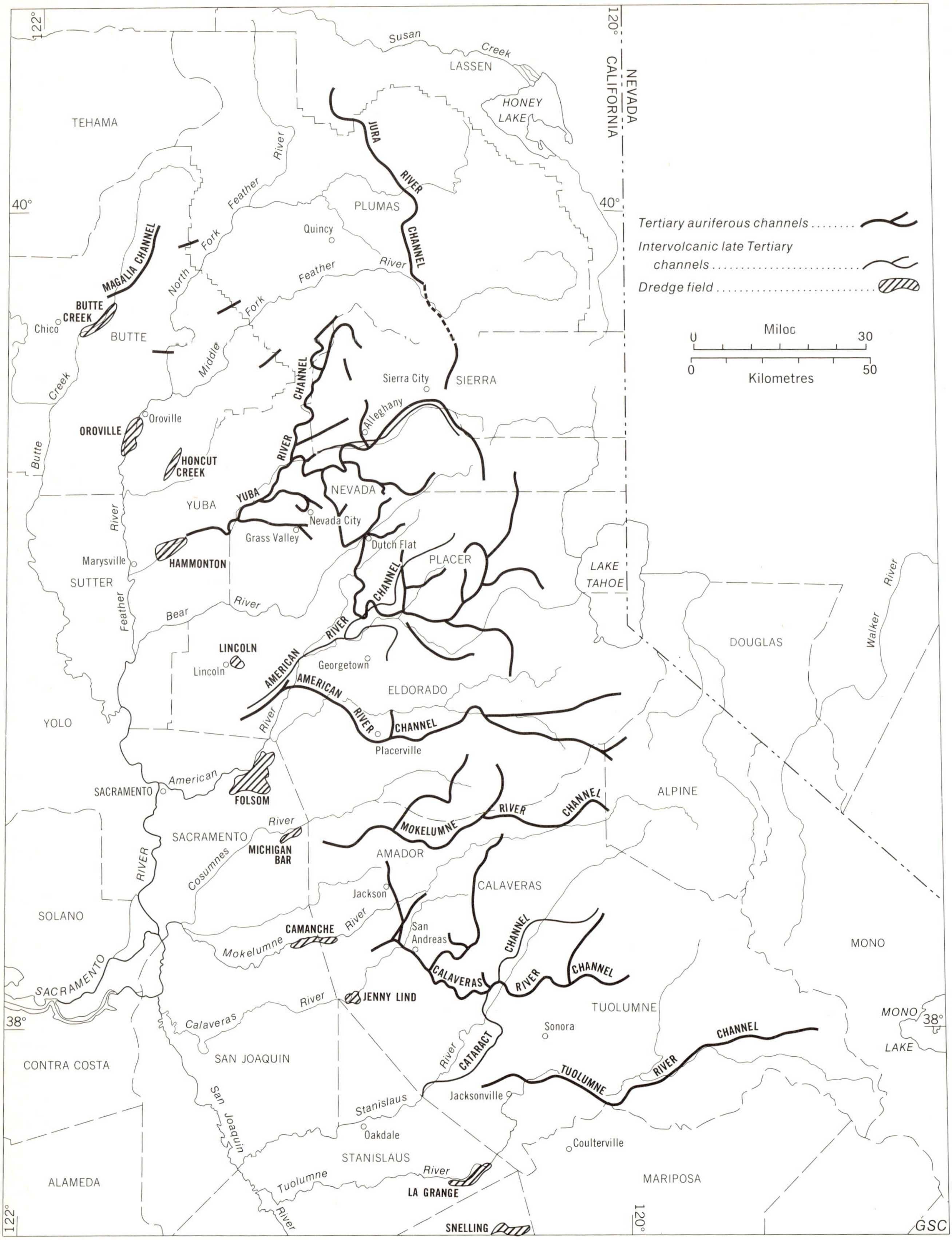

Figure 83. Map of Tertiary channels and dredge fields, Sierra Nevada, California (after Lindgren, 1911; and Clark, 1970).

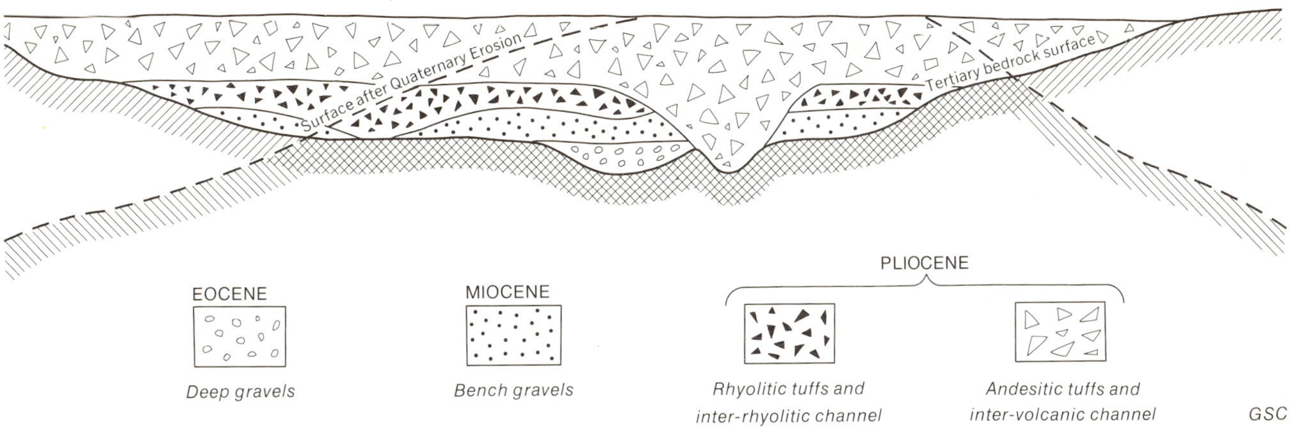

Figure 84. Schematic representation of the four principal epochs of Tertiary gravels in the Sierra Nevada, California (after Lindgren, 1911).

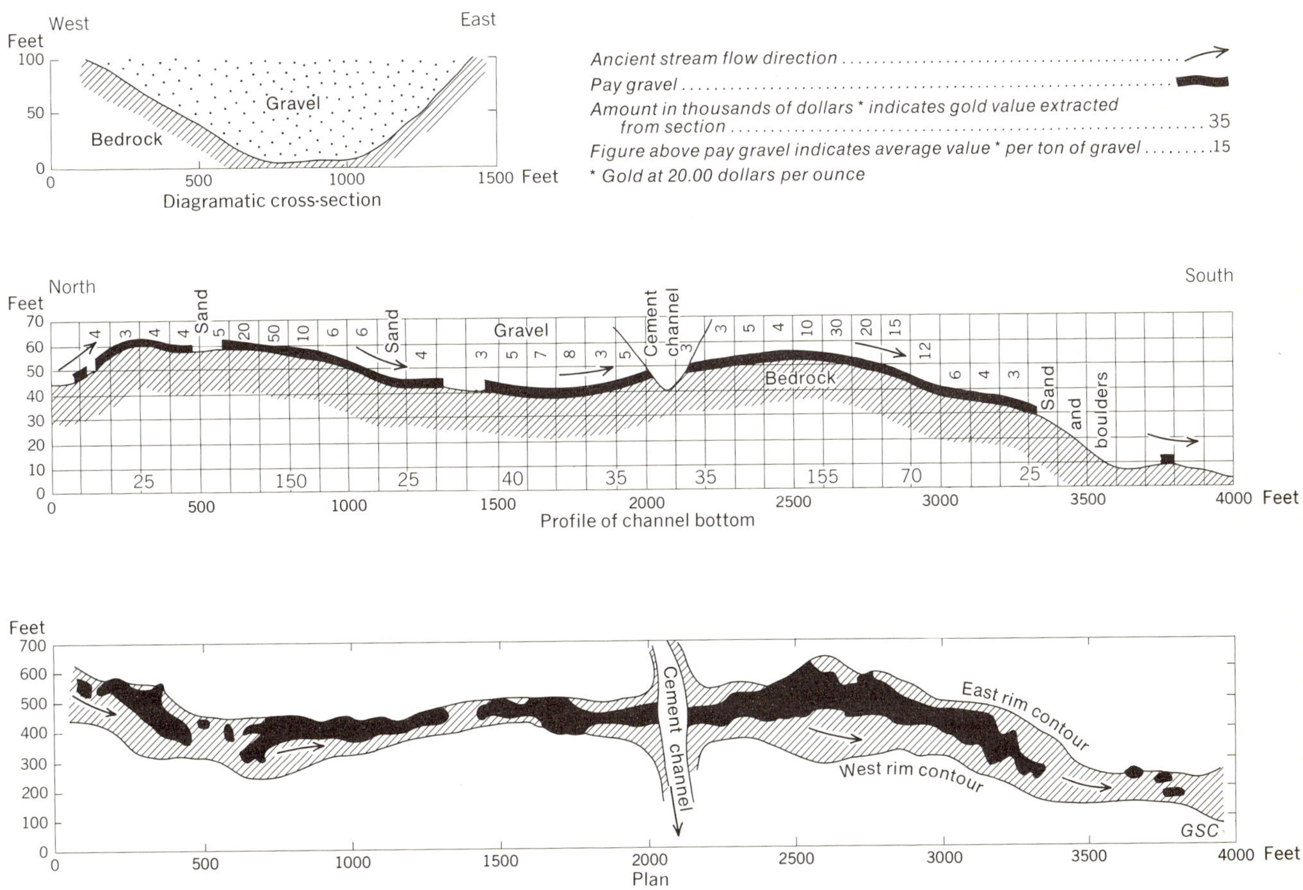

Figure 85. Section, profile and plan of part of the Mayflower channel, Forest Hill divide, Placer County, California (after Lindgren, 1911).

The Chocó is covered by thick forest typical of the tropical rain belt. The geological terrane is mainly Tertiary, consisting essentially of conglomerates, sandstones, shales and various types of volcanics and serpentinized basic intrusives. Quartz veins containing gold, silver, copper, lead and zinc are widespread in the contact zone where Tertiary granite intrudes andesites, greenstones and metasediments. The placers are river and bench gravels that yield abundant gold and platinum. In some placers the Au/Pt ratio is 100:1, in others 1:1 and in still others 1:3. Numerous other districts in Colombia also yield gold and platinum mainly from placer deposits, but some districts also contain lode gold deposits. There is a good summary of these deposits in the book by Emmons (1937) and in the paper by Meyer (1941).

The deep leads of Victoria, Australia Alluvial gold was discovered in Victoria at Clunes in March 1850, and there followed in 1851 the great rush that covered most of the state. The first workings were in the present streams and at the heads and shallow tributaries of the deep leads. By deep leads are meant, according to the legal definition of the Mining Development Act of Victoria, 1896 "any water-course or gutter below the surface of the earth containing alluvial deposits, at depth of not less than 100 feet from such surface."

The bedrock of the goldfields of Victoria is tightly folded Ordovician and Silurian shales, slates, greywackes and sandstones intruded by granitic rocks probably of Devonian age. The gold deposits are quartz saddle reefs interbedded veins and crosscutting veins containing native gold, pyrite, arsenopyrite and a few other sulphides. The principal mining areas, and the sites of many of the deep leads, are Stawell, Glenorchy, Ararat, Moyston, Raglan, Avoca Valley, Ballarat, Bendigo, Daylesford, Castlemaine, Clunes, Malmsbury, Rutherglen and many others.

The Paleozoic rocks and their contained gold deposits were deeply weathered and eroded, and the debris deposited in old river channels mainly during Tertiary time according to the fossil evidence. The deep leads so formed were deposited by rivers trending north or south from a watershed that extended from east to west across the country, coinciding in places with the great dividing range of Victoria. There followed in late Tertiary time further infilling by alluvium, and in places where volcanism was active the valleys were capped by basalt.

The deep leads of Victoria are described in great detail by Hunter (1909), who classifies them as alluvium and basaltic covered deep leads. Their nature is shown in Figure 86. Some of the placers lie close to the original veins in weathered debris and appear to be buried eluvial deposits. In places recent streams have cut down through alluvium and basalt to expose the deep leads. Where this occurs the downstream gravels of the recent streams may be highly auriferous (Fig. 75).

The deep leads are composed mainly of disintegrated country rock, quartz, heavy minerals, flour gold, and nuggets and slugs of the metal. The width of the pay streaks in the channels varies from 100 to 1000 ft or more. The grade varied from 20 to 65 oz Au/running foot in the richer parts. Recent work on the deep lodes of Victoria and a general bibliography of the deposits is given by Bowen and Whiting (*in* Knight, 1975).

Lena, Aldan and Amur drainage systems, Siberia, U.S.S.R. Gold placers are widespread in Siberia. Many are now largely exhausted, but some in the Lena, Aldan, Amur and other drainage systems are still productive and are the largest river placers known. Production from these placers probably exceeds 40 million ounces of gold since their discovery in 1829. Descriptions of the placers of Siberia are given by De Batz (1899), Maclaren (1908), Mushketov (1930), Obroutchev and Guérassimov (1929), Shakhov (1969), Kartashov (1971), Borodaevskaya and Rozhkov (1974) and numerous others (*see* the Selected bibliography).

The placers of Siberia occur mainly near the headwaters of the Ob, Yenisey, Lena, Aldan, Kolyma, Amgun and Amur rivers or near the headwaters of tributaries to these great systems (Fig. 87). Nearly all are in deeply dissected terrains marked by ranges of low hills and an irregular network of streams. They owe their origin to the weathering and erosion of uplifted zones of considerable extent that are marked by gold-bearing deposits of many types in rocks of both sedimentary and volcanic origin. Most placers in Siberia carry one pay streak on or near the bedrock that commonly varies in width from 100 to 1000 ft or more and ranges in thickness from 3 to 25 ft. The grade varies widely, the placers of western Siberia being leaner than in the eastern part of the region. The Yenisey placers are said to have ranged in value from 0.02 to 0.05 oz Au/yd^3; those in east Transbaikal averaged about 0.10 oz Au/yd^3. The Lena placers are much richer in places, and

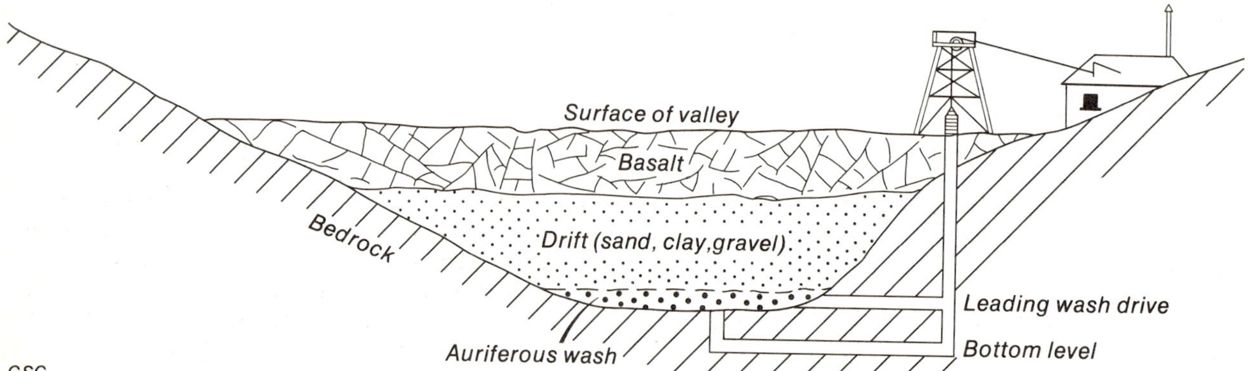

Figure 86. Diagrammatic section showing the nature of occurrence of the 'deep leads' of Victoria, Australia (after Hunter, 1909).

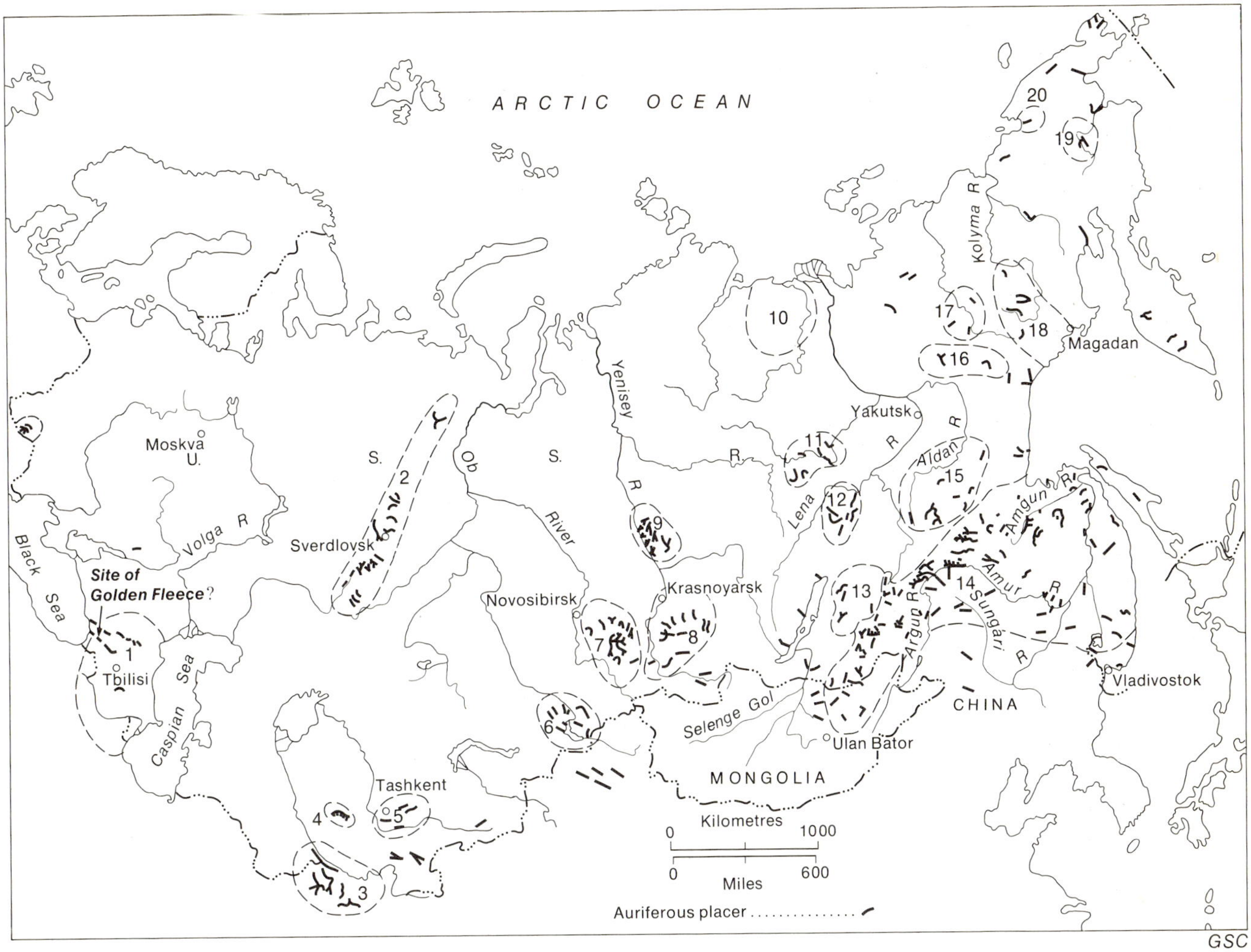

INDEX TO PLACER DISTRICTS

1. Ancient Colchis (placers known)
2. Urals
3. Amu Darya (the ancient Oxus)
4. Nuratau Ridge
5. Tashkent
6. Lake Zaysan
7. Kuznetsk - Alatau
8. Krasnoyarsk (Sayan)
9. Yenisey Range
10. Anabar - Olenek (placers reported)
11. Vilyuy
12. Vitim - Patomsk Highland
13. Vitim River
14. Amur - Argun - Amgun rivers
15. Aldan River
16. Verkhoyansk Mountains
17. Elgi River
18. Upper Kolyma
19. Belaya Gora
20. Chaun River

Figure 87. Map showing the general distribution of the principal auriferous placer districts in U.S.S.R., Mongolia and northern China (Manchuria) (compiled from various sources). Many of the placers shown are exhausted, others are small and probably have produced only limited amounts of gold. Most of the gold production from placers in U.S.S.R. now comes from eastern Siberia.

they frequently carry up to three or more stacked pay streaks.

Most of the placers of Siberia are of late Tertiary and Quaternary age. Many are in the permafrost zone, and a number contain remains of the wooly mammoth (*Elephas primigenius*) and other animals. Some 20 000 wooly mammoths are said to have been found in the placers, a number with the fleshy parts of the bodies intact, a feature suggestive of a catastrophic onset of the frigid climate of the glacial period.

The bedrocks of the Lena placer district are mainly highly folded and faulted crystalline and metamorphic rocks of Precambrian (mainly Proterozoic) and Cambrian age. They comprise mainly sediments such as slates, greywackes, quartzites, phyllites and derived quartz-mica schists and gneisses intruded by quartz-feldspar porphyries and small granitic intrusions, all enclosing quartz veins and stringers of low gold tenor and auriferous pyrite impregnations (Buryak, 1967*a,c*; Polikarpochkin, 1973). Mushketov (1930) stated that the placers of the Lena are located principally in valleys, rich in pyrite impregnations, that are cut into the Precambrian. He also evidently believed that the gold in the preglacial gravels was derived from these pyritic bodies and was chemically precipitated from groundwaters flowing beneath the drift in postglacial times. Research carried out by Sementchenko and others on the pyrite of the Lena placers, which was mentioned by Maclaren (1908), shows this mineral to be rich in gold. This has led to the general conclusion by a number of investigators that the alluvial gold came from the pyrite in the schists of the country rock.

The general topography of the Lena placer district is that of a deeply dissected terrain with low hills and an irregular network of streams. The placers occur mainly in the basins of the Bodaibo, Vacha, Juia and Khomolkho rivers and are of three types: (1) deep alluvial preglacial gravels in old river channels and on benches buried deeply by glacial and interglacial deposits; (2) relatively shallow interglacial gravels in old river valleys; and (3) surface and near surface modern alluvial sands and gravels.

The first are of prime economic importance, the second of some importance, and the third are spotty and lean. The pay streaks of the preglacial placers are a few to several feet thick and wide, 100 to 1000 ft or more, generally much wider than the present river channels. The second and third types of placers were apparently derived from the preglacial gravels. The gold in the preglacial channels is nuggety; some is angular, some crystallized (octahedra and cubes) some rough, and some is well worn. The fineness ranges from 810 to 930, with an average of about 900 according to the old and some of the new records. The average grade seems to be about 6 ppm or 0.17 oz Au/ton of gravel. The normal heavy minerals are magnetite, ilmenite, pyrite, arsenopyrite, native copper, galena, native bismuth, garnet, rutile, tourmaline, zircon, etc.

The extensive placers of the Bodaibo River, a right tributary of the Vitim River, are typical of the Lena drainage system. Most of the gold in these placers is thought to have originated from wide and extensive zones of slightly auriferous quartz veins and sulphide impregnations. Eight benches (terraces) varying in width from 20 to 250 m above the bedrock notch or bottom (thalweg) have been recognized in the Bodaibo Valley. The oldest complex of alluvial deposits in the bottom and succeeding five benches is represented by extensively decomposed and weathered, bright yellowish brown and greyish brown, clayey pebble and boulder beds and gravels. These are considered to be of Early and Middle Quaternary age and are overlain by morainal boulder clay assigned to the maximal Samarov glacial period. The youngest complex comprises grey and yellowish grey, sandy and clayey gravels, boulder beds and sands of the upper three benches and the bottom (thalweg) of the present Bodaibo River. The gravels and sands on the benches are regarded as of lacustrine and fluvial origin and are assigned to the Kazantsev interglacial period; the latest alluvium is considered to be of Late Quaternary and Recent age.

The richest placers of the Bodaibo are those on the second bench above the bottom, but the first bench and others below the glacial clay have also been productive, especially the fourth bench, which is the most extensive. The dimensions of individual pay streaks on the benches vary greatly being from 10 to 200 m wide and up to several hundred metres in length. Most average about 1 m in thickness, but some are up to 5 m thick. The depth of overburden to the deepest pay streaks averages 25 m but may be up to 150 m in some sections. In places gold penetrates into the bedrock (mainly sandstones and shales) to depths of 1 m or more. Most of the gold grains in the Bodaibo placers average about 3mm in size. Nuggets are particularly common, some weighing several kilograms or more. This type of nuggety gold usually has a high fineness, 870 to 930. A second variety of gold, occurring locally in the western part of the basin, has complex shapes, is spongy, drusy and dendritic, and the grains are relatively low in fineness, 730 to 820.

The Amur River (Chinese, Hei-lung Chiang) forms part of the frontier between the Soviet Union and the People's Republic of China. It rises on the northern border of northwest China and Inner Mongolia, flows generally eastward and empties into Tatarskiy Strait separating Siberia from Sakhalin Island. Along its course the Amur and its tributaries, the Zeya, Bureya, Sungari, Ussuri and Amgun, traverse a vast territory with rocks ranging in age from Precambrian to Recent. The topography of this enormous drainage system ranges from mountainous to subdued deeply weathered to lowland. The mineral deposits in the area range from auriferous quartz veins to auriferous sulphide bodies of many types to mercury- and tin-bearing deposits.

The alluvial placers of the Amur drainage system include creek, gulch, river and bench types. Some are worked in a small way; others are extensively dredged. In total the placers of the Amur provide a large proportion of the gold production of the Soviet Union. The details of the placers are much the same as those described for the Lena system and need not be repeated at any length.

Permafrost is comparatively rare in the maritime placer areas but becomes more persistent in the mid-reaches and head waters of the Amur and its tributaries. Frozen ground is uncommon along the courses of the main rivers, making dredging especially favourable. In most of the placer areas the alluvium is auriferous throughout with particularly rich sections at or near or in the first few feet of bedrock. In most sections the ratio of the gold-bearing alluvium to the overburden (turf) averages about 6:1. The alluvium, which comprises both sands and gravels, ranges from less than 0.04 oz Au/ton to values greater than 0.15 oz Au/ton. The average gold content of the valley gravels along the larger rivers appears to be about 0.02 oz/ton and the average of the alluvium for most of the placer districts is 0.05 oz/ton.

China and Mongolia. Much of the modern and ancient gold production of China has come from placers of both eluvial and alluvial origin (Juan, 1946). The bedrocks of China are chiefly Precambrian metasediments and volcanics, Paleozoic and Mesozoic sedimentary rocks and smaller areas of Tertiary sediments and volcanics. Vast areas of the country are covered by Pleistocene and Recent alluvium. Gold deposits comprise principally quartz veins and stockworks, silicified pyritic bodies, polymetallic lodes in limestone and auriferous conglomerates, the first three types being related to granitic intrusion and metamorphic events of Proterozoic (Algonkian), late Paleozoic and Cretaceous-Tertiary age (Emmons, 1937; Juan, 1946).

Placers are associated with many of the auriferous deposits mentioned above. These include ancient (Tertiary?) valley gravels, Pleistocene and Recent high level (terrace) gravels,

Recent stream gravels and marine placers. The last are relatively insignificant. The principal producing rivers and streams are in Shantung, Kiangsu, Fukien, Chihli, Hupeh (Han and Yangtze rivers), Hunan, Kwangsi, Szechuan, Kansu, Tsinghai, Yunnan and Heilungkiang, Kirin and other regions of Manchuria. The Manchurian placers seem to be the most productive, and they are said to have large reserves. The important placer drainage systems include those of the Sungari, Tumen Ho, Nonni, Amur and Yalu rivers (Fig. 87). Some 200 000 oz of gold per annum are won from these placers.

Alluvial gold placers are widespread in Mongolia particularly in the area north of Ulan Bator near the U.S.S.R. border. There, auriferous gravels have long been worked along the Selenge Gol, Orhon Gol, Tula Gol, Hereleng Gol and their tributaries. The source of the gold is mainly scattered quartz veins in various schists and gneisses intruded by granitic stocks and dykes. These placers and deposits are on the southwestern extension of those in the Transbaikal region of U.S.S.R.

Modern beach placers

Modern beach placers carrying economic amounts of gold are relatively rare. The classic example is the beach placers at Nome in the Seward Peninsula, Alaska. Other examples, all now exhausted are known along the Oregon and California coasts, in Nova Scotia, British Columbia, along some of the beaches in Chile, along the coasts of Westland and Southland in New Zealand and elsewhere. Those at Nome demonstrate most of the features of gold-bearing beach placers.

The beach placers at Nome and the geology of the area are described in detail in the publications by Collier *et al.* (1908), Smith (1908), Gibson (1911), Moffit (1913), Hummel (1962*a,b*) and Cobb (1973). Briefly, the bedrock in the vicinity of Nome is composed of highly folded and faulted metamorphic rocks, mainly of Paleozoic age, consisting of various sediments now largely schists, graphitic quartzites, phyllites, limestones and graphitic slates, all cut in places by granitic rocks and greenstones. Veins and small lenticular masses of quartz and calcite are widely distributed in all of the metamorphic rocks but most abundantly in the schists of the Nome Group near the localities rich in placer gold. Most of the veins, especially the larger ones that crosscut the foliation, carry free gold and sulphides. Silver-lead veins are also known from Seward Peninsula, and there are quartz veins bearing cassiterite and a little gold in the far western part (Moffit, 1913; Cathcart, 1922).

The Nome area is characterized by a crescent-shaped, flat, alluvial coastal plain or tundra that intervenes between the sea and the highlands which rise to 1000 ft or more some 5 mi inland. The alluvium of the coastal plain is predominantly fine, usually sand and sandy clay, with some coarser gravel layers and frequent beds of clay. Most of it is slightly auriferous. These sediments represent late Tertiary and Quaternary deltaic deposits laid down in the sea by streams and rivers bringing material from the auriferous highlands. The gravels and sands are up to 100 ft or more in thickness in places and are covered with muck to varying depths. Inland, the alluvial deposits are river, stream and lake gravels and sands. The Nome area was extensively glaciated during Pleistocene time, and parts are underlain by permafrost.

The placers of the Nome area (Fig. 88) include the following types:

1. *Eluvial placers* mainly on the slopes of hills represented by those on Rock Creek, Pioneer Gulch, Boer Creek and Nekula Gulch: The gold is coarse and angular and occurs on or in the upper part of disintegrated schists. These placers have been worked in a small way.
2. *Gulch, creek, stream* and *river placers* represented by those in the Snake River basin, Dry Creek and Newton Gulch, Nome River basin, Hastings Creek, Fred Gulch, Hazel Creek and others: These have been extensively worked; characteristic heavy minerals are scheelite, bismuth, pyrite, arsenopyrite, magnetite, ilmenite and limonite. The gold is fine to coarse grained, and nuggets are common.
3. *High-bench placers* are stream placers laid down in former drainage systems. Several have been worked mainly at the head of Dexter Creek in the low saddles in the watershed between the Snake and Nome rivers.
4. *Gravel-plain placers* are scattered and do not occur as well defined concentrations but as irregular accumulations disseminated through the gravels. Most are small and lean. Their origin is uncertain. Most appear to be related to river deposits.
5. *Beach placers* include modern, buried and raised types. They are the subject of the discussion that follows.

Work done up to 1970 indicated that there were 12 recognizable beaches in the Nome tundra, including offshore, modern, submarine, second, intermediate, Monroeville, third and fourth as indicated in Figure 89. The beaches represent successive periods of Quaternary and older coastal uplift that extended over a considerable length of time.

On the present beach the gold was concentrated on clay or micaceous sandy false bottoms. The pay streaks were 250 to 300 ft wide and about 2 to 18 in. thick above the clay beds, the latter being some 45 ft above bedrock. The pay streaks were a series of discontinuous stratified lenses containing essentially quartz and granite sand and pebbles with schist and limestone shingle in which gold was plentiful and associated with magnetite, garnet, limonite, hematite, ilmenite, scheelite and pyrite. Most of the gold was fine grained and well worn.

The second beach lies about ½ mi inland and some 40 ft above sea level. The gold streak, some 6 in. thick and 25 to 100 ft wide, lies on a false bottom of sandy and micaceous sediment and is buried under 15 ft or so of overburden. Most of the gold was a bright yellow colour, flaky and fine grained; some was about the size of gourd seeds. The pay streak in places ran up to 0.1 oz/pan.

The submarine beach lies almost directly under the second beach and some 56 ft lower. Where the bedrock is high the pay streak lies on it, elsewhere it lies on a false bottom of clay, sand and gravel at an elevation equal to that of the top of the rises in the bedrock. The pay streak lies under 60 ft or so of overburden and is 1 ft to less than 3 ft in thickness and from 300 to 400 ft wide. The gold is uniformly distributed through it, and the heavy minerals are somewhat different than those of other beaches. They include abundant pyrite and arsenopyrite with garnets, magnetite and ilmenite. The gold was bright yellow and fairly coarse; the grade of the pay streak ranged from 0.2 to 0.5 oz Au/yd^3.

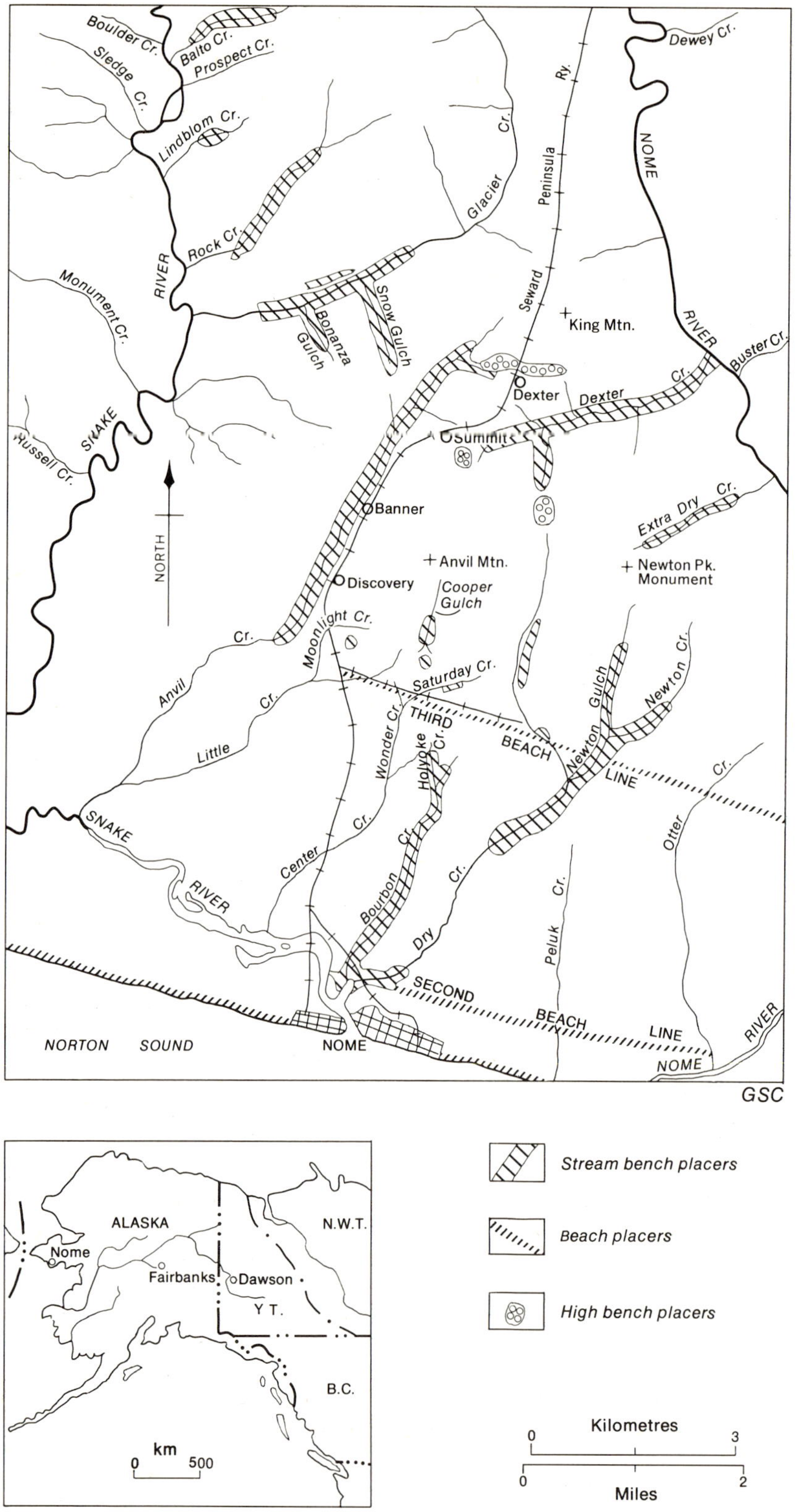

Figure 88. Map showing the distribution of the various types of auriferous placers in the Nome district, Alaska (after Collier *et al.*, 1908).

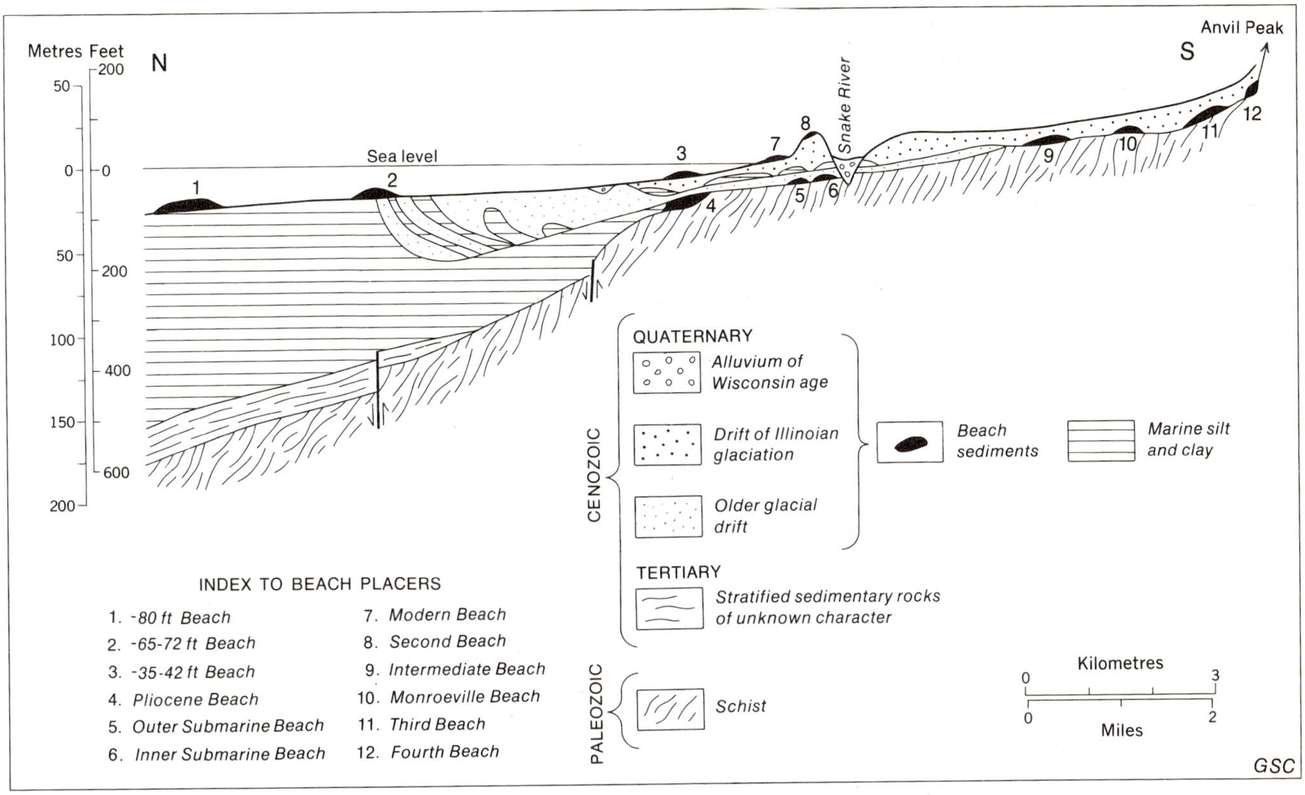

Figure 89. Geological section across the coastal plain area at Nome, Alaska showing the various auriferous beaches (after Nelson and Hopkins, 1972).

The intermediate beach lies about 1¾ mi inland and rests on bedrock some 20 ft above sea level. The pay streak is 1 to 3 ft thick, 400 to 600 ft wide, lies on or in the first foot or so of the bedrock, and the gold tenor is 0.25 to 0.75 oz Au/yd^3. The gold is bright yellow and fine grained, accompanied by abundant pyrite, arsenopyrite, magnetite and ilmenite. A characteristic feature of this beach is the presence of abundant mollusk shells, hence the name 'Clamshell Beach'.

The third beach, which is the richest of all, lies 70 ft above sea level and much of it rests on bedrock. The pay streak is mainly on the bedrock, 5 to 12 ft thick, 100 to 600 ft wide and covered by 20 to 120 ft of overburden. The gold was fine grained, but some nuggets were also found. Some parts of the pay streak were fabulously rich, in places 1000 oz/yd^3 or more. Much of the gravel of the pay streak ran 0.25 oz/yd^3. Gibson (1911) remarks that pay streaks in naturally thawed areas (thaliks) are not nearly as rich as those in the areas of permafrost. The reason for this he says is unclear. In Yukon there are indications of a similar phenomena in places. The writer thinks that the reason is due to leaching of the gold from these thawed areas by the groundwater. That gold (and silver) is capable of migrating in the permafrost is beyond doubt as shown by the writer (Boyle, 1951,1965*a*); under warmer conditions where groundwaters are moving, the effect is reinforced with time.

The Monroeville beach is considerably different than the others. The gravel is coarser, and the gold is much coarser than in the other pay streaks and is mainly in the shattered bedrock. There is also little garnet but plenty of pyrite and arsenopyrite. The pay streak is low grade and 300 to 500 ft wide being composed of 1 ft of gravel and 2 to 3 ft of bedrock. It is overlain by 50 ft or so of overburden and is 33 ft above sea level.

The fourth beach marks an ancient shoreline, but apparently there is no rich pay streak associated with this beach.

Extensive areas of the Nome beaches have been dredged after thawing of the gravels.

Recent work (since 1970) in the Nome area described by Nelson and Hopkins (1972) has revealed the presence of other beaches offshore and on bedrock as shown in Figure 89. In addition gold has been found over a widespread area in the sediments of the northern Bering Sea. (*See* further under the section on glacial and postglacial placers.)

The origin of the auriferous Nome beaches is of interest. Originally the gravels were of coastal plain deltaic origin and relatively lean in gold. During coastal uplift each successive beach has seen the winnowing and concentration of gold from the gravel plains by wave action. There may also have been contributions of gold by streams debouching their sedimentary load in the shallows along the beaches.

Small marine beach placers have been worked along the coasts of Oregon and northern California. Those in Oregon have been described by Pardee (1934). Recent work on the gold content of sediments along the coasts of Oregon and California by Clifton (1968) and Moore and Silver (1968) revealed some anomalous areas requiring further investigation.

Some of the black sands of Graham Island, Queen Charlotte Islands, British Columbia are auriferous. These occur along the beaches and originated from the reworking of extensive morainic deposits by wave and wind action (Mandy, 1934). The sands consist essentially of magnetite, garnet, epidote and olivine. The gold occurs as fine colours and is said to be present in amounts up to ⅓ oz/yd^3 or more in some of the magnetite-rich lenses. Overall, however, the beach sands are characteristically low grade.

In Nova Scotia small gold placers, mostly of academic interest, occur along the beaches of the Atlantic where the gold-bearing Meguma Group rocks are eroded by waves. One occurrence at the Ovens contained some 2000 oz of gold as small flakes and nuggets in the sand, the metal apparently being derived from nearby lean gold-bearing quartz veins and auriferous pyrite- and arsenopyrite-bearing slate and greywacke (Malcolm and Faribault, 1929). Offshore there may be placers of similar origin and these could be relatively rich and large where the Meguma Group rocks have been extensively eroded. These are discussed subsequently.

In Chile beach sands occur at Cucao and at Carelmapu, the latter in southern Chile. Some of these placers were very rich in places.

Auriferous beach placers of Recent age have a widespread distribution along the coastline of Westland and Nelson, New Zealand. They have been sluiced, dredged and worked as deep leads extensively in the past at Bruce Bay, Gillespies Beach, Addisons Flat, Okarito, Saltwater Beach, Barrytown, Charleston and elsewhere. Three types of beach placers can be recognized: those developed along the present strandline; those developed along former strandlines that are now submerged; and those developed along former strandlines that are elevated and now appear as raised terraces (beaches) inland from the coast. Well sorted and rounded pebble gravels with sandy intervals constitute most of the placer materials; some are cemented by ferruginous oxides, mainly limonite. The gold is extremely fine grained and is generally accompanied by ilmenite, magnetite, garnet, zircon, monazite, rutile, tourmaline, hornblende and epidote. Some scheelite and thorite occurred in the more southern placers. The thickness of the beach gravels ranges widely; some are up to 50 ft and contain stacked gold streaks that tend to parallel the present coastline. The source of their gold would appear to be multiple depending on their location, including auriferous Pleistocene gravels, older auriferous Tertiary and Cretaceous conglomerates, Precambrian(?) Greenland and Waiuta rocks and their gold-bearing lodes and the Alpine Schists. In Westland and Nelson these placers and beachsands represent a large reserve of blacksand from which gold could be won as a byproduct. Their details and locations are discussed by Williams (1974).

Glacial and postglacial placers

Glaciation generally destroys preexisting placers and scatters the gold far and wide. Few glacial deposits such as till sheets, kames, moraines, eskers, terraces, stream gravels and outwash deposits contain concentrations of gold. However, a few of the water-sorted materials associated with glaciers of both mountain and continental type may contain small placer concentrations. The placers of the Chaudière basin, Quebec described in some detail above probably owe their origin at least in part to concentration in glacial sands. Similarly, some of those in the Cariboo and in other parts of British Columbia, Yukon and Alaska commonly have a glacial sand and gravel component, and some of the gold has been reworked from older gravels into pay streaks, lean in most places. Similar conditions seem to have prevailed in the placer districts of Siberia and northern Europe.

It should be stressed that glaciation does not preclude the presence of gold placers. The existence of the Cariboo, Chaudière, Nome, Aldan, Lena and many other placers in regions that have been extensively glaciated should dispel any thought that glaciation is inimical to the preservation of Tertiary and Quaternary placers and the formation of new placers from these during glacial periods. In this respect it is worthwhile considering that deep Tertiary(?) or Pleistocene valleys now buried under great thicknesses of glacial clay and till may contain valuable placers in Canada, Europe, Asia and elsewhere.

Postglacial concentrations of gold in the gravels and sands of river systems and in oceanic sediments are common in Canada and many other countries. A few examples will serve to illustrate the nature of these placers.

Economic gold placers of glacial derivation have been worked extensively in Westland and Nelson provinces, New Zealand. In northern Westland these placers are floored or bottomed by the 'Blue Bottom', an upper Miocene formation of marine clay with glauconite. It is nonauriferous and is succeeded by an early Pleistocene formation known as the 'Old Man Gravels', consisting of loosely consolidated, rusty well rounded pebbles of greywacke, granite and schist. This formation has been recognized from the Grey River in the north south to the Mikonui River at Ross; it is slightly auriferous, but the source of the gold is uncertain, being probably partly derived from the Precambrian(?) Greenland and Waiuta rocks, which contain auriferous lodes and partly from the geosynclinal schists east of the Alpine Fault. The Old Man Gravels have been worked at several places but apparently seldom with profitable results. Their chief importance is that they were probably extensively reworked and are an important source of gold for the younger Pleistocene and Recent placers.

Moraines and glaciofluvial deposits were laid down during Pleistocene time over large areas in Westland Province by piedmont and valley glaciers that descended from the Alps to the present Westland coastline. The deposits of interest, many of which are auriferous and apparently of several generations, include esker, kame, stream and outwash plain gravels and sands that extend from north of Reefton to Ross. They have been extensively worked at Stafford, Goldsborough, Kumara, Greenstone, Dunganville, Stillwater Valley, Notown, Red Jack's and elsewhere. The auriferous materials constitute gravels and sands laid down below the glaciers and winnowed from their frontal moraines, mainly the latter. The source of the constituents of the gravels and sands and the gold is varied, some was derived from the Precambrian Greenland and Waiuta greywackes and their auriferous lodes, some from the Alpine Schists and some from the reworking of older auriferous early Pleistocene, Tertiary and Cretaceous conglomerates and sands. The gold in most of the glacial gravels is flaky, fine grained with an occasional small nugget. In most areas the quality of gold in the gravels was noted to diminish progressively with distance from the ancient ice edges. The Pleistocene gravels have, however, been extensively reworked yielding Quaternary and Recent placers along many of the rivers, streams and beaches of Westland from the Buller River south to the Mikonui River. In the past these have also been extensively sluiced and are largely exhausted, but one dredge still operates on gravels in the Taramakau River northwest of Kumara, producing some 1000 oz Au/month. Details of the auriferous Pleistocene and Recent gravels are given by Park (1908), Morgan (1908) and Williams (1974).

Gold has been known in the North Saskatchewan River since about 1859 and has been won by hand methods and dredging (Tyrrell, 1915). Gold is also present in a number of other rivers in western Canada including the Red Deer, Athabasca and McLeod (Fig. 79). Most of the gold in the North Saskatchewan River is present as very minute flakes and pellets in coarse gravel on the top few inches or few feet of the bars in the river all the way from Rocky Mountain House to Prince Albert, a distance of some 500 mi or more. The richest parts begin near Goose encampment, where the big coal seam at the top of the Cretaceous Edmonton Formation crosses the river, continue past Edmonton and cease in the vicinity of Beaver Lake Creek. Throughout the whole of this distance the coal-bearing Edmonton Formation forms the banks of the North Saskatchewan Valley. Tyrrell says that some 28 000 oz of gold have been washed from the river. He also believes that a small amount of platinum is also associated with the gold.

The auriferous pay streaks in the North Saskatchewan River are certainly of postglacial origin. The source of the gold, however, is problematical. Some have thought that the gold originated in undiscovered lodes in the Rockies; others have considered that the metal was washed out of the glacial till, clay, sand and gravel spread over the foothills and prairie; and still others think the gold came from the Tertiary sedimentary formations or from the Cretaceous rocks of the Edmonton Formation. Tyrrell (1915) was among the advocates of a source in the Edmonton Formation, since he found traces of gold in the coal ashes and clinkers resulting from the in situ burning of the seam. It seems probable that the coal-bearing parts of the Edmonton Formation are indeed the source of the gold since pyrite is common in these rocks, and pyrite in coal is commonly slightly auriferous. (*See also* the section on gold in coal in Chapter II.)

The Vermilion River in Ontario has long been known to contain gold. In the writer's opinion the Vikings may have found gold there, but the recent discovery of the gold-bearing sands was made in 1896. Meyn (1970) summarizes the available information on the occurrence of gold in the river basin. Earlier reports were written by Gracey (1898), Coleman (1901) and Prest (1949). The gold is very fine grained and disseminated throughout the sands and gravels. There appear to be no pay streaks, although some of the gravels are richer than others. Occasionally one can recover a piece of gold the size of a pin head in places, but according to all reports nuggets are rare. The average value of most of these gravels is about 0.005 oz Au/yd^3 or less, although some have been reported to run up to 0.02 oz Au/yd^3. The source of the gold is uncertain. Coleman (1901) thought that it probably came from the auriferous deposits in the Huronian rocks to the north and northeast; others have considered the Porcupine and Shining Tree areas as the source. Probably much of the gold has come from the winnowing of glacial till, gravel and sand by postglacial and modern streams and rivers in the Vermilion River system. If this is so, the gold has a widespread origin, coming from rocks and deposits of many types.

Other rivers and areas where placer gold can be panned from reworked glacial materials in Ontario include Wanapitei River, Meteor Lake, Lake Manitou, Savant Lake, Grassy River (south of Timmins) and various glacial sands in Munro, McCool, Montrose, Timmins and other townships. In Northwest Territories there are numerous eskers, outwash sand deposits and reworked glacial sands that yield a colour or two of gold with persistent panning. None of these are economic.

Gold concentrations in modern river sands and gravels, similar to those just described, occur in many of the river systems of the world of which the most celebrated must be the Rhine, immortalized in Wagner's prelude "Das Rheingold" to the great opera "Der Ring des Nibelungen".

Gold concentrations in ocean sediments on continental shelves occur in various parts of the world. Most are low grade and up to the present have not been mined. Numerous examples of these concentrations can be cited. Two offshore concentrations from the Ovens in Nova Scotia have been described by Libby (1969). Here, the gold appears to occur in ancient buried stream channels, although some probably also occurs in old beach deposits. The tenor of the sands and gravels is said to range from a few cents per cubic yard to more than four dollars per cubic yard (0.001 to 0.11 oz/yd^3). The volume of gravel available in one deposit is 6 million and in the other 36 million yd^3.

Pilkey and Bornhold (1970) examined the heavy mineral fractions of sediments from the Carolina continental margin United States, offshore from the deposits in the Piedmont and Blue Ridge gold provinces. Most of the measurable concentrations of gold were very low (0.7 to 7.0 ppb) in the heavy mineral fractions and restricted to the general vicinity of Cape Lookout and northward.

A number of submarine areas in Alaska have been investigated for detrital gold. Reimnitz *et al.* (1970) found no gold in the bottom mud of Nuka Bay, a fiord system along the northern Gulf of Alaska fed by numerous streams carrying gold. A small amount of gold was detected in two samples of relict gravel and in relatively unsorted coarse material dumped into the head of the fiord by glacier. The investigators concluded that the gold apparently moves with the coarse fraction of the sediment supply and is retained in deltas and moraines where there is little chance of finding an economic concentration.

In the northern Bering Sea, Nelson and Hopkins (1972) found that glaciers had moved auriferous debris as much as 5 km beyond the present shoreline of Seward Peninsula and nearly to the centre of the Chirikov Basin from Siberia's Chukotka Peninsula. The richest concentrations (920 ppb) and coarsest particles (1 mm or larger) of gold were found in seafloor relict gravels that mantle glacial drift lobes in the Nome nearshore region and in gravel patches over bedrock in the Sledge Island area. These bodies of relict gravel are thought to have formed during transgression and regression of the shoreline when eustatic changes of sea level occurred in

Pleistocene time. Except along the present shoreline, Holocene (Recent) sands and muds throughout most of the southern Bering Sea were generally found to contain only sparse fine gold flakes (0.25 mm in diameter or smaller). The total background gold content of Chirikov Basin sediments is not well defined, but analysis of several whole samples suggests an average of about 10 ppb.

Fossil gold placers

Fossil gold placers may be of residual, alluvial or aeolian origin and include ancient eluvial, alluvial fan, piedmont fan, talus, gulch, creek, stream, river flat and channel, deltaic, beach, coastal plain and marine types. On uniformitarian principles we assume that these deposits originated in a manner similar to that described above for each of the modern types. Their detailed history, however, may be complex as pointed out by Fayzullin (1968). There may have been a number of intermediate collectors of gold between its primary source and its final site of deposition. This is especially true where multiple erosion cycles are a feature of the geological history of a district. Furthermore, this recycling process may partly explain why the gold in fossil placers is generally much finer grained than that which we commonly see in modern (first cycle) placers near the source of the gold. Most fossil gold placers are in transgressive rock sequences and commonly occur on or near ancient weathered and eroded surfaces, i.e., unconformities, although some may be stratigraphically higher than the basal beds at unconformities. In age fossil gold placers range from Precambrian to Tertiary. Two general types can be recognized – one type is extensively lithified and contains much pyrite and often other sulphides, a few deposits of this type contain hematite (or ilmenite) as the principal iron mineral; another type is only moderately to weakly lithified and contains little pyrite and other sulphides but often plenty of iron oxides (limonite, hematite). To the first belongs the quartz-pebble conglomerate deposits, the largest and most productive gold placers known; the second category includes many gold-bearing sandstones, quartzites and conglomerates throughout the world. No attempt will be made here to describe any of the latter type in any detail, but some specific examples will be briefly mentioned.

Auriferous, pyritic quartz-pebble conglomerates are especially characteristic of the Precambrian and are generally of Lower Proterozoic age. They most commonly lie immediately above the Archean-Proterozoic boundary as emphasized by Derry (1961) and seem to have derived most of their gold from auriferous deposits in Archean rocks. The most productive of these deposits have been described above, the references to other occurrences of a similar type throughout the western world are given there. In U.S.S.R. there are numerous pyritic quartz-pebble conglomerates of Precambrian age, but so far as is now known none of these are productive of gold. Most apparently contain only from 0.1 to 1 ppm Ag. They include occurrences in Karelia, Krivoy Rog, Kursk magnetic anomaly, Voronez syneclise, in the district west of Lake Baikal, Vitim-Patom Highlands, the eastern and western portions of the Aldan Shield, around the rim of the Anabar Shield and elsewhere. These conglomerates and methods of their study are the subject of monographs by Smirnov (1969) and Ivensen *et al.* (1969).

Pyritic quartz-pebble conglomerates are not restricted solely to the Precambrian. On the contrary, occurrences of this kind are known in Paleozoic and Mesozoic rocks and may also occur in Tertiary rocks. None of the younger occurrences are, however, economic, as far as the writer can ascertain.

The economic quartz-pebble conglomerates have had a long and complex chemical and metamorphic history, and their origin is controversial. It is suggested in the last part of this chapter that these deposits are essentially placers in which the various elements, including gold, have been extensively redistributed.

The second type of fossil placer occurs in sandstones, conglomerates and quartzites that are marked by the presence of little pyrite but often contain much hematite or limonite. This type is usually only moderately lithified, and the gold is in most cases easily recognized as detrital. There are a number of occurrences of these types of placers in Canada.

Near Gays River, Nova Scotia, Carboniferous (Horton Group) conglomerates, consisting of pebbles of slate and quartz in a gritty matrix of quartz, slate particles and iron oxides, overlie auriferous slates and greywackes of the Meguma Group for 4 mi (Malcolm and Faribault, 1929; Stevenson, 1959). The basal beds and the underlying slate, where it is fractured or jointed, are auriferous. The gold, bright and of a deep yellow colour, occurs as thin flakes and small nuggets in the matrix of the conglomerate and in the fractures of the slate. It is associated with considerable amounts of iron oxides in places. The source of the gold appears to be auriferous quartz veins in Meguma rocks to the southeast.

Analyses of the various types of gold-bearing conglomerates are given in Table 70. It will be noted that the manganese and iron contents of some of the facies are relatively high. No systematic correlation could be found between the gold contents and the iron and manganese contents, except that in some places the iron-rich varieties of conglomerates seem to contain a little more gold. Of the other elements only arsenic appears to be higher than normal, being probably concentrated mainly in the iron oxides.

The gold contents of the various facies of the conglomerates are low according to Geological Survey analyses (<10–35 ppb) (<0.0002–0.001 oz Au/ton). Some of the heavy concentrates made from the conglomerates are rich in gold (up to 18.5 ppm; 0.54 oz Au/ton) as well as in elements such as Ag, Ti, Mn, Ba, Cr, V, Ni, Cu, Zr, Ce, Y, Nd, Nb and Th, mostly reflecting the presence of heavy minerals such as magnetite, ilmenite, wad, barite, zircon and monazite.

According to Malcolm and Faribault (1929), 1878 oz of gold were won from 11 877 tons of crushed conglomerate between the years 1873 to 1881. This gives an average of 0.158 oz Au/ton (5.4 ppm).

Pan concentrates from the sediments of streams near the conglomerates are enriched in gold (up to 8.5 ppm). This feature suggests a method of tracing the auriferous conglomerates in drift-covered areas. Careful assessment of the results will be necessary, however, to differentiate gold anomalies due to weathering of the conglomerate from those due to weathering of auriferous deposits in the Meguma rocks.

In New Brunswick Carboniferous conglomerates are slightly auriferous near Nauwigewauk (Ingall, 1893). The conglomerate beds consist of interstratified coarse sandstones

Table 70. Analyses of Carboniferous auriferous conglomerates, Gays River area, Nova Scotia

Constituent	Manganiferous conglomerate	Hematitic conglomerate	Grey conglomerate
		(%)	
SiO_2	73.9	74.2	79.6
Al_2O_3	9.7	10.1	10.0
Fe_2O_3	3.5	3.5	2.5
FeO	0.8	1.6	<0.1
CaO	2.4	2.3	0.3
MgO	0.9	1.1	0.4
Na_2O	0.8	0.5	0.4
K_2O	2.8	2.7	2.6
H_2O (total)	1.6	1.6	1.5
TiO_2	0.29	0.33	0.31
P_2O_5	0.10	0.08	0.08
MnO	1.06	1.00	0.38
CO_2	2.6	2.6	0.2
S	0.13	0.12	<0.02
F	<0.02	0.10	<0.02
Cl	0.08	0.04	<0.02
C	0.16	0.09	nf
O ≡ Cl, F, S	0.07	0.10	0
Total	100.8	101.9	98.3
		(ppm)	
Cu	27	15	33
Ag	0.14	0.12	0.08
Au	<0.01	0.01	<0.01
Sr	76	70	55
Ba	430	310	300
Zn	19	19	16
Cd	<0.6	1.3	1.3
Hg (ppb)	10	23	12
B	143	145	163
Ga	12.8	13.3	13
Tl	<1.0	<1.0	<1.0
Sc	6.1	8.8	8.2
U	0.4	<0.1	1.4
Ge	2.4	2.4	3.1
Sn	3.3	3.3	3.3
Pb	12	12	12
Zr	120	110	109
As	25	13	18
Sb	2	<2	2
Bi	0.6	0.53	<0.50
V	40	37	31
Cr	21	26	25
Mo	1	3	2
W	<2	<2	<2
F	<0.02	0.10	<0.02
Cl	0.08	0.04	<0.02
Co	23	24	18
Ni	28	28	28

Note: nf–not found.

and pebble beds, the pebbles being mainly of quartz, shale and limestone. It is said that the gold is free and very fine and apparently contained in the quartz pebbles that make up part of the conglomerate. The more pebbly beds are said to carry up to 0.2 oz Au/ton.

The Edmonton Formation along the North Saskatchewan River is slightly auriferous as noted in the previous section. Some of the Tertiary rocks overlying the Edmonton Formation may also be slightly auriferous judging from old reports.

The Tertiary (Eocene) conglomerates southwest of the Indian River (McKinnon and Bismark creeks) in the Klondike district of Yukon are slightly auriferous (Fig. 76). These conglomerates are composed of well rounded white quartz pebbles and cobbles and schist particles cemented by fine-grained quartz and schist fragments. A few grains of pyrite occur in places, usually near the base of the series. The conglomerates are not unlike the auriferous 'White Channel Gravels' of the various creeks of the Klondike except that they are more strongly lithified. In gold content the conglomerates are low, but when panned a few fine colours usually appear. Maclean (1914) reports nil or only traces of gold from the old workings; others report from trace to 0.1 oz Au/ton. As far as the writer is aware, little has been done on the conglomerates except some diamond drilling since Maclean's report. Mertie (1940) mentions similar auriferous conglomerates in Alaska south of the Yukon River in a narrow belt that extends from the Internatonal Boundary west-northwest for more than 100 mi.

Elsewhere in the world gold-bearing conglomerates, sandstones and quartzites appear to be relatively common. A few examples will suffice to indicate their nature.

The Paleozoic Deadwood Formation of the Black Hills region of South Dakota has long been known to be auriferous, and parts of the formation (Cambrian Potsdam Formation of Devereux) have been mined (Devereux, 1882; Irving *et al.*, 1904; Connolly, 1927; Norton, 1974). One of the richest auriferous parts of the Deadwood Formation is near the large Homestake gold deposit where the basal conglomerate, breccia and sandy beds rest on Precambrian schists, amphibolites, etc. (Fig. 90). The auriferous conglomerate ('cement ore') is composed mainly of quartz pebbles and boulders, schist shingle and pebbles of hematite (iron-formation) cemented partly by pyrite, oxidized to various iron oxides in the oxidized zone. Pockets of iron oxide or pyritic material in depressions and channels in the underlying Precambrian rocks were especially rich in gold. Some of the gold occurs in well rounded small flattened nuggets on or near the Precambrian bedrock. Another type of gold occurs as films and flakes along schist planes and in fractures and crevices in the Precambrian schistose bedrock to depths of 10 ft or so in places. Small amounts of gold are also present in, or associated with, the pyrite that may have been introduced in Tertiary time. Yet another type of gold occurs some miles distant in quartzites intruded by porphyry dykes of Tertiary age. This gold occurs in the native form in a very fine state of division mainly in cavities lined with quartz crystals in the quartzite.

Devereux (1882), Irving *et al.* (1904) and Connolly (1927) consider that the gold is essentially detrital, derived mainly from the great Homestake lode and deposited in a nearby beach placer. Connolly lists a variety of evidence, which he thinks confirms the placer hypothesis. Among the various features are the facts that the gold in the conglomerates has all the physical characteristics of placer gold, and that it is invariably finer than the gold in the Homestake lode. The pyrite is said by Connolly (1927) to have been introduced during Tertiary time. This may be the case, although it seems more probable that the pyrite had an origin similar to that of the Rand, namely bacterial reduction of sulphates followed by fixation of available iron by the H_2S generated.

It has been said (Hosted and Wright, 1923) that the particulate gold and the gold in films and seams in the conglomerates and quartzites is of Tertiary age, related to Tertiary dykes and stocks that invade the Paleozoic rocks. There are certainly deposits of gold-tungsten, gold-silver and

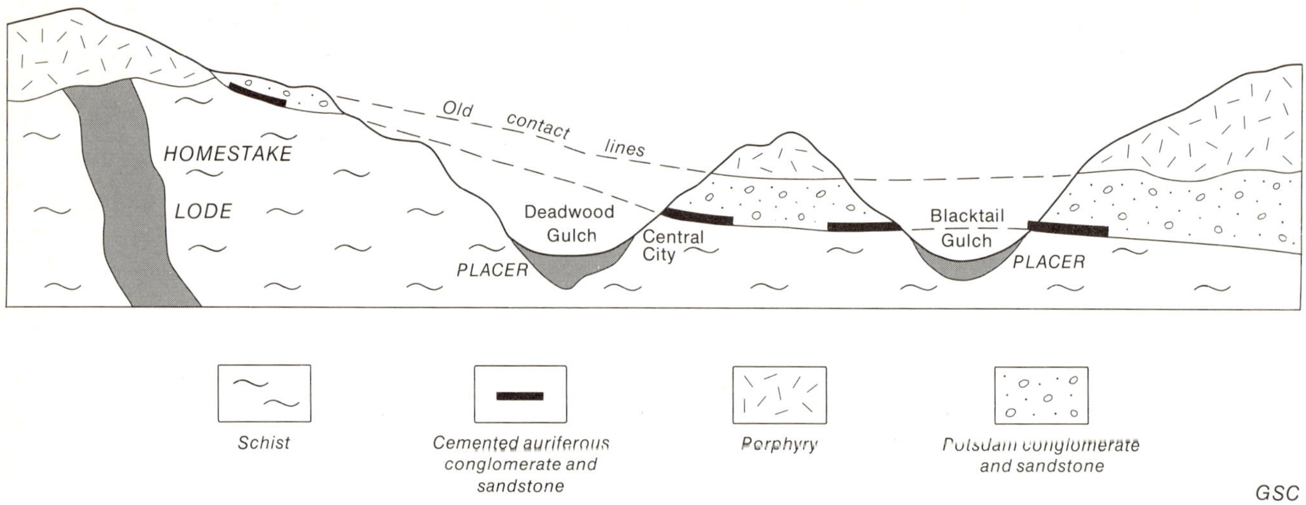

Figure 90. Generalized geological cross-section in the vicinity of the Homestake Mine, Lead, South Dakota (after Devereux, 1882). (No scale given on original.)

lead-silver that are of Tertiary age in higher parts of the Deadwood Formation, particularly in dolomites, in the general vicinity of the Homestake lode, but it is unlikely that the particulate gold and the gold in seams in the conglomerates and quartzites are of the same age. It appears more probable that the detrital gold has been partly remobilized by the intrusion of the Tertiary dykes. Devereux (1882) indicates that this has happened in some places, since he noted that the gold of the fossil placer had been partly dissolved out near a dyke. Contrast this with the same phenomenon noted in the Rand (page 324). One also wonders if the source of the gold in the undoubted Tertiary deposits is not to be found in the Paleozoic placers and in the Precambrian rocks rather than in Tertiary magmas. This theme will be developed more fully in the last section of this chapter.

Some of the sediments of Lower Eocene age in Texas contain gold (Dumble, 1913). Most of the auriferous beds are associated with the lignitic stage of the Eocene and include bedded clays and sands, the latter often containing pyrite and some iron oxides. Again we see a connection between gold and coal such as that which Tyrrell (1915) observed in the Edmonton Formation of Alberta as noted above. In fact the occurrence of auriferous sandstones and conglomerates with coal measures in various parts of the world is rather remarkable.

The thick clay beds (claystones) of the Shinarump Group (Triassic) in southern Utah are auriferous in places according to Lawson (1913). The gold, about 0.0025 oz/ton (0.085 ppm) is apparently very finely divided and is disseminated rather evenly throughout the colloidal clay materials (bentonite), which swell enormously when wet. Elsewhere in Utah, gold in ancient sediments seems to be widespread. Butler *et al.* (1920) mention the presence of gold in the sandstones in the San Juan River region and in the Triassic sandstones in the vicinity of the Henry Mountains.

Antweiler and Love (1967) investigated the gold content of Cretaceous and Tertiary sandstones and conglomerates in northwest Wyoming. Slightly modified, their conclusions are as follows:

1. Gold occurs in all areas of Cretaceous and Tertiary quartzite-bearing conglomerates that have been sampled in northwest Wyoming. The bulk rock analyses of the various formations are as follows (all values in parts per billion): Harebell Formation (latest Cretaceous), 65; Pinyon Conglomerate (Paleocene), 86; Fort Union Formation (Paleocene), 35; conglomerate of earliest (?) Eocene age, 94; Wind River Formation (early Eocene), 222; Pass Peak Formation of Eardley and others (middle(?) Eocene), 47; Miocene (?) conglomerate, 65; Quaternary deposits, 103.

2. The volume of low grade gold-bearing strata is so vast (50 mi^3 or more) that development of new mining and recovery methods designed for low-cost handling of huge quantities of relatively poorly consolidated material may eventually be justified.

3. Most of the gold was not derived from adjacent mountains but came from Precambrian and possibly Paleozoic quartzite in a now-buried uplift northwest of the Teton Range.

4. To analyze for gold in low concentrations, a method combining cyanide extraction and atomic absorption was found most reliable.

5. Analyses of 754 panned concentrates demonstrate some of the problems of recovering gold by mechanical and chemical methods and of evaluating various types of analytical data.

The detailed petrology and study of the paleocurrents of the Harebell Formation, Pinyon conglomerate and other coarse clastic deposits in northwestern Wyoming has been published by Lindsey (1972).

Turner (1903) describes a Cretaceous auriferous conglomerate from the Cottonwood Mining district, Siskiyou County, northern California. The conglomerate is part of a marine sandstone series and contains some fossil (shells) layers near the top. It has a maximum thickness of about 75 ft. Turner considered the gold to be of local origin derived from quartz veins in the metamorphic series all along the outcrop of the basal conglomerate. The conglomerate is said to have contained up to 0.175 oz/ton of conglomerate. The auriferous Cretaceous conglomerates have given rise to Pleistocene and younger placers in the Cottonwood valley. This is a case of an intermediate collector in the system – primary lode deposit → alluvial placer.

Attempts to mine a basal Cambrian conglomerate (Deadwood Formation) near Bald Mountain, Bighorn Mountains, Wyoming, were made in the early part of the century (Darton, 1906).

Barker (1969), searching for fossil placers, found only low contents of gold in the Precambrian clastic rocks of southwestern Colorado and northern New Mexico. In Colorado, the Vallecito Conglomerate and conglomerates of the Uncompahgre Formation of the Needle Mountains were sampled. In New Mexico, the Ortega Quartzite, the Big Rock Conglomerate Member, the Jawbone Conglomerate Member and the upper quartzite member of the Kiawa Mountain Formation in the Tusas Mountains and the lower quartzite member of the Ortega Quartzite and the conglomerate member of the Vadito Formation in the Picuris Range were sampled. All of these rocks contained less than 0.1 ppm gold; they apparently were derived from a terrane of quartzite, jasper, argillite and iron-formation that contained virtually no gold.

Auriferous sandstone, grit and conglomerate are widespread in the South Island of New Zealand, mainly in Otago and northeast Southland provinces. These include basal Cretaceous greywacke-quartz-schist conglomerates, basal early Tertiary sandstone, grit and quartz-pebble conglomerates and late Tertiary to early Pleistocene loosely cemented gravel conglomerates.

Several types of fanglomerates and cement conglomerates of Cretaceous age have been worked at a number of places in Otago, the most important being at Bluespur (Gabriels Gully), Wetherstons, Waitahuna Gully and Forsyth, where wedges of a cement conglomerate have been intermittently preserved by infaulting along a prominent fault (Tuapeka Fault). The workings of Gabriels Gully at Lawrence (the site of Gabriel Read's pioneer discovery of an economic gold deposit in South Island, N.Z.) are extensive and have produced some 200 000 oz of gold. Viewed at a distance the conglomerate has a bluish cast (hence the name Bluespur), but at close range it is seen to be greenish in colour due to a chloritic cement, which in places is also siliceous. The pebbles comprise schist, greywacke, quartzite and quartz of coarse gravel size with occasional cobbles up to 1 ft across. The sorting is relatively good, at least in the remnants of the conglomerate seen by the writer, and there is a fair stratification and some sandy and shaly laminae. Near the base the conglomerate contains some secondary pyrite. Most of the gold was evidently finely divided, and the pay streak was mainly localized near the bottom. The grade of the lower part of the conglomerate, as calculated from old records, was apparently about 1 ppm. Details of this conglomerate are given by Marshall (1918) and Williams (1974).

The basal early Tertiary sandstones, grits and quartz-pebble conglomerates rest mainly on the basement auriferous Otago (Haast Group) Schists of Triassic-Jurassic age and locally on Cretaceous fanglomerates. The age of the early Tertiary quartz-pebble conglomerates is said to be Oligocene, and most appear to be of fluvial origin. Some are up to 300 ft thick and commonly contain (secondary?) pyrite near their base. In places the gold appears to have been relatively evenly distributed through the conglomerate, but more generally the metal occurs over the schist basement in more or less oriented pay streaks. The gold is generally flaky and fine, the colours ranging from 0.5 mm down to paint and microscopic particles with occasional nuggets on the schist bottom. The fine division of the gold is considered by Williams (1974) to be the result not only of long attrition but also due to an origin from the weathering of auriferous pyrite from the lodes of the Otago Schist. Most of these conglomerates were worked where drawn up along schist blocks during late Tertiary orogenic movements. The grade of the early Tertiary conglomerates appears to have been about 0.07 ppm. Details of these conglomerates are given in Macpherson (1933) and Williams (1974).

Fanglomerates and various water-sorted beds of cream to reddish coarse gravels, loosely cemented and consisting mainly of schist and quartz pebbles were laid down along rising fault scarps and outward from these scarps during the late Miocene to early Pleistocene as a result of the Late Tertiary Kaikoura Orogeny, which affected much of the terrain underlain by the Otago Schists. These conglomerates were known to the early miners of the last century as the 'Maori Bottom' since they floor many of the auriferous Quaternary alluvials. In places these conglomerates are gently inclined to steeply dipping; most are auriferous to some degree. The richest sections have been worked at Nasby and in the Nevis Valley. Details and references concerning these auriferous conglomerates are given by Williams (1974).

Auriferous conglomerates, some worked extensively in the past, also occur in northwestern and southeastern Nelson. The former are considered to be of Oligocene age, rest mainly on Lower Paleozoic rocks containing gold-quartz lodes in places and consist mostly of quartz and slate pebbles and small boulders cemented in places by limonite and other ferruginous cement. In southeastern Nelson the basal Tertiary (Eocene) conglomerates are only slightly auriferous. Along the west coast certain Cretaceous and lower Tertiary conglomerates and sandstones are slightly auriferous in places. The Cretaceous Hawks Crag Breccia (a complex of poorly sorted coarse sandstone, siltstone, conglomerate and sedimentary breccia) is apparently leanly auriferous in places. This breccia is also pyritiferous and in places uraniferous as in the Buller Gorge area, some of the uranium being in hydrocarbon. Williams (1974) has concluded that the uranium minerals are epigenetic and were probably introduced during the lower Cretaceous orogeny. Others (noted by Williams, 1974) have opted for a syngenetic-remobilization origin or for multiple mineralizations. The basal conglomerates of the Tertiary coal-measures in the Greymouth and Buller districts and other Eocene conglomerates in the area of the Grey and Inangahua rivers are slightly auriferous and in places pyritiferous. Details and a bibliography of all these conglomerates and breccias are given by Williams (1974).

Slightly auriferous conglomerates of early Pleistocene age have a widespread distribution in northern Westland occurring intermittently northwards from the Mikonui River at Ross to the Grey River. These auriferous conglomerates, which consist essentially of rusty greywacke, granite and schist pebbles, were known as the 'Old Man Gravels' and rest over large areas on a barren upper Miocene clay-glauconite formation known to the old placer miners as the 'Blue Bottom'. The source of the gold in the gravels is uncertain, some probably coming from the Precambrian(?) Greenland and Waiuta formations, which contain auriferous lodes, and some from the geosynclinal schists east of the Alpine Fault. The Old Man Gravels have been worked in places but apparently unprofitably. Their main importance is that they were probably one of the principal sources of gold for the younger Pleistocene and Recent gravels. Details of the Old Man Gravels are given by Suggate (1950) and Williams (1974).

Over the terrain underlain by the Otago Schists and their equivalents in Nelson and in the region in front of the glacial moraines that spread westward from the Southern Alps in Westland, New Zealand, several generations of auriferous conglomerates and gravels can be distinguished including those of Cretaceous, early Tertiary, late Tertiary to early Pleistocene and Quaternary age. The first three generations are described briefly above, and the last is mentioned in the previous section. The origin of the gold in these various generations of conglomerates and gravels is of interest. Williams (1974) thinks that most of the gold was ultimately derived from the weathering of the auriferous lodes of the Otago (Haast Group) Schists, the very finely divided variety in particular, coming from the breakdown of the auriferous pyrite that these lodes contain. Alexander McKay (1903), however, considered that the lodes in the Otago Schists were not abundant enough to provide the large amounts of gold won from the alluvials and conglomerates and thought that the bulk of the gold ultimately came mainly from the quartz laminae that are profusely developed in the schists. These are auriferous in places. Thus McKay concluded, "The gold of Otago must therefore be regarded as mainly having origin in the schists, and in near association with the quartz foliae that characterize the middle and lower divisions of the schists, and is but little due to the quartz reefs occurring in the same rocks."

These differences of opinion concernir. the origin of the gold in placer fields arise from time to time among investigators, and there is as yet no certain solution or consensus. As mentioned later there is considerable evidence to suggest that at least some of the gold came from lodes, but that probably the bulk came from the weathering of the country rocks. With respect to the Otago deposits this multiple hypothesis would seem to resolve the problem. It might be added that in addition to the gold derived from the quartz laminae in the schists can probably also be added that weathered from pyrite, which is abundantly distributed through the graphitic phyllites and schists and other similar rocks of the Otago Schists.

The recycling of gold during sedimentation processes is markedly illustrated by the auriferous conglomerates and gravels of Otago. The initial auriferous rocks and lodes are those in or hosted by the Triassic and Jurassic Otago Schists. Their weathering initially gave Cretaceous conglomerates and sandstones locally enriched in placer gold. There followed then the recycling of gold from these conglomerates through Tertiary and Quaternary sediments with probable contributions at each stage from the auriferous Otago Schists and their contained lodes.

Numerous gold-bearing conglomerates, sandstones and quartzites ranging in age from Cambrian to Tertiary, occur in U.S.S.R. Ivensen *et al.* (1969) describe a number of them and give comparisons with similar deposits in other parts of the world. Most of the lithified types are low grade and have not been mined. The Lower Cambrian Khuzhir auriferous conglomerates and sandstones in East Sayan near Cheremkhovo rest on an Archean basement containing scattered gold-bearing quartz and quartz-sulphide stringers. The conglomerates are composed essentially of pebbles of quartz, quartzite, gneiss, granite, etc. heavily stained with iron oxides. They and their associated sandstones and siltstones represent alluvial river bed deposits. The gold is present as individual, fine grained, rounded particles and in quartz pebbles; some crystals and dendrites have also been noted. The general grade of the conglomerates appears to range from 0.2 to 0.4 ppm.

A great belt of auriferous conglomerates of Neogene age occurs along the southern flanks of the North Pamir and North Kunlun foredeeps of the continental molasse in U.S.S.R. and China respectively. They are the sources of most of the gold in the Quaternary placers of this region, which have been worked as far back as the 10th century. The source of the gold in the conglomerates is thought to be primary deposits in the Upper Paleozoic volcanic and sedimentary strata of the North Pamirs. Nearly all of the conglomerates are very low grade, ranging from 0.0075 to 0.750 ppm. Most of the gold is in the beds and lenses with large pebbles (boulder beds).

In summarizing this section on fossil gold placers we can state certain general paleogeologic and paleogeographic conditions that must exist in order to give rise to such deposits. These include, with brief notes of explanation:

1. An auriferous hinterland that has undergone gradual uplift, weathering, erosion and denudation. The primary source of the gold may have been (a) epigenetic gold-bearing deposits of many types; (b) disseminated gold in rocks especially those rich in pyrite and arsenopyrite, the gold being an intimate constituent of these sulphides; and (c) a combination of (a) and (b).

2. A favourable paleoenvironment for the concentration of gold in eluvial and alluvial deposits. The paleosedimentary conditions for the formation of fossil eluvial, alluvial, beach and aeolian placers have probably remained constant through time and hence are similar to those observed for modern placers as described above. By a study of the modern conditions one should be able to predict with some assurance the possible occurrence of fossil placers in any particular sedimentary column. Predictions may be difficult, however, where multiple erosion cycles have predominated in the geological history of a district.

Ivensen *et al.* (1969) have drawn the following conclusions with respect to the position and origin of ancient placers in the sedimentary and tectonic framework of an area.

1. Fossil auriferous placers are in most cases related neither to buried nor to relict paleoreliefs. Rather, they are related to various facies and belong to diverse formations. Most of the old placers are essentially part of the platform-type and orogenic, mainly continental, sedimentary formations. Economic gold deposits of the fossil placer type are known in various continents in geosynclinal formations of the Upper Archean and Lower Proterozoic periods, in the geosynclinal (?) and orogenic formations of the Lower and Middle Proterozoic periods that form part of the composition of the upper structural stage of the protoplatform, on the floor of the

platform mantle covering the Epikarelian platform, in continental autochthonous and allochthonous formations deposited at the late stage of the evolution of a stable platform region, in the orogenic sedimentary formations deposited at the late stage of development of mobilized portions of a platform and in the orogenic formation of foredeeps and intermontane troughs.

2. All of the deposits (i.e., the so-called fossil placers) are of sedimentary origin, and their gold has passed through various katagenetic (near surface) transformations, although the original pattern of the gold distribution has generally been retained. These gold-bearing deposits (placers) can be subdivided into two major groups: clastogene (placers in the proper sense of the word) and the group in which gold has been precipitated from finely dispersed suspensions mainly in a deltaic and marine littoral environment. The largest known Precambrian auriferous conglomerates, namely those of the Witwatersrand, belong to the second group. Ancient deltaic and marine littoral environments should, therefore, be scrutinized carefully for fossil gold deposits during prospecting.

Problems concerning the origin of placer gold

Despite the fact that the origin of gold in placers seems at a cursory glance to be relatively simple there still remain a number of mystifying features about gold in placers that have perplexed geologists for a century or more. Many outstanding investigations on gold in placers have been completed, but some of the problems remain unsolved. A few of the interesting problems with respect to the origin of gold in placers include the following: (1) the precise details of the origin of the gold in placers; (2) the reasons for the common occurrence of gold on the bedrock or on false bottoms; (3) the origin of nuggets; and (4) the reasons for the formation of a large number of productive gold placers in the Cenozoic era, particularly during Tertiary time in all of the climatic zones of the earth.

1. The precise details of the origin of the gold in placers has long been a problem in certain placer districts. In some districts one can point to gold-bearing deposits of various types as a source for the gold, but in others large auriferous deposits are lacking. In their place large quantities of widely distributed slightly auriferous quartz stringers, blows, boudins and irregular quartz masses are commonly found. These are said to have provided the gold to the placers. Examples of this feature are widespread, and many have been mentioned in preceding sections of the text. The localization of the principal placers of the Klondike within the confines of the Klondike Schist (Fig. 76) is a striking example. Another is found in the Bendeleben Quadrangle, Seward Peninsula, Alaska where practically all of the placers of the quadrangle occur in areas underlain by the York Slate, which is marked by intrusive bodies of acid to intermediate composition, small base metal veins carrying gold and major fault zones that have localized a plethora of auriferous quartz veinlets (Sainsbury, 1974). Finally, in still other districts it is evident from our analyses that the pyrite and arsenopyrite in graphitic schists and other pyritiferous rocks could have supplied large amounts of gold to placers during weathering of these rocks. A number of possible origins for the gold in placers are, therefore, evident—auriferous deposits, widely distributed slightly auriferous quartz bodies, various types of rocks, especially those bearing abundant pyrite and combinations of all three. Strange as it may seem, considering how long placers have been exploited, we have at present no certain idea of the contribution of each of the possible three sources of gold in the placers of any district. This is a problem that merits intensive research, for its solution could be particularly useful in geochemical prospecting as will be evident to anyone searching for economic primary gold deposits. If for instance the gold in placers or stream sediments is wholly or at least partly derived from primary quartz or sulphide deposits, use of the element as an indicator of such deposits is valid. If, however, the sources are widely distributed quartz blows, stringers, etc. or the country rocks, the use of gold as an indicator for the location of economic gold-bearing primary quartz or sulphide deposits is invalid. This problem is discussed further in Chapter V.

2. The reasons for the occurrence of pay streaks of gold on or near the bedrock or on false bottoms has intrigued many students of placers. Why should this be so, considering the great quantities of sand and gravel that often overlie the pay streaks? One would suspect on casual examination that the gold would be rather evenly distributed throughout the gravels and sands rather than concentrated in well-defined streaks.

Cheney and Patton (1967) mention that most economic geology text books do not mention this problem or treat it rather vaguely. These authors consider two general processes that might produce pay streaks on bedrocks or false bottoms—downward chemical movement of gold, which they dismiss; or downward physical movement of particles of gold through unconsolidated sediments, which they also consider improbable. They view the streaks on the bedrock as the results of extensive scouring (sluicing) of the whole sediment column in stream and river channels during floods of unusually large magnitude at irregular intervals of decades or even centuries.

Gunn (1968) commenting on these views concludes that there can be no all-embracing explanation for bedrock placers. From his experience, however, he believes that there is a downward physical movement of gold through the unconsolidated materials in placers. The source of the agitation (or jigging) to produce this movement is microseisms, one activating agency of these being moving water. Gunn (1968) also noted that in some placers the barren gravels may have been deposited under quite different conditions than those containing the pay streaks. Tuck (1968), after mentioning that bedrock concentrations are not the sole occurrence of gold in placer deposits, there being also concentrations on various types of false bottoms and as disseminations throughout some gravels, generalizes as follows:

> Commercial bedrock concentrations, in the case of stream deposits, are generally the result of a long period of erosion during which the stream is dominantly downcutting. Assuming a source of gold, weathering and erosion are separating and concentrating the gold from the country rock into the bottom of the valley. All during this period, which may represent several thousand feet or more of vertical erosion, gold as it is liberated is on or near bedrock since the stream is essentially flowing on bedrock. The action has been likened to that of a sluice box on a large scale, and there is no better analogy. Once gold reaches bedrock it generally remains

there. All during this period the stream valley is essentially youthful. Eventually a baselevel is reached, maturity is attained, and dominantly the stream begins to widen its valley.

Characteristic of many of the important bedrock concentrations of gold is a narrow rich concentration meandering through a wider valley in which the bedrock of the bordering area may or may not have about the same elevation as that of the rich bedrock concentration. The bordering bedrock area may also have a concentration of gold, but it is of much lesser value than the narrow rich streak. In many cases, the narrow rich channel was high-grade enough to be hand-mined while later it and the bordering low-grade areas were dredged. The most logical explanation for this narrow rich channel is that it marks the bottom of the valley when the stream ceases dominantly to downcut and begins to cut laterally having reached a mature stage. The narrow rich channel is the concentration of gold from a volume of country rock very large relative to that which furnished the lower grade concentrations in the bordering area. It also logically follows that generally the narrow rich channel contains much coarser gold, and the particles have a higher gold-silver ratio, than the bordering area. Floods have their greatest importance in carrying gold to bedrock principally in the stage when the valley has reached maturity and is being widened with aggradation and degradation alternately occurring. At a still later stage aggradation becomes dominant, and the bedrock concentration formed during the youthful and early mature stages becomes covered with alluvium.

Tuck (1968) also reminds us that the history of placer streams and beach placers is much more complex than this simple exposition, pointing out among other factors, the necessity for a favourable erosion history and a suitable bedrock source of gold in the free state in particle sizes large enough to form commercial placers.

Krook (1968) claims that the climate is an important factor, a point also stressed by Mertie (1940). Krook says that the transition from vertical to lateral erosion and aggradation in a region is caused by a change in climate. As an illustration he mentions the case in some regions of the world where during Pleistocene time lateral erosion and aggradation took place during the glacial periods under periglacial conditions while during the warm interglacial intervals downcutting was extensive, giving rise to a series of river terraces. Under such conditions gold is highly concentrated in the lower terrace deposits and stream bottoms of a valley since with each phase of downcutting the gold is concentrated because it is transported only short distances horizontally, whereas the bulk of the sediment is swept far away downstream.

Kolesov (1975*b*), in a recent contribution to the problem of why placer gold tends to find the bedrock or false bottoms, maintains that the laws of hydrodynamic sorting in a flowing aqueous medium must be taken into consideration. He develops a number of equations which indicate that large gold particles will tend to accumulate in the bottom of the placer and that the finer particles will be distributed relatively uniformly throughout the cross section of the placer. Details should be sought in the original paper.

3. The origin of gold nuggets has long been a subject of discussion amongst students of gold placers. Two general theories prevail: one holds that the nuggets are formed mainly by chemical accretion processes; the other maintains that they are detrital in origin, had essentially the same weight as they now possess, but their shape and features are due to the rolling and hammering they have received as they have been moved along in the gravels. A third theory is a compromise holding that nuggets are partly detrital and partly chemical in origin.

There has been no consensus about the origin of nuggets for centuries. Albertus Magnus (1200–1280) mentions the problem in his *Book of Minerals* ((Wychoff, 1967, p.184) and concludes, "But gold which is formed in sands as a kind of grains, larger or smaller, is formed from a hot and very subtle vapour, concentrated and digested in the midst of the sandy material, and afterwards hardened into gold. For a sandy place is very hot and dry; but water getting in closes the pores so that the vapour can not escape; and thus it is concentrated upon itself and converted into gold. And therefore this kind of gold is better (finer)." From this we can assume that Albertus considered that alluvial gold is formed in situ. He was one of the first to advocate a chemical accretion process. His theory is repeated in the *Bergwerk-und Probierbüchlein* (1500) where reference is made to "gold generated in rivers and born in streams" (Sisco and Smith, 1949). In his *De la Pirotechnia* published in 1540, Vannoccio Biringuccio disagreed with the chemical theory and states that the gold of placers is merely transported by streams from oxidized and disintegrated bedrock deposits (Smith and Gnudi, 1959). Georgius Agricola, in his *De re metallica*, published in 1556, also adheres to this view (Hoover and Hoover, 1912).

Egleston (1881) considered that nuggets, and indeed most placer gold, were of chemical origin, the gold having been deposited from surface waters and waters that circulated through the placers. He explained the gold values on bedrocks in this manner, stating that as the gold-bearing solutions reach bedrock they can go downward no farther and precipitate their gold around nuclei due to the presence of reductants such as lignite, fossil wood, pyrite, etc. Gold in the upper few feet of bedrock in placer deposits was similarly explained.

Liversidge (1893*c*) reviewed the literature and theories on the origin of gold nuggets back to 1860. He mentions that A.R.C. Selwyn was the first to propound a chemical accretion theory for the origin of nuggets in the Victoria placers, Australia. Selwyn thought that the nuggets were chemically accreted about small nuclei of gold in the placers from large quantities of saline and acid thermal waters, which obtained their gold by percolation through the great thicknesses of gold-bearing gravels and sands. This took place during the period of great volcanic activity, which produced the basalts that cover the deep leads in some districts of Victoria. It should be mentioned in passing that a similar theory was advocated by some investigators for the deep leads in California.

Liversidge (1893*c*) carried out many experiments on the precipitation of gold on gold nuclei and conceded that gold nuggets could be formed in this way. However, he concluded that large nuggets have not been formed in situ in this manner. He stated that they had been set free from reefs and that any small additions of chemically precipitated gold were immaterial. In the case of gold dust and gold grains he thought the situation might be different for these expose a large surface and thus chemical accretion ("electroplating" he called it) may have had an appreciable effect in increasing the amount of this type of gold.

Maclaren (1908) reviewed all the evidence available about the origin of gold nuggets up to the turn of the century and concluded that chemical accretion played a large part in the formation of the nuggets and other gold particles in placers. Lindgren (1933) doubted that chemical accretion played any part in the formation of placer gold. Fisher (1935), who carried out an extensive investigation of the nature and structure of alluvial gold in the Morobe goldfield, New Guinea, concluded that the nuggets and flakes were derived mechanically from denuded gold-quartz veins or lodes. McConnell (1905) and Johnston and Uglow (1926) advocated the same origin for the placer gold of the Klondike and Cariboo respectively. Mustart (1965) also concluded that most of the placer gold he examined from the central Yukon was essentially detrital. Desborough (1970) and Desborough *et al.* (1970) have inferred that the placer gold they investigated from the western United States is detrital in origin. Boyle (1968*b*) has advocated a multiple hypothesis for the origin of placer gold, facets of which are amplified subsequently.

A number of criteria have been used by the proponents of the detrital and chemical hypotheses to support their contentions. These are listed with brief comments.

Criteria suggesting a detrital origin for nuggets and placer gold:

1. Gold placers are formed predominantly near the source of epigenetic gold deposits of various types. If chemical factors were involved one might expect that placers would be formed both near the source of the gold and at any distance where the chemical factors are favourable. Those who advocate a chemical mechanism might, however, retort that the presence of fine-grained gold near the source (epigenetic deposits) is the main reason for chemical accretion since the gold dust particles provide nuclei for precipitation.

2. Gold placers commonly contain nodules of quartz with masses of undoubted primary gold. Also large nuggets often have quartz attached to them in a manner similar to that found in primary quartz veins.

3. In many gold placers coarse gold is found near the source and increasingly finer grained gold downstream. This is suggestive of increasing abrasion and comminution of detrital material varying with distance travelled.

4. Gold in placers is commonly accompanied by a suite of heavy resistant minerals that accompany the gold in primary deposits, examples being scheelite, barite, pyrite and arsenopyrite. These minerals are obviously detrital and have no secondary chemical history. Gold probably behaved in a similar manner.

5. In a number of placers the gold contains essentially the same trace elements or trace minerals as those found in the gold of the primary deposits. This criterion is tenuous since most of the trace elements, having a siderophile (metallic) tendency, such as Hg, Ag, Cu, etc., may have precipitated with the gold on the nuclei of the nuggets. The traces of included minerals in nuggets are more difficult to explain chemically, but during accretion processes it is not uncommon for fine-grained particles of sulphides, tellurides, quartz, etc. to be incorporated mechanically in various types of pyrite nodules, iron concretions, etc. The same may take place in gold nuggets.

6. Gold in placers exhibits a crystalline nature similar to that of vein gold. There is commonly also a thin rind of relatively pure gold on some nuggets that has been interpreted as due to leaching of silver and other metals because these have a much higher solubility than gold. It should perhaps be noted in passing that not all gold nuggets, flakes and dust have a rind of purer gold, a feature noted by Stumpfl and Clark (1965) and many others. The crystallinity argument is not particularly valid. We assume that gold in epigenetic deposits, such as veins, is deposited from dilute solutions over a long period, and certainly this process gives rise to a good crystalline structure. There is no reason to assume that nuggets accreting gold from equally dilute surface solutions should not, likewise, have a good crystalline structure. The rind of relatively pure gold may be the surface expression of the precipitation mechanism. With time and growth of the nugget this rind is gradually incorporated into the crystalline fabric as a result of diffusion and recrystallization processes. Crook (1939) has suggested that recrystallization of gold nuggets is probably common. He noted that placer nuggets from 'new' or recent deposits show definite strain lines while nuggets from 'old' deposits where the gold has remained at rest for long periods of time have equiaxed (recrystallized) at normal temperatures due to cold working.

Criteria suggesting a chemical origin for nuggets and placer gold:

1. Gold and silver are constituents of most natural waters being relatively enriched in waters leaching primary gold and silver deposits. Gold and silver are also capable of migrating in permafrost (Boyle, 1951, 1965*a*) by a mechanism that is not entirely understood. Both elements seem to migrate in the permafrost while attached to water vapour molecules. Many years ago Freise (1931) pointed out that the *pinta que paga,* i.e., the 'paying spark' in gold placers in Brazil reappears after a number of years particularly when the exhausted goldfield has been hidden by the sun, by vegetation or other means. This he attributed to the transport of gold in humic (black) waters and its redeposition in the stream bottoms. As noted in Chapter II his theory of transport as gold humates has been verified by modern methods of analysis and experiment.

2. Nuggets usually have a mammillary appearance, which is suggestive of chemical accretion processes. Some have said, however, that this appearance is due to abrasion and rolling about in the placers. The microstructures of some mammillary nuggets in alluvial placers are zonal, and some are concentric, features that bespeak chemical accretion. Similar structures are often seen in gold in oxidized zones and in eluvial placers. In these situations the gold is undoubtedly accretionary in origin.

3. Gold crystals are often present in placers. They are relatively rare in most primary gold deposits. In placers the salient edges of the crystals are often sharp and well defined, indicating absence of agents of attrition or of solution. In other words many of these crystals grew in the placers.

4. Some nuggets and irregular particles of gold have crystal protuberances on their ends or on small projections. Such structures are rare in primary deposits. In the Urals, some rounded gold nuggets in the placers have developed crystal faces (Shcherbina, 1956*b*). Even more significant is the common occurrence of thin deposits and films of high purity gold on rounded particles of osmiridium (sysertskite) and

platinum in the gold placers of the Kuznetsk Alatau and Gornaya Shoria in U.S.S.R. (Syrovatskii *et al.,* 1969). These authors also refer to the work of a number of others who have noted chemical migration of gold in the Western Siberian placers. Voronin and Goldberg (1972) figure a number of occurrences of assymetrical secondary growths of gold on placer gold. They attribute this phenomenon to electrochemical processes in the placers, the small gold nuclei acting as collecting agencies.

5. Films, threads, wires, dendrites and paint gold commonly occur on stones in placers, on magnetite, attached to nuggets or limonite cement and especially in or near organic material such as buried wood, that has exerted a reducing action. 'New' or redeposited gold has been repeatedly observed in the Lena alluvial deposits, U.S.S.R. Nikolaeva (1958), however, considers that the amount of redeposited gold is very small and of no commercial importance.

6. Pyrite that has replaced fossil wood in some placers is invariably auriferous. Some say, however, that this is due to entrapment of flour or dust gold by the pyrite.

7. A number of investigators have shown experimentally that particles of gold accrete dissolved gold where reducing matter or minerals such as pyrite are available, as probably is always the case in placers. Goni *et al.* (1967) have demonstrated experimentally that gold nuggets can be formed from colloidal solutions of gold by compaction of the colloidally precipitated gold.

8. It has repeatedly been observed that there is an increase in the fineness of gold with increasing distance from the source. Pliny at the beginning of our era commented on this phenomenon. Thus, the fineness increases consistently in the order: primary deposit – eluvial placer – alluvial placer – marine placer all things being equal. Aberrations in the pattern are common, but in general the phenomenon appears to hold in many places according to Maclaren (1908), Lindgren (1933), Mertie (1940) and many others. With respect to the various elements present in placer gold there is a general decrease in the silver content and in some placers a gradual decrease in elements such as Fe, Mn, Cu, Ni, Pb, Bi, As, Sb and Te with distance from the source. Gorshkov *et al.* (1971) traced the fate of gold from the Khualazinsk primary deposit through eluvial and alluvial placers along the Tinkan River to a beach deposit along the Bay of Tinkan, far eastern southern maritime province, U.S.S.R. They noted a regular increase in the roundness of the gold particles, a general decrease in size and a fairly regular increase in fineness (800–860 in the primary deposit to 930 in the beach placer). They also observed a regular decrease in the content of Pb, Bi, Sb and Te in the gold in proportion to its distance from the source. Cu remained constant while mercury increased slightly. Internal structural changes were also noted with distance from the source, the most interesting being that the rind of purer gold on the particles increased in width, and there was an increase in the degree of recrystallization (60–75% recrystallized in the marine placers). The thicker rind in the marine placers was attributed to a greater degree of electrolytic corrosion by the seawater compared with the river water. The explanation of the increase in fineness of placer gold with increasing distance from the source has varied among those who have worked on the problem. Most investigators, including McConnell (1907), who first provided conclusive evidence for a rind of relatively pure gold on nuggets, have suggested that silver is progressively leached out of the gold because of the differential solubility of the two metals. Others have considered the explanation to be increasing precipitation of gold on small gold nuclei (dust) in the placers. Mertie (1940), who examined the problem carefully, considered the solution of silver from gold nuggets improbable. He pointed to the difficulty of removing silver from gold-silver alloys by the action of very dilute cold surface waters when concentrated inorganic acids were incapable of doing the job when the ratio of gold to silver is greater than 1:1.5 (400 fine) (the lower limit of gold for the removal of silver by the process of inquartation). Since most placer gold is at least 600 fine the amount of silver removed should be infinitesimal. Mertie and others have also pointed to the volume relationship of the so-called silver leached rinds compared to the total volume of gold. These rinds are very narrow (0.002 mm) and cannot have much overall effect in the fineness of the gold, especially where the gold is coarse (>1 mm). Furthermore, gold from which silver is leached is porous; microscopically a marked surficial porosity has been observed only rarely on particles of placer gold.

Mertie (1940) pictures the pay streak as a series of overlapping wedges of gold-bearing alluvium, which were progressively deposited upstream during a single erosional cycle. According to his view the downstream end of a pay streak is the oldest, and the upstream end the youngest, if the deposition has taken place during a single cycle of erosion. He also advanced the hypothesis that the apical part of a gold lode contains gold of higher fineness than its lower horizons. Taken together, he thought these two concepts lead directly to the conclusion that the gold of highest fineness is likely to be found in the downstream end of a pay streak, and that the fineness of the gold may diminish progressively upstream. This relationship was found to exist in some placer pay streaks in Alaska, but in others the fineness was observed to change erratically, or not at all.

On the other hand there is some experimental evidence that the fineness of placer gold can be increased by leaching and mechanical treatment. Koshman and Yugay (1972) were able to increase the fineness of gold by leaching with HF and H_2SO_4 and by grinding the gold. The first procedure removed both mineral and chemical impurities and the second mainly mineral impurities. Campbell *et al.* (1973) have reported similar findings with respect to acid leaching of placer gold. They noted that Ag, Cu, Hg, Pb, Bi and a number of other elements were removed and the fineness increased. How valid these experiments are with respect to the natural systems is a moot question. The leaches were far beyond any of the concentrations one might expect to find in natural waters.

9. Some gold nuggets in eluvial and alluvial placers appear to be greatly enriched in mercury compared with the gold particles in the primary deposits from which the nuggets presumably came. The reason for this may be that as the gold nuggets accrete in the oxidized zones of auriferous deposits they avidly collect the mercury liberated during the oxidation of minerals such as tetrahedrite and sphalerite. More data than are now available should be acquired on this aspect of the formation of nuggets since the enrichment of mercury may

assist in solving the perennial problem of the origin of gold nuggets.

The writer can add little to the vexed question of the origin of nuggets in placers. In the Yukon there is undoubted evidence that the gold in the oxidized zones is chemically accreted from very finely divided or lattice gold in pyrite and arsenopyrite. There, and elsewhere, there is positive evidence that the coarse gold in the upper parts of oxidized zones is finer than that in the primary deposits. What happens to the gold as it passes from the oxidized zones into the eluvium and from there into the alluvial placers and thence downstream is a problem that has not been adequately researched in the writer's opinion. There is certainly positive evidence that the soil solutions and stream waters contain both gold and silver, and all of the chemical evidence points to the probability that gold will be precipitated on any available gold nuclei provided reductants are present, which is nearly always the case in eluvium and alluvium. There is also considerable evidence that the gold of placer deposits has been extensively recrystallized, and the weight of the chemical evidence with respect to the relatively pure gold rinds on nuggets is that the rinds in most cases represent chemically precipitated gold and some silver rather than that the silver was leached from the surface layer. There may, however, be cases where the silver is leached from the surface of nuggets, but how this leaching could have proceeded throughout a nugget the size of a hen's egg, thus increasing its fineness considerably in excess of that found for the gold in the primary source, is difficult to understand.

These considerations have led the writer to conclude that the gold of placers is of both detrital and chemical origin. In some placers the detrital component may predominate; in others the chemical is more important. One should view the formation of gold nuggets in a dynamic sense, the agents forming them being both chemical and mechanical and their action being concomitant. As gold deposits weather down their particles of gold pass into the oxidized zones and then into the eluvium, in both of which they slowly accrete dissolved or colloidal gold (and some silver) from the oxidizing solutions in the veins and from the soil solutions. As the weathering cycle proceeds the material in eluvial deposits may be further decomposed, eroded, transported and redeposited. This can be effected by gravity, wind, streams, rivers, ice or by the surf of the sea, resulting in aeolian, alluvial and beach placers. During these movements the gold-silver particles may be hammered, comminuted and reduced to a flour, or, as is more often the case, especially in an aqueous environment, the small particles may be partly dissolved, the gold reprecipitated on the larger particles, and these may be cold worked and recrystallized as they are moved along in the placers. Since silver is more mobile (more soluble) than gold the chemical processes of solution and reprecipitation generally lead to an increase in the fineness of the gold flakes and nuggets with increasing distance from the source.

There is a tradition among placer miners that is centuries old, namely that gold 'grows' in placers. There is probably much truth in the adage.

Under certain conditions it seems possible that the gold in placers may be simply comminuted by mechanical abrasion, there being little or no chemical action. Yeend (1975) asked the question – What happens to a sand-size particle of gold as it travels along the bed of a river? – and sought to answer it by a laboratory experiment in which gold particles were tumbled in a mill with sand, sand and cobbles and only cobbles in a manner simulating a high-energy river environment. The results of his experiment are of interest: with sand and gravel, the common bedload of many placer rivers, gold was abraded at rates of 0.015 to 0.007 per cent (by weight)/h of travel (at 0.5 mi/h or 0.845 km/h). Cobbles, rather than sand, were responsible for most of the physical changes and abrasion of the gold. Ten gold fragments tumbled for 120 h with cobbles and water (no sand) were broken down to 68 recoverable fragments and lost about 25 per cent of their weight to particles smaller than could be recovered using conventional panning techniques. Gold, tumbled for 120 h with sand and water, lost less than 1 per cent of its weight. Gold was abraded faster by wet sand than by dry sand. Velocity appeared to be more important as a factor in the abrasion of gold than travel distance – a fourfold increase in velocity produced a tenfold increase in hourly abrasion rates of gold. Shilo *et al.* (*in* Petrovskaya, 1974) also carried out a number of simulative experiments on the abrasive action of ground quartz on placer gold. They noted that the gold grains were changed in form by erosive action and impact deformation. The average wear of the placer particles was 0.05 to 0.06 per cent/h.

4. The reasons for the formation and preservation of large numbers of productive gold placers throughout the world during Cenozoic time, particularly in the Tertiary period, as compared with earlier periods, has received little mention in the literature. But, before continuing the discussion of why these placers are so prevalent some of the general features of the distribution and nature of gold placers with time should be briefly reviewed. Large economic placerlike deposits occur in the Precambrian (e.g., Witwatersrand), but they are of restricted occurrence, of controversial origin, and if originally placers have been highly modified. The sedimentary rocks of the Paleozoic and Mesozoic contain a few fossil placers, but all are low grade, the gold being finely divided and generally widely disseminated throughout the conglomerates and sandstones. In Cenozoic sediments, particularly those of Tertiary age, gold placers are widespread occurring in practically every country. Many are or were phenomenally rich. An interesting feature about gold placers is the relationship of the size of the gold particles with time. Precambrian placers are all characterized by very fine-grained gold, often in the range 50 to 100μ. Paleozoic and Mesozoic gold placers are invariably also marked by fine-grained gold, but in places small nuggets a few millimetres in diameter can be found. They are, however, relatively rare. Tertiary and younger placers often contain coarse gold, nuggets weighing 1 oz or more being relatively common in places.

The Tertiary was a period of great volcanic activity and gold deposition in many parts of the world as noticed in the first part of this chapter (Fig. 7). But this abundant primary source of gold cannot be the sole parameter, or indeed a vital parameter, in the production of the many placers in the Tertiary and younger periods, for older eras were also marked by an abundant primary gold deposition, particularly the Precambrian.

The Tertiary was marked by extensive tectonic uplift in most parts of the world, and the paleoclimatologists tell us that the climate was generally hot and humid over almost the whole face of the earth, from the poles to the equator. Great forests of tropical to subtropical type also covered large areas of the land. The weathering, mainly lateritic, was deep and the downcutting by stream and river systems due to uplift was prolonged and extensive. Certainly these were optimum conditions for the formation of gold placers. But they are not unique in the geological history of the earth. Similar conditions probably existed in late Precambrian time, and certainly prevailed during parts of Paleozoic and Mesozoic time.

On close scrutiny it appears that neither source, tectonics nor climatic conditions are the decisive factors which gave rise to the multitude of placers in Tertiary and later times. One factor alone seems to be responsible, namely the generation of the sediments in which the placers occur. In nearly all Tertiary and younger gold placers the sediments are first generation or in some placers locally second generation. In other words the natural milling process has been relatively limited, and the gold has not been comminuted and dispersed. In older rocks most of the sediments represent many generations (cycles), and the natural milling has been extensive and pervasive. The result has been the wholesale comminution of gold and its dispersal throughout a vast volume of sandstones, conglomerates, etc. This is apparently why there is a diminution in the size of gold particles in placers with time. Recognition of these features is important in prospecting for fossil placers. First or second generation sediments in the sedimentary column are most likely to contain placers; only rarely in certain basins, such as the Witwatersrand, has the dispersed gold been reconcentrated into payable quantities by secondary (diagenetic) processes as outlined in a later section.

Miscellaneous sources of gold

Gold is won or can be won from a great variety of deposits not specifically mentioned in the above descriptions. These include the following:

1. *Nickel-copper ores associated with basic intrusions – Sudbury type:* The content of gold (and silver) in these ores is very low. In his studies Hawley (1962) found native gold (electrum) in only one of the deposits at Sudbury (the Frood deposit), and there it was relatively rare. The gold was rich in Ag, Pd and Ir and was predominantly associated with chalcopyrite. In the Noril'sk ores Razin and Yurkina (1971) found four gold minerals electrum, porpezite, argentocuproaurite and native gold. In the impregnation ores these minerals, all relatively high in Pd, are commonly intergrown with various platinoid minerals. Paragenetic relationships suggest that argentocuproaurite was the earliest to crystallize followed by porpezite, native gold and finally electrum.

In a study of the gold content of the Cu-Ni deposits in the Noril'sk region Vakhrushev *et al.* (1972) found great differences in the gold contents of the various sulphides. The highest contents (up to 9 ppm) were found in pyrrhotite, cubanite, chalcopyrite and pentlandite, and the lowest amounts in pyrite ($<$0.05 ppm) and sphalerite. They thought that this was good evidence for a difference in origin of the two suites of minerals, the gold-rich suite being essentially early (primary magmatic) and the gold-poor suite late post-magmatic (hydrothermal). They also made the interesting observation that pyrite related to trap (basic) intrusions commonly has less gold than that in skarn, polymetallic and vein deposits.

Keays and Crocket (1970) carried out an extensive investigation of the gold and platinoid content of the minerals of the Sudbury ores. Their averages in ppm for gold in the various minerals are: pyrrhotite (0.012), chalcopyrite (0.078), magnetite (0.0014), pentlandite (0.038) and gersdorffite (0.021). From these figures it is apparent that gold is associated mainly with chalcopyrite and pentlandite. In the deep ore zone of the Strathcona Mine at Sudbury, Chyi and Crocket (1976) found gold concentrated in pentlandite (up to 0.333 ppm). Pyrrhotite, chalcopyrite and magnetite contained up to 0.0277, 0.0351 and 0.011 ppm respectively. The general averages for pyrrhotite, chalcopyrite, pentlandite and magnetite were 0.0142, 0.0198, 0.1566 and 0.0056 ppm respectively.

A review of the literature on the distribution of gold in the Sudbury, Noril'sk and similar ores indicates that gold is commonly highest in content in the copper-rich (chalcopyrite, cubanite, etc.) ores and lowest in the pyrrhotite- and pyrite-rich ores, particularly the latter. The same relationship holds for the ores at Insizwa (Scholtz, 1936). In most nickel-copper ores of the Sudbury type, therefore, gold tends to follow copper and nickel and is commonly associated with the platinoids. Silver has a similar coherence particularly with copper. Most of the gold, at least about 75 per cent, occurs in the nickel-copper ores of the Sudbury type as native metal or as an alloy with platinoids, Ag and Cu. The remainder of the gold is probably present as an isomorphic mixture or in a finely divided state in the sulphides, particularly chalcopyrite, pentlandite, pyrite, arsenopyrite and pyrrhotite.

Our analyses and calculations of the content of gold and silver in the Sudbury ores as mined indicate average values of the order of: Au, 0.08 ppm; Ag, 2.3 ppm. Some of the so-called 'offset orebodies' are relatively rich in gold and silver; the chalcopyrite ores of the Vermilion deposit for instance average 1.9 and 100 ppm Au and Ag respectively. Other ores in these 'offset orebodies' that are rich in pyrrhotite have less gold and silver than the average quoted above.

The deposits of the Sudbury type yield considerable amounts of gold and silver throughout the world. The deposits at Sudbury produce some 2 million oz of silver and 50 000 oz of gold annually.

2. *Massive sulphide deposits containing essentially Fe, Cu, Pb and Zn sulphides in volcanic and sedimentary terranes – Noranda-Bathurst type:* In these deposits the contents of gold and silver are generally low. Most deposits rarely contain more than 0.05 oz/ton Au (1.7 ppm) and 5 oz/ton Ag (171 ppm). The highest gold contents found in deposits of this type appear to occur in the Kurokô ores of Japan (up to 1.5 oz Au/ton; 51.3 ppm). These are discussed separately below. Locally, small high grade massive sulphide deposits in many parts of the world may contain up to 2 oz Au/ton or more. The Au/Ag ratio in the Fe, Cu, Pb and Zn massive sulphide deposits appears to range widely from 0.006 to 1; the general average is about 0.025. Most of the gold in these deposits is generally present as the native metal or as tellurides; aurostibite has not been reported from massive sulphide deposits.

Some gold is also probably present in lattice sites in various sulphosalts, particularly tetrahedrite-tennantite and in chalcopyrite, bornite, pyrite and arsenopyrite. In many of these deposits the gold is associated with the copper-rich parts; in others with the siliceous pyritic or arsenopyritic zones. In certain deposits the gold appears to have been introduced during the galena-sphalerite stage. In this respect the gold mineralization behaves much like that discussed for gold-quartz veins. In the Phoenix Mine, Boundary district, British Columbia, Sinclair and Percy (1969) found, by multiple regression analysis of assay data for 220 samples, that there was a correlation of high gold and silver values with high chalcopyrite and pyrite content of the ore. In addition, a negative correlation between gold and lime-bearing minerals was observed. In the copper ores of the Cobar region, New South Wales the native gold is commonly associated with bismuth minerals and often with the sphalerite-galena stage of mineralization (Rayner, 1969). In the Bathurst, New Brunswick massive sulphide ores there is a positive correlation of gold with lead and zinc (galena-sphalerite stage) in some deposits; in others there is a correlation with copper; and in still others the gold appears to have been enriched at the pyrite-arsenopyrite stage. Some deposits with bornite are rich in gold, for example, the Urup deposit in the North Caucasus where up to 39.4 ppm Au in finely dispersed particles occurs in the bornite (Vikhter *et al.*, 1968). Others rich in arsenopyrite are, likewise, high in gold content. Such a deposit was the great Boliden deposit in Sweden (Grip and Wirstam, 1970). The ores were complex in this deposit and included arsenopyrite-rich bodies, gold-bearing quartz-tourmaline veins and large bodies composed mainly of pyrite and chalcopyrite. During the life of the deposit some 8 341 550 tons of sulphide ore were mined, containing 15.2 ppm Au, 49 ppm Ag, 1.42 per cent Cu, 25.1 per cent S and 6.91 per cent As. Much of the gold was microscopic and was present in the arsenopyrite; coarser gold occurred in the quartz-tourmaline veins.

In the massive sulphide bodies in volcanic and sedimentary terranes the silver is generally associated with the higher grade lead parts of some deposits and with the high-grade copper zones in others. Most of the primary silver is in either galena or tetrahedrite; some may be associated with sphalerite or chalcopyrite. In some deposits, much silver is contained in argentite, native silver, electrum, gold, sylvanite, altaite, stromeyerite and other silver-bearing sulphides, sulphosalts, tellurides and selenides. There is commonly no particular relationship between the silver and gold contents in the massive sulphide ores. As an extreme example one may mention the Kidd Creek deposit near Timmins, Ontario, which contains some 5 oz Ag/ton, whereas the gold content of the ores is nil or only traces. The native silver in this deposit, likewise, is devoid of gold in more than trace amounts. Supergene silver in the massive sulphide deposits occurs as the native metal, in chalcocite, covellite and acanthite, and in chlorargyrite, argentojarosite, etc. Supergene gold is invariably present as the native metal. Some of the oxidized zones of the massive sulphide bodies in volcanics and sediments are enriched in gold (*see also* Chapter IV).

In some deposits of this type there is commonly a complex relationship between gold, gold tellurides and the other constituents of the massive sulphide bodies. The classic example is the Horne Mine at Noranda described in great detail by Price (1934). He recognized three types of gold mineralization in this great orebody – gold in the massive sulphide deposits, gold in the flux deposits (silicified zones associated with the mineralization) and gold in the chloritic (alteration) deposits. Briefly the relationships of the gold in these three deposits are as follows according to Price:

(a) *Gold mineralization in the massive sulphide deposits:*

1) Gold and tellurides (especially the former) forming minute veinlets, irregular areas and dots in idiomorphic pyrite crystals.
2) Veinlets (predominantly gold) along the borders of pyrite crystals, between the pyrite and the adjacent sulphide.
3) Gold-telluride areas replacing pyrrhotite, the free gold particles being surrounded or 'sheathed' by tellurides.
4) Gold replacing chalcopyrite.
5) Gold and tellurides in veinlets and angular patterns replacing minute chlorite aggregates in the massive sulphides.
6) Quartz veinlets carrying gold and tellurides, cutting the earlier sulphides.
7) Type 5 repeated at the contacts of highly chloritized syenite porphyry where such material occurs in the midst of a large sulphide mass.

(b) *Gold mineralization in the flux deposits:*

1) Gold and tellurides in the forms given under Types 1 and 2 for the massive sulphides – in, and around, the borders of pyrite crystals.
2) Gold and tellurides in quartz veinlets cutting pyrite crystals.
3) Gold-telluride-quartz veinlets cutting the siliceous country rock.
4) Gold and tellurides replacing chloritized areas in and around isolated pyrite aggregates.

(c) *Gold mineralization in the high-grade chloritic deposits:*

1) The usual pyrite associations and textures given under (a) and (b).
2) Replacement of chalcopyrite by the tellurides and gold.
3) Replacement of the chlorite by tellurides in angular patterns.
4) Extremely minute gold and telluride veinlets in quartz aggregates in the massive chlorite.
5) Tellurides and gold replacing original magnetite.
6) Gold and telluride veinlets cutting chloritized metadiabase.

Throughout the world considerable amounts of gold and silver are won as a byproduct of the smelting and metallurgical treatment of the ores won from the massive sulphide bodies in volcanic and sedimentary terranes.

3. *Polymetallic vein and lode deposits containing essentially Fe, Cu, Pb and Zn sulphides in volcanic and sedimentary terranes:* These are widespread throughout the mineral belts of the world. Some are rich in gold (up to 1 oz/ton or more); others contain only traces. The silver content grades all the way from a few tenths of an ounce of silver per ton to 40 oz/ton or more. There is commonly no particular relationship between the gold and silver contents of the ores. Each deposit or group of deposits appears to be individualistic with respect

to the Au/Ag ratio, but statistically the majority of the deposits have a very low ratio.

Most of the gold in these deposits occurs in the native form, intimately mixed with the sulphides or as tellurides. Aurostibite has not been reported from these deposits. Some gold is also present as a lattice constituent of sulphosalts, particularly tetrahedrite-tennantite and of pyrite, arsenopyrite, bornite and chalcopyrite. Vikhter *et al.* (1968), who studied the precise form of gold in the pyritic polymetallic deposits in the northern Caucasus, U.S.S.R., found the metal to be present as flaky, spongy, streaky, dendritic and often filmy particles and aggregates. Gold tends to favour those deposits with a quartz gangue or with large amounts of arsenopyrite and pyrite, although this is by no means an invariable feature. Much of the silver in these deposits occurs in galena or in tetrahedrite-tennantite. Some of the deposits contain silver sulphides, sulphosalts and native silver. Many of the oxidized zones of these deposits are greatly enriched in gold, mainly in the native state, and in silver as the native metal or in chalcocite, covellite, acanthite, chlorargyrite, argentojarosite, etc. (*see also* Chapter IV).

Throughout the world large amounts of gold and silver are won as a byproduct of the smelting and metallurgical treatment of the ores from these deposits.

4. *Kurokô (black ore) deposits of Japan:* These deposits may belong partly in the massive sulphide category (2 above) and partly in the polymetallic category (3 above). They are, however, somewhat different from both and merit a brief description since they are commonly rich in both gold and silver, which are won mainly as a byproduct of the smelting of the Pb-Zn-Cu ores.

The Kurokô deposits are characteristically developed along the inner zone of Tertiary sediments and effusive rocks in northeast and southwest Japan. The geology near the numerous mines is complex and differs from district to district. Briefly, the host rocks are a volcanic sequence of tuffs, breccias, conglomerates, black shales, eruptive liparites, rhyolites, dacites and andesites and intrusive dolerites, andesites, diorites and quartz porphyries. The best general descriptions of the deposits are those by Kinoshita (1931), Horikoshi (1951–1952; 1969), Ishikawa *et al.* (1962), Urashima and Sato (1968), Matsukuma and Horikoshi (1970), Tatsumi and Watanabe (1971), Lambert and Sato (1974) and Ishihara *et al.* (1974).

Most of the deposits occur in the tuffs and shales and consist of veins, stockworks, irregular lenses, masses and nodules and conformable beds. They consist of an intimate mixture of very fine-grained sphalerite, galena and barite, usually accompanied by some chalcopyrite and often large masses of pyrite. In some areas large bodies of gypsum occur near the sulphide bodies. The name 'kurokô' derives from the black colour of the ores, but the colour as well as the texture and mineral composition varies considerably and a number of varieties are distinguished. These include massive 'kurokô', brecciated 'kurokô' banded 'kurokô', ôkô or yellow ore, which is a crystalline mass of pyrite and chalcopyrite accompanied by some quartz, and 'keikô' or siliceous ore, which is composed of silicified liparite or tuff heavily impregnated with pyrite, sphalerite, galena and other ore minerals. All of these varieties may occur individually or they may be intimately related in a single deposit.

The country rocks of the kurokô deposits are affected by marked alterations in places. These include propylitization, albitization, kaolinization and silicification. In some mines these alteration effects appear to be related spatially to the deposits; in others the relationships are not clear since the alteration effects spread throughout the country rocks.

The grade of the deposits is variable. Some deposits yield essentially lead, zinc and copper; others yield mainly copper or zinc. All types yield some silver and gold. The antimony-arsenic–rich ores containing fahlerz (tetrahedrite-tennantite) tend to have the highest contents of silver (up to 100 oz Ag/ton or more) and gold (up to 1.0 oz Au/ton or more). Most published analyses of these ores show that silver is always in excess of gold and that the Au/Ag ratio ranges from 0.005 to 0.05. Most of the gold in the kurokô-type deposits is evidently present as electrum, which according to Matsukuma and Horikoshi is generally concentrated in kurokô ore. The hypogene silver is largely present in tetrahedrite-tennantite and in a number of other silver minerals, including argentite, sternbergite, stromeyerite, jalpaite, polybasite, pyrargyrite, electrum and native silver.

A number of the deposits are deeply oxidized and present a vertical profile consisting of a surface gossan zone, a leached zone, an oxide enriched zone and a secondary sulphide enriched zone. The gossan is composed mainly of iron oxides and hydrated oxides and barite. The components of the sulphides are nearly completely leached out. The leached zone is depleted in copper and zinc and enriched in silver and gold. The oxide enriched zone may contain iron oxides, marcasite, pyromorphite, native copper, linarite, malachite, cerussite, anglesite and native silver. Copper and silver are greatly enriched in some of the deposits. The secondary sulphide enriched zone contains chalcocite, bornite, covellite, argentite, etc. It is enriched in copper, silver and a variety of other elements. The secondary sulphide zone is generally less than 5 m thick.

There has been considerable discussion on the origin of the kurokô ores. Some Japanese geologists consider them to be epigenetic replacement deposits derived by hydrothermal processes related to Tertiary magmatism. Certain investigators, noting the conformable banded nature of some of the ores, consider them to be of sedimentary (volcanic-exhalative) origin. Judging from the detailed descriptions of most of the deposits and from my own observations the weight of the evidence appears to favour an epigenetic origin.

The kurokô ores are interesting in that they are probably young equivalents of the massive sulphide deposits found in older rocks. Had they gone through a period of metamorphism and recrystallization they would probably present the features we now see in the older massive sulphide bodies found in Precambrian, Paleozoic and Mesozoic volcanic terranes. In this respect we may note that the Buchans lead-zinc-copper-silver-gold deposits in Newfoundland are remarkably similar to the kurokô bodies. The Buchans deposits, described in some detail by Relly (1960) and Swanson and Brown (1962) occur in the western part of the central volcanic belt of Newfoundland, which consists of a complex series of Ordovician andesites, dacites, rhyolites, volcanic breccias, tuffs and arkose. The orebodies consist of great lenses and irregular bodies of massive and disseminated sulphides in volcanic breccias and tuffs. The principal minerals are sphalerite, galena, chalcopyrite, tetrahedrite and pyrite. The main gangue minerals are barite and quartz. The wallrock alteration effects are those of sericitization, silicification, pyritization and chloritization. The grade of the ore is about 15.5 per cent Zn, 7.8 per cent Pb, 1.4 per cent Cu, 3.5 oz Ag/ton and 0.05 oz Au/ton. The silver occurs in the native form, in argentite, in galena and probably also in the chalcopyrite and tetrahedrite. There are no obvious correlations of the two precious metals with Cu, Pb or any other element in the orebodies.

5. *Disseminated deposits – porphyry Cu-Mo type:* These deposits generally contain only low contents of gold and silver. Overall the deposits average about 0.01 oz Au/ton and 0.1 oz Ag/ton, although certain heavily mineralized sections of some deposit may contain 2 to 5 times this content of the precious metals. Many of the deposits are deeply oxidized, and both the oxidized zones and secondary sulphide zones are commonly enriched in both gold and silver (*see also* Chapter IV).

The nature of gold in porphyry Cu-Mo deposits has received little study. Probably much of the gold is present in the native state and as tellurides in the hypogene ores. In the supergene sulphides the same states may predominate, although there may be considerable amounts of gold in lattice sites in minerals such as chalcocite and covellite. In the oxidized zones most of the gold is present in the native state. In a Transcaucasian Cu-Mo deposit apparently of a porphyritic nature, Karapetyan (1962) found the gold to be present in the primary ores in the following forms: (1) coarse-grained native gold in the interstices of the sulphides and other associated minerals; (2) finely dispersed and submicroscopic gold in the early sulphides (pyrite, chalcopyrite, bornite, etc.); and (3) gold tellurides. In the oxidized zones most of the gold is present in the native form. Most of the gold in a complex copper porphyry deposit in the Almalyk region U.S.S.R. was deposited during two major stages of mineralization according to Meshchaninov and Azin (1972). These are indicated in the five stages represented in the deposit as follows: (1) pre-ore; (2) Cu-Mo (gold-bearing); (3) Pb-Zn; (4) Au ore in stockworks and veins; and (5) rare metal.

The Island Copper Mine, Port Hardy, Vancouver Island, British Columbia is characterized mainly by the introduction of pyrite, chalcopyrite and molybdenite into extensive fractured and brecciated zones. Much of the gold in the porphyry-type deposit occurs as ovoid and hackly particles of free metal with an average size of 15μ. The greater part of this gold is associated with the pyrite and chalcopyrite. Our studies indicate that there is no particular correlation between the gold content and that of Cu, As, Sb, Bi, Se and Te in the deposit. The grade of the ore in the Island Copper Mine, as calculated from production records, is 0.004 oz Au/ton and 0.02 oz Ag/ton. Similar grades occur in certain of the other porphyry copper deposits in British Columbia, e.g., Granisle deposit – Au–0.004 oz/ton; Ag–0.04 oz/ton, whereas in other deposits the grades are very low in gold, e.g., Brenda deposit – Au–0.0004 oz/ton; Ag–0.03 oz/ton.

Gold and silver, like molybdenum, correlate closely with the cupriferous zone above the Kalamazoo porphyry copper deposit, San Manuel district, Pinal County, Arizona (Chaffee, 1976). The highest concentrations of both gold and silver are mostly in the potassic alteration zone, and gold is also concentrated to a certain degree in the propylitic zone. Chaffee concluded that Cu, Mo, Au and Ag shared a common genesis, with gold having a broader dispersion with respect to the 0.1 per cent copper boundary and the pyrite aureole.

In the great Bougainville porphyry copper orebody in New Guinea, chalcopyrite and pyrite are the principal ore minerals, with a little bornite. The tenor of the gold varies directly with that of copper. Reserves are estimated to be 900 million tons containing 0.48 per cent Cu and 0.018 oz Au/ton (Knight *et al.*, 1973).

The production of gold and silver from porphyry Cu-Mo deposits is considerable. In fact a large proportion of the gold and silver production of the United States comes from these deposits. For instance some 200 million oz of silver and 11 million oz of gold have been won from about 1 billion tons of ore in the Bingham Canyon deposit in Utah (Peters, *et al.*, *in* Titley and Hicks, 1966). Some deposits evidently produce much more gold than others since Kesler (1973) has shown that porphyry copper deposits are divisible into geologically real copper-molybdenum deposits with relatively little gold and copper-gold deposits enriched in the precious metal.

Much of the gold in porphyry-type copper and molybdenum deposits is associated with or contains small amounts of palladium.

6. *Pyritiferous shales and schists and cupriferous shales and schists – Kupferschiefer type:* Most pyritic shales contain only traces of gold, the highest contents recorded being about 1 ppm Au. Oil shales, are, likewise, generally low in gold content. Commercial recovery from these types of rocks is not feasible at present, but may become so in the future when such rocks are mined for their bitumen, uranium and base metal content. Commercial recovery of gold and silver from oil shales and tarry residues is suggested in a patent by Petty (1968).

The copper shales and schists are commonly enriched in silver (Boyle, 1968*b*) but generally contain only traces of gold. Our analyses of samples from the White Pine, Michigan deposits show them to contain only $<$0.005 to 0.020 ppm Au, the average being $<$0.005 ppm. The silver contents range from $<$0.4 to 33.3 ppm, the average being about 4.5 ppm. The ores of the Kupferschiefer average about 0.5 ppm Au and 180 ppm Ag. The Zechstein cupriferous black shales of the Fore Sudeten Monocline in Poland are, likewise, low in gold, containing usually from 0.01 to 0.3 ppm (Salamon, 1976). The silver content by comparison averages about 150 ppm with local concentrations up to 5700 ppm or more. Some of the gold is present in electrum, the remainder in native silver and a great variety of silver and copper sulphides and sulphosalts.

Some pyritic carbonaceous slates and schists are slightly auriferous where they contain extensive seams, blows and boudins of quartz. The gold is epigenetic in situ, although it may have been concentrated in the quartz or adjacent schist from the sedimentary beds during metamorphism. Such types of deposits are widespread in many terranes of sedimentary rocks, but few contain more than 1 ppm Au when bulk sampled. One deposit at Blue Hill, Tallapoosa County, Alabama has been studied by the U.S. Bureau of Mines (Sullivan *et al.*, 1969). The average of analyses of the graphitic schist was

0.08 oz Au/ton (2.7 ppm). Most of the gold was found to be free and a concentrate assaying 24.3 oz Au/ton was obtained. Jigging, tabling and amalgamation recovered 86.5 per cent of the gold from the schist.

7. *Cupriferous and argentiferous sandstones – Red-bed type:* Some of these deposits are markedly enriched in silver (Boyle, 1968*b*), but most contain only traces of gold. The writer's analyses of these deposits from various parts of the world show that the ores generally contain less than 0.002 oz Au/ton (0.06 ppm). Kobe (1960) reports that the red-bed ores at Negra Huanusha in central Peru average 3.21 per cent Cu; 2.92 oz/ton Ag; and 0.006 oz/ton Au. The writer's limited research on the red-bed deposits of New Brunswick and Nova Scotia indicates that chalcocite is the principal carrier of gold and silver. Contents up to 0.08 ppm Au and 445 ppm Ag have been recorded from pure chalcocite from the Dorchester Mine and elsewhere in New Brunswick. Tarr (1910) reports up to 31.35 oz Ag/ton (1083 ppm) and 0.1 oz Au/ton (3.4 ppm) in the chalcocite from the red-bed deposits of Oklahoma. Shpora *et al.* (1972) comment on the relatively low content of gold in the copper-bearing sandstones of Naukat, U.S.S.R. They remark that the gold content in the sandstones is not anomalously high in comparison with the gold content of other sedimentary rocks.

8. *Carbonatites:* These complexes contain small amounts of silver (up to 4.5 ppm), but their gold content is commonly very low (<0.01 ppm). Some complexes such as those that produce copper (Phalaborwa, South Africa) contain native gold and/or electrum and/or native silver (Forster, 1958).

9. *Platinum deposits:* Reefs and sulphide segregations associated with basic and ultrabasic rocks such as the Merensky Reef in South Africa are slightly auriferous, and some gold is won from treatment of the platinoid concentrates, which as a whole average about 0.25 oz/ton of rock. In some samples from the Merensky Reef Crocket *et al.* (1976) found that Au and Pt did not show a clear enrichment in separated sulphides and oxides, but the two metals exhibited a maximum content (Au–3.4 ppm; Pt–15.0 ppm) immediately above the marker chromite seam and coincident with maxima for sulphur, nickel and cobalt. There is a considerable reserve of gold in the platiniferous nickel-copper sulphide segregations of the Great Dyke of southern Rhodesia. Panned concentrates from some of these segregations run 0.6 per cent gold (Worst, 1960) (*see also* 1 above).

10. *Black sands:* Certain black sands in many parts of the world contain small amounts of gold which can be won as a byproduct of their mining for magnetite, ilmenite, rutile, zircon, etc. Such sands occur on Graham Island, Queen Charlotte Islands, British Columbia (Mandy, 1934), along the west coast of South Island, New Zealand (Williams, 1974) and elsewhere.

11. *Sediments of the sea:* All sediments of the sea are very slightly auriferous, averaging about 0.005 ppm Au. Some such as those in parts of the Red Sea are slightly auriferous and argentiferous containing an average of 0.66 ppm Au and 28.4 ppm Ag as compiled from the available data (*see* Table 14). These sediments are also enriched in Cu and Zn. Several methods of winning the base and precious metals have been suggested (Degens and Ross, 1969; Lassiter, 1971). The subsea gold resources of the world are discussed by McKelvey and Wang (1970).

12. *The sea:* Seawater contains an average of 0.02 ppb Au and 0.2 ppb Ag (*see also* the section on natural waters in Chapter II). Both elements are also present in marine manganese nodules and other marine sediments as noted above. The trick is how to win these precious metals economically. The economics of winning metals from the sea are discussed in detail by Mero (1965).

> The observations among the gold-bearing rocks of the Southern States seem to show that the precious metal was originally deposited in the beds of various sedimentary rocks, such as slates, quartzites and limestones, and that by subsequent process, it has been, in some instances, accumulated in the veins which intersect these rocks.
>
> –Sir William E. Logan, 1860

Origin of gold deposits

In the foregoing descriptions of the various types of gold deposits, probable origins for the principal types have been suggested. It now remains to summarize these and to amplify some of the concepts. The origin of typical epigenetic deposits such as gold-quartz veins is discussed first, and this is followed by an outline of the probable origin of the enigmatic auriferous quartz-pebble conglomerates, typified by the world's largest gold deposits, namely those of the Witwatersrand.

Epigenetic gold deposits

Epigenetic gold deposits include those developed in skarns; in gold-quartz lenses and silicified zones in fractures, faults and shear zones; and in various replacement and disseminated bodies in iron-formations, limestones and other chemically receptive rocks. The essential features of all of these deposits is that they were dependent upon the presence of preexisting structures or chemically receptive rocks within which some medium or other has deposited the gold and its accompanying gangue.

The source of gold and its associated metallic and gangue elements in epigenetic deposits has long been the subject of debate among geologists. Four distinct possibilities have been considered since the time of Agricola. These can be conveniently tabulated as follows:

(1) The gold and its accompanying elements came from a deep source in the earth. The mantle is often mentioned as a possibility, but even deeper regions have been suggested. An origin based on these sources is commonly referred to as an abyssal origin.

(2) The gold and its accompanying elements came from a crystallizing magmatic body. Basic magmas are favoured by some investigators (e.g., Bilibin and others); more acid magmas such as those that give rise to granodiorites and granites appeal to others (e.g., Emmons, Lindgren and others). This source is the basis of the magmatic hydrothermal theory.

(3) The gold and its accompanying elements came from the country rocks that enclose the deposits or from rocks that lie above, below or lateral to the deposits. This is the basis of the various secretion (lateral, metamorphic, etc.) theories.

(4) The gold and its associated elements came from a weathered surface and were carried into favourable structures by descending meteoric waters. This mode of origin can, in general, be dismissed for practically all types of epigenetic gold deposits. Such a process seems quite incapable of producing the varied types of epigenetic gold deposits that occur in mineralized volcanic and sedimentary terranes. Furthermore, the regional zonation of certain minerals and elements in epigenetic gold deposits, the marked differences in elemental precipitation as the mineralization progressed in many areas, the characteristic wall-rock alteration effects and the remarkably low solubility of elements such as gold and lead in normal groundwaters all militate against the weathering hypothesis. For these reasons we shall not consider it further.

Abyssal theories of origin

These theories generally postulate that the constituents of ores in veins and other epigenetic deposits came directly from deep zones within the earth in the form of diffusing elements, as vapours, in a molten state or in various types of aqueous solutions. It is important to distinguish the details of this mechanism clearly, since ultimately all crustal rocks and the constituents of ores can be traced to deep sources within the earth. Since René Descartes first suggested the idea in 1644, abyssal sources have long been advocated for gold and other metallic ores, the mantle appealing to many as the most probable zone from which the metals, sulphides and gangue were extracted. The precise nature of the mother rock or magma is usually referred to in the vaguest of terms, although some seem to think that it is basic or ultrabasic in composition. Brown (1948), however, considers that a metal sulphide layer exists in the upper part of the mantle, ready to be tapped when channelways such as faults and fractures are opened by tectonism in the crust above. Quiring (1954) postulates a chalcosphere in the earth from which the sulphides and metals are derived. Kennedy (1948) considered that ore deposits with plutonic associations were derived from magmas that resulted from remelting of the deep parts of the granitic layer of the earth. The granitic layer he envisaged as consisting of granitic igneous rocks and metamorphic rocks, mainly of sedimentary origin, that were enriched in particular elements in certain zones. Remelting of parts of these zones, attended by concentration of the elements in orebodies, he thought could explain the phenomenon of metallogenic provinces.

All of the abyssal theories put the responsible source rocks or mother substance well out of reach of any modern techniques of examination and analysis; hence, they remain highly speculative, unsupported by any facts. There are, however, some manifestations that may suggest a deep source for gold and its accompanying elements. For instance, many epigenetic gold deposits are closely associated with volcanic rocks, examples being particularly common in Precambrian and Tertiary terranes. Since it is generally assumed that the volcanic rocks were derived from magmas originating deep within the earth, probably the mantle, the gold and its accompanying elements may have had a similar origin. The evidence of the association of gold deposits with volcanism can, however, be interpreted in other ways as will be discussed later. Here, it is sufficient to note that there is generally a considerable time interval between the stages of volcanism and mineralization. This interval is marked in most places by metamorphic, granitization and/or igneous injection events, usually of a complex nature. The other feature that could be taken to indicate a deep source for gold and its accompanying elements is the common occurrence of gold deposits in or near great fault and fracture systems that run for many tens of miles, examples being the Kirkland Lake–Larder Lake–Cadillac 'break' of Ontario and Quebec; the Giant-Campbell System at Yellowknife, Northwest Territories; the Mother Lode System of California; and the Champion System of the Kolar goldfield in India. The great linear extent of these systems suggests that they cut deeply into the earth, perhaps to the mantle, from which they may have derived the constituents of their gold lodes. This possibility was examined in the Yellowknife district by the writer (Boyle, 1961*a*) where the Giant-Campbell shear zone system (Archean age) some 15 mi in extent is cut by a great series of steeply dipping late faults (Proterozoic age) many of which can be traced for more than 50 mi (Fig. 17). If the mantle or some other deep zone were the source of gold and its accompanying elements, one would expect both systems of faults to have been mineralized with gold since both evidently cut deeply into the earth. Such is not the case: the Giant-Campbell System is highly auriferous, the late faults, even where they are extensively mineralized with quartz and some pyrite, are essentially barren. The reasons for this difference in mineralization are complex but can be related directly to metamorphic events. The auriferous Giant-Campbell and its subsidiary systems were formed during the late stage of regional metamorphism, granitization and granitic intrusion of the greenstone belt. As the systems dilated, the constituents of the gold-quartz lodes, mobilized from the greenstone pile, were drawn into and precipitated in favourable structural sites in the systems. When the late faults were formed the processes of regional metamorphism, granitization and granitic intrusion were not operative, and hence gold was not mobilized and concentrated from the host rocks.

Magmatic differentiation theories

The basis of these theories is that the gold and its accompanying elements came from a crystallizing magmatic body injected into or derived from crustal rocks. Two general concepts have been developed. One envisages differentiation of the magmatic body to give an ore magma composed of fused silica, carbonates, sulfides, metals, etc. from which the vein constituents crystallized. The other postulates that as the magmatic bodies crystallize they yield hydrothermal extracts containing the metals, sulphides and gangue elements such as silica, all of which are deposited as constituents in the epigenetic gold deposits. The precise origin of the magmatic bodies, which are said to give rise to the ore constituents, is rather vague in the literature, especially when specific mineral belts are discussed. Following R.A. Daly (1914) most of the older investigators apparently thought that the magmatic bodies were of subcrustal origin, derived from the differentiation of primary basaltic magma representing a fused part of the mantle or some other deeper zone in the earth. In recent years granitization and/or anatectic processes involving crustal rocks have been more popular as the source of magmas. The rocks most commonly thought to give rise to granitic magmas by these processes are pelitic sediments, although piles of such sediments and volcanics have also been suggested as the original materials. The phenomenon of granitization is discussed at length by Read (1957) and Reynolds (1946, 1947) and its relationship to ore genesis by Guimaraes (1947), Sullivan (1948) and MacGregor (1951). The last investigator summed up his views as follows: "The conclusion is reached that the origin of gold is the Sima where it is diffused in the glassy or crystalline basaltic rock. From time to time the Sima in a molten state ascends fissures through the Sial and flows out in vast fields of lava. In belts of mountain-building the granite of the Sial fuses and the more fluid parts of it ascend to fill domelike chambers formed by the subsidence of the overlying rocks. Sinking fragments of basalt dissolve or disintegrate in the molten granite. Metallic sulphides and carbon dioxide from the sunken rocks react with silica, water and other fluxes in the granite to form solvents of gold. As the granite consolidates, the solutions are ejected upwards to form mineral deposits in the upper, already crystallized portion of the granite and in the overlying rock."

In recent papers Smirnov (1968, 1976) outlined current views in U.S.S.R. on the sources of ore-forming material. Three principal groups of ore-forming source material are assumed: (1) juvenile sources associated with deep-seated subcrustal magmas of basaltic composition; these give rise to a basaltophilic group of metals including Fe, Mn, V, Ti, Cr, platinoids, Cu and Zn; (2) assimilative sources in shallow palingenetic (anatectic) magmas within the predominantly granitoid portion of the crust of the earth; from these derive a granitophilic group of elements including Sn, W, Be, Li, Nb and Ta; and (3) filtration sources in which ore-forming material is leached along the circulation pathways of aggressive hydrothermal solutions that differ widely in origin (magmatic, metamorphic or exogenetic). Gold deposits can evidently derive from all three source materials. Thus one finds in the voluminous Soviet literature on mineral deposits a number of papers that deal specifically with gold deposits in accord with these conceptual ideas. Investigators supporting the magmatic derivation of gold include among others Rozhkov (1968*a*, 1970), who advocates a paragenetic relationship of gold deposits with magmas of both subcrustal (simatic origin) and intracrustal (granitic and andesitic) derivaton; Il'enok (1970), who relates epigenetic gold deposits of various types to a variety of magmas; Shcherbakov (1970), who maintains that both basaltic and granitic magmas are important sources of gold; Volarovich and Shilin (1971), who connect gold deposits with the crystallization of gabbro-granodioritic magmas and maintain that the gold content of rocks from these calc-alkaline magmas increases with the differentiation sequence gabbro to aplite; Karelin *et al.* (1974), who concluded that the source of gold is deep magma chambers; and Marakushev (1977), who on thermodynamic grounds relates endogenic gold deposits to diorites, granodiorites, plagiogranites and granites. Among those supporting granitization, remobilization, redistribution and regeneration theories are a number of investigators who describe the gold deposits of the Soviet far east in the book edited by Radkevich (1966). Some of these have the idea that the gold deposits are regenerated by assimilation of ancient metamorphic horizons by anatectic magmas. In other publications, Moiseenko *et al.* (1970) consider that much of the gold in deposits is redistributed as a result of the destruction of minerals containing trace amounts of gold during postmagmatic alteration processes; Shcherbakov (1970), Shcherbakov *et al.* (1972) and Shcherbakov *et al.* (1973) maintain that endogenous gold deposits may be derived not only from magmas but also from anatectic granitoids and from piles of rocks with higher contents of gold while they are undergoing hydrothermal and contact metamorphism; and Perelygina (1973) advocates a Mesozoic rejuvenation of the gold in the Aldan Shield for the origin of the deposits in the cover rocks of this massif.

The ore magma is an old concept, originally postulated by James Hutton in 1795. Since then it has been considerably amplified and used to explain auriferous quartz deposits in many parts of the world. Thomas Belt (1861), enquiring into the origin of the auriferous quartz veins in Australia, concluded "that in these veins the distribution of gold, and the structure and arrangement of the quartz, are explained by the theory that they are fissures that have been filled with molten silica containing entangled metallic vapours." Belt later worked in Nova Scotia and entertained a similar origin for the veins and saddle reefs in the Meguma Group. J.E. Spurr (1923) postulated a molten origin for gold-quartz bodies in his classic treatise on the ore magmas. Farmin (1938, 1941) has explained the gold-quartz veins of Grass Valley, California by injection of a siliceous fluid rich in CO_2, and Bichan (1944, 1947) advocated deposition of gold from silicothermal melts for most epigenetic gold deposits.

Originally, the derivation of gold-quartz veins from siliceous melts seems to have gained favour as a result of the mistaken idea that silica (quartz) is insoluble in water and from the general observation that many gold-quartz veins are closely associated with igneous bodies especially dykes, some features of which resemble those found in veins, viz. chilled dyke contacts, which are somewhat analogous to fine-grained quartz borders in veins. Later the presence of unsupported inclusions of wall rock in gold-quartz veins was taken as evidence that the depositing medium was dense enough to float the fragments or sufficiently viscous to support them. Finally, the sharpness of the contact between the vein material and wall rock and the absence of replacement of the wall rock was assumed to indicate that the depositing medium was too viscous to penetrate discontinuities such as grain boundaries in the wall rock.

The ore magma theory of the origin of gold-quartz veins has not been favourably received from its inception. Kirwan (1799), criticizing Hutton's ideas, showed that silica is soluble in hot water. The observation that quartz veins show a close association with igneous bodies in some mineralized areas has stood the test of time, but most investigators doubt that quartz veins are magmatic offshoots of such bodies. The unsupported wall-rock inclusions and sharp vein contacts as evidence of a melt injection have also been questioned, even at Grass Valley where such phenomena are widespread (Johnston, 1940). It is considered more probable that the first is due to repeated movement and deposition of quartz along the veins, during which process residuals of unsupported wall rock are left within the resultant quartz bodies. The sharp contact relationships have, likewise, been reinterpreted as due to similar physico-chemical relationships existing between the wall rocks and the quartz-depositing medium.

A magmatic source for the metallic and gangue elements in hydrothermal deposits has been favoured by many geologists since Elie de Beaumont and others first suggested the possibility more than 125 years ago. The modern theory envisages a crystallizing magma during its last differentiating stages giving off water and other volatiles that transport the metals and gangue elements to their depositional sites in fractures, faults, shear zones and other favourable locales. The variations on this theme are many, and the physico-chemical mechanisms conjured up to explain the various manifestations of the process would fill many volumes. Some recent concepts are outlined in the book edited by Barnes (1967). In the present century most epigenetic gold deposits have been explained by the magmatic hydrothermal theory particularly by such astute observers as Pošepný (1902), Lindgren (1933), Emmons (1940), Bruce (1937*a,b*), Moore (1940), Bateman (1950) and a number of U.S.S.R. geologists including Bilibin, Betekhtin, S.S. Smirnov and others (Shilo, 1971; Ivankin and Rabinovich, 1972). As noted in a previous part of this chapter, the auriferous quartz-pebble conglomerates of the Witwatersrand have been similarly explained by Graton (1930). Emmons (1924, 1937) in his assessment of the space relations of gold deposits to granite batholiths laid great stress on the area of exposure of the magmatic bodies. He stated that no deposits may be expected in the main batholiths where they were exposed over areas of more than 12 to 15 mi in diameter. The tops of batholiths, represented by cupolas, domes and other projections, and the nearby invaded rocks, were the favourable sites where rising hydrothermal solutions had deposited their mineral load.

A thorough critical review of the shortcomings of the magmatic hydothermal hypothesis cannot be attempted here, but a few pertinent points suggesting that magmas may not be the source of metals and gangue elements in epigenetic gold deposits can be briefly mentioned.

Many epigenetic gold deposits occur in areas where there are no observable synchronous igneous rocks from which the metals and gangue elements could have come. This is especially true in the Canadian Shield and in many other Precambrian terranes as noted by Bruce (1937*a,b*) and numerous other investigators. A similar situation exists in younger rocks in many gold belts. In most gold belts the evidence appears to indicate that the proximity of epigenetic gold deposits to exposed igneous rocks is due only to structural factors and has no genetic significance whatsoever. Perhaps this criticism is not a valid argument against the magmatic hypothesis since deep-seated igneous rocks could conceivably lie far below the roots of the deposits. For many areas such a proposition, however, is pure speculation and is 'grasping at straws' to support the hypothesis. It is more logical and more in accord with the facts to assume on the available evidence that the elements in such deposits were not derived from a magma.

It is generally considered that the metals and gangue elements were exuded from the magma during its final stages of differentiation. If we assume that the magma differentiated in the line – basic-intermediate-acid components – there is considerable difficulty on geochemical grounds in deriving elements such as copper, zinc, iron, calcium and sulphur, all of which accompany gold in its deposits, by such a process. These and many other metallic and gangue elements are enriched in basic rocks, not in acidic rocks. If there is any differentiation of such elements during magmatic processes, it is to reduce the amount of such elements toward the close of differentiation, that is during the formation of granites. Certain elements such as lead and uranium do, however, show an enrichment with differentiation and are strongly concentrated in acidic phases where they substitute in various minerals, e.g., lead in feldspars and uranium in various accessory minerals. Most of these minerals are, however, generally undersaturated with respect to lead and uranium, a feature suggesting that little if any of these elements would be drawn off in hydrothermal solutions.

A number of investigators have linked gold mineralization to gold abundance in nearby associated igneous rocks. They claim that higher (as well as lower) than background gold values in the igneous rocks indicate that gold deposits were derived from the magmas from which these rocks crystallized. The various views and arguments are summarized in Chapter II, where it was concluded that there is probably no particular relationship between the gold content of most igneous bodies such as batholiths, stocks and dykes, and nearby associated gold mineralization. Certain pyritic porphyries may be an exception to this statement, but even where these slightly auriferous rocks occur they are cut and commonly altered by the gold-bearing quartz deposits. In a recent paper Tilling *et al.* (1973) have concluded from extensive analytical data that gold becomes depleted in residual silicate melts in differentiating calc-alkalic magmas. They also observed no particular relationship between the gold content of igneous bodies and the presence of gold mineralization.

Also pertinent to the argument against magmatic hydrothermal solutions as a source for the elements in epigenetic gold deposits is the oft quoted observation that the deposits show no close time relationship to granitic or other igneous bodies, and that where the two are found together, the deposits, especially veins, cut and offset both the granitic bodies and other closely related rocks such as porphyries. This means that the granitic rocks were fully consolidated, and any process of differentiation that might have produced hydrothermal solutions had ceased when the veins were formed. Again the supposed igneous source could be sought at depth, but this is an artifice to circumvent the facts. In any event, in the writer's experience gold veins cut the igneous rocks, to which they are often said to be genetically related, down to the deepest levels of mining. Furthermore, if crystallization differentiation and formation of hydrothermal solutions is a feature of igneous rocks, one must assume on physico-chemical grounds that the outer parts of batholiths and stocks would cool first and the central parts would be the last to consolidate. The outer parts of the igneous bodies should, therefore, contain numerous hydrothermal veins, and these should increase in number and richness in gold and other metals inward and downward in the bodies. Such is not the case in nature; the central parts of batholiths and stocks are practically devoid of veins and other types of deposits in most mineralized areas.

Another similar problem for the magmatist is found in granitic bodies where several phases of granitic material, ranging from intermediate to acidic to pegmatitic composition, are present with the more acidic material being successively younger in age. These phases, in the writer's experience, show no enrichment of gold or any of the metals (molybdenum and bismuth excluded), sulphur, carbon dioxide or other volatiles with increasing acidity and youth. The pegmatites are generally devoid of gold and most sulphides (molybdenite and bismuthinite excluded) suggesting that the last extracts that gave rise to the pegmatites (assuming a magmatic origin for these bodies) were notably deficient in gold and the metallic elements common to veins. There is, therefore, a break in the lineage – granite-pegmatites-veins – which the magmatist has not yet satisfactorily answered. Could it not be that pegmatites have an origin other than by magmatic solutions? Certainly many pegmatites seem to be of metamorphic secretion origin as suggested by numerous writers since T. Sterry Hunt first postulated such an origin more than 100 years ago. Other granitic pegmatites that cut their granitic host rocks, seem to have been derived by diffusion of their constituents from the granitic rocks that were at relatively high temperature but fully crystallized and consolidated.

Another point not favouring the derivation of metals such as gold and gangue elements from magmatic hydrothermal solutions is the fact that the rocks adjacent to the supposed parent igneous bodies are not characterized by extensive hydrated, carbonated or sulphidized zones – features that should prevail if differentiation products of the igneous magmas, such as water, carbon dioxide and sulphur, were expelled outwards into the wall rocks of the stocks and batholiths. One cannot expect that all of these components would only escape up fractures or faults. On the contrary they should, when expelled, diffuse outward into the wall rocks. However, such is not the case at Yellowknife (Boyle, 1959, 1961*a*) at Keno Hill (Boyle, 1965*a*) or other localities where the writer has examined this problem carefully. If one can judge from the chemical effects in wall rocks adjacent to batholiths and stocks, it is certain that the magmas that gave rise to these rocks did not exude any quantity of volatiles, metals or gangue elements during the final stages of their differentiation.

Contact metamorphic and certain skarn bodies have frequently been quoted as evidence for the migration of metals, silica and volatiles from adjacent solidifying magmatic bodies. This could conceivably be the case in some places, but more generally it is apparent that the source of the epigenetic elements is the invaded country rocks. As suggested by the writer (Boyle, 1969*a*) these elements have moved by diffusion towards and have been precipitated in thermally and chemically active dilatant sites (usually contorted limestone beds), commonly near the borders of granitization or injected igneous bodies.

Finally, the process by which acidic and intermediate igneous rocks are formed is in considerable doubt. There are good field and physico-chemical reasons for suspecting that granite, granodiorite, monzonite and other similar rocks, which are thought to be parent sources of the constituents of epigenetic gold deposits, are not derived by igneous processes but by granitization or ultra-metamorphism. Even if magmas do arise by these more logical processes it is almost certain that they are greatly deficient in water, carbon dioxide, sulphur and the various metals such as zinc, copper, silver, gold, etc., because these elements are driven out of zones undergoing high temperature regional metamorphism early in the process, and few if any remain in any quantity in the fused materials to be differentiated at a late crystallization stage as hydrothermal solutions.

Secretion theories

General. Secretion theories, based on the assumption that the country rocks enclosing the epigenetic deposits or that lie below, above or lateral to the deposits are the source of the constituents of the epigenetic deposits, have had a long history. Agricola, writing in his famous classics, *De Ortu et Causis Subterraneorum* and *De Re Metallica* published in 1546 and 1556 respectively, concluded that veins and epigenetic lodes were deposited by circulating underground waters and reasoned that surface waters percolated downward, became heated, dissolved mineral matter, rose again and deposited this mineral matter in what he called "canales" (fissures). Agricola was, therefore, one of the first meteoric water secretionists, and it is a tribute to his genius to note that modern research has shown that certain mineral deposits, including some Tertiary gold deposits, probably originated in the manner he suggested.

More than 300 years were to pass before further progress was made in elucidating many aspects of the secretion theories. In 1885, J.G. Forchhammer, by painstaking analysis, demonstrated that ordinary rocks contain traces of many of the metals and other elements commonly found in veins. This gave a great impetus to the various secretion theories, since the presence of the metals in country rocks was held to be conclusive proof of the source of the constituents of veins and other types of epigenetic deposits. As early as 1845 K. Gustav Bischof of Bonn, the distinguished geochemist, supported the secretion theory in his *Textbook of chemical and physical geology*, and similar support can be found in the writings of G.A. Daubrée, T. Sterry Hunt, J.A. Phillips, F. Sandberger, S.F. Emmons, G.F. Becker and C.R. Van Hise during the latter half of the 19th century and the early part of the present century. All of these investigators ascribed the source of the metals in veins to the country rocks containing them but differed in their opinion as to whether ascending or descending meteoric water was the responsible transporting medium. Of all these investigators R. Daintree, writing in his *Geology of the Ballan District, Victoria, 1866*, seems to have appreciated the geochemical cycle of gold most clearly and its importance in the formation of gold-quartz reefs. He said: "I had long come to the conclusion that most, if not all, the gold in the quartz reefs was derived from the rocks in which these reefs occur. That the strata themselves received their supply of gold at the period of their deposition from the ocean in which they were deposited. That the organic matter and the gases generated therefrom on decomposition, sulphuretted hydrogen, etc. were the cause of the precipitation; and that the amount of metallic deposit was in proportion to the amount of organic matter deposited with the organic sediment."

The secretion theories enjoyed considerable popularity in Europe and in America during the latter part of the 19th century but were strongly opposed during its closing years by A.W. Stelzner of Freiberg, F. Pošepný of Przibram, J.R. Don of Otago and L. De Launay of Paris. These investigators were the first advocates of the magmatic hydrothermal theory. They demonstrated that the metals could not have come from the rocks immediately adjacent to veins as F. Sandberger had suggested in his restricted lateral secretion theory, but they drew the entirely erroneous conclusion that the metals in the country rocks were all introduced from the fissures and faults in which the veins occur. They, therefore, failed to appreciate the painstaking analytical work of Bischof, Forchhammer and Sandberger and the important geochemical considerations bearing on the migration and concentration of the elements, such as precipitation of various metals during sedimentation processes, elucidated by Sterry Hunt.

During the first 50 years of the 20th century the magmatic hydrothermal theory gained in popularity, and secretion theories were practically forgotten. Nearly all epigenetic gold deposits known up to 1950 were, therefore explained in terms of the hydrothermal theory. Since 1950, the major advances in geochemistry and in our knowledge of the migration and concentration of the elements have caused many to question this theory, and one now sees a decided swing of the pendulum of thought toward secretion theories for the origin of many epigenetic gold deposits.

The secretion theories of the origin of epigenetic deposits embrace a number of concepts and mechanisms. Some of these are briefly outlined below with some general criticisms; the details of the mechanisms are described and discussed subsequently.

Meteoric water secretion theories. Most of the early investigators considered meteoric water, either descending or ascending or both, to be the collecting and transporting medium for the metals, sulphides and gangue minerals found in epigenetic deposits. The source of these constituents was considered by most geologists to be the local rocks in the general vicinity of the deposits; Sandberger (1882–1885), however, concluded from his analytical researches that the vein elements were derived from the wall rocks immediately adjacent to the veins. This view is in general incorrect, but as we shall see later, many of the metals and compounds found in gold-bearing veins and lodes, such as Fe, Ca, Mg and SiO_2 may have been derived under certain conditions from the adjacent wall rocks.

Recent research utilizing $^{18}O/^{16}O$, D/H and other isotopic ratios strongly suggests that gold deposition in many of the Tertiary and younger deposits took place essentially from meteoric water (including in places sea water) set into convective circulation by interaction with rising igneous intrusions. Examples are widespread including Tonopah, Goldfield and Comstock Lode, Nevada (Taylor, 1973); Bodie, California and elsewhere (O'Neil *et al.*, 1973; O'Neil and Silberman, 1974); Creede, Colorado (Bethke *et al.*, 1973); Sunnyside Mine, Eureka mining district, San Juan County Colorado (Casadevall, 1976); Tui Mine, Hauraki goldfield, New Zealand (Robinson, 1974); Kurokô deposits, Japan (Sakai and Matsubaya, 1974; Ohmoto and Rye, 1974); and epithermal Au-Ag districts, United States (White, 1974). There is evidence that some of the epithermal gold deposits in the northeastern U.S.S.R. had a component of meteoric water (with dissolved air) in the solutions from which they were precipitated. This is surmised from the work of Goncharov and Shugurova (1973), who studied the gaseous composition of gas-liquid inclusions in the constituent minerals. They found a ratio for O/N = 0.25–0.26, approximately equal to that in the atmosphere (0.27). From this they concluded that atmosphere constituents played a part in the origin of these deposits. In general, therefore, it appears that meteoric waters have dominated during mineralization processes in epithermal gold-silver belts, but some investigators consider that magmatic water also played a part (Robinson, 1974; O'Neil and Silberman, 1974; White, 1974) or was entirely responsible for the precipitation of gold as in the Tertiary Megradzorsk gold deposits in Armenian S.S.R. (Akopyan *et al.*, 1976). It has never been clear to the writer how one can isotopically characterize magmatic water from metamorphic water sweated out of the rocks. Until this problem is resolved the term 'magmatic water' should remain in abeyance.

While the source of the water seems at least partly solved, there still remain problems with respect to the source of the base metals, sulphur, gold and silver. Some investigators have concluded that the magmas with which the meteoric water has interacted are the most reasonable source; others consider the volcanic sedimentary pile or the underlying basement rocks more likely sources. The last two appear more probable. In passing it is interesting to note that the so-called epithermal deposits were those with which the early meteoric water secretionists were most familiar.

While meteoric water seems the most probable medium of transport for gold and other elements in what are called epithermal or near-surface deposits, the situation may be different for deep-seated, high-temperature deposits. In their formation, if water was involved as a collecting and transporting medium, it was probably mainly derived by dehydration of minerals such as micas, etc. during metamorphism (metamorphic water).

Metamorphic secretion theories. Metamorphic secretion theories as applied to epigenetic gold deposits assume that the gold, silver and gangue elements, initially present in the country rocks, were mobilized and concentrated in available faults, shear zones, fractures and chemical traps (carbonate rocks) during metamorphic events. There are many variations on this theme, some of which are discussed subsequently. In recent years many gold deposits have been explained by metamorphic secretion theories.

Boyle (1955,1959,1961*a*) and Wanless *et al.* (1960) explained the Yellowknife gold-quartz deposits in Precambrian greenstones and (greywacke-slate) sediments in terms of mobilization, migration and concentration of elements during regional metamorphic events involving granitization. The gold-quartz deposits in the argillite-graphitic schist-quartzite terrane of the Yukon Group in the Keno Hill area, Yukon were explained in a similar manner (Boyle, 1965*a*), as were also the gold-quartz deposits in the slate-greywacke-quartzite assemblage of the Meguma Group in Nova Scotia (Boyle, 1966). The papers on the Yellowknife deposits have been criticized by Ames (1962) and McConnell (1964*a*), who claim that the data presented favour a hydrothermal origin. Replies to the criticisms have been published (Boyle, 1964; Boyle *et al.*, 1963*b*).

In 1953 Knight (1957) postulated the "source bed concept" for the origin of sulphide orebodies. By this he suggested "that all sulfide orebodies of the majority of fields are derived from sulfides that were deposited syngenetically at one particular horizon of the sedimentary basin constituting the field, and that the sulfides subsequently migrated in varying degree under the influence of rise in temperature of the rock environment." He gave a few examples of auriferous deposits (Rand, Canadian Shield) whose origin he thought was encompassed by the concept.

Nifontov (1957) attributes a number of the gold-quartz deposits of U.S.S.R. to metamorphic processes, especially those occurring in sedimentary rocks. Chaikovskii (1963) and Badalov (1965*b*, 1966) have, likewise, stated that the gold in some of the deposits of U.S.S.R. was extracted from the country rocks.

Davies (1963,1964) studied a number of gold deposits in the Rice Lake–Wanipigow River area of Manitoba and concluded that the concentration of gold is due to a redistribution or diffusion and migration of the element in the country rocks into fractures, faults and shear zone in zones of lower grade metamorphism. This view is corroborated by the work of Stephenson (1971) and Stephenson and Ehmann (1971) in the same area.

Moiseenko and co-workers attach considerable importance to the processes of metamorphism in the formation of gold deposits. In a series of papers, Moiseenko (1965), Moiseenko *et al.* (1970), Shcheka and Moiseenko (1970), Moiseenko and Fat'yanov (1972) and Moiseenko and Neronskii (1973) reached the general conclusion that the primary sources of gold in the earth's crust are basic and ultrabasic magmatic rocks in which the element is present in a dispersed form. During granitization and metamorphism gold is released, is transferred to melts or solutions, migrates, and is deposited in local tectonic structures. Repeated mobilization and deposition of gold is frequently necessary for the formation of economic epigenetic gold deposits according to these investigators.

Shcherbakov (1967,1970) considers that the source of gold in endogenous deposits may be basaltic magma, granitoid magma resulting mainly from the fusion of sedimentary rocks or volcanic-sedimentary piles from which the gold is mobilized during hydrothermal metamorphism and collected in local structures such as faults and fractures. Buryak (1967*a*; 1975*a*) and Buryak *et al.* (1972) also considered that gold was mobilized during metamorphic and granitization events. During these processes in the Vitim-Patom Upland in particular gold was decreased in amount as the rocks passed from shales to gneisses to granitic rocks. The authors calculated that the loss of gold during metamorphism and granitization processes amounted to 3.3 tons/km^3.

Metamorphic secretion theories have been favoured in recent years as an explanation of the origin of epigenetic gold deposits in Precambrian terranes comprising greenstones, iron-formations and metasediments. Anhaeusser *et al.* (1969), Vilijoen *et al.* (1969,1970) and Anhaeusser (1976) consider that the gold in the deposits of the greenstone belts of southern Africa (Barberton Mountain Land, Rhodesia, etc.) and elsewhere was initially present in the mafic and ultramafic lavas and associated sediments and was mobilized and concentrated in dilatant zones during metamorphic events attendant upon granitic intrusion. In a more recent paper Saager (1973) advances a similar origin for epigenetic gold deposits in Precambrian rocks and suggests that the ultimate source of the gold in the vein deposits of the Swaziland greenstone belts is the 'primitive' peridotitic and basaltic komatiites of the Onverwacht Group rather than the intruding granites. The komatiites are relatively rich in gold (0.02 ppm). In more recent papers Anhaeusser *et al.* (1975) and Anhaeusser (1976) note that the komatiites contain from 0.001 to 0.0015 ppm, a value substantially lower than the worldwide tholeiite average of 0.0025. They think two interpretations of these new data are possible. One is that these are primary concentrations, and the other is that the lavas contained significantly higher amounts of gold, but that later metamorphism and hydrothermal activity has erased all record of this fact. Judging from the petrographic data available on these rocks and from the data of numerous studies in U.S.S.R., the latter appears most probable. For the Kirkland Lake–Larder Lake deposits in Ontario, Ridler (1970) has postulated a syngenetic-metamorphic origin for the gold deposits, the gold being derived mainly from iron-formation. Belevtsev (1970) considers that the source of the metals in a number of Precambrian deposits is the host rocks and thinks that they were mobilized and concentrated mainly by the action of metamorphism aided greatly by metamorphic water. The gold-quartz bodies in the greenstones of the Ramagiri goldfield in Andhra Pradesh, India have been similarly explained by Ghosh *et al.* (1970), but their metamorphogenic origin has been questioned by Rao and Srinivasan (1971), who consider the veins to be typically hydrothermal, controlled by structures and tectonism. Iyer (1970), however, considers these veins and others in the greenstone belts of southeast Mysore to be of metamorphic secretion origin. Deposits in other Precambrian terranes have also been explained by metamorphic secretion theories. Here we may mention the Kalgoorlie goldfield in Western Australia (Travis *et al.*, 1971); the Ashanti Gold Mine, Obuasi, Ghana (Wilson, 1971); the Kilo-Moto deposits, district of Mongbwalu, Zaire (Lavreau, 1973); the Homestake deposit, South Dakota (Sawkins and Rye, 1971,1974; Rye and Rye, 1972; Rye *et al.*, 1974*a*); and Precambrian gold deposits in general (Henley, 1971).

A number of deposits of variable age and type have been explained by metamorphic secretion, remobilization or regeneration theories in recent years. Morávek and Rus (1970) consider that the quartz-carbonate veins enriched in gold, uranium and selenium in the Sedlcany-Krásná Hora area of Czechoslovakia owe their origin to regeneration or remobilization of gold from earlier auriferous veins in Variscan deposits associated with the Central Bohemian pluton. At Tilkerode, Harz Mountains, Germany, Tischendorf (1970) has shown that a number of the elements in the carbonate-hematite gold selenide vein were mobilized and concentrated from the country rock. The specific mobility of the elements in decreasing order was found to be as follows: S, Mo, Cu, Se, Co, Au, Ni, V, Zn, Pb, Cr and Sn. Keays and Kirkland (1972) found that gold was leached during alteration of the Thomson River dyke, Victoria, Australia. The dyke was originally an ultramafic differentiate with segregated copper-nickel sulphides. The leaching of the gold and its redeposition in quartz stringers with chalcopyrite and/or pyrite in a nearby dilatant zone took place mainly during faulting, shearing and extensive alteration by carbonate-bearing solutions. The extraction of the gold from the pre-existing sulphides in the ultrabasic body is clear as shown by depletion haloes of gold about the orebody. The platinoids, on the other hand, were not affected to the same extent as the gold. In the Urals gold province Zhelobov (1972) suggests that the gold is extracted from ultramafic rocks by granitization and magmatic assimilation processes along deep-seated fault systems, and Vilor and Sarapulova (1971) consider that epigenetic gold deposits in general are derived mainly from sedimentary rocks, the gold in these rocks being leached out during progressive metamorphism and redeposited in fractures and faults under conditions of changed pressure and temperature. Even crystallized granitoids may be a source of gold. Anoshin and Potap'yev (1966) found that the element is lost from granitoids that are metamorphosed or hydrothermally altered. They postulated that the gold so lost is concentrated in quartz veins. Shcherbakov *et al.* (1973) examined the gold content of granites and other rocks and their accessory magnetites in the Altai-Sayan gold belt. Intensive redistribution of gold was noted in the regions of hydrothermal metamorphism from which they concluded that much gold was leached from these parts. Belyaev *et al.* (1972) are of the opinion that metamorphic processes are responsible for the gold-sulphide mineralization in the amphibolites in the Olenii Ridge Lower Proterozoic sedimentary-volcanic formations, U.S.S.R.

In a recent contribution on the Kalgoorlie goldfield, Western Australia, Tomich (1974) postulates that the Golden Mile is an Archean volcanogenic-type deposit, modified and redistributed along later developed structures.

Li and Shokhina (1974*a*, *b*) demonstrated that the initial gold content in the rocks of the Yenisey Range is not changed substantially during regional metamorphism either in the low temperature greenschist and epidote-amphibolite facies or in the high temperature amphibolite and granulite facies. Radical changes in mineral composition under the conditions of high temperature metamorphism are, however, accompanied by redistribution and dispersion of gold within the rocks. Mushkin *et al.* (1974) think that the source of gold in metamorphic terranes is due to eclogitization processes operating in the lower part of the crust. These processes are attended by release of Si, Na, K and Au ($\sim$0.015 g/m^3). They drew this conclusion from their study of the eclogitization of gneisses in the granulite facies of the southern Tien-Shan.

In a more recent contribution Moore (1975) ascribes the origin of the Proterozoic gold deposits of the Saudi Arabian Shield to mobilization of the gold from the country rocks and concentration in suitable structures during the period of regional metamorphism. Similarly Pyke (1975,1976), noting that many of the gold deposits in the Timmins (Porcupine) area of Ontario show a close spatial relationship of Archean ultramafic metavolcanics, has postulated that these rocks are the principal source of gold in the deposits, and that the gold was extracted from them and concentrated in dilatant zones through a process of lateral secretion when the area was subjected to dynamothermal metamorphism and the ultramafic volcanics suffered extensive carbonatization. Hutchinson (1975) has also advocated a metamorphic secretion theory for the origin of Precambrian and Phanerozoic lode gold deposits. He postulates that initially the gold was deposited in sulphide bearing interflow sediments (iron-formations and tuffs) in eugeosynclinal volcanic terranes, the metal being subsequently concentrated, probably by lateral secretion, in lodes, veins and stockworks, as a result of deformation, recrystallization and redistribution during later metamorphism.

Roslyakova and Roslyakov (1975) document numerous researches in U.S.S.R. where it is evident that most, if not all, of the gold in epigenetic deposits has come from country rocks by processes of metamorphism or hydrothermal activity. Most of these researches have already been mentioned in the text above; one by A.T. Bendik, not available to the writer, is of interest. Bendik claims that the gold in the great Muruntau deposits in Uzbek S.S.R. was initially present in the sedimentary host rocks (carbonaceous siltstones and sandstones) and was mobilized and concentrated as a result of hydrothermal metamorphism. Badalov *et al.* (*in* Petrovskaya, 1974) remark that in the Almalyk auriferous region metasomatism of middle-upper Paleozoic effusive-intrusive complexes has liberated large amounts of silica and suggest that the ubiquitous coincident occurrence of gold and silica (quartz) indicates that both were derived from the same source.

A study of the sulphur isotopes in the auriferous Wood's Point dyke swarm, Victoria, Australia (Fig. 53) by Green and Keays (1976) suggests that the bulk of the sulphides associated with the gold could not have been derived from the adjacent sedimentary country rocks or from the copper-nickel sulphides in the dykes. The more mafic dyke bulges, copper-nickel sulphides and gold mineralization are all concentrated along two lineations paralleling the fold axes of the host sediments. Green and Keays considered that these lineations may represent deepseated fractures that controlled the upward migration of gold-bearing hydrothermal solutions from depth. The source of the gold, they thought, could have been various rock types present at depth, including copper-nickel sulphides and Lower Paleozoic sediments.

An investigation of the gold distribution and foliation development in the Lower Paleozoic rocks of central Victoria, Australia and its bearing on gold mineralization by Glasson (1975) has shown that one possible source for gold is the black pyritic Cambrian shales whose gold contents (av. 0.068 ppm) are well above those of the greywacke, quartzite and other sediments (av. 0.003 ppm). Furthermore, during foliation development within the Clunes goldfield, pyrite and other minerals such as albite are dissolved and their constituents rearranged. During this rearrangement Glasson considered that some of the constituents of the pyrite and other minerals, including gold, may have contributed to the filling of the gold-quartz reefs.

Henley *et al.* (1976) and C.J. Paterson (pers. commun., 1976) have made an extensive study of the gold-quartz and scheelite-gold-quartz lodes in the Mesozoic Haast Schist, Central Otago, New Zealand. They have concluded that the source of the elements are rocks deep in the schist pile, and that during metamorphism, uplift and hydraulic fracturing gold, tungsten and their accompanying elements were concentrated in available fractures and faults. There is an interesting historical summary in the publication by Williams (1974) of the various genetic theories of origin of the auriferous Otago lodes; these range from meteoritic water secretion to magmatic hydrothermal theories.

Few younger (Tertiary) deposits associated with volcanism have been examined carefully from the viewpoint of secretion theories. In the old literature, Becker (1882) claimed that the hard, fresh undecomposed diabase (andesite) hosting the Comstock Lode contained twice as much gold and silver as the highly altered (propylitized) rocks in the immediate vicinity of the lode. From these data he drew the conclusion that the filling of the lode was by lateral secretion, the precious metals and the gangue being derived essentially from the propylitized rocks. The same idea was held by B. von Inkey, F.W. Hutton, J.M. Maclaren and others for the origin of the Hauraki gold-quartz veins (Finlayson, 1909; Williams, 1974), although later investigators, including J.R. Don, A.M. Finlayson and W. Lindgren, considered the auriferous deposits to be of magmatic hydrothermal origin.

In the more recent literature lateral secretion theories have again been considered. De Roever and Lodder (1967) postulate that the Rodalquilar deposits in Almeria, southeastern Spain are of syngenetic origin. The deposits are veins, consisting mainly of banded fine-grained quartz and alunite, some pyrite, marcasite, pyrrhotite, covellite, delafossite, jarosite, hematite, geothite and native gold, in highly altered ignimbrites of Neogene age. De Roever and Lodder consider that gold-bearing volatiles emanated from the ignimbritic materials and deposited their gold along the walls of fumaroles. The deposits are, therefore, fossil fumaroles that are enriched in gold.

At Guanajuato, Mexico, Gross (1975) attributes the source of vein materials in the gold-silver deposits in the Sierra, Veta Madre and La Luz fault systems to the underlying Mesozoic sedimentary rocks. Alkaline groundwater in the rocks was heated to temperatures above 350°C during the period of volcanism in the middle Oligocene. The hot solutions leached vein-forming materials from the surrounding rocks and escaped along faults formed by crustal subsidence following volcanism. The solutions moved upward along the faults, and the deposition of vein material during the upper Oligocene was caused primarily by a decrease in temperature estimated to be approximately 10°C/100m.

Casadevall (1976) and Casadevall and Ohmoto (1977) have made an extensive study of the Sunnyside Mine, Eureka mining district, San Juan County, Colorado by means of fluid inclusions and various isotopes. The gold and base metal ores are of Miocene age (13.0–16.6 m.y ago) and are localized in faults and fracture fillings associated with the San Juan–Uncompahgre calderas. Sulphur isotope studies suggest that the source of the sulfur in the ores was upper Paleozoic marine evaporites; isotopic studies of lead and rubidium-strontium indicate that the metals in the ores were scavenged from both Tertiary volcanic rocks and the underlying Precambrian basement rocks. The hydrothermal fluids were predominantly deep circulating meteoric water.

Capacity of country rocks to supply gold

The gold content of most rock types is discussed in considerable detail in Chapter II, and the average gold content of 1 mi^3 and 1 km^3 of several common rock types is shown in Table 71. It is immediately apparent that practically all rock types are adequate sources of gold for concentration in most types of auriferous deposits. Among the igneous rocks those of ultrabasic and basic character followed by the intermediate types contain the most gold. This may partly account for the marked frequency of gold deposits in greenstone and propylitic belts originally constituted of ultrabasic flows and sills, basalts, andesites, dacites, etc. It may also explain why in auriferous belts there is often an association of gold deposits with highly altered (serpentinized and carbonated) ultrabasic bodies. Among the various clastic sedimentary rocks the sandstones and conglomerates hold the most gold followed by the greywackes and subgreywackes. Again there is a correlation of the frequency of occurrence of auriferous deposits with types of rocks, the metamorphosed greywacke-sandstone terranes in many parts of the world being characterized by the presence of numerous gold-quartz veins, saddle reefs, etc. Further correlations are evident with respect to auriferous deposits and the presence of the relatively gold-rich sulphidic schists and pyritic slates. Metamorphosed volcanic and sedimentary terranes containing such rocks are frequent hosts for various types of gold deposits.

Mechanisms of migration and concentration of gold in secretion theories

General. Secretion of gold from the country rocks and its deposition in structural traps such as faults, fractures and shear zones or in chemical traps, mainly carbonate beds, involves various mechanisms for the mobilization, migration and concentration of the element. Few of these are clearly understood in a physicochemical sense since it is practically impossible to experiment on the scale necessary to elucidate the mechanisms.

One can assume on reasonably good grounds (*see* Chapter II) that much of the gold in the common rocks is present in pyrite with smaller amounts in magnetite and various other minerals. There is probably also an interstitial component of the element loosely bound in an unknown manner along the grain boundaries of the rock and along various other structural

Table 71. Gold content of a cubic mile and cubic kilometre of various types of igneous and sedimentary rocks*

Rock type (Gold content in ppm in brackets)		Gold content/mi^3		Gold content/km^3	
		($g \times 10^6$)	($oz \times 10^6$)	($g \times 10^6$)	($oz \times 10^6$)
Igneous type rocks					
Ultrabasic	(0.0045)	56.273	1.809	13.500	0.434
Basic	(0.0072)	90.036	2.895	21.600	0.694
Intermediate	(0.0047)	58.774	1.890	14.100	0.453
Acid	(0.0027)	33.764	1.086	8.100	0.260
Sedimentary rocks					
Sandstone, arkose, conglomerate, etc.	(0.0263)	328.882	10.574	78.900	2.537
Shale, mudstone, argillite, etc.	(0.0039)	48.770	1.568	11.700	0.376
Sulphidic schists, pyritic black shale, etc.	(0.0143)	178.822	5.749	42.900	1.379
Tuffs	(0.0023)	28.762	0.925	6.900	0.222
Limestone, dolomite	(0.0034)	42.517	1.367	10.200	0.328
Greywacke and subgreywacke	(0.0076)	95.03	3.055	22.800	0.733

*Data compiled from Tables 13 and 15.

discontinuities such as microfractures. When a pile of rocks with gold so disposed is subjected to the conditions of regional metamorphism (P = 10kbar; T = 500°C) or affected by intrusion of igneous bodies a great variety of complex reactions take place. None of these are known in any detail, but it is reasonable to assume that water and other volatiles, silica and the various metals such as Cu, Pb, Zn, Ag and Au are mobilized. That this is the case is readily shown by heating powdered pyritiferous shales or volcanic rocks in a current of steam and condensing the effluent in a cold trap. Experiments such as this were carried out more than 70 years ago by Gautier and others in their researches on gases in rocks (Clarke, 1924). They have been recently revived by Walker and Buchanan (1969), who obtained a highly reactive gaseous mixture when black shale was heated at temperatures ranging from 300 to 400°C. They considered that the gaseous mixture, which included H_2O, HCl, CO_2, CH_4, H_2S, S_2 and higher polymers, is likely to be important in the transport of the metals.

Similar experiments by Belevtsev *et al.* (1972), utilizing various aqueous acid, neutral and alkaline solutions to leach rocks at 300°C and 300 atm, showed that both the ore and petrogenic elements pass into solution rather readily especially from unmetamorphosed sedimentary rocks. When the rocks are metamorphosed (e.g., gneisses) removal of the various elements is more difficult and restricted. Details should be sought in the original publication. The authors concluded that the experimental data confirm the great mobility of ore- and rock-forming elements during metamorphism and ultrametamorphism.

Once released from its bonded sites in rocks gold may migrate either by mass transport, diffusion or a combination of both mechanisms. Movement by mass transport generally involves solution of the gold in liquid or gaseous water, migration of the solution or gas along structural discontinuities such as fractures, faults and shear zones and precipitation in favourable sites as a result of various chemical interactions or a decrease in temperature and/or pressure in the system. The various modes of solution and precipitation of gold in aqueous solutions and gases are discussed in Chapter II and need not be repeated. It is easy to visualize a model in which gold and its associated gangue elements follow the consecutive courses; dissolution (mobilization) from deeply buried rocks under high temperature and pressure, migration of the gold and other elements in solution upward along a fault system under a driving gas pressure and precipitation of the quartz, sulphides and gold in dilatant zones in a region where the temperature and pressure are greatly reduced compared with that where the gold was dissolved (mobilized). Such a model is applicable in a number of auriferous belts, particularly those where the deposits were formed near the surface as in certain Tertiary terranes. For deposits formed under deep-seated conditions, as in Precambrian terranes, such a model is not applicable or is applicable only in a restricted sense. Under these conditions diffusion processes are more probable.

Diffusion is the spontaneous statistical process whereby atoms, ions and molecules migrate through rocks, natural solutions and gases, as a result of differences in free energy (chemical potential) between any two points in these media. Since diffusion is essentially dependent on the kinetic properties of atoms, ions and molecules the nature of the binding of the medium through which diffusion takes place largely determines the rate. In gases diffusion is rapid because the mean free path, which is equivalent to the average distance between atomic, ionic or molecular collisions, is long. In liquids it is short and diffusion is less rapid, and in solids it is very short resulting in slow rates. Increase of temperature and pressure respectively increase and decrease the rates in all media. Shearing, however, increases the rates in solids.

In rocks diffusion may take place in the following ways:

1. Migration of constituents (SiO_2, CO_2, H_2O, S, As, Cu, Ag, Au, etc.) in a gaseous state or as ions and molecules along grain boundaries and fractures and through pores and other discontinuities in rocks. The migration is accomplished in part by surface diffusion, but where gases are involved there may also be some mass transport. This mechanism is probably operative in regional metamorphism, formation of skarn deposits and other relatively high temperature phenomena. It has been suggested by the writer (Boyle, 1959, 1961*a*, p. 94) as an explanation for the distribution of various volatiles in the regional metamorphic zonation in the Yellowknife gold belt, Northwest Territories. Confirmation of the process seems evident from sulphur isotopic data (Wanless, *et al.*, 1960).

2. Migration of ions, hydrated ions and molecules through a relatively stationary water gas or other natural gas such as carbon dioxide that pervades the grain boundaries, pores and other discontinuities in rocks. This mechanism is likewise a relatively high temperature phenomenon. Various constituents such as SiO_2, K, Na, Zn, Cu, Ag, Au, Fe, etc. may be transferred in this manner judging from the data of Morey (1957). The writer (Boyle, 1955, 1961*a*, p. 94, 120, 136) has explained the metamorphic zonation and the transfer of silica and other constituents in the wall-rock alteration zones of the gold-quartz veins and lenses at Yellowknife by this mechanism. Confirmation seems evident from sulphur isotopic data (Wanless *et al.*, 1960).

3. Migration of ions, hydrated ions and molecules through stationary water that occupies all fractures, breccia zones, pores and other discontinuities in rocks. Many near-surface gold veins in (Tertiary) volcanic terranes probably owe their origin to this mechanism. Support for this mechanism is found in the remarks by Ewers (1969) with respect to the diffusion of gold and the constituents of quartz and pyrite through stationary solutions during hydrothermal experiments.

4. Diffusion of ions, atoms and molecules through the crystal lattices of the constituent minerals of rocks. At low temperatures the rate is extremely slow. At high temperatures and where the crystal lattices are under shear the rate is greatly increased. This mechanism is probably operative in the formation of gold-bearing contact metamorphic deposits, in certain shear zone deposits and in the formation of high temperature gold-quartz veins in intensely metamorphosed rocks.

The mechanisms of formation of epigenetic gold deposits appear to have differed little with time, the processes that gave rise to Precambrian deposits recurring throughout successive geological eras. An outline of the mechanisms thought to have operated in the formation of Precambrian deposits can, therefore, serve, with modifications, as a basis for the discussions on the formation of deposits in the Paleozoic, Mesozoic and Cenozoic eras.

Precambrian (mainly Archean) deposits occur principally in volcanic or sedimentary terranes or in granitic bodies that invade the volcanics or sediments. Most of the deposits in volcanic terranes are localized in faults, fractures or shear (schist) zones and comprise quartz veins and lenses, stockworks, or pyritized and silicified bodies of rock usually enveloped by extensive alteration zones marked by chloritization, carbonatization, pyritization and sericitization. Deposits in sedimentary terranes are mainly quartz veins and lenses, saddles or irregular bodies developed along bedding planes, in drag folds, or in local shears and faults. Wall-rock alteration effects are absent or minimal. Deposits in granitic rocks are commonly veins, stockworks or pyritized and silicified bodies that may or may not be accompanied by marked alteration zones.

There are a number of features in gold-bearing Precambrian greenstone belts that have a direct bearing on the origin of the gold deposits in these belts. Most of these have been mentioned previously in various parts of the text and will now be summarized briefly.

Most greenstone belts are composed essentially of basalts, andesites, dacites, gabbroic-dioritic intrusive plugs and dykes and subsidiary rhyolite. Ultrabasic flows and intrusive plugs and dykes are generally local. None of these rocks are particularly enriched in gold – the initial average for the rocks of greenstone piles probably being about 0.007 ppm (*see* Table 13). Within greenstone belts, however, there are usually three other rock types that are commonly enriched in gold (and silver). These include various sulphide schists, iron-formations and pyritiferous quartz-feldspar porphyry dykes and bosses. These rocks vary widely in their gold (and silver) contents as shown in Tables 13, 15 and 16. A general average for these rocks of 0.02 ppm seems reasonable. Such rocks were probably the main source of the gold that has been concentrated in epigenetic deposits in greenstone belts, although the volcanics and basic and ultrabasic intrusives may have made sizeable contributions as we shall see later.

All Precambrian greenstone belts have been folded, extensively metamorphosed, granitized and/or injected by granitic bodies. Contemporaneous with these events, and in places somewhat later, the rocks have been extensively faulted and sheared. During these processes at least three mechanisms could conceivably have given rise to epigenetic gold deposits.

The first mechanism for consideration involves granitization of enormous volumes of volcanic rocks and their associated sediments. During this process large amounts of volatiles (H_2O, CO_2, S, etc.) and metals, including gold and silver, originally present in the volcanic pile may possibly have been dissolved in the resultant granitic melt, which upon cooling and crystallizing ejected solutions containing silica, volatiles, gold, etc., and these found their way into available fractures, faults and shear zones where they were partitioned and ultimately crystallized as quartz, carbonates, pyrite, arsenopyrite, gold and other minerals.

There can be little doubt that granitization of volcanic piles releases vast quantities of volatiles and metals, as can be readily seen by comparing the average contents of the rocks before and after granitization (Table 72). When one considers that most Precambrian greenstone belts have been granitized over many miles, and hence that many cubic miles of rock have been transformed, there is no need to look further than the volcanic piles of rocks for the source of the gold and its associated elements. An estimate by the writer of the gold that could have been released by granitization of the Yellowknife greenstone belt amounts to some 30 million oz.

This attractive hypothesis has, however, a number of defects from a chemical and thermodynamic viewpoint. Chief among these is the production of silica (quartz) one of the most abundant components of practically all types of epigenetic gold deposits. During granitization of intermediate and basic volcanic rocks silica is required in considerable amounts at the sites of transformation as shown by analyses of rocks in granitized sequences (Boyle, 1961*a*, p. 72). This precludes the possibility of excess amounts of silica at the end stages of granitization in the writer's opinion. Another flaw in the hypothesis is the thermodynamic argument of the apparent impossibility of retaining the volatiles (the carriers of the metals) in magmatic melts during the granitization process. Since the system involved in granitization is invariably open, the volatiles and hence the metals will migrate by diffusion away from the sites of transformation mainly during the early stages of the process, not at the end stages. Profiles of volatiles outward from the centres of granitization suggest that this is indeed the case (Boyle, 1961*a*, p. 85).

Because the granitization process, as envisioned with a final magmatic phase differentiating volatiles and metals, appears improbable the writer has suggested a granitization-metamorphic-dilation mechanism for the origin of gold deposits in Precambrian volcanic belts (Boyle, 1961*a*). The essentials of this mechanism can be described as follows.

During the orogenic events, which gave rise first to complex folding and then granitization of volcanic piles and associated sediments, marked temperature gradients existed outward from the foci of granitization leading to the formation of the zone facies of metamorphism in greenstone belts. At this time much water, carbon dioxide, sulphur and probably some of the chalcophile elements were mobilized and migrated toward the cooler parts, that is, down the temperature gradient at right angles to the isothermal lines as shown by the vectors in Figure 91*a*.

If no structural breaks were formed in the rocks, the result would be a simple metamorphic halo about a granitized centre. Such unmineralized belts occur in some Precambrian terranes. In many belts, however, great systems of faults and shear zones were formed near the end of the orogeny and formation of the granitic bodies. As a consequence of these structural events the migration vectors were radically changed by the dilatancy of the shear zones as shown by Figure 91*b*. The shear zones literally sucked the mobile carbon dioxide, water, sulphur and other elements from the country rocks and channelled them toward the surface. Because of the great extent and depth of most of the shear zone and fault systems the effects of dilatancy were felt over great volumes of the country rocks, and mobile elements were probably contributed from points thousands of feet horizontally and vertically from the shear-zone and fault systems. Extensive structures

	(%)			(ppm)										
Content in volcanic (greenstone) belts prior to granitization	1.80	0.262	0.094	7.0	1.0	80	6	93	123	50	240	0.007	0.12
Content in derived granodiorite rocks	0.97	0.084	0.024	2.0	0.2	10	12	55	10	5	5	0.003	0.05
Gain (+) or loss (–) in short tons (or troy ounces) X10[6] during transformation of a cubic mile of greenstone rocks to granodiorite*	–114.4	–24.5	–9.65	–0.069	–0.011	–0.965	+0.082	–0.524	–1.558	–0.620	–3.239	–1.608 (oz)	–28.14 (oz)
	(–24.9)	(–5.34)	(–2.1)	(–0.015)	(–0.0024)	(–0.210)	(+0.018)	(–0.114)	(–0.339)	(–0.135)	(–0.705)	(–12.0g)	(–210.g)
Content in pelitic (greywacke and slate) sedimentary belts prior to granitization	3.13	0.30	0.250	20	1.0	65	15	110	70	26	135	0.012	0.20
Content in derived granitic rocks	0.97	0.084	0.024	2.0	0.2	10	12	55	10	5	5	0.003	0.05
Gain (+) or loss(–) in short tons (or troy ounces) X10[6] during transformation of a cubic mile of pelitic rocks to granite†	–297.7	–29.77	–31.15	–0.248	–0.011	–0.758	–0.041	–0.758	–0.827	–0.289	–1.792	–3.618 (oz)	–60.32 (oz)
	(–64.8)	(–6.48)	(–6.78)	(–0.054)	(–0.0024)	(–0.165)	(–0.009)	(–0.165)	(–0.180)	(–0.063)	(–0.390)	(–27.0g)	(–450.0g)
Content in ultrabasic rocks prior to granitization	7.80	0.67	0.047	4.0	0.4	80	4	97	1285	135	2000	0.005	0.08
Content in derived granodiorite rocks	0.97	0.084	0.024	2.0	0.2	10	12	55	10	5	5	0.003	0.05
Gain (+) or loss (–) in short tons (or troy ounces) X10[6] during transformation of a cubic mile of ultrabasic rocks to granodiorite‡	–941.46	–80.78	–3.170	–0.028	–0.003	–0.965	+0.110	–0.579	–17.575	–1.792	–27.500	–0.80 (oz)	–12.06 (oz)
	(–204.9)	(–17.58)	(–0.690)	(–0.006)	(–0.0006)	(–0.210)	(+0.024)	(–0.126)	(–3.825)	(–0.390)	(–5.985)	(–6.0g)	(–90.0g)

*Numbers in parentheses refer to metric tons or grams X10[6] gained or lost during transformation of a cubic kilometre of greenstone rocks to granodiorite.
†Numbers in parentheses refer to metric tons or grams X10[6] gained or lost during transformation of a cubic kilometre of pelitic rocks to granite.
‡Numbers in parentheses refer to metric tons or grams X10[6] gained or lost during transformation of a cubic kilometre of ultrabasic rocks to granodiorite.

that impose such marked dilatant effects on the country rocks may be termed 'first-degree' dilatant zones.

During the formation of the large fault and shear zones, carbon dioxide, water and sulphur reached a high concentration along them. As a consequence the chemical equilibrium was strongly displaced, and chloritization, carbonatization, pyritization, etc. of enormous tonnages of volcanic rock took place. This led to the liberation of silica, potassium, calcium, iron, etc., in addition to gold, silver and other metallic elements present in the rock affected by alteration. These liberated compounds and elements, together with those added to the shear zones by diffusion from the country rocks, migrated laterally and vertically to 'second-degree' low-pressure dilatant zones at shear-zone junctions and other structural locales where contorted zones or openings were formed. In these sites, secondary reactions, promoted by the low-pressure and lower-chemical potential, resulted in the precipitation of quartz, carbonates, auriferous pyrite and arsenopyrite, various other sulphides and sulphosalts and native gold, all of which form the familar gold-quartz-sulphide bodies common in volcanic (greenstone) belts.

In this model the source of the gold is assumed to be the pile composed of various volcanic rocks, intrusive (?) pyritic porphyries and interbedded sediments, especially the iron-formations and sulphide schists, some of which are relatively enriched in gold (Tables 15 and 16). A more local source for the gold (and quartz), however, seems probable in some Precambrian greenstone belts.

Many greenstone belts are marked by great zones of shearing and schisting running for many miles in which the rocks are intensely altered to chlorite schists, chlorite-talc-carbonate schists and in places to great masses of carbonate (ankerite). Some of the gold deposits in greenstone belts are located within the confines of these great carbonated and chloritized zones. Examples are the orebodies of the Giant and Con mines in the Giant-Campbell System, Yellowknife, Northwest Territories and the orebodies of the Kerr Addison and Chesterville mines at Larder Lake, Ontario. Other deposits are in fractures, faults and shear zones that are subsidiaries of the great carbonated zones. Examples are widespread throughout the world, being present at Yellowknife Northwest Territories (Negus-Rycon System subsidiary to the Giant-Campbell System); at Kirkland Lake, Ontario (Kirkland Lake fault subsidiary to the Kirkland Lake–Larder Lake Break); at Ramore, Ontario (Ross Mine where the orebodies are adjacent to extensive carbonated zones); at Porcupine (Timmins) Ontario where belts and patches of carbonatization are widespread both in the productive vein zones and well outside these zones; at the McWatters and other mines in the Rouyn area of Quebec which are in or adjacent to the carbonate-chlorite-talc schist zone that marks the Cadillac Lake Fault; in the Kalgoorlie gold belt of Australia (Golden Mile zone); in Rhodesia; South Africa; and elsewhere.

When rocks are extensively chloritized and/or carbonatized, such as happens when water and carbon dioxide are introduced along great shear systems, there is a major loss of silica as shown by the analyses in Tables 48 and 73. This silica is transferred to dilatant sites in the great shear (schist) zone and fault systems or to similar sites in subsidiary shear zones and faults where it is precipitated mainly as quartz. It is also

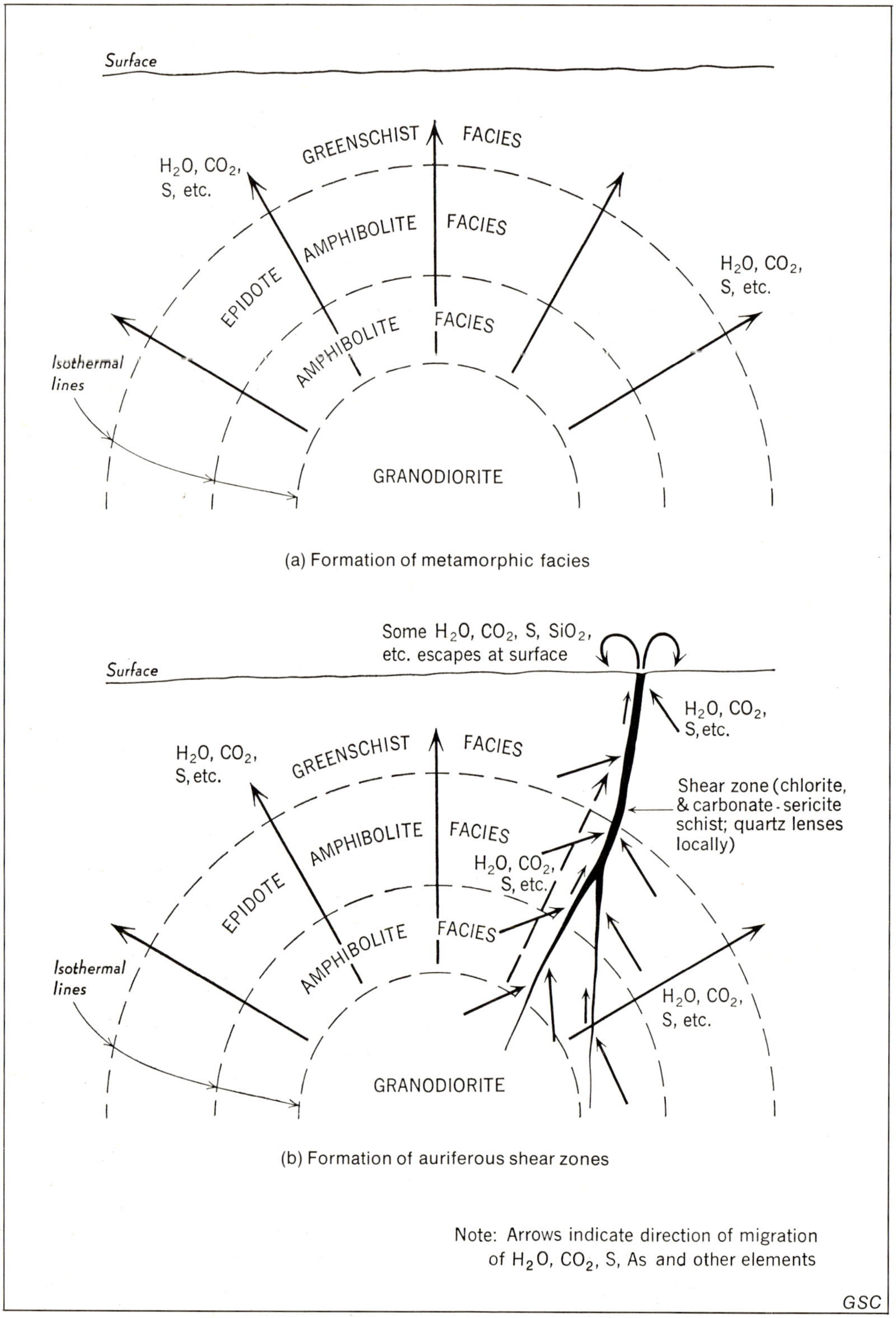

Figure 91. Schematic diagrams illustrating the formation of metamorphic facies and auriferous shear zones.

Table 73. Gain or loss of some constituents during extensive carbonatization of a cubic mile and cubic kilometre of Archean greenstone belts and ultrabasic rocks

Constituent	CO_2	SiO_2	S	As	Sb	Cu	Pb	Zn	Ni	Co	Cr	Au	Ag
	(%)							(ppm)					
Content in volcanic (greenstone) belts prior to carbonatization	0.26	53	0.094	7.0	1.0	80	6	93	123	50	240	0.007	0.12
Content in carbonated greenstone rocks*	8.00	15	0.040	2.0	0.5	25	2	50	80	30	210	0.003	0.05
Gain (+) or loss (–) in short tons (or troy ounces) X10^6 during carbonatization of a cubic mile of greenstone rocks†	+1066.9 (+232.2)	–2067.64 (–450.0)	–7.444 (–1.620)	–0.0689 (–0.0150)	–0.0069 (–0.0015)	–0.758 (–0.1650)	–0.0551 (–0.0120)	–0.593 (–0.1290)	–0.593 (–0.1290)	–0.276 (–0.0600)	–0.414 (–0.0900)	–1.608 (oz) (–12.0g)	–28.14 (oz) (–210.0g)
Content in ultrabasic rocks prior to carbonatization	0.67	40.1	0.047	4.0	0.4	80	4	97	1285	135	2000	0.005	0.08
Content in carbonated ultrabasic rocks*	8.00	30	0.040	2.0	0.2	30	2	50	200	35	1000	0.002	0.05
Gains (+) or loss (–) in short tons (or troy ounces) X10^6 during carbonatization of a cubic mile of ultrabasic rocks‡	+1010.4 (+219.9)	–1392.2 (–303.0)	–0.9649 (–0.210)	–0.0276 (–0.006)	–0.0028 (–0.0006)	–0.6892 (–0.150)	–0.0276 (–0.006)	–0.6479 (–0.141)	–14.956 (–3.255)	–1.3784 (–0.300)	–13.784 (–3.00)	–1.206 (oz) (–9.00g)	–12.06 (oz) (–90.0g)

*These contents are rough estimates from selected carbonated zones. With increase in carbonatization most of the metals decrease significantly to low values, usually only a few parts per million.
†Numbers in parentheses refer to metric tons (or grams) X10^6 gained or lost during carbonatization of a cubic kilometre of greenstone rocks.
‡Numbers in parentheses refer to metric tons (or grams) X10^6 gained or lost during carbonatization of a cubic kilometre of ultrabasic rocks.

probable that much of the gold contained in the mafic silicates, pyrite, oxides (magnetite, etc.) and other minerals in the zones affected by chloritization and carbonatization is liberated and migrates together with the silica into available dilatant zones. It is difficult to prove this contention, although careful work at Yellowknife (Boyle, 1961*a*, p.167) suggests that probably much of the gold in the economic shoots came from the great shear (schist) zones as a result of chloritization and carbonatization. Numerous other elements are also liberated (Table 73), and these probably contribute substantially to the mineral composition of the gold deposits. In recent years the results of investigations by others support the idea that much of the gold in quartz veins and other auriferous deposits may have been locally derived as a result of various alteration processes.

Roslyakova *et al.* (1970) in their investigation of the behaviour of gold in the primary halos of some gold veins (Berikul' and Darasun deposits) noticed that highly altered zones of carbonatization and sericitization constituted negative gold halos, that is the gold in these zones appeared to have been extracted and probably contributed to that present in the veins. They refer to other cases of this phenomenon in the auriferous deposits of U.S.S.R.

Laipanov and Sher (1971) in their study of the wall-rock alteration effects near gold-bearing quartz veins of the Byoryozovsky deposit observed that the carbonatization and chloritization of amphibolites and serpentinites reduced the gold content from 0.015 to 0.008 and 0.004 ppm respectively. They concluded that the gold thus removed was transferred to the veins.

Keays and Kirkland (1972) examined the mobilization of gold at the Thomson River Copper Mine, Victoria, Australia. There, a hornblendite dyke has been extensively sheared, faulted, altered and carbonatized. Cu, Pd, Ir, Au, and to a lesser extent Co and Ni in the main portion of the dyke were shown to have migrated from the sheared and altered rocks and to have been deposited in quartz stringers in a previously segregated body of magmatic sulphides trapped along the footwall of the dyke. Of the various metals noted above, the greatest redistribution was shown by gold.

It should be emphasized in concluding this part of the discussion that not all chloritized and carbonated shear zones are depleted in gold. On the contrary many are slightly enriched in the element as noted in a previous section in this chapter. Many situations actually exist. The most common are shear zones that are slightly depleted in gold throughout much of this extent; other shear and schist zones, however, are depleted in gold in certain parts, either along strike or on their dip, and enriched in others. The latter is the case at Yellowknife where the writer (Boyle, 1961*a*) has shown that silica, and probably also gold, are extracted at depth and have migrated to higher levels where they now form the economic ore shoots. One must think and work in three dimensions on problems of ore genesis.

Considering now the Precambrian sedimentary terranes we see from Tables 15 and 71 that pyritic graphitic shales, argillites, and certain types of sandstones (quartzites) and greywackes are relatively rich in gold (and silver). These rocks were probably the main source of gold in the epigenetic deposits in the Precambrian metasediments. During metamorphism and granitization of the sedimentary piles, silica, gold, silver, sulphur and various base metals were mobilized and drawn into available dilatant zones in drag-folded parts of the beds, along opened bedding planes, along faults, along sheared axes of anticlines and synclines, and along the crests of anticlines (saddle reefs). In all these sites the various mobile constitutents were precipitated as quartz, carbonates, sulphides and native gold. The model suggested for the formation of gold deposits in the sedimentary terranes is essentially the same as that for volcanic (greenstone) belts. In both cases migration of the elements now present in the gold deposits was keyed to the advancing wave of metamorphism and granitization, and the precipitation of the elements depended essentially on structural events. The latter made available dilatant zones in which the elements were precipitated in various mineral forms mainly as a result of differential chemical potentials, the dilatant zones being sites of low pressure and low chemical concentration for specific elements compared with conditions in the stressed and heated host rocks. Where carbonate members occur in the volcanic and sedimentary piles dilatancy was only partly responsible for the precipitation of skarn-type orebodies in most places. More important were the complex chemical reactions between the carbonate minerals and the elements and compounds in the diffusion currents associated with the vanguard of the advancing wave of metamorphism and granitization. In other words the carbonate rocks were chemical traps (chemical potential sinks) that immobilized constituents such as silica, sulphur, base metals, silver and gold, thus precipitating various Ca-Mg-silicates, pyrite, pyrrhotite, arsenopyrite, base metal sulphides and native gold – minerals stable under the P-T conditions prevailing during the formation of skarns.

In some Precambrian (and younger) sedimentary terranes (and in certain mixed greenstone and sedimentary terranes) the precise relationships of the auriferous deposits to metamorphic events, granitization and granitic bodies invading the sedimentary (and/or volcanic) terranes are complex. In many regions the metamorphic, granitic and mineralization events are coeval or nearly so as outlined in the previous paragraph; in other regions, however, some of the quartz deposits appear to be truncated by the granitic bodies, and xenoliths of the quartz veins occur in the granitic rocks. In a few places (e.g., Yellowknife) some of the auriferous quartz veins are cut by pegmatites. Such relationships are common where there are two ages of granitic rocks in the terrane as in the Precambrian Yellowknife Supergroup and in the Ordovician Meguma Group of Nova Scotia. Generally the first phase is a biotite granite, whereas the second is a muscovite granite usually with a plethora of associated pegmatites. Where these geological features exist the sequence of events seems to have been:

1. Initial stage of folding of greywacke and slate (and volcanics) with the concomitant effects noted in 2 below.

2. Rise of the geotherms through the sedimentary (and/or volcanic) pile due to metamorphism and granitization at depth. This is attended by the migration of silica, sulphur, arsenic, gold and other constituents into available dilatant zones to form the saddle reefs, leg reefs, interbedded veins, stockworks, lodes and other auriferous bodies with their contained auriferous sulphides and native gold.

3. Continued folding resulting in the corrugation of the quartz veins, formation of quartz boudins, ptygmatic veins and a general crushing and recrystallization of the quartz. During this period remobilization and reconcentration of the initial gold into dilatant sites (shoots), commonly as the native metal, may take place in the quartz bodies. In addition further increments of gold, sulphur, arsenic, silica and other constituents may be provided by the prevailing diffusion currents.

4. Continued granitization at depth followed by the injection of high level granitic stocks and small batholiths. The development of pegmatites commonly takes place during this stage.

5. Late stage minor folding and faulting.

The epigenetic gold deposits formed during Paleozoic, Mesozoic and Cenozoic times are essentially similar in most repects to those of Precambrian age.

There are some differences, but these are minor. For instance, great zones of shearing, schisting, chloritization and carbonatization are not common in rocks younger than Mesozoic, but their place is taken by enormous irregular zones of pervasive propylitization. It is from these zones that some of the silica, sulphur, base metals and gold in the veins may have come, having been liberated from the andesites, pyroclastics, etc. during hydration and carbonatization processes (Fig. 92).

Another model is possible for the origin of certain gold deposits in rocks lying in older geological terranes (e.g., Paleozoic rocks on Precambrian terranes; Tertiary andesites and pyroclastics lying on Mesozoic rocks). In this model (Fig. 92) the source rocks for much of the gold and its associated elements may be underlying pyritic graphitic schists, volcanics, etc., and the receptacle rocks for the deposits may be overlying faulted and fractured andesites, pyroclastics, limestone, etc. During igneous activity the heat wave associated with intrusive bodies or granitization fronts liberates water, carbon dioxide and other volatiles from the basement rocks, and these in turn promote the migration of silica, sulphur, base metals, silver and gold. These rise into the available dilatant zones such as faults, fractures and breccia pipes where they are precipitated as quartz, sulphides and native gold. Such an ore genesis model as this appears to answer many of the problems associated with the formation of Tertiary epigenetic gold deposits. In this respect it is of interest to note that many of the basement rocks underlying Tertiary deposits in many parts of the world are Mesozoic or older graphitic, pyritic schists, greywackes and slates and volcanic assemblages, rocks that are commonly enriched in gold, silver, sulphur and the various base metals now present in the Tertiary gold deposits.

Our present knowledge of the migration of gold and its associated elements during metamorphic events, or as a result of extensive alteration effects such as chloritization and carbonatization, is meagre, and this is one of the weak points in all metamorphic secretion theories. In a previous section diffusion in its many manifestations has been considered as the most probable process by which gold is transferred from source rocks to the deposits. At Yellowknife it cannot be shown conclusively that the gold and its accompanying elements came from the country rocks, although there is a strong suggestion that much gold was liberated within the shear zone systems during carbonatization and chloritization

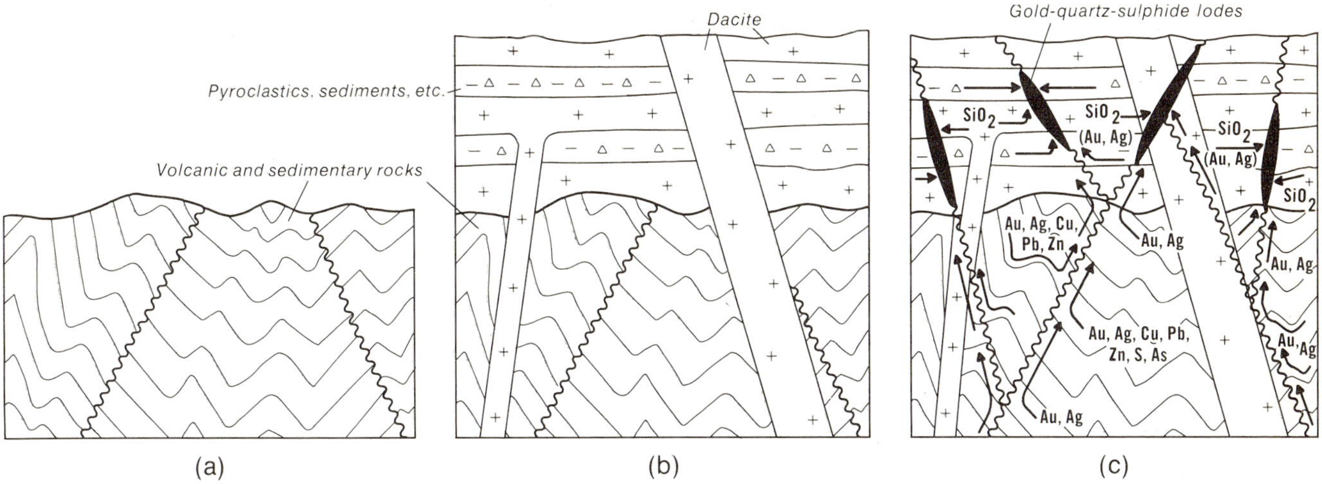

Metamorphosed, folded and faulted volcanic-sedimentary basement complex

Basement complex is overlain by andesites, pyroclastics and sediments, and the whole is intruded by dacite porphyries, etc. This is accompanied by a rise in the geotherms through the basement complex

Late faulting and fracturing coeval with or followed by propylitization of the volcanic pile; mobilization and migration of SiO_2 from zones of propylitization; contemporaneous mobilization, migration and concentration of gold, silver, base metals, etc. into lodes, the source of these elements being the rocks of the basement complex as well as the zones of propylitization in places

Figure 92. Schematic diagrams illustrating the sequence of events in the formation of certain gold-silver epithermal deposits.

and migrated into local dilatant zones. The chemical profiles for water, carbon dioxide and sulphur across the metamorphic zones of the greenstone belt are, likewise, highly indicative of a process whereby these constituents were driven out of the high temperature zones (amphibolite facies) into the low temperature zones (greenschist and chloritized shear zones) (Boyle, 1959; 1961*a*, p. 85). Confirmatory evidence is provided by sulphur isotope studies done on samples from traverses across the metamorphic facies in the greenstone belt (Wanless *et al.* 1960).

Others have observed a variety of phenomena that suggest gold is relatively mobile during the metamorphism of rocks. Some of these involve the injection of dykes into auriferous deposits. At Yellowknife and in the Little Long Lac–Sturgeon River gold belt of Ontario, particularly the latter, gold is extensively mobilized near basic dykes that cut across the orebodies. For example, at the Northern Empire Mine east of Beardmore (Fig. 46), a flat diabase sheet cutting across the gold ore shoots is accompanied by an alteration halo some 50 ft or more thick in which gold and much quartz has been leached and dissipated (Benedict and Titcomb, 1948). Similar phenomena occur in other parts of the world and are notable near dykes cutting the bankets of the Rand in places (Sharpe, 1955; De Kock, 1964). While these examples are microphenomena compared to regional metamorphic events they do, however, show that gold is relatively easily mobilized where a thermal gradient prevails.

The behaviour of a variety of elements during the metamorphism of various types of rocks has been studied in the Patom Highland of U.S.S.R. by Petrov and Makrighina (1971) and Petrov *et al.* (1972). In the first paper the authors found a number of petrogenic and rare elements such as Li, Rb, Ba, Sr, etc. to behave inertly whereas Fe, Co, Ni and Au showed considerable migration. Aided by volatiles these elements moved down the temperature gradient and accumulated in the lower temperature facies (epidote-amphibolite and greenschist). In the second paper the Au, U, Th and K contents were determined in meta-pelites, calcareous mudstones, marls, limestones and sandstone. During metamorphism gold was found to exhibit significant changes in content in meta-pelites and siltstones, i.e., in rocks that were initially rich in water. Furthermore, the meta-pelites lost gold from the amphibolite facies with a consequent gain in the epidote-amphibolite and greenschist facies. U, Th and K were found to be largely immobile during metamorphism.

These tendencies, however, do not appear to be unequivocal. Li (1974*a,b*) and Li and Shokhina (1974*a,b*) for example noted no significant change in the gold content during progressive metamorphism of Precambrian rocks on the Yenisei Ridge, Siberia, U.S.S.R. Even under very high temperature conditions (granulite facies) the only result was a local redistribution and dispersal of gold.

Petrov (1974), in an exhaustive study of the genesis of the gold-quartz veins of the Yenisei region, concluded that the gold mineralization resulted from three primary factors: (1) the gold content of the sedimentary-metamorphic rocks; (2) the development of fissure zones in these rocks, which was connected with folding and dragging associated with extensive faulting; and (3) the development of 'chemical differentiation' of the rock in the zones of fracturing and fissuring. By 'chemical differentiation' Petrov indicated that the gold and other elements can quite easily be derived from the host rocks. He defined 'chemical differentiation' as a typically ore-producing 'hydrothermal' process, similar to metamorphic differentiation in the kinetics of chemical reactions and the transfer of matter, but separated from metamorphic differentiation in time. 'Ore-producing' chemical differentiation manifests itself before or after metamorphism and metamorphic differentiation, and it is also possible for it to appear in the initial and final stages of these processes but not together with them. When considering the genesis of quartz veins alone and particularly in the North Yenisei region, Petrov (1974) clarified the definition of 'chemical differentiation' as follows:

> During the formation of quartz veins, chemical differentiation is that process of the diffusional rearrangement of rock matter as a result of which there occurs a concentration of silica in some volume of that rock and the complete concentration of some ore components earlier disseminated in that volume more or less uniformly in the forms of veins, veinlike masses and deposits of composite form. 'Ore-producing' chemical differentiation is not directly linked with magmatic activity; its closeness to metamorphic differentiation is due solely to their common relationship to tectonic stresses, which serve as a kind of trigger for both processes.

In another part of this most interesting work Petrov (1974) carried out experiments on the geological diffusion of silicon under stress (tectonic) conditions. When tension fissures formed under compression were available the rate of diffusion of silicon was of the order of 1 to 3 $\times$ 10^{-5} cm^2/s, which is 10^5 times more than the diffusion coefficient of silicon commonly quoted in the literature. These values agree reasonably well with those calculated and observed for silver by the writer many years ago, (Boyle, 1968*b*). As Petrov (1974) indicates these rates mean "that quartz veins with a thickness of about one millimetre may arise in nature in less than a year."

Detailed mechanisms of the formation of epigenetic gold deposits

General characteristics of epigenetic gold deposits. In the previous section the generalized mechanisms of the formation of epigenetic gold deposits were outlined. This section will deal with the details of the mechanisms, specifically with parameters such as localizing structures and chemically favourable host rocks, dilatancy, migration and redistribution of the ore and gangue elements, and the temperature and pressure conditions under which the deposits are thought to have formed.

Following the views of the structural geologists, one recognizes two extreme zones in the crust of the earth where auriferous veins, lenses and other types of deposits may form – the near surface zone of fracture and the deep zone of flowage and metamorphism where open fractures and cavities are largely inhibited. An intermediate zone combines the features of both these zones to a degree dependent upon the prevailing tectonic and geochemical conditions. These conceptions are fundamental, and they influence greatly the types of gold deposits that will form.

No particular figures can be ascribed to the lower limit of the low load zone of fracture in tectonic environments, since many factors are involved depending upon the types of rocks and the nature of the applied forces. For normal rock piles (volcanics and sediments) the lower limit of the fracture zone is probably at 4000 m. Below this the intermediate zone continues to 8000 m or slightly more, below which the great load would inhibit the general presence of cavities and open fractures for any length of time.

In the zone of fracture, especially under near-surface conditions, open structures such as fractures, faults, breccia zones and porous zones are formed and preserved, and these become filled with groundwater, connate water, metamorphic water[29], or mixtures of these. Under certain conditions this water will form a static medium through which ore and gangue constituents derived from the country rocks will migrate in the form of ions, hydrated ions, molecules and gases to sites of deposition determined by differences in the chemical potential of constituents in the system. Under other conditions the water may be in motion, and mass transport and subsequent deposition of dissolved constituents gathered from the country rocks will take place. Even under these conditions diffusion processes are at work, but they are of a short-range nature, whereas in the first situation they are of a relatively long-range order. In rare cases the movement of constituents in the fractures, faults, etc. may be entirely in the gaseous state.

In the zone of flowage and high grade metamorphism the exact state of water cannot be accurately defined. Below the critical temperature (273°C) it probably exists as a gas, above this point as a dense gas. In both cases it would pervade all grain boundaries and other discontinuities in the rocks thus forming a more or less stationary medium through which diffusion of gold, silver, base metals and gangue constituents may take place. In other circumstances, as in rocks undergoing very high grade metamorphic changes, water may be largely expelled from the rocks, thus necessitating a dry diffusion of constituents along grain boundaries and other discontinuities.

Gold deposits formed in the zone of fracture are usually of two types: veins and lodes generally of the gold-quartz type deposited in faults and fractures; and disseminated deposits developed near faults and fractures in chemically reactive or porous and permeable rocks, usually limestones, dolomites, argillaceous/arenaceous carbonate rocks or a variety of carbonate-bearing sandstones, arkose, etc. The structures, textures, and paragenesis of the mineral aggregates of these two types of deposits are characteristic.

In the veins and lodes deposited in the zone of fracture open cavities and vugs are abundant and open space filling manifested by crustified growth banding, comb structures, cockade structures, delicate colloform banding and intricate lamellar quartz aggregates are characteristic (Pls. 4, 12 and 13). Quartz with massive structure is common in some deposits, but book and ribbon structures in quartz are relatively rare. Early precipitated minerals such as calcite and barite are often replaced, sometimes pseudomorphously, by quartz and/or adularia. Such phenomena suggest major changes with time in the composition of the mineralizing media or in the conditions of deposition. Replacement of fault breccia and gouge, represented by silicification, sericitization, carbonatization, pyritization, alunitization, etc. is common, but the vein matter is generally frozen to the walls. Wall-rock alteration zones are characteristically developed in some rocks, subdued or practically absent in others. (*See also* the section on wall-rock alteration in this chapter.) In cafemic rocks (andesites, chlorite schists, etc.) the zone developed adjacent to the veins and lodes is characterized essentially by carbonatization, sericitization and pyritization. This grades outward through a chloritized-carbonatized zone into a weakly chloritized zone and ultimately into fresh rock. In acidic

[29]Metamorphic water is here defined as that water expelled from deep zones as a result of the dehydration of shales and other hydrous rocks undergoing regional or thermal metamorphism.

rocks (granite, rhyolite, gneiss, etc.) the alteration zones are generally less distinct but commonly include a sericitized and pyritized zone adjacent to the veins and lodes and an outer sericitized or argillized zone, which grades imperceptibly into the fresh rock. Silicification may occur in places, commonly in the zone nearest the veins or lodes. Argillization is widespread in all zones in some deposits as is also alunitization. Most alteration zones are marked by an increase in the K_2O/Na_2O ratio as the veins are approached. Propylitization, alunitization and silicification, particularly the first two, may assume a regional aspect in some areas. Loss of silica is marked where hydration, carbonatization, propylitization and alunitization processes are manifest. The silica is evidently delivered to the fissures and faults where it crystallizes as quartz.

The distribution of the gold in the vein deposits formed in the zone of fracture is characteristic. Most of the precious metal occurs within the vein filling or in seams along the contact between the vein filling and the wall rocks. The characteristic wall-rock alteration zones generally contain only minor amounts of gold, and the gold values drop off rapidly with distance from the vein. Only rarely can the vein filling and wall-rock alteration zones be profitably mined as a whole.

The disseminated deposits in the zone of fracture are rudely tabular, commonly stratiform, occasionally pipelike and usually have a compact to moderately porous, fine-grained to coarse texture, the gold and auriferous minerals being disseminated, or occasionally arranged in rude bands throughout a groundmass of various minerals, some residual and some resulting from alteration and mineralization processes. In the lower temperature deposits, developed mainly in carbonate rocks, argillaceous/siliceous carbonate rocks or carbonate-bearing sandstone, quartzite, etc., the alteration effects are mainly silicification, sericitization, argillization, chloritization, alunitization and pyritization, together with the introduction of a variety of elements, mainly Au, Ag, S, As, Sb, Ba, Zn, Cu, Pb, Hg, W, Mo, Se and B. There is generally a major abstraction of CO_2 (carbonate) and an introduction of K, the latter usually resulting in an increase in the K_2O/Na_2O ratio in the ore zones compared with the unaltered host rocks. The higher temperature disseminated deposits are developed in similar host rocks, and the elemental introductions and abstractions are essentially similar. The resulting groundmass minerals are, however, different. Instead of silicification and argillization the deposits are marked essentially by the development of a compact skarn or hornfels manifest by the presence of a variety of Ca-Mg-Fe-silicates such as garnet, epidote, etc. Skarn-type gold deposits formed under near surface conditions are, however, relatively rare.

In both the vein and lode deposits formed in the zone of fracture successive bands or crusts of ore minerals commonly permit a general paragenesis to be determined, but numerous studies by many investigators show that the succession is commonly complicated, contemporaneous deposition of minerals, incomplete sequences, minor reversals, omissions and partial repeated sequences punctuated by fracturing being frequently encountered. In the compact disseminated deposits a paragenetic sequence of the ore and gangue minerals can generally be made out by microscopic methods. Disregarding details and various complications the common paragenetic sequence in both types of deposits is (1) quartz, carbonates, sericite, feldspars, pyrite, pyrrhotite and/or arsenopyrite followed by (2) sphalerite, (3) chalcopyrite, tetrahedrite, tennantite, enargite, (4) galena, (5) acanthite, (6) stibnite and various Fe, Cu, Zn, Pb, Ag antimony sulphosalts, and (7) gold and gold tellurides. A more detailed paragenesis is given later.

The position of gold in the paragenetic sequence is of interest. In some deposits the early pyrite and/or arsenopyrite are often auriferous, the metal being present in lattice sites or as finely divided submicroscopic particles in the sulphides. In nearly all deposits the gold and gold tellurides are the last minerals to crystallize. In massive quartz veins, and even in some of those that exhibit comb structures and depositional banding, the quartz and early pyrite and/or arsenopyrite are crushed, fractured, sheared or brecciated, the gold and tellurides being deposited often with clear limpid quartz and carbonates in the fractures, crushed zones, shears and interstices between the breccia fragments. In the literature mention of this feature is repeated *ad infinitum* and is discussed at length by Bain (1930) and White (1943), and figured in many beautiful specimens by Ebbutt (1948).

Deposits formed deep within the earth in the zone of flowage have many features in common with those formed in the zone of fracture, but there are major differences as well.

The deposits are of two types: veins, lenses, saddles and irregular bodies in shear zones, schist zones and fractures; and replacement bodies in chemically favourable rocks. The structures and textures of the mineral aggregates in these two types of deposits are characteristic.

In the veins, lenses, saddles, etc., the quartz, which generally makes up more than 90 per cent of the bodies, commonly has a compact texture and is medium to coarse grained but locally highly mylonitized with innumerable recrystallized sections and phantom veinlets. There may be a rude banding in the quartz, but depositional banding of quartz and other minerals and comb structures, cockade arrangements and vugs are relatively rare. A characteristic feature is the presence of book structure (parallel quartz sheets sandwiched between slabs of wall rock) and ribbon quartz (Pl. 3), the origins of which have been long discussed and variously interpreted (McKinstry and Ohle, 1949; Chace, 1949; Stillwell, 1950, 1953*b*). The chief hypotheses that have been suggested to explain book structure are: accretion of quartz along parallel lines of weakness, replacement processes along similar lines, inflation of lines of weakness by invading solutions and the force of growing crystals. For the origin of ribbon quartz the alternative hypotheses are the same with the addition of post-quartz shearing processes. Probably all of these mechanisms have operated in the formation of deep-seated quartz veins, lenses and saddle reefs; the argument is really over which is the most important mechanism. To the writer the activating agency that results in the formation of book and ribbon structure is dilation, both macro and micro, attendant upon shearing, faulting and the opening of the crests of folds promoted mainly by obliquely directed compressive forces. According to this concept quartz lenses are the result of repeated formation of incipient openings in shear zones, faults and saddles, where sheeting, buckling, dragging and mashing of the wall rock was prevalent. These openings and contorted zones are sites of low pressure and low chemical potential into which mineral matter (mainly silica) migrated and was deposited. Large composite quartz lenses and veins marked by ribbon quartz evidently grew mainly by accretion of numerous small quartz lenses filling dilation zones. Once these small lenses were formed, repeated warping and dragging resulted in recrystallization of the quartz and further dilation, drawing in more silica. This action, together with some replacement, tended to obliterate the seams and patches of wall rock, mainly schist, surrounding the small lenses, and the final result was ribboned quartz lenses or bodies of massive quartz containing trains of sericite, chlorite, carbonate, etc., which mark the sites of former ribbons.

The deep-seated auriferous skarn-type deposits are generally irregular bodies or bunches of ore but some are tabular and stratiform. Their texture is compact and coarse, characteristics of replacement processes under high temperatures and pressures. Banding, crustification and vugs may occur, but they are infrequent and local. The characteristic minerals, mainly Ca-Mg-Fe silicates and the various oxides and sulphides, are essentially similar to those formed in skarn deposits developed near the surface.

The wall-rock alteration effects in deep-seated auriferous veins, lodes and skarn-type deposits are similar to those mentioned for the near surface deposits. The alteration effects are, however, commonly more marked in the deeper deposits. Alunitization and argillization are rare in the deep deposits whereas tourmalinization, biotitization and ankeritization are particularly common. Epidotization is, likewise, common in some deep-seated deposits, and minerals like garnet, andalusite and amphiboles make their appearance in some of the alteration sequences. Gold (and silver) is commonly present in the alteration zones of these deposits and may contribute substantially to the worth of the ore, the ore boundaries being determined in many cases by assay. Occurrences are even known where the bulk of the gold is in the altered rock whereas the vein materials are relatively barren.

The paragenetic sequence of the minerals in the deep-seated deposits is essentially similar to that described above for the deposits formed in the zone of fracture under near surface conditions.

It is well known that gold is rarely if ever distributed uniformly throughout its deposits; more generally the element is concentrated in specific zones referred to as ore shoots. Mention of these has been made at various places in the preceding text, and it now remains to summarize the features of these phenomena and suggest reasons for their occurrence.

Ore shoots can be defined as zones with definite boundaries within deposits that can be mined at a profit under normal conditions of mining. As such they depend, therefore, essentially on the content of useful elements they contain; in the case of most auriferous deposits this generally means the content of gold, although in certain deposits closely associated elements such as Cu, Ag, Pb, Zn, Bi, Te, etc. may be won as a byproduct. The delineation or blocking out of ore shoots is usually done by sampling and assaying, which in the case of auriferous deposits requires considerable experience and care because of the camouflaged occurrence of the element (e.g., gold in pyrite or arsenopyrite) and the particulate and dispersed nature of native gold. Novices evaluating gold deposits and organizing sampling programs for the delineation of ore shoots should consult textbooks on the subject such as those by Forrester (1946) and McKinstry (1948) before commencing work.

Ore shoots are regular to highly irregular in outline; commonly elongate and pipelike or chimneylike in form and large or small in tonnage. Their geometry is defined by their width or thickness, their breadth, their stope length and their pitch length. Deposits may contain only one ore shoot, although most contain several separated by barren (uneconomic) mineralized material. In origin ore shoots may be of primary (hypogene) or supergene derivation. The general nature and causes of ore shoots are discussed at considerable length in Lindgren (1933), Newhouse (1942) and Bateman (1950). In the following text, discussion is restricted to hypogene ore shoots in auriferous epigenetic deposits such as veins, stockworks and replacement deposits. Localization of ore shoots in the quartz-pebble conglomerate type of deposit is discussed in the previous section, and an outline of the causes of auriferous supergene ore shoots is given in Chapter IV.

The distribution of ore shoots in auriferous deposits is not fortuitous, although in some cases their causes may be elusive and due to factors that are not readily determined. The localization of ore shoots in auriferous deposits may be due to tectonic (structural) features, physico-chemical (pressure, temperature, concentration) factors, or a combination of both. As discussed subsequently tectonism produces a variety of dilatant features that control the migration and deposition of gold and its accompanying elements in deposits and more specifically in ore shoots. Various physico-chemical factors such as the type of wall rock and its porosity, the mingling of solutions and diffusion currents, and changes in temperature and pressure as solutions, gases and diffusion currents rise toward the surface also exert an influence on the deposition of gold in ore shoots.

The causes of localization of auriferous ore shoots in deposits can usually be ascertained by careful geological, mineralogical, chemical and assay work, paying particular attention to types of primary and secondary structures, types of wall rocks and the type and degree of wall-rock alteration. When the various geological and mineralogical features and the assay and chemical data are plotted and the latter contoured on plans and sections, particularly longitudinal sections, of the deposits certain relationships generally emerge between the location of enrichments of gold (ore shoots) and features such as the curvature of faults or shear zones, impermeable barriers and favourable wall rocks. Once an ore shoot control is established the search for other shoots is commonly facilitated.

Ore shoots controlled by secondary or superimposed structural conditions. Structural distortion of rocks related to shearing, faulting, fracturing, brecciation and folding produces various types of dilatant features that are sites of low chemical potential where gold and its associated elements concentrate according to physico-chemical principles discussed subsequently. Some of these dilatant features include:

1. Open zones and discontinuous cavities caused by strike slip or dip slip movements of opposite walls along a curved or irregular fissure, fault or shear zone: Ore shoots localized in such zones are generally lenselike in form and commonly have

extensive pitch dimensions. Many of the auriferous ore shoots in the Yellowknife Greenstone Belt owe their origin to their type of dilatancy (Boyle, 1961*a*), and a number of the ore shoots in the Mother Lode System of California are considered by Knopf (1929) to have had such an origin.

2. Open spaces, brecciated zones, and contorted, dragged, mashed and shattered zones at the intersections or junctions of fissures, faults or shear zones: In addition to dilatancy such sites may also produce a large surface area of comminuted rock for chemical interaction with solutions or diffusion currents. Ore shoots localized in such sites may be pipelike, saddlelike, lenselike or irregular in form. Most have limited strike dimensions but extensive pitch dimensions. Many of the auriferous ore shoots in the Yellowknife Greenstone Belt are at shear zone junctions (Fig. 18). Most of the ore shoots in the Grass Valley veins in California are at vein intersections.

3. Dragged, crenulated, severely contorted and dilated sites in wide schist or shear zones: Such zones in addition to producing dilatancy along myriad contorted sheetlike openings also provide a large surface area of comminuted rock for chemical reaction with migrating solutions or diffusion currents. Many of the auriferous ore shoots in greenstone belts at Yellowknife, Timmins, Kirkland Lake (Kerr Addison) and western Quebec are localized in such sites.

4. Tension fractures in wide schist and shear zones: Where these are closely spaced they become filled with gold-quartz veins and stringers that constitute irregular stockworks or en echelon ore shoots. Some of the orebodies in the Con Mine at Yellowknife and the Kerr Addison and Chesterville mines at Larder Lake (Fig. 23) are of this nature.

5. Dilatant zones created by slippage, shearing or faulting along the walls of quartz veins and lenses: These zones originate as the result of late movements and commonly contain significant enrichments of gold in many auriferous belts.

6. Open spaces along faults, shears and fractures and brecciated, dragged and contorted zones in competent rocks below or otherwise associated with impervious barriers such as shale, slate, igneous rocks, fault gouge: These barriers being athwart the dilatant zones, pond solutions and otherwise obstruct diffusion currents, giving rise to impounded ore shoots. Some of the auriferous lodes at Keno Hill, Yukon are of this nature (Boyle, 1965*a*).

7. Dilatant zones created by tension fractures, shrinkage joints, or small faults and shears in dykes, sills, competent lava flows and massive sedimentary beds: These give rise to ladder veins (shoots) and en echelon veins and lodes. Examples are numerous in the Beresovsk gold belt of U.S.S.R. and elsewhere (Figs. 19, 52, 53 and 54).

8. Dilatant zones associated with pipes and chimneys: These have various origins including brecciation and comminution of rocks at the intersection of faults, fractures, shears or joint systems; throats of old volcanoes; diatreme breccia pipes or volcanic explosion vents; and solution channels and callapse structures in carbonate-bearing rocks. Ore shoots in these sites generally have short plan dimensions but extensive pitch dimensions. Examples are common in auriferous belts (e.g., Cresson Blowout, Cripple Creek, Colorado, Figs. 35 and 36; Sunbeam Kirkland auriferous pipe in the Falcon Lake stock, Manitoba, Fig. 51).

9. Extensive zones of fracturing, faulting, crushing and brecciation developed in dykes, sills, competent lava flows and massive sedimentary beds, intrusive plugs, stocks and small batholiths: Such dilatant zones commonly encompass very large volumes of rock that constitute the ore shoots. In some cases, however, enriched zones in great volumes of fractured and altered rock embody individual randomly distributed ore shoots. Deposits in this category include the so called 'porphyry gold deposits' by analogy with the 'porphyry copper deposits'. Examples of low grade porphyry gold deposits are widespread. They are described in a previous section. (*See especially* Figs. 49 and 50.)

10. Dilatant zones related to openings on the crests of anticlines and less frequently in the troughs of synclines: During regional folding these zones give rise to the classical saddle reefs (Figs. 38, 39, 40 and 44) so common in terranes of greywackes and quartzites interbedded with slates and phyllites. In saddle reefs the richest ore shoots are predominantly localized in the caps of the reefs and regionally on domes.

11. Dilatant zones in tension fractures, faults and shear zones along the sheared or faulted axes of anticlines and synclines. Shoots in these sites tend to be lenselike in form; most have a short strike dimension but commonly extend to considerable depth.

12. Dilatant zones created during regional folding by drag folding, shearing, brecciation and contortion of incompetent sedimentary beds, tuffs and volcanics. Such zones give rise to leg reefs (shoots) in isoclinally folded strata (Figs. 38, 40 and 41), lenselike and irregular lenselike shoots in tuffs and other rocks (Fig. 55) and irregular ore shoots in rolls, drag folds and other contortions in iron-formations (Figs. 56 and 57).

Ore shoots controlled by chemical conditions. It is axiomatic to state that all ore shoots are precipitated in sites of low chemical potential; in certain cases decreases in pressure and temperature are the promoting agencies for precipitation, whereas in other cases changes in the concentration (stability) factor of elements (or their complexes) in solutions or diffusion currents are paramount in causing widespread precipitation due to interaction with reactive wall rocks, entry into porous zones, and so on. Some of the conditions where this takes place include:

1. Shoots controlled by steep temperature and pressure gradients: As mineral laden solutions or diffusion currents rise from depth they eventually enter zones of lower temperature and pressure where precipitation takes place as a result of the P-T factor in the chemical potential. In some conduits (faults, shears, fracture zones) the P-T gradients are steep and a marked vertical zoning may result as in certain auriferous Tertiary terranes (Hauraki goldfield, New Zealand; Translyvania goldfield, Romania); in other conduits the gradients are so flat that precipitation takes place over a long vertical interval as in certain auriferous deep seated deposits (Kolar, India; Kirkland Lake, Ontario).

2. Ore shoots controlled by intermediate, basic and ultrabasic wall rocks: When mineralizing solutions and diffusion currents laden with H_2O, CO_2, S and As react with ultrabasic, basic and intermediate rocks such as serpentinites, basalts and andesites extensive changes take place resulting in the binding of Fe, Ca, Mg and Mn in the rocks by CO_2, H_2O,

S and As to form carbonates, chlorite, pyrite and arsenopyrite. Silica, which is released, contributes to the vein fillings. The alkali-bearing sulphide, sulphide-arsenide and sulphide-antimonide gold complexes are broken down, the alkalies (K and Na) forming sericite and/or albite and adularia, and the gold precipitating in pyrite, arsenopyrite or as the native metal. These various processes are discussed at length in the sections on natural waters (Chapter II) and wall-rock alteration (Chapter III).

3. Ore shoots controlled by carbonaceous wall rocks: These are the auriferous ore shoots controlled by the so-called 'indicators', which are numerous in certain gold belts but particularly at Ballarat, Victoria and Gympie, Queensland. The action of the so-called indicators is probably due to the marked reduction and precipitation effect that carbonaceous matter has on gold solutions. (*See also* the section on gold in natural waters.)

4. Ore shoots controlled by iron sulphide– and iron oxide–rich rocks: Pyritiferous and pyrrhotiferous iron-formations in many parts of the world have a marked precipitating effect on gold solutions, examples being the Central Patricia and Pickle Crow mines in Ontario (Figs. 20 and 21). The rich auriferous ore shoots in quartz veins adjacent to highly pyritiferous amphibolites at Ophir in California are further examples of the phenomenon. The function of sulphide– and iron oxide–rich rocks in precipitating gold has been established experimentally. (*See* the section on gold in natural waters.)

5. Ore shoots controlled by carbonate-bearing rocks: Carbonate-rich rocks such as limestones, dolomites and calcareous shales notably control the location of auriferous ore shoots under both high and low temperature conditions. The reason for this is evidently the ease with which these rocks can be replaced and the neutralizing effect they have on neutral to slightly acid ore solutions and diffusion currents. Under high temperature conditions skarns with auriferous shoots in pyritic and arsenopyritic zones are formed (e.g., Nickel Plate and French mines at Hedley, British Columbia); at low temperatures siliceous zones with ore shoots enriched in gold, silver, pyrite and a variety of sulphides and sulphosalts result as at Carlin Nevada (Figs. 58 and 59) and Salsigne, France (Fig. 42).

6. Ore shoots controlled by primary (initial) porosity conditions: Such conditions exist initially in porous sandstones, grits and conglomerates; in volcanic breccias, volcanic explosion vents, diatremes, vesicular lava flows and certain tuffaceous rocks; and as a result of solution phenomena during diagenesis that give rise to solution cavities, collapse breccia structures and so on. These porosity conditions are sites of low concentration and have given rise to auriferous ore shoots in many gold belts, examples being Salsigne, France (Fig. 42) where some of the ore shoots occur in initially porous grits and phyllitic quartzites and Cripple Creek, Colorado (Cresson Blowout, Figs. 35 and 36) and Central City district, Colorado (The Patch, Fig. 48) where fabulously rich ore shoots occurred in volcanic breccia pipes.

In concluding this discussion on auriferous ore shoots it should be emphasized that conditions for rich and highly productive shoots (bonanzas) in deposits are most propitious when a conjunction of both structural and chemical factors are present. Another particularly favourable condition prevails when quartz veins and other auriferous deposits have been intensely sheared, brecciated and recrystallized resulting in the remobilization of gold from sulphides in the deposits followed by its concentration mainly as the native metal in well defined shoots marked by tensional dilation cracks, fissures and discontinuities in gangue minerals and along such features as the contacts between veins and their wall rocks. Such conditions are most commonly found in deep seated (Precambrian and Paleozoic) deposits; in shallower deposits such as those in Tertiary terranes similar phenomena may obtain but more generally brecciation and reopening of veins due to recurrent movements results in pulsations of mineralization from sources exotic to the deposits, and these precipitate gold and other minerals in the new fractures, brecciated zones or along the contacts of veins and wall rocks. The subject of the remobilization of gold and multiple generations of gold is discussed below.

Temperature, pressure and pH conditions of gold deposition in epigenetic deposits. Estimates of the temperature-pressure conditions of formation of auriferous epigenetic deposits have been made for many years utilizing mineral intergrowths and thermal stabilities, pyrite geothermometer, sphalerite geothermometer, decrepitation, isotopic data and liquid inclusion data. What follows is a general summary of a few of the multitudinous investigations extant in the literature.

The techniques of studying liquid inclusions in minerals and decrepitation methods are described at length by Scott (1948), Smith and Peach (1949*a,b*) and Peach (1949). There is an extensive review and bibliogrpahy on both subjects by Smith (1953), and the more recent investigations are covered in the works by Kashiwagi *et al.* (1955), Takenouchi (1962), Yermakov *et al.* (1965) and in the Proceedings of COFFI (Roedder, 1968–1974).

A brief summary of the fundamentals of liquid inclusion and decrepitation research in determining temperatures and other features of ore deposition as it pertains to gold follows.

In general, quartz contains two types of liquid inclusions. These have been called primary inclusions and secondary inclusions depending upon their nature and location in the quartz crystals and grains. Primary liquid inclusions occur along growth planes, interlineage boundaries and other crystal discontinuities and are thought to have resulted from the trapping of some of the fluid from which the quartz grew. Secondary liquid inclusions occur along healed fractures, around grain boundaries and in small, clear crosscutting (phantom) veinlets. In most thin sections of vein quartz, trains of secondary inclusions cut across grain boundaries and hence appear to have formed by the trapping of some liquid from solutions migrating along grain boundaries and fractures that formed after the deposition of the quartz crystals and grains. Inclusions in the minerals of gold deposits may be simple or complex. Some contain only gas at room temperature; others contain at least one liquid phase at room temperature, generally water plus a gas bubble. Salt crystals of various types are common in some inclusions, and hydrocarbons (petroleum) have been noticed in a few inclusions from certain, usually young, deposits. Studies of the salt content of primary inclusions are useful in understanding the chemistry of ore-bearing solutions, provided that certain assumptions are valid.

In the ideal case at the temperature and pressure of formation of both primary and secondary liquid inclusions the trapped liquid would just fill the inclusions. During the drop in temperature and pressure to that prevailing in the rocks where the quartz is now found, the liquid contracts and small vapour bubbles (vacuoles) form within the inclusions. These inclusions with their vapour bubbles can be seen under high magnification in nearly all quartz grains and crystals.

If the quartz is slowly heated, the liquid in the inclusions expands to refill the inclusion and the vacuole disappears (the filling temperature). Further heating produces pressure in excess of the tensile strength of the quartz, and the liquid breaks the quartz with a 'pop', that is, the quartz decrepitates. The temperature at which the quartz decrepitates is known as the decrepitation temperature. The difference between the filling temperature and decrepitation temperature is called the over-shoot in temperature.

Three methods have been developed to determine the filling temperature:

1. The original optical method of Sorby (1858), consisting of measuring the diameter of the inclusion cavity and vacuole, converting the relative diameters to relative volumes, thus giving the degree of filling, and utilizing tables of pressure-temperature-volume of water and salt solutions to estimate the filling temperature.

2. A direct optical method utilizing a heating-stage mounted on a high-power microscope through which an operator observes the disappearance of the vacuoles in the inclusions, measuring the filling temperature by thermometer or thermocouple.

3. The decrepitation geothermometer consisting of an electrically heated oven in which a specific quantity of clean crushed (–40+80 mesh) quartz is placed and heated at a controlled constant rate. The miniature explosions or 'pops' of the decrepitating quartz are picked up by a sensitive microphone, amplified, fed to an electronic integrating mechanism and then to a recorder that draws a curve of decrepitation rate vs. time. From this curve the decrepitation temperatures of both primary and secondary inclusions can be read. To determine the filling temperature from these temperatures what is called the over-shoot in temperature must be subtracted. This usually amounts to a few degrees, generally less than 20.

It should be pointed out that the filling temperature of the liquid inclusions does not represent the mineral crystallization temperature that prevailed during the formation of the liquid inclusions. To obtain the crystallization temperature, either an estimated pressure correction or a pressure correction determined by some other geothermal method (e.g., pyrite geothermometer, Smith, 1947) isotopic ratios, etc. must be applied to the filling temperature.

In gold deposits the precise relationship of the gold to liquid inclusions of both primary and secondary origin in associated minerals, mainly quartz, is fundamental to the interpretation of the temperature and other features of deposition of the gold in ore shoots as determined by liquid inclusions or decrepitation methods. This is a problem for which there are no certain answers as all will recognize who have studied the detailed relationships of gold in its deposits. Some general assumptions, however, appear to be valid. In some gold-quartz deposits the gold and the quartz appear to have crystallized nearly contemporaneously. Such deposits are common in younger rocks, but certain (usually paragenetically late) veins in rocks as old as Precambrian also show this feature. In such deposits the temperature and other features of deposition of the quartz as revealed by liquid inclusions and decrepitation probably apply with only slight modification to the deposition of the gold. In many types of gold-quartz deposits of all ages native gold occurs in fractures, crushed zones and small vuggy parts of the quartz. This gold appears to have been deposited during a late stage of the mineralization process. This type of gold is discussed in more detail in an earlier section in this chapter dealing with the associated minerals of gold. Studies of this late gold and the liquid inclusions in the bulk of the quartz indicate that there is probably no relationship between the primary inclusions and the gold; there may, however, be a relationship between the secondary inclusions and the gold, but whether the gold was deposited from the solutions trapped within the secondary inclusions is not certain as the inclusions may have been formed earlier or later than the gold. The writer's (Boyle, 1954) interpretation has been that where gold occurs along fractures now marked by planes of secondary liquid inclusions the gold probably crystallized from the residual solutions trapped in these inclusions.

Machairas (1970*b*) examined the relationship of gold and liquid inclusions in quartz from Guyana; Timmins, Ontario; St Yrieix, France; and elsewhere. He did not distinguish between primary and secondary inclusions, but his figures suggest that both types are present. Nevertheless, he noted a genetic association of gold particles with fluid inclusions containing mainly water and CO_2. In some specimens of quartz he observed gold particles in the liquid inclusion cavities, and further observed that the content of CO_2 in the inclusions in the quartz was greatest in auriferous quartz. As a generality he stated that the concentrations of fluid inclusions and gold in quartz are proportional.

An early investigation of the ore deposition temperature and pressure by decrepitation of quartz and use of the pyrite geothermometer at the McIntyre Mine, Timmins, Ontario by Smith (1948) indicated two separate stages of mineralization in the gold-quartz bodies in greenstones and quartz porphyry, the first extending over the range of 630 to 400°C at a depth of 10 km and the second at 150°C at a much shallower depth. Blais (1954) working with the gold-bearing quartz veins at the O'Brien Mine, northwestern Quebec found by decrepitation methods that the auriferous quartz decrepitated in the temperature range 75 to 120°C, whereas the barren quartz decrepitation began above 130°C. In a recent paper on the O'Brien deposit Krupka *et al.* (1977) optically observed four different types of fluid inclusions: type I – pregold mineralization fluids, CO_2 rich; type II – gold mineralizing fluids showing evidence of boiling at temperature of 380–300°C; type III – postgold mineralization fluids; vapour-liquid two phase inclusions; and type IV – postgold mineralization fluids, halite bearing. Using data on the degree of filling, P-V-T data on the H_2O-NaCl system, and estimates of the depth of mineralization the authors calculated the filling temperatures of the inclusions trapped during the gold mineralization stage (type II) to range from 380 to 300°C. Using similar data the filling temperatures of the other types of inclusions were also estimated: 420 to 250°C for type I inclusions (pregold mineralizing fluids; 75 ± 5% filling); 330 to 170°C for type III inclusions (postgold mineralizing fluids; 85 ± 5% filling); 270 to 140°C for type IV inclusions postmineralizing fluids with NaCl crystals, 90 ± 5% filling). The authors noted that these temperatures are the minimum temperatures of hydrothermal fluids, because the fluids were trapped at pressures greater than those at boiling conditions and pressure corrections must be applied. Furthermore, the uncertainty in the filling temperatures of type I inclusions is probably larger than those shown above, because these inclusions contain liquid CO_2, but they too were assumed to be simple H_2O-NaCl fluids. Despite these uncertainties the authors concluded that the temperatures of hydrothermal fluids at the O'Brien gold deposit gradually decreased toward the later stages of mineralization.

The various results recorded on the geothermal history of the O'Brien deposit mentioned above, point out the great uncertainties in liquid inclusion work. One method (decrepitation) gave an apparent gold deposition temperature of 120 to 75°C, whereas another (degree of filling of inclusions) recorded temperatures of 380 to 300°C. There are in addition uncertainties with respect to leakage of inclusions, pressure corrections, relationship of the liquid inclusions to the actual process of gold deposition and so on. It is obvious that much more sustained research is necessary on the geothermometry of epigenetic gold-quartz veins in order to establish the true temperature limits of gold deposition. The complexities and uncertainties will become even more evident from the results of the following researches.

Boyle (1954) examined the quartz of two gold-bearing shear zone systems (Campbell and Negus-Rycon) by decrepitation methods at Yellowknife, Northwest Territories over a vertical extent of some 1400 ft. Two general ranges of decrepitation were noted 350° to 180°C and 180° to 90°C, the first apparently related to the initial temperature of deposition of the vein quartz and first stage auriferous pyrite and arsenopyrite, and the second probably related to a late stage introduction or to the remobilization and precipitation of gold concentrated during the initial stage of mineralization. More recently Chary (1971), after making a great many assumptions, concluded that the temperature of deposition of the gold-quartz veins in the Con-Rycon Mine at Yellowknife ranged from 350 to 300°C based on sulphur isotope and fluid inclusion studies. The maximum pressure of deposition was estimated at 850 to 900 atm.

At the Lamaque Mine, Bourlamaque, Quebec, Smith (1954) examined the quartz of the No. 6 vein with a view to determining the direction of flow of late stage solutions that probably deposited much of the gold. He found decrepitation temperatures ranging between 80 and 124°C for the quartz.

In recent years Sidorov *et al.* (1971) and Shilo *et al.* (1971) have carried out extensive pressure-temperature research on various types of gold deposits in the northeast U.S.S.R mainly by liquid inclusion methods. Depending on the types of deposits the temperature interval of mineralization ranged from 450 to 50°C, the most productive periods being in the interval 310 to 150°C. The pressure ranged from 5 to 280 atm. Shilo *et al.* give two tables that are of interest to those dealing with various types of gold and gold-silver deposits in plutonic and volcanic terranes. These tables have been slightly modified and are reproduced as Tables 74 and 75.

The so-called epithermal (mainly Tertiary) gold-bearing deposits in New Zealand and United States have recently received attention by means of liquid inclusions and isotopic studies. Robinson (1974) outlined four stages in the mineral paragenesis of the Tui deposit in the Hauraki goldfield of the Coromandel Peninsula: (1) pyrite I + chalcopyrite I + hematite + quartz; (2) galena + sphalerite + quartz, followed by minor pyrite II + chalcopyrite II; (3) minor hematite, barite, cinnabar, kaolinite and carbonates; and (4) cerussite, smithsonite and other supergene minerals. The fractionation of sulphur isotopes between coexisting pairs of minerals indicated temperatures of formation of 400 ± 80°C for pyrite I/chalcopyrite I and 320 ± 60°C for sphalerite/galena. Temperatures derived from quartz, carbonate and barite/water oxygen isotope fractionation showed a gradual decrease from about 250°C at the end of Stage 2, to about 60°C at the end of Stage 3 and about 30°C for the formation of the supergene carbonates. Taylor (1973) concluded from $^{18}O/^{16}O$ isotopic analyses that the Tertiary gold deposits in the Tonopah and Comstock Lode mining districts of Nevada were deposited mainly from heated meteoric waters with temperatures in the range 250 to 300°C; in the Goldfield district much lower temperatures, 125 to 200°C were suggested by the isotopic data. Similar conclusions were drawn for the Tertiary deposits in the Bodie mining district, Mono County, California by O'Neil *et al.* (1973) using oxygen and hydrogen isotope data. They determined that the temperatures of deposition of the gold-silver mineralization fell in the range 215 to 245°C.

Nash and Cunningham (1973) have examined the conditions of deposition of fluorite and gold in the Jamestown district of Colorado. There, fluorite deposits and gold, gold-telluride and lead-silver ores of Tertiary age occur in close proximity. The gold deposits are mainly veins; the fluorite deposits occur principally as disseminations, in breccia zones, stockworks and pipe-shaped bodies in and adjacent to a sodic granite stock. The gold and gold telluride mineralization occurs with and without fluorite in peripheral veins. Liquid inclusions associated with the main fluorite stage were found to have salinities of 26 to 50 per cent and filling temperatures of 250 to 350°C. Two fundamentally different types of fluids were considered to be present in the gold-bearing veins. Inclusions in pregold quartz and fluorite were found to have filling temperatures ≤375°C and high salinities, suggesting that early quartz and fluorite of the gold-bearing veins were roughly contemporaneous with the main period of fluorspar deposition. Gold and telluride deposition occurred at <300°C. Fluid inclusions most directly associated with gold deposition have filling temperatures ranging from 205 to 270°C and have ~4 per cent salinity, suggesting that the Au-bearing fluids either evolved from the fluorite-bearing fluids or were later unrelated hydrothermal pulses introduced into reopened structures. The occurrence of boiling and the heterogeneity of CO_2 contents were considered to indicate that

Table 74. Characteristics of the gold deposits of the northeastern U.S.S.R.

Association	Type of deposit	Depth of formation	Relationship to magmatic events	Mineral composition	Sulphide content	Temperature interval of mineralization	Temperatures of deposition of productive associations
		(km)			(%)	(°C)	
Plutonic	Gold-rare metal	1.5–2	Genetic with granitoids of orogenic volcano-plutonic associations	Quartz, topaz, tourmaline, wolframite, cassiterite, molybdenite, arsenopyrite, lollingite, minerals of bismuth and cobalt, tellurides of gold and silver	up to 5	450–130	310–160
	Gold-quartz	2–3	Coeval and coextensive with early-orogenic intrusions	Quartz, arsenopyrite, pyrite, sphalerite, galena, gold	up to 1	360–50	250–190
Volcanic	Gold-sulphide	1.5–2	Coextensive with volcanics of andesite composition	Quartz, albite, chalcopyrite, pyrite, galena, sphalerite, tetrahedrite, gold	from 5 to 50	270–150	200–150
	Gold-silver	0.5–1.5	Coextensive with volcanics of liparite-ignimbrite and andesite-basalt composition	Quartz, adularia, calcite, rhodonite, rhodochrosite, dolomite, argentite, sulphosalts of silver, galena, sphalerite, pyrite, marcasite, gold	from 1 to 10	410–50	300–230

Source: Shilo *et al.* (1971).

Table 75. Temperature of formation of gold-silver deposits of northeastern U.S.S.R.

Temperature of formation	Type of deposits	Mineral composition	Content of ore minerals	Temperature interval of mineralization	Temperatures of deposition of productive associations
			(%)	(°C)	
High-temperature	Gold-argentite (with rhodonite)	Quartz, rhodonite, rhodochrosite, argentite, gold, galena, sphalerite, pyrite	up to 1	410–250	up to 400
Medium-temperature	Gold-sulphosalt-argentite	Quartz, calcite, adularia, sphalerite, galena, pyrargyrite, argentite, gold	up to 0.5	325–130	320–290
	Gold-galena-sphalerite	Quartz, calcite, galena, sphalerite, pyrite, marcasite, chalcopyrite, tetrahedrite, argentite, pyrargyrite, gold	up to 10	360–50	300–210
Low temperature	Gold-argentite (with adularia)	Quartz, adularia, fluorite, argentite, aguilarite, galena, sphalerite, pyrite, chalcopyrite, gold	up to 1	240–80	220
	Gold-sulphosalt-stibnite	Quartz, adularia, pyrargyrite, stephanite, polybasite, tetrahedrite, argentite, stibnite	from 1–2 to 50 (in stibnite veins)	270–80	240–220

Source: Shilo *et al.* (1971).

fluid pressures were low and variable, ~150 bar. The fluids were thought to have emanated from an intrusive body at depth.

Nash (1975) concluded from fluid inclusion studies of quartz and sphalerite in vein, pipe and replacement deposits in the northwestern San Juan Mountains, Colorado (Silverton Caldera) that deposition of base metals and gold in the Argentine and adjoining veins (Idarado Mine) essentially took place within a relatively narrow range (280–290°C; corrected for pressure, 305–315°C) from solutions with a low salinity (less than 2 wt. %). The ore fluids appeared to be physically homogeneous liquids over the 3700 ft (1100 m) vertical and 11 200 ft (3400 m) lateral range studied. Other details are elaborated in the original paper.

An investigation of the gold-base metal mineralization at the Sunnyside Mine, Eureka mining district, San Juan County, Colorado by Casadevall (1976) and Casadevall and Ohmoto (1977) revealed six periods of mineralization with their approximate volumetric abundances as follows: (1) pyrite-quartz ores (5%); (2) banded quartz-sulphide ores (30%); (3) massive galena-sphalerite-chalcopyrite-bornite-hematite ores (40%); (4) gold-telluride-quartz ores (1–2%); (5) manganese ores (20%); and (6) quartz-fluorite-carbonate-sulphate ores (5%). These ores fill faults and fractures associated with the upper Oligocene

San Juan–Uncompahgre calderas and were emplaced between 13.0 and 16.6 m.y. ago. Heating studies of fluid inclusions in quartz, fluorite, and rhodochrosite indicated temperature ranges of 250 to 320°C for period (1) to (5) mineralization and 170 to 245°C for period (6) mineralization. Evidence of boiling of period (6) fluids at 240°C was observed from samples collected at 3800 m elevation. These data together with the P-V-T data of dilute salt solutions and a maximum erosion rate of 17cm/1000y indicated P_{H_2O} of between 110 and 220 bars for periods (1) to (5) and about 35 bars for period (6). The ore-forming fluids were essentially water with a small proportion of CO_2 in which were dissolved K, Na, Ca, Cl, oxidized sulphur species such as HSO_4^-, KSO_4^- and $NaSO_4^-$, and the metals including Cu, Fe, Zn and Mn at concentrations ranging from less than 10 to 1000 ppm. The pH of periods (1) to (5) fluids ranged from 4.3 to 5.9. In the last part of their paper Casadevall and Ohmoto, drawing on isotope date for O, H, S and C, conclude that the mineralizing solutions were essentially meteroric water and that the sulphur and metals came from the country rocks in the vicinity of the deposit (Mesozoic and/or Paleozoic marine sediments provided the various salts and sulphur and the Tertiary volcanic rocks and Precambrian basement rocks the metallic elements such as lead, zinc, gold, etc.).

In U.S.S.R. a number of studies of the temperature of deposition of epithermal (volcanic class) gold-silver deposits have been done. The work of Naiborodin and Goncharov (1970) on the Agatovskoye deposit in Cretaceous-Paleogene volcanics in the southern part of the Okhotsk-Chukotka volcanic belt may be mentioned here. The mineralization is in veins carrying essentially quartz, carbonates, barite, galena, sphalerite, chalcopyrite, pyrite, marcasite, arsenopyrite, argentite, pyrargyrite and gold (electrum). The veins are of four types: (1) quartz, (2) carbonate-quartz, (3) sulphide-carbonate-quartz, (4) sulphide-carbonate stage and (5) barite-opal-chalcedony stage. The study of gaseous-liquid inclusions indicated that the Agatovskoye deposit formed within a temperature range of 365 to as low as 50°C. The deposition of ore minerals took place between 300 to 200°C. Native gold was deposited between 250 and 230°C. The temperature changed sharply during the process of mineralization, which was accompanied by temperature inversions of 100 to 150°C between the stages. These conditions were explained by the authors as due to internal changes of thermodynamic conditions, mixing of high temperature fluids with meteoric water and as a result of the effect on the temperature regime by subvolcanic intrusions.

The temperature of formation of the so-called subvolcanic gold-silver deposits in central Kamchatka has been studied by Andrusenko and Shchepotyev (1974) by means of liquid inclusion thermometry. The results indicated that the following types of mineralization formed at the given temperatures: (a) copper-molybdenum mineralization precipitated from highly concentrated essentially chloride solutions within the temperature range of 340 to 70°C; (b) gold-silver mineralization precipitated from alkaline chloride-sulfate-bicarbonate solutions with carbon dioxide in the ranges of 400 to 300°C (early gold), 320 to 240°C (main gold concentrations), 260 to 115°C (late segregations); and (c) mercury mineralization in the temperature range 260 to 105°C.

In Japan, Hattori (1975) carried out a detailed geochemical investigation of the Yatani Mine, a representative of the epithermal precious and base metal deposits of Tertiary age that occur mainly in Neogene volcanics in the so called Green Tuff region. The Yatani deposits occur as fissure fillings and are divided into two types – lead-zinc veins and gold-silver veins. All of the veins exhibit breccia, cockade or crustified banding structures with occasional druses. The major constituent minerals of the lead-zinc veins are sphalerite, galena, chalcopyrite, pyrrhotite, marcasite, quartz, iron chlorite, muscovite, adularia, Mn-silicate and manganiferous carbonate with minor amounts of arsenopyrite, stannite, electrum and argentite. The gold-silver veins are Ginguro bands composed essentially of sphalerite, galena, chalcopyrite, electrum, argentite, pyrite, quartz and adularia with minor amounts of cassiterite, stannite, marcasite, montmorillonite, iron chlorite and manganiferous carbonate. Using various geochemical parameters Hattori estimated the environmental factors of the two types of veins as follows: temperature of formation was 250 to 200°C for the Au-Ag veins and approximately the same for the Pb-Zn veins; the sulphur fugacity was 10^{-11} atm at 250°C to 10^{-13} atm at 200°C for the Au-Ag veins and 10^{-12} to 10^{-14} atm at 250°C or 10^{-14} to 10^{-16} atm at 200°C for the Pb-Zn veins; and the oxygen fugacity was $10^{-34.5}$ to 10^{-39} atm at 250°C or $10^{-37.5}$ to 10^{-42} atm at 200°C for the Au-Ag veins and 10^{-37} to 10^{-40} atm at 250°C or 10^{-40} to $10^{-45.5}$ atm at 200°C for the Pb-Zn veins. Regarding the chemical characteristics of the ore-forming solutions Hattori (1975) concluded that they had an NaCl equivalent concentration of 0.1 to 0.2 mol/L; a pH, about 6 or more for the Au-Ag veins and 6 or less for the Pb-Zn veins; a concentration of total dissolved sulphur equal to 10^{-3} to 10^{-2} mol/L; the dominant dissolved sulphur species for the Pb-Zn veins was hydrogen sulphide, whereas the Au-Ag veins were formed near the boundary separating dominant sulphate ion from hydrogen sulphide.

Hattori (1975) further concluded that the gold was mainly transported as gold-sulphur complexes rather than as gold-chloride complexes, and that the precipitation of gold was mainly caused by an oxidation process. Coprecipitation of sphalerite and galena with gold he thought would be expected in neutral or weakly alkaline solution with low salinity and total dissolved sulphur of 10^{-2} mol/L when the fugacity of oxygen increased as a result of the rising of the ore solution and when the breakdown of the metal-sulphur complexes occurred.

Czamanske *et al.* (1973) have used annealing history phenomena (interdiffusion in the Au-Ag system) in natural gold to estimate upper limits of gold deposition. Concentration changes of up to 30 weight per cent Ag indicate probable deposition temperatures of less than 300°C for electrum from Copper Basin, Arizona and Alder Gulch, Montana. The gold in the Homestake Mine, South Dakota probably was formed at temperatures well below 400°C.

The sphalerite geothermometer has been used by a few investigators to estimate the temperature of deposition of sphalerite-pyrrhotite in gold deposits, but the results have been quite erratic. At Keno Hill the temperatures obtained seemed to be at variance with the geological situation, and the method was deemed not to be applicable to the various deposits (Boyle and Jambor, 1963).

According to Clark (1960) and Kretschmar (1973) the composition of arsenopyrite varies with temperature. This is the basis of the arsenopyrite geothermometer, and precise temperatures can generally be obtained if an estimate of the sulphur fugacity prevailing at the site of deposition can be made from various coexisting minerals. Since arsenopyrite is a common associate of gold, the composition (as measured from certain interplanar spacings by X-ray diffraction) of the sulphide-arsenide may be useful in estimating the formation temperatures particularly of the early stage of deposition of some gold deposits (the gold-arsenopyrite-pyrite stage noted in the next section dealing with paragenesis). The temperature of formation of later stages can also be estimated, if arsenopyrite is present in the mineral assemblage.

The temperature of formation of arsenopyrite in the gold-quartz deposits of the Yenisei Ridge has been investigated by Li *et al.* (1971) by means of X-ray diffraction measurements. The temperatures obtained ranged from 350 to 575°C. A correlation was noted between the temperature of formation of the arsenopyrite, its morphology and the distance of its host deposit from a large granitic intrusion; in general higher temperatures were noted for those deposits near the intrusive and lower temperatures for those distant from the intrusive.

Summarizing the data on gold-quartz deposits we can say that the deposition of the bulk of the quartz and sulphide mineralization took place within the interval 500 to 150°C. There is considerable evidence to support the thesis that much of the gold in many deposits was deposited (or mobilized from pre-existing auriferous sulphides, etc. and redeposited) within the interval 150 to 75°C. In some deposits higher temperatures for the deposition of gold seem probable, usually in the range 300 to 150°C. The range of pressure appears to be from about 2 bars for near surface gold-quartz deposits to 300 bars for deep seated deposits.

Relatively little work has been done on the geothermometry and geobarometry of auriferous skarn type deposits. Our limited results based on decrepitation and liquid inclusions suggest temperatures of formation of skarn minerals to range from about 350 to 700°C and the temperature of deposition of quartz associated with gold in these deposits to range from 100 to 350°C. Shcherbakov *et al.* (1972) investigated the skarn-type gold deposits in the Sinyukhinsk district of the Altai and found that the quartz subjected to skarnification gave a temperature range of 420 to 780°C by liquid inclusion methods; the garnet registered 420 to 700°C. Quartz associated with the gold stage gave temperatures ranging from 88 to 350°C by liquid inclusion methods. The pressures at which skarns and their associated gold mineralization form must vary widely depending on their depth of formation; probably the pressures vary from a few hundred bars to 5 kbars or more for most deposits.

The pH conditions of the migration and deposition of gold are discussed at length in Chapter II. There it was concluded that the solutions that deposited gold in epigenetic deposits were mainly neutral to slightly alkaline, a situation supported by data on natural waters that are now depositing gold and by data on wall-rock alteration effects noted in many gold-quartz deposits. Acid solutions may have been responsible for the deposition of gold where minerals such as alunite were developed.

Paragenesis and zoning in epigenetic gold deposits. The paragenesis of the various minerals in epigenetic gold deposits as determined by microscopic methods and the observations of crosscutting veinlets is reasonably constant, although there are commonly reversals in the sequences and other complications that need not concern us here. The major elemental and mineral paragenesis (oldest to youngest stages) in most epigenetic gold deposits are given below and diagrammed in Figure 93. Not all stages or elements are present in any one deposit. Elements and minerals in parentheses are rare in the respective stage.

1. Fe, (Ni), (Co), S, As, SiO_2, Ca, Mg, Mn, K, Na, CO_3, C, Au and Ag represented mainly by pyrite and/or arsenopyrite essentially coeval with quartz and carbonates. Marcasite may occasionally accompany or take the place of pyrite in some deposits, especially those formed near the surface. Pyrrhotite is commonly a later mineral in this stage but is generally rare in most deposits except those formed under deep-seated conditions or in iron-formations in which case it takes the place of pyrite. In some deposits pyrrhotite accompanies or follows Stage 3. Gold and silver are lattice constituents of pyrite and/or arsenopyrite, or the two precious metals occur in submicroscopic and microscopic particles of native gold in the sulphide minerals of this stage. Where Ni-Co arsenides are present, they appear to be about the same age as arsenopyrite. Carbon (graphite) is usually introduced at this stage, although in places the substance is late occurring along late fractures and slips. The bulk of the feldspars and sericite are generally formed during this stage; some chlorite is introduced.
2. W. Mo, (Sn) and S represented by wolframite, scheelite and (cassiterite). In some deposits these minerals are contemporaneous or nearly so with those in Stage 1. Molybdenite is a late mineral of this stage in some deposits.
3. Mo, S represented by molybdenite. In most deposits molybdenite appears to be an early mineral, but in some deposits there are late generations of the mineral.
4. Bi, S represented by native bismuth, bismuthinite, etc. Two generations of these minerals are apparent in gold deposits. In some deposits only an early generation is present; in others only a late generation (Stage 11); and in still others both generations.
5. Zn, Cd, (Au), (Ag), S represented by sphalerite. Au and Ag are usually only minor constituents of this phase.
6. U, Th and rare earths represented by uraninite (pitchblende) and/or brannerite. In some deposits uraninite is coeval with Stage 7; in others the uraniferous mineralization follows Stage 13.
7. Cu, S, As, Sb, Au and Ag represented mainly by chalcopyrite, tennantite, tetrahedrite, enargite, cubanite and bornite. The paragenesis of these copper minerals is often complicated. In some deposits tetrahedrite precedes chalcopyrite; in others the reverse is true. Gold and silver are lattice constituents of the copper minerals or occur in submicroscopic and microscopic particles of gold in the sulphides and sulphosalts.
8. Pb, Ag, Au, (Bi), S represented by galena. Silver is a lattice constituent of galena or occurs in a variety of exsolved (?) silver minerals. The position of gold is uncertain; it may be a lattice constituent or may occur as exsolved (?) native gold

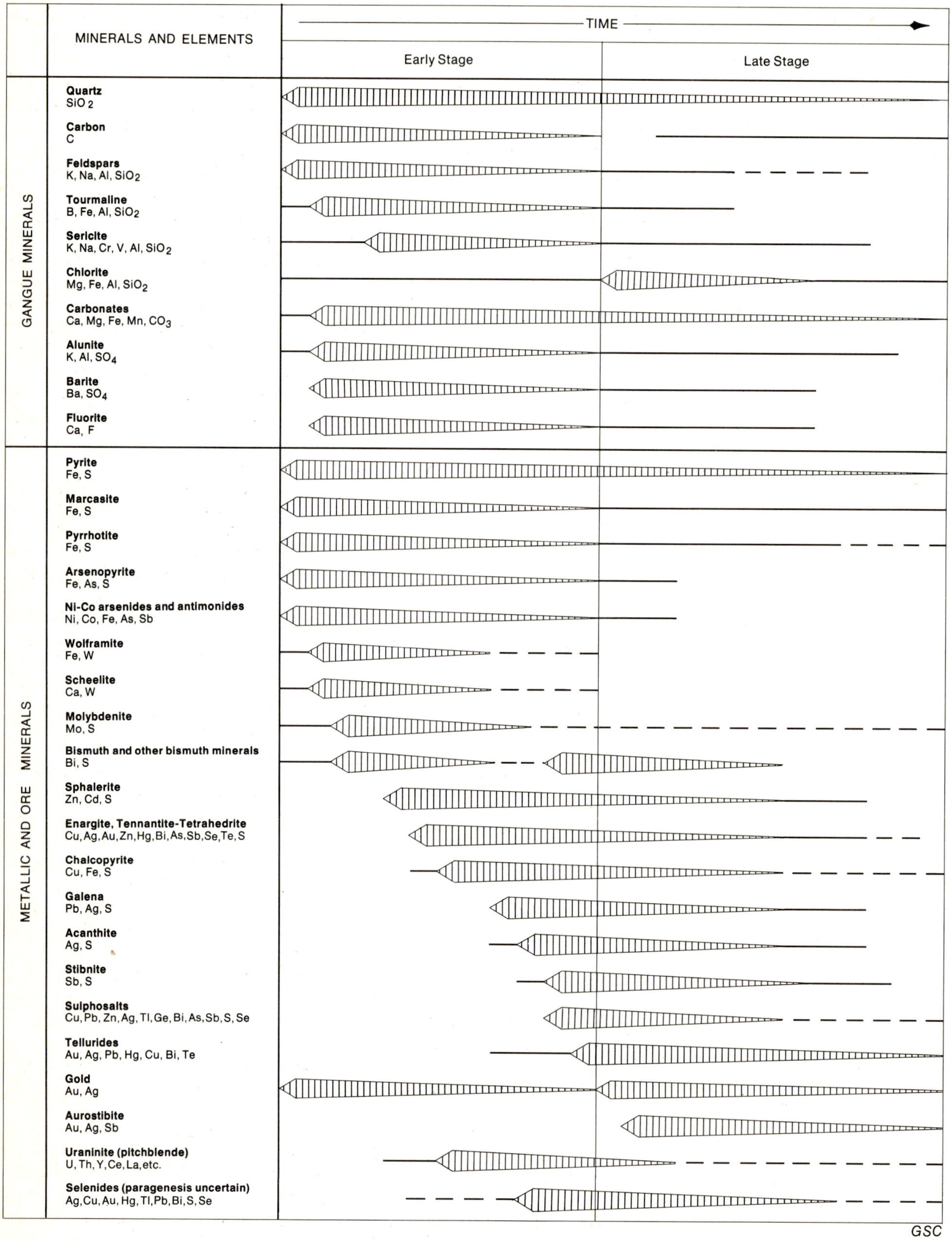

Figure 93. Generalized paragenetic sequence of principal minerals and elements in epigenetic gold deposits. Minerals and elements shown include only those in veins, lodes, etc.; minerals and elements in wall-rock alteration zones are excluded.

or gold tellurides in the galena.

9. Ag, S represented by acanthite (argentite).

10. Sb, (As), S, (Se), Fe, Cu, Zn, Pb, Bi, Ag, Au represented by stibnite and various sulphosalts, e.g., pyrargyrite. Gold and silver may be lattice constituents of these minerals or may occur in native gold as inclusions in the minerals. The paragenesis of these minerals is often complicated, but stibnite commonly precedes the sulphosalts in many deposits. In other deposits it is coeval with the antimony sulphosalts.

11. Bi, Ag, Au, S represented by native bismuth and bismuthinite. The gold and silver may occur in native gold intimately intergrown with native bismuth or the precious metals may occur in the mineral maldonite.

12. Au, Ag, Pb, Hg, Cu, Bi, Te represented by the precious and base metal tellurides. Commonly the tellurides are paragenetically older than native gold, but in some deposits they all appear to be coeval.

13. Au, Ag represented by native gold and electrum.

14. Au, Ag, Sb represented by aurostibite.

15. Hg, As, S, (Te) represented by cinnabar, orpiment and realgar and in places by mercury tellurides. The exact paragenesis of orpiment and realgar in gold deposits is uncertain. Both are late minerals, and in some veins they may slightly precede the gold. More generally, however, these minerals are paragenetically coeval with gold or younger.

The exact position of the mercury minerals is uncertain; they are, however, invariably late in the paragenetic sequence. The position of gold is not constant; it appears in a number of the stages commonly being precipitated throughout all stages in some deposits. In many deposits there is much gold in the early pyrite and arsenopyrite and some gold in the sphalerite, chalcopyrite, galena and sulphosalts. In most deposits, however, the largest amounts of free gold, electrum and gold tellurides are the last minerals to crystallize. Aurostibite is generally closely associated with native gold and in some deposits is paragenetically later than the metal, rimming or veining it. In other deposits aurostibite and native gold appear to be coeval. Aurostibite appears to occur only in deposits that are relatively enriched in antimony, and in some deposits the mineral appears to accompany Stage 10. There is commonly a close paragenetic relationship between gold and native bismuth and bismuthinite as well as with a great variety of other bismuth-bearing minerals. In most deposits the gold minerals are slightly later than the bismuth minerals or intimately mixed with them indicating nearly contemporaneous deposition. In some gold deposits bismuth follows lead and is precipitated in galena apparently as a lattice constituent in some cases. Galena may also contain bismuth in the form of minute inclusions of minerals such as matildite (Ag,Bi,S_2) and aramayoite ($Ag(Bi,Sb)S_2$). There are reports in the older literature of the occurrence of bismuthinite following sphalerite (Stage 5), but I have not been able to verify this microscopically. Molybdenite presents a varied habit in gold deposits. In some deposits it appears as an early mineral; in others it is late occurring along late slips. The tungsten minerals are invariably early minerals. Scheelite commonly appears to be about the same age as the carbonates in skarn deposits and gold-quartz veins. This apparent contemporaneity may, however, be due to the fact that the tungsten radical has replaced the carbonate radical ($CaCO_3 + WO_4^{2-} = CaWO_4 + CO_3^{2-}$) in which case the scheelite is paragenetically later than the carbonates. Wolframite is generally later than pyrite, pyrrhotite and arsenopyrite and earlier than the Zn, Cu, Pb and Ag sulphides and sulphosalts and the late gold minerals. Cassiterite, when present in deposits, is invariably paragenetically older than native gold and other gold minerals. The position of barite and fluorite in the sequence is uncertain and seems to be variable from deposit to deposit. Commonly the two minerals accompany the intermediate quartz and carbonate stages. In some deposits, however, they are relatively older than native gold and other gold minerals. Alunite exhibits a similar behaviour, although it is usually an early mineral and seems to accompany the pyrite stage in some deposits. Feldspars are generally early minerals in the paragenesis usually being mainly deposited during Stage 1; some deposits have several ages of feldspar. Quartz and the various carbonates are invariably early minerals in the paragenetic sequence, and it is notable that they continue to be deposited continuously or intermittently throughout the paragenetic stages manifest in epigenetic gold deposits. Sericite and chlorite may be deposited throughout the paragenesis but usually only in small amounts after Stage 1. Excepting some auriferous skarns, magnetite is a relatively rare gangue mineral in most deposits. In a few vein deposits magnetite can apparently take the place of pyrite and/or pyrrhotite as in the Ust'-Kara gold belt in eastern Transbaikal, U.S.S.R. (Kolosova and Onishchuk, 1971). In these deposits the gold and magnetite, which are paragenetically early minerals in a second stage of mineralization, are coeval or nearly so. Certain of the auriferous veins in Chile are also rich in magnetite and specularite, which appear to be early minerals in the paragenesis. Most of the auriferous deposits containing magnetite and/or hematite occur in granitic intrusives or in high grade metamorphic terranes. Those at Tennant Creek, Northern Territory, Australia are, however, in relatively low grade metasediments (Large, 1975). In some deposits uraninite (pitchblende) and brannerite are distinctly earlier in their paragenesis than gold and gold tellurides. Their maximum deposition appears to precede chalcopyrite but in some deposits is coeval with this mineral. In other deposits veins carrying pitchblende are distinctly later than auriferous veins in the same lode system (e.g., Box Mine, Goldfields, Saskatchewan). The position of the various selenides is uncertain. These minerals are generally late in the paragenetic sequence and appear to overlap the deposition of chalcopyrite and galena with the maximum deposition taking place during the sulphosalt–telluride stages.

In conclusion it should be recognized that the successive stages noted above are rarely sharply demarkated either in time or space. There is much overlapping, telescoping, repetition and omission of some of the stages. A more detailed discussion of the paragenetic relationships of minerals to native gold and other gold minerals is given in previous sections on the elemental and mineral associates of gold.

A sequence (zonation) of elements and minerals horizontally outward from intrusive granitic bodies or granitization centres is commonly observed as shown in Figure 94 and Table 76. This zonation was noted by a number of early investigators and elaborated upon by Spurr (1923) and

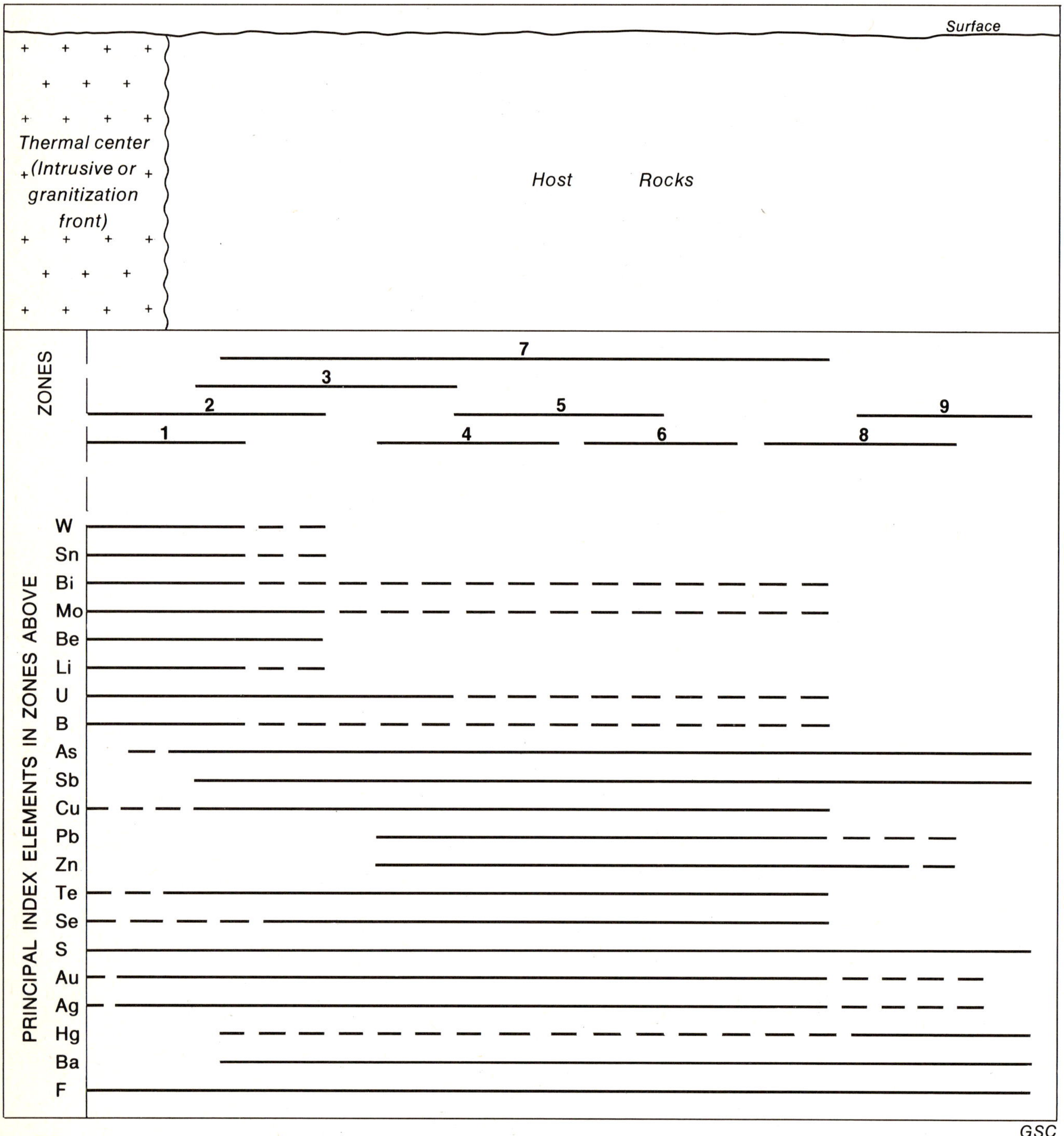

Figure 94. Generalized horizontal endogenic zonation of elements outward from a thermal centre. Elements shown include only those in veins, lodes, etc.; elements in wall-rock alteration zones are excluded (*see also* Table 76).

Table 76. Generalized horizontal endogenic zonation of elements and minerals outward from a thermal centre*

Zone	1	2	3	4	5	6	7	8	9
Persistent elements	SiO_2, Fe, S	SiO_2, Fe, Ca, Mg, Mn, S	SiO_2, Fe, Ca, Mg, Mn, S	SiO_2, Fe, Ca, Mg, Mn, S	SiO_2, Fe, Ca, Mg, Mn, S	SiO_2, Fe, Ca, Mg, Mn, S	SiO_2, Fe, Ca, Mg, Mn, S	SiO_2, Fe, Ca, Mg, Mn, S	SiO_2, Fe, Ca, Mg, Mn, S
Index elements	W, Sn, Mo, Bi, Be, Ta, Nb, Li, F, RE†, Th, Zr, U, B, As, S, (Cu)	Fe, S, As (Sb), (Te), Au, Ag, Cu, Mo	Cu, S, As, Sb Au, Ag, Te, U	Zn, Cd, S (Au), (Ag)	Pb, Ag, Au, S, Te	Ag, S, (Au)	Sb, As, S, Cu, Zn, Pb, Bi, Ag, Au, Fe, Te, Se	As, Sb (Au), (Ag), (H_g)	Hg, (Sb), (As)
Au/Ag ratio	High	High		Intermediate	Intermediate		Low	Low	
Persistent minerals	Quartz, feldspar, micas, pyrite, pyrrhotite, (magnetite and/or hematite may take the place of pyrite in some deposits)	Quartz, feldspar, pyrite, pyrrhotite, carbonates. (Magnetite and/or hematite may take the place of pyrite in some deposits)	Quartz, feldspar, pyrite, pyrrhotite, carbonates	Quartz, pyrite, pyrrhotite, carbonates	Quartz, pyrite, carbonates	Quartz, pyrite, marcasite, carbonates	Quartz, pyrite, marcasite, carbonates	Quartz, pyrite, marcasite, carbonates	Quartz, chalcedony, pyrite, marcasite, carbonates
Index minerals	Wolframite, scheelite, cassiterite, arsenopyrite, molybdenite, bismuth, bismuthinite, topaz, beryl, tantalite-columbite, uraninite and various other rare metal minerals	Arsenopyrite Ni-Co arsenides and antimonides, molybdenite, native gold, scheelite	Chalcopyrite, tennantite-tetrahedrite, enargite, barite, fluorite, uraninite (pitchblende)	Sphalerite, barite, fluorite	Galena, tellurides, barite, fluorite	Acanthite	Stibnite, sulphosalts, barite, fluorite	Realgar, orpiment, stibnite, barite, fluorite	Cinnabar, realgar, orpiment, stibnite, barite, fluorite
Remarks	Mineralization commonly in the intrusives and their border zones. Deposits are generally not auriferous. Includes pegmatites	Contact metamorphic deposits may carry copper minerals such as bornite, chalcocite, molybdenite, etc. Some desposits may be in intrusive thermal centre (porphyry Cu-Mo deposits)	May overlap zone 2	May coalesce with zone 5	May coalesce with zone 4	May coalesce with zones 4 and 5	May overlap zones 3–6	May overlap zone 7. Some veins and replacement deposits carry gold	May overlap zone 8. Rarely auriferous

*See also Figure 94.
†RE = rare earths.

Emmons (1924). Perused carefully the zonation sequence from granitic body outwards is essentially the same as the elemental and mineralogical paragenetic sequence of oldest to youngest. Yet another type of zonation is found in some deposits in fractures and faults that cut deeply into the earth. This is a vertical zonation of elements and minerals (Fig. 95) noted in a general way by a number of early investigators and later discussed in detail by Spurr (1923), Emmons (1924), Lindgren (1927, 1937) and Graton (1933). This type of vertical zonation is manifest most clearly in Tertiary deposits; it is not marked in Precambrian deposits, some of these extending to depths of 10 000 ft (Kolar) or more without decrease in gold content and radical change in the mineralization or with a very gradual change from pyrite to pyrrhotite and diminution of minerals such as stibnite and sulphosalts. The reason for this difference is not clear; one can speculate that the Tertiary deposits were precipitated under a steep thermal gradient, whereas the thermal gradient during the deposition of the Precambrian deposits would appear to have been gentle or practically nonexistent. In general, the vertical zonation in gold deposits where present is similar to the paragenetic sequence and lateral zonation.

It should not be assumed, however, that all auriferous deposits exhibit a regular vertical zonation as shown in Figure 95. Most deposits show a general tendency toward the ideal, but some diverge widely from it. Even some Precambrian deposits exhibit a complex vertical zonation. Large (1975) describing the zonation at the Proterozoic Juno Mine in the Tennant Creek goldfield, Central Australia gives a complex vertical sequence in this small but rich gold lode. Both the gangue minerals and ore minerals are vertically zoned.

The nature of the gold in deposits commonly exhibits differences with respect to proximity to igneous bodies. Numerous deposits near granitic bodies have relatively little free gold, the element being intimately bound in pyrite or arsenopyrite as a lattice constituent or in a very finely divided form. On the other hand deposits at a distance from the intrusives generally have increasing quantities of free crystallized gold. There are, however, numerous exceptions to these generalizations. The Au/Ag ratio of deposits as a whole may show a pattern, the deposits next to granitic and other igneous bodies having a higher ratio that decreases with distance from the igneous bodies. This is obviously due in some places to the increase in silver minerals with increasing distance from the igneous bodies. The Au/Ag ratio (fineness) of native gold, likewise, appears to show the same relationship in some belts, although there are many aberrations. Commonly, there is no pattern, or a complex or erratic one exists. However, in certain deposits formed near the surface (Tertiary epithermal deposits), the early gold in pyrite and arsenopyrite is frequently silver deficient (850 or more fine), whereas late crystallized gold is usually silver rich (800 fine or less) in the same deposit. Deep seated deposits do not generally show this tendency; in fact in certain Precambrian (Archean) deposits the reverse obtains. The reason for this may be subsequent recrystallization of the early auriferous pyrite and arsenopyrite resulting in the migration of gold and silver (*see below*). Vertical zonations in the Au/Ag ratio of deposits have also been noted. (*See also* the section on Au/Ag ratios in this chapter.) Probably, in the majority of deposits the Au/Ag ratio increases with depth, a circumstance due, it is thought by a number of investigators, to the relatively high mobility of silver, which compared with gold, tends to migrate (diffuse) out of the higher temperature zones.

The trace elemental content of gold in deposits may exhibit zonal variations. Thus, in some deposits elements such as antimony and mercury may be much more abundant in the free gold near the surface than at depth. This is especially true in certain Tertiary deposits and seems to be true in some Precambrian deposits. Similarly the free gold in deposits in and near thermal centres (intrusive granites, etc.) is commonly impoverished in elements such as antimony and mercury compared with the gold in deposits distant from the centres.

There are many exceptions to the statements made above about the fineness of native gold and the trace element content of the metal. The cause of some of these exceptions is not readily ascertained in some deposits; in others the cause seems to be predicated on the type of wall rock. As noted in an earlier section ore shoots in acidic wall rocks tend to have much lower Au/Ag ratios, and in some cases the gold also appears to have a lower fineness than in shoots developed in basic rock. In addition, subsequent processes of recrystallization and migration of gold and silver after the initial deposition of the two metals obviously has an effect in some, generally deep-seated, deposits. All of these features require further detailed research.

The precise causes of the relatively consistent paragenetic sequences and zonations has eluded the geologist and chemist for more than a century due to the enormous chemical complexity of the ore-forming processes. Early investigators concluded that the phenomena were due predominantly to temperature effects; later investigators assumed a single ore solution or mineralization stage and considered that the determining factors were due to hardness of minerals, power of crystallization of minerals and abundance of mineral constituents (Gilbert, 1924); solubilities of ore and gangue minerals (Newhouse, 1928; Verhoogen, 1938); progressive change in anion/cation ratio as mineralization proceeded (Bandy, 1940); impedance of structures (MacKay, 1946); volatility of ore constituents (Brown, 1948); metallic melting point (Sullivan, 1954); relative amounts of ionic and covalent bonding in sulphides (Ringwood, 1956); increase in chemical potential of sulphur (McKinstry and Kennedy, 1957); relative stability of covalent-bonded complex ions (Barnes, 1962); chalcophile element abundances (Taylor, 1963), polar-covalent bond stabilities (Pasquali and Bisque, 1975); and chemical affinities and acidity–basicity indicators (Marakushev, 1977). Polyascendent solutions have also been assumed. A number of investigators now think that zoning and the paragenetic relationships are the result of pulsation processes of mineralization punctuated in places by structural activities such as the reopening of fractures or formation of faults and fractures with different orientations or at successively greater distances from intrusive or granitization centres. There are also the effects of wall rocks on zoning to consider. These effects are partly structural and partly chemical. It is evident in some areas that certain rocks form extensive dilatant zones whereas others do not; similarly there is commonly a differential effect on the precipitation of elements exerted by different types of host rocks. For example, acidic solutions are more profoundly

ZONE	SURFACE	Au/Ag RATIO	PERSISTENT ELEMENTS	INDEX ELEMENTS	PERSISTENT MINERALS	INDEX MINERALS	REMARKS
1			Si0₂, Fe, Ca, Mg, Mn, S	Ba, Sr, F	Quartz (chalcedony), pyrite, marcasite, carbonates	Barite, fluorite, (cinnabar)	This zone is rare in most veins. Generally barren of gold and silver
2			Si0₂, Fe, Ca, Mg, Mn, S	Hg, Sb, As, Ba, Sr, F	Quartz (chalcedony), pyrite, marcasite, carbonates	Cinnabar, realgar, orpiment, stibnite, barite, fluorite	In certain veins this zone may be auriferous and argentiferous
3		Low	Si0₂, Fe, Ca, Mg, Mn, S	Ag, Au, As, Sb, Cu, Zn, Pb, Bi, Te, Se	Quartz (chalcedony), pyrite, carbonates, galena, sphalerite, chalcopyrite, gold (electrum), (molybdenite)	Acanthite, gold, stibnite, sulphosalts, tellurides, selenides, bismuthinite and other bismuth minerals, adularia, alunite, Mn-silicates, Mn-carbonates, fluorite, barite, (scheelite), (wolframite)	Constitutes bonanza section of most epithermal gold-silver deposits. Passes with decrease in gold and silver into zone 4. Generally absent in deep-seated deposits.
4			Si0₂, Fe, Ca, Mg, Mn, S	Cu, Zn, Pb	Quartz, pyrite, carbonates, galena, sphalerite, chalcopyrite	Chalcopyrite, pyrite, sphalerite, galena, barite	Commonly low or barren in gold and silver, Represents the bottoms of many epithermal Au-Ag veins.
5	VEIN OR LODE	Medium	Si0₂, Fe, Ca, Mg, Mn, S	Au, Ag, Cu, Pb, Zn, As, Sb, Te,	Quartz, pyrite, arsenopyrite, carbonates, galena, sphalerite, chalcopyrite, gold, (molybdenite)	Galena, sphalerite, tetrahedrite-tennantite, (acanthite), sulphosalts, (scheelite), (wolframite), tellurides, gold	Sphalerite commonly increases with depth.
6			Si0₂, Fe, Ca, Mg, Mn, S	Au, Ag, Cu, As, Sb	Quartz, pyrite, arsenopyrite, carbonates, gold, (molybdenite)	Chalcopyrite, tetrahedrite-tennantite, enargite; uraninite (pitchblende); scheelite, wolframite, tellurides, gold, (tourmaline)	Merges into zone 7
7			Si0₂, Fe, Ca, Mg, Mn, S	Au, Ag, As, Te, Cu, W	Quartz, pyrite, arsenopyrite, carbonates, feldspar, gold, (molybdenite)	Arsenopyrite, chalcopyrite, scheelite, wolframite, tourmaline, tellurides, gold	Merges into zone 8
8		High	Si0₂, Fe, Ca, Mg, Mn, S	Au, Ag, Cu, W	Quartz, pyrite, pyrrhotite, feldspar, carbonates, tourmaline, gold, (molybdenite)	Pyrite, pyrrhotite, chalcopyrite, arsenopyrite, scheelite, wolframite, gold, tourmaline	
9			Si0₂, Fe, S	W, Sn, Mo, Bi, B, F, As	Quartz, feldspar, pyrite, pyrrhotite, molybdenite	Pyrite, pyrrhotite, bismuth, molybdenite, arsenopyrite, bismuthinite, wolframite, scheelite, cassiterite, tourmaline, topaz	Generally barren of gold and silver

GSC

Figure 95. Idealized vertical endogenic zonation of elements and minerals in epigenetic gold deposits. Minerals and elements shown include only those in veins, lodes, etc.; minerals and elements in wall-rock alteration zones are excluded.

affected (neutralized), and the processes of ore formation more radically changed by reaction with limestones than they are with schists and shales. With respect to gold deposits there is also a marked effect on the Au/Ag ratio exerted by acidic and basic wall rocks as mentioned in a previous section in this chapter.

Probably all of the factors mentioned above are involved to some extent in zoning, but we shall not pursue any of them in detail here. General reviews of the various suggested mechanisms of zoning can be found in the papers by Lindgren (1927), Park (1955), Barnes (1962) and Petrovskaya *et al.* (1974) and in the volumes edited by Kutina and Stemprok (1963–1965) and Sokolov (1974). Finally, to conclude this review of the literature, it is of interest to note that Ewers and Keays (1977) observed a crude metalliferous zoning in present day auriferous hydrothermal systems in the Broadlands geothermal field, New Zealand. They noted that the elements As, Sb, Au and Tl are enriched in the near surface parts of the system whereas Ag, Se, Te, Bi, Pb, Zn, Cu and Co are concentrated mainly at depth. They concluded that the deposition of Tl occurs primarily in response to decreasing solution temperatures, while the other elements are precipitated through the combined effects of decreasing temperatures and boiling of the hydrothermal solutions within permeable horizons. Oxidation reactions and changes in pH resulting from solution/wall-rock interactions were considered unlikely to have influenced deposition to any great extent. For those elements believed to be transported as thio complexes (i.e., As, Au and Sb), the decrease in pressure as solutions are discharged at the surface and resulting losses of H_2S through boiling reduced the stabilities of these complexes and contributed to the surface enrichment of As, Au and Sb.

In considerations of zoning of mineral deposits both solution and diffusion effects must be taken into account. Where the systems are relatively simple and mineral deposition has apparently taken place at low temperatures from solutions that permeate faults and fractures, it appears that the paragenetic (deposition and replacement) sequence can be related in a crude way to the solubilities (solubility products) of carbonates, sulphates and sulphides as shown by the writer (Boyle, 1972) for the barite-sulphide mineralization at Walton, Nova Scotia. When gold deposits are considered, however, there are many more factors involved in paragenesis and zoning than solubilities in aqueous solutions. In auriferous deposits the role of diffusion of constituents is important as are also extensive wall-rock alteration reactions and remobilization processes involving gold initially deposited in sulphides subjected to post-depositional structural deformation (crushing, shearing and recrystallization). Each of these complex problems are dealt with in turn.

The migration of ore and gangue constituents in any form is the result of differences in chemical potential (μ) of a constituent (component) between its source and site of deposition. For the migration of a constituent to take place in thermodynamic terms the following condition must obtain:

$$\mu \text{ (deposition site)} < \mu \text{ (source)}$$

and at equilibrium when no migration takes place:

$$\mu \text{ (deposition site)} = \mu \text{ (source)}$$

The first is the fundamental thermodynamic law governing the flow of constituents (components) by diffusion, gaseous flow or other modes of transport. If we could evaluate the chemical potentials of ore and gangue constituents there would be no difficulty in explaining the zonal and paragenetic features of mineral deposits. Unfortunately, at our state of knowledge, this cannot be done, and a moment's reflection on the thermodynamic principles of chemical potentials and endogenic geological phenomena will indicate why.

Simply stated the chemical potential of a constituent (component), μ, is a function of the concentration (activity), C, of a constituent, the volume of the constituent and hence the external pressure on the system, P, and the temperature, T. This is usually stated as:

$$\mu = f(C, P, T)$$

In any endogenic system it is sometimes possible to estimate the temperature between source and site of deposition; the pressure, likewise, may be estimated from the load of superincumbent rocks, or differences in pressure can be estimated from the fact that the pressure commonly approaches zero in a dilatant zone, whereas outside the zone it is equivalent to the load of superincumbent rocks, neglecting the effect due to compressive forces in tectonic zones. The latter can cause very large errors in the pressure estimates. Estimation of the concentration factor in endogenic processes is only rarely possible mainly because we do not generally know the form in which the constituent migrated. For instance the particular element in question may have migrated as a diffusing ion, in a gaseous form, or as some soluble species such as a sulphide complex, carbonate complex, etc. For each of these forms the concentration factor in the chemical potential is different. In addition the form of migration may change, being a diffusive ion at great depths and a simple or complex soluble species as the element on its upward migration enters near-surface fractures filled with water. All of these imponderables make any estimation of the chemical potential of a constituent virtually impossible at our state of knowledge. If metamorphic or granitization processes are responsible for the formation of epigenetic gold deposits as many now suppose, there are yet other energy factors for consideration. One of these, the most important, is the energy required to mobilize an element or its compounds from a bound state in the rocks. If the element or its compounds are mobilized from the lattices of minerals, much more energy is generally required than that for mobilization of the same element or its compounds from grain boundaries. In addition it is also axiomatic that each of the elements and their compounds have specific mobilization energies from their bound states.

To summarize briefly we see that the mobilization of an element or its compounds from rocks, its migration and its precipitation in a new site such as a fracture is a most complicated process in an energy context. Not only differential mobilization energies prevail for the different elements, but a whole array of factors affect the chemical potentials. We cannot yet put numbers on the mobilization energies nor on the chemical potentials in endogenic processes. Until we can we shall not understand mineral zonation and paragenesis in a quantitative sense.

Nevertheless, it is possible to assign a qualitative index to the migration capacity of an element from our knowledge of its geochemistry. Thus, certain elements possess high migration energies, whereas others have low migration energies. The former migrate reluctantly under conditions of ore formation, or stated another way they require a large amount of thermal, solution or other energy to maintain them in a mobile state. They have low migration capacity indexes. Other elements migrate readily and have high migration capacity indexes, and elements with all grades of intermediate migration capacity lie between these extremes. An example of an element (compound) that has a high migration capacity index is silicon (silica). It is found in practically all mineral deposits from the highest temperature to the lowest and occurs in deposits in all grades of metamorphism. An element with a low migration capacity index is titanium, its minerals being generally concentrated only under relatively high temperature conditions. Zirconium has a similar tendency. It should be noted, however, that these general tendencies are not rigid; there are for instance situations where zirconium, as a complex, is relatively mobile in carbonated waters.

Based on their thermodynamic properties, diffusivity, binding energies in lattices, ability to form soluble complexes, solution properties under different Eh/pH conditions and ability to form colloids one can arrange the common elements in groups having migration capacities as follows:

Gangue elements:

$Si > B > K > Na > Ca = Mn > Mg > Co > Fe = Ni > Al$

Gaseous compounds:

$CO_2 = S = Se > As > Sb > Te$

Base metals:

$Hg > Zn > Cd > Cu > Pb > Mo = Bi = Sn = W$

Precious metals:

$Ag > Au > Pt$

This sequence is somewhat similar to that proposed by Udodov and Shvartsev (1965) for the coefficients of migration of certain trace elements in natural subsurface waters. These authors do not include silicon or the various gaseous elements in their table, but they indicate that gold has the highest coefficient of migration, i.e., the metal is the most mobile of the elements considered.

The migration capacity index of gold (and silver) in epigenetic processes is essentially the same as silicon (silica) as judged by the universal occurrence of both precious metals with quartz. Similarly gold has about the same migration capacity index as sulphur, arsenic and antimony. With respect to the base metals it would seem to fall between mercury and zinc. The reason for the relatively high migration capacity of gold during endogenic processes would seem to be partly due to the fact that in rocks gold is not usually in isomorphic replacement sites in mineral lattices, but rather the metal is present in microfractures and other discontinuities in minerals and rocks from which it is readily mobilized. That this is so is proven by the fact that gold is relatively easily leached from rocks and minerals as Davletov *et al.* (1973) have shown. This feature is also indicated by the long known fact that practically every quartz vein and stringer in zones of medium to high grade metamorphism contain small but detectable amounts of gold. There is also a considerable array of experimental data (*see* the section on gold in natural waters in Chapter II), which shows that gold is a highly mobile element under a great variety of chemical conditions.

The high migration capacity of gold has interesting consequences in the reworking, recrystallization or metamorphism of epigenetic deposits following the initial deposition of gold with early formed pyrite and arsenopyrite. The problem of the various generations of gold in deposits was analyzed by the writer many years ago (Boyle, 1961*a*). Briefly, the situation at Yellowknife, as well as in many other gold deposits in Precambrian greenstone belts, is that there are at least two ages of gold. At Yellowknife there are some four ages of gold and silver. These are as follows, the oldest first:

1. Early gold and silver contained in pyrite and arsenopyrite in the alteration zones and in quartz lenses. Much of this gold and silver is present in pyrite and arsenopyrite in lattice sites or in an extremely finely divided submicroscopic form. To free these two elements, roasting followed by cyanidation is commonly necessary.

2. Native gold containing silver; gold and silver in solid solution in sulphosalts in fractures and crushed zones in the quartz lenses and their alteration zones. Four types of native gold are present. One variety possesses a light yellow colour and is generally rich in silver. A more abundant type has a rich golden colour and occurs as blebs, sprigs and irregular masses in fractures in quartz and in granulated pyrite and arsenopyrite; also in sulphosalts, stibnite, chalcopyrite, etc. It is often accompanied by aurostibite. A third variety, of limited occurrence, is a reddish coloured gold that is often associated with chalcopyrite. The fourth variety occurs on slickensided surfaces of chlorite and sericite schist as thin films and gives the appearance of being painted on the surfaces.

3. Native gold containing silver; gold and silver in solid solution in freibergite, bournonite and other sulphosalts, and chalcopyrite. The native gold occurs in calcite and ankerite filling vugs and crushed parts in crosscutting quartz-carbonate veinlets. It is generally accompanied by the sulphides and sulphosalts mentioned above.

4. Native gold containing silver in late fractures cutting the orebodies. This gold is generally associated with pyrite and scalenohedral calcite and occurs as splendent leaf gold, spongiform gold and distorted octahedral crystals. In one case in the Giant System the leaf and sponge gold were observed in an ice lens (Boyle, 1951).

The presence of successive ages of gold and silver is usually explained by separate introductions from magmatic solutions. Just why this should take place has, however, not been satisfactorily answered. It would seem that a more plausible hypothesis, having regard to the geochemical facts at our disposal, is that the succeeding ages of gold and silver are simply due to the redistribution of the two elements orginally in the first generation of sulphides. The process would probably be as follows: gold and silver together with other chalcophile elements, sulphur, arsenic and antimony, migrated into and through the schist of the shear zones. When they found a site of marked dilatancy, early high-temperature pyrite and arsenopyrite were formed, which incorporated the available gold and silver in their lattices. As the deposit cooled and successive structural adjustments took place with crushing and partial recrystallization of the early sulphides, some gold and silver were exsolved together with some sulphur, antimony and other chalcophile elements. These migrated into nearby fractures where the secondary sulphides were precipitated together with native gold. At this stage aurostibite ($AuSb_2$) was probably formed by reaction of gold and antimony, and much gold and silver were incorporated in the various sulphosalts which formed at this time. The third age of gold and silver in the crosscutting quartz-carbonate veins probably had a similar origin. The fourth age of gold may owe its origin in part to supergene processes and was precipitated from underground waters in late fractures and faults.

Others have noted the phenomenon of remobilization of gold after its initial deposition in deposits. Bichan (1947) comments on the fact that gold has rarely remained at the original point of precipitation and thinks that the distinctive concentration of the metal into shoots, separated by zones of relatively barren vein material, is the result of lateral or vertical migration of gold from zones of extreme pressure or compression into nearby zones of tensional dilation. Moiseenko (1965) and Moiseenko *et al.* (1971) in studies of the gold deposits of the Pacific Belt, U.S.S.R. and particularly of the metamorphism of the Priamur gold deposits, noted that the fineness of gold increased and the form and internal structure of the native metal were changed from that initially deposited. Experimentally they noted that when gold-bearing ores were heated the small gold particles recrystallized and formed larger aggregates of greater fineness than the original particles. During the heating Au is more inert than Hg, As, Sb, Zn and Ag; the latter elements partly migrated out of the gold particles. At 700°C in the system Au-Ag the rate of diffusion of Ag was found to be 2.5 times greater than that for Au. Numerous details on the diffusivity of gold are given in the second paper. Petrovskaya (1970) noted redeposition of both quartz and gold during plastic deformation of early quartz veins of the Yenisei Ridge, U.S.S.R. And finally, Sakharova *et al.* (1972) heated polished specimens of auriferous minerals including pyrite, pyrrhotite, chalcopyrite and arsenopyrite from the Darasun deposit, U.S.S.R. This deposit is noted for its invisible and microscopic gold in the sulphides. During the heating tests some of the fine dispersed gold grew into larger particles, and some was completely dispersed. The migration of the gold began in the range 250 to 300°C. The investigators concluded from their experiments that initially deposited gold may undergo redistribution over a long time interval down to temperatures as low as 150°C.

It should be emphasized that not all auriferous deposits have multiple ages of gold due to the reworking or the regeneration processes described above. In many deposits it is evident that the different generations of gold are due to separate introductions of the metal during the mineralization process, each generation being keyed to dilation (fracturing) of the deposits. Some of the deposits in sediments (saddle reefs) and many of those formed under near surface conditions during the Tertiary appear to fall into this category.

Finally, if we accept the migration capacity index concept during endogenic mineralization it is relatively easy to construct a model of the regional zonation and many other features of the separation of elements during the mineralization history of a simple area. Thus, near granites or granitization centres only those elements such as tin and tungsten with low migration capacities (high mobilization energies) can be mobilized in the enclosing host rocks and concentrated in available dilatant structures. Elements with high migration capacities (low migration energies) will be set in motion in these environments, but they cannot be precipitated; hence, they will migrate into cooler zones where they are subsequently precipitated in available dilatant structural sites or chemical traps. Similarly those elements with low migration capacity indexes are precipitated first in the paragenetic sequence, followed successively by those with increasingly higher migration capacity indexes.

Where the situation is complex with a number of pulsations or epochs of mineralization the model is necessarily more involved. The situation in the Keno Hill area of Yukon, Canada will serve as an example (Boyle, 1965*a*). There, the country rocks are mainly folded and faulted phyllites, schists and quartzites, and the first phase of mineralization is marked by early barren quartz veins and boudins; the second by scheelite-bearing skarn bodies, a cassiterite replacement lode, and small pegmatites and coarse-grained quartz veins and lenses carrying scheelite and wolframite; the third by quartz-pyrite-arsenopyrite-sulphosalt-gold veins in extensive faults; and the last by one or more generations of siderite-quartz-galena-sphalerite-freibergite veins in extensive faults. All stages are clearly marked in time being punctuated by tectonic events resulting in a variety of suitable dilatant zones. Gold only appears in the third stage; the last stage is phenomenally rich in silver. With respect to thermal centres manifested in the area by intrusive granitic bodies the following zonation is evident. The first stage quartz bodies occur throughout the area and show no zoning relationship to the intrusives; the tin-tungsten bodies are localized in or near the granitic stocks; the quartz-pyrite-arsenopyrite-sulphosalt-gold veins are widespread but mainly without the intrusives; and the siderite-galena-sphalerite-freibergite lodes lie well outside the intrusives, generally midway between two major granitic stocks.

The mode of formation of these various ages of deposits at Keno Hill has been explained by the writer as follows: During the initial period of folding and regional low-grade metamorphism numerous small dilatant zones were formed in the country rocks, and these were filled with the more mobile constituents in the country rocks (e.g., those elements and compounds with high migration capacity indexes, mainly SiO_2, CO_2, Ca, etc.) to form the myriad stringers and lenses of quartz containing a few carbonates in places. During the main period of regional metamorphism and granitic intrusion, water and carbon dioxide were mobilized as a result of the rise of the geotherms through the sedimentary pile. Mobilization of SiO_2, S, As, Sb, Zn, Pb, Ag, Au, etc., was thereby facilitated, but little migration of any of these constituents, including H_2O and CO_2, took place until the vein faults were formed. Once this occurred the more mobile constituents, viz., SiO_2 and some S, As, Sb, Fe and Au, were drawn into the vein faults and precipitated in the more dilatant parts to form the quartz-pyrite-arsenopyrite-gold lodes. Following this period of mineralization, increased temperatures at depth coupled with renewed tectonic activity and the formation of more extensive dilatant zones brought larger quantities of CO_2 and S into the diffusion system. The CO_2 strongly depressed the mobility of silica, as mentioned in a previous section in Chapter II, but at the same time this constituent, together with sulphur, greatly enhanced the mobility of Fe, Mn, Mg and the chalcophile elements, Zn, Pb, Ag, etc. All these constituents were drawn into the extensive dilatant parts of the vein faults to form the siderite-galena-sphalerite-freibergite lodes.

The formation of the various types of deposits through a long time interval has depended, therefore, on the timing of regional metamorphism, granitic intrusion and tectonic activity and ultimately on the migration capacity index of the various elements involved in ore formation.

General synthesis of the metamorphic secretion-dilation theory of the formation of epigenetic auriferous deposits

The generalized mechanisms implicit in the metamorphic secretion-dilation theory of the formation of auriferous deposits have been outlined in the various sections above. It now remains only to summarize these mechanisms and amplify them with a few details.

Three principal types of epigenetic auriferous deposits are known – deposits in skarn, deposits in fractures and faults exhibiting little if any wall-rock alteration and deposits marked by extensive alteration zones. The first is readily explained on the 'chemical trap' principle. As diffusing ions (or solutions) of the metals, set in motion from the country rocks by granitization or by the introduction of granitic bodies, migrate outwards from the thermal centres and reach chemically contrasting rocks such as limestones or calcareous shales reactions occur that result in the precipitation of silica, sulphides and the metals. Later dilatant structures developed in the skarns may draw in further increments of the migrating elements from the country rocks or from extensive skarnified rocks, thus forming local enriched shoots. This is the only part played by dilatancy in the formation of skarn orebodies.

Deposits, mainly auriferous quartz bodies without extensive alteration zones, formed in fractures or in openings on the crests of folds (saddle reefs) are due mainly to the effects of dilation of the structures. Thus, during the metamorphism and granitization of shale-greywacke sequences, silica, the metals, sulphur, arsenic and antimony are mobilized and migrate essentially by diffusion into available dilatant zones where the various elements are partitioned into the minerals we now find in the deposits. Silica, being one of the most mobile of thc constituents, and being amply available in the shale and greywacke rocks, fills most of the dilatant zones as quartz, which generally contains some gold.

Where alteration is marked the chemistry of the formation of gold deposits is more complex. In Precambrian and younger volcanic belts where the gold-quartz deposits are often associated with great linear carbonated zones and have well-marked alteration zones the origin of the deposits can best be illustrated by a short historical synthesis as shown in Figures 91 and 96.

Following extensive outpouring of marine lavas and intermittent sedimentation most of the belts were folded, metamorphosed and partly granitized. During these orogenic processes, marked temperature gradients existed outward from the foci of granitization leading to the formation of the zone facies of metamorphism in greenstone belts and their associated sedimentary rocks. At this time much water, carbon dioxide, sulphur and probably some of the chalcophile elements were mobilized and migrated toward the cooler parts, that is, down the temperature gradient at right angles to the isothermal lines as shown by the vectors in Figure 91*a*. If there had been no structural breaks in the rocks the result would have been a simple metamorphic halo about a granitized centre.

However, in many greenstone belts a great system of shear zones was formed near the end of the orogeny and formation of the granitic bodies. As a consequence the migration vectors were radically changed by the dilatancy of the shear zones as shown by Figure 91*b*. The shear zones literally sucked the mobile carbon dioxide, water, sulphur and other elements from the country rocks and channelled them toward the surface. Because of the great extent and depth of the shear zones the effects of dilatancy were undoubtedly felt over great volumes of the country rocks, and mobile elements were probably contributed from points thousands of feet horizontally and vertically from the shear-zone systems. Extensive structures, which impose such marked dilatant effects on the country rocks, may be termed 'first-degree' dilatant zones (Fig. 96).

During the formation of the large shear zones, carbon dioxide, water and sulphur reached a high concentration along them. As a consequence the chemical equilibrium was strongly displaced, and chloritization, carbonatization, pyritization, etc. of enormous tonnages of volcanic rock took place. This led to the liberation of silica, potassium, calcium, iron, etc. in addition to gold, silver and other metallic elements present in the rock affected by alteration. These liberated compounds and elements, together with those added to the shear zones by diffusion from the country rocks, migrated laterally and vertically to 'second-degree' low-pressure dilatant zones at shear-zone junctions and other structural locales where contorted zones or openings were formed. In these sites, secondary reactions promoted by the low-pressure and lower-chemical potential, resulted in the precipitation of quartz, carbonates and gold- and silver-bearing pyrite and arsenopyrite, forming quartz-sulphide lenses with adjacent alteration haloes. The details of the formation of auriferous quartz lenses and veins in the second degree dilatant zones (Fig. 96) are complex but are of great importance and merit some description.

As noted above, the shear-zones when first formed constituted first-degree dilatant zones into which the volatiles were drawn from the surrounding rocks. Once in the structures

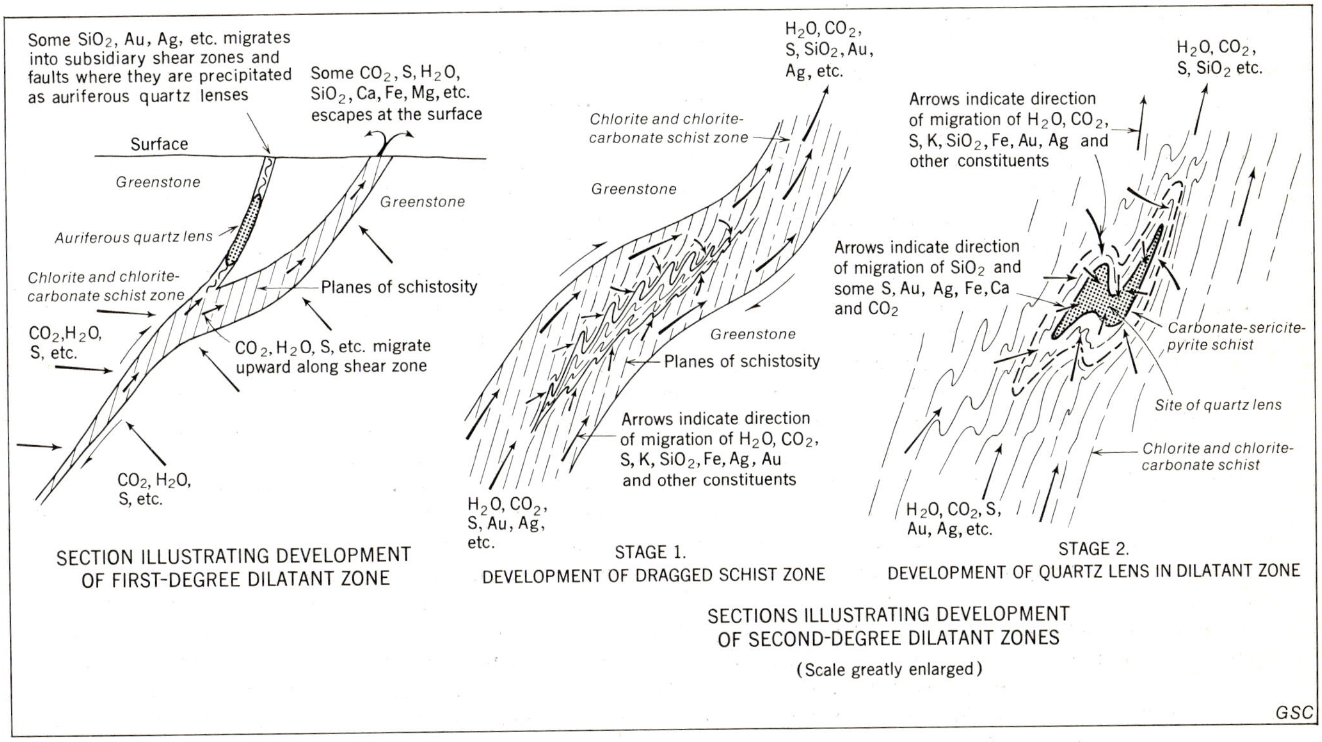

Figure 96. Diagrams illustrating development of first- and second-degree dilatant zones, alteration zones and auriferous quartz lenses.

the volatiles would tend to migrate upward toward the surface that would be the region of lowest pressure, and hence there would be a constant flux of volatiles migrating through the shear zones toward the surface. During this process, the chemical equilibrium was severely displaced, and carbonization and hydration of the silicates took place transforming enormous volumes of rock and releasing large amounts of silica as well as much gold, silver and base metals. The gold together with the silver, base metals, sulphur and arsenic migrated upward and may have been partly dispersed at the surface. This interaction of both chemical and structural factors produced the extensive chlorite and chlorite-carbonate schist zones containing only scattered small quartz lenses. The great linear carbonated zones in many greenstone belts are manifestations of this process. These are frequently either low in gold or leached of some of their gold and other elements. The writer estimates that in some of these carbonated zones at least half of the gold in the rock has been mobilized and moved either into the second-degree dilatant zones noted below or into fractures and faults that are subsidiary to the great carbonated zones.

During the course of the formation of some of the first-degree shear zones as outlined above, particular parts were subjected to intense contortion and dragging at structural sites such as shear zone junctions and flexures. These severely contorted sites marked second-degree dilatant zones further disturbing the chemical equilibrium within the shear zones. The migrating volatiles together with released potash and silica from the chlorite zones streamed toward these sites where the following reactions took place.

(Ca,Mg,Fe silicates) + CO_2 = ankerite + silica

(Ca,Mg,Fe silicates) + S = pyrite + silica + Ca + Mg + other constituents

(Ca,Mg,Fe silicates) + S + As = arsenopyrite + silica + Ca + Mg + other constituents

Chlorite + K^+ = sericite + other constituents

As the volatiles were delivered to the contorted sites, dilatancy (volume increase) must have proceeded concomitantly, otherwise the reactions would not have run from left to right, because one of the principal products, silica, must be withdrawn. Once dilatancy ceased the reactions stopped, and equilibrium was restored. The net result, therefore, was the formation of quartz lenses in the more dilatant parts surrounded by desilicated carbonate-sericite-pyrite-arsenopyrite zones grading into desilicated chlorite-carbonate zones.

The formation of the quartz lenses and their alteration zones, as suggested above, is actually the operation of Reicke's principle on a grand scale whereby various mobile constituents are transferred from zones of directed high pressure to those of low pressures. The thermodynamics of the process are complex and not given to a simple exposition here. It is sufficient to state simply that the mineralogical changes in the shear zones merely reflect the chemical adjustments attendant upon the chemical potential differences induced by dilatancy. The transfer of the various constituents was essentially down the pressure gradient, and the migration vectors have in general been oriented toward the low-pressure zones.

The state in which the mobile constituents were transferred through the shear zones can only be inferred from our present inadequate knowledge of the physical chemistry of diffusion of matter in solids. It is probable that most of the constituents were transferred in a gaseous or ionized state because many of the constituents such as water, carbon dioxide, sulphur and arsenic pass readily into the gaseous state. Furthermore it is well known that gases at high temperatures and pressures diffuse more readily through solids than do liquids. Another important consideration is that silica and other oxides have a marked solubility in gaseous water at high temperatures as shown by numerous investigations.

Water was undoubtedly the main catalyst in the processes whereby numerous constituents were transferred through both the country rocks and shear zones. Under the deep-seated conditions of formation of many of the gold-quartz lenses in greenstone belts the state of the water was probably gaseous, as noted above, and it may have been partly ionized as H^+, H_3O^+ and OH^-. Carbon dioxide probably existed as the ionized components of carbonic acid, H^+, HCO_3^- and CO_3^{2-}, and sulphur was probably present as H_2S or the ionized components of this gas, H^+ and HS^-. Silica was probably hydrated as $Si(OH)_4 \cdot nH_2O$ but may also have migrated as hydrated sodium and potassium complexes. Numerous other constituents were probably transferred as hydrated complexes.

There seems to be little justification for assuming that the migration of the constituents through the shear zones required the movement of large volumes of water. Ions, especially when hydrated, diffuse readily through gaseous water at high temperatures. This mechanism, therefore, provides an adequate means of transport, and in the writer's opinion the migration of the elements took place through a nearly stationary flux of gaseous water that permeated both the country rocks and shear zones. The density of the hydrous flux was probably controlled in part by the thermal gradient and also by the dilatancy of the structures. Such a mechanism of transport is actually analogous to the diffusion of ions in liquid water, a phenomenon well known to chemists and shown to be applicable to mineralization processes many years ago by Duffell (1937).

The relative rates of migration of the various constituents through shear and schist zones is a problem that requires much experimental data for an adequate solution. The main factors bearing on the problem can, however, be stated and an attempt can be made at interpreting some of the chemical features of the shear and schist zones.

Shearing was undoubtedly one of the most important factors. It is well known that shearing strains mineral lattices and promotes diffusion (Mott, 1955). In addition, extreme contortion also comminutes the minerals increasing chemical reaction. Furthermore, the effects of directed pressure, such as shearing, on the mobility of the elements, must also be considered. Bridgman (1931) has shown that all atoms can be compressed at high pressures, and that the compressibilities of the alkali metals are greater than those of other elements. It can also be shown from lattice theory that the greater the compressibility of an element the weaker is its bonding energy. It follows, therefore, that under great shearing stress, the more compressible an element the easier it will be detached from its lattice. As Bugge (1945) has suggested, this phenomena is a plausible explanation for the relatively high mobility of the alkalis compared with calcium, magnesium, etc., in zones of intense shearing. In the shear zones, excepting the volatiles, soda and potash commonly exhibit the largest percentage gains and losses, attesting to their higher mobility compared with other constituents. (*See for example* Tables 46, 47 and 48.) This higher mobility can probably be attributed in part to the action of shearing in view of the theoretical concepts mentioned above.

When water is present in the system, consideration must also be given to the effect of the ionic potential[30] on the migration of the elements. Elements such as potassium, sodium, calcium and magnesium have a relatively low ionic potential and form easily soluble cations that are not readily precipitated by changes in pH. Elements like iron and aluminum have an intermediate ionic potential, are readily hydrolyzed, and their solubilities are largely pH dependent. Silicon, sulphur, phosphorus, etc., with high electrostatic charges and high ionic potentials, form oxy-acid anions whose solubilities may depend in part on the pH of the environment.

As noted above, sodium and potassium have low ionic potentials, and their cations are readily soluble. Transfer of these elements through a pervasive water gas would, therefore, be relatively easy and hence their mobility in the shear zones was high. Calcium and magnesium, on the other hand, have higher ionic potentials and their mobility was lower as shown rather conclusively by their smaller percentage gains and losses in many shear zones as compared to sodium and potassium (Tables 46, 47 and 48).

Silica forms complex ions whose solubilities vary only slightly with pH near the neutral point; at high and low pH's their solubilities increase. Referring to the tables just mentioned one sees that the percentage gains and losses for silica in many shear zones are not high suggesting that the pH of the environment was close to the neutral point. For alumina the percentage gains and losses are relatively low indicating a low mobility for this constituent. Aluminum is an amphoteric element with a relatively high ionic potential, and alumina has solubility maxima at pH 4 and 10. The iron oxides show intermediate values in their percentage gains and losses indicating a moderate mobility. This means that the pH of the environment during the migration of iron must have been on the acid side because the mobility of iron above pH 7 is relatively low. Correlating these observations on silica, alumina and iron, it seems logical to conclude that a slightly acidic to neutral condition prevailed during the main stages of the alteration processes.

The effects of carbon dioxide, sulphur and arsenic have not been considered. Their presence has without question modified the course of the reactions and the mobility of many elements. It would be expected that carbon dioxide would have increased the acidity and hence promoted the migration of Ca, Fe, Mg and Mn and perhaps K and Na. The presence of sulphur and arsenic, especially in a gaseous hydrated form, would certainly have promoted the migration of many of the trace elements such as Cu, Zn, Au, Ag, etc. As the volatile constituents were used up, the pH of the environment probably became alkaline permitting a greater migration of silica and perhaps alumina.

[30]The ionic potential is the charge of an ion divided by its radius and is a measure of the behaviour of an ion towards water.

It is apparent, therefore, that the mobility and transfer of the elements in shear zones in greenstone belts has depended upon several interacting factors. Chief among these was shearing which comminuted the rock, permitting extensive chemical reaction to proceed. It also promoted diffusion, and because of the intrinsic feature of compressibility certain elements such as Na and K were more readily detached from their constituent minerals than others. Once free the elements migrated in accordance with their ionic potentials. Those with low ionic potentials migrated freely; those with high potentials less so. The pH of the environment was critical for the migration of some elements especially aluminum and iron. The influence of carbon dioxide, sulphur and arsenic was marked not only in promoting the migration of Ca, Mg, Fe, etc., but also the metallic trace elements and finally in fixing many of the elements in the carbonate and sulphide minerals of the shear zones. The schist of the shear zones and the early quartz lenses were thus formed essentially contemporaneously as the result of intense shearing, which triggered and promoted the chemical effects characteristic of many of the auriferous deposits in greenstone belts.

During cooling of the deposits and further structural adjustments within the shear zones, much gold, silver, antimony and other elements were exsolved from the early sulphides and found their way into local dilatant zones such as fractures in quartz lenses and local tension fractures and small faults in the shear zones. Some silica and the components of the late carbonates did likewise. This process gave rise to the successive later generations of native gold, sulphosalts, pyrite, quartz and carbonates in the shear zones of most greenstone belts.

In deposits of all ages the sequence of events and the details of the geochemistry of the formation of the auriferous deposits were probably much the same as those described above. Certain deposits in rocks overlying basement complexes, such as many of those of Tertiary age as well as some of those of older vintage, may have had a slightly different history. Many of these deposits seem to have derived some or all of their constituents from the rocks of the basement complexes, particularly from pyritiferous graphitic shales, greywacke and quartzites that underlie many of the auriferous belts. The mechanism of concentration of the quartz, the gold and silver, and the various other gangue and base metal minerals differs only in detail from what has been described above. In some cases the quartz, and indeed some of the gold and other metals, may have originated mainly from the pervasive propylitization of andesites and other rocks. In other cases the thermal wave associated with deep-seated volcanism probably mobilized the vein constituents in the basement rocks, and these were drawn into the extensive dilatant fractures and faults formed in the superincumbent andesites, breccias, dacites, etc. A schematic representation of the probable process has already been illustrated in Figure 92.

In passing it is interesting to note that only certain Tertiary volcanic districts are auriferous despite the fact that basic to acidic volcanics are present and the basement rocks are favourable sediments and/or volcanics. One can contrast the auriferous Tertiary volcanics of Romania and New Zealand and the barren volcanics of Scotland and eastern Australia in this respect. In my experience the difference between productive and nonproductive auriferous Tertiary volcanic districts lies in the phenomenon of propylitization. In Romania, New Zealand and in other productive Tertiary auriferous belts propylitization is extensive; in Scotland, Ireland, eastern Australia and elsewhere the Tertiary rocks are relatively fresh, and there are no gold deposits despite the fact that the strata are commonly highly faulted. Whatever has produced the extensive propylitization has also produced the gold deposits. More research on this phenomena, why it is present in some areas and absent in others, is obviously needed.

Origin of the auriferous and uraniferous quartz-pebble conglomerates

These deposits, the largest gold producers in the world, have had a controversial history as regards their origin since they were first discovered in the last part of the 19th century. As noted in the section describing the deposits in detail, two principal ideas have been suggested for their origin – a placer or modified placer origin for the gold and uranium, and a hydrothermal origin for the two metals. The pros and cons regarding each hypothesis are discussed in detail under the description of each deposit or group of deposits and need not be repeated. In the following the writer will briefly review some of the salient geological and geochemical characteristics of these deposits and inject his own thoughts into a final synthesis of how he thinks the auriferous and uraniferous quartz-pebble conglomerates originated.

The rocks in which the deposits occur are clastic sediments, principally conglomerates and quartzites, laid down in great subsiding basins on basement rocks of varied types. The stratigraphic sequence in some basins contains volcanics, mainly the basic variety. Conglomerates and quartzites in basins overlying terranes marked by the presence of auriferous deposits in greenstone and associated sedimentary belts that are invaded by granitic rocks are generally gold bearing (Rand, Tarkwa) and some are also uraniferous (Rand). Those overlying terranes marked mainly by granitic rocks and gneisses are not generally auriferous but are commonly enriched in uranium, thorium and rare-earths (Blind River–Elliot Lake). From these observations we can conclude that the provenance of the sedimentary materials is an important parameter in assigning a source to the gold, uranium, rare-earths, etc.

The broad geologic setting of the various basins is of interest. Considering the Rand and the Blind River–Elliot Lake deposits it seems probable that both lie in basins flanked by extensive faults that probably mark great graben structures. Volcanism is evident in both basins, although the volcanics are not present in the immediate Blind River–Elliot Lake area. Great basic intrusives are present somewhat removed from the basins (Bushveld complex and Sudbury eruptive), but within the general geological framework. In the Rand, dykes of Bushveld age intrude the rocks of the basin, and in the Blind River–Elliot Lake basin dykes of Sudbury eruptive age, likewise, intrude the basin. And finally, there is a distinctive type of mineralization related in some manner or other to these basic intrusives, which in both South Africa and Canada includes gold veins, copper deposits of several types and nickel-cobalt-platinoid mineralization. Broadly speaking these features are also evident in the vicinity of the Tarkwa and Jacobina basins, although the late stage of mineralization associated with the basic bodies is weak or absent. How these various features are to be integrated metallogenetically cannot at present be deciphered, but they should be kept in mind by future researchers dealing with the origin of these deposits. It is of interest to note the similarity in the setting of the auriferous conglomerates and the copper conglomerates and quartzites in the Keweenaw Peninsula, Michigan. Jolly (1974) suggested that the copper in the conglomerates was derived by metamorphism (dehydration) of underlying volcanics rather than derivation of the copper by magmatic hydrothermal solutions as is generally supposed. In other words the copper was redistributed within the sedimentary-volcanic pile as a result of diagenesis and metamorphism – exactly the same process that has been suggested by many for the postplacer history of the gold in the Rand and other similar deposits.

The mineral composition of the quartz-pebble conglomerate deposits is variable. The principal iron minerals are pyrite and iron oxides, both being apparently mutually exclusive in individual deposits. Deposits that contain abundant pyrite are generally uraniferous, whereas those that contain only the iron oxides do not seem to be uraniferous. The occurrence of gold is independent of the presence of pyrite or iron oxides, the precious metal occurring in both types of deposits (Rand, Tarkwa) in economic amounts.

Some of the pyrite in the quartz-pebble conglomerates is detrital, but the largest amount is authigenic. The iron oxides in the deposits appear to be partly detrital and partly authigenic. The sulphur in the pyrite of the various deposits, so far as is now known, does not show the marked fractionation effects usually attributed to biogenic (bacterial) activity, although some of the authigenic pyrite in the Rand exhibits a slight shift towards enrichment in ^{32}S that may be due to bacterial transformations.

Deposits that contain abundant pyrite generally also have small amounts of a number of other base metal sulphides, particularly pyrrhotite, sphalerite, chalcopyrite and galena. Nearly all of these sulphides are authigenic minerals. It is of interest to note that many conglomerates and quartzites at or near unconformities are frequently enriched in authigenic base metal sulphides; few of these are even slightly auriferous, uraniferous or thoriferous. Here, we may mention the Laisvall deposits in Sweden; the Maubach deposits near Aachen, West Germany; the Silver Mine deposits near Sydney, Nova Scotia; and the disseminated base metal mineralization in the Cobalt quartzites at Cobalt, Ontario. The source of the constituents in the base metal sulphides in these deposits, as in the pyritiferous auriferous quartz-pebble conglomerates, is uncertain.

The gold in the auriferous quartz-pebble conglomerates is invariably very fine grained (~ 0.05 mm) and has a definite authigenic appearance in all deposits. Nuggets of variable size such as occur in Tertiary and Quaternary placers do not occur in the quartz-pebble conglomerates. The fineness of the gold in most of the quartz-pebble conglomerates averages about 925, a relatively low fineness (high silver content) for placer deposits. The fineness is commonly uniform in the deposits, another feature that contrasts with modern placers. As Horwood (1917) pointed out many years ago, the purity of detrital gold is generally in inverse proportion to the size of the grains. Considering the size of the gold in the Rand and other similar deposits one would expect the gold to have a very high fineness if the metal was simply detrital in origin.

The radioactive minerals in the quartz-pebble conglomerates, including uraninite and brannerite, occur in discrete grains, in tiny veinlets and in thucholite as fine disseminations. Most of the radioactive minerals are probably authigenic in the writer's opinion; some may be detrital, but the amount is small. Even the detrital component appears to have been highly modified.

Thucholite and other types of carbonaceous substances are universally present in the Rand deposits but appear to occur only in small amounts in the other deposits. The mineraloid occurs in small globules 'fly speck carbon', in thin columnar seams, in small veinlets and as small isolated patches. The thucholite encloses minute particles of uraninite and small irregular sprigs and particles of gold. Some seams of thucholite in the Rand (Carbon Leader) are exceptionally rich in platinoid minerals, native silver, Ni-Co arsenides and various sulphides. Gold exhibits a particularly close association with the thucholite. Many suggestions have been ascribed for the origin of the thucholite. Some say it is a polymerized hydrocarbon and others that it represents the organic remains of algae, lichenlike plants and so on. The writer is of the opinion that nearly all of the thucholite and other similar solid hydrocarbons are polymerized residues of petroleumlike materials derived in some manner, as yet unknown, from the humification of algae, bacteria and other low forms of life. There is a strong suggestion that these petroleumlike substances are capable of transporting gold, uranium, thorium and various other metals probably in a chelated form. As they set (polymerize) in a gellike mass, hence the ovoid and columnar form of many of these hydrocarbons, the gold, uranium and other components are split out, appearing as minute inclusions or along cracks in the hardened gel. Some of the columnar varieties of thucholite in the Rand appear to be lichen mats as suggested by Hallbauer and van Warmelo (1974), but the ovoid nodules and specks containing gold and uraninite are exactly the same as the thucholite, also with gold and uraninite, in the Richardson skarn gold deposit at Eldorado, Ontario, described by Boyle and Steacy (1973). In the latter, ovoid fungal hyphae or plant debris are ruled out, although the constituents of the thucholite probably derived ultimately from primitive organic matter.

The trace element content of the quartz-pebble conglomerate deposits reflects the abundance of the various allogenic and authogenic minerals. Notable are the relatively high contents of arsenic and mercury in some of the deposits. The signficance of this is uncertain. Both elements are commonly enriched in pyritic sediments, and hence they may be syngenetic in the deposits. On the other hand these elements are characteristic of epigenetic (hydrothermal) gold deposits. The very low contents of tin and tungsten in the deposits are difficult to explain by any hypothesis. If the deposits were originally placers one would have thought that the heavy concentrates would be somewhat enriched in tin and tungsten minerals. Cassiterite is, however, not generally abundant in Precambrian (Archean) terranes, and this may be the reason for its paucity in deposits such as the Rand. Tungsten minerals, especially scheelite, are common in auriferous Precambrian deposits and being heavy and resistant should have collected together with gold in the placers. However, the amount of scheelite in Precambrian gold deposits is low, and dissipation throughout the vast volume of quartz-pebble conglomerates would probably have reduced the tungsten content below the detection limit by ordinary analysis. If one postulates a hydrothermal origin for the deposits, the absence of tungsten seems unusual considering that nearly all other types of Precambrian gold deposits contain detectable amounts of the element, and some are considerably enriched in scheelite. The content of zirconium in some of the reefs of the Rand and Elliot Lake (Table 35) seems low for most placers (average about 100 ppm), but this may be a feature of the provenance of the resistate zirconium minerals, mainly zircon. The presence of platinoids in the Rand is of interest. It is said that some of these elements (e.g., Pt, Pd, Ir, Os) are detrital, but those in the Carbon Leader cannot be detrital in the form in which they now occur. In this reef they are present as sperrylite, braggite, cooperite, and the metallic compounds platiniridium, platinum and osmiridium, associated with pyrrhotite, chalcopyrite and a variety of Ni-Co arsenides and sulphides. Such a mineralization suggests some connection with basic and ultrabasic rocks not necessarily as a result of magmatic differentiation processes but more probably related to metamorphic secretion from basic dykes or volcanics (Ventersdorp system and/or Bushveld intrusives). Whatever the origin of this mineralization, it should be kept in mind in future research.

The source and chemical history of the principal constituents of the matrix of the quartz-pebble conglomerates are of interest in the matter of the origin of these deposits. Microscopic examination suggests that the quartz in the matrix is probably only partly detrital, a considerable amount of silica having probably been initially present in a dissolved or colloidal state in the waters in which the conglomerates and quartzites were laid down. The same is probably true of the other components, particularly alumina, that now constitute sericite, chlorite, chloritoid, pyrophyllite and the other micaceous minerals. In addition to the hydrolysates, silica and alumina, it seems highly probable that a number of other hydrolysate elements followed these two main constituents of the matrix; here may be mentioned especially the rare-earths, thoria, zirconia, urania and titania, although the last (as rutile, leucoxene, etc.) may have been partly derived as residues of the sulphidization of black sand (ilmenite) if such a process went on during diagenesis and metamorphism. It seems probable that iron was delivered to the basins probably as ferrous sulphate, $FeSO_4$. Oxidation and hydrolysis within the sedimentary basin ultimately precipitated the iron as the hydrous and anhydrous oxides. The sulphate released was then available for transformation by bacteria to H_2S, which in most basins fixed much of the hydrous iron oxides as pyrite during sedimentation and diagenetic processes. In some basins (Tarkwa) the iron was fixed as the oxide and remained as such probably because of the absence of bacterial activity. The various reactions to produce pyrite from the colloidal iron oxides (limonite) are complex and not readily set down in equation form. The hydrolysis reactions are probably as follows:

$$12FeSO_4 + 3O_2 + xH_2O \rightarrow 4Fe_2(SO_4)_3 + 2Fe_2O_3{\cdot}xH_2O$$
$$Fe_2(SO_4)_3 + 3H_2O + xH_2O \rightarrow Fe_2O_3{\cdot}xH_2O + 3H_2SO_4$$

Splitting out of water from the hydrous iron oxides yields a limonitic product FeO·OH.

$$Fe_2O_3 \cdot xH_2O \rightarrow 2FeO \cdot OH + yH_2O$$

Bacterial action involving the sulphate available from the hydrolysis reactions yields H_2S.

$$3H_2SO_4 \xrightarrow{\text{Bacteria}} 3H_2S + 6O_2$$

The H_2S reacts with the limonitic product to give iron sulphide, which evidently recrystallizes or polymerizes in a complicated manner to yield pyrite.

$$4\,FeO \cdot OH + 4H_2S \rightarrow 4\,FeS + 6H_2O + O_2$$

$$FeS + S \rightarrow FeS_2$$

All of these reactions are apparently concomitant during sedimentation and diagenesis. In fact bacterial H_2S may act directly on any available $FeSO_4$ to produce iron sulphide directly and this recrystallizes or polymerizes to yield pyrite.

$$FeSO_4 + H_2S \rightarrow FeS + H_2SO_4$$

$$FeS + S \rightarrow FeS_2$$

It will be noted in the above discussion on iron that oxygen is required in the hydrolysis reaction. It has often been stated that the quartz-pebble conglomerates were accumulated under anoxogenic conditions. This seems improbable having regard to the presence of abundant iron-formations in older (Keewatin) and coeval rock sequences. In addition it appears probable that free oxygen was one of the elements entering the primitive oceans as a result of the early degassing of the earth (Boyle, 1976*a*). The presence of detrital pyrite in the quartz-pebble conglomerate and the possible presence of detrital uraninite in these rocks are not indications of the absence of oxygen in the atmosphere since abundant pyrite commonly occurs in placers in cold climates. It would seem that temperature is more significant than the presence of oxygen in the weathering of pyrite under certain conditions.

The foregoing preamble permits a synthesis of the possible origin of the auriferous and uraniferous quartz-pebble conglomerates and quartzites. These deposits are essentially sediments laid down in a subsiding basin, all of their constituents being either accumulated by detrital or chemical processes. The quartz pebbles and part of the quartz, sericite and other silicate matrix minerals are detrital, derived from the weathering of ancient granitic-greenstone terranes, some marked by the presence of auriferous deposits. Much of the silica and alumina in the matrix was probably delivered to the basin in solution or as colloids, as were also thoria, zirconia, the rare-earths, urania and much titania. Iron reached the basin most probably as dissolved ferrous sulphate or in some cases perhaps as colloidal hydrous oxides. Various hydrolysis reactions precipitated the silica and alumina as colloidal silica (quartz) and Al-Si complexes (clay minerals), and these coprecipitated and carried down most of the thoria, zirconia, the rare-earths, urania and titania as hydrous oxides or adsorbed constituents. The iron was precipitated as gelatinous hydrous oxides (limonite) as a result of hydrolysis, leaving the sulphate in a dissolved form in the interstitial waters of the sediments.

The form in which the gold reached the basin is problematical. It may have been delivered in solution or as a colloid. It seems more probable, however, that the gold was thoroughly milled and reached the basin as very fine flour or skim gold similar to that found in the sediments of many deltas. Whatever the form it appears that the gold was initially widely spread throughout the sediments, especially the conglomerates.

After the initial sedimentation processes a long interval of diagenesis prevailed followed by an extensive period of low-grade metamorphism. During diagenesis anaerobic sulphur reducing bacteria acting on interstitial sulphate-saturated waters produced much H_2S, and this reacted with available iron to give ferrous sulphide and ultimately pyrite or pyrrhotite. The residues of the bacteria gave rise to various oily substances that accumulated in the sediment. Where bacteria were absent the iron oxides recrystallized to give hematite as in the Tarkwa deposits. During the metamorphic period that followed considerable recrystallization, migration and formation of new minerals took place. The matrix of the conglomerates was recrystallized to yield sericite, pyrophyllite and secondary quartz. Ferrous sulphide gave rise to pyrite or pyrrhotite by polymerization and recrystallization processes, and there was considerable local migration of the constituents of the iron sulphides to give secondary or authigenic pyrite and pyrrhotite. At the same time constituents such as lead, zinc, arsenic, cobalt and nickel, originally incorporated in the initial ferrous sulphide complex, were split out and combined to give the various base metal sulphides, arsenides, etc. Under certain conditions as in the Rand, some of the gold and uranium were mobilized as metallo-organic complexes involving the hydrocarbon residues of bacteria and algae. These complexes then migrated and were precipitated in the more porous zones, particularly in the conglomerates, in the form of hydrocarbon gels from which native gold and uraninite were split out. Under other conditions much of the gold and uranium probably migrated in complex sulphide solutions from which pyrite, arsenopyrite, the base metal sulphides, uraninite and native gold were precipitated in favourable sites, mainly porous zones in the conglomerates and quartzites. During this stage of the redistribution of constituents in the sediments, but after penetration of the sediments by basic dykes and the deposition of basic flows in the Witwatersrand system, it seems probable that nickel, cobalt, the various platinoids and gold were mobilized from the sediments and especially from the basic dykes and flows and were precipitated in reducing environments particularly in the Carbon Leader (thucholite), which seems to have originated at least in part from organic (lichen) mats during sedimentation.

The last phase of the mineralization process in the auriferous and uraniferous quartz-pebble conglomerate deposits was the formation of quartz veins with sulphides and gold in fractures and small faults. These probably represent late secretion veins, their constituents being derived locally from the conglomerates and quartzites.

Thus, the formation of the auriferous and uraniferous quartz–pebble conglomerate deposits has involved processes of both a sedimentary and metamorphic nature. Constituents that were deposited both detritally and chemically were later mobilized during diagenesis and low grade metamorphism and concentrated after relatively local migration in favourable porous sites mainly in the conglomerates. The results of such processes can be seen today in an embryonic stage in certain older Mesozoic-Cenozoic fossil placers in the Urals (Barannikov *et al.*, *in* Petrovskaya, 1974) and in numerous young (Tertiary) placers in California, Australia and elsewhere.

Es thut kein Gang so gut
Er hat einen eisernen Hut.

There is no lode like that,
Which has an iron hat.
–Old German mining proverb

Chapter IV. Oxidation and secondary enrichment of gold deposits

General principles

Gold deposits, subjected to oxidation, may become enriched near the surface both as a result of chemical processes involving the migration of gold and by the removal of soluble gangue and sulphides. Where silver and lead occur in quantity in auriferous deposits in argentite, tetrahedrite, galena or other sulphides, there is also a parallel enrichment of these elements as a result of oxidation processes. In contrast to copper and silver, however, gold exhibits less tendency to be enriched in secondary sulphide zones in veins and other deposits.

The mobility, migration, concentration or dispersion of gold during the oxidation of gold deposits are strongly influenced by the following factors:

1. The type of deposit and its mineral composition. The nature of the gangue, types of sulphides and primary gold-bearing minerals are the main considerations. The first two control to a large degree the nature and presence of solubilizing and precipitating constituents in the oxidizing waters. They also control to a varying degree the various reactions between the solid components of the veins and the oxidizing waters.

2. Types of enclosing rocks, specifically their mineral and chemical composition. These have some control on the nature and presence of solubilizing and precipitating constituents in the oxidizing waters. They also control to a varying degree the various reactions between the oxidizing waters and the solid components in the wall rocks.

3. Degree of fracturing, crushing or shearing of the deposit and its enclosing rocks.

4. Climatic factors.

5. Position of the water table.

6. Organic agencies. Gold is soluble as certain organic compounds, or it may be complexed and rendered soluble by the products of these agencies, e.g., H_2S produced by bacteria that gives rise to $[AuS]^-$.

All these factors interact to produce a variety of complex chemical systems that are impossible to follow in detail here. The discussion will be restricted to some of the general features of oxidized gold deposits and to a relatively simple view of the chemical processes involved. Data on the oxidation of gold minerals and deposits may be found in the publications by Skey (1871), Weed (1901, 1902), Emmons, S.F. (1901), Rickard (1902), Lenher (1904, 1912, 1918), Stokes (1906), Brokaw (1910, 1913), Soper (1911), Palmer and Bastin (1913), Grout (1913), Emmons, W.H. (1917), Schneiderhöhn (1924), Freise (1931), Lindgren (1933), Fetzer (1934), MacKay (1944), Chukhrov (1947), Smirnov (1951), Al'bov (1952, 1960), Morris and Lovering (1952), Shcherbina (1956*b*), Kreiter *et al.* (1959), Sato (1960), Machairas (1967, 1970*a*), Radtke *et al.* (1972*a*), Perel'man (1972), Roslyakov *et al.* (1971*a*, 1972), Rusakova and Povarennyk (1973), Bezverkhnii and Baumshtein (1973), Nesterov *et al.* (1974*a,b*), Nikolaeva and Badalova (1974), Borodaevskaya and Rozhkov (1974), Petrovskaya (1974), Kolotov *et al.* (1975*b*), Pitul'ko (1976), Roslyakov (1976) and Nestorenko *et al.* (1976). Other references to the migration of gold in natural waters and soils and its presence in biological materials will be found in Chapter II. The present chapter should be read in conjunction with Chapter IV in the writer's monograph on silver (Boyle, 1968*b*).

In this chapter we shall deal mainly with the course of gold during oxidation processes. The fate of various elements associated with gold in its deposits during oxidation processes is given in the section on elemental associates of gold in Chapter III.

The principal hypogene gold-bearing minerals are native gold, aurostibite, fischesserite and the various tellurides, e.g., sylvanite. Auriferous pyrite, pyrrhotite and arsenopyrite should also be included because although they are not gold minerals *per se* they provide a large proportion of the gold in a number of districts.

There is only one supergene mineral of gold namely the native metal. In some auriferous lodes there is a suggestion that aurostibite and some of the tellurides may be of supergene origin, but the evidence is equivocal. Supergene native gold occurs in a great variety of forms ranging from irregular slugs, nuggets, platelets, dendrites, wires, sprigs, tufts, hairs, filaments, filligrees, spongy masses, mossy aggregates and foils to very finely divided metal commonly called paint or mustard gold. Much of the gold in the oxidized zones of auriferous deposits occurs in relatively discrete particles and aggregates perched on, attached to or coating the various products of oxidation or in some cases intimately intergrown with minerals such as secondary calcite, gypsum, sulphur, kaolinite and opal. In addition there is commonly much microscopic and submicroscopic gold in mineral aggregates such as limonite and wad. Some of this exists in discrete microcrystals and masses; the remainder is in an adsorbed or chemically combined form.

Native gold is relatively inert and in many of its deposits, especially those with a paucity of sulphides and sulphosalts, the mineral passes into the oxidized zones in essentially the same condition as it is present in the primary ore. Under other conditions it is dissolved or migrates in a variety of forms discussed in some detail subsequently.

The fate of gold during oxidation processes is greatly affected by the nature and size of the gold particles in the primary deposits. As noted in Chapter II and elsewhere in the text, the solubility of gold is much greater when the element is released in an ionic form from the lattices of pyrite, arsenopyrite, etc. or when the gold is released in a submicroscopic (colloidal) form from these host minerals. Much of this gold evidently migrates either in solution, as a colloid or in suspension. When the gold is present in particles greater than about 100 μ in diameter, its solubility is greatly restricted. Most of this gold migrates in a physical form as platelets, spangles or small nuggets. Gold in this form also tends to restrict the migration of ionic gold, since it forms readily available nuclei for precipitation of the dissolved gold. Al'bov (1952) has claimed that the size of the gold basically determines its behaviour in the zone of oxidation.

Oxidation of gold tellurides may yield soluble gold and silver or finely divided native gold (containing some silver) and tellurite or a variety of tellurites and tellurates. Native tellurium may also be of supergene origin in the oxidized zones of some auriferous telluride deposits, although it is often difficult to prove that this is in fact the case. In the zones of reduction of certain auriferous telluride deposits, however, there is little doubt that at least some of the native tellurium is supergene in origin. In some deposits tellurite and/or paratellurite is a product of the oxidation of sylvanite, calaverite, etc.

$$AuTe_2 + 2O_2 \rightarrow Au + 2TeO_2$$

In others, insoluble iron tellurites such as emmonsite, mackayite and blakeite are precipitated mainly by reaction of Fe^{3+} with the tellurous acid formed during the oxidation of the tellurides.

$$AuTe_2 + 2O_2 + 2H_2O \rightarrow Au + 2H_2TeO_3$$

In still other places, tellurates may be precipitated by reactions involving telluric acid formed by the oxidation of the tellurides under very high oxidation potentials.

$$AuTe_2 + 3O_2 + 6H_2O \rightarrow Au + 2H_6TeO_6$$

Relatively little is known about the oxidation of aurostibite, $AuSb_2$. My observations indicate that the common products of oxidation are spongy masses of gold and antimony ochres, probably the oxides cervantite, senarmontite and valentinite although they give no X-ray patterns. The oxidation reaction can be written empirically as:

$$2AuSb_2 + 3O_2 \rightarrow 2Au + 2Sb_2O_3$$

Presumably under certain conditions the gold may be liberated in a soluble form. Similarly the Sb (III) and Sb (V) oxides may be dissolved under alkaline conditions and may precipitate various antimonites and antimonates. The oxidation of gold-bearing tetrahedrite is somewhat similar to that for aurostibite.

The oxidation of the silver-gold selenide, fischesserite, probably yields native gold enriched in silver and various selenites and/or selenates depending on the magnitude of the oxidation potential.

The oxidation of auriferous pyrite, pyrrhotite and arsenopyrite yields various soluble iron and arsenic compounds and gold. The latter may be released in a soluble form or as finely divided, spongy or mustard gold.

The fate of primary gold and finely divided native gold released by the oxidation of auriferous pyrite, pyrrhotite, arsenopyrite, chalcopyrite, stibnite, various other sulphides and sulphosalts, tellurides, aurostibite and fischesserite is varied and depends on many factors operating in the oxidized zone. My observations show that much of the primary gold and finely divided gold released from sulphides is incorporated into the cementing minerals of the gossans where it is inextricably intermixed with limonite, wad and other mineral aggregates. Some of this gold also finds its way into the eluvium and ultimately into alluvial placers. The very finely divided gold (flour gold), however, may retain considerable mobility in the water and be moved downstream in the drainage system for great distances. Where the oxidation zones are highly porous and loose, much of the fine gold moves physically downward simply by gravity or in downward percolating waters. Finally, the gold may migrate in a variety of chemical forms under certain conditions. We do not know the actual chemical forms of gold in the oxidizing waters of gold deposits for no one has measured these directly. However, we can infer certain possible forms from the known chemistry of the element and from the chemistry of groundwaters and mine waters. Some of the mechanisms of chemical migration follow. The details of these mechanisms are discussed more fully in the section on natural waters in Chapter II.

During oxidation processes in gold-bearing deposits gold may migrate in the oxidizing waters:

1. As the metal Au° apparently in solution or in a dispersed state as a colloid protected by a variety of other colloids including silica, hydrous iron oxides, hydrous manganese oxides, etc. Extremely finely divided gold and lattice gold released as a result of the oxidation of auriferous pyrite, pyrrhotite, arsenopyrite, chalcopyrite, stibnite, tetrahedrite, the various gold tellurides and aurostibite would seem to be especially susceptible to migration in these forms.

2. As hydroxocomplexes of the type $[Au(OH)]$, $[Au(OH)_2]^-$, $[Au(OH)_4]^-$ and $[Au(HS)(OH)]^-$. Roslyakov *et al.* (1972) claim that some of these complexes are relatively stable in oxidized zones.

3. As various dissolved sulphur (thio) species. During the oxidation of sulphides and sulphosalts the sulphur component may yield a variety of species, including sulphide ion, thiosulphate, sulphite, polythionate and sulphate depending upon the Eh and pH.

$$\frac{\text{MS}}{\text{Metal sulphide}} \rightarrow S^{2-} \rightarrow (S_2O_3)^{2-} \rightarrow (SO_3)^{2-} \rightarrow (S_nO_6)^{2-} \rightarrow (SO_4)^{2-} \quad (n = 2\text{–}6)$$

A number of other complexes may also be formed such as HS^-, HSO_3^- and probably many other H-S-O species, making the oxidation of a sulphide a most intricate process. A number of these complexes render gold (and silver) soluble, in the neutral and alkaline pH range including particularly HS^-, $(S_2O_3)^{2-}$ and $(SO_3)^{2-}$. The complexes formed are of the types $[AuS]^-$, $[Au(HS)_2]^-$, $[Au(S_2O_3)_2]^{3-}$ and $[Au(SO_3)_2]^{3-}$. Where arsenic and antimony are abundant in gold-bearing sulphide deposits gold arsenothio and gold antimonothio complexes of the type $[Au(AsS_3)]^{2-}$ and $[Au(Sb_2S_4)]^-$ may be responsible for the migration of gold. That such complexes are probable is suggested by analogy with silver, the presence of secondary (supergene) pyrargyrite and proustite being relatively common in some silver deposits.

Gold (III) salts of oxy-anions are not very stable, but complex auric sulphates of the type $[Au(SO_4)_2]^-$ are known. Such complexes may be present where the oxidation potential is high in oxidizing sulphide zones and where H_2SO_4 and an oxidant such as MnO_2 are present. This mechanism may partly account for the relatively high migration capacity of gold in some oxidizing sulphide bodies. (*See also* the section on gold in natural waters in Chapter II.) Roslyakov *et al.* (1972) consider that the solubility of gold as sulphate complexes is improbable from thermodynamic considerations, the standard electrode potential for the formation of the $[Au(SO_4)_2]^-$ complex being unfavourable, even under strongly oxidizing conditions.

It seems probable that much of the migration of gold in sulphide deposits is the result of solution of gold by sulphur, arsenic and antimony complexes particularly the thiosulphate, sulphite and sulphide species. Complex auric sulphates may also be a factor in the mobility of gold since we have observed that finely divided gold is slightly soluble in ferric sulphate solutions (Boyle *et al.*, 1975).

Listova *et al.* (1966) carried out a number of experiments involving solutions formed during the oxidation of various natural Pb, Zn and Fe sulphides. They found that gold was dissolved in these solutions, especially when $CaCO_3$ was present and concluded that the metal was complexed under weakly alkaline conditions by thiosulphates and polythionates formed during reaction of carbonates with the products of the oxidizing sulphides.

4. As various dissolved halide species, mainly chloride complexes. Gold has long been known to be soluble as chloro complexes:

$$2Au^\circ + 2H^+ + 4Cl^- \rightarrow 2[AuCl_2]^- + H_2$$

Provided an oxidant such as Fe^{3+} is present, gold may have a relatively high mobility in oxidized zones where chloride occurs in the waters. It has been considered for many years that the generation of the reactive chlorine that solubilizes the gold is due to the following reaction.

$$MnO_2 + 2Cl^- + 4H^+ = Mn^{2+} + 2H_2O + Cl_2$$

Since all of these reactants may be present in some oxidation zones it seems probable that some gold migrates as the soluble chloride complexes. There have, however, been critics of the MnO_2 mechanism mainly because of the lack of this compound or because of unfavourable pH conditions. Eddingfield (1913*a*) for instance notes that the influence of manganese applies mainly to noncalcareous deposits. In the Philippines where abundant manganiferous calcite occurs in the veins the oxidizing waters are neutral or alkaline, and this prevents the formation of any free chlorine. He attributes the near-surface enrichment in the gold deposits of the Philippines mainly to the removal of gangue elements. Other critics have doubted that all of the necessary reactants are brought together at the most propitious time to effect solution of gold. Another possibility obtains, however, in oxidized zones, and this would seem to be the most important. Ferric salts such as the sulphate solubilize gold in the presence of dissolved chlorides under acid conditions. The ferric ion maintains the gold in an oxidized state that is then capable of binding with chloride to form soluble complexes. The simplified ionic reactions probably run as follows in an acidic and complexing environment:

$$Fe^{3+} + Au^\circ \rightarrow Fe^{2+} + Au^+$$

$$Au^+ + 2Cl^- \rightarrow [AuCl_2]^-$$

Since ferric sulphate is one of the most common compounds in oxidizing zones containing sulphides and dissolved salts such as NaCl, KCl, etc. are present in some oxidizing waters, it follows that gold may be solubilized as suggested.

Lakin (1969*b*) observed that when NaCl was added to synthetic leach solutions approximating the composition of solutions resulting from the commercial leaching of dumps of porphyry copper deposits gold was dissovled in the presence of manganese dioxide. The test solution of ferric sulphate and cupric sulphate (pH 1.5) contained 0.5 wt. per cent NaCl. Leaf gold was added to two 200 ml portions of solution, and 0.5 g MnO_2 was added to one of these portions. Air was bubbled through both solutions for 3 days, and the solutions were then filtered. The filtrate of the portion that had been in contact with MnO_2 contained 400 μg (micrograms) of gold, the other portion only 4 μg.

Supergene gold is frequently found in association with the silver halides, chlorargyrite, bromargyrite and iodargyrite in a manner suggesting that the two precious metals were carried as the chloride, bromide or iodide in the supergene waters.

5. As various organic complexes. Gold is readily soluble as cyanide and thiocyanate complexes of the type $[Au(CN)_2]^-$, $[Au(CN)_4]^-$ and $[Au(CNS)_4]^-$. Such complexes may be present in oxidized zones where large amounts of organic matter are present in the overlying soils and eluvium as suggested in Chapter II. Humic substances may also solubilize or carry gold in an adsorbed form under similar conditions. Actually we have few data on the efficiency of these particular types of organic compounds to carry gold in oxidizing zones. Cyanide complexes if present would seem to be only transitory because of the relatively low pH in zones where sulphides are undergoing oxidation. Similarly, the stability of humic complexes would not seem to be particularly high under acidic conditions.

6. Adsorbed to various organic and inorganic colloids. The possibility of this mode of transport by humic colloids has been alluded to above and is discussed at length in Chapter II. It may be important where highly humic conditions prevail. Transport as an adsorbed constituent on inorganic colloids such as hydrous iron and manganese oxides and silica is highly probable in the oxidized zones of most types of gold-bearing deposits. In these substances (colloids and gels) we have repeatedly found enriched amounts of gold (and silver) in gold-silver deposits.

The pH and Eh of the oxidizing solutions and environment affect the mobility and precipitation of gold in a number of ways. Many of the soluble complexes of gold including the various sulphide species such as $[Au(HS)]^-$, $[Au_2(HS)_2S]^{2-}$, the thiosulphate, sulphite, cyanide and the thioarsenic and thioantimony species are stable only in neutral or alkaline solutions. Acidification of such solutions breaks up the complexes and native gold is precipitated. On the other hand the chloro complexes are stable only in acid and near neutral media. Increase in pH precipitates native gold. The pH and Eh also affect the mobility of gold in other ways, the principal ones being the effect on the ferrous-ferric and manganous-manganese dioxide couples which in turn markedly influence the solubility and precipitation of gold as we shall see.

The pH of oxidizing solutions in gold deposits is controlled mainly by the presence of various acids, especially H_2SO_4, which derives directly from the oxidation of pyrite and other sulphides or the hydrolysis of certain sulphates, mainly ferric sulphate, and by the neutralizing effect of carbonate rocks or gangue. Thus, where the gangue is mainly quartz, pyrite and other sulphides and the wall rocks unreactive such as slate, greywacke, quartzite, gneiss or schist the conditions are commonly acid, and the waters are only very slowly neutralized. If chlorides are present in the environment, gold may show considerable migration as the chloro complexes. In addition there is less tendency for absorption of gold on hydrous iron oxides (limonite) and hydrous manganese oxides (wad). In fact, the latter may not be precipitated at all because of the acidic conditions. On the other hand where the gangue or wall rocks are carbonate bearing the solutions are quickly neutralized, and the alkaline complexes of gold are stable thus imparting a relatively high mobility to the element. However, abundant hydrous iron and manganese oxides are formed under slightly acid, neutral and alkaline conditions, and these tend to adsorb gold strongly in this pH range (5–8). The mobility of gold is, therefore, a rather complicated phenomenon, particularly so when one encounters various mixtures of gangue minerals and a variety of host rocks. Gold deposits tend to be individual, and the factors that have led to the supergene mobility and precipitation of gold have to be worked out for each deposit. To generalize one can say that gold is particularly mobile in an acidic or strongly alkaline environment, and hence the most favourable conditions for secondary enrichment are weakly acid, neutral and weakly alkaline conditions. This is also the general view of Roslyakov (1976).

With respect to silver and the base metals, gold generally has a relatively low mobility during oxidation processes. In the absence of humic complexing, studies carried out by the writer indicate that the generalized sequence of mobility is $Zn > Cd = Hg > Ag > Cu \geq Mo > Co \geq Ni > Au > Pb > Sn \geq W \geq Bi$. Morris and Lovering (1952) found a somewhat similar sequence in their study of the dispersion of heavy metals in the Tintic district, Utah. As with all generalized sequences there are commonly local modifying influences. When humic matter is present in the system the order of the mobility of the various elements may be greatly modified, depending on the type of humic complexes present and other factors. Gold appears to have a relatively high mobility in humic waters as noted in Chapter II; its mobility is probably about the same as that of Ag, Cu, Hg and Mo.

A number of mechanisms for the precipitation of gold from downward migrating supergene solutions are possible. Some of these are:

1. Increase or decrease in the pH of the solutions due to oxidation of pyrite with the consequent production of H_2SO_4, hydrolysis of ferric sulphate which produces H_2SO_4 and reactions with gangue minerals and wall rocks, which tend to neutralize or make the solutions alkaline. Increase in pH destroys the chloro complexes of gold and precipitates the native metal. Decrease in pH has a similar effect on the alkaline complexes such as $[AuS]^-$, $[Au(HS)_2]$, thiosulphate, sulphite, etc., and native gold is precipitated.

2. Precipitation of native gold from solution by ferrous ion. Where gold-bearing solutions encounter ferrous ion as in zones where a low oxidation prevails native gold is precipitated.

$$Fe^{2+} + Au^+ \rightleftharpoons Au^\circ + Fe^{3+}$$

Commonly this reaction is characteristic of the deeper zones of oxidizing deposits, but in some veins and lodes the phenomenon is marked in the near-surface zones. This mechanism of precipitation seems to be one of the most important in deposits containing abundant pyrite, pyrrhotite, arsenopyrite, siderite and other readily oxidized iron minerals. It is probably the main mechanism of precipitation of gold in many gossans; also in the zone of reduction (zone of supergene sulphides).

Experimentally, Machairas (1967) has shown that auriferous pyrite and arsenopyrite yield secondary gold as a result of reduction by $FeSO_4$. He recognized three stages in the process: (1) oxidation of pyrite and arsenopyrite to yield H_2SO_4 + $Fe_2(SO_4)_3$, which solubilizes the primary gold; (2) reduction of the dissolved gold by $FeSO_4$ to give secondary gold; and (3) hydrolysis reactions, which precipitate the iron oxides (limonite). These results agree with those carried out in our laboratories and reported elsewhere (Boyle *et al.*, 1975).

3. Precipitation of gold from solution by manganous ion in feebly acid, neutral or alkaline solutions:

$$2Au^+ + Mn^{2+} + 4OH^- \rightarrow 2Au^\circ + MnO_2 + 2H_2O$$

This commonly takes place in the deeper parts of gossans and oxidized gold deposits. The native gold is commonly adsorbed or absorbed by the hydrous manganese oxide in a very finely divided form and is rarely visible (*see* 4 below).

4. Adsorption and/or coprecipitation of negatively-charged gold complexes and colloids by positively charged gels such as hydrous ferric oxides (limonite). This mechanism appears to be particularly effective since most gelatinous iron oxides in or near gold-bearing deposits are generally enriched in gold (Table 22). Gelatinous manganese oxides, likewise, adsorb and/or coprecipitate gold, but the mechanism is not entirely clear. Since the gold complexes and colloids are negatively charged they should be repelled by the negative hydrous manganese oxides. However, these gels invariably contain considerable iron and other elements that may reverse the overall charge giving the complex a positive charge which is effective in precipitating the negative gold complexes and colloids. Alternatively, anionic gold complexes may be adsorbed on the positively charged hydrous manganese (II) oxides. (*See also* the section on gold in wad in Chapter II.)

The writer's observations on natural gold- and silver-bearing hydrous iron and manganese oxide gels suggest that initially the reaction is essentially an adsorption phenomenon, that is, the gels exist as a random arrangement of micelles of the oxides with adsorbed gold and silver ions. No gold or silver can be observed in these dried gels even under the highest powers available with the optical microscope. With aging of the gel some of the gold and silver may be desorbed and split out of the complex yielding small spangles and filaments of native gold in the vicinity of the oxides (limonite and wad). In other cases much of the gold and silver is retained by the oxides as an adsorbed phase or in an extremely finely divided form inextricably mixed with the oxides.

Clay minerals and gelatinous silicate complexes also tend to markedly concentrate gold. One frequently finds seams, bunches and patches of clay minerals in the oxidized zones of gold deposits greatly enriched in gold. Gouge and puggy clay along fractures and faults, likewise, tends to concentrate gold in the oxidized zones. The reason for this is evidently largely due to adsorption processes as explained in the section on natural waters in Chapter II.

Numerous other natural gellike substances tend to

precipitate gold readily. Among these may be mentioned humic gels, humic-limonitic gels, humic-limonite-wad-alumina gels, bismuth ochres, tellurium ochres and amorphous antimony and arsenic ochres. The precipitation mechanism is evidently largely due to adsorption on these substances.

5. Coagulation and/or precipitation of gold colloids by various charged ions, sols and gels. Metallic gold colloids carry a negative charge and hence positively charged clay minerals, hydrous iron oxide and various other positively charged ions, sols and gels should effectively coagulate or precipitate the gold. The importance of this mechanism is unknown since we have no data on the transport of gold as a colloid in the oxidized zones of gold deposits.

6. Precipitation due to the presence of various natural reductants. Various field and laboratory observations show that native gold is readily precipitated from gold-bearing solutions by practically all sulphides, silicates, carbonates, native metals, clay minerals, carbonaceous substances, and a great variety of supergene hydrous oxides, arsenates, antimonates, etc. The gold is only rarely found coating these various substances; more generally the dissolved metal is precipitated on nearby minute nuclei of native gold, and these continue to grow to give relatively coarse plâtelets, wires, sprigs, flakes, slugs and nuggets in the oxidized zones. In some veins, however, there is a general precipitation of gold in the form of finely divided particles throughout such materials as disintegrated sulphides, limonite, wad, puggy clay and various ochres in the oxidized zones.

7. Coprecipitation and/or adsorption by numerous supergene minerals. Among these may be mentioned bismutite, chlorargyrite, the jarosites, beudantite, bindheimite, malachite, azurite, gypsum, native sulphur, pyromorphite, scorodite, and various other phosphates, vanadates, arsenates and antimonates. The exact nature of gold in some of these materials is unknown. Some gold is evidently in lattice positions in these minerals, but most is probably present in a very finely divided form (dust). In some occurrences in chlorargyrite and the analogous bromide and iodide the gold may be coarse and often well crystallized.

8. Precipitation of gold by H_2S. Hydrogen sulphide precipitates native gold from solution in the absence of solubilizing agents such as the alkali carbonates. Such a mechanism may operate where pockets of H_2S occur in the oxidized zones due to the oxidation of various sulphides at low Eh, due to the action of H_2SO_4 on sphalerite and other sulphides, and possibly due to the bacterial reduction of sulphates. In the zones of reduction (supergene sulphides) the precipitating effect of H_2S may also be partly responsible for the enrichment of gold in chalcocite, covellite and other supergene sulphides.

9. Precipitation of gold by reaction with supergene and hypogene sulphides. Gold is commonly enriched in supergene sulphide zones and also in the upper parts of sulphide orebodies just below the zone of oxidation, even where the latter are lacking in supergene sulphides. As mentioned above part of the precipitation of gold in these zones of reduction may be due to the effect of H_2S. On the other hand electrochemical and replacement effects are probably also involved. Gold is the lowest metal in the electro-chemical series and its sulphide is probably less soluble than the sulphides of copper, zinc and iron (pyrite). Hence gold should replace these elements in their sulphide combinations. Whether this is actually what takes place or not is a question. However, it is a fact that many chalcocite zones in copper sulphide deposits are enriched in gold (and silver). Much of the gold is submicroscopic or extremely finely divided. Only occasionally does one see visible gold in these zones that is obviously of supergene origin.

Examples of gold enrichment in deposits

The supergene enrichment of gold in deposits may take place by a simple mechanical mechanism, by chemical means or by a combination of these. First the enrichment may be primarily the result of the chemical removal of a large part of the gangue minerals, in which case the gold migrates slowly downward by gravity or is left behind as a residual component. Such enrichments require no chemical movement of the gold; the lodes simply weather down and the gold collects mainly by gravity in the accumulated residue. Secondly, the gold is enriched as a result of chemical migration of gold. In most deposits the movement of the gold is usually downward, although some exhibit chemical enrichments that indicate a degree of lateral movement of the solutions. In the oxidized zones of gold deposits both mechanisms have generally operated, and there are rarely sufficient criteria to distinguish which mechanism is the dominant one.

Furthermore, the supergene enrichment of gold in deposits is often highly irregular and not all deposits exhibit the same features. Practically every deposit is a study in itself. There are, however, certain general regularities that deserve mention, and these are best outlined by dealing individually with the types of deposits in which gold occurs. Before proceeding, however, there is one generality with respect to the vertical zoning of oxidized gold deposits that can be mentioned. Many deposits exhibit a surface-oxidized zone or gossan in which there is frequently, but not always, an enrichment of gold often in the free state. This generally passes imperceptibly into the primary gold-quartz or gold-sulphide ores that are leaner in gold. This common near-surface enrichment must be taken into account when examining surface prospects lest grave errors be made in estimating the grade of the primary ores. Where conditions were such as to produce supergene sulphides in the zone of reduction at depth, enrichments of gold may also be encountered in this zone. The ideal zonation in an oxidized gold lode is, therefore, a moderately gold-enriched near-surface gossan or oxidized zone, a highly gold-enriched secondary sulphide zone at or near the present or former water table and a lean gold-bearing primary zone that extends to depth. Considerable divergence from this ideal zoning arrangement is common.

For a summary of the chemistry of the various elements that are enriched or depleted in the various zones noted above see the section on associated elements in Chapter III.

The oxidation effects displayed by gold-bearing pegmatites, coarse-grained granitic rocks and porphyries are essentially the same as those described for gold-quartz veins. In general the degree of enrichment of gold in these deposits depends in large measure on the amount of pyrite and arsenopyrite that they contain.

Skarn-type gold deposits usually contain much auriferous pyrite, pyrrhotite and arsenopyrite and considerable amounts of carbonates. Oxidation of the sulphides yields abundant ferrous sulphate that is rapidly oxidized to ferric sulphate. This undergoes hydrolysis and hydrous iron oxides are precipitated. Some H_2SO_4 may be produced by direct oxidation of sulphides and by the hydrolysis of sulphates, but this is soon neutralized by the carbonates. In general the oxidized zones and gossans on skarn-type gold deposits are deep and extensively developed in unglaciated areas, and most are enriched in gold compared with the primary ores. Furthermore, the gold is in a freer state being much more amenable to amalgamation and cyanidation, having been released from its host minerals, particularly pyrite and arsenopyrite. Where lead and silver are abundant in the ores these two elements may exhibit a parallel enrichment with gold. Copper, zinc, cadmium and tellurium are commonly leached from the gossans and upper parts of the oxidized zones. Some arsenic invariably remains in the gossans adsorbed on limonite or present as scorodite and other arsenates; much arsenic is, however, leached in most places.

A good example of a deeply leached gold-bearing skarn type deposit is provided by the Rosita Mine in Nicaragua described by Bevan (1973). The mine is situated in a tropical forest plain where the annual rainfall is around 110 in. The orebodies are in garnet-epidote skarns developed in Cretaceous calcareous sediments (now marbles) mainly on the southern flank of a small Tertiary dioritic intrusion. The primary ore minerals in the skarn are chiefly massive and disseminated chalcopyrite, pyrite, pyrrhotite and magnetite. The metals won from the ores are copper, silver and gold.

Three zones are present, from the surface downward – an oxidized zone, a zone of secondary sulphides and the primary ore. The oxidized zone contained mainly malachite, with some azurite, chrysocolla, chalcanthite, tenorite, cuprite and native copper. The copper grade was over 5 per cent; the content of gold and silver were respectively 0.019 and 0.890 oz/ton as calculated from the production records. In the secondary sulphide zone, chalcocite was the principal economic mineral with minor amounts of covellite, bornite and 'grey coppers'. The copper grade was 4 per cent, and the content of gold and silver were respectively 0.063 and 0.534 oz/ton. As mentioned above the principal copper mineral in the primary ore is chalcopyrite. The copper grade is 1.65 per cent and the gold and silver grades respectively 0.028 and 0.196 oz/ton. These figures show a general depletion of gold in the oxidized ores and an enrichment in the secondary sulphide zone as compared with the primary ore. Silver on the other hand is markedly enriched in the oxidized zone and slightly enriched in the secondary sulphide zone compared with the primary zones.

The gold-bearing skarn-type deposits in the Bau mining district Sarawak described in Chapter III are deeply weathered to depths of 20 ft or more (Wolfenden 1965; Pimm 1967). The weathered materials are multicoloured auriferous clays and sandy clay with mineralized rubble and blocks of limestone and skarn. They average about 0.2 oz Au/ton and range from 0.05 to 1 oz Au/ton. In some the silver is up to 1 oz/ton. The gold is extremely finely divided in most of the clays. Some of the clays contain abundant black manganese oxides; these are enriched in gold, up to 1.5 oz/ton.

The gold-bearing skarn-type deposits in the glaciated parts of Canada exhibit only shallow oxidation effects. At the Nickel Plate Mine, Hedley Mining District, British Columbia the oxidized zone, developed on the skarn containing auriferous arsenopyrite, pyrite, pyrrhotite and sphalerite, was only a few feet thick and consisted of a deep red limonitic and hematitic earth in which resided a rubble of sulphide-skarn material with occasional wisps of erythrite. The gold in the red earth is visible as small ragged particles, which may have been exsolved from the sulphides or which may have been accreted from the finely divided and lattice distributed metal in the auriferous arsenopyrite. The latter seems more probable since free gold was rarely if ever seen in the Nickel Plate ores. According to Camsell (1910) there was a slight surface enrichment of gold, but this was not of any great importance economically.

There are innumerable examples of the oxidation of gold-quartz veins, gold-bearing polymetallic lodes and auriferous massive sulphide deposits throughout the gold districts of the world. Most of the good examples occur in unglaciated terrains, but a few in glaciated areas exhibit features that are of interest. The literature on oxidation of these types of deposits is so voluminous that no attempt will be made to discuss any of the investigations in any detail. The interested reader will find the general reviews and summaries by Maclaren (1908), Lindgren (1933), Emmons (1917) and Smirnov (1951) invaluable if he wishes to pursue the subject in depth. Most of the oxidized and enriched zones described in these works were long mined out before the summaries appeared in print. Likewise, today there are few good examples of oxidized gold deposits available for study.

Gold-quartz veins and gold-bearing polymetallic deposits exhibit oxidation characteristics that can be summarized as follows:

1. Some of these deposits exhibit the three zones commonly found in oxidized deposits – a deep zone of oxidation marked by the presence of oxides, carbonates, sulphates, arsenates, etc. grading into a thin zone of secondary (supergene) sulphides at or near the present or former water table, followed downward by the primary gold-quartz or sulphide ores. Other deposits exhibit no zone of supergene sulphides; the oxidized zone merely passes imperceptibly into the primary zone. Still others show no marked development of an oxidized zone, the primary gold-quartz and sulphide ores being capped only by a few inches or feet of limonitic quartz-sulphide rubble.

2. Many of these deposits show no marked enrichment or depletion of gold as a result of oxidation. The gold is, however, in a freer state, generally coarser, often finer and more amenable to cyanidation and amalgamation in the oxidized zones. There are innumerable references to these particular features in the literature. The reasons for little change in the gold tenor during oxidation in some deposits is not clear. Evidently when a particle of gold is released from its gangue under such conditions it simply remains in situ and accretes any nearby soluble gold, thus growing coarser. Lack of oxidants such as MnO_2 and complexing agents such as chloride and thiosulphate are said by some to be the reasons for the immobility of gold during oxidation processes. The

writer, however, has observed that the state of subdivision of the released gold is a major factor, the finer the subdivision the greater the mobility of the metal in the oxidized zones.

3. In some gold-quartz veins and gold-bearing polymetallic lodes the gossans and oxidized rubble near the surface are extensively leached of gold, the metal being precipitated farther down in the oxidized zones, in the zones where the oxidation zones and primary zones interfinger or in the zones of supergene sulphides if such are present. In the oxidized zones rich pockets of supergene gold are frequently encountered as are gold-rich patches of clay and disintegrated gouge. The zones of secondary sulphides usually have patchy rich streaks or pockets, although in some of these zones the gold values are relatively uniformly distributed.

4. In certain gold-quartz veins and auriferous polymetallic lodes the gossans and near surface oxidized rubble are enriched in gold whereas the lower parts of the oxidized zones and secondary sulphide zones have a normal tenor of gold. In many of these gossans the gold is extremely finely divided and inextricably mixed with the limonite, wad and other products of oxidation.

In Canada there are few examples of deep oxidation of gold-quartz veins and gold-bearing polymetallic deposits mainly because glaciation has removed the oxidized zones. In the Keno Hill–Dublin Gulch area, Yukon the gold-bearing quartz-pyrite-arsenopyrite veins are deeply oxidized, in places down to depths of 600 ft (Boyle, 1965*a*). Unfortunately, from the point of view of research these veins commonly occupy the same fissures and faults as later pyrite-siderite-galena-sphalerite-freibergite lodes, making it difficult if not impossible to separate the effects resulting from oxidation of each of these two types of veins. Nevertheless, some general tendencies can be discerned with respect to the gold veins. As the auriferous pyrite and arsenopyrite are oxidized their gold is released and collects in small slugs, rough wires and dust in the limonite-scorodite-quartz rubble. This gold is much coarser than that which is seen only rarely in the veins. In some veins in the Keno Hill–Dublin Gulch area there is some leaching of gold (and silver) as shown in Table 77, but overall there is no marked migration of gold in most of the veins.

Table 77. Gold and silver contents of gold-quartz veins and lodes and their oxidized zones, Keno Hill–Dublin Gulch area, Yukon

Description of sample	Au	Ag	Powder density
	(oz/ton)		
No. 6 vein Keno Hill			
Composite sample of highly oxidized part of quartz-pyrite-arsenopyrite vein	0.015	0.71	2.89
Composite sample of fresh quartz-pyrite-arsenopyrite vein	0.215	3.36	3.28
Dublin Gulch area			
Composite sample of limonite-scorodite material in ozidized zone of veins	0.30	1.0	2.90
Composite sample of quartz-pyrite-arsenopyrite-jamesonite veins	0.57	1.0	3.25

The Tertiary Mount Nansen gold-silver deposits in the Dawson Range, Yukon are systems of narrow steeply dipping veins in metasediments and volcanics of Precambrian to Paleozoic age, andesitic volcanics and granitic rocks of Mesozoic age and Cretaceous to early Tertiary dacite porphyries (Saager and Bianconi, 1971). The hypogene mineralization is essentially pyrite, arsenopyrite, stibnite, sphalerite, chalcopyrite, galena, freibergite, jamesonite, boulangerite, bournonite and gold. In the intensely oxidized zones and partly oxidized zones that extend to depths of 200 ft or more the common supergene minerals are limonite, scorodite, arsenobismite, arsenolite, covellite, anglesite, bindheimite, beudantite and native silver. Very little migration of silver and gold has taken place, and there are no supergene enrichment zones.

In the Yellowknife district the gold-quartz-pyrite-arsenopyrite-sulphosalt veins and lodes in greenstones are only superficially oxidized down to depths of 1 ft or so in most places. The oxidized materials are a rubble of quartz and disintegrated sericite schist in pulverulent iron oxides, mainly limonite. Some of these oxidized zones are greatly enriched in gold as shown in Table 78. On the other hand gossanous material, composed mainly of disintegrated sericite schist cemented by limonite on certain gold-bearing zones, is only slightly enriched in gold.

Table 78. Gold and silver contents of gold-quartz veins and lodes and their oxidized zones, Yellowknife, Northwest Territories

Description of sample	Au	Ag
	(oz/ton)	
Crestaurum shear zones		
Composite sample of iron oxide rich material in oxidized part of shear zones.	1.16	0.20
Composite sample of quartz-pyrite-arsenopyrite-sulphosalt ore	0.5	0.10
Lynx schist zone		
Composite sample of gossan material	0.005	0.03
Composite sample of silicified and pyritized shear zone	0.001	0.03

The gold-antimony ores of West Gore, Hants County, Nova Scotia occur in Meguma slates and greywackes of Ordovician age. The mineralization is mainly quartz and some calcite in zones of crushed slate. In this gangue there is much pyrite, arsenopyrite, stibnite and some native antimony, all with associated native gold. Near the surface the ores are oxidized to shallow depths probably not exceeding 10 ft in most places. In this zone there is a rubble of unoxidized ore particles and lumps highly stained in places by limonite and antimony ochres. Locally kermesite, Sb_2S_2O, is present, and there is some supergene (?) reddish stibnite, Sb_2S_3, in places in the zone of reduction. The gold in the oxidized parts of the veins is mainly free.

The older literature contains a number of good examples of secondary gold enrichment. At Battle Mountain, Colorado, Guiterman (1890) describes exceptional enrichments of the pyritiferous veins in quartzites down to depths of 250 ft. The pay ore averaged 7 oz Au/ton and 50 oz Ag/ton. Twisted and lumpy nuggets of gold intimately mixed with abundant chlorargyrite are described mainly in oxidized material rich in clay. Guiterman ascribes the enrichment of gold to the solution effect of ferric sulphate derived from the slow oxidation of pyrite.

At Tintic, Utah, Lindgren (1915) described rich shoots of gold ore in the oxidized zones that are undoubtedly due to the migration and enrichment of gold. The primary ores of Tintic are of two types – veins in (Tertiary) monzonite and replacement bodies in (Ophir) limestones and dolomite of Paleozoic age. The former contain essentially quartz, barite, pyrite, enargite, galena and a few other base metal sulphides. The replacement deposits are marked by silicification (jasperoid) and contain quartz, barite, galena, enargite, sphalerite, some pyrite and a little chalcopyrite, tetrahedrite and famatinite. The deposits are oxidized to the unusual depths of 1650 to 2400 ft. The oxidized zones were constituted mainly of limonite with abundant residuals of primary ore, in addition to supergene minerals such as anglesite, cerussite, plumbojarosite, smithsonite, calamine and hydrozincite in the predominantly lead-zinc ores and a variety of copper arsenates, malachite and more rarely cuprite and native copper in the cupriferous deposits. The secondary silver minerals were argentite, cerargyrite and native metal; some rich oxidized ores showed native gold. Most of the oxidized ores were greatly enriched in silver.

Some of the oxidized copper ores contained gold running from 0.5 to 0.75 oz/ton, but in addition certain parts of shoots were very rich, assaying 1 to 5 oz/ton. In these shoots the gold was a deep yellow colour and of high purity. Its secondary nature was apparent from its deposition between quartz grains or as thin flakes on joints. Manganese was always present in the rich shoots of gold, and Lindgren (1915) concluded that sodium chloride, furnished by the winds from the Salt Lake deserts in conjunction with H_2SO_4 and manganese dioxide in the deposits, was responsible for the migration and secondary concentration of the gold.

A secondary sulphide zone is mentioned as being present in one of the mines. It was phenomenally rich in supergene silver (100–3000 oz/ton) contained mainly in pearceite, which accompanies galena, spherulitic sphalerite and marcasite. There is no mention of the gold content of this secondary sulphide zone.

Lovering (Morris and Lovering, 1952) studied the supergene mobility and migration of gold, silver, copper, lead and zinc in the Tintic district. Under the conditions of oxidation that have existed over some 25 years he found that lead had not moved perceptibly, gold had migrated only a few inches, copper had moved a few score feet at most and only in acid solutions, and zinc had travelled hundreds of feet but not as far as silver in some cases. From the data he collected Lovering concluded that gold has a relatively high mobility in acid ferric sulphate solutions with iron just starting to hydrolyze. Silver, he found, may travel far in the oxidized zone, but if it comes into contact with sulphides its migration is quickly halted.

The Creede district in the San Juan Mountain of Colorado is noted for its Tertiary silver-copper-lead-zinc-gold ores that occur mainly in veins and mineralized breccias in faults representing subsidence structures in a large cauldron complex of Tertiary volcanic rocks and associated tuffs (Steven, 1968). The principal primary minerals in the veins and mineralized fault breccias are clay, chlorite, amethystine to white quartz, galena, sphalerite, chalcopyrite, pyrite and minor amounts of argentite, fluorite, barite, rhodochrosite and ankerite. Open-space filling has dominated in the formation of the deposits. Beds of tuff containing fragments of carbonized wood are mineralized with exceptional quantities of silver in places. Apparently none of these are of economic value.

In the older literature on Creede (Emmons and Larsen, 1923; Larsen, 1930), mention is made of the extensive oxidation effects impressed on the veins during which minerals such as limonite, hematite, jarosite, wad, chalcedony, kaolin, gypsum, anglesite, cerussite, oxidized nodules of galena, smithsonite, goslarite, chrysocolla and malachite are developed. Near the surface (150–300 ft) the veins were generally not productive, having been leached of much of their metal content. The most productive (enriched) zone lay at depths of 150 to about 1000 ft. According to Emmons and Larsen (1923) this was the zone of maximum precipitation of both gold and silver. Unoxidized ore carried about 0.1 oz Au/ton whereas the enriched ores carried from 0.5 to 5 oz Au/ton or more. Much of the secondary gold, which was dark yellow and of high fineness, occurred in particles about the size of mustard seed in manganese dioxide (wad). Studies of the annual production showed a noteworthy increase in gold content with depth. Silver on the other hand exhibited enrichment near the surface and a decrease with depth due mainly to the formation of chlorargyrite and native metal in the near-surface parts of the veins.

Some of the Appalachian gold deposits in the southeastern United States were enriched as a result of weathering. Kinkel and Lesure (1968) and Lesure (1971) describe some of these deposits, mainly in Georgia. At the Calhoun Mine, Lumpkin County, the primary ore is auriferous quartz stringers and veins in mica schist. Weathering of the schist and auriferous veins has yielded a saprolite some 20 to 50 ft thick in which there has been much leaching of silica, alumina, total iron and other major constituents; concomitantly there has also been an enrichment of gold, arsenic and a number of other elements including B, Be, Co, La, Nb, Ni, Pb, Zn and Zr. Freshly deposited limonite coating the wall of an adit was found to contain 2.9 ppm Au, indicating that some gold is being transported with iron during weathering.

The gold-quartz and other types of deposits in Brazil, many associated with various facies of iron-formation, are extensively oxidized, and some show marked enrichments near the surface. One unusual type 'jacutinga' occurs as thin (inches to a few feet) lines or bands in itabirite and is a decomposition, or more accurately a surface chemical disintegration, product of the iron-formation. It is composed essentially of powdery ferric oxide (limonite and hematite), manganese oxides, clay minerals and talc in which nuggets, plates and threads of native gold are present. Some of the 'jacutinga' were rich, up to 0.5 oz Au/ton or more, and some extended to depths of 700 ft or more. Oliveira (1931, 1932) thinks that as pyrite is oxidized to form H_2SO_4 it reacted with NaCl, which is found in every mine, to form HCl. The latter attacked the MnO_2 present to give nascent Cl that dissolved the gold. The acidic solution of $AuCl_3$ then reached an area where there was no MnO_2, and there it encountered Fe^{2+} in $FeSO_4$ that precipitated the gold. Alternatively the solution may have reacted with organic compounds, H_2S, carbon, sulphides, tellurides or carbonates resulting in the precipitation of the gold.

Many of the gold deposits of Chile are deeply oxidized and some are enriched. Most of these deposits occur in the Coast Range and in the low hills of the Central Valley, commonly in granodiorite intrusions and their nearby sediments and volcanics. According to Benitez (1959) the primary minerals are essentially quartz, pyrite, hematite and small amounts of the base metal sulphides. The pyrite and sulphides are extensively altered to limonite, chrysocolla, malachite, azurite, atacamite, copper oxides and calcanthite. In the oxidized zones there is also much kaolin, clay and chlorite. The gold is mainly in the limonite in a very finely divided form and much of it is 'rusty' making it difficult to float.

A number of the rich gold veins in the Waihi district, Hauraki goldfield, New Zealand were extensively enriched during oxidation processes (Morgan, 1924). The primary minerals in this field were essentially quartz, calcite, valencianite (adularia), pyrite, galena, sphalerite, chalcopyrite, argentite, pyrargyrite and native gold. Oxidation was pronounced and deep, up to 1500 ft in some of the lodes, yielding limonite, wad, azurite, malachite, chrysocolla, cuprite and native silver locally. Some veins were very nearly completely oxidized to a rubble of disintegrated and platy quartz and the other supergene minerals noted above to depths of several hundred feet; others were only partly oxidized. Morgan (1924) asserts that there was only a minor enrichment due to removal of gangue elements. Most of the enrichment he attributes to the solution and redeposition of gold and silver by descending meteoric waters. In some of the Waihi lodes there was some leafy supergene gold on joint surfaces near the surface, but much of the dissolved gold and silver was carried to deeper levels, some into the primary sulphide zones at depth where the metals were precipitated giving material whose assays averaged 1 to 2 oz Au/ton and 30 to 60 oz Ag/ton.

With respect to the Hauraki goldfield in New Zealand Williams (1974) maintains that the effects of secondary sulphide enrichment were not entirely understood, and that this led to a general conflict in the literature on the origin of the lodes. His view is of interest.

> Unfortunately, Morgan failed to set out the mineralogical form of secondary sulphide enrichment in Waihi mine – the only deep mine-workings then accessible – in fact it is clear that he did not fully understand the practical meaning of secondary enrichment, for he described rich ore comprising pyrite, sphalerite, galena and chalcopyrite as secondary material formed by the action of descending waters. He more correctly ascribed the pyrargyrite and proustite in the Waihi and Grand Junction mines to secondary enrichment.
>
> Generally then, the Hauraki goldfields were worked for about three-quarters of a century with little awareness of the meaning of secondary enrichment processes. Yet the list of minerals set out by the various writers contains many which are characteristic of secondary enrichment. A clear instance of secondary sulphide enrichment seems to be presented by the exceedingly rich ore in the stopes on 12-level in the Bonanza section of Maria mine–where covellite was reported with a little bornite. While describing Waihi, Maclaren (1908) pointed out that "The uppermost sulphide zone at the base of the zone of oxidation was very rich, averaging from 1 to 2 ounces gold and 30 to 69 ounces silver per ton, while exceedingly rich portions assayed 25 ounces gold and 1,000 ounces silver per ton." As already noted, Fraser and Adams (1907) recorded that "It would appear that proximity to the original surface has been favourable to the deposition of the high-grade silver ores." To the present writer the description of the silver-bearing reefs at Komata and Maratoto seem clearly to indicate secondary enrichment.
>
> Although pyrite pervades the mineralization, there is some doubt as to whether it was sufficiently abundant in the lodes to initiate a strong zone of secondary enrichment. For instance in the Tui mine, pyrite is inconspicuous – hence there is no strong gossan and no evidence of a strong secondary enrichment. Moreover, even though sulphides are consistently referred to at Waihi, according to Maclaren (1908) they constitute only 3 per cent of the vein material.

Kraft *et al.* (1974) have carefully examined the effects of oxidation on some gold quartz deposits in the Central Mongolian Peoples Republic. These deposits contain essentially quartz and native gold with about 5 per cent of sulphides and tellurides comprising arsenopyrite, pyrite, enargite, sphalerite, tennantite, tetrahedrite, chalcopyrite, bournonite, galena, hessite, altaite and petzite. The principal oxidation minerals are limonite (goethite, lepidocrocite, hematite), covellite, chalcocite, malachite, azurite, cerussite, scorodite, romeite and smithsonite. In the oxidized zones an increase in the gold content as a whole was noted; the primary native gold in the oxidized ore exhibited a decrease in silver; the silver content of the oxidized ore as a whole was reduced; and the Au/Ag ratio of the oxidized ore increased compared with the primary ore. In addition the gold particles in the oxidized zones exhibited a pronounced coarsening of grain size. Kraft *et al.* (1974) considered that colloidal dissolution and colloidal migration of gold were the dominant factors in the mobility of the gold. Because of the low stability of gold colloids the metal does not travel far in oxidized zones. Local solution and reprecipitation of gold over long periods of time are thought to give rise to the coarsening of grain of the gold particles during oxidation and to lead to important metal (gold, silver) displacements during oxidation processes.

Anhaeusser (1966) describes supergene gold enrichment at the Lily Mine in the Barberton Mountain Land in South Africa. This mine worked quartz veins and a plunging pipelike brecciated shoot filled with quartz, both types of deposits being in the Precambrian Fig Tree series in shear zones comprising shale, banded iron-formation, chert, greywacke and their metamorphic equivalents. The primary ores are composed mainly of quartz, pyrrhotite, arsenopyrite, chalcopyrite, magnetite and gold. The oxidized ores extend to depths of 150 to 200 ft and contain much limonite, goethite, hematite and melnikovite pyrite in a rubble of quartz and partly oxidized sulphides. The oxidized ore averaged between 0.15 and 0.25 oz Au/ton. Bonanzas of rich gold ore were found in the pipelike brecciated body and at intervals along some of the quartz veins. The ore in the oxidized zones was free milling, that in the fresh or sulphide-bearing rock required roasting. Anhaeusser (1966) concluded from this fact, from the fact that the fineness of the gold decreased with depth, from the secondary nature of the gold and from the occurrence of rich pockets of gold in favourable structural traps that much of the gold was of secondary origin, derived from the oxidized protore and concentrated by supergene and residual processes of enrichment in the oxidized and semioxidized sections of the mine.

Some of the auriferous deposits in Ghana were deeply

weathered and slightly enriched in gold. Those in the Marlu gold belt are of particular interest (Pickering, 1935; Thomas, 1937). This belt lies about 60 mi north of Takoradi. The deposits occur entirely in Precambrian Birrimian phyllites, ash flows and tuffs. The hypogene ores occur in large argillized, silicified and otherwise altered crush and breccia zones along faults mineralized with fine-grained pyrite, fine needle arsenopyrite, graphitic material and gold. The graphitic material is said to have largely controlled the deposition of the gold. The surface parts of the orebodies down to about 20 ft or so were lateritized and highly oxidized. This ore was deep red ochre in colour and contained large amounts of colloidal size hydrous ferric oxides and aluminum oxides. The gold was present apparently mainly in the finely divided free state in this lateritized material. Below the lateritized zone the phyllites and ore material were highly oxidized but not to the extent of that in the laterites. The gold in this zone was also free. This oxidized zone passed gradually into the primary ore. The average grade of the lateritized ore was 0.26 oz/ton, that of the relatively fresh ore 0.4 oz/ton indicating a general depletion of gold during oxidation, although the metal in the oxidized parts was more amenable to metallurgical treatment.

Deposits like this are particularly difficult to discover because little is left in the oxidized zones but lateritic material and gold. The metal is also often so finely divided that it does not remain in the pan. Such deposits do not generally yield placers, another feature that makes their discovery difficult. Geochemical soil methods using gold, arsenic or antimony as indicators are particularly useful in their discovery and delineation during development.

The supergene geochemistry of tellurium, the element with which gold is so commonly associated, is discussed in Chapter III in the section on associated elements. There it was stated that in general tellurium exhibits a marked mobility under some conditions of oxidation; under others the mobility of the element is restricted and supergene minerals of tellurium such as tellurite, paratellurite, emmonsite (durdenite), etc. are formed. Supergene tellurides, particularly the base metal varieties, such as altaite and tetradymite have been reported from certain oxidized zones, but it is commonly difficult to differentiate these so-called 'supergene tellurides' from those that are paragenetically late in the primary mineral sequence.

Most of the gold-telluride types of deposits appear to exhibit relatively little enrichment of gold during oxidation. At Cripple Creek, Colorado, Lindgren and Ransome (1906) found little enrichment of gold. In a number of the veins the primary minerals of the sheeted zones, essentially quartz, fluorite, pyrite, sphalerite, galena, molybdenite and gold-silver tellurides, were oxidized to depths ranging from 300 to 900 ft to a brownish clayey material composed mainly of limonite and kaolinite in which the original vein structure was lost. The fluorite was dissolved and dispersed while the tellurides were reduced to dark brown powdery gold. Some of the tellurium was carried away in solution, and some remained bound as tellurite and ferric tellurites such as emmonsite (durdenite). There was little or no enrichment of gold in the oxidized zone, but a pronounced leaching of the small amounts of silver originally contained in tellurides and tetrahedrite took place in some veins. It should be noted that Lindgren and Ransome (1906) found that some of the tellurides near the surface, even in the oxidized zones, were relatively unoxidized. This feature is apparently common in some gold districts where the deposits contain tellurides since these minerals are often found in placers.

Stillwell (1953*a*), describing the gold-silver tellurides at Kalgoorlie, Western Australia says that the gold-bearing tellurides are decomposed in the oxidation zone with the liberation of free 'mustard gold'. Iron tellurates have been identified in this field, and it is probable that much of the tellurium remains behind in this form in the oxidized zones. Stillwell further states that occurrences of silver chloride are recorded, but that probably the bulk of the silver migrated downward in solution. Evidence of the downward percolation of silverbearing solutions, he thinks, is disclosed in isolated specimens by the dissociation of the richer gold tellurides into sylvanite, petzite and hessite.

Little definite information exists concerning the oxidation processes in the classical Tertiary gold telluride deposits in Romania (Transylvania). These famous deposits were worked by the Romans, and all of the oxidized portions of the veins have long since been removed. By piecing together bits and pieces of information from a great many old reports and from personal observations in eastern Europe a general summary of the effects of oxidation can be constructed.

The primary minerals in the Romanian (Transylvanian) gold deposits are essentially quartz, calcite, rhodochrosite, pyrite, galena, sphalerite, chalcopyrite, bournonite, alabandite, tetrahedrite, stibnite, realgar, stephanite, pyrargyrite, free gold and various gold tellurides including sylvanite, nagyagite, petzite and krennerite. In general the tenor of these ores ranged from about 0.25 to 0.75 oz Au/ton. The silver contents were highly variable from a few to a few hundred ounces per ton. The veins were oxidized at the surface and down to depths of several hundred feet in places. In these oxidized zones the principal minerals were limonite, wad, rubbly quartz, leached calcite, malachite, anglesite and residuals of oxidized sulphides. Tellurite and native tellurium are mentioned as being present in some of these zones. The gold tenor of these zones and of the zones where they interfinger with the primary zones was much higher than that in the primary zones. Assays of ounces per ton are mentioned for numerous vein systems. The gold apparently occurred relatively free in these zones and was easily won by washing and simple metallurgical practices. It was also of relatively high fineness, 850+ in most veins, whereas the fineness of the gold in the primary ores was usually about 500 to 750 fine. Leafy gold was common in some veins in the oxidized zones and in the upper part of the primary zones. Gold particles of high purity (800–900 fine), perched on pyrite and nests of gold in vugs and open spaces just below the oxidized zones, were common. All these features combined with the observation repeated many times in the literature that the veins decreased in gold tenor with depth suggest extensive enrichment of gold as a result of oxidation processes. Some investigators have attributed this enrichment to extensive removal of the gangue, but some of the facts bearing on the fineness and distribution of gold suggest considerable migration and reprecipitation of the gold. Little detailed information is available in the old reports concerning the oxidation of the tellurides. Some geological papers mention that at Nagyág native gold was rarely seen and when visible is said to have been secondary and to have resulted from the decomposition (oxidation) of the tellurides. In old mineralogical reports there are frequent references to the alteration (oxidation) of sylvanite, nagyagite and other Au-Ag tellurides to mustard gold, tellurite and other tellurites.

The oxidation of the telluride ores of Boulder County, Colorado are discussed in some detail by Kelly and Goddard (1969). There, the primary telluride veins are composed of an interlacing network of pyritic or marcasitic horn quartz seams in which the ore minerals are quite sparse and irregularly distributed. The primary ore and gangue minerals are numerous (about 67), the most important being quartz, roscoelite, ankerite, calcite, fluorite, barite, pyrite, marcasite, galena, sphalerite, sylvanite, petzite, hessite, calaverite, krennerite, various other tellurides, native tellurium and native gold.

The telluride veins have not been deeply weathered and the residual enrichment of gold is correspondingly slight. Partially oxidized ore contains abundant jarosite, limonite and tellurium oxides and in places some supergene native tellurium, mercury, hessite and the copper tellurides. Fine spongy gold in limonite ('rusty gold') is common in the outcrops and is in places associated with native silver and the silver halides. The geochemical behaviour of the principal metals, gold, silver, tellurium and iron is discussed in terms of acidities, oxidation potentials and chloride ion activities in the oxide zone. The mobility of gold has been generally low in the deposits; on the other hand silver has been extensively leached close to the surface, and only minor amounts were reprecipitated at depth. Tellurium was highly mobile in the oxide zone and has been extensively or completely leached from most of the vein outcrops. Eh-pH diagrams for the systems Te-H_2O, Ag-Te-S-H-O-Cl, and for the relations in the near surface telluride ores are given, as is also an equilibrium oxidation potential-chloride ion activity diagram for the system Ag-Te-S-H-O-Cl. Kelly and Goddard (1969) place considerable emphasis on the role of chloride in the various reactions involving the supergene history of gold, silver and tellurium.

Relatively little has been published about the oxidation of auriferous selenide-bearing deposits. Presumably in oxidized zones of these deposits gold appears mainly in the native form and the element selenium, as native selenium, seleniferous sulphur and as various selenites or selenates such as chalcomenite and kerstenite.

The oxidized zones of massive and disseminated base metal sulphide bodies are frequently enriched in gold (and silver) in many parts of the world. Much of this enrichment is due to removal of the gangue components, but some is also the result of chemical migration and redeposition of gold. These deposits commonly exhibit three well-defined zones in ideal cases – a gossan and oxidized zone in which the primary sulphides are leached and altered to minerals such as limonite, wad, jarosite, anglesite, etc.; a secondarily enriched sulphide zone at the base of the oxidized zone at or near the present or former water tables where sulphides such as chalcocite, marcasite, melnikovite pyrite, covellite and more rarely sphalerite, galena, pyrargyrite and proustite are developed; and a primary zone of pyrite, pyrrhotite and various base metal sulphides in massive aggregates or disseminated through intensely fractured, brecciated or faulted rocks of practically all types. Among these deposits we recognize the Ni-Cu orebodies associated with basic igneous rock complexes and the Cu-Zn orebodies commonly found in sedimentary and volcanic terranes. The gold content of the primary ores of these bodies varies widely, but only rarely as in the Kurokô ores of Japan and in a few other deposits are assays greater than 0.1 oz/ton recorded. The silver content varies between wide extremes; some bodies contain only a few tenths of an ounce to 1 oz/ton or so, others average several ounces or tens of ounces per ton. As an example of the variations in gold and silver the great Kidd Creek deposit at Timmins may be mentioned. The main massive sulphide orebodies contain practically no gold (<0.01 oz/ton) whereas the silver content, principally in the form of native silver, ranges up to 5 oz/ton or more. In most massive and disseminated sulphide bodies and in all porphyry copper deposits the Au/Ag ratio is generally much less than 1. The nature and distribution of the gold in the primary ores of these bodies is poorly known despite the importance of the subject. In some, finely divided native gold or electrum is present; in others there is much gold in telluride form and in minerals such as chalcopyrite, bornite, tetrahedrite and other sulphosalts; and in still others pyrite and arsenopyrite are the principal auriferous minerals. In the writer's experience the gold content of these bodies often varies directly with the copper, arsenic and antimony contents, and locally in some deposits with the lead and zinc contents; the silver content generally follows the copper or lead content of the ores. These generalizations are commonly not realized in local samples and even in some orebodies as a whole. In this category also belong the so-called 'porphyry' copper deposits that occur in rocks of many types but commonly in highly fractured granitic rocks such as granodiorite, monzonite and quartz-feldspar porphyries. Their primary gold content is low (~ 0.005 oz/ton) and their Au/Ag ratio is likewise low (~ 0.01). Much of the gold occurs in the native state in these bodies, but some is closely associated with chalcopyrite, bornite, enargite, tetrahedrite-tennantite and pyrite as submicroscopic particles or lattice constituents. Further details on the tenor and distribution of gold (and silver) are given in the following descriptions of oxidized massive sulphide and porphyry copper deposits.

Many of the massive sulphide deposits in Canada were marked by gossans and shallow oxidation zones, but few have been studied. At Sudbury some of the Ni-Cu orebodies had extensive gossans essentially composed of limonite-cemented wall-rock fragments and highly weathered residuals of the massive sulphide ore. Few of these gossans are available for study today, and the remains of those that are present have been considerably disturbed or removed by pitting, mining and road building making any reasonable attempt at adequate sampling virtually impossible. Nevertheless, some bulk grab sampling of some of these gossans has given some interesting results, as shown in Table 79. There is a slight leaching of gold, a marked enrichment of silver, a general enrichment of platinum, a leaching of palladium and general enrichments of selenium and tellurium in most gossans compared with the estimated averages of the primary ores.

All of the massive sulphide orebodies in the Bathurst-Newcastle area of New Brunswick were marked by gossans, some extensive and up to 10 ft or more in thickness. These

Table 79. Precious metal contents of gossans, Sudbury area, Ontario

Description	Au	Ag	Pt	Pd	Se	Te
	(ppm)					
Murray gossan						
Limonite and limonite-cemented weathered sulphides (average of 3 bulk samples)	0.06	6	0.260	0.135	35	2.0
Sudbury primary ores, estimated average	0.08	2.3	0.17	0.27	35	0.1
Vermilion Mine						
Massive limonite and limonite-cemented rubble (average of 2 bulk samples)	0.545	145	67.6	55.0	250	160
Fresh ore, mainly chalcopyrite and pyrite (average of 2 bulk samples)	1.88	100	35.0	108.0	165	115
Worthington Mine area						
Gossan; mainly limonite and rubble of country rock (average of 2 bulk samples)	0.025	3.5	1.25	0.045	76	1.0
Sudbury primary ores; estimated average	0.08	2.3	0.17	0.27	35	0.1

have been studied in some detail by the writer and his colleagues (Boyle and Davies, 1964; Presant, 1971; Davies *et al.*, 1969). Only a brief summary of the features with respect to gold and silver will be attempted here. The primary ores of the massive sulphide orebodies are composed, in decreasing order of abundance, of pyrite, sphalerite, galena, pyrrhotite and chalcopyrite. Less common are arsenopyrite, bornite and tetrahedrite. Small amounts of stannite, boulangerite, cassiterite and domeykite are also present. The distribution and nature of the primary gold in the various deposits is not known. In some deposits finely divided free gold is probably present, but considerable amounts of the metal seem to be either submicroscopic or lattice constituents of minerals such as pyrite, arsenopyrite, chalcopyrite, sphalerite and tetrahedrite. Primary silver tends to be enriched in the zones rich in lead and zinc, particularly where galena is abundant. The gossans are limonite-cemented conglomerates and are composed mainly of limonite with some wad in places both of which cement schist particles, till and stones. Below the limonitic cap there is often an oxidized zone and a residual type of gossan containing limonite, highly leached till, leached quartz, jarosite, scorodite, beudantite, anglesite, cerussite, pyromorphite, barite and highly leached and disintegrated residuals of pyrite and other sulphides. Below this again and along faults in some deposits there is an enriched secondary sulphide zone consisting mainly of pyrite and other sulphides with supergene covellite and chalcocite. The total thickness of all of these various zones only rarely exceeds 20 ft; commonly it is only a few feet. Much of the residual gossan material appears to have formed before Pleistocene time and is preglacial. The remainder is of post-Pleistocene age, for glacial boulders and stones of granitic and other composition are cemented by the limonite and other secondary minerals.

The results in Table 80 show that gold is enriched in most of the gossan zones developed on the massive sulphide bodies.

Table 80. Gold and silver contents of gossans, secondary sulphide zones and primary ores, Bathurst-Newcastle district, New Brunswick

Description	Au	Au	Ag	Ag
	(ppm)	(oz/ton)	(ppm)	(oz/ton)
Brunswick Mining and Smelting No. 6 Deposit				
Gossan, limonite cemented conglomerate (average of 14 bulk samples)	0.18	0.005	30.9	0.90
Gossan, leached zone and residual gossan below above (average of 6 bulk samples)	0.02	0.0006	4.3	0.12
Gossan, beudantite-rich residual material	0.42	0.012	107	3.13
Sulphide zone with some supergene covellite, chalcocite and pyrite (average of 2 bulk samples)	3.1	0.09	262	7.66
Primary sulphide zone (average assay of 3 bulk samples)	0.26	0.007	47.2	1.4
Heath Steele Mine				
Gossan, mainly limonite and wad (average of 4 bulk samples)	1.5	0.043	142.9	4.17
Sulphide zone with supergene covellite and chalcocite (average of 4 bulk samples)	0.5	0.014	95.9	2.80
Sulphide zone (average of primary ore)	0.06	0.002	23.1	0.7
Anaconda				
Gossan, mainly limonite and some wad (average of 2 bulk samples)	20	0.58	171	5.0
Sulphide zone, supergene covellite and chalcocite (average of 2 bulk samples)	2.8	0.08	109	3.2
Sulphide zone (average of primary ore)	0.03	0.001	42.0	1.2
Leger prospect				
Gossan, mainly limonite (average of 2 bulk samples)	1.9	0.05	86.8	2.5
Sulphide zone (average of primary ore)	0.68	0.02	36.5	1.06
Nigadoo veins				
Gossan, mainly limonite (average of 4 bulk samples)	0.181	0.005	220.1	6.4
Gossan, beudantite-rich (1 bulk sample)	0.41	0.01	314.6	9.1
Sulphide zone (average of primary ore)	0.095	0.0027	150.5	4.4

The enrichment appears to be greatest where wad is developed with limonite, or where beudantite and other supergene minerals are present in the gossans. Most of the supergene sulphide zones also exhibit an enrichment in gold. Silver shows a trend similar to gold in the various parts of the gossans.

An excellent example of the enrichment of gold in the oxidized zones of a massive and disseminated sulphide deposit is provided by Mount Morgan in Queensland, Australia. This great deposit is a breccia pipe with numerous offshoots primarily mineralized with pyrite, pyrrhotite, magnetite, chalcopyrite, sphalerite, quartz and gold (Staines, 1953; Staff, Mount Morgan Limited, 1965; Cornelius, 1969). The host rocks are tuffs, pyroclastics, rhyolite, andesite and other volcanic rocks intruded by gabbro, diorite, quartz diorite, and rhyolite and quartz latite bodies. Virtually all of the orebodies occur in the intrusive rhyolite and the tuff and lava units.

The grade of the primary ore averages about 0.8 per cent copper and 0.112 oz Au/ton. Some of the gold is native, and some occurs in calaverite. It is also stated that considerable amounts of gold are intimately associated with pyrite, about 30 per cent in solid solution according to I.W. McPheat quoted by Cornelius.

The deposit was deeply weathered during Lower Jurassic time and a deep and extensive gossan (100 acres) and oxidized zone were developed down to a depth of 362 ft. The gossan and oxidized zone were then covered by Lower Jurassic sediments (Cornelius, 1967). Judging from old reports the principal minerals in the gossan and oxidized zone were limonite, hematite, kaolin, wad, cellular quartz, wisps of chalcanthite, chrysocolla, malachite and azurite, disintegrated pyrite, native copper, chalcocite, telluric ochres and gold. The change from oxidized to unoxidized ore was abrupt, and there was no enriched supergene copper sulphide zone, although minor amounts of chalcocite and covellite were noted a few feet below the gossan. Parts of the gossan and nearly all of the deep oxidized zone were greatly enriched in gold. Entire benches in the great open pit ran 5 oz Au/ton, and some large blocks of gossanous ore assayed 800 oz/ton. Silver was notably absent, the fineness of the gold averaging about 997. In the primary ore the gold/silver ratio is about 3:1. Some 2.5 million oz of gold were recovered alone from the gossan and oxidized ore. In some of the old descriptions rich silver-bearing siliceous kaolinitic patches are mentioned, but in general most of the copper and silver were extensively leached.

Cornelius (1969) gives some interesting facts about the migration of gold in the Mount Morgan orebody in the last 75 years. During this time gold has been taken into solution and deposited in stalactites of goethite and melanterite. Assays of these stalactites show them to contain from a trace to 0.161 oz Au/ton. Measurements of pH and Eh values of mine water dripping from the stalactites range from 2.3 to 3.0 and from +0.555 to +0.845 volts respectively.

Mount Morgan was probably near the sea during Lower Jurassic time, and one would expect that considerable amounts of wind-blown chlorine would have been present in the oxygenated meteoric waters leaching the orebodies. This component would complex with gold and silver rendering them mobile. Gold being less mobile than silver in the form of these complexes in the presence of $FeSO_4$ and other reductants would be more easily precipitated whereas silver would be dispersed. This natural refining action has probably led to the fabulous bonanza of gold in the gossan and oxidized zone of the deposit. In this respect it is interesting to note that Dunn (1905) thought that sea water was responsible for the enrichment of gold in the oxidized zone.

Many of the rich copper and gold-copper lodes of the Cobar mining field in New South Wales, Australia were oxidized to varying depths (Rayner, 1969). Most of the deposits are massive and disseminated deposits containing essentially pyrite, magnetite, chalcopyrite, pyrrhotite, sphalerite and galena with smaller amounts of cubanite, vallerite, bornite, tetrahedrite, arsenopyrite, bismuthinite, cassiterite, gold, and a number of other minerals as the primary constituents. The oxidation minerals derived from the above minerals are numerous and include limonite, malachite, azurite, chrysocolla, cuprite, atacamite, covellite, chalcocite, wad, cerargyrite, embolite, iodyrite, cerussite, anglesite, scorodite, erythrite and others. Some of the gossans and siliceous cappings were relatively rich in gold, native silver, cuprite, native bismuth and the bismuth ochres (oxides). Many of the gossans were leached of much of their copper, and some were also leached of their lead and zinc. A number of the deposits were secondarily enriched in supergene copper minerals, mainly chalcocite, at and near the present or past water tables. The oxidation zoning was generally as follows: a near surface siliceous-limonite zone, followed downward by a copper carbonate and oxide zone which gave over to an enriched chalcocite zone at or near the water table, below which the ores were mainly primary. In some deposits this oxidation zonation extended to depths of 500 ft or more. Gold appears to have been present in all of the oxidation zones, enriched in places and depleted in others.

According to a variety of old reports and papers, the great Rio Tinto pyritic copper deposits in Huelva, Spain were capped by extensive hematitic and limonitic gossans up to 100 ft thick. These contained practically no copper and only traces of silver and gold. Immediately below the gossan a clearly demarcated earthy zone was present, and this was greatly enriched in both gold and silver. Below this zone the massive pyrite was leached for 3 ft or more, and this was followed by an enriched chalcocite zone, also somewhat enriched in gold and silver. Below lay the immense pyritic lenses containing a little chalcopyrite, galena and sphalerite in which only traces of gold were present. Most of the mining has been in the pyritic lenses. Recently, however, mining of the 15 million tons of slightly auriferous gossan at Rio Tinto and an underlying enriched copper zone has been in progress (Pryor *et al.*, 1972). Most of these gossans and the underlying leached and enriched zones flank the massive pyritic lenses, now largely mined out, and overlie irregular chimney-shaped pyrite- and chalcopyrite-bearing stockworks. The gossan, composed mainly of hematite, goethite, jarosite and clay is considerably enriched in Au, Ag, Pb, As, Hg and $BaSO_4$. The underlying rich secondary sulphide zone contains abundant chalcocite. The grade of the gossan is 2.5 and 45 ppm gold and silver respectively (0.07 and 1.3 oz/ton). The gold and silver grades of the enriched sulphide zones and primary sulphide stockworks are not given. Much of the gold and silver in the gossan is evidently free since they can be removed by cyanide treatment. Pryor *et al.* (1972) discuss methods of sampling and data processing suitable for estimating grades and tonnages of auriferous gossans and similar oxidized ores that should be useful in dealing with these slightly enriched deposits.

At the Pueblo Viejo Gold Mine in the Dominican

Republic large bodies of gold-bearing oxide and sulphide ore are being prepared for mining.[31] The orebody covers an area of approximately 1200 m by 500 m, the host rocks being Early to Middle Cretaceous acid flows and pyroclastics, cherts and carbonaceous siltstones of the Los Ranchos Formation. These rocks are extensively silicified and pyropylitized in the vicinity of the primary mineralization that consists principally of pyrite, with lesser amounts of sphalerite, enargite, tennantite, covellite and gold-silver minerals in veins and disseminations. The weathered (oxidized) zone is some 30 m thick and is completely leached of sulphides leaving only gold and silver in a lateritic residuum and gossan. There has apparently been relatively little supergene enrichment of gold and silver although the latter metal does show a slight enrichment at the oxide-sulphide interface. The published grades of the ore reserves are as follows:

Weathered zone: over 20 000 000 tons assaying 0.15 oz Au/ton and 0.76 oz Ag/ton.

Sulphide zone: over 17 000 000 tons assaying 0.13 oz Au/ton and 1.12 oz Ag/ton; also 2.19 per cent Zn and 0.25 per cent Cu.

The so-called 'porphyry copper' deposits throughout the world provide a large production of gold and silver. Despite this, our knowledge of the details of the distribution and nature of the two precious metals in these deposits is largely unknown. Why this should be so is uncertain since nearly every other aspect of these bodies has been intensively investigated. Since both gold and silver are good indicators in geochemical prospecting for porphyry copper deposits an accurate knowledge of their chemistry, especially in the oxidized zones would be invaluable. The notes that follow are based on the analysis of grab samples from some deposits, a most unsatisfactory procedure, and on bits and pieces of information gleaned from personal communications and the extensive literature on the deposits. For the deposits in the United States and Mexico there is some general information on the distribution of gold and silver in the monograph edited by Titley and Hicks (1966).

The overall content of gold and silver in porphyry copper bodies is low, probably near the values 0.01 oz Au/ton and 0.1 oz Ag/ton. This compares with an average copper content of about 0.5 per cent in most of the bodies as a whole.

Many porphyry copper deposits exhibit three zones from the surface downward – an oxidized zone, a supergene sulphide zone and a primary sulphide zone. Some lack a supergene sulphide zone, whereas others lack both the oxidized zone and supergene sulphide zone, the primary mineralization being capped only by slightly oxidized and leached material a few feet in thickness. The latter type of deposit generally occurs only in glaciated terrains or at high altitudes; in arid, semiarid and humid tropical regions there is generally considerable oxidation, leaching and supergene enrichment of the bodies. In arid and semiarid regions the oxidized zones are often deeply leached leaving an iron-stained porous siliceous mass; in tropical regions a gossanous lateritic cap develops. In both cases, but particularly in the latter, there may be some residual accumulation of gold and silver as noted above for the Pueblo Viejo Mine in the Dominican Republic. Most such zones rarely contain more than 0.1 oz Au/ton and 0.5 oz Ag/ton although there are cases where the grades were evidently higher. Thus, it is said that the oxidized zones at Bingham were considerably enriched in gold and silver, as were also

[31]Information provided by the New York and Honduras Rosario Mining Co. of New York and the J.R. Simplot Company of Boise, Idaho.

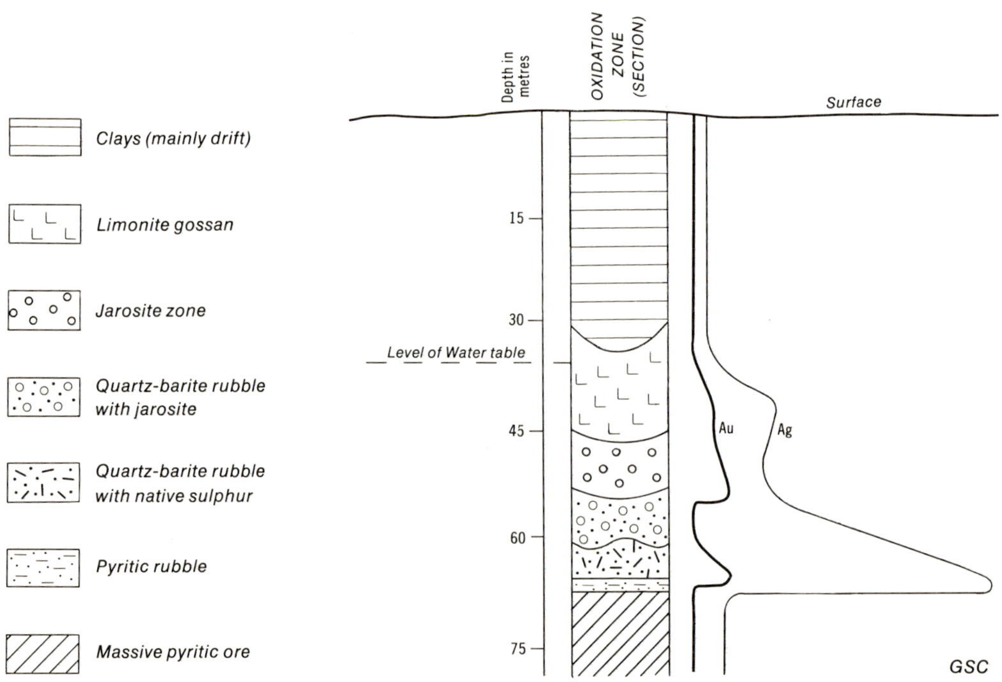

Figure 97. Localization of gold and silver in the oxidized zone of the Maikain 'S' deposit, northern Kazakhstan, U.S.S.R. (after Borodaevskaya and Rozhkov, 1974).

those at Butte (Emmons, 1917). In actual fact it would appear from the records that the surface materials (leached outcrops) on the Bingham, Utah and Ely, Nevada porphyry copper deposits, as well as others in the southwestern United States, were worked first for gold. Later at depth these deposits blossomed into enormous copper orebodies of both secondarily enriched and primary types. In the oxidized zones of certain porphyry copper deposits, copper carbonates and silicates may be extensively developed. These generally carry only a little gold and silver (0.01–0.2 ppm Au; 0.34–6.8 ppm Ag). The supergene sulphide zones, marked by the development of chalcocite and covellite, are enriched in gold and silver in most deposits. Few assays of such zones are available, but most appear to carry the two precious metals in amounts around 0.03 oz Au/ton and 0.5 oz Ag/ton (1.02 and 17.1 ppm respectively). This compares with an average of about 2 per cent copper for these zones. Much of the gold and silver appears to be lattice or submicroscopic constituents of chalcocite, but other modes of occurrence are also possible such as the native metal.

In U.S.S.R. it appears that many of the gold deposits are deeply oxidized both in the Urals and in the various gold belts of Siberia. Most of the deposits exhibit features similar to those described above, and hence the details can be omitted. Few of the descriptions in the U.S.S.R. literature give grades and tonnages, and it is most difficult to ascertain the enrichment (or depletion) factors with respect to gold and silver and their accompanying elements. In a paper by Roslyakov *et al.* (1972) describing the general oxidation phenomenon in a large number of oxidized gold deposits in eastern Transbaikal and the Kuznetskii Alatau, the authors state that the ratio of the mean gold content in the oxidized zones to that in the primary ores fluctuates in the range 1.1 to 1.5. It is also stated that the oxidation zones of various metasomatic deposits exhibit less enrichment in gold than other types. Other general discussions of the oxidation phenomena in gold-silver deposits in U.S.S.R. may be found in the publications by Chukhrov (1947), Smirnov (1951), Al'bov (1952,1960), Kreiter *et al.* (1959), Berzon and Bushlyakov (1971), Roslyakov *et al.* (1971*a*, 1972), Petrovskaya (1973), Rusakova and Povarennyk (1973), Nesterov *et al.* (1974*a,b*), Borodaevskaya and Rozhkov (1974), Nestorenko *et al.* (1976) and Roslyakov (1976). In the publication by Borodaevskaya and Rozhkov (1974), the localization of gold and silver in the Maikain 'S' deposit is given. This deposit is a massive pyritic deposit with small amounts of sphalerite, chalcopyrite, galena, tetrahedrite, enargite and bornite. Quartz and barite are the principal gangue minerals. The gossan and oxidized zones provide a gold-silver ore, the two elements being concentrated in the jarositic lower parts of the oxidized zone as shown in Figure 97.

Gold is where you find it.
—Anonymous

Chapter V. Prospecting for gold deposits

Introduction

Gold is won from deposits specifically mined for the metal and from a great variety of polymetallic deposits in which it is a minor byproduct. Gold is, therefore, a suitable indicator not only of its own deposits but frequently also of a large number of other types, particularly those carrying copper, silver, lead and zinc. The converse of the old Chilean prospecting proverb, "If thou findest copper thou hast gold" is true for many metals with which gold is associated. In the following we shall deal principally with methods applicable to those deposits in which gold is an important constituent and only briefly with deposits in which the element is a byproduct.

Selection of areas

Those searching for gold-bearing deposits should study carefully the types of deposits and their geological setting as outlined in Chapter III. Particular note should also be taken of the modes of oxidation of the various types of gold deposits as discussed in Chapter IV. Since gold and silver usually occur together the chapter on prospecting in the writer's monograph on silver (Boyle, 1968*b*) will complement parts of the present chapter.

There are no known bodies of granitic, intermediate or basic rocks, pegmatites or porphyries where indigenous (syngenetic) amounts of gold and silver are enriched in sufficient quantities to constitute orebodies. A few granitic and intermediate stocks and quartz-feldspar porphyry bodies are known, however, in which higher than average amounts of gold and silver are present. Most of these are greatly enriched in indigenous pyrite with which the gold and silver are associated. Conceivably, bodies such as these could exist that have commercial quantities of indigenous gold and silver. They should be sought in volcanic terranes, particularly those with abundant tuffs and interbedded pyritic sediments, and in sedimentary belts rich in pyritic schists, slates and argillites. The smaller granitic, dacitic and quartz-feldspar porphyry, stocks, irregular bodies and their associated dykes appear to be the most favourable for primary enrichments of gold and silver. Such rocks are also frequently marked by higher than average contents of Cu, Mo, As and Sb.

Bodies of igneous rocks ranging from parts of batholiths to dykes are frequently the hosts for gold deposits. In these the gold is epigenetic, that is introduced into favourable structural sites such as fractures, faults and shear zones. These types of deposits are considered subsequently.

Gold-bearing skarn and contact metamorphic deposits require little comment. They occur mainly in the contact aureoles of granitic and allied rocks and in medium- and high-grade metamorphic terranes. In these environments nearly all deposits are developed in limestones, dolomites, calcareous sandstones or calcareous shales. Metamorphosed sedimentary piles of argillites, black schists and the various calcareous rocks seem to be particularly favourable as do piles containing metavolcanic rocks, tuffs and calcareous rocks. The thought should be kept in mind that the pyritiferous black shales and argillites and the volcanic rocks may be the source of the gold and silver and other elements, and that these were mobilized during metamorphism and granitization and chemically trapped in the calcareous rocks where suitable primary (porosity) and secondary structural conditions prevailed.

A great many of the metamorphic and granitized terranes in Canada are susceptible to mineralization of the skarn and contact metamorphic types. The metamorphic terranes of the Appalachian Belt, the Grenville, the vast metamorphic terranes of the Canadian Shield and the numerous metamorphic belts and contact zones of the western Cordillera are examples of broad select areas in which to prospect for gold-bearing deposits of the skarn type. The Grenville rocks of Ontario and Quebec should receive more attention than they have as favourable sites for the deposition of gold-silver deposits.

Among the various vein, lode, stockwork and disseminated types of deposits categorized in Chapter III there are two general types – the gold-quartz type in which gold is the principal economic constituent and the polymetallic type containing lead, zinc, copper and other base metals with gold and silver as valuable and sometimes indispensable byproducts. There appear to be no general rules governing the localization of these deposits, a feature that should be kept in mind by the prospector. There are, however, certain patterns of occurrence that seem worthy of note.

On a statistical basis the gold-quartz and gold-bearing polymetallic veins and lodes show two characteristic modes of occurrence with respect to the various types of country rocks. A large number of these types of deposits occur in volcanic piles especially those containing abundant interbedded tuffs and pyritiferous sediments. The second greatest concentration of these deposits is in metamorphosed sedimentary piles, especially those containing greywackes, quartzites and pyritiferous argillites and slates. Neither of these modes of occurrence are at all new to the seasoned prospector – the greenstones and propylites of many mineral belts and the greywacke-slate assemblages with their abundant quartz bodies have long been known as favourable auriferous areas. The philosophy of prospecting in these two environments has, however, been that the metals were derived from local granites or some other suitable igneous rock. Perhaps we should take a more careful look at the country rocks themselves as the source of the metals. There is abundant geochemical evidence to suggest that the pyritiferous schists, slates, iron-formations and tuffaceous rocks intercalated with the older greenstones, or in the case of Tertiary volcanic piles lying below them, may be the source of the gold and silver, and that the andesites, basalts, trachytes, etc. as well as the intrusive porphyries may also have made their contributions. Likewise, it is also highly probable that sulphide-bearing black slates, schists, phyllites, etc. are the source of the noble metals in the deposits localized in the greywacke-slate assemblages. As these various rocks were structurally deformed and metamorphosed, numerous dilatant zones were formed along faults, shear zones, sheared beds and on the crests of anticlines. These low pressure zones became the depositional sites for the more mobile constituents from the rocks, including quartz, the components of pyrite and arsenopyrite, gold and silver and other constituents. (*See also* the last section of Chapter III.)

The philosophy in prospecting for these two types of gold-silver deposits should be to seek out areas containing the favourable rocks mentioned above, paying particular attention to sulphide-bearing sediments. From this point careful attention should be directed to areas that were structurally deformed during the metamorphic history of the favourable areas. It should be mentioned that the black pyritiferous schists, phyllites and argillites are usually incompetent rocks and do not generally contain the deposits. They probably provided the elements to the deposits; more competent rocks such as the greywackes, greenstones and local intrusive stocks and dykes are generally the sites of the structural traps.

The old volcanic belts (greenstones) of Precambrian to Mesozoic age inclusive are generally folded, faulted and extensively sheared. The gold-quartz and auriferous polymetallic deposits in these belts tend to favour the lower grade epidote amphibolite and greenschists facies, but this is not an invariable rule. Numerous deposits are known in the high grade amphibolite facies, in batholithic granitic bodies near their contact with the greenstones or in stocks, plugs, dykes and sills of various igneous rock types in the higher grade metamorphic facies. Many deposits are in great carbonated shear zones or in subsidiary structures. During shearing, chloritization, hydration and carbonatization of the shear zones, SiO_2, various metals, sulphur, arsenic and antimony were released and migrated into dilatant zones as outlined in the last section of Chapter III. A number of auriferous deposits in greenstone belts are in tuffs or iron-formations, and others are disseminations and stockworks in quartz-feldspar porphyry, dacite and diorite dykes or stocks that frequently carry much pyrite and pyrrhotite. These types of deposits are mentioned later.

In some of the old volcanic (greenstone) belts the gold deposits may show a preference for certain rock sequences. In the Birch Lake–Uchi Lake area of Ontario Goodwin (1965*b*) found by detailed stratigraphic studies that the volcanic components are arranged in superimposed sequences or cycles. Each cycle exhibits a progression from predominantly mafic effusives below to predominantly felsic extrusives above. Sedimentary rocks are preferentially associated with the felsic extrusives. Two volcanic cycles are represented, and the gold occurrences are preferentially distributed with respect to them. Of 50 recorded occurrences, 39 are located in the upper felsic division of the Lower volcanic cycle and 40 are associated with closely interbanded dacite-rhyolite pyroclastic rocks, andesitic lavas and siliceous sediments. The gold-bearing quartz veins either lie at the contacts of enclosing lithologic units or occupy crosscutting fractures in one or more adjacent units.

No preferential distribution of gold deposits within the volcanic sequence is present in the Yellowknife Greenstone Belt nor is any immediately obvious in the Red Lake district. At Timmins, however, there is a slight tendency for the deposits to lie within certain stratigraphic members. The story of the relationships of gold deposits in the old greenstone belts is, therefore, complex. The often repeated assertion that gold (and sulphide) deposits exhibit a relationship to felsic volcanic members may reflect structural rather than volcanogenic conditions. In volcanic sequences the rhyolites, trachytes and dacites appear to fail structurally more readily than basalts and andesites. Failure of the felsic members is also commonly by brittle fracture, whereas the mafic members shear or form schist zones. Rocks that fail by brittle fracture seem also to be more susceptible to mineralization in some belts compared with those that shear or form schist.

In the recent literature some stress has been placed on the association of gold deposits and ultramafic rocks, these rocks being considered as possible sources for the gold. Viljoen *et al.* (1969,1970) noted a possible relationship between ultramafics and gold deposits in the Barberton Mountain Land, South Africa, and Pyke (1976) has drawn attention to a similar relationship in the Timmins area of northeastern Ontario. The latter author also pointed to the coincidence of ultramafics and gold deposits in other Archean greenstone belts, including the Holtyre area, Kirkland Lake–Larder Lake area and the Val d'Or area, all in the Abitibi Greenstone Belt. Gold deposits are also commonly concentrated in or near ultramafic bodies in a number of other places in the writer's experience. Examples are widespread in the Mother Lode System of California (Knopf, 1929) where the system cuts serpentinites, in the Bridge River area of British Columbia and in a number of greenstone belts in Western Australia, particularly Coolgardie (McMath *et al.*, 1953) where the relationships of many of the veins, including Bayley's Reward, and ultrabasic rocks (flows and/or sills) are well displayed. McMath *et al.* underline the fact that "ultrabasic rocks are the most favoured host rocks of the Coolgardie district."

The reasons for the association of gold deposits and ultrabasic rocks are not entirely clear. Some investigators consider the rocks to be the source of the gold; others think that structural considerations are more important. Probably both are responsible agents. As noted in Chapter III carbonatization and hydration during shearing and faulting of ultrabasic rocks may deliver large amounts of gold and other elements to migrating solutions (Table 73) that ultimately deposit these in suitable dilatant zones such as fissures and faults. Ultrabasic rocks, as is well known, are particularly susceptible to hydration and carbonatization and probably release of gold. Not all ultrabasic rocks are enriched in gold as shown by the researches of Anhaeusser *et al.* (1975) and the compilation recorded in Table 13. One must conclude from these facts that both source and tectonic effects are important considerations.

It bears emphasizing that the old greenstone belts as well as many of the younger volcanic terranes, and also certain sedimentary sequences, are marked by extensive developments of carbonatization and hydration (chloritization, sericitization, etc.). These are manifest in some terranes by random and irregular zones comprising large volumes of rock; more generally, however, the zones are linear or anastomosing and follow lines of extensive faulting, shearing, crushing and drag folding. Gold may have been released from such zones as noted in Chapter III and quantified in Table 73. Examples are numerous of the close association of gold deposits and extensive zones of hydration and carbonatization (Boyle, 1976). Prospectors finding such zones should consider them particularly favourable indicators and should search diligently for nearby dilatant zones (open fissures, faults, shear zones, drag folds, etc.) in which the gold and gangue elements have been concentrated.

At this point it should be mentioned that Ridler (1976) has also emphasized the relationship of many gold deposits in Archean rocks to extensive carbonate zones. His interpretation of these zones differs, however, from that of the author; he considers these zones, e.g., the so-called Larder Lake Break, to be carbonate exhalites associated with the volcanism of the greenstone belts.

From the above discussion it follows that those searching for auriferous deposits in old volcanic belts should seek out favourable source rocks such as interbedded pyritiferous sediments, tuffs, iron-formation, pyritic porphyries and zones of extensive carbonatization and hydration in the vicinity of favourable structures such as faults, fracture zones, schist zones, shear zones, breccia pipes, drag folds, etc. Frequently these structures are marked by depressions or ridges depending on their weathering characteristics and are often revealed by lineaments, circular and S-shaped lakes, marshes and bogs and irregular topographically high or low features. In the Canadian Shield and elsewhere many favourable gold-bearing structures are marked by lineaments on the ground and on air photos. Such lineaments mark depressions and are produced mainly by the differential weathering of shear zones, faults, dykes, tuff beds, etc. that have been highly carbonated and pyritized. Many of these features have been accentuated by the abraiding action of glaciation. An excellent example of the use of lineaments in prospecting for gold deposits is afforded by the Yellowknife Greenstone Belt. In this gold belt there are three types of lineaments. One type marks diabase dykes, another late Proterozoic faults, and the third, and most important, marks the carbonated chlorite schist zones in which the gold-quartz veins and lodes occur (Fig. 17). During early exploration in this belt the late Dr. A.S. Dadson of Ventures recognized the significance of the latter type of lineament in prospecting, and his recommendations for diamond drilling along one of the prominent linear valleys of the area led directly to the discovery of the great orebodies of the Giant Mine. It should be pointed out that not all gold-bearing structures are marked by depression features. On the contrary, in some volcanic belts the mineralized zones may be heavily silicified and may weather as linear, curved or irregular ridges.

With regard to prospecting in old Precambrian shield areas (Canadian Shield, Indian Shield, African Shield, Australian Shield, etc.), Emmons (1932) laid great stress on the relationship of auriferous deposits to granitic intrusions and classified their position with respect to these intrusions as follows, the least productive first:

1. Granite areas 1 mi or more from a contact with invaded (greenstone, sedimentary) rocks. Lodes are numerous but generally not workable. Few deposits are productive.

2. Granite areas less than 1 mi from their contact with greenstones or sediments. A few valuable lodes have been worked in these areas, but most are small and only moderately productive compared with the lodes of the next two groups.

3. Invaded (greenstones, sediments) rocks of the 'island' or 'roof pendant' areas. These contain numerous deposits, many of which have been profitably mined.

4. Areas of the invaded (greenstones, sediments) rocks near small stocklike intrusives. These contain most of the largest gold lode mining districts in the shields.

The flows, tuffs and associated sediments in young (Tertiary) volcanic belts are usually relatively flat-lying but are generally extensively faulted. Great volumes of these rocks are highly propylitized (basic rocks) or sericitized (rhyolites) often on a regional scale. Silicification and argillization are two other effects that are often widespread, and in some districts the rocks may be alunitized. The mineralized zones commonly show little direct relationship to these alteration effects, although most occur within the general confines of the regionally altered zones. In a few regions irregular belts or bands of propylitic, sericitic, silicic or argillic alteration flank the mineralized belts and are fairly reliable indicators. Similarly, bands, belts and patches of alunitization may also indicate the presence of veins within, flanking or underlying the belts and patches. An example of this type of relationship is shown in Figures 32 and 33.

Most of the gold-quartz veins and gold-bearing polymetallic deposits in the young volcanic belts occur in faults or in brecciated zones. These cut all types of rocks but are commonly concentrated in linear belts or in curved patterns in the peripheral zones of rhyolite or dacite necks, dykes or sills. Some of the faults are marked by pronounced scarps or lineaments. Particularly favourable loci for the occurrence of auriferous deposits in the young volcanic belts are where andesites, rhyolites, etc. overlie pyritiferous sediments and volcanics of older vintage, the young andesites and other volcanics being the receptacle rocks and the older sediments, volcanics, etc. the source of the gold and other metals (*see* Fig. 92).

Deposits in sedimentary belts are commonly concentrated in greywacke-quartzite-slate assemblages; also in argillite-shale-sandstone-black schist piles containing limestone, limy shale or limy sandstone members. Particularly favourable are greywacke-slate assemblages isoclinally folded and then cross-folded with the formation of abundant pitching folds. The structural traps in these piles of rock are provided by dilation zones along the crests of folds (saddle reefs), along bedding planes (bedded-veins) and in the myriad small faults, drag folds and contorted zones that result from the folding and faulting of the rocks. Chemical traps are provided by the fractured and contorted carbonate rocks. The source of the gold and its accompanying elements, including silicon (quartz), is probably the sedimentary pile, particularly the pyritiferous argillites, slates and schists. The migration of the elements was promoted in part by dilation of the various structures, but also by the thermal fronts associated with regional metamorphism, granitization and granitic injection. In this respect, most of the deposits appear to have been controlled by the thermal fronts since they occur predominantly in the low-grade (relatively unmetamorphosed) facies. This is well marked in Nova Scotia (Newhouse, 1936) and at Yellowknife (Boyle, 1961*a*). A few are clustered along the isograd line between the relatively unmetamorphosed greywackes, slates and argillites and their metamorphic equivalents the knotted (andalusite, garnet, cordierite) quartz-mica schists. Where structures are the controlling features and are formed late in the metamorphic history of a sedimentary belt, deposits may occur in the high-grade facies and in any enclosed granitic bodies. Gold and polymetallic deposits localized in the carbonate members of sedimentary piles show no particular relationship to metamorphic facies; some, usually the skarn-type deposit, are formed in the high-grade facies whereas others, the highly silicified types, are localized mainly in the low-grade facies.

Disseminated and stockwork types of gold deposits in igneous dykes, sills and small stocks should be sought in both volcanic and sedimentary terranes, generally though not always in the higher grade metamorphic facies. In the younger volcanic terranes the intrusive host rocks of these types of deposits are often highly sericitized or propylitized. All dykes, sills and small stocks that contain pyrite and/or stockworks of quartz or carbonate veinlets should be carefully analyzed for gold (and silver). Similarly, all igneous bodies of this nature that exhibit alteration effects such as chloritization, sericitization, carbonatization, alunitization, etc. should be analyzed for gold. Even those surface zones that report only trace amounts of the metal should be considered favourable since they may be the upward extension of bonanzas at depth. Where this is suspected deep vertical holes should be drilled to investigate the conditions at depth in the stocks, sills and dykes.

Disseminated gold deposits in sedimentary and volcaniclastic beds are of two general types as mentioned in Chapter III – those in tuffs and associated iron-formations and those in carbonate-bearing beds (Carlin type). The former are restricted mainly to volcanic belts and are usually in tuffs and iron-formations interbedded with andesites, basalts, etc. (greenstones), although a few occur in sedimentary belts. Most are marked by abundant developments of pyrrhotite, pyrite and often arsenopyrite; others contain these minerals in addition to abundant quartz and carbonates in veinlets or disseminated throughout the tuffs and iron-formations. The sulphides tend to weather readily, and the bands, belts and zones that are gold-bearing frequently weather low and form marked linear depressions. Others, commonly the highly silicified types, weather high and form prominent ridges. Nearly all are heavily iron stained, and some have marked gossans. The disseminated gold deposits in carbonate-bearing beds (Carlin type) seem to occur mainly in sedimentary belts, although there is no reason to assume that they should not also occur in volcanic belts. There are few criteria to distinguish their precise localization judging from the occurrence of those that are known in United States and U.S.S.R. Most of these deposits seem, however, to be localized in tectonically disturbed zones near regional faults of some magnitude. Most of these deposits are easily located by geochemical methods utilizing rocks, stream sediments and soils as the sampling media.

The presence of certain dyke rocks and stocklike bodies generally of igneous origin within volcanic and sedimentary belts has long been considered a favourable indicator of the possible occurrence of gold (and certain types of sulphide) deposits. Here the author may mention quartz-feldspar porphyries that occur frequently in the Precambrian gold belts of the world, porphyritic dacites that are common in most Tertiary gold belts, and various albite aplites and porphyries (keratophyres, albitites, alaskites) that occur in gold belts of various ages. There appears to be some prospecting value in these observations especially where the various types of porphyries are enriched in pyrite. The reason for the association of gold with quartz-feldspar and albite porphyries and dacites may be one of source as mentioned in Chapter III, it may be one of structural competency to hold fractures, or it may be the result of both these factors.

The association of gold deposits with albite porphyry dykes and stocks variously named keratophyres, albitites and alaskites has been noticed by Gallagher (1940) and Ward (1958). The latter detailed the types of deposits genetically related to the albite porphyries as follows: (1) auriferous silicified albite porphyries; (2) auriferous quartz veins in albite porphyries; (3) auriferous quartz veins in the country rocks in the vicinity of the porphyries; and, (4) auriferous sulphide lodes in the country rocks in the vicinity of the porphyries.

These deposits are said to be characterized by a high Au/Ag ratio (generally more than 9) and a simple mineral association comprising gold, pyrite, pyrrhotite, arsenopyrite, chalcopyrite, sphalerite, galena, and small amounts of scheelite and gold tellurides.

Ward (1958) hypothesized that a simple mineral association and a high Au/Ag ratio point to a genetic association with albite porphyries. Once this relationship is established the possibility of the four different types of gold ore deposits stated above can be suspected to occur in a district. In alluvial gold areas, exposures of albite porphyries similarly advise the search for more than one type of gold ore deposit.

Bilibin (1947) considers that many auriferous deposits are associated with small diorite intrusives that are the precursors of batholithic bodies in some gold belts (Kolyma region of U.S.S.R.). In other belts he noted that many gold deposits are confined to zones marked by small postbatholithic intrusives as in northern Kazakhstan, the Kalba and Kuznetsk Alatau regions and elsewhere. Other Soviet geologists have also noted the association of auriferous deposits with small intrusives particularly in the Urals and eastern Transbaikal. Ivankin and Rabinovich (1972) review the evidence for the association of gold deposits with small intrusives and list the following combination of factors, which they think leads to the formation of granitic magmatic systems that yield auriferous solutions in the early magmatic stage, during the formation of the batholiths proper and during the postbatholithic stage.

1. High basicity of the palingenic magma, whose composition is shifted from predominantly granodioritic toward dioritic and gabbro-dioritic due to the magmatic replacement of different iron-magnesian sedimentary-volcanogenic and

other rocks – all products of the basaltic magmas of previous cycles.

2. Complete differentiation of these specific magmas with the appearance of contrasting petrographic types of rocks, such as gabbros and plagioclase granites, diorites, syenites, etc.

3. Considerable upward movement of the magmas in the mesobyssal and hypabyssal zones of the earth's crust, facilitated by the segregation of volatile components and differentiation of the magmas. This is followed by the accumulation of mineralizing agents in the outer (frontal) parts of the magmatic columns and by the divergence at different levels and convergence of the hydrothermal solutions. Immobile magmas even with a high primary gold content evidently do not give rise to ore-magmatic systems.

4. The assimilation of carbonate rocks by the magma as one of the causes of its differentiation.

5. The interaction of the mobile structural-tectonic elements with the magmas during the period of the latter's formation, migration and differentiation. In many respects this determines the morphology of the frontal zones of the plutons and the development of the ore-magmatic systems.

Borodaevskaya and Rozhkov (1974) remark on the paragenetic link between minor intrusions and gold deposits in numerous auriferous regions. They distinguish the most common minor intrusions with which gold deposits are associated as follows: (1) minor intrusions of the middle stages of orogenic development of an area, and (2) minor intrusions of the late orogenic stage and of activation zones in regions where the folding has been completed. They point out that a characteristic feature of the interrelationship of minor intrusions and auriferous mineralization is the frequent occurrence of deposits in the intervals between injection of igneous rocks of different ages – a feature that they think emphasizes the close link between ore- and magma-generating sources.

Certain unusual rocks have often been thought to be signs of the presence of auriferous deposits. There is an adage among some gold prospectors that lamprophyres are good indicators of the presence of gold deposits. Such seems to be the case in some belts (McLennan, 1915), but in others there is no relationship whatsoever in the writer's opinion. The possibility of a genetic connection should, however, be kept in mind in further research work in gold belts.

It bears repeating that all regional and local zones and patches of propylitization, alunitization, pyritization, arsenopyritization, carbonatization, chloritization, sericitization and silicification in all ages and types of rocks should be carefully scrutinized for enrichments of gold and silver. Such zones are often revealed by a bleached appearance on the weathered surface; some are reddish due to heavy iron staining, and others are black due to the presence of manganese oxides. All such zones should be analyzed for gold and silver and their indicator elements such as arsenic. Even where the surface zones are relatively low in these elements, the thought should always be entertained that such zones are the upward manifestations of deeper bonanzas. Where this is suspected deep drilling is suggested utilizing the concepts of primary halos as outlined subsequently.

Those seeking areas containing modern (nonlithified) gold placers should be guided by the principles discussed in the section on placers in Chapter III and should consult the references given in that section, particularly the works by Gardner and Johnson (1934–1935), Jenkins (1946), West (1971) and Wells (1973). In applying the basic principles governing the formation of placers it must be constantly kept in mind that the climatic and geomorphological conditions when the placers were formed in any particular area were in most cases much different than those that obtain today.

One must think and construct models in terms of paleoclimatic and paleogeomorphic conditions and reconstruct the history of erosion and deposition of the auriferous sediments. This will be appreciated by examining Figure 72 where it will be seen that the old stream and placer deposits bear no particular relationship to the present day topographic or drainage patterns.

The occurrence of gold placers in an area is dependent on four general conditions – a source for the gold, a mechanism for the release of the metal from its source, a concentration mechanism and favourable conditions for preservation of the deposits. Each of these conditions need only be reviewed briefly here since the details are given in Chapter III.

The source of the gold may be one or more of the following: epigenetic gold-quartz deposits; auriferous disseminated and massive sulphide bodies; auriferous polymetallic deposits, slightly auriferous quartz stringers, blows and veins in schists, gneisses and various other rocks; various slightly auriferous minerals such as pyrite and other sulphides in graphitic schists and other rocks; slightly auriferous conglomerates, quartzites and other rocks; and old placers. Intermediate sources such as the last two in the system – primary source–alluvial placer – must always be kept in mind as reiterated in Chapter III.

Weathering is the principal method of release of gold from its deposits, although glacial action may be effective in glacial terrains. The erosive action of wind is an important method of liberation of gold in some desert regions.

Concentration of gold in placers is accomplished mainly by physical means, but chemical processes also play a large part in some cases as outlined in Chapter III. The principal concentrating agents are: gravity (eluvial placers), water (gulch, stream, river, flood plain, delta and beach placers), wind (eolian placers) and glaciers (glacial placers).

The favourable conditions for preservation of gold placers are: abandonment of stream and river courses containing pay streaks as a result of change of course or rapid downward cutting by streams and rivers; regional uplift, which mainly affects alluvial deposits especially marine placers raising them above the active surf; and burial, which may be by landslide materials, glacial materials, flood plain gravels, alluvial fans, desert outwash detritus, or volcanic flows and tuffs.

Throughout the world areas containing modern placers are generally marked by a 'kindly' terrain that is characterized by a subdued topography marked by broad, often terraced intrenched valleys and rounded deeply weathered hills commonly with nearly accordant summit levels. The paleogradients have usually been relatively low, approximately 30 ft/mi. Few extensive placers were formed in terrains marked by sharp alpine features and high gradient V-shaped valleys; similarly excessively flat terrains far from mountain systems and their foothills have yielded few productive placers. The rocks in placer districts are invariably metamorphosed, and there has usually been considerable igneous activity both extrusive and intrusive. Deep secular weathering is a feature of all placer areas.

Fossil placers occur predominantly in sedimentary terranes generally in conglomerates or quartzites or sequences of both these rocks. Occurrences in greywackes are not common, and there are no known occurrences of fossil placers in shales and limestones. Conglomerates are of many origins including marine, lacustrine, estuarine, fluviatile (fanglomerate) and glacial. Sandstones have similar origins. Both types of rocks generally mark periods of deep secular weathering that is favourable for the production of gold placers.

It is not possible *a priori* to determine where fossil placers are likely to occur. As noted above sedimentary belts should be sought out and the various conglomerates, quartzites and sedimentary breccias examined carefully with the pan or by chemical or fire assay analyses for gold. Zones in sandstones and conglomerates containing magnetite or ilmenite should receive particular care in sampling. Likewise, all conglomerates and sandstones, in fact any sedimentary rock or tuff carrying pyrite, pyrrhotite and/or arsenopyrite, should be carefully tested for gold, platinum, uranium, rare earths and base metals.

The location of fossil auriferous conglomerates and quartzites in sedimentary sequences is difficult to predict. Many lie near the base of the sequences, but others lie well above the base in some districts. Those seeking such conglomerates and quartzites should carefully study the sections in Chapter III, dealing with fossil (lithified) placers and auriferous quartz-pebble conglomerates. Sufficient examples are given there to illustrate practically all possible occurrences of these types of gold deposits. Surface indications of auriferous conglomerates and quartzites are commonly present in regions containing these deposits, being often manifest by large amounts of pyrite in the rocks; excess quantities of quartz pebbles or sands in streams and soils; large concentrations of SO_4 in ground and surface waters due to the oxidation of pyrite; abundant iron staining and gossans, both residual and transported; anomalous radioactivity in the rocks, soils and waters; and the presence in the environment of solid hydrocarbons enriched in one or more of Th, U, rare earths and Au.

The report of the working group on uranium in quartz-pebble conglomerates (International Atomic Energy Agency, Vienna, 1974; Formation of uranium ore deposits; Proceed. Symp. Athens, p. 707) is of interest in prospecting for auriferous deposits in the quartz-pebble conglomerates. They state:

> Uranium-bearing, oligomictic, quartz-pebble conglomerates are epicontinental sediments laid down in Precambrian intracratonic basins. Most of these conglomerates are found in the sedimentary section prior to the development of the first limestones. None have been found of age younger than the first red beds and we believe that none younger will be found. We suggest that reports of young conglomerates of this type, such as a Cambrian conglomerate described in Soviet literature, are in error.
>
> The deposition of uranium-bearing minerals in the known conglomerates appears closely related to sedimentary features. This is particularly true in the Witwatersrand and at Elliot Lake, where the presence of uranium in young, cross-cutting conglomerate beds clearly relates to scour of older beds.
>
> Where the uraniferous mineral is structurally ordered thorian uraninite, we believe it to be of detrital origin. However, pitchblende occurs in some of the conglomerates and, in our opinion, this must have been deposited from solution even though the uranium has not moved far in dissolved form. In the main, such pitchblende is associated with carbonaceous material normally referred to as thucholite. The available evidence suggests that this material is of organic origin.
>
> The conglomerates do not normally contain magnetite or ilmenite. On the other hand they always contain abundant pyrite. We are forced to believe that most pyrite originates by sulphidization of detrital magnetite and ilmenite. The sulphur originated from volcanogenic processes, active at the early stages of crustal evolution, which pertain to the time of deposition of the conglomerates. Some of the pyrite so formed may have entered the conglomerates in detrital form.
>
> Conglomerates and contained minerals ultimately formed by collection, concentration and distribution of detritus by major stream systems.
>
> The formation of an economic deposit relates both to material available at source and to the degree to which the sorting and cleaning process produces a heavy mineral concentrate in the ultimate conglomerate. Known conglomerates differ widely in this regard so that uraniferous conglomerates may contain a variety of other valuable minerals. At present, gold is the significant example of this.
>
> Age determinations suggest that the conglomerates form between about 2800 m.y. and 2200 m.y. As noted earlier, this is prior to the development of the first red beds of the sedimentary section. Prospection for such ores should utilize this as a working hypothesis.
>
> In the opinion of the Working Group, those encountering conglomerates of this kind should prepare heavy mineral concentrates from reasonable rock volumes to facilitate study and understanding of the origin of uranium-bearing minerals and other heavy minerals in the rocks. Except for South Africa, where extensive work of this kind has been done, there is regrettably little data available. The Working Group recommends radiometric logging of all drill holes as the Rand-type thin conglomerates can easily be missed if core recovery is poor.

The writer would urge caution with respect to the conclusions drawn about the age of the quartz-pebble conglomerates. There may well be much younger representatives than those now known. Finally, the statement about magnetite and ilmenite may not be valid as regards the auriferous types, as the Tarkwa deposits in Ghana so clearly testify. The statements that the conglomerates formed earlier than the first limestones is questionable. Archean rocks on which the productive quartz-pebble conglomerates were laid down contain notable amounts of sedimentary carbonates (limestones in places) (Boyle, 1976*a*).

Mordvin *et al.* (1973) have investigated the specific characteristics of auriferous Precambrian conglomerates in eastern Siberia and have concluded the following, which is of interest in prospecting for these deposits: (1) the auriferous conglomerates are confined to the margins of ancient shields; (2) most are restricted to Precambrian rocks formed in depressions that underwent slow subsidence; (3) all are only gently folded and metamorphosed; (4) the ore minerals are confined to the cement of monomictic quartzites or oligomictic conglomerates with well packed average size pebbles; (5) auriferous sections have large intervals of sedimentary hiatus; (6) the stratification in the auriferous conglomerates is better developed than that in basal conglomerates, indicating their formation as quartzose terrigenous rocks which were products of deep chemical weathering; (7) auriferous conglomerates occur in regions marked by preexisting epigenetic gold deposits; and (8) many of the auriferous conglomerates occur in strata enriched in iron.

To conclude this section it is worth mentioning that the prospector should prospect in old geological reports, ancient treatises and the narratives of explorers. Gold has been sought by man for thousands of years, and he has documented its occurrence since the invention of writing. Old gold workings, prospects and chance occurrences of the precious metal are described in the most ancient writings, some now long forgotten. Perusal of these descriptions is often valuable in selecting auriferous areas for further work not only in the search for gold but also for silver, copper and other base metals with which gold is commonly associated. As an example one can point to the Pontic goldfield in Turkey, mentioned in ancient Greek and Roman writings and then largely forgotten. Today this field (Trabzon) is one of Turkey's important sources of copper, lead, zinc and silver.

> To properly prospect for gold deposits one must have a knowledge of the geochemistry of the primary and secondary processes of concentration of the precious metal.
>
> –A.E. Fersman, 1939

Geochemical methods

General

Gold has not been used extensively as an indicator element in geochemical surveys mainly because of difficulties in its estimation in field laboratories. It is, however, a very valuable indicator of various types of deposits as will be shown presently and for this reason should be employed more frequently. In addition to indicating deposits in which it is a major constituent, gold is also useful in detecting various types of silver deposits, polymetallic lead-zinc-copper deposits and certain types of copper deposits.

Indicator elements

In its deposits gold is associated with a large number of elements as detailed in Chapter III and discussed in another publication by the writer (Boyle, 1974*a*). Most epigenetic auriferous deposits contain essentially the same suite of elements, but the contents range widely from mere traces to minor amounts. The gold-bearing quartz-pebble conglomerates contain a suite of elements that differ only slightly from other types of gold deposits, and there are some differences in the elemental associates of gold in modern and fossil placers.

The elements greatly enriched in epigenetic gold deposits such as skarns, veins and stockworks include Al, Ca, Mg, Fe, Mn, Na, K and B and those bound in the quartz, carbonate and sulphide gangue minerals. In addition there are a large number of other elements that exhibit enrichments, being present either in the gangue minerals, the accompanying metallic minerals, sulphides, sulphosalts, etc., or in the native gold and various other gold minerals. These include: Li, Rb, Cs, Cu, Ag, Sr, Ba, Zn, Cd, Hg, B, Ga, In, Tl, Sc, Y, La, rare earths, Th, U, Ge, Sn, Pb, Ti, P, As, Sb, Bi, V, Se, Te, Cr, Mo, W, F, Re, Co, Ni and the Pt metals. Most of these elements are useful only as general indicators of the presence of gold and should be only considered as such in all types of geochemical surveys. Some such as Li, K, Rb, Cs, Ti, V and Cr are useful indicators principally in geochemical surveys utilizing rocks (primary halos) as a sampling medium. The elements most commonly accompanying gold in epigenetic deposits and those with the highest specificity as indicators include the following in their approximate order of importance: Si(SiO_2, quartz), S (FeS_2, pyrite), Ag, As, Sb, Te, B, Bi, Hg, Mo, W, Cu, Zn, Pb, Cd, Tl, Ba, Sr, Mn, Se, F, Cl, U and Th. Silver, arsenic and antimony are nearly universally enriched in all types and ages of epigenetic deposits. Tellurium and selenium enrichments are usually restricted to specific mineral belts and where found may serve usefully in geochemical surveys. Selenium tends to be enriched mainly in Tertiary deposits. Boron is characteristic of epigenetic gold deposits in sedimentary piles but may also be enriched in various types of auriferous deposits in volcanic terranes. Boron is more characteristic of older deposits than those in Tertiary and younger rocks. Bismuth is enriched in a number of epigenetic gold deposits of various ages and types. Mercury exhibits the highest enrichments in the young deposits, falling off generally in amount with increasing age of the deposits. Other factors than age, however, appear to control the enrichment of mercury in some deposits. Thus, veins and lodes that are enriched in sphalerite, stibnite, tetrahedrite and other sulphosalts are usually, likewise, high in mercury content regardless of their age. This is emphasized by certain oreshoots in the Precambrian Yellowknife Greenstone Belt that contain nearly as much mercury as Tertiary deposits in Japan and New Zealand. Molybdenum is common in many types of epigenetic gold deposits but rarely in significant amounts to be useful as an indicator in geochemical surveys except in certain belts. Tungsten is a frequent element in relatively deep-seated epigenetic gold deposits and occurs in some of the younger shallower veins as well. The significance of tungsten as an indicator for epigenetic gold deposits is largely unknown. Because of its low abundance its appearance in higher than average amounts in soils, stream sediments and vegetation may signal the presence of auriferous mineralization. Copper, zinc, lead, cadmium and barium are useful but not specific indicators of epigenetic gold-bearing deposits. They tend to signal the presence of polymetallic deposits, although lead and barium are frequently useful in soil analyses and stream sediment surveys for gold deposits because of their low mobility and their tendency to be concentrated in these media near the deposits. Manganese is particularly abundant in Tertiary deposits in certain regions. Its presence in anomalous amounts in water, soils, stream sediments and vegetation may be a general indicator of both gold and polymetallic deposits of this age, as well as those of older vintage, in an area. Fluorine is particularly characteristic of certain Tertiary deposits (e.g., Cripple Creek) and should be a good indicator of these deposits under most conditions. The element is present only in small amounts in Precambrian, Paleozoic and Mesozoic deposits. In these deposits the efficacy of fluorine as an indicator is limited except where extensive developments of sericite, other micas, tourmaline or minerals such as topaz are present. Chlorine may be enriched in the oxidized zones of certain auriferous deposits, the element being bound mainly in silver and copper halides such as chlorargyrite and boleite.

Thallium is enriched in some of the younger auriferous deposits, particularly those of Tertiary age; its effectiveness as an indicator in geochemical prospecting is unknown. Rubidium and cesium are enriched in certain wall-rock alteration zones associated with auriferous deposits as mentioned in Chapter III. At the Ogofau Gold Mine in southern Wales Al-Atia and Barnes (1974) found a well-developed primary rubidium halo in the shaly country rocks enclosing the auriferous veins and lodes. The rubidium aureoles were better developed than those for Pb, Zn, Cu and Hg. Uranium and thorium occur in various types of auriferous skarns, veins and lodes but generally only within specific mineral belts characterized metallogenetically by these two elements. Uranium, in particular, may be a good indicator of gold-bearing deposits in such belts.

In the quartz-pebble conglomerate type of auriferous deposits the elements greatly enriched include those bound in the gangue minerals, particularly Si(SiO_2), Al, Fe and S. In addition there are a large number of others that exhibit enrichments, often only to a slight degree. These elements occur in the gangue elements, in various metallic minerals, in oxides and complex substances such as uraninite and thucholite, or in the native gold. These include Ag, U, Th, rare-earths, As, Cu, Pb, Zn, Hg, Co, Ni and platinoids. Some deposits (e.g., Tarkwa) are essentially devoid of U, Th, rare-earths and the base metals. Of the elements associated with gold in the quartz-pebble conglomerates the most indicative of the precious metals would appear to be Ag and As as well as U, Th, rare-earths, Hg and the base metals in particular areas.

The heavy minerals associated with gold in eluvial and alluvial placers are familiar and need not be considered in any detail here. The most frequent associated heavy minerals are magnetite, hematite, ilmenite, pyrite, arsenopyrite, garnet, zircon and monazite; in addition cassiterite, platinoids, native bismuth, amalgam, native copper, cinnabar, scheelite, wolframite, barite, galena, stibnite and sulphosalts are present in some areas. None of these minerals are specific indicators of gold, but some, such as arsenopyrite, native bismuth, scheelite, stibnite, sulphosalts and platinoids, are suggestive of the presence of the precious metal.

The elements in the heavy concentrates from stream sediments, soils, eluvial placers, etc., most indicative of gold, are Ag, Pt metals, Te, As, Sb, Bi, W, B, Ba, Hg, Cu, Zn, Cd, Pb, Tl, Se, Sn, Ni, Co, U, Th, rare-earths, Zr and P in about that order of specificity.

Gold differs in its chemistry and the resistate character of its minerals from most of its indicator elements (*See* the section on associated elements in Chapter III.) Consequently, the relationship of the primary and secondary halos and dispersion trains of gold and those of its indicator elements is complex. In some terranes and surficial environments the halos and dispersion trains and fans of both gold and its indicators are coincident; in others the halos and dispersion trains and fans of the indicators may be displaced with respect to those of gold. From this it follows that each gold belt is more or less individual, and pilot studies to determine the exact nature and relationship of the gold halos and dispersion trains and fans and those of its indicators should precede detailed surveys. This applies particularly to surveys utilizing surficial materials (water, soils, vegetation, etc.); less so to lithogeochemical surveys.

Lithogeochemical methods

Relatively little detailed work has been done using the gold or indicator element content of fresh unsheared and unaltered country rocks to outline geochemical or metallogenic provinces containing gold deposits; hence our knowledge of this aspect of broad regional geochemical prospecting for gold is rather limited. Judging from the trace element studies done by the writer at Yellowknife (Boyle, 1961*a*), Keno Hill (Boyle, 1965*a*), and in the Bathurst-Newcastle district of New Brunswick there are no definite elemental patterns in the unaltered and unsheared country rocks that would suggest the presence of gold (and silver) deposits in these areas. Certain of the rocks, especially the pyritic black schists, pyritic tuffs and iron-formation in these areas contain higher than normal amounts of gold (and silver), but this seems to be true also of areas where no gold (and silver) deposits are known. No particular patterns in the gold content of the various granitic intrusives in these districts were noted; all were very low in gold content (<0.005 ppm). Some quartz-feldspar porphyries in the Yellowknife Greenstone Belt were found to be somewhat enriched in what is thought to be indigenous gold (0.10 ppm). Most of these are spatially related to the gold deposits.

Seeland (1973*a,b*) carried out a broad scale reconnaissance for gold in the sedimentary and metamorphic rocks of the United States south of the Great Lakes region, the aim being to discover fossil placers in clastic rocks on the southward dipping paleoslope draining the auriferous areas in the Canadian Shield north of the Great Lakes. In 799 samples gold was detected in only 36. Four samples yielded 0.1 to 0.4 ppm Au. Seeland (1973*a,b*) concluded from the data that the frequency with which gold was detected in the rocks of the region suggested the possibility that certain undetermined syngenetic chemical processes have concentrated anomalous amounts of gold in rocks of chemical origin (Grenville marble, glauconitic sandstone, gypsum, etc.). He thought that these chemical processes may be more effective concentrators of gold than the expected placer-forming physical processes, even in a region relatively close to a source of clastic gold. In addition he speculated that the general lack of gold and the small grain size of the gold that was found (less than 5% of the hydraulic equivalent size in the nonglauconitic sandstone-quartzite category) are results of postdepositional leaching.

Finally, he postulated that the apparent abundance of chemically or biochemically precipitated gold and the relative lack of clastic gold in a region close to a major gold source could also be explained by the presence of an east-west late Precambrian–early Paleozoic continental divide just north of the Great Lakes separating a known south- and southeast-trending paleocurrent system from a probably north-trending system. Thus, he concluded that most Canadian Shield gold deposits would lie north of this divide, and the logical place to look for fossil placers would then be in the basal Paleozoic clastic rocks near James Bay.

Wrücke and Armbrustmacher (1973) determined the gold contents of bedrock samples collected during a geochemical reconnaissance in the Gold Acres–Tenabo areas, northern Shoshone Range, Nevada. Included in this report are also analytical data for 30 other elements. No conclusions were drawn from the data.

Roslyakova and Roslyakov (1975) found complex patterns in the regional gold and indicator element halos associated with auriferous fields in U.S.S.R. Their research is summarized at the end of this section.

Summarizing we can say that much more research comparing the trace element content of fresh rocks in mineralized belts with those in barren areas is necessary before an answer to the problem of whether or not the unsheared and altered country rocks reflect definite geochemical gold provinces or belts in which it is more probable to find gold deposits.

On a more limited scale, however, considerable data are now available on the distribution of gold and its indicator elements in the country rocks, mainly intrusives, in the vicinity of gold deposits. Many years ago Emmons (1937) emphasized the intimate relationship of some gold deposits to the apices of granitic stocks and small batholiths, the philosophy being that as the differentiated metal-bearing hydrothermal solutions streamed out of the crystallizing magma they would concentrate in the fractures and faults in the country rocks around the apices and apophyses of the stocks. Gross (1950) tested this general philosophy in his study of the spatial relation of gold deposits to intrusive bodies in Manitoba and northwestern Ontario. He used indicators that he considered to be late crystallization products of the intrusives, namely silica, zircon, uranium and thorium. Four intrusives, the Dome and McKenzie–Gold Eagle stocks of Red Lake, Ontario; part of a batholith in the Rice Lake district of Manitoba; and the Froghead Bay stock in the Eagle Lake district of Ontario were investigated. The first three have associated gold deposits; the fourth has no known associated gold deposits. In the first three the concentrations of the indicators increased in the direction of the gold deposits; in the fourth there was no pattern. This was said to indicate a migration of the late solutions towards the known ore deposits. It is obvious that such patterns could be particularly useful in prospecting for gold deposits in the vicinity of stocks and small batholiths.

A number of investigations of the gold content of stocks and small batholiths in gold belts have now been completed and attempts have been made to relate the data obtained to the presence or absence of associated gold deposits. The results have been mixed to say the least. Some investigators find that intrusives in the vicinity of gold-bearing deposits exhibit enrichments of gold; others have found no such relationships. The various investigations dealing with this feature of geochemical prospecting that have come to the notice of the writer are summarized in Chapter II and need not be repeated here. The writer's experience in the matter of intrusives in gold belts is that some small stocks and irregular porphyry bodies, usually enriched in indigenous pyrite or pyrrhotite, often contain higher than normal amounts of gold in the vicinity of gold deposits. This is true in the greenstone belts at Yellowknife, at Matachewan, at Timmins and elsewhere. Moorhouse (1942) has commented on the common occurrence of auriferous albitite dykes and albite porphyries in the Canadian Shield and elsewhere. He gave evidence that he thought pointed to the magmatic differentiation origin for the gold. Others have considered that the higher than normal content of gold is due to secondary processes attendant upon fracturing or shearing. This is certainly true in many cases; in others the gold enrichments appear to be primary.

Wolfe (1975) gives interesting data in this respect from six Early Precambrian felsic intrusions in Ontario. These range in size from 4 to 88 km^2. Four (Dome stock, Red Lake area; Larder Lake stock; Lebel stock, Kirkland Lake area; and Cairo stock, Matachewan area) have associated gold deposits and two (Kabenung Lake and Coyle Lake) are barren. Wolfe observed that the gold levels in unmineralized stocks are consistently below the 5 ppb detection limit of the analytical method used. In contrast, analyses of samples from stocks with associated gold mines within their borders and in nearby metavolcanic-metasedimentary host rocks show a strongly skewed trace element gold population with a substantial percentage of gold values exceeding 5 ppb and ranging as high as 320 ppb. From these data he hypothesized that (1) orestage hydrothermal metasomatism has produced widespread gold enrichment in Precambrian intrusions that host epigenetic vein-type gold deposits, and (2) geochemical gold analysis of samples collected from 30 to 50 randomly distributed sites can be used to estimate the gold exploration potential of stock-sized bodies and to outline internal patterns of gold variation that may be useful in directing exploration to particular contacts or portions of a stock. He cautioned that his results *do not* conclusively indicate that the presence of auriferous deposits is signalled by unusually high gold contents in primary unaltered igneous rocks.

When relating gold deposits to igneous rock bodies such as dykes, stocks and small batholiths one must be certain that the igneous bodies are either older than or coeval with the gold mineralization. In the first case the mineralization may be associated as a result of metamorphic secretion processes; in the second the mineralization may be related genetically via differentiation processes.

The concentration of certain indicator elements in intrusive stocks and small batholiths may signal the presence of auriferous deposits in some districts. In this respect Garrett (1974) has shown from extensive data on the mean mercury content of granitoid plutons in the McQuesten area of the central Yukon that a positive correlation exists between plutons with a high mercury content and the presence of gold deposits.

Dvornikov (1964) working in the southeastern Donets Basin of the U.S.S.R. also noted a regional relationship between the mercury content of the rocks and the presence of gold deposits. He found mercury dispersion aureoles in the country rock, and in soils and alluvial deposits. Two types were distinguished (a) primary, formed in sandstones, clay shales, limestones and coal simultaneously with the lead-zinc and gold deposits and occurrences, and (b) secondary, formed in soils, loose alluvial deposits and waters subsequent to the formation of the deposits.

Both types of mercury dispersion aureoles fringe the districts of lead-zinc and gold deposits of the Nagol'nyy Range. The mercury dispersion aureoles were found to be usually several times as wide as the zones of mineralization, deposits and ore-bearing belts. The mercury content of the rock within the contours of the dispersion aureoles was determined to be tens of times higher than the geochemical background.

Cornwall *et al.* (1967) examined the Ag, Hg and Au contents of residual soil, rock from mine workings and bedrock over a large area in Nevada marked by two bonanza silver-gold districts (Comstock and Tonopah). They observed that geochemical anomalies for silver clearly delineate areas of principal silver mines in both districts, and mercury anomalies show the same pattern at Tonopah. In the Comstock district, however, mercury values were found to be highest outside the main silver-gold mining area a feature that they thought suggested horizontal zoning from a central silver-gold area outward to a mercury area. They speculated that if similar zoning occurs in the vertical plane, several mercury anomalies in the districts may be underlain by silver-gold mineralization.

Analyses of mineral separates in the country rocks may be used to indicate the presence of gold deposits. The best mineral separates for this type of work in our experience appear to be pyrite and arsenopyrite, particularly the latter. Pyrite and arsenopyrite isolated from the greywacke and slate wall rocks adjacent to gold-quartz veins in the Meguma Group of Nova Scotia are invariably enriched in gold (and silver) compared with these minerals taken from rocks distant from the veins. Similar relationships were noted at Keno Hill in Yukon. Not enough work was done, however, to establish if there is a gradual increase in gold content towards the veins. This approach should be tried on long drill cores where they are available. Magnetite appears to concentrate gold in some rocks as noted in Chapter II; it could be a useful mineral in studies of this type.

With respect to pyrite, Sher and Demchenko (1962) comment on the use of the mineral as a guide to concentrations of gold and remark on the frequent enrichment of pyrite in and near gold-bearing deposits. In the Lena Gold Belt, U.S.S.R., they noted two distinct types of pyritization, the gold mineralization being associated with late pyritization characterized by a major development of pentagonal dodecahedral crystals. In the Philippines, Santos and Walters (1971) found that pyrite, chalcopyrite and sphalerite in deposits and prospects of various types were gold collectors and noted that these minerals were consistently high in the precious metal in auriferous districts. On this basis they concluded that the three sulphides are useful in evaluating prospective gold-bearing belts. They considered that a lower threshold limit of 1 ppm gold distinguished gold-productive belts from those containing essentially base metal deposits.

Vakrushev and Tsimbalist (1967) examined the gold content of sulphide minerals in a number of gold-bearing skarns, gold-quartz deposits and nongold-bearing skarn in the Altai-Sayan, U.S.S.R. All sulphides, particularly pyrite, were enriched in gold in the first two types of deposits; in the last the sulphides were characterized by minimal gold contents. From these data they concluded that the gold content of pyrite gives a direct indication of commercial gold mineralization in evaluating skarn deposits.

In a similar study on auriferous and barren skarns in the Kommunarovsk and Ulen-Tuimsk regions of the Altai-Sayan, Okhapkin and Bozin (1971) found that the ore magnetites contain lower amounts of Ti, V and Cr than the accessory (barren) magnetites. The authors considered that the low content of Cr in the auriferous region could be used as a prospecting criterion.

Many types of gold deposits are enveloped by primary halos of various types, involving both major and trace elements. Some of the halos are of the additive type[32], that is they exhibit an increase in the content of particular constituents as the veins are approached; others are of subtractive nature and show depletion or decrease of a particular constituent as the veins are approached.

The major constituents that seem to be of the greatest use in geochemical prospecting for gold using primary halos appear to be SiO_2, K, Na, CO_2, S and H_2O. The behaviour of these constituents during wall-rock alteration processes and the mineralization of most types of gold deposits is outlined in Chapter III and need not be repeated in any detail here.

Silica exhibits major decreases as gold-quartz veins and lodes are approached particularly in ultrabasic, basic and intermediate volcanic and intrusive host rocks. The effect is not marked in most acidic host rocks and sediments of the greywacke-slate type and their metamorphic equivalents. Silicification, or increase in silica content as the veins and lodes are approached, is a feature mainly of limy host rocks, certain acidic rocks and occasionally sediments of the greywacke-slate type and their metamorphic equivalents. Silica profiles may be of interest in locating gold-quartz deposits in sheared greenstones and in the various other types of rocks mentioned above. The method is probably best used on drill core although grids of surface rock samples may also be employed. It is imperative to ascertain the exact nature of the profiles to be expected before employing the method; in this respect one or more pilot profiles across known gold deposits will give the required information as to whether subtractive, additive or uniform profiles prevail for the types of deposits sought. The effectiveness of using silica profiles alone in prospecting for gold deposits is relatively unknown. Probably decreases or increases in silica content provide no more than general information that mineralization processes have been active in the particular shear zone, fault or bed. Combined with information on the distribution of other major and trace elements in halos about gold deposits, silica profiles, however, provide another arrow in the quiver of the ore hunter.

---

[32]The terms additive and subtractive are meant only in a descriptive sense. They are not intended to infer that the constituents were added to the wall rocks from the vein channels (fractures, faults, etc.) or similarly removed from the wall rock to the vein channels. Primary halos are usually complex and difficult to analyze with respect to the original direction of flow of constituents. In homogeneous, unfractured wall rocks an increase in a constituent toward a vein site may indicate movement of constituents from the vein channels into the wall rocks; or alternatively it may indicate the frozen profile of a smooth buildup of diffusing ions at a precipitation front. Where the wall rocks are extensively fractured an increase in constituents may indicate an increase in the number of small fractures as the vein site (the main locus of fracturing) is approached. In this case again the constituents may have migrated from the main channel into all the subsidiary channels; or alternatively diffusion of constituents toward and into the vein site may have resulted in the filling of the principal fractures as well as all of the small subsidiary ones. The subtractive halos may indicate that constituents moved toward and into the vein sites; or alternatively that they were widely dispersed into the wall rock beyond the halo. The genesis of halos is really not the problem in geochemical prospecting, although it is advantageous to have an idea as to how each halo was produced if possible. Empirically it is known that certain gold deposits have particular types of halos, whereas others have different types. *See also* the discussion on wall-rock alteration on Chapter III and the outline of the origin of epigenetic gold deposits in the same chapter.

Potash and soda commonly show an antithetical arrangement in alteration profiles as gold-quartz and other auriferous deposits are approached (*see* Tables 45 to 59). This relationship is also reflected in their ratios. Thus, as the deposits are approached potash commonly increases consistently whereas there is a simultaneous decrease in soda; the K_2O/Na_2O ratio, therefore, generally increases as the veins are approached. This is of considerable importance in surveys utilizing drill core or grids of surface rock samples, since if a progressive increase in the ratio K_2O/Na_2O is observed, the indication is that gold mineralization is being approached (Boyle, 1974*b*). Similar phenomena have been observed by the writer around polymetallic deposits of various types. Actually the phenomena is only a reflection of potash metasomatism associated with mineralization. The use of the ratio K_2O/Na_2O is, however, more sensitive and indicative.

Borcos and Stanciu (1974) observed a genetic relationship between the gold-silver mineralization and the wall-rock alteration effects, manifested principally by intense sericitization of the host Tertiary andesites, in the Hanes deposit in the metalliferous mountains of Romania. When the K_2O content of the alteration zones exceeds 3 per cent proximity to economic orebodies is indicated.

Carbon dioxide (carbonate) is particularly indicative of gold-quartz mineralization, since nearly all types of host rocks exhibit an increase in CO_2 as the auriferous deposits are approached. In skarn and silicified bodies in limy rocks the reverse is true; there is a major decrease in CO_2 due to skarnification or silicification. It is imperative, therefore, to ascertain the types of profiles to be expected in an area before detailed prospecting using this constituent is attempted. In this respect a study of Tables 45 to 59 is instructive. In the aureoles of most gold-quartz deposits CO_2 is introduced into the wall rocks concomitant with removal of SiO_2. The ratio SiO_2/CO_2, therefore, decreases as mineralization is approached in most cases. In limy rocks that are skarnified or silicified, on the other hand, the ratio SiO_2/CO_2 usually increases consistently toward the ore zones. Analyses for CO_2 and SiO_2 of drill cores or of rock samples on grids should be useful in assessing an approach to gold orebodies, but it must be stressed that careful scrutiny of the results and a knowledge of the profiles to be expected in an area are necessary for success. Keyed with trace elemental surveys of the same rock samples the information obtained from SiO_2/CO_2 ratios may be invaluable.

Fakhry (1974) has shown that infrared spectroscopy is a useful means for evaluating the ratio of carbonate to silicate minerals in rock samples from the alteration zones of gold deposits, particularly at the Fawakhir Mine, eastern desert, Egypt. The ratio of absorbance of the C-O vibration absorption band of carbonates and the Si-O vibration absorption band of silicates (C-O/Si-O) defines a characteristic parameter (δ) that ranges from 0 in rocks with no carbonate to some high positive value where carbonatization is marked. This ratio, therefore, increases with carbonatization, that is toward auriferous veins, where this type of alteration prevails.

Barkhudaryan and Grebenchikov (1974) described a method utilizing infrared spectroscopy for evaluating the gold content in auriferous quartz veins.

Konstantinova *et al.* (1973) state that the composition of carbonates may be used as an indicator of the intensity of hydrothermal gold mineralization in the Sukholozhsky prospect in the Lena region of U.S.S.R. Thus, superore zones are characterized by carbonates with high contents of Ca and Sr, and there is a general increase in these elements and an increase in Fe and Mg in the carbonates with depth, that is toward and into the auriferous ore zones. Gormasheva *et al.* (1973) by approximate extraction methods of the components of carbonates with acetic acid have been able to differentiate auriferous propylite zones from those that are barren in the Evensky Gold Belt, U.S.S.R. The auriferous variety associated with the gold deposits is characterized by the presence of Fe-Mn carbonates whereas the barren ones contain mainly calcite and dolomite.

Sulphur commonly increases in the host rocks of gold deposits as the ores are approached. All profiles studied by the writer in nearly all types of gold deposits show this effect except those in sulphide-rich iron-formations and certain pyrite-rich rocks such as pyritiferous shales and porphyries. In these any pattern is usually swamped by the large amount of sulphur present in indigenous pyrite and pyrrhotite. Combined with studies of the distribution of heavy metals and especially of the semimetals As, Sb, Se and Te, the distribution of sulphur may be particularly useful in exploration work involving drill cores and surface rock samples on grids.

The water content of host rocks due to hydration processes adjacent to gold-quartz and other types of gold-bearing deposits is commonly erratic and dependent upon certain equilibrium relationships inherent in the formation of chlorite, sericite, clay minerals and other hydrous minerals. These relationships are commonly difficult to ascertain without detailed petrographic work. The water content of rocks is, therefore, not particularly useful in geochemical surveys using primary halos. A more useful index is the ratio $SiO_2/(CO_2+H_2O+S)$. In most types of gold deposits, excepting skarn and silicified deposits, this ratio consistently decreases as the orebodies are approached. This is easily understood from the fact that as the silicates are carbonated, hydrated or sulphidized, silica is released and this usually migrates into the dilatant zones to form quartz.

$$(Ca,Fe,MG)\text{ silicates} + CO_2 \rightarrow (Ca,Mg,Fe,)CO_3\text{ (ankerite)} + SiO_2$$

$$\text{Mg-Fe-silicate} + H_2O \rightarrow \text{chlorite} + SiO_2$$

$$\text{Fe-silicate} + S \rightarrow \text{pyrite} + SiO_2$$

Where arsenic and boron occur in quantity in the alteration halos these elements should be added to the total of the volatiles.

Gold is associated with a large number of trace elements that can serve under particular conditions as indicators in geochemical prospecting surveys using grids of surface rock samples and drill cores. The chemistry of these elements and their relationships to gold in hypogene deposits and their wall-rock alteration zones are discussed in detail in the sections in Chapter III on associated elements and wall-rock alteration phenomena. The elements that should serve as the best indicators in lithogeochemical work include Rb, Cs, Cu, Ag, Ba, Zn, Hg, B, U, Pb, As, Sb, Bi, Se, Te, Tl, Cr, Mo, W, Mn, Co and Ni. Of these the most specific indicators are Cu,

Ag, Zn, Hg, B, As, Sb, Bi and Te. As, Sb and Te are nearly universally present in hypogene gold deposits. Of these three, As is the best indicator; Sb and Te are normally very low in content in many gold deposits, and hence if they are found in amounts higher than their normal abundance (Sb–0.7 ppm; Te–0.05 ppm) this should signal the probable presence of gold (and silver) mineralization. Fluorine may be useful in certain types of deposits where tourmalinization is well developed or where fluorite is a gangue mineral as in some Tertiary deposits (e.g., Cripple Creek). The efficiency of chlorine, bromine and iodine as indicators of gold deposits in lithogeochemical surveys is relatively unknown. Judging from the general erratic behaviour of chlorine in some of the alteration halos examined by the writer it would appear that these elements have only a limited usefulness under the most favourable conditions.

Many of the indicator elements of gold exhibit a much smoother behaviour in their distribution compared with the precious metal. This is often manifested in both regional profiles and in those across primary halos associated with auriferous deposits. The elements most often exhibiting this desirable behaviour include As, Sb, B and Zn. Many factors influence the zoning of indicator elements associated with the focus of auriferous mineralization, particularly the lithology, degrees of porosity and permeability of the host rocks (principally the amount of fracturing, shearing and contortion to which the rocks have been subjected; also the initial porous and permeable character of the rock), and the penetrative nature (gas, liquid) of the mineralizing medium. From my general observations I conclude that the mobility of the indicator elements with respect to gold and silver in primary halos follows the general sequences:

Gangue elements:

$Si > B > K > Na > Ca = Mn > Mg > Co > Fe = Ni > Al$

Gaseous compounds:

$CO_2 = S = Se > As > Sb > Te$

Base metals:

$Hg > Zn > Cd > Cu > Pb > Mo = Bi = Sn = W$

Precious metals:

$Ag > Au > Pt$

In primary and leakage halos, therefore, one tends to find that compounds such as CO_2, Hg, B, S, As, etc. have travelled far from the focus of Au-Ag mineralization. They commonly occur in greater concentrations well up-dip from the auriferous orebodies and generally have broad lateral halos, whereas the other constituents such as Cu, Pb, Mo, Bi, Ag and Au have relatively few up-dip manifestations and restricted lateral halos.

A number of investigations utilizing gold and its indicator elements in geochemical surveys using rocks to locate gold-bearing deposits are described in the literature. Most of these are based on trace elements in primary halos.

Goloubinoff (1937) seems to have been one of the first to suggest the use of rock samples in prospecting for gold deposits. He noted that rocks in gold-bearing districts have a higher than average gold content (clarke). He gave as the average clarke for gold a value of 0.10 ppm, and found a micronorite from Haut Tonkin to be eight times the clarke, a pegmatite from Annam 25 times, and a diabase from the Belgian Congo (Zaire) and a norite from French Guinea (Guinea) to be 10 times, all in auriferous areas.

Sokoloff (1950) carried out geochemical surveys utilizing rock and drill cores in the vicinity of the gold deposits of Kalgoorlie, Western Australia. The greenstones showed a number of gold positive bands that the author interpreted as trace analogues of economic lodes, localized by microstructures in the country rock. These appeared to bear no particular relationship to the main lodes. On continuous gold halos around the lodes the author commented as follows: "Diffusion of traces of gold into walls of the lode channel would result in a reasonably continuous anomaly in the rock that could be termed a gold halo. Suggestions of such diffusion are indeed present locally. The linear extent of the diffusion is very small, however. Some lodes are entirely devoid of the halos. Rock within a few inches of the lode may be gold-negative."

In an abstract Robertson (1955) mentions the interesting fact that geological structures can be traced by geochemical means in some places. At Maryville, Montana, a contact of granodiorite with hornfelsed Belt sediments is conspicuously marked by concentrations of Au and Ag in the granodiorite.

In recent years a considerable number of investigations have been carried out on primary trace element halos associated with gold deposits in an effort to use the data obtained as a guide to the location of the deposits, especially those that are blind or deeply buried. Those noticed in this compilation deal mainly with vein, lode and replacement type gold-bearing deposits and include investigations by Kutyukhin (1960), Boyle, (1961*a*), Khramyshkin (1962), Kvyatkovskii (1963), Mukanov and Romashin (1964), Rakhmatullaev and Gamaleev (1964), Mazzucchelli (1965), Fakhry (1965), Polikarpochkin *et al.* (1965), Palei *et al.* (1967), Mahrholz and Slaughter (1967), Gapontsev and Polikarpochkin (1969), Kitaev *et al.* (1968), Khramyshkin (1968), Potapenko *et al.* (1968), Grigor'ev (1968), Eleyeva and Rusinova (1969), Bezirganov (1969), Baumshtein and Grigor'ev (1969), Zashchinskii *et al.* (1969), Obraztsov (1969), Stevens *et al.* (1969), Kitaev *et al.* (1970*a,b,c*), Mel'nikov and Mel'nikova (1970), Roslyakova *et al.* (1970), Laipanov and Sher (1971), Grigoryan and Zubov (1971), Wilson (1971), Ovchinnikov and Grigoryan (1971), Sears (1971), Bolter and Al-Shaieb (1971), Gapontsev (1971), Voin (1971), Orazbekov (1971), Polikarpochkin and Kitaev (1971), Bowen (1972), Prosnyakov *et al.* (1972), Atabek'yants (1972), Obrazbekov (1972), Mercer and Crocket (1972), Ovchinnikov *et al.* (1972), Aristov *et al.* (1972), Al-Shaieb (1972), Jae (1973), Kitaev and Chumakin (1973), Chibisov *et al.* (1973), Korobeinikov (1974*a*), Gustavson and Neathery (1975) and Steed *et al.* (1976).

The results and conclusions drawn from these investigations have been varied and are not readily systematized. In a general way it can be said that the primary trace element halos associated with most types of gold veins, lodes and replacement bodies are marked. Widths, extents and up-dip projections vary widely depending upon the types of host rocks, localizing structures, degree of wall-rock alteration, post-mineralization fracturing and other geological factors. Each gold belt and commonly each deposit has its individual characteristics. The most satisfactory indicators for use in primary halo surveys in the search for gold deposits appear to be Au, Ag, As, Sb and Hg. Other elements such as Zn, Pb, Cu, B, Ba and W are applicable in certain belts. All investigators agree that As and Sb, particularly the former, are the best universal indicators. At Yellowknife, enrichments of these two elements in the shear zones can be noticed hundreds of feet up-dip from underlying orebodies. Laterally, the schist and weakly mineralized material in shears and fractures are enriched in both elements up to distances of 500 ft or more from known orebodies. At Norseman in central Western Australia, Mazzucchelli (1965) noted strong leakage halos of arsenic above lenticular gold-quartz reefs in inclined zones of reverse shearing. These persisted up the shear zones as much as 2000 ft up-dip from blind orebodies and could be detected in soils over the barren suboutcrops of the ore-bearing structures.

The details of the dispersion of various trace elements including gold in primary halos associated with auriferous mineralization are complex. Some observers find the lateral dispersion limited to a few feet and even to a few inches in some gold districts; in others the lateral dispersion may exceed 300 ft. The vertical dispersion, up-dip of orebodies in shear zones, fracture zones or replaced beds is commonly much greater than the lateral dispersion for reasons of permeability and porosity. In most districts the leakage halos can be detected at distances measured in hundreds and in places in a few thousands of feet above blind orebodies. Multiple element analyses of primary halos have been done in a number of gold districts in U.S.S.R. especially at the Baleisky and Onokhovsky gold deposits in Transbaikal and the Belaya Gora deposit in the Far East (Polikarpochkin *et al.*, 1965; Polikarpochkin and Kitaev, 1971). These deposits are clusters of epithermal gold-quartz veins in a variety of sediments and volcanics that exhibit marked chloritized, silicified, kaolinized and pyritized zones adjacent to and between the veins. The minerals in the chalcedonic quartz veins are essentially native gold of low fineness, stibnite, pyrite and arsenopyrite with varying amounts of pyrargyrite, galena, sphalerite, chalcopyrite, cinnabar, fluorite, realgar and orpiment. The details of the elemental distributions in the halos of the various deposits are complex and should be sought in the original works. In general it can be stated that the halos have a multicomponent nature exhibiting enrichments of Au, Ag, As, Sb, Cu, Pb, Zn and Hg. Normally the halos are narrow (30 m) near the roots of the deposits, widen out (in places to 300 m) in the rocks flanking the productive parts of the deposits and continue so for distances up to 300 m above the orebodies. A vertical zonation of elements in the halos is also apparent; maxima for Au and Ag occur adjacent to the zones of productive mineralization whereas the maxima for As, Sb and Hg are displaced higher up the structures. The Au/Ag ratios also show vertical variations, the zones of maximum mineralization (maximum gold deposition) being distinguished by a maximal value of the ratio, whereas the halos above and below are characterized by decreased values of the ratio.

Polikarpochkin and Kiteav (1971) conclude from their research on the primary halos associated with epithermal gold deposits that:

> The marked lateral extent of the halos aids in the lateral search for orebodies, and the projection of the halos upward above the orebodies for a considerable distance is particularly helpful in locating blind orebodies. In the latter respect, the search for deep orebodies is best done by using the halos for arsenic, antimony, mercury, and copper. The halos for gold and silver, on the other hand, are restricted in their vertical extent and are thus not particularly useful.
>
> Prospecting for blind orebodies can be guided by the elemental composition of the halos and the nature of the change of the concentration of the elements with depth. If the halos exhibit high concentrations of only arsenic, mercury, and antimony without high concentrations of gold and silver, this indicates that one is high above the zone of optimal gold mineralization. Copper concentrations in the halos decrease markedly as one approaches this zone of optimal mineralization. The presence of high contents of gold and silver indicates a near approach to the zone of maximal mineralization, especially when a decrease in the Ag/Au ratio is apparent. The vertical interval over which these elemental variations are manifest is equivalent to the vertical range of the supra-ore halos and is approximately 200 to 300 meters. A decrease in the content of gold and silver with depth and an increase in the Ag/Au ratio means that the zone of optimal mineralization is higher than the level of sampling or that it has been eroded. The identification of sub-ore halos is particularly important in prospecting. They can generally be recognized by their narrow width and a progressive decrease in their content of gold with depth. The silver contents in these halos are frequently high, and there may also be higher lead contents in places.

Much more work is required on the distribution of gold and silver in primary halos associated with gold deposits to define their exact nature. This is especially true where gold-quartz veins and lodes occur in wide schist zones as at Yellowknife and elsewhere. In such zones the wide shear (schist) zones of chlorite schist appear in places to be depleted in gold and silver with respect to the country rocks, the two metals being probably added to the quartz veins and their adjacent intensely altered carbonate-sericite schist zones (Boyle, 1961*a*). Others have noted this phenomenon (*see* the section on origin of gold deposits in Chapter III). In particular, Roslyakova *et al.* (1970) remark on negative gold halos associated with the Darasun gold deposits in eastern Transbaikal and the Berikul' deposit in the Kuznetskii Alatau Mountains, both deposits being steeply dipping quartz veins, the former in granodiorite and the latter in plagioclase hornblende porphyries and tuffs respectively. The negative gold halos correspond to the zones where the most intense wall-rock alteration effects are manifest, in granodiorite they may be up to 2 m wide adjacent to the narrow veins. Roslyakova *et al.* (1970) conclude that the gold (and other elements such as Ag, As, Pb, Zn, Cu, Ni, Co, etc.) were transferred to the vein sites from the wall rocks. Bolter and Al-Shaieb (1971) have also noticed negative anomalies in gold and other metals in the immediate vicinity of veins in andesites in the Searchlight district of Nevada. In general, it appears that gold and silver tend to show a regular increase in quantity towards deposits in some places, whereas in others the general increase in the concentration curve may be interrupted by negative (i.e., leached out) sections (Fig. 98).

Leakage halos associated with gold mineralization and manifest by pervasive alteration of country rocks (propylitization, pyritization, arsenopyritization, silicification, alunitization, etc.) within mineral belts or between vein systems; weakly mineralized schist, mylonite, gouge, etc. in satellite and subsidiary shears, fractures and faults; patches and linear streaks of silicified rock; and rocks exhibiting other evidence of epigenetic introduction of elements such as silicification of limestones, development of skarn and impregnation of rocks by pyrite and arsenopyrite are considerable aids in the location of gold deposits buried deep within rocks or hidden

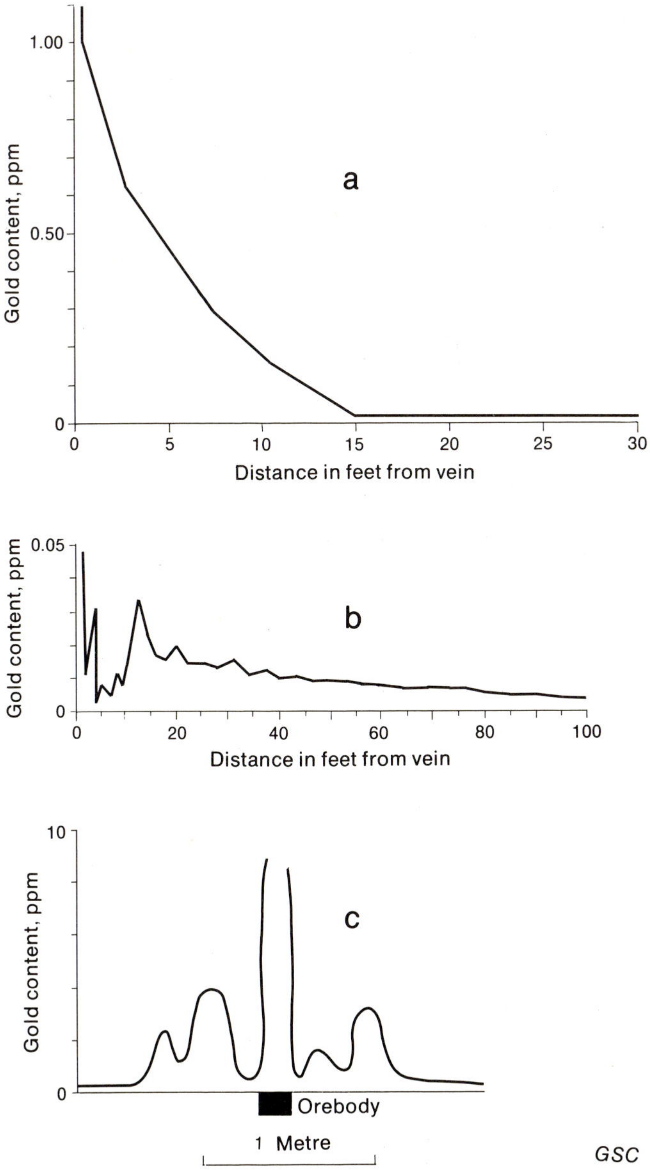

Figure 98. Positive and negative primary profiles in auriferous deposits. (a) General distribution of gold in the primary halos at Yellowknife, N.W.T. (b) Distribution of gold at the Chief of the Hill Mine, Searchlight district, Nevada (after Bolter and Al-Shaieb, 1971). (c) Distribution of gold at one of the Klyuchi orebodies, eastern Transbaikal, U.S.S.R. (after Roslyakova and Roslyakov, 1975).

by overburden especially in valleys. Analyses of the above materials for gold and silver and their indicator elements, particularly As, Sb, Te, Hg, etc. when plotted on geological maps of an area and analyzed in three dimensions may indicate the presence of various types of gold deposits. Thus, the large gold-quartz lodes of the Yellowknife area, Northwest Territories, most of which lie in drift-filled valleys, are clearly indicated by satellitic and subsidiary shears and fractures weakly mineralized with Au, Ag, As and Sb in the rocks flanking the valleys. Similar conditions exist in the Red Lake Camp and elsewhere in the Canadian Shield according to the writer's observations. Disseminated deposits of the Carlin, Nevada type are particularly susceptible to discovery by surveys utilizing analyses of the materials of narrow quartz veins, calcite veins and shear zones in partially silicified limestone for As, Sb, W, Hg, Au, etc. Thus, the anomaly in the Cortez district of Nevada outlined by Erickson *et al.* (1966) led directly to the discovery of over 3 million tons of open pit gold ore averaging 0.3 oz Au/ton (U.S. Geol. Surv. Circ. 560, 1968, p. 1).

The use of analyses of the materials of leakage halos and of pervasive alteration zones cannot be over-emphasized in prospecting for covered and deeply buried, blind gold-bearing orebodies. When carrying out a lithogeochemical survey based on leakage halos or pervasive mineralization effects using either rock samples or drill cores the following points should be kept in mind. In addition to analyzing the rock and core samples, all shear zones, fractures, contorted zones and altered zones should, likewise, be sampled and analyzed for gold and its indicator elements. A detailed geological map showing all these features as well as any small veins, no matter what size, should be plotted and the gold and indicator element values entered at the appropriate sites. Where drilling is done, sections with all of this detail should also be prepared. Only in this way is it possible to observe patterns in the primary gold dispersion in the rocks and from these patterns to predict the locus of large deposits. It should be constantly borne in mind that most large deposits have a halo of smaller satellites developed in subsidiary or parallel fractures or in favourable sites in porous, permeable and chemically replaceable rocks. Trace-element work on small shear zones, fractures, etc. increases our ability to differentiate smaller and smaller satellites. It may well be that the data when plotted will show an increase in the gold or indicator element content of alteration zones in a certain direction or that the number of gold-bearing fractures, etc. increases toward a certain valley or draw beneath which lies a major shear or fault zone containing deposits. It is also advisable to contour the results since this method often brings out zones that should be trenched or drilled.

Palei *et al.* (1967) have investigated the geochemistry of gold and some of its indicator elements in alteration zones in the vicinity of the Sultanuizdag deposit in Uzbekistan, U.S.S.R. Some of their observations are of interest. They found that the nearby Dzhamansai intrusion is the most enriched in gold (0.066 ppm Au) in the area. Among the various alterations in the intrusive rocks only sericitization and silicification displayed a direct connection between the degree of alteration and gold content. A direct positive correlation was observed between the concentrations of Au and Ba. The relation between Au and Pb or Ga was inverse. In the Aktausk intrusion a direct relation was noticed between the Au concentration and the degree of muscovitization and silicification and a direct relation was traceable between the concentrations of Au and Ba. For Au and Cu an inverse ratio obtained. Mica-plagioclase-quartz schists form a hydrothermally but slightly affected belt at a distance of 1 to 2 km from the Au-bearing zone. The Au concentration of the belt averages 0.0036 ppm. As the mining area is approached hydrothermal alterations intensify, Au concentrations rise to 0.009 ppm, and within 10 to 20 m from the orebodies, to some tenths of a ppm. In order to determine the relationships between Au and the accessory elements the correlation ratios were calculated. The results showed that the Ba, Mn, As and Be concentrations rise in direct proportion to that of Au. The reverse applies for Pb. A marked relationship prevails also between Au and the elements Ga, Zr and Y. There was no observed correlation between Au and Cu or Zn. The main amount of Au was deposited during the post-magmatic stage and is due to epigenetic processes.

Geochemical surveys utilizing leakage halos and zones characterized by pervasive mineralization effects associated with gold-bearing deposits have been carried out by the United States Geological Survey for a number of years mainly in the western States, but also in Alaska and elsewhere. Investigations noticed during this completion include those by Erickson *et al.* (1964, 1966), Gott and McCarthy (1966), Gott *et al.* (1969), Drewes (1967), Cornwall *et al.* (1967), Lovering *et al.* (1968), Wrücke *et al.* (1968), Elliott and Wells (1968), Whitebread and Hoover (1968), MacKevett and Smith (1968), Gott and Zablocki (1968), McCarthy *et al.* (1969*a*), Akright *et al.* (1969), Wells and Elliott (1971), Chaffee (1972), Bailey and Williams (1975), Ashley and Albers (1975) and Ashley and Keith (1976). In the areas in Nevada containing disseminated gold deposits in limestone (e.g., Carlin, Cortez, Getchell, Gold Acres), oxidized materials along fractures, small quartz and calcite veins, silicified zones, mineralized material in shear zones and breccia zones, altered rocks, gossanous materials and jasperoid[33] have proven to be excellent sampling materials. In these Au, As, Sb, Hg, W, Ag and Bi appear to be the best indicators of gold mineralization. A number of other elements including B, Ni, Cu, La, Y, Ga, Sc, Pb, Mo, Sn, Zn, Cu and Te are also present in anomalous amounts in the sampling materials and in the known orebodies. In the Drum Mountains, Utah, jasperoid is gold bearing and commonly contains anomalous amounts of Ag, Bi, As, Sb, Sn, Pb, Cu, Hg and Y. Silver in altered rocks and other materials clearly marks the gold-silver-bearing lodes and zones in the Comstock Lode and Tonopah areas, Nevada, and mercury appears to be particularly useful for indicating gold-silver mineralization at depth in both areas. In the southern Santa Rita Mountains, Arizona, sampling of altered rock revealed large zones of granitic rocks enriched in Cu, Pb, Zn, Ag and Au. At Cripple Creek the volcanic rocks between the productive veins contain an average of about 0.6 ppm Au, and a large area of rocks marked by a gold anomaly averaging 2.5 ppm Au was located near the Cresson Mine. This anomaly also contains enriched amounts of Ag, Te and Hg. In the Ely porphyry copper district of Nevada, jasperoid, gossanous material, silicified fracture fillings and dolomites exhibiting silica boxworks were found to be good sampling materials. Gold, silver, tellurium and mercury in these materials are distributed in zones clearly related to centres of alteration that are in turn related to the major copper deposits. The four elements form halos around a copper-rich core. The Ely district as a whole is greatly enriched in tellurium, large volumes of rock containing of the order of 100 ppm Te (2000 times the average abundance of Te in igneous rocks). Tellurium and mercury are also enriched in certain jasperoid samples (Lovering *et al.* 1966). They found that the 'favourable type' of jasperoid commonly associated with metalliferous ores generally contained >1 ppm Te and >5 ppm Hg, whereas the 'unfavourable type' found in barren zones contained only low amounts of these elements.

In the Taylor Mining District and vicinity near Ely, Nevada, Lovering and Heyl (1974) found analyses of jasperoid to be particularly indicative of precious metal mineralization. They observed certain characteristics of the jasperoid which appear to be most useful as guides to the focus of silver-gold mineralization as follows: (1) a change from dominantly brown to dominantly grey to dark grey or nearly black, (2) a coarsening in texture of the matrix jasperoid, (3) an increase in the number of vugs and (4) an increase in the ratio of copper to chromium. They noted that all the metals including silver and gold, that have been produced from the Taylor district exhibit strong anomalies in jasperoid samples from the central part of the district – in fact, the ores are actually mineralized jasperoid – but the metals show no consistent zonal distribution relative to each other. They also observed that the metals tend to vary considerably in concentration among samples from the same locality and tend to increase and decrease simultaneously as shown by samples collected across a mineralized body. However, they found a variety of patterns in the regional anomalies of the metals that are detailed in the original paper.

Two recent studies in the Goldfield area of Nevada, one on the silicified rocks of the district (Ashley and Keith, 1976) and the other on the distribution of gold near orebodies in the oxidized zone (Ashley and Albers, 1975) are of interest in the problems related to prospecting for these deposits. Briefly two correlations of elements were noted in the silicified rocks – the first Au, Ag, Pb, Bi, As and Cu represents a relict hypogene association preserved in spite of oxidation, and the second As, Cu, Mo and Fe is a supergene association produced during oxidation when varying amounts of these elements were mobilized and precipitated with limonite. For the second investigation 278 samples of argillized and silicified dacite were collected in 1966 from excavations at the Combination and January mines, which once yielded gold in commercial quantities. Semiquantitative analyses show that Au, Ag, Pb, B, Hg and As are notably enriched in rocks of the cuts. All these elements except lead and mercury formed conspicuous ore minerals. Geochemical maps and one geochemical profile across strike show that relatively high concentrations of all these elements are restricted to silicified zones. This low-tenor metallization dispersed through silicified zones does not extend into adjacent clay-bearing rocks. During oxidation, arsenic, copper, molybdenum and zinc were

[33] Jasperoid refers to an epigenetic rock body formed mainly by fine-grained, chertlike siliceous replacement of a pre-existing rock. Most jasperoid bodies range in size from small pods to masses more than 1000 ft wide. Most are structurally controlled and occur along fault and fracture systems; some are chemically controlled by the presence of limy rocks, the silica replacing the carbonates over large volumes or along specific linear zones probably marked by faults and other porous zones. Many of the individual masses of jasperoid are brecciated, the fragments being cemented by younger quartz or less frequently by coarse calcite. The jasperoid rock is generally aphanitic to very fine grained and exhibits a variety of colours including shades of grey-black, grey, greyish brown, light brown, yellowish brown and reddish brown. Much jasperoid is marked by micro vugs, especially in the highly mineralized belts. In most mineralized belts the ore metals, the jasperoid and certain rhyolite porphyry dykes all appear to be genetically related. The ore metals appear to have been introduced penecontemporaneously with the late stages of the silicificiation that gave rise to the jasperoid bodies, by solutions moving upward along fault and fracture systems.

leached from the silicified zones and the orebodies within them, but these metals did not form distinct supergene halos in the surrounding argillized rocks. From the semiquantitative data available, the average amount of gold in silicified vein material was calculated to be between 2 and 4 ppm.

The mineralogical and chemical criteria for assessing what may happen to the precious metal values in auriferous deposits with depth has intrigued geologists and mining engineers for centuries. In the writer's opinion there are few sure and certain criteria that can be considered as signals that the values will continue unchanged, increase or peter out with depth. The changes with depth manifest in horizontal and vertical trace element halos and their relationship to continuance of gold values are discussed above, and the essentials of vertical elemental and mineral zoning in auriferous deposits are discussed in the last part of Chapter III and need not be repeated. A discussion of the use of the Au/Ag ratio in predicting conditions at depth in auriferous deposits is also given in Chapter III and the effect of supergene enrichment, one of the factors in the variation of gold values with depth, is outlined in Chapter IV. In many auriferous areas the persistence of the structural elements localizing the orebodies is an important consideration in the depthwise continuation of gold values, although it is well known that in many belts the structures may continue to depth, but the gold content may diminish or cut-off abruptly. To generalize, one can say that in the shallow, commonly Tertiary, deposits the precious metal values tend to diminish with depth. Few such deposits are productive of gold below 2000 ft. On the other hand the deep-seated, generally older than Tertiary deposits, may extend to depths greater than 10 000 ft (e.g., Kolar).

Anderson (1935) gave a number of geological guides that he considered could be helpful in examining gold-quartz mines with respect to depth continuation and other matters in the North American Cordilleran region. Indications of petering out of gold-quartz orebodies considered were: (1) persistent increase in the base metals. This may be gradual or abrupt; (2) marked decrease in the fineness of gold. This indicates an approach to the bottom of the ore zone; (3) presence of a dyke (the most acidic dyke in the particular igneous assemblage of rocks in the district) from which the gold-quartz orebodies segregated. The dyke(s) can often be recognized by the presence of finely disseminated primary pyrite; and (4) presence or association of pegmatite or feldspar with the gold-quartz veins. The last two criteria are based on Anderson's (1935) idea that gold-quartz veins differentiate directly from the most acid dykes of any igneous assemblage. Hence, when the gold-quartz veins pass into the dykes they have reached their source and must terminate. These ideas have been criticized by Dougherty (1936) who pointed out that in many areas gold-quartz orebodies continue downward to considerable depth in, along or near the most felsic (acidic) dykes in the belt, such as dykes of alaskite, albitite, quartz porphyry and feldspar porphyry.

The recent book by Roslyakova and Roslyakov (1975) on the subject of the endogenic (primary) halos of gold deposits merits close attention since it embraces a large amount of very careful work. Only a brief summary can be given here; the details should be sought in the original publication, a translation of which is available from the Geological Survey of Canada library.

Roslyakova and Roslyakov review past work on endogenic halos associated with gold deposits and give brief descriptions of the types of gold deposits covered by their investigations. These include mainly quartz vein and disseminated types, auriferous sulphide bodies and auriferous skarns. The deposits include the Darasun, Klyuchi, Natilii, Muruntau and many others that are described in the present volume. The narrative continues with a discussion of the background contents of gold in the rocks (sedimentary, igneous, metamorphic) in the various regions studied. Samples were taken of fresh rock well removed from the mineralized centres (50–80 km). The general averages obtained in ppm were as follows: sedimentary–0.003, with values up to 0.0065 for conglomerates and sulphidic sediments; volcanics – 0.0005–0.005 with values up to 0.010, the basic varieties being highest in gold; acidic intrusive igneous rocks – 0.001–0.005; basic intrusive igneous rocks – 0.001–0.006; metamorphic rocks – 0.001–0.0054. From their data Roslyakova and Roslyakov (1975) concluded that each region in the U.S.S.R. has its own characteristic regional gold background, determined mainly by the development of specific rock types. On the whole they concluded that for most regions the background (in the rocks) ranged from 0.0018 to 0.0034 ppm and the distribution generally follows the lognormal law. They found no particular correlation of gold with other elements in the rocks, although in sediments and their metamorphic equivalents they detected a correlation between the abundance of the precious metal and organic matter in certain regions only. (Compare the data and discussion on gold in rocks given in Chapter II.)

The characteristics of endogenic halos associated with the various types of auriferous deposits follow. This section is preceded by a general summary that can be paraphrased as follows:

Anomalous concentrations of gold and its associated elements are distributed unevenly as distance increases from individual orebodies and then as distance increases from the auriferous ore zones. There is, however, a definite pattern common to all the ore fields, auriferous ore zones and individual orebodies.

At a distance of 10 km or more from the auriferous ore zones anomalous contents of gold, and in certain cases also of arsenic, lead, molybdenum and copper, are encountered in restricted zones. Anomalies of this kind are not strongly marked, either in terms of gold content or of its dispersion. These anomalies are caused by the deviation from the norm to a greater or lesser extent of the contents of carbonaceous matter, dark-coloured minerals and accessory magnetite and sulphides in the country rock.

The gold contents begin to vary more and more on approaching the auriferous ore zones and then on approaching individual orebodies; their average values also increase. Several zones, i.e., halos, can be identified by reference to the content of gold and the nature of its distribution, its concentrator minerals and correlations with other elements. At the present stage of research, halos of three orders can be distinguished with confidence; (1) ore field halos; (2) ore zone halos; and (3) individual orebody halos.

The endogenic ore field halos are in reality local sectors

of the earth's crust with anomalous amounts of gold as compared with the background or clarke of a large region. Most of these halos are extensive according to the authors, the area of the Tertiary Oganchensk Goldfield in Kamchatka being about 50 km^2. The ratio of its width to its length is approximately 1:2. The halo conforms with the orientation of the main tectonic structures of the goldfield, i.e., with the great zones of fracturing and brecciation. Compared with the regional background the gold content in the ore field halo is some 4 times higher in amount. The fluctuation in particular precious metal values is within 2 to 3 orders of magnitude, whereas the fluctuation in the regional background is within one order of magnitude.

Similar features were noted in the other goldfields studied by Roslyakova and Roslyakov (1975). The endogenic ore field halo of the Kozlovskoe field is irregular in shape but some 150 km^2 in area; that for the Klyuchi field more than 12 km^2 in area; and that for the Darasun field about 600 km^2 in area. Gold values in these various halos exceed at least twice the regional averages. Summarizing, the authors make the following statements with respect to ore field halos:

1. The ore field halo serves as the local background to the auriferous zones; the larger these zones, the larger the scales and the higher the average contents and dispersion of gold in the ore field halo. The halos of large ore fields are hundreds of square kilometres in area, those of comparatively small fields cover tens of square kilometres, and those of individual gold occurrences extend to a few kilometers only.

2. The ore field halos have a complex morphology, which is dependent on the basic structures of the region. Their average gold contents are three or more times greater than the regional background. Their particular values vary within 3 to 4 orders of magnitude, and often, especially in the vicinity of deposits, the gold contents are one order of magnitude lower than the regional background. It is, therefore, vital when defining ore field halos to show both the increased and the reduced gold contents as well as the nature of the gold distribution in sectors of the ore field that differ in their composition and in the extent of their alteration.

3. Disseminated sulphides, accessory magnetite, light-coloured minerals and micas are the gold concentrators in ore field endogenic halos. There is no reliable correlation between the gold in the halos and other primary ore elements. In places the gold is directly correlated with barium and copper.

4. Increased gold concentrations (tens of parts per billion) and a distinct irregularity in the gold distribution over large areas are vital geochemical prospecting criteria for gold mineralization. The presence of anomalously low gold contents against an increased background may be a reliable sign of the near approach to a deposit.

Extensive data for auriferous ore zone halos are given in the following section, and these are discussed in great detail. It is stated that these halos are in general complex, gold and other associated elements being often distributed in an irregular and often erratic manner. Ore zone halos are usually revealed against the local background by a large spread of from four to five orders of magnitude in the gold contents, and copper, lead, zinc, arsenic and other chalcophile elements are found in increased amounts. Both positive and negative (leached out) gold halos occur, and these vary in size according to the scale of the mineralization. In some ore zones the gold contents in the halos equal the regional background. Pyrite and arsenopyrite and certain skarn minerals are the principal gold carriers in ore zone halos. Gold correlates with numerous elements within the halos, but the patterns are often complex, and each ore zone has its own characteristics. The most common correlations of gold are with copper, lead, zinc, silver, cobalt, nickel, barium and arsenic.

Individual orebody halos are described at great length and discussed in the light of wall-rock alteration effects. Most orebody halos are complex and each vein, impregnated zone, sulphide lens, etc. has its own characteristics, although within an ore zone there is much regularity. Individual orebody halos may be marked by positive or negative halos or combinations of these. The main concentrator mineral of gold in positive halos is generally pyrite.

Much detail is given with respect to the zonality and correlation of the various chalcophile and other elements with gold in individual orebody halos. Only a brief summary can be attempted here as follows:

The horizontal distribution of trace elements in altered rocks near auriferous veins such as Ag, Pb, Zn, Bi, Mo, Y, Sc and Hg has much in common with the distribution of gold. Ni, Co, Sn, Cr, Zr and V behave differently. All these elements may, however, form positive, external negative and exocontact negative halos near the orebodies. The positive halo maxima and the negative halo minima for the various elements often fail to coincide. The discrepancy with the gold halo maximum and minimum is particularly great in the case of Ni, Co, Sn, Cr, Zr and V halos, which are considerably more extensive. Anomalously high arsenic and antimony contents can be detected only in the vicinity of the ores, encompassing both the positive and negative halos of Au, Ag, Pb and the other associated elements. All these factors make the correlation between gold and trace elements in the country rocks in the vicinity of the orebodies extremely complicated. Calculation of the coefficients of correlation between the contents of elements and distances from the orebodies commonly reveals a low inverse relationship between them. This situation is thought to be due to the alteration of negative and positive halos, and especially by the presence of exocontact negative halos.

Despite the complexity of the relationship between elements in hydrothermally altered rocks horizontally outward from auriferous orebodies, the elements are commonly distributed according to their concentration maxima as follows: As, Sb-Au, Ag, Pb, Zn, Cu, Bi, Mo, Hg-Ni, Co, Sn, V. Variations in this sequence are normal in some auriferous deposits, especially those with very rich ore shoots.

Vertical zonality in the distribution of elements in halos is apparent from the occurrence of the highest concentrations of certain halo elements in the upper parts of some orebodies, of others in the central parts, and of yet others in the deep parts. In particular, for the Darasun deposit, the following vertical zonality is normal: Pb, Zn, As, Ag and Sb are concentrated in the halos of the upper parts of orebodies, and Cu, Bi, and more rarely Cr, Ni and Co, concentrate in the lower parts. The vertical zonality in the Muruntau deposit exhibits a predominance of Ag, Pb and Cu in the rocks above the ore and of Au, Co, Mo and W in the lower parts of the ore zone. These examples indicate that there is no universal vertical zonality in the halos associated with auriferous deposits.

Regarding changes in the width of gold halos with depth, it can be said in some cases that there is no change, in others an increase and in yet other cases a decrease. Some investigators have established that endogenic gold halos are most marked in the vicinity of the zone of optimum gold mineralization, and that all other conditions being equal the width and productivity (an index of the nearness to ore shoots) of gold halos decrease as the distance increases from the ore shoots upward and downward along the dip.

In the last section of the book on prospecting for gold deposits using endogenic halos the authors emphasize the high cost incurred in prospecting for auriferous deposits using endogenic halos, a feature that makes it imperative to have criteria that can be used to concentrate attention on a limited number of promising sectors. Prior to the development of highly sensitive methods for the estimation of gold, indicator elements (Cu, Pb, As. etc.) have been employed extensively in geochemical prospecting for gold. Roslyakova and Roslyakov think a change in this approach is necessary. They advocate the use of gold directly as an indicator now that the analytical methods are available for its precise determination. They point out that while arsenic is a good pathfinder for gold, it is not infallible and mention a number of deposits where the element is ineffective. They further consider that even the Muruntau deposit, which was found by using arsenic, could have been better and faster located by using gold directly as an indicator. Mercury is, likewise, not always a good indicator of gold deposits as shown by the Darasun Goldfield. Here again gold is the better indicator. Their final statement is worthy of quotation.

"Our observations indicate that gold should be determined at each stage of the geochemical investigations. Gold deposits should be sought in terms of gold. Only this approach permits a correct assessment of the potential of an area and the discovery within it of blind deposits at least cost."

Within the confines of auriferous deposits in shear zones, faults, stockworks, other dilatant zones and chemically favourable rocks a number of geochemical techniques can be utilized to determine the trend and location of the pathways of mineralization, to outline the zonation of the mineralization, to predict the possible continuation of known ore on the strike and pitch of the oreshoots, to predict the vertical extent of mineralization and to determine a variety of other features useful in the development and mining of orebodies. This is a large subject that lies outside the scope of this bulletin, and only a brief listing of the methods can be considered here. These include:

1. Plotting single metal values, the sum of metal values and/or metal ratios on plans and sections, particularly on longitudinal sections, of the deposits and their associated primary halos. Such plots permit various comparisons, indicate which metals are associated, the probable pathways of mineralization, changes of mineralization with type of wall rock, zonation parameters, the extent of secondary enrichment or depletion due to oxidation and reduction processes, the nature of the primary halos and the possible vertical extent of primary mineralization. The Au/Ag ratio of native gold and of the ore shoots as a whole in auriferous deposits is of considerable use in this type of work. (*See* the section on Au/Ag ratios of deposits in Chapter III.)

2. Plotting of mineralogical data, including mineral types, asymmetric growth banding of crystals and textures of mineral intergrowths in the ore. Such data may indicate the direction and extent of pathways of mineralization ('the plumbing system') and may give information on zoning. In a number of deposits, particularly those of Tertiary age (epithermal deposits), there is a consistent mineralogical zonation as discussed at length in the last part of Chapter III. In these deposits the precious metals are enriched near the surface with base metals at depth. Deep seated deposits do not generally exhibit this type of zonation, although in some auriferous districts there may be consistent changes in the type and texture of the ore and gangue minerals that are significant.

3. Plotting data obtained from liquid inclusions and from various mineralogical geothermometers and geobarometers on plans and sections of the deposits. The methods and some of the results of these techniques are discussed in a subsequent section.

4. Plotting isotopic data on ore and gangue elements on plans and sections of the deposits. Such data may indicate pathways of mineralization and may give information on zoning (*see further* the section on other methods in this chapter).

Further details on some of the techniques mentioned above with appropriate bibliographies are given by McKinstry (1948) and Goodell and Petersen (1974) and in various papers recorded in the Selected bibliography.

Pedogeochemical methods

The materials used in pedogeochemical surveys include soils, weathered residuum, calcrete, silcrete, ferricrete, and various glacial materials including till, sand, gravel, etc. These materials may be analyzed directly for gold and its indicator elements; alternatively samples of the materials can be panned and a heavy (or light) concentrate obtained for mineralogical and/or chemical analyses. The latter method is particularly effective since gold tends to collect as the metal in soils, weathered debris and glacial materials in the vicinity of its deposits. The halos and trains developed in soils, tills and allied materials are of a chemical (hydromorphic) and/or resistate (mechanical) nature. Total analyses of soils and allied materials for gold will outline the combined chemical and resistate halos and trains; selective extraction techniques generally outline only the chemical halos and trains; and heavy and light concentrates the resistate halos and trains.

Pedogeochemical surveys should be conducted on a closely spaced grid (25 to 50 ft centres) because of the relatively narrow width of most types of gold-bearing deposits. For deposits of the quartz-pebble conglomerate, stockwork, mineralized pipes and disseminated types the spacing can be increased to 100 ft centres or more in most cases.

Gold may be the best indicator of its deposits in some districts whereas some of its associated elements may be more effective in others. The selection of the best indicator(s) will depend on the type of deposit and the conditions of oxidation, soil formation, glaciation, etc. Those contemplating work in glaciated terrains should read the papers by Forgeron (1971) and Nichol and Bjorkland (1973). Pilot studies should be carried out if possible before conducting extensive surveys. During the pilot stage the elements associated with gold in the types of deposits in the district should be ascertained, as should also the effectiveness of these elements as indicators in the soils, tills, etc. In our experience, and judging from the investigations of others, the best indicators of gold deposits utilizing soils, weathered residuum and glacial materials are Au, Ag, Hg, Sb and As, particularly the last element. In some districts Mo, W, Cu, Pb, Zn, Cd and U may be useful if these elements accompany gold in its deposits; Te, Se, Bi, Ba, Ni, Co, Sn, F and B appear to have only a restricted use in districts where the primary deposits are enriched in these elements. For surveys based on analyses of heavy mineral separates of soils, weathered residuum and glacial materials Au and Ag are generally the best indicators; others include As, Sb, W, Mo, Cu, Pb, Zn, Hg, Cd, platinoids, Te, Bi, Sn, Tl, Ba, Sr, Ni, Co, U, V, Cr, Sc, rare-earths and B. Pilot studies to determine which of the latter indicators are effective should be carried out before embarking on extensive surveys.

Selecting the horizon or horizons from which to obtain the soil samples, weathered residuum or glacial materials usually requires some knowledge of the soils and their parentage or of the glaciation in the area in which the survey is to be carried out. Here again pilot studies should indicate the best horizons to sample and in addition give information on the backgrounds of the various indicator elements in the different horizons. Generally the B horizon is the best for sampling, although the A horizon may be effective in some areas. The latter often exhibits an enrichment in gold, silver, arsenic, antimony and other indicator elements, and the anomalies may have greater contrast than those in other horizons. In some areas only the C horizon close to bedrock gives satisfactory results. This is especially true when the weathered residuum or glacial deposits are thick. Heavy and light mineral surveys are best carried out by using materials as close to the bedrock as possible, although in some places near surface samples of soils, tills, etc. may be satisfactory.

Deciding what are anomalous gold contents in soils, weathered residuum and glacial materials depends on the district and should be determined by preliminary surveys. The normal gold content of soils is usually less than 0.005 ppm and the normal silver content ranges from 0.1 to 0.5 ppm (Table 19). The range for the other two common indicators of gold in background soils, weathered residuum and glacial materials is As – 7 ppm; Sb – 0.7 ppm. Means for some other indicators in normal soils, weathered residuum and glacial materials are as follows: Hg – 0.06; Mo – 2; B – 8; W – <4; Cu – 20; Pb – 15; Zn – 75; Cd – 0.2; Te – 0.01; Se – <0.2; U – 1.0; Bi – <0.5; Sn – 2; Tl – 0.1; Ba – 400; Sr – 100; Ni – 15; Co – 5; and F – 250; all in parts per million. The writer has found that values above 0.01 ppm Au and 0.7 ppm Ag are generally anomalous and should prompt the prospector to investigate the cause. Anomalous values of the indicator elements cannot be stated with any assurance since the dispersion and enrichment characteristics of the various elements vary so widely. However, consistent values 2 or 3 times the average abundance of the figures given above should receive attention.

A number of geochemical prospecting investigations utilizing soils, weathered residuum and glacial materials have been described in the literature. Those noticed in this compilation, and dealing specifically with the detection of gold deposits, include the following:

In Canada the first attempt to use geochemical methods for tracing gold mineralization through glacial overburden is that described by Chisholm (1950). He was able to successfully trace a gold-bearing zone in the Kenora district of Ontario under glacial overburden ranging from a few inches to 5 ft in depth by utilizing copper, lead and zinc as indicators.

Sokoloff (1950) found that analyses of mature residual soils in Western Autralia showed a correlation between gold anomalies in the soil and known lodes. He cautioned that soil anomalies are difficult to interpret and cannot distinguish between a large underlying deposit or one where only the remnants remain after deep erosion. In this situation the level of erosion with respect to the potential target is important to determine where possible. In Japan, Shima (1953*a,b*) used copper and zinc successfully as indicator elements in soil surveys carried out to locate gold-silver veins. James (1957) investigated the metal dispersion patterns in soils over arsenical and antimonial gold deposits in three Rhodesian areas. The results showed that anomalous arsenic values are commonly found in all depths of the soil in the immediate vicinity of mineralization. Anomalous quantities of antimony also occurred, but copper and zinc showed no increase in content. The retention of As and Sb in soils was highest in Fe-rich soils. Narayanaswami (1960) found arsenic to be a particularly good indicator in soil surveys for locating arsenopyrite-bearing gold lodes in the Kolar Goldfields, India. Martinet (1959) also found that the arsenic content of soils was a good indicator of gold deposits in French West Africa. The gold and arsenic anomalies shown on the profiles are mainly coincident. Webb (1958) and Mather (1959) also report that arsenic (and antimony) are good pathfinder (indicator) elements for locating gold deposits utilizing soil surveys in Rhodesia and Sierra Leone. Webb makes reference to the common observation that arsenic is strongly retained in Fe-rich soils. He also notes that there is relatively restricted leaching of arsenic during weathering where such soils are formed, and such lateral dispersion as does take place is probably largely mechanical.

Chapman (1959) found that anomalous amounts of arsenic and/or lead in the soil, particularly in the C horizon are the best indicators of gold- and silver-bearing galena veins in the Kantishna area of Alaska. Zinc was also found in anomalous amounts at some sites, but copper rarely showed more than a background amount. Similar results were found by Boyle (1965*a*) in the Keno Hill–Galena Hill–Dublin Gulch area of Yukon. There, arsenic is a particularly good indicator in residual soils for detecting quartz-pyrite-arsenopyrite-gold-silver veins. Over or slightly downhill from these veins marked arsenic anomalies with a high contrast were obtained; a similar phenomenon was noted for antimony. In areas covered by glacial materials deep sampling near the bedrock is the only satisfactory way of using pedogeochemical methods.

Cu, Pb, Zn, Hg and As have been favoured indicators utilizing pedogeochemical methods in a variety of terrains, throughout the world. Bayley and Janes (1961) found arsenic to be a suitable indicator in the Atlantic district of Wyoming, and Brotzen and Obial (1963) found that Cu, Zn and Pb were suitable indicators for a number of types of deposits, including those carrying gold, in the Philippines. Granier *et al.* (1963) noted a parallelism between the contents of Cu and Au in deeply weathered lateritic layers overlying metadiorite on Mount Flotouo (Ivory Coast). The gold content fluctuated from 1 to 8 ppm and the copper content from 400 to 1500 ppm. In Japan, Kishimoto *et al.* (1963) found mercury to be an excellent indicator of Tertiary (Neogene) Au-Ag quartz veins of the epithermal type. An intimate relationship between the mercury dispersion in the soil profiles and underlying hypogene gold-silver deposits was found.

Arsenic in soils has proven to be an excellent indicator of gold deposits in the Basin and Range Province of the western United States according to Cavender (1963). Near the Getchell Gold Mine in northern Nevada soils samples were collected along 50 profiles across potentially gold-bearing structures and areas, and the arsenic concentrations were statistically evaluated. The results were as follows: (1) arsenic anomalies in residual soils showed a direct relation to gold-arsenic mineralization and potential gold-bearing areas were delineated, (2) unknown and mineralized high-angle faults were uncovered, (3) arsenic mineralization was found in low-angle thrusts – structures previously considered unmineralized, and (4) the recognition and interpretation of local fault structures was augmented.

Dvornikov *et al.* (1963) observed that dispersion aureoles of mercury in the A horizon of the soils in the Nagol'nyi Range coincide with a belt of polymetallic and gold deposits and with fault zones, crests of anticlines and zones of quartz veins. The broadest and most strongly enriched aureoles were located over sulphide and quartz veins. The mercury was found to occur as soluble (chloride), oxide, sulphide, metallic and metal-organic compounds. The clay fraction was relatively richer in mercury than the bulk composition or the coarse fraction. The investigators considered that the high concentrations form as a result of the mechanical decomposition of orebodies and by the accumulation of mercury in the A horizon by the decay of plants whose roots tapped mercury-bearing soil solutions. They concluded that high mercury concentrations in the A soil horizon are an important prospecting clue for new mineralization zones.

An extensive integrated geochemical survey involving analyses of plants, soils and waters has been carried out by Razin and Rozhkov (1963) over the Kuranakh type gold deposits in the permafrost region of the Aldan Shield, Yakutia, U.S.S.R. These deposits are developed in a karst terrane in a mountainous taiga environment and are described in some detail in the section on placers. The soils are mainly soddy-forest loams. The gold content overlying the mineralized zones ranged from 0.1 to 5.7 ppm and increased down the soil profile, the largest amount being present in the C horizon. The highest gold content in the true soils (up to 2.8 ppm Au) was recorded in the B_1 horizon. Enrichments of gold were found only in the humic horizons of podzols in amounts ranging from 0.1 to 5.5 ppm. A clear relationship between the gold and Fe_2O_3 contents was noted in the soils and a less definite relationship between the gold and Al_2O_3 contents. Increased contents of Fe_2O_3 were usually marked by an enhancement of gold, especially in podzols. Razin and Rozhkov (1963) concluded that pedogeochemical methods were effective for locating gold deposits in permafrost regions.

Baxter and Poet (1964) found that Cu, Pb, Zn and As in soils gave a negative response over veins in the Cripple Creek area of Colorado, although they concluded that anomalous amounts of these elements may indicate favourable zones for general exploration. Tellurium was found to be more effective as an indicator of the veins and mineralized zones. The tellurium contents of the soil ranged from 0.25 to 12.5 ppm, the higher values being associated with veins and zones mineralized with gold tellurides.

Hawkes (1965) carried out a geochemical survey using arsenic as indicator for gold mineralization in the Marudi Mountain area, Guyana. The gold occurs as eluvial and alluvial deposits and also in arsenical quartz veins. Two hundred and fourteen soil samples from a depth of 2 ft were analyzed for their arsenic content. The results showed a high frequency of samples containing between 0 and 39 ppm arsenic. The mean of these values, which is 10 ppm, was taken as the regional background. The value for the threshold (130 ppm) was calculated using the mean (background) plus twice the standard deviation. A maximum to background ratio of 30:1 represented the greatest contrast recorded in the area. Five main and six smaller areas of anomalous arsenic values were located, and in two instances these apparently marked the extension of known gold veins.

Mazzucchelli (1965) and Mazzucchelli and James (1966) examined both the primary and secondary dispersion of arsenic and other elements related to gold mineralization in a belt of auriferous rocks, some 300 mi in length, in central Western Australia (Coolgardie-Kalgoorlie districts of the Australian Precambrian Shield. They observed that anomalies related to arsenical gold mineralization are detectable in the various types of residual soils occurring in the area studied, including relict lateritic soils. Arsenic was found to be concentrated mainly in secondary iron oxides in the coarse (+80 mesh) sand and silt fractions of the soils. Appreciable concentrations of arsenic were also found to be associated with the clays, particularly in the alkaline near-surface horizon of the solonized brown soils. The distribution of arsenic in the soil profile appeared to be dependent on pH and the abundance of secondary iron oxides. Lateral dispersion processes under the present semiarid climatic conditions were found to be dominantly mechanical, although appreciable hydromorphic dispersion of arsenic apparently took place during the development of the lateritic soils in the Pliocene.

Bateson (1965) emphasizes the point that in Guyana gold is a particularly good pathfinder for base metals. There, despite intense weathering and leaching, gold is relatively immobile and stable remaining behind in the residual clays and soils. The presence of gold in the weathered residuum, therefore, indicates mineralized zones which may contain base metals. The prospecting procedure employed is to identify the gold-bearing areas and follow this by soil surveys using Cu, Pb and Zn as indicators in the search for base metal mineralization.

Since 1965 there has been a general increase in the use of soils and allied materials on prospecting for gold throughout the world. Komov (1966) describes extensive aurimetric surveys using soils in the Yenisei Range, U.S.S.R. where the deposits are mainly gold-quartz veins. Both gold and arsenic proved useful as indicators. Gold exhibited the least migration and hence marked the location of the deposits most accurately. The zones mineralized with gold were indicated by dispersion halos in the soils with a gold content of 0.05 ppm or higher. Kurbanayev (1966) found that the content of thallium in the secondary dispersion aureoles in eluvium, alluvium and weathered materials of the Maykain gold-barite-polymetallic deposit (northern Kazakhstan) is a clue to the existence of sericitized rocks beneath the surface. Such rocks, containing thallium as a primary constituent, are often indicative of hydrothermal gold mineralization. In places fivefold enrichments of thallium in the eluvium, alluvium and weathered materials were found when compared with the bedrock (1.0 ppm Tl compared to 0.2 ppm respectively).

Anthony (1967) relates the history of the discovery of the Keystone Gold Mine, Cleary Hill area, Fairbanks district, Alaska by geochemical soil methods. The C horizon just above the uppermost stone stripe layer was sampled at a depth of about 33 in. Dithizone analyses for total soluble Cu, Pb and Zn were done on the spot, and the results plotted. The geochemical anomaly revealed as the result of the work marked a high grade gold-quartz vein some 2 ft in width and more than 2000 ft in length.

In the Tavua Basin, Fiji, Johnson (1967) found that arsenic was the most satisfactory pathfinder for gold in soils. There, the deposits are epithermal, occurring in shears and fractures that are associated with a Tertiary caldera (*see also* Chapter III).

A number of investigations have been carried out in recent years in various parts of the world mainly to determine the applicability and efficiency of soil sampling methods in prospecting for gold deposits of various types. Those noticed in the literature include investigations by Kurbanaev and Atchibaev (1968), Nason *et al.* (1968), Potapenko *et al.* (1968), Obraztsov (1969), Banister (1970), Reed and Miller (1971), Evangulov *et al.* (1971), Kartsov (1971), Watson (1972), Leonard (1973), Belogolova and Shibanov (1974), Lakin *et al.* (1974), Pasquali and Bisque (1975), Gustavson and Neathery (1975), Rosylakov (1976), Nestorenko *et al.* (1976), Watterson *et al.* (1976), and Kokkola and Pehkonen (1976). In these investigations the best indicator elements for gold-bearing deposits have proved to be Au, Ag, As, Sb, Pb, Cu, Zn, Ni, Co, Bi and Mo, particularly the first four. Hg has proved useful in some districts and Te in others. Kurbanayev (1966) found thallium to be enriched in the soils and eluvium overlying the auriferous polymetallic-barite Maikain deposit, U.S.S.R. and concluded that it was a useful pathfinder for such deposits. Steed *et al.* (1976) observed that anomalous arsenic and rubidium contents in soils best outlined the ore zones at the Ogofau gold mines, Dyfed, Wales.

In the West End Creek area, Yellow Pine district, Valley County, Idaho, Leonard (1973) found a gold anomaly near the Yellow Pine Mine by soil sampling. The gold anomaly is accompanied by a silver anomaly and by conspicuous though minor mercury, antimony, arsenic and tungsten anomalies. The gold content of 128 soil samples ranged from <0.05 to 8 ppm, the median value being 0.70 ppm. The Tertiary mineralization at the Yellow Pine Mine (Stibnite area) consists of disseminated bodies associated with faults and shear zones and containing essentially auriferous pyrite and arsenopyrite, gold, scheelite and stibnite, in a gangue of quartz, feldspar, sericite and carbonate. There are mercury prospects and mines in the area; these contain cinnabar as the principal mercury mineral.

Watson (1972) and others (*in* the Chamber of Mines Journal, Rhodesia, 1968, 1969*a,b*) describe an interesting gold anomaly in Kalahari sand associated with termite mounds in Rhodesia. He concluded that the enriched amounts of gold (0.05 ppm) (background 0.02 ppm) in the soil and termite mounds may have been carried up by termites from gold-bearing fissures in the Basement Complex to their mounds. Repeated erosion of the termite mounds probably dispersed the gold in the surrounding soil, thus producing an anomaly that has a width of 100 m or more in places. (*See also* the discussion on termite mounds in the section on Biogeochemistry in Chapter II and in the subsequent section on Biogeochemical methods.)

Humus-rich forest soil (mull) as a sampling medium in geochemical exploration for gold has been fairly extensively employed in the western United States. In the Empire district, Clear Creek County, Colorado, Curtin *et al.* (1968) and Lakin *et al.* (1974) found that the gold content of pine and aspen mulls is more representative of gold deposits in bedrock beneath colluvial and glacial cover than is the gold content in float pebbles, cobbles and soil below the mull. Similarly, in the Orogrande district, Idaho, the gold and silver contents of ashed samples of conifer needle humus samples were successfully used to outline a low-grade gold prospect (Rice, 1970). Using the same technique several gold-antimony anomalies were outlined near Stibnite, Idaho by Banister (1970). In a later publication Curtin *et al.* (1971) state that in the Empire district, gold deposits in bedrock beneath the colluvial and glacial cover are best delineated by the gold and, to a lesser extent, by the copper and bismuth contents of the mull ash. The distribution of anomalously high amounts of the last two metals in the soil below the mull corresponds to the gold deposits in bedrock only in places, and many of the high anomalies in soil probably reflect the inclusion of vein material in otherwise barren colluvial or morainal cover. They consider that the results obtained in the Empire district indicate that mull may be an effective geochemical sampling medium for gold, silver, bismuth, lead and zinc in forested areas blanketed by colluvium or glacial drift where the transported material offers no clue to the nature of the underlying bedrock. In a final summary of their work on mull Lakin *et al.* (1974) indicate the usefulness of the mull horizon (A or humic horizon) in prospecting for gold deposits in the western United States because of the general enrichment of gold in this horizon near auriferous deposits. They also note that the gold content of soil profiles in some areas increases near the bottoms of the profiles at or near bedrock. They suggest that this may be due to the release of gold from disintegrating fragments of vein material and/or to the downward migration of gold particles within the soil profiles during downslope creep of the soil.

Aferov *et al.* (1968) noted that the subhorizon A_0 of soils in the vicinity of the Darasun deposits is enriched in gold, and Talipov (1972) observed that the A (humus) horizon of the soils over auriferous deposits containing abundant arsenopyrite in central Kyzl-kum, U.S.S.R. is rich in arsenic. C.F. Gleeson (pers. commun., 1974) states that the samples from the A (humus) horizon developed on till in the Duparquet area of Quebec are much more useful than those from other horizons in prospecting for gold-quartz deposits. The contrast of the gold anomalies compared with the background is marked.

There are few published investigations dealing with prospecting for auriferous conglomerates utilizing soils or weathered residuum as a sampling medium. Zagoskin and Zagoskina (1971) state that the Khuzhir occurrence of auriferous conglomerates in a taiga-forest environment is indicated by a dispersion halo of gold in soils some 60 to 200 m wide. The halos are apparently complex due to the wide disperson of gold in soils near the conglomerates.

The gold content of soils and weathered residuum has been used effectively in the search for porphyry-type deposits. In Yukon, Canada, Archer and Main (1971) relate the history of the discovery of the Casino copper-molybdenum porphyry deposit by stream sediment geochemistry and the outlining of the principal mineralized zones by soil geochemistry, utilizing in both types of surveys Cu, Mo, Pb, W, Au and Ag as indicators. Over the deposit the gold content of the soils (B+C horizons) exceeded 0.5 ppm in places compared with a background of 0.05 ppm or less. Learned and Boissen (1973) also found gold to be a useful pathfinder element in the search for porphyry copper deposits in the Rio Vivi copper district of Puerto Rico. There, many of the deposits in the copper belt are characterized by intensely leached cappings and covered by residual soils and weathered residuum. Within the copper belt the residual soils overlying five major deposits were found to contain strongly anomalous amounts of gold. The gold content of the soils was uniform from deposit to deposit and was about equivalent to the gold content of the primary deposits (0.20 ppm). The general background content of gold in the soils remote from the deposits was generally less than 0.02 ppm.

The calcium carbonate (calcite) content of soils as determined by a gas-volumetric method was applied in prospecting for gold deposits in the Aldan mining district (Koltykan deposits) by Borovitskii and Shemyakin (1965). The Koltykon deposits are in fault, fracture and crushed zones that cut Archean gneisses, schists and marbles overlain by Cambrian dolomites, limestones, marly shales and sandstones all intruded by Cimmerian (Mesozoic) granitic and porphyritic bodies. The gold deposits are post-Cimmerian and consist of veins and lenses of pyrite, quartz, ankerite and Cu-Pb-Zn sulphides. The ores are deeply oxidized and converted to a loose argillo-limonitic mass containing fine-grained gold in amounts up to 300 ppm. The drift cover is some 10 to 15 m thick.

The reason given for applying the carbonate survey in the Koltykon field was that late calcite terminated the deposition of the ores. Soil samples taken at 15 cm depth and analyzed for $CaCO_3$ (calcite) showed marked anomalies over the known deposits, even though the overburden was 15 m thick. Borovitskii and Shemyakin (1965) admit that the mechanism of enrichment of calcium carbonate in soils over dislocations that control the gold mineralization is obscure, but stress that this circumstance is a fact regardless of the presence of carbonate host rocks in the surveyed area.

The writer suggests that Borovitskii and Shemyakin (1965) appear to have found salt halos in the soils overlying the faults and fractures. These halos are probably due to the fixation of calcium in the soils by carbon dioxide that has risen along the fracture system from depth. Another possible mechanism is the fixation of calcium (or elements such as Mg, Fe, Mn) in the soil by carbon dioxide derived from the (bacterial) oxidation of methane which has migrated up faults from depth. Another mineral that is commonly found in soils above fractures and faults is gypsum; derived by the fixation of the available calcium in the soil by SO_4 resulting from the (bacterial) oxidation of hydrogen sulphide that has risen along the fractures. Carbon dioxide, methane and hydrogen sulphide are common gaseous constituents of fault systems, some containing gold deposits, even in the Precambrian (Boyle, 1961*a*). Salt halos in soils over mineralized fault and fracture zones, marked by the presence of minerals such as calcite, gypsum, marcasite, etc., should receive much more detailed attention from those doing research in geochemical prospecting methods.

Analyses of heavy mineral concentrates obtained from soils, weathered residuum or glacial materials by panning, rocking, sluicing or other methods are a most effective method of outlining secondary dispersion trains and halos related to gold deposits. Several techniques can be applied in heavy concentrate studies, including: (1) the heavy concentrates can be studied mineralogically and the frequency of gold colours, pyrite, arsenopyrite, native bismuth, barite, limonite and other heavy mineral associates of gold determined and plotted; (2) the whole sample of heavy concentrates or separated fractions such as the magnetic and/or nonmagnetic fractions can be assayed or analyzed for gold (and silver) and the results plotted; or (3) the heavy concentrates or separate mineral fractions such as pyrite, magnetite and limonite can be analyzed for gold and the elemental indicators of the element such as As, Sb, Bi, Te, Zn, Cu, Pb, Ba, etc., and the results plotted. In all of these techniques the sampling media can be collected on a grid or on lines as the terrain dictates. The work should be quantitative – a standard quantity of soil, glacial materials, etc. should be collected from which all of the heavy minerals are separated and the whole or an aliquot analyzed. The mineralogical and chemical results on the heavy concentrates should then be uniform, but if different samples sizes are used in the survey the mineralogical and chemical results should be calculated to a standard sample size. In practice, the sample size for obtaining the optimum amount of heavy concentrates for mineralogical and chemical work has to be determined in the field since the content of heavy minerals varies widely in soil, eluvium, and glacial materials. Any of the horizons of the soil can be chosen for sampling, but generally the B and C horizons are the best. In glacial terrains samples taken near the bedrock (in the basal till, sand, gravel, etc.) are generally the most effective in outlining dispersion trains and fans.

It should be emphasized that if the soil or glacial materials are sieved before panning for heavy minerals or gold the best sieve size must be ascertained to give the optimum contrasting anomalies for the heavy minerals employed. This can only be done by pilot surveys in an area. This feature is critical for gold particles since in any given area these tend to a relatively uniform size. If too small a sieve size is used, the majority of the gold particles may be screened out with the result that the analytical results will show spotty and often spurious anomalies that have no bearing on the actual occurrence of gold.

Crude panning methods are effective if gold mineralization alone is sought. One simply pans the soil, eluvium, weathered residuum and glacial materials on a grid or on lines as dictated by the geology and the terrain. The colours of gold are counted and recorded on a plan of the grid or lines. Increase in the number of colours in a dispersion train or fan heralds the approach to the primary sources. It need not be emphasized that this method has led to the discovery of countless gold deposits throughout the world, especially in countries that are covered by thick residual deposits. With the advent of more sophisticated analytical methods of geochemical prospecting the gold pan has been somewhat neglected, a circumstance that is unfortunate because there is no better instrument for tracing fans and trains of heavy minerals (gold, cinnabar, barite, cassiterite, platinum, etc.) in soils, eluvium, weathered residuum and stream sediments as emphasized by Mertie (1954), Theobald (1957) and Wells (1973). In glacial terrains the use of the pan is somewhat restricted when applying pedogeochemical methods except for materials on or near the bedrock.

Light mineral associates of gold obtained from soils, eluvium and glacial materials can also be used in tracing secondary dispersion trains and fans related to primary gold deposits. Particularly applicable in this respect is the presence of vein quartz in the soils and drift. By plotting the frequency of vein quartz nodules and particles it is sometimes possible to locate the source of the veins. Many gold deposits were found by the old prospectors using this technique. Lee (1963) provides a modern example by utilizing vein quartz, chloritic fragments and other vein materials as indicators in glacial fans at Kirkland Lake, Ontario. Other minerals such as ankerite, limonite, wad and fuchsite can also be used as indicators. Before attempting work on light minerals, however, it is desirable to study the nature of the vein quartz, carbonates and other associated light vein minerals in the known gold veins of a district in order that proper comparisons can be made between the minerals in the primary deposits and those in the secondary dispersion trains and fans.

Only a few investigations of heavy minerals in soils in the secondary dispersion trains and fans associated with gold deposits have been published. Those noticed in this review include papers by Dvornikov *et al.* (1964) and Bolotnikova (1969). In the soils and friable bedrocks of the Nagol'nyi Ridge the first investigators found native Au, an Ag-Hg alloy and native Pb, Zn and Cu near zones of Pb-Zn and Au mineralization. Native lead was exceptionally abundant. These native metals and alloys are considered to be good indicators of the deposits on the ridge. Bolotnikova deals mainly with the particle size distribution of gold in the secondary dispersion halos over the gold-sulphide deposits in the Maikainsk district. She recommended that the fractions -0.5 to $+0.1$ mm of the residual materials should be used as the initial sampling media during prospecting for gold-sulphide deposits in the district.

Gurev (1970) describes mineralogical-geochemical heavy concentrate surveys carried out in the Kyzyl-kum gold province. There, as a result of oxidation processes part of the gold is fixed in the iron hydroxides (limonite), which is pseudomorphous after the pyrite. Analyses of these hydroxides separated from the heavy concentrates of soils, residual materials, etc., in conjunction with other chemical and mineralized work, have proved useful in the metallometric surveys.

Certain auriferous porphyry copper deposits in Melanesia contain abundant rutile in the ore zones, five times more than in the host rocks (L.J. Lawrence, pers. commun., 1975). On weathering, this rutile may be contributed to the soils and ultimately to the stream sediments and alluvium. Excess amounts of rutile in the heavy separates of these materials could be a valuable indicator of these types of auriferous porphyries.

Analyses for gold, silver and other common associated elements in sulphide, gossan and vein quartz boulders in glacial materials may be a useful approach for outlining secondary dispersion trains and fans related to gold-bearing deposits in the glaciated regions of Canada and other countries. The method is an old one and was generally referred to as tracing float by the old prospectors. An early reference to this type of prospecting is found in the paper by Moore (1925), who relates the presence of highly auriferous quartzite boulders north of Goudreau Lake in Ontario, the source of which had not yet been found at his time of writing. Float tracing, or boulder tracing as it is now called, has been used successfully in Sweden and Finland (Grip 1953; Okko and Peltola, 1958; Hyvärinen *et al.*, 1973). Similar studies have been carried out by Lee (1963, 1965), Cachau-Herreillat and La Salle (1971) and Shilts (1973, 1974) using vein quartz fragments, chloritic fragments, trace elements and heavy minerals including gold in the glacial materials of basal tills, eskers, frost boils and buried valleys in the vicinity of the Kirkland Lake gold-silver deposits, in the auriferous belts of Quebec and in mineralized belts of Northwest Territories. Boulder tracing and closely allied methods involve a close study of the glacial features followed by the plotting of the location of the boulders and chemical analyses on geological maps in order to discover the apex or source of the fan or train that marks the outcrop sites. Details of sampling and concentrating equipment are discussed at length in the papers by Lee and Shilts. Lee (1971) reviews progress in float and boulder tracing in the Canadian Shield in Canada and notes that the discovery of some seven major gold deposits in this geological terrane was assisted by float and heavy clast surveys.

In recent years overburden drilling has been employed extensively in the search for gold and many other types of mineral deposits. The techniques are discussed at length by Van Tassel (1969), Gleeson and Cormier (1971) and Wennervirta *et al.* (1971) and are particularly applicable to permafrost terrains, deeply covered glacial terranes, lateritic terranes, filled valleys and buried eluvial and alluvial placer fields. Overburden surveys are conducted on grids or selected lines or traverses in a manner similar to that employed in pedogeochemical surveys. Samples for analyses may be collected from any level but in general are usually obtained from the zone immediately above bedrock. It is advisable, particularly in placer fields, to sample the near-surface bedrock as well.

Geochemical prospecting for auriferous deposits in deserts and other arid terrains where much of the cover consists of wind-blown materials presents a number of unique problems that have been discussed by Bugrov (1974) and Bugrov and Shalaby (1975). They found that sieving-out of the –0.25 mm fraction to eliminate the dilution effect of aeolian sands enhanced anomaly contrasts in surveys utilizing eluvium and colluvium. Other details are given in the original papers.

Hydrogeochemical methods

The hydrogeochemical methods discussed include those based on ground and surface waters, precipitates at springs and in water courses, stream, river and lake sediments and heavy mineral concentrates in spring, stream, river and lake sediments as sampling media. In practice, surveys may utilize only one of the media, but generally two or more are often used. For thorough work all media should be employed, especially where the geology and types of mineralization are poorly known.

The normal content of gold in natural waters is very low (of the order of 0.00003 ppm [0.03 ppb] or less). In gold-bearing districts the gold content of spring, stream, river and lake waters is, likewise, low, generally in the range 0.001 to 1.0 ppb Au (*see* Table 21). It appears doubtful, therefore, if hydrogeochemical surveys using gold as indicator would be effective using any simple field method of estimation of the element in natural waters. More sophisticated analytical techniques utilizing coprecipitation, carrier, resin collection and solvent extraction methods followed by colorimetric, atomic absorption, spectrographic or neutron activation analysis are suitable for the determination of gold in natural waters, but they are slow and costly and require a well established laboratory. If such methods are employed, care must be taken to analyze the water samples as soon after collection as possible to avoid, among other problems, the extensive adsorption of the traces of gold on the containers. Analyses of evaporated residues of large amounts of water for gold, silver and other associated elements may be useful in certain districts to delimit the stream systems receiving soluble gold from deposits. In the writer's experience, however, the method is tedious and subject to numerous errors.

Numerous investigations have been published on the content of gold in natural waters as an indicator of its deposits. The results have been mixed, some investigators claiming success and others registering failure with the method. (*See also* the section on gold in natural waters in Chapter II).

Kropachev (1935) seems to have been the first to use gold as an indicator in water surveys. He concluded that most gold-bearing areas are marked by waters with gold contents greater than 0.06 ppb.

Konovalov (1941) found that the gold content of river and stream waters of the Sutar River basin was variable (0.000–0.10 ppb) on an areal basis as well as with respect to position in the water course and with time. He concluded that the general physicochemical factors of the migration of gold in natural waters are complex and largely unknown. The gold content of the waters cannot, therefore, be taken as a criterion of the gold content of the soils, rocks or deposits of a given drainage basin. On the other hand Albul and Miller (1959) considered that the gold content of natural groundwaters was effective in prospecting for gold-bearing deposits. In the groundwaters of a district in Transbaikal they recorded normal contents up to 0.05 ppb Au; in and near the gold deposits the groundwaters ranged from 0.2 to 80 ppb Au. Increased amounts of Ag (2.6), Sb (600), As (250) and U (7.8) were found to accompany the gold in the groundwaters in or near the gold deposits, all values being in parts per billion.

Razin and Rozhkov (1963, 1966) carried out an extensive investigation of the gold content of the various ground and surface waters associated with the Kuranakh-type gold deposits in the Aldan Shield, Yakutia region, U.S.S.R. (*see also* the section on placers in Chapter III.) The deposits are in the permafrost region in a mountain taiga terrain. The water types differentiated in the area include those in unconsolidated recent formations above the permafrost; those above and interpermafrost in the Lower Cambrian rocks; and those below the permafrost in the Lower Cambrian and Archean rocks. The authors noted that the content of gold in the various waters depended on the pH and in particular on the content of ferric ions in aqueous solution. The gold is said to occur in both a dissolved form and as a suspension, with the latter predominating. Mildly acid waters leaching the most productive parts of the karst deposits contained an average of 0.001 ppm dissolved gold and 0.009 ppm suspended gold. The authors considered that hydrogeochemical methods utilizing gold as an indicator were suitable for prospecting for the Kuranakh (karst) type of gold deposits in permafrost areas.

Sal'e (1965*b*) also dealt with hydrogeochemical methods of prospecting in the area of the Kuranakh deposits. He found that the subsurface and surface waters contained from traces to 1 ppb Au and were commonly enriched in Tl, As and Sb. The Tl and As were mainly restricted to dispersion halos in the waters near the gold-bearing deposits, whereas Sb had a much wider dispersion. Sal'e concluded that Tl, As and Sb were good indicators of the Kuranakh-type deposits in hydrogeochemical surveys utilizing water as the sampling medium.

Borovitskii *et al.* (1966) also noted increased amounts of As, Sb and Tl in the Kuranakh area waters. In the surface stream waters near the deposits they found less than 0.03 ppb Au, and in the groundwaters in the mines and near the deposits they recorded gold contents ranging from 0.03 to 10 ppb, maximum concentrations being present in those waters high in dissolved manganese and iron.

Vsevolozhskaya (1966) used semiquantitative spectral analyses of dried residues from the subsurface waters of the Aldan area in U.S.S.R. in hydrogeochemical surveys. She noted the presence of some 20 trace elements in the residues and found that Ag, Sb and As were particularly good indicators for metasomatic gold deposits. She noted some factors that decreased the efficiency of hydrogeochemical methods based on groundwater analyses in permafrost areas including: high organic content in the waters; dilution due to heavy precipitation of rain water; and dilution due to thawing of the permafrost.

Goleva (1968, 1970) found that the best indicators of gold deposits in the waters of the Balei district, eastern Transbaikal, U.S.S.R. are Au, Ag, As, Zn, Pb and Sb. Auxiliaries under certain conditions include Bi, Cu, and SO_4^{2-}. In the groundwaters in fractures and fissures the backgrounds were Au<0.07;

Zn<20; As<50; Bi<0.1; and Ag<5, all in ppb. Anomalous groundwaters near gold-bearing deposits contained the following amounts of the various elements; Au>0.15; Zn>30; As>60; Bi>1.5; and Ag>10, all in ppb. In alluvial and surface waters the backgrounds were Au<0.05; Zn<15; As<2; Bi<0.05; and Ag<3, all in ppb. Anomalous alluvial and surface waters contained the following amounts of the various elements: Au>0.1; Zn>20; As>5; Bi>1.0; and Ag>6, all in ppb. During detailed hydrogeochemical surveys Goleva recommended that Au be determined in addition to the other indicators because water anomalies with a similar association of elements may be created by sulphide and tourmaline veins that are barren of gold in the area.

Gosling *et al.* (1971), using neutron activation, investigated the gold content of natural waters draining gold-barren and gold-enriched rocks in the Colorado Front Range. They found a relatively small range in the total gold content of the waters, from nondetectable to 0.150 ppb. 'Solute' or dissolved gold concentrations (average 0.010 ppb) were generally higher than particulate gold concentrations (average about 0.003 ppb), and both were found to be unrelated to rock or water types. The authors concluded that hydrogeochemical prospecting using gold as an indicator in waters is unpromising.

Since gold is commonly only present in natural waters in the parts per billion range and is difficult to determine accurately, it is probably advisable to use other indicators of its deposits. These indicators depend on their concentration in the gold-bearing deposits and their mobility in the ground and surface waters. For any specific area the effective indicators may be quite different from those applicable in another area. It thus follows that a pilot survey should be run in an area before extensive surveys based on water analyses are undertaken. During the actual surface survey all springs and stream waters should be analyzed carefully and the results plotted. Analyses of groundwaters from drillholes, wells and other sources can be effectively applied to outline hydrogeochemical halos associated with gold deposits of practically all types. Some of the techniques useful in this type of work are outlined by Boyle *et al.* (1971).

In my experience and judging from investigations reported in the literature the most effective indicators in natural waters of most types of gold deposits are Zn, Ag, As, Sb and Cu. Ni and Co may be useful indicators where the gold deposits contain these elements; similarly with U and Hg.

Cavender (1963) found the arsenic content of springs in the area of the Getchell Gold Mine in northern Nevada effective in indicating areas of potential gold mineralization and corroborated similar findings from soil sampling. He also observed that the arsenic concentration in an individual sample was helpful in classifying the type or mineralization source of the other trace metals, found in the water sample.

Adilov *et al.* (1971) used antimony as an indicator in subsurface water surveys of a gold belt in the Kuraminsk Range, U.S.S.R. They noted that the Sb content is 10 times higher in waters of the nonoxidized zone than in waters of the oxidized zone that contained 0.8 to 6.0 ppb Sb. The investigators concluded that Sb can be used as an element-indicator in the search for deep-seated auriferous deposits. Makarov (1972) used Sb, As, Au and sulphate ion as the indicators in hydrogeochemical prospecting for gold-antimony mineralization in northeastern Yakutia, U.S.S.R., and Loginova (1972) noted that waters associated with gold deposits in Bashkiria contain more silver than those associated with copper deposits. Skryabin (1968) found Mo a useful indicator in waters, Sal'e (1965*a*) noted that Tl was enriched in waters near the Kuranakh-type deposits, and Goleva (1970) concluded that Bi could be used under certain conditions as an auxiliary indicator in the Balei gold district, eastern Transbaikal.

Waters leaching gold deposits enriched in Li, K, Rb and Cs may have high contents of these elements because of their marked solubilities as the sulphates and other salts. Se and Te may occur in increased amounts in natural waters leaching deposits containing enrichments of these elements. Finally, the SO_4^{2-} content of waters may be an indication of the presence of sulphides in which gold may occur. However, great care must be exercised in interpreting SO_4^{2-} results since the sulphate may derive from quite barren pyritic bodies, pyritic schists, gypsum bodies, etc.

In a recent study Kolotov *et al.* (1975*a*) have indicated that zones of potassic alteration (metasomatism) associated with auriferous deposits are reflected by higher contents of K in the hydrologic regime. Their section shows that most zones enriched in gold give coincident anomalies in the water that are high in potassium.

In the auriferous Lena region of U.S.S.R. Polikarpochkin *et al.* (1974) found that sulphate ion in surface waters could be used in prospecting for gold deposits. The anomalies have a fair contrast and the sulphate dispersion is extensive in this region; hence, sampling at widely spaced intervals (1–3 km) is effective. They also noted that the As and Mo contents of water were suitable indicators of auriferous mineralization under certain circumstances. In the waters Au, Zn and Cu were not particularly useful as indicators of the gold deposits.

Jonasson and Allan (1973) have proposed snow as a sampling media in hydrogeochemical prospecting in temperate and permafrost regions. They noted that such metals as Hg, Zn, Cu, Pb, Ni, Cd and Mn occur in snow samples overlying buried mineralization and that the dispersion aureoles are usually quite broad, especially near massive sulphide deposits. The dispersion pattern of Cu and Zn in snow samples showed anomalous trends in the vicinity of the Joe Lake Gold Mine in southeastern Ontario.

Stream sediment surveys have proven their usefulness in prospecting for base metal deposits of all types in most terrains. In recent years these types of surveys, including those based on lake sediments, have been applied in the search for precious metal deposits, particularly those bearing gold and silver. The literature is replete with examples of stream sediment surveys executed under practically all conditions, geological and climatic, but no effort will be made here to review any of the work in detail. Some of the stream sediment investigations carried out mainly in gold-bearing districts that have come to notice during compilation of this chapter include those by Safronov *et al.* (1960*a,b*), Petrov (1962), Osokin (1962), Fedorenko (1962), James (1965), Mazzucchelli (1965), Sal'e (1965*b*), Nichol *et al.* (1966), Gleeson *et al.* (1966–1968), Garrett and Nichol (1967), Jasper (1967), Smith (1968), Eakins (1968), Hale and Govett (1968), Nickerson (1972), Brosgé and Reiser (1972), Bell (1973), Plant and Coleman (1973) and Gundobin and Kolesnikov (1974). Some of these investigators used gold as a direct indicator of its deposits; others employed various other indicators particularly Cu, Ag, Zn, Tl, Pb, As, Sb, Bi, Ni, Co, Mo and W. My experience suggests that of the last group of indicators the most specific for gold-bearing deposits are As, Sb and Ag.

A number of other indicators may be useful in stream, river, pond and lake sediment surveys where the primary deposits contain them in enriched amounts. They include Ba, U, Ni, Co, Cd, Hg, B, rare-earths, Se, Te and F. Increased concentrations of boron in stream sediments reflect the presence of gold-quartz veins and Pb-Zn-Ag lodes in the Keno Hill–Galena Hill area Yukon (Gleeson *et al.*, 1968). Boron is in fact a most useful indicator of nearly all types of gold deposits since the element tends to be enriched either in the deposits or in the associated wall-rock alteration zones (Boyle, 1971). However, care should be taken in interpreting geochemical surveys based on boron as an indicator of gold deposits since the boron from country rocks containing tourmaline, dumortierite, etc., resulting mainly from metamorphism or igneous intrusion, may confuse or mask the dispersion patterns from the deposits. Manganese minerals are commonly abundant in epithermal (Tertiary) gold-silver deposits. Warren and Delavault (1959) suggest the use of manganese as a pathfinder for these deposits in stream sediments, soil and biogeochemical surveys.

While conducting surveys based on analyses of natural waters or stream, river, pond and lake sediments the geochemist should be constantly on the alert for precipitates of all kinds at the orifices of springs, on stream bottoms or on the stones and rocks in the streams. These precipitates should be carefully studied and samples taken for analysis. Limonite- or wad-cemented conglomerates, composed of stones, sand, rock chips and vegetable remains should receive detailed attention. Where found they should be carefully mapped and sampled and their location analyzed with respect to springs and the groundwater systems of the area. In many places these bodies occur near springs issuing from faults that either contain auriferous deposits or intersect such deposits along their course.

Limonite, wad and other precipitates are commonly enriched in gold near gold-bearing deposits (*see* Table 22). Similarly they may also be enriched in the elements that accompany gold in its deposits. Particularly indicative of gold deposits are enrichments of the following elements in manganese- and iron-rich precipitates: Li, Cu, Ag, Sr, Ba, Ra, rare-earths, Zn, Cd, Hg, B, Tl, U, Pb, As, Sb, Bi, Se, Te, Mo, W, F, Co and Ni. Of these the best indicators appear to be Ag, Zn, Hg, As and Sb. The other elements are generally only effective indicators if present in appreciable concentrations in the primary deposits. Factors such as the mobility of the particular elements under the prevailing conditions of Eh and pH, organic activity, etc. are also significant. Kanurkov and Stefanov (1968) found that gold in limonite was a reliable alternative indicator element to Se or Te in certain auriferous deposits in Bulgaria. Limonite derived from auriferous pyrite and chalcopyrite averaged 1.2 ppm Au whereas that arising from oxidation of iron carbonates averaged only 0.016 ppm. Siliceous precipitates (Table 22) are often enriched in gold and in many of the elements accompanying it in its primary deposits. Such precipitates are frequently associated with hot springs and are hence not necessarily related to primary gold deposits. However, gold-bearing siliceous precipitates should be looked upon favourably since the waters from which they derived may have deposited gold along the underground channels by which the water reached the surface. Carbonate, barite and other types of precipitates (Table 22) rarely contain above normal concentrations of gold, but some may coprecipitate many of the elements that accompany gold in its deposits. All should be tested.

Frequently hydrous manganese and iron oxides, particularly the former, coating stream, river, pond and lake sediments cause anomalies due to adsorption and/or coprecipitation of the metals from stream, river, pond and lake waters. In areas where some streams are manganiferous and others not, this may be a disadvantage since the manganiferous streams tend to concentrate the metals from a normal background, giving false anomalies. On the other hand in areas that are uniformly manganiferous the concentrating action may be advantageous, and one can make good use of manganiferous and ferriferous sediments. In this respect it is desirable, therefore, when carrying out stream sediment surveys to take note of the quantity of manganese and iron oxides present and devise a qualitative index for each stream or part of a stream surveyed. If spectrographic or atomic absorption analyses are to be carried out on the stream sediments, both iron and manganese should be determined and plotted. Comparison of heavy metal maps with those of manganese and iron will reveal the effects of these two elements in concentrating the metals.

In gold-bearing areas manganese oxides (wad) coating rocks, etc. tend to collect gold (and silver). Iron oxides may do likewise if there is some arsenic or antimony in the environment. Analyses of these oxide coatings may, therefore, indicate the presence of gold mineralization in an area. The same can be said for humic precipitates in some areas and for aluminous and siliceous precipitates in others.

Heavy mineral surveys of surface drainage systems are particularly effective in prospecting for gold. The methods are essentially the same as those discussed for soils. The sediments of gulches, creeks, streams, rivers, lakes, seas, oceans and beaches are panned, and the heavy concentrates are collected and analyzed mineralogically or chemically. The work should be made quantitative by taking either a standard weight of sediment (a full pan or two for example) from which all or an aliquot of the heavies are analyzed, or by taking a standard weight of heavy minerals for analysis and referring the result back to the amount of gravel or sand from which the sample was obtained. Care must be taken in panning to retain the fine (flour) gold, otherwise the values obtained on analysis will be low. (*See also* the discussion on panning in a subsequent section on analytical methods.) The age-old technique of counting colours is often satisfactory, if gold alone is used as the indicator. In practice the sample size for obtaining the optimum amount of heavy concentrates for mineralogical and chemical work has to be determined in the field since the content of heavy minerals varies widely in stream sediments, lake sediments, beach sands, etc.

It should be emphasized that if the stream sediment is sieved before panning for heavy minerals or gold the best sieve size to give the optimum contrasting anomalies for the heavy minerals employed must be ascertained. This can only be done by pilot surveys in an area. This feature is critical for gold particles since in any given area these tend to a relatively uniform size. If too small a sieve size is used, the majority of the gold particles may be screened out with the result that the analytical results will show spotty and often spurious anomalies that have no bearing on the actual occurrence of gold.

Where moss abounds along stream courses and spring flooding occurs, it is often advantageous to obtain samples of the moss at regular intervals since this plant tends to mechanically collect fine (skim) gold. In many cases the gold can be washed out of the moss when dry and the colours counted; alternatively the moss can be ashed and the gold determined chemically or by fire assay.

If mineralogical work is done on the heavy concentrates, several mineral indicators of gold deposits can be used in following the dispersion trains, particularly native gold, platinoids, arsenopyrite, stibnite, native bismuth, scheelite, barite, limonite, etc. Excess amounts of rutile in stream and lake sediments may be an indicator of certain types of auriferous porphyry copper deposits (*see also* the section on Pedogeochemical methods). The discussion on the mineral associates of gold in Chapter III will suggest others. A good textbook, such as that by Raeburn and Milner (1927), and the mineral atlas by Guigues and Devismes (1969), both describing alluvial minerals, are recommended, and where possible X-ray methods of identification should be used to confirm visual determinations whenever possible. The frequency of occurrence should be determined for each indicator mineral at each sample site. An increase in the frequency of occurrence of the indicator mineral or minerals generally signals an approach to the source.

When analyses of the heavy concentrates are employed several techniques can be used such as analysis of the whole heavy mineral suite, analysis of magnetic and/or nonmagnetic fractions, analysis of specific mineral fractions such as pyrite, magnetite or limonite or those that lie within certain specific gravity ranges and so on. Gold is generally the best indicator, but other elements such as Ag, As, Sb, Cu, Pb, Zn, Bi, Se, Te, etc. can often be successfully employed. A knowledge of the elemental and/or mineral associates of gold in the deposits of an area is desirable in ascertaining which indicators are the most likely to yield positive results.

The light mineral associates of gold in stream sediments, lake sediments, beach sands, etc. can also be used in tracing secondary dispersion trains in drainage systems. Vein quartz, chloritic pebbles, carbonates, limonite, wad and the other light associates can all be used. Before attempting work on light minerals, however, it is desirable to study the nature of the vein quartz, carbonates, oxidation products and so on in the known gold deposits of a district in order that proper comparisons can be made between the minerals in the primary deposits and those in the secondary dispersion trains.

Placers of all types, except those deeply buried and not being eroded, can generally be located by employing stream sediment and heavy concentrate surveys. The best indicator of these secondary deposits is of course gold, but some of the other more specific indicators that may give a clue to the presence of gold placers are platinoids, Ba (barite), Hg (cinnabar), As (arsenopyrite), Sb (stibnite, sulphosalts), Bi (bismuth, bismuthinite), Te (tellurides), W (wolframite, scheelite), Cr (chromite) and Ti (ilmenite). Some of these minerals (elements) are present in much greater abundance than gold, and hence their trains have a relatively high contrast with the background and are easier to follow.

With respect to prospecting for placers it should perhaps be mentioned in passing that not all areas containing gold lodes contain placers and *vice versa*. In many districts the placer production varies inversely to that from lodes; the causes are manifold; for example in some districts the placers that were formed have been destroyed by glaciation or other geological forces and in other districts the veins and lodes have not been eroded deeply enough to yield their gold. Some deposits, the great Witwatersrand bankets for instance, yield no placers for reasons that are obscure. Finally, telluride deposits yield few placers as they are eroded. In both these cases the liberated gold may be of such a fine-grained nature (flour gold) that it is swept away and dispersed through great volumes of sediment.

Two problems frequently arise in the interpretation of gold anomalies found in drainage sediments. One is the distance that the gold has travelled and the other is the source of the gold. The first has no certain answer since the distance travelled by native gold is highly variable and depends on many factors such as particle size, gradient of stream, frequency of flash flooding and a host of other parameters all of which are discussed in some detail in the section on placers.

Flour or flood gold, being very fine grained (1000 to 5000 colours worth 1 cent) may travel for hundreds of miles. It tends to collect in the upstream parts of river and stream bars particularly those on the inside of curves. There and elsewhere in drainage systems and in deltas where there is slack water, gold anomalies can often be expected in gold-bearing regions. In fact these 'skim bars', so called by the old prospectors because they are replenished practically each year and can be skimmed of their gold quite easily, led to many a disappointment in the early days before their nature was clearly understood. They are common along a 400 mi stretch of the Snake River in western Wyoming and southern Idaho (Hill, 1915; Hite, 1933*a*, *b*), all along the Fraser River in British Columbia and along a great stretch of the North Saskatchewan River in Alberta. Flour gold is a troublesome feature in interpreting geochemical prospecting surveys using drainage sediments. Its source can rarely be ascertained since its migration is so extensive. Furthermore, it may cause innumerable anomalies that are veritable pitfalls for the enthusiastic geochemist. One may think one has discovered Eldorado whereas what one really has found is an enrichment of gold particles that may have come from a source 100 mi or more away. It is difficult to obviate the problem of flour gold, but there are two techniques which may prove useful. One is that flour gold generally has a very high fineness (>950) an indication that it has travelled far from its source or that it is not derived from primary deposits such as veins and lodes (*see below*). Determination of the Au/Ag ratio of the fine-grained gold may, therefore, be of some assistance in the interpretation stage of stream sediment and heavy mineral surveys in ascertaining the nature and source of the gold. The other technique that can be used to sort out anomalies due to flour gold far from its source is to analyze the stream sediment and/or heavy concentrates for the common associated elements in gold deposits, particularly As, Sb, Te, Bi, Ag, B, etc. If these do not support the gold anomalies by being present in higher than normal amounts, the gold has probably travelled far from its source.

Coarse gold (>1 mm) on the other hand generally remains relatively close to its source in veins and lodes. Since the turn of the last century there have been many estimates of how far coarse gold migrates from its source. Most agree that 5000 to 20000 ft is about the length of the highly anomalous paystreak from the primary source of the gold under normal conditions. In the Kolyma River basin where the natural rock riffles (fissility) are aligned perpendicular to the stream flow, Gorbunov (1959) estimates that the distance the maximum amounts of gold have travelled from their primary source average from 700 to 1500 m (2300–4920 ft). Tuck (1968) states that studies indicate that during some 5000 ft of erosion, coarse gold (>1 mg in weight) liberated at the beginning of the erosion cycle does not travel more than 10 000 to 15 000 ft horizontally. These distances agree reasonably well with my investigations in the Yukon and elsewhere, and I conclude that the downstream vector of the movement of coarse gold derived from primary vein and lode deposits is small in most places, measurable in miles rather than in tens of miles. The immobility of gold is emphasized in some places. There are many records of the gold in Tertiary placers being not far removed from its primary source, commonly only a few thousand feet. Later (Quaternary) streams that have downcut through the Tertiary placers almost at right angles are, likewise, enriched within a few thousand feet of the secondary (Tertiary) concentration of gold.

The problem of the precise source of gold in stream sediments and other drainage deposits does not enter into the geochemical considerations if only gold placers are sought. It is, however, of fundamental concern if gold is to be used as an indicator of its primary deposits. As discussed in the section on placers there are two sources from which the gold in stream sediments may come – one is primary veins and lodes; the other is widespread disseminations in quartz veinlets, blows and in pyrite and other minerals in large volumes of rocks. In the first case, use of gold as an indicator of its deposits is entirely valid; in the second, suites of rocks that are only slightly auriferous will be outlined.

There is probably no certain way to differentiate gold derived from the two sources mentioned above. Some clues may be obtained from the fineness of the gold. In general gold derived from veins and lodes decreases in fineness (the ratio Au/Ag of the native gold decreases) as the source is approached. This is generally not the case with fine-grained gold derived from weathered pyrite and from widely distributed small quartz stringers and blows in various rocks. This type of gold is commonly high in fineness throughout the sedimentary system eluvium-alluvium. Gold derived from the weathering of tellurides and aurostibite in veins and lodes and from massive sulphide bodies where the gold is held as microscopic or lattice constituents of the sulphides is, however, also commonly very high in fineness throughout the sedimentary system, a complicating feature that should be borne in mind by the geochemist. The physical nature of the gold may also give a clue to its origin. Gold derived from quartz veins and lodes containing the element mainly as native gold exhibits an increase in the porosity, roughness and hackly appearance as the source is approached. The size of the gold particles generally also shows an increase towards the source. Gold derived from pyrite in rocks or in disseminated small quartz stringers, blows, etc. is generally very fine grained with a smooth appearance, except where extensive accretion into large grains and nuggets has taken place. Gold derived from sulphide bodies where the element is held mainly as lattice constituents of sulphides and sulphosalts or as microscopic particles in these and other minerals may be very fine grained or may be accreted and be similar to that liberated from disseminations in large volumes of rock. These features complicate the picture, but they should be clearly recognized by the geochemist. Commonly accretion nuggets and grains derived from disseminations in rocks or from sulphide bodies are much smoother and more gnarled than the rough hackly nuggets or particles liberated mechanically from their source in gold-quartz veins and lodes in which the metal is present mainly in the native state. Further clues as to the source of the gold may be provided by visual and microscopic examination of the gold and its associated minerals. For example flakes and small nuggets of gold with attached vein quartz, carbonates or limonite-wad (gossan) generally indicate a nearby vein source. The type and nature of the heavy mineral associates of gold may indicate the type of source. Thus, the presence of abundant relatively angular particles or crystals of arsenopyrite, stibnite, sulphosalts, tellurides, bismuth, bismuthinite, scheelite, wolframite, tourmaline, limonite-wad (gossan), etc. commonly indicates a nearby lode source. On the other hand well rounded particles of these mineral associates indicate a lode source at some distance from the sample point. Similarly, microscopic inclusions of pyrite, quartz, arsenopyrite, tellurides, bismuthides, etc. in the placer gold generally indicate a lode source, although care must be taken to differentiate when possible minerals included during nugget accretion processes.

Probably the best way of differentiating the two types of gold that may produce anomalies in stream sediment and heavy mineral surveys is to analyze the stream sediment and/or heavy concentrates for the commonly associated elements in gold deposits, particularly As, Sb, Te, Bi, Ag, B, W, etc. If these do not support the gold anomalies by being present in higher than average amounts, the gold has either travelled far from its source (either primary lodes or disseminations in rocks) or the gold has been derived from a large volume of rock through which it is widely disseminated.

The trace and minor element content of native gold in stream sediments and placers may be a clue to the primary source of the gold. Warren and Thompson (1944) spectrographically analyzed a wide variety of native gold samples, mainly from British Columbia and Yukon, but also from other parts of Canada and other countries and found that the amounts of various elements such as Ag, Cu, Fe, Mn, V, Ti, Hg, Pb, Bi, Te, As, Sb, Zn, Cd, Sn, Pd and Pt varied widely. From this they concluded that the differences are more related to metallogenetic zones or provinces than to any difference in the type of deposit in which the gold is found. They thought that by detailed analytical work gold could be used as a pathfinder and that it should be possible to determine the metallogenetic province or zone, and in some cases even the deposit from which a particular sample of gold originates.

Fisher (1945, 1950) in two classic works on the fineness of gold states that the uses to which a knowledge of gold fineness can be put are manifold. Thus, the relative importance as sources of gold of the various areas drained by different streams can be assessed with some accuracy; similarly sudden changes in the fineness of gold along a stream may indicate the presence of important new sources of primary gold; and in areas where the fineness of gold associated with each type of intrusive has been well established, the fineness of the native gold not only gives one a definite lead towards its source, but may also indicate its probable geological associations and the type of deposit to be searched for.

Mustart (1965) carried out a study of trace and minor elements in the alluvial gold of the Dawson and Mayo (Dublin Gulch–Keno Hill–Highet Creek) areas, Yukon. The elements determined semiquantitatively by spectrographic methods in the placer gold included Ag, Cu, Fe, Sn, Hg, Pb, As, Sb, Bi, Zn and Mn. He found that certain assemblages of trace elements particularly Zn, Bi, As and Sb show a characteristic distribution and can often be correlated with the presence of known mineralization in the area. Thus, Zn, As and Sb were markedly concentrated in the Mayo samples indicating the characteristic Zn, As and Sb mineralization of this area. In the Dawson area these elements were present in only low amounts again indicating the general weak Zn, As, Sb mineralization of the Klondike.

Desborough (1970) and Desborough *et al.* (1970) examined placer gold grains from numerous localities in the western United States and Alaska for their silver and copper content and their mineral inclusions, mainly sulphides, by electron microprobe. Their work is detailed and merits close study. Briefly, they noted that the variation in the copper and silver content among grains from a particular locality is large, but that the mean silver content of the interior of the placer gold grains from each locality and (or) the variation in copper content may be of value in distinguishing lode sources and gold mining districts. Thus, microprobe analysis shows that the interior of gold grains is independent of chemical actions that affect the border of placer gold grains during their transport history, and it is evident that distinct compositional groups of different lode sources may be identified even in a single sample – information that may aid in recognizing the existence of concealed lodes that once contributed to a placer environment.

Antweiler and Campbell (1976) noted that the composition of native gold, i.e., its 'signature', may be a useful guide in mineral exploration since the mineral is associated with a great variety of ore deposits and has characteristic signatures for each of several types of deposits. These signatures comprise distinctive contents of Au, Ag and Cu, a unique suite of trace elements that includes Pb, Bi, Sb, As, Te, Zn, Sn, Cr and Ni, and characteristic Au/Ag and Au/Cu ratios. The authors list a number of ore deposits in the United States in which the gold has characteristic signatures, including porphyry copper, pyrometasomatic, contact metamorphic and hydrothermal vein deposits. They also noted that signatures of gold from the Central City district, Colorado correlate with zoning patterns. In this district the silver content of gold was found to increase from a low of 5 to 10 per cent in the central, productive high-temperature zone to a high of 35 per cent in the nonproductive low-temperature zone, whereas the copper content of the same gold samples decreases from a high of 500 to 1000 ppm to less than 100 ppm. From this the authors concluded that the silver and copper contents of gold may identify the zonal position of a vein and thereby help to predict its potential productivity. Finally, Antweiler and Campbell (1976) noted that another application of gold compositional data is in tracing placer gold to its bedrock source. As an example of this they mentioned the Tertiary quartz monzonite Montgomery Gulch stock in the Tarryall district in Colorado. There, the silver content of placer gold in the district differed from that of nearly all of the bedrock sources of gold found by early prospectors. However, one scarcely prospected area peripheral to the Montgomery Gulch stock was found to contain gold with a silver content similar to that of the placer gold. This area is thought by the authors to be the most likely source of gold in the productive placers and is a potential exploration target.

Comparisons of the internal structure of gold as observed microscopically following etching or other treatments may be a clue to the source of the stream or placer gold if sufficient control samples of both primary (lode) and secondary (placer) gold are available from a specific area. If control samples of the lode gold are not available comparisons cannot be made, and hence interpretation of the source may be difficult if not impossible since the internal structure of stream (placer) gold depends on many factors that are discussed in Chapters II and III. Those contemplating work on the internal structural characteristics of gold should consult the papers by Fisher (1935), Valpeter and Davidenko (1970) and Petrovskaya (1973).

The results of a number of heavy mineral surveys carried out in auriferous regions have been described in the literature. These are briefly summarized in the following:

Overstreet and Bell (1960) carried out a geochemical and heavy mineral reconnaissance of the Concord SE quadrangle, Cabarrus County, North Carolina. The sample points were widely spaced, generally one sample per stream tributary. Gold placers and vein deposits are known in and adjacent to the quadrangle. Gold and scheelite were used as indicators during the survey, gold being identified visually and scheelite by its fluorescence (ultraviolet radiation of 2500 Å wavelength). The authors concluded that the geochemical and heavy mineral reconnaissance provides a rapid and inexpensive method of locating high-background areas as a first step in the exploration for ore deposits.

Heavy minerals collected from the stream sediments in the Chinkuashih area of Taiwan by rocker were used by Li-Ping and Fang-Sung (1968) in prospecting for gold and copper deposits. The indicators chosen were native gold, barite and pyrite, which occur in the primary deposits. The distribution pattern of the native gold indicated an anomalous area of about 10 km^2, including all the known gold and copper deposits in the Chinkuashih area. No native gold was found in the heavy mineral samples outside the known or potential mineralized areas. The copper content of pyrite in the vicinity of copper deposits was found to be high (1100 to $>$1700 ppm), relatively low in and near gold deposits (40–480 ppm) and low in and near coal beds (100–120 ppm).

Analyses of nonmagnetic parts of concentrates panned from arroyo sediments near the Monticello Box, northern Sierra Cuchillo, Socorro County, New Mexico were used by Griffitts and Alminas (1968) to indicate possible concealed mineral deposits, mainly those of gold and silver in Tertiary volcanic rocks and in Paleozoic limestone. Concentrates southeast of Monticello Box were found to be anomalous, containing as much as 2 per cent Pb, at least 0.2 per cent each of Mo, Zn and Cu, and anomalous amounts of Ba and Ag. Te was found in the part of the original samples that passed through an 80-mesh sieve, but this material was found to be less useful than the concentrates in outlining the anomalies.

Fischer and Fisher (1968) experimented with panned concentrates from streams in the northwestern part of the San Juan Mountains, Colorado. The concentrates were analyzed for gold by the fire assay–atomic absorption method. Three types of areas where chosen for study (1) 'barren' areas, where gold mineralization might be geologically possible, but no deposits are known; (2) slightly mineralized areas that contain only a few known veins and prospects and small mines; and (3) well-mineralized areas that contain numerous veins and some very productive mines.

The results of the survey indicated that replicate analyses of large samples were consistent enough to permit placing considerable confidence in the results obtained for smaller samples on which only one analysis was made. For general field practice, it was necessary to pan enough sand and gravel to yield about 15 g of concentrate. The analytical results obtained were also quantitatively compatible with known geologic relations and indicated that a few samples from a stream are adequate to distinguish between 'barren' and mineralized areas and to determine the relative amount of gold in mineralized areas.

Guigues and Devismes (1969) carried out an extensive heavy mineral survey of the Precambrian Massif Armoricain in France, one of the indicator minerals being gold. They display a map showing marked concentrations of alluvial gold in Brittany in the Fosse Centrale Amoricaine and in Normandy, the native metal being localized both at the border and some distance from granitic bodies. The Paleozoic Châteaulin Basin is also shown as having concentrations of alluvial gold. This report is a fine example of how to carry out heavy mineral surveys, and in addition contains an excellent alluvial (heavy) mineral atlas.

Panned concentrates have been used extensively for many years in U.S.S.R. apparently with considerable success (Sigov, 1939). Two recent investigations of interest based on gold and associated heavy minerals include those by Gurev (1970) and Zavorotnykh (1971, 1972). Komov (1966) mentions that panning and heavy mineral studies have been used effectively together with other geochemical and geophysical methods in prospecting for primary gold deposits and placers in the Yenisei Range, U.S.S.R.

The gold content of the heavy mineral concentrates of stream sediments in the Keno Hill area, Yukon was investigated by Boyle and Gleeson (1972). The concentrates were analyzed for gold by the fire assay–atomic absorption method. The results of the survey showed that the gold contents in the heavy mineral concentrates reflect the presence of four northeast-trending vein systems, two being known and two being suspected from the data obtained. These comprise an early gold-bearing pyrite-arsenopyrite assemblage and late pyrite-galena-sphalerite-freibergite lodes. In some systems the former predominate; in others the latter are well developed. All of the known gold placers of the area were indicated by high gold contents in the sediments, and some streams, not known to contain placers, were found to have relatively high gold values.

A heavy mineral survey was carried out by Good *et al.* (1973) in two areas (one mineralized and the other barren) in the northeast striking gold–base metal belt in Goochland, Fluvanna and Louisa counties of the Central Piedmont of Virginia, United States. The heavy mineral fraction obtained by sieving and panning 20 lb samples of stream sands was analyzed for gold by atomic absorption. Samples of concentrate from the mineralized auriferous region yielded values greater than 100 ppm Au in the immediate vicinity of the Shannon Hill–Caledonia cluster of mines and 10 to 100 ppm Au up to 1 mi downstream. Farther downstream the values were less than 1 ppm Au in the concentrate. In the barren area only 10 samples out of 128 reported gold contents greater than 1 ppm, the highest being 7 ppm. The heavy mineral survey proved more effective for delineating gold mineralization than reconnaissance surveys of soils, mull (humus), leaf ash or ordinary sieved stream sands. Further details of this survey are given in a later paper by Good *et al.* (1977).

Ocean sediments were tested for gold along the southeast coast of Nova Scotia in 1968 (Libby, 1969). Samples were obtained by a unique drilling method employing a high pressure water jet stream pipe and an air lift pipe for vacuuming the hole cuttings and conveying them to the ship through hoses. On shipboard the samples were passed over riffles in a sluice box, and the gold and heavy minerals were collected for study and evaluation. The survey outlined three significant marine placers off the Ovens Peninusla with a combined total volume of 42 million yd^3 of auriferous material. Other investigations concerned with recent marine placers are outlined in the section on placers in Chapter III.

Gold placers and stream sediments enriched in gold may indicate the presence of various types of metalliferous deposits, particularly massive sulphide bodies, polymetallic lodes and porphyry copper deposits. There are numerous instances of these circumstances in the Canadian Cordillera, one striking example being the Casino porphyry copper deposit in Yukon (Archer and Main, 1971). In the western United States and Mexico several of the great porphyry copper desposits had associated gold placers, a good example in Utah being the placers in Bingham Canyon and tributary drainages, which were derived mostly from oxidized copper and lead-zinc-silver ores (Johnson, 1973*b*). Learned and Boissen (1973) have pointed out that gold is a useful pathfinder element in the search for porphyry copper deposits in Puerto Rico. One could go on at great length relating the presence of gold placers as indicators of metalliferous deposits but this seems unnecessary. Figure 67 shows clearly that the distribution of gold placers marks metallogenic zones throughout the world where glaciation has not been severe. Even in glaciated areas the gold content of till, stream sediments, etc. often indicates the presence of metalliferous deposits.

Biogeochemical methods

The biogeochemistry of gold is discussed in some detail in Chapter II where it is stated that the gold content of most plants, animals and their fossil residues is low. For plants the normal range is 0.005 to 0.10 ppm in the ash; for animals <0.0001 to 0.02 ppm in the dry matter; and for coal, petroleum and other fossil organic materials <0.05 ppm in the ash. In general it can be said that the gold contents of flora and fauna tend to exceed the norm by a factor of 2 or more in the vicinity of gold deposits.

Prior to 1950 the analytical determination of gold in organic materials was plagued by many difficulties, and the results are semiquantitative at best. In recent years the determination of gold in organic materials by neutron activation and fire assay–atomic adsorption methods is more reliable but still beset with a number of problems. Because of sampling problems and analytical difficulties biogeochemical methods have not been widely used in prospecting for gold deposits.

The techniques employed in biogeochemical surveys are varied. One technique employs the indicator qualities of certain plants and animals; another utilizes analyses of plants and animals for gold or its indicator elements in the search for deposits; and still others employ analyses of fossil residues of plants and animals, canine sniffing and so on. The methods and procedures employed in biogeochemical prospecting are discussed at length by Brundin (1939), Thyssen-Bornemisza (1942), Viktorov (1955), Cannon (1960), Malyuga (1964), Nesvetaylova (1970) and Brooks (1972).

There are probably no indicator plants that are specific for gold (or silver), although some plants may be specific indicators of soils or waters rich in As, Sb, S, Se, Te, Cu, Mn and other elements that commonly accompany gold in its deposits. A discussion of these indicator plants, however, lies outside the scope of the present treatise. The interested reader should consult the works of Sykora (1959), Chikishev (1965), Nesvetaylova (1970) and Brooks (1972).

Lidgey (1897) and Dorn (1937) mention a number of plants that are reputed to be indicators of gold and silver-rich areas. In Brazil *Crecropia laetivirens* Hub.; *C. palmate* Willd; *C. lyratiloba*; *Alpina speciosa* Schum.; *Typha latifolia* L. (cattail) and *T. dominguensis* Kunth, all containing gold in their ashes, are reported to be gold indicators, and it is said that *Eriogonum ovalifolium* (buckwheat family) is a silver indicator, which led to the discovery of several silver deposits in Montana. Sykora (1959) made brief reference to a number of plant indicators for gold and silver, and Carlisle and Cleveland (1958) gave a list of sampler plants for silver and other elements. There are also some very interesting references to indicator plants for silver and gold in old Chinese writings (Needham, 1954, 1959). For instance, it is stated that the *tshung* plant (the ciboule onion, *Allium fistulosum*) is an indicator for silver, and the *hsiai* plant (*Allium bakeri*) indicates gold. Sykora (1959) quoting others says that certain psammophytic (silica-loving) plants may be indirect indicators of gold deposits. He further remarks that the early prospectors in California were guided in their search for placer fields by certain psammophytic shrubs, which in spring are especially visible from great distances on account of their white flowers. He does not name the shrubs.

There are no references in the literature to chlorotic or other effects, if any, induced in plants by soils enriched in gold (or silver).

Since 1950 a number of investigations using analyses of plants in the search for gold-bearing deposits have been described in the literature. Some have employed gold directly as the indicator element; others have used arsenic and other elements that commonly accompany gold in its deposits.

Warren and Delavault (1950) examined the gold and silver content of some trees and horsetails in British Columbia; a gold-bearing area (north fork of Watson Creek) and a number of areas barren of gold were chosen for the study. In the former, the ash of horsetails, Douglas fir, aspen, willow and dwarf juniper contained from 0.17 to 1.02 ppm Au; in the latter the recorded gold values did not exceed 0.03 ppm. The corresponding silver values were: positive area, nil to 0.39 ppm; negative areas, 0.10 to 1.40 ppm. Some of the negative areas (i.e., barren of gold) contain silver-rich deposits, e.g., Sullivan Mine. More recent work by Warren and Hajek (1973) in an auriferous zone in the Stirrup Creek area (a northern tributary of Watson Bar Creek) has shown that stems of white pine and Douglas fir contain from 0.50 to 0.60 ppm Au in the ash. In addition Mountain Phacelia (*Phacelia sericea*) contains from <0.01 to 1.20 ppm Au in oven-dried material. This plant also reports tellurium in amounts ranging from <0.05 to 3.5 ppm in oven-dried material.

Razin and Rozhkov (1963) investigated the gold content in the ash of a number of whole plants and roots over the Kuranakh ore deposits, Yakutia (*see* the section on placers in Chapter III for a description of these deposits). The plants chosen for study were varied and included larch, pine, birch, alder, poplar, rosemary, blueberry, horsetails, mosses, etc. Outside the mineralized zones gold was not generally detected in the ash of these plants. Within the confines of the ore zones the average gold content of the plant ash was about 0.45 ppm. Details should be sought in the original publication. The authors concluded that analyses of plants for Au (and Cu and Zn) are effective for outlining zones in the rocks enriched in gold. (*See also* the section on pedogeochemical and hydrogeochemical methods where the results of these surveys in the district of the Kuranakh deposits are described.)

Warren *et al.* (1964) used arsenic in the ash of Douglas fir (*Pseudotsuga menziesii*) as an indicator of Au, Sb, Pb and Zn mineralization in British Columbia. They noted that in unmineralized areas the ash of Douglas fir seldom carried more than 2 ppm As; in the vicinity of mineralization arsenic contents in the ash ranged from 7 to 10 000 ppm, the values being higher in the present year twigs. In addition, most of the biogeochemical anomalies were more marked than the pedogeochemical. In a later paper, on the same subject Warren *et al.* (1968) found that Douglas fir commonly contained 10 to 100 times more As in its ash than any of the trees and lesser plants normally associated with it. An analysis of all of their data indicated that in areas where there is no known arsenical mineralization, the ash of first year Douglas fir twigs will generally contain less than 100 ppm As. In the vicinity of mineralization the ash contains from 100 to >1000 ppm As. First year needles and all older growth generally carry much smaller amounts of arsenic than first year twigs.

In India, Dekate (1971) has carried out a number of

investigations of gold in soil and plants in areas of gold mineralization. He found gold enrichments in humus-rich soils developed on zones of gold mineralization and noted that plants producing hydrocyanic acid such as *Sorghum saccharatum Pers* are capable of preferentially absorbing gold from the soil, at times, to the extent of depleting the gold in the soil in the proximity of plant roots (*see also* the section on biogeochemistry in Chapter II). Dekate (1971) concluded that cyanogenic plants could be used advantageously in biogeochemical surveys mounted to prospect for gold in areas of gold mineralization.

A number of surveys utilizing analysis of plant ashes for gold and its indicator elements have been carried out in the arid desert areas (Tamdytau, Kyzylkum, etc.) of Uzbekistan, U.S.S.R. in the search for gold deposits (Khamrabaev *et al.*, 1965; Aripova and Talipov, 1966; Talipov, 1968; Talipov *et al.*, 1968; Talipov *et al.*, 1969; Talipov *et al.*, 1971; Glushchenko *et al.*, 1973; Talipov and Khotamov, 1973, 1974). The region is one where the gold deposits contain abundant arsenic and where the surface is covered with from 3 to 5 m of residual materials and loose rock. The vegetation includes many desert varieties principally wormwood (*Artemisia*) perennial grasses, saltwort (*Salsola*), *Lagochilus intermedius* and giant fennel (*Ferula*). The ashes of *Artemisia* proved to be the best sampling medium, but other plants were useful in certain areas. Over buried gold deposits up to 85 ppm Au was found in the ash of *Artemisia*, and this herb was also commonly enriched in As and Sb. In general the ashes of the desert plants were manyfold higher in gold in gold-bearing areas than the corresponding plants in background areas, most plant ashes contained from 40 to 150 times the amount of gold in the surface soils and residuum. Elevated contents of Sb, Ag, Pb, Cu, Zn, Cr, Ni, Sn and Mo were detected in a variety of plant ashes in a number of areas containing gold mineralization, and As was found to be much higher in content in the plants in gold-bearing areas compared with its normal abundance in plants. In a number of places a direct correlation between the As and Au and Sb and Au contents of the ash of plants was noted with the presence of gold-bearing deposits. Most authors concluded that the biogeochemical method is superior to the lithogeochemical methods of prospecting in the deserts of Uzbekistan and Kazakhstan. The best time for sampling was found to be in April and May while the plants were blooming. Later in the summer, and in September and October, the conditions were unfavourable for sampling and biogeochemical surveys in general. Other details should be sought in the original papers.

A number of investigations involving the determination of gold in plants by neutron activation have been carried out in Austria, Hungary and Czechoslovakia by Schiller *et al.* (1971*a*) and Kaspar *et al.* (1972). At Bockhard, Austria, *Aconitum napellus* growing near a tunnel in an old gold mine contained 60 ppb Au (in the dry plant), whereas samples taken from an area about 2 km distant had only 15 ppb Au. *Allium victoralis* (garlic) collected in the same localities contained 20 and 6.5 ppb Au respectively. In the Jilové area, some 20 km from Prague, gold occurs in gold-quartz veins and silicified zones in mica schists near a granodiorite pluton of Paleozoic age. Profiles of plant samples across the veins showed that the plants contained 8 ppb Au (in the dry plant) over the vein zone and from 1 to 3 ppb Au at distances ranging from 1 to 4 km from the zone. Shallow rooted plants and shrubs proved to be the best sampling media; deep rooted trees were unsuitable. In a later communication Schiller *et al.* (1973) reported that studies on maize and grass plants showed that the highest gold collecting efficiency was in the spring, probably caused by factors such as intensive uptake of elements by plants during this season, a higher rainfall and a more intensive movement of subsurface waters. Finally, these investigators stress the importance of further research in order to place biogeochemical surveys for gold on a quantitative basis since it is still uncertain how and under what conditions gold is taken up by plants.

In the Donets Basin of U.S.S.R., Dvornikov *et al.* (1973) examined the biogeochemical dispersion halos of a number of chalcophile elements (Pb, As, Ga, Ge and Tl) in 10 species of plants growing on auriferous deposits. The concentrations in fescue (*Festuca sulcata* Hack), *Euphorbia esula* and *Artemisia campestris* L. were higher than in any of the other seven species examined. The authors recommended that *F. sulcata* and *A. campestris* could be utilized as indicator plants in prospecting for auriferous deposits.

In the permafrost regions of U.S.S.R., Pitul'ko (1973) states that the analyses of tree leaves or aqueous extracts from these leaves can be used in prospecting for tin and gold placers. The best indicators proved to be Mo, B, Ag and Ti.

Quin *et al.* (1973) investigated the efficacy of plant and soil indicators of quartz reefs, some auriferous and argentiferous, in the Maratoto Valley, New Zealand, the species analyzed being *Berilschmiedia tawa, Brachyglottis repanda* and *Dysoxylum spectabile.* They established that the best indicators of the reefs were elevated silver concentrations in the A horizon soils; a 'negative mercury anomaly' of less than 0.23 ppm in the B horizon; low values of chromium and magnesium in C horizon soils; and elevated copper concentrations in all three species studied. The high copper levels they thought, were associated with the presence of this element in silver ores in the area and the negative mercury anomaly may be due to loss of this element at the temperature of formation of the reef. The low values of chromium and magnesium probably reflect the lower concentrations of these elements in the lodes than in the andesitic country rocks.

The recent paper by Talipov *et al.* (1975) on the use of gold in plants as a prospecting aid in the Chatkalo-Kuraminsk region, Uzbek S.S.R. contains much of interest. They noted that the accumulation of gold (and a number of other elements) in plants depends to a large degree on the concentration of the element(s) in the groundwaters. They also observed that some elements have an antagonistic effect on the uptake of certain elements in plants; thus high calcium contents in groundwaters decreased the migration of gold and hence its accumulation in plants.

In a further paper on the gold and antimony distribution in plants in the auriferous Chatkal-Kuraminsk region Talipov *et al.* (1976) observed a direct proportionality between the gold and antimony contents, a relation that they concluded could be used in surveying the potential of gold deposits in the eastern Almalyk area. Certain plants such as *Spiraea* and honeysuckle were found to accumulate large amounts of gold when growing in the vicinity of the auriferous deposits. Also in the Almalyk area (Talipov *et al.* 1974) a linear correlation was observed between the chemical composition of plants, soil and water and the content of Au, Ag, Cu, Zn and Pb. Soil, water and plant ashes from the auriferous deposits had 2 to 10 times the Au content of the auriferous zones of the Cu-Mo and complex ore deposits. The highest Cu, Ag and Pb concentrations were observed in southern Almalyk plants. The authors concluded that the good correlation between plant composition and metal distribution indicates that a biogeochemical method can be used in prospecting for Au and Cu-Mo ore deposits on the northern slope of the Karamin Range.

At the Ogofau Gold Mines, Dyfed Wales the gold orebodies are quartz veins containing auriferous pyrite and arsenopyrite in Silurian and Upper Ordovician shales and siltstones. The veins also contain abundant hydromuscovite, minor galena and sphalerite and rare cookeite (Li silicate). Six species of vegetation (birch, oak, hazel, ash, hawthorn and gorse) from the Ogofau area were sampled by Steed *et al.* (1976) and analyzed for Pb, Zn, Rb, Mn, Cu, Ni and Fe during a pilot survey. The results showed a marked dependence on the species and to a lesser extent on the type of plant material (twigs, leaves, bark), aspect and seasonal effects. Two year old twigs were the most satisfactory sampling materials. Birch trees gave the greatest anomaly contrast and magnitude, and these were used in the main survey, being supplemented by oak trees to cover areas where birch trees were absent. Rubidium reflected the pattern in the soil and proved to be the most useful pathfinder for the auriferous mineralization. Further details should be sought in the original paper.

The gold content of plant juice is recommended as an efficient method of prospecting for auriferous deposits by Kyuregyan and Burnutyan (1972). In the Zodsk gold belt, Armenian S.S.R., they found that the gold content of the juice from several plants was significantly higher than that in aqueous extracts of the plants or soil. The technique for obtaining and analyzing the plant juice is given in the original paper.

Krendelev and Pogrevniak (1977) have suggested another novel way of prospecting for gold deposits utilizing the analyses of birch tree sap. One of the advantages of the method over conventional biogeochemical methods cited by the authors is that the results of the sampling require no corrections for the time of sampling since the flow of the sap in the spring begins and terminates at the same times over large regions. Another advantage cited is the widespread occurrence of birch forests in U.S.S.R., a feature that should also apply to other countries in the northern latitudes such as Canada, Finland, Sweden and Norway. Further advantages are the simplicity of obtaining and processing samples and the fact that the sap is a ready-made natural metal-bearing solution that can be subjected to activation analysis and atomic absorption analysis without any preliminary preparation. Samples of sap obtained by tapping the birch *Betula platyphylla* were taken during the spring sap run along profiles across a pyritic gold deposit in western Transbaikal. The gold content of the sap varied from 1.5 to 133.0 $\times$ 10^{-9} g/l, the higher values clearly coinciding with anomalous gold contents present in the secondary lithochemical (soil) halos over the deposit.

The utilization of sap as a sampling medium should be applicable to regions that have a widespread distribution of certain species of tree. The maple and various conifers come to mind in Canada, and these and other species should be considered as sampling media elsewhere.

Organic deposits such as bogs, muskegs and marshes are commonly enriched near gold-bearing deposits in many of the elements associated with the gold, particularly Cu, Ag, Ba, Zn, Hg, Pb, As, Sb, U, Se, Te, Mo, W, Co and Ni. Gold may show slight enrichments in these organic deposits, but so far as my experience goes they are not marked. Where bogs are found enriched in metals and other elements commonly found in gold deposits, the source of the metals should be sought by tracing their dispersion in the surface or underground water courses that debouch into the organic deposits. The bogs may overlie the auriferous deposits or their distribution pattern may take the form of halos or partial halos about mineralized zones.

In rocks containing coal fragments, carbonized plants or thin coal seams the general presence of auriferous mineralization may be indicated by higher than normal amounts of gold in the coaly materials, but more generally by the indicator elements found in auriferous deposits. During surveys using these carbonaceous materials samples of the coaly fragments, beds, etc. should be collected whenever found in outcrop or drill core, ashed and analyzed for gold, silver and their indicator elements. The results should then be plotted on plans and sections of the geology of the area in which the work is being carried out. Such representation may reveal general zones of mineralization or stratigraphic members that enclose auriferous deposits. It should be cautioned in passing that coaly materials tend to concentrate numerous chalcopile elements and other indicators of gold from groundwaters traversing relatively barren rocks. In general, coaly fragments in barren areas are not as highly enriched in the various indicator elements as those in mineralized areas. By using pilot surveys and one's experience it is usually possible by judicious screening to eliminate those samples that indicate mineralization from those that are ordinary natural collectors of stray amounts of As, Sb, Te and other indicator elements from circulating stratal waters.

Bitumen and various solid hydrocarbons particularly anthraxolite and thucholite can be used as sampling media in the manner just described above for coaly fragments. It should not be forgotten that the great bulk of the gold of the enormous Witwatersrand deposits is either in or closely associated with thucholite.

Two novel ways of utilizing biological agents in prospecting for gold have recently been advocated. One involves the use of termite mounds in Rhodesia and the other is concerned with 'man's best friend' the dog.

'Termite prospecting' has received considerable attention in Rhodesia, particularly in areas covered by Kalahari sand (Chamber of Mines Journal, Rhodesia, 1968, 1969*a,b*). The termites are thought by some to bring particles of mineralization up from depth and deposit them in their mounds. Analyses of the mounds for gold and its indicator elements may, therefore, reveal anomalies that are related to auriferous deposits. It appears, however, from the work by Watson (1970) that termites do not bring up mineralized material rich in zinc from depth, but rather that this element, as well as copper and lead, move upward in the soil solutions in Kalahari sand. Watson concluded from this that it is possible to prospect for hidden deposits by standard geochemical techniques, and that no advantage is gained by sampling termite mounds. In a more recent article concerned with gold in the Rhodesian termite mounds Watson (1972) concluded that: "The passage of termites through a gold-bearing fissure in the Basement Complex at 23 m depth and the presence of anomalous concentrations of gold in the 3 m thick overburden of Kalahari sand suggest that termites may have carried material containing gold from gold-bearing fissures in the Basement Complex to their mounds at the surface whilst excavating galleries to the water table at about 27 m depth. Subsequent erosion of mounds would have dispersed the gold in the surrounding soil. However, the difference between the mean concentrations of gold in termite mounds and soils is small and not statistically significant. It is, therefore, probable that the gold in existing termite mounds was derived mainly from soil carried by termites from less than 3 m depth in the course of constructing their mounds."

A further contribution to geochemical prospecting using the metal content of termite mounds is provided by d'Orey (1975). In central Mozambique he found that the copper and nickel contents in termite mounds of the *Macrotermitidea* family were enriched above a group of known copper deposits that are hidden by deep transported soil and were not revealed by geophysical and normal soil sampling surveys. d'Orey noted that sampling of termite mounds is also excellent for discriminating between different types of geological formations and, in the area investigated, for distinguishing between extrusive and intrusive serpentinites. He concluded that the sampling of termite mounds is a valuable supplement to established geological methods of prospecting, and recommended further investigation of the technique in places where there are deep-penetrating termites.

Dogs have long been known to have an acute sense of smell and were used to sniff out such objects as mines and other explosives during World War II. It is only natural that the scenting ability possessed by man's best friend should be turned to more peaceful pursuits, but it is not clear who first suggested the use of dogs to sniff out float boulders of oxidizing sulphides. Apparently the idea developed in Russia and Finland about the same time, probably in the early 60's. In Finland Dr. Aarno Kahma of the Geological Survey of Finland conceived the idea in 1962 that dogs should be able to scent ore boulders and thereby aid in outlining ore boulder trains and fans. Subsequently, dogs were trained in Finland, Sweden and Canada for this purpose. The method of training and the problems involved in using 'prospecting dogs' are outlined in some detail by Orlov *et al.* (1969), Brock (1972) and Nilsson (1973). It would appear that the use of dogs in prospecting for auriferous deposits is most effective where the deposits contain abundant sulphides.

Finally, Pagliuchi (1925) records an interesting case of the biogeochemical association of a bird, a particular tree and the occurrence of gold in southeastern Venezuela. He called it a "Feather guide to gold" and says:

> While engaged in the examination of the above-described mines and prospects, I constantly heard the sharp cry of a bird which is seldom seen, as its habitat is in the dense foliage of tall trees. The bird is gray in colour and about the size of a robin. It is called "El Minero" (the Miner). The natives informed me that this bird is always found in the vicinity of gold mines or whenever quartz is abundant. Thereafter, as I rode through the country, whenever I heard the cry of El Minero either I saw a quartz ledge extending across the trail or was able to find one by looking about a little. Apparently the natives were correct. Why the affinity which seemed to exist between bird and quartz ledges? My investigations and attempts to solve the mystery left me still puzzled. Finally a miner told me that if I wanted gold I must always look for the Mora tree. He pointed out to me some trees which grew next to the quartz ledge that I was examining at that moment. Then the solution of the puzzle dawned on me. The Mora tree grows and thrives only in siliceous-ground, and El Minero feeds on the berries of this tree; hence whenever his cry is heard, one is assured of finding a quartz vein in the vicinity. Of course the quartz may or may not be auriferous, but when prospecting in a country known to be auriferous, it must be admitted that the bird is a helpful guide to the prospector.

Atmogeochemical methods

The ascent of warm exhalations from fissures in the earth, often containing mineral deposits, has been observed for centuries. Agricola writing in his classic, *De re metallica* says: "Further, we search for the veins by observing the hoar-frosts, which whiten all herbage except that growing over the veins, because the veins emit a warm and dry exhalation which hinders the freezing of the moisture, for which reason such plants appear rather wet than whitened by the frost."[34]

Gases associated with mineral deposits form both mobile and static gaseous aureoles. Four principal origins for the gases can be recognized:

1. Those that form mobile gaseous aureoles due to radioactivity, oxidation of mineral constituents or in certain places to present day metamorphism of mineral constituents. The gases produced are He, Rn and Ar in the case of deposits containing the radio-elements Th, U and K, and SO_2, H_2S, other thiocompounds, CO_2 and various volatile metallic organic and inorganic compounds in the case of oxidation and metamorphism.

2. Those migrating along faults and other structures some of which contain mineral deposits. The source of these gases generally lies in the rocks at depth, and their origin is commonly unknown. The gases most often encountered are H_2S, CO_2, N_2, CH_4 and the inert gases.

3. Those that form mobile gaseous aureoles and are related to present day mineralizing processes as in hot springs and fumaroles. The gases involved are generally H_2S, CO_2, H_2 and various volatile B, As, Sb, F and Cl compounds.

4. Those that form static gaseous aureoles and are related to past mineralizing, diagenetic and metamorphic processes in the rocks. These gases are trapped in inclusions in the lode and wall-rock minerals, along grain boundaries, and in small faults, fractures and other discontinuities in the lodes and country rocks. The gases and volatile compounds encountered are many and varied including H_2S, CO_2, N_2, CH_4, H_2, inert gases and so on (*see also* the discussion on liquid inclusions in the next section of this chapter).

Sampling materials utilized in atmogeochemical surveys

[34]H.C. Hoover and L.H. Hoover; Translation of Georgius Agricola, *De re metallica*, 1912, p. 37.

may include rocks and lode materials, soils and glacial materials, surface and underground waters and the air (atmosphere). Specialized techniques are required to extract and analyze the gases in all of these materials. Discussion of these lies outside the scope of this monograph; those wishing to pursue the subject further should consult the reference listing.

Gold forms no common inorganic volatile compounds that occur in nature in concentrations that are readily measurable. The element does, however, form a number of organic compounds, including alkyls (methylated compounds), organic arsenic compounds and chelates, some of which have a high volatility. None of these, so far as is now known, have been recorded in nature. One would expect such compounds to be present in highly organic environments where bacteria are particularly active. Some of the common elements associated with gold yield volatile compounds or gases by various processes. Here, we may mention particularly Hg, As and Sb, which form a number of volatile methylated compounds; in addition mercury as the metal has a relatively high volatility, and arsenic and antimony form volatile hydrides (arsine and stibine). Sulphur is perhaps the most common associate of gold that forms abundant gaseous compounds, often in some concentration as a result of oxidation processes, bacterial activity, metamorphism, etc. (e.g., SO_2, H_2S, various inorganic thiospecies and a host of organic sulphur compounds such as the mercaptans). Other volatile compounds of interest among the Group VIA elements include H_2Se, SeO_2 and various other inorganic and organic seleniferous species where high selenium contents are present in auriferous ores; also perhaps organic tellurium compounds in telluriferous ores. Where uranium and thorium are associated with gold, helium and radon gases are evolved. Zinc-rich ores containing gold may give rise to volatile methylated zinc compounds where bacteria are especially active. Fluorine and chlorine occur in some abundance in certain auriferous deposits, and these elements or their gaseous compounds may be of interest as atmophile indicators.

From the literature there appears to have been only a few atmogeochemical surveys carried out particularly to locate gold deposits. Frederickson *et al.* (1971) describe an airborne and truck-borne mass spectrometer–computer technique for regional exploration utilizing the F, Cl, Br, H_2S, SO_2, AsH_3 and Hg content of near surface air as a sampling medium. Only the briefest details are given, but when the system was tried in the auriferous Cripple Creek area of Colorado the results were successful. A total of 31 stations were occupied by the truck-mounted unit in 1 day. In addition eight stations were occupied up to five times each to demonstrate the repeatability of results. Because of the fairly extensive road network in this area, good areal coverage was obtained. Anomalies three or more times greater than background were obtained for fluorine, chlorine, hydrogen sulphide and (questionably) mercury. Comparison of the results with the more extensive ground geochemical survey by Gott *et al.* (1969) showed good correlation of anomalies. Furthermore, comparison of the fluoride results with unpublished fluorine data in the area also showed good correlation.

Other investigations include one by Eremeev *et al.* (1973) that outlines one profile of a high contrast helium anomaly over a gold deposit in northern Kazakhstan. The writer has noted low concentrations of CH_4, CO_2 and H_2S rising along various structures that contain auriferous bodies in the Canadian Shield; one of these is well documented (Boyle, 1961*a*). The source of these gases is unknown. One can only say in summary that it seems probable that atmogeochemical surveys utilizing rocks, soils, waters and the air will prove useful in the future for locating structures in which auriferous deposits may be found. In some cases perhaps atmogeochemical surveys using the volatile compounds of gold and its associated elements may actually prove useful in locating auriferous deposits directly. Those seeking further information on atmogeochemical surveys should consult the recent papers by Dyck (1969), Bristow and Jonasson (1972), McCarthy (1972), Ovchinnikov *et al.* (1973), Eremeev *et al.* (1973) and Meyer and Peters (1973).

Other methods

A number of other methods may be useful in locating gold deposits. These include plotting the distribution of various mineral indicators, panning of gossans, radioactive surveys, isotope variations and certain other geochemical-geophysical methods.

Quartz, one of the most resistant of minerals to chemical and mechanical agencies, is a universal associate of gold in nearly all types of auriferous deposits. These facts have been appreciated by prospectors for thousands of years, and it is no exaggeration to say that most of the great gold mines of the world have been found by tracing quartz float to its source in gold-quartz veins or other types of auriferous siliceous deposits. All types of quartz may be auriferous with perhaps the exception of rose quartz, which seems to occur only in quartz pegmatites. However, certain varieties of quartz are more indicative of gold mineralization than others. Prospectors tracing quartz float in Precambrian terranes should pay particular attention to black, grey and mottled quartz and to crushed and recrystallized white quartz since these are the most common auriferous varieties. In Paleozoic and Mesozoic terranes milky and grey varieties predominate in gold-quartz veins. Amethystine, crustified and chalcedonic varieties of quartz are particularly indicative of Tertiary gold deposits. All quartz floats exhibiting limonitic (iron oxide) staining or containing sulphides should be considered as a favourable indication of aurificity and should be assayed. In fact where quartz occurs relatively abundantly in any terrane a number of floats should be assayed on a routine basis.

More details on the varieties of auriferous quartz are given in the section on associated minerals in Chapter III, and the more sophisticated uses of quartz as a mineral indicator are mentioned in other contexts below.

Gold is associated in many deposits with galena, which oxidizes slowly to anglesite and cerussite. Often these two secondary minerals form only crusts on the galena preventing further oxidation. These partly oxidized nodules of galena collect in the soil, eluvium and in places in the stream beds. The panning of such stream beds, eluvium or soil may indicate the presence of gold (and silver) deposits, which can be traced by following the small particles of galena to their source. In some areas the oxidized galena may occur in relatively large masses, enabling one readily to follow the train of float in the soil, rock talus, etc. In other areas the galena may be finely comminuted, requiring microscopic and spectrographic or other chemical examination of the pannings. Often a simple map indicating the location of each piece of float, the grain count of comminuted galena, or the chemical analysis of the soil pannings for lead, gold and silver will reveal the float train and help in locating the source. By the same token a stream map of the area with similar data plotted on it may indicate the locale from whence the galena, gold and silver came. Numerous other heavy resistant minerals can be employed in a similar manner. Some of these include auriferous pyrite and arsenopyrite, oxidized stibnite particles, various sulphosalts, cinnabar, barite, scheelite, wolframite and chlorargyrite, the last being common in the oxidized zones of some gold-silver veins in arid environments (*see also* the sections on pedogeochemical and hydrogeochemical methods in this chapter).

The panning of gossans during prospecting for auriferous deposits is an ancient technique mentioned by Pliny (*Historia naturalis,* Book XXXIII). He describes how the Romans used the method in Spain; by panning *segullum,* a red earthy (limonitic) residuum in the oxidized parts of gold veins they were able to estimate the approximate worth of the (oxidized) veins. It is interesting to note that *segulla* is still the term used by Castillian prospectors for oxidized residuum on auriferous deposits.

Gossans of all types should be panned carefully for gold. Particular care should be paid to the oxidation products of indigenous gossans, and a search should be made for the oxidation products of minerals that accompany gold in its deposits such as anglesite, cerussite, jarosites, bindheimite, scorodite, antimony ochers, silver halides (chlorargyrite) and malachite (*see also* Chapter IV). Assays and spectrographic analyses are useful in determining the presence of gold and silver and should be done on a routine basis on all gossans. Gold and silver and other metals may be nearly completely removed from some indigenous gossans leaving them barren except for traces of gold, silver and other metals. In such cases deep trenching or diamond drilling may reveal the presence of economic concentrations of gold and silver at depth. Gossans resulting from the transportation of iron, manganese and other elements should be analyzed by spectrograph or other geochemical methods for silver, zinc, copper, lead and other metals. Often such transported (or pseudo) gossans result from the oxidation of pyrite in slates, iron-bearing carbonates in sediments, etc. and are not related to metallic deposits. In other cases they show a direct relationship to metallic deposits, yet they may not be significantly enriched in the metals present in the primary ores. The reason for this behaviour is not clear, but it seems to be related to the colloidal behaviour of hydrous iron and manganese oxides and the recrystallization processes that go on to make limonite and geothite. During such processes any adsorbed metals seem to have been leached out in a number of transported gossans the writer has investigated. It is, therefore, not possible with our present knowledge to tell which transported gossans are related to metallic deposits and which are not. Until further research elucidates this problem the cardinal rule with respect to gossans of all types is that they should not be left until all geological and chemical possibilities concerning their origin and possible relationship to metallic deposits have been exhausted.

Certain gold-bearing deposits and their alteration halos contain uranium (and thorium). Many of these deposits occur in Tertiary rocks (e.g., Central City district, Colorado), but others are known in rocks of all ages including Precambrian (e.g., Witwatersrand, South Africa; Box Mine, Lake Athabasca district, Saskatchewan; Richardson Mine, Ontario; Tennant Creek, Northern Territory, Australia). Such deposits can often be detected by geiger, scintillometer or X-ray spectrometer surveys or by using uranium (and thorium) as an indicator element in rock, soil, stream sediment or water surveys. At Tennant Creek for instance, Large (1975) observed an umbrella-shaped halo of uranium values that capped one of the auriferous magnetite-chlorite orebodies. Contents greater than 80 ppm U coincided with the outer edge of the auriferous zone.

Many types of gold deposits, particularly those in basic and intermediate volcanic and intrusive rocks, are characterized by enrichments of potassium in their wall-rock alteration zones (*see* the section on wall-rock alteration in Chapter III). Since the potassium isotopic species contains a radioactive member, ^{40}K, it should be possible using suitable detection apparatus to outline positive anomalous potash-rich zones which may be related to gold mineralization. Alternatively, potash metasomatism and other types of pervasive alteration may have removed elements such as U, Th and Ra from the rocks leaving negative or subnormal gamma-emitting aureoles. There appear to have been few surveys carried out in North America, Europe or Africa with the aim of discovering gold deposits using highly sensitive gamma-ray spectrometers either on the ground or airborne, although such methods have been used extensively to locate other types of deposits (Darnley and Grasty, 1971; Bennett 1971; Cook *et al.,* 1971; Johnson, C.H., 1971). Eight references to these types of surveys were noticed in the literature on gold deposits in U.S.S.R. Ostrovskiy *et al.* (1970) studied the possibility of using an aerial survey of radioactivity along the 3500 km long Okhotsk-Chukotka Mesozoic-Cenozoic volcanic belt where there are a number of gold-silver occurrences. In this belt the gold-silver mineralization occurs mainly in volcanics and associated sediments and is associated with adularization resulting from potassium metasomatism, which is accompanied by a decrease in radioactivity due to loss of Th, U and Ra. On the other hand, the radioactivity is higher outside the adularized rocks. The profiles and plans presented are most interesting; over the adularized zones 'highs' are recorded for potassium, and these are matched by corresponding lows for thorium and radium (Fig. 99). Details should be sought in the original paper. Gamma-ray spectrometric surveys carried out by Blyumentsev *et al.* (1974) at the surface and in diamond-drill holes near gold deposits in a similar geological environment exhibited patterns almost identical to those reported by Ostrovskiy *et al.* (1970). Both groups of investigators concluded that surveys employing gamma-ray spectrometry are useful in outlining areas of adularization associated with gold deposits. In the Darasun Goldfield Zashchinskii *et al.* (1970) used a potassium radiometric survey to outline tectonic fracture systems to which the auriferous deposits are related, and Portnov *et al.* (1971) used aerial γ-ray spectrometric and hydrogeochemical methods to define zones of adularization that mark Au-Ag mineralization at the Karamkensk deposit, localized in paleovolcanic caldera structures. In the alteration zone of a near-surface gold deposit in the U.S.S.R. Grebenchikov *et al.* (1974) noted that uranium and thorium formed a negative halo whereas potassium was found to increase towards the ore; sodium was leached. Somewhat similar results were noted by Zhokhov *et al.* (1975) in the auriferous Kokpatas field. There, in comparison with the volcanic-sedimentary host rocks, they found that the hydrothermally altered rocks differed sharply in their potassium content but did not differ noticeably in their uranium and thorium content. In the near-surface zone of hydrothermally altered rocks, however, an increase in uranium was observed, being caused by the sorption of uranium by iron hydroxides and other oxidation products. The authors concluded that the gamma-ray spectrometric method was useful in mapping the zones of hydrothermally altered rocks in which gold-sulphide mineralization was localized. Fel'dman *et al.* (1975) also found gamma-ray spectrometry of use in localizing auriferous deposits. Using the (U+Th)/K ratio they noted that the auriferous lodes were marked by minima; on the graph for the K content the positions of the gold lodes are marked by increased values. Krendelev *et al.* (1976) used gamma-ray spectrometric methods for contouring auriferous ore zones in the Transbaikal region, U.S.S.R. The method was capable of locating and rapidly outlining ore zones that formed with the redistribution of the natural radioactive elements in the host rocks. Zones of high gold content (quartz-pyrite-disseminated gold) were distinguished by a minimum K content and high U content. High U and occasional Th and K contents were commonly noted in rocks close to ore zones in tectonic structures (faults, shear zones).

Gross (1952), noting that some auriferous deposits are located in and around the boundaries of igneous intrusives such as small stocks and batholiths, examined the radioactivity (alpha activity) of nine intrusives including the Athona stock, Goldfields, Saskatchewan; the Falcon Lake stock, Manitoba; the Rice Lake intrusive, Manitoba; the McKenzie stock, Red Lake, Ontario; the Dome stock, Red Lake, Ontario; the Eagle Lake intrusive, Ontario; the Turnbull Township intrusive, Ontario; the Pearl Lake porphyry, Timmins, Ontario; and the Larder Lake intrusive, Ontario. The results of these surveys showed that stocks and batholiths of igneous or sedimentary (granitization) origin have a zone of higher-than-normal radioactivity in the vicinity of ore structures. Intrusives with no important structures close at hand have no such concentration. The determination of the distribution of the radioactive elements, can therefore, be used as a general guide to ore structures. Furthermore, the distribution of zirconium, silica and the heavy metals can be used in much the same way as radioactivity. Zirconium, however, appears to be less reliable as a structural indicator in granitized sediments, whereas in more basic rocks, silica is more critical than radioactivity. (*See also* the review of another paper on this subject by Gross in the section on lithogeochemistry.) Gross (1952) considered the radioactive technique to be of practical value as it enables the prospector to restrict his detailed work to relatively small areas around an intrusive body. Furthermore, if no major concentrations of radioactivity, zirconium, etc. are found, the intrusive may well be eliminated from detailed prospecting.

Gold has only one isotope, and as noted in a previous publication on silver (Boyle, 1968*b*), there are no variations of any consequence in the isotopic abundance of this most common associate of gold. There does not seem, therefore, to be any way of using either silver or gold isotopes in prospecting at the present time. Two common elemental associates of gold are lead and sulphur, both of which exhibit differences in their isotopic ratios depending on the mode of formation of their minerals and other factors. These two elements and a few others offer some possibilities in prospecting for gold-bearing deposits.

Neglecting details, metallic deposits, containing lead minerals tend to have a distinctive lead isotopic composition for all of the epigenetic lead minerals. For instance, lead minerals in uranium deposits tend to have a high proportion of uranium-lead, that is they are enriched in ^{206}Pb and ^{207}Pb the derivatives of ^{238}U and ^{235}U; in thorium-rich deposits a high enrichment of thorium-lead, ^{208}Pb can be expected in the

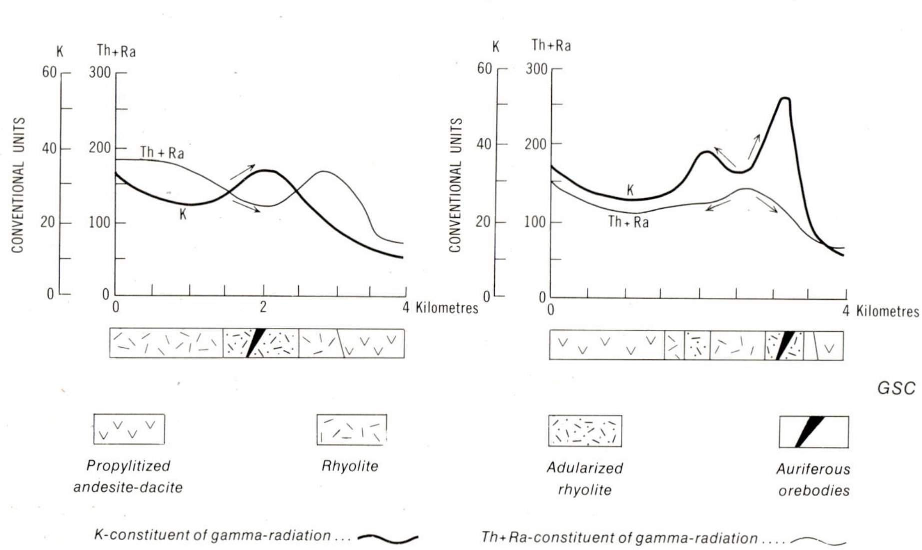

Figure 99. Typical radiation aeroprofiles over auriferous adularized zones in unaltered Tertiary volcanics, Okhotsk-Chukotka Belt, U.S.S.R. (after Ostrovskiy *et al.*, 1970).

lead minerals. In ordinary lead-bearing deposits the lead minerals have a component of original (primal) lead isotopes (^{204}Pb, ^{206}Pb, ^{207}Pb, ^{208}Pb), plus varying amounts of radiogenic lead (^{206}Pb, ^{207}Pb, ^{208}Pb) depending among other factors on the age of the deposit. And finally there are deposits in which the lead minerals have isotopic ratios that are unusual or anomalous (the J-lead of the Mississippi Valley deposits, but also of such deposits as Keno Hill, Yukon and elsewhere). The reasons for the variable isotopic composition of lead in deposits is extremely complex and certainly not yet understood. We need to understand the processes involved in the migration and concentration of lead isotopes before our theories can be placed on a firm basis (Boyle, 1959). These problems notwithstanding, it is possible to 'fingerprint' lead deposits and minerals derived from lead-bearing deposits by means of their isotopic ratios. This particular feature can be used in prospecting for gold- and silver-bearing deposits containing lead as Boyle (1968*b*), Angino *et al.* (1971), Cannon *et al.* (1971) and Antweiler *et al.* (1972) have suggested. The last two papers are particularly instructive in this matter. Cannon *et al.* (1971) observed evidence for a lead isotope gradient in the Leadville district, Colorado that appeared to point toward the focus of mineralization and also found favourable prospecting indications in the Mississippi Valley district, which do not concern us here. They concluded from their work that nearly all of the lead isotopic analyses presented in their paper represented galenas collected from mines or prospects in bedrock, and that the field is wide open for the use of isotopic analyses to evaluate geochemical lead anomalies in soils and waters.

In the Hayns Peak area of Colorado Antweiler *et al.* (1972) examined the lead isotope composition of galena in deposits, feldspar in rocks and placer gold in an effort to determine the origin of the placer gold, whose source has been variously attributed to rocks or disseminated mineralization of Precambrian, Mesozoic and Cenozoic (Tertiary) age. Briefly, the geology of the area involves a setting of Precambrian basement rocks overlain by late Paleozoic, Mesozoic and Tertiary sediments, all of which are intruded by a Tertiary (Pliocene) rhyolite porphyry stock and sills. The known bedrock mineralization includes Pb-Zn-Cu-Ag deposits and small widely distributed gold-quartz veins in a variety of rocks. The data from the lead isotope study are complex, but briefly the trace amounts of lead extracted from the placer gold is very close in isotopic composition to that in the galena associated with the Tertiary intrusions, and it lies on a compositional line joining the galena-leads with the feldspar-lead in the Tertiary porphyry. Furthermore, the isotopic composition of the lead in the placer gold has nothing in common with ore leads from the Precambrian sulphide deposits. The investigators concluded that this evidence implies that most of the placer gold was derived from the epigenetic Tertiary mineralization.

These examples suggest that lead isotopic ratios in placer gold, as well as in various minerals associated with gold such as pyrite, galena and lead sulphosalts, in soils, eluvium and stream sediments may be used to 'fingerprint' the source of the lead and may be an ancillary aid in various types of geochemical surveys carried out to locate gold deposits.

A number of detailed sulphur isotopic studies of wall rocks and ores in gold and polymetallic sulphide deposits have been done in recent years mainly to elucidate the nature of the ore-forming processes. Some of the results suggest that sulphur isotopes may be useful in detailed prospecting for gold-bearing deposits. Wanless *et al.* (1960) observed in a series of profiles across country rock, alteration zones and gold-quartz lenses at Yellowknife, Northwest Territories that as the ores are approached there is a progressive increase in ^{34}S content of sulphides (mainly pyrite and arsenopyrite) with a maximum enrichment in the gold-quartz lenses. Similar enrichments in the ^{34}S content of sulphides with approach to ore were noted by Shima and Thode (1971) in polymetallic deposits in Japan and Canada. Such phenomena are, however, not invariable as Boyle *et al.* (1970) found in the Keno Hill–Galena Hill area, Yukon. There, the sulphides in the gold-quartz and Pb-Zn-Ag-Cd lodes were lower in ^{34}S content compared with the host sedimentary rocks. From this it follows that before isotopic ratios can be used in estimating approach to mineralization the nature of the sulphur isotopic dispersion profiles about the deposits must be known. When this is clear, sulphur isotopic ratios in sulphides from drill cores may be a useful ancillary procedure in conjunction with lithogeochemical ratios for estimating nearness to ore (*see also* the section on lithogeochemical methods).

In a more general way sulphur isotopic ratios may be an aid in prospecting for gold-bearing deposits. Jensen (1971) reviewed the state of sulphur isotopic work in prospecting and concluded that:

> Measurements with high-precision, dual-collector mass spectrometers of the stable sulphur isotope ratio $^{34}S/^{32}S$ have enabled the economic geologist to determine if sulphide minerals have formed at low temperatures (bacteriogenically – generally syngenically) or at high temperatures (epigenetically). Primary sulphates can readily be distinguished from secondary sulphides and from primary sulphides in hydrothermal deposits by the differences in $^{34}S/^{32}S$ ratios of their respective sulphur. In some deposits, primary sulphates are associated with primary metal values and secondary sulphates are not. In addition the distinctions between a magmatic hydrothermal and a sedimentary origin of sulphide deposits can generally be resolved by sulphur isotopic measurements.

The differentiation between primary and secondary (supergene) sulphates by sulphur isotopic analyses has been clearly shown by Field (1966) for a number of deposits in the western United States and by Jensen *et al.* (1971) at Goldfield, Nevada. In all of these deposits there is a measurably distinct isotopic dissimilarity between the primary sulphates, which commonly are greatly enriched in ^{34}S with respect to associated sulphides, and supergene sulphates, which usually have ratios similar to the primary sulphides from which they were derived. With respect to alunite related to gold mineralization it is advantageous to be able to distinguish the primary alunite (which is associated with primary gold mineralization) from secondary alunite that may not bear any particular relationship to the period of gold deposition. At Keno Hill (Boyle *et al.*, 1970) the sulphur isotope ratio of supergene sulphates in the oxidized zones of the deposits reflects the ratio in the hypogene sulphides and is quite different (enriched in ^{32}S) from that in hypogene barite. Similarly, the ratio in the sulphate of the meteoric waters leaching mineralized zones reflects the ratio in the ores, a fact that can be used in hydrogeochemical surveys to 'fingerprint' sulphur derived from ores from that originating from the country rocks.

The isotopic ratios of various elements may also be of value in prospecting for gold-bearing deposits. Of interest may be the ratios of various elements so commonly associated with gold, particularly Cu (2 natural isotopes), Sb (2 natural isotopes), Te (8 natural isotopes), Cd (8 natural isotopes), Tl (2 natural isotopes), B (2 natural isotopes), Zn (5 natural isotopes), Hg (7 natural isotopes) and so on. There are no data on the distribution of the isotopes of these various elements in gold deposits. Jensen (1971) remarks that $\delta^{18}O$ surveys of rock and mineral samples may be of use in prospecting, permitting the plotting of isopleths, i.e., lines of equal $\delta^{18}O$ values, indicating increasing temperatures that existed as the source of the mineralizing solution is approached. Figure 100 illustrates this approach in an idealized manner.

Akopyan *et al.* (1976) studied the oxygen isotope constitution of quartz in ores, wall-rock alteration zones and host intrusive rocks of the Megradzorsk gold deposit, Armenia, U.S.S.R. They observed that the $\delta^{18}O$ values of the intrusives relative to SMOW (Standard mean ocean water) increase from early to late differentiates (6.25 to 10.60 °/oo). $\delta^{18}O$ values within the narrow range from 11.95 to 12.25 °/oo characterize the quartz of the hydrothermal stage with which increased contents of gold are associated. Furthermore, the quartz of the productive veins and their associated wall-rock alteration zones have the same or similar $\delta^{18}O$ values, a feature that led the authors to conclude that this fact is useful in detecting productive parts of the deposit enriched by circulating hydrothermal solutions.

Menke (1951) describes an isotopic method of geochemical prospecting based on the identification of ^{198}Hg. This isotope results from fast neutron bombardment of gold such as might occur where the precious metal occurs in association with uranium and thorium minerals. The mercury so formed may be rendered mobile and may hence be used as a travelling indicator of gold deposits. In using the method it is necessary that the small amounts of mercury not be masked by very much larger amounts of ordinary mercury (which contains some 10% of ^{198}Hg).

Reference has been made in previous sections to the use of gold fineness in surveys based on soils, eluvium and stream

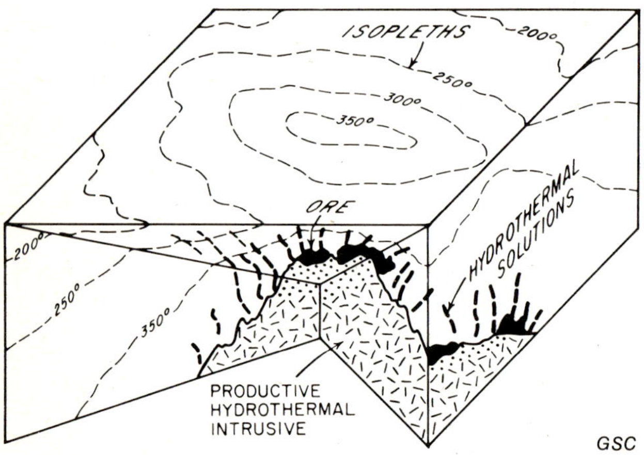

Figure 100. Idealized potential application of $\delta^{18}O$ studies to hydrothermally altered contact zones near a mineralizing fluid source (after Jensen, 1971).

sediments. The fineness of gold may also be used in a practical way in developing deposits already discovered. Thus, a detailed study of the fineness of gold at various levels of a mine may furnish useful data with regard to the amount of leaching and secondary enrichment that has taken place (Fisher, 1945,1950). Similarly, according to the same author, the fineness may be of value in determining the possible extension in depth of an orebody. For further details the interested readers should consult the section on Au/Ag ratios in Chapter III.

Numerous studies of the quartz of gold-bearing deposits have been carried out with a view to using the data in prospecting. Microscopically and macroscopically the quartz of gold deposits is extremely varied. There are, however, numerous features that distinguish gold-bearing vein quartz from other types of quartz, and these can be used in detailed lithogeochemical, pedogeochemical and stream sediment surveys if samples of the quartz can be obtained. *A priori* knowledge of the types and the features of vein quartz in an area are advisable before attempting to use the mineral as an indicator. Some of the specific characteristics of auriferous quartz are outlined in Chapter III.

Vertushkov *et al.* (1970) discuss various ways of distinguishing orebearing from barren quartz in gold districts in the Urals and elsewhere. Their method is based on statistical evaluation of such features as texture of the quartz aggregates, translucency, losses in heating and composition of gas-liquid inclusions. The last expressed as a 'water index' (V_{H_2O}/V_{CO_2}) is said to vary according to the content of gold in the quartz. For instance the following results were obtained from the Byn'gi deposit

'Water index'	Au content (ppm)
8.7	0.6 (barren quartz)
6.0	3.5
2.5	9.2 (ore quartz)

Maps are made of the quartz occurrences and indices for the various features are plotted. It is said that these maps aid in prospecting for gold deposits by differentiating areas of gold-bearing quartz from those that are barren. Details of the methods of computation of the indices should be sought in the original paper.

Pyrite and arsenopyrite are almost universal minerals in auriferous deposits, and both are excellent indicators of gold. Many studies of these two sulphides in auriferous deposits have been done (*see* the section on associated minerals in Chapter III). Certain crystallographic, X-ray, trace element and electrical properties sometimes characterize and 'fingerprint' the pyrite and arsenopyrite from specific auriferous deposits or belts. Pyrite or arsenopyrite with this 'fingerprinting' found in alteration zones, eluvium, soils and stream sediments can often give a clue to the presence of unsuspected deposits in unprospected or heavily overburdened parts of specific auriferous belts.

Studies based on liquid inclusions in minerals, especially quartz and the decrepitation characteristics of minerals in gold deposits may be of interest in prospecting for deposits or extending knowledge of those already known. The techniques of studying liquid inclusions in minerals and decrepitation methods are described at length by Scott, H.C., (1948), Smith and Peach (1949*a*, *b*) and Peach (1949). There is an extensive review and bibliography on both subjects by Smith (1953), and numerous investigations of gold-bearing quartz in a variety of deposits are described in Smith *et al.* (1949–1951). More recent work on liquid inclusions and decrepitation can be found in the works by Kashiwagi *et al.* (1955), Takenouchi (1962), Yermakov *et al.* (1965), and in the Proceedings of COFFI (Roedder, 1968–1974).

A general outline of the basis of liquid inclusion and decrepitation methods has been given in Chapter III. Here, we shall deal briefly with a few of the investigations extant in the literature that bear on prospecting for epigenetic gold deposits.

Since pressure and temperature largely control the deposition of gold in vein systems, and since these parameters can be estimated by liquid inclusion and decrepitation measurements in conjunction with other physico-chemical methods and calculations, it has been considered possible, in ideal situations, to predict the direction of flow of mineralizing solutions, the location of ore shoots and other features of gold mineralization. It is on this basis that a number of investigations of gold deposits have been carried out using decrepitation and other methods.

An early investigation of the ore deposition temperature and pressure by decrepitation of quartz and use of the pyrite geothermometer at the McIntyre Mine, Timmins, Ontario by Smith (1948) indicated two separate stages of mineralization in the gold-quartz bodies in greenstones and quartz porphyry, the first extending over the range of 630 to 400°C at a depth of 10 km and the second at 150°C but at a much shallower depth. Variations of temperature of deposition of pyrite were found to be unrelated to depth in the mine and to individual veins but were found to be related to the distance from the contact of the porphyry and the wall rock. A gradient of the mean temperature of deposition was found to be 14°C per 100 ft, from 500 ft within the porphyry to 500 ft outside the porphyry intrusive.

In a later paper Smith (1951) used the combined decrepitation and pyrite geothermometer methods to predict the maximum depth of the gold deposition in the Archean Ontario-Quebec gold belt comprising the McIntyre Mine at Schumacher (Porcupine Belt), the Kerr Addison Mine east of Kirkland Lake and the Malartic Gold Fields Mine in western Quebec. It was calculated that the vertical extent of the epithermal (late stage low temperature) mineralization was at least 9000 ft below the base of the Proterozoic Cobalt sediments (the datum level).

Blais (1954) investigated the auriferous quartz veins at the O'Brien Mine, northwestern Quebec by decrepitation methods. There, the ore shoots have a long vertical extension and are localized in or near vein rolls, vein intersections, vein deflections and graphitic shears where concentrated fracturing and shattering of the vein quartz afforded minute openings for the migration of auriferous solutions and deposition of late gold. The observed relationships indicated a connection between late gold and secondary liquid inclusions. Auriferous quartz was found to decrepitate in the temperature range 75 to 120°C, whereas barren quartz decrepitation began above 130°C. Many trend lines of low decrepitation temperature were found to correspond to ore shoots.

Boyle (1954) examined the quartz of two gold-bearing shear zone systems (Campbell and Negus Rycon) by decrepitation methods at Yellowknife, Northwest Territories over a vertical extent of some 1400 ft. There, the ore shoots occur in quartz veins and lenses some of which have been intensely crushed and recrystallized. Much of the free gold occurs in late fractures and recrystallized parts of the quartz veins, and some gold is intimately associated in a submicroscopic form with arsenopyrite and sulphosalts that are respectively early and late in the mineralization sequence. Curves of decrepitation temperature (135–250°C) and intensity (1–5) where drawn on the ore shoots and examined for existing patterns. The details are complex, but the following generalities are evident:

1. The quartz of the ore shoots is not characterized by consistently higher or lower decrepitation temperatures than that outside the ore shoots, although there is a tendency for higher decrepitation temperatures to occur in the ore shoots.

2. The gold in the orebodies is markedly concentrated in the crushed and recrystallized quartz, and in this respect the zones of higher decrepitation temperatures and lower decrepitation intensities mark the ore shoots.

3. In the Campbell System there is a decrease in decrepitation temperature in both early and late quartz with depth; however, this relationship is not pronounced in the Negus-Rycon System.

4. The decrepitation temperature and intensity apparently are controlled to a large extent by the proportion of microcrystalline quartz to medium-grained quartz in the sample; thus the greater the proportion of microcrystalline quartz the higher the decrepitation temperature and the lower the decrepitation intensity. Detailed petrographic study indicates that the microcrystalline quartz is probably developed by crushing and recrystallization of medium-grained quartz. It seems logical that during the crushing and brecciation large numbers of liquid inclusions would be destroyed; at the same time it is possible that the tensile strength of the quartz would be increased in a similar manner to that of a metal that has been work-hardened. The reduction in the number of liquid inclusions accounts for the decrease in decrepitation intensity and the probable increase in the tensile strength may produce the higher decrepitation temperatures. Thus, the variation of decrepitation temperatures and intensities is related to some extent to the amount of crushing and recrystallization that the quartz has undergone. The instrumental effect must be considered, however, when there are only a few liquid inclusions in the sample. In this case the slope of the decrepitation curve may be so low, that the point of break is difficult to determine, in which case there is a tendency to select a decrepitation temperature higher than the true value.

5. The significant decrease of the decrepitation temperature with depth of the early and late quartz in the Campbell System may be caused by decreasing crushing and recrystallization of the quartz at depth; that is, the control may be structural. On the other hand, the decrease in decrepitation temperature with depth may be explained by the temperature and pressure gradient existing at the time of formation of the secondary liquid inclusions. Petrographic and structural studies tend to substantiate the first of these conclusions.

It was concluded that there are certain poorly developed

patterns in the decrepitation results that may be significant in the analysis of the ore-forming process, but none appeared to be clear-cut enough to determine the direction of flow of mineralizing solutions or predict the possible location of ore shoots. The reason for the poorly developed patterns is the complexity introduced by extreme crushing and recrystallization of the quartz. In this respect the quartz in the Yellowknife orebodies is not suitable for decrepitation studies.

At the Lamaque Mine, Bourlamaque Quebec, Smith (1954) examined the quartz of the No. 6 vein with a view to determining the direction of flow of late solutions that probably deposited much of the gold. Briefly, the vein is composed of a coarse, glassy to white quartz with some ankerite, tourmaline, pyrite and native gold. The vein cuts across a zoned granodiorite-diorite plug and greenstone. When heated the quartz decrepitates between the range 80 to 124°C. High grade ore specimens have a relatively rapid rate of decrepitation; low grade specimens have a relatively slow rate. The measured decrepitation temperatures when plotted in the plane of the vein, and contoured, show that the lower values are near the central part of the vein, where it crosses the principal granodiorite-greenstone contact, and the higher values are around the outer limits of ore in the vein. Smith (1954) concluded that if the assumptions are made that the measured decrepitation is due to filling of secondary liquid inclusions trapped during the stage of gold deposition, and that the measured variation of decrepitation is due to variation of pressure during that stage, the contoured data can be taken to mean that the later gold-bearing solutions entered the quartz vein at the principal wall-rock contact and moved up and down the dip, depositing gold during and possibly due to, the drop in pressure.

In recent years an enormous amount of work has been done on liquid inclusions by all methods of study. Most of this work has involved geothermal projects and elucidation of the nature of the ore-bearing solutions. Only scattered references deal with prospecting for gold deposits. Abstracts and summaries of the various investigations can be consulted in the proceedings of COFFI (Roedder, 1968–1974).

It is difficult at present to evaluate the effectiveness of methods based on liquid inclusions in prospecting for gold deposits. It would seem that up to now the methods have been of limited practical value in detailed prospecting but have given considerable information on the temperature, pressure and nature of solutions from which gold has been deposited. Perhaps we cannot expect more at this stage of the research.

One approach to using inclusions and decrepitation in prospecting for gold may lie in 'fingerprinting' quartz and other minerals. Thus, the auriferous quartz of a mineralized belt is usually marked by specific liquid inclusion and decrepitation characteristics. If such quartz is found in soils, stream sediments, etc., it signals the fact that auriferous quartz veins lie uphill or within some local drainage basin.

The thermoelectric properties of pyrite have been investigated as a means of detailed prospecting for deposits containing the sulphide and gold. The pyrite geothermometer developed by Smith (1947) is based on the assumption that crystals of pyrite (an electronically conducting mineral) deposited at a high temperature have a more positive thermoelectric potential than crystals deposited at a low temperature, and that the thermoelectric potential varies continuously between any given limits of the temperatures of deposition.

In the McIntyre Mine, Schumacher Ontario, Mutch (1952) found that there were distinct differences in readings obtained by pyrite geothermometer in the ore-bearing parts of the No. 25 vein as compared with those obtained outside the vein. In the ore-bearing parts positive readings were obtained whereas outside only negative readings were found. He reasoned that since the area of the vein containing the pyrite giving negative readings is in excess of the ore section, determination of thermoelectric potential of pyrite may have an application in locating and outlining gold orebodies. Rozova and Gavrilov (1970) also noted a zoning of thermo-e.m.f. changes in pyrite in the Kokpatas gold deposit, central Kyzul-kum U.S.S.R. There, the pyrite occurs in a disseminated gold-sulphide deposit in metamorphosed carbonaceous sediments and volcanic rocks. Mineral zoning consists of an intensely sulphidized zone in the centre (mainly arsenopyrite) surrounded by zones with increasing amounts of pyrite. The thermo-e.m.f. changes in the pyrite outlined five separate zones with the thermo-e.m.f. increasing toward the periphery of the deposit. Corresponding changes in the specific electrical resistivity of pyrite were also found. A relationship between the thermo-e.m.f. of pyrite and the arsenic content of the mineral was noted.

Thermoluminescence of minerals has been proposed as a detailed prospecting tool, and some work has been done on auriferous quartz and in alteration halos around various types of mineral deposits. Thermoluminescence (thermostimulated phosphorescence) is the property of a mineral or rock to emit light (photons) when heated below red heat (below about 400°C). The phenomenon is due to unstable electronic conditions (trapped electrons) within the crystal lattice. These conditions can be induced by radioactive bombardment, stress, heat absorption and a variety of other causes not yet well understood. Details should be sought in the book edited by McDougall (1968).

MacDiarmid (1963) observed a general erratic distribution of thermoluminescence in the vicinity of hydrothermal deposits. On the other hand, McDougall (1964,1966) considers that his investigations indicate that thermoluminescence should have at least limited application as an exploratory tool. He observed a number of patterns adjacent to orebodies, some carrying gold, as follows (1) no thermoluminescence is observed; (2) thermoluminescence is present, but no distinctive change is observed as the zone is approached; (3) thermoluminescence is low or nil immediately adjacent to the ore zone and rises to a higher background level at some distance away; (4) thermoluminescence is low or nil immediately adjacent to the ore zone, rises to a higher value at some distance away, and then drops to a more or less uniform background (or to nil values) at a still greater distance; and (5) thermoluminescence is high immediately adjacent to the ore zone and decreases away from it to a more or less uniform background (or to nil values). The profiles figured by McDougall are remarkedly similar to those obtained for trace elements, including gold, in the primary halos in rocks surrounding orebodies.

Grishin *et al.* (1971) examined the thermoluminescence of quartz from a number of primary gold deposits of northern

Kazhakhstan, U.S.S.R. They found that quartz with visible gold and quartz devoid of gold showed no difference in their thermoluminescence. There was also no particular relationship with any specific trace element in the quartz. They did, however, note significant differences in the glow curves of the quartz with different typomorphic characteristics (e.g., cryptocrystalline, fine grained, coarse grained, columnar, sugary, etc.) that they related to the conditions of formation of the quartz. In this respect the authors thought that thermoluminescence of quartz could be useful in ascertaining the conditions of formation of the auriferous quartz and thereby gain some knowledge on the possible distribution of orebodies within a district.

The possibility of applying neutron activation analyses of rock, soil, sediments, etc. in situ for locating gold deposits has been considered by a number of investigators. Senftle *et al.* (1969) have suggested the use of ^{252}Cf as the source of neutrons and a Fe(Li) crystal as the gamma-ray detector. This technique was used to obtain spectra of marine manganese nodules and gold ore (0.3 oz/ton) in a simulated marine environment. Siegel (1971) details the techniques and problems of prospecting by in situ neutron activation apparatus in the marine environment.

Fire is the test of gold
–Seneca

Analytical methods for estimating gold

Gold can be determined in geological materials by a variety of methods, including fire assay, colorimetric analyses using mainly organic reagents, inorganic analytical methods, spectrography, combined assay and spectrography, atomic absorption spectrography, spot tests, neutron activation and panning. Only the last can be employed as a field test as will be discussed subsequently.

The analytical method employed to determine gold depends essentially on the type of geological material to be analyzed and the level of the gold content. It should be borne in mind that the gold contents of normal rocks, soils, glacial materials, stream sediments and spring precipitates usually lie within the range 0.0005–0.01 ppm. In normal natural waters the average is 0.00003 ppm (0.03 ppb) and in normal biological materials (dry weight) the gold content only rarely exceeds 0.01 ppm. Ores range widely in their gold contents but generally between 5 and 35 ppm. Placers, likewise, range widely in their gold contents; most carry less than 1 ppm Au. *See also* the extensive data given in Chapter II.

There are a number of treatises, textbooks and papers dealing with practically all aspects of the sampling procedures for analysis, analytical chemistry and fire assaying of gold, including those by Fulton and Sharwood (1929), Dillon (1955), Sandell (1959), Furman (1962), Stanton and McDonald (1964), Wise (1964), Herz (1966), Chow (1966), Beamish (1966), Huffman *et al.* (1967), Thompson *et al.* (1968), Van Sickle and Lakin (1968), Popova (1971), Green *et al.* (1971), Hildon and Sully (1971), Beamish and Van Loon (1972), West (1973), Purushottam *et al.* (1973), Brown and Hilchey (1975), Rajan and Raju (1975), Sighinolfi and Santos (1976*b*), Beamish and Van Loon (1977) and Haffty *et al.* (1977). Numerous mineralogical texts and treatises outline the mineralogical and chemical methods for the determination of native gold and its minerals. The reader is referred particularly to those by Short (1940), Palache *et al.* (1944,1951), Ford (1958), Ramdohr (1960), Hey (1962,1963), Chukhrov (1960–1965), Berry and Thompson (1962), Uytenbogaardt and Burke (1971) and Galopin and Henry (1972).

Fire assay methods using lead, tin or other collectors are best adapted to ores or rocks enriched in gold. With proper attention being paid to fluxes and procedures fire assay methods for gold are accurate and highly reproducible. Fire assay methods can also be combined with a number of other methods for the estimation of gold. One method involves the optical spectrographic determination of the gold in the assay beads; others involve solution of the gold-silver beads by aqua regia or cyanide followed by determination of the gold by colorimetric and other wet chemical methods such as atomic absorption spectroscopy or neutron activation. The combination methods are adaptable to practically all geological materials except water.

Colorimetric and other types of wet chemical methods are not generally satisfactory for the direct determination of gold in geological materials because of the relatively low amounts of the metal present. If combined with enrichment or preconcentration (collector) techniques such as fire assay, solvent extraction, ion exchange or addition of tellurium or mercury compounds as collectors the wet chemical methods are satisfactory but tedious and are not suitable for rapid determination of gold in large numbers of samples.

The limit of detection of gold in the DC arc is about 10 ppm, and hence direct optical spectrographic methods are generally not suitable for the determination of the metal in most types of geological materials. Even for ores the spectrographic method is unreliable because of the erratic and often particulate nature of gold. X-ray spectrographic methods are, likewise, unsuitable for the determination of gold in most geological materials because of the low detection limit of the methods (~10 ppm). The detection limits for trace analysis of gold by spark-source mass spectroscopy are, likewise, high being 0.01 to 0.1 ppm (Beske, 1974). Suitable enrichment or collector techniques such as fire assay, ion exchange, solvent extraction, etc. followed by spectrographic determination of the gold in the beads, (ashed) resins, (ashed) solvent extracts, etc. have been employed with success, but the methods are tedious and the reproducibility is commonly low.

Atomic absorption spectroscopy for the determination of gold following collection of the metal by suitable means including fire assay, solvent extraction, ion exchange, tellurium and mercury compounds, etc. have found wide usage in geochemistry (Angino and Billings, 1972). A thorough review of the methods applicable for determination of gold in rocks, soils, waters and biological materials with a bibliography up to August 1972 is given by Sen Gupta (1973).

Spot tests are not generally suitable for the qualitative determination of gold in most geological materials because of poor sensitivity. They can be employed, however, for the qualitative determination of gold in minerals such as the tellurides, sulphosalts, etc. Feigl (1958) lists a number of specific tests for gold in solutions.

Neutron activation methods are extremely sensitive and

particularly adaptable for the determination of gold in most geological materials. The methods are costly and tedious, but in recent years they have been successfully adapted to geochemical prospecting surveys where large numbers of samples are involved. They are also especially useful in fundamental geochemical studies on the distribution of gold in geological and biological materials. Most of the methods involve a preconcentration step, especially when dealing with water. Details are discussed in the books and papers by Lobanov *et al.* (1966,1967*a,b*), Barnett *et al.* (1968), Rowe and Simon (1968), Simon and Millard (1968), Fritze and Robertson (1969), Nikanorov *et al.* (1971), Schiller *et al.* (1971*b*), Santos and Walters (1971), Leushkina *et al.* (1971), Stoyanka and Dancheva (1972), Rehman (1972), Plant and Coleman (1973) and Nadkarni and Morrison (1974).

Various collectors and carriers of gold are employed in analytical methods including among others activated charcoal, metal sulphides, mercury and tellurium compounds in conjunction with various reductants and various resins. Among the last is an ion exchange resin with high selectivity for gold (Koster and Schmuckler, 1967) as well as other noble metals. This resin, a styrene-divinyl-benzene copolymer with a resonating amino group was found by Green and Law (1970) and Green *et al.* (1971) to quantitatively collect gold from acid solutions over a wide range of acid concentrations. The common metals were not collected by the resin nor did they seriously interfere with the collection of gold. They proposed X-ray spectrographic and neutron activation analytical methods for gold based on the use of paper disks containing this resin.

Native gold is readily recognized by its golden colour, its hardness (2.5), its high specific gravity (19), its marked malleability and its resistance to attack by all common acids. It is, however, readily soluble in aqua regia and in alkali cyanide solutions. It is rarely misidentified, although pyrite and chalcopyrite may sometimes be mistaken for gold. The brittleness of these two sulphides and their solution in nitric acid distinguish them readily from gold.

The gold tellurides, aurostibite and fischesserite are not easily identified by eye. Most gold tellurides have a brilliant metallic lustre and are steel grey to silver white in colour. Some incline to a slight yellowish tinge. When treated with nitric acid they are decomposed with the formation of a residue of rusty coloured gold. Aurostibite is metallic grey in colour with a bright to dull lustre. Treatment with nitric acid yields a residue of rusty coloured gold which may be obscured by white and yellow ochers of antimony oxide. Fischesserite resembles the tellurides, especially petzite. In polished sections it is isotropic under crossed nicols and has a pink colour in reflected light. If tellurides, aurostibite or fischesserite are suspected in ores, the minerals should be X-rayed to establish their positive identification.

The presence of gold in sand, gravel, soil and in ores of various types can be readily ascertained by the age old method of panning. The method can be used directly where the materials are unconsolidated. For ores a given weight of the material should be crushed, ground to a fine powder and roasted in a tin can or some other convenient receptacle in the camp fire overnight. The oxidized material is then carefully panned. By counting colours one can estimate fairly accurately the value of placer sands or various types of gold-bearing ores.

Numerous types of gold pans are in use throughout the world, including the gourd, the common gold pan, the batea and others. The operation of panning for gold cannot be adequately described; it must be seen and practised but can be easily learned by anyone. One can wet-pan if water is available, or dry-pan where there is no water by simply allowing the material to fall a variable distance onto a ground sheet the dross being removed by a suitable wind. Those wishing more details about panning should consult the three excellent papers on the subject by Mertie (1954), Theobald (1957), and Wells (1973). The techniques described in these works are also particularly applicable to geochemical methods using heavy minerals as indicators or as a sampling medium. When employing panning it should be emphasized that if the soils, glacial materials or sediments are sieved one must ascertain the best sieve size to give the optimum contrasting anomalies for the heavy minerals employed. This feature is critical for gold particles, since in any given area these tend to a relatively uniform size. If too small a sieve size is used the majority of the gold particles may be screened out with the result that significant anomalies for gold may be missed.

In addition to the pan a number of primitive apparatus can be employed in the field to separate gold from the dross. Here we need only mention the rocker, the sluice and the dry washer. The latter is often used in the desert and other arid regions. A description of this machine is given by West (1971).

When small samples are to be tested for gold a miner's spoon or oxhorn spoon should be used to separate the precious metal from the dross.

The touchstone, a piece of grey or black chert, has long been used to test gold. Agricola used it and touch-needles (a series of strips of gold alloys of graded composition) for examining bullion, coins and jewelery. The golden streak left on the touchstone is unaffected by nitric or hydrochloric acid but is removed by aqua regia.

> All is not golde that hath a glistering hiew,
> But what the touchstone tries and findeth true.
> —Times Whistle

Nature herself makes it clear that the production of gold is laborious, the guarding of it difficult, the zest for it very great, and its use balanced between pleasure and pain.

–Diodorus Siculus, 1st Century B.C.

Chapter VI. Production and uses of gold

Production and consumption

Gold mining in one form or another goes back into antiquity at least for more than 4000 years. During the earliest times one can imagine that the winning of gold was a haphazard activity, nuggets of the native metal being plucked from some auriferous stream, the oxidation zones of gold or sulphide deposits or from residuum near such deposits. Production was probably erratic and generally unorganized. By the time of the Pharaohs, however, organized gold mining in the Arabian and Egyptian shields appears to have been well established, and gold was seemingly relatively plentiful compared with silver for in the Code of Menes, the first pharaoh, who reigned in Egypt about 3500 B.C., it was decreed that "one part of gold is equal to two and one-half parts of silver in value." Since that time up to the fall of the Roman Empire gold appears to have been mined almost continuously mainly from placers but also from the oxidized zones of deposits in all organized societies throughout the world. There is evidence of ancient gold workings and placers ranging from Japan, through China, India, U.S.S.R., Turkey, the Mediterranean islands and lands, central Europe, Spain, France, Britain, Mexico, Central America and South America. The early Minoan goldsmiths had a source of gold for they produced some of the finest pieces of early gold work known to man (Frontispiece). The Greeks mined gold extensively throughout their empire and appear to have prospected for it far and wide judging from the legend of Jason and the Golden Fleece. The Romans, likewise, sought gold throughout their empire obtaining it from placers and mines in Spain, France, Britain, Germany, Central Europe and the Middle East. The yearly production of gold in the Roman Empire at the beginning of our Era can only be roughly estimated; a reasonable figure seems to be about 250 000 oz. One of the great gold placer areas in the Middle East was the river Pactolus in Lydia (a tributary of the present Gediz in Anatolia in Turkey). Another was the Pontic field in Turkey, on the southern shore of the Black Sea, possibly the site of the Garden of Eden and probably one of the sites of the Golden Fleece sought by Jason.

Nearly all of the gold mined by the ancients was wrought into idols, shrines, bowls, vases, flasks, cups and jewellery. Later, about 1000 B.C., gold and silver came into general usage for coins as a medium of exchange, certainly in all countries between the Indus and the Nile and probably on a much more widespread geographical basis. By Philip of Macedon's time (356 B.C.) gold and silver were generally valued in the ratio of 10:1. This value ratio did not change much through the centuries that followed for we find that in England the ratio in 1464 was 11:1 according to the recorded mint price. Since then there has been considerable fluctuation in the value ratio as shown in Table 81. There is an interesting comment in the Engineering and Mining Journal (vol. 95, p. 1163, 1913) with regard to the origin of the ratio of value between gold and silver that merits quotation:

> "I have endeavored to discover as to how and when the ratio of value between gold and silver commenced, and as to what was its origin", said Bedford McNeill, in his presidential address before the Institution of Mining and Metallurgy. "Sir David Barbour has called my attention to an interesting theory of Professor D'Arcy Thompson, 'that the origin of the ratio may have been astronomical, gold being associated with the sun, and silver with the moon; the period of the earth's revolution round the sun being 365¼ days, and that of the moon round the earth 27.32 days, it will be found that the ratio between these figures is very nearly 13⅓, and this figure 13⅓ was the ratio that was fixed between gold and silver during the Babylonian Empire (2000 B.C.) down to the Lydian Empire (500 B.C.).' Professor D'Arcy Thompson has also pointed out that the Lydians made coins of what was called 'electrum', and reckoned one electrum coin as being of the same value as 10 silver coins of the same weight. It was supposed that the Lydians were unable to separate the gold from the silver, but when we find by assay that the proportions in these old coins of gold to silver were approximately as 73 and 27, calculating gold as 13⅓ times as valuable as silver, and we find that an electrum coin would be worth exactly 10 silver coins of the same weight, we cannot help but admit that the Lydian assayers must have been well versed in the science and practice of their art, and that such figures cannot be mere coincidences."

Since the earliest times gold has been hoarded by kings, states, popes and individuals, the prime example being Croesus of the Mermnadae, last King of Lydia (560–546 B.C.). It is said that most of his fortune in gold came from the placers of the Pactolus, but it seems more likely that his wealth had a more widespread origin since the Lydians were renowned for their trading pursuits. How much gold was hoarded in ancient times is unknown, but the practice seems to have increased at the beginning of our era, so much so that during most of the Dark and Middle Ages both of the precious metals were scarce. Some of the scarcity was undoubtedly due to the exhaustion of most of the placers in Europe and the Middle East and also to a general decline in mining, the economic limits, in terms of depth, of profitable ore having been reached in many districts mainly because of engineering problems (ventilation and flooding by groundwater). It is estimated that the total annual production of gold during the Dark and Middle Ages probably averaged only about 100 000 oz.

Geographical exploration brought an abrupt change in this situation, the manifestations of which affect us to this day. With the rediscovery of America by the Spaniards in 1492 came the discovery of great stores of both gold and silver first in the islands of the West Indies, then in Mexico and finally in Central and South America. At first, great quantities of gold were looted from the natives, but as this source declined active placering and mining by slave labour provided

Table 81. Value ratio of gold to silver through the ages*

Year	Value ratio	Remarks, authority, etc.†
Prior to 3500 B.C.	1 or less	No records
c. 3500	2.5	In Egypt. Code of Menes
708	13.5	In Assyria. Cuneiform inscriptions at Nineveh
500	13.0	In Persia. Herod III
400	12.0	In Greece. Plato
404–336	12.0	In Greece, Peloponnesian war to time of Alexander
300	10.0	In Greece. Fall in gold value due to influx of Alexander's spoil
207	13.7	In Rome
186	10.0	In Rome
58–49	8.9	In Rome. Fall in gold value due to influx of Caesar's spoil
1–37 A.D.	10.9	In Rome. Reigns of Augustus and Tiberius
54–68	11.8	In Rome. Reign of Nero
81–96	11.3	In Rome. Reign of Domitian
312	14.4	In Byzantium. Reign of Constantine
438	14.4	In Rome and Byzantium. Theodosian code
864	12.0	In Europe. Edictum Pistense
1344–1482	11.1	In England. Mint returns
1482–1492	12.2	In Europe
1492	11.0	In Europe. Discovery of America
1497	10.7	In Spain
1550	12.0	In Europe
1641	14.0	In Europe. Moran and John Locke on money
1690	16.0	In Europe. Sir Isaac Newton
1730	16.0	In Europe. Kelly's "Cambist"
1760	14.3	In London. Del Mar
1800	15.7	In Europe & U.S.A. EMJ
1850	15.7	In Europe & U.S.A. EMJ
1886	22.0	In Europe & U.S.A. Discovery of the Rand
1890	22.1	In Europe & U.S.A. EMJ
1900	34.4	In Europe & U.S.A. EMJ
1910	38.2	In Europe & U.S.A. EMJ
1915	38.2	In Europe & U.S.A. 1st Great War, EMJ
1920	20.8	In Europe & U.S.A. EMJ
1930	60.8	In Europe & U.S.A. EMJ
1935	45.5	In Europe & U.S.A. EMJ
1940	50.0	In Europe & U.S.A. 2nd Great War, EMJ
1950	41.2	In Europe & U.S.A. EMJ
1960	38.5	In Europe & U.S.A. EMJ
1970	21.1	In Europe & U.S.A. EMJ
1975 (January)	41.9	In Europe & U.S.A. EMJ

* The figures given are calculated value ratios in the currency of the time and all should be read as follows: 2.5 indicates that at the stated time gold was two and one-half times the value of silver; 12 that it was twelve times, and so on.

† Sources of information include various documents in the British Museum and British Library, London; National Museum, Athens, Greece; Vatican Library, Rome; Jacob (1831), and Del Mar (1880). The values since 1800 were obtained mainly from the Engineering and Mining Journal (EMJ) annual volumes. No attempt was made to authenticate all of the values shown, and no adjustments have been made to compensate for differences between the various currencies of different countries. For a thorough discussion of the value ratios prior to 1880 see Del Mar (1880).

a large and continuous influx of gold and silver to Spain. Despite an attempt by the Spanish government to keep the precious metals to itself, they soon filtered over the whole of Europe, increasing the supply of money and inflating prices. It is estimated that from 1492 to 1600 more than 10 million oz of gold came from the Spanish Americas, which was about 40 per cent of the world production at that time. Marked increases in gold production from South America, particularly from the placers of Colombia, were registered during the period 1600–1800; for example some 48 million oz were won during the 18th century, accounting for about 80 per cent of the world production. Two further increases in the world gold production followed during the periods 1820–1880 and 1890–1920. The first marked the discoveries of the great placer and lode mining areas of Siberia, California, Australia and New Zealand; the second followed the discoveries of placers and bedrock deposits in Alaska, Yukon, central Canada and on the Witwatersrand of South Africa. Since 1920 there has been only one period, 1933–1939, when the production of gold increased significantly. During and since World War II a general decline in the production of gold in nearly all countries except South Africa and the Soviet Union has taken place. In South Africa the average annual production for the last 5 years has been about 27.5 million oz; in the Soviet Union the annual production for the last 5 years has probably averaged 10 million oz, although this figure is only a rough estimate, the true value being a closely guarded state secret.

It is estimated that the total amount of gold won by man from the earth to 1975 is about 3 billion (3×10^9) oz. Of this amount some 2 per cent was produced prior to 1492; 8 per cent during the period 1492–1800; 25 per cent during the interval 1801–1900; and 65 per cent from 1901–1975, all figures being rough estimates.

Approximately 88 per cent of the annual world production of gold is obtained from primary deposits and placers (fossil and modern) mined essentially for gold. The remaining 12 per cent comes as a byproduct of base metal and silver mining. The average annual world production during the period 1971–1975 inclusive was about 49 million oz. The principal producing countries are shown in Table 82.

Table 82. World gold production

Country	Average annual production for years 1971–1975 inclusive
	(oz x 10^6)
South Africa	27.5
U.S.S.R.	10.5
Canada	1.9
United States	1.5
Australia	0.8
Ghana	0.7
Others	6.0
Total	48.9

Sources: U.S. Department of the Interior, Minerals Yearbook (1971–1975); Mining Annual Review (1971–1975); Canadian Mining Journal, Annual Mineral Industry Review (February issue) (1971–1975); Department of Energy, Mines and Resources, Canadian Minerals Yearbook (1971–1975). The figure given for the U.S.S.R. is estimated.

Uses

The historical uses of gold are of interest: From the dawn of civilization until about 1000 B.C. the uses were restricted mainly to ornamentation, decoration and the display of princely power. This has often been called the decorative or ornamentative stage in the history of the precious metal. After 1000 B.C. gold came into general use as money and freely circulated until about 1916. This was the monetary stage in the history of the metal, although considerable amounts continued to be used for ornamental purposes. After World War I the movement of gold and its use as money was greatly curtailed and eventually restricted in many countries, most of the metal, except that permitted for the manufacture of jewellery, going straight from the mints into the vaults of most countries. This period continues to the present day, although there has been a relaxation in a number of countries that permit their citizens to own and trade in the metal. Since 1950 there has been increased industrial use of gold – sufficient it would seem to suggest that we are entering the industrial age in the history of the metal.

Specifically, the uses of gold depend essentially on its traditional role as a monetary measure by governments and central banks in the settlement of international balances; on its intrinsic quality as the most beautiful of metals; and on its chemical inertness, great malleability and high electrical and thermal conductivity. In its international monetary role gold is utilized mainly in the form of high purity bars, tablets or more rarely as coins with a specified gold content. In its other roles the metal is employed in the pure form or alloyed with other metals such as silver, copper and the platinum metals.

It is difficult to state accurately the apportionment of the annual world production of gold among its various uses. There are several reasons for this, chief of which is the tendency for certain financial institutions and individuals to add to their hoards of the metal. Rarely is the magnitude of these hoards made public nor are their annual increase or decrease known. The purchase of gold, for uses other than the manufacture of jewellery, industrial purposes, etc., is prohibited by law in many countries, but despite this hoarding is traditional in many Asiatic and European countries and is probably also common in North and South America. It is estimated that about 20 per cent of the annual world production of gold goes into the vaults of governments and central banks as a monetary reserve; some 5 per cent probably passes into the hands of private and corporate hoarders; and the remainder about 75 per cent, some 36 million oz, is used in fabricated articles. Of the last some 60 per cent is consumed in dentistry and the jewellery and coin industry; the remainder is utilized in a great variety of electronics and other industries.

Gold is the royal and aristocratic ornamental metal par excellence with a seductive natural colour, lustre and satinlike texture. Since ancient times this most beautiful of metals has traditionally enjoyed a use in articles of personal adornment, particularly in rings and a variety of costume jewellery. Considerable amounts of gold are also used in fountain pens, medallions, gold watches and in gold fillings in dentistry. Some gold has also been utilized since antiquity in the fabrication of chalices, cups, vases, shrines and a great variety of other articles cherished for their intrinsic value. In all of these uses gold is generally alloyed with copper or silver, the fineness of the gold alloy being expressed in 'carats'. Pure gold is '24 carat gold'.

Pure gold or that containing a low percentage of silver or copper can be hammered into extremely thin foil, about 0.000005 in. in thickness. In the foil or leaf form gold has been used for a great variety of decorative purposes on buildings, statues and other articles since pre-Biblical times. Thin layers of gold welded to base metals such as nickel, copper or brass can be rolled or drawn into complex forms without rupturing the thin gold layers. Material of this nature is called gold-filled plate and is used extensively in various types of jewellery, watch cases, spectacle rims, etc.

The industrial uses of gold depend essentially on its softness, the ease with which it alloys with silver, copper or platinum metals, its great ductility and malleability, its great electrical and thermal conductivity and its chemical inertness and hence corrosion-resistance to oxygen, sulphur, most chemical compounds and nearly all of the single common acids. Most of the gold used in industry is consumed by the electronic and electrical engineering industries, the products ranging widely from coatings of vacuum tubes, specialized electrical contacts, electrical precision drawn wire leads, gold electroplate high frequency conductors, to printed circuits in computers, radios and television sets. Considerable amounts of gold are also used by industry in the fabrication of high temperature brazing alloys, linings for specialized chemical and nuclear plants, platinum-gold alloy spinnerets for viscose rayon plants, infrared and thermal reflectors in aircraft and space vehicles and heat shields for jet and rocket engines. Gold-coated window glass for buildings in hot climates and electrically heated windscreens having a thin transparent and conductive layer of gold are finding an increasing use in cars, aeroplanes, ships and locomotives throughout the world. Some gold is also consumed by the printing and furniture industries in the form of gold paints and by the ceramic industry in the form of organic 'liquid golds' for application in pottery and glassware. On firing or other treatment the organic gold compounds are reduced leaving a thin film of gold tightly bonded to the ceramic ware or glass. Small amounts of gold are also used for colouring glass, and gold salts are used to a small extent in certain photographic and medical preparations.

It seems probable that the industrial use of gold will increase substantially in the future as will also its use in jewellery and other articles of adornment. The amount of the annual increase is, however, conjectural, perhaps some 2 per cent of the present yearly production or about 1 million oz. One can also predict that there will be a substantial yearly increase in hoarding although the amounts cannot be reliably estimated. Presumably it will vary with the confidence that individuals place in their respective country's paper notes and base metal coinage. The fate of gold as a monetary standard in the settlement of international balances can only be conjectured. Should Utopia descend upon us we will have no need of an objective, tangible, enduring, weighable, readily transportable medium such as gold in our international financial transactions. But since that day appears far removed it would seem that gold will be that medium which will continue to keep all men honest in their trading and financial manipulations despite views to the contrary by many of our modern economists, stimulated it would seem by the thoughts of John Maynard Keynes, who referred to the position of the metal in the modern world as that "barbarous relic". Gold is the discipline that curbs the power of the printing press. Stephen Leacock (1932) touched upon this point in a memorable address on gold:

> We have to have – I would say 'stable prices'; you cannot have a world with all the prices stable; you can only have a world in which the price level moves slowly with a regular and intermittent oscillation. But you can have and must have a fixed exchange. You cannot have a foreign trade based on monetary speculation; you have got to have something absolutely stable for your dealings with foreign nations, and I tell you, gentlemen, I am convinced that there is nothing that will take the place of some absolutely metallic standard, some standard of things, and not of opinions.

The price of a troy ounce of gold at the time of writing (December, 1977) on the free market was approximately $US 170; the equivalent price of silver was $US 5.00 giving an Au:Ag value ratio of 34, a considerable increase since the time of Pharaoh Menes.

Those who wish to pursue the details of the production, uses and economics of gold should consult the informative works by Malcolmson (1907), Keynes (1930), Altman (1960), Badone and Spence (1961), Morgan (1965), Kavanagh (1968, 1976), Green (1970), Weil and Davidson (1970), West (1975), Hogan (1975), Collins (1975), Fahlgren (1975), Vilar (1975), Fells (1975) and Fells and Glynn (1976).

As we look in retrospect at the last 5000 years of the history of gold we see that the noblest and most beautiful of metals has played a remarkable and sometimes dominant role in human experience and progress, first in an ornamental way, next in coinage, then as an international medium of exchange, and now as an indispensable element in industry. We see also that gold has elicited both good and evil works in man, as have also most other materials of this earth. On the good side of the coin, avarice for the metal has given us great discoveries both chemical and geographical; on the evil side the *auri sacra fames* has led to conquest, enslavement of nations, civil contention and the vilest treatment of men ever devised.

As we look in prospect we perceive an industrial vista where gold will play an ever increasing role in the production of high speed computers, telecommunications, space vehicles, pharmaceuticals and a thousand other artifacts of future civilizations. We perceive also that man will lose none of his fascination for the metal that he has long admired for its natural beauty and enduring qualities.

Long have I laboured in the realms of gold
And much have I learned.

Selected bibliography

Abdukhayumov, A.A.
1973: Effect of the geological environment on chemical characteristics of beresitization of rocks with gold sulphide mineralization (as illustrated by the Terekkan and Left-Bank Taldy-Bulak deposits); Tr. Tashk. Politekh. Inst., v. 95, p. 102–103. (Chem. Abstr., v. 83, 63540k.)

Abdulkabirova, M.A., Ankinovich, O.S., Grishin, V.M., Kayupov, A.K. and Kudaibergenova, N.K.
1971: Gold ore fields of northern Kazakhstan; Nauka Kaz. S.S.R., Alma-Ata, U.S.S.R., 168 p. (Chem. Abstr., v. 77, 64926d.)

Abreu, Sílvio Fróis
1973: Recursos minerais do Brasil; vols. I and II. Editora, Edgard Blücher Ltda; Sao Paulo, Brasil.

Adams, F.D.
1938: The birth and development of the geological sciences; The Williams & Wilkins Co., Baltimore, 506 p.

Adams, S.F.
1920: A microscopic study of vein quartz; Econ. Geol., v. 15, p. 623–664.

Addie, G.
1964: Silver-gold ratios in some Witwatersrand conglomerates; Econ. Geol., v. 59, p. 325.

Adilov, V.B., Shvartseva, N.M. and Lyan, D.
1971: Antimony in subsurface waters of some gold-ore deposits of the Kuraminsk Range; Vop. Gidrokhim. Prir. Vod. Aridnoi Zony SSSR, N.N. Khodzhibaev, ed., p. 91–102. (Chem. Abstr., v. 81, 94129m.)

Aferov, Yu. A., Zvyagin, V.G., Roslyakova, N.V., Roslyakov, N.A., Shabynin, L.L. and Epov, I.N.
1968: Gold in rocks, plants, and waters of the Darasun deposit; Vop. Geol. Pribaikal. Zabaikal., no. 3, p. 146–149. (Chem. Abstr., v. 70, 30889j.)

Ahlfeld, F.
1937: Typen bolivianischer Goldlagerstätten; Zentralbl. Mineral. Geol. Palaeontol., Abt. A, p. 240–255.

Aho, L.
1975: Ore mineralization at Ritovuori, Pihtipudas, central Finland; Geol. Surv. Finl., Bull. 275, 21 p.

Aicard, P.
1957: Le Precambrien du Togo et du nord-ouest du Dahomey; Gouv. Gen. Afr. Occid. Fr., Bull. Dir. Fed. Mines Geol. Dakar, Grand Imprimère Africaine, no. 23, 221 p.

Aitken, J.D.
1959: Atlin map-area, British Columbia; Geol. Surv. Can., Mem. 307, 89 p.

Akhmedov, A.M., Gavrilenko, B.V. and Predovskii, A.A.
1974: Distribution of gold in carbonaceous sulfide schists of the Pechenga complex; Vop. Geol. Metallog. Kol'sk Poluostrova, v. 5, pt. 2, p. 231–235. (Chem. Abstr., v. 84, 92837 p.)

Akopyan, M.S., Iskandaryan, G.G. and Dontsova, E.I.
1976: Isotopic oxygen study of the Megradzorsk gold ore deposit; Geokhimiya, no. 2, p. 283–288.

Akright, R.L., Radtke, A.S. and Grimes, D.J.
1969: Minor elements as guides to gold in the Roberts Mountains Formation, Carlin gold mine, Eureka County, Nevada; *in* Int. Geochem. Explor. Symp., F.C. Canney, ed.; Colo. Sch. Mines, Q., v. 64, no. 1, p. 49–66.

Al-Atia, M.J. and Barnes, J.W.
1974: Rubidium: a primary dispersion pathfinder at Ogofau Gold Mine, southern Wales; Geochemical exploration 1974; I.L. Elliot and W.K. Fletcher, eds., Assoc. Explor. Geochem., Spec. Publ. no. 2, Elsevier Co., Amsterdam, p. 341–352.

Al-Shahristani, H. and Al-Atyia, M.J.
1973: Determination and geochemical significance of trace elements in Iraqi crude oils using instrumental neutron activation analysis; J. Radioanal. Chem., v. 14, no. 2, p. 401–413.

Al-Shaieb, Z.
1972: Geochemical anomalies in the igneous wall rock at Mayflower mine, Park City district, Utah; Univ. Microfilms, Ann Arbor, Mich., 113 p. (Chem. Abstr., v. 80, 5689t.)

Alabina, A.A., Roslayakov, N.A., Sotnikov, V.I., Kharitonov, I.I. and Tsimbalist, V.G.
1969: Distribution characteristics of gold and molybdenum in rocks of the Davendinsk-Klyucheva ore field; Zap. Zabaikal. Filiala Geogr. Obshchest. SSSR, no. 36, p. 44–49. (Chem. Abstr., v. 77, 116734m.)

Albers, J.P. and Kleinhampl, F.J.
1970: Spatial relation of mineral deposits to Tertiary volcanic centers in Nevada; U.S. Geol. Surv., Prof. Pap. 700-C, p. 1–10.

Al'bov, M.N.
1952: Forms of the migration of gold in the oxidized zone of mineral deposits; Akad. Nauk SSSR, Izv., Ser. Geol., no. 4, p. 41–52.
1960: Determination of zone of secondary enrichment in vein gold deposits; Tr. Sverdlovsk. Gorn. Inst., no. 35, p. 58–64. (Chem. Abstr., v. 60, 1460b.)

Albul, S.P. and Miller, A.D.
1959: Prospecting for gold-ore deposits according to hydrogeochemical zones of dispersion; Novoe Methodike Tekh. Geologorazvedoch. Rabot, Leningrad Sb., v. 2, p. 133–139. (Chem. Abstr., v. 55, 6273c.)

Alcock, F.J.
1936: The gold deposits of Lake Athabaska; Can. Inst. Min. Metall., Trans., v. 39, p. 531–546.

Aleksandrov, A.L., Olovyashnikov, V.M. and Potorochenko, A.N.
1975: Florencite in rocks of the Bodaibo gold ore region as an indicator of hydrothermal processes; Akad. Nauk SSSR, Dokl., v. 224, no. 5, p. 1152–1155. (Chem. Abstr., v. 84, 76878x.)

Aliskerov, A.A.
1974: Acidity-alkalinity of hydrothermal mineral formation processes in the Avachinsk-Kitkhoisky uplift zone; Gidroterm. Mineraloobraz. Rastvory Obl. Akt. Vulkanizma, p. 200–203. (Chem. Abstr., v. 82, 173654n.)

Allen, E.T. and Day, A.L.
1935: Hot springs of the Yellowstone National Park; Carnegie Inst. Wash. 525 p.

Aller, L.H.
1961: The abundance of the elements; Interscience Publishers Inc., New York, 283 p.

Allman, R. and Crocket, J.H.
1972: Gold; *in* Handbook of geochemistry, v. II/3, K.H. Wedepohl, ed., Springer-Verlag, Berlin, Chap. 79.

Alminas, H.V., Watts, K.C. and Siems, D.F.
1972: Trace metal distribution in the Winston, Chise, and Priest Tank quadrangles, Sierra County, New Mexico; U.S. Geol. Surv., Misc. Field Study Map MF-400 (Ag and Au).

Altman, O.
1960: The role of gold in international liquidity; Mines Mag., v. 50, ño. 10, p. 39–42.

Alvir, A.D.
1930: Antamokite, a new gold-silver telluride; Philipp. J. Sci., v. 41, p. 137–139.

Amarskii, V.G.
1972: Distribution of gold and mercury mineralization in the central part of the Stanovoi Ridge; Zap. Zabaikal. Filiala Geogr. O-va. SSSR, v. 86, p. 68–76. (Chem. Abstr., v. 82, 62013h.)

Ames, H.G.
1948: Perron Mine; *in* Structural geology of Canadian ore deposits, v. 1, Can. Inst. Min. Metall., Montreal, p. 893–898.

Ames, R.L.
1962: The origin of the gold-quartz deposits, Yellowknife, N.W.T.; Econ. Geol., v. 57, p. 1137–1140.

Amiryan, Sh. O.
1960: Mineralogy of gold-deposits; Akad. Nauk Arm. SSR, Dokl., v. 31, no. 1, p. 43–48; Akad. Nauk Arm. SSR, Izv., Geol. Geograf. Nauki, v. 13, nos. 3–4, p. 47–55. (Chem. Abstr., v. 55, 8194b.)
1974: Mineralogy and geochemistry of ores of the Laligyukh gold-sulphide deposit; Akad. Nauk Arm. SSR, Izv. Nauki Zemle, v. 27, no. 2, p. 57–61. (Chem. Abstr., v. 82, 46347q.)

Amosov, R.A., Chuvikina, N.G., Bochek, L.I. and Epov, I.N.
1971: Hydrothermal smythite from the Darasun gold-ore deposit; Akad. Nauk SSSR, Dokl., v. 197, no. 3, p. 674–677. (Chem. Abstr., v. 75, 8352k.)

Amosov, R.A., Chuvikina, N.G., Novozhilov, Yu.I. and Gureev V.F.
1975: Orientation cocrystallization of native gold and pyrite; Zap. Vses. Mineral. Obschest. 5, p. 630–634.

Amosov, R.A. and Gureev, V.F.
1970: Formation of gold on sulphides; Mater. Konf. Molodykh Nauch. Sotrudnikov Tsent. Nauch. Issled. Gornorazved. Inst. Tsvet., Redk. Blagorod. Metal, 1969, p. 12–15. (Chem. Abstr., v. 80, 123167u.)

Amukun, S.E.O. and Turnock, A.C.
1971: Composition of gold-bearing quartz vein rocks, Bissett area, Manitoba; *in* Geology and geophysics of the Rice Lake region, southeastern Manitoba (Project Pioneer); Manitoba Mines Br., Publ. 71–1, p. 325–336.

Anasenko, A.V.
1970: Gold deposits of the Adycha-Indigirka region; Geol. Rud. Mestorozhd., no. 5, p. 105–108.

Anders, E., Ganapathy, R., Keays, R.R., Laul, J.C. and Morgan, J.W.
1971: Volatile and siderophile elements in lunar rocks: comparison with terrestrial and meteoritic basalts; Proc. 2nd Lunar Sci. Conf., Geochim. Cosmochim. Acta, Suppl. 2, v. 2, p. 1021–1036.

Anderson, J.C.
1935: Essential criteria in examining gold-quartz mines; Eng. Min. J., v. 136, p. 445–449.

Anderson, J.S.
1945. Chemistry of the earth; R. Soc., N.S.W., J. Proc., v. 76, p. 329–345.
1953: Observations on the geochemistry of indium; Geochim. Cosmochim. Acta, v. 4, p. 225–240.

Anderson, R.E.
1969: Geology and geochemistry Diana lakes area, western Talkeetna Mountains, Alaska; State Alaska Geol. Rep., no. 34, 27 p.

Andrade Ramos, J.R. de and Fraenkel, M.O.
1974: Uranium occurrences in Brazil; *in* Formation of uranium ore deposits; Int. Atomic Energy Agency—SM—183/24, p. 637–658.

Andreev, V.V., Evtushenko, S.I. and Timofeeva, T.S.
1970: The geochemistry of gold and silver in ores of the Kuru-Tegerek deposit in Kirghizia; Zap. Uzb. Otd. Vses. Mineral. Obshchest., no. 23, p. 169–171.

Andrew, A.R.
1910*a*: The detection of minute traces of gold in country rock; Inst. Min. Metall., Trans., v. 19, p. 276–281.
1910*b*: The geology of the Dolgelley gold belt, north Wales; Geol. Mag., Decade V, v. VII, p. 159–171, 201–211, 261–271.

Andrews-Jones, D.A.
1968: The application of geochemical techniques to mineral exploration; Colo. Sch. Mines, Mineral. Ind. Bull., v. 11, no. 6, 31 p.

Andrianarisoa, R. and Raharinosy, M.
1972: Evaluation of indications of cinnabar, cassiterite, and gold in the Amilobe Region; Malagasy, Rapp. Annu. Serv. Geol., 1972, p. 95–104. (Chem. Abstr., v. 82, 127635j.)

Andrusenko, N.I. and Shchepotyev, Yu. M.
1974: Temperature conditions of the formation and stages of subvolcanic gold-silver deposits of central Kamchatka; Geokhim., no. 2, p. 179–186.

Angino, E.E. and Billings, G.K.
1972: Atomic absorption spectrometry in geology, 2nd ed.; Elsevier Publ. Co., Amsterdam, 191 p.

Angino, E.E., Goebel, E.D. and Waugh, T.C.
1971: Lead isotopes and metallic sulphides as exploration guides in Mid-Continent Paleozoic rocks; *in* Geochemical exploration, R.W. Boyle and J.I. McGerrigle, eds., Can. Inst. Min. Metall., Spec. Vol. 11, p. 453–456.

Anhaeusser, C.R.
1966: Supergene gold enrichment in the Barberton Mountain Land with particular reference to the Lily Mine; Univ. Witwatersrand, Econ. Geol. Res. Unit, Inform. Circ. no. 29, 16 p.
1976*a*: The nature and distribution of Archaean gold mineralization in southern Africa; Miner. Sci. Eng., v. 8, p. 46–84.
1976*b*: Archean metallogeny in southern Africa; Econ. Geol., v. 71, p. 16–43.

Anhaeusser, C.R. and Button, A.
1974: Review of southern Africa stratiform ore deposits. Their position in time and space; Univ. Witwatersrand, Econ. Geol. Res. Unit, Inform. Circ. no. 85, p. 1–48. (Chem. Abstr., v. 81, 172937z.)

Anhaeusser, C.R., Fritze, K., Fyfe, W.S. and Gill, R.C.O.
1975: Gold in "primitive" Archaean volcanics; Chem. Geol., v. 16, p. 129–135.

Anhaeusser, C.R., Mason, R., Viljoen, M.J. and Viljoen, R.P.
1969: A reappraisal of some aspects of Precambrian Shield geology; Geol. Soc. Am., Bull., v. 80, no. 11, p. 2175–2200.

Anhaeusser, C.R., Roering, C., Viljoen, M.J. and Viljoen, R.P.
1968: The Barberton Mountain Land: a model of the elements and evolution of an Archean fold belt; Geol. Soc. S. Afr., Trans. (Annex.), v. 71, p. 225–254.

Ankudovich, P.M. and Roseeva, A.N.
1974: Gold distribution in metasomatic rocks of the Kuramin subzone (Uzbek S.S.R.); Metallog. Geokhim. Uzb., p. 167–172. (Chem. Abstr., v. 82, 142820j.)

Anonymous
1933: Gold; a reprint of the special number of "The Times"; The Times Publishing Co., Ltd., Printing House Square, London, 238 p.

Anoshin, G.N.
1977: Gold in magmatic rocks; Trans. Inst. Geol. Geophys., no. 289, Acad. Sci. USSR Siber. Br., "Nauka" Publ. House, Novosibirsk, 207 p.

Anoshin, G.N., Bushlyakov, I.N. and Potap'ev, V.V.
1970: Distribution of gold in rocks and minerals of the Altai, Transbaikalia, and Urals granitic rock massifs; Izv. Tomsk. Politekh. Inst., v. 239, p. 39–44. (Chem. Abstr., v. 75, 111760r.)

Anoshin, G.N. and Emel'yanov, E.M.
1969: Gold in magmatic rocks of the Altantic Ocean (Radioactivation analysis data); Akad. Nauk SSSR, Dokl., v. 189, no. 5, p. 1107–1110.

Anoshin, G.N., Emel'yanov, E.M. and Perezhogin, G.A.
1969: Gold in the modern sediments of the northern part of the Atlantic Ocean Basin; Geochem. Int., v. 6, no. 5, p. 897–905.

Anoshin, G.N. and Kepezhinskas, V.V.
1972: Petrochemical features related to gold distribution for the Cenozoic volcanic rocks of the Kuril-Kamchatka province; Geochem. Int., v. 9, no. 4, p. 618–629.

Anoshin, G.N. and Potap'yev, V.V.
1966: Gold in granites of the Kolyvan' (Altai) and Khangilay-Shilinskiy (Transbaikalia) massifs (according to radioactivation analysis data); Geokhim., no. 9, p. 1070–1074. *Also* Geochem. Int. 1966, v. 3, no. 5, p. 850–854.

Anoshin, G.N., Zolotarev, B.P. and Tsimbalist, V.G.
1971: Gold in rocks of tholeiite-basalt formations; Akad. Nauk SSSR, Dokl., v. 201, no. 3, p. 700–703. (Chem. Abstr., v. 76, 48206t.)

Anthony, K.M.
1967: The discovery of the Keystone gold mine, Cleary Hill area, Fairbanks district, Alaska—a geochemical prospecting case history; *in* Proc. Symp. Geochem. Prospecting, Ottawa, April 1966, E.M. Cameron, ed., Geol. Surv. Can., Pap. 66–54, p. 3–12.

Antipin, V.S., Kuz'min, M.I., Kozlov, V.D. and Glazunova, A.D.
1969: Relative average contents and distribution of molybdenum and gold in eastern Transbaikalia granitic rocks of various ages; Zap. Zabaikal. Filiala Geogr. Obschest. SSSR, v. 35, p. 104–108. (Chem. Abstr., v. 75, 89975r.)

Antrobus, E.S.A.
1956: The origin of the auriferous reefs of the Witwatersrand system; Geol. Soc. S. Afr., Trans., v. 59, p. 1–15, disc. p. 17–22.

Antweiler, J.C. and Campbell, W.L.
1976: Application of gold compositional analyses to mineral exploration in the United States; 25th Int. Geol. Congr., Sydney, Australia, Abstr., v. 2, p. 433–434.

Antweiler, J.C., Doe, B.R. and Delevaux, M.H.
1972: Lead isotope and other evidence on the bedrock source of placer gold at Hahns Peak, Colorado; Econ. Geol., v. 67, p. 302–314.

Antweiler, J.C. and Love, J.D.
1967: Gold-bearing sedimentary rocks in northwest Wyoming—a preliminary report; U.S. Geol. Surv., Circ. 541, 12 p.

Antweiler, J.C. and Sutton, A.L.
1970: Spectrochemical analysis of native gold samples; U.S. Clearinghouse Fed. Sci. Tech. Inform., P B Rep. 1970, no. 194809, 32 p. (Chem. Abstr., v. 75, 8625b.)

Anufriev, Yu. N., Stupakov, G.P. and Moskalyuk, A.A.
1973: Characteristic properties of quartz from the Kochkar gold-ore deposit; Akad. Nauk SSSR, Izv., Ser. Geol., no. 3, p. 68–81. *Also* Int. Geol. Rev., v. 16, p. 405–416, 1974.

Apel'tsin, F.E., Erofeev, B.N. and Lugov, S.F.
1968: Some features of gold metallogeny in the region of the Okhotsk-Chukotka volcanic belt; 23rd Int. Geol. Congr., Prague, 1968, Abstr., Sec. 7, p. 174.

Aprelkov, S. Ye. and Kharchenko, Yu. I.
1968: Gold polymetallic and gold-silver occurrences in southern Kamchatka; Geol. Rud. Mestorozhd., v. 10, p. 109–112. *Also* Int. Geol. Rev., v. 13, p. 740–743, 1970.

Arapov, V.A., Bragin, I.K., Gar'kovets, V.G., Malakhov, A.A., Matsokina, T.M., Palei, L.Z., Petrov, V.M., Rakhmatullaev, Kh.R. and Khamrabaev, I.Kh.
1968: Metallogenic mapping of Uzbekistan for gold and type classification of gold-bearing areas; *in* Rud. Form. Osn. Cherty Metallogen. Zolota Uzb., I.Kh. Khamrabaev, ed., Izd. "Fan" Uzb. SSR, Tashkent, p. 303–329. (Chem. Abstr., v. 73, 37336x.)

Arapov, V.A., Bragin, I.K., Kustarnikova, A.A., Malakhov, A.A., Matsokina, T.M., Palei, L.Z. and Khamrabaev, I.Kh.
1968: Geological factors in the distribution of endogenic gold ore formations; Rud. Form. Osn. Cherty Metallogen. Zolota Uzb., I.Kh. Khamrabaev, ed., p. 337–346. (Chem. Abstr., v. 73, 37335w.)

Archer, A.R. and Main, C.A.
1971: Casino, Yukon—a geochemical discovery of an unglaciated Arizona-type porphyry; *in* Geochemical exploration, R.W. Boyle and J.I. McGerrigle, eds., Can. Inst. Min. Metall., Spec. Vol. 11, p. 67–77.

Aripova, Kh. and Prikhid'ko, P.L.
1965: Determination of gold in plants and soils during biogeochemical studies; Uzb. Geol. Zh., v. 9, no. 4, p. 50–53. (Chem. Abstr., v. 64, 1341a.)

Aripova, Kh. and Talipov, R.M.
1966: Concentration of gold in soils and plants in the southern part of the Tamdyn Mountains; Uzb. Geol. Zh., v. 10, no. 3, p. 45–51. (Chem. Abstr., v. 66, 48364m.)

Aristov, V.V., Bezirganov, B.G. and Shevyrev, I.A.
1972: The question of the origins of gold in auriferous deposits and its significance for the solution of prospecting problems; Izv. Vyssh. Ucheb. Zaved. Geol. Razved., no. 12, p. 61–67.

Arkhipov, Yu. V. and Bilanenko, V.A.
1973: Antimony mineralization in eastern Yakutia; Razved. Okhr. Nedr., no. 10, p. 8–13. (Chem. Abstr., v. 80, 85687d.)

Armstrong, E.G. and Armstrong, K.F.
1931: The glycosides; Longmans, Green and Co., Ltd., London, 123 p.

Armstrong, H.S.
1943: Gold ores of the Little Long Lac area, Ontario; Econ. Geol., v. 38, p. 204–252.

Arsen'ev, V.P., Baranov, V.A. and Kotkin, V.V.
1975: Gold-bearing capacity of carbonaceous terrigenous rocks of the Precambrian and early Paleozoic in Sayan-Baikal fold region; Geol. Geofiz., no. 6, p. 3–11. (Chem. Abstr., v. 83, 196320k.)

Ashley, R.P.
1974: Goldfield mining district, Nevada; Nev., Bur. Mines, Rep., v. 19, p. 49–66.

Ashley, R.P. and Albers, J.P.
1975: Distribution of gold and other ore-related elements near ore bodies in the oxidized zone at Goldfield, Nevada; U.S. Geol. Surv., Prof. Pap. 843-A, 48 p.

Ashley, R.P. and Keith, W.J.
1976: Distribution of gold and other metals in silicified rocks of the Goldfield mining district, Nevada; U.S. Geol. Surv., Prof. Pap. 843-B, 17 p.

Ashley, R.P. and Silberman, M.L.
1976: Direct dating of mineralization at Goldfield, Nevada by potassium-argon and fission-track methods; Econ. Geol., v. 71, p. 904–924.

Aslaner, G. and Ottemann, J.
1972: Pure gold in the zinkenite ore veins in Toktamis of Ordu, northeast Anatolia; Bull. Miner. Res. Explor. Inst. Turk., no. 78, p. 55–58. (Chem. Abstr., v. 79, 33631r.)

Atabek'yants, K.P.
1972*a*: Primary haloes of gold sulphide deposits; Sov. Geol., no. 1, p. 128–135.
1972*b*: Zoning of primary haloes and their relation to formation characteristics of the Yuzhnoe deposit (Uzbekistan); Sov. Geol., no. 9, p. 98–109. (Chem. Abstr., v. 78, 18857e.) *Also* Int. Geol. Rev., v. 15, no. 12, p. 1394–1402.

Atanasov, V.A.
1975: Argentian mercurian tetrahedrite, a new variety, from the Chiprovtsi ore deposit, western Stara-Planina mountains, Bulgaria; Mineral. Mag., v. 40, p. 233–237.

Atkinson, S.
1619: The discoverie and history of the gold mynes in Scotland; G.L. Meason, ed., (1825) for the Ballantyne Club; James Ballantyne and Co., Printer, Edinburgh, 119 p.

Aubert, G., Duong, P.K. and Geffroy, J.
1964: Sur la localisation de l'or dans le mineral du Châtelet (Creuse), Soc. Fr. Mineral. Cristallogr., Bull., 87, p. 623–624.

Auger, P.E.
1941: Zoning and district variations of the minor elements in pyrite of Canadian gold deposits; Econ. Geol., v. 36, p. 401–423.

Augustithis, S.S.
1967: On the textures and paragenesis of the gold-quartz-tourmaline veins of Ondonoc, W-Ethiopia; Miner. Deposita (Berl.), v. 3, p. 48–55.

Avdonin, V.N. and Vertushkov, G.N.
1955: Amethyst from the gold ore deposits of Berezovsk (Ural); Tr. Gorno-Geol. Inst., Akad. Nauk SSSR, Ural. Filiala, no. 26, p. 228–229. (Chem. Abstr., v. 51, 5645d.)

Averill, C.V.
1946: Placer mining for gold in California; Calif. Div. Geol. Mines, Bull., 135, 377 p.

Averin, Yu. A., Babaev, K.L., Markova, E.A. and Palei, L.Z.
1968: Types and distribution regularities of gold ore fields in central Asia; Osn. Probl. Metallogen. Tyan-Shanya, Tr. Vses. Metallogen. Soveshch., 5th, 1968, M.M. Adyshev, ed., Izd. "Ilim", Frunze, U.S.S.R., p. 400–405. (Chem. Abstr., v. 77, 22922u.)

Aver'yanov, G.S., Dyshchuk, Yu. I. and Peregudov, V.V.
1974: Characteristics of gold ore manifestations in the western part of the southern Pamir gold-bearing belt; Akad. Nauk. Tadzh. SSR, Dokl., v. 17, no. 5, p. 45–47. (Chem. Abstr., v. 82, 5973x.)

Babaev, K.L. and Abdukayumov, A.A.
1968: Genetic types of gold ore deposits and their distribution regularities in central Asia; Osn. Probl. Metallogen. Tayn-Shanya, Tr. Vses. Metallogen. Soveshch., 5th, 1968, M.M. Adyshev, ed., Izd. "Ilim", Frunze, U.S.S.R., p. 406–412. (Chem. Abstr., v. 77, 22944c.)
1974: Gold ore paraberesites in quartz-biotite schists in Terekkan; Zap. Uzb. Otd. Vses. Mineral. O-va, v. 27, p. 138–141. (Chem. Abstr., v. 83, 100872t.)

Babaev, K.L., Kochnev, E.A., Palei, L.Z. and Troitskii, Z.I.
1968: Formation conditions and character of distribution of placer deposits in Uzbekistan; Rud. Form. Osn. Cherty Metallogen. Zolota Uzb., I. Kh. Khamrabaev, ed., Izd. "Fan" Uzb. SSR, Tashkent, U.S.S.R., p. 276–290. (Chem. Abstr., v. 73, 37358f.)

Babaev, K.L. and Palei, L.Z.
1968: Gold placer deposits in Uzbekistan; Rud. Form. Osn. Cherty Metallogen. Zolota Uzb., I. Kh. Khamrabaev, ed., Izd. "Fan" Usb. SSR, Tashkent, U.S.S.R., p. 263–276. (Chem. Abstr., v. 73, 37357e.)

Babička, J.
1943: Gold in living organisms; Mikrochemie Ver. Mikrochim. Acta, v. 31, p. 201–253 (in German).

Babička, J., Komárek, J.M. and Némec, B.
1945: Gold in animal bodies; Acad. Tcheque Sci., Bull. Int., Cl. Sci., Math., Natur. Med., v. 45, p. 131–137.

Babina, E.A.
1972: Gold ore mineralization in skarns of the Ampalyk deposit; Izv. Tomsk. Politekh. Inst., no. 201, p. 9–11. (Chem. Abstr., v. 80, 123226n.)

Babkin, P.V., Goncharov, V.I. and Sidorov, A.A.
1973: Physicochemical analysis of the formation of volcanic (ore) deposits; Tr. Sev.-Vost. Kompleks. Inst., Dal' nevost. Tsentra, Akad. Nauk SSSR, no. 55, p. 185–191. (Chem. Abstr., v. 84, 7597t.)

Babkin, P.V., Kalinin, A.I. and Struzhkov, F.E.
1974: A new type of gold-silver orebody; Akad. Nauk. SSSR, Dokl., v. 216, p. 619–621.

Backman, O.L.
1948: Broulan mine; *in* Structural geology of Canadian ore deposits, v. 1, Can. Inst. Min. Metall., Montreal, p. 554–558.

Bacon, W.R.
1975*a*: Lode gold deposits in western Canada; (Abstr.), Can. Inst. Min. Metall., Bull., v. 68, no. 762, p. 51.
1975*b*: Lode gold deposits in western Canada; *in* Proc. 5th Gold and Money Session and Gold Technical Session, Pacific Northwest Metals and Minerals Conference, Portland, Oregon, Oreg. Dept. Geol. Miner. Ind. publ., p. 139–163.

Badalov, S.T.
1961: The geochemical relationship between indium and silver in zinc-silver-lead deposits; Geochemistry, no. 10, p. 1005–1009.
1965*a*: On the role of predominant components in the geochemistry of minor and rare elements of ore deposits; Geochem. Int., v. 2, no. 5, p. 857–860.
1965*b*: Some considerations on the genesis of gold-bearing quartz veins; Symp. Problems of Postmagmatic Ore Deposition, vol. 2 append., Acad. Sci. Czech., Prague, p. 104–106.
1966: Country rocks as a possible source of gold in endogenic quartz-gold deposits; Mineral. Geokhim. Sul'fidnykh Mestorozhd. Uzb., Akad. Nauk Uzb. SSR, Inst. Geol. Geofiz., p. 81–89. (Chem. Abstr., v. 66, 97395f.)
1972: The origin of gold concentrations in sulphides and its causes; Uzb. Geol. Zh., no. 2, p. 53–56.

Badalov, S.T. and Badalova, R.P.
1967: Some regularities of distribution of gold and silver in the principal ore minerals of hypogene deposits of Karamazar and western Kazakhstan; Geochem. Int., v. 4, no. 4, p. 660–668.

Badalov, S.T. and Rabinovich, A.V.
1966: The geochemistry of indium and thallium in the Karamazar ore region (the Uzbek and Tadzhik republics); Geochem. Int., v. 3, p. 1095–1101.

Badalov, S.T. and Terekhovich, S.L.
1966: Geochemistry of elements of the Pt group in the Almalk ore region; Akad. Nauk. SSSR, Dokl., v. 168, p. 1397–1399.

Badalova, R.P.
1966: Main supergene minerals in the Uzbekistan gold-ore deposits and occurrences; Nauch. Tr., Tashkent. Gos. Univ., no. 273, p. 35–40. (Chem. Abstr., v. 67, 13727a.)

Badalova, R.P. and Badalov, S.T.
1967: The standard of gold from hypogene ore deposits and occurrences of Uzbekistan; Akad. Nauk SSSR, Dokl., v. 173, no. 4, p. 914–916. (Acad. Sci. USSR, Dokl., Earth Sci. Sec., v. 173, nos. 1–6, p. 194–197.)

Badalova, R.P., Bakieva, S.A., Meshchaninov, Ye.Z., Meshchaninova, G.S., Nikolaeva, L.A. and Yakovleva, L.I.
1971: The composition and internal structure of gold from the Amalyk area; Nauch. Zap. Tashkent. Univ., no. 405, p. 27–31.

Badalova, R.P., Borozdin, V.G. and Nikolaeva, L.A.
1973: Relation between the behaviour of gold and its depth in ore manifestations of the Karzhantau Ridge; Nauch. Tr., Tashkent. Gos. Univ., no 429, p. 49–52. (Chem. Abstr., v. 80, 50496p.)

Badalova, R.P., Lebedeva, S.I., Yakovleva, L.I., Bakieva, S.A. and Nedoboi, V.F.
1973: Density of minerals in the gold-silver series; Zap. Uzb. Otd. Vses. Mineral. Obshechest., no. 26, p. 103–104. (Chem. Abstr., v. 80, 50368y.)

Badalova, R.P. and Palei, L.Z.
1965: Main features of the metallogeny of gold in western Uzbekistan; Geol. Rud. Mestorozhd., v. 7, no. 5, p. 38–46. (Engl. Rev. *in* Econ. Geol., v. 63, p. 852.)

Badalova, R.P., Popenko, G.S., Augustovskii, V.B. and Abramov, Yu. F.
1972: Mineral composition of gold-bearing Quaternary placers of the Pskem River basin; Nauch. Tr., Tashkent. Gos. Univ., no. 429, p. 53–57. (Chem. Abstr., v. 80, 85654r.)

Badalova, R.P., Popenko, G.S. and Tsypkina, P.E.
1970: Morphology of native and placer gold illustrated by deposits in western Uzbekistan; Nauch. Tr., Tashkent. Gos. Univ., no. 372, p. 31–34. (Chem. Abstr., v. 74, 144514w.)

Badone, L. and Spence, N.S.
1960: Physical metallurgy and uses of gold; Can., Dep. Mines Tech. Surv., Mines Br., Inform. Circ., 116, 86 p.
1961: Physical metallurgy and uses of gold; Can., Dep. Mines Tech. Surv., Mines Br., Inform. Circ., 129, 192 p.

Baedecker, P.A.
1967: The distribution of gold and iridium in meteoritic and terrestrial materials; U.S. At. Energy Comm. (Pub.) ORO-2670-17, 110 p. (Ph.D. thesis, Univ. Kentucky, Lexington, Ky.)

Baedecker, P.A., Chou, C.-L., Grudewicz, E.B. and Wasson, J.T.
1973: Volatile and siderophilic trace elements in Apollo 15 samples: geochemical implications and characterization of the long-lived and short-lived extralunar materials; *in* Proc. 4th Lunar Sci. Conf., Geochim. Cosmochim. Acta, Supp. 4, v. 2, p. 1177–1195.

Baedecker, P.A., Chou, C.-L., Sundberg, L.L. and Wasson, J.T.
1972: Extralunar materials in Apollo 16 soils and the decay rate of the extralunar flux 4.0 GY ago; Earth Planet. Sci. Lett., v. 17, p. 79–83.
1974: Volatile and siderophilic trace elements in the soils and rocks of Taurus-Littrow; *in* Proc. 5th Lunar Sci. Conf., Geochim. Cosmochim. Acta, Supp. 5, v. 2, p. 1625–1643.

Baedecker, P.A. and Ehmann, W.D.
1965: The distribution of some noble metals in meteorites and natural materials; Geochim. Cosmochim. Acta, v. 29, p. 329–342.

Bagdavadze, N.V., Barbakadze, L.V., Ginturi, E.N., Kuchava, N.E., Mosulishvili, L.M. and Kharabadze, N.E.
1965: Radioactivation method for determination of gold in blood; Akad. Nauk Gruz. SSR, Soobshch. (Bull. Acad. Sci. Georgian SSR), v. 39, pt. 2, p. 287–294. (Chem. Abstr., v. 64, 5439b.)

Bailes, A.H.
1971: Geology and geochemistry of the Pilot-Smuggler Shear Zone, Rice Lake region, southeastern Manitoba; *in* Geology and geophysics of the Rice Lake region, southeastern Manitoba (Project Pioneer); Man. Mines Br., Publ. 71-1, p. 229–311.

Bailey, G.B. and Williams, K.L.
1975: The genesis and economic potential of gold-bearing jasperoids in the central Drum Mountains, Utah; Econ. Geol., v. 70, p. 1317 (abstr.).

Baimukhamedov, Kh. N., Karimov, Kh. K. and Protasevich, L.N.
1975*a*: Gold- and scheelite-bearing skarn-like formations in the sedimentary-metamorphic Precambrian strata in the central Kyzylkum (south of Tamdytau); Uzb. Geol. Zh., v. 19, no. 4, p. 3–6. (Chem. Abstr., v. 84, 20419f.)
1975*b*: Rare metal epidote-pyroxene-amphibolic metasomatites of the gold-ore deposit of Muruntau; Akad. Nauk Uzb. SSR, Dokl., v. 32, no. 5, p. 48–49. (Chem. Abstr., v. 84, 108660p.)

Baimukhamedov, Kh. N., Yusupov, R.G., Batyrmukhamedov, A.K. and Borodin, Yu, V.
1972: Metallogenic aspects of gold-containing granitic rocks; Uzb. Geol. Zh., v. 16, p. 21–25. (Chem. Abstr., v. 78, 162100v.)

Bain, G.W.
1930: Structure of gold-bearing quartz in northern Ontario and Quebec; Am. Inst. Min. Metall. Eng., Tech. Publ., no. 327, 44 p.
1933: Wall-rock mineralization along Ontario gold deposits; Econ. Geol., v. 28, p. 705–743.
1960: Patterns of ores in layered rocks; Econ. Geol., v. 55, p. 695–731.

Bain, H.R., ed.
1911: Types of ore deposits; Mining Scientific Press, San Francisco and Mining Magazine, London, 378 p.

Baker, J.W. and Geological Staff.
1957: Kerr-Addison mine; *in* Structural geology of Canadian ore deposits, v. 2, Can. Inst. Min. Metall., Montreal, p. 392–402.

Baldock, J.W., Hepworth, J.W. and Marengwa, B.S.
1976: Gold, base metals, and diamonds in Botswana; Econ. Geol., v. 71, p. 139–156.

Bandy, M.C.
1940: A theory of mineral sequence in hypogene ore deposits; Econ. Geol., v. 35, p. 359–381, 546–570.

Banister, D.P.
1970: Geochemical investigations for gold, antimony, and silver at Stibnite, Idaho; U.S. Bur. Mines, Rep. Invest. 7417, 7 p.

Barannikov, A.G. and Shub, I.Z.
1970: Relation of placer gold with native sources as exemplified by the southern Urals deposits; Tr. Sverdlovsk. Gorn. Inst., no. 74, p. 9–14. (Chem. Abstr., v. 78, 126877v.)

Baranov, I. Ya. and Stativkin, E.V.
1972: Auriferous mineral associations on the Tyrnyauz ore field; Mineral-Petrogr. Geokhim. Issled. Sev. Kavkase Donbasse, I.A. Shamrai, ed., Izv. Rostov. Univ., Rostov-on-Don, U.S.S.R., p. 57–69. (Chem. Abstr., v. 81, 138602b.)

Barbieri, D.
1935: Experimental researches on the gold-bacteria and gold-toxin complexes; Boll. Soc. Int. Microbiol. Sez. Ital., v. 7, p. 452–456. (Chem. Abstr., v. 30, 3851–2.)

Barbosa, A.F., Abrahao, B. and Arroyo, A.
1948: Notas sobre o minerio da mina de Passagem, MG; Min. Metal., v. 13, no. 74, p. 101–110.

Barbosa, A.L.M.
1968: Geologia do Quadrilatero Ferrifero; XXII Congr. Brasil. Geol. Belo Horizonte, Roteiros das Excursoes, p. 2–6.

Barbosa, A.L.M., Grossi Sad, J.H. and Dorr, J.V.N.
1974: The "consanguineous" origin of a tourmaline-bearing gold deposit: Passagem de Mariana (Brazil); Econ. Geol., v. 69, p. 416–419.

Bariand, P. and Pelissier, G.
1972: Origine de l'or de Zarehshuran (Iran occidental). Paragenèse à pyrite aurifère avec orpiment et sulfures d'antimoine; Soc. Fr. Minéral. Cristallogr., Bull., v. 95, p. 625–629.

Barker, F.
1969: Gold investigations in Precambrian clastic and pelitic rocks, southwestern Colorado and northern New Mexico; U.S. Geol. Surv., Bull. 1272-F, 22 p.

Barkhudaryan, N.B. and Grebenchikov, A.M.
1974: Evaluation of the gold content in gold-bearing quartz veins; U.S.S.R. patent 438,888. (Chem. Abstr., v. 82, 67884c.)

Barkovskaya, M.B.
1963: Some features of the formation of economic grade heavy minerals concentration in basins of deposition; Geol. Rud. Mestorozhd. v. 5, p. 50–64. *Also* Econ. Geol., v. 58, p. 1357–1358, 1963.

Barnes, H.L.
1962: Mechanisms of mineral zoning; Econ. Geol., v. 57, p. 30–37.

Barnes, H.L. ed.
1967: Geochemistry of hydrothermal ore deposits; Holt, Rinehart and Winston Inc., New York, 670 p.

Barnett, P.R., Skinner, D.L. and Huffman, C.
1968: Determination of gold, platinum, and palladium by a combined fire-assay, ion-exchange, and spectrochemical technique: U.S. Geol. Surv., Prof. Pap. 600-C, p. C161–C163.

Barrass, P.F.
1975: Platinoid minerals in the gold reefs of the Witwatersrand Basin; Inst. Min. Metall., Trans., v. 84 (Bull. no. 822), p. B-70 (abstr.).

Barrett, R.W. and Johnston, A.W.
1948: Central Patricia mine; *in* Structural geology of Canadian ore deposits, v. 1, Can. Inst. Min. Metall., Montreal, p. 368–372.

Barsanov, G.P.
1945: The structures of quartz and ore composition of the auriferous veins of the Miass region of the southern Urals; Akad. Nauk SSSR, Izv., Ser. Geol., no. 2, p. 74–86. (Chem. Abstr., v. 39, 4817.)

Barton, M.D., Kieft, C., Burke, E.A.J. and Oen, I.S.
1978: Uytenbogaardtite, a new silver-gold sulphide; Can. Mineral., v. 16, p. 651–657.

Barton, P.B.
1971: The Fe-Sb-S system; Econ. Geol., v. 66, p. 121–132.

Bartram, G.D. and McCall, G.J.H.
1971: Wall-rock alteration associated with auriferous lodes in the Golden Mile, Kalgoorlie; Geol. Soc. Aust., Spec. Publ., no. 3, p. 191–199.
Baskakov, M.P. and Chebotarev, G.M.
1972: Possible forms of gold fixation in placers; Akad. Nauk Uzb. SSR, Dokl., v. 29, p. 40–42. (Chem. Abstr., v. 78, 74705r.)
Baskakova, M.P.
1970: Organic matter in gold-ore quartz veins in the Kyzyl-Kum sands; Akad. Nauk Uzb. SSR, Dokl., v. 27, no. 11, p. 47–48. (Chem. Abstr., v. 75, 8417k.)
Bastin, E.S.
1915: Experiments with colloidal gold and silver; Wash. Acad. Sci., J., v. 5, p. 64–71.
1922: Bonanza ores of the Comstock Lode, Virginia City, Nevada; U.S. Geol. Surv., Bull. 735, p. 41–63.
Bastin, E.S. and Laney, F.B.
1918: The genesis of the ores at Tonopah, Nevada; U.S. Geol. Surv., Prof. Pap. 104, 50 p.
Bate, L.C. and Dyer, F.F.
1965: Trace elements in human hair; Nucleonics, v. 23, p. 74–81.
Bateman, A.M.
1950: Economic mineral deposits, 2nd ed.; John Wiley & Sons, Inc., New York, 916 p.
Bateman, J.D.
1940: Rock alterations in the Uchi gold area; Econ. Geol., v. 35, p. 382–404.
1952: Some geological features at Giant Yellowknife; Geol. Assoc. Can., Proc., v. 5, p. 95–107.
1958: Uranium-bearing auriferous reefs at Jacobina, Brazil; Econ. Geol., v. 53, p. 417–425.
Bateson, J.H.
1965: Geochemical breakthrough in mineral prospection in British Guiana; Can. Min. J., v. 86, no. 12, p. 71–73.
Battiston, U. and Moauro, A.
1974: Determination of the total content of silver, arsenic, gold, cobalt, rhenium, antimony, uranium and zinc in natural waters by neutron activation analysis; Com. Naz. Energ. Nucl., RT-CHI 74(8), 31 p. (Chem. Abstr., v. 82, 47566x.)
Bauer, F.K. and Schermann, O.
1971: Über eine Peckblende-Gold-Paragenese aus dem Bergbau Mitterberg, Salzburg (ein Vorbericht); Sond. Verhand. Geol. Bund., Heft 4, s. A97–A100.
Bauer, R. and Schaudy, R.
1970: Activation analytical determination of elements in meteorites, 3. Determination of manganese, sodium, gallium, germanium, copper and gold in 21 iron meterorites and 2 mesosiderites; Chem. Geol., v. 6, p. 119–131.
Baumshtein, R.A. and Grigor'ev, N.G.
1969: Characteristics of primary dispersion haloes in the Kozlovsk gold ore deposit; Zap. Zabaikal. Filiala Geogr. Obshchest. SSSR, no. 36, p. 92–97. (Chem. Abstr., v. 77, 116687y.)
Baur, E.
1942: Sur la signification et le dosage de l'or des eaux marines; Bull. Inst. Oceanogr., no. 830.
Baxter, R.A. and Poet, S.E.
1964: Geochemical prospecting for gold telluride ores; Mines Mag., v. 54, p. 27–28.
Bayley, R.W. and Janes, W.W.
1961: Geochemical surveying for gold veins in the Atlantic district, Wyoming; U.S. Geol. Surv., Prof. Pap. no. 424-D, p. 332–333.
Bayramgil, O.
1945: Mineralogische Untersuchung der Erzlagerstätte von Isikdag (Türkei); Schweiz. Mineral. Petrog. Mitt., Bd. 25, H. 1, p. 23–113.
Beamish, F.E.
1966: The analytical chemistry of the noble metals; Pergamon Press, Oxford, 608 p.
Beamish, F.E. and Van Loon, J.C.
1972: The present status of methods of separating and determining the noble metals; Miner. Sci. Eng., v. 4, p. 3–17.
1977: Analysis of noble metals, overview and selected methods; Academic Press, New York, 327 p.
Beard, R.C. and Garratt, G.L.
1976: Gold deposits of the Kenora–Fort Frances area; Ont. Div. Mines, Miner. Deposit Circ. 16, 46 p.
Beck, R.
1905: The nature of ore deposits, vols. 1 and 2; Eng. Min. J., New York and London.
Becker, G.F.
1882: Geology of the Comstock Lode and the Washoe district; U.S. Geol. Surv., Monogr. 3, 422 p.
1887: Natural solution of cinnabar, gold and associated sulfides; Am. J. Sci., Ser. 3, v. 33, p. 199–210.
1895: Gold fields of the southern Appalachians; U.S. Geol. Surv., 16th Annu. Rep., 1894–1895, p. 251–331.
1898: Reconnaissance of the gold fields of southern Alaska, with some notes on general geology; U.S. Geol. Surv., Annu. Rep., no. 18, pt. 3, p. 1–86.
Beda, V.D. and Kucherova, M.A.
1970: Gold content in Kamchatka rocks; Vop. Geol., Geokhim. Metallogen. Sev.-Zap. Sekt. Tikhookean. Poyasa, Mater. Nauch. Sess. 1969, I.N. Govorov, ed., Dal'nevost. Geol. Inst., Vladivostok, U.S.S.R., p. 210–212. (Chem. Abstr., v. 75, 79221c.)
Beevers, J.R.
1967: A method for the determination of the gold content of samples by solvent extraction and atomic absorption spectrophotometry; Econ. Geol., v. 62, no. 3, p. 426–428.
Béland, J.
1957: St. Magloire and Rosaire–St. Pamphile areas; Que. Dep. Mines, Geol. Surv. Br., Geol. Rep. 76, 49 p.
Beleva-Naumova, S. and Dancheva, R.
1972: Atomic-adsorption method for determination of gold; Rudodobiv, v. 5., p. 25–28. (Chem. Abstr., v. 80, 55535k.)
Belevtsev, Ya. N.
1970: Sources of metals of metamorphic-hydrothermal deposits; *in* Problems of hydrothermal ore deposition, Z. Pouba and M. Stemprok, eds., Int. Union Geol. Sci., ser. A, no. 2, Schweizerbart'sche, Stuttgart, p. 30–35.
Belevtsev, Ya. N., Fomenko, V. Yu., Kucher, V.N. and Kuzenko, S.V.
1972: Mobilization of metals from sedimentary and metamorphic rocks by aqueous solutions; Geol. Zh., v. 32, no. 3, p. 42–51.
Bell, A.M.
1948: Hallnor mine; *in* Structural geology of Canadian ore deposits, v. 1, Can. Inst. Min. Metall., Montreal, p. 547–553.
Bell, H.
1960: A synthesis of geological work in the Concord area, North Carolina; U.S. Geol. Surv., Prof. Pap. 400-B, p. B 189–B 191.
1973: Some results of geochemical sampling in McCormick County, South Carolina; U.S. Geol. Surv., Bull. 1376, 22 p.
Bell, J.M.
1906: The salient features of the economic geology of New Zealand; Econ. Geol., v. 1, p. 735–750.
1924: Carbonaceous matter at Porcupine (1) — its geological occurrence; Inst. Min. Metall., v. 33, p. 459–464.
Bell, J.M. and Fraser, C.
1912: The geology of the Waihi-Tairua subdivision, Hauraki division; N.Z., Geol. Surv., Bull. 15 (new ser.), 192 p.
Bell, J.P.
1961: The gold-bearing deposits of the Potaro River down stream of Tumatumari, British Guiana; Proc. 5th Inter-Guiana Geol. Conf., p. 267–272.

Bell, L.V.
1948: Caribou mine; *in* Structural geology of Canadian ore deposits, v. 1, Can. Inst. Min. Metall., Montreal, p. 927–936.

Belogolova, G.A. and Shibanov, V.A.
1974: On the secondary dispersion halos of deposits in the eastern Lena gold district and the use of the drill UPB-25; Ezheg. Inst. Geokhim., Sib. Otd., Akad. Nauk SSSR, 1973, L.V. Tauson, ed., p. 251–253.

Belokon, V.G., Blank, M. Ya., Nikitin, N.M., Sidenko, O.G., Trushin, I.A. and Yasyrev, A.P.
1974: Placer gold of Meso-Cenozoic formations of the northern outlying regions of the Donets Basin; Geol. Zh., v. 34, no. 4, p. 109–113. (Chem. Abstr., v. 82, 19496f.)

Belt, T.
1861: Mineral veins: an enquiry into their origin founded on a study of the auriferous quartz veins of Australia; John Weale Publisher, London, 52 p.

Belyaev, K.D., Karpov, R.V. and Yakovlev, I.A
1972: Gold content of amphibolites in the Olenii Ridge and characteristics of geochemical prospecting for gold; Nauch, Osn. Geokhim. Metod. Poiskov Mestorozhd. Polez. Iskop. Ots. Potentsialnoi Rudonos. Magmat. Metamorf. Kompleks. p. 170–3 (in Russian) L. N. Ovchinnikov, ed. (Chem. Abstr., v. 81, 80490v.)

Bendik, A.T. and Nesterova, N.P.
1971: Gold-silver mineralization in the Muruntau deposit and in the Muruntau region; Uzb. Geol. Zh., v. 15, no. 4, p. 75–77. (Chem. Abstr., v. 77, 8217m.)

Benedict, P.C.
1948: Island Mountain mine; *in* Structural geology of Canadian ore deposits, v. 1, Can. Inst. Min. Metall., Montreal, p. 149–162.

Benedict, P.C. and Titcomb, J.A.
1948: Northern Empire mine; *in* Structural geology of Canadian ore deposits, v. 1, Can. Inst. Min. Metall., Montreal, p. 389–399.

Beneš, P.
1967: On the state of manganese and gold traces in aqueous solutions; J. Inorg. Nucl. Chem., v. 29, p. 2889–2898.

Beneš, P. and Riedel, A.
1967: Adsorption and desorption of manganese and gold on semiconductor silicon; Collect. Czech. Chem. Commun., v. 32, p. 2547–2556.

Beneš, P. and Singer, J.
1968: Sorption of manganese and gold on paper; Collect. Czech. Chem. Commun., v. 33, p. 1508–1521.

Benitez, F.
1959: Floating Chile's oxidized gold ores (in 2 parts); Eng. Min. J., v. 160, nos. 5–6, p. 106–112, 116–122.

Bennett, R.
1971: Exploration for hydrothermal mineralization with airborne gamma-ray spectrometry; *in* Geochemical exploration, R.W. Boyle and J.I. McGerrigle, eds., Can. Inst. Min. Metall., Spec. Vol. 11, p. 475–478.

Bensusan, A.J.
1929: Auriferous Jacutinga deposits; Inst. Min. Metall., Trans. (London), v. 38, p. 450–483.
1942: Two unusual gold occurrences; Chem. Eng. Min. Rev., v. 34, no. 402, p. 178.

Berg, G.
1929: Vorkommen und Geochemie der mineralischen Rohstoffe; Leipzig, 414 p.
1932: Das Vorkommen der chemischen Elemente auf der Erde; Leipzig, J.A. Barth, 204 p.

Berg, H.C. and Cobb, E.H.
1967: Metalliferous lode deposits of Alaska; U.S. Geol. Surv., Bull. 1246, 254 p.

Berg, R.
1928: Das allgemeine Vorkommen von Gold in Nahrungsmitteln und Organen; Biochem. Z., v. 198, p. 424–427.

Bergendahl, T.J.
1975: The oxidation states of gold; J. Chem. Educ., v. 52, p. 731–732.

Berger, B.R.
1975: Trace element variations associated with disseminated gold mineralization at the Getchell mine, Humboldt County, Nevada; Econ. Geol., v. 70, p. 1318 (abstr.).

Berger, B.R., Silberman, M.L. and Koski, R.A.
1975: Discussion on K-Ar relations of granodiorite emplacement and tungsten and gold mineralization near the Getchell mine, Humboldt County, Nevada — a reply; Econ. Geol., v. 70, p. 1487–1491.

Berger, V.I. and Kuz'min, V.G.
1975: Age-related position of gold-antimony mineralization of the Yana-Kolyma belt; Akad. Nauk SSSR, Dokl., v. 224, p. 165–167. (Chem. Abstr., v. 84, 108859k.)

Beridze, G.I., Macharashvili, G.R. and Mosulishvili, L.M.
1969: Quantitative determination of the gold content in certain wines by the neutron activation method; Sov. Radiochem., v. 11, no. 6, p. 713–715 (transl. by Consultants Bureau, New York).

Berman, Yu. S. and Gorelyshev, A.V.
1974: Gold-silver ratios exemplified by a gold-silver deposit of the northeastern part of the U.S.S.R.; Geokhim., no. 11, p. 1613–1621. (Chem. Abstr., v. 82, 142826r.)

Berman, Yu. S., Volarovich, G.P. and Kazarinov, A. I.
1973: Mineralogical-geochemical characteristics of gold deposits developed at shallow depths; Int. Geol. Rev., v. 15, p. 784–790.

Berndt, A.F. and Cummins, J.D.
1970: The crystal structure of the Au_2Hg phase; Acta Crystallogr., v. B26, pt. 6, p. 864–867.

Bernhard, J.
1966: Die Mitterberger Kupferkieslagerstätte Erzführung und Tektonik; Jahrb. Geol. B.A., Bd. 109, s. 1–90, Vienna.

Bertolani, M. and Sighinolfi, G.P.
1964: Metalliferous mineralization of the gold deposit of Cani in the valley of Anzasca (Novara), Italy; Rend. Soc. Mineral. Ital., v. 20, p. 71–82. (Chem. Abstr., v. 62, 1457b.)

Bertrand, G.
1932: Peut-on compter l'or parmi les éléments de la matière vivante? Bull. Soc. Chim. Fr., 4e sér., t. 51, p. 564–568.

Berry, L.G. and Thompson, R.M.
1962: X-ray powder data for ore minerals: The Peacock Atlas; Geol. Soc. Am., Mem. 85.

Berzon, R.O.
1973: Rare minerals from the Byn'govsk gold ore deposit; Issled. Obl. Rud. Miner., p. 48–60. (Chem. Abstr., v. 80, 62017w.)

Berzon, R.O. and Bushlyakov, I.N.
1971: Gold/silver ratios in ores of the Ben'govskoye ore deposit in the central Urals; Geol. Rud. Mestorozhd., v. 13, no. 6, p. 96–97.

Berzon, R.O. and Karyshev, Yu. M.,
1973: Selenium and tellurium in gold ore deposits of the Urals; Geokhim., no. 11, p. 1615–1619. (Chem. Abstr., v. 80, 62143j.)

Beske, H.E.
1974: Trace analysis of gold and platinum by spark mass spectroscopy; Eurisotop Off. Inform. Booklet no. 72, p. 657–695. (Chem. Abstr., v. 82, 164419n.)

Bethke, P.M., Barton, P.B., Lanphere, M.A. and Steven, T.A.
1976: Environment of ore deposition in the Creede mining district, San Juan Mountains, Colorado: II. Age of Mineralization; Econ. Geol., v. 71, p. 1006–1011.

Bethke, P.M., Barton, P.B. and Rye, R.O.
1973: Hydrogen, oxygen and sulfur isotopic compositions of ore fluids in the Creede district, Mineral County, Colorado; Econ. Geol., v. 68, p. 1205 (abstr.).

Beus, A.A. and Grigorian, S.V.
1977: Geochemical exploration methods for mineral deposits; A.A. Levinson, ed., transl. R. Teteruk-Schneider, Applied Publishing Ltd., Wilmette, Illinois, 287 p.

Bevan, P.A.
1973: Rosita mine — a brief history and geological description; Can. Inst. Min. Metall., Bull. v. 66, no. 736, p. 80–84.

Beyschlag, F., Vogt, J.H.L. and Krusch, P. (transl., S.J. Truscott)
1914: The deposits of the useful minerals and rocks, v. 1; MacMillian and Co., Ltd., London, 514 p.
1916: The deposits of the useful minerals and rocks, v. 2; MacMillian and Co., Ltd., London, p. 515–1262.

Bezirganov, E.G.
1969: Use of endogenic geochemical haloes of arsenic in prospecting for gold lodes (illustrated by a gold deposit of the Lesser Caucasus); Uch. Zap., Erevan. Gos. Univ., no. 1, p. 169–176. (Chem. Abstr., v. 73, 79376r.)

Bezverkhnii, M.P. and Baumshtein, R.A.
1973: Characteristics of the comparative mobility of elements in supergene haloes of rare metal and gold ore deposits; Zap. Zabaikal. Filiala Geogr. O-va. SSSR, v. 89, p. 39–40. (Chem. Abstr., v. 85, 66065m.)

Bichan, W.J.
1944: Gold deposition; Econ. Geol., v. 39, p. 234–241.
1947: The ubiquity of gold values and their relation to zones of tensional dilation; Econ. Geol., v. 42, p. 396–403.

Bilibin, Yu. A.
1938: The principles of placer geology; G.O.N.T.I. (State Joint Scientific and Technical Publishing House) Moscow-Leningrad, 505 p.
1947: Metallogensis of gold; Zap. Vses. Mineral. Obshchest., Ser. 2, v. 76, p. 77–91 (in Russian).
1950: On geochemical types of orogenic zones; 18th Int. Geol. Congr., Great Britain, Pt. II, Sec. A., p. 22–28.
1955: Metallogenic provinces and metallogenic epochs; Gosgeoltehk., Moscow, 85 p. (in Russian).

Billingsley, P.
1926: Geology of the Zaruma gold district of Ecuador; Am. Inst. Min. Eng., Trans., v. 74, p. 255–275.

Billingsley, P. and Hume, C.B.
1941: The ore deposits of Nickel Plate Mountain, Hedley, B.C.; Can. Inst. Min. Metall., Trans., v. 44, p. 524–590.

Bindermann, N.N. and Grabeklis, R.V.
1964: Relation of gold and molybdenum mineralizations in south-western part of the Transbaikal gold-molybdenum belt; Akad. Nauk. SSSR, Dokl., v. 159, no. 6, p. 1290-1292. (Chem. Abstr., v. 62, 6270d.)

Bishopp, D.W.
1962: Metallogenic provinces and maps—a first approach; Geol. Surv., S. Afr., Annals, v. 1, p. 163–168.
1964– The mode of origin of banket deposits, discussion; Inst.
1965: Min. Metall., Trans., v. 74, no. 704, p. 658–659.

Blackadar, R.G.
1967: Kodlunarn Island and Frobisher's "Gold"; Arctic Circular, v. 17, no. 1, p. 1–12.

Blagonadezhdin, B.I.
1972: Kalana gold ore deposit in the Republic of Mali; Tr. Voronezh. Gos. Univ., v. 86, p. 71–74. (Chem. Abstr., v. 81, 155820p.)
1975: Gold mineralization in relation to tectonics and magmatism, as in the northwestern margin of Birrim gold belt, west Africa; Int. Geol. Rev., v. 17, p. 1080–1090.

Blais, R.A.
1954: A petrologic and decrepitometric study of the gold mineralization at the O'Brien mine, northwestern Quebec; unpubl. Ph.D. thesis, Univ. Toronto, Toronto, 292 p.

Blainey, G.
1969: The rush that never ended, 2nd ed.; Melbourne University Press, Melbourne, Australia, 389 p.

Blake, W.P.
1897: Gold in granite and plutonic rocks; Am. Inst. Min. Eng., Trans., v. 26, p. 290–298.

Blakemore, K.
1971: The book of gold; Stein and Day Publishers, New York, 224 p.

Blanchard, R.
1933: Chemical migration: Post mine phenomena in New Guinea; Eng. Min. J., v. 134, p. 365–368, 425–428.

Blatchford, A.
1953: Geology of the Tavua goldfield, Viti Levu, Fiji; Proc. Aust. Inst. Min. Metall., nos. 168–169, p. 13–51.

Blockley, J.G.
1974: Notes on the Paterson Range gold prospects; Geol. Surv. W. Aust., Annu. Rep. 1973, p. 71–73.

Blokhina, N.A.
1972: Occurrence of gold in skarn deposits of the southern Gissar Range; Akad. Nauk. Tadzh. SSR, Dokl., v. 15, no. 10, p. 38–40. (Chem. Abstr., v. 78, 162083s.)
1974: Boron mineralization in skarns of gold sulphide deposits of the Tarórsk group in central Tadzhikistan; Akad. Nauk. Tadzh. SSR, Dokl., v. 17, no. 8, p. 47–50. (Chem. Abstr., v. 82, 173624c.)

Blomfield, A.L., Rood, H.S., Crocker, B.S. and Williamson, C.L.
1936: Milling-investigations into the ore as occurring at the Lake Shore mine; Can. Inst. Min. Metall., Trans., v. 39, p. 279–434.

Blyumentsev, A.M., Khrust, A.R. and Chepizhnaya, E.A.
1974: Radioactive elements as indicators of gold in effusives; Yad. Geol., p. 187–197. (Chem. Abstr., v. 83, 100886a.)

Bobkov, A.I.
1969: Geochemical criteria for determination of initial rock composition in metasomatites of Kuranakh ore field; Rudoobrazov. Ego Svyaz Magmat., p. 160–161. (Chem. Abstr., v. 73, 37362c.)

Bobrievich, A.P. Gritsik, E.P. and Gritsik, V.V.
1971: Gold content of the Chivchinsk Mountains and Ciscarpathia; Tr. Inst. Miner. Res., Akad. Nauk Ukr. RSR, no. 4, p. 75–81. (Chem. Abstr., v. 79, 68594h.)

Boericke, W.F.
1933: Prospecting and operating small gold placers; John Wiley & Sons, Inc., New York, 136 p.

Boev, N.I.
1971: Hydrogeochemical criteria for exploration work in the Beregova district of the Trans-Carpathian region; Geol. Zh., v. 31, no. 1, p. 68–78.

Bogach, V.A.
1973: Gold ore mineralization associated with tourmaline-quartz-amphibole metasomatites; Mater. Respubl. Nauch. Teor. Konf. Molod. Geol. Kaz. SSR 3rd, p. 51–52. (Chem. Abstr. 83, 208413h.)

Bogdanovich, V.A.
1974: Primary zonality of Soviet gold bearing deposits and its practical significance; Tr. Sib. Nauch. Issled. Inst. Geol., Geofiz. Miner. Syr'ya, v. 145, p. 88–93. (Chem. Abstr., v. 83, 82754z.)

Bolarovich, G.P.
1972: Prospects for revealing subsurface gold-ore deposits related to volcanogenic formations; Sov. Geol., no. 11, p. 74–85.

Bolotnikova, I.V.
1969: Particle size distribution of gold in secondary dispersion haloes over gold-sulfide deposits in the Maikainsk ore field; Metod. Tekh. Razved, no. 64, p. 91–92. (Chem. Abstr., v. 72, 113684f.)

Bolter, E. and Al-Shaieb, Z.
1971: Trace-elements anomalies in igneous wall rocks of hydrothermal veins; *in* Geochemical exploration, R.W. Boyle and J.I. McGerrigle, eds., Can. Inst. Min. Metall., Spec. Vol. 11, p. 289–290.

Bonev, P.K.
1969–1970: Quelques changements métasomatiques et minéralisations prémétallifères dans les gisements or-polymétallique Bakadjik, près de Jambol; Ann. Ecole Sup. Mines Geol., Sofia Fasc. II. Géologie, Tom. XVI, p. 27–38.
1974: Ore formation and origin of gold polymetallic deposits in Bakadzhik, Bulgaria; Abstracts of Papers, Int. Assoc. Genesis Ore Deposits (IAGOD), 4th Symp., p. 85–86.
1976: Ore formation and genesis of the Bakadjik gold-polymetallic deposit, south-eastern Bulgaria; *in* Problems of Ore Deposition, v. 1, Volcanogenous ore deposits, 4th IAGOD Symp., Varna, B. Bogdanov, ed., Bulg. Acad. Sci., Sofia, p. 226–232.

Bonham, H.F. and Garside, L.J.
1974: Tonopah mining district and vicinity, Nevada; Nev., Bur. Mines, Rep., v. 19, p. 42–48. (Chem. Abstr., v. 82, 173551b.)

Boorman, R.S. and Abbott, D.
1967: Indium in co-existing minerals from the Mount Pleasant tin deposit; Can. Mineral., v. 9, p. 166–179.

Borchers, R.
1961: Exploration of the Witwatersrand System and its extensions; Geol. Soc. S. Afr., Trans. (Annex.), v. 64, p. LXVII–XCVIII.
1964: Exploration of the Witwatersrand system and its extensions; *in* The geology of some ore deposits in southern Africa, vol. I, S.H. Haughton, ed., Geol. Soc. S. Afr., Johannesburg, p. 1–23.

Borchert, H.
1935: Neue Beobachtungen an Tellurerzen; Neues Jahr. Mineral. Geol. Palaeontol., v. 69, p. 460–477.

Borcos, M., Gheorghita, I., Lang, B. and Russo, D.
1972: Metallogenetic activity associated with quartz andesites in the southern part of the Gutii Mountains; Stud. Teh. Econ., Inst. Geol. (Rom), ser. 1, v. 6, p. 89–120. (Chem. Abstr., v. 80, 17450w.)

Borcos, M. and Stanciu, C.
1974: Hydrothermal alteration of quartz-bearing andesite in the Hanes deposit (Metalliferous mountains); Bull. 6th Congr. Assoc. Geol. Carpatho-Balk., 1963, v. 4, p. 813–828. (Chem. Abstr., v. 85, 23678w.)

Bordonosov, V.P.
1974: Prospecting for metasomatic gold ores in the contact zone of Kalamin plutonic rocks; Tr. Sib. Nauch. Issled. Inst. Geol., Geofiz. Miner. Syr'ya, v. 145, p. 124–126. (Chem. Abstr., v. 83, 82755a.)

Borisenko, L.F.
1961: Scandium: principal features of the geochemistry, mineralogy, and genetic types of its deposits; Moscow, Izdatel'stvo AN SSSR, 130 p. (in Russian). (Transl. from the Russian by Consultants Bureau, New York, 78 p.

Borisenko, L.F., Miller, A.D. and Fisher, E.I.
1972: The abundance of gold in ultra-basic rocks; Geochem. Int., v. 1972, nos. 1–2, p. 121–128.

Borisenko, L.F. and Uskov, E.D.
1971: Gold content of ultrabasic rocks of the Gusevogorsk massif (Urals); Akad. Nauk SSSR, Dokl., v. 201, no. 2, p. 450–452. (Chem. Abstr., v. 76, 48193m.)

Borisenok, L.A.
1970: Geochemistry of gallium; MGU Press, Moscow (in Russian.)

Borodaevskaya, M.E. and Rozhkov, I.S.
1974: Gold deposits; *in* Ore deposits of the U.S.S.R., V.I. Smirnov, ed., v. 3, p. 5–77, "Nedra" Moscow.

Borodaevskii, N.I.
1962*a*: Regularities in the distribution, genesis, and types of gold-bearing shows of Kalba; Tr. Tsent. Nauch. Issled. Gornorazved. Inst., no. 41, p. 87–100. (Chem. Abstr., v. 58, 9968h.)
1962*b*: Genetic types of gold deposits; Tr. Tsent. Nauch. Issled. Gornorazved. Inst., no. 50, p. 3–15. (Chem. Abstr., v. 61, 5371a.)

Borodaevskii, N.I. and Kytyukin, P.I.
1939: Origin of vein fractures of the Berezov gold-ore deposits; Sov. Geol., v. 9, no. 2, p. 81–84.

Borodin, O.P. and Taraseich, Yu. N.
1972: Placer gold in the Cenozoic formations of Sakhalin; Vop. Geol. Rud. Mestorozhd. Dal'nego Vostoka, G.I. Osipova, ed., Akad. Nauk SSSR Dal'nevost. Geol. Vladivostok, U.S.S.R. p. 86–98. (Chem. Abstr., v. 82, 19486c.)

Borovitskii, V.P., Miller, A.D. and Shemyakin, V.N.
1966: Determination of minute amounts of gold in natural waters of the Aldan region; Geochem. Int., v. 3. no. 2, p. 371–377.

Borovitskii, V.P. and Shemyakin, V.N.
1965: Carbonate survey practise as applied in prospecting for gold deposits in the Aldan mining district; Sov. Geol., no. 10, p. 145–146. *Also* Int. Geol. Rev., v. 8, no. 9, p. 1037–1038.

Bostock, H.H.
1967: Geological notes Itchen Lake map-area, District of Mackenzie; Geol. Surv. Can., Pap. 66–24, 13 p.
1968: Gold-arsenopyrite-loellingite-pyrrhotite deposits in amphibolite, Itchen Lake–Contwoyto Lake area, District of Mackenzie; *in* Report of Activities, Pt. B; Geol. Surv. Can., Pap. 68–1, Pt. B, p. 72–76.

Bostock, H.S.
1942: Ogilvie, Yukon Territory; Geol. Surv. Can., Map 711A.

Bothwell, S.A.
1938: Geology of the Pickle Crow gold mine; Can. Inst. Min. Metall., Trans., v. 41, p. 132–140.

Botinelly, T., Neuerburg, G.I. and Conklin, N.M.
1973: Galkhaite, (Hg,Cu,Tl,Zn) (As,Sb) S_2, from the Getchell mine, Humboldt County, Nevada; U.S. Geol. Surv., J. Res., v. 1, no. 5, p. 515–517.

Bouška, V., Havlena, V. and Sůlcek, Z.
1963: Geochemistry and petrography of the Cenomanian coals from Bohemia and Moravia; Cesk. Akad. Ved, Rozpr. Rada Mat. Prirdn. Ved, v. 73, no. 8, p. 3–78 (in Czech).

Bowen, H.J.M.
1968: The uptake of gold by marine sponges; J. Mar. Biol. Assoc. U.K., v. 48, p. 275–277.

Bowen, K.G.
1972: Arsenic as a guide to gold mineralization at the Wattle Gully and Sambas mines; Min. Geol. J. (Mines Dep. Victoria), v. 7, no. 2, p. 5–15.

Bowie, A.J.
1879: Hydraulic mining in California; Am. Inst. Min. Eng., Trans., v. 6, p. 27–100.

Bowie, S.H.U.
1964–1965: The mode of origin of banket orebodies, discussion; Int. Min. Metall., Trans., v. 74, no. 702, p. 492–497.

Boydell, H.C.
1924: The role of colloidal solutions in the formation of mineral deposits; Inst. Min. Metall., Trans. (London), v. 34, pt. 1, p. 145–337.

Boyer, F. and Picot, P.
1963: Sur la présence de maldonite (Au_2Bi) à Salsigne (Aude); Soc. Fr. Minéral. Cristallogr., Bull. 86, p. 429.

Boyer, F., Touray, J.C. et Vogler, M.
1967: Présence d'inclusions à CO_2 liquide dans des quartz du district aurifère de Salsigne; C.R. Acad. Sci. Paris, Ser. D, t. 265, p. 11–13.

Boyes, W.T. and Harrison, J.M.
1948: Squall Lake property; *in* structural geology of Canadian ore deposits, v. 1, Can. Inst. Min. Metall., Montreal, p. 302–304.

Boyle, R.W.
1951: An occurrence of native gold in an ice lens, Giant Yellowknife Gold Mines, Yellowknife, Northwest Territories; Econ. Geol., v. 46, p. 223–227.
1953: On the colour of black and grey quartz from Yellowknife, Northwest Territories, Canada; Am. Mineral., v. 38, p. 528–535.
1954: A decrepitation study of quartz from the Campbell and Negus-Rycon shear zone systems, Yellowknife, Northwest Territories; Geol. Surv. Can., Bull. 30, 20 p.
1955: The geochemistry and origin of the gold-bearing quartz veins and lenses of the Yellowknife Greenstone Belt; Econ. Geol., v. 50, p. 51–66.
1959: The geochemistry, origin, and role of carbon dioxide, water, sulphur and boron in the Yellowknife gold deposits, N.W.T., Canada; Econ. Geol., v. 54, p. 1506–1524.
1960: Some geochemical considerations on lead-isotope dating of lead deposits; Econ. Geol., v. 54, p. 130–135 (Geol. Surv. Can. Reprint 15).
1961*a*: The geology, geochemistry, and origin of the gold deposits of the Yellowknife district; Geol. Surv. Can., Mem. 310, 193 p.
1961*b*: Native zinc at Keno Hill; Can. Mineral, v. 6, pt. 5, p. 692–694.
1963: Diffusion in vein genesis; Symp. Problems of Postmagmatic Ore Deposition, Prague, Czech., v. 1, p. 377–383.
1964: Origin of Yellowknife gold deposits—a reply; Econ. Geol., v. 59, p. 1176–1177.
1965*a*: Geology, geochemistry, and origin of the lead-zinc-silver deposits of the Keno Hill–Galena Hill area, Yukon Territory; Geol. Surv. Can., Bull. 111, 302 p.
1965*b*: The geochemistry of cadmium in the lead-zinc-silver deposits of the Keno Hill–Galena Hill area (Yukon, Canada); Akad. Nauk SSSR, Probl. Geokhim, Izd. "Nauka", Moscow, p. 220–231.
1966: Origin of gold and silver in the gold deposits of the Meguma Series, Nova Scotia; (abstr.) Progr. Geol. Assoc. Can. and Mineral Assoc. Can. Annu. Meet., Halifax, N.S.
1968*a*: Fahlbands, sulfide schists, and ore deposition; Econ. Geol., v. 63, p. 835–838.
1968*b*: The geochemistry of silver and its deposits; Geol. Surv. Can., Bull. 160, 264 p.
1968*c*: Conditions of formation of rhombohedral (3R) molybdenite; Akad. Nauk SSSR, Dokl., v. 183, p. 332–333.
1969*a*: The source of elements in epigenetic mineral deposits—a metallogenic concept for prospecting; Can. Min. J., v. 90, p. 104–107.
1969*b*: Elemental associations in mineral deposits and indicator elements of interest in geochemical prospecting; Geol. Surv. Can., Pap. 68–58, 45 p.
1969*c*: Hydrothermal transport and deposition of gold; Econ. Geol., v. 64, p. 112–115.
1970: Regularities in wall-rock alteration phenomena associated with epigenetic deposits; *in* Problems of hydrothermal ore deposition, Int. Union Geol. Sci. A., no. 2, Z. Pouba and M. Stemprok, eds., Schweizerbart'sche, Stuttgart, 396 p.
1971: Boron and boron minerals as indicators of mineral deposits; *in* Geochemical exploration, R.W. Boyle and J.I. McGerrigle, eds., Can. Inst. Min. Metall., Spec. Vol. 11, p. 112.
1972: The geology, geochemistry, and origin of the barite, manganese, and lead-zinc-copper-silver deposits of the Walton-Cheverie area, Nova Scotia; Geol. Surv. Can., Bull. 166, 181 p.
1974*a*: Elemental associations in mineral deposits and indicator elements of interest in geochemical prospecting (revised); Geol. Surv. Can., Pap. 74–45, 40 p.
1974*b*: The use of major elemental ratios in detailed geochemical prospecting utilizing primary halos; J. Geochem. Explor., v. 3, p. 345–369.
1976*a*: Mineralization processes in Archean greenstone and sedimentary belts; Geol. Surv. Can., Pap. 75–15, 45 p.
1976*b*: Native lead at Keno Hill, Yukon; Can. Mineral., v. 14, p. 577.
1977: Cupriferous bogs in the Sackville area, New Brunswick, Canada; J. Geochem. Explor., v. 8, p. 495–527.

Boyle, R.W., Alexander, W.M. and Aslin, G.E.M.
1975: Some observations on the solubility of gold; Geol. Surv. Can., Pap. 75–24, 6 p.

Boyle, R.W., Chamberlain, J.A., Holman, R.H.C. and Smith, A.Y.
1963*a*: Eight papers on regional geochemistry in Canada; Geol. Surv. Can., Pap. 63–23, 46 p.

Boyle, R.W. and Dass, A.S.
1971*a*: The geochemistry of the supergene processes in the native silver veins of the Cobalt–South Lorrain area, Ontario; Can. Mineral., v. 11, pt. 1, p. 358–390.
1971*b*: The origin of the native silver veins at Cobalt, Ontario; Can. Mineral., v. 11, pt. 1, p. 414–417.

Boyle, R.W., Dass, A.S., Church, D., Mihailov, G., Durham, C., Lynch, J. and Dyck, W.
1969: Research in geochemical prospecting methods for native silver deposits, Cobalt area, Ontario, 1966; Geol. Surv. Can., Pap. 67–35, 91 p.

Boyle, R.W. and Davies, J.L.
1964: Geology of the Austin Brook and Brunswick No. 6 sulphide deposits, Gloucester County, New Brunswick; Geol. Surv. Can., Pap. 63–24, 23 p.

Boyle, R.W. and Gleeson, C.F.
1972: Gold in the heavy mineral concentrates of stream sediments, Keno Hill area, Yukon Territory; Geol. Surv. Can., Pap. 71–51, 8 p.

Boyle, R.W., Hornbrook, E.H.W., Allan, R.J., Dyck, W. and Smith, A.Y.
1971: Hydrogeochemical methods—application in the Canadian Shield; Can. Inst. Min. Metall., Bull., v. 64, no. 715, p. 60–71.

Boyle, R.W. and Jambor, J.L.
1963: The geochemistry and geothermometry of sphalerite in the lead-zinc-silver lodes of the Keno Hill–Galena Hill area, Yukon Territory; Can. Mineral., v. 7, pt. 3, p. 479–496.

Boyle, R.W. and Jonasson, I.R.
1973: The geochemistry of arsenic and its use as an indicator element in geochemical prospecting; J. Geochem. Explor., v. 2, p. 251–296.

Boyle, R.W. and Steacy, H.R.
1973: An auriferous radioactive hydrocarbon from the Richardson mine, Eldorado, Ontario; *in* Report of Activities, April to October, 1972, Geol. Surv. Can., Pap. 73–1, Pt. A, p. 282–285.

Boyle, R.W., Wanless, R.K. and Lowdon, J.A.
1963*b*: The origin of the gold-quartz deposits, Yellowknife, N.W.T.—Discussion; Econ. Geol., v. 58, p. 804–807.

Boyle, R.W., Wanless, R.K. and Stevens, R.D.
1970: Sulfur isotope investigation of the lead-zinc-silver-cadmium deposits of the Keno Hill–Galena Hill area, Yukon, Canada; Econ. Geol., v. 65, p. 1–10.

Bradshaw, P.M.D., ed.
1975: Conceptual models in exploration geochemistry; The Canadian Cordillera and Canadian Shield; J. Geochem. Explor., v. 4, no. 1 (Spec. Iss.), 213 p.

Bragg, J.G.
1943: Rock alteration at the San Antonio Mine; Can. Min. J., v. 64, p. 553–556.

Brammall, A.
1926: Gold and silver in the Dartmoor granite; Mineral. Mag., v. 21, no. 112, p. 14–20.

Brammall, A. and Dowie, D.L.
1936: The distribution of gold and silver in the crystalline rocks of the Malvern Hills; Mineral. Mag., v. 24, no. 152, p. 260–264.

Brammall, A. and Harwood, H.F.
1924: The occurrence of gold-bearing pegmatite on Dartmoor; Mineral Mag., v. 20, no. 105, p. 201–211.

Brannock, W.W., Fix, P.F., Gianella, V.P. and White, D.E.
1948: Preliminary geochemical results at Steamboat Springs, Nevada; Am. Geophys. Union, Trans., v. 29, no. 2, p. 211–226.

Bray, A.
1928: Notes on the banket reefs of the Gold Coast Colony; Inst. Min. Metall., Trans. (London), v. 38, p. 21–69.

Bridgman, P.W.
1931: The physics of high pressure; G. Bell and Sons, Ltd., London, 398 p.

Brinck, J.W.
1956: Gold deposits in Surinam; Leidsche Geol. Meded., v. 21, p. 1–246.

Bristow, Q. and Jonasson, I.R.
1972: Vapour sensing for mineral exploration; Can. Min. J., v. 93, n. 15, p. 39–47, 85.

Brock, B.B. and Pretorius, D.A.
1964*a*: Rand Basin sedimentation and tectonics; *in* The Geology of some ore deposits in southern Africa, S.H. Haughton, ed., Geol. Soc. S. Afr., v. 1, p. 549–599.
1964*b*: An introduction to the stratigraphy and structure of the Rand goldfield, *in* The Geology of some ore deposits in southern Africa, S.H. Haughton, ed., Geol. Soc. S. Afr., v. 1, p. 25–61.

Brock, J.S.
1972: The use of dogs as an aid to exploration for sulphides; Western Miner, v. 45, no. 12, p. 28–32.

Brock, R.W.
1904: Original native gold in igneous rocks; Eng. Min. J., v. 77, p. 511.

Broderick, T.M.
1945: Geology of the Ropes gold mine, Marquette County, Michigan; Econ. Geol., v. 40, p. 115–128.

Brokaw, A.D.
1910: The solution of gold in the surface alterations of ore bodies; J. Geol., v. 18, p. 321–326.
1913: The secondary precipitation of gold in ore bodies; J. Geol., v. 21, p. 251–267.

Brooks, H.C. and Ramp, L.
1968: Gold and silver in Oregon; Oreg., Dep. Geol. Miner. Ind., Bull. 61, 337 p.

Brooks, R.R.
1960: The use of ion-exchange enrichment in the determination of trace elements in sea-water; Analyst, v. 85, p. 745–748.
1961: Apparent geochemical association of bismuth and thallium; Nature, v. 189, no. 4768, p. 910–911.
1972: Geobotany and biogeochemistry in mineral exploration; Harper & Row, Publishers, New York, 290 p.

Brosgé, W.P. and Reiser, H.N.
1972: Geochemical reconnaissance in the Wiseman and Chandalar districts and adjacent region, Southern Brooks Range, Alaska; U.S. Geol. Surv., Prof. Pap. 709, 21 p.

Brotzen, O. and Obial, R.
1963: Geochemical prospecting applied in the Philippines; Philipp. Geol., v. 17, no. 1, p. 26–40.

Broughton, P.L.
1973: Economic geology of gold-silver distribution in the Snare Group sediments (Proterozoic) at Norris Lake, Northwest Territories; Mountain Geol., v. 10, no. 2, p. 45–51.

Brown, B.W. and Hilchey, G.R.
1974: Sampling and analysis of geochemical materials for gold; Geochemical exploration 1974, I.L. Elliott and W.K. Fletcher, eds.; Assoc. Explor. Geochem., Spec. Publ. no. 2, Elsevier Sci. Publ. Co., Amsterdam, p. 683–690.

Brown, C.E.G. and Dadson, A.S.
1953: Geology of the Giant Yellowknife Mine; Can. Inst. Min. Metall., Trans., v. 56, p. 69–86.

Brown, C.E.G., Dadson, A.S. and Wrigglesworth, L.A.
1959: On the ore-bearing structures of the Giant Yellowknife Gold Mine; Can. Inst. Min. Metall., Trans., v. 62, p. 107–116.

Brown, G.F. *et al.*
1951: Geologic reconnaissance of the mineral deposits of Thailand; U.S. Geol. Surv., Bull. 984, 183 p.

Brown, J.B.
1960: Solubility of gold in mercury; J. Chem. Educ., v. 37, no. 8, p. 415–416.

Brown, J.C.
1935: Gold in Burma and the Shan States; Min. Mag., v. 52, p. 9–20, 82–92.

Brown, J.C. and Dey, A.K.
1955: India's mineral wealth; Oxford Univ. Press, Inc., New York, 761 p.

Brown, J.S.
1948: Ore genesis; Hopewell Press, Hopewell, N.J., 204 p.

Brown, R.A.
1948: O'Brien mine; *in* Structural geology of Canadian ore deposits, v. 1., Can. Inst. Min. Metall., Montreal, p. 809–816.

Brown, R. and Wolstenholme, W.A.
1964: Analyses of geological samples by spark source mass spectrometry; Nature, v. 201, no. 4919, p. 598.

Brown, W.L.
1948: Jerome mine; *in* Structural geology of Canadian ore deposits, v. 1, Can. Inst. Min. Metall., Montreal, p. 438–441.

Browne, P.R.L.
1969: Sulfide mineralization in a Broadlands geothermal drill hole, Taupo volcanic zone, New Zealand; Econ. Geol., v. 64, p. 156–159.

Brownell, G.M.
1941: Geology of the Falcon Lake stock, southeastern Manitoba; Can. Inst. Min. Metall., Trans., v. 44, p. 230–250.

Bruce, E.L.
1933: Mineral deposits of the Canadian Shield; Macmillan, Toronto, 428 p.
1937*a*: Geological relations of some major gold deposits of the Canadian Shield; Am. Inst. Min. Metall. Eng., Tech. Publ. no. 807, 13 p.
1937*b*: Geological relations of the major gold deposits of the Canadian Shield; Finl., Comm. Geol., Bull., v. 19, no. 115, p. 165–177.
1939: Structural relations of some gold deposits between Lake Nipigon and Long Lake, Ontario; Econ. Geol., v. 34, p. 357–368.
1941*a*: Albite and gold; Econ. Geol., v. 36, p. 455–458.
1941*b*: Concentrated saline water from the Sturgeon River gold mines; R. Soc. Can., Proc. Trans., ser. 3, v. 35, sec. 4, p. 25–59.
1942: Geology of the Goudreau-Lochalsh area; Ont. Dep. Mines, Annu. Rep., 1940, vol. 49, pt. 3, 50 p.
1943: Gold-silver ratios in certain Ontario gold mines; R. Soc. Can., Trans., v. 37, sec. 4, p. 23–31.
1944: Geology of the Rennie-Leeson area; Ont. Dep. Mines, Annu. Rep., 1942, v. 51, pt. 8, p. 1–26.
1948: Renabie mine; *in* Structural geology of Canadian ore deposits, v. 1, Can. Inst. Min. Metall., Montreal, p. 436–438.

Bruemmer, F.
1966: Kodlunarn Island's gold rush; Can. Geogr. J., v. 72, no. 2, p. 48–51.

Brundin, N.
1939: Method of locating metals and minerals in the ground; U.S. Pat. Off., 2,158,980, 3 p.

Brunfelt, A.O., Heier, K.S. and Steinnes, E.
1971: Determination of 40 elements in Apollo 12 materials by neutron activation analysis; Proc. 2nd Lunar Sci. Conf., Geochim. Cosmochim. Acta, Supp. 2, v. 2, p. 1281–1290.

Brunton, S.
1926*a*: The gold deposits of Nova Scotia: an analysis of the history and present status and a hypothesis concerning the structural features of the province in relation to the deposition of gold; Can. Inst. Min. Metall., Bull., v. 29, p. 424–497.
1926*b*: The gold deposits of Nova Scotia; a new hypothesis concerning the structural features of the province; Inst. Min. Metall., Trans., v. 35, p. 356–389.

Brückner, W.
1955: The mantle rock ("laterite") of the Gold Coast and its origin; Geol. Rundsch., v. 43, p. 307–327.

Bryner, L.
1969: Ore deposits of the Philippines—an introduction to their geology; Econ. Geol., v. 64, p. 644–666.

Bueno, J.A.
1955: Deposits of uranium and other metals in the La Baja region, Municipao Calif., Dep. Santander (Columbia); Ministerio Mines Petrol., Inst. Geol. Nacl., Bol. Geol., no. 3, p. 1–83, anexo p. 1–21. (Chem. Abstr., v. 51, 152f.)

Buffam, B.S.W.
1948*a*: Moneta Porcupine mine; *in* Structural geology of Canadian ore deposits, v. 1, Can. Inst. Min. Metall., Montreal, p. 457–464.
1948*b*: Aunor mine; *in* Structural geology of Canadian ore deposits, v. 1, Can. Inst. Min. Metall., Montreal, p. 507–515.

Buffam, B.S.W. and Allen, R.B.
1948: Chesterville mine; *in* Structural geology of Canadian ore deposits, v. 1, Can. Inst. Min. Metall., Montreal, p. 662-671.

Bugge, J.A.W.
1945: The geological importance of diffusion in the solid state; Av. Nor. Vidensk. Akad. Oslo I. Mat. Naturvidensk. K1., no. 13, 59 p.

Bugrov, V.
1974: Geochemical sampling techniques in the eastern desert of Egypt; J. Geochem. Expl. v. 3, p. 67–75.

Bugrov, V.A. and Shalaby, I.M.
1975: Geochemical prospecting in the eastern desert of Egypt; *in* Geochemical exploration 1974, I.L. Elliot and W.K. Fletcher, eds.; Elsevier Sci. Publ. Co., Amsterdam, p. 523–530.

Bujor, D.I.
1948: Synthetische Versuche über die Bildung der Golderzlagerstätten zu Brad in Rumanien; Ark. Kemi, Mineral. Geol., v. 26A, no. 3, p. 1–7. (Chem. Abstr., v. 44, 7196d.)

Bulatova, L.Sh.
1972*a*: Characteristics of gold ore mineralizations in the Kassan region (southern Kirgizia); Geol. Rus. Raionov Mestorozhd. Srednei Azii. Metody Ikh, p. 82–89. (Chem. Abstr., v. 80, 98353s.)
1972*b*: Mineralogical-geochemical characteristics of gold ore mineralization in the Kassan region; Geol. Petrologiya, Mineral. Endogennykh Mestorozhd. Srednei Azii, L.A. Bykov, ed., "Nedra" Moscow, USSR, p. 32–41. (Chem. Abstr., v. 80, 29204g.)

Bulgatov, A.N.
1973: Tectonic structures and distribution regularities of gold in the Barguzinsk taige; Geol. Geofiz., no. 12, p. 29–37. (Chem. Abstr., v. 80, 98381z.)

Bulynnikov, A.Ya.
1963: Genetic types of gold mineralization in the Kuznetsk Ala-Au and Salair; Tr. Tomsk. Gos. Univ., Ser. Geol. v. 164, p. 75–85. (Chem. Abstr., v. 60, 15608e.)

Burbank, W.S.
1930: Revision of geologic structure and stratigraphy in the Ouray district of Colorado, and its bearing on ore deposition; Colo. Sci. Soc., Proc., v. 12, no. 6, p. 151–232.

Burbank, W.S. and Luedke, R.G.
1968: Geology and ore deposits of the western San Juan Mountains, Colorado; *in* Ore deposits of the United States, Graton-Sales Vol., J.D. Ridge, ed., v. I, Am. Inst. Min. Metall., Pet. Eng. Inc., New York, p. 714–733.

Bürg, G.H.
1930: Die Sichtbarmachung des feinverteilten Goldes in goldhoffingen Erzen und ihre wirtschaftliche Bedeutung; Metall. Erz., v. 27, p. 333.
1935: Natur des in den Pyriten nicht sichtbar enthaltenen Goldes; Z. Prakt. Geol., v. 43, pt. 2, p. 17–26.

Burgess, J.A.
1911: The halogen salts of silver and associated minerals at Tonopah, Nevada; Econ. Geol., v. 6, p. 13–21.

Burn, R.G.
1973: Geochemical exploration in northeastern Nicaragua; Inst. Controam. Invest. Tecnol. Ind. Publ. Geol., v. 4, p. 17–29. (Chem. Abstr., v. 82, 75648u.)

Burnham, C.W.
1959: Metallogenic provinces of the southwestern United States and northern Mexico; N.M., Bur. Mines Miner. Resour., Bull. 65, 76 p.
1962: Facies and types of hydrothermal alteration; Econ. Geol., v. 57, p. 768–784.

Burnol, L., Autrau, A., Bonnici, J.-P. and Geoffroy, J.
1974: Acid granites and associated mineralization in the northwestern part of the French central masse; Int. Geol. Correlation Prog. Metallization Associated With Acid Magmatism, Excursion C Guidebook, Bur. Rech. Geol. Minières (Fr.), Paris, 206 p.

Burrows, A.G.
1925: The Porcupine gold area; Ont. Dep. Mines, Annu. Rep., 1924, v. 33, pt. 2, 112 p.
1929: The wire gold discovery south of the Porcupine District, Ontario; Eng. Min. J., v. 127, p. 241.

Burrows, A.G. and Hopkins, P.E.
1925: Kirkland Lake gold area; Lebel and Gauthier townships; Ont. Dep. Mines, Annu. Rep., 1923, v. 32, pt. 4, 96 p.

Buryak, V.A.
1964: Effect of regional metamorphism on the development of the gold-sulphide mineralization of the central part of the Lena gold fields; Fiz. Khim. Usloviya Magmat. Metasomatoza, Inst. Geol. Geofiz. Sibirsk, Otd. Akad. Nauk SSSR, p. 184–189. (Chem. Abstr., v. 63, 6726c.)
1967*a*: Gold ore formations of the Lena gold-bearing region; Int. Geol. Rev., v. 9, no. 12, p. 1553–1559.
1967*b*: A new type of gold mineralization in the Lena gold province; Akad. Nauk SSSR, Dokl., v. 169, no. 1, p. 163–166, 1966 (Akad. Sci. USSR, Dokl., Earth Sci. Sec., v. 169, nos. 1–6, p. 14–16).
1967*c*: Hypogene zoning of mineralization in the old gold provinces of Siberia; Acad. Sci. USSR, Dokl., Earth Sci. Sec., v. 173, p. 212–214 (Am. Geol. Inst. transl.).
1975*a*: Metamorphogenic-hydrothermal type of gold mineralization; Geol. Rud. Mestorozhd., v. 17, no. 2, p. 37–46.
1975*b*: Criteria for prospecting for gold mineralization of metamorphogenic-hydrothermal type; Sov. Geol., no. 10, p. 35–48. (Chem. Abstr., v. 84, 76947u.)
1976: Volcanic-sedimentary and hydrothermal-sedimentary minerogenesis and the role it plays in the formation of gold deposits in black (carbonaceous) shales; Acad. Sci. USSR, Dokl., v. 226, p. 907–910.

Buryak, V.A., Kashcheyeva, T.V. and Khemelevskaya, N.M.
1966: Influence of host rock and composition on the development of Precambrian gold mineralization (Patom Highlands); Acad. Sci. USSR, Dokl., Earth Sci. Sec., v. 169, nos. 1–6, p. 44–47.

Buryak, V.A., Lisii, V.A., Popov, N.P. and Khrenov, P.M.
1974: Gold mineralization in ancient strata of the Sayan-Baikal folded region; Razved. Okhr. Nedr., no. 2, p. 6–11. (Chem. Abstr., v. 81, 66383f.)

Buryak, V.A., Sizykh, Yu. I., Chernenko, A.I., Rusina, L.D., Klyachina, R.M., Volynets, T.P. and Yelizar'yeva, T.I.
1972: Behaviour of gold during granitization and pegmatitization in the Vitim-Patom highlands; Acad. Sci. USSR, Dokl., Earth Sci. Sec., v. 207, p. 230–233 (Am. Geol. Inst. transl.).

Buschendorf, F.
1926: Betrachtungen über die Gangkomponenten, sowie über das Vorkommen und die Verteilung des Goldes in der Primarzone alter Goldquarzgänge; Z. Prakt. Geol. Jahrg. 34, p. 1–11.
1931: Die primären Golderze des Hauptganges bei Brandholz im Fichtelgebirge unter besonderer Berücksichtigung ihrer Paragenesis und Genesis; Neues Jahrb. Miner., Geol. Palaeontol., Beil. Bd. 62, Abt. A, p. 1–50.

Buschendorf, F. and Hüttenhain, H.
1931: Über das Vorkommen von Gold und Wismuterzen in den Siegerländer Spateisensteingängen; Neues Jahrb. Miner., Geol. Palaeont., Beil. Bd. 62, Abt. A, p. 51–56.

Busev, A.I. and Ivanov, V.M.
1973: Analytical chemistry of the elements: analytical chemistry of gold; Nauka, Moscow, U.S.S.R., 263 p. (Chem. Abstr., v. 80, 33598f.)

Bushliakov, I.N.
1971: The distribution of gold in the granitoids of the Verkh-Isetsk Massif of the Urals; Geokhim., no. 12, p. 1442–1448.

Busschau, W.J.
1949: The glamour of gold, its attraction through the ages, pt. 2; S. Afr. Min. Eng. J., v. 60, pt. 1, no. 2941, p. 545–547.

Butler, B.S., Loughlin, G.F., Heikes, V.C., *et al.*
1920: The ore deposits of Utah; U.S. Geol. Surv., Prof. Pap. 111, 672 p.

Buturlinov, N.V. and Latysh, I.K.
1970: Gold and silver in magmatic rocks of the Donets Basin; Geol. Zh. (Kiev), v. 30, no. 5, p. 69–75.

Byers, A.R.
1940: Geology of the Nighthawk Peninsular gold mine; Econ. Geol., v. 35, p. 996–1011.
1948: Porcupine Peninsular mine; *in* Structural geology of Canadian ore deposits, v. 1, Can. Inst. Min. Metall., Montreal, p. 565–569.

Byers, A.R. and Gill, J.E.
1948: Sladen Malartic mine; *in* Structural geology of Canadian ore deposits, v. 1, Can. Inst. Min. Metall., Montreal, p. 858–864.

Cabri, L.J.
1965*a*: Phase relations in the Au-Ag-Te system and their mineralogical significance; Econ. Geol., v. 60, p. 1569–1606.
1965*b*: The occurrence of telluride minerals at the Acupan gold mine, Mountain Province, Philippines; Econ. Geol., v. 60, p. 1080–1082.
1972: The mineralogy of the platinum-group elements; Miner. Sci. Eng., v. 4, no. 3, p. 3–29.

Cabri, L.J., Harris, D.C. and Stewart, J.M.
1970: Paracostibite (CoSbS) and nisbite ($NiSb_2$), new minerals from the Red Lake area, Ontario, Canada; Can. Mineral., v. 10, pt. 2, p. 232–246.

Cabri, L.J. and Laflamme, J.H.G.
1974: Rhodium, platinum, and gold alloys from the Stillwater complex; Can. Mineral., v. 12, p. 399–403.

Cabri, L.J. and Rucklidge, J.C.
1968: Gold-silver tellurides: relation between composition and X-ray diffraction data; Can. Mineral., v. 9, pt. 4, p. 547–551.

Cachau-Herreillat, F. and La Salle, P.
1971: The utilization of eskers as ancient hydrographic networks for geochemical prospecting in glaciated areas; *in* Geochemical exploration, R.W. Boyle and J.I. McGerrigle, eds., Can. Inst. Min. Metall., Spec. Vol. 11, p. 121.

Cadigan, R.A.
1972: Geochemical anomalies and alteration in the Moenkopi Formation, Skull Creek, Moffat County, Colorado; U.S. Geol. Surv., Prof. Pap. 761, 21 p.

Caillere, S. and Dietrich, J.E.
1966: Maucherite, gold, and niccolite in the Aghbar mine, Bou-Azzer region, (Anti-Atlas, Morroco); Morroco, Serv. Geol., Notes Mem., no. 188, p. 139–141. (Chem. Abstr., v. 71, 62905y.)

Caley, E.R.
1964: Analysis of ancient metals; Macmillan and Co., Ltd., New York, 176 p.

Callahan, W.H.
1974: Occurrence of lorandite, thallium arsenide sulfide, $TlAsS_2$, at the Carlin gold deposit, Nevada; Econ. Geol., v. 69, p. 121–124.

Callow, K.J.
1967: The geology of the Thanksgiving mine, Baguio district, Mountain Province, Philippines; Econ. Geol., v. 62, p. 472–481.

Callow, K.J. and Worley, B.W.
1965: The occurrence of telluride minerals at the Acupan gold mine, Mountain Province, Philippines; Econ. Geol., v. 60, p. 251–268. *Also* Cabri, L.J., Econ. Geol., v. 60, p. 1080–1082.

Calvert, J.
1853: The gold rocks of Great Britain and Ireland and a general outline of the gold regions of the world with a treatise on the geology of gold; Chapman & Hall, Ltd., London, 324 p. and appendix and index.

Cameron, A.G.W.
1959: A revised table of abundance of the elements; Astrophys. J., v. 129, p. 676–699.
1968: A new table of abundance of the elements in the solar system; *in* Origin and distribution of the elements, L.H. Ahrens, ed., Pergamon Press, Oxford, p. 125–143.

Cameron, E.N., Jahns, R.H., McNair, A.H. and Page, L.R.
1949: Internal structure of granitic pegmatites; Econ. Geol., Monogr. 2, 115 p.

Cameron, G.W. and Hood, P.J.
1975: Residual aeromagnetic anomalies associated with the Meguma Group of Nova Scotia and their relation to gold mineralization; Geol. Surv. Can., Pap. 75-1, Pt. C, p. 197–211.

Campbell, J.D.
1953: The structure of the Kalgoorlie goldfield; *in* Geology of Australian ore deposits, v. I, A.B. Edwards, ed., 5th Emp. Min. Metall. Congr., Australia and New Zealand, p. 79–93.
1965: Gold ore deposits of Australia; *in* 8th Commonw. Min. Metall. Congr., Australia and New Zealand, v. 1, Geology of Australian ore deposits, 2nd ed., J. McAndrew, ed., p. 31–38.

Campbell, W.L., Mosier, E.L. and Antweiler, J.C.
1973: Effects of laboratory treatments on silver and other elements in native gold; J. Res., U.S. Geol. Surv., v. 1, no. 2, p. 211–220.

Camsell, C.
1910: The geology and ore deposits of Hedley mining district, British Columbia; Geol. Surv. Can., Mem. 2, 218 p.
1913: Geology and mineral deposits of the Tulameen district, B.C.; Geol. Surv. Can., Mem. 26, 188 p.

Cannon, H.L.
1960: Botanical prospecting for ore deposits; Science, v. 132, no. 3427, p. 591–598.

Cannon, H.L., Shacklette, H.T. and Bastron, H.
1968: Metal absorption by *Equisetum* (horsetail); U.S. Geol. Surv., Bull. 1278-A, 21 p.

Cannon, R.S., Pierce, A.P. and Antweiler, J.C.
1971: Suggested uses of lead isotopes in exploration; *in* Geochemical exploration, R.W. Boyle and J.I. McGerrigle, eds., Can. Inst. Min. Metall., Spec. Vol. 11, p. 457–463.

Carlisle, D. and Cleveland, G.B.
1958: Plants as a guide to mineralization; Calif. Dep. Nat. Resour. Div. Mines, Spec. Rep. 50, 31 p.
Carnegie, D.W.
1898: Spinifex and sand: a narrative of five years pioneering and exploration in western Australia; C. Arthur Pearson Ltd., London (Penguin Colonial facsimiles, Penguin Books Australia, Ltd., Ringwood, Victoria, Australia, 1973, 454 p.)
Carpenter, J.A.
1941: An investigation as to the presence of commercial quantities of mercury and gold in the dry lakes of Nevada; Univ. Nevada, Bull., v. 35, no. 4, Geol. Min. Ser. no. 35, 20 p.
Carter, J.W.
1968: Gold mineralization in central Guyana; Trans. 4th, Caribbean Geol. Conf., Trinidad, v. 1965, p. 407–414.
Carter, O.F.
1948: Coniaurum mine; *in* Structural geology of Canadian ore deposits, v. 1, Can. Inst. Min. Metall., Montreal, p. 497-503.
Cartwright, A.P.
1962: The gold miners; Purnell & Sons (SA) Pty Ltd., Johannesburg, 340 p.
Casadevall, T.
1976: Sunnyside mine, Eureka mining district, San Juan County, Colorado: geochemistry of gold and base metal ore formation in the volcanic environment; 25th Int. Geol. Congr., Sydney, Australia, Abstr., v. 3, p. 805.
Casadevall, T. and Ohmoto, H.
1977: Sunnyside mine, Eureka mining district, San Juan County, Colorado: geochemistry of gold and base metal ore deposition in a volcanic environment; Econ. Geol., v. 72, p. 1285–1320.
Cathcart, S.H.
1922: Metalliferous lodes in southern Seward Peninsula; U.S. Geol. Surv., Bull. 722, p. 163–261.
Cavender, W.S.
1963: Arsenic in geochemical gold exploration; Min. Eng., v. 15, no. 11, p. 60 (abstr.).
Cerveira, A.
1952: Relationship between hypogene deposits of gold and tungsten in Portugal; Port., Soc. Geol., Bol., v. 10, p. 133–144 (in Portuguese).
Cevales, G.
1961: Ore deposits of the Gran Paradiso (Graian Alps); Rend. Soc. Mineral. Ital., v. 17, p. 193–217. (Chem. Abstr., v. 56, 2179i.)
Chace, F.M.
1949: Origin of the Bendigo saddle reefs with comments on the formation of ribbon quartz; Econ. Geol., v. 44, p. 561–597.
Chae, S.H. and Kim, R.C.
1972: Distribution of gold deposits in the northwestern region of Korea; Chijil Kwa Chiri, v. 12, no. 4, p. 30–34. (Chem. Abstr., v. 79, 148095m.)
Chaffee, M.A.
1972: Distribution and abundance of gold and other selected elements in altered bedrock, Empire mining district, Clear Creek County, Colorado; U.S. Geol. Surv., Bull. 1278-C, 23 p.
1976: The zonal distribution of selected elements above the Kalamazoo porphyry copper deposit, San Manuel district, Pinal county, Arizona; J. Geochem. Explor., v. 5, p. 145–165.
Chaika, V.M.
1969: Precambrian metal-bearing conglomerates and fundamentals for evaluation of Precambrian placer deposits; Probl. Metallonosnosti Drevnikh Konglomeratov Territ. SSSR Mater. Soveshch., 1966, V.I. Smirnov, ed., Izd. "Nauka", Moscow, U.S.S.R., p. 29–52. (Chem. Abstr., v. 72, 34222a.)
1970: Problems of placer deposit evolution; Sostoyanie Zadachi Sov. Litol., Dokl. Zased. Vses. Litol. Soveshch., 8th, 1968, no. 2, p. 149–158. (Chem. Abstr., v. 74, 101646y.)
Chaika, V.M. and Nezhdanova, G.E.
1969: The migration of gold during the metamorphism of the sulphide-bearing graphitoid-kyanite shales of the Keyv Series of the Kola Peninsula; Akad. Nauk SSSR, Dokl., v. 188, no. 2, p. 438–441.
Chaikovskii, V.K.
1963: Gold in the sedimentary-metamorphic rocks and its study; Geokhim. Petrogr. Mineral. Osad. Obrazov., Akad. Nauk SSSR Gos. Kom. SSSR, p. 381–394. (Chem. Abstr., v. 61, no. 2, 1620f.)
Chakraborty, A.R.
1971: Bibliography of gold in India; Indian Miner., v. 25, no. 1, p. 82–92.
Chakraburtty, A.K., Stevens, C.M., Rushing, H.C. and Anders, E.
1964: Isotopic composition of silver in iron meteorites; J. Geophys. Res., v. 69, no. 3, p. 505–520.
Chakravorty, P.S.
1971: Mineragraphic studies of the gold-bearing quartz veisn around Kundrakocha, Singhbhum district, Bihar; Inst. Miner. Assoc., Pap. Proc. 7th Gen. Meet., 1970, no. 3, p. 418–424.
Chamber of Mines, Rhodesia
1968: Antheaps—a guide to mineral deposits? Chamber Mines J., v. 10, no. 12, p. 29.
1969*a*: Anthill theory winning many adherents in Rhodesia; Chamber Mines J., v. 11, no. 3, p. 16.
1969*b*: "Termite prospecting" going well; Chamber Mines J., v. 11, no. 9, p. 19.
Chamber of Mines, South Africa
1968–: Gold bulletin, a quarterly review of research on gold and its applications in industry; Chamber Mines S. Afr., 5 Hollard Street, Johannesburg.
Chance, H.M.
1900: The discovery of new gold districts; Am. Inst. Min. Eng., Trans., v. 29, p. 224–230.
Ch'ang ta
1963: The geology of China; U.S. Dep. Commer. OTS: 63-21820 (transl.).
Chapman, R.M.
1959: Geochemical exploration in the Kantishna area, Alaska; Geol. Soc. Am., Bull. v. 69, no. 12, pt. 2, p. 1751 (abstr.).
Charlewood, G.H.
1964: Geology of deep developments on the Main Ore Zone at Kirkland Lake; Ont. Dep. Mines, Geol. Circ. no. 11, 49 p.
Chary, K.N.
1971: An isotopic and geochemical study of gold-quartz veins in the Con-Rycon mine, Yellowknife, N.W.T.; Unpubl. M.Sc. thesis, Univ. Alberta, Edmonton, 90 p.
Chasovitin, M.D. and Filimanov, M.V.
1973: On the geochemistry of gold in early Cretaceous intrusive rocks of Chukotka; Geol. Geofiz., no. 7, p. 110–113.
Cheglokov, S.V.
1972: Spatial distribution characteristics of gold in the Balei ore field; Probl. Obrazov. Rud. Stolbov, Mater. Simp., 1969, F.N. Shakhov, ed., "Nauka" Sib. Otd., Novosibirsk, U.S.S.R., p. 89–92. (Chem. Abstr., v. 79, 55827m.)
Chekalova, K.A.
1968: Gold in ores of the Sokol'noe deposit; Geol., Geokhim. Mineral. Zolotorud. Raionov Mestorozhd. Kaz., no. 1, p. 119–121. (Chem. Abstr., v. 73, 5926y.)
Chemodanov, N.I.
1958: Morphological and chemical characteristics of placer gold in the Chaun-Uchkotsk region; Vses. Nauch. Issled. Inst. Zolota Redkikh Metal., Geol. no. 34, p. 27–36. (Chem. Abstr., v. 54, 19327c.)
Chen, P.-Y. and Lee, C.W.
1974: Fuchsite in gold-bearing rock from Laochi, Hualien; Acta Geol. Taiwan., no. 17, p. 7–11. (Chem. Abstr., v. 82, 19354h.)

Cheney, E.S. and Patton, T.C.
1967: Origin of the bedrock values of placer deposits; Econ. Geol., v. 62, p. 852–853.

Cherepnin, V.K., Mikov, A.D. and Gribanov, A.P.
1973: Secondary zoning in the gold ore deposits of Kuznetsk Alatau; Geol. Geofiz., no. 11, p. 43–47. (Chem. Abstr., v. 81, 15480c.)

Cherkasov, R.F.
1973: Archean strata of Aldan as a possible source of ore material in Mesozoic activation deposits; Metallog. Obl. Tektono-Magmat. Akt., Tezisy Dokl. Vses., M.M. Odintsov, ed., Akad. Nauk. SSSR, Sib. Otd., Inst. Zemnoi Kory, Irkutsk, U.S.S.R., p. 135–136. (Chem. Abstr., v. 82, 12656m.)

Chernyayev, A.M., Chernayeva, L.E., Yeremeyeva, M.N. and Andreyev, M.I.
1969: Hydrogeochemistry of gold; Geochem. Int., v. 6, no. 2, p. 348–358.

Chesnokov, B.V.
1972: Causes of the size fluctuations of the unit cell of pyrite from the Berezov gold ore deposit in the Urals; Zap. Vses. Mineral. Obshchest., v. 101, no. 4, p. 492–495.

Chesnokov, B.V. and Popova, V.I.
1971: Crystallomorphology of pyrite from the Berezovo gold ore deposit in the Urals; Mineral. Kristallogr., Tr. Soveshch. 1969, G.N. Vertushkov, ed., p. 43–47. (Chem. Abstr., v. 79, 128015g.)

Chetvertkov, S.S.
1972: Origin of gold-bearing placers; Probl. Izuch. Chetvertich. Perioda, Mater. Vses. Soveshch., 3rd 1968, S.M. Tseitlin, ed., "Nauka": Moscow, U.S.S.R., p. 357–362. (Chem. Abstr., v. 79, 81613h.)

Chibber, H.L.
1934: The mineral resources of Burma; The Macmillan Company, London, 320 p.

Chibisov, N.P., Kozlov, G.G. and Koksharskii, M.G.
1973: The correlation of primary and secondary geochemical halos of a gold-antimony deposit (northeast Yakutsk); Geol. Rud. Mestorozhd., v. 15, no. 4, p. 80–83. (Chem. Abstr., v. 80, 85685b.)

Chikishev, A.G.
1965: Plant indicators of soils, rocks, and subsurface waters; Consultants Bureau, New York, 210 p.

Chisholm, E.O.
1950: A simple chemical method of tracing mineralization through light non-residual overburden; Can. Inst. Min. Metall., Bull., v. 43, p. 64–68.

Chou, C.-L., Baedecker, P.A., Bild, R.W. and Wasson, J.T.
1974: Volatile-element systematics and green glass in Apollo 15 lunar soils; *in* Proc. 5th Lunar Sci. Conf., Geochim. Cosmochim. Acta, Suppl. 5, v. 2, p. 1645–1657.

Choubert, B.
1974: Le Précambrien des Guyanes; Mem. Bur. Rech. Geol. Minières (Fr.), no. 81, 213 p.

Chow, A.
1966: The determination of gold in ores; Unpubl. Ph.D. thesis, Univ. Toronto, Toronto, 130 p.

Chow, A. and Beamish, F.E.
1967: A comparison of methods for determining gold in ores; Can. Min. J., v. 88, p. 66–69.

Christie, A.M.
1953: Goldfields—Martin Lake map-area, Saskatchewan; Geol. Surv. Can., Mem. 269, 126 p.

Chukarov, E.V., Finkel'shtein, Yu. V. and Tabachkov, V.A.
1974: Geological-mineralogical characteristics of gold occurrence in the Kattaicha River basin (western Uzbekistan); Zap. Uzb. Otd. Vses. Mineral. O-va, v. 27, p. 53–56. (Chem. Abstr., v. 83, 118738p.)

Chukhrov, F.V.
1947: The migration of gold in the zone of oxidation; Akad. Nauk. USSR, Izv., Ser. Geol., no. 4, p. 117–125.

Chukhrov, F.V., ed.
1960–1965: Mineral reference book, 2 vols; Izdatl'stvo Akad. Nauk, SSSR, Moscow.

Chukhrov, F.V., Arest-Yakubovitch, R.E. and Kozlova, N.A.
1940: On the composition of jarosites from the deposits of central Kazakhstan; Akad. Nauk, Dokl., v. 28, no. 9, p. 829–831.

Church, B.N.
1973: Geology of the White Lake basin; B.C. Dep. Mines, Bull. 61, 120 p.

Chyi, L.L. and Crocket, J.H.
1976: Partition of platinum, palladium, iridium, and gold among coexisting minerals from the deep ore zone, Strathcona mine, Sudbury, Ontario; Econ. Geol., v. 71, p. 1196–1205.

Cioflică, G., Savu, H., Borcos, M., Stafan, A. and Istrate, G.
1973: Alpine volcanism and metallogenesis in the Apuseni Mountains; Symp. Volcanism Metallogenesis, Bucharest, Guideb. No. 13, 70 p., Published by Geol. Inst. Bucharest.

Cirkel, F.
1911: Alluvial gold deposits in Quebec; Eng. Min. J., v. 92, p. 1035-1038.

Cissarz, A.
1965: Einführung in die allgemeine und systematische, Lagerstättenlehre; E. Schweizerbart'sche Verlagsbuchhandlung, Stuttgart, 228 p.

Clapp, C.H.
1915: Alunite and pyrophyllite in Triassic and Jurassic volcanics at Kyuquot Sound, British Columbia; Econ. Geol., v. 10, p. 70–88.

Clappison, R.J.S.
1953: The Morning Star mine, Wood's Point; *in* Geology of Australian ore deposits, vol. 1, A.B. Edwards, ed.; 5th Emp. Min. Metall. Congr., Australia and New Zealand, p. 1077–1081.

Clark, E.S.
1955: The crystal structure of gold (III) chloride; Unpubl. thesis, Univ. Calif., Berkeley; U.S. At. Energy Comm., UCRL 3190, 48 p.

Clark, E.S., Templeton, D.H. and MacGillavry, C.H.
1958: The crystal structure of gold (III) chloride; Acta Crystallogr., v. 11, pt. 4, p. 284–288.

Clark, L.A.
1960: The Fe-As-S system: phase relations and applications; Econ. Geol. v. 55, p. 1345–1381, 1631–1652.

Clark, S.H.B., Condon, W.H., Foster, H.L. and Hoare, J.M.
1969: Geochemical data in the Sikonsina Pass area (Alaska), *in* Some shorter mineral resource investigations in Alaska; U.S. Geol. Surv., Circ. 615, p. 21–25.

Clark, W.B.
1970: Gold districts of California; Calif. Div. Mines Geol., Bull. 193, 186 p.

Clarke, E. de C. and Ellis, H.A.
1939: Metasomatism of country near ore-bodies and its possible economic significance; Econ. Geol., v. 34, p. 777–789.

Clarke, F.W.
1924: The data of geochemistry, 5th ed.; U.S. Geol. Surv., Bull. 770, 841 p.

Clarke, F.W. and Washington, H.S.
1924: The composition of the earth's crust; U.S. Geol. Surv., Prof. Pap. 127, 117 p.

Clayton, D.D. and Fowler, W.A.
1961: Abundances of heavy nuclides; Ann. Phys. (New York), v. 16, p. 51–68.

Clevenger, G.H. and Caron, M.H.
1925: The treatment of manganese-silver ores; U.S. Bur. Mines, Bull. 226, 110 p.

Clifford, T.N.
1966: Tectono-metallogenic units and metallogenic provinces of Africa; Earth Planet. Sci. Lett., v. 1, no. 6, p. 421–434.

Clift, D.S.L.
1975: Gold dredging in New South Wales; Geol. Surv. N.S.W., Min. Res. 41, 148 p.
Clifton, H.E.
1968: Gold distribution in surface sediments on the continental shelf off southern Oregon: a preliminary report; U.S. Geol. Surv., Circ. 587, 6 p.
Clifton, H.E., Hubert, A. and Phillips, R.L.
1967: Marine sediment sample preparation for analysis for low concentrations of fine detrital gold; U.S. Geol. Surv., Circ. 545, 11 p.
Clifton, H.E., Hunter, R.E., Swanson, F.J. and Phillips, R.L.
1969: Sample size and meaningful gold analysis; U.S. Geol. Surv., Prof. Pap. 625-C, p. C1–C17.
Cloke, P.L.
1963: Synthetic and natural phases in the system Au-Ag-Te; Econ. Geol., v. 58, p. 1163–1166.
Cloke, P.L. and Kelly, W.C.
1964: Solubility of gold under inorganic supergene conditions; Econ. Geol., v. 59, p. 259–270.
Cobb, E.H.
1962: Lode gold and silver occurrences in Alaska; U.S. Geol. Surv., Min. Invest. Resour. Map MR-32.
1964: Placer gold occurrences in Alaska; U.S. Geol. Surv., Min. Invest. Resour. Map MR-38.
1973: Placer deposits of Alaska; U.S. Geol. Surv., Bull. 1374, 213 p.
Cobb, J.C.
1967: A trace-element study of iron meteorites: J. Geophys. Res. v. 72, no. 4, p. 1329–1341.
Cockfield, W.E.
1932: The geology of placer deposits; Can. Inst. Min. Metall., Trans. v. 35, p. 58–64.
Coetzee, F.
1965: Distribution and grain-size of gold, uraninite, pyrite and certain other heavy minerals in gold-bearing reefs of the Witwatersrand Basin; Geol. Soc. S. Afr., Trans., v. 68, p. 61–88.
Cohen, A.J.
1959: Moldavites and similar tektites from Georgia, U.S.A.; Geochim. Cosmochim. Acta, v. 17, p. 150–153.
Cohen, E.M.
1962: Revised geology of the Tavua Goldfield, Fiji; Proc. Australas., Inst. Min. Metall., no. 204, p. 135–160.
Coleman, A.P.
1901: The Vermilion River placers; Ont. Bur. Mines, v. 10, p. 151–159.
Coleman, L.C.
1953: Mineralogy of the Yellowknife Bay area, N.W.T.; Am. Mineral., v. 38, p. 506–527.
1957: Mineralogy of the Giant Yellowknife Gold Mine, Yellowknife, N.W.T.; Econ. Geol., v. 52, p. 400–425.
Colin, L.L.
1946: Gold fineness in relation to geology: considerations of the Macequece field (Mozambique); S. Afr. Min. Eng. J., v. 57, pt. I, p. 279–283.
Collender, F.D.
1964: The geology of the Cam and Motor Mine, southern Rhodesia; *in* The geology of some ore deposits in southern Africa, vol. II, S.H. Haughton, ed.; Geol. Soc. S. Afr., Johannesburg, p. 15–27.
1965: The mode of origin of banket orebodies, discussion; Inst. Min. Metall., Trans., v. 74, p. 497–500.
Collier, A.J., Hess, F.L., Smith, P.S. and Brooks, A.H.
1908: The gold placers of parts of Seward Peninsula, Alaska; U.S. Geol. Surv., Bull. 328, 343 p.
Collins, G.E.
1902: The relative distribution of gold and silver values in the ores of Gilpin County, Colorado; Inst. Min. Metall., Trans. (London), v. 12, p. 480–499.
Collins, R.S.
1975: Gold; Miner. Resour. Consultative Comm., Miner. Doss. no. 14, Her Majesty's Stationery Office, London, 66 p.
Collins, W.H.
1925: North shore of Lake Huron; Geol. Surv. Can., Mem. 143, p. 160.
Colombo, U.P., Sironi, G., Fasolo, G.R. and Malvano, R.
1964: Systematic neutron activation technique for the determination of trace metals in petroleum; Anal. Chem., v. 36, no. 4, p. 802–807.
Conn, E.E.
1969: Cyanogenic glycosides; J. Agr. Food Chem., v. 17, no. 3, p. 519–526.
Connolly, J.P.
1927: The Tertiary mineralization of the northern Black Hills; S. Dakota Sch. Mines, Bull. no. 15, 130 p.
Connor, J.J., Shacklette, H.T. *et al.*
1975: Background geochemistry of some rocks, soils, plants, and vegetables in the conterminous United States; U.S. Geol. Surv., Prof. Pap. 574-F, 168 p.
Cook, B., Duval, J. and Adams, J.A.S.
1971: Progress in the calibration of airborne gamma spectrometers for geochemical exploration; *in* Geochemical exploration, R.W. Boyle and J.I. McGerrigle, eds., Can. Inst. Min. Metall., Spec. Vol. 11, p. 480–484.
Cook, E.W.
1973: Elemental abundance in Green River oil shale; Chem. Geol., v. 11, no. 4, p. 321–324.
Cooke, A.
1964: Canada's first gold rush; The Beaver, Summer, Hudson's Bay Co., p. 24–27.
Cooke, H.C.
1943: Gold: the unrivalled medium of exchange; Can. Inst. Min. Metall., Bull., no. 374, p. 280–293.
1946: Canadian lode gold areas, summary account; Geol. Surv. Can., Econ. Geol. Ser., no. 15, 86 p.
Cooke, H.C. and Johnston, W.A.
1932: Gold occurrences of Canada, summary account; Geol. Surv. Can., Econ. Geol. Ser., no. 10, 61 p.
Cooke, H.R.
1947: The Original Sixteen to One gold-quartz vein, Alleghany, California; Econ. Geol., v. 42, p. 211–250.
Cooper, M.
1971: Selected annotated bibliography on the geochemistry of gold, U.S. Geol. Surv., Bull. 1337, 63 p.
Cooper, R.A.
1923: Mineral constituents of Rand concentrates; J. Chem. Metall. Min. Soc. S. Afr., v. 24, p. 90–95, 264–266.
Cooper, W.G.G.
1934: The geology of the Prestea goldfield; Gold Coast Geol. Surv., Mem. no. 3, 20 p.
Corcoran, R.E. and Zwick, H.W. (co-chairmen)
1975: Proceedings 5th Gold and Money Session and Gold Technical session; Pacific Northwest Metals and Minerals Conference, Portland, Oregon, Oreg. Dep. Geol. Miner. Ind. publ., 201 p.
Cordani, U.G., Amaral, G. and Kawashita, K.
1973: The Precambrian evolution of South America; Geol. Rundsch., v. 62, no. 2, p. 309–317.
Corking, W.P.
1948: Pickle Crow mine; *in* Structural geology of Canadian ore deposits, vol. 1, Can. Inst. Min. Metall., Montreal, p. 373–376.
Cormie, J.M.
1948: East Malartic mine; *in* Structural geology of Canadian ore deposits, vol. 1, Can. Inst. Min. Metall., Montreal, p. 865–867.
Cornelius, K.D.
1967: Breccia pipe associated with epigenetic mineralization, Mount Morgan, Queensland; Econ. Geol., v. 62, p. 282–285.
1969: The Mount Morgan mine, Queensland — a massive gold-copper pyritic replacement deposit; Econ. Geol., v. 64, p. 885–902.

Cornwall, H.R., Lakin, H.W., Nakagawa, H.M. and Stager, H.K.
1967: Silver and mercury geochemical anomalies in the Comstock, Tonopah, and Silver Reef districts, Nevada-Utah; U.S. Geol. Surv., Prof. Pap. 575-B, p. 10–20.

Cotton, F.A. and Wilkinson, G.
1967: Advanced inorganic chemistry, 2nd ed.; Interscience Publishers, New York, 1136 p.

Courtis, W.M.
1890: Gold quartz; Am. Inst. Min. Eng., Trans., v. 18, p. 639–644.

Cousins, C.A.
1956: The value distribution of economic minerals with special reference to the Witwatersrand gold reefs; Geol. Soc. S. Afr., Trans., v. 59, p. 95–113.
1973: Platinoids in the Witwatersrand system; J. S. Afr. Inst. Min. Metall., v. 73, no 6, p. 184–199.

Coveney, R.M.
1971: Hydrogen and serpentinite — their roles in the localization of gold ores at the Oriental Mine, Alleghany, California; Econ. Geol., v. 66, p. 1265–1266 (abstr.).

Cox, D.P.
1967: Regional environment of the Jacobina auriferous conglomerates, Brazil; Econ. Geol., v. 62, p. 773–780.

Crampton, F.A.
1937: Occurrence of gold in stream placers; Mining J., v. 20, no. 16, p. 3–4, 33–34.

Crane, W.R.
1908: Gold and silver; John Wiley & Sons, Inc., New York, 727 p.

Cranton, G.E. and Heyding, R.D.
1968: The gold/selenium system and some gold selenotellurides; Can. J. Chem., v. 46, p. 2637–2640.

Crawford, A.R.
1970: Continental drift and un-continental thinking; Econ. Geol., v. 65, p. 11–16.

Crocket, J.H. and Chyi, L.L.
1972: Abundances of Pd, Ir, Os and Au in an alpine ultramafic pluton; 24th Int. Geol. Congr., Montreal, Sec. 10, Geochemistry, p. 202–209.

Crocket, J.H., Keays, R.R. and Hsieh, S.
1967: Precious metal abundances in carbonaceous and enstatite chondrites; Geochim. Cosmochim. Acta, v. 31, p. 1615–1623.
1968: Determination of some precious metals by neutron activation analysis; J. Radioanal. Chem., v. 1, p. 487–507.

Crocket, J.H., Macdougall, J.D. and Harriss, R.C.
1973: Gold, palladium and iridium in marine sediments; Geochim. Cosmochim. Acta, v. 37, p. 2547–2556.

Crocket, J.H., Teruta, J. and Garth, J.
1976: The relative importance of sulfides, spinels, and platinoid minerals as carriers of Pt, Pd, Ir, and Au in the Merensky Reef at Western Platinum Limited, near Marikana, South Africa; Econ. Geol. v. 71, p. 1308–1323.

Crocket, J.H., Vincent, E.A. and Wager, L.R.
1958: The distribution of gold in some basic and ultra-basic igneous rocks and minerals (abstr.);Geochim. Cosmochim. Acta, v. 14, p. 153–154.

Crohn, P.W.
1965: Tennant Creek gold and copper field; *in* 8th Commonw. Min. Metall. Congr., Australia and New Zealand, Geology of Australian ore deposits, vol. 1, 2nd ed., J. McAndrew, ed., p. 176–182.

Crook, W.J.
1939: Preliminary spectrographic and metallographic study of native gold; Am. Inst. Min. Metall. Eng., Tech. Publ. no. 998, 14 p.

Crouse, C.S.
1925: The precious metal content of the black Devonian shales of Kentucky; *in* Oil Shales of Kentucky; Kentucky Geol. Surv. ser. 6, v. 21, p. 49–58.

Csajka, M., Revina, L.D., Yukina, L.V. and Lavrukhina, A.K.
1974: Activation analysis of iron meteorites; Magyar Kemiai Folyoirat, v. 80, nos. 10–11, p. 484–487. (Chem. Abstr., v. 82, 10779t.)

Cumenge E. and Robellaz, F.
1898: L'or dans la nature; P. Vicq-Dunod et Cie, Éditeurs, Paris, 106 p.

Curle, J.H.
1899: The gold mines of the world, containing concise and practical advice for investors; Waterlow and Sons Ltd., London, 317 p.
1902: The gold mines of the world, 2nd ed.; Waterlow and Sons Ltd., London, 380 p.
1905: The gold mines of the world, 3rd ed.; George Routledge & Sons Ltd., London, 308 p.

Curtin, G.C., Lakin, H.W. and Hubert, A.E.
1970: The mobility of gold in mull (forest humus layer); U.S. Geol. Surv., Prof. Pap. 700-C, p. 127–129.

Curtin, G.C., Lakin, H.W., Hubert, A.E., Mosier, E.L. and Watts, K.C.
1971: Utilization of mull (forest humus layer) in geochemical exploration in the Empire district, Clear Creek, County, Colorado; U.S. Geol. Surv., Bull. 1278-B, 39 p.

Curtin, G.C., Lakin, H.W., Neuerburg, G.J. and Hubert, A.E.
1968: Utilization of humus-rich forest soil (mull) in geochemical exploration for gold; U.S. Geol. Surv., Circ. 562, 11 p.

Cusack, F.
1973: Bendigo, a history; Heinemann, Melbourne, 262 p.

Cushing, C.E. and Rancitelli, L.A.
1972: Trace element analyses of Columbia River water and phytoplankton; Northwest Sci., v. 46, no. 2, p. 115–121.

Czamanske, G.K., Desborough, G.A. and Goff, F.E.
1973: Annealing history limits for inhomogeneous, native gold grains as determined from Au-Ag diffusion rates; Econ. Geol., v. 68, p. 1275–1288.

Dadson, A.S.
1949: The Giant Yellowknife; Western Miner, v. 22, no. 10, p. 82–90.

Dadson, A.S. and Bateman, J.D.
1948: Giant Yellowknife mine; *in* Structural geology of Canadian ore deposits, v. 1, Can. Inst. Min. Metall., Montreal, p. 273–283.

Daly, R.A.
1914: Igneous rocks and their origin; McGraw-Hill Book Co., Inc., New York, 563 p.

Darnley, A.G. and Grasty, R.L.
1971: Mapping from the air by gamma-ray spectrometry; *in* Geochemical exploration, R.W. Boyle and J.I. McGerrigle, eds., Can. Inst. Min. Metall., Spec. Vol. 11, p. 485–500.

Darton, N.H.
1906: Geology of the Bighorn Mountains; U.S. Geol. Surv., Prof. Pap. 51, 129 p.

Das Sarma, B., Sen, B.N. and Chowdhury, A.N.
1965: Platinum and gold contents of granite G-1 and diabase W-1; Econ. Geol., v. 60, p. 373–374.

Dautov, A.I.
1973: Distribution of gold in dark coloured minerals in Koshrabadsk and Yakhtonsk intrusive rocks in western Usbekistan; Akad, Nauk Uzb. SSR, Dokl., v. 30, p. 58–60. (Chem. Abstr., v. 81, 15462y.)

Davidenko, N.M.
1970: Zoning of auriferous ore bodies in the Keperveem area of the Chukota Peninsula; Geol. Rud. Mestorozhd., no. 2, p. 103–106.
1973*a*: Mineralogical-geochemical characteristics of gold in Malo-Anyuisk region of western Chukotka; Geol. Geofiz., no. 3, p. 133–139.
1973*b*: Isometric crystals and dendrites of gold on a nugget; Priroda, no. 11, p. 123–124.
1975: On the paragenesis of native gold and rock crystal on Chukotka; Miner. Paragenezisy Miner. Endog. Mestorozhd; P.M. Tatarinov and D.V. Rundkvist, eds., Pub. "Nauka" Leningrad Otd., Leningrad, U.S.S.R., p. 122-125. (Chem. Abstr., v. 85, 81051x.)

Davidson, C.B.
1967: Geochemical prospecting abstracts, January 1961–December 1962; U.S. Geol. Surv., Open File Rep. 955.

Davidson, C.B., Markward, E.L. and Cieslewicz, J.
1967: Geochemical prospecting abstracts, July 1957–December 1960; U.S. Geol. Surv., Open File Rep. 954.

Davidson, C.F.
1953: The gold-uranium ores of the Witwatersrand; Mining Mag., v. 88, p. 73–85.
1957: On the occurrence of uranium in ancient conglomerates; Econ. Geol., v. 52, p. 668–693.
1962: On the cobalt: nickel ratio in ore deposits; Mining Mag., v. 106, p. 78–85.
1964–1965: The mode of origin of banket orebodies; Trans. Inst. Min. Metall., v. 74, pt. 6, p. 319–338. *Also* v. 74, pt. 12, p. 844–857.
1965: A possible mode of origin of strata-bound copper ores; Econ. Geol., v. 60, p. 942–954.

Davidson, C.F. and Bowie, S.H.U.
1951: On thucholite and related hydrocarbon-uraninite complexes, with a note on the origin of the Witwatersrand gold ores; Geol. Surv. G.B., Bull., no. 3, p. 1–18.

Davidson, S.
1948: Beattie mine; *in* Structural geology of Canadian ore deposits, v. 1, Can. Inst. Min. Metall., Montreal, p. 692–701.

Davidson, S. and Banfield, A.F.
1944: Geology of the Beattie Gold Mine, Duparquet, Quebec; Econ. Geol., v. 39, p. 535–556.

Davies, D.N.
1964: The nickel-tungsten-gold deposits at Forbes Reef, Hhohho District; Swaziland Geol. Surv. Mines Dep., Bull., no. 4, p. 45–50.

Davies, D.N. and Hunter, D.R.
1964: The gold deposits of the Barberton Mountain Land in Swaziland; *in* The geology of some ore deposits in southern Africa, v. 2; S.H. Haughton, ed.; Geol. Soc. S. Afr., Johannesburg, p. 59–75.

Davies, J.F.
1963: Geology and gold deposits of the Rice Lake–Wanipigow River area, Manitoba; unpubl. Ph.D. thesis, Univ. Toronto, Toronto.
1964: Mineral deposits related to major structures in the Precambrian of Manitoba; Can. Inst. Min. Metall., Bull., v. 57, no. 623, p. 313 (abstr.).
1977: Structural interpretation of the Timmins mining area, Ontario; Can. J. Earth Sci., v. 14, p. 1046–1053.

Davies, J.L., Tupper, W.M., Bachinski, D., Boyle, R.W. and Martin, R.
1969: Geology and mineral deposits of the Nigadoo River–Millstream River area, Gloucester County, New Brunswick; Geol. Surv. Can., Pap. 67–49, 70 p.

Davies, O.
1935: Roman mines in Europe; Clarendon Press, Oxford.

Davletov, I.K.
1970: Estimation of the average gold content in the main rock forming minerals of intrusive rocks; Akad. Nauk SSSR, Dokl., p. 1448–1450.
1974: Distribution of gold in rocks of an intrusive complex in the Chatkal zone of the Tien-Shan; Geokhim., no. 4, p. 635–638. *Also* Geochem. Int., v. 11, p. 456–458. (Chem. Abstr., v. 81, 124280h.)

Davletov, I.K. and Dzhakshibayev, Sh.
1970: The mineral balance and the behavior of gold during the emplacement of an intrusive body; Geochem. Int., v. 7, no. 6, p. 997–1005.

Davletov, I.K., Dzhenchruaeva, R.D. and Satykeev, D.S.
1972: Types of skarns and their gold content in the Chatkal ore region; Zap. Kirg. Otd. Vses. Mineral. Obshcest., no. 8, p. 27–33. (Chem. Abstr., v. 79, 94773k.)

Davletov, I.K. and Popkov, N.N.
1970: Major characteristics of gold metallogeny in the Kirghiz S.S.R.; *in* Osn. Probl. Metallogen. Tyan-Shanya, Tr. Vses. Metallogen. Soveshch. 5th 1968, M.M. Adyshev, ed., p. 417–424. (Chem. Abstr., v. 77, 22960e.)

Davletov, I.K., Shchelkov, E.E. and Solov'eva, V.V.
1973: Results of an experimental study of the form of a gold find in intrusive rocks; Akad. Nauk SSSR, Dokl., v. 212, no. 1, p. 202–204. (Chem. Abstr., v. 80, 17475h.)

Davydov, A.S. and Goroshko, G.G.
1970: Comparative characteristics of gold from placer deposits situated near the Pyrkanaisk granitic massif, western Chukota; Vop. Geol., Geokhim. Metallogen. Sev.-Zap. Sekt. Tikhookean. Poyasa, Mater. Nauch. Sess. 1969, I.N. Govorov, ed., Dal'nevost. Geol. Inst., Vladivostok, U.S.S.R., p. 271–275. (Chem. Abstr., v. 75, 89986v.)

Day, F.H.
1963: The chemical elements in nature; Reinhold Publ. Corp., New York, 372 p.

De Batz, R.
1898: Les gisements aurifères de Sibérie; Chamerot et Renouard, Paris, 176 p.
1899: The auriferous deposits of Siberia; Am. Inst. Min. Eng., Trans., v. 28, p. 452–467.

De Grazia, A.R. and Haskin, L.
1964: On the gold content of rocks; Geochim. Cosmochim. Acta, v. 28, p. 559–564.

De Jongh, A.C.
1930: Gold in the Dutch East Indies; *in* Gold resources of the world; 15th Int. Geol. Cong., Pretoria, p. 107–122.

De Kock, W.P.
1940: The Ventersdorp contact reef; Geol. Soc. S. Afr., Trans., v. 43, p. 85–108.
1964: The geology and economic significance of the West Wits line; *in* The geology of some ore deposits in southern Africa, v. 1, S.H. Haughton, ed., Geol. Soc. S. Afr., Johannesburg, p. 323–386.

De Kun, N.
1965: The mineral resources of Africa; Elsevier, Amsterdam, 740 p.

De Launay, L.
1913: Gîtes minéraux et métallifères; tomes 1, 2 et 3, C.H. Béranger, Editeur, Paris et Liège.

De Oliveira, E.P.
1930: Gold in Brazil; *in* Gold resources of the world; 15th Int. Geol. Congr., S. Afr., p. 57–61.
1932: Genesis of the deposits of auriferous jacutinga; Econ. Geol., v. 27, p. 744–749.

De Roever, W.P. and Lodder, W.
1967: Indications of syngenetic origin of gold ore and ignimbrites near Rodalquilar (S.E. Spain); Bull. Volcanol., t. 30, p. 35–40.

De Villiers, J.E.
1957: The mineralogy of the Barberton gold deposits; S. Afr. Dep. Mines, Bull., 24, 60 p.

Degens, E.T. and Ross, D.A., eds.
1969: Hot brines and recent heavy metal deposits in the Red Sea: a geochemical and geophysical account; Springer-Verlag, New York, 600 p.

Dekate. Y.G.
1971: Absorption of gold by *Sorghum saccharatum Pers*; J. Indian Geosci, Assoc., v. 13, p. 75–78.

Delitsyn, L.M.
1974: Experimental data on solubility, transport, and deposition of gold; Tr. Tsent. Nauch. Issled. Geologorazved. Inst. Tsvet, v. 112, p. 94–106. (Chem. Abstr., v. 83, 45803u.)

Del Mar, A.
1880: A history of the precious metals; Bell and Sons, Ltd., London, 373 p.

Demina, N.N.
1969: Gold and titanomagnetite of Kamchatka marine placers and their extraction; Tr. Tsent. Nauch. Issled. Gornorazved. Inst., no. 82, p. 128–130. (Chem. Abstr., v. 73, 37659e.)

Denholm, L.S.
1967: Lode structures and ore shoots at Vatukoula, Fiji; Proc. Australas. Inst. Min. Metall., no. 222, p. 73–83.

Denisov, S.V.
1973: The source-province of the marine near-shore gold placer deposits; Akad. Nauk. SSSR, Izv., Ser. Geol., no. 8, p. 73–78.

Denisov, S.V., Koshman, P.N. and Yugay, T.A.
1966: Trace elements in gold from some districts of the Amur region; Geochem. Int., v. 3, p. 904–908.

Dennen, W.H. and Puckett, A.M.
1972: On the chemistry and color of amethyst; Can. Mineral., v. 11, p. 448–456.

Derby, O.A.
1884: Peculiar modes of occurrence of gold in Brazil; Am. J. Sci. Ser. 3, v. 28, p. 440–447.
1903: Notes on Brazilian gold ores; Am. Inst. Min. Eng., Trans., v. 33, p. 282–287.
1911: On the mineralization of the gold-bearing lode of Passagem, Minas Geraes, Brazil; Am. J. Sci., Ser. 4, v. 32, p. 185–190.

Derry, D.R.
1951: A 3000 year old gold mine; Can. Min. J., v. 72 (pt. 2), no. 8, p. 68–72.
1961: Economic aspects of Archaean-Proterozoic boundaries; Econ. Geol., v. 56, p. 635–647.

Derry, D.R. and Herz, H.C.
1948: Canadian Malartic mine; *in* Structural geology of Canadian ore deposits, v. 1, Can. Inst. Min. Metall., Montreal, p. 853–857.

Derry, D.R., Hopper, C.H. and McGowan, H.S.
1948: Matachewan Consolidated mine; *in* Structural geology of Canadian ore deposits, v. 1, Can. Inst. Min. Metall., Montreal, p. 638–643.

Desborough, G.A.
1970: Silver depletion indicated by microanalysis of gold from placer occurrences, western United States; Econ. Geol., v. 65, p. 304–311.

Desborough, G.A., Heidel, R.H., Raymond, W.H. and Tripp, J.
1971: Primary distribution of silver and copper in native gold from six deposits in the western United States; Miner. Deposita (Berl.), v. 6, p. 321–334.

Desborough, G.A., Raymond, W.H. and Iagmin, P.J.
1970: Distribution of silver and copper in placer gold derived from the northeastern part of the Colorado mineral belt; Econ. Geol., v. 65, p. 937–944.

Deul, M.
1958: Biochemical and geochemical origins of ash-forming ingredients in coal; Am. Chem. Soc. Meet., Chicago, Div. Gas Fuel Chem., prepr. p. 169–182.

Devereux, W.B.
1882: The occurrence of gold in the Potsdam Formation, Black Hills, Dakota; Am. Inst. Min. Eng., Trans., v. 10, p. 465–475.

Dias, De, A.
1969: Gondomar antimony-gold mine; Port., Republ. Min. Econ. Dir. General Minas Serv. Geol., v. 19, fascs. 1–2, p. 91–170.

Dickson, F.W., Radtke, A.S., Weissberg, B.G. and Heropoulos, C.
1975: Solid solutions of antimony, arsenic, and gold in stibnite (Sb_2S_3), orpiment (As_2S_3), and realgar (As_2S_2); Econ. Geol., v. 70, p. 591–594.

Diller, J.S.
1914: Mineral resources of south-western Oregon; U.S. Geol. Surv., Bull. 546, 147 p.

Dillon, V.S.
1955: Assay practice on the Witwatersrand; Transvaal and Orange Free State Chamber of Mines, Cape Times Ltd., 603 p.

Diman, E.N.
1976: On minimum purity of gold in association with pyrite; Geokhim., no. 3, p. 404–412.

Dimitrov, R., Kuikin, S. and Andreev, M.
1973: Geochemical methods used in prospecting for nonferrous and noble metals in the Bulgarian People's Republic; Sostoyanie Puti Dal'neishego Soversh. Metodov Poiskov Skrytykh Rudn. Tel Svintsovo-Tsinkovykh, Zoloto-Serebryanykh Mednykh, Olovyannykh Vol'framouykh Mestorozhd., Dokl. Semin., 1971, M.B. Borodaevskaya, ed., p. 149–163. (Chem. Abstr., v. 82, 142823n.)

Distanov, E.G., Stebleva, A.T. and Obolenskii, A.A.
1975: Genesis of the Uderei gold-antimony deposit in the Enisei Ridge area; Geol. Geofiz., no. 8, p. 19–27. (Chem. Abstr., v. 84, 47318j.)

Ditmar, G.V.
1966: Sampling gold of various ages in the northern part of the Stanovoy Range; Geol. Rud. Mestorozhd., v. 8, p. 92–96. *Also* Econ. Geol., v. 64, p. 359.

Dobretsova, I.L. and Melnikova, R.D.
1977: Gold from ophiolite and glaucophane schist rock; Geol. Geophys. no. 1, p. 143–150 (in Russian).

Dodin, A.L.
1941: Gold content of the Usin district of the Kuznetsk Alatau; Sov. Geol., no. 3, p. 108–109. (Chem. Abstr., v. 36, 6455, 7.)

Dodin, D.A., Solomina, L.M. and Shatkov, V.A.
1973: Some aspects of the geochemistry of platinum minerals and gold, and their application in prospecting for the Norilsk-type copper nickel ore deposits; *in* The North Siberian nickel-bearing region and its economic outlook, N.N. Urvantsev *et al.*, eds., Ministry Geol. USSR, Leningrad, p. 108–115.

Dodwell, C.R.
1971: Gold metallurgy in the twelfth century; Gold Bull., Chamber Mines S. Afr., v. 4, no. 3, p. 51–55.

Doelter, C. *et al.*
1911–1931: Handbuch der Mineralchemie; Verlag von Theodor Steinkopff, Dresden and Leipzig, 4 vols.

Dolmage, V. and Brown, C.E.G.
1945: Contact metamorphism at Nickel Plate Mountain, Hedley, B.C.; Can. Inst. Min. Metall., Trans., v. 48, p. 27–85.

Dolzhenko, B.H.
1973: A new type of gold mineralization in the Tien Shan; Akad. Nauk SSSR, Dokl., v. 210, no. 5, p. 1191–1192.

Dolzhenko, V.N.
1975: On manifestation of gold mineralization in serpentinites; Acad. Sci. USSR, Dokl., v. 224, p. 1174–1176.

Domaszewska, T.
1964: Occurrence and exploitation of gold in Lower Silesia; Przegl. Geol., v. 12, no. 4, p. 180–185.

Dominian, L.
1912: History and geology of ancient gold-fields in Turkey; Am. Inst. Min. Eng., Trans., v. 42, p. 569–589.

Don, J.R.
1898: The genesis of certain auriferous lodes; Am. Inst. Min. Eng., Trans., v. 27, p. 564–668.

d'Orey, F.L.C.
1975: Contribution of termite mounds to locating hidden copper deposits; Trans. Inst. Min. Metall., v. 84, p. B150–B151.

Dorn, P.
1937: Pflanzen als Anzeichen für Erzlagerstätten; Der Biol. (Munich), v. 6, p. 11–13.

Dorr, J.V.N.
1969: Physiographic, stratigraphic, and structural development of the Quadrilátero Ferrifero, Minas Gerais, Brazil; U.S. Geol. Surv., Prof. Pap. 641-A, 110 p.
1971: Review: U.S.G.S. Prof. Pap. 641-A on the Quadrilátero Ferrifero, Minas Gerais, Brazil; Econ. Geol., v. 66, p. 351–352.

Dorr, J.V.N. and Miranda Barbosa, A.L.
1963: Geology and ore deposits of the Itabira district, Minas Gerais, Brazil; U.S. Geol. Surv., Prof. Pap. 341-C, 110 p.

Dougherty, E.Y.
1935: Geologic problems of the Canadian pre-Cambrian gold fields; Econ. Geol., v. 30, p. 879–889.
1936: Essential criteria in examining gold-quartz mines; Eng. Min. J., v. 137, p. 44–45.

Douglas, G.V.
1948: Structure of the gold veins of Nova Scotia; *in* Structural geology of Canadian ore deposits, v. 1, Can. Inst. Min. Metall., Montreal, p. 919–926.

Douglas, R.J.W., ed.
1970: Geology and economic minerals of Canada; Geol. Surv. Can., Econ. Geol. Rep. no. 1 (5th ed.), 838 p.

Dowie, D.L.
1964– The mode and origin of banket orebodies, discussion;
1965: Inst. Min. Metall., Trans., v. 74, p. 500–503.

Dresser, J.A.
1908: Report on a recent discovery of gold near Lake Megantic, Quebec; Can., Dep. Mines, Geol. Surv. Br. Publ., v. 2, no. 1028, 13 p.

Dresser, J.A. and Denis, T.C.
1949: Geology of Quebec, v. 3, Economic geology; Que. Dep. Mines, Geol. Rep. 20, 562 p.

Drewes, H.
1967: A geochemical anomaly of base metals and silver in the southern Santa Rita mountains, Santa Cruz county, Arizona; U.S. Geol. Surv., Prof. Pap. 575-D, p. 176–182.

Drysdale, C.W.
1915: Geology and ore deposits of Rossland; Geol. Surv. Can., Mem. 77, 317 p.

Du Bois, G.C.
1903: Beitrag zur Kenntnis der surinamischen Laterit — und Schutzrinden—bildungen; Tschermak's Mineral. Petrog. Mitt., v. 22, p. 1–61.

Du Toit, A.L.
1940: Developments on and around the Witwatersrand; Econ. Geol., v. 35, p. 98–108.

Dudey, N.D., Ross, L.E. and Noshkin, V.E.
1969: Application of activation analysis and Ge(Li) detection techniques for the determination of stable elements in marine aerosols; U.S. Natl. Bur. Stand., Spec. Publ. 312, pt. 1, p. 55–61.

Duffell, S.
1937: Diffusion and its relation to ore deposition; Econ. Geol., v. 32, p. 494–510.

Dugas, J.
1966: The relationship of mineralization to Precambrian stratigraphy in the Rouyn-Noranda area, Quebec; Geol. Assoc. Can., Spec. Pap. 3, p. 43–55.

Dunbar, W.R.
1948: Structural relations of the Porcupine ore deposits; *in* Structural geology of Canadian ore deposits, v. 1, Can. Inst. Min. Metall., Montreal, p. 442–456.

Dumble, E.T.
1913: The occurrence of gold in the Eocene deposits of Texas; Am. Inst. Min. Eng., Trans., v. 44, p. 588–591.

Dunn, E.J.
1905: The Mount Morgan gold mine, Queensland; R. Soc. Victoria, Proc. (New Series), v. 17, pt. 2, p. 341–355.
1929: Geology of gold; Charles Griffin & Co. Ltd., London, 303 p.

Dunn, P.G.
1976: Zonation of hydrothermal minerals at the Juno mine, Tennant Creek goldfield, central Australia — a discussion; Econ. Geol., v. 71, p. 1614.

Duong, P.K.
1965: Enquête sur l'or dans les roches, origins de l'or des gisements; Chronique des mines et de la recherche minière, 33 année, no. 343, p. 175–188.

Dvornikov, A.G.
1964: Types of mercury dispersion aureoles in the southeastern part of the Donets Basin; Akad. Nauk SSSR, Dokl., v. 154, no. 5, p. 1110–1112 (Acad. Sci. USSR, Dokl., Earth Sci. Sec., v. 154, nos. 1–6, p. 158–160).

Dvornikov, A.G., Ovsyannikov, L.B. and Sidenko, O.G.
1973: Biogeochemical dispersion haloes of chalcophile elements in gold ore deposits of the Nagol'nyi Ridge (Donets Basin); Dopov. Akad. Nauk Ukr. RSR, Ser. B., v. 35, no. 6, p. 490–494. (Chem. Abstr., v. 79, 145290k.)

Dvornikov, A.G., Tkach, B.I., Shtanchenko, M.S. and Antonov, V.M.
1964: Minerals of the native element group in loose rocks and soils of the Nagol'nyi Ridge; Dopov. Akad. Nauk Ukr. RSR, no. 9, p. 1226–1229. (Chem. Abstr., v. 62, 1449g.)

Dvornikov, A.G., Vasilevskaya, A.E. and Shcherbakov, V.P.
1963: Mercury dispersion aureoles in the soils of the Nagol'nyi Range; Geochemistry, no. 5, p. 501–509.

Dyachenko, A.P., Legendza, V. Ya., Shelekhov, A.N. and Radionova, Ye. S.
1970: Auriferous ore occurrences in Precambrian rocks of the Stary Oskol region of the Kursk magnetic anomaly; Geol. Rud. Mestorozhd., no. 3, p. 92–96. *Also* Akad. Nauk SSSR, Dokl., v. 193, no. 2, p. 415–417.

Dyck, W.
1969: Development of uranium exploration methods using radon; Geol. Surv. Can., Pap. 69–46, 26 p.
1971: The adsorption and coprecipitation of silver in hydrous oxides of iron and manganese; Geol. Surv. Can., Pap. 70–64, 23 p.

Dyer, W.S.
1936: Geology and ore deposits of the Matachewan-Kenogami area; Ont. Dep. Mines, v. 44, pt. 2, p. 1–55.

Dzhantuganov, N.I.
1973: Geological-petrologic characteristics of the Sultansary gold ore deposit (northern Tien-Shan); Uzb. Geol. Zh., v. 17, no 2, p. 56–59. (Chem. Abstr., v. 80, 39277x.)
1974: Morphology of the native gold from several types of rocks and ores of the regions of northern Tien Shan; Uzb. Geol. Zh, v. 18, no. 3, p. 35–38.

Dzhukebaev, I.K. and Marichev, K.I.
1975: Endogenic gold and rare metal mineralization and magmatism of the southern part of the Sarysu-Teniz uplift; Akad. Nauk Kaz. SSR, Izv., Ser. Geol., v. 32, no. 2, p. 24–31. (Chem. Abstr., v. 83, 135090p.)

Eakins, G.R.
1968: A geochemical investigation of the Wood River–Tikchik Lakes area, southwestern Alaska; Alaska, Dep. Nat. Resour., Div. Mines, Miner., Geochem. Rep., no. 17, p. 1–31.

Eales, H.V.
1961: Fineness of gold in some southern Rhodesian gold mines; Inst. Min. Metall., Trans., v. 71, no. 660, p. 49–73.
1962: The occurrence of aurostibite in the Gwanda District, southern Rhodesia; Geol. Soc. S. Afr., Trans., v. 65, pt. 2, p. 79–83.
1968: Determining fineness variation characteristics in gold ores by reflectometry; Econ. Geol., v. 63, p. 688–691. Discussion by E.F. Stumpfl, Econ. Geol., v. 64, p. 341–342.

Eales, H.V. and Viljoen, E.A.
1973: Determination of silver content of natural gold alloys by reflected-light microscopy; Inst. Min. Metall., Trans., Sec. B, v. 82, p. 47–50.

Earl, T.C.
1915: The testing of alluvials; privately printed, 86 p.

Ebbutt, F.
1948: Relationships of minor structures to gold deposition in Canada; *in* Structural geology of Canadian ore deposits; v. 1, Can. Inst. Min. Metall., Montreal, p. 64–77.

Echegoyén, S.J., Romero, M.S. and Velázquez, S.S.
1972: Geology and mineral deposits of the central part of the Guanajuato mining district, Mexico; Mex., Cons. Recursos Natur. Renov., Bol. no. 75. (Chem. Abstr., v. 80, 50549h.)

Eddingfield, F.T.
1913*a*: Alteration and enrichment in calcite quartz-manganese gold deposits in the Philippine Islands; Philipp. J. Sci., v. 8, p. 125–134.
1913*b*: Gogo, *Entada Scandens* Bentham, and its effect on gold and gold solutions; Philipp. J. Sci., v. 8, p. 135–139.

Edwards, A.B.
1953: Geology of Australian ore deposits, v. 1, 5th Emp. Min. Metall., Congr., Australia and New Zealand, 1290 p.
1958: The mineral composition of the Maude and Yellow Girl gold ore, Glen Wills, Victoria; F.L. Stillwell Anniv. Vol., Melbourne, Australas. Inst. Min. Metall., p. 105–132.

Efendiev, G.Kh. and Babaeva, E.E.
1970: Distribution of silver, bismuth, and gold in the sulphide minerals of complex ore deposits of the Lesser Caucasus (Azerbaidzhan S.S.R.); Issled. Obl. Neorg. Fiz. Khim, p. 18–26, I.A. Orudzheva, ed., "Elm", Baku, Azerb. SSR. (Chem. Abstr., v. 76, 74702w.)

Egleston, T.
1881: The formation of gold nuggets and placer deposits; Am. Inst. Min. Eng., Trans., v. 9, p. 633–646.

Ehmann, W.D.
1971: Gold, *in* Handbook of elemental abundances in meteorites, B. Mason, ed.; Gordon and Breach Science Publishers, New York, p. 479–485.

Ehmann, W.D. and Baedecker, P.A.
1968: The distribution of gold and iridium in meteoric and terrestrial materials; *in* Origin and distribution of the elements, L.H. Ahrens, ed., Pergamon Press, Oxford, p. 301–311.

Ehmann, W.D., Baedecker, P.A. and McKown, D.M.
1970: Gold and iridium in meteorites and some selected rocks; Geochim. Cosmochim. Acta, v. 34, p. 493–507.

Ehmann, W.D. and Gillum, D.E.
1972: Platinum and gold in chondritic meteorites; Chem. Geol., v. 9, no. 1, p. 1–11.

El Boushi, I.M.
1972: Geology of the Gebeit gold mine, Democratic Republic of the Sudan; Econ. Geol., v. 67, p. 481–486.

El Shazly, E.M.
1957: Classification of Egyptian mineral deposits; Egypt. J. Geol., v. 1, no. 1, p. 1–20.

Eleyeva, I.V. and Rusinova, O.V.
1969: Genesis and prospecting importance of quartz-kaolinite formations of upper horizons of the Balei deposit; Geol. Rud. Mestorozhd., v. 11, no. 4, p. 88–95.

Ellert, R.
1974: Gold mineralization associated to volcanism in the Quadrilatero ferrifero ("Iron Quadrangle"), Minas Gerais, Brazil, Abstracts of Papers, Int. Assoc. Genesis Ore Deposits (IAGOD), Varna, Bulgaria, p. 441–443.

Elley, H.W.
1915: Precipitants of gold and silver (discussion); Econ. Geol., v. 10, no 6, p. 580–582.

Elliott, J.E. and Wells, J.D.
1968: Anomalous concentrations of gold, silver, and other metals in the Mill Canyon area, Cortez Quadrangle, Eureka and Lander counties, Nevada; U.S. Geol. Surv., Circ. 606, 20 p.

Ellis, A.J.
1967: The chemistry of some explored geochemical systems; *in* Geochemistry of hydrothermal ore deposits, H.L. Barnes, ed., Holt, Rinehart and Winston, Inc., New York, p. 465–514.

Emel'yanov, E.L. and Perfil'ev, V.V.
1973: New types of endogenous rare metal and gold mineralization in the Paleozoic activation zone of a region in eastern Siberia; Metallog. Obl. Tekt. Magmat. Akt., Tezisy Dokl. Vses. Metallog. Soveshch., 7th, M.M. Odintsov, ed., Akad. Nauk SSSR, Sib. Otd., Inst., Zemnoi Kory, Irkutsk, U.S.S.R., p. 177–178. (Chem. Abstr., v. 82, 158713n.)

Emery, J.F. and Leddicotte, G.W.
1961: The radiochemistry of gold; Natl. Acad. Sci., Natl. Res. Counc. Nucl. Sci. Ser. NAS-NS 3036, 34 p.

Emmons, S.F.
1886: Geology and mining industry of Leadville, Colo.: U.S. Geol. Surv., Monogr. 12, v. 2, p. 585–608.
1901: The secondary enrichment of ore deposits; Am. Inst. Min. Eng., Trans., v. 30, p. 177–217.
1903: Platinum in copper ores in Wyoming; *in* Contributions to economic geology, 1902, U.S. Geol. Surv., Bull. 213, p. 94–97.

Emmons, W.H.
1917: The enrichment of ore deposits; U.S. Geol. Surv., Bull. 625, 530 p.
1924: Primary downward changes in ore deposits; Am. Inst. Min. Eng., Trans., v. 70, p. 964–997.
1932: Prospecting in the shield area; Eng. Min. J. v. 133, p. 139–140.
1937: Gold deposits of the world; McGraw-Hill Book Company Inc., New York, 562 p.
1940: The principles of economic geology, McGraw-Hill Book Company Ltd., New York, 529 p.

Emmons, W.H. and Calkins, F.C.
1913: Geology and ore deposits of the Philipsburg quadrangle, Montana; U.S. Geol. Surv., Prof. Pap. 78, 271 p.

Emmons, W.H. and Larsen, E.S.
1923: Geology and ore deposits of the Creede district, Colorado; U.S. Geol. Surv., Bull. 718, 198 p.

Enikeev, A.M.
1972: Geological characteristics of the Daugyztau deposit; Nauch, Tr., Tashkent, Gos. Univ., no. 429, p. 331–334. (Chem. Abstr., v. 80, 29214k.)

Entin, A.R.
1975: Presence of gold in Archean iron ore deposits of the central part of the Aldan Shield; Akad. Nauk SSSR, Dokl., v. 223, no 3, p. 722–725. (Chem. Abstr., v. 84, 20401u.)

Epshtein, Yu. A.
1969: Gneissic dome in southern Primor'ye and its effect on the distribution pattern of gold; Sov. Geol. no. 1, p. 116–118, 1968. *Also* Int. Geol. Rev., v. 11, no. 1, p. 135–137.
1973: Senonian magmatism of the southern maritime territory and gold mineralization related to it; Vop. Magmat., Metamorf. Orugeneniya Dal'nego Vostoka, Dal'nevost. Petrogr. Reg. Soveshch. (Mater.), 2nd 1972 S.S. Zimin, ed., Akad. Nauk SSSR, Dal'nevost. Nauch. Tsent., Vladivostok, U.S.S.R., p. 232–234. (Chem. Abstr., v. 81, 155902s.)

Erämetsä, O.
1954: The distribution and transport of gold in the sulfate pulping process and mineral synthesis which occur in the sulfate process smelter furnace; Suom. Kemistilehti 27A, p. 204–207. (Chem. Abstr., v. 51, 14264i.)

Eremeev, A.N., Sokolov, V.A., Solovov, A.P. and Yanitskii, I.N.
1973: Application of helium surveying to structural mapping and ore deposit forecasting; *in* Geochemical exploration, 1972, M.J. Jones, ed., Inst. Min. Metall. (London), p. 183–192.

Eremin, R.A.
1970: Relation of adularization to gold-silver mineralization in the Finish ore occurrence, Okhotsk-Chukotka volcanic belt; Geol. Geofiz., no. 2, p. 34–40. (Chem. Abstr., v. 73, 27538a.)

Erenburg, A.M. and Peshchevitskiy, B.I.
1969: On the stability of iodide complexes of gold (I) in aqueous solutions; Zh. Neorg. Khim., v. 14, no. 3, p. 932–935.

Erikson, J.E.
1957: Geochemical prospecting abstracts, July 1952–December 1954; U.S. Geol. Surv., Bull. 1000-G, p. 357–395.

Erickson, R.L. and March, S.P.
1971: Geochemical, aeromagnetic, and generalized geologic maps showing distribution and abundance of mercury, arsenic, antimony, tungsten, gold, copper, lead and silver, Golconda and Iron Point quadrangles, Humboldt County, Nevada; U.S. Geol. Surv., Misc. Field Studies Maps MF-312, MF-313, MF-314 and MF-315.
1973: Geochemical, aeromagnetic and generalized geologic maps showing distribution and abundance of lead, silver, gold, copper, mercury, arsenic, antimony, tungsten, molybdenum, and zinc, Goldrun Creek quadrangle, Humboldt County, Nevada; U.S. Geol. Surv., Misc. Field Studies Maps MF-506, MF-507 and MF-508.
1974: Geochemical, aeromagnetic, and generalized geologic maps showing distribution and abundance of mercury, arsenic, antimony, tungsten, gold, copper, molybdenum, lead, bismuth, and silver, Brooks Spring quadrangle, Humboldt County, Nevada; U.S. Geol. Surv., Misc. Field Studies Maps MF-563, MF-564, MF-565, MF-566 and MF-567.

Erickson, R.L., Marranzino, A.P., Oda, U. and Janes, W.W.
1964: Geochemical exploration near the Getchell mine, Humboldt County, Nevada; U.S. Geol. Surv., Bull. 1198-A, 26 p.

Erickson, R.L., Van Sickle, G.H., Nakagawa, H.M., McCarthy, J.H. and Leong, K.W.
1966: Gold geochemical anomaly in the Cortez district, Nevada; U.S. Geol. Surv., Circ. 534, 9 p.

Ershov, A.D.
1974: On the relation of the deposits belonging to nickel-cobalt-gold (silver)-bismuth-uranium (five metal) formation to tectonic activization; Geol. Rud. Mestorozhd., no. 5, p. 86–89.

Ershov, V.M. and Shcheglova, A.I.
1958: Germanium in mine waters of the Kizelov coal basin; Geochemistry, no 4, p. 490–492.

Ershov, Sh. E., ed.
1971: The gold fields of North Kazakhstan; "Nauka" Publ. House, Acad. Sci. Kazakh SSR, Alma Ata, 167 p.

Ertl, R.F., Niedermayr, G. and Seemann, R.
1975: Tauerngold; Veroeff. Naturhistorischen Mus., Neue Folge 10, Verlag Naturhistorisches Mus., Wien, Erscheinungsort Wien, 31 p.

Esenov, Sh.E. *et al.*
1968: Main characteristics of the Kazakhstan endogenic metallogeny; Metallogen. Tyan-Shanya, p. 224–225. (Chem. Abstr., v. 71, 62970r.)
1970: Main characteristics of Kazakhstan metallogeny; Zakonomer. Razmeshcheniya Polez. Iskop., no. 9, p. 145–161. (Chem. Abstr., v. 74, 144509y.)

Eshkin, V.Yu.
1965: Nugget gold in rock crystal-bearing quartz vein from the Circumpolar Urals; Zap. Vses. Mineral. Obshchest., v. 94, no. 2, p. 203–204. (Chem. Abstr., v. 63, 2751c.)

Evangulov, B.B., Zysin, A.M., Arskii, Yu.M. and Faizullin, R.M.
1971: Variability of metal content and the optimum number of prospecting network points in one of the gold deposits of Yakut, A.S.S.R.; Zap. Leningrad. Gorn. Inst., v. 60, no. 2, p. 102–111. (Chem. Abstr., v. 76, 61737m.)

Evans, T.G.
1975: Gold in Texas; Tex., Bur. Econ. Geol., Miner. Res. Circ. 56, 36 p.

Evseev, Yu.P., Grabeklis, R.V., Zinov'eva, G.N. and Polokhov, V.P.
1973: Major ore formations of the Kara gold-bearing region; their relation to magmatism and prospecting features; Geol. Nek. Rudn. Raionov Polei Zabaik, F.I. Vol'fson, ed., p. 107–114. (Chem. Abstr., v. 84, 62601q.)

Ewers, G.R. and Keays, R.R.
1977: Volatile and precious metal zoning in the Broadlands geothermal field, New Zealand; Econ. Geol., v. 72, p. 1337–1354.

Ewers, W.E.
1969: Hydrothermal transport and deposition of gold; Econ. Geol., v. 64, p. 575.

Fahlgren, J.E.J.
1975: An analysis of gold mining in the context of present economic conditions; Can. Inst. Min. Metall., Bull., v. 68, p. 113–115.

Fainshtein, G. Kh.
1974: Weathering crusts and their role in the alluvial ore formation in the mantle of the Siberian platform; Rudonos. Kory Vyvetrivaniya, V.I. Smirnov, ed., "Nauka" Moscow, U.S.S.R., p. 271–277. (Chem. Abstr., v. 82, 75593x.)

Fairbairn, H.W. *et al.*
1951: A cooperative investigation of precision and accuracy in chemical, spectrochemical, and modal analysis of silicate rocks; U.S. Geol. Surv., Bull. 980, 71 p.

Fakhry, A.A.
1965: Zoning of primary dispersion haloes over a gold deposit and the migration capacity of the elements; Vest. Mosk. Univ., Ser. IV, Geol., v. 20, no. 5, p. 31–43. (Chem. Abstr., v. 64, 10938.)
1974: Geochemical studies at the Fawakhir gold mine, Eastern Desert, Egypt, 2. Infrared absorption spectroscopy as a means of characterising the primary dispersion haloes of gold deposits; J. Geochem. Explor., v. 3, p. 265–275.

Faribault, E.R.
1926: The gold measures of Nova Scotia and deep mining; N.S. Dep. Pub. Works Mines, Halifax, N.S., Repr. 15 p.

Farmin, R.
1938: Dislocated inclusions in gold-quartz veins at Grass Valley, California; Econ. Geol., v. 33, p. 579–599.
1941: Host-rock inflation by veins and dikes at Grass Valley, California; Econ. Geol., v. 36, p. 143–174.

Fawley, A.P.
1947: An electrum-ruby silver deposit at East gold mine, B.C.; Can. Inst. Min. Metall., Bull. no. 426, p. 460–470.

Fayzullin, R.M.
1967: Lithological-facies characteristics and the gold content of a series of conglomerates of the Kamensk deposit; Vop. Rudonos. Vost. Zabaikal., F.I. Vol'fson, ed., Izd. "Nedra"; Moscow, U.S.S.R., p. 126–140. (Chem. Abstr., v. 70, 108202b.)
1968: Problem of intermediate collectors of gold; Int. Geol. Rev., v. 10, no 12, p. 1383–1385.

Fayzullin, R.M. and Tolmacheva, T. Ya.
1971: Nature of native gold from the placers of the Chikoi-Menzinsk interfluve (western Transbaikalia); Vop. Reg. Geol. Metallogen. Zabaikal., no 6, p. 39–48. (Chem. Abstr., v. 77, 154966w.)

Fayzullin, R.M. and Turchinova, D.M.
1974: On relationships between gold placers and primary sources or intermediate collectors; Akad. Sci. USSR, Dokl., Earth Sci. Sec., v. 212, p. 242–243 (Am. Geol. Inst. Transl.).

Fazal-Ur-Rehman
1972: The determination of gold by radiochemical and non-destructive neutron activation analyses in some sulphide ore samples from Pakistan; Punjab. Univ., Geol. Bull., no. 9, p. 37-42.

Fazal-Ur-Rehman, Brunfelt, A.O. and Finstad, K.G.
1974: Gold, silver, and mercury in material from some Norwegian sulphide mines; Norg. Geol. Unders., no. 311, p. 17–24.

Feather, C.E. and Koen, G.M.
1975: The mineralogy of the Witwatersrand reefs; Min. Sci. Eng., v. 7, no 3, p. 189–224.

Fedorenko, Ya.D.
1962: Geochemical methods for prospecting for gold deposits in the Transbaikal region; Byull. Nauch. Tekh. Inform. Min. Geol. Okhr. Nedr SSSR, no. 1, p. 41–42. (Chem. Abstr., v. 60, 11760h.)

Fedorov, B.M.
1947: Mesozoic placers of the Uchalin region in the southern Urals; Razved. Okhr. Nedr., v. 13, no. 1, p. 8–11. (Chem. Abstr., v. 41, 5824a.)

Feigl, F.
1958: Spot tests in inorganic analysis, 5th ed.; Elsevier, Amsterdam, 600 p.

Fel'dman, A.A.
1966: Geophysical and geochemical research in prospecting for gold ore deposits of the Kuranach type; Min. Geol. SSSR, Moscow, 52 p. (Chem. Abstr., v. 67, 92929u.)

Fel'dman, A.A., Slepnev, P.V. and Kul'kov, B.N.
1975: Use of gamma spectrometry in prospecting for gold ore deposits near the surface; Razved. Okhr. Nedr, no. 10, p. 59–60. (Chem. Abstr., v. 84, 138431g.)

Feldtmann, F.R.
1916: Contributions to the study of the geology and ore deposits of Kalgoorlie, Geol. Surv. W. Aust., Bull., no. 69, pt. 3, 152 p.

Felippe, J.C. and Martins, T.
1926: Radioactivity and oligodynamic action of minerals; Compt. Rend. Soc. Biol. v. 95, p. 1488–1489. (Chem. Abstr., v. 22, 1928, p. 250–251.)

Fells, P.D.
1975: Gold in 1975; Consolidated Gold Fields Limited, London, 48 p.

Fells, P.D. and Glynn, C.
1976: Gold 1976; Consolidated Gold Fields Limited, London, 26 p. plus appendices.

Ferguson, H.G.
1913: Lode deposits of the Alleghany district, California; U.S. Geol. Surv., Bull. 580, p. 153–182.
1917: Placer deposits of the Manhattan district, Nevada; U.S. Geol. Surv., Bull. 640-J, p. 163–193.

Ferguson, H.G. and Gannett, R.W.
1929: Gold-quartz veins of the Alleghany District, California; Am. Inst. Min. Metall. Eng., Tech. Publ. no. 211, 40 p.

Ferguson, R.B.
1950: Red gold from the San Antonio gold mine, Bissett, Manitoba; Am. Mineral., v. 35, p. 459–460.

Ferguson, S.A.
1966: The relationship of mineralization to stratigraphy in the Porcupine and Red Lake areas, Ontario; Geol. Assoc. Can., Spec. Pap. 3, p. 99–119.

Ferguson, S.A., Groen, H.A. and Haynes, R.
1971: Gold deposits of Ontario, Part 1. Districts of Algoma, Cochrane, Kenora, Rainy River, and Thunder Bay; Ont. Dep. Mines North. Affairs, Mineral. Res. Circ. no. 13, 315 p.

Fernandes, L.L.
1961: The gold and diamond deposits of British Guiana; Proceed. Inter-Guiana Geol. Conf., 5th, p. 273–283.

Fersman, A.E.
1934–1939: Geochemistry, vols. 1–4; Leningrad.
1939: Gold; *in* Geochemistry, v. 4, Leningrad, p. 262–271.

Fetzer, W.G.
1934: Transportation of gold by organic solutions; Econ. Geol., v. 29, p. 599–604.
1946: Humic acids and true organic acids as solvents of minerals; Econ. Geol., v. 41, p. 47–56.

Field, C.W.
1966: Sulfur isotopic method for discriminating between sulfates of hypogene and supergene origin; Econ. Geol., v. 61, p. 1428–1435.

Finch, J.W. *et al.*, ed.
1933: Ore deposits of the western States (Lindgren Vol.); Am. Inst. Min. Metall. Eng., New York, 797 p.

Finkelshtein, Yu.V.
1970: Concerning the relation between the mercury and the gold mineralization in Uzbekistan; Razved. Okhr. Nedr., no. 6, Tashkent, U.S.S.R., p. 7–9.

Finkelstein, N.P. and Hancock, R.D.
1974: A new approach to the chemistry of gold; Gold Bull. (Natl. Inst. Metall., Johannesburg, S. Afr.), v. 7, no. 3, p. 72–77.

Finlayson, A.M.
1909: Problems in the geology of the Hauraki gold fields, New Zealand; Econ. Geol., v. 4, p. 632–645.

Finucane, K.J.
1953: The Hill 50 gold mine, Boogardie; *in* Geology of Australian ore deposits, v. 1, 5th Emp. Min. Metall. Congr., Australia and New Zealand, p. 231–235.
1965: Ore distribution and lode structures in the Kalgoorlie goldfield, *in* Geology of Australian ore deposits, v. 1, 2nd ed., J. McAndrew, ed., 8th Commonw. Min. Metall. Congr., Australia and New Zealand, p. 80–86.

Finucane, K.J. and Jensen, H.E.
1953: Lode structures in the Kalgoorlie goldfield; *in* Geology of Australian ore deposits, v. I., A.B. Edwards, ed., 5th Emp. Min. Metall. Congr., Australia and New Zealand, p. 94–111.

Firsov, L.V.
1956: Conditions for the occurrence of albite in gold quartz veins; Zap. Vses. Mineral. Obshchest. v. 85, p. 228–232. (Chem. Abstr., v. 51, 2963c.)
1963: Gold in the Yana-Kolyma folded belt; Geol. Metallogeni. Sov. Sekt. Tikhookean. Rud. Poyasa, Akad. Nauk SSSR, Sibirsk. Otd., Dal'nevost. Filial, Dal'nevost. Geol. Inst., p. 290–304. (Chem. Abstr., v. 61, 4075e.)
1967: Gold-quartz ore deposits of the Yana-Kolyma belt; *in* Endogennyye rudnyye formatsii Sibiri i Dal'nego Vostoka (Engogene ore formations of Siberia and the Far East); Moscow, Nauka, p. 132–144. *Also* Int. Geol. Rev., v. 9, no. 12, p. 1544–1552.

Fischer, K.W.
1958–1959: Zur geochemie der edelmetalle; Wiss. Z, Hochsch. Arkitekt. Bauw. Weimar, v. 6, no. 2, p. 85–91.
1966: Edelmetalle in der Saale und in ihrem Einzugsgebiet; Geol. Jahr. 15, Heft 4/5, p. 550–561.

Fischer, R.P. and Fisher, F.S.
1968: Interpreting pan-concentrate analyses of stream sediments in geochemical exploration for gold; U.S. Geol. Surv., Circ. 592, 9 p.

Fisher, E.I. and Fisher, V.L.
1970: Prospecting for gold ore deposits in northeastern Transbaikalia based on their relation with secondary dispersion haloes and alluvial gold placers; Vop. Reg. Geol. Metallogen. Zabaikal., no. 5, p. 108–114. (Chem. Abstr., v. 78, 100429a.)

Fisher, E.I., Fisher, V.L. and Miller, A.D.
1974: Nature of the interaction of natural organic acids with gold; Sov. Geol., no. 7, p. 142–146. (Chem. Abstr., v. 82, 19513j.)

Fisher, J.R.
1976: Gold in the search for the Americas; Gold Bull., Chamber Mines S. Afr. v. 9, no. 2, p. 58–63.

Fisher, M.S.
1935: The origin and composition of alluvial gold, with special reference to the Morobe goldfield, New Guinea; Inst. Min. Metall., Bull. (London), no. 365, p. 1–46.

Fisher, N.H.
1945: The fineness of gold with special reference to the Morobe goldfield; New Guinea; Econ. Geol., v. 40, p. 449–495, 537–563.
1950: Application of gold fineness to the search for ore; Australas. Inst. Min. Metall., Proc., nos. 156–157, p. 185–190.

Fisher, N.H., ed.
1973: Metallogenic provinces and mineral deposits in the southwestern Pacific; Aust. Bur. Miner. Resour., Geol. Geophys., Bull. 141, 225 p.

Fitzgerald, A.C., Graham, R.J., Gross, W.H. and Rucklidge, J.C.
1967*a*: The application of gold-silver ratios to a fault displacement problem, Bourlamaque, Quebec; Can. Inst. Min. Metall. Prog., Ottawa, p. 11 (abstr.).
1967*b*: The application and significance of gold-silver ratios at Val D'Or, Quebec; Econ. Geol., v. 62, p. 502–516.

Flanagan, F.J.
1969: U.S. Geological Survey standards—II. First compilation of data for the new U.S.G.S. rocks; Geochim. Cosmochim. Acta, v. 33, no. 1, p. 81–120.
1973: 1972 values for international geochemical reference samples; Geochim. Cosmochim. Acta, v. 37, no. 5, p. 1189–1200.

Flawn, P.T.
1966: Mineral resources; Rand McNally & Co., Chicago, 406 p.

Fleischer, E.B. and Laszlo, A.
1969: Synthesis of a gold porphyrin; Inorg. Nucl. Chem. Lett. v. 5, no. 5, p. 373–376.

Fleischer, M.
1969: U.S. Geological Survey standards—I. Additional data on rocks G-1 and W-1, 1965–1967; Geochim. Cosmochim. Acta, v. 33, no. 1, p. 65–79.

Fleischer, M. and Harder, J.O.
1955: Geochemistry of germanium; U.S. At. Energy Comm., TID-5212, p. 93–105.

Fleischer, R. and Routhier, P.
1973: The "Consanguineous" origin of a tourmaline-bearing gold deposit: Passagem de Mariana (Brazil); Econ. Geol., v. 68, p. 11–22.
1974: The "Consanguineous" origin of a tourmaline-bearing gold deposit: Passagem de Mariana (Brazil)—a reply; Econ. Geol., v. 69, p. 419–422.

Flores, T.
1920: Estudio geologico-minero de los distritos de El Oro y Tlalpujahua; Inst. Geol. Mex., Bol. 37, 85 p.

Foley, L.L.
1960: Gold and silver in manganese ore, Polk Co., Arkansas; Econ. Geol., v. 55, p. 1757.

Folinsbee, R.E.
1942: Zone facies of metamorphism in relation to the ore deposits of the Yellowknife-Beaulieu region, Northwest Territories; unpubl. Ph.D. thesis, Univ. Minnesota, Minneapolis, Minn.
1971: Precambrian metallogenic epochs—atmospheric or centrospheric? Int. Geochem. Congr., Moscow, Abstr. II, p. 962–963.

Folinsbee, R.E., Kirkland, K., Nekolaichuk, A. and Smejkal, V.
1972: Chinkuashih—a gold-pyrite-enargite-barite hydrothermal deposit in Taiwan; *in* Studies in mineralogy and Precambrian geology, Doe, B.R. and Smith, D.K., eds., Geol. Soc. Am., Mem. 135, p. 323–326.

Fomin, D.P., Glotov, A.M. and Gromova, E.I.
1973: Gold mineralization of Cimmerian age in western Uzbekistan; Akad. Nauk SSSR, Dokl., v. 213, no. 1, p. 176–177. (Chem. Abstr., v. 80, 110816e.)

Fominykh, V.G. and Znamenskii, N.D.
1974: Regularities of distribution of platinum group metals and gold in magnetites from Uralian granitoids of different genetic types; Akad. Nauk SSSR, Dokl., v. 217, p. 1185–1186.

Forbes, R.J.
1964: Gold; *in* Studies in ancient technology, Chapter 5, p. 151–192, E.J. Brill, Leiden, Holland.

Ford, W.E.
1958: A textbook of mineralogy, 4th ed.; John Wiley & Sons, Inc., New York, 851 p.

Forgeron, F.D.
1971: Soil geochemistry in the Canadian Shield; Can. Inst. Min. Metall. Bull., v. 64, no. 715, p. 37–42.

Forrester, J.D.
1946: Principles of field and mining geology; John Wiley & Sons, Inc., New York, 647 p.

Forster, I.F.
1968: Paragenetical ore mineralogy of the Loolekop-Phalaborwa carbonate complex, eastern Transvaal; Geol. Soc. S. Afr., Trans., v. 61, p. 359–363.

Forsythe, D.L.
1968: Calaverite at Vatukoula, Fiji; Econ. Geol., v. 63, p. 694–695.
1971: Vertical zoning of gold-silver tellurides in the Emperor gold mine, Fiji; Australas Inst. Min. Metall., Proc., no. 240, p. 25–31.

Fortier, Y.O.
1946: Preliminary Map, Yellowknife-Beaulieu region, Northwest Territories, Geol. Surv. Can., Pap. 46–23.
1947: Preliminary map, Ross Lake, Northwest Territories; Geol. Surv. Can., Pap. 47–16.

Fortin, N.
1969: Etude géostatistique de la minéralisation aurifère à la mine Sullivan, Comté d'Abitibi-Est, Province de Québec; unpubl. M.Sc. thesis, Ecole Polytechnique, Montreal, 94 p. and appendix.

Forward, F.A. and Mackiw, V.N.
1955: Chemistry of the ammonia pressure process for leaching Ni, Cu, and Co from Sherritt Gordon sulphide concentrates; J. Metals, no 3, p. 457–463.

Foshag, W.F. and Fries, C.
1942: Tin deposits of the Republic of Mexico; U.S. Geol. Surv., Bull. 935-C.

Foss, G.V.
1966*a*: Gold; *in* Metals in sedimentary rock series: precious metals; radio-active, trace and rare earth elements; L.V. Pustovalov, ed., "Nauka", Moscow, p. 2–32.
1966*b*: Geochemistry and minerals of gold; Metal. Osad, Tolshchakh, Blagorod. Metal. Radioaktiv. Rasseyan. Redkozemel. Elem. Akad. Nauk SSSR, p. 3–50.

Foster, R.L.
1968: Potential for lode deposits in the Livengood gold placer district, east-central Alaska; U.S. Geol. Surv., Circ. 590, 18 p.
1971*a*: Cation exchange capacity and metal deposition: a suggestion; Econ. Geol., v. 66, p. 200–201 (Disc.).
1971*b*: Gold deposits at Slate Creek, northern Nye County, Nevada; Econ. Geol., v. 66, p. 965–966.

Foster, R.P.
1977*a*: Solubility of scheelite in hydrothermal chloride solutions; Chem. Geol., v. 20, p. 27–43.
1977*b*: The geochemistry of gold with reference to Archean mineralization in Rhodesia; Univ. Rhodesia Inst. Min. Res. Rep. 25, 36 p.

Fouché, K.E. and Smales, A.A.
1966: The distribution of gold and rhenium in iron meteorites; Chem. Geol., v. 1, no. 4, p. 329–339.
1967: The distribution of trace elements in chondritic meteorites. 2. Antimony, arsenic, gold, palladium and rhenium; Chem. Geol., v. 2, no. 2, p. 105–134.

Frankel, J.J.
1940: Notes on some of the minerals in the Black Reef Series; Geol. Soc. S. Afr., Trans., v. 43, p. 1–8.

Fraser, C.
1910: The geology of the Thames subdivision, Hauraki, Auckland; N.Z. Geol. Surv., Bull. 10 (New Series), 136 p.

Fraser, C. and Adams, J.H.
1907: The geology of the Coromandel Subdivision, Hauraki, Auckland; N.Z. Geol. Surv., Bull. 4, 154 p.

Fraser, J.A.
1954: Gold dish and kava bowl; J.M. Dent & Sons Ltd., London, 262 p.

Frederickson, A.F., Lehnertz, C.A. and Kellogg, H.E.
1971: Mobility, flexibility highlight a mass spectrometer-computer technique for regional exploration; Eng. Min. J. v. 172, no. 6, p. 116–118.

Freise, F.W.
1931: The transportation of gold by organic underground solutions; Econ. Geol., v. 26, p. 421–431.

Frens, G.
1973: Controlled nucleation for the regulation of the particle size in monodisperse gold suspensions; Nature, v. 241, no. 105, p. 20–22.

Fricker, A.G.
1976: Placer gold—measurement and recovery; Symp. on Sampling Practices in the Mineral Industries; Australas. Inst. Min. Metall., Parkville, Australia, p. 115–127.

Friedensburg, F.
1953: Die Metallischen Rohstoffe, Heft 3, Gold; Ferdinand Enke Verlag Stuttgart, 234 p.

Friedrich, G.
1960: Petrographische und erzmikroskopische Beobachtungen an der Goldlagerstätte Rodalquilar Prov. Almeria/Spanien; Neues Jahrb. Mineral., Abh., 94, p. 208–227.

Friedrick, K.
1906: Untersuchungen über den Goldgehalt von Gebirgsproben und solen deutscher Salzlagerstätten; Metallurgie, v. 3, no. 2, p. 627–630.

Friend, J.N. and Allchin, J.P.
1939: Colour of celestine; Nature, v. 144, p. 633.
1940: Blue rocksalt; Nature, v. 145, p. 266–267.

Fries, C. and Schmitter, E.
1948: Tin-bearing placers near Guadalcázar, State of San Luis Potosí, Mexico, U.S. Geol. Surv., Bull. 960-D, 149 p.

Fripp, R.E.P.
1976: Stratabound gold deposits in Archean banded iron-formation, Rhodesia; Econ. Geol., v. 71, p. 58–75.

Fritze, K. and Robertson, R.
1969: Precision in the neutron activation analysis for gold in standard rocks G-1 and W-1; *in* Modern trends in activation analysis; Nat. Bur. Stand., Spec. Publ. 312, p. 1279–1283.

Frohberg, M.H.
1936: The gold (ore) deposits of the Michipicoten area; Ont. Dep. Mines Annu. Rep. 1935, pt. 8, p. 39–83.

Frondel, C.
1938: Stability of colloidal gold under hydrothermal conditions; Econ. Geol., v. 33, no. 1, p. 1–20.
1962: The system of mineralogy, 7th ed., v. 3, Silica minerals, John Wiley & Sons, Inc., New York, 334 p.

Frondel, C. and Ito, J.
1957*a*: Crystal chemistry and geochemistry of scandium; Geol. Assoc. Can., Prog., p. 27 (abstr.).
1957*b*: Geochemistry of germanium in the oxidized zone of the Tsumeb mine, South-West Africa; Am. Mineral., v. 42, p. 743-753.
1967: Crystal chemistry and geochemistry of scandium; Can. Mineral., v. 9, pt. 2, p. 289.

Fuchs, E. and De Launay, L.
1893: Traité des gîtes minéraux et métallifères, Tomes 1 et 2, Paris.

Fuchs, W.
1935: Rare elements in German brown-coal ashes: Ind. Eng. Chem., v. 27, p. 1099–1100.

Fujii, T. and Haramura, H.
1976: Gold chloride complex in silica solution at low temperature and pressure; 25th Int. Geol. Congr., Sydney, Australia, Abstr., v. 2, p. 563–564.

Fukai, R. and Meinke, W.W.
1962: Activation analyses of vanadium, arsenic, molybdenum, tungsten, rhenium, and gold in marine organisms; Limnol. Oceanogr., v. 7, p. 186–200.

Full, R.P. and Grantham, R.M.
1968: Ore deposits of the Republic mining district, Ferry County, Washington; *in* Ore deposits of the United States, J.D. Ridge, ed., Publ. Am. Inst. Min. Metall. Pet. Eng. Inc., New York, p. 1481–1494.

Fulton, C.H.
1916: The ores of copper, lead, gold and silver; U.S. Bur. Mines, Tech. Pap. 143, 45 p.

Fulton, C.H. and Sharwood, W.J.
1929: A manual of fire assaying, 3rd ed.; McGraw-Hill Book Co., Inc., New York, 268 p.

Furman, N.H., ed.
1962: Standard methods of chemical analysis, 6th ed.; D. Van Nostrand Co., Inc., Princeton, New Jersey, 1401 p.

Furnival, G.M.
1935: The large quartz veins of Great Bear Lake, Canada; Econ. Geol., v. 30, p. 843–859.

Furse, G.D.
1948: McIntyre mine; *in* Structural geology of Canadian ore deposits, v. 1; Can. Inst. Min. Metall., Montreal, p. 482–496.

Fyfe, W.S. and Henley, R.W.
1973: Some thoughts on chemical transport processes, with particular reference to gold; Mineral. Sci. Eng., v. 5, no. 4, p. 295–303.

Fyles, J.T., Harakal, J.E. and White, W.H.
1973: The age of sulfide mineralization at Rossland, British Columbia; Econ. Geol., v. 68, p. 23–33.

Gaarden, J.
1940: Gold nuggets of the world; Gaarden Publishing Co., Hollywood, Calif., 149 p.

Gabrielse, H., Souther, J.G. and Roots, E.F.
1962: Dease Lake, British Columbia; Geol. Surv. Can., Map 21-1962.

Gadd, N.R.
1964: Surficial geology, Beauceville map-area, Quebec; Geol. Surv. Can., Pap. 64-12, 3 p.

Gage, M.
1948: The geology of the Reefton quartz lodes; N.Z., Geol. Surv., Bull. 42, 69 p.

Gair, J.E.
1962: Geology and ore deposits of the Nova Lima and Rio Acima quadrangles, Minas Gerais, Brazil; U.S. Geol. Surv., Prof. Pap. 341-A, 67 p.

Galakhov, I.N., Mushkin, I.V. and Chernyavskii, Yu. A.
1974: Gold mineralization of western Uzbekistan; Akad. Nauk SSSR, Dokl., v. 217, no. 1, p. 169–171. (Chem. Abstr., v. 82, 19497g.)

Galkiewicz, T.
1972: Metallogeny in Bulgaria; Rudy Metale Niezelaz, v. 17, no. 12, p. 590–591. (Chem. Abstr., v. 79, 7967k.)

Galkin, M.A.
1967: Mineral associations in mercury deposits of northeastern Yakutia; Akad. Nauk SSSR, Dokl., v. 169, no. 2, p. 438–440. *Also* Acad. Sci. USSR, Dokl., Earth Sci. Sec., v. 169, nos. 1–6, p. 165–167.

Gallagher, D.
1940: Albite and gold; Econ. Geol. v. 35, p. 698–736.
1941: A microscopic study of some ores of the Lupa goldfield, Tanganyika Territory, East Africa; Econ. Geol., v. 36, p. 306–323.

Galopin, R. and Henry, N.F.M.
1972: Microscopic study of opaque minerals; W. Heffer and Sons Ltd., Cambridge, 322 p.

Gamaleev, I.E., Khamrabaev, I. Kh., Muninov, Sh. A. and Baranov, V.V.
1967: Picrite porphyrites from southern Fergana and northern Nuratau; Uzb. Geol. Zh., v. 11, no. 1, p. 3–9. (Chem. Abstr., v. 68, 89038t.)

Gamyanin, G.N.
1974: Types of gold ore deposits of eastern Yakutia; Vop. Rudonos. Yakutii, G.N. Gamyanin, ed., p. 5–34. (Chem. Abstr., v. 84, 76923h.)

Gamyanin, G.N. and Solov'ev, V.I.
1969: Eastern Yakutia gold-ore formations; Str. Zemnoi Kory Yakutii Zakonomer. Razmeshcheniya Polez. Iskop. Yu. P. Ivensen, ed., Izd. "Nauk" Moscow, U.S.S.R., p. 256–274. (Chem. Abstr., v. 73, 27539b.)

Ganapathy, R., Keays, R.R. and Anders, E.
1970: Apollo 12 Lunar samples: trace element analysis of a core and the unconformity of the regolith; Science, v. 170, no. 3957, p. 533–535.

Ganapathy, R., Keays, R., Laul, J.C. and Anders, E.
1970: Trace elements in Apollo 11 lunar rocks: implications for meteorite influx and origin of moon; Proc. Apollo 11 Lunar Sci. Conf., Geochim. Cosmochim. Acta, Suppl. 1, v. 2, p. 1117–1142.

Ganapathy, R., Morgan, J.W., Higuchi, H. and Anders, E.
1974: Meteoric and volatile elements in Apollo 16 rocks and in separated phases from 14306; Proc. 5th Lunar Sci. Conf., Geochim. Cosmochim, Acta, Suppl. 5, v. 2, p. 1659–1683.

Ganapathy, R. Morgan, J.W., Krähenbühl., U. and Anders, E.
1973: Ancient meteoric components in lunar highland rocks: clues from trace elements in Apollo 15 and 16 samples; Proc. 4th Lunar Sci. Conf., Geochim. Cosmochim. Acta, Suppl. 4, v. 2, p. 1239–1261.

Ganchev, P. and Ninov, I.
1971: Relation between the gold and silver contents in some veins of the Govezhda deposit and its application in practice: Rudodobiv, iss. 6, p. 1–4. (Chem. Abstr., v. 80, 72759d.).

Ganiev, A.G., Sil'vanovich, Y.A. and Rakhimov, Kh.R.
1971: Neutron-activation determination of precious metals in some basic rocks and minerals connected with them; Neitron. Aktiv. Anal. p. 139–145. (Chem. Abstr., v. 76, 148471u.)

Gannett, R.W.
1919: Experiments relating to the enrichment of tungsten ores; Econ. Geol., v. 14, p. 68–78.

Gapeeva, M.M.
1973: Gold ore manifestations in metamorphic rock masses of the northwestern part of the Vitim-Olekma Interfluve; Metamorf. Komplek. Vost. SSSR, A.M. Smirnov, ed., p. 226–233. (Chem. Abstr., v. 82, 61983n.)

Gapon, A.E.
1970: Gold concentration in the black clays of the Dogaldinsky suite (Vitim-Patom Highland); Acad. Sci. USSR, Siberian Div., Inst. Geochem. Yearb. 1969, L.V. Tauson, ed., Irkutsk, p. 239–240.

Gapon, A.E. and Voloskova, S.E.
1976: The relationship of gold with organic matter in Pre-Cambrian sedimentary rocks (Vitim-Patom Highland); Ezheg. Inst. Geokhim. Sib. Odt., Akad. Nauk SSSR, 1974, p. 208–211.

Gapontsev, G.P.
1971: Determination of the productive depth of the auriferous veins of the Balei-type according to geochemical data; Geol. Rud. Mestorozhd., v. 13, no. 1, p. 106–110. *Also* summary in Econ. Geol., v. 63, p. 1150–1160.

Gapontsev, G.P. and Polikarpochkin, V.V.
1969: Distribution of elements accompanying gold in the veins of the Baleisk ore field on the basis of multiple analyses; Ezheg. Inst. Geokhim. Sib. Otd., Akad. Nauk SSSR, 1968, p. 217–221. (Chem. Abstr., v. 73, 37346a.)

Garces, H.
1942: Solubilidad del oro en acides humicos; Anal. Prim. Congr. Panamericano Ing. Min. Geol., v. 3, p. 1135–1138.

Gardiner, A.H.
1914: The map of the gold mines in a Ramesside papyrus at Turin; Cairo Sci. J. v. 8, p. 42–46.

Gardner, E.D.
1950: Guide to prospecting for lode gold; U.S. Bur. Mines, Inform. Circ. 7535 (pt. 1), p. 1–12.

Gardner, E.D. and Allsman, P.T.
1938: Power shovel and dragline placer mining; U.S. Bur. Mines., Inform. Circ. 7013, 68 p.

Gardner, E.D. and Johnson, C.H.
1934–1935: Placer mining in the western United States; Pt. I, General information, hand shoveling, and ground sluicing; U.S. Bur. Mines, Inform. Circ. 6786, 74 p.: Pt. II, Hydraulicking, treatment of placer concentrates, and marketing of gold; U.S. Bur. Mines, Inform. Circ. 6787, 89 p.: Pt. III, Dredging and other forms of mechanical handling of gravel, and drift mining; U.S. Bur. Mines, Inform. Circ. 6788, 82 p.

Garlick, W.G.
1953: Reflections on prospecting and ore genesis in northern Rhodesia; Inst. Min. Metall., Trans. (London), v. 63, no. 563, p. 9–20.

Garrels, R.M. and Christ, C.L.
1965: Solutions, minerals, and equilibria; Harper & Row, Publishers, New York, 450 p.

Garrett, R.G.
1974: Mercury in granitoid rocks of the Yukon and its relation to gold-tungsten mineralization; J. Geochem. Explor., v. 3, p. 277–289.

Garrett, R.G. and Nichol; I.
1967: Regional geochemical reconnaissance in eastern Sierra Leone; Inst. Min. Metall. Trans., Sec. B, v. 76, p. B97–B112; discussion, v. 77, p. B78–B79, 1968; authors' reply, v. 78, p. B54–B55.

Gather, B. and Blacknik, R.
1974: Das System Gold-Wismut-Tellur; Z. Metallk., Bd. 65, p. 653–656.

Gavrikov, S.I.
1962: Occurrence of gold in sedimentary rocks in the basin of the Elga; Geol. Rud. Mestorozhd., no. 2, p. 114–115.

Gavrikova, S.N., Shevyrev, I.A., Alkin, V.S. and Ivanov, V.N.
1973: Elements of vertical zoning in beresites and ores of the Itakinsk ore field (eastern Transbaikal); Geol. Rud. Mestorozhd., v. 15, no. 3, p. 117–123.

Gavrilenko, B.V., Belolipetski, A.P., Bolotov, V.I. and Vorobets, A.A.
1974: Geochemistry of gold in the metamorphic complex of Kolmozero-Voronye (Kola Peninsula); Geokhim, v. 1974, no. 1, p. 132–138.

Gavrilenko, B.V., Goryainov, P.M. and Evdokimov, B.N.
1976: Distribution of gold in geological formations of the ferruginous-siliceous deposits of the central part of the Kola Peninsula; Dokl. Acad. Nauk. USSR, v. 231, p. 159–161. (Chem. Abstr., v. 86, 109330j.)

Gavrilov, A.M.
1971: Syngenesis with sulfide nature of finely dispersed gold in the Bakyrchik gold ore deposit (eastern Kazakhstan); Tr. Tsent. Nauch. Issled. Gornorazved. Inst., no. 96 (pt. 1), p. 159–166. (Chem. Abstr., v. 77, 91152e.)

Gavrilov, A.M., Alysheva, E.I. and Shevchenko, V.N.
1974: Nature of submicroscopic gold in arsenopyrite of a gold-arsenic deposit; Tr. Tsent. Nauch. Issled. Geol. Inst. Tsvet, v. 112, p. 47–52. (Chem. Abstr., v. 83, 63656c.)

Gay, N.C.
1963: A review of the geochemical characteristics of gold in ore deposits; Univ. Witwatersrand, Econ. Geol. Res. Unit, Inform. Circ. 12, 70 p.
1964: The composition of gold from the Barberton Mountain Land; Univ. Witwatersrand, Econ. Geol. Res. Unit, Inform. Circ. 19, 53 p.

Gayev, I.A. and Gayeva, N.M.
1971: Gold in Precambrian rocks of the middle Dnieper region; Acad. Sci. USSR, Dokl., Earth Sci. Sec., v. 198, p. 127–128.

Gebler, I.V. and Stramkovskaya, K.K
1957: On the absorption by lignite of heavy metals from weak solutions of their salts; Izv. Filiala Akad. Nauk SSSR, no. 12, p. 78–82.

Gedansky, L.N. and Hepler, L.G.
1969: Thermochemistry of gold and its compounds; Engelhard Ind. Tech. Bull., v. 10, no. 1, p. 5–9.

Geffroy, J.
1963: La brannérite du filon aurifère de La Gardette (Isère) et sa signification métallogénique; Bull. Soc. Fr., Minéral. Cristallogr., t. 86, p. 129–132.

Geier, B.
1933: Die Goldlagerstätte, El Chivato, bei Talca (Chile); Z. Prakt. Geol., v. 41, p. 61–69.

Gelman, M.L.
1976: On the role of regional metamorphism in gold mineralization of the South-East of the USSR; Acad. Sci. USSR Dokl., v. 230, p. 1406–1409.

Generalov, P.P., Kostyuk, B.F., Gavrilyuk, I.V. and Zasyad'ko, G.N.
1972: Geological and geomorphological premises of the placer gold content in the Severo-Sosvinsk and Lyapinsk Urals; Tr., Zapad. Sib. Nauch. Issled. Geologorazved. Neft. Inst., no. 52, p. 6–24. (Chem. Abstr., v. 78, 113925j.)

Geological Survey of South Africa
1959: The mineral resources of the Union of South Africa, 4th ed.; Gov. Printer, Pretoria, 614 p.

Gerglelchev, V.N.
1974: Concentric zoning in the Madzharovo; Spis. Bulg. Geol. Druzh., v. 35, no. 2, p. 218–223. (Chem Abstr., v. 82, 114185n.)

Gesner, A.
1867: Gold and its separation from other minerals; N.S. Inst. Nat. Sci., v. 1, p. 54–60.

Gevers, T.W.
1961: Outline of the geology of Southern Africa; Trans. 7th. Commonw. Min. Metall. Congr., Pap. Disc., v. 1, S. Afr. Inst. Min. Metall., p. 25–37.

Ghitulescu, T.P. and Socolescu, M.
1941: Étude géologique et minière des Monts Métallifères; An. Inst. Geol. Rom., v. 21, p. 185–464.

Ghosh, D.B., Sastry, B.B.K., Rao, A.J. and Rahim, A.A.
1970: Ore environment and ore genesis in Ramagiri Gold Field, Andhra Pradesh, India; Econ. Geol., v. 65, p. 801–814.

Gibbins, W.
1970: Geology of the Falcon Lake Stock, southeastern Manitoba; Geol. Assoc. Can., Annu. Meet., Winnipeg (abstr.), p. 26.

Gibson, F.H. and Selvig, W.A.
1944: Rare and uncommon chemical elements in coal; U.S. Bur. Mines, Tech. Pap. 669, 23 p.

Gibson, J.C. and Stockwell, C.H.
1948: San Antonio mine; *in* Structural geology of Canadian ore deposits, v. 1, Can. Inst. Min. Metall., Montreal, p. 315–321.

Gibson, T.M.
1911: Pay-streaks at Nome; Mining and Scientific Press, v. 102, p. 424–427, 462–467.

Gilbert, G.
1924: The relation of hardness to the sequence of the ore minerals; Econ. Geol., v. 19, p. 668–673.
1948: Rossland Camp; *in* Structural geology of Canadian ore deposits, v. 1, Can. Inst. Min. Metall., Montreal, p. 189–196.

Gilbert, R.D.
1940: The gold deposits of the Paracale-Mambulao district; Eng. Min. J., v. 141, no. 6, p. 45–50.

Gill, J.E.
1948: Wasa Lake mine; *in* Structural geology of Canadian ore deposits, v. 1, Can. Inst. Min. Metall., Montreal, p. 730–734.

Gill, J.E. and Byers, A.R.
1948: Surf Inlet and Pugsley mines; *in* Structural geology of Canadian ore deposits, v. 1, Can. Inst. Min. Metall., Montreal, p. 99–104.

Gillett, T.E.
1964–1965: The mode of origin of Banket orebodies, discussion; Inst. Min. Metall. Trans., v. 74, p. 571.

Gillum, D.E. and Ehmann, W.D.
1971: Noble metal determination in meteorites; Radiochim. Acta, v. 16, p. 123–128.

Ginzburg, A.I. *et al.*
1959: Minerals of germanium and its deposits; *in* Geology of the deposits of rare metals, no. 5, Gosgeoltekhizdat, Moscow, 112 p.

Ginzburg, A.I., Zhuravleva, L.N., Ivanov, I.B. and Shcherbina, V.V.
1959: Rare elements and their deposits; *in* Geology of the deposits of rare metals, no. 3, Gosgeoltekhizdat, Moscow, 125 p.

Ginzburg, I.I.
1960: Principles of geochemical prospecting; Pergamon Press, Oxford, 311 p.

Ginzburg, V.L. and Rogover, G.B.
1961: Regularities in the distribution of nonferrous and precious metals in principal ore minerals and silicates of the Noril'sk deposits; Sov. Geol., no. 3, p. 48–60. (Int. Geol. Rev., v. 3, no. 10, p. 917–926.)

Giuscă, D., Borcos, M., Lang, B. and Stan, N.
1973: Neogene volcanism and metallogenesis in the Gutăi Mountains; Symp. Volcanism and Metallogenesis, Bucharest, Guidebook no. 11, 50 p., published by Geol. Inst. Bucharest.

Giuscă, D., Cioflica, G. and Udubasa, G.G.
1968: Metallogenesis associated with Neogene volcanism of the Rumanian Carpathians; 23rd Int. Geol. Congr., Prague, Abstr., p. 190–191.

Gladkov, N.A.
1972: Relation of Salaira gold-bearing placers to natural electrical and geochemical fields; Geol. Geofiz., no. 3, p. 86–91. (Chem. Abstr., v. 77, 91156j.)

Glasson, M.J.
1975: Gold distribution and foliation development in Lower Proterozoic rocks in central Victoria: bearing on gold mineralization; unpubl. M.Sc. thesis, Univ. Melbourne, Melbourne, Austr., 164 p.

Gledhill, T.L.
1927*a*: Goudreau-Lochalsh gold area, District of Algoma; Ont. Dep. Mines, Annu. Rep., v. 36, pt. 2, p. 50–86.
1927*b*: Michipicoten gold area, District of Algoma; Ont. Dep. Mines Annu. Rep., v. 36, pt. 2, p. 1–49.

Gleeson, C.F.
1970: Heavy mineral studies in the Klondike area, Yukon Territory; Geol. Surv., Can., Bull. 173, 63 p.

Gleeson, C.F. *et al.*
1965: Heavy metal content of stream and spring waters and stream and spring sediments, Mayo Lake, Keno Hill, Mt. Haldane, Rackla River, Scougale Creek, McQuesten Lake, and Dublin Gulch, Yukon Territory; Geol. Surv., Can., Maps 18-1964 to 31-1964 inclusive.
1966–1968: Lead, silver, zinc, arsenic, antimony, copper, molybdenum, tungsten and tin, nickel, cobalt, manganese, and boron content of stream and spring sediments of Keno Hill area, Yukon Territory; Geol. Surv., Can., Maps, Prelim. Ser., nos. 45-1965 (lead) to 56-1965 (boron) inclusive.

Gleeson, C.F. and Cormier, R.
1971: Evaluation by geochemistry of geophysical anomalies and geological targets using overburden sampling at depth; *in* Geochemical exploration, R.W. Boyle and J.I. McGerrigle, eds.; Can. Inst. Min. Metall., Spec. Vol. 11, p. 159–165.

Gleizer, L.M.
1970: Distribution of gold in the Lower Paleozoic formations of the Pskemsk Ridge (Tien-Shan); Zap. Uzb. Otd. Vses. Mineral. Obshchest., no. 22, p. 206–207. (Chem. Abstr., v. 74, 114513c.)

Globa, V.A. and Narseev, V.A.
1973: Evolutionary series of gold mineralization of volcanic belts in Kazakhstan; Evol. Vulkanizma Istor. Zemli, Mater. Vses. Paleovulkan, p. 284–285. (Chem. Abstr., v. 84, 20363h.)

Glukhov, R.G.
1971: Akkarginsk gold deposit; Sov. Geol., no. 9, p. 130–137.
1974: Near ore metasomatites of gold ore deposits; Akad. Nauk Kaz. SSR, Izv., Ser. Geol. v. 31, no 1, p. 74–81. (Chem. Abstr., v. 81, 80605m.)

Glushchenko, V.M., Talipov, R.M., Nishanov, P.N., Lunin, A.A., Samigdzhanova, M.A. and Arilova, Kh.A.
1973: Comparative data on the determination of the trace content of gold, by different methods in the ash of plants and water; Uzb. Geol. Zh., v. 17, no. 4, p. 77–79. (Chem. Abstr., v. 80, 66425a.)

Gmelin
1950–1954: Gmelins Handbuch der anorganischen Chemie, System-nummer 62, Gold, pt. 1 and 2, Weinheim/Bergstrasse, Verlag Chemie, GMBH, 406 p.

Godlevskii, M.N., Razin, L.V. and Konkina, O.M.
1970: Gold content in the Norilsk-type differentiated intrusions; Tr. Tsent. Nauch. Issled. Gornorazved. Inst. Tsvet., Redk. Blagorod Metal., no. 87, p. 42–57. (Chem. Abstr., v. 74, 56300t.)

Goldberg, E.D.
1957: Biogeochemistry of trace metals; *in* Treatise on marine ecology and paleoecology, v. 1, J.W. Hedgpeth, ed., Geol. Soc. Am., Mem. 67, p. 345–357.

Goldberg, E., Uchiyama, A. and Brown, H.
1951: The distribution of nickel, cobalt, gallium, palladium, and gold in iron meteorites; Geochim. Cosmochim. Acta, v. 2, no. 1, p. 1–25.

Goldberg, I.
1964: Notes on the relationship between gold deposits and structure in southern Rhodesia; *in* The geology of some ore deposits in southern Africa, v. 2, S.H. Haughton, ed., Geol. Soc. S. Afr., Johannesburg, p. 9–13.

Goldschmidt, V.M.
1933*a*: Über die Anreicherung seltener Elemente in Steinkohlen; Gesell. Wiss. Göttingen, Nachr., Math. Phys. Kl., no. 4, p. 371–386.
1933*b*: Grundlagen der quantitativen Geochemie; Fortsch. Mineral. Kristallogr. Petrog., v. 17, p. 112–156.
1934: Drei Vortäge über Geochemie; Geol. Foeren. Stockh. Foerh., v. 56, p. 385–427.
1935: Rare elements in coal ashes; Ind. Eng. Chem., v. 27, p. 1100–1102.
1937: Geochemische Verteilungsgesetze der Elemente. IX. Die Mengenverhaltnisse der Elemente und der Atom-Arten; Nor. Vidensk. Oslo, Skr., Mat. Natur. Kl., no. 4, 148 p.
1954: Geochemistry; Clarendon Press, Oxford, 730 p.

Goldschmidt, V.M., Krejci-Graf, K. and Witte, H.
1948: Trace metals in sediments; Nachr. Akad. Wiss. Göttingen, Math. Phys. Kl., Math. Physik. Chem., Abt. 35–52. (Chem. Abstr., v. 47, 11095c.)

Goldschmidt, V.M. and Peters, Cl.
1932: Zur Geochemie der Edelmetalle; Gesell. Wiss. Göttingen, Nachr., Math. Phys. Kl., no. 4, p. 377–401.
1933: Über die Anreicherung seltener Elemente in Steinkohlen; Gesell. Wiss. Göttingen Nachr., Math. Phys. Kl., no. 4, p. 371–386.

Goleva, G.A.
1968: Hydrogeochemical prospecting for hidden ore deposits; "Nedra" Publ. House, Moscow, U.S.S.R., 291 p. (in Russian).
1970: Hydrogeochemistry of the gold deposits of the Balei district; Int. Geol. Rev., v. 12, no. 2, p. 195–203.

Goleva, G.A. and Chitaeva, N.A.
1973: The geochemistry of cadmium in the zone of hypergenesis of ore deposits; Litol. Polez. Iskop., no. 3, p. 92–101.

Goleva, G.A., Krivenko, V.A. and Gutz', Z.G.
1970*a*: Geochemical regularities of the distribution and migration forms of gold in natural waters; Geochem. Int., no. 6, p. 744–757.
1970*b*: Geochemical trends in the occurrence and migration forms of gold in natural waters; Geochem. Int., v. 7, no. 3, p. 518–529.

Goleva, G.A. and Vorob'yeva, I.N.
1967: The migration of germanium in the ground waters of ore deposits; Geochem. Int., v. 4, p. 809–817.

Goloubinoff, V.
1937: Sur la prospection géochimique différentielle des gîtes minéraux; Comptes rendus, Acad. Sci. Paris, t. 204, p. 1075–1077.

Golovnya, S.V.
1974: Peculiarities of gold distribution in eclogites of the Salnyne Tundras (Kola peninsula); Geokhim., no. 4, p. 638–640. (Geochem. Int. v. 11, no. 2, p. 459–461.)

Goncharenko, A.I.
1970: Auriferous listwanites as a new type of mineralization in the northern part of the Kuznetsk Alatau; Izv. Tomsk. Politekh. Inst., v. 239, p. 110–114. (Chem. Abstr., v. 75, 111707d.)

Goncharov, V.I.
1973: Chemical composition of hydrothermal solutions of gold-silver deposits; Tr. Sev. Vost. Kompleksn. Inst., Dal'nevost. Tsent., Akad. Nauk SSSR, no. 55, p. 174–184. (Chem. Abstr., v. 84, 7596s.)

Goncharov, V.I., Naiborodin, V.I. and Sidorov, A.A.
1972: Temperature characteristics in the formation of subvolcanic gold-silver deposits of the northeastern U.S.S.R.; Rud. Sreda Vklyuch. Miner., V.I. Smirnov, ed., "Nauka" Moscow, p. 124–131. (Chem. Abstr., v. 78, 149674r.)

Goncharov, V.I. and Shugurova, N.A.
1973: The composition of the volatile constituents of hydrothermal solutions of near-surface type (exemplified by the gold-silver deposits of the northeast U.S.S.R.); Geochemistry, no. 10, p. 1583–1585.

Goni, J., Guillemin, C. and Sarcia, C.
1967: Géochemie de l'or exogène. Etude expérimentale de la formation des dispersions colloidales d'or et de leur stabilité; Mineral. Deposita (Berl.), v. 1, p. 259–268.

Good, R.S., Fordham, O.M. and Halladay, C.R.
1973: Geochemical exploration for gold in the central Piedmont of Virginia (abstr.); Min. Eng. v. 25, no. 8, p. 31-32.
1977: Geochemical reconnaissance for gold in the Caledonia and Pendleton quadrangles in the Piedmont of central Virginia; Virginia Miner., v. 23, no. 2, p. 13–22.

Goodchild, W.H.
1918: The evolution of ore deposits from igneous magmas; Min. Mag., v. 19, no. 4, p. 188–199.

Goodell, P.C. and Petersen, U.
1974: Julcani mining district, Peru: a study of metal ratios; Econ. Geol. v. 69, p. 347–361.

Goodwin, A.M.
1965*a*: Mineralized volcanic complexes in the Porcupine–Kirkland Lake–Noranda region, Canada; Econ. Geol., v. 60, p. 955–971.
1965*b*: Volcanism and gold deposition in the Birch-Uchi Lakes area; Can. Inst. Min. Metall., Bull., v. 58, no. 635, p. 304–314.

Goossens, P.J.
1972: Metallogeny in Ecuadorian Andes; Econ. Geol., v. 67, p. 458–468.

Gorbunov, E.Z.
1959: On the distances travelled by alluvial gold from its native sources; Sov. Geol., no. 6, p. 98–105.

Gordon, J.B.
1977*a*: Gold Deposits of Ontario, northwestern sheet, District of Kenora (Patricia portion); Ont. Geol. Surv. Prelim. Map P. 1226, Mineral Deposits Ser., scale 1:1,013,760 or 1 in.: 16 mi. Compilation 1974, 1975, 1976.
1977*b*: Gold Deposits of Ontario, east central sheet, Districts of Thunder Bay, Algoma, Cochrane, Sudbury, Timiskaming and Nipissing; Ont. Geol. Surv. Prelim. Map P. 1228, Mineral Deposits Ser., scale 1:1,013,760 or 1 in.: 16 mi. Compilation 1974, 1975, 1976.
1977*c*: Gold Deposits of Ontario, west central sheet, Districts of Kenora (Patricia Portion), Thunder Bay, Algoma and Cochrane; Ont. Geol. Surv. Prelim. Map P.1227, Mineral Deposits Ser., scale 1:1,013,760 or 1 in.: 16 mi. Compilation 1974, 1975, 1976.
1977*d*: Gold Deposits of Ontario, southern sheet, southern Ontario and District of Nipissing; Ont. Geol. Surv. Prelim. Map P.1229, Mineral Deposits Ser., scale 1:1,013,760 or 1 in.: 16 mi. Compilation 1974, 1975, 1976.

Gorman, W.A.
1955: Preliminary report on St. Georges–St. Zacharie area, Beauce and Dorchester counties; Que. Dep. Mines, Geol. Surv. Br. Prelim. Rep. no. 314, 5 p.

Gormasheva, G.S., Zakharov, M.N. and Sanin, B.P.
1973: Separation of ore-bearing and barren propylite zones by approximate phase analyses of carbonate components of propylite andesites (Illustrated by gold-ore deposits of the Evensk ore node); Ezheg., Inst. Geokhim. Sib. Otd., Akad. Nauk SSSR, 1972, p. 315–319. (Chem. Abstr., v. 81, 138581u.)

Gornostaev, N.N.
1937*a*: Genetic types of the gold deposits of the U.S.S.R.; 17th Int. Geol. Congr., U.S.S.R., Abstract of Papers, p. 178.
1937*b*: The auriferous petrographic complexes of the U.S.S.R., 17th Int. Geol. Congr. U.S.S.R., Abstract of Papers, p. 171.

Gorshkov, L.S., Nikolaeva, L.A. and Epshtein, Yu.A.
1971: Changes in gold during river and marine placer deposit formation in the Khualaztza ore field, southern Maritime Territory; Izv. Vyssh. Ucheb. Zaved., Geol. Razved., v. 14, no. 6, p. 54–60. (Chem. Abstr., v. 75, 79067g.)

Gorzhevskii, D.I., Fogel'man, N.A. and Alektorova, Ye.A.
1968: Transbaikal gold-molybdenum belt; Acad. Sci. USSR, Dokl., Earth Sci. Sec. v. 178, p. 46–47.

Gorzhevskii, D.I., Rozhkov, I.S. and Sheglov, A.D.
1971: Neogene volcanism and gold ore deposits of the Apuseni Mountains; Probl. Geol. Sess. Mezhdunar. Geol. Kongr., 23rd year, A.P. Vinogradov, ed., "Nauka" Moscow, U.S.S.R., p. 317–324. (Chem. Abstr., v. 77, 22977r.)

Gosling, A.W., Jenne, E.A. and Chao, T.T.
1971: Gold content of natural waters in Colorado; Econ. Geol., v. 66, p. 309–311.

Gott, G.B. and McCarthy, J.H.
1966: Distribution of gold, silver, tellurium, and mercury in the Ely mining district, White Pine County, Nevada; U.S. Geol. Surv., Circ. 535, 5 p.

Gott, G.B., McCarthy, J.H., VanSickle, G.H. and McHugh, J.B.
1969: Distribution of gold and other metals in the Cripple Creek district, Colorado; U.S. Geol. Surv., Prof. Pap. 625-A, 17 p.

Gott, G.B. and Zablocki, C.J.
1968: Geochemical and geophysical anomalies in the western part of the Sheep Creek Range, Lander County, Nevada; U.S. Geol. Surv., Circ. 595, 17 p.

Gottesmann, W.
1976: Material aspects of endogenous gold accumulation; Z. Angew. Geol. v. 22, no. 6, p. 251–258 and v. 22, no. 9, p. 406–412.

Gottfried, D. and Greenland, L.P.
1972: Variation of iridium and gold in oceanic and continental basalts; 24th Int. Geol. Congr., Montreal, Sec. 10, p. 135–144.

Gottfried, D., Rowe, J.J. and Tilling, R.I.
1972: Distribution of gold in igneous rocks; U.S. Geol. Surv., Prof. Pap. 727, 42 p.

Gottfried, D., Rowe, J.J., Tilling, R.I. and Dodge, F.W.
1969: Geochemical behaviour of gold during magmatic differentiation; Geol. Soc. Am., Abstr. with programs, pt. 7, Annu. Meet., Atlantic City, N.J., p. 277–278.

Goubeau, J. and Birckenbach, L.
1938: Untersuchung des Edelmetallgehaltes von Kalisalzlagerstätten; Z. Anorg. Allg. Chem., Bd. 236, p. 37–44.

Goulden, P.D., Afghan, B.K. and Brooksbank, P.
1972: Determination of nanogram quantities of simple and complex cyanides in water; Anal. Chem., v. 44, no. 11, p. 1845–1850.

Grabezhev, A.I.
1973: Similar temperature interval of formation of the Ural gold ore and rare metal hydrochemical deposits and the corresponding beresites and greisens; Ezheg., Inst. Geol. Geokhim., Akad. Nauk SSSR, Ural Nauch. Tsent. 1972, p. 154–157. (Chem. Abstr., v. 81, 52456v.)
1974: Metasomatic processes of rocks in the Kochkar gold ore deposit; Tr. Inst. Geol. Geokhim., Ural, Nauch. Tsent. Akad. Nauk SSSR, v. 108, p. 3–22. (Chem. Abstr., v. 84, 62728 m.)

Gracey, A.H.
1898: Placer gold on Vermillion River; Ont. Bur. Mines, v. 7, pt. 3, p. 256–259.

Graham, A.R.
1931: Sturgeon Lake gold area, Districts of Kenora and Thunder Bay; Ont. Dep. Mines, Annu. Rep., 1930, v. 39, pt. 2, p. 36–50.

Graham, A.R. and Kaiman, S.
1952: Aurostibite, $AuSb_2$—a new mineral in the pyrite group; Am. Mineral., v. 37, p. 461–469.

Grange, L.I.
1937: The geology of the Rotorua-Taupo Subdivision, Rotorua and Kaimanawa Divisions; N.Z. Geol. Surv., Bull. no. 37 (New Series), p. 118–119.

Granier, C.
1937: Introduction à la Prospection Géochimique des Gîtes Métallifères; Masson et Cie, Paris, 143 p.

Granier, C., Lajoinie, J.-P. and Vitali, C.
1963: Géochimie de l'or et du cuivre dans les formations latéritiques argileuses du Mont Flotouo (Ity, Côte d'Ivoire); Soc. Fr. Minéral. Cristallogr., Bull., v. 86, p. 252–258.

Grantham, D.R.
1928: Lupa gold field; Geol. Surv., Tanganyika Territory, Short Pap., no. 2, 6 p.
1933: The eastern extension of the Lupa goldfield (Ipogola-Sengambi-Shoga); Geol. Sur., Tanganyika Territory, Short Pap., no. 11, 9 p.

Graton, L.C.
1930: Hydrothermal origin of the Rand gold deposits; Econ. Geol., v. 25, Suppl. to no. 3, 185 p.
1933: The depth-zones in ore deposition; Econ. Geol., v. 28, p. 513–555.

Gravrilenko, B.V., Belolipetskiy, A.P., Bolotov, V.I. and Vorobets, A.A.
1974: Geochemistry of gold in the Kolmozero-Voron'ya metamorphic complex (Kola Peninsula); Geochem. Int., v. 11, no. 1, p. 111–117.

Grazhdantsev, N.G.
1971: Statistical comparative analyses of the similarity of the epithermal Dzhungara-Balkhash region gold-silver deposits and the Balei deposit; Mat. Metody Geol., no. 2, p. 96–99. (Chem. Abstr., v. 82, 33376y.)

Grebenchikov, A.M., Khrust, A.R. and Chepizhnaya, E.A.
1974: Radiogeochemical characteristics of near-surface gold-ore deposits and their prospecting value; Izv. Vyssh. Ucheb. Zaved., Geol. Razved., v. 17, no. 10, p. 44–49. (Chem. Abstr., v. 82, 88555m.)
Grechishnikov, N.P. and Sakhatskii, I.I.
1973: Conditions of gold accumulation in the conglomerates of Krivoi Rog; Geol. Zh., v. 33, no. 3, p. 110–114.
Green, A.H. and Keays, R.R.
1976: Geology and gold mineralization of the Woods' Point dyke swarm, Victoria, Australia; Geol. Assoc. Can., Annu. Meet., Edmonton, Abstr., p. 57.
Green, J.
1959: Geochemical table of the elements for 1959; Geol. Soc. Am., Bull., v. 70, p. 1127–1184.
Green, L.
1977: The gold hustlers; Alaska Northwest Publishing Co., Anchorage, Alaska, 339 p.
Green, L.H. and Roddick, J.A.
1972: Dawson, Yukon Territory; Geol. Surv., Can., Map 1284A.
Green, T.
1970: The world of gold; Simon and Schuster, Inc., New York, 254 p.
1973: The world of gold today; Walker and Company, New York, 287 p.
Green, T.E., Campbell, W.J. and Law, S.L.
1971: Application of a combined ion exchange paper-X-ray spectrographic method to geochemical exploration; *in* Geochemical exploration, R.W. Boyle and J.I. McGerrigle, eds., Can. Inst. Min. Metall., Spec. Vol. 11, p. 561 (abstr.).
Green, T.E. and Law, S.L.
1970: Properties of an ion exchange resin with high selectivity for gold; U.S. Bur. Mines, Rep. Invest. 7358, 9 p.
Green, T.E., Stephen, L.L. and Campbell, W.J.
1970: Use of selective ion exchange paper in X-ray spectrography and neutron activation, application to the determination of gold; Anal. Chem., v. 42, no. 14, p. 1749–1753.
Greenberg. R.
1964–1965: The mode of origin of banket orebodies, discussion; Inst. Min. metall., Trans., v. 74, no. 703, p. 574–576.
Greenland, L.
1967: The abundances of selenium, tellurium, silver, palladium, cadmium, and zinc in chondritic meteorites; Geochim. Cosmochim. Acta, v. 31, no. 5, p. 849–860.
Greig, J.A.
1976: Claymore gold placers, Ladue River area, Yukon-Alaska; Geol. Assoc. Can. Cordilleran Sec., Geomorphology of the Canadian Cordillera and its bearing on mineral deposits, Prog. Abstr., Vancouver, p. 21.
Gribnitz, K.H.
1964: Notes on the Barberton goldfield; *in* The geology of some ore deposits in southern Africa, v. 2, S.H. Haughton, ed.; Geol. Soc. S. Afr., Johannesburg, p. 77–90.
Griffis, A.T.
1962: A geochemical study of the McIntyre mine; Can. Inst. Min. Metall., Bull., v. 55, no. 598, p. 76–83.
Griffith, S.V.
1960: Alluvial prospecting and mining, 2nd. ed.; Pergamon Press, Inc., New York, 245 p.
Griffitts, W.R. and Alminas, H.V.
1968: Geochemical evidence for possible concealed mineral deposits near the Monticello Box, northern Sierra Cuchillo, Socorro County, New Mexico, U.S. Geol. Surv., Circ. 600, 13 p.
Griffitts, W.R., Alminas, H.V. and Mosier, E.L.
1971: Trace metal distribution in the Vicks Peak, Steel Hill, and Black Hill quadrangles, Socorro County, New Mexico; U.S. Geol. Surv., Open File Rep. Temp. 1436 (Nb and Au).
Grigor'ev, A.M.
1968: Geochemical underground prospecting for gold in northern Kazakhstan; Glubinnye Poiski Rud. Mestorozhd., v. 2, p. 267–289. (Chem. Abstr., v. 73, 79378 t.)
Grigoryan, S.V. and Zubov, M.A.
1971: Endogeneous geochemical haloes of gold ore deposits; Vop. Prikl. Geokhim., no. 2, p. 44–52. (Chem Abstr., v. 77, 77800h.)
Grimanis, A.P., Pantazis, G., Papadopoulas, C. and Tsanos, N.
1965: Determination of trace elements in the Greek lakes by neutron-activation analysis; Proc. 3rd Int. Conf. Peaceful Uses Atomic Energy, Geneva 1964, v. 15, p. 412–419.
Grip, E.
1953: Tracing of glacial boulders as an aid to ore prospecting in Sweden; Econ. Geol., v. 48, p. 715–725.
Grip, E. and Wirstam, A.
1970: The Boliden sulphide deposit; Sver. Geol. Unders., Ser. C, no. 651, v. 64, no. 8, 68 p.
Grishin, V.M. and Zlomanova, Z.G.
1971: Morphogenetic differences in vein quartz of gold ore deposits of northern Kazkhstan; Tr. Inst. Geol. Nauk, Akad. Nauk Kaz. SSR, v. 31, p. 63–72. (Chem. Abstr., v. 80, 123151j.)
Grishin, V.M., Izyumskiy, S.I. and Gapouov, T.G.
1971: Thermoluminescence of quartz from primary gold deposits of northern Kazhakstan; Zap. Vses. Mineral. Obshchest., v. 100, no. 2, p. 153–162.
Gritsik, V.V., Gritsik, E.P. and Matkovskii, O.I.
1969: Native gold from the upper reaches of the Belyi Cheremosh River; Mineral. Sb. (Lvov), v. 22, no. 4, p. 386–393. (Chem. Abstr., v. 72, 45981y.)
Gross, J. and Scott, J.W.
1927: Precipitation of gold and silver from cyanide solution on charcoal; U.S. Bur. Mines Tech., Pap. 378, 78 p.
Gross, W.H.
1950: A study of the spatial relation of gold ore to intrusive bodies in northwestern Ontario; Geol. Assoc. Can., Proc., v. 3, p. 123–139.
1952: Radioactivity as a guide to ore; Econ. Geol., v. 47, p. 722–742.
1968: Evidence for a modified placer origin for auriferous conglomerates, Canavieiras mine, Jacobina, Brazil; Econ. Geol., v. 63, p. 271–276.
1975: New ore discovery and source of silver-gold veins, Guanajuato, Mexico; Econ. Geol. v. 70, p. 1175–1189.
Grosse, E.
1932: Zur Kenntnis der Gold-Silberlagerstätten von Titiribi; Z. Prakt. Geol., v. 40, p. 44–45.
Grossman, L.
1973: Refractory trace elements in Ca-Al-rich inclusions in the Allende meteorite; Geochim. Cosmochim. Acta, v. 37, p. 1119–1140.
Grout, F.F.
1913: On the behaviour of cold acid sulphate solutions of copper, silver and gold with alkaline extracts of metallic sulphides; Econ. Geol., v. 8, p. 407–433.
1946: Microscope characters of vein carbonates; Econ. Geol., v. 41, p. 475–502.
Gruner, J.W.
1933: The solubilities of metallic sulphides in alkali sulphide solutions; Econ. Geol., v. 28, p. 773–777.
Grutterink, J.A.
1950: The origin of the residual gold deposits of Surinam (Dutch Guyana); Geol. Mijnbouw, no. 3, p. 89–93.
Guigues, J. et Devismes, P.
1969: La prospection minière à la batée dans le Massif Armoricain; B.R.G.M., Mem., no. 71, 167 p.
Guigues, J. et Machairas, G.
1972: Les diverses minéralisations aurifères du Massif Armoricain, en relation avec le volcanisme, la sédimentation et les orogénies; 24th Int. Geol. Congr. Montreal, Sec. 4, Mineral deposits, p. 188–195.

Guiguet, M.
1961: Cariboo gold quartz; Western Miner, v. 34, no. 7, p. 37–44.
Guild, P.W.
1972: Metallogeny and the new global tectonics; 24th Int. Geol. Congr., Montreal, Sec. 4, Mineral deposits, p. 17–24.
Guimaraes, D.
1947: Mineral deposits of magmatic origin; Econ. Geol., v. 42, p. 721–736.
Guiterman, F.
1890: Gold deposits in the quartzite formation of Battle Mountain, Colorado; Colo. Sci. Soc., Proc., v. 3, pt. 3, p. 264–268.
Gümüs, A.
1970: Türkiye metalojenisi; Maden Tetkik ve Arama Enstitüsü Yayinlarindan, no. 144, Ankara, 30 p.
Gundobin, G.M. and Kolesnikov, V.N.
1974: Dispersion trains of subsurface gold-silver deposits; Ezheg. Inst. Geokhim., Sib. Otd., Akad. Nauk SSSR, 1973, L.V. Tauson, ed., p. 274–278.
Gunn, C.B.
1968: Origin of the bedrock values of placer deposits; Econ. Geol., v. 63, p. 86.
Gunning, H.C.
1948: Privateer mine; *in* Structural geology of Canadian ore deposits, v. 1, Can. Inst. Min. Metall., Montreal, p. 86–88.
Gurev, V.F.
1970: Mineralogical-geochemical heavy concentrate method of prospecting for gold deposits; Izv. Tomsk, Politekh. Inst., no. 239, p. 411–415. (Chem. Abstr., v. 75, 120080w.)
Gurov, L.P.
1972: Gold-bearing deposits of the upper Amur River region; Vop. Geol. Rud. Mesterozhd. Dal'nego Vostoka, G.I. Osipova, ed., Akad. Nauk SSSR, Dal'nevost Geol., Vladivostok, U.S.S.R., p. 66–72. (Chem. Abstr., v. 82, 19286n.)
Gurulev, S.A., Feofilaktov, G.A. and Truneva, M.F.
1972: Genetic properties of the gold-bearing pyrite complex-metal mineralization of the Kamennyi region in the northern Muisk Range; Mater. Geol. Polez. Iskop. Buryat. ASSR, v. 15, p. 88–94. (Chem. Abstr., v. 82, 127549j.)
Gustafson, J.K.
1933: Metamorphism and hydrothermal alteration of the Homestake gold-bearing formation; Econ. Geol., v. 28, p. 123–162.
Gustavson, J.B. and Neathery, T.L.
1975: Geochemical prospecting for gold in Alabama; Am. Inst. Min. Eng., Annu. Meet., New York, Prepr. no. 75-L-54, 31 p.
Haber, F.
1927: Das Gold im Meerwasser; Z. Angew. Chem., v. 40, p. 303–314.
1928: Das Gold im Meere; Z. Ges. Erdkunde Berlin Erg.-Heft Nr. 3, p. 3–12.
Hackett, J.P. Jr. and Bischoff, J.L.
1973: New data on the stratigraphy, extent, and geologic history of the Red Sea geothermal deposits; Econ. Geol., v. 68, p. 553–564.
Haffty, J., Riley, L.B. and Goss, W.D.
1977: A manual on fire assaying and determination of the noble metals in geological materials; U.S. Geol. Surv. Bull. 1445, 58 p.
Hagen, J.C.
1954: Some aspects of the geochemistry of platinum, palladium, and gold in igneous rocks with special reference to the Bushveld Complex, Transvaal; unpubl. Ph.D. thesis, Mass. Inst. Technol., Cambridge, Mass., 300 p.
Hale, R.C.
1974: Gold deposits of the Coker Creek district, Monroe district, Tennessee, Tenn., Div. Geol., Bull., v. 72, p. 1–93. (Chem. Abstr., v. 82, 114266q.)
Hale, W.E. and Govett, G.J.S.
1968: Applied geochemistry in exploration for selected mineral occurrences in the Philippines; Econ. Geol., v. 63, p. 87 (abstr.).
Halet, R.A.
1948: Malartic Gold Fields mine; *in* Structural geology of Canadian ore deposits, v. 1, Can. Inst. Min. Metall., Montreal, p. 868–875.
1957: Quesabe mine; *in* Structural geology of Canadian ore deposits, v. 2, Can. Inst. Min. Metall., Montreal, p. 413–415.
Haley, C.S.
1923: Gold placers of California; Calif. St. Min. Bur., Bull. 92, 167 p.
Hall, A.L.
1930: The gold resources of the Union of South Africa — Murchison Range; *in* The gold resource of the world, 15th Int. Geol. Congr., Pretoria, p. 335–337.
Hall, G.W.
1977: The gold mines of Merioneth; Griffin Publications, Gloucester, U.K., 120 p.
Hallbauer, D.K.
1975: The plant origin of the Witwatersrand carbon; Mineral. Sci. Eng., v. 7, no. 2, p. 111–131.
Hallbauer, D.K. and Joughin, N.C.
1972: Distribution and size of gold particles in the Witwatersrand reefs and their effects on sampling procedures; Inst. Min. Metall. Trans., Sec. A, v. 81, p. 133–142.
Hallbauer, D.K. and van Warmelo, K.T.
1974: Fossilized plants in thucholite from Precambrian rocks of the Witwatersrand, South Africa; Precambrian Res., v. 1, p. 199–212.
Hallberg, J.A.
1972: Geochemistry of Archean volcanic belts in the Eastern Goldfields region of western Australia; J. Petrol., v. 13, no. 1, p. 45–56.
Hamaguchi, H., Kuroda, R., Tomura, K., Osawa, M. *et al.*
1961: Values for trace elements in G-1 and W-1 with neutron activation analysis; Geochim. Cosmochim. Acta, v. 23, p. 296–299.
Hammond, E.M.
1889: The auriferous gravels of California; 9th Annu. Rep., State Mineral., Calif. St. Min. Bur., p. 105–138.
Hamrabaev, I.H., Rahmatullaev, H.R. and Masdiev, R.A.
1972: Some specific features of the distribution of Au, Pb, U, and Th in granitoid rocks concealed beneath other rocks in the territory of the Kyzyl Kum desert; *in* Geol. Mineral. Geokhim. Uzb., Tashkent, "FAN" Publishing House, p. 180–184.
Hamrabaev, I.H. and Urunbaev, K.U.
1971: On the geochemistry of gold in the magmatic process; Int. Geochem. Congr., Moscow, U.S.S.R., Abstracts of Reports I, p. 173–174.
Hancock, R.D. and Finkelstein, N.P.
1971: π- and σ-bonding in the stabilities of d^{10} metal-ion complexes in aqueous solution; Inorg. Nucl. Chem. Lett., v. 7, p. 477–484.
Hanna, M.A. and Wolf, A.G.
1941: Gold, silver, and other elements in salt-dome cap rocks; Am. Assoc. Petrol. Geol., Bull., v. 25, p. 750–752.
Hanson, G.
1935: Barkerville gold belt, Cariboo district, British Columbia; Geol. Surv. Can., Mem. 181, 42 p.
Hanson, G. and McNaughton, D.A.
1936: Eagle-McDame area, Cassiar district, British Columbia; Geol. Surv. Can., Mem. 194, 16 p.
Harbaugh, J.W.
1953: Geochemical prospecting abstracts through June 1952; U.S. Geol. Surv., Bull. 1000-A, p. 1–50.

Hargraves, R.B.
1960: A bibliography of the geology of the Witwatersrand System; Univ. Witwatersrand, Econ. Geol. Res. Unit, Inform. Circ., no. 1, 41 p.
1961: Silver content of gold in Witwatersrand conglomerates; Univ. Witwatersrand, Econ. Geol. Res. Unit, Inform., Circ., no. 3, 25 p.
1963: Silver-gold ratios in some Witwatersrand conglomerates; Econ. Geol., v. 58, p. 952–970.

Hargraves, R.B. and Winter, H. de la R.
1964: A bibliography of the geology of the Witwatersrand system; *in* The geology of some ore deposits in southern Africa, v. 1, S.H. Haughton, ed.; Geol. Soc. S. Afr., Johannesburg, p. 601–616.

Harriman, F.G.
1965: Madsen Red Lake Gold Mines Limited, mine geology, Private Rep., 7 p.

Harris, J.F.
1961: Summary of the geology of Tanganyika, pt. 4: economic geology; Tanganyika Geol. Surv., Mem., no. 1, 143 p.

Harris, M. and Radtke, A.S.
1974: Relation of statistical findings to the geochemistry and genesis of the Carlin gold deposit, Nevada; Econ. Geol., v. 69, p. 1180–1181 (abstr.).
1976: Statistical study of selected trace elements with reference to geology and genesis of the Carlin gold deposit, Nevada; U.S. Geol. Surv., Prof. Pap. 960, 21 p.

Harrison, H.L.H.
1962: Alluvial mining for tin and gold; Mining Publications, Ltd., London, 313 p.

Harrison, J.B.
1905: Report on the petrography of the Cuyuni and Mazaruni districts, and of the rocks at Omai, Essequibo River, with some notes on the geology of part of the Berbice River: Georgetown, British Guiana (Rep. Gov. Geol.), 71 p.
1908: The geology of the goldfields of British Guiana; Dulau & Co., London, 320 p.

Harrison, J.M.
1948: Nor-Acme mine; *in* Structural geology of Canadian ore deposits, v. 1, Can. Inst. Min. Metall., Montreal, p. 304–306.

Harriss, R.C., Crocket, J.H. and Stainton, M.
1968: Palladium and gold in deep-sea manganese nodules; Geochim. Cosmochim. Acta, v. 32, p. 1049–1056.

Hart, E. A. and Gill, J.E.
1948: Arntfield mine; *in* Structural geology of Canadian ore deposits, v. 1, Can. Inst. Min. Metall., Montreal, p. 711–719.

Hart, R.C., Harper, H.G. and Algom field staff.
1957: Quirke Lake trough; *in* Structural geology of Canadian ore deposits, v. 2, Can. Inst. Min. Metall., Montreal, p. 316–324.

Hartmann, A. and Sangmeister, E.
1972: Zur Erforschung urgeschichtlicher Metallurgie; Angew. Chem., Bd. 84, no. 14, p. 668–678.

Harvey, R.D. and Vitaliano, C.J.
1964: Wall-rock alteration in the Goldfield District, Nevada; J. Geol., v. 72, no. 5, p. 564–579.

Hastings, J.B.
1908: Primary gold in a Colorado granite; Am. Inst. Min. Eng., Trans., v. 39, p. 97–103.

Hatch, F.H. and Corstorphine, G.S.
1909: The geology of South Africa, 2nd ed.; Macmillan and Co. Ltd., London.

Hatschek, E.
1919: An introduction to the physics and chemistry of colloids, 3rd ed.; J. & A. Churchill, London, 116 p.

Hatschek, E. and Simon, A.L.
1912: Gels in relation to ore deposition: Inst. Min. Metall. Trans., v. 21, p. 451–480.

Hattori, K.
1975: Geochemistry of ore deposition at the Yatani lead-zinc and gold-silver deposit, Japan; Econ. Geol. v. 70, p. 677–693.

Haugh, I.
1962: A petrographic study of the Falcon Lake stock, eastern Manitoba; unpubl. M.Sc. thesis, Univ. Manitoba, Winnipeg, 85 p.

Haughton, S.H., ed.
1964*a*: The geology of some ore deposits in southern Africa, v. 1; Gold deposits of the Witwatersrand Basin, Geol. Soc. S. Afr., Johannesburg, 625 p.
1964*b*: The geology of some ore deposits in southern Africa, v. 2; Geol. Soc. S. Afr., Johannesburg, 739 p.

Hausen, D.M. and Kerr, P.F.
1968: Fine gold occurrence at Carlin, Nevada; *in* ore deposits of the United States 1933–1967, v. 1, J.D. Ridge, ed., Am. Inst. Min. Metall. Pet. Eng. New York, p. 908–940.

Hauser, H.
1901: L'or; Librairie Nony et Cie, Paris, 363 p.

Hawkes, D.D.
1965: A note on the geochemical dispersion of arsenic in relation to gold mineralization at Marudi Mountain; Br. Guiana Geol. Surv., Rec., v. 3, p. 31–33.

Hawkes, H.E.
1957: Principles of geochemical prospecting; U.S. Geol. Surv., Bull. 1000-F, 355 p.

Hawkes, H.E. and Webb, J.S.
1962: Geochemistry in mineral exploration; Harper and Row, New York, 415 p.

Hawkins, D.B. and Roy, R.
1963: Experimental hydrothermal studies on rock alteration and clay mineral formation; Geochim Cosmochim. Acta, v. 27, p. 1047–1054.

Hawley, J.E.
1932: The Siscoe gold deposit; Can. Min. Metall., Bull., v. 25, p. 368–386.
1939: The association of gold, tungsten, and tin at Outpost Islands, Great Slave Lake; Univ. Toronto Studies, Geol. Ser. No. 42, Contrib. Can. Mineral,. p. 53–57.
1948: Francoeur mine; *in* Structural geology of Canadian ore deposits, v. 1, Can. Inst. Min. Metall., Montreal, p. 701–710.
1952: Spectrographic studies of pyrite in some eastern Canadian gold mines; Econ. Geol., v. 47, p. 260–304.
1962: The Sudbury ores: their mineralogy and origin; Can. Mineral., v. 7, pt. 1, 207 p.

Hawley, J.E. and Hart, R.C.
1948: Preston East Dome mine; *in* Structural geology of Canadian ore deposits, v. 1, Can. Inst. Min. Metall., Montreal, p. 528–538.

Hawley, J.E. and Nichol, I.
1961: Trace elements in pyrite, pyrrhotite, and chalcopyrite of different ores; Econ. Geol., v. 56, p. 467–487.

Haycock, M.H.
1937: The role of the microscope in the study of gold ores; Can. Inst. Min. Metall., Trans., v. 40, p. 405–414.

Head, R.E.
1934: Form and occurrence of gold in pyrite from a metallurgical standpoint; U.S. Bur. Mines, Rep. 3226, p. 27–31.
1935*a*: Form and occurrence of gold in pyrite; U.S. Bur. Mines, Rep. 3275, p. 39–46.
1935*b*: Form and occurrence of gold in pyrite-coated gold; Can. Min. J., v. 56, p. 517–521.

Head, R.E. and Miller, V.
1928: The occurrence of jarosite minerals in oxidized lead ores as a factor in metal losses; U.S. Bur. Mines, Rep. Invest., no. 2870, 13 p.

Healy, J. *et al.*
1963: Waiotapu geothermal field; N.Z. Dep. Sci. Ind. Res., Bull. 155, 141 p.

Healy, J.F.
1978: Mining and metallurgy in the Greek and Roman world; Thames and Hudson, London, 316 p.

Heatherington, A.
1868: Gold fields of Nova Scotia; John Lovell, Montreal, 170 p.

Hegemann, F. and Leybold, C.
1954: Eine Methode zur quantitative spektrochemischen Analyse von Pyrite; Z. Erzbergbau Metallhuttenw., v. 7, p. 108–113.

Heinen, H.J. and Eisele, J.A.
1974: Recovery of gold from ores; U.S. Patent 3,825,651. (Chem. Abstr., v. 81, 12464b.)

Heinz, W.F.
1975: Prospecting for gold; Pegasus Press, Christchurch, 80 p.

Helgeson, H.C. and Garrels, R.M.
1968: Hydrothermal transport and deposition of gold; Econ. Geol., v. 63, p. 622–635.

Helke, A.
1934: Die Goldtellurerzlagerstätten von Săcărâmb (Nagyág) in Rumänien; Neues Jahrb. Mineral., Geol. Paläeontol., v. 68, pt. A, p. 19–85.

Hemley, J.J., Hostetler, P.B., Gude, A.J. and Mountjoy, W.T.
1969: Some stability relations of alunite; Econ. Geol., v. 64, p. 599–612.

Henderson, J.
1917: The geology and mineral resources of the Reefton subdivision; N.Z. Geol. Surv., Bull. 18, 232 p.
1930: The gold resources of New Zealand; *in* The gold resources of the world, 15th Int. Geol. Congr., Pretoria, S. Afr., 1929, p. 259–263.

Henderson, J. and Bartrum, J.A.
1913: The geology of the Aroha subdivision, Hauraki, Auckland; N.Z. Geol. Surv., Bull. 16 (new ser.), 127 p.

Henderson, J.F.
1948*a*: The relation of gold deposits to structure, Yellowknife, N.W.T.; Precambrian, v. 21, no. 7, p. 6–11.
1948*b*: Structural control of ore deposits in the Canadian Shield between Great Slave and Great Bear lakes, Northwest Territories; *in* Structural geology of Canadian ore deposits, v. 1, Can. Inst. Min. Metall., Montreal, p. 238–243.

Henderson, J.F. and Brown, I.C.
1950: Structure of the Yellowknife Greenstone Belt, Northwest Territories; Can. Inst. Min. Metall., Trans., v. 53, p. 415–422.
1966: Geology and structure of the Yellowknife Greenstone Belt, District of Mackenzie; Geol. Surv. Can., Bull. 141, 87 p.

Henderson, J.F. and Fraser, N.H.C.
1948: Camlaren mine; *in* Structural geology of Canadian ore deposits, v. 1, Can. Inst. Min. Metall., Montreal, p. 269–272.

Henderson, J.F. and Jolliffe, A.W.
1939: Relation of gold deposits to structure, Yellowknife and Gordon Lake areas Northwest Territories; Can. Inst. Min. Metall., Trans., v. 42, p. 314–336.

Hendricks, R.L., Reisbick, F.B., Mahaffey, E.J., Roberts, D.B. and Peterson, M.N.A.
1969: Chemical composition of sediments and interstitial brines from the Atlantis II, Discovery and chain deeps; *in* Hot brines and recent heavy metal deposits in the Red Sea, E.T. Degens and D.A. Ross, ed., Springer-Verlag New York Inc. 1969, p. 407–440.

Henley, R.W.
1971: Geochemistry and genesis of Precambrian gold deposits; Ph.D. thesis, Univ. Manchester. (abstr., Int. Min. Metall. Trans., sec. B, v. 81, 1972, p. 126. *Also* Min. Sci. Eng., v. 5, no. 4, 1973, p. 295–303.)
1973: Solubility of gold in hydrothermal chloride solutions; Chem. Geol., v. 11, no. 2, p. 73–87.

Henley, R.W., Norris, R.J. and Paterson, C.J.
1976: Multistage ore genesis in the New Zealand geosyncline: a history of post-metamorphic lode emplacement; Miner. Deposita (Berl.), v. 11, p. 180–196.

Henley, K.J.
1975: Gold ore mineralogy and its relation to metallurgical treatment; Mineral. Sci. Eng., v. 7, p. 289–312.

Herman, H.
1914: Economic geology and mineral resources of Victoria; Victoria (Australia) Geol. Surv., Bull. 34, 36 p.

Herz, N.
1966: Gold; *in* Treatise on analytical chemistry, I.M. Kolthoff, P.J. Elving and E.B. Sandell, eds., pt. 2, v. 4, p. 71–105, Interscience Publishers, New York.

Hess, F.L.
1917: Tungsten minerals and deposits; U.S. Geol. Surv., Bull. 652, 85 p.

Hewett, D.F.
1931: Zonal relations of the lodes of the Sumpter Quadrangle; Am. Inst. Min. Metall. Eng., Trans., Gen. Vol., p. 305–346.

Hewett, D.F. and Rove, O.N.
1930: Occurrence and relations of alabandite; Econ. Geol., v. 25, p. 36–56.

Hewett, D.F., Fleischer, M. and Conklin, N.
1963: Deposits of the manganese oxides: supplement; Econ. Geol., v. 58, p. 1–51.

Hey, M.H.
1962: An index of mineral species and varieties arranged chemically, 2nd ed; Br. Mus., 728 p.
1963: Appendix to the second edition of an index of mineral species and varieties arranged chemically; Br. Mus., 135 p.
1966: Catalogue of meteorites, 3rd ed.; Br. Mus., Publ. no. 464, London, 637 p.

Hey, M.H. and Embrey, P.G.
1974: A second appendix to the second edition of an index of mineral species and varieties arranged chemically; Br. Mus., Publ. no. 725, 168 p.

Hildebrand, F.A. and Gott, G.B.
1974: Coloradoite, acanthite, and jarosite from the Cripple Creek district, Teller County, Colorado; U.S. Geol. Surv., J. Res., v. 2, no. 3, p. 339–340.

Hildon, M.A. and Sully, G.R.
1971: The determination of gold in the p.p.b. and p.p.m. ranges by atomic absorption spectrometry; Anal. Chim. Acta, v. 54, p. 245–251.

Hill, J.M.
1915: Notes on the fine gold of Snake River, Idaho; U.S. Geol. Surv., Bull. 620-L, p. 271–294.

Hillebrand, J.R.
1954: A suggested source of the thiophile elements of the Tertiary ore deposits of the southwest; Econ. Geol., v. 49, p. 863–876.

Hiller, J.E.
1962: Die mineralischen Rohstoffe; E. Schweizerbart'sche Verlagsbuchhandlung, Stuttgart, 359 p.

Hills, E.S.
1947: The metalliferous geochemical zones of Australia; Econ. Geol., v. 42, p. 478–491.
1953: Tectonic setting of Australian ore deposits; *in* Geology of Australian ore deposits, v. 1, A.B. Edwards, ed., 5th Emp. Min. Metall. Congr., Australia and New Zealand, p. 41–61.

Hintenberger, H., Jochum, K.P. and Seufert, M.
1973: The concentrations of the heavy metals in the four new Antarctic meteorites Yamato (a), (b), (c), and (d) and in Orgueil, Murray, Allende, Abee, Allegan, Mocs, and Johnstown; Earth Planet. Sci. Lett., v. 20, no. 3, p. 391–394.

Hintze, C.
1889–1968: Handbuch der Mineralogie; 6 vol., Berlin and Leipzig.

Hirst, T.
1938: The geology of the Tarkwa goldfield and adjacent country; Gold Coast Geol. Surv., Bull. no. 10, 24 p.
1942: The geology of the Konongo gold belt and surrounding country; Bull. Imp. Inst., v. 40, no. 1, p. 36–61.

Hite, T.H.
1933*a*: Fine gold and platinum of Snake River, Idaho; Econ. Geol., v. 28, p. 256–265.
1933*b*: Special features of fine gold from Snake River, Idaho; Econ. Geol., v. 28, p. 686–691.

Ho, C.S.
1953: Mineral resources of Taiwan; Taiwan Geol. Surv., 313 p.

Hobbs, F.
1943: Gold, the real ruler of the world; The Business Foundation Publishers, Chicago, Ill., 265 p.

Hobson, B.
1971: Historic gold coins of the world; Doubleday & Company, Inc., Garden City, New York, 192 p.

Hobson, G.V. and Robinson, A.M.
1943: The Gold Coast banket: some aspects of its geology in relation to mining; Inst. Min. Metall., Bull., no. 460, p. 1–23.

Hodgson, J.N.
1968: The optical properties of gold; J. Phys. Chem. Solids, v. 29, p. 2175–2181.

Hoefs, J., Nielsen, H. and Schidlowski, M.
1968: Sulfur isotope abundances in pyrite from the Witwatersrand conglomerates; Econ. Geol., v. 63, p. 975–977.

Hoefs, J. and Schidlowski, M.
1967: Carbon isotope composition of carbonaceous matter from the Precambrian of the Witwatersrand System; Science, v. 155, p. 1096–1097.

Hoffman, A.
1947: Free gold, the story of Canadian mining; Rinehart, New York and Toronto, 420 p.

Hoffman, F.
1931: Die Gold-Silverlagerstätte von Titiribi (Kolumbien); Z. Prakt, Geol., v. 39, p. 1–13, 19–26.

Höfler, H. and Sorantin, H.
1967: Application of nondestructive activation analysis to meteorites: determination of aluminum, vanadium, manganese, and gold in stony and iron meteorites; Chem. Geol., v. 2, no. 4, p. 273–278.

Hofmann, F.
1965: Untersuchungen über den Goldgehalt der Oberen Marinen Molasse und des Stubensandsteins in der Gegend von Schaffhausen; Schweiz. Mineral. Petrogr. Mitt., v. 45, no. 1, p. 131–137. (Chem. Abstr., v. 53, 14553c.)

Hogan, J.
1975: Gold; Can. Min. J., v. 96, no. 2, p. 109–112.

Hogg, N.
1957: Nor-Acme mine; *in* Structural geology of Canadian ore deposits, v. 2, Can. Inst. Min. Metall., Montreal, p. 262–275.

Hoggins, F.E. and Brooks, R.R.
1973: Natural dispersion of mercury from Puhipuhi, Northland, New Zealand; N.Z. J. Mar. Freshwater Res., v. 7, nos. 1 and 2, p. 125–132.

Holland, S.S.
1944: Lode-gold deposits, north-eastern British Columbia and Cariboo and Hobson Creek areas; British Columbia Dep. Mines, Bull. no. 20, pt. 6, 16 p.
1950: Placer gold production of British Columbia; B.C. Dep. Mines, Bull. 28, 89 p.

Holmes, S.W.
1957: Pronto Mine; *in* Structural geology of Canadian ore deposits, v. 2, Can. Inst. Min. Metall., Montreal, p. 324–339.

Holmes, T.C.
1948: Dome mine; *in* Structural geology of Canadian ore deposits, v. 1, Can. Inst. Min. Metall., Montreal, p. 539–547.

Hoover, H.C.
1899: The superficial alteration of Western Australian ore deposits; Am. Inst. Min. Eng., Trans., v. 28, p. 758–765.

Hoover, H.C. and Hoover, L.H., translators
1912: De re metallica by G. Agricola (1556); Mining Mag., London, 637 p.

Hopkins, H.
1950: Geology of the Wright-Hargreaves mine; Ont. Dep. Mines, v. 57, pt. 5, p. 161–176.

Hopkins, P.E.
1922: Ontario gold deposits; Ont. Dep. Mines, Annu. Rep., 1921, v. 30, pt. 2, 73 p.
1923: Lebel and Gauthier townships; Ont. Dep. Mines, v. 32, pt. 4, p. 53–88.
1925*a*: Larder Lake area; Ont. Dep. Mines, Annu. Rep., 1924, v. 33, pt. 3, p. 1–26.
1925*b*: Night Hawk Lake gold area; Ont. Dep. Mines, Annu. Rep., 1924, v. 33, pt. 3, p. 27–36.

Hopper, S.A.J.
1948: Golden Manitou mine; *in* Structural geology of Canadian ore deposits, v. 1, Can. Inst. Min. Metall., Montreal, p. 891–893.

Horikoshi, E.
1969: Volcanic activity related to the formation of the Kurokô-type deposits in the Kosaka district, Japan; Miner. Deposita (Berl.), v. 4, p. 321–345.

Horikoshi, Y.
1951–1952: Fundamental studies of the Kurokô deposits, with special reference to the Hanaoka mine, I–III; Min. Geol. (Japan), v. 1, no. 1, p. 1–11 and v. 1, no. 2, p. 69–78.

Horn, M.K. and Adams, J.A.S.
1966: Computer-derived geochemical balances and element abundances; Geochim. Cosmochim. Acta, v. 30, p. 279–297.

Horwood, C.B.
1917: The gold deposits of the Rand; Charles Griffin & Company Ltd., London, 400 p.

Horwood, H.C.
1938: Geology of the Casummit Lake area and the Argosy mine; Ont. Dep. Mines, Annu. Rep., 1937, v. 46, pt. 7, 33 p.
1945: Geology and mineral deposits of the Red Lake area; Ont. Dep. Mines., Annu. Rep., 1940, v. 49, pt. 2, 231 p.
1948*a*: General structural relationships of ore deposits in the Little Long Lac–Sturgeon River area; *in* Structural geology of Canadian ore deposits, v. 1, Can. Inst. Min. Metall., Montreal, p. 377–384.
1948*b*: General structural relationships of ore deposits in the Red Lake area; *in* Structural geology of Canadian ore deposits, v. 1, Can. Inst. Min. Metall., Montreal, p. 322–328.
1948*c*: Gold Eagle mine; *in* Structural geology of Canadian ore deposits, v. 1, Can. Inst Min. Metall., Montreal, p. 346–351.
1948*d*: McKenzie Red Lake mine; *in* Structural geology of Canadian ore deposits, v. 1, Can. Inst. Min. Metall., Montreal, p. 352–356.
1948*e*: Howey and Hasaga mines; *in* Structural geology of Canadian ore deposits, v. 1, Can. Inst. Min. Metall., Montreal, p. 340–345.

Hosted, J.O. and Wright, L.B.
1923: Geology of the Homestake orebodies and the Lead area of South Dakota; Eng. Min. J., v. 115, p. 793–799, 836–843.

Houtz, R.E. and Phillips, K.A.
1963: Interim report on the economic geology of Fiji; Fiji, Geol. Surv. Dep., Econ. Rep. no. 1, 36 p.

Hovorka, D. and Jaros, M.
1974: Gold abundances in West Carpathian ultramafic rocks; Geol. Zb. (Slov. Akad. Vied), v. 25, no. 2, p. 355–362.

Howe, E.
1924: The gold ores of Grass Valley, California; Econ. Geol., v. 19, p. 595–622.

Howie, F.H. and Veale, C.R.
1966: Low temperature synthesis of gold telluride; J. Inorg. Nucl. Chem., v. 28, p. 1149–1154.

Hrytsyk, V.V., Hrytsyk, K.P. and Plavshudin, V.G.
1969: First discovery of native gold in earthy manganese ores of the Chyvchyn Mountains; Rep., Acad. Sci. Ukr. SSR, Ser. B, Geol., Geophys. Chem., Biol., v. 31, no. 6, p. 493–496.

Huang, C.K.
1955: Gold-copper deposits of the Chinkuashih mine, Taiwan, with special reference to the mineralogy; Acta Geol. Taiwan., no. 7, p. 1–20.
1960: Gold deposits of the Wutanshan area, Taiwan, and their relation to the Chinkuashih deposits; Acta Geol. Taiwan., no. 8, p. 13–25.
1972: Characteristic features of the gold-copper deposits of the Chinkuashih area, and prospect of copper production in Taiwan; Geol. Soc. China, Proc., no. 15, p. 5–6.
1973: Minor gangue minerals of the Chinkuashih gold-copper deposits, Taiwan; Acta Geol. Taiwan., no. 16, p. 31–38. (Chem. Abstr., v. 80, 62060e.)

Hughes, I.G.
1973: Mineral resources of Jamaica; Jamaica Geol. Surv. Dep., Bull. 8, 89 p.

Huffman, C., Mensik, J.D. and Riley, L.B.
1967: Determination of gold in geologic materials by solvent extraction and atomic-absorption spectrometry; U.S. Geol. Surv., Circ. 544, 6 p.

Hulin, C.D.
1929– Metallization from basic magmas; Dep. Geol., Univ.
1930: Calif., Bull., v. 18, no. 9, p. 233–274.
1930: A Mother Lode gold ore; Econ. Geol., v. 25, p. 348–355.

Hummel, C.H.
1962*a*: Preliminary geologic map of the Nome D-1 quadrangle, Seward Peninsula, Alaska; U.S. Geol. Surv., Min. Invest. Field Studies Map MF-248.
1962*b*: Preliminary geologic map of the Nome C-1 quadrangle, Seward Peninsula, Alaska; U.S. Geol. Surv., Min. Invest. Field Studies Map MF-247.

Hummel, R.W.
1957: Determination of gold in sea water by radioactivation analysis; Analyst, v. 82, p. 483–488.

Hunahashi, M.
1973: Quartz fabrics in an epithermal ore vein; Contrib. Mineral. Petrol., v. 39, no. 2, p. 157–170.

Hunt, T.S.
1879: Chemical and geological essays, 2nd ed.; Scientific Publ. Co., New York and London, 489 p.

Hunter, S.
1909: The deep leads of Victoria; Victoria (Australia) Geol. Surv., Mem. 7., 142 p.

Hurst, M.E.
1927: Arsenic-bearing deposits in Canada; Geol. Surv. Can., Econ. Geol. Ser. no. 4, 181 p.
1931: Pickle Lake-Crow River area, District of Kenora (Patricia Portion); Ont. Dep. Mines, Annu. Rep., 1930, v. 39, pt. 2, p. 1–35.
1935: Vein formation at Porcupine, Ontario; Econ. Geol., v. 30, p. 103–127.

Hussak, E.
1898: Der goldführende, kiesige Quarzlagergang von Passagem in Minas Geraes, Brasilien; Z. Prakt. Geol., v. 5, p. 345–357.

Hutchinsón, R.W.
1973: Volcanogenic sulfide deposits and their metallogenic significance; Econ. Geol., v. 68, p. 1223–1246.
1975: Lode gold deposits; the case for volcanogenic derivation; *in* Proc. 5th Gold and Money Session and Gold Technical Session, Pacific Northwest Metals and Minerals Conference, Portland, Oregon, Oreg. Dep. Geol. Miner. Ind. publ., p. 64–105.

Huttl, J.B.
1940: Working a gossan gold ore with modern earth-moving equipment; Eng. Min. J., v. 141, p. 44–47.

Hytonen, K. and Kallio, P.
1974: Heavy minerals of the Lemmenjoki alluvial gold-bearing deposit, Inari, northern Lapland; Geologi, v. 26, p. 103.

Hyvärinen, L., Kauranne, K. and Yletyinen, V.
1973: Modern boulder tracing in prospecting; *in* Prospecting in areas of glacial terrain, M.J. Jones, ed.; Inst. Min. Metall., p. 87–95.

Ianovici, V. *et al.*
1969: Evolutia geologica a muntilor metaliferi; Editura Acad. Republ. Soc. Romania, Bucuresti, 741 p.

Ibbotson, P.
1962: The geology of the Tavua area, Viti Levu; Fiji, Geol. Surv. Dep., Bull. 8, 25 p.
1967: Petrology of the Tertiary caldera Tavua goldfield; Fiji, Geol. Surv. Dep., Mem. 3, 59 p.

Idriess, I.L.
1933: Prospecting for gold, 4th ed.; Angus and Robertson, Ltd., Sydney, Australia, 293 p.
1934: Gold-dust and ashes; the romantic story of the New Guinea goldfields; Angus and Robertson, Ltd., Sydney, Australia, 285 p.

Igumnova, N.S.
1974: Gold in endogenous mineral associations of the Khangalas deposit; Vop. Rudonos. Yakutii, G.N. Gamyanin, ed., p. 47–52. (Chem. Abstr., v. 84, 76925k.)

Ikeda, N.
1955*a*: The hot springs of Nasu IX; Nippon Kagaku Zasshi, v. 76, p. 833–836.
1955*b*: Chemical studies on the hot springs of Nasu XI; Nippon Kagaku Zasshi, v. 76, p. 1071–1073.

Il'enok, S.S.
1970: Genetic relation of mineralization with intrusions; *in* Geol. Zolotorud. Mest. Sib., F.M. Shakhov, ed., "Nauka", Sib. Ord., Novosibirsk, U.S.S.R., p. 3–30. (Chem. Abstr., v. 74, 24190r.)

Il'jin, K.B.
1968: Map of the major metallogenic zones of the U.S.S.R.; U.S.S.R. Min. Geol. (VSEGEI).
1974: Regional metallogeny of the U.S.S.R.; "Nedra" Publ. House, Moscow, 237 p.

Indolev, L.N., Koshik, I.M. and Zhadanov, Yu. Ya.
1974: Structure and mineral composition of the Sarylakh gold-antimony deposit; Mineral. Endog. Mestorozhd. Yakutii, p. 92–108. (Chem. Abstr., v. 82, 114223y.)

Indolev, L.N. and Zhadanov, Yu. Ya.
1975: Aurostibite from the gold antimony deposits of Yakutia; Zap. Vses. Mineral. O-Va, v. 104, p. 586–588.

Ingall, E.D.
1893: The precious metals—New Brunswick; Geol. Surv. Can., Annu. Rep. 1891 (new ser.), v. 5, pt. 2, p. 147ss.

International Geological Congress
1930: The gold resources of the world; 15th Int. Geol. Congr., Pretoria, South Africa, 457 p.

Ipat'eva, I.S.
1968: Accessory minerals in granitic rocks of the Upper Indigirka region (eastern Yakut A.S.S.R.); *in* Aktsessoryne Miner, Izverzhennykh Porod, I.E. Smorchkov, ed., Izd. "Nauk", Moscow, p. 173–180. (Chem. Abstr., v. 71, 63163s.)

Irving, J.D., Emmons, S.F. and Jaggar, T.A.
1904: Economic resources of the northern Black Hills; U.S. Geol. Surv., Prof. Pap. 26, 222 p.

Ishihara, S. *et al.*
1974: Geology of Kurokô deposits; Min. Geol. Spec. Issue no. 6, Publ. Soc. Min. Geol. Japan, 435 p.

Ishii, E.
1974: Statistical analysis of the veins of the Ohkuchi Mine, Kagoshima Prefecture, Japan; Kozan Chishitsu, v. 24, no. 3, p. 213–222. (Chem. Abstr., v. 81, 155890m.)

Ishikawa, H., Kuroda, R. and Sudo, T.
1962: Minor elements in some altered zones of "Kurokô" (Black Ore) deposits of Japan; Econ. Geol., v. 57, p. 785–798.

Ishmurzin, F.Ya. and Volkov, Yu.Ya.
1974: Association of gold with bismuth telluride at the Kumbel'skoe ore field in northern Kirgizia; Uzb. Geol. Zh., v. 18, no. 3, p. 82–83. (Chem. Abstr., v. 82, 61963f.)

Ismagilov, M.I.
1969: Primary zoning in Maisk gold complex ore deposit; Ural'skoe Petrografischeskoe Soveshchanie, Tr., 2nd, Sverdlovsk, 1966, no. 5, p. 178–184.

Ismagilov, M.I. and Ismagilova, M.Z.
1974: Geochemistry and mineralogy of trace elements of the Baimak gold deposits in southern Urals; Mineral., Geokhim. Metamorf. Polez. Iskop. Bashk. Zaural'ya, p. 62–68. (Chem. Abstr., v. 83, 208418p.)

Itsikson, M.I.
1974: Metallogenic systems of the Pacific segment of the earth; Geol. Rud. Mestorozhd., v. 16, no. 2, p. 10–28.

Ivankin, P.F. and Rabinovich, K.R.
1972: Gold-bearing ore magmatic systems of the granitoid series; Int. Geol. Rev., v. 14, no. 9, p. 1002–1007.

Ivanov, T., Shalaby, I.M. and Hussein, A.A.A.
1973: Metallogenic characteristics of South-Eastern Desert of Egypt; Egypt, Geol. Surv., Annals, v. 3, p. 139–166.

Ivanov, V.V.
1959: On the genetic types of endogenous thallium-bearing deposits; Trans. Inst. Min. Geochem. Crystallogr. Chem. no. 2, Acad. Sci. USSR, Moscow, p. 230–241.

Ivanov, V.V. and Meituv, G.M.
1970: On the problem of the source of some trace elements in hydrothermal ores; *in* Problems of hydrothermal ore deposition, Z. Pouba and M. Stemprok, ed., Int. Union Geol. Sci., Ser. A, no. 2; E. Schweizerbart'sche Verlagsbuchhandlung Stuttgart and Akademiai Kiado, Budapest, p. 100–102.

Ivanov, V.V., Poplavko, E.M. and Gorokhova, V.N.
1972: The geochemistry of rhenium; Int. Geol. Rev., v. 14, no. 2, p. 1–105.

Ivensen, Yu.P., Amuzinskii, V.A., Korobitsyn, A.V., Kikhtinskii, G.G. and Trunilina, V.A.
1974: Nature of the distribution of gold in magmatic rocks; Akad. Nauk. SSSR, Dokl., v. 216, no. 6, p. 1387–1389. (Chem. Abstr., v. 82, 33463z.)

Ivensen, Yu. P., Levin, V.I. and Nuzhnov, S.V.
1963: Placer deposits. Gold-bearing conglomerates; Str. Zemnoi Kory Yakutii Zakonomer. Razmeshcheniya Polez. Iskop., Yu. P. Ivensen, ed., Izd. "Nauka", Moscow, U.S.S.R., p. 325–330. (Chem. Abstr., v. 73, 27541w.)
1969: The types of fossil gold placers in various formations and methods of prospecting for them; USSR Acad. Sci., Sib. Div., Yakutsk Branch, "Nauka" Publishing House, Moscow, 210 p.

Ivensen, Yu. P., Stepanov, A.A. and Chaikovskii, B.K.
1963: Gold-bearing conglomerates; Razved. Okhr. Nedr., v. 29, no. 2, p. 1–7. (Chem. Abstr., v. 59, 3653a.)

Ivkin, I.I. and Sokolov, V.A.
1971: Quantitative and paragenetic interrelations of gold and silver in metasomatites of secondary quartzite formations; Mat. Metody Geol., no. 2, p. 37–43. (Chem. Abstr., v. 82, 46356s.)

Iwano, S. and Kuwano, N.
1974: Miargyrite and other minerals from the Hojo mine, Kagoshima Prefecture, Japan; Chigaku Kenkyu, v. 25, no. 1–6, p. 94–101. (Chem. Abstr., v. 84, 7591m.)

Iyer, G.V.A.
1970: Geochemistry of the alkali feldspars and origin of gold-quartz veins in the Precambrian of S.E. Mysore; unpubl. Ph.D. thesis, Indian Inst. Sci., Bangalore, 224 p.

Jack, T. and Powell, J.
1972: Copper, silver and gold: annual survey covering the year 1970; Organometal. Chem. Rev., Sec. B., v. 10, p. 281–300.

Jacob, W.
1831: History of the precious metals, London.

Jae, H.K.
1973: A geochemical study on mercury element in the environments of gold-silver ore deposits; J. Korean Inst. Mineral. Min. Eng., v. 10, no. 4, p. 273–279 (Engl. abstr.).

Jaenicke, J.
1935: Habers Forschungen über das Goldvorkommen im Meerwasser; Naturwissenschaften 23, p. 57–63.

Jahns, R.H.
1955: The study of pegmatites; Econ. Geol., 50th Anniv. Vol., pt. 2, p. 1025–1130.

James, C.H.
1957: Applied geochemical studies in southern Rhodesia and Great Britain; unpubl. Ph.D. thesis, Univ. London, R. Sch. Mines. *Also* Imperial College Sci. Technol. Geochem. Prospecting Res. Cen., Tech. Comm. no. 12, p. 1–237.
1962: A review of the geochemistry of mercury (excluding analytical aspects) and its application to geochemical prospecting; Univ. London, Imperial College, Sci. Technol., Geochem. Res. Cen., Tech. Comm. no. 41, 42 p.
1964: The potential role of mercury in modern geochemical prospecting; Min. Mag., v. 110–111, p. 23–32.

James, L.D.
1965: Regional geochemical reconnaissance in the northern and southern sections of the Sula Mountains schist belt, Sierra Leone; unpubl. Ph.D. thesis, Univ. London, Imperial College, R. Sch. Mines.

James, W.F.
1948: Siscoe mine; *in* Structural geology of Canadian ore deposits, v. 1, Can. Inst. Min. Metall., Montreal, p. 876–882.

Janssen, E.M.V., Pohlmann, F. and Wiegers, G.A.
1976: The phase diagram of the gold-chlorine system; J. Less-Common Metals, v. 45, p. 261–273.

Jasper, M.W.
1967: Geochemical investigations, Willow Creek southerly to Kenai Lake region, south-central Alaska; Alaska, Div. Mines, Miner., Geochem. Rep. no. 14, p. 1–47.

Jenkins, O.P.
1964: Geology of placer deposits, pt. 1–9; Miner. Inform. Serv., Calif. Div. Mines, Geol. v. 17.

Jenkins, O.P. and Wright, W.Q.
1934: California's gold-bearing Tertiary channels; Eng. Min. J., v. 135, no. 11, p. 497–502.

Jenks, W.F., ed.
1956: Handbook of South American geology; Geol. Soc. Am., Mem. 65, 378 p.

Jenne, E.A., Chao, T.T. and Heppting, L.M.
1968: Use of mercurous chloride to recover trace amounts of gold from waters; Econ. Geol., v. 63, p. 420–421.

Jenny, W.P.
1903: The chemistry of ore deposition; Am. Inst. Min. Eng., Trans., v. 33, p. 445–498.

Jensen, M.L.
1971: Stable isotopes in geochemical prospecting; *in* Geochemical exploration, R.W. Boyle and J.I. McGerrigle, eds., Can. Inst. Min. Metall., Spec. Vol. 11, p. 464–468.

Jensen, M.L., Ashley, R.P. and Albers, J.P.
1971: Primary and secondary sulfates at Goldfield, Nevada; Econ. Geol., v. 66, p. 618–626.

Jewitt, W.G. and Gray, S.
1940: The Box mine; Can. Inst. Min. Metall., Trans,. v. 43, p. 447–467.

Johan, Z., Picot, P., Pierrot, R. and Kvacek, M.
1971: La fischesserite, Ag_3AuSe_2, premier séléniure d'or, isotype de la petzite, Soc. Fr. Mineral. Cristallogr. Bull., t. 94, p. 381–384.

Johansen, P.G. and Buchanan, A.S.
1957: An application of the microelectrophoresis method to the study of the surface properties of insoluble oxides; Aust. J. Chem., v. 10, no. 4, p. 398–403.

John, Y.W.
1973: Mineralization of alaskite gold veins; Kwangsan Hakhoe Chi, v. 10, no. 1, p. 5–9. (Chem. Abstr., v. 79, 44345b.)

Johnson, B.F.G.
1971: The chemistry of gold; Gold Bull., S. Afr. Chamber Mines, v. 4, no. 1, p. 9–11.
1976: Recent advances in organogold chemistry; Gold Bull., S. Afr. Chamber Mines, v. 9, p. 46–49.

Johnson, B.F.G. and Davis, R.
1973: Gold, *in* Comprehensive inorganic chemistry, J.C. Bailar *et al.*, eds., Pergamon Press, Inc., Oxford, p. 129–186.

Johnson, C.H.
1971: Recent advances in sensitivity and data analyses of airborne gamma-ray spectrometry for mineral exploration (abstr.); *in* Geochemical exploration, R.W. Boyle and J.I. McGerrigle, eds., Can. Inst. Min. Metall., Spec. Vol. 11, p. 501.

Johnson, I.R.
1967: Geochemical exploration for gold in the Tavua basin, Fiji; N.Z., J. Geol. Geophys., v. 10, no. 5, p. 1186–1187 (abstr.).

Johnson, M.G.
1972*a*: Placer gold deposits of New Mexico; U.S. Geol. Surv., Bull. 1348, 46 p.
1972*b*: Placer gold deposits of Arizona; U.S. Geol. Surv., Bull. 1355, 103 p.
1973*a*: Placer gold deposits of Nevada; U.S. Geol. Surv., Bull. 1356, 118 p.
1973*b*: Placer gold deposits of Utah; U.S. Geol. Surv., Bull. 1357, 26 p.
1974: Placer gold deposits in the southwest; Min. Eng., v. 26, no. 3, p. 20–23.

Johnston, H.L. and Leland, H.L.
1938: The solubility of gold hydroxide in alkali and equilibria in the saturated solutions: J. Am. Chem. Soc., v. 60, p. 1439–1445.

Johnston, W.A. and Uglow, W.L.
1926: Placer and vein gold deposits of Barkerville, Cariboo district, British Columbia; Geol. Surv. Can., Mem. 149, 246 p.

Johnston, W.D.
1940: The gold quartz veins of Grass Valley, California; U.S. Geol. Surv., Prof. Pap. 194, 101 p.

Johnston, W.G.Q.
1973: Preliminary geochemical data of Southend, Saskatchewan; Sask. Dep. Min. Res., N.T.S. 64 D/6.

Jolliffe, A.W.
1945: Yellowknife—faults and orebodies; Precambrian, v. 18, no. 5, p. 7–11.

Jolly, W.T.
1974: Behaviour of Cu, Zn, and Ni during prehnite-pumpellyite rank metamorphism of the Keweenawan basalts, northern Michigan; Econ. Geol., v. 69, p. 1118–1125.

Jonasson, I.R. and Allan, R.J.
1973: Snow: a sampling medium in hydrogeochemical prospecting in temperate and permafrost regions; *in* Geochemical exploration 1972, M.J. Jones, ed., Inst. Min. Metall., p. 161–176.

Jonasson, I.R. and Boyle, R.W.
1972: Geochemistry of mercury and origins of natural contamination of the environment; Can. Inst. Min. Metall., Bull., v. 65, no. 717, p. 32–39.

Jonasson, I.R. and Sangster, D.F.
1975: Variations in the mercury content of sphalerite from some Canadian sulphide deposits; Geochemical exploration 1974, I.L. Elliott and W.K. Fletcher, eds., Assoc. Explor. Geochem., Spec. Publ. no. 2., Elsevier Sci. Publ. Co., Amsterdam, p. 313–332.

Jones, L.H.P.
1957: The solubility of molybdenum in simplified systems and aqueous soil suspensions; J. Soil. Sci., v. 8, no. 2, p. 313–327.

Jones, O.A.
1947: Ore genesis in Queensland; R. Soc. Queensland, Proc., v. 59, pt. 1, 91 p.

Jones, R.S.
1968: Gold in meteorites and in the earth's crust; U.S. Geol. Surv., Circ. 603, 4 p.
1969: Gold in igneous, sedimentary and metamorphic rocks; U.S. Geol. Surv., Circ. 610, 28 p.
1970: Gold content of water, plants and animals. U.S. Geol. Surv., Circ. 625, 15 p.

Jones, R.S. and Fleischer, M.
1969: Gold in minerals and the composition of native gold; U.S. Geol. Surv., Circ. 612, 17 p.

Jones, W.A.
1944: Mineralogy of the Ross veins, Ramore, Ontario; Can. Inst. Min. Metall., Trans., v. 47, p. 55–70.
1948*a*: Hollinger mine; *in* Structural geology of Canadian ore deposits, v. 1, Can. Inst. Min. Metall., Montreal, p. 464–481.
1948*b*: Ross mine; *in* Structural geology of Canadian ore deposits, v. 1, Can. Inst. Min. Metall., Montreal, p. 570–579.

Joralemon, P.
1951: The occurrence of gold at the Getchell mine, Nevada; Econ. Geol., v. 46, p. 267–310.
1975: K-Ar relations of granodiorite emplacement and tungsten and gold mineralization near the Getchell mine, Humboldt County, Nevada; Econ. Geol., v. 70, p. 405–406.

Joubin, F.R.
1948: Bralorne and Pioneer mines; *in* Structural geology of Canadian ore deposits, v. 1, Can. Inst. Min. Metall., Montreal, p. 168–177.

Joubin, F.R. and James, D.H.
1957: Algoma uranium district; *in* Structural geology of Canadian ore deposits, v. 2, Can. Inst. Min. Metall., Montreal, p. 305–316.

Jouravsky, G.
1952*a*: Or, *in* Géologie des gîtes mineraux Marocains; 19th Congr. Géol. Int., Rabat, p. 223–232.
1952*b*: Sur la présence d'une paragénèse nouvelle à molybdénite dans les filons 7 et 5 de la région mineralisée de Bou Azzer (Sud Marocain); C. R., Acad. Sci., Paris, t. 234, p. 230–231.

Juan, V.C.
1946: Mineral resources of China; Econ. Geol., v. 41, p. 399–471.

Junner, N.R.
1921: The geology of the gold occurrences of Victoria, Australia; Econ. Geol., v. 16, p. 79–123.
1932: The geology of the Obuasi goldfield; Gold Coast, Geol. Surv., Mem. 2, 43 p.

Junner, N.R. and Hirst, T.
1944: The Gold Coast Banket: some aspects of its geology in relation to mining, Disc.; Inst. Min. Metall., Bull. no. 466, p. 23–29.

Junner, N.R., Hirst, T. and Service, H.
1942: The Tarkwa Goldfield; Gold Coast, Geol. Surv., Mem. 6, 75 p.

Juve, G.
1974: Ore mineralogy and ore types of the Stekenjokk deposit, central Scandinavian Caledonides, Sweden; Sver. Geol. Unders., ser. C, no. 706, Arsbok 68, no. 13, 162 p.

Kachalovskaya, V.M.
1974: Gold in Nova-Shirokinsk deposit complex metal ores; Nauch. Tr. NII Tsvet. Met. no. 35, p. 138–140. (Chem. Abstr., v. 83, 119089q.)

Kalenov, A.D.
1958: Geochemistry of scandium in the supergene zone; Geochemistry, no. 2, p. 171–175.

Kalitkina, N.A.
1971: Consolidation and disintegration of gold in pyrite and arsenopyrite; Vest. Mosk. Univ. Geol., v. 26, no. 6, p. 107–110. (Chem. Abstr., v. 76, 143394m).

Kal'nichenko, S.S.
1969: Gold ore mineralization in relation to Lower Paleozoic magmatism and its role in the formation of gold-containing placers of the Vitim plateau and eastern Tuva; Zap. Zabaikal. Filiala Geogr. Obshchest. SSSR, no. 36, p. 35–39. (Chem. Abstr., v. 77, 116685w.)

Kang, P.K. and Hwang, C.S.
1974: Size and shape of natural gold grains in Bujun gold deposit; Chijil Kwa Chili, v. 14, no. 2, p. 39–42. (Chem. Abstr., v. 82, 142850u.)

Kanungo, S.C.
1972: Gold mineralization in Koraput district, Orissa; Indian Miner., v. 26, no. 2, p. 61–64.

Kanurkov, G. and Stefanov, G.
1968: Determination of gold in limonite by the neutron activation method; Geol. Inst., Bulg. Acad. Sci., Jubilee Geol. Vol., V. Tsankov, ed., Izd. "BAN", Sofia, Bulg., p. 323–326. (Chem. Abstr., v. 74, 24201v.)

Karajian, H.A.
1920: Mineral resources of Armenia and Anatolia; Armen Technical Book Co., New York, 211 p.

Karapetyan, A.I.
1962: Character of gold mineralization in a copper molybdenum deposit; Int. Geol. Rev., v. 4, no. 8, p. 925–928.

Karelin, V.V., Maksimov, E.P., Nikitin, V.N. and Uyutov, V.I.
1974: Gold in Mesozoic igneous rocks of the eastern part of the Aldan Shield; Geokhim, no. 4, p. 554–561. (Chem. Abstr., v. 81, 124279q.)

Kartashov, I.P.
1971: Geological features of alluvial placers; Econ. Geol., v. 66, p. 879–885.

Kartsov, A.I.
1971: The estimation of reserves of gold from secondary dispersion halos of associated elements; Razved. Okhr. Nedr., no. 5, p. 34–37.

Kashirin, K.F., Konstantinova, I.M. and Polikarpochkin, V.V.
1975: Gold in granitoids and contact aureoles of granitoid intrusions of the Lena gold-bearing region; Geokhim., no. 10, p. 1571–1575.

Kashiwagi, T., Nishio, S. and Imai, H.
1955: On the formation temperature of minerals by thermal microscope (heating microscope stage) and decrepitation method; Mining Geol. (Japan), v. 5, no. 17, p. 155–161.

Kashkai, M.A.
1964: Metasomatic zoning and chemistry of listwanites; Fiz. Khim. Usloviya Magmat. Metasomatoza, Inst. Geol. Geofiz., Sibirsk. Otd. Akad. Nauk SSSR, p. 198–214. (Chem. Abstr., v. 64, no. 7, 8048d.)

Kaskmenskaya, O.V. and Khvorostova, Z.M.
1969: Main stages of relief development and formation of gold placer deposits in the upper Kolyma and Indigirka river basin during Cenozoic time; Probl. Chetvertich. Geol. Sib., V.N. Saks, ed., Izd. "Nauka", Moscow, U.S.S.R., p. 131–142. (Chem. Abstr., v. 73, 17303f.)

Kaspar, J., Hudec, I., Schiller, P., Cook, G.B., Kitzinger, A. and Wölfl, E.
1972: A contribution to the migration of gold in the biosphere of the humid mild zone; Chem. Geol., v. 10, no. 4, p. 299–305.

Katakura, Y.
1965: Neutron activation analysis in the determination of colloidal gold and lutecium in biological materials; Tohoku Univ., Sci. Rep., Ser. C12, p. 223–239.

Kato, T., Shibata, H. and Nakamoto, A.
1933: Telluriferous gold-silver vein at the Rendaizi mine, Idu, with special reference to tellurite; J. Geol. Soc. Tokyo, v. 40, p. 233–241. (Chem. Abstr., v. 29, 4703.) *Also* Mineral. Abstr., v. 5, p. 479.

Kavanagh, P.M.
1968: Have 6000 years of gold mining exhausted the world's gold reserves? Can. Inst. Min. Metall., Bull., v. 61, no. 672, p. 553–558.
1976: Gold reserves of the world; Geosci. Can., v. 3, no. 3, p. 151–155.

Kazakevich, Yu.
1972: Conditions of formation and preservation of composite buried gold placers; Izd. "Nedra" Publ. House, Moscow, 215 p.
1976: Specific features of formation of gold placers in rift zones; 25th Int. Geol. Cong., Sydney, Australia, Abstr., v. 1, p. 213.

Kazanskii, V.I.
1962: Problem of the primary source of gold in Malyi Khingan; Geol. Rud. Mestorozhd., v. 4, no. 2, p. 109–113. (Engl. abstr., *in* Econ. Geol., v. 58, p. 468.)

Kazanskii, V.I., Krupennikov, V.A. and Rozanov, Yu. A.
1970: Localization conditions of the Mesozoic gold-bearing metasomatites in the crystalline basement of the central Aldan area; Akad. Nauk. SSSR, Izv., Ser. Geol., no. 6, p. 5–16. (Chem. Abstr., v. 73, 89992c.)

Kazarinov, A.I.
1969: Origin of gold-ore deposits of the Kuranakh type; *in* Zolotorud. Form. Dal'nego Vostok., E.A. Radkevich, ed., "Nauka", Moscow, p. 125–135. (Chem. Abstr., v. 74, 5450j.)

Kazarinov, A.I., Mikhailova, M.S. and Fomin, P.S.
1970: Relation of gold and gold-silver mineralization with propylites, secondary quartzites, and quartz metasomatites (in a group of subvolcanic deposits in the eastern part of the USSR); Izv. Tomsk. Politekh. Inst., v. 239, p. 50–54. (Chem. Abstr., v. 75, 111608x.)

Keays, R.R. and Crocket, J.H.
1970: A study of precious metals in the Sudbury nickel eruptive ores; Econ. Geol., v. 65, 438–450.

Keays, R.R., Ganapathy, R., Laul, J.C., Anders, E., Herzog, G.F. and Jeffery, P.M.
1970: Trace elements and radioactivity in lunar rocks: implications for meteorite infall, solar wind flux, and formation conditions of moon; Science, v. 167, no. 3918, p. 490–493.

Keays, R.R. and Kirkland, M.C.
1972: Hydrothermal mobilization of gold from copper-nickel sulfides and ore genesis at the Thomson River Copper mine, Victoria, Australia; Econ. Geol., v. 67, p. 1263–1275.

Keays, R.R. and Scott, R.B.
1976: Precious metals in ocean-ridge basalts: implications for basalts as source rocks for gold mineralization; Econ. Geol., v. 71, p. 705–720.

Keele, J.
1905: The Duncan Creek mining district, Stewart River, Yukon Territory; Geol. Surv. Can., Sum. Rep. 1904, p. 18–42.

Keesing, N.
1976: History of the Australian gold rushes; Angus and Robertson, Melbourne, 412 p.

Kelley, K.K.
1960: Contributions to the data on theoretical metallurgy, XIII. High temperature heat content, heat capacity, and entropy data for the elements and inorganic compounds; U.S. Bur. Mines, Bull. 584.

Kelly, W.C. and Cloke, P.L.
1961: The solubility of gold in near-surface environments; Papers, Mich. Acad. Sci., Arts, Lett., v. 46, p. 19–30.

Kelly, W.C. and Goddard, E.N.
1969: Telluride ores of Boulder County, Colorado; Geol. Soc. Am., Mem. 109, 237 p.

Kemp, J.F.
1902: The geological relations and distribution of platinum and associated metals; U.S. Geol. Surv., Bull. 193, 95 p.
1905: Ore deposits of the United States and Canada; Eng. Min. J., New York, 481 p.

Kempthorne, H.R.
1957: Bevcon mine; *in* Structural geology of Canadian ore deposits, v. 2, Can. Inst. Min. Metall., Montreal, p. 416–419.

Kennedy, W.Q.
1948: Crustal layers and the origin of ore deposits; Schweiz. Mineral. Petrogr. Mitt., v. 28, p. 222–229.

Kenny, E.J.
1924: Gold; N. S. W., Geol. Surv. Bull., no. 7, 60 p.

Kenyon, W.
1975: "All is not golde that shineth"; The Beaver, Summer Outfit 306:1, p. 40–46.

Kepezinskas, K., Anoshin, G.N. and Tsimbalist, V.G.
1970: Chemical composition of the rocks of the kyanite schist facies and the distribution of gold in them (illustrated by Bolshoi and Malyi Khodakan rivers, Vitim-Patom Upland); Probl. Petrologii Genet. Mineral., v. 2, p. 98–107. (Chem. Abstr., v. 73, 111750w.)

Keppie, J.D.
1976: Structural model for the saddle reef and associated gold veins in the Meguma Group, Nova Scotia; Can. Inst. Min. Metall. Bull., v. 69, no. 774, p. 103–116.

Kerr, W.C.
1881–1882: Some peculiarities in the occurrence of gold in North Carolina; Am. Inst. Min. Eng., Trans., v. 10, p. 475–476.

Kesler, S.E.
1972: Crustal environment and silver-gold-molybdenum abundances in magmatic-hydrothermal ore deposits; 24th Int. Geol. Congr., Montreal, Abstr., p. 131–132.
1973: Copper, molybdenum and gold abundances in porphyry copper deposits; Econ. Geol., v. 68, p. 106–112.

Keynes, J.M.
1930: A treatise on money, v. 1, 363 p. and v. 2, 424 p.; Harcourt, Brace and Company, New York.

Keys, M.R.
1948: Berens River mine; *in* Structural geology of Canadian ore deposits, v. 1, Can. Inst. Min. Metall., Montreal, p. 365–368.

Khairullin, B.G., Yusupov, R.G., Abdurakhmanov, G.A., Bubnov, L.S., Chistaya, G.I. and Panasenko, F.N.
1968: Gold specialization of Middle-Upper Carboniferous low-depth granodiorite-adamellite batholithic intrusions in the Angara Plateau (Kuramin Ridge); Metallogen. Tyan-Shanya, Frunze, "Ilim", p. 278–281. (Chem. Abstr., v. 71, 115117b.)

Khamrabaev, I. Kh., ed.
1969: The gold ore deposits and principal features of the metallogeny of gold in Uzbekistan; "Fan" Publ. Office, Uzb. S.S.R., Tashkent, 396 p.

Khamrabaev, I. Kh., Chebotorev, G.M., Mansurov, M., Tillyaev, Kh. S., Dautov, A.I. and Khrenov, V.A.
1971: Geological-mineralogical characteristics of the Charmitan gold deposit in western Uzbekistan; Uzb. Geol. Zh., v. 15, no. 3, p. 3–7. (Chem. Abstr., v. 75, 99913h.)

Khamrabaev, I. Kh., Ismailov, M.I. and Rakhmatullaev, Kh. R.
1971: Main types of gold and rare metal ore formations in Tien-Shan and the Donets Basin; Geol. Zh., v. 32, no. 5, p. 28–33. (Chem. Abstr., v. 76, 143366h.)

Khamrabaev, I. Kh, Rakhmatullaev, Kh. R., Kasymov, A.K. and Aripova, Kh.
1965: Auriferous rocks in the southern part of the Tamdytau; Uzb. Geol. Zh., v. 9, no. 1, p. 15–19. (Chem. Abstr., v. 63, 1599a.)

Khamrabaev, I. Kh., Rakhmatullaev, Kh. R. and Magdiev, R.A.
1972: Distribution characteristics of gold, lead, uranium, and thorium in granitic rocks of enclosed territories in central Kyzylkums; Zap. Uzb. Otd. Vses. Mineral. Obshchest., no. 25, p. 148–151. (Chem. Abstr., v. 78, 60662n.)

Khamrabaev, I. Kh. and Uruibaev, K.U.
1972: Geochemistry of gold in magmatic processes; Mezhdunar. Geokhim. Kongr. Dokl. 1st, 1971, pt. 1, p. 594–605. (Chem. Abstr., v. 81, 5297j.)

Kharechenko, Yu. I.
1970: Possible parental rock sources of gold alluvial placer deposits in central and southern Kamchatka; Probl. Geol. Rossypei, Soveshch., (Dokl.) 3rd., 1969, V.I. Smirnov, ed., Sev. Vost. Kempleks. Inst. Magadan, U.S.S.R., p. 138–146. (Chem. Abstr., v. 75, 38848u.)

Kharkar, D.P. and Turekian, K.K.
1971: Analyses of Apollo 11 and Apollo 12 rocks and soils by neutron activation; Proc., 2nd Lunar Sci. Conf., Geochim. Cosmochim. Acta, Suppl. 2, v. 2, p. 1301–1305.

Khetagurov, G.V.
1969: Distribution of silver and gold in ores, minerals and enrichment products of the lead-zinc deposits of the Greater Caucasus; Geochem. Int., v. 6, p. 1070–1077.

Khitarov, N.I.
1957: The chemical properties of solutions arising as a result of the interaction of water with rocks at elevated temperatures and pressures; Geokhim., no. 6, p. 481–492.

Khorvat, V.A.
1972: Talcosed serpentinites in the gold-sulfide deposits of the Balyska ore field in northern Kirgizia; Geol. Rud. Raionov Mestorozhd., Srednei Azii, Metody Ikh. Izuch. Razved., C.K. Abadullaev, ed., p. 93–97. (Chem. Abstr., v. 80, 98432s.)

Khotamov, Sh., Lobanov, E.M. and Kist, A.A.
1966: The problem of the concentration of gold in organs of plants within ore fields; Akad. Nauk Tadzhik. SSR, Dokl., v. 9, no. 11, p. 27–30.

Khramyshkin, P.V.
1962: Geochemical methods for prospecting for gold and mercury in Uzbekistan; Byul. Nauch. Tekhn. Inform. Min. Geol. Okhr. Nedr., SSSR, no. 1, p. 56–60. (Chem. Abstr., v. 60, 3881d.)
1968: Geophysical and geochemical prerequisites for detection of gold-ore deposits in western Uzbekistan; Rud. Form. Osn. Cherty Metallogen. Zolota Uzb., I. Kh. Khamrabaev, ed., Izd. "Fan" Uzb. SSR, Tashkent, U.S.S.R. p. 347–355. (Chem. Abstr., v. 73, 37339a.)

Khripkov, A.V.
1971: Formation dynamics and source of littoral-marine placers in the Ukraine; Vop. Geol. Mineral. Rud. Mestorozhd. Ukr., no. 4, p. 248–258. (Chem. Abstr., v. 80, 62137k.)

Khripkov, A.V. and Zaitsev, A.A.
1973: Presence of gold in the weathering crust in the central Dnieper River region; Geol. Zh., v. 33, no. 4, p. 132–134. (Chem. Abstr., v. 80, 5708y.)

Khryachkov, M.A.
1972: The origin of gold in the central part of Sakhalin Island; Geol. Geofiz., no. 12, p. 120–123. (Chem. Abstr., v. 78, 126912c.)

Kibanov, G.A.
1968: Gold content in middle Vitim region conglomerates; Geol. Razved. Mestorozhd. Polez. Iskop. Zabaikal., p. 50–51. (Chem. Abstr., v. 70, 79888x.)

Kieft, C. and Oen, I.S.
1973: Ore minerals in the telluride-bearing gold-silver ores of Salida, Indonesia, with special reference to the distribution of selenium; Miner. Deposita (Berl.), v. 8, no. 4, p. 312–320.
1974: Ore mineral parageneses in the Mn-Sn-Ag-Au-bearing veins of Mangani near Padang, Sumatra, Indonesia; Abstr. Int. Assoc. Genesis Ore Deposits (IAGOD), 4th Symp., Varna, Bulgaria, p. 302–303.

Kieft, C. and Oen, I.S.
1977: Ore mineral parageneses in Mn-Sn-Ag-Au-Se-bearing veins of Mangani, Sumatra, Indonesia; *in* Problems of ore deposition, v. 2, 4th IAGOD Symp., Varna, Bulgaria, B. Bogdanov, ed., Bulgarian Acad. Sci., Sofia, p. 295–302.

Kim, J.H.
1973: Geochemical study of mercury in gold-silver ore deposits; Taehan Kwangsan Hakhoe Chi, v. 10, no. 4, p. 273–279 (in Korean).

Kim, Ki Chan
1972: Possible forms of gold migration on the basis of thermodynamic stabilities; Chijil Kwa Chiri, v. 12, no. 6, p. 36–39. (Chem. Abstr., v. 79, 138574c.)

Kim, O.J.
1971: Metallogenic epochs and provinces of South Korea; Chijil Hakhoe Chi, v. 7, no. 1, p. 129–155. (Chem. Abstr., v. 82, 173549g.)

Kimura, K., Fujiwara, S. and Yasuda, Y.
1950: Chemical prospecting IV. Chemical prospecting at the southern part of the Sado gold mine. The distribution of minute amounts of silver and copper in rocks; J. Chem. Soc. Japan, Pure Chem. Sec., v. 71, p. 467–471 (in Japanese). (Chem. Abstr., v. 45, no. 6130i.)

Kindle, E.D.
1936: Gold occurrences of Ontario, east of Lake Superior; Geol. Surv., Can., Mem. 192, 167 p.
1964: Copper and iron resources, Whitehorse copper belt, Yukon Territory; Geol. Surv. Can., Pap. 63-41, 46 p.

Kindle, L.F.
1932: Kowkash-Ogoki gold area, District of Thunder Bay; Ont. Dep. Mines, Annu. Rep., 1931, v. 40, pt. 4, p. 55–104.

King, F.A.
1955: When Martin Frobisher searched for gold; Can. Min. J., v. 76, pt. 2, no. 12, p. 67–69.

King, P.B.
1969: The tectonics of North America — a discussion to accompany the tectonic map of North America, Scale 1:5,000,000; U.S. Geol. Surv., Prof. Pap. 628, 95 p.

Kingdom of Saudi Arabia
1965: Mineral resources of Saudi Arabia, Bull. no. 1, 73 p.
1969: Mineral resources research, 1967–68, 112 p.
1970: Mineral resources research, 1968–69, 209 p.

Kingsbury, J.M.
1964: Poisonous plants of the United States and Canada; Prentice-Hall, Inc., Englewood Cliffs, N.J., 626 p.

Kinkel, A.R.
1948: Buffalo Ankerite mine; *in* Structural geology of Canadian ore deposits, v. 1, Can. Inst. Min. Metall., Montreal, p. 515–519.

Kinkel, A.R. and Lesure, F.G.
1968: Residual enrichment and supergene migration of gold, southeastern United States; U.S. Geol. Surv., Prof. Pap. 600-D, p. 174–178.

Kinoshita, K.
1931: On the "Kurokô" (Black Ore) deposits; Jap. J. Geol. Geog., v. 8, no. 4, p. 281–352.

Kirillov, G.A.
1972: Facies-dynamic factors in the formation and distribution of gold placers in the Cretaceous formations of western Uzbekistan; Nauch. Tr., Tashkent. Gos. Univ. v. 431, p. 162–165. (Chem. Abstr., v. 81, 155806p.)

Kirillov, V.P., Legedza, V. Ya. and Sidorov, V.A.
1970: Experimental study of the possibility of formation of auriferous iron disulfide at normal temperatures under atmospheric pressure; Akad. Nauk. SSSR, Dokl., v. 195, no. 4, p. 941–943.

Kirkham, R.V.
1968: A mineralogical and geochemical study of the zonal distribution of ores in the Hudson Bay Range, British Columbia; unpubl. Ph.D. thesis, Univ. Wisconsin, Madison, Wis.

Kirwan, R.
1799: Geological essays, London, 502 p.

Kishimoto, F., Kato, K., Takashima, K., Tono, T. and Katsume, K.
1963: Geochemical prospecting for gold-silver deposits, an example of mercury indication at the Okuchi mine; Kozan Chishitsu, v. 13, no. 5, p. 243–252. (Chem. Abstr., v. 62, 10231c.)

Kisielov, W. and Gregorowicz, Z.
1955: Mineral substances in Polish crude oils; Nafta, Miesiecznik Posweizony Nause, Technica, Statystyce Poland, v. 11, p. 258–259 (in Polish).

Kisterov, K.V.
1976: Flatness of gold particles and mechanisms of their migration in the formation process of alluvial placers; Geol. Rud. Mestorozhd., v. 18, no. 2, p. 129–131. (Chem. Abstr., v. 85, 127209k.)

Kitaev, N.A.
1973: Primary halos of epithermal gold-ore deposits; Geol. Geokhim. Prognoznaya Ots. Rud. Raionov Mestorozhd. Zabaikal., F.I. Vol'fson, ed., p. 102–112. (Chem. Abstr., v. 84, 62586p.)

Kitaev, N.A. and Bulanov, V.A.
1972: Types of hydrothermal mineralization and mineral associations of the Lubavinsky-Haverchinsky gold-ore deposit (Transbaikalia); Ezheg. Inst. Geokhim., Sib. Otd., Akad. Nauk. SSSR, 1971, L.V. Tauson, ed., p. 218.

Kitaev, N.A., Bulanov, V.A. and Chumakin, V.N.
1973: Formation particularities of Lyubavinsk gold ores; Ezheg. Inst., Geokhim., Sib. Otd., Akad. Nauk SSSR, 1972, L.V. Tauson, ed., p. 320–324. (Chem. Abstr., v. 81, 138580t.)

Kitaev, N.A. and Chumakin, V.N.
1973: Haloes of chemical element association of the Lyubavinsk ore field; Ezheg. Inst. Geokhim., Sib. Otd., Akad. Nauk SSSR, 1972, L.V. Tauson, ed., p. 243–247. (Chem. Abstr., v. 81, 138579z.)

Kitaev, N.A., Naryshev, A.M. and Polikarpochkin, V.V.
1968: Primary dispersion haloes of mercury in the Baleisk goldfield; Vop. Metall. Rtuti, V.A. Kuznetsov, ed., Izd. "Nauka" Moscow, U.S.S.R., p. 228–232. (Chem. Abstr., v. 71, 93440x.)

Kitaev, N.A., Nesterova, A.A. and Sarapulova, V.N.
1970*a*: Characteristics of accessory element distribution in pyrites and total sulphide fraction from haloes of the Baleisk goldfield; Izv. Tomsk. Politekh. Inst., v. 239, p. 225–232.

Kitaev, N.A., Novikov, V.M., Naryshev, A.M. and Polikarpochkin, V.V.
1970*b*: Comparative data on element composition and zoning of primary haloes over Transbaikalia and Far East epithermal gold deposits; Ezheg. Inst. Geokhim., Sib. Otd., Akad. Nauk. SSSR, 1969, L.V. Tauson, ed., p. 223–226. (Chem. Abstr., v. 74, 101672d.)

Kitaev, N.A., Sarapulova, V.N. and Korotaeva, I. Ya.
1970*c*: Silver/gold ratio in primary haloes of the Baleisk goldfield; Geol. Geofiz., no. 9, p. 117–123. (Chem. Abstr., v. 74, 33455h.)

Kitami, M.
1973: Recent exploration and development of the Takatama gold and silver mine, Fukushima Prefecture; Mining Geol. (Japan), v. 23, p. 191 (abstr.).

Kitson, A.
1930: Gold resources of the Gold Coast, British West Africa; *in* Gold resources of the world; 15th Int. Geol. Congr., Pretoria, 1929, p. 161–168.

Kitson, A.E. and Felton, W.J.
1930: Minerals of concentrates from stream-gravels, soils and crushed rocks of the Gold Coast; Gold Coast, Geol. Surv., Bull. no. 6, 50 p.

Kittl, E.
1931– 1932: Los yacimientos auríferos de la República Argentina, su génesis y posicíon geologica; Rev. Minera t. III, nos. 4, 5, 6, 7; t. IV no. 1., Repr. Publ., Mus. Nac. Hist. Nat. "Bernardino Rivadavea", Bernardo de Irigoyen 331, Buenos Aires, 142 p.
1965: Zur entstehung der Golderzlagerstätten Argentiniens; Freiberg. Forschungsh., C186, p. 137–143.

Klempert, S. Ya., Sakhor, D.A. and Khasanov, R.
1973: Submicroscopic gold in Terek deposit pyrites; Akad. Nauk Uzb. SSR, Dokl., v. 30, no. 2, p. 42–43. (Chem. Abstr., v. 81, 15473c.)

Knight, C.L.
1957: Ore genesis—the source bed concept; Econ. Geol., v. 52, p. 808–817.

Knight, C.L., ed.
1975: Economic geology of Australia and Papua, New Guinea: 1—Metals; Australas. Inst. Min. Metall., 1126 p.

Knight, C.L., Fraser, R.B. and Baumer, A.
1973: Geology of the Bougainville copper orebody, New Guinea; Aust., Bur. Min. Resour., Geol. Geophys., Bull. 141, p. 123–133.

Knight, C.W.
1925: Lightning River area; Ont. Dep. Mines, Annu. Rep., 1924, v. 33, pt. 3, p. 41–49.

Knight, O.W.
1911: Research work on certain Nova Scotian gold ores; Can. Min. J., v. 32., p. 622–626, 657–660.

Knopf, A.
1915: A gold-platinum-palladium lode in southern Nevada; U.S. Geol. Surv., Bull. 620A, 18 p.
1924: Geology and ore deposits of the Rochester district, Nevada; U.S. Geol., Surv., Bull. 762, 78 p.
1929: The Mother Lode system of California; U.S. Geol. Surv., Prof. Pap. 157, 88 p.

Kobe, H.W.
1960: Cu-Ag deposits of the red-bed type at Negra Huanusha in central Peru; Schweiz. Mineral Petrogr. Mitt., v. 40, pt. 1, p. 163–176.

Koch, S. and Grasselly, Gy.
1952: The minerals of the sulphide ore-deposit of Nagybörzsöny; Acta Univ. Szegediensis, Acta Mineral. Petrog. Tomus VI, p. 1–21 (in English).

Koděra, M.
1963: Polymetallic subvolcanic deposits of Banská Stiavnica and Hodruša; *in* Some ore deposits of the western Carpathians, guide to excursion, Symposium on Problems of Postmagmatic Ore Deposition; Publ. House Czech. Acad. Sci., Praha, p. 64–79.

Koeberlin, F.R.
1934: An hypothesis as to the origin of gold in volcanic ash; Eng. Min. J. v. 135, p. 394–396, 400.
1938: Sedimentary copper, vanadium-uranium, and silver in southwestern United States, Econ. Geol., v. 33, p. 458–461.

Koen, G.M.
1958: The attrition of uraninite; Geol. Soc. S. Afr., Trans., v. 61, p. 183–196.

Koga, A.
1960: Chemical studies on the hot springs of Beppu. I. Trace elements in the Beppu hot springs; Onken Kiyo, v. 12, p. 52–60.
1961: Chemical studies on the hot springs of Beppu, XXV, Distribution of gold; Nippon Kagaku Zasshi, v. 82, p. 1476–1478, VII, Distribution of silver; Nippon Kagaku Zasshi, v. 78, p. 1717–1721.

Kogen, V.S.
1969: Mineral associations of the central Apsakanskoye ore field (Stanovy Range); Int. Geol. Rev., v. 11, no. 8, p. 860–866.
1971: Characteristics of native gold from the Stanovoi Ridge ore formations and placer deposits; Akad. Nauk. SSSR, Izv. Ser. Geol., no. 8, p. 74–84. (Chem. Abstr., v. 76, 5645t.)

Kogen, V.S., Naprasnikova, L.A. and Ryabtseva, G.I.
1974: Distribution and origin of local beach concentrations of gold as exemplified by a bay in the southern maritime territory; Izv. Vyssh. Ucheb, Zaved. Geol. Razved, v. 17, no. 1, p. 54–60. (Chem. Abstr., v. 81, 66370z.)

Kokkola, M. and Pehkonen, E.
1976: Kangaskyla: gold in till; J. Geochem. Explor., v. 5, no. 3, p. 239–244.

Kolesov, S.V.
1975*a*: Subsidence of gold grains in allochthonous alluvial and coastal marine placers; Int. Geol. Rev., v. 17, no. 8, p. 915–953.
1975*b*: Flattening and hydrodynamic sorting of placer gold; Int. Geol. Rev., v. 17, no. 8, p. 940–944.

Kolonin, G.R. and Aksenova, T.P.
1971: Effect of temperature on complex formation of some ore elements in chloride solutions; Int. Geochem. Congr., Abstracts of Reports I, Moscow, U.S.S.R., p. 274–275.

Kolosova, T.B. and Onishchuk, Yu. V.
1971: A new type of gold-ore mineralization in eastern Transbaikal; Int. Geol. Rev., v. 13, no. 7, p. 1027–1035.

Kolotov, B.A., Boldyrev, M.A., Wagner, B.B., Minacheva, L.I., Spasskaya, T.S. and Sokalovskaya, L.A.
1975*a*: About zones of potassic metasomatosis being promising for gold-silver mineralization which are singled out according to hydrochemical data; Geokhim., no. 2, p. 306–307.

Kolotov, B.A., Spasskaya, T.S., Minacheva, L.I. and Wagner, B.B.
1975*b*: Acid sorptive barrier for gold in the zone of hypergenesis of ore deposits; Geokhim., no. 12, p. 1898–1900.

Komov, I.L.
1966: Gold prospecting methods in Yenisey Range; Int. Geol. Rev., v. 8, no. 10, p. 1237–1240.
1969: New type of gold-ore mineralization in the Yenisey Range; Sov. Geol., no. 4, p. 143–144, 1968. *Also* Int. Geol. Rev., v. 11, no. 3, p. 298–299.

Kondakov, L.A. and Perunov, V.I.
1973: Origin of Tanzanian ferruginous quartzites illustrated by the Lupa gold field; Akad. Nauk SSSR, Izv. Ser. Geol. no. 12, p. 136–138.

Koneev, R.I.
1974: Mineralogical characteristics of the Katranga gold ore occurrence (Chatkal Ridge); Zap. Uzb. Otd. Vses. Mineral. O-va, v. 27, p. 41–43. (Chem. Abstr., v. 83, 100867v.)

Konkina, O.M.
1970: Distribution of gold in rocks of the Zlatogorsk ultrabasic differentiated massif (northern Kazakhstan); Mater. Konf. Molodykh Nauch. Sotrudnikov Tsent. Nauch. Issled. Gornorazved. Inst. Tsvet., Redk. Blagorod. Metal. 1969, p. 16–19. (Chem. Abstr., v. 80, 123168v.)

Konovalov, E.P.
1941: On the hydrochemical method of prospecting for gold; Sov. Geol., no. 2, p. 114–117.

Konovalov, I.V.
1973: Wallrock zonation of a gold deposit (Lena region); Int. Geol. Rev., v. 15, no. 10, p. 1180–1181.
1975: Dependence of the temperature of formation of gold-ore deposits in the Lena region on facies of metamorphism; Akad. Nauk SSSR, Dokl., v. 220, no. 3, p. 694–697. (Chem. Abstr., v. 83, 13448r.)

Konstantinov, M.M.
1971: Altered wallrocks from some Transcaucasian gold-ore deposits; Tr., Tsent. Nauch. Issled. Gornorazved. Inst., no. 96, pt. 2, p. 59–75. (Chem. Abstr., v. 78, 149694x.)
1973*a*: Structures of Transcaucasian auriferous fields and main factors governing their location; Akad. Nauk SSSR, Izv., Ser. Geol., no. 5, p. 68–81.
1973*b*: Prospecting criteria for gold ore deposits in Transcaucasus; Razved. Okhr. Nedr., no. 5, p. 11–16. (Chem. Abstr., v. 79, 81633q.)
1974: Structures of gold fields in Transcaucasia and fundamental relationships in their positions; Int. Geol. Rev., v. 16, no. 3, p. 269–280.
1977: Geochemical types of gold mineralization in the volcanic regions, Akad. Sci. USSR, Dokl., v. 234, p. 180–183.

Konstantinov, M.M. and Grushin, V.A.
1972: Structural conditions of localization of the Zodsk gold ore deposit; Izv. Vyssh. Ucheb. Zaved., Geol. Razved., v. 15, no. 5, p. 60–67. (Chem. Abstr., v. 77, 77766b.)

Konstantinov, M.M. and Strizhov, V.P.
1974: Isotope composition of sulphur of sulphides from Transcaucasian gold ore deposits; Tr., Tsent. Nauch. Issled. Geol. Inst. Tsvetn. Blagorodn. Met., v. 112, p. 112–115. (Chem. Abstr., v. 83, 82748a.)

Konstantinov, M.M., Tomson, I.N., Cheglokev, S.V. and Andreyeva, M.G.
1967: The character of zoning of mineralization in some orebearing areas of the eastern Transbaikal region; Geol. Rud. Mestorozhd., v. 9, no. 3, p. 41–52. *Also* Econ. Geol., v. 64, p. 587.

Konstantinova, I.M., Zontov, V.N. and Polikarpochkin, V.V.
1973: Carbonate composition of gold ore occurrences in the Lena region and the use of carbonate analyses for prospecting deep mineralization; Ezheg. Inst. Geokhim. Sib. Otd., Akad. Nauk SSSR, 1972, L.V. Tauson, ed., p. 325–329. (Chem. Abstr., v. 81, 138582v.)

Köppel, V.H. and Saager, R.
1973: Lead isotope evidence for the detrital origin of Witwatersrand pyrites and its bearing on the provenance of the Witwatersrand gold; Univ. Witwatersrand, Econ. Geol. Res. Unit, Inform. Cir. no. 79, 17 p.
1974: Lead isotope evidence on the detrital origin of Witwatersrand pyrites and its bearing on the provenance of the Witwatersrand gold; Econ. Geol., v. 69, p. 318–331.

Koptev-Dvornikov, V.S., Yemel'yanenko, P.F., Spiridonov, E.M. and Fel'dman, V.I.
1969: Petrographic features of gold-bearing intrusive masses of the Stepnyak type (northern Kazakhstan); Int. Geol. Rev., v. 11, no. 12, p. 1392–1398.

Korobeinikov, A.F.
1971: Morphological features of gold from some Siberian gold ore deposits; Mineralog. Mineral. Kristallogr., Tr., Soveshch. 1969, G.N. Vertushkov, ed., Gorn. Inst., Sverdlovsk, U.S.S.R., p. 33–34. (Chem. Abstr., v. 79, 128012d.)
1974*a*: Behaviour of gold during the near-vein metasomatism of rocks of some Siberian gold-ore deposits; Geokhim., no. 6, p. 831–844. (Chem. Abstr., v. 82, 33492h.) *Also* Geochem. Int., v. 11, p. 592–605.
1974*b*: Geochemistry of hydrothermal solutions of gold-ore deposits according to the gas-liquid inclusions in minerals; Tr. Sib. Nauch. Issled. Inst. Geol., Geofiz. Miner. Syr'ya, v. 144, p. 88–106. (Chem. Abstr., v. 84, 76955v.)
1976: On the geochemistry of gold in contact-metasomatic processes of granitoid intrusions, Akad. Sci. SSSR, Dokl., v. 227, p. 1224–1227.

Korobeinikov, A.F. and Oleynikov, B.V.
1976: The gold in magnetite of traps, granitoid and contact metamorphic rocks from Siberia; Akad Sci. USSR, Izv. Geol. Ser. no. 1, p. 83–97.

Korobushkin, I.M.
1970: The form in which "finely dispersed" gold is found in pyrite and arsenopyrite; Akad. Nauk SSSR, Dokl., v. 192, no. 5, p. 1121–1122.

Korolev, N.N., Timofeyev, T.S. and Romanenko, D.Ya.
1971: Gold particle shapes in the sulphide endogenous deposits of central Asia; Zap. Vses. Mineral. Obshchest., v. 100, p. 49–54.

Koroleva, N.N.
1968: Telluride pyrite-gold ore formation in the Burgundin ore field; Rud. Form. Osn. Cherty Metallogen. Zolota Uzb., I.Kh Khamrabaev, ed., p. 70–80. (Chem. Abstr., v. 73, 37344y.)
1970: Küstelite from a quartz-gold-silver ore occurrence in the Karamazar; Acad. Sci. USSR, Dokl., Earth Sci. Sec., Transl., v. 195, p. 131–133 (Am. Geol. Inst. transl.).
1971: Relation between Ag:Au ratio and mineral associations; Geochem. Int., v. 8, no. 2, p. 200–205.

Koroleva, N.N., Korolev, V.A. and Overchkin, V.V.
1966: Gold-sulfide mineralization in carbonate rocks—a new type of gold mineralization in Middle Asia; Geol. Rud. Mestorozhd., v. 8, no. 5, p. 30–44. (Engl. rev. *in* Econ. Geol., v. 64, p. 354–355.)

Koroleva, N.N., Markova, E.A. and Klempert, S.Y.
1971: Behaviour of gold, bismuth, zinc, lead, and antimony in hydrothermal process; Int. Geochem. Congr., Moscow, Abstr. I, p. 368–369.

Korotayeva, I.Ya.
1971: The geochemistry of gold in Mesozoic continental depositions of the Undino-Dainsky depression, eastern Transbaikalia; Ezheg. Inst. Geokhim., Sib. Otd., Akad. Nauk SSSR, 1970, L.V. Tauson, ed., p. 299.

Korotayeva, I.Ya. and Polikarpochkin, V.V.
1969: Gold content in the Paleozoic sedimentary rocks of the Nerchinsk Zavod area (eastern Transbaikalia); Geokhim., no. 11, p. 1386–1391. (Chem. Abstr., v. 72, 34346u.)
1970: Gold in organic remnants in ferruginous sediments of recent river sediments, eastern Transbaikalia. Ezheg. Inst. Geokhim., Sib. Otd., Akad. Nauk SSSR 1969, L.V. Tauson, ed., p. 227–230. (Chem. Abstr., v. 74, 114445h.)

Korzhinskii, D.S.
1965: Abriss der metasomatischen Prozesse; Akademie-Verlag, Berlin.

Koschmann, A.H. and Bergendahl, M.H.
1962: Gold in the United States, exclusive of Alaska and Hawaii; U.S. Geol. Surv., Mineral. Invest. Resour., Map MR-24.
1968: Principal gold-producing districts of the United States; U.S. Geol. Surv., Prof. Pap. 610, 283 p.

Koshman, P.N. and Yugay, T.A.
1972: The causes of variation in fineness levels of gold placers; Geochem. Int., v. 9, no. 3, p. 481–484.

Koster, G. and Schmuckler, G.
1967: Separation of noble metals from base metals by means of a new chelating resin; Anal. Chim. Acta, v. 38, p. 179–184.

Koulomzine, T.
1948; Consolidated Central Cadillac mine; *in* Structural geology of Canadian ore deposits, v. 1, Can. Inst. Min. Metall., Montreal, p. 816–821.

Kovalenko, V.I., Kuz'min, M.I. and Koval, P.V.
1973: General scheme of Mesozoic metallogeny in Mongolia; Ezheg., Inst. Geokhim., Sib. Otd., Akad. Nauk SSSR, 1972, L.V. Tauson, ed., p. 109–114. (Chem. Abstr., v. 81, 155855d.)

Kowalski, H. and Reimers, P.
1974: Nondestructive analysis of medieval gold coins; Eurisotop Off. Inform. Bookl., v. 72, p. 391–415. (Chem. Abstr., v. 83, 8341w.)

Kozlovskaya, A.A., Kozlovskii, G.M. and Kosyak, E.A.
1974: Mineralogical properties of the ores from the gold-skarn Sary-Adyr deposit in central Kazakhstan; Akad. Nauk SSSR, Izv. Ser. Geol., v. 31, no. 4, p. 67–73. (Chem. Abstr., v. 82, 19469z.)

Kraft, M., Taubert, P. and Ungethüm, H.
1974: Influence of supergene processes on the distribution of gold illustrated by hydrothermal gold deposits of the Central Mongolian Peoples Republic, Pt. I and II; Z. Angew. Geol., Bd 20, p. 458–468, 495–502. (Chem. Abstr., v. 82, 127630d.)

Krähenbühl, U., Ganapathy, R., Morgan, J.W. and Anders, E.
1973: Volatile elements in Apollo 16 samples: implications for highland volcanism and accretion history of the moon; *in* Proc., 4th Lunar Sci. Conf., Geochim. Cosmochim. Acta, Suppl. 4, v. 2, p. 1325–1348.

Krainov, S.R.
1967: Geochemistry of germanium in the thermal carbonate waters (illustrated by examples from the Pamirs and Greater Caucasus); Geoçhem. Int., v. 4, p. 309–320.

Krainov, S.R. and Loginova, L.G.
1968: Notes on the geochemistry of rare earth elements (Ce, La, Yb) and yttrium in alkaline waters of the Lovozero massif of intrusive rocks; *in* Lithology and useful minerals, Acad. Sci. USSR, p. 107–110.

Kraskov, V.V. and Evdokimov, E.I.
1971: Geomorphological prospecting for tin and gold placers in the central Chukotka valleys; Uch. Zap., Tsent. Nauch. Issled. Inst. Olovyannoi Prom., no. 2, p. 16–20. (Chem. Abstr., v. 79, 68662d.)

Kraskov, V.V. and Lobanov, V.V.
1973: Morphometric factors of endothermal gold concentration in Chaun-Chukotka placers; Geomorfol., no. 4, p. 29–34. (Chem. Abstr., v. 80, 135867k.)

Krasnikov, V.I., ed.
1957: Geochemical exploration for ore deposits; Geol. Lit. Publ. House, Moscow, U.S.S.R., 467 p.
1968: Evaluation of gold content in ore deposits and dispersion haloes from accessory elements under eastern Transbaikalia conditions; Geol. Nekot. Rud. Mestorozhd., Zabaikal, p. 274–296. (Chem. Abstr., v. 72, 45982z.)

Krasnova, V.N.
1960: Age of the Kumak schist zone; Mater. Geol. Polez. Iskop. Yuzh. Urala, Yuzhno-Ural'sk. Geol. Upr. no. 2, p. 40–42. (Chem. Abstr., v. 60, 310e.)

Krauskopf, K.B.
1951: The solubility of gold; Econ. Geol., v. 46, p. 858–870.
1956: Factors controlling the concentrations of thirteen rare metals in sea water; Geochim. Cosmochim. Acta, v. 9, p. 1–32B.

Kravchenko, G.L. and Saskhatskii, I.I.
1975: Presence of gold and silver in copper-bearing sandstones of the Donets Basin; Geol. Zh., v. 35, no. 1, p. 119–123. (Chem. Abstr., v. 83, 13442j.)

Kreiter, V.M., Aristov, V.V., Volynskii, I.S., Krestovnikov, A.B. and Kuvichinskii, V.V.
1959: The behaviour of gold in the zone of oxidation of auriferous sulfide deposits; Moscow, Gosgeoltekhizdat, 268 p. (Engl. rev. by O. Zvyagintsev *in* Geochemistry, no. 6, p. 683–685, and by C.F. Davidson *in* Econ. Geol., v. 55, no. 8, 1761.)

Krendelev, F.P.
1974: Zonation of gold-sulphide mineralization in sedimentary-volcanogenic series of the Enisei Ridge; Abstracts of Papers, Int. Assoc. Genesis Ore Deposits (IAGOD), 4th Symp., Varna, Bulgaria, p. 106–107.

Krendelev, F.P., Mironov, A.G. and Gofman, A.M.
1976: Use of gamma-spectrometric methods in contouring ore zones in the trans-Baikal region; Geol. Geofiz, no. 8, p. 67–75. (Chem. Abstr., v. 86, 19674v.)

Krendelev, F.P. and Pavlov, A.L.
1972: The causes of the joint occurrence of gold and uranium in ore provinces; *in* Radioaktivnye elementy v. gornykh porodakh (Radio-active elements in rocks), pt. 2, Novosibirsk, p. 22–23. *Also* Ref. Zh. Geol., no. 11, 1972, 11B23.

Krendelev, F.P. and Pogrevniak, Iu. F.
1977: Possibility of searches for ore manifestations of gold and zinc according to their contents in birch sap; Akad. Sci. USSR, Dokl., v. 234, p. 184–187.

Krendelev, F.P., Zozulenko, L.B. and Orlova, L.M.
1970: Temperature of homogenization and composition of gases in gas-fluid inclusions in pebbles from ancient conglomerates of sulphide type; Fluid Inclusion Res., Proc. COFFI, v. 3, p. 101–107.

Kretschmar, U.
1973: Phase relations involving arsenopyrite in the system Fe-As-S and their application; unpubl. Ph.D. thesis, Univ. Toronto, Toronto, Ont., 146 p.

Krieger, P.
1932: An association of gold and uraninite from Chihuahua, Mexico; Econ. Geol., v. 27, p. 651–660.

Krieger, P. and Hagner, A.F.
1943: Gold-nickel mineralization at Alistos, Sinaloa, Mexico; Am. Mineral., v. 28, p. 257–271.

Kripkov, A.J.
1969: Graphical plotting of the distribution of useful components in placers; Vop. Geol. Mineral., Rud. Mestorozhd. Ukr., no. 3, p. 147–151. (Chem. Abstr., v. 77, 22967n.)

Krook, L.
1968: Origin of the bedrock values of placer deposits; Econ. Geol., v. 63, p. 844–846.

Kropachev, C.K.
1935: The distribution of gold in nature; Sovet. Zolotoprom., no. 8, p. 46–47. (Chem. Abstr., v. 32, 6980f.)

Kruglov, G.P., Li, L.V. and Nelyubov, V.A.
1969: New type of gold ore mineralization in the Enisei Ridge; Mater. Geol. Polez. Iskop. Krasnoyarsk. Kraya no. 6, p. 114–119. (Chem. Abstr., v. 77, 22969q.)

Krupka, K.M., Ohmoto, H. and Wickman, F.E.
1977: A new technique in neutron activation analyses of Na/K ratios of fluid inclusions and its application to the gold-quartz veins at the O'Brien Mine, Quebec, Canada; Can. J. Earth Sci., v. 14, p. 2760–2770.

Kryukov, V.D. and Tarasov, A.V.
1971: Prospecting for noble metal placer occurrences in the Noril'sk area; Geol. Polez. Iskop. Noril'sk. Raiona, N.N. Urvantsev, ed., Nauch. Issled. Inst. Geol. Arktiki, Leningrad, U.S.S.R., p. 108–112. (Chem. Abstr., v. 77, 128993h.)

Krylova, V.V.
1975: First find of native lead at a gold-silver deposit in the northeastern U.S.S.R.; Akad. Nauk. SSSR, Dokl., v. 221, no. 2, p. 445–446. (Chem. Abstr., v. 83, 82743v.)

Ksenofonotov, O.K. and Leonova, E.P.
1973: Akkargina gold deposit (western Turgai); Vop. Geol. Genezisa Polez. Iskop., v. 3, p. 54–99. (Chem. Abstr., v. 82, 61966j.)

Kubaschewski, O. and von Goldbeck, O.
1975: The thermochemistry of gold; Gold Bull., v. 8, p. 80–85.

Kucha, H.
1973: Organic gold compounds in copper-bearing shale of the Fore-Sudetic monocline; Rudy Metale Niezelaz., v. 18, no. 6, p. 302–303. (Chem. Abstr., v. 80, 29298r.)

Kudaibergenova, N.K.
1969: Near vein alterations in northern Kazakhstan gold deposits; Akad. Nauk Kaz. SSR, Izv., Ser. Geol., v. 26, no. 6, p. 66–68. (Chem. Abstr., v. 72, 102515a.)

Kudryavteseva, A.I.
1969: Regular characteristics in the distribution of gold in skarn minerals of the Tardansk deposit; Mater. Geol. Tuvinskoi USSR, no. 1, p. 68–71. (Chem. Abstr., v. 75, 51408v.)

Kuikin, S.
1974: Distribution of gold ore and gold content of endogenic deposits in Bulgaria; Abstracts of Papers, Int. Assoc. Genesis Ore Deposits (IAGOD), 4th Symp., Varna, Bulgaria, p. 329–330.

Kulikova, M.F.
1962*a*: Trace elements in the oxidized zone of polymetallic deposits of eastern Transbaikalia; Geochemistry, no. 2, p. 176–185.
1962*b*: Behaviour of indium in the oxidized zones of some polymetallic deposits of eastern Transbaikalia; Geochemistry, no. 7, p. 707–714.
1964: Distribution of trace elements in the oxidized zone of some lead-zinc deposits of middle Asia; Geochem. Int., nos. 1–6, p. 877–891.
1966: Geochemistry of gallium and indium in the oxidized zone of lead-zinc deposits in Soviet Central Asia; Geochem. Int., v. 3, p. 982–995.

Kulikova, N.N.
1972: Geochemistry of gases of gold-ore deposits of the Transbaikal region; "Nauka", Moscow, U.S.S.R., 115 p. (Chem. Abstr., v. 78, 114261b.)

Kunaev, I.V.
1963: On the zonal distribution of the rare-metal and gold metallization in the Northern Sikhote-Alin and the lower Amur region; Symposium Problems of Postmagmatic Ore Deposition, v. 1, Publ. House, Czech. Acad. Sci., p. 194–199.

Kurauti, G.
1941: Synthetic study of gold-containing pyrite; Suiyokwai-Si, v. 10, p. 419–424. (Chem. Abstr., v. 35, 3563.)

Kurbanayev, M.S.
1966: Thallium in the dispersion aureoles of the Maykain gold-barite-polymetallic deposit; Geochem. Int., v. 3, p. 568–572.

Kurbanayev, M.S. and Atchibaev, K.A.
1968: Problems of geological efficiency of aurimetric surveys; Geol. Geokhim. Mineral. Zolotorud. Rainov Mestorozhd. Kaz., no. 1, p. 125–128. (Chem. Abstr., v. 73, 27524t.)

Kuryliw, C.J.
1957: Cochenour Willians mine; *in* Structural geology of Canadian ore deposits, v. 2, Can. Inst. Min. Metall., Montreal, p. 295–304.

Kushmuradov, O.K.
1970: Notes on the gold content of sedimentary, metamorphic, and granitoid rocks of the North Nuratau Mountains; Uzb. Geol. Zh., v. 14, no. 1, p. 8–11.

Kushnarev, P.I. and Kashcheev, L.P.
1974: Substantiation of parameters for sampling during the exploration of littoral-marine gold placers; Izv. Vyssh. Ucheb. Zaved. Geol. Razved, v. 17, no. 1, p. 88–93. (Chem. Abstr., v. 81, 66372b.)

Kutina, J. and Stemprok, M., ed.
1963–1965: Symposium—Problems of postmagmatic ore deposition; vols. I and II, and Appendices I and II, Publisher Geol. Surv. Czech., Czech. Acad. Sci., Prague.

Kutyukhin, P.I.
1960: Significance of country rock alteration in exploration for blind gold-bearing veins in the Dzhetygara deposit; Tr. Sver. Gorn. Inst., no. 35, p. 27–40. (Chem. Abstr., v. 60, 1460e.)

Kuznetsov, Yu. A.
1973: New type of gold mineralization in the Donets Basin; Mineral. Sb. (Lvov), v. 27, no. 3, p. 247–249. (Chem. Abstr., v. 82, 61948e.)

Kuznetsov, Yu. A., Kuznetsova, S.V., Metnik, M.A. and Sidenko, O.G.
1973*a*: Florencite as an associated mineral of gold in the Bobrik ore manifestation in the Donets Basin; Akad. Nauk SSSR, Dokl., v. 210, no. 3, p. 682–684. (Chem. Abstr., v. 79, 68674j.)

Kuznetsov, Yu. A., Kuznetsova, S.V., Samoilovich, L.G., Sidenko, O.G., Sharkin, O.P. and Yushin, A.A.
1973*b*: New data on chemical composition and some physical properties of gold from the Nagol'nyi Ridge (Donbas); U.S. Geol. Surv. Transl. from Konst. Svoistva Mineral., v. 7, p. 117–119.

Kuznetsov, K.R. and Panifilov, R.V.
1974: Silver deposits; *in* Ore deposits of the U.S.S.R., V.I. Smirnov, ed., v. 3, Nedra Press, Moscow, p. 78–95.

Kvyatkovskii, E.M.
1963: Primary dispersion haloes of quartz-gold veins in the Lyubovinsk ore field (east Transbaikal); Zap. Leningrad. Gorn. Inst., v. 45, no. 2, p. 10–15. (Chem. Abstr., v. 59, 11113f.)

Kyuregyan, E.A. and Burnutyan, R.A.
1972: Gold in the juice of plants and a method for its detection; Akad. Nauk Arm. SSR, Izv., Nauki Zemle, v. 25, no. 2, p. 83–85. (Chem. Abstr., v. 81, 130520c.)

Kyuregyan, E.A. and Eksuzyan, Ts. O.
1974: Content of gold in the subsurface waters of some regions in the Armenian SSR; Zap. Arm. Otd. Vses. Mineral. O-Va, v. 7, p. 111–119. (Chem. Abstr., v. 84, 35116d.)

Lacomme, A. and Fontan, F.
1974: Auriferous alluviums of Ariege on the Salat; C.R. Congr. Natl. Soc. Savantes, Sec. Sci. v. 96, p. 669–675. (Chem. Abstr., v. 84, 92852q.)

Lacroix, A.
1922–1923: Mineralogie de Madagascar; 3 vols., A. Challamel, éditeur, Paris.

Laffitte, P., coordinator
1968–1970: Carte métallogénique de l'Europe; Int. Union Geol. Sci. and Int. Geol. Congr., Publ. by UNESCO and Bur. Rech. Géol. Minières (Fr.).

Laffitte, P. et Rouveyrol, P.
1964: Carte minière du globe sur fond tectonique; Annal. des Mines, no. 12, p. 25–51.

Laipanov, Kh. Kh. and Sher, S.D.
1971: Contribution to the study of metasomatic alterations of rocks of various composition near gold-bearing quartz veins; Int. Geochem. Congr., Moscow, Abstr. II, p. 542–543.

Lajoinie, J.P. et Fonteilles, M.
1968: Une gîte de skarns latéritisés: le gisement aurifère D'Ity (Côte-D'Ivoire); Chron. Mines, no. 373, p. 143–153.

Lakes, A.
1905: Geology of western ore deposits; Kendrick Book and Stationary Co., Denver, 438 p.

Lakin, H.W.
1969*a*: Trace elements in supergene sulfate minerals; U.S. Geol. Surv., Circ. 622, p. 4.
1969*b*: Solution of gold by porphyry copper leach solutions; U.S. Geol. Surv., Circ. 622, p. 4.

Lakin, H.W., Curtin, G.C. and Hubert, A.E.
1971: Geochemistry of gold in the weathering cycle; *in* Geochemical exploration, R.W. Boyle and J.I. McGerrigle, eds.; Can. Inst. Min. Metall., Spec. Vol. 11, p. 196.

Lakin, H.W., Curtin, G.C., Hubert, A.E., Shacklette, H.T. and Doxtader, K.G.
1974: Geochemistry of gold in the weathering cycle; U.S. Geol. Surv., Bull. 1330, 80 p.

Lakin, W.H. and Nakagawa, H.M.
1965*a*: Simplified spectrophotometric determination for gold; Eng. Min. J., v. 166, no. 10, p. 108–110.
1965*b*: A spectrophotometric method for the determination of traces of gold in geological materials; U.S. Geol. Surv., Prof. Pap. 525-C, p. C168–C171.

Lamb, J.
1957: The French mine, Hedley, B.C.; *in* Structural geology of Canadian ore deposits, v. 2, Can. Inst. Min. Metall., Montreal, p. 47–50.

Lamb, J., Bush, J.B. and Williams, C.T.
1957: Nickel Plate mine, Hedley, B.C.; *in* Structural geology of Canadian ore deposits, v. 2, Can. Inst. Min. Metall., Montreal, p. 42–46.

Lambert, I.B. and Sato, T.
1974: The Kurokô and associated ore deposits of Japan: a review of their features and metallogenesis; Econ. Geol., v. 69, p. 1215–1236.

Lamey, C.A.
1966: Metallic and industrial mineral deposits; McGraw-Hill Book Co., New York, 567 p.

Lancaster, F.H.
1973: The gold content of sea-water. Its extraction an unrealisable dream; Gold Bull., v. 6, no. 4, p. 111.

Landsberg, A. and Hoatson, C.L.
1970: The kinetics and equilibria of the gold-chlorine system; J. Less-Common Metals, v. 22, p. 327–339.

Lang, A.H.
1933: Gold prospecting—Rouyn-Bell River region, Quebec; Can. Min. J., v. 54, p. 267–272.
1948: Regional structure of Barkerville district; *in* Structural geology of Canadian ore deposits, v. 1, Can. Inst. Min. Metall., Montreal, p. 141–149.
1970: Prospecting in Canada, 4th ed; Geol. Surv. Can., Econ. Geol. Rep. No. 7, 308 p.

Lang, A.H., Griffith, J.W. and Steacy, H.R.
1962: Canadian deposits of uranium and thorium; Geol. Surv. Can., Econ. Geol., Ser. 16 (2nd ed.), 324 p.

Langford, G.B.
1930: Beardmore-Nezah gold area, Ontario; Econ. Geol., v. 25, p. 251–269.

Langford, G.B. and Hancox, E.G.
1936: Hypogene anhydrite from McIntyre mine, Porcupine district, Ontario; Econ. Geol., v. 31, p. 600–609.

Langille, E.G.
1948*a*: Premier mine; *in* Structural geology of Canadian ore deposits, v. 1, Can. Inst. Min. Metall., Montreal, p. 121–124.
1948*b*: Geology of the Gold Drop mine, Portland Canal district; Western Miner, v. 21, no. 11, p. 48–50.

Lantsev, I.P., Nikolaeva, L.A., Badalova, R.P. and Denisova, L.K.
1971: Distribution of trace elements in native gold from different deposits; Tr. Tsent. Nauch. Issled. Gornorazved. Inst., no. 96 (pt. 1), p. 130–137. (Chem. Abstr., v. 77, 78005q.)

Lapo, M.S.
1971: Distribution features of gold in the effusive rocks of the Bolshaya Sitsa and Valenku River basins; Vop. Geol. Rud. Mestorozhd. Dal'nego Vostoka, G.I. Osipova, ed., Akad. Nauk. SSSR, Dal'nevost Geol., Vladivostok, U.S.S.R., p. 99–109. (Chem. Abstr., v. 82, 19487d.)

Large, R.R.
1974: Hydrothermal mineral zonation at the Juno mine, Tennant Creek goldfield, central Australia; Econ. Geol., v. 69, p. 1183 (abstr.).
1975: Zonation of hydrothermal minerals at the Juno mine, Tennant Creek Goldfield, central Australia; Econ. Geol., v. 70, p. 1387–1413.
1976: Zonation of hydrothermal minerals at the Juno mine, Tennant Creek goldfield, central Australia—a reply; Econ. Geol., v. 71, p. 1615–1617.

Large, R.R. and Mumme, W.G.
1975: Junoite ,"Wittite", and related seleniferous bismuth sulfosalts from Juno mine, Northern Territory, Australia; Econ. Geol., v. 70, p. 369–383.

Larsen, E.S.
1930: Recent mining developments in the Creede district, Colorado; U.S. Geol. Surv., Bull. 811, p. 89–112.

Lassiter, J.B.
1971: Some chemical aspects of the evolution and utilization of the Red Sea heavy metal deposits; Am. Chem. Soc. Div. Water, Air, Waste Chem., Gen. Pap., v. 11, no. 2, p. 114–121.

Latimer, W.M.
1952: Oxidation states of the elements and their potentials in aqueous solutions, 2nd ed.; Prentice-Hall Inc., Englewood Cliffs, N.J., 392 p.

Latulippe, M.
1966: The relationship of mineralization to Precambrian stratigraphy in the Matagami Lake and Val d'Or districts of Quebec; Geol. Assoc., Can., Spec. Pap. 3, p. 21–42.

Latysh, I.K. and Sidenko, O.G.
1967: Gold of the Bobrikovsk deposit (Nagolny Ridge, Donbas); Konst Svoistva Miner., Akad. Nauk Ukr. SSR, Respub. Mezhvedom. Sb., no. 2, p. 116–119. (Chem. Abstr., v. 68, 61644p.)

Latysh, I.K., Shirinbekov, N.K., Frolova, A.N., Solovits'kii, V.N., Vozgrin, B.R., Podanev, G.A. and Braevich, I.M.
1971: Gold content in the northwestern part of the Ukrainian crystalline shield; Akad. Nauk Ukr. RSR, Dopov, Ser. B, v. 33, no. 4, p. 305–306. (Chem. Abstr., v. 75, 51603e.)

Latysh, I.K., Ryabokon, V.V. and Kostyuchenko, V.S.
1974: Manifestation of gold in rocks of the Sob River Basin; Geol. Zh, v. 34, no. 1, p. 148–149.

Lau, J.W.E.
1970: Mineralogical study of the arsenical gold ore, Bau mining district, Sarawak, Malaysia; unpubl. B.Sc. thesis, Carleton Univ., Ottawa, 60 p.

Laul, J.C., Case, D.R., Wechter, M., Schmidt-Bleek, F. and Lipschutz, M.E.
1970: An activation analysis technique for determining groups of trace elements in rocks and chondrites; J. Radioanal. Chem., v. 4, p. 241–264.

Laul, J.C., Ganapathy, R., Morgan, J.W. and Anders, E.
1972: Meteoritic and non-meteoritic trace elements in Luna 16 samples; Earth Planet. Sci. Lett. v. 13, no. 2, p. 450–454.

Laul, J.C., Hill, D.W. and Schmitt, R.A.
1974: Chemical studies of Apollo 16 and 17 samples; *in* Proc., 5th Lunar Sci. Conf., Geochim. Cosmochim. Acta, Suppl. 5, v. 2, p. 1047–1066.

Laul, J.C., Morgan, J.W., Ganapathy, R. and Anders, E.
1971: Meteoritic material in lunar samples: characterization from trace elements; Proc., 2nd Lunar Sci. Conf., Geochim. Cosmochim. Acta, Suppl. 2, v. 2, p. 1139–1158.

Laul, J.C. and Schmitt, R.A.
1973*a*: Chemical composition of Apollo 15, 16, 17 samples; *in* Proc., 4th Lunar Sci. Conf., Geochim. Cosmochim. Acta, Suppl. 4, v. 2, p. 1349–1367.
1973*b*: Chemical composition of Luna 20 rocks and soil and Apollo 16 soils; Geochim. Cosmochim. Acta, v. 37, no. 4, p. 927–942.

Laur, F.
1906: Sur la présence de l'or et de l'argent dans le Trias de Meurthe-et Moselle; C. R., Acad. Sci., v. 142, p. 1409–1412.

Lawn, J.G.E.
1924: Presidential address on the subject of silver in Witwatersrand ores; Geol. Soc. S. Afr., Proc., v. 27, p. 19–31.

Lawson, A.C.
1913: The gold of the Shinarump at Paria; Econ. Geol., v. 8, p. 434–448.

Lavreau, J.J.
1973: New data about the Kilo-Moto gold deposits (Zaire); Miner. Deposita (Berl.), v. 8, no. 1, p. 1–6.

Lay, D.
1940: Fraser River Tertiary drainage-history in relation to placer-gold deposits; B. C. Dep. Mines, Bull., no. 3, 30 p.

Laznicka, P.
1973: Development of nonferrous metal deposits in geological time; Can. J. Earth Sci., v. 10, p. 18–25.

Leacock, S.
1932: If gold should cease to be "gold"; Can. Inst. Min. Metall., Bull., v. 25, no. 244, p. 430–439.

Learned, R.E. and Boissen, R.
1973: Gold—a useful pathfinder element in the search for porphyry copper deposits in Puerto Rico; *in* Geochemical exploration 1972, M.J. Jones, ed., Inst. Min. Metall. (London), p. 93–103.

LeBourdais, D.M.
1957: Metals and men; McClelland & Stewart, Ltd., Toronto, 416 p.

Lee, H.A.
1963: Glacial fans in till from the Kirkland Lake Fault: a method of gold exploration; Geol. Surv. Can., Pap. 63-45, 36 p.
1965: 1. Investigation of eskers for mineral exploration; 2. Buried valleys near Kirkland Lake, Ontario; Geol. Surv. Can., Pap. 65-14, 20 p.
1971: Mineral discovery in the Canadian Shield using the physical aspects of overburden; Can. Inst. Min. Metall., Bull., v. 64, no. 715, p. 32-36.

Lee, J.K.
1974: Gold placers of the Sangjeon River basin, Gwangchon district, Chijil Kwangmul Chosa Yongu Pokoso, v. 2, no. 2, p. 161-170. (Chem. Abstr., v. 83, 135156q.)

Lee Moreno, J.L.
1974: Geochemical prospecting for epithermal precious metal veins in the vicinity of the Pachuca-Real del Monte (Mexico) (abstr.); Mining Eng., v. 26, no. 8, p. 52.

Legedza, V.Ya.
1966: Occurrences of chemogenic gold in sedimentary rocks; Genet. Osob. Obshch. Zakonomer. Razved. Zolotoi Miner. Dal'nego Vostoka, Dal'nevost. Geol. Inst. Sib. Otd. Akad. Nauk SSSR, p. 55-66. (Chem. Abstr., v. 68, 23513n.)
1967: Role of diagenesis in the accumulation of gold in sulfides of certain sedimentary rocks; Acad. Sci. USSR, Dokl., v. 172, p. 206-208 (Am. Geol. Inst. transl.).

Leie, Yu. A. and Teplov, V.P.
1973: Neogene tectonomagmatic activation in the Transcarpathians and basic features of its metallogeny; Metallog. Obl. Tektono-Magmat. Akt., Tezisy Dokl. Vses. Metallog. Soveshch., 7th, M.M. Odintsov, ed., Akad. Nauk. SSSR, Sib., Otd., Inst. Zemnoi Kory: Irkutsk, U.S.S.R. (Chem. Abstr., v. 82, 127568q.)

Leigh, R.W.
1964: The geology of the Dalny mine, southern Rhodesia; *in* The geology of some ore deposits in southern Africa, v. 2, S.H. Haughton, ed.; Geol. Soc. S. Afr., Johannesburg, p. 29-40.

Lemos, J.C.
1974: Uranium and gold of the Serra da Jacobina; Bol. Com. Nac. Energ. Nucl. (Brazil), v. 6, p. 1-24. (Chem. Abstr., v. 84, 153313p.)

Lenher, V.
1904: Solubility of gold in certain oxidizing agents; J. Am. Chem. Soc., v. 26, p. 550-554.
1909: Some observations on the tellurides; Econ. Geol., v. 4, p. 544-564.
1912: The transportation and deposition of gold in nature; Econ. Geol., v. 7, p. 744-750.
1918: Further studies on the deposition of gold in nature; Econ. Geol., v. 13, p. 161-184.

Leo, G.W., Cox, D.P. and Carvalho, J.P.P.
1964: Geologia da parte sul da Serra de Jacobina, Bahia, Brasil; Brasil Ministerio Minas e Energia, Dep. Nac. Prod. Miner., Bol. 209, 87 p.

Leo, G.W., Rose, H.J. and Warr, J.J.
1965: Chromian muscovite from the Serra de Jacobina, Bahia Brazil; Am. Mineral., v. 50, p. 392-402.

Leonard, B.F.
1973: Gold anomaly in soil of the West End Creek area, Yellow Pine district, Valley County, Idaho; U.S. Geol. Surv., Circ. 680, 16 p.

LeRoy, L.W. and Koksoy, M.
1962: The lichen—a possible plant medium for mineral exploration; Econ. Geol., v. 57, p. 107-111.

Lesure, F.G.
1971: Residual enrichment and supergene transport of gold, Calhoun mine, Lumpkin County, Georgia; Econ. Geol., v. 66, p. 178-186.

Letnikov, F.A. *et al.*
1975: Evolution of fluids during formation of gold-bearing metamorphogenetic quartz veins and possible forms of gold transfer; Geokhim,. no. 12, p. 1827-1836.

Leushkina, G.V., Lobanov, E.M., Dutov, A.G. and Matveeva, N.P.
1971: Determination of gold in biogeochemical samples based on a seven-second gold-197 isotope; Neitronaktiv. Anal., p. 27-33. (Chem. Abstr., v. 77, 13605r.)

Leutwein, F.
1939: Über das Vorkommen einiger seltener Elemente in metamorphen Gesteinen des Schwarzwaldes; Zentralbl. Mineral., Geol. Palaeontol., Abt. A, no. 4, p. 123-128.
1951: Über das Vorkommen von gold in Thüringischen Alaun- und Kieselschiefern; Hallesches Jahrb. Mitteldeut. Erdgeschich., Bd. 1, no. 2, p. 83-85.

Levat, E.D.
1895: L'or en Sibérie orientale, Tomes 1 et 2, Paris, Édouard Rouveyre, Éditeur.
1903: Richesses Minérales des possessions Russes en Asie Centrale; Ann. Mines, 10e Sér., v. 3, p. 181-354.

Levington, K.R.
1972: Ore deposits and mines of the Charters Towers 1:250,000 sheet area, north Queensland; Geol. Surv. Queensl., Rep. 1972, no. 57, p. 1-103.

Levinson, A.A.
1974: Introduction to exploration geochemistry; Applied Publishing Ltd., Calgary, 612 p.

Li, L.V.
1974*a*: Gold-ore mineralization of southern Enisei Ridge; Tr. Sib. Nauch Issled. Inst. Geol., Geofiz. Miner. Syr'ya, v. 144, p. 82-87. (Chem. Abstr., v. 84, 76954u.)
1974*b*: Relation between formation of gold ore deposits and the processes of progressive regional metamorphism in the Enisei Ridge; Tr. Sib. Nauch. Issled. Inst. Geol., Geofiz. Miner. Syr'ya, v. 145, p. 102-103. (Chem. Abstr., v. 83, 63547t.)

Li, L.V. and Datsenko, V.M.
1973: The gold distribution in the granitoids of Proterozoic intrusions in the South Yenisey Ridge; Geochem. Int., v. 10, no. 2, p. 649-655.

Li, A.F. and Kop'eva, V.N.
1961: Rare elements in gold ores of eastern Siberia; Sb. Nauch. Tr., Irkutskii Nauch. Issled. Inst. Redk. Metallov., no. 9, p. 35-41. (Chem. Abstr., v. 57, 13435i.)

Li, L.V. and Kornev, T.Y.
1972: Gold content in ultrabasic rocks of the Yenisey Ridge; Geokhim., no. 11, p. 1397-1400. (Geochem. Int., v. 9, p. 949-951.)

Li, L., Mikheev, V.G. and Rukosuev, M.N.
1971: Probable formation temperature of arsenopyrite in gold deposits on the Yenisei Ridge according to data from x-ray investigations; Geol. Rud. Mestorozhd., v. 13, no. 4, p. 108-112.

Li, L.V., Nevolin, V.A. and Mikheev, V.G.
1974: Geological-mineralogical patterns and genetic features of a gold ore deposit of the Enisei Ridge; Tr. Sib. Nauch. Issled. Inst. Geol., Geofiz. Miner. Syr'ya, v. 145, p. 69-88. (Chem. Abstr., v. 83, 135086s.)

Li, L.V., Okhaphin, N.A., Shebanin, S.P. and Mikheev, V.G.
1974: Composition of element impurities in some deposits of the Enisei Ridge and other gold-bearing regions in the U.S.S.R.; Tr. Sib. Nauch. Issled. Inst. Geol., Geofiz. Miner. Syr'ya, v. 144, p. 108-115. (Chem. Abstr., v. 84, 75957x.)

Li, L.V. and Shokhina, O.I.
1974*a*: Behaviour of gold during processes of progressive regional metamorphism of Precambrian rocks of the Yenisei Mountain Ridge; Geokhim., no. 3, p. 402-410.
1974*b*: Gold distribution in Proterozoic sedimentary-metamorphic rocks of the Enisei Ridge; Tr. Sib. Nauch. Issled. Inst. Geol., Geofiz. Miner. Syr'ya, no. 145, p. 27-37. (Chem. Abstr., v. 83, 118885j.)

Li, Tung and Yio, Chi-Lung
1966: The abundance of chemical elements in the earth's crust and its major tectonic units; Sci. Sin., v. 15, no. 2, p. 258-272.

Li-Ping, T. and Fang-Sung, Y.
1968: Heavy-mineral reconnaissance for gold and copper deposits of the Chinkuashih area, Taiwan; Acta Geol. Taiwan., no. 12, p. 41–57. (Abstr. *in* UNESCO Min. Res. Dev. Ser., no. 38, 1970, p. 162.)

Libby, F.
1969: Gold in the sea; Sea Frontiers, v. 15, no. 4, p. 232–241.

Lidgey, E.
1897: Some indications of ore deposits; Australas. Inst. Min. Eng., Trans., v. 4, p. 110–120.

Liebenberg, W.R.
1955: The occurrence and origin of gold and radioactive minerals in the Witwatersrand System, the Dominion Reef, the Ventersdorp Contact Reef, and the Black Reef; Geol. Soc. S. Afr., Trans., v. 58, p. 101–254.

Liffa, A.
1965: Geology of the metallic deposits near Telkibánya; Magy. Allami Földt. Intérz. Évk., v. 42, p. 211–259 (in Hungarian; French summary).

Lightfoot, B.
1930: Gold resources of southern Rhodesia; *in* The Gold resources of the world; 15th Int. Geol. Congr., Pretoria, 1929, p. 351–366.

Lincoln, F.C.
1911*a*: Types of Canadian gold deposits; Eng. Min. J., v. 91 p. 470–472.
1911*b*: Certain natural associations of gold; Econ. Geol., v. 6, p. 247–302.

Lindahl, T.
1970: The crystal structure of Au_6Hg_5; Acta Chem. Scand., v. 24, p. 946–952.

Lindgren, W.
1894: An auriferous conglomerate of Jurassic age from the Sierra Nevada; Am. J. Sci., ser. 3, v. 47, p. 275–280.
1895: Characteristic features of California gold-quartz veins; Geol. Soc. Am., Bull., v. 6, p. 221–240.
1896: The gold-quartz veins of Nevada City and Grass Valley districts, California; U.S. Geol. Surv., 17th Annu. Rep., pt. 2, p. 1–262.
1901: Metasomatic processes in fissure-veins; Am. Inst. Min. Eng., Trans., v. 30, p. 578–692.
1902: Test for gold and silver in shales from western Kansas; U.S. Geol. Surv., Bull. 202, 21 p.
1905– 1906: Metasomatic processes in the gold deposits of western Australia; Econ. Geol., v. 1, p. 530–544.
1906: Occurrence of albite in the Bendigo veins; Econ. Geol., v. 1, p. 163–166.
1909: Metallogenetic epochs; Econ. Geol., v. 4, p. 409–420.
1910: The hot springs of Ojo Caliente and their deposits; Econ. Geol., v. 5, p. 22–27.
1911: The Tertiary gravels of the Sierra Nevada of California; U.S. Geol., Surv., Prof. Pap. 73, 226 p.
1915: Processes of mineralization and enrichment in the Tintic mining district; Econ. Geol., v. 10, p. 225–240.
1917: Gold and silver deposits in North and South America; Am. Inst. Min. Metall. Eng., Trans., v. 55, p. 883–909.
1920: Vein filling at Bendigo, Victoria; Econ. Geol., v. 15, p. 312-314.
1927: Magmas, dikes, and veins; Am. Inst. Min. Metall. Eng, Trans., v. 74, p. 71–126.
1933: Mineral deposits, 4th ed.; McGraw-Hill Book Co., Inc., New York, 930 p.
1937: Succession of minerals and temperatures of formation in ore deposits of magmatic affiliations; Am. Inst. Min. Metall. Eng., Trans., v. 126, p. 356–376.

Lindgren, W. and Bancroft, H.
1914: Republic (Eureka) district; U.S. Geol. Surv., Bull. 550, p. 133–166.

Lindgren, W. and Ransome, F.L.
1906: Geology and gold deposits of the Cripple Creek district, Colorado; U.S. Geol. Surv., Prof. Pap. 54, 516 p.

Lindsey, D.A.
1972: Sedimentary petrology and paleocurrents of the Harebell Formation, Pinyon conglomerate, and associated coarse clastic deposits, northwestern Wyoming; U.S. Geol. Surv., Prof. Pap. 734-B, 68 p.

Linn, T.A. and Moore, C.B.
1968: Neutron activation determination of gold in iron meteorites and inclusions; Earth Planet. Sci. Lett., v. 3, no. 5, p. 453–456.

Lipman, P.W., Fisher, F.S., Mehnert, H.H., Naeser, C.W., Luedke, R.G. and Steven, T.A.
1976: Multiple ages of mid-Tertiary mineralization and alteration in the western San Juan Mountains, Colorado; Econ. Geol. v. 71, p. 571–588.

Lishmund, S.R.
1973: Upper Hunter gold field; Geol. Surv. N.S.W., Rec., v. 15, no. 2, p. 109–162.

Listova, L.P., Vainshtein, A.Z. and Ryabinina, A.A.
1966: Dissolution of gold in media forming during oxidation of some sulphides; Metallogen. Osad. Osad. Metamorf. Porod, Akad. Nauk SSSR, Lab. Osad. Polez. Iskop., p. 189–199. (Chem. Abstr., v. 68, 88967h.)

Little, H.W.
1959: Tungsten deposits of Canada; Geol. Surv. Can., Econ. Geol. Ser., no. 17, 251 p.
1960: Nelson map-area, west half, British Columbia; Geol. Surv. Can., Mem. 308, 205 p.

Little, H.W., Smith, E.E.N. and Barnes, F.Q.
1972: Uranium deposits of Canada; 24th Int. Geol. Congr., Montreal, Field Excursion C67 Guidebook, 64 p.

Littlepage, J.D. and Bess, D.
1938: In search of Soviet gold; Harcourt, Brace and Co., New York, 310 p.

Litvinov, V.L., Lyakhov, Yu.V. and Popivnyak, I.V.
1970: Paleotemperature zoning in the Kariyskoye gold deposit; Geol. Rud. Mestorozhd., no. 5, p. 96–101.

Liversidge, A.
1893*a*: On the origin of moss gold; J. R. Soc. N.S.W., v. 27, p. 287–298.
1893*b*: On the condition of gold in quartz and calcite veins; J. R. Soc. N.S.W., v. 27, p. 299–303.
1893*c*: On the origin of gold nuggets; J. R. Soc. N.S.W., v. 27, p. 303–343.
1893*d*: On the crystallization of gold in hexagonal forms; J. R. Soc. N. S. W., v. 27, p. 343–346.
1897*a*: Presence of gold in natural saline deposits and marine plants; J. Chem. Soc. (London), v. 71, p. 298–299.
1897*b*: The crystalline structure of gold and platinum nuggets and gold ingots; J. Chem. Soc. (London), v. 71, p. 1125–1131.

Lloyd-Jacob, D.
1971: The role of gold in industry; Gold Bull., v. 4, no. 2, p. 25–29, Publ., Chamber Mines S. Afr., Johannesburg.

Lobanov, E.M., Khatamov, Sh. and Khamidova, R.V.
1967*a*: Determination of gold in biological and geological objects by neutron activation without a chemical decomposition; Aktiv Anal. Gorn. Porod Drugizk Ob'ektov p. 147–157. (Chem. Abstr., v. 69, 631f.)

Lobanov, E.M., Khatamov, Sh. and Kist, A.A.
1967*b*: Radiochemical and preliminary extraction of gold in neutron activation analyses from samples of plants and geological materials; Aktiv. Anal. Gorn. Porod Drugikh. Ob'ektov, p. 158–166. (Chem. Abstr., v. 68, 119123m.)

Lobanov, E.M., Khatamov, Sh. and Talipov, R.M.
1966: Efficiency of neutron activation logging during biogeochemical prospecting for gold ore deposits (in central Kazakhstan); Uzb. Geol. Zh., v. 10, p. 49–54. (Chem. Abstr., v. 69, 29220g.)

Lock, A.G.
1882: Gold; its occurrence and extraction; E. & F.N. Spon Publishers, London, 1229 p.

Lodder, W.
1966: Gold-alunite deposits and zonal wall-rock alteration near Rodalquilar, S.E. Spain; Univ. Amsterdam Geol. Inst., Mededling, no. 318, 93 p.

Logan, W.E., Murray, A., Hunt, T.S. and Billings, E.
1863: Geology of Canada; Dawson Brothers, Montreal, 983 p.

Loginov, V.N.
1972: Minerals of the oxidation zone of a gold-antimony deposit in the southern Transurals; Tr. Sver. Gorn. Inst., v. 86, p. 101–105. (Chem. Abstr., v. 83, 13370j.)

Loginova, L.A.
1972: Hydrogeochemical criteria for ore prospecting; Gidrogeol. SSSR, no. 15, A.B. Sidorenko, ed., "Nedra", Moscow, p. 261–264.

Lombardi, O.W.
1964: Di-β-naphthylthiocarbazone (Dinaphthizone) compared with dithizone as an analytical reagent for the determination of trace metals in natural waters; Anal. Chem., v. 36, no. 2, p. 415–418.

Longley, C.S. and Lazier, T.A.
1948: Paymaster mine; *in* Structural geology of Canadian ore deposits, v. 1, Can. Inst. Min. Metall., Montreal, p. 520–528.

Longridge, C.C.
1913: Gold and tin dredging; Min. J., London, 425 p.

Lonsdale, J.T.
1927: Geology of the gold-pyrite belt of the Northeastern Piedmont, Virginia; Va., Geol. Surv., Bull. 30, 110 p.

Lopez, J.S. *et al.*
1968: La rodalquilarite, chlorotellurite de fer, une nouvelle espèce minérale; Soc. Fr. Minéral. Cristallogr., Bull., v. 91, p. 28–33.

Lord, C.S.
1941*a*: The Tompson-Lundmark gold mines; The Miner, v. 14, no. 12, p. 31–33.
1941*b*: Mineral industry of the Northwest Territories; Geol. Surv. Can., Mem. 230, 136 p.
1949: Firth River gold; Arctic, v. 2, no. 2, p. 125–126.
1951: Mineral industry of District of Mackenzie, Northwest Territories; Geol. Surv. Can., Mem. 261, 336 p.

Loshchinin, V.P.
1971: Distribution of gold in the Precambrian formations of Taskazgan rhythmic suites in Taskazgan (Tamdytau); Uzb. Geol. Zh., v. 15, no. 5, p. 62–65.

Loughlin, G.F. and Koschmann, A.H.
1935: Geology and ore deposits of the Cripple Creek District, Colorado; Colo. Sci. Soc., Proc., v. 13, no. 6, 435 p.

Louis, H.
1934: Mineral deposits; E. Benn., Ltd., London, 384 p.

Lounamaa, J.
1956: Trace elements in plants growing wild in different rocks in Finland; Annu. Bot. Soc. Zoo. Bot. Fenn. 'Vanamo', v. 29, no. 4, p. 1–195.

Lovering, T.G. and Heyl, A.V.
1974: Jasperoid as a guide to mineralization in the Taylor mining district and vicinity near Ely, Nevada; Econ. Geol., v. 69, no.1 , p. 46–58.

Lovering, T.G., Lakin. H.W. and Hubert, A.E.
1968: Concentration and minor element association of gold in ore-related jasperoid samples; U.S. Geol. Surv., Prof. Pap. 600-B, p. B112–B114.

Lovering, T.G., Lakin, H..W and McCarthy, J.H.
1966: Tellurium and mercury in jasperoid samples; U.S. Geol. Surv., Prof. Pap. 550-B, p. 138–141.

Lovering, T.S.
1950: The geochemistry of argillic and related types of rock alteration; Colo. Sch. Mines, Q., v. 45, no. 1B, p. 231–260.

Lovering, T.S. *et al.*
1949: Rock alteration as a guide to ore—East Tintic District, Utah; Econ. Geol., Monogr. 1, 64 p.

Lovering, T.S. and Goddard, E.N.
1950: Geology and ore deposits of the Front Range, Colorado; U.S. Geol. Surv., Prof. Pap. 223, 318 p.

Low, A.P.
1906: The cruise of the Neptune; Gov. Print. Bur., Ottawa, 355 p.

Lowell, W.R.
1942: The paragenesis of some gold and copper ores of southwestern Oregon; Econ. Geol., v. 37, p. 557–595.

Lucio, A. and Gaines, R.V.
1973: Minerals of the Morro Velho gold mine, Brazil; Mineral. Rec., v. 4, no. 5, p. 224–229. (Chem. Abstr., v. 80, 39223b.)

Lunde, G.
1927: Über das Vorkommen des Platins in norwegischen Gesteinen und Mineralen; Z. Anorg. Allg. Chem., Bd. 161, p. 1–20.

Lunde, G. and Johnson, M.
1928: Vorkommen und Nachweis der Platinmetalle in norwegischen Gesteinen, II; Z. Anorg. Allg. Chem., Bd. 172, p. 167–195.

Lunev, B.S. and Utkin, R.E.
1973: Explorations for fine gold; Uch. Zap., Permsk. Gos. Univ. v. 266, p. 116–130. (Chem. Abstr., v. 81, 138570q.)

Lungwitz, E.E.
1900*a*: The lixiviation of gold deposits by vegetation; Eng. Min. J., v. 69, p. 500–502.
1900*b*: Der geologische Zusammenhang von Vegetation und Goldlagerstätten; Zeit. Prakt. Geol., v. 8, p. 71–74.

Lupander, K. and Räisänen, W.K.
1954: The Haveri mine; *in* The mines and quarries of Finland, E. Aurola, ed., Geol. Tutkimuslaitos Geotek. Julk. no. 55, p. 47–52.

Luttrell, G.W.
1966: Base-and-precious-metal and related ore deposits of Virginia: Va., Div. Miner. Resour., Miner. Resour. Rep. 7, 167 p.

Lyakhov, Yu. V.
1975: Temperature zoning of the Darasun deposit; Geol. Rud. Mestorozhd., v. 17, no. 2, p. 28–36.

Lyalikova, N.N. and Kulikova, M.F.
1969: Bacterial oxidation of germanium-bearing sulfides; Acad. Sci. USSR, Dokl., v. 184, Earth Sci. Sec., p. 189–191.

Lyalikova, N.N. and Mokeicheva, L.Ya.
1969: The role of bacteria in gold migration in deposits; Microbiology, v. 38, no. 5 (transl. by Consultants Bureau, N.Y.), p. 682–686.

Maag, R.
1975: History of gold; Mineralienfreund, v. 13, no. 2, p. 17–23.

Macadam, P.
1931: The distribution of gold and carbon in the Witwatersrand bankets; Geol. Soc. S. Afr., Trans. (Annex.), v. 34, p. 81–88.
1936: The heavier metals and minerals in the Witwatersrand banket; Geol. Soc. S. Afr., Trans., v. 39, p. 77–79.

MacArthur, D.M.
1972: Gold reduction and oxidation in aqueous solutions; J. Electrochem. Soc., v. 119, no. 6, p. 672–677. (Chem. Abstr., v. 77, 28197h.)

MacDiarmid, R.A.
1963: The application of thermoluminescence to geothermometry; Econ. Geol., v. 58, p. 1218–1228.

MacDougall, J.B.
1946: Two thousand miles of gold; from Val d'Or to Yellowknife; McClelland & Stewart, Ltd., Toronto, 234 p.

MacGregor, A.M.
1928: The geology of the country around the Lonely Mine, Bubi District: Geol. Surv. S. Rhodesia, Bull. no. 11, 96 p.
1951: The primary source of gold; S. Afr., J. Sci., v. 47, no. 6, p. 157–161.
Machairas, G.
1963: Etude des phénomènes de migration chimique de l'or. Cas de la Guyanne Française et d'Ity en Côte-d'Ivoire; Soc. Fr. Minéral. Cristallogr., Bull., v. 86, p. 78–80.
1967: Dissolution et recristallisation de l'or primaire pendant l'oxydo-réduction des sulfures aurifères; Bur. Rech. Geol. Min. (Fr.), Bull., no. 5, p. 111–121.
1970*a*: Contribution à l'étude, minéralogique et métallogenique de l'or; Bur. Rech. Geol. Min. (Fr.), Bull. (2e Sér.), Sec. 2, no. 3, p. 1–109.
1970*b*: Association des inclusions fluides et des particules d'or dans le quartz aurifère; Schweiz. Mineral. Petrog. Mitt. v. 50, pt. 1, p. 167–171.
MacKay, B.R.
1921: Beauceville map-area, Quebec; Geol. Surv. Can., Mem. 127, 105 p.
MacKay, R.A.
1946: The control of impounding structures on ore deposition; Econ. Geol., v. 41, p. 13–46.
MacKay, R.A.C.
1944: Purity of native gold as a criterion for secondary enrichment; Econ. Geol., v. 39, p. 56–68.
MacKevett, E.M. and Smith, J.G.
1968: Distribution of gold, copper and some other metals in the McCarthy B-4 and B-5 quadrangles, Alaska; U.S. Geol. Surv., Circ. 604, 25 p.
Maclaren, J.M.
1906: The source of the waters of geysers; Geol. Mag., v. 3, p. 511–514.
1908: Gold, its geological occurrence and geographical distribution; Mining J., London, 687 p.
Maclean, T.A.
1914: Lode mining in Yukon; Can. Dep. Mines, Mines Br., Publ. no. 222, 205 p.
Maconachie, R.J.
1940: Lode-gold deposits, Upper Lemon Creek area and Lyle Creek–Whitewater Creek area, Kootenay district; B. C. Dep. Mines, Bull. no. 7, 50 p.
Macpherson, E.O.
1933: Gold-bearing conglomerates of Central Otago; N.Z. Dep. Sci. Ind. Res., Bull. 40, p. 262–274. *Also* N.J., J. Sci. Technol., v. 14, p. 262–274.
Magakian, I.G.
1952: Metallogenic specializations in tectono-magmatic complexes; Zap. Vses. Mineral. Obshchest., v. 81, p. 169–174. (Chem. Abstr., v. 47, 452d.)
1972: The complexes (series) of ore formations in different types of U.S.S.R. ore provinces; 24th Int. Geol. Congr., Sec. 4, Mineral deposits, Montreal, p. 60–64.
Mahrholz, W.W. and Slaughter, A.L.
1967: Research on geochemical exploration underground; Min. Congr. J., v. 53, p. 42–46, 48–49.
Makarov, V.N.
1972: Hydrogeochemical prospecting for gold-antimony mineralization (as exemplified by northeastern Yakutia); *in* Nov. Dannye Metallogen. Geol. Str. Yano-Kolymskogo Mezhdurech'ya, M.I. Rabkin, ed., p. 141–143. (Chem. Abstr., v. 79, 148086j.)
Malakhov, A.A.
1970: Tien-Shan endogenous gold ore formations; *in* Osn. Probl. Metallogen. Tyan-Shanya, Tr. Vses. Metallogen. Soveshch. 5th, 1968, M.M. Adyshev, ed., p. 413–416. (Chem. Abstr., v. 77, 22844v.)
Malcolm, W.
1912: Gold fields of Nova Scotia; Geol. Surv. Can., Mem. 20-E, 324 p.
Malcolm, W. and Faribault, E.R.
1929: Gold fields of Nova Scotia; Geol. Surv. Can., Mem. 156, 253 p. *Also* Geol. Surv. Can., Mem. 385 (Mem. 156 repr.).
Malcolmson, J.W.
1907: The history of gold and silver; Eng. Min. J., v. 84, p. 1021–1023.
Malissa, H., Hermann, F., Kluger, F. and Kiesel, W.
1972: Chemical and microprobe investigations of the Allende-meteorite; Mikrochim. Acta, pt. 3, p. 434–450.
Mallet, J.W.
1887: On the occurrence of silver in volcanic ash from the eruption of Cotopaxi of July 22nd and 23rd, 1885; Proc., R. Soc., v. 42, p. 1–3.
1890: On a second case of the occurrence of silver in volcanic dust, namely in that thrown out in the eruption of Tunguragua in the Andes of Ecuador, January 11th, 1886; Proc., R. Soc., v. 47, p. 227–281.
Malouf, S.E. and Hinse, R.
1957: Campbell Chibougamau mines; *in* Structural geology of Canadian ore deposits, v. 2, Can. Inst. Min. Metall., Montreal, p. 441–449.
Malouf, S.E. and Thorpe, W.
1957: Chibougamau Explorers mine; *in* Structural geology of Canadian ore deposits, v. 2, Can. Inst. Min. Metall., Montreal, p. 449–454.
Maltizov, A.G. and Tyutrin, I.I.
1971: Geological-geomorphological premises of gold placer deposits and development trends in exploration and prospecting in Sakhalin; Geol. Polez. Iskop. Sakhalina Kuril'skikh Ostrovov, I.I. Tyutrin, ed., Sakhalin. Territ. Geol. Upr., Yuzhno-Sakhalinsk, U.S.S.R., p. 59–68. (Chem. Abstr., v. 81, 66450a.)
Malyuga, D.P.
1964: Biogeochemical methods of prospecting; Acad. Sci. USSR, 205 p. (Engl. transl. by Consultants Bur. Enterprises, Inc., New York.)
Mandy, J.T.
1934: Gold bearing black-sand deposits of Graham Island, Queen Charlotte Islands; Can. Inst. Min. Metall., Trans., v. 37, p. 563–572.
Manskaya, S.M. and Drozdova, T.V.
1968: Geochemistry of organic substances; Pergamon Press, Oxford, 345 p.
Mantei, E., Bolter, E. and Al Shaieb, Z.
1970: Distribution of gold, silver, copper, lead and zinc in the productive Marysville stock, Montana; Miner. Desposita (Berl.), v. 5, p. 184–190.
Mantei, E.J. and Brownlow, A.H.
1967: Variation in gold content of minerals of the Marysville quartz diorite stock, Montana; Geochim. Cosmochim. Acta, v. 31, p. 225–235.
Marakushev, A.A.
1977: Geochemical properties of gold and conditions of its endogenic concentration; Miner. Deposita (Berl.), v. 12, p. 123–141.
Marinenko, J. and May, I.
1968: Fluorometric determination of gold in rocks with rhodamine B; Anal. Chem., v. 40, no. 7, p. 1137–1139.
Markham, N.L.
1960: Synthetic and natural phases in the system Au-Ag-Te; Econ. Geol., v. 55, p. 1148–1178, 1460–1477.
Markova, E.A.
1967: Occurrence of volynskite in the gold ore deposit of central Asia; Zap. Vses. Mineral. Obshchest., v. 96, no. 3, p. 324–326. (Chem. Abstr., v. 67, 92662b.)
Markward, E.L.
1961: Geochemical prospecting abstracts, January 1955–June 1957; U.S. Geol. Surv., Bull. 1098-B, p. 57–160.
Marmo, V.
1960: On the sulphide and sulphide-graphite schists of Finland; Finl., Comm. Géol., Bull. no. 190, 80 p.

Marques de Almeida Rolff, P.A.
1944: Gold in pegmatites of northeast Brazil; Min. Metal., v. no. 8, p. 61–62. (Chem. Abstr., v. 39, 256.)

Marshall, C.E.
1964: The physical chemistry and mineralogy of soils; John Wiley & Sons, Inc., New York, 388 p.

Marshall, P.
1918: The geology of the Tuapeka District; N.Z., Geol. Surv., Bull., no. 19, 79 p.

Martinet, R.
1959: Dosages semi-quantitatifs par confined spot de traces de métaux dans les sols As-Ni-Cu-Pb-Zn-Au; Symp. Explor. Geoquimica v. 2, 20th Int. Geol. Congr., Mexico, 1956, p. 403–413.

Martinet, R. and Cuper, J.
1961: Dosage rapide de l'or dans les sols, les minérais et les eaux; Dosage rapide de l'or dans les latérites; Bur. Rech. Géol. Min. (Fr.), Bull. 2, p. 47–66.

Martynov, Yu. T., Semenov, G.S., Sulin, V.V., Shatsov, A.H. and Shorikov, S.I.
1972: Application of nuclear physical methods for the study of rock, ore, and mineral samples. Estimation of the titanium, zirconium, and hafnium contents in samples of recent placers of the Baltic Sea by gamma-activation methods; Tr. Vses. Nauch. Issled. Inst. Yad. Geofiz. Geokhim., no. 13, p. 32–38. (Chem. Abstr., v. 80, 10045j.)

Maslenitsky, I.
1944: On some cases of formation of disperse gold segregations in iron sulphides, C. R. Acad. Sci. USSR, Dokl., v. 14, no. 9, p. 385–388.

Maslov, Yu. S.
1974: Geological-geomorphological premises of the formation of gold- and tin-bearing marine placer deposits of the Chuan Bay; Sov. Geol., no. 8, p. 117–122. (Chem. Abstr., v. 82, 62036t.)

Mason, B.
1962: Meteorites; John Wiley & Sons, Inc., New York, 274 p.
1966: Principles of geochemistry, 3rd ed.; John Wiley & Sons, Inc., New York, 329 p.

Mason, B., ed.
1971: Handbook of elemental abundances in meteorites; Gordon and Breach Science Publishers, New York, 555 p.

Masson, D.L.
1962: Plants for geobotanical prospecting—indicator and accumulator plants used for sampling for geochemical prospecting; Wash. St. Inst. Technol., Circ. 1, 9 p.

Mather, W.B.
1937: Geology and paragenesis of gold ores of the Howey mine, Red Lake, Ontario: Econ. Geol., v. 32, p. 131–153.

Mather, A.L.
1959: Geochemical prospecting studies in Sierra Leone; D.I.C. thesis, Univ. London, Imperial College Sci. Technol.

Matheson, A.F.
1948*a*: Bankfield and Tombill mines; *in* Structural geology of Canadian ore deposits, v. 1, Can. Inst. Min. Metall., Montreal, p. 401–406.
1948*b*: Jellicoe mine; *in* Structural geology of Canadian ore deposits, v. 1, Can. Inst. Min. Metall., Montreal, p. 399–401.
1956: The St. John Del Ray Mining Company, Limited, Minas Geraes, Brazil; Can. Inst. Min. Metall. Bull., v. 49, no. 525, p. 37–43.

Matheson, A.F. and Douglas, J.H.
1948: Hard Rock mine; *in* Structural geology of Canadian ore deposits, v. 1, Can. Inst. Min. Metall., Montreal, p. 406–413.

Mathews, W.H.
1944: Lode-gold deposits, south-eastern British Columbia; B.C. Dep. Mines, Bull. no. 20, pt. 2, 19 p.

Matkovskii, O.I., Yasinskaya, A.A. and Razumeeva, N.N.
1974: New type of gold ore mineralization in the Transcarpathians; Akad. Nauk. Ukr. RSR, Dopov., Ser. B, v. 36, no. 12, p. 1074–1077. (Chem. Abstr., v. 82, 142835t.)

Matsukuma, T.
1962: Gold and silver deposits and their ores in Kyushu, Japan; J. Min. College Akita Univ., Ser. A, v. 2, p. 20–59.

Matsukuma, T. and Horikoshi, E.
1970: Kurokô deposits in Japan, a review; *in* Volcanism and ore genesis, T. Tatsumi, ed., Univ. Tokyo Press, p. 153–179.

Matveenko, V.T.
1973: Regional regularities in the formation in time and distribution in space of gold ore mineralization in the northwestern part of the Pacific Ocean Tectonic Belt; Zakonomer. Razmeshcheniya Polez. Iskop., v. 10, p. 213–229. (Chem. Abstr., v. 80, 72783g.)

Mauer, W. and Pose, H.
1943: Neutronenemission des Urankerns als Folge seiner spontanen Spaltung; Z. Phys., Bd. 121, p. 285–292.

Mawdsley, J.B.
1938: Late gold and some of its implications; Econ. Geol., v. 33, p. 194–210.
1948: Red Lake Gold Shore mine; *in* Structural geology of Canadian ore deposits, v. 1, Can. Inst. Min. Metall., Montreal, p. 334–340.

Mazzucchelli, R.H.
1965: Geochemical dispersion related to gold mineralization, Western Australia; unpubl. Ph.D. thesis, R. Sch. Mines, Imperial College, Univ. London, London, U.K.

Mazzucchelli, R.H. and James, C.H.
1966: Arsenic as a guide to gold mineralization in laterite-covered areas of Western Australia; Inst. Min. Metall., Trans., Sec. B, v. 75, p. B286–B294; Disc., v. 75, p. B127–B129; reply by authors, v. 76, p. B218.

McAndrew, J.
1965: Gold deposits of Victoria; *in* 8th Commonw. Min. Metall. Congr., Australia and New Zealand, v. 1, Geology of Australian ore deposits, 2nd ed. J. McAndrew, ed., p. 450–456.

McCann. W.S.
1922: The gold-quartz veins of Bridge River District, B.C. and their relationship to similar ore-deposits in the Western Cordilleras; Econ. Geol., v. 17, p. 350–369.

McCarthy, J.H.
1972: Mercury vapour and other volatile components in the air as guides to ore deposits; J. Geochem. Explor., v. 1, p. 143–162.

McCarthy, J.H., Learned, R.E., Botbol, J.M., Lovering, T.G., Watterson, J.R. and Turner, R.L.
1969*a*: Gold-bearing jasperoid in the Drum Mountains, Juab and Millard counties, Utah; U.S. Geol. Surv., Circ. 623, 4 p.

McCarthy, J.H., Vaughn, W.W., Learned, R.E. and Meuschke, J.L.
1969*b*: Mercury in soil gas and air—a potential tool in mineral exploration; U.S. Geol. Surv., Circ. 609, 16 p.

McCauley, C.K. and Butler, J.R.
1966: Gold resources of South Carolina; South Carolina Div. Geol., Bull. 32, 78 p.

McConnell, G.W.
1964*a*: Yellowknife gold-quartz deposits; Econ. Geol., v. 59, p. 328–330.
1964*b*: Notes on similarities between some Canadian gold deposits and the Homestake deposits of South Dakota; Econ. Geol., v. 59, p. 719–720.

McConnell, R.G.
1905: Report on the Klondike gold fields; Geol. Surv. Can., Annu. Rep., pt. B, v. 14, p. 1–71. *Also* Geol. Surv. Can. Mem. 284, p. 64–113.
1907: Report on gold values in the Klondike high level gravels; Geol. Surv., Can. Rep. No. 979, 34 p. *Also* Geol. Surv. Can., Mem. 284, p. 217–238.

McDougall, D.J.
1964: A review of thermoluminescence as applied to mineral exploration and rock mechanics; Can. Min. J., v. 85, no. 3, p. 50–54.
1966: A study of the distribution of thermoluminescence around an ore deposit; Econ. Geol., v. 61, p. 1090–1103.
1968: Thermoluminescence of geological materials; Academic Press, London and New York, 678 p.

McGerrigle, H.W.
1936: Gold placer deposits of the Eastern Townships; Que. Bur. Mines, Annu. Rep., 1935, pt. E, 65 p.

McIlroy, R.J.
1951: The plant glycosides; Edward Arnold & Co., London, 138 p.

McIntosh, R.
1948: Madsen Red Lake Mine; *in* Structural geology of Canadian ore deposits, v. 1, Can. Inst. Min. Metall., Montreal, p. 328–334.

McKay, Alexander
1903: Gold deposits of New Zealand; John Mackay, Government Printer, Wellington, 75 p.

McKay, G.A., Cormie, A.M. and Coulson, C.J.
1948: Leitch mine; *in* Structural geology of Canadian ore deposits, v. 1, Can. Inst. Min. Metall., Montreal, p. 385–389.

McKee, E.H. *et al.*
1974: Guidebook to the geology of four Tertiary volcanic centers in central Nevada; Nev. Bur. Mines, Geol. Rep. 19, 72 p.

McKelvey, V.E. and Wang, F.F.H.
1970: World subsea mineral resources; U.S. Geol. Surv., Map 1-632.

McKinney, J.S. *et al.*
1964: Geology of the Anglo American group mines in the Welkom Area, Orange Free State Goldfield; *in* The geology of some ore deposits in southern Africa, v. 1, S.H. Haughton, ed.; Geol. Soc. S. Afr., Johannesburg, p. 451–506.

McKinstry, H.E.
1948: Mining geology; Prentice-Hall, Inc., New York, 680 p.
1957: Source of iron in pyritized wall rocks; Econ. Geol., v. 52, p. 739–754.

McKinstry, H.E. and Kennedy, G.C.
1957: Some suggestions concerning the sequence of certain ore minerals; Econ. Geol., v. 52, p. 379–390.

McKinstry, H.E. and Ohle, E.L.
1949: Ribbon structure in gold-quartz veins; Econ. Geol., v. 44, p. 87–109. (Disc. by E.Y. Dougherty, Econ. Geol., v. 45, p. 177–179.)

McLaren, D.C.
1944: A story of gold; Can. Min. J., v. 65, no. 7, p. 447–452.

McLaughlin, D.H.
1931: The Homestake enterprise—ore genesis and structure; Eng. Min. J., v. 132, no. 7, p. 324–329.
1937: A comparison of the gold deposits of Australia and North America; Proc., R. Can. Inst., Ser. IIIA, v. 11, p. 53–62.

McLellan, R.R., Berkenkotter, R.D., Wilmot, R.C. and Stahl, R.L.
1974: Drilling and sampling Tertiary gold-bearing gravels at Badger Hill, Nevada County, Calif.; U.S. Bur. Mines., Rep. Invest. 7935, 50 p.

McLennan, J.F.
1915: Quartz veins in lamprophyre intrusions; Eng. Min. J., v. 99, p. 11–13.

McLeod, I.R., ed.
1965: Australian mineral industry: the mineral deposits; Aust., Bur. Min. Resour., Geol. Geophys., Bull. No. 72, 690 p.

McMath, J.C., Gray, N.M. and Ward, H.J.
1953: The geology of the country about Coolgardie, Coolgardie Goldfield, W.A.; Geol. Surv. West. Aust., Bull. 107, 365 p. and suppl. atlas.

McMurchy, R.C.
1948: Powell Rouyn mine; *in* Structural geology of Canadian ore deposits, v. 1, Can. Inst. Min. Metall., Montreal, p. 739–747.

McPheat, I.W., Gooden, J.E.A. and Townend, R.
1969: Submicroscopic gold in a pyrite concentrate; Australas. Inst. Min. Metall., Proc., no. 231, p. 19–25.

Means, A.H.
1914: Tourmaline bearing gold quartz veins of the Michipicoten District, Ontario, Econ. Geol., v. 9, p. 122–135.

Mehliss, A.T.M.
1964: The Dawn gold deposit, Queens area, Bulawayo, southern Rhodesia; *in* The geology of some ore deposits in southern Africa, v. 2, S.H. Haughton, ed., Geol. Soc. S. Afr., Johannesburg, p. 41–58.
1968: The geology of the Indarama gold-antimony mine, Sebakwe area, Que Que, Rhodesia; Geol. Soc. S. Afr., Trans. (Annex.), v. 71, p. 135–146.

Meikle, B.K.
1970: Camflo Mines Limited—geology and mining; Can. Inst. Min. Metall., Bull., v. 63, no. 704, p. 1406–1410.

Mellor, E.T.
1916: The conglomerates of the Witwatersrand; Inst. Min. Metall., Trans., v. 25, p. 226–348.
1931: The origin of the gold in the Rand banket: discussion on Professor Graton's paper; Geol. Soc. S. Afr., Trans. (Annex.), v. 34, p. 55–69.

Mellor, J.W.
1923: A comprehensive treatise on inorganic and theoretical chemistry; v. 3, Cu, Ag, Au and the alkaline earths; Longmans, Green and Co., Ltd., London.

Mel'nichuk, E.V.
1974: Gold content of the northeastern part of the Ukrainian Shield in relation to crust formation processes; Rudonos. Kory Vyvtrivaniya, V.I. Smirnov, ed., "Nauka" Moscow, U.S.S.R., p. 183–185. (Chem. Abstr., v. 82, 75731r.)

Mel'nikov, V.D. and Mel'nikova, L.V.
1970: Regular characteristics of gold and trace element distribution in hydrothermally altered volcanic rocks of the Belaya Gora deposit, Lower Amur area; Isv. Tomsk. Polotekh. Inst., v. 239, p. 45–49. (Chem. Abstr., v. 75, 111759x.)

Menke, J.R.
1951: Method for locating deposits; U.S. Patent 2,551,449.

Mennell, F.G.S.
1905: The banket formation of Rhodesia; Geol. Soc. S. Afr., Trans., v. 8, p. 82–87.

Men'shikov, V.I., Malykh, V.D. and Shestakova, T.D.
1974: Direct determination of gold in solids by an atomic-absorption method; Zh. Anal. Khim., v. 29, no. 11 p. 2132–2136. (Chem. Abstr., v. 82, 179889q.)

Mercer, W. and Crocket, J.
1972: Gold and palladium in host rocks and ores from the Heath Steele B-1 deposit, New Brunswick; 24th Int. Geol. Congr., Montreal, Sec. 10, Geochemistry, p. 180–185.

Mering, J., Mathieu-Sicaud, A. and Perrin-Bonnet, I.
1953: Fixation de particules d'or colloidal sur la kaolinite; Cong. Géol. Int. 19ieme, Alger, Fasc. 18, p. 103–107.

Mero, J.L.
1965: The mineral resources of the sea; Elsevier Publishing Co., Amsterdam, 310 p.

Merrill, G.P.
1896: An occurrence of free gold in granite; Am. J. Sci., v. 1, p. 309–311.

Mertie, J.B.
1921: Lode mining in the Juneau and Ketchikan districts; *in* Mineral resources of Alaska; U.S. Geol. Surv., Bull. 714, p. 105–128.
1940: Placer gold in Alaska; J. Wash. Acad. Sci., v. 30, p. 93–124.
1954: The gold pan: a neglected geological tool; Econ. Geol., v. 49, p. 639–651.
1969: Economic geology of the platinum metals; U.S. Geol. Surv., Prof. Pap. 630, 120 p.

Meshchaninov, E.Z. and Azin, V.N.
1972: Distribution of gold in a copper porphyry deposit in the Almalyk region; Izv. Vyssh. Ucheb. Zaved, Geol. Razved. v. 15, no. 8, p. 46–50. *Also* Int. Geol. Rev., v. 15, no. 5, p. 660–663.

Meshchaninova, G.S. and Meshchaninov, E.Z.
1974: Gold distribution in mineral complexes of the Kaulda deposit; Zap. Uzb. Otd. Vses. Mineral. O-va, v. 27, p. 38–40. (Chem. Abstr., v. 83, 100866u.)

Meshcheryakov, S.S.
1963: Distribution of indium in ore deposits of tin and gold; Zap. Vses. Mineralog. Obshchest., v. 92, no. 4, p. 466–470. (Chem. Abstr., v. 60, 2682c.)

Meyer, A.
1941: In the Choco, Colombia; Eng. Min. J., v. 142, p. 35–39.

Meyer, C. and Hemley, J.J.
1967: Wall rock alteration; *in* Geochemistry of hydrothermal ore deposits, H.L. Barnes, ed., Holt, Rinehart and Winston Inc., New York, p. 166–235.

Meyer, W.T. and Peters, R.G.
1973: Evaluation of sulphur as a guide to buried sulphide deposits in the Notre Dame Bay area, Newfoundland; *in* Prospecting in areas of glacial terrain, M.J. Jones, ed., Inst. Min. Metall. (London), p. 55–56.

Meyn, H.D.
1970: Geology of Hutton and Parkin townships; Ont. Dep. Mines, Geol. Rep. 80, 78 p.

Michel, A. and Hunt, T.S.
1866: Reports of Mr. A. Michel and Dr. T.S. Hunt on the gold region of Canada; Geol. Surv. Can., Progr. Rep., 1863–1866, p. 49–90.

Middlehurst, B.M. and Kuiper, G.P.
1963: The moon, meteorites, and comets; Univ. Chicago Press, Chicago, 810 p.

Miholić, S.
1954: Genesis of the Witwatersrand gold-uranium deposits; Econ. Geol., v. 49, p. 537–540.

Mikhailova, Yu. V.
1970: Characteristics of gold distribution in sulfide occurrences in Chakchar and Yakkabag mountains, southern Uzbekistan; Zap. Uzb. Otd. Vses. Mineral. Obshchest., no. 22, p. 196–201. (Chem. Abstr., v. 74, 114505b.)

Mikhalev, G.P. and Shamshina, E.A.
1969: Placer deposits. Diamond placer deposits in Yakutia; Str. Zemnoi Kory Yakutii Zakonomer. Razmeshcheniva Polez. Iskop., Yu. P. Ivensen, ed., Izd. "Nauka", Moscow, U.S.S.R., p. 342–352. (Chem. Abstr., v. 73, 27537z.)

Mikov, A.D.
1970: Thermal and x-ray diffraction studies of limonites from Kuznetsk Alatau gold deposits; Izv. Tomsk. Politekh. Inst., no. 239, p. 446–454. (Chem. Abstr., v. 75, 111586p.)

Mikshevich, V.N., Moiseev, V.E. and Antonov, V.A.
1971: Neutron-activation determination of gold, iridium, and platinum in pyroxenites; Aktiv. Anal., Mater. Vses. Soveshch., 2nd, 1968, p. 127–130. (Chem. Abstr., v. 77, 28465u.)

Miller, A.D. and Fisher, E.I.
1973: Dissolution of gold during oxidation by MnO_2; Geochem. Int., v. 10, no. 3, p. 656–663.
1974: Dissolution of gold in $FeCl_3$ solutions and its precipitation on pyrite (laboratory model); Geokhim., no. 3, p. 411–417.

Miller, B.L. and Singewald, J.T.
1919: The mineral deposits of South America; McGraw-Hill Book Co., Inc., New York, 598 p.

Miller, T.P. and Ferrians, O.J.
1968: Suggested areas for prospecting in the central Koyukuk River region, Alaska; U.S. Geol. Surv., Circ. 570, 12 p.

Miller, W.G.
1902: The eastern Ontario gold belt; Ont. Bur. Mines Rep. for 1902, p. 186–207.

Miller, W.G. and Knight, C.W.
1914: The Precambrian geology of southeastern Ontario; Ont. Bur. Mines Rep., v. 22, pt. 2, 151 p.

Mills, J.W.
1948: Mic Mac mine; *in* Structural geology of Canadian ore deposits, v. 1, Can. Inst., Min. Metall., Montreal, p. 803–808.
1954: Vertical zoning at the O'Brien gold mine, Kewagama, Quebec; Econ. Geol., v. 49, p. 423–430.

Milton, C. and Milton, D.J.
1958: Nickel-gold ore of the Mackinaw Mine, Snohomish County, Washington; Econ. Geol., v. 53, p. 426–447.

Mincheva-Stefanova, Y. and Stefanov, G.
1972: Use of gold content in sphalerite as a typochemical indicator; Spis. Bulg. Geol. Druzh. (Rev. Bulgarian Geol. Surv.), v. 33, no. 3, p. 271–281 (Engl. summary).

Mineev, G.G.
1976: Participation of organisms in the geochemical cycle of gold migration and concentration; Geokhim. no. 4, p. 577–583.

Minigulov, I.M.
1975: Finely dispersed gold of the Lebedin deposit (central Aldan); Akad. Nauk SSSR, Dokl., v. 225, p. 918–919. (Chem. Abstr., v. 85, 35547r.)

Minguzzi, C.
1947: Dosatura spettrografica dell oro in piriti italiane; Soc. Toscana Sci. Nat. Atti., Mem., v. 54, p. 210–243.

Minnett, R.C.A., Button, A. and Kable, E.J.D.
1973: The gold contents of pre-Malmani argillaceous sediments in the Transvaal Supergroup, northeastern Transvaal; Univ. Witwatersrand, Econ. Geol. Res. Unit, Inform. Circ. 82, 11 p.

Minter, W.E.L.
1970: Gold distribution related to sedimentology of a Precambrian Witwatersrand conglomerate, South Africa, as outlined by moving-average analyses; Econ. Geol., v. 65, p. 963–969.
1976: Detrital gold, uranium, and pyrite concentrations related to sedimentology in the Precambrian Vaal Reef placer, Witwatersrand, South Africa; Econ. Geol., v. 71, p. 157–176.

Mints, M.V.
1973: Gold in igneous rocks of the southwestern part of the Ulinsk superposed trough (Okhotsk -Chukotsk volcanic belt); Akad. Nauk. SSSR, Izv. Ser. Geol., no. 11, p. 50–58. (Chem. Abstr., v. 80, 110706u.) *Also* Int. Geol. Rev., v. 17, p. 604–611.

Mirkhodzhaev, I.M. and Lisogor, L.N.
1972: Age of commercial gold ore mineralization in Karamazar; Nek. Itogi Petrometallogenicheskikh Issled. Uzb., p. 171–198. (Chem. Abstr., v. 84, 7620v.)

Miroshnichenko, L.A., Kalinin, S.K. and Zamyatina, G.M.
1971: Germanium in the skarns of central Kazakhstan; Geochem. Int., v. 8, no. 3, p. 371–374.

Mirzaev, M.R. and Khabirov, N.A.
1972: Determination of trace impurities in natural diamonds by the activation method; Akad. Nauk Uzb. SSR, Izv., Ser. Fiz. Mat. Nauk, v. 16, p. 91–93. (Chem. Abstr., v. 77, 13592j.)

Modnikov, I.S., Skvortsova, K.V. and Chesnokov, L.V.
1974: Correlations of uranium-molybdenum, fluorite, and gold ore deposits in regions of continental volcanism; Sov. Geol., no. 12, p. 48–61. (Chem. Abstr., v. 82, 142833r.)

Moffit, F.H.
1913: Geology of the Nome and Grand Central quadrangles, Alaska; U.S. Geol. Surv., Bull. 533, 140 p.

Mogarovskii, V.V.
1961: Geochemistry of thallium in the oxidized zone of the Daraiso sulfide deposit (Middle Asia); Geochemistry, no. 9, p. 848–852.

Moiseenko, V.G.
1965: Metamorfizm Zolota Mestorozhdenii Priamur'ya (Metamorphism of gold in deposits of the Amur River region); Khabarovsk: Kn. Izd., 127 p. (Chem. Abstr., v. 65, 3608c.)

Moiseenko, V.G. and Fat'yanov, I.I.
1972: Geochemistry of gold; 24th Int. Geol. Congr., Montreal, Sec. 10, Geochemistry, p. 159–165.

Moiseenko, V.G., Nechkin, G.S. and Fat'yanov, I.I.
1970: Geochemical conditions of redistribution of gold associated with magmatic rocks; *in* Problems of hydrothermal ore deposition, Z. Pouba and M. Stemprok, eds., Int. Union. Geol. Sci., Ser. A, no. 2, p. 115–120.

Moiseenko, V.G. and Neronskii, G.I.
1973: Confinement of gold mineralization to metamorphic complexes; Metamorf. Kompleksy Vostoka SSSR, A.M. Smirnov, ed., Akad. Nauk. SSSR, Dal'nevost Nauch. Vladivostock, p. 219–225. (Chem. Abstr., v. 82, 61982m.)

Moiseenko, V.G., Shcheka, S.A., Fat'yanov, I.I. and Ivanov, V.S.
1971: The geochemical peculiarities of the distribution of gold in the rocks of the Pacific Belt; Izd. "Nauka" Publ. House, Moscow, 207 p.

Montgomery, J.H., Sinclair, A.J., Symonds, D.F. and Giroux, G.H.
1977: Geostatistical study of the Ladner Creek gold deposit of Carolin Mines, Ltd.; Western Miner, v. 50, no. 3, p. 11–22.

Moore, E.S.
1912: Hydrothermal alteration of granite and the source of vein-quartz at the St. Anthony Mine; Econ. Geol., v. 7, p. 751–761.
1925: Gold in quartzite in the Goudreau District, Ontario; Can. Min. J., v. 46, p. 142.
1932: Goudreau and Michipicoten gold areas, District of Algoma; Ont. Dep. Mines, Annu. Rep., 1931, v. 40, pt. 4, p. 1–54.
1940: Genetic relations of gold deposits and igneous rocks in the Canadian Shield; Econ. Geol., v. 35, no. 2, p. 127–139.

Moore, G.W. and Silver, E.A.
1968: Gold distribution on the sea floor off the Klamath Mountains, California; U.S. Geol. Surv., Circ. 605, 9 p.

Moore, J. McM.
1975: Regional metamorphism as an ore-forming process in the Saudi Arabian Shield; Inst. Min. Metall., Trans., Sec. B., v. 84, p. B59–B62.

Moorhouse, W.W.
1942: Gold mineralization in minor igneous intrusions, Econ. Geol., v. 37, no. 4, p. 318–329.

Morachevskii, D.E. and Nechaeva, A.A.
1960: Characteristics of migration of rhenium from molybdenites; Geochemistry, no. 6, p. 648–651.

Morávek, P.
1963: A contribution to the study of zonal structure of gold-bearing veins; *in* Symp. Problems of Postmagmatic Ore Deposition; v. 1, p. 118–125, Geol. Surv. Czech., Prague.
1971: Ore deposits, structure, and mineralization of the Jilové gold-mining district; Sb. Geol. Ved. Loziskova Geol. Rada Lg Svazek 13, 170 p.

Morávek, P. and Rus, V.
1970: Regeneration of gold in the central Bohemian pluton area, Czechoslovakia; *in* Problems of hydrothermal ore deposition, Z. Pouba and M. Stemprok, eds., Int. Union Geol. Sci., Ser. A., no. 2, p. 120–125.

Mordvin, P.P. and Levin, V.I.
1969: Khudzhir auriferous conglomerates in the eastern Sayans; Form. Tipy Drevnykh Zolotonosnykh Rossypei Metody Ikh Poiskov, N.V. Cherskii, ed., Izd. "Nauka", Moscow., U.S.S.R., p. 160–174. (Chem. Abstr., v. 72, 81383p.)

Mordvin, A.P., Stanikov, V.A. and Fainshtein, G.Kh.
1973: Problem of gold occurrence in Precambrian conglomerates of eastern Siberia and ways of solving it; Litol. Osad. Polez. Iskop. Dokembr. Sib. Dal'nego Vostoka, Mater. Soveshch. Sib. Otd. Kom. Osad. Porodam Otd. Geol., Geokhim. Geofiz. Akad. Nauk. SSSR, 8th, 1972, Yu.P. Kazanskii, ed. (Chem. Abstr., v. 84, 138455t.)

Morey, G.W.
1957: The solubility of solids in gases; Econ. Geol., v. 52, p. 225–251.

Morey, G.W. and Ingerson, E.
1937: The pneumatolytic and hydrothermal alteration and synthesis of silicates; Econ. Geol., v. 32, suppl. to no. 5, p. 607–761.

Morgan, E.V.
1965: A history of money; Pelican Books Inc., Baltimore, Md., 237 p.

Morgan, J.W., Ganapathy, R., Higuchi, H., Krähenbühl, U. and Anders, E.
1974: Lunar basins: tentative characterization of projectiles, from meteoritic elements in Apollo 17 boulders; *in* Proc. 5th Lunar Sci. Conf., Geochim. Cosmochim. Acta. Suppl. 5, v. 2, p. 1703–1736.

Morgan, J.W., Ganapathy, R. and Krähenbühl, U.
1975: Meteoritic trace elements in lunar rock 14321,184; Geochim. Cosmochim. Acta, v. 39, p. 261–264.

Morgan, J.W., Krähenbühl, U., Ganapathy, R. and Anders, E.
1972: Trace elements in Apollo 15 samples: Implications for meteorite influx and volatile depletion on the moon; Proc. 3rd Lunar Sci. Conf., Geochim. Cosmochim. Acta, Suppl. 3, v. 2, p. 1361–1376.
1973*a*: Luna 20 soil: abundance of 17 trace elements; Geochim. Cosmochim. Acta, v. 37, no. 4, p. 953–961.
1973*b*: Trace element abundances and petrology of separates from Apollo 15 soils; *in* Proc. 4th Lunar Sci. Conf., Geochim. Cosmochim. Acta, Suppl. 4, v. 2, p. 1379–1398.

Morgan, J.W., Laul, J.C., Krähenbül, U., Ganapathy, R. and Anders, E.
1972: Major impacts on the moon: characterization from trace elements in Apollo 12 and 14 samples; Proc. 3rd Lunar Sci. Conf., Geochem. Cosmochim Acta, Suppl. 3, v. 2, p. 1377–1395.

Morgan, J.W., Rebagay, T.V., Showalter, D.L., Nadkarni, R.A., Gillum, D.E., McKown, D.M. and Ehmann, W.D.
1969: Allende meteorite: some major and trace element abundances by neutron activation analysis; Nature, v. 224, p. 789–791.

Morgan, P.G.
1908: The geology of the Mikonui subdivision, New Westland; N.Z., Geol. Surv., Bull., no. 6 (New Series), 175 p.
1924: The geology and mines of the Waihi district, Hauraki goldfield, New Zealand; N.Z., Geol. Surv., Bull. 26 (New Series), 218 p.

Morrell, W.P.
1968: The gold rushes; Adam and Charles Black, London, 427 p.

Morris, H.L.
1918: A study of the chemistry of gold at high temperatures and pressures; J. Am. Chem. Soc., v. 40, p. 917–927.

Morris, H.T. and Lovering, T.S.
1952: Supergene and hydrothermal dispersion of heavy metals in wall rocks near ore bodies, Tintic district, Utah; Econ. Geol., v. 47, p. 685–716.

Mott, N.F.
1955: Physics of the solid state; Advancement Sci., v. 12, no. 46, p. 148–154.

Mueller, G.
1971: Organic inclusions within hydrothermal minerals from S.W. Africa and elsewhere; Advances in Organic Geochemistry, p. 707–716, Pergamon Press, Oxford.

Mukaiyama, H. and Izawa, E.
1964: The gold-silver ores from the Kushikino mine, Kagoshima prefecture, Japan, with special reference to the occurrence of naumannite; Min. Geol. (Japan), v. 14, p. 191–199.

Mukanov, K.M. and Aubakirova, R.B.
1969: The geochemistry of rhenium in the oxidation zones of sulphide deposits; Akad. Nauk SSSR, Dokl., v. 184, no. 2, p. 445–447.

Mukanov, K.M. and Romashin, S.S.
1964: Primary geochemical halos at the Akbastau-Kusmurun ore field; Geol. Rud. Mestorozhd., v. 6, no. 5, p. 90–93.

Mukhin, L.
1974: Evolution of organic compounds in volcanic regions; Nature, v. 251, p. 50–51.

Muller, M.C.
1974: Compounds with gold-antimony bonds; Gold Bull., S. Afr. Chamber Mines, v. 7, no. 2, p. 39–40.

Mumme, W.G.
1975: Junoite, $Cu_2Pb_3Bi_8(S, Se)_{16}$, a new sulfosalt from Tennant Creek Australia: its crystal structure and relationship with other bismuth sulfosalts; Am. Mineral., v. 60, p. 548–558.

Mumme, W.G. and Watts, J.A.
1976: Pekoite, $CuPbBi_{11}S_{18}$, a new member of the bismuthinite-aikinite mineral series: its crystal structure and relationship with naturally-and synthetically-formed members; Can. Mineral. v. 14, p. 322–333.

Murchison, R.I.
1867: On the original introduction of gold into the earth's crust and its subsequent distribution in debris over various parts of the earth; *in* Siluria, 4th ed., John Murray Publisher, London, p. 448–475.

Murdock, J. and Webb, R.W.
1966: Minerals of California—Centennial Volume (1866–1966), with sections by Ian Campbell and Eleanor M. Learned; Calif. Div. Mines Geol., Bull. 189, 559 p.

Murovtsev, A.V.
1968: Gold in sedimentary-volcanic strata of a central Asia gold field; Nauch. Tr., Tashkent, Gos. Univ., no. 337, p. 59–63. (Chem. Abstr., v. 72, 69142b.)

Murthy, M.K.
1964: Geology and gold-bearing lodes of Hutti gold mines, Raichur district, Mysore State; 22nd Int. Geol. Congr., New Delhi, India, pt. 5, Genetic problems of ores, p. 87–101.

Mushketov, D.
1930: Sketch of the gold ore deposits of U.S.S.R.; *in* The gold resources of the world; 15th Int. Geol. Congr., Pretoria, 1929, p. 305–319.

Mushkin, I.V. and Yaroslavskii, R.I.
1974: Gold in alkaline basaltoids and some types of abyssal inclusions of the southern Tien-Shan; Geokhim., no. 7, p. 1041–1044.

Mushkin, I.V., Yaroslavskii, R.I. and Kotsyuba, A.S.
1974: Behaviour of gold during eclogitization of the granulite facies gneisses exemplified by xenoliths in dikes of alkaline basaltic rocks of the southern Tien-Shan; Geokhim., no. 8, p. 1172–1178.

Musin, V.N.
1973: Relation of gold mineralization in the Amur Region to magmatic formations; Vop. Magmat., Metamorf. Orudeneniya Dal'nego Vostoka, dal'nevost. Petrogr. Reg. Soveshch., (Mater.), 2nd 1972, S.S. Zimin, ed., Akad. Nauk SSSR, Dal'nevost. Nauch. Tsent., Vladivostok, U.S.S.R., p. 238–241. (Chem. Abstr., v. 81, 155903t.)

Mustart, D.A.
1965: A spectrographic and mineralographic investigation of alluvial gold from the central Yukon; unpubl. B.Sc. thesis., Univ. British Columbia, Vancouver, 46 p.

Musylev, S.A., ed.
1966: Map of mineral resources of Africa; Min. Geol., U.S.S.R. (2 sheets).

Mutch, A.D.
1952: Variation of thermoelectric properties of pyrite in association with gold ore; Min. Eng., v. 4, p. 880–883.

Myers, W.B.
1971: Precambrian pyritic gold- and uranium-bearing conglomerates; Econ. Geol., v. 66, p. 980 (abstr.).

Myerscough, L.
1973: The nuclear properties of gold; Gold Bull., S. Afr. Chamber Mines, v. 6, no. 3, p. 62–68.

Mysnik, A.M.
1973: Trace elements in pyrites of Kalba; Tr. Inst. Geol. Nauk, Akad. Kaz. SSR, v. 33, p. 151–155. (Chem. Abstr., v. 81, 172973h.)

Nadkarni, R.A. and Morrison, G.H.
1974: Determination of noble metals in geological materials by neutron activation analysis; Anal. Chem., v. 46, no. 2, p. 232–236.

Nadson, G.A. and Stern, C.A.
1934: The biological action of metals at a distance; Compt. Rend., v. 198, p. 282–284. (Chem. Abstr., v. 28, 2387-8.)

Nahai, L.
1958: The mineral industry of Turkey, U.S. Bur. Mines, Inform. Circ. 7855, 140 p.

Naiborodin, V.I. and Goncharov, V.I.
1970: Temperatures prevailing during the formation of the Agatovskoye gold-silver deposit; Geol. Rud. Mestorozhd., v. 12, no. 6, p. 46–51. (Summary *in* Econ. Geol., v. 69, p. 730.)

Naiborodin, V.I. and Sidorov, A.A.
1973: Ore-formation series of gold deposits in the Okhotsk-Chukotka volcanic belt; Zakonomer. Razmeshcheniya Polez. Iskop., no. 10, p. 240–249. (Chem. Abstr., v. 80, 72784h.)

Narayanaswami, S.
1960: Geochemical prospecting for gold-bearing lodes in the Kolar gold fields, India; Symp. Explor. Geoquimica., v. 3, 20th Int. Geol. Congr., Mexico City, 1956, p. 541–553.

Narayanaswami, S., Ziauddin, M. and Ramachandra, A.V.
1960: Structural control and localization of gold-bearing lodes, Kolar gold field, India; Econ. Geol., v. 55, p. 1429–1459.

Nash, J.T.
1975: Fluid inclusion studies of vein, pipe, and replacement deposits, northwestern San Juan Mountains, Colorado; Econ. Geol., v. 70, p. 1448–1462.

Nash, J.T. and Cunningham, C.G.
1973: Fluid-inclusion studies of the fluorspar and gold deposits, Jamestown district, Colorado; Econ. Geol., v. 68, p. 1247–1262.

Nasibov, T.N., Orudzhev, G.Yu and Melikov, N.E.
1972: New type of mercury deposit in Terter River middle reaches basin; Akad. Nauk. Azerb. SSR, Dokl., v. 28, no. 8, p. 48–52. (Chem. Abstr., v. 79, 94760d.)

Nason, A.P., Quesada, A., Lopez, V.M., Dennen, W.H. and Roth, J.
1968: Geochemical prospecting for gold in the Callao district, Venezuela; Geos. (Caracas) Esc. Geol. Minas, Met., v. 18, p. 24–40. (Chem. Abstr., v. 72, 14584s.)

Naumenko, V.V. and Skarzhinskii, V.I.
1970: The endogenic ore deposits of Neogene age in Transcarpathia; Geol. Rud. Mestorozhd., no. 1, p. 83–87.

Naumov, V.N., Pachadzhanov, D.N. and Burichenko, T.I.
1971: Behaviour of bismuth, silver, lead, and zinc in the process of secondary sulfide enrichment; Geochem. Int., v. 8, no. 3, p. 459.

Naz'mova, G.N. and Shalaev, Yu. S.
1969: On age of gold mineralization in north Kazakhstan, Akad. Nauk. SSSR, Dokl., v. 188, no. 6, p. 1355–1356.
1971: Mineral associations of gold ore deposits in northern Kazakhstan; Vest. Mosk. Univ. Geol., v. 26, no. 2, p. 98–103. (Chem. Abstr., v. 75, 79045y.)

Naz'mova, G.N., Spiridonov, E.M. and Shalaev, Yu. S.
1975: Aurostibite of the Bestyube field in northern Kazakhstan, first find in the U.S.S.R.; Akad. Nauk SSSR, Dokl., v. 222, no. 3, p. 687–689. (Chem. Abstr., v. 84, 33618v.)

Nedachi, M.
1974: Mineralization of the Kohoku gold, silver, and copper ore deposits, Miyagi prefecture, Japan; Tohoku Univ., Sci. Rep., Ser. 3, v. 12, no. 3, p. 331–394.

Needham, J.
1954: Prospection geóbotanique en Chine médiévale; J. Agric. Trop. Bot. Appl., v. 1, nos. 5–6, p. 143–147.
1959: Science and civilization in China, v. 3; Mathematics and sciences of the heavens and the earth; Cambridge Univ. Press, London, 877 p.

Negurey, E.P. and Temel, V.K.
1936: Tellurides in the placers of Krasny Creek in the Upper Angora region; Prob. Sov. Geol., v. 7, p. 62–66. *Also* Rev. Geol., v. 17, p. 497. (Chem. Abstr., v. 32, 6182,9.)

Nekrasov, I.Ya.
1973: Correlation of gold and tin mineralization in northeastern U.S.S.R., Geol. Rud. Mestorozhd., v. 15, no. 3, p. 16–28. (Chem. Abstr., v. 80, 5725b.)

Nekrasov, Ye. M.
1973: Structural types of gold-ore deposits localized in diversely deformed sandstone-slaty series; Akad. Nauk. SSSR, Izv., Ser. Geol., no. 12, p. 57–71.
1974: Distribution of gold in veined bodies intersecting sandshale strata; Akad. Nauk. SSSR, Izv., Ser. Geol., no. 12, p. 81–94. (Chem. Abstr., v. 82, 158727v.)

Nel, L.T.
1960: The genetic problem of uraninite in the South African gold-bearing conglomerates; 21st Int. Geol. Congr., Norden, pt. XV, p. 15–25.

Nelson, C.H. and Hopkins, D.M.
1972: Sedimentary processes and distribution of particulate gold in the northern Bering Sea; U.S. Geol. Surv., Prof. Pap. 689, 27 p.

Nelson, T.R.H.
1944: Gold mining in South Wales; Mines Quarry Eng., v. 9, p. 3–10, 33–38, 55–60.

Němec, B., Babička, J. and Oborský, A.
1936: Über das Vorkommen von Gold in den Schachtelhalmen (*Equisetum palustre* u. *arvense*); Bull. Int. Acad. Sci. Bohême, v. 37, p. 1–6.

Němec, B., Babička, J. and Smoleř, I.
1937: Spektroskopische und chemische Untersuchungen über Gold in den Pflanzen; Acad. Tcheque Sci. (Prague), Bull. Int., v. 38, p. 40–45.

Němec, D.
1974: Gold in pyrites in some Czechoslovakian deposits; Mineral. Slovaca, v. 6, no. 2, p. 151–154. (Chem. Abstr., v. 82, 75734u.)

Neronskii, G.I., Dobraya, V.T. and Fedulov, V.G.
1972: Distribution and properties of placer gold in prospecting for gold ore deposits. Vop. Geol. Rud. Mestorozhd. Dal'nego Vostoka, G.I. Osipova, ed., Akad. Nauk S.S.S.R., Da'nevost. Geol. Vladivostok, U.S.S.R., p. 79–85. (Chem. Abst., v. 82, 19485b.)

Nesmeyanov, A.N., Perevalova, E.G., Grandberg, K.I. and Lemenovskii, D.A.
1974: Gold organic complexes; Akad. Nauk. SSSR, Izv., Ser. Khim., no. 5, p. 1124–1137.

Nestorenko, G.V.
1973: Lower Cretaceous oxidation zone at the Itakina gold ore deposits in northeastern Transbaikalia; Geol. Geofiz., no. 10, p. 66–76. (Chem. Abstr., v. 80, 39269w.)

Nestorenko, G.V., Vorotnikov, B.A. and Osintsev, S.R.
1976: Behaviour of gold and accessory elements in the processes of oxidation of deposit outcrops (as exemplified by N-E Trans. Baikal and South Yenisei taiga); *in* Gold and rare elements in geochemical processes, Yu. G. Shcherbakov, ed.; "Nauka" Publ. House, Siberian Br., Novosibirsk, p. 140–165.

Nesterov, N.V.
1974: Relation of gold ore deposits in the Soviet part of the Pacific Ocean belt to weathering crusts; Rudonos. Kory Vyvetrivaniya, V.I. Smirnov, ed., "Nauka", Moscow, U.S.S.R., p. 186–191. (Chem. Abstr., v. 82, 46298z.)

Nesterov, N.V., Levin, V.I. and Igumnova, N.S.
1974*a*: Geochemical conditions of the oxidation of ore bodies of the Tokursk deposit; Geol. Geofiz., no. 4, p. 79–84. (Chem. Abstr., v. 81, 52477c.)
1974*b*: Formation of enriched sections of oxidation zones in gold ore deposits, as illustrated by deposits of the Far East; Vop. Rudonos. Yakutii, p. 67–77. (Chem. Abstr., v. 84, 108597y.)

Nesterova, N.P. and Chebotarev, G.M.
1969: The mineralogical and geochemical features of the ores of the Muruntau deposit; *in* The gold ore formations and principal features of the metallogeny of gold in Uzbekistan, I. Kh. Khamrabaev, ed., "Fan" Publ. Off. Uzb. SSR, Tashkent, p. 176–189.

Nesvetaylova, N.G.
1970: Botanical prospecting for ores; Izd. "Nedra" Publ. House, Moscow, 97 p.

Neubauer, W.H.
1962: Geologie der Goldagerstätte von El Sid in Oberägypten mit einem Beitrag zur Geologie der zentralen Arabischen Wüste; Geol. Jahrb., v. 80, p. 117–159.

Nevolin, V.A.
1971: Relation of placers to gold ore deposits in the Yenisei Ridge; Tr. Sib. Nauch. Issled. Inst. Geol., Geofiz. Miner. Syr'ya, no. 114, p. 76–78. (Chem. Abstr., v. 79, 81656z.)

Newhouse, W.H.
1928: The time sequence of hypogene ore mineral deposition; Econ. Geol., v. 23, p. 647–659.
1936: A zonal gold mineralization in Nova Scotia; Econ. Geol., v. 31, p. 805–831.

Newhouse, W.H., ed.
1942: Ore deposits as related to structural features; Princeton Univ. Press, Princeton, N.J. 280 p.

Newhouse, W.H. and Zuloaga, G.
1929: Gold deposits of the Guayana Highlands, Venezuela; Econ. Geol., v. 24, p. 797–810.

Nice, R.W.
1971: Recovery of gold from active carbonaceous ores at McIntyre; Can. Min. J., v. 92, no. 6, p. 41–44, 49.

Nickel, E.H.
1977: Mineralogy of the "green leader" gold ore at Kalgoorlie, Western Australia; Proc., Aust. Inst. Min. Metall., no. 263, p. 9–13.

Nichol, I. and Bjorklund, A.
1973: Glacial geology as a key to geochemical exploration in areas of glacial overburden with particular reference to Canada; J. Geochem. Explor., v. 2, p. 133–170.

Nichol, I.I., James, L.D. and Viewing, K.A.
1966: Regional geochemical reconnaissance in Sierre Leone; Inst. Min. Metall., Trans., Sec. B, v. 75, p. B146–B161; Disc., v. 76, p. B69–B72, 1967; authors' reply, v. 78, p. B52–B54.

Nickerson, D.
1972: An account of a lake sediment geochemical survey conducted over certain volcanic belts within the Slave structural province of the Northwest Territories during 1972; Geol. Surv. Can., Open File Rep. 129, 22 p., append. and maps.

Nicolini, P.
1970: Gitologie des concentrations minérales stratiformes; Gauthier-Villars Editeur, Paris, 792 p.

Nifontov, R.V.
1957: Sur la minéralisation aurifère homogène des roches sédimentaires de quelques régions minéralisees, (en Russe); *in* Method. Isslen. Miner. Syr'ja, Gosgeoltekhizdat, Moscow, p. 90–97. *Also* Chron. Mines, 1962, no. 313, p. 320–321.)
1960: Chemogenic gold in some sedimentary rocks and its role in formation of placer deposits; Zakonomer. Razmeshcheniya Polez. Iskop., Akad. Nauk. SSSR, Otd., Geol. Geograf. Nauk, no. 4, p. 58–62. (Chem. Abstr., v. 56, 185c.)

Nikanorov, G.S., Kotsuba, A.S., Kovalenko, V.V. and Sil'vanovich, Yu. A.
1970: Simultaneous determination of clarke values of gold, mercury, and antimony in rocks by using the neutron activation method; Aktiv. Anal. Blagorod. Metal., p. 113–114. (Chem. Abstr., v. 76, 10043z.)

Nikanorov, G.S., Samoilenko, Yu. N. and Chulyakov, A.S.
1971: Neutron activation determination of gold and rhenium in geochemical studies; Yadernofiz. Metody. Anal. Veshchestva, p. 42–46. *From* Ref. Zhur., Khim. 1972, Abstr., no. 7G129. (Chem. Abstr., v. 78, 11229t.)

Nikitin, N.M. and Yasyrev, A.P.
1974: On the average gold content in sedimentary rocks, Geokhim., no. 8, p. 1237–1239.

Nikolaeva, L.A.
1958: "New" gold in the Lena alluvial deposit; Tr. Tsent. Nauch. Issled. Gornorazved. Inst., v. 2, no. 25, p. 119–122. (Chem. Abstr., v. 54, 18216h.)
1973: Changes in the structure of gold during the metamorphism of ores in some central Asian deposits; Zap. Uzb. Otd. Vses. Mineral. Obshchest., no. 26, p. 105–109. (Chem. Abstr., v. 80, 50473d.)

Nikolaeva, L.A. and Badalova, R.P.
1974*a*: Secondary gold as an index of the sulfide content of ores; Nauch. Tr., Tashkent, Gos. Univ., no. 438, p. 45–47. (Chem. Abstr. v. 85, 35586c.)
1974*b*: Typomorphic importance of the composition of native gold; Tr. Sib. Nauch. Issled. Inst. Geol., Geofiz. Miner. Syr'ya, v. 144, p. 106–108. (Chem. Abstr., v. 84, 76956w.)

Nikol'skii, N.S.
1963: Some features of the distribution of ore mineralization on Sakhalin; Tr. Sakhalin. Komplesk. Nauch. Issled. Inst., Akad. Nauk. SSSR, Sib. Otd., no. 15, p. 14–21. (Chem. Abstr., v. 63, 12892f.)

Nilsson, G.
1973: Nickel prospecting and the discovery of the Mjövattnet mineralization, northern Sweden: a case history of the use of combined techniques in drift-covered glaciated terrain; *in* Prospecting in areas of glacial terrain, M.J. Jones, ed.; Inst. Min. Metall. (London), p. 97–109.

Nishanov, P. Kh., Akbarov, U. and Talipov, R.M.
1975: Volatility of gold when ashing plant samples; Uzb. Geol. Zh., v. 19, no. 3, p. 92–93. (Chem. Abstr., v. 83, 128446f.)

Nishiwaki, C., Matsukuma, T. and Urashima, Y.
1971: Neogene gold-silver ores in Japan; Soc. Min. Geol. Japan, Spec. Issue 3, p. 409–417.

Noble, J.A.
1950: Ore mineralization in the Homestake gold mine, Lead, South Dakota; Geol. Soc. Am. Bull., v. 61, p. 221–252.

Noddak, I. and Noddack, W.
1930: Die Häufigkeit der chemischen Elementen; Naturwissenschaften, v. 18, p. 757–764.
1931: Die Geochemie des Rheniums; Z. Phys. Chem., v. 154A, p. 207–244.
1934: Die geochemischen Verteilungskoeffizienten der Elemente; Sv. Kem. Tidskr., v. 46, no. 8, p. 173–201.
1939: Die Häufigkeiten der Schwermetalle in Meerestieren; Ark. Zool., Bd. 32A, no. 4, p. 1–35.

Nolan, T.B.
1930: The underground geology of the western part of the Tonopah mining district, Nevada; Univ. Nevada Bull., v. 24, no. 4, p. 5–35.
1933: Epithermal precious-metal deposits; *in* Ore deposits of the western States, Lindgren Vol., Am. Inst. Min. Metall. Eng., New York, p. 623–640.
1935: The ground geology of the Tonopah mining district, Nevada; Univ. Nevada, Bull., v. 29, no. 5, 49 p.

Noldart, A.J. and Threader, V.M.
1965: Gold deposits of Tasmania; *in* 8th Commonw. Min. Metall. Congr., Australia and New Zealand, v. I, Geology of Australian ore deposits, 2nd ed., J. McAndrew, ed., p. 518–521.

Nomura, K., Mikami, A., Kato, T. and Oka, Y.
1970: The determination of scandium and gold in meteorites, tektites and standard rocks by neutron activation analyses with an internal-reference method; Anal. Chim. Acta, v. 51, p. 339–408.

Norman, G.W.H.
1948*a*: Major faults, Abitibi region; *in* Structural geology of Canadian ore deposits, v. 1, Can. Inst. Min. Metall., Montreal, p. 822–839.
1948*b*: The Malartic-Haig section of the southern Gold Belt of western Quebec; *in* Structural geology of Canadian ore deposits, v. 1, Can. Inst. Min. Metall., Montreal, p. 839–845.

Normand, D. and Phan, K.D.
1968: Répartition géochimique de l'or dans le district aurifère de St. Yrieix (Massif Central); 23rd Int. Geol. Congr., Prague, Proc., sec. 7, Prague, p. 367–382.

North, H.H. and Allen, C.C.
1948: Young-Davidson mine; *in* Structural geology of Canadian ore deposits; v. 1, Can. Inst. Min. Metall., Montreal, p. 633–637.

Norton, J.J.
1974: Gold in the Black Hills, South Dakota, and how new deposits might be found; U.S. Geol. Surv., Circ. 699, 22 p.

Novgorodova, M.I.
1976: Platinum, palladium and gold in chalcopyrite ores of the southern Urals; Akad. Nauk SSSR, Dokl., v. 226, p. 942-944.

Novgorodova, M.I. and Neronskii, G.I.
1975: Distorted forms of gold crystals from placers of the Amur river region; Zap. Vses. Mineral. O-va, v. 104, no. 4, p. 449–452. (Chem. Abstr., v. 84, 62423h.)

Novgorodova, M.I. and Tsepin, A.I.
1976: On the phase composition of copper gold; Akad. Nauk SSSR, Dokl., v. 227, p. 184–187.

Nowlan, J.P.
1948: Cochenour Willans mine; *in* Structural geology of Canadian ore deposits, v. 1, Can. Inst. Min. Metall., Montreal, p. 357–365.

Nozhikin, A.D., Krendelev, F.P., Gavrilenko, V.A., Mironov, A.G., Bobrov, V.A. and Tsimbalist, V.G.
1972: The distribution of gold and radioactive elements in the Chingansk Series of the Uvolzhsk graben (Yenisei Range); Akad. Nauk SSSR, Sib. Otd., Tr. Inst. Geol. Geofiz. Geokhimiya Usloviay Obrazovaniya Rud. Zolota Redkikh Metallov, no. 149, p. 171–187.

Nuffield, E.W.
1954: Brannerite from Ontario, Canada; Am. Mineral., v. 39, p. 520–522.

Obolenskii, A.A. and Obolenskaya, R.V.
1972: Gold-antimony and mercury ore formations in Yakutia; *in* Geol. Genezis Endogennykh Rud. Form. Sib., V.A. Kuznetsov, ed., "Nauka", Moscow, p. 53–64. (Chem. Abstr., v. 78, 18855c.)

Obrazbekov, G.T.
1972: Coefficients of correlation between gold and associated elements in the primary haloes of the Zhanalyk deposit (northern Kazakhstan); Akad. Nauk Kaz. SSR, Izb., Ser. Geol., v. 29, no. 2, p. 77–80. (Chem. Abstr., v. 77, 22875f.)

Obraztsov, A.I.
1969: Characteristics of prospecting for blind gold-antimony deposits; Metod. Tekh. Razved, no. 64, p. 56–58. (Chem. Abstr., v. 72, 113683e.)
Obroutchew, W.A. et Guérassimov, A.P.
1929: Carte géologique de la région aurifère de la Léna, 172 p., Feuilles VI-1 et VI-2; Comm. Geol. Leningrad. *Also* Geol. Zentrabl. Abstr., v. 43, no. 6, 1931, p. 379.
Oelsner, O.
1961: Atlas of the most important ore mineral parageneses under the microscope; Pergamon Press, Oxford, 311 p.
Oen, I.S. and Kieft, C.
1974: Nickeline with pyrrhotite and cubanite exsolutions, Ni-Co-rich loellingite and an Au-Cu-alloy in Cr-Ni ores from Beni-Bousera, Morocco; Neues Jahrb. Mineral. Monatsh., no. 1, p. 1–8.
Ogryzlo, S.P.
1935: Hydrothermal experiments with gold; Econ. Geol., v. 30, p. 400–424.
Ogura, T.
1967: Geology and mineral resources of the Far East, vol. 1; Univ. Tokyo Press.
1969: Geology and mineral resources of the Far East, vol. 2; Univ. Tokyo Press.
Ohmoto, H. and Rye, R.O.
1974: Hydrogen and oxygen isotopic compositions of fluid inclusions in the Kurokô deposits, Japan; Econ. Geol., v. 69, p. 947–953.
Oka, Y., Kato, T. and Sasaki, M.
1964: Determination of gold in seawater by neutron activation analysis; Nippon Kagaku Zasshi, v. 85, no. 10, p. 643–647.
Okhapkin, N.A. and Bozin, A.V.
1971: Trace elements as possible indicators of gold mineralization in the contact-metasomatic magnetite deposits of the Altai-Sayan folded region; Tr. Sib. Nauch. Issled. Inst. Geol. ,Geofiz. Miner. Syr'ya, no. 114, p. 16–23. (Chem. Abstr., v. 79,68678p.)
Okko, V. and Peltola, E.
1958: On the Outokumpu boulder train; Comm. Geol., Finl. Bull., no. 180, p. 113-134.
Oleinikov, B.V.
1974: Geochemical trends in the behaviour of copper, lead, and gold during the evolution of tholeiite melts in intermediate-depth chambers; Akad. Nauk. SSSR, Dokl., v. 218, p. 1204–1206. (Chem. Abstr., v. 82, 75665x.)
Oleinikov, B.V. and Korobeinikov, A.F.
1974: Chief geochemical tendencies of gold in the evolution of basic rock melts under plutonic conditions; Vop. Rudonosn. Yakutii, G.N. Gamyanin, ed., p. 78–89. (Chem. Abstr., v. 84, 92919s.)
Oleinikov, B.V., Vasil'ev, Yu.R., Korobeinikov, A.F., Nikishov, K.N. and Chesnokov, V.L.
1973: Gold in some rocks of mantle origin; *in* Geol. Geokhim. Bazitov Vost. Chasti Sib. Platformy, V.V. Koval'skii, ed., "Nauka", Moscow, p. 226–230. (Chem. Abstr., v. 80, 85637n.)
Oliveira, E.P. de
1931: Origin of Jacutinga auriferous deposits; Ann., Acad. Brasil. Cienc., v. 3, p. 151–157. (Chem. Abstr., v. 26, 3755.)
1932: Genesis of the deposits of auriferous Jacutinga; Econ. Geol., v. 27, p. 744–749.
1937: A politica do ouro; Brazil, Serv. Geol. Mineral., Bol. No. 73, 46 p.
Oliver, T.A.
1949: Ore minerals of the Berens River mine; Can. Min. J., v. 70, p. 83–86.
Ol'shevskiy, V.M.
1973: On the occurrence of aquamarine in gold-quartz veins of Karalveem deposit (west Chukotka); Akad. Nauk. SSSR, Dokl., v. 212, p. 1208–1209. (Chem. Abstr., v. 80, 39191q.)
1975: Dependence of the relationship Ag/Au in near-surface gold-silver deposits on the composition of ore-enclosing rocks; Akad. Nauk SSSR, Dokl., v. 224, no. 4, p. 925–928. (Chem. Abstr., v. 84, 76815z.)
Olson, J.C., Shawe, D.R., Pray, L.C. and Sharp, W.N.
1954: Rare-earth mineral deposits of the Mountain Pass District, San Bernadino County, California; U.S. Geol. Surv., Prof. Pap. 261, 75 p.
O'Neil, J.J. and Gunning, H.C.
1934: Platinium and allied metal deposits of Canada; Geol. Surv. Can., Econ. Geol. Ser., no. 13, 165 p.
O'Neil, J.R. and Silberman, M.L.
1974: Stable isotope relations in epithermal Au-Ag deposits; Econ. Geol., v. 69, p. 902–909.
O'Neil, J.R., Silberman, M.L., Fabbi, B.P. and Chesterman, C.W.
1973: Stable isotope and chemical relations during mineralization in the Bodie mining district, Mono County, California; Econ. Geol., v. 68, p. 765–784.
Ong, H.L. and Swanson, V.E.
1969: Natural organic acids in the transportation, deposition, and concentration of gold; Colo. Sch. Mines, Q., v. 64, no. 1, p. 395–425.
Ong, H.L., Swanson, V.E. and Bisque, R.E.
1970: Natural organic acids as agents of chemical weathering; U.S. Geol. Surv., Prof. Pap. 700-C, p. 130–137.
Onishi, H.
1959: Spectrophotometric determination of traces of gold with rhodamine B; Mikrochim. Acta, Heft 1, p. 9–17.
Orazbekov, G.T.
1971: Structure and zonality of quartz-barite type primary haloes in northern Kazakhstan gold mineralization; Akad. Nauk Kaz. SSSR, Izv., Ser. Geol., v. 27, no. 3, p. 64–65. (Chem. Abstr., v. 75, 66116x.)
Orlov, A.P., Robonen, V.I. and Kirilenko, G.M.
1969: Geological prospecting with ore-searching dogs; "Nedra", Moscow, 47 p.
Orlova, Z.V.
1972: Heavy mineral concentrates and features of gold in the alluvial placers of west Chukotka; Tr. Sev. Vost. Kompleks. Inst., Dal'nevost. Tsent. Aka., v. 44, p. 130–134. (Chem. Abstr., v. 81, 155818u.)
Osipov, A.P.
1971: Distribution characteristics of the gold-silver deposits in the Okhotsk-Chukotsk volcanic belt; Izv. Vyssh. Ucheb. Zaved., Geol. Razved., v. 14, no. 12, p. 71–76. (Chem. Abstr., v. 76, 74763s.)
Osipov, A.P. and Sidorov, A.A.
1973: Features and perspectives of gold-rare metal formation; Tr. Sev. Vost. Kompleks. Inst., Dal'Nevost. Tsent., Akad. Nauk SSSR, no. 55, p. 163–173. (Chem. Abstr., v. 84, 7595r.)
Osokin, P.V.
1962: Results of metallometric prospecting for ores in the China-Amalat interfluvial in the Vitim plateau; Mater. Geol. Polez. Iskop. Buryat. ASSR, Buryatsk. Geol. Upr. no. 8, p. 61–76. (Chem. Abstr., v. 59, 11113c.)
Ostrovskiy, E. Ya, Portnov, A.M. and Drabkin, I.E.
1970: Gamma ray radiation aerial surveys in prospecting for near-surface gold deposits; Geol. Rud. Mestorozhd. v. 12, no. 3, p. 87–91. *Also* Int. Geol. Rev., v. 14, no. 7, p. 688–691.
Ottemann, J. and Augustithis, S.S.
1967: Geochemistry and origin of "platinum-nuggets" in lateritic covers from ultrabasic rocks and birbirites of W. Ethiopia; Miner. Deposita (Berl.), v. 1, p. 269–277.
Ovichinnikov, L.N., ed.
1972: The ore resources of the Urals; Izd. "Nauka" Publ. House, Moscow, 185 p.
Ovchinnikov, L.N., Baranov, E.N., Grigoryan, S.V. and Vostroknutov, G.A.
1972: Geochemical prospecting methods for ore deposits and their application in the Urals; *in* Rud. Baza Urala, L.N. Ovchinnikov, ed., "Nauka", Moscow, p. 67–91. (Chem. Abstr., v. 79, 68614q.)

Ovchinnikov, L.N. and Grigoryan, S.V.
1971: Primary halos in prospecting for sulphide deposits; *in* Geochemical exploration, R.W. Boyle and J.I. McGerrigle, eds., Can. Inst. Min. Metall., Spec. Vol. 11, p. 375–380.

Ovchinnikov, L.N., Sokolov, V.A., Fridman, A.I. and Yanitskii, I.N.
1973: Gaseous geochemical methods in structural mapping and prospecting for ore deposits; *in* Geochemical exploration, 1972, M.J. Jones, ed., Inst. Min. Metall. (London), p. 177–182.

Overstreet, W.C. and Bell, H.
1960: Geochemical and heavy-mineral reconnaissance of the Concord S.E. Quadrangle, Cabarrus County, North Carolina; U.S. Geol. Surv., Mineral Invest. Field Studies, Map MF-235.

Ozerova, N.A. *et al.*
1969: Recent mercury mineralization in the Mendeleev volcanic (Kurile Islands); Geol. Rud. Mestorozhd., v. 11, no. 5, p. 17–33. (Chem. Abstr., v. 73, 57956c.)

Paar, W.
1976: Telluride der Gold-Nasturan-Paragenese von Mitterberg, Salzburg (Osterreich); Neues Jahrb. Mineral. Monatsh., Heft 5, S. 193–202.

Pabst, A. and Stinson, M.C.
1960: Brannerite with gold from Plumas County, California; Geol. Soc. Am., Bull., v. 71, p. 2071–2072.

Page, R.W. and McDougall, I.
1972: Ages of mineralization of gold and porphyry copper deposits in the New Guinea Highlands; Econ. Geol., v. 67, p. 1034–1048.

Pagliuchi, F.D.
1925: Gold in southeastern Venezuela; Eng. Min. J., v. 120, p. 485–491.

Pain, S.A.
1960: Three miles of gold: the story of Kirkland Lake; Ryerson Press, Toronto, 109 p.

Palache, C., Berman, H. and Frondel, C.
1944: The system of mineralogy, v. 1, Elements, sulfides, sulfosalts, oxides; John Wiley & Sons, Inc., New York, 834 p.
1951: The system of mineralogy, v. 2, Halides, nitrates, etc.; John Wiley & Sons, Inc., New York, 1124 p.

Palandzhyan, S.A. and Razin, L.V.
1974: Gold distribution in alpine-type massifs of ultrabasic and basic rocks of the Sevan ridge; Akad. Nauk Arm. SSR, Izv., Nauki Zemle, v. 27, no. 3, p. 25–37. (Chem. Abstr., v. 81, 124243y.)

Palei, L.Z.
1968: Age and distribution characteristics of gold ores in central Asia; Int. Geol. Rev., v. 10, no. 7, p. 741–744.
1971: Distribution conditions of endogenous ores. Quantitative accounting for mineralogical-geochemical ore indicators in medium- and large-scale geological exploration forecasting; Metody Detal'nogo Kolichestvennogo Prognozirovaniya Endogennykh Rud. Polei Mestorozkd., S.I. Ibadullaev, ed., Nauch. Issled. Inst. Geol. Miner. Syr'ya, Tashkent, U.S.S.R., p. 60–62. (Chem. Abstr., v. 79, 138714c.)

Palei, L.Z., Murovtsev, A.V. and Borozenets, N.I.
1967: Geochemistry of gold in Sultanuizdag; Uzb. Geol. Zh., v. 11, no. 6, p. 50–55. *Also* Geochem. Int., v. 4, no. 6, p. 1197–1201. (Chem. Abstr., v. 69, 4228x.)

Palmer, C. and Bastin, E.S.
1913: Metallic minerals as precipitants of silver and gold; Econ. Geol., v. 8, p. 140–170.

Palmer, W.S.
1935: Gold in petrified wood; Min. Metall., v. 16, p. 335.

Pampura, V.D. and Trufanova, L.G.
1968: Lithium and cesium in zones of argillic alteration, Baleisk gold deposit (eastern Transbaikal); Geochem. Int., v. 5, no. 4, p. 825–827.

Pankov, V.P. and Chukhrov, F.V.
1962: Mineralogy of gold tellurium ores from the Manka deposit in the southern Altai; Mineral. Sb. L'vovsk. Geol. Obshchest. pri L'vovsk. Gos. Univ. 16, p. 195–209.

Papish, J. and Stilson, C.B.
1930: Gallium IV: occurrence of gallium in zinc minerals; Am. Mineral., v. 15, p. 521–527.

Papon, A. and Lemarchand, R.
1973: Geology and mineralization of southwestern Ivory Coast; Ivory Coast, Dir. Mines Geol., Bull., v. 6, p. 1–286. (Chem. Abstr., v. 82, 5962t.)

Pardee, J.T.
1933: Placer deposits of the western United States; *in* Ore deposits of the western states; Am. Inst. Min. Eng., Lindgren Vol., p. 419–450.
1934: Beach placers of the Oregon Coast; U.S. Geol. Surv., Circ. 8, 41 p.

Pardee, J.T. and Park, C.F.
1948: Gold deposits of the Southern Piedmont; U.S. Geol. Surv., Pap. 213, 156 p.

Parès, Y. et Martinet, R.
1964: Intervention des bactéries dans le cycle de l'or. Etude biologique de phénomène; Bur. Rech. Geol. Min. (Fr.,) Bull., no. 3, p. 1–29.

Parilov, Yu. S. and Netaliyeva, T.D.
1973: Occurrence of silver in the form of halides in sphalerites of the lead-zinc deposits; Geochem. Int. v. 10, p. 336–338.

Park, C.F.
1955: The zonal theory of ore deposits; Econ. Geol., 50th Anniv. Vol., p. 226–248.
1970: Review of U.S. Geol. Surv. Prof. Pap. 641-A, Physiographic, stratigraphic and structural development of the Quadrilatero Ferrifero, Minas Gerais, Brazil; Econ. Geol., v. 65, p. 523–524.

Park, C.F. and MacDiarmid, R.A.
1970: Ore deposits; W.H. Freeman and Company, San Francisco, 522 p.

Park, J.
1897: The geology and veins of the Hauraki goldfields; N.Z., Inst. Min. Eng., Session 1897, 105 p.
1908: The geology of the Cromwell subdivision, western Otago division; N.Z., Geol. Surv., Bull. No. 5, (New Series) 92 p.
1927: A textbook of mining geology; Charles Griffin and Co. Ltd., London.

Park, W.C.
1966: Geochemistry of scandium, a brief review and its problems; J. Geol. Soc., Korea, v. 2, no. 2, p. 39–40.

Parker, B.H.
1974: Gold placers of Colorado; Colo. Sch. Mines, Q., v. 69, no. 3, 268 p. and no. 4, 224 p.

Parker, R.L.
1967: Composition of the earth's crust; U.S. Geol. Surv., Prof. Pap. 440-D, 19 p.

Parr, R.M. and Taylor, D.M.
1963: Determination of gold in human liver by thermal neutron activation analysis; Physics in Medicine and Biology, v. 8, p. 43–50.

Parsons, F.L.
1907: The coal-mining situation in northern Wyoming; Eng. Min. J., v. 84, p. 930–935.

Parsons, G.E.
1948: Bidgood Kirkland mine; *in* Structural geology of Canadian ore deposits, v. 1, Can. Inst. Min. Metall., Montreal, p. 653–658.

Pascal, P.
1957: Nouveau traité de Chimie Minérale; t. 8, Masson et Cie, Paris.

Pasquali, Z.J. and Bisque, R.E.
1975: Interpretation of soil geochemical expression of mineralization within El Callao gold mining district, Venezuela; Colo. Sch. Mines, Q., v. 70, no. 1, p. 83–107. (Chem. Abstr., v. 83, 13493b.)

Patterson, D.O. and Thomson, J.O.
1970: A study of the intermetallic compounds of gold and manganese through the use of the ^{197}Au Mössbauer effect at 4.2°K. and as a function of pressure; Phys. Rev. B., v. 2, p. 2440–2448.

Paule, R.C. and Mandel, J.
1972: Analyses of interlaboratory measurements on the vapour pressure of gold; J. Pure Appl. Chem., v. 31, no. 3, p. 371–394.

Paulo, A.J. and Salamon, W.
1973: Native gold in ore veins of the western part of Gory Kaczawskie Mountains (west Sudeten); Mineral. Pol., v. 4, p. 85–91. (Chem. Abstr., v. 82, 173663q.)

Pavlov, A.L. and Pavlova, L.K.
1971: Thermodynamic parameters of gold occurrence during ore formation; Fiz. Fiz.-Khim. Rudoobrazuyushchikh Protessov, G.L. Pospelov, ed., "Nauka", Sib. Otd. Novosibirsk, U.S.S.R., p. 121–147. (Chem. Abstr., v. 76, 143437c.)

Pazdzerskii, V.A.
1974: Localization of gold mineralization at the Yuzhnoe gold-sulphide deposits (Kokpatas ore field), Metallog. Geokhim. Uzb., p. 57–61. (Chem. Abstr., v. 82, 142806j.)

Pazenkova, N.I.
1967: Germaniferous limonite-forming mechanism; Acad. Sci. USSR, Dokl., Earth Sci. Sec., v. 173, p. 204–206 (Am. Geol. Inst. Transl.).

Pchelintseva, N.F. and Fel'dman, V.I.
1973: Gold in metamorphic rocks of Koktchetav Highland; Geokhim., no. 12, p. 1846–1854. *Also* Geochem. Int., v. 10, no. 6, p. 1357–1365.

Peach, P.A.
1949: A decrepitation geothermometer; Am. Mineral., v. 34, nos. 5–6, p. 413–421.

Pegg, W.C.
1950: A contribution to the geology of the West Rand area; Geol. Soc. S. Afr., Trans., v. 53, p. 209–227.

Peirson, D.H., Cawse, P.A., Salmon, L. and Cambray, R.S.
1973: Trace elements in the atmospheric environment; Nature, v. 241, p. 252–256.

Pelletier, R.A.
1940: Pyrrhotite as an enriching factor in Rand ores; J. Chem. Metall. Min. Soc. S. Afr., v. 40, no. 11, p. 353–365.
1964: Mineral resources of south-central Africa; Oxford Univ. Press, Cape Town, 277 p.

Penning, W.H.
1888: The South African goldfields; J. R. Soc. Arts, v. 36, p. 433–444.

Percy, J. and Smith, R.
1854: On the detection of gold in lead and its compounds, Phil. Mag., v. 7, p. 126–130.

Pereira, J. and Dixon, C.J.
1964– Evolutionary trends in ore deposition; Inst. Min.
1965: Metall., Trans., v. 74, pt. 9, p. 505–527.

Perel'man, A.I.
1972: The geochemistry of elements in the zone of supergenesis; Izdat. "Nedra", Moscow, 287 p.

Perelyaev, A.P.
1953: Composition and structure of native gold; Zap. Vses. Mineral. Obshchest., v. 82, p. 196–206. (Chem. Abstr., v. 48, 1904b.)

Perelygina, A.I.
1972: Major types of gold ore deposits and occurrence in the Aldan Shield; Izv. Vyssh. Uch. Zaved., Geol. Razved., v. 15, no. 4, p. 56–64. (Chem. Abstr., v. 77, 8190x.)
1973: Main types of gold-ore deposits and occurrences of the Aldan Shield; Int. Geol. Rev., v. 15, no. 1, p. 19–24.

Pereyaslov, V.P.
1970: Reliable bulk sampling during prospecting for placer gold; Mater. Metod. Tekh. Geologorazved. Rob., no. 2, p. 76–78. (Chem. Abstr., v. 77, 142305s.)

Perez, Re. J.
1969: Investigacienes auríferas en la provincia de Cáceres; Bol. Geol. Minero, v. 80, no. 2, p. 131–145.

Peshchevitskii, B.I., Anoshin, G.N. and Erenburg, A.M.
1965: Chemical forms of gold in seawater; Akad. Nauk SSSR, Dokl., v. 162, no. 4, p. 915–917. (Engl. transl. *in* Acad. Sci. USSR, Dokl., Earth Sci. Sec., v. 162, nos. 1–6, p. 205–207.)
1970: Chemical forms of gold in sea water and problems of the redox potential of sea water; Khim. Resur. Morei Okeanov, Nauka Moscow, U.S.S.R., S.V. Bruevich, ed., p. 141–144.

Peterson, D.W., Yeend, W.E., Oliver, H.W. and Mattick, R.E.
1968: Tertiary gold-bearing channel gravel in northern Nevada County, California; U.S. Geol. Surv., Circ. 566, 22 p.

Peterson, U.
1964: Synthetic and natural phases in the system Au-Ag-Te; Econ. Geol., v. 59, p. 1390–1394.
1970: Metallogenic provinces in South America; Geol. Rund., v. 59, no. 3, p. 834–897.

Petrov, B.V., Krendelev, F.P., Bobrov, V.A., and Tsimbalist, V.G.
1972: Behaviour of radioactive elements and gold during metamorphism of the Patomsk Highland sedimentary rocks; Geokhim., no. 8, p. 947–956. *Also* Geochem. Int., v. 9, no. 4, p. 647–655.

Petrov, B.V., Krendelev, Ph. P., Bobrov, V.A. and Tsimbalist, V.T.
1970: The behaviour of radioactive elements and gold during sedimentary rock metamorphism in the Patom Highland; Ezheg. Inst. Geokhim., Sib. Otd., Akad. Nauk SSSR, 1969, L.V. Tauson, ed., p. 115–119.

Petrov, B.V. and Makrighina, V.A.
1971: The geochemistry of the regional metamorphism in high pressure conditions; Int. Geochem. Congr., Moscow, Abstracts of Reports II, p. 446–447.

Petrov, P.A.
1962: Evaluation of the gold content in a region from the alluvium; Tr. Sib. Nauch. Issled. Inst. Geol., Geofiz., Mineral'n Syr'ya, no. 25, p. 195–197. (Chem. Abstr., v. 59, 1386d.)

Petrov, V.G.
1974: Environment of gold ore deposits in the northern part of the Yenisei Range; Acad. Sci. USSR, Sib. Br., Trans. Inst. Geol. Geophys. No. 69, "Nauka" Publ. House Siberian Div. Novosibirsk, 138 p.

Petrov, Yu. I.
1972: Topogeochemistry of the Muruntau gold ore deposit in central Asia; Lithogeokhimicheskie Metody Poiskakh Skrytogo Orudeneniya, L.N. Ovchinnikov, ed., p. 38–41. (Chem. Abstr., v. 81, 80405w.)

Petrovskaya, N.V.
1955: Productive mineral associations in gold deposits; Reps. All Soviet Min. Soc. No. 3, p. 290–307 (in Russian).
1960: Character of auriferous mineral associations and formation of gold ores in U.S.S.R.; 21st Int. Geol. Congr., Copenhagen, Genetic Prob. Ores, Contrib. Sov. Geol. p. 145–158.
1968: Auriferous formations of Uzbekistan; Geol. Rud. Mestorozhd., v. 10, no. 3, p. 3–16. *Also* Econ. Geol., v. 65, p. 75.
1970: Redeposition of gold and quartz during formation of gold deposits; Izv. Tomsk. Politekh. Inst., v. 239, p. 133–139. (Chem. Abstr., v. 75, 111606v.)
1971: Growth and subsequent changes in gold crystals; Mineral. Soc. Japan, Spec. Pap. 1, p. 116-123.
1973: Native gold; Izd. "Nauka", Moscow, 347 p.

Petrovskaya, N.V., ed.
1974: Mineralogy and geochemistry of gold; Tezizy Dok. Simp. Mineral. Geokhim Zolota, vols. 1 and 2, Akad. Nauk SSSR Dal'Nevost Nauch. Tsent., Vladivostok, U.S.S.R.

Petrovskaya, N.V., Bernshtein, P.S., Mirchink, S.G. and Andreeva, M.G.
1961: Geological structure, mineralogy, and characteristics of genesis of gold ore deposits of the Balei ore field (eastern Transbaikal); Tr. Tsent. Nauch. Issled. Gornorazved. Inst., no. 45, pt. I, 97 p. (Chem. Abstr., v. 58, 360g.)

Petrovskaya, N.V., Elinson, M.M. and Nikolaeva, L.A.
1971*a*: Composition and formation conditions of gas inclusions in native gold; Int. Geochem. Congr., Moscow, U.S.S.R., Abstracts of Reports I, p. 326–327.

Petrovskaya, N.V. and Fastalovich, A.I.
1955: Changes in the internal texture of native gold buried in placers; Vop. Geol. Azii, no. 2, p. 245–256.

Petrovskaya, N.V., Konstantinov, R.M. and Sirotinskaya, S.V.
1971*b*: Quantitative study of similarity in mineral composition of gold deposits; Sov. Geol., v. 14, no. 4, p. 1–22. (Chem. Abstr., v. 75, 23810b.)

Petrovskaya, N.V., Lyuttsau, V.G., Frolova, K.E., Laputina, I.P. and Vasil'eva, G.V.
1971*c*: Heterogeneity in native gold; Vop. Odnorodnosti Neodnorodnosti Miner., F.V. Chukhrov, ed., "Nauka", Moscow, U.S.S.R., p. 44–62. (Chem. Abst., v. 77, 91087n.)

Petrovskaya, N.V., Novgorodova, M.I., Frolova, K.E. and Gorshkov, A.I.
1976: New data on the composition of phases in heterogeneous segregations of native gold; Izv. Akad. Nauk SSSR, Ser. Geol. no. 3, p. 67–73. (Chem. Abstr., v. 85, 49355p.)

Petrovskaya, N.V., Novikov, V.M., Fomin, P.S., Frolova, K.E. and Evstignyeva, T.L.
1971*d*: Specific characteristics of native gold as the indicator of subsurface Tertiary deposits in the Lower Amur region; Geol. Rud. Mestorozhd., v. 13, no. 1, p. 17–27.

Petrovskaya, N.V., Timofeevskii, D.A. and Borodaevskii, N.I.
1974: Endogenic zoning of gold ore fields and deposits; Zon. Gidroterm. Rud. Mestorozhd. (Zoning of hydrothermal ore deposits), v. 2, G.A. Sokolov, ed., "Nauka", Moscow, U.S.S.R., p. 86–122.

Petrulian, N.S.
1934: Gisement aurifère de Rosia-Montana et du gisement de plomb et de zinc de Herja (Transylvanie, Roumanie); M.O., Imprimerie Nat., Bucarest, 83 p.

Petty, J.W.
1968: Recovery of precious metals from oil shale; U.S. Patent 3,383,201. (Chem. Abstr., v. 69, 20963u.)

Peyton, A.L. and Lynch, V.J.
1953: Investigation of the Brewer topaz deposit Chesterfield County, S.C.; U.S. Bur. Mines, Rep. Invest. 4992, 19 p.

Phan, K.D.
1965: Enquête sur l'or dans les roches; Chron. Mines Rech. Minière, no. 343, p. 175–188.

Phaup, A.E.
1964: Gold mines in southern Rhodesia—introduction; *in* The geology of some ore deposits in southern Africa, vol. 2, S.H. Haughton, ed., Geol. Soc. S. Afr., Johannesburg, p. 1–7.

Phillips, J.A.
1896: A treatise on ore deposits; MacMillan and Co., Ltd., London, 943 p.

Pickering, T.
1935: Mineralized crush zones in Gold Coast Colony; Mining Mag., v. 53, no. 6, p. 329–336.

Pienaar, P.J.
1963: Stratigraphy, petrology, and genesis of the Elliot Group, Blind River, Ontario, including the uraniferous conglomerate; Geol. Surv. Can., Bull. 83, 140 p.

Pilkey, O.H. and Bornhold, B.D.
1970: Gold distribution on the Carolina continental margin—a preliminary report; U.S. Geol. Surv., Prof. Pap. 700-C, p. C30–C34.

Pimm, A.C.
1967: Bau Mining District, West Sarawak, East Malaysia, pt. 2 Krokong; Malaysia, Geol. Surv., Borneo Region, Bull. 7, pt. 2, 97 p.

Piper, D.Z. and Graef, P.A.
1974: Gold and rare-earth elements in sediments from the East Pacific Rise; Mar. Geol., v. 17, p. 287–297.

Pisemskii, G.V. and Pletnev, A.G.
1972: Metallogenesis of gold of the Ukrainian crystalline Shield; Sov. Geol., no. 10, p. 67–79.

Pisemskii, G.V., Sher, S.D., Zhadnova, T.P. and Ganzha, L.M.
1974: Types of gold ore manifestations and prospecting criteria of gold mineralization in the basement of the Russian platform; Sov. Geol., no. 7, p. 47–55. (Chem. Abstr., v. 82, 19509n.)

Pitul'ko, V.M.
1973: Distribution of immobile and mobile forms of chemical elements in biogeochemical dispersion haloes of ore deposits in the subarctic region; Pochvy Rastit Merzlotnykh Raionov SSSR, Mater Vses. Simp. "Biol. Probl. Sev.", 5th, 1972, p. 364–368. (Chem. Abstr., v. 82, 75587y.)

Pitul'ko, V.M.
1976: Behaviour of gold in oxidation zone of deposits of the Soviet Far North; Geokhim., no. 4, p. 569–576.

Plaksin, I.N. and Troitskii, A.V.
1949: Central Siberia—base for the production of ferrous and nonferrous metals; Gorn. Zh., v. 123, no. 1, p. 5–8. (Chem. Abstr., v. 44, 88b.)

Plant, J. and Coleman, R.F.
1973: Application of neutron activation analysis to the evaluation of placer gold concentrations; *in* Geochemical exploration 1972, M.J. Jones, ed., Inst. Min. Metall. (London), p. 373–381.

Plecash, J., Hopper, R.V. *et al.*
1963: Operations at La Luz mines and Rosita mines, Nicaragua, Central America; Can. Inst. Min. Metall., Bull., v. 56, no. 616, p. 624–641.

Pletnev, A.G.
1972: Gold content in the Likhmanovsk syncline ferruginous-siliceous formation of the Krivoi Rog area; Geol. Zh., v. 32, no. 2, p. 139–141. (Chem. Abstr., v. 76, 143402n.)

Pliny the Elder (Gaius Plinius Secundus)
79 A.D.: Historia naturalis, Book XXXIII, Aurum, Roma.

Pokrovskii, E.V.
1968: The gold content of Middle-Upper Cambrian sediments of southwestern part of the Siberian Platform; Geol. Geofiz., no. 11, p. 115–120.

Pokrovskii, P.V. and Berzon, R.O.
1975: Composition of copper and silver gold from the deposit Zolotaya Gora; Ezheg., Inst. Geol. Geokhim., Akad. Nauk SSSR, Ural. Nauchn. Tsent., 1974, p. 94–97. (Chem. Abstr., v. 86, 7236r.)

Polański, A.
1948: A new essay of evaluation of the chemical composition of the earth; Soc. Amis Sci. Lett. Poznań Bull., Ser. B., v. 9, P. 25–46.

Polikarpochkin, V.V.
1973: The interrelation of types of zonalities of gold-bearing mineralization in the Lena region; Ezheg. Inst. Geokhim., Sib. Otd., Akad. Nauk SSSR, 1972, L.V. Tauson, ed., p. 287–288.

Polikarpochkin, V.V. and Kitaev, N.A.
1971: Endogenic haloes of epithermal gold-bearing deposits; *in* Geochemical exploration, Can. Inst. Min. Metall., Spec. Vol. 11, R.W. Boyle and J.I. McGerrigle, eds., p. 381–383.

Polikarpochkin, V.V., Kitaev, N.A., Gapon, A.E. and Zontov, V. I.
1971: The geochemical particularities of formation processes of gold-bearing quartz-sulphide veins and zones; Int. Geochem. Congr., Moscow, Abstracts I, p. 376–377.

Polikarpochkin, V.V., Kitaev, N.A. and Sarapulova, V.N.
1965: Structure and vertical zonation of the primary dispersion aureoles at the Balei gold deposits; Geokhim., no. 8, p. 1024–1037. *Also* Geochem. Int., v. 2, no. 4, p. 741–753.

Polikarpochkin, V.V. and Korotayeva, I.Ya.
1976: Gold distribution in sedimentary rocks of the Undino-Dainsk depression (eastern Transbaikal region); Geokhim., no. 3, p. 396–403.

Polikarpochkin, V.V., Korotayeva, I.Ya. and Gapon, A.E.
1969: The concentration of chemogenic gold in sedimentary formations; Ezheg. Inst. Geokhim., Sib. Otd., Akad. Nauk SSSR, 1968, L.V. Tauson, ed., Irkutsk, p. 240–245.

Polikarpochkin, V.V., Voloskova, S.E., Konstantinova, I.M. and Korotaeva, I.Ya.
1974: The use of the hydrogeochemical method for gold ore deposit prospecting in the Lena region; Ezheg. Inst. Geokhim., Sib. Otd., Akad. Nauk. SSSR, 1973, L.V. Tauson, ed., p. 263–266.

Polkanov, Yu. A. and Yalovenko, I.P.
1970: Finds of gold in coastal and estuary sands on the northwestern shore of the Black Sea; Akad. Nauk SSSR, Dokl., v. 191, no. 4, p. 905–908. (Chem. Abstr., v. 73, 27540v.)

Poltorikhin, P.I.
1974*a*: Gold content of pyrites in the Kalba gold belt; Int. Geol. Rev., v. 16, no. 2, p. 154–155.
1974*b*: Geochemical characteristics of the northwestern part of the Kalba gold-ore belt; Geokhim., no. 9, p. 1331–1341. *Also* Geochem. Int., v. 11, no. 5, p. 927–936.

Poltorikhin, P.I. and Mukanov, K.M.
1970: Geochemical prerequisites in prospecting for gold deposits in western Kalba; Akad. Nauk. Kaz. SSR, Izv. Ser. Geol., v. 27, no. 3, p. 25–29. (Chem. Abstr., v. 73, 79382q.)

Polyakov, V.D.
1959: Formation of foam and analysis of slime in the Kara-Bogaz-Gol Bay; Prob. Kompleks. Ispol'zovan Mineral. Bogatstv Kara-Bogaz-Gola, Acad. Nauk. Turkmen. SSR, p. 325–328. (Chem. Abstr., v. 56, 7062d.)

Polyanitsyn, A.V.
1973: Model of gold-ore deposit dispersion flow; Ezheg., Inst. Geokhim., Sib. Otd., Akad. Nauk SSSR, 1972, L.V. Tauson, ed., p. 330–334. (Chem. Abstr., v. 82, 33369y.)

Pomârleanu, V., Movileanu, A., Mihalka, J. and Mihalka, S.
1972: Beitrag zum Studium der hydrothermal en Erzlagerstätte Văratic (Baia Mare, Rumänien); Chem. Erde, Bd. 31, p. 275–285.

Popenko, G.S.
1970*a*: Internal and submicrostructures of gold in Quaternary formations of the Muruntau deposit; Zap. Uzb. Otd. Vses. Mineral. Obshchest, no. 23, p. 62–65.
1970*b*: Comparative characteristics of gold from placer and hard-rock deposits in Tamdyn Mountains, western Uzbekistan; Zap. Uzb. Otd. Vses. Mineral Obshchest., no. 21, p. 63–68. (Chem. Abstr., v. 74, 101602f.)

Popenko, G.S., Tsypkina, P.E. and Nikolaeva, L.A.
1973: Composition and internal structure of gold in the Quaternary placers of the Chirchik River basin; Zap. Uzb. Otd. Vses. Mineral. Obshchest., no. 26, p. 117–119. (Chem. Abstr., v. 80, 50476g.)

Popkov, N.N.
1961: Gold mineralization in skarns of the northern Tien-Shan; Zap. Kirgizsk. Otd. Vses. Mineralog. Obshchest., no. 2, p. 47–64. (Chem. Abstr., v. 58, 13619c.)

Popov, V.M.
1972: Gold-pyrite-complex metal deposits in the Baimak region of Bashkiria and some characteristics of their localization; Mater. Geol. Polez. Iskop. Yuzhn, Urala, v. 5, p. 177–182. (Chem. Abstr., v. 82, 173643h.)
1975: On the geochemistry of gold in the Upper-Jurassic halogen formation of south Central Asia; Akad. Nauk SSSR, Dokl., v. 224, no. 4, p. 929–932.

Popova, L.M.
1971: Application of spectrochemical methods for the determination of gold during hydrogeochemical investigations; Geol. Geofiz., no. 5, p. 135–138. (Chem. Abstr., v. 76, 88538j.)

Popova, V.I.
1972: Evolution of growth forms of pyrite crystals from the Berezovsk gold ore deposit in the Ural Mountains; Tr. Sverdlovsk. Gorn. Inst., v. 86, p. 109–111. (Chem. Abstr., v. 83, 19438j.)

Popova, V.S. and Mushchinskaya, N.A.
1973: Concentration of antimony and gold rich ores; Nauch. Tr. Sredneaz. Proekt. Tsvet. Metall., no. 7, p. 52–56. (Chem. Abstr., v. 82, 158921d.)

Popovic, A.
1954: Noble metals in the ash of some coals of the Timok basin; J. Chem. Soc., Belgrade, v. 19, p. 305–307 (Engl. summary).

Portnov, A.M., Ostnovskii, E.Ya. and Kolotov, B.A.
1971: Prospecting for near-surface gold-silver deposits by aerial gamma-spectrometric and hydrogeochemical methods; Razved. Okhr. Nedr., v. 37, no. 11, p. 13–16. (Chem. Abstr., v. 76, 61759v.)

Pošepný, F.
1902: The genesis of ore deposits, 2nd ed.; Am. Inst. Min. Eng., New York, 806 p.

Potapenko, G.M., Yakubov, O.K. and Kislyi, B.I.
1968: Deep-seated aurimetric survey in composite prospecting for gold in northern Kazakhstan; Geofiz. Issled., p. 284–291. (Chem. Abstr., v. 72, 14579u.)

Potap'ev, V.V., Anoshin, G.N. and Malikova, I.N.
1968: Gold and tantalum in ore-bearing and ore-free granitic massifs; Metallogen. Magmat. Tyan-Shanya, p. 277–278. (Chem. Abstr., v. 72, 14735s.)

Pourbaix, M. *et al.*
1963: Atlas D'équilibres électrochimiques; Gauthier-Villars & Cie., Paris, 644 p.

Povarennykh, A.S. and Rusakova, L.D.
1973: The new mineral kafehydrocyanite; Geol. Zh., v. 33, p. 24–30.

Poznyak, V.O.
1968: Variscite in a gold ore deposit of western Chukotka; Izv. Vyssh. Ucheb. Zaved., Geol. Razved., v. 11, no. 4, p. 145–147. (Chem. Abstr., v. 69, 29188c.)

Prashnowsky, A.A. and Schidlowski, M.
1967: Investigation of Precambrian thucholite; Nature, v. 216, p. 560–563.

Presant, E.W.
1971: Geochemistry of iron, manganese, lead, copper, zinc, arsenic, antimony, silver, tin, and cadmium in the soils of the Bathurst area, New Brunswick; Geol. Surv. Can., Bull. 174, 93 p.

Prest. V.K.
1949: The Pleistocene geology of the Vermillion River system near Capreol, District of Sudbury, Ontario (with special reference to placer gold); Ont. Dep. Mines, Prelim. Rep., 1949-2, 8 p.

Pretorius, D.A.
1964: The geology of the Central Rand goldfield; *in* The geology of some ore deposits in southern Africa, v. 1, S.H. Haughton, ed.; Geol. Soc. S. Afr., Johannesburg, p. 63–111.
1966: Conceptual geological models in the exploration for gold mineralization in the Witwatersrand Basin; Univ. Witwatersrand, Econ. Geol. Res. Unit, Inform. Circ., no. 33, 39 p.
1974*a*: Nature of the Witwatersrand gold-uranium deposits; Univ. Witwatersrand, Econ. Geol. Res. Unit, v. 86, p. 1–50.
1974*b*: Gold in the Proterozoic sediments of South Africa. Systems, paradigms, and models; Univ. Witwatersrand, Econ. Geol. Res. Unit, v. 87, p. 1–22.
1975: The depositional environment of the Witwatersrand goldfields: a chronological review of speculations and observations; Miner. Sci. Eng., v. 7, no. 1, p. 18–47.

Price, P.
1934: The geology and ore deposits of the Horne mine, Noranda, Quebec; Can. Inst. Min. Metall., Trans., v. 37, p. 108–140.
1948: Horne mine *in* Structural geology of Canadian ore deposits, v. 1, Can. Inst. Min. Metall., Montreal, p. 763–772.
Price, P. and Bray, R.C.E.
1948: Pamour mine; *in* Structural geology of Canadian ore deposits, v. 1, Can. Inst. Min. Metall., Montreal, p. 558–565.
Prochnau, J.F.
1971: Distribution and mode of occurrence of gold in the Chibougamau district, Quebec; unpubl. M.Sc. thesis, McGill Univ., Montreal, 134 p.
Prokopenko, N.M.
1941: Principal stages in the history of indium in the earth's crust; Acad. Sci. USSR, Dokl., v. 31, p. 903–906.
Prokuronov, P.V.
1975: Alluvial gold placers in the high-mountain deeply-dissected topographical environments. Transportation range of gold-particles in the instance of North Caucasus; Akad. Nauk SSSR, Izv., Ser. Geol., no. 5, p. 137.
Prosnyakov, M.P., Chetyrbotskaya, I.I. and Berengilova, V.V.
1972: Ratio of gold and silver as an important prospecting feature during an appraisal of a gold-ore mineralization; Razved. Okhr. Nedr, v. 38, no. 3, p. 8–14. (Chem. Abstr., v. 77, 8259b.)
Proust, G.P.
1920: L'or—prospection, gisement, extraction; Gauthier-Villars et Cie, Éditeurs, Paris, 319 p.
Pryor, R.N., Rhoden, H.N. and Villalón, M.
1972: Sampling of Cerro Colorado, Rio Tinto, Spain; Inst. Min. Metall., Trans., Sec. A, v. 81, p. 143–159.
Pryor, T.
1923–1924: The underground geology of the Kolar gold field; Inst. Min. Metall., Trans., v. 33, p. 95–115.
Purington, C.W.
1903: Ancient gold mining; Eng. Min. J., v. 75, p. 437.
Purkayastha, B.C. and Das, N.R.
1965: On the study of recovery of gold from sea water; Sci. Cult., v. 31, no. 8, p. 403–407.
Purushottam A., Naidu, P.P. and Lal, S.S.
1973: Solvent extraction-cum-atomic absorption spectrophotometric determination of traces of gold in rocks, minerals and ores; Curr. Sci. v. 42, no. 10, p. 367.
Putnam, G.L.
1953: The gold content of sea water; J. Chem. Educ., v. 30, p. 576–579.
Putzer, H.
1976: Metallogenetische Provinzen in Südamerika; E. Schweizerbart'sche, Verlagsbuchhandlung, Stuttgart, 316 p.
Pye, E.G.
1952: Geology of Errington Township, Little Long Lac area; Ont. Dep. Mines, Annu. Rep., 1951, v. 60, pt. 6, 140 p.
Pyke, D.R.
1975: On the relationship of gold mineralization and ultramafic volcanic rocks in the Timmins area; Ont. Div. Mines, Misc. Pap. 62, 23 p.
1976: On the relationship between gold mineralization and ultramafic volcanic rocks in the Timmins area, northeastern Ontario; Can. Inst. Min. Metall., Bull., v. 69, no. 773, p. 79–87.
Quin, B.F., Brooks, R.R. and Boswell, C.R.
1973: Plant and soil indicators of quartz reefs in the Maratoto Valley; N.Z., J. Sci., v. 16, no. 3, p. 737–746.
Quinn, H.A.
1964: Geology and mineral resources of Ethiopia; World Mining, v. 17, no. 2, p. 34–41 and v. 17, no. 3, p. 31–35.
Quiring, H.
1954: Schalenbau der Erde und sphärogene Erze; C. R. Congr. Geol. Int. 19e, Alger, Sec. 13, pt. 3, Fasc. 15, p. 431–438.
Rabenau, A. and Rau, H.
1968: Hydrothermale Züchtung von Goldkristallen; Naturwissenschaften, v. 55, no. 7, p. 336–339.
Rabenau, A., Rau, H. and Rosenstein, G.
1970: Telluride halides of gold; J. Less-Common Metals, v. 21, p. 395-401.
Rabinovich, K.R., Akchurina, V.N. and Bulynnikov, V.A.
1972: Ore shoots and zoning in gold ore deposits; Probl. Obrazov. Rud. Stolbov. Mater. Simp. 1969, F.N. Shakhov, ed., "Nauka" Sib. Otd. Novosibirsk U.S.S.R., p. 75–81. (Chem. Abstr., v. 79, 33575a.)
Radanova, T.G. and Karadzhova, B.
1971: On the distribution of lithium, rubidium and cesium in the propylites and secondary quartzites from Spahievo ore field, Bulgaria; Int. Geochem. Congr., Moscow, Abstracts of Reports I, p. 411–412.
Radkevich, Ye. A.
1961: On the types of metallogenic provinces and ore districts; Int. Geol. Rev., v. 3, no. 9, p. 759–783.
Radkevich, Ye. A., ed.
1966: Genetic characteristics and general relationships in the development of gold mineralization in the Far East; "Nauka", Moscow, 204 p.
1969: The gold ore formations of the Far East; Izd. "Nauka" Publ. House, Moscow, 139 p.
Radkevich, Ye. A., Usenko, S.F. and Chebotarev, M.V.
1965: On the relations of tin and gold mineralization to the main structural elements of the southern part of the far east; Geol. Geofiz., no. 3, p. 25–38.
Radtke, A.S. and Dickson, F.W.
1974: Controls on the vertical position of fine-grained replacement-type gold deposits; Abstracts of Papers, Int. Assoc. Genesis Ore Deposits (IAGOD), 4th Symp., Varna, Bulgaria, p. 68–69.
1975: Carlinite, Tl_2S, a new mineral from Nevada; Am. Mineral., v. 60, p. 559–565.
1976*a*: Genesis of Carlin-type gold deposits in the framework of tectonic evolution of the Basin and Range province; Econ. Geol., v. 71, p. 702 (abstr.).
1976*b*: Genesis and vertical position of fine-grained disseminated replacement-type gold deposits in Nevada and Utah, U.S.A.; *in* Problems of ore deposition, v. 1, Volcanogenous ore deposits, 4th IAGOD Symp., Varna, B. Bogdanov, ed., Bulgarian Acad. Sci., Sofia, p. 71–78.
Radtke, A.S., Heropoulos, C., Fabbi, B.P., Scheiner, B.J. and Essington, M.
1972*a*: Data on major and minor elements in host rocks and ores, Carlin gold deposit, Nevada; Econ. Geol., v. 67, p. 975–978.
Radtke, A.S. and Scheiner, B.J.
1970*a*: Influence of organic carbon on gold deposition at the Carlin and Cortez deposits, Nevada; Econ. Geol., v. 65, p. 739–740 (abstr.).
1970*b*: Studies of hydrothermal gold deposition (I). Carlin gold deposit, Nevada, The role of carbonaceous material in gold deposition; Econ. Geol., v. 65, p. 87–102.
1971: Cation exchange capacity and metal deposition: a suggestion—a reply; Econ. Geol., v. 66, p. 201.
Radtke, A.S., Taylor, C.M. and Christ, C.L.
1972*b*: Chemical distribution of gold and mercury at the Carlin deposit, Nevada; Econ. Geol., v. 67, p. 1009 (abstr.).
Radtke, A.S., Taylor, C.M., Dickson, F.W. and Heropoulos, C.
1974*a*: Thallium-bearing orpiment, Carlin gold deposit, Nevada; U.S. Geol. Surv., J. Res., v. 2, no. 3, p. 341–342.
Radtke, A.S., Taylor, C.M., Erd, R.C. and Dickson, F.W.
1974*b*: Occurrence of lorandite, $TlAsS_2$, at the Carlin gold deposit, Nevada; Econ. Geol., v. 69, p. 121–123.
Radtke, A.S., Taylor, C.M. and Frost, J.E.
1967: Bismuth and tin minerals in gold-and silver-bearing sulfide ores, Ohio mining district, Marysvale, Utah, U.S. Geol. Surv., Prof. Pap. 575–D, p. 127–130.

Radtke, A.S., Taylor, C.M. and Heropoulos, C.
1973: Antimony-bearing orpiment, Carlin gold deposit, Nevada; J. Res., U.S. Geol. Surv., v. 1, no. 1, p. 85–87.

Raeburn, C. and Milner, H.B.
1927: Alluvial prospecting; Thomas Murby & Co., London, 478 p.

Raicevic, D. and Cabri, L.J.
1976: Mineralogy and concentration of Au-and Pt-bearing placers from the Tulameen River area in British Columbia; Can. Inst. Min. Metall., Bull., v. 69, p. 111–119.

Rajan, S.C.S. and Raju, N.A.
1975: Titrimetric determination of gold by precipitation with hydroquinone; Talanta, v. 22, p. 185–189.

Rakhmatullaev, Kh. R.
1970: Relation of Hercynian gold and mercury ore formations in southern Tien-Shan; Uzb. Geol. Zh., v. 14, no. 4, p. 77–82 (Chem. Abstr., v. 74, 33484s.)
1972: Caledonian and Hercynian epochs of gold ore mineralization in western Uzbekistan (as illustrated by the Kokpatas ore field); Nek. Itogi Petrometallogenicheskikh Issled. Uzb., p. 154–170. (Chem. Abstr., v. 84, 7619b.)

Rakhmatullaev, Kh.R. and Gamaleev, I.E.
1964: Elements indicating gold mineralization in western Uzbekistan; Akad. Nauk Uzb. SSR, Dokl., v. 21, no. 6, p. 38–41. (Chem. Abstr., v. 62, 340c.)

Rakhmatullaev, Kh.R. and Sher, S.D.
1969: The metasomatic wall rock alteration in the Muruntau ore deposits; *in* The gold deposits and principle features of the metallogeny of gold in Uzbekistan, I. Kh. Khamrabaev, ed.; "Fan" Publ. Off., Uzbek S.S.R., Tashkent, p. 173–176.

Ramdohr, P.
1953*a*: Maldonit; Sitzungsber. Deut. Akad. Wiss. Berlin, Kl. Math. Naturwiss., no. 5, p. 3–8.
1953*b*: Über Metamorphose und sekundäre Mobilisierung; Geol. Rundsch., v. 42, pt. 1, p. 11–19.
1955: Neue Beobachtungen an Erzen des Witwatersrandes in Süd Afrika und ihre genetische Bedeutung; Abh. Deut. Akad. Wiss. Berlin, v. 5, p. 1–55.
1958*a*: New observations on the ores of the Witwatersrand in South Africa and their genetic significance; Geol. Soc. S. Afr., Trans., (Annex.), v. 61, p. 1–51.
1958*b*: Die uran-und goldlagerstätten Witwatersrand–Blind River District–Dominion Reef–Serra de Jacobina; Deut. Akad. Wiss. Berlin, Abh., 35 p.
1960: Die Erzmineralien und ihre Verwachsungen; 3rd ed. Akademie Verlag. Berlin, 1089 p. (Transl.—The ore minerals and their intergrowths, Pergamon Press, Oxford, 1969, 1174 p.)
1965: Rheingold als Seifenmineral; Jahrb. Geol. Landesamt Baden-Würtemberg, v. 7, p. 87–95.
1967: The wide-spread paragenesis of ore minerals originating during serpentinization; Geol. Rud. Mestorozhd., no. 2, p. 32–43.
1973: The opaque minerals in stony meteorites; Elsevier Publishing Co., Amsterdam, 245 p.

Ramdohr, P. and Udubasa, G.
1973: Frohbergite in gold ore deposits of Sacaramb and Fata Baii (Romania); Miner. Deposita (Berl.), v. 8, no. 2, p. 179–182. (Chem. Abstr., v. 79, 116903g.)

Ramović, M.
1968: Principles of metallogeny; Geogr. Inst., Nat. Sci. Fac., Univ. Sarajevo, Yugoslavia, 271 p.

Ramsay, W.R.H. and Kobe, H.W.
1974: Great Barrier Island silver-gold deposits, Hauraki Province, New Zealand; Miner. Deposita (Berl.), v. 9, p. 143–153.

Randall, J.A.
1973: Silver-gold ratios related to copper mineralization at the Anglo-Rouyn mine, Saskatchewan; Program and Abstracts, Geol. Assoc. Can.–Can. Soc. Pet. Geol. and Mineral Assoc. Can., Annu. Meet.—Arctic Symp.—Cretaceous Colloq., May 23–26, Univ. Sask., Saskatoon, p. 89 (abstr.).

Rankama, K., ed.
1963–1970: The Precambrian; 4 vols., Interscience Publishers, New York.

Rankama, K. and Sahama, Th.G.
1950: Geochemistry; Univ. Chicago Press, Chicago, 912 p.

Ransome, F.L.
1901: A report on the economic geology of the Silverton Quadrangle, Colorado; U.S. Geol. Surv., Bull. no. 182, 265 p.
1909: The geology and ore deposits of Goldfield, Nevada; U.S. Geol. Surv., Prof. Pap. 66, 258 p.
1910: Geology and ore deposits of the Goldfield district, Nevada; Econ. Geol., v. 5, p. 301–311, 438–470.

Rao, B.K. and Srinivasan, R.
1971: Ore environment and ore genesis in Ramagiri goldfield, Andhra Pradesh, India; Econ. Geol., v. 66, p. 1083–1084.

Rast, N. and Grant, R.
1973: Transatlantic correlation of the Variscan-Appalachian orogeny; Am. J. Sci., v. 273, p. 572–579.

Rayner, E.O.
1969: The copper ores of the Cobar region, New South Wales; N.S.W., Geol. Surv., Mem. 10, 131 p.

Raynor, G.V.
1976: The alloying behaviour of gold; Gold Bull., v. 9, pt. 1, p. 12–19, pt. 2, p. 50–54.

Razenkov, N.I. and Galaktionova, G.F.
1962: On the mode of occurrence of gallium in the oxidized zone of sulfide deposits; Geochemistry, no. 1, p. 104–108.

Razin, L.V. and Begizov, V.D.
1973: Gold-silver mineralization in the Talnakh and Oktyabr'sk ore deposits; Geol. Rud. Mestorozhd., v. 15, no. 6, p. 32–45. (Chem. Abstr., v. 81, 124213p.)

Razin, L.V., Gomonova, A.I., Bykov, V.I. and Meshchankina, V.I.
1971: A new natural gold, copper, and palladium intermetallic compound from the ores of the Talnakh deposit; Zap. Vses. Mineral. Obshest., USSR, v. 100, no. 1, p. 66–76.

Razin, L.V. and Rozhkov, I.S.
1963: The geochemistry of gold in the zone of weathering and in the biosphere in the permafrost zone of the Aldan Shield; Acad. Sci. USSR, Siberian Br., Yakutia Sec., Trans. Geol. Ser. no. 16, p. 5–22.
1966: Geochemistry of gold in the crust of weathering and the biosphere of gold-ore deposits of the Kuranakh type; Nauka, Moscow, 254 p. (Rev. in Econ., Geol., v. 62, p. 437–438.)

Razin, L.V. and Yurkina, K.V.
1971: The gold minerals in the impregnation ore of the Noril'sk deposit; Geol. Rud. Mestorozhd., v. 13, no. 1, p. 93–97

Read, H.H.
1957: The granite controversy; Thomas Murby & Co., London, 430 p.

Redmayne, R.A.S., ed.
1922: The mineral industry of the British Empire and foreign countries, Gold, Her Majesty's Stationery Office, London, 366 p.

Reed, B.L. and Miller, R.L.
1971: Orientation geochemical soil survey at the Nixon Fork mines, Medfra quadrangle, Alaska; U.S. Geol. Surv., Bull. 1312-K, 21 p.

Reed, G.W.
1963: Heavy elements in the Pantar meteorite; J. Geophys. Res., v. 68, p. 3531–3535.

Reed, H.W.
1911: Secondary enrichment of gold deposits; Min. Sci. Press, v. 102, no. 24, p. 825.

Reeves, T.J.
1971: Gold in Ireland; Irel., Geol. Surv., Bull. 1, p. 73–85.

Rehman, F.U.
1972: The determination of gold by radiochemical and non-destructive neutron activation analyses in some sulphide ore samples from Pakistan; Punjab Univ., Geol. Bull., no. 9, p. 37–42.

Reid, J.A.
1904– The structure and genesis of the Comstock Lode; Univ.
1906: Calif., Dep. Geol., Bull., v. 4, no. 10, p. 177–199.
1941: Albite and gold; Econ. Geol., v. 36, p. 217–219.
1945: The Hardrock "Porphyry" of Little Long Lac; Econ. Geol., v. 40, p. 509–516.

Reider, R. and Wänke, H.
1969: Study of trace element abundance in meteorites by neutron activation; *in* Meteorite research, P.M. Millman, ed., Springer-Verlag New York, Inc., New York, p. 75–86.

Reimer, T.O.
1975: Age of the Witwatersrand system and other gold-uranium placers. Implications on the origin of the mineralization; Neues Jahrb. Mineral. Monatsh., no. 2, p. 79–98. (Chem. Abstr., v. 83, 13689v.)

Reimnitz, E., Von Huene, R. and Wright, F.F.
1970: Detrital gold and sediments in Nuka Bay, Alaska; U.S. Geol. Surv., Prof. Pap. 700-C, p. C35–C42.

Reinecke, L.
1927: The location of the payable ore-bodies in the gold-bearing reefs of the Witwatersrand; Geol. Soc. S. Afr., Trans., v. 30, p. 89–119.
1930: Origin of the Witwatersrand system; Geol. Soc. S. Afr., Trans., v. 33, p. 111–133.

Reinecke, L. and Stein, W.G.A.
1930: Ore bodies of the Pilgrims Rest goldfield; Geol. Soc. S. Afr., Trans., v. 32, p. 65–88.

Relly, B.H.
1960: The geology of Buchans Mine, Newfoundland; unpubl. Ph.D. thesis, McGill Univ., Montreal.

Reuning, E.
1925: Die Natasmine in Südwest-Afrika, eine pegmatitisch-pneumatolytisch-hydrothermale Übergangs-lagerstätte mit Scheelit, Molybdänglanz, Kupfererzen und Gold; Neues Jahr. Mineral., Geol. Palaeontol. Beil. Bd. 52, p. 192–264.

Reynolds, D.G.
1965: Geology and mineralization of the Salsigne gold mine, France; Econ. Geol., v. 60, p. 772–791.

Reynolds, D.L.
1946: Geochemical changes leading to granitization; Geol. Soc. London, Q.J., v. 102, pt. 3, p. 389–446.
1947: The granite controversy; Geol. Mag., v. 84, no. 4 p. 209–223.

Rice, H.M.A.
1960: Geology and mineral deposits of the Princeton map-area, British Columbia; Geol. Surv. Can., Mem. 243, 136 p.

Rice, H.R.
1949: Renabie mines; Can. Min. J., v. 70, p. 61–67.

Rice, R.
1964– The mode of origin of banket orebodies, discussion;
1965: Inst. Min. Metall., Trans., v. 74, no. 702, p. 503–504.

Rice, W.L.
1970: Investigation of a low-grade gold deposit in the Orogrande district Idaho; U.S. Bur. Mines, Rep. Invest. 7425, 14 p.

Rich, R.L. and Taube, H.
1954: The induced exchange of Cl^- and $AuCl_4^-$; evidence for Au(II); J. Phys. Chem., v. 58, p. 6–11.

Richards, F.
1948: Cariboo gold quartz mine; *in* Structural geology of Canadian ore deposits, v. 1, Can. Inst. Min. Metall., Montreal, p. 162–168.

Richards, G.G., Christie, J.S. and Wolfhard, M.R.
1976: Specogna: a Carlin type gold deposit, Queen Charlotte Islands, British Columbia; Can. Inst. Min. Metall. Bull., v. 69, no. 773, p. 64 (abstr.).

Richardson, W.G.
1974: A survey of Canadian mining history; Can. Inst. Min. Metall., Spec. Vol. 14, 115 p.

Richter, D.H.
1970: Geology and lode-gold deposits of the Nuka Bay area, Kenai Peninsula, Alaska; U.S. Geol. Surv., Prof. Pap. 625-B, 16 p.

Richter, D.H. and Matson, N.A.
1968: Distribution of gold and some base metals in the Slana area, Eastern Alaska Range, Alaska; U.S. Geol. Surv., Circ. 593, 20 p.

Rickaby, H.C.
1932: Bannockburn gold area; Can. Inst. Min. Metall., Bull., no. 245, p. 501–511.

Rickard, M.J.
1966: Reconnaissance geology of Vanua Levu; Fiji, Geol. Surv. Dep., Mem. no. 2, p. 1–81.

Rickard, T.A.
1896: The Enterprise mine, Rico, Colorado; Am. Inst. Min. Metall. Eng., Trans., v. 26, p. 906–980.
1899: The alluvial deposits of western Australia; Am. Inst. Min. Metall. Eng., Trans., v. 28, p. 490–537.
1902: The formation of bonanzas in the upper portions of gold-veins; Am. Inst., Min. Eng., Trans., v. 31, p. 198–220.
1912: The domes of Nova Scotia; Inst. Min. Metall., Trans. (London), v. 21, p. 506–566.
1925: The geologic distribution of gold; Eng. Min. J., v. 119, p. 486–487.
1932: Man and metals, vols. I and II; McGraw-Hill Book Co., Inc., New York, 1068 p.
1934: The primitive use of gold; Inst. Min. Metall., Bull. no. 360, p. 1–28, no. 361, p. 1–5 and no. 364, p. 33–37.
1942: The gold digging of British Columbia; Can. Min. J., v. 63, no. 10, p. 640–646.
1944: The romance of mining; Macmillan Company of Canada Ltd., Toronto, 450 p.

Ridge, J.D., ed.
1968: Ore deposits of the United States, 1933–1967; Graton-Sales Vol.; Publ. Am. Inst. Min. Metall. Pet. Eng. Inc., New York, vols. 1 and 2.
1972: Annotated bibliographies of mineral deposits in the western hemisphere; Geol. Soc. Am., Mem. 131, 681 p.

Ridland, G.C.
1941: Mineralogy of the Negus and Con mines, Yellowknife, Northwest Territories, Canada; Econ. Geol., v. 36, p. 45–70.

Ridler, R.H.
1970: Relationship of mineralization to volcanic stratigraphy in the Kirkland–Larder Lakes area, Ontario; Geol. Assoc. Can., Proc., v. 21, p. 33–42.
1976: Stratigraphic keys to the gold metallogeny of the Abitibi Belt; Can. Min. J., v. 97, no. 6, p. 81–87.

Ries, H.
1930: Economic geology, 6th ed.; John Wiley & Sons, Inc., New York, 860 p.

Ringsleben, W.C.
1935: Hollinger enterprise, geology; Can. Min. J., v. 56, p. 364–372.

Ringwood, A.E.
1955: The principles governing trace element distribution during magmatic crystallization; Geochim. Cosmochim. Acta, v. 7, p. 189–202, 242–254.
1956: Surface tensions of molten heavy metal iodides and their relation to sulphide paragenesis; Aust. Inst. Min. Metall., Proc., no. 180, p. 55–74.

Ritter, C.J.
1971: Trace elements of gold-bearing quartz veins of the Lamaque mine, Bourlamaque, P.Q., Canada; Univ. Microfilms, Ann. Arbor, Mich. (Chem. Abstr., v. 76, 130120u.)

Roberts, B.
1971: The Carolina gold rush; McNally and Loftin, Charlotte, N.C., 80 p.

Roberts, R.J. and Morris, J.H.
1974: The structural geology of the Upper Beaver Mine, Kirkland Lake, Ontario: a volcanogenic copper-gold deposit in Archean mafic volcanic rocks; Geol. Assoc. Can.-Miner. Assoc. Can, Annu. Meet., Abstr., p. 75.

Roberts, R.J., Radtke, A.S., Coats, R.R., Silberman, M.L. and McKee, E.H.
1971: Gold-bearing deposits in north-central Nevada and southwestern Idaho; Econ. Geol., v. 66, p. 14–33.

Robertson, D.S.
1974: Basal Proterozoic units as fossil time markers and their use in uranium prospection; *in* Formation of uranium ore deposits; Int. At. Energy Agency—SM—183/35, p. 495-512.

Robertson, D.S. and Steenland, N.C.
1960: On the Blind River uranium ores and their origin; Econ. Geol., v. 55, p. 659-694.

Robertson, F.
1955: Tracing geological structures by geochemical means; Econ. Geol., v. 50, p. 100 (abstr.)

Robertson, I.D.M.
1972: Mercury in Rhodesia; Rhodesia Geol. Surv. Miner. Resour. Ser., no. 17, 17 p.

Robinson, A.H.A.
1932: Gold in Canada; Can. Dep. Mines, Mines Br. Publ. No. 730, 92 p.
1933: Gold in Canada; Can. Dep. Mines, Mines Br. Publ. No. 734, 92 p.
1935: Gold in Canada; Can. Dep. Mines, Mines Br. Publ. No. 769, 127 p.

Robinson, B.W.
1974: The origin of mineralization at the Tui mine, Te Aroha, New Zealand, in the light of stable isotope studies; Econ. Geol., v. 69, p. 910–925.

Robinson, S.C.
1955: Mineralogy of uranium deposits, Goldfields Saskatchewan; Geol. Surv. Can., Bull. 31, 128 p.

Roedder, E., ed.
1968–1974: Fluid inclusion research, Proc. Commission on ore-forming fluids in inclusions (COFFI), vols. 1–7; privately published, Washington, D.C. and University of Michigan Press, Ann Arbor, Mich.

Rogers, A.F.
1912: Lorandite from the Rambler Mine, Wyoming; Am. J. Sci., Ser. 4, v. 33, p. 105–106.

Rolfe, C. and Hume-Rothery, W.
1967: The constitution of alloys of gold and mercury; J. Less-Common Metals, v. 13, p. 1–10.

Romanowitz, C.M., Bennett, H.J. and Dare, W.L.
1970: Gold placer mining: Placer evaluation and dredge selection; U.S. Bur. Mines, Inform. Circ. 8462, 56 p.

Roscoe, S.M.
1959: Monzonite as an ore mineral in Elliot Lake uranium ores; Can. Min. J., v. 80, p. 65–66.
1969: Huronian rocks and uraniferous conglomerates in the Canadian Shield; Geol. Surv. Can., Pap. 68-40, 205 p.

Roscoe, S.M. and Steacy, H.R.
1958: On the geology and radioactive deposits of Blind River Region; Proc., 2nd U.N. Conf. on Peaceful Uses of Atomic Energy, v. 2, Pap. 222, p. 475–483.

Rose, T.K. and Newman, W.A.C.
1937: The metallurgy of gold, 7th ed.; Charles Griffin and Co., London, 561 p.

Rosenbaum, J.B., May, J.T. and Riley, J.M.
1969: Gold in seawater—fact or fancy; Mines Mag., v. 59, no. 9, p. 14–17.

Rosenthal, E.
1970: Gold! Gold! Gold! The Macmillan Company, London, 372 p.

Roslyakov, N.A.
1969: Distribution of gold and mercury in the Darasun ore field; Zap. Zabaikal. Filiala Geogr. Obshchest. SSSR, v. 36, p. 56–62. (Chem. Abstr., v. 77, 116718j.)
1976: Gold in the zone of hypergenesis of gold ore deposits of East Trans-Baikal, and gold behaviour in the hypergenic processes of an oxidizing series; *in* Gold and rare elements in geochemical processes, Yu.G. Scherbakov, ed., "Nauka" Publ. House, Siberian Br., Novosibirsk, p. 113–140.

Roslyakov, N.A., Nepeina, L.A., Tsimbalist, V.G., Ushakov, G.D. and Reshchevitsky, B.I.
1972: Form of deposition and migration of gold in the weathering crust of gold-sulphide deposits; Tr. Inst. Geol. Geofiz., Akad. Nauk SSSR, Sib. Otd., v. 149, p. 125–138.

Roslyakov, N.A., Peshevitsky, B.I., Nepeyna, L.A. and Tsimbalist, V.G.
1971a: Geochemistry of gold in the processes of weathering crust formation; Int. Geochem. Congr., Moscow, Abstracts of Reports II, p. 719–720.

Roslyakova, N.V. and Roslyakov, N.A.
1975: Endogenic aureoles of gold deposits; Izd. "Nauka" Sib. Otd., no. 182, 132 p.

Roslyakova, N.V., Roslyakov, N.A., and Zvyagin, V.G.
1970: The behaviour of gold in the primary haloes of some vein gold deposits; Izv. Tomsk. Politekh. Inst., v. 239, p. 208–215.

Roslyakov, N.A., Zvyagin, V.G., Roslyakova, N.V., Epov, I.N., Kharitonov, I.I. and Alabina, A.A.
1971b: Data on the feed source of gold-bearing placer deposits in Darasun rivers; Zap. Zabaikal. Filiala Geogr. Obshchest. SSSR, no. 52, p. 116–125. (Chem Abstr., v. 78, 32554k.)

Ross, C.P.
1933: The lode deposits in the Boise Basin, Idaho; Econ. Geol., v. 28, p. 329–343.

Routhier, P.
1963: Les gisements métallifères; t. I and II, Masson et Cie., Paris.

Rowe, J.J.
1969: Fractionation of gold in a differentiated tholeiitic dolerite; Chem. Geol., v. 4, p. 421–427.

Rowe, J.J. and Simon, F.O.
1968: The determination of gold in geologic materials by neutron-activation analysis using fire assay for the radiochemical separations; U.S. Geol. Surv., Circ. 599, 4 p.

Rowe, R.C.
1948: Gold—its past and its future; Can. Min. J., v. 69, no. 4, p. 63–70.

Roy, B.C.
1963: Investigations for gold in India by the Geological Survey of India; Geol. Soc. India, Mem. 1, p. 31–38.

Roy, B.C., coordinator
1963: Mineral distribution map of Asia and the Far East; United Nations, New York.

Rozhkov, I.S.
1940: Auriferous sand deposit on Leba River in the Urals; Zolotaya Prom., v. 12, no. 9, p. 24–25. (Chem. Abstr., v. 38, 3223,8.)
1966: The Kolar gold deposits; Geol. Rud. Mestorozhd., v. 8, no. 6, p. 3–15.
1967: Proterozoic auriferous conglomerates of the Tarkwa district, Ghana; Geol. Geofiz., no. 1, p. 60–74.
1968*a*: The genetic types of gold deposits and their position within geotectonical structures; Geol. Geofiz., no. 7, p. 3–16.
1968*b*: The characteristic features of gold deposits of mobile belts and platforms; 23rd Int. Geol. Congr., Prague, Endogeous Ore Deposits, sec. 7, v. 7, p. 313–322.
1970: The conditions of origin of auriferous deposits of gold telluride type; *in* Problems of postmagmatic ore deposition, Z. Pouba and M. Stemprok eds.; Int. Union Geol. Sci. Ser. A, no. 2, p. 126–129.
1971*a*: Gold content of Krivoy Rog iron ore; Acad. Sci. USSR, Dokl., Earth Sci. Sec., v. 196, p. 92–95 (Am. Geol. Inst. Transl.)
1971*b*: Near surface gold deposits; Geol. Rud. Mestorozhd., v. 13, no. 3, p. 3–14.
1971*c*: Gold deposits of moderate depth; Geol. Rud. Mestorozhd., v. 13, no. 5, p. 23–37 (in Russian).
1971*d*: Structural conditions of the formation of near-surface gold deposits; Soc. Min. Geol. Japan, Spec. Issue 3, p. 126–131.

Rozhkov, I.S. *et al.*
1970: Nonuniform distribution of gold in rocks and minerals as determined by neutron activation analysis; Akad. Nauk SSSR, Dokl., v. 191, no. 4, p. 927–930. (transl. Am. Geol. Inst.), Acad. Sci. USSR, Dokl., Earth Sci. Sec., v. 191, no. 1–6, p. 200–203.)
1971: The presence of gold in the Krivoi Rog iron ores; Acad. Nauk. USSR, Dokl., v. 196, no. 4, p. 923–926.

Rozhkov, I.S., Kaminsky, F.V. and Franzesson, E.V.
1973: Gold in kimberlites and ultrabasic inclusions; Geokhim., no. 12, p. 1898–1903. *Also* Geochem. Int. v. 10, no. 6, p. 1385–1389.

Rozhkov, I.S. and Pisemskey, G.V.
1966: The problem of gold in ancient conglomerates; Geol. Rud. Mestorozhd., no. 5, p. 116–118.

Rozhkov, I.S. and Zhadnova, T.P.
1972: Gold content of the Russian Platform; Akad. Nauk SSSR, Dokl., v. 203, no. 2, p. 434–437. (Chem. Abstr., v. 76, 143396p.)

Rozova, E.V. and Gavrilov, A.M.
1970: Use of pyrite electrical properties in studying zoning in the Kokpatas gold deposit central Kyzyl-Kum; Tr. Tsent. Nauch. Issled. Gornorazved. Inst., no. 93, p. 122–129. (Chem. Abstr., v. 75, 23830h.)

Ruiz, C.F. and Ericksen, G.E.
1962: Metallogenic provinces of Chile, S.A.; Econ. Geol., v. 57, p. 91–106.

Rundall, W.H.
1927–1928: Note on the occurrence of galena in clay, South Shan States, Burma; Inst. Min. Metall., Trans., v. 37, p. 27–49.

Rusakova, L.D.
1974 Supergene zoning in gold-pyrite deposits; Rudonos. Kory Vyvetrivaniya, V.I. Smirnov, ed., "Nauka", Moscow, U.S.S.R., p. 197–202. (Chem. Abstr., v. 82, 61994s.)

Rusakova, L.D. and Povarennykh, A.S.
1973: Secondary enrichment of gold deposits; Geol. Zh., v. 33, no. 3, p. 20–31.

Rush, P.M.
1976: Prospecting methods for gold-quartz vein deposits; unpubl. M.Sc. thesis, Imperial College, Sci. Technol., Univ. London, London, 139 p. and append.

Rusinov, V.L.
1968: Propylites and mineralization; Geol. Rud. Mestorozhd., v. 10, no. 6, p. 75–86. (Review in Econ. Geol., v. 65, p. 230.)

Russell, A.
1929: On the occurrence of native gold at Hope's Nose, Torquay, Devonshire; Mining Mag., v. 22, p. 159–162.

Ruzicka, V. and Steacy, H.R.
1976: Some sedimentary features of conglomeratic uranium ore from Elliot Lake, Ontario; Geol. Surv. Can., Pap. 76-1A, p. 343–346.

Ryabinin, A.I., Romanov, A.S., Khatamov, Sh. and Khamidova, R.
1974: On the question of gold in the waters of the World Ocean; Geokhim., no. 1, p. 158–162.

Ryan, C.W.
1960: A guide to the known minerals of Turkey; Min. Res. Explor. Inst. Turkey, Ankara, 196 p.

Ryan, G.R., ed.
1976: Mining centres of northern Australia; 25th Int. Geol. Congr., Sydney, Australia, Excursion Guide no. 49AC, 52 p.

Rye, D.M., Doe, B.R. and Delevaux, M.H.
1974*a*: Homestake gold mine, South Dakota: II Lead isotopes, mineralization ages, and source of lead in ores of the Northern Black Hills; Econ. Geol., v. 69, p. 814–822.

Rye, R.O., Doe, B.R. and Wells, J.D.
1974*b*: Stable isotope and lead isotope study of the Cortez, Nevada, gold deposit and surrounding area; U.S. Geol. Surv., J. Res., v. 2, no. 1, p. 13–23.

Rye, D.M. and Rye, R.O.
1972: The origin of the Homestake gold deposit, South Dakota in the light of stable isotope studies; Econ. Geol., v. 67, p. 1010 (abstr.).
1974: Homestake gold mine, South Dakota: I. Stable isotope studies; Econ. Geol., v. 69, p. 293–317.

Rytuba, J.J. and Dickson, F.W.
1977: Reaction of pyrite + pyrrhotite + quartz + gold with NaCl-H_2O solutions, 300–500°C, 500–1,500 bars, and genetic implications; *in* Problems of ore deposition, v. 2, 4th IAGOD Symp., Varna, Bulgaria, B. Bogdanov, ed., Bulgarian Acad. Sci., Sofia, p. 320–326.

Ryzhov, B.V.
1971: Structure of placer deposits of cassiterite, wolframite, and gold in small valleys of the Shilka-Argun interfluve; Izv. Geol. Razved., v. 14, no. 8, p. 68–73. (Chem. Abstr., v. 75, 120072v.)
1972: Structure and formation conditions of placer deposits in the northern Sosva gold-bearing region; Sov. Geol., v. 15, no. 5, p. 94–103. (Chem. Abstr., v. 77, 77803m.)

Saager, R.
1968: Newly observed ore minerals from the Basal Reef in the Orange Free State Goldfield in South Africa; Econ. Geol., v. 63, p. 116–123.
1969: The relationship of silver and gold in the Basal Reef of the Witwatersrand, South Africa; Miner. Deposita (Berl.), v. 4, p. 93–113.
1970: Structures in pyrite from the Basal Reef in the Orange Free State Gold-Field; Geol. Soc. S. Afr., Trans., v. 73, pt. 1, p. 29–46.
1973: Metallogenese präkambrischer Gokdvorkommen in den vulkano-sedimentären Gesteinskomplexen (greenstone belts) der Swaziland-Sequenz in Südafrika; Geol. Rundsch., v. 62, p. 888–901.
1975: New geochemical observations on the determination of the primary source of gold in the Witwatersrand deposit, South Africa; Fortsch. Mineral., v. 53, no. 1, p. 70. (Chem. Abstr., v. 84, 33731b.)

Saager, R. and Bianconi, F.
1971: The Mount Nansen gold-silver deposit, Yukon Territory, Canada; Miner. Deposita (Berl.), v. 6, p. 209–224.

Saager, R. and Mihálik, P.
1967: Two varieties of pyrite from the Basal Reef of the Witwatersrand System; Econ. Geol., v. 62, p. 719–731.

Sadovskii, I.A., Narseev, V.A., Bugaetz, A.N., Lopatnikov, V.V., Pyshko, E.P. and Uvarov, V.V.
1971: The behaviour of tantalum and gold in magmatic process in formation of conjugate rare-metal and gold-bearing belts; Int. Geochem. Congr., Moscow, Abstract of Reports I, p. 167–168.

Safiannikoff, A.
1972: Le gisement primaire aurifère de Twangitza; Ann. Soc. Geol. Belg., v. 95, p. 311–321.

Safronov, N.I., Polikarpochkin, V.V. and Trushkov, Yu. N.
1960*a*: Combined method of gold exploration; Sov. Geol., no. 4, p. 92–110. (Chem. Abstr., v. 55, 1307b.)

Safronov, N.I., Polikarpochkin, V.V. and Utgof, A.A.
1960*b*: Spectrographic aurometric surveying as a method of prospecting for gold ore deposits not accompanied by mechanical haloes (placers); *in* Novoye Metodike Tekhnike Geologorazvedochnykh Rabot. Sbnoronik 1, Leningrad, Gostoptekhizdat, p. 100–108, 1958. (Engl. transl. *in* Int. Geol. Rev., v. 2, no. 3, p. 254–258.)

Saqui, C.L.
1928: The ancient mining works of Cassandra, Greece; Econ. Geol. v. 23, p. 671–680.

Sahli, E.W.
1961: Antimony in the Murchison Range of north-eastern Transvaal; Trans. 7th. Commonw. Min. Metall. Congr., Johannesburg, South Africa, v. 1, p. 181–199.

Sainsbury, C.L.
1974: Geologic map of the Bendeleben Quadrangle, Seward Peninsula, Alaska; Report prepared in cooperation with U.S. Bur. Mines, U.S. Geol. Surv., and the Mapmakers, Anchorage, Alaska by Air Samplex, Golden, Colorado, 31 p.

Sakai, H. and Matsubaya, O.
1974: Isotopic geochemistry of the thermal waters of Japan and its bearing on the Kurokô ore solutions; Econ. Geol., v. 69, p. 974–991.

Sakharova, M.S.
1969*a*: Study of the composition of gold by X-ray spectral microanalysis; Akad. Nauk SSSR, Dokl., v. 186. no. 2, p. 430–433.
1969*b*: Mineralogy of Darasun gold field in eastern Transbaikalia; Int. Geol. Rev., v. 11, no. 1, p. 45–59.
1970: On the distribution with depth of sulfosalts and gold in a hydrothermal ore deposit; *in* Problems of hydrothermal ore deposition, Z. Pouba and M. Stemprok, eds., Int. Union Geol. Sci., Ser. A, no. 2, p. 129–134.
1972: Hydrothermal ore deposits. Typomorphism of associations of bismuth and tellurium minerals in gold ore deposits of eastern Transbaikalia; Tip. Miner. Ego Prakt. Znachenie, F.V. Chukhrov, ed., "Nedra" Moscow, p. 233–240. (Chem. Abstr., v. 80, 123182v.)

Sakharova, M.S., Appollonov, V.N., Kalitkina, N.A. and Krivitskaya, N.N.
1972: Early gold and its possible redistribution in sulphides; Tr. Sev. Vost. Kompleks., Inst. Dal'nevost. Tsent., Akad. Nauch. SSSR, no. 44, p. 64–69.

Sakharova, M.S., Batrakova, Yu.A. and Ryakhovskaya, S.K.
1975: Investigation of interaction between sulfides and gold-containing solutions; Geokhim., no. 5, p. 740–746.
1976: Influence of the anionic composition of the solution upon the process of gold and silver coprecipitation on sulphides; Geokhim., no. 6, p. 871–878.

Sakharova, M.S. and Demidov, V.G.
1972: Gold to silver ratio in the Darasun deposit; Tr. Mineral. Muz., Akad. Nauk SSSR, no. 21, p. 127–132. (Chem. Abstr., v. 77, 154986c.)

Sakharova, M.S. and Krivitskaya, N.N.
1970: Mineralogical and geochemical characteristics of the lead-cinnabar-bismuth sulphosalts in the gold deposits of the eastern Transbaikal region; Geol. Rud. Mestorozhd., no. 4, p. 56–70.
1972: Mineralogical-geochemical characteristics of lead-antimony-bismuth sulfosalts in gold deposits in eastern Transbaikalia; Int. Geol. Rev., v. 14, no. 8, p. 811–822.

Sakharova, M.S. and Lobacheva, I.K.
1967: Electrochemical investigation of processes of gold deposition on sulphides; Geol. Rud. Mestorozhd., v. 9, no. 4, p. 46–55. (Engl. Rev. *in* Econ. Geol., v. 64, p. 591–592.)

Sakharova, M.S. and Nekrasov, E.M.
1964: The mineralogical and geochemical characteristics and conditions of localization of new types of gold-sulfides ores in the Darasun deposit; Geol. Rud. Mestorozhd., v. 6, p. 39–55. *Also* Econ. Geol., v. 60, p. 642–643.

Saks, S.E.
1974: Hydrodynamic differentiation in flow and its effect on the variability of the metal content in placers; Izv. Vyssh. Ucheb. Zaved., Geol. Razved, v. 17, no. 1, p. 73–81. (Chem. Abstr., v. 81, 66371a.)
1975: Dynamics of marine placer formation; Izv. Vyssh. Ucheb. Zaved., Geol. Razved. v. 18, p. 81–89. (Chem. Abstr., v. 84, 124511w.)

Sal'e, O.V.
1965*a*: Thallium in prospecting for gold deposits in the Aldan territory; Izv. Vyssh. Ucheb. Zaved., Geol. Razved., v. 8, no. 12, p. 141–143. (Chem. Abstr., v. 64, 7883g.)
1965*b*: Hydrochemical method in prospecting for gold in the Aldan area; Zap. Leningrad. Gorn. Inst., v. 48, no. 2, p. 77–80. (Chem. Abstr., v. 63, 12868g.)

Salamon, W.
1976: Noble metals in black shales of the Fore Sudeten Monocline; Rudy Metale Niezelaz., v. 21, no. 12, p. 472–477 (in Polish).

Sales, R.H. and Meyer, C.
1948: Wall rock alteration at Butte, Montana; Am. Inst. Min. Metall. Eng., Trans., v. 178, p. 9–35. *Also* A.I.M.M.E. Tech. Publ. no. 2400.
1950: Interpretation of wall-rock alteration at Butte, Montana; Colo. Sch. Mines, Q., v. 45, p. 261–273.

Salmon, J.H.M.
1963: A history of goldmining in New Zealand; R.E. Owen, Government Printer, Wellington, 309 p.

Salop, L.I.
1964: Precambrian geochronology and some features of the early stage of the geological history of the earth; Rep. 22nd Session, Int. Geol. Congr., India, Pt. X, Proc., Sec. 10, Geochemistry, p. 131–139.

Salotti, C.A., Heinrich, E.W. and Giardini, A.A.
1971: Abiotic carbon and the formation of graphite deposits; Econ. Geol., v. 66, p. 929–932.

Samsonov, G.V., ed.
1968: Handbook of the physicochemical properties of the elements; IFI/Plenum, New York.

Samusikov, V.P., Sergeenko, A.I., Trushkov, Yu.N. and Tskhurbaev, F.I.
1969: Placer deposits. Eastern Yakutia gold placer deposits; Str. Zemnoi Kory Yakutti Zakonomer. Razmescheniya Polez. Iskop., Yu. P. Ivensen, ed., Izd. "Nauka", Moscow, U.S.S.R., p. 330–339. (Chem. Abstr., v. 73, 27542x.)

Samusikov, V.P. and Tsabul, L.N.
1972: Chemical and granulometric composition of placer gold in the Allakh-Yun'sk region; Rossypi Zolata Ikh Svyazi Korennymi Mestorozhd. Yakutii, Yu. N. Trushkov, ed., Yakutsk, Knizhnoe Izd., Yakutsk, U.S.S.R., p. 122–144. (Chem. Abstr. v. 78, 162174x.)

Sandberger, F.
1882–1885: Untersuchungen über Erzgänge; Wiesbaden Kreidel, Part 1 (1882) p. 1–158, Part 2 (1885) p. 159–431.

Sandefur, B.T.
1942: The geology and paragenesis of the nickel ores of the Cuniptau mine, Goward, Nipissing District, Ontario; Econ. Geol. v. 37, p. 173–187.

Sandell, E.B.
1948: Colorimetric determination of traces of gold; Anal. Chem., v. 20, no. 3, p. 253–256
1959: Colorimetric determination of traces of metals, 3rd ed., Interscience Publ. Inc., New York, 1032 p.

Sanderson, L. ;
1964: Gold, its properties, occurrences and extraction; Can. Min. J., v. 85, no. 4, p. 113–116 and no. 5, p. 81–85.

Santos, G.G. and Walters, L.J.
1971: Gold provinces in the Philippines defined by activation analysis; Nucl. Tech. Mineral. Explor. Exploit., Int. Atom. Energy Agency, Vienna, p. 143–156.

Saprykin, A.A. and Yablokova, S.V.
1970: Internal structural characteristics of gold from the Amur ancient placer deposits; Izv. Tomsk Poltiekh. Inst., v. 239, p. 390–397. (Chem. Abstr., v. 75, 111631z.)

Sarafian, P.G. and Furbish, W.J.
1965: Solubilities of natural and synthetic ferrimolybdite; Am. Mineral., v. 50, p. 223–226.

Sato, M.
1960: Oxidation of sulfide ore bodies; 1. Geochemical environments in terms of Eh and pH; Econ. Geol., v. 55, p. 928–961.

Saukov, A.A.
1963: Geochemical methods of prospecting for useful mineral deposits; State Univ., Moscow, 247 p.

Savage, E.M.
1934: Prospecting for gold and silver; McGraw-Hill Book Co., Inc., New York, 307 p.

Savage, W.S.
1964: Mineral resources and mining properties in the Kirkland Lake–Larder Lake area; Ont. Dep. Mines, Min. Res. Circ. no. 3, 108 p.

Sawkins, F.J. and Rye, D.M.
1971: On the relationships of certain Precambrian gold deposits to iron formation; Econ. Geol., v. 66, p. 981 (abstr.).
1974: Relationship of Homestake-type gold deposits to iron-rich Precambrian sedimentary rocks; Inst. Min. Metall., Trans., Sec .B., v. 83, p. B56–B59. *Also* discussion in Trans., Sec. B. v. 84, 1975, p. B37–B38.

Sawyer, A.R.
1894: The gold-fields of Mashonaland; John Heywood, Manchester, 99 p.

Saxby, J.D. and Stephens, J.F.
1973: Carbonaceous matter in sulphide ores from Mount Isa and McArthur River: an investigation using the electron-probe and the electron microscope; Miner. Deposita (Berl.), v. 8, no. 2, p. 127–137.

Sazomov, V.N.
1969: Gold-bearing granitic rocks and secondary quartzites in the Berezovsk ore field; Geol. Polez. Iskop. Urala, no. 2, p. 62–64. (Chem. Abstr., v. 72, 92147r.)

Schaudy, R., Kiesl, W. and Hecht, F.
1967: Activation and analytical determination of elements in meteorites; Chem. Geol., v. 2, no. 4, p. 279–287.
1968: Activation analytical determination of elements in meteorites, 2; Chem. Geol., v. 3, no. 4, p. 307–312.

Scheiner, B.J., Lindstrom, R.E. and Henrie, T.A.
1971: Oxidation process for improving gold recovery from carbon-bearing gold ores; U.S. Bur. Mines, Rep. Invest. 7573, 14 p.

Schidlowski, M.
1965: Probable life forms from the Precambrian of the Witwatersrand-System (South Africa); Nature, v. 205, p. 895–896.
1966: Mineralbestand und Gefügebilder in Faseraggregaten von kohliger Substanz (Thucholith) aus den Witwatersrand-Konglomeraten; Contrib. Mineral. Petrol., v. 12, p. 365–380.
1968: The gold fraction of the Witwatersrand conglomerates from the Orange Free State goldfield (South Africa); Miner. Deposita (Berl.), v. 3, p. 344–363.

Schiller, P. and Cook, G.B.
1971: Determination of trace amounts of gold in natural waters by nondestructive activation analysis using preconcentration before irradiation; *in* Nuclear techniques for mineral exploration and exploitation; Int. Atom. Energy Agency, Vienna, p. 137–142.

Schiller, P., Cook, G.B. and Beswick, C.K.
1971*a*: A contribution to geobotanical and geochemical prospecting for gold by activation analysis; *in* Nuclear techniques for mineral exploration and exploitation, Proc. Panel Nucl. Tech. Miner. Resour. (Cracow, 1969), Int. Atom. Energy Agency, Vienna, p. 129–135.
1971*b*: Determination of gold by nondestructive activation analysis for geochemical and geobotanical prospecting; Mikrochim. Acta, no. 3, p. 420–428.

Schiller, P., Cook, G.B., Kitzinger-Skálová, A. and Wölfl, E.
1973: The influence of the season variation for gold determination in plants by neutron activation analysis; Radiochem. Radioanal. Lett., v. 13, nos. 5–6 p. 283–286.

Schindewolf, U. and Wahlgren, M.
1960: The rhodium silver, and indium content of some chondritic meteorites; Geochim. Cosmochim. Acta, v. 18, p. 36–41.

Schmeisser, K.
1898: The gold-fields of Australasia; MacMillan and Co., Ltd., London, 254 p.

Schmid, K.
1973: Gold content of the streams and sediments of the Miocene Molasse of the northwestern Napf area (Canton Lucerne, Switzerland); Schweiz. Mineral. Petrogr. Mitt., v. 53, no. 1, p. 125–156. (Chem. Abstr., v. 79, 7948e.)

Schmidt, A.
1919: Goldhaltiges Grünbleierz von Neu-Bulach im Schwarzwald; Z. Prakt. Geol., v. 27, p. 157–159.

Schmidt, J.W. and Conn, K.
1971: Abatement of water pollution in the base metal mining industry; Can. Min. J., v. 92, no. 3, p. 57–63.

Schmitt, H.
1950: Origin of the "Epithermal" mineral deposits; Econ. Geol., v. 45, p. 191–200.

Schneiderhöhn, H.
1924: Die oxydations-und Zementationszone der sulfidischen Erzlagerstätten; Fortsch. Mineral. Kristallogr. Petrog., v. 9, p. 469–562.
1929: The mineragraphy and spectrography of the sulfide platinum ores of the Bushveld complex, chap. XVII, p. 206–246, *in* Wagner, P.A., The platinum deposits and mines of South Africa, Oliver and Boyd, Edinburgh and London.
1934: Die Ausnutzungsmöglichkeiten der deutschen Erzlagerstätten; Metallwirtschaft, 13, p. 151–157.
1958: Die Erzlagerstätten der Erde; Gustav Fisher Verlag, Stuttgart, vols. 1, 2 and 3.

Schneiderhöhn, H. and Moritz, H.
1931: Spektrographische Untersuchungen über die Verteilung der Platinmetalle in den Mineralien der sudafrikanischen Platinlagerstätten; Festschr. Plantinschmelze G. Siebert, Hanau, p. 257–287.

Scholtz, D.L.
1936: The magmatic nickeliferous ore deposits of East Griqualand and Pondoland; Univ. Pretoria, Pretoria, Ser. II, Nat. Sci., no. 1, pt. 1, 210 p.

Schonwandt, H.K.V.
1974: Gold from Flat; Nor. Geol. Tidsskr., v. 54, no. 1, p. 63–68.

Schutz, D.F. and Turekian, K.K.
1965: The investigation of the geographical and vertical distribution of several trace elements in sea water using neutron activation analysis; Geochim Cosmochim. Acta, v. 29, no. 4, p. 259–313.

Schwartz. G.M.
1944: The host minerals of native gold; Econ. Geol., v. 39, p. 371–411.
1947: Hydrothermal alteration in the "porphyry copper" deposits; Econ. Geol., v. 42, p. 319–352.
1950: Problems in the relation of ore deposits to hydrothermal alteration; Colo. Sch. Mines, Q., v. 45, p. 197–208.
1955: Hydrothermal alteration as a guide to ore; Econ. Geol., 50th Anniv. Vol., pt. 1, p. 300–323.
1959: Hydrothermal alteration; Econ. Geol., v. 54, no. 2, p. 161–183.

Schweigart, H.
1965: Solid solution of gold in sulfides; Econ. Geol., v. 60, p. 1540–1542.

Schweitzer, G.K. and Bishop, W.N.
1953: Low concentration chemistry, VI: Some properties of tracer gold in solution; J. Am. Chem. Soc., v. 75, p. 6330–6332.

Scott, H.C.
1948: The decrepitation method applied to minerals with fluid inclusions; Econ. Geol., v. 43, p. 637-654.

Scott, J.S.
1948: Quemont mine; *in* Structural geology of Canadian ore deposits, v. 1, Can. Inst. Min. Metall., Montreal, p. 773–776.

Searls, F.
1948: A contribution to the published information on the geology and ore deposits of Goldfield Nevada; Univ. Nevada Bull., v. 42, no. 5, Geol. Min. Ser., no. 48, 24 p.
1952: Karst ore in Yunnan; Econ. Geol., v. 47, p. 339–346.

Sears, W.P.
1971: Mercury in base metal and gold ores of the province of Quebec; *in* Geochemical exploration, R.W. Boyle and J.I. McGerrigle, eds., Can. Inst. Min. Metall., Spec. Vol. 11, p. 384–390.

Seeland, D.A.
1973*a*: Geochemical reconnaissance for gold in the sedimentary rocks of the Great Lakes region, Minnesota to New York; U.S. Geol. Surv., Bull., no. 1305, 16 p.
1973*b*: Geochemical reconnaissance for gold in sedimentary and metamorphic rocks of the Great Lakes region. Tabulated data on 799 samples; U.S. Natl. Tech. Inform. Serv., PB Rep., No. 222892/2, 60 p. (Chem. Abstr., v. 80, 50676x.)

Seiranyan, V.B., Sarkisyan, S. Sh. and Zav'yalov, E.N.
1974: Zeolites from the ores of the Armaniss gold-complex metal deposit (northern Armenia); Geol. Rud. Mestorozhd., v. 16, no. 6, p. 112–117. (Chem. Abstr., v. 83, 63465q.)

Segerstrom, K. and Ryberg, G.E.
1974: Geology and placer-gold deposits of the Jicarilla Mountains, Lincoln County, New Mexico; U.S. Geol. Surv., Bull. 1308, 25 p.

Selwyn, A.R.C.
1972: Notes and observations on the gold fields of Quebec and Nova Scotia; Halifax, N.S., 16 p.

Semenenko, N.P. and Latysh, I.K.
1973: On purity of the native gold of the Ukraine; Akad. Nauk Ukr. RSR, Dopov., Ser. B., v. 35, p. 132–135.

Sen, B.
1959: A new spot test for gold; Mikrochim. Acta, p. 513–515.

Sen Gupta, J.G.
1973: A review of the methods for the determination of the platinum-group metals, silver, and gold by atomic-absorption spectroscopy; Miner. Sci. Eng., v. 5, no. 3, p. 207–218.

Senftle, F.E., Duffey, D. and Wiggins, P.F.
1969: Mineral exploration of the ocean floor by *in situ* neutron absorption using a Californium-^{252}Cf source; J. Mar. Tech. Soc., v. 3, no. 5, p. 9–16.

Sestini, G.
1970: Sedimentological study of the Tarkwaian gold deposits; Ghana; Univ. Leeds, Res. Inst. Afr., Geol. Dep. Earth Sci., v. 14, p. 23–26.
1973: Sedimentology of a paleoplacer. The gold-bearing Tarkwaian of Ghana; *in* Ores in sediments, 8th Int. Sedimentol. Congr., 1971, G.C. Amstutz and A.J. Bernard, eds., p. 275–305.

Sevensma, P.H.
1941: Les gisements d'or de la région de Saint-Yrieix; Georg & Cie S.A., Genêve, 122 p.

Seward, T.M.
1973: Thio complexes of gold and the transport of gold in hydrothermal ore solutions; Geochim. Cosmochim. Acta, v. 37, no. 3, p. 379–399.

Shabanin, M.A.
1972: Facial-paleogeographical and geomorphological conditions of placer formation in Fergana; Nauch. Tr., Tashkent. Gos. Univ., no. 429, p. 304–306. (Chem. Abstr., v. 80., 50504q.)

Shabynin, L.L.
1966: On the migration of gold in natural waters; Izvest. Tomsk Polytekh. Inst., U.S.S.R., v. 151, p. 273–280.
1970: Some experimental data on the migration of gold under supergene conditions; *in* Geology of goldbearing deposits of Siberia, V.N. Shakhov, and S.S. Il'enok, eds., Acad. Sci. USSR, Siberian Br., p. 137–141.

Shacklette, H.T.
1965: Element content of bryophytes: U.S. Geol. Surv., Bull. 1198-D, 21 p.

Shacklette, H.T., Lakin, H.W., Hubert, A.E. and Curtin, G.C.
1970: Absorption of gold by plants; U.S. Geol. Surv., Bull. 1314-B, 23 p.

Shadrin, A.I. and Lokotko, V.V.
1972: Talatuisk gold ore deposit; Zap. Zabaikal. Filiala Geogr. Obshchest. SSSR, no. 69, p. 57–59. (Chem. Abstr., v. 80, 98375a.)

Shah, K.R., Filby, R.H. and Haller, W.A.
1970: Determination of trace elements in petroleum by neutron activation analysis; J. Radioanal. Chem., v. 6, p. 413–422.

Shahine, Salah A. and Mahmound, Rafaa, M.
1976: A selective and sensitive spot-test for gold; Mikrochim. Acta, v.1, no. 1, p. 89–92.

Shakhov, F.N.
1961: Major trends in scientific investigations in the auriferous districts of Siberia; Geol. Geofiz., no. 10, p. 89–101. (Engl. transl. *in* Int. Geol. Rev., 1964, v. 6, no. 2, p. 202–211.)

Shakhov, F.N., ed.
1969: The geology of placer deposits in the southern part of western Siberia; "Nauka" Publ. House, Moscow, 192 p.

Shakhov, F.N. and Il'enok, S.S., eds.
1970: Geology of the Siberian gold deposits; "Nauka" Novosibirsk, U.S.S.R., 147 p.

Sharma, N.N.
1972: Concentration of gold in sea water; Indian J. Mar. Sci., v. 1, no. 2, p. 151–152. (Chem. Abstr., v. 79, 139578u.)

Sharpe, J.W.N.
1949: The economic auriferous bankets of the Upper Witwatersrand beds and their relationship to sedimentation features; Geol. Soc. S. Afr., Trans., v. 52, p. 265–300.
1955: Discussion: Occurrence and origin of gold and radioactive minerals in the Witwatersrand system, the Dominion Reef, the Ventersdorp Contact Reef and the Black Reef; Geol. Soc. S. Afr., Trans., v. 58, p. 229–230.

Sharwood, W.J.
1911*a*: Notes on tellurium-bearing gold ores; Econ. Geol., v. 6, p. 22–36.
1911*b*: Notes on tellurium-bearing gold ores; Min. Sci. Press, v. 102, no. 12, p. 434–435.
1911*c*: Analyses of some rocks and minerals from the Homestake mine, Lead, South Dakota; Econ. Geol., v. 6, p. 729–789.

Shaw, D.M.
1957: The geochemistry of gallium, indium, thallium—a review; *in* Physics and chemistry of the earth; v. 2, L.H. Ahrens *et al.*, eds., Pergamon Press, London, p. 164–211.

Shcheka, S.A. and Moiseenko, V.G.
1970: Regular characteristics of gold distribution in basic and ultrabasic rocks; Izv. Tomsk. Politek. Inst., v. 239, p. 37–38.

Shcheka, S.A., Moiseenko, V.G. and Fominykh, V.G.
1971: Patterns of gold distribution in intrusive basic and ultrabasic rocks; Akad. Nauk SSSR, Dokl., v. 201, no. 2, p. 461–464.

Shcherbakov, I.B. and Latysh, I.K.
1972: Geochemistry of gold in rocks of the Ukrainian Shield central region; Akad. Nauk Ukr. RSR, Dopov., Ser. B, v. 34, no. 2, p. 127–129.

Shcherbakov, Yu.G.
1967: Gold ore provinces and formations; Int. Geol. Rev., v. 9, p. 1537–1543.
1968: Distribution of gold in igneous rocks and specialization of gold-bearing intrusions; Geochem. Int., v. 5, p. 1226–1228.
1969: Distribution and conditions of gold concentrations in ore provinces; "Nauka" Moscow, 268 p., 1967. (Engl. rev. *in* Int. Geol. Rev., v. 11, no. 4, p. 511–512.)
1970: Source of gold in endogenous deposits; *in* Problems of hydrothermal ore deposition, Z. Pouba and M. Stemprok, eds., Int. Union. Geol. Sci., Ser. A, no. 2, p. 134–136.
1973: Geochemical characteristics of gold ore deposition in regions of autonomous activation; Metallog. Obl. Tektono-Magmat. Akt., Tezisy Dokl. Vses. Metallog. Soveshch., 7th, M.M. Odintsov, ed., Akad. Nauk SSSR, Sib. Otd., Inst. Zemnoi Kory, Irkutsk, U.S.S.R. p. 180–182. (Chem. Abstr., v. 82, 46363s.)
1974: Geochemistry of gold ore deposits in the Kuznetsk-Alatau and Altai mountains; F.N. Shakhov, ed., Akad. Nauk SSSR, Sib. Otd., Tr. Inst. Geol. Geofiz. V'pusk, 56, 278 p.
1976: Gold and rare elements in geochemical processes; "Nauka" Publ. House, Siberian Branch, Novosibirsk, 311 p.

Shcherbakov, Yu.G., Pavlova, L.K., Mel'nikova, R.D. and Tsimbalist, V.G.
1972: Geochemistry of gold in the Sinyukha skarn-ore field (Gornyi Altai); Tr. Inst. Geol. Geofiz., Akad. Nauk SSSR, Sib. Otd., no. 149, p. 7–20. (Chem. Abstr., v. 79, 106793g.)

Shcherbakov, Yu.G. and Perezhogin, G.A.
1963: Geochemical relation between gold mineralization, intrusives, and the enclosing rocks in western Siberia; Geochem. Int., no. 9, p. 882–890.
1964: Geochemistry of gold; Geokhim., no. 6, p. 518–528. *Also* Geochem. Int. 1964, no. 3, p. 489–496.

Shcherbakov, Yu.G., Roslyakova, N.V., Pavlova, L.K. and Mel'nikova, R.D.
1973: Geochemistry of gold in the formation of hydrothermal gold deposits; Mezhdunar. Geokhim. Kongr., (Dokl.) 1st, 1971, pt. 2, A.P. Vinogradov, ed., p. 131–142. (Chem. Abstr., v. 81, 94060s.) (Engl. abstr. *in* Int. Geochem. Congr., Moscow, Abstracts of Reports I p. 238–239.)

Shcherbina, V.V.
1956*a*: Geochemical significance of quantitative Ag-Au ratios; Geokhim., no. 3, p. 65–73. *Also* Geochemistry, no. 3, p. 301–311.
1956*b*: Complex ions and the transfer of elements in the supergene zone; Geochemistry, no. 5, p. 486-493.
1960: Features of the geochemistry and types of occurrences of scandium; Geology of occurrences of rare elements, No. 8, All Sov. Sci. Res. Inst. Miner. Raw Materials (VIMS), Moscow, 58 p.
1964: Do thiosulphates exist in hydrothermal solutions? Geol. Rud. Mestorozhd., no. 3, p. 110–112.
1972: Principles of geochemistry; "Nedra", Moscow, 295 p.

Shcherbina, V.V. and Zar'yan, R.N.
1964: Paragenesis of silver and gold tellurides as solid phases in the system Ag-Au-Te; Geochem. Int., no. 4, p. 653-657.

Shchepot'ev, Yu.M. and Andrusenko, N.I.
1975: Genetic features of gold and mercury deposits near the surface on Kamchatka; Sov. Geol., no. 6, p. 62–71. (Chem. Abstr., v. 83, 196317q.)

Shenon, P.J. and Reed, J.C.
1934: Geology and ore deposits of the Elk City, Orogrande, Buffalo Hump, and Tenmile districts, Idaho County, Idaho; U.S. Geol. Surv., Cir. 9, p. 30.

Sher, S.D.
1966: Quantitative estimate of gold content in geotectonic provinces; Int. Geol. Rev., v. 8, p. 1329–1333.
1970: Wall-rock alteration in one of the gold deposits of Middle Asia; *in* Problems of hydrothermal ore deposition, Z. Pouba and M. Stemprok, eds., Int. Union Geol. Sci., Ser. A, no. 2, p. 312–315.
1972: Metallogeny of gold (North America, Australia, and Oceania); "Nedra", Moscow, U.S.S.R., 295 p. (Chem. Abstr., v. 78, 100788s.)

Sher, S.D. and Demchenko, A.V.
1962: Applications of crystallography to prospecting for gold deposits; Geol. Rud. Mestorozhd., no. 4, p. 84–96.

Sher, S.D. and Timofeevskii, D.A.
1974: Vertical prolongation of gold ore deposits, localization and differentiation of geotectonic provinces; Abstracts of Papers, Int. Assoc. Genesis Ore Deposits (IAGOD), 4th Symp., Varna, Bulgaria, p. 56–57.

Shergin, B.V.
1973: Gold in sedimentary rocks of the Lena gold-bearing region; Ezheg. Inst. Geokhim. Sib. Otd., Akad. Nauk SSSR, 1972, L.V. Tauson, ed., p. 261–265. (Chem. Abstr., v. 81, 155872g.)

Shevkalenko, V.L.
1971: Practical importance of the silver-gold ratio in ores of the Shkol'noe deposit (northern Tadzhikistan); Akad. Nauk Uzb. SSSR, Dokl., v. 28, no. 3, p. 40–42. (Chem. Abstr., v. 80, 17468h.)

Shilin, N.L.
1968: Gold in the rocks of central Kamchatka; Akad. Nauk SSSR, Izv., Ser. Geol., no. 11, p. 69–76.

Shilo, N.A.
1956: Some pecularities of placer formation in the zone of permafrost development; Sov. Geol., v. 53, p. 102–117.
1960: The role of subpolar climate in the formation and disposition of placer deposits; *in* The principles of the disposition of useful minerals IV, Placer deposits; U.S.S.R. State Sci. Tech. Publ. House Lit. Mining, p. 20–29.
1963: Some features of the composition of alluvial placers in the Yana-Kolyma belt; Tr. Sev. Vost. Kompleks. Nauch. Issled. Inst. no. 3, p. 87–105. (Chem. Abstr., v. 61, 9304e.)
1968: Certain problems of auriferous lodes and placers; Int. Geol. Rev., v. 10, no. 11, p. 1241–1246.
1970*a*: Placer-forming ore formation and their relation to placer deposits Probl. Geol. Rossypei, Soveshch., (Dokl.) 3rd, 1969, V.I. Smirnov, ed., Sev. Vost. Kompleks. Inst.; Magadan, U.S.S.R., p. 13–24. (Chem. Abstr., v. 75, 23807f.)
1970*b*: Placer-forming minerals and placer deposits; Pacific Geol., no. 2, p. 29–33. (Chem. Abstr., v. 74, 14825n.)
1971: The problems of the geology of gold; Earth-Sci. Rev., v. 7, p. 215–225.
1973: Metal content of placers of the northwestern Pacific area; Aust., Bur. Min. Resour., Geol. Geophys. Bull., no. 141, p. 221–222.
1974: Gold and gold-silver mineralization in the Okhotsk-Chukotka volcanic belt and some problems of ore formation; Zap. Leningrad. Gorn. Inst., v. 67, no. 2, p. 20–34. (Chem. Abstr., v. 83, 82774f.)

Shilo, N.A., ed.
1972: The present problems of the geology of gold in the far eastern part of the U.S.S.R.; Trans., Far East Multipurpose Inst., Magadan, no. 44, 202 p.

Shilo, N.A., Belyi, V.F. and Sidorov, A.A.
1974: Volcanic belts of East Asia and their bearing on the problems of tectonics, magmatism and metallic ore generation; Geol. Geofiz., no. 5, p. 70–88.

Shilo, N.A. *et al.*
1975*a*: Pre-Mesozoic gold content of the Omolonsky massif; Acad. Sci. USSR, Dokl., v. 225, p. 1165–1167.
1975*b*: Auriferous formation of Oloi zone; Akad. Nauk SSSR, Geol. Geofiz., no. 10, p. 3–10.
1975*c*: A new model of the tectonic and metallogenic zonality of the northeastern U.S.S.R.; Acad. Nauk USSR, Dokl., v. 223, no. 4, p. 961–964.

Shilo, N.A., Gorodinskii, M.E. and Gulevich, V.V.
1975: Gold ore formations of the Oloi zone; Geol. Geofiz., no. 10, p. 3–10. (Chem. Abstr., v. 84, 124530b.)

Shilo, N.A., Sakharova, M.S., Batrakova, Yu.A. and Ryakhovskaya, S.K.
1974: Investigation of gold and silver coprecipitation from aqueous gold-silver solutions; Akad. Nauk SSSR, Dokl., v. 217, no. 6, p. 1398–1400. (Chem. Abstr., v. 81, 173108s.)

Shilo, N.A. and Shumilov, Yu.V.
1970: New experimental data on the behaviour of gold particles in an aqueous medium; Akad. Nauk. SSSR, Dokl., v. 195, no. 1, p. 193–196. (Chem. Abstr., v. 74, 44401p.)
1976: Mechanisms of behaviour of gold during placer formation processes in the northeast of the U.S.S.R.; 25th Intl. Geol. Congr., Sydney, Australia, Abstr., v. 1, p. 224.

Shilo, N.A., Sidorov, A.A., Goncharov, V.I. and Naiborodin, V.I.
1971: The temperature conditions and the depths of formation of gold-ore deposits; Soc. Min. Geol. Japan, Spec. Iss. 3, p. 356–359. (Chem. Abstr., v. 76, 74717e.)

Shilo, N.A., Sidorov, A.A., Naiborodin, V.I. and Goncharov, V.I.
1969: Gold-ore associations of the northeastern U.S.S.R.; Acad. Sci. USSR, Dokl., Earth Sci. Sec., v. 188, p. 97–99 (Am. Geol. Inst. transl.).

Shilo, N.A., Sidorov, A.A. and Zhelnin, S.G.
1972: Problems of gold metallogeny and prospects for gold ore deposits in the northeastern U.S.S.R.; Tr. Sev. Vost. Kompleks, Inst. Dal'nevost. Tsent. Akad. Nauk SSSR, no. 44, p. 27–34. (Chem. Abstr., v. 82, 88423z.)

Shilts, W.W.
1973: Drift prospecting; geochemistry of eskers and till in permanently frozen terrain: District of Keewatin; Northwest Territories; Geol. Surv. Can., Pap. 72–45, 34 p.
1974: Drift prospecting in the Ennadai–Rankin Inlet greenstone belt, District of Keewatin; Geol. Surv. Can., Pap. 74–1, pt. A, p. 259–260.

Shima, M.
1953*a*: Geochemical prospecting for gold, silver, copper, lead, and zinc ore deposits; Pt. 1—Prospecting for gold and silver ores; Reps. Sci. Res. Inst. (Japan), v. 29, p. 111–115. (Chem. Abstr., v. 47, 12145b.)
1953*b*: Geochemical prospecting for gold, silver, copper, lead, and zinc ores: Pt. 4—Gold; Reps. Sci. Res. Inst. (Japan), v. 29, p. 245–248. (Chem. Abstr., v. 48, 1207h.)
1970: Geochemical prospecting methods, 2nd ed.; Kyoritsu Publ. Co. Ltd., Tokyo, 273 p.

Shima, M. and Thode, H.G.
1971: A geochemical prospecting method using stable isotopes; *in* Geochemical exploration, R.W. Boyle and J.I. McGerrigle, eds., Can. Inst. Min. Metall., Spec. Vol. 11, p. 469–472.

Shklyar, K.T. and Kakovskii, R. Sh.
1972: Preparation and structure of gold amalgams; Izv. Vyssh. Ucheb. Zaved., Tsvet. Met., v. 15, no. 1, p. 57–60.

Shmidt, A.I.
1961: Age relations of pyrite and gold-lead-zinc mineralization in the Kurosan ore field; Geol. Rud. Mestorozhd., v. 3, no. 6, p. 27–40.

Shmotov, A.P.
1974: Tectonic deformations and the accompanying hydrothermal and metasomatic transformations of the country rocks (the Lena gold region); Akad. Nauk. SSSR, Dokl., v. 218, no. 1, p. 178–181. (Chem. Abstr., v. 82, 101330f.)

Short, M.N.
1940: Microscopic determination of the ore minerals, 2nd ed.; U.S. Geol. Surv., Bull. 914, 314 p.

Shpora, L.D., Martynov, V.V. and Mesitskii, L. Yu.
1972: Gold and silver content of the copper-bearing sandstones of Naukat; Nauch. Tr., Tashkent Gos. Univ., no. 432, p. 206–212. (Chem. Abstr., v. 82, 19446q.)

Shpunt, B.R.
1969: Exploration sampling Quaternary placer deposits in Anabar-Olenek interfluve containing gold, diamonds, and platinum; Uch. Nauch. Issled. Inst. Geol. Arktiki, Reg. Geol., no. 15, p. 171–172. (Chem. Abstr., v. 73, 37361b.)
1973: Placer gold occurrences in Cenozoic formations of the Lena-Anabar interfluve; Rossypnaya Almazonosnost Stredni Sib., A.P. Puminov, ed., Nauch. Issled. Inst. Geol. Arktiki, Leningrad, U.S.S.R., p. 31–35. (Chem. Abstr., v. 79, 148079j.)
1974: Typomorphic characteristics and genesis of placer gold deposits in the northern Siberian platform; Geol. Geofiz., no. 9, p. 77–88. (Chem. Abstr., v. 82, 114186p.)

Shvartsev, S.L.
1974: Gold in ground waters of the West-Siberian artesian basin; Vopr. Geokhim. Podzemn. Vod Svyazi. Poiskami Rudn. Polezn. Iskop., p. 191–193 and 205–211. (Chem. Abstr. v. 86, 7334w.)

Sidgwick, N.V.
1950: The chemical elements and their compounds; v. 1, p. 103–192, Oxford.

Sidorov, A.A.
1965: Epithermal gold and silver occurrences of Sopka Rudnaya; Geol. Rud. Mestorozhd., v. 7, no. 2, p. 91–96. (Engl. rev. *in* Econ. Geol., v. 63, p. 577.)
1966: Gold-silver mineralization of Central Chukotsk; Akad. Nauk SSSR Magadan. Sev. Vost. Kompleks. Nauch. Issled. Inst. Tr., vyp. 14, 146 p.
1974: On the problem of origin and evolution of a gold-silver formation; Evol. Vulkan. Istor. Zemli Tr. Vses. Paleovulkanol. Simp. 1st 1973, I.V. Luchitskii and G.M. Fremd, eds.; p. 424–430. (Chem. Abstr. v. 85, 8297c.)

Sidorov, A.A., Gontcharov, V.I. and Naiborodin, V.I.
1971: Thermobarogeochemistry of the deposits of gold-silver formation; Int., Geochem. Congr., Moscow, Abstract of Reports I, p. 382–383.

Sidorov, A.A., Naiborodin, V.I., Eremin, R.A. and Gontcharov, V.I.
1973: Peculiar features of vertical zonality of gold-silver deposits of the north-east of the U.S.S.R.; Akad. Nauk. SSSR, Dokl., v. 209, no. 2, p. 440-443.

Sidorov, A.A. and Yugai, T.A.
1972: Information about Mesozoic-Cenozoic gold-silver deposits; Tr. Sev. Vost. Kompleks. Inst. Dal'nevost. Tsent. Akad. Nauk SSSR, v. 44, p.53–63. (Chem. Abstr., v. 82, 101237f.)

Siegel, F.R.
1971: Marine geochemical prospecting—present and future; *in* Geochemical exploration, R.W. Boyle and J.I. McGerrigle, eds., Can. Inst. Min. Metall., Spec. Vol. 11, p. 251–257.
1974: Applied geochemistry; John Wiley & Sons, Inc., New York, 353 p.

Siegl, W.
1972: Uranium paragenesis of Mitterberg (Salzburg, Austria); Tschermak's Mineral. Petrogr. Mitt., Bd. 17, Heft 4, p. 263–275. (Chem. Abstr., v. 77, 22836u.)

Sighinolfi, G.P. and Gorgoni, C.
1977: Gold distribution in Ivrea-Verbano gabbroic complex; Chem. Geol., v. 20, no. 2, p. 99–107.

Sighinolfi, G.P. and Santos, A.M.
1976*a*: Geochemistry of gold in Archean granulite facies terrains; Chem. Geol. v. 17, p. 113–123.
1976*b*: Determination of gold in geological samples at parts per milliard levels by flameless atomic-absorption spectroscopy; Mikrochim. Acta II, no. 1-2, p. 33–40.

Sigit, S., Purbo-Hadiwidjojo, M.M., Sulasmoro, B. and Wirjosudjono, S.
1969: Minerals and mining in Indonesia; Ministry Mines, Djakarta, Indonesia, 123 p.

Sigov, A.P.
1939: Prospecting by heavy mineral studies; Trans. Ural Sci. Res. Inst. Geol. Prosp. Econ. Mineral. Sverdlovsk Mining Inst. No. 4, 64 p.

Sigov, A.P. *et al*
1972: Placers of the Urals – their formation, distribution and elements of geomorphic prediction; Sov. Geogr., v. 13, p. 375–387.

Sikka, D.B., Royce, J. and Nehru, C.E.
1964: Geosynclines in the Precambrian of India and their potential economic significance; Mineral Markets, v. 3, no. 6, p. 15–33.

Silberman, M.L. and Ashley, R.P.
1970: Age of ore deposition at Goldfield, Nevada, from potassium-argon dating of alunite; Econ. Geol., v. 65, p. 352–354.

Silberman, M.L., Berger, B.R. and Koski, R.A.
1974: K-Ar age relations of granodiorite emplacement and tungsten and gold mineralization near the Getchell mine, Humboldt County, Nevada; Econ. Geol., v. 69, p. 646–656.

Silberman, M.L., Chesterman, C.W., Kleinhampl, F.J. and Gray, C.H.
1972: K-Ar ages of volcanic rocks and gold-bearing quartz-adularia veins in the Bodie mining district, Mono County California; Econ. Geol., v. 67, p. 597–604.

Silichev, M.K.
1970: The geology and structure of the Nezhdaninskoye gold deposit; Geol. Rud. Mestorozhd., no. 2, p. 96–102.

Silichev, M.K. and Belozertseva, N.V.
1973: Gold distribution in the sediments and igneous rocks around the Nezhdana gold-ore deposit; Geokhim., no. 6, p. 931–935.

Sillén, L.G. and Martell, A.E.
1964: Stability constants of metal-ion complexes; Chem. Soc., Spec. Publ. no. 17, London, 754 p.

Simon, F.O. and Millard, H.T.
1968: Determination of gold in rocks by neutron activation analysis using fire-assay preconcentration; Anal. Chem., v. 40, no. 7, p. 1150–1152.

Simonin, L.
1880: L'or et l'argent; Paris, 296 p.

Simons, F.S. and Prinz, W.C.
1973: Gold: *in* United States mineral resources, Brobst D.A. and W.P. Pratt, eds., U.S. Geol. Surv., Prof. Pap. 820, p. 263–275.

Simpson, E.S. and Gibson, C.G.
1912: Contributions to the study of the geology and ore deposits of Kalgoorlie, East Coolgardie goldfield (in 7 parts); Geol. Surv. West. Aust., Bull. 42, 198 p.

Simpson, P.R. and Bowles, J.F.W.
1977: Uranium mineralization of the Witwatersrand and Dominion Reef systems; Phil. Trans. R. Soc. London, Ser. A, v. 286, p. 527–548.

Sims, P. K., Drake, A.A., and Tooker, E.W.
1963: Economic geology of the Central City district, Gilpin County, Colorado; U.S. Geol. Surv., Prof. Pap. 359, 231 p.

Sinclair, A.J. and Percy, G.R.
1969: Multiple regression analyses of precious metal values, Phoenix mine, British Colombia, Canada; Econ. Geol., v. 64, p. 822–824.

Singer, C.
1954: Some early goldwork; Endeavour, v. 13, no. 50, p. 86–93.

Singewald, Q.D.
1950: Mineral resources of Colombia; U.S. Geol. Surv., Bull. 964-B, 204 p.

Sinyakov, V.I.
1974: Propylitization in skarn-magnetite deposits in Gornaya Shoriya; Tr. Inst. Geol. Geofiz., Akad, Nauk SSSR, Sib. Otd., v. 55, p. 138–147. (Chem. Abstr., v. 82, 173659t.)

Sinyugina, E.Ya. and Strepetova, Z.V.
1970: Lithological characteristics of gold-bearing layers in Siberian alluvial placer deposits; Sostoyanie Zadachi Sov. Litol., Dokl. Zased. Vses. Litol. Soveshch., 8th, 1968, no. 2, p. 207–212. (Chem. Abstr., v. 74, 101660y.)

Sirin, N.A.
1963: Main stages of magmatic and metallogenic occurrences on the eastern slopes of the Urals; Magmat. Metamorfizin, Metallogen. Urala, Sverdlovsk, Sb., v. 1, p. 111–120. (Chem. Abstr., v. 61, 10461b.)

Sisco, A.G. and Smith, C.S., translators
1949: Bergwerk- und Probierbüchlein (1500); Am. Inst. Min. Metall. Eng. New York, 196 p.

Skerl, A.C.
1948: Geology of the Cariboo Gold-Quartz mine, Wells, B.C., Econ. Geol., v. 43, p. 571–597.

Skey, W.
1871: On the reduction of certain metals from their solutions by metallic sulphides, and the relation of this to the occurrence of such metals in a native state; Trans., Proc. New Zealand Inst., v. 3, p. 225–231.

Skryabin, A.E.
1972: Relation of placer gold dispersion to the location of the original deposits in the Taryn-El'ginsk area (Yano-Kolymsk gold-bearing belt); Rossypi Zolota Ikh. Suyazi Korennyi Mestorozhd. Yakutii, Yu. N. Trushkov, ed., p. 46-103. (Chem. Abstr., v. 79, 7955e.)

Skryabin, A.I.
1974: Types of gold fields of the Taryno-El'Ginskaya and Selerikanskaya zones, the Yana-Kolyma belt; Vop. Rudonos. Yakutii, G.N. Gamyanin, ed., p. 175–186. (Chem. Abstr., v. 84, 76928p.)

Skryabin, V.F.
1968: Hydrogeochemical prerequisite in prospecting for gold mineralization under mountain conditions of the Chatkal-Kurama territory; *in* Rud. Form. Osn. Cherty Metallogen. Zolota Uzb., I. Kh Khamrabaev, ed., Izd. "Fan" Uzb. SSSR, Tashkent., p. 358–362. (Chem. Abstr., v. 73, 37340u.)

Slaughter, A.L.
1968: The Homestake mine; *in* Ore deposits of the United States, 1933–1967; J.D. Ridge, ed., v. 2, p. 1436–1459.

Smales, A.A.
1955: Some trace element determinations in G-1 and W-1 by neutron activation; Geochim. Cosmochim. Acta, v. 8, nos. 5/6, p. 300.

Smales, A.A., Mapper, D. and Fouché, K.F.
1967: The distribution of some trace elements in iron meteorites as determined by neutron activation; Geochim. Cosmochim. Acta, v. 31, p. 673–720.
1968: A comparison of trace element distribution in the metal phase of chondrites and in iron meteorites; *in* Origin and distribution of the elements, L.H. Ahrens, ed., Pergamon Press, Oxford, p. 329–344.

Smales, A.A., Mapper, D., Webb, M.S.W., Webster, R.K., Wilson, J.D. and Hislop, J.S.
1971: Elemental composition of lunar surface material (part 2); Proc., 2nd Lunar Sci. Conf., Geochim. Cosmochim. Acta, Suppl. 2, v. 2, p. 1253–1258.

Smirnov, F.L. and Entin, A.R.
1975: Geochemical bond between gold and phosphorus and its significance in the estimation of the prospects of the Aldan Shield gold content; Geokhim., no. 1, p. 129–131.

Smirnov, S.S.
1951: The oxidation zone of sulphide deposits; Akad. Nauk SSSR, Izv., Moscow, 335 p.

Smirnov, V.I.
1959: Essai de subdivision metallogénique du territoire de l'U.R.S.S.; Soc. Geol. Fr., Bull., Ser. 7, v. 1, no. 5, p. 511–526.
1965: Geology of economic minerals; "Nedra" Moscow, 590 p.
1968: The sources of ore-forming material; Econ. Geol., v. 63, p. 380–389.
1976: Geology of mineral deposits; Mir Publishers, Moscow, 520 p.

Smirnov, V.I., ed.
1969: Problem of the occurrence of metals in ancient conglomerates in the territory of the Soviet Union; "Nauka" Moscow, 192 p.
1974: Ore deposits of U.S.S.R., vols. 1, 2, and 3, "Nedra" Moscow.

Smirnov, V.N.
1972: Geological-geomorphological conditions for the formation of alluvial gold in the Taigonos peninsula; Tr. Sev. Vost. Kompleks. Inst., Dal'nevost. Tsent., Akad. Nauk SSSR, v. 44, p. 125–130. (Chem. Abstr., v. 81, 173044t.)

Smith, A.
1948*a*: Tulsequah area; *in* Structural geology of Canadian ore deposits, v. 1, Can. Inst. Min. Metall., Montreal, p. 112–121.
1948*b*: Surf Point and Edye Pass mines; *in* Structural geology of Canadian ore deposits, v. 1, Can. Inst. Min. Metall., Montreal, p. 94–99.

Smith, C.S. and Gnudi, M.T., translators
1959: *De la Pirotechnia* by Vannoccio Biringuccio. (1540); Am. Inst. Min. Metall. Eng., New York, 477 p.

Smith, F.G.
1943: The alkali sulphide theory of gold deposition; Econ. Geol., v. 38, p. 561–590.
1947: The pyrite geo-thermometer; Econ. Geol., v. 42, p. 515–523.
1948: The ore deposition temperature and pressure at the McIntyre mine, Ontario; Econ. Geol., v. 43, p. 627–636.
1951: Gold deposition temperature—pressure gradients in the Ontario-Quebec mining region; Can. Inst. Min. Metall., Trans., v. 54, p. 44-48.
1953: Historical development of inclusion thermometry; Univ. Toronto Press, Toronto, 149 p.
1954: Direction of flow of late stage solutions in the Lamaque No. 6 vein; Econ. Geol., v. 49, p. 530–536.

Smith, F.G. *et al.*
1949–1951: Temperature pressure research of hydrothermal mineral deposits; vols. 1, 2 and 3, privately printed.

Smith, F.G. and Peach, P.A.
1949*a*: Recent advances in the laboratory study of ore; Can. Inst. Min. Metall. Bull., no. 447, p. 351–353.
1949*b*: Apparatus for the recording of decrepitation in minerals; Econ. Geol., v. 44, p. 449–451.

Smith, P.S.
1908: Investigations of the mineral deposits of Seward Peninsula; U.S. Geol. Surv., Bull. 345, p. 206–267.
1930: The gold resources of Alaska; Econ. Geol., v. 25, p. 176–196.

Smith, W.H.
1968: Geochemical investigation of a portion of the Fortymile district, east-central Alaska; Alaska, Dep. Nat. Res., Div. Mines Miner., Geochem. Rep., no. 16, p. 1–17.

Smolin, A.P.
1970: Native gold ores of the Urals; "Nedra", Moscow, U.S.S.R., 144 p.
1975: Structural documentation of gold deposits; "Nedra", Moscow, U.S.S.R., 240 p.

Snelgrove, A.K.
1935: Geology of gold deposits of Newfoundland; Newfoundland, Dep. Nat. Resour., Geol. Sec., Bull. 2, 46 p.

Sobolev, S.F., Anoshin, G.N. and Perezhogin, G.A.
1974: Distribution of gold in rocks of ultrabasic and basic formations in Ural; Int. Geol. Rev., v. 16, p. 800–809.

Sobolevskii, V.I.
1939: Investigation of tellurides containing gold from the Krasnyi Brook deposits (Verkhnyaya Angora River, Baikal Lake region, Siberia); Sov. Geol., no. 4–5, p. 95–103. (Chem. Abstr., v. 34, 5791,5.)
1970: Noble metals, Gold (Series Science of the Earth, No. 3); "Znanie", Moscow, U.S.S.R., 48 p. (Chem. Abstr. v. 77, 129258j.)

Sobotka, J.
1954: Aurostibite, $AuSb_2$, in the ore veins of Krásná Hora and Milešov (first macroscopical occurrence); Rozpr. Cesk. Akad. Ved, v. 64, no. 7, p. 43–60.

Sobotovich, E.V., Rudenko, A.K. and Smirnova, A.V.
1971: Distribution of gold in kamacite of the Sikhote-Alin meteorite; Probl. Kosmokhimii Meteoritiki, Rosshir, Yubileinyi plenum, p. 164–169.

Soeharto, O.
1976: Comparative study of low grade gold ore analysis; 25th Int. Geol. Congr., Sydney, Australia, Abstr., v. 2, p. 459.

Sokoloff, V.P.
1950: Dispersion of gold in rock and soil; Chem. Eng. Min. Rev., v. 43, no. 1, p. 7–10.

Sokolov, G.A., ed.
1974: Zoning of hydrothermal ore deposits; vols. 1 and 2, Izd. "Nauka" Publ. House, Moscow, U.S.S.R.

Solodilova, V.V.
1972: Geology of the northern Balkhash Rivers region. Secondary quartzites. Bosaga Massif; Geol. Metallog. Sev. Pribalkhash'ya, Sh. E. Esenov, ed., "Nauka" Kaz. SSR, Alma-Ata, U.S.S.R., p. 112–117. (Chem. Abstr., v. 81, 155905v.)

Sonstadt, E.
1872: On the presence of gold in sea water; Chem. News, v. 26, no. 671, p. 159–161.

Soper, E.K.
1911: The genesis of ore deposits—II; Eng. Min. J., v. 92, p. 947–949.

Sorby, H.C.
1858: On the microscopical structure of crystals, indicating the origin of minerals and rocks; Q. J. Geol. Soc. London, v. 14, pt. 1, p. 453–500.

Sorokin, V.N.
1971: Chemogenic and hydrothermal gold content of the Bakyrchik deposit (eastern Kazakhstan); Tr., Tsent. Nauch. Issled. Gornorazved. Inst., no. 96. (pt. I), p. 148–158. (Chem. Abstr., v. 77, 91151d.)
1973: On the probable form of gold transportation in hydrothermal solutions; Geokhim., no. 12, p. 1891–1894.

Souther, J.G.
1967: Cordilleran volcanic study; *in* Report of Activities, Pt. A, 1966, Geol. Surv. Can., Pap. 67-1, pt. A, p. 89–92.

Spevackova, V. and Krivanek, M.
1974: Determination of gold in platinum by neutron activation analysis; J. Radioanal. Chem., v. 21, no. 2, p. 485–488. (Chem. Abstr., v. 82, 79998f.)

Spiridonov, E.M.
1971: Gold and copper in the rocks of the Stepnyak and Bestyubinsk gold ore field; Geol. Geofiz., v. 141, no. 9, p. 124–130.

Spridonov, E.M., Sokolova, N.F. and Gapeev, A.K.
1974: Mineral associations of the Zhana-Tyube gold telluride ore deposit (northern Kazakhstan); Geol. Rud. Mestorozhd., v. 16, no. 1, p. 54–65.

Spurr, J.E.
1903: Gold in the dioritic rock from Mashonaland; Eng. Min. J., v. 76, p. 500.
1904: Native gold original in metamorphic gneisses; Eng. Min. J., v. 77, p. 198–199.
1905: Geology of the Tonopah mining district, Nevada; U.S. Geol. Surv., Prof. Pap. 42, 295 p.
1906: Ore deposits of the Silver Peak quadrangle, Nevada; U.S. Geol. Surv., Prof. Pap. 55, 174 p.
1923: The ore magmas—a series of essays on ore deposition (in 2 vols.); McGraw-Hill Book Co., Inc., New York, 915 p.

Squair, H.
1965: Geology and mineralization at the Tundra gold mine, Northwest Territories, Canada; unpubl. Ph.D. thesis, Univ. London, London, 323 p. and maps.

Srinivasan, B., Alexander, E.C. and Manuel, O.K.
1972: Te^{130}–Xe^{130} age determination of tellurium minerals; Econ. Geol., v. 67, p. 592–596.

Staines, H.R.E.
1953: Mount Morgan copper and gold mine; *in* Geology of Australian ore deposits, v. 1, 5th Emp. Min. Metall. Congr., Australia and New Zealand, Melbourne, Aust., p. 732–750.

Staff, Mount Morgan Limited
1965: Copper-gold ore deposit of Mount Morgan; *in* 8th Commonw. Min. Metall. Congr., Australia and New Zealand, v. 1, Geology of Australian ore deposits, 2nd ed., J. McAndrew, ed., p. 364–369.

Stanciu, C.
1973: Hydrothermal alteration of Neogene volcanic rocks from ore deposits in Gutîi Mountains (East Carpathians); Rev. Roum. Geol., Geophys., Geog., Ser. Geol., v. 17, no. 1, p. 43–62 (Engl.).

Stanikov, V.A.
1974: Role of early gossans in the formation of gold mineral placers; Rudonos. Kory Vyvetrivaniya, p. 192–196. (Chem. Abstr., v. 82, 127475g.)

Stanton, R.E. and McDonald, A.J.
1964: The determination of gold in soil with brilliant green; Analyst, v. 89, no. 1065, p. 767–770.

Stark, W.
1943: Über die Goldführung der Meere; Helv. Chim. Acta, v. 26, p. 424–441.

Steacy, H.R., Boyle, R.W., Charbonneau, B.W. and Grasty, R.L.
1973: Mineralogical notes on the uranium occurrences at South March and Eldorado, Ontario; Geol. Surv. Can., Pap. 73–1, pt. B, p. 103–105.

Steacy, H.R., Plant, A.G. and Boyle, R.W.
1974*a*: Notes on the association of brannerite and native gold at the Richardson mine, southeastern Ontario; Geol. Surv. Can., Pap. 74–1, pt. B., p. 175.
1974*b*: Brannerite associated with native gold at the Richardson mine, Ontario; Can. Mineral., v. 12, p. 360–363.

Steed, G.M., Annels, A.E., Shrestha, P.L. and Tater, P.S.
1976: Geochemical and biogeochemical prospecting in the area of the Ogofau gold mines, Dyfed, Wales; Inst. Min. Metall., Trans. v. 85, B109–B117.

Steelink, C.
1963: What is humic acid? J. Chem. Educ., v. 40, no. 7, p. 379–384.

Steiner, A.
1953: Hydrothermal rock alteration at Wairakei, New Zealand; Econ. Geol., v. 48, p. 1–13.
1970: Genesis of hydrothermal K-feldspar (adularia) in an active geothermal environment at Wairakei, New Zealand; Min. Mag., v. 37, no. 292, p. 916–922.

Stepanov, G.A.
1959: The distribution of native gold in the Berezovo gold deposits (Urals); Tr. Gorno-Geol. Inst., Akad. Nauk SSSR, Ural. Filial., no. 40, pt. 1, p. 159-163. (Chem. Abstr., v. 56, 11262c.)

Stephenson, J.F.
1971: Gold deposits of the Rice Lake-Beresford Lake greenstone belt, Southeastern Manitoba; *in* Geology and geophysics of the Rice Lake region, southeastern Manitoba (Project Pioneer); Man. Mines Br. Publ. 71-1, p. 337–374.

Stephenson, J.F. and Ehmann, W.D.
1971: Neutron activation analysis of gold in Archean igneous and metamorphic rocks of the Rice Lake–Beresford Lake area, southeastern Manitoba; Econ. Geol., v. 66, p. 933–939.

Steven, T.A.
1968: Ore deposits in the central San Juan Mountains, Colorado; *in* Ore deposits of the United States, Graton-Sales Vol., J.D. Ridge, ed., v. 1, Am. Inst. Min. Metall. Pet. Eng. Inc., New York, p. 706–713.
1969: Possible relation of mineralization to thermal springs in the Creede district, San Juan Mountains, Colorado: a discussion; Econ. Geol., v. 64, p. 696–698.

Steven, T.A. and Lipman, P.W.
1976: Calderas of the San Juan volcanic field, southwestern Colorado; U.S. Geol. Surv., Prof. Pap. 958, 35 p.

Stevens, D.N., Bloom, D.N. and Bisque, R.E.
1969: Evaluation of mercury vapour anomalies at Colorado Central Mines, Clear Creek County, Colorado; *in* Int. Geochem. Explor., Symp., F.C. Canney, ed., Colo. Sch. Mines, Q., v. 64, no. 1, p. 513–514.

Stevenson, I.M.
1959: Shubenacadie and Kennetcook map-areas, Colchester, Hants and Halifax counties, Nova Scotia; Geol. Surv. Can., Mem. 302, 88 p.

Stevenson, J.S.
1948*a*: Twin "J" mine; *in* Structural geology of Canadian ore deposits, v. 1, Can. Inst. Min. Metall., Montreal, p. 88–93.
1948*b*: Localization of ore in Zeballos area; *in* Structural geology of Canadian ore deposits, v. 1, Can. Inst. Min. Metall., Montreal, p. 78–85.

Stewart, G.O.M.
1975: The Jewel mine; Western Miner, v. 48, no. 9, p. 39–42.

Stillwell, F.L.
1918: Replacement in the Bendigo quartz veins and its relation to gold deposition; Econ. Geol., v. 13, p. 100–111.
1949: Occurrence of tellurides at Vatukoula, Fiji; Aust. Inst. Min. Metall., Proc., no. 154–155, p. 3–27.
1950: Origin of the Bendigo saddle reefs; Econ. Geol., v. 45, p. 697–701.
1952: The occurrence of gold antimonide in the antimony veins of Costerfield, Victoria; Aust. Inst. Min. Metall., Proc., nos. 164–165, p. 77–91.
1953*a*: Tellurides in western Australia; *in* Geology of Australian ore deposits, v. 1, A.B. Edwards, ed., 5th Emp. Min. Metall. Congr., Australia and New Zealand, p. 119–127.
1953*b*: Formation of Bendigo quartz reefs; *in* Geology of Australian ore deposits, v. 1, A.B. Edwards, ed., 5th Emp. Min. Metall. Congr., Australia and New Zealand, p. 1028–1033.

Stillwell, F.L. and Edwards, A.B.
1941: Coated gold from Cobar, New South Wales; Aust. Inst. Min. Metall., Proc., new ser., v. 121, p. 1–10.
1942: The mineral association of Tennant Creek gold ores; Aust. Inst. Min. Metall., Proc., new ser., no. 126, p. 139–144.
1946: An occurrence of sub-microscopic gold in the Dolphin East Lode, Fiji; Aust. Inst. Min. Metall., Proc., new ser., no. 141, p. 31–46.

Stiopol, V. and Draghici, I.
1973: Gold and silver distribution in the Bucium deposit (Metaliferi Mountains); Buchar. Univ., An. Ser. Geol., no. 22, p. 39–50 (Engl. summary).

Stockwell, C.H.
1948: Structural control of mineral deposits in southeastern Manitoba; *in* Structural geology of Canadian ore deposits, v. 1, Can. Inst. Min. Metall., Montreal, p. 306–314.
1970: Geology of the Canadian Shield; *in* Geology and economic minerals of Canada; Geol. Surv. Can., Econ. Geol. Rep. no. 1, R.J.W. Douglas, ed., p. 45–54.
1973: Revised Precambrian time scale for the Canadian Shield; Geol. Surv. Can. Pap. 72–52, 4p.

Stockwell, C.H. and Harrison, J.M.
1948: Structural control or ore deposits in northern Manitoba; *in* Structural geology of Canadian ore deposits, v. 1, Can. Inst. Min. Metall., Montreal, p. 284–291.

Stokes, H.N.
1906: Experiments on the solution, transportation and deposition of copper, silver, and gold; Econ. Geol., v. 1, p. 644–650.

Stoll, W.C.
1961: Tertiary channel gold deposits at Tipuani, Bolivia; Econ. Geol., v. 56, p. 1258–1264.
1962: Notes on the mineral resources of Ecuador; Econ. Geol., v. 57, p. 799–808.

Stone, J.B.
1937: The structural environment of the Bendigo goldfield; Econ. Geol., v. 32, p. 867–895.

Stone R.W.
1912: Coal near the Black Hills, Wyoming–South Dakota; U.S. Geol. Surv., Bull. 499, 66 p.

Stoyanka, B.-N. and Dancheva, R.
1972: Atomic-absorption method for determination of gold; Rudodobiv, no. 5, p. 25–28. (Chem. Abstr., v. 80, 55535k.)

Strasheim, A. and Jackson, P.F.S.
1971: Analytical survey of raw bullion from South African gold mines by mass spectrography; J.S.Afr. Inst. Min. Metall., v. 72, no. 5, p. 121–138. (Chem. Abstr., v. 76, 143385j.)

Streckeisen, A. and Ghitulescu, T.P.
1930: Les ressources d'or de Roumanie; *in* The gold resources of the world; 15th Int. Geol. Congr., Pretoria, 1929, p. 295–303.

Street, N. and Buchanan, A.S.
1956: The zeta-potential of kaolinite particles; Aust. J. Chem., v. 9, no. 4, p. 450–466.

Stringham, B.
1952: Fields of formation of some common hydrothermal-alteration minerals; Econ. Geol., v. 47, p. 661–664.

Ströhl, G.W., Von
1966: Der Einfluss von Salzböden und Goldbergwerken auf die chemische Wasserqualität des Vaal zwischen Vaalstaudamm und Kimberley; Gesundh. Ing., v. 87, no. 8, p. 230–235.

Strominger, D., Hollander, J.M. and Seaborg, G.T.
1958: Table of isotopes; Rev. Mod. Phys., v. 30, no. 2, pt. 2, p. 585–904.

Stumpfl, E.F.
1969: Determining fineness variation characteristics in gold ores by reflectometry; Econ. Geol., v. 64, p. 341–342.
1970: New electron probe and optical data on gold tellurides; Am. Mineral., v. 55, p. 808–814.
1974: The genesis of platinum deposits: further thoughts; Minerals Sci. Eng., v. 6, no. 3, p. 120–141.

Stumpfl, E.F. and Clark, A.M.
1965: Electron-probe microanalysis of gold-platinoid concentrates from southeast Borneo; Inst. Min. Metall., Trans., v. 74, p. 933–946.

Suess, H.E. and Urey, H.C.
1956: Abundances of the elements; Rev. Mod. Phys., v. 28, no. 1, p. 53–74.

Suggate, R.P.
1950: Quartzose coal measures of west Nelson and north Westland; N.Z., J. Sci. Technol., v. 31, p. 1–14.

Suleimenov, K.D.
1973: Distribution regularities of the gold ore mineralization in the eastern part of the Arganatinsk uplift; Akad. Nauk Kaz. SSR, Izv., Ser. Geol., v. 30, no. 6, p. 70–72. (Chem. Abstr., v. 81, 66416u.)

Suleimanova, N.T.
1974*a*: Distribution of gold and silver in rocks and minerals of the Almalyk syenite-diorite massif; Uzb. Geol. Zh., v. 18, no. 5, p. 31–33. (Chem. Abstr., v. 82, 127762y.)
1974*b*: Gold distribution in the Almalyk syenite-diorite massif (Dal'nee deposit); Zap. Uzb. Otd. Vses. Mineral. O-va, v. 27, p. 44–45. (Chem. Abstr., v. 83, 100868w.)

Sullivan, C.J.
1948: Ore and granitization; Econ. Geol., v. 43, p. 471–498.
1954: Metallic melting point and ore deposition; Econ. Geol. v. 49, p. 555–574.

Sullivan, C.J. and Ivanac, J.F.
1953: The Tennant Creek Goldfield; *in* Geology of Australian ore deposits; 5th Emp. Min. Metall. Congr., Australia and New Zealand, p. 305–316.

Sullivan, G.V., Browning, J.S. and Sanders, S.J.
1969: Recovering gold from a graphitic schist from Tallapoosa County, Ala.; U.S. Bur. Mines, Rep. Invest. 7251, 11 p.

Sultanmuratov, Sh., Gamaleev, I.E. and Dadadzkanov, B.
1974: Case of the occurrence together of gold and cinnabar in a region of the western continuation of the southern-Fergana mercury-antimony belt; Uzb. Geol. Zh., v. 18, no. 2, p. 88–91. (Chem. Abstr., v. 82, 5941k.)

Summers, R.
1969: Ancient mining in Rhodesia; Nat. Mus. Rhodesia, Mem. 3, Salisbury, 236 p.

Sundell, I.G.
1936: Fineness and composition of alluvial gold from the Ivalojoki, Finnish Lapland; Finl., Comm. Geol., Bull. no. 115, p. 115–160.

Surgai, V.T.
1973: Regional geochemical conditions of the distribution of gold in the Tien Shan; Akad. Nauk Kirg. SSR, Geol. Inst., Izd. "Ilim", Frunze, 76 p.

Sutherland, C.H.V.
1969: Gold; its beauty power and allure, 2nd rev. ed.; McGraw-Hill, Inc., New York, 196 p.

Sutherland Brown, A.
1957: Geology of the Antler Creek area, Cariboo district, British Columbia; B.C. Dep. Mines, Bull No. 38, 105 p.

Sutherland Brown, A. and Holland, S.S.
1957: The northeastern part of the Cariboo district; *in* Structural geology of Canadian ore deposits, v. 2, Can. Inst. Min. Metall., Montreal. p. 78–84.

Sutton, A.L., Havens, R.G. and Sainsbury, C.L.
1973: A spectrochemical method for determining the composition of native gold; U.S. Geol. Surv., J. Res., v. 1, no. 3, p. 301–307.

Svanholm, J.
1975: Gold in Honduras; World Min., v. 28, p. 30–31, 61.

Sviridoff, A.P.
1923: Present status of gold mining in eastern Siberia; Eng. Min. J., v. 116, p. 621–628.

Svistunov, N.V.
1970: Mechanism of the dissolution of gold in chlorine water; Publ. Higher Educ. Inst., Metallurgy of non-ferrous metals; v. 13, no. 5, p. 69–71.

Swaine, D.J.
1955: The trace-element content of soils; Commonw. Bur. Soil Sci. (Gt. Brit.), Tech. Comm. 48, 157 p.
1962: The trace-element content of fertilizers; Commonw. Bur. Soils, Harpenden; Tech. Comm. 52, 306 p.

Swanson, E.A. and Brown, R.L.
1962: Geology of the Buchans orebodies; Can. Inst. Min. Metall., Trans., v. 65, p. 288–296.

Swiegers, J.U.
1940: Gold, carbon, pyrite and other sulphides in the Black Reef; Geol. Soc. S. Afr., Trans., v. 42, p. 35–45.

Sykora, L.
1959: Plants in geological research; Czech. Acad. Sci., Prague, 322 p.

Syrovatskii, V.V., Bal'ter, B.L. and Manaeva, A.V.
1969: Authigenic gold intergrown with osmium-iridium and platinum in Kuznetsk Alatau placer deposits; Zap. Vses. Mineral. Obshchest., v. 98, no. 3, p. 338–340. (Chem. Abstr., v. 71, 83412a.)

Syrovatskii, V.V. and Zimoglyadov, B.N.
1969: Relation of gold placer deposits to gabbro-dunite-pyroxenite intrusive formations in western Siberia; Geol. Rossypei Yuga Zapad. Sib., V.A. Kryzhanovskii, ed., Izd. "Nauka", Moscow, U.S.S.R., p. 155-159. (Chem. Abstr., v. 72, 45984b.)

Szádeczky-Kardoss, E.
1958: Hydrated ionic radii and hydrothermal ore genesis from the point of view of the geochemical potentials; Acta Geol., Acad. Sci. Hung., Tomus 5, Fasc. 3–4; p. 351–357.

Szeky-Fux, V.
1964: Propylitization and potassium metasomatism; Acta Geol. (Hungary), v. 8, nos. 1–4, p. 97–117.

Taber, S.
1913: Geology of gold belt in the James River Basin, Virginia; Va., Geol. Surv., Bull., no. 7, 271 p.
1948: Gold crystals from the southern Appalachians; Am. Mineral., v. 33, p. 482–488.

Taguchi, S. and Watanabe, T.
1973: Clay minerals associated with gold ores from the Fuke mine, Kagoshima Prefecture, with special reference to chlorite-saponite interstratified minerals; Kyushu Daigaku Rigakubu Kenkyu Hokoku, Chishitsugaku, v. 11, no. 2, p. 243–250. (Chem. Abstr., v. 84, 138387x.)

Takashima, K. and Igarashi, T.
1973: Metallogenic map of Japan; Geol. Surv. Japan, 1:2,000,000 map ser. no. 14.

Takenouchi, S.
1962: Study of temperatures of mineral formation in hydrothermal ore deposits by the liquid inclusion methods; Mining Geol. (Japan), v. 12, no. 55, p. 282–293.

Takeuchi, T. and Abe, H.
1967: Alteration processes of volcanic rocks under hydrothermal conditions (hydrothermal studies on the wall-rock alteration I); Sci. Rep., Tohoku Univ., Ser. 3, v. 10, no. 1, Sendai, Japan, p. 151–172.

Talipov, R.M.
1968: Biogeochemical prerequisites in prospecting for gold mineralization under desert conditions; *in* Rud. Form. Osn. Cherty Metallogen. Zolota USSR, p. 355–358. (Chem. Abstr., v. 73, 47379c.)
1972: Distribution characteristics of copper, gold, arsenic, and silver in soils of gold ore occurrences in central Kyzyl-Kum; Zap. Uzb. Otd. Vses. Mineral. Obshchest. v. 25, p. 203–205. (Chem. Abstr., v. 78, 32562m.)

Talipov, R.M., Adylov, B.B. and Enikeev, N.I.
1972: Geochemistry of gold, silver, and arsenic in subsurface waters of central Kyzyl-Kum; Uzb. Geol. Zh., v. 16, no. 4, p. 59–62. (Chem. Abstr., v. 78, 162326y.)

Talipov, R.M., Aripova, Kh., Karabaev, K.K., Khatamov, Sh. and Akhunkhodzhaeva, N.
1968: Possible use of arsenic in biogeochemical prospecting for gold-ore deposits (as illustrated by Tamdytau); Uzb. Geol. Zh., v. 12, no. 5, p. 43–47. (Chem. Abstr., v. 71, 5374a.)

Talipov, R.M., Aripova, Kh., Karabaev, K.K. and Yusupov, P.G.
1969: Results of biogeochemical surveys in central Kyzyl-Kum; Biogeokhim. Poiski Rud. Mestorozhd., p. 177-182. (Chem. Abstr., v. 75, 90154d.)

Talipov, R.M., Glushchenko, V.M. and Lezhneva, N.D.
1975: Correlation of gold content in plants and waters of several ore fields in the Karaminskie Mountains; Uzb. Geol. Zh., v. 19, no. 4, p. 21–26. (Chem. Abstr., v. 84, 33795s.)

Talipov, R.M., Glushchenko, V.M., Nishanov, P. Kh. and Samigdzhanova, M.
1974: Distribution of gold and its accessory elements in subsurface waters of Almalyk; Zap. Uzb. Otd. Vses. Mineral. O-va, v. 27, p. 49-50. (Chem. Abstr., v. 83, 118941z.)
1976: Some principles of gold distribution and trace elements accompanying it in plants of the Almalyk ore region; Akad. Nauk. Uzb. SSR, Dokl., v. 31, no. 1, p. 53–54. (Chem Abstr., v. 85, 35551n.)

Talipov, R.M., Glushchenko, V.M., Tverskaya, K.L. and Nishanov, P.
1976: Some characteristics of gold and antimony distribution in plants in ore deposits of the Chatkal-Kurama region; Uzb. Geol. Zh., no. 3, p. 65–69. (Chem. Abstr., v. 85, 119581k.)

Talipov, R.M., Karabaev, K.K. and Akhunkhodzhaeva, N.
1971: Results of biogeochemical studies in northern Tamdytau (western Uzbekistan); Zap. Uzb. Otd. Vses. Mineral. Obshchest., no. 24, p. 197–203. (Chem. Abstr., v. 77, 37514m.)

Talipov. R.M. and Khotamov, Sh.
1973: Biogeochemical studies in the northern part of the central Kyzyl-Kum; Uzb. Geol. Zh., v. 17, no. 6, p. 26–31. (Chem. Abstr., v. 80, 132019z.)
1974: Distribution of trace elements in plants of the Tamdynsk Mountains (central Kyzyl-Kum); Uzb. Geol. Zh, v. 18, no. 1, p. 23–27.

Tananaeff, N.A.
1925: Über die Gleichgewichtsreaktion, $Fe^{2+} + Ag^{+} \leftrightharpoons Fe^{3+} + Ag$; Z. Phys. Chem., v. 114, p. 49–58.

Taneda, S. and Mukaiyama, H.
1970: Gold deposits and Quaternary volcanoes in Southern Kyushu; Guidebook 11, Excursion B8, Int. Assoc. Genesis Ore Deposits (IAGOD), Tokyo-Kyoto Meet., 31 p.

Tarr, W.A.
1910: Copper in the "red beds" of Oklahoma; Econ. Geol., v. 5, p. 221–226.

Tatarinov, P.M., ed.
1971: Map of useful minerals of the world's continents (1: 15,000,000 scale); All Union Sci. Res. Geol. Inst. (VSEGEI), Ministry Geol., U.S.S.R.

Tatsch, J.H.
1973: Mineral deposits; Tatsch Associates, Sudbury, Mass., 264 p.
1975: Gold deposits, origin, evolution and present characteristics; Tatsch Associates, Sudbury, Mass., 320 p.

Tatsumi, T., ed.
1970: Volcanism and ore genesis; Univ. Tokyo Press, Tokyo, 420 p.

Tatsumi, T. and Watanabe, T.
1971: Geological environment of formation of the Kurokô-type deposits; Soc. Min. Geol. Japan, Spec. Iss. 3, p. 216–220.

Taylor, C.M. and Radtke, A.S.
1967: New occurrences and data on nolanite; Am. Mineral., v. 52, p. 734–743.

Taylor, F.C. and Schiller, E.A.
1966: Metamorphism of the Meguma Group of Nova Scotia; Can. J. Earth Sci., v. 3, p. 959–974.

Taylor, H.K.
1964–1965: The mode of origin of banket orebodies, discussion; Inst. Min. Metall., Trans., v. 74, no. 705, p. 801.

Taylor, H.P.
1963: Importance of chalcophile element abundances in determining the sequence of sulfide mineral deposition from monoascendent ore-forming solutions; *in* Symposium Problems of Post Magmatic Ore Deposition, v. 1, J. Kutina, ed., Czech., Acad. Sci., Prague, p. 267–272.
1973: O^{18}/O^{16} evidence for meteoric-hydrothermal alteration and ore deposition in the Tonopah, Comstock Lode, and Goldfield Mining districts, Nevada; Econ. Geol., v. 68, p. 747–764.

1974: The application of oxygen and hydrogen isotope studies to problems of hydrothermal alteration and ore deposition; Econ. Geol., v. 69, p. 843–883.
Taylor, R.B.
1948: Delnite mine; *in* Structural geology of Canadian ore deposits, v. 1, Can. Inst. Min. Metall., Montreal, p. 504–507.
Taylor, S.R.
1964: Abundance of chemical elements in the continental crust; a new table; Geochim. Cosmochim. Acta, v. 28, p. 1273–1285.
Taylor, S.R. and Sachs, M.
1960: Trace elements in australites; Nature, v. 188, p. 387–388.
Tayurskii, D.N.
1973: Evaluation of placer gold occurrence according to native sources; Landshafty Yuga Dal'nego Vostoka, p. 123–125. (Chem. Abstr., v. 83, 63513d.)
Tegengren, F.R.
1963: A historical review of gold in the Philippines; Philipp. J. Sci. v. 92, no. 4, p. 551–600.
Tenyakov, V.A., Rakovskiy, E. Ye. and Filippova, T.P.
1970: Abundance and some general geochemical features of gold in bauxite (as shown by neutron activation analysis); Akad. Nauk. SSSR, Dokl., v. 195, p. 194–197.
Terziev, G.
1966: Kostovite, a gold-copper telluride from Bulgaria; Am. Mineral., v. 51, p. 29–36.
Theobald, P.K.
1957: The gold pan as a quantitative geologic tool; U.S. Geol. Surv., Bull. 1071-A, 54 p.
Thomas, B.I.
1973: Gold-lode deposits, Fairbanks mining district, central Alaska; U.S. Nat. Tech. Inform. Serv., PB Rep. 1973, no. 222610/8, 20 p. (Chem. Abstr., v. 80, 50450u.)
Thomas, D.E.
1953: Mineralization and its relationship to the geological structure of Victoria; *in* Geology of Australian ore deposits, v. 1, A.B. Edwards ed.; 5th Emp. Min. Metall. Congr., Australia and New Zealand, p. 971–985.
Thomas, F.L.
1937: Marlu gold-mining areas; Mining Mag., v. 56, p. 329–341.
Thomas, G.
1954: Electrode potentials and the dissolution of gold; Can., Dep. Mines Tech. Surv., Mines Br., Tech. Pap., no. 9, 14 p.
Thomas, T.M.
1961: The mineral wealth of Wales and its exploitation; Oliver and Boyd, Edinburgh, p. 209–213.
Thompson, C.E., Nakagawa, H.M. and Van Sickle, G.H.
1968: Rapid analysis for gold in geologic materials; U.S. Geol. Surv., Prof. Pap. 600-B, p. 130–132.
Thompson, R.M.
1949: The telluride minerals and their occurrence in Canada; Am. Mineral., v. 34, p. 342–382.
Thomson, E.
1922: The occurrence of tellurides in Ontario; Univ. Toronto Studies, Geol. Ser. no. 14, p. 91–98.
1928: A new telluride occurrence in Quebec; Univ. Toronto Studies, Geol. Ser. no. 27, p. 11–14.
Thomson, J.E.
1936: Gold deposits of the belt extending from Manitou Lake to Lake of the Woods, Ontario; Can. Inst. Min. Metall., Trans., v. 39, p. 686–701.
1938: Structure of gold deposits in the Crow River area, Ontario; Can. Inst. Min. Metall., Trans., v. 41, p. 358–374.
1939*a*: The Crow River area; Ont. Dep. Mines, Annu. Rep., 1938, v. 47, pt. 3, p. 1–65.
1939*b*: The Uchi Lake area; Ont. Dep. Mines, Annu. Rep., 1938, v. 47, pt 3, p. 68–82.
1948*a*: Regional structure of the Kirkland Lake–Larder Lake area; *in* Structural geology of Canadian ore deposits, v. 1, Can. Inst. Min. Metall., Montreal, p. 627–632.
1948*b*: Omega mine; *in* Structural geology of Canadian ore deposits, v. 1, Can. Inst. Min. Metall., Montreal, p. 658–662.
1950: Geology of Teck Township and the Kenogami Lake area, Kirkland Lake gold belt; Ont. Dep. Mines, 57th Ann. Rep., 1948, v. 57, pt. 5, p. 1–53.
Thomson, J.E. *et al.*
1950: Geology of the main ore zone at Kirkland Lake; Ont. Dep. Mines, 57th Ann., Rep., 1948, v. 57, pt. 5, p. 54–188.
Thorne, W.E.
1926: Testing and estimating alluvials; Mining Publ. Ltd., London, 52 p.
Thorpe, R.I.
1967: Controls of hypogene sulphide zoning, Rossland, British Columbia; unpubl. Ph.D. thesis, Univ. Wisconsin, Madison, Wis., 131 p.
Thorpe, R.I. and Little, H.W.
1973: The age of sulfide mineralization at Rossland, British Columbia; Econ. Geol., v. 68, p. 1337–1340.
Thyssen-Bornemisza, S.V.
1942: Geochemische und pflanzenbiologische Zusammenhänge im Luchte der angewandten Geophysik; Beitr. Geophys., v. 10, p. 35–84 (Engl. summary).
Tihor, L.A. and Crocket, J.H.
1976: Origin and distribution of gold-bearing carbonate zones of the Kirkland Lake–Larder Lake area, Ontario; Geol. Surv. Can., Pap. 76-1A, p. 407–408.
Tilling, R.I., Gottfried, D. and Rowe, J.J.
1973: Gold abundance in igneous rocks: bearing on gold mineralization; Econ. Geol., v. 68, p. 168–186.
Time-Life Books
1974: The Forty-niners; text by W.W. Johnson, Time-Life Books, New York, 240 p.
Timofeev, V.I.
1969: Placer deposits. Western Yakutia gold placer deposits; Str. Zemnoi Kory Yakutii Zakonomer. Rozmeshcheniya Polez. Iskop., Yu.P. Ivenisen, ed., p. 330–339. (Chem. Abstr., v. 73, 27542x.)
Timofeeva, T.S.
1972: Mineralogical-geochemical characteristics of some gold ore occurrences in Central Asia; Geol., Petrologiya, Mineral. Endogennykh Mestorozhd. Srednei Azii, L.A. Bykov, ed., "Nedra", Moscow, U.S.S.R., p. 42–47. (Chem. Abstr., v. 80, 29156t.)
Timofeevskii, D.A.
1971: Formation, classification, mineral types, and gold mineral associations of the gold ore deposits of the U.S.S.R.; Tr., Tsent. Nauch. Issled. Gornorazved. Inst., no. 96, pt. 1, p. 5–32. (Chem. Abstr., v. 77, 64681v.)
Timofeevskii, D.A. and Adrianova, S.I.
1974: On the physico-chemical conditions of gold deposit formation as distinct from ore formation of the U.S.S.R; Abstracts of Papers, Int. Assoc. Genesis Ore Deposits (IAGOD), 4th Symp., Varna, Bulgaria, p. 265–266.
Tischendorf, G.
1959: Zur Genesis einiger Selenidvorkommen, insbesondere von Tilkerode im Harz; Freiberg. Forschungsh., C69, p. 1–168.
1970: The influence of the country rock on the formation of the selenide paragenesis at Tilkerode, Harz Mountains; *in* Problems of hydrothermal ore deposition, Z. Pouba and M. Stemprok, eds., Int. Union Geol. Sci., Ser. A, no. 2, p. 316–321.
Tishchenko, E.I. and Tishchenko, M.D.
1974: Coefficient of flatness of gold in placers; Razved. Okhr. Nedr., no. 3, p. 52–54. (Chem. Abstr., v. 81, 66397p.)
Titley, S.R. and Hicks, C.L., eds.
1966: Geology of the porphyry copper deposits, southwestern North America; Univ. Arizona Press, Tucson, Arizona, 287 p.
Tolbert, G.E.
1964: Geology of the Raposos gold mine, Minas Gerais, Brazil; Econ. Geol., v. 59, 775–798.

Tollon, F.
1970: Le district aurifère de Salsigne—Aude; Société des Mines et Produits Chimiques de Salsigne (Aude), JF—Editions-Impressions, 14, rue Idrac, Toulouse, 176 p.

Tolmachev, I.I., Kalinin, S.K., Terekhovich, S.L., Syromyatnikov, N.G. and Zaravnyayeva, V.D.
1971: Distribution of Pd, Pt and Au in the Riphaean and Cambrian igneous rocks of the Boshchekul'sk region, northeastern Kazakhstan; Geokhim., no. 3, p. 325–331. *Also* Geochem. Int., v. 8, no. 2, p. 194–199.

Tolman, C.
1936: Lake Etchemin map-area, Quebec; Geol. Surv. Can., Mem. 199, 20 p.

Tomich, S.A.
1974: A new look at Kalgoorlie Golden Mile geology; Proc. Australas., Inst. Min. Metall., no. 251, p. 35.
1976: Further thoughts on the application of the volcanogenic theory to the Golden Mile ores at Kalgoorlie; Proc. Australas., Inst. Min. Metall., no. 258, p. 19–29.

Tooker, E.W.
1963: Altered wall rocks in the central part of the Front Range mineral belt, Gilpin and Clear Creek counties, Colorado; U.S. Geol. Surv., Prof. Pap. 439, 102 p.

Topley, W.
1887: Gold and silver: their geological distribution and their probable future production; Br. Assoc. Advanced Sci., Rep., p. 510–535.

Tourtelot, H.A. and Riley, L.B.
1973: Size and shape of gold and platinum grains; *in* Ore in sediments, 8th Int. Sedimentol. Congr. 1971, p. 307–319. (Chem. Abstr., v. 79, 138557z.)

Townsley, B.F.
1935: Mine-Finders, the history and romance of Canadian mineral discoveries; Saturday Night Press, Toronto, 246 p.

Traill, R.J.
1954: A preliminary account of the mineralogy of radioactive conglomerates in the Blind River region, Ontario; Can. Min. J., v. 75, p. 63–68.
1963: A rhombohedral polytype of molybdenite; Can. Mineral., v. 7, pt. 3, p. 524–526.
1970: A catalogue of Canadian minerals; Geol. Surv. Can., Pap. 69-45, 649 p.

Travin, Yu. A.
1972: Structural features of alluvial placer deposits of recent river valleys in the northeastern U.S.S.R.; Tr. Sev. Vost. Kompleks, Inst., Dal'nevost. Tsent. Aka, v. 44, p. 118–125. (Chem. Abstr., v. 81, 155817t.)

Travis, G.A.
1966: Geostatistics of primary gold deposits and a study of the gold and telluride mineralization at Kalgoorlie, Western Australia; D.I.C. thesis, Imperial College Sci. Technol., London.

Travis, G.A. and Woodall, R.
1975: A new look at Kalgoorlie Golden Mile geology; Proc. Australas. Inst. Min. Metall., no. 256, p. 33–40.

Travis, G.A., Woodall, R. and Bartram, G.D.
1971: The geology of the Kalgoorlie goldfield; Geol. Soc. Aust., Spec. Publ. no. 3, J.E. Glover, ed., p. 175–190.

Trdlicka, Zd. and Zyka, V.
1964: A contribution to the geochemistry of gold; Sb. Ved. Pr. Vys. Sk. Bánské Ostrave Rocnik X., Cislo 1–2, Rada Hornicko-Geologická, Clánek c. 93, p. 11–12.

Tremblay, L.P.
1966: Contwoyto Lake map-area, District of Mackenzie; Geol. Surv. Can., Pap. 65–21, 17 p.

Trenholme, L.S.
1948: Belleterre mine; *in* Structural geology of Canadian ore deposits, v. 1, Can. Inst. Min. Metall., Montreal, p. 796–803.

Trenina, T.I. and Shumilov, Yu.V.
1970: Natural gold amalgam in some placer deposits of the Bilibinsk area; Kolyma, no. 2, p. 40–41. (Chem. Abstr., v. 74, 44165q.)

Trofimov, V.S.
1964: Basic factors controlling the formation and location of mineral placers; Lithol. Miner. Resour., no. 6, p. 5–18.
1969: The origin of gold in certain ancient conglomerates; *in* The problem of the occurrence of metals in ancient conglomerates in the U.S.S.R.; "Nauka" Publ. House, Moscow-Leningrad, p. 58–63.
1971: Two varieties of economic mineral placers of littoral-marine origin; Litol. Polez. Iskop., no. 6, p. 44–52. (Chem. Abstr., v. 76, 88378g.)

Troshin, Yu.P., Mikheeva, Z.I. and Alekseeva, N.N.
1961: Distribution of gold, silver, tellurium, and selenium in the sulfide minerals of Transbaikalian hydrothermal deposits of diverse origin; Zap. Zabaikal. Filiala Geogr. Obschchest. SSSR, no. 36, p. 97–102. (Chem. Abstr., v. 77, 103957m.)

Tsonev, D.
1972: Gold and mercury from the Radka deposit; God. Sofii. Univ., Geol. Geogr. Fak. 1972–1973, v. 65, p. 157–165. (Chem. Abstr., v. 79, 106740.)

Tsukagoshi, S., Hashmito, M. and Hibi, F.
1974: Gold-silver ore deposit of the Tochibora Mine, Kamioka, Gifu Prefecture; Kozan Chishitsu, v. 24, no. 2, p. 111–118. (Chem. Abstr., v. 81, 124237z.)

Tsypukov, Yu.P.
1976: The early Mesozoic gabbro-granite association and problems of gold deposits in the northwest Khentay (Mongolia); Ezheg. Inst. Geokhim. Sib. Otd., Akad. Nauk SSSR, 1974, p. 86–91.

Tsypukov, Yu.P. and Lombo, D.
1973: On the goldbearing mineralized zones in the northwest Khentay (Mongolia); Ezheg. Inst. Geokhim. Sib. Otd., Akad. Nauk SSSR, 1972, L.V. Tauson, ed., p. 260.
1974: Gold in the magmatic rocks of the northwestern Khentei in the Mongolian People's Republic; Ezheg. Inst. Geokhim. Sib. Otd., Akad. Nauk SSSR p. 75–78. (Chem. Abstr., v. 84, 47528c.)

Tuck, R.
1968: Origin of the bedrock values of placer deposits; Econ. Geol., v. 63, p. 191–193.

Tully, D.W.
1963: The geology of the Upper Canada mine; Can. Inst. Min. Metall., Bull., v. 56, no. 609, p. 24–34.

Tunell, G.
1954: The crystal structures of the gold-silver tellurides (final report); Office of Naval Research, Contract N6 onr 275, Task Order 10, Research Project NR-081-105, Univ. California, Dep. Geol., 68 p.

Tung, Li and Chi-Lung, Yio
1966: The abundance of chemical elements in the earth's crust and its major tectonic units; Sci. Sinica, v. 15, no. 2, p. 258–272.

Tunyan, G.A.
1974: Conditions for formation of Tei-Lichkvaz gold-ore mineralization; Zap. Arm. Otd. Vses. Mineral. O-va, v. 6, p. 13–21. (Chem. Abstr., v. 83, 63526k.)

Turekian, K.K. and Kharkar, D.P.
1970: Neutron activation analysis of milligram quantities of Apollo 11 lunar rocks and soil; Proc., Apollo 11 Lunar Sci. Conf., Geochim. Cosmochim. Acta, Suppl. 1, v. 2, p. 1659–1664.

Turkevich, J., Stevenson, P.C. and Hillier, J.
1951: A study of the nucleation and growth processes in the synthesis of colloidal gold; Disc. Faraday Soc., no. 11, p. 55–75.

Turneaure, F.S.
1955: Metallogenetic provinces and epochs; Econ. Geol., 50th Anniv. Vol., pt. 1, p. 38–98.

Turner, H.W.
1895: Auriferous gravels of the Sierra Nevada; Am. Geol., v. 15, p. 371–379.
1903: The Cretaceous auriferous conglomerate of the Cottonwood Mining District, Siskiyou County, California, Eng. Min. J., v. 76, p. 653–654.

Turner, J.K.
1915: Mining in the shale at Goldfield; Min. Sci. Press, v. 110, p. 995–996.

Turner, S.
1933: Economic aspects of gold and silver; U.S. Bur. Mines, Inform. Circ. 6740, 13 p.

Turovskii, S.D., Makarov, V.A. and Nosyrev, I.V.
1962: Discovery of ore gravel in the Lower Carboniferous conglomerates of the Boordu region (northern Tien-Shan); Akad. Nauk SSSR, Dokl., v. 147, p. 210–211. (Chem. Abstr., v. 58, 5389f.)

Tvalchrelidze, T.A.
1959: Gold-bismuth-tellurium associations in the Damblud and Zod deposits; Geol. Sb. Kavkaz. Inst. Miner. Syr'ya, p. 65–70. (Chem. Abstr., v. 55, 24404f.)
1961: Rare-metal mineralization in the copper pyrrhotite formation on the southern slope of the Caucasus Mountains; Tr. Kavkaz. Inst. Miner. Syr'ya, Min. Geol. Okhr. Nedr. SSSR, no. 3, p. 15–18. (Chem. Abstr., v. 60, 3882b.)

Tylecote, R.F.
1962: Metallurgy in archaeology; Edward Arnold Ltd., London, 368 p.

Tyrrell, J.B.
1912: The law of the pay-streak in placer deposits; Inst. Min. Metall., Trans., v. 21, p. 593–613.
1915: Gold on the North Saskatchewan River; Can. Min. Inst., Trans., v. 18, p. 160–173.

Tyurin, N.G.
1963: On the composition of hydrothermal solutions; Geol. Rud. Mestorozhd., no. 4, p. 24–42.
1965*a*: On finely dispersed gold in pyrite; Geol. Rud. Mestorozhd. v. 7, no. 5, p. 70–75. (Engl. rev. *in* Econ. Geol., 1968, v. 63, p. 853.)
1965*b*: Time of precipitation of gold from metal-bearing solutions; Akad. Nauk SSSR, Izv., Ser. Geol., no. 8, p. 40–44.

Tyurin, N.G. and Kakovskii, I.A.
1960: The behaviour of gold and silver in the zone of oxidation of sulphide deposits; News of higher educational institutions; Metallurgy of non-ferrous metals, no. 2, p. 6–13 (in Russian).

Udodov, P.A. and Shvartsev, S.L.
1965: Regular characteristics of trace components migration in subsurface waters of the supergene zone; *in* Proc., Inter-Univ. Conf. on the Hydrogeochemistry and Paleo-Hydrogeology Methods of Prospecting for Deposits of Useful Minerals, Tomsk, Siberia, p. 39–43.

Uglow, W.L. and Osborne, F.F.
1926: A gold-cobaltite-lodestone deposit, British Columbia, with notes on the occurrence of cobaltite; Econ. Geol., v. 21, p. 285–293.

Ukhanov, A.V. and Pchelintseva, N.F.
1972: Gold contents of peridotite and eclogite xenoliths from a kimberlite pipe; Geokhim., no. 2, p. 247. *Also* Geochem. Int., v. 9, no. 1, p. 157.

Uklonskii, A.S.
1970: On the geochemistry of gold; Zap. Uzb. Otdel. Vses. Mineral Obshchest. Acad. Sci., Uzb. SSR, Tashkent, no. 23, p. 127–131.

Umitbayev, R.B., Sindeyev, A.S. and Zhupakhin, E.N.
1968: Prospecting for epithermal deposits of gold in north-eastern U.S.S.R.; Int. Geol. Rev., v. 10, no. 7, p. 745–748.

Umpleby, J.G.
1910: Geology and ore deposits of Republic mining district; Wash. Geol. Surv., Bull. no. 1, 65 p.

U.N.E.S.C.O.
1970: Proceedings of the second seminar on geochemical prospecting methods and techniques (with special reference to their applicability in tropical humid zones); UNESCO Miner. Resour. Dev. Ser., no. 38, 413 p.

United Nations
1972: Mineral resources of the Lower Mekong Basin and adjacent areas of Khmer Republic, Laos, Thailand and Republic of Viet-Nam; Miner. Resour. Ser., no. 39, 148 p.

Urashima, Y.
1956: "Bosa" quartz veins, especially the fine-grained quartz aggregates, of the Konomai mine in Hokkaido, Japan; J. Fac. Sci. Hokkaido Univ., Ser. 4, Geol. Mineral., v. 9, no. 3, p. 371–380.

Urashima, Y. and Sato, J.
1968: Gold-bearing ores and native gold in the Shakanai Kurokô deposits of Japan; Min. Geol. (Japan), v. 81, p. 11–24.

U.S. Bureau Mines, Field Staff
1967: Production potential of known gold deposits in the United States; U.S. Bur. Mines, Inform. Circ. 8331, 24 p.

U.S. Geological Survey
1968: U.S. Geological Survey heavy metals program, progress report 1966–67; U.S. Geol. Surv., Circ. 560, 24 p.
1969*a*: U.S. Geological Survey heavy metals program, progress report 1968—field studies; U.S. Geol. Surv., Circ. 621, 34 p.
1969*b*: U.S. Geological Survey heavy metals program, progress report 1968—topical studies; U.S. Geol. Surv., Circ. 622, 19 p.
1969*c*: Geochemistry of water; gold in hot spring deposits, Wyoming and Nevada; U.S. Geol. Surv., Prof. Pap. 650-A, p. A121.

Usenko, I.S., Kravchenko, G.L. and Sakhatskii, I.I.
1973: Distribution of gold in ferruginous-siliceous and some other crystalline rocks of the Sea of Azov region; Geol. Zh., v. 33, no. 5, p. 60–68. (Chem. Abstr., v. 80, 110902e.)

Utting, E.P.
1953: The Hannan's North gold mine; *in* Geology of Australian ore deposits, v. 1, A.B. Edwards, ed., 5th Emp. Min. Metall. Congr., Australia and New Zealand, p. 112–118.

Uvad'ev, M.D., Gertman, Yu.L. and Markina, V.M.
1973: Nature of metasomatic processes in the Kaul'dy gold ore deposit (Uzbek S.S.R.); Zap. Uzb. Otd. Vses. Mineral. Obshchest., no. 26, p. 10–12. (Chem. Abstr., v. 80, 62209k.)

Uvarov, V.V., D'yachkov, B.A., Burmistrov, V.R., Stepanov, V.M. and Madiyanov, T.
1972: Geochemistry of gold in the magmatic complexes of the Kalba-Narymsk region (eastern Kazakhstan); Geokhim., no. 9, p. 1130–1132. *Also* Geochem. Int., v. 9, no. 5, p. 774–776.

Uytenbogaardt, W. and Burke, E.A.J.
1971: Tables for microscopic identification of ore minerals, 2nd ed.; Elsevier Publ. Co., Amsterdam, 430 p.

Vakrushev V.A.
1972: Mineralogy, geochemistry and origin of deposits of the gold-bearing skarn type; "Nauka" Publ. House, Siberian Div., Novosibirsk, U.S.S.R., 237 p.

Vakhrushev, V.A. and Dorosh, V.M.
1966: Selenium and tellurium distribution in sulfide ore deposits of the Altai-Sayan region; Geokhim., no. 11, p. 1349–1354. *Also* Geochem. Int., v. 3, no. 6, p. 1076–1081.

Vakhrushev, V.A., Ryabov, V.V. and Tsimbalist, V.G.
1972: Gold in some sulfide minerals in the Noril'sk region; Geochem. Int., v. 9, no. 3, p. 474–477.

Vakhrushev, V.A. and Tsimbalist, V.G.
1967: Distribution of gold in the sulfides of the Altai-Sayan skarn deposits; Geokhim., no. 10, p. 1076–1081. *Also* Geochem. Int., v. 4, no. 5, p. 972–977.
1972: Gold in terrestrial ultramafic rocks; Geokhim., no. 9, p. 1128–1130. *Also* Geochem. Int., v. 9, no. 5, p. 772–773.

Vakhrushev, V.A., Tychinskii, A.A. and Tsimbalist, V.G.
1971: Gold content of sulfides in complex ore deposits of eastern Transbaikalia; Geol. Geofiz., no. 12, p. 130–133. (Chem. Abstr., v. 76, 156670y.)
1973: Gold potential of sulfides in some polymetallic deposits of eastern Transbaikal; Int. Geol. Rev., v. 15, no. 7, p. 778–780.

Valpeter, A.P. and Davidenko, N.M.
1970: Criteria of placer deposit relation to parental rocks; Probl. Geol. Rossypei, Soveshch., (Dokl.) 3rd 1969, p. 116–124. (Chem. Abstr., v. 75, 38853s.)

Van Aubel, R.
1934: Géochimie de l'or; Ann. Soc. Geol. Belg., t. 57, p. B131–B150.
1936: Origin of graphite in the auriferous veins of Kivu; C.R. Soc. Géol. Fr., p. 236–237. (Chem. Abstr., v. 31, 342.)
1939: Sur l'importance dans les minerals d'or—du calibre des particules; Ann. Mines (Paris), 13e Sér., t. 16, p. 155–161.

Van Bemmelen, R.W.
1970: The geology of Indonesia, v. 2, Economic Geology; Martinus Nijhoff, The Hague, 267 p.

Van Eeden, O.R., Partridge, F.C., Kent, L.E. and Brandt, J.W.
1939: The mineral deposits of the Murchison Range east of Leydsdorp; Union S. Afr., Dep. Mines, Geol. Surv. Div., Mem. 36, 163 p.

Van Hise, C.R.
1904: A treatise on metamorphism; U.S. Geol. Surv., Monogr., v. 47, 1286 p.

Van Sickle, G.H. and Lakin, H.W.
1968: An atomic-absorption method for the determination of gold in large samples of geologic materials; U.S. Geol. Surv., Circ. 561, 4 p.

Van Tassell, R.E.
1969: Exploration by overburden drilling at Keno Hill Mines, Limited; Colo. Sch. Mines, Q., v. 64, no. 1, p. 457–478.

Vanderbilt, A.T.
1888: Gold; Swan Sonnenschein and Co., Paternoster Square, London, 155 p.

Vanderburg, W.O.
1936: Placer mining in Nevada; Univ. Nevada Bull., v. 30, no. 4, 178 p.

Varley, T.
1922: Bureau of mines investigates gold in oil shales and its possible recovery; U.S. Bur. Mines., Rep. Invest., Ser. no. 2413, 10 p.

Vasil'ev, B.D.
1970: The distribution of gold in the skarns of the Natal'evka deposits; Izv. Tomsk. Politekh. Inst., v. 239, p. 216–219.

Vasilevskii, M.M.
1970: Correlation of propylitization and mineralization; Probl. Metasomatizma, Tr. Konf. Okolorundnomu Metasomatizmu 2nd, 1966, p. 115–121. (Chem. Abstr., v. 74, 144521w.)
1972: Propylites and ore formation; Problemy Metasomatizma. (Tr. II Konf. Okolorundnomu Metasomatizmu) Trans. All-Union Sci. Res. Geol. Inst. (VSEGEI), p. 115–121.

Vasudev, V.N. and Naganna, C.
1973: Mineragraphy of gold-quartz-sulphide reefs of Hutti gold mines, Raichur District, Mysore State; J. Geol. Soc. India, v. 14, no. 4, p. 378–383.

Vasyunina, E.D.
1963: Some regularities in the relation of placer and rock gold; Sb. Nauch. Tr. Chelyab. Nauch. Issled. Inst. Gorn. Delta, no. 2, p. 296–304. (Chem. Abstr., v. 61, 5372b.)

Vaughan, B.E.
1975: Suspended particle interactions and uptake in terrestrial plants; Rep. 1974, BNWL-SA-5132, 22 p. (Chem. Abstr., v. 82, 151435q.)

Veizer, J. and Demovic, R.
1969: Geochemistry of sedimentary carbonate rocks (mantle series of the high Tatra mountains and middle Triassic of the Slovak karst region); Geol. Zb., Geol. Carpathica, v. 20, no. 2, p. 303–332.

Vemban, N.A., Subramanian, K.S. and Sundaram, S.M.
1972: Gold occurrences near Kotagiri, Nilgiri district, Tamil Nadu; Indian Miner., v. 26, no. 2, p. 37–41.

Vennor, H.G.
1870: Geological Survey of Canada, Report of Progress from 1866–1869, p. 165–171.

Verhoogen, J.
1938: Thermodynamical calculation of the solubility of some important sulphides, up to 400°C; Econ. Geol., v. 33, p. 34–51.

Vertushkov, G.N., Kainov, V.I. and Chesnokov, B.V.
1972: Cosalite from the Berezovsk and Kochkarsk gold ore deposits; Tr. Sverdlovsk. Gorn. Inst., v. 86, p. 106–109. (Chem. Abstr., v. 83, 13371k.)

Vertushkov, G.N., Sokolov, Yu. A. and Popov, V.A.
1970: Vein quartz of some gold deposits of the Urals; Acad. Sci. USSR Ural's Br.; Minerals of magmatic and metamorphic rocks of the Urals. Contrib. Geol. Geochim. Inst., no. 86, Sverdlovsk, U.S.S.R., p. 38–43.

Vetluzhskikh, V.G. and Frolov, V.I.
1974: Nature of the presence of gold and causes of coal metamorphism in the Chul'man depression; Akad. Nauk SSSR, Dokl., v. 217, no. 2, p. 409–411. (Chem. Abstr., v. 81, 138626n.)

Vikhter, B.Ya., Razumova, R.V., Khol', F.I. and Kurbanov, A.Sh.
1968: The form of gold in pyrite deposits; Acad. Sci. USSR, Dokl., Earth Sci. Sec., v. 178, p. 152 (Am. Geol. Inst. transl.).

Viktorov, S.V.
1955: The use of the geobotanical method in geological and hydrogeological investigations; Acad. Sci. USSR, Moscow (Engl. transl. *in* U.S. Atom. Energy Comm., AEC-tr-3968, 219 p.).

Vilar, P.
1975: A history of gold and money, 1450–1920; New Left Books, London, 360 p.

Viljoen, E.A.
1971: An electron-microprobe analysis of gold in the Witwatersrand banket and in ores from the Barberton Mountain Land; Natl. Inst. Metal., S. Afr., Rep., no. 1361, 7 p.

Viljoen, R.P., Saager, R. and Viljoen, M.J.
1969: Metallogenesis and ore control in the Steynsdorp Goldfield, Barberton Mountain Land, South Africa; Econ. Geol., v. 64, p. 778–797.
1970: Some thoughts on the origin and processes responsible for the concentration of gold in the Early Precambrian of southern Africa; Miner. Deposita (Berl.), v. 5, p. 164–180.

Vilor, N.V. and Kazmin, L.A.
1973: Investigations of the complexing of the chloride hydrothermal system "gold and quartz"; Ezheg. Inst. Geokhim., Sib. Otd., Akad. Nauk SSSR, 1972, L.V. Tauson, ed., p. 360–364.

Vilor, N.V. and Sarapulova, V.N.
1971: Migration capacity of gold during the endogenous process; Tr. Soveshch. Eksp. Tekh. Mineral. Petrogr., 8th 1968, V.V. Lapin, ed., "Nauka", Moscow, U.S.S.R., p. 90–91. (Chem. Abstr., v. 76, 102139p.)

Vilor, N.V. and Shkarupa, T.A.
1971: Fine-disperse gold dissolution in hydrothermal solutions at high temperatures and pressures; Int. Geochem. Congr., Abstracts of Reports I, Moscow, U.S.S.R., p. 280–281.

Vincent, E.A. and Smales, A.A.
1956: The determination of palladium and gold in igneous rocks by radioactivation analysis; Geochim. Cosmochim. Acta, v. 6, p. 154–160.

Vincent, E.A. and Crocket, J.H.
1960: Studies in the geochemistry of gold—I. The distribution of gold in rocks and minerals of the Skaergaard intrusion, east Greenland. II. The gold content of some basic and ultrabasic rocks and stone meteorites; Geochim. Cosmochim. Acta, v. 18, nos. 1–2, p. 130–148.

Vinogradov, A.P.
1953: The elementary chemical composition of marine organisms; Engl. transl. by Sears Found. Mar. Res., Mem. 2, 647 p.
1956: Regularity of distribution of chemical elements in the earth's crust; Geochemistry, no. 1, p. 1–43.
1959: The geochemistry of rare and dispersed chemical elements in soils, 2nd ed.; Consultants Bureau Inc., New York, 209 p.
1962: Average contents of chemical elements in the principal types of igneous rocks of the earth's crust; Geochemistry, no. 7, p. 641–664.
1973: Preliminary data on lunar soil collected by the Luna 20 unmanned spacecraft; Geochim. Cosmochim. Acta, v. 37, no. 4, p. 721–729.

Vinogradov, A.P., Lavrukhina, A.K., Ganiyev, A.G., Sil'vanovich, Yu.A. and Rakhimov, Kh.R.
1972: Partition of platinum metals and gold between phases in meteoric material, pt. 1; Geochem. Int., v. 9, no. 6, p. 967–974.
1973: Distribution of platinoids and gold between various phases of the meteoritic matter, pt. 2; Geokhim., no. 7, p. 963–974.

Vinogradov, V.I.
1957: Solubility of the secondary molybdenum minerals in weak solutions of H_2SO_4 and Na_2CO_3; Geochemistry, no. 3, p. 279–286.

Vishnevsky, A.S.
1958: New data on the distribution of indium in minerals of the oxidized zone; Geochemistry, no. 7, p. 851–855.

Visnapuu, A., Marek, B.C. and Jensen, J.W.
1971: Dissociation and vaporization of gold chlorides and gold bromides; U.S. Bur. Mines, Rep. Invest. RI 7513, 22 p.

Visser, D.J.L., ed.
1968: The symposium on the Rhodesian basement complex; Geol. Soc. S. Afr. (Annex.), v. 71, 279 p.

Vivaldi, J.L.M., Sierra, J. and Leal, G.
1971: Some aspects of the mineralization and wall-rock alteration in the Rodalquilar goldfield, S.E. Spain; Soc. Min. Geol. Japan, Spec. Iss. 2, p. 145–152.

Vlasov, K.A., ed.
1966–1968: Geochemistry and mineralogy of rare elements and genetic types of their deposits; Vols. I, II, and III, Isr. Progr. Sci. Transl., Jerusalem.

Vlasov, K.A., Kuz'menko, M.Z. and Es'kova, E.M.
1966: The Lovozero alkali massif; Oliver & Boyd, Edinburgh and London, 627 p.

Vlasov, A.S. and Prigorovskii, V.I.
1971: Classification of permafrost alluvial gold deposits and unified methods of drilling prospecting; Tr. Vses. Nauch. Issled. Inst. Zolota Redk. Metal., no. 31, p. 3–29. (Chem. Abstr., v. 80, 23209j.)

Vobecký, M., Frána, J., Řanda, Z., Benada, J. and Kunciř, J.
1971: Analytical possibilities of reactor neutron activation method to nondestructive analysis of meteorites; Radiochem. Radioanal. Lett., v. 6, no. 4, p. 237–247.

Voin, M.I.
1971: Near-surface primary geochemical halos over the Khakandzhin gold deposit and their use in detection of blind ore bodies; *in* Nauch. Osn. Geokhim. Metod. Poiskov Glubokozalegayushchikh Rud. Mestorozhd., Simp. 1970, v. 2, V.V. Polikarpochkin, ed., p. 60–68. (Chem. Abstr., v. 78, 32513w.)

Voin, M.I. and Kazak, A.P.
1973: Zonality of rare-metal and gold ore mineralization in the Kumak-Kotansunskoi zone of warping of the Orsk Transurals; Tr. Vses. Nauch. Issled. Geol. Inst., no. 185, p. 102–111. (Chem. Abstr., v. 82, 46387c.)

Voinkov, D.M., Grinenko, L.N. and Davidenko, N.M.
1976: On the sources of the gold ore manifestations in the Chukchi folded region (from sulfur isotopic data); Izv. Vyssh. Ucheb. Zaved. Geol. Razved. v. 19, no. 2, p. 74–80. (Chem. Abstr. v. 85, 8327n.)

Voitovich, V.S. and Khenkina, S.B.
1971: Structural characteristics and gold-bearing features of the Voitsk and Shuezersk deposits in east Karelia; Geol. Rud. Mestorozhd., v. 13, no. 1, p. 102–105.

Vokes, F.M.
1958: Metallogenic provinces and epochs in Norway; Saertrykk av Tidsskrift for Kjemi, Bergvesen Og Metallurgi 4, p. 47–55.

Volarovich, G.P.
1972: Prospects for revealing subsurface gold-ore deposits related to volcanogenic formations; Sov. Geol., no. 11, p. 74–85.

Volarovich, G.P., Andrusenko, N.I. and Shchepot'ev, Yu. M.
1974: Conditions of formation of near surface gold and mercury mineralization in Kamchatka; Abstracts of Papers, Int. Assoc. Genesis Ore Deposits (IAGOD), 4th Symp., Varna, Bulgaria, p. 423–424.

Volarovich, G.P. and Mikhailova, M.S.
1973: Metasomatism and gold mineralization in young volcanic belts in the eastern U.S.S.R.; Evol. Vulkanizma Istor. Zemli, Mater. Vses. Paleovulkan, p. 277–278. (Chem. Abstr., v. 84, 20362g.)

Volarovich, G.P. and Shilin, N.G.
1971: Gold behaviour during crystallization of granitoid magma on the example of the Miocene gabbro-granodiorite formation in central Kamchatka; Int. Geochem. Congr., Moscow, U.S.S.R., Abstracts of Reports II, p. 664–665.

Volarovich, G.P. and Shokhor, F.A.
1970: Relation of gold placer deposits with parental sources in the Far East; Probl. Geol. Rossypei, Soveshch. (Dokl.) 3rd 1969, V.I. Smirnov, ed., Sev. Vost. Kompleks. Inst., Magadan, U.S.S.R., p. 147–154. (Chem. Abstr., v. 75, 38849v.)

Volchenko, Yu.A.
1970: Composition of platinoids and gold in the Gusevogorsk gabbro-pyroxenite massif (central Urals); Ezheg. Inst. Geol. Geokhim. Akad. Nauk SSSR, Ural. Filial 1969, p. 188–192. (Chem. Abstr., v. 78, 162055j.)

Vol'fson, N.B., Chembarisov, Sh. A. and Chistyi, B.I.
1962: Geochemical and geophysical methods for prospecting for gold ore mineralization in Uzbekistan; Byull. Nauch. Tekh. Inform. Min. Geol. Okhr. Nedr. SSSR, no. 1, p. 52–56. (Chem. Abstr., v. 60, 3881c.)

Volkov, G.A. and Shakhbazova, L.V.
1973: Data on the geochemistry of gold in natural waters of the Beskes ore field; Izv. Vyssh. Ucheb. Zaved. Geol. Razved., v. 16, no. 9, p. 92–98. (Chem. Abstr., v. 80, 72894u.)

Volodin, V.F.
1966: Precambrian gold-bearing conglomerates of U.S.S.R.; Int. Geol. Rev., v. 8, no. 7, p. 874–880.

Von Cotta, B.
1870: A treatise on ore deposits; D. Van Nostrand, New York, 574 p.

Von Palfy, M.
1911–1912: Geologische Verhältnisse und Erzgänge der Bergbaue des Siebenbürgischen Erzgebirges; Mitt. Jahrb. Kgl. Ung. Geol., Bd. 18, p. 229–526.

Von Rahden, H.V.R.
1965: Apparent fineness values of gold from two Witwatersrand gold mines; Econ. Geol., v. 60, p. 980–997.

Von Rahden, H.V.R. and Urli, G.L.P.
1969: The association of gold and uranium with other minerals in different Witwatersrand reefs; Natl. Inst. Metal., S. Afr., Res. Rep. no. 837, 40 p.

Von Veimarn, P.
1913: Precipitation of gold in orebodies; Z. Chem. Ind. Kolloide, v. 11, p. 287. *Also* Sci. Press, 1913, v. 107, p. 309.

Vorob'ev, V.P. and Chernysheva, V.S.
1972: Mineralogical study of the heavy fractions of marine sediments from the shelves of the Sea of Japan; Izv. Vyssh. Ucheb. Zaved., Geol. Razved., v. 15, no. 1, p. 42–50. (Chem. Abstr., v. 76, 88679f.)

Vorob'ev, V.P. and Kolesov, S.V.
1975: Stream transport of placer gold; Lithol. Miner. Resour., v. 10, no. 3, p. 378–380 (Transl. Consultants Bureau, New York).

Vorob'yev, V.P. and Krapivner, R.B.
1975: Preliminary classification of coastal-marine placers; Int. Geol. Rev., v. 17, no. 7, p. 823–829.

Voronin, D.V. and Goldberg, I.S.
1972: Electrochemical processes in placer deposits of native metals; Akad. Nauk SSSR, Dokl., v. 207, no. 1, p. 190–192.

Vorotnikov, V.A., Nikolaeva, N.M. and Pirozhkov, A.V.
1973: Forms of deposition of gold in waters of the southern Enisei region; Geol. Geofiz., no. 11, p. 37–42. (Chem. Abstr., v. 80, 147749m.)

Voskresenskaya, N.T.
1967: Geochemical behaviour of thallium and rubidium in geological processes and their petrogenetic significance; *in* Chemistry of the Earth's Crust, vol. 2, A.P. Vinogradov, ed., Isr. Progr. Sci. Transl., Jerusalem, p. 401–415.

Voskresenskaya, N.T. and Zvereva, N.F
1968: Geochemistry of gold and its distribution in the magmatic complexes of northern Kazakhstan; Geokim., no. 4, p. 422–431. *Also* Geochem. Int., v. 5, no. 2, p. 373–381.

Voskresenskaya, N.T., Zvereva, N.F., Konkina, O.M. and Feldman, V.I.
1970: Behaviour of gold in the process of basic magma differentiation; Geokhim., no. 12, p. 1438–1445.

Vsevolozhskaya, M.A.
1966: Use of hydrochemical methods of prospecting in areas with insular distribution of permafrost; Merzolotnye Issled., Geol. Fak., Kafedra Merzlotoved. Mosk. Univ., Sb. Statei no. 6, p. 175–183. (Chem. Abstr., v. 66, 12933g.)

Wagner, P.A.
1926: Occurrence of platinum metals in South Africa; Econ. Geol., v. 21, p. 109–134, 243–270.
1927: A new gold discovery in the Rustenburg district, Transvaal; Geol., Soc. S. Afr., Trans., v. 29, p. 95–108.

Wagoner, L.
1901: The detection and estimation of small quantities of gold and silver; Am. Inst. Min. Eng., Trans., v. 31, p. 798–810.
1908: The presence of gold and silver in deep-sea dredgings; Am. Inst. Min. Eng., Trans., v. 38, p. 704–705.

Wakabayashi, K., Sukeshita, M. and Uemura, K.
1973: Geology and exploration of Kushikino mine, Kagoshima Prefecture; Min. Geol. (Japan), v. 23, p. 190 (abstr.).

Walker, A.L. and Buchanan, A.S.
1969: Geochemical processes in ore formation, Pt. 1. The production of hydrothermal fluids from sedimentary sequences; Econ. Geol., v. 64, p. 919–922.

Walker, T.L.
1921: "Allemontite" from Atlin, B.C.; Am. Mineral., v. 6, no. 6, p. 97–99.

Wang, Y.
1973*a*: Wall rock alteration of Late Cenozoic mineral deposits in Taiwan: Geologic settings and field relations; Geol. Soc. China, Proc., no. 16, p. 145–160.
1973*b*: Wall rock alteration of Late Cenozoic mineral deposits in Taiwan. Mineralogical and physicochemical aspects; Acta Geol. Taiwan., no. 16, p. 1–29. (Chem. Abstr., v. 80, 62157s.)

Wänke, H., Baddenhausen, H., Dreibus, G., Jagoutz, E., Kruse, H., Palme, H., Spettel, B. and Teschke, F.
1973: Multielement analyses of Apollo 15, 16, and 17 samples and the bulk composition of the moon; *in* Proc. 4th Lunar Sci. Conf., Geochim. Cosmochim. Acta, Suppl. 4, v. 2, p. 1461–1481.

Wänke, H., Begemann, F., Vilcsek, E., Rieder, R., Teschke, F., Born, W., Quijano-Rico, M., Voshage, H. and Wlotzke, F.
1970: Major and trace elements and cosmic-ray produced radioisotopes in lunar samples; Science, v. 167, no. 3918, p. 523–525.

Wänke, H., Palme, H., Baddenhausen, H., Dreibus, G., Jagoutz, E., Kruse, H., Spettel, B., Teschke, F. and Thacker, R.
1974: Chemistry of Apollo 16 and 17 samples: bulk composition, late stage accumulation and early differentiation of the moon; *in* Proc., 5th Lunar Sci. Conf., Geochim. Cosmochim. Acta, Suppl. 5, v. 2, p. 1307–1335.

Wänke, H., Wlotzka, F., Baddenhausen, H., Balacescu, A., Spettel, B., Teschke, F., Jagoutz, E., Kruse, H., Quijano-Rico, M. and Rieder, R.
1971: Apollo 12 samples: chemical composition and its relation to sample locations and exposure ages, the two component origin of the various soil samples and studies on lunar metallic particles; Proc., 2nd Lunar Sci. Conf., Geochim. Cosmochim. Acta, Suppl. 2, v. 2, p. 1187–1208.

Wanless, R.K., Boyle, R.W. and Lowdon, J.A.
1960: Sulfur isotope investigation of the gold-quartz deposits of the Yellowknife district; Econ. Geol., v. 55, p. 1591–1621.

Ward, F.N., ed.
1975: New and refined methods of trace analysis useful in geochemical exploration; U.S. Geol. Surv., Bull. 1408, 105 p.

Ward, H.J.
1958: Albite porphyries as a guide to gold ore; Econ. Geol., v. 53, p. 754–756.

Ward, J.H.W.
1968: Geological aspects of the Patchway gold mine, Rhodesia; Geol. Soc. S. Afr. (Annex.), v. 71, p. 147–158.

Ward, W. *et al.*
1948: The gold mines of Kirkland Lake; *in* Structural geology of Canadian ore deposits, v. 1, Can. Inst. Min. Metall., Montreal, p. 644–653.

Warren, H.V.
1936: A gold-bismuth occurrence in British Columbia; Econ. Geol., v. 31, p. 205–211.

Warren, H.V. and Cummings, J.M.
1937*a*: The relationship between gold and metallic minerals in British Columbia; Can. Inst. Min. Metall., Trans., v. 40, p. 1–4.
1937*b*: Textural relations in gold ores of British Columbia; Am. Inst. Min. Metall. Eng., Tech. Publ. no. 777, 15 p.

Warren, H.V. and Delavault, R.E.
1950: Gold and silver content of some trees and horsetails in British Columbia; Geol. Soc. Am., Bull., v. 61, pt. 1, p. 123–128.
1959: Pathfinding elements in geochemical prospecting; Symp. Explor. Geoquim., v. 2, 20th Int. Geol. Congr., Mexico, 1956, p. 255–260.

Warren, H.V., Delavault, R.E. and Barakso, J.
1964: The role of arsenic as a pathfinder in biogeochemical prospecting; Econ. Geol., v. 59, p. 1381–1385.
1968: The arsenic content of Douglas fir as a guide to some gold, silver and base metal deposits; Can. Inst. Min. Metall., Bull., v. 61, no. 675, p. 860–866.

Warren, H.V. and Hajek, J.H.
1973: An attempt to discover a "Carlin-Cortez"; Western Miner, v. 46, no. 10, p. 124–134.

Warren, H.V. and Matthews, W.H.
1942: Some notes on a gold-tungsten occurrence in the Bridge River district; Miner, v. 15, no. 5, p. 31–34.

Warren, H.V. and Thompson, R.M.
1944: Minor elements in gold; Econ. Geol., v. 39, p. 457–471.

Warren, R.G.
1972: A commentary on the metallogenic map of Australia and Papua, New Guinea; Aust. Bur. Miner. Resour., Geol. Geophys., Bull. 145, 83 p.

Washington, R.A. and Holman, R.H.C.
1966: A rapid and sensitive method for determining gold in rocks and other geological materials; Geol. Surv. Can., Pap. 65-7, 18 p.

Wasson, J.T. and Baedecker, P.A.
1970: Ga, Ge, In, Ir and Au in lunar, terrestrial and meteoritic basalts; *in* Proc., Apollo 11 Lunar Sci. Conf., Geochim Cosmochim. Acta, Suppl. 1, v. 2, p. 1741–1750.

Watanabe, M.
1952: Modes of occurrence of tellurium-bearing minerals in Japan; Tohoku Univ., Sci. Reps. ser. 3, v. 4, no. 2, p. 45–80.

Watanabe, T.
1943: Geology and mineralization of the Suian district, Tyosen (Korea); J. Fac. Sci., Hokkaido Univ., ser. 4, Geol. Min. v. 6, nos. 3–4, p. 205–303.

Watkins, T.H.
1971: Gold and silver in the west; American West Publ. Co., Palo Alto, Calif., 287 p.

Watson, J.P.
1970: Contribution of termites to development of zinc anomaly in Kalahari sand; Inst. Min. Metall., Trans., v. 79, sec. B, p. 53–59.
1972: The distribution of gold in termite mounds and soils at a gold anomaly in Kalahari sand; Soil Sci., v. 113, no. 5, p. 317–321.

Watson, J.V.
1973: Influence of crustal evolution on ore deposition; Inst. Min. Metall., Trans., v. 82, sec. B, p. 107–113.

Watterson, J.R., Gott, G.B., Neuerburg, G.J., Lakin, H.W. and Cathrall, J.B.
1976: Tellurium dispersion patterns as guides to mineral deposits; 25th Int. Geol. Congr., Abstr., Sydney, Australia, v. 2, p. 462.

Way, H.J.R.
1965: The mode of origin of banket orebodies, discussion; Inst. Min. Metall., Trans., v. 74, no. 704, p. 571–574.

Weatherbe, D.
1907: Dredging for gold in California; Mining Scientific Press, San Francisco, 217 p.

Webb, J.S.
1958: Observations on geochemical exploration in tropical terrains; Symp. Explor. Geoquim., v. 1, 20th Int. Geol. Congr., Mexico, 1956, p. 143–173.

Webber, B.N.
1935: Bajada placers of the arid southwest; Am. Inst. Min. Metall., Eng. Tech. Publ., no. 588, 16 p.

Weber, W. and Stephenson, J.F.
1973: The content of mercury and gold in some Archean rocks of the Rice Lake area; Econ. Geol., v. 68, p. 401–407.

Wedepohl, K.H.
1971: Geochemistry; Holt, Rinehart and Winston, Inc., New York, 231 p.

Weed, W.H.
1901: The enrichment of gold and silver veins; Am. Inst. Min. Eng., Trans., v. 30, p. 424–448.
1902: The enrichment of gold and silver veins; *in* Genesis of ore deposits, by Franz Pošepný, 2nd ed., Am. Inst. Min. Eng., New York, p. 473–497.
1904*a*: Original native gold in igneous rocks; Eng. Min. J., v. 77, p. 440–441.
1904*b*: Free gold in basic igneous rocks in southeastern Russia; Eng. Min. J., v. 77, p. 522–523.
1905: Economic value of hot springs and hot-spring deposits; U.S. Geol. Surv., Bull. 260, p. 598–604.

Weeks, M.E.
1968: Discovery of the elements, 7th ed.; publ. by Journal Chemical Education, Easton, Pa., 806 p.

Weil, G.L. and Davidson, I.
1970: The gold war; Secker and Warburg, London, 245 p.

Weinig, A.J.
1960: Gold forms in ores; Mines Mag., v. 50, p. 27–32.

Weiser, H.B.
1933: Inorganic colloid chemistry, v. 1, The colloidal elements; John Wiley & Sons, Inc., New York, 389 p.
1935: Inorganic colloid chemistry, v. 2, The hydrous oxides and hydroxides; John Wiley & Sons, Inc., New York, 429 p.

Weiss, H.V. and Lai, M.G.
1963: Cocrystallization of ultramicro quantities of elements with 2-mercaptobenzimidazole: determination of gold in sea water; Anal. Chim. Acta, v. 28, p. 242–248.

Weissberg, B.G.
1969: Gold-silver ore-grade precipitates from New Zealand thermal waters; Econ. Geol., v. 64, p. 85–108.
1970: Solubility of gold in hydrothermal alkaline sulfide solutions; Econ. Geol., v. 65, p. 551–556.

Weissberg, B.G. and Wodzicki, A.
1970: Geochemistry of hydrothermal alteration and origin of sulphide mineralization at the Tui mine, Te Aroha, New Zealand; N.Z. J. Sci., v. 13, no. 1, p. 36–60.

Wells, J.D.
1971: Association of gold and arsenic with pyrite, Cortez and Carlin mines, Nevada; Econ. Geol., v. 66, p. 1270–1271 (abstr.).

Wells, J.D. and Elliott, J.E.
1971: Geochemical reconnaissance of the Cortez-Buckhorn area, southern Cortez Mountains, Nevada; U.S. Geol. Surv., Bull. 1312-P, 18 p.

Wells, J.D. and Mullens, T.E.
1973: Gold-bearing arsenian pyrite determined by microprobe analysis, Cortez and Carlin gold mines; Econ. Geol., v. 68, p. 187–201.

Wells, J.D., Stoiser, L.R., and Elliott, J.E.
1969: Geology and geochemistry of the Cortez gold deposit, Nevada; Econ. Geol., v. 64, p. 526–537.

Wells, J.H.
1973: Placer examination-Principles and Practice; U.S. Dep. Interior, Bur. Land Management, Tech. Bull. 4, 204 p.

Wennervirta, H., Bölviken, B. and Nilsson, C.A.
1971: Summary of research and development in geochemical exploration in Scandinavian countries; *in* Geochemical exploration, R.W. Boyle and J.I. McGerrigle, eds.; Can. Inst. Min. Metall., Spec. Vol. 11, p. 11–14.

Wernecke, L.
1932: Alaska Juneau—geology of the ore zones; Eng. Min. J., v. 133, p. 493–499.

West, J.M.
1971: How to mine and prospect for placer gold; U.S. Bur. Mines, Inform. Circ. 8517, 43 p.
1975: Gold; U.S. Dep. Interior, Bur. Mines, Miner. Yearb., v. 1, 1974, p. 603–626.

West, W.F.
1973: Geochemical field assaying made easy; Rhodesia, Chamber Mines J., v. 15, no. 2, p. 33–35.

White, C.E., Ross, R.H. and Campbell, N.
1949: The Con-Rycon Mine, Yellowknife, N.W.T.; Can. Inst. Min. Metall., Trans., v. 52, p. 133–147.

White, D.E.
1955: Thermal springs and epithermal ore deposits; Econ. Geol., 50th Anniv. Vol., pt. 1, p. 99–154.
1974: Diverse origins of hydrothermal ore fluids; Econ. Geol., v. 69, p. 954–973.

White, D.E., Anderson, E.T. and Grubbs, D.K.
1963*a*: Geothermal brine well: mile deep hole may tap ore-bearing magmatic water and rocks undergoing metamorphism; Science, v. 139, p. 919–922.

White, D.E., Hem, J.D. and Waring, G.A.
1963*b*: Chemical composition of sub-surface waters; *in* Data of geochemistry, 6th ed., U.S. Geol. Surv., Prof. Pap. 440-F, 67 p.

White, D.E. and Waring, G.A.
1963: Volcanic emanations; *in* Data of geochemistry, 6th ed., U.S. Geol. Surv., Prof. Pap. 440-K, 29 p.

White, M.G.
1956: Uranium in the Serra de Jacobina State of Bahia, Brazil; *in* Proc. U.N. Int. Conf. Peaceful Uses of Atomic Energy, Geneva, v. 6, p. 140–142.
1961: Origin of uranium and gold in the quartzite-conglomerate of the Serra de Jacobina, Brazil; U.S. Geol. Surv., Prof. Pap. 424-B, p. 88–89.

White, W.H.
1943: The mechanism and environment of gold deposition in veins; Econ. Geol. v. 38, p. 512–532.

Whitebread, D.H. and Hoover, D.B.
1968: Preliminary results of geological, geochemical, and geophysical studies in part of the Virginia City quadrangle, Nevada; U.S. Geol. Surv., Circ. 596, 19 p.

Whiting, F.B.
1959: Structural belts and mineral deposits of northwestern Argentina; Econ. Geol., v. 54, p. 903–912.

Whitmore, D.R.E., Berry, L.G. and Hawley, J.E.
1946: Chrome micas; Am. Mineral., v. 31, p. 1–21.

Whittaker, E.J.W. and Muntus, R.
1970: Ionic radii for use in geochemistry; Geochim. Cosmochim. Acta, v. 34, p. 945–956.

Whittle, A.W.G.
1965: The scope for mineral exploration in New Zealand; Mining Chem. Eng. Rev. (Australia), v. 57, no. 2, p. 19–23.

Wicks, C.E. and Block, F.E.
1963: Thermodynamic properties of 65 elements—their oxides, halides, carbides and nitrides; U.S. Bur. Mines, Bull. 605.

Wiebols, J.H.
1955: A suggested glacial origin for the Witwatersrand Conglomerates; Geol. Soc. S. Afr., Trans., v. 58, p. 367–382.

Willden, R. and Hotz, P.E.
1955: A gold-scheelite-cinnabar placer in Humboldt County, Nevada; Econ. Geol., v. 50, p. 661–668.

Williams, G.J.
1974: Economic geology of New Zealand; Australas. Inst. Min. Metall. Monog. Ser. 4, 2nd Ed., 490 p.

Williams, M.
1918: Absorption of gold from colloidal solutions by fungi; Ann. Bot., v. 32, p. 532–534.

Williams, S.A.
1972: Embreyite, a new mineral from Berezov, Siberia; Mineral. Mag., v. 38, p. 790–793.

Williamson, H.C. and Barr, D.J.
1965: Gold mineralization in the Yilgarn goldfield; *in* Geology of Australian ore deposits, 2nd ed., J. McAndrew, ed., 8th Commonw. Min. Metall. Congr., p. 87–94.

Willimott, C.W.
1884–1885: Minerals of the Ottawa district; Ottawa Field-Naturalists' Club, Trans., no. 6, vol. 2, no. 2, p. 189–210.

Willis, J.L.
1973: Mining history of gold deposits of the Far South Coast, New South Wales; N.S.W., Geol. Surv., Bull., v. 24, p. 1–117. (Chem. Abstr., v. 82, 114276t.)

Wilson, E.D. and Fansett, G.R.
1961: Arizona gold placers; small scale gold placering; *in* Gold placers and placering in Arizona, Ariz., Bur. Mines, Bull. 168, p. 11–117.

Wilson, H.D.B.
1944: Geochemical studies of the epithermal deposits at Goldfield, Nevada; Econ. Geol., v. 39, p. 37–55.

Wilson, H.S.
1948: Lamaque mine; *in* Structural geology of Canadian ore deposits, v. 1, Can. Inst. Min. Metall., Montreal, p. 882–891.

Wilson, I.R.
1971: Progress report on a study on geochemical variations in wall rock, Ashanti gold mine, Obuasi, Ghana; Univ. Leeds, Dep. Earth Sci., 15th Annu. Rep., Res. Inst. Afr. Geol., p. 39–43.
1972: Carbonaceous matter in wall rock, Ashanti gold mine, Obuasi, Ghana; Univ. Leeds Res. Inst. Afr. Geol. Dep. Earth Sci. Annu. Rep. Sci. Results, no. 16, p. 33–34.

Wilson, M.E.
1948*a*: Structural features of the Noranda-Rouyn area; *in* Structural geology of Canadian ore deposits, v. 1, Can. Inst. Min. Metall., Montreal, p. 672–683.
1948*b*: McWatters mine; *in* Structural geology of Canadian ore deposits, v. 1, Can. Inst. Min. Metall., Montreal, p. 783–789.
1965: The Deloro stock and its mineralized aureole; Econ. Geol., v. 60, p. 163–167.

Wilson, M.E., Hopper, R.V. and Trenholme, L.S.
1948: Rouyn Merger mine; *in* Structural geology of Canadian ore deposits, v. 1, Can. Inst. Min. Metall., Montreal, p. 789–796.

Wilson, M.E. and Lee, A.C.
1948: Senator-Rouyn mine; *in* Structural geology of Canadian ore deposits, v. 1, Can. Inst. Min. Metall., Montreal, p. 735–739.

Wilson, M.E. and McQuarry, W.R.
1948: Stadacona mine; *in* Structural geology of Canadian ore deposits, v. 1, Can. Inst. Min. Metall., Montreal, p. 776–782.

Wilson, N.W.
1948: Iran and its mineral resources; Mining Mag., v. 74, p. 277–289.

Wise, E.M., ed.
1964: Gold: recovery, properties, and applications; D. Van-Nostrand Co. Inc., Princeton, New York, 367 p.

Wisser, E.
1941: Albite and gold; Econ. Geol., v. 36, p. 658–663.

Wiwchar, M.B.
1957: Consolidated Discovery Yellowknife mine; *in* Structural geology of Canadian ore deposits, v. 2, Can. Inst. Min. Metall., Montreal, p. 201–209.

Wlotzka, F., Spettel, B. and Wänke, H.
1973: On the composition of metal from Apollo 16 fines and the meteoritic component; *in* Proc. 4th Lunar Sci. Conf., Geochim. Cosmochim. Acta, Suppl. 4, v. 2, p. 1483–1491.

Wolfe, W.J.
1975: Gold in early Precambrian plutonic rocks; the relation between geochemical abundance and concentration to exploitable levels; Soc. Mining Eng., Am. Inst. Min. Eng., prepr. 75-L-5, 17 p.

Wolfenden, E.D.
1965: Bau Mining district, West Sarawak, Malaysia, Pt. 1: Bau; Malaysia, Geol. Surv., Borneo region, Bull. 7, pt. 1, 147 p.

Wong, W.H.
1919: The mineral resources of China; Geol. Surv. China, Ser. B, no. 1, 270 p.

Wood, E.D.
1971: Gold in sea water; unpubl. Ph.D. thesis, Univ. Alaska, College, Alaska.

Wood, J.M.
1974: Biological cycles for toxic elements in the environment; Science, v. 183, p. 1049–1052.

Woodall, R.W.

1965: Structure of the Kalgoorlie goldfield; *in* 8th Commonw. Min. Metall. Congr., Australia and New Zealand, v. 1, Geology of Australian ore deposits, 2nd ed., J. McAndrew, ed., p. 71–79.

1975*a*: Gold resources of Australia; Private Rep., Western Mining Corp. Ltd., Kalgoorlie, Western Australia, 24 p.

1975*b*: Gold resources of Australia; *in* Proceedings 5th Gold and Money Session and Gold Technical Session, Pacific Northwest Metals and Minerals Conference, Portland, Oregon, Oreg. Dep. Geol. Miner. Ind. publ., p. 106–138.

Woodtli, R.

1961*a*: Gold impregnation deposits in the Moto area (central Africa); Econ. Geol., v. 56, p. 603–607.

1961*b*: Relationships of general structure to gold mineralization in the Kilo area (central Africa); Econ. Geol., v. 56, p. 584–591.

1977: Geologic factors in looking for hidden gold deposits; *in* Problems of ore deposition, v. 2, 4th IAGOD Symp., Varna, Bulgaria, B. Bogdanov, ed., Bulgarian Acad. Sci., Sofia, p. 409–412.

Worst, B.G.

1960: The great dyke of southern Rhodesia; South. Rhodesia Geol. Surv. Bull. 47, 234 p.

Worthington, J.E. and Kiff, I.T.

1970: A suggested volcanigenic origin for certain gold deposits in the Slate Belt of the North Carolina Piedmont; Econ. Geol., v. 65, p. 529–537.

Wright, K.

1969: Textures from some epigenetic mineral deposits of Tennant Creek–central Australia; *in* Remobilization of ores and minerals, P. Zuffardi, ed., Cagliari, Italy, p. 219–251.

Wright, L.B.

1947: Geologic relations and new ore bodies of the Republic district, Washington; Am. Inst. Min. Metall. Eng., Tech. Publ. no. 2197, 19 p.

1976: Fifty six years of exploration: a search for gold; Craftsman Printers, Lubbock, Tex., 234 p.

Wrücke, C.T. and Armbrustmacher, T.J.

1973: Geochemical analyses of bedrock samples collected in the Gold Acres and Tenabo areas northern Shoshone Range, Nevada; U.S. Natl. Tech. Inform. Serv., PB Rep., no. 220645/6, 136 p. (Chem. Abstr., v. 79, 138693r.)

Wrücke, C.T., Armbrustmacher, T.J. and Hessin, T.D.

1968: Distribution of gold, silver and other metals near Gold Acres and Tenabo, Lander County, Nevada; U.S. Geol. Surv., Circ. 589, 19 p.

Wybergh, W.

1930: Goldfields of Pilgrims Rest and Sabie; *in* The gold resources of the world, 15th Int. Geol. Congr., Pretoria, p. 320–332.

Wychoff, D., translator

1967: Book of minerals by Albertus Magnus; Clarendon Press, Oxford, 309 p.

Yablokova, S.V.

1972: New morphologic variety of gold and its origin; Acad. Sci. USSR, Dokl., Earth Sci. Sec., v. 205, p. 143–146 (Am. Geol. Inst. transl.).

Yablokova, S.V. and Polkanov, Yu.A.

1974: Gold in titanium-bearing placers of the Ukraine; Akad. Nauk. Ukr. RSR, Dopov., Ser. B, v. 36, no. 8, p. 701–703. (Chem. Abstr., v. 82, 62042s.)

Yablokova, S.V. and Ryzbov, B.V.

1972: Role of ancient gold in the feeding of Quaternary placers of the Mari Taiga; Izv. Vysch. Ucheb. Zaved., Geol. Razaved., v. 15, no. 10, p. 60–65. (Chem. Abstr., v. 78, 74712r.) *Also* Int. Geol. Rev., v. 15, no. 10, p. 1182–1185.

Yakovlev, V.I. and Nyuppenen, T.I.

1974: Vertical zoning of mineralization at the Nezhdaninsk gold deposit; Zap. Leningrad. Gorn. Inst., v. 64, no. 2, p. 37–42. (Chem. Abstr. v. 84, 47310a.)

Yampol'skaya, M.Ya., Savchenko, L.A. and Vtorina, L.S.

1975: Extraction of gold from coastal-marine sands; Tsvet. Met., no. 3, p. 88–89. (Chem. Abstr., v. 83, 119011h.)

Yang, Min-chih, Ni, Chi-Tsung and Tai, Feng-Fu

1974: Geochemistry of precious metals in skarns and hydrothermal copper deposits from a certain district in China; Geochimica (Peking), no. 3, p. 166–168 (Engl.).

Yassoglou, N.J. and Nobeli, C.

1968: The distribution of gold in the silt fraction of some soil profiles and its genetical significance; Soil Sci. Soc. Am., Proc., v. 32, no. 5, p. 705–708.

Yasuda, M.

1927: The estimation of gold and silver in sea water; Chem. Soc. Japan, Bull., v. 2, no. 12, p. 321–323.

Yasyrev, A.P.

1967: Gold and silver in phosphorites of the Russian Platform; Tr. Tsent. Nauch. Issled. Gornorazved. Inst., no. 32, p. 200–203. (Chem. Abstr., v. 67, 83871s.)

1969: Gold in Mesozoic nodular phosphate deposits of the Russian Platform, Acad. Sci. USSR, Dokl., Earth Sci. Sec., v. 185, p. 183–186.

1971: Concretion phosphorites of the Russian Platform as an intermediate collector of gold during placer formation; Akad. Nauk SSSR, Dokl., v. 199, no. 2, p. 452–455. (Chem. Abstr., v. 75, 131658s.)

Yatsenko, A.S.

1972: Gold localization in subhorizontal quartz veins imbedded in the granite bodies of the central Vitim highlands; Mater. Geol. Polez. Iskop. Buryat. ASSR, v. 15, p. 95–100. (Chem. Abstr. v. 82, 127555h.)

Yazkov, N.M.

1972: Geology and endogenous mineralization of the Lomamsk gold-bearing region; Zap. Zabaikal. Filial. Geogr. O-va. SSSR, v. 86, p. 53–61. (Chem. Abstr., v. 82, 62012g).

Yeend, W.E.

1974: Gold-bearing gravels of the Ancestral Yuba River, Sierra Nevada, California; U.S. Geol. Surv., Prof. Pap. 772, 44 p.

1975: Experimental abrasion of detrital gold; J. Res., U.S. Geol. Surv., v. 3, no. 2, p. 203–212.

Yeleyeva, I.V. and Rusinova, O.V.

1969: Genesis of quartz-kaolinite rock alteration in the upper levels of the Balei deposits and its importance in prospecting; Geol. Rud. Mestorozhd., v. 11, no. 4, p. 88–95.

Yen, T.F., ed.

1975: The role of trace metals in petroleum; Ann Arbor Science Publ. Inc. Ann Arbor, Mich., 221 p.

Yermakov, N.P. *et al.*

1965: Research on the nature of mineral-forming solutions with special reference to data from fluid inclusions; Pergamon Press, Oxford, 743 p.

Youchev, U., Dimitrov, C., Bogdanov, B., Zhelyaskova-Panayotova, M., Dokov, R., Manolov, N. and Rashkov, R.

1974: Genetic types of ore deposits in Bulgaria; Int. Assoc. Genesis Ore Deposits (IAGOD), Sofia, Bulgaria, 47 p.

Young, G.A.

1909: A descriptive sketch of the geology and economic minerals of Canada; Can. Mines Dep., no. 1085, 151 p.

Young, R.B.

1917: The banket of the South African Goldfields; Gurney and Jackson, London.

1931: Discussion on the origin of the gold in the Witwatersrand system, The genesis of the gold in the Rand banket; Geol. Soc. S. Afr., Trans. (Annex), v. 34, p. 1–14.

Yudalevich, Z.A., Sandomirskii, G.G. and Nikanorov, G.S

1973: Gold content in granitic rocks and quartz veins of northern Nurutau and Bakantau; Uzb. Geol. Zh., v. 17, no. 5, p. 17–22. (Chem. Abstr., v. 80, 98500n.)

Yudin, I.M.
1971: Finely dispersed sulfide mineralization of sedimentary-metamorphic country rocks of the Muruntau deposit and its genesis; Geol. Rud. Mestorozhd., v. 13, no. 2, p. 43–51. (Chem. Abstr., v. 75, 23847u.)

Yudin, S.S., Yudina, V.M. and Shilin, N.L.
1972: Presence of gold in volcanogenic series of the central part of the Okhotsk volcanogenic belt; Akad. Nauk SSSR, Dokl., v. 207, no. 1, p. 157–160.

Yugai, T.A.
1967: Genetic groups of Tertiary occurrences of gold in volcanic formations of the lower reaches of the Amur River; Geol. Rud. Mestorozhd., v. 9, no. 3, p. 79–81. (Rev. *in* Econ. Geol., v. 64, p. 588.)
1970: Gold placer deposits in the area of the Belaya Gora epithermal deposit; Probl. Geol. Rossypei, Soveshch., (Dokl.) 3rd 1969, V.I. Smirnov, ed., Sev. Vost. Kompleks. Inst., Magadan, U.S.S.R., p. 155–161. (Chem. Abstr., v. 75, 38847t.)
1971: Change in ore gold standards dependent on depth; Akad. Nauk SSSR, Dokl., v. 200, no. 2, p. 446–448. (Chem. Abstr., v. 75, 131642g.)

Yushmanov, V.V.
1972: Genesis of gold placers in the western part of the Stanovoi Ridge; Zap. Zabaikal. Filiala Geogr. O-Va. SSSR, v. 86, p. 104–107. (Chem. Abstr., v. 82, 62014j.)

Zagoskin, V.A. and Zagoskina, E.I.
1971: Prospecting for auriferous conglomerates by geochemical methods; Razved. Okhr. Nedr., v. 37, no. 5, p. 30–33. (Chem. Abstr., v. 75, 79054a.)

Zagruzina, I.A.
1970: The age of the gold mineralization of the northeastern part of the Soviet Union; Geol. Rud. Mestorozhd., no. 1, p. 79–82.
1972: Age of gold mineralization in the Northeast; Int. Geol. Rev., v. 14, no. 4, p. 372–374.

Zagruzina, I.A. and Polazan'ev, V.P.
1975: Paleozoic gold mineralization in the Omolon massif; Geol. Rud. Mestorozhd., v. 17, no. 1, p. 74–80. (Chem. Abstr., v. 83, 118757u.)

Zaitseva, M.L., Ivanovskii, M.D. and Larina, N.K.
1973: Sorption properties of carbonaceous substances during the cyaniding of gold ores; Tsvet. Metal., no. 1, p. 77-79. (Chem. Abstr., v. 79, 56200p.)

Zakharov, M.N., Gundobin, G.M. and Glazunova, A.D.
1974: Geochemical study and gold content of intrusions in the Evensk ore node in the northeastern U.S.S.R.; Ezheg. Inst. Geokhim., Sib. Otd., Akad. Nauk SSSR, p. 52–55. (Chem. Abstr., v. 84, 47527b.)

Zakharov, N.N., Gundobin, G.M. and Andrulaitis, L.D.
1976: Regularities of gold distribution in Upper Cretaceous and Paleogene igneous rocks of the Turomchinsky imposed depression (Okhotsko-Chutotsky volcanic belt); Ezheg. Inst. Geokhim., Sib. Otd., Akad. Nauk SSSR, 1974, p. 80–85.

Zakharov, Ye. Ye.
1971: Auriferous Mesozoic metasomatites in crystalline base of the central Aldan district; Int. Geol. Rev., v. 13, no. 7, p. 1024–1026.

Zakharov, Ye. Ye., Novikov, V.P. and Pilipenko, G.N.
1969: The occurrence of gold in Mesozoic metasomatites in the crystalline basement of the central Aldan region; Geol. Rud. Mestorozhd., v. 11, no. 2, p. 85–88. *Also* Int. Geol. Rev., v. 13, no. 7, 1971, p. 1024–1026.

Zashchinskii, L.A., Kortunov, V.A., Nomokonova, G.G. and Turysheva, T.I.
1970: Use of potassium radiometric survey during study of ore-controlling factors in the Darasan gold field; Izv. Tomsk. Politekh. Inst., no. 239, p. 436–440. (Chem. Abstr., v. 75, 120042k.)

Zashchinskii, L.A., Miko, A.D. and Turysheva, T.I.
1969: Use of composite density and boronometric surveys for outlining zones of gold-containing xenoliths; Geol. Geofiz., no. 12, p. 130–133. (Chem. Abstr., v. 72, 69147g.)

Zaslow, M.
1975: Reading the rocks: the story of the Geological Survey of Canada 1842–1972; Macmillan, Toronto 599 p.

Zavaritskii, A.N.
1955: Igneous rocks; Akad. Nauk SSSR, Moscow, 479 p. (in Russian).

Zavorotnykh, P.R.
1971: Heavy concentrate and geochemical prospecting of ore deposits by using water flow dispersion haloes; Vop. Reg. Geol. Metallogen. Zabaikal'ya, no. 6, p. 82–88. (Chem. Abstr., v. 77, 167143a.)

Zavorotnykh, I.R.
1972: Possibilities of detecting the sources for gold-bearing placers according to geochemical data; Zap. Zabaikal. Filiala Geogr. Obschest. SSSR, no. 69, p. 90–92. (Chem. Abstr., v. 80, 98377c.)

Zhelnin, S.G.
1973: Characteristics of gold-placer forming formations in the northeast (U.S.S.R.); Tr. Sev. Vost. Kompleks. Inst., Dal'nevost. Tsent., Akad. Nauk. SSSR, no. 55, p. 239–271. (Chem. Abstr., v. 84, 7599v.)

Zhelnin, S.G. and Travin, Yu. A.
1976: Comparative features of placer geology in the modern structure of the northeastern U.S.S.R.; 25th Int. Geol. Congr., Sydney, Australia, Abstr., v. 1, p. 229–230.

Zhelnin, S.G., Valpeter, A.P. and Pruss, Yu. V.
1972: Role of weathering crusts in the formation of placer deposits in the northeast U.S.S.R.; Tr. Sev. Vost. Kompleks. Inst. Dal'nevost. Tsent. Aka, v. 44, p. 101–117. (Chem. Abstr., v. 81, 155816s.)

Zhelobov, P.P.
1972: The role of conjugations and intersections of deep-seated faults in the location of goldbearing areas of the Urals; *in* Tr. Inst. Geol. Geofiz. Geokh. Uslov Obr. Rud.Zolota Redk. Met., no. 149, "Nauka" Publ. House, Siberian Br., Novosibirsk, U.S.S.R., p. 37–43.

Zhirnov, A.M.
1972: Hypogene colloidal gold in the Kaul'da gold ore deposit (central Asia); Uzb. Geol. Zh., no. 1, p. 93–95.
1973: Metacolloidal gold in the Kaul'dy gold ore deposit (Kuramin Ridge); Zap. Uzb. Otd. Vses. Mineral. Obshchest., no. 26, p. 114–116.(Chem. Abstr., v. 80, 50475f.)

Zhirnov, A.M. and Badalov, S.T.
1972: New data on the time of deposition of gold in hydrothermal deposits (as illustrated by Almalyk); Uzb. Geol. Zh., no. 6, p. 47–50. (Chem. Abstr., v. 79, 44316t.)

Zhokhov, G.A., Klimov, V.I., Karimov, Kh.K. and Abbasov, I.M.
1975: Radioactive elements in country rocks and hydrothermally altered rocks of the Kokpatas ore field; Uzb. Geol. Zh., v. 19, p. 32–36. (Chem. Abstr., v. 84, 138643c.)

Zhukova, R.I.
1974: The relationship of Au, Pb, Zn with organic matter and water-soluble elemental forms in secondary dispersion halos; Ezheg. Inst. Geokhim., Sib. Otd., Akad. Nauk SSSR, 1973, L.V. Tauson, ed., p. 247–250.

Ziauddin, M.
1967: Gold in India, Retrospect and prospect; Indian Miner., v. 21, no. 4, p. 329–338.
1970: Notes on the geochemistry of gold; Indian Miner., v. 24, no. 1, p. 28–41.

Zlobin, V.A. and Sovetov, Yu. K.
1975: Copper, gold, and radioactive elements in formations of the Taseev series of the Enisei Ridge; Geol. Geofiz, no. 8, p. 147–152. (Chem. Abstr., v. 84, 62823p.)

Zontov, V.N. and Sarapulova, V.N.
1974: Element content and zonality of one gold deposit; Ezheg. Inst. Geokhim., Sib. Otd., Akad. Nauk SSSR, 1973 L.V. Tauson, ed., p. 206–209.

Zsigmondy, R., and Thiessen, P.A.
1925: Das kolloide Gold; Akad. Verlagsgesellschaft M.B.H. Leipzig, 226 p.

Zulfugalry, D.I.
1960: Rasprostranenie mikroelementov v kaustobiolitakh, organizmakh, osadochnykh porodakh i plastovykh vodakh; Acad. Nauk SSSR Inst. Geol. Razrabot. Iskop. Azerbaidzhan. Gos. Univ., Baku, 291 p. 1964; German transl. (Distribution of trace elements in caustobioliths, organisms, sedimentary rock, and ground water; Leipzig, VEB Deut. Verlag Grundstoffindustrie, 291 p.) 1963 (Engl. abstr. *in* Bibliography and index of geology exclusive of North America, v. 28, p. 633).

Zverev, K.S., Levchenko, V.M. and Miller, E.I.
1947: Determination of gold in Matzesta waters; Hydrochemical essays, v. 13, p. 258–260, Acad. Sci. USSR, Hydrochem. Inst. Novocherkassk.

Zvereva, N.F. and Gavrilenko, B.V.
1971: Gold in the rock forming minerals of the Krykkuduk intrusive complex (northern Kazakhstan); Geokhim., no. 1, p. 114–118.

Zvyagintsev, O.E.
1941: Geochemistry of gold; Izd. Akad. Nauk SSSR, Moscow-Leningrad, 114 p. (in Russian).
1959: Review of book by V.M. Kreiter *et al.*, Migration of gold in the oxidized zone of gold-bearing sulphide deposits; Geochemistry, no. 6, p. 683–685.

Zvyagintsev, O.E. and Paulsen, I.A.
1938: Concerning the action of minerals on gold solutions; Acad. Sci., USSR, C.R. (Dokl.), v. 21, no. 4, p. 176–178.
1940: Contribution to the theory of formation of vein gold deposits; Acad. Sci. USSR, C.R. (Dokl.), v. 26, p. 647–651.

Index

146
2151 3 9